Concise
Dictionary
of Modern
Medicine

Notice

Concise Dictionary of Modern Medicine

Joseph Segen, MD, FCAP
Department of Histopathology
Chase Farm Hospital
Enfield, United Kingdom

McGraw-Hill

New York   Chicago   San Francisco   Lisbon   London   Madrid   Mexico City
Milan   New Delhi   San Juan   Seoul   Singapore   Sydney   Toronto

The *McGraw-Hill* Companies

Cataloging-in-Publication Data is on file with the Library of Congress.

Concise Dictionary of Modern Medicine

2 3 4 5 6 7 8 9 0   DOC/DOC   0 1 0 9 8 7 6

ISBN 0-8385-1535-5

This book was printed on acid-free paper.

It was set by the McGraw-Hill Professional's composition unit in Hightstown, N.J. RR Donnelley, The Lakeside Press, was printer and binder.

For Amy, Art, Cindy, Big David

Publisher's Preface

This concise compendium of current medical terminology has been designed for physicians, medical residents, medical students, nurses, and all other health professionals—as well as the curious lay public. The focus is on the real-world language of practice. The goal has been to provide, in a convenient format, readable comprehensible definitions of terminology, jargon, acronyms, neologisms, and casual speech heard in today's health care environment. Many of the approximately 20,000 terms are not included in ordinary medical dictionaries, or are only briefly and inadequately defined. Readers will find entries in alternative medicine, forensics, and sports medicine in addition to classical areas of medicine, such as cardiology and endocrinology. Important concepts in basic science encountered in practice also are covered.

The *Concise Dictionary of Modern Medicine* aims to be learned enough for professional users, yet written in clear, simple language that will appeal to students and general readers wanting to explore the exciting new terminology of contemporary medicine.

Mark D. Licker
Publisher

Author's Preface

Today's world differs vastly from that in which the standard medical dictionaries were compiled at the dawn of the 20th century. Then, a well-educated physician was versed in the classic languages–Greek and Latin; specialization in a particular field was uncommon and consisted of a one-to-two-year apprenticeship in a handful of disciplines at one of the great medical centers in Europe; subspecialization was nonexistent. Some disciplines– eg, virology, genetics, and immunology–were primitive. Others–eg, molecular biology, informatics, diagnostic imaging, and endoscopic surgery–were a half-century in the future and the stuff of science fiction. Equally inconceivable were issues of medical malpractice, regulatory oversight, and third-party payers. Our predecessors' diagnostic tools–a microscope, a stethoscope, and rudimentary X-ray devices–seem now quaintly primitive. Disease mechanisms were poorly understood, infections often fatal, trauma to a limb the prelude to amputation. Little effective therapy was available. Some diseases didn't exist; others weren't recognized. The average person lived less than 55 years.

The world, medicine, physicians, and their patients have changed considerably in the ensuing years. Decades have passed since Latin (or classic Greek) was a prerequisite for studying medicine in the US or in other civilized countries. Specialization and subspecialization are expected of today's physicians and require up to 8 years post-med school. English has become the lingua franca of medicine and of the world at large. Patients are now 'clients' and have average lifespans of nearly 80 years. Disease mechanisms are dissected by molecular biologists, immunologists, and other scientists using a bewildering array of techniques. Today's cornucopia of diagnostic modalities–MRI, ultrasonography, PET, endoscopy, and in situ hybridization– to name but a few, and therapeutic options–laparoscopic surgery, gamma knife, laserectomies and -otomies, genetic manipulation, and so on–would have left our predecessors breathless. Medical information doubles every 7 to 10 years. Computers and communication gadgets have become fixtures in the workplace. Physicians now share decisions with patients, bureaucrats, and ethicists, and the consequences of those decisions–outcomes–with peer reviewers and possibly with attorneys. The changes in medicine have engendered a new vocabulary; a medical dictionary should have its finger on the pulse of those changes. This is a compilation of terms, many of recent vintage, that are integral to the language of modern medicine, a language replete with acronyms, jargon, neologisms, and the argot of new disciplines, diseases, their diagnoses, and therapies.

In the 13 years since the publication of the author's *Dictionary of Modern Medicine*, countless thousands of terms have been been added to the working medical parlance, reflecting quantum leaps in the understanding and management of obesity, Alzheimer's disease, HIV-related diseases, oncology, to name but a few of the areas in which basic research has been translated into effective therapies. More people embrace alternative forms of health care despite the paucity of valid efficacy data.

Multinational, multicenter megatrials on the effects of various interventions are now common–eg, ASTRONAUT, CABG patch, DINAMIT, SUPREMO, TURBO, etc, etc–as are meta-analyses–eg, the Cocharan collaborations–that distill the data into meaningful 'nuggets' for clinical practice. Issues related to physician-assisted suicide, euthanasia, and other permutations of the formerly unthinkable have crept towards center stage, as evidenced by the Terry Schiavo case in 2005. Mobile phones and laptop computers have transitioned from toys to tools. The human genome has been sequenced...ahead of schedule.

The Internet is for some a necessity, and for others, an obsession–and yes, Internet addiction disorder has been described. In less than 10 years, the Internet has become a driving force in communication, and impacted on the way in which scientific data is published (sic) and medicine is practiced. Its potential as a vehicle for disseminating scientific information has borne the fruit of e-journals, which publish research data online months before the print version arrives by 'snail mail.' The ever faster throughput of data on the Internet has raised the profile of the nascent fields of telemedicine–and its 'twiglets,' teleradiology, telepathology, teletceteraology–which have had multiple 'ripple' effects on public access to health care information, the 'place' where medicine is practiced, and competition for patients. In a sense, the present work is a 'poster child' for the effect that the Internet has had on medical informatics, as virtually all notes regarding term usage, preferred names, and source material for the *Concise Dictionary of Modern Medicine* were obtained via the Internet.

In these same 13 years, the world has become smaller–CNN and the BBC bombard us with issues of human rights, global economies, global warming, natural and human-made disasters. As 2004 closed, we were reminded of our helplessness in stopping the forces of nature as the Boxing Day tsunami swept 300,000 victims to their deaths. Terrorism has added to the stress of life in the 21st century–the most recent wave of hatred began on September 11th, 2001, and shows no sign of abatement, as shown by the July 2005 bombings in London. Such drastic circumstances create new vocabulary and new uses for old terms, as the medical community is called upon to care for a new genre of victims in the borderless new order.

The author's earlier dictionary, published in early 1992, had 12,000 entries. Now, in 2005, the database from which the *Concise Dictionary of Modern Medicine* has been carved has over 150,000 entries. The reader might be relieved to know that he or she is lugging around only 20,000 or so of the gems that the author thought most useful for those in clinical training.

Introduction

Dr. Samuel Johnson, who compiled the first comprehensive dictionary of the English language said, *'Dictionaries are like watches, the worst is better than none, and the best cannot be expected to go quite true.'* I hope that after over 20 years of development, the present work goes as true as possible.

The purpose of a language is to transmit information or ideas The purpose of a dictionary is to catalog the structural units–words and phrases–used in a particular language. The formulation of a language's vocabulary is an imperfect exercise validated by colloquial usage and the passage of time–rarely by logical parameters. In a dynamic language such as medicine, new terms are constantly added; others change meaning; still others disappear without due notice. A key indicator of a contemporary medical dictionary's usefulness is whether it contains the terminology used in the working parlance. The present work attempts to catalog and define new terms that have entered the modern medical lexicon, while including those in continued use.

Since the first editions of the Dorland's and Stedman's dictionaries were compiled at the dawn of the 20th century, the language of medicine has undergone enormous changes related to:

A shift in linguistic roots The major medical dictionaries used in English-speaking countries were compiled at the beginning of the 20th century, when knowledge of the classic languages (Latin and Greek) was expected of undergraduates, and fluency in German and/or French required for those who studied 'cutting edge' medicine. Few of today's physicians have more than a passing familiarity with the classics. English is essentially the only international language of medical communication.

Specialization After finishing medical school, every physician undergoes the further rigors of at least 4 years of training in a specialty–and often a subspecialty–of medicine, surgery, or other field of health care. Each specialty and subspecialty has its own vocabulary.

'Players' The practice of medicine is no longer that of a simple doctor-patient relationship. Many parties are now involved in patient management, including allied health professionals, administrators, bureaucrats, lawyers, educators, researchers, ethicists, information managers, and others. Each new player, his tools, and methods add a colorful swatch to the quiltwork of the medical lexicon.

Turf The domain encompassed by health care has expanded, as have the ethical obligations of its practitioners. The country doctor of the early 20th century who maintained a solo practice, exchanging services for the produce from a patient's farm in a small town–a norm at the turn of the century–is an anachronism with little relevance to the health care environment of the 21st century. More than half of all new physicians practice in some form of a group. Physician services are reimbursed by third-party payers. Medical practice in academic and/or research environments requires political savvy that would tax the Borgias. The Internet, still in its infancy as an environment for practicing medicine, has added, and will continue to add, terms to the working medical lexicon. The physician's role as an ethical bastion has expanded with global communication. The images of genocide, starvation, environmental, and human-made catastrophes are, courtesy of the media, only an f-stop away. A parochial mindset is no longer an option for the physician of the 21st century.

Technology Any advancing technology has two effects on the working vocabulary in a particular field: new terms are added and other terms are retired. Medicine is a prime example of this two-step process. The use of increasingly sophisticated tools for studying disease mechanisms has led to a greater understanding of biomedical phenomena and improvements in both diagnostic and therapeutic modalities. New data often invalidate older classification schema based on observations by light microscopy, a primitive tool by today's standards, and clinical observations.

'Democracy' The explosive growth of the personal computer industry has changed the rules of communication, and created an environment in which anyone with a modem has access to medical information and can coin terminology that suits his/her needs. With the Internet, Pandora's Box has been opened and the 'right to write'–long controlled by editors who demanded quality of writing–has become a fundamental freedom.

Criteria for selection The main criterion is active use–if the author has seen the term or phrase in a journal or current text or heard it in rounds, in meetings with colleagues, at conferences, or in other medical venues, it is 'fair game' for inclusion. The terms included are those which are new, of heightened interest to biomedical scientists, or interesting. Interesting terms include (1) Droll–Jogger's nipple, Parting of the Red Seas syndrome, Toxic sock syndrome (sic), Tiger Wood syndrome, Twinkie™ defense; (2) Fascinating–Entomophagy, Quackery, Placentophagy, Typhoid Mary; (3) Disturbing–Chemical warfare, Drive-by shooting, Landmines, Nazi science, 9-11, the Rape of Nanking; (4) Of general interest–Anthropology, Ballistics, El Niño, Extreme sports, Internet, Nobel Prizes, Smog; and (5) American slang*–Cadillac version, Couch potato, Go sour, Postal worker syndrome, Shrinkage. British medical slang will be incorporated in the next edition (Sorry, Pete).

The *Concise Dictionary of Modern Medicine* targets students of clinical medicine, including residents, registrars, SHOs, interns, medical students on the wards, as well as physician assistants and nurse practitioners. Because the *Concise* is intended to be a companion to the on-hands patient experience, it lacks the literature references found in the unabridged version of this dictionary which is still in production–all definitions are culled from primary sources–eg, *New England Journal of Medicine, Nature, Science,* etc. I apologize to those authors who might mistakenly believe I've commandeered their words as my own–this is not the case, merely a reflection of the size constraints placed on this work by the defining adjective, *Concise.*

---

*The insinuation of spoken 'American' into the working medical lexicon is hardly surprising, given that the US is the planet's largest English-speaking country. And Americans tend to borrow or bend a word to fit a need and, if one is not available, extemporaneously create one. A compiler's role is to document, not pass sentence on the validity of the neologistic activities of others.

On a Personal Note

In April 2005, I jumped the pond and now work for the NHS north of London. It's been said that the UK and the US are two countries separated by a common language,[1] one of many truths told in jest. American is to English what Brazilian is to Portuguese–they're written the same, but the spoken word differs. Thus, as with most Americans, English is my second language. I once thought I could splice together a phrase as well as the next guy. I was wrong. I received the first of many spankings even before I arrived when I took the IELTS[2] exam, which assesses a person's aptitude in English…and failed. The average Brit's command of the Queen's tongue has made me more than once reflex to Mark Twain's advice, *"it's better to keep your mouth shut and be thought a fool, than to open it and remove all doubt."* Much time will pass before I stop wondering if, after finishing a monologue, I've removed all doubt in the listener.

In the UK not only did I have to relearn English, I've had to relearn medicine. It would be hard to imagine a more challenging career move–from a sleepy "Band-Aid" hospital in the hills of Appalachia, to a 1000-bed two-campus NHS trust with a soup-to-nuts array of diagnostic dilemmas. The professionalism of my new clinical and pathology colleagues and the quality of medicine practiced here are nothing less than intimidating. I often feel like the dog in a Far Side cartoon who is desperately trying to understand quantum physics.

The frustration of a lexicographer is that he knows the book will never be finished. Although I've heard countless 'pearls,' eg, smash-and-grab operation, medical defense, to "top and tail" a patient, stopping the clock, fit, chap/bloke/wanka, and others, in the past few months that warrant inclusion, this edition is too close to the bindery to incorporate any of the terms, not to mention that the production department at McGraw-Hill would organize a proper lynch mob if I suggested any changes at this juncture–so stay tuned for the next edition.

Many thanks go to Dr. Peter Kench–head of the department of histopathology, histopathology consultants Erich Langner, Ines Miselevich, Hugh Reid, Anupam Joshi, and Khurram Chaudhary, also of the department, all for their tolerance and patience in teaching an old dog new tricks. Thanks also to Richard von Witt, emeritus chair, who pushed to have a token American in the department, Ed James, current pathology chair, Marilyn Treacy, CMG chair, T.F. Chan of the department, one of the Trust's "can-do" people, and Richard Harrison, Medical Director of the Barnet-Chase Farm Hospitals Trust. Further sucking up is warranted to clinical colleagues whom I see during the weekly MDTs,[3] including Mssrs[4] Al-Dubaisi, Bernhardt, Bunce, Gellister, Elton, Jackson, Jaffe, Mathur, Mitchell, Murray, Pearson, Ward, Warren, Webster,

---

[1] You say toMAHtoe, I say toMAYtoe, or in medicine, you say cerVIcal, I say CERvical.

[2] International English Language Testing System.

[3] Multidisciplinary team meetings. In the UK, all new malignancies (with the exception of basal cell carcinomas) require a team approach to care management.

[4] Surgeons are all misters–women are miss or missus, which stems from the English art of understatement. I witnessed a most astounding example of said art at the Enfield Pageant of Motoring shortly after arriving. Nestled among the highly polished pride-and-joy–but essentially worthless–MGs, Morrises, and Minis, was a rusty, ratty D-type Jaguar (yeah, a real one). I waited an hour for the owner, who was hunting cheap shock absorbers among the vendors. I sidled up to him and asked how much he'd take for the car, he said £750. I knew he meant thousand. To save a few quid on shock absorbers and use a nearly $2 million-dollar car as a daily driver. …that, THAT, dear reader, is understatement on this side of the puddle!

ON A PERSONAL NOTE

Yanni, and Doctors Isaacs, Karp, Kaplan, Lotzof, Melcher, Ostler, MDT coordinators Allison, Anne, Barbara, and Denise, clinical attachment Nilmini, pathology registrars Mohamed Ben-Gashir, Hiruni Dhanapala, John Du Parcq, and Rashpal Flora. I'd be remiss if I forgot my new bench colleagues at arms, including senior scientists, Rowena Berliner, Hasit Patel, Hillary Riches, Bina Welch, and Iain Bowman, and a host of hey-yous and hoi polloi, including Anita, Delwyn, Janet, Jennie, the Jos–Big Jo, Little Jo, NewJo–John, Karen, Michael, Orla, Val, Wanda, Zina, the typing pool–Anne and Anne and Elaine–and Jeanette. Last, but certainly not least, to Elwira Rumniak, for sticking her thumb in the dike/dyke[5] when I realized errors had slipped through the copy editing and I needed the extra hands to finish this before the next Ice Age.

Joseph C. Segen, MD
Department of Histopathology
Barnet-Chase Farm Hospitals Trust
Enfield EN2 8JL, MIDDX, UK
joseph.segen@bcfh.nhs.uk

---

[5]Dike refers to an embankment that prevents flooding; dyke has another meaning.

Abbreviations Used in Text

For convenience, the listing is repeated on page 761.

1° Primary
2° Secondary
2-D Two-dimensional
3° Tertiary
3-D Three-dimensional
aa Amino acid
ABG Arterial blood gases
ACE Angiotensin-converting enzyme
ACEI Angiotensin-converting enzyme inhibitor
AD Autosomal dominant
ADD Attention deficit (hyperactivity) disorder
ADH Antidiuretic hormone (vasopressin)
AF Atrial fibrillation
AFB Acid-fast bacillus
AFP Alpha-fetoprotein
AIDS Acquired immunodeficiency syndrome
aka Also known as
alk phos Alkaline phosphatase
ALL Acute lymphocytic (lymphoblastic) leukemia
ALS Amyotrophic lateral sclerosis
ALT Alanine aminotransferase (formerly GPT)
Am American
AMA American Medical Association
ama Against medical advice
AMI Acute myocardial infarction
AML Acute myelocytic (granulocytic, myeloid, myelogenous) leukemia
AMN American Medical News
ANA Antinuclear antibody
ANLL Acute nonlymphocytic leukemia
ANP Atrial natriuretic protein
AP Anteroposterior
APLM Archives of Pathology and Laboratory Medicine
apo Apolipoprotein
aPTT Activated partial thromboplastin time
AR Autosomal recessive
ARDS 1. Acute respiratory distress syndrome 2. Adult respiratory distress syndrome
ASAP As soon as possible
ASCUS Atypical squamous cells of undetermined significance
ASD Atrial septal defect
ASHD Atherosclerotic heart disease
Ass Association
AST Aspartate aminotransferase (formerly GPT)
ATP Adenosine triphosphate
AV 1. Arteriovenous 2. Atrioventricular
AW Atomic weight
BAL Bronchoalveolar lavage
BBB 1. Bundle branch block 2. Blood-brain barrier
BCC Basal cell carcinoma
β-blocker Beta adrenergic receptor blocking agent
BFP Biological false positive

BID Twice a day
BM 1. Bone marrow 2. Basement membrane 3. Bowel movement
BMJ British Medical Journal
BMT Bone marrow transplantation
bp Base pairs
BP Blood pressure
BPH Benign prostatic hypertrophy
BR Bilirubin
BUN Blood urea nitrogen
Bx Biopsy
CA Cancer, malignancy
CABG Coronary artery bypass graft
CAD Coronary artery disease
C albicans *Candida albicans*
cAMP Cyclic adenosine monophosphate
CAP College of American Pathologists
CBC Complete blood count
CCB Calcium channel blocking agent
CDC Centers for Disease Control & Prevention—based in Atlanta
cDNA Complementary DNA
CEA Carcinoembryonic antigen
CF Complement fixation
CHD Coronary heart disease
CHF Congestive heart failure
CIE Counter-immunoelectrophoresis
CIN Cervical intraepithelial neoplasia
CIS Carcinoma-in-situ
CK Creatinine (phospho)kinase
CLL Chronic lymphocytic (lymphoblastic, lymphoid) leukemia
CML Chronic myelocytic (granulocytic, myelogenous, myeloid) leukemia
CMV Cytomegalovirus
CNS Central nervous system
CO₂ Carbon dioxide
COD Cause of death
Coll College
Cons Disadvantages
COPD Chronic obstructive pulmonary disease
CPR Cardiopulmonary resuscitation
CPT Current Procedural Terminology, cocktail party trivia
CRH Corticotropin-releasing hormone
C-section Cesarean section
CSF Cerebrospinal fluid
CT Computed tomography
C trachomatis *Chlamydia trachomatis*
CVA Cerebrovascular accident
c/w Consistent with
CXR Chest X-ray
DAD Diffuse alveolar damage
DEA Drug Enforcement Administration, US Department of Justice
DIC Disseminated intravascular coagulation
DiffDx Differential diagnosis

DKA Diabetic ketoacidosis

DM Diabetes mellitus

D melanogaster *Drosophila melanogaster*

DNA Deoxyribonucleic acid

DOA Dead on arrival

DOE Dyspnea on exertion

DRG Diagnosis-related group

DSM-IV Diagnostic and Statistical Manual, 4th edition

DVT Deep vein thrombosis

DWI Driving while intoxicated

Dx Diagnosis

EBV Epstein-Barr virus

EC Enzyme Commission

E coli *Escherichia coli*

EDTA Ethylene diamine tetraacetic acid

EEG Electroencephalogram, electroencephalographic

eg *exempli gratia*, for example

EGF Epidermal growth factor

EIA Enzyme immunoassay

EKG Electrocardiography

ELISA Enzyme-linked immunosorbent assay

EM Electron microscopy, ultrastructure

EMG Electromyography

EMS Emergency medical services

EMT Emergency medical technician

ENT Ears, nose, and throat, otorhinolaryngology

EPA Environmental Protection Agency

ER 1. Emergency room, emergency ward 2. Endoplasmic reticulum

ERCP Endoscopic retrograde cholangiography

ERT Estrogen replacement therapy

ESR Erythrocyte sedimentation rate

ESRD End-stage renal disease

et al And others

etc Etcetera; yada, yada, yada

FDA United States Food and Drug Administration

FDP Fibrinogen degradation product(s)

FFA Free fatty acid

FISH Fluorescence in situ hybridization

FNA Fine-needle aspiration (biopsy or cytology)

FSH Follicle-stimulating hormone

FTT Failure to thrive

FUO Fever of unknown origin

GABA Gamma-aminobutyric acid

GC Gas chromatography

GC-MS Gas chromatography-mass spectroscopy

G-CSF Granulocyte-colony stimulating factor

GERD Gastroesophageal reflux disease

GFR Glomerular filtration rate

GGT Gamma-glutamyl transferase

GH-RH Growth hormone-releasing hormone

GI Gastrointestinal

GLC Gas-liquid chromatography

GM-CSF Granulocyte-macrophage colony-stimulating factor

GMS Gomori-methenamine-silver

GN Glomerulonephritis

GNP Gross National Product

GNR Gram-negative rods (bacilli)

G6PD Glucose-6-phosphate dehydrogenase

GRH Gonadotropin-releasing hormone

GTP Guanosine triphosphate

GTT Glucose tolerence test

GU Genitourinary

GVHD Graft-versus-host disease

H Hour

H⁺ Hydrogen ion

HAV Hepatitis A virus

Hb Hemoglobin

HBV Hepatitis B virus

hCG Human chorionic gonadotropin

HCFA Health Care & Financing Administration

HCl Hydrochloric acid

HCV Hepatitis C virus

HDL High-density lipoprotein

HDL-C High-density lipoprotein-cholesterol

HDN Hemolytic disease of the newborn

H&E Hematoxylin & eosin

hGH Human growth hormone

HHS US Department of Health and Human Services

HHV Human herpesvirus (HHV-1, HHV-etc)

H influenzae *Haemophilus influenzae*

HIV Human immunodeficiency virus

HLA Human leukocyte antigen (the major histocompatibility complex of humans)

HMO Health maintenance organization

H₂O₂ Hydrogen peroxide

HPLC High-performance liquid chromatography

HPV Human papillomavirus

H pylori *Helicobacter pylori*

HR Heart rate

HSV Herpes simplex virus

HTLV-I Human T cell leukemia/lymphoma virus

HUS Hemolytic-uremic syndrome

HVZ Herpes varicella-zoster

Hx History

IA Intraarterial

IBD Inflammatory bowel disease

ICP Intracranial pressure

ICU Intensive care unit

ID Identification, identify; infectious disease

IDDM Insulin-dependent diabetes mellitus

ie *id est*, that is to say

IFA Indirect fluorescent antibody

IFN Interferon

Ig Immunoglobulin

IGF-I Insulin-like growth factor I

IL Interleukin

IM Intramuscular

ImPx Immunoperoxidase

INR International normalized ratio

IQ Intelligence quotient

IR Infrared

IRMA Immunoradiometric assay

ISH In situ hybridization

ITP Idiopathic thrombocytopenic purpura

IU International Units

IUBMB International Union of Biochemistry and Molecular Biology

IUD Intrauterine (contraceptive) device

IV Intravenous

IVDU Intravenous drug use/user

IVP Intravenous pyelogram

JACC Journal of the American College of Cardiology

JAMA Journal of the American Medical Association

JCAHO Joint Commission of Accredited Hospitals Organization

JVP Jugular venous pulse

K⁺ Potassium

kb Kilobase

kbp Kilobasepair

kD Kilodalton

KS Kaposi sarcoma

LBW Low birth weight

LC Liquid chromatography

L&D Labor and delivery

LDH Lactate dehydrogenase

LDL Low-density lipoprotein

LDL-C Low-density lipoprotein-cholesterol

LES Lower esophageal sphincter

LGV Lymphogranuloma venereum

LH Luteinizing hormone

LH-RH Growth hormone-releasing hormone

LLQ Left lower quadrant

LM Light micoscopy

LMP Last menstrual period

LN Lymph node

LOC Loss of consciousness

LP Lumbar puncture

LT Leukotriene

LUQ Left upper quadrant

LVEF Left ventricular ejection fraction

MAb Monoclonal antibody

MAC *Mycobacterium avium* complex

MAOI Monoamine oxidase inhibitor

MCP Mayo Clinic Proceedings

MCTD Mixed connective tissue disease

MD Physician, medical doctor

MEN Multiple endocrine neoplasia

µg Microgram

mg Milligram

MHC Major histocompatibility complex

MI Myocardial infarction

MIM Mendelian Inheritance in Man

min Minute (time)

M&M Morbidity and mortality

mo/ma Monocyte/macrophage (tissue histiocyte)

MPS Mucopolysaccaride(s), mucopolysaccharidosis

MRI Magnetic resonance imaging

mRNA Messenger RNA (ribonucleic acid)

MRSA Methicillin-resistant *Staphylococcus aureus*

MS Multiple sclerosis

M tuberculosis *Mycobacterium tuberculosis*

MTX Methotrexate

MVA Motor vehicle accident

MW Molecular weight

NA Nomina Anatomica

Na⁺ Sodium

N/C ratio Nuclear/cytoplasmic ratio

N-CAM Neuronal-cell adhesion molecule

NCI National Cancer Institute (US)

NEJM New England Journal of Medicine

NG Nasogastric

NGF Nerve growth factor

N gonorrheae *Neisseria gonorrbeae*

NYHA New York Hear Association

NHL Non-Hodgkin's lymphoma

NIDDM Non-insulin-dependent diabetes mellitus

NIH National Institutes of Health

NK cell Natural killer cell

NMDA *N*-methyl-D-aspartic acid, *N*-methyl-D-aspartate

NO Nitric oxide

NO₂ Nitrogen dioxide

NOS Nitric oxide synthase; not otherwise specified

NPO nothing by mouth

NSAID Nonsteroidal anti-inflammatory drug

NY New York

NYC New York City

O₂ Oxygen

Ob/Gyn Obstetrics and gynecology

OC Oral contraceptive(s)

OCD Obsessive-compulsive disorder

OD Overdose, optical density

O&P Ova and parasites

OR Operating room, operating suite

OSHA Occupational Safety and Health Administration

OTC Over-the-counter

PAF Platelet-activating factor

PAS Periodic acid-Schiff

PBC Primary biliary cirrhosis

PC Personal computer; politically correct

PCBs Polychlorinated biphenyls

PCP 1. Phencyclidine 2. *Pneumocystis carinii* pneumonia

PCR Polymerase chain reaction

PCTA Percutaneous transluminal angioplasty

PDA Patent ductus arteriosus

PDGF Platelet-derived growth factor

PEEP Positive end-expiratory pressure

PEF Peak expiratory flow

PET Positron emission tomography

PG Prostaglandin

PID Pelvic inflammatory disease

PKU Phenylketonuria

PMN Polymorphonuclear neutrophil or leukocyte, segmented neutrophil

PMS Premenstrual syndrome

PNS Parasympathetic nervous system, peripheral nervous system (see context)

PO By mouth

POC Products of conception

ppb Parts per billion

ppm Parts per million

PP&M Pharmacodynamics, pharmacokinetics, and metabolism

ppt Parts per trillion

PPID GL Mandell et al, Principles & Practice of Infectious Diseases, 4th ed, Churchill-Livingstone, 1995

ppm Parts per million

pron Pronounced

Pros Advantages

PRN *pro re nata*, as needed

PROM Premature rupture of membranes

PSA Prostate-specific antigen

PT Prothrombin time

Pt Patient

PTE Pulmonary thromboembolism

PTH Parathyroid hormone

aPTT (activated) Partial thromboplastin time

PUD Peptic ulcer disease

PVC Premature ventricular contraction

Px prognosis

Q Every

Q 4 H Every four hours

QA Quality assurance

QC Quality control

QD Every day

QID Four times a day

QOL Quality of life

r Recombinant

RAA Renin-aldosterone-angiotensin (system)

RA Rheumatoid arthritis

RAST Radioallergosorbent test

RBCs Red blood cells, erythrocytes

RBRVS Resource-based relative value scale

R&D Research & development

RDS Respiratory distress syndrome

re Regarding, about, concerning, in reference to

REM sleep Rapid eye movement sleep

RFLP Restriction fragment length polymorphism

rh Recombinant human

rH Rhesus factor

RIA Radioimmunoassay

RLQ Right lower quadrant

RN Registered nurse

R/O Rule out, exclude from diagnostic consideration

ROM Range of movement

RR Relative risk

rRNA Ribosomal RNA (ribonucleic acid)

RSV Respiratory syncytial virus

RT Radiation therapy, reverse transcriptase

RTI Respiratory tract infection

RUQ Right upper quadrant

Rx Therapy

S aureus *Staphylococcus aureus*

SCID Severe combined immunodeficiency

SD Standard deviation

sec Second (time)

SI International System (of units)

SIADH Syndrome of inappropriate secretion of ADH

SIDS Sudden infant death syndrome

SIL Squamous intraepithelial lesion

SIV Simian immunodeficiency virus

SLE Systemic lupus erythematosus

SNS Sympathetic nervous system

SO$_2$ Sulfur dioxide

SOB Shortness of breath

SOM School of medicine

SPECT Single photon emission-computed tomographic

SPEP Serum protein electrophoresis

S pneumoniae *Streptococcus pneumoniae*

spp Species

ssp Subspecies

STD Sexually transmitted disease

SVT Supraventricular tachycardia

Sx Symptoms

T$_3$ Triiodothyronine

T$_4$ Thyroxin

TAH-BSO Total abdominal hysterectomy with bilateral salpingo-oophorectomy

TB Tuberculosis

TDM Therapeutic drug monitoring

TdT Terminal deoxytransferase

TG Triglyceride

TGF-β Transforming growth factor-β

TIA Transient ischemic attack

TIBC Total iron-binding capacity

TID Three times a day

TLC Thin-layer chromatography

TNF Tumor necrosis factor

TNM Tumor, Node Metastasis classification

TMJ Temporomandibular joint

TOP Termination of pregnancy

tPA Tissue plasminogen activator

TPN Total parenteral nutrition

tRNA Transfer RNA (ribonucleic acid)

TRH Thyrotropin-releasing hormone

T-S Trimethoprim-sulfamethoxazole

TSH Thyroid-stimulating hormone, thyrotropin

TTP Thrombotic thrombocytopenic purpura

TURP Transurethral resection of the prostate

TX Thromboxane

Tx Tissue, pathology

U 1. Unit 2. University

UK United Kingdom

UN United Nations

URI Upper respiratory tract infection

US 1. Ultrasound, ultrasonographic 2. United States

USDA United States Department of Agriculture

usw *und so weiter*, etcetera

UTI Urinary tract infection

UV Ultraviolet

VDRL Venereal disease research laboratory (test) for syphilis

VEGF Vascular endothelial growth factor

VIP Vasoactive intestinal polypeptide

VLDL Very low density lipoportein

V/Q Ventilation/perfusion

VNTR Variable number of terminal repeats

VRE Vancomycin-resistant enterococcus

vs Versus, in contrast to, in comparison with

VSD Ventricular septal defect

VZV Varicella-zoster virus

WBCs White blood cells, leukocytes

WHO World Health Organization

WNL Within normal limits

WPW Wolff-Parkinson-White syndrome

WSJ Wall Street Journal
X-R X-linked recessive

SYMBOLS

• Bullet point
< Less than
<<< Much less than
≤ Less than or equal to
? Uncertain
> More than
>>> Much more than

≥ More than or equal to
→ Leads to, results in
↔ Reversible
↓ Decrease, decreased, decreases, decreasing, reduces
↑ Increase, increased, increases, increasing
♀ Female, women
♂ Male, men
± About, approximately, circa
‡ See there
† Deceased, dead, mortality

Concise Dictionary of Modern Medicine

A Symbol for: 1. Actin 2. Adenine 3. Adrenaline 4. Adriamycin 5. Alanine 6. Ampere 7. Amphetamine 8. Anaphylaxis 9. Angstrom 10. Anion 11. Anterior 12. Aqueous 13. Area 14 Asparagine 15. Atrium 16. Auricular 17. Axis

a Symbol for: 1. Absorptivity 2. Acceleration 3. Chemical activity 4. Arterial blood gas 5. *atto-*

a- absent, deficient, lack of–eg, atrophy

AA 1. Acute appendicitis 2. Alcoholics Anonymous, see there 3. Amino acid, see there, also written 'aa' 4. Amyloid-associated 5. Aplastic anemia 6. Arachidonic acid 7. Ascending aorta 8. Ascorbic acid 9. Atomic absorption 10. Autoanalyzer

aa of each

A antigen A major blood group–ABO antigen that defines the blood type A, which assumes the codominant allele at the ABO locus is either blood group A or H. Cf B antigen, Bombay phenotype, H antigen

AAA 1. Abdominal aortic aneurysm 2. Aminoadipic acid 3. Aneurysm of ascending aorta

the three As IMMUNOLOGY A term that simplistically addresses management of Pts with insect allergy: adrenaline, avoidance, allergist

4 As The four As–house officer GRADUATE EDUCATION A popular term for those qualifications–ability, affability, attitude, availability that indicate the potential for a successful house officer–resident in training

a wave A component of a normal jugular phlebogram produced by retrograde transmission of the pressure pulse, corresponding to atrial systole; the a wave begins before the first heart sound peaking at the moment the first sound begins; abnormalites of the a wave indicate cardiopulmonary disease; it may disappear in A Fib, be 'swallowed' in the *v-y* descent of a prolonged P-Q interval, it may be very large–giant when the atrium is contracting against resistance–eg tricuspid valve stenosis or atresia, pulmonary HTN or pulmonary edema or bear a presystolic 'notch'–pulmonary edema. Cf Cannon wave

A4 Amyloid β peptide

A68 A neuronal antigen–recognized by a monoclonal antibody Alz-50, which is present in the fetus by 32 wks of development, and disappears by age 2; it is present in brain tissue and CSF, is a major subunit of paired helical filaments of Alzheimer's dementia and is derived from LMW tau protein–possibly Tau 69, but is more phosphorylated

AAASPS CARDIOLOGY A clinical trial–African-American antiplatelet stroke prevention study–sponsored by the National Stroke Association, designed to assess stroke prevention treatment; specifically, the study compares the effectiveness of aspirin to Hoffmann-La Roche's Ticlid®–ticlopidine in preventing recurrent stroke in African-American Pts

AABB American Association of Blood Banks A professional, non-profit organization established in 1947 and dedicated to the education, formulation of standards, policy and other facets of transfusion medicine

AAD American Association of Dermatology

a-ADCO₂ ARTERIAL BLOOD GASES A value determined by capnography which corresponds to the difference between the arterial and alveolar–end-tidal CO_2, which is normally 2–3 mm Hg; the a-ADCO₂ can be widened by incomplete alveolar emptying, sampling error, V/Q abnormalities

AAFP American Association of Family Practice

AAN American Association of Neurology

A&Ox3 Alert & oriented to person, place, & time

AAO 1. American Association of Ophthalmology 2. American Association of Orthodontists 3. American Academy of Osteopathy, see there 4. American Academy of Otolaryngology.

AAOS American Academy of Orthopaedic Surgery

AAP 1. American Academy of Pediatrics 2. American Academy of Pedodontics 3. American Academy of Periodontists.

AAPCC Adjusted average per capital cost MANAGED CARE The funds a managed care plan receives from the CMS, formerly HCFA, to cover costs. See Capitation

Aarskog syndrome Aarskog-Scott syndrome, Faciodigitogenital syndrome An X-R or AD condition characterized by short stature, facial abnormalities–ocular hypertelorism, broad nose, long philtrum, hypoplastic maxilla, low-set floppy ears, anteverted nostrils, broad upper lip, curved linear dimple below the lower lip, ptosis, down-slanted palpebral fissures, ophthalmoplegia, strabismus, hyperopic astigmatism, large cornea, lop-ears, cleft lip/palate, cryptorchidism, proximal finger joint hyperextensibility, flexed distal finger joints, stretchable skin, shawl and/or saddle-bag scrotum, widow's peak hair, brachydactyly, digital contractures, clinodactyly, mild syndactyly, transverse palmar crease, lymphedema of the feet, ligament laxity, osteochondritis dissecans, genu recurvatum, flat feet, cervical spine hypermobility, odontoid anomaly, macrocytic anemia, hemochromatosis, hepatomegaly, portal cirrhosis, imperforate anus, rectoperineal fistula, interstitial lung disease, sternal deformity

Aase-Smith syndrome An AD condition characterized by hypoplastic BM and thumbs with 3 rather than 2 phalanges, which may be accompanied by leukopenia, radial hypoplasia, narrow shoulders, retinopathy, late closure of fontenels, mild growth retardation, cleft lip/palate, hepatosplenomegaly, and cardiac anomalies–eg, VSD

AASK CARDIOLOGY A clinical trial–African American Study of Kidney Disease and Hypertension Pilot Study that studied the relationship between anti-hypertensive therapy and progression of renal disease See Hypertension

Ab IMMUNOLOGY Portion of antigen that interacts with an MHC molecule

ab- off, away from: abduct

abacavir Ziagen™, zintavir, 1592 AIDS An HIV nucleoside reverse transcriptase inhibitor–NRTI/integrase inhibitor similar to AZT ADVERSE EFFECTS Life-threatening allergic reactions–fever, dyspnea, rashes, intraoral blistering, nausea, headache. See AIDS, Nucleoside reverse transcriptase inhibitor, Reverse transcriptase

abandonment MALPRACTICE A physician's unilateral severance–or 'negligent termination' of a professional relationship with a Pt–so that the Pt may continue care with another physician, either without the Pt's consent, or without reasonable or adequate notification, and/or under circumstances in which the need for continuing medical care remains. See Malpractice SOCIAL MEDICINE The desertion of a child by a parent or adult caretaker with no provisions for reasonable child-care or apparent intention to return; a child may be considered abandoned if left alone or with siblings or nonrelated and unsuitable individuals; abandonment is considered a form of physical neglect

abasia NEUROLOGY Lack of motor control or muscle coordination while walking

abbreviated injury scale EMERGENCY MEDICINE A numerical scoring system for rating organ damage sustained during trauma, which is based on physical examination, operative reports, and autopsy results. See Injury Severity Score

abbreviated new drug application PHARMACOLOGY An application made in the US by a pharmaceutical company requesting authority to market a 'new' drug for which both its therapeutic indications and formulation were previously approved by the FDA in another similar drug. See Me too drug

ABC 1. Aneurysmal bone cyst, see there 2. Antigen-binding capacity-immunology 3. Apnea, bradycardia, cyanosis 4. ATP-binding cassette 5. Avidin-biotin complex 6. Axiobuccocervical-dentistry

ABC sequence EMERGENCY MEDICINE The first level of life support measures used in CPR according to the 'American school' of cardiology, is the simple mnemonic of 'airway, breathing, circulation'; advanced life support then continues as DEF–'drugs, EKG, fibrillation'. See C-A-B

ABC/DEF SEQUENCE

AIRWAY	Ensure airway patency–clear bronchotracheal tree
BREATHING	Breathing by intermittent positive pressure ventilation
CIRCULATION	Compress chest at 60/minute
DRUGS/FLUIDS	Place IV line
EKG	Examine/monitor EKG
FIBRILLATION	Defibrillate ppp

ABC study CARDIOVASCULAR DISEASE A series of trials initiated by the ABC–Association of Black Cardiologists to study angiotensin II receptor blockers in hypertensive African-Americans–eg, the efficacy and tolerability of candesartan cilexetil–Atacand® See Hypertension

ABCD DERMATOLOGY A simple mnemonic for the clinical features of early melanomas, where A refers to asymmetry, B to border irregularity, C to variegation of color, and D to a diameter ≥ 6 mm

ABCD trial CARDIOVASCULAR DISEASE 1. A prospective, randomized, blinded, clinical trial–Appropriate Blood Pressure Control in Diabetes–which compared, in type 2–NIDDM diabetics, 1. The efficacy of nisoldipine, a CCB, with enalapril, an ACE inhibitor, as 1st-line antihypertensive agents and 2. The effects of moderate BP control–target–80-89 mm Hg vs 'tight' BP control–target–75 mm Hg on cardiovascular M&M. See Enalapril, Nisoldepine 3. A clinical trial–Alternans Before Cardioverter-Defibrillator

abciximab ReoPro® CARDIOLOGY A proprietary–Eli Lilly monoclonal antibody directed against platelet glycoprotein IIb/IIIb, as an adjunct for PTCA or atherectomy to ↓ coronary artery ischemia, in Pts with unstable angina not responding to conventional medical therapy See EPILOG

abdominal abscess A localized abdominal suppuration, caused by perforation or postop complications MANAGEMENT Percutaneous or open surgical drainage

abdominal adhesions Fibrous intraabdominal scars 2° to surgery, C-sections, chronic inflammation

abdominal angina Chronic mesenteric ischemia, see there

abdominal aorta ANATOMY The portion of the aorta that begins below the diaphragm, extends to the bifurcation of the iliac arteries, and supplies blood to the abdominal viscera, pelvic organs and legs BRANCHES Inferior phrenic, lumbar, celiac trunk, superior mesenteric, inferior mesenteric, middle adrenal, renal, testicular and ovarian arteries

abdominal aortic aneurysm A focal aortic dilation of ≥ 50% ↑ in diameter, accompanied by distension and weakened aortic wall EPIDEMIOLOGY Incidence is rising 12/10⁵–1951; 36/10⁵–1980; ♂:♀ ratio 2:1; results in 15,000 deaths/yr–US 1988, population ± 230 x 10⁶ CLINICAL Sudden, severe abdominal pain radiating to back DIAGNOSIS 100% accuracy with ULTRASONOGRAPHY–CONS Difficult in obese subjects, or with excess bowel gas, periaortic disease, technique does not document proximal or distal ends of aneurysm for surgery TREATMENT ≥ 5 cm in diameter AAAs should be repaired; therapy of smaller AAAs is controversial CONTRAINDICATIONS TO ELECTIVE RECONSTRUCTION MI in last 6 months, intractable angina or CHF, severe pulmonary insufficiency with dyspnea at rest, severe renal insufficiency, life expectancy of ≤ 2 yrs PROGNOSIS Treatment mortality is ± 6%; 25–40% of untreated cases rupture in 5 yrs, with a 90% mortality if ≥ 5 cm

abdominal bath RADIOTHERAPY A treatment field from the diaphragm to the pelvis, which is used for abdominal lymphomas or ovarian CA

abdominal cavity ANATOMY A body space bounded superiorly by the diaphragm, laterally by the abdominal wall, inferiorly by the pelvis; the AC is arbitrarily separated from the pelvic cavity by an imaginary plane that passes across the superior opening of the pelvis CONTENT GI tract except esophagus and anus, kidneys, spleen, adrenal glands

abdominal colic CLINICAL MEDICINE A condition characterized by intense cramping or colicky pain, which may be accompanied by nausea and vomiting ETIOLOGY Urinary stones, far less commonly, heavy metal–arsenic, lead, thallium, mushroom, or organophosphate pesticide poisoning, or drug withdrawal. See Colic

abdominal examination CLINICAL MEDICINE A hands-on evaluation of the abdominal cavity to identify abnormalities, if any, based on any change in size, shape, consistency, or sound on percussion of the organs found therein. See Abdomen

abdominal guarding CLINICAL MEDICINE A spasm of the anterior abdominal wall muscles detected on physical examination, which is a 'protective response' to intraabdominal inflammation seen in Pts with appendicitis, diverticulitis or peritonitis

abdominal hernia Ventral hernia SURGERY A protrusion of loops of intestine, fat, or fibrous tissue through a defect or weakened region of the abdominal wall–eg, an umbilical hernia

abdominal hysterectomy GYNECOLOGY Surgical removal of the uterus through an incision in the abdominal wall. See Total abdominal hysterectomy. Cf Vaginal hysterectomy

abdominal-jugular reflux Abdominojugular reflux, hepatojugular reflux

CARDIOLOGY An ↑ in JVP that follows 10–30 secs of pressure on the periumbilical region, which is due to an ↑ in flow of blood from the abdominal veins into the right atrium

abdominal lipectomy Abdominal panniculectomy COSMETIC SURGERY The excision of fat and excess skin from the lower abdomen in Pts with major weight loss and loose abdominal skin folds. See Abdominoplasty, Liposuction

abdominal muscles CLINICAL ANATOMY The large muscles of the anterior abdominal wall–external oblique, internal oblique, rectus abdominalis, which help in breathing, support spinal muscles while lifting, and help maintain abdominal organs and GI tract in their normal position

abdominal obesity Androgenous obesity, truncal obesity PUBLIC HEALTH A clinical form of obesity which is more typical of ♂; those with AO waists > 40 inches had a 3 fold > risk of high cholesterol, were 4 times more likely to be in poor physical condition, and had a 7-fold ↑ risk of DM

abdominal-perineal resection SURGICAL ONCOLOGY The surgical excision of the lower rectum and anus with loss of the anal sphincter and a need for colostomy. See Colostomy

abdominal pregnancy OBSTETRICS An ectopic pregnancy in which a fertilized egg implants on the peritoneal surface of the abdominal cavity. See Ectopic pregnancy

abdominal tap Belly tap, paracentesis CRITICAL CARE MEDICINE A method for differentiating a surgical abdomen, ie that requiring surgery, from a 'non-surgical' abdomen, avoiding an unnecessary laparotomy INDICATIONS Blunt abdominal trauma alone or in combination with injuries to the head, thorax or the extremities or with concomitant substance abuse, acute pancreatitis, post-operative peritonitis, or peritonitis in children with a 2ⁿᵈ disease. Cf Pericardiocentesis

abdominal thrust maneuver EMERGENCY MEDICINE A maneuver used in drowning victims to remove fluids from the upper respiratory tract; there are no controlled studies that support its efficacy See Drowning. Cf Heimlich maneuver

abdominoplasty Tummy tuck COSMETIC SURGERY The excision of a large horizontal ellipse of skin and fat from the anterior wall of the lower abdomen, and an upper abdominal flap is stretched to the suprapubic incision and sewn in place; the umbilicus is exteriorized through an incision in the flap at the proper level RAISON D'ETRE Vanity COST $4500. See Cosmetic surgery

abduct *verb* To move away from the body

abducted Distal angulation of an extremity away from the midline of the body in a transverse plane and away from a sagittal plane passing through the proximal aspect of the foot or part, or away from some other specified reference point

abduction NEUROLOGY Movement of an extremity on a transverse plane away from the axis or midline. Cf Adduction

ABECB Acute bacterial exacerbation of acute bronchitis

ABER Auditory brainstem evoked response

aberration MEDTALK A defect, deviation, or irregularity PSYCHIATRY See Mental disorder

abetalipoproteinemia Bassen-Kornzweig syndrome A rare AR condition most common in Askanazi Jews CLINICAL Cerebellar ataxia, peripheral neuropathy, retinitis pigmentosa, fat malabsorption, steatorrhea, chronic diarrhea, anemia, FTT LAB Acanthocytosis, ↓ VLDL-cholesterol, ↓ LDL-cholesterol, absent apoB, when homozygous MANAGEMENT Medium-chain TGs, water-miscible vitamin E

ABG 1. Arterial blood gas 2. Axiobuccogingival–dentistry

ABI Ankle/brachial blood pressure index, see there

ability See Athletic ability, Mental ability

ability to practice medicine MEDICAL MALPRACTICE The cognitive capacity to make appropriate clinical diagnoses and exercise reasonable medical judgements, keep abreast of medical developments, refer Pts to appropriate providers, prescribe appropriate medications and other substances, maintain and review Pts' charts, and provide coverage for Pts. See Impaired health care worker

ablate *verb* To remove; excise

ablation CARDIOLOGY Ablation therapy The destruction, isolation or removal of an endometrial region linked to a particular arrhythmia INDICATIONS Atrial flutter/fibrillation, AV nodal reentry, AV reentry, ventricular reentry. See Balloon ablation, Radiofrequency catheter ablation ENDOCRINOLOGY The pharmacologic or surgical elimination of hormone-producing tissue. See Androgen ablation, Endometrial ablation GYNECOLOGY The stripping out of a tissue. See Cryoablation, Endometrial ablation MEDTALK Removal, excision SURGERY Removal, excision

ablation therapy See Ablation–cardiology

ablutophobia PSYCHOLOGY Fear of washing or bathing. See Phobia

ABN Advance beneficiary notice, see there

abnormal *adjective* Not normal, contrary to a usual structure, position or behavior

abnormal presentation OBSTETRICS Any position in which a part of the baby other than the crown of its head (cephalic presentation), emerges first EXAMPLES Breech, face, shoulder. Cf Breech presentation, Brow presentation, Face presentation, Posterior presentation, Shoulder presentation

ABO antibody See ABO system

ABO incompatibility TRANSFUSION MEDICINE A type of blood incompatibility, in which certain recipients–eg genotypes AO, BO, OO of donated blood products-usually understood to mean packed RBCs, has antibodies to antigens–A or B on the surface of RBCs being transfused. See ABO system

ABO system ABO blood group system TRANSFUSION MEDICINE The major alloantigen system in humans, based on 3 carbohydrate antigens expressed on RBCs, encoded on chromosome 9q34 SIGNIFICANCE ABO blood groups determine who can donate blood to, or accept blood from, whom; type A or AB blood causes an immune reaction in type Bs; type B and AB blood cause an immune reaction in type As; conversely, type O blood has no A or B antigens–type Os are universal donors; since AB blood already produces both antigens, type ABs can accept any of the other blood types without suffering an immune reaction, and are thus known as universal recipients, as he/she forms antibodies to neither A nor B antigens. See Blood group antigens, Lewis antigen, Transplantation, Universal donor, Universal recipient

abortifacient PHARMACOLOGY An agent that induces the expulsion of an embryo or fetus; a drug, herb other chemical that dilates of the cervix and causes the uterus to contract resulting in a spontaneous termination of pregnancy

abortion OBSTETRICS The premature expulsion of the products of conception–POCs from the uterus of the embryo or of a nonviable fetus CLINICAL Uterine contractions, uterine hemorrhage, softening and dilatation of cervix, presentation or expulsion of all or part of the POCs STATISTICS Rate––0.5% of ♀ age 15–44 Netherlands; 1.4% in UK; 2.7% US; 6% Cuba; 18% Russia;

where abortions are illegal, the rate of complications are much higher. See Complete abortion, Criminal abortion, Early abortion, Elective abortion, Habitual abortion, Incomplete abortion, Induced abortion, Inevitable abortion, Late abortion, Late-term abortion, Medical abortion, Missed abortion, Partial birth abortion, Prostaglandin-induced abortion, Recidive abortion, Recurrent abortion, Saline abortion, Septic abortion, Spontaneous abortion, Threatened abortion, Urea abortion, Vacuum abortion

abortion pill See Contragestive, Oral contraceptive, RU-486

abortion ratio OBSTETRICS The number of spontaneous and induced abortions/100 live births/yr. See Abortion

abortion trauma syndrome PSYCHOLOGY '...*a medical syndrome that does not exist.*' ABS is a 'nonentity' presumed to have been created by 'pro-life'–anti-abortion activists, and alleged to occur in women who have undergone abortion, who allegedly suffer deleterious physical and emotional consequences after abortion. See Abortion

above-the-knee amputation AKA SURGERY An 'elective' procedure used for severe–gangrenous peripheral vascular disease, which is commonly required in older diabetics; AKA is preferred to below-the-knee amputation in treating peripheral vascular disease if the gangrene extends above the malleoli, as the AKA has a higher healing rate–85-100%, better rehabilitation with a prosthesis, and fewer complications. Cf Below-the-knee amputation, Potato chip operation

ABP Doxorubicin, bleomycin, prednisone ONCOLOGY A combination chemotherapy regimen used to treat Pts with NHL. See Combination chemotherapy

ABPA Allergic bronchopulmonary aspergillosis

abracadabra therapy A type of placebo effect, which may effect a cure in certain conditions in mentally impressionable subjects. See Placebo effect

abrasion DENTISTRY The wearing away of enamel DERMATOLOGY A scrape; superficial injury to a mucocutaneous surface caused by rubbing or scraping from a sharp object, resulting in an area of body surface denuded of skin or mucous membrane PATHOLOGY The wearing away of a substance or structure–such as the skin or the teeth through a mechanical process

abreaction PSYCHIATRY Emotional release or discharge associated with remembering and resolving repressed mental trauma experienced and repressed in childhood. See False memory

abruptio placentae Ablatio placentae, abruptio, premature separation of placenta OBSTETRICS The premature separation of the placenta from its site of implantation in the endometrial before the delivery of the fetus; some degree of AP occurs in 1:85 deliveries; severe AP with total separation of the placenta is an obstetric emergency which occurs about 1 in 500-750 deliveries, and is accompanied by fetal death CLINICAL Constant abdominal and/or back pain, irritable, tender or hypertonic uterus, vaginal bleeding—seen in most; 30% are asymptomatic PREDISPOSING FACTORS Preeclampsia-eclampsia, chronic HTN, DM, chronic renal disease; mechanical causes are rare–1-5% and include transabdominal trauma, sudden decompression, as occurs in the delivery of a 1st twin or rupture of membranes in hydramnios, or traction of a short placenta MANAGEMENT Expectant therapy if fetus is immature and bleeding limited, treat shock if present, vaginal delivery if possible, C-section COMPLICATIONS DIC, acute cor pulmonale, renal cortical and tubular necrosis, uterine apoplexy, transfusion hepatitis

abscess INFECTIOUS DISEASE A local accumulation of pus in tissues, organs or confined spaces, almost invariably due to an infection MICROBIOLOGY The bacteria cultured from an abscess are largely a function of the region affected. See Abdominal abscess, Alveolar abscess, Amebic abscess, Apical abscess, Areolar gland abscess, Brain abscess, Brodie's abscess, Cold abscess, Collar button abscess, Collar-stud abscess, Crypt abscess, Intracranial abscess, Kogoj's abscess, Microabscess, Munro's microabscess, Perforating abscess, Peritonsillar abscess, Ring

abscess, Satellite abscess, Stellate abscess, Walled abscess

abscess scan A nuclear scan in which a Pt's own WBCs are tagged with radioactive indium, then injected into the bloodstream; the course of the WBCs can then be mapped using a gamma camera to provide an image that shows the location of the radioactive WBCs–infection or inflammation; AS is used to detect a hidden source of bacterial infection

absence Absence attack, absence seizure, petit mal epilepsy NEUROLOGY A common form of childhood epilepsy characterized by episodic arrest of sensation and voluntary activity CLINICAL Transient loss of contact with the environment-eg brief staring spell, minimal motor manifestations, ± decline in school performance DIAGNOSIS 3-mins of hyperventilation test may elicit an 'absence' MANAGEMENT Trimethadione, ethosuximide. Cf Grand mal seizure

absent periods Amenorrhea, see there

absolute alcohol LABORATORY Ethanol that contains ≤ 1% water by weight; it is a critical working reagent in certain areas of the clinical laboratory, in particular histopathology, where it is required in certain steps of embedding tissues in paraffin

absolute contraindication DECISION-MAKING A reason for *not* performing a particular therapeutic intervention which is so compelling or carries such a grave risk that its performance would be reasonably regarded as constituting malpractice. See Contraindication, Patient selection

absolute risk EPIDEMIOLOGY The observed or calculated probability of occurrence of an event, X, in a population related to exposure to a specific hazard, infection, trauma; the number of persons suffering from a disease when the exposed population is known with certainty. Cf Relative risk

absorbable suture see Catgut, Suture, Synthetic absorbable suture

absorbed dose RADIATION PHYSICS The energy imparted by ionizing radiation per unit mass of irradiated material–eg, tissue, defined as an SI unit, the gray–Gy, which corresponds to the 100 rads, the old unit for absorbed dose

absorption The process of taking in IMMUNOLOGY Agglutinin absorption A lab technique consisting of either removal of an antibody from serum by adding its cognate antigen, or removal of an antigen by adding its cognate antibody; absorption allows an antiserum to be purified by removing unwanted immunoglobulins, or may be used to 'fish' for an antigen or antibody of interest MEDTALK The uptake of material across a surface–eg, epidermis, GI mucosa, renal tubules PHARMACOLOGY The process by which a drug enters the body and is available for therapeutic activity; agents administered IV are absorbed completely; agents administered extravascularly are usually absorbed by passive diffusion of the nonionized drug fraction; a drug's concentration in the blood is a function of the ratio of absorption to elimination. See Accumulation

absorptive dressing WOUND CARE A dressing designed to absorb drainage from a wound

abstinence PSYCHOLOGY Self-denial; voluntary refraining from indulgences–eg, eating, drinking inebrients, sexual activity. See Continence, Fast

abulia NEUROLOGY ↓ action and thought coupled with indifference or lack of concern about the consequences of action PSYCHIATRY Lack of will or motivation, often expressed as inability to make decisions. See Anhedonia

abuse PUBLIC HEALTH A behavior defined as '...*the willful infliction of physical pain, injury, or mental anguish, or willful deprivation by a caretaker of services necessary for physical and/or mental well being.*' See Battered child, Battered wife syndrome, Child abuse, Child sexual abuse, Domestic violence, Elder abuse, Emotional abuse, Institutional abuse, Medical student abuse, Physical abuse, Psychological abuse, Sexual abuse, Solvent abuse, Substance abuse, Super Bowl Sunday abuse, Verbal abuse

abuse-liable THERAPEUTICS Pharmacologics, both illicit–eg, heroin and legal–eg, methamphetamines with potential for abusive dependency. See Drug seeking, Gateway drug

abusive behavior PUBLIC HEALTH Any of various behaviors–aggressive, coercive or controlling, destructive, harassing, intimidating, isolating, threatening–which a batterer may use to control a domestic partner/victim. See Domestic violence

ABVD Doxorubicin, bleomycin, vinblastine, dacarbazine ONCOLOGY A 'salvage' chemotherapy regimen used for Pts who have had disease–eg, lymphoma relapse after 1° RT or chemotherapy. See CHOP, MOPP, Salvage chemotherapy

AC joint Acromio-clavicular joint ORTHOPEDICS The point where the acromium process of scapula and the distal end of the clavicle meet, which is the location of most frequent shoulder separations.

aca LAB MEDICINE The very first automated random-access analyzer. See Random access analyzer

ACADEMIC CARDIOLOGY A clinical trial–Azithromycin in Coronary Artery Disease: Elimination of Myocardial Infection with Chlamydia

academic detailing THERAPEUTICS The use of educational 'props' by pharmaceutical companies and representatives–drug 'reps' to improve drug prescribing practices. Cf Detailing

academic medical center ACADEMIA A health care organization that is often linked to a medical school and hospital complex Missions: teaching of medical students and physicians in training; research; 3° Pt care in close affiliation or as part of a degree-granting university. See Ivory tower

academic progress SOCIAL MEDICINE The constellation of school achievements of children < age 18

academic question Moot question VOX POPULI An issue that does not require an answer or adjudication because it is irrelevant to the central issue being considered

acalculia NEUROLOGY Loss of a facility with arithmetic calculation

acalculous cholecystitis SURGERY Gallbladder inflammation without gallstones ULTRASONOGRAPHIC FINDINGS Sludge, gallbladder distension, wall thickening, wall lucencies, localized pericholecystic fluid

acanthamebiasis *Acanthamoeba* infection INFECTIOUS DISEASE An infection by *Acanthamoeba* spp, characterized by pustules, vasculitis, granulomatous amebic encephalitis, and granulomas of skin; it is common in weak or immunocompromised subjects, often leading to death HIGH-RISK GROUPS AIDS, alcohol abuse, DM, immunosuppression-transplantation-related, cancer–leukemia, lymphoma, malnutrition TREATMENT Uncertain; a case of *A rhysodes* responded to topical chlorhexidine and ketoconazole, systemic pentamidine–IV and oral itraconazole

Acanthamoeba A genus of free-living pathogenic amoebas SOURCES tap water, dust, soil, sewage, air conditioning units. See Acanthamebiasis. Cf *Leptomyxid, Naegleria*

acanthocyte Acanthrocyte HEMATOLOGY An RBC with a irregularly spaced, pointed spicules, of varying shape and number–Greek, ακαντηο, spike; acanthocytes are seen in abetalipoproteinemia–hereditary acanthocytosis, as well as in alcoholic cirrhosis with hemolytic anemia, hemolytic anemia due to pyruvate kinase deficiency, in neonatal hepatitis, after heparin administration or splenectomy. See Red blood cells

acanthocytosis 1. Abetalipoproteinemia, see there 2. A general term for the presence of acanthocytes in the peripheral blood. See Acanthocytes

acantholysis DERMATOLOGY The pathologic disruption of intercellular bridges between keratinocytes in the squamous epithelium of mucocutaneous surfaces, resulting in intraepithelial vesicles DISEASES WITH Pemphigus vulgaris, benign familial pemphigus (Hailey-Hailey disease), Darier's disease, staphylococcal scalded-skin syndrome. See Pemphigus vulgaris, Tombstone appearance

acanthosis nigricans DERMATOLOGY A condition characterized by hyperkeratosis and papillomatosis FORMS Malignant–associated with CA–eg, gastric CA, but also rarely Hodgkin's disease, or bone CA; inherited; endocrine–associated with pituitary tumors, polycystic ovary syndrome, insulin resistance, idiopathic

acapnia Hypocapnia; ↓ CO_2 in blood. Cf Hypercapnia

Acapulco gold DRUG SLANG Marijuana from southwestern Mexico See Marijuana

ACAS CARDIOLOGY A clinical trial–Asymptomatic Carotid Atherosclerosis Study which evaluated the 5-yr risk of fatal and non-fatal stroke-primary outcome in Pts with asymptomatic but severe carotid atherosclerosis. See Carotid stenosis

AC-BC gap AUDIOLOGY A normal finding in pure tone audiometry, in which the threshold for hearing sounds conducted through the air–air conduction, AC is lower than the threshold for sounds conducted through the bones of the skull–bone conduction, BC; the relationship changes dramatically in conductive hearing loss, where the threshold for BC is lower than that for AC. See Pure tone audiometry

accelerated fractionation RADIATION ONCOLOGY The delivery of radiation at a rate of accumulation that is up to 50% faster than that of standard fractionation, which substancially ↓ duration of therapy, and ↓ potential for tumor repopulation between fractions. See Fractionation. Cf Hyperfractionation

accelerated leukemia A progressive phase of myelogenous leukemia, in which immature, abnormal WBCs in the BM and blood are higher than in chronic leukemia, but not as high as in a blast phase

accelerated phase ONCOLOGY A progressive phase of CML characterized by immature, defective WBCs in BM and peripheral blood which is higher in the chronic phase, but less than in the blast phase. See Blast crisis

acceleration-deceleration injury EMERGENCY MEDICINE Brain injury due to blows, MVAs, etc, which is a major cause of cerebral morbidity, related to abrupt movement and deformation of the brain in the cranial cavity

acceptable daily intake NUTRITION An estimate of the amount of a food that a person can ingest daily over a lifetime humans without appreciable health risk, divided by an average person's lifespan. See Tolerable daily intake

access HEALTH CARE 1. The ability of an individual or group of individuals to obtain health insurance 2. The ability of an individual to obtain adequate or appropriate health care services; the availability of medical care to a Pt, which can be determined by location, transportation, type of medical services in the area, etc. See Direct access, Health care access INTENSIVE CARE See Vascular access

ACCESS TO HEALTH CARE, FACTORS IN

GEOGRAPHIC OR LOGISTIC FACTORS, eg rural communities have poor access to medical attention, due to a relative lack of providers

FINANCES or ability to pay for services and

OTHER FACTORS: ethnic, social and psychiatric aspects of the person seeking health care

access device INTENSIVE CARE Any device used to obtain access to the body's fluid compartments—eg, Hickman catheter, central line, PICC, Port-a-Cath, shunts and other devices. See Vascular access

accessible SURGERY *adjective* Reachable; a term used in the context of a lesion that can be managed by a standard surgical technique, as the lesion is not deep in the brain or adjacent to vital structures—ie, inaccessible

accession LAB MEDICINE *noun* An ordered test or group of tests on a particular; a specimen that has been formally received by a laboratory or other health care service and received an accession number *verb* To input the demographics for a particular Pt's analysis into an information system; to log or document receiving a specimen in the lab

accessory placenta Succenturiate, placenta, supernumerary placenta OBSTETRICS An extra dollop of placental tissue separate from the main placenta

accessory spleen *splen accessorius* Any of a number of small aggregates or masses of encapsulated splenic tissue located adjacent to the spleen or along the gastrosplenic ligament. See Spleen

accident MANAGED CARE An event that is unforeseen, unexpected, and unintended NEUROLOGY See Cerebrovascular accident PUBLIC HEALTH An unintentional and/or unexpected event or ocurrence that may result in injury or death; MVAs are the most common cause of accidental death in developed countries. See Hit&run accident, Motor vehicle accident

accident-prone *adjective* Referring to a person's real or percieved tendency to suffer from accidents of various types

accidental *adjective* Unexpected; by chance

accidental bodily injury HEALTH INSURANCE Physical injury sustained from an accident

accidental hypothermia EMERGENCY MEDICINE '...*an unintentional decline in the core (body) temperature below 35°C*,' a temperature at which thermoregulatory systems begin to fail, as compensatory responses intended to reduce heat loss through conduction, convection, evaporation, radiation, and respiration are limited ETIOLOGY Alcoholism, drug addiction, mental illness, cold water immersion, winter sports/occupations CLASSIFICATION Mild, body temperature 32.2–35°C; moderate, 28–32.2°C; severe, < 28°C CLINICAL Disorientation, aphasia, amnesia, prolonged coma MANAGEMENT Rewarming—passive if mild; active rewarming if moderate or severe. See Rewarming

acclerated hypertension Malignant hypertension, see there

Accolate® Zafirkulast, see there

accommodation OPHTHALMOLOGY The automatic adjustment of the lens curvature, resulting in a change in the focal length of the eye, which brings images of objects from various distances into focus on the retina; the ability of the eye to focus at various distances, by changing lens shape. See Reasonable accommodations

accommodation list LAB MEDICINE A list of commercial in non-FDA-approved in vitro tests—IVTs, which are regarded as gold standards and used to guide clinical decisions. See Gold standard, Home brew product, Standard of care test

accountability A condition in which individuals who exercise power are constrained by external means and internal norms MEDICAL ETHICS The extent to which a person is answerable to a higher authority

accountable health partnership MANAGED CARE A competing economic or for-profit unit—eg physicians considered as an aggregate, or competing hospitals, that was proposed as a key component of any other 'universal' health plan. See Clinton Plan, Managed competition

accreditation GRADUATE EDUCATION The process required of a hospital or medical center for it to act as an 'official'—ie accredited training program for interns, residents, and fellows

ACCT CARDIOLOGY A clinical trial—Amlodipine Cardiovascular Community Trial—that evaluated the effect of sex and age on response to the antihypertensive, amlodipine. See Amlodipine, Antihypertensive, Hypertension

Accu-Chek® system POC TESTING A blood glucose monitoring system used for home monitoring of glucose with menu-driven screens and analysis of lifestyle data—eg, exercise, which may affect health and glucose control. See Diabetes mellitus, Glucose

accumulation A build-up, assembly; the process of accumulating, collecting together THERAPEUTICS The amount of drug in the body relative to the amount lost by elimination, which is a function of ratio of the dose interval and the drug's half-life; accumulation continues until the amount eliminated per dosage interval is equal to the amount administered during that time. Cf Absorption

accumulation period HEALTH INSURANCE The period during which an insured incurs eligible medical expenses to satisfy a deductible. See Deductible

Accupril® Quinapril, see there

AccuProbe® UROLOGY A cryotherapy device for treating prostate CA, and possibly, breast CA. See Cryotherapy

accuracy LAB MEDICINE The extent to which a value from a test reflects or agrees with the reference value of the analyte being tested, measured statistically by standard deviations; the proportion of correct outcomes of a method—often used interchangeably with concordance. See Coefficient of variation, Diagnostic accuracy, Inaccuracy, Two-by-two table. Cf Precision

Accutane Isotretinoin, see there

Accutane baby syndrome Vitamin A embryopathy, see there

Accuzyme™ WOUND CARE An enzymatic ointment for debriding acute and chronic wounds. See Wound care

ACE Angiotensin-converting enzyme, see there. Cf ACE inhibitors, also 1. Acute Care for Elderly 2. Adriamycin, cyclophosphamide, etoposide 3. Alcohol, chloroform, ether—obsolete anesthetic cocktail 4. American College of Epidemiology

Ace bandage Ace wrap ORTHOPEDICS A proprietary elastic bandage used to ↓ swelling and protect contused joints; if placed too tightly, may ↓ circulation and cause pain and paresthesia

ACE inhibitor Angiotensin-converting enzyme inhibitor PHARMACOLOGY Any of a family of drugs that are used to manage essential HTN, ↓ CHF-related M&M PROS ACEIs are cardioprotective and vasculoprotective; cardioprotective effects include improved hemodynamics and electric stability, ↓ SNS activity and ↓ left ventricular mass; vasculoprotective benefits include improved endothelial function, vascular compliance and tone, and direct antiproliferative and antiplatelet effects; ACEIs also stimulate PG synthesis, ↓ the size of MIs, ↓ reperfusion injury and complex ventricular arrhythmias; ACEIs are the treatment of choice in CHF with systolic dysfunction; they are vasodilators which ↓ preload and afterload; ACEI-induced ↓ in angiotensin II inhibits the release of aldosterone, which in turn ↓ sodium and water retention which, by extension, ↓ preload; ACEIs improve hemodynamics of CHF by ↓ right atrial pressure, pulmonary capillary wedge pressure, arterial BP, as well as pulmonary and systemic vascular resistances; ACEIs ↑ cardiac and stroke indices by the left ventricle and ↓ the right ventricular end-diastolic volumes, thereby resulting in ↑ cardiac output, while simultaneously ↓ cardiac load and myocardial O_2 consumption; ACEIs also downregulate the

SNS, which is linked to the pathogenesis of CHF ADVERSE EFFECTS Idiopathic–eg, rashes, dysgeusia, BM suppression; class-specific–eg, hypotension, renal impairment, hyperkalemia, cough, angioneurotic edema, the latter 2 of which are mediated by small vasoactive substances, eg, bradykinin, substance P, and PG-related factors

ACE INHIBITOR EFFECTS IN HEART DISEASE

CARDIOPROTECTIVE EFFECTS

- Restores balance between myocardial O_2 supply & demand
- Reduces left ventricular preload and afterload
- Reduces left ventricular mass
- Reduces sympathetic stimulation

VASCULOPROTECTIVE EFFECTS

- Antiproliferative & antimigratory effects on smooth muscle & inflammatory cells
- Antiplatelet effects
- Improved arterial compliance and tone
- Improved and or restored endothelial function
- Antihypertensive
- Possibly, antiatherosclerotic effect

After Lonn et al, 1994

ACE2 Angiotensin-converting enzyme 2 THERAPEUTICS An ACE-like enzyme expressed in vascular endothelial cells of the heart and kidney PHYSIOLOGY ACE2 appears to be a critical regulator of heart function and may counteract ACE activity. See ACE

acebutolol Sectral® CARDIOLOGY A β_1-blocker for managing HTN and tachydysrhythmias, effective in most other clinical uses of beta blockers ADVERSE EFFECTS Antinuclear antibodies; SLE-like disease is rare. See β-blocker

acellular vaccine IMMUNOLOGY A vaccine consisting of immunogenic parts of pathogens, but not whole cells. See Vaccine

acenocoumarol A fast-acting vitamin K-antagonist anticoagulant

acentric chromosome A chromosome lacking a centromere, which is essential for division and retention of the chromosome in the cell

aceruloplasminia METABOLIC DISEASE A condition characterized by progressive extrapyramidal signs, cerebellar ataxia, dementia, and DM associated with ↓ plasma ceruloplasmin due to a mutation in the ceruloplasmin gene; Pts have accumulation of iron in glia and neurons, especially in the basal ganglia and dentate nucleus, liver cells, pancreatic islets MANAGEMENT Chelation with deferoxamine; possibly administration of ceruloplasmin. See Ceruloplasmin. Cf Wilson's disease

ACES CARDIOLOGY A clinical trial–Azithromycin & Coronary Events Study NUTRITION An acronyn for the major dietary antioxidants–vitamin A and beta carotene, vitamin C, vitamin E, and selenium

acetaldehyde TOXICOLOGY The major metabolic product of ethanol, which is generated by ethanol dehydrogenase and subsequently metabolized to acetate by aldehyde dehydrogenase, and responsible for drinking-alcohol's toxic effects

acetaminophen Paracetamol–UK, Tylenol® PAIN MEDICINE An OTC analgesic–headache, muscle, joint pain and antipyretic which lacks anti-inflammatory activity THERAPEUTIC RANGE 10-25 mg/L CRITICAL VALUE ≥ 300 mg/L TOXICITY Overdose can cause fatal liver failure

acetazolamide Dazamide®, Diamox® THERAPEUTICS A heterocyclic sulfonamide used to manage respiratory acidosis by inhibiting renal carbonic anhydrase, which ↑ renal excretion of Na⁺, K⁺, and bicarbonate, and ↓ ammonia excretion; acetazolamide is also used to ↓ fluid retention in CHF, control 2° glaucoma and preoperatively in acute angle-closure glaucoma, epilepsy EFFECT ↓ Serum pH; ↑ urine pH THERAPEUTIC RANGE 10–15 μg/mL

TOXIC RANGE ≥ 20 μg/mL

acetic acid PHARMACOLOGY A pharmaceutical necessity containing 36% $C_2H_4O_2$ by weight; it is bactericidal, spermatocidal, and used in vaginal douches for managing *Trichomonas, Candida, Haemophilus* infections

acetic acid solution Lugol's solution, see there

acetone ENDOCRINOLOGY A ketone body *normally* present in scant amounts in the urine and serum of normal individuals produced by oxidation of fats; ketones ↑ in DM, DKA, starvation. See Ketone body

acetowhite lesion GYNECOLOGY A whitish patch on the uterine cervix when it is 'painted' with 5% acetic acid–vinegar, the whiter the lesion, the greater the hyperkeratosis. See Colposcopy, HPV

acetyl-CoA Acetylcoenzyme A METABOLISM A coenzyme derivative in the metabolism of glucose and fatty acids that contributes substrates to the Krebs cycle; acetyl CoA provides the acetyl unit for multiple biochemical reactions and plays a central role in intermediary metabolism–synthesis, catabolism, or use of nutrients for energy production and growth. See Citric acid cycle

acetylcholine NEUROPHYSIOLOGY An acetic acid ester of choline-a substance that functions as a major cholinergic neurotransmitter released from vertebrate neurons into the synaptic spaces after stimulation from the PNS ACTION Vasodilation, cardiac inhibition, GI peristalsis; it is involved in the control of thought, mood, sleep, muscles, bladder, sweat glands

acetylcholine receptor antibody AChR antibodies, motor end plate antibody CLINICAL IMMUNOLOGY A group of antibodies that are reactive with epitopes other than the binding site for acetylcholine or α-bungarotoxin; AChR-binding antibodies wax and wane as a function of disease severity, and block and destroy acetylcholine receptors, causing muscle weakness; although serum levels of ARAs correlate poorly with severity of weakness, MG may improve with immunosuppressive therapy SPECIMEN Serum at room temperature REF RANGE < 0.03 nmol/L METHOD Radioreceptor assay COMMENTS Failure to maintain specimen at room temperature interferes with results; Pts undergoing thymectomy, thoracic duct drainage, immunospressive therapy or plasmapheresis may have ↓ levels. See Anti-striated muscle antibody, Myasthenia gravis

acetylcholinesterase NEUROPHYSIOLOGY A hydrolase that metabolizes acetylcholine to acetyl and choline in the synaptic cleft, restoring it to a ground state, in preparation for the next nerve impule; acetylcholinesterase is found in the CNS, RBCs, motor endplates

acetylcoenzyme A See Acetyl-CoA

acetylcysteine Mucomyst THERAPEUTICS A mucolytic used to ↓ the viscosity of lung secretions, and is thought to improve O_2 delivery and consumption by replenishing glutathione stores; acetylcysteine is also used per os or IV as an antidote, and minimizes hepatocellular necrosis in Pts with fulminant liver failure; it has also been used for treating dry eye–keratoconjunctivitis sicca, and as an enema for managing bowel obstruction caused by meconium ileus. See Cystic fibrosis

acetylsalicylic acid Aspirin THERAPEUTICS An odorless, white, bitter drug which is water and alcohol-soluble and melts at 132-136°C used to ↓ pain, fever, inflammation, and to prevent blood clotting TOXICOLOGY Gastritis, GI bleeding, children taking aspirin during viral infections are at risk for Reye syndrome. See Reye syndrome

ACG 1. Acycloguanosine–antiviral 2. Ambulatory care group 3. Angiocardiography 4. Angle closure glaucoma 5. Apex cardiogram, see there

ACGME Accreditation Council for Graduate Medical Education

achalasia GI DISEASE An idiopathic dysmotility disorder, characterized by

Achard-Thiers syndrome

a loss of peristalsis in the distal ⅔ of the esophagus accompanied by impaired relaxation of the lower esophageal sphincter CLINICAL Dysphagia, chest pain, vomiting, heartburn IMAGING Plain films, air-fluid level in an enlarged esophagus; barium swallow, bird's beak tapering and, if prolonged, sigmoid esophagus MANAGEMENT Pneumatic dilation of esophagus; medication–eg nitroglycerin, CCBs–eg, nifedipine; surgery–eg, Heller myotomy. See GERD

Achard-Thiers syndrome ENDOCRINOLOGY A condition affecting post-menopausal ♀ characterized by DM, hirsutism and masculinization, which is caused by overproduction of androgens

acheilia Liplessness

acheiria Handlessness

achievement by proxy PSYCHIATRY The vicarious experiencing, by a parent or coach, of the success of a young athlete. Cf Munchausen by proxy

Achilles' reflex Ankle reflex, see there

Achilles tendinitis SPORTS MEDICINE A condition characterized by pain and swelling along the tendon sheath proximal to the calcaneus CLINICAL Stiffness with ankle movement, tenderness, crepitus IMAGING Usually nada, rarely, soft tissue thickening MANAGEMENT Physical therapy or brief immobilization, antiinflammatory agents

Achilles tendon *tendo calcaneo* ANATOMY The tendon that attaches the calf muscles to the calcaneous bone; the long common tendon of both venters–bellies of the calf muscles–gastrocnemius and soleus, which is the strongest tendon in the body; the AT begins below the posterior calf and is attached distally at the midline at the posterior aspect of the calcaneus CONDITIONS AFFECTING Insertional tendinitis, rupture

achiral *adjective* Referring to a molecule or material which is superimposable on its mirror image, ie does not display 'handedness'. See Chiral

achlorhydria Absence of HCl in gastric juice

achondroplasia PEDIATRICS An AD –80% are due to random mutations–short-limbed dystrophy characterized by upper arms and thighs that are disproportionately smaller than forearms and legs FEATURES Bowing of extremities, waddling gait, limited ROM of major joints, frontal bossing, short stubby fingers, moderate hydrocephalus, depressed nasal bridge, lumbar lordosis IMAGING Short, thick tubular bones, irregular epiphyseal plates; broad, cupped diaphyses; delayed epiphyseal ossification MANAGEMENT Osteotomies for severe deformities; Ilizarov procedure COMPLICATIONS Disk herniation may lead to acute paraplegia. See Frontal bossing, Ilizarov procedure, Osteotomy

achoo syndrome Photic sneeze reflex An AD condition affecting ± 25% of the population, which is characterized by paroxysms of sneezing when passing from dark to bright light or direct sunlight–aka helicociliosternutatogenic reflex

achromotrichia Absence or loss of pigment in hair

acid CHEMISTRY A chemical that can accept a pair of electrons or donate a proton *adjective* 1. Relating to an acid, acidic 2. Sour in taste *noun* Any usually water-soluble compound that donates a hydrogen ion–H⁺ or proton in a chemical reaction, or can accept a pair of electrons and combine with metals to form salts. See Acetylsalicylic acid, Alpha-lipoic acid, Amino acid, Arachidonic acid, Azelaic acid, Benzoic acid, Betulinic acid, Bile acid, Binary acid, Cis fatty acid, Conjugated linoleic acid, Deoxyribonucleic acid, Decosaenoic acid, Delta-aminolevulinic acid, Dextromethorphan acetic acid, Diethylenetriaminepentaacetic acid, Dimethylxanthenone acetic acid, Domoic acid, DMSA acid, Eicosapentaenoic acid, Ellagic acid, Essential amino acid, Essential fatty acid, Excitatory amino acid, Fatty acid, acid, Free-form amino acid, Fibric acid, Folic acid, Formic acid, Gamma-linolenic acid,

Glacial acetic acid, Homogentisic acid, Homovanillic acid, Hyaluronic acid, Hydrofluoric acid, 5-Hydroxyindole acetic acid, Hydrochloric acid, Ibotenic acid, Kainic acid, Lactic acid, Linoleic acid, Linolenic acid, Lipid-associated sialic acid, Methylmalonic acid, Mevalonic acid, n-3 fatty acid, Nalidixic acid, Nicotinic acid, Nitrilotriacetic acid, Nordihydroguaiaretic acid, Nucleic acid, Okadaic acid, Organic acid, Orotic acid, Oxolinic acid, Pangamic acid, Pantothenic acid, Paraaminobenzoic acid, Phosphoenolpyruvic acid, Phytanic acid, Picric acid, Polyunsaturated fatty acid, Retinoic acid, Ribonucleic acid, Saturated fatty acid, Uncoded amino acid, Uric acid. Cf Base DRUG SLANG Popular for LSD, see there

acid-fast stain MICROBIOLOGY A generic term for any of a number of special histologic stains–Ziehl-Neelsen, Kinyoun, hot carbolfuschin, et al, used to identify *Mycobacterium* spp

acid hemolysis test Ham test

acid infusion test Bernstein test

acid loading test NEPHROLOGY A test used to diagnose renal tubular acidosis, in which a Pt takes ammonium chloride tablets for 3 days to ↓ blood pH, followed by collection of serum and urine; RTA is classified based on GFR, serum K⁺, proximal H⁺ secretion, urine pH, and urinary anion gap

acid phosphatase A group of enzymes with broad specificity which transphosphorylates; APs are present in the prostate gland, semen, liver, spleen, RBCs, BM, platelets REF RANGE 0-1.1 Bodansky U ↑ Prostate CA or infarction, Paget's disease, Gaucher's disease, multiple myeloma. See PSA

acid-reflux disorder Any of a number of conditions caused by reflux of gastric secretions, on the esophageal mucosa–GERD, and laryngeal mucosa–reflux laryngitis and lungs–'reflux pneumonitis' DIAGNOSIS Postprandial heartburn relieved by antacids, endoscopy–± biopsy, pH monitoring MANAGEMENT-MEDICAL Change lifestyle-eg raise head of bed, ↓ bedtime snacks, ↓ fatty foods, smoking cessation, ↓ alcohol, use of antacids or alginic acid; histamine-receptor blockers–eg, cimetidine, ranitidine, omeprazole MANAGEMENT-SURGICAL Nissen fundoplication-widely preferred, Hill gastropexy, Belsey fundoplication. See Barrett's esophagus, GERD, Reflux pneumonitis

acid-base balance Acid-base homeostasis NEPHROLOGY A physiologic state defined by the pH in the serum, which normally is a tad alkaline; acids CO_2 and molecules containing H⁺, which are buffered by alkaline substances, primarily HCO_3, with some buffering by phosphate, proteins, Hb, and other substances; ↑ CO_2 or ↓ HCO_3 causes acidosis; conversely, ↓ CO_2 or ↑ HCO_3 causes alkalosis; if the defect is caused by changes in CO_2, it is a respiratory acidosis or respiratory alkalosis; if the defect is caused by changes in HCO_3 levels, it is a metabolic acidosis or metabolic alkalosis; the lungs regulate the levels of CO_2 through changes in the breathing rate; CO_2 is lost with faster breathing and ↑ when breathing slows, which provides fast but temporary regulation of body pH; major, long-term regulation of body pH occurs in the kidneys, which excrete acids, and excrete or create HCO_3 for use in the body

acidified serum lysis test Ham test

aciduria The presence of acid in urine. See Orotic aciduria

Acinetobacter spp BACTERIOLOGY A widely distributed bacterium found in moist hospital environments, which may establish itself in the respiratory flora and on the skin of Pts with prolonged hospitalization, often via contaminated medical instruments–eg, catheters and IV lines which introduce *Acinetobacter* to normally sterile sites; infections are generally nosocomial, occur in warmer seasons and involve the GU tracts and respiratory tract, wounds, and soft tissues

acini Plural of acinus, eg, milk-producing glands of breast

acinic cell carcinoma A low-grade salivary gland malignancy EPIDEMIOLOGY ACC comprises 1–3% of all salivary gland tumors; ♂ predominance; peaks in 3rd decade MANAGEMENT Usually adequately treated with *wide* local excision PROGNOSIS 5-yr survival, 90%; 20-yr survival, 55%

ACIP CARDIOLOGY A clinical trial–Asymptomatic Cardiac Ischemia Pilot Study that evaluated 3 therapeutic strategies[2] for ↓ myocardial ischemia during exercise testing. See Angina, Coronary artery disease, Myocardial infarction, Silent ischemia IMMUNOLOGY Advisory Committee on Immunization Practices

Aciphex® Rabeprazole, see there

acitretin THERAPEUTICS A retinoid used for CA prevention and psoriasis. See Retinoids

ACL 1. Anterior cruciate ligament, see there 2. Automated coagulation laboratory

ACL injury See Anterior cruciate ligament injury

ACLS certified EMERGENCY MEDICINE A person who is trained in providing advanced cardiovascular life support

ACME CARDIOLOGY A clinical trial–Angioplasty Compared to Medicine that compared outcomes of angioplasty vs. medical therapy on M&M in Pts with single-vessel coronary artery disease. See Angina, Angioplasty, Coronary artery disease, Percutaneous transluminal coronary angioplasty

acne Pimples, 'zits' DERMATOLOGY A general term for a primarily pubertal condition characterized by local inflammation of the pilosebaceous unit of the face, upper back and shoulders due to keratinaceous plugging of the sebaceous glands at the base of hair follicles. See Acne arthritis, Acne vulgaris, Cystic acne

acne rosacea Rosacea DERMATOLOGY An idiopathic skin disorder affecting light-skinned–of Celtic descent, middle-aged–♂ more severely, ♀ more commonly, characterized by chronic inflammation of cheeks, nose, chin, forehead, eyelids, often accompanied by erythema or acne-like eruptions, prominent subcutaneous blood vessels, tissue edema ASSOCIATIONS Migraines, other skin disorders–acne vulgaris, seborrhea, blepharitis, keratitis. Cf Acne

acne vulgaris Acne, common acne DERMATOLOGY A condition caused by chronic sebaceous gland inflammation characterized by comedones, papules and pustules of sebaceous areas–face, chest, back and resolving with scarring reaction; AV is the most common disease seen by dermatologists, affecting ± 5% of Americans TREATMENT Comedolytics–eg, retinoic acid, benzoyl peroxide; antibiotics–eg, clindamycin, erythromycin, tetracycline COMPLICATIONS Nonresponsive AV may evolve to cystic acne. See Cystic acne, Isoretinoin

ACOG American College of Obstetricians & Gynecologists

acoustic *adjective* Referring to sound

acoustic neuroma Cerebellopontine angle tumor, vestibular schwannoma NEUROSURGERY A usually benign tumor of the 8th cranial nerve, which arises from the Schwann cells that ensheath the acoustic nerve EPIDEMIOLOGY 1:100,000; 8% are bilateral; 16% are associated with type 2 von Recklinghausen's disease CLINICAL Hearing loss, tinnitus, vertigo DIAGNOSIS Auditory evoked potentials, MRI, CT MANAGEMENT Surgical removal by dissection, cautery, laser obliteration. See Malignant schwannoma

acoustic rhinometry ENT A simple reproducible technique for measuring nasal airflow, which is used to identify fixed lesions–eg, septal deviations, or alterations in cross-sectional area induced by allergens or drugs See Nasal compliance

acoustic tumor Acoustic neuroma, see there

ACP American College of Physicians

acquired *adjective* New, not inherited

acquired aplastic anemia See Aplastic anemia

acquired deafness See Noise-induced hearing loss

acquired immunity IMMUNOLOGY 1. Adaptive immunity Any immune response to exogenous antigens or immunogens 2. Secondary immunity Any compromise in immune function unrelated to inherited defects in the immune system. See AIDS 3. Immunity in which non-self antigens trigger an antiself immune reaction after a sensitization period.

acquired immunodeficiency syndrome See AIDS

acquired leukoderma DERMATOLOGY A condition characterized by regions of otherwise normal skin of nonpigmented white patches of varied sizes, often symmetrically distributed and usually bordered by hyperpigmented areas. Hair in the affected areas is usually, but not always, white

acquired platelet function defect Acquired qualitative platelet disorders HEMATOLOGY Any non-hereditary defect in platelet function ETIOLOGY Polycythemia vera, leukemia, myelofibrosis, renal failure, multiple myeloma, drugs–penicillins, salicylates, phenothiazines CLINICAL Defective coagulation, easy bruising, bleeding gums, nosebleeds, vaginal bleeding, rectal bleeding, hematemesis, epistaxis, hematuria DIAGNOSIS Bleeding time, PT, PTT, platelet count

acquisition IMAGING The obtaining of an image of a dynamic process or flow through a vascular lumen. See Real-time imaging

ACR American College of Radiology

acral lentiginous melanoma A rare, flat, palmoplantar or subungual melanoma, more common in non-whites, average 5-yr survival < 50%, unrelated to actinic exposure, but possibly related to ectopic pigmentation. See Melanoma

ACRE CARDIOLOGY A clinical trial–Appropriateness of Coronary Revascularization. See Revascularization PUBLIC HEALTH Advisory Committee on Releases to the Environment

acrocephalosyndactyly PEDIATRICS A family of AD conditions due to premature closure of cranial sutures resulting in a peaked head and facial dysmorphia IMAGING Skull film MANAGEMENT Surgery to correct skull and facial abnormalities. See Bird face

acrocyanosis Raynaud sign CLINICAL MEDICINE An acquired condition marked by symmetrical cyanosis of the extremities, with persistent, blue and/or red mottling of the skin of the digits, wrists and ankles, accompanied by profuse sweating and cold extremities. See Raynaud's phenomenon

acrodermatitis enteropathica Congenital zinc deficiency An AR condition characterized by dermatitis and diarrhea; the dermatitis affects the cheeks, elbows, knees, periorbital and perianal regions, and is accompanied by hair loss on the scalp, eyebrows and lashes; AE is associated with short stature, anemia, skin hyperpigmentation, hepatosplenomegaly, hypogonadism, poor wound healing, immune deficiency, recurrent bacterial and fungal infections due to immune defect LAB ↓ serum zinc due to impaired uptake MANAGEMENT Oral zinc

acromegaly ENDOCRINOLOGY A disease of adults due to excess hGH secretion of anterior pituitary or extrapituitary origin, or due to excess secretion of GH-RH by hypothalamic tumors or ectopic hGH production by small cell carcinoma of the lungs, carcinoids, islet cell tumors, adrenal adenomas or other 'endocrine' tumors CLINICAL Coarsened, enlarged facies, lips, nose, jaw, hands, feet, and frontal bones, widely spaced teeth, bone proliferation in extremities, soft tissue thickening, hyperhidrosis, macroglossia, headache, amenorrhea, impotence, somnolence, moodiness, glucose intolerance, HTN, heart disease, carpal tunnel syndrome, sleep apnea. See Giantism. See Acromegaloidism

acromioclavicular joint injury ORTHOPEDICS A disruption of the articulation formed between the acromion process and the clavicle, which may be accompanied by tearing of ligaments SPORTS Football, wrestling, equestrianism, hockey CLINICAL Pain at the top of the shoulder, ↓ ROM, splinting with arm held to the side; severe ASs may be accompanied by a 'lump' on the top of the shoulder MANAGEMENT Ice is nice; grade I, II, sling for pain; grade III controversial-some advocate open reduction and internal fixation; others say, *'if it ain't (really) broke, don't fix it.'*

acromioclavicular sprain Acromioclavicular joint injury, see there

acronym VOX POPULI A neologism created from the first letter of the each of the words in a particular phrase. See Acroeponym, Trial

acrophobia PSYCHOLOGY An abnormal or morbid fear of heights. See Phobia

acrosin test REPRODUCTION MEDICINE A test that assesses the enzymatic activity of the sperm, which is required to penetrate the outer layer of egg, resulting in fertilization

acrosomal reaction REPRODUCTIVE MEDICINE A necessary and irreversible step in fertilization which is triggered by sperm receptor, in which the sperm's and ovum's plasma membranes fuse and the sperm penetrates the zona pellucida. Cf Capacitation

acrosome REPRODUCTION BIOLOGY A membrane-bound cap-like structure at the head of a spermatozoon that contains lytic enzymes-acrosin hyaluronidase which degrade the outer coat of the egg and allow the penetration of one sperm. See Acrosome reaction. Cf Capacitation

acrylamide NUTRITION A substance found in ↑ concentrations in fried foods-eg, potato chips, French fries, and regarded by the WHO as a probable human carcinogen

acrylic DENTISTRY A plastic used to bond false teeth, retainers, and other dental devices

acrylonitrile TOXICOLOGY A flammable, toxic, carcinogenic and volatile liquid, which is absorbed through the skin, lungs and GI mucosa USES Monomer for acrylic fibers, resins, and plastics for home appliances, boats, etc; acrylonitrile may be teratogenic in rats at levels that are maternally toxic; reproductive risk is uncertain

ACT CARDIOLOGY Two clinical trials, 1. Angioplasty Compliance Trial. See Angioplasty 2. Attacking Claudication with Ticlopidine. See Ticlopidine PSYCHOLOGY ACT–A Controlled effectiveness Trial–Consumer vs Non-consumer Assertive Case Management Teams and Usual Care funded by the SAMHSA, which evaluated the use of services by Pts with serious and chronic mental illness, according to different types of case management

act of smoking factor PUBLIC HEALTH Any act which includes fumbling for a cigarette or a 'light', flicking ashes from clothing, tossing the cigarette butt out the window, etc, all of which ↑ the risk of an MVA. See Motor vehicle accident

ACT-UP AIDS Coalition To Unleash Power AIDS A NY-based organization of AIDS activists which aggressively pursue legislation favoring improved treatment for Pts with AIDS or HIV infection. See AIDS

ACTG 076 AIDS Clinical Trial Group Study 076 AIDS A study that validated the prophylactic use of AZT in HIV-infected pregnant ♀–HIPW, orally during pregnancy, IV during labor, and subsequently to newborns ↓ by ⅓ HIV-positivity in infants, saving 1 in 7 infants born to HIPW; ACTG 076 is the standard of care for these ♀. See AIDS, HIV

ACTH Adrenocorticotrophic hormone ENDOCRINOLOGY A 39-amino acid hormone produced by the anterior pituitary gland, which regulates adrenocortical activity, stimulating release of cortisol, mineralocorticoids, and adrenal androgens; pituitary ACTH is in turn regulated by corticotropin-releasing hormone–CRH; ACTH is ↑ in pituitary-dependent Cushing syndrome DAILY PATTERN ACTH levels vary over the space of a 24 hr day, peaking at 6–8 am and 6–11 pm; ACTH suppression or stimulation testing may be necessary to confirm the diagnosis of changes in adrenal function; ACTH is ↑ in Addison's disease, congenital adrenal hyperplasia, Cushing's disease, ectopic ACTH-producing tumors, Nelson syndrome; ACTH is ↓ in 2° adrenocortical insufficiency, adrenal carcinoma, adenoma. See Cushing syndrome

ACTH suppression test Dexamethasone suppression test, see there

Acticoat™ WOUND CARE A silver nitrate-based antibacterial burn dressing for wound protection. See Pressure ulcer

Actigall® Ursodiol, ursodeoxycholic acid An agent used to dissolve gallstones as an alternative to cholecystectomy in Pts with radiolucent, noncalcified gallstones ≤ 20 mm in diameter, and to prevent gallstone development in Pts undergoing rapid weight loss

acting internship Externship, junior internship, senior clerkship, subinternship GRADUATE MEDICAL EDUCATION A clinical rotation undertaken in a specific area of medicine–eg cardiology, infectious disease, emergency medicine by a 4th-yr medical student, which often indicates the student's specialty interest See warm-up rotation; Cf Internship

acting out CHILD PSYCHIATRY Self-abusive, aggressive, violent and/or disruptive behavior PSYCHIATRY A form of displacement in which a Pt's behavior is a response to a current situation. See Displacement

actinic cancer A skin cancer attributable to excess exposure to solar radiation, most common in the head & neck, followed by the legs and trunk. See Actinic keratosis, Melanoma

actinic keratosis Senile keratosis, solar keratosis DERMATOLOGY A premalignant lesion of sun-exposed skin that is overly sensitive to the effects of UV light–sunlight, and more common in the fair-skinned or elderly CLINICAL Discrete scaly or gritty erythematous plaques located on a sun exposed surface DIAGNOSIS Biopsy MANAGEMENT Cryosurgery or electrocautery; some topicals may be used to promote peeling PROGNOSIS ± 20% of AKs develop into SCC PREVENTION Broad-spectrum sunscreens with a sun protection factor of 17 may ↓ new lesions; avoid photosensitizing drugs–eg, tetracyclines. See Squamous cell carcinoma

Actinomyces MICROBIOLOGY A genus of slow-growing, facultatively anaerobic, nonmotile, nonspore-forming, gram-positive bacteria, which form branching, irregularly staining filaments, filamentous microcolonies, and metabolize by fermentation of certain sugars. See Actinomycosis, Sulfur granules. Cf *Bacteroides* spp, *Nocardia* spp

actinomycin D Cosmegen®, dactinomycin ONCOLOGY A relatively toxic antibiotic produced by *Streptomyces* spp, used as a chemotherapeutic MECHANISM Inhibits DNA transcription by RNA polymerase INDICATIONS Rhabdomyosarcoma, Wilms' tumor ADVERSE EFFECTS Anorexia, N&V, myelosuppression with pancytopenia, diarrhea

actinomycosis INFECTIOUS DISEASE A chronic local or systemic granulomatous infection by *Actinomyces israelii*, a filamentous, gram-positive bacterium CLINICAL Weight loss, weakness, fever, local pain, indolent suppurative lesions of face & neck–40-60% of cases, lungs & chest–15%, abdomen and other regions, often accompanied by draining sinus tracts/abscesses containing yellow aggregates–'sulfur granules' TREATMENT IV penicillin. See *Actinomyces*, Sulfur granules

action potential CARDIOLOGY The constellation of changes in electric potential generated by myocardial cell membranes after stimulation PHYSIOLOGY The sequential, electrochemical polarization and depolarization that traverses the membrane of a neuron in response to mechanical stimulation–eg, touch, pain, cold, etc. See Depolarization

Actiq® Fentanyl citrate A lollipop delivery system developed to alleviate breakthrough pain in cancer Pts. Cf Breakthrough pain

Activase® A proprietary thrombolytic, which may ↑ survival in TIA victims by 33% MECHANISM Activation of the body's fibrinolytic system by stimulating conversion of plasmin from plasminogen, and several other clotting factors. See TIA, tPA

activated charcoal Medicinal charcoal TOXICOLOGY AC is used for early management of oral poisoning and is effective against most toxins except mercury, iron, lithium, cyanide; if the drug has an enterohepatic cycle, as do barbiturates, glutethimide, morphine, and other narcotics and tricyclic antidepressants, AC may be repeated for up to 24 hrs; adult dose 50–100 g THERAPEUTICS A carbon residue of destructive distillation–heated to 200°C-to remove volatile gases of organic materials–eg, bone, wood, resulting in carbonaceous granules–charcoal; AC is effective in removing various substances from a fluid or gas of interest

activated macrophage IMMUNOLOGY A mononuclear phagocyte that has been 'turned on'–ie has enhanced activity by cytokines; AMs are bigger–twice the size of resting macrophages and 'badder'–↑ lysozymes and surface expression of MHC class II antigens, and are pivotal in defending against microorganisms that grow well in histiocytes/tissue-based macrophages and other cells–eg, *Listeria* spp, and *Salmonella* spp

activated partial thromboplastin time aPTT HEMATOLOGY A test that evaluates the clotting factors of the intrinsic pathway–except VII and XIII, by measuring the time required to form a fibrin clot; aPTT is used to screen for bleeding tendencies and to monitor heparin therapy; it is ↑ in coagulation factor deficiencies–factors V, VIII, IX, X, XI, XII, DIC, Hodgkin's disease, hypofibrinogenemia, leukemia, cirrhosis, vitamin K deficiency, von Willebrand's disease, and in drug therapy–eg, heparin and aspirin REF RANGE 30-40 secs. See Prothrombin time

activated protein C resistance APC resistance HEMATOLOGY A condition caused by an inherited defect in the anticoagulant response to APC and clinically characterized by ↑ venous thrombosis; it is responsible for 20-50% of DVT PATHOGENESIS Protein C, a key regulator of coagulation, circulates in an inactivated form and is activated by the binding of thrombin to thrombomodulin receptors on vascular endothelial cells; once activated, protein C lyses coagulation factors Va and VIIIa; APCR may be due to a selective defect in factor V coagulant function

activated thrombin HEMATOLOGY A key clot promoting enzyme that converts fibrinogen to fibrin and protects against fibrinolysis by activating thrombin-activatable fibrinolysis inhibitor. See Fibrin, Fibrinolysis, TAFI

activation HEMATOLOGY See Plasminogen activation IMMUNOLOGY Lymphocyte activation, see there. See Complement activation, Macrophage activation NEUROLOGY The stimulation of a nerve to a level high enough to kick over its action potential

activation product RADIATION BIOLOGY An unstable nucleus/radioisotope formed during activation. See Activation-radiation biology

Activa™ system An implantable device to deliver mild electrical stimulation to block signals that cause involuntary tremor. See Essential tremor, Parkinson's disease

active *adjective* Not passive, not expectant

active arm Treatment arm, see there

active euthanasia MEDICAL ETHICS The practice of injecting a Pt with a lethal dose of medication with the primary intention of ending the Pt's life. Cf Active euthanasia

active fixation lead CARDIAC PACING A pacing lead with a mechanism at the tip–eg a corkscrew which is anchored in the endocardium, ↓ the risk of displacement. See Lead

active immunotherapy The administration of substances to evoke a protective immune response in the form of specific antibodies. Cf Passive immunotherapy, Vaccine

active life expectancy Active life expectancy at age *x* EPIDEMIOLOGY The average number of yrs of life remaining in an independent state–ie free from significant disability for a population of individuals, all of age *x*, all of whom are subject for the remainder of their lives to the observed age-specific risks of disability. See Life expectancy

active metabolite THERAPEUTICS A drug metabolite with therapeutic activity similar to the parent compound, which must be considered in therapeutic pharmacokinetics

active placebo STATISTICS A placebo with side effects similar to those of a therapeutic agent which would otherwise allow a Pt to identify whether he is receiving drug or placebo–eg, dry mouth is associated with chlorpromazine. See Placebo

active staff HOSPITAL PRACTICE 1. A medical staff status defined by the minimum number of Pts a practitioner admits to the hospital during an average month; AS members are eligible to vote, hold office, and serve on medical staff committees. See Medical staff 2. A physician or other health care provider who admits a determined number of Pts to a hospital each yr. See Staff

activity The ability to produce some effect; the extent or intensity of a function or action CARDIAC PACING Body movement which affects metabolic demand and pacing rate ENDOCRINOLOGY The functional effect of a hormone or hormone-like substance. See HCG-like activity, Melanoma growth stimulatory activity, NSILA, Plasma renin activity EPIDEMIOLOGY See Surveillance activity SEXOLOGY Sexuoerotic rubbing of orogenital mucosae. See Safe sexual activity, Sexual activity VOX POPULI Doing stuff. See Hyperactivity, Instrumental activities of daily living activity, Major life activity, Physical activity

activity of daily living MEDTALK A core physical function–eg, self-care, bathing, dressing, grooming, ambulation, food preparation, shopping, housekeeping, work, leisure. See ADL scale

activity threshold CARDIOLOGY The level of physical activity above which activity-seeking rate responsive pacemakers ↑ the pacing rate. See Pacemaker

Actonel® Risedronate, see there

Actos® Pioglitazone, see there

actual charge HEALTH INSURANCE The actual amount charged by a physician for medical services rendered

actualization PSYCHIATRY The realization of one's full potential

acute infection of bladder Acute cystitis. See Cystitis

Acucare model EMERGENCY MEDICINE A format for ↓ ER overutilization, in which medical intensivists triage ER Pts, oversee emergency care, monitor utilization and authorize treatment or discharge

acuity AUDIOLOGY Clarity of hearing EMERGENCY MEDICINE Acuteness, intensity, level of urgency. See Triage OPHTHALMOLOGY Sharpness of vision, expressed as a ratio–eg, 20/100, where a subject can see at 20 meters what a person with perfect vision can see at 100 meters

Acumed™ suture ORTHOPEDICS An implantable soft-tissue-to-bone anchor used in rotator cuff and soft tissue repair of the foot and ankle. See Rotator cuff

acumen Astuteness, perception, perspicacity

acute *adjective* Of abrupt onset, or short duration, usually of hrs or days in duration, used in reference to a disease or symptoms. Cf Chronic, Subacute

acute abdomen A relatively nonspecific symptom complex, in which a Pt is first seen in a 'toxic' state, complaining of incapacitating abdominal pain, variably accompanied by fever, and leukocytosis; AA may also be defined as an acute intra-abdominal inflammatory process that may require surgical intervention; appendicitis is the most common cause of an AA; nearly 100 other conditions may present in a similar fashion, in particular, ruptured ectopic pregnancy in a fallopian tube, ruptured acute diverticulitis and acute mesenteric lymphadenitis

ACUTE ABDOMEN ETIOLOGY

INFECTION Amebiasis, hepatitis, falciparum malaria, pneumococcal pneumonia, rheumatic fever, salmonella gastroenteritis, staphylococcal toxemia, syphilis in 'tabetic crisis,' trichinosis, TB, typhoid fever, viral enteritides, herpes zoster, infectious mononucleosis, Whipple's disease

INFLAMMATION Appendicitis, cholangitis, cholecystitis, Crohn's disease, diverticulitis, gastroenteritis, hepatitis, SLE, mesenteric lymphadenitis, pancreatitis, peritonitis due to organ perforation, perinephric abscesses, pyelonephritis, ulcerative colitis, intestinal obstruction, rheumatoid arthritis, polyarteritis nodosa, Henoch-Schönlein disease

INTOXICATION Black widow spider bite, heavy metals, mushrooms

ISCHEMIA Renal infarction, mesenteric arterial thrombosis

MALIGNANCY Pain due to organ infarction, Hodgkin's disease ('classically' associated with alcohol ingestion), leukemia, lymphoproliferative disorders

METABOLIC DISEASE Adrenal insufficiency (Addisonian crisis), DKA, familial hyperlipoproteinemia, familial Mediterranean fever, hemochromatosis, hereditary angioneurotic edema, hyperparathyroidism, hyperthyroidism, acute intermittent porphyria, uremia, substance abuse withdrawal

OB/GYN Twisted ovarian cyst, ectopic pregnancy, endometriosis, PD

REFERRED PAIN Pneumonia, MI, pleuritis, pericarditis, myocarditis, hematomata of the rectal muscle, renal colic, peptic ulcer, nerve root compression

TRAUMA PERFORATION/RUPTURE–AORTIC ANEURYSM, SPLEEN, BLADDER

acute adrenal crisis Acute adrenal insufficiency ENDOCRINOLOGY A life-threatening condition caused by an abrupt ↓ cortisol AT RISK PTS Those taking corticosteroids for prolonged periods–wks to months–eg with Addison's disease, especially if steroids are stopped abruptly CLINICAL Shock, hypotension, weakness, headache, vomiting, fever, chills, tachycardia, sweating MANAGEMENT Vasopressors, steroids PREVENTION Taper steroid withdrawal

acute AIDS syndrome Acute HIV syndrome, see there

acute angle closure glaucoma Angle closure glaucoma OPHTHALMOLOGY An abrupt block in the fluid circulation in the eye, resulting in ↑ intraocular/anterior chamber pressure and potential damage of the optic nerve and blindness which occurs when the intensity of iris bombé is sufficient to occlude the anterior chamber angle–ACA; AACG is more common in Pts with preexisting narrowing of the ACA, which occurs in hypermetropes CLINICAL Sudden onset of blurred vision, severe ocular or facial pain, halos, N&V, ↑ intraocular pressure, shallow anterior chamber, steamy cornea, fixed–nonreactive dilated pupil, ciliary injection MANAGEMENT IV/oral acetazolamide, topical beta blockers, hyperosmotic agents, pilocarpine or intraocular pressure; after intraocular pressure is ↓, peripheral iridectomy to establish a permanent communication between anterior and posterior chambers; if there is no response to medical management, emer-

gency trabeculectomy, laser sclerostomy. See Chronic angle closure glaucoma, Open angle glaucoma

acute bacterial meningitis See Bacterial meningitis

acute bilateral obstructive uropathy UROLOGY A condition of relatively abrupt onset caused by urethral obstruction ETIOLOGY Men–BPH, women–bladder cystocele, bladder tumors, prostate CA, tumors or other structures of the bladder neck or urethra, kidney or bladder stones CLINICAL Irritability, urgency, bladder incontinence, urine stasis, UTIs, urine reflux into ureters and kidney, leading to pyelonephritis and hydronephrosis

acute blood loss INTERNAL MEDICINE A state of vascular instability caused by external or internal hemorrhage CLINICAL Supine tachycardia > 100/min, ↑ pulse > 30/min, supine or postural hypotension. See Dehydration, Volume depletion

acute bronchitis PULMONOLOGY A lower RTI–up to 95% of which are viral–that causes reversible bronchial inflammation CLINICAL Cough, fever, sputum, wheezing, rhonchi DIFFDx Asthma, aspergillosis, occupational exposure, chronic bronchitis, sinusitis, pneumonia MANAGEMENT Antibiotics rarely shorten the course of disease; bronchodilators–eg, albuterol may provide symptomatic relief. See Asthma

acute care CRITICAL CARE Care administered to a Pt with an acute–eg burns, trauma, cardiac arrest, or other condition, which may be administered in ambulances, ERs, CCUs, and elsewhere MANAGED CARE Skilled, medically necessary health care provided by medical and nursing personnel to restore a person to good health

acute chest syndrome HEMATOLOGY A complex seen in Pts with sickle cell anemia–SCA CLINICAL Fever, tachycardia, chest pain, leukocytosis, and pulmonary infiltrates; it is the most common cause for hospitalization in SCA and is due to vascular occlusion and/or infection; in children, ACS is often due to bacterial pneumonia, especially *S pneumoniae*–preventable by pneumococcal vaccine, *Mycoplasma pneumoniae*, and others. See Sickle cell anemia

acute cholangitis GASTROENTEROLOGY Acute bile duct inflammation ETIOLOGY Stricture or obstruction of bile ducts, infection, pancreatic reflux, drugs, chemicals CLINICAL Right upper quadrant pain that may radiate to the shoulder, variable severity, N&V, constipation, episodic chills, fever, slow pulse, Murphy sign, anorexia, weight loss; spontaneous remission is rare IMAGING ERCP, percutaneous transhepatic cholangiography LAB ↑ WBCs, ↑ BR, urobilinogen in urine MANAGEMENT Cholecystectomy

acute confusional state Acute brain syndrome NEUROLOGY A state of severe confusion or an abrupt change in mental status, due either to psychiatric or organic disease CLINICAL Lethargy, agitation, confusion, disorientation, delirium. Cf Dementia

acute coronary syndrome CARDIOLOGY A term that encompasses the permutations of acute ischemic heart disease, which is a heterogeneous constellation of clinical symptoms associated therewith DIAGNOSIS Careful clinical Hx, PE, resting 12-lead and serial EKG–marked symmetrical T-wave inversion in precordial leads, various cardiac markers; for Pts unlikely to have CAD, AHRQ guidelines recommend a treadmill test without imaging, recording an EKG both during Sx and after Sx relief; recurrent Sx or a change in clinical status; ST-elevation-ACS–which requires immediate reperfusion–must be rapidly excluded RISK FACTORS DM, smoking, HTN, ↑ cholesterol TYPES ST-elevation ACS is esentially the same as what was formerly termed Q-wave infarction, non-ST-elevation ACS encompasses a broader range of disease–eg, rest angina, new-onset angina, increasing angina, postinfarction angina, variant angina, and non-Q-wave infarction EARLY MANAGEMENT Bed rest, control of precipitating factors, initiation of medical therapy PROGNOSIS Most ACS patients stabilize; 10% undergo in-hospital MI; some Pts die suddenly; Pts with cardiomegaly and non-responder to initial medical therapy have a higher risk for subsequent cardiac events; randomized clinical trials have provided data on the incidence of subsequent cardiac events in non-ST-elevation ACS patients with unstable angina; ESSENCE, GUSTO II, PARAGON, PURSUIT trials reported death rates between 2.9% and 4.7%, reinfarction rates of 5% to 12%, and stroke rates of 0.4% to 0.9%; other risk factors associated with adverse outcomes include male sex, older age, HTN, left ventricular hypertrophy, poor performance on exercise stress testing, and extensive disease on angiography or nuclear testing; antagonists–thrombogenic stimuli–ie, ulcerated plaques and intimal tears can persist for prolonged periods, and the hemostatic system may remain activated for months after ACS ↓ M&M in Pts treated with GP IIb/IIIa inhibitors–abciximab, eptifibatide, tirofiban, lamifiban; agents that directly inhibit thrombin, including hirudin and bivalirudin, are being explored as alternatives to unfractionated heparin in ACS and during percutaneous coronary interventions THEORETICAL ADVANTAGES Inhibit clot-bound thrombin, ↓ variability in dosing, eliminate risk of HIT and thrombocytopenic purpura NEXT GENERATION TREATMENTS VEGF, fibroblast growth factor; laser revascularization, external counterpulsation; gene therapy with various molecules, IV GP IIb/IIIa. See GUSTO 2, Myocardial infarction, Reperfusion-eligible acute MI

acute disease EPIDEMIOLOGY Any condition—eg infection, trauma, pregnancy, fracture, with a short, often < 1 month clinical course; ADs usually respond to therapy; a return to a state of complete–pre-morbid health is the rule. Cf Chronic disease

acute disseminated encephalitis NEUROLOGY An acute complication of viral infection–1:1000 cases of measles or vaccination–1:10⁶ measles vaccinations, involving the entire brain and spinal cord or focally affecting a nerve or cord root CLINICAL Meningial signs and, if serious, coma and death TREATMENT None

acute diverticulitis GI DISEASE Acute inflammation of a diverticulum of the large–less commonly, small–intestine STAGING See Hinchey staging CLINICAL Pain in hypogastrium, which localizes to the left lower quadrant; altered bowel habits–diarrhea > constipation, dysuria, urinary frequency, urgency–if affected segment is close to bladder IMAGING CT is the diagnostic method of choice; it is safe, cost-effective, correlates with diagnosis of AD; false negative rate–2-21% MANAGEMENT First attack-liquid diet, 7-10 days of broad-spectrum antibiotics; for refractory AD, 'triple therapy' has been advocated, although newer broad-spectrum antibiotics–eg, pipericillin, tazobactam, may be as effective; CT-guided drainage is advocated for peridiverticular abscesses of > 5 cm³; ±20% of Pts with AD require surgery; in Pts who respond to medical management, and in whom diverticular abscesses can be controlled by percutaneous drainage, a single-stage procedure with 1° anastomosis is appropriate, and can be performed by laparoscopy; in emergencies, a 2-stage procedure with resection and colostomy in stage one, and reanastomosis in stage 2 is preferred See Diverticular disease

ACUTE DIVERTICULITIS–PATHOGENESIS
↑ INTRALUMINAL PRESSURE (IP ↓ stool bulk → ↑ GI transit time → ↑ IP; hypersegmentation and ↑ IP result in herniation of colonic mucosa at weakened areas adjacent to the points of penetration of the vasa recta in the intestinal wall–which explains why diverticula are arranged in role
WEAKENED BOWEL WALL Once diverticula develop, undigested food may become entrapped; obstruction of the neck of the diverticulum sets the stage of distension as a result of mucus secretion; diverticula may consist solely of mucosa and thus are susceptible to vascular compromise

acute effects of overexposure OCCUPATIONAL SAFETY Any adverse effects evident at the time of, or shortly after, exposure to a hazardous material

acute episode A period when an injury is most intense, usually immediately following the injury or flare-up

acute fatty liver of pregnancy Fatty liver of pregnancy, see there

acute febrile illness A nonspecific term for an illness of sudden onset accompanied by fever

acute glomerulonephritis Acute nephritic syndrome NEPHROLOGY A nonspecific term for acute glomerular inflammation ETIOLOGY Streptococcal infection, SLE, syphilis, bacterial endocarditis, sepsis, vasculitis, Goodpasture syndrome, typhoid fever, Henoch-Schönlein purpura, hepatitis, viral infection–eg, mumps, measles, infectious mononucleosis

acute gouty arthritis Acute gout RHEUMATOLOGY An abrupt gouty attack, which may be precipitated by overeating, alcohol, surgery, emotional stress, infection, antibiotics, insulin CLINICAL Crushing pain of a joint–most often the great toe–which is swollen, hot, shiny. See Gout, Gouty arthritis

acute hepatitis CLINICAL MEDICINE Liver inflammation of abrupt onset, which may be due to a viral infection–eg HAV or toxins CLINICAL Low-grade fever, anorexia, N&V, fatigue, malaise, headache, photophobia, pharyngitis, cough; later, dark urine, light stool, jaundice, hepatomegaly, ± splenomegaly, ± lymphadenopathy DIAGNOSIS Liver biopsy; ↑ transaminases–ALT, AST, ↑ virus-specific IgG or IgM PATHOLOGY Diffuse lymphocytic and plasmacytic inflammation, patchy–piecemeal necrosis, and liver cell degeneration in the form of either lytic–ballooning or coagulative–acidophilic degeneration; AH may be accompanied by a collapse of the central reticulin fibers surrounding the hepatocytes; activated Kupffer cells may be filled with phagocytic debris. Cf Chronic hepatitis

acute HIV syndrome Acute HIV infection, primary symptomatic HIV infection A transient, flu-like early response to HIV-1, that occurs 1-6 wks after exposure to HIV-1 in 50-70% of those with 1° HIV infection; AHS presents as an acute infectious mononucleosis-like complex and is accompanied by viremia and an immune response to HIV within 1 wk to 3 months CLINICAL Fever, severe fatigue, sore throat, lymphadenopathy, anorexia, N&V, and a maculopapular rash, less commonly, diarrhea, lightning-like pain, major weight loss, abdominal cramping, palmoplantar desquamation, myalgia, arthralgia, headache, photophobia, lymphadenopathy, pruritic maculopapular or urticarial rash, lymphocytic meningoencephalitis and peripheral neuropathy, all Sx remit, and reappear as AIDS after a latency of up to several yrs LAB Mild leukopenia, occasionally inversion of the CD4:CD8 ratio, antibodies to HIV products–gp120, gp160, p24 and p41 first appear ≥ months after infection. See AIDS, CD4:CD8 ratio, HIV-1

acute idiopathic polyneuritis Guillain-Barré syndrome, see there

acute intermittent porphyria HEMATOLOGY An AD condition caused by a deficiency of porphobilinogen deaminase, resulting in overproduction of δ-aminolevulinic acid CLINICAL Recurrent abdominal colic, constipation, fever, leukocytosis, postural hypotension, peripheral neuritis, polyneuropathy, paraplegia, urinary retention, respiratory paralysis, behavioral changes and episodic organic psychosis, photosensitivity; worse with barbiturates LAB ↑ δ-aminolevulinic acid, urine porphobilinogen

acute interstitial nephritis Acute allergic nephritis NEPHROLOGY Renal inflammation characterized by cellular—primarily mononuclear—and fluid exudates, often with epithelial degeneration TYPES Idiopathic, 2° to drugs or infections FINDINGS-HYPERSENSITIVITY Fever, rash, sore throat, malaise, arthralgia, myalgia, hepatitis, splenomegaly, eosinophilia, lymphoadenopathy, thrombocytopenia, autoimmune hemolysis FINDINGS-ANAPHYLAXIS Respiratory tract–laryngeal edema, asthma; skin–urticaria, angioedema; GI tract–N&V, diarrhea; cardiovascular–capillary leak, hypotension LAB-PROXIMAL TUBULE Glycosuria, aminoaciduria, tubular proteinuria–usually < 1 g/24 h, hypouricemia DISTAL TUBULE Distal renal tubular acidosis, ↑ K⁺, isosthenuria, Na⁺ wasting MANAGEMENT Corticosteroids *may* be effective PROGNOSIS From mild renal dysfunction to acute renal failure

acute kidney failure Acute renal failure, see there

acute leukemia HEMATOLOGY A rapidly progressive malignancy of sudden onset, characterized by an uncontrolled 'clonal' proliferation of immature WBCs which replace BM and spill into the peripheral circulation; untreated AL may be fatal in wks to months. See Acute lymphocytic leukemia, Acute myelocytic leukemia

acute low back pain REHABILITATION MEDICINE A nonspecific symptom of abrupt onset or an exacerbation of chronic low back pain of < 3 wks in duration MANAGEMENT Continuing normal activities within the limits allowed by the pain leads to more rapid recovery than either bedrest or back-mobilizing exercises See Low back pain

acute lymphocytic leukemia Acute lymphoblastic leukemia, ALL A malignant lymphoproliferative process that commonly affects children and young adults affecting ± 1800/yr–US; ± 650/yr–UK ETIOLOGY ALL has a hereditary component; it is 20 X ↑ Pts with Down syndrome; it is linked to benzene exposure, RT in ankylosing spondylitis; it is unrelated to exposure to magnetic fields CLINICAL Abrupt onset, often ± 3 month Hx of fatigue, fever, hemorrhage from multiple sites, lymphadenopathy, hepatomegaly, splenomegaly PROGNOSIS 9% long-term disease-free survival–1962-66; 71%–1984-88–St Jude's Children's Hospital; 90-95% achieve remission; improved cure rate is attributed to prophylaxis for meningeal leukemia and more intense systemic chemotherapy

acute massive pulmonary embolus An '...*obstruction or significant filling defect involving two or more lobar pulmonary arteries, or the equivalent amount of emboli in smaller or other arteries*' See Pulmonary embolism

acute mitral regurgitation See Mitral regurgitation

acute mountain sickness WILDERNESS MEDICINE A condition caused by prolonged exposure to high altitude CLINICAL Dry cough, SOB, poor exercise tolerance, dizziness, headache, sleep difficulty, anorexia, confusion, fatigue, tachycardia MANAGEMENT Move to low altitude PROPHYLAXIS Acetazolamide–Diamox. See Mountain sickness, acute

acute myelocytic leukemia Acute myelocytic (myelogenous, myeloid, nonlymphocytic) leukemia ONCOLOGY A rapidly progressing form of leukemia which is characterized by the proliferation of immature WBCs–blasts in peripheral circulation EPIDEMIOLOGY Primarily in adults, infants < age 1; 2,000 new cases are diagnosed/yr, UK COMPLICATIONS Bleeding, ↑ infections CLINICAL Fatigue, weight loss, fevers, weakness, pallor, bone & joint pain, bleeding gums, nosebleeds, bruising, lymphadenopathy MANAGEMENT Chemotherapy and/or BMT. See Leukemia

acute myocardial infarction CARDIOLOGY The abrupt death of heart muscle due to acute occlusion or spasm of the coronary arteries EPIDEMIOLOGY ±1.5 million MIs/yr–US, 75,000 AMI follow strenuous physical activity, of whom $\frac{1}{3}$ die; ± $\frac{1}{4}$ of all deaths in the US are due to AMIs; > 60% of the AMI-related deaths occur within 1 hr of the event; most are due to arrhythmias, in particular ventricular fibrillation TRIGGERS Heavy exertion in ±5% of Pts, which is inversely related to Pt's habitual physical activity ETIOLOGY Occlusion of major coronary artery–CA, in a background of ASHD, due primarily to the plugging of the vessel with debris from an unstable plaque–see Uncomplicated plaque CLINICAL Main presenting symptom–retrosternal chest pain accompanied by tightness, discomfort, & SOB; cardiac pain often radiates to the arm & neck, and less commonly to the jaw; the pain of AMI generally is *not* relieved with nitroglycerin, in contrast to esophageal pain, which is often identical in presentation, and may respond, albeit slowly, to nitroglycerin; the characteristic clinical picture notwithstanding, there is a high rate of false negative diagnoses of AMIs DIAGNOSIS Clinical presentation, physical examination, EKG–sensitivity in diagnosing AMI is 50–70%, and is lower in lateral MIs than in anterior and inferior MIs; CXR may demonstrate left ventricular failure, cardiomegaly ECHOCARDIOGRAPHY M-mode, 2-D & Doppler RADIOISOTOPIC STUDIES Radionuclide angiography, perfusion scintigraphy, infarct-avid scintigraphy, & PET can be used to detect an AMI, deter-

mine size & effects on ventricular function, and establish prognosis; a radiopharmaceutical, ^{99m}Tc-sestamibi, has become the perfusion imaging agent of choice, given its usefulness for measuring the area of the myocardium at risk for AMI, and for recognizing the myocardium salvaged after thrombolytic therapy OTHER IMAGING TECHNIQUES–eg, CT, and MRI LAB CK-MB, troponin I DIFFDX AMI is *the* most common cause of acute chest pain in older adults, other conditions must be excluded–PREVENTION ↓ Smoking, ↓ cholesterol, ↓ HTN; ↑ aerobic exercise; influence of other factors–eg maintaining normal body weight, euglycemic state in diabetes, estrogen-replacement therapy, mild-to-moderate alcohol consumption, effect of prophylactic low-dose aspirin-on incidence of AMI is less clear. See AIMS, ASSET, EMERAS, EMIP, GISSI, GISSI-2, GUSTO-1, INJECT, ISIS-2, ISIS-3, LATE, MITI-1, MITI-2, RAPID, TAMI-5, TAMI-7, TEAM-2, TIMI-2, TIMI-4, Trial

---

DIFFERENTIAL DIAGNOSIS OF ACUTE MYOCARDIAL INFARCTION

ARM PAIN Myocardial ischemia, cervical/thoracic vertebral pain, thoracic outlet syndrome

EPIGASTRIC PAIN Myocardial ischemia, GI tract–esophagus, peptic ulcers, pancreas, liver disease–cholecystitis, hepatic distension, pericardial pain, pneumonia

RETROSTERNAL PAIN Myocardial ischemia, aortic dissection, esophageal pain, mediastinal lesions, pericardial pain, PTE

SHOULDER PAIN Myocardial ischemia, cervical vertebra, acute musculoskeletal lesions, pericardial pain, pleuritis, subdiaphragmatic abscess, thoracic outlet syndrome

---

acute necrotizing ulcerative gingivitis Trench mouth ORAL PATHOLOGY A condition characterized by progressive necrosis of intraoral tissues, seen in those with poor oral hygiene and suboptimal nutrition CLINICAL Pain, edema, punched-out ulcers, pseudomembrane formation, halitosis; anaerobic flora–eg, *Fusobacterium* spp, spirochetes ± related to *Treponema pallidum*, which also cause Vincent's angina–which affects the soft palate and tonsils, cancorum oris, upper respiratory abscesses DIFFDX Erythema multiforme, lichen planus, pemphigus, pemphigoid TREATMENT H_2O_2, antibiotics–eg, tetracycline, if fever or lymphadenopathy, saline mouth rinse, local anesthetics. Cf Periodontal disease

acute neurologic illness NEUROLOGY A condition defined by the Natl Childhood Encephalopathy Study–UK as 1. Acute or subacute encephalitis, encephalomyelitis, encephalopathy–eg, postinfectious encephalitis but not pyogenic infection 2. Unexplained loss of consciousness 3. Convulsions lasting > $\frac{1}{2}$ hr, or followed by coma lasting ≥ 2 hrs or by paralysis or other neurologic signs not previously present 4. Infantile spasms 5. Reye syndrome See Encephalopathy

acute otitis media Acute ear infection ENT A bacterial or viral middle ear inflammation that is most common in children, which presents with a rapid onset of pain, irritability, anorexia, or vomiting RISK FACTORS Children with siblings, day care centers with ≥ 7 children, ♂, Native Americans, Eskimos, tobacco smoke, family history of atopy; AOM is ↓ in summer and in breast-fed infants MICROBIOLOGY *S pneumoniae*, *H influenzae*, *Moraxella catarrhalis*, RSV, rhinovirus, adenovirus, influenza virus; OM is more common in children as their eustachian tubes are shorter, narrower, and more horizontal than adults RISK FACTORS Recent illness, crowded or unsanitary living conditions, heredity, high altitude, cold climate, bottle feeding allowing fluid to pool in the throat at the eustachian tube TREATMENT 1st-line, amoxicillin; 2nd-line, amoxicillin-clavulanate, TS, erythromycin-sulfisoxazole See Otitis externa, Otitis media

acute pain PAIN MANAGEMENT A normal physiologic and usually time-limited response to an adverse (noxious) chemical, thermal or mechanical stimulus, associated with surgery, trauma, and acute illness and historically responsive to opioid therapy. See Pain; Cf Chronic pain

acute pancreatitis Inflammation of the pancreas of abrupt onset, often with gallstones and alcohol ingestion EPIDEMIOLOGY 109,000 hospitalizations, 2251 deaths–US; 10-fold ↑ from 1960s to 1980s–reason unclear; ? alcohol abuse; ? widened diagnostic criteria; ± 250 admissions/10^6 population/yr, higher in certain populations–eg, 4-22% in AIDS Pts ETIOLOGY

Obstruction, toxins or drugs, trauma, metabolic defects, infection, vascular defects, idiopathic DIAGNOSIS Abdominal pain, ↑ amylase, ↑ lipase, ultrasonography, contrast-enhanced CT MANAGEMENT Supportive, bowel rest with parenteral nutrition PROGNOSIS Ranson's criteria, modified Glasgow criteria, APACHE II PROGNOSIS 25% have complications, 9% die of pancreatitis, sepsis, pulmonary failure, etc. See Chronic pancreatitis

acute phase response A constellation of nonspecific host responses to cytokines released in response to tissue injury, infection, inflammation and rarely malignancy—eg, Hodgkin's disease, renal cell CA, and causes functional liver changes–↑ synthesis of acute phase proteins, endocrine system–abnormal glucose tolerance, ↑ gluconeogenesis, thyroid dysfunction, altered lipid metabolism, immune system–left shift leukocytosis, hypergammaglobulinemia, metabolic system–↓ albumin synthesis, energy consumption, ↑ ceruloplasmin, ↓ iron and zinc levels, and CNS–lethargy; the most measured molecule in the response is the highly nonspecific CRP, which may ↑ 10- to 1000-fold within hrs from a normal of 100 µg/L

acute poststreptococcal glomerulonephritis Post-infectious glomerulonephritis NEPHROLOGY Glomerular inflammation which follows streptococcal infection–eg strep throat with group A beta-hemolytic streptococci EPIDEMIOLOGY More common in ♂ age 3 to 7 CLINICAL Edema of face, extremities, oliguria, proteinuria, HTN, joint pain LAB Hematuria, proteinuria, ↑ ASO MANAGEMENT Symptomatic; bed rest, antibiotics; antihypertensives. See Strep throat

acute promyelocytic leukemia ONCOLOGY A type of leukemia that comprises 10% of AML, which has a poor prognosis in children CLINICAL Presents with bleeding diathesis—fatal in 8-47%, due to chemotherapy-induced thrombocytopenia and hypofibrinogenemia MANAGEMENT Induction with all-*trans*-retinoic acid, followed by conventional chemotherapy, commonly an anthracycline plus cytarabine. See Retinoic acid syndrome

acute prostatitis UROLOGY An inflammation of the prostate of abrupt onset, caused by bacterial infection–eg, E coli, but also STD bacteria–*N gonorrhoeae, U urealyticum, Trichomonas vaginalis*; AP may also follow urethral instrumentation—eg, catheterization or cystoscopy, trauma, bladder outlet obstruction, or systemic infection CLINICAL Chills, fever, lower abdominal discomfort or perineal pain, burning with urination, difficulty urinating, painful defecation and ejaculation ASSOCIATIONS Epididymitis, orchitis, men age 20 to 35 yrs old, multiple sexual partners, high-risk sexual behaviors

acute pulmonary eosinophilia Loeffler's syndrome PULMONOLOGY A self-limiting inflammation in the lungs accompanied by eosinophils ETIOLOGY Unknown CXR Diffuse infiltrates LAB ↑ Eosinophils MANAGEMENT APE may respond to corticosteroids COMPLICATIONS Restrictive cardiomyopathy due to pericardial fibrosis PROGNOSIS Often resolves without therapy. See Eosinophilia

acute pulmonary coccidioidomycosis San Joaquin Valley fever INFECTIOUS DISEASE Lung infection due to *Coccidioides immitis*, endemic in southwestern US, which follows a 10 to 30 day incubation period and more commonly affects immune compromised hosts. See Coccidiodomycosis

acute pulmonary histoplasmosis INFECTIOUS DISEASE An acute URI caused by exposure to *H capsulatum* spores, which is usually accompanied by mild flu-like Sx. See Histoplasmosis

acute radiation injury syndrome Atomic bomb disease A complex described in the radiation-exposed survivors of Hiroshima and Nagasaki, as well as nuclear reactor accidents CLINICAL 5-25 centiGray–cGy; the symptoms are a function of level of exposure

acute renal failure Acute kidney failure NEPHROLOGY An abrupt decline in renal function, triggered by various processes—eg, sepsis, shock, trauma, kidney stones, drug toxicity-aspirin, lithium, substances of abuse, toxins, iodi-

nated radiocontrast. Cf Chronic kidney failure

acute self-limited colitis GI DISEASE A condition with features that overlap idiopathic IBD; ASLC is a diagnosis of exclusion–absence of histologic criteria for diagnosing IBD–eg distorted crypt architecture, ↑ round cells and PMNs in lamina propria, villous surface, epithelioid granuloma(s), basal lymphoid aggregates, basal giant cells DIFFDx Diversion colitis, preps for scoping–eg, Fleet's enema, herbal cleansers. See Inflammatory bowel disease

acute silicoproteinosis PULMONOLOGY A rare, fulminant respiratory complication of intense exposure to respirable silica, usually in the context of sandblasting, coal mining, or any other job in which there is heavy exposure to fine particulate silica; in contrast to classic silicosis, AS appears within 6 months to 3 yrs of initial exposure and leads to severe respiratory impairment, death IMAGING–CXR Multifocal ground glass appearance and air space consolidation. Cf Coal workers' pneumoconiosis

acute skin failure BURN CARE Potentially fatal systemic consequences of skin damage, as occurs in extensive 2nd degree–deep partial and 3rd degree–full-thickness burns or in toxic epidermal necrolysis. See Toxic epidermal necrolysis

acute spinal injury NEUROSURGERY Any trauma to the spinal cord or cauda equina that may result in long-term neurologic deficit INITIAL MANAGEMENT Airway maintenance, CPR, spinal immobilization, drugs-metaraminol for hypotension, methylprednisolone to minimize motor loss, ancillary–adjunctive therapy–eg indwelling urinary catheter, NG tube-except where contraindicated as in facial injuries, gastroduodenal ulcer prophylaxis with H₂-blockers, PTE prophylaxis with low-dose heparin, pneumatic compression boots

acute stroke NEUROLOGY A stage of stroke starting at the onset of Sx and last for a few hrs thereafter. See Stroke, Transient ischemic attack

acute toxicity PHARMACOLOGY Illness caused by a single exposure to a toxic substance

acute tubular necrosis NEPHROLOGY A pathologic change of acute renal failure due to shock, crush injuries, hemoglobinuria, toxic nephrosis, sepsis, drugs-aminoglycosides, amphotericin B, cyclosporine, radiocontrast, ischemia in transplanted kidneys PREDISPOSING CONDITIONS DM, liver disease; ATN in transplanted kidneys is managed in an expectant fashion as renal function may resume spontaneously in 2-4 wks

acute urethral syndrome UROLOGY A condition characterized by dysuria with 'sterile urine'–many Pts have mild *E coli, Staphylococcal saprophyticus, Chlamydia trachomatis* infection MANAGEMENT Antibiotics

acute urinary tract infection See Urinary tract infection

acute vascular occlusion CARDIOVASCULAR DISEASE Any abrupt closure of any arterial or venous blood vessel which may be due to vasospasm or plugging by thromboemboli

acyclovir Zovirax INFECTIOUS DISEASE A nucleoside analogue used to manage viral infections in Pts with BMTs, chemotherapy-induced or acquired immunosuppression–eg, AIDS INDICATIONS HSV-1, HSV-2, HVZ ADVERSE EFFECTS Upset stomach, headache, nausea; hair loss from prolonged use. Cf Foscarnet, Gancyclovir

acylated plasminogen streptokinase complex APSAC, see there

AD 1. Adenovirus 2. Admitting diagnosis 3. Alcohol dehydrogenase 4. Alzheimer's disease 5. Androstenedione 6. Anxiety disorder 7. Autosomal dominant, see there

ad hoc *adjective* For the purpose at hand

ad hoc committee A committee formed with the purpose of addressing a specific issue or issues, which theoretically is disbanded once its raison d'etre is finished

ad- toward, to

ADA 1. Adenosine deaminase, see there 2. American Dental Association 3. American Diabetes Association 4. Americans with Disabilities Act, see there

AdalatCC® Nifedipine, see there

ADAM CARDIOLOGY A clinical trial–Amsterdam Duration of Antiretroviral Medication

ADAM complex Amniotic band syndrome, see there

Adam complex PSYCHIATRY A guilt complex of a person who breaks a parental 'law' that the subject, often a child, did not previously know was forbidden–eg, incestuous relations

Adam Ecstasy, see there

Adapin® Doxepin, see there

adaptation OPTHALMOLOGY The ability of the eye to adjust to variations in light intensity PSYCHOLOGY The fitting of behavior to the environment by modifying one's impulses, emotions, or attitudes. See Social adaptation

adaptation energy The energy a person expends when subjected to a severe emotional or physical trauma. See Burnout

adaptation response See Adaptation, psychology

adaptic WOUND CARE A clear occlusive covering used for dressing wounds of varying size including decubital ulcers, epidural insertion, arterial line insertion, where there is an ongoing need to monitor the site for infection and inflammation

ADC 1. Adrenal cortex 2. AIDS dementia complex, see there 3. Analog-to-digital computer 4. Average daily census

ADD 1. Adenosine deaminase 2. Attention deficit disorder, see there

add-on test Add test, add-on LAB MEDICINE An order for a test that is added to a previously collected specimen–eg, blood. Cf Reflex testing

addict SUBSTANCE ABUSE A person vulnerable to compulsive heavy consumption of substances with abuse potential. See Controlled drug substance, –holic, Substance abuse, Therapy addict

addiction PSYCHIATRY A preoccupation with and compulsive use of a substance despite recurrent adverse consequences; addiction often involves a loss of control and ↑ tolerance, and may be associated with a biological predisposition to addiction. See Sexual addiction. Cf Dependence SUBSTANCE ABUSE 1. A physiologic, physical, or psychological state of dependency on a substance–or pattern of compulsive use, which is characterized by tolerance, craving, and a withdrawal syndrome when intake of the substance is reduced or stopped; the most common addictions are to alcohol, caffeine, cocaine, heroin, marijuana, nicotine–the tobacco industry argues that nicotine's addictive properties are unproven, amphetamines RISK OF ADDICTION Cocaine, amphetamines > opiates & nicotine > alcohol, benzodiazepine, barbiturates > cannabis, hallucinogens, caffeine 2. A disorder involving use of opioids wherein there is a loss of control, compulsive use, and continued use despite adverse social, physical, psychological, occupational, or economic consequences. See Substance abuse 3. A neurobehavioral syndrome with genetic and environmental influences that results in psychological dependence on the use of substances for their psychic effects; addiction is characterized by compulsive use despite harm. See Pain. Cf Chronic pain

addiction medicine SUBSTANCE ABUSE The health field that addresses the needs of individuals addicted to substances of abuse including alcohol and illicit drugs–eg, cocaine, marijuana, heroin, and others; AM focuses on prevention and treatment and mental health

addiction specialist Substance abuse specialist, addictionologist, addictologist A health professional–eg, a psychiatrist, who manages a Pt with dependence on various substances of abuse–eg, alcohol, cocaine, opiates, tobacco SALARY $79K + 17% bonus

Addison's disease Chronic adrenal insufficiency ENDOCRINOLOGY A endocrinopathy characterized by ↓ production of aldosterone and cortisol ETIOLOGY Trauma, hemorrhage, TB of adrenal gland, pituitary gland destruction CLINICAL Weakness, hypotension, easy fatigability, weight loss, N&V, abdominal pain, muscle and joint pain, amenorrhea, bronzing of the skin and hyperpigmentation-especially at skin folds LAB Anemia, neutropenia, eosinophilia, lymphocytosis, hypoglycemia, ↓ Na⁺, ↓ cortisol, ↑ Ca²⁺, ↑ K⁺, ↑ BUN MANAGEMENT Hydrocortisone, fludrocortisone WARNING Pts should wear medical alert bracelet. See Polyglandular autoimmune syndrome

address *pronounced* uh-DRESS PROFESSIONAL COMMUNICATION A speech, oration, or written statement directed to a particular group of persons; to be differentiated from a lecture, which may have a didactic end. See Keynote address, National Library of Medicine

adductor Adductor muscle ANATOMY Any muscle that pulls toward the midline–medial plane of the body

ADE Adverse drug event, see there

adefovir dipivoxil VIROLOGY A nucleotide analogue effective against the polymerases of hepadnaviruses, retroviruses, herpesviruses, used to treat hepatitis Be antigen in adults with evidence of active viral replication, ↑ LFTs, or histologically active liver disease ADVERSE EFFECTS Renal damage requiring monitoring, diarrhea, nausea; in Pts with HBeAg-positive chronic hepatitis B, 48 wks of adefovir dipivoxil resulted in histologic liver improvement, reduced serum HBV DNA and alanine aminotransferase levels, and ↑ rates of HBeAg seroconversion. See AIDS, Hepatitis B

ADEK NUTRITION An acronym for a multivitamin containing fat-soluble vitamins A, D, E and K indicated for Pts with malabsorption

aden- *prefix*, Latin, pertaining to a gland

adenectomy SURGERY The surgical removal of all or part of a gland

adeno- *prefix*, Latin, pertaining to a gland

adenocarcinoma AdenoCA ONCOLOGY A carcinoma that arises in a secretory epithelium or mucosa; it is the malignant counterpart of an adenoma. See Clear cell adenocarcinoma, Minimum deviation adenocarcinoma of the endocervix. Cf Adenoma

adenocarcinoma of renal cells See Renal cell carcinoma

adenocarcinoma of bronchus See Lung cancer

adenocarcinoma of uterus Endometrial carcinoma, see there

adenofibroma Fibroadenoma, see there

adenoid *adjective* 1. Gland-like 2. Relating to the adenoids *noun* See Adenoids

adenoid cystic carcinoma Cylindroma SURGICAL PATHOLOGY An uncommon carcinoma characterized by a cribriform or sieve-like pattern when viewed by low-power LM SITES SIGHTED Salivary glands, as well as breast, cervix, lung and in the head & neck; in the salivary glands, ACC is indolent, but malignant with a tendency to recur and invade perineurial spaces. See Salivary gland tumor

adenoid hypertrophy Adenoidal hypertrophy, see there

adenoidal hypertrophy Adenoid hypertrophy ENT Chronic enlargement of the adenoids, usually accompanied by recurrent infections which, if deemed excessively frequent, is an indication for adenoidectomy, see there

adenoidectomy ENT Surgical removal of adenoids INDICATIONS Upper airway or nasal obstruction, sleep apnea, chronic otitis media. See Tonsils and adenoids

adenoiditis Inflammation of the adenoids

adenoids ANATOMY 1. Adenoid glands, pharyngeal tonsil Lymphoid tissue that forms a prominence on the wall of the pharyngeal recess of the nasopharynx 2. A popular term for adenoidal hypertrophy, see there

adenoids & tonsils See Tonsils and adenoids

adenomatous polyp GI DISEASE A premalignant glandular tumor arising in the GI tract mucosa, which is more common in older persons MANAGEMENT Polypectomy, followup every 3 yrs

adenomatous polyposis coli Familial adenomatous polyposis, see there. See APC gene, APC protein

adenomyosis Stromal endometriosis GYNECOLOGY A condition characterized by extension of endometrial glands into the myometrium, often accompanied by diffuse and symmetric enlargement of the uterus RISK POPULATION More common in multiparous ♀ > age 30, possibly exacerbated by OCs CLINICAL Prolonged or heavy menstrual bleeding, painful menses with cramping MANAGEMENT Analgesics; hysterectomy

adenopathy A general term for a lymph node disease. See Lymphadenopathy

Adenoscan® CARDIOLOGY A sterile formulation of adenosine used as an adjunct to thallium in cardiac imaging to evaluate CAD in Pts unable to adequately exercise and take the treadmill component of related tests. See Treadmill test

adenosine CARDIOLOGY An endogenous nucleoside composed of adenine linked to D-ribose, which results from the hydrolysis of adenylic acid; adenosine is a therapeutic alternative to verapamil, in treating both narrow- and wide-complex SVT; it ↓ heart rate, followed by a reflex ↑ in sinus discharge. See Calcium-channel blocker, Supraventricular tachycardia

adenosine deaminase deficiency ADA deficiency A uniformly fatal AD disease, which consitutes 40% of Pts with SCID CLINICAL Cellular immune dysfunction, oral candidiasis, intractable diarrhea, FTT, severe diaper rash, pseudoachondrodysplasia, death by age 2 LAB ↓ Lymphocytes < 0.5 x 10^9/L–US, < 500 mm³, especially T cells; eosinophilia; ↑ adenosine and deoxyadenosine in serum and urine TREATMENT Gene therapy; rarely BMT. See Cartilage-hair syndrome

adenosis BREAST Any hyperplastic process that primarily involves the glands of the breast. See Blunt duct adenosis, Sclerosing adenosis

adenotonsillectomy ENT Simultaneous excision of the tonsils and adenoids

adenovirus Any of a family of > 40 icosahedral–20-sided nonenveloped DNA viruses with a 36 kb genome, which has 'early' genes that encode regulatory proteins, and 'late' genes that encode structural proteins GENE THERAPY Adenovirus is '…*particularly attractive as a vector* (for gene therapy), *as it can produce large amounts of highly purified recombinant virus, which efficiently infects differentiated cells*'; the virus is engineered to ensure that it remains in the desired location by replacing the early genes with the genes of interest–eg, *CFTR*–cystic fibrosis transmembrane conductance regulator gene VIROLOGY Adenovirus can cause common cold, noninfluenzal acute respiratory disease, pneumonia, epidemic keratoconjunctivitis, acute febrile pharyngitis, acute hemorrhagic cystitis DIAGNOSIS Indirect methods–antibod-

ies; direct methods–adenovirus antigens by immune–antigen-antibody reactions; viral culture is too time-consuming to be practical. See EIA

Adentri™ CVT-124 CARDIOLOGY An adenosine A1 receptor antagonist in clinical trials in Pts with CHF. See Congestive heart failure

Adepril® Amitriptyline, see there

ADEPT PSYCHOLOGY A study–Adolescent Depression Prevention & Treatment–funded by the Natl Inst Mental Health, examining the cost-effectiveness of a 'service' model for treating children of depressed adults enrolled in HMOs. See Depression

ADFR therapy Activate, depress, free, repeat A therapeutic modality–phosphorus, followed by calcitonin and calcium, administered in 3 month cycles, used to attenuate post-menopausal bone loss in ♀ for whom estrogen therapy is contraindicated or unacceptable

ADG Ambulatory diagnostic group

ADH 1. Alcohol dehydrogenase, see there 2. Antidiuretic hormone, see there

ADHD Attention deficit/hyperactivity disorder. See Attention-deficit disorder

adherence to treatment Compliance THERAPEUTICS The following of a recommended course of treatment by taking all prescribed medications for the length of time necessary

adhesiolysis SURGERY The surgical lysis of adhesions, usually by laparoscopy

adhesion The stable joining of parts to each other, or the union of 2 opposing tissue surfaces, which may be normal or abnormal HEMATOLOGY See Platelet adhesion SURGERY Synechia A collagen-rich fibrous band, scar, or stricture, that forms after an intervention in a surgical field, classically in the peritoneal cavity after abdominal surgery or laparotomy; adhesions may be related to a focal ↓ in plasminogen activator in the mesothelial lining or to local inflammation or infection; gentle manipulation of the organs and removal of blood minimizes adhesive band formation, which may be severe enough to cause intestinal obstruction; nothing effectively prevents adhesions. See Violin string adhesions

adhesion filter Adsorption filter TRANSFUSION MEDICINE A '3rd-generation' blood component filter that removes 99-99.9% of WBCs–by adhesion in units of packed red cells and platelets. See Blood filters, Leukocyte reduction

adhesive capsulitis ORTHOPEDICS A condition caused by prolonged immobility of the shoulder joint CLINICAL Shoulder is painful, tender, ↓ passive and active ROM MANAGEMENT Physical therapy, corticosteroid injection; surgery if unresponsive. See Range of motion

adiadochokinesia NEUROLOGY Inability to perform rapid alternating movements of one or more extremity

adipose *adjective* Pertaining to fat cells or tissue

adjunct MEDTALK A thing joined or added to another thing but which is not an essential part thereof–eg, RT is an important adjunct to surgery and may represent appropriate adjuvant therapy

adjunct staff MANAGED CARE A body of physicians, dentists, osteopaths, or other medical practitioners, who have achieved a certain level of professional distinction, and participate in educational or research programs in a hospital/health care center, but in general do not provide or supervise provision of direct Pt management or medical care—eg by residents or fellows, and don't admit Pts. See House staff, Medical staff

adjunct test NIHSPEAK A test that provides information that adds to or helps interpret the results of other tests, and provides information useful for risk assessment

adjunctive therapy MEDTALK A therapeutic maneuver(s) with an ancillary role in treating a disease by ↓ M&M, but not part of the immediate therapy required to stabilize the Pt. Cf Adjuvant therapy

adjustment MANAGED CARE See Case-mix adjustment PSYCHIATRY Functional, often transitory, alteration or accommodation by which one can better adapt to the immediate environment and inner self. See Adaptation

adjustment disorder CHILD PSYCHIATRY A constellation of extreme reactions in adolescents to social demands for establishing personal identity and independence from family

adjuvant IMMUNOLOGY Any nonspecific immune enhancer—eg Freund's adjuvant, BCG vaccine, consisting of a particulate-rich oily substances, which promotes protein aggregation; adjuvant mixed with an antigen acts as a tissue depot, slowly releasing antigen and activating the immune system ONCOLOGY The addition of chemotherapy to a traditional therapeutic modality to ↓ M&M PHARMACOLOGY A substance which, when added to a medication, enhances its pharmacologic effect. See Neoadjuvant

adjuvant analgesic PAIN MANAGEMENT An ancillary agent with independent or additive analgesic properties, which allows a ↓ in the amount of analgesics needed to relieve symptoms in Pts with CA, AIDS, and other dread disease

adjuvant therapy Any therapy that ↑ a primary treatment's efficacy ONCOLOGY A treatment—eg, chemotherapy, RT, or hormone therapy, used after a primary treatment of a tumor, to prevent metastases, or for residual malignancy after excision; AT is used after one or more of the conventional therapeutic arms—surgery, chemotherapy, RT, has failed. See IL-2/LAK cells. Cf Adjuvant chemotherapy THERAPEUTICS Therapy that enhances an primary therapy; auxiliary therapy.

ADL scale Activity of daily living scale CLINICAL MEDICINE Any of a number of instruments used to assess physical functions—eg, self-care, ambulation, food preparation, shopping, housekeeping, etc. See Barthel index, Framingham Disability scale, Instrumental ADL scale, Katz ADL scale, Kenny Self-Care scale, Performance test of ADL, Timed manual performance

ADME parameters DRUG DEVELOPMENT The critical—absorption, distribution, metabolism, elimination pharmacokinetic—'hoops' that a drug must jump through before FDA approval for clinical use. See Pharmacokinetics

administer PHARMACOLOGY *verb* To apply a substance—by injection, inhalation, ingestion or by other means, to the body of a Pt or research subject by either a health practitioner or his authorized agent and under his direction, or by the Pt or research subject himself. Cf Dispense MEDTALK → VOX POPULI Give

administration MANAGED CARE The sum total of management and direction of health care organizations—personnel, budgets, and logistics, and the implementation of health care policy. See Management tools

administrative cost MANAGED CARE A cost incurred by the 'business' end of a health care facility or university—eg, staffing and personnel costs, nursing home and hospital administration, insurance, and overhead expenses. See Indirect costs

administrative intervention DIAGNOSTIC MEDICINE Any intervention on the part of an administrative body—eg in a hospital or other health care facility, which is intended to influence a physician's pattern of practice—eg, to ↓ overordering of tests for Pt management

administrative responsibility Any task or duty related to managing an institution; non-Pt management-related responsibilities of physicians include chart review, participation in the tumor board or tissue committee, etc. Cf Clinical responsibility

administrator BUREAUCRACY A white-collar worker, usually a nonphysician, who directs and/or manages operational aspects of a business or bureaucracy. See Scientific review administrator, State fluoridation administrator, Third-party administrator

ADMIRAL CARDIOLOGY A clinical trial—Abciximab before Direct Angioplasty & stenting in Myocardial Infarction Regarding Acute and Long-term followup. See Abciximab, Angioplasty, Myocardial infarction, Stenting

admission HOSPITAL PRACTICE An episode of in-hospital health care. See Frequent flyer, Nth admission, Readmission

admission hyponatremia LAB MEDICINE Serum Na+ levels of < 130 mmol/L which, in older persons–> age 64 is associated with 2-fold ↑ in mortality

admit *noun* A popular term for a Pt who has been admitted to a hospital or ward *verb* To arrange for a person's admission into a hospital

admits HOSPITAL CARE The number of admissions to a hospital including outpatient and inpatient facilities. See Admission

admitting Admitting department MEDTALK A popular term for the hospitalization process which begins in the ER and/or outpatient registration and consists of taking Pt demographics, insurance information, and assigning a room. Cf Discharge, Office area

admitting privilege MANAGED CARE The right, by virtue of membership on a hospital's medical staff, to admit private Pts in a particular medical center or hospital, and to render specific diagnostic or therapeutic services in that hospital. See Staff privileges

adnexae Fallopian tubes and ovaries

adnexitis GYNECOLOGY Inflammation of uterine adnexae

Adofen Fluoxetine, see there

adolescence ADOLESCENT MEDICINE A period that begins with the onset of 2° sexual characteristics—puberty and ends with the cessation of growth—adulthood, in the vernacular, teenagehood, generally between 13 and 18. Cf Adulthood, Childhood

ADOLESCENCE—SIGNS OF PSYCHOLOGICAL PROBLEMS

PROBLEM BEHAVIORS, eg substance abuse, sexual promiscuity, delinquency

INTERPERSONAL ISOLATION from friends and family

COGNITIVE DYSFUNCTION, eg decline in academic performance, irregularities in expressive or receptive language

To be normal during adolescence is, by itself abnormal—Anna Freud

adolescent A person in the adolescent period. See Homeless adolescent, Teenage pregnancy

adolescent crisis PSYCHOLOGY The constellation of relatively abrupt and profound changes that the physical and emotional rigors of adolescence place on their 'victim'. Cf Midlife crisis

adolescent gynecomastia INTERNAL MEDICINE A condition that affects ±15% of ♂ at puberty, caused by growth of glandular tissue and enlargement of the breasts in response to the hormones; typically the enlargement is minimal and self-correcting, but may be severe enough to require corrective mammoplasty

adolescent medicine Ephebiatrics A subspecialty of internal medicine or pediatrics, dedicated to managing and maintaining the health issues of older children. See Pediatrics

adolescent pregnancy See Teenage pregnancy

ADOPT ENDOCRINOLOGY A 4-yr phase IV trial–A Diabetes Outcome Progression Trial–comparing early treatment of Pts with type 2 DM with Avandia®, metformin, or glyburide–a sulfonylurea in glucose control, delay or prevention of DM complications–eg, kidney disease, etc

ADOPT-A CARDIOLOGY A clinical trial–Atrial Dynamic Overdrive Pacing Trial-A. See Pacemaker

adopted person Adoptee SOCIAL MEDICINE A person who has been legally adopted by one or two parents. See Adoption

adoption SOCIAL MEDICINE The act of lawfully assuming the parental rights and responsibilities of another person, usually a child under the age of 18; the care and nurturing of a child by a non-blood-related adult who assumes the roles, rights, and obligations of a natural parent; 2% of children < age 18–US are adopted–± 1 million. See Cooperative adoption, Designated adoption, Independent adoption, Infant adoption, Informal adoption, Open adoption, Relative adoption, Semiadoption, Simple adoption, Traditional adoption, Transracial adoption, Wrongful adoption, Zygote adoption

adoptive parents SOCIAL MEDICINE Persons who lawfully adopt children, who are generally married couples but may be single persons, including homosexuals; most APs are married

ADPKD Autosomal dominant polycystic kidney disease, see there

ADR Adverse drug reaction, see there

adrenal antibodies See Anti-adrenal antibodies

adrenal crisis Addisonian crisis, acute adrenal insufficiency ENDOCRINOLOGY Acute life-threatening adrenocortical insufficiency with ↓ serum cortisol, seen in Pts with Addison's disease TRIGGERS Infections, trauma, hemorrhage, TB, surgery, dehydration with salt deprivation, destruction of pituitary gland or evoked by replacing thyroid hormone in Pts with hypothyroidism of hypothalamic or pituitary origin and underlying mild ACTH deficiency CLINICAL Hypotension, shock, fever, dehydration, anorexia, weakness, apathy, headache, vomiting, chills, tachycardia, sweating LAB ↓ Na⁺, ↑ K⁺, ↑ WBCs, eosinophilia, hypoglycemia MANAGEMENT Pharmacologic doses of IV hydrocortisone, BP support

adrenal hyperplasia Diffuse enlargement of the adrenal glands. See Congenital adrenal hyperplasia

adrenal insufficiency Adrenal gland insufficiency, adrenocortical insufficiency, Addison's disease ENDOCRINOLOGY A condition characterized by a marked ↓ in adrenal function CLINICAL Weakness, hypotension, easy fatigability, weight loss, N&V, abdominal pain, muscle and joint pain, amenorrhea, bronzing of the skin and hyperpigmentation-especially at skin folds MANAGEMENT Hydrocortisone, fludrocortisone

1⁰ ADRENAL INSUFFICIENCY

ETIOLOGY

- **ABRUPT ONSET** Adrenal hemorrhage, necrosis, thrombosis, sepsis, coagulopathy, warfarin therapy, antiphospholipid syndrome
- **SLOW ONSET** Autoimmune disease, TB, adrenomyeloneuropathy, systemic fungal infection, metastatic cancer, AIDS-related-eg Kaposi sarcoma, CMV

CLINICAL

- **SYMPTOMS SHARED WITH 2⁰ ADRENAL INSUFFICIENCY** Tiredness, weakness, mental depression, anorexia, weight loss, vertigo, orthostatic hypotension, nausea, vomiting, diarrhea, hyponatremia, hypoglycemia, mild anemia, increase WBCs, eosinophilia
- **1⁰ ADRENAL INSUFFICIENCY** Hyperpigmentation, ↑ K⁺, vitiligo, autoimmune thyroid disease, CNS Sx if adrenomyeloneuropathy

DIAGNOSIS

ADRENAL INSUFFICIENCY–EXCLUSION OF Basal cortisol at 8-9 am–≤ 3 µg/dL–normal, 6-24 µg/dL; corticotropin stimulation test-little ↑ after stimulation–normal, plasma cortisol ≥ 20 µg/dL

- **1⁰ ADRENAL INSUFFICIENCY** Corticotropin stimulation test-no ↑ after stimulation–normal, plasma cortisol ≥ 20 µg/dL; plasma cortisol in low-normal range; plasma corticotropin always ≥ 100 pg/mL

2⁰ ADRENAL INSUFFICIENCY

ETIOLOGY

- **ABRUPT ONSET** Postpartum pituitary necrosis, necrosis or hemorrhage into pituitary adenoma, pituitary or adrenal surgery
- **SLOW ONSET** Pituitary–1⁰ or metastatic tumor, craniopharyngioma–± accompanied by pituitary surgery or radiation, sarcoidosis, Langerhans cell histiocytosis–histiocytosis X, empty sella syndrome, hypothalamic tumors, long-term glucocorticoid therapy

CLINICAL

- **SYMPTOMS SHARED WITH 1⁰ ADRENAL INSUFFICIENCY** Tiredness, weakness, mental depression, anorexia, weight loss, vertigo, orthostatic hypotension, nausea, vomiting, diarrhea, hyponatremia, hypoglycemia, mild anemia, increased WBCs, eosinophilia
- **2⁰ ADRENAL INSUFFICIENCY** Pallor of skin, amenorrhea, ↓ libido, ↓ axillary and pubic hair, small testicles, 2° hyperthyroidism, defect in prepubertal growth, delayed puberty, headache, defects in vision, diabetes insipidus

DIAGNOSIS

- **ADRENAL INSUFFICIENCY (RULE OUT)** Basal cortisol at 8-9 am–≤ 3 µg/dL–normal, 6-24 µg/dL; corticotropin stimulation test-little ↑ after stimulation–normal, plasma cortisol ≥ 20 µg/dL
- **2⁰ ADRENAL INSUFFICIENCY** Insulin-induced hypoglycemia-little or no ↑ in plasma corticotropin and cortisol in 2° adrenal insufficiency; short metaprone test–insufficient ↑ in plasma corticotropin and cortisol; low-dose corticotropin stimulation test-little ↑ after stimulation–normal, plasma cortisol ≥ 20 µg/dL

NEJM 1996; 335:1206nv

adrenal medullary scan IMAGING A nuclear scan of the adrenal glands after IV injection of a radioactive tracer, used to detect pheochromocytomas, in particular, extraadrenal tumors

adrenal medullary transplantation Brain graft surgery, see there

adrenal tumor 1. Adrenocortical adenoma, see there 2. Adrenocortical adenocarcinoma, see there

adrenalectomy Removal of the adrenal glands

adrenaline British for epinephrine, see there

adrenergic *adjective* Referring to 1. Neural activation by catecholamines–eg, epinephrine, norepinephrine, dopamine. See Biogenic amines, Neurotransmitters, Sympathetic nervous system 2. Sympathetic nerve fibers that liberate epinephrine or norepinephrine into a synapse *noun* Any agent with adrenergic–agonist activity

adrenergic bronchodilator PHARMACOLOGY An agent that dilates bronchial lumina and ↓ airway resistance, used for asthmatic Pts. See Asthma, Wheezing

adrenergic receptor NEUROPHYSIOLOGY Any of a family of cell membrane receptors that receive neuronal impulses from postganglionic adrenergic fibers from the sympathetic nervous sytem, which are divided into α receptors, which results in an excitatory response of smooth muscle cells to catecholamines, and β receptors, which result in an inhibitory response to catecholamines; the GI tract is an exception, in that either α or β receptor stimulation results in relaxation

adrenocortical *adjective* Referring to the adrenal cortex

adrenocortical carcinoma ONCOLOGY A rare–≤ 2 in 10⁶/yr tumor of the adrenal cortex; average age 46; ♂:♀ ratio 1:2.5 CLINICAL ± 70% of Pts with AC have endocrine symptoms due to excess secretion of glucocorticoids and/or androgens, less commonly due to excess mineralocorticoids or estrogens MANAGEMENT Surgical resection, mitotane. See Mitotane

ADRENOCORTICAL CANCER

STAGE I CA < 5 cms; no spread to periadrenal tissues

STAGE II CA > 5 cms; no spread to periadrenal tissues

STAGE III CA spread to periadrenal tissues or lymph nodes

STAGE IV CA spread to regional tissues or organs, to lymph nodes around adrenal cortex, or metastasized

adrenocortical hormone Any hormone–eg, cortisol, mineralocorticoids, estrogen, etc–secreted by the adrenal cortex

adrenogenital syndrome Congenital adrenal hyperplasia, see there

adrenoleukodystrophy Melanodermic leukodystrophy An X-linked peroxisomal disease in which impaired oxidation of saturated long-chain fatty acids is associated with adrenal insufficiency–Addison's disease and neurologic impairment, and in late-onset cases, adrenomyeloneuropathy. See Leukodystrophy

adrenolytic PHARMACOLOGY *adjective* Referring to 1. Inhibition of epinephrine–adrenaline activity or secretion 2. Sympathetic nerve fibers that block epinephrine or norepinephrine activity at a synapse *noun* Any agent with adrenolytic–antagonist activity

β_2-adrenoceptor agonist β_2-agonist A family of antiasthmatics-albuterol, formoterol, salmeterol, terbutaline, that induce bronchodilation

adriamycin Doxorubicin HCl ONCOLOGY An anthracycline antibiotic with antineoplastic activity, used to treat both leukemia and solid tumors; its efficacy is hampered by cardiotoxicity that $\uparrow$ as the cumulative dose passes 500 mg/mm^3

adult immunization The administration of vaccines to prevent clinical infection in adulthood; '*The contrast between the impact of vaccine-preventable diseases of adults compared with those of children is striking. Each yr, < 500 persons in the U.S. die of vaccine-preventable diseases of childhood … 50,000 to 70,000 adults die of influenza, pneumococcal infections, HBV.* Cf Childhood immunization

ADULT IMMUNIZATION

INFECTIONS	ANNUAL DEATHS	VACCINE EFFICACY	CURRENT VACCINE USE	PREVENTABLE DEATHS
PNEUMOCOCCI	40,000	60%	20%	19,200
INFLUENZA	20,000	70%	41%	8,260
HBV	5000	90%	10%	4,050
MMR‡	<30	95%	variable	<30
TETANUS-DIPHTHERIA	<25	99%	40%	<15

‡Measles-mumps-rubella. From JAMA 1994; 272:1133

adult obesity PUBLIC HEALTH Overweight in an adult, defined as an average body-mass index of $\geq$ 27.8 in $\male$ and 27.3 in $\female$. See Morbid obesity, Obesity. Cf Childhood obesity

adult-onset diabetes mellitus Type 2 diabetes mellitus, see there

adult-onset Still's disease RHEUMATOLOGY Still's disease–acute febrile onset of arthritis in adults CLINICAL Multiple remissions and exacerbations, loss of wrist extension, carpal ankylosis, distal interphalangeal joint involvement; other features include daily fever spikes, loss of neck motion, evanescent pink macules most prominent when febrile, pericarditis, pleural effusions, severe abdominal pain LAB Normal rheumatoid factor, ANA, complement See Rheumatoid arthritis, Still's disease

adult respiratory distress syndrome PULMONARY MEDICINE A clinical condition consisting of a diffuse infiltrative process in the lung, which affects ±150,000/yr–US characterized by acute pulmonary edema

and respiratory failure, poor oxygenation, $\uparrow$ functional residual capacity, and $\downarrow$ compliance; ARDS may accompany various medical and surgical conditions and may be associated with interstitial pneumonitis–usual, desquamative and lymphoid types ETIOLOGY Gram-negative sepsis, pneumonia, shock, gastric acid aspiration, trauma, drug overdose, toxic gas–chlorine, NO$_2$, smoke exposure, severe metabolic derangement, pancreatitis CLINICAL A 6-24 hr latency period is followed by hypoxia, $\downarrow$ aeration, dyspnea, severe SOB, 'stiff' lungs, ie $\downarrow$ pulmonary compliance RADIOLOGY Extensive, diffuse bilateral fluffy infiltrates MANAGEMENT Nitric oxide–NO, 18 ppm in one study inhalation therapy results in a $\downarrow$ mean pulmonary artery pressure–37 to 30 mm Hg, $\downarrow$ intrapulmonary shunting–36% to 31%, $\uparrow$ ratio of partial pressure of arterial O$_2$ to the fraction of inspired O$_2$–PaO$_2$/FiO$_2$, an index of arterial oxygenation efficiency–±152 to ±199 PEEP, prayer PROGNOSIS A function of underlying cause MORTALITY ± 60%, the cause of death has shifted from hypoxia to multiorgan failure

adult T-cell leukemia-lymphoma ONCOLOGY A rapidly-progressive lymphoproliferative malignancy of mature T lymphocytes, commonly associated with infection by the retrovirus, HTLV-I, first described in southeastern Japan, also seen in the Caribbean, Africa and in blacks in the southeastern US, in whom the disease is aggressive with skin lesions, hypercalcemia, rapid enlargement of hilar, retroperitoneal and peripheral lymph nodes with mediastinal sparing, invasion of CNS, lungs, GI tract, opportunistic infections–eg, *Pneumocystis carinii*. See T-cell lymphoma

ADULT T-CELL LEUKEMIA-LYMPHOMA–CLINICAL FORMS

ACUTE Median age 52, lymphadenopathy, hepatosplenomegaly, cutaneous lesions, up to a 20-year latency, often resistant to chemotherapy with poor prognosis following disease onset LAB $\uparrow$ Ca^{++}, WBCs 10-500 × 10^9–US: 10-500 000/mm^3, Sezary-like cells with CD3, CD4, CD2, and Tac+ surface antigens, causing a chronic, smoldering lymphoma

CHRONIC Clinically between acute and smoldering disease

SMOLDERING Characterized by erythematous skin nodules filled with lymphocytes that may undergo 'blast transformation' to the typical acute T cell leukemia

CRISIS When either 2. or 3. transform to ATL

LYMPHOMA Most common in US; affects blacks with hypercalcemia, leukemia, hepatosplenomegaly, erythematous skin lesions, lytic bone lesions

adult teeth Permanent teeth, see there

adult temper tantrum PSYCHIATRY A prolonged anger reaction which, unlike temper tantrums in toddlers who are merely trying to get their own way, is distinctly abnormal. Cf Temper tantrum

adulteration PHARMACOLOGY The substitution of one material or substance for another, such that a manufactured product is incorrectly labeled and/or dosage information is not in accordance with the FDA requirements, which $\downarrow$ potency or value, or adds unnecessary ingredients

adulthood VOX POPULI 1. The period of physical maturity that follows adolescence 2. Twenty-one yrs of age–in the US, by act of Congress, childhood officially ends at the 18th birthday–Public Law 98-292, The Child Protection Act of 1984

advance directive Advance medical directive, self-determination MEDICAL ETHICS Instruction(s) that provide a mentally competent person with a

ADVANCE DIRECTIVE TYPES

LIVING WILL, in which the person outlines–usually in writing, specific treatment guidelines to be followed by health care providers

HEALTH CARE PROXY Power of attorney for healthcare decision making, proxy to make the health care decisions. The person designates a trusted individual to make medical decisions in the event of inability to make such decisions

vehicle for directing his/her own treatment in the event of serious illness and/or loss of mental ability to communicate those wishes; in an AD, the person indicates in advance, how treatment decisions are to be made with regard to the use of artificial life support. See DNR orders, Durable powers of attorney, Euthanasia, Living will

advance medical directive Advance directive, see there

advanced beneficiary notice MANAGED CARE A document signed by a Pt accepting responsibility for paying for a test or diagnostic service which the Pt's primary care thinks is appropriate, which Medicare may not under Medicare's 'reasonable and necessary' standard, and therefore not pay the party performing the test. See Reasonable and necessary

Advanced Cardiac Life Support See ACLS

advanced life support EMERGENCY MEDICINE A generic term for resuscitation efforts that extend beyond basic CPR

ADVANCED LIFE SUPPORT METHODS

ADVANCED VENTILATORY SUPPORT, eg tracheal intubation, pharyngotracheal lumen airway, esophageal obturator airway, transtracheal catheter ventilation

IV ACCESS

CORRECTION OF ACIDOSIS

DIAGNOSTIC TESTS, eg EKG, arterial blood gases

TREAT ARRHYTHMIAS, either with drugs or electroconversion

ADVANCED PERFUSION SUPPORT. Cf Basic life support

---

advanced-practice nurse A registered nurse with specialty training and education–usually a masters degree, ie, 6+ yrs of formal college or university education in primary care–eg nurse practitioner, midwife, or acute care of inpatients–eg clinical nurse specialist, intensive care specialist who performs various examinations–eg, flexible sigmoidoscopic examination

adverse action MEDICAL MALPRACTICE Any of a number of professional reprimands or sanctions. See Incompetence, National Practitioner Data Bank, Staff privileges

adverse drug reaction Adverse drug event PHARMACOLOGY Any noxious, undesired, or unintended response to a drug, which occurs at dosages used for prophylaxis, diagnosis, therapy, or modification of physiologic functions; ADEs do not include therapeutic failures, poisoning, or intentional overdoses; ADEs occur in 1-15% of all drug administrations, but are rarely fatal

adverse effect PHARMACOTHERAPY An undesirable and unintended, although not necessarily unexpected, result of therapy or other intervention–eg, headache following spinal tap or intestinal bleeding associated with aspirin therapy TOXICOLOGY An abnormal or harmful effect on an organism due to exposure to a chemical or noxious substance; AEs cause functional or anatomic damage, irreversible changes in homeostasis, or ↑ an organism's susceptibility to other chemical or biologic stress CLINICAL Change in food or water consumption, body or organ weight, enzyme activity, visible illness or death

adverse event MALPRACTICE An injury caused by medical management–rather than by the underlying disease–which prolongs hospitalization, produces a disability at the time of discharge, or both ETIOLOGY Drug effects, wound infections, technical complications, negligence, diagnostic mishaps, therapeutic mishaps, and events occurring in the emergency room. See Malpractice, Misadventure, Negligence. Cf Adverse effect

adverse selection MANAGED CARE 1. A stance adopted by health care insurers, which fiercely compete among themselves to insure the healthiest and wealthiest segment of a particular population, and thus adversely select the population which they target for selling insurance policies. See 'Safety net' hospital 2. A health plan, whether indemnity or managed care, is selected over other plans by enrollees who are more likely to file claims and use services, causing an inequitable proportion of enrollees requiring more medical services in that plan

advertising The public notification of a product's availability and related activities for its promotion MEDICAL COMMUNICATION A public notice, usually in the form of a paid announcement, which may be printed or broadcast. See 'Coming soon advertising' Direct-to-consumer advertising, Institutional advertising, Introductory advertising, Remedial advertising, Reminder advertising, Sexist advertising,, Tobacco advertising

Advil Ibuprofen, see there

adviser GRADUATE EDUCATION A person formally affilitated with an academic institution who counsels one or more students and–ideally–acts as the student's advocate

advocate ETHICS *noun* (pron. ad´ ve ket) A person who acts on the behalf of or speaks for another–eg, for a cause or plea. See Amicus curiæ *verb* (pron. ad ve kát) To act or speak for another person or group of person

AECB Acute exacerbation of chronic bronchitis. See Chronic bronchitis

AECD Automatic external cardioverter-defibrillator A device that monitors cardiac rhythm and automatically defibrillates

AED Automated external defibrillator, see there

aerate PHYSIOLOGY *verb* To add air or O_2 into a liquid. See Waste treatment

aerial *adjective* Referring to that which is found in or takes place in the air

aerobic *adjective* SPORTS MEDICINE Referring to exercise in which energy is supplied by O_2 and is required for sustained periods of generally 20 mins or more with a generally high pulse rate at ± 80% of maximum. Cf Anaerobic

aerobic bacteria Bacteria that grow in the presence of O_2, which are the most common causes of clinical infection. Cf Anaerobic bacteria

aerobic fitness CLINICAL MEDICINE A value obtained from exercise testing, which is expressed as either $V_{O_2,peak}$–O_2 consumption at peak exercise, or W_{peak}–peak work capacity or tolerance, expressed as a percentage of normal. See Exercise testing, Physical fitness

aerodigestive tract SURGICAL ANATOMY A term that encompasses the oral cavity, sinonasal tract, larynx, pyriform sinus, pharynx, and esophagus

aerogenic Gas-producing–eg, aerogenic fermentation

Aeromonas **spp** MICROBIOLOGY A genus of gram-negative, facultatively anaerobic, nonspore-forming bacilli, which have been isolated from various foods, including dairy products, meats and vegetable; *Aeromonas* spp may contaminate skin wounds and cause intestinal and extra-intestinal infections including meningitis and sepsis in immunocompromised hosts

aerophagia Aerophagy The excessive swallowing of air, which is usually an unconscious act that occurs during normal eating or drinking or is associated with other conditions CLINICAL Abdominal bloating, belching, flatulence ETIOLOGY Anxiety, rapid eating–'shoveling' or drinking–'guzzling,' chewing gum, smoking, or poorly fitted dentures. See Pneumatosis cystoides intestinalis

aerosol EPIDEMIOLOGY A particulate suspension of infectious agents which may remain pathogenic for long periods of time; a fine mist or spray containing minute particles from soil, clothes, bedding or floors when moved, cleaned or blown by wind; aerosolized material includes fungal spores–infective agents themselves–bits of infected feces, or particles of dirt or soil that have been contaminated with a pathogen. See Droplet nuclei OCCUPATIONAL MEDICINE A suspension of particles in air. See Acid aerosol, Airborne transmission, Inhalant, Flammable aerosol, Sclerosol® (talc powder) intrapleural aerosol THERAPEUTICS A suspension or dispersion of fine particles, ranging 10^{-6}-10^{-9} in diameter, of a solid or liquid in a gas, which is atomized into a fine mist for inhalation therapy, or for a nebulizer and inhaled

aerospace medicine A field of medicine that studies the effects of atmospheric & space flight on human physiology and health TYPES 1. Aviation medicine, see there 2. Space medicine, see there

aerotolerant *adjective* Referring to the ability to survive in the presence of O_2, as in facultative anaerobes, which don't require O_2 for survival, but are not harmed by its presence

Aesculapian staff MEDICAL HISTORY The symbol of medicine and the art of healing, which is depicted by a *single* serpent coiled around a rough knotty staff. Cf Caduceus

AFB Acid-fast bacillus, also 1. Aflatoxin B 2. Aorto-femoral bypass

AFB₁ Aflatoxin B1, see there

afebrile *adjective* Feverless

affair SEXOLOGY A long-term sexual liaison, usually understood to mean one in which one or both partners are committing adultery VOX POPULI As commonly used, a major whoops–faux pas, boo-boo, etc between 2 or more people. See Baltimore affair, Burt affair, Cantekin affair, Cleveland affair, Slutsky affair, Swine influenza vaccine affair

affect PSYCHIATRY 1. The observed emotional state of a Pt, which may be modified by such adjectives as blunted, dramatic, labile, sad 2. The subjective experience of emotion accompanying an idea or mental representation; affect is loosely synonymous with feeling, emotion, or mood. See Emotion, Flat affect, Inappropriate affect, Mood

affective disorder PSYCHIATRY Any mood disorder, including depressive and bipolar illness. See Bipolar disorder, Depression

afferent arteriole ANATOMY A blood vessel in the kidney that supplies blood to glomeruli

afferent *adjective* 1. Conveying or transmitting 2. Referring to the movement of blood or nerve impulses centrally–eg, through vessels toward the heart, or nerves to the brain, Cf Efferent

afferent loop syndrome Blind loop syndrome A type of postgastrectomy syndrome, caused by partial or complete obstruction of the proximal portion–loop of a Billroth II anastomosis between the stomach and jejunum CLINICAL Pain, bloating, abdominal tenderness, with bacterial overgrowth and malabsorption

affidavit FORENSIC MEDICINE Deposition; a sworn pledge, oath, or statement by a person about certain facts. See testimony

affiliation MEDICAL PRACTICE A professional relationship, including an employment relationship, a position as an independent contractor or granting of privileges by a health care facility or HMO See Hospital affiliation

affirmative action SOCIAL MEDICINE The removal of artificial barriers to the employment of ♀ and disadvantaged minorities, or admission of same to highly selective institutions of higher learning; AA refers to any effort to recruit and hire members of previously disadvantaged groups, as a means of erasing past inequities. See *Bakke v Regents of the University of California*, Reverse discrimination; Cf Glass ceiling

affirmative defense MEDICAL LIABILITY A defense used by a physician in a lawsuit that is based on adherence to the local standards of care–which may be established by Ob/Gyns, emergency room specialists, and anesthesiologists. See Medical malpractice

affluent diet Western diet CLINICAL NUTRITION A diet characterized by a

↑↑↑ fat, saturated fat, cholesterol, and calories, which is commonly consumed in wealthier nations. Cf Mediterranean diet, Ornish diet

afibrinogenemia HEMATOLOGY A rare AR condition characterized by complete incoagulability of blood, which may be first identified by excess bleeding from the umbilical stump; other findings include bone and liver lesions, and spontaneous rupture of spleen

AFO Ankle-foot orthosis

AFP α-fetoprotein, see there

African American MULTICULTURE A person having origins in any of the black racial groups of Africa. See Race

after-baby blues Post-partum depression, see there

afterbirth OBSTETRICS Placenta and fetal membranes that are normally expelled from the uterus after delivering the infant, ergo, 'after birth'. See Placenta

afterload CARDIOLOGY The amount of hemodynamic pressure–peripheral vascular resistance downstream from the heart–which ↑ in heart failure 2° to aortic stenosis and HTN. Cf Preload PHYSIOLOGY The tension produced by heart muscle after contraction

agalactia Agalactosis Absence or failure to secrete milk

agammaglobulinemia 1. Swiss agammaglobulinemia, see there 2. X-linked agammaglobulinemia, see there

age MEDTALK *noun* The time period elapsed since birth. See Maternal age effect, Paternal age *verb* To grow old, senesce

age-based rationing MANAGED CARE, MEDICAL ETHICS A proposed form of rationing of publicly-funded health care services, in which limits would be placed on the type and amount of such services that would be freely available to persons above a certain age. See Rationing

age class MEDTALK A term defined in ecology as a group of individuals of a species with the same age

age-related macular degeneration See Macular degeneration

age-specific mortality rate EPIDEMIOLOGY A mortality rate limited to a particular age group, in which the numerator is the number of deaths in that age group, and the denominator the number of persons in that age group in the population

age spot DERMATOLOGY A flat, pigmented macule commonly seen on the skin of older folks; ASs are benign and of merely cosmetic concern LOCATION, Sun exposed areas–eg, forehead, back of hands PREVENTION Avoid sun; sunscreen–SPF of 15 *may* ↓ incidence MANAGEMENT Creams and lotions to bleach skin; cryotherapy may be required. See Sun protection factor

ageism GERIATRICS A bias or belief that may be held by a health care provider that depression, forgetfulness, and other disorders are a normal part of aging and that older individuals will not benefit from treatment of mental disorders. See elderly. PSYCHIATRY Systematic stereotyping of and discrimination against the elderly to distance oneself from their social plight and skirt personal fears of aging and death. See Elder abuse. Cf Gerontophobia

agenesis EMBRYOLOGY The incomplete or nondevelopment of a body part, organ or tissue. See Anal agenesis, Renal agenesis

agent CLINICAL PHARMACOLOGY An authorized person who acts on behalf of or at the direction of a manufacturer, distributor, or dispenser, which does not include a common or contract carrier, public warehouseman, or employee of the carrier warehouseman CHOICE IN DYING An adult appointed by the declarant under an advance directive, executed or made in accordance with the legal provisions, to make health care decisions for the declarant. See Declarant EPIDEMIOLOGY A factor, such as a microorganism, chemical substance, or form of radiation, whose excessive presence, or in deficiency diseases, relative absence, is essential for the occurrence of a disease MEDTALK A thing capable of producing an effect. See Biological agent, Challenge agent, Controlled drug substance Scheduled agent, Cytoprotective agent, Cytotoxic agent, Dirty agent, Gene transfer agent, Intercalating agent, Nerve agent, Radiopaque contrast agent, Reducing agent, Reversal agent, Schedule I agent, Schedule II agent, Thrombolytic, Vesicant/blistering agent PHARMACOLOGY Any substance capable of producing a physical, chemical or biologic effect. See Alkylating agent, Antidiabetic agent, Antimitotic agent, Antineoplastic agent, Antiplatelet agent, Antipsychotic agent, Chemotherapeutic agent, Depolarizing agent, Inotropic agent, Keratolytic agent, Negative inotropic agent, Nondepolarizing agent, Positive inotropic agent VIROLOGY An unidentified virus or pathogen. See Creutzfeldt agent, Hawaii agent, Norwalk agent, Pittsburgh pneumonia agent, TORCH agent, TWAR agent

AGENT CARDIOLOGY A clinical trial–Angiogenic Gene Therapy

ageusia Loss of a sense of taste–eg, paisley with plaid MANAGEMENT Queer Eye for the Straight Guy

agglutination LAB MEDICINE The clumping of aggregates of antigens or antigenic material-eg bacteria, viruses, with antibodies in a solution. See Latex agglutination test REPRODUCTIVE BIOLOGY The conjoining of 2 organisms of the same species for sexual reproduction, which may be mediated by a carbohydrate on one organism and a protein on the other, thereby forming a glycoprotein

aggravation MEDTALK An ↑ in seriousness, intensity or severity of disease; worsening VOX POPULI An annoyance

aggression PSYCHIATRY Forceful physical, verbal, or symbolic action which may be appropriate and self-protective–eg, healthy self-assertiveness, or inappropriate–eg, hostile or destructive behavior; aggression may be directed toward the environment, another person/personality, or toward the self–eg, depression

aggressive *adjective* Referring to 1. A clinical stance in which the treatment is peremptory, and intended to eradicate a particular lesion or process, as in aggressive chemotherapy, radiotherapy, or surgery 2. A rapid-growing or metastatic tumor which usually has a poor prognosis 3. Violent behavior. See Sexually aggressive

aging A multifaceted process in which bodily structures and functions undergo a negative deviation from the optimum; aging phenomena include ↓ memory, muscle strength and mass, manual dexterity, cardiac output, auditory and visual acuity, loss or thinning of hair, ↑ body fat, ↑ risk of CA, DM, infections, osteoarthritis, osteoporosis accompanied by ↓ in height due to ↓ intervertebral space. See Geriatrics, Life-extending diet DERMATOLOGY Changes in the skin and subcutaneous tissues of those whose future is shorter than their past; aging effects result from intrinsic and extrinsic processes See Aging, Aging skin, Sagging face

aging face syndrome COSMETIC SURGERY The dynamic, cumulative, and degenerative effects of aging on the superficial and structural integrity of the face CONTRIBUTING FACTORS Latitude, occupation, recreational activities, genetic substrate, general health COMPONENTS Skin, soft tissues, structural landmarks See Aging, Aging skin, Sagging face

aging phenomena GERIATRICS The constellation of changes of aging

AGING

INTRINSIC AGING The immutable effects of chronologic aging, eg atrophy-attenuation of epidermis, retration of rete pegs, ↓ number of Langerhans' cells and melanocytes, general decay of structural dermal and epidermal components

EXTRINSIC AGING Effects of external factors, eg sunlight, smoking, gravity and gravidity, keratinocytic dysplasia, solar elastosis, and possibly carcinogenesis; intrinsic & extrinsic aging are intimately linked and thus not subdivided

agism Ageism, see there

agitation NEUROLOGY A state of restless anxiety CLINICAL ↑ nonpurposeful motor activity, usually associated with internal tension–eg, ants in pants, fidgeting, pacing, pulling of clothes

agliophobia PSYCHOLOGY Fear of pain. See Phobia

aglossia Tonguelessness–ergo, speechlessness

agnogenic myeloid metaplasia Myelofibrosis with myeloid metaplasia HEMATOLOGY A chronic progressive condition–panmyelosis and variable BM fibrosis, massive splenomegaly 2° to extramedullary hematopoiesis and leukoerythroblastic anemia with dysmorphic RBCs, circulating normoblasts, immature WBCs, atypical platelets CLINICAL Pts are often > age 50, insidious weight loss, anemia, abdominal discomfort due to splenomegaly, often with hepatomegaly; 80% have nonspecific chromosome defects DIAGNOSIS BM Bx MANAGEMENT No specific therapy; packed RBCs for anemia, androgens may ↓ transfusion requirements, but are poorly tolerated in ♀; recombinant erythropoietin PROGNOSIS Survival ± 5 yrs, often → acute leukemia. See Pseudonym syndrome

agnosia NEUROLOGY An inability to recognize sensory stimuli–objects, people, sounds, shapes or smells, common in parietal lobe tumors; agnosias are classified according to the sense affected–eg, touch–tactile agnosia, hearing—auditory agnosia, sight—visual agnosia, smell–olfactory agnosia, taste–gustatory agnosia. See Gustatory agnosia, Spatial agnosia, Visual agnosia

agonal *adjective* Relating to that which occurs just before death

agonist PHARMACOLOGY A substance that promotes a receptor-mediated biologic response, often by competing with another substance at the same receptor. Cf Antagonist

agoraphobia PSYCHOLOGY Fear of open spaces or of being in crowded, public places like markets; fear of leaving a safe place. See Phobia. Cf Monophobia, PAD syndrome

agrammatism NEUROLOGY A condition characterized by a dissolution of spoken syntax, which becomes laborious and reduced to scattered substantives, and short phrases; in agrammatism, 'function' words are omitted or confused with each other more often than 'content' words See Jargon agrammatism

agranulocytosis Granulocytopenia, granulopenia HEMATOLOGY A marked ↓ in PMNs < 500/mm³ CLINICAL Fever, malaise, mucocutaneous ulcers–throat, GI tract, skin ETIOLOGY Acquired due to adverse response to prescription drugs–chloramphenicol, clozapine, nitrous oxide, procainamide, sulfonamides, thiazide diuretics. See Infantile genetic agranulocytosis

agraphia NEUROLOGY A form of aphasia, characterized by a loss in ability to write, which is most commonly seen in Pts with tumors of the parietal lobe which involve the dominant cerebral hemisphere

Agrylin® Anagrelide, see there

AH interval CARDIOLOGY A period measured by electrophysiologic studies of the heart, equal to the time between the onset of the first rapid atrial deflection and the His bundle deflection; as the lower right atrium and the His bundle delineate the anatomic boundaries of the AV node, the AHI–usually 55-130 msec is essentially equivalent to the AV nodal conduction time; the AHI is ↓ by atropine and isopreterenol, and ↑ by adenosine, digitalis, some class I antiarrhythmics–eg, moricizine, propranolol, rapid or premature atrial pacing, vagal maneuvers, and verapamil. Cf HV interval

AI 1. Allergy index 2. Angiogenesis inhibitor 3. Angiotensin I 4. Apical impulse 5. Artificial insemination 6. Artificial intelligence, see there

aichmophobia PSYCHOLOGY Fear of needles or pointed objects. See Phobia

AID 1. Acute infectious disease 2. Artificial insemination by donor. See Artificial reproduction 3. Automatic implantable defibrillator

AIDS Acquired immunodeficiency syndrome A condition defined by CDC criteria, which is intimately linked to infection by a retrovirus, human immunodeficiency virus–HIV-1; long-term survival after HIV infection is possible; once clinical AIDS develops, it is fatal, despite temporary response to various therapies. See ARC, 'Dominant dozen', gp120, gp160, Hairy leukoplakia, HIV-1, HIV-2, *Isospora belli*, Nonprogressive HIV infection Patient zero, *Pneumocystis carinii*, VLIA–virus-like infectious agent, Walter Reed classification

---

AIDS CDC SURVEILLANCE CASE DEFINITION

AIDS DIAGNOSED DEFINITIVELY W/O NEED TO CONFIRM HIV INFECTION

Candidiasis of esophagus, trachea, bronchi, lungs

Cryptococcosis, extrapulmonary

Cryptosporidiosis > 1 month duration

CMV infection of any organ EXCEPT liver, spleen, or lymph nodes in Pts > 1 month of age

Herpes simplex infection, mucocutaneous > 1 month duration and/or of esophagus, bronchi, lungs

Kaposi sarcoma < age 60

Primary CNS lymphoma < age 60

Lymphoid interstital pneumonitis and/or pulmonary lymphoid hyperplasia < age 13

Mycobacterium avium complex or *M kansasii* disseminated

Pneumocystis carinii pneumonia

Progressive multifocal leukoencephalopathy

Toxoplasmosis of the brain in Pts > 1 month of age

AIDS DIAGNOSED DEFINITIVELY WITH CONFIRMATION OF HIV INFECTION

Multiple or recurrent pyogenic bacterial infections

Coccidioidomycosis, disseminated

Histoplasmosis, disseminated

Isopora spp infection, > 1 month duration

Kaposi sarcoma, any age

Mycobacterium (not *M tuberculosis*), disseminated

Mycobacterium tuberculosis–extrapulmonary

Non-Hodgkin's lymphoma (small noncleaved cell, Burkitt or non-Burkitt, immunoblastic sarcoma)

Primary CNS lymphoma, any age

Salmonella septicemia, recurrent

AIDS DIAGNOSED PRESUMPTIVELY WITH CONFIRMATION OF HIV INFECTION

Candidiasis of esophagus

CMV retinitis

Disseminated mycobacterial infection–culture not required

HIV encephalopathy

HIV wasting syndrome

Kaposi sarcoma

Lymphoid interstital pneumonitis and/or pulmonary lymphoid hyperplasia < age 13

Pneumocystis carinii pneumonia

Toxoplasmosis of the brain in Pts > 1 month of age

AIDS-defining disease A disease which, when accompanied by evidence of HIV infection, fulfills the criteria necessary to diagnose AIDS PCP, MAC, AIDS dementia complex, AIDS wasting syndrome, Kaposi's sarcoma, CMV retinitis. See AIDS dementia complex, AIDS wasting syndrome, CMV retinitis, Kaposi sarcoma, *Mycobacterium avium* complex, *Pneumocystis carinii* pneumonia

AIDS dementia complex AIDS An insidious–30% of asymptomatic HIV-positive subjects have EEG abnormalities, progressive cognitive, motor, behavioral dysfunction, which affects up to ⅔ of AIDS Pts; ADC may be complicated by infections—eg, *Toxoplasma gondii*, CMV, lymphomas CLINICAL Inability to concentrate, loss of memory, gait incoordination, dysgraphia, slowing of psychomotor functions and eventually, apathy PATHOLOGY Degeneration of subcortical white matter and deep gray matter, white matter vacuolization in the lateral and posterior columns of the spinal cord

AIDS disclosure law AIDS Any law that requires an HIV-infected person to inform sexual significant other(s) about HIV status. See Partner notification

AIDS embryopathy AIDS An HIV-induced complex in children born to IVDU mothers, which is characterized by craniofacial defects, including microcephaly, hypertelorism, box-like head, saddle nose, long palpebral fissures with blue sclera, a triangular philtrum, and patulous lips

AIDS enteropathy AIDS An AIDS-related condition, defined by the *absence* of microorganisms which, when seen in AIDS-related complex, may presage clinical AIDS CLINICAL Diarrhea, often worse at night, weight loss, occasionally fever and possibly malnutrition with impaired D-xylose absorption PATHOLOGY Partial villous atrophy with crypt hyperplasia in the small intestine, viral inclusions, ↓ plasma cells and ↑ intraepithelial lymphocytes in the large and small intestine See Gay bowel disease

AIDS fraud HEALTH FRAUD Any form of health care fraud committed against Pts with AIDS. Cf AIDS quackery, FLV 23/A, Unproven methods for cancer management

AIDS pattern EPIDEMIOLOGY A demographic pattern of those who have a greater risk of becoming infected with HIV HEMATOPATHOLOGY A constellation of pathologic findings in AIDS BMs, which includes hypercellularity, ↑ reticulin fibers and lymphocytes and eosinophilic and megakaryocytic hyperplasia. See Bone marrow biopsy

AIDS precautions Those activities intended to minimize exposure to contaminated blood–and potentially bloody fluids, including extreme care with infected sharps, hand washing, use of face masks, double-gloving. See Universal precautions

AIDS quackery The use of an agent, in particular herbal formulations that have been anecdotally reported to be of some benefit in treating AIDS; agents used include acemannan, bitter melon and its protein extract, curcumin, glycyrrhizin, MAP-30, megadoses of vitamins, and Chinese herbal formulas. See Acemannin, AIDS fraud, Bitter melon, CanCell, Curcumin, Glycyrrhizin, Hydrogen peroxide therapy, MAP-30, Ozone treatment. Cf AIDS fraud

AIDS-related complex A pre-AIDS condition with prodromal manifestations of AIDS, where the criteria for defining a case as AIDS are not yet present; ARC is a 'Chinese menu disease', requiring 2 or more clinical features and 2 or more abnormal lab results; the Pts may also have non-specific lymphadenopathy. See AIDS, Follicle lysis, HIV-1

AIDS test LAB MEDICINE Any test performed on a standard venipuncture blood specimen which detects HIV antibodies–ELISA, or antigens–eg, Western blot, or viral nucleic acid–eg, viral load by RNA. See Western blot

AIDS therapy HIV treatment may be: preventive-eg to prevent in utero infection of HIV-positive mothers; prophylactic-eg to prevent opportunistic infections when CD4 levels fall below certain level; based on efficacy. See AIDS fraud, AIDS quackery, AIDS vaccine

AIDS vaccine A hypothetical vaccine intended to either prevent HIV infection or ensure that those infected will not fall victim to AIDS; the most promising vaccine is that using a naked DNA plasmid, reported by Letwin et al in 20/10/00 Science; as of early 2001, there is no effective vaccine. See gp160 vaccine, HGP-30, HIV, V3 loop, Zagury

AIL 1. Angiocentric immunoproliferative lesion 2. Angioimmunoblastic lymphadenopathy

AILA Angioimmunoblastic lymphadenopathy, see there

AIN Anal intraepithelial neoplasia. See Intraepithelial neoplasia

AIP Acute intermittent porphyria

air ambulance EMERGENCY MEDICINE A helicopter or, less commonly, a fixed wing aircraft, used to evacuate a person who requires immediate medical attention that cannot be provided at his/her current location

air-conditioner lung PULMONARY MEDICINE A type of hypersensitivity pneumonitis seen in those who work with air conditioning and climate control systems and are exposed to *Aspergillus* spp and thermophilic organisms. See Hypersensitivity pneumonitis

air embolus Air embolism The presence of gas in blood vessels, which can cause an interruption of normal blood flow; AE is of greatest importance in the coronary and cerebral arteries. See Embolism

air meniscus sign IMAGING A crescent-shaped radiolucency bordering a mass lesion, classically seen in pulmonary hydatid cysts, where air enters the cyst forming a radiolucency between the outer layer–host and the inner layer–hydatid membrane of *Echinococcus granulosus:* the most common AMS in ultrasonography is caused by *Aspergillus fumigatus*

air sickness A permutation of motion sickness, which occurs during ascent and/or descent in an airplane. See Airline food

air travel disease TRAVEL MEDICINE Any condition related to the unique conditions of air travel INFECTIONS The potential for infection by an airborne pathogen from another inmate in the confined quarters of commercial flights is understudied. See Airline food, Air sickness

---

AIR TRAVEL & DISEASE

AEROTITIS MEDIA Middle ear trauma

BAROSINUSITIS Pain, bleeding of sinuses; flying with sinusitis may cause a spread to the CNS

DEEP VEIN THROMBOSIS 'Economy class' syndrome

HEART ATTACK Related to anxiety, sprinting to connecting flight, ↓ cabin O₂ ergo hypoxia, ↓ cabin humidity

INFECTIONS eg, TB and other communicable diseases

MUSCULOSKELETAL Low back pain and injuries

NAUSEA

SYNCOPE/VERTIGO-Possibly related to ↓ cabin O₂ leading to hypoxia

SEPARATION-RELATED Panic, anxiety, depression

TRAUMA Due to objects tumbling from overhead–avoid the aisle seat

---

airborne allergen Aeroallergen A substance that is light enough to be carried through air currents, and capable of evoking an immune response EXAMPLES Pollens, fungal spores, and algae, which make miserable the lives of those who mount an immune response thereto

airborne transmission EPIDEMIOLOGY The transmission of pathogens by aerosol, which enter the body by the respiratory tract. See Aerosol

AIRE CARDIOLOGY A clinical trial–The Acute Infarction Ramipril Efficacy Study that evaluated the effect of antihypertensive therapy with ramipril, a long-acting ACE inhibitor on M&M in Pts with post-MI heart failure. See ACE inhibitor, Acute myocardial infarction, Heart failure, Ramipril

airway resistance LUNG PHYSIOLOGY A measure of the resistance–in cm H₂O to the flow–in L/min of air in upper airways, the result of natural recoil–resiliency of anatomic structures–oro- and nasopharynx, larynx, and nonrespiratory portions of the lungs–trachea, bronchi, and bronchioles through which air passes on the way to the alveoli; assessment of AR evaluates airway responsiveness, provocation testing–eg bronchial challenge,

evaluation of sites of airflow resistance or closures, and characterization of the type of lung disease; airway resistance is ↑, either focally or globally in asthma, COPD, and smokers. See Airway responsiveness, Asthma, COPD

airways ANATOMY The 'pipes'–trachea, bronchi, bronchioles–through which air passes to and from the alveoli. See Small airways

AIS 1. Abbreviated Injury Scale A classification of severity of MVA injury formulated in 1985 by the American Association of Automotive Medicine. See Injury Severity Score 2. Adenocarcinoma in situ 3. Anal carcinoma in situ 4. Androgen-insensitivity syndrome, see there

AJCC American Joint Committee on Cancer

AK 1. Above-the-knee 2. Actinic keratosis

AKA/aka 1. Also known as 2. Above-the-knee amputation

akathisia Antsiness NEUROLOGY Motor restlessness ranging from a feeling of inner disquiet to inability to sit still or lie quietly, accompanied by a sensation of muscular quivering, and an urge to be in constant motion, a common extrapyramidal effect of neuroleptics/antipsychotics. See Extrapyramidal syndrome

akinesia NEUROLOGY Absent or ↓ voluntary movement PHARMACOLOGY Temporary paralysis of a muscle by procaine injection PSYCHIATRY Hysterical paralysis, see there

ALA 1. Alanine 2. delta–Aminolevulinic acid, see there

alalia ENT Speechlessness; mutism; loss of speech due to an impaired articulatory apparatus

alanine aminotransferase Glutamine pyruvic transaminase, GPT CLINICAL CHEMISTRY An enzyme found primarily in the liver, with lesser amounts in the kidneys, heart and skeletal muscles; low levels of ALT are normal in the circulation; after liver damage, ALT is released into the bloodstream before more obvious clinical findings of liver damage–eg jaundice, occur; ↑ ALT is an early indicator of acute liver damage; ALT is measured as part of a panel of blood chemistry tests REF RANGES ♂ 10-32 U/L; ♀ 9-24 U/L; children 2 times > adults; AA is ↑ in viral hepatitis, drug-induced hepatitis, infectious mononucleosis, chronic hepatitis, intrahepatic cholestasis, cholecystitis, active cirrhosis, acute MI. See Aspartate amino transferase, Gamma-glutamyl transferase

ALARA As Low As Reasonably Achievable RADIATION SAFETY A generic stance regarding radiation exposure

alarm clock headache NEUROLOGY A migraine of abrupt onset, which often occurs at the same time each day

alb- Latin, white

albendazole PARASITOLOGY An inexpensive benzimidazole-type scolicidal anthelmintic, with a broad range of anti-parasitic activity against nematodes–roundworms, cestodes–flatworms, protozoa, and possibly, microsporidiosis; it is not FDA-approved; albendazole can be used in conjunction with percutaneous drainage for managing hepatic hydatid cysts SUSCEPTIBLE ORGANISMS *Ascaris lumbricoides*, *Giardia lamblia*, hook-worms–*Ancylostoma duodenale* and *Necator americanus*, *Hymenolepis nana*, *Opisthorchis viverrini*, *Strongyloides stercoralis*, *Taenia solium*, *Trichuris trichiuria* ADVERSE EFFECTS Abdominal pain, headaches, vertigo, N&V, alopecia, ↑ aminotransferase, neutropenia. See Benzimidazole, Hydatid cyst disease

albinism A group of hereditary and congenital, often AR diseases that share a metabolic defect in the production of mature melanin, which translates clinically into hypopigmentation of skin, hair, and eyes–iris See

Oculocutaneous albinism type I, Oculocutaneous albinism type II, Yellow albinism

ALBINISM, MAJOR GROUPS

GENERALIZED (OCULOCUTANEOUS) ALBINISM All 6 subtypes are AR; the most common, type IA is due to tyrosinase deficiency, which may be due to a missense mutation; GA also occurs in Chediak-Higashi, Hermansky-Pudlak, & Cross syndromes

PARTIAL ALBINISM An AD condition with a focal white patch, similar to Waardenburg syndrome

OCULAR ALBINISM An X-linked recessive condition

albino A person with albinism

Albright syndrome McCune-Albright syndrome, see there

albumin CLINICAL CHEMISTRY A 66 kD heat-coagulable, acid-precipitable, water-soluble protein produced by the liver, which acts as an osmotic regulator, stabilizer, a nutritive substrate for tissues, binding and transport protein and, experimentally, a growth media supplement; albumin is the major–± 60% total–plasma protein; 40% of albumin in adults–± 125 g is intravascular; it is responsible for 75-80% of intravascular oncotic pressure; daily turnover is 10-16%; it is a major transport protein for large organic anions–eg, fatty acids, BR, drugs, enzymes, and hormones–eg, cortisol and thyroxine, when their specific binding globulins are saturated; serum albumin levels serve as a surrogate marker for liver disease; albumin is ↑ in dehydration and ↓ in liver disease, protein malnutrition, chronic disease, neoplasia, thyroid disease, burns, active inflammation, renal disease See Glycosylated albumin, Prealbumin, Proalbumin, Sonicated albumin, Timed collection THERAPEUTICS Albumin is available as plasma protein fraction-50g/L; it carries *no* risk of hepatitis INDICATIONS Acute blood loss, burns, hypo– or analbuminemia CONTRAINDICATIONS Malnutrition. See Colloids, Crystalloids. Cf Albumen, Ovalbumin REF RANGE Serum ♂, 4.2-5.5 g/dL; ♀ 3.7-5.3 g/dL; urine, 3.9-24.4 mg/24 hrs; CSF, 15-45 mg/dL TRANSFUSION MEDICINE A colloid-type volume expander consisting of albumin in 2 standard concentrations–5% and 25% in a physiologic solution; albumin is prepared by heating to 60°C for 10+ hrs, which inactivates both HBV and HIV, and is used in surgical blood management by hemodilution. See Hemodilution, Surgical blood management

albuminuria NEPHROLOGY 1. The spilling of albumin into the urine, which may be associated with immune complex nephritis 2. A rarely used synonym for proteinuria

ALC 1. Alcohol 2. Alternate level of care, see there

Alcaligenes faecalis Acaligenes odorans, Pseudomonas odorans BACTERIOLOGY An environmental bacterium which colonizes moist areas in hospitals and may transiently colonize the skin MODE OF TRANSMISSION Contaminated medical devices and fluids–eg, IV, hemodialysis, irrigation, disinfectants; often a contaminant, *A faecalis* may be isolated from blood, sputum, and urine, especially in immunocompromised hosts

alcaptonuria Alkaptonuria, see there

Alcelam Alprazolam, see there

alcohol CHEMISTRY Any of a broad category of organic chemicals containing one or more hydroxyl—OH groups with a minimal tendency to ionize; alcohols can be liquids, semisolids or solids at room temperature COMMON ALCOHOLS Ethanol or CH_3CH_2OH/'drinking' alcohol, methanol–CH_3OH, wood/grain alcohol, which can cause blindness and CNS damage, propanol–$(CH_3)_2CHOH$, 'rubbing' alcohol. See Absolute alcohol, Perillyl alcohol CLINICAL MEDICINE The commonly ingested alcohol, ethyl alcohol–ethanol, once consumed, peaks in the blood in ±30 mins; ±1 hr is needed to eliminate each 10g of alcohol ingested; blood alcohol levels reliably indicate the amount of ethanol 'on board'; alcohol consumption may relieve anxiety for several hrs, but, long-term, may aggravate or provoke anxiety and panic disorders HEALTH BENEFITS OF As little as 1 drink/wk ↓ risk of stroke REF RANGE Negative; serum levels of > 0.05% is sufficient to cause impairment

alcohol consumption PUBLIC HEALTH Moderate alcohol consumption–1-2 drinks/day is associated with ↓ cardiovascular mortality–RR = 0.5, and an improved lipid profile, due to ↑ HDL_2 and HDL_3; there is a direct positive association between alcohol consumption and plasma levels of endogenous tissue-type plaminogen activator–tPA, which averages 10.9 ng/mL for those who consume alcohol daily, 9.7 ng/mL–wkly 9.1 ng/mL–monthly and 8.1 ng/mL–never; these data suggest that changes in the fibrinolytic potential may be a mechanism whereby alcohol consumption ↓ the risk of cardiovascular disease; moderate AC is associated with ↓ mortality in ♀, in particular those at ↑ risk for CAD

alcohol profile LAB MEDICINE A screening test for identifying alcohols–methyl, ethyl, isopropyl, and acetone-in blood SPECIMEN Blood in gray top tube LOWEST DETECTABLE VALUE Acetone 5 mg/dL; ethanol 10 mg/dL; isopropanol 5 mg/dL; methanol 5 mg/dL. See Alcohol

alcohol-related birth defects Any birth defect–eg, pre– or postnatal growth retardation, facial dysmorphia–thin upper lip, poorly-developed philtrum, short nose, and eye openings, CNS defects with mental retardation; when multiple ARBDs are present, the term fetal alcohol syndrome is used. See Fetal alcohol syndrome

alcohol withdrawal A constellation of clinical findings that result when a person with habitual long-term or heavy alcohol use history goes 'on the wagon' and abstains from alcohol. See Delirium tremens

alcoholic *adjective* Referring to any condition induced by prolonged exposure to ethanol *noun* A person suffering from alcoholism, see there

alcoholic cardiomyopathy A clinicopathologic state induced by chronic alcoholism, and a major cause of dilated cardiomyopathy, characterized by severe left ventricular dysfunction, associated with a 40-80% 3-yr mortality, and it may present as sudden death or ventricular fibrillation

alcoholic cirrhosis Alcoholic liver disease, Laennec's cirrhosis, nutritional cirrhosis, portal cirrhosis A clinicopathologic entity which usually evolves from the early lesion of alcoholic fatty liver, in which the liver is 'greasy' and weighs up to 3 kg, with intact lobules and veins, portal infiltration by PMNs and prominent Mallory body formation to the end-stage–irreversible alcoholic cirrhosis CLINICAL Liver fibrosis leads to portal HTN, which evokes bleeding esophageal varices and hepatic encephalopathy. Cf Micro-micronodular cirrhosis

alcoholic encephalopathy Wernicke's encephalopathy, see there

alcoholic fatty liver A liver with acute and subacute, ie precirrhotic, changes induced by alcohol, a toxin that interferes with fatty acid oxidation, impairing the tricarboxylic acid cycle, resulting in incomplete β-oxidation products from fatty acids. See Alcoholic cirrhosis. Cf Fatty liver of pregnancy

alcoholic hepatitis Inflammation of the liver which occurs in a chronic drinker after an episode of exceptionally heavy alcohol consumption; AH ranges from mild–abnormal laboratory tests are the only indication of disease, to severe liver dysfunction with jaundice, hepatic encephalopathy, ascites, bleeding esophageal varices, abnormal blood clotting and coma PATHOLOGY Ballooning degeneration of hepatocytes, inflammation with PMNs and sometimes Mallory bodies–aggregates of intermediate filament proteins; AH is reversible if the patient stops drinking, but it usually takes several months to resolve; it can lead to liver scarring and cirrhosis, and often occurs in alcoholics with cirrhosis. See Wernicke syndrome

alcoholic ketoacidosis Ketoacidosis 2° to alcohol abuse, in which ketone bodies–eg, acetone, accumulate in the circulation, which is evoked by conditions that encourage fat metabolism. See Diabetic ketoacidosis

alcoholic liver disease HEPATOLOGY A general term for any of a number of clinical conditions caused by chronic excess of alcohol consumption, including alcoholic cirrhosis and alcoholic fatty liver. See Alcoholic hepatitis, Cirrhosis

alcoholic neuropathy Pseudotabes, alcoholic polyneuropathy NEUROLOGY A nutritional neuropathy 2° to chronic alcohol abuse, exacerbated by specific nutritional deficiencies–eg, thiamin–vitamin B_1 and vitamin B_{12} deficiency CLINICAL Burning pain and weakness of legs with atrophy and fasciculations, paresthesia, usually acral in distribution, ↓ tactile and position sense, ataxia, ulcers of immobile parts; sensory defects, muscle weakness and cramps, heat intolerance, impotence, dysuria, dysphagia, speech impairment, diarrhea, constipation TREATMENT Diet, vitamin B_{12}

alcoholic paranoia Othello syndrome, see there

alcoholic polyneuropathy Alcoholic neuropathy, see there

alcoholic type I 'Maintenance-type' alcoholic An anxiety-prone or passive-dependent person who drinks to alleviate problems; the onset in either sex is after age 25; AT-Is have a high reward dependence and avoid harmful or novel situations; most have minimal–if at all—antisocial tendencies

alcoholic type II 'Binge' drinker An alcoholic who enjoys the novelty of drinking; AT-IIs may be antisocial, have a history of violence, may be the biological son of an alcoholic father, and become alcoholic by age 25

alcoholism SUBSTANCE ABUSE A condition characterized by a pathologic pattern of alcohol use causing a serious impairment in social or occupational functioning; also defined as a '...*primary, chronic, disease with genetic, psychosocial, and environmental factors influencing its development and manifestations. The disease is often progressive and fatal. It is characterized by ... distortions in thinking, most notably denial*'; alcoholism is characterized by the regular intake of ≥ 75 g/day of alcohol CHRONIC EFFECTS Co-morbidity due to portal HTN, hepatic failure, hyperestrogenemia, infections–especially pneumonia, which may be due to alcohol-induced suppression of various immune defenses, psychosocial disruption, transient hyperparathyroidism with ↓ Ca^{2+}, ↓ Mg^{2+}, osteoporosis. See Blood alcohol levels, Standard drink

alcopop SUBSTANCE ABUSE A cider, juice or other beverage which has been commercially 'spiked' with a low amount of alcohol–eg, ±5%; some authorities believe alcopops may lead to future heavier alcohol use. See Alcohol, Wine cooler

Alder-Reilly anomaly GENETICS An AD condition characterized by large, intensely azurophilic granules in the cytoplasm of PMNs, lymphocytes, and monocytes, accompanied by defects in eosinophilic and basophilic granules

aldicarb TOXICOLOGY A lipid-soluble, sulfurated anticholinergic insecticide/nematocide TOXICITY Abdominal cramps, nausea, diarrhea, dizziness, sweating, muscle spasms, blurred vision MANAGEMENT Atropine

aldolase CLINICAL CHEMISTRY A serum enzyme that cleaves fructose 1,6-diphosphate to dihydroxyacetone phosphate and glyceraldehyde-3-phosphate in the muscles to produce energy; aldolase is distributed in all tissues and ↑ in skeletal muscle disease or injury, metastatic CA, CML, megaloblastic anemia, hemolytic anemia, or infarction; it is measured in suspected myopathies, as the intensity of ↑ reflects the severity of disease; aldolase may also be ↑ early in Pts who later develop muscular dystrophy and can also be used to monitor steroid therapy in inflammatory myopathies REF RANGE 3.1-7.5 U/L; aldolase is ↑ in Duchenne's muscular dystrophy, dermatomyositis, polymyositis, trichinosis

Aldoril® THERAPEUTICS A proprietary combination of methyldopa and hydrochlorothiazide used to manage HTN

aldose reductase An enzyme normally present in the eye and elsewhere, which converts excess glucose–as occurs in DM——into sorbitol; excess sor-

bitol in the eye, nerves, and kidneys can lead to retinopathy, neuropathy, and nephropathy. See Aldose reductase inhibitors

aldose reductase inhibitor CLINICAL THERAPEUTICS Any of a family of compounds–eg, epalrestat, ponalrestat, sorbinil, tolrestat, which block aldose reductase; ↑ intracellular glucose leads to ↑ sorbitol, which competitively inhibits glomerular and neural synthesis of *myo*-inositol; the ↓ in *myo*-inositol synthesis depresses phosphoinositide metabolism, and ↓ in Na^+K^+-ATPase activity; ARIs do not prevent the complications of type 1 DM See Aldose reductase

aldosterone ENDOCRINOLOGY An adrenocortical mineralocorticoid hormone that controls the body's electrolyte and water homeostasis by regulating reabsorption of Na^+ and Cl^- in exchange for K^+ and H^+ ions, and maintaining BP and blood volume; aldosterone secretion is controlled by the RAA system and by concentrations of K^+ in the circulation, which if ↑, evokes secretion of aldosterone; ↓ Na^+ evokes renin release, which stimulates aldosterone secretion; aldosterone may be measured when evaluating HTN; aldosterone is ↑ in adrenocortical adenoma or CA, bilateral adrenal hyperplasia, renovascular HTN, liver disease, CHF, cirrhosis, nephrotic syndrome, pregnancy–3rd trimester; it is ↓ in 1° hypoaldosteronism, salt-losing syndrome, toxemia of pregnancy, Addison's disease REF RANGE Serum, ≤ 20 mg/dL; ≤ 20mg/24hrs, urine. See Hypertension, Pseudoaldosterone, Timed collections.

Aldrich syndrome An X-R condition characterized by thrombocytopenia, eczema, bloody diarrhea–melena, ↑ risk of bacterial infection; death is due to hemorrhage or overwhelming sepsis, usually by age 10

alendronate Fosamax® ENDOCRINOLOGY A biphosphonate used to treat postmenopausal osteoporosis and Paget's disease of bone. See Osteoporosis. Cf Salmon calcitonin

Alepam® Oxazepam, see there

alert NEUROLOGY *adjective* 1. Referring to a state in which an individual is awake and appropriately answers all questions 2. Attentive; quick to think or act VOX POPULI On-the-ball, 'with it,' cooking with gas, smelling the coffee, firing on all 12–Italy, 10–Chrysler, 8–US, 6–Europe, 5–Sweden, 4–Japan, 3–France, cylinders

ALERT NEPHROLOGY A clinical trial–Assessment of Lescol® in Renal Transplantation

alert & oriented x 3 PHYSICAL EXAMINATION-MENTAL Alert & oriented to person, place & time

ALERT system CARDIOLOGY An external generator and catheter system for providing electrical impulses directly into the heart to convert a Pt from A Fib to a normal sinus rhythm. See Defibrillation

alert value Panic value, see there

Alesse® GYNECOLOGY A combination oral contraceptive containing the lowest hormone doses–levonorgestrel/ ethinyl estradiol–now on the market while still maintaining greater than 99% efficacy when used as directed. See Birth control

aleukemic leukemia A variant of acute leukemia where the peripheral blood reveals pancytopenia and few discernible blast cells on peripheral blood smears; leukemic cells are seen only by a BM Bx. See Leukemia

alexia NEUROLOGY 1. Inability to understand printed words or sentences 2. Loss of a previously possessed reading facility that is unexplained by defective visual acuity PSYCHIATRY Loss of the ability to grasp the meaning of written or printed words and sentences. See Dyslexia

alfentanil ANESTHESIOLOGY An opioid analgesic that is 10-fold more potent than morphine CONS Profound respiratory depression and skeletal muscle rigidity. See Opioid anesthetic. Cf Muscle relaxant

ALG Antilymphocyte globulin, see there

alglucerase Ceredase® A monomeric 497 AA glycoprotein and modified form of β-glucerebrosidase, which catalyzes the hydrolysis of glucocerebroside in reticuloendothelial system lysosomes, which is produced by genetic engineering; alglucerase is the most effective therapy for type 1 Gaucher disease, resulting in ↓ hepatospenomegaly, hematologic defects, ↑ bone mineralization, and reversal of cachexia CosT Very high. See Gaucher disease, 'Orphan drug'

algophobia PSYCHIATRY A morbid fear of pain. See Phobia

algor mortis The cooling of a body after death. Cf Rigor mortis

algorithm DECISION-MAKING A logical set of rules for solving a specific problem, which assumes that all of the data is objective, that there are a finite number of solutions to the problem, and that there are logical steps that must be performed to arrive at each of those solutions NIHSPEAK A step-by-step procedure for solving a problem; a formula. See Back-propagation, Critical pathway, Genetic algorithm, Risk of ovarian cancer algorithm

algorithmic testing Reflex testing, see there

alien hand Alien limb phenomenon NEUROLOGY A clinical finding in which there is awkward asymmetric involuntary movement of the hand or limb, which is interpreted by the internal sensors as 'alien'; the limb moves without being controlled by the Pt. Cf Progressive supranuclear palsy

alienation PSYCHIATRY 1. The sensation that one has been removed from friends, family or one's usual social setting; cultural estrangement. See Depersonalization 2. The sense of being removed from one's own emotions—alienation of affect

alimentary *adjective* Gastrointestinal, pertaining to food, nutrition, or the organs of digestion

aliphatic *adjective* Referring to organic compounds in which the carbons are arranged in straight 'open' chains—eg, alkanes, alkenes, alkynes

ALIVE AIDS A clinical trial—AIDS Linked to IV Drug Use which periodically assessed the seroconversion rate of a community-based cohort of 2960 IV drug users recruited in Baltimore in 1988-1989. See AIDS, CD4, HIV CARDIOLOGY 1. A series of trials—Azimilide Postinfarct Survival Evaluation—designed to test the efficacy and safety of azimilide—Stedicor® in preventing sudden death due to cardiac arrhythmia in post-MI Pts. See ASAP. 2. A clinical trial—Adenosine Lidocaine Infarct Zone Viability Enhancement

alkaline phosphatase 'Alk phos' CLINICAL CHEMISTRY A 69 kD homodimeric metalloenzyme with broad specificity, which is widely distributed in nature, has an optimal activity at a high—± 10 pH, and occurs in multiple forms—isoenzymes in the liver, intestines, and bone; AP hydrolyzes phosphate esters, yielding alcohol and phosphate; because ↑ total AP indicates either liver or bone disease, additional clinical studies—eg, evaluation of AP isoenzymes, are required to determine the cause of the elevation PHYSIOLOGIC ↑ IN AP: During bone growth, infants, children and adolescents-levels are 3-fold > than adults, pregnancy, hepatobiliary disease—eg viral hepatitis, severe biliary obstruction, biliary cirrhosis, intrahepatic cholestasis, Paget's disease of bone, osteomalacia, osteogenic sarcoma, bone metastasis, hyperparathyroidism, infectious mononucleosis, vitamin D deficiency rickets; ↓ in hypoposphatasia, protein deficiency, magnesium deficiency REF RANGE AP ranges differ according to method and lab. See Alkaline phosphatase isoenzyme, Immunoperoxidase

alkaline phosphatase isoenzyme CLINICAL CHEMISTRY Any of the organ-specific alkaline phosphatase isoenzymes: liver, bone, intestine and placenta; intestinal AP occurs almost exclusively in individuals with blood group B or O and is ↑ 8 hrs after a fatty meal; placental AP first appears in the 2nd trimester of pregnancy, accounts for half of all AP in the 3rd trimester and drops to normal levels 1 month post partum; the Regan isoenzyme resembles the placental AP and is present in some Pts with cancer. See Alkaline phosphatase

alkalosis PATHOPHYSIOLOGY A clinical state due to either an accumulation of bases or loss of acids—↓ H⁺, resulting in ↑ pH. See Contraction alkalosis, Metabolic acidosis, Respiratory alkalosis. Cf Acidosis

alkaptonuria Black pain disease, black urine disease, alcaptonuria, alcaptonuric ochronosis An AR defect in tyrosine and phenylalanine metabolism, more common in ♂, due to homogentisic acid oxidase—HAO deficiency; metabolic pathway of phenylalanine and tyrosine → ring opening of homogentisic acid → malylacetoacetic acid; alkaptonuria is first recognized by the mother who cannot clean the children's diaper as the urine oxidizes to a pitch black color upon exposure to air CLINICAL Arthritis due to homogentisic acid deposition in cartilage, tendons, as well as in the sclera, viscera, skin; when severe, pigment deposition can compromise cardiac, renal, or pulmonary function, and spill into the urine as a melanin-like substance. Cf Blue diaper syndrome

alkyl group CHEMISTRY A functional group lacking an H⁺

alkylating agent MOLECULAR BIOLOGY An organic compound able to transfer an alkyl group to a nucleotide ONCOLOGY A generic term for any of a family of chemotherapeutics that cause irreversible damage to tumor cells and apoptotic destruction ROUTE OF ADMINISTRATION IV, oral ADVERSE REACTIONS Stomatitis, N&V, diarrhea, skin rash, anemia, alopecia; with cyclophosphamide, hemorrhagic cystitis, cardiac toxicity. Cf Antimetabolite, Plant alkaloid, Topoisomerase inhibitor

ALL 1. Acute lymphocytic leukemia, see there 2. Allergy

all-or-none phenomenon All-or-nothing principle, all-or-none response CARDIAC PHYSIOLOGY The property of cardiac muscle in which stimulation from a single myocyte travels to the atrium and ventricle before contracting, resulting in a coherent and coordinated pump activity. See Action potential PHYSIOLOGY A rule applied to the activation of individual muscle or nerve cells, where the response to stimuli–depolarization only occurs above a certain threshold, usually −55 mV, after which a complete action potential occurs which is maximum in intensity, ie the strength of the nerve impulse is not dependent on or a function of the strength of the stimulus PSYCHOLOGY A 'soft' rule in behavioral studies, which refers to the observation that a behavioral stimulus will produce either a complete response or none at all

all-nighter GRADUATE EDUCATION A long, intense study session before an exam, intended to help a student maximize her grade in a particular subject, by studying all night; research suggests that memory retention is greatest after a good night's sleep than with an all-nighter. See Sleep hygiene

all-payer system MANAGED CARE A proposed health care system in which all insurers would use the same fee schedule. See Health care reform

Allegra-D® Fexofenadine/pseudoephedrine ALLERGY MEDICINE A nonsedating antihistamine with a decongestant for relief of seasonal allergic rhinitis. See Allergic rhinitis, Antihistamine

Allegron® Nortryptyline, see there

allele GENETICS An alternate form of a gene, which results in different gene products; any one of 2 or more variants of a gene that occupy the same position—locus on a chromosome, which may differ in nucleotide sequence, but not substantively in function or effect. See Amorphic allele, Pseudoallele allele, *Reeler* allele, Wimp allele

Allen's test REHABILITATION MEDICINE A test used to determine patency of the ulnar or radial artery; the hand is clenched to force blood out; if the blood does not flow back into the hand rapidly, one or more arteries are stenosed or occluded–eg, due to throacic outlet syndrome

allergen IMMUNOLOGY A substance–eg, pollen, dander, mold, which can evoke an immediate-type hypersensitivity–allergic reaction, triggering a release of histamine. See Airborne allergen, Cockroach allergen, Feline allergen, Immunogenic allergen

allergen immunotherapy Desensitization, hyposensitization, immunotherapy ALLERGY MEDICINE A modality that attempts to ↓ IgE-mediated hypersensitivity to various substances, by administering ever-increasing amounts of an antigen–eg urushiol in poison ivy, sumac, and pollen, to evoke formation of blocking antibodies. See Elimination diet, Immune tolerance; Cf Bee venom therapy

allergic *adjective* Pertaining to, caused or affected by an allergy. Cf Allergenic

allergic alveolitis PULMONARY MEDICINE Lung inflammation 2° to exposure to a chemical, organic dust, fungi, molds which, if chronic, can lead to interstitial lung changes on CXR CLINICAL Cough, fever, SOB, wheezing. See Bird-handler's disease, Farmer's lung, Hypersensitivity pneumonitis

allergic asthma CLINICAL IMMUNOLOGY A condition characterized by bronchoconstriction and SOB CLINICAL Wheezing, dyspnea—especially exhaling, chest tightness EXACERBATED BY Abrupt changes in temperature or humidity, allergies, URIs, exercise, stress, cigarette smoke MANAGEMENT Bronchodilators; steroids if unresponsive; RhuMAb-E25, a humanized monoclonal antibody may be used in severe AA. See Asthma, Status asthmaticus

allergic bronchopulmonary aspergillosis PULMONOLOGY A condition characterized by episodic–asthmatiform pulmonary obstruction, immune response to *Aspergillus fumigatus*–eg, by precipitating *A fumigatus* antibodies, positive skin reactivity to *A fumigatus* antigens CLINICAL Cough–expectoration of brown plugs, fever, dyspnea, wheezing, malaise, chest pain, diaphoresis, hemoptysis IMAGING Central bronchiectasis, atelectasis, transient or fixed pulmonary infiltrates, mucus plugging, hyperinflation LAB Eosinophilia, ↑ IgE, *A fumigatus* in sputum MANAGEMENT Corticosteroids

allergic conjunctivitis OPHTHALMOLOGY Allergy-induced conjunctival inflammation which may accompany hay fever CLINICAL Itchy, red, tearing eyes MANAGEMENT Topical cromolyn, vasoconstrictors, cold compresses, oral antihistamines, air-conditioning. See Hay fever

allergic contact dermatitis Allergic dermatitis DERMATOLOGY A condition caused by cell-mediated immunity due to contact with haptens–eg, nickel, chromates, ursodiols in poison ivy and poison oak, synthetic chemicals, drugs, cosmetics, jewelry, neomycin ointment, etc, and may affect any body part CLINICAL Intense pruritus, erythema, intercellular edema, papulovesicles, which with continued exposure, are followed by vesiculation, rupture and oozing dermatitis MODIFYING FACTORS–HOST Genetic predisposition, age, route of 1° sensitization, and concomitant disease–eg, immunocompromise as in AIDS MODIFYING FACTORS–ENVIRONMENT Physicochemical modulation. DIAGNOSIS Patch test MANAGEMENT Topical corticosteroids. See Contact dermatitis, Patch test

allergic polyp Inflammatory polyp, see there

allergic reaction IMMUNOLOGY Any response to an allergic stimulus, which can be localized or systemic CLINICAL Rash, itching, hives, swelling, dyspnea, ↓ BP LAB ↑ IgE, ↑ mast cells, basophils, which release histamine, PGs, LTs, kinins, et al. See Asthma, Hypersensitivity reaction

allergic rhinitis Hay fever CLINICAL IMMUNOLOGY An inflammatory response in the nasal passages to allergens, which is the most common form of atopic–allergic disease, which affects 5-20% of the general population; AR

is initiated by exposure of the nasal mucosa to airborne antigens, evoking IgE production; upon repeated re-exposure to the allergen–eg, ragweed pollen, histamine, leukotrienes C_4, D_4, E_4, B_4, PGD_2, kinins, kininogen, serotonin are released CLINICAL Paroxysms of sneezing, nasal congestion, nasal and ocular pruritus, tearing, rhinorrhea, anosmia, ageusia, postnasal drip which may cause coughing, partial or total obstruction of airflow, throat clearing, and allergic 'shiners' DIAGNOSIS Skin testing with appropriate inhalant allergens is of greater use than measuring serum IgE MANAGEMENT Avoid allergens, antihistamines, espeially H_1 receptor antagonists, sympathomimetic amines, anticholinergic agents, corticosteroids, decongestants, cromolyn sodium, immunotherapy. See H receptors, Hay fever, Immunotherapy, Sensitization

allergist IMMUNOLOGY A physician, who is often trained in both internal medicine and clinical immunology and who manages Pts with allergies MEAT & POTATOES DISEASES Allergy-induced asthma, stings, bites, sinusitis SALARY $137K + 12% bonus

allergy IMMUNOLOGY 1. A state of hypersensitivity induced by exposure to a particular antigen/allergen, resulting in adverse immune reactions on subsequent re-exposure to the allergen. See Anaphylactic shock, Cross allergy, Food allergy, Hypersensitivity reaction, Latex allergy, Peanut allergy, Pseudoallergy 2. The medical specialty dedicated to diagnosing and managing allergic disorders

allergy desensitization See Desensitization therapy

allergy shots See Desensitization therapy .

allergy skin test Patch test, see there

allergy testing See Patch testing, RAST, Skin testing

ALLHAT CARDIOLOGY An ongoing randomized, open label, multicenter trial evaluating whether antihypertensive therapy reduces M&M in CAD, and to determine whether lipid-lowering pravastatin therapy in moderately hypercholesteremic Pts reduces heart-related M&M. See Antihypertensive, Hypertension, Pravastatin

allied health personnel Paramedical personnel All health care personnel who are 1. Not physicians, dentists, podiatrists, and registered nurses and 2. Have received specialized training and require special licensure; these workers include chiropractors, dieticians, laboratory technologists, medical illustrators, technicians and transcriptionists, medical records technicians, occupational therapists, optometrists, phlebotomists, physicians' assistants, podiatrists, practical nurses

alligator forceps Alligators SURGERY A type of forceps with sharp teeth

allocated benefits HEALTH INSURANCE Payments authorized for specific purposes with a maximum specified for each; in hospital policies, there may be scheduled benefits for X-rays, drugs, dressings, and other specified expenses. See Benefits

allocation VOX POPULI The distribution of a thing. See Nonrandomized allocation, Resource allocation

allogeneic *adjective* Relating to that which is genetically dissimilar, but of the same species

allogeneic transplant IMMUNOLOGY The transplantation of an organ or tissue from a genetically matched relative or other donor

allograft Allogeneic graft IMMUNOLOGY A graft–organ, tissues, or cells donated from a genetically distinct–in humans, allogeneic at one or more MHC loci—individual of the same species. See Renal transplantation. Cf Autograft, Xenograft

allograft immunity IMMUNOLOGY Any immune response by the host against a transplanted tissue. See Graft-versus-host disease, Rejection

alloimmune *adjective* Relating to alloimmunization or a nonself immune reaction

allopurinol Zyloprim® THERAPEUTICS A xanthine oxidase inhibitor which lowers high uric acid and thus is effective for both 1°–eg gout and 2°–eg due to hematologic disorders or neoplasa–hyperuricemia PHARMACOLOGY Allopurinol and its 1° metabolite, oxypurinol, inhibit xanthine oxidase; oral absorption; peak plasma at < 1 hr; $T_{1/2}$ 2-3 hrs ADVERSE EFFECTS Hypersensitivity reaction

allotype An inherited set of determinants or sequence of amino acids–eg, on an Ig–and other proteins demonstrating heterogeneity, which is specific for an individual and more common in a racial group. Cf Idiotype, Isotype

allowable charge MANAGED CARE The lesser of the actual charge, the customary charge and the prevailing charge, which is the amount on which Medicare bases Part B payments

allowable error Allowable analytical error STATISTICS A systemic error that is 'acceptable', both statistically and analytically–eg, 95% limit of error. See Standard deviation

Allscripts THERAPEUTICS A handheld electronic prescribing product

Almazine® Lorazepam, see there

Alopam® Doxepin, see there

alopecia Baldness DERMATOLOGY 1. Loss or absence of hair on the scalp 2. Baldness, see there See Hair replacement, Hot comb alopecia, Moth-eaten alopecia

ALOPECIA TYPES

MALE PATTERN On the front and top–blame mother

PATCHY Alopecia areata–blame mother, angry lover

PERMANENT Related to RT–blame radiation oncologist

TOTAL Alopecia capitis totalis–blame mother

TRANSIENT Due to chemotherapy—cyclophosphamide, cytosine arabinoside, doxorubicin–blame oncologist

alopecia areata DERMATOLOGY A noncicatricial, presumed autoimmune form of transient patchy baldness, that affects up to 2% of the population, peak age, 30-50 ETIOLOGY Uncertain; it has been attributed to anxiety, stress, celiac disease PATHOLOGY Hair may have an 'exclamation mark' appearance MANAGEMENT Intralesional glucocorticoids, topical immunotherapy, anthralin, biological response modifiers–eg, minoxidil PROGNOSIS Spontaneous remission & recurrence is common

ALOS Average length of stay, see there

Alpers disease Poliodystrophy An AR condition characterized by premature closure of cranial sutures resulting in a peaked skull and abnormal facies CLINICAL accompanied by seizures, incoordination, mental deterioration, spasticity, cortical blindness, cortical deafness MANAGEMENT Surgery to correct skull and facial defects PROGNOSIS Poor, death is the norm within 3 yrs of onset

α Symbol for A band in serum electrophoresis STATISTICS The probability of committing a Type I or false-positive error

alpha activity SLEEP DISORDERS The presence of alpha waves or alpha rhythm in an EEG. See Alpha rhythm

alpha-1 antitrypsin deficiency An inherited condition–frequency, ±1:10,000, characterized by low or absent production of alpha-1 antitrypsin, an enzyme which is critical to tissue remodeling CLINICAL The PiZZ phenotype

is characterized by early-onset emphysema and liver-related symptoms—jaundice, cholestasis, fatigue, cirrhosis, liver failure, ascites, mental changes, GI bleeding, and ↑ risk of liver CA TREATMENT IV or nebulized prolastin if COPD, for direct delivery to lungs; alpha-1 proteinase inhibitor; liver transplant; the gene for A1AT may be transferred via adenoviruses to the lung epithelium; following transfer, A1AT mRNA is expressed as functioning A1AT MANAGEMENT Prolastin, O_2, antibiotics, phenobarbital or cholestyramine for jaundice and itching, liver transplant

alpha carotene NUTRITION A carotenoid abundant in carrots, sweet potatoes, cantalope, which has vitamin A and immunostimulatory activity; AC consumption is linked to ↓ risk of lung CA and cancer cell growth in vitro. See Carotenoid

alpha effect ENDOCRINOLOGY The metabolic, hemodynamic and modulatory effects of epinephrine and norepinephrine reflect the concentration of adrenergic receptors on the α and β cells of the pancreatic islets; α effects–epinephrine acting on β cells include glycogenolysis, gluconeogenesis, inhibition of insulin-stimulated glucose uptake in skeletal muscle and vasoconstriction

alpha-fetoprotein Fetoglobulin CLINICAL CHEMISTRY A 70 kD glycoprotein, synthesized by the embryonic yolk sac, fetal GI tract and liver, which has 40% homology with albumin, which peaks at 13 wks of fetal age; AFP's role in fetal development is unclear; AFP levels are measured in pregnancy to screen for open neural tube defects–incidence, 1-2/1000 births and for Down syndrome, and in adults to detect liver cancer and germ cell tumors REF RANGE Non-pregnant adults < 30 ng/mL; maternal serum–13-16 wks, < 1.0-4.4 µg/dL; amniotic fluid–13-16 wks, 0.9-4.1 mg/dL AFP levels in fetal serum are 150-fold > amniotic fluid, which in turn are 200-fold > maternal serum; maternal serum levels are 300-400 µg/L in the 3rd trimester; AFP levels in fetal serum and amniotic fluid peak at 13 wks, while the maternal levels peak at 30 wks; AFP in pregnancy is > in twins and higher multiple pregnancies; if the fetal neural tube fails to close completely, large quantities of AFP enter the amniotic fluid, resulting in ↑ levels in the mother's serum; confirmatory tests such as amniocentesis and/or ultrasonography are used to identify neural tube defects is elevated; after the immediate postnatal period, ↑ serum AFP levels occur only with conditions of abnormal cell multiplication; although AFP measurement is not FDA-approved for cancer screening, in practice, it is used to both detect and monitor therapy in liver cell cancer and germ cell tumors of gonadal, retroperitoneal, or mediastinal origin. See Liver cell carcinoma, Triple screen

alpha heavy chain disease Seligmann's disease The most common heavy chain disease–paraproteinemia, in which there is an excess production of an incomplete IgA₁–partial heavy chain, no light chain; AHCD affects Sephardic Jews, Arabs, Mediterranean rim, South America, Asia CLINICAL Onset in childhood or adolescence as a lymphoproliferative disorder in the respiratory or GI tracts with severe diarrhea, malabsorption, steatorrhea, weight loss, hepatic dysfunction, lymphadenopathy, and marked mononuclear infiltration, which may evolve to lymphoma; AHCD may remit spontaneously, respond to antibiotics or, if monoclonal, require combination chemotherapy LAB ↑ Alk phos, ↓ Ca^{2+} MANAGEMENT Antibiotics, chemotherapy PROGNOSIS May cause death by age 20-30. See IPSID, Mediterranean lymphoma

alpha-2-macroglobulin Alpha-2-M, antiplasmin An 800 kD protein, which maps to chromosome 12 which inhibits proteases–trypsin, chymotrypsin, elastase, thrombin, plasmin REF RANGE 125–420 mg/dL, higher in ♀ ↑ A2M Liver disease, DM, nephrotic syndrome, neural tube defects, ↑ estrogen, pregnancy, neonates and children, ataxia-telangiectasia ↓ A2M Protein-losing gastroenteropathy, protein malnutrition, DIC, fibrinolytic therapy

alpha rhythm Alpha activity, alpha frequency NEUROLOGY A type of electrical activity in adults detected by EEG in the posterior brain, which may be abolished with visual stimulation and attenuated by thinking; AR is seen in relaxed adults with closed eyes; AR has a frequency of 8-13 Hz over the occipital lobe and has bihemispheric asynchrony, where the non-dominant hemisphere has a greater wave amplitude; focal CNS disease is accompanied by focally altered α rhythm, which becomes diffuse in coma SLEEP DISORDERS The EEG pattern that corresponds to the awake state; present in most, but not all, normal individuals; most consistent and predominant during relaxed wakefulness, particularly with reduction of visual input

alphavirus Group A arbovirus VIROLOGY A genus of the family Togaviridae characterized by 50-60 nm virions containing a single-stranded linear RNA PATHOGENIC ALPHAVIRUSES Eastern, western, and Venezuela equine encephalitis viruses, Sindbis virus and Semliki Forest viruses–which may be used as vectors for expressing heterologous genes, Ross River, o'nyong-nyong virus EPIDEMIOLOGY Pathogenic alphaviruses of the Western hemisphere occur primarily in the summer CLINICAL Headache, fever, chills, N&V, mental confusion, somnolence LABORATORY Lymphocytosis and ↑ protein in CSF MANAGEMENT Supportive, intensive nursing care SEQUELAE Neurologic effects–eg, mental retardation, convulsions, paralysis in 30–70% of survivors

alpha wave NEUROLOGY A superficial brain wave with a frequency of 8–13 cycles/sec–Hz, an amplitude of 50 μV, which is normally seen by EEG in the occipital and parietal lobes. See Alpha rhythm, Brain waves, Silva mind control. Cf Mantra, Meditation, Theta waves, Yoga

Alphagan Brimonidine OPHTHALMOLOGY An alpha-2 agonist for treating open-angle glaucoma. See Open angle glaucoma

Alport syndrome A heterogeneous group of conditions with variable patterns of heredity–most are X-R but also AD and AR frequency 1 in 50,000 CLINICAL Glomerulonephritis, hematuria, proteinuria, evolving to HTN, nephrotic syndrome, and end-stage renal disease, variably accompanied by sensorineural deafness, colored urine, swelling, cough, poor vision TREATMENT Kidney transplant may be successful in absence of anti-glomerular basement membrane and anti-tubular antibodies

Alpram Alprazolam, see there

alprazolam NEUROPHARMACOLOGY A benzodiazepine used to manage cholecystokinin related anxiety-panic disorders and CNS depression ADVERSE EFFECTS Drowsiness, loss of coordination, mood swings TOXIC RANGE > 75 μg/L

alprostadil UROLOGY A formulation of PGE-1 which relaxes the corpus cavernosum, allowing engorging during erection. See Erectile dysfunction, Orgasm

Alprox-TD® Alprazolam, see there

ALRI Acute Lower Respiratory Infection

ALS Amyotrophic lateral sclerosis, see there, also 1. Advanced Life Support 2. Alternate lifestyle, see there 3. Antilymphocyte serum

Alström syndrome Alström-Hallgren syndrome, retino-otodiabetic syndrome OPHTHALMOLOGY An AR condition characterized by progressive blindness, type 1 DM, obesity, deafness, normal mental capacity

ALT 1. Alanine aminotransferase-glutamate pyruvate transaminase, GPT 2. Acquisition lead time 3. Antilymphocyte therapy, see there 4. Argon laser trabeculoplaty 5. Autolymphocyte therapy

Altace® Ramipil VASCULAR DISEASE An ACE inhibitor antihypertensive ADVERSE EFFECTS Headache, fatigue, dizziness, dry cough, angioedema CONTRAINDICATIONS Hx of angioedema. See ACE inhibitor

alteplase Activase® CARDIOLOGY A thrombolytic used to manage and prevent pulmonary embolism. See Pulmonary embolism, Thrombolytic therapy

altered immune response A reaction in the immune system caused by an allergen or irritant.

alterego Someone who is perceived to be like oneself

alternans test CARDIOLOGY A noninvasive test for detecting T-wave alternans and identifying Pts at high risk for life-threatening cardiac arrhythmias and sudden cardiac death.

Alternaria spp MYCOLOGY A species of aeroallergenic fungi that causes hypersensitivity pneumonitis or woodworker's disease and IgE-mediated allergies, which are common in Pts with atopic dermatitis, see there

alternate delivery system HEALTH INSURANCE Health services that are more cost-effective than in inpatient, acute care hospitals, such as skilled and intermediary nursing facilities, hospice programs, and in-home services

alternate level of care Pt care–eg, hospice, at-home nursing, in which efforts are no longer made to cure or aggressively treat a disease, but rather prepare for death and alleviate pain, as needed by those who are terminally ill with cancer or AIDS. See Hospice, Terminal illness syndrome

alternate site testing LAB MEDICINE The performance of an array of tests in sites other than a traditional lab environment. See Point-of-care testing

alternating hemiplegia PEDIATRIC NEUROLOGY A rare condition defined by an early onset–≤ 18 months of age, repeated attacks of hemiplegia involving both sides of the body, oculomotor abnormalities during the attack, autonomic disturbances, and mental and neurologic defects EEG Usually normal; some ↑ slow wave activity MANAGEMENT Possibly flunarazine

alternative *adjective* Referring to that which is not conventional, mainstream, or traditional, which is used primarily in the context of alternative–complementary medicine. Cf Fringe, Integrative, Mainstream *noun* One of two or more choices; option. See Reduction alternative, Refinement alternative, Replacement alternative, Therapeutic alternative

alternative birthing center OBSTETRICS An obstetrics unit or facility that provides a pleasant, 'friendly' atmosphere for ♀ who are expected to have an uncomplicated vaginal delivery. See Alternative obstetrics, Doula, Lamaze, Midwife, Natural childbirth, Natural family planning

alternative complement pathway Properdin pathway IMMUNOLOGY A route of complement activation that occurs independently of complement-fixing antibodies; the ACP is more complex than the classic complement pathway; it requires a 'priming' C3 convertase–C3,Bb, and an 'amplification' C3 convertase–C3b,Bb; in the presence of properdin, C3 convertase is stabilized, activating later complement components, which leads to opsonization, leukocyte chemotaxis, ↑ vascular permeability, and cytolysis; ACP is activated by properdin, IgA, IgG, lipopolysaccharide, and snake venom; both pathways are stimulated by trypsin-like enzymes. Cf Complement

alternative healing Natural healing A philosophical stance based on alternative medicine principles, in which a person is returned to a state of well-being through a therapy that is not 'mainstream' in nature. See Alternative medicine

alternative hypothesis EPIDEMIOLOGY A hypothesis to be adopted if a null hypothesis proves implausible, where exposure is linked to disease. See Hypothesis testing. Cf Null hypothesis

alternative lifestyle Alternate lifestyle, queer SOCIAL MEDICINE A generic, 'politically sensitive' term for any form of living arrangements with a 'significant other' in which sexual orientation differs from the usual male-female dyad–eg, a male or female homosexual dyad

alternative medicine *…a heterogeneous set of practices 'that are offered as an alternative to conventional medicine, for the preservation of health and the diagnosis and treatment of health-related problems; its practitioners are often called healers'*; alternative health care practices constitute a vast array of treatments and ideologies, which may be well-known, exotic, mysterious, or even dangerous, and are based on no common or consistent philosophy; the practitioners range from being sincere, well-educated, and committed to their form of healing, to charlatans, deprecatingly known as 'quacks'. See Fringe medicine, Holistic medicine, Integrative medicine, Office of Alternative Medicine. Cf Unproven methods for cancer management

ALTERNAIVE THERAPIES, TYPES OF

ALTERNATIVE (FORMAL) SYSTEMS Acupuncture, ayurvedic medicine, Chinese herbal medicine, homeopathy, osteopathy

BODY AWARENESS Exercise and movement therapies, eg dance therapy, martial arts, yoga

MANIPULATIVE THERAPIES Chiropractic, Hellerwork, Rolfing®

MENTAL THERAPIES Humanistic psychology, hypnosis

NATURAL REMEDIES Diet, eg macrobiotics, naturopathy

SENSORY THERAPIES Art, color, and music therapy

alternative site provider MANAGED CARE A provider of health care other than a traditional hospital, medical practice, and outpatient settings–eg, home care, specialty services, birthing centers, ambulatory surgery centers, outpatient mental health, etc

alternobaric trauma A transient vestibular or auditory dysfunction seen in fliers and divers during ascent, attributed to elevated and probably asymmetric middle ear pressure as a function of eustachian tube opening pressures; vertigo, hearing loss, and tinnitus usually resolve in 10 to 15 mins

altitude anoxia Hypoxia which is directly correlated with a person's altitude

altitude decompression sickness WILDERNESS MEDICINE A permutation of decompression sickness–which almost invariably affects scuba divers–that affects those ascending above 5500 meters MANAGEMENT Hyperbaric oxygen. See Hyperbaric oxygen therapy, Mountain sickness. Cf (the) Bends, Decompression sickness

altitude sickness Mountain sickness, see there

altretamine Hexalen ONCOLOGY An alkylating anticancer agent. See Chemotherapy

aluminum Aluminium A metallic element–atomic number 13; atomic weight 26.98 TOXICOLOGY Changes of aluminum toxicity include vitamin D-refractory osteodystrophy with ↓ mineralization, ↓ bone formation, hypercalcemia, anemia, progressive encephalopathy, dementia MANAGEMENT Chelation with deferroxamine

Alupram Diazepam, see there

alveolar *adjective* Pertaining to an alveolus

alveolar abscess DENTISTRY A pus pocket adjacent to a tooth root which is related to plaque and calculus deposition CLINICAL Tooth pain, tenderness, which may be accompanied by facial swelling, fever TREATMENT Antibiotics, drainage, deep cleaning. See Apical abscess

alveolar macrophage PHYSIOLOGY A free mononuclear cell of the lower respiratory tract, which has a high phagocytic capacity, and is responsible for clearing inhaled particles and lung surfactant; AMs are transiently attached by pseudopodia to the surface of alveolar epithelium; presence of AMs in sputum submitted for cytologic evaluation are markers of specimen adequacy; AMs are generally submerged in a film of phospholipids, and thus form part of the surface lining of the aveoli See Dust cell, Dust macule

alveolitis 1. Acute or chronic accumulation of inflammatory–PMNs, lymphocytes and immune effector–macrophages cells in the lung alveoli. See Allergic alveolitis, Fibrosing alveolitis, Pulmonary alveolitis 2. Inflammation of an alveolus of a tooth

Alzheimer's disease Alzheimer's dementia NEUROLOGY A progressive, neurodegenerative disease, which is the most common cause of dementia, and characterized by progressive mental deterioration accompanied by disorientation, memory and language defects, confusion, leading to progressive dementia, which may be accompanied by dysphasia, and apraxia; in the DSM-IV, AD '...is ...a diagnosis of exclusion, and all other causes (of) cognitive defects ... must first be ruled out.' AD affects 3% in those aged 65–74; 18% aged 75–84; 47% > age 85; AD has been linked to a form of apolipoprotein E, those with 1 defective *APOE* ε4 gene–located on chromosome 19 are at an ↑ risk of AD, and those with 2 copies have AD of early onset; A68 protein is present in AD brain homogenates, and linked to formation of neuritic plaques and neurofibrillary tangles–68 kD; one Alzheimer's amyloid precursor–APP may inhibit serine proteases, and is generated by alternative enzyme splicing, releasing an intact β fragment, possibly explaining the deposition of these fibrils in AD brains MANAGEMENT Tetrahydroaminoacridine-THA, aka tacrine, an acetylcholinesterase inhibitor, may improve the quality of life in AD Pts. See Tacrine

Alzheimer's disease-associated protein Any of a group of 3 major protein subunits–eg, A68, identified by the ALZ-50 antibody, present in neuritic–senile plaques and neurofibrillary tangles, the classic histologic markers for Alzheimer's disease

AMA 1. American Medical Association, see there 2. Aerospace Medical Association 3. Against medical advice, see ama 4. Alternative Medical Association 5. Antimitochondrial antibodies, see there

ama Against medical advice The self-discharge of a Pt from a health care facility, contrary to what his/her physician(s) perceive to be in the Pt's best interests. See Good Pt

amalgam Dental amalgam DENTISTRY A silver-copper-tin alloy with varying amounts of mercury to fill carious teeth. See Biological dentistry, Cremation, Fluoridation, Mercury

Amanita phalloides TOXICOLOGY A mushroom which, with related species–*A bisporigera, A verna, A virosa* are the most common cause of fatal mushroom poisonings CLINICAL After a 12-hr latency, N&V, abdominal colic, severe watery diarrhea; this is followed by a 24-hr latency period, then–if the amount ingested was significant—by fatal hepatitis and renal failure MANAGEMENT Symptomatic-rehydration, IV glucose, instillation of 100 g of activated charcoal per os, mannitol to prevent oliguria; 50% of late-treated Pts die. See Poisonous mushroom

Amantadine™ THERAPEUTICS An antiviral that prevents the release of viral nucleic acid into the host cells, which is most effective against influenza virus; in parkinsonism, amantadine ↑ presynaptic dopamine release, blocks dopamine reuptake into the presynaptic neurons, and has anticholinergic effects ADVERSE EFFECTS Nausea, vertigo, confusion, hallucinations, anxiety, restlessness, depression, irritability, peripheral edema, orthostatic hypotension, psychotic reaction

Amaryl® Glimepiride ENDOCRINOLOGY An agent which may be used in combination with metformin–Glucophage® as 2nd-line therapy for type 2 diabetes when monotherapy with either Amaryl or metformin fails to achieve adequate blood glucose control.

amatoxin TOXICOLOGY Any of a family of potentially lethal toxins present in certain mushrooms—*Amanita phalloides, Lepiota chlorophyllum* and others, which may cause accidental poisoning in amateur mycophagists CLINICAL 12-hr latency, then N&V, abdominal pain, and diarrhea, which may be followed by an asymptomatic period before acute hepatic dysfunction and death TREATMENT Charcoal hemoperfusion

amaurosis Blindness, see there

amaurosis fugax Transient retinal ischemia, see there

amaurotic familial idiocy Batten-Mayou disease, neuronal ceroid-lipofuscinosis MOLECULAR MEDICINE An AR inborn error of metabolism characterized by neuronal deterioration ONSET Age 5 to 10, rapid deterioration of vision and slower deterioration of intellect, accompanied by seizures and psychotic behavior

ambidexterity NEUROLOGY The ability to perform tasks requiring manual dexterity with either hand

Ambien® Zolpidem tartrate NEUROLOGY A hypnotic used for short-term management of insomnia See Osteoprotegerin

ambiguous genitalia ENDOCRINOLOGY ♂ or ♀ external genitalia that are undifferentiated, indistinct or discordant with the genotype. See Hermaphroditism, Intersexuality

ambiguous sexuality PSYCHOLOGY Acquired sexual discordance in which an individual's phenotype and genotype are ♂–or ♀, but his/her 'psychotype' is ♀–or ♂, and thus requires transsexual conversion. See Transsexuality

Ambisome® Liposomal amphotericin B INFECTIOUS DISEASE A formulation of amphotericin B for treating presumed fungal infection in febrile, neutropenic Pts; treatment of Pts with *Aspergillus*, *Candida*, *Cryptococcus* infections refractory to amphotericin B deoxycholate or in Pts with renal impairment or where toxicity precludes use of standard amphotericin B, visceral leishmaniasis, and cryptococcal meningitis See Amphotericin B

ambivalence PSYCHIATRY The coexistence of contradictory emotions, attitudes, ideas, or desires vis-á-vis a particular person, object, or situation

amblyopia OPTHALMOLOGY Impaired vision without an organic eye lesion

AMBRI ORTHOPEDICS An acronym for shoulder joint instability which is Atraumatic, Multidirectional, often Bilateral, requires Rehabilitation as first-line therapy, Inferior capsular shift as the best alternative (surgical) therapy. See Shoulder instability

Ambu bag EMERGENCY MEDICINE A self-refilling bag-valve-mask unit with a 1-1.5 liter capacity, used for artificial respiration which, although suboptimal for the non-intubated Pt, is effective for ventilating and oxygenating intubated Pts, allowing both spontaneous and artificial respiration

ambulance EMERGENCY MEDICINE A vehicle for transporting a Pt to or from a hospital or medical center, which is equipped with supplies to render emergency medical care, and manned or womanned by one or more individuals formally trained in providing such care. See Air ambulance, Emergency medical srvice, EMT

ambulatory 1. Ambulant 2. Not stationary. See Ambulatory care, Ambulatory surgery

ambulatory blood pressure monitoring CARDIOLOGY A noninvasive technique by which multiple indirect blood pressure readings can be obtained automatically for 1 to 3-day periods with minimal intrusion into a Pt's daily activities. See Holter monitor, Hypertension, Hypotension

ambulatory care Outpatient care MEDICAL PRACTICE Any form of medical care—including diagnosis, observation, treatment, rehabilitation, which is provided on an outpatient basis—ie, does not require an overnight admission, to those who are not confined to a hospital but who are 'ambulatory'. See Outpatient surgery. Cf Inpatient care

ambulatory care center Walk-in clinic MEDICAL PRACTICE A free-standing facility that provides non-emergent medical, or less commonly, dental services

ambulatory electrocardiographic monitoring Holter monitoring, see there

ambulatory monitoring CARDIAC PACING Monitoring of a body function–eg, heart rhythm, via a portable electronic device; the person usually participates in normal activities while wearing the device. See Holter monitor

ambulatory patient group MANAGED CARE A prospective payment system, analogous to inpatient DRGs, for ± 300 hospital outpatient technical component services. See Ambulatory payment classification, DRG

ambulatory surgery Outpatient surgery MEDICAL PRACTICE Surgery performed on an outpatient basis in otherwise healthy and clinically stable persons in a facility which may be either free-standing or a suite inside the hospital PROS ↓ Cost; return to daily routine is faster; procedure is less invasive, less debilitating CONS Cost-control pressures, less invasive and more convenient care may ↑ utilization and costs; ↑ family costs and stress

ambulatory surgery center A free-standing center that performs various types of surgery

AMD Age-related macular degeneration, see there

amebiasis PARASITOLOGY Infection with the protozoan, *Entamoeba histolytica*, a pathogen associated with poor sanitary conditions CLINICAL Anorexia, N&V, diarrhea DIAGNOSIS O&P in stool TREATMENT Diloxanide furoate, paromomycin, metronidazole plus a luminal agent. See Amebic abscess, Amebic dysentery

amebic *adjective* Pertaining to 1. Amebiasis, see there 2. Amoeba, see there

amebic dysentery PARASITOLOGY A clinical form of amebiasis–due to *Entamoeba histolytica*, which is characterized by diarrhea, and accompanied by ulcerative inflammation, which mimics ulcerative colitis. See Amebiasis

amebic liver abscess PARASITOLOGY An area of liquefaction necrosis in the liver due to amebic infection. See Amebiasis

ameboid *adjective* Amoeba-like

amelanotic melanoma DERMATOLOGY A rare, poorly differentiated form of melanoma, which occurs in Pts with a previous pigmented melanoma. See Melanoma, Regression

amelia Limblessness. See Phocomelia

ameloblastoma Adamantinoma A locally-aggressive, but almost invariably benign tumor that arises from the odontogenic epithelium in a fibrous stroma of the mandible or maxilla, which is most common in ♂ in the 4th decade TREATMENT Wide local excision; may recur

amendment Amended application, revised application CLINICAL TRIALS Any protocol change which occurs after activation of a trial GOVERNMENT A change in an existing law. See Doggett amendment, Helms amendment, Social Security Amendments of 1983, Synar amendment

amenorrhea GYNECOLOGY The absence or abnormal cessation of menses, which may be 1°–no menstrual period by age 16, or 2°–menses stops for ≥ 6 months in a ♀ in whom normal menstruation was established or for 3 normal intervals in a ♀ with oligomenorrhea SCOPE 1° amenorrhea affects 2.5% of US ♀; 2° amenorrhea affects 3% of the population; it is most common in Pts under extreme stress IMPACT Amenorrheic ♀ do not ovulate, which makes them inconceivable; amenorrhea accompanied by absence of estrogen is linked to genital atrophy and osteoporosis; amenorrhea with minimal estrogen is linked to endometrial hyperplasia; ♀ with 1° amenorrhea may suffer psychosocial and psychosexual problems. See Post-pill amenorrhea

America's Best Hospitals MEDIA & HEALTH An annual 'report card' on the quality of care received in US hospitals published by US News & World

Report, that is either proudly quoted by those who are rated or dismissed by those who are not

American Academy of Osteopathy The professional organization for osteopathic physicians, who actively use osteopathic manipulation in their practice, and adhere to a philosophy that follows the original precepts of osteopathy, as espoused by Andrew Taylor Still, the founder of osteopathy. See Osteopathy, Still. Cf American Osteopathic Association

American Cancer Society ONCOLOGY A national organization that disseminates information about cancer, and funds cancer research

American Law Institute Formulation FORENSIC MEDICINE Section 401 of the ALI's Model Penal Code states that '*a person is not responsible for criminal conduct if at the time of such conduct as a result of mental disease or defect he lacks the substantial capacity either to appreciate the wrongfulness of his conduct or to conform his conduct to the requirements of law*' See Temporary insanity

American National Standards Institute See ANSI

American Osteopathic Association The primary professional organization for osteopathic physicians, which promotes education about osteopathy, and acts in the interests of its practitioners. See Osteopathy. Cf American Academy of Osteopathy

American Society of Anesthesiology Classification A system used by anesthesiologists to stratify severity of patients' underlying disease and potential for suffering complications from general anesthesia

AMERICAN SOCIETY OF ANESTHESIOLOGY PATIENT CLASSIFICATION STATUS

ASA I	Normal healthy Pt
ASA II	Pt with mild systemic disease; no functional limitation–eg, smoker with well-controlled HTN
ASA III	Pt with severe systemic disease; definite functional impairment–eg, DM and angina with relatively stable disease, but requiring therapy
ASA IV	Pt with severe systemic disease that is a constant threat to life–eg, DM + angina + CHF; Pts have dyspnea on mild exertion and chest pain
ASA V	Unstable moribund Pt who is not expected to survive 24 hours with or without the operation
ASA VI	Brain-dead Pt whose organs are removed for donation to another
E	Emergency operation of any type, which is added to any of the 6 above categories, as in ASA II E

American trypanosomiasis American sleeping sickness, agent *T cruzi*–Chagas' disease VECTOR Reduviid–kissing bug CLINICAL The acute form most commonly affects infants, causing malaise, fever, hepatosplenomegaly; the chronic or asymptomatic form is more subtle and is accompanied by altered cardiac conduction–the most common cause of CHF in South America, megaesophagus, megacolon MANAGEMENT Melarsoprol, a toxic agent used for end-stage meningoencephalitic disease or propoxydecanoic acid, a myristic acid analogue which is highly toxic to, and specific for trypanosome, eflornithine

Americans with Disabilities Act Legislation passed in the US in 1990 that was intended to remove the physical barriers and biases in places of public access and in the workplace, that had previously prevented those with physical and mental disabilities–handicaps, 'challenges' from enjoying the full benefits of freedoms guaranteed by the US Constitution. See Barriers, Disabilities. Cf Impairment

amethopterin Methotrexate, see there

AMI Acute myocardial infarction

Amicen Amitryptiline, see there

amikacin THERAPEUTICS The broadest of the broad-spectrum aminogly-

cosides, which is resistant to aminoglycoside-inactivating enzymes INDICATIONS Nosocomial gram-negative bacterial infections–eg, with *Enterobacter* spp, *Klebsiella* spp, *Proteus* spp, *Pseudomonas aeruginosa*, *Serratia* spp, *E coli*, possibly MAC ADVERSE EFFECTS Ototoxicity, nephrotoxicity

Amilit Amitryptiline, see there

amiloride NEPHROLOGY A drug that blocks the Na^+/H^+ antiporter, used clinically as a potassium-sparing diuretic. See Diuretics, Potassium-sparing diuretic PULMONOLOGY An aerosolized sodium channel blocker that may slow the progression of pulmonary dysfunction in cystic fibrosis. See Diuretics, Potassium-sparing diuretic

amimia PSYCHIATRY A language disorder characterized by inability to gesticulate or understand the significance of gestures

amine precursor uptake & decarboxylation See APUD cell, APUDoma

Amineurin® Amitryptiline, see there

amino acid drink Protein drink, see there

aminoglutethimide ONCOLOGY A nonsteroidal aromatase inhibitor which inhibits estrogen production and suppresses estrogen-dependent tumor growth. See Estrogen-receptor

aminoglycoside THERAPEUTICS Any of a family of broad-spectrum antibiotics–amikacin, gentamicin, kanamycin, netilmicin, neomycin, framycetin, streptomycin, tobramycin, which are used primarily against aerobic gram-negative bacteria PHARMACODYNAMICS Poorly absorbed per os, poor penetration of CNS–BBB, rapid excretion if kidneys are normal TYPES Gentamicin, streptomycin, tobramycin TOXICITY Dose-related–kidneys, vestibular, auditory, and neuromuscular systems, minor skin rash, drug fever, $\downarrow Mg^{2+}$, $\downarrow Ca^{2+}$, $\downarrow K^+$; it is common practice to monitor Pts receiving AGs. See Therapeutic drug monitoring

aminolevulinic acid ONCOLOGY A drug used in photodynamic therapy which is absorbed by tumor cells and activated when exposed to light, killing the cells. See Photodynamic therapy

aminopeptidase Any of the hydrolases–enzymes, which catalyze the removal of amino terminal amino acids or dipeptides from a protein or peptide

aminophylline THERAPEUTICS The ethylenediamine salt of theophylline, administered IV to Pts with acute asthma Sx MECHANISM Inhibits cAMP phosphodiesterase ACTION Relaxes upper airway and pulmonary vessel smooth muscle, resulting in broncho- and vasodilation; it is also diuretic, coronary vasodilator, cardiac and cerebral stimulant ADVERSE EFFECTS GI irritation–anorexia, N&V, epigastric pain, restlessness, insominia, headache CONTRAINDICATIONS Peptic ulcers, seizures, hypersensitivity to ethylenediamine. See Theophylline

aminopterin ONCOLOGY A folic acid antagonist and inhibitor of dihydrofolate reductase; it is a potent cytotoxic agent fomerly used to manage leukemia; it has been replaced by MTX. See Methotrexate

amiodarone Cordorone® CARDIOLOGY A class III agent that prolongs the duration of the action potential; amiodarone is used to treat refractory ventricular and supraventricular tachycardia, A Fib, conduction block ADVERSE EFFECTS Pulmonary fibrosis, which occurs in ±6% of Pts taking the drug, hypothyroidism, hyperthyroidism, deposits in cornea—causing photosensitivity and/or skin-blue/gray skin pigmentation, leukocytoclastic vasculitis, hepatitis, ↑ digitoxin levels, neurotoxicity, GI toxicity

amiodarone-induced thyrotoxicosis CARDIOLOGY Thyroid disease caused by amiodarone, an iodine-rich antiarrhythmic, which occurs in up to

±10% of amiodarone recipients living in areas of moderate iodine deficiency. See Thyrotoxicosis

Amiprol Diazepam, see there

AMISTAD CARDIOLOGY A clinical trial–Acute Myocardial Infarction Study of Adenosine to determine the efficacy of IV adenosine in combination with either streptokinase or tPA in limiting the size of AMIs See Acute myocardial infarction

Amitril® Amitryptiline, see there

amitriptyline Elavil NEUROPHARMACOLOGY A tricyclic antidepressant, with sedative and anticholinergic properties, which may be used for peripheral neuropathy ADVERSE EFFECTS Rash, nausea, weight gain/loss, drowsiness, nervousness, insomnia, confusion, seizures, coma, orthostatic hypotension

AML Acute myelocytic leukemia, see there

amlodipine besylate Norvasc® CARDIOLOGY A CCB used to treat HTN and angina pectoris. See ACT, Calcium channel blocker, PRAISE

ammonia NH₃ PHYSIOLOGY NH₃ is produced in the liver, intestine, and kidneys as endproduct of protein metabolism; the liver converts ammonia into urea, which is then excreted by the kidneys; in liver disease this conversion is diminished, resulting in ↑ serum ammonia; serial measurement of ammonia is used to follow the progression of hepatic encephalopathy in Reye syndrome and other conditions REF RANGE 15-49 µg/dL ABNORMAL VALUES ↑ Hepatic coma, Reye syndrome, severe CHF, GI hemorrhage, erythroblastosis fetalis, drugs–eg, diuretics and antibiotics. See Hepatic encephalopathy

amnesia NEUROLOGY A pathologic impairment or lack of memory, which is often temporally linked to a traumatic event TYPES Anterograde–no memory for that occurring *after* the event; retrograde–no memory for that occurring *before* the event; amnesia may be organic, emotional, mixed origin, or time-limited. See Anterograde amnesia, Hysterical amnesia, Retrograde amnesia, Selective amnesia, Sensorimotor amnesia, Source amnesia

amniocentesis OBSTETRICS A procedure in which fluid is obtained by an ultrasonographically-guided needle from the amniotic cavity–usually between wks 15-17 of pregnancy to detect fetal abnormalities by karyotyping or chemical analysis; accuracy is reported to be 99.4%; complication rate < 0.5% above the background pregnancy loss of 2-3%; fetal loss minimal DEFECTS IDENTIFIED BY AMNIOCENTESIS Cultured amniotic cells can be used for cytogenetic studies, DNA analysis–karyotyping, and enzyme assays. Cf Chorionic villus biopsy

AMNIOCENTESIS—INDICATIONS FOR

Maternal age > 35

3+ spontaneous abortions

Previous child with chromosome defect, metabolic disease, neural tube defect

Parent, or family Hx of chromosome defect(s)

Possible carrier of X-linked disease

amnioinfusion Intra-amniotic infusion OBSTETRICS The injection of any fluid into the amniotic cavity, either to 1. Replenish intra-uterine volume after rupture of membranes to facilitate fetal manipulation before a vaginal deliver, or 2. To induce labor. See Saline abortion, Prostaglandin abortion

amniotic band 'syndrome' OBSTETRICS A triad of amnion-denuded placenta, fetal attachments to, or entanglement by, amniotic remnants, and fetal deformation, malformation, or limb disruption. See ADAM complex, Sequence

amniotic epithelium See Amnion

amniotic fluid *liquor amnioticus* OBSTETRICS The fluid that surrounds the fetus in the amniotic sac, which cushions it from injury and plays a central role in fetal development; osmolality, sodium, creatinine, and urea content

of AF and maternal serum are virtually identical VOLUME ± 800 mL at birth; specific gravity, ±1.008; pH, 7.2

amniotic fluid analysis LAB MEDICINE A series of tests performed on the fluid obtained by amniocentesis. See Amniocentesis

amniotic fluid embolism OBSTETRICS A condition resulting from a traumatic delivery and 'injection' of amniotic fluid containing lanugo, squames, mucus and debris into the opened maternal circulation, which communicates with the amniotic fluid INCIDENCE 1:80,000 deliveries ETIOLOGY Idiopathic, predisposed to by the high intrauterine pressure that allows amniotic fluid to pass into the maternal venous circulation, where the meconium is toxic to the mother, potentially causing DIC MORTALITY ±80%. See Embolism

amniotic sac *amnion*, bag of waters, membranes OBSTETRICS An amniotic fluid-filled thin membraned sac which surrounds the embryo of amniote vertebrates, and provides a fluid cushion, prevents dehydration, acting as a waste receptacle during embryonic development. See Amnion

amniotomy Breaking the bag of waters, membrane rupture OBSTETRICS The deliberate rupture of the amniotic sac to induce or hasten labor

amobarbital Amytal® THERAPEUTICS An intermediate-acting sedative barbiturate

Amoeba PARASITOLOGY A genus of amebas of the order Amoebida

amoeba Ameba An imprecise name for several types of free living unicellular phagocytic organisms; the pathogenic amebas have been reclassified as *Entamoeba* spp, *Endolimax* spp, and others

amorphous *adjective* Lacking a fixed shape; shapeless

amotivational syndrome SUBSTANCE ABUSE A condition linked to chronic marijuana abuse, which most commonly affects young, learning disabled, and emotionally immature individuals. See 'Gateway' drugs, Marijuana

amoxapine NEUROPHARMACOLOGY A tricyclic antidepressant ADVERSE EFFECTS Tardive dyskinesia, sedation, postural hypotension, cholinergic effects—eg dry mouth, blurred vision, constipation, urinary retention, weight gain, neuroleptic malignant syndrome, cardiovascular effects—EKG, slow AV conduction; withdrawal symptoms accompany abrupt withdrawal

amoxicillin INFECTIOUS DISEASE A broad-spectrum semisynthetic penicillin with activity similar to that of ampicillin

AMPA receptor NEUROPHYSIOLOGY Any of a family of distinct ionotropic glutamate–excitatory post-synaptic receptors widely expressed in the CNS, which are the 1° memory receptors. See Excitatory amino acid receptor channel, Glutamate receptor

AMPA-kainate receptor A glutamate receptor that mediates the fast component of excitatory post-synaptic potentials; the AMPA-kainate receptors mediate a large fraction of excitatory transmission by neurotransmitters, and are regulated by cAMP-dependent protein kinase and phosphates

ampakine NEUROPHARMACOLOGY Any compound that facilitates synaptic transmission by binding to glutamate and enhance long-term potentiation, a critical component in long term memory, by enhancing the activity of AMPA receptor-rich neurons which enhance AMPA receptor activity by amplifying neural signals generated by glutamate binding; ampakines improve memory in early trials in Alzheimer's disease, schizophrenia See AMPA receptor, Long-term potentiation

amphetamine PHARMACOLOGY A CNS stimulant, anorexiant and drug of abuse USED FOR Hyperactivity; narcolepsy; obesity. See Drug Screening, Therapeutic drug monitoring

Amphotec® Amphocil INFECTIOUS DISEASE A proprietary antifungal formulation in which toxic amphotericin B is wrapped inside a liposome envelope. See Amphotericin B, Liposome

amphotericin B INFECTIOUS DISEASE A heptaene macrolide antibiotic produced by the bacteria, *Streptomyces nodosus*, administered IV EFFECTIVE AGAINST *Blastomycosis dermatitidis*, *Candida*, *Coccidioides immitis*, *Cryptococcus neoformans*, *H capsulatum*, *Paracoccidioides braziliensis*, *Torulopsis glabrata* ADVERSE EFFECTS Severe–fever, azotemia, nephrotoxicity, chills, headache, anorexia, N&V, diarrhea, kidney damage, BM suppression LAB Hypochromic normocytic anemia, neutropenia See Amphotec, Liposome

ampicillin INFECTIOUS DISEASE An analogue of penicillin, which inhibits crosslinking of peptidoglycan chains in the eubacterial cell wall. Ampicillin is comparable to penicillin G against gram-positive bacteria, but more active against gram-negative bacilli ADVERSE EFFECTS Rash, especially if Pt also receiving allopurinol or actively infected with EBV

Amplicor A proprietary–Roche Diagnostics PCR-based test used to monitor HIV-1 viral load

amplification MOLECULAR BIOLOGY An in vivo–as in the fragile X syndrome or in vitro–as in cloning or PCR ↑ in the number of copies of a specific gene or DNA sequence of interest. See Cloning, DNA amplification, Exon amplification, Extreme amplification, Gene amplification, Linked linear amplification, PCR, PCR amplification of specific alleles, Solid phase amplification

amplification system PHYSIOLOGY A generic term for any group of proteins that function in coordinated sequences, forming positive feedback loops for expanding the response to a low intensity signal

AMPLIFICATION SYSTEMS

COAGULATION, eg factor Xa activating factor 'X' in the presence of factor VIII, Ca^{2+}, and phospholipid

COMPLEMENT Augments B-cell response. See Alternative and Classic pathways

CYTOKINES Amplifies T-cell response, ILs, kinins, lipid mediators and mast cell products

amplitude CARDIAC PACING The maximum absolute value attained by an electrical waveform, or any quantity that varies periodically; pacemaker amplitudes express the value of the potential difference in volts or current flow in amperes; pacemaker output pulses have typically averaged 5 volts and 10 milliamps. See Relative fusional vergence

amprenavir AIDS A protease inhibitor in clinical trials for treating HIV, used in combination with other protease inhibitors–eg, nelfinavir, indinavir, or saquinavir ADVERSE EFFECTS N&V, diarrhea, headache, perioral paresthesias, stomach discomfort, rash; other effects include hyperglycemia, DM, acute hemolytic anemia, spontaneous bleeding in hemophiliacs, and fat redistribution. See AIDS, Combination therapy, HIV, Protease inhibitor

ampule Ampoule-British MEDTALK A small hermetically sealed glass or plastic container which contains a sterile solution–eg, lidocaine, etc to be administered parenterally–IM, IV, subcut

ampulla A saclike enlargement of a duct or tube 1. Ampulla of Vater, formally, hepatopancreatic ampulla 2. Ampule

amputation SURGERY The partial or total surgical excision of a limb, appendage or digit. See Above-the-knee amputation, Below-the-knee amputation, Forequarter amputation, Guillotine amputation, Hemipelvectomy, Translumbar amputation

amrinone Inocor® CARDIOLOGY An agent which improves cardiac output by vasodilatory and positive inotropic activity ADVERSE EFFECTS Dose-related thrombocytopenia

AMS 1. Acute maxillary sinusitis. See Sinusitis 2. Acute mountain sickness, see there

Amsterdam criteria ONCOLOGY Criteria for diagnosing hereditary non-polyposis colorectal cancer, see there

AMSTERDAM CRITERIA

FAMILY HISTORY Presence of histologically verified colorectal cancer in ≥ 3 relatives–one of whom is a 1st-degree relative of the other 2

'VERTICAL' HISTORY Presence of disease in ≥ 2 successive generations

AGE OF ONSET Age < 50 in > 1 affected relative

EXCLUSION Hereditary polyposis syndromes have been excluded

amusia NEUROLOGY Inability to recognize or produce music. See Tone-deafness

amygdalin A β-cyanogenic glycoside structurally related to the semisynthetic laetrile, derived from the seeds of certain fruits. See Laetrile

amylase CLINICAL CHEMISTRY An enzyme synthesized in the pancreas and salivary glands and secreted in the GI tract, which digests starch and glycogen; amylase is measured in Pts with suspected pancreatitis; serum and urine levels peak 4-8 hrs after onset of acute pancreatitis, and normalize within 48-72 hrs; parotitis due to mumps or radiation therapy also ↑ serum amylase; in cases of ↑ serum amylase without pancreatitis or parotitis, requires quantification of amylase isoenzymes REF RANGE Varies by laboratory; 25-90 U/L, serum; 4-30 U/2 hrs, urine; amylase is ↑ in acute pancreatitis, obstruction of common bile duct, pancreatic duct or ampule of Vater, pancreatic injury from perforated peptic ulcer, acute salivary gland disease; amylase is ↓ in chronic pancreatitis, pancreatic CA, cirrhosis, hepatitis, eclampsia. See Macroamylase

Amyline Amitryptiline, see there

amyloid β-fibrillosis A homogeneous, extracellular glycoprotein with a fibrillary ultrastructure, derived either from 1. The N-terminal of lambda or kappa Ig light chains–amyloid of immune origin, a 5-18 kD glycoprotein produced by a single clone of plasma cells or 2. Amyloid of unknown origin, from serum amyloid A–SAA, an acute phase protein that ↑ sharply during inflammation; all amyloids have a common molecular theme, that of the β-pleated protein sheet, demonstrable by X-ray crystallography and responsible for amyloid's Congo red staining and resistance to proteolytic digestion See Beta amyloid. Cf Alzhemier's disease

amyloid plaque NEUROLOGY A pathologic lesion of Alzheimer's brains characterized by aggregated amyloid staining material. See Alzheimer's disease, Neurofibrillary tangle

amyloidosis Systemic amyloidosis INTERNAL MEDICINE A group of diseases in which amyloid protein is deposited in specific organs–localized amyloidosis or throughout the body–systemic amyloidosis deposition of amyloid, which can be 1°, which usually affects nerves, skin, tongue, joints, heart, or liver, or 2° to other conditions–eg, TB, CA, leprosy, and accompanied by immune system changes, which affect the spleen, kidneys, liver, and adrenal glands. See Cardiac amyloidosis, Cerebral amyloidosis, Hereditary amyloidosis

AMYLOIDOSIS

AL AMYLOIDOSIS 1° amyloidosis An uncommon condition associated with plasma cell dyscrasias, where the accumulated amyloid fibers correspond to fragments of Ig light chains; the median survival in one study of 153 Pts with primary AL amyloidosis as 20 months, less if they had signs of renal or cardiac amyloidosis and up to 40 months if neither system was involved

AA AMYLOIDOSIS Reactive amyloidosis, 2° amyloidosis A group of conditions seen in chronic inflammation, in which the accumulated fibers derive from a circulating acute-phase lipoprotein, known as serum protein A

Both types of amyloidosis contain amyloid P component, a non-fibrillary glycoprotein found in the circulation; scintigraphy after injection of [125I]-labelled serum amyloid P component can locate tissue deposition of amyloidosis and if the deposition is active, cytotoxic therapy may be instituted; focal amyloidosis commonly occurs in those organs most susceptible to aging, eg the heart and brain, and only is symptomatic if the amyloid deposition is significant

amyotrophic lateral sclerosis Lou Gehrig's disease NEUROLOGY A chronic, progressive degenerative, motor neuron disease, characterized by upper limb weakness, atrophy and focal neurologic signs EPIDEMIOLOGY Incidence, 0.5-1.5/10⁵, more common in ♂, usually > age 50; occurs randomly thoughout the world with local clustering on the Kii Peninsula, Japan and Guam, where it is associated with dementia, parkinsonism, Alzheimer's CLINICAL Loss of fine motor skills, triad of atrophic weakness of hands and forearms; leg spasticity; generalized hyperreflexia MANAGEMENT Possible riluzole. See Motor neuron disease

amyotrophy NEUROLOGY Muscle wasting. See Hereditary neuralgic amyotrophy

Amytal® Amobarbital, see there

ANA Antinuclear antibodies, see there, also, 1. American Neurological Association 2 Antiviral nucleoside analogue, see there

ANA-negative systemic lupus erythematosus SLE characterized by absence of antinuclear antibodies–seen in 5% of SLE, photosensitivity, features of Sjögren syndrome, a low incidence of lupus nephritis and lupus psychosis, presence of rheumatoid factor and antibodies to Ro/SSA antigen and La/SSB antigens and single-stranded DNA. See Antinuclear antibodies, Lupus erythematosus. Cf Antiphospholipid syndrome

anabolic MEDTALK *adjective* Relating to anabolism–stimulating synthesis or building up, in particular of tissue, especially muscle

anabolic steroid Anabolic-androgenic steroid ENDOCRINOLOGY A drug or hormone-like substance, chemically or pharmacologically related to 17-α-alkylated testosterone that promotes muscle growth, which is commonly abused by athletes LIPID CHANGES BY ASs ↓ HDL-C–especially HDL₂ ↑ hepatic TG lipase–HDL catabolism INDICATIONS Children, adolescents with delayed puberty, ↓ growth, small penis, hypogonadism, testosterone deficiency, osteoporosis management, aplastic anemia, endometriosis, angioedema, sports performance enhancement—no longer legal, relief and recovery from common injuries, rehabilitation, weight control, anti-insomnia, regulation of sexuality, aggression, cognition ROUTE Oral, parenteral METABOLIC EFFECTS ↑ Protein sythesis and amino acid consumption, androgenesis, catabolism ADVERSE EFFECTS–♂ Breast enlargement, testicular atrophy, sterility, sperm abnormalities, impotence, and prostatic hypertrophy; ♀ Clitoral hypertrophy, beard growth, baldness, deepened voice, decreased breast size; ♀/♂ Aggression and antisocial behavior, 'roid rage'; ↑ risk of cardiovascular disease, peliosis hepatis, hemorrhage, jaundice, acne, accelerated bone maturation, resulting in short stature, liver tumors–hepatic adenomas and CA, which may regress with abstinence LAB ASs are detectable to 1 ppb 4 days after last use if the hormone is water-soluble, or 14 days after use in lipid-soluble compounds FDA STATUS ASs are 'schedule III' drugs per the Controlled Substances Act; AS abusers are ↑ risk for HIV transmission, given the common practice of injecting ASs and 'fraternal' sharing of needles. See 'Roid rage, 'Stacking'

anabolism Synthesis; a phase of metabolism involving chemical reactions within cells that result in the production of larger molecules from smaller ones. Cf Catabolism

anaclitic depression NEONATOLOGY A response in infants who have been separated from their mothers for prolonged periods of time, resulting in a disruption of the mother-child dyad. See Genie, Holtism, Social isolation; Cf Bonding, Companionship, Infant massage

anaerobe MICROBIOLOGY Any organism, usually a bacterium, capable of living without air; anaerobic pathogens obtain their energy from fermentation; nonpathogenic anerobes in nature obtain their energy from anaerobic

respiration in which nitrate or sulfate serves as electron acceptors; oropharynx, skin, colon, vagina harbor up to 10¹¹ anaerobes/cm³; anaerobes are common causes of infection, and may be associated with aerobic flora in infections and abscesses of the oral cavity, upper respiratory tract, colon, genital tract, skin, and brain; factors controlling anaerobes' virulence are uncertain TREATMENT Penicillin for supradiaphragmatic anaerobic infections; clindamycin, metronidazole, chloramphenicol, or cephoxatin, if the infection is below the diaphragm. See Aerotolerant anaerobe, Facultative anaerobe, Microaerophile, Obligate aerobe, Obligate anaerobe

anaerobic *adjective* 1. Referring to an anaerobe 2. Lacking O₂

anaerobic exercise SPORTS MEDICINE A general term for exercise consisting of slow rhythmic movements against a force–eg, calisthenics–push-ups, sit-ups, weight lifting, which evoke a minimal ↑ in heart rate; AE strengthens muscles, ↑ joint mobility, and ↓ risk of musculoskeletal injury. Cf Aerobic exercise

Anafranil Clomipramine, see there

anagrelide HEMATOLOGY An agent used to manage essential thrombocythemia and thrombocythemia due to myeloproliferative disorders–eg, CML, polycythemia vera—to ↓ platelets, risk of thrombosis and other Sx

anal *adjective* Pertaining to the anus

anal agenesis A condition caused by a defective development of the anus; the anal dimple is present; perianal stimulation of a neonate results in puckering, which indicates the presence of the external sphincter CLINICAL Complete intestinal obstruction in absence of a perineal fistula. See Anorectal anomalies

anal atresia Imperforate anus, see there

anal canal The terminal tubular part of the large intestine that opens via the anus

anal cancer ONCOLOGY A malignancy of the anal canal, which has been linked to anal intercourse PATHOLOGY Adenocarcinoma is more common in the proximal anus; the transition zone is associated with various malignancies–eg, carcinoma, lymphoma, melanoma; squamous cell carcinoma is most common in the distal anus, and may be accompanied by HPV infection. See Anal intraepithelial neoplasia

ANAL CANCER STAGING

I	CA spread beyond anal mucosa; < 2 cm
II	CA spread beyond anal mucosa; > 2 cm
IIIA	CA spread to perirectal lymph nodes or to adjacent organs–eg, vagina, bladder
IIIB	CA spread to abdominal or inguinal lymph nodes or to both nearby organs and perirectal lymph nodes
IV	CA spread to distant lymph nodes within the abdomen or metastasized

anal character PSYCHIATRY A personality type that manifests excessive orderliness, miserliness, and obstinacy; in psychoanalysis, a pattern of behavior in an adult that may originate in the anal phase of infancy, between age 1 and 3. See Psychosexual development

anal fissure SURGERY A common condition, consisting of a tear or superficial laceration in the anal mucosa, extending from the anal verge toward the dentate line; 90% of 1° AFs are posterior CLINICAL Pain, spasms, bright red blood in stool MANAGEMENT Surgery–which permanently weakens the internal sphincter; conservative-chemical denervation with botulinum toxin by injection; topical application of nitroglycerin See Botulinum toxin

anal intercourse Insertion of an erect penis per anus in a fashion analogous to that of vaginal intercourse. See Anal epithelial lesion, Anal intraepithelial neoplasia, Gay bowel syndrome

anal intraepithelial neoplasia ONCOLOGY A carcinoma in situ of the anorectal mucosa, which most commonly affects immunosuppressed ♂ homosexuals who are often infected by various HPV types; AIN is histologically characterized by irregular nuclear shape and chromatin and hyperchromasia. See CIN, HPV, Intraepithelial neoplasia. Cf Anal epithelial lesion

anal personality See Anal-retentive

anal-retentive PSYCHOLOGY *adjective* Referring to a person with an 'anal personality' who, according to classic Freudian psychoanalysis, has traits that arose in the anal phase of psychosexual development, in which defecation constituted the 1° source of pleasure, and retention of feces is viewed as a manifestation of defiance to a parent figure *noun* A popular term for an anal-retentive person, who is obstinate, rigid, meticulous, compulsive, and overly conscientious, a behavioral profile typical of many physicians and scientists

anal stage PSYCHIATRY The period of pregenital psychosexual development, usually from age 1 to 3, in which the child has particular interest and concern with the process of defecation and the sensations connected with the anus; the pleasurable part of the experience is termed anal eroticism. See Anal character

anal tag SURGERY Swollen skin at the peripheral end of an anal fissure, often accompanied by pain on defecation and fresh bleeding. See Anal fissure

analgesia NEUROLOGY A state of insensitivity to pain, which is the result of 1. Pharmacotherapy with an analgesic. See Oligoanalgesia, Patient-controlled analgesia, Preemptive analgesia 2. Derangement of sensation

analgesic *adjective* Referring to ↓ pain *noun* PAIN MANAGEMENT Painkiller An agent that ↓ perception of pain–nociceptive stimulation without causing loss of consciousness or anesthesia. See Adjuvant analgesic, Anesthesiology

analgesic drug 'ladder' CLINICAL PHARMACOLOGY An algorithm for managing cancer pain, according to the levels of intensity of therapy, delineated by the WHO

ANALGESIC LADDER–PER WHO

1ST STEP	Non-opioids with/without adjuvants–neurolytic blockage, cordotomy, chemical hypophysectomy and others, using choline magnesium trisalisylate, salsalate
2ND STEP	Weak opioids, with/without non-opioids and/or adjuvants, eg codeine, hydrocodone, often produced in combination with aspirin or acetaminophen
3RD STEP	Strong opioids, with/without non-opioids and/or adjuvants, eg morphine sulfate, available in various forms, eg tablets, elixir, suppository

analgesic nephropathy Analgesic-associated nephropathy, phenacetin nephritis A form of kidney damage caused by overexposure to certain analgesics, including phenacetin, acetaminophen, aspirin, and NSAIDs CLINICAL Hematuria, renal colic, pyelonephritis IMAGING Intravenous pyelogram reveals 'ring' shadows and multiple characteristic cavitations COMPLICATIONS ↑ Risk of transitional cell carcinoma, acute renal failure. See Interstitial nephritis

analgesic tolerance PAIN MANAGEMENT An ↑ need for opioids to achieve the same level of analgesia. Cf Addiction

analingus SEXOLOGY Stimulation of the anus with tongue and lips; not a paraphilia; it is either part of normophilic sexuoerotic activity or part of the sadomasochistic repertory. See Anal intercourse

analog INFORMATICS *adjective* Referring to data in the form of continuously variable–non-discrete physical quantities, the mode in which most laboratory instruments produce information, where data is generated as non-discrete signals, as AC or DC current, voltage changes or pulse amplitudes. Cf Analogue, Digital

analogue PHARMACOLOGY A therapeutic agent with structural or chemical similarity to another substance, or which mimics the effects of another agent, but has a different chemical structure. See Antinucleoside analogue, Base analogue, Nicotine analogue, Nucleoside analogue, Purine analogue, Cf Analog

analysis LAB MEDICINE The quantification or other form of assessment of the constituents or elements of a substance of interest. See Activation analysis, Amniotic fluid analysis, ANCOVA, ANOVA, Aura analysis, Base sequence analysis, Bayesian analysis, Best interests analysis, Blood gas analysis, BRAC analysis, Breakeven analysis, Canonical correlation analysis, Capillary ion analysis, Character analysis, Chimerism analysis, Chinese hand analysis, Chromosome analysis, Clinical decision analysis, Clonal analysis, Clot waveform analysis, Combustion analysis, Coordination analysis, Cost-benefit analysis, Cost-effectiveness analysis, Cost of failure analysis, Cost of illness analysis, Cost minimization analysis, DF extremes analysis, Decision analysis, Deletion analysis, Discriminant analysis, Distributive and synthesis analysis, DNA analysis, DNA ploidy analysis, DNA sequence analysis, Dual parameter analysis, Energy-dispersive-x-ray analysis, Error rate-bound analysis, Expression analysis, Factor analysis, Foot analysis, Gap analysis, Gastric analysis, Gel shift analysis, Genetic linkage analysis, Hair analysis, Head space analysis, Heteroduplex analysis, High-resolution chromosome analysis, Image analysis, Information for Management analysis, Intent to treat analysis, Interim analysis, Kaplan-Meier analysis, Karyotype analysis, Laban movement analysis, Linkage analysis, Live cell analysis, Markovian analysis, Markovian texture analysis, Meta-analysis, Microanalysis, Microarray analysis, Microsatellite analysis, Model-fitting analysis, Molecular genetic analysis, Monte Carlo analysis, Movement analysis, Multivariate/logistic-regression analysis, Network analysis, Neutron activation analysis, Orthogonal regression analysis, Paradigm analysis, Perturbation analysis, Planning analysis, Ploidy analysis, Quantal analysis, Quantitative trait loci association analysis, RAPD analysis, Replacement analysis, Restriction analysis, Risk analysis, ROC analysis, Root cause analysis, S phase analysis, Semen analysis, Sensitivity analysis, Signature pattern analysis, Single-strand conformation polymorphism analysis, Sister chromosome exchange analysis, Slot blot analysis, Spectral analysis, Survival analysis, SWOT analysis, Synovial fluid analysis, Task analysis, Temporal analyisis, Time series analysis, Tree-structured survival analysis, Univariate analysis, VNTR analysis, Volume-outcome analysis, Workflow analysis MEDTALK The assessment of the individual components of a process or substance PSYCHIATRY Psychoanalysis, see there. See Ego analysis, Freudian analysis, Jungian psychoanalysis analysis, Psychoanalysis, Psychometric analysis, Transactional analysis

analysis bias Deviation of results or inferences from the truth resulting from flaws in the analysis or interpretation of results. See Bias

analysis of variance ANOVA STATISTICS An analytical method used for continuous variables, which determine whether the source of variability among 3 or more data sets is due to true differences in the sets or due to random variations or 'statistical noise'; ANOVA compares the means of several random variables, assuming that each has a normal distribution with the same variance; these algorithms are quite complex and are usually computer-based. See Statistics

analytic sensitivity LAB MEDICINE The concentration at which the mean response is statistically beyond the noise limits of the signal at zero concentration. See Sensitivity. Cf Functional sensitivity

analytic epidemiology EPIDEMIOLOGY The aspect of epidemiology concerned with identifying health-related causes and effects; AE uses comparison groups, which provide baseline data, to quantify the association between exposures and outcomes, and test hypotheses about causal relationships

analytical psychology PSYCHIATRY The name given by psychoanalyst Carl Jung for his system, which minimizes the influences of sexual factors in emotional disorders and stresses mystical religious influences and a belief in the collective unconscious. See Freudian psychology.

anamnesis PSYCHIATRY The developmental history of a Pt and of his/her illness; the act of remembering

anaphylactic reaction Anaphylaxis CLINICAL IMMUNOLOGY An antigen-induced, IgE-mediated release–and production—of chemical mediators, the target of which is blood vessels and smooth muscle PATHOGENESIS An AR is a hypersensitivity reaction that follows re-exposure to an antigen to which the body has previously formed an IgE antibody; within seconds of exposure to the antigen(s), which may be proteins, polysaccharides, and haptens, IgE molecules cross-link on the surface of mast cells and basophils, stimulating vesicle degranulation and release of LBW mediators of anaphylaxis; in the 1° response, preformed molecules are released, including eosinophil chemotactic factor, and vasoactive substances—eg, heparin, histamine, serotonin, and various enzymes; in the 2° response, acute phase reactants are produced and released; fatal and near-fatal ARs in children are commonly evoked by peanuts > nuts > eggs, milk, fish, and others CLINICAL Bronchospasm, dyspnea, hypotension, edema, shock, and possibly death MANAGEMENT Epinephrine ASAP. See Acute phase reactants. Cf Anaphylactic shock, Anaphylactic reaction

anaphylactic shock IMMUNOLOGY An extreme manifestation of an allergic reaction, which is characterized by ↓ blood pressure, shock–poor tissue perfusion and difficulty breathing. See Anaphylactic reaction

anaphylactoid purpura Henoch-Schönlein purpura, see there

anaphylactoid reaction IMMUNOLOGY An anaphylaxis-like reaction that occurs without an allergen-IgE antibody event, which is caused by a nonimmune release–eg reaction to radiocontrast, chymopapain, aspirin, of vasoactive and inflammatory mediators, including histamine. Cf Anaphylactic reaction

anaplastic carcinoma An often aggressive epithelial malignancy that lacks the histologic criteria required to confirm its embryologic lineage; ACs lack architectural landmarks and are classified based on the cell type, divided into small cell, intermediate cell, giant cell, spindle cell and mixed cell types; ACs occur in the lungs, thyroid and rarely elsewhere PROGNOSIS Poor–anaplasia implies that the tumor is 'primitive'; ACs are often aggressive and have a variable response to excision, chemotherapy, radiotherapy PANCREAS A rare variant of 'garden variety' pancreatic ductal adenocarcinoma, which affects ♀ > age 50 PROGNOSIS Poor, < 6 months THYROID A rare, rapid growing and invasive form of thyroid cancer, which affects those > age 60 ETIOLOGY Unknown, possibly linked to radiation exposure CLINICAL Hoarse voice, cough, hemoptysis, tracheal obstruction; physical exam may reveal nodules in thyroid LAB Thyroid function is usually normal DIAGNOSIS Biopsy TREATMENT Surgery ± RT

Anapsique Amitryptiline, see there

anaptic *adjective* Referring to an impaired sense of touch

anastomosis SURGERY 1. Any opening between 2 normally separate spaces, lumina, or organs, regardless of the manner–surgical, traumatic or pathological—in which the opening was created 2. The surgical connection between 2 tubular structures–eg, end-to-end anastomosis of the colon or rectum after a cancerous segment has been excised, or end-to-side anastomosis of a saphenous vein during a CABG

anastomotic *adjective* Referring to a surgical connection of 2 luminal–tubular structures

anastrozole Arimidex® ONCOLOGY An aromatase inhibitor, which converts androstenedione to estradiol in peripheral fat INDICATIONS Postmenopausal ♀ with advanced estrogen-dependent breast CA that does not respond to tamoxifen; unlike tamoxifen, anastrozole is not associated with ↑ risk of endometrial CA. See Breast cancer. Cf Tamoxifen

anatomic snuffbox ANATOMY A triangular depression on the dorsal pollicar aspect of the hand when the thumb is fully extended

anatomical imaging Structural imaging, see there

anatomical position An erect body stance with the eyes directed interiorly, the arms at the sides, palms of the hands facing interiorly, and fingers pointing straight down; anatomists and clinicians use the AP to build hypothetical biomechanical models of normalcy in which to describe movement of the center of gravity

anatomy The study of bodily structures and their relationships with each other. See Gross anatomy, Regional anatomy, Surgical anatomy

ANCA Antineutrophil cytoplasmic antibody, see there

anchor suture SURGERY A small suture used for ligament repair of small joints. See Suture

anchovy paste appearance PARASITOLOGY A fanciful descriptor for the olive-brown or creamy white grumous material seen in hepatic and cerebral amebiasis—*Entamoeba histolytica*, composed of autolyzed necrotic debris and hemorrhage. See Amebiasis

ancillary service INPATIENT CARE Any of a broad range of services–eg phone, TV–that are outside of the routine room and board charges incidental to a hospital stay OUTPATIENT CARE Health services–eg, in-office lab testing, imaging–mammography and other support services, which are not procedures or medical services, which relate to medical practice See Added-value service, Ambulatory surgery, Room & board

ancillary support GRADUATE EDUCATION Any form of support that eases Pt management workload–eg, hospital information system, Pt and specimen transport services, nursing support, consulting services, etc. See Call schedule

anconeus *musculus anconeus* ANATOMY ORIGIN Lateral epicondyle of humerus and adjacent capsule of elbow joint INSERTION Olecranon and posterior surface of ulna. ACTION Extends elbow and abducts ulna in pronation NERVE Radial

ANCOVA Analysis of covariance STATISTICS The use of grouped regression analysis, which ensures that comparisons of the variates between 2 groups are not be confounded by possible differences in covariates

Ancylostoma duodenale PARASITOLOGY The only hookworm found in the US, which enters the body as a 3rd-stage larva through breaks in the skin—eg bare feet in contact with contaminated soil; once in the circulation, larvae migrate to the pulmonary alveoli, where they are coughed up, swallowed, then enter the duodenum and attach themselves to villi CLINICAL Intense pruritus, erythema, and a vesicular rash at the site of larval penetration, anemia, malnutrition MANAGEMENT Mebendazole; iron for anemia

ANDA Abbreviated new drug application, see there

Andersen disease Glycogen storage disease IV, see there

Anderson clamp MD Anderson clamp, see there

androgen Androgenic hormone ENDOCRINOLOGY Any natural or synthetic 'male' hormone–eg, testosterone, methyltestosterone, fluoxymesterone, danazol, which promotes the development and maintenance of ♂ sex characteristics; natural androgen is produced chiefly by the testis, but also by the adrenal cortex and, in small amounts, by the ovary THERAPEUTIC INDICATIONS Manage androgen deficiency, delayed puberty in ♂ and some forms of breast CA. See Androgen replacement therapy, Dihydrotestosterone, Testosterone

androgen ablation Androgen suppression ONCOLOGY The therapeutic ↓ of circulating androgens-testosterone and 5α-dihydrotestosterone by either orchiectomy or by an LHRH agonist; AA is commonly used in metastatic prostate CA and results in a response, albeit short-lived–12-18 months of up to 80% of Pts; AA can be augmented by flutamide, an androgen receptor antagonist that blocks the effect of androgens produced by the adrenal gland.

See Androgen-independent prostate cancer

androgen deficiency ENDOCRINOLOGY A relative or absolute paucity of androgens. See Androgen replacement therapy

ANDROGEN DEFICIENCY

PREPUBERTAL AD Delayed onset of puberty; caused by cryptorchidism, bilateral torsion of the testes, orchitis, vanishing testis syndrome, orchidectomy, 1° testicular failure–eg Klinefelter syndrome, anorchia, and other conditions

ADULT AD Regression of androgen-dependent activities, fatigue and loss of energy, somatic changes, eg relative ↓ in lean body mass and skeletal muscle, ↑ fat, sexual behavior–impotence, erectile dysfunction, ↓ sexual desire, and regression of 2° sexual characteristics

androgen-independent prostate cancer ONCOLOGY A form of prostate CA characterized by metastases, aggressive clinical behavior, and a poor response to androgen ablation; most AIPCs express high levels of androgen receptor gene transcripts, which may be due to androgen receptor mutations. See Androgen ablation, Orchiectomy, Prostate cancer

androgen-induced hermaphroditism ENDOCRINOLOGY A hermaphroditic sexual defect induced in a 46, XY gonadal ♀ fetus by excess masculinizing hormone transmitted from the mother through the placenta

androgen insensitivity syndrome Incomplete testicular feminization, partial androgen resistance, testicular-feminizing syndrome UROLOGY A common–1/500 births–condition characterized by a genotypic–46,XY ♂ and a phenotypic ♀ identical to genotypic ♀; AIS have groin masses corresponding to testes or absent/sparse pubic and axillary hair PATHOGENESIS Cells are unresponsive to androgens, not to estrogen; masculine internal development is incomplete; externally, genitalia are ♀, except for a blind vagina, which can be made functional by either dilation or surgical lengthening; these genotypic ♂ are amenorrheic and infertile, but have phenotypic ♀ breasts

androgen replacement therapy ENDOCRINOLOGY The administration of androgens–eg, testosterone or its congeners in a ♂ with hypogonadal androgen-deficiency, which may be 1° or 2°, and either congenital or acquired; the intent of ART is to restore normal physiologic effects of testosterone, which depend on the stage of sexual development; in androgen-deficient boys, the goal of ART is to initiate and maintain androgen-dependent activities including somatic development–eg, development of skeletal muscle and ↑ strength, long bone growth, redistribution of body fat, and erythropoiesis, sexual behavior–↑ libido and potency, and development of 2° sexual characteristics–eg, ♂ pattern of hair growth, penile and scrotal enlargement, laryngeal enlargement and thickening of the vocal cords and deepening of the voice. See Androgen ablation, Orchiectomy, Prostate cancer

androgen resistance–complete Testicular feminization, see there

androgen resistance–partial Androgen insensitivity syndrome, incomplete testicular feminization ANDROLOGY An X-R endocrinopathy that affects a heterogeneous group of X,Y individuals CLINICAL Ambiguous external genitalia, hypoplastic testes, ↓ 'male' hair LAB ↑ LH, testosterone, estrogen, estradiol; ± ↑ FSH PATHOGENESIS Peripheral androgen resistance due to quantitative or qualitative defects in the androgen receptor

androgeic alopecia Common baldness, female pattern hair loss, male pattern baldness DERMATOLOGY Hereditary thinning of hair induced by androgens in genetically susceptible ♂ and ♀, which begins between age 12 and 40 in ± 50% of the general population MECHANISM Dihydrotestosterone binds to androgen receptor of susceptible scalp hair follicles, activating genes that gradually transform large terminal hair follicles to miniature follicles, producing finer hair in shorter hair cycles; dihydrotestosterone is formed by peripheral conversion of testosterone by one of 2 isoforms of 5α-reductase which, with other enzymes, regulate specific steroid transformations in skin; those with AA have ↑ 5α-reductase, ↑ androgen receptors, ↓ cytochrome P-450 aromatase, which converts testosterone to estradiol in hair follicles in the frontal scalp MANAGEMENT Finasteride, minoxidil. See Hair replacement therapy, Male-pattern baldness, Terminal hair

androgenital syndrome Congenital adrenal hyperplasia, see there

androgyny Androgeny PSYCHIATRY A combination of ♀ and ♂ corporal, mental or behavioral characteristics in a person. See Transvestite

andrologist A physician, often a urologist, who treats ♂ infertility and sexual dysfunction. See Urology

andrology A medical specialty dedicated to treating male infertility and sexual dysfunction

andropathy UROLOGY Any condition that affects only ♂–ie involves the prostate, testes, penis, and ancillary structures

andropause Male climacteric, male menopause ENDOCRINOLOGY A constellation of changes that occur in older ♂, including ↓ libido, sexual performance, ↓ sperm quantity and quality, erectile dysfunction, frailty, ↓ muscle and bone mass, and ↑ body fat CLINICAL Hot flashes, insomnia, mood swings, irritability, weakness, lethargy, ↓ lean body and bone mass, and impotence MANAGEMENT Some data suggest that androgen replacement therapy may ameliorate andropause. See Androgen replacement therapy, Midlife crisis

androstenedione An androgenic steroid, less potent than testosterone, which is produced by the adrenal cortex and gonads, and converted to estrone in fat and the liver; in ♂, overproduction of androstenedione may cause feminization; androstenedione is ↑ in Cushing syndrome, ovarian, testicular, or adrenocortical tumors, adrenal hyperplasia, polycystic ovary disease; it is ↓ in hypogonadism, Addison's disease

androsterone ENDOCRINOLOGY A ♂ sex hormone, derived from progesterone, which has 15% the strength of testosterone; it is found in ♂ and ♀ plasma and urine

anecdotal *adjective* Unsubstantiated; occurring as single or isolated event. See Fringe, Unproven. Cf Blinding

anemia HEMATOLOGY A condition characterized by ↓ RBCs or Hb in the blood, resulting in ↓ O_2 in peripheral tissues CLINICAL Fatigability, pallor, palpitations, SOB; anemias are divided into various groups based on cause–eg, iron deficiency anemia, megaloblastic anemia–due to ↓ vitamin B_{12} or folic acid, or aplastic anemia–where RBC precursors in BM are 'wiped out'. See Anemia of chronic disease, Anemia of investigation, Anemia of prematurity, Aplastic anemia, Arctic anemia, Autoimmune hemolytic anemia, Cloverleaf anemia, Congenital dyserythropoietic anemia, Dilutional anemia, Dimorphic anemia, Drug-induced immune hemolytic anemia, Fanconi anemia, Hemolytic anemia, Idiopathic sideroblastic anemia, Immune anemia, Iron-deficiency anemia, Juvenile pernicious anemia, Macrocytic anemia, Megaloblastic anemia, Microcytic anemia, Myelophthisic anemia, Neutropenic colitis with aplastic anemia, Nonimmune hemolytic anemia, Pseudoanemia, Refractory anemia with excess blasts, Sickle cell anemia, Sideroblastic anemia, Sports anemia

GENERAL GROUPS OF ANEMIA

MORPHOLOGY

MACROCYTIC

Megaloblastic anemia

    Vitamin B_{12} deficiency

    Folic acid deficiency

MICROCYTIC HYPOCHROMIC

Iron-deficiency anemia

Hereditary defects

Sickle cell anemia

Thalassemia

Other hemoglobinopathies

NORMOCYTIC

Acute blood loss

Hemolysis

BM failure

Anemia of chronic disease

Renal failure

ETIOLOGY

DEFICIENCY

Iron

Vitamin B$_{12}$

Folic acid

Pyridoxine

CENTRAL—DUE TO BM FAILURE

Anemia of chronic disease

Anemia of senescence

Malignancy

BM replacement by tumor

Toxicity due to chemotherapy

Primary BM malignancy, eg leukemia

PERIPHERAL

Hemorrhage

Hemolysis

---

anemia of chronic disease HEMATOLOGY A form of anemia that accounts for $\frac{1}{4}$ of all anemias in hospitalized Pts; it is the predominant form of hypoproliferative anemia, and seen in Pts with arthritis, chronic infections, and malignancy, which interferes with RBC production and shortens RBC lifespan CLINICAL Findings reflect the underlying disease LAB Mild-to-moderate anemia, often microcytic ± hypochromic; iron stores may be low/normal; RBC indices may be normal PATHOGENESIS Unknown; ? related to IFN-γ produced by activated macrophages TREATMENT Transfusion, erythropoietin

anemia of investigation Iatrogenic anemia, nosocomial anemia LAB MEDICINE Anemia 2° to multiple phlebotomies, most common in ICU Pts

anemia panel LAB MEDICINE A group of laboratory parameters regarded as cost-efficient, sensitive, and specific in evaluating a Pt with anemia COMPONENTS CBC with indices and automated WBC–'diff' and reticulocyte counts. See Organ panel

anemia of prematurity NEONATOLOGY A condition characterized by ↓ erythrocyte mass, which is most common in low- and very-low-birth weight infants LAB ↓ Reticulocytes, ↓ erythropoietin production

anemic infarct White infarct A localized area of ischemic necrosis in a solid organ, caused by an abrupt arterial occlusion; by 1-2 days post-insult, the region is pale yellow-white; AIs are characteristic of occlusions in organs with a single supply of blood–eg kidney, spleen, heart. See 'White' graft; Cf Hemorrhagic–red infarct

anencephaly NEONATOLOGY A lethal malformation consisting of congenital partial or complete absence of the cranial vault accompanied by absence of overlying tissues, including the brain and cerebral hemispheres, skull and scalp; anencephaly develops in the 1st month of gestation and affects 0.14-0.7/1000 live births; the 1° abnormality is failure of cranial neurulation, the embryologic process separating the forebrain precursors from the amniotic fluid; since neural tissue is exposed, cerebral tissue is hemorrhagic, fibrotic, gliotic without functional cortex ETIOLOGY Usually idiopathic, possibly multifactorial or polygenic in origin. See Uniform Determination of Death Act

anergy IMMUNOLOGY Depression or absence of an immune response to an antigen to which the host was previously sensitive; it is characterized by ↓/absent lymphokine secretion by viable T cells when the T cell receptor is engaged by an antigen; it can be tested by loss of delayed hypersensitivity–eg, to PPD, *Candida* antigens, or DCNB. See Clonal anergy, Deletion. Cf Allergy

anergy testing IMMUNOLOGY The administration of various substances to determine anergy; those who do not react to any substances, including tuberculin/PPD, after 48 to 72 hrs–ie, < 3 mm of induration to skin tests, are considered anergic

anesthesia 1. Loss of normal sensation; numbness 2. Loss of pain sensation, as intentionally induced by drugs or medication COMPLICATIONS N&V, aspiration pneumonitis, renal failure, liver dysfunction. See Anesthesiology, Augment anesthesia, General anesthesia, Glove & stocking anesthesia, Infiltration anesthesia, Local anesthesia, One lung anesthesia, Tumescent anesthesia, Vocal anesthesia

anesthesiologist A physician trained in administering anesthetics during surgery and caring for anesthetized Pts OCCUPATIONAL HAZARDS Chronic exposure to anesthetics, infectious diseases, substance abuse, radiation exposure. Cf Anesthetist SALARY $165K + 13% bonus

anesthesiology MEDICAL PRACTICE The branch of medicine dedicated to relieving pain and inducing anesthesia during surgery and other medical procedures; the scope of anesthesiology includes non-surgery-related pain management; management of painful syndromes; monitoring, restoring, and maintaining hemostasis; teaching CPR; evaluating and applying respiratory therapy

anesthetic *adjective* 1. Producing, referring or pertaining to, or characterized by anesthesia 2. Characterized by a loss of sensation or awareness; numbness *noun* An agent or drug that abolishes the sensation of pain or awareness of surroundings. See General anesthetic, Inhalation anesthetic, Local anesthetic, Opioid anesthetic

anesthetist A specialist who administers all forms of anesthesia–general, spinal block, local, regional. Cf Anesthesiologist

aneuploid *adjective* Referring to aneuploidy, see there

aneurysm SURGERY A weakened or attentuated site on the wall of an artery, a vein or the ventricle, which has stretched or ballooned and filled with blood or, in an artery, resulting in a splitting of the wall, leading to pooling of blood in the vessel wall CLINICAL Pulsating mass, bruit–aneurysmal bruit over the swelling, ± symptoms from pressure on contiguous parts. See Abdominal aortic aneurysm, Aortic aneurysm, Berry aneurysm, Cerebral aneurysm, Mycotic aneurysm, Ruptured aortic aneurysm

aneurysmal bone cyst ORTHOPEDICS A tumor-like osteolytic lesion often located in the metaphysis of long bones of young Pts IMAGING Physalliferous–soap bubble-like osteolytic cortical expansion TREATMENT Curette bone defects, fill with bone chips

aneurysmectomy SURGERY The surgical removal of an aneurysm, see there

angel dust PCP, see there

angiitis Vasculitis, see there. See Cerebral granulomatous angiitis, Cutaneous leukocytoclastic angiitis

angina Angina pectoris CARDIOLOGY A condition characterized by 'suffocatingly' intense pain, exertional substernal pressure radiating to the jaw/left shoulder/left arm, associated with N&V, dyspnea, diaphoresis, lasting < 30 min, which is relieved with nitroglycerin and rest, and classically accompanied by a sense of impending doom DIFFDX Aortic dissection, costochondritis, GERD, intercostal neuritis, MI, peptic ulcer disease, PTE, tension pneumothorax, unstable angina DIAGNOSIS Clinical history, EKG at rest and during activity, thallium stress test, coronary artery angiography EKG ST segment depression, T wave flattening RISK FACTOR MODIFICATION Smoking cessation, treat HTN, DM, ↓ cholesterol MANAGEMENT Initial therapy assumes acute MI–ergo aspirin, O₂, beta-blockers, heparin, telemetry monitoring MEDICAL THERAPY Nitroglycerin, long-acting nitrates, beta-blockers, CCBs SURGICAL THERAPY PCTA, CABG, stenting. See Canadian Cardiovascular Society functional classification, Chronic stable angina, High-risk unstable angina, Intestinal angina, Prinzmetal's angina, Recurrent angina, Unstable angina, Variant angina. Cf Prinzmetal's angina

anginal *adjective* Relating to or characterized by angina

anginal equivalent CARDIOLOGY A general term for Sx of myocardial ischemia other than angina per se–eg, SOB, faintness, fatigue, belching. See Angina

Angio-Seal™ INTERVENTIONAL CARDIOLOGY A mechanical and biochemical seal used to close arterial puncture wounds created during diagnostic and therapeutic catheterization procedures–eg cardiac angiographies and percutaneous transluminal coronary angioplasty–PTCA, peripheral angiography, coronary and peripheral stent placement. See Interventional cardiology

angiocentric lymphoproliferative lesion HEMATOLOGY A family of primarily extranodal immunoproliferative lesions that affect the sinonasal region, lungs, skin, intestine, and brain. See Angiocentric lymphoma, lymphomatoid granulomatosis

angiodysplasia Colonic angiodysplasia GASTROENTEROLOGY An abnormal aggregate of blood vessels–vascular ectasias usually in the right colon in persons > age 60; angiodysplasia accounts for 40% of the cases of recurrent or chronic lower GI bleeding in the elderly

angioedema 1. A general term for a vascular reaction of the deep dermis, subcutaneous or submucosal tissues, which corresponds to localized edema 2° to vasodilation and ↑ capillary permeability 2. Angioneurotic edema, see there

angiofibroma Juvenile angiofibroma ENT A benign nasopharyngeal tumor which is most common in adolescent/adult ♂ CLINICAL Repeated epistaxis, nasal congestion and discharge, hearing loss IMAGING Skull film, CT of head MANAGEMENT Excision if lesion is enlarging or blocking airway

angiofollicular lymphoid hyperplasia Castleman's disease, see there

angiogenesis The sprouting of new blood vessels and capillary beds from existing vessels, which plays a fundamental role in embryonic development, tissue and wound repair, resolution of inflammation, and onset of neoplasia; angiogenesis is linked to certain pathologies–eg, cancer, diabetic retinopathy, rheumatoid arthritis

angiogenesis inhibitor ONCOLOGY A chemotherapy adjuvant which inhibits the angiogenesis required for tumor growth and survive, especially for metastastatic tumors See CAI, CM101, IFN-alpha, IL-12, Marimastat, Pentosan polysulfate, Platelet factor 4, Thalidomide, TNP-470

angiogenic *adjective* Relating to angiogenesis

angiogenic factor Any of a group of substances present in the circulation–most of which are polypeptides–eg, angiogenin, fibroblast growth factor, transforming growth factors, some lipids which help form blood vessels; AFs ↑ after myocardial ischemia

angiogenic gene therapy CARDIOLOGY An investigational therapy that delivers a growth factor gene to diseased coronary arteries evoking growth factor production and stimulating angiogenesis and formation of blood vessels that bypass occluded coronary arteries

angiogram INTERVENTIONAL CARDIOLOGY A film or digital image of a radiocontrast-filled blood vessel

angiography IMAGING A radiographic technique for evaluating vascular defects in the form of arterial or venous obstruction, in which radiocontrast is injected into a vessel of interest in order to evaluate anatomy INDICATIONS When initial non-invasive procedures–CT, MRI fail to reveal the cause of a suspected defect. See Cardiac catherization, Cerebral angiography, Coronary angiography, Digital subtraction angiography, Directional color angiography, Equilibrium radionuclide angiography, Fluorescein angiography, Lymphangiography, Magnetic resonance angiography, Magnetic resonance coronary angiography, Percutaneous transluminal cerebral angiography, Percutaneous transluminal coronary angiography, Pulmonary angiography, Transluminal angiography

angioid streak OPHTHALMOLOGY Any of number of minute interruptions of the elastic tissue of the retina, which appear as peripapillary, gray–in non-whites to red-brown–whites, linear striations radiating from the optic fundus along stress lines or toward the equator in an abnormal Bruch's membrane; ½ occur in Pts with pseudoxanthoma elasticum; ASs also occur in fundi rendered brittle by calcium–tumor-related calcinosis or heavy metal–eg, lead; ASs also occur in acromegaly, abetalioproteinemia, DM, hemochromatosis, hemolytic anemia, hypercalcinosis, hyperphosphatasia, ITP, myopia, neurofibromatosis, Paget's disease of bone, senile elastosis, sickle cell anemia, Sturge-Weber syndrome, tuberous sclerosis

angioma A tumor of blood–hemangioma or lymph–lymphangioma blood vessels

angioneurotic edema Angioedema, Quincke's disease NEUROLOGY A chronic and potentially fatal condition characterized by episodic localized subcutaneous, periorbital, periocular and laryngeal edema, abdominal pain ETIOLOGY Absent C1 esterase inhibitor CLINICAL Recurrent edema, abdominal pain, laryngeal edema which can compromise breathing DIAGNOSIS Hx of recurrent angioedema, absent C1 esterase inhibitor in blood TRIGGERING FACTORS Allergies–eg foods, pollen, insect bites, drugs–eg ACE inhibitors, salicylates, stress, exposure to cold, water, sunlight, heat MANAGEMENT Epinephrine, antihistamines, corticosteroids (androgens). See Episodic angioedema, HANE

angiopathy A disease of blood vessels–arteries, veins, capillaries. See Cerebral amyloid angiopathy, Diabetic angiopathy, Microangiopathy

angioplasty Any interventional repair of a blood vessel. See Balloon angioplasty, Deferred adjunctive coronary angioplasty, Immediate adjunctive coronary angioplasty, Kissing balloon coronary angioplasty, Percutaneous transluminal coronary angioplasty, Primary coronary angioplasty, Rescue adjunctive coronary angioplasty, Rescue renal angioplasty

angioscopy The visualization of a blood vessel lumen with a narrow bore flexible endoscope. See Coronary angioscopy

angiotensin Any of a family of vasoconstricting peptides

angiotensin I An N-terminal decapeptide produced when renin acts on angiotensinogen; A-I is the precursor of angiotensin II, but itself has no known physiologic role. See Angiotensinogen

angiotensin-converting enzyme A key enzyme in the RAA system, which converts the inactive decapeptide angiotensin I to the octapeptide, angiotensin II, a potent vasoconstrictor that also stimulates aldosterone secretion; ACE is also involved in metabolizing bradykinin REF RANGE 18–67 U/L, > age 20; those < 20 have higher levels; ACE is ↑ in sarcoidosis, Gaucher disease, leprosy, histoplasmosis, cirrhosis, asbestosis, berylliosis, DM, Hodgkin's disease, hyperthyroidism, amyloidosis, PBC, idiopathic pulmonary fibrosis, PE, scleroderma, silicosis, TB; ACE ↓ in response to prednisone therapy for sarcoidosis. See Renin/angiotensin/aldosterone system

angiotensin II receptor antagonist PHARMACOLOGY Any of a family of agents-eg losartan and valsartan, which block the binding of angiotensin II–A-II to its cognate cell membrane receptors–AT_1, AT_2, and others; 1st generation ARAs included the sartan family of agents, which only block AT_1, interacting with the amino acids in the transmembrane domains, blocking the binding of A-II to AT_1; an alternative to ACEI therapy for Pts with CHF; unlike ACEIs, ARAs do not interfere with bradykinin and prostaglandin metabolism, interference which has been linked to some of the adverse effects of ACEI therapy, particularly to cough and angioedema.

angiotensinogen Renin substrate A 60 kD glycoprotein of the α_2 globulin fraction of plasma proteins, which is synthesized and released from the liver, and cleaved in the circulation to form the biologically inactive, angiotensin I, a decapeptide split from the N-terminal by renin, a proteolytic enzyme. See Angiotensin, Renin-angiotensin system

angle closure glaucoma Acute angle closure glaucoma, see there

angle hook GYNECOLOGY A device for grasping and manipulating the cervix, fundus, and intravaginal soft tissue during gynecologic procedures

angry back syndrome Excited skin syndrome ALLERGY MEDICINE A generalized erythema seen on the back of Pts undergoing skin patch testing, caused by the high concentration of the allergens, resulting in an irritation–but not immune-mediated–response to the allergen; such a response may result in false-positivity to all tested allergens. See Patch testing

angular cheilitis Perlèche DERMATOLOGY A condition characterized by inflammation, exudation, maceration and fissuring at the angles of the lips ETIOLOGY AC is most often linked to candidiasis, but is also related to the ↓ vertical dimension of lower face in edentulous elderly with loss of alveolar bone, 'sagging' of cheeks due to myotonia, sialorrhea, ariboflavinosis–with glossitis, keratitis, and seborrhea-like dermatitis, malnutrition, streptococcal infection.

angular gyrus NEUROPHYSIOLOGY A region of the parietal lobe which links visual input with the stored sounds of language. See Dyslexia

anhedonia PSYCHIATRY Absence of pleasure from activities that are normally or had previously been pleasurable. See Sexual anhedonia. Cf Hedonism

anhidrosis DERMATOLOGY The lack of appropriate sweat production in response to thermal or pharmacologic stimulation, which may become a medical emergency with hyperthermia, heat exhaustion, heatstroke, and death; anhidrosis may affect the entire body or be segmental in distribution and is divided into structural–eg, anhidrotic ectodermal dysplasia, sweat gland necrosis, or functional defects, often related to autonomic or thermoregulatory control which involve the central or peripheral nervous systems

animal PHARMACOLOGY Any nonhuman animate being endowed with the power of voluntary action. See Cat, Cow, Dog, Fish, Horse, Monkey, Pig, Sentinel animal, Snake. VOX POPULI Etc.

animal dander See Dander

animal fat CLINICAL NUTRITION A general term for saturated fatty acids of animal–beef, pork, lamb origin, which are associated with an ↑ risk of colorectal––CA. See Fatty acid, Saturated fatty acid. Cf Unsaturated fatty acid

anion gap LAB MEDICINE A mathematic approximation of the difference between unmeasured anions–PO_4^-, SO_4^-, proteins and organic acids, and unmeasured cations–Ca^{2+}, Mg^{2+}, which normally exceed unmeasured cations; the AG is the difference between the sum of the most abundant measured serum anions–Cl^- and HCO_3^- and serum cations–Na^+ and K^+; urinary AG is calculated as $Na^+ + K^+ - Cl^-$ and is a crude index of the levels of urinary ammonium and used to evaluate hyperchloremic metabolic acidosis REF RANGE 8-16 mEq/L; AG is ↑ in renal failure due to defective renal tubu-

lar acidification with an ↑ in phosphate and sulfate, starvation-related DKA due to an accumulation of acetoacetate and β-hydroxybutyrate or alcohol abuse, in disorders of amino acid metabolism and hyperglycemic nonketotic coma due to various organic acids, in lactic acidosis, overdose or poisoning––eg salicylates, methanol ethylene glycol antifreeze or paraldehyde; AG is ↓ in hypermagnesemia, GI loss of bicarbonate, in nephrotic syndrome due to a loss of albumin which is anionic at a physiological pH, after lithium ingestion, and multiple myeloma and Waldenström's macroglobulinemia, due to an ↑ in cationic proteins

aniridia OPHTHALMOLOGY A congenital AD or AR condition characterized by a virtually complete absence of the iris, which may be accompanied by other ocular defects–eg, congenital cataracts, corneal dystrophy, foveal hypoplasia, and therapeutically refractory glaucoma PROGNOSIS Poor

anisakiasis PARASITOLOGY Infestation of the upper GI tract–stomach, small intestine mucosae by larvae of the family Anisakidae, which are common ascaroid parasites of marine fish; human infestation results from eating raw fish–eg, sushi; the larval stage of *Anisakis simplex* and *Phocanema–Pseudoterranova decipiens* account for all US cases of anisakiasis; another anasakine, *Contracecum*, may rarely cause anisakiasis CLINICAL *A simplex*–abdominal pain, N&V, diarrhea, eosinophilia with occult blood in stool, if gastric anisakiasis; leukocytosis sans eosinophilia if intestinal anisakiasis; *P decipiens*––few significant Sx, as it does not penetrate gastric or intestinal wall MANAGEMENT-GASTRIC ANISAKIASIS–endoscopic removal of larva INTESTINAL ANISAKIASIS–surgical excision of involved intestine. See Sushi

anismus GASTROENTEROLOGY An idiopathic form of anorectal outlet obstruction that affects older ♀ and young boys, characterized by inappropriate contraction and typical EMG changes in the pelvic floor muscles and external anal sphincter CLINICAL Constipation, perineal pain, defecatory dysfunction PATHOGENESIS Anismus may be a functional defect, as no organic cause has been identified MANAGEMENT Biofeedback training

anistreplase See APSAC

ankle/brachial blood pressure index Ankle-arm index, ABI CARDIOLOGY A simple clinical test of a Pt's risk of future MI and cardiovascular 'reserve'; in the ABI, BP is measured in the upper arm and at the ankle. See Blood pressure

ankle jerk Ankle reflex, see there, aka Achilles tendon reflex

ankle reflex Achilles tendon reflex, Ankle jerk NEUROLOGY An abrupt plantar jerk of the ankle evoked by tapping the Achilles tendon with an unrestricted forefoot. See Achilles tendon

ankle sprain ORTHOPEDICS A stretching of the ankle ligaments and/or muscles with swelling

ankyloglossia Tongue-tie, see there

ankylosing spondylitis RHEUMATOLOGY A polyarthritis of the vertebral column, which is characterized by progressive, painful stiffening of the joints and ligaments, which primarily affects young ♂ RISK FACTORS AS is linked to HLA B27; individuals with B27 have a 300-fold ↑ risk for AS CLINICAL ↓ ROM of back, ↓ chest expansion; transient–50% or permanent peripheral arthritis; uveitis in 25% IMAGING Diagnostic changes in sacroiliac joints LABORATORY ↑ ESR, normal ANAs, rheumatoid factors; 90% of AS Pts have HLA B27 MANAGEMENT NSAIDs, physical therapy

ankylosis ORTHOPEDICS A fusion of bones across a joint, which may be a complication of chronic inflammation. See Ankylosing spondylitis

anlage EMBRYOLOGY Any early embryonic structure that differentiates into a more complex structure. See Retinal anlage.

ANLL Acute nonlymphocytic leukemia

44 **Ann Arbor classification** A system for staging Hodgkin's disease and guiding therapy; clinical data is added based on A–absence or B–presence of associated symptoms–eg, night sweats, fever or weight loss of > 10%; other Sx of Hodgkin's disease include lethargy, fatigability, anorexia, pruritus. See Hodgkin's disease

ANN ARBOR CLASSIFICATION

I   A single involved lymphoid region, organ or site

II 2+ involved lymphoid regions, or one extralymphoid site and a lymphoid region on the same side of the diaphragm

III Lymphoid regions involved on both sides of the diaphragm, ± accompanied by local ized involvement of extralymphatic organs or spleen

IV Disseminated involvement of 1+ extralymphatic organs or tissues, with/without asso ciated lymphadenopathy

annamycin ONCOLOGY A liposomal anthracycline for treating CAs–eg multiple-drug-resistant tumors–eg, refractory breast CA, refractory/relapsed AML, ALL, blast crisis of CML. See Anthracycline, Liposomes

anogenital *adjective* Relating to the anus and external genitalia

anomaly An abnormal thing PEDIATRICS A marked deviation from the norm or a standard, especially due to a congenital–birth or hereditary defect. See Alder-Reilly anomaly, May-Hegglin anomaly, Pelger-Huët anomaly, Pseudo-Chediak-Higashi anomaly, Pseudo-Pelger-Huët anomaly

anomic Amnestic aphasia NEUROLOGY Loss of the ability to name objects

anonymous sex PUBIC HEALTH Any sexual activity in which the partners' identities are unknown–often intentionally to each other at the time of the activity's occurrence. See Bathhouse, Glory hole, Sex club

anonymous testing PUBLIC HEALTH The testing of an individual for certain infections, in particular, HIV, providing the results to public health departments without identifying that person by name, but rather by a number. Cf Named reporting

anorchia Anorchism UROLOGY Congenital absence of one or both testes. Cf Cryptorchism

anorectal abscess An abscess formed adjacent to the anus CLINICAL Perianal tenderness, swelling, pain on defecation

anorectal fistula SURGERY An acutely inflamed 'tract' which connects the anus and the rectum ETIOLOGY Trauma, abscesses, inflammatory processes–eg, CA, Crohn's disease CLINICAL Chronic purulent discharge from a para-anal opening which, with a probe, leads to the rectum, stool strain pain PATHOLOGY Inflammation MANAGEMENT Small AFs heal spontaneously; large ones require surgery

anorectic *adjective* Referring to anorexia *noun* A person with anorexia nervosa See Anorexia nervosa, Anorexic

anorexia nervosa PSYCHOLOGY An eating disorder primarily–>95% of young ♀, characterized by an extreme aversion to food, and attributed to a misperception of body image–anorectics believe they are overweight even when significantly underweight CLINICAL Average 25% below normal weight:height ratio, absence of ≥ 3 menstrual periods, cold intolerance, hypothermia, constipation, hypotension, bradycardia CLINICAL Anemia, leukopenia, electrolyte abnormalities, ↑ BUN and creatinine, ↑ cholesterol, ↓ LH, FSH DIFFDX Panhypopituitarism, Addison's disease, hyperthyroidism, DM, Crohn's disease, CA, TB, CNS tumors MANAGEMENT Psychotherapy, hospitalization. See Bulimia, Eating disorders

anorexic *adjective* Lacking a normal appetite. See Anorexia nervosa. Cf Anorectic

anoscope A metal tube inserted into the anus for viewing the anal mucosa. Cf Colonoscope, Rectoscope

anoscopy The use of an anoscope to directly visualize the anorectal mucosa, and detect hemorrhoids, polyps, anal fissures and rectal bleeding of uncertain etiology

anosmia Anosphrasia, olfactory anaesthesia Absence of the sense of smell, a symptom typical of frontal lobe brain tumors

ANOVA Analysis of variance, see there

anovulatory Anovular *adjective* Unassociated with discharge of an egg from the ovary

anovulatory bleeding See Dysfunctional uterine bleeding

anovulatory cycle Infertile cycle GYNECOLOGY A cycle in which an egg is not released from the ovary

anoxia Hypoxia, see there

ANP Atrial natriuretic peptide, see there

Anpress Alprazolam, see there

anserine bursa ANATOMY A bursa of the medial knee which is at the convergence of the tendons of the gracilis, sartorius, semitendinosus and tibial collateral ligaments. See Anserine bursitis

anserine bursitis ORTHOPEDICS Inflammation of the anserine bursa, which occurs in those who heavily exercise the knee–joggers or in obese older ♀ with osteoarthritis of the knee or valgus angulation of knee CLINICAL Pain on walking and lying with the knees together, tenderness over medial tibia at anserine bursa MANAGEMENT NSAIDs, ice after activity, heat while resting, corticosteroid injection; if valgus angulation of knee while standing, insert a lateral wedge in the shoe to ↓ tension on anserine bursa

ANSI American National Standards Institute INFORMATICS A nonprofit organization belonging to the ISO, which is involved in establishing electronic data standards in the US. See International Committee for Standardization

Ansial Buspirone HCl, see there

Ansiolin Diazepam, see there

Ansiopax® Alprazolam, see there

Anspor® Cephradine, see there

Antabuse® Disulfuram, see there

antacid PHARMACOLOGY A basic agent that neutralizes acid in the gastric lumen; the final pH achieved is usually a function of the amount of antacid administered GI EFFECTS Antipeptic, ↑ in acid secretion, ↑ GI motility, ↑ mucus secretion; antacids vary in absorption–eg, $NaHCO_3$ and sodium citrate and are completely absorbed, which may result in metabolic acidosis

antagonistic *adjective* Referring to any combination of 2 or more drugs, which results in a therapeutic effect that is less than the sum of each drug's effect. Cf Additive, Synergism

antagonistic effect The negative effect that one chemical or family of chemicals has on other chemicals

antalgic gait OCCUPATIONAL MEDICINE A gait pattern specifically modified to reduce the amount of pain a person is experiencing; the term is usually applied to a rhythmic disturbance in which as short a time as possible is spent on the painful limb and a correspondingly longer time is spent on the healthy side. See Gait

ante- before, in front of

antebrachium Forearm, see there

antecedent *adjective* Existing or occurring before, which may be linked to subsequent events. See Preexisting

Anten Doxepin, see there

Antenex® Diazepam, see there

anterior compartment syndrome A condition in which swelling within the anterior compartment of the lower leg jeopardizes the functionality of the muscles, nerves, and arteries that serve the foot

anterior cord syndrome NEUROLOGY A post-traumatic spinal cord symptom complex characterized by a loss of voluntary motor function, pain, and temperature sense, and intact distal position, vibration, and light touch sense, dysfunctional anterior and lateral columns and intact posterior columns ETIOLOGY Trauma to relatively mobile cervical vertebrae, common in whiplash injury–spinal hyperextension

anterior cruciate ligament injury SPORTS MEDICINE An injury most common in sports characterized by abrupt changes of direction–eg, football, skiing, tennis, soccer CLINICAL Swelling, tenderness of knee MANAGEMENT ACL reconstruction via arthroscopy

anterior drawer test ORTHOPEDICS A test for evaluating anterior cruciate ligament integrity. See Anterior cruciate ligament

anterior dynamized system ORTHOPEDICS A spinal intervertebral body fixation orthosis, consisting of dynamized closed screws, closed transverse fixator connector, anterior rods and closed blocker

anterior horn disease NEUROLOGY A group of conditions that affect the anterior horn of the spinal cord–eg, Werdnig-Hoffmann disease–aka infantile spinal muscle atrophy, the classic 'floppy infant' syndrome, is characterized by hypotonia, symmetric, arreflexic weakness with death by age 2; amyotrophic lateral sclerosis is a distal AHD of adults with fasciculations and wasting of 'bulbar' muscles and poliomyelitis, characterized by fever, asymmetric distal involvement, later becoming generalized, with muscle weakness of respiratory and bulbar muscles

anterior presentation Vertex presentation, see there

anterior uveitis Iritis, nongranulomatous uveitis OPHTHALMOLOGY Inflammation of the anterior eye classically associated with autoimmune disease–eg, rheumatoid arthritis or ankylosing spondylitis

anterograde amnesia NEUROLOGY Amnesia which occurs from the moment of physical or mental trauma; AA is characterized by an inability to form new memories of life events. Cf Retrograde amnesia

anterolateral cordotomy Spinothalamic tractomy PAIN MANAGEMENT A type of neuroablative procedure in which a lesion is created in the spinothalamic tract as a means of interrupting pain transmission; open cordotomies require laminectomies, but result in excellent analgesia and can be done in the thoracic region for leg pain. See Neuroablative procedure

anthracosis PULMONOLOGY A generic term for blackening of tissues, often understood to mean carbon dust deposition in the lung and lymph nodes, which does not itself cause disease, and is usually present in urban dwellers, and in those working in certain occupations–eg, coal mining. See coal workers' pneumoconiosis

anthracycline ONCOLOGY A chemotherapeutic used in leukemia to prevent cell division ADVERSE EFFECTS Dose-dependent stomatitis, N&V and BM suppression which 'maxes' at ± 10 days; alopecia is normal; extravasation at IV injection site causes severe, protracted ulcers and necrosis. See Chemotherapy, Daunorubicin, Doxorubicin, Idarubicin

anthrax Greek, *anthrax*, a burning coal, charbon, milzbrand INFECTIOUS DISEASE An often fatal bacterial infection which occurs when *Bacillus*

anthracis endospores, primarily of grazing herbivore–cattle, sheep, horses, mules–origin enter via skin abrasions, inhalation, or orally PATHOGENESIS Anthrax endospores germinate within macrophages, become vegetative bacteria, multiply within the lymphatics, enter the bloodstream and cause massive septicemia CLINICAL URI-like symptoms, followed by high fever, vomiting, joint pain, SOB, internal and external hemorrhage, hypotension, meningitis, pulmonary edema, shock sudden death; intestinal anthrax is caused by ingestion of contaminated meat; cutaneous anthrax is rare DIAGNOSIS ELISA for capsule antigens–95+% sensitivity, for protective antigen–72% sensitivity; detection of exotoxins in blood is unreliable PREVENTION Prophylaxis–6 wks with doxycycline or ciprofloxacin; vaccination, with anthrax vaccine absorbed; decontamination with aerosolized formalin MANAGEMENT Penicillin, doxycycline; if allergic to penicillin, chloramaphenicol, erythromycin, tetracycline, ciprofloxacin See *Bacillus anthracis*, Cutaneous anthrax, Industrial anthrax, Inhalation anthrax

ANTHRAX, CLINICAL FORMS

PULMONARY Almost universally fatal–due to inhalation of anthrax spores which germinate and produce toxins resulting in pleural effusions, hemorrhage, cyanosis, SOB, stridor, shock, death

INHALATION Anthrax pneumonia, inhalational anthrax, pulmonary anthrax An almost universally fatal form due to inhalation of 1 to 2 μm pathogenic endospores which are deposited in alveoli, engulfed by macrophages and germinate en route to the mediastinal and peribronchial lymph nodes, produce toxins CLINICAL Mediastinal widening, pleural effusions, fever, nonproductive cough, myalgia, malaise, hemorrhage, cyanosis, SOB, stridor, shock, death, often accompanied by mesenteric lymphadenitis, diffuse abdominal pain, fever

CUTANEOUS Once common among handlers of infected animals, eg farmers, woolsorters, tanners, brushmakers and carpetmakers in an era when brushes were from animals CLINICAL Carbuncle–a cluster of boils, that later ulcerates, resulting in a hard black center surrounded by bright red inflammation; rare cases which become systemic are almost 100% fatal

GASTROINTESTINAL After ingesting contaminated meat–2 to 5 days; once ingested spores germinate, causing ulceration, hemorrhagic and necrotizing gastroenteritis CLINICAL Fever, diffuse abdominal pain with rebound tenderness, melanic stools, vomit, fluid and electrolyte imbalances, shock; death is due to intestinal perforation or anthrax toxemia

OROPHARYNGEAL Uncommon, follows ingestion of contaminated meat CLINICAL Cervical edema, lymphadenopathy–causing dysphagia, respiratory difficulty

ANTHRAX MENINGITIS A rare, usually fatal complication of GI or inhalation anthrax with death occurring 1 to 6 days after onset of illness CLINICAL Meningeal symptoms, nuchal rigidity, fever, fatigue, myalgia, headache, N&V, agitation, seizures, delirium, followed by neurologic degeneration and death

anthrax vaccination A series of 6 shots over 6 months and booster shots annually, given routinely to veterinarians, livestock workers, military personnel in the US, UK, Russia. See Anthrax, Biological warfare, Sverdlosk

anthropometry MEDTALK 1. The measurement of a person's physical parameters–height and weight 2. The field that deals with the physical dimensions, proportions, and composition of the human body, as well as the study of related variables that affect them

antiacid Antacid, see there

antiallergic *adjective* Countering allergy or an allergic state

antiandrogen ENDOCRINOLOGY A hormone or other agent–eg, megestrol acetate, spironolactone, flutamide, nilutamide, and cimetidine, which interferes with androgen function by competitively inhibiting androgen binding to cognate receptors at the target organ and is either biologically inert or functionally very weak; these compounds are used to manage androgen-dependent CAs–♂ breast and prostate, hirsutism, acne

antianginal *adjective* Referring to an agent or mechanism that counters anginal pain *noun* An agent that prevents or ameliorates anginal pain

antiangiogenesis ONCOLOGY Prevention of the growth of new blood vessels into a solid tumor. See angiogenesis

antiarrhythmic *adjective* Referring to an agent or mechanism that counters cardiac arrhythmias *noun* An agent that prevents or ameliorates cardiac arrhythmias

antiarrhythmic therapy CARDIOLOGY The use of therapeutic agents to counteract potentially life-threatening and refractory ventricular arrhythmias, and ↓ mortality; the Vaughn Williams/Harrison classification divides antiarrhythmic agents into 5 groups. Cf Proarrhythmic effect

antiasthmatic *adjective* Referring to an agent or mechanism that counters asthma *noun* An agent that ameliorates asthma

antibacterial *adjective* Referring to an agent or effect that suppresses or inhibits bacterial reproduction *noun* A general term for any agent that suppresses bacterial growth or destroys bacteria

antibacterial ointment A topical bactericidal or bacterostatic ointment, used for acne, or eye infections

antibacterial soap A bactericidal agent used to clean the skin–eg, Betadine, pHisoHex

antibiogram MICROBIOLOGY The profile of an organism's susceptibility/resistance to a panel of antibiotics, which can be used to determine genetic relatedness of various bacteria Cf Molecular strain typing

antibiotic *adjective* Relating to the destruction of living things *noun* MEDTALK 1. An agent obtained directly from a yeast or other organism which is used against a bacterial infection 2. Any agent used to kill or reduce the growth of any infectious agent, including viruses, fungi and parasites. See Drug resistance, Macrolide antibiotic, Polyene antibiotic MOLECULAR BIOLOGY A substance that interferes with a particular step of cellular metabolism, causing either bactericidal or bacteriostatic inhibition; sometimes restricted to those having a natural biological origin

antibiotic assay Sensitivity testing, see there

antibiotic-associated colitis Pseudomembranous colitis, see there

antibiotic-associated diarrhea Antibiotic-associated colits, gastroenteritis Diarrhea caused by *Clostridium difficile*, most often seen in a Pt taking antibiotics; many persons infected with *C difficile* are asymptomatic; in others, a *C difficile* toxin causes diarrhea, abdominal pain, severe colitis, fever, ↑ WBCs, vomiting, dehydration and, with time, develop pseudomembranous colitis and perforation MANAGEMENT *Lactobacillus GG* therapy

antibiotic-associated enteritis Pseudomembranous colitis, see there

antibiotic bonding A technique that pretreats plastics used for indwelling devices–eg, intravascular catheters, in order to prevent their being coated by bacterial glycocalyx, which counteracts host opsonization, and renders systemic antibiotics ineffective

antibiotic-induced diarrhea See Antibiotic-associated diarrhea

antibiotic ointment Any of a number of topical antibacterial ointments or creams

antibiotic resistance INFECTIOUS DISEASE The relative or complete ability of an organism–bacterium, fungus to counteract the desired bactericidal or bacteriostatic effect of one or more antimicrobial agents

antibiotic-resistant *Streptococcus pneumoniae* Any of a number of strains of *S pneumoniae* which are resistant to one or more antibiotics. See *S pneumoniae*

antibiotic sensitivity & identification A test performed in the microbiology lab after a pathogenic bacterium has been isolated by culture

antibody IMMUNOLOGY An immunoglobulin produced by plasma cells, which has a specific amino acid sequence and specifically binds to the antigen(s)–eg, foreign proteins, microbes or toxins, that induced its synthesis; antibodies may bind to closely related antigens. See Acetylcholine receptor antibody, Anticardiolipin antibody, Anticentromere antibody, Anti-double-stranded DNA antibody, Anti-epidermal antibody, Anti-extractable nuclear antibody, Antigliadin antibody, Antihistone antibody, Anti-idiotype antibody, Anti-insulin antibody, Anti-islet cell antibody, Anti-Jo-1 antibody, Anti-LANA antibody, Antimicrosomal antibody, Antimitochondrial antibody, Antimyelin antibody, Antimyeloperoxidase antibody, Antineuronal-nuclear antibody, Antineutrophil cytoplasmic antibody, Antinuclear antibody, Antiparietal cell antibody, Anti-platelet antibody, Anti-PRP antibody, Anti-Purkinje cell antibody, Anti-receptor antibody, Anti-reticulin antibody, Anti-ribosomal antibody, Anti-Ro/SSA antibody, Anti-single-stranded DNA antibody, Anti-striated muscle antibody, Anti-thyroglobulin antibody, Antithyroid antibody, Antithyroid peroxidase antibody, Anti-tumor necrosis factor-α monoclonal antibody, Autoantibody, Bexxar radiolabeled monoclonal antibody, Secondary antibody, Binding antibody, Blocking antibody, Catabolic antibody, Catalytic antibody, Chimeric antibody, Core antibody, Cross-reactive antibody, Designer antibody, Enhancing antibody, Fluorescent treponemal antibody, Functional antibody, GAD antibody, Glutamic acid decarboxylase autoantibody, HAMA antibody, Heterophile antibody, HIV antibody, HTLA antibody, Humanized antibody, Immunoglobulin, Insulin receptor antibody, Intrinsic factor antibody, Islet antibody, Isotypic control antibody, Ku antibody, LW antibody, Miniantibody, MOC-31 antibody, Monoclonal antibody, Neutralizing antibody, Parietal cell antibody, Plantibody, PM-1 antibody, Polyclonal antibody, Primary antibody, Purkinje cell antibody, RANA antibody, Scleroderma antibody, Sjögren antibody, Sm antibody, Smooth muscle antibody, Sperm antibody, Thyroid stimulating hormone receptor antibody, Trichinosis antibody, Warm antibody

antibody therapy CLINICAL IMMUNOLOGY Any therapeutic intervention in which a monoclonal or other concentrated antibody is used to manage a condition–eg, cancer or severe infection

antibody titer The amount of a specific antibody present in the serum, usually as a result of an acquired infection; titers for IgM usually rise abruptly at the time of infection–acute phase and fall slowly; during the 'convalescent' phase, IgG ↑ and is elevated for life. See Seropositivity

anticancer diet CLINICAL NUTRITION A popular term for a series of dietary guidelines promulgated by the American Cancer Society, NCI, and American Institute for Cancer Research, which are intended to ↓ a person's risk of cancer. See Folk cures for cancer, Unproven methods for cancer management

ANTI-CANCER DIET–GUIDELINES TO ↓ CANCER RISK

↓ Fat consumption, especially saturated fats

↓ Consumption of smoked, salt-cured, and nitrate-cured foods, eg bacon, ham, cheese, corned beef, luncheon meats

↓ Alcohol

↑ High fiber foods, including whole grains, legumes, fruits, vegetables

↑ Foods rich in vitamins A and C, carotenoids, eg leafy green vegetables, citrus fruits, and red, orange, and yellow fruits & vegetables

↑ Cruciferous vegetables, eg broccoli and cabbage

OTHER RECOMMENDATIONS

Avoid tobacco

Avoid refined sugar and salty foods

Lose weight

Exercise regularly and vigorously

anticancer drug Chemotherapeutic, see there

anticardiolipin antibody CLINICAL IMMUNOLOGY Any of a family of antibodies that recognize an epitope on the cardiolipin molecule; ACA are associated with immune-mediated disease, strokes, fetal wastage, syphilis, but are also found in healthy subjects without disease, represent natural antibodies, and indicate a normally functioning immune system NORMAL VALUES IgA < 22 APL–IgA Phospholipid Units–PLU; IgG < 22 GPL–IgG PLUs; IgM < 11 MPL–IgM PLUs. See Antiphospholipid syndrome

anticholinergic Parasympatholytic *adjective* Referring to an agent or effect that suppresses or inhibits acetylcholine activity *noun* Any agent that inhibits parasympathetic activity by blocking the neurotransmitter, acetylcholine; anticholinergics are used for asthma, COPD, diarrhea, N&V, Parkinson's disease, and to ↓ smooth muscle spasms–eg, in the urinary bladder; anticholinergics may be antimuscarinic, ganglionic blockers, and neuromuscular blockers

anticholinesterase PHARMACOLOGY An agent–eg, certain nerve gases, which blocks nerve impulses by inhibiting anticholinesterase EXAMPLES Insecticides–eg, parathion, and nerve gas agents–eg, sarin, soman, tabun; AChEs can be reversible or irreversible ACTION Eyes–hyperemia and pupillary constriction, GI tract—↑ GI contractions and secretion of gastric acid

anticipatory anxiety PSYCHIATRY Anxiety caused by an expectation of anxiety or panic in a particular situation. See performance anxiety

anticipatory side effect PSYCHOLOGY Any adverse effect linked to a particular clinical maneuver–eg, administration of chemotherapy, which evokes nausea; such side effects are caused by an uncontrolled autonomic response to anxiety and stress MANAGEMENT Relaxation training, systematic desensitizaton, distraction

anticoagulant A general term for any substance that prevents coagulation of blood HEMATOLOGY Anticoagulants administered to prevent or treat thromboembolic disorders include heparin, a parenteral agent which inactivates thrombin and other clotting factors and oral anticoagulants–warfarin, dicumarol et al, which inhibit the hepatic synthesis of vitamin K-dependent clotting factors LAB MEDICINE Anticoagulants used to prevent clotting of blood specimens for laboratory analysis: heparin and substances that make Ca^{2+} unavailable for clotting–eg, EDTA, citrate, oxalate, fluoride MEDTALK Blood thinner TRANSFUSION MEDICINE Anticoagulant solutions used to preserve stored whole blood and blood fractions: ACD–acid citrate dextrose, CPD–citrate phosphate dextrose, CPDA-1–citrate phosphate dextrose adenine and heparin

ANTICOAGULANTS–CATEGORIES & USES

• COUMADIN® Prevent blood clot formation
• HEPARIN Prevent blood clot formation
• THROMBOLYTICS, eg tPA, streptokinase–dissolve blood clots

anticoagulant therapy HEMATOLOGY The use of anticoagulants to prevent intravascular clot formation, or dissolve clots that have already formed INDICATIONS DVT/thrombophlebitis, CAD, TIA/stroke, dysrhythmia, prosthetic heart valve, cancer MONITORING Serial measurement of PT, PTT. See Heparin, tPA, Warfarin

anticonvulsant *adjective* Related to preventing seizures *noun* Any agent used to prevent, reduce or stop seizures or convulsions. See Epilepsy

antidepressant *adjective* Relieving depression *noun* An agent used to manage depression, anxiety, panic disorders. See Depression, Prozac/fluxetine, Tricyclic antidepressant

antidiabetic *adjective* Relating to an effect or agent that counters DM *noun* Antidiabetic agent Any agent that improves glucose control–eg, efficiency of glucose utilization. See Diabetes mellitus, Insulin, Triglitazone

antidiuretic hormone Arginine vasopressin ENDOCRINOLOGY An octapeptide hormone synthesized in the hypothalamus and released from the posterior pituitary, which promotes renal tubular reabsorption of water by kidneys in response to ↑ osmolality/↓ plasma volume with ↑ sodium and solutes; ↓ osmolality–water excess results in ↓ secretion of ADH, thereby ↑ excretion of water to maintain fluid balance; ADH is ↑ in bronchogenic carcinoma, acute porphyria, hypothyroidism, Addison's disease, cirrhosis, hepatitis, hemorrhage, shock, CHF; ADH is ↓ with diabetes insipidus, viral infection, metasta-

tic CA, sarcoidosis, TB, Langerhans cell histiocytosis/Hand-Schuller-Christian disease, syphilis, head trauma THERAPEUTIC EFFECTS ADH is an antidiuretic and vasopressor, and used for diabetes insipidus

anti-dumping law HEALTH LAW Any legislation enacted to prevent the inappropriate transfer of Pts who are medically unstable–eg, in early labor, or with impending rupture of aortic aneurysm, to other health care facilities. See Dumping

antiemetic *adjective* Countering emesis, vomiting *noun* An agent–eg, odansetron, granisetron, which prevents or alleviates nausea and vomiting associated with chemotherapeutics–eg, cisplatin, cyclophosphamide, dacarbazine, etc

antiepileptic *adjective* Referring to an agent or effect that suppresses or inhibits seizures *noun* A general term for any agent that counters epilepsy, anticonvulsive therapy. See Convulsion, Epilepsy, Seizure disorder

antiestrogenic *adjective* Referring to an effect that suppresses or inhibits estrogenic activity *noun* A general term for an agent that counters estrogenic activity

antifolate chemotherapy Antifolate therapy ONCOLOGY The use of an antimetabolite–eg, MTX, to compete with folate, inhibiting dihydrofolate reductase, the enzyme responsible for → inactive dihydrofolate into an active tetrahydrofolate; this block causes a buildup of toxic dihydrofolate, shutting down synthesis of purine nucleotides and thymidylate; AC is used to treat malignancy. See Leukovorin rescue. Methotrexatre

antifreeze TOXICOLOGY A fluid with a high concentration of ethylene glycol or methanol, which is added to car, truck, and other cooling systems for internal combustion engines to substantially ↓ the freezing point of the coolant; antifreeze has long had currency as a surrogate inebriant, despite its being associated with acidosis and renal shutdown–ethylene glycol and blindness–methanol CLINICAL In low amounts, antifreeze simulates alcohol metabolism; at high levels, it is associated with N&V, seizures, coma, death, at levels above 100 ml of ethylene glycol. See Leukovorin rescue. Methotrexatre

antifungal *adjective* Referring to an effect that kills or inhibits fungal growth *noun* An agent that kills or inhibits fungal growth

antigen A molecule that usually has a molecular weight of > 1 kD, is a protein, often foreign–ie, non-self, which is capable of evoking a specific immune response, antibody production. See A antigen, Acquired B antigen, Alloantigen antigen, Antigenicity, Australia antigen, Autoantigen, B antigen, Bladder tumor antigen, Cancer-associated antigen, CENP antigen, Cromer-related antigen, Early antigen, Eclipsed antigen, Epithelial membrane antigen, Epstein-Barr nuclear antigen, Epstein-Barr viral capsid antigen, Extractable nuclear antigen, GAS, H antigen, Hepatitis surface antigen, High frequency antigen, H-Y antigen, Leukocyte common antigen, Lewis Y antigen, Low-frequency antigen, Mls antigen, Myeloid antigen, O antigen, Oncofetal antigen, P antigen, p24 antigen, Platelet antigen, Private antigen, PSA-prostate-specific antigen, Proliferating cell nuclear antigen, PRP antigen, Public antigen, Rh antigen, S antigen, Self antigen, Sm antigen, Soluble antigen, Squamous cell carcinoma antigen, SS-A antigen, SS-B antigen, Superantigen, TA-90, TAG-72, Targett antigen, Tumor-associated antigen, U antigen, Vi antigen, Xg^a antigen. Cf Antibody, Hapten

antigenicity Immunogenicity, see there

antigliadin antibody IMMUNOLOGY Gliadins are a class of proteins found in the gluten of wheat and rye grains; in genetically susceptible individuals, α-gliadins evoke production of IgA and IgG antibodies in Pts with celiac disease, resulting in a malabsorption syndrome–celiac sprue or gluten-sensitive enteropathy–GSE, and possibly dermatitis herpetiformis–DH REF RANGE IgA < 31 AU, negative; 32–39 AU, equivocal; > 39 positive; IgG < 89 AU, negative; 90–110 AU, equivocal; > 110 positive. See Celiac disease, Gluten

anti-glomerular basement membrane antibody Anti-GBM antibody IMMUNOLOGY An antibody that is usually ↑ in Pts with Goodpasture syndrome which may be serially measured to monitor response to therapy; anti-GBM antibodies may be measured in conjunction with anti-neutrophil cytoplasmic antibodies in Wegener's granulomatosis REF RANGE Negative < 10 U; positive > 20 U. See Goodpasture syndrome

antihemophiliac factor A Factor VIII, see there

antihemophiliac factor B Factor IX, see there

antihistamine Antihistaminic PHARMACOLOGY An agent that counteracts the effects of histamine released during allergic reactions by blocking histamine–H_1 receptors ADVERSE EFFECTS Dry mouth, drowsiness, urine retention in ♂, tachycardia. See Histamine receptor

antihormone ENDOCRINOLOGY A natural or synthetic analogue of a hormone–eg, FSH, LH, which binds to the hormone's receptor, blocking its activity. See B/I ratio

antihypertensive *adjective* Referring to an agent or mechanism that reduces HTN *noun* An agent used to manage HTN

anti-IgE antibody ALLERGY MEDICINE An antibody directed against IgE. See Allergy

anti-inflammatory *adjective* Referring to an agent or mechanism that ↓ inflammation *noun* An agent–ice, aspirin, ibuprofen–that attenuates inflammation. See Nonsteroidal antiinflammatory drugs

anti-insulin antibody A serum antibody which may be present in Pts with type 1 DM and insulin resistance. See Anti-islet cell antibody

anti-islet cell antibody DIABETOLOGY Any of a number of antibodies found at the onset of type 1–childhood-onset DM–IDDM, which 'recognize' components of the islet cell's cytoplasm and surface

antikickback law HEALTH CARE LAW Legislation enacted to prevent industries from having an unfair advantage in obtaining government contracts, where a company 'X' would offer a illegal 'kickback' fee to a person instrumental in ensuring that 'X' would get a lucrative contract for a service performed or product sold. See Kickback

antileukotriene Any of a class of agents that either interfere with leukotriene synthesis or antagonize leukotriene receptors, which are as effective as cromolyn or theophylline, and may ↓ the amount of inhaled steroids needed to control inflammation See Asthma, Leukotrienes, Zafirlukast, Zileuton

ANTILEUKOTRIENES

LTD$_4$ RECEPTOR ANTAGONISTS Zafirkulast-Accolate® Benefits LTD$_4$-induced bronchoconstriction, early and late responses, exercise challenge, cold-induced asthma, chronic asthma

5-LIPOXYGENASE INHIBITOR Zileutron-Zyflo® Benefits asthma induced by exercise, cold, aspirin, bronchial hyperresponsiveness

FLAP INHIBITORS None are FDA-approved Benefits early and late responses and cold-induced asthma

antilymphocyte globulin Antilymphocyte serum CLINICAL IMMUNOLOGY A polyclonal antiserum 'raised' in one species of animals against the lymphocytes—in particular T cells of another—ie xenogeneic species, which, upon injection, causes profound immunosuppression and lymphocytopenia

antimetabolite Antimetabolic agent PHARMACOLOGY Any agent–eg, MTX, 6-mercaptopurine, thioguanine, 5-FU, gemcitabine that is a structural analogue of a native cell metabolite, which either inhibits the enzymes of a particular metabolic pathway or is incorporated during synthesis to produce defective product or prevents replication; anti-metabolites are used as chemotherapeutics, antivirals and as immunosuppressants, and include analogues of purines–eg azathioprine, pyrimidines, folic acid–eg aminopterin, MTX See Azathioprine, Methotrexate

antimicrobial *adjective* Referring to an agent or mechanism that kills or inhibits the growth or reproduction of microbes *noun* An agent that attenuates, kills or inhibits the growth or reproduction of microbes

antimicrosomal antibodies Antithyroid microsomal antibodies ENDOCRINOLOGY Antibodies directed against components of thyroid microsomes, in particular peroxidase; AMAs are the most useful of antithyroid antibodies as they are often present in thyroid disease, and in higher titers than those for antithyroglobulin antibodies, especially in younger Pts; AMAs are present in Hashimoto's thyroiditis-99% positive, Graves' disease-80%, hypothyroidism, atrophic thyroiditis, and are ↑ in the elderly. See Antithyroglobulin antibodies, Antithyroid antibodies

anti-mitochondrial antibody Any antibody that reacts to various mitochondrial antigens; one such antigen is M2, which evokes the production of an AMA present in 90-95% of Pts with PBC, some of whom also have scleroderma; AMA is directed against the ATPase lipoprotein present on the inner mitochondrial membrane

antimitotic agent Antimicrotubule agent, mitotic inhibitor ONCOLOGY An agent that inhibits cancer growth by stopping cell division PHARMACOLOGY A drug that inhibits or blocks mitosis in a particular phase of the cell cycle–eg, metaphase arrest caused by colchicine and vinca alkaloids. See Chemotherapy

antimongoloid slant Antimongolic fissure PEDIATRICS A downward slant of the eyelid in the horizontal plane that is opposite that of individuals with trisomy 21–'mongolism'; the finding is nonspecific and may be seen in various congenital syndromes–eg, Treacher-Collins syndrome and Franceschetti–oculomandibulofacial syndrome, and is accompanied by a bird-like facies

antimony TOXICOLOGY A metallic element–atomic number 51; atomic weight 121.75 CLINICAL-ACUTE N&V, bloody diarrhea, hepatitis, kidney failure CLINICAL-CHRONIC Itching, conjunctivitis, laryngitis, headache, anorexia, weight loss, anemia, jaundice, kidney failure LETHAL DOSE 100 to 200 mg MANAGEMENT Supportive; gastric lavage/activated charcoal; dimercaprol to accelerate excretion; transfusions for hemolysis REF RANGE < 10 μg/L

antimyelin antibody NEUROLOGY Any of a number of antibodies associated with demyelinating disorders–eg, multiple sclerosis in particular multiple sclerosis

antimyocardial antibody CARDIOLOGY An autoantibody associated with Dressler's post MI syndrome, pericarditis, acute rheumatic fever, myocarditis, post coronary bypass surgery Pts. See Dressler syndrome

antineoplastic *adjective* Referring to an antineoplastic agent or mechanism *noun* Chemotherapeutic agent, see there

antineoplastic antibiotic Anticancer antibiotic, antitumor antibiotic ONCOLOGY Any of a group of anticancer drugs that block cell growth by interfering with DNA. See Daunomycin, Vinblastin, Vincristin

antineutrophil cytoplasmic antibody ANCA IMMUNOLOGY Any autoantibody directed against certain components of granulocytes, myeloid-specific lysosomal enzymes; ANCAs are most commonly found in systemic vasculitides–eg, necrotizing vasculitis, active generalized Wegener's granulomatosis–WG, 84-100% are positive, polyarteritis nodosa, inflammatory conditions of the lung and kidney–eg, crescentic glomerulonephritis, unexplained renal failure, Churg-Strauss syndrome, HIV infection, IBD, drug-induced lupus, SLE, rheumatoid arthritis, and others

antinuclear antibody ANA IMMUNOLOGY Any of a number of circulating antibodies that are directed against various antigens in the nucleus, including histone, double- and single-stranded DNA, and ribonucleoprotein; ANAs are often present in serum of Pts with SLE and other connective tissue diseases. See Speckled pattern

antinucleoside analogue THERAPEUTICS Any of a family of compounds that resemble natural nucleotide bases-the building blocks of DNA, which are used as antivirals; ANAs selectively interfere with viral DNA replication, without interfering with cellular DNA replication ADVERSE EFFECTS Short-term, minimal; long-term, adverse systemic effects on oxidative phosphorylation, which has features of inherited mitochondrial diseases, resulting in myopathy, cardiomyopathy, neuropathy, lactic acidosis, failure of exocrine pancreas, liver, BM

antioxidant NUTRITION Any agent—eg, vitamins A, C, and E, selenium, and others, that is capable of reducing highly histotoxic O_2 reduction products and reactive O_2 species—eg hydroxyl radical, which derive from superoxide anion–O_2·$^-$ and H_2O_2, the univalent and bivalent reduction products of O_2, generated during the normal intermediate metabolism of the respiratory chain; other antixodants include glutathione, α-tocopherol–vitamin E, bilirubin. See Catalase, Ceruloplasmin, Free radical, Glutathione, Peroxidase, Superoxide dismutase, Transferrin

antioxidant therapy THERAPEUTICS A general term for the use of any agent—eg, antioxidant vitamins, glutathione reductase, superoxide dismutase, to 'scavenge' O_2 free radicals–OFRs or excited O_2 molecules, which are by-products of normal metabolic reactions; excess OFRs have been linked to cancer $2°$ to OFR-induced DNA damage, and to cardiovascular disease $2°$ to OFR-induced oxidation of LDL-cholesterol to a more atherogenic form; in cardiology, AT attempts to block the oxidative modification of LDL, which is thought to be an early step in fatty streak development, atherogenesis and ASHD-related pathologies–eg, CAD and CVAs. See Antioxidant, Antioxidant vitamin

antioxidant vitamin NUTRITION Any vitamin–eg, beta carotene–provitamin A, ascorbic acid–vitamin C, and alpha-tocopherol–vitamin E with antioxidant activity. See Antioxidant, Antixoxidant therapy

anti-pancreatic islet cell antibody See Anti-islet cell antibody

antiparietal cell antibody IMMUNOLOGY An autoantibody that occurs in megaloblastic–pernicious anemia–PA, severe atrophic gastritis; in autoimmune thyroiditis, Pts with PA may also have antithyroid antibodies; APCA is also found in 2 to 3% of the normal population and its presence $\uparrow$ with age

antiphospholipid antibody Either of 2 antibodies: 1. Anticardiolipin antibody, see there 2. Lupus anticoagulant, see there. See Antiphospholipid antibody syndrome

antiphospholipid antibody syndrome Antiphospholipid syndrome, circulating lupus anticoagulant syndrome IMMUNOLOGY The association of recurrent thromboses–cerebral, repeated spontaneous abortions and renal disease often in ANA-negative SLE Pts, which may be accompanied by repeated fetal wastage and IgM gammopathy; APA has been defined as a ' *thrombophilic disorder in which venous or arterial thrombosis, or both, may occur. The serologic markers are antiphospholipid antibodies–anticardiolipin antibodies, the lupus anticoagulant, or both'*; APAS is characterized by the presence of circulating antiphospholipid antibodies–APA, in particular against cardiolipin–ACA, here used interchangeably, which overlap with lupus anticoagulants. Cf Lupus anticoagulant

antiplatelet agent THERAPEUTICS Any agent–eg, aspirin, that $\downarrow$ platelet clumping and clotting

antiplatelet antibody HEMATOLOGY An auto- or alloantibody directed against platelet antigens, which may be measured in thrombocytopenia or in Pts who are refractory to platelet transfusions. See Platelet transfusion

antiplatelet therapy HEMATOLOGY Any therapy that inhibits platelet adhesion and/or aggregation, both of which $\uparrow$ complications of ASHD. See Aspirin, Thrombolytic therapy

antipromoter ONCOLOGY A substance that blocks the action of a promoter molecule in carcinogenesis, potentially acting at any stage in a neoplastic transformation sequence EXAMPLES Dietary fiber, vitamins A, C and E, selenium, indoles, flavones, isothiocyanates. See Tumor promoter

antipruritic *adjective* Relieving or preventing itching *noun* An agent that prevents or relieves itching

antipsychotic agent Major tranquilizer, neuroleptic NEUROPHARMACOLOGY Any drug that attenuates psychotic episodes AGENTS Phenothiazines, thioxanthenes, butyrophenones, dibenzoxazepines, dibenzodiazepines, diphenylbutylpiperidines INDICATIONS Management of schizophrenic, paranoid, schizoaffective and other psychotic disorders, acute delirium, dementia, manic episodes–during induction of lithium therapy, control of movement disorders—in Huntington's disease, Tourette syndrome, ballismus, intractable hiccups, severe nausea and vomiting by blocking the medulla's chemoreceptor trigger zone ADVERSE EFFECTS Extrapyramidal effects–dystonia, akathisia, parkinsonism, tardive dyskinesia due to blocking of basal ganglia; sedation and autonomic side effects–orthostatic hypotension, blurred vision, dry mouth, nasal congestion and constipation are due to blocking of histamic, cholinergic and adrenergic receptors

antipyretic Antifebrile *adjective* Referring to an antifebrile agent or effect *noun* An agent that relieves or reduces fever

anti-receptor antibody IMMUNOLOGY A generic term for an autoantibody directed against a substrate's receptor, which is capable of altering the cell's response to that substrate; ARAs are pathogenically linked to endocrine disorders, either $\uparrow$ or $\downarrow$ hormonal activity, and are directed against receptors for corticotropin, H_2–histamine, parathyroid, islet β cells, insulin, thyrotropin–TSH, gastrin, and FSH. See Autoimmunity

antiretroviral VIROLOGY *adjective* Referring to an agent or effect that counters a retrovirus *noun* A drug that counters or acts against a retrovirus, usually understood to be HIV; FDA-approved antiretrovirals include reverse transcriptase inhibitors, nucleoside analogues and protease inhibitors See Antiretroviral, Non-nucleoside reverse transcriptase inhibitor

anti-Scl-70 Anti-scleroderma or anti-topoisomerase I antibody IMMUNOLOGY An autoantibody that occurs in Pts with scleroderma. See Scleroderma

antisepsis INFECTION CONTROL A general term for procedures or chemical treatments intended to kill or inhibit microbial growth

antiseptic *adjective* Referring to an agent or effect that counters microbial growth *noun* MEDTALK A substance that arrests or prevents the growth of microorganisms by inhibiting their activity without necessarily destroying them. Cf Aseptic, Disinfectant, Germicide

antiserum IMMUNOLOGY A serum that contains Igs against specified antigens, used therapeutically

antisocial personality disorder Dyssocial personality disorder, psychopathy, sociopathy PSYCHIATRY A disorder affecting an individual with complete disregard for the rights of others, who engages in antisocial behavior without remorse; APD begins in childhood or early adolescence and continues into adulthood PREVALENCE 3% ♀; 1% ♂; more common with substance abuse or in prison, or forensic settings. See Conduct disorder, Personality disorder

---

ANTISOCIAL PERSONALITY DISORDER

A. **PERVASIVE PATTERN OF DISREGARD FOR & VIOLATION OF RIGHTS OF OTHERS OCCURRING FROM AGE 15, INDICATED BY 3 + OF FOLLOWING**

1. Failure to conform to social norms with respect to lawful behaviors that is grounds for arrest

2. Deceitfulness as indicated by repeated lying, use of aliases, manipulating others for personal profit or pleasure

3. Impulsivity or failure to plan ahead

4. Irritability & aggressiveness, indicated by repeated physical assault

5. Reckless disregard for safety of self or others

6. Consistent irresponsibility, indicated by inconsistent work behavior or not honoring financial commitments

7. Lack of remorse, indicated by indifference to, or rationalizing having hurt, mistreated, or stolen from another

B. AGE ≥ 18

C. EVIDENCE OF A CONDUCT DISORDER–SEE THERE–BEFORE AGE 15

D. APD-DEFINING BEHAVIOR DOESN'T OCCUR EXCLUSIVELY DURING COURSE OF SCHIZO-PHRENIA OR MANIC EPISODE

antispasmodic PHARMACOLOGY An agent that relieves spasm. See Spasmolytic

anti-sperm antibody Any of a group of antibodies produced against 4 distinct components of the sperm; although ASAs are linked to infertility, their presence is not associated with a particular disease; ASAs may be found in the circulation, either free or as immune complexes, in seminal fluid and/or attached to the sperm surface REF RANGE/NORMAL IgA, IgG, IgM < 15% binding

antistenotic therapy CARDIOLOGY Any therapy or device designed to prevent complete occlusion of a stenosed coronary artery at risk of acute ischemia. See Balloon angioplasty, Stenting

Antistreplase® APSAC, see there

antistreptolysin O assay ASO test LAB MEDICINE A serologic test that monitors group A β-hemolytic streptococcal infection, ie 'strep throat'–90% positive, which may progress to acute streptococcal glomerulonephritis–± 25% are ASO positive, scarlet fever, and rheumatic fever, and thus used to monitor response to therapy. See Strep throat

antisubstitution law HEALTH LAW Any legislation or regulation that prevents pharmacists from substituting generic–almost invariably cheaper, alternative drugs, for those prescribed by a physician when he/she specifies '*dispense as written*'. See Generic drug

Antitampering Act A US federal law that criminalizes changing the labeling or content of non-prescription consumer products. See Cyanide, Tylenol® incident

antithrombin III HEMATOLOGY A 58 kD α_2-glycoprotein with a single polypeptide chain that inactivates serine proteases–thrombin and other coagulation proteins including factor Xa, IXa, kallikrein and others by an irreversible heparin-dependent reaction FUNCTION AT III dissolves blood clots that normally form within the circulation; heparin's anticoagulant activity hinges on activation of AT-III; AT-III-deficient individuals do not benefit from heparin therapy; ↓ AT-III may be a congenital AD condition, or acquired, occurring in DIC–due to 'consumption' or in liver disease–due to ↓ AT-III production, resulting in an ↑ risk of coagulation; AT III is ↓ in congenital deficiency, liver transplant, DIC, nephrotic syndrome, cirrhosis, chronic liver disease, carcinoma, mid-menstrual cycle; AT III is defective in 0.14% to 0.5% of the general population. See Hereditary thrombophilia, Recombinant human antithrombin III

antithrombin III deficiency Congenital antithrombin III deficiency An AD condition characterized by ↑ risk of DVT and PE. See Antithrombin III

antithrombotic HEMATOLOGY *adjective* Referring to an agent or effect that prevents or reduces thrombosis *noun* An agent that prevents or interferes with thrombosis. See Heparin, Warfarin. Cf Thrombolytic

anti-thyroglobulin antibody IMMUNOLOGY An antibody that is ↑ in autoimmune thyroiditis, Hashimoto's thyroiditis, thyrotoxicosis, hypothyroidism, thyroid CA, myxedema, Sjögren syndrome and autoimmune hemolytic anemia, SLE, rheumatoid arthritis. See Thyroiditis

antithyroid antibody Thyroid antibody Any antibody directed against 'self' antigens, which may be either cellular components–eg antimicrosomal antibodies or proteins–eg thyroglobulin, of thyroid origin, that are often present in autoimmune diseases; ATAs damage the thyroid and occur in 40-90% of Pts with chronic thyroiditis, Graves' disease or thyrotoxicosis, Hashimoto's thyroiditis, hypothyroidism; ATAs may also be ↑ in pernicious anemia, rheumatoid arthritis, and SLE REF RANGE Titers < 1:32-antithyroglobulin, and < 1:100-antimicrosomal are negative. See Antimicrosomal antibodies, Antithyroglobulin

antithyroid drug PHARMACOLOGY Any of a family of therapeutic agents–eg, propylthiouracil and methimazole, containing thiourea's thiocarbamide radical, which inhibits thyroid hormone 'organification'; ADs are used to treat children, young adults and pregnant ♀ with Graves' form of hyperthyroidism, but are not generally used for hyperfunctioning tumors or for Hashimoto's disease ADVERSE EFFECTS Occur in 1-5% and include rash, fever, urticaria, arthritis, transient leukopenia, agranulocytosis–0.5%, toxic hepatitis, and aplastic anemia

antithyroid microsomal antibody Antimicrosomal antibody, see there

antithyroid peroxidase antibody ENDOCRINOLOGY An antibody in thryoid autoimmune disease–TAD, eg, Hashimoto's thyroiditis, Graves' disease, post-partum TAD, and other autoimmune disorders. See Thyroiditis

antitoxin IMMUNOLOGY An antibody-rich serum from an animal stimulated with specific antigens or bacterial toxins–eg, botulinus, tetanus or diphtheria, which is used to provide passive immunity. See Passive immunity

antitrust law GOVERNMENT Legislation that limits the ability of organizations or groups of individuals to monopolize a service–or product, thereby controlling and restricting free trade. See Safe harbor rules

antituberculosis drug INFECTIOUS DISEASE Any drug–eg, isoniazid, rifampin, ethambutol, streptomycin, pyrazinamide, ethionamide, para-aminosalicylic acid, kanamycin, cycloserine, capreomycin, ciprofloxacin, amikacin, used to manage TB; multidrug-resistant isolates of *M tuberculosis* are most often resistant to isoniazid and rifampin. See Drug-resistant tuberculosis

antitumor antibiotic Antineoplastic antibiotic, see there

antitussive *adjective* Referring to an agent or mechanism that relieves or prevents cough *noun* An agent that relieves or prevents cough

antivenom Antivenin TOXICOLOGY A vehicle that contains an antibody or other substance that binds specifically to a toxin, deactivating it

antiviral *adjective* Referring to an agent or mechanism that counters viral effects or infection *noun* Antiviral agent, see there

antiviral agent Antiviral INFECTIOUS DISEASE An agent that prevents viral invasion or replication, treats an infection, or thrashes the virus into latency; antivirals may be specific–see below or nonspecific–eg, IFNs, which stimulate host defenses against viruses, making cells more resistant to viral genes, enhancing cellular immune responses or suppressing viral replication TYPES Inhibitors of viral DNA replication, transcription, or translation; topical surfactants; those that reduce viral infectivity by altering the lipid envelope; photodynamic inactivation with dyes which, when activated with UV light, are incorporated into the viral DNA, causing reading errors, IFN and IFN inducers, immune modulators and vaccines; AAs should not be used prophylactically, as their use selects for drug-resistant strains of virus

Antivirogram™ VIROLOGY A diagnostic system that evaluates HIV resistance to antiretrovirals, allowing optimal treatment of HIV/AIDS Pts. See Antiviral

Antizol® Fomepizole A proprietary injectable antidote for ethylene glycol–antifreeze poisoning. See Antifreeze

ANUG Acute necrotizing ulcerative gingivitis, see there

anuria Anuresis NEPHROLOGY The complete lack of urine production ETIOLOGY Acute renal failure, which may be due to shock or dehydration, fluid-ion imbalance, bilateral ureteral obstruction

Anxicalm Diazepam, see there

Anxiedin Lorazepam, see there

anxiety PSYCHOLOGY The emotional component of biological responses to imagined danger, linked to intrapsychic conflict CLINICAL-PHYSICAL Tachycardia, dyspnea, trembling, cognitive difficulties, hypersensitivity, dizziness, weakness, dysrhythmia, sweating, fatigue CLINICAL-MENTAL Sense of impending doom, powerlessness, apprehension, tension TYPES External stress–exogenous anxiety, internal stress–endogenous anxiety; it is pathological when it interferes with effectiveness in living, achieving of goals or satisfaction, or reasonable emotional comfort. See Anxiety disorder, Anticipatory anxiety, Castration anxiety, Free-floating anxiety, Monitor anxiety, Performance anxiety, Separation anxiety, Signal anxiety

anxiety attack PSYCHIATRY An episode of extremely uncomfortable anxiety; severe AAs are called panic attacks. See Anxiety, Anxiety disorder, Panic attack

anxiety disorder CHILD PSYCHIATRY Exaggerated or inappropriate responses to the perception of internal or external dangers PSYCHIATRY A general term that encompasses a broad range of conditions attributed to a state of heightened mental stress, anxiety and panic disorders, both exogenous and endogenous STATISTICS ADs affects 10–15% of Americans, and commands ±10% of the mental health bill ETIOLOGY Mental or physical trauma–eg, immune response to infections, substance abuse, genetic components TYPES OF AD Panic attack, agoraphobia, specific phobia(s), social phobia, obsessive-compulsive disorder, post-traumatic stress disorder, acute stress disorder, AD associated with medical conditions, substance abuse, or not otherwise specified

anxiety reaction PSYCHOLOGY An acute, transient episode of anxiety often accompanied by systemic changes–eg, hyperventilation-induced changes—periorbital and fingertip tingling, tachypnea, syncope

Anxiolan Buspirone, see there

anxiolytic *adjective* Pertaining to an anxiolytic *noun* A drug that reduces anxiety or tension

Anxionil Diazepam, see there

Anxira Lorazepam, see there

anything new, fear of PSYCHOLOGY Neophobia. See Phobia

Anzemet™ Dolasetron mesylate An antiemetic for postoperative and chemotherapy-related N&V

Anzepam Lorazepam, see there

AOA 1. American Orthopaedic Association 2. American Osteopathic Association, see there

AΩA ACADEMIA An honor society whose members are really–really, really smart See the Match

aortic aneurysm CARDIOLOGY A fusiform swelling of the aorta, which is linked to ASHD, ±75% of which occur in the abdominal aorta CLINICAL Most

are asymptomatic and emerge as an incidental finding during physical examination–PE; if pain is present, it is gnawing, relieved by positional changes, and often pulsating in character DIAGNOSIS PE, CT, US, MRI, digital subtraction angiography, aortic angiography NATURAL HISTORY AAs enlarge with time; 15% to 20% rupture in 1 yr if < 6 cm in diameter; 50% rupture if ≥ 6 cm MANAGEMENT Elective replacement of AA with synthetic vascular graft if > 4 cm and Pt is stable; for large AAs, Creech's intrasaccular approach, which places the graft inside of the aneurysm is associated with the least morbidity. See Aortic dissection

aortic arch syndrome 1. Takayasu's arteritis, see there 2. Subclavian steal syndrome, see there

aortic coarctation Coarctation of aorta CARDIOLOGY A congenital heart defect which occurs in 7% of Pts with CHD; ♂ ♀ ratio, 2:1; AC is characterized by narrowing of the aortic lumen, often distal to the origin of the left subclavian artery at the site of the aortic ductal attachment–the ligamentum arteriosum; extensive collateral arterial circulation develops though the internal thoracic, intercostal, subclavian and scapular arteries to supply the rest of the body in Pts with AC CLINICAL Rarely asymptomatic; headaches, dizziness, fainting, nosebleeds, ↓ pulses, rib notching and muscle cramps in legs with activity, palpitations; may be associated with gonadal dysgenesis–Turner syndrome, PDA, VSD, bicuspid aortic valve, mitral stenosis or regurgitation or aneurysms of the circle of Willis PE Systolic arterial pressure higher in arms than legs; diastolic pressure the same; femoral pulse is weak, delayed; possible findings include systolic thrill palpable in the suprasternal notch, LV hypertrophy, systolic ejection click due to bicuspid aortic valve, harsh systolic ejection murmur along the left sternal border, and a systolic murmur heard in the back EKG LV hypertrophy IMAGING CXR-notching of ribs due to ↑ collateral flow through the IC arteries; pre– and poststenotic dilatation of the aorta yields the reverse E or '3' sign; other imaging modalities include echocardiography, Doppler, CT, MRI, contrast aortography COMPLICATIONS Left ventricular failure, premature CAD, infective endocarditis, CVAs due to rupture of intracranial aneurysms MANAGEMENT Surgical repair, if transcoarctation pressure gradient is > 30 mm Hg; the earlier the repair, the less likely is the Pt to have persistent or recurrent HTN. See Aortic stenosis

aortic dissection Aortic dissection of hematoma, dissecting aneurysm of aorta CARDIOVASCULAR DISEASE A condition which affects ±2000 people/yr–US, characterized by the presence of a second–false lumen in the aortic wall; dissection occurs when blood flows into a torn tunica intima of an aortic wall weakened by cystic medial necrosis, usually in a background of chronic HTN, which follows degeneration of the elastic tissue and collagen of the tunica media HIGH RISK GROUPS Older ♂, pregnant ♀, blacks, HTN, connective tissue diseases, vasculitis or congenital aortic coarctation; AD may be induced by the trauma of the cardiac catheterization and exercise CLINICAL Hypotension with loss of consciousness due to compromise of the brachiocephalic vessels, reactive HTN, pulse-blood pressure dissociation and focal neurologic defects due to involvement of the spinal arteries RADIOLOGY Abnormal chest films with widening of the mediastinal silhouette, an abnormal aortic contour, left-sided pleural effusion, a disparity of the luminal diameter between the ascending and descending aorta and separation of intimal calcification ≥ 0.5 cm from the outer edge of the soft tissue border of the aorta–'calcium sign'; other defects may be seen by aortic angiography, echocardiography, CT, MRI MANAGEMENT Sodium nitroprusside, β-blockers, CCBs; blood flowing into a dissection may re-flow into the aortic lumen if there is a 2nd intimal tear or may rupture into the pleural or pericardial cavities, with potential fatal consequences MORTALITY 21% of ruptured AD die in the first 24 hrs; 90% within 3 months. See Cystic medial necrosis

aortic ectasia CARDIOLOGY A condition characterized by aortic root dilatation associated with HTN and aging, resulting in aortic regurgitation. See Aortic regurgitation

aortic occlusion CARDIOLOGY The partial or complete occlusion of the aorta along its path, an event which, given its hemodynamic consequences, is incompatible with long-term viability of the tissues 'south' of the occlusion

aortic regurgitation Aortic insufficiency CARDIOLOGY The reflow of blood back from the aorta into the left ventricle due to incompetency of the aortic valve ETIOLOGY Congenital or acquired valve defect of either the aortic leaflets–infectious endocarditis, rheumatic fever, or the aortic root–annuloaortic ectasia, Marfan syndrome, aortic dissection, collagen vascular disease, syphilis CLINICAL DOE, orthopnea, fatigue, ± angina, ↑ pulse pressure, systolic HTN with ↑ afterload on left ventricle EXAMINATION ↑ pulse pressure–Corrigan's pulse, Hill sign, Musset sign, Quincke's pulse; systolic murmur, diastolic rumble–Austin Flint murmur over cardiac apex WORKUP Doppler echocardiography to estimate severity of AR, confirmed by aortography MANAGEMENT Early valve replacement surgery, guided by the '55 rule'–performed when the ejection fraction is < 55% and/or the end-systolic dimension is ≥ 55 mm. Cf Aortic stenosis

aortic stenosis CARDIOLOGY Narrowing of the aortic annulus caused by degeneration and calcification of the valve leaflets; AS is more common and occurs earlier in Pts with underlying valve defects–eg, bicuspid valves; AS in previously normal valves develops after age 60 and is associated with HTN and hypercholesterolemia ETIOLOGY Rheumatic fever, congenital heart disease, idiopathic sclerosis CLINICAL Angina, syncope, CHF HEMODYNAMICS Chronic stenosis leads to LV enlargement, CHF EXAMINATION Systolic ejection murmur radiating to the neck; in mild stenosis, the murmur peaks early in systole and is often associated with a thrill; with ↑ severity the murmur peaks progressively later in systole, and may become softer as cardiac output ↓ DIAGNOSIS Doppler echocardiography indicates ↑ flow rates and ↓ total flow; catheterization, coronary angiography MANAGEMENT If asymptomatic, none needed except for prophylactic antibiotics to cover for infective endocarditis; valve replacement surgery; balloon valvotomy is only palliative. Cf Aortic regurgitation, Pulmonary stenosis

aortic ultrasound CARDIOLOGY The use of ultrasonography to evaluate aorta size, shape, adequacy of blood flow, and detect abdominal aortic aneurysms. See Aortic aneurysm

aortic valve ANATOMY A 3-leaflet valve at the base of the aorta which opens during systole–left ventricular contraction, then shuts during distole–atrial contraction preventing the backwash of oxygenated blood from the aorta into the ventricle

aortic valve stenosis Aortic stenosis, see there

aortic window IMAGING A space below the inferior margin of the aortic arch and the superior margin of the left pulmonary artery seen on a plain AP chest film

aortitis CARDIOLOGY Inflammation of the aorta. See Syphilitic aortitis

AP diameter Anterior-posterior diameter

AP film Anteroposterior film IMAGING A radiologic image in which the beams pass front-to-back. Cf PA film

APACHE III Acute Physiology & Chronic Health Evaluation INTENSIVE CARE A 'third-generation' system for estimating the risk of hospital death in adult ICU Pts based on physiological assessments of most severely affected values during the first 24 hrs in the ICU and subjecting the results to logistic regression modeling techniques. See Medisgroups, Prognostic scoring systems

apareunia SEXOLOGY Total coital impairment. Cf Dyspareunia

apathetic hyperthyroidism ENDOCRINOLOGY A masked hyperthyroidism, more common in depressed older Pts CLINICAL Hypermetabolic state manifest by weight loss, CHF, supraventricular tachyarrhythmia

apathy PSYCHOLOGY Indifference, numbness. See Erotic apathy

APC 1. Activated protein C, see there 2. Adenomatous polyposis coli, see there 3. Antigen-presenting cell, see there 4. Atrial premature contraction, see there 5. CLINICAL TRIALS Adenoma Prevention with Celecoxib 6. MANAGED CARE Ambulatory payment classification, see there

APC resistance Activated protein C resistance, see there

apex Latin, summit Any tip of a pyramidal or rounded structure LUNG The superior-most part of the lung HEART The heart formed by the left ventricle-its inferior-most portion TEETH The tip of a tooth root

apex beat Apical thrust, cardiac impulse CARDIOLOGY A cardiac pulsation that corresponds to the impact of the apex–left ventricular as it rotates forward during systole; the AB is palpable or visible on the anterior chest wall in the 5th left intercostal space, and may be displaced by heart–eg enlarged right ventricle or aortic aneurysm or lung–eg dilated pulmonary artery disease. See Apexcardiography

apexcardiography CARDIOLOGY A technique that graphically records movements of the chest wall caused by the beating heart, which is used to evaluate left ventricular function, MI, aneurysms, ischemia, and pericarditis; it provides ancillary information and is used with EKG, phonocardiography, and carotid or jugular vein pulse tracings ABNORMAL VALUES Absent or abnormal waves occur in AFib, mitral and aortic valve stenosis, HTN, hypertrophic cardiomyopathy, ventricular aneurysm, ischemia, MI, CAD

APGAR score NEONATOLOGY, OBSTETRICS A bedside test for evaluating a neonate's post-partum status and potential for survival, based on an acronym of Virginia Apgar's name; the higher the score, the better the infant will fare during the neonatal period

APGAR SCORE

APPEARANCE Color–0 for blue, 2 for pink

PULSE Heart rate–0 for none, 1 for <100/min, 2 for > 100/min

GRIMACE Reflex–0 for none, 1 for grimace, 2 for cough/sneeze

ACTIVITY Muscle tone–0 for limp, 2 for full flexion

RESPIRATORY EFFORT 0 for absent, 2 for strong crying

aphakia Lenslessness

aphasia Dysphasia NEUROLOGY Partial or total inability to understand or create speech, writing, or language due to damage to the brain's speech centers; loss of a previously possessed facility of language comprehension or production unexplained by sensory or motor defects or diffuse cerebral dysfunction ETIOLOGY Stroke, brain disease, injury; anomia–nominal or amnesic aphasia and impaired ability to communicate by writing-agraphia are usually present in all forms of aphasia. See Broca's/Motor aphasia, Sensory/Wernicke's aphasia, Tactile aphasia

APHASIA

MOTOR Broca's aphasia A primary deficit in language output or speech production, which ranges in severity from the mildest, cortical dysarthria, characterized by intact comprehension and ability to write, to a complete inability to communicate by lingual, phonetic, or manual activity

SENSORY Wernicke's aphasia Pts with sensory aphasia are voluble, gesticulate, and totally unaware of the total incoherency of their speech patterns; the words are nonsubstantive, malformed, inappropriate–paraphasia Sensory aphasia is characterized by 2 elements:

Impaired speech comprehension–due largely to an inability to differentiate spoken and written phonemes–word elements-due to either involvement of the auditory association areas or separation from the 1° auditory complex

Fluently articulated but paraphasic speech, which confirms the major role played by the auditory region in regulating language

TOTAL Global aphasia, complete aphasia A form of aphasia caused by lesions that destroy significant amounts of brain tissue, eg occlusion of the middle cerebral or left internal carotid arteries, or tumors, hemorrhage, or other lesions; total aphasia is characterized by virtually complete impairment of speech and recognition thereof; afflicted Pts cannot read, write, or repeat what is said to them; although they may understand simple words or phrases, rapid fatigue and verbal and motor perseverence, they fail to carry out simple commands; total aphasia of vascular origin is almost invariably accompanied by right hemiplegia, hemianesthesia, hemianopia of varying intensity

aphonia ENT Complete speechlessness resulting from an inability to produce normal sounds due to organic—eg, laryngeal disease or mental cause. See Stroke. Cf Alalia, Spasmodic dysphonia

aphthae Coldsores, see there; herpetic ulcers

aphthous ulcer Canker sore ORAL PATHOLOGY A small, often painful mouth ulcer due to injury to the oral mucosal, viral infection or vitamin deficiency

apical abscess DENTISTRY An abscess near a tooth root's apex caused by deep caries; AAs may extend to the tooth's pulp chamber, destroying nerves, blood vessels MANAGEMENT Root canal therapy—endodontics, tooth extraction

apical cap sign RADIOLOGY 1. A subtle blush or shadow on a plain chest film in the left lung apex caused by blood leaking from a traumatically ruptured aorta into the pleura, indicating a need for emergency aortic aortography and surgery 2. Dunce cap sign A finding on a plain abdominal film in pheochromocytomas; the enlarged rounded tumor located in the adrenal medulla forms the 'head' and the triangular dunce cap representing the normal residual cortex which may be pushed superiorly

apical tumor A tumor at the top of an organ, often the lung

apicoectomy Root resection DENTISTRY Amputation of the apex of a tooth root, usually to treat an infection

Apis mellifera Honeybee IMMUNOLOGY A major cause of life-threatening anaphylaxis in sensitized individuals CLINICAL Fever, chills, light-headedness, hives, joint and muscle pain, bronchial constriction, SOB, hypotension, pulmonary edema, shock, and possibly, death. See Hymenopteran

Aplacasse Lorazepam, see there

aplasia Absence of tissue or organ development. See Pure red cell aplasia, Pure white cell aplasia. Cf Atrophy, Hypoplasia

aplastic anemia HEMATOLOGY Anemia due to failure in BM production of RBCs, often accompanied by failure in WBC and platelet production—ie, granulocytopenia, thrombocytopenia ETIOLOGY Toxins—benzene, toluene, insecticides; RT; drugs—carbamazepine, chemotherapy, chloramphenicol, gold salts, phenytoin, phenylbutazone, quinacrine, sulfonamides, tolbutamide; autoimmunity, SLE, post-hepatitis, pregnancy, PNH PROGNOSIS AA is often unresponsive to specific therapy

aplastic crisis HEMATOLOGY A transient cessation of BM activity seen in sickle cell anemia, which may be triggered by infection and/or folic acid deficiency, and characterized by disappearance of reticulocytes from peripheral circulation and abrupt worsening of anemia. See Sickle cell anemia

Apligraf™ Graftskin WOUND CARE A proprietary manufactured human skin equivalent/artificial skin containing living cells used to cover venous ulcers, burn wounds, diabetic and decubitus ulcers. See Artificial skin

apnea SLEEP DISORDERS A cessation of airflow of ≥ 10 seconds GRADING Mild-obstructive apnea-hypopnea index of ≤ 20; moderate 20–40; severe ≥ 40. See Obstructive apnea-hypopnea index

apnea monitor PEDIATRICS An impedance-type device that monitors both the respiratory and heart rate of an infant, and sounds an alarm alerting care-givers of a possible need to perform CPR in the event of either apnea or a marked ↑ or ↓ in heart rate. See Central sleep apnea, SIDS

apneic *adjective* Referring to apnea, breathless

apneustic center ANATOMY A wad of nuclei in the brain stem that participate in the rhythmic control of breathing

apo- 1. A prefix indicating a protein component in a conjugated molecule—eg, apoferritin, apolipoprotein, see there 2. Apolipoprotein, see there

apo A-IV A gene that encodes apo A-IV; allelic polymorphism in *apo A-IV* results in the substitution of histidine for glutamine at position 360 near the COOH terminus, generating the isoform apo IV-2, which attenutates the hypercholesterolemic response to dietary cholesterol. See Apolipoprotein A-IV

APOE ε4 MOLECULAR NEUROLOGY The type 4 allele of the apolipoprotein E gene locus located on chromosome 19, which may ↑ the risk of late-onset Alzheimer's disease, and has been associated with ↓ cerebral parietal metabolism; possession of an *APOE* ε4 allele is a strong predictor for cognitive decline. See Alzheimer's disease

apoenzyme The protein part of an enzyme that requires a cofactor/coenzyme. See Apoprotein

apolipoprotein Any of a family of small proteins on the surface of lipoprotein complexes, which bind to specific enzymes or transport proteins across cell membranes, which correspond to the protein moiety of lipoproteins. See HDL-cholesterol, Kringle domain

apolipoprotein A-I A 28 kD single chain protein that comprises 75% of the ApoA in HDL; apoA-I is the major protein constituent of lipoproteins, and participates in cholesterol transport by activating LCAT-lecithin:cholesterol acyltransferase; apoA-I levels are inversely proportional to the risk of CAD

apolipoprotein B-100 A 550 kD protein synthesized in the liver that is the major component in lipoproteins of endogenous origin—LDL, VLDL, IDL, provides the recognition signal targeting LDL to the LDL—apoB, E receptor and is considered to be the l'enfant terrible of ASHD REF RANGE 40 to 125 mg/dl ABNORMAL VALUES ↑ in familial combined hyperlipidemia, acquired hyperlipidemia, in acute angina and MI

apolipoprotein B-48 A 250 kD protein that is a major chylomicron component synthesized in the intestine; ApoB-48 has an obligatory role in the synthesis of chylomicrons and is essential for intestinal absorption of dietary fats and fat-soluble vitamins

apolipoprotein C-II deficiency An AR condition characterized by recurring pancreatitis, DM LAB Absent apoC-II, 50% ↓ apoA-I, ↓ A-II, ↓ apoB, ↓ LDL, ↓ HDL; ↑ TGs, ↑ cholesterol; ↑ chylomicrons, ↑ VLDL, ↑ apoE TREATMENT Diet

apolipoprotein C-III An 8.7 kD protein and major constituent of chylomicrons and VLDL; it is a cofactor for sphingomyelinase, an inhibitor of lipoprotein lipase, may activate lecithin; cholesterol acyltransferase and modulate uptake of TG-rich remnants by hepatic receptors; Pts with apoC-III deficiency have an ↑ risk of CAD

apolipoprotein E A 34-kD cholesterol-binding glycoprotein, which comprises 15% of VLDL; apoE maps to chromosome 19, is secreted by macrophages that mediate the uptake of lipoproteins—VLDL, HDL, LDL and cholesterol esters into cells via distinct binding domains for each receptor; apo-E has a central role in the metabolism of TG-rich lipoproteins, and mediates the uptake of chylomicrons and VLDL by the liver

apolipoprotein E deficiency A rare cause of type III hyperlipoproteinemia, caused by specific mutations in apoE which prevent the binding of chylomicrons and VLDL to the LDL-receptor; AED is characterized by ↑↑↑ serum cholesterol, ↑ TGs, accumulation of β-migrating remnants, and premature ASHD. See Apolipoprotein E

Aponal® Doxepin, see there

aponeurosis ANATOMY A flat sheet of fibrotendinous tissue which forms the site of attachment of flat muscles or corresponds to a zone of separation of flat muscles. See Bicipital aponeurosis, Plantar aponeurosis

aponeurosis plantaris Plantar aponeurosis, see there

apophysitis of tibial tubercle Osgood Schlatter disease, see there

apoplexy NEUROLOGY An imprecise term for an acute devastating intracerebral event–eg stroke, intracranial hemorrhage, often associated with loss of consciousness and paralysis. See Abdominal apoplexy, Pituitary apoplexy

apoptotic cell CELL BIOLOGY A dense, eosinophilic, pyknotic cell surrounded by a thin clear space, often lying within epithelium, which is due to apoptosis

Apozepam Diazepam, see there

APP Amyloid β protein precursor A membrane-spanning glycoprotein of unknown function, which is expressed in many mammalian tissues, the gene for which maps to chromosome 21. See Alzheimer's disease, β-amyloid

apparent agency Ostensible agency MEDICAL MALPRACTICE A relationship in which a person–eg, a physician, has a relationship of responsibility with an entity–eg, a hospital. See Captain of the ship doctrine, Malpractice

apparent volume of distribution PHARMACOLOGY The ratio of the total amount of drug in the body to the concentration of the drug in the plasma, or the 'apparent' volume necessary to contain the entire amount of a drug, if the drug in the entire body were in the same concentration as in the plasma. Cf Therapeutic drug monitoring

appearance VOX POPULI The way something looks or appears. See Ace of spades appearance, Anchovy paste appearance, Bag-of-worms appearance, Bearskin rug appearance, Beaten brass/silver appearance, Blocked pipe appearance, Blood & thunder appearance, Blueberry muffin appearance, Boiled lobster appearance, Bone spicule appearance, Bread crumb appearance, Brick dust appearance, Bulls-eye appearance, Candle flame appearance, Candlestick appearance, Carpet tack appearance, Chain of lakes appearance, Chair rung appearance, Champagne glass appearance, Chinese character appearance, Clear glass appearance, Cluster of grapes appearance, Cobblestone appearance, Cocktail sausage appearance, Coffee bean appearance, Coffee grounds appearance, Coiled spring appearance, Copper wire appearance, Coral reef appearance, Corduroy cloth appearance, Cornflake appearance, Cornrow appearance, Coup de sabre appearance, Crazy pavement appearance, Cream cheese appearance, Crow's feet appearance, Crumpled tissue paper appearance, Crushed cranberry appearance, Cupid's bow appearance, Cushingoid appearance, Dewdrop on a rose petal appearance, Dilapidated brick wall appearance, Dinner fork appearance, Dog-ear appearance, Double ring appearance, Elephant ear appearance, Envelope appearance, Erlenmeyer flask appearance, Figure 8 appearance, Finger-in-glove appearance, Flea-bite appearance, Foamy appearance, Fried egg appearance, Frog neck appearance, Garden hose appearance, Gasoline pump appearance, Goblet appearance, Gothic arch appearance, Grape-like appearance, Groomed whisker appearance, Ground glass appearance, Hairbrush appearance, Hair-on-end appearance, Halo appearance, Heaped up appearance, Holly leaf appearance, Lacquer crack appearance, Lacy appearance, Leafless tree appearance, Light bulb appearance, Link sausage appearance, Lizard skin appearance, Lollipop appearance, Lollipop tree appearance, Maltese Cross appearance, Mariner's wheel appearance, Medusa head appearance, Mickey Mouse appearance, Milk of calcium appearance, Molar tooth appearance, Motor oil appearance, Mulberry appearance, Mushroom appearance, Nipple appearance, Onion skin appearance, Owl eye appearance, Party wall appearance, Peach pit appearance, Peau d'orange appearance, Philadelphia cream cheese appearance, Picture frame appearance, Pigskin appearance, Pizza pie appearance, Plucked chicken appearance, Porcelain white appearance, Pruned tree appearance, Railroad track appearance, Retort tube appearance, Rib-within-a-rib appearance, Roman bridge appearance, Rugger jersey spine appearance, Safety pin appearance, Salt & pepper appearance, School of fish appearance, Scrambled egg appearance, Sea fan appearance, Sealed envelope appearance, Serpentine appearance, Shaving cream appearance, Shepherd's crook appearance, Shiny coin appearance, Shredded appearance, Shredded casette appearance,

Silver wire appearance, Slapped cheek appearance, Smiling face appearance, Snow ball appearance, Stacked coin appearance, Thorn apple appearance, Thrush breast appearance, Tin tack appearance, Tissue culture appearance, Toy balloon appearance, Tramtrack appearance, Tree bark appearance, Treefrog hand appearance, Veiled appearance, Watered silk appearance, Wedding ring appearance, Wheat sheaf appearance, Whetstone appearance, Whiskering appearance, Windblown appearance, Wooly appearance, Worn tennis ball appearance

appendectomy SURGERY The removal of an appendix by a conventional–McBurney abdominal–incision or by laparoscopy ANESTHESIA General HOSPITAL STAY 1-3 days, shorter if laparoscopic. See Appendicitis

appendicitis SURGERY Inflammation of the vermiform appendix which is most common in children CLINICAL Right lower quadrant pain of acute onset, rebound tenderness over McBurney's point in right hypogastrium, fever, anorexia, constipation, diarrhea, N&V LAB ↑ WBCs, left shift of WBCs, ↑ ESR DIAGNOSIS Hx, PE, ultrasound, CT MANAGEMENT Appendectomy COMPLICATIONS Rupture, purulent peritonitis; untreated, death. See Appendix. Cf Left-sided appendicitis

apperception PSYCHIATRY Perception modified by personal emotions, memories, biases

appetite center Feeding center

apple POPULAR NUTRITION The fruit of the tree *Malus sylvestris*, which has a long tradition as a 'healthy food'; the components responsible for the apple's effect include its high pectin content, soluble fiber, and polyphenols. See Healthy foods

apple shape CLINICAL MEDICINE A popular term for truncal obesity, defined as a body-mass index of ≥ 27 and a waist-to-hip ratio of ≥ 0.88, often associated with ASHD and an ↑ risk of acute MI. Cf Pear shape

appliance GI DISEASE A disposible pouch attached to the skin of a stoma–eg, colostomy, into which waste material is discharged. See Ostomy

application PHARMACEUTICAL INDUSTRY A formal request for FDA review of a stage in the regulatory process of bringing a drug to market. See Abbreviated new drug application, IND application, NDA application, Premarket approval application, Request for application, Supplemental application VOX POPULI 1. A specific use 2. A formal request, usually in writing for a position, service, or right

applied genomics Molecular genetics MOLECULAR MEDICINE The use of genetic information to diagnose, prognosticate and manage disease, based on the techniques of PCR, real time-PCR, DNA sequencing, and gene rearrangement studies for leukemia and lymphoma. See Gene rearrangement, Genetics, In situ hybridization, PCR

applied research CLINICAL RESEARCH Research that applies results to a specific problem–eg studying the effects of different methods of law enforcement on crime rates

appointment HOSPITAL PRACTICE The admission of a physician to a hospital's staff, based on his/her credentials, work Hx, malpractice claims Hx, and reputation MEDICAL PRACTICE A scheduled timeslot in which a person is examined by a physician during his/her normal office hrs

appose MEDTALK *verb* To contact, juxtapose

approach-avoidance conflict PSYCHOLOGY An intrapersonal conflict characterized by both attraction towards and repulsion from something

appropriate care HEALTH CARE POLICY A health care service which is–per RAND investigators—'worth doing'. See Health care rationing

appropriate for gestational age NEONATOLOGY *adjective* Referring to an infant whose gestational age and weight are synchronous according to standardized age and growth curves. See Low birthweight

appropriate period of observation END OF LIFE The minimum time which must pass between performing the 1st examination to determine whether brain death has occurred and a 2nd corroborative examination. See Attending physician, Brain death, Corroborating physician

appropriateness HEALTH CARE The use of a resource or service in the most suitable, or efficient manner possible. See Inappropriate. Cf Practice guidelines

approved charge HEALTH INSURANCE The monies paid under Medicare as the maximum fee for a covered service

approved program GRAD EDUCATION An education program which is approved by a overseeing body–eg, a licensing or professional board or governmental agency

approximate 1. Not exact, thereabouts, more or less, sort of, approximal MEDTALK → VOX POPULI *verb* To make closer

apraxia NEUROLOGY A disorder of motor function characterized by an inability to execute a skilled or learned or purposeful motor act, despite normal muscle function, unrelated to paralysis or lack of comprehension ETIOLOGY Cortical lesions. See Construction apraxia, Ideational apraxia, Ideokinetic apraxia, Ideomotor apraxia, Neurapraxia

APRES CARDIOLOGY A clinical trial–Angiotensin-converting enzyme inhibition Post REvascularization Study

Apresoline® Hydralazine, see there

APRICOT CARDIOLOGY A clinical trial–Aspirin vs Coumadin in Prevention of Reocclusion & Recurrent Ischemia after Successful Thrombolysis which compared 2 strategies to ↓ risk of recurrent CAD

APSAC Acylated plasminogen streptokinase complex, Antistreplase®, Eminase® CARDIOLOGY A thrombolytic prepared from streptokinase and human plasminogen, the active site of which is acylated to block activation by other plasma proteins, while retaining fibrin-binding capacity. See Thrombolytic therapy. Cf Streptokinase, tPA

APSGN Acute poststreptococcal glomerulonephritis, see there

APUD system Amine Precursor Uptake & Decarboxylation system A morphologic and functional subgroup of the endocrine system, encompassing the C cells of the thyroid and ultimobranchial body, type I cells of the carotid body and paraganglia, norepinephrine and epinephrine-producing cells of the adrenal medulla, melanoblasts, pineal gland, and posterior pituitary cells: APUD cells take up tryptophan, convert it to 5-HT–serotonin, which is converted to MAO, 5HIAA

APUDoma ONCOLOGY A tumor that produces small peptide hormones of the APUD system EXAMPLES Small–oat cell carcinoma of lung, carcinoids of lung, thymus, GI tract and prostate, medullary CA of thyroid, pancreatic islet cell tumor, malignant melanoma, ganglioneuroma CLINICAL Heterogeneous; reflects functional nature of tumor and/or metastases. See MEN

Aquacel® Hydrofiber® WOUND CARE A highly absorbant, nonadherent wound dressing/packing ConvaTec. See Absorptive dressing

aquatic therapy Water therapy REHAB MEDICINE The exercising of muscle groups under water, which increases range-of-motion and light resistance for rehabilitation. See Rehabilitation medicine

aque- Water: aqueous

A/R Accounts receivable, see there

AR Autosomal recessive, also 1. Access recirculation, see there 2. Acoustic reflex 3. Active resistance–rehabilitation medicine 4. Adrenergic receptor 5. Allergic

reaction 6. Allergic rhinitis 7. Aortic regurgitation 8. Apical rate

ara-C Cytarabine, see there

arabinose BIOCHEMISTRY A pentose that occurs in D and L configurations

arabinosyladenine THERAPEUTICS An antiviral adenosine analogue

arabinosylcytosine THERAPEUTICS An antibiotic that blocks DNA synthesis, derived from cytosine by replacing ribose with arabinose

arachidonate A salt, ester, or anion of arachidonic acid

arachnic injury TOXICOLOGY Any lesion cause by biting arachnids including spiders, scorpions, tarantulas, et al. See Black widow spider, Brown recluse spider, Scorpion sting, Tarantula

arachnidism ENTOMOLOGY Intoxication by arachnid venom, especially from a black widow spider bite. See Necrotizing arachnidism

arachnodactyly 1. Long fingers/toes 2. Obsolete for Marfan syndrome

arachnoid *adjective* Like a cobweb, covered with or consisting of entangled soft fibers or hair-like material *noun* A member of the arachnoid family

arachnoiditis NEUROPATHOLOGY Inflammation of the arachnoid membrane and subarachnoid space

arachnophobia PSYCHIATRY A morbid fear of spiders. See Phobia

Aralen® Chloroquine phosphate, see there

Arato v. Avedon FORENSIC MEDICINE A decision from the California Supreme Court on informed consent that revolves around the Pt's right to a truthful prognosis. See Truth-telling

Arava™ Leflunomide RHEUMATOLOGY An agent used to manage rheumatoid arthritis in adults and slow structural damage as evidenced by bone erosion and joint-space narrowing. See Rheumatoid arthritis

arbovirus VIROLOGY A large, heterogeneous group of single stranded RNA viruses with an envelope surrounding the capsid, which are so named as most are transmitted by arthropod bites, ie are ARthropod-BOrne VECTORS Mosquitoes, sandflies, ticks CLINICAL Most arboviral infections are mild; clinical syndromes include hemorrhagic fever, encephalitis, systemic fever complex, hepatitis FAMILIES Arenavirus, Bunyaviridae, Flaviviridae, Reoviridae, Rhabdoviridae, Toagaviridae See California encephalitis, Eastern equine encephalitis, St Louis encephalitis, Yellow fever, Western equine encephalitis

ARC AIDS-related complex, also 1. American Red Cross, see there 2. Association pour Recherche du Cancer, see there

arc of motion Range of motion, see there

Arcanobacterium haemolyticum INFECTIOUS DISEASE A bacterial pathogen of adolescents and young adults that causes an exanthem similar to that caused by viruses, toxins, and drugs CLINICAL From mild pharyngitis to diphtheria-like with septicemia TREATMENT Antibiotics

arch wire A stainless steel orthodontic wire attached to braces or brackets to move the teeth in dentistry, or stabilize the jaw in LeFort fractures. See LeFort fracture

architectural barrier PUBLIC HEALTH Any structure or design feature that makes a building inaccessible to a person with a disability–eg, lack of ramps, narrow elevator doors. See Americans with Disabilities Act, Service dog

archiving INFORMATICS The storage of data in archives. See Mirroring, Optical disk archiving

arcitumomab CEA scan IMMUNOLOGY A 99mTc-labelled murine IgG monoclonal Fab fragment which attaches in vivo to CEA and can be used as a tumor marker to detect GI tract CAs using immunoscintigraphy See Monoclonal antibody

arctic anemia Polar anemia HEMATOLOGY An idiopathic anemia 2º due to cold exposure LAB Microcytic anemia, then normocytic anemia. See Anemia

arcus senilis GERIATRICS A cloudy, opaque ring around the cornea and iris which, in those younger than 50, often indicates hypercholesterolemia, but in the elderly has no significance

ardeparin A formulation of LMW heparin. See Low-molecular-weight heparin

ARDS Adult respiratory distress syndrome, see there

area under the curve CLINICAL PHARMACOLOGY The amount of a therapeutic agent that is present in the circulation in a determined time period, often 24 hrs. See Therapeutic drug monitoring

Aredia® Pamidronate, see there

areflexia NEUROLOGY An absence of reflexes

Arenavirus VIROLOGY The single genus of the family Arenaviridae, which includes Lassa virus, lymphocytic choriomeningitis virus and hemorrhagic fever—eg Junin, Machupo—viruses VECTOR Rodents. See Arenaviridae

areolar gland abscess An abscess of areolar glands in the breast CLINICAL Tender, painful lump, which may drain pus MANAGEMENT Warm wet compresses, antibiotics; incision and drainage may be required

ARESLD Alcohol-related end-stage liver disease

argatroban Novastan® HEMATOLOGY An anticoagulant used to treat Pts with heparin-induced thrombocytopenia–HIT, heparin-induced thrombocytopenia & thrombosis syndrome–HITTS, which may be used with thrombolytics–eg, tPA and streptokinase in managing acute MI

argentaffinoma ONCOLOGY A tumor located anywhere in the GI tract—± 90% in appendix, remainder are in the ileum, stomach, colon or rectum or lungs, which secretes excess serotonin, evoking a carcinoid syndrome. See APUD system, Carcinoid syndrome

Argentine hemorrhagic fever A viral illness caused by the Junin arenavirus EPIDEMIOLOGY Transmitted by contact with rodent urine; 23 outbreaks have been recorded, in the maize-producing region of Argentina RODENT VECTORS *Akodon arenicola, Calomys laucha, C musculinus* CLINICAL 1-2 wk incubation, followed by mucocutaneous hemorrhage, fever, anorexia, N&V, fluid loss and oliguria, hypotension, shock, severe myalgia, leukopenia, thrombocytopenia, transient hypocomplementemia MANAGEMENT Rehydration; high–15-45% mortality may ↓ to 1-4% with convalescent serum-specific Junin virus immune plasma

arginase deficiency METABOLIC DISEASE Congenital deficiency of arginase, which results in hyperargininemia and episodic hyperammonemia, leading to mental retardation and spasticity

arginine BIOCHEMISTRY A 'facultatively' essential amino acid that contains a guanido group with a pKa > 12, which carries a positive charge at physiological pH; it becomes an essential amino acid when the body is under stress or injured SOURCES Turkey, chicken and other meats. See Unproven methods for cancer management

arginine deficiency A condition caused by ↓ growth due to ↓ dietary arginine, which affects babies born with a phosphate synthetase deficiency MOLECULAR IMMUNOLOGY AD reduces early B-cell maturation

arginine test GH stimulation test PEDIATRIC ENDOCRINOLOGY A test that measures hGH secreted by the pituitary gland in response to arginine, which

is used to evaluate infants with growth retardation and adults with pituitary tumors ABNORMAL VALUES Failure to ↑ hGH may indicate hypopituitarism or dwarfism. See Growth hormone

arginine vasopressin ADH-antidiuretic hormone, see there

argument MEDTALK The reason(s) advanced for a particular thing's existence. See Drug-baby argument, Health freedom argument, Particular person argument

Argyll-Robertson pupil OPHTHALMOLOGY A pupil that reacts to accommodation but not to light, a finding typical of 3º syphilis

argyria TOXICOLOGY The deposition of silver salts in skin 2º to acute excess or chronic exposure to low levels CLINICAL-ACUTE Burning of mouth and throat, vomiting, diarrhea, collapse, coma, death CLINICAL-CHRONIC Permanent blue-bronze discoloration of skin and mucosa, especially of sun-exposed zones

ARI Acute respiratory infection, see there

Aricept® Donepezil NEUROLOGY An agent that transiently slows the mental decline in Alzheimer's disease. See Alzheimer's disease

Arilvax® VACCINOLOGY A live-attenuated yellow fever vaccine. See Yellow fever

Arimidex® Anastrozole ONCOLOGY A nonsteroidal 3ʳᵈ-generation selective aromatase inhibitor used to manage postmenopausal ♀ whose advanced breast CA is resistant to tamoxifen. See Breast cancer

Aripax Lorazepam, see there

arithmophobia PSYCHOLOGY Fear of numbers. See Phobia

Arizona MICROBIOLOGY A genus of gram-negative bacilli of *Salmonella* tribe, of the Enterobacteriaceae family; infection is uncommon, and usually an extension of colonization CLINICAL Usually GI–nausea, diarrhea

arm ONCOLOGY The 'side' on which a Pt in clinical trial is placed, which is usually either a treatment arm or a placebo arm, and assigned in a random fashion. See Control arm, Treatment arm, Q arm

armamentarium THERAPEUTICS A general term commandeered from the military for the battery of therapies used to battle a particular condition VOX POPULI Weaponry

ARMD Age-related macular degeneration, see there

Armed Forces Institute of Pathology A section of the US military which provides consultations, reference atlases and educational programs for pathologists

Armonil Diazepam, see there

Arnett v. Dal Cielo PEER REVIEW A lawsuit initiated by the D Arnett, director of the Medical Board of California against Alameda Hospital to obtain records from an anesthesiologist who admitted drug abuse, and after undergoing treatment, was to be monitored by the hospital for competence and to determine whether he represented a risk to Pts

Arnold-Chiari malformation NEUROLOGY A congenital compression deformity of the 'lower brain' characterized by herniation of the cerebellar tonsils and distal medulla oblongata through the foramen magnum into the spinal canal; ACM is accompanied by compression of the 4ᵗʰ ventricle and firm adhesions at the cisterna magna

AROM Active range of movement. See Range of motion

aromatic *adjective* Referring to a substance with a fragrant–usually due to the presence of volatile oils *noun* A general term for an herbal medicine with a fragrant odor, many of which are said to be mild stimulants. See Aromatics®, Aromatherapy, Herbal medicine

Aropax® Paroxetine, see there

arousable Capable of being aroused–from a stuporous state

arousal NEUROLOGY A state of behavioral or physiologic activation SEXOLOGY See Sexual arousal SLEEP DISORDERS An abrupt change from a 'deep' stage of NREM sleep to a 'lighter' stage, or from REMS to awake, with the possibility of awakening as the final outcome; arousal may be accompanied by ↑ tonic EMG activity and heart rate, as well as body movements; the interruption of sleep by continuous alpha activity and ↑ EMG activity of > 3 secs. See Sleep apnea

arousal disorder SLEEP DISORDERS Any parasomnia disorder attributed to an abnormal arousal mechanism with frequent and/or prolonged stress AROUSAL DISORDERS Sleepwalking, sleep terrors, confusional arousals RISK FACTORS Sleep apnea, heartburn, or periodic limb movement during sleep COMPLICATIONS Injury, violence, excess eating, disturbance of bedpartner or family MANAGEMENT Prescription drugs, behavior modification–eg, hypnosis or relaxation/mental imagery. See Adaptation response, Fight-or-flight response. Cf Relaxation response

ARP-1 Apolipoprotein, AI, apoAI regulatory protein-1 A member of the steroid receptor superfamily, which has various effects on lipid metabolism and cholesterol homeostasis; ARP-1 binds to DNA, down-regulating the apoAI gene, and in addition binds to the thyroid hormone-responsive element, and the regulatory regions of apoB, apoCIII, and insulin gene

ARPKD Autosomal recessive polycystic kidney disease

ARREST CARDIOLOGY A clinical trial–AngioRad™ Radiation for Restenosis–to evaluate the safety of US Surgical's AngioRad™ gamma system in ↓ restenosis in native vessels after angioplasty with/without stents. See Angiorad

arrest CARDIOLOGY *noun* Cardiac arrest, see there *verb* To stop, a term referring to the ceasing of all activity of an organ. See Hypothermic circulatory arrest

arrhenoblastoma Sertoli-Leydig cell tumor

arrhythmia CARDIAC PACING Any rhythm in the heart that falls outside the norm with respect to rate, regularity, and propagation sequence of depolarization wave. See Atrial arrhythmia, Cardiac arrhythmia, Proarrhythmia, Reentrant arrhythmia, Ventricular arrhythmia CARDIOLOGY Any variation in the normal rate or periodicity of heart beats

arrhythmia mapping CARDIOLOGY The use of hardware and software to ID the source of tachyarrhythmias; radiofrequency ablation is then directed to the specific arrhythmogenic site for potentially curative treatment. See Tachycardia

arrhythmogenic right ventricular dysplasia Right ventricular dysplasia, see there

arsenic intoxication Arsenic poisoning A toxic trace metal that is a key component of herbicides, insecticides, rodenticides, wood preservatives, and used in manufacturing glass, and paints; the usual fatal dose is 100–200 mg; there are ± 1900 arsenic intoxications/yr–US, 85% of which are accidental by children < age 6; the rest are adult suicides CLINICAL Vague GI–N&V and neurologic–apprehension and SOB symptoms, and a classic sign, 'garlic' breath, followed by dysphagia, tachycardia, severe abdominal pain and bloody diarrhea, then by renal and cardiac failure and circulatory collapse TREATMENT Dimercaprol–BAL. See Heavy metals

arsenic trioxide ONCOLOGY An anticancer drug that induces apoptosis in certain cancer cells

Arsitran Chlordiazepoxide, see there

ART 1. Antiretroviral therapy, see there 2. Assisted reproductive technology, see there

arterial anastomosis CARDIOLOGY Any end-to-end or end-to-side joining of one artery to another

arterial blood oximetry CRITICAL CARE A clinical test that quantifies carboxyhemoglobin–COHb and methemoglobin, and calculates the content of O_2 bound to Hb; ABO is indicated in Pts with anemia or polycythemia, for evaluating altered Hbs, O_2 content of the blood, and to correlate the findings of arterial blood gas analysis SPECIMEN Arterial blood; because the analyte, carbon monoxide, is unstable, the specimen must be sent to the lab ASAP on ice and analyzed STAT REF VALUE COHb < 1.5%; Heavy smokers ≥ 5%

arterial blood gas CRITICAL CARE Analysis of arterial blood for O_2, CO_2, bicarbonate content, and pH, which reflects the functional effectiveness of lung function and to monitor respiratory therapy REF RANGE pO_2, 75-100 mm Hg; pCO_2, 35-45 mm Hg; pH: 7.35-7.42, O_2 content: 15-23%; O_2 saturation, 94-100%; HCO_3^-, 22-26 mEq/L. See Metabolic acidosis, Metabolic alkalosis, Respiratory acidosis, Respiratory alkalosis

arterial catheterization CARDIOLOGY The insertion of an indwelling catheter in order to monitor a particular hemodynamic parameter in the circulation in 'real time'

ARTERIAL CATHETERIZATION–INDICATIONS

MONITOR Hemodynamically unstable Pts who may be hypo– or hypertensive, currently treated with vasopressor or vasodilator agents

SAMPLE BLOOD on a regular basis to assess blood gases–oxygen, carbon dioxide in a mechanically ventilated Pt

DETERMINE CARDIAC OUTPUT

arterial embolism CARDIOLOGY An abrupt interruption in arterial blood flow caused by a blot clot or atherosclerotic plaque that has migrated through the circulation, resulting in ischemia and necrosis distal to the occlusion; therapeutic AE can be performed therapeutically to block the flow of blood to a tumor RISK FACTORS A Fib, vascular injury, ↑ platelets CLINICAL Pain, coolness of extremity and absent/diminished pulse MANAGEMENT Thrombolytic therapy–eg tPA, streptokinase; surgery

arterial gas embolism Air embolism EMERGENCY MEDICINE A bolus of gas or air within the blood vessels, which may be caused by overinflation of the lungs ETIOLOGY Traumatic chest injury, 2° to distension and barotraumatic rupture of alveoli due to trapped gases in scuba divers, mechanical ventilation, placement of a central venous catheter, cardiothoracic surgery, hemodialysis, orogenital sex during pregnancy MANAGEMENT Hyperbaric O_2. See Decompression sickness, Hyperbaric oxygen therapy

arterial plethysmography A test that measures the systolic BP of a normal extremity compared to an upper extremity, detecting an arterial block in the lower extremities REF RANGE < 20 mm Hg difference in systolic BP between the upper and lower extremities ABNORMAL VALUE > 20 mm Hg occurs in arterial occlusive disease or arterial blockage. See Plethysmography

arterialization PHYSIOLOGY A mechanical and ventilatory process in which venous blood passes through the pulmonary circulation and, through the exchanges of gases, undergoes ↑ O_2 and ↓ CO_2

arteriogram Arteriograph RADIOLOGY The film image obtained by arteriography, see there

arteriography IMAGING The visualization of arteries by injecting a radiocontrast into an artery suspected of having significant stenosis or occlusion–eg, affected by ASHD which, if untreated, may lead to infarction of a limb or organ, long-term disability, death SITES PERFORMED Carotid, coronary, femoral arteries; each may require surgical or medical—ie non-surgical therapy, if the artery is severely occluded.

arteriolar narrowing Atherosclerosis, see there

arteriolar nephrosclerosis Arteriolonephrosclerosis NEPHROLOGY A destructive form of ASHD involving the renal blood vessels 2° to severe—malignant HTN which, if prolonged, leads to renal failure. See Atherosclerotic heart disease

arteriole ANATOMY A teeny weeny artery that connects arteries with capillaries

arteriolosclerosis CARDIOLOGY A condition characterized by sclerosis and thickening of the walls of arterioles

arteriopathy MEDTALK Any pathology arising in an artery. See Plexogenic pulmonary arteriopathy

arteriosclerosis ASHD, hardening of the arteries CARDIOLOGY ASHD's early effects are in the lower extremities, with subtotal occlusion and decreased ability to withstand exercise without frequent rest periods; atherosclerosis is a generic term for arterial 'hardening'—calcium deposition, sclerosis, and thickening by fibrous tissue with loss of elasticity forms of arteriosclerosis including atherosclerosis—in which there is lipid deposition, Mönckeberg sclerosis, arteriolosclerosis; it is a common disorder usually affecting > age 50 and refers to any of a group of diseases characterized by thickening and hardening of the artery wall and in the narrowing of its lumen RISK FACTORS Personal or family history of coronary artery or cerebrovascular disease, DM, HTN, kidney disease involving hemodialysis, smoking, or obesity. See Atherosclerosis, Hyaline arteriosclerosis

ARTERIOSCLEROSIS

ARTERIOLOSCLEROSIS

• Benign—associated with hyaline arteriolosclerosis

• Malignant—associated with myofibroblast hyperplasia, 'onion-skinning' of endothelial basement membrane and deposit of fibrinoid material in vascular wall

ATHEROSCLEROSIS Formed by cholesterol and cholesterol esters, covered by a fibrous plaque which, with time becomes calcified, ulcerated and causes thromboembolism in coronary artery disease—strokes, MIs, leg ischemia-especially in DM, ischemia of large intestine

MÖNCKEBERG SCLEROSIS Idiopathic and often asymptomatic annular calcified bands occurring in the muscular media of medium to small blood vessels of the extremities that have been fancifully likened to a goose's neck

arteriosclerotic heart disease See Atherosclerosis, Atherosclerotic heart disease

arteriovenous fistula SURGERY 1. Arteriovenous shunt, see there 2. A post-traumatic anastomosis between arteries and veins, which bypasses the capillary bed

arteriovenous malformation NEUROLOGY A potentially fatal congenital intracranial anomaly with large arteries feeding in a mass of communicating vessels which empty into large draining veins filled with 'arterialized' blood; large AVMs produce a shunt large enough to ↑ cardiac output LOCATION Brain, medulla, spinal cord CLINICAL Headaches, seizures, subarachnoid hemorrhage IMAGING CT, MRI, endovascular angiography ASAP

arteriovenous shunt Arteriovenous fistula THERAPEUTICS The surgical joining of an artery and a vein under the skin to create a hemodialysis access port COMPLICATIONS Large AVSs cause cardiac overload as arterial blood passes to the venous circulation without delivering nutrients and O$_2$ to tissues

arteritis Inflammation of an artery. See Granulomatous arteritis, Small vessel arteritis

arthralgia Pain in the joints

arthritis Inflammation of joints of which there are ±100 clinical forms ETIOLOGY Infectious, autoimmune, idiopathic, metabolic, traumatic CLINICAL Stiffness, warmth, swelling, redness, pain. See Acne arthritis, Ankylosing spondylitis, Bacterial arthritis, Degenerative arthritis, Familial histiocytic dermatoarthritis arthritis, Gonococcal arthritis, Gouty arthritis, Lupus erythematosus, Lyme arthritis, Osteoarthritis, Pseudogout, Rheumatoid arthritis, Sexually acquired reactive arthritis, Systemic-onset juvenile rheumatoid arthritis, Tuberculous arthritis, Viral arthritis

arthritis-dermatitis syndrome INFECTIOUS DISEASE An arthropathy caused by *N gonorrhoeae*, characterized by tenosynovitis with skin rash and fever, often with a positive blood culture; the arthralgia of go nococcal infection is most often an asymmetric polyarthralgia; ¹/₄ of Pts develop a monoarthralgia only; asymmetric joint involvement frequently affects the knee, elbow, wrist, metacarpophalangeal, ankle joints. See Gonococcal arthritis

arthritis diet Anti-arthritis diet NUTRITION A series of dietary guidelines promulgated by the Arthritis Foundation—US, which *may* ↓ a person's risk of arthritis; the recommendations include eating a variety of foods, maintaining an optimal weight, avoiding excess cholesterol, fat, sugar, sodium, ↓ alcohol consumption, and consumption of adequate starch and fiber. See Diet, Dong diet, No-nightshade diet; Cf 'Anti-cancer diet'

arthritis panel LAB MEDICINE A panel of lab tests—eg, uric acid, antinuclear antibodies, rheumatoid factor, sed rate used to evaluate possible causes of arthritis

arthrocentesis Closed joint aspiration, joint tap ORTHOPEDICS A procedure in which a needle is introduced into a joint and fluid removed for diagnostic or therapeutic purposes DIAGNOSTIC INDICATIONS Idiopathic joint effusions or arthritic complaints, suspected infectious arthritis, suspected gout, monitoring of antibiotic response in Pt with septic arthritis THERAPEUTIC INDICATIONS Decompress painful joint effusion, evacuate pus from septic joint, remove inflammatory infiltrate and crystals in Pt with gout, inject corticosteroids

arthrodesis Joint fusion ORTHOPEDICS The creation of a bony union across a joint, which can be spontaneous or surgical; best arthrodesis joints: ankle, knee, shoulder, hip INDICATIONS Differ according to the joint; ankle arthrodesis is performed for osteoarthritis, rheumatoid arthritis, and posttraumatic arthritis; arthrodesis of the knee is usually performed as a last resort, given the compromise in mobility that it causes—eg sitting on airplanes or in movies, for infections—eg, TB or neuropathic joint 2° to DM or syphilis. See Chamley arthrodesis

arthrography ORTHOPEDICS A test in which air or radiocontrast medium is injected into various joints—eg shoulder, wrist, knee, ankle to allow visualization of the articular space ABNORMAL IN Osteochondral disruption and fractures, osteochondritis dissecans, rheumatic diseases, synovial defects, tears of rotator cuff, joint capsule, ligaments, etc

arthropathy A general term for any disease of the joints. See Neuropathic arthropathy, Silicone arthropathy, Traumatic arthropathy

arthroplasty Surgical repair of a joint. See Implantation arthroplasty, Total hip replacement, Total knee replacement, WOMAC

arthroscope ORTHOPEDICS A thin fiberoptic endoscope with 3 channels which is introduced into a joint space via a small incision

arthroscopic knee surgery Arthroscopic knee repair ORTHOPEDICS A minimally invasive arthroplastic procedure, usually performed to repair one or more of the major knee ligaments—eg, anterior and posterior cruciate ligaments, lateral or medial meniscus, lateral collateral or medial collateral ligaments

arthroscopy ORTHOPEDICS The direct examination of a joint—eg, shoulder, wrist, knee, ankle, with an arthroscope INDICATIONS Diagnose and/or biopsy lesions of the meniscus, synovium, and extrasynovial tissues, and manage intraarticular lesions–eg, torn ligaments and cartilage ABNORMAL FINDINGS Baker's cyst, chondromalacia, chondromatosis, fractures, osteochondral disruption, fractures, osteochondritis dissecans, rheumatic disease, synovial defects, capsular or ligament tears COMPLICATIONS Nerve damage related to portal placement, postoperative joint infection, soft tissue or bone infections. See Arthroscope, Arthroplasty

arthrosis 1. A general term for any joint disease 2. Pseudoarthrosis, see there

Arthrotec® Diclofenac/misoprostol RHEUMATOLOGY An enteric-coated analgesic used in Pts with rheumatoid arthritis or osteoarthritis who are at ↑ risk for NSAID-induced ulcers ADVERSE EVENTS Diarrhea, dyspepsia, N&V

articular cartilage A hyaline cartilaginous covering over the opposing/articulating surface of the bones of synovial joints

articulate DENTISTRY The conforming of the upper to the lower teeth, especially when adjusting prostheses, bridgework, and crowns to the 'natural' apposing surface SPEECH To speak concisely

articulation NEUROLOGY 1. Speech 2. The ability to produce intelligible speech, through the appropriate interaction of the lips, tongue and palate

articulation deficiency 1. Articulation disorder, see there; speech defect 2. A general term for a mechanical disparity between 2 joints

articulation disorder AUDIOLOGY An inability to correctly produce speech sounds–phonemes because of imprecise placement, timing, pressure, speed, or flow of movement of the lips, tongue, or throat

articulatory technique High velocity-low amplitude technique OSTEOPATHY A technique in which joints are adjusted rapidly, often accompanied by a popping or snapping sound; AT results in a transient stretching of joint capsules, and is said to reset the position of the spinal cord and nerves, allowing the nervous system to function optimally and improve the body's biomechanical efficiency. See Osteopathy

artifact Artefact A structure not normally present, but produced by some external action; something artificial; the distortion of a substance or signal, which interferes with or obscures the interpretation of a study, or a structure that is not representative of a specimen's in vivo state, or which does not reflect the original sample, but rather the result of an isolation procedure, its handling or other factors; artifacts in electronic readout devices–eg, EEG, EKG, and EMG, may be due to loose leads or electrical contacts CARDIAC PACING An electrical impulse of noncardiac origin which is recorded as a vertical spike on an EKG or other ECG monitor–eg a pacemaker pulse; electrical signals from muscle contractions, or myopotentials, are called muscle artifacts IMAGING The artifact seen depends on the procedure–eg, barium enema, where zones of inconstant segmental contractions of the colon may be confused with organic constrictions or anatomic variations, due to mucosal or intramural tumors, or a metal surgical clip that obscures an anatomical structure. See Beam-hardening artefact, Edge artefact, Mosaic artefact, Ring artefact

artificial blood Artificial oxygen carrier TRANSFUSION MEDICINE A synthetic or semisynthetic substance used to transport O_2; usually a blood loss of 20-25% is well tolerated and crystalloids–eg, dextrose, are adequate to raise the blood volume to acceptable levels

artificial heart CARDIOVASCULAR SURGERY A biomechanical device intended to replace the heart's hemodynamic function. See Jarvik-7, Penn State heart, Ventricular assist device

artificial heart valve CARDIOVASCULAR SURGERY A synthetic-mechanical or natural–porcine–valve surgically placed into the heart to replace a defective or malfunctioning valve; the aortic and mitral valves are the most frequently replaced with AHVs

artificial insemination REPRODUCTIVE MEDICINE The instillation of sperm-bearing semen in the vagina, cervix, or endocervical canal to fertilize an egg; ± 75,000 ♀/yr–US undergo AI. See Assisted reproduction, Intracervical insemination, Intrauterine insemination, Intratubal insemination

artificial intelligence INFORMATICS The study of intelligence using ideas and methods of computation whose central goal is to understand the principles that make intelligence possible; a format of computer programming that attempts to simulate human 'intelligence.' See Bayesian network, Expert system, Machine learning, Neural networking, Symbolic reasoning

artificial liver BIOTECH A cartridge with cloned human liver cells, through which blood flows to facilitate removal of waste products. See Extracorporeal liver assist device, Liver dialysis

artificial skin CRITICAL CARE Any synthetic material designed to have the physicochemical properties of skin–eg, optimal 'wetting' and 'draping,' leading to adherence, ↓ bacterial invasion and fluid loss, eliciting cellular and vascular invasion, synthesizing a dermal matrix while biodegrading the artificial graft. See Burns, Dermagraft-TC, Integra Artificial Skin. Cf Split-thickness graft, Spray-on-skin

artificial tears A solution containing 0.5% carboxymethyl cellulose or 5% polyvinyl alcohol, used to treat dry eye–xerophthalmia, often associated with Sjögren syndrome, which may also be due to sarcoidosis, senile lacrimal gland atrophy, acute or chronic infectious dacryoadenitis–eg, gonococcal and trachoma or tumors–eg, lymphomas, pseudolymphoma, 1° or metastatic carcinoma

ARTISTIC CARDIOLOGY A clinical trial–AngioRad™ Radiation Therapy for In-Stent Restenosis Intracoronary Trials–to evaluate the AngioRad™ gamma system in stented Pts See Stent

artistic temperament PERFORMING ARTS MEDICINE A personality 'profile' well-described in writers, artists, and composers which, in the extreme case, borders on a mental illness

arylsulfatase A deficiency Metachromatic leukodystrophy, see there

arylsulfatase B deficiency Mucopolysaccharidosis VI/ Maroteaux-Lamy disease, see there

Arzepam Diazepam, see there

as needed prn. See prn order.

AS-PCR Allele-specific polymerase chain reaction, see there

ASA 1. Acetylsalicylic acid. See Aspirin 2. Adams-Stokes attack 3. Allele specific amplification. See PCR amplification of specific alleles 4. American Society of Anesthesiologists 5. Aminosalicylic acid 6. Anterior septal artery 7. Argininosuccinic acid 8. Arysulfatase A

Asacol® Mesalamine, see there

ASAP As soon as possible A generic term for a level of urgency less urgent than 'stat' CARDIOLOGY A clinical trial–Azimilide Supraventricular Arrhythmia Program–on the efficacy of daily azimilide/Stedicor® for managing A Fib and other supraventricular arrhythmias. See ALIVE

asbestos ENVIRONMENT Any finished natural product containing a type of incombustible mineral fiber; the US has used 30 billion tons of asbestos since 1900; it is a component of ± 3000 manufactured products; maximum exposure levels–1976 OSHA standards = 2 fibers/ccm^3/8 hr period

asbestosis PULMONARY MEDICINE A condition characterized primarily by pleural fibrosis 2° to chronic inhalation of asbestos fibers PATHOGENESIS Asbestos fibers are phagocytosed by alveolar macrophages but, given their insoluble nature, the phagocytic vesicle bursts, releasing enzymes, killing the would-be killer CLINICAL Asbestosis is graded by severity of peribronchial fibrosis; fibers may not be found in the plaques. See Asbestos, Mesothelioma. Cf Coal workers' pneumoconiosis

ASC Ambulatory surgery center, see there

ascariasis INFECTIOUS DISEASE Infection by a nematode, *Ascaris lumbricoides* EPIDEMIOLOGY Infection occurs after ingesting eggs in contaminated food or more commonly, carried to mouth by hands after contact with contaminated soil; after an early pulmonary phase–larval migration, worms stay in the intestine CLINICAL Asymptomatic; pneumonitis at time of transpulmonary migration; diarrhea, abdominal colic MANAGEMENT Mebendazole, pyrantel pamoate

ascending *adjective* Rising; upward

ascending tick paralysis Tick paralysis, see there

ASCENT INTERVENTIONAL CARDIOLOGY A clinical trial–ACS Stent Clinical Equivalence in de Novo lesions Trial

ascites GI DISEASE A pathologic accumulation of serous fluid in the peritoneal–abdominal cavity, common in decompensated–advanced–liver disease, that develops in 50% of those with cirrhosis; Pts with cirrhosis who develop ascites have a 50% 2-yr survival ETIOLOGY-HEPATIC Cirrhosis, alcoholic hepatitis, massive metastases to liver, fulminant hepatic failure, vascular compromise–cardiac failure, Budd-Chiari syndrome, portal vein thrombosis, veno-occlusive disease, fatty liver of pregnancy EXTRAHEPATIC ORIGIN Peritoneal carcinomatosis, peritoneal TB, biliary or pancreatic ascites, nephrotic syndrome, serosal inflammation CLINICAL Abdominal distension which, if extreme, causes SOB, portal HTN, water and Na$^+$ retention LAB Hypoalbuminemia, specific gravity < 1.010, protein content of ≤ 3% TREATMENT Paracentesis, ↓ Na$^+$ in diet, diuretics, liver transplant, peritoneal shunt, transjugular intrahepatic portosystemic shunt–TIPS, extracorporeal ultrafiltration and reinfusion. See Dialysis ascites, Malignant ascites

ASCUS Atypical squamous cell of undetermined significance CYTOLOGY A cell seen in a pap smear of the cervix that fulfills some of the criteria–eg nuclear enlargement and irregularity, cytoplasmic clearing, and thickened cell membranes, that define cells typical of either a condyloma or a neoplasm. See CIN, HSIL, LSIL. Cf AGUS, SIL

ASCVD Arteriosclerotic cardiovascular disease. See Atherosclerosis, Artherosclerotic heart disease

ASD Atrial septal defect, see there

aseptic *adjective* Freed of infection, microorganisms, sepsis; sterile. Cf Antiseptic

aseptic bursitis Noninfectious inflammation of a bursa

aseptic meningitis INFECTIOUS DISEASE Nonpurulent meningeal inflammation, which is more common in those < age 30 ETIOLOGY Viruses, especially Coxsackievirus and echovirus, circumscribed bacterial infections, hemorrhage, neoplasia–eg leukemia and lymphoma, chemical or mechanical irritants CLINICAL Headache, stiff neck or back, N&V, fever, photophobia, pain on eye movement; more common in late summer LAB CSF pleocytosis, ↑ lymphocytes MANAGEMENT Corticosteroids may be effective. Cf Purulent meningitis NEUROLOGY A progressive life-threatening form of meningovascular neurosyphilis which may occur in untreated syphilis CLINICAL Headaches, mental changes, ↓ nerve function–eg, vision, movement or sensation, vascular Sx, eg stroke. See Neurosyphilis, Syphilis

aseptic necrosis Avascular necrosis, osteonecrosis ORTHOPEDICS Death of bony tissue, usually due to ischemia. See Necrosis

asexual *adjective* Sexless; occurring without sex; unrelated to sex

ash leaf lesion DERMATOLOGY An oblong hypopigmented cutaneous macule seen by Wood's lamp in children with tuberous sclerosis; ≥ 3 lesions is diagnostic for TS. See Tuberous sclerosis. Cf Oak leaf configuration, Neurofibromatosis

ASHD Arteriosclerotic heart disease, see there

Asherman syndrome OBSTETRICS A condition, most common in the postpartum period which is characterized by intrauterine adhesions, which may result in amenorrhea TREATMENT D&C

Ashman phenomenon Ashman beats CARDIOLOGY An EKG finding in Pts with A Fib and aberrant beats due to shortened refractory periods; unlike PVCs, which AP mimics, specific management is unnecessary. See Atrial fibrillation

ASIA motor score American Spinal Injury Association motor score A clinical tool used to evaluate neuromuscular dysfunction in Pts with spinal cord injury

Asian or Pacific Islander MULTICULTURE A person with origins in any of the peoples of the Far East, Southeast Asia, Indian subcontinent, Pacific Islands–eg China, India, Japan, Korea, the Philippine Islands and Samoa

ASIST CARDIOLOGY A clinical trial–Atenolol Silent Ischemia Study that evaluated the effect of atenolol on M&M in Pts with CAD and/or silent myocardial ischemia. See Atenolol, Coronary artery disease, Silent ischemia

ASK CARDIOLOGY A clinical trial–African American Study of Kidney Disease and Hypertension Pilot Study–that studied the relationship between anti-hypertensives and progression of renal disease See Hypertension

asminyl® Dyphilline, see there

ASO 1. Administrative services organization, see there 2. Allele-specific–oligonucleotide hybridization 3. Anti-streptolysin O, see there

ASO titer Anti-streptolysin O titer, see there

ASP 1. Alternative site provider, see there 2. Application service provider, see there

L-asparaginase ONCOLOGY A chemotherapeutic used for lymphomas and lymphocytic leukemia ADVERSE EFFECTS Vomiting, hepatic dysfunction, allergic reaction, lethargy

asparagine BIOCHEMISTRY A nonessential amino acid which is the β-amide of aspartic acid–AA; asparaginine assists in the neural metabolism, and when the extra amino group is removed, the resulting AA acts as an excitatory transmitter and allows it to be used interchangeably with AAs in protein building

aspartame Nutrasweet® An artificial sweetener/ester of aspartic acid and phenylalanine; it may be safer than saccharin except in Pts with phenylketonuria ADVERSE REACTIONS Rare, with large amounts–mild depression, headaches, insomnia, loss of motor control, nausea, seizures, etc, and possibly brain cancer. See Artificial sweeteners. Cf Aspartate, Cyclamate

aspartate aminotransferase GOT, glutamate oxaloacetate transaminase AST A cytoplasmic and mitochondrial transaminase enzyme that catalyzes the reaction of aspartate and 2-oxoglutarate yielding glutamate and oxaloacetate; the transport of amino acids is central to protein buildup-anabolism or breakdown-catabolism; AST is ↑ in hepatic, myocardial, renal and cerebral infarction, and in hepatic and skeletal muscle disease

aspartate BIOCHEMISTRY A nonessential amino acid that has a central role in transferring amino groups by aspartate aminotransferase in the liver; in proteins, aspartate takes the form of its amide, asparagine. See Aspartame, Aspartate aminotransferase

ASPECT CARDIOLOGY A Dutch clinical trial–Anticoagulation in the Secondary Prevention of Events in Coronary Thrombosis which evaluated the efficacy of an oral anticoagulant-coumadin in preventing recurrent MIs, and determined the dose needed to ↓ risk of reinfarction and hemorrhage in elderly post-MI Pts. See Myocardial infarction, Warfarin

Asperger disorder PEDIATRIC PSYCHIATRY A neurobiological condition characterized by autistic-like behaviors with severe defects in social and communication skills, which may be coupled with high intelligence and hyperfocusing on one particular area of interest EPIDEMIOLOGY More common in boys, especially, "nerds". See Autism, Pervasive personality disorder, Sequential thinking

aspergillosis INFECTIOUS DISEASE Infection by *Aspergillus* spp, which evokes granulomatous lesions of lungs, auditory canal, skin or ocular, nasal, or urethral mucosa EPIDEMIOLOGY Nosocomial aspergillosis is linked to hospital construction and contaminated ventilation systems HIGH-RISK DISEASES Asthma–pulmonary aspergillosis, AIDS CLINICAL Cough, haemoptysis, weight loss, wheezing, fever, chills, hematuria, ↓ urine output MANAGEMENT Amphotericin B alone or with flucytosine, surgical excision of invasive aspergillosis of the brain, paranasal sinuses, and noninvasive sinus colonization PROGNOSIS *Aspergillus* endocarditis has a poor prognosis; otherwise conservative management may be adequate. See Allergic bronchopulmonary aspergillosis

Aspergillus MYCOLOGY A genus of fungi of the class Deuteromycetes, which are soil dwellers, with ± 300 species, some of which are human pathogens. See *Aspergillus fumigatus*

Aspergillus fumigatus MICROBIOLOGY The fungal species that is the most common cause of human aspergillosis, which may infect the lungs, invade blood vessels, or disseminate to various organs. See Aspergillosis

aspermatogenesis UROLOGY The lack of production of sperm

asphyxia PHYSIOLOGY 1. Impaired breathing 2. A pathological state caused by the inadequate intake of O_2, with accumulation of CO_2 and hypoxia. See Autoerotic asphyxia, Sexual asphyxia

asphyxiant A gas or vapor that compromises availability of O_2 for breathing, by either displacing O_2–eg, CO_2 or by replacing O_2–eg, carbon monoxide

aspirate To suck in CHEST MEDICINE *noun* Fluid withdrawn from a cyst *verb* To inhale foreign fluid or semi-fluid material, in particular gastric content into the upper respiratory tract, resulting in aspiration pneumonia CLINICAL MEDICINE The drawing of a fluid under negative pressure from a joint–eg, to ↓ pressure LAB MEDICINE The drawing of a fluid under negative pressure into a receptacle for storage—eg syringe, or for analysis— as in a laboratory instrument that aspirates material being tested on a batch analyzer

aspiration 1. Aspiration 2. The withdrawal of fluid from a body cavity or a mass–eg a cyst with a needle and a syringe by suction or siphonage, such as a syringe. See Gastric aspiration, Paracentesis

aspiration biopsy Fine-needle aspiration biopsy, see there

aspiration cytology CYTOLOGY A term referring to diagnostic cytologic material from internal organs–eg, breast, liver, lymph node, prostate, salivary gland, thyroid, and other relatively inaccessible sites, obtained by fine needle aspiration. Cf Exfoliative cytology

aspiration pneumonia Aspiration pneumonitis PULMONOLOGY A condition characterized by the inhalation or inappropriate passage of highly acidic gastric content–food, gastric acid, vomitus-into the respiratory tract, a clinical event most common in the comatose; after the insult, there is progressive respiratory depression, hypoxia, tachypnea, and tachycardia; the tracheobronchial tree 'sweats' thin frothy fluid; lung parenchyma is acutely inflamed, hemorrhagic and edematous with atelectasis and necrosis MORTALITY Up to 70%. See Gastric aspiration, Pneumonia

aspirator DENTISTRY A negative suction device used clear saliva

aspirin Acetylsalicylate, acetylsalicylic acid THERAPEUTICS A widely used analgesic, antipyretic, antiinflammatory, and antiplatelet agent; it is one of the safest drugs currently prescribed ADVERSE EFFECTS GI upset, occult bleeding; high doses inhibit cyclooxygenase and can produce tinnitus, tachycardia, dizziness, deep rapid breathing, hallucinations, convulsions, coma USED FOR Pain, fever, rheumatic complaints; acute rheumatic fever; TIAs, acute MI. See Controlled-release aspirin, NSAID, Superaspirin

ASR Age/sex rate, see there

assassin bug Any of the cone-nosed arthropods of the hemipteran family Reduviidae, order Hymenoptera, which includes true bugs, the trivial name refers to their insect predatory activity; those of the subfamily Triamtominae are vectors for *Trypanosoma cruzi*–Chagas' disease agent

assault FORENSIC MEDICINE The unlawful placing of an individual in apprehension of immediate bodily harm without his/her consent. See Sexual assault, Nonsexual genital assault

assault & battery FORENSIC MEDICINE ASSAULT The unlawful placing of an individual in apprehension of immediate bodily harm without his/her consent BATTERY The unlawful touching of another individual without his consent. See Malpractice, Negligence

assault/interpersonal violence A social act involving a serious abuse of power, consisting of the exertion of physical force and power over another individual with the intent of controlling, disempowering and/or injuring that individual. See Domestic violence

assay LAB MEDICINE The quantification of a substance of interest by a specific chemical, enzymatic, immunologic or radioimmune-mediated reaction

ASSENT I CARDIOLOGY A clinical trial–Assessment of the Safety of a New Thrombolytic, which evaluated the rates of intracranial hemorrhage–bleeding strokes and mortality in Pts treated with TNK-tPA. See TNK-tPA

assent MEDICAL ETHICS Agreement by a minor or other person not competent to give legally valid informed consent–eg, a child or cognitively impaired person–to participate in research. Cf Informed consent

assertive behavior PSYCHIATRY Bold and/or insistent communication of suggestions or actions to others. See Assertiveness training, Passive behavior. Cf Passive-aggressive behavior

assertiveness training PSYCHIATRY A procedure in which subjects are taught appropriate interpersonal responses involving frank, honest, and direct expression of their feelings, both positive and negative

assessment MEDTALK Evaluation. See Exposure assessment, Functional assessment, Nutritional assessment, Probabalistic safety assessment, Process assessment, Quality assessment, Quantitative risk assessment, Risk assessment

ASSET CARDIOLOGY A clinical trial–Anglo-Scandinavian Study of Early Thrombolysis that compared the mortality in acute MI Pts randomized to tPA or placebo. See Acute myocardial infarction, tPA. Cf GISSI, TIMI-2

assignment of benefits HEALTH INSURANCE A method where the person receiving the medical benefits assigns the payment of those benefits to a physician or hospital. See Dual assignment, Mandatory assignment, Participation, Random assignment

assimilation PSYCHIATRY A person's comprehension and integration of new experiences

assisted mechanical ventilation Mechanical ventilation, see there

assisted reproduction Artificial reproduction REPRODUCTIVE MEDICINE Noncoital and/or non-natural manipulation of reproductive processes such that one–or rarely, both of the child's genetic parents are not the rearing parent(s) TYPES In vitro fertilization, gamete intrafallopian transfer–GIFT, zygote intrafallopian transfer, embryo cryopreservation, egg or embryo donation, surrogate birth. See Baby M, in vitro fertilization, Surrogate motherhood

assisted suicide 1. Physician-assisted suicide, see there. See Double effect 2. Any suicide with a 2ⁿᵈ party participant

assistive device PUBLIC HEALTH Any device designed or adapted to help people with physical or emotional disorders to perform actions, tasks, and activities. See Americans with Disabilities Act, Architectural barriers, Assistive technology

assistive technology Enabling technology DISABILITIES A technology designed to improve the quality of life a person with disabilities and function in the most optimal possible fashion. See American with Disabilities Act

Assival Diazepam, see there

association *adjective* NEUROLOGY Referring to an area of the brain that links motor and sensory cortical areas. See Visual association *noun* EPIDEMIOLOGY A statistical relationship between two or more events, characteristics, or other variables–eg, an association between an exposure to X and a health effect, Y, which does not necessarily imply cause and effect PSYCHIATRY 1. Any connection between the conscious and the unconscious 2. A relationship between ideas and emotions by contiguity, continuity, or similarity. See Clang association, Direct association, Free association, Loosening of associations, Strength of association VOX POPULI An organized group of similarly-minded individuals. See AALAS, American Association of Acupuncture & Oriental Medicine, American Association of Naturopathic Physicians, American Osteopathic Association, Blue Cross & Blue Shield Association, Independent practice association, International Association of Cancer Victims & Friends, National Rifle Association, World Medical Association

association neuron Interneuron, internuncial neuron A nerve cell confined to the CNS, which conveys impulses in an arc from sensory to motor neurons

associative conditioning Pavlovian conditioning, see there

assumption-of-risk doctrine *Volenti non fit injuria* MEDICAL MALPRACTICE A doctrine that states that an individual who knowingly exposes him/herself to hazards with potential for bodily harm, cannot hold others liable if harm occurs; under the AORD, a person who consents to a medical procedure–or alternatively, decides to forego a physician-recommended therapy–with knowledge that injury is a foreseeable—albeit uncommon—result, waives the right to a future complaint that any 'foreseeable' injury was caused by the physician's negligence, assuming medical treatment was performed with proper care, and *res ipsa loquitur* cannot be evoked. See 'Blood shield' statutes, Informed consent, *Res ipsa loquitur*. Cf Contributory negligence

assurance VOX POPULI A commitment to do a thing. See Cooperative oncology group assurance, General assurance, Multiple project assurance, Quality assurance, Single project assurance

AST 1. Aspartate aminotransferase, see there 2. Assisted reproductive technology, see there

astemizole Histamal® THERAPEUTICS A non-sedating H₁-receptor antagonist-antihistamine used for seasonal rhinitis–hay fever, hives ADVERSE EFFECTS Minimal sedation, anticholinergic effects

astereognosis NEUROLOGY Inability to recognize familiar objects by touch that cannot be explained by a defect of elementary tactile sensation

asterixis Flapping tremor NEUROLOGY An exaggerated muscle tremor characterized by involuntary jerking of the hand in flexion and extension, classically seen in advanced liver disease and hepatic encephalopathy

asthenia NEUROLOGY Loss of strength and energy, weakness

asthma INTERNAL MEDICINE per the American Thoracic Society, 1987–*Asthma is a* [chronic]*clinical syndrome characterized by increased responsiveness of the tracheobronchial tree to a variety of stimuli. The major symptoms of asthma are paroxysms of dyspnea* [particularly exhaling air, accompanied by chest tightness], *wheezing, and cough, which may vary from mild and almost undetectable to severe and unremitting (status asthmaticus). The primary physiological characteristic of this hyperresponsiveness is variable airways obstruction. This can take the form of spontaneous fluctuations in the severity of obstruction, substantial improvements in the severity of obstruction following bronchodilators or corticosteroids, or increased obstruction caused by drugs or other stimuli. Histologically, patients with fatal asthma have evidence of mucosal edema of the bronchi, infiltration of the bronchial mucosa or submucosa with inflammatory cells, especially eosinophils, and shedding of epithelium and obstruction of peripheral airways with mucus* EXACERBATING FACTORS Rapid changes in temperature or humidity, allergies, URIs, exercise, stress or cigarette smoke MORTALITY 18.8/million blacks; 3.7/million whites–US, age 15-24 CLINICAL Wheezing, tachypnea, tachycardia, bronchiolitis, prolonged expiration, inter- & subcostal retraction, nasal flaring DIFFDX Aspiration, bronchitis, bronchopulmonary dysplasia, cystic fibrosis, GERD, vascular rings, pneumonia WORKUP ABGs–hypoxia, respiratory acidosis; CXR–hyperinflation; PFTs— ↓ vital capacity, ↑ functional residual capacity, ↑ residual volume SPIROMETRY–↓ FEV₁ LAB Eosinophilia, ↑ Hct if dehydrated PATHOLOGY Bronchial and bronchiolar occlusion by plugs of thick, tenacious mucus, accompanied by Curschman spirals, Charcot-Leyden crystals, thickening of bronchial epithelium, edema and inflammation with abundant eosinophils, ↑ size of submucosal glands, hypertrophy of bronchial wall muscle MANAGEMENT Bronchodilators–in particular β₂-adrenoreceptor agonists, corticosteroids; anti-T-cell agents, phosphodiesterase inhibitor, K⁺-channel activators, thromboxane antagonists See Allergic asthma, Cardiac asthma, Exercise-induced asthma, Occupational asthma, Status asthmaticus, Yokohama asthma. Cf COPD

ASTHMA, GUIDELINES FOR TREATMENT/MANAGEMENT

SYMPTOMS–Sx	FEV₁/PEFR variability	MANAGEMENT STRATEGY
MILD INTERMITTENT ASTHMA		
Symptoms ≤ 2 X/wk	> 80%/< 20%	Inhaled β₂ agonists PRN
Nocturnal Sx < 2 X/month		
MILD PERSISTENT ASTHMA		
Symptoms > 2 X/wk; < 1 X/day	> 80%/20–30%	Inhaled corticosteroids (200–800 μg/day) or
Nocturnal Sx > 2 X/month		cromolyn or nedocromil; zafikulast or zileuton
		if Pt > age 12; long-acting β₂ agonist, sustained
		release theophylline, especially for nocturnal Sx
		Inhaled β₂ agonists PRN
MODERATE PERSISTENT ASTHMA		
Symptoms daily	60–80%/>30%	Inhaled corticosteroids (800–2000 μg/day)
Nocturnal Sx > 1 X/wk		Long-acting β₂ agonist or sustained release
Daily β₂ agonists required		theophylline, especially for nocturnal Sx
		Inhaled β₂ agonists PRN
SEVERE PERSISTENT ASTHMA		
Continuous Sx	<60%/>30%	Inhaled corticosteroids (800–2000 μg/day)
Frequent nocturnal Sx		Long-acting β₂ agonist or sustained release
Frequent exacerbations		theophylline, especially for nocturnal Sx
Limitation of physical activity		Inhaled β₂ agonists PRN, oral corticosteroids

FEV₁ Forced expiratory volume in 1 second; PEFR: Peak expiratory flow rate (% of predicted)

PEFR variability, defined as evening PEFR—morning PEFR X 100/50% (evening PEFR + morining PEFR)

From the 1997 Guidelines for the Management of Asthma (National Asthma Education Program)

asthma triad A triad of Sx associated with asthma: bronchospasm, nasal polyps, ↑ sensitivity to aspirin–acetylsalicylic acid, but not sodium salicylate, *and* other exogenous agents–eg, indomethacin, aminopyrine, yellow food additives–eg, tartrazine yellow, FD&C yellow #5. See Asthma

astigmatism OPHTHALMOLOGY A visual disturbance due to irregularity in curvature of the refractive surfaces, which produces a blurred image on the retina and multiple focal points or lines

astrapophobia PSYCHOLOGY Fear of thunder and lightning. See Phobia

ASTRID CARDIOLOGY A clinical trial–Atrial Sensing Trial to prevent Inappropriate Detections

astringent *adjective* Causing local contraction after topical application *noun* PHARMACOLOGY A topical agent–eg, aluminum-based compounds, used to precipitate proteins, as topical hemostatics, to ↓ mucosal inflammation, toughen skin, promote healing, as antiseptics, and as an antiperspirant

astronaut diet SPACE MEDICINE A fiber-free food fare for ↓ stool bulk, intended to minimize or eliminate the problems inherent in solid waste management systems in zero gravity. See Diet. Cf Airline food

asymmetry Lacking symmetry NEUROLOGY A proportional difference/discordance between the right/left hemispheres, especially vis-`a-vis function

asymptomatic *adjective* Lacking obvious signs or symptoms of disease–usually understood to mean–despite the presence of a pathologic condition

asymptomatic bacteriuria UROLOGY The presence of bacteria in the urine at a level indicating infection without Sx; in ♀ w/ DM, antimicrobial therapy does not ↓ complications. See Cystitis, Too numerous to count

asymptomatic hematuria NEPHROLOGY Hematuria not associated with pain or other Sx. See Hematuria

asymptomatic HIV infection AIDS A state in which HIV is present in a person without signs of clinical disease; AIDS may follow infection by HIV by up to 10 yrs. See AIDS, HIV

asymptomatic proteinuria NEPHROLOGY Proteinuria not associated with pain or other Sx. See Proteinuria

asynchronous pacemaker CARDIAC PACING A pacemaker which fires at a fixed, preset rate, independent of the heart's intrinsic electrical and/or mechanical activity

asystole Cardiac standstill; cardiac arrest; heartbeatlessness; cor sans systole

AT/NC Atraumatic, normocephalic

AT LAST AIDS A clinical trial–Antiretroviral Trial Looking at Sex & Treatment. See AIDS

'at-risk' pregnancy OBSTETRICS A pregnancy at risk for spontaneous abortion–an event occurring in 20-60% of all pregnancies. See Abortion

ATAC Arimidex, tamoxifen and combination therapy

Atacand® Candesartan cilexitil CARDIOLOGY An angiotensin II receptor blocker used alone or with other agents for HTN. See ABC, CANDLE

atavism Any of a number of normally dormant traits–eg, the presence in humans of multiple nipples, appearance of vestigial hind limbs in whales, or possibly hereditary hypertrichosis in humans

ataxia NEUROLOGY Muscle incoordination and gait unsteadiness due to cerebellar dysfunction, and compromise in regulating limb movement. See Cerebellar gait, Friedreich's ataxia, Hereditary cerebellar ataxia, Spinocerebellar ataxia VOX POPULI Wobbling

ataxia-telangiectasia Louis-Bar syndrome NEUROLOGY An AR condition associated with sinopulmonary infections, choreoathetosis—wobbly gait, slurring of speech, muscular atrophy, 'red eyes' LAB ↓ IgG4 and IgA2, ±↓ IgE, ↑ AFP CLINICAL Progressive cerebellar ataxia, oculocutaneous telangiectasia, thymic aplasia or hypoplasia–cellular defect, immune deficiency, autoimmune phenomena, ↑ susceptibility to radiation-induced chromosomal damage–breakage and rearrangement, and ↑ in lymphoid, breast, and other CA, and rearrangements due to defective DNA repair; immune complex deposits in glomeruli, choroid plexus, heart valves, synovium; ↑ in vasoactive amines–histamine, serotonin, IgE, platelets; presence of macrophages; antigen and antibody valences DIAGNOSIS Clinical findings, ↑ AFP, cell culture assay for radiosensitivity and atypical radioresistant DNA synthesis; cell culture methods can be used for prenatal diagnosis. See Phakomatosis

ataxic gait NEUROLOGY Awkward, uncoordinated ambulation/walking. See Gait

ATCC American Type Culture Collection, see there

Ateben Nortryptyline, see there

atelectasis Partial lung collapse CHEST MEDICINE The partial or total collapse of lung parenchyma, due to obstruction of a bronchus by a mucus plug, infection or CA; it is common post general anesthesia CLINICAL Low-grade fever, dry cough, chest pain, SOB MANAGEMENT Mild atelectasis is treated with deep breathing exercises and respiratory therapy; atelectasis 2° to CA may be managed by therapeutic bronchoscopy. See Plate atelectasis

atenolol Tenormin® CARDIOLOGY A β-blocker used to treat HTN, angina. See β-blocker, Myocardial infarction

Atensine Diazepam, see there

ATF MOLECULAR VIROLOGY Activating transcription factor A cellular protein that stimulates transcription of adenovirus E4 transcription unit, which acts early in infection at any of several 'enhancer' binding sites

ATG 1. Angiotensinogen 2. Antithymocyte globulin, see there 3. Antithyroglobulin–antibodies

Atgam® Antithymocyte globulin, see there

athelia Nipplelessness

atherectomy The removal of potentially occlusive atheromatous plaques from the lumen of a major artery–eg, coronary, carotid, or arteries of the limbs, by coronary balloon angioplasty, lasers, drill-tipped catheter, or other methods. See Directional coronary atherectomy

atherogenic *adjective* Referring to the ability to initiate or accelerate atherogenesis—the deposition of atheromas, lipids, and calcium in the arterial lumen

atherosclerosis Hardening of the arteries HEART DISEASE A common type of arteriosclerosis found in medium and larger arteries in which raised areas in the tunica intima are formed from smooth muscle cells, cholesterol, and other lipids; the progressive narrowing and hardening of arteries due to intramural deposition of LDL, and calcium, 2° to exposure of smooth muscle to lipid, resulting in platelet-induced smooth muscle proliferation, and ↑ risk of future stroke and MI CLINICAL Atherosclerosis is symptomatic when the narrowing of an artery ↓ the blood flow to a particular tissue enough to cause pain in the tissues supplied by that artery DIAGNOSIS Clinically significant atherosclerosis is identified by angiography and Doppler ultrasonography studies FACTORS HTN 'HARD' RISK FACTORS HTN > 160/95 mm Hg, ↑ LDL-C–total cholesterol > 265 mg/dL, smoking > 1 pack/day, DM 'SOFT' RISK FACTORS ♂, family Hx of ASHD, obesity, ↑ apoB, ↑ apoC-III, ↑ total cholesterol, ↑ TGs, ↓ HDL-C, ↑↑↑ homocysteine, a highly reactive amino acid, which is toxic to vascular endothelium and may potentiate the autooxidation of LDL-C, promoting thrombosis COMPLICATIONS IN LESIONS Aneurysms–dissecting and fusiform of arterial wall subjacent to atheroma, bleeding into plaque, calcification, thrombosis MEDICAL MANAGEMENT Lifestyle changes–eg, exercise, vegetarian diet, fish better than meat and eggs, 1-2 alcoholic drinks/day–red wine may be best, possibly biofeedback modalities–eg, yoga, Cholesterol-lowering drugs—statins SURGICAL MANAGEMENT Balloon angioplasty–PCTA, plaque scraping or 'grinding'—largely abandoned due to poor outcomes, stenting, CABG. See Atherosclerotic plaque

atherosclerotic heart disease CARDIOLOGY A general term for the progressive narrowing and hardening of coronary arteries, due to atheroma deposition which, with time undergo calcification and ulceration RISK OF PROGRESSION ↑ Cholesterol, HTN, smoking, DM, family Hx of ASHD. See Atherosclerosis, Atherosclerotic plaque

atherosclerotic plaque CARDIOLOGY The core lesion of atherosclerosis; it begins as a fatty streak, an ill-defined yellow lesion–fatty plaque, which develops well-demarcated edges that evolve to fibrous plaques, whitish lesions with a grumous lipid-rich core which, with time, becomes a complicated plaque composed of WBCs, smooth muscle cells and extracellular matrix in large artery intimas. See Complicated plaque

atherosclerotic renal disease Arteriolar nephrosclerosis, see there

atherosclerotic vascular disease Atherosclerosis, see there

athetosis Mobile spasm NEUROLOGY Constant, slow involuntary writhing movements, most severe in hands; A CNS disorder characterized by continual uncoordinated movements of the limbs

athetotic gait Dyskinetic gait PHYSICAL EXAM Ambulation typical of cerebral palsy–CP in which a person with CP attempts to compensate for impaired voluntary muscle control through bizarre twisting motions and exaggerated posturing to bridge the fragmented motor control. See Gait

athlete SPORTS MEDICINE A person who participates in an interscholastic, intercollegiate, or intramural athletic activity being conducted by an educational institution, or a professional athletic activity. See Athletic trainer, Athletic training

athlete drug testing SPORTS MEDICINE Testing of athletes for the presence of substances that enhance or stimulate performance, or substances of abuse. See Anabolic-androgenic steroids, Blood doping, Gender verification, 'Roid rage

athlete's foot Tinea pedis A malodorous dermatophytosis affecting moist, warm toe webs and soles of feet; AF often occurs in athletes, especially in adolescent ♂, and causes maceration, flaking, peeling, erosion, pruritus RISK FACTORS Poor hygiene, occlusive footwear, prolonged moisture of skin, minor skin or nail injuries AGENTS *Trichophyton rubrum, T mentagrophytes, Epidermophyton floccosum* TREATMENT Drying, if recalcitrant, haloprogin, tolnaftate; if refractory, griseofulvin. See Jock itch, Tinea pedis, Tinea cruris

athlete's heart Athletic heart syndrome SPORTS MEDICINE A heart typical of trained athletes, and characterized by ↑ left ventricular diastolic volume and ↑ thickness of the left ventricular wall, as seen by 2-D echocardiography; arrhythmias seen in athletes' hearts are usually benign and include sinus bradycardia, wandering pacemaker, cardiac blocks, nodal rhythm, atrial fibrillation, ST segment and T-wave changes, ↑ P wave amplitude, right ventricular hypertrophy. See Sudden unexplained nocturnal death

athletic training SPORTS MEDICINE The practice of physical conditioning and reconditioning of athletes and prevention of injuries incurred by athletes. See Athlete, Athletic trainer

Ativan® Lorazepam, see there

Atkins' diet POPULAR NUTRITION A carbohydrate-poor, fat-rich 'fad' diet developed by Dr Robert Atkins in which 73% of the caloric content is fat; the basis of the diet is the deliberate induction of ketosis, in which stored fat is burned for energy. See Fad diet, Diet

ATLANTIC CARDIOLOGY A clinical trial–Angina Treatments–Lasers And Normal Therapies In Comparison

atlantoaxial subluxation ORTHOPEDICS An upper cervical spine subluxation which is either rotatory or anterior; rotatory AAS is more common in children, often due to apparently trivial injury; torticollis may be seen IMAGING On an open mouth film, the odontoid is not equidistant between lateral masses of C1 MANAGEMENT Reduce subluxation and brace; recurrent or chronic cases can be treated with C1-C2 fusion; anterior AAS is due to traumatic disruption of transverse ligament IMAGING Abnormal atlanto-dental interval; if anterior AAS is suspected, a CT rules out fracture; an MRI is used to evaluate the transverse ligament which, if disrupted, requires fusion; a cervical brace suffices for fracture without disruption. See Atlantodental interval

atlantodental interval NEURORADIOLOGY The distance between the dens and anterior ring of C1, which is ≤ to 3 mm in adults, and ≤ 5 mm in children, which is altered in Pts with atlantoaxial subluxation. See Atlantoaxial subluxation

ATLAS CARDIOLOGY A five-yr, randomized, double-blind trial–Assessment of Treatment with Lisinopril & Survival–which evaluated the effect of high vs low doses of Zestril®/lisinopril on M&M in Pts with CHF; high doses appear to be effective. See Lisinopril SPORTS MEDICINE A 5-yr controlled trial–Athletes' Training and Learning to Avoid Anabolic Steroids begun in 1993 and funded by the National Institute of Drug Abuse. See Anabolic steroids

atlas fracture C1 fracture, see there

AT LAST AIDS A clinical trial–Antiretroviral Trial Looking at Sex & Treatment. See AIDS, Safe sex

atmospheric inner ear barotrauma ENT Injury to the inner ear which is often permanent due to inner ear damage, most common in divers CLINICAL Sensorineural hearing loss and tinnitus, less commonly, vertigo, nystagmus and ataxia; divers usually report Sx during descent, with unilateral difficulty clearing negative pressure MANAGEMENT Conservative–eg bed rest, head elevation, monitoring of auditory and vestibular Sx; surgical–eg tympanotomy and middle ear exploration–reserved for progressive hearing loss or persistent vestibular Sx

ATN Acute tubular necrosis, see there

atonia Atony NEUROLOGY Lack of normal tone or strength

atopic dermatitis Atopic eczema, infantile eczema, neurodermatitis IMMUNOLOGY An immune-mediated dermatopathy affecting 1-3% of children, which is characterized by severe pruritus of early–usually in infancy onset, and a familial tendency; it may be associated with IgE-mediated skin reactions, allergic rhinitis and/or asthma ETIOLOGY Idiopathic in children; in adults, hypersensitivity to chemicals–eg detergents or soaps, metals–eg nickel or plants–eg poison ivy, poison oak CLINICAL In infancy, AD tends to be a weeping, papulovesicular—which rupture and ooze, and intensely pruritic inflammation of cheeks and inguinal region; in later childhood, it is more lichenified and is more prevalent over antecubital, popliteal, and collar regions EXACERBATING FACTORS Anxiety, stress, depression LAB Eosinophilia MANAGEMENT Avoid known allergens; reduce skin dryness scratching, and inflammation; antihistamines SIDE EFFECT Burning sensation; the ↑ susceptibility of Pts with AD to infection may be due to a ↓ in expression of antimicrobial peptides See Endogenous antimicrobial peptide, Tacrolimus

atopy CLINICAL IMMUNOLOGY A state of ↑ sensitivity to common antigens–eg, house dust, animal dander, pollen, with ↑ production of allergen-specific IgE; atopy may have a hereditary component as there is ↑ susceptibility to hay fever, asthma, and eczematoid dermatitis. See Asthma, Atopic dermatitis, Hay fever

atorvastatin Lipitor® CARDIOLOGY A cholesterol-lowering HMG-CoA reductase inhibitor or statin which improves the lipid profile—↓ 35% to 60% LDL-C; ↓TGs 20–37%;↑ 5–9% HDL-C ADVERSE EFFECTS Liver dysfunction, rhabdomyolysis with acute renal failure 2° to myoglobinuria, constipation, flatulence, dyspepsia, abdominal pain, myalgia, weakness. See AVERT, Statin

atovaquone Mepron INFECTIOUS DISEASE An agent used for treating PCP in Pts with HIV infection and for toxoplasmosis and microsporidiosis PROS Less toxic, more tolerable than T-S CONS Less effective than T-S ADVERSE EFFECTS Headache, nausea, diarrhea, fever and rashes. See AIDS, *Pneumocystis carinii* pneumonia, Trimethoprim–Sulfamethoxazole

ATPase Adenosine triphosphatase, ATP synthase An enzyme that catalyzes the conversion of phosphate + ADP ↔ ATP during oxidative phosphorylation in mitochondria See Na⁺/K⁺ ATPase

ATR Achilles tendon reflex, see Ankle reflex

atracurium besylate ANESTHESIOLOGY An injectable intermediate-acting nondepolarizing muscle relaxing anesthetic used with general anesthetics in Pts undergoing surgery or mechanical ventilation where skeletal muscle relaxation is required. See Cisatracurium, Muscle relaxant, Nondepolarizing agent. Cf Depolarizing agent

Atragen™ ONCOLOGY A lipid-based IV formulation of all-trans retinoic acid–ATRA for acute promyelocytic leukemia and other hematologic malignancies, solid tumors. See Promyelocytic leukemia

atransferrinemia MOLECULAR MEDICINE An AR condition characterized by a complete absence of transferrin CLINICAL Severe hypochromic anemia, hemosiderosis of heart and liver, heart failure MANAGEMENT Parenteral transferrin. See Transferrin

atraumatic *adjective* Without injury

atrazine TOXICOLOGY A nonphytoestrogenic herbicide. See Phytoestrogen

atresia EMBRYOLOGY Closure or absence of an orifice or tubular structure. See Anal atresia, Biliary atresia, Choanal atresia, Duodenal atresia, Esophageal atresia

atria ANATOMY Plural of atrium, see there

atrial *adjective* Relating to an atrium, see there, usually understood to be the atrium of the heart

atrial arrhythmia CARDIOLOGY Any irregularity in rate and rhythm which arises in the atrium

atrial fibrillation CARDIOLOGY The most frequent sustained cardiac arrhythmia, which is very common in older individuals and characterized by disorganized electrical conduction in the atria, resulting in ineffective pumping of blood into the ventricle; ± 1 x 10⁶ in US have AF; 80% also have heart disease ETIOLOGY HTN, heart failure, valve disease–mitral valve prolapse, rheumatic heart disease, especially when associated with clinically silent mitral stenosis, dilated cardiomyopathy, pericarditis, cardiothoracic surgery, hyperthyroidism, alcohol use or withdrawal, acute illness–eg, pneumonia, decompensated COPD, sepsis, and other conditions, PTE, sympathetic triggers–eg, cocaine, amphetamines, atrial myxoma, tachycardia-bradycardia syndrome, lone AF RISK FACTORS DM, MI, HTN, ↑ age–peak at age 75-79; untreated AF Pts suffer strokes at a rate of 4.5%/yr SEQUELAE Loss of coordinated electromechanical activity, resulting in blood stasis and atrial thrombosis; AF precipitates heart failure as it results in the loss of the so-called atrial kick, which accounts for 5-30% of cardiac output, and a shorter diastole CLINICAL Palpitations, dizziness, dyspnea, angina, syncope EKG Narrow QRS complex < 120 msec with no interval between the QRS complexes; absent P waves in leads II, III, aVF, and V₁₋₂ WORKUP Hx, especially alcohol or drug abuse, thyroid studies, CBC, chemical profile, CXR for pulmonary disease and/or heart failure, EKG, echocardiogram DIAGNOSIS Transesophageal

echocardiography identifies Pts with atrial emboli requiring short-term anticoagulation with heparin before cardioversion. See Linkage analysis MANAGEMENT-ACUTE Adenosine, digoxin, magnesium sulfate, CCBs–eg, verapamil, diltiazem MANAGEMENT-LONG TERM Cardioversion, anticoagulation MEDICAL CONVERSION OF AF Class IA antiarrhythmics–eg, quinidine as well as disopyramide, procainamide, performed as an inpatient; class IC agents–eg, flecanide, propafenone, or class III agents–eg, amiodarone, sotalol are increasingly popular in acute conversion of AF to a sinus rhythm, as is Ibutilide INTERVENTIONAL CONVERSION OF AF Failure of medical conversion of AF to a sinus rhythm, accompanied by ventricular rates unresponsive to antiarrhythmics, requires AV node-His bundle ablation with implantation of a rate-adaptive VVI pacemaker, or preferably, atrial or dual chamber pacing, as the incidence of AF is lower than with VVI pacing. See Atrial kick, Atrial flutter

atrial flutter CARDIOLOGY Rapid, well organized contraction of the atrium at a rate of 250-350 beats/min; at atrial beats < 200/min, the ventricles can respond in a 1:1 fashion–at higher atrial rates, they respond with a 2:1, 3:1, or 4:1 block EKG Sawtooth waves and evidence of continued electric activity, best seen in II, III, aV₁ or V₁ ETIOLOGY Paroxysmal AF may occur in normal hearts; chronic–persistent AF is often associated with underlying heart disease–eg, rheumatic or ischemic heart disease, cardiomyopathy, or PTE, mitral or tricuspid stenosis or regurgitation, thyrotoxicosis, alcoholism, pericarditis CLINICAL Asymptomatic, flutter in jugular venous pulse MANAGEMENT Synchronous DC–direct current cardioversion at low energy–< 50 joules; CCBs–eg, verapamil and diltiazem

atrial gallop Fourth heart sound, see there

atrial kick Atrial systole CARDIOLOGY The contraction of the atrium, which accounts for 5-30% of cardiac output; it appears as an abrupt notch in the pressure curve in the ventricular outflow tract, and is typical of hypertrophic cardiomyopathy–formerly idiopathic hypertrophic subaortic stenosis

atrial lead CARDIOLOGY A pacing lead designed to be used in–endocardial, or on–epicardial, the atria; endocardial atrial leads are often J-shaped to facilitate placement in the atrial appendage. See Lead. Cf Ventricular lead

atrial myxoma CARDIOLOGY The most common 1° cardiac neoplasm, age of onset 25-55 CLINICAL Symptoms may be obstructive–right-sided congestion—most are right-sided, ± ascites, constitutional–fever, fatigue, arthralgias, myalgias, weight loss, Raynaud's phenomenon, skin rash, clubbing of digits, and related to embolism–SOB, orthopnoea, paroxysmal nocturnal dyspnea, fainting, palpitations, dizziness and syncope, pleuritic chest pain, hemoptysis DIAGNOSIS 2-D, transesophageal echocardiography MANAGEMENT Excision PROGNOSIS Excellent after excision

atrial premature depolarization Premature atrial complex, see there

atrial septal defect CARDIOLOGY An acyanotic congenital heart disease–CHD, which is common–¹/₃ of congenital heart defects found in adults, 2 to 3-fold more common in ♀, due to nonclosure of the foramen ovale at birth, resulting in a defect between the atria; ASDs are usually asymptomatic until the 3ʳᵈ or 4ᵗʰ decades of life SEVERITY < 0.5 in diameter, no consequences; > 2.0 cm, substantial hemodynamic consequences; because the Rt ventricle–RV is more compliant, blood shunts from Lt to Rt atrium, causing ↑ pulmonary blood flow, dilation of both atria, RV and pulmonary arteries; as the RV fails or loses compliance, L→R shunt ↓ and R→L shunting may occur; with a large ASD, an RV or pulmonary arterial impulse may be palpable HEART SOUNDS Normal 1ˢᵗ HS, wide fixed splitting 2ⁿᵈ HS, a systolic ejection murmur, heard in the 2ⁿᵈ IC space, peaking in midsystole and ending before the 2ⁿᵈ HS, is usually so soft that it is mistaken for an innocent murmur EKG Rt axis deviation and incomplete Rt BBB; normal rhythm for 1ˢᵗ 3 decades, then–AFib, SVT CLINICAL SOB, fatigue on exertion, palpitations, flowed by SVTs, Rt heart failure, paradoxical embolism, recurrent lung infections DIAGNOSIS Catheterization to localize ASDs IMAGING CXR-prominent pulmonary arteries and peripheral vascular pattern–shunt vascularity; transthoracic echocardiography–dilation of atria and RV MANAGEMENT Clinically significant–defined as the ratio of pulmonary to system blood flow of > 1.5, which requires surgical closure to prevent RV dysfunction; Pts with irreversible pulmonary vascular disease and pulmonary HTN are poor surgical candidates. See Atrium, Congenital heart disease. Cf Ventricular septal defect

atrial synchronous VAT CARDIAC PACING *adjective* Referring to a dual-chamber pacemaker which senses atrial activity and paces only in the ventricle; the rate of ventricular stimulation is directly synchronized to sensed atrial activity. See Pacemaker

atrial synchronous, ventricular inhibited VDD CARDIAC PACING *adjective* Referring to a dual-chamber pacemaker that senses in the atrium and paces in the ventricle; the pacemaker's response above the programmed lower rate depends upon and is synchronized to the patient's atrial rhythm. See Pacemaker

atrial systolic failure Atrial failure A marked decline in late diastolic atrial transport, classically associated with primary amyloidosis, with most atrial filling occurring in early ventricular diastole; ASF has a characteristic trans-mitral Doppler spectral velocity profile and may presage a poor prognosis in Pts with 1° amyloidosis, possibly due to thromboembolism. Cf Atrial standstill

atrial tachycardia CARDIOLOGY Tachycardia triggered by a focus in the atrium, which beats at 160 to 190 bpm. See Atrial fibrillation, Atrial flutter

Atridox™ Doxycycline hyclate DENTISTRY A subgingival anti-infective for periodontal disease, which is placed in the periodontal pocket, solidifies, then is released over 7 days See Periodontal disease

atrioventricular *adjective* Pertaining to cardiac atrium and a ventricle

atrioventricular block A-V block CARDIOLOGY A delay in conduction or failure of the electrical impulse to reach the ventricular conducting system, which may occur at the atrium, at the A-V node, in the bundle of His, or in the bundle branches DRUGS CAUSING Clonidine, methyldopa, verapamil CLINICAL Asymptomatic or may require a pacemaker. See Complete heart block, Sick sinus syndrome

ATRIOVENTRICULAR BLOCKS

1ST-DEGREE A-V BLOCK P-R intervals > 0.2 sec; all P waves are conducted to ventricle, ie, followed by a QRS complex–usually benign

2ND-DEGREE A-V BLOCK

MOBITZ I–AKA WENCKEBACH BLOCK P-R intervals ↑ in length until a beat is 'dropped'; PR interval after a blocked beat is prolonged; Mobitz I blocks often follow an MI–usually benign

MOBITZ II P-R interval of conducted beat is constant–conduction ratio ≥ 3:1; occasionally P fails to conduct an impulse–may have serious clinical implications

3RD-DEGREE A-V BLOCK Dissociation of atrial and ventricular activity; atrial rate is > ventricular rate; permanent pacemaker is required

atrioventricular bundle Bundle of His, AV bundle ANATOMY A group of specialized cardiac fibers that conduct impulses from the AV node to the ventricle

atrioventricular canal defect CARDIOLOGY A VSD located near the junction of the mitral and tricuspid valves. See Congenital heart disease, Ventricular septal defect

atrioventricular conduction defect CARDIOLOGY Any derangement-block or delay in electric conduction through the AV node or bundle of His above the bifurcation of the AV fascicle; an ACD is localized by a 12-lead EKG. See Atrioventricular block

atrioventricular dissociation AV dissociation CARDIOLOGY The independent depolarization of the atria and ventricles, where atrial rhythm is controlled by one pacemaker and the ventricular rhythm is controlled by another, accompanied by a loss of A-V synchrony. See Third-degree block

atrioventricular nodal reentrant tachycardia AVNRT CARDIOLOGY The most common form of paroxysmal SVT TREATMENT Antiarrhythmics are rarely successful; catheter ablation of the fast P wave causes complete heart block in 10%, requiring pacemaker placement; selective catheter ablation of the atrial end of the slow PW using radiofrequency current may eliminate AVNRT with little risk of AV block. See Supraventricular tachycardia

atrophic gastritis GASTROENTROLOGY A condition that is the end result of chronic gastritis, characterized by mucosal atrophy, which may be a precursor of gastric CA. See *Helicobacter pylori*, Stomach cancer

atrophic vaginitis GYNECOLOGY A condition characterized by post-menopausal inflammation of vaginal mucosa 2° to thinning, with loss of elasticity of the vaginal wall, common in postmenopausal ♀ due to ↓ endogenous estrogen CLINICAL Pruritus ± burning sensation, ↓ in vaginal secretions, dyspareunia, post-coital bleeding, ±bacterial infection TREATMENT Topical and/or oral estrogen–contraindicated in Pts with prior breast or endometrial CA. See Lichen sclerosis

atrophy Wasting away; a ↓ in size of a cell, tissue, organ or part, due to defective or failed nutrition. See Blue atrophy, Brown atrophy, Disuse atrophy, Fat pad atrophy, Gastric atrophy, Geographic atrophy, Group atrophy, Multiple system atrophy, Pseudoatrophy of brain, Spinal muscle atrophy, White atrophy. Cf Dystrophy GYNECOLOGY A thinning of the ♀ genital mucosa due to ↓ estrogen in menopause APPEARANCE Smooth, thin epithelium, prominent blood vessels, ↑ risk of trauma DIFFDX Severe dysplasia, invasive CA MANAGEMENT Exogenous HRT

atropine THERAPEUTICS A racemic mixture of toxic alkaloids extracted from *Atropa belladonna*, a competitive antagonist of anticholinergics EFFECTS Tachycardia, ↓ salivation, GI motility–constipation, sweating, cycloplegia, mydriasis, urinary retention, bronchodilation THERAPEUTIC USE Bronchodilator, reverses effects of organophosphate pesticides, nerve gases CONTRAINDICATIONS Narrow-angle glaucoma, synechiae between iris and lens, GI obstruction, obstructive uropathy, megacolon, GERD, unstable cardiovascular disease and acute hemorrhage, tachycardia, myasthenia gravis

Atrovent® Ipratropium Br, see there

Atruline Sertraline, see there

ATSDR Agency for Toxic Substances & Disease Registry

attached patient PATIENT CARE A Pt known to have a regular attending physician when he/she is registered/seen in an ER or other 'neutral' health care setting. See Attending physician, Flipping. Cf Unattached patient

attachment PSYCHIATRY The behavior of an organism that relates in an affiliative or dependent manner to another object which develops during critical periods of life and can be extinguished by lack of opportunity to relate; if separation occurs before maturation can provide for adaptive adjustment, personality deviation can occur

attachment disorder CHILD PSYCHIATRY A condition characterized by difficulty in forming loving, lasting, intimate relationships SOCIAL MEDICINE A condition characterized by psychologic and physical distancing from adults

attack VOX POPULI An episode or event of abrupt onset. See Crack attack, Drop attack, Fatty food attack, Gallbladder, Panic attack, Sleep attack, Transient ischemic attack

attack rate CLINICAL EPIDEMIOLOGY An expression of incidence, defined as the proportion of persons at risk for an infection who become infected over a period of exposure. See Incidence rate

attending physician MEDICAL PRACTICE The physician who is on the medical staff of a hospital or health care facility, and legally responsible for the care given to a particular Pt while in the hospital. See Private physician TERMINAL CARE The physician–or designee who is primarily responsible for the care and treatment of the individual upon whom a declaration of brain death is to be made See Appropriate period of observation, Brain death, Corroborating physician

attention deficit disorder The most widely used term for the condition which is officially–per the DSM-IV—known as Attention deficit-hyperactivity disorder, see there

attention deficit-hyperactivity disorder Attention deficit disorder PSYCHIATRY An inability to control behavior due to difficulty in processing neural stimuli, resulting in ↑ motor activity, ↓ attention span EPIDEMIOLOGY ADHD is the most common neurobehavioral disorder of childhood, with a point-prevalence of 2% to 18%—depending on defining criteria; in children, the ♂:♀ ratio is 2:1; in adults, 1:2, characterized by impulsiveness, distractibility, variably accompanied by hyperactivity and/or aggressiveness, immaturity and emotional lability; although ADHD is considered idiopathic, neurochemistry and genetics may play a role DIAGNOSIS ADHD is a diagnosis of exclusion IMAGING NMR, ADHD brains may be 5% smaller in the anterior frontal region, the right caudate and globus pallidus; by PET, ADHD Pts maintain blood flow through temporal region, in contrast to controls, where the blood flow is ↓ MANAGEMENT Stimulants-methylphenidate-Ritalin®, dextroamphetamine, pemoline; antidepressants–desipramine, imipramine, bupropion; alpha-adrenergic agonists–clondine DIFFDX Sensory deficit, receptive language disorder, specific learning disability, seizures, emotional problems, oppositional defiant disorder, conduct disorder, problems of parent-child interaction, mental retardation See Breuning affair

attentional distraction PSYCHOLOGY Any of a number of techniques–eg use of video games, playing with pets, headphones with music, intended to block a Pt's perception of an unpleasant event–eg, anticipatory nausea associated with chemotherapeutics, dentist drilling. See Anticipatory side effects. Cf Relaxation training, Systematic desensitizaton

attenuated *adjective* MEDTALK 1. Reduced in virulence or infectivity 2. Thin, minimal, markedly reduced or decreased

attenuated virus A functionally effete virus that is no longer virulent, which could be used a live virus vaccine

attenuation A generic term for a reduction or diminution of activity, intensity, power, or virulence of a reaction, effect, or organisms ability to grow and/or multiply MICROBIOLOGY ↓ virulence of a microorganism–eg, that of bacillus Calmette-Guerin–BCG, a strain of *Mycobacterium bovis* that has been weakened by multiple–238–subcultures on a bile-glycerine medium; the resulting bacterium is immunogenic, ie capable of eliciting antibody formation, but non-virulent; live attenuated organisms are used to produce the poliomyelitis vaccine but these may revert to a wild type RADIATION BIOLOGY A process by which a beam of radiation is ↓ in intensity when passing through material, due to absorption and scattering processes, leading to a ↓ in flux density of the beam when projected through matter

attestation MANAGED CARE A document signed by a physician, stating that he/she performed the diagnostic or therapeutic procedure on a Pt for which a bill is being submitted

at the elbow standard GRADUATE EDUCATION A standard for billing which requires that a physician billing for services rendered to Pts treated in a teaching hospital be at the resident's or intern's elbow

attitude PSYCHOLOGY "…*the tendency towards a mode of response, toward the object in question.*" See Abstract attitude

ATTRACT RHEUMATOLOGY A clinical trial–Anti-TNF Trial in Rheumatoid Arthritis with CA2 Treatment–which evaluated the effects of humanized anti-TNF–cA2–treatment on Pts with RA. See Remicade®, Rheumatoid arthritis

attributable proportion EPIDEMIOLOGY A measure of the public health impact of a causative factor; the proportion of disease in a group exposed to a particular factor attributable to exposure to that factor

attributable risk EPIDEMIOLOGY Any factor which ↑ the risk of suffering a particular condition. See Relative risk, Risk factor. Cf Nonattributable risk STATISTICS The rate of a disorder in exposed subjects that is attributable to the exposure derived from subtracting the rate–incidence or mortality of the disorder of a nonexposed population from the rate in an exposed population

atychiphobia PSYCHOLOGY Fear of failure. See Phobia

atypical *adjective* Not normal, irregular, abnormal, not conforming to a type, abnormal in presentation, morphology, appearance or behavior

atypical adenomatous hyperplasia–prostate SURGICAL PATHOLOGY A lesion of the transition zone of the prostate, distinguished from well-differentiated CA by a relative lack of nuclear or nucleolar enlargement, infrequent crystalloids, and a fragmented but partially intact basement membrane; AAH is more common in older Pts with larger prostates, more nodular hyperplasia, more cancer, and, if malignant, with higher Gleason scores. See Gleason score; PIN Cf Benign prostatic hypertrophy, Prostate cancer

atypical carcinoid An intermediate form of neuroendocrine tumor between low-grade malignant–typical carcinoid and high-grade small cell carcinoma. See Carcinoid, Small cell carcinoma

atypical depression Panic disorder, see there

atypical GERD INTERNAL MEDICINE An atypical presentation of GERD which affects up to 30% of Pts with classic GERD CLINICAL 1. Lungs–asthma, cough, chronic bronchitis, pulmonary fibrosis, pneumonia; 2. ENT–laryngitis, sinusitis, hoarseness, vocal nodules, globus hystericus; 3. Others–noncardiac chest pain, dental erosion, hiccups DIAGNOSIS Dx of exclusion; Hx of severe chest pain–postprandial or nocturnal; endoscopic esophagitis, < 50% of Pts; esophageal pH monitoring–85% sensitivity; test of therapy MANAGEMENT Unlike classic GERD, atypical GERD requires longer treatment, is less responsive to therapy; high-dose proton pump inhibitor therapy for 12 wks; nonresponders may require fundoplication. See GERD, Noncardiac chest pain

atypical lipomatous tumor Atypical lipoma, well-differentiated liposarcoma PATHOLOGY A soft tissue tumor, most often retroperitoneal, also seen in the skin PROGNOSIS ALTs often recur locally, but rarely metastasize

atypical lymphocyte Downey cell HEMATOLOGY An enlarged dysmorphic lymphocyte characterized by an often irregular monocyte-like nucleus which may stretch the length of the cell, with chromatin strands that parallel the length of the nucleus; nucleoli and azurophilic granules may be present; ALs are seen in various non-neoplastic conditions, classically in infectious mononucleosis, toxoplasmosis, CMV infection, and viral hepatitis. See Azurophilic granule

atypical mycobacterial infection INFECTIOUS DISEASE Clinical infection with mycobacteria other than those causing TB or lepra Risk factors Immune compromise, AIDS CLINICAL Abscesses, septic arthritis, osteomyelitis; AMIs include those with *M avium* complex, associated with AIDS, and *M marinum*, seen in the Chesapeake Bay region, affecting fishermen and aquarium keepers and *M ulcerans*, endemic to the banks of the upper Nile MANAGEMENT Difficult due to multidrug resistance. See Atypical mycobacterium

atypical mycobacterium Any *Mycobacterium* spp exclusive of *M lepra*, *M tuberculosum*, or *M bovis*–the latter 2 of which cause 'typical' TB; AM are so designated because they grow more rapidly, produce no niacin, don't

reduce nitrates, produce heat-stable catalase,and are usually resistant to isoniazid. See Atypical mycobacterial infection, Multidrug-resistant tuberculosis

atypical nevus DERMATOLOGY A pigmented lesion with a clinical
appearance that differs from a 'garden variety' mole, which may be larger,
have irregular borders, lack uniformity of color; it can be flat or raised above
the skin surface. See Nevus

atypical pneumonia CHEST MEDICINE A clinically 'atypical' form of
pneumonia, which lacks the classic signs and Sx of pneumonia TYPES
Chlamydia pneumonia, *Mycoplasma*, influenza A or B, adenovirus, *Legionella*
MANAGEMENT Antibiotics if bacterial PROGNOSIS Sx often improve in < 2 wks

atypical squamous cell of undetermined significance See ASCUS

atypical transformation zone GYNECOLOGY A transformation
zone–the interface between the cervical and endocervical mucosa, which is
characterized by various patterns–eg, leukoplakia, acetowhite epithelium
and abnormal vascular patterns, changed epithelial thickness, nuclear density, vascular proliferation; 90% of cervical neoplasias–CIN and invasive CA
arise in the transformation zone. See Transformation zone

atypical vessels OB/GYN Blood vessels associated with malignant
epithelium with ↑ metabolic needs, accompanied by asymmetrical vascular
proliferation, often associated with progressively smaller blood vessels; with
invasion, AVs may have sharp angulations with cork-screw, hockey-stick or
spaghetti-like patterns; the distance between vessels is often ↑, resulting in
bizarre patterns

AU 1. Allergy unit, see there 2. Antitoxin unit

AUA American Urological Association, see there

AUC$_{0-∞}$ ANTIBIOTICS Area under the concentration time-curves from time
zero to infinity; which corresponds to a calculation of mean concentration
levels of a therapeutic agent in the body See Pharmacokinetics

audio-CASI Audio-enhanced, computer-assisted self-interviewing CLINICAL
STUDIES A method used to improve the quality of survey data from questionnaires, intended to overcome respondents' limited literacy and incentives
to misstate the answers to questions that may be regarded as sensitive

audiogram AUDIOLOGY A test in which hearing is measured over a range
of sound frequencies. See Pure tone audiometry

audiologist AUDIOLOGY A non-MD health care professional trained to
identify and measure hearing impairment and related disorders–eg, balance
or vestibular disorders and tinnitus, and rehabilitate Pts with impaired hearing and related disorders; audiologists use various tests and procedures to
assess hearing and balance SALARY $52K + 6% bonus. See Speech pathology.

audiology The study of hearing

audiometric zero AUDIOLOGY A value arbitrarily assigned to 0
dB–deciBel hearing level, the average hearing acuity for a normal population, which corresponds to 24.5 dB SPL–sound pressure level at 250 Hertz

audiometry The measurement of hearing. See Play audiometry, Pure tone
audiometry, Speech threshold audiometry

AUDIT Alcohol Use Disorders Identification Test A screening tool used to identify
alcoholics. See Alcoholism

audit trail INFORMATICS A software tracking system used for data security,
which is attached to a file each time it is opened so an operator can identify
who and when a file has been accessed. See Internal audit

audition rotation Away externship, away rotation GRADUATE EDUCATION A
clinical rotation by a 4th yr medical student interested in a specific residency

program and/or target specialty in a location other than that of his/her medical school. See Match

auditory agnosia NEUROLOGY An inability to recognize sounds, a common finding in parietal lobe tumors. See Agnosia

auditory association PSYCHOLOGY The ability to associate spoken
words in a meaningful fashion

auditory brainstem response test AUDIOLOGY A test used for assess
hearing in infants and young children, which involves attaching electrodes
to the head to record electrical activity from the auditory and other regions
of the brain; it is also used to test cerebral activity in neurologically unresponsive Pts. See Hearing test

auditory evoked potentials test Brainstem auditory evoked potentials
NEUROLOGY A test that measures brain and brainstem response to auditory stimulation, which serves to evaluate neurologic integrity and hearing
APPLICATION Evaluate hearing in neonates, suspected stroke, acoustic neuroma, Meniere's disease, multiple sclerosis

auditory memory The ability to remember words and sounds. See Memory

auditory perception NEUROLOGY The ability to identify, interpret, and
attach meaning to sound

auditory prosthesis NEUROLOGY A device that substitutes or enhances
hearing ability. See Hearing aid

auditory vocal automatic Grammatic closure

augment anesthesia ANESTHESIOLOGY The use of a systemic
agent–eg, a hypnotic in addition to local anesthesia to enhance analgesic
effects rather than subjecting a Pt to general anesthesia. See Local anesthesia

augmentation Augmentation mammoplasty, see there

augmentation cystoplasty UROLOGY Reconstructive surgery in which
a segment of the bowel is removed and used to replace part of the diseased
bladder, in Pts with severe urge incontinence COMPLICATIONS Bowel obstruction, blood clots, infection, pneumonia, urinary fistulae, UTIs, difficulty urinating, and rarely tumors

augmentation mammoplasty COSMETIC SURGERY Any surgical procedure which ↑ breast size COMPLICATIONS Capsule contraction–20% of cases,
hematoma, infection, implant exposure, deflation or rupture of implant,
breast asymmetry, external scars. See Mammoplasty. Cf Reduction mammoplasty

augmentative device AUDIOLOGY Any device that facilitates communication by those with limited or absent speech EXAMPLES Communication
boards; pictographs–symbols that look like the things they represent; ideo-
graphs–symbols representing ideas. See Americans with Disabilities Act

Augmentin® INFECTIOUS DISEASE A combination antibiotic containing
amoxicillin and clavulinic acid INDICATIONS Ampicillin-resistant *S aureus*, *B
fragilis*, β-lactam-producing Enterobacteriaceae

Aunt Millie approach Pattern recognition MEDTALK An unsound–albeit
often applied—philosophy for clinical decision-making, from the quip:

HOW DO I KNOW IT'S AUNT MILLIE?
BECAUSE IT LOOKS LIKE AUNT MILLIE

Pattern recognition is the traditional model for teaching pathology and radiology, both of which are visual 'arts'; although it allows molding of future generations in an accepted paradigm, exceptions to any 'rules' are generally relegated
to wastepaper basket categories, as exceptions cannot be explained or understood
in an arbitrarily-defined context. Cf Heuristic method, Stochastic process

aura NEUROLOGY A subjective—illusionary or hallucinatory. or objective–motor event marking the onset of an epileptic attack, grand mal seizure, or a migraine. See Migraine, Seizure PARANORMAL An energy field said to envelop the human body, and correspond to the soul See Chakra, Cf Vital force

aural *adjective* 1. Pertaining to the ear 2. Relating to perception by the ear, as an aural stimulus 3. Referring to an aura

aural polyp Otic polyp ENT A benign polyp in the external ear canal that can be attached to the tympanic membrane

aural rehabilitation AUDIOLOGY Any technique used for the hearing-impaired to improve their speech and communication. See Speech therapy

Aurorix® Moclobemide, see there

auscultate *verb* PHYSICAL EXAMINATION To listen with a stethoscope to internal organs–heart, lungs, GI tract–for sounds of diagnostic portent

Auspitz' sign DERMATOLOGY Removal of a 'virgin' yellow-white, sharply demarcated plaque of psoriasis, results in pinpoint hemorrhage

Austin Flint murmur CARDIOLOGY A sign of aortic regurgitation, characterized as a diastolic/presystolic rumble heard best over the cardiac apex. See Aortic regurgitation

Austrian syndrome A term used for Pts–usually alcoholics–with pneumococcal pneumonia, meningitis, and endocarditis and ruptured aortic valve, who present with bacteremia and, despite adequate antibiotic therapy, have a high–80% mortality

autacoid ENDOCRINOLOGY A locally acting hormone-like substance–eg, histamine, serotonin, angiotensin, eicosanoids

authentication BIOMETRICS The use of a personal feature, eg, face, hand, fingerprint, signature, voice, iris, or other highly specific indicator to verify a person's identity, and restrict access to information to authorized persons; confirmation of a user's identity, generally through user name and password or biometric characteristics. See Biometrics. Cf Authorization

author JOURNALISM A person involved in writing a manuscript

author 'inflation' The growth in number of people receiving authorship credit on published reports in biomedical sciences. See Mega-author paper

authority An expert in a particular field

authority figure PSYCHIATRY A projected or real person in a position of power or authority–eg, a parent, spouse

authorization MANAGED CARE Formal approval by the gatekeeping arm of a third-party payer for payment of a requested procedure. See Gatekeeper

authorized person LAB MEDICINE A person–eg a physician, who orders tests and receives test results on persons for whom payment is sought under Medicare. See CLIA 88

authorized user RADIATION PHYSICS A person who, having satisfied the applicable training and experience requirements, is granted authority to order radioactive material and accepts responsibility for its safe receipt, storage, use, transfer and disposal

authorship SCIENCE JOURNALISM The state of being an author. See Author, Author misconduct, CV-weighing, Darsee affair, Honorary authorship, Mega-author paper, Slutsky affair, Spurious authorship, Unearned authorship

autistic disorder Autism, autistic psychopathy PEDIATRICS A pervasive developmental disorder, which affects 1:2500 children with a 3-4:1 ♂:♀ ratio, onset usually by age 3; AD is characterized by profound introversion, self focus, lack of reality sense, withdrawal and developmental delays and deficits in social interaction, communication, creative/imaginative play, behavior, interpersonal relationships CLINICAL Autistic behaviors–eg, whirling, flapping, self-mutilation, body rocking, toe walking MANAGEMENT Facilitated communication, drugs—clomipramine, haloperiodol, fluvoxamine maleate–a potent serotonin reuptake inhibitor, naltrexone to control self-mutilation; intensive behavioral therapy PROGNOSIS Poor; < 20% are gainfully employed as adults; < 20% function in sheltered environments; > ⅔ require permanent supervision, support

autoagglutination HEMATOLOGY The aggregation of RBCs, often caused by a cold-reacting antibody, as seen in cold agglutinin disease or other autoimmune hemolytic anemias

autoantibody IMMUNOLOGY Any antibody produced by an organism against one of its own–self antigens. See Antibody, Hashimoto's thyroiditis, Lupus erythematosus, Myasthenia gravis, Rheumatoid arthritis

autoantigen IMMUNOLOGY Any self antigen–eg, glomerular basement membrane, mitochondria, muscle, parietal cells, thyroglobulin and others, which may evoke production of autoantibodies. See Antigen

Autocyte PREP® CYTOLOGY A monolayer slide preparation and discrete-staining lab product used for cytologic specimens–eg, Pap smears. See Pap smear

autodigestion PATHOLOGY The release of enzymes from macrophages, resulting in tissue destruction linked to cell and tissue death in acute pancreatitis

autoerotic *adjective* Referring to sexuoerotic self-stimulation–eg masturbation. See Masturbation

autoerotic death SEXOLOGY 1. Death from self-strangulatory asphyxia or electrical self-stimulation as part of a paraphilic masturbatory ritual in which a ligature of some type is placed around the neck 2. Death from asphyxiation or electrocution, as a miscalculation in a paraphilic sexuoerotic ritual involving self-strangulation or self-applied electric current. See Electrocutophilia

autogenous *adjective* Of self origin; self-produced

autogenous control The regulation of a system or process–eg, enzymes, gene production, by its own products

autograft SURGERY A tissue obtained from one site on the body, which is 'donated'–engrafted–to another

autoimmune *adjective* Referring to an immune response to self antigens

autoimmune chronic active hepatitis Lupoid hepatitis, see there

autoimmune disease CLINICAL IMMUNOLOGY Any condition in which the body recognizes its own tissues as foreign and directs an immune response against them; AD is linked to production of antibodies against self antigens, which affects ± 5% of adults–⅔ are ♀ in Western nations

autoimmune hemolytic anemia HEMATOLOGY A condition in which IgG autoantibodies against a component of the Rh factor coat, the RBC surfaces ETIOLOGY Idiopathic–±50%, SLE, CLL, lymphomas CLINICAL Anemia of rapid onset, fatigue with angina or CHF, jaundice, splenomegaly LABORATORY Anemia, spherocytosis, nucleated RBCs, positive direct Coombs test MANAGEMENT Prednisone, IV immune globulin, splenomegaly PROGNOSIS Good. See Autoimmunity, Hemolysis

autoimmune hepatitis Lupoid hepatitis A type of chronic active hepatitis attributed to various circulating autoantibodies, which may be linked to other autoimmune diseases–eg, thyroiditis, DM, ulcerative colitis, Coombs-positive hemolytic anemia, proliferative glomerulonephritis, Sjögren syndrome

autoimmune thrombocytopenia purpura Idiopathic thrombocytopenic purpura, see there

autoimmune thyroiditis Hashimoto's disease, see there

autoimmunity Self immunity IMMUNOLOGY The reaction of an organism's immune system to self antigens as if they were non-self or foreign; like alloimmunity, autoimmunity is characterized by the activation of T cells, clonal expansion, and antibody production. See Anti-nuclear antibodies, Anti-receptor antibodies, Clonal anergy, Superantigen

auto-Lancet™ PHLEBOTOMY A finger-lancing device, with push-button trigger, used to obtain capillary blood samples

autologous MEDTALK Self TRANSPLANT BIOLOGY *adjective* Referring to a tissue that comes from the same person, in contrast to that donated by another person. Cf Heterologous, Homologous

autologous bone marrow transplantation TRANSPLANTATION MEDICINE The administration to an individual 'X' of his/her own BM, often to a leukemic Pt in relapse who, because a suitable HLA-matched donor is not available, would otherwise die of the disease. See Bone marrow

autologous chondrocyte transplantation ORTHOPEDIC SURGERY A procedure for treating defects of articular cartilage BACKGROUND Defects of articular cartilage often follow joint trauma and, if large and deep enough, lead to pain and joint dysfunction followed by osteoarthritis, eventually requiring replacement of the joint by an artifical prosthesis; total knee replacement is commonly performed on those > 60 yrs of age, but is increasingly problematic in younger Pts, as mechanical joints have a limited lifespan. See Total knee replacement

autologous graft Autologous transplant TRANSPLANT MEDICINE The transfer of a tissue from one site to another in the same individual

autologous transfusion Autologous blood transfusion TRANSFUSION MEDICINE The collection and re-infusion of the Pt's own blood and/or blood products; the volume of blood available for AT can be ↑ with recombinant erythropoietin and iron supplements; AT is most often used in orthopedic–total hip replacement and cardiovascular surgery. See Intraoperative 'autologous' blood transfusion

autologous unit TRANSFUSION MEDICINE A unit of RBCs or other blood product to be transfused into the donor at the time of elective surgery. See Autologous transfusion

autolymphocyte therapy EXPERIMENTAL ONCOLOGY A form of immunotherapy for treating metastatic renal carcinoma, in which a Pt's WBCs are removed, stimulated by monoclonal antibodies, which causes the WBCs to produce and secrete cytokines; the cytokine supernatant is then removed and readministered with an aliquot of the Pt's own WBCs. Cf Immunoaugmentive therapy

automated cytology device CYTOPATHOLOGY Any device that automates any process in cytology, which includes computer-aided imaging–eg Papnet, AutoPap, screening, process control, and automated preparation–eg, ThinPrep. See AutoPap, Computer-aided imaging device, Papnet, ThinPrep

automated external defibrillator EMERGENCY MEDICINE A portable device designed for use by first-response personnel for out-of-hospital emergency treatment of Pts suffering from cardiac arrest. See First-response personnel

automatic antitachycardia pacemaker CARDIAC PACING A pulse generator capable of detecting tachycardias and delivering stimuli automatically, which stops when the tachycardia terminates

automatic epilepsy NEUROLOGY Seizures with elaborate and multiple sensory, motor, and/or psychiatric components CLINICAL Clouded consciousness, amnesia, complex behaviors–eg, bursts of anger or emotion, anxiety, automatisms EEG Spike discharges in temporal lobe during sleep. See Complex partial seizure

automatic external defibrillator Smart defibrillator CARDIOLOGY A device designed to monitor the heart's electric activity and, if ventricular fibrillation is identified, deliver an electric shock. See Defibrillator, Ventricular fibrillation

automatic negative thought PSYCHIATRY An unpleasant anxiety-provoking thought automatically triggered in a person by a particular situation. See Cognitive therapy

automatic suspension HOSPITAL PRACTICE The immediate suspension of a practitioner from the medical staff due to activities of such egregious nature as to warrant said suspension–eg, revocation of the practitioner's state license to practice medicine. Cf Summary suspension

automaticity CARDIAC PACING The inherent property of individual myocardial cells to depolarize spontaneously NEUROLOGY Automatism, see there

automatism Monophasia, recurring utterances, verbal stereotypy NEUROLOGY A form of motor aphasia, characterized by stereotyped utterances repeatedly repeated, as if by compulsion; an involuntary compulsion to perform a motor act ASSOCIATIONS Psychomotor epilepsy, catatonic schizophrenia, psychogenic fugue, complex partial seizure, post-traumatic automatism, etc. See Aphasia, Motor aphasia PSYCHIATRY Automatic and apparently undirected nonpurposeful behavior that is not consciously controlled. See Automatic behavior

automobile blindness OPHTHALMOLOGY A term for vision that is so poor that the responsible licensing agency in a particular state will not issue a driver's license. See Blind spot, Legally blind

autonomic *adjective* 1. Autonomous; self controlled; functionally independent 2. Referring to the autonomic nervous system, see there

autonomic dysreflexia NEUROLOGY A potentially life-threatening ↑ in BP, sweating, and other autonomic reflexes in reponse to various stimuli–eg, bowel impaction. See Autonomic failure

autonomic failure NEUROLOGY A condition characterized by sympathetic and often parasympathetic failure; CLINICAL Syncope, orthostatic hypotension; AF can be divided into 1°, 2°, and drug-induced forms with considerable overlap ETIOLOGY AF may be evoked by or associated with aging, alcohol imbibition, carcinomatous autonomic neuropathy, CNS disease, dopamine β-hydroxylase deficiency, familial dysautonomia or Riley-Day syndrome, idiopathic orthostatic hypotension, Parkinson's disease, Shy-Drager syndrome. See Baroreflex failure

autonomic hyperreflexia NEUROLOGY An exuberant autonomic response to various insults ETIOLOGY Upper spinal cord injury, cystoscopy or distention of bladder or colon CLINICAL HTN, sweating, bradycardia, severe headache

autonomic neuropathy NEUROLOGY A symptom complex caused by damage to autonomic nerves ETIOLOGY DM, alcohol use, traumatic nerve injury, anticholinergics CLINICAL Abdominal bloating, heat intolerance, N&V, impotence, diarrhea, constipation, orthostatic vertigo, dysuria, urinary incontinence

autonomy VOX POPULI Personal capacity to consider alternatives, make choices, and act without undue influence or interference of others. See Functional autonomy, Physician autonomy

autophobia PSYCHIATRY A morbid fear of being alone or in solitude

Autoplex® T HEMATOLOGY An agent used to treat hemophilia A Pts who are refractory to factor VIII replacement. See Hemophilia

autoprothrombin C Factor X, see there

autoprothrombin I Factor VII, see there

autopsy Postmortem examination, necropsy PATHOLOGY A post-mortem examination of a body, which helps determine cause of death and identify any diseases that had not been detected while the Pt was alive, or confirms the presence of conditions that had been diagnosed before the Pt died. See Forensic autopsy, Hospital autopsy, Organ-limited autopsy. Cf Psychologic autopsy

---

AUTOPSY TYPES

BIOPSY ONLY A minimalist post-mortem in which the prosector examines the organs, but only samples small fragments–'biopsies' for histologic examination

CHEST ONLY An autopsy in which only the lungs and heart are examined; findings in a 'chest only' autopsy is to ID an occluding thrombus in the coronary arteries, massive PTE, or evaluate a person for compensation under the Black Lung Compensation act of 1969

COMPLETE An autopsy in which the thoracic, abdominal, and cranial cavities are examined

HEAD ONLY An autopsy in which the pathology of interest is presumed to reside entirely in the cranial cavity

NO HEAD An autopsy examining the chest and abdominal cavity without cranial cavity

---

autopsy permit PATHOLOGY A legal document required for a nonforensic autopsy, which contains a core of basic information about the decedent, particulars of the person authorizing the autopsy, and extent of the authorized autopsy. See Autopsy

autosexual SEXOLOGY *adjective* Characterized by self-sex contact, usually as a genital act (masturbation), with or without an accompanying erotic fantasy or ritual. See Masturbation

autosomal dominant GENETICS Referring to a mode of inheritance, in which the presence of only one copy of a gene of interest on one of the 22 autosomal–non-sex chromosomes, will result in the phenotypic expression of that gene; the likelihood of expressing an autosomal gene in progeny is 1:2; ♂ and ♀ are affected equally. Cf Autosomal recessive, X-linked recessive

autosomal dominant polycystic kidney disease ADPKD A common–1:400-1:1000 AD condition, which causes 6-9% of ESRD in developed countries CLINICAL Acute or subacute onset of azotemia and HTN, due to ↑ activity of the RAA system, possibly related to the ischemic pressure induced by the expanding cysts; ADPKD first appears in adults with upper quadrant tenderness; extrarenal disease is due to defective extracellular matrix, with hepatic cysts, diverticulosis, berry and abdominal aneurysms, annuloaortic ectasia, valvular regurgitation, anemia, ↑ ESR, ↑ WBCs DIAGNOSIS Ultrasonography. See Polycystic kidneys

autosomal recessive GENETICS Referring to a mode of inheritance, in which the phenotypic expression of a gene of interest requires its presence on both paired chromosomes–ie, a homozygous state; if both parents have the recessive gene of interest, the likelihood of expression in progeny is 1:4; ♂ and ♀ are affected equally. Cf Autosomal dominant, X-linked recessive

autosome CYTOGENETICS Any chromosome in an organism's complement–22 pairs in humans–other than sex chromosomes, X and Y See Chromosome, Diploid, Genome. Cf Sex chromosome

autosuggestion Autohypnosis, self-hypnosis POP PSYCHOLOGY The use of hypnosis on oneself. See Hypnosis, Hypnotherapy

autozygote GENETICS Homozygote in which the 2 alleles are identical by descent–ie they are copies of an ancestral gene. See Allele, Allozygote, Gene

AUTS Adult Use of Tobacco Survey TOBACCO CONTROL A survey conducted by the CDC. See Tobacco

A-V Atrioventricular, also 1. Anteversion 2. Aortic valve 3. Arteriovenous 4. Audiovisual

A-V block Atrioventricular block, see there

A-V conduction defect Atrioventricular conduction abnormality. See Atrioventricular block

A-V dissociation Atrioventricular dissociation, see there

A-V fistula Arteriovenous fistula, see there

A-V interval CARDIAC PACING The length of time in a dual-chamber pacemaker between an sensed or paced atrial event and a paced ventricular event. See Pacemaker

A-V node Atrioventricular node, see there

A-V sequential CARDIAC PACING A dual-chamber pacemaker that paces at a programmed rate in the atrium and *senses* and paces in the ventricle

A-V synchrony CARDIAC PACING A heart activation sequence in which first the atria then, after an appropriate delay, the ventricles contract; dual-chamber pacemakers are designed to stimulate this normal heart sequence

A-V valve Atrioventricular valve, see there

availability bias RISK ANALYSIS A bias in risk assessment in which a Pt overestimates the risk of an adverse outcome based on the notoriety of the risk–eg breast CA in ♀. See Bias. Cf Anchoring bias

Avakine™ Remicade®, see there

avalanche GEOMEDICINE A natural disaster in which a massive block of snow cascades down a steep incline STATISTICS In North America, ± 15 die thereof/yr; in Europe, 150/yr; most are recreational deaths–eg, snowmobilers, mountaineers, backcountry skiers AVALANCHE RISKS ↓ Snow stability; slope angles > 35° COD Crush injury, asphyxia. See Geological disaster

Avalide® CARDIOLOGY A tablet containing irbesartan/Avapro®, an angiotensin II receptor blocker, and hydrochlorothiazide, a thiazide diuretic for hypertensive Pts whose BP is poorly controlled by either agent as a monotherapy See Hypertension

Avandia® Rosiglitazone DIABETES An agent used with conventional therapy as a first-line monotherapy for type 2 diabetes. See Type 2 diabetes

Avapro® Irbesartan CARDIOVASCULAR DISEASE A once-daily angiotensin II receptor antagonist used for treating HTN. See Hypertension

avascular necrosis of hip Legg-Perthes' disease, see there

Aventyl® Nortriptyline, see there

average Arithmetic average, arithmetic mean STATISTICS A measure of central tendency, calculated by the sum of all data points dividing by the number–n of data points. See Time-weighted average

average cost Average benefit MANAGED CARE The cost per unit of output, calculated as the cost or benefit of all units of output divided by the number of units

average length of stay MANAGED CARE The total number of Pt days divided by the number of admissions and discharges during a specified period of time, which results in an average number of days in the hospital for each person admitted

average payment rate MANAGED CARE The amount of money that the HCFA could conceivably pay an HMO or CMP for service to Medicare recipients under a risk contract. See Risk contract

aversion therapy PSYCHOLOGY A behavior therapy technique in which stimuli associated with undesirable behavior—eg alcohol, drug, or tobacco use, are paired with a painful or unpleasant stimulus, resulting in suppression of the undesirable behavior. See Behavioral therapy, Encounter group therapy, Flooding, Image aversion therapy, Systematic desensitization

AVERT CARDIOLOGY 1. A clinical trial—Atorvastatin Versus Revascularization Treatments—which evaluated the efficacy of aggressive lowering of cholesterol in Pts at risk of future cardiovascular events 2. A clinical trial–Artificial Valve Endocarditis Reduction Trial

aviation medicine TRAVEL MEDICINE IN-FLIGHT EMERGENCIES Federal Aviation Agency regulations require that an 'enhanced' medical kit–stethoscope, sphygmomanometer, airway tube, syringes, epinephrine, nitroglycerin, 50 ml 50% dextrose, diphenhydramine injectable be carried on any airplane with > 30 seats IN-FLIGHT EMERGENCIES Syncope 29%, cardiac/chest pain 16%, asthma/shortness of breath 10%, allergic reactions 5%, in-flight deaths DISEASE TRANSMISSION Resurgence of TB has made the aerosol transmission through aircraft ventilation systems a health hazard of unknown epidemiologic significance STATISTICS 0.31 deaths/10^6 passengers—regardless of flight length; 125 deaths/10^9 passenger kilometers; 25 deaths/10^6 departures; the average victim was ♂, age 53.8, physicians were available in 43% of cases CAUSE OF DEATH Cardiac 56%, terminal cancer 8%, respiratory 6%, miscellaneous and no cause, the remainder. See Traveler medicine

aviator's astralagus ORTHOPEDICS A type of fracture and fracture-dislocations of the talus, in which the point of impact is a centrally located transverse ridge impacting on the foot. Cf Cavalry fracture

AVID CARDIOLOGY A clinical trial–Antiarrhythmics Versus Implantable Defibrillators that compared the effect of implantable defibrillators vs the best medical therapy–antiarrhythmics for survivors of MI or those with nonsustained ventricular tachycardia on the morbidity and mortality in Pts resuscitated from near-fatal ventricular arrhythmias. See AVID, CABG, MADIT

avidity IMMUNOLOGY The degree of stability of an antibody's binding with its antigen, which is a function of the number of shared binding sites and the binding energy of an antibody and antigen–the sum of the binding affinities of the combining sites on the antibody. Cf Affinity

aviophobia PSYCHOLOGY Fear of flying. See Phobia

AVM 1. Acute viral meningitis, see there 2. Arteriovenous malformation, see there

avoidance PSYCHIATRY The intentional skirting of new interpersonal contacts to the extent of impairing social function

avoidant personality disorder PSYCHOLOGY A pervasive pattern of social inhibition, feelings of inadequacy, and hypersensitivity to negative evaluation, which begins by early adulthood, and is present in various contexts

---

AVOIDANT PERSONALITY DISORDER-DSM 301.82 Requires 4 + of the following

AVOIDS OCCUPATIONAL ACTIVITIES involving significant interpersonal contact, due to fear of criticism, disapproval, or rejection

UNWILLING TO GET INVOLVED with people unless certain of being liked

RESTRAINT IN INTIMATE RELATIONSHIPS due to fear of ridicule or shame

PREOCCUPIED with social criticism or rejection

INHIBITED IN NEW INTERPERSONAL SITUATIONS due to feelings of inadequacy

NEGATIVE SELF IMAGE Views self as socially inept, unappealing, or inferior to others

RELUCTANT TO TAKE RISKS or engage in activites that have the potential for being embarrassing

Modified from DSM-IV, Am Psych Asso, Washington DC, 1994

---

avulsion MEDTALK The tearing away, as may occur with a nerve or part of a bone

avulsion fracture Strain fracture ORTHOPEDICS A fracture that occurs when a fragment of bone is yanked out by tightened ligaments due to a strain or intense muscle contraction against resistance

awake Fully aroused SLEEP DISORDERS A polysomnographically defined state not equal to any NREM stages or REMS: characterized by alpha and beta waves, tonic EMG, voluntary REMS, and eye blinks

awake, alert & oriented PHYSICAL EXAM A status—the Pt is up & at em, bright-eyed & bushy-tailed and knows which side of bed he/she currently is on; AAO is determined by physical examination and generally indicates that the Pt is mentally appropriate

Aware® A rapid point-of-care HIV screen which yields results from a fingerstick blood sample. See AIDS, HIV, Western blot

AWD Alive with–disseminated and/or metastatic disease

Axcis™ percutaneous myocardial revascularization system INTERVENTIONAL CARDIOLOGY A minimally invasive system for treating CAD by accessing the beating heart through the femoral artery, creating 15 to 20 channels by laser catheter in the left ventricle to revascularize the damaged heart. See PTMR, TMR

axial compression ORTHOPEDICS A type of force, especially of the foot and vertebral column, in which body weight falls centrally on a particular bone. See Compression fracture

axial-load teardrop fracture SPORTS MEDICINE A vertebral body and posterior neural archburst fracture caused by severe compression; the unstable fracture encroaches on the spinal cord, resulting in paralysis. Cf Teardrop fracture

Axid® Nizatidine, see there

axillary line PHYSICAL EXAM A tissue landmark which extends southward from the axilla, under which lies mammary tissue. See Breast

axillary lymph node dissection SURGERY The excision of the lymph nodes in the armpit, a procedure commonly performed with mastectomy for breast CA. See Breast cancer

axillary lymphadenopathy CLINICAL MEDICINE A general term for clinically obviously enlargement of lymph nodes in the axillary region, which is a zone of lymphatic drainage from the arm and breast

axillary nerve dysfunction Axillary nerve neuropathy NEUROLOGY A general term for shoulder dysfunction related to axillary nerve injury. See Neuropathy

axillary tail Axillary fat pad SURGERY The fibroadipose tissue in the axilla which contains 5 to 25 lymph nodes and breast and arm lymphatics

axis MEDTALK A central or straight line between two structures. See Brain-gut axis, Hypothalamic-pituitary-adrenal axis

axis fracture C2 fracture, see there

Axis I PSYCHIATRY A classification dimension used with DSM-IV, which includes clinical disorders and syndromes and/or other areas of concern. See DSM-IV, Multiaxial system

Axis II PSYCHIATRY A dimension used with DSM-IV, which includes personality disorders: paranoid, schizoid, schizotypal, antisocial, borderline, histrionic, narcissistic, dependent, obsessive-compulsive, personality "NOS" and mental retardation. See DSM-IV, Multiaxial system

Axis III PSYCHIATRY A dimension used with DSM-IV for conditions which may impact emotions–infections, neoplasia, endocrine, nutritional, metabolic & immunity, hematologic, neurologic, circulatory, respiratory, digestive, genitourinary, pregnancy/childbirth, skin, musculoskeletal, connective disorders, congenital anomalies, perinatal problems, ill-defined conditions, injury/poisoning. See DSM-IV, Multiaxial system

Axis IV PSYCHIATRY A dimension used with DSM-IV for psychosocial stressors–death, divorce, loss of job, etc in the form of problems; primary support group problems, social environment problems, educational problems, occupational problems, housing problems, economic problems, problems with access to health care, problems related to legal system/crime. See DSM-IV, Multiaxial system

Axis V PSYCHIATRY A dimension used with DSM-IV for factors that affect a person's mental functions–eg, psychologic, social, and occupational factors, impairment from physical or environmental limitations, "trait" measure of functioning–eg, best in past yr, etc. See DSM-IV, Multiaxial system

axis of motion An axis that is perpendicular to the plane in which the joint motion occurs; the closer the axis of the motion is to the body plane, the less movement there is in that body plane

Axokine® Ciliary neurotrophic factor, see there

azacitidine Vidaza® HEMATOLOGY A nucleoside analogue that may be used to treat β-thalassemia as it stimulates fetal globin production and myelodysplastic syndrome SIDE EFFECTS Neutropenia, thrombocytopenia, renal failure, liver failure

azathioprine Imuran® IMMUNOLOGY An immunosuppressant used to prevent rejection of heart, kidney, lung and other allografts, acting primarily on T-cells; it is used in rheumatoid arthritis, myasthenia gravis ADVERSE EFFECTS BM suppression–leukopenia, thrombocytopenia, macrocytic anemia, GI tract–N&V, diarrhea, fever, malaise, myalgia, LFT abnormalities, hepatotoxicity, pancreatitis; should not be given with allopurinol. See Heart transplant, Kidney transplant, Lung transplant, Transplantation

Azedipamin Diazepam, see there

azimilide Stedicor® CARDIOLOGY An antiarrhythmic used for A Fib, supraventricular arrhythmias, and to prevent sudden cardiac death in post-MI Pts. See ALIVE, ASAP

azithromycin Zithromax® INFECTIOUS DISEASE A broad-spectrum once-daily advanced-generation macrolide antibiotic used to manage community-acquired and other pneumonias caused by *S pneumoniae, H influenzae, Moraxella catarrhalis, C pneumoniae, M pneumoniae,* chlamydial infections, UTIs, MAC, possibly also toxoplasmosis and cryptosporidiosis ADVERSE EFFECTS Diarrhea, N&V, abdominal pain, vertigo, photosensitivity, hearing loss, injection site pain RELATIVE CONTRAINDICATIONS Cystic fibrosis, nosocomial infections, bacteremia, elderly, debilitated Pts. See Community-acquired pneumonia

azole INFECTIOUS DISEASE One of a family of synthetic broad-spectrum antifungal antibiotics–including ketoconazole, fluconazole, itraconazole INDICATIONS Various azoles have established efficacy in treating blastomycocosis, candidiasis, coccidioidomycosis, cryptococcosis, histoplasmosis, paracoccidoidomycosis and sporotrichosis ADVERSE EFFECTS N&V, anorexia, rash, pruritus, nonspecific ↑ liver enzymes, anemia, leukopenia, thrombocytopenia, headache, etc DRUG INTERACTIONS Plasma levels of azoles are ↓ by antacids, H₂-receptor antagonists, sucralfate, isoniazid, phenytoin, rifampin. See Fluconazole, Itraconazole, Ketoconazole

Azolid® Phenylbutazone, see there

AZOOR Acute zonal occult outer retinopathy, see there

azoospermia UROLOGY Complete absence of sperm in the semen, with resultant infertility. See Oligospermia

Azopt® Brinzolamide OPHTHALMOLOGY A topical carbonic anhydrase inhibitor used to manage ↑ intraocular pressure in Pts with ocular HTN or open-angle glaucoma. See Open-angle galucoma

azotemia NEPHROLOGY A higher than normal blood urea–BUN or other nitrogen-containing compounds in the blood; ↑ BUN may be: (1) prerenal, due to ↓ renal blood flow–with ↓ glomerular filtration rate–GFR and/or excess urea production, seen in dehydration, shock, ↓ blood volume, and CHF; (2) renal, with ↓ GFR due to acute or chronic renal failure; (3) postrenal, due to urinary tract obstruction or perforation with extravasation of urine; ↓ BUN occurs in pregnancy–due to ↑ GFR, malnutrition, high fluid intake, severe liver disease–↓ protein production. See Uremia

AZT 3'-azido-3'-deoxythimidine, zidovudine, Retrovir AIDS A nucleoside analogue, used with other anti-HIV agents to manage AIDS and HIV infection ADVERSE EFFECTS Long-term AZT use is associated with muscle loss, nausea, anemia, myelosuppression, oral ulcers, BM damage, headache. See AIDS, HAART

Azurogen Lorazepam, see there

B Symbol for: 1. *Bacillus* 2. Bel 3. Blood 4. 5-Bromouridine 5. Bronchodilator 6. Brucella 7. Buccal 8. Magnetic induction, expressed in teslas

b Symbol for: 1. Base 2. Bed 3. Inconvertible enzyme 4. Magnetic flux density

B antigen TRANSFUSION MEDICINE A major blood group–ABO antigen which defines the blood type B, and assumes that the codominant allele at the ABO locus is B or H; B antigens are highly immunogenic; when a B unit of packed red cells is transfused into an A or O recipient, the natural antibodies present in the recipient are capable of evoking a severe or fatal hemolytic transfusion reaction. Cf A antigen, Bombay phenotype, H antigen

B cell B lymphocyte HEMATOLOGY One of the 2 major classes of lymphocytes, which comprises 30% of circulating lymphocytes and is concentrated in the follicular zones of lymphoid tissue–in contrast T cells is located in the deep cortex; B cells are responsible for antibody production, a transformation evoked by interaction with the appropriate CD4 T-helper cells; Igs are responsible for specific defense against viruses and bacteria, immune surveillance–cytolysis of potentially malignant 'self' cells, mediation of antibody-dependent cell cytotoxicity, allergic reactions, formation of antigen-antibody complexes, and production of cytokines; surface and cytoplasmic antigens indicate the degree of B cell maturation and function; cytoplasmic IgM is present in pre-B cells and surface Ig in mature B lymphocytes and plasma cells; complement receptors are seen in mature cells; B cell 'markers' include CD9, CD10, CD19, CD20, CD24, Fc receptor, B1, BA-1, B4 Ia. See Pan B cell markers, Cf T cells

B-cell lymphoproliferative syndrome An uncommon, life-threatening complication of BM or organ transplantation caused by profound immunosuppression, which may be induced by EBV; BLS occurs in < 0.5% of the recipients of HLA-identical BM, but is common in Pts who suffered severe GVHD and were treated with anti-CD3 antibodies CLINICAL Ranges from self-limited, spontaneously-resolving infectious mononucleosis to oligoclonal or monoclonal proliferations and aggressive lymphomas PROGNOSIS 80-90% mortality in BM recipients; 60% survival in those receiving other organs. See EBV-related associated lymphoproliferative disorder

B-CHOP ONCOLOGY A chemotherapy regimen consisting of Bleomycin Cytoxan, hydroxydaunomycin, Oncovin, prednisone. See CHOP

B chromosome GENETICS A supernumerary–'extra' segment of DNA present in many species, which may be driven to self-duplicate, as BCs are transmitted at higher rates than otherwise expected from classic mendelian genetics

B complex vitamin NUTRITION The water-soluble B vitamins: B_1–thiamin, B_2–riboflavin, B_3-niacin, B_6-pyridoxine, B_{12}-cobalamin, and folate

B-DOPA Bleomycin, dacarbazine, Oncovin-vincristine, prednisone, Adriamycin-doxorubicin ONCOLOGY A 'salvage' chemotherapy regimen used for treating Pts who have had disease–eg, lymphoma relapse after 1° radiotherapy or chemotherapy. See Chemotherapy, Salvage chemotherapy

B mode See Ultrasound

B-MOPP ONCOLOGY A chemotherapy regimen consisting of bleomycin, mechlorethamine, Oncovin-vincristine, procarbazine, prednisone. See Chemotherapy

B symptom ONCOLOGY Any manifestation of systemic disease associated with leukemia and lymphoma, including significant fever–in Hodgkin's disease, the classic, but uncommon Pel-Ebstein fever, night sweats and unintentional weight loss of > 10%; 'A' symptoms correspond to an absence of clinical manifestations of malignancy

B1 therapy ONCOLOGY A therapy for chemotherapy-refractory low-grade and transformed low-grade NHLs. See Non-Hodgkin's lymphoma

B-type natriuretic peptide See BNP

B-type virus B virus, see there

B value Self-actualization need, see there

B virus B-type virus, herpesvirus simiae A virus enzootic to macaques and Old World monkeys that is closely related to herpes simplex. See C-type virus, Mouse mammary tumor virus

B19 VIROLOGY A parvovirus that infects humans, causing fifth disease-erythema infectiosum, hydrops fetalis–due to intrauterine infection, pure red cell aplasia and chronic anemia in immunocompromised Pts. See Fifth disease

babbling NEUROLOGY Quasi-random vocalizations in infants that precede language acquisition. See Lalling stage

babbling stage NEUROLOGY The stage of language development that follows cooing, which begins ± 6 months after birth, and consists of vowel-consonant-labial and nasogutteral combinations. Cf Cooing stage, Lalling stage

Babesia microti The babesiosis agent, an intraerythrocytic protozoan parasite endemic in rodents, *Peromyscus* spp and *Microtus* spp, transmitted to humans via a tick, *Ixodes dammini* DIAGNOSIS LM of thick smears of peripheral blood

babesiosis INFECTIOUS DISEASE A systemic infection caused by *Babesia* spp, in particular *B microti*; in Nantucket, an endemic region, ±40% of *I dammini* have *B microti* sporozoites in their saliva CLINICAL 1-4 wk incubation, fever, shaking chills, malaise myalgias, fatigue, hemoglobinuria TREATMENT Clindamycin, quinine. See *Babesia microti*

Babinski sign Babinski's reflex, plantar reflex NEUROLOGY A reflex movement of the big toe upward instead of downward when the plantar aspect of the foot is stroked, a maneuver used to test injury to, or diseases of, the upper motor neurons

baby VOX POPULI A popular term for an infant, or the youngest person in a family, from birth to toddlerhood–circa age 2. See Blue baby, Blueberry muffin baby, Bollinger baby, Cloud baby, Collodion baby, Crack baby, Designer baby, Juicy baby, Test tube baby

baby aspirin THERAPEUTICS A popular term for a formulation that contains 81 or less mg of aspirin, used to ↓ blood coagulability. See Aspirin

baby batter Ejaculate, see there

baby bottle PEDIATRICS A surrogate mammary that may be filled with various liquids including baby formula CONS Sweetened formula may result in cavities, loss of deciduous teeth

baby bottle syndrome PEDIATRICS Severe caries of the deciduous teeth, caused by prolonged use of milk or juice bottles as a sleeping aid for infants

Baby Doe law Public Law 98-457 HEALTH & LAW Legislation passed in 1984, which requires states to establish mechanisms in their child-protection services responsive to reported medical neglect of disabled children. See Baby Doe, Baby Jane Doe

Baby Doe regulations HEALTH & LAW Federal regulations promulgated in 1985 for implementing the 'Baby Doe Law', which require that disabled infants with life-threatening conditions receive the '…*appropriate nutrition, hydration, and medication, which in the treating physician's…reasonable medical judgement will be most likely to be effective in ameliorating or correcting all such conditions*'. See Baby Doe, Baby–Jane Doe

baby fat COSMETIC SURGERY A popular term for 2 poorly-studied clinical forms of ♀ fat deposition–1. The fat that marks a girl's prepubescent body and face contour, ie baby *fat* 2. The extra tonnage/'lipo-bulk' that accumulates during pregnancy, ie *baby* fat, ne'er to lose. See Cellulite

baby shortage SOCIAL MEDICINE A term referring to the disparity between the number of people interested in adopting infants from the US and Canada and the number of infants in need of adoption. See Adoption

baby selling Black market adoption SOCIAL MEDICINE The illegal act of selling an infant to adoptive parents or other persons by the birthparents or by an intermediary. See Adoption. Cf Independent adoption

baby teeth Deciduous teeth DENTISTRY The 20 teeth of the 1st dentition, which begin to calcify at the 4th month. Cf Permanent teeth

babysitter A person, often an intelligent family member, who stays by the bedside of a Pt requiring mechanical ventilation, and guards for equipment malfunctions or other problems

bacillary angiomatosis Epithelioid angiomatosis INFECTIOUS DISEASE A distinct vascular proliferative disorder of the skin and lymph nodes seen in HIV-positive subjects ETIOLOGY *Rochalimaea quintana* is the most common cause of BA; *R henselae* is less common, but may also cause bacteremia, as well as bacillary peliosis hepatis, and splenitis CLINICAL Erythematous papules and nodules, fever, bacteremia TREATMENT Erythromycin, other antibiotics. See AIDS, Peliosis hepatitis

Bacillus anthracis INFECTIOUS DISEASE A gram-positive organism which causes often fatal infections when its endospores–resistant to heat, drying, UV light, gamma radiation, and many disinfectants–enter the body and cause septicemia MILITARY MEDICINE *B anthracis* has been touted as a viable biological weapon; it was used only once, by the Japanese army in Manchuria in the 1940s. See Anthrax, Biological warfare

back extensor strength BES GERIATRICS A parameter used to evaluate elderly Pts with lower back pain and osteoporosis; it is measured by using a back isometric dynamometer

back labor OBSTETRICS A popular term for the location–ie, the lower back, of pain and discomfort due to uterine contractions, which most commonly occurs with posterior presentation. See Labor

back pain See Low back pain

backbone VOX POPULI Vertebral column, see there

background information SOCIAL MEDICINE Data provided to prospective adoptive parents on a child and his/her biological family

background radiation RADIATION The amount of ionizing-electromagnetic radiation to which a person is exposed from natural sources including terrestrial radiation due to natural radionuclides in the soil–eg, radon, cosmic radiation, and fallout in the environment from anthropogenic sources. See Radon

background retinopathy Nonproliferative retinopathy OPHTHALMOL-OGY Early stage of diabetic retinopathy; usually does not impair vision

backup INFORMATICS A function that allows users to copy one or more files to a storage pool to protect against data loss. See Incremental backup, Mirroring, Selective backup. Cf Restore

backup contraception GYNECOLOGY The use of 2° contraception in the event of failure or suboptimal 1° contraception. Cf Emergency contraception

backward failure CARDIOLOGY Heart failure attributed to ↑ filling pressure of the ventricles, due to obstruction, as occurs with mitral or tricuspid stenosis, which causes ↑ venous pressure with congestion, ie backward failure. Cf Forward failure

backwash ileitis GASTROENTEROLOGY The contiguous mucosal involvement of the terminal ileum, extending proximally, seen in 10-15% of cases of ulcerative colitis–UC as a 'spillover' phenomenon; unlike Crohn's disease which may involve the entire GI tract, UC is usually seen in the colon and rectum and occasionally, 'backwashed' ileum IMAGING Ulceration and irregularity; irregularity and stringiness may be seen in the mucosal pattern of postevacuation films; the mucosa may have a pseudopolypoid appearance with cobblestoning. See Ulcerative colitis

baclofen Lioresal® NEUROLOGY A GABA antagonist used to ↓ recalcitrant spinal spasticity in Pts with multiple sclerosis, spinal-cord injury, and possibly also Parkinson's disease

BACOD ONCOLOGY A chemotherapy regimen–bleomycin, Adriamycin-doxorubicin, cyclophosphamide, Oncovin–vincristine, dexamethasone

m-BACOD ONCOLOGY A 2nd-generation combination chemotherapy regimen–MTX, bleomycin, Adriamycin-doxorubicin, cyclophosphamide, Oncovin–vincristine, dexamethosone

BACOP ONCOLOGY A 3rd-generation combination chemotherapy regimen–bleomycin, Adriamycin-doxorubicin, cyclophosphamide, Oncovin–vincristine, prednisone, used for NHL. See CHOP

bacteremia The presence of bacteria in blood

bacteremic shock INFECTIOUS DISEASE Shock induced by bacteria actively growing in the circulation, equivalent to septic shock. See Septic shock, Shock

bacterial arthritis Pyogenic arthritis RHEUMATOLOGY An acute arthropathy characterized by painful swelling of a joint, fever, ↑ WBCs, local heat and inability to move the joint; early, the joint is distended with pus, which may be accompanied by aseptic necrosis of subchondral bone; if untreated, the synovial space may be replaced by granulation tissue and fibrosis, resulting in bony ankylosis AGENTS *Staphylococcus, Streptococcus, Gonococcus*.

bacterial count PUBLIC HEALTH The concentration of coliform bacteria in water, a quantity that loosely correlates with the level of contamination of drinking and recreational waters. See Public water

bacterial meningitis Acute bacterial meningitis NEUROLOGY Meningeal inflammation caused by bacteria which, if untreated, is often fatal, or associated with significant sequelae EPIDEMIOLOGY 60% are community-acquired–CM, 40% nosocomial–NM PREDISPOSING FACTORS Recent neurosurgery/use of neurosurgical devices–60% NM, immune dysfunction–20.5% NM; 7.5% CM, CSF leak–8.6% NM, 3% CM, head injury–8.6% NM, 3.5% CM,–acute or chronic otitis media–< 1% NM, 10% CM, sinusitis, pneumonia, endocarditis, DM, alcoholism CLINICAL Fever–95% at presentation, nuchal rigidity–88% at presentation, neurologic signs—confusion, lethargy, seizures, papilledema MICROBIOLOGY *H influenzae* > *S pneumoniae* > *N meningitidis* MANAGEMENT Antibiotics and dexamethasone MORTALITY ± 25%, highest in those > age 60, obtunded at presentation, or with seizures in past 24 hrs

bacterial overgrowth GI DISEASE The multiplication of opportunistic bacteria in the lower GI tract, often due to antibiotic therapy. See Pseudomembranous colitis LAB MEDICINE The multiplication of contaminating bacteria in a specimen–eg, blood, urine, due to inadequate fixation or delay in processing the specimen in a timely fashion

bacterial skin infection Cellulitis, see there

bactericidal Bacteriocidal *adjective* Referring to that which kills bacteria, or is capable of killing bacteria. Cf Bacteriostatic

bacteriological specimen INFECTIOUS DISEASE Any body fluid, secretion, or tissue–eg, blood, sputum, urine, CSF, biopsy material, sent to a laboratory, from which bacterial cultures are performed

bacteriostatic BIOLOGY *adjective* Referring to inhibition of bacterial growth and/or reproduction *noun* An agent that inhibits bacterial growth and/or reproduction. Cf Bactericidal

Bacteroides fragilis MICROBIOLOGY A gram-negative anaerobe which is a frequent isolate in anaerobic cultures; it is associated with abscesses, aspiration pneumonia, endocarditis, septic arthritis and empyema

bacteruria NEPHROLOGY Bacteria in urine. See Too numerous to count, Urine culture

Bactrim® Trimethoprim/sulfamethoxazole. See TMP/SMX

Bactroban® Mupirocin cream A topical antibiotic for treating traumatic skin lesions with 2° infection by susceptible strains of *Staphylococcus aureus*, methicillin-resistant *S aureus*–MRSA, *Streptococcus pyogenes*

bad baby OBSTETRICS A popular term for a baby with unexpected birth-related defects or injuries, who has a multiproblem clinical course post-partum

bad baby law Neurologic birth injury compensation plan MALPRACTICE Any legislative act that limits the liability that obstetricians have with birth-related injuries to the spinal cord or brain of a live-born infant weighing ≥ 2500 grams at birth. See Bad baby

bad cholesterol LDL-cholesterol CARDIOVASCULAR DISEASE Cholesterol transported in the circulation by low-density lipoprotein, the elevation of which is directly related to the risk of CAD and cholesterol-related morbidity See LDL-cholesterol. Cf Good cholesterol

bad debt MEDICAL PRACTICE Any bill submitted for payment by a third-party payer or Pt, which is not paid in full

bad eicosanoid A popular term for an eicosanoid arising from the metabolism of dihomogamma-linolenic acid by the enzyme delta-5-desaturase into arachidonic acid, which is converted into thromboxane A$_2$—which ↑ platelet aggregation, PGE$_2$—which promotes pain and depresses the immune system, and leukotrienes—which are linked to allergic responses and skin disease. See Zone-favorable diet. Cf Good eicosanoid

bad habit Unhealthy habit CLINICAL MEDICINE A patterned behavior regarded as detrimental to physical or mental health, which is often linked to a lack of self-control. Cf Good habit

bad trip SUBSTANCE ABUSE A hallucinogenic drug-induced experience in which the pattern of time-space disorientation causes an intense adverse neuropsychiatric response in the person taking the 'trip'. See Flashbacks, 'High'

Baecke questionnaire A survey of habitual physical–eg, sports-activity, ranging from 1–lowest activity to 5—highest activity, which is used in epidemiological studies

BAER Brainstem auditory evoked response, see there

bag DRUG SLANG Street slang for a container for drugs

bag of waters See Amniotic sac

bag of worms appearance A lesion or density that has a multivermiform pattern, or one in which the activity is likened to a quivering mass of live elongated organisms NEUROLOGY Asynchronous, involuntary quivering, and squirming of multiple skeletal muscle fascicles of the tongue, which disappear during sleep and may be suppressed with rest, sedation, or volition, classically seen in Sydenham's chorea UROLOGY The tactile sensation of a varicocele lying within the scrotal sac

bag price Price per accession HEALTH REIMBURSEMENT A 'unit' price charged to a client of a commercial lab, and used to calculate costs, profit margins, commissions for salespersons, and so on

bagassosis OCCUPATIONAL MEDICINE A hypersensitivity pneumonitis seen in cane sugar workers sensitive to *Thermoactinomyces sacchari*, *T vulgaris*, a fungus that grows well in sugar cane pressings. See Farmer's lung

baggage PSYCHOLOGY A popular term for a constellation of mental issues based on memories and experiences that any person has when entering into an intimate relationship

bagging DRUG SLANG A 'street' term for the use of inhalant abuse substances EMERGENCY MEDICINE Manual respiration for a Pt with dyspnea, using a handheld squeeze bag attached to a face mask

bags Bags under the eyes COSMETOLOGY Loose suborbital skin caused by inflammation and edema, associated with vasodilation–the blood vessels impart the dark color CAUSES Sleep deprivation, smoking, aerosolized irritants, allergic 'shiners', various dermatopathies, aging

Bairnesdale ulcer INFECTIOUS DISEASE A necrotizing skin ulcer caused by *Mycobacterium ulcerans*, first described in Bairnesdale, Australia. See Buruli ulcer

Bak *Bak* A member of the *bcl-2* family expressed in a wide range of cells which, when overexpressed in NGF-deprived sympathetic neurons accelerates apoptosis, and counteracts Bcl-2's apoptosis-protecting effects. See Bcl-2

BAK procedure ORTHOPEDICS A surgical procedure that uses the BAK interbody fusion system to stabilize and fuse spinal vertebrae in Pts with chronic low back pain 2° to degenerative disk disease. See Low back pain

Baker's cyst Synovial cyst of popliteal space ORTHOPEDICS A localized post-traumatic swelling of a bursa sac behind the patella corresponding to a cyst which consists of a membrane-lined sac filled with synovial fluid that has escaped from the joint. See Synovial fluid

Baker scale COSMETIC SURGERY A system–from grade 1 to 4–used by plastic surgeons to quantify the degree of capsular contraction surrounding a breast implant in Pts with augmentation mammoplasty

Bakke v Regents of the University of California MEDICAL EDUCATION A lawsuit initiated by a white student in California, in response to an affirmative action in 1973 by the U of California Medical School, Davis, which admitted minority students who had lower grades and test scores than white students. See Affirmative action, Project 3000 by 2000, Reverse discrimination, Underrepresented minorities

BAL 1. Blood alcohol level, see there 2. Bronchoalveolar lavage, see there

Baladi inverter HEART SURGERY A device inserted into an aortic incision which is inverted to form a bowl, providing a blood-free workspace during aortic grafting procedures, as an alternative to clamping in conventional and beating heart procedures. See Beating heart surgery

balance NEUROLOGY Equilibrium, see there PHYSIOLOGY Homeostasis. See Fat balance, Health balance, Nitrogen balance, Water balance

balance billing MANAGED CARE The practice of billing Pts in excess of the amount approved for payment by a health plan, Medicare, or private fee-for-service insurance. See Allowable charge, Nonparticipating physician

balance disorder AUDIOLOGY A disturbance in equilibrium due to a disruption of the labyrinth. See Equilibrium

balanced diet CLINICAL NUTRITION A diet with proportionate amounts of foods which are optimal for good health. See Food groups, Food pyramid

balanitis Balanoposthitis UROLOGY Inflammation of the glans penis (and prepuce), usually linked to phimosis. See Phimosis

balantidiasis PARASITOLOGY Infection with the intestinal ciliate, *Balantidium coli* EPIDEMIOLOGY Human infection–balantidiosis–is linked to pig farming and slaughtering CLINICAL Generally asymptomatic; heavy trophozoite loads results in bloody dysentery, severe dehydration or death; some Pts have intestinal ulcers, mesenteric lymphadenitis with extension to the liver, lung etc MANAGEMENT Tetracycline, metronidazole

bald tongue Complete atrophy of lingual papillae, seen in pernicious or iron-deficiency anemias, pellagra–bald tongue of Sandwith, syphilis CLINICAL Pain, burning, beefy red color

ball VOX POPULI 1. A sphere. See Birthing ball, Retraction ball 2. A testicle. See Blue ball syndrome

ball & socket Ball-in-socket *adjective* Referring to a morphology in which there is a sharply circumscribed round mass surrounded by a clear or lucent space followed by a density similar to the first central 'ball' IMAGING An appearance of the sequelae of epiphyseal fractures with a primary fusion of the central portion of the epiphysis, appearing as a ball within a relatively radiolucent 'socket', seen in infantile scurvy, battered child syndrome, acromelic dwarfism RHEUMATOLOGY A descriptor for the radiologic appearance of compressive erosions in the interphalangeal joint of rheumatic diseases, where a central sclerotic zone is surrounded by an osteoporotic 'ring'

ball & socket joint The most freely movable type of synovial joint–eg, shoulder joint, hip joint

ball valve obstruction CHEST MEDICINE A partial endobronchial obstruction which allows facile entry but not egress of air, resulting in a build-up of pressure in the terminal airways and potential rupture of alveoli; air percolates into soft tissues, causing interstitial emphysema COMPLICATIONS Extreme BVO leaks may cause tension pneumothorax with ↑ positive pressure in the hemithorax, shifted mediastinum and compromised circulation

ballistic stretching Bouncing stretching SPORTS MEDICINE Rapid, jerking movements in which a body part is moved with a momentum that would stretch the muscles to a maximum; during the bouncing motion, the muscle responds by contracting, to protect itself from overstretching

balloon angioplasty CARDIOLOGY A minimally invasive procedure in which a catheter with an inflatable balloon tip is inserted through the femoral or brachiocephalic artery and 'snaked' to a previously identified–by angiography zone of arterial stenosis or occlusion. See Atherosclerosis, Cholesterol, Percutaneous transluminal coronary angioplasty. Cf Stenting

balloon dissector SURGERY An instrument used as an alternative to blunt dissection; the BD is inserted uninflated through an incision between tissue layers, then filled with air or saline, resulting in separation of the planes

balloon tamponade EMERGENCY MEDICINE A hemostatic procedure for upper GI bleeding using a Sengstaken-Blakemore tube inflated in the stomach, which anchors the device in the desired location INDICATIONS Bleeding esophageal varices or any persistent esophageal hemorrhage. See Sengstaken-Blakemore tube

balloon valvoplasty A method used to treat stenotic cardiac valves, including 1. Pulmonic valve, where balloon valvoplasty is considered by some to be an optimal therapy 2. Mitral valve The results may be suboptimal if the valve ring is extensively calcified, but good if the valve is pliable or 3. Aortic valve Valvoplasty is considered the optimal therapy only for frail, elderly Pts who are otherwise poor surgical candidates; recurrence of symptoms, restenosis and death occur in 50%. See Inoue balloon

ballooning IMAGING A term for the biconcave compression of the end-plate of a vertebral body, caused by ↑ pressure in the intervertebral discs, usually in the lumbar spine, which may be seen in osteoporosis. Cf Fish vertebrae

ballotable *adjectice* PHYSICAL EXAM Referring to an intraabdominal mass which is palpable and which can be "bounced" back and forth

balsalazide Colazal® INTERNAL MEDICINE An agent that delivers concentrated 5-aminosalicylic acid to the colonic mucosa in Pts with ulcerative colitis. See Ulcerative colitis

BALT Bronchiole-associated lymphoid tissue, see MALT

BAMI CARDIOLOGY A study–Biochemical markers for Acute Myocardial Infarction that compared serum troponin I levels with CK-MB levels for diagnosing acute MI. See Cardiac profile guideline, Troponin I

banana form Crescent form PARASITOLOGY A term for the morphology of macrogametocytes, the ♀ sexual intraerythrocytic form of *Plasmodium falciparum* which also has compact chromatin

banana sign IMAGING An ultrasonographic finding seen when a major neural tube defect accompanies Arnold-Chiari malformation APPEARANCE Cerebellar tonsils and midbrain structures herniate into the foramen magnum, causing ventriculomegaly 2° to compression of the outflow from the 3rd and 4th ventricles; the 'banana' corresponds to the compressed cerebral hemisphere

bancroftian filariasis TROPICAL MEDICINE Infection with *Wuchereria bancrofti*, which causes elephantiasis, hydrocele, and regional economic loss TREATMENT Diethylcarbamazine, ivermectin, albendazole. See Diethylcarbamazine, Filariasis

band LAB MEDICINE An aggregate of a particular protein or group of proteins on an electrophoresis of serum protein

Band-Aid™ hospital MEDTALK A popular term for a health care facility which provides for minimal care of significance

Band-Aid™ surgery A popular term for laparoscopic surgery, so called as the size of the incision is so small that once sutured, the wound can be covered with a small adhesive strip.

band keratopathy OPHTHALMOLOGY A broad deposit of opaque calcium phosphate in vertical lines parallel to, within and often lateral to the limbus on Bowman's membrane, seen by slit lamp examination

bandage scissors WOUND CARE Scissors used to remove bandages and other objects taped or adherent to the skin surface, which is characterized by having a flat plate perpendicular and outside of one or both cutting blades. See Scissors

banding CARDIAC PATHOLOGY Zonal changes of cardiac muscle due to ischemia, which are characterized by opaque transverse bands within myocytes adjacent to an intercalated disc, accompanied by shortening and scalloping of the sarcomere, fragmentation of Z bands, distortion of myofibrils and displacement of mitochondria away from intercalated disc. See Contraction band necrosis. Cf Wavy changes

BANF Bilateral acoustic neurofibromatosis. See Neurofibromatosis, type 2

bank A central repository of something of value, for future withdrawal or retrieval. See Blood bank, Brain bank, National Practitioner Data Bank, Organ Bank, Sperm bank, Tissue bank

Bankart lesion ORTHOPEDICS Shoulder instability due to detachment of the inferior glenohumeral ligament complex from the inferior glenoid, which is often accompanied by stretching of the remaining fibers, leading to shoulder laxity. See Position. Cf Beach chair position

banking To place something of value in a repository for future withdrawal or retrieval. See Sperm banking

BAO Basal Acid Output, see there

BAP ONCOLOGY Bleomycin, Adriamycin-doxorubicin, prednisone–chemotherapy regimen

BAPP ONCOLOGY A chemotherapy regimen consisting of bleomycin, Adriamycin-doxorubicin, cisplatin, prednisone

bar-reading test NEUROLOGY A test of binocular and stereoscopic vision in which a ruler is held midway between the eyes and printed material

Bárány's sign NEUROLOGY Nystagmus on the opposite side of an ear irrigated with cold water.

barbell tumor Dumbbell tumor, see there

barber's itch Barber's rash, sycosis DERMATOLOGY A staphylococcal infection of the hair follicles in the bearded area of the face, usually the upper lip, which is aggravated by shaving; tinea barbae is similar to barber's itch, but is caused by a fungus

barbiturate Downer NEUROPHARMACOLOGY Any sedative/hypnotic; they are derived from barbituric acid, are divided based on their duration of action into ultrashort, short, intermediate and long-acting–eg phenobarbital; barbiturates are the suicide drug of choice, or anticonvulsant, and among street drugs used as a 'downer'

barbiturate coma MEDICAL ETHICS The use of barbiturates–eg, pentobarbital, to induce a loss of consciousness, which has a 2°–'double' effect of inducing hypoxia and respiratory failure. See Euthanasia, Physician-assisted suicide

bare See Go bare

BARI CARDIOLOGY A clinical trial–Bypass Angioplasty Revascularization Investigation/Intervention–which compared the efficacy of CABG and PCTA as methods of revascularization in Pts with multivessel CAD. See Coronary artery bypass graft surgery, Percutaneous coronary angioplasty

bariatric medicine Obesitology CLINICAL MEDICINE The formal study of obesity and its treatment. See Morbid obesity, Obesity, Overweight

barium enema Lower GI series IMAGING Examination of the large intestine, by instilling barium sulfate, either alone–single-contrast, or with air–double contrast, per rectum until the colon is filled; any regional or zonal defects in the 'column' of barium may indicate inflammation, polyps, or tumors; visible pouches of barium can indicate diverticuli INDICATIONS Pts with Hx of altered bowel habits, lower abdominal pain, or passage of blood, mucus, or pus in the stool; abnormal in CA, diverticulitis, chronic ulcerative colitis, Crohn's disease, polyps, intussusception, gastroenteritis, stenosis, irritable colon, vascular injury, rarely in acute appendicitis CONTRAINDICATIONS Tachycardia, ulcerative colitis, toxic megacolon, suspected perforation of colon

barium peritonitis A rare complication of a barium enema due to colonic perforation with leakage/spillage of radiocontrast material, rarely accompanied by feces into the peritoneum, release of histamine and vasoactive substances with activation of coagulation pathways, resulting in fibrinous peritonitis MORTALITY High–± 50%

barium swallow IMAGING A technique in which a radiocontrast 'milkshake' of barium sulfate is swallowed to detect benign or malignant lesions of the pharynx, esophagus, stomach, and small intestine, and evaluate the integrity of the swallowing mechanism; the progress of the barium is followed radiographically to detect filling 'defects'–places where normal outline of barium should be seen but is not INDICATIONS Detection of foreign bodies, strictures, tumors, Barrett's esophagus, fistulas, reflux; definitive diagnosis of lesions requires endoscopic biopsy

BARN Bilateral acute retinal necrosis OPHTHALMOLOGY A condition characterized by herpes virus-induced anterior and posterior uveitis, papillitis with retinal detachment 1-3 months after onset, of which only 50% are bilateral. See Herpes

BarOn test BarOn Emotional Quotient Inventory PSYCHOLOGY A proprietary test of 'emotional intelligence'. See Marshmallow test

baroreflex failure PHYSIOLOGY A constellation of clinical findings characterized by marked lability of BP with systolic/diastolic HTN and tachycardia, ± headache, diaphoresis, emotional lability, and refractoriness of heart rate in response to exogenous vasoactive substances alternating with hypotension and bradycardia MANAGEMENT Clonidine suppresses pressor and tachycardic surges, diazepam, reduces stress. Cf Autonomic failure

barotitis externa ENT Pressure-related inflammation of the external ear canal

barotrauma AUDIOLOGY Middle ear injury caused by ↑ air pressure; trauma to the inner ear 2° to atmospheric pressure alteration, which occurs while flying or deep water diving, resulting in ↓ visual and proprioceptive cues due to ↓ vestibular input CLINICAL Disequilibrium, disorientation, N&V SPORTS MEDICINE Tissue injury due to the failure of a gas-filled body space–eg, lungs, middle ear, sinuses, to equalize internal pressure to ambient pressure; barotrauma often results from rapid or extreme changes in external pressure–eg, explosions. See Atmospheric inner ear barotruma, Pulmonary barotrauma

barrel chest CHEST MEDICINE A descriptor for a broad chest with hyperinflated, poorly aerated lungs, typical of emphysema; a similar short, but broadened chest of a different etiology occurs in Morquio syndrome, aka mucopolysaccharidosis, type IV

barren *adjective* GYNECOLOGY Infertile, sterile, fruitless, inconceivable

Barrett's esophagus GASTROENTEROLOGY A condition estimated to occur in ± 2 million Americans, which develops in Pts with GERD DEFINITION Replacement of normal stratified squamous epithelium with metaplastic, premalignant intestinal columnar epithelium in the distal esophagus, ± accompanied by peptic ulceration, typically a sequel to chronic reflux; the degree of dysplasia correlates with aneuploidy by flow cytometry ENDOSCOPY

BE changes include a proximal migration of the squamocolumnar Z-line, and patchy areas corresponding to single layered columnar cells in intimate contact with underlying blood vessels; although most Pts are adults, BE may affect children, suggesting BE has a congenital component; BE carries a ± 35-40-fold ↑ risk of esophageal adenoCA, which is almost invariably accompanied by dysplasia, and has a prognosis similar to that of epidermoid carcinoma–14.5% 5-yr survival MANAGEMENT-RESTORE NORMAL SQUAMOUS EPITHELIUM Electrocoagulation, argon plasma coagulation, laser therapy, laser + antireflux surgery, photodynamic therapy; thermal ablation of metaplastic esophageal mucosa has fallen into disfavor LOW-GRADE DYSPLASIA Follow-up, possibly β-carotene HIGH-GRADE DYSPLASIA Esophagectomy; endoscopic mucosal ablation, photodynamic therapy

barrier PHYSIOLOGY A physical or functional hurdle which a substance or cell must surmount or circumvent to have free access to a tissue or site in the body. See Blood-brain barrier, Bone marrow barrier SOCIAL MEDICINE An impediment in access to a service or activity, defined in the context of the Americans with Disabilities Act–ADA, which may be architectural–eg requiring widened doors, wheelchair ramps, and others or communication-related–eg linguistic barrier, vision defects. See Americans with Disabilities Act, Architectural barrier, Cultural barrier, Disability, Inequitable barrier, Reasonable accommodations

barrier-free *adjective* Pertaining or referring to structural or architectural design that does not impede use by individuals with special physical needs. See American with Disabilities Act, Architectural barrier, Barrier

barrier method REPRODUCTION MEDICINE An intervention that counters the formula, boy + girl = baby, with mechanical–eg condom, diaphragm—barriers. See Condom, Contraception, Contraceptive sponge, Diaphragm, Pearl index, Spermicide

barrier precautions INFECTION CONTROL A general term referring to any method or device used to ↓ contact with potentially infectious body fluids, including facial masks, doubled gloves and fluid-resistant gowns. See Isolation, Reverse isolation, Universal precautions

Barthel Index CLINICAL MEDICINE An interviewer-based instrument similar to the Kenny Self-Care Scale, used to assess physical functions, specifically self-care and ambulation–eg, stair-climbing. See ADL scale

Bartholin's abscess GYNECOLOGY An abscess due to a bacterial infection in Bartholin's glands, see there

Bartholin's cyst GYNECOLOGY A cyst that develops when the duct of a Bartholin gland is obstructed

Bartholin's gland *Glandula vestibularis major*, vulvovaginal gland ANATOMY Either of the paired glands on each side of the vaginal orifice which are homologues of the ♂ bulbourethral glands

bartholinitis GYNECOLOGY Inflammation of Bartholin's glands, which is usually bacterial

Bartonella henselae *Rochalimaea henselae* INFECTIOUS DISEASE A slender, fastidious coccobacillary bacterium of the normal flora of cats associated with bacteremia, endocarditis, cat-scratch disease, bacillary angiomatosis, peliosis hepatis; it may affect HIV-infected and immunocompetent Pts, causing persistent or relapsing fever; *B henselae* and *B quintana* have been linked to bacillary angiomatosis. See Bacillary angiomatosis

Bartonella quintana *Rochalimaea quintana* INFECTIOUS DISEASE A slender, fastidious coccobacillary bacterium found in the normal flora of small rodents transmitted by body lice, which causes trench fever, bacillary splenitis, bacteremia, endocarditis, cat-scratch disease, and cutaneous bacillary angiomatosis, relapsing fever, chronic lymphadenopathy in alcoholics, and in the homeless MANAGEMENT Empirical–nafcillin, ceftriaxone etc. See Trench fever

Bartter syndrome ENDOCRINOLOGY A rare endocrinopathy, usually of younger Pts, and may be seen in dwarfism; it is characterized by potassium-wasting, normotensive hyperreninemia, hyperaldosteronism, and vascular hyporesponsiveness to endogenous pressors–eg, norepinephrine, angiotensin II–attributed to ↑ PGE CLINICAL Proximal muscle weakness, failure to thrive, short stature, convulsions, tetany, cramps, ileus, gout, mental retardation, enuresis, nocturia, polyuria, salt craving, constipation, insensitivity to angiotensin II pressor effects, normal BP LAB Metabolic alkalosis, ↓ serum K+, ↑ urine K+, ↑ renin, ↑ aldosterone, DIFFDX Prolonged nasogastric suctioning, habitual vomiting, diuretic abuse MANAGEMENT Potassium loading, spironolactone to block aldosterone effects

basal acid output BAO Production of gastric H+ under baseline conditions, normal: 0-10 mmol/hr; BAO serves to evaluate the completeness of vagotomy; Pts with Zollinger-Ellison syndrome have a ratio of basal to maximal acid output–BAO/MAO of > 60%; BAO is also ↑ in pernicious anemia, gastric CA, myxedema, rheumatoid arthritis. Cf MAO, PAO

basal body temperature REPRODUCTION MEDICINE The lowest possible *normal* body temperature, which usually occurs in the morning before getting out of bed; measurement of BBT is a popular method for determining whether ovulation has occurred. See Anovulatory cycle

basal cell carcinoma Basal cell cancer, Ronald Reagan tumor DERMATOLOGY A usually indolent skin cancer most common in the sun-exposed regions of the head, neck, and upper body in older individuals TREATMENT Local excision; if areas where tissue border may compromise function or appearance–eg, angle of eyes, eyelids, nose, Mohs surgery may be indicated

basal cell nevus syndrome Nevoid basal cell carcinoma syndrome, basal cell carcinoma syndrome, Gorlin-Goltz syndrome A rare AD condition characterized by childhood onset of multiple nevoid basal cell carcinomas accompanied by skin defects, 'pits' in the hands and feet, in the form of 2-3 mm in diameter 'dells' occasionally filled with carcinoma, milia, sebaceous cysts, lipomas, fibromas, lymphomesenteric cysts, CNS disease–mental retardation, EEG abnormalities, calcification of dura, medulloblastoma, schizophrenia, endocrine system–ovarian cysts or fibroma, ♂ hypogonadism, ♀ escutcheon, scanty facial hair, eyes–canthal dystopia, hypertelorism, coloboma, congenital blindness, typical facies–hypertelorism, lateral displacement of medial canthi, mandibular prognathism, accentuated supraorbital ridges, jaw cysts and a broad nasal root, skeleton–spina bifida occulta, fused, absent or cervical ribs, kyphosis, scoliosis, cervical and thoracic vertebral fusion, bridging of sella turcica, frontal and temporoparietal bone 'bossing', shortened 4th-5th metacarpals, epithelial jaw cysts

basal energy expenditure The amount of oxygen consumed while resting and fasting, extrapolated to 24 hrs, roughly equivalent to 25 kcal/kg; BEE is used to determine an individual's caloric needs; BEE is multipled by level of activity–eg, 1.2 for bed rest; 1.4 for moderate activity and by an injury factor–1.2 for pneumonia, surgery, 1.3 for major injury, 1.5–1.6 for severe sepsis. See Basal metabolic rate

basal metabolic rate Basic metabolic rate, BMR A baseline rate of metabolism expressed as oxygen consumption or heat production under resting or basal conditions, usually long after eating

basal skull fracture ORTHOPEDICS A fracture involving the base of the cranium CLINICAL May be asymptomatic; raccoon eyes, Battle sign, hemotympanum, CSF rhinorrhea IMAGING Plain skull film may not reveal fracture; CT or MRI is more reliable. See LeFort fracture

base gait Stride width, walking base BIOMECHANICS The side to side distance–measured in millimeters–between the line of the 2 feet, usually measured at the mid point of the heel, but sometimes the center of the ankle joint

BASE jumping SPORTS MEDICINE An extreme sport in which participants jump–with parachute-from 4 types of structures that constitute the acronym: Buildings, Antennas, Spans–bridges, Earth–cliffs

baseline characteristic MEDICAL PRACTICE An initial finding or value in a Pt, before any formal intervention

base of tongue SURGICAL ANATOMY An area defined by the Am Joint Committee on Cancer, as extending from the line of the circumvallate papillae to the junction of the base of the epiglottis–valleculae, including the pharyngoepiglottis and glossoepiglottic folds See Oropharynx, Oropharynx proper

base pairing MOLECULAR BIOLOGY The specific complementary hydrogen bonding of the bases–purines and pyrimidines—in a double stranded nucleic acid; BP results in formation of a double helix from 2 complementary single strands; in DNA the pairs are adenine + thymine and guanine + cytosine; in RNA, the pairs are adenine + uracil and guanine +cytosine. See Hybridization

baseball finger Mallet finger SPORTS MEDICINE A flexion deformity at a 30° angle of the distal phalanx, produced by a blow to the tip of the finger, which in US is often associated with catching a baseball thrown at high speed, resulting in the forced flexion of the distal phalanx and a separation–by rupture or avulsion–fracture of the common extensor tendon from its insertion in the base of the distal phalanx, accompanied by an inability to extend the fingertip

baseball stitch SURGERY A type of surgical repair used to close the uterus in a classic cesarean section incision, an incision more cephalad than the now-preferred lower uterine segment incision, as it is associated with greater morbidity

baseline examination CLINICAL PRACTICE A physical exam which is part of an initial Pt-physician contact, and designed to assess a Pt's eligibility for enrollment in a clinical trial and produce requisite baseline data.

baseline test CLINICAL PRACTICE Any test than measures current or pre-treatment parameters, including chemistries, cell counts, enzyme levels and so on, against which response(s) to therapy, if any, is evaluated

bash SOCIAL MEDICINE *verb* To speak or refer to another in a disparaging fashion. See Gay bashing, Physician bashing

basic *adjective* Referring to a base; acting as a base in a salt; alkaline

basic benefits package MANAGED CARE A 'stripped-down' health care package offered by some health insurance policies, which may include preventive screening–eg pap smears and limited number of allowed inpatient and outpatient days

basic DRG payment rate MANAGED CARE The payment a hospital receives for a Medicare Pt in a particular DRG; the rate is calculated by adjusting the standardized amount to reflect wages and non-wage cost of living differences in the hospital's geographic area and cost of the DRG. See DRG, Standardized amount

basic life support EMERGENCY MEDICINE The constellation of emergency procedures needed to ensure a person's immediate survival, including CPR, control of bleeding, treatment of shock and poisoning, stabilization of injuries and/or wounds, and basic first aid. Cf Advanced life support

basic pacing rate Basic pacing interval CARDIAC PACING The rate at which an implanted pulse generator emits pacing stimuli in the absence of intrinsic cardiac activity

basic sciences Preclinical medicine GRADUATE EDUCATION The 2-3 yr period of medical school, that precedes clinical instruction and training, which provides a core of basic knowledge required for success in clinical medicine–for the student's rotations though surgery, internal medicine, gynecology, pediatrics, and the other fields of clinical medicine; BSs include anatomy, physiology, biochemistry, molecular biology, pharmacology, microbiology, pathology

basilar artery migraine Basilar migraine, see there

basilar migraine NEUROLOGY A type of classical migraine that first appears in adolescence, primarily in young ♀, often associated with the menstrual cycle, which is linked to a vasomotor defect of a major brain artery (carotid, vertebrobasilar); in BMs, aura symptoms are thought to arise in the brainstem or occipital lobes CLINICAL Loss of consciousness, drop attacks, vertigo, alternating hemiplegia, and confusion, ± accompanied by ataxia, incoordination, diplopia, dysarthria, and nystagmus, as well as pallor, vomiting, and photophobia, possibly followed by polydipsia, polyuria, irritability. See Classical migraine

basiliximab Simulect® NEPHROLOGY A high-affinity, monoclonal antibody used to minimize acute rejection post renal transplant, which complements antirejection drugs–eg, Neoral®. See Kidney transplant

basket case Train wreck VOX POPULI A derogatory term for a Pt with a dread disease or a terminal illness; a person to be pitied

basket cell HEMATOLOGY Fragile cell, smudge cell A fragmented and degenerated WBC in a peripheral blood smear, with a bare nucleus partially surrounded by a coarse network of splayed, red-purple nucleoplasm, which may be seen in normal subjects and ↑ in atypical lymphocytosis, CLL, AML, and CML, thus being similar in origin to 'smudge' cells

basophil Basophilic granulocyte HEMATOLOGY A type of granular WBC with large distinctly basophilic/"blue" 2° granules containing heparin, histamine, PAF and other mediators of the immediate hypersensitivity response, which are released when IgE cross-links to the high affinity Fc receptors on the cell surface

basophilia Basophilic leukocytosis HEMATOLOGY An absolute basophil count of > 100/mm³

BAT Blunt abdominal trauma. See Blunt trauma

bathing suit pattern Bathing suit distribution DERMATOLOGY A pattern of truncal skin involvement in rheumatic heart disease–erythema marginatum—characterized by flattened, nonblanchable erythema slowly enlarging maculopapules that undergo central healing DIFFDX X-linked lipidosis, Fabry's disease, glycosphingolipidosis, fucosidosis, sialidosis RISK FACTORS Lymphopenia, immobility, dry skin, low body weight, activity limited to bed or chair PUBLIC HEALTH A term for the pattern of wounds seen in abused ♀, which are those that would be covered by a one-piece bathing suit. See Domestic violence, Wife-abuser

bathroom privileges NURSING The allowing of postoperative Pts sufficient autonomy to address personal elimination needs. See Bed-ridden

bathtub-related electrocution FORENSIC MEDICINE An electrocution that occurs when a "live" electrical appliance falls in an occupied water-filled bathtub. See Electrocution

batimastat ONCOLOGY An anticancer angiogenesis inhibitor of matrix metalloproteinase

Batista procedure Left ventricular diameter reduction, left ventriculectomy, ventricular reduction CARDIOVASCULAR SURGERY A radical surgical procedure for treating some forms of end-stage heart failure, in particular CHF due to dilated cardiomyopathy. Cf Cardiomyoplasty

batophobia PSYCHOLOGY Fear of heights or being close to high buildings. See Phobia

batrachophobia PSYCHOLOGY Fear of amphibians, such as frogs, newts, salamanders, etc. See Phobia

battered child Abused child, battered baby PEDIATRICS A young child, often under age 3, who has been repeatedly and severely neglected by caretakers; BCs have signs of multiple episodes of trauma–eg, subdural hematomas, fractures, bruises in various stages of healing, often with FTT and chronic malnutrition See Child abuse

battered wife syndrome PUBLIC HEALTH A general term for a constellation of signs and symptoms that are typical of a ♀ wife/common-law partner who is physically abused by a husband/boyfriend. See Domestic violence, Spousal abuse. Cf Superbowl syndrome

battery CARDIAC PACING One or more power cells, usually chemical, that serve as a source of electrical power LAB MEDICINE A panel of tests. See Panel, Test battery MEDICAL MALPRACTICE The unauthorized touching of another person. See Assault, Ghost surgery, Informed consent, Malpractice

battle fatigue Posttraumatic stress disorder, see there

Battle sign PHYSICAL EXAM Postauricular ecchymoses of the head & neck, especially over the mastoid bone, a finding typical of basilar fracture of the skull

Baycol® Cerivastatin VASCULAR DISEASE A cholesterol-lowering, HMG-CoA reductase inhibitor–statin–for managing hypercholesterolemia and mixed dyslipidemia–it ↑ HDL-C by 4% and ↓ LDL-C by 36%. It was withdrawn fronm the market in mid-200/. See Statin

Bayesian analysis A decision-making analysis that '...*permits the calculation of the probability that one treatment is superior based on the observed data and prior beliefs...subjectivity of beliefs is not a liability, but rather explicitly allows different opinions to be formally expressed and evaluated.*' See Algorithm, Critical pathway, Decision analysis

Bayesian logic A type of reasoning in which the likelihood of an event occurring can be described in quantitative—ie probabilistic terms. See Artificial intelligence, Computer-assisted diagnosis

Bayley Scales of Infant Development PSYCHOMETRIC TESTING A revised standardization of the California First-Year Mental Scale, used in early stages of the Berkeley Growth Study, Bayley scales are applicable from birth to 15 months, measure varying stages of growth at each age level, supplemented by extensive longitudinal data on groups of infants

Baylisascaris procyonis Raccoon roundworm PARASITOLOGY An intestinal parasite of raccoons, the eggs of which may be ingested by humans, hatch in the intestine and migrate through organs and muscles CLINICAL Nausea, fatigue, hepatomegaly, loss of muscle coordination, blindness, encephalitis, death MANAGEMENT Possibly, albendazole PROGNOSIS Poor, profound neurologic impairment, partial paralysis, cortical blindness, developmental delay, etc

bayonet hand ORTHOPAEDICS A deformity of hereditary multiple exostosis–diaphyseal aclasis, which is characterized by ulnar deviation of the carpus and subluxation of the radius

bayonet incision An elongated S-shaped incision used to treat lacerations and for access in reconstructive surgery to the wrist bones

BBVP-M ONCOLOGY A chemotherapy regimen consisting of BCNU, bleomycin, VePesid, prednisone, methotrexate

BCAP ONCOLOGY A chemotherapy regimen consisting of BCNU, cyclophosphamide, Adriamycin-doxorubicin, prednisone

BCC 1. Basal cell carcinoma 2. Benign cellular changes, see there

bCG bacillus Calmette-Guerin ALTERNATIVE ONCOLOGY A strain of *Mycobacterium bovis* that has been grown multiple generations on potato bile glycerine agar to a point where it is is used to nonspecifically stimulate the immune response in Pts with certain malignancies–eg, melanoma. See Immune booster; Cf Coley's toxin, Ubenimex

BCMF ONCOLOGY A chemotherapy regimen: Bleomycin, cyclophosphamide, MTX, 5-fluorouracil

BCNU Bis-chlorethylnitrosourea, carmustine A chemotherapeutic related to lomustine–CCNU and semustine, which partially overlaps the activity/toxicity of alkylating agents INDICATIONS Hodgkin's disease, possibly NHL, melanoma, myeloma, brain tumors, GI carcinomas; BCNU crosses the blood-brain barrier and may be used for meningeal leukemia and brain tumors SIDE EFFECTS N&V, ↓ platelets, ↓ WBCs, 2° leukemia, pulmonary fibrosis, renal failure

BCOP ONCOLOGY A chemotherapy regimen: BCNU, cyclophosphamide, Oncovin-vincristine, prednisone

BCVPP ONCOLOGY A chemotherapy regimen: BCNU, cyclophosphamide, vinblastine, procarbazine, prednisone

BD Behavioral disorder, see there

beach chair position ORTHOPEDICS One of 2 positions–the other is lateral decubitus—for shoulder arthroscopy. See Position. Cf Lateral decubitus position

beading CARDIOLOGY Luminal irregularity of arteries supplying regions affected by electrical injury; Pts with beaded vessels are at ↑ risk for subsequent thromboses ORTHOPEDICS Multiple post-fracture tumefactions of the ribs, typical of osteogenesis imperfecta, type II. See Accordion, Rosary

beaking BONE IMAGING A finding in a lateral spine film in Pts with mucopolysaccharidoses and mucolipidoses, characterized by a bird-beak-like tapering of the anteroinferior or anterosuperior margin of the lumbar vertebrae; a 'beak' is also described in the medial aspect of the proximal tibia at the epiphyseal plate in Blount's disease or coxa vara–functional correction may require osteotomy

Beano™ GASTROENTROLOGY A deflatulent with simethicone added to beans deemed hyperflatulogenic; Beano's enzymes digests raffinose and stachyose, carbohydrates for which humans have no enzymes. See Beans, Flatulence

BEAT CARDIOLOGY A clinical trial–Bucindolol Evaluation in Acute Myocardial Infarction Trial being conducted in Europe to evaluate the effect of Bextra™bucindolol–on survival in Pts with ↓ left ventricular systolic function within 7 days of an acute MI

beat *noun* CARDIOLOGY A pulse, stroke of the heart. See Apex, Escape beat, Fusion beat, Pseudofusion beat, Pseudopseudofusion beat *verb* To strike, to punish by blows

beat knee Prepatellar bursitis caused by prolonged kneeling often associated with trauma and/or infection, either acute, associated with serous effusions or chronic with hemorrhage, loose bodies, and calcifications

beating heart surgery CARDIOLOGY Any cardiovascular procedure, often performed endoscopically, in which the heart is not stopped. See CABG

beats per minute CARDIAC PACING The unit of measure for the frequency of heart depolarizations or contractions each minute–or pulse rate

Beck Depression Inventory Beck Hopelessness Scale PSYCHIATRY A questionnaire that assesses the severity of depression by evaluating self-dissatisfaction, indecisiveness, work difficulty, fatigability, suicidal ideation. See Depression, Personal Attributes Questionnaire. Cf Beck Depression Inventory

Becker's muscular dystrophy NEUROLOGY An X-linked condition characterized by slowly progressive muscle weakness of the legs and pelvis, difficulty walking, mental retardation, fatigue and pseudohypertrophy of calf muscles

Beckwith-Wiedemann syndrome PEDIATRICS An AD condition characterized by overgrowth with visceromegaly–± asymmetric, resulting in hemihypertrophy, macroglossia, macrosomia, omphalocele, seizures, hyperinsulinemic hypoglycemia, mental retardation, microcephaly, renal medullary disease, pancreatic and renal hyperplasia, and ↑ benign–adrenal adenoma, hamartoma of heart and malignant–nephroblastoma, adrenocortical carcinoma, and embryonal tumors–eg, Wilms' tumor, hepatoblastoma, rhabdomyosarcoma. See Wilms' tumor

beclomethasone dipropionate Vanceril™ A new steroid molecule with enhanced receptor binding activity, but more rapid presystemic metabolism, which is used as inhalation antiasthmatic therapy. See Asthma,

bed HOSPITAL CARE A unit of 24-hr Pt occupancy in a hospital or other inpatient health facility, which indicates hospital size. See Certified bed, Managed care, Observation bed, 'Swing' bed, Water bed

bed-wetting Enuresis PEDIATRICS The involuntary nocturnal passage of urine in children after ages 4 or 5 yrs; BW is 2-fold more common in ♂ MANAGEMENT Behavior modification, supportive attitude by parents, limit fluids at bedtime; alarms–a sleep pad that triggers an alarm when wet, vasopressin nasal spray has been used in some children

bedbug Chinch A blood-sucking arthropod TYPES Cosmopolitan–eg, *Cimex lectularius*, tropical–eg, *C hemipterus*; bedbugs cause pruritus and in sensitive individuals, urticaria, vesiculobullous lesions, arthalgia, asthmatic Sx

bedside glucose monitoring ENDOCRINOLOGY A format for measuring glucose, in which blood is obtained from the Pt and measured immediately; it has been claimed that BGM results in ↓ hospital stays and hospitalization costs for Pts with DKA. See Point-of-care testing

bedside manner MEDTALK A popular term for the degree of compassion, courtesy, and sympathy displayed by a physician towards Pts in a clinical setting. See Doctor-patient game

bedside spirometry Portable spirometry A procedure for evaluating and monitoring Pts with potential respiratory problems, those recuperating from acute illness, and monitoring weaning from mechanical ventilation

bedside test LAB MEDICINE Any evaluation of analytes close to a Pt who may be a relatively critical state; devices used for BTs may be less accurate than those used in a hospital's laboratory, but have the advantage of short 'turn-around' time–eg, 2 minutes, facilitating therapy using minimal volumes–eg, 0.25–0.5 mL, of specimens; BT may be used for pH, PO_2, PCO_2, Na^+, K^+, Hct, glucose, Ca^{2+}, and chloride. Cf Stat testing

bedsore Pressure ulcer, see there, aka decubital ulcer

bedtime SLEEP DISORDERS The time when one attempts to fall asleep–as distinguished from the time when one gets into bed

bee sting IMMUNOLOGY A sting from bees and other stinging insects including yellow jackets, hornets and wasps can trigger allergic reactions of variable severity; avoidance and prompt treatment are essential; in selected cases, allergy injection therapy is highly effective

bee sting kit EMERGENCY MEDICINE An emergency kit carried by those with a Hx of sudden and severe allergic reactions to bee stings; the BSK includes an antihistamine and an epinephrine autoinjector

beep Bleep, British *verb* To contact by portable pager–eg, *Dr. Kildare was beeped on rounds*

beeper A portable paging device that allows a person to contact–'beep' or page its wearer by calling a specific telephone number; depending on the device's sophistication and capacity, a person contacting the person wearing the beeper may send a message of various length at a range of up to 100 kilometers–62 miles from the 'home base'

beer NUTRITION A fermented liquor made from a malted grain, commonly barley, with hops or some other substance to impart a bitter flavor. See Alcoholism, Purple grapes

behavior Conduct, bearing, demeanor, manner PSYCHOLOGY Manner of behaving—good or bad; mode of conduct; comportment. See Affective behavior, Catatonic behavior, Compensatory behavior, Dyssocial behavior, Eusocial behavior, High-risk behavior, Homosexual behavior, Novelty-seeking behavior, Preening behavior, Purging behavior, Sexual behavior, Suicidal behavior, Symbolic behavior, Withdrawing behavior

behavior management PSYCHOLOGY Any nonpharmacologic maneuver–eg contingency reinforcement–that is intended to correct behavioral problems in a child with a mental disorder–eg, ADHD. See Attention-deficit-hyperactivity syndrome

behavior therapy PSYCHIATRY Any treatment that focuses on modifying observable and, at least in principle, quantifiable behavior by means of systematic manipulation of the environment and behavioral variables thought to be functionally related to the behavior EXAMPLES Operant conditioning, shaping, token economy, systematic desensitization, aversion therapy, flooding. See Biofeedback

behavioral disorder PSYCHIATRY A disorder characterized by displayed behaviors over a long period of time which significantly deviate from socially acceptable norms for a person's age and situation

behavioral health professional PSYCHOLOGY A person who is licensed by the state, whose professional activities address a client's behavioral issues; examples include psychiatrists, psychologists, social workers, psychiatric nurse practitioners, marriage and family counselors, professional clinical counselors, licensed drug/alcohol abuse counselors and mental health professionals. See Mental health worker

behavioral intervention Behavior modification, behavior 'mod', behavioral therapy, behaviorism PSYCHIATRY The use of operant conditioning models, ie positive and negative reinforcement, to modify undesired behaviors–eg, anxiety. See Aversion therapy, Encounter group therapy, Flooding, Imaging aversion therapy, Systematic desensitization

behavioral medicine A medical discipline that integrates behavioral, psychosocial, and biomedical approaches to improving health and reducing illness. See Mind/body medicine

behavioral symptom NEUROLOGY In Alzheimer's disease, any of the Sx that relate to action or emotion, such as wandering, depression, anxiety, hostility, sleep disturbances. See Alzheimer's disease

behavioral technique PSYCHIATRY Any coping strategy in which Pts are taught to monitor and evaluate their behavior and to modify their reactions to pain

Behçet's disease Behçet syndrome An idiopathic multisystem condition most common in the Middle East; ♂:♀ ratio 2:1, aged 15–40 CLINICAL Pustules, aphthous stomatitis–90%, genital ulcers; vasculitis; erythema nodosa–80%, meningoencephalitis, synovitis, uveitis LAB ↑ ESR, ↑ Igs, anemia MANAGEMENT Symptomatic, corticosteroids

Behçet's triad A trio of symptoms: recurring oral–aphthous ulcers, genital ulcers, uveitis. See Behçet's disease

being buried alive, fear of PSYCHOLOGY Taphephobia or taphophobia. See Phobia

BEIR studies Biological effects of exposure to low levels of ionizing radiation EPIDEMIOLOGY A series of studies from the Natl Research Council–UK that periodically analyze cancer data from Japanese atomic bomb blast survivors and from those with long-term exposure to low levels of radiation

Bell's palsy Bell's paralysis, facial nerve palsy NEUROLOGY A condition affecting the 7th–facial nerve, resulting in unilateral facial paralysis; BP can be differentiated from a central–stroke deficit by inability to raise eyebrow on affected side. See Facial nerve

la belle indifference Conversion disorder, see there

belly button MEDTALK Umbilicus, navel

belly tap Abdominal tap, see there

belonephobia Aichmophobia PSYCHOLOGY A morbid fear of needles, pins and other sharp objects. See Sharps

below the knee amputation SURGERY An 'elective' procedure often required for peripheral vascular disease; while preferred by the Pt–since the sense of loss is lessened, the BKA often requires re-amputation, an important factor related to mortality; furthermore, the prosthetic device fits less satisfactorily to the BKA than to an above-the-knee amputation. See Phantom limb, Potato chip operation

belt EPIDEMIOLOGY A popular term for any broad geographical region with an ↑ incidence of a particular disease. See AIDS belt, Asian esophageal cancer belt, Cancer seat, Goiter belt, Lymphoma belt, Stroke belt PUBLIC HEALTH See Seat belt

Benadryl® Diphenhydramine ALLERGY MEDICINE An antihistamine which, like other older sedating agents, has been associated with fatal MVAs. See Antihistamine

benazepril Lotensin A nonsulfhydryl ACE inhibitor and antihypertensive used to manage CHF, which protects against progression of renal failure in Pts with renal disease–eg, glomerulopathies, interstitial nephritis, nephrosclerosis, polycystic kidney disease, diabetic nephropathy, etc See ACE inhibitors, Angiotensin-converting enzyme

Bence Jones protein An abnormal dimer of light-chain Ig derived from the clonal expansion of plasma cells, found in the urine of 50-80% of Pts with myeloma and Waldenstrom's macroglobulinemia; these proteins are small enough to be excreted by the kidney

the bends Acute decompression sickness, the chokes INTERNAL MEDICINE A clinical complex caused by rapid whole body decompression, with intravascular 'boiling' of nitrogen, and resultant M&M in scuba divers and high-altitude pilots or workers in high-pressure environments–eg, caissons–in chronic decompression sickness CLINICAL Headache, N&V, vertigo, tinnitus, dyspnea, tachypnea, convulsions, shock, joint and abdominal pain; nitrogen gas in the brain causes 'boxcar' air bubbles in meningeal vessels separating the blood 'column', potentially causing death. Cf Caisson disease

benecol CLINICAL NUTRITION A soy based plant sterol-rich proprietary margarine available in Finland, which ↓ serum cholesterol. Cf Isoflavone, Nutraceutical, Soy, Textured vegetable protein

Beneficat Trazodone, see there

beneficiary HEALTH INSURANCE A person(s) other than the member of an insurance or pension plan who has been designated to receive benefits–eg proceeds of an accident insurance policy or pension plan in the event of an employee's death MANAGED CARE A person eligible for or receiving benefits under an insurance policy or plan, Medicare or Medicaid programs

benefits MEDTALK Services provided under the terms of a health insurance policy. See Allocated benefits, Ancillary benefits, Catastrophic benefits, Hospital benefits, Mandated benefits, Marginal benefits, Net benefits, Spousal benefits

benefits package HEALTH INSURANCE Services covered by a health insurance contract or plan and the financial terms of such coverage, including cost sharing and limitations on amounts of services. See Cost sharing

BeneFix® Recombinant coagulation factor IX HEMATOLOGY A recombinant factor IX for treating spontaneous bleeds, prophylaxis, and preventing bleeds during surgery and postoperatively in Pts with hemophilia B

benign *adjective* Not cancerous; not malignant; referring to a nonmalignant lesion or tumor that does not invade or metastasize, for which surgical excision is curative. Cf Malignant

benign breast disease See Fibroadenoma, Fibrocystic disease, Mastopathy

benign exertional headache NEUROLOGY A headache evoked by running, lifting, coughing, sneezing, or bending.

benign faint Syncope as a physiologic response to stress, especially emotional CLINICAL Lightheadedness, nausea, flushing, sensation of warmth, transient loss of consciousness

benign glandular inclusion SURGICAL PATHOLOGY A nonpathologic displaced ductular and acinar structure in a perineural space. See Benign metastasis, Perineural invasion

benign headache NEUROLOGY 1. A headache unrelated to an identifiable 'malignant' pathology–eg, tumor, infection, stroke, vasculopathy 2. Tension headache, see there

benign lymphadenopathy Any non-malignant regional or generalized enlargement of lymph nodes. See Lymph node necrosis

benign lymphoepithelial lesion Mikulicz disease A lesion of the salivary and lacrimal glands, clinically related to Sjögren syndrome, which may be autoimmune in nature. Cf 'Eskimoma'

benign 'metastasis' A popular term for the presence of nonmalignant nonlymphoid tissue in lymph nodes–eg, thyroid follicles in cervical lymph nodes, which may be confused with carcinoma. See Lymph node inclusions

benign neglect DECISION-MAKING A stance of nonintervention that a clinician may adopt in the face of lesions and clinical conditions which have an uncertain or stable clinical course. Cf Watchful waiting

benign paroxysmal positional vertigo Cupulolithiasis NEUROLOGY A form of transient vertigo caused by utricular degeneration which liberates otoconia; otoconia drift into the lower part of the vestibule, the ampulla of the posterior semicircular canal; once there, the otoconia alter the cupola's specific gravity, changing its response characteristics from a purely angular acceleration detector to one that is stimulated by linear movements and gravity INCIDENCE BPPV is a common form of vertigo which is more common in older adults ETIOLOGY Closed head injury, vestibular neuronitis, infections, post-stapedectomy DIAGNOSIS History–single bouts of severe vertigo of < 1 min in duration, after a change in head position, often more severe on one side, when bending, when looking to take an object off a shelf, tilting the head back; the episodes are clustered in time and separated by remissions lasting months or more; most BPPV resolves spontaneously within several months of onset, especially following head injury; persistent, near disabling Sx may mandate surgery: singular neurectomy, vestibular neurectomy, or posterior semicircular canal occlusion

benign persistent hematuria Benign recurrent hematuria, persistent microscopic hematuria NEPHROLOGY A heterogeneous condition characterized by microscopic hematuria MANAGEMENT Depends on the cause

benign prostatic hypertrophy Benign prostatic hyperplasia UROLOGY A benign enlargement of the prostate, which is normal after age 50, 2° to ♂ hormones; BPH pushes against the urethra and the bladder, blocking urine flow CLINICAL Bladder-outlet obstruction, seen in 50% of ♂ ≥ age 60; excess enlargement may obstruct the urethra, causing urinary retention; 30+% require surgery. See Finasteride, TURP, Urinary retention

benign tumor ONCOLOGY A nonmalignant tumor that neither invades locally not metastasizes, which is often surrounded by a fibrous capsule or pseudocapsule; BT's characteristics generally include slow growth by expansion and enclosure in a fibrous capsule

Bennett fracture ORTHOPEDICS An intraarticular fracture/dislocation of the ulnar aspect of thumb/1st metacarpal, in which a small fragment of the metacarpal continues to articulate with the trapezium MECHANISM Forced thumb abduction COMPLICATIONS Pseudoarthrosis PROGNOSIS Better than Rolando fracture. See Fracture. Cf Rolando fracture

benoxaprofen THERAPEUTICS An oral NSAID. See NSAIDs

Benpine Chlordiazepoxide, see there

bentiromide test GASTROENTEROLOGY A 'tubeless' pancreatic function test in which bentiromide–N-benzoyl-L-tyrosyl-p-aminobenzoic acid is cleaved by chymotrypsin, yielding p-aminobenzoic acid which is measured in urine FALSE POSITIVITY Liver disease, renal failure, malabsorption. See Pancreolauryl test, Tubeless test

Benton verbal fluency test PSYCHOLOGY A test in which a subject is asked to generate a list of words in 60 seconds, beginning with the letters F, A, and S, with a rest period between each letter; a score of 31-44 is normal; heavy–abuse of marijuana is associated with a ↓ in Benton VFT scores

benzedrine Amphetamine sulfate PHARMACOLOGY A vasoconstricting non-narcotic stimulant, first marketed as an OTC inhalation stimulant to treat congestion. See Amphetamine

benzodiazepam See Benzodiazepine

benzodiazepine PHARMACOLOGY A class of widely prescribed and often overdosed sedative-hypnotics EFFECTS Sedation, hypnotic, ↓ activity, muscle relaxation, anxiolytic, anticonvulsant ADVERSE EFFECTS Physical and psychological dependence

benztropine Cogentin® NEUROLOGY An anticholinergic used to manage parkinsonism ADVERSE EFFECTS Usually dose related–nervousness, impaired memory, numbness, listlessness, depression, confusion, excitement, hallucinations with high doses; GI–dry mouth, constipation, N&V, blurred vision, mydriasis, hyperthermia, anidrosis, urinary retention, dysuria, weakness, rash, tachycardia

benzoylecgonine TOXICOLOGY The major metabolite of cocaine which is detectable in the urine. See Cocaine

Beraprost® Epoprostenol A formulation of prostacyclin in trials for pulmonary vascular disease 2° to treatment with diet drugs

bereavement PSYCHOLOGY Feelings of deprivation, desolation, and grief at the loss of a loved one; the bereaving or mourning the loss of a 'significant other' or other loved one, which is often accompanied by a transient–usually < 6 months period of depression; it may be misdiagnosed as a mental disorder including major depression. See Complicated bereavement, Depression

Berger's disease IgA nephropathy, see there

beri-beri NUTRITION A disease caused by deficiency of thiamine–vitamin B₁, which affects the heart–cardiac dilatation, fatty degeneration, peripheral nerves—myelinolysis and axonal degeneration, subcutaneous tissues–edema and vascular congestion and serosal linings–effusions CLINICAL Fatigue, apathy, irritability, depression, muscular atrophy, paresthesias, abdominal pain and if extreme, ↑ intracranial pressure and coma. See Shoshin beri-beri

---

BERI-BERI

DRY BERI-BERI Affects central and peripheral nervous systems CLINICAL, CHILDREN Plethoric with pallor, hoarseness due to nerve paralysis, apathy, dyspnea, tachycardia, hepatomegaly

WET BERI-BERI More cardiocentric CLINICAL The afflicted children are malnourished with pallor, edema, tachycardia, dyspnea, renal failure, right-sided cardiac failure TREATMENT Thiamine-rich foods, eg milk, vegetables, cereals, fruits

---

Bernard-Soulier syndrome Congenital hemorrhagic thrombocytic dystrophy HEMATOLOGY An AR or codominant condition with mucocutaneous and visceral hemorrhage due to deficiency of glycoprotein Ib, the receptor for von Willebrand factor–vWF, and GP1's—glycocalicin, both are involved in the interaction between vWF and platelet membrane, which is critical for normal platelet adhesion in the early phases of 1° hemostasis CLINICAL Major mucocutaneous hemorrhage of early onset including epistaxis, ecchymoses/purpura, nosebleeds, gingival, GI, and menorrhagia, and bleeding after minor trauma DIFFDX ITP, giant platelet syndrome, May-Hegglin anomaly, Epstein syndrome, Montreal syndrome, platelet-type von Willebrand syndrome–pseudo-von Willebrand's disease with macrothrombocytopenia LAB Variable thrombocytopenia, morphologically abnormal giant platelets on peripheral blood smears, defective prothrombin consumption ↑ bleeding time due to poor platelet adhesion to subendothelium, no platelet aggregation with ristocetin PLATELET DEFECTS ↑ size, basophilia of membrane, aggregation–or absence of cytoplasmic granules, pseudopod formation, and cytoplasmic vacuolization PREVENTION Avoid aspirin, anti-inflammatories or other prohemorrhagic agents. Cf Glanzmann's thrombasthenia

berry aneurysm NEUROLOGY A 0.2-0.5 cm saccular dilation of arteries at the base of the brain, at or near the circle of Willis–95% of BAs or at the vertebrobasilar arteries–5%, due to a developmental or congenital weakness in the medial muscle layer of the cerebral arteries; 30% of BAs are multiple, located at bifurcations, often anterior and appear in 1-2% of all autopsies; most BAs rupture at > 1 cm in diameter. See Circle of Willis

berry picking operation SURGICAL ONCOLOGY A popular term for the removal of multiple metastases from an organ or body cavity; for colorectal CA metastatic to the liver, most authors believe resection of up to 4 metastases, ideally < 5 cm in diameter is reasonable. Cf Cherry picking

BERT CARDIOLOGY A clinical trial–Beta Energy Restenosis Trial—designed to evaluate the safety and feasibility of Novoste's Beta-Cath system at 3 radiation doses. See Beta-Cath System, Coronary artery disease

beryllium poisoning Berylliosis TOXICOLOGY Inhalation of beryllium dusts, salts or fumes.

BERYLLIUM POISONING

ACUTE BP Causes chemical acute pneumonitis–caused by short exposures to high concentrations of beryllium, due to lung toxicity

CHRONIC BP Causes chronic interstitial, sarcoid-like granulomatous lung disease due to beryllium levels over periods from months to years

---

BES 1. Back extensor strength, see there 2. Balanced electrolyte solution

BESS Brain edema severity score. See Hepatic encephalopathy

BEST CARDIOLOGY A clinical trial–Beta-blocker Evaluation of Survival Trial—on the effect of adding bucindolol-Bextra™ to standard therapy on mortality in

Pts with CHF; BEST was conducted by the National Heart, Lung, and Blood Institute/NIH and the VA

Best's disease Best's macular degeneration, vitelliform dystrophy OPHTHALMOLOGY An AD retinal degeneration of variable penetration and early onset FINDINGS Fundoscopic appearance ranges from mild pigment defects to a vitelliform–egg yolk–appearance in the central macula; subsequent degeneration may result in subretinal neovascularization and hemorrhage, and severe macular scarring MOLECULAR PATHOLOGY Mutation of *bestrophin* gene, which maps to chromosome 11. See Bestrophin

beta β The second letter of the Greek alphabet; STATISTICS The probability of a Type II–false-negative error. See Type II error. Cf Alpha

beta-amyloid A4, β–amyloid NEUROLOGY A 4 kD polypeptide encoded on chromosome 21 arising from altered processing of amyloid precursor protein, an integral membrane glycoprotein secreted as a truncated carboxyl-terminal molecule; BA forms plaques in the brains of Pts with Alzheimer's disease–AD, Down syndrome, infectious encephalopathy, cerebral amyloid angiopathy; BA is found in skin, intestine, adrenal gland. See Alzheimer's disease, Amyloid, Presenilin. Cf Alpha2-macroglobulin

beta blocker Beta-adrenergic blocking agent PHARMACOLOGY Any of a class of agents that blocks β_1 and/or β_2 adrenergic receptors in the nervous system EFFECT ↓ Heart rate, ↓ BP, ↓ anxiety INDICATIONS Angina, arrhythmias, HTN, mitral valve prolapse, tachycardia, etc

beta-gamma bridge IMMUNOLOGY A 'spanning' of the usually well-defined peaks in the β and γ regions in an SPEP which occurs in chronic liver disease, classically in alcoholic liver disease, as well as in chronic infections and connective tissue disease, and is due to polyclonal production of proteins that migrate in the region, which may 'bury' small monoclonal expansions. See serum protein electrophoresis

beta rhythm Beta activity, beta waves SLEEP DISORDERS An EEG frequency in the range of 13-35 Hz; when the predominant frequency, BR is usually associated with alert wakefulness or vigilance and is accompanied by a high tonic EMG

beta testing The testing of a vendor's product–a device, equipment, hardware or software by applications-knowledgeable users, who are familiar with the product's use but not with its design. Cf Alpha testing

Betadine® Povidone-iodine NURSING A popular topical antiseptic used as a surgical scrub

betamimetic PHARMACOLOGY *adjective* Stimulating, mimicking or referring to beta adrenergic receptors of the sympathetic nervous systems *noun* A betamimetic agent

Betapace® Sotalol, see there

Betapam Diazepam, see there

Betaseron® Recombinant IFN-β1b, see there

betaxolol CARDIOLOGY A β-blocker used to treat HTN and glaucoma. See β-blocker

bethanechol chloride Urecholine A cholinergic agent with muscarinic effects, which stimulates smooth muscle activity of the GI and GU tracts INDICATIONS Gi and GU atony–eg, neurogenic bladder, atony of stomach after vagotomy, postoperative gastric retention, acute postoperative and postpartum urinary retention, reflux esophagitis, GERD

the Bethesda System CYTOLOGY A system for reporting results from pap smears, which provides a uniform format for cervical and vaginal cytologic specimens, classifying noninvasive lesions and standardizes the lexicon

for cervical/vaginal cytology reports, providing clinically relevant information. See Cervical intraepithelial neoplasia, High-grade squamous epithelial lesion, HPV, Low-grade squamous intraepithelial lesion. Cf Papanicolaou system

BETTER CARDIOLOGY A clinical trial–Beta Radiation Trial to Eliminate Restenosis. See BRITE, In-stent restenosis

Bextra® CARDIOLOGY Bucindolol An investigational nonselective beta-blocker with vasodilating properties in phase III clinical trials for treating CHF, removed from the marketplace. See BEAT, BEST PAIN MANAGEMENT Valdecoxib A next-generation of a proprietary Cox-2-specific inhibitors

bezoar GASTROENTEROLOGY A mass of foreign material in the stomach–food, mucus, vegetable fiber, hair, or other indigestible material, facilitated by partial or complete gastrectomy, as acid hydrolysis of gastric content is ↓; undigested bezoars cause discomfort or pain, halitosis, gastric erosion or ulceration and potentially peritonitis, hemorrhage, obstruction, N&V; the mass is more easily palpable in tricho- than in phytobezoars

Bezold-Jarisch triad ENT A Sx triad said to be more common in Pts with otosclerosis: 1. ↓ perception of low tones; 2. ↓ bone conduction or sound; 3. Negative Rinne's test. See Rinne test

BF 1. Barrier filter 2. Biceps femoralis 3. Blocking factor 4. Blood flow

BFP Biological false positive, see there

BHAT CARDIOLOGY A clinical trial–Beta-blocker Heart Attack Trial—which evaluated the effect of β-blockers on post-MI Pts. See Congestive heart failure

Biafine® WOUND CARE A topical preparation to ↓ the risk of and treat skin reactions to radiation therapy/radiodermatitis, as well as treatment and management of full and partial-thickness dermal wounds, 1st- and 2nd-degree burns. See Radiation therapy

bias EPIDEMIOLOGY Deviation of results or inferences from the truth, or processes leading to such systematic deviation; any trend in the collection, analysis, interpretation, publication, or review of data that can lead to conclusions that are systematically incorrect

Biaxin® Clarithromycin, see there

bicarbonate HCO_3 NEPHROLOGY A general term for any salt containing the anion HCO_3^-, which is the most important buffer in the blood; bicarbonate is regulated by the kidney, which excretes it in excess and retains it when needed; it is ↑ in ingestion of excess antiacids, diuretics, steroids; it is ↓ in diarrhea, liver disease, renal disease, chemical poisoning. See Blood gases

bicarbonate loading SPORTS MEDICINE Administration of sodium bicarbonate before competing in a sports event, to neutralize lactic acid produced during anaerobic metabolism of muscles. Cf Carbohydrate loading, Phosphate loading

biceps reflex Elbow flexion NEUROLOGY A reflex contraction of the biceps brachii

bicipital tendinitis RHEUMATOLOGY Tendinitis of the biceps brachii ETIOLOGY ↑ Activity of biceps or shoulder, especially if repetitious CLINICAL Shoulder pain aggravated by shoulder movement or resisted flexion of biceps muscle MANAGEMENT Rest, NSAIDs, RICE

biclonal gammopathy ONCOLOGY The presence of 2 clonal expansions–IgG > IgA > IgM, in the same Pt; $\frac{2}{3}$ of BGs are of undetermined significance; $1/3$ are associated with myeloma or lymphoproliferative disorders, including lymphomas and leukemias. Cf Monoclonal gammopathy of undermined significance

biclonality The rare occurrence of an uncontrolled expansion of 2–or more clones of neoplastic cells, as in the biclonal expansion of 2 B cell lines or B and T cell lines. Cf Composite lymphoma

bicornuate uterus INFERTILITY A congenital malformation of the uterus where the upper portion–horn is duplicated. See Uterus didelplus

bicuspid aortic valve CARDIOLOGY A congenital heart defect in which the aortic annulus has 2 instead of 3 semilunar valves, seen in 3% of the population; ♂:♀, 4:1; 20% of those with a BAV have other cardiovascular disease–eg, PDA or aortic coarctation; the valve is subjected to abnormal hemodynamic stress, leading to leaflet thickening, calcification and aortic stenosis. See Aortic stenosis

bicycling SPORTS MEDICINE MAIN INJURIES Pain, numbness and injuries, primarily of the knee, but also wrists, buttocks, pelvis, carpal tunnel and ulnar nerve, skin breakdown of the buttocks due to overtraining, pelvic nerve compression, transient sexual dysfunction due to long rides. See Water therapy

big-ticket item MANAGED CARE A popular term for an expensive therapeutic or diagnostic procedure

bilateral cancer ONCOLOGY A cancer that occurs in both paired organs–eg, both breasts, kidneys, ovaries, etc

bilateral hemothorax Blood in both pleural cavities

bilateral hydronephrosis UROLOGY Enlargement of the renal pelvis and calyces of both kidneys, which is associated with conditions that interfere with the drainage of urine from kidneys to the ureters and bladder ETIOLOGY Acute and chronic bilateral obstructive uropathy, vesicoureteric reflux, ureteropelvic junction obstruction, neurogenic bladder, bladder outlet obstruction, prune belly syndrome

bilateral locked facets ORTHOPEDICS A lower cervical spine injury resulting in subluxation with locking of both facets due to hyperflexion; 90% of Pts with BLFs have a neurologic injury; $2/3$ of Pts have a complete spinal cord injury, $1/4$ have myelopathy. Cf Unilateral locked facet

bilateral mastectomy SURGERY The excision of both breasts usually for CA. See Mastectomy, Prophylactic mastectomy

bilateral obstructive uropathy Damage to both kidneys caused by obstruction of urine flow, which may lead to HTN and/or renal failure

bilateral oophorectomy GYNEOLOGY The removal of two or more ovaries. See TAH-BSO

bilateral symmetry *adjective* Referring to a form divisible into equal mirror halves in one plane only

bile duct stricture Biliary stricture SURGERY An abnormal narrowing of the common bile duct, due to local scarring, which may cause biliary obstruction ETIOLOGY Prior local surgery, pancreatitis, trauma, gallstones CLINICAL Jaundice, fever, chills, abdominal pain MANAGEMENT Endoscopic surgery

bile duct cancer Cholangiocarcinoma, see there

bile nephrosis Hepatorenal syndrome, see there

bile peritonitis Biliary peritonitis CLINICAL MEDICINE Peritoneal inflammation caused by leakage of bile into the peritoneal cavity

biliary atresia GASTROENTEROLOGY A rare condition caused by an abnormal development of intrahepatic or extrahepatic bile ducts; obstruction

of bile flow from the liver can lead to cirrhosis if untreated CLINICAL Jaundice in 2nd to 3rd wk of life, clay-colored stools. See Neonatal jaundice, Newborn jaundice.

biliary calculus Gallstone, see there

biliary disease SURGERY Any pathology that affects the gallbladder and its conduits, commonly cholecystitis, cholelithiasis, cholesterolosis, etc

biliary dyskinesis Sphincter of Oddi dysfunction GASTROENTEROLOGY An idiopathic syndrome linked to structural and/or functional abnormalities of Oddi's sphincter; it may cause recurrent abdominal pain in middle-aged ♀ after cholecystectomy, even in absence of objective evidence of biliary or pancreatic disease. See ERCP

biliary obstruction Bile duct obstruction CLINICAL MEDICINE A blockage of the bile ducts with accumulation of bile in the liver ETIOLOGY Stone, tumor, pancreas–tumors or pancreatitis, cholecystitis, bile duct cysts, trauma, bile duct stricture, enlarged lymph nodes CLINICAL Colic, jaundice. See Gall bladder disease

biliary stent Biliary endoprothesis SURGERY A tube inserted into a stenosed or obstructed narrowed or blocked bile duct to improve bile flow

bilirubin BR LAB MEDICINE A yellow-red Hb breakdown product derived from catabolized RBCs, present in bile transported from the liver to the gallbladder to the intestines; BR gives bile its color and is normally passed in stool; ↑ production or ↓ excretion of BR results in jaundice of the skin and ocular sclera; BR that has *not* been metabolized in the liver is indirect bilirubin–BR_I, and is attached to albumin in the circulation; after hepatic metabolism, it is no longer bound to proteins, and is called direct bilirubin–BR_D; the sum of BR_D and BR_I is total bilirubin–BR_T, which is usually measured as part of a routine chemistry profile, and in liver profiles; if BR_T is ↑, the laboratory automatically measures BR_D; BR_I is a calculated value of $BR_T − BR_D$ REF RANGE BR_T, umbilical cord <2.0 mg/dL; 0-1 day of life <6.0 mg/dL; 1-2 day <8.0 mg/dL; 3-5 day <12 mg/dL, thereafter <0.2-1.0 mg/dL; BR_D < 0.3 mg/dL; BR_I 0.2–1.3 mg/dL; BR_T 0.3–1.6 mg/dL; BR_I is ↑ in liver damage, hemolytic anemia, congenital enzyme deficiencies; BR_D is ↑ in biliary obstruction; BR_T is ↑ in continuous hemolysis, biliary obstruction with hepatic damage URINE Negative AMNIOTIC FLUID 28 wk <0.075 mg/dL; 40 wk <0.025 mg/dL See Conjugated bilirubin, Delta bilirubin

bilirubin encephalopathy Kernicterus, see there

billable test LAB MEDICINE A unit of productivity in clinics and hospitals–where it is an 'ordered' laboratory test. Cf Workload unit

billing MANAGED CARE The submission to a payer–Pt or his/her health insurance carrier, a bill for services rendered or products tendered. See Balance billing, Central billing, Consolidated billing, Direct billing, Indirect billing, Insurance only billing

Billroth I operation Billroth I anastomosis SURGERY The excision of the pylorus with an end-to-end anastomosis of the stomach and the duodenum, which was once used in managing peptic ulcer disease, and continues to be used for gastric CA. See Dumping syndrome

Billroth II operation Billroth II anastomosis SURGERY A procedure consisting of subtotal gastrectomy and gastroenterostomy linking the gastric pouch to the jejunum distal to the ligament of Trietz. See Dumping syndrome

bilocular *adjective* Having or referring to 2 cavities

bimalleolar fracture ORTHOPEDICS A fracture of the lower tibia which affects the internal and external malleolus

bimanual Bimanous *adjective* Referring to or having 2 hands

bimanual pelvic examination GYNECOLOGY The use of both

hands–2 fingers from one hand in the vagina, the other on the abdomen, which allows evaluation of the position, size, firmness, mobility of the uterus See Rectal examination

Bimaran Trazodone, see there

binaural fusion NEUROPHYSIOLOGY The process by which the brain compares auditory information received in each ear and translates the differences into a unified perception of a single sound coming from a specific region of space.

binge PSYCHIATRY A period of consumption of a very large quantity of food during a short period, with lack of control over eating; binges are often followed by self-induced emesis or 'purging' COMPLICATIONS Gastric rupture-Mallory-Weiss syndrome, vascular compression, pancreatitis, aspiration pneumonia, ipecac-induced myocarditis, heart failure, refeeding edema, hypokalemia, hypochloremia, metabolic alkalosis. See Bulimia nervosa

binge drinking An early phase of chronic alcoholism, characterized by episodic 'flirtation' with the bottle by binges of drinking to the point of stupor, followed by periods of abstinence; BD is accompanied by alcoholic ketoacidosis–accelerated lipolysis and β-hydroxybutyric acid production due to impaired insulin secretion, ↓ food consumption and recurrent vomiting. See 'Eyeopener'

binge eating disorder PSYCHIATRY A condition characterized by recurrent binge eating–at least 2 days/wk for 6 months; absence of anorexia nervosa; no recurrent purging, excessive exercise or fasting marked distress with 3+ of following: Eating very rapidly; eating until uncomfortably full; eating when not hungry; eating alone; feeling disgusted or guilty after a binge See Eating disorder

binge-purge syndrome See Bulimia

bioabsorbable pin ORTHOPEDICS A bioprosthetic used for short-term fixation of an unstable bone, which is later eliminated from the body

bioaccumulation factor TOXICOLOGY The concentration of a chemical in tissue divided by its concentration in the diet

bioanalytical laboratory LAB MEDICINE A place, establishment, or institution organized and operated primarily for performing chemical, microscopic, serologic, parasitologic, bacteriologic, or other tests, by the practical application of one or more of the fundamental sciences, to material originating from the human body, for the purpose of obtaining scientific data which may be used as an aid to ascertain the state of health See CLIA, Laboratory

bioavailability CLINICAL PHARMACOLOGY The degree to which a drug is available to a target tissue after administration which, for oral drugs, reflects the rate and extent of GI tract absorption; BA is the in vivo presence of a substance in a form that allows it to be metabolized, serve as a substrate, bind a specific molecule, or participate in biochemical reactions. See Oral bioavailability

Biobypass® CARDIOLOGY A gene-based drug delivery product that promotes angiogenesis in tissues with poor blood flow–eg, in Pts with CAD and and peripheral vascular disease. See Angiogenesis

bioceramic ORTHOPEDIC SURGERY A calcium phosphate–eg, hydroxyapatite or tricalcium phosphate, used to coat the gliding surfaces of artificial joints and ↓ wear and tear in prosthetic joints. See Total hip replacement, Total knee replacement

biochemical hepatitis LAB MEDICINE Any minor ↑ in liver enzymes–ALT, AST, related to aspirin use in Pts with rheumatic fever, juvenile rheumatoid arthritis, SLE, Reiter syndrome. See ALT, AST, Hepatitis

biochemical mechanism Any chemical reaction or series of reactions, often enzyme-mediated, which result in a physiologic effect

biocompatibility 1. The extent to which a foreign, usually implanted, material elicits an immune or other response in a recipient 2. The ability to coexist with living organisms without harming them

biocycle 1. Biorhythm, see there 2. A cycle of biological phenomena that occurs in a particular organism–eg, menstruation in humans, the daily sleep cycle, the disturbance of which results in jet lag

biodegradable *adjective* Referring to a substance–eg, a chemical, which is degradable by natural systems or components thereof–eg, soil bacteria, weather, plants or animals, to a simpler form

biodynamics The formal study of vital forces, physiological interactions and behavior

bioelectricity A general term for the low-power electric currents that normally flow within nerves and muscles

bioengineered tissue IMPLANTATION SURGERY Any tissue created by the techniques of cell biology–eg biochemistry and tissue culture, and materials science, which replaces failed or failing tissue. See Biomaterial

bioengineering The science of developing and manufacturing artificial replacements for organs, limbs and tissues. See Biomaterial

bioequivalent allergy unit ALLERGY MEDICINE A unit based on quantitative skin testing, which uses standardized extracts of various allergens. See Desensitization, Skin testing

biofeedback 1. A process in which a Pt learns to influence physiologic responses which are either involuntary or usually well-regulated, but the regulation has broken down because of trauma or disease 2. A method of controlling a living system by informing it of its previous performance. See Biofeedback device

biofeedback device Any instrument that measures physiologic parameters eg electromyographic activity, galvanic–electrodermal skin resistance, muscle tension, BP, and others; some mainstream physicians believe BDs may be used to control tachycardia, HTN, fecal incontinence, and other conditions. See Electromyograph feedback, Galvanic skin response

Biofix absorbable rods & screws ORTHOPEDICS Proprietary bioorganic fixation devices for repair of non-load bearing fractures of cancellous bone

biogenic Biogenous *adjective* Arising or originating from a biologic process

biogenic amine theory PSYCHIATRY A theory that explains major depression–MD; according to the BAT, MD is due to ↓ CNS levels of amine neurotransmitters–ANs–eg, dopamine, norepinephrine, serotonin; the BAT also explains the efficacy of antidepressants, which ↑ ANs that bind to post-synaptic receptors on neurons. See Major depression

Bioglass™ ORTHOPEDICS A product made from calcium salts, phosphorous, sodium salts and silicon–essential elements of mineralized bone which bonds with bone and promotes bone regeneration INDICATIONS Mandibular augmentation, repair of damaged middle ear

biographer CLINICAL MEDICINE A popular term for a Pt who describes his/her own medical history

biography CLINICAL MEDICINE A Pt's written medical history

biohazardous waste PUBLIC HEALTH Waste products–eg, body fluids and tissues, which may carry human pathogens; BW often originates from health care facilities and/or research laboratories, and places a relatively small or confined group of people at up ↑ risk for infection during the time necessary for the infectious agent to dry or otherwise become inactive. See Regulated waste, Sharps

biohybrid organ THERAPEUTICS A device that marries biologic units–eg, cells or tissues, to a delivery vehicle to circumvent–through immunoisolation–immune attack on transplanted–non-self tissues. See Artificial liver, Artificial pancreas, Bioengineering

bioinformatics INFORMATICS The use of information technology to acquire, store, manage and analyze biological data. See Genomics, Informatics. Cf Chemoinformatics

biologic transmission EPIDEMIOLOGY Indirect vector-borne transmission of an infectious agent in which the agent undergoes biologic changes within the vector before transmission to a new host

biological *adjective* Referring to biology *noun* THERAPEUTICS Biologic Any of a number of FDA-regulated agents–eg antitoxins, antisera, vaccines, and blood plasma products prepared from donor pools or obtained directly from various living organisms–often mammals; they are *not* amenable to the chemical or physical standardization steps required of pharmaceuticals; they are impure chemically; safety cannot be assumed. See Antisera, Antitoxins, Vaccines

biological agent MILITARY MEDICINE A euphemism for a pathogenic microorganism or virus, or other toxic biological material, intended for use as a weapon of mass destruction.

biological false positive A lab result that is positive in a subject who is known to be a true negative for the substance being measured. See Ulysses syndrome

biological family SOCIAL MEDICINE A family related by blood or marriage. See Family, Nuclear family

biological half-life $T_{1/2}$ BIOLOGY The time required for ½ of the total amount of a particular substance in a biologic system to be degraded by biological processes when the rate of removal is nearly exponential RADIATION PHYSICS The length of time required for $1/2$ of a radioactive substance to be biologically eliminated from the body

biological parent Birth parent, genetic parent, natural parent SOCIAL MEDICINE The ♀ and ♂ who conceive a child

biological response modifier IMMUNOLOGY Any of a broad family of natural or synthetic molecules that up- or down-regulate, or restore immune responsiveness TYPES OF BRMs Interferons, ILs, CSFs, TNFs, B-cell growth factor, B-cell differentiating factors, eosinophil chemotactic factor, lymphotoxin, macrophage chemotactic factor, macrophage activating factor, macrophage inhibiting factor, osteoclast-activating factor, and others; BRMs are generated after a T cell recognizes an antigen present on the surface of a self antigen-presenting cell which, once activated, produces a multiple lymphokines–cytokines THERAPEUTIC EFFECTS OF BRMs 1. Regulation and/or increased immune response; 2. Cytotoxic or cytostatic activity against tumor cells; 3. Inhibition of metastasis, or cell maturation; 4. Stimulation of BM stem cells, required for recuperation from cytotoxic insult secondary to chemotherapy. See B cell, Colony-stimulating factor, Interferon, Interleukin, T cell, Tumor necrosis factor

biological variability LAB MEDICINE The variability in a lab parameter due to physiologic differences among subjects–interindividual BV, and in the same subject over time–intraindividual BV

biology The formal study of living organisms. See Aerobiology, Building biology, Cell biology, Conservation biology, Cryobiology, Developmental biology, Molecular biology, Population biology, Psychobiology, Sociobiology, Structural biology, Topobiology

biomagnetism MAINSTREAM MEDICINE Biomagnetic therapy The formal study of magnetic fields associated with life activities; mainstream biomagnetism is an intradisciplinary hybrid with roots in quantum mechanics, superconductivity, and bioelectricity

biomarker DIAGNOSTIC MEDICINE Any relatively specific biochemical parameter–eg, PSA, Hb_{1a} or troponin I which, when found in ↑ amount in the blood, other body fluids, or tissues, suggest the presence of certain diseases or cancers. See Cardiac marker, Tumor marker

biomaterial 1. Any synthetic material or device–eg implant or prosthesis-intended to treat, enhance or replace an aging or malfunctioning–or cosmetically unacceptable—native tissue, organ or function in the body. See Bioengineering, Breast implants, Hybrid artificial pancreas, Shiley valve, Teflon, Total hip replacement 2. A biomaterial used for its structural, not biological, properties–eg, collagen in cosmetics, carbohydrates modified by biotechnology to be used as lubricants for biomedical applications or as bulking agents in food manufacture

biomechanical dysfunction ORTHOPEDICS An imbalance in the musculoskeletal system resulting faulty movement patterns of the body. Injuries that are not properly identified and rehabbed lead to these dysfunctions.

biomechanics ORTHOPEDICS The application of mechanical laws to living structures, especially to the musculoskeletal system and locomotion; biomechanics addresses mechanical laws governing structure, function, and position of the human body

biomedical engineering TECHNOLOGY The application of engineering principles and methods to medical problems–eg, manufacture of prosthetic limbs and organs

biomedical test A test intended to evaluate qualitative or quantitative derangements in physiology and predict impaired health due to exposure to hazardous materials

biometrics HEALTH INFORMATICS Electronic capture and analysis of biological characteristics–eg, fingerprints, facial structure or patterns in the eye. See Authentication

biomicroscopy 1. Examination of tissues by microscopy 2. Slit-lamp microscopy

Bionx screw ORTHOPEDICS An absorbable, cannulated screw used for screw fixation of distal radius–Colles fractures

biophysical profile OBSTETRICS The measurement of 5 physical parameters in fetuses at ↑ risk for poor outcome, a non-stress test which evaluates the reactivity of the heart rate to various stresses, other parameters are measured simultaneously by dynamic ultrasound imaging: Fetal breathing movement–a non-stress test, fetal muscle tone, gross body movement, and amniotic fluid volume

bioprosthesis An implanted device of natural–ie, nonsynthetic origin–eg, porcine tissue designed to replace a defective body part–eg, a heart valve. See Heart-valve prosthesis

bioprosthetic valve SURGERY An implanted device of nonsynthetic origin designed to replace a defective heart valve; porcine BV, are less thrombogenic than mechanical valves, but are more prone to structural degeneration, limiting their durability. See Heart-valve prosthesis

biopsy *noun* A term for 1. A surgical procedure in which a small piece of tissue is removed from a Pt 2. The tissue itself; the changes in the biopsy are interpreted by a pathologist, usually under a microscope, who renders a diagnosis based on relatively standard morphologic criteria. See Abdominal wall fat pad biopsy, Agonal biopsy, Aspiration biopsy, Biochemical biopsy, Blastocyst biopsy, Blind biopsy, Bone marrow aspiration & biopsy, Breast biopsy, Cervical biopsy, Chorionic villus biopsy, Cleavage stage biopsy, Cold cone biopsy, Cone biopsy, Core biopsy, Endobronchial biopsy, Endometrial biopsy, Endomyocardial biopsy, Endoscopic biopsy, Excisional biopsy, Fine needle aspiration biopsy, Guided wire open biopsy, Heart biopsy, Incisional biopsy, Jumbo biopsy, Metabolic biopsy, Microbiopsy, Mirror image biopsy, Muscle biopsy, Needle biopsy, Nerve biopsy, Open biopsy, Open lung biopsy, Pleural biopsy, Polar body biopsy, Prostate biopsy, Punch biopsy, Renal biopsy, Salivary gland biopsy, Saucerization biopsy, Sentinel lymph node biopsy, Sextant biopsy, Shave biopsy, Skin biopsy, Skinny biopsy, Skinny needle biopsy, Small intestinal biopsy, Stereotactic biopsy, Stereotactic needle biopsy, Transbronchial needle biopsy, Transbronchial biopsy, Wedge biopsy, Wire-guide excisional biopsy

biopsy forceps SURGERY A disposable forceps used during minimally invasive GI and urologic endoscopy for collecting biopsies

biopsy needle SURGERY A thin–'skinny' needle passed percutaneously into an organ, often liver and kidney to obtain tissue for evaluation by light microscopy

biosafety level EPIDEMIOLOGY A classification for the degree of caution required when working with specific groups of pathogens. See Maximum containment facility

BioScrew® ORTHOPEDICS A bioabsorbable interference screw and delivery system for tibial and femoral repair; instruments included: BioScrew tri-lobe driver, tap, guides, and tunnel notcher

biosynthesis PHARMACOLOGY See Combinatorial biosynthesis

biosynthetic human insulin Humulin, see there

Biot breathing PULMONOLOGY A form of recurrent apnea characterized by regular volleys of breathing with a fixed tidal volume, punctuated by periods of apnea, which may be associated with either hypo- or hyperventilation. Cf Cheyne-Stokes breathing

bioterrorism GLOBAL VILLAGE A hypothetical scenario in which a hostile individual, organization or nation threatens the use of biologic weapons as a vehicle for extortion. See Anthrax, Ecoterrorism, Smallpox

biotin deficiency CLINICAL NUTRITION A condition caused by a lack of biotin characterized by anemia, dry skin, enteritis, and hypercholesterolemia; BD is rare, as biotin is formed by intestinal flora; it occurs in those who consume raw eggs.

biparental inheritance GENETICS Inheritance of a maternal and a paternal allele of a gene, ie normal mendelian inheritance

bipedal *adjective* Capable of locomotion on 2 feet

biphasic basal body temperature REPRODUCTION MEDICINE A pattern typical of ovulation and formation of a corpus luteum, which secretes progesterone, which $\uparrow$ BBT ± 0.5°C during the 2nd half of the menstrual cycle. See Basal body temperature

biphasic insulin ENDOCRINOLOGY An insulin formulation consisting of a mixture of intermediate- and fast-acting insulin. See Diabetes mellitus, Insulin

biphosphonate ENDOCRINOLOGY Any of a number of carbon-substituted analogues of pyrophosphate, an endogenous inhibitor of bone mineralization INDICATIONS Malignant hypercalcemia, Paget's disease, breast and prostate CA, myeloma; the most potent is zoledronic acid, which can be given at intervals of up to one yr ADVERSE EVENTS Oral–GI intolerance; IV–myalgia, pyrexia. See Osteoporosis

biplane cath lab INTERVENTIONAL CARDIOLOGY A place in a specialized cardiac catheterization suite where TEE is performed. See Transesophageal echocardiography

biplane transesophageal echocardiography Biplane intraoperative transesophageal echocardiography, BTEE CARDIAC IMAGING A type of TEE in which there are 2–longitudinal and transverse-imaging planes allowing real-time detection of residual cardiac lesions during the perioperative management of Pts with congenital heart disease. See Transesophageal echocardiography

bipolar *adjective* Having 2 poles CARDIAC PACING Having 2 electrodes, both of which are external to the pulse generator, usually in the heart. See Pulse generator

bipolar disorder Bipolar disease, bipolar illness, manic-depressive disease/illness, manic depression PSYCHIATRY A condition characterized by episodic mania-euphoria, alternating with bouts of depression, which affects 1% of the general population; BD first appears by age 30; ½ of Pts have 2-3 episodes during life, each from 4-13 months in duration CLINICAL Mood swings in BD may be dramatic and rapid, but more often are gradual; manic episodes are characterized by disordered thought, judgment, and social behavior, unwise business or financial decisions may be made when an individual is in a manic phase TREATMENT Lithium; if manic episode is unresponsive, electroconvulsive therapy may be effective

BIPOLAR DISORDER

BIPOLAR I DISORDER is characterized by a occurrence of one or more manic episodes or mixed episodes, and one or more major depressive episodes, and an absence of episodes better accounted for by schizoaffective, delusional, or psychotic disorders

BIPOLAR II DISORDER Recurrent major depressive episodes with hypomanic episodes Bipolar II is characterized by one or more major depressive episodes, one or more hypomanic episodes, and an absence of manic or mixed episodes or other episodes better accounted for by schizoaffective, delusional, or psychotic disorders

Famous manic-depressives: Paul Gauguin, Ernest Hemingway, Herman Hesse, Gustav Mahler, Edgar Allan Poe, Franz Schubert, Mark Twain, Vincent van Gogh, Tennessee Williams, Virginia Woolf

bipolar electrogenerator LAPAROSCOPIC SURGERY A device used to coagulate tissue during laparoscopic surgery, which is preferred to unipolar systems, as electical current is generated only between the 2 electrodes being used Cf Unipolar electrogenerator

bipolar trait A personality trait that represents extreme opposites of expression–eg, dominance-submission, extroversion-introversion, passive-aggression. See Bipolar disorder

bird beak sign IMAGING A descriptor for GI tract findings by barium studies COLON Ace of spades appearance A sharply defined voluptuously-curved, cut-off of the enema column in a volvulus of the sigmoid colon; if the barium passes proximally, 2 kissing 'bird beaks' are seen, known as an Omega loop ESOPHAGUS Sigmoid esophagus The over-distended esophagus of achalasia tapers into a pointed beak, which corresponds to the non-relaxing lower esophageal sphincter, seen by an upper GI radiocontrast study–barium 'swallow'

bird handler's disease Bird breeder's disease, bird fancier's lung, parrot fever PULMONARY MEDICINE A type of hypersensitivity pneumonitis–extrinsic allergic alveolitis caused by allergy to bird components* or bird droppings, avian serum or gut proteins AT RISK BIRDS Budgerigars, chickens, ducks, parakeets, pigeons, turkeys. See Hypersensitivity pneumonitis

bird leg A popular term for a leg greatly $\downarrow$ in diameter due to muscular atrophy. See Champagne bottle leg

birth 1. The act of being born, as in the birth of a child 2. Lineage; extraction; descent; noble extraction. See Husband-coached birth, Multiple birth, Natural childbirth, Primal–scream therapy, Rebirthing, Stillbirth, Virgin birth

birth canal Parturient canal A popular term for the passage through which the fetus travels the travails of parturition, to wit, the uterus, the vagina, and the vulva

birth control pill Oral contraceptive, see there

birth defect NEONATOLOGY 1. Any distortion of a part or general disfigurement of the body 2. Congenital malformation, see there 3. An abnormality of structure, function or body metabolism present at birth that may result in a physical and/or mental disability or is fatal

birth mother Biological mother, genetic mother, natural mother REPRODUCTIVE MEDICINE A ♀ who carries a baby to term, who she plans to give up for adoption. See Baby M, Surrogacy SOCIAL MEDICINE The ♀ who, with the birth father conceived a child, carried the pregnancy to term and delivered, then subsequently placed the child for adoption

birth parent Biological parent, see there

birth rate Crude birth rate The number of live births divided by the average population, or the population at midyr. Cf Mortality rate

birth weight The weight of a newborn child which, in the US, averages 3.2 kg at 37-wk/term. See Low-birth weight, Very low-birth weight

birthing room OBSTETRICS A hospital room designed to attend to the "creature comfort" needs of woman expected to have an uncomplicated vaginal L&D and recovery STUFF IN BR Room for family & friends, Martha Stewartoid decor, TV, magazines PROS Homey, happy customers CONS Can't sterilize, unsafe, not amenable to emergent care. See Alternate birthing center

birthmark DERMATOLOGY A popular term for a painless and benign, red or purple vascular skin marking–eg, cavernous hemangioma, which often disappears or fades by school age MANAGEMENT Local steroid injections may ↓ size. See Hemangioma, Nevus

bisacodyl Dulcolax An agent used to manage constipation

bisexual *adjective* 1. Having both ♂ and ♀ sexual characteristics or orientation 2. Of both sexes; hermaphroditic; as a flower with stamens and pistil, or an animal with ovaries and testes *noun* A person who engages in both homo– and heterosexual activity. Cf Heterosexual, Homosexual

bisferiens pulse CARDIOLOGY A double-beat pulse palpated over the carotid or brachial arteries, which is characteristic of obstructive cardiomyopathy and aortic regurgitation; the ascending limb–percussion wave initially arises rapidly and forcefully, producing a systolic pulse peak, followed by a dip or trough, and a second slower and broader positive–tidal wave; in some Pts, the pulse is intermittent, and may be evoked by mechanical–Valsalva or pharmacologic–nitroglycerin or catecholamines–maneuvers. Cf Spike-and-dome

bite *verb* To seize with the teeth; to lacerate, crush, or wound with teeth INFECTIOUS DISEASE A chomp from a dentated mammal; dog bites are relatively clean; monkey bites often contain pathogens; human bites more so; 10-20% of human bites are on the face, neck, breasts, or genitals, and occur during sexual activity. See Closed bite, Closed fist injury, Live bite, Snake bite PEDIATRICS See Stork bite

bivalency The state of being bivalent

bizarre *adjective* Referring to a strange thing; bizarre is used by pathologists for highly abnormal cells or patterns

BK virus INFECTIOUS DISEASE A small human polyomavirus which, like the JC virus, can transform infected cells in culture; 1° BKV infection is usually subclinical, is most common in early childhood, persists in renal epithelium and may be reactivated, causing severe tubulointerstitial nephritis, and cystitis in BM transplant recipients, or in the immunocompromised Pts–eg, those with hyper-IgM immunodeficiency syndrome; BKV may cause encephalitis DIAGNOSIS Serology–IgM, IgG antibodies; PCR, T_2-weighted MRI. See Polyomavirus. Cf JC virus

BKA Below the knee amputation, see there

black death Black plague. See Plague

black gallstone INTERNAL MEDICINE A gallstone composed of bilirubin polymers mixed with mucin glycoprotein matrix; BGs are most common in a background of cirrhosis and chronic hemolysis, as occurs in sickle cell anemia or thalassemia; BGs comprise 10-90% of all gallstones, depending on the population being studied, and have more calcium, carbonate and unmeasured residue, but less cholesterol and fat. See Gallstone. Cf Brown gallstone

black hairy tongue CLINICAL MEDICINE A condition characterized by hyperkeratosis of darkened, hyperplastic filiform papillae on the dorsum of the tongue with 2° hemosiderin deposition; BHT is anterior to the circumvallate papillae, due to prolonged antibiotic therapy, resulting in overgrowth of chromogenic bacteria, or related to food, coffee, or tobacco or oral bismuth therapy MANAGEMENT Brush tongue with a toothbrush ILLUSTRATION New England Journal of Medicine 1997, volume 337:897. See Hairy tongue. Cf Hairy leukoplakia

black lung disease 1. Anthracosis, see there 2. Coal workers' pneumoconiosis, see there

black mold *Stachybotrys chatarum* PUBLIC HEALTH A fungus found in moist environments–eg, schools, etc, which may cause nasal congestion, eye irritation, fever, wheezing, SOB

black palm Tâche noire SPORTS MEDICINE A condition affecting young adults, and seen on the thenar eminence of those participating in various sports–eg, climbing, gymnastics, golfing, tennis, weight lifting, and others; BP is caused by lateral shearing stress of epidermis against the rete pegs of the papillary dermis, resulting in rupture of capillaries, intraepidermal, and intracorneal hemorrhage which translate into multiple symmetrical black spots See Sports dermatology, Talon noir

black patch delirium GERIATRICS A transient clinical complex common in elderly Pts who have had both eyes patched after eye surgery CLINICAL Restlessness, anxiety, disorientation in time and space, persecutory delusions, hallucinations, suicidal ideation TREATMENT Passage of time

black tarry stool Black stool Stool that is viscid and blackened by oxidized blood, a finding often caused by cancer-related occult hemorrhage; BTSs appear with as little as 50-75 mL of blood, usually above the ligament of Treitz–oxidation of heme by combined action of enzymes and bacteria requires several hrs DIFFDX Drugs–eg, salicylates, steroids, rauwolfia, phenylbutazone, indomethacin, which may cause GI blood loss in normal persons and in those with underlying GI tract anomalies; BS may also be iatrogenic–charcoal, iron, bismuth

black widow spider *Latrodectus mactans* TOXICOLOGY A venomous spider indigenous to North America–most bites occur in California, more common in summer–which is 13 mm in length with a leg spread of 40 mm and which bites with anterior fangs and may be fatal in the very young or old CLINICAL Abrupt sharp, cramping and/or burning pain, dizziness, weakness, spreading to the entire body, activation of autonomic nervous system–N&V, sweating, salivation, tremors, muscle cramping and spasms, twitching, paresthesias, eyelid edema, ptosis; severe cases are accompanied by SOB, tachypnea, respiratory grunting, tachycardia, systolic HTN, acute nephritis and hemoglobinuria in small children MANAGEMENT Supportive, mild sedation, bed rest; for muscle spasms, calcium gluconate, antihypertensives; antivenin-Lyovac may be indicated in some Pts

blackhead Open comedo A blocked sebaceous gland which is open to air, where the secretions oxidize, turning black. Cf Whitehead

blackout NEUROLOGY A sign of early chronic alcohol or other substance abuse, characterized as an episode of total amnesia lasting from hrs to days after a period of intense drinking or alcohol binge; blackouts may be due to alterations in central serotoninergic neurotransmission, as these Pts have ↓ plasma levels of tryptophan

bladder cancer Cancer of urinary bladder UROLOGY A malignant epithelial neoplasm of the urinary bladder; it is the 5th most common cancer of ♂ in developed nations STATISTICS 52,300 new cases/1993–US; 90% 5-yr survival; 9% if distant metastasis when diagnosed CLINICAL TYPES Superficial—80% of total—throughout their clinical course; invasive *ab initio* RISK FACTORS Tobacco, occupational exposure to petrochemicals–benzene, exhaust fumes and carcinogens in rubber, chemical and leather industries, schistosomiasis DIAGNOSIS History, PE, urine cytology, imaging; confirm by cystoscopy & biopsy CLINICAL Hematuria, ↑ urinary frequency MANAGEMENT Depends on growth, size, location of tumor; for superficial UC, close followup is appropriate; for invasive BC, cystectomy, cystoprostatectomy, and radical cystectomy may be needed; RT, neoadjuvant–preemptive chemotherapy, or bCG may be used

BLADDER CANCER STAGING

STAGE I Cancer spread to bladder mucosa but not to muscular wall

STAGE II Cancer spread to muscular wall of bladder

STAGE III Cancer spread through muscular wall to peritoneum and/or to the nearby reproductive organs

STAGE IV Cancer spread to abdominal wall, pelvis, to nearby lymph nodes, or it has metastasized

bladder compliance UROLOGY Distensibility of the urinary bladder when filled. See Cystometrogram.

bladder-conserving therapy UROLOGY Any treatment for muscle-invading urinary bladder cancer that is designed to retain bladder function; BCT consists of complete transurethral "stripping" of transitional–urothelial epithelium, followed by induction chemoradiotherapy–cisplatin with 40 + Gy of RT, or preceded by MTX, cisplatin, vinblastine PROGNOSIS Good. Cf Radical cystectomy

bladder infection 1. Cystitis, see there 2. Urinary tract infection, see there

bladder outlet obstruction UROLOGY A general condition for any difficulty in the passage of urine from the bladder to the urethra which is more common in ♂, and due to BPH. See Benign prostate hypertrophy

bladder polyp UROLOGY A nonspecific pedunculated tumor that arises from the bladder mucosa. See Papillary urothelial cell carcinoma, Transitional cell carcinoma

bladder stone A concrement in the urinary tract. See Kidney stone

bladder training A type of biofeedback therapy used in Pts with urinary incontinence, a condition that affects ⅓ of ♀ ≥ age 60; BT is based on a combination of behavior modification, scheduling of voluntary micturition, and Pt education that emphasizes neurologic control of lower urinary tract function. See Biofeedback training, Behavioral modification

bladder ultrasonography Bladder capacity ultrasound, bladder scan UROLOGY The noninvasive determination of postvoid residual urine volume using an ultrasound device, which automatically calculates and displays bladder volume INDICATIONS Lower urinary tract dysfunction, urinary incontinence and/or neurogenic bladder. See Intermittent catheterization, Postvoid residual urine volume

blade plate ORTHOPEDICS A plate commonly seen on postoperative films, usually shaped at an oblique or right angle for subtrochanteric or supra-condylar fractures of the femur, occasionally used to bridge a femoral osteotomy; one arm has a chisel-shaped end driven into the bone, bridging the fracture; the other arm is used as a side plate and anchored to the bone with multiple screws

bland diet CLINICAL NUTRITION A mechanically soft and nonirritating diet commonly prescribed for Pts with IBD and peptic ulcer disease, despite its uncertain efficacy. See Diet. Cf Spicy foods

bland food NUTRITION A food that is not chemically or mechanically irritating–eg, toast, crackers, pasta, low fat meats. Cf Soft food

blast *noun* HEMATOLOGY A general term for a primitive blood cell. See Blast cell, Blast crisis

blast cell Blast HEMATOLOGY An immature cell in the BM's proliferative compartment, and earliest identifiable precursor of a cell line–erythroblast, lymphoblast, megakaryoblast and myeloblast; BCs are large–15-20 μm, have high N:C ratios, nuclei with fine, lacy to granular chromatin that contain one or more prominent nucleoli, basophilic, agranular cytoplasm, and abundant RNA, and actively synthesize DNA; BCs represent up to 5% of BM cells

blast crisis Blast phase, blast transformation ONCOLOGY The abrupt conversion of a chronic, relatively indolent leukemia usually CML into an acute decompensated, accelerated phase, with a marked–30+% of WBCs—↑ proportion of blasts and number of lymphocytes or myelocytes in circulation and BM CLINICAL Lymphadenopathy, hepatosplenomegaly, spleen and bone pain, fever, thrombosis LAB Leukocytosis, thrombocytosis or thrombocytopenia, anemia MANAGEMENT Response to BCs is usually short-lived; myeloblast transformations are commonly treated with hydroxyurea; ¼ respond to prednisone with vincristine. Cf Blast transformation, Relapse

blast injury TRAUMA MEDICINE An injury due to explosions or rapid decompression, the severity of which is a direct function of the intensity of the blast wave; death is caused by exsanguination from ruptured pulmonary vessels with hemorrhage, hemoptysis, air embolism, hypoxia, and respiratory failure; other lesions include cardiac contusion, causing arrhythmia, rupture of hollow organs, cerebral injuries–parenchymal hemorrhage and air embolism, and rupture of tympanic membranes TREATMENT Supportive; if air embolism is present, hyperbaric oxygen is indicated; other facets of blast injuries include impinging flying objects and whether the subject was submerged or free-standing at the time of the explosion. Cf Nuclear war

blastomycosis Chromoblastomycosis MYCOLOGY A suppurative granulomatous infection caused by *Blastomyces dermatitidis* EPIDEMIOLOGY ± 4/10⁵ symptomatic, many more asymptomatic–US CLINICAL-SYSTEMIC Usually acquired by inhalation, may produce dermatitis, pneumonitis or present as a systemic disease in the skin, lungs, bones, kidney, liver, spleen, CNS; usually begins as a respiratory infection, accompanied by cough, pleuritic chest pain, ARDS, chills, malaise, anorexia, weight loss MANAGEMENT Amphotericin B

bleed *noun* SURGERY Traditionally used as a verb, 'bleed' has acquired nominative status in the highly popular synonym for an 'episode of hemorrhage' *verb-active* To emit blood; to lose blood, eg, the arm bleeds, to bleed at the nose *verb-passive* To let blood; as, Dr. Q. Wack bleeds Pts with fever. See Herald bleed

bleeder SURGERY *noun* 1. An actively hemorrhaging Pt 2. A Pt whose BM has been ravaged by chemotherapy, and is thus highly susceptible to bleeding due to marked thrombocytopenia–eg, < 10,000 platelets/mm³ 3. An overtly hemorrhaging blood vessel in an operative or endoscopic field. Cf 'Clotter' VOX POPULI Phlebotomist, see there

bleeding CLINICAL MEDICINE Emitting blood; hemorrhaging; bloodletting GYNECOLOGY See Breakthrough bleeding, Dysfunctional uterine bleeding, Postmenopausal bleeding, Postpartum bleeding, Varicial bleeding

MEDTALK Hemorrhagic diathesis, see there; expressing anguish or compassion

bleeding disorder Coagulopathy, see there

blemish DERMATOLOGY A cosmetic skin defect–eg, birthmarks and lesions caused by aging or exposure to the sun EXAMPLES Port-wine stains, scars, acne scars, birth-related stretch marks MANAGEMENT Green-light lasering. See Facial resurfacing

Blenoxane Bleomycin, see there

BLEO-MOPP ONCOLOGY A chemotherapy regimen: bleomycin, nitrogen mustard, Oncovin, procarbazine, prednisone. See Chemotherapy

bleomycin Blenoxane, BLEO, BLM ONCOLOGY A chemotherapeutic antibiotic, which consists of a mixture of low MW–±1.5 kD metal and oxygen-dependent DNA cleavers, isolated from *Streptomyces verticillus verticellus*, which is used to treat lymphomas, Hodgkin's disease, Kaposi sarcoma, testicular CA, and other germ cell tumors MECHANISM OF ACTION Induction of single- and double-stranded breaks in DNA ADVERSE EFFECTS Fever, stomatitis, skin lesions, BM suppression, fever, chills, and hair loss

blepharitis Blear eye CLINICAL MEDICINE Chronic inflammation of the eyelids, with secretion of sebaceous material

blepharoplasty PLASTIC SURGERY 1. The surgical removal of the 'baggy' parts of the upper and lower lid 2. Surgical repair of an eyelid; often performed for ptosis

blepharospasm NEUROLOGY A focal dystonia consisting of the involuntary twitching of one or both eyelids; the spasms may completely close the eyelids, causing functional blindness even if the vision is normal ETIOLOGY Related to emotional stress, sleep deprivation or use of stimulants–eg, amphetamines, caffeine, nicotine, decongestants

blind OPHTHALMOLOGY *adjective* 1. Referring to the inability to see 2. Not well marked or easily discernible; hidden; unseen; concealed 3. Lacking openings for light or passage; open only at one end; as in the 'blind gut' or cecum *noun* Inability to see; sightless *verb* 1. To make blind; to deprive of vision or discernment 2. To darken; to obscure to the eye or understanding; to conceal; to deceive

blind biopsy PATHOLOGY An endoscopic biopsy of tissue that is grossly–macroscopically normal; BBs may be diagnostically positive—especially if a submucosal lesion is suspected, and if negative, may establish 'clear' margins for a surgical resection. Cf Biopsy

blind loop syndrome Afferent loop syndrome, stagnant loop syndrome SURGERY A complication of Billroth II subtotal gastroenterostomy–end-to-side enteroenteric anastomosis that may arise long after surgery; the afferent loop consists of duodenum and a variable portion of jejunum, a loop that is a temporary reservoir for 1-1.5 liters of biliary and pancreatic secretions; after a fatty meal, the contents of a partially obstructed afferent loop ↑ and 'explosively' enter the stomach and may be regurgitated as greenish bilious fluid; other Sx include intermittent diarrhea due to disaccharidase deficiency, abdominal 'colic', hemorrhage, vitamin deficiency and neurologic Sx; with prolonged partial obstruction, the stool is steatorrheic–bulky, gray, and greasy, accompanied by weight loss; complete blind loop obstruction may be a medical emergency with rapid deterioration, shock, and perforation peritonitis TREATMENT Antibiotics–eg, T-S, loop shortening, afferent-to-efferent or Roux-en-Y anastomoses or gastrojejunostomy

blind spot OPHTHALMOLOGY 1. A small area of the retina where the optic nerve enters the eye; occurs normally in all eyes 2. Any gap in the visual field corresponding to an area of the retina where no visual cells are present; associated with eye disease; optic disk, see there PSYCHIATRY An area of a person's personality of which he is totally unaware, since recognition would cause painful emotions PUBLIC HEALTH A physical space behind a driver's outer shoulder between that seen by the rear-view mirrors mounted on the inside and outside of a car

blinding CLINICAL RESEARCH The concealment of group assignment–to either the treatment or control group from the knowledge of Pts and/or investigators in a clinical trial; while blinding is intended to remove bias and subjectivity, it is not always practical–eg when comparing surgery to drug treatment, but should be used when possible and compatible with optimal Pt care. See Double blinded study, 'Nocebo', Placebo, Triple blinded study. Cf Control

blindness An inability to see effectively. See Blue color blindness, Legal blindness, Night blindness, Occupational blindness, Snowblindness, Transient monocular blindness

blister A skin vesicle filled with serous fluid, caused by burns, trauma, or by a vesicatory; a collection of serous fluid

bloating VOX POPULI A lay term for post-prandial abdominal fullness or swelling

Blocadren® Timolol, see there

block ANESTHESIOLOGY Anesthesia intended to ameliorate locoregional pain. See Caudal block, Celiac plexus block, Epidural block, Field block, Hemiblock, Nerve block, Neurolytic block, Neurolytic celiac plexus block, Nonneurolytic block, Paracervical block, Pudendal block, Ring block, Spinal block CARDIOLOGY Impeded electromechanical activity. See Atrioventricular block, Bundle branch block, Exit block, Heart block, Left bundle-branch block, Möbitz type I block, Möbitz type II block, Mucosal block, Right bundle-branch block GASTROENTEROLOGY An obstruction NEUROLOGY A mental impasse. See Writer's block PUBLIC HEALTH See Sunblock

blockage Wax, see there

blocking CLINICAL TRIALS The process of establishing defined groups, as in a treatment schedule designed to ensure a specific allocation ratio

blood HEMATOLOGY A circulating tissue composed of a fluid portion–plasma, suspended formed elements–RBCs, WBCs, platelets, and other components, including CO_2, O_2, proteins, glucose, cholesterol and other fats, which circulates in a closed system–the heart, arteries, capillaries and veins, and is charged with transporting O_2 and nutrients to cells, and removing CO_2 and waste products to the appropriate sites. See Artificial blood, Bad blood, Deoxygenated blood, Euroblood, Fecal occult blood, Frozen blood, Leukocyte-poor blood, Occult blood, Safe blood, Strawberry cream blood, Umbilical cord blood, Whole blood, Yellow blood. Cf Snake blood

blood alcohol concentration Blood alcohol level CLINICAL CHEMISTRY The serum levels of ethanol in an individual who has imbibed same; U.S. federal mandate has defined 80 mg/dL as legal intoxication

blood blister DERMATOLOGY A collection of blood in a skin blister caused by minor skin trauma–eg, a pinch or crushing injury

blood-borne pathogen A generic term for pathogenic microorganism(s) present in blood including viruses–eg HIV, HBV, HCV, CMV, and others, and parasites–eg malaria, *Leishmania*, *Babesia* PROPHYLAXIS After HBV exposure, HBV hyperimmune globulin may be used prevent clinical disease. See Bad Bug Book. Cf Water-borne pathogen

blood-brain barrier PHYSIOLOGY A selectively permeable structural and functional barrier that exists between the capillaries and the brain; water, O_2 and CO_2 readily cross the BBB, glucose is slower, Na^+, K^+, Mg^{++}, Cl^-, HCO_3^- and HPO_4^- require 3-30-fold more time to equilibrate with the CSF than with other interstitial fluids; urea penetrates very slowly; catecholamines and bile salts essentially do not cross the BBB–kernicterus is due to accumulation of bile salts in the brains of neonates whose BBB is yet immature; integrity of the BBB is impaired in hepatic encephalopathy

blood cell A lay term for any of the 3 major cells in the peripheral circulation–eg, RBCs–see red cells, WBCs–see white cells, and platelets, see there

blood clot HEMATOLOGY Thrombus, see there; the conversion of blood from a liquid form to solid through coagulation. See Atherosclerotic plaque, Coagulation, Embolism, Thrombus

blood clotting factor Coagulation factor, see there

blood component therapy Component therapy The therapeutic use of specific portions–components of blood–eg, factor VIII concentrates, packed red cells, or platelets rather than whole blood

blood count Complete blood count, see there

blood filter TRANSFUSION MEDICINE A device attached to a unit of blood or components designed to retain cells, blood clots, debris. See Leukocyte reduction

blood gas CLINICAL MEDICINE 1. The major gases–CO_2 and O_2 in blood 2. Blood gas analysis, see there

blood gas analysis CLINICAL MEDICINE The evaluation of arterial blood for O_2, CO_2, and bicarbonate content as well as blood pH, to determine oxygenation and acid base status. See Anion gap, Metabolic acidosis, Metabolic alkalosis, Respiratory acidosis, Respiratory alkalosis

blood glucose DIABETOLOGY The principal sugar produced by the body from food–especially carbohydrates, but also from proteins and fats; glucose is the body's major source of energy, is transported to cells via the circulation and used by cells in the presence of insulin. See Glucose

blood glucose monitoring Sugar monitoring LAB MEDICINE The periodic testing of serum glucose in Pts known to have DM. See Bedside glucose monitoring, Beta cell implants, Diabetes, Glucometer, Glycosylated hemoglobin, Non-Invasive glucose monitoring

blood group TRANSFUSION MEDICINE A popular term for any of the more than 20 antigen systems including ABO, Duffy, Kelly, Kidd, Lewis, Rh and other antigen systems, which are found primarily, on RBCs See ABO Cartwright, Chido/Rodgers, Cromer-related, Do/Gy/Hy/Jo, Duffy, Gerbich, H, Indian, JMH, Kell, Kidd, Knops, Kx, Lewis, Lutheran, LW, MN, OK^k, P$_1$, Rh, Type & cross, Type & screen, Xg^a

blood in urine VOX POPULI Hematuria

blood pressure CARDIOLOGY The force that blood in the circulation exerts on arterial walls, 2° to myocardial contraction in response to various demands–eg, exercise, stress, sleep, which is divided into systolic–due to heart contractions and diastolic–relaxation phases; BP varies with age and sex RULE OF THUMB FOR NORMAL SYSTOLIC BP–ADULTS BP = 100 + age; CHILDREN BP = 2 x age + 80; DIASTOLIC BP should be ± $^2/_3$ NORMAL BP 120/80 mm Hg. See Hypertension, Hypotension, Sphygmomanometer–blood pressure cuff

blood product A general term for a biopharmaceutical purified from human blood–eg factor VIII, fresh frozen plasma–FFP

blood shortage TRANSFUSION MEDICINE A relative deficiency in the supply of blood and blood products available for transfusion in a particular region. See Euroblood

blood storage lesions TRANSFUSION MEDICINE The reversible changes that occur in packed RBCs during storage at 4°C: ↑ K$^+$, ↑ inorganic phosphate; ↓ pH, ↓ Na$^+$, ↓ 2,3 DPG. Transfusion

blood sugar See Glucose, Hyperglycemia, Hypoglycemia

blood test LAB MEDICINE Any test performed on the blood, in contrast to urine or other body fluid. See Bilirubin, BUN, Calcium, CBC, Cholesterol, Metabolic panel, Glucose, Phosphorus, Protein, Uric acid

blood thinner VOX POPULI Anticoagulant, see there

blood transfusion The transfer of blood or blood products from a donor into a recipient, usually to replace red cells or blood products lost through severe bleeding. See Autologous transfusion, Blood conserving therapy, Directed donation

blood urea nitrogen BUN NEPHROLOGY A metabolic byproduct from the breakdown of blood, muscle and protein which is a measure of the urea level in blood; ↑ BUN often indicates early kidney damage, as well as dehydration, CHF, GI bleeding, starvation, shock or urinary tract obstruction–by tumor or prostate gland; ↓ BUN may indicate liver disease, malnutrition or a low protein diet

blood warmer A device that warms blood stored at 4°C to body temperature. See Cold antibody

blood work A popular term referring to any diagnostic testing performed on the fluid or cells of peripheral blood

bloodless surgery center Bloodless medical and surgical center HOSPITAL PRACTICE A free-standing affiliate of a hospital, which offers a range of medical and surgical procedures that eliminate or markedly ↓ blood use. See Transfusion medicine. Cf Autologous transfusion

Bloodloc system TRANSFUSION MEDICINE A proprietary system that uses a mechanical barrier with dedicated transfusion identification to prevent improper transfusion. See Transfusion medicine

bloodshot VOX POPULI *adjective* Red and inflamed; infused with blood, or having turgid blood vessels; especially, inflamed or irritated conjunctivae

bloody show OBSTETRICS A blood-tinged mucus cervical plug that filled the cervical canal during pregnancy, which is a classic sign of impending labor; BS also refers to the beginning of menstruation

blowout fracture HEAD A fracture of the floor and medial walls of the orbit caused by a blunt object–eg, baseball, fist, rock–with herniation of the orbital contents–fat, inferior oblique, and inferior rectus into maxillary sinus, accompanied by diplopia, enophthalmos, and limitation of upward gaze. See LeFort fractures HIP Dashboard fracture, see there

blowout metastasis ONCOLOGY An osteolytic bone metastasis, which is an expansile, marginated, trabeculated lesion typical of thyroid and renal CAs, also seen in liposarcoma, melanoma, pheochromocytoma, lung, breast CAs

blue baby An infant with neonatal cyanosis, most often due to congenital heart malformation–eg, transposition of great vessels and in tetralogy of Fallot MANAGEMENT Surgery

blue balls Lover's nuts SEXOLOGY A popular term for testicular pain caused by prolonged sexual stimulation without ejaculation

blue belly Cullen sign Periumbilical bluish discoloration due to intraperitoneal hemorrhage, seen in acute pancreatitis

blue bloater A popular term for the appearance of a Pt with COPD with Sx of chronic bronchitis, normal to ↓ lung capacity, ↑ residual volume with air-trapping, ↓ expiratory flow, and characteristic arterial blood gas parameters–↓ PO$_2$, ↑ PCO$_2$, despite normal diffusing capacity, cyanosis and right heart failure, due to sleep apnea and progressive chronic pulmonary HTN; with time, it becomes indistinguishable from other forms of COPD. See Chronic obstructive pulmonary disease, 'Pink puffer'

Blue Cross MANAGED CARE A not-for-profit US bureaucracy involved in hospital reimbursement. Cf Blue Shield

94 **Blue Cross & Blue Shield Association** The Blues MANAGED CARE A federation of locally governed, regulated, and not-for-profit licensed entities that disburses monies to health care providers in the US. See Blue Cross, Blue Shield, 'Major medical', Wrap-around policy

blue domed cyst BREAST A component of fibrocystic breast disease seen in perimenopausal ♀, which consists of cysts filled with light brown–posthemorrhagic fluid with benign stromal and epithelial hyperplasia; tenderness is ↑ before menstruation. See Fibrocystic disease of breast

blue nevus DERMATOLOGY A sharply circumscribed black-blue intradermal nevus present at birth or which develops with time; it is most common on acral parts and buttocks of ♀, seen as elevated sebaceous yellow, smooth plaques on the scalp, face, < 1 cm in diameter, which may also be seen in the oral cavity, uterine cervix and prostate

Blue Shield A US not-for-profit health care insurer that is a reimbursement intermediary for physicians. Cf Blue Cross

blue spell PEDIATRICS An episode of hypoxia, cyanosis, dyspnea, restlessness and syncope with acutely ↓ arterial PO_2 and ↓ of already compromised pulmonary blood flow, seen in Fallot's tetralogy, but also in other cyanotic congenital heart malformations. See Shunts–right-to-left

blue toe syndrome CARDIOLOGY Atherothrombotic microembolism of lower extremities due to recurrent cholesterol embolic 'showers' with painful cyanotic discoloration of toes and embolism elsewhere, resolving between attacks; despite the gangrenous appearance, BTs may respond to conservative therapy without amputation DIAGNOSIS Arteriography TREATMENT-MEDICAL Dypyridamole plus aspirin SURGICAL Thromboendarterectomy of aorta. See Black toe disease

blue top tube CLINICAL CHEMISTRY A 5 ml tube with sodium citrate as an anticoagulant, used to collect specimens for coagulation factor assays, fibrinogen, glucose-6-phosphate dehydrogenase, PTT, PT, thrombin time

blue valve 'syndrome' CARDIOLOGY A descriptor for the bluish, vaguely translucent, hypermobile cardiac valves removed from Pts with Marfan syndrome, which have mucoid degeneration. See Marfan syndrome

the Blues Blue Cross & Blue Shield Association, see there

the blues PSYCHIATRY Transient mental depression, often related to exogenous events. See Christmas blues, Cocaine blues, Postpartum blues

blunt dissection SURGICAL TECHNIQUE The separation of tissues–dissection along fascial planes with a blunt instrument; BD preserves locoregional architecture and structural integrity of nerves, vessels, and lymph nodes. See Dissection

blunt injury A traumatic injury effected by a blunt object or force, in which the skin was not penetrated; usually results from assaults, abuse, accidents or resuscitative measures

blunt trauma MOLECULAR Any injury sustained from blunt force, which may be related to MVAs, or mishaps, falls or jumps, blows or crush injuries from animals, blunt objects or unarmed assailants. Cf Penetrating trauma

BM Bone marrow, also 1. Basal metabolism 2. Body mass 3. Bowel movement 4. Buccomesial

BMI Body mass index, see there

BMP 1. Basic metabolic panel, see there 2. Bone morphogenetic protein, see there ONCOLOGY A chemotherapy regimen: BCNU, MTX, procarbazine

BMR Basal metabolic rate, see there

BMT Bone marrow transplant, see there

BNP B-type natriuretic peptide, brain natriuretic peptide PHYSIOLOGY A 32-residue peptide hormone produced predominantly in the ventricles, secreted in response to fluid overload–eg, CHF. See Atrial natriuretic peptide

BO 1. Basioccipital 2. Body odor, see there 3. Bowel obstruction 4. Buccoocclusal

BOAP ONCOLOGY A chemotherapy regimen: bleomycin, Oncovin, Adriamycin, prednisone. See Bleomycin, Chemotherapy

board ADMINISTRATION An authorized assembly or meeting, public or private of persons appointed or elected to manage or direct a business or trust, as a board of directors, trustees. See Institutional review board, National Board of Medical Examiners, Professional board, Specialty board, State board of medicine INFORMATICS Healthcare Informatics Standards Board

board-certified *adjective* Referring to a US or Canadian physician who has 1. Completed 4-8 yrs-of post-medical school residency training, ie a physician who is 'board-eligible' and 2. Passed an examination, commonly called the 'Boards' that tests theoretical knowledge in an area of specialization. Cf Board-eligible *noun* A designation that a physician has successfully completed an approved educational program and evaluation process by the American Board of Medical Specialties which includes an examination designed to assess the knowledge, skills, and experience required to provide quality Pt care in a given specialty

board-eligible *adjective* Referring to a North American physician who has completed 4-8 yrs of post-medical school residency training, and is eligible to sit for an examination–the 'boards' that tests theoretical knowledge in the area of specialization. Cf Board-certified

board of health MANAGED CARE A body of a municipal or state government which is responsible for coordinating public health activities in a specified city, county, or state. See Centers for Disease Control

board-like rigidity CLINICAL MEDICINE Spastic rigidity of abdominal wall muscles induced by acute peritonitis, classically elicited by a perforated ulcer–any fulminant peritonitis may elicit the same reaction. See Hippocratic facies

board of medical examiners CLINICAL PRACTICE A body recognized by a state government, which validates a health professionals, credentials, to determine the individual's ability to practice medicine in a particular state; the board is also authorized to suspend, place a physician on probation, or revoke his/her license to practice medicine

board of trustees POLITICS The posse of thugs who oversee an institution's administration. See Board of directors

boards See State boards

BOAT CARDIOLOGY An ongoing study–Balloon Angioplasty vs Optimal Atherectomy on the efficacy of aggressive atherectomy as a means of improving the short-term results of angioplasty. Cf CAVEAT

body burden THERAPEUTICS A popular term for the total amount of a particular agent or chemical in the body; for some chemicals, the BB is high, because the agent is stored in fat or bone or is eliminated very slowly

body dysmorphic disorder Beauty hypochondria, dysmorphophobia PSYCHIATRY A psychiatric condition in which the Pt has a profound negative distortion of his or her body image, which may hinge on a perceived craniofacial flaw; Pts with BDD have obsessive body-image concerns, leading to compulsive checking of appearance in the mirror, intense self-consciousness, social avoidance and isolation, and depression. See Body image. Cf Anorexia nervosa

body fluid 1. Any fluid in the body including blood, urine, saliva, sputum, tears, semen, milk, or vaginal secretions 2. BF is often used with specific reference to those fluids to which health care workers might reasonably be exposed including blood, urine, saliva, and semen

body image PSYCHOLOGY The concept of one's body or self as presented to others; the BI is core aspect of 'selfness,' which incorporates personal attitudes and perceptions concerning one's appearance. See Hair replacement

body language PSYCHOLOGY An informal often culture-independent form of communication in which emotions, feelings, motives, and thoughts are expressed by changes in facial expressions, gestures, posture, body positions, and other nonverbal signs. See Kinesics

body lice VOX POPULI Pediculosis humanis corporis. See Louse

body mass index PHYSIOLOGY A calculated value that correlates with body fat, which is used to define obesity; according to the WHO, ↑ risk of various obesity-related conditions occur at BMIs of ≥ 25. See Fat balance

body mechanics Correct positioning of the body for a given task, such as lifting a heavy object or typing. See Biomechanics

body odor A malodorous body scent. Cf Flatulance, Halitosis

body odors, fear of PSYCHOLOGY Osmophobia, osphresiophobia. See Phobia

body packing SUBSTANCE ABUSE A method for smuggling narcotics in which a 'courier' swallows condoms or other containers filled with pure cocaine or heroin to escape detection by customs agents and specially trained dogs; the body packer 'syndrome' consists of a constellation of medical complaints–eg, intestinal obstruction and rupture of bags, the latter is almost always fatal DIAGNOSIS Scout films. See 'Double condom' sign, Drug mules

body piercing BODY IMAGE A disruption of a mucocutaneous surface with jewelry or dangling artifices. See Tattoos

body rocking NEUROLOGY Monotonous rhythmic movements, which can be seen in various conditions–eg, autism, mental retardation, psychiatric disorders, and in prolonged institutionalization

body sculpting SURGERY A highly popular term for cosmetic surgery intended to change the contours of a person's body to achieve what he or–more commonly–she perceives to be a perfect physique. See Cosmetic surgery

body substance isolation A system of precautionary measures taken to ↓ nosocomial transmission of pathogens in health care, which requires that gloves be worn for any contact with mucous membranes and moist body substances–eg, oral secretions, stool, etc–BSI ignores air-borne pathogens. Cf Universal precautions

Boerhaave syndrome Traumatic rupture of the lower esophagus after major blunt chest trauma, CPR or forceful protracted vomiting; BS is more common in Pts with pre-existing esophageal disease–eg, reflux esophagitis CLINICAL Abrupt chest pain, which may radiate to the neck, accompanied by shock, sepsis, death within 48 hrs if untreated

boil INFECTIOUS DISEASE *noun* A painful, inflamed, circumscribed, often staphylococcal skin infection with pus and blood and a central fibrous mass of dead tissue, aka core; if multiple, a carbuncle; so-called blind boils suppurate imperfectly, or fail to come to a head CLINICAL Tender pea-sized or greater, red nodule, which may ooze pus or weep MANAGEMENT Warm, wet compresses; oral or topical antibiotics COMPLICATING FACTORS CA, DM, immunosuppressants. See Dehli boil *verb* To heat to a boiling point, or cause ebullition; as, to boil water

Bolivian hemorrhagic fever VIROLOGY An arenavirus infection similar to Argentine HF; BHF is endemic to the grain-producing province of Beni in Amazonian Bolivia AGENT Machupo virus VECTOR Excreted in urine of the rodent vector, *Calomys callosus* CLINICAL BHF affects the cardiovascular, hematopoietic, renal systems, and the CNS; early fever, anorexia, N&V, myalgia, neurologic signs—50% have intention tremor, 25% convulsive encephalopathy MORTALITY 10-20%, especially children. See Machupo virus. Cf Haverhill fever

bolus 1. Any concentrate given as a single dose to achieve an immediate effect 2. Any mass or blob–eg, masticated food, in transit through a tube ENDOCRINOLOGY An extra boost of insulin given to cover expected rise in blood glucose–sugar such as the rise that occurs after eating THERAPEUTICS A large IV dose of a drug given "all at once" at the beginning of treatment, which raises the concentration in the body to a therapeutic level

bolus thrombolytic CARDIOLOGY *noun* A thrombolytic–eg, tPA, administered as a bolus–ie, to ↓ time to treatment and ↓ mortality in acute MI See ASSENT I, Bolus, TNK-tPA

bomb MILITARY MEDICINE A device designed to cause physical damage to a specified area, by exploding on impact or when a particular event occurs–eg, being moved. See B-61, Dirty bomb, Genetic bomb POPULAR PSYCHOLOGY See Time bomb

bonding NEONATOLOGY The formation of emotional ties between an infant and mother or other caregiver that occurs in the early post-partum. See Companionship, Infant massage. Cf Anaclitic depression, Inner bonding, Male bonding, Social isolation PSYCHIATRY The attachment and unity of 2 people whose identities are significantly affected by mutual interactions PUBLIC HEALTH See Antibiotic bonding

bone ANATOMY A solid, rigid, ossified connective tissue forming an organ of the skeletal system; any of the 206 bones in the body. See Basisphenoid bone, Blue bone, Carpal bone, Cancellous bone, Compact bone, Cuboid bone, Cuneiform bone, Disappearing bone, Dumbbell bone, Endochondral bone, Facial bone, Frontal bone, Funny bone, Halbard bone, Hetereotopic bone, Hungry bone, Hyoid bone, Innominate bone, Lacrimal bone, Membranous bone, Moth-eaten bone, Nasal bone, Navicular bone, Peppermint stick candy bone, Ping pong bone, Red bone, Rider's bone, Shooter bone, Spongy bone, Wormian bone, Woven bone, Zygomatic bon

bone cancer ONCOLOGY Any malignancy of bone, which is usually 2° to, ie metastatic from primary tumors of prostate, breast, kidneys, etc; 1° BCs include osteosarcoma, chondrosarcoma CLINICAL Intense–"gnawing"–bone pain DIAGNOSIS Hx, PE, x-rays, Bx LAB ↑ Alk phos TREATMENT Depends on type and mitotic activity, location, size, extent of tumor

bone densitometry The measurement of bone mass or density; the current methods—single-photon absorptiometry, dual-energy photon absorptiometry, dual-energy X-ray absorptiometry, are based on tissue absorption of photons derived from either a radionuclide or an X-ray tube; the latter are more accurate with shorter scan time. See Osteoporosis

bone density The amount of trabecular bone in a certain volume of bone, which can be measured by quantitative computed tomography

bone glue ORTHOPEDICS A calcium/phosphate goo that hardens into dahllite 5 mins post mixing, acting as 'instant bone' for fractures. See Dahllite

bone graft ORTHOPEDIC SURGERY Sterilized bony tissue, often of cadaveric origin, used to fill and/or 'sculpt' bone defects INDICATIONS Spinal fusion, revision of failed articular prostheses, filling traumatic or malignant bone defects, or periodontal defects. See Tissue bank

bone marker LAB MEDICINE Any protein degradation product that indicates bone breakdown. See N-telopeptides

bone marrow ablation ONCOLOGY The destruction of BM with radiation or drugs

bone marrow aspiration & biopsy HEMATOLOGY A diagnostic procedure in which tissue is obtained from BM INDICATIONS Chronic anemia, leukopenia, thrombocytopenia, aplastic anemia, leukemias and lymphomas, polycythemia vera, myelodysplastic syndromes, myeloma; BMA&B may be performed on an ambulatory basis by a hematologist or oncologist and used to follow efficacy of therapy

bone marrow harvesting HEMATOLOGY The collecting of disease-free BM for storage and future use in BM transplant

bone marrow suppression ONCOLOGY A reduction of BM stem cells, a side effect of chemotherapeutics and antivirals–eg, AZT; BMS leads to ↓ WBCs, RBCs and platelets, ergo, anemia, bacterial infections and spontaneous or excess bleeding

bone marrow transplantation HEMATOLOGY The infusion of hematopoietic precursors or stem cells to re-establish BM function after chemotherapy or RT damage, or defective BM COMPLICATIONS GVHD–35-60%, interstitial pneumonia, idiopathic, or viral–CMV, HSV, varicella-zoster; post-BMT leukemic relapse or AML from donor into an HLA-identical sibling for CML CAUSE OF DEATH Opportunistic infections–eg, CMV pneumonia–85% mortality; prophylactic gancyclovir ↓ this complication; other causes of interstitial pneumonia include *Pneumocystis carinii* or RT CLINICAL GVHD with skin rash, hepatic defects, diarrhea, infections, autoimmunity TYPES Allogeneic, autologous INDICATIONS BMT is appropriate and often successful in leukemia and aplastic anemia, but less so in treating lymphoma, thalassemia major, osteopetrosis, inborn errors of metabolism–adrenoleukodystrophy, Hurler syndrome, osteopetrosis, metachromatic leukodystrophy, congenital immune deficiencies–eg, Wiscott-Aldrich disease, SCID TYPES Autologous–Pt's own BM; allogeneic–BM from someone else; syngeneic–BM from identical twin PROGNOSIS Aplastic anemia–5-yr survival-80%; ANLL-50%; ALL-25%; BMT survival in leukemia depends on chronicity and cell type; BMT survival in non-neoplastic conditions 63-75+%; minor HLA mismatching is OK. See Autologous BMT, Chemotherapy, Relapse, Remission

bone metastases ONCOLOGY Cancer that has spread from a primary tumor to the bone

bone mineral density A measurement of bone mass, expressed as the amount of mineral–in grams divided by the area scanned in cm². See Bone densitometry

bone modeling See Modeling

bone paste 1. Fragmented bone and fibro-osseous tissue removed by various forms of surgery to the foot 2. A mineral-rich filler used in open bone and fracture repair

bone remodeling See Remodeling

bone scan Bone scintigraphy NUCLEAR MEDICINE A method in which a radioactive compound–eg, ⁹⁹ᵐTc IDA, is administered and its distribution in the body analyzed by a scintillation camera for ↑ or ↓ uptake in bone, an indicator of infection or malignancy. Cf Skeletal survey

bone sclerosis Eburnation IMAGING An ↑ bone density, such that the bone is 'whiter' than normal

bone spur Lipping IMAGING A small osteophyte seen at the margin of a joint's articular surface

bone tumor ONCOLOGY A generic term encompassing both malignant and benign tumors in bone; most cancer in bone tissue is 2° to metastasis from a distant 1°s–eg, from breast or prostate; 1° bone CA–eg, osteogenic sarcoma is rare. See Osteoma, Osteosarcoma

bone/joint panel LAB MEDICINE A group of tests once regarded as a cost-effective, sensitive, specific method for evaluating Pts with bone and joint complaints. See Organ panel. Cf N-telopeptide, Pyridinium cross-links

Bonefos® A proprietary form of clodronate, see there

BoneSource A hydroxyapatite cement/paste used to repair cranial defects

Bonnevie-Ullrich syndrome A genetic disorder in ♀ characterized by lack of an X chromosome CLINICAL Webbing of neck, short stature, retarded development of 2° sex characteristics, absent menses, infertility, coarctation of aorta, low hairline, eye defects–drooping eyelids, skeletal deformities MANAGEMENT Estrogen supplements at puberty; hGH replacement; surgical correction of coarctation

bonus Any benefit of a job or a lump-sum of cash given to an employee, which is a tangible expression of the employer's gratitude for the employee's performance. See Immune bonus, Marriage bonus, Signing bonus

bony *adjective* Consisting of bone/bones; full of bones; pertaining to bones

boob job A popular term for breast augmentation, see there

booked procedure SURGERY A surgical procedure that has been scheduled for surgery and time allocated in a particular OR

bookmark ONLINE A popular term for the noting of a person for future reference–a metaphor borrowed from Web browsers

BOOP Bronchiolitis obliterans organizing pneumonia A disease once considered a form of interstitial pneumonia ETIOLOGY Obscure; ? associated with toxic fumes, infection, connective tissue disease CLINICAL Cough, dyspnea, 'flu' symptoms, 50% recovery, 12% BOOPs eventually die of disease, many develop usual interstitial pneumonia; obstructive symptoms are limited to smokers, most of whom have restrictive disease and impaired diffusing capacity, remission may follow use of corticosteroids RADIOLOGY Patchy ground-glass appearance on AP chest film DIFFDX Organization in infections, pneumonias associated with other conditions–eg COPD, cystic fibrosis, bronchiectasis, distal obstruction, aspiration diffuse alveolar damage, eosinophilic pneumonia, collagen vascular disease, allergic reactions–eg extrinsic allergic alveolitis, vasculitis, and reactive to other processes–abscesses, infarcts, tumors

boosted reaction TUBERCULOSIS A positive reaction to a tuberculin skin test, due to a boosted immune response from a skin test given up to a year earlier. See Booster phenomenon, Two-step testing

booster Booster dose IMMUNOLOGY The deliberate re-exposure of an individual to an antigen as a means of inducing a 2° immune response months to yrs after a primary response to a vaccine or other antigen. See Immune

booster phenomenon TUBERCULOSIS An ↑ in size of a tuberculin reaction–> 6 mm, or from < to > 10 mm after a 2ⁿᵈ PPD skin test for TB; the BP is attributed to enhanced immunologic 'recall', due to either 1. Previous infection with *M tuberculosis* or 2. Infection with a non-TB mycobacterium; BP is most common in elderly Pts with inactive *M tuberculosis*, who rarely convert to active disease

booster shot IMMUNOLOGY A 2ⁿᵈ immunization dose, administered after an appropriate time interval, allowing the body to mount an immune response; when a person is exposed–eg, to 'dirty' wounds, or plans potential exposure–eg, travel to regions endemic for certain infectious agents, the BS provides a rapid anamnestic response that outpaces the development of disease–eg, tetanus

boot INFORMATICS *verb* To load the operating system–eg, Windows, OS X into the computer's RAM or main memory, after which the computer can run applications. See Random access memory

BOP ONCOLOGY A chemotherapy regimen: 1. BCNU, Oncovin, prednisone 2. Bleomycin, Oncovin, Platinol

BOPP ONCOLOGY A chemotherapy regimen: BCNU, Oncovin, procarbazine, prednisone. See Chemotherapy CLINICAL RESEARCH Boronated porphyrin An investigational photosensitizing drug; clinical trials will assess BOPP's safety for treating malignant brain tumors

borborygmus GASTROENTEROLOGY A rumbling, gurgling, tinkling–stomach "growling" noises heard on auscultation of the abdomen in conditions of ↑ intestinal peristalsis

borderline An adjectival expediency widely used in medicine for any condition that cannot be neatly placed in one of usually 2 categories, each of which has a distinct clinical significance, therapy and prognosis

borderline hypertension That range of systolic and diastolic BPs in which there is no unequivocal benefit obtained by therapy

borderline isolated systolic hypertension A subtype of HTN–systolic BP between 140 and 159 mm Hg, diastolic BP < 90 mm Hg; BISH is the most common type of untreated HTN in adults ≥ age 60; after 20 yrs of followup, 80% had progressed to definite HTN–vs 45% of normotensives with ↑ risk of cardiovascular disease

borderline personality disorder PSYCHIATRY A disorder of adult onset, which is characterized by instable interpersonal relationships, self-image, and affect, impulsivity in various contexts, and fluctuations in intense moods

BORDERLINE PERSONALITY DISORDER

1. Frantic efforts to prevent real or imagined abandonment
2. A pattern of intense and unstable interpersonal relationships swinging between extremes of idealization and devaluation
3. Unstable self-image
4. Impulsivity in 2+ areas & self-destructive potential, eg binge eating, driving, gambling, sexual relations, substance abuse
5. Recurring suicidal or self-mutilating gestures or behaviors
6. Marked lability of moods and affect
7. Chronic feeling of 'emptiness'
8. Inappropriate anger and inability to control anger
9. Transient stress-related paranoid ideation or dissociative symptoms

borderline tumor SURGICAL PATHOLOGY A term used for neoplasms with many histologic criteria of malignancy, for which the future behavior is uncertain. See Stump.

Bordetella pertussis MICROBIOLOGY A small, aerobic, gram-negative bacillus, causative organism of whooping cough; B *pertussis* produces various toxins including a dermonecrotizing toxin, an adenyl cyclase, an endotoxin and pertussis toxin, as well as surface components such as fimbrial hemagglutinin DIAGNOSIS Culture, serology, clinical Rx

bore VOX POPULI 1. The inside diameter of a tubular structure. 2. A person who bores

Borg scale CHEST MEDICINE A system for scoring the perception of dyspnea, consisting of a linear scale ranking the degree of difficulty in breathing, ranging from none–0 to maximum–10

boron therapy Boron neutron capture therapy RADIATION ONCOLOGY A form of RT in which a cancer Pt ingests a boron compound and then is exposed to a stream of neutrons from a nuclear reactor; the compound concentrates in the tumor, and itself becomes radioactive, delivering the highest radiation directly into the tumor

Borrelia burgdorferi The spirochete agent of Lyme disease, which contains several outer membrane proteins and a highly immunogenic flagellar protein which may be important in the diagnosis and treatment of Lyme disease

DIAGNOSIS Culture, serology, ID in tissue, PCR of synovial fluid. See Lyme disease, Tick

borrowed servant doctrine MALPRACTICE A principle under which the party usually liable for a person's actions–eg, a hospital responsible for a nurse, is absolved of that responsibility when that person is asked to do something–eg, by a surgeon, which is outside of the bounds of hospital policy. See 'Captain of the Ship' doctrine, Malpractice, Respondeat superior

bosentan VASCULAR DISEASE A mixed ET$_A$–endothelin and ET$_B$ receptor antagonist, used as a long-term antihypertensive, which ↓ BP without reflex neurohormonal activation. See Endothelin ET$_A$ receptor, Endothelin ET$_B$ receptor

bossing Frontal bossing, boxhead, caput quadratum Rounded prominence of the frontal and parietal bones in an infant's cranial vault, due to various causes–eg, untreated vitamin D-induced rickets, causing a thickened outer table with permanent enlargment of the head, congenital anemia–eg, thalassemia with massive hematopoiesis in an expanded marrow space, aka 'hot cross bun skull', and others–eg, conditions associated with 'gargoyle' facies, acromesomelic dysplasia of Maroteaux, anhydrotic ectodermal dysplasia, craniometaphyseal dysplasia of Pyle, frontodigital syndrome, Kenney's tubular stenosis syndrome, basal cell nevus syndrome, Taybi syndrome, thanatophoric dwarfism, AIDS embryopathy. See Hot cross bun skull

Boston exanthem INFECTIOUS DISEASE An exanthematous roseoliform salmon-colored maculopapular face and chest rash caused by echovirus 16, preceded by a high, 1-2 day fever which subsides with rash onset; both last a wk, are more common in children in the late summer; epidemic or sporadic

botanical toxicity HERBAL MEDICINE Any adverse effect caused by herbs or other plants. See Poisonous plants. Cf Herbal medicine

botox Botulinum toxin A, see there

botryoid sarcoma Myxoid variant of embryonal rhabdomyosarcoma A tumor of hollow organs–urinary bladder, vagina of children DIFFDX Intramuscular myxoma, myxoid liposarcoma, myxoid MFH, botryoid sarcoma–myxoid embryonal rhabdomyosarcoma MANAGEMENT Excision, chemotherapy, RT PROGNOSIS Poor without aggressive therapy; early metastases

botryomycosis Bacterial ball INFECTIOUS DISEASE A chronic bacterial infection that presents as a hardened fibrotic mass with draining sinuses and purulent granular debris; it usually affects the skin and subcutaneous tissue, but may involve the nasal cavity and sinuses, liver, lung, kidneys, brain, GI tract, lymphoid tissue DIFFDX Aggressive, malignant tumors MANAGEMENT Wide surgical evacuation

bottom edge of stomach VOX POPULI Greater curvature. See Stomach

botulinum toxin A Oculinum NEUROLOGY One of several toxins produced by *C botulinum*, of which the 150 kD type A toxin has been purified and used to treat various neuromuscular junction disorders including strabismus, blepharospasm, spasmodic torticollis, other dystonias, hemifacial spasms, and anal incontinence

botulism A paralyzing disease caused by a potent toxin produced by *C botulinum* under anaerobic conditions, which is either foodborne or is acquired via wounds CLINICAL Progressive dizziness, blurred vision, slurred speech, dysphagia, nausea, ↓ gag reflex CLINICAL FORMS Infant botulism, food-borne botulism, wound botulism, idiopathic TREATMENT Botulism antitoxin

bougie SURGERY A long, flexible instrument, introduced into the urethra, esophagus, etc, to remove obstructions THERAPEUTICS A slender rod consisting of medicine in a vehicle–wax, gelatin that melts at body temperature, which is introduced via the urethra

bougienage GASTROENTEROLOGY A technique for treating corrosive burns of the esophagus in which a bougie is passed to the point of obstruction and dilated. See Self-bougienage

bounceback MEDTALK A popular term for a Pt who returns to the ER with the same complaint shortly after being released. Cf Frequent flier

boutique hospital Luxury hospital HOSPITAL CARE A hospital that specializes in high-volume, big-ticket, inpatient surgery–eg CABG and total hips, which provides more personalized care to Pts and discharges them earlier than standard hospitals. See Hotel services

bovine spongiform encephalopathy Bovine spongiform encephalitis, BSE, mad cow disease A disease of cattle, characterized by high-stepping or staggering gait, anxiety, ↑ sensitivity, and kicking while being milked, less commonly, frenzy, aggression; BSE was first described in the UK in cows fed with sheep offal. See Prions

bowel A general term for the small and large intestines; intestine

bowel disimpaction Manual removal of impacted fecal matter from the rectum. See Bobbing for apples

bowel infection 1. Gastroenteritis, see there 2. Colitis, see there

bowel obstruction GI DISEASE Blockage of the GI lumen CLINICAL Constipation, bloating, abdominal pain MANAGEMENT IV fluids, bowel rest, nasogastric suction, possibly surgery

bowel perforation Gastrointestinal perforation GI DISEASE Complete penetration of the intestinal wall resulting in bacterial contamination of the abdominal cavity/peritonitis

bowel prep A preoperative enema, often used with prophylactic oral and parental antibiotics to purge the colon of food and feces before intestinal surgery; the most popular agent for preparing the large intestine for surgery is polyethylene glycol for mechanical cleansing; surgeons also use conventional enemas, dietary restriction and cathartics

bowel sound PHYSICAL EXAM-ABDOMEN Any sound caused by contractions of the large intestine as it propels stool forward TYPES Normoactive, hyperactive, hypoactive, high-pitched, inaudible, tympanitic, decreased, markedly diminished

Bowen's disease Bowenoid dysplasia, intraepithelial cancer Carcinoma in situ arising in, and confined to, the epithelium

bowenoid *adjective* Pertaining to Bowen's disease

bowenoid dysplasia Bowen's disease, or carcinoma in situ of the skin

bowleg Genu varum ORTHOPEDICS External deviation of the knee(s); a certain degree is normally present in infants, and corrects itself with bipedal ambulation; when excessive, rickets is considered–vitamin D-induced osteomalacia allows bending of the femoral shaft bearing the mechanical brunt of ambulatory kinetics; when combined with anterior curvature of the tibia and fibula, the children have a 'saddle-sore' stance; anterior or antero-lateral bowing of the tibia may occur in neurofibromatosis with fractures, which may be complicated by pseudoarthrosis

bowler hat sign GI RADIOLOGY Spreading of barium contrast around an adenomatous or villotubular colon polyp, at the base of which there is a recess between the stalk and the polyp 'body', forming the hat's 'crown' DIFFDX Benign-leiomyoma, lipoma, hyperplastic polyp, premalignant; malignant–leiomyosarcoma, lymphoma, melanoma

bowler's thumb SPORTS MEDICINE A nerve compression syndrome due to pressure of the ulnar and radial digital nerves of the thumb on the edge of the bowling ball's thumb hole CLINICAL Numbness, paresthesias, hypesthesia of skin distal to the nerve; BT may lead to perineural fibrosis and the formation of a painful nodule TREATMENT Rest, thumb guard, change grip, round-

ing off edge of hole in bowling ball; surgery is not indicated

boxer's fracture SPORTS MEDICINE A fracture of the 5th metacarpal neck after a direct blow on the 5th metacarpal head with the fist clenched, causing dorsal angulation of the fracture line and volar displacement of the head

boxers Boxer shorts REPRODUCTIVE MEDICINE A type of undergarment that leaves the "guys" swinging in the breeze; according to popular culture, sperm quality and successful insemination ↑ dramatically with prolonged use thereof. Cf Briefs

boxers' encephalopathy Dementia pugilistica SPORTS MEDICINE Boxing causes major neuropsychologic defects in its long-term practitioners when tested by the Wechsler and Bender Gestalt tests, causing variable organic mental disease and impaired short-term memory, dysarthria, nystagmus, reasoning ability, and motor skills; CT evidence of cortical atrophy ACUTE BOXING INJURIES Cerebral edema, ischemia, and temporal or uncal herniation. See Blood sport, Broughton rules, Marquis of Queensbury rules, 'Punch-drunk' syndrome

boxing SPORTS MEDICINE A contact sport in which 2 latter-day gladiators pummel each other with gloved fists until one falls to his knees or floor and the match is formally ended

boy A male child, from birth to age of puberty. See Bubble boy, Little boy, Tomboy

bp 1. Base pair, see there 2. Blood pressure, see there 2. Boiling point

BPD Bipolar disorder, see there

BPH Benign prostatic hypertrophy, see there

BPPV Benign paroxysmal positional vertigo, see there

brace ORTHOPEDICS A device that shores biomechanically weakened body parts. See Milwaukee brace

brachial plexopathy Brachial plexus injury, see there

brachial plexus injury OBSTETRICS The squashing of the brachial plexus, almost always due to a shoulder dystocia in a vaginal delivery, which is often associated with transient paralysis See Operative vaginal delivery.

brachio- Latin *brachium*, arm Relating to the arm

brachium Upper arm; segment of forelimb between the shoulder and elbow

brachy- Greek root for short

brachycephaly Brachycephalia, brachycephalism A disproportionately short head

brachydactyly Brachydactylia Abnormally short fingers

brachytherapy Implant radiation, internal radiation, intracavitary therapy, interstitial radiation therapy RADIATION ONCOLOGY RT in which radioactive materials–iridium-192, radium-226, and other radioisotopes, sealed in needles, seeds, wires, or catheters are placed in direct contact with certain carcinomas to deliver locally intense ionizing radiation–eg, as needed in invasive CA of the uterine cervix. See Intravascular brachytherapy

bracket DENTISTRY A metal or ceramic part glued onto a tooth that serves as a means of fastening the arch wire

brady- Greek root for slow

bradyarrhythmia A heart rhythm in which there is an arrhythmia combined with a rate < 60 beats/min

bradycardia CARDIOLOGY Slow heart rate; commonly defined as a rate < 60 bpm or a rate that is too slow to be physiologically appropriate for the person and/or activity; alternatively< 45 beats/min in ♂; <50 beats/min in ♀ CLINICAL Sx may be specific–syncope, or nonspecific–dizziness, fatigue, weakness, heart failure MANAGEMENT Beta blockers–pindolol, pacemaker. See Cardiac output

bradycardia-tachycardia syndrome Brady-tachy syndrome CARDIOLOGY A form of sinus-node dysfunction–sick sinus syndrome characterized by alternating periods of atrial tachyarrhythmias and bradycardias. See Sick sinus syndrome

bradycardic *adjective* Referring to bradycardia

bradykinetic NEUROLOGY *adjective* Referring to abnormally slow movement

bradylalia Bradyglossia NEUROLOGY Abnormally slow speech

Braille PUBLIC HEALTH Alphanumeric writing designed for the vision impaired; characters are encoded and typed in relief so properly trained fingers can "read" written communication. Cf Americans with Disabilities Act, Service dog

brain abscess NEUROLOGY A localized intracranial infection filled with pus due to a bacterial infection

brain aneurysm Cerebral aneurysm NEUROLOGY A dilated and weak segment of a cerebral artery, often located in the circle of Willis at the base of the brain, which is susceptible to rupture; BAs may be caused by birth defects or follow poorly controlled HTN CLINICAL "Thunderclap headache" often associated with N&V, ↓ consciousness

brain attack NEUROLOGY A lay term for an abrupt, nonpsychogenic change in mental status, unrelated to infection or trauma; BAs correspond to either a TIA or CVA. See Cerebrovascular accident, Stroke, Transient ischemic attack

brain concussion Cerebral concussion NEUROLOGY An acute change in level of neurologic function which is usually linked to trauma to the head CLINICAL Prolonged unconsciousness, memory loss, vision defects, loss of equilibrium, dyspnea, pupillary dilation

brain contusion NEUROLOGY A head injury often associated with a concussion, which is of sufficient force to bruise the brain surface and cause extravasation of blood without rupturing the pia-arachnoid. See Coup, Contrecoup

brain death END OF LIFE The irreversible cessation of all functions of the entire brain, including the brainstem. See Appropriate period of observation, Attending physician, Corroborating physician, Harvard criteria, Multiorgan donation

HARVARD CRITERIA FOR BRAIN DEATH

- Unreceptivity and unresponsiveness
- No movement or breathing
- No reflexes
- Flat electroencephalogram (confirmatory)

IN ADDITION, THE FOLLOWING MUST BE PRESENT

- Body temperature ≥ to 32° C
- Absence of CNS depressants

brain edema Cerebral edema NEUROLOGY Fluid accumulation outside the vascular compartment of the brain–ie, within cerebral tissue. Cf Brain edema

brain-graft surgery NEUROSURGERY A procedure for treating Parkinson's disease, in which chromaffin adrenal cells are transplanted into the caudate nucleus. See Parkinson's disease

brain herniation NEUROLOGY A pressure-induced prolapse of part of the brain into adjacent spaces, which occurs when the brain is under very high pressure CLINICAL Coma, paralysis, unilateral dilated pupil ETIOLOGY Head injury, 1° or metastatic brain tumor, bacterial meningitis, brain abscess TYPES Cerebellar herniation, uncal–temporal herniation, transtentorial herniation of the brain

brain imaging IMAGING Any technique that permits the in vivo visualization of the substance of the CNS–eg, CT, PET, MRI, yielding a series of 2-D images/"slices" of brain regions of interest which may be manipulated by computer to generate 3-D simulations; other techniques–eg, ultrasound, angiography, radionuclide scans, regional cerebral blood flow measurements, brain electrical activity mapping and its variants, and even the now-obsolete pneumoencephalogram provide images of aspects of the CNS, but are limited in the structures visualized or degree of resolution, or other parameters. See PET-CT scan

brain injury NEUROLOGY A condition in which a person before, during or after birth suffered trauma or encephalitis, which compromises normal learning process

brain metastases ONCOLOGY Cancer that has spread from a primary tumor, most commonly, small cell carcinoma of the lung, to the brain

brain scintigraphy Brain scan IMAGING A procedure in which a radioisotope is administered IV and its distribution monitored with a gamma camera; BS informs on adequacy of blood flow to the cerebral cortex, and detects abnormalities in the brain before they can be detected by CT or MR scans INDICATIONS Evaluation of strokes, TIAs, seizure activity, organic brain disease, tumors, especially if combined with SPECT

brain tumor NEUROLOGY A neoplasm affecting the brain which may be 1°–brain or meninges, or 2°–ie metastatic to the brain; malignant gliomas account for 2.5% of all cancer-related deaths; BTs are the 3rd most common CA in ages 15-34; 35,000 BTs occur/yr–US; 1st-degree relatives of children with brain tumors have a 5-fold ↑ in the risk of CNS tumors, leukemia, and other childhood tumors in the affected family CLINICAL Seizures, vision or hearing loss, hemiparesis, double vision, headache, weird behavior, N&V, memory loss IMAGING MRI without and with contrast, CT MANAGEMENT Surgery, gamma knife radiotherapy are often effective; chemotherapy, immunotherapy are not. See Gamma knife

BRAIN TUMORS/MASSES

PRIMARY

NONNEOPLASTIC Craniopharyngioma, colloid cysts

PRIMARY–BENIGN Meningioma, pituitary adenoma, acoustic neuroma, epidermoid tumors, choroid plexus papilloma

PRIMARY–LOW GRADE Pilocytic astrocytoma, astrocytoma, hemangioblastoma, oligodendroglioma, ganglioglioma

PRIMARY–MALIGNANT Anaplastic astrocytoma, glioblastoma multiforme, ependymoma, lymphoma, medulloblastoma, primitive neuroectodermal tumor, germ cell tumor, pineal cell tumor, chordoma, choroid plexus carcinoma

METASTATIC Carcinoma, meningeal carcinomatosis

brain waves NEUROLOGY Oscillations/sec that correspond to various types of cerebral activity, as measured on an EEG. See Electroencephalogram

Brainerd diarrhea EPIDEMIOLOGY An epidemic outbreak of chronic watery diarrhea of unknown origin characterized by acute onset, prolonged duration, marked urgency, and no systemic symptoms

brainstem auditory evoked response BAER, Brainstem auditory evoked potential NEUROPHYSIOLOGY A method for evaluating hearing using clicking sounds and recording the responses–known as auditory evoked potentials with EEG electrodes placed on the scalp; the early responses reflect electrical activity at the cochlea, cranial nerve VIII and brainstem; late responses are due to cortical activity; after each click, 7 consecutive potentials are recorded, designated as waves I through VII, each of which corresponds to a specific relay point in the brainstem; BAER is an objective means of diagnosing and localizing early lesions of the auditory system INDICATIONS BAER is used to diagnose acoustic neuroma, intrinsic brainstem lesions–eg, multiple sclerosis, infarction, gliomas, degenerative disease–eg, Charcot-Marie-Tooth disease, Meniere's disease, olivopontocerebellar degeneration, stroke, Wilson's disease, ↑ intracranial pressure, brain death

brainstem glioma NEUROLOGY A tumor of the pons and medulla which comprises ±15% of pediatric brain tumors CLINICAL Double vision, facial weakness, vomiting, gait ataxia MANAGEMENT RT, chemotherapy MORTALITY Up to 30%

brainstem implant AUDIOLOGY An auditory prosthesis that bypasses the cochlea and auditory nerve, used in Pts who don't benefit from a cochlear implant because of auditory nerve defects

brake drug SPORTS MEDICINE An anabolic steroid used by ♀ athletes to delay–ie to 'brake' body growth and minimize body fat, preventing a shift in the center of gravity SIDE EFFECTS Premature closure of epiphyseal plate, ↓ adult height. See Anabolic-androgenic steroids

bran CLINICAL NUTRITION A byproduct of milled wheat, which contains ± 20% indigestible cellulose, acting as a bulk laxative; it has been recommended for cardiovascular disease, constipation, diarrhea, diverticulosis, hemorrhoids, and IBD. See Dietary fiber, Oat bran. Cf Water-soluble fiber

branchial cleft cyst Branchial cyst A cyst-like embryologic rest–remnant present at birth, which arises from branchial clefts, usually the 2nd, at the end of the trachea and branch into the lungs; BCCs often present as asymptomatic unilateral fluctuant masses posteroinferior to the angle of the jaw and anterior to the sternocleidomastoid muscle; it is characterized by having cartilage as part of the supporting wall structure; BCCs may not be recognized until adolescence when enlarged; BCCs may develop a sinus or drainage path to the skin surface from which mucus can be expressed MANAGEMENT Total surgical excision; recurrence is rare

branchial cyst Branchial cleft cyst, see there

branching snowflake test A crude bedside test using a glass slide for differentiating amniotic fluid–positive from maternal urine–negative, based on a radiating snowflake-like appearance of dried amniotic fluid

branchio-oto-renal syndrome An AD condition characterized by mixed hearing loss accompanied by a Mondini-type cochlear malformation, bilateral renal dysplasia with abnormalities of collecting ducts, and bilateral branchial clefts and/or cysts; BORS may be linked to breakpoint mutations on chromosome 8q

brandy A strong alcoholic liquor distilled from wine

Branham sign Nicoladoni-Branham sign INTERNAL MEDICINE Bradycardia following compression of an AV fistula

brassy cough CHEST MEDICINE A non-productive 'metallic' cough heard in children with acute bacterial or viral laryngotracheitis, often accompanied by inspiratory stridor and respiratory distress. Cf Whooping cough

BRAT(T) diet Bananas, rice, apples, toast (tea) CLINICAL NUTRITION A bland diet prescribed for viral gastroenteritis which, with water, replenishes liquids and electrolytes lost in pediatric diarrhea; other foods appropriate for diarrhea include plain chicken, crackers, and potatoes. See Diet

BRAVO CARDIOLOGY A clinical trial–Blockade of the GP IIB/IIIA Receptor to Avoid Vascular Occlusion– which evaluated lotrafiban in preventing strokes and acute MI. See GP IIB/IIIA

brawny edema VASCULAR MEDICINE A change typical of chronic

venous insufficiency, characterized by thickening, induration, liposclerosis and non-pitting edema; the brawny color is due to hemosiderin from lysed RBCs; with chronic ischemia, the skin undergoes atrophy, necrosis, and stasis ulceration, surrounded by a rim of dry, scaling, and pruritic skin

Braxton Hicks contraction OBSTETRICS Any of the painless irregular uterine contractions that begin after the 3rd month of pregnancy

BRCA-related gynecologic cancer ONCOLOGY Any cancer of the ♀ genital tract which occurs more commonly in those with mutations in BRCA1 or BRCA2 BRGCs Breast–60 to 85% lifetime risk; ovarian–15 to 65% lifetime risk. See BRCA1, BRCA2

BRCA1 MOLECULAR ONCOLOGY A large tumor suppressor gene on chromosome 17 which is linked to breast, ovarian, prostate and other CAs; Pts with BRCA1 represent 5% of all breast CAs; ♀ with BRCA1 have an 85% chance of developing breast CA before age 65; mutations of BRCA1 are common in Ashkenazi Jews LAB BRCA1 and BRCA2 are part of some commercial diagnostic labs' genetic services. See Tumor suppressor gene

BRCA2 MOLECULAR ONCOLOGY A tumor suppressor gene linked to breast and ovarian CA See Tumor suppressor gene

bread & butter disease A catch-phrase for routine conditions seen in each medical specialty, which comprise most of a physician's work-load and present little intellectual challenge

breaking EMERGENCY MEDICINE A maneuver used in drowning and near-drowning victims to clear the pharynx of water and vomitus; the victim is turned prone, hands of person rendering CPR are locked together under abdomen, fluid-filled body is lifted to expel water and gas. Cf Flake maneuver, Heimlich maneuver

breaking the bag of waters Amniotomy, see there

breakthrough bleeding GYNECOLOGY A term applied to various gynecologic "bleeds," usually refers to mid-cycle bleeding in OC users, and is attributed to insufficient estrogens; the term is not applied to *abnormal* bleeding in OC users

breakthrough infection IMMUNOLOGY An infection by the same organism that a vaccine is designed to protect against; BIs may be caused by exposure to the organism before the vaccine took effect, or before the entire vaccine was administered. See Booster

breakthrough pain ONCOLOGY Severe intermittent cancer pain that does not respond to the usual modalities–eg, opiates; BP may require anesthetic procedures–eg, nerve blocks, intraspinal infusions of analgesic medication, or neuroablation

breast abscess Submammary abcess GYNECOLOGY An acute infection of the breast, which is particularly common during breast-feeding

breast atrophy PHYSICAL EXAMINATION Loss of tissue in the ♀ breast, a typical postmenopausal change

breast cancer ONCOLOGY An uncontrolled growth of abnormal breast tissue, usually epithelial in nature STATISTICS BC is the most common CA in ♀; occurs in 12% of all US ♀, killing 3.5%–it is the leading COD in ♀ age 40-55; in Japan, BC is ⅕ as common as the US, but there has doubled from the mid-1930s to 1990s–± 55 to 105/10⁵ STATISTICS Deaths/Newly diagnosed 44.5/185.7 DIAGNOSIS Self- and physician examination, mammography, ultrasonography, biopsy RISK FACTORS Breast feeding–≥ 6 months ↓ risk of BC; lactation and breast feeding before age 20 is associated with a RR of 0.54 CHEMOTHERAPY Maximum effect with early treatment and maximum tolerable doses with cyclophosphamide, doxorubicin, fluorouracil SURVEILLANCE Early detection, monthly breast self-examination, mammography at age-appropriate intervals THERAPEUTIC TRENDS 1985-1993 Breast-conserving therapy, 31%→54%; axillary node dissection 52%→40%; RT 38%→54%. See *BRCA1, BRCA2*

BREAST CANCER STAGING

STAGE I Cancer ≤ 2 cms; no spread outside the breast

STAGE II One of following– cancer ≤ 2 cm–spread to the axillary lymph nodes, *or* cancer of 2 to 5 cm that may/may not have spread to axillary lymph nodes, *or* cancer ≥ 5 cm without spread to axillary lymph nodes

STAGE III

A Defined by either 1. ≤ 5 cm and + axillary lymph nodes, which have grown into each other or into other structures and are attached to them, *or* 2. ≥ 5 cm and spread to axillary lymph nodes

B Cancer has spread to tissues near the breast–skin, chest wall, including the ribs and the muscles in the chest, *or* has spread to lymph nodes inside the chest wall along the breast bone

STAGE IV Cancer has metastasized, most often to bone, lungs, liver, or brain *or* has spread locally to skin and lymph nodes inside the neck, near the collarbone

breast cancer gene(s) See *BRCA1, BRCA2*

breast-conserving surgery SURGICAL ONCOLOGY An operation to remove the breast CA but not the breast TYPES Lumpectomy, quadrantectomy, segmental mastectomy. See Breast reconstruction, Lumpectomy, Quadrantectomy, Segmental mastectomy

breast feeding PEDIATRICS The provision of a neonate and infant with liquified lacteal products 'on tap'; lactation and BF–≥ 6 months before age 20 is associated with a relative risk of 0.54–ie ↓ risk of subsequent development of breast cancer in BF mothers. See Breast milk, La Leche League, Natural family planning

breast implant PLASTIC SURGERY An inert sac, once filled with silicone, now saline; some are covered by polyurethane foam, inserted under anterior chest wall skin to augment or restore cosmetically the ♀ contour or for post-mastectomy reconstruction; BIs may be associated with connective-tissue diseases–see there, or other disease–eg, autoimmune disorders–eg, Hashimoto's thyroiditis, PBC, sarcoidosis, CA COMPLICATIONS Capsule contraction, implant rupture, hematoma, wound infection. See Biomaterials, Breast cancer, Cosmetic surgery, Silicone

breast infection Mastitis, see there

breast lump Breast mass, see there

breast marker LAB MEDICINE An antigen variably present in breast tissue used to either differentiate between benign or malignant breast lesions, or identify the breast as the origin of metastases. See BRCA, Immunoperoxidase

breast mass Breast lump SURGICAL ONCOLOGY A breast lump that is either benign–eg, breast abscess, fat necrosis, fibroadenoma, fibrocystic disease, or malignant–CA. See Breast biopsy

breast milk NEONATOLOGY Human milk is similar to cow milk in the water content–88%, specific gravity, 1.030, fat content–3.5%, energy value–0.67 kcal/ml and type of sugar––lactose. See Breast-feeding, La Leche League; Cf Certified milk, Humanized milk, Raw milk, White beverages

breast milk jaundice NEONATOLOGY Jaundice caused by an ↑ in BR in the late postnatal period, attributed to enterohepatic cycling of bile pigments See Breast feeding

breast pain VOX POPULI → MEDTALK Mastodynia

breast prosthesis SURGICAL ONCOLOGY An artificial breast worn under clothing. See Mammoplasty

breast pump PEDIATRICS A tubular mechanical device that provides gentle suction for milk extraction, used when breasts are engorged or when direct infant feeding is not possible for various physical–eg prematurity or logistic–eg mommy has to work–reasons

breast reconstruction SURGICAL ONCOLOGY A surgical revision of the anterior chest wall after mastectomy for breast CA, which may be performed during initial therapy or as a 2nd procedure. See Mammoplasty

breast self-examination PUBLIC HEALTH The periodic—eg, once monthly palpation by a ♀ of her own breast, to detect new growths; although there is little evidence of the effectiveness of BSE, it is the only means by which neoplasms can be detected between exams by health care personnel

breast surgeon A general surgeon specialized in breast surgery

breast tumor profile LAB MEDICINE A battery of tests used to evaluate breast cancer aggressiveness after diagnosis; tests in a BTP include estrogen and progesterone receptor assays, DNA index/cell cycle analysis by flow cytometry or by image analysis and Ki-67 localization, cathepsin D, c-*erb*B-2/*HER-2/neu* oncoprotein analysis. Cf Breast cancer profile

breast ultrasound IMAGING A test which uses ultrasonic waves to scan the breast and evaluate fibrocystic breast disease, implants, masses, and cysts

breath sound CHEST MEDICINE A pulmonary sound heard through a stethoscope over the chest

breath test CHEST MEDICINE Any of a number of clinical tests used to evaluate malabsorption, in which a food containing a substance emitting low levels of radioactivity is ingested and, if malabsorbed, is exhaled through the lungs. See ²H-lactose test, ¹⁴C-xylose test GI DISEASE See Urea breath test LAB MEDICINE The testing of gases exhaled from a Pt for the presence of volatile organic compounds–VOCs

breathalyzer PUBLIC HEALTH A device used to detect alcohol on a suspected drunk driver's breath; see DWI

breathe To inhale and exhale during respiration; to respire

Breathe Right™ SPORTS MEDICINE A proprietary Band-Aid™-like nasal strip worn on the nasal bridge, said to improve breathing by opening the nasal passages INDICATIONS Temporary relief of breathing difficulties due to deviated nasal septum, ↓/elimination of snoring, temporary relief of nasal congestion

breathing CHEST MEDICINE Respiration; the act of inhaling and exhaling air. See Biot breathing, Cheynes-Stokes breathing, Noisy breathing, Rescue breathing

breathing market ALLERGY MEDICINE A popular term for products designed to ↓ inhaled allergens–eg, air filters, dehumidifiers, animal dander wipes, washable toys, inhaler covers, etc. See Allergies

breathlessness EMERGENCY MEDICINE A lack of breathing, detected by a rescuer by looking for chest movements, listening for air escaping during exhalation, and feeling for air flow. See Rescue breathing, Shortness of breath

breech delivery OBSTETRICS Extraction or expulsion of the fetus feet or buttocks first

breech presentation OBSTETRICS Buttocks position at the time of delivery, a position that carries a 3-6-fold ↑ mortality due to complications–umbilical cord prolapse, tentorial tearing, cerebral hemorrhage of after-coming head; in the US, BPs are usually delivered by C-section. See Cesarean section

Brennen mesh SURGERY A biosynthetic mesh used for hernia repair where connective tissue has ruptured, to support the bladder neck in treating female urinary incontinence secondary to urethral hypermobility or sphincter deficiency

brick-red sputum Popular term for a thickened mixture of blood, bacteria, necrotic lung tissue and mucus, characteristic of *Klebsiella pneumonia*

bridge DENTISTRY A fixed partial denture; a prosthetic replacement of missing teeth cemented or attached to abutment teeth or implants adjacent to the space; removable partial denture is a prosthetic replacement of missing teeth on a framework that can be removed by a Pt PHYSICAL THERAPY An exercise in which a person lays on his/her back with bended knees, while lifting the pelvis, placing thighs, back and pelvis in a straight line, strengthening abdominal, lower back, gluteus and hamstring muscles TRANSPLANTATION MEDICINE An organ surrogate that carries out a particular physiologic function and "buys time" for a Pt awaiting a donor organ for transplantation

bridging CARDIOLOGY A term for the systolic narrowing of the left anterior descending coronary artery seen by angiography as an isolated finding during cardiac catheterization or in Pts with CAD, left ventricular hypertrophy or hypertrophic cardiomyopathy. Cf Rattail, Sawfish patterns TRAUMATOLOGY The 'spanning' of breaks in the skin by blood vessels, seen after a blunt object strikes tightened skin, rupturing the epidermis and dermis while blood vessels–being more mobile–remain intact, 'bridging' the gap

Brief Pain Inventory NEUROLOGY A brief, relatively simple, self-administered questionnaire for evaluating pain, which addresses the relevant aspects of pain–history, intensity, timing, location, and quality and the pain's ability to interfere with the Pt's activities. See Pain

brief reactive psychosis PSYCHIATRY A psychotic episode that lasts from hrs to 1 wk; BRP is evoked by ↑ stress–eg, death of loved one CLINICAL Delusions, hallucinations, disordered thinking, impaired speech, bizarre social activities

briefs White tighties REPRODUCTIVE MEDICINE A type of undergarment that keeps the "guys" close to the body; according to popular culture, use of briefs ↓ sperm quality and successful insemination. Cf Boxers

bright light therapy Light therapy, phototherapy The use of the visible range of the electromagnetic spectrum as a therapeutic modality; light acts on the hypothalamus and releases neurotransmitters and factors after receiving neural impulses from retina; BLT is the therapy of choice for SAD; modalities of LT include the use of full spectrum light–eg sunlight, bright light–10,000 lux, UV light, colored light, hemoirradiation. See Heliotherapy, Seasonal affective disorder, Spring fever

BRITE CARDIOLOGY A clinical trial–Beta Radiation to Reduce In-Stent Restenosis–evaluating the efficacy of the RDX™ Catheter Radiation Delivery System in reducing the risk of restenosis after intracoronary intervention for ASHD. See BETTER, In-stent restenosis

British Thermal Unit A unit of energy–equal to 252 calories or 1055 Joules needed to raise one pound of water by one degree Fahrenheit

brittle bones 1. Osteogenesis imperfecta Bones with ↑ osseous fragility, a phenomenon seen in osteogenesis imperfecta, due to genetic defects–eg, point mutation in collagen, type I 2. Osteoporosis, see there

brittle diabetes ENDOCRINOLOGY 1. Labile diabetes, unstable diabetes Type 1 DM in which serum glucose fluctuates widely, swinging rapidly from the hypoglycemia to hyperglycemia despite frequent 'titration' of insulin doses 2. Type I DM, see there

BRM Biological response modifier, see there

broad-spectrum *adjective* Referring to any agent, usually understood to be

an antibiotic, which is effective against a wide range of microorganisms

Broca's aphasia Motor aphasia NEUROLOGY Loss of the ability to produce spoken and usually written language with retained comprehension See Aphasia

Brodie's abscess ONCOLOGY A localized abscess that occurs in young ♂ age 10–20 in the metaphysis of long leg bones CLINICAL Aching, boring pain of variable intensity, swelling and localized tenderness

bromazepam Lexotan® PHARMACOLOGY A benzodiazepine

bromhidrosis MEDTALK Body odor–eg, stinky feet

bronch A popular verb for performing bronchoscopy, as in 'bronching'–pronounced bronking–a Pt

bronchial brushings Brush biopsy A procedure in which cells from the mucosa of the upper airways–trachea, bronchi, bronchioles are obtained for cytologic evaluation under direct bronchoscopic visualization of suspicious mucosal lesions. Cf Bronchial washings

bronchial challenge Inhalation of antigens to detect allergic reactions; in a positive BC, the Pt's temperature and WBC count rise, and the FEV_1, and FVC fall within 4–12 hrs of inhalation

bronchial hyperresponsiveness Exaggerated bronchial constriction most common in asthma, in response to nonspecific provocation, inhalation of various bronchoconstrictors, but also to physical challenges–eg, exercise, dry or cold air, hypertonic or hypotonic aerosols

bronchial veins Veins that begin at the root of the lung and receive blood from vessels near the bronchial arteries.

bronchial washing A procedure in which isotonic saline is instilled through a bronchoscope and fluid containing cells, microorganisms, or other material from the upper airways–trachea, bronchi, bronchioles is aspirated into a trap; the material is then centrifuged to concentrate the cells, stained, and examined by microscopy, or cultured if infection is suspected. Cf Bronchial brushings

bronchiectasis CHEST MEDICINE Progressive dilatation of bronchi or bronchioles due to pneumonitis, pneumonia, obstruction, tumors or congenital disease–eg, cystic fibrosis CLINICAL Halitosis, paroxysmal coughing, expectoration of mucopurulent matter; it may affect bronchioles uniformly–cylindric bronchiectasis, occur in irregular pockets–saccular bronchiectasis or dilated bronchi may have terminal bulbous enlargements–fusiform bronchiectasis; rarely congenital, it is usually acquired in childhood

bronchiole-associated lymphoid tissue BALT Lymphoid tissues of the hilar and peribronchiolar region, which produces IgA specific for inhaled antigens. See Gut-associated lymphoid tissue, Mucosa-associated lymphoid tissue, Skin-associated lymphoid tissue

bronchiolitis CHEST MEDICINE Inflammation of the bronchioles, usually 2° to viral infection–eg, RSV

bronchioloalveolar carcinoma ONCOLOGY A type of adenocarcinoma–representing ±4% of non-small cell lung CAs–that spreads widely throughout the lungs. See Lung cancer

bronchitis CHEST MEDICINE Inflammation, often with edema and erythema, of one or more bronchi, usually 2° to infection CLINICAL Coughing, pain, SOB. See Chronic bronchitis, Plastic bronchitis

bronchoalveolar lavage CYTOLOGY A 'wash' of the upper respiratory tract to obtain cells for evaluating inflammation or cancer of lungs; BAL material is used for 1. Cytologic analysis 2. Analysis of CD4:CD8 ratio and, rarely 3. To obtain cells for gene rearrangement, ie Southern blot hybridization, to diagnose lymphoma

bronchoalveolitis Bronchopneumonia, see there

bronchocentric granulomatosis CHEST MEDICINE A condition characterized by a granulomatous replacement of mucosa with eosinophilia and sometimes angiitis, due either to a fungal–allergic bronchopulmonary aspergillosis–or TB reaction or hypersensitivity to various antigens AGE GROUPING Young–± age 22 yrs, typical Sx of asthma and eosinophilia; older group with neither CLINICAL Progressive dyspnea, cough, hemoptysis, malaise, fever; cavities may result from lung abscesses, cavitated granulomas, distended bronchi; 75% of BGs are unilateral, and more common in upper lobes IMAGING Lobar and segmental consolidation and atelectasis, irregular masses, linear opacities, shadows due to abnormal bronchi, mucoid impaction. See Allergic bronchopulmonary aspergillosis

bronchoconstriction Narrowing of a bronchus, bronchostenosis

bronchodilator CHEST MEDICINE Any agent that dilates airway lumina, to ↑ airflow, given to wheezing Pts EXAMPLES Theophylline, aminophylline, epinephrine, Alupent, metaproterenol, isoproterenol, Ventolin, Proventil, salmeterol, pirbuterol, albuterol. See Adrenergic bronchodilator, Asthma, Wheezing

bronchogenic cancer Bronchogenic carcinoma ONCOLOGY A cancer, usually an SCC, which arises in bronchiolar mucosa/epithelium. See Lung cancer, Squamous cell carcinoma. Cf Small cell carcinoma

bronchophony CHEST MEDICINE A change in the character of the spoken voice, which is higher-pitched and less muffled than normal; it is due to an ↑ transmission of sound 2° to consolidation, infarction, atelectasis, or compression of lung tissue See Egophony

bronchopneumonia Bronchopneumonitis CHEST MEDICINE Lung inflammation that usually begins in terminal bronchioles, which become clogged with mucopurulent exudate, forming consolidated patches in adjacent lobules; it is often 2° to URIs and debilitation, affecting infants, elderly and immunocompromised Pts

bronchopulmonary *adjective* Pertaining to the lungs, bronchi, and bronchioles

bronchopulmonary dysplasia A chronic lung disease affecting ± 7000 premature infants/yr–US, treated for respiratory distress syndrome with supplemental O_2 and mechanical ventilation for ≥ 1 wk, with Sx of persistent respiratory distress, who have rounded radiolucencies on a plain chest films LAB ↑↑ Leukotrienes C_4, D_4, E_4 in lavage fluid PROGNOSIS 40% die of pulmonary dysfunction in later life, by airway obstruction, airway hyperreactivity and hyperinflation. See Hyaline membrane disease

bronchopulmonary segment A subdivision of a lobe of a lung based on its connection to a segmental bronchus

bronchoscope PULMONOLOGY A thin, flexible, lighted endoscope used to examine the upper airways, vocal cords, and tracheobronchial tree to the 4th to 6th division, obtain diagnostic material, ie biopsies, brushings, washings, and instill medicine, and mechanics for easy guidance through the tree. See Bronchoscopy

bronchoscopy Fiberoptic bronchoscopy The use of a flexible endoscope to directly examine the upper airways, vocal cords, and the tracheobronchial tree to the 4th to 6th division; bronchoscopy is used to evaluate suspected malignancy or infections, hemoptysis, persistent coughing, and occasionally to take biopsies–transbronchial and cytology–eg, bronchial washings, specimen culture and to remove foreign bodies in the upper airways; bronchoscopy should not be performed unless absolutely necessary in Pts with asthma, severe hypoxia, unstable angina pectoris, or recent MI. See Bronchoalveolar lavage, Laser bronchoscopy

BRONCHOSCOPY COMPLICATIONS

INSTRUMENT-RELATED Common–hemorrhage, pneumothroax, pulmonary infiltrates; rare–air embolism, mediastinal emphysema

MEDICATION/PREMEDICATION Common–respiratory depression, laryngospasm; rare–bronchospasm, excitability, seizures, hypotension, syncope, cardiorespiratory arrest

PROCEDURAL Common–laryngospasm, bronchospasm, hypoxia; uncommon–arrhythmia, fever, bacteremia, pneumonia

TECHNICAL Rare–airway, vascular trauma, instrument breakage (Fishman 2nd ed, 1989, p458)

bronchospasm CHEST MEDICINE Spasmodic contraction of bronchial smooth muscle, as in asthma. See Asthma, Exercise-induced bronchospasm

bronchostaxis Bronchial hemorrhage

bronze baby syndrome PEDIATRICS A complication of phototherapy in infants with clinical jaundice and indirect hyperbilirubinemia CLINICAL Dark gray-brown skin discoloration that may persist for many months, ↑ direct bilirubin, mixed hyperbilirubinemia, and obstructive liver disease. See Breast milk jaundice

bronze diabetes A clinically distinct form of DM associated with hemochromatosis, in which there are yellow-red hemosiderin deposits in the skin, liver and heart and other organs, eventually leading to renal failure, death CLINICAL DM-like Sx, orange skin MANAGEMENT Repeated phlebotomy

Broviac® catheter A tunneled central venous catheter

brown fat Brown adipose tissue A special form of fat that generates heat by non-shivering thermogenesis, so designated as it is rich in mitochondria, which imparts a pardous hue; BF is rich in sympathetic nerve endings and vessels and its metabolic activity and development is regulated by norepinephrine, and it is normally located in the axillary, subscapular, and interscapular regions, around the large thoracoabdominal vessels, heart, kidneys, and adrenal glands; BF is ↑ in Chagas' disease, CHF, Duchenne's muscular dystrophy, malignancy, pheochromocytoma, SIDS, and in malnutrition

brown spider *Loxosceles reclusa*–warmer parts of the US and *L laeta*–South America, more dangerous than *L reclusa* are secretive, seek secluded sites and bite when bothered; they are identified by the presence of a characteristic violin-like marking on the dorsal surface LOCAL SYMPTOMS Cyanosis, necrosis, pustule or bullae, rimmed by ischemia and erythema expanding to 20 cm over wks and months, accompanied by lymphadenitis SYSTEMIC SYMPTOMS Fever, chills, headache, N&V, dizziness, purpura, myalgia, arthralgia, DIC, and potentially fatal intravascular hemolysis with hemoglobulinuria and acute renal failure MANAGEMENT Surgical excision of bite site ASAP, which limits the spread of toxins, split thickness skin grafting, corticosteroids, antihistamines; with advanced lesions, tetanus immunization is required, bacterial culture, and broad-spectrum antibiotics

brown tumor Osteitis fibrosa cystica NEPHROLOGY A hyperparathyroidism-induced tumor-like mass of bony tissue characterized by fibrosis, cyst formation, marked osteoclastic resorption, multinucleated giant cells, and hemosiderin–which imparts the brown color deposits and rounded, cyst-like radiologic defects; BTs also occur in 2° hyperparathyroidism and may be the first sign of renal osteodystrophy in Pts with ESRD who are kept alive by renal dialysis, and have enough time to develop the osseous reaction. See End-stage renal disease

brownish urine UROLOGY Dark urine due to excretion of various substances, which can be 'physiological'—beets, drugs–azo gantricin, methylene blue, aniline dyes, fuschin, menses, or 'pathologic'–dipyrrole–from unstable Hb, RBCs, Hb, homogentisic acid, melanin, myoglobin, porphyrin

browser Web browser INFORMATICS A program used to search various resources on the Internet. See Client, Homepage, Netscape Mosaic, URL, WWW

broxuridine Broxine, BUdR ONCOLOGY A radiation sensitizer used to treat malignant gliomas and other brain tumors

104 **Bruce protocol** CARDIOLOGY A treadmill exercise protocol used to classify a Pt's functional–NYHA status. Cf Cornell protocol

brucellosis Bang's disease, Malta fever, undulating fever INFECTIOUS DISEASE Infection by *Brucella* spp, primarily *B abortus*, less commonly, *B melintensis* and *B suis* EPIDEMIOLOGY Primarily affects veterinarians, farmers, wool sorters, dairy workers, who are occupationally exposed to infected animals, meats, spore-laden wool CLINICAL Fever, sweating, malaise, aches, meningitis, abscesses of brain, liver, spleen, cholecystitis, endocarditis, arthritis, spondylitis, osteomyelitis, erythema nodosum, inhalation pneumonitis LAB Agglutination positive MANAGEMENT Doxycycline, rifampin for 6+ wks

Brudzinski sign NEUROLOGY A physical sign of meningitis, which is evoked by either passive flexion of one leg resulting in a similar movement on the opposite side, or if the neck is passively flexed, flexion occurring in the legs

Brugada syndrome CARDIOLOGY A condition in which people with no known heart problems or defects suffer sudden cardiac death or aborted sudden cardiac death EKG Right bundle branch block, persistent ST-segment elevation in V1 to V3 unexplained by electrolyte disturbances, ischemia, structural heart disease TREATMENT Implantable defibrillator. See Long Q-T syndrome

bruise A contusion 2° to traumatic injury of the soft tissues which interrupts capillaries and causes leakage of RBCs; in the skin it appears as a reddish-purple discoloration which does not blanch when pressed upon; when it fades it becomes green and brown as the body metabolizes the RBCs in the skin MANAGEMENT Local ice packs after injury

bruit CARDIOLOGY An audible swishing sound or murmur heard over an arterial 'thrill' caused by atherosclerosis; when auscultated over the carotid arteries, bruits predict future CVAs; it is unclear whether surgical correction improves the ultimate outcomes, as the ischemic event often occurs at a distance from the identified 'danger zone'; artery or vascular channel; indicates ↑ turbulence often caused by a partial obstruction. See Carotid endarterectomy

brunneroma Antral gland hyperplasia, brunner gland adenoma, heterotopic adenomatous polyp, GI DISEASE A polypoid proliferation of antral glands with normal overlying foveolae, which occur in small clusters separated by bundles of smooth muscle

brush CYTOLOGY A disposible with synthetic 'whiskers', used to scrape cells from mucosal surfaces. See Endocervical brush

brush border PHYSIOLOGY A specialized portion of the free or apical aspect of certain epithelial cells–eg, of the GI tract, consisting of closely packed microvilli that facilitate absorption; brush border cells line the luminal surface of the intestine and proximal convoluted tubules of the kidneys; the BB contains absorptive microvilli and glycocalyx–which is rich in hydrolytic enzymes

brushing CYTOLOGY The scraping of loose cells from the cervix, endocervix and vaginal wall. See Differential brushing DENTISTRY The cleaning of teeth with a tooth brush. See Bass brushing

bruxism PSYCHIATRY Compulsive grinding or clenching of teeth, which occurs unconsciously while awake or during stage 2 sleep. May be secondary to anxiety, tension, or dental problems

BS 1. Bachelor of Science 2. Basisphenoid 3. Bile salt 4. Blood sugar 5. Bowel sounds 6. Breath sounds

BSE 1. Bilateral, symmetrical, equal 2. Bovine spongiform encephalopathy, see there 3. Breast self-examination, see there

BSL Biosafety level, see there

BSO Bilateral salpingo-oophorectomy. Excision of both ovaries

BSS Balanced salt solution An ophthalmologic irrigating solution

BT CLINICAL MEDICINE Body temperature LAB MEDICINE Bleeding time PSYCHIATRY Behavioral therapy

BU 1. Bodansky Unit 2. Bromouracil 3. Busulfan 4. Burn unit

bubble stability test Foam stability test, see there

bubo INFECTIOUS DISEASE A tender, swollen, 4-5 cm in diameter, purplish lymph node most commonly seen in the inguinal or the axilla; bubos are classically associated with lymphogranuloma venereum, but may be associated with 1° syphilis–when it accompanies a chancre, gonorrhea, plague–hence the name bubonic plague, TB, et al. See Pseudobubo

bubonic plague Black death, black plague INFECTIOUS DISEASE A rare bacterial infection due to *Yersinia pestis*; in its full-blown fulminant form–explosive *Y pestis* growth–may kill in 24 hrs, by destroying normal tissues; after 3 days of incubation, high fever, black blotchy rashes–DIC plus petechial hemorrhage, delirium; bursting of a bubo–a massively enlarged lymph node–is painful enough to 'raise the dead' CLINICAL Painful, enlarged lymph nodes, fever, headache, prostration, pneumonia, sepsis EPIDEMIOLOGY *Y pestis* is transmitted by Oriental rat fleas–*Xenopsylla cheopis*, which bite the rat, ingesting *Y pestis*; these rapidly reproduce in the flea, forming a 'plug' of obstructing bacteria in the flea's gut, making the flea ravenously hungry, which goes into a feeding frenzy, repeatedly biting the rat and regurgitating *Y pestis*; once the usual hosts–rats–die, the fleas becomes less discriminating and attack any mammal; in humans, aerosol is the common mode of transmission INCUBATION 2-10 days MORTALITY Without antibiotics, nearly 100%; with antibiotics, 5%. See *Yersinia pestis*

buccal tube DENTISTRY A small metal part welded on the outside of a molar bank, which contains a slots to hold archwires, lip bumpers, facebows and other devices used to move the teeth

buccopharyngeal *adjective* Referring to the cheek/mouth and pharynx

bucindolol Bextra™ CARDIOLOGY An investigational nonselective beta-blocker with vasodilating properties; removed from marketplace

Buck's traction ORTHOPEDICS An apparatus for applying longitudinal traction on the leg by contact between the skin and adhesive tape, for maintaining the proper alignment of a leg fracture; friction between the tape and skin permits application of force through a cord over a pulley, suspending a weight; elevation of the foot of the bed allows the body to act as a counterweight; a type of traction in which a nonconstricting boot with weights is worn by the Pt to maintain proper alignment. See Traction

bucket handle fracture FORENSIC RADIOLOGY Fragmentation of the distal end of one or both femurs appearing at the bone margins as a crescent-shaped osseous density paralleling the metaphysis, which is seen on x-rays when the growth plate is oblique to the radiographic beam; incomplete 'bucket handles' are characteristic of child abuse-related injuries, which may also be associated with subperiosteal neoosteogenesis–trauma of recent origin, distal transverse dense lines–previous growth disturbance and cortical thickening–evidence of remote trauma. See Battered infant syndrome, Child abuse

bucking RESPIRATORY THERAPY Violent resistance by a Pt to intubated ventilation that may cause asynchronous breathing, ergo V/Q mismatching and risk of barotrauma, cardiac arrhythmia, and ↑ intracranial pressure; the newer ventilatory support devices rarely evoke this reaction, but Pts may still require sedation, or narcotics. See V/Q mismatch

Bucky® IMAGING A cassette film in a Potter-Bucky diaphragm

Budd-Chiari syndrome Hepatic vein obstruction HEPATOLOGY A loosely defined term for hepatic venous outflow obstruction, which has been used by various authors for both thrombotic and nonthrombotic obstruction of the large hepatic veins and the inferior vena cava ETIOLOGY Intravascular lesions including malignancy, myeloproliferative disorders, PNH, infection–eg schistosomiasis, intravascular webs; BCS is characterized by clotting of blood in the hepatic vein due to coagulation 2° to polycythemia vera or linked to OC use CLINICAL Ascites LAB Abnormal LFTs DIFFDX Hepatic failure, GI bleeding, starvation MANAGEMENT Unsatisfactory

buddy taping ORTHOPEDICS Immobilization of fingers or toes adjacent to one that is fractured or otherwise injured; buddy taping is used for injuries of the proximal interphalangeal joint with incomplete collateral ligament tearing, where the affected finger is taped to an adjacent 'buddy' and immobilized for 2 wks or more if there is also distal dislocation

budesonide Rhinocort PHARMACOLOGY A nebulizable synthetic glucocorticosteroid with high topical antiinflammatory activity, low systemic activity, and enhanced receptor binding activity; nebulized budesonide is used for croup and control of mild asthma, for inflammatory colitides–eg, ulcerative colititis and Crohn's disease, and seasonal or perennial allergic rhinitis. See Allergic rhinitis

buffalo hump Gibbus A popular term for a mass of adipose tissue present at the lower cervical and upper thoracic vertebrae, characteristically seen in Cushing's disease/syndrome; a gibbus may also accumulate in the thoracolumbar regions of black South Africans with achondroplasia–possibly related to their practice of carrying infants on their backs, mucopolysaccharidoses, chronic osteomyelitis induced by coccidioidomycosis and *M tuberculosis*

buffer CHEMISTRY A chemical system that minimizes the effects, in particular the pH, of changes in the concentration of a substance

buffered PHARMACOLOGY Referring to pills coated with a special substance that neutralizes stomach acid; drugs are buffered to ↓ stomach upset or ↑ absorption by the intestines

buffy coat LAB MEDICINE The flavescent yellow-white band of cells and debris present between the upper layer of plasma and the lower layer of RBCs, when whole blood is spun at 5000 RPM, which corresponds to WBCs

bug MEDICAL ENTOMOLOGY Any of a number of insects that are bloodsucking–eg, bed bugs–*Cimex lectularius* and/or act as vectors for disease–eg, reduviid bugs, carriers of trypanosomiasis. See Assassin bug, Kissing bug, Red bug, Reduviid bug MICROBIOLOGY A popular synonym for bacteria. See Superbug

bugeyes ENDOCRINOLOGY Popular for marked bilateral exophthalmos with proptosis, caused by the infiltrative ophthalmopathy, which is seen in more than ½ of Graves' thyrotoxicosis, but which is less common in non-Graves' hyperthyroidism. See Graves' disease

bugs in the rug A regionally popular term for pubic lice

bulbous nose Rhinophyma, see there

bulging disk NEUROSURGERY A condition caused by protrusion, herniation, or prolapse of a vertebral disc from its normal position in the vertebral column; the displaced disc may exert force on a nearby nerve root causing the typical neurologic symptoms of radiating pain–to an extremity, numbness, paresthesias, weakness; recurrent back pain is common MANAGEMENT NSAIDs, corticosteroids, rest; if advanced, laminectomy, microdisc surgery

bulimia Bulimia nervosa PSYCHIATRY A compulsive eating disorder characterized by binge eating, frequent fasting, laxative use, induced vomiting, inappropriate compensation to prevent weight gain; BN usually affects ♀ at a slightly later–age 17-25 onset than anorexia nervosa, but shares its preoccupation with food; bulimics may consume enormous quantities of food, in a 'binge', followed by self-induced emesis, a 'purge'; bulimia is either 1°, or a component of other diseases–eg, schizophrenia, OCs, Klüver-Bucy, and

Kleine-Levin syndromes; bulimics may have concomitant impulsive behavior–alcohol and drug abuse, poor peer and parental relations; sexual promiscuity, prostitution, and stealing may be required to financially support the eating 'addiction'; bulimia affects 1.3% of ♀ and 0.1% of ♂; bulimics may have imparied cholecystokinin secretion which may response to tricyclic antidepressants. See Binge. Cf Scarlet O'Hara 'syndrome'

bulk-forming laxative GASTROENTEROLOGY Bulk cathartic, bulk laxative, bulking agent Any laxative–eg psyllium, calcium polycarbophil, methylcellulose—that absorbs water and expands, ↑ in stool volume, facilitating passage of stools, ↑ motility. See Laxative; Cf Stool softener, Stimulant laxative

bulking up SPORTS MEDICINE A term used in weight training for the intentional ↑ of muscle mass often using anabolic steroids. See Anabolic steroids, Weight training

bulky disease ONCOLOGY A term for a CA with a considerable tumor burden–'bulk'; BD generally has a poorer prognosis independent of the histologic grade; a major priority in oncology is to debulk a tumor to optimize chemo- or RT. See Debulking operation

bull neck INFECTIOUS DISEASE A popular term for prominent and acute cervical lymphadenopathy associated with pitting edema, which is warm, tender and affects children over age 6; the change may be so extensive as to cover the sternocleidomastoid muscle's border; it is typically seen in children with epiglottitis who present with agitation, a muffled cry, SOB, cyanosis, drooling and dysphagia, due to *H influenzae* type B, *Corynebacteria diphtheriae* and rarely, *S pneumoniae* and *Staphylococcus aureus*

bulla plural, bullae ANATOMY A rounded thin-walled bony prominence DERMATOLOGY A large bleb or vesicle filled with serous fluid which may correspond to a separation of the epidermal-dermal junction. See Pemphigus PULMONOLOGY See Emplysema

bullet FORENSIC MEDICINE A charge which, when issued from a firearm, causes damage to person or property that correlates with its mass and velocity. See Ballistics, Black talon bullet, Dum-dum bullet, Hollow tip, Magic bullet, Retained bullet

bullous *verb* Pertaining to or characterized by bullae–as in bullous pemphigoid, or blistering, as in 2nd-degree burns

bullous disease DERMATOLOGY Any of a number of conditions characterized by the formation of bullae covering large portions of the skin surface; many BDs are immune-mediated and can be separated based on the location of the lesions. See Bullous pemphigoid, Pemphigus vulgaris

bullous myringitis ENT An infectious condition resulting in painful blisters on the surface of the tympanic membrane, which may occur in a background of otitis media caused by *Mycoplasma*

bullous pemphigoid DERMATOLOGY A disease characterized by tense blister–bullae formation on mucocutaneous surfaces, caused by circulating antibodies–usually IgG to the basement membrane of stratified epithelium CLINICAL BP may be chronic and mild without affecting the general health MANAGEMENT Topical corticosteroids–may require high doses of systemic cortisone. Cf Pemphigus vulgaris

bulls-eye lesion DERMATOLOGY The morphology of erythema multiforme, which is mediated by circulating immune complexes, elicited by infections, drugs, connective tissue disease; the skin may also have erythematous plaques and vesiculobullous lesions; BELs are also a classic lesion of Lyme disease seen 1-3 wks before the onset of the arthritic symptoms; mucosal involvement has been designated as Stevens-Johnson syndrome ENDOSCOPY A finding in the GI lumen, in which edematous folds surround a central depression–mass lesion with a central ulcer, non-specific finding seen in solitary amebomas, actinomycosis, amyloidosis, appendiceal disease, TB; single BELs include submucosal carcinoid, primary carcinoma, KS, leiomyoma, leiomyosarcoma, lipoma, and lymphoma; multiple BELs occur in metastases from primary breast, lung, and renal carcinoma, lymphoma, and mastocytosis GI RADIOLOGY BELs are similar to those seen by endoscopy, are caused by centrally ulcerated lesions, suggestive of malignancy, where the central zone is hypodense, ie necrotic, implying a rapidly proliferating lesion that has outgrown its vascular supply; multifocal BELs are suggestive of melanoma; unifocal BELs occur in carcinoma metastatic to the GI tract, benign or malignant smooth muscle tumors of the intestinal wall, KS and other sarcomas, eosinophilic granuloma, ectopic pancreatic tissue

bumetanide Bumex® CARDIOLOGY A loop diuretic with a rapid onset but short duration of action indicated for Pts with HTN and CHF CONTRAINDICATIONS Not indicated due to hypovolemia and fetal hypoxia. See Diuretic, Loop diuretic

bumper fracture FORENSIC MEDICINE A compression fracture of the lateral tibial plateau, with separation of the plateau's margin or depression of the central articular surface; BFs occur in a leg abruptly abducted by an automobile bumper, which strikes the lateral aspect of an extended leg and fixed foot, where the valgus stress forces the 2 bones into a close contact; BFs are often bilateral and involve both the tibia and fibula

BUN Blood urea nitrogen NEPHROLOGY A measure of serum urea, a by-product of the breakdown of blood, muscle and protein; urea is cleared by the kidney, and diseases which compromise the function of the kidney will frequently lead to ↑ blood levels NORMAL RANGE 7–20 mg/dL ↑ IN Renal disease, dehydration, CHF, GI bleeding, starvation, shock, urinary tract obstruction–by tumor or prostate gland ↓ IN Liver disease, malnutrition or a low protein diet REF RANGE ♀ 6–20 mg/dL; ♂ 8–23 mg/dL; SI expresses nitrogen as urea–normal adults, 2.9-8.2 mmol/L; BUN/creatine ratio normally is 20:1. See Azotemia, Uremia

bundle branch block CARDIOLOGY Abnormal conduction through one of the conduction branches which normally supplies the right and left ventricles, often resulting in delayed conduction though either the right or left bundle branches. See Left bundle branch block, Right bundle branch block

bundling MANAGED CARE The consolidation of 2 or more services–bundled service, supplies, drugs or other resources into fewer categories for payment–bundled payment. See Ancillary packaging, Significant procedure packaging. Cf Unbundling

bunion PODIATRY A localized, often painful swelling or deformity at the first metatarsal head of the great toe, which may be associated with bursitis or degenerative joint disease (osteoarthritis)

bunionectomy SURGERY The surgical excision of a bunion, most often from the base of the great toe, followed by immobilization in a plaster cast for 1-2 months

Bunyaviridae VIROLOGY A family of single-stranded RNA enveloped arboviruses composed of > 200 viruses, including genera Bunyavirus, Hantavirus, Phlebovirus, Nairovirus, Uukuvirus; they infect vertebrates and arthropods and some genera cause serious disease–eg, hemorrhagic fever

bupivacaine Marcaine® ANESTHESIOLOGY A potent local anesthetic with a 4-8 hr duration of action, doubled by adding epinephrine. Cf Ropivacaine

buprenorphine ADDICTION MEDICINE An agent that may be superior to methadone in ↓ heroin and cocaine abuse SIDE EFFECTS Sedation, constipation. Cf Methadone

bupropion Zyban® PSYCHIATRY An antidepressant SIDE EFFECTS Dry mouth, insomnia CONTRAINDICATIONS Seizures, bulimia

burbulence GI DISEASE A popular coinage for 'gassy' Sx–eg, Belching, bloating, flatulence. See Flatulence

Burkholderia cepacia Pseudomonas cepacia BACTERIOLOGY A bacterium found in the environment–eg, plants, water, soil, and in hospital environment, which may colonize the respiratory tract of Pts with cystic fibrosis; transmitted by direct physical contact from contaminated environmental sources, fluids and medical devices and, by aerosol, person-to-person CLINICAL May cause life-threatening URIs in Pts with cystic fibrosis, chronic granulomatous disease; also associated with UTIs and may infect sterile body sites

Burkholderia pseudomallei Pseudomonas pseudomallei BACTERIOLOGY A Pseudomonas-like bacterium CLINICAL Ranges from asymptomatic to melioidosis; skin infection and multifocal abscesses, URI, septicemia and shock

Burkitt lymphoma ONCOLOGY An aggressive extranodal B-cell NHL of children and young adults associated with EBV infection and a characteristic translocation between chromosomes 8 and 14 . See Lymphoma, Lymphoma belt, WHO classification

burn An injury, or effect due to fire or intense heat INFORMATICS A popular term for the engraving of 'master' compact disk–CD, or DVD GI DISEASE See Heartburn PUBLIC HEALTH A major health problem in which various thicknesses of skin are injured or destroyed by fire or intense heat; in 2002, 1 million Pts required medical attention for burns; burn depends on the depth, area and location of the burn; burn depth is generally categorized as 1st, 2nd, 3rd-degree SURGERY Cauterize, see there

burn cancer A squamous cell carcinoma that arises in a background of burns–especially in radiant energy-induced burns

burn center HOSPITAL CARE A specialized unit in a medical center dedicated to managing Pts with deep burns. See Burn

burned out mucosa GASTROENTEROLOGY A finding in severe, long-standing ulcerative colitis, with virtually complete mucosal denudation and scattered residual pseudopolyps

burned-out germinal center HEMATOLOGY A morphologically distinct germinal center that is typical of angioimmunoblastic lymphadenopathy; BOGCs are composed of loose aggregates of pale histiocytes and scattered immunoblasts or epithelioid cells mixed with amorphous eosinophilic and PAS-positive intercellular material, and thus resemble granulomas. See Germinal centers

burned-out phase 'Spent' phase HEMATOLOGY A desirable end-stage of polycythemia vera, where hyperproduction of erythrocytes settles down and much of the BM is replaced by reactive fibroblasts which produce the collagen typical of myelofibrosis

burner Acute, transient brachial plexus injury SPORTS MEDICINE A typically transient traction neurapraxia of younger athletes caused by direct cervical spine trauma–common in football occurring in younger athletes which may occur with spear tackling and accompanied by shoulder depression and lateral neck deviation away from the side of injury CLINICAL Burning or tingling sensation extending into the shoulder or arm when the neck is forced beyond the normal ROM MANAGEMENT Cowboy collar to prevent extreme hyperextension and lateral bending of the cervical spine, neck and shoulder exercise; athletes may return to collision activities when asymptomatic–normal strength, neurologically normal, and full range of painless cervical motion. See Spear tackling

burnout DRUG SLANG 1. A heavy abuser of drugs 2. Street argot for heavy abuse of drugs GYNECOLOGY See Cervical burnout PSYCHIATRY A stress reaction developing in persons working in an area of unrelenting occupational demands CLINICAL ↓ work performance, fatigue, insomnia, depression, ↑ susceptibility to physical illness, reliance on alcohol or other drugs of abuse for temporary relief. See Flight-or-fight response, Old Soldier syndrome. Cf Adaptation response, Alarm stage

burr A drill bit used to cut hard tissues–enamel, bone in dentistry or orthopaedics

burr cell HEMATOLOGY RBCs with regular spines or bumps on the surface, most commonly associated with snake bites, uremia, pyrokinase deficiency

bursa aspiration ORTHOPEDICS Insertion of a needle in a bursa to withdraw fluid–for diagnostic purposes or inject drugs–corticosteroids or a local anesthetics. See Bursitis

bursectomy Surgical drainage and removal of a bursa

bursitis Inflammation of a bursa, which may be accompanied by calcification of the supraspinatus tendon, or of the subdeltoid bursa CLINICAL Inflammation, pain, limited movement, ↓ ROM ETIOLOGY Idiopathic, chronic overuse, trauma, rheumatoid arthritis, gout, infection SITES Shoulder, knee, elbow, Achilles tendon, first metatarsal of the foot–bunion, etc MANAGEMENT–NON-INFECTIOUS Rest, ice, NSAIDs, analgesics INFECTIOUS Antibiotics, aspiration, surgery. See Anserine bursitis, Aseptic bursitis, Calcific bursitis, Knee bursitis, Septic bursitis, Shoulder bursitis

burst & taper INTERNAL MEDICINE A popular term for the therapeutic strategy used for treating Pts with corticosteroids, where large ('pharmacologic') doses are given early to bring an inflammatory process under control, then reduced to minimize the adverse steroidal effects

burst pacing CARDIAC PACING The delivery–eg, atrial BP–of rapid, multiple electrical stimuli, typically used to interrupt tachcardia. See Pacemaker

burst therapy THERAPEUTICS A relatively high dose administered over a short period, of a drug which has untoward side effects when therapy with the agent is prolonged

Büschke-Loewenstein tumor Giant condyloma acuminatum A verrucous carcinoma of the ♀ anogenital region TREATMENT Early lesions require radical local excision, while recurrent or metastasizing lesions require abdominoperineal resection

buserelin Suprefact® INFERTILITY A long-acting gonadotropin-releasing hormone analogue which downregulates the pituitary-gonadal axis, block testosterone production in the testes; it has been used to treat metastatic prostate CA. See Disease flare

business DRUG SLANG Injection paraphernalia

Buspar® Buspirone, see there

buspirone Buspar® NEUROPHARMACOLOGY A psychotropic agent for managing anxiety and ADD. See Attention deficit disorder

busulfan BU, Myleran® ONCOLOGY An alkyl sulfonate, the sole antineoplastic activity of which is myelosuppression; at low doses, it suppresses granulocytopoiesis; at higher doses, evokes pancytopenia INDICATIONS CML TOXICITY Confined to BM RESPONSE RATE 85–90%

butabarbital Pyridium® PHARMACOLOGY A sedative/hypnotic of intermediate duration of action

butalbital Fiorinal® PHARMACOLOGY A sedative/ hypnotic of intermediate duration of action, which may be used for tension headaches

Butazolidin® Phenylbutazone, see there

Butisol® Butabarbital, see there

butterfly fragment ORTHOPEDICS A popular term for a wedge-shaped fragment of bone split from the main fragments, seen in a comminuted–multifragmented fracture, usually of long bones

butterfly needle PHLEBOTOMY A short needle with flexible plastic handles that fold for insertion and lay flat for stabilization with tape, aka scalp vein needles, as they are the most practical and commonly used IV needles for infants

butterfly pattern IMAGING The fine diffuse infiltrates seen by a plain chest film that radiates bilaterally from the hilum to the lung periphery, characteristic of pulmonary alveolar proteinosis

butterfly rash INTERNAL MEDICINE An often photosensitive facial rash typical of SLE, which consists of an erythematous blush or scaly reddish patches on the malar region, extending over the nasal bridge–the facial 'seborrheic' region, potentially becoming bullous and/or secondarily infected; 'butterfly' region rashes have been described in AIDS, ataxia-telangiectasia, Bloom syndrome, Cockayne syndrome, dermatomyositis, erysipelas, pemphigus foliaceus, pemphigus erythematosus, riboflavin deficiency, tuberous sclerosis with smooth, red-yellow papules–adenoma sebaceum, appearing by age 4 at the nasolabial fold. See Systemic lupus erythematosus

button hole stenosis Fishmouth stenosis, see there

button sequestrum ORTHOPEDICS A preserved island of bone–sequestrum lying in a 'punched-out' osteolytic lesion of the diploë of the skull, well-described in eosinophilic granuloma, and seen in infections–TB, staphylococcal infection, metastatic CA, myeloma, radiation necrosis, meningioma, benign bone tumors, and ventriculoatrial shunts

buttress plate ORTHOPEDICS A plate used to support bone that is unstable in compression or axial loading; BPs are often used in the distal radius and tibial plateau to hold impacted and depressed fragments in position once they have been elevated. See Plate

buttressing effect FORENSIC PATHOLOGY The ↓ in size of a projectile's exit wound caused by compression or 'shoring' of the skin by clothing, seating materials, or any other externally placed deformable object

butylated hydroxyanisole BHA A preservative used in the food processing industry to prevent fats and oils from becoming rancid, it also added to packaged foods; a BHA-rich diet in pregnant mice is alleged to significantly reduce cholinesterase activity in offspring, and impact on the animals' level of aggressiveness, sleep patterns, and weight. See Alternative medicine, BHT

buy-In MEDICAL PRACTICE The purchasing of a portion of a group practice as a vehicle for becoming one of the partners in the practice

buzzword VOX POPULI A generic name for a term that has been recently incorporated into the argot of a particular field

BVAP A chemotherapy regimen consisting of BCNU, vincristine, Adriamycin-doxorubicin, prednisone

BVPP A chemotherapy regimen consisting of BCNU, vincristine, procarbazine, prednisone

bypass CARDIOVASCULAR SURGERY A surgical procedure in which a cardiovascular surgeon creates a new pathway for the flow of body fluids; bypass graft. See Cardiopulmonary bypass, Coronary artery bypass graft MANAGED CARE The re-routing of a Pt to be admitted to facility A–eg, to an ER, pediatric ICU, or other, to an equivalent facility B, when A is filled to capacity. Cf Anti-dumping laws, Bed, Dumping

bypass graft SURGERY A surrogate blood vessel used to reroute blood; BGs may be synthetic–Dacron, or autologous–vein from the Pt's own leg, to substitute for diseased vessel

bypass obstruction A partial occlusion of a mainstem bronchus resulting in hyperdistension of an entire lobe, with potential for interstitial emphysema and rupture, in a fashion identical to that of the ball-valve phenomenon

bypass surgery Coronary artery bypass graft, see there

byssinosis Brown lung OCCUPATIONAL MEDICINE A lung disease, secondary to inhalation of airborne dust from cotton, hemp, and linen; the early stages of disease are attributed to endotoxin CLINICAL Coughing, wheezing, airway obstruction; > 10 yrs, chronic bronchitis, emphysema, and interstitial lung disease, long-term disability MANAGEMENT Bronchodilators, change of occupation. See Farmer's lungs

byte COMPUTERS A set of 8 binary bits, a unit for measuring computer memory and storage capacity. See Bit, Computers, RAM

C-section Cesarean section, see there

C spine Cervical spine, see there

c wave CARDIOLOGY The wave in the normal jugular phlebogram, which corresponds to an early ventricular systole caused by bulging of the tricuspid valve into the right atrium

C1 Atlas ANATOMY A unique ring-shaped vertebrae without a vertebral body and no adjacent disc, which has narrow anterior and posterior arches connected to stout wedge-shaped lateral masses; C1 transitions a rigid skull and a mobile cervical spine, acting like a 'washer', buffering various forces; transverse atlantal ligament and alar ligaments stabilize C1 on C2; atlantoaxial joint ROM is 40° rotation, 10° flexion and 10° lateral bending. See C1 fracture

C1 fracture Atlas fracture ORTHOPEDICS A fracture which occurs with axial loading or axial loading + neck flexion, extension, lateral bending, or axial rotation; C1Fs are most common in the 2nd decade of life with a 2:1 ♂:♀ ratio, account for 4-15% of all cervical fractures and linked to MVAs and falls; 40% are associated with C2 injuries CLINICAL C1F victims either die at the scene or present without neurologic deficit–neck stiffness, limited neck movement, suboccipital pain, muscle spasm, headache IMAGING Open mouth odontoid views may show displacement of the lateral masses and are more useful than lateral views; *up to 25% of C1 fractures are missed on plain radiographs*; a CT provides detailed visualization of fracture(s) TYPES OF C1FS Posterior arch fracture, Jefferson fracture. See C1

C1 posterior arch fracture ORTHOPEDICS A stable fracture of the posterior arch of C1, usually caused by marked cervical hyperextension, effectively treated with a cervical collar. See C1

C2 fracture Axis fracture ORTHOPEDICS Fractures of C2 account for 17% of all cervical spine fractures TYPES Odontoid fracture–55%, followed by hangman's fracture–23%, other C2 fractures–22%. See Hangman's fracture, Odontoid, Odontoid fracture

CA 1. Cancer 2. Celiac axis 3. Chronologic age 4. Coarctation of the aorta 5. Cold agglutination 6. Common antigen 7. Computer-assisted 8. Condyloma acuminatum, see there 9. National Cancer Institute

CA-125 A cell surface glycoprotein expressed on the cell membrane of normal ovarian tissue, ovarian, cervical, endometrium, GI tract, and breast CAs; rising levels indicate a poor prognosis, but low levels are of little clinical utility; CA-125 may also be ↑ in liver disease, acute pancreatitis, renal failure, occasionally in normal ♀, lymphoma

CAA Cerebral amyloid angiopathy, see there

C-A-B sequence The order of emergency life support, where C–Chest compression is followed by A–Airway maintenance and B–Breathing. See A-B-C sequence

CABADAS CARDIOLOGY A clinical trial–Prevention of Coronary Artery Bypass Graft Occlusion by Aspirin, Dipyridamole and Acenocoumarol/Phenoprocoumon Study–that evaluated various strategies for preventing restenosis after CABG in Pts with CAD. See Angioplasty, CABG, Coronary artery disease, Low-dose aspirin

cabergoline Dostinex® A selective, potent, long-acting dopamine antagonist that suppresses prolactin secretion and restores gonadal function in ♀ with hyperprolactinemic amenorrhea; it is more effective and better tolerated than bromocriptine

CABG Coronary artery bypass graft, see there

CABG Patch trial CARDIOLOGY A clinical trial that evaluated prophylactic use of implanted cardiac defibrillators in Pts at high risk for ventricular arrhythmias after CABG. See AVID, Coronary artery bypass graft surgery, MADIT

C Symbol for: 1. Calorie 2. Carbon 3. Caudal 4. Centigrade 5. Chest 6. Cholesterol 7. Clearance rate 8. Closure 9. Cocaine–drug slang 10. Complement 11. Complex 12. Concentration 13. Constant region in immunoglobulin 14. Contracture 15. Contralateral 16. Cysteine 17. Cyanosis 18. Methcathinone; amphetamine–drug slang 19. Velocity of sound–blood, in Doppler ultrasonography

c Symbol for: 1. calorie–$^{1}/_{1000}$ of a Calorie 2. centi- 3. Complementary–molecular biology 4. *cum*–Latin, with 5. Molar concentration

C-peptide Connecting peptide ENDOCRINOLOGY A biologically inactive moiety of proinsulin produced endogenously in the pancreas, and stored in secretory granules in a 1:1 ratio with insulin; unlike factitious hypoglycemia, which is induced by insulin of exogenous origin, an ↑ in C-peptide–≥ 0.2 nmol/L, as well as ↑ insulin, ≥ 42 pmol/L is characteristic of insulinoma; C-peptide quantification is used to detect factitious insulin injection and diagnose insulin-secreting tumors in diabetics, where > 7 ng/ml of C-peptide after induced hypoglycemia supports a diagnosis of insulinoma. See Diabetes, Insulin

C-peptide suppression test ENDOCRINOLOGY A test which may be used to identify the causes of hypoglycemia, which '...*is based on the observation that beta cell secretion (as measured by levels of C peptide) is suppressed during hypoglycemia to a lesser degree in persons with insulinomas than in normal persons.*' The test requires that the serum glucose be ≥ 60 mg/dL and that the data be adjusted for body-mass index and age

C reactive protein LAB MEDICINE A 120 kD polypeptide of the pentraxin family which is produced by the liver during inflammation and detectable in serum in various conditions particularly during acute immune responses, named for its ability to bind the C polysaccharide of the *S pneumoniae* cell wall; CRP is an 'acute phase reactant' and is a biological marker for inflammation and necrosis FUNCTION Complement activation, T-cell binding, inhibition of clot retraction, suppression of platelet and lymphocyte function, enhancement of phagocytosis by PMNs; CRP is more predictive of cardiovascular events than LDL cholesterol; CRP is used to monitor early deterioration or development of complications of therapy–eg, empyema after pneumonectomy. See Acute phase reactant

C&R See Control & Restraint

cabin fever Relapsing fever, see there

CABRI CARDIOLOGY A clinical trial–Coronary Angioplasty versus Bypass Revascularization Investigation—comparing the outcome of PCTA vs CABG in Pts with multivessel coronary artery disease. See Angina, Angioplasty, CABG, Percutaneous transluminal angioplasty

Cache Valley virus VIROLOGY A common Bunyamwera virus, isolated in Utah in 1956 and recovered primarily from mosquitos–genera *Culiseta*, *Aedes*, *Anopheles*, and occasionally from vertebrates; in livestock, CVS may cause congenital malformations–musculoskeletal and CNS defects CLINICAL Rare in humans; CVS infection may be linked to severe encephalitis and multiorgan failure. See Bunyavirus, Encephalitis

cachexia CLINICAL MEDICINE A state of severe weight lose and tissue wasting 2° to underlying disease–eg, AIDS, terminal CA, anorexia nervosa, or malnutrition. See Tumor necrosis factor

cacophobia PSYCHOLOGY Fear of ugliness. See Phobia

cactus buttons DRUG SLANG A popular term for mescaline. See Mescaline

CAD 1. Cold agglutinin disease, see there 2. Computer-assisted design 3. Coronary artery disease, see there

cadaveric organ TRANSPLANT SURGERY An organ transplanted after its owner's death. See Harvesting, Mandated choice, Presumed consent

cadaveric transplant A transplant from a deceased and usually unrelated donor. See Cadaveric organ

CADILLAC A clinical trial–Controlled Abciximab/ReoPro® & Device Investigation to Lower Late Angioplasty Complications

cadmium poisoning TOXICOLOGY A condition associated with industrial exposure to cadmium CLINICAL renal tubule disfunction–aminoaciduria, glucosuria, hyperphosphatemia, hepatic fibrosis, emphysema and COPD, osteomalacia accompanied by bone pain. See Cadmium, Itai-itai disease, Toxic metal

caduceus MEDICAL HISTORY The time-honored, yet incorrect, symbol of medicine and art of healing, depicted by *two* serpents coiled around a winged staff. Cf Æsculapian staff

CAESAR AIDS study A non-US–Canada, Australia, Europe, South Africa AIDS trial that evaluated lamivudine with zidovudine, indicating slowing of disease progression and improved survival

café coronary Vallecular dysphagia EMERGENCY MEDICINE Complete and abrupt upper airway obstruction by a bolus of food, often meat, which occludes the esophagus and larynx, so named as the sudden onset of Sx simulates acute MI; CC victims are speechless, breathless and, without help–eg Heimlich maneuver–lifeless AT-RISK GROUPS Inebriated, bedentured, mentally retarded, demented, gluttonous CLINICAL Violent coughing, cyanosis, collapse, death. Cf Steakhouse syndrome, Sushi syncope

caffeine PHARMACOLOGY A methylxanthine that is *the* most widely used psychoactive substance; it is present in coffee, tea, maté, soft drinks, cocoa, Excedrin, NoDoz, kola nuts, guarana products; low doses–20–200 mg produce positive subjective effects, feelings of well-being, alertness, energy; higher doses have adverse effects–eg, nervousness, anxiety. See Coffee

caffeine dependence syndrome SUBSTANCE ABUSE A clinical complex defined by fulfilling 3 of 4 generic criteria for substance dependence from the DSM-IV. See Coffee, Tea. Cf Caffeine withdrawal syndrome

CAFFEINE DEPENDENCE SYNDROME–DEPENDENCY CRITERIA

TOLERANCE Criterion 1, DSM-IV substance dependence, seen in 75% of cohort

SYMPTOMS OF WITHDRAWAL in absence of the use of caffeine–Criterion 2, seen in 94%

PERSISTENT DESIRE or unsuccessful attempt to reduce or control use of caffeine products–Criterion 4, seen in 81%

CONTINUED USE DESPITE KNOWLEDGE of a persistent or recurrent problem that is likely to have been caused or exacerbated by substance abuse–Criterion 7, seen in 94%

caffeine withdrawal syndrome A complex associated with cessation of caffeine consumption; CWS is a defining criterion for the caffeine dependence syndrome CLINICAL Headache, lethargy, muscle pain or stiffness, ↓ performance, dysphoric mood changes–eg, depression, ±N&V. See Caffeine, Caffeine-dependence syndrome, Coffee

Caffey disease Infantile cortical hyperostosis PEDIATRICS An AD condition characterized by fever, inflammation,. swelling of affected bones–tibia, mandible, ulna and facial bones, restlessness

CAG 1. Chronic atrophic gastritis 2. Coronary angiography, see there

CAH Congenital adrenal hyperplasia, see there

Cain complex PSYCHIATRY A destructive sibling rivalry, in which one of the sibs resents the other for perceived favoritism from a parental figure

'caine family CLINICAL PHARMACOLOGY A popular term for a group of essentially interchangeable local anesthetics, which includes carbocaine, lidocaine, xylocaine, and others

caisson disease Chronic decompression sickness OCCUPATIONAL MEDICINE A condition caused by long-term whole body decompression, with repeated 'boiling' of nitrogen and resultant M&M of workers in high-pressure environments–eg, caissons CLINICAL Dysbaric–ischemic osteonecrosis with medullary infarcts of femoral, humoral, and tibial heads and rarely, CA–malignant fibrous histiocytoma arising in bone infarcts. Cf the Bends

calamine lotion A lotion containing zinc oxide–98% of weight mixed with iron oxide, or zinc carbonate; it is mildly astringent and protective, and used for various skin conditions–eg, poison ivy and sunburn

calcaneal apophysitis Seever's disease ORTHOPEDICS A condition characterized by heel pain in adolescents preceding fusion of the calcaneal apophysis, attributed to overuse

calcaneal spur Heel spur ORTHOPEDICS A bony outgrowth of the calcaneus, which may cause recurrent heel pain

calcaneal valgus Calcaneovalgus, rearfoot valgus ORTHOPEDICS An everted rearfoot See Valgus

calcaneal varus Rearfoot varus ORTHOPEDICS An inverted rearfoot. See Varus

calcaneodynia MEDTALK Heel pain

calcific tendonitis ORTHOPEDICS Inflammation of a tendon accompanied by focal calcium deposits, common in the supraspinatus tendon of shoulder joint CLINICAL Pain, ↓ ROM MANAGEMENT NSAIDs, needle aspiration, surgery, pulse ultrasound. See Tendinitis

calcification MEDTALK The deposition of calcium in tissues

calcinosis cutis DERMATOLOGY Circumscribed subcutanous deposition of calcium, which may occur in a background of chronic inflammation–eg, in connective tissue diseases–eg, scleroderma, SLE MANAGEMENT Medical; intralesional steroid injection; etidronate disodium, a diphosphonate, may inhibit biomineralization; surgery is of questionable efficacy

calcitonin Thyrocalcitonin PHYSIOLOGY A 32 residue polypeptide–plasma levels 100 pg/mL hormone produced by the parafollicular or 'C'–ultimobranchial cells of the thyroid; calcitonin is rapidly secreted in response to ↑ plasma Ca²⁺, pentagastrin, glucagon, β-adrenergics, and alcohol; it is produced by several neoplasms, especially, medullary CA of thyroid–MCT, which may also be associated with other endocrine tumors ↑ IN MCT, small cell of the lung, breast CA. See C cells

calcium A metal, atomic number 20, atomic weight 40.08, which is a divalent cation abundant in the body, especially in bone and teeth CALCIUM METABOLISM Ca²⁺ is the most critical mineral in bone–added by osteoblasts; removed by osteoclasts, Ca²⁺ maintains metabolic processes–eg, muscle contraction, neural transmission, cardiac activity, coagulation and inhibition of cell destruction; serum Ca²⁺ levels are controlled by a balance between PTH and calcitonin–produced by the thyroid's C or parafollicular cells; proper absorption of Ca²⁺ hinges on appropriate gastric acidity, presence of vitamin D, and a balance of other minerals–eg, phosphorus and mangesium; PTH ↑ serum Ca²⁺ levels by ↑ bone resorption via osteoclasts and mobilizing Ca²⁺, and indirectly ↑ GI absorption of Ca²⁺ by ↑ vitamin D production; PTH also ↑ phosphate excretion in the urine; calcitonin ↓ serum Ca²⁺ and phosphate levels by inhibiting bone resorption DAILY REQUIREMENT ± 400–1000 mg/day REF RANGE Infant to 1 month: 7.0-11.5 mg/dL; 1 month to 1 yr: 8.6-11.2 mg/dL >1 yr: 8.2-10.2 mg/dL; chronic abuse of laxatives, excess transfusions and various drugs can ↓ Ca²⁺ levels; Ca²⁺ is ↑ in hyperparathyroidism, parathyroid tumors, Paget's disease, myeloma, metastatic CA, multiple Fx, prolonged immobilization, renal disease, adrenal insufficiency, ↑ Ca²⁺ ingestion, antacid abuse; Ca²⁺ is ↓ in Cushing syndrome, hypoparathyroidism, malabsorption, acute pancreatitis, renal failure, peritonitis. See Hypercalcemia, Hypocalcemia, Ionized calcium

calcium alginate dressing WOUND CARE A wound dressing used to manage exudates in partial to full-thickness wounds, providing a moist environment for healing. See Wound care

calcium channel blocker Calcium channel antagonist PHARMACOLOGY An agent that blocks entry of Ca²⁺ into cells, thereby preventing cell dealth and loss of function due to excess Ca²⁺; CCBs are used to manage heart conditions and stroke, may be useful for Alzheimer's disease

calcium deficiency 1. Acute CD–see Hypocalcemia 2. Chronic calcium deficiency A long-term deficit of calcium, resulting in poor mineralization of bones, in adults, soft bones–osteomalacia and osteoporosis; in children, rickets, impaired growth. See Hypocalcemia

calcium metabolism The constellation of ionic checks & balances that maintain Ca²⁺ homeostasis in the blood and tissues. See Calcium

calcium pyrophosphate deposition disease A disorder characterized by polymorphous arthropathies CLINICAL Pseudogout–acute crystal-induced synovitis or knee and/or wrist, arthropathy–osteophytes of 2ⁿᵈ and 3ʳᵈ metacarpal heads, subchondral cysts of carpal bones; chondrocalcinosis of symphysis pubis, knee, annulus fibrosus of knee, nucleus fibrosus of intervertebral disk. Cf Gout

calcium supplementation METABOLISM The addition of Ca²⁺ to the diet, usually in the form of calcium carbonate

calculus Plural, calculi DENTISTRY Tartar Indurated, yellow–brown/black deposits on teeth formed by bacteria in dental plaques from mineralized calcium salts in saliva and subgingival transudates KIDNEYS A stone in the urinary tract. See Mulberry calculus, Staghorn calculus

Caldwell-Luc operation Intraoral antrostomy ENT An operation intended to relieve chronic sinusitis by improving maxillary sinus drainage, creating an opening into the upper jaw above a 2ⁿᵈ molar; once drained, the neofistula is allowed to heal closed

caliber The diameter of a tube or cylinder FORENSIC PATHOLOGY The diameter of a bullet measured in hundredths of an inch–in the US and in English-speaking countries COMMON BULLET CALIBERS .22, .25, .30, .32, .38, and .45; in other–'metric' countries, calibers are measured in millimeters–eg, .38 caliber corresponds to 9 mm. See Ballistics, Gauge, Yaw

caliceal stone A renal calculus the forms in the renal pelvis or calices

calicivirus VIROLOGY A ubiquitous small RNA virus with a characteristic Star of David morphology of Picornaviridae, which may be related to the Norwalk agent, of which 5 pathogenic strains are recognized CLINICAL Gastroenteritis with diarrhea and vomiting which affects young children and the elderly. See Gastroenteritis, Norwalk virus, Rabbit caliciviral disease

California encephalitis A viral infection by any of 4 California Bunyaviridae; most CE is by the LaCrosse virotype, which occurs in the summer in north central US; in endemic regions, CE causes 20% of acute childhood meningitides with a very low mortality rate; the infectious cycle is maintained in the mosquito vector, *Aedes triseriatus* CLINICAL Non-specific viral prodrome, followed by a 3-8 day meningismus with spontaneous resolution PROGNOSIS Excellent; rare psychological residua

call The 'after hours' responsibility that a physician has for evaluating already hospitalized Pts or admitting new Pts to a particular service or to the hospital per se. See Call schedule, Courtesy call, Home/beeper call, House call, Jeopardy call, Long call, Oncall, Short call, Weekend call. Cf Close call

call schedule GRADUATE EDUCATION A time period in which a resident is responsible for Pts, usually understood to be the time span after normal working hours. See Home/beeper call, House officer, Jeopardy call, Long call, Nightfloat, Short call

Call syndrome An acquired vasospastic disorder condition most common in ♀ with migraine, tends to occur during puerperium, and characterized by reversible segmental vasoconstriction of multiple cerebral arteries CLINICAL Headache, seizures, transient or persistent multifocal brain signs, accompanied by brain edema and ↑ intracranial pressure DIAGNOSIS Alternating zones of vasoconstriction and vasodilation

call to stool PHYSIOLOGY The neural signal that follows the arrival of ordnance in the rectum, telling the pilot that Dresden is in the sights, the bombay doors can be opened and the 500-pounders released. See Defecation

callosity A bony bump, callus

callus Callosity, tyloma DERMATOLOGY Localized hyperkeratosis of skin 2° to repeated friction or pressure ORTHOPEDICS A mass of indurated bony trabeculae and cartilage formed by osteoblasts early in healing fractures

caloric restriction The deliberate ↓ in caloric intake to levels up to 30% below a 'usual' diet See Diet, Methuselah factor. Cf Protein restriction

caloric test NEUROLOGY A test of vestibular function in which the ear canal is irrigated with cold and hot water, which often identifies an impairment or loss of thermally induced nystagmus on the involved side; the use of thermic stimuli to induce endolymphatic flow in the horizontal semicircular canal and horizontal nystagmus by creating a temperature gradient between the 2 sides of the canal

calorie CHEMISTRY A unit of measurement defined as 4.184 absolute joules–the amount of energy required to raise the temperature of 1 gram of water from 15° to 16°C NUTRITION Food calories equal to 1,000 calories–ie, 1 food calorie = 1 kilocalorie. See Empty calorie, Exchange list, Meal plan

Calot's triangle Cystohepatic triangle ENDOSCOPY The region in the liver bed bounded by the cystic artery, cystic duct and common hepatic duct, all of which must be ID'd and protected during laparoscopic cholecystectomy. See Laparoscopic cholecystectomy

Cambridge classification GI DISEASE A classification of pancreatitis based on time of presentation–acute vs chronic, severity–mild, moderate, severe–if multisystem failure and/or complications occurred in the form of abscess or pseudocyst

camel back curve INFECTIOUS DISEASE A fever curve with double daily febrile 'spikes'–eg, gonococcal endocarditis, measles, visceral leishmaniasis, Rift Valley fever. Cf 'Saddleback' curve

CAMELOT CARDIOLOGY A clinical trial–Comparison of Amlodipine versus Enalapril/Lipitor® to Limit Occurrences of Thrombosis. See Enalapril

Campath® ONCOLOGY A humanized monoclonal antibody to leukocyte antigen CD52, used to manage CLL refractory to standard therapies; it is expressed on lymphocytes and monocytes, but not hematopoietic stem cells

campus The physical environment of a large medical center with multiple buildings; areas or sections that are located in the on-site complex of buildings are 'on campus', off-site affilitates of the medical center are said to be 'off campus'. See 'In-house', Off-campus

Campylobacter A genus of gently-curved gram-negative rods that are common zoonotic commensals found in the GI tracts of wild and domesticated animals, and cause 3 types of human disease: enteric–eg, diarrhea, typically by *C jejuni*, extraintestinal, most often by *C fetus*, and gastric, due to *C pylori*, re-classified as *Helicobacter pylori*; human infections are attributed to contaminated water or food. See *Helicobacter pylori*

campylobacter enteritis INFECTIOUS DISEASE A water-borne gastroenteritis caused by *C jejuni*, a cause of travelers' diarrhea EPIDEMIOLOGY Linked to ingestion of contaminated eggs, poultry, water; 2-4 day incubation period CLINICAL Abdominal pain, ± bloody, watery diarrhea, fever. See *Campylobacter jejuni*

Campylobacter fetus* ssp *fetus *Campylobacter fetus* ssp *intestinalis* A subspecies of *C fetus* that causes opportunistic infections in humans–eg, bacteremia, meningitis, endocarditis, rarely diarrhea

Campylobacter jejuni *Vibrio jejuni, Campylobacter fetus* ssp *jejuni* A curved or spiral gram-negative bacillus with a single polar flagellum EPIDEMIOLOGY Linked to contact with domestic and farm animals, unpasteurized milk, primates, day care centers; peaks in summer CLINICAL Acute gastroenteritis of abrupt onset; malaise, myalgia, headache; Sx may be accompanied by abdominal colic, N&V, anorexia, tenesmus MANAGEMENT Erythromycin, tetracycline, gentamicin, furazolidine, chloramphenicol

Canadian Cardiovascular Society functional classification CARDIOLOGY A system used to stratify the severity of angina pectoris. See APSAC. Cf New York Heart Association classification

---

CANADIAN CARDIOVASCULAR SOCIETY–FUNCTIONAL CLASSIFICATION SEVERITY OF UNSTABLE ANGINA

CLASS I   Usual physical activity, eg walking or climbing stairs, does not cause angina; angina is evoked by strenuous and/or rapid work or recreation

CLASS II   Slight limitation of ordinary activities, eg after walking 2 blocks, climbing one flight of steps, under normal circumstances, after meals, in the cold, wind, in the morning, or when under emotional stress

CLASS III   Marked limitation of ordinary activities, eg walking 1-2 blocks or climbing stairs under normal circumstances

CLASS IV   Inability to carry out any physical activity without discomfort–angina may be present at rest

Circulation 1976, 54:522; JAMA, 1999; 281:1258-1260

---

canaliculate *adjective* Relating to a longitudinal groove, canal, or channel

canalplasty ENT A procedure used to repair defects of the external auditory canal by removing abnormal bone growth–eg, anterior overhang or exostosis, removal and replacement of intractably infected mucocutaneous tissue in the canal and enlargement or straightening of a stenotic or tortuous external auditory canal

Canavan disease Spongy degeneration of CNS An early onset AR condition caused by a defect or deficiency of aspartoacylase resulting in accumulation of N-acetylaspartic acid in brain, primarily in Jews CLINICAL Atonia of neck muscles, hyperextension of legs, flexion of arms, blindness, severe mental retardation, megacephaly, death by 18 months MANAGEMENT Nada; CD is a candidate for gene transfer therapy. See *Lorenzo's Oil*

cancellous bone Spongy bone, see there

cancellous screw ORTHOPEDICS A screw with relatively coarser thread and often with a smooth, unthreaded portion, which allows it to act as a lag screw and to anchor in soft medullary bone. See Screw

cancer Malignancy A malignancy of any embryologic origin, defined by WH Clark, Jr as a '…*population of abnormal cells showing temporally unrestricted growth preference (continually increasing number of cells in the population) over normal cells. Such abnormal cells invade surrounding tissues, traverse at least one basement membrane zone, grow in the mesenchyme at the primary site, and may metastasize to distant sites. It is the totality of properties that determines whether a given lesion should be designated as a cancer.*' See Actinic cancer, Anal cancer, Apoptosis, Bilateral cancer, Bladder cancer, Bone cancer, Brain cancer, BRCA-related gynecologic cancer, Breast cancer, Burn cancer, Cervical cancer, Chimney sweeps' cancer, Coelomic epithelial cancer, Colorectal cancer, Distant cancer, Early stage breast cancer, Endocrine cancer, Endometrial cancer, Environmental cancer, Epithelial ovarian cancer, Esophagus cancer, Extensive-stage small cell lung cancer, Familial cancer, Gastric cancer, Gynecologic cancer, Head & neck cancer, Hereditary nonpolyposis colorectal cancer, In situ cancer, Infiltrating cancer, Inflammatory breast cancer, Invasive cancer, Interval cancer, Invasive cervical cancer, Islet cell cancer, Kang cancer, Kangri cancer, Khaini cancer, Kidney cancer, Laryngeal cancer, Limited-stage small cell lung cancer, Liver cancer, Localized cancer, Locally advanced cancer, Lung cancer, Major cancer, Meningeal cancer, Microfocal cancer, Minimal breast cancer, Minor cancer, Nonmelanoma cancer, Oral cancer, Nonsmall cell lung cancer, Osteophilic cancer, Ovarian cancer, Pancreatic cancer, Pelvic cancer, Penile cancer, Prostate cancer, Pitch workers' cancer, Radiation-induced cancer, Recurrent cancer, Refractory cancer, Regional cancer, Residual cancer, Scar cancer, Second cancer, Skin cancer, Small cell lung cancer, Solid cancer, Solid cancer, Spontaneous regression of cancer, Terminal cancer, Testicular cancer, Thyroid cancer, Unresectable cancer, Uterine cancer, Vaginal cancer, Virally induced cancer, Yang cancer, Yin cancer, Yin and yang cancer. Cf Carcinoma, Neoplasm, Tumor

cancer, fear of Cancerphobia, see there

cancer cachexia ONCOLOGY A complex, multifactorial syndrome characterized by anorexia and/or unintended loss of appetite, accompanied by generalized host tissue wasting, skeletal muscle atrophy, immune dysfunction, and metabolic derangements. See Cachexia, Malnutrition

cancer center NIHSPEAK An institution designated by NCI as a comprehensive or clinical cancer center and eligible to conduct IND drug studies

cancer cluster EPIDEMIOLOGY A cancer that occurs in a group of people living or working in a geographically defined region who may share one or more environmental factors–eg, DES, and a characteristic lesion–eg, vaginal adenoCA, in common. See Clusters

cancer detection Any maneuver intended to identify cancer in an asymptomatic person. See Cancer screen, Cancer screening guidelines

---

CANCER DETECTION

SYMPTOMS Changes in bowel or bladder habits, nonhealing sores, unusual bleeding or discharges, new lumps in the breast or any other part of the body, indigestion or difficulty swallowing, change in the appearance of a pigmented lesions, persistent cough or hoarseness

EXAMINATION Palpation of breast, lymph nodes, rectal examination for prostate

TESTS Various endoscopies, imaging–CXR in high-risk Pts, pap smear, fecal occult blood

---

cancer effect level TOXICOLOGY The lowest concentration of a chemical in study or group of studies capable of producing a significant ↑ in the incidence of cancer–or tumors above that of a control population

cancer fallacy FRINGE ONCOLOGY Scientifically invalid reasoning about the nature of cancer, which may be used by purveyors of unsound cancer therapies to promote products or services. See Unproven methods for cancer management. Cf Cancer myth

cancer incidence ONCOLOGY The frequency with which cancer occurs, statistics of interest, US; estimated new cancer cases by sex, per the American Cancer Society

CANCER INCIDENCE BY SEX, 1998

MALES		FEMALES	
Prostate	185,400	Breast	178,700
Lung	91,400	Colorectum	68,900
Colorectum	66,000	Lung	80,100
Bladder	39,500	Uterus	36,100
Lymphoma	34,800	Lymphoma	27,700
Oral	20,600	Ovary	25,400
Melanoma	24,300	Melanoma	17,300
Kidney	17,600	Pancreas	14,900
Leukemia	16,100	Bladder	14,900
Stomach	14,300	Leukemia	12,600
Pancreas	14,100	Kidney	12,300
Larynx	9,000	Oral	9,700
All sites	627,900		600,700
All sites-both sexes			1,228,600

Cancer Information Service ONCOLOGY A service of the NCI that provides information on prevention, treatment options, clinical trials, newly approved anticancer drugs, supportive care, etc (☎ 800.422.6237; rex.nci.nih.gov)

cancer of kidney & ureter See Transitional cell carcinoma

cancer of mouth Oral cancer, see there

cancer myth Any popular belief about the nature of malignancy, which may sharply differ from available scientific evidence. See Cancer fallacy, Unproven forms of cancer therapy

cancer of unknown primary origin ONCOLOGY A cancer identified in one part of the body, although primary site cannot be found

cancer phobia Cancerophobia ONCOLOGY An excessive fear of suffering the ravages of malignancy; CP more commonly affects those who have directly cared for a loved one who suffered marked pain or disfigurement for a protracted period before death

cancer risk The likelihood of developing malignancy, the combined results of lifestyle–eg, low fiber diet, smoking, chronic alcohol abuse, exposure to carcinogens, and family Hx of cancer

cancer screening guideline Any guideline promulgated by an authoritative organization–eg Am Cancer Society, for early detection of a malignancy common in a particular population, the diagnosis of which, if caught early, results in a complete cure or improved long-term survival. See Cancer screening, Cancer screening test; Cf Unproven methods for cancer management

CANCER SCREENING GUIDELINES

BREAST CA Self-breast examination on a monthly basis, a baseline mammogram at age 40 and mammography every 1-2 years thereafter, depending on risk factors

COLORECTAL CA Per NCI, Am Cancer Society, and Am College of Physicians–annual fecal occult blood test > age 40 and flexible sigmoidoscopy every 3-5 years > age 50 (NEJM 1991; 325:37)

PROSTATE CA Annual digital rectal examination after age 40, and measurement of prostate-specific antigen or acid phosphatase in the serum

UTERINE CERVIX Annual Pap smear and pelvic examination after initiation of sexual activity; after 3 normal years, the test may be reduced in frequency at the discretion of the Pt's physician

cancer screening test Cancer screen A measurable clinical or lab parameter used to detect early malignancy; although these tests are relatively nonspecific, they are highly sensitive, and detect many persons who are 'abnormal' for the parameter being measured EXAMPLES Occult blood in stool–colorectal CA; mammography–to detect microcalcifications and densities for breast CA. See Cancer screening, Cancer screening guideline

cancer surgery SURGICAL ONCOLOGY Surgical excision of a malignancy. See Operable cancer, Resectable cancer

cancer symptom Any change attributable to a malignancy–eg, marked and rapid weight loss, changed bowel or bladder habits, nonhealing ulcers, abnormal bleeding or discharge, lumps in breast, dyspepsia, dysphagia, recent changes in a pigmented lesion, persistent cough

cancer-to-cancer metastasis A rare event in which there is metastatic penetration of one malignancy into another– 'DONOR' CANCERS Bronchogenic 33%, breast 10%, GI 10%, prostate 10%, thyroid 10% 'RECIPIENT' CANCERS Renal cell cancer 60%, lymphoproliferative processes 12%, and others

cancer virus Oncogenic virus, see there

cancerization ONCOLOGY Carcinogenesis, see there

CancerLit ONCOLOGY The NCI's bibliographic database of published research that contains compilations of select citations and brief reviews on cancer topics: wwwicic.nci.nih.gov/canlit/canlit.htm

CancerNet ONCOLOGY Information from NCI–eg, the PDQ database, sorted for the general public, health care providers, researchers

Candida albicans A dimorphic fungus that is a major opportunistic pathogen in immunocompromised Pts; *C albicans* is part of the normal GI flora, and commonly causes vaginal candidiasis MANAGEMENT Nystatin, ketoconazole, fluconazole

Candida hypersensitivity syndrome Candida syndrome, yeast syndrome A controversial condition attributed to an overgrowth of *Candida albicans* on mucosae, especially of vagina and GI tract ETIOLOGY CHS has been linked to various agents, ranging from overuse of antibiotics to OCs, and pregnancy CLINICAL Chronic GI complaints–eg bloating, heartburn, constipation, diarrhea, anxiety, depression, loss of memory, poor concentration, fatigue, infections, irritability, depression, headache, mood swings, nasal congestion, weight gain DIAGNOSIS Based on a questionnaire–!!! rather than blood or tissue studies TREATMENT Nystatin. See Candidiasis, Crook diet, Yeast connection

candidiasis Moniliasis, candidosis INFECTIOUS DISEASE Infection with *Candida*, especially of mucocutaneous surfaces and caused by *C albicans* CLINICAL Dysphagia, oral lesions DIAGNOSIS Endoscopy, cytology, cultures MANAGEMENT Ketoconazole, fluconazole

candidiasis of oral cavity Oral thrush, see there

candidiasis of skin Cutaneous candidiasis, see there

CANDLE CARDIOLOGY A clinical trial–Candesartan versus Losartan Efficacy Comparison–to compare the antihypertensive effect of candesartan cilexetil–Atacand® with losartan. See Candesartan cilexetil

candle flame appearance CARDIOLOGY A finding by Doppler color flow imaging–the central blue color that corresponds to a zone of high velocity, surrounded by a yellow-orange blush, corresponding to a turbulent zone of lower blood flow; the CFA is typical of mitral stenosis–MS, other Doppler changes in MS have scimitar, mushroom and bifid jet shapes

candle wax drippings A descriptor for an appearance likened to melted candle wax gutterings OPHTHALMOLOGY Fundoscopic changes in which the vascular sheath is surrounded by pretetinal inflammatory exudates, appearing as bulbous perivascular dilatations, a finding suggestive of sarcoidosis

candlestick appearance BONE IMAGING A sharply marginated 'cutoff' of the terminal phalanges with a central depression, seen in burns, DM, gout, leprosy, malabsorption, vasoocclusive disease, porphyria, psoriasis, Raynaud's disease, rheumatoid arthritis, scleroderma. Cf Penciling

Canesten® Clotrimazole INFECTIOUS DISEASE A topical used for *Candida* spp and/or trichomonas infection. See *Candida albicans*

canker sore Aphthous ulcer, cold sore, fever blister One or more painful, recurrent ulcers of the oral cavity or vermilion border, preceded by tenderness and pruritus, beginning as an erythematous indurated papule that rapidly erodes, leaving an ulcer covered with grayish exudate–time from onset to healing ± 10-14 days; most CSs are idiopathic and may be due to allergy, autoimmunity, drugs, endocrinopathy, infections–viral, nutritional–vitamin deficiency, stress or trauma, with 2° bacterial infections that respond to topical tetracycline; CSs may respond poorly to treatment, and require oral rinses with benadryl, xylocaine and corticosteroids

cannabinoid receptor THC receptor, see there

cannabis intoxication See Amotivational syndrome, Marijuana

Cannabis sativa Marijuana, see there

cannon 'a' wave CARDIOLOGY An abnormal jugular venous pressure curve with an accentuated 'a' wave of sufficient intensity to cause the earlobes to 'flap', due to ↓ right ventricular compliance, tricuspid stenosis or an arrhythmia in which the atrium contracts against a closed or stenosed tricuspid valve; a less 'explosive' but still prominent 'a' wave may by associated with pulmonary HTN; CAWs may be regular, as are AV junctional rhythms, where a CAW occurs every 2nd beat in a 2:1 block, or irregular, which are more common and may occur in complete heart blocks without A Fib, V tach, AV dissociation

cannonball metastases One or more large, well-circumscribed metastatic nodules in the lungs, classically in renal cell carcinoma, but also in choriocarcinoma

cannula SURGERY A tube inserted into a duct, cavity or other space; during insertion, its lumen is occluded by a trocar. See Endoscopy, Nasal cannula, QuickDraw™ venous cannula, Trocar

cannulated screw ORTHOPEDICS A screw with a hollow shaft, a low pullout strength compared to conventional screws, which can be placed with reasonable precision and minimal trauma using Kirscher ("K") wires COMPLICATION CSs may perforate the joint if placed in a bone with their tips close to the subchondral bone. See Screw

canthaxanthin FOOD INDUSTRY A synthetic carotenoid which in humans cannot be converted to vitamin A, which has been marketed as a tanning agent under various names; it has been implicated in aplastic anemia. See Artificial dye, Carotenoid, Tanning, Unproven methods for cancer management

canyoneering Canyoning SPORTS MEDICINE An 'extreme sport' in which the participants slide, jump or rappel down waterfalls and mountain streams through a canyon, clad in wetsuits, helmets, life jackets, and climbing harnesses, combining the skills of caving, climbing, rappeling, and 'reading' the river. See Extreme sport

CAP 1. College of American Pathologists. See CAP inspection, CAP survey

2. Community-acquired pneumonia, see there 3. Cyclophosphamide, Adriamycin, prednisone–an oncology regimen

cap DENTISTRY A popular term for a crown, see there GYNECOLOGY See Cervical cap MANAGED CARE A limit on reimbursement for a health care service imposed by an insurance company or governmental agency. See Capitation OBSTETRICS See Cradle cap

capacitation REPRODUCTIVE MEDICINE The postrelease maturation in sperm, which are capable of penetrating the zona pellucida and fertilizing an oocyte after exposure to secretions in the ♀ genital tract. Cf Acrosome reaction

capacity PATIENT RIGHTS The capability of a person to function as an autonomous unit. See Testamentary capacity

CAPARES CARDIOLOGY A clinical trial–Coronary Angioplasty Amlodipine Restenosis Study. See Amlodipine

CAPD Continuous/chronic ambulatory peritoneal dialysis. See Dialysis, Peritoneal dialysis

CAPE CARDIOLOGY A clinical trial–Circadian Anti-Ischemia Program in Europe that evaluated the effect of an anti-ischemic, amlodipine, on M&M in Pts with CAD. See Amlodipine, Coronary artery disease, Silent ischemia

capillaritis Inflammation of capillary vessels. See Necrotizing alveolar capillaritis, Vasculitis

capillary hemangioma DERMATOLOGY A painless benign vascularized red–purple skin lesion that develops shortly after birth PROGNOSIS Many CHs disappear in early childhood MANAGEMENT Local steroid injections may ↓ the CH's size

capillary leakage syndrome Progressive dyspnea and pericarditis, which may occur in therapy with GM-CSF, which has been used to treat metastatic CAs; a similar response may occur in IL-2 therapy, where large amounts of fluid–10-20 liters are held hostage in peripheral tissues, and accompanied by high fever, confusion, and disorientation. See GM-CSF

capillary specimen Fingerstick specimen, see there

capitation Capitated payment MANAGED CARE 1. A method of payment for health services in which a physician or hospital are paid a fixed amount is paid per enrollee to cover a defined scope of services for a defined population set–aka covered lives for a defined period of time, regardless of actual number or nature services provided; capitation may be used by purchasers to pay health plans or by plans to pay providers; 2. capitation payment is made on the basis of the number of members each month. See Bundling, Fee for service, HMO, Medicare risk contract, Per diem, PMPM, PPO, Rate setting. Cf Fee-for-service

Caplan lesion Caplan syndrome, rheumatoid pneumoconiosis PULMONARY MEDICINE A large > 4 cm in diameter silicotic nodule, with smooth borders, peripheral palisading of macrophages, concentric internal lamination, and central necrosis; CLs may represent juxtaposition of silicotic and rheumatoid nodules. See Rheumatoid arthritis, Silicosis

***Capnocytophaga canimorsus* sp nov** CDC group DF-2, DF2, dysgonic fermenter-2 INFECTIOUS DISEASE A fastidious gram-negative bacillus linked to dog-bite infections CLINICAL Sx range from mild and localized to bacteremia, necrosis, cellulitis of soft tissue and skin, potentially serious septicemia with Waterhouse-Friderichsen-type adrenal cortical collapse, DIC, endocarditis, gangrene, malar purpura–a finding quasi-pathognomonic of generalized Schwartzman reaction, fulminant sepsis in cirrhosis or postsplenectomy; other infections include pneumonia, endocarditis, meningitis. See Dog

capnography Capnometry PULMONOLOGY The measurement and graphic display of CO_2 levels in the airways, which can be performed by infrared spectroscopy; capnography facilitates Pt management by providing

1. Continuous and noninvasive monitoring of ventilation in critically ill Pts and 2. Early detection of clinically significant changes in respiratory status by displaying changes in the amount of CO_2 and abnormal CO_2 waveforms; capnography can be used to detect a wide range of clinical conditions–eg, extubation of respiratory support, hypotension or massive blood loss, emphysema, COPD, pulmonary embolism, and others. See Infrared capnography

Capoten® Captopril, see there

CAPPP CARDIOLOGY A randomized, open label, placebo-controlled, multicenter trial designed to compare the effectiveness of conventional therapy vs captopril on cardiovascular M&M in Pts with HTN. See Antihypertensive, Captopril, Hypertension

capreomycin INFECTIOUS DISEASE An antibiotic used to treat MAC and TB SIDE EFFECTS Anorexia, thirst, excess urination, hematuria, anemia, hearing loss. See AIDS

CAPRIE CARDIOLOGY A clinical trial–Clopidogrel Versus Aspirin in Patients at Risk of Ischemic Events which compared the efficacy of the antiplatelet agent, clopidogrel, with aspirin therapy as a stroke prophylaxis. See Clopidogrel

CAPS CARDIOLOGY A clinical trial–Coronary Arrhythmia Pilot Study which evaluated the effect of antiarrhythmic therapy on survival in Pts with non-sustained ventricular tachycardia and PVCs

caps Abbreviation for capsule DRUG SLANG 1. Crack 2. Capsules, pills in general 3. Heroin, see there 4. Psilocybin/psilocin

capsaicin NEUROLOGY Capsaicine A chemical from red hot chili peppers that may be used for painful dysesthesias of herpes and DM; topical capsaicin triggers release of the neuropeptide, substance P from type C nociceptive fibers, opens Ca^{2+} and Na^+ channels causing the initial pain associated with 'hot' foods; substance P is not replenished, thus pain sensation is ↓ after the initial pain; capsaicin binding is relatively strong and attributed to its lipophilic side chain MANAGEMENT Casein, a lipophyilic phosphoprotein acts like a detergent and strips the capsaicin from the receptors in the oral cavity; topical capsaicin may ↓ the symptoms of painful diabetic neuropathy. See Blister beetle, Scoville unit, Spicy foods

capsular contracture BREAST SURGERY Tightening of the capsule around a breast implant, which may be accompanied by discomfort, implant distortion

capsule PHARMACOLOGY A solid dosage form of a drug in which the drug is enclosed in a hard or soft soluble container or shell of an appropriate gelatin. See Microcapsule

capsulectomy COSMETIC SURGERY The operative removal of an entire capsule and implant–especially of the breast

capsulitis Inflammation of the capsule of a joint. See Adhesive capsulitis

capsulotomy BREAST SURGERY Closed capsulotomy Manual release of the capsule–Pt's own scar tissue–around a breast implant NEUROSURGERY Anterior capsulotomy A form of psychosurgery in which the presumed fronto-thalamic connections of the anterior limb of the internal capsule of the brain are interrupted–by thermocoagulation using a bipolar electrode system–at the point where the connections pass between the head of the caudate nucleus and putamen INDICATIONS Obsessive neurosis–ie, OCD, depression COMPLICATIONS Transient confusion, nocturnal incontinence, seizures, depression, excess fatigue, poor memory, slovenliness. See Psychosurgery ORTHOPEDICS An incision into the capsule of a ball joint–femur, humerus—to relieve pressure and ↓ risk of hematomas and bone necrosis. See Hip capsule

CapSure™ continence shield UROGYNECOLOGY An external suction cuplike device for managing dribbling in ♀ with stress incontinence, which fits over the meatus;. See Urinary incontinence

Captain Kangaroo MEDICAL SLANG A popular term for the chairman of a pediatrics department. See Medical slang

captain of the ship doctrine MEDICAL MALPRACTICE An adaptation from the 'borrowed servant rules', as applied to an operating room, which arose in *McConnell* v *Williams,* holding the person in charge–eg, a surgeon responsible for all under his supervision, regardless of whether the 'captain' is directly responsible for an alleged error or act of alleged negligence, and despite the assistants' positions as hospital employees. See 'Borrowed servant', Respondeat superior

CAPTIN CARDIOLOGY Either of 2 clinical trials–1. Captopril before Reperfusion in Acute Myocardial Infarction 2. Captopril Plus Tissue Plasminogen Activator following Acute Myocardial Infarction

Captopril Capoten® CARDIOLOGY A short-acting ACE inhibitor used for HTN, in type 1 DM, heart failure, and post-MI Pts with left ventricular dysfunction; long-term administration of captopril is associated with ↑ survival, and ↓ cardiovascular M&M, possibly attenuating ventricular dilatation and remodeling WARNING Angioedema of upper airways ADVERSE EFFECTS Neutropenia, agranulocytosis, proteinuria, rash, pruritus, hypotension, dysgeusia. See ACE inhibitor

capture CARDIAC PACING Depolarization of the atria and/or ventricles by an electrical stimulus delivered by an artificial pacemaker; one-to-one capture occurs when each electrical stimulus causes a corresponding depolarization. See Stimulation threshold

caput medusae Medusa head HEPATOLOGY A term for the engorged veins that radiate from a recanalized umbilical vein–falciparum ligament in portal HTN, most common in advanced alcoholic cirrhosis, and accompanied by ascites, hepatosplenomegaly, patent hepatic veins

caput succedaneum NEONATOLOGY A yarmulka-like edematous bulge that forms on that portion of the infant scalp immediately overlying the cervical os in a vertex presentation; CS is most prominent in prolonged labor and an incompletely dilated cervix and disappears shortly after birth

car sickness Motion sickness, see there

carapace pattern NEUROLOGY A broad sensory loss affecting very short nerves of the upper body–trunk and thorax, with a distribution likened to the carapace of a shellfish

carbachol A parasympathomimetic formed by substituting an acetyl with a carbamyl group on acetylcholine, which acts on muscarinic and nicotinic receptors

carbamazepine Carbatrol®, Tegretol®, divalproex NEUROLOGY An anticovulsant used as a monotherapy for treating partial seizures–eg, generalized tonic-clonic seizures in children; in contrast to valproate which is as effective as carbamazepine in treating generalized tonic-clonic seizures, carbamazepine is associated with better seizure control and seizure-rating score, ↓ seizures, and time to first seizure; it may be used for manic episodes ADVERSE EFFECTS Rash, hair loss, tremor, myelosuppression; carbamazepine induces enzymes that metabolize warfarin, clonazepine, phenytoin $T_{1/2}$ 15 hrs THERAPEUTIC RANGE 4–10 mg/L TOXIC RANGE ≥ 15 mg/L. See Convulsions, Seizures. Cf Valproate

carbofuran N-methyl-carbamate OCCUPATIONAL MEDICINE A broad-spectrum anticholinergic pesticide/insecticide sprayed by air to protect various crops–eg alfalfa, rice, grapes, and cotton TOXICITY Most commonly affecting farm workers—headache, N&V, muscle weakness, tearing, salivation, bradycardia, diaphoresis, miosis

carbohydrate

carbohydrate NUTRITION An abundant organic compound, which is one of the 3 main classes of foods and a principal source of energy; ingested carbohydrates are sugars and starches, which are metabolized into glucose, or assembled into glycogen, and stored in the liver and muscle for future use. See Complex. Cf Fats, Protein

carbohydrate drink SPORTS MEDICINE A sports drink that contains glucose polymers, intended to replenish the reserves of energy during and after exercise. See Sports drink

carbohydrate loading Pasta loading SPORTS MEDICINE The ingestion of a low-fat, low-protein, low-fiber, carbohydrate-rich meal–after previous depletion of the hepatic stores of glycogen, with the intent of maximizing glycogen storage in muscle. Cf Bicarbonate loading, Phosphate loading

carbon dioxide CO_2 PHYSIOLOGY A metabolic byproduct of carbohydrate metabolism; it accumulates in tissues, is released to the blood in veins, and is eliminated via the lungs

carbon dioxide content CO_2 content ARTERIAL BLOOD GASES A measure of the relative blood concentration of CO_2, measured using pH electrodes, by enzymes, or based on changes in pH REF RANGE < age 2–18-28 mmol/L; > 2 yrs–venous 22-26 mmol/L; > 2 yrs–arterial 22-32 mmol/L ↓ IN Respiratory alkalosis–DKA, lactic acidosis, alcoholic ketoacidosis, hyperventilation, metabolic acidosis, kidney disease, renal failure, diarrhea, Addison's disease, ethylene glycol poisoning, methanol poisoning ↑ IN Severe vomiting, gastric drainage, hypoventilation–eg, emphysema or pneumonia, hyperaldosteronism, Cushing syndrome. See Arterial blood gases

CO_2 laser ablation GYNECOLOGY The excision of part of the uterine cervix for CIN, focal CIS, condylomas

carbon monoxide CO CLINICAL TOXICOLOGY A byproduct of combustion, which is a tasteless, odorless gas that outcompetes O_2 for Hb binding–CO has a 200-fold > affinity for active heme sites than O_2 CLINICAL-EARLY Headache, nausea CLINICAL-LATE Coma, cardiovascular collapse CLINICAL-TOO LATE Death

carbon monoxide poisoning EMERGENCY MEDICINE Intoxication caused by excess of carbon monoxide in ambient air, due to either smoke inhalation or suicide attempts CLINICAL Acute Sx occur at 20% concentration–of CO in blood, severe symptoms at 30%, headache and confusion at 40–50%, unconsciousness and seizures as 60–70%, and ≥60% can be fatal; levels in normal nonsmokers = 0.25–2.1%, with smokers and certain industrial-exposed individuals having up to 10% MANAGEMENT Hyperbaric oxygen. See Hyperbaric oxygen therapy

carbon tetrachloride CCl_4 TOXICOLOGY A volatile liquid used in dry cleaning and fire extinguishers

carboplatin ONCOLOGY A chemotherapeutic for advanced ovarian and other CAs ADVERSE EFFECTS Cytopenias, nausea, diarrhea, hair loss, pain, neurologic complaints

carboxyhemoglobin COHb Hb in which carbon monoxide–CO is irreversibly bound

carbuncle DERMATOLOGY A deep skin infection, often by staphylococci, involving several interconnected hair follicles CLINICAL A red nodular pus pocket, fever, malaise MANAGEMENT Hot wet compresses, drainage, antibiotics PROGNOSIS Worse with immune compromise or underlying illness–eg AIDS, cancer, DM, pernicious anemia, dermatitis

carcinoembryonic antigen See CEA

carcinogen ONCOLOGY Any physical or chemical agent or substance which, when administered by an appropriate route, ↑ incidence of tumors when compared to unexposed control population. See Cocarcinogen, Complete carcinogen, Natural carcinogen, Proximal carcinogen

carcinogenesis Transformation, tumourigenesis ONCOLOGY A series of genotypic and phenotypic changes that result in a cell being identified as malignant by virtue of its metastatic potential. See Cancerization, Chemical carcinogenesis, Inducer, One-hit/two-hit model, Proto-oncogenes, Tumor promoter, Tumor suppressor

carcinoid syndrome ONCOLOGY A symptom complex caused by carcinoids, which arise from the enterochromaffin system, a system most concentrated in the midgut; carcinoids release vasoactive substances–eg, serotonin, and its metabolite, 5-HIAA, as well as bradykinin, histamine, PG, substance P; serotonin's precursor is tryptophan, used by the body to produce niacin and some proteins; carcinoid Sx are most common in metastases, often from a primary ileal carcinoid metastatic to the liver CLINICAL Episodic skin flushing on the face, spreading to the trunk, then telangiectasia, precipitated by alcohol, food, stress, or liver palpation; also diarrhea or GI obstructive symptoms, bronchospasm, wheezing, pleural, peritoneal, retroperitoneal, endocardial fibrosis of the right valves and right ventricular wall, which may cause failure, carcinoid myopathy and atrophy of type II muscle fibers LAB ↑ Serotonin, 5-HIAA in serum, urine MANAGEMENT Resection. See APUD system, Atypical carcinoid

carcinoma ONCOLOGY A malignant neoplasm of epithelial and occasionally neuroepithelial origin; carcinomas are divided according to tissue of origin–eg, glands–adenoCA, squamous epithelium–SCC, and bladder epithelium–transitional cell carcinoma; carcinomas may metastasize to bone, liver, lung, brain. See Acinic cell carcinoma, Adenocarcinoma, Adrenocortical carcinoma, Anaplastic carcinoma, Anaplastic carcinoma of pancreas, Anaplastic carcinoma of thyroid, Basal cell carcinoma, Basaloid carcinoma, Carcinoma in situ, CASTLE, Chromophobe cell carcinoma, Clear cell carcinoma, Collecting duct carcinoma, Colloid carcinoma, Ductal carcinoma in situ, Duct cell carcinoma, Embryonal carcinoma, Endometrial carcinoma, Epithelial carcinoma, Epithelial-myoepithelial carcinoma, Fibrolamellar carcinoma, Follicular carcinoma, Giant cell carcinoma, Glassy cell carcinoma, Hürthle cell carcinoma, Inflammatory carcinoma, In situ carcinoma, Intraductal carcinoma, Intramucosal carcinoma, Juvenile carcinoma, Krebs' carcinoma, Large cell carcinoma, Large cell undifferentiated carcinoma of lung, Laryngeal carcinoma, Lobular carcinoma in situ, Medullary carcinoma, Merkel cell carcinoma, Microinvasive carcinoma, Minimal deviation adenocarcinoma of cervix, 'Murky cell' carcinoma, Nasopharyngeal carcinoma, Non-small cell carcinoma of lung, Oat cell carcinoma, Ovarian small cell carcinoma–hypercalcemic type, Pleomorphic carcinoma, Pleomorphic lobular carcinoma, Renal cell carcinoma, Sarcomatoid carcinoma, Scirrhous carcinoma, Secretory carcinoma, Small cell carcinoma, Spindle cell carcinoma, Squamous cell carcinoma, Stump carcinoma, Superficial spreading carcinoma, Terminal duct carcinoma, Transglottic carcinoma, Transitional cell carcinoma, Tubular carcinoma, Undifferentiated carcinoma, Verrucous carcinoma. Cf Cancer

carcinoma of breast See Breast cancer

carcinoma of cervix See Cervical cancer

carcinoma of endometrium See Endometrial cancer

carcinoma in situ SURGICAL PATHOLOGY A carcinoma in which all of the cytologic and pathologic criteria used to define malignancy have been met, but which has yet to invade; CIS may regress or may be stable for long periods; although the cervix was one of the first locations where CIS was recognized, other epithelia in the body have CIS lesions. See Cervical intraepithelial neoplasia, Ductal carcinoma in situ, Intraepithelial neoplasia, Lobular carcinoma in situ, Microinvasive carcinoma. Cf Borderline tumors

carcinomatosis Carcinosis ONCOLOGY An end-stage condition characterized by disseminated, fulminant extension of metastatic CA. See Metastatic carcinoma

CARDIA CARDIOLOGY A clinical trial–Coronary Artery Risk Development in Young Adults. See Coronary artery disease

cardiac *adjective* Pertaining to the 1. Heart 2. Cardia, a part of the stomach, see there

cardiac arrest Asystole CARDIAC PACING Complete cessation of the heart's normal and rhythmic electrical and/or mechanical activity

Cardiac Arrhythmia Suppression Trial See CAST, CAST-2

cardiac arrhythmia CARDIOLOGY Any defect in the heart's electrical activity, which may be detectable only by EKG, or manifest as an abnormality in rate, rhythm, or sequence of cardiac events CLINICAL From asymptomatic to palpitations, syncope, etc

cardiac blood pool imaging CARDIOLOGY A noninvasive test using radionuclides to delineate cardiac chambers and major vessels, ID acute MIs, ASD, VSD, dilated cardiomyopathy, CHF, mitral stenosis, CAD, superior vena cava syndrome

cardiac catheterization CARDIOLOGY A procedure in which a flexible catheter is inserted in a peripheral blood vessel, usually a leg–femoral or arm—antecubital vein, passed through the inferior vena cava and, under fluoroscopic guidance, placed in the region(s) of interest COMPLICATIONS Arrhythmias, embolism–cerebral, pulmonary, MI, pericardial tamponade. See Doppler echocardiography

---

CARDIAC CATHETERIZATION INDICATIONS

Evaluate heart valves and detect stenosis and regurgitation

Determine regional BP and detect pulmonary HTN

Obtain blood samples to evaluate oxygenation of blood

Inject dye and evaluate heart function in 'real time'–cardiac angiography and assess patency of the coronary arteries–coronary angiography.

RIGHT-SIDED CARDIAC CATHETERIZATION Evaluate tricuspid and pulmonary valve function, and measure pressures of and take blood samples from the right atrium and ventricle, and pulmonary artery ABNORMALITIES, RIGHT SIDE: Pulmonary HTN, pulmonary valve stenosis, tricuspid valve stenosis, atrial and ventricular septal defects

LEFT-SIDED CARDIAC CATHETERIZATION Evaluate function of mitral and aortic valves, and coronary arteries ABNORMALITIES, LEFT SIDE Aortic valve regurgitation, CAD–stenosis or occlusion, mitral valve stenosis or regurgitation, ventricular hypertrophy or aneurysm

---

cardiac 'cirrhosis' A hepatopathy characterized by liver cell atrophy, centrilobular necrosis and extensive fibrosis–the end stage is virtually identical to posthepatitis cirrhosis, caused by repeated and/or prolonged CHF with ↑ venous pressure and ↓ hepatic blood flow. See Nutmeg liver

cardiac concussion See Steering wheel injury

cardiac 'cripple' A person whose innocent cardiac murmur or normal physiologic variant of an EKG pattern–eg ST segment deviation, such as an early repolarization was misinterpreted as indicating cardiac disease or failure, whose physical activity is subsequently restricted and/or who receives various cardiac drugs. See Innocent murmurs

cardiac cycle CARDIOLOGY A cycle of pumping–systole and refilling–diastole. See Electrophysiology

cardiac enzymes LAB MEDICINE A group of 3 enzymes–AST, total CK, and LD, once used to diagnose and monitor suspected MI. See β enolase, Cardiac markers, CK-MB, Flipped pattern, Troponin

cardiac event Coronary event CARDIOLOGY Any severe or acute cardiovascular condition including acute MI, unstable angina, or cardiac mortality

cardiac failure See Congestive heart failure, Heart failure, Left heart failure, Right heart failure

cardiac glycoside PHARMACOLOGY A drug that blocks the Na$^+$/K$^+$ pump

cardiac hypertrophy Cardiac enlargement Compensatory enlargement of the

heart, which may be pathologic, due to underlying cardiac disease–eg, CHF, valve disease, HTN, or physiologic, as in athletes. See Athlete's heart syndrome, Congestive heart failure

cardiac index CARDIOLOGY The cardiac output of blood–L/min/m^2 surface area LAB MEDICINE The ratio of CK-MB to total CK, an indicator of myocardial ischemia, and ↑ risk for acute MI. See Cardiac profile guideline, CK-MB, Troponin I

cardiac injury panel An abbreviated battery of parameters for evaluating Pts who may have suffered an acute MI, measuring LD, CK, CK isoenzymes. See Cardiac marker, Organ panel

cardiac insufficiency Heart failure, see there

cardiac marker CARDIOLOGY An analyte measured in serum that reflects myocardial damage or acute cardiovascular disease–eg, acute MI EXAMPLES CK-MB, myoglobin, myosin light chains, troponin. See Cardiac enzymes, CK-MB, Troponin I

cardiac output CARDIOLOGY A measure of blood flow through the heart to the systemic circulation, expressed as volume of blood/unit time or L/min, calculated as left ventricular forward stroke X heart rate; CO is calculated by Fick method–O$_2$ consumption divided by the AV O$_2$ difference, or by thermodilution, with Swan-Ganz catheterization. See Bradycardia, Tachycardia

cardiac pacemaker A device that delivers a small electric shock to the heart to effect cardiac contraction at a pre-determined rate

cardiac profile guideline CARDIOLOGY An algorithm based on clinical–Pt Hx and EKG and lab data–troponin and total CPK, used to decide whether to admit a Pt to observation, if there is a low risk of an acute MI. See Cardiac index

cardiac reserve The ability of the heart to respond to ↑ demand beyond its usual workload

cardiac risk evaluation panel A group of tests used to stratify a person according to the risk for suffering ASHD-related morbidity; CREP measures cholesterol, triglycerides, HDL-cholesterol, glucose. See Organ panel

cardiac tamponade Interference with the venous return of blood to the heart 2° to accumulation of fluids or blood in pericardium, resulting in ↑ mean right atrial pressure and near-equalization with intrapericardiac pressure, which has a wide range of clinical and hemodynamic effects ETIOLOGY 2° to dissecting aneurysm, HTN, post-MI, renal failure, pericarditis, hypothyroidism, autoimmune disease–eg, SLE, chest trauma, CA DIAGNOSIS Echocardiogram MANAGEMENT Pericardiocentesis, ie needle aspiration, pericardial window

cardiogenic *adjective* Originating in the heart

cardiogenic shock Cardiac shock CARDIOLOGY The inability of the heart to deliver sufficient blood–O$_2$ to the tissues to meet resting metabolic demands due to pump failure EPIDEMIOLOGY CS complicates 7–10% cases of acute MI and is the leading cause of death in Pts hospitalized with acute MI; when hemodynamic monitoring is available, CS is defined by a systolic BP < 30 mm Hg–or < 80 mm Hg in absence of hypovolemia, ↑ arteriovenous O$_2$ difference–> 5.5 ml/dL, and ↓ cardiac index–< 2.2 L/min/m^2 body surface, with an ↑ pulmonary capillary wedge pressure–>15 mm Hg ETIOLOGY MI, cardiomyopathy, overwhelming infection, heart attack or disease, hormonal insufficiency, hypoglycemia, hypothermia, allergic reaction, drugs, spinal cord injury CLINICAL Cold extremities, cyanosis, persistent oliguria, CHF MORTALITY > 80% MANAGEMEMT Emergency revascularization–CABG or angioplasty, fluid restriction, diuretics, vasopressors–eg, dopamine to maintain BP; IV agents–eg, dobutamine, to ↑ inotropism

–cardiogram *suffix* A recording of cardiac activity, as in apex cardiogram, Doppler echocardiogram, echocardiogram, electrocardiogram, phonocardiogram, vectorcardiogram

cardiologist Heart specialist An internist specialized in diseases of the heart and cardiovascular system MEAT & POTATOES DISEASES HTN, arrhythmias, chest pain, acute MI SALARY $221K + 13% bonus

cardiology The medical field dedicated to diagnosing and treating diseases of the heart and cardiovascular system. See Interventional cardiology, Nuclear cardiology

cardiomegaly Enlargement of the heart

cardiomyopathy CARDIOLOGY A general term for a disease of the heart, defined by WHO as a *primary disease process of heart muscle in absence of known underlying etiology*. See Alcoholic cardiomyopathy, Beer drinkers cardiomyopathy, Dilated cardiomyopathy, Hypertrophic cardiomyopathy, Idiopathic dilated cardiomyopathy, Ischemic cardiomyopathy, Peripartum cardiomyopathy, Restrictive cardiomyopathy, Secondary cardiomyopathy

cardiopathy Any disease of the heart, due to inflammation, ASHD, fatty degeneration–thiamin deficiency, HTN, renal disease, thyrotoxicosis, toxicity–eg adriamycin, alcohol, cadmium, cobalt, cardiopathy, valve disease

cardioprotection CARDIOLOGY Therapy to prevent heart disease–eg, use of beta blockers to ↓ mortality and recurrent coronary events after acute MI, rather than to treat angina or HTN. See Myocardial infarction

cardiopulmonary bypass CARDIOVASCULAR SURGERY A procedure in which the flow of blood to the heart is diverted to a heart-lung machine–a pump-oxygenator before returning it to the arterial circulation, used in modern open heart surgery; aortic cannulation is used for arterial inflow; a single right atrial cannula is used for venous return to the pump; after the ascending aorta is clamped, cold potassium cardioplegia solution is infused into the aortic root, which arrests and protects the heart during the heart surgery

cardiopulmonary murmur CARDIOLOGY An innocent murmur related to movement of the heart, which disappears when breath is held. See Murmur

cardiopulmonary resuscitation EMERGENCY MEDICINE The restoration of cardiopulmonary function after cardiac arrest COMPONENTS Compression of anterior chest wall to stimulate blood flow through the heart, artificial ventilation–eg, mouth-to-mouth breathing, defibrillation. See ABC method, CAB method

cardiopulmonary sleep study Polysomnography, see there

cardioselective *adjective* Relating to a therapeutic or other effect that is greater on the heart than on another tissue

cardiotonic *adjective* Referring to a tonic effect on the heart *noun* An agent with said effect–eg, digitalis

cardiotoxicity A deleterious effect on the heart. See Cocaine cardiotoxicity

cardiovascular disease Cardiovascular disorder INTERNAL MEDICINE Any of the diseases–ASHD, CAD, cerebrovascular disease, HTN–that primarily impact the heart and blood vessels

cardiovascular exercise SPORTS MEDICINE Any vigorous aerobic exercise, which near-maxes the heart rate–eg, basketball, bicycling, cross-country skiing, dancing, hiking, jogging, race-walking, racquetball, running, skating, soccer, stair-climbing, volleyball. See Aerobic exercise, Exercise, Vigorous exercise

cardiovascular fitness Fitness A benchmark of a subject's cardiovascular and respiratory 'reserve', assessed by exercise testing; improved CF ↓ risk of acute MI. See Aerobic exercise, Exercise, MET, Thallium stress test, Vigorous exercise. Cf Anaerobic exercise

cardiovascular surgery HEART SURGERY An operation for repairing structural defects of the cardiovascular system EXAMPLES CABG, repair of congenital heart defects, varicose veins, aortic aneurysms, ventricular remodeling, transmyocardial

cardioversion INTERVENTIONAL CARDIOLOGY The conversion of a cardiac arrhythmia, usually a tachyarrhythmia to a normal sinus rhythm; CV is most effective in terminating tachycardias due to defective reentry–eg, atrial flutter, A Fib, AV nodal entry, WPW syndrome, V tach with a pulse, flutter; the electric shock depolarizes all excitable myocardium, prolongs refractoriness, interrupts reentry circuits, and establishes electrical homogeneity; it is attempted in Pts with AFib in order to improve cardiac function, relieve symptoms, and ↓ risk of thrombus formation; transesophageal echocardiography identifies Pts with atrial emboli requiring short-term anticoagulation with heparin before cardioversion. See Chemical cardioversion, Direct current cardioversion, Single-pulse cardioversion. Cf Defibrillation

cardiovocal syndrome Hoarseness due to compression of the left laryngeal nerve between the aorta and dilated pulmonary artery, which is often first seen in infancy CLINICAL Sx reflect the underlying cardiac disease–eg, lesions of the aortic arch and congenital cardiac defects, mitral valve stenosis, CAD, HTN

carditis CARDIOLOGY Inflammation of the heart. See Endocarditis, Myocarditis, Pericarditis

Cardizem® Diltiazem, see there

Cardura® Doxazosin mesylate, see there

care MEDTALK Patient-related health services; management. See Alternate level of care, Ambulatory care, Appropriate care, Collaborative care, Comfort care, Complementary cancer care, Cooperative care, Critical care, Custodial care, Emergency care, Episode of care, Foster care, Front-end health care, Futility care, Health care, Home care, Home health care, Inappropriate care, Inpatient care, Intensity of care, Intermediate care, Kangaroo care, Kinship care, Level of care, Managed care, Medicare, Monitored anesthesia care, National health care, Opt-out managed care, Palliative care, Pharmaceutical care, Potentially ineffective care, Preventive care, Primary care, Quality of care, Referred care, Replicare care, Respite care, Scar care, Secondary care, Skilled nursing care, Supportive care, Tertiary care, Uncompensated care, Unproven care, Urgent care, Wound care

CARE A series of clinical trials CARDIOVASCULAR DISEASE 1. Calcium Antagonist in Reperfusion Trial. See Reperfusion injury 2. Carvedilol Atherectomy Restenosis trial 3. A clinical trial–Cholesterol And Recurrent Events Trial–of the potential benefits of lipid-lowering agents–eg pravastatin on M&M in normocholesterolemic ♂ survivors of acute MIs. See Calgary-CARE, Lipid-lowering therapy, Pravastatin. Cf 4S, WOSCOPS

care coordination MANAGED CARE 1. The brokering of services for Pts to ensure that needs are met and services are not duplicated by the organizations involved in providing care 2. The process in which an HMO's gatekeeper–often a primary care physician, who sends a Pt for specialist services, continues to manage the Pt, assure followup, and continuity of care

care proxy Power of attorney for health-care decision-making CHOICE IN CARE/DYING A form of advance medical directive–AMD, in which the person designates another to make medical decisions in the event that he/she becomes too incapacitated to make such decisions. See Advance medical directives, Living will

care team MANAGED CARE A gallimaufry of HCWs–eg physicians, physician extenders, nurses, medical assistants and those providing ancillary and diagnostic services–eg, radiology and lab technologists, physical therapists, nutritionists, psychotherapists, massage therapists. See Managed care

CARE-HD NEUROLOGY A clinical trial–Coenzyme Q10 And Remacemide: Evaluation in Huntington's Disease

career adviser GRADUATE EDUCATION An academic 'guru'–eg, department, junior faculty member, or other at a medical school, charged with guiding a medical student in his/her choice of specialty. See Internship–the Match

caregiver HEALTH CARE The person–eg, a family member or a designated HCW–who cares for a Pt with Alzheimer's disease, other form of dementia or chronic debilitating disease requiring provision of nonmedical protective and supportive care

caries DENTISTRY Cavity, tooth decay The destruction of tooth enamel and dentin, which is linked to infection by *Streptococcus mutans* and microaerophilic organisms that thrive when protected by a layer of hardened dental plaque; caries is most common in the young with refined carbohydrate-rich diets, especially in 'snackers' who have ↑ oral pH; caries may affect older Pts with DM, CA, or immunodeficiencies. See Fluoridation, Periodontal disease, Plaque, Tartar

CARISA CARDIOLOGY A clinical trial–Combination Assessment of Ranolazine In Stable Angina

carisoprodol Rela® PHARMACOLOGY A muscle relaxant related to meprobamate. See Meprobamate

CARMEN Cryoablative reduction of menstruation GYNECOLOGY A procedure for treating abnormal uterine bleeding

carmustine BCNU ONCOLOGY An anticancer alkylating nitrosourea

carnal *adjective* Referring to the flesh, to baser instincts, often referring to sexual "knowledge"

Carney's complex CARDIOLOGY A variable clinical complex of mucocutaneous, visceral, and endocrine disorders, defined by ≥ 2 of the following

---

CARNEY'S COMPLEX

CARDIAC MYXOMA

CUTANEOUS MYXOMA

MAMMARY MYXOID FIBROMA

SPOTTY MUCOCUTANEOUS PIGMENTATION Lentigines, blue, junctional or compound nevi

PRIMARY PIGMENTED NODULAR ADRENOCORTICAL DISEASE

TESTICULAR TUMOR–large cell calcifying Sertoli cell tumors, often bilateral/multicentric

PITUITARY hGH-SECRETING TUMOR–Acromegaly or gigantism (MCP 1986; 61:165)

---

carnitine deficiency METABOLIC DISEASE A condition caused by ↓ carnitine palmityl transferase, failure of carnitine transport in kidney, muscle, fibroblasts or cobalamin deficiency, electron transfer flavoprotein deficiency, Fanconi syndrome, isovaleric acidemia, medium chain acylCoA dehydrogenase deficiency, methylmalonic and propionic acidemias; Pts receiving valproate for seizure disorders may develop a toxicity syndrome with associated carnitine deficiency CLINICAL Myoglobinuria, renal failure, hypoglycemia, hypotonia, hepatomegaly, hepatic coma, CHF, neurologic changes–progressive myasthenia, encephalopathy, lethargy, coma, death, cardiomegaly, cardiac arrest, impaired growth and development

carnivore A meat-eater. Cf Omnivore, Vegetarian

carnosinase deficiency Carnosinemia An AR condition characterized by severe psychomotor retardation, myotonic and grand mal seizures TREATMENT Low-protein diet

Caroli's disease HEPATOLOGY A condition characterized by a diffuse distribution of intrahepatic biliary cysts and other cystic lesions–eg, medullary sponge kidneys CLINICAL Childhood onset–eg, jaundice, episodic fever, pain

COMPLICATIONS Hepatic fibrosis, cirrhosis, portal HTN, esophageal bleeding, death in childhood

carotenemia Transient yellowing of skin due to excess dietary carotene, seen in infants fed too much carrots or adults consuming mucho carrots or beta carotene

carotenoid NUTRITION A vitamin A precursor with antioxidant activity; although beta carotene is the best known of the group, 600 carotenoids have been identified; 40 are common in fruits and vegetables; high carotenoid consumption is associated with ↓ risk of bladder, colon, lung, skin CAs and growth of CA cells. See Beta carotene, Vitamin A

carotid artery occlusion Subclavian steal syndrome, see there

carotid bruit CARDIOLOGY A systolic murmur heard at the base of the neck–over a carotid artery evoked by turbulence 2° to intravascular stenosis. See Stroke

carotid endarterectomy NEUROLOGY Removal of atherosclerotic plaque by "scraping" the vascular intima of the carotid arteries to ↓ risk of CVAs and TIAs. See Stroke, Transient ischemic attack

carotid sinus syncope Carotid sinus syncope CARDIOLOGY A syncope resulting from overstimulation of the carotid sinus, either spontaneous or after by pressure on the carotid sinus CLINICAL Convulsions

carotid sinus syndrome Carotid sinus syncope, see there

carotid stenosis CARDIOVASCULAR DISEASE The partial occlusion of one or both carotid arteries, which is linked to an ↑ risk of strokes & CVAs. See Stroke

carotid ultrasound IMAGING The use of ultrasound to evaluate blood flow through the carotid arteries in the neck in Pts at risk of CVAs or TIAs

carpal tunnel release SURGERY Relief of pressure on median nerve entrapped in the carpal tunnel by incision or endoscopic repair

carpal tunnel syndrome Median nerve dysfunction, pinched nerve syndrome OCCUPATIONAL MEDICINE A syndrome of painful paresthesiae of the median nerve, affecting hands and fingers, in particular index, middle and ring fingers, 2° to fibrous adhesions in the CT; it is most common in ♀; if prolonged, the CTS may result in partial atrophy of the lateral half of the thenar eminence with weakness of abduction and opposition of the thumb; it is either a chronic idiopathic flexor tenosynovitis, repetitive stress syndrome, or part of systemic conditions–eg, acromegaly, amyloidosis, DM, granulomatous disease, hypothyroidism, mucopolysaccharidosis I-S, myxedema, obesity, pregnancy, or local processes, including ganglion cyst, bone dislocations, lipoma, callus formation in wrist fractures–eg, Colles or Smith fractures, gout, pseudogout, and rheumatoid arthritis causing flexor tenosynovitis EXAMINATION Phalen sign, Tinel sign CLINICAL Most symptomatic at night, hypesthesia of median nerve region, paresthesias, thenar muscle atrophy, slowed nerve conduction and EMG studies; the median nerve is a 'mixed' nerve and sensory loss precedes motor dysfunction, affecting sensation in the palmar aspect of the radial 3½ fingers DIFFDx Proximal sites of nerve entrapment–eg, Pancoast tumor, pronator teres syndrome TREATMENT, EARLY Wrist splint, NSAIDs, corticosteroid injection–eg, β-methasone, yoga TREATMENT, LATE Longitudinal section of epineurium and flexor retinaculum for thenar atrophy DIAGNOSIS EMG is the 'gold standard' test, which detects ↓ conduction velocity

Carpentier-Edwards® valve HEART SURGERY A bioprosthetic heart valve designed for small hearts without requiring post-op anticoagulation. See Heart valve

carpopedal spasm MEDICINE Spasms of hands and feet due to ↓ Ca²⁺ or hyperventilation

CARPORT CARDIOLOGY A clinical trial–Coronary Artery Restenosis Prevention on Repeated Thromboxane-Antagonism Study that evaluated thromboxane A2-receptor blockade in preventing restenosis after PCTA in Pts with CAD. See Angioplasty, Coronary artery disease, Percutaneous transluminal angioplasty

carrier EPIDEMIOLOGY A person or animal without apparent disease who harbors a specific pathogen can transmit it to others; the carrier state may occur in a person with an asymptomatic infection–asymptomatic carrier, or during the incubation period, convalescence, and postconvalescence of a person with clinically recognizable disease; the carrier state may be of short or long duration–transient or chronic. See Latent carrier, Silent carrier GENETICS A state in which a person has a gene known to be linked to a particular condition, who does not manifest the disease; in humans the classic carrier state is that of a ♀ with a defective gene on the X chromosome, which does not manifest itself in ♀ with 2 X chromosomes–one of which is presumed to be normal–for a particular condition INFECTIOUS DISEASE A person infected with a bug, who can act as a 'vector' and transmit the infection to others but is asymptomatic TYPES Silent carriers–eg with TB, retain infectiousness; latent carriers–eg those with HSV are *not* infectious. See Typhoid Mary MANAGED CARE 1. An organization–eg, an insurance company, with an HCFA contract to administer claims processing and make Medicare payments to health care providers for Medicare Part B benefits. See Fiscal Intermediary, Part B 2. A private contractor that administers claims processing and payment for Medicare Part B services. See Supplementary Medical Insurance OBSTETRICS A surrogate mother who is carrying a gestational product to term. See Gestational carrier PHARMACOLOGY A peptide, protein, or other substance that binds to a therapeutic agent, and transports it in the circulation. See Vehicle

carrier test A direct test–detects a mutant gene–eg, Tay-Sachs gene, or indirect test–detects a specific effect–eg, sickle cell or thalassemia trait designed to identify carriers of a gene for an X-linked or AR disorder

cart Cot, gurney, stretcher, trolley MEDTALK A narrow movable bed used to transport Pts or maintain them in a horizontal position in emergency or critical care settings VOX POPULI A non-self-propelled multiwheel structure designed to carry a load. See Crash cart, Shopping cart

carteolol A β-blocker used to treat HTN. See Beta-blocker

Cartia® Diltiazem HCl CARDIOLOGY A generic Cardizem® used for angina and HTN. See Angina, Hypertension

Carticel® Cartilage-cell culturing service ORTHOPEDICS A technique in which a Pt's own cartilage cells are removed, grown, reimplanted to fill a cartilage defect of the knee and, at the surgeon's discretion, glued with fibrin glue for watertightness. See Cell culture

cartilage-hair hypoplasia syndrome Metaphyseal chondrodysplasia, McKusick type An AR form of short-limbed dwarfism, common in the Amish, of German descent CLINICAL Bone dysplasia, bradycarpia, redundant skin, sparse hair, defective pigment, malabsorption; death by age 20 IMMUNE DEFECTS Neutropenia with defective T-cell immunity, potentially fatal vaccinia, progressive vaccine-related poliomyelitis, defective antibody production and severe combined immunodeficiency, related to defective adenosine deaminase activity TREATMENT Antibiotics, IFN, BM transplantation. See Adenosine deaminase deficiency

cartoon MEDICAL COMMUNICATION A schematic diagram or illustration, usually in color, representation of a biological process or molecule, presented either in a journal or in a lecture. Cf 'Nettergram'

caruncle ANATOMY A fleshy bump of mucocutaneous tissue–eg, the carunculae hymenales, the residua of a torn hymen; carucula lacrimalis, the reddish bump on the medial angle of the eye; carunculae sublinguaes, the bumps on either side of the frenulum, at the tops of which are the orifices of the sublingual salivary gland

carve-out MANAGED CARE *adjective* Referring to 1. A program that excludes certain services—usually from an organization's capitated rate—and tends to focus on one disease in depth 2. An arrangement in which some benefits–eg mental health are removed from coverage provided by an insurance plan, but are provided through a contract with a separate set of providers 3. A population subgroup for whom separate health care arrangements are made

carve-out coverage MANAGED CARE The provision of certain benefits–'coverage' by an insurance plan through a contract with a separate set of providers. See Coverage

carving out MANAGED CARE *adjective* Referring to the practice of allowing healthy persons in small employer groups to buy lower cost health insurance policies, while workers who are sicker must buy more expensive high-risk pool coverage

cascade PHYSIOLOGY A molecular system capable of self-propagation or amplification, especially of a weak signal; once a cascade is initiated, it may continue to be amplified through positive feedback loops and pathways, until down-regulated by local mechanisms–eg, by proteolytic enzymes. See Complement cascade, Ischemic cascade, Metastatic cascade

CASCADE CARDIOLOGY A clinical trial–Cardiac Arrest in Seattle: Conventional vs Amiodarone Drug Evaluation–that compared conventional therapy vs amiodarone, an antiarrhythmic on M&M in survivors of refractory out-of-hospital ventricular arrhythmias. See Amiodarone, Ventricular arrhythmia. Cf BASIS

cascade testing LAB MEDICINE The sequencing of tests to diagnose a disease or process–eg, fetal lung maturity; CT is appropriate when a 'gold standard' method is technically demanding and/or costly and the diagnosis can usually be established by simpler or more cost-effective strategies. Cf Reflex testing

CASE NEUROLOGY A clinical trial–Canadian Activase for Stroke Effectiveness. See Thrombolytic therapy

case EPIDEMIOLOGY A countable instance in the population or study group of a particular disease, health disorder, or condition under investigation; sometimes, an individual with the particular disease. See Reportable case FORENSIC MEDICINE A civil or criminal action or event. See Ayala case, Detroit case, Helga Wanglie case, Index case, LifeNet case, Newport Hospital case, Pappas case MEDTALK A popular term for a person who presents with a particular set of findings. See Good intern case, Great case VOX POPULI A person who is unique or special, as in "he's a case". See Basket case

case definition EPIDEMIOLOGY A set of standard criteria for diagnosing a particular disease or health-related condition, by specifying clinical criteria and limitations on time, place, and person

case-fatality ratio EPIDEMIOLOGY A value calculated as 100 cases of a disease 'X', divided by the number of persons with the disease who died in a given period of time; the resulting ratio is equal to the rate of a disease's occurrence. See Cause-fatality ratio

case management MANAGED CARE The creation of a coordinated, ongoing and personalized strategy for Pts with a variety of health care needs–eg, the elderly and those with long-term illnesses; primary care physicians act as case managers, planning specialist referrals and provide continuity within the separate services delivered. See Disease management SOCIAL MEDICINE A service that assists clients to obtain and coordinate community resources such as income assistance, education, housing, medical care, treatment, vocational preparation, and recreation

case mix MANAGED CARE The characteristics–age, gender and health status–of the population served by a health system or physician's office in a given period of time, which are classified by disease, diagnostic or therapeutic procedures performed, method of payment, duration of hospitalization, and intensity and type of services provided; in the US, a hospital's CM is based on the diagnosis-related groups. See Demographics, DRGs

case rate INFECTIOUS DISEASE The number of cases of a particular infection or exposure during a unit of time, divided by the population during that period; CRs are often expressed in terms of a population of 100,000

case study Anecdotal report, single case report EPIDEMIOLOGY An uncontrolled–prospective or retrospective observational study involving an intervention and outcome in a single Pt

case-to-infection ratio Case-to-infection proportion EPIDEMIOLOGY The number of cases of clinical disease in humans compared to the number of infections with the agent that causes the disease in humans

CASH INTERVENTIONAL CARDIOLOGY A clinical trial–Cardiac Arrest Study of Hamburg that compared the effect of device–implantable defibrillator therapy vs the best medical therapy for survivors of MI OB/GYN A clinical trial–Cancer and Steroid Hormone study which evaluated the effects of prolonged use of OCs on the risk of breast cancer. See Breast cancer

cash out MEDICAL PRACTICE A popular term referring to the translation of intangible and/or inaccessible assests into cash at current prices, as may occur when a physician sells his office practice VOX POPULI To sell any interest in an enterprise

CASS CARDIOLOGY, CARDIOVASCULAR SURGERY A randomized, open label, multicenter trial that compared the outcomes of CABG vs. medical therapy on M&M in Pts with coronary artery disease after an MI. See Angina, CABG, Silent ischemia

Cassadan Alprazolam, see there

cast MEDTALK A mold of formed structure or wall. See Decidual cast NEPHROLOGY A translucent proteinaceous mold of the renal tubules seen in the urine; casts are ↑↑↑ in renal disease, and the types are crude indicators of the type of renal disease; the matrix of all casts is composed of albumin, small–< 50 kD globulins and glycoprotein secreted by Henle's ascending loop and the distal tubule, forming Tamm-Horsfall protein, aka uromodulin. See Granular cast, Telescoped cast, Waxy cast ORTHOPEDICS A temporary plaster or fiberglass device used to immobilize a Fx. See SPICA cast, Walking cast

CAST CARDIOLOGY A clinical trial that examined the effects of suppressing asymptomatic/mildly symptomatic ventricular ectopic beats and arrhythmias in post-MI Pts with CAD using encainide and flecainide. See Arrhythmias, Encainide, Flecainide. Cf BASIS, CASCADE, CAST II, Moricizine

CAST 2 CARDIOLOGY A trial–Cardiac Arrhythmia Suppression Trial–of the effect of moricizine, an antiarrhythmic, on ventricular arrhythmia in Pts with CAD; moricizine ↑ mortality and was withdrawn from the market. See CAST, Moricizine. Cf BASIS, CASCADE, Encainide, Flecainide

castigophobia PSYCHOLOGY Fear of punishment. See Phobia

CASTLE CARDIOLOGY A clinical trial–Candesartan Amlodipine Study of Tolerability & Efficacy

Castleman's disease A heterogeneous group of idiopathic, multicentric lymphoproliferative disorders primarily of young adults CLINICAL Hepatosplenomegaly, lymphadenopathy LAB ↑ IL-6 in serum and tissues, anemia, hypoalbuminemia, hypergammaglobulinemia MANAGEMENT Surgical resection

castor oil An oil cold-pressed from the kernel of *Ricinus communis* seeds, which contains glycerides of ricinoleic and isoricinoleic acids–eg, dihydroxystearin, isoricinolein, palmitin, and triricinolein; it has been used externally as an emollient and internally as a laxative. See Castor bean, Ricin

castration The removal of gonads; castration occurs in various contexts: 1. Forensic–a controversial procedure intended to ↓ sexual drives in repeat sex offenders 2. Pathologic–due to infection, which renders both testes functionally inactive–eg, with filariasis in elephantiasis 3. Therapeutic–to remove the source of androgens in Pts with androgen-sensitive prostate CA–which can also be accomplished by RT. See Chemical castration, Medical castration, Surgical castration PSYCHIATRY The fantasized loss of the genitals; a mental state of impotence, powerlessness, helplessness, or defeat UROLOGY Surgical removal of testes in ♂ with androgen-dependent prostate CA

castration anxiety PSYCHIATRY Anxiety due to fantasized danger or injuries to the genitals and/or body, precipitated by everyday events with symbolic significance which appear threatening, such as loss of a job, loss of a tooth, or an experience of ridicule or humiliation

casualty A victim of an accident or mass disaster. See Mass disaster

CAT 1. Catalase 2. Catalyst 3. Catecholamine 4. Chloramphenicol acetyl transferase, a 'reporter' gene 5. Choline acetyl transferase 6. Cholesterol acyl transferase 7. Computerized axial tomography, see there

CAT fund Catastrophic liability fund MALPRACTICE A professional liability fund that exists in some states–eg Pennsylvania, into which physicians and hospitals pay for malpractice insurance

CAT scan Computer tomography, see there

cat *Felis catus* A mammal of medical interest that is a model for some human diseases, eg, dermatosparasix, and a vector for bacteria, fungi, and parasites

CATS, IMPACT ON MEDICINE

MODEL SYSTEMS-human diseases, eg dermatosparasix, a defect in converting type I procollagen to collagen; mannosidosis–affects shorthair cats; Niemann-Pick disease, type I–affects Siamese cats

VECTORS FOR DISEASE

- **BACTERIA** *Bartonella (Rochalimaea) henselae, Bergeyella (Weeksella) zoohelcum, Brucella suis*–anthrax, *Campylobacter jejuni, Capnocytophaga canimorsus*, CDC group NO-1, *Chlamydia psittaci*–feline strain, *Dipylidium caninum, Francisella tularensis, Neisseria canis, Pasteurella multocida*, Q-fever, *Rickettsia felis*, salmonellosis, *Yersinia pestis*–plague, *Yersinia pseudotuberculosis*
- **FUNGI**, eg *Microsporum canis*–dermatophytosis, *Sporothrix schenckii*
- **PARASITES**, eg *Ancylostoma braziliense, A caninum, Brugia pahangi*, Clonorchis sinensis, Cryptosporidium*, cutaneous larva migrans, *Dipylidium caninum, Dracunculiasis medinensis*, Echinococcus vogeli, E multilocularis, Gnathostoma spinigerum, Isospora belli, Leptospira* spp–leptospirosis, *Opistorchis felineus, Sarcoptes scabiei*–scabies, *Toxoplasma gondii, Trypanosoma cruzi**, Trichinosis, visceral larva migrans, *Wuchereria bancrofti*. See Cat scratch disease
- **VIRUSES** Cowpox, poxvirus, rabies medicine.bu.edu/dshapiro/zoocat.htm

ALLERGIES Some individuals are highly allergic to cats, which is attributed to the Fel dl antigen, see there

*Parasites that have part of their life cycle in humans

cat scratch disease Cat scratch fever A self-limited regional lymphadenitis of children/adolescents, caused by close contact with or being scratched by household pets; 95% are due to cats–especially kittens, 5% to dogs EPIDEMIOLOGY ± 22,000/yr–US 2,000 hospitalizations; most cases occur in those under age 20, more common in ♂, often in fall and winter; fleas may be vectors CLINICAL Erythematous papules at the inoculation sites, eg hands and forearms, anorexia, malaise, fever, parotid swelling, maculopapular rashes, lymphadenopathy, splenomegaly, encephalopathy DIAGNOSIS Hx, indirect fluorescence for antibodies to *Rochalimaea henselae* TREATMENT Gentamycin, ciprofloxacin. Cf Bacillary angiomatosis,

CATS CLINICAL TRIALS 1. A randomized, double-blind, placebo-controlled clinical trial–Canadian-American Ticlopidine Study which assessed the effect of ticlopidine in ↓ CVAs, MIs, or vascular death in Pts with thromboembolic strokes 2. A clinical trial–Captopril And Thrombolysis Study

cats, fear of PSYCHOLOGY Aclurophobia, ailurophobia, elurophobia, felinophobia, galeophobia, gatophobia. See Phobia

cat's elbow Lichen planus actinicus DERMATOLOGY A type of lichen planus, seen in the Middle East, on sun-exposed parts, especially the face CLINICAL Annular lesions with pigmented centers, centrally thinned epidermis, well-demarcated pale, raised margins

catabolic disease Catabolic illness METABOLIC DISEASE A condition characterized by rapid weight loss and loss of fat and skeletal muscle mass, which may occur in a background of either an acute, self-limited disease–eg, injury, infection or a chronic condition–eg DKA, multisystem organ failure, AIDS, advanced CA, chemotherapy, RT ASSOCIATED CLINICAL EVENTS Immunosuppression, muscle weakness, predisposition to PTE, thrombophlebitis, altered stress response MANAGEMENT Aggressive nutritional support in CI may not prevent protein loss; specialized nutrition–eg, glutamine, arginine, branched-chain amino acids, n-3–omega-3 fatty acids, growth hormone, possibly IGF-I and EGF, may ↓ protein loss, hospitalization

catabolic pathway A series of metabolic reactions leading to breakdown of a complex organic molecule to a simpler ones, with release of energy. Cf Anabolic pathway

catalepsy PSYCHIATRY A state of ↓ responsiveness with a trancelike states, which occurs in organic or psychologic disorders, or under hypnosis

cataplexy NEUROLOGY An abrupt ↓/loss of muscle tone either limited to muscle groups, or generalized, leading to muscle weakness, paralysis or postural collapse; cataplexy in an awake person is pathognomonic of narcolepsy, and is triggered by emotional stimuli or stress, which may cause knee-buckling; cataplectic attacks are dangerous for machinists, house painters, construction workers MANAGEMENT Imipramine, protripyline, IMAOs. See Narcolepsy

cataract Lens opacity OPHTHALMOLOGY Partial or complete opacification of the ocular lens or capsule, which impairs vision or causes blindness CLASSIFICATION Morphology–size, shape, location; etiology–eg DM, corticosteroids, trauma, und so weiter; time of occurrence–eg in elderly CLINICAL Double or blurred vision; ↑ sensitivity to light, glare DIAGNOSIS Slit lamp microscopy TREATMENT Phacoemulsion, surgery. See Black cataract, Congenital cataract, Snowflake cataract, Sunflower cataract

cataract surgery Cataract extraction EYE SURGERY A procedure that restores function and improves visual acuity in ±90% of Pts with cataracts. See Phacoemulsion, Posterior capsulotomy

catarrh INFECTIOUS DISEASE Inflammation of the nasopharyngeal mucosa with fluid discharge

catastrophe policy Major medical, see there

catastrophic illness A morbid condition that results in health care costs that exceed a person's income, or which compromise financial independence, reducing him/her to subsistence or near-poverty levels; CIs are usually life-threatening and may leave significant residual disability–eg, AIDS, major burns, trauma with residual paralysis or coma, and terminal CA

catastrophic limit MANAGED CARE A ceiling–eg $1000 on the amount of money that a person must pay out-of-pocket for the health care expenses incurred by a catastrophic illness–eg AIDS, burns, CA, MVA, etc, before the insurer pays bills

catatonia NEUROLOGY A psychomotor disturbance characterized by periodic muscle rigidity, fixed posture, inability to move or talk and unresponsiveness, finding typical of schizophrenic disorders. See Cataplexy, Schizophrenia

catchment area MEDTALK A region served by a health care facility or health plan, and delineated by population distribution, geography, or transportation patterns. See Demographics

catecholamine ENDOCRINOLOGY A biogenic amine from tyramine/phenylalanine which contains a catechol nucleus EXAMPLES Epinephrine–adrenaline in UK, norepinephrine–noradrenaline and dopamine, which act as hormones and neurotransmitter in the peripheral and central nervous system; catecholamines are produced by sympathetic nervous system activation ACTIVITY Autonomic arousal, fight-or-flight stress response, reward response. See Biogenic amine, Dopamine, Epinephrine, Indolamine, Norepinephrine

category X drug PHARMACOLOGY A therapeutic with known teratogenicity which is contraindicated for use during pregnancy–eg, vitamin A congeners–for severe acne–which cause a characteristic syndrome in neonates. See Vitamin A embryopathy

catgut SURGERY An absorbable suture material from the submucosa of bovine intestine. Cf Silk

catharsis Cathartic method PSYCHIATRY Any psychoanalytic technique in which the client is led to recognize the underlying basis for underlying mental issues, and release associated suppressed or forgotten emotions by talking them out; catharsis is integral to primal therapy and Reichian therapy. See Primal therapy, Psychoanalysis, Reichian therapy, Repression

cathartic adjective Referring to 1. Catharsis, see there 2. An evacuative therapeutic agent MAINSTREAM MEDICINE noun A purgative, laxative

cathartic colon GI DISEASE A colon affected by long-term abuse of cathartics, laxatives CLINICAL Intractable diarrhea with hypokalemia, protein-losing enteropathy, cachexia, hypogammaglobulinemia, finger-clubbing, and potentially ↑ constipation, with vicious cycle of ↑ use of purgatives IMAGING Barium studies may be normal, have complete loss of haustral markings, or mimic ulcerative colitis

catheter CLINICAL PRACTICE A thin, flexible tube placed in a vascular or other lumen–eg, bladder, and used to deliver fluids, nutrients or therapeutic, withdraw fluids, or obtain samples–eg, of blood, urine; the ideal IV catheter is made from a material that both inhibits the formation of thrombi–Silastic — and resists adherence of microorganisms–Teflon/polytetrafluoroethylene, or polyurethane. See Central venous catheter, Dilatation catheter, Dual-lumen catheter, Foley catheter, Groshong catheter, Guardian™ catheter, Hickman catheter, Indwelling catheter, Intrauterine pressure catheter, Lynx over-the-wire balloon catheter, Nitrofuran delivery catheter, PTCA balloon catheter, Pigtail catheter, Swan-Ganz catheter. Cf Phlebotomy

catheter ablation therapy CARDIOLOGY A nonsurgical technique in which an electrode catheter is introduced into the veins of the arms or legs and advanced into the heart; if an abnormal electric pathway of depolarization is identified in the conduction system–as is commonly the case with SVT or WPW syndrome–it is 'zapped' with a high intensity–> 25 joules radiofrequency wave to neutralize it

catheter-associated UTI Catheter-associated urinary tract infection UROLOGY A UTI linked to use of a catheter–1% to 5% of Pts develop a UTI after a single catheterization, 100% after an indwelling catheter is in place for 4+ days AGENTS E coli, Proteus, enterococcus, Pseudomonas, Enterobacter, Serratia, Candida. See Urinary tract infection

catheterization The placement of a catheter in a lumen. See Right heart catheterization, Subclavian vein catheterization, Swan-Ganz catheterization

cathode ray tube See Video display unit

cation A positively-charged ion–eg, Na⁺, K⁺, Ca²⁺, Mg²⁺

cattle PUBLIC HEALTH Ruminants–most commonly, cows–bred for food; cows are important carriers of various pathogens Actinomyces pyogenes, anthrax, bovine spongiform encephalitis, brucellosis, campylobacteriosis, cowpox, cryptosporidiosis, *E coli* O157:H7, European tick-borne encephalitis, foot & mouth disease, giardiasis, leptospirosis, *Mycobacterium bovis*, pseudocowpox, Q-fever, rabies, salmonellosis, *Streptococcus zooepidemicus*, *Taenia saginata*, *Yersinia enterocolitica*

Caucasian POPULATION BIOLOGY *adjective* White, usually understood to be of non-Hispanic European descent. Cf Asian, Black, Hispanic

cauda equina syndrome Acute cauda equina syndrome NEUROSURGERY A condition caused by compression of multiple lumbosacral nerve roots in the spinal canal due to an abrupt prolapse of the lumbar disk CLINICAL *CES is a medical emergency* characterized by bilateral sciatica in the lower back and upper buttocks, saddle anesthesia, urinary retention, bowel dysfunction DIAGNOSIS Myelography, CT, MRI MANAGEMENT Standard laminotomy-diskectomy–98% success; microsurgical diskectomy–86%, percutaneous diskectomy–67%

caudal block ANESTHESIOLOGY Extradural anesthesia which is administered as a single injection or as a continuous drip; CBs are technically demanding and require more anesthetic than other forms of locoregional anesthesia. See Block

cauliflower ear SPORTS MEDICINE An external ear–pinnae deformity caused by inadequate treatment– through-and-through stitches–of an otohematoma, with interruption of contact between the perichondrium and the underlying cartilage, resulting in a thickened ear ETIOLOGY Contact sports–wrestling, boxing-related trauma in adolescents, or falls in the elderly and children, the natural evolution of hematomas in this site is towards granulation tissue and exuberant fibrosis; primary CE occurs in polychondritis

causal factor MEDTALK A factor linked to the causation of a disease or health problem

causalgia Caulsalgia syndrome PAIN MEDICINE A sensation of persistent severe burning pain of either organic–direct or indirect trauma to a sensory nerve, accompanied by trophic changes–or psychologic origin. See Headache, Somatiform disorder

causation Cause & effect LAW & MEDICINE 1. In the context of disability evaluation, where a particular condition might be linked to the workplace; medical definition of causation requires valid scientific proof; legal definition requires either a probability of > 50% or that the event was more likely than not to be causative. See Pulmonary function test MALPRACTICE The establishment of a cause-and-effect relationship between an allegedly negligent act and the purported injuries. See Malpractice, Negligence

cause LAW & MEDICINE That which creates a condition or results in an effect. See Immediate cause of death, Necessary cause, Proximate cause, Sufficient cause, Underlying cause of death

cause of cancer ONCOLOGY Any factor etiologically linked to the development of malignancy; > ½ of CAs are linked to external factors–eg, sun/skin CA; smoking/lung CA; sexual activity/cervical CA; uncircumcised penis/penile CA

cause of death FORENSIC MEDICINE The reason or event that precipitates death TYPES Proximate COD, immediate COD STATISTICS-'TOP TEN' US–ASHD/HTN, CA, CVA/stroke, COPD, accidents–falls, fires, drowning, etc, pneumonia & influenza, DM, HIV, suicide, chronic liver disease & cirrhosis 'TOP TEN' WORLDWIDE: CAD, acute RTIs in children, CVA/stroke, diarrhea in children, COPD, TB, malaria, accidents–falls, fires, drowning, measles, other heart disease. See Death, Mechanism of death, World health

CAUSE OF DEATH

PROXIMATE COD The most important, immediate, direct or actual cause, or last event or act that occurred before the chain of events leading to death

IMMEDIATE COD The concluding or final event that actually produces death; other CODs include natural causes, HIV risk factors, injury and poisoning, dementia, periprocedural deaths associated with medical treatment or diagnostic/therapeutic procedures and devices, perinatal death (CAP Today 12/95, p57)

cause-specific mortality rate EPIDEMIOLOGY The mortality rate from a specified cause for a population; the numerator is the number of deaths attributed to a specific cause during a specified time interval; the denominator is the size of the population at the midpoint of the time interval

cauterization Diathermy, electrodiathermy The use of heat to destroy tissues or close minute bleeding vessels METHODS Thermal, electrical, cryocauterization, laser, chemicals

cauterize *verb* To stop bleeding with high heat or a chemical–eg, silver nitrate

cavalry fracture A crushing plantar-flexion fracture of the tarso-metatarsal joint, which occurs when a violent force is applied to the heel along the axis of the foot when the toe is fixed; tarso-metatarsal fractures are associated with MVAs, stepladders, or misguided steps. Cf Aviator's astragalus

CAVEAT CARDIOLOGY A trial–Coronary Angioplasty Versus Excisional Atherectomy Trial–that compared PCTA vs. atherectomy outcomes for managing Pts with CAD. See Angioplasty, Atherectomy, BOAT, Coronary artery disease

CAVEAT II CARDIOLOGY A trial–Coronary Angioplasty Versus Excisional Atherectomy Trial–that compared outcomes of PCTA vs. directed coronary atherectomy for managing CAD. See Angioplasty, Atherectomy, Coronary artery disease

CaverMap™ UROLOGY A device that maps and locates the cavernosal nerve filaments during radical prostatectomy, helping preserve the nerve bundles, when feasible, in an attempt to maintain Pt potency and continence. See Prostate cancer, Radical prostatectomy

cavernous hemangioma Cavernoma, stork bites, strawberry mark DERMATOLOGY A benign, painless, red-purple vascular skin lesion that develops after birth and usually disappears in early childhood MANAGEMENT Local steroid injections may ↓ size; surgery

cavernous sinus syndrome NEUROLOGY A condition caused by masses in the cavernous sinus and/or parasellar region resulting in pressure on the 3rd, 4th, 6th, and part of the 5th cranial nerves CLINICAL Oculomotor nerve paralysis–ophthalmoplegia, lid edema, ptosis, mydriasis, and anesthesia of the eyeball, characterized by loss of the corneal reflex in the ophthalmic branch of the trigeminal nerve, often accompanied by blindness and cortical analgesia DIAGNOSIS CT, MRI. See Tolosa-Hunt syndrome

CAVH Continuous arteriovenous hemofiltration, see there

cavity BALLISTICS The permanent space created along a bullet's trajectory caused by the bullet per se. See Ballistics DENTISTRY Caries A hole in a tooth that may be confined to the enamel, or penetrate into the dentin and pulp CLINICAL If deep, pain and ↑ sensitivity to changes in temperature TUBERCULOSIS A hollow space in the lung, visible on a CXR, which may contain a gazillion tuberculous bacilli, especially in persons with severe pulmonary TB

cavovalgus Talipes cavovalgus ORTHOPEDICS A type of clubfoot, characterized by an overly high longitudinal arch and outwardly turned heel. See Clubfoot

cavovarus Talipes cavovarus ORTHOPEDICS A type of clubfoot characterized by an overly high longitudinal arch and an inwardly turned heel. See Clubfoot

CBC Complete blood count LAB MEDICINE The automated analysis of certain parameters of the cells in the circulation, as performed by an automated cell counter–eg RBCs, WBCs, platelets, Hb, Hct, mean corpuscular Hct, Hb, RBC volume and WBC differential count, dividing them into lymphocytes, monocytes, granulocytes–neutrophils, basophils, eosinophils. See Platelet count, Red cell count, White cell count

CC Complications, see there

cc Cubic centimeter(s)

CCB Calcium channel blocker, see there

CCD TELEMEDICINE A semiconducting device–charge-coupled device in a solid-state camera or scanner which contains photosensitive (optical) cells; each pixel of an image requires a cell–the more CCDs in a scanner, the higher an image's resolution. See CCD camera

C/C/E Cyanosis, clubbing, edema An abbreviation used in physical examination of the extremities, which is a crude indicator of the adequate oxygenation of blood. See Pink puffer

CCNU Cyclonexyl-chloroethyl-nitrosourea, lomustine ONCOLOGY A chemotherapeutic used in adjuvant therapy of high-grade astrocytomas, also used in NHL, melanoma, myeloma, GI CAs ADVERSE EFFECTS N&V, ↓ platelets, ↓ WBCs, 2° leukemia, pulmonary fibrosis, renal failure. See BCNU, Nitrogen mustards

CCPD Continuous cycling peritoneal dialysis, see there

CCR-1 RANTES–Regulated upon Activation Normal T-cell Expressed & Secreted AIDS, MOLECULAR MEDICINE A chemokine for CCR-5 receptor which blocks the entry of M-tropic strains of HIV, which preferentially infect macrophages; it is secreted by CD8 T cells, platelets, renal epithelial, and mesangial cells, and is chemoattractive for basophils, eosinophils, monocytes, and T cells; it activates basophils, inducing histamine release. See AIDS, CCR-3, CCR-5, Chemokine, HIV

CCR-3 CMKBR3, CC CKR3, eotaxin R CELL BIOLOGY A 41 kD coreceptor for HIV encoded on chromosome 3q21.3 LIGANDS Eotaxin, Eot-2, Eot-3, MCP-2, MCP-3, MCP-4, MCP-4, MIP-1δ, RANTES. See Coreceptor, RANTES

CCR-5 CMKBR2, CC CKR2, ChemR13 CELL BIOLOGY A 40 kD receptor encoded on chromosome 3q21; CCR-5 is a coreceptor for SIV and HIV which, with CD4, provides the portal for HIV entry into macrophages LIGANDS MIP-1a, MIP-1b, RANTES. See AIDS, Coreceptor, HIV

CCR-5 AIDS, MOLECULAR MEDICINE The gene that encodes CCR-5 which, when mutated, encodes a defective receptor protein incapable of being infected with HIV. See AIDS, CXCR4, HIV

CCU Critical care unit, see there

CD 1. Cluster designation, cluster of differentiation IMMUNOLOGY A nomenclature system for surface antigens of human leukocytes, macrophages, platelets, and other cells, characterized by monoclonal antibodies, allowing these cells to be categorized by lineage; most cells have > 1 surface antigen; 'w'–workshop designations indicate an incompletely characterized antigen; also 1. Chemical diabetic 2. Chief of division 3. Childhood disease 4. Common duct 5. Communicable disease 6. Compact disk 7. Complicated delivery 8. Contact dermatitis 9. Controlled drug 10. Convulsive disorder 11. Corneal dystrophy 12. Crohn's disease 13. Cystic duct 14. Left posterior descending coronary artery, see there

CD-ROM Compact disk-read only memory INFORMATICS A compact disk, capable of holding ± 750 MB of permanent data, which is recorded or burned only once. Cf CD-RW

CD-RW Compact disk-read-write INFORMATICS A rewritable compact disk, capable of holding ± 750 MB of data; CD-RW is an active storage medium, allowing repeated data additions. Cf CD-ROM

CD1 A family of MHC class I-like molecules expressed on the surface of immature thymocytes, Langerhans cells, and certain B cells; CD1 is a ligand for T cell subpopulation, is present in intestinal epithelium adjacent to GALT, possibly involved in epithelial immunity. See Gut-associated lymphoid tissue

CD3 T3 A complex of 5 invariable polypeptide chains, which are part of a larger complex that includes the T cell receptor; CD3 is 'pan-T' cell antigen–ie, present on most mature T cells–60-85% of peripheral T cells, 20-85% of thymocytes. See Pan-T cell markers

CD4 A transmembrane glycoprotein present on T helper/inducer cells, which participates in adherence of T cells to target cells, and is involved in thymic maturation and transmission of intracellular signals during T cell activation by the class II MHC; CD4 has inducer or helper activity for T cell, B cell, and macrophage interactions, and evokes T cell proliferation in response to soluble antigens or autologous non-T cells, providing appropriate signals for B cell proliferation and differentiation into Ig-secreting cells; CD4 is also a high-affinity receptor for HIV-1's gp120, binding 42–55 of the NH_2 terminal domain amino acids, and has an Ig-like fold similar to the complementarity-determining region of the kappa light chain; CD4 also binds Igs independently of the Fc receptor, and is an accessory to the T-cell receptor

CD4 cell CD4+ lymphocyte A circulating T cell with a 'helper' phenotype; in AIDS Pts, the levels of CD4+ cells is a crude indicator of immune status and susceptibility to certain AIDS-related conditions; these Pts may suffer KS as CD4+ cells fall below 0.3 x 10^9/L–US: 300/mm³, NHL < 0.15 x 10^9/L–US: 150/mm³, *Pneumocystis carinii* < 0.1 x 10^9/L–US: 100/mm³, MAC < 0.05 x 10^9/L–US: 50/mm³. See CD4 count, Kaposi sarcoma

CD4 count Absolute T4 count AIDS The number of 'helper' T cells/mm³ in Pts with HIV, which declines as infection progresses; the absolute CD4 count was the standard lab parameter used to monitor the immune suppression in persons with HIV; HIV RNA levels are more valid markers of immune responsiveness, and are the gold standard for Pts with AIDS/HIV infection

CD4 T cells Helper T cells, see there

CD4/CD8 coexpression HEMATOLOGY The expression on the surface of a T cell of both CD4–the 'helper' T cell antigen and CD8–the 'supressor' T cell antigen; it is common in Crohn's disease, myasthenia gravis, rheumatoid arthritis, lymphoproliferative disorders–eg, AML, B-cell lymphoma. See CD4, CD8

CD4/CD8 ratio AIDS The ratio of circulating T cells with the 'helper cell' determinant–CD4 on the cell surface to T cells with 'suppressor cell' determinant–CD8. See CD4, CD8, CD4/CD8 coexpression

CD8 T8 antigen A disulfide-linked heterodimeric protein, for which each monomer is 32–34 kD; CD8 is a marker for T cells with suppressor and cytotoxic activity; it is a co-receptor with class I MHC antigens on antigen-presenting cells, and is physically associated with a p56 tyrosine kinase, which phosphorylates adjacent proteins; NK cells may weakly express CD8. See Cytotoxic T cells

CD8 cells T cells with CD8 on the surface, which are immunosuppressive and suppress mitogen-induced and antigen-specific antibody production, and require CD4 cell cooperation

CDE Certified Diabetes Educator A health care professional qualified by the Am Assn of Diabetes Educators to teach diabetics how to manage their disease. See Diabetes mellitus

CDS Controlled Drug Substance, see there

CEA NEUROSURGERY Carotid endarterectomy, see there ONCOLOGY Carcinoembryonic antigen A fetal gut glycoprotein normally present in nanogram amounts in the circulation, which is ↑ in up to 30% of colorectal, lung, liver, pancreas, breast, head & neck, bladder, cervix, prostate and medullary thyroid CAs, and may be ↑ in lymphoproliferative disorders, melanoma, heavy smokers, and Pts with IBD; ↑ CEA is not a reliable cancer screening tool, but is useful for monitoring recurrent colorectal CA; some experts believe that a 35% ↑ of CEA above a Pt's post-resective surgery base-line levels warrants a 'second look' operation to rule out metastases

CEA scan Arcitumomab, see there

CEDARS CARDIOLOGY A clinical trial–Comprehensive Evaluation of Defibrillators And Resuscitative Shock. See Defibrillation, Resuscitation

Cedax® Ceftibuten, see there

cefazolin Kefzol® INFECTIOUS DISEASE A 3rd generation broad-spectrum cephalosporin, active against *E coli* and *Klebsiella* but not against most Enterobacteriaceae

cefepime Maxipime® INFECTIOUS DISEASE A 4th-generation cephalosporin INDICATIONS RTIs–eg bronchitis, pneumonia; cellulitis and other skin and soft-tissue infections; UTIs–eg pyelonephritis, febrile neutropenia, PID, gon-orrhea, septicemia, intra-abdominal infections, meningitis, surgical prophy-laxis. See Cephalosporin

ceftibuten Cedax® A once-daily broad-spectrum cephalosporin used for acute bacterial otitis media, acute bacterial exacerbation of chronic bronchitis, pharyngitis/tonsillitis. See Cephalosporin

ceftriaxone Rocephin® INFECTIOUS DISEASE An advanced generation cephalosporin used for RTIs, UTIs, skin infections, PID, gonorrhea, sep-ticemia, intra-abdominal infections, meningitis, neurosyphilis, surgical pro-phylaxis SIDE EFFECTS Diarrhea, liver dysfunction

ceiling VOX POPULI A maximum, upper limit. See Glass ceiling, Nondisclosure ceiling

Celebrex® Celecoxib PAIN MANAGEMENT A COX-2 inhibitor NSAID approved for rheumatoid arthritis, osteoarthritis and other pain PROS Much lower incidence of GI bleeding ADVERSE EFFECTS Abdominal pain, diarrhea, dyspepsia. See Nonsteroidal anti-inflammatory drug. Cf Celexa™, Cerebyx®

celery stick sign PEDIATRIC RADIOLOGY A descriptor for a disorganized diaphysis and metaphysis with osteitis of long bones, seen in the distal femur and proximal tibia, where longitudinal radiolucent striations alternate with sclerotic bands; 'celery stick' changes are typical of children with rubella embryopathy, especially if infected in the 1st trimester DIFFDX Other transpla-cental infections–eg, CMV, HSV, syphilis, toxoplasmosis–see TORCH, malig-nancy–eg, leukemia, neuroblastoma, systemic disease, scurvy, hypervita-minosis D, osteogenesis imperfecta, osteoporosis

Celexa® Citalopram PHARMACOLOGY An SSRI for treating major depres-sion SIDE EFFECTS Nausea, sleep disturbances, dizziness. See Major depression, Selective serotonin reuptake inhibitor. Cf Celebrex®, Cerebyx®

celiac disease Celiac sprue, celiac syndrome, gluten-sensitive enteropathy GI DISEASE A malabsorptive syndrome caused by hypersensitivity of intes-tinal mucosa to α-gliadin, a gluten extract composed of glutamine and proline-rich proteins, present in wheat, barley, rye, and oats CLINICAL Diarrhea, copious fatty stools, abdominal distension, weight loss, anemia, hemorrhage, osteopenia, muscle atrophy, peripheral neuropathy, CNS and spinal cord demyelination–sensory loss, ataxia, amenorrhea, infertility, edema, petechiae, dermatitis herpetiformis, especially if an HLA B27 hap-lotype LAB Transaminases–ALT, AST are ↑ in ± 40% of Pts with CD, which usually normalizes with gluten-free diet; occult bleeding occurs in ± ½ of Pts with CD and may be severe enough to cause iron deficiency anemia; Pts may also have alopecia areata, antigliadin and antiendomysial anti-bodies DIAGNOSIS Small intestine Bx MANAGEMENT Eliminate gliadin from diet; without treatment 10-15% develop lymphoma–eg, immunoblastic

lymphoma, less commonly, T cell lymphoma; CD predisposes to GI lym-phoma, and carcinoma of the oral cavity, esophagus; the otherwise rare small intestinal adenoCA is 80-fold more common in CD. Cf Tropical sprue

celiac plexus block PAIN MANAGEMENT A nerve block for severe, often cancer-related pain, the most commonly performed neurolytic nerve block, used early in managing pain due to pancreatic and other upper abdominal malignancies ADVERSE EFFECTS Diarrhea, orthostatic hypotension, pneumo-thorax, neurologic sequelae–eg, L-1 root neuralgia or paraplegia. See Nerve block, Neurolytic block, Nonneurolytic block

celibacy SEXOLOGY A state of being unmarried and abstinent from sexual relationships; vows of celibacy may be taken by clerics, and required by their religious order

cell block CYTOLOGY A paraffin-embedded specimen derived from dried mucus, sputum, or debris found in fluids of pleural, pericardial, endo-bronchial and other sites that cannot be processed in the usual fashion for cytologic analysis

cell phone Cellular telephone A portable device that sends and receives analog or digital telecommunication signals; CP use while driving carries a 4.3-fold ↑ risk of MVAs; use of CPs while driving is prohibited by law in New York

cell Samaritan MEDTALK A person who calls 911 on a cell phone to report a public safety occurrence without physically contributing any useful aid to the resolution of the problem; CSs assuage their consciences by doing some-thing without effort or involvement or direct assistance

cell surface marker A molecule usually found on the plasma membrane of a specific cell type or a limited number of cell types

cellophane tape procedure Scotch Tape™ test, see there

cellulite DERMATOLOGY A popular term for cosmetically undesirable sub-cutaneous layered fat that cause peau d'orange-like dimpling of the skin sur-face; cellulite is not an accepted medical term, either clinically or patholog-ically, as histologic examination reveals 'garden variety' fat and adipocytes. See Obesity

cellulitis DERMATOLOGY An acute, diffuse and boggy, suppurative inflam-mation of deep subcutaneous tissues and sometimes muscle ETIOLOGY Infection of wounds 2° to surgery, trauma, burns or other skin lesions–eg, erysipelas, by bacteria, especially group A streptococci and *Staphylococcus aureus*; it also occurs in immunocompromised Pts CLINICAL Edema, warmth, tenderness, indistinct margins, tendency to spread along tissue planes RISK FACTORS Insect bites/stings, animal/human bites, skin wounds, peripheral vascular disease, DM, ischemic ulcers; recent heart, chest, dental surgery, use of steroids, immunosuppressants MANAGEMENT Antibiotics. See St Helenian cel-lulitis, Synergistic necrotizing cellulitis. Cf Erysipelas

Celsius scale Centigrade A system used to measure temperature in Europe and other intellectually advanced countries NORMAL BODY TEMPERATURE 37°C = 98.6°F CONVERSION Fahrenheit → Celsius: C° = (F°–32) x ⁵/₉; to convert Celsius → Fahrenheit: F° = (C° x ⁹/₅) + 32

cement DENTISTRY Any of a number of bonding materials used in cavities and restorations

Cenestin® GYNECOLOGY A plant-derived synthetic estrogen product for treating vasomotor Sx of menopause, possibly of use for osteoporosis. See Menopause

census HOSPITAL CARE 1. The number of inpatients–ie occupied beds in a service, unit, ward, or in the entire hospital or health care facility, exclusive of newborns 2. A list of Pts in a service, unit, ward, or the entire hospital. See Bed. Cf Bypass

center ANATOMY A region in the body where something occurs. See Burned-out germinal center, Cell center, Ejaculation center, Erection center, Feeding center, Germinal center, Ossification center, Regressively transformed germinal center, Satiation center, Vasomotor center VOX POPULI A place where something occurs. See Academic medical center, Alternative birthing center, Ambulatory care center, Ambulatory surgery center, Burn center, Cancer center, Centers for Disease Control & Prevention, Center of excellence, CEPH center, Community mental health center, Day care center, Diagnostic center, Fox Chase Cancer Center, Imaging center, Level one trauma center, Medical center, MTOC, Multicenter, Poison control center, Research center, Residential treatment, Secondary care center, Surgicenter, Tertiary care center, Urgent care center, Yerkes Regional Primate Research center

center of excellence TERTIARY CARE A hospital designated by Medicare as one in which it will reimburse expenses for a particular procedure–eg, liver transplantation, based on that center's higher than average rate of success

Centers for Disease Control & Prevention CDC EPIDEMIOLOGY The premier epidemiologic agency of the world which operates under the HHS, located in Atlanta, Georgia; its mission is to promote health and quality of life by preventing and controlling disease, injury and disability; it is nonregulatory and has 11 centers, offices and institutes

Centers for Medicare & Medicaid Services See CMS, formerly HCFA

central auditory processing disorder AUDIOLOGY The inability to differentiate, recognize, or understand sounds in a person with normal hearing and intelligence

central cord syndrome NEUROLOGY A post-traumatic condition affecting the cervical spinal cord, in which necrosis spreads from the central gray matter peripherally to the myelin, resulting in focused damage to the corticospinal tracts; voluntary myelinated motor fibers to the arms are more central and those to the legs more peripheral; in CCS, lower motor neuron changes occur in the arms and are accompanied by leg spasticity; sensory defects reflect the degree of anterolateral and posterior column destruction, often accompanied by altered pain and temperature sensation in hands; CCS of acute onset may be accompanied by urinary retention and incontinence. Cf Central spinal cord syndrome

central core myopathy Central core disease of muscle, Shy-Magee disease An AD myopathy of neonatal onset, causing hypotonia in infancy–'floppy baby' syndrome CLINICAL Slowly progressive muscle weakness, post-exercise cramping, and sparing of the cranial muscles GENETICS The defect is located on chromosome 19 and involves the ryanodine receptor-1

central dogma MOLECULAR BIOLOGY The pedagogical tenet that translation of a protein invariably follows a chain of molecular command, where DNA acts as the template for both its own replication and for the transcription to RNA–and with subsequent maturation, to mRNA, which then serves as a template for translation into a protein. See DNA, Nucleic acid, Protein, Reverse transcription, RNA, RNA polymerase. Cf Prion

central hypothyroidism ENDOCRINOLOGY Hypothyroidism caused by pituitary or hypothalamic disease in the guise of tumors or vascular defects

central line Central venous catheter, see there

central nervous system leukemia Involvement of the CNS by leukemia

CNS LEUKEMIA–PER 'MODIFIED' ROME WORKSHOP CRITERIA
GROUP 1 No blasts detectable by cytocentrifuge
GROUP 2 < 5 leukocytes/mL, blasts detectable by cytocentrifuge
GROUP 3 > 5 leukocytes/mL, blasts, cranial nerve palsies

central nervous system prophylaxis ONCOLOGY Chemotherapy or RT to the CNS as preemptive therapy for killing CA cells that may be lurking

in the brain and spinal cord

central obesity Abdominal obesity, truncal obesity Obesity defined by an ↑ waist-to-hip ratio, waist-to-thigh ratio, waist circumference, and sagittal abdominal diameter, and linked to an ↑ risk of cardiovascular events. See Body mass index, Obesity

central pontine myelinolysis NEUROPATHOLOGY A condition characterized by softening of the base of the brain at the pons with damage to the myelin sheath, related to aggressive correction of hyponatremia; first identified in alcoholics–Wernicke-Korsakoff syndrome, CPM also occurs in AIDS, infection, lymphoproliferative disorders–eg, AML, malnutrition, and venous obstruction CLINICAL Weakness, double vision, muscle spasms, speech defects, delirium, sleep disorders, hallucinations, tremors and uncontrolled eye movements IMAGING Defects are seen by MRI PREVENTION Slow correction of electrolytic imbalance

central sleep apnea SLEEP DISORDERS A type of life threatening sleep apnea due to defective responses to O_2 and CO_2 in the circulation MECHANISM Possibly ↓ sensitivity to CO_2. See Sleep apnea syndrome

central spinal cord syndrome NEUROSURGERY A condition affecting the cervical spinal cord, which follows spinal injury with acute hyperextension of the neck in older Pts often with congenital or acquired cervical spinal stenosis due to spondylosis; in CSCS, the motor defect is more severe in the upper extremity, and most marked in the intrinsic muscles of the hand; severity of sensory deficits and bowel and bladder dysfunction vary; although spontaneous improvement of neurologic function is the rule, residual defects are common, and may reflect the severity of the initial injury. Cf Central cord syndrome

central venous catheter Central line, central venous access catheter, Hickman catheter HOSPITAL CARE A thin flexible device inserted into a large vein and used to establish a diagnosis–by drawing blood or administering therapy

Centrazepam Diazepam, see there

centrifugation LAB TECHNOLOGY The process of separating fractions of systems in a centrifuge. See Axial centrifugation, Density gradient centrifugation, Ultracentrifugation

centronuclear myopathy Myotubular myopathy, see there

cephalad march CRITICAL CARE MEDICINE The progression of weakness and loss of sensory perception seen in an acute epidural hematoma

cephalexin Keflex® INFECTIOUS DISEASE A 3rd generation broad-spectrum cephalosporin

cephalic presentation OBSTETRICS Any position of the baby in which the head presents, which may be the vertex (normal)–the easiest presentation to deliver; abnormal cephalic presentations include the brow, face, and posterior presentations at the time of delivery. See Cephalic version, Cesarean section

cephalic version OBSTETRICS The manipulative or spontaneous flipping of the fetus in the uterus so that the head presents to the opening of the birth canal. See External cephalic version

cephalohematoma NEONATOLOGY A hematoma under the scalp 2° to minor birth trauma, which is resorbed and rarely requires specific intervention

cephalopelvic disproportion OBSTETRICS A structural and functional disparity between the fetus and the birth canal which makes vaginal delivery difficult or impossible. See Birth canal, Dystocia

cephalosporin INFECTIOUS DISEASE Any of a family of broad-spectrum tetracyclic triterpene antibiotics derived from *Cephalosporium* spp, which are similar chemically and in mechanism of action to penicillin INDICATIONS Skin and soft tissue infections, RTIs, UTIs, STDs, meningitis, endocarditis, septicemia of unknown portal of entry, anaerobic infections, polymicrobial infections PHARMACOKINETICS Absorbed orally, excreted by the distributed in tissues ADVERSE REACTIONS Hypersensitivity, rash, serum sickness, acute tubular damage. See Fourth-generation cephalosporin

CEPHALOSPORIN GENERATIONS

1ST   IV agents–Cephalothin, cephaloridine

2ND   Oral agents, which had a longer half-life

3RD   ↑ antibacterial spectrum to include *H influenzae*, anaerobic bacteria, eg *Bacteroides fragilis*

4TH   Not all authors use the term; further; ↑ antibacterial spectrum to include *Pseudomonas* species

cephalothin Cefalotin® INFECTIOUS DISEASE A parenteral semisynthetic derivative of cephalosporin C, and 3ʳᵈ generation broad-spectrum cephalosporin

-cephaly A Greek root for head. See Macrocephaly, Microcephaly

cephradine Anspor®, Velosef® A semisynthetic oral/IV cephalosporin excreted unchanged in the urine which accumulates in serum; it is contraindicated in renal failure

Ceprate® ONCOLOGY A product line that selectively removes cell lines–eg, stem cells from BM and peripheral blood before tumor-purging therapy, or removes T cells to ↓ GVHD in allograft recipients. See Marrow ablation

cerclage SURGERY The operative encircling of a part, especially of an incompetent cervix, tightening it with sutures. See Cervical incompetence

cerclage wire ORTHOPEDICS A wire placed around bone to approximate fracture fragments

cerebellar ataxia NEUROLOGY A condition characterized by a usually abrupt onset of unsteady gait, nystagmus, and dysarthria, which in children may persist in the form of residual movement or behavioral disorders. See Ataxia

cerebellar herniation Foramen magnum herniation, uncal herniation NEUROLOGY The prolapse of the cerebellum, when subjected to ↑ pressure, forcing it down through the foramen magnum, compressing the brainstem and compromising the respiratory center, causing coma, paralysis, unilateral dilated pupil, or death ETIOLOGY Head injury, brain tumors or abscesses, bacterial meningitis

cerebellopontine angle tumor Acoustic neuroma, see there

cerebr–, cerebri– *prefix*, Latin, pertaining to the brain

cerebral amyloid angiopathy Cerebral amyloidosis, cerebral congophilic angiopathy NEUROLOGY A sporadic disease of the elderly, present in 5-10% of primary nontraumatic cerebral hemorrhage, which may be found incidentally at autopsy, often with the histologic changes of Alzheimer's disease and/or dementia, characterized by deposition of amyloid in arterial walls; CAA is idiopathic or may be associated with human hereditary cerebral hemorrhage with amyloidosis, Dutch type CLINICAL Dementia, Sx of lobar hemorrhage due to ↑ vascular fragility DIAGNOSIS Cerebral edema by CT; leukoencephaly and petechial hemorrhages by MRI. See Amyloidosis, HCHWA-D, Lobar intracerebral hemorrhage

cerebral aneurysm Brain aneurysm NEUROLOGY A dilated and weakened, rupture-prone segment of a cerebral artery, which may be a birth defect or develop 2° to poorly controlled HTN; ±5% of general population has an aneurysm; rupture occurs in $4/10^5$/yr CLINICAL-RUPTURE Severe 'thunderclap' headache, weakness, numbness, N&V, neurologic defects, ↓ consciousness. See Aneurysm

cerebral angiography IMAGING The injection of radiocontrast into the carotid artery, to evaluate the need for endarterectomy due to ASHD

cerebral atrophy 1. Alzheimer's disease, see there 2. Pick's disease, see there

cerebral cortex NEUROLOGY The outer portion of the brain, the neocortex consisting of gray-colored layers of nerve cells, and the interconnecting neural circuitry, which is intimately linked to cognition. See Limbic system

cerebral cysticercosis PUBLIC HEALTH A CNS infestation by intermediate-stage larva–metacestode of pork tapeworm, *Taenia solium* EPIDEMIOLOGY Oral ingestion of *T solium* eggs shed in feces of human carriers of pork tapeworms, and thus may occur in those who have not ingested or had contact with pork CLINICAL Aphasia, seizures, late-onset epilepsy, hydrocephalus DIAGNOSIS ID of cysts by CT, MRI, serum antibody by ELISA MANAGEMENT Praziquantel, albendazole

cerebral edema MEDTALK Brain swelling NEUROLOGY Swelling of cerebral interstitial tissue ETIOLOGY CE may be seen in bacterial meningitis in children, after trauma and in alcoholics CLINICAL Absent oculocephalic response, fixed eye deviation, coma MANAGEMENT IV mannitol, decompression

cerebral granulomatous angiitis An inflammatory reaction of multiple etiologies, which is defined histologically by giant cell or epithelioid cell granulomatous inflammation of cerebral blood vessels; CGA is heterogeneous with regard to caliber, location, and type of vessels involved, degree of necrosis, presence of systemic vasculitis

cerebral hematoma A blood clot in the brain. See Epidural hematoma, Subdural hematoma

cerebral hemorrhage Brain bleed NEUROLOGY Abrupt bleeding into cerebral tissue, which may be 2° to HTN, ASHD malformations or trauma. See Arteriovenous malformation, Berry aneurysm, Cerebrovascular accident, Stroke, Subdural hematoma

cerebral herniation The displacement of part of the brain due to a marked ↑ in intracranial pressure, forcing the brain in the direction of least resistance, ie down the foramen magnum CLINICAL Paralysis, coma, unilateral dilated pupil ETIOLOGY Head injury, brain tumors, abscess, bacterial meningitis TYPES Cerebellar, uncal, transtentorial

cerebral hypoxia ↓ O₂ in brain; depending on the duration and severity, Sx range from mild–eg, lethargy to serious neurologic damage–eg, coma, seizures, death

cerebral laceration NEUROLOGY Tearing of brain tissue by a sharp object; CLs may occur when the falx lacerates the corpus callosum after trauma of the brain. Cf Extradural hemorrhage, Subdural hemorrhage

cerebral lymphoma Brain lymphoma, primary lymphoma of CNS ONCOLOGY A brain tumor characterized by a massive infiltrate of lymphoid cells, which is more common in immunocompromised hosts–eg, AIDS Pts, post-transplants Pts on long-term immunsuppression, etc CLINICAL Seizures, neurological and personality changes. See Lymphoma

cerebral palsy Little's disease NEUROLOGY A persistent, nonprogressive SX complex affecting up to 0.4% of term births, attributed to intrauterine and/or perinatal hypoxia; characterized by defective motor development, with spastic paraplegia, gait defects, incoordination, seizures, ataxia, speech defects, ±mental retardation

cerebral performance category scale EMERGENCY MEDICINE A scale that grades Pt response to CPR, on a scale of 1 to 5. See Glasgow coma scale, Harvard criteria

cerebral vasculitis

CEREBRAL PERFORMANCE CATEGORY SCALE
CPC 1 A return to normal cerebral function and normal living
CPC 2 Cerebral disbility but sufficient function for independent activities of daily living
CPC 3 Severe disability, limited cognition, inability to carry out independent existence
CPC 4 Coma
CPC 5 Brain death

cerebral vasculitis Temporal arteritis, see there

cerebritis Inflammation of the brain

cerebrospinal fluid Spinal fluid NEUROLOGY A clear, colorless fluid that contains small quantities of glucose and protein, which surrounds the brain, spinal cord, ventricles, subarachnoid space, and the central canal of the spinal cord, provides nutrients, and acts as a shock absorber; CSF analysis is accomplished by lumbar puncture; WBCs or bacteria in the CSF indicate bacterial–septic meningitis. See Lumbar puncture

cerebrospinal fluid analysis LAB MEDICINE The in vitro evaluation of the CSF. See Lumbar puncture

cerebrospinal fluid leak CSF leak NEUROLOGY The inappropriate loss of fluid from the otherwise sealed CSF space ETIOLOGY Trauma to head–eg CSF rhinorrhea, CSF otorrhea, cranial base surgery DIAGNOSIS Suspicious post-op nasal or ear drainage, β-transferrin assay, endoscopic fluorescein technique IMAGING MRI, CT, CT cisternography, radioisotope scans TREATMENT Soft tissue and mucosal flaps, biomaterial grafting

cerebrovascular adjective Relating to cerebral blood vessels

cerebrovascular accident Stroke, cerebral hemorrhage NEUROLOGY Sudden death of brain cells due to ↓ O_2 2° to vascular obstruction or ruptured cerebral artery CLINICAL Abrupt unilateral weakness, paralysis DIAGNOSIS CT, MRI PREVENTION Control HTN, DM PREVENTION Carotid endarterectomy ↓ risk of future stroke; in asymptomatic Pts with stenosed carotids; CVA risk ↓ with aspirin and ticlopidine–Ticlid™, which have antiplatelet activity. See Stroke, Transient ishemic attack

cerebrovascular congestion Malignant cerebral edema NEUROLOGY Brain edema caused by intracranial vasodilation 2° to autoregulatory dysfunction. Cf Brain edema

cerebrovascular disease NEUROLOGY Any vascular disease affecting cerebral arteries–eg ASHD, diabetic vasculopathy, HTN, which may cause a CVA or TIA with neurologic sequelae–speech, vision, movement of variable duration. See Cerebrovascular accident, Transient ischemic attack

Cerebyx® Fosphenytoin PHARMACOLOGY An IV agent for status epilepticus, which may replace IV Dilantin–phenytoin. Cf Celexa

Ceredase® A recombinant glucocerebrosidase used for Gaucher's disease

Cerepax® Temazepam, see there

Ceresine® NEUROLOGY An agent for preventing CNS ischemia in closed head injuries, stroke and children with congenital lactic acidosis. See Cerebrovascular accident

Ceretec® IMAGING Technetium 99m–^{99m}Tc imaging agent commonly used with SPECT imaging to detect CVAs, which may be useful for diagnosing Alzheimer's disease. See Alzheimer's disease

cerivastatin Baycol® CARDIOLOGY Cholesterol-lowering, HMG-CoA reductase inhibitor/statin for managing hypercholesterolemia and mixed dyslipidemia; it ↑ HDL-C and ↓ LDL-C; withdrawn from the market as it was linked to rhabdomyolysis. See Statin

ceroid A complex of alcohol-insoluble, oxidized polyunsaturated lipid pigment(s) resulting from the peroxidation of unsaturated lipids that are similar or identical to lipofuscin; ceroid accumulates in macrophages of the heart, liver, GI tract, and brain in the elderly and is thus termed 'wear and tear' pigment; it has been inculpated in age-related organ dysfunction–see 'garbage can' hypothesis, hypovitaminosis E, cathartic colon, and hereditary conditions–eg, Batten's disease, sea-blue histiocytosis

certifiable adjective Capable of being certified, as in either 1. An infectious disease, rarely used in this context, or 2. A mental disorder, the latter being by far the most common use in American 'English'

certificate VOX POPULI An official document generated by the appropriate federal, state, or other authorizing body attesting to a certain fact. See Birth certificate, Death certificate, Digital certificate, Employee certificate of insurance, Group certificate, Security certificate

certified bed A 'legal' bed in a health care facility approved by authorities for use by Pts on a permanent basis, and which the governing body–usually the state board of health–deemed to have sufficient staffing to support its unqualified use. See Bed. Cf 'Swing' bed

certified nurse midwife Nurse midwife OBSTETRICS A registered state-licensed registered nurse who, by virtue of added knowledge and skill gained through an organized program of study and clinical experience, is qualified to manage the care of women and/or newborns during the antepartum, intrapartum, and postpartum periods, and to provide expressly limited well-woman health care. See Midwife

certifying death The act of confirming that a person is dead after another person (physician) has pronounced or determined that the individual is dead. Cf Pronouncing death

ceruloplasmin Ferroxidase CLINICAL CHEMISTRY A dehydrogenase produced in the liver involved in copper detoxification and possibly in mopping up excess oxygen radicals or superoxide anions ↑ IN Neoplasms, inflammation, liver disease, rheumatoid arthritis, SLE, pregnancy, OCs, estrogen therapy ↓ IN Wilson's disease, Menkes' kinky hair syndrome, hepatitis, cirrhosis, nephrotic syndrome, sprue, scleroderma of small intestine. See Copper

cerumen Earwax ENT A waxy secretion of the hair follicles and glands of the external auditory canal which protects the ear by trapping dust, microorganisms, and foreign particles, preventing them from entering and damaging the ear. See Wet cerumen

ceruminous gland A specialized gland that secretes cerumen into the external auditory canal

cervical adjective Referring to the 1. Uterine cervix 2. Neck

cervical adenopathy Cervical lymphadenopathy, see there

cervical biopsy-cytology correlation CYTOLOGY A part of quality assurance in cytology, which consists of determining whether the results of the pap smear–based on The Bethesda System reflect the diagnosis of a subsequent biopsy. See Cervical cancer, Pap smear

cervical biopsy A biopsy of the uterine cervix usually performed days to wks after a pap smear reveals changes–especially epithelial cell abnormalities, warranting further evaluation COMPLICATIONS Discomfort, bleeding; the tissue obtained is placed in formalin and submitted to a pathologist for interpretation. See Cervical intraepithelial neoplasia, Colposcospy, Pap smear

cervical canal Endocervical canal The canal between the uterine cervix and the endometrium, consisting of the endocervix, and lined by endocervical glands; mucus plugs the cervical canal and normally prevents foreign materials from entering the endometrium and uterus

cervical cancer ONCOLOGY Invasive malignancy of the uterine cervix EPIDEMIOLOGY Incidence is 2.5-fold > in black ♀ PREVENTION Regular pelvic exams, pap smears CLINICAL Abnormal bleeding PATHOLOGY SCC–85%, adenoCA–10% MANAGEMENT Cryosurgery, cauterization, laser surgery PROGNOSIS Poor if advanced; CC-related deaths are ugly; death is accompanied by uremia, and a typical 'funky' urinary odor. See Conization, LEEP. Cf Cervical intraepithelial neoplasia

CERVICAL CANCER STAGING

STAGE I No spread to nearby tissues

- IA–minimal microscopic CA found in deep cervical tissues
- IB–larger amount of CA found in deep cervical tissues

STAGE II Extension of tumor, but confined to pelvis

- IIA–spread beyond cervix to upper two thirds of the vagina
- IIB–spread to pericervical tissue

STAGE III Spread throughout pelvic area, eg to lower vagina, blockage of ureters

STAGE IV Metastases

- IVA–spread to bladder or rectum
- IVB–spread to distant organs, eg lungs

cervical carcinoma in situ GYNECOLOGY An SCC of the cervix, that usually arises in a CIN, which invades < 5 mm in greatest dimension. See Cervical cancer, CIN, Pap smear

cervical conization See Conization

cervical cord neurapraxia SPORTS MEDICINE A transient injury to the cervical spine resulting in neurologic sequelae; CCN are generally benign, but Pts should be counseled about recurrence, depending on the spinal canal/vertebral body ratio

cervical cord neurapraxia SPORTS MEDICINE A clinical entity due to developmental narrowing of the AP diameter of the cervical canal and acute mechanical deformation of the spinal cord; the typical CCN involves an athlete who has an acute transient neurologic episode of cervical cord origin CLINICAL Sensory changes–eg, burning pain, numbness, or tingling in both arms, both legs, all 4 extremities, or an ipsilateral arm and leg; in absence of instability or structural deficiency of the cervical spine, CCN is clinically benign, with complete neurologic recovery and no permanent morbidity in Pts who return to contact sports

cervical disk syndrome NEUROSURGERY A condition caused by cervical cord or root displacement or compression–often at multiple levels, characterized by radicular–suboccipital, cervical, interscapular, and thoracic pain that radiates into the upper extremities, aggrevated by movement, accompanied by paresthesias, and dysesthesias in cervical dermatomes; CDS may be accompanied by muscle fasciculations, spasms, and atrophy, lower extremity spasticity with extensor plantar sign, and spastic bladder DIAGNOSIS Narrowed disk spaces, osteophytes, or spinal stenosis by radiology, myelography, CT, MRI TREATMENT-MEDICAL Immobilization, traction in a neutral position, transcutaneous injection of enzymes–eg collagenase, to dissolve cartilage TREATMENT-SURGICAL Decompression with laminotomy, foraminotomy, total diskectomy via bilateral laminotomy followed by interbody fusion

cervical dysplasia Cervical intraepithelial neoplasia–CIN, see there

cervical dystonia Spasmodic torticollis, see there

cervical erosion Cervical ulceration GYNECOLOGY A partial or complete loss of the cervical mucosa ETIOLOGY CA, trauma–eg, coitus–especially with multiple partners, tampon insertion, infection, chemicals–eg, spermicidal creams or foams, douche CLINICAL Vaginal or post-coital bleeding DIAGNOSIS Colposcopy, Bx PROGNOSIS Spontaneous healing without intervention

cervical eversion Ectropion GYNECOLOGY The outward migration of columnar endocervical glands, the T zone or transition zone into the squamous epithelium of the uterine cervix; CE is usually asymptomatic and is associated with an ↑ in baseline estrogens as seen in pregnancy, at puberty or with exogenous hormones, as in OC use. See Transition zone

cervical incompetence Incompetent cervical os GYNECOLOGY A condition characterized by dilation and effacement of uterine cervix in later pregnancy which may be associated with spontaneous abortion and prematurity FREQUENCY 1-2% of all pregnancies; causes 20-25% of 2nd trimester abortions CLINICAL Vaginal bleeding or spotting; sensation of lower abdominal weightRISK FACTORS Prior cervical surgery or previous difficult vaginal delivery, cervical malformation, DES exposure, multigestation MANAGEMENT Cerclage. See Cerclage

cervical intraepithelial neoplasia Cervical dysplasia, CIN GYNECOLOGY Precancerous change of uterine cervical epithelium SCREENING Pap smears, colposcopy and pelvic exam PEAK AGE 25 to 35 RISK FACTORS Multiple sexual partners, early onset of sexual activity–< age 18, early childbearing–< age 16, Hx of STDs–eg, genital warts, genital herpes, HIV; CIN represents a continuum of histologic changes ranging from CIN 1–formerly, mild dysplasia, to severe dysplasia/carcinoma in situ, CIN 3; the lesion arises at the squamo-columnar cell junction at the transformation zone of the endocervical canal, with a variable tendency to develop invasive SCC, a tendency that is enhanced by concomitant HPV infection, of which HPV 6 and 11 are associated with the 'garden variety', ie benign condylomas; HPV types 16, 18 occur in CIN 3; types 31, 33, 35, 52, 56 also occur in CIN; 78% of ♀ who are positive for HPV, especially HPV 16 and 18, eventually develop CIN 2-3 TREATMENT Cone biopsy, laser vaporization or excision, loop electrosurgical excision, cryotherapy. See Carcinoma-in-situ, Cervical cancer, Intraepithelial neoplasia, Dysplasia, Low-grade intraepithelial neoplasia, High-grade intraepithelial neoplasia

PROGRESSION OF CERVICAL INTRAEPITHELIAL NEOPLASIA

	CIN 1	CIN 2	CIN 3
Regression	60%/50%	40%/43%	33%/–
Persistence	30%/41%	40%/48%	55%/–
Progression to CIN 3	10%/9%	20%/9%	NA
Progression to SCC	± 1%	5%	≥ 12%

From Int J Gynecol Pathol 1993; 12:186/Modern Pathol 1990; 3:679

cervical laminectomy NEUROSURGERY A procedure for relieving pressure on the spinal cord or nerve root 2° to a slipped or herniated disk of the cervical spine; CL entails removing part of the vertebral bone

cervical lymphadenopathy Cervical adenopathy, cervical lympadenitis ENT Enlarged and inflamed LNs of neck ETIOLOGY Viral, bacterial–eg, *Streptococcus* infection, or tonsillitis, pharyngitis, dental infections CLINICAL Pain, tenderness, lymphadenopathy

cervical mucus GYNECOLOGY A viscous fluid that plugs the cervical os, and prevents sperm and bacteria from entering the uterus; at midcycle, under estrogenic influence, CM becomes thin, watery, and stringy, and allows free passage of sperm into the uterus. See Cervical mucus method, Cervix

cervical os ANATOMY The opening of the uterine cervix which is covered by squamous epithelium. See Cervical incompetence, Transformation zone

cervical osteoarthritis Cervical spondylosis, degenerative joint disease of the cervical spine ORTHOPEDICS A degenerative disorder of the cervical spine 2° to progressive erosion of cartilage covering weight-bearing joints; bone deposit or spur formation may ↓ the diameter of the foramina of exiting nerve roots resulting in compressive Sx or nerve irritation. See Herniated disk

cervical polyp GYNECOLOGY A usually benign 'prolapse' of endocervical tissue from the cervix, with inflammation, congestion, ↑ estrogens; CPs are more common in ♀ > age 20 who have had children; some may cause bleeding and interfere with the menstrual cycle DIAGNOSIS Pelvic exam, colposcopy, pap smears–cervical CA may be polypoid TREATMENT Excision

cervical radiculopathy NEUROLOGY Irritation of nerve roots of the neck due to a herniation or prolapse of a intervertebral disk from its normal position, which impinge on nearby nerves resulting in pain and neurologic Sx. See Cervical disk syndrome, Prolapsed disk

cervical rib A uni- or bilateral congenital anomaly of the 1st thoracic rib or 7th cervical rib, affecting up to 0.5–1% of the population, 15% of whom have an associated thoracic outlet syndrome–see there–due to associated neural and arterial defects; cervical ribs range from type I, a short bar extending from the transverse process to type IV, a complete extra rib articulating with the sternum

cervical ripening OBSTETRICS The stromal response of the cervix in phase 1 of parturition, which precedes the onset of labor, in which the cervix becomes softer and more dilatable. See Cervical effacement

cervical smear Pap smear, see there

cervical spine CLINICAL ANATOMY The region of the vertebral column encompassing C1 through C7

cervical spine fusion ORTHOPEDICS The operative fusing of 2 or more cervical vertebrae due to trauma to the spine and/or bone degeneration

cervical spine injury ORTHOPEDICS A posttraumatic injury to the cervical spine, resulting in vertebra displacement; horizontal displacement of > 3.5 mm; rotation > 11° is an absolute contraindication to future participation in contact sports–eg, football, rugby. See Burner, Cervical cord neurapraxia, Spear tackler spine

cervical spine trauma ORTHOPEDICS A traumatic event, especially common in contact sports, resulting in cervical spine injury, see there

cervical stenosis GYNECOLOGY A block of the cervical canal due to a congenital defect or complications of surgery–eg, cryosurgery. See Cervix

cervical traction ORTHOPEDICS A type of continuous or intermittent traction in which a head halter with weights is worn by the Pt to maintain proper alignment of a fracture of the cervical spine. See Traction

cervicitis Cervical erosion GYNECOLOGY 1. A nonspecific inflammatory response to infection and non-infectious agents, characterized by hyperemia and a vascular pattern similar to punctation by colposcopy; unlike punctation, changes of cervicitis extend beyond the transformation zone. See Colposcopy. Cf Mucopurulent cervicitis 2. Inflammation of the cervix by *N gonorrhoeae* or *Chlamydia trachomatis*, which may spread to the fallopian tubes causing PID. See Cervicitis, Pelvic inflammatory disease

cervicovaginal junction CLINICAL ANATOMY The virtual line on the mucosa between the superior and external part of the cervix and the dome of the vaginal, forming a landmark for vaginal hysterectomies. Cf Fornix

cervix ANATOMY A narrow necklike portion of an organ; the neck, connecting the head and trunk or a constricted part of an organ GYNECOLOGY Cervix uterine, neck of uterus A doughnut-shaped structure directly contiguous with the uterus; the lower and narrow end of the uterus, between the isthmus and ostium uteri. See Barrel cervix, Cockscomb cervix, Ripe cervix, Strawberry cervix, Uterus

cervix mark NEONATOLOGY A fluid-filled vesicle of no clinical significance, that lies atop the caput succedaneum-the edematous 'mass' at the presenting aspect of the head in a vaginally-delivered infant

CES Cranial electrotherapy stimulation, see there

cesarean section C-section, abdominal delivery OBSTETRICS The delivery of a fetus via an abdominal incision and surgical extraction of a product of conception; CSs are the most commonly performed invasive procedure in many nations INDICATIONS Failure of progress, fetal distress, cephalopelvic disproportion, placenta previa, placental abruption, placental insufficiency, breech presentation, active genital herpes, multiple gestation, preeclampsia and excessive scarring from previous surgeries STATISTICS Brazil has the world's highest rate of C-sections–$^{32}/_{100}$ hospital deliveries; Japan and Czechoslovakia the lowest– $^{7}/_{100}$. See Vaginal delivery

cessation VOX POPULI The stopping of a thing. See Smoking cessation

cestode Tapeworm A ribbon-shaped segmented worm that inhabits the GI tract of vertebrates including humans EPIDEMIOLOGY Cestode infestations are more common in developing nations; US cases are linked to infected dogs or cats; *Taenia solium*–pork tapeworm and *T saginata*–beef tapeworm occur in those who eat undercooked meats; live up to 20 yrs and reach 10 m in length CLINICAL Unexplained weight loss, pernicious anemia, eggs or ribboned segments in stools

cetirizine Zyrtec® THERAPEUTICS A potent, antiinflammatory H1-receptor antagonist used for respiratory allergies; it has little CNS reactivity or cardiotoxicity and doesn't interact with other medicines. See Antihistamine

CF 1. Central field 2. Complement fixation, see there 3. Conversion factor–health care management 4. Coronary flow .5 Cystic fibrosis

CFR 1. Case-fatality ratio 2. Catastrophic failure rate 3. Code of Federal Regulations 4. Constant flow rate

CFS Chronic fatigue syndrome, see there

CGD Chronic granulomatous disease, see there

Chaddock sign NEUROLOGY Great toe extension by stroking the skin over the external malleolus, an indicator of corticospinal tract lesions–eg, MS. Cf Babinski sign

Chadwick sign OBSTETRICS A bluish discoloration of cervix and vagina, classically associated with pregnancy

Chagas' disease American trypanosomiasis PARASITOLOGY An infection caused by a protozoan, *Trypanosoma cruzi*, which is a major cause of M&M in Latin American EPIDEMIOLOGY Circa 17 x 10⁶ have chronic CD in South America; *T cruzi* is transferred in the feces of hematophagous triatomine insects–kissing or reduviid bugs which may contaminate the bite itself, the conjunctiva or a mucosal surface; less common routes of transmission include blood transfusion–20,000 transfusion-related cases/yr, Brazil, maternofetal–vertical, breast milk, accidental–lab workers CLINICAL-ACUTE CD Mild with 5% mortality; fever, malaise, headache, anorexia, edema of face and lower extremities, conjunctivitis, myocarditis, lymphadenopathy, hepatosplenomegaly; rarely muscle or CNS invasion; the acute phase lasts 2 to 3 months; infants may develop meningitis and heart involvement CHRONIC CD 10-30 yrs latency; severe cardiomyopathy with biventricular enlargement, thinning of ventricular walls, mural thrombi, interstitial fibrosis, conduction defects–eg, right BBB, or complete AV block, CHF; GI disease with megaesophagus and/or megacolon due to either local denervation or possibly to an autoimmune mechanism LAB-ACUTE Parasites in peripheral blood smear LAB-CHRONIC Serology for IgG by ELISA, CF, indirect immunofluorescence TREATMENT Therapies, benznidazole, nifurtimox are inadequate; cures in 50% at high toxicity PROGNOSIS Poor in Pts with CHF, left ventricular aneurysm or dysfunction

chain of custody Chain of evidence FORENSIC MEDICINE A formally documented continuity of possession, and proof of integrity of evidence collected, which establishes each person having custody/being in possession of the evidence; Note: The CoC is the path that objects–eg bullets, knives or clinical specimens–eg semen specimen in rape must take for these materials to be legally accepted as evidence in court. See Admissible evidence, Rape, SART

chair ACADEMENTIA 1. See Endowed professorship 2. A non-sexist sobriquet for the chairman, -woman or -person, ie the presiding officer in an organization; in academics, the 'chair' is often a full professor responsible for the academic, clinical, administrative and research activities of the department. Cf Endowed chair, Professor

chairman 1. Chairperson, see there 2. Chief of staff, see there

chart

131

chairperson Chairman The head of an academic department. See 'Chair', Cf Chief

chalasia PEDIATRICS Gastroesophageal reflux (GERD) in infants which, if severe, may be associated with ↑ vomiting, crying, discomfort, possibly ↑ risk of GERD as adults. See GERD

chalazion Meibomian cyst, meibomian gland lipogranuloma, tarsal cyst OPHTHALMOLOGY A cyst in the eyelid caused by plugging of the meibomian gland MANAGEMENT Warm wet compresses; surgical drainage if unresponsive to conservative measures

challenge IMMUNOLOGY The administration of an antigen or allergen to a person who has been previously exposed to the antigen, specifically to evoke an immune response

challenged *adjective* Compromised, disabled, disadvantaged, handicapped. See Americans with Disabilities Act, Impaired, Politically correct

champagne bottle legs NEUROLOGY Marked distal peroneal muscle atrophy of the legs with tapering of the distal extremities and hypertrophy of the proximal muscles, imparting an inverted champagne bottle or 'stork leg' appearance; CBLs are typical of advanced Charcot-Marie-Tooth type of chronic familial peripheral neuropathy; pes cavus may be the only early finding, which is followed by foot drop–leading to a high 'steppage' gait; walking is difficult given the combination of sensory ataxia and muscular weakness EMG ↓ conduction velocity, ↑ distal latency. See Onion bulb, Piano playing

CHAMPUS Civilian Health & Medical Program for Uniformed Services A health care plan for military dependents and retirees operated by the DoD TYPES OF SERVICE HMO, PPO, and fee-for-service, through a single health plan known as TriCare

chance STATISTICS The likelihood that a thing has happened unpredictably without human intervention or observable cause. See Monte Carlo simulation, Random

chancre STD The classic 1° skin lesion of syphilis, which consists of a painless 1–2 cm ulcer; the 1° chancre is highly contagious, contains zillions of spirochetes, and heals as a papule; chancres outside the vagina or on the scrotum render condoms useless in disease-preventing systems. See Kissing chancre, Syphilis

chancroid STD A shallow, painful ulcer caused by the sexually transmitted *Haemophilus ducreyi*, which clinically mimics syphilitic chancres; usually located on the penis or vulva, it is classically associated with tender lymphadenopathy MANAGEMENT Ceftriaxone

chandelier sign GYNECOLOGY A popular term for pain hypersensitivity in ♀ with PID; pelvic examination evokes pain of such intensity that the Pt seemingly leaps out of the examination stirrups, reaching 'for the chandelier'. See Pelvic inflammatory disease

change An alteration in circumstances or events or status quo. See Age change, Benign cellular changes, Chemical change, Conformational change, Fatty change, Sex change, Structural change, Wavy changes

change of life Menopause, see there

channeling MEDICAL PRACTICE The directing of a Pt to a specific health care or rehabilitation facility

CHAOS Cambridge Heart Anti-Oxidant Study CARDIOLOGY An on-going study that concluded that vitamin E supplements ↓ the risk of acute MI in Pts with known CAD. See Angina, Coronary artery disease, Silent ischemia

Chapel Hill Consensus Conference INTERNAL MEDICINE An international panel that adopted names and definitions for systemic vasculi-

tides–SVs; the CHCC divided SVs by size and shunned terms–arteritis, arteriolitis, venulitis, capillaritis, and others, which are difficult for pathologists to document. See Systemic vasculitis

chaplain HOSPITAL CARE A member of the clergy who provides for the spiritual needs of Pts and families. See Prayer

chapping VOX POPULI Drying of skin

character PSYCHIATRY The sum of a person's relatively fixed personality traits and habitual modes of response. See Metacharacter, Personality

characteristics VOX POPULI The features of a thing. See Demand characteristics, Operational characteristics

characterization VOX POPULI The delineation of a thing's features. See Activated characterization, Risk characterization, Secondary sex characterization

Charcot's joint Neuropathic arthropathy ORTHOPEDICS A joint characterized by ↓ pain or position sense due to tabes dorsalis, diabetic neuropathy, amyloidosis or leprosy, resulting in destruction of joints and soft tissue

Charcot-Marie-Tooth disease Hereditary peroneal nerve dysfunction NEUROLOGY A clinically heterogenous and most common–1:2500 of the inherited peripheral neuropathies CLINICAL Slowly progressive atrophy of distal muscles, especially those innervated by the peroneal nerve, leading to muscular weakness and atrophy of hands, feet, and legs with pes cavus deformity, claw-hand, stork-leg appearance, with foot drop and a slapping gait EMG ↓ nerve conduction velocity, due to destruction of nerves with degeneration of myelin sheath MANAGEMENT Nada

charge MANAGED CARE The posted price of a service provided by a hospital or health care facility; Medicare requires hospitals to apply the same schedule of charges to all Pts, regardless of the expected sources or amount of payment. See Fees

CHARGE complex A disease of probable neural crest origin, diagnosed when infants have 4 of the acronym's 6 components: **C**oloboma, **H**eart defects–conotruncal or septal, **A**tresia of nasal choanae, **R**etarded growth and development, **G**enital hypoplasia, **E**ar anomalies and/or deafness; other findings: facial palsy, renal anomalies, cleft lip/palate

charge explosion MANAGED CARE A feature of a billing computer system whereby a single charge item is expanded into numerous line-item charges–in computer-ese, a macroexpansion

charge nurse NURSING A nurse with supervisory responsibility for the nurses on a shift in a particular ward or unit

charity care MANAGED CARE Free or reduced fee care provided due to financial situation of Pts. See Pro bono, Stewardship

Charley DRUG SLANG A street term for heroin. See Heroin

Charley Horse A popular term for leg pain caused by a contusion to the thigh involving the quadriceps muscle; without immediate compression, the ensuing hematoma and pain can be substantial; the term is often applied to painful injuries to other muscle groups

CHARM CARDIOLOGY A clinical trial–Candesartan cilexitil or Atacand™ in Heart Failure Assessment of Reduction in Mortality & Morbidity.

Charnley arthrodesis ORTHOPEDICS An operation consisting of trochanteric osteotomy with cemented acetabular and femoral components which is used in younger Pts with ankylosing spondylitis. See Ankylosing spondylitis

chart *noun* A formal document that includes relevant data, records, reports–eg radiology, pathology, of a person's medical Hx. See Bar chart,

Hospital chart, Organized chart, Run chart, Subway chart *verb* To place orders, progress notes or data in a person's chart

chart war A popular term for a Pt management-related skirmish between 2 physicians or services that occurs in writing on the pages of a Pt's chart, with attacks and ripostes in the form of progress notes, with each party questioning the other's justification for performing a particular procedure. See Chart

chasing the dragon Dragon chasing SUBSTANCE ABUSE Inhalation of heated heroin–or crack, associated with progressive spongiform leukoencephalopathy CLINICAL progressive bradykinesia, ataxia, slurred speech, abulia, decorticate posture, spasticity, quadriplegia, dysmetria, broad-based gait IMAGING MRI—diffuse, symmetric area of white matter hyperintensity, most prominent in the cerebellum; posterior brain, splenium of the corpus callosum, and posterior limb of internal capsule MANAGEMENT None PROGNOSIS 25% mortality. See Cocaine, Heroin

chastity SEXOLOGY An abstinence from unlawful sexual activity, either outside or inside marriage; total abstinence, celibacy. See Celibacy

chat room ONLINE A virtual environment on the Internet, in which participants with similar interest 'talk' via keyboard

CHD 1. Congenital heart disease 2. Coronary heart disease

check LAB MEDICINE See Delta check SPORTS MEDICINE See Body check VOX POPULI An evaluation to verify something

check-list method PEDIATRIC PSYCHOLOGY An objective method used to evaluate children, in which the observer checks the behavior of a child according to a specific list of traits and reactions

check sample PUBLIC HEALTH A distribution water sample sent to a state lab or to a state-approved lab for analysis

checkup See Yearly checkup

Chediak-Higashi syndrome Chediak-Higashi-Steinbrink syndrome, hereditary leukomelanopathy MOLECULAR MEDICINE A rare AR condition characterized by giant lysosomes and ↑ susceptibility to infections CLINICAL Partial albinism, lymphadenopathy, hepatosplenomegaly, debilitating neuropathy, photophobia, purulent infections; without BMT most Pts die by age 3 LAB Giant granules in WBCs and platelets, ↓ chemotaxis, bactericidal activity MANAGEMENT BMT, steroids PROGNOSIS Death by age 10

CHEESE AIDS A clinical trial–Comparative Trial of HIV-Infected Patients Evaluating Efficacy & Safety of Saquinavir-Enhanced Oral Formulation and Indinavir given as part of a Triple Drug Therapy. See AIDS, Indinavir, Saquinavir

cheilitis DERMATOLOGY Inflammation and/or cracking of the lips which may be 2° to a vitamin C or B complex, mineral deficiency, sunburn, cosmetics, dermatitis

chelation therapy The use of a chemical to bind and thereby inactivate a substance. TOXICOLOGY The only approved indication for EDTA chelation is for treating heavy metal–eg lead, mercury poisoning

chelator A chemical–eg, EDTA that binds metal ions from solutions. See Chelation therapy

Chem-6 LAB MEDICINE A battery of 6 serum analytes commonly ordered as part of a routine 'stat' profile: Na^+, K^+, Cl^-, CO_2, glucose, urea nitrogen

Chem-7 LAB MEDICINE A battery of 7 serum analytes often ordered together–chloride, CO_2 content, creatinine, glucose, K^+, Na^+, urea. See Chem-6

Chem-12 LAB MEDICINE A battery of 12 clinically important tests on serum analytes, which are often ordered together–albumin, alk phos, AST, calcium, cholesterol, glucose, LD, phosphate, total BR, total protein, urea nitrogen–BUN, and uric acid. Cf Chem-6, Chem-7

chematology lab LAB MEDICINE Consolidated hematology and chemistry departments in a clinical lab, enabled by the automation in each

chemical burn OPTHALMOLOGY A topical eye injury evoked by toxic fluids; untreated CBs can rapidly lead to permanent blindness; alkalines are more dangerous than acids as the high pH causes saponification of membranes with cell disruption and cell death; acids coagulate the superficial proteins, limiting the toxic agent's penetration AGENTS Ammonia derivatives–cleaning agents, fertilizers, refrigerants, lime products–plaster, mortar MANAGEMENT Irrigate with water, saline or any neutral pH liquid ASAP X ≥ 10 mins

chemical cardioversion CARDIOLOGY The use of antiarrhythmics to convert an arrhythmia into a normal pattern. See Antiarrhythmics. Cf Direct current cardioversion

chemical castration Pharmacologic castration PUBLIC HEALTH The treatment of ♂ with paraphilia with methoxyprogesterone acetate, which inhibits gonadotropin secretion. See Chemical castration, Megan's law, Pedophilia

chemical colitis Soap colitis An acute inflammatory colitis after a 'cleansing' enema; although classically associated with soapsuds, acute CC may also occur after rectal instillation of a wide range of 'insulting fluids' including herbal concoctions, H_2O_2, vinegar, potassium permanganate, sodium diatrizoate–Hypaque, glutaraldehyde, Fleet's Phospho-Soda enema or inappropriate fluids, which may have hypertonic, detergent, or directly toxic effects CLINICAL Pain, cramping, anaphylaxis, serosanguineous diarrhea, hypovolemia, hemoconcentration; if the mucosal damage is severe, bacteria may penetrate the mucosa, cause sepsis, hypokalemia, pseudomembrane formation, hemorrhagic necrosis, intestinal gangrene, and acute renal failure. See Colon therapy, Detoxification therapy, High colonic

chemical denervation THERAPEUTICS The use of any agent that blocks neural transmission in a relatively complete fashion for a prolonged period. See Botulinum toxin

chemical dependency SUBSTANCE ABUSE A condition involving continued misuse of chemical substances. See Substance abuse

chemical diabetes mellitus A sub/preclinical form of DM characterized by DM-like response curves to GTT or other provocation tests; benefits of treating these Pts are uncertain. See Glucose tolerance test

chemical eye injury OPHTHALMOLOGY An eye injury caused by a fluid with pH that is either too high–alkali injury, base or too low–acid injury; alkali CEIs cause liquefaction necrosis; acid CEIs cause coagulation necrosis MANAGEMENT Copious irrigation

chemical imbalance PSYCHOLOGY A popular term of uncertain utility, which refers to a belief that many, if not all, mental disorders are attributable to a disequilibrium of one or more neurotransmitters

chemical injury TOXICOLOGY A caustic burn related to direct contact of a strong acid or base with a mucocutaneous surface–eyes, skin, respiratory tract, upper GI tract. See Chemical eye injury

chemical matricectomy PODIATRY The use of a caustic chemical–eg, phenol, to selectively destroy part of the nail bed in Pts with ingrown toenails. See Matricectomy

chemical peel Chemexfoliation DERMATOLOGY A technique in which phenol or trichloroacetic acid–TCA is 'painted' on elderly sun-exposed skin covered by extensive premalignant actinic keratosis. See Actinic keratosis. Cf Laser resurfacing, Mohs' surgery

chemical pneumonia Chemical pneumonitis PULMONOLOGY Chronic lung inflammation 2° to inhaled toxins–eg, phosgene or chlorine, organic dust, fungi or spores CXR Chronic interstitial lung changes, pulmonary edema CLINICAL Cough, fever, SOB, wheezing, ↓ O_2 exchange–acute with edema, hypoxia; chronic with interstitial fibrosis, possibly respiratory failure

AGENTS Bleach, cleaners, beryllium–in old fluorescent lightbulbs, methylene chloride–paint strippers. See Animal house fever, Aspiration pneumonia, Chemical warfare, Pneumonia

chemical spill PUBLIC HEALTH An inadvertent release of a liquid chemical regarded as hazardous to human health which in a workplace is identified with hazardous materials labels. See Material Safety Data Sheets

chemical substance PHARMACOLOGY *Any* substance including alcohol, drugs, or medications–eg, those taken pursuant to a valid prescription for legitimate medical purposes, and in accordance with the prescriber's direction, as well as those taken illegally. See Substance abuse

chemically-modified hemoglobin TRANSFUSION MEDICINE A stabilized Hb solution designed to carry O_2 or serve as a blood substitute; CMHs remain in the intravascular space for 12 hrs to 4 days and may interfere with measurement of Hb, BR, and lipids, but do not interfere with antibody testing. See Blood substitute, Perfluorocarbon

chemiclave SURGERY A machine that sterilizes surgical instruments with high-pressure, high-temperature water, alcohol, formaldehyde vapor

chemonucleolysis NEUROSURGERY The injection of an enzyme–eg, chymopapain, as an alternative to laminectomy in certain cases of intervertebral disc rupture. See Laminectomy

chemoprevention NUTRITION The use of drugs, chemicals, vitamins or other substance in the diet or to prevent or ↓ the incidence of a disease–eg, CA; chemopreventive agents–supported by 'soft' epidemiologic data are vitamins A, C, E, bran, other dietary fibers, and cruciferous vegetables. See Antioxidant, Antioxidant therapy. Cf Chemoprophylaxis, Unproven methods for cancer management

chemoprophylaxis The use of a chemical, usually a therapeutic, to prevent the occurrence of a disease, usually infectious–eg, malaria or TB. Cf Chemoprevention

chemoradiation Chemoradiotherapy ONCOLOGY The sequential use of chemotherapeutics–eg, 5-FU, cisplatin, etoposide, then RT, to treat CA

chemoreceptor trigger zone Area postrema NEUROANATOMY The neural center for emesis, located in the floor of the 4th ventricle, which receives vagal afferents or stimulated directly by apomorphine, cardiac glycosides, ergot compounds, chemotherapeutics, staphylococcal enterotoxin, salicylate and nicotine and other circulating chemicals; in contrast to the adjacent but distinct vomiting center, the CTZ does not respond to electrical stimulation

chemoresistance 1. Resistance of cells to the action of a specific therapeutic agent 2. Resistance of a particular tumor to chemotherapy; treatment refractory

chemoresponsive *adjective* Referring to tumor responsiveness to chemotherapy

chemosensitivity 1. Sensitivity of chemoreceptors to hypoxia and hypercapnea; ↓ chemosensitivity to hypoxia has been linked to fatal asthma attacks 2. Sensitivity of a tumor to chemotherapy

chemosensory dysfunction Chemosensory disorder ENT A defect of smell and/or taste, which has subtle influences on a person's quality of life ETIOLOGY Head trauma, URI, chronic nasal and paranasal sinus disease; psychological depression and thyroid dysfunction are associated with CD; estrogens ameliorate the CD seen in postmenopausal Pts; a specific type of CS, the burning mouth syndrome, is associated with extreme weight loss

chemosis Edema of bulbar conjunctiva around the iris

chemosurgery Mohs' micrographic technique PLASTIC SURGERY A technique for excising superficial, locally invasive, tumor microfingerlets of primary, nonmelanoma skin CA, yielding a cure rate of > 98% for broad, super-

ficial skin CAs, best for > 1-2 cm recurrent tumors, recurrence-prone sites–nose, eyes, ears, aggressive histologic subtypes–eg, morphea-like or metatypical BCC. See Basal cell carcinoma. Cf Chemical peel

chemotactic factor Chemoattractant, chemotaxin IMMUNOLOGY Any small molecule that acts as a chemical stimulus along a concentration gradient, attracting macrophages and other cells–eg, to a site of inflammation. See Chemotaxis

chemotherapeutic *adjective* Referring to a chemotherapeutic agent, effect or regimen *noun* Chemotherapeutic agent, see there

chemotherapeutic agent An agent used to treat CA, administered in 'regimens'–one or more 'cycles' that combine 3 or more agents over wks; CAs are toxic to any cell with a high rate of proliferation–the CA itself, the GI tract–causing N&V, BM–cytopenias, and hair–alopecia EXAMPLES Busulphan, cisplatin, cyclophosphamide, MTX, daunorubicin, doxorubicin, melphalan, vincristine, vinblastine, chlorambucil. See Immunosuppressive agents

chemotherapy Chemotherapeutics ONCOLOGY The use of various agents, most of which are toxic to dividing cells, to induce tumor cell lysis; successful CT is a function of tumor responsiveness, which most predictably occurs in lymphoproliferative malignancy–eg, leukemia and lymphoma, and small cell–undifferentiated carcinoma; because CT is most effective against rapidly proliferating cells, collateral damage to dividing cells in skin and hair, BM, GI tract, are common, resulting in reversible hair loss, myelosuppression, N&V; a late effect of CT, especially in pediatric leukemia, is induction of a 2nd malignancy, which may be refractory to therapy. See Adjuvant chemotherapy, Antifolate chemotherapy, Combination chemotherapy, Combined modality therapy, Consolidation chemotherapy, Damocles' syndrome, Electrochemotherapy, High-dose chemotherapy, Hyperfractionation trans-C chemotherapy, Induction chemotherapy, Intensification chemotherapy, Intraperitoneal chemotherapy, Intraperitoneal hyperthermic chemotherapy, Intrathecal chemotherapy, Maintenance chemotherapy, Microdose chemotherapy, Neoadjuvant chemotherapy, Photochemotherapy, Preemptive chemotherapy, Regional chemotherapy, Salvage chemotherapy

chemotherapy-induced emesis Chemotherapy-induced nausea & vomiting ONCOLOGY A side effect of many chemotherapeutic agents which, while often the most anxiety-provoking of the toxic effects of chemotherapy, is self-limited and rarely life-threatening HIGHLY EMETOGENIC Cisplatin, carmustine, dacarbazine, dactinomycin, mechlorethamine–nitrogen mustard, streptozocin MANAGEMENT Dopamine D_2 high-dose metoclopramide, serotonin–5-HT_3 receptor antagonists–eg, ondansetron. See Ondansetron

chemotherapy-induced leukemia ONCOLOGY A leukemia linked to prior chemotherapy, especially with alkylating agents, which induce ANLL, often with mutations of chromosomes 5 and 7. See Secondary malignancy

chemotherapy-resistant disease ONCOLOGY A cancer that responds to a chemotherapeutic regimen for ≤ 1 month. See Relapse. Cf Remission

Chérambault-Kandinsky syndrome PSYCHIATRY A false conviction by a Pt that his/her life is controlled by an unattainable loved one who secretly reciprocates the love

cherry angioma Senile angioma DERMATOLOGY A benign, ruby red, 1-3 mm in diameter papule, surrounded by a pale halo; the CA is commonly seen on the trunk and extremities of older adults, located in the superficial corium, and consists of dilated, thinned capillaries that cause superficial bumps MANAGEMENT Excision by cryotherpapy for cosmetic purposes

cherry red color A descriptor for the bright red, mucocutaneous discoloration, classically associated with CO poisoning; 'cherry red' also refers to the oropharyngeal discoloration seen in acute epiglottitis. See Carbon monoxide intoxication

cherry spot METABOLIC DISEASE A bright red macule in the optic fundus of Pts with Tay-Sachs disease–GM2-gangliosidosis type 1, Niemann-Pick diseases, Sandhoff's disease–GM2-gangliosidosis type 2, generalized gangliosidosis–GM1-gangliosidosis type 1, cherry red spot myoclonus syndrome or sialidosis, type 1, sialidosis type 2, Goldberg syndrome, mucolipidosis type 1, metachromatic leukodystrophy, retinal vasculopathy

cherubism Familial intraosseous fibrous swelling of jaws An AD condition, 100% penetration in ♂ with a cherub-like physiognomy, first recognized by age 5 CLINICAL Puffed-out cheeks, agenesis of permanent teeth, dental dysgenesis, exophthalmos, progressive, bilateral soap-bubble expansile lesions of bone, at the mandibular angle, submandibular lymphadenopathy

CHESS DECISION ANALYSIS Comprehensive Health Enhancement Support System A comprehensive interactive computer-based information/support tool which helps Pts understand and make decisions about diagnostic and treatment options in dread or debilitating diseases–eg breast and prostate CA, AIDS, menopause and heart disease. See Decision-making, Empowerment PHARMACOLOGY Cornell High-Energy Synchrotron Source A powerful x-ray device housed in Ithaca, NY, used to analyze crystallized molecules. See Drug development

chest Thorax A popular term for the region between the neck and abdomen, which contains the heart and lungs plumbing and wiring. See Barrel chest, Dirty chest of Simon, Flail chest

chest examination The evaluation of the chest, in which the examiner listens to breath and heart sounds, location and nature of pain, to identify disease

CHEST EXAMINATION

FINDING	BREATH SOUND*	PERCUSSION	FREMITUS
Pleural effusion	↓	Dull	↓
Pneumonia	↓	Dull	↑
Pneumothorax	↓	Hyper	↓

*BS Breath sounds; Hyper Hyperresonant; PC Percussion; TF Tactile fremitus

chest film Chest X-ray IMAGING A plain AP radiograph of the chest, the most commonly ordered X-ray, to evaluate lungs, heart, aorta, thoracic bones

chest pain A general term for any dull, aching pain in the thorax, usually referring to that of acute onset, which is often regarded as being myocardial in origin unless proven otherwise

chest tube Thoracostomy A tube, placed under negative pressure in the pleura cavity, to drain fluid

chest wall pain Chest pain of noncardiac cause, which may be due to musculoskeletal inflammation

chest X-ray Chest film, see there

Cheung BPH system UROLOGY A microwave heating + balloon catheterization system to shrink obstructing prostatic tissue in BPH and create a biological stent to maintain patency. See Benign prostatic hypertrophy

chew Chewing tobacco. See Smokeless tobacco

chewing gum diarrhea A type of osmotic diarrhea caused by an excess of sugarless chewing gum, which contains hexitols–eg, sorbitol and mannitol; hexitols are major constituents of sugar-free dietary foods, are not metabolized by bacteria, remain in the GI lumen, and ↑ intraluminal osmotic pressure, resulting in diarrhea

chewing tobacco Smokeless tobacco, see there

Cheyletiella blakei ENTOMOLOGY A mite that may transiently burrow into the skin of humans belonging to a cat, evoking pruritus and innumerable papular urticarial lesions that heal by scabbing. See Cats, Mites

Cheyletiella yasguri MEDICAL ENTOMOLOGY A dog mite that may transiently burrow into a dog owner's skin and evoke pruritus, causing self-resolving papulourticarial lesions. See Dogs, Mites

Cheyne-Stokes breathing INTERNAL MEDICINE A form of recurrent apnea seen in Pts with neurologic and heart disease, characterized by regular volleys of apnea followed by regular 'crescendo-decrescendo' fluctuations in respiratory rate and tidal volume, punctuated by periodic apnea, hypo- or hyperventilation; CSB is triggered by ↑ arterial PCO_2, sedatives, opiates; it is more common during sleep and at high altitudes, and seen in comas 2° to a severe neurologic insult. Cf Biot breathing

CHF Congestive heart failure, see there

chi squared test X^2 STATISTICS A test of association between 2 or more variables, used to identify a linkage between a factor or attribute and an outcome, and determine if the observed values of a variable are significantly different from those expected based on a null hypothesis. See Null hypothesis

chicken soup Chicken broth FOLK MEDICINE Jewish penicillin A fowl broth with a long tradition as a home remedy for URIs, which may be a nasal decongestant, inhibit growth of pneumococci in vitro, and stimulate immune responsiveness in WBCs MAINSTREAM MEDICINE A popular term for a drug, maneuver, or device with little or no benefit, which may be used in absence of an efficacious product. See Band-Aid® therapy, 'Homeopathic', Placebo

chickenpox Varicella, human herpesvirus type 3 Acute HHV-3 infection, most common before age 10; 3.5 million cases & 50 children die/yr of chickenpox–US; 9000 are hospitalized CLINICAL 2-wk incubation, then a scarlatiform prodromal rash, low-grade fever, weakness, sore throat, cough, anorexia, malaise, crops of reddish papules that become intensely pruritic vesicles, which ↑ in number for 3-4 days; itching and excoriation may cause extensive (permanent) scarring COMPLICATIONS Otitis, pneumonia, 2° bacterial rashes and infections, encephalitis–5-15% mortality, 15% with permanent neurologic sequelae, ataxia, palsies, Reye syndrome, a potentially fatal complication, viral pneumonia–1:400 require hospitalization, thrombocytopenia, purpura fulminans, myocarditis, glomerulonephritis, hepatitis, myositis; after resolution of clinical disease, HHV-3 becomes latent, integrating its DNA into the dorsal root ganglion cells MANAGEMENT Acyclovir may shorten duration of disease. See Chickenpox vaccine, Shingles

chickenpox vaccine PEDIATRICS A one-shot live varicella virus vaccine that confers 6+ yrs of protection in 70–90% ADVERSE EFFECTS Injection site pain, erythema CONTRAINDICATIONS Hx of anaphylactoid reactions to neomycin or gelatin, current fever, pregnancy, immunocompromise; pregnancy be avoided for 3 months post-vaccination. See Chickenpox

chief *adjective* Principal, main, see Chief complaint *noun* 1. Chief of service, see there 2. Chief of staff, see there

chief complaint The principal reason for a person seeking medical attention; CCs include acute abdomen, colicky pain, crushing chest pain, hip fracture, etc. See Complaint, Minor complaint

chief of service MEDICAL PRACTICE Head of a department or section of a clinically oriented service in a hospital or health care facility

chief of staff MEDICAL PRACTICE The chief administrative officer of the medical staff; the physician or other health care professional who is in charge of the medical staff in a hospital or health care organization ORIENTAL MEDICINE The liver. See Twelve vital organs

chief syndrome A phenomenon that may occur when a 'very important person'–V.I.P.–is admitted to a major medical center or university hospital. See V.I.P. 'syndrome'

chigger A popular term for the dermatitis caused by mite larvae–chiggers CLINICAL Intense pruritis, red rash on the waist, ankle and skin folds MANAGEMENT Usually resolve spontaneously; ± antihistamines for itching. See Chigger

chiggers Harvest mites, red mites DERMATOLOGY Larvae of the family Trombiculidae, genus *Eutrombicula*–southern US, *Trombicula*–Europe which causes skin infestation HABITAT Berry patches, tall grass, weeds, woods. See Chiggers

child A person who has not attained the legal age for consent to treatment or procedures involved in the research, as determined under the applicable law of the jurisdiction in which the research will be conducted MEDTALK Pediatric patient. See Adopted-in child, Adopted-away child, Battered child, Chosen child, FLK, Latchkey child, Puppet child, Wednesday's child, *The Wild Child*

child abuse Battered child syndrome, trauma 'X' PUBLIC HEALTH A tragedy that claims 2000-5000 lives/yr and causes countless injuries in the US; CA is often first recognized by characteristic radiologic findings–eg metaphyseal fragmentation, incomplete 'bucket handles', old fractures, sub-periosteal hematomas–with epiphyseal dislocations, metaphyseal cupping, shortening of long bone shafts, and a ball-and-socket configuration, pelvic fractures, fractures of posterior ribs, spine, and sternum, a post-mortem radiologic survey may be required to convict the caretaker/parent of manslaughter SOCIAL MEDICINE Behavior by a parent or guardian that causes significant negative emotional or physical consequences in a child TYPES Physical abuse, emotional abuse, sexual abuse, neglect. See Batttered child, Child maltreatment, Domestic violence, Emotional abuse, Neglect, Physical abuse, Sexual abuse, Shaken-baby syndrome

child abuser PUBLIC HEALTH A person who mentally or physically abuses a child TYPICAL CA PROFILE Age < 30, slightly more likely to be ♀, whose mother was unemployed/employed part time as a manual laborer TYPICAL VICTIM Young children, teens. See Child abuse, Domestic violence

child labor PUBLIC HEALTH Gainful employment of a child–usually, < age 15 in a workplace that employs adults functioning in equal or similar capacities. See Sweatshops. Cf Terms of Engagement

child maltreatment '...*intentional harm or threat of harm to a child by someone acting in the role of a caretaker, for even a short time...* CATEGORIES *Physical abuse, sexual abuse, emotional abuse, neglect...*', the last being most common. See Child abuse, Child abuser

child protective services SOCIOLOGY A state or county agency that addresses issues of child abuse and neglect

child psychiatrist PSYCHIATRY A psychiatrist specialized in mental, emotional, or behavior disorders of children and adolescents; CPs are qualified to prescribe medications

child psychologist PSYCHOLOGY A mental health professional with a PhD in psychology who administer tests, evaluates and treats children's emotional disorders, but can't prescribe medications

CHILD syndrome An X-linked congenital lethal complex that is fatal in ♂–♂:♀ ratio, 19:1 CLINICAL Unilateral ichthyosis, limb malformation, accompanied by ipsilateral hypoplasia of paired organs–eg, lung, thyroid, psoas muscle, CNS, and cranial nerves

child welfare agency CHILD PSYCHIATRY An administrative organization providing protection to children, and supportive services to children and their families

Child-Pugh score HEPATOLOGY A scoring system used in Pts undergoing TIPS, which describes a range of severity of liver disease. See TIPS

childbirth NEONATOLOGY The constellation of processes that result in the birth of an infant. See Natural childbirth

childbirth, fear of PSYCHOLOGY Lockiophobia, Maleusiophobia, Parturiphobia, or Tocophobia. See Phobia

childhood PEDIATRICS The period of development between infancy and puberty FORENSIC DEFINITION The period of development until the 18th birthday

childhood immunization Children's vaccination, childhood vaccination In the US, it is recommended that all children receive vaccination against Diphtheria, tetanus, pertussis, HBV, *H influenzae* type B–HIB, poliovirus, measles, mumps, rubella, varicella zoster virus–chickenpox. Cf Adult immunization varicella/zoster

childhood obesity PUBLIC HEALTH Overweight in a child, an average BMI of ≥ 85% for age and sex; ≥ 95% for age and sex is *very obese*. See Body-mass index, Obesity. Cf Adult obesity

childhood volvulus PEDIATRICS A condition characterized by lower GI obstruction, and variable compromise in vascularity CLINICAL N&V, bloody stool, abdominal pain, constipation, shock MANAGEMENT Surgical reduction. See Paralytic ileus

chill CLINICAL MEDICINE 1. A sensation of coldness often accompanied by shivering, chattering of teeth, goosebumps–gooseflesh, and skin pallor, which may follow exposure to a cold, damp environment, or precede or occur at the same time as a cold; chills are a response to an abrupt disparity between the set point of the hypothalamic thermostat and the blood temperature

chin-lift maneuver See Head-tilt/chin-lift maneuver

China flu INFECTIOUS DISEASE A strain of influenza A, H3N2, that caused outbreaks in China and the US

Chinese restaurant syndrome Monosodium glutamate allergy, MSG allergy An abrupt allergic reaction, the susceptibility to which is AR, caused by sensitivity to monosodium glutamate–MSG, a seasoning used in Chinese restaurants and in soy sauce CLINICAL Abrupt onset of severe headaches, heartburn, numbness, palpitations, vertigo–especially with chronic MSG exposure, thirst, abdominal and chest pains/heartburn, sweating, and flushing. Onset 30 min after meal, lasting up to 12 h MANAGEMENT Nada, usually self-limited; antihistamines in some Pts. See MSG

chip fracture A small fracture through a bone region that serves as the site of a ligamentous attachment. See Fracture

chipmunk face A descriptor for expanded globular maxillae, with BM hyperexpansion into facial bones, combined with prominent epicanthal folds, characteristic of severe β thalassemia; CF may also refer to soft tissue swelling–eg, diffuse parotid gland swelling, accompanied by xerostomia and reddened eyes, seen in Sjögren syndrome. See β-thalassemia

chipper DRUG SLANG An occasional user of illicit drugs. See Recreational drug use TOBACCO A popular term for a person who smokes < 5 cigarettes/day, who may be resistant to nicotine dependence or addiction, and often born to non-smoking parents

chiropody Podiatry, see there

chiropractic ALTERNATIVE MEDICINE Referring to a system of health care which is based on the belief that the nervous system is the most important determinant of a person's state of health; according to chiropractic theory, most diseases are the result of 'nerve interference,' caused by spinal subluxations, which are said to respond to spinal manipulation; abnormal nerve function may result in musculoskeletal derangements and aggravate pathologic processes in other body regions or organs. See Chiropractor, Medically-oriented chiropractic, Mixed chiropractic, Network chiropractic, Straight chiropractic, Subluxation-based chiropractic. Cf Massage therapy, Osteopathic medicine

PRINCIPLES OF CHIROPRACTIC

VITALISM The body has an intrinsic ability to heal itself; the chiropractor's role is to facilitate the body's ability to restore the vital or life force–termed innate intelligence, to its optimum level, and therefore be allowed to heal itself

HOLISM All organs and systems function as one interconnected unit; anything that affects the nervous system has widespread effects elsewhere in the body

CORRECTION OF SUBLUXATION Subluxation is defined as a malalignment of the vertebrae that causes pressure on the spinal cord, nerve roots, and nerves; chiropractics have labelled this subluxation-induced pressure on nerves 'nerve reflex'–which has a different connotation for neurologists

chiropractic therapy Chiropractic treatment consists of adjustment and manipulation–*chiro*–, Greek, hand of the vertebral column and extremities, which some chiropractors supplement with physical therapy, nutritional support, and radiography–for diagnostic purposes only CONDITIONS THAT MAY BE REGARDED AS THE THERAPEUTIC DOMAIN OF A CHIROPRACTOR Arthritis, pain in neck, shoulder, back, arms, legs, pain described as 'pins and needles' and numbness, sports injuries, whiplash, migraine, sprains, strains, insomnia, nerve entrapment, muscle cramps, stiffness, occupational injuries. See Chiropractic, Chiropractor

chiropractor A health professional trained in chiropractic; chiropractors do not perform surgery or prescribe drugs; of 50,000 licensed chiropractors in the US, many practice 'straight' chiropractic, ie lumbar manipulation, believed–a belief shared by some mainstream physicians–to benefit Pts with low back pain. See Chiropractic, Chiropractic therapy

chisel fracture An incomplete fracture of the head of the radius, in which the fracture line extends distally from the center of the articular surface

Chlamydia pneumoniae *C psittaci* TWAR A pathogen that causes pneumonia, asymptomatic RTIs, pharyngitis, otitis media

Chlamydia psittaci The agent that causes psittacosis, see there

Chlamydia trachomatis STD A human pathogen, similar to gonorrhea in transmission and disease; it is found in the cervix and urethra and survives in the throat or rectum EPIDEMIOLOGY It is the most common STD agent in the US–causing ± 4.5 million cases/yr; it is present in 1-3% of all ♂ and 15-40% of ♀ in STD clinics CLINICAL Inclusion conjunctivitis, lymphogranuloma venereum, urethritis, epididymitis and proctitis in ♂, mucopurulent cervicitis, endometritis, salpingitis–*C trachomatis* is implicated in 50% of salpingitis and PID, bartholinitis, and acute urethral syndrome in ♀ and conjunctivitis and pneumonia in neonates; infection may be asymptomatic DIAGNOSIS Direct fluorescent antibody staining, solid phase immunoassay, ELISA, cell culture, nucleic acid probe, PCR COMPLICATIONS Fallopian tube destruction, ±infertility, ectopic/tubal pregnancy, preterm delivery, severe PID MANAGEMENT Doxycycline, azithromycin. Cf *Mycoplasma pneumoniae*

chlamydial urethritis STD A common STD of the ♂ urethra caused by *C trachomatis*. See *Chlamydia trachomatis*

chloasma Melasma, cloasma hepaticum DERMATOLOGY A term for melasma-like facial hyperpigmentation, which may be a skin manifestation of pregnancy or chronic liver disease

chloracne OCCUPATIONAL DERMATOLOGY A condition caused by halogenated organic compounds–eg, chlorinated chemicals

chlorambucil Leukeran™ ONCOLOGY An alkylating chemotherapeutic of the nitrogen mustard family, used for lymphomas and other CAs ADVERSE EFFECTS BM suppression, leukemia, infertility, convulsions

chloramphenicol INFECTIOUS DISEASE A broad-spectrum antibiotic effective against gram-positive cocci–eg *Staphylococcus aureus* and gram-negative coccobacilli–eg *Brucella abortus* COMPLICATIONS Aplastic anemia

chlordane TOXICOLOGY A viscous, chlorinated organic compound once used as an insecticide and fumigant, now EPA-limited to control termites and non-food plants; it is absorbed through the skin, causes liver and kidney damage, possibly CA CLINICAL Convulsions, depression, hyperexcitability, incoordination, death

chlordiazepoxide NEUROPHARMACOLOGY A psychoactive benzodiazepine sedative, hypnotic, muscle relaxant, anticonvulsant EFFECTS Undetermined limbic CNS depressant effects USED FOR Anxiety, anxiety disorder CONS Physical and psychologic dependence similar to barbiturates. See the Little Yellow Pill

chlorination PUBLIC HEALTH Addition of chlorinated compounds to drinking water as disinfectants. Cf Ozonation

chlorine A toxic gaseous element–atomic number 17, atomic weight 35.45, used as a bleaching agent; although clorine is the critical for metabolism, it is present as chloride, which has a valence of –1. See Chloride

Chlorophyllum molybdites *Lepiota morgenii* TOXICOLOGY A mushroom that causes non-fatal poisoning CLINICAL Nausea, cramping, diarrhea for up to 3 days–the active components of *C molybdites* are GI irritants MANAGEMENT Supportive, as for severe gastritis. See Poisonous mushroom

chlorpheniramine An antihistamine. See Allergy

chlorpromazine Thorazine® An aliphatic phenothiazine, neuroleptic, antipsychotic, with antiadrenergic, anticholinergic, antiemetic, central antihypertensive activity

chlorpropamide ENDOCRINOLOGY An oral sulfonylurea hypoglycemic, structurally related to tolbutamide, which ↑ release of insulin from beta islet cells in Pts with type 2 DM

chlorpryrifos PUBLIC HEALTH The most widely used US pesticide used against cockroaches, termites, fleas. See Cockroaches, Pesticide bomb

choanal atresia Posterior choanal atresia ENT A partial or complete obstruction of bony–90% of cases or membranous passages from the nasal cavity to the nasopharynx, due to a persistent nasopharyngeal or buccopharyngeal membrane MANAGEMENT Transpalatal removal of obstruction and part of the posterior nasal septum, to prevent scarring; if membranous, probe perforation, laserization, cauterization may suffice.

chocolate A comestible prepared from ground and roasted beans of the cacao plant, *Theobroma cacao*, native to South America; it is composed of cocoa butter, a substance high in stearic acid, which is converted in vivo to oleic acid, which may lower cholesterol; ⅓ of cocoa butter is palmitic acid which ↑ cholesterol. See Chocolate craving, Cf Carob DRUG SLANG A regional street term for opium; amphetamine; marijuana

chocolate cyst Endometrioma GYNECOLOGY A periadnexal or ovarian cyst filled with thick inspissated, old and unclotted blood, grossly simulating chocolate, seen in endometriosis. See Endometriosis

choke CLINICAL MEDICINE *verb* To suffer a sensation of obstruction of the upper airways FORENSIC PATHOLOGY *verb* To intentionally obstruct the upper airways of another person by external compression, at the level of the trachea. See Choke hold, Strangulation

choked disk OPHTHALMOLOGY Papilledema with swelling of the optic nerve head, caused by ↑ intracranial pressure with edema-induced blurring of the disk margins and obliteration of the optic cup, elevation of the nerve head, capillary congestion, hyperemia, venous engorgement, loss of venous pulse, peripapillary exudates, retinal wrinkling, and punctate nerve fiber layer hemorrhage; if pressure is reduced, fundus returns to normal without loss of vision; ↑ intracranial pressure is due to meningoencephalitis, hemorrhage, metabolic disease, toxins, trauma, tumors. See Pseudotumor cerebri

the 'chokes' Sudden onset of respiratory distress in caisson's disease, which is associated with pulmonary edema, hemorrhage, atelectasis, and emphysema; the 'chokes' may be due to an ↑ in platelet adhesion to gas bubbles that release vasoconstrictors, platelet factor 3, causing coagulopathy. See the Bends, Caisson's disease

cholangiectasis GI DISEASE Dilatation of bile ducts, due to ↓ muscle

cholangiocarcinoma Bile duct cancer, cancer of bile ducts A rare–5/100,000/yr cancer of intrahepatic bile ducts seen > age 60 ETIOLOGY Anabolic steroids, Thorotrast, *Clonorchis sinensis*, possibly ulcerative cholitis, cholecystitis; *not* associated with alcohol abuse CLINICAL Jaundice—71%, abdominal pain-49%, weight loss—44% PROGNOSIS 53% 1-yr survival, 9% 3-yr, 4% 5-yr. Cf Hepatocellular carcinoma

cholangiography IMAGING Oral, IV or percutaneous administration of radiocontrast excreted into the bile tract, to detect gallstones or visualize bile ducts. See Percutaneous transhepatic cholangiography

cholangioma 1. Any tumor of the bile ducts 2. Cholangiocarcinoma, see there

cholangiopathy Any disease of the bile ducts. See AIDS-associated sclerosing, Infantile obstructive

cholangitis HEPATOLOGY Inflammation of a bile duct. See Acute cholangitis, Primary sclerosing cholangitis, Recurrent pyogenic cholangitis

cholecystectomy SURGERY The surgical removal of the gallbladder INDICATIONS Gallstones, cholecystitis, cancer. See Bile, Cholecystitis, Gallstones, Laparoscopic cholecystectomy

cholecystitis GI DISEASE Inflammation of the gallbladder. See Acalculous cholecystitis, Acute cholecystitis, Chronic cholecystitis, Emphysematous cholecystitis

CHOLECYSTITIS

ACUTE CLINICAL Right upper quadrant pain, variable severity, sudden onset, rigidity and rebound tenderness with peritonitis, N&V, constipation, fever, slow pulse, Murphy sign, spontaneous remission is not uncommon IMAGING Stones by ultrasound, plain films, CT MANAGEMENT Cholecystectomy

CHRONIC 'Classically' occurs in the '4 F' group—female, fat, fertile, and forty; cholecystitis is also associated with oral contraceptives and heredity CLINICAL Some Pts are asymptomatic; otherwise the symptoms of acute inflammation wax and wane IMAGING Bilirubin or cholesterol stones by ultrasound, plain films, CT MANAGEMENT Conservative therapy to dissolve stones

choledocholithiasis Common bile duct stone(s) GI DISEASE The presence of gallstones in the common bile duct, which occurs in ±15% of Pts with gallstones CLINICAL Abdominal pain, N&V; if severe, fever, chills and jaundice. See Bile tract, Gallstone

choledochotomy Incision into the common bile duct

cholelithiasis The presence of gallstones. See Gallstones

cholera INFECTIOUS DISEASE A severe form of gastroenteritis, caused by *V cholera*, often acquired through contaminated water supplies; cholera may be seasonal in developing regions–eg, Africa, Asia, South America CLINICAL Copious watery stools, abdominal colic and eventual collapse from dehydration MANAGEMENT Aggressive hydration. See Bengal cholera, Cholera toxin, WHO solution

cholera toxin INFECTIOUS DISEASE A heat-sensitive multimeric enterotoxin produced by *Vibrio cholera*, which transfers ADP-ribose to a G protein, locking adenyl cyclase in an 'on' position by ADP ribosylation of a G_s protein

Cholestagel® INTERNAL MEDICINE A nonabsorbed cholesterol reducer, that ↓ LDL-C in hypercholesterolemia. See Cholestatin

cholestasis HEPATOLOGY An intra– or extrahepatic block of bile flow or bile ducts resulting in ↑ serum BR, ergo jaundice

cholesteatoma DERMATOLOGY A benign plug of keratinized squamous epithelium and cholesterol in the middle ear post acute otitis media CLINICAL Hearing loss, ear fullness, pain

cholesterol BIOCHEMISTRY A precursor of steroid hormones and bile acids; it is an integral component of cell membranes and plasma lipoproteins, and found in animal fats, bile, blood, brain, milk, egg yolk, myelin

sheaths, liver, kidneys, and adrenal glands; it is absorbed from the diet *and* synthesized in the liver; diets high in saturated–animal fats ↑ cholesterol levels; it is the main component of the most common type of gallstones, and integral to arterial atheromas, in cysts and in malignancy; it is a precursor of bile acids and important in the synthesis of steroid hormones; diets low in saturated fats ↓ cholesterol levels, as does exercise; ↑ cholesterol is associated with ASHD, CAD, ↑ risk of death due to acute MIs and strokes; total cholesterol–TC is measured in routine chemistry panels; cholesterol is transported in the circulation by carrier proteins, which are classified according to their density–high-density lipoprotein—HDL, LDL, VLDL, based on density gradient ultracentrifugation; HDL-C is metabolized efficiently, and thus is 'good' cholesterol; 'bad cholesterol,' LDL-C, is inefficiently metabolized HIGH RISK FOR ASHD TC > 6.21 mmol/L–US > 240 mg/dL; LDL-C > 160 mg/dl, HDL-C < 35 mg/dl BORDERLINE RISK 5.17-6.18 mmol/L–US: 200-239 mg/dL LOW RISK < 5.17 mmol/L–US 200 mg/dL, LDL-Cl < 130 mg/dL, HDL-C > 55 mg/dL ↑ IN Hypercholesterolemia, nephrotic syndrome, hypothyroidism, biliary cirrhosis ↓ IN Malnutrition, hyperthyroidism, colorectal CA MANAGEMENT, ↑ CHOLESTEROL Diet–↓ saturated fats, weight loss, regular exercise, medications. See Bad cholesterol, Fish, Good cholesterol, HDL, Hypercholesterolemia, LDL, RLP, VLDL

cholesterol embolization syndrome Atheroembolic renal disease VASCULAR DISEASE A condition affecting ± 4/10,000, more common in ♂ ≥ age 60, in which showers of cholesterol and debris from atherosclerotic plaques embolize to renal arterioles, blocking blood flow, resulting in renal infarction and possibly renal failure RISK FACTORS ASHD, DM, HTN, smoking, obesity, hyperlipidemia. See Atherosclerotic heart disease

cholesterol-lowering drug THERAPEUTICS Any of a family of agents that ↓ serum cholesterol; the most cost-effective agents for lowering LDL-C are nicotinic acid and lovastatin; the most efficient for ↑ HDL-C are nicotinic acid and gemfibrozil

cholesterol-raising fatty acid CARDIOLOGY A dietary fat that ↑ total and/or LDL-C EXAMPLES Lauric acid, myristic acid, palmitic acid, *trans* monosaturated fatty acids. See Cholesterol, Tropical oils

cholesterosis Cholesterolosis A condition characterized by deposition of cholesterol in tissues, most commonly in the gallbladder as aggregates of cholesterol-laden macrophages below the mucosa

cholesteryl ester storage disease Cholesterol ester storage disease MOLECULAR MEDICINE An AR defect in lysosomal acid lipase activity, resulting in major tissue accumulation of cholesterol esters and TGs

cholestipol CARDIOLOGY A bile acid sequestrant that ↑ conversion of cholesterol into bile acids, and stimulates the synthesis of LD receptors PROS ↓ CAD, safe CONS Expensive, inconvenient, GI side effects. See Cholesterol

cholestyramine CARDIOLOGY A bile acid sequestrant that ↑ conversion of cholesterol into bile acids, and secondarily stimulates the synthesis of LD receptors PROS ↓ risk of CAD, generally safe CONS Expensive, inconvenient, GI side effects. See Cholesterol

choline acetyltransferase Choline acetylase An enzyme which controls acetylcholine production in synapses of the autonomic nervous system, muscle, and SNS, which may be depleted in Alzheimer's disease

cholinergic PHARMACOLOGY *adjective* Referring to a parasympathomimetic effect, specifically on cholinergic receptors, evoking acetylcholine release *noun* A chemical or drug–eg, bethanechol, that stimulates acetylcholine release from parasympathetic nerve endings. See Parasympathetic nervous system. See Cholinergic crisis. Cf Adrenergic

cholinesterase There are 2 cholinesterases: one is synthesized in the liver and present in the serum, and the other–now formally known as acetylcholinesterase—is synthesized in the RBCs; both are used to determine the extent of organophosphate exposure; the serum form is more useful in detecting acute toxicity while acetylcholinesterase better reflects chronic exposure; some people have genetic variants of cholinesterase, which act more slowly on substrates than the normal enzyme, and they may experience prolonged apnea after anesthesia with suxamethonium-type muscle relaxants; these variant enzymes can be detected by screening before undergoing anesthesia. See Dibucaine number. Cf Acetylcholinesterase

cholinesterase inhibitor PHARMACOLOGY Any agent–eg, pyridostigmine, ambenonium, neostigmine, that inhibits acetylcholinesterase, the enzyme which breaks down acetylcholine, thereby preventing transmission of nerve impulses to a muscle

chondralgia Pain in/around cartilage; chondral pain

chondritis Inflammation of cartilage

chondrocalcinosis Pseudogout RHEUMATOLOGY A chronic gout-like recurrent arthritis characterized by calcium pyrophosphate deposition in multiple joints

chondrocranium ANATOMY The bones of the base of the brain which are formed by endochondral ossification

chondrodysplasia PEDIATRIC ORTHOPEDICS A defect in development of cartilage especially of long bones, resulting in arrested growth and dwarfism

chondrodysplasia punctata A heterogeneous group of bone dysplasias, all of which have epiphyseal stippling in infancy

chondrodystrophy A disturbance in the development of cartilage, primarily of the long bones, resulting in arrested growth and dwarfism

chondroid *adjective* Referring to cartilage *noun* Hyaline cartilage, see there

chondromalacia Progressive erosion of cartilage

chondromalacia patellae A condition characterized by progressive erosion of knee cartilage, more common in younger persons CLINICAL Pain with climbing, grinding sensation in knees

chondropathy A nonspecific term for any disease of cartilage

CHOP ONCOLOGY A 'first-generation' combination chemotherapy regimen used for low and intermediate-grade NHLs. See Chemotherapy, Non-Hodgkin's lymphoma, Remission

chordee UROLOGY A fixed downward curvature of the penis, which may be congenital–eg, hypospadias or acquired–eg, due to gonorrhea

chorea NEUROLOGY A condition characterized by involuntary but seemingly well-coordinated, rapid, complex, spastic movements. See Sydenham's chorea

choreoathetosis NEUROLOGY Involuntary slow, twisting, aimless muscle movements

chorioallantoic membrane COMPARATIVE ZOOLOGY An extraembryonic membrane formed in birds and reptiles by the apposition of the allantois to the inner face of the chorion; the CM is highly vascularized

chorioamnionitis OBSTETRICS Infection of the amniotic sac and villi CLINIC High fever, prolonged rupture of membranes, prematurity, and ↑ fetal M&M

choriocarcinoma Chorioblastoma, chorioepithelioma, gestational choriocarcinoma, malignant mole An aggressive rapidly growing malignancy arising from the placental trophoblast, which may arise in a hydatidiform mole, or de novo EPIDEMIOLOGY Most common in Asian ♀, predominantly in ♂ with a peak incidence CLINICAL Continued vaginal bleeding in ♀ with a recent hydatidiform mole, abortion or term pregnancy MANAGEMENT MTX CAUSE OF DEATH Hemorrhage, pulmonary insufficiency, drug toxicity, sepsis, renal failure

chorionic gonadotropin Human chorionic gonadotrophin A hormone secreted by the anterior pituitary gland, which supports pregnancy COMPONENTS α-hGH 14.5 kD; β-hGH 22.2 kD. See FSH, Gonadotropin, LH

CHORIONIC GONADOTROPIN LEVELS—POST CONCEPTION	
WEEKS	**IU/L**
1 wks	5-50
2 wks	40-1000
3 wks	100-4000
4 wks	800-10,000
5-6 wks	5000-100,000
8 wks	20,000-200,000
9-12 wks	10,000-200,000
2nd trimester	8,000-50,000
3rd trimester	6,000-50,000

chorionic villus biopsy Chorionic villus sampling OBSTETRICS A method for early–1st trimester diagnosis of fetal genetic defects and other diseases; tissue is obtained at 8-12 wks–vs 16th wk for amniotic fluid analysis from developing placenta by ultrasound-guided transcervical catheter aspiration Bx; tissue obtained is from the chorion frondosum, the layer from which chorionic villi develop DIAGNOSTIC YIELD CVB, 97.8% vs amniocentesis at 16 wks, 99.4%. See Amniocentesis

choristoma Non-specialized tissue that develops in utero, which corresponds to normal cells and tissues located in abnormal sites–eg, ectopic breast, liver, or other tissues. See Neuromuscular choristoma

choroid plexus tumor NEUROPATHOLOGY A rare brain cancer arising in the ventricles, seen in children < age 2

choroideremia Tapetochoroidal dystrophy A form of X-linked hereditary retinal degeneration–other hereditary retinal degenerations include Refsum's disease, gyrate atrophy and abetaipoproteinemia–characterized by a centripetal loss of visual fields due to a gene mutation in chromosome Xq21

choroidoretinitis See Uveitis

Chow technique ORTHOPEDICS A technique for endoscopic carpal ligament release through an open-slotted cannula, performed under local anesthesia without a tourniquet. See Carpal tunnel syndrome

chrematistophilia SEXOLOGY A paraphilia in which sexuoerotism hinges on being forced to pay, or being robbed by, the partner for sexual service

Christian-Weber disease Relapsing non-suppurative panniculitis A condition characterized by focal painful aggregates of subcutaneous fat necrosis with erythematous, ulcerating and eventually atrophic skin, seen in chubby middle-aged ♀ CLINICAL CWD may be acute or chronic, fulminant or transient, systemic or confined to the skin, and variably associated with polyserositis and fever ETIOLOGY Unknown; possibly related to trauma, cold, drugs, chemicals; it may occur in Pts with SLE, rheumatoid arthritis, DM, sarcoidosis, after corticosteroid withdrawal, in acute and chronic pancreatitis and in pancreatic carcinoma LAB ↑ Lipase, ↑ amylase

Christmas disease Hemophilia B HEMATOLOGY An X-R hemophilia, seen in 1/50,000 ♂, due to a defect in coagulation factor IX CLINICAL Easy bruising, nosebleeds, bleeding gums, muscle hematomas, hemarthrosis MANAGEMENT Factor IX concentrates. See Factor IX deficiency

chromatic aberration OPTICS The differences in the focal points when multiple wave lengths–colors of light–eg, those of white light, pass through a lens system

chromatin GENETICS The stainable material of interphase nuclei corresponding to chromosomes; chromatin consists of nucleic acids–DNA and associated histone protein, which are packed into nucleosomes; euchromatin is loosely packed and accessible to RNA polymerases. See Salt & pepper chromatin. Cf Heterochromatin

chromatin body Barr body A dense X chromosome in ♀ cells, best recognized in buccal mucosa scrapings

chromium deficiency A rare condition characterized by ↓ weight, glucose intolerance, insulin resistance, ↓ respiratory quotients, peripheral neuropathy. See Chromium

chromium intoxication Poisoning due to toxic levels of chromium Sources Industrial–chromium-laden fumes and dusts in electroplating, manufacture of steel, dyes, chemicals, leather tanning, photography Clinical Acute CI causes allergic reactions, conjunctivitis, dermatitis, edema; chronic CI causes GI symptoms, hepatitis, and ↑ risk of lung CA. See Chromium

chromoblastomycosis Chromomycosis Infectious disease An infection by various dematiaceous fungi: *Fonsecaea pedrosoi*, *F compacta*, *Phialophora verrucosa*, *Cladosporium carrionii*, *Rhinocladiella aquaspersa* Epidemiology Most US infections occur in whites, age 30–50, who are farmers or manual laborers Clinical Verrucoid, ulcerated, crusted lesions with pseudoepitheliomatous hyperplasia, keratolytic microabscesses. Cf Blastomycosis, Phaeohyphomycosis

chromogen Chromagen A chemical or compound that reacts to produce a colored end-product, used to detect the presence of a substance of interest; chromogens are critical detectors in immunoenzymatic reactions. See Porter-Silber chromogen

chromosome Genetics *adjective* Etymologically incorrect, but widely preferred *noun* Any of a number of paired units of the self-replicating genetic material in the eukaryotic nucleus, the 'master genetic database' containing the complete information present in a cell or virus which results in the palette of phenotypic expression of the individual; human chromosomes consist of 23 long–100-300 million bp, each–paired DNA or, in some organisms, RNA molecules that in humans are associated with RNA and histone proteins, and most readily recognized during mitosis as they align themselves on the metaphase plate; chromosomes are divided into structurally similar groups based on length from the centromere: group A–chromosomes 1-3; B–chromosomes 4, 5; C–chromosomes 6–12, X chromosome; D–chromosomes 13–15; E–chromosomes 16–18; F–chromosomes 19, 20; G–chromosomes 21, 22, Y chromosome. See Acentric chromosome, Accessory chromosome, Autosomal chromosome, B chromosome, Bacterial artificial chromosome, C banding, Christchurch chromosome, Eukaryote, Flow cytometry, G banding, Gene, Harlequin chromosome, Honorary chromosome, Homologous chromosome, Human genome project, Isochromosome, Lampbrush chromosome, Marker chromosome, Minichromosome chromosome, Nucleotide, Philadelphia chromosome, Ploidy analysis, Polytene chromosome, Protein, Q Banding, Ring chromosome, Sex chromosome, Translation, Transcription, Unbanded chromosome, X chromosome, Y chromosome

chromosome analysis Genetics A procedure in which cells–usually of fetal origin are obtained, either in the 1st trimester by chorionic villus biopsy, or later in pregnancy by amniocentesis, and grown in a tissue culture, to detect major chromosome defects Indications Mixed congenital anomalies with mental or growth retardation, infertility, in cryptorchidic testes, ambiguous genitalia, repeated neonatal death, advanced maternal age, and in adults, analysis of neoplasia Specimens Amniotic fluid–prenatal; blood in green top tube; BM if leukemia is suspected; maintain at room temperature. See Banding. Cf DNA hybridization, FISH, PCR

chromosome breakage syndrome Any of a group of inherited diseases in which chromosomes are more ↑ fragile–eg, ataxia-telangiectasia, Bloom syndrome, Fanconi syndrome, and xeroderma pigmentosum, resulting in ↑ susceptibility to certain Cas

chromosome complement The entire set of chromosomes in a species–in humans, 46 chromosomes

chromosome disorder Chromosome aberration, chromosome defect Genetics Any abnormality of a chromosome number or structure due to loss, duplication, or rearrangement of DNA. See Chromosome

chronesthesy Administration time of a drug based on the circadian or other bioperiodic rhythm, related to the rhythm in receptor number or conformation and rate-limiting steps in metabolic pathways. See Circadian rhythm

chronic *adjective* Referring to a condition lasting ≥ 3 months–per US Natl Center for Health Statistics; persisting or occurring over a long period of time. Cf Acute

chronic active Epstein-Barr virus infection Pathology A condition characterized as a specific immunodeficiency for EBV Clinical Prolonged fever, hepatosplenomegaly, lymphadenopathy, liver dysfunction, which may be accompanied by aneurysms and vasculitis. See Epstein-Barr virus

chronic active hepatitis 1. Obsolete term. See Chronic hepatitis 2. Chronic viral hepatitis

chronic atrophic pyelonephritis Reflux nephropathy, see there

chronic bilateral obstructive uropathy Obstructive uropathy Urology A disorder characterized by prolonged and/or progressive blockage of urine flow from both kidneys due to urethral obstruction Etiology ♂–BPH, prostate CA, ♀–bladder cystocele, bladder tumors, tumors or other structures of the bladder neck or urethra, kidney, bladder stones Clinical Irritability, urgency, bladder incontinence, urine stasis, UTIs, urine reflux into ureters and kidney, leading to pyelonephritis and hydronephrosis; CBOU leads to HTN and/or renal failure

chronic bronchitis Pulmonary medicine A common condition that is more common in middle-aged men, which is often related to cigarette smoke, less often to air pollutants–eg, NO_2, SO_2; CB is often complicated by bacterial–eg *H influenzae*, *S pneumoniae* or viral–eg RSV infections. See COPD, Emphysema. Cf Panlobular emphysema

chronic cholecystitis Chronic inflammation of the gallbladder, a condition caused by repeated episodes of acute cholecystitis, which may lead to fibrosis of the gallbladder wall. See Cholecystitis, Gallstones

chronic disease Medtalk Any condition–eg connective tissue disease, CA, ASHD, HTN, Alzheimer's disease that has a protracted–usually ≥ 6 months clinical course; CDs require long-term therapy; response is suboptimal; return to a state of complete or pre-morbid normalcy is the exception, not the rule Risk factors for chronic disease Cigarette smoking, sedentary lifestyle, obesity. Cf Acute disease

chronic eosinophilic pneumonia A condition characterized by certain clinical and histologic features that overlap those of bronchiolitis obliterans and organizing pneumonia

chronic exertional compartment syndrome Sports medicine A condition that usually arises in the anterior compartment of the lower leg, characterized by cramping, pain, and tightness, often with numbness and tingling in the foot. Cf Medial tibial stress syndrome

chronic fatigue syndrome Chronic fatigue & immune dysfunction syndrome A chronic, idiopathic debilitating condition that may follow viral infections–eg herpes, hepatitis, CMV, EBV; CFS is probably a real entity; linked to a combination of CMV and EBV infection Clinical Unexplained fatigue, weakness, muscle pain pain, lymphadenopathy and malaise; CFS is a diagnosis of exclusion; there are no valid tests, treatment consists of relief of symptoms, life style changes; some cases may resolve with time Management None; the 'cures' reported may be mere placebo effect or spontaneous remission. See Fibromyalgia syndrome

CHRONIC FATIGUE SYNDROME–CDC CASE DEFINITION

MAJOR CRITERIA (required)

1. Recent onset of debilitating or recurring fatigue of > 6 months duration and
2. Exclusion of clinically similar conditions

MINOR CRITERIA (eight of ten required)

1. Low-grade fever (< 38.6° C) or chills
2. Sore throat (or pharyngitis)
3. Painful anterior and/or posterior cervical and axillary lymphadenopathy
4. Unexplained muscular weakness
5. Myalgia
6. Generalized fatigue of > 24 hours after previously tolerated exercise
7. Severe generalized headache
8. Migratory arthralgia
9. Neuropsychological, eg photophobia, irritability, poor concentration, depression
10. Sleep disturbance

chronic glaucoma OPHTHALMOLOGY A disorder caused by ↑ intraocular pressure, 2° to blockage of the circulation of the aqueous humor, which may damage the optic nerve and cause blindness CLINICAL ↓ vision, halos around lights–worse at night, mild headaches MANAGEMENT Beta-blocker eyedrops

chronic glomerulonephritis Chronic nephritis NEPHROLOGY An advanced kidney disease characterized by inflammation and slowly progressive renal failure ETIOLOGY Idiopathic–¹⁄₄, amyloidosis, diabetic nephropathy, focal segmental glomerulosclerosis, IgA nephropathy, lupus nephropathy, myeloma, rapidly progressive glomerulonephritis. See Diabetic nephropathy, Focal segmental glomerulosclerosis, IgA nephropathy, Lupus nephropathy, Rapidly progressive glomerulonephritis

chronic granulomatous disease IMMUNOLOGY Any of a heterogeneous group of inherited (⅔ are X-linked; ⅓ AR) phagocytic defects, characterized by recurrent and potentially fatal pyogenic infections of early onset; PMNs ingest pathogens, but don't produce superoxide and microbicidal oxygen intermediates–the respiratory 'burst'–due to defective NADPH oxidase, resulting in recurrent infections by catalase-bacteria–eg, *S aureus*, *Enterobacteriaceae*, *Aspergillus* spp, and fungi with sinusitis, pneumonia, abscess formation; ♂:♀ ratio 4:1 MANAGEMENT IFN-γ may ↓ serious infection by 'boosting' the immune system

chronic hepatitis HEPATOLOGY Inflammation of the liver that continues without improvement for ≥ 6 months. See Cirrhosis. Cf Acute hepatitis

chronic idiopathic diarrhea Chronic nonspecific diarrhea A self-limited condition defined as the presence of persistently loose stools for ≥ 4 wks, without identifiable cause, and occurring in absence of systemic illness; the diarrhea is 2° in nature and does not respond to antibiotics, but resolves in several months. See Brainerd diarrhea

chronic inflammatory polyneuropathy NEUROLOGY A common, slowly progressive, often bilateral disorder characterized by recurrent inflammation of multiple nerves, accompanied by ↓ movement and sensation MANAGEMENT Systemic corticosteroids or chemotherapy for immune suppression PROGNOSIS Variable. See Guillain-Barré syndrome

chronic leukemia HEMATOLOGY A malignancy of lymphoid and myeloid stem cells. See Chronic lymphocytic leukemia, Chronic myelocytic leukemia

chronic Lyme disease A predominantly neurologic condition ranging from mild–eg, fatigue, paresthesia, arthralgia, memory loss, mood swings, and dysomnia, to severe–eg, spastic paraparesis, tetraparesis, ataxia, chorea, cognitive impairment, bladder dysfunction, cranial nerve deficits, myelitis, brainstem encephalitis and demyelination Lyme disease triad: Lymphocytic meningitis, cranial neuritis–especially of 7th and 8th cranial nerves and radi-

culitis DIAGNOSIS IgG antibody–'Western' immunoblotting to *B burgdorferi* that may disappear with time; persistent, ie treatable infection should be ruled out by a specific T-cell lymphoblastic response assay. See Lyme disease

chronic lymphocytic leukemia Chronic lymphoblastic lymphoma HEMATOLOGY A slowly progressive form of leukemia more often seen in older adults, which is characterized by ↑ mature lymphocytes CLINICAL Variable Sx; CLL may be diagnosed fortuitously during a routine CBC for an unrelated illness, before clinical disease Cf Chronic myelocytic leukemia

CHRONIC LYMPHOCYTIC LEUKEMIA–STAGES

0 Lymphocytosis in blood, without other Sx of leukemia
I Lymphocytosis; lymphadenopathy
II Lymphocytosis; lymphadenopathy; hepatosplenomegaly
III Lymphocytosis; lymphadenopathy; hepatosplenomegaly, anemia
IV Lymphocytosis, lymphadenopathy; hepatosplenomegaly, thrombocytopenia ± anemia

chronic lymphocytic thyroiditis Hashimoto's disease, see there

chronic mesenteric ischemia Abdominal angina A condition characterized by intermittent severe ischemia, resulting in abdominal colic, beginning 15-30 mins post-prandially, lasting 1-2 hrs, and appearing when 2 or all 3–superior and inferior mesenteric and celiac major abdominal arteries have severe ASHD; since the intestine's O₂ demand ↑ with meals, Pts avoid the pain by not eating, and thus lose weight; malabsorption may occur because absorption is O₂-dependent MANAGEMENT Bypass, endarterectomy, vascular reimplantation, percutaneous transluminal angioplasty

chronic motor tic disorder Chronic vocal tic disorder NEUROLOGY A Tourette-like condition, characterized by recurrent motor activity and vocalizations. See Chronic vocal tic disorder, Tourette syndrome

chronic myelocytic leukemia Chronic granulocytic leukemia, chronic myelogenous leukemia, chronic myeloid leukemia HEMATOLOGY An indolent leukemia, characterized by an ↑ in mature granulocytes in the peripheral circulation TREATMENT Imatinib mesylate induces major cytogenetic responses in 60% of Pts with chronic-phase CML and complete hematologic responses in 95% of CML Pts. See Imatinib mesylate

chronic nephritis Chronic glomerulonephritis, see there

chronic nonspecific diarrhea See Chronic idiopathic diarrhea

chronic obstructive lung disease See COPD

chronic osteomyelitis CLINICAL MEDICINE Osteomyelitis with bone necrosis due to compromised vascular supply, which may persist for yrs RISK FACTORS Recent trauma, DM, hemodialysis, IV drug abuse. See Osteomyelitis

chronic otitis externa Otitis externa ENT A condition of young adults, characterized by inflammation, irritation or infection of the external auditory canal, caused by mechanical trauma or chemical irritation. Cf Otitis media

chronic otitis media Middle ear infection ENT Recurrent middle ear inflammation due to mechanical trauma, chemicals, allergies, pathogens; COM may follow persistent acute OM, extend from mastoiditis or is linked to fluid build-up with tympanic membrane rupture or damage to auditory ossicles

chronic pain disorder Somatiform pain disorder PAIN MANAGEMENT A nonmalignant condition characterized by nonspecific aches and pain, accompanied by chronic anxiety, depression and, often, drug dependency. See Pain management

chronic pancreatitis Chronic relapsing pancreatitis GI DISEASE Recurrent pancreatitis linked to alcohol abuse or hemochromatosis, which may worsen with time. See Pancreatitis

chronic phase chronic myelogenous leukemia HEMATOLOGY Early CML–which may last from months to yrs, where the number of immature, abnormal WBCs in the BM and peripheral blood is higher than normal, but lower than in accelerated/blast phase. See Blast crisis

chronic prostatitis UROLOGY Inflammation of the prostate due to a bacterial infection, associated with UTI, cystitis, urethritis, epididymitis, or acute prostatitis RISK FACTORS Alcoholism, perineal injury, and certain sexual practices, especially anal intercourse. See Benign prostatic hypertrophy

chronic pulmonary coccidioidomycosis INFECTIOUS DISEASE Lung infection due to *Coccidioides immitis*, endemic in southwestern US; it may follow a 20-yr latency period and is more common in immunocompromised hosts. See Coccidioidomycosis

chronic pulmonary histoplasmosis INFECTIOUS DISEASE Histoplasmosis in a person exposed to bird droppings who has COPD or a compromised immune system. See Histoplasmosis

chronic recurrent root neurapraxia SPORTS MEDICINE An injury of older football players due to compression of the nerve root in the intervertebral foramina 2° to hyperextension and ipsilateral deviation of the head and neck, often with foraminal stenosis and/or degenerative disk change in combination with developmental cervical stenosis DIAGNOSIS Spurling's maneuver

chronic renal failure Chronic kidney failure NEPHROLOGY A slow decline in renal function, which may be 2° to chronic HTN, DM, CHF, SLE, or sickle cell anemia and, if extreme, leads to ESRD, mandating kidney dialysis; an abrupt decline in renal function may be triggered by acute intercurrent processes–eg, sepsis, shock, trauma, kidney stones, kidney infection, drugs–aspirin or lithium, toxins, abuse substances, or injection of iodinated radiocontrast LAB Fluid retention, uremia MANAGEMENT Low-protein diet to conserve renal function; transplantation if ESRD

chronic sinusitis Chronic sinus infection ENT Inflammation of the sinuses that empty into the nasal cavity ETIOLOGY Allergic rhinitis, nasal obstruction, deviated nasal septum, tooth abscesses, URIs

chronic stable angina CARDIOLOGY The most common form of angina, characterized by chest discomfort due to myocardial ischemia, and unaccompanied by myocardial necrosis; the cause of pain is uncertain, possibly substances released during transient ischemia–eg, adenosine, histamine, bradykinin, serotonin, as well as acidosis, ↑ K+

chronic stress A state of prolonged tension from internal or external stressors, which may cause various physical manifestions–eg, asthma, back pain, arrhythmias, fatigue, headaches, HTN, irritable bowel syndrome, ulcers, and suppress the immune system. See Fight-or-flight response

chronic symptomatic HIV infection Middle stage HIV disease AIDS HIV infection with non-life-threatening signs and Sx of HIV infection–eg, oral thrush, gingivitis, seborrheic dermatitis, molluscum contagiosum, fevers, fatigue, lymph node swelling, malaise, weight loss. See AIDS, HIV

chronic toxicity TOXICOLOGY A condition caused by repeated or long-term exposure to low doses of a toxic substance

chronic urethritis Urethral syndrome UROLOGY A condition characterized by painful urination and urinary frequency, which may be caused by bacterial infections–eg, STDs or linked to structural defects

chronic urinary tract infection Chronic UTI, recurrent urinary tract infection UROLOGY Repeated episodes–> 2 times in 6 months, or prolonged bacterial infection of bladder and urethra accompanied by repeated cystitis, or UTI that does not respond to the usual treatment or that lasts > 2 wks; in young ♀, recurrent UTIs may indicate obstructive urinary tract defects–eg, ureterovesical reflux, and should be evaluated. See Urinary tract infection

chronic vascular syndrome DIABETOLOGY A term for the often fatal consequences of type 1 DM on the large and small blood vessels, where there is ↑ in thickness and ↑ 'leakiness' of vessels. See Diabetes mellitus

chronic venous insufficiency Venous insufficiency VASCULAR DISEASE A condition characterized by poor flow of venous blood, especially in the leg veins CLINICAL Leg swelling, pain, cramps, risk of DVT

chronicity Characterized by long duration; the state of being chronic

chronobiology PHARMACOLOGY The formal study of the effects of circadian rhythms on the timing of illness, and therapy PHYSIOLOGY The formal study of circadian rhythms on physiologic and pathologic events. See Biorhythm, Chronotherapy, Circadian rhythm

chronopharmacology PHARMACOLOGY The study of the interactions of biologic rhythms with medications; chronopharmacology is focused on 2 areas: 1. Biologic rhythm dependencies of medications and underlying mechanisms 2. Effect of timing pharmacotherapy on biologic time structure and relationships among rhythms. See Circadian rhythm, Pharmacokinetics

chronotherapy Any therapy based on the timing of physiologic and pathologic event. See Bright light therapy, Chronobiology, Circadian rhythm, Seasonal affective disorder ONCOLOGY The adminstration of chemotherapy doses synchronized to the body's circadian rhythm; CT may ↑ allowable doses of chemotherapeutics, ↓ tumor burden, and chemotherapy-related side effects

chronotropic response CARDIAC PACING Change of heart rate over time in response to stimuli–eg, exercise

Chryseobacterium meningosepticum *Flavobacterium meningosepticum*, CDC group IIa BACTERIOLOGY A bacterium of soil, water, plants, foods, hospital water–incubators, tap water, hemodialysis systems, pharmaceuticals MODE OF TRANSMISSION Uncertain, but probably linked to contaminated medical devices and fluids; occasionally transmitted via the birth canal to neonate CLINICAL Bacteremia, neonatal meningitis, rarely adult meningitis; may be found in mixed wound infections, ocular infections, URIs, UTIs, sinusitis, endocarditis, peritonitis, fasciitis

Chrytemin Imipramine, see there

CHS CARDIOLOGY Any of a number of clinical trials 1. Cardiovascular Health Study 2. Charleston Heart Study 3. Community Health Study 4. Congenital Heart Surgeons Society Study 5. Coronary Heart Study

chubby puffer syndrome A sleep apnea complex of obese, prepubescent ♂, which resembles that of the Pickwick syndrome, caused by primary alveolar hypoventilation; temporarily ameliorated by tonsillectomy, CPS is more central in the nervous system, possibly arising in the reticular activating system MANAGEMENT Weight loss, TCAs. See Pickwick syndrome., Sleep apnea syndrome, Tricyclic antidepressants. Cf Pink puffer

Churg-Strauss syndrome Allergic granulomatosus & angiitis INTERNAL MEDICINE A small vessel vasculitis characterized by eosinophil-rich and granulomatous inflammation of the respiratory tract, formation of ANCA and necrotizing vasculitis of small-to-medium-sized vessels associated with asthma, edema. See Small vessel vasculitis, Systemic vasculitis

Chvostek sign NEUROLOGY Unilateral spasm due to muscle tetany which is induced by tap over the facial nerve which occurs in severe hypocalcemia

chylomicronemia syndrome Hyperchylomicronemia A condition characterized by marked chylomicronemia with plasma TG levels > 225 mmol/L–US: 2000 mg/dL CLINICAL Abdominal and chest pain, pancreatitis, memory defects, carpal tunnel-like paresthesiae, hepatosplenomegaly, chronic eruptive xanthomata, and insulin-resistant DM, possibly related to marked hypertriglyceridemia; Sx are exacerbated by alcohol, β blockers, diuretics, estrogens, glucocorticoids

chyme The mass of partially digested food goo that passes from the pylorus to the duodenum

chymotrypsin PHYSIOLOGY A GI tract serine protease synthesized in the pancreas as a prohormone, which cleaves proteins at hydrophobic amino acids–leucine, phenylalanine, tryptophan, tyrosine

CI 1. Cardiac index, see there 2. Confidence Interval

CIC Circulating immune complexes. See Immune complexes

cicatrix Scar, see there

CID Cytomegalic inclusion disease, see there

cidofovir Forvade, GS-504, HPMPC, Vistide AIDS A nucleotide analogue effective against a broad spectrum of herpesviruses INDICATIONS CMV retinitis, CMV-induced blindness, genital warts ADVERSE EFFECT Nephrotoxicity

CIDS CARDIOLOGY A trial–Canadian Implantable Defibrillatory Study–that compared a device–implantable defibrillator therapy vs the best medical therapy in survivors of acute MI

cigarette TOBACCO A cylinder of cut blonde or black tobacco ensheathed in a tube of usually white paper, often with a filter to ↓ inhaled tars, which is lighted and inhaled at the lips to achieve a pleasurable level of nicotine. See Clove cigarette, Fetal tobacco syndrome, Fire-safe cigarette, Low-tar cigarette, Passive smoking, Safe cigarette, Smokeless cigarette, smoking, Tobacco

ciguatera poisoning NUTRITION The ciguatera, a coral reef fish that secretes ichthyosarcotoxin–ciguatoxin, a substance produced by the reef dinoflagellate, *Gambierdiscus toxicus*, and concentrated, unchanged up the food chain by herbivores and carnivores; CP is the most common marine intoxication in the US; 400 spp of fish are implicated–eg, barracuda, grouper, red snapper, amberjack, surgeonfish, sea bass and–unlike scombroid poisoning, may cause morbidity regardless of the form of preparation CLINICAL Onset 6-12 hrs after ingestion; N&V, cramping, diarrhea, paresthesias, reversal of temperature sense, arthralgias, myalgias, cranial nerve palsies, pruritus with alcohol ingestion, chills, hypotension, bradycardia, respiratory paralysis or death, average duration 8 days DIAGNOSIS RIA, ELISA TREATMENT IV mannitol reverses Sx. See Fish, Sushi. Cf Scombroid poisoning

ciliary processes OPHTHALMOLOGY The extensions or projections of the ciliary body that secrete aqueous humor

cilostazol Pletal® CARDIOLOGY An antiplatelet vasodilator used to manage intermittent claudication due to peripheral vascular disease by ↑ blood flow to affected limbs, per improved ankle/brachial index CONTRAINDICATIONS CHF

cimetidine Tagamet™ CLINICAL THERAPEUTICS An H_2-receptor antagonist used to treat peptic ulcer disease ADVERSE REACTIONS Diarrhea, headache, drowsiness, fatigue, muscle pain, constipation; rarely mental confusion, agranulocytosis, gynecomastia, impotence, allergic reactions, myalgias, tachycardia, arrhythmias, interstitial nephritis, mild ↑ in creatinine. See Histamine receptor antagonists

CIN Cervical intraepithelial neoplasia, see there

Cincinnati Prehospital Stroke Scale NEUROLOGY A 3-item–arm weakness, defects in speech, facial droop–scale for stratifying stroke victims as candidates for thrombolytics. See NIH Stroke Scale, Stroke, Thrombolytics

Cinderella delusion A popular term for the fantasy of some medical students that they will, upon graduation from medical school '…*suddenly be able to develop succinct differential diagnoses, make accurate diagnostic and treatment plans, and sign medication orders …without a senior physician's countersignature controlling for error*'. See July phenomenon

cineangiography INTERVENTIONAL CARDIOLOGY Real-time angiography recorded on film for documentation

CIO Chief information officer. See Information systems

ciprofloxacin Cipro® ANTIBIOTICS A broad-spectrum–GNRs, staphylococci, fluoroquinolone with limited activity against streptococci, anaerobes INDICATIONS Acute sinusitis, acute exacerbation of chronic bronchitis, UTI, acute cystitis in ♀, bacterial prostatitis, intra-abdominal infection, skin, bone and joint infection, infectious diarrhea, typhoid fever, gonorrhea ADVERSE EFFECTS GI pain, N&V, diarrhea, seizures, rash, photosensitivity. See Fluoroquinolone

circadian pacemaker A cluster of neurons, the activity of which fluctuates in ± 24 hr cycles; the CP resides in the pineal gland, weighs 100-180 mg, and derives embryologically from the ependyma at the roof of the 3rd ventricle; the CP influences the pineal gland, which produces melatonin at night. See Bright light therapy, Chronotherapy, Circadian rhythm, Melatonin, Seasonal affective disorder, Zeitgeber

circadian rhythm Diurnal rhythm, ultradian rhythm PHYSIOLOGY An innate, daily, fluctuation of physiologic and behavioral functions–eg sleep waking, generally tied to the 24 hr day-night cycle; the diurnal cadence in humans without cyclical cues provided by natural light, is 25.4 hrs; CR affects drug metabolism, serum levels of various substances–eg, ACTH, physiologic activities–eg, BP, myocardial blood flow and O_2 demand, psychosomatic disease and sleep cycles, cell division, hematopoiesis, NK cell activity. See Biorhythm, Circadian pacemaker, Insomnia, Jet lag, Melatonin, Shift work

circle of Willis *circulus arteriosus cerebri* ANATOMY A conduit of anastomosed arteries that encircle the optic chiasm and hypophysial region at the base of the brain, consisting of parts of each internal carotid, anterior, middle, and posterior cerebral arteries, and anterior and posterior communicating arteries. See Berry aneurysm

circling the drain FTD–fixing to die, near extremis, pre-code MEDTALK Referring to a Pt whose future prospects of life are dim

circulating nurse A nurse who participates in a surgical procedure, coordinating, planning and implementing all the nurse-related activities during an operation, but who has not scrubbed with the surgical team itself. See Scrub nurse

circulation MEDTALK Blood stream. See Enterohepatic recirculation, Fetal circulation, Pulmonary circulation, Systemic circulation

circulatory system The cardiovascular 'plumbing,' heart, arteries, capillaries, veins, which transport oxygenated blood from the lungs and heart to the general circulation, and return deoxygenated systemic blood to the lungs and heart. See Heart

circumcision UROLOGY Surgical removal of the foreskin/prepuce, either by an obstetrician, or as a part of a religious rite–eg, a bris, performed by a rabbi in Judaism at birth and Moslems in preadolescence. See Sunna circumcision. Cf Female circumcision

circumscribed *adjective* Having distinct borders

circumstantial homosexuality SEXUALITY The practice of homosexual acts when there are no opportunities for heterosexual activity–eg, adolescents in detention centers, in correctional facilities–prisons, and residential treatment centers, ±accompanied by sadism and sexual exploitation. See Sadism. Cf Latent homosexuality

circumstantial speech NEUROLOGY Speech in which there are tightly linked associations, interspersed with nonessential associations, thus taking a circuitous route before reaching its goal ETIOLOGY Bipolar–manic-depressive disorder, organic brain disease. Cf Tangentiality

circus movement Reentry, reciprocal movement CARDIOLOGY Aberrant electrical impulses that flow through the cardiac conduction system, which form the basis for some–if not all–SVT

cirrhosis Greek, orange-yellow 1. A term rarely used for chronic interstitial inflammation of non-liver organs–eg, cardiac cirrhosis HEPATOLOGY The final irreversible stage of chronic inflammation and/or cell injury of the liver, which results in a scarred, contracted, and functionally effete organ ETIOLOGY Alcohol, viral hepatitis–HAV, HBV, HCV, HDV, PBC, hemochromatosis, Wilson's disease, alpha-1-antitrypsin deficiency, intestinal bypass, venous outflow obstruction CLINICAL Jaundice, itching, fatigue and, with failure, end-stage liver disease DIAGNOSIS Physical exam, labs, liver Bx LAB ↑ Transaminases–ALT, AST, ↓ albumin COMPLICATIONS Confusion, coma–due to hepatic encephalopathy, fluid accumulation–ascites, internal bleeding, kidney failure MANAGEMENT Conservative–eg, diuretics, alcohol detoxification–'drying out'; liver transplantation may be an option for Pts with advanced cirrhosis. See Alcoholic cirrhosis, Cardiac cirrhosis, Indian childhood cirrhosis, Micro-micronodular cirrhosis, Primary biliary cirrhosis

cirrhous ONCOLOGY *adjective* Hardened, indurated

CIS Carcinoma-in-situ, see there

***cis* fatty acid** NUTRITION A natural fatty acid, in which the carbon moieties lie on the same side of the double bond; natural fats and oils contain only *cis* double bonds–eg, oleic acid, a monounsaturated fatty acid with a *cis* configuration. See Fatty acid; Cf *Trans* fatty acid

cisapride Propulsid® GI DISEASE An agent that ↑ lower esophageal sphincter pressure and for nocturnal therapy of GERD ADVERSE EFFECTS Headache, diarrhea, abdominal pain, nausea, rhinitis, constipation. See GERD

cisatracurium ANESTHESIOLOGY A stereoisomer of atracurium, a nondepolarizing muscle relaxing anesthetic preferred in multiorgan failure. See Muscle relaxant, Nondepolarizing agent. Cf Depolarizing agent

cisplatin Cis-platinum, diaminedichloride, Platinol A chemotherapeutic that forms covalent bonds and crosslinks with DNA, which is used for head & neck, ovarian, testicular, bladder CAs, lymphomas ADVERSE EFFECTS Nephrotoxicity, nausea, neurotoxicity

cisplatin neuropathy NEUROLOGY A dose-related peripheral neuropathy similar to vitamin E toxicity caused by cisplatin CLINICAL Numbness and tingling in fingers and toes

CISS Continuous-flow isotonic solution system GYNECOLOGY A system used in hysteroscopic procedures utilizing safer isotonic fluids, compared to present usage of hypotonic fluids, which if absorbed can cause serious complications, including death. See Hysteroscopy

citalopram Celexa® NEUROLOGY A serotonin reuptake inhibitor for managing major depression ADVERSE EVENTS Nausea, insomnia, ejaculation disorder, diaphoresis, fatigue CONTRAINDICATIONS MAOI therapy

citation INFORMATICS The record of an article, book, or other report in a bibliographic database that includes summary descriptive information–eg, authors, title, abstract, source and indexing terms. See Report

citric acid urine test LAB MEDICINE A test that measures citric acid in urine ↓ in renal tubular acidosis or kidney stones; ↑ in high carbohydrate diets, estrogens, vitamin D therapy

Citrobacter diversus MICROBIOLOGY A *Citrobacter* in soil, water, sewage and food, and an opportunistic pathogen EPIDEMIOLOGY *C diversus* is an important cause of bacterial meningitis, brain abscesses, and may cause endocarditis and hospital-acquired bacterial infections

Citrobacter freundii MICROBIOLOGY A *Citrobacter* opportunistic pathogen MANAGEMENT Cephalothin, aminoglycosides

citrullinemia METABOLIC DISEASE An AR condition caused by a defect in argininosuccinate synthase, resulting in an accumulation of citrulline in serum, CSF, and urine CLINICAL Severe vomiting, mental retardation, and early death in most Pts; onset may be delayed–late Sx include enuresis, delayed menarche, insomnia, sleep reversal, night sweats and terrors, diarrhea, tremors, episodic post-prandial confusion, hallucinations, coma, bizarre behavior misdiagnosed as mental disorder LAB Orotic aciduria, hyperammonemia

civil commitment Involuntary hospitalization, see there

CJD Creutzfeldt-Jakob disease, see there

CK Creatinine phosphokinase, see there

CK-MB Creatine phosphokinase MB isoenzyme CARDIOLOGY A CK isoenzyme usually ↑ in acute MI; CK-MB may be ↑ in muscular dystrophy, polymyositis, myoglobinuria, malignancy–eg, lung CA. Cf Troponin I, Troponin T

Cl Chemical symbol for chloride

clade Cladus, subtype GENETICS A branch of biological taxa or species that share features inherited from a common ancestor; a single phylogenetic group or line. See Inheritance, Species

Cladosporium carrionii *Cladophialophora ajelloi* INFECTIOUS DISEASE The dematiaceous aeroallergenic fungal agent of chromoblastomycosis. See Chromoblastomycosis

claim A demand for compensation MANAGED CARE A written request by an insured or assignee–eg, provider for payment of benefits covered by an insurance policy; a bill for healthcare service(s) sent by a provider to the Pt's insurance or health plan, which may review the claim for validity before paying benefits. See Aberrant claim, Electronic billing, Unassigned claim MALPRACTICE A formal statement by a plaintiff alleging that a civil wrong has been committed by a defendant. See Cross claim

claims-made policy MALPRACTICE A type of medical malpractice or professional liability insurance policy for a physician or other health care professional in which coverage is provided for any claim that occurs *only while the policy is in force*, regardless of when the alleged malpractice, negligence, or injury occurred. Cf Nose coverage, Occurrence policy, Tail coverage

clamp SURGERY A surgical device that closes an opened channel. See Clark clamp, Koala vascular, Tension clamp

clang association NEUROLOGY A shift in a conversation or flow of ideas, based on the sound of the words being used, not the content ETIOLOGY Schizophrenia, manic phase of bipolar disorder. See Bipolar disorder

clap STD Sobriquet for *N gonorrhoeae* infection. See Gonorrhea

Clarion® cochlear implant AUDIOLOGY An implant which bypasses ear damage, sending electric signals directly to the auditory nerve, interpreted by the brain as sounds. See Hearing aid

clarithromycin Biaxin® INFECTIOUS DISEASE A broad-spectrum semi-synthetic macrolide antibiotic used for acute exacerbation of chronic bronchitis, acute maxillary sinusitis, acute otitis media by *H influenzae*, *M catarrhalis*, and *S pneumoniae*, community acquired pneumonia by *S pneumoniae*, *Mycoplasma pneumoniae*, *C pneumoniae*, as well as *S aureus*, *M catarrhalis*, MAC, combined with other agents or prophylactically with omeprazole, for *H pylori* ADVERSE EFFECTS Diarrhea, N&V, dyspepsia, abdominal pain, headache, dysgeusia. See Chronic bronchitis

Claritin® ALLERGY MEDICINE A nonsedating sympathomimetic antihistamine containing loratidine and pseudoephedrine, used for seasonal allergies CONTRAINDICATIONS MAOI therapy, narrow-angle glaucoma, urinary retention, HTN, CAD. See Antihistamine

Clark clamp SURGERY A clamp used for dissecting and tunneling deep pelvic tissues

144 **CLAS** 1. Cholesterol-Lowering Atherosclerosis Study A study using colestipol and niacin in ♂ with previous CABG surgery 2. Circulating lupus anticoagulant syndrome. See Antiphospholipid antibody syndrome, Lupus anticoagulant

CLASP CARDIOLOGY A clinical study–Collaborative Low-Dose Aspirin Study in Pregnancy which evaluated the effects of low-dose aspirin–used to prevent pregnancy-related HTN—on intrauterine growth

clasp knife phenomenon NEUROLOGY A manifestation of corticospinal spasticity, in which there is ↑ tone in either flexion or extension with sudden relaxation, as the muscle continues to be stretched, imparting a sensation likened to that of an opening clasp knife; the CKP is often accompanied by weakness of the affected extremity, ↑ tendon reflexes, and a Babinski sign. Cf Cogwheel phenomenon, Gegenhalten

class BIOLOGY A taxonomic division of a phylum which is in turn divided into orders. See Genus, Order, Phylon VOX POPULI A grouping of any type. See Age class, Inhalation class, Management class

CLASS NEUROLOGY A clinical trial–Clomethiazole Acute Stroke Study RHEUMATOLOGY A clinical trial–Celecoxib/Celebrex Long-term Arthritis Safety Study, which compared a proprietary Cox-2 inhibitor to standard NSAIDs

class action lawsuit MEDICAL LIABILITY A legal action by one or more plaintiffs on their own behalf, and on the behalf of other persons with an identical interest in an alleged wrong. See Agent Orange, Breast implants, Dalkon shield, Shiley valve

class I device REGULATORY AFFAIRS A 'low risk' medical device–eg, elastic bandages, tongue depressors, which is minimally regulated by the FDA

class II device REGULATORY AFFAIRS A 'medium risk' medical device–eg, hearing aids, syringes, approved by the FDA for use in humans. See 510(K)

class III device REGULATORY AFFAIRS A highly regulated 'high risk' medical device–eg, life-support or life-sustaining devices–eg, pacemakers and heart valves, approved by the FDA for use in humans; CIIIDs are also defined as those which pose a potentially unreasonable risk of illness or injury. See Medical device

class recall REGULATORY AFFAIRS The recall of an FDA-regulated medical device or product, the use of or exposure to which a reasonable probability exists for adverse health effects or death

class switching PHYSIOLOGY A step in the normal maturation of Hb during fetal development that requires switching from embryonal zeta chain, which is structurally similar to the α chain, and occurs on chromosome 16 to mature α chain production; β chain production is a 2-stage maturation process with early loss of the ε chain and the presence during fetal and early post-natal life of a γ chain, which ↓ as β chain expression is ↑. See Introduction to this volume: note

classic MEDTALK *adjective* Referring to a classic; traditional, well-known, typical, widely used *noun* A thing that is widely known, traditional, timeless. See Citation classic

classic complement pathway IMMUNOLOGY The usual route of activation of the complement cascade, initiated by C1q binding to either IgM or to 2 adjacent IgG molecules; the resulting conformational change of C1q autoactivates C1r2, in turn activating C1s2, cleaving C4–C4b followed by C2–C2a; C4b,2a–the C3 convertase of CCP–initiates opsonization, leukocyte chemotaxis, ↑ vascular permeability, and cytolysis; CCP is activated by IgG, IgM, DNA, staphylococcal protein A and C-reactive protein; both classic and alternate pathways are stimulated by trypsin-like enzymes. See Common pathway. Cf Alternate pathway

classic migraine Migraine with aura NEUROLOGY An episodic headache that accounts for up to 20% of all migraines, lasts 4 to 72 hrs, is associated with N&V, photo- and phonophobia, and often follows an aura, which may not occur on the same side as the migraine; Pts are asymptomatic between episodes; in most, the headaches are unilateral, pulsating and of moderate to severe intensity; there may be a family history of migraines and of travel sickness. See Migraine. Cf Common migraine

classification Any systematic arrangement of similar entities organisms, disease processes, etc, which are separated based on specific types of differences. See Ambulatory payment classification, Ann Arbor classification, Bethesda classification, Black's classification, BLEED classification, Bormann's classification, Broders' classification, Caldwell-Molloy classification, Cambridge classification, Canadian Cardiovascular Society functional classification, Cladistic classification, Deafness classification, DeBakey classification, Denis classification, Dukes classification, FAB–French-American-British classification, FDA classification of devices, FDA Classification of Teratogenicity, Gustilo classification, Hamilton classification, Hazard classification, Hinchey grading classification, International Classification of Diseases–9th edition, Clinical Modification classification, International Workshop classification for chronic lymphocytic leukemia, ILO classification, Kiel classification, Killip classification, Lovejoy's classification, Ludwig classification, Lukes-Collins classification, Marseille classification, McKusick classification, New York Heart Association classification, Norwood classification, Obesity Task Force classification, Papanicolaou classification, Pesaro classification, Physical status classification, Quebec classification, Rappaport classification, REAL classification, Rosenthal classification, Savary-Miller classification, Shandall classification, TNM classification, Ulcerative colitis classification, Vaughan Williams classification, WHO classification, Wolfe classification, Working Formulation

claudication CARDIOLOGY Walking-induced pain in one or both legs that does not disappear with continued walking, and is relieved only by rest; claudication is present in 15% to 40% of Pts with peripheral arterial disease and associated with a ↓ ability to perform daily tasks MANAGEMENT Exercise training. See Peripheral arterial disease, Exercise training ORTHOPEDICS Limping. See Intermittent claudication, Pseudoclaudication .

clause VOX POPULI A phrase in a legal document that binds the contracting parties to do a thing. See Benefits not provided clause, Conscience clause, Consent to settle clause, Consultation clause, Delaney clause, Gag clause, Grandfather clause, Hold harmless clause, Incontestable clause, Indemnification clause, Insuring clause, Most favored nation clause, Noncompete clause, Recurring clause

claustrophobia PSYCHIATRY An abnormal/morbid/irrational fear of closed spaces–eg elevators, tunnels. See Phobia

clavicle strap Figure of 8 strap ORTHOPEDICS A backpack-like device that facilitates healing of a fractured clavicle, by aligning Fx fragments

clavicular *adjective* Pertaining to the clavicle

claw foot ORTHOPEDICS A deformity caused by atrophic paralysis of intrinsic foot muscles, which allows the long toe extensors to dorsiflex the proximal phalanges, and the long flexors to shorten the foot, heighten the arch and flex the distal phalanges, pulling the foot into talipes equinus; CF occurs in chronic polyneuropathies, and is typical of Charcot-Marie-Tooth disease. Cf Lobster claw deformity

claw hand Main en griffe The hand deformity characterized by flexion and atrophy, which may follow the claw foot deformity of Charcot-Marie-Tooth disease, where the atrophy is usually confined to the distal arm; clawhand deformity may also occur in Dejerine-Sottas' hypertrophic polyneuropathy, leprosy, and Refsum's disease; a similar 'stiff hand' occurs in mucopolysaccharidosis, type I-S. Cf Lobster claw deformity

clean *adjective* Free of dirt or pollution MEDICAL LIABILITY Referring to a malpractice 'virgin,' ie a physician without any lawsuits, past or present SUBSTANCE ABUSE Drug-free when examined

clean-catch urine LAB MEDICINE A specimen of 'midstream' urine for bacterial culture that is obtained from ♂ easily, and from ♀ after cleansing the perineum and meatus with povidone-iodine wipes

clean wound A superficial wound produced by uncontaminated sharp objects, either electively–eg, surgical procedure or by accident, being cut by sharp glass or metal–eg, broken glass. Cf Clean-contaminated wound, Dirty wound

clear EMERGENCY MEDICINE A widely used imperative, used to indicate to participants in a 'code blue' resuscitation that an operator is about to turn on the 'juice' in the defibrillation paddles. See Code blue

clear-glass appearance IMAGING A descriptor for the 'empty' holes, spaces and clefts seen in the x-ray of osteoporotic bone, which is similar to the ground-glass graininess characteristic of osteomalacia

clear up DERMATOLOGY Disappearance–eg of a rash. See Rash

clearance PHARMACOLOGY The elimination of a drug, therapeutic agent, or other substance from the body or other biologic system; clearance is expressed as a hypothetical volume that is completely removed in a given unit of time; in terms of pharmacokinetics, clearance is the product of the volume of distribution and the elimination rate constant; much of a drug's elimination is via the kidneys and clearance is commonly expressed in mL/min or L/hr. See Hepatic clearance, Renal clearance, Therapeutic drug monitoring, Total body clearance PHYSIOLOGY 1. The removal of a substance from the blood by metabolism or excretion. See Nasal mucociliary clearance 2. A quantitative measure of item 1

cleft lip EMBRYOLOGY A congenital defect characterized by a failure in the fusion of the upper lip; often associated with cleft palate. See Cleft palate

cleft palate EMBRYOLOGY A congenital defect characterized by a failure in the fusion of the hard and/or soft palate, often associated with cleft lip. See Cleft lip

cleido- Greek, clavicle *adjective* Related to the clavicle

cleidocranial dysostosis Cleidocranial dysplasia PEDIATRICS An AD condition characterized by partial or complete absence of clavicle, dental defects, joint laxity and a facies–heavy brow, protruding jaw, wide nasal nasal bridge, malaligned teeth, characteristic facies–frontal bossing

clenched fist sign Levine sign CARDIOLOGY The holding of a clenched fist over chest to describe constricting, pressing pain typical of angina pectoris, see there

clenched fist syndrome A condition caused by a traumatic laceration, typically over the 3rd and 4th metacarpophalangeal joints, which occurs when 'A' strikes 'B' in the teeth with a clenched fist; this results in a cutaneous abscess often infected with *Eikenella corrodens*, part of the normal oral flora; improper management may result in osteomyelitis MANAGEMENT Debridement, broad-spectrum antibiotics to 'cover' for anaerobic bacteria

clenoliximab IMMUNOLOGY An anti-CD4 immunotherapeutic for treating rheumatoid arthritis, possibly also severe asthma and psoriasis. See Monoclonal antibody

CLIA Clinical Laboratory Improvement Amendments of 1988 Congressional legislation that promulgated quality assurance practices in clinical labs, and required them to measure performance at each step of the testing process from the beginning to the end-point of a response to a test result. See HCFA, POLs, Waived tests

click-murmur syndrome Mitral valve prolapse, see there, aka Barlow's syndrome

client PSYCHOLOGY Patient Any person who is voluntarily or involuntarily receiving mental health services or substance abuse services from any mental health service provider

climacteric GYNECOLOGY *noun* 1. The perimenopausal period of ovarian involution, characterized by vasomotor lability–eg, hot flashes, dysmenorrhea, redistribution of fat, dyspareunia, early osteoporosis, ↓ skin elasticity, hormonal changes–eg, ↑ FSH and LH, ↓ PGE$_2$, estrogen, and progesterone; postmenopausal ovaries continue to secrete androgens, which are converted into estrogens 2. Analogous age period in men *adjective* Menopausal

Climara® GYNECOLOGY A transdermal ERT patch for menopausal ♀, which ↓ risk of osteoporosis, vasomotor menopausal Sx, vulval and vaginal atrophy, hypoestrogenism. See Estrogen replacement therapy, Menopause

climatologic disaster PUBLIC HEALTH Any natural disaster caused by climatologic disturbances, which may be linked to global phenomena–eg El Niño, La Niña that impact on local weather EXAMPLES Floods, earthquakes and storms–often land-based–tornadoes or water-based–cyclones, hurricanes, typhoons. See Earthquake. Cf Geological disaster

climax Orgasm, see there

climbing up on oneself Gower sign NEUROLOGY A clinical sign for the manner in which children with well-developed Duchenne's muscular dystrophy–DMD, rise from a sitting to a standing position, by grasping and pulling on body parts from the knees to hips until they are in an erect position

clinafloxacin ANTIBIOTICS A broad-spectrum–gram-negative bacilli, staphylococci fluoroquinolone antibiotic with ↑ activity against streptococci and anaerobes; it is also active against PCP and toxoplasmosis ADVERSE EFFECTS Diarrhea due to *Clostridium difficile* overgrowth. See Fluoroquinolone

clindamycin INFECTIOUS DISEASE An antibiotic combined with pyrimethamine to treat and prevent toxoplasmosis, PCP and, topically, for acne vulgaris ADVERSE EFFECTS Diarrhea, dysgeusia. See AIDS

clinic MEDICAL PRACTICE A site where Pts are seen on an outpatient basis, either as a first-time visit, or as a follow-up to some form of previous evaluation or therapy. See Ambulatory care, Betty Ford Clinic, Free clinic, Pain clinic, University health clinic PSYCHIATRY A place where Pts are treated for a specific kind of disorder; either medical or mental

clinical *adjective* 1. Pertaining to a clinic or to the bedside; that which can be observed in Pts 2. Pertaining to or based on observation and management of Pts, in contrast to theoretical or basic sciences

clinical alert MEDTALK Tickler A message–electronic, snail mail, etc–that reminds a health care provider of a particular Pt, of management related issues germane to that person's management

clinical cooperative group RESEARCH A group–eg SWOG, ECOG–of health care institutions or investigators that collaborate to develop and implement common therapeutic protocols

clinical decision analysis A quantitative approach to complex decisions, first used by the military and industry, and increasingly popular in medicine, as a vehicle for creating useful decision-making models. See Algorithm, Critical pathway, Decision analysis

clinical depression PSYCHIATRY Persistent sadness or loss of interest in activities for ≥ 2 wks in absence of external precipitants DIFFDX Grief, bereavement CLINICAL Changed eating habits, insomnia, early morning wakening, ↓ interest in normal activities, depressed mood, fatigue, suicidal ideation. Cf Bereavement

clinical diagnosis MEDTALK A working hypothesis based on collected symptom data, both subjective and objective, which are used to consider potential cause-and-effect relationships

clinical director MEDTALK A locally defined term that variously translates into lab director, medical director, chief of service, etc. See Laboratory director, Medical director

clinical etiquette

clinical etiquette Professional comportment MEDICAL PRACTICE The components of medical practice which, in addition to ethics and competence, define what it is to be a physician

CLINICAL ETIQUETTE

BEDSIDE MANNER Avoid easy familiarity, be attentive of Pts needs, do not eat on rounds

DRESS Conservative & appropriate

GROOMING Clean, neat, unobtrusive

LANGUAGE Respectful, at level of audience, non-use of vulgar vernacular or demeaning appellations, discretion regarding others' condition (JAMA 1988; 260:2559)

clinical evaluation MEDTALK An evaluation of whether a Pt has symptoms of a disease, is responding to treatment, or is having adverse reactions to therapy

clinical faculty GRADUATE EDUCATION An unpaid member of a medical school staff who has a private practice, and devotes < 50% of his time to the institution; the responsibilities of 'voluntary track' faculty include participation in teaching efforts, administrative functions and/or research efforts; clinical faculty members receive the titles of clinical instructor, clinical assistant professor, clinical associate professor, and clinical professor, at increasing levels of accomplishments

clinical indicator PATIENT CARE An objective measure of the clinical management and outcome of Pt care

clinical laboratory test A generic term for any test regarded as having value in assessing health or disease states

CLINICAL LABORATORY TEST PURPOSES

DEFINE RISK OR DISEASE, eg detect hyperglycemia, hypercholesterolemia

STRATIFY A PERSON INTO A DISEASE OR NONDISEASE STATE, in which the population is bimodal with overlapping parameters, which include above normal without disease and low normal with disease

MONITOR A CONDITION, eg glucose for DM, cholesterol for ASHD (CAP Today, 1994; 8:34)

clinical nurse specialist A registered nurse with an advanced degree in a particular area of Pt care–eg, neurosurgery nurse specialist

clinical pathology The field of pathology dedicated to measuring and/or identifying substances, cells, or microorganisms in body fluids AREAS Clinical microbiology–bacteriology, mycology, parasitology, virology; immunology; chemistry; hematology; immunohematology–blood banking. Cf Anatomic pathology, Surgical pathology

clinical pathway Critical pathway, treatment pathway CLINICAL MEDICINE A standardized algorithm of a consensus of the best way to manage a particular condition MODALITIES USED Teletherapy, brachytherapy, hyperthermia and stereotactic radiation. See Oncology, Surgical oncology MEDTALK A multidisciplinary set of prescriptions and outcome targets for managing the overall care of a specific type of Pt–from pre-admission to post-discharge for Pts receiving inpatient care; CPs are intended to maintain or improve quality of care and ↓ costs for Pts, in particular DRGs

clinical practice guidelines Clinical policies, practice guidelines, practice parameters, practice policies MEDTALK Systematically developed statements to assist practitioner and Pt decisions about appropriate health care for specific clinical circumstances. See Psychology

clinical presentation The constellation of physical signs or symptoms associated with a particular morbid process, the interpretation of which leads to a specific diagnosis

clinical preventive service MANAGED CARE A health care service delivered in clinical settings for the purpose of preventing the onset or progression of a health condition or illness

clinical proteinuria LAB MEDICINE Urinary protein excretion > 0.5 g/24 hr, a level detectable by ordinary dipstick testing; CP corresponds to an albumin excretion of > 300 mg/24 hr; CP is a typical finding in early DM. See Diabetes mellitus

clinical psychology PSYCHOLOGY A field that applies psychologic principles to assessing, preventing, ameliorating and rehabilitating mental distress, disability, dysfunctional behavior, and enhance psychologic well-being. See Psychology

clinical responsibility Any task or duty involving the professional component of medical practice, which requires the exercise of clinical judgement with respect to Pt care. Cf Administrative responsibility

clinical rotation MEDICAL EDUCATION A period in which a medical student in the clinical part of his/her education passes through various 'working' services[3] in 1-4 month blocks

clinical significance MEDTALK A conclusion that an intervention has an effect of practical meaning to Pts and health care providers; even though an intervention is found to have a statistically significant effect, this effect might not be clinically significant

clinical study PHARMACOLOGY A study designed to measure the safety, efficacy, and appropriate dosage of a new drug or biological; CSs may be placebo-controlled or, less commonly, compared with a 'gold standard' therapy. See Phase I, II, III study

clinical suspicion A working hypothesis about a Pt's diagnosis, which is then tested with appropriately targeted tests to arrive at a definitive diagnosis; a CS is based on a constellation of findings in a Pt that suggests to the physician a limited palette of possible diagnoses

clinical trial Clinical medical trial, clinical research trial RESEARCH A controlled study involving human subjects, designed to evaluate prospectively the safety and effectiveness of new drugs or devices or behavioral interventions. See Drug discovery, IND, Phase I, II, and III studies

clinician A health care professional–physician, physician assistant, or nurse–involved in active Pt management. See Primary mental health care clinician, Staff clinician. Cf Academician

clinicopathological conference A formal discussion of a Pt's clinical, radiologic, and lab data, usually in front of a large group of junior and senior colleagues. Cf Professorial rounds

Clinoril® Sulindac, see there

clipping NEUROSURGERY A definitive therapy for intracranial aneurysms, which consists of the direct obliteration of the aneurysmal neck with a clip of the proper strength, shape, and size, that prevents the flow of blood into the dome of the aneurysm, while preserving the parent artery. Cf Interventional neuroradiology, Trapping

Clitocybe TOXICOLOGY A genus of poisonous mushrooms CLINICAL Abrupt onset-15 mins after ingestion; profuse sweating, nausea, abdominal colic, blurred vision, salivation, rhinorrhea, dizziness MECHANISM Stimulation of parasympathetic system MANAGEMENT Atropine. See Poisonous mushroom

clitoridectomy HUMAN RIGHTS A type of ♀ circumcision. See Female circumcision. Cf Infibulation

CLITORIDECTOMY TYPES

I Partial or total clitoris removed-likened to penis amputation

II Clitoridectomy and partial labia minora excision

The wound is closed with thread, grass, or other suture materials, or with a poultice

Clitoridectomy is deeply rooted in the culture of certain African countries, and symbolizes societal control over a woman's sexuality (NEJM 1994; 331:712a)

clitoroplasty GYNECOLOGY An operation in which an overlarge phallus of an intersexual is pared to that of a typical clitoris, with excision of part of the corpus cavernosus and shaping of the foreskin and scrotum to create labia majora and minora. See Female circumcision, Intersex, John/Joan

CLL 1. Chronic lymphocytic leukemia 2. Cholesterol-lowering lipid

CLO test® Rapid urease test LAB MEDICINE A diagnostic test performed during a gastric Bx to detect *H pylori*. See Helicobacter pylori

cloacogenic carcinoma Basaloid carcinoma, see there

cloasma Melasma, 'pregnancy mask' OBSTETRICS A rash 2° to the hormonal effects of pregnancy or OCs, exacerbated by sunlight CLINICAL Irregular, flat, symmetrical, light brown areas on the malar region, cheeks, and forehead, which often fade with delivery or cessation of birth control pills

clodronate Ostac®, Bonefos® METABOLISM A biphosphonate used to manage osteoporosis, Paget's disease of bone, cancer-induced hypercalcemia in Pts with prostate CA, myeloma and other CAs by regulating osteoclastic activity THERAPEUTIC EFFECT ↓ pain, risk of Fx, new bone metastases. See Osteoporosis

clofazimine Lamprene® THERAPEUTICS A lipophilic rhimophenazine used to manage leprosy and, with other drugs, atypical *Mycobacterium* infections—eg, MAC, discoid lupus, pyoderma gangrenosum ADVERSE EFFECTS GI upset, skin discoloration, rash

clomiphene citrate Clomid® OBSTETRICS An ovulation-inducing agent–'fertility drug', which acts by releasing gonadotropins from the pituitary; CC ↑ multiple gestations, and ↑ the risk–RR = 2.3 of borderline or invasive ovarian tumors

clomipramine NEUROPHARMACOLOGY A tricyclic antidepressant analogue of imipramine, used for OCD ADVERSE EFFECTS Sedation, anticholinergic effects, orthostatic HTN, tremor, nausea, sweating, seizures, arrhythmias, sexual dysfunction

clonality A state of proliferation determined by the cell(s) of origin; daughter cells arising from multiple cells–ie polyclonal are reactive in nature, while those arising from a single cell–ie monoclonal are neoplastic; determining clonality is important in treating a disease, and understanding its physiopathology, and can be assessed in ♀ by molecular analysis of the patterns of X chromosome inactivation. See Clonal analysis

clonazepam PHARMACOLOGY A benzodiazepine sedative and anticonvulsant used to treat petit mal seizures ADVERSE EFFECTS Drowsiness, ataxia, behavioral changes, hyperactivity, restlessness, irritability, hypersalivation CONTRAINDICATIONS Severe liver disease, narrow-angle glaucoma

clone A population of cells derived from a single parent cell and thus genetically identical; genetic differences in clonal population may arise from random spontaneous mutations during cell growth

clonic *adjective* Pertaining to clonus, see there

clonidine suppression test The inability of clonidine, an α_2-agonist, to suppress catecholamine secretion, a finding suggestive of a pheochromocytoma. See Pheochromocytoma

clonus NEUROLOGY 1. A volley of muscle contractions and relaxations, occurring in rapid succession 2. An abrupt transient muscle contraction

clopidogrel Plavix® CARDIOLOGY An antiplatelet agent that ↓ the risk of ischemic stroke, MI, and vascular death in Pts with prior cardiovascular disease CONTRAINDICATIONS Active bleeding–eg, PUD, intracranial hemorrhage, or Pts at risk for bleeding due to trauma, surgery or other drug therapy ADVERSE EVENTS Dyspepsia, purpura, diarrhea, rash, intracranial hemorrhage, neutropenia. See CAPRIE. Cf Aspirin

clorazepate PHARMACOLOGY An anxiolytic and sedative hypnotic prodrug for nordiazepam

close call MEDTALK A popular term for a serious medical error that does not result in harm to the Pt. See Medical error

close contact EPIDEMIOLOGY Any person who has been in intimate contact for a period of time with a person with infectious TB

close contact infection An infection transmitted by close, nonsexual, contact between susceptible and infectious individuals

closed angle glaucoma Acute glaucoma, see there

closed bite Deep bite DENTISTRY A dental malocclusion in which there is overbite–upper teeth cover the lower teeth due to a ↓ occlusal vertical dimension

closed comedo Whitehead, see there

closed fist injury SPORTS MEDICINE The tearing of the skin on knuckles, which traumatically impact on the teeth of an opponent, often in the context of non-gloved fisticuffs; CFIs are commonly associated with infections requiring systemic antibiotics.

closed fracture Simple fracture ORTHOPEDICS A fracture in which the skin is not broken. See Fracture

closed loop system CRITICAL CARE MEDICINE An electronic system used in ICUs to monitor multiple Pt parameters–eg, pulse and respiratory rate, and regulate and control all by computer, including the rate of infusion of IV drugs and fluids. See ICU

closed order book GRADUATE EDUCATION A practice in teaching hospitals in which only the intern–a physician in the first yr of post-medical school training—can write therapeutic orders for hospitalized Pts in his/her care. See Libby Zion, 405 regulations

clostridial myonecrosis Gas gangrene, necrotizing subcutaneous infection EMERGENCY MEDICINE A rapidly progressive, life-threatening form of gangrene that is a rare complication of 'dirty' traumatic wounds, which are infected with *Clostridium* spp ETIOLOGY War wounds, bee stings, venipuncture; CM is due to production of toxins, especially alpha toxin, leading to tissue necrosis, shock MANAGEMENT Surgical decompression, excision of necrotic tissue, penicillin, hyperbaric oxygen. See Hyperbaric oxygen therapy, Necrotizing fasciitis

Clostridium botulinum MICROBIOLOGY A gram-positive, spore-forming anaerobe which produces a potent neurotoxin. See Botulism

Clostridium difficile A common cause of bacterial colitis; it is the causative agent in 99% of pseudomembranous colitis, and 20-30% of antibiotic-associated diarrhea

***Clostridium difficile* colitis** INFECTIOUS DISEASE Colonic infection by *C difficile* CLINICAL Some are asymptomic and become *C difficile* carriers; more commonly, diarrhea, abdominal pain, colitis, fever, vomiting dehydration; if severe, pseudomembranous colitis, with possible perforation DIAGNOSIS Stool—cytotoxicity assay is sensitive–≥ 94% and specific–≥ 99%, but expensive; ELISAs are more cost-effective but less sensitive LAB ↑ WBC TREATMENT Metronidazole–98% response rate, vancomycin–96%, bacitracin–83%, cholestyramine–68%. See Pseudomembranous colitis

Clostridium perfringens INFECTIOUS DISEASE An anaerobic gram-positive spore-forming rod, widely distributed in nature and present in the intestine of humans and other mammals. *C perfringens* type A accounts for ±15% of outbreaks of food poisoning in the US

clot VOX POPULI *noun* An intravascular coagulum. See Blood clot, Hard clot, Sentinel clot *verb* To coagulate

clot buster Thrombolytic (agent), see there

148 **clot-dissolving drug** Thrombolytic (agent), see there

clotrimazole INFECTIOUS DISEASE An antifungal imidazole used topically for candidiasis ADVERSE EFFECTS Nausea, dizziness, diarrhea, changed LFTs

clotter A popular term for 1. A hematologist specialized in disorders of coagulation 2. A Pt with ↑ coagulation

clotting factor Coagulation factor. See Factor

cloud baby A popular term for an infant with an infection that spreads by aerosol, who releases 'clouds' of viral or bacteria-rich material into the ambient air, and is a vector for miniepidemics of URIs

cloxacillin INFECTIOUS DISEASE A beta-lactam penicillin that is *not* hydrolyzed by staphylococcal penicillinases, ie most isolates of *S aureus* and some coagulase-negative staphylococci are susceptible to cloxacillin; it is also useful for streptococci

clozapine Clozaril® PSYCHIATRY A dibenzodiazepine antipsychotic and sedative, and effective neuroleptic for acutely psychotic and treatment-resistant schizophrenics, without extrapyramidal effects–dystonia, parkinsonism, tardive dyskinesia or ↑ prolactin levels ADVERSE EFFECTS Hypotension, hypersalivation, sedation, agranulocytosis–1% of Pts, tachycardia, weight gain. See Bundling, Positive and Negative Symptom Scale, Schizophrenia. Cf Haloperidol

club drug Any 'recreational' drug–eg Ecstasy, Herbal Ecstasy, Rohypnol, GHB, Ketamine, LSD, widely used by high school and college students, young adults who frequent night clubs and all-night rave parties RISKS, CHRONIC USE Dropping out–school, college, loss of job, crime

Club Med dermatitis A photodermatitis consisting of linear, mirror-image, hyperpigmented tender patches with scalloped edges, located on the inner thighs

clubbing of fingers Expansion–broadened and/or thickened fingertips or toes, with ↑ lengthwise curvature of the nail and a ↓ in the angle between the cuticle and the nail, often accompanied by underlying hypertrophic osteitis associated with chronic 'central'–pulmonary hypoxia–↓ blood O_2 due to infection, emphysema, congenital heart disease–tetralogy of Fallot, transposition of great vessels, cystic fibrosis, bronchogenic CA, TB, bronchiectasis, pneumoconiosis, interstitial pulmonary fibrosis DIFFDX Nonpulmonary disease–eg, atrial myxomas, Hodgkin's disease, Crohn's disease, ulcerative colitis, hepatitis, cirrhosis, thyroid acropachy, Graves' disease

clubfoot Calcaneal valgus, calcaneovalgus, metatarsus varus, talipes calcaneus, talipes equinovarus, talipes equinus, talipes valgus, talipes varus, valgus calcaneus ORTHOPEDICS The most common congenital deformity of the foot, in which the foot is turned downward and inward; clubfoot is formed by a combination of equinus–plantar flexion of forefoot; calcaneus–dorsiflexion of forefoot, the calcaneus forms the plantar prominence; varus–heel and forefoot are inverted, plantar surfaces shifted medially; and valgus–eversion of forefoot and lateral facing of the plantar surface; a general term for a group of deformities of the ankles and/or feet that are usually present at birth; the defect may be mild or severe and may affect one or both ankles and/or feet TYPES Talipes equinovarus–foot is turned inward and downward; calcaneal valgus–foot is angled at heel with toes pointing upward and outward; metatarsus varus–front of foot is turned inward; uncorrected CF may result in gait defects ETIOLOGY Hereditary and environmental factors; isolated, or due to various underlying disorders–eg, muscular dysplasia, anomalous insertion of tendons, arthrogryposis, congenital constriction bands, in utero compression, CNS disease–spina bifida, poliomyelitis, Friedrich's ataxia, Sheldon-Freeman 'whistling face' syndrome, Moebius syndrome MANAGEMENT Orthopedic splints or casts to correct the foot position

clumsy child syndrome Dyspraxia PEDIATRIC NEUROLOGY A child whose fine and/or gross motor skills are immature, erratic, slow, imprecise;

neurologic exam is normal; children may have cognitive or perceptual problems due to learning disabilities TYPES Constructional, dressing, ideational, ideomotor, oromotor dyspraxias. Cf Cerebral palsy

cluster headache Histamine headache, migrainous neuralgia NEUROLOGY A chronic recurring headache which is more common in younger ♂, attributed to histamine release; CHs are intense, but short-lived–½-2 hrs, unilateral, often periorbital headache with a 'clock-setting' predictability, which often occurs with spring to fall seasonality, over 3-8 wks, disappears for months to yrs, and may begin within 2-3 hrs of falling asleep CLINICAL Knifelike intranasal or retrobulbar pain; unlike migraines in which the Pts prefer to lie still in a darkened room, CH victims restlessly pace, bang their heads against the wall and have suicidal ideation; CHs may be of vascular origin and accompanied by a blocked, runny nose, tearing, facial flushing and swelling, ptosis, pupil constriction MANAGEMENT Prevention–ergotamine tartrate, methysergide is more effective than analgesics once acute attack begins

cluster suicides PSYCHIATRY Multiple suicides, usually among adolescents, in a circumscribed period of time and area, which have a 'contagious' element. See Suicide

cluttering SPEECH PATHOLOGY A condition characterized by an excessive rate of speech with an irregular rhythm, collapsing of sounds and words, and loss of syllables; cluttering can range in severity from garbled, but generally intelligible, to virtually unintelligible, and may co-exist with stuttering TREATMENT Bethanechol may be effective. See Stuttering

Clutton's joint RHEUMATOLOGY A painful symmetric arthritis of the knees, occasionally also elbows and ankles, due to congenital syphilis CLINICAL Acute or insidious onset of pain and swelling, with synovial thickening and effusion IMAGING Tibial periostitis

C_{max} PHARMACOLOGY The peak serum concentration of a therapeutic drug. See Pharmaco-kinetics, T_{max}. Cf Trough

CME Continuing medical education GRADUATE EDUCATION Education for health professionals that follows completion of formal post-medical school specialty training–eg, residency, fellowships FORMATS Lectures, seminars, refresher courses, workshops, audio- and video-tapes SPONSORS Medical schools, professional organizations, hospitals. See Graduate medical education, On-line CME. See Sponsored symposia. Cf Libby Zion

CMF regimen ONCOLOGY An adjuvant combination–cyclophosphamide, MTX, fluorouracil–regimen administered after modified radical mastectomy in lymph node-positive breast CA. See Breast cancer

CMI MANAGED CARE Case Mix Index, see there PSYCHIATRY Chronic mental illness

C_{min} Trough serum concentration, see there

CML Chronic myelocytic leukemia, see there

CMO MANAGED CARE Chief medical officer, see there

CMR Crude mortality rate, see there

CMS Centers for Medicare and Medicaid Services, formerly HCFA, Health Care Financing Administration. See Managed care

CMV Cytomegalovirus, see there

CN Certified nurse practitioner, see there

CNS Central nervous system, see there

CNS abscess Brain abscess NEUROSURGERY A space-occupying cerebral pus pocket 2° to a bacterial or, less commonly, fungal infection ETIOLOGY Direct–epidural abscess, chronic ear infection or sinusitis, dental care, open head trauma RISK FACTORS Congenital heart disease–eg, tetralogy of Fallot, defective vessel walls—eg, Osler-Rendu-Weber syndrome CLINICAL Headache, neurologic defects—eg, motor or sensory loss MANAGEMENT Surgical drainage

CNS leukemia Central nervous system leukemia, see there

CNS metastases Central nervous system metastases, see there

CNS tumor ONCOLOGY Any tumor of the CNS, which may be primary—eg brain stem glioma, craniopharyngioma, medulloblastoma, meningioma, or metastatic or secondary. See Brain tumor

co-sleeping Bed-sharing PEDIATRICS The sleeping of an infant or child in a parent's bed PROS Intimate contact with parent during critical formative period of infancy CONS Risk of death—±60 occur/yr in the US—due to suffocation, strangulation in bed clothing, or overlying. See Overlying

coactivated T cells IMMUNOLOGY T cells coated with monoclonal antibodies to enhance their ability to kill tumor cells. See Activated T cell

coagulation HEMATOLOGY Clot formation SURGERY The physical disruption of tissue to form an amorphous residuum, as in electrocoagulation and photocoagulation. See Coagulopathy, Interstitial laser coagulation

coagulation disorder See Coagulopathy

coagulation factor Blood clotting factor HEMATOLOGY Any of a number of serum proteins—factors I-XIII, designated by Roman numerals, which act in a coordinated fashion to clot blood

coagulation factor I Fibrinogen, see there

coagulation factor II Prothrombin, see there

coagulation factor III Thromboplastin, a heterogeneous group of phospholipids—tissue factors present in the brain, lung, placenta. See Tissue factor

coagulation factor IV Obsolete for ionized calcium. See Calcium

coagulation factor V Factor V, see there

coagulation factor VI Obsolete for accelerin, factor Va

coagulation factor VII Factor VII, see there

coagulation factor VIII Factor VIII, see there

coagulation factor IX Factor IX, see there

coagulation factor X Factor X, see there

coagulation factor XI Factor XI, see there

coagulation factor XII Factor XII, see there

coagulation factor XIII Factor XIII, see there

coagulation factor assay Any test that measures levels or functional activity of one or more coagulation factors; the one-stage extrinsic CFA is abnormal in the face of defects of one or more of the following factors: II, V, VII, or X; the one-stage intrinsic CFA evaluates factors VIII, IX, XI, XII. See Extrinsic pathway, Factor(s), Intrinsic pathway. Cf Coagulation panel

coagulation panel Coagulation profile A battery of tests that identifies coagulation defects, based on time required for blood to clot; the CP includes prothrombin time, activated partial thromboplastic time, platelet count, bleeding time. Cf Organ panel

coagulopathy HEMATOLOGY A clotting defect in which bleeding does not stop in the usual time period ETIOLOGY Hemophilia, drug-induced defects—eg, aspirin, thrombocytopenia, liver disease, Von Willebrand's disease. See Consumption coagulopathy, DIC, Leukemic coagulopathy

coal dust OCCUPATIONAL MEDICINE Finely particulate carbonaceous dust, often mixed with silicates; CD is a known lung irritant intimately linked to coal workers' pneumoconiosis—black lung disease. See Coal dust macule, Coal workers' pneumoconiosis

coal miner's elbow Student's elbow Olecranon bursitis caused by prolonged pressure, friction or trauma to the elbow. Cf Beat knee

coal tar THERAPEUTICS A blackish, semisolid byproduct of the destructive distillation of bituminous coal, which contains benzene, naphthalene, phenols, and other organic compounds; CT is used to treat psoriasis, and may induce clearance of psoriatic plaques CONS Unpleasant odor, carcinogenic; CT is usually combined with UVB light. See Psoriasis

coal workers' pneumonoconiosis Anthrasilicosis LUNG MEDICINE An occupational lung disease affecting those with prolonged—> 10 yrs exposure to coal dust, which accumulates in the macrophages of peribronchiolar tissues CXR Scarring, nodules in lungs with advanced disease CLINICAL Early pneumoconiosis is often a 'pathologist's disease'—ie, established by histology, followed, with time, by wheezing, chronic cough, dyspnea, pulmonary HTN, respiratory failure IMAGING 2 radiologic patterns of CWP: simple—nodular, which in turn is divided into 3 stages, and complicated TREATMENT None. See Anthracosis

coaptive thermocoagulation ENDOSCOPIC SURGERY A procedure used to ablate bleeding ulcers, in which 2 vessel walls are compressed and "fried"

coarctation A stricture or point of contraction. See Aortic coarctation

coarse rales Coarse crackles PULMONARY MEDICINE A loud and low-pitched, discontinuous, 'explosive' lung sound. Cf Fine crackles

coated tongue 1. Oral candidiasis, see there 2. Strawberry tongue, see there 3. A popular term for a film of bacteria and debris covering the tongue in absence of clinical disease

Coats' disease OPHTHALMOLOGY A usually unilateral AR condition, characterized by exudates and retinal detachment, affecting children from 18 months to 18 yrs of age; CD may be accompanied by hearing loss, muscle weakness, myopathic facies, mental retardation DIFFDX Retinoblastoma

cobalamin deficiency See Vitamin B_{12} deficiency

cobalt knife Cobalt Scalpel™ RADIATION ONCOLOGY A focused RT system for noninvasive stereotactic CA treatment. See Radiation therapy

COBALT CARDIOLOGY A clinical study—Continuous Infusion vs Double-Bolus Administration of Alteplase—designed to compare the effect on M&M of Pts undergoing AMI of accelerated—which has now become the standard administration of alteplase—over 90 mins, with 2 bolus doses of alteplase given 30 mins apart. See Alteplase. Cf GUSTO

cobblestone appearance Cobblestone pattern A widely used term for multiple, equally-sized rounded densities that project from a single linear surface, when the image is 2-D or that rise above a flattened plane when viewed in 3-D, a pattern which has been likened to roads paved by multiple similarly-sized 'cobbled' stones GI DISEASE A characteristic radiologic and gross appearance of the intestinal mucosa in Crohn's disease, due to submucosal involvement; to the endoscopist, cobblestoning refers to the uniform nodules—due to the submucosal edema, while the pathologist refers to severe ulcerative disease with crisscrossing of the ulcers through inflamed but intact mucosa; intestinal 'cobblestoning' may also occur in ulcerative colitis—where ulcers alternate with regenerating mucosa, ischemic colitis, lymphoid hyperplasia of CVID, amyloidosis, mucoviscidosis, pneumatosis cystoides intestinalis, multiple lymphangiomas, and polyposis coli; in the intestine, the mucosal rugosities may correspond to polyps, or be filled with air, lymphoid tissue, or amyloid ORAL DISEASE Multiple, closely-set intraoral papilloma-like fibromas that impart a pebbly tactile sensation in Cowden's premalignant multiple hamartoma syndrome

cocaine Coke, snow, C, girl, stardust, crack, rock SUBSTANCE ABUSE An alkaloid powder–benzoylmethyl-ecgonine $C_{17}H_{21}NO_4$–derived from *Erythroxylon coca*; cocaine is used in medicine as a topical anesthetic; it is a potent psychostimulant, and evokes intense physical and psychological addiction STATISTICS 3 million Americans use cocaine regularly–600,000 use heroin, 10-15% of US population has tried cocaine–40% of those between age 25-30; 10-15% become addicted; cocaine is implicated in 5/1000 deaths in ages 25-30; up to 25% of MVA fatalities in drivers aged 15-45 in NYC had cocaine in their system CLINICAL Neurologic Sx–agitation, disorientation, convulsions, TIAs, CVAs, mood swings, delirium; cardiovascular Sx–arrhythmia–eg V tach, V fib, myocarditis, acute MI; cocaine's adverse effects on myocardial O_2 supply are exacerbated by cigarette smoking, ↑ demand for O_2, ↓ diameter of coronary arteries; other features include fetal damage, sudden death, hyperpyrexia, cerebral vasculitis, nasal changes, ↓ O_2-diffusing capacity, spontaneous pneumomediastinum, eating disorders–eg bulimia and anorexia, and madarosis–singeing of eyebrows and eyelashes due to hot vapors associated with smoking crack; chronic abuse is associated with ↑ sensitivity–'reverse tolerance' to cocaine's non-euphorigenic effects–eg hyperactivity and anesthesia CHRONIC ABUSE Seizures are common in habitual abusers, who have diffuse cortical atrophy, by CT and diffuse slowing of waves on EEG DESIRED EFFECTS Stimulates CNS, causing intense euphoria–±45 mins in duration, ↓ hunger, indifference to pain and fatigue, illusions of great physical strength and mental ability ADVERSE EFFECTS Cardiovascular, malnutrition, anemia DRUG TESTING Plasma $T_{1/2}$ 90 mins; urine and serum are screened for benzoylecgonine, cocaine's major metabolite, by enzyme-labeled competitive immunoassay–see EMIT, and confirmed for legal purposes by the 'gold standard' method, GC-MS PATHOLOGY-BRAIN Subarachnoid hemorrhage, cortical atrophy GI TRACT GI ischemia due to adrenergic vasoconstriction HEART Mononuclear cell inflammation of myocardium with myocytic necrosis LUNGS Congestion, spontaneous pneumothorax ORAL CAVITY Cocaine's pH of 4.5 causes dental erosion, and ↑ caries in cocaine abusers, although this may be related to altered eating habits PREGNANCY Abruptio placentae, premature labor, vaginal bleeding after injection of cocaine, placental vasoconstriction, spontaneous abortion, congenital malformation, ↑ perinatal mortality, neurologic and behavior defects, tachycardia, HTN PSYCHIATRIC DISORDERS Dysphoria, paranoid psychosis, depression SEXUAL DYSFUNCTION Cocaine's myth as an aphrodisiac derives from its ability to delay ejaculation and orgasm, causing temporary mood elevation, and heightened sensory awareness, a myth that may be undeserved; chronic cocaine abuse evokes impotence, subfertility–eg, ↓ sperm counts, ↓ motility, ↑ abnormal sperm forms MANAGEMENT Cocaine-induced acute rhabdomyolysis may respond to dantrolene, a drug used in malignant hyperthermia; pharmacologic management of cocaine dependence and withdrawal has failed with antidepressants–eg desipramine and flupenthixol, dopaminergics–eg bromocriptine, mazindol, flupentixol, opiate antagonists–eg naltrexone and mixed agonists-antagonists–eg buprenorphine, anxiolytics–eg buspirone and anticonvulsants–eg carbamazepine; psychotherapy and behavior therapy have been equally ineffective CAUSE OF DEATH Arrhythmia due to ventricular fibrillation and cardiovascular collapse, respiratory arrest with pulmonary edema, cerebrovascular insults associated with HTN. See Binge, Cocaine nose, Crack, Freebase, Kindling

cocaine blues A popular term for depression after extended cocaine use

cocaine cardiotoxicity A condition caused by long-term cocaine use CLINICAL Chest pain, tachycardia, HTN, acute MI, ventricular arrhythmia, asystole, CHF, sudden death due to cardiovascular collapse. See Cocaine

cocaine nose A constellation of findings in chronic intranasal–snorting cocaine abusers CLINICAL Frequent rubbing of nose, chronic rhinitis, rhinorrhea, loss of sense of smell and, due to collapse of nasal mucosa and matrix, a nose to smell with RHINOSCOPY From nada to perforation SURGICAL COMPLI-

CATIONS Localized septal collapse, poor mucosal healing, inadequate correction of septal deflection TREATMENT Rhinoplasty on highly selected Pts–results are often poor; submucosal resection and septoplasty should be avoided

cocaine rhabdomyolysis syndrome SUBSTANCE ABUSE A potentially fatal complication of cocaine abuse characterized by renal failure, profound hypotension, hyperpyrexia, ↑↑↑ CK; in those who die, DIC is almost always present. See Cocaine

cocaine withdrawal syndrome Cocaine abstinence syndrome SUBSTANCE ABUSE Abstinence from cocaine results in a triphasic response consisting of 1. 'Crash' with paranoia, hypersomnia followed by hyperphagia 2. Withdrawal with prolonged anhedonia and cocaine craving 3. Extinction and possibly major depression

co-carcinogen ONCOLOGY An agent–eg, a chemical, radiation, with a 'helper' role in carcinogenesis; co-carcinogens differ from tumor promotors in that they must be present at the same time as the carcinogen. See Tumor promoter

coccidioidomycosis Coccidiosis INFECTIOUS DISEASE An infection by spores of a soil fungus, *Coccidioides immitis* CLINICAL 60% of those infected are asymptomatic or have disease indistinguishable from URI, identified only by skin testing; in the rest, there is a 1-3 wk asymptomatic period, followed by a lower RTI, with fever, sweating, anorexia, weakness, arthralgias, cough, chest pain, weight loss, erythema nodosum COMPLICATIONS Pleural effusion, dissemination, possibly meningitis PROGNOSIS Poor if disseminated TREATMENT IV amphotericin B; oral fluconazole, itraconazole, ketoconazole

cochlear implant AUDIOLOGY A multicomponent electronic prosthetic device for those with severe hearing loss, whose sensory neurons have been damaged, but not completely destroyed, and for whom conventional hearing aids are inadequate; CIs bypass damaged structures in the inner ear and directly stimulate the auditory nerve. See Cochlea, Hearing aid

Cochran Collaboration JOURNALISM A project being coordinated in 10 centers worldwide, and carried out by thousands of volunteers, that searches the world medical literature on randomized control trials, and together with all the unpublished trials that can be located, publish the findings in an electronic form. See Meta-analysis

Cockayne syndrome An AR condition characterized by dwarfism, microcephaly, 'salt and pepper' choroidoretinitis, optic atrophy, cerebral calcifications, mental retardation, intention tremor, tottering gait, deafness, small trunk, long extremities, ↓ subcutaneous fat, sexual infantilism, hepatosplenomegaly, ASHD, early death. Cf Hutchinson-Gilford syndrome

cockroach ENTOMOLOGY A largely nocturnal insect, Order Blattaria; most common US roaches are *Periplaneta americana*, *Blattella germanica* and *B orientalis*, and are of medical interest as potential vectors for bacteria–eg, *Salmonella* spp; other organisms cultured from cockroaches include *Shigella*, *Proteus*, *Mycobacterium* spp, *E coli*, *Klebsiella pneumoniae* and *Pseudomonas aeruginosa*

cocktail party syndrome Chatter-box syndrome A descriptor for the behavior of children with arrested hydrocephalus, who may be sociable, gabby, pseudointelligent, speaking in a seemingly erudite fashion on subjects about which they have no true understanding, scanning dysrhythmic speech. See Williams syndrome

cocktail therapy A mixture of drugs used to produce a particular effect

cocoon A descriptor for the fibrotic encasement of the entire small intestine in sclerosing peritonitis, a spontaneous idiopathic process in young ♀, which follows peritoneovenous shunting, practolol therapy, peritoneal dialysis, chemotherapy, where unknown toxins may stimulate fibroblastic proliferation, reactive fibrosis

co-counseling Peer counseling PSYCHOLOGY A therapeutic philosophy that originated in the 1960s in Seattle; in a typical co-counseling session, each person has an allotted time to act as a counselee, while the peer counselor helps him/her recognize hidden or blocked emotions; once one person's allotted time ends, the roles are switched. See Encounter groups, Humanistic psychology

COD Cause of death

cod-liver oil NUTRITION An oil from cod liver, rich in n-3 fatty acids, in particular eicosapentaenoic acid–EPA; 10 g of CLO contributes 1 g of EPA to the diet. See Eicosapentanoic acid, Fish oil

code *noun* EMERGENCY CARE True code A widely used, highly popular term for 1. A cardiopulmonary arrest or other emergency requiring resuscitation, or 2. A call for personnel over the hospital's PA system to respond to such an emergency. See Failed code, Slow code ETHICS A set of rules or principles. See Code of Hammurabi, Nuremburg code, Hunter-killer code MANAGED CARE A system for classifying medical or surgical procedures for payment by third-party payers. See A code, Activity code, Barcode, -GB code, J code, K code, NBG code, Uniform billing code, V code VOX POPULI A system for organizing large amounts of information, in which each block of data is designated by an alphanumeric. See Barcode *verb* To suffer a cardiac arrest in a hospital environment

code black MEDTALK Public address system jargon for a deceased person admitted to an ER. See Code

code blue Code EMERGENCY MEDICINE A message announced over a hospital's public address system, indicating that a cardiac arrest requiring CPR is in progress; to be 'coded' is to undergo CPR

code orange Code four HOSPITAL A message announced over a hospital's public address system warning the staff of 1. a bomb threat 2. a radioactive spill 3. person with mental issues is loose in the halls of the hospital

code pink HOSPITAL A message announced over a hospital's public address system warning the staff of an infant abduction

code purple VIP-patient plan A contingency plan announced over a hospital's public address system indicating that 1. a VIP or 2. violent person or Pt is in house. See VIP syndrome. Cf Chief syndrome

code red PUBLIC HEALTH A message announced over a hospital's public address system indicating that 1. A fire is occurring in the facility, or 2. There are weather conditions that are potentially hazardous to health by dint of pollutants and/or high air temperatures

code status The formally indicated–by signed documents—status of a Pt in a hospital, with respect to desire for resuscitative–ie CPR efforts, should the need arise; unless the Pt specifically requests *not* to be resuscitated, ie DNR (do not resuscitate) status, CPR is performed. See Advance directive, DNR

code yellow HOSPITAL A message announced over a hospital's public address system alerting the staff about, and the need to prepare for, a pending emergency or external disaster–eg, multitrauma, major effects of storm, etc

codec Coder/decoder TELEMEDICINE A device that uses hardware and/or software to translate analog transmissions into digital transmissions and compresses data from 10- to 100-fold; codecs are used to transmit information in telemedicine. See Asynchronous transfer mode switching, T-1, Telemedicine

codeine TOXICOLOGY An oral opioid analgesic with antitussive activity EFFECTS Blocks pain messages to brain; ↓ central response USED FOR Pain; antitussive. See Opioids

co-dependency SUBSTANCE ABUSE 1. The activity of an enabler, a person who allows an alcoholic or other substance abuser to continue with the addiction. See Enabler 2. The concomitant presence of 2 or more states of drug dependency–eg, alcohol and substance–marijuana, heroin, barbiturates

coding MANAGED CARE The placement of HCPCS codes–eg, 99251 for an initial inpatient consultation–on bills submitted to Medicare and other 3rd-party payers, identifying and defining the complexity of physician or other health care provider services. See Centralized coding, Computerized coding, CPT, Decentalized coding, Downcoding, Dynamic coding, Huffman coding, Static coding, Upcoding PSYCHOLOGY The process in which a person acquires an image of his/her sexual self. See Gender coding, Gender crosscoding

Codman's triangle RADIOLOGY A wedged elevation of periosteum seen on a plain film of the long bones, a 'classic' finding of Ewing sarcoma, also seen in osteosarcoma, bone metastases, hematomas, syphilis, TB

coefficient VOX POPULI A variable or factor which allows the calculation of a property or quantity of a substance under various conditions. See Absorption coefficient, Activity coefficient, Adsorption coefficient, Attenuation coefficient, Dice coefficient of similarity, Inbreeding coefficient, Intraclass correlation coefficient, Mass attentuation coefficient, Mass energy absorption coefficient, Octanol-water partition coefficient, Spearman's rank (order) correlation

coefficient of retraction PULMONOLOGY The ratio of the maximal static recoil pressure to total lung capacity, a sensitive indicator of elastic recoil; ↓ COR is typical of severe emphysema, and results in a left shift in volume-pressure curve. See Elastic recoil, Lung-reduction surgery

coefficient of variation LAB MEDICINE The standard deviation divided by the mean, expressed as a percentage, used to evaluate and compare methodologies and instruments. See Mean, Standard deviation

coercion PUBLIC SAFETY Threat of kidnapping, extortion, force or violence to be performed immediately or in the future or use of parental, custodial, or official authority over a child < age 15; the use of some form of force to compel a person into therapy, most commonly psychiatric–eg, child psychiatry, or treatment of substance abuse

coffee LIFESTYLE A beverage made from dried, roasted beans of the coffee tree–*Coffea arabica*, a moderate stimulant causing mild physical dependence

coffee grounds appearance A descriptor for the color and consistency of bleeding in benign and malignant gastric ulcers; 'coffee grounds vomitus' is characteristic of yellow fever

Cogan syndrome An autoimmune condition characterized by interstitial keratitis and bilateral, rapidly progressive audiovestibular dysfunction CLINICAL Vasculitis of CNS, aorta, heart, pericardium, lungs; ocular complaints–photophobia, blurred vision, lacrimation, pain; Meniere's disease-like Sx–vertigo, ataxia vegetative Sx, with progression to complete absence of vestibular function, manifested by ataxia and oscillopsia, bilateral hearing loss MANAGEMENT High dose steroids, immunosuppression

Cognex® NEUROLOGY An agent that transiently slows the mental decline in Alzheimer's Pts. See Alzheimer's disease

cognitive behavioral therapy PSYCHIATRY Therapy that seeks to alleviate specific conditions–eg, phobias, by modifying thought and behavior EFFICACY Uncertain. See Psychotherapy

cognitive function NEUROLOGY Any mental process that involves symbolic operations–eg, perception, memory, creation of imagery, and thinking; CFs encompasses awareness and capacity for judgment

cognitive sciences The areas of medicine that study the nature and processes of mental activity–eg, neurology, psychiatry, psychology

cognitive skill PSYCHOLOGY Any of a number of acquired skills that reflect an individual's ability to think; CSs include verbal and spatial abilities, and have a significant hereditary component

cognitive therapy Cognitive treatment PSYCHIATRY Therapy in which a person is taught to cope with internal conflicts–anxiety, stress, guilt, phobias, and emotional negativity by consciously changing the way of thinking; CT is used for depressive disorders, ADD, and emphasizes rearranging maladaptive processes of thinking, perceptions, and attitudes. See Attention-deficit-hyperactivity syndrome, Distorted thinking

cogwheel phenomenon NEUROLOGY A circular jerking rigidity in flexion and extension in a background of tremor, which continues throughout an entire range of movement, a finding typical of parkinsonism; it may also be 'smoothly' rigid, and termed lead pipe rigidity. Cf Clasp knife phenomenon, Gegenhalten

coherent speckle ULTRASONOGRAPHY A deterioration in contrast resolution that occurs in pulse-echo ultrasonography, due to phase interference of echoes returning simultaneously from different tissue regions to the receiving transducer. See Speckle pattern

cohort CLINICAL TRIALS A group of persons with a common characteristic, set of characteristics or exposure, who are followed for the incidence of new diseases or events, as in a cohort for a prospective study. See Birth cohort, Cluster, Inception cohort

coiled spring appearance RADIOLOGY A descriptor for a finding in radiocontrast studies of the upper GI tract, caused by 1. Posttraumatic intramural hematoma of the 3rd segment of the duodenum, where the folds over the mass are stretched; a similar finding occurs in intussusception. See Stacked coin appearance 2. Distension of small intestinal loops, especially the jejunum

coin lesion Solitary pulmonary nodule RADIOLOGY A rounded, circumscribed nodule measuring < 4 cm that may be surrounded by well-aerated pulmonary parenchyma, which often appears as an incidental finding in an otherwise unremarkable plain CXR; ±50% of CLs in the US DIAGNOSIS Age, smoking Hx, geographic location, Hx of previous malignancy ETIOLOGY Infection–abscesses, aspergilloma, bacteria, coccidioidomycosis, echinococcal cysts, *Dirofilaria immitis*, histoplasmosis, TB, benign masses–bronchial adenoma, chondroma, diaphragmatic hernia, benign mesothelioma, neurogenic tumor, sarcoidosis, sclerosing hemangioma, Wegener's granulomatosis, rheumatoid nodules; malignant masses–$1°$ lung CA, which comprise $\geq 35\%$, metastases $\pm 10\%$, sarcoma, myeloma, Hodgkin's disease, choriocarcinoma

coinfection The simultaneous presence of 2 or more infections, which may $\uparrow$ disease severity and duration

co-insurance MANAGED CARE A cost-sharing requirement in many health insurance policies or health plans, in which the insured and insurer share payment of an approved charge/fee for covered services in a specified ratio, after an annual deductible is paid, up to a pre-set limit or maximum liability. See Allowed Charge, Copayment, Cost Sharing, Deductible, 80/20, Maximum liability

coital fantasy Copulation fantasy; intercourse fantasy SEXOLOGY Imagery of erotically stimulating content that precedes sexual intercourse, which may be essential to attaining an orgasm. See Masturbation fantasy

coital pain Painful sexual intercourse in ♀. See Dyspareunia

coitus Coition, copulation SEXOLOGY Sexual union between a ♂ and ♀ involving insertion of the penis into the vagina

coitus interruptus A form of natural family planning in which the man's orgasm and ejaculation are an out-of-body experience. See Contraception, Natural family planning, Pearl index

Coke® bottles OPHTHALMOLOGY A popular term for thick glasses, which have been fancifully likened to the bottoms of the 'classic' bottles of Coca-Cola™

colchicine An alkaloid isolated from autumn crocus-*Colchicum autumnale* used for long term control of gout MECHANISM Colchicine blocks mitosis at metaphase by binding to the tubulin heterodimer, interfering with microtubule assembly INDICATIONS Management of gouty arthritis; used in research to arrest cells during mitosis–by disrupting the spindle, to visualize chromosomes

cold Common cold, see there

cold abscess A well-circumscribed focus of acute inflammation without the usual signs of a 'hot' abscess–eg, calor, rubor, tumor, dolor–per Celsus; CAs 'classically' occur in paravertebral TB, and may be seen in normal persons infected by an attenuated organism, or in Pts with hyperimmunoglobulin E syndrome given inadequate antibiotic therapy. See Job syndrome

cold agglutinin disease Cold agglutinin syndrome CLINICAL IMMUNOLOGY An immune disorder characterized by IgM autoantibodies that optimally agglutinate RBCs at very low temperatures–eg, 4°C; low titers–< 1:32 of cold agglutinins–CAs are detectable in many normal subjects; polyclonal CAs $\uparrow$ after certain infections–eg, mycoplasma, CMV, EBV, trypanosomiasis, and malaria, peak in 2-3 wks and are insignificant if non-hemolytic; CAs may occur in any Pt with acquired hemolytic anemia and a positive direct Coomb's test; certain antibodies have been implicated–eg, anti-I, -i, -Pr, -Gd, Sd^d. See Autoantibody. Cf 'Room temps'

cold compress ORTHOPEDICS CCs are usually applied intermittently to acutely injured muscle, joints or bone, up to 48 hrs after the initial trauma

cold cone biopsy Cold knife conization GYNECOLOGY The excision of a conical piece of uterine cervix, to confirm the Dx of HPV, CIN or CA based on the results of a pap smear, treat same, since the pathology is usually confined to the cone. See CIN, HPV. Cf LEEP

cold hypersensitivity An excess autonomic nervous system reaction to low ambient temperature, characterized by bradycardia and a local wheal-and-flare reaction; when familial, the reaction is termed 'cold urticaria'

cold injury A general term for any injury induced by low ambient temperatures–eg, chilblain, trench foot, frostbite, which may be accompanied by tissue freezing RISK GROUPS Infants, young children, elderly, alcohol use

cold laser Excimer laser, see there

cold nodule IMAGING A focus of $\downarrow$ radioisotope uptake on a ^{123}I or ^{99m}Tc scintillation scan of the thyroid, seen in cystic or solid, often nonfunctional lesions; when the mass is solid by ultrasound, a biopsy is warranted as most carcinomas are cold nodules. Cf Hot nodule, Warm nodule

cold remedy POPULAR PHARMACOLOGY Any OTC product for relief of one or more common cold symptom TYPES Antihistamines, decongestants PROS CRs provide some relief by partially suppressing nasal congestion, runny nose, cough CONS CRs are not antimicrobial, don't enhance the immune system, or alter duration. See Common cold

cold scan IMAGING A round non-imaging region in a solid organ–eg, liver, seen by scintigraphy. See Cold nodule. Cf Hot nodule

cold sore DERMATOLOGY A popular term for a perioral infection by HSV 1, resulting in a blister and a weeping ulcer; Pts should avoid direct–oral, orogenital mucosal contact with others, to prevent spread of infection; herpes labialis may recur, exacerbated by stress, sunlight, fever or illness MANAGEMENT Antiviral creams, oral–acyclovir; pre-treatment with acyclovir before certain exposures may $\downarrow$ exacerbations. Cf Canker sore

cold turkey method SUBSTANCE ABUSE A cessation technique used in narcotic or nicotine addicts, where withdrawal Sx are treated as little as possible, with the hope that once the addict has passed through the catharsis of withdrawal–and its unpleasant Sx, he/she would be unwilling to re-initiate drug abuse

cold urticaria-angioedema Cold-induced urticaria-angioedema Autonomic nervous system reactivity to low temperatures CLINICAL FORMS, ACQUIRED After cold exposure, the Pt suffers a pruritic urticarial eruption, potentially evolving into angioedema, accompanied by headaches and wheezing; if the whole body is cooled–swimming in winter, hypotension, collapse, possibly death HEREDITARY, IMMEDIATE-TYPE Erythematous maculopapules accompanied by a burning sensation, pyrexia, arthralgias, ↑ WBCs HEREDITARY, DELAYED-TYPE Erythematous swelling 9-18 hrs after a cold 'challenge'

colectomy GI DISEASE The surgical removal of various lengths of the colon; partial colectomies–with a rim or margin of healthy Tx are motivated by CA and acute diverticulitis; more complete colectomies–hemi-colectomy, total colectomy are indicated for long-standing ulcerative colitis, familial adenomatous polyposis, premalignant conditions which eventuate into cancer. See Open colectomy

colic CLINICAL MEDICINE Pain that crescendos to a peak of severity and then slowly subsides. See Abdominal colic, Renal colic PEDIATRICS Infantile colic A common Sx of infants, in which paroxysmal abdominal pain of presumed GI origin is accompanied by irritability, tensing of muscles, severe crying, lasting until the infant is completely exhausted ETIOLOGY Uncertain, possibly air swallowing, overfeeding, cow's milk allergy, undiluted juices, and emotional stress; in absence of organic causes—strangulated hernia, intussusception, pyelonephritis, others—no therapy gives consistent relief; infants often outgrow colic by 6 months of age

coliform bacteria Coliforms MICROBIOLOGY Any of a number of small, gram-negative, rod-shaped, facultative anaerobic bacteria which are used to monitor water pollution by feces

colitis GI DISEASE Inflammation of the large intestine, which is divided into structural and/or functional subtypes—ulcerative, Crohn's, infectious, pseudomembranous, spastic CLINICAL Rectal bleeding, abdominal colic, diarrhea DIAGNOSIS Visualization by colonoscopy, sigmoidoscopy; barium enema, virtual colonoscopy MANAGEMENT Differs by type, and includes medical or surgical management. See Chemical colitis, Crohn's colitis, Collagenous colitis, Diversion colitis, Enterocolitis, Indeterminant colitis, Irritable bowel syndrome, Ischemic colitis, Neutropenic colitis with aplastic anemia, Pseudomembranous colitis, Ulcerative colitis

collaboration PSYCHIATRY A helping relationship between a family member and a mental health professional who share responsibility for a child with an emotional disorder

collagen disease & arthritis panel LAB MEDICINE A battery of serum tests designed to establish the diagnosis of a rheumatic disease in a most cost-effective manner; the CD&AP measures ESR, rheumatic factor–by latex agglutination, uric acid levels, ANAs, C-reactive protein. See Organ panel

collagen vascular diseases Connective tissue diseases, see there

collagenous colitis GI DISEASE A condition characterized by watery diarrhea and the case-defining, but patchy, histologic finding of ↑ collagen in the upper lamina propria of the intestinal mucosa; the collagen may obscure the lower border of the epithelial basement membrane, and be accompanied by a mononuclear inflammation of the lamina propria; fecal WBCs occur in 55% MANAGEMENT Uncertain, possibly 5-aminosalicylic acid, cholestyramine, corticosteroids, sulfasalazine

collapse A state of extreme prostration and depression, with circulatory failure. See Volitional collapse

collar button abscess An advanced abscess of the finger that forms on the palmar and dorsal surfaces, commonly at the metacarpophalangeal joint TREATMENT Palmar and dorsal incision and drainage

collar button lesion GI DISEASE Any of a number of small deep ulcers with narrow necks, which parallel a barium-filled large intestinal lumen,

affected by ulcerative colitis; the 'button' surface corresponds to an eroded ulcer base, and the 'button neck' corresponds to the sides of the islands of preserved mucosa; a similar phenomenon occurs in colonic amebiasis ulcers. See Flask lesion

collateral *adjective* Referring to that which occurs in addition to a desired effect, is located adjacent to, or on the radius of a circle, secondary, or accessory. See Collateral damage

collateral circulation CARDIOLOGY Blood flow that pursues a channel or system of vessels that is alternative to, or develops in substitution for, a major vascular pathway

COLLATERAL CIRCULATION CORONARY ARTERY

GRADE 0 No flow in the collateral

GRADE 1 Collateral is barely apparent; dye is present in at least 3 consecutive frames

GRADE 2 Collateral is moderately opaque but present throughout at ≥ 75% of the cardiac cycle; there is antegrade motion of the dye rather than diffuse filling.

GRADE 3 Collateral is well opacified and the column of dye is well defined–ie, > 0.5 mm in diameter, but < 0.7 mm wide over majority of length; collateral has clear antegrade dye motion

GRADE 4 Collateral is well opacified, fills antegrade, and is very large; > 0.7 mm in diameter over entire length (Am Heart J 1999; 137:169)

collateral damage SURGERY A popular term for any undesired but unavoidable co-morbidity associated with a therapy–eg, chemotherapy-induced CD to the BM and GI tract as a side effect of destroying tumor cells

collecting station LAB MEDICINE A facility specifically designed to collect specimens and send them to a central lab; CSs perform no tests. See CLIA, Laboratory. Cf Drawing station

collection agency MANAGED CARE An enterprise that demands payment for a product sold or a service rendered, after an appropriate period of time has passed and the purchaser has not paid the seller. See Dun

collective attitude PSYCHOLOGY A general term for those attitudes that develop due to social pressure, from family, peers, mentors or the media

Colles fracture Dinner fork deformity ORTHOPEDICS A fracture of the distal radius at the wrist due to a fall on an outstretched hand, where the distal Fx fragment is angled upwards–dorsal angulation, imparting a fork-like appearance

collimation RADIATION PHYSICS The formal process in which a beam or field of radiation is reduced with a lead diaphragm, tube, or cone

collision OBSTETRICS A mechanical obstruction to the birth of twins, such that the lay of one fetus impedes the engagement of the other; the most extreme collision is known as interlocking, see there PUBLIC HEALTH Motor vehicle accident. See MVA

colloid goiter Endemic goiter, see there

colloid solution FLUID THERAPY A suspension of particles that are so small–1 nm to 1 μm in diameter—that they don't settle out of solution without external force–eg, centrifugation TRANSFUSION MEDICINE A balanced solution–eg, Dextran 40, Dextran 70, used for volume expansion PROS Composed of HMW substances retained in vessels, unassociated with peripheral edema CONS Allergic reactions; ↑ cost. Cf Crystalloid solution

colloquialism VOX POPULI A term of ordinary everyday speech, conversational. See Medical slang

co-localization The presence of 2 or more substances in the same site

colon cancer See Colorectal cancer

colon cut-off sign IMAGING Gaseous distension of the right and trans-

colon polyp

154

verse colon with ↓ or absent air beyond the splenic flexure, a finding characteristic of acute pancreatitis; the inflamed transverse mesocolon–which is attached to the anterior pancreas 'cuts off' the barium flow, causing mesenteric arterial thrombosis and ischemic colitis

colon polyp GASTROENTEROLOGY A hereditary or acquired pedunculated neoplasm arising from the colonic mucosa; small CPs are usually benign, but may become malignant; like colorectal CA, CPs may present with occult bleeding. See Polyp, Colorectal cancer

COLON POLYP

ACQUIRED POLYPS Adenomatous or tubular–villous in morphology, ↑ frequency with age; although often asymptomatic, larger polyps are often announced by bleeding or changed bowel habits; if very large, APs may form a leading 'front' of an intussusception; distinction between adenomatous polyps–'tight' round glands and villous adenomas–finger-like fronds of elongated glands has little practical importance–both have malignant potential; periodic colonoscopy and polypectomy yields a 3-fold ↓ in subsequent cancer; hyperplastic polyps are also acquired but are non-neoplastic

HEREDITARY POLYPS

• **FAMILIAL ADENOMATOUS POLYPOSIS** A premalignant, AD condition presenting in early adulthood with 100s to 1000s of colonic polyps, related to a loss of the normal repression of DNA synthesis in the entire colonic epithelium; adenocarcinoma occurs in 70–100% of Pts, prevented by prophylactic colectomy

• **GARDNER SYNDROME** A rare AD condition with premalignant polyps of the entire GI tract, which is identical to FAP, but has, in addition, extraintestinal tumors; most Pts develop colon carcinoma; other neoplasms in GS Pts include bile duct carcinoma, osteomas of mandible, skull, and long bones, soft tissue tumors–fibromas, lipomas, sebaceous cysts, and rarely, thyroid and adrenal gland cancers

colon resection SURGERY The segmental or subtotal surgical removal of colon INDICATIONS Colorectal cancer, angiodysplasia, ulcerative colitis, acute diverticulitis COMPLICATIONS Anastomic dehiscence, infection, necrosis. See Anterior resection, Hemicolectomy, Sigmoidectomy

colonic irrigation SURGERY An intraoperative procedure for antegrade cleansing of the large intestine in an emergency colon resection, which can be used in elective left-sided colonic surgery in Pts who are clinically stable, circumventing the need for a temporary colostomy. See Bowel preparation.

colonoscope GI DISEASE A flexible lighted fiberoptic endoscope inserted per rectum to perform colonoscopy, which is equipped with optics to allow the examiner to see and photograph the mucosa, and channels for obtaining a biopsy, or to spritz air through the colonoscope to improve visualization. See Colonoscopy. Cf Virtual colonoscopy

colonoscopic polypectomy Removal of any polyp–hyperplastic, adenomatous, villous adenoma, or polypoid–eg lymphoid hyperplasia, lesion by colonoscopy. See Colon polyp

colonoscopy GI DISEASE Visualization of the large intestine with a colonoscope; lesions found during colonoscopy are biopsied and examined by a pathologist to determine whether they are benign or malignant ABNORMAL RESULTS GI bleeding, diverticuli, polyps, stricture, tumor, IBD–eg, ulcerative colitis, Crohn's disease COMPLICATIONS Rarely, intestinal perforation. See Virtual colonoscopy

color blindness Colorblind The partial or, rarely, complete inability to distinguish colors, which affects up to 10% of men; most CB is X-linked–90% occurs in ♂ and is tested by the Ishihara pseudoisochromic charts

color flow Doppler Angiodynography IMAGING A diagnostic procedure that uses ultrasound to image arteries–which show as red and veins–blue, to identify vascular obstruction

Colorado tick fever American mountain fever, mountain fever, mountain tick fever INFECTIOUS DISEASE A rare acute tick-borne–vector, *Dermacentor andersoni* RNA orbiviral infection occurring in the early spring in the Rocky Mountains CLINICAL 3 to 6 day incubation, then chills, sweats, joint pain,

biphasic–'saddleback' fever, myalgias of the back and legs, headache, retroorbital pain, photophobia, malaise, N&V, rash, weakness MANAGEMENT Tick removal, acetaminophen for fever PROGNOSIS Excellent; self-limited with little residua

colorectal cancer Colon cancer ONCOLOGY A malignant epithelial tumor arising from the colonic or rectal mucosa, which is the 3rd leading cause of cancer in ♂, 4th ? in ♀ in the US; risk of CC is ↓ with a low fat, high fiber diet EPIDEMIOLOGY 152,000 new cases, 57,000 deaths–1993, US SURVEILLANCE Annual Fecal occult blood testing is reported to ↓ mortality by 33% PREDISPOSITION Adenomatous polyps, family Hx–highest if 1st-degree family member–parents, siblings or children had CC and even higher if < age 55, ulcerative colitis SCREENING Most colorectal cancers develop from polyps; colon polypectomy ↓ CC; colon polyps and early cancer may be asymptomatic; screening is recommended every 3 yrs CLINICAL Rectal bleeding, occult blood in stools and, in advanced cases, bowel obstruction and weight loss DIAGNOSIS Colonoscopy with biopsy, CT, barium enema PATHOLOGY Most CCs are adenocarcinomas; 'raromas' include lymphomas, neuroendocrine carcinomas, and sarcomas MOLECULAR PATHOLOGY CCs develop as genetic alterations accumulate–eg, K-*ras* oncogene on chromosome 12, and tumor-suppressor genes on chromosomes 5, 17p–which encodes p53, and 18q–*DCC* gene MANAGEMENT Surgery; cure likely if CA is confined to intestine. See Colorectal adenoma

COLORECTAL CANCER–TNM CLASSIFICATION

STAGE I Tumor invades muscularis propria, but has not spread to nearby lymph nodes

STAGE II Tumor spread into the subserosa and/or perirectal tissues with up to 3 regional lymph nodes, or directly invades adjacent tissues without lymph node involvement

STAGE III Any depth of tumor invasion with four or more positive lymph nodes, *without* distant metastases

STAGE IV Any depth of tumor involvement; any number of involved lymp nodes, *with* distant metastases

colorectal surgeon A surgeon specialized in managing colorectal disease–eg, hemorrhoids, cancer

colostomy SURGERY A hole in the abdominal wall at which the colon communicates directly with the outside; colostomies are created as an artificial anus and required when surgery mandates temporary or permanent loss of the rectum, anal function INDICATIONS Diverticulitis, Crohn's disease, ulcerative colitis, for diverting fecal stream in colon cancer, intestinal obstruction, anorectal defects. See Ileostomy, Urostomy

colostrum OBSTETRICS A sticky yellow-white fluid secreted by the breasts from late pregnancy to several days after birth, but before breast milk is produced. See Breast milk

colposcopy GYNECOLOGY A technique in which a lighted magnifying device, a colposcope, is used to evaluate and/or assist in biopsying lesions of the uterine cervix and upper vagina, previously identified by a pap smear. See Acetowhite lesion, Cervical biopsy, CIN, HPV

colposuspension needle UROLOGY A malleable tipped needle used in Stamey bladder neck suspension procedures. See Bladder suspension

coma NEUROLOGY A state of unarousable unconsciousness; a sleep-like state; not conscious ETIOLOGY Head injury, neurologic disease, acute hydrocephaly, intoxication or metabolic derangement or hypoglycemia. See Barbiturate coma, Diabetic coma, Induced pentobarbital coma, Pseudocoma

coma panel Coma profile LAB MEDICINE A battery of tests used to ID the cause of a coma PARAMETERS MEASURED Alcohol, ammonium, calcium, creatinine, glucose, lactic acid, osmolality, phenobarbital, toxicology screen of blood and urine. See Organ panel

comatose NEUROLOGY *adjective* Referring to a state in which an individual is unresponsive even to painful stimuli; not conscious Cf Semicomatose 2. In a coma

combat fatigue Battle fatigue, combat shock PSYCHIATRY A condition that affects soldiers after long tours of combat duty, which is characterized by a loss of self-esteem, anxiety, tremulousness, depression, extreme emotional lability, dyspepsia, dyspnea. See 'Burn-out syndrome', Old soldier's heart, Post-traumatic stress disorder

combination therapy PHARMACOLOGY Cocktail therapy The use of multiple agents to treat a clinical condition–eg, cancer, AIDS REPRODUCTIVE MEDICINE The use of estrogen and progestin as an OC. Cf Chemotherapy, Combined modality therapy, Polypharmacy

CombiPatch® Estalis® GYNECOLOGY A dermal patch containing estradiol/norethindrone acetate used to manage the vasomotor Sx of menopause; vulvar and vaginal atrophy; hypoestrogenism due to hypogonadism, castration, or primary ovarian failure

Combivent® Ipratropium Br/albuterol PULMONOLOGY A combination used to improve pulmonary function tests in COPD

co-medication PSYCHIATRY The use of a second medication to alleviate the side effects of another medication

comedo Blackhead, plural comedones DERMATOLOGY A plug of keratin and sebum within the dilated orifice of a hair follicle, which may contain bacteria–eg, *Propionibacterium acnes*, *P granulosum*, *Staphylococcus albus*, or yeast, *Pityrosporon ovale*. See Acne. Cf Comedocarcinoma

comfort care TERMINAL CARE Palliative and supportive treatment for Pts who are suffering from a terminal illness–eg AIDS, cancer, or who have refused life-sustaining treatment; CC is aimed at relieving symptoms, enhancing the quality of remaining life, and easing the dying process. See DNR, Euthanasia, Hospice, Initiative 119, Kevorkian, Palliative therapy, Physician-assisted suicide

coming out SEXOLOGY Gay vernacular, from the phrase, "coming out of the closet", referring to acknowledging to oneself and others, homophilic sexuoerotic orientation. See Closet

'coming soon' advertising 'Teaser' advertising which indicates the name of a drug, without claims for potential indications, safety, or effectiveness. See Advertising

command hallucination PSYCHIATRY A hallucination in which a person perceives spoken orders or commands from an 'entity' within; CHs in a schizophrenic should trigger an ↑ in monitoring of Pts to prevent violent behavior or suicide

commando operation SURGERY A term attributed to Hayes Martin, a pioneer in head & neck surgery, for the en bloc removal of an advanced 1° malignancy of the oral cavity, usually SCC–lymphoma is amenable to RT or chemotherapy; the CO is a very aggressive procedure, and entails partial removal of the mandible, floor of the mouth and/or tongue, accompanied by a radical neck dissection. See Heroic surgery, Radical neck dissection

comminuted fracture ORTHOPEDICS A fracture with multiple pieces of broken bone. See Fracture

commission BUSINESS & HEALTH A type of finder's fee set by insurance brokers or agents for selling health plans, built into the premiums paid by the insured. See Biological Stain Commission, Commission on Dietary Supplement Labels, Commission E, Enzyme Commission, Joint Commission on Accreditation of Healthcare Organization, National Commission for Certification of Acupuncturists, National Commission for the Protection of Human Subjects of Biomedical & Behavioral Research, Nuclear Regulatory Commission, Pepper Commission, Pew Commission

commissurotomy SURGERY The surgical splitting of a commissure–eg in the brain or cusps of heart valves. See Percutaneous transvenous mitral commissurotomy

commitment FORENSIC PSYCHIATRY An order or process by which a court or magistrate directs its officer(s) to take a mentally ill person to a mental health facility or penal institution; commitment proceedings can be civil or criminal, voluntary or involuntary, and usually require a court or judicial proceeding. See Emergency psychiatric commitment

committee HOSPITAL PRACTICE A group of health professionals affiliated with a hospital who meet regularly to address an area–eg, infection control, safety, etc that needs a multidisciplinary approach to ensure issues are addressed and quality of service is maintained. See Accredited Standards Committee, Blue ribbon committee, Committee of 10,000, Committee for Freedom of Choice in Cancer Therapy, Credentials committee, Disaster committee, Ethics committee, Executive committee, Hospital committee, Interagency Research Animal committee, International Committee for Standardization, Microbiology & Infectious Diseases Research Committee, NAAIDC subcommittee, National Committee for Quality Assurance, Patient care committee, Political action committee, Radioactive drug research committee, Radiation safety committee, Recombinant DNA Advisory committee, Safety committee, Search committee, Seattle committee, Steering committee, Tissue review committee, Utilization review committee

commodity test LAB MEDICINE Any lab test that can be performed in terms of 'economy of scale,' ie the higher the volume of tests of a particular type performed, the lower the cost per unit; in general, CTs do not require a rapid turnaround, and the price is reduced to a minimum

common cold Acute nasopharyngitis INFECTIOUS DISEASE A popular term for a viral URI, 2° to inflammation of the nasal mucosa–rhinitis and coryza, which is usually spread by aerosol, and caused by any of a number of viruses AGENTS Rhinovirus has 111 serotypes,–caused 15-40% of CCs, coronavirus, 10-20%; viral causes of CC include influenza A, B, C, parainfluenza, RSV, adenovirus, rarely coxsackie and enterovirus; group A β-hemolytic streptococci cause 2-10% of CCs; in 30-50% no etiologic agent is identified IMMUNITY Because the CC is due to many differnt viruses, the body never builds up resistance–immunity to all of them, and CCs are a common and recurring problem FREQUENCY Preschool children ±9 CCs/yr; kindergarten ±12 colds/yr; adolescents and adults, ± 7 colds/yr OLD WIVES TALE DEBUNKED Exposure to cold weather probably has no effect on the spread of CCs INCIDENCE The CC affects 41/100 population/yr–US; it results in 23 million days of lost work, and a loss of $12 billion CLINICAL Influenza-like syndrome MANAGEMENT Zinc gluconate lozenges may ↓ incidence by 50% SUSCEPTIBILITY Psychologic stress may ↑ susceptibility to CCs

common migraine Migraine without aura NEUROLOGY An episodic headache that accounts for ±80% of migraines, and lasts between 4 and 72 hrs, associated with N&V; it is a common type of chronic headache, more common in ♀ between age 10 and 46, and not associated with typical migraine prodromes–eg, aura; Pts are asymptomatic between episodes; in ⅔, the headaches are unilateral, pulsating and of moderate to severe intensity, and may be accompanied by photophobia and phonophobia; there may be a family history of migraines and also of travel sickness. See Migraine. Cf Basilar migraine, Classical migraine

common pathway Any final route in a molecular 'cascade' in which there is a complex interplay among enzymes, substrates, activators, inactivators, and a relatively small signal is 'amplified' by a positive feedback loop to produce an effect COAGULATION A CP is initiated by either the extrinsic or intrinsic pathway, either of which activates factor X–Xa, which in turn activates factor II, converting it into thrombin in the presence of factor V, Ca^{2+}, and membrane phospholipid, and activate prothrombinase producing thrombin from prothrombin; prothrombin then converts fibrinogen into fibrin, forming a blood clot, that becomes irreversible when factor XIII is activated

common peroneal nerve dysfunction Common peroneal nerve neuropathy, peroneal nerve injury NEUROLOGY A peripheral neuropathy caused by damage to the myelin sheath or nerve per se, caused by trauma or injury to the knee, fibula, compressive casts, excess pressure on the posterior knee, DM, surgery, polyarteritis nodosa. See Peripheral neuropathy

common source outbreak EPIDEMIOLOGY An outbreak that results from a group of persons being exposed to a common pathogen or toxin PUBLIC HEALTH A mass infection from a single contaminated source

common variable immunodeficiency A heterogeneous, often AR group of primary immune dysfunctions, which affects 20-90/10⁶ live births, characterized by a ↓ in most immunoglobulin isotypes–B-cell precursors are present, but don't differentiate into plasma cells and T cell defects without major defects in cell-mediated immunity; CVID is usually limited to intrinsic B-cell defects, but in most Pts, there are also defects of T-cell activation and ↓ secretion of IFN-γ, IL-2, IL-4, IL-5, and B-cell differentiation factor CLINICAL Average age at diagnosis is 12 yrs, by which time many Pts have suffered recurrent bacterial infections, chronic otitis, sinusitis, bronchiectasis, pneumonitis, diarrhea, malabsorption, sprue-like enteritis, achlorhydria, pernicious anemia, cholestasis, and giardiasis, which affect up to 50% of Pts LAB ↓↓↓ Igs, ↓ serum IgG, IgA, and usually IgM, impaired antibody response to antigens, ↓ 5´-nucleotidase in lymphocytes RISKS CVID Pts have a 50-fold ↑ risk for gastric CA, 70% of whom have ↓ gastrin secretion in response to bombesin–a clinical marker for CVID Pts at risk for gastric CA MANAGEMENT IV gammaglobulins

communicable disease An infection by a pathogen that is easily spread by close or relatively close contact

communicating hydrocephalus Normal pressure hydrocephalus NEUROLOGY An enlargement of cerebral ventricles due to ↑ secretion and/or ↓ production and absorption of CSF, where the ventricular conduits are open, and the fluid moves freely into the spinal subarachnoid space, but is blocked by obliteration of the subarachnoid cisterns around the brainstem or subarachnoid spaces over the cerebral convexities ETIOLOGY Arnold-Chiari malformation, infections–eg, bacterial meningitis, toxoplasmosis, CMV, or other viral meningitides, subarachnoid hemorrhage, ↑ production of CSF–eg, choroid plexus papilloma, Hurler syndrome due to fibrosis in the subarachnoid space, or hypervitaminosis A

community-acquired pneumonia Pneumonia caused by an infection currently present in the community; CAP is the most common cause of infectious death–US, and number 6 killer overall; of the 57% of CAPs in which a pathogen is identified, *S pneumoniae* causes 60%, *H influenzae* 15%, *Legionella pneumophila* 10-15%, *Staphylococcus aureus* 2-10% CLINICAL Atypical pneumonia with delayed recognition; early SX are distinctly non-pulmonary–eg, dry cough, myalgia, arthralgia DIAGNOSIS Pathogens in pleural fluid or blood cultures, *Pneumocystis carinii* in sputum or BAL, a 4-fold ↑ in *Mycoplasma pneumoniae* antibody titers, isolation of *L pneumophila*, or a 4-fold ↑ in *L pneumophila* antibody titers, or positive direct fluorescent antibody test for legionella, *S pneumoniae* antigen in serum, urine RISK FACTORS Alcoholism, seizure disorders, smoking, immunosuppression

community medicine An informal division of general medicine, defined as the care of Pts in the context of their daily lives. See Family practice

community mental health center CLINICAL PSYCHOLOGY An institution that provides mental health services required by §1916(c)(4) of the Public Health Service Act and is certified by the appropriate state authorities as meeting such requirements. Cf Clinical psychologist

community rating MANAGED CARE The setting of insurance premiums based on the community's demographics and claims experience, without adjusting for the group's median age, gender, and health conditions. See Premium, Experience Rating

community support system PSYCHIATRY An organized system of care to assist adults with long-term psychiatric disabilities to meet their needs and develop their potentials without being unnecessarily isolated or excluded from the community

comorbidity The simultaneous presence of 2+ morbid conditions or diseases in the same Pt, which may complicate a Pt's hospital stay; in the US health care system, comorbidity carries considerable weight in determining the reasonable length of hospitalization under the DRG classification of diseases. See DRGs

companion SOCIAL MEDICINE An individual with whom a person has a close personal relationship EXAMPLES Spouses, lovers, children, parents, friends, pets and others, who provide an individual with a sense of belonging and of being needed. See Companionship

comparison film IMAGING An X-ray taken of the side opposite of the side thought to have a lesion, to compare the film quality and texture. See Old film

compartment syndrome Compressive syndrome ORTHOPEDICS A symptom complex caused by ischemia, trauma–fractures, inflammation or infection of a closed anatomic space, resulting in compression of nerves, blood vessels, or tendons that traverse the space CLINICAL Numbness, paresthesias, pain or loss of movement of an extremity MANAGEMENT Early therapy–fasciotomy is crucial as end-stage disease requires major reconstructive surgery to salvage function. See Carpal tunnel syndrome, Tarsal tunnel syndrome

compassionate use PHARMACOLOGY The use of an agent to treat Pts for whom conventional therapies have failed, or for whom no other drug exists; CU refers to the use of an agent on humanitarian grounds before it has received regulatory–FDA–approval

compatible *adjective* PHARMACOLOGY Referring to the lack of adverse interactions between medications, which allows simultaneous administration without nullification or aggravation of either's effect POP PSYCHOLOGY Referring to the ability of 2 adults to function as a couple TRANSFUSION MEDICINE Referring to a blood transfusion without a transfusion reaction in the recipient

Compazine® Prochlorperazine, see there

compensation ORTHOPEDICS A change of structure, position or function of a part in an attempt by the body to adjust to or neutralize the abnormal force of a deviation of structure, position or function of another part PSYCHIATRY 1. An unconscious defense mechanism in which one attempts to compensate for real or perceived defects 2. A conscious process in which one strives to compensate for real or perceived defects of physique, performance skills, or psychological attributes; often the 2 types merge. See Individual psychology, Overcompensation

compensation package The combination of salary, retirement fund contributions, payment for continuing medical education activities, books, journals, bonuses, licensing fees, payment of malpractice fees, health insurance and hospitalization, disability insurance, and other components with monetary value–eg, car leasing, memberships to professional organizations, clubs, and others, that comprise the financial arrangement offered to a physician or other professional, for providing services to a hospital, HMO, or group practice

compensatory mechanisms CARDIAC PACING Physiologic responsiveness of cardiovascular system whereby it changes its function and characteristics to ↑ or ↓ cardiac output. See Cardiac output

competence PATIENT'S RIGHTS A legal term for the capacity of a person to act on his/her own behalf; the ability to understand information presented, to appreciate the consequences of acting–or not acting–on that information, and to make a choice. See Autonomy. Cf Incapacity, Incompetence PSYCHOLOGY A constellation of abilities possessed by a person for adequate decision-making; competency is a measure of a person's autonomy and ability to give permission for diagnostic tests or for dangerous, but potentially life-saving procedures. Cf Autonomy VOX POPULI Skill, ability. See Cultural competence

complaint A Sx of which a person is aware or which causes discomfort, generally described from a Pt's perspective–eg, loss of weight, crushing chest pain, FUO, and is often the principal reason for seeking medical attention; in the working parlance in the US, complaints are divided into chief–major complaints and minor complaints. See Chief complaint, Minor complaint

complement IMMUNOLOGY *adjective* Pertaining to the complement system *noun* 1. Any protein of the complement system 2. Complement system The term was first used for a heat labile factor in serum that caused immune cytolysis of antibody coated cells; it now refers to the entire functionally related system comprising ± 25 distinct serum proteins, which mediate the non-specific inflammatory response to various antigens through a complex sequence of enzymatic cleavages; complement is thus the effector not only of immune cytolysis but also of other biologic functions; it is activated by 2 routes, the classic and alternative pathways. See Alternative pathway, Classic pathway, Complement activation MEDTALK *noun* A set of something. See Chromosome complement

complement C1q deficiency IMMUNOLOGY The absolute or relative lack of complement C1q, 90% of the affected have SLE; C1q apparently clears apoptotic cells, which may stimulate the immune system, ↑ autoantibody production. See Lupus erythematosus

complement C3 Beta-1-C-globulin, factor A A 180 kD serum protein; normal concentration 70–150 mg/dL, which is ↑ in chronic inflammation, bile obstruction, amyloidosis, severe protein malnutrition; ↓ in complement activation, immune complex disease, immunologic disease, bacteremia, genetic disease

complement cascade IMMUNOLOGY A complex, multimolecular biologic system with ± 25 different proteins that self-assemble on cell surfaces, functioning in concert with the specific immune systems to mediate host reactions and anti-microbial defense; the coup de grâce for the hapless target cell is the polymerization of C9 on its surface, which forms a transmembrane 'doughnut' that facilitates the egress of ions, cell lysis and death. See Complement. Cf Alternate pathway, Classic pathway, Common pathway

complement deficiency A state in which any of the complement proteins is subnormal

COMPLEMENT DEFICIENCIES–ASSOCIATED DISORDERS
C1

C1r SLE, renal disease, repeated infections

C1s SLE

C2 SLE, vasculitis, membranoproliferative glomerulonephritis, dermatomyositis

C3 Repeated infections

C4 SLE

C5 SLE, gonococcal disease

C6 Relapsing meningococcal meningitis, gonococcal infection

C7 Raynaud's disease, chronic renal disease, gonococcal infection

C8 SLE, gonococcal infection

complementation PSYCHOLOGY The process of functioning differently than, and in reciprocation to, someone else, by responding to that person's activities, behavior, and reactions, especially vis-á-vis differentiation of gender-identity/role. Cf Identification

complete abortion OBSTETRICS An abortion or miscarriage in which all tissues have been expelled; an abortion may be completed by curettage to eliminate necrotic decidual tissue in the uterus, which might act as a nidus for infection. See Abortion

complete blood count See CBC

complete fracture ORTHOPEDICS A fracture in which the bone fragments are completely separated. Cf Fracture, Incomplete fracture

complete hysterectomy Total hysterectomy GYNECOLOGY Complete surgical removal of the uterus and cervix, but not the adnexae–ovaries and fallopian tubes. See TAH-BSO

complete linkage GENETICS An inheritance pattern for 2 gene loci on the same chromosome, in which the observed crossover frequency between the loci is zero. See Chromosome, Crossing over, Gene, Inheritance, Linkage, Locus, Nonlinkage, Partial linkage

complete remission Complete response ONCOLOGY Disappearance of all signs and symptoms of disease–eg, cancer, multiple sclerosis, with normalization of all biochemical and radiologic parameters, as well as a negative repeat biopsy–pathologic remission. Cf Partial remission

completed stroke NEUROLOGY A stroke in which there is no further ischemia or loss of functional activity. See Cerebrovascular accident, Transient ischemic attack

complex *adjective* Complicated, not simple MEDTALK A bunch of related stuff. See Activated complex, AIDS dementia complex, AIDS-related complex, Antigen-antibody complex, B complex, Carney's complex, CHARGE complex, CREST complex, H-2 complex, HIV-associated cognitive motor complex, Immune complex, K complex, Lambda complex, Laryngeal complex, LBW complex, Lymphoid granular complex, MAIS complex, Medea complex, Membrane adaptor complex, *Mycobacterium avium* complex, Nasal complex, Oriental flush complex, Parachute valve complex, Polymyositis/dermatomyositis complex, QRS complex, Ribosomal-lamellar complex, Sarcoglycan complex, Sequestration complex, Sicca complex PSYCHIATRY A group of associated ideas with common, strong emotional tone, which is largely unconscious and significantly influences attitudes and associations. See Cain complex, Diana complex, Electra complex, Faust complex, Giving up-given up complex, God complex, Icarus complex, Inferiority complex, Jocasta complex, Joseph complex, King Lear complex, Mother Superior complex, Oedipus complex, Phaedra complex, Pygmalion complex, Superiority complex

complex closure SURGERY Any skin closure involving layering, debridement or advanced repair–plasty. See Debridement, Pedicle, Plasty

complex disease Complex trait, polygenetic disease, polygenic disease GENETICS A medical condition that arises from a intricate interaction of inherited–'nature' and environmental–'nurture' factors EXAMPLES Cancer, CAD, DM, HTN, bipolar disorder, obesity, schizophrenia. Cf Complicated disease, Mendelian genetics

complex partial seizure Complex seizure, partial complex seizure NEUROLOGY A brief, temporary seizure with multiple elaborate sensory, motor, or psychic components; rare in toddlers ETIOLOGY Regional necrosis due to hypoxia, trauma, tumors or discrete brain lesions CLINICAL Clouding of consciousness, event amnesia, bursts of anger or emotion, anxiety, automatisms EEG Temporal lobe spikes during sleep, abnormal electrical activity in the nerve cells of a discrete region of the brain. See Seizures

complex regional pain syndrome Reflex sympathic dystrophy INTERNAL MEDICINE A condition characterized by pain and tenderness associated with vasomotor instability, skin changes, and rapid development of bony demineralization–eg, osteoporosis often following localized trauma, stroke, or peripheral nerve injury TYPE CRPS Type 1–no definable nerve lesion; CRPS type 2, aka causalgia, nerve lesion is present

complex test Any test–eg, CT, IVP, endoscopy, immunoperoxidase staining, performed after simpler tests have not provided a diagnosis. Cf Simple test

complexion PHYSICAL EXAMINATION The color/appearance of the facial skin, which may be described as pale, flushed–transiently reddish, ruddy–reddish over a prolonged period. See Muddy complexion

compliance The capacity or ability to yield to a pressure or force without disruption or dysfunction; compliance is a measure of tissue distensibility–eg, of an air- or fluid-filled organ CLINICAL MEDICINE A measure of the extent to which Pts follow a prescribed treatment plan–eg, take drugs, undergo a medical or surgical procedure, exercise or quit smoking. See Patient compliance. Cf Noncompliance MANAGED CARE The adherence of a particular organization to statutes or mandates from regulatory agencies—governing agencies or bodies—or to an official mandate or obligatory standard. See HCFA 1500, UB92

complicated disease MEDTALK Any condition that has a stormy clinical course or which 'goes badly,' eg, advanced DM accompanied by diabetic retinopathy, renal failure, gangrene of lower extremity, etc. Cf Complex disease

complicated plaque VASCULAR DISEASE An advanced or fully developed lesion of atherosclerosis, defined as one or more of the following: rupture or fracture of the fibrous cap of the fibrous plaque, hemorrhage into the plaque, mural thrombosis, prominent fibrosis. See Atherosclerotic plaque

complication 1. Any adverse medical response to a procedure or therapy; in drug therapy, aka, adverse–side effects, see there 2. The simultaneous presentation of 2 or more diseases PATIENT MANAGEMENT Any adverse or undesired result of disease management. See Contraindication, Indication

component therapy Blood component therapy, see there

composite cultured skin DERMATOLOGY A skin substitute used to cover burn wounds and epidermolysis bullosa. See Artificial skin

composite/foam SURGERY A wound dressing that combines a thin nonadherent layer of hydrocolloid with a polyurethane sponge. See Dressing

composite lymphoma A rare lymphoma composed of 2 or more malignant cell lines in the same lymph node. See Lymphoma

composition VOX POPULI 1. That of which a thing is composed or formed. See Base composition 2. La-la-la-la-la-la etc, á la Mozart

Composix·Mesh® SURGERY A biosynthetic material used in ventral hernia repair; comprised of Bard mesh on one side for tissue ingrowth, bonded to ePTFE on the other side to prevent adherence to other tissues and adhesion formation. See Gore-Tex®

compound CLINICAL PHARMACOLOGY *verb* To combine two or more active pharmacologics to produce a single preparation, often referred to as a dosage form. See Formulation

compound fracture Open fracture ORTHOPEDICS A fractured bone exposed to potential bacterial contamination through a laceration in the soft tissues overlying the fracture site

comprehensive metabolic panel LAB MEDICINE A battery of analytes–albumin, alk phos, AST, BUN, calcium, chloride, glucose, potassium, sodium, total protein–which are measured to establish a baseline and detect metabolic disorders. See Panel

compress A pad of folded gauze, which may be applied with pressure to an area of skin and held in place for a period of time–eg, to cover an open wound or stop bleeding; compresses can be cold or hot; moist or dry. See Cold compress, Hot compress

compression CLINICAL MEDICINE Squeezing; pressure on a body or vital structure by 2 or more external forces, where the force is exerted by the bodies in a linear direction toward each other TELEMEDICINE A process used to transmit graphic images by eliminating extraneous data and packaging the file into lean block of data that is de-compressed by the receiver TYPES Lossless, lossy. See Lossless compression, Lossy compression

compression dressing Compression bandage WOUND CARE A bandage designed to provide pressure to a particular area

compression fracture Compression axial fracture, crush fracture ORTHOPEDICS 1. A lower cervical spine fracture caused by axial loading

CLINICAL. Ranges from minor–little displacement of bone, managed with external immobilization, to severe–bursting injury of the vertebral body and retropulsion of bone into the spinal canal or with significant comminution, best managed with cervical corpectomy and bone grafting 2. A fracture of a vertebral body due to axial compression, with loss of height of the vertebral body on X-ray, common in the lumbar spine in postmenopausal ♀ with osteoporosis

compression plate ORTHOPEDICS A plate used to stabilize fractures, used with lag screws to provide dynamic compression on the tension side of bone; dynamic CPs are the most common types of plates, and have special oval screw holes with a beveled floor and an inclined surface, to approximate the ends of the bone as the screws tighten. See Plate

compressive ear dressing ENT A thick dressing of soft gauze applied over the ear, covered by a circular wrap; CEDs *may* prevent post-traumatic ear swelling which causes cauliflower ear. See Cauliflower ear

compulsion PSYCHIATRY A behavior or mental act which is repetitive–eg, handwashing, double-checking, or mental–eg, repeating words silently, which a person feels compelled to perform in response to an obsession, or according to rules that must be applied strictly *or* behaviors or mental acts aimed at preventing or ↓ distress or preventing some dreaded event or situation, which are not realistically connected with what they are intended to neutralize or prevent, or behaviors that are clearly excessive. See Repetition compulsion. Cf Obsession, Obsessive-compulsive disorder, Obsessive-compulsive personality disorder

compulsive personality disorder PSYCHOLOGY A mental disorder characterized by avoidance of feelings, emotion and intimacy, accompanied by strict adherence to external or internalized rules of order. See Obsessive-compulsive personality disorder

computed tomography Computed axial tomography, computerized tomography, CT scanning IMAGING A special radiographic diagnostic technique in which multiple X-rays are taken from different angles in a single plane and a series of 2-D images–'slices' of the different tissue densities–eg fat, muscle, bone, etc, are constructed by computer; in contrast to conventional radiology, CT results in a 1000-fold ↑ in image resolution, and can pinpoint lesions < 2 mm in greatest dimension; dyes may be injected IV to ↑ resolution of abnormal tissue and blood vessels, or radiocontrast can be used for dynamic testing. See Hounsfield units, Multislice CT, PET/CT, Quantitative CT, Spiral CT, TACT, Ultrafast CT. Cf Magnetic resonance imaging, Ultrasonography

computer-assisted diagnosis Any use of computer algorithms to arrive at a clinical diagnosis. See Algorithm, Artificial intelligence, Cyberdoc, Expert system. Cf Computer-assisted therapeutics management

computer-assisted screening CYTOLOGY The use of optical scanning devices and computer algorithms to render a diagnosis on a pap smear; Neuromedical Systems–Papnet, and AutoPap–Neopath; both companies went belly-up

computer-assisted therapeutics management MEDICAL INFORMATICS Any computerized decision-support system for selecting appropriate medications, optimizing dosages, and duration of therapy, while minimizing adverse effects and hypersensitivity reactions and reducing costs. See Computer-assisted diagnosis

computer-based patient record Electronic medical record HEALTH INFORMATICS A 'personal health library' providing access to all resources on a Pt's health history and insurance information

computer vision syndrome OCCUPATIONAL MEDICINE A condition linked to prolonged computer monitor use; persons viewing computer screens tend to blink less and open their eyes more widely, resulting in dryness, fatigue, burning, difficulty in focusing, headaches MANAGEMENT Moisturizing drops. See CRT, Internet addiction disorder

computers, fear of PSYCHOLOGY Cyberphobia. See Phobia

COMT inhibitor Catechol-o-methyltransferase inhibitor NEUROLOGY A drug that enhances the effectiveness of levodopa by blocking COMT, an enzyme that metabolizes levodopa before it reaches the brain; ↑ levodopa bioavailability in the brain results in steady and continuous dopaminergic stimulation, and more consistent and effective symptom relief in Parkinson's disease

Comtan® Entacapone NEUROLOGY A COMT inhibitor which may improve motor performance in Parkinson Pts receiving levodopa/carbidopa. See Parkinson's disease

con- Latin, with, together

concentrate *verb* To ↑ a substance's strength by evaporation or removal of solvent NEUROLOGY *verb* To focus on a particular mental task

concentration EPIDEMIOLOGY The density of a population in a particular demographic region LAB MEDICINE The proportion of one molecule or substance to its diluent. See Blood alcohol concentration, Critical concentration, Critical dissolved oxygen concentration, Derived air concentration, Minimum bactericidal concentration, Minimum effective concentration, Minimum inhibitory concentration, Passive concentration, Peak serum concentration, Potentially toxic concentration, Serum concentration, Steady-state serum concentration, Total drug concentration, Trough serum concentration NEUROLOGY A general term for the degree of mental focus required to carry out a task TOXICOLOGY The ratio of a mass or volume of a solute to the mass or volume of a solvent. See Lethal concentration, Median lethal concentration, Minimum detectable concentration

concentration-effect curve PHARMACOLOGY A graph of the relationship between exposure concentration of a drug or other chemical and the graded effect that it produces. See Area under the curve concentration

concentric contraction SPORTS MEDICINE Muscle contraction that occurs while the muscle is shortening as it develops tension and contracts to move a resistance. Cf Eccentric contraction

conception REPRODUCTION BIOLOGY The onset of pregnancy, marked by implantation of a blastocyst in the endometrium, and formation of a viable zygote

conceptus The embryo/fetus and extraembryonic tissues from fertilization to birth

concierge fee HEALTH CARE A fee charged by some physicians–eg, cardiologists, for providing expanded personal access and services. See Boutique hospital

concomitant *adjective* Accompanying, accessory, joined with another

concordance LAB MEDICINE A measure of test performance corresponding to the proportion of substances tested that are correctly classified as positive or negative, which depends on the prevalence of positives in the population. See Accuracy, Two-by-two table

concrement Stone, see there

concrete thinking PSYCHIATRY Cognition that reflects experience, rather than abstraction, typical of those who are unable to generalize

concupiscence Horniness, see there

concussion NEUROLOGY A severe brain injury, due to a violent blow or shaking, which is characterized by immediate and transient impairment of brain function and one or more of the following: loss of consciousness, amnesia, seizure, or a change in mental status, which may be accompanied by defects in cognition, vision, equilibrium. See Transient ischemic attack. Cf Contusion SPORTS MEDICINE A traumatically induced alteration in mental status; impairment of brain function caused by injury to the head; speed and degree of recovery depend on severity of the brain injury. See Labyrinthine concussion

---

CONCUSSION–PER COLORADO MEDICAL SOCIETY

GRADE I MILD CONCUSSION 'Having the bell rung', 'head ding' Confusion without amnesia or loss of consciousness; difficult to detect MANAGEMENT Remove from game, examine immediately for amnesia, postconcussive symptoms; return to game in no less than 20 minutes; three grade I concussions should end player's season; no contact sports for ≥ 3 months

GRADE II MODERATE CONCUSSION Confusion with amnesia, no loss of consciousness; posttraumatic and retrograde amnesia may occur in severe cases; athlete should be evaluated for 24 hours MANAGEMENT Remove permanently from game; examine repeatedly for intracranial problems; return to practice ≥ 1 week; return to playing ≥ 1 month or end athlete's season; season ends if ≥ 3 grade II concussions or abnormality by CT or MRI

GRADE III SEVERE CONCUSSION Loss of consciousness MANAGEMENT Transport to hospital with cervical spine immobilization; full neurologic examination; return to practice ≥ 2 weeks; return to play ≥ 1 month or end athlete's season; season ended if ≥ 2 grade II concussions or any abnormality by CT or MRI

---

condition *noun* A state, mode, or state of being; the physical status of the body as a whole or of one of its parts, usually indicates abnormality. See Medical condition, Permissive condition, Preexisting condition Pregnancy-related conditions, Qualifying condition, Restrictive condition, Stress-related condition *verb* To subject a person or organism to a set of circumstances that ↑ functionality SPORTS MEDICINE Endurance training, see there

conditioning PSYCHIATRY The establishing new behavior through psychologic modifications of responses to stimuli. See Operant conditioning, Respondent conditioning, Work conditioning

condom REPRODUCTION MEDICINE A diaphanous sleeve, often produced from latex, which fits snugly over the penis and is used to prevent pregnancy and STDs. See Contraception, Natural family planning, Pearl index. Cf Female condom

condom therapy REPRODUCTION MEDICINE Therapy prescribed to ↓ sperm antibodies in a ♀ by using a condom during intercourse for ≥ 6 months and by the ♀ refraining from all skin contact with the partner's sperm

conduct disorder PSYCHIATRY A repetitive and persistent pattern of behavior in which the basic rights of others or major age-appropriate social norms or rules are violated–criterion A GROUPINGS Aggressive conduct that causes or threatens physical harm to other people or animals; nonaggressive conduct that causes property loss or damage; deceitfulness or theft; serious violations of rules PREVALENCE 3 to 9% ♀; 1% to 16% ♂. See Oppositional defiant disorder

conduction CARDIAC PACING The passage of an electrical charge; the active propagation of a depolarization wave in the heart PHYSIOLOGY The transmission of nerve impulses. See Retrograde conduction

conduction system *systema conducens cordis* ANATOMY A network of specialized nerve fibers innervated by both the sympathetic–SNS and parasympathetic nervous systems–PNS; PNS stimulation, eg, ↑ vagal stimulation, ↓ sinus node automaticity and ↓ AV nodal conduction; SNS stimulation ↑ automaticity, enhances conduction. See Atrioventricular node, Sinus node

conductive hearing loss Conductive hearing impairment ENT Hearing loss due to a defect in the outer ear–eg a block of the external ear canal or middle ear–ossicles or eardrum, in the face of a normal auditory nerve

condyloma acuminatum Genital warts, venereal warts, verruca acuminata STD A sexually-transmitted HPV-induced papilloma on the external urogenital ♂ or ♀ epithelium MANAGEMENT Cryotherapy, surgical excision, IFN-α-2a PROGNOSIS Usually benign; ↑ risk of cervical cancer

cone GYNECOLOGY Cone biopsy, see there NEUROPHYSIOLOGY 1. A color receptor cell in the retina of the eye 2. Growth cone, see there UROGYNECOLOGY See Vaginal cone

160 **cone biopsy** Conization The surgical excision of the cone-shaped uterine cervix, which encompasses the ectocervix and endocervical portion of the uterine cervix; it is performed as definitive therapy for CIN 1 to 3, and has been used for circumscribed carcinoma in situ, microinvasive SCC, and rarely, condylomas, confined to the cervix, while preserving the uterus COMPLICATIONS Bleeding, infection, cervical stenosis, cervical incompetence. See Cervix, CIN, HPV, HSIL, LSIL. Cf LEEP

confabulation NEUROLOGY The falsification of plausible experiences and events in response to questions about situations or events that are not recalled, as a means of 'covering' gaps in memory, a finding typical of Wernicke's encephalopathy. Cf False memory

conference effect MEDICAL PRACTICE A popular term for a transient shift in a health care provider's pattern of practice, and the conditions diagnosed, based on information acquired at a recently-attended conference or symposium

confidence interval STATISTICS A range of values for a variable of interest–eg, a rate, constructed so that the range has a specified probability of including the true value of the variable. See Confidence limits

confidence limits MEDICAL PRIVACY The endpoints–minimum or maximum value–of a confidence interval, the range over which it can be stated with a given probability or 'degree of confidence', that a parameter of interest–eg, a mean or standard deviation, is present. See Confidence interval, Standard deviation

confidentiality PSYCHIATRY The ethical principle that a physician may not reveal any information disclosed in the course of medical care. See Anne Sexton, Bennett-Leahy bill, Doctor-patient relationship, Hippocratic Oath, Malpractice, Privilege, Privileged communication.

confined spaces, fear of PSYCHOLOGY Claustrophobia. See Phobia

confinement OBSTETRICS See EDC/EDL–expected date of confinement or expected date of labor

confirmation Verification of the results obtained from a screening method, using a gold-standard technique, or so-called confirmation method

conflict NEUROLOGY See Visual-vestibular conflict PSYCHIATRY A mental struggle that arises from the simultaneous operation of opposing impulses, drives, external–environmental or internal demands TYPES Intrapsychic–between forces within the personality; extrapsychic–between the self and the environment. See Approach-avoidance conflict VOX POPULI 1. Collision, clash. See Feto-maternal conflict 2. War, battle. See Man-made disaster

confrontation PSYCHIATRY Bellicose communication that deliberately invites a Pt or client to self-examine some aspect of his/her behavior in which there is a discrepancy between words and deeds

confusion NEUROLOGY Disorientation with respect to time, space–place, or person, which may be accompanied by disordered consciousness. See Nocturnal confusion

congenital *adjective* Referring to a condition present at birth, regardless of causation

congenital adrenal hyperplasia Adrenogenital syndrome, congenital lipoid adrenal hyperplasia, congenital virilizing adrenal hyperplasia ENDOCRINOLOGY A group of AR conditions characterized by a partial or complete defect of enzymes–most commonly steroid 21-hydroxylase, that synthesize cortisol in the adrenal cortex; the lack of cortisol results in ↑ ACTH–due to lost feedback inhibition, leading to adrenal cortical hypertrophy and hyperplasia, ↑ production of mineralocorticosteroids—aldosterone and androgens CLINICAL Virilization, salt loss, HTN, defects in spermatogenesis, ovulation; in ♀, dysmenorrhea, deep voice, hirsutism, ambiguous genitalia due to excess production of androgens; in ♂, enlarged penis, small testes, early development

of 2° ♂ characteristics; untreated, mortality rate is ±100% for both sexes MANAGEMENT Glucocorticoids and mineralocorticoids prevent untimely and, in girls, incongruous postnatal virilization; plastic surgery is needed to feminize the genitalia; with appropriate therapy, survival and good physical and mental health is the norm. See StAR

congenital agranulocytosis Congenital neutropenia, see there

congenital cataract NEONATOLOGY Clouding of cornea at birth ETIOLOGY Cerebrohepatorenal syndrome, congenital rubella, Conradi-Huhnermann syndrome, Down syndrome, ectodermal dysplasia, galactosemia, Hallerman-Streiff syndrome, Lowe syndrome, Marinesco-Sjögren syndrome, Pierre-Robin syndrome, Sieman syndrome, trisomy 13 MANAGEMENT Cataract removal and insertion of artificial lens. See Cataract

congenital dyserythropoietic anemia PEDIATRIC HEMATOLOGY A group of inherited defects of erythropoiesis CLINICAL Lifelong mild-to-moderate anemia and ineffective erythropoiesis

congenital epulis ORAL PATHOLOGY A benign mesenchymal mass of the gingiva of infant ♀, identical to granular cell tumor MANAGEMENT Excision

congenital gout *gutta*, Latin, droplet A condition linked to 2 X-linked enzymes–hypoxanthine-guanine phosphoribosyl transferase, defective in Lesch-Nyhan disease, and 5-phosphoribosyl-1-pyrophosphate LAB ↑ serum uric acid, often ≥ 410 μmol/L–US: 7.0 mg/ml, a level found in a significant minority of ♂–90% of gout occurs in ♂; family Hx of gout is present in ± ½ of cases; ½ present with the classic 'podagra'–1st metatarsophalangeal joint; up to 90% suffer podagra at some time during their disease MANAGEMENT–ACUTE Colchicine, NSAIDs, corticosteroids INTERVAL Diet CHRONIC Allopurinol, sulfinpyrazone, salicylates

congenital heart disease A congenital malformation–eg, coarctation of aorta, VSD, ASD, tetralogy of Fallot–of the heart or great blood vessels, which may or may not have clinical consequences. See Baby Faye heart, Shunt

CONGENITAL HEART DISEASE

RIGHT→LEFT SHUNT Cyanotic shunt Tetralogy of Fallot, transposition of the great vessels, trucus arteriosus, tricuspid valve atresia

LEFT→RIGHT SHUNT Acyanotic shunt Patent ductus arteriosus, atrial septal defect, ventricular septal defect, aortic stenosis, pulmonary stenosis, aortic coarctation (NEJM 2000; 342:256RV)

congenital herpes Intrauterine herpes INFECTIOUS DISEASE A perinatal or prenatal HSV infection in which HSV is capable of causing chorioretinitis, microophthalmia, vesicular rashes, lethargy, fever, seizures, SOB, hemorrhagic diathesis, encephalitis, cerebral edema, DIC. See Herpes simplex, TORCH

congenital hip dislocation Congenital hip dysplasia PEDIATRIC ORTHOPEDICS A hip joint malformation present at birth, thought to have a genetic component CLINICAL Hip dislocation, asymmetry of legs and fat folds, and ↓ movement on the affected side; CHD may be asymptomatic and must be diagnosed by physical examination

congenital hypernatremia Idiopathic neonatal hypernatremia of Ballard A potentially life-threatening condition of perinatal onset with ↑↑↑ serum sodium, often > than 134-146 mmol/L–US: 310-340 mg/dl CLINICAL CNS dysfunction, seizures, neuromuscular spasms, production of low volumes of concentrated urine, weight loss, hypotension, ↑ BUN; Sx are exacerbated by concomitant sepsis, asphyxia or CNS hemorrhage ETIOLOGY ↑ Na⁺ loss, as in diabetes insipidus, either ↓ ADH secretion or ↓ sensitivity to ADH, osmotic diuresis, diaphoresis, diarrhea; ↓ water intake, due to disordered thirst sensation, ↑ salt ingestion without adequate fluid, electrolytic imbalance, ↑ Ca²⁺, ↑ K⁺

congenital hypothyroidism Congenital myxedema, cretinism Hypothyroidism of neonatal onset, which occurs in 1/±7000 births, more commonly in ♀, characterized by mental and physical retardation due to inappropriate thyroid development or inadequate maternal intake of iodine during gestation. See Myxedema

congenital leukemia PEDIATRIC HEMATOLOGY A leukemia that develops in the first month of life, usually acute non-lymphocytic leukemia; CL is often associated with congenital malformations and chromsome defects, especially trisomy 21/Down syndrome, but also trisomy 13, mosaic trisomy 9, monosomy 11, Turner syndrome, etc. See Leukemia

congenital malformation Congenital defect A heterogenous group of structural defects, which are usually identified at birth MAJOR CMs, US PDA, hypospadias, clubfoot, ventricular septal defect, hydrocephalus, Down syndrome, hip dislocation, valve stenosis and/or atresia, pulmonary artery stenosis, microcephalus, cleft lip ± cleft palate, spina bifida, rectal atresia and stenosis, polydactyly, supernumerary nipples, branchial clefts, abdominal wall defects

congenital megacolon 1. Hirschsprung's disease 2. Megacolon, see there

congenital platelet function defect HEMATOLOGY Any inherited condition–eg, Bernard-Soulier disease, Glanzmann's thrombasthenia, gray platelet syndrome, storage pool disease, etc, characterized by defective platelet function. See Bernard-Soulier disease, Glanzmann's thrombasthenia, Gray platelet syndrome

congenital rubella syndrome A malformation complex in a fetus infected in utero with rubella; the defects reflect the embryologic stage at the time of infection, with developmental arrest affecting all 3 embryonal layers, inhibiting mitosis, causing delayed and defective organogenesis; maternal infection in the 1st 8 wks of pregnancy causes embryopathy in 50-70% of fetuses; the susceptible period extends to ±20th wk; infection in late pregnancy carries little fetal morbidity CLINICAL Cardiac defects–eg PDA, pulmonary valve stenosis, VSD, hepatosplenomegaly, interstitial pneumonia, LBW, congenital cataracts, deafness, microcephaly, petechia, purpura, CNS Sx–eg, mental retardation, lethargy, irritability, dystonia, bulging fontanelles, ataxia LAB Viral isolation, IgM antibodies in fetus by hemagglutination inhibition VACCINATION Attenuated live virus vaccine between 15 months and puberty; effective antibodies develop after immunization in 95% of Pts. See Extended rubella syndrome

congenital syphilis Congenital lues, fetal syphilis NEONATOLOGY Transplacental infection with *Treponema pallidum* CLINICAL Early–hepatomegaly, irritability, FTT, fever, perioral and genital rash–condyloma lata, nasal discharge or snuffles and saddle nose; late–Hutchinson's teeth, saber shins, neurologic impairment, deafness, blindness LAB ↑ liver enzymes, anemia, monocytosis DIAGNOSIS Serologic tests performed at birth may be negative. See Syphilis

congenital toxoplasmosis A transplacental infection with the protozoan *Toxoplasma gondii* affecting ± ⅓ of fetuses of ♀ with acute acquired toxoplasmosis, most severe if the infection occurs in 1st trimester; children are often normal at birth, followed by blindness, mental retardation; it may affect the brain, lung, heart, eyes, liver CLINICAL Hydrocephalus, microcephaly, cerebral calcifications, atrophy, chorioretinitis, uveitis, vitritis, convulsions, hyperbilirubinemia, hepatomegaly DIAGNOSIS PCR of amniotic fluid to detect 35-fold repeat of the *B1* gene of *T gondii* Sensitivity 97%–vs 90% with conventional parasitology LAB IgM immunoassay MANAGEMENT Pyrimethamine, sulfadiazine, leukovorin. See TORCH

congested ENT *adjective* Referring to a boggy blood-filled tissue. See Nasal congestion

congestive *adjective* Referring to an abnormal accumulation of blood in tissue

congestive cardiomyopathy Dilated cardiomyopathy, see there

congestive heart failure Congestive heart disease CARDIOLOGY '...a complex clinical syndrome characterized by abnormalities of left ventricular function and neurohormonal regulation, which are accompanied by effort intolerance, fluid retention, and reduced longevity'; an impairment of cardiac function in which failing ventricles cannot adequately perfuse tissue to meet metabolic demands; CHF usually develops over a long period, but may be abrupt in onset EPIDEMIOLOGY CHF is a major health problem which affects 2-3 million, US; 400,000 new cases are diagnosed/yr CLINICAL Low-output 'forward CHF'–weakness, fatigue, lethargy, light-headedness, and confusion; in decompensated CHF, cardiac cachexia ensues, characterized by exhaustion and loss of lean muscle mass; low-output backward CHF–pulmonary congestion–fluids accumulate in lungs, causing dyspnea, initially only on exertion; also seen, peripheral and pedal edema, rales, S3 gallop, sinus tachycardia, hypotension, ↑ jugular venous pressure, and abdominojugular–hepatojugular reflux HIGH-OUTPUT HEART FAILURE–'NON-CARDIAC' CHF Albright's disease–polyostotic fibrous dysplasia, anemia, carcinoid syndrome, arteriovenous fistulas–trauma, Paget's disease of bone, hemangiomatosis, glomerulonephritis, hemodialysis, liver disease–alcohol-related thiamin deficiency, ↓ peripheral arterial resistance, hyperkinetic heart syndrome, polycythemia vera, thyrotoxicosis CHF PRECIPITANTS Alcohol, cor pulmonale, drug-related–inappropriate medications, non-compliance, ↑ fluid and/or sodium intake, fever, hypothyroidism, hypoxia, infection, obesity, pregnancy, pulmonary embolism, renal failure, uncontrolled HTN WORKUP-EKG–to exclude myocardial ischemia or infarction, and/or arrhythmia WORKUP-LAB CBC–to exclude anemia, BUN and creatinine–to assess renal function, electrolytes–K+ and magnesium, liver enzymes, cardiac markers including enzymes–eg, LDH isoenzymes and proteins–e.g., troponin T–to exclude recent MI or ischemia, thyroid function tests–to exclude thyrotoxicosis, a major cause of high-output heart failure RADIOLOGY Cardiac enlargement, interstitial and/or alveolar edema, and pulmonary vascular redistribution in acute CHF, findings which are *less* common in chronic CHF ECHOCARDIOGRAPHY–M-mode, 2-D, & Doppler to determine the left ventricular ejection fraction; CHF typically has concentric left ventricular hypertrophy MANAGEMENT OF PRECIPITATING FACTORS Eliminate noncardiac factors–eg, alcohol, drugs, excess fluid and/or sodium intake, fever, hypothyroidism, hypoxia, infection, renal insufficiency, HTN, and other factors; control precipitating factors, which may eliminate the signs and symptoms of CHF ACUTE MANAGEMENT Acute CHF with extreme respiratory distress is a medical emergency that requires immediate treatment to ↓ volume–preload and myocardial O₂ demand and ↑ forward blood flow LONG-TERM MANAGEMENT Once CHF develops, it requires continuous therapy to ↓ M&M–ie, 1. non-pharmacologic maneuvers–eg, salt restriction in the form of a 'no salt added' regimen, reduction of alcohol intake, exercise as tolerated and, for Pts with impaired renal function or psychogenic polydipsia, fluid restriction 2. drug therapy with multiple agents–diuretics, ACE inhibitors, digoxin, nitrates, CCBs, beta-blockers, inotropic agents that ↑ intracellular sodium–eg, vesnarinone–have a narrow therapeutic range PROGNOSIS Poor; 5-yr mortality rate for CHF after the onset of Sx-per Framingham Heart data, is 62% for ♂; 42% for ♀; 200,000 deaths attributable to CHF occur/yr–US; ↑ norepinephrine levels direct correlates with hemodynamic severity and poor prognosis. See Beta-blockers, Calcium channel blockers, Nitrates

---

CONGESTIVE HEART FAILURE–ETIOLOGY

DILATED CARDIOMYOPATHY Congestive cardiomyopathy

Infectious–eg viral, parasites

Toxic–eg alcohol abuse, adriamycin, cyclophosphamide

Nutritional–eg carnitine, selenium, thiamin, or protein deficiency

Connective tissue disease

HYPERTROPHIC CARDIOMYOPATHY Concentric left ventricular hypertrophy

INFILTRATIVE CARDIOMYOPATHY Amyloid, hemochromatosis, sarcoidosis

HYPERTENSIVE HEART DISEASE

MYOCARDIAL ISCHEMIA/INFARCTION Accompanied by left ventricular dysfunction

VALVULAR HEART DISEASE

---

conicotine A nicotine metabolite detectable in serum up to a wk after the last cigarette. See Nicotine, Smoking

conization Cone biopsy, see there

conjugate movement OPHTHALMOLOGY The simultaneous movement of both eyes in the same direction

conjugated bilirubin Direct bilirubin Bilirubin chemically bound to a glucuronide in the liver, which is excreted in bile by the liver and stored in the gallbladder or transferred to the duodenum REF RANGE Direct BR: 0–0.3 mg/dL ↑ IN Bile duct obstruction, cirrhosis, Crigler-Najjar syndrome, Dubin-Johnson syndrome, hepatitis. See Bilirubin

conjugation The act of being joined together or conjugated

conjunctivitis Pink eye OPHTHALMOLOGY Conjunctival inflammation, the single most common eye disease ETIOLOGY Infection—bacterial, viral, fungal, parasitic; immunologic–hypersensitivity or autoimmune; chemical/irritative–occupational, iatrogenic; 2° to systemic diseases–Reiter syndrome, dermatitis herpetiformis, epidermolysis bullosa, Kawasaki disease, gout, thyroid disease, carcinoid; mechanical issues–eg, canaliculitis, dacryocystitis CLINICAL Conjunctival hyperemia, fluid discharge, tearing, exudate, pseudoptosis, papillary hypertrophy, chemosis MANAGEMENT Antibiotics, management of systemic disease, vasoconstrictors, cold compresses; corticosteroids *may* be indicated, but are linked to glaucoma, cataracts COMPLICATIONS Corneal ulceration. See Allergic conjunctivitis, Keratoconjunctivitis, Neonatal conjunctivitis, Shipyard conjunctivitis

connection VOX POPULI A link of one form or another. See Africa connection, AIDS-malaria connection, Cross-connection, Monkey connection, Mosquito connection

connective tissue disease Autoimmune disease, collagen-vascular disease Any of the diseases affecting connective tissues, with an autoimmune component, and immunologic/inflammatory defects CLINICAL Arthritis, connective tissue defects, endocarditis, myositis, nephritis, pericarditis, pleuritis, synovitis, vasculitis LAB ANAs, direct antiglobulin–Coombs' test, hemolytic anemia, leukopenia, thrombocytopenia, ↑ or ↓ Igs, rheumatoid factors, BFP for syphilis TYPES Ankylosing spondylitis, dermatomyositis, IBD-related arthritis, polychondritis, polymyalgia rheumatica, polymyositis, psoriatic arthritis, rheumatoid arthritis, Sjögren syndrome, SLE, systemic sclerosis, vasculitis

consanguineous *adjective* Referring to a blood relationship–ie, descendent from a common ancestor

conscious *adjective* NEUROLOGY Awake, alert. See Conservatorship, Unconscious *noun* PSYCHIATRY The content of mind or mental functioning of which one is aware

conscious sedation Moderate sedation ANESTHESIOLOGY Minimally depressed consciousness in which a Pt retains the ability to independently and continuously maintain an open airway and a regular breathing pattern, and to respond appropriately and rationally to physical stimulation and verbal commands; CS may be induced by parenteral or oral medications or combination thereof. Cf General anesthesia

consensual light reflex OPHTHALMOLOGY The ability of both pupils to react to light which is tested by shining a beam of light in one eye, and assessing the response of the other eye, which normally constricts

consent MEDTALK A voluntary yielding of a person's free will to another. See Informed consent, Presumed consent

conservative therapy PATIENT CARE Management of a clinical condition with the least aggressive of available therapeutic options; CT is often equated with 'medical' vs 'surgical' treatment–eg, drug, diet, and lifestyle management of severe, but asymptomatic ASHD, in contrast to management of the same condition with CABG. See Benign neglect, Palliative therapy. Cf 'Heroic' therapy, Radical surgery

consistent with C/W CLINICAL DECISION MAKING A phrase used by practitioners of the 'visual arts' of medicine, ie pathology and radiology, in which a diagnosis is based on a subjective interpretation of a particular pattern in a tissue, organ, or body region. See Aunt Millie approach, Defensive medicine

consolidation IMAGING An ↑ in radiologic density of an air-filled space, due to accumulation of fluid and WBCs, as occurs in the lungs in acute pneumonia. See Hepatization

consolidation chemotherapy Consolidation treatment ONCOLOGY A phase of chemotherapy–CT intended to liquidate residual malignant cells in Pts with leukemia or other lymphoproliferative malignancies in remission; CC follows 'intensification' phase of CT, which in turn is followed by a maintenance phase. Cf Adjuvant chemotherapy, Induction chemotherapy, Intensification chemotherapy, Maintenance chemotherapy, Salvage chemotherapy

constipation GI DISEASE Infrequent and/or incomplete evacuation of stool ETIOLOGY IBS, diverticulosis, medications–eg overuse of laxatives, colorectal CA DIAGNOSIS Barium enema, endoscopy and Bx MANAGEMENT Bulk laxatives, high-fiber diets. Cf Diarrhea

constitution PSYCHIATRY A person's intrinsic physical and psychologic endowment; sometimes used more narrowly to indicate physical inheritance or intellectual potential

constitutional evaluation MEDICAL PRACTICE A part of a physical exam, required for CPT-related documentation of a physician's evaluation and management services; CE includes recording of BP–sitting, standing, supine, pulse rate or regularity, respiration, temperature, weight, height, as well as evaluation of physical appearance vis-à-vis development, nutrition, body habitus, deformities, attention to grooming. See Evaluation and management services

constrictive pericarditis CARDIOLOGY A condition characterized by a chronic inflammation, fibrosis and scarring of a pericardium contracted to the point of compromising normal diastolic filling of the ventricles ETIOLOGY Infection, connective tissue disease, malignancy, trauma, metabolic disorders–eg uremia, RT, sarcoidosis, asbestosis, previous MI, TB, viral infection, cardiac surgery CLINICAL Right-sided CHF, ↓ cardiac output COMPLICATIONS Cardiac tamponade, pulmonary edema MANAGEMENT Surgical incision. See Pericardium

construction apraxia NEUROLOGY An acquired inability to draw 2-D objects or forms or copy 3-D arrangements of forms or shapes, in absence of motor defects, due to a defect in the non-dominant parietal lobe

consult *noun* 1. A formal consultation from another health care professional on the diagnosis, therapy, or prognosis of a disease. See Consultation 2. A physician's office *verb* To engage in a formal tête-à-tête with another health care professional

consultation Consult A service type provided by a physician whose opinion or advice regarding evaluation and/or management of a specific problem requested by another physician or other appropriate source; a formal or, less commonly, informal, 2nd opinion obtained from a consultant, which may be sought by an attending physician regarding a diagnosis or therapeutic plan being considered in the context of Pt management. See Second opinion MEDICAL PRACTICE The performance by another physician or health professional of a specific diagnostic or therapeutic task, without transfer of responsibility for the Pt's care or even for ongoing management of the Pt's problem. See Consultant, Episodic consultation, Health consultation, Pathology consultation, Teleconsultation, Video consultation, Virtual consultation. Cf Referral

consumable cost ADMINISTRATION Those necessary expenses borne by the lab or other hospital service which includes reagents, disposables, and other supplies, as well as maintenance and lease contracts. See Disposables

consumption coagulopathy Disseminated intravascular coagulopathy HEMATOLOGY A condition characterized by ↑ platelet consumption with coagulation factor depletion–prolonged PT, PTT, ↑ fibrinolysis–generation of fibrin split products ETIOLOGY Sepsis, extensive burns, trauma, retained dead fetus, heat stroke, mismatched blood transfusion, metastatic CA, some forms of leukemia CLINICAL Severe bleeding, bruising

contact dermatitis DERMATOLOGY A dermatitis evoked by a substance in regular direct contact with the skin TYPES Irritant CD–accounts for 80% of occupational dermatitides and damages skin by nonimmune mechanisms–eg detergents, solvents, adhesives, ointments, etc; allergic CD is due to an immune skin response, and is more common in a nonoccupational setting and caused by allergy to various substances–eg, jewelry, rubber, latex, poison ivy, neomycin, etc. See Allergic contract dermatitis, Irritant contact dermatitis, Latex allergy

contact lens OPHTHALMOLOGY Crystalline ocular lens A transparent soft or rigid device placed directly on the cornea to correct refractory errors

contact rate EPIDEMIOLOGY The rate at which susceptibles meet infecteds, measured as individuals per unit time

contact sensitivity IMMUNOLOGY An allergic response to contact with irritant, usually hypersensitivity. See Contact dermatitis

contact sport SPORTS MEDICINE A sport in which body contact either is an integral component of the sport–eg, boxing, football, martial arts, rugby, wrestling, or commonly occurs while engaged in the sport–eg, basketball, hockey, lacrosse. See Blood sports

contagious EPIDEMIOLOGY *adjective* Infectious; capable of being transmitted from one person to another by contact or close proximity

containment PUBLIC HEALTH The confining or prevention of further dissemination of a potentially hazardous–eg biologic, radioactive or toxic–agent. See Biological containment, Biosafety levels, Regulated waste

contaminant An impurity, any substance or material that enters a system–the environment, human body, food, etc. where it is not normally found CLINICAL CHEMISTRY A substance in a Pt specimen that may skew diagnostic results MICROBIOLOGY A microorganism not found in a Pt, which may grow in culture and provide erroneous results PHARMACOLOGY An undesired adulterant SURGICAL PATHOLOGY See Floater–surgical pathology

contaminated sharp Any object that is capable of penetrating mucocutaneous surfaces including, but not limited to, needles, scalpels, broken glass, broken capillary tubes, and exposed ends of dental wires, which is contaminated by blood and/or pathogens

contamination Pollution by an inferior material INFECTIOUS DISEASE Introduction of organisms in a wound. See Cross contamination PUBLIC HEALTH The presence of any foreign or undesired material in a system–eg, toxic contamination of the ground water in an ecosystem or untreated sewage into a stream RADIATION PHYSICS The deposition of radioactive material in any place where it is not wanted. See Radioactive contamination

contamination with germs, fear of PSYCHOLOGY Misophobia, mysophobia. See Phobia

continence Self control vis-à-vis defecation, sexual activity, urination

continent reservoir SURGERY A pouch formed from small intestine to hold urine after the bladder has been remove. See Ileostomy, Urostomy

contingency fee LAW & MEDICINE An attorney fee based on a percentage of the money recovered in a lawsuit

continuing medical education See CME

continuity of care MEDTALK Uninterrupted health care for a condition from the time of first contact–eg, to the point of resolution or long-term maintenance

continuous ambulatory peritoneal dialysis See Peritoneal dialysis

continuous filtration NEPHROLOGY A device that automatically collects platelets while removing leukocytes from donated blood. See Hemofiltration

continuous positive airway pressure CRITICAL CARE A type of artificial ventilation, in which the lung pressure is maintained above atmospheric pressure during the entire respiratory cycle. See PEEP

continuous quality improvement MANAGED CARE An approach to health care based on evaluation of a product or the outcome(s) of a process, and on understanding the needs and expectations of the consumers of these products or processes. See Quality assurance, Total quality management

continuous venovenous hemofiltration NEPHROLOGY A renal replacement therapy consisting of continuous removal of ultrafiltrate at > 5 ml/min, with fluid replacement, which ultrafilters plasma, removing small and medium-sized molecules and replacing electrolytes; CVVH is used for renal failure; it is inadequate for Pts with frank renal failure. See Hemodialysis

continuous-loop event recorder CARDIOLOGY An ambulatory EKG monitoring system, which continuously records data, but saves data only when it is manually activated by the Pt PROS More cost-effective and efficient than a Holter monitor. Cf Holter monitor

contraception The prevention of conception or impregnation, including natural family planning, oral contraceptives, 'morning-after' pill, spermicidal foam, RU486 or devices–eg condoms, diaphragms, IUDs. See Back-up contraception, Breast feeding, Coitus interruptus, Contraceptive, Natural family planning, Pearl index, Rhythm method

contraceptive OBSTETRICS *adjective* Relating to contraception *noun* Any device or method for preventing fertilization, or a term product of conception TYPES Barrier methods–condoms, diaphragms, hormone combinations, spermicides, implantable hormonal devices, RU-486, etc. See Contraception, Dalkon shield, IUD, 'Litogen', Lunelle, Mirena, Nuvaring, Oral contraceptive, Ortho Evra, Pearl index, RU-486, Seasonale, Sequential oral contraceptive, Wrongful birth

contract MANAGED CARE A health care policy or plan in which a provider offers certain services delineated in writing, to which the purchaser–Pt agrees by signature. See Guaranteed renewable contract, Provider risk contract, Subscriber contract

contraction CARDIOLOGY A heart beat. See Premature ventricular contraction GI DISEASE The shortening of the muscularis propria of the GI tract, resulting in peristalsis. See Giant peristaltic contraction OBSTETRICS The shortening of myometrial cells, resulting in ↑ intrauterine tension. See Braxton Hicks contraction PHYSIOLOGY A ↓ in muscle length, accompanied by ↑ tension. See Concentration contraction, Isometric contraction, Isotonic contraction

contraction alkalosis FLUID BALANCE An ↓ in extracellular fluids with ↑ extracellular bicarbonate and ↑ bicarbonate resorption in the renal tubules, a finding typical of furosemide therapy MANAGEMENT Discontinue therapy, substitute acetazolamide–carbonic anhydrase inhibitor until metabolic derangement is corrected. Cf Metabolic alkalosis, Respiratory alkalosis

contractions OBSTETRICS Volleys of tightening and shortening of myometrium–uterine muscle, which occur during labor, cause dilatation and thinning of the cervix and aid in the descent of the infant in the birth canal. See Labor. Cf Decelerations

contracture ORTHOPEDICS A fixed resistance to passive movement of a musculoskeletal unit or joint, usually due to local fibrosis, often caused by prior ischemic insult. See Capsular contracture, Fibromyalgia, Volkman's ischemic contracture. Cf Contraction

contragestion REPRODUCTIVE MEDICINE Any contraceptive method that specifically prevents the gestation of a fertilized egg–eg, the 'morning after' pill or RU-486, either by making the implantation site uninhabitable, or by promoting the fertilized product's expulsion. See Contraceptive, RU-486

contraindication CLINICAL DECISION-MAKING Any medical reason for not performing a particular therapy; any condition, clinical symptom or circumstance indicating that the use of an otherwise advisable intervention in some particular line of treatment is improper, undesirable, or inappropriate. See Absolute contraindication, Patient selection, Relative contraindication. Cf Indication

contralateral *adjective* Referring to an opposite side. Cf Ipsilateral

contrast bath SPORTS MEDICINE A bath with fluctuating temperatures to induce alternating vasodilation and vasoconstriction, or reduce subacute or gravity-dependent edema. See Cold bath, Hot bath

contrast echocardiography CARDIOLOGY Enhancement of echocardiography with contrast media–eg, with sonicated albumin, which is used to detect coronary disease by IV injection of fluorocarbons. See Harmonic ultrasound imaging, Sonicated albumin, Transient response imaging

contrast medium Contrast, contrast agent IMAGING A substance with a density–eg, a dye or signal differing from that of the organ or structure being imaged, which allows delineation of contour abnormalities; CMs that are more radiopaque–eg with barium or iodine—than the organ or structure being analyzed are known as positive CM, while those that are less radiopaque–eg, with air, are known as negative CM. See Radiopaque contrast

contrast imaging CARDIOLOGY The use of a microaggregate or other ultrasonography contrast agent to ↑ sensitivity of real-time myocardial perfusion imaging. See Ultrasonography

contrast study IMAGING An imaging procedure in which a contrast medium is introduced to enhance the image of a particular body region or structure. See Contrast medium

contrecoup injury TRAUMATOLOGY A brain 'bruise' diametrically opposite the site of an impacting blow to the cranium, where the head is in motion and the brain lags behind by a split-second; a blow to the back of the head results in lesions of the frontal lobes and horns of the temporal lobes; a blow to the top of the skull results in contrecoup lesions to the hippocampus and the corpus callosum. See Boxing. Cf Coup

contributory negligence Contributory neglect MALPRACTICE A Pt's or plaintiff's conduct while under a physician's care that falls below that which a reasonable person would exercise for his or her own protection, thereby contributing to the physician's alleged negligence. See Medical malpractice, Negligence, Reasonable person standard

control CLINICAL RESEARCH Control subject A nontreated or 'negative' individual in a study who serves as a reference. See Concurrent nonrandomized control, Control population, Control subject, Historical control EPIDEMIOLOGY In a case-control study, a comparison group of persons without disease LAB MEDICINE A specimen with known or standardized values for an analyte, that is processed in tandem with an unknown specimen; the 'control' specimen is either known to have the substance being analyzed, ie 'positive' control or known to lack the substance of interest, ie 'negative' control. See Negative control, Positive control, Quality control PSYCHOLOGY The degree to which a person can limit or modify verbal or physical responses to external stimuli. See Administrative control, Impulse control

controllable life-style specialty 'Regulated' specialty MEDICAL PRACTICE A specialty that can be scheduled in terms of work hours. Cf Noncontrollable life-style specialty

controlled drug substance Any drug or therapeutic agent–commonly understood to include narcotics, with a potential for abuse or addiction, which is held under strict governmental control, as delineated by the Comprehensive Drug Abuse Prevention & Control Act passed in 1970

CONTROLLED DRUG SUBSTANCES

SCHEDULE I DRUGS High abuse potential, no accepted medical use in US–Acetorphine, acetyl methadol, allyprodine, α–acetylmethadol, bufotenine, dextromoramide, diethyltryptamine, dimethyltryptamine–DMT, etorphine, heroin, ibogaine, ketobemidone, LSD–N,N-diethyl-D-lysergamide or lysergic acid diethylamide, marijuana, mescaline, PCP–phencyclidine, peyote, phenadoxone, phenampromide, racomoramide, tetrahydrocannibol

SCHEDULE II High abuse potential, potentially leading to severe psychologic or physical dependence; schedule II agents have acceptable medical uses, eg narcotics–alphaprodine, anileridine, cocaine, codeine, diphenoxylate, diprenorphine, etorphine HCl, ethymorphine, hydrocordone, hydromorphone, levorphanol, meperidine, methadone, morphine, oxymorphone, poppy straw concentrates, powdered opium, raw opium, thebaine and non-narcotics–amphetamine, amobarbital, methaqualone, methamphetamine, methaqualone, pentobarbital, percodan, phencyclidine, phenmetrazine, secobarbital

SCHEDULE III High abuse potential, moderate to low physical dependence, and high psychologic dependence potential, with acceptable medical uses, which may be narcotic–eg nalorphine, paregoric, or nonnarcotic–eg aprobarbital, benzphentamine, butabarbital, chlorphentermine, chlortermine, glutethimide, mazindol, methyprylon, phendimetrazine, probarbital, talbutal, thiamylal, thiopental, vinbarbital

SCHEDULE IV Minimal abuse potential, limited physical or psychological dependence potential, nonnarcotic, eg barbital, chloral hydrate, chlordiazepoxide, clonazepam, chlorazepate, dextropropoxyphene, diazepam, diethylpropion, ethchlorvynol, ethinamate, fenfluramine, lorazepam, mebutamate, methobarbital, meprobamate, methohexital, oxazepam, paraldehyde, phenobarbital, phentermine, prazepam

SCHEDULE V Very low abuse/dependence potential–eg brown mixture–opium, some codeine preparations, diphenoxylate preparations–Lomotil, ethylmorphine-Cidicol, opium–Donnagel-PG, terpin hydrate, or non-narcotic, eg loperamide

controlled drug substance Schedule VI agent PHARMACOLOGY Any compound, mixture, or preparation containing any stimulant or depressant exempted from Schedules III, IV, V, or any drug not in Schedules I–V, which, because of potential toxicity must be prescribed by a physician.

controlled trial CLINICAL RESEARCH A clinical study in which one group of participants receives an experimental drug while the other receives either a placebo or an approved–'gold standard' therapy. See Blinding, Double-blinded

contusion DERMATOLOGY A bruise, an injury without a break in the skin, in which subcutaneous blood vessels rupture, resulting in ecchymotic patches, often due to a blow from a blunt object. See Brain contusion, Cerebral contusion, Cortical contusion, Hip-pointer contusion

conundrum A problem with no satisfactory solution; a dilemma

conus medullaris ANATOMY The inferior, tapering portion of the spinal cord. See Spinal cord

convalescent serum Serum from a person who has recuperated from a particular infection–eg, scarlet fever, which may be of use in treating a person with the same infection; while acute-phase serum has ↑ IgM antibodies, CS has ↓ IgM, and ↑ IgG antibodies

convergence INFORMATICS The melding of disparate technologies into a single user modality, such as multipurpose fax/printers, internet telephony, Web phones, cable modems and voice-over-IP OPTICS The degree of alignment of images or pictures of the 3 primary colors—red, green and blue—in video cameras and computer monitors

convergent therapy Convergent combination therapy THERAPEUTICS The use of 2 or more drugs that target the same protein, in particular of a highly mutable virus–eg, HIV

conversion PSYCHIATRY An unconscious defense mechanism by which anxiety caused by intrapsychic conflict is converted and expressed in a somatically symbolic fashion CLINICAL Paralysis, pain, sensory loss

conversion disorder Histrionic personality disorder, hysteria, hysterical neurosis PSYCHIATRY A group of psychiatric reactions in which the Pt 'converts' mental problems into a physical manifestation, with an inappropriate lack of concern about their disabilities EXAMPLES Sensation of a thing stuck in the throat–'globus hystericus', recurrent abdominal pain without physical findings, hysterical blindness, gait defects, paralysis, sensory loss, seizures, urine retention. See 'la Belle indifference', Factitious disease, Hysterical neurosis, Post-traumatic stress disorder

conversion reaction Conversion hysteria, conversion symptom PSYCHIATRY A constellation of Sx that suggest organic brain disease, which affect voluntary muscles and sensory organs CLINICAL Paralysis, blindness, dysphagia, deafness, aphonia, SOB, spells, anesthesia, incoordination, amnesia, unconsciousness. Cf Hysterical reaction

conversion sickness AIDS A constellation of clinical findings classically associated with the seroconversion to production of anti-HIV antibodies PSYCHOLOGY Conversion reaction

converting enzyme Angiotensin converting enzyme, see there

convolution An elevation on the surface of a structure and an infolding of the tissue upon itself

convulsion NEUROLOGY Violent involuntary contraction(s) of skeletal muscle. See Salaam convulsion

cooing murmur Musical murmur CARDIOLOGY A pure frequency murmur/heart sound with a tonal quality most commonly heard in elderly Pts with calcified aortic stenosis and heard primarily over the apex

cooing stage NEUROLOGY The earliest stage of linguistic development which begins several wks after birth and consists of phonemes of variable duration. Cf Babbling stage, Lalling stage

cookie cutter border DERMATOLOGY A descriptor for the gross morphology of the scalloped ulcerated margins of SCC. See Squamous cell carcinoma

cool down SPORTS MEDICINE A period after vigorous exercise, during which the heart rate returns to normal; the CD period may be associated with a paradoxic ↑ in physiologic stress, as the abrupt withdrawal of catecholamines is coupled to a surge in vagal tone, evoking A Fib

Coombs' test Antiglobulin test, Coombs reaction IMMUNOLOGY A lab test for detecting autoantibodies on RBCs

cooperative *noun* VOX POPULI An organization whose members share purchasing clout, resources or other materials. See Health insurance purchasing

cooperative learning EDUCATION THEORY A student-centered teaching strategy in which heterogeneous groups of students work to achieve a common academic goal–eg, completing a case study or a evaluating a QC problem. See Problem-based learning, Socratic method

coordination NEUROLOGY The harmonious functioning of the motor apparatus of the brain and particular groups of muscles for performing adaptive responses. See Hand-eye coordination

COP 1. Capillary osmotic pressure 2. Colloid oncotic pressure 3. Cyclophosphamide, Oncovin–vincristine, prednisone

cop-killer bullet PUBLIC HEALTH A popular term for a bullet capable of penetrating bullet-proof vests worn by law enforcement officers–cops. See Ballistics, Soft body armor

Copaxone® Glatiramer acetate, copolymer-1 THERAPEUTICS A noninterferon, nonsteroidal agent that ↓ relapses in Pts with multiple sclerosis. See Multiple sclerosis

co-payment MANAGED CARE That portion of a claim or medical expense that a health plan member must pay out-of-pocket for specific med-

ical services–eg, hospital care, drugs, office visits, etc; the insurer pays the remaining portion

COPD Chronic obstructive pulmonary disease PULMONOLOGY An umbrella term for a group of usually progressive lung disorders with overlapping signs and symptoms, including asthma, bronchiectasis, chronic bronchitis, and emphysema; COPD, usually associated with a long Hx of cigarette smoking, is the 5th most common COD–65,000 deaths/yr, US, the 3rd most common–after heart diseases and schizophrenia–cause of chronic disability of older individuals, and the most common cause of pulmonary HTN and cor pulmonale in the US; the major COPD lesions, chronic bronchitis and emphysema, commonly coexist; the former is responsible for the alveolar hypoxia, ↓ PO_2, ↑ CO_2, and ↓ pH that lead to pulmonary HTN, which is seen in 65% of ♂ at autopsy and 15% of ♀, and is due to the unopposed effect of elastases in the lungs CLINICAL SOB, wheezing, chronic cough DIAGNOSIS Clinical Hx, PE, pulmonary function tests COMPLICATIONS Bronchitis, pneumonia, lung cancer; Pts with COPD have been divided into type A with emphysema, fancifully known as 'pink puffers' and type B with chronic bronchitis–'blue bloaters'; respiratory function and dyspnea in severe COPD may improve with theophylline, which improves respiratory-muscle function MANAGEMENT Bronchodilators, O_2 for advanced disease PREVENTION Smoking cessation, ↑ dietary n-3 polyunsaturated fatty acids may protect against COPD, possibly by interfering with the production of inflammatory mediators, including leukotrienes, platelet-activating factor, IL-1 and TNF. See Emphysema

---

MANAGEMENT OF COPD

MINIMIZE AIRFLOW RESTRICTION

Reduce production of secretions

  ↑ Eliminate secretion

Bronchodilatation

Sympathomimetic agents, eg inhaled β_2-adrenoreceptor agonists

Anticholinergic agents, eg ipratropium, nebulized atropine

Theophylline

Corticosteroids–maximum benefit in 1st 2 wks of therapy (NEJM 1999; 340:1941oA)

CORRECT 2º PHYSIOLOGIC ALTERATIONS

Hypoxemia–O_2 administration

Pulmonary hypertension and cor pulmonale

Hypercapnia

OPTIMIZE FUNCTIONAL CAPACITY

Exercise conditioning

Upper extremity training

Respiratory muscle training

Respiratory muscle rest

Dyspnea

Nutrition

Physical and occupational therapy

Psychosocial rehabilitation

OTHER ISSUES OF MANAGEMENT

α_1-antitrypsin augmentation

Bullectomy

Lung transplantation

Antibiotics with exacerbations

Smoking cessation

Copelandia TOXICOLOGY A mushroom with significant quantities of psilocybin, a hallucinogen CLINICAL 20–60 mushrooms are ingested to obtain the desired effect; small amounts of *Copelandina* spp evoke euphoria, accompanied by dizziness, and weakness; larger amounts alter temporal and spatial perception and induce visual disturbances including hallucinations MANAGEMENT Supportive; effects are transient. See Poisonous mushroom

166 **coping mechanism** PSYCHIATRY Any conscious or unconscious mechanism of adjusting to environmental stress without altering personal goals or purposes

copper A metallic element–atomic number 29; atomic weight 63.56; it is an essential trace mineral, and required in certain metabolic reactions–eg, iron absorption and metabolism, and formation of RBCs, nerves

copper sulfate test TRANSFUSION MEDICINE A rapid test for specific gravity of blood, which is an indirect measurement of hematocrit, and is used to determine blood donor acceptability. See Blood donor

coprolalia Scatologia The use of words, phrases, and 'colorful' language that is regarded as obscene which may be a manifestation of paraphilia or Tourette syndrome. Cf Telephone scatologia

coproporphyria An AD condition that primarily affects ♀, which is often latent until puberty CLINICAL ⅔ of Pt are asymptomatic; Sx occur as acute attacks of photosensitivity or mental dysfunction, precipitated by alcohol, barbiturates, diphenylhydantoin, estrogens, griseofulvin, mephenytoin, meprobamate, progestins, sulfonamides, valproic acid, and other drugs LAB ↑ coproporphyrin III in feces, accumulation in the liver; during attacks, coproporphyrin, porphobilinogen, δ-aminolevulinic acid–ALA are ↑ in urine MANAGEMENT Morphine, phenothiazine, chlorpromazine; if no improvement in 24 hrs, IV ferric heme to suppress ALA synthase activity

cor biloculare CARDIOLOGY A congenital heart defect characterized by defect of the atrial and ventricular septae, resulting in a 2 chambered heart with a single atrium, a single ventricle and a common AV valve

cor pulmonale CARDIOLOGY Right ventricular enlargement–ventricular wall ≥ 5 mm or autopsy weight of the right ventricle of > 65 g 2° to pulmonary HTN ETIOLOGY 1° lung disease–eg, pulmonary vascular disease, parenchymal defects–eg, emphysema, bronchiectasia, lungs with an abnormal ventilatory drive, and defects in the thoracic cage; in chronic CP, cardiac hypertrophy is combined with dilatation and with time evolves to CHF; in acute CP, there has only been time sufficient for cardiac dilatation; in older Pts, chronic CP is the 3ʳᵈ most common cardiac disorder after ASHD and HTN; given its relation to cigarette smoking, CP is more common in ♂ MEDICAL MANAGEMENT Supplementary O₂, corticosteroids, anticoagulants, vasodilators, other therapy for underlying lung disease SURGICAL MANAGEMENT Some Pts are candidates for lung and heart-lung transplantation

coral reef appearance GI DISEASE A popular term for the endoscopic appearance of the tortuous, ectatic submucosal veins that may rupture in colonic angiodysplasia, see there

cord ENT See Fixed vocal cord, Spinal cord, Vocal cord OBSTETRICS See Umbilical cord

cord blood stem cells Umbilical cord blood HEMATOLOGY A therapeutic 'agent' containing concentrated hematopoietic stem cells for Pts with BM depleted–'wiped out' by disease; they are transfused into HLA-identical siblings and used to reconstitute the BM of Pts with Fanconi's anemia, aplastic anemia, leukemia, X-linked lymphoproliferative disorder, severe thalassemia and chemotherapy and RT effects. Cf Cord cells

cord strangulation OBSTETRICS The wrapping of the umbilical cord around the baby's neck, which is more common with abnormal presentations, especially breech and should presentation. See Prolapsed cord, Umbilical cord

cordocentesis 1. Fetal blood sampling, see there 2. Percutaneous umbilical blood sampling, see there

cordotomy PAIN MANAGEMENT Surgery to cut fibers of the spinal cord; used to ↓ pain in Pts with intractable cancer-related pain

core decompression ORTHOPEDIC SURGERY A procedure for non-traumatic osteonecrosis, in which the BM is decompressed by removing a core of medullary bone, which is then reinserted to support the weakened cortical bone. See Osteoporosis

core window INFECTIOUS DISEASE That timespan in which an HBV-infected Pt has detectable hepatitis core antigen–HBc in the serum, but has yet to produce detectable levels of HB surface antibody; thus 'core window' period in HB differs from the 'window' period. See Hepatitis serology, Window period

Corgard® Nadalol, see there

coring reamer ORTHOPEDICS A device for harvesting autograft bone cylinders and creating tibial tunnels during ACL repair. See Anterior cruciate ligament repair

corkscrew esophagus The radiocontrast image of an esophagus with periodically spaced, high-amplitude spastic peristaltic contractions of the lower esophagus, due to ↑ responsiveness to neurotransmitters or hormones CLINICAL 'Corkscrew esophagus triad': retrosternal pain, ↑ intraluminal pressure, uncoordinated muscle contractions, and dysphagia and weight loss; it may be asymptomatic; in the lower esophagus, the contractions may ↑ intraluminal pressure, producing transient pseudodiverticuli

corkscrew vessels IMAGING Blood vessels with tight tortuosity seen in advanced micronodular, often alcoholic, cirrhosis, due to collapse of the hepatic parenchyma, without loss of vascular length, resulting in 'crimping' of vessels

corn PODIATRY A small, hard, conical hyperkeratosis caused by friction and pressure; the corn's apex may rub against subcutaneous nerve fibers causing significant pain TYPES Hard, soft MANAGEMENT Paring with a scalpel blade, appropriate footwear, padding–eg, hammer toe splint or corn pads. See Hard corn, Soft corn

corneal abrasion OPHTHALMOLOGY The mechanical scraping of the cornea due to sand, metal dust, hard contact lenses or other foreign bodies

corneal graft A cadaveric cornea used to treat herpetic scars, keratoconus, corneal edema. See Tissue bank

corneal neurotrophic ulcer OPHTHALMOLOGY A corneal ulcer 2° to a loss of sensory innervation of the cornea, leading to ↓ number of corneal stem cells, ↓ rate of metabolism and mitotic activity, resulting in ↑ corneal permeability, and ↓ acetylcholine and acetyltransferase ETIOLOGY 5ᵗʰ nerve palsy, viral infections, chemical burns, corneal surgery, abuse of topical anesthetics, neurotrophic keratitis, DM, multiple sclerosis MANAGEMENT Nerve growth factor may be effective; other modalities–eg, covering the cornea with a patch or a soft contact lens, tarsorrhaphy with conjunctival flap are largely ineffective

corneal reflex Corneal response NEUROLOGY Irritation of the cornea causing lid closure

corneal transplant OPHTHALMOLOGY The replacement of a damaged cornea with a cadaveric–healthy donor cornea; CTs are indicated in severe corneal injury or for corneal ulcers with residual scarring

Cornell protocol CARDIOLOGY A treadmill exercise protocol for classifying a person's functional–NYHA status. See NY Heart Association classification. Cf Bruce protocol

coronary *adjective* Referring to the blood vessels, nerves, and ligaments related of the myocardium *noun* A popular term for an acute MI. See Café coronary, Coital coronary

coronary angiography INTERVENTIONAL CARDIOLOGY A diagnostic technique in which a radiocontrast is injected directly into the coronary arteries, allowing visualization and quantification of stenosis and/or obstruction. See Deferred adjunctive coronary angioplasty, Immediate adjunctive coronary angioplasty, Percutaneous transluminal coronary angioplasty, Primary coronary angioplasty, Rescue coronary angioplasty

coronary artery bypass graft CABG *pron.* cabbage CARDIOLOGY A procedure in which vascular grafts, usually saphenous veins–legs, rarely cephalic veins, are anastomosed end-to-side to the internal mammary arteries, bypassing atherosclerotically stenosed coronary arteries; the internal mammary artery might be a better donor, given its relative resistance to collapse; internal mammary procedures are associated with ↑ survival, better long-term patency, ↓ rate of reoperation STATISTICS 800,000/yr–US survive acute MI–200,000 do not; up to 350,000/yr undergo CABG, ⅔ of whom may benefit from the procedure INDICATIONS Pts with coronary lesions that are unsuitable for, or do not respond to, catheter-based intervention; perioperative death and MI are 4% and 10% higher, respectively, when CABG is performed urgently rather than electively; thus efforts should be made to stabilize the Pt pharmacologically before attempting revascularization; an intra-aortic balloon pump–IABP can serve as a bridge to catheterization or revascularization as long as the Pt does not have severe peripheral vascular disease, significant aortic insufficiency, or known severe aortoiliac disease, including aortic aneurysm; Pts who do not stabilize after IABP placement should be reevaluated to confirm the diagnosis of ischemic heart disease and considered for emergency catheterization MEDICAL THERAPY VS CABG; It is unclear whether medical therapy–diet, exercise, and cholesterol-lowering drugs is more beneficial than surgery; medical therapy is preferred in those without evidence of myocardial ischemia, and in Pts with 1- or 2-vessel disease without significant LADS; CABG is indicated in Pts with chronic stable angina who are medical 'failures', Pts with 2-vessel disease and left anterior descending coronary artery stenosis–LADS; CABG is also indicated in 2- or 3-vessel disease; 'triple bypass' CABG is indicated for unstable angina and ischemia detected by an exercise stress test PRESURGERY WORKUP EKG, stress test, echocardiography, coronary angiography COMPLICATIONS Progression of ASHD, recurrent angina, arrhythmia, sudden death, within 5 yrs in ± 2% of surgically treated and ± 6% of medically treated Pts RECOVERY ± 7-10 days. See ACME, Angina, CABG patch trial, CABRI, CASS, EAST, ERACI, GABI, RITA, Thallium imaging, Treadmill exercise test. Cf Percutaneous transluminal angioplasty; Stenting

coronary artery disease Coronary heart disease CARDIOLOGY Atherosclerosis of the coronary arteries which ↑ a person's risk of cardiovascular M&M. See Atherosclerosis, Myocardial infarction

coronary artery vasospasm Acute coronary insufficiency, anterior chest wall syndrome, coronary artery spasm, coronary vasoconstriction, Prinzmetal's angina, variant angina, vasospastic angina CARDIOLOGY A condition characterized by sudden and usually transient vasoconstriction of a coronary artery with ↓ myocardial O_2 CLINICAL Severe, crushing, chest pain at rest and ST segment deviation PRECIPITATING FACTORS Emotional stress, medications, street drugs–cocaine, cold exposure, stress, vasoconstrictors DIAGNOSIS Normal angiogram, ST segment ↑ EKG ST segment deviation–usually ↑ without preceding ↑ in heart rate or BP MANAGEMENT Nitrates, nifedipine or diltiazem with aspirin; slow CCBs–eg, verapamil, may contribute to AV block. Cf Angina pectoris

coronary event See Cardiac event

coronary perfusion pressure A pressure gradient between aortic and right atrial pressures during the relaxation phase in CPR; CPP correlates well with myocardial blood flow and predicts outcome during cardiac arrest; a minimum pressure of 15 mm Hg is required for spontaneous return of circulation. See Perfusion

coronary revascularization procedure CARDIOLOGY Any procedure–eg, CABG, PTCA, stenting, used to ↑ coronary artery blood flow. See Balloon angioplasty, Coronary artery bypass graft, Coronary artery stent

coronary steal CARDIOLOGY A condition characterized by shunting of all relatively well oxygenated blood from a critical area of low perfusion, to an area of higher perfusion; it is unique as it may be iatrogenic and occur in pharmacologic stress imaging–see there, using dipyridamole to induce vasoconstriction; this causes a fall in blood flow to the subendocardium distal to the site of the stenosed coronary artery. See Steal. Cf Subclavian steal

coronary stent Intracoronary stent CARDIOLOGY An expandable tubular device which can be inserted percutaneously, and left within a coronary artery lumen to maintain its patency PROS Clinical and angiographic outcomes are better with intracoronary artery stent implantation than with standard balloon angioplasty, ↓ rate of restenosis and ↓ need for revascularization procedures CONS ↑ vascular complications, ↑ hospital stay. Radiation Cf Balloon angioplasty, Coronary artery bypass graft

coronary thrombosis CARDIOLOGY A thrombus in a coronary artery lumen, which may cause an acute MI; thrombi form when microscopic cracks in the atherosclerotic coronary artery wall expose collagen and trigger platelet adhesion and thrombus formation

coronary vasodilator reserve Coronary vasodilator response The ratio of maximal to basal coronary blood flow, which can be used as a functional index of the severity of coronary artery stenosis; the CVR also correlates with the geometry of the stenosis by quantitative arteriography

Coronaviridae VIROLOGY A family of single-stranded pleomorphic myxoma-like RNA viruses which cause acute respiratory disease or acute gastroenteritis

coronavirus VIROLOGY The single genus of the family Coronaviridae, which have a corona or halo-like appearance by EM; coronaviruses cause RTIs , common cold, and severe acute respiratory syndrome. See SARS

coroner FORENSIC MEDICINE An elected–less commonly appointed–public official whose chief responsibility is to investigate and provide official interpretation regarding the manner and possible cause(s) of unexplained deaths; in contrast to a medical examiner, coroners are usually not required to be medical doctors, although the requirements depend on the laws governing the jurisdiction. See Forensic pathology. Cf Medical examiner

corporal punishment PUBLIC HEALTH The use of physical punishment–beating or other form of bodily injury to discipline children and control misbehavior. See Domestic violence. Cf Capital punishment

corporate liability HEALTH LAW A doctrine that hospital and health care centers have an responsibility to Pts that extends beyond that of merely furnishing facilites for treatment

corps VOX POPULI A cadre of similarly minded coves or chaps. See National Health Service Corps

corpses, fear of PSYCHOLOGY Necrophobia. See Phobia

corpus callosum agenesis-chorioretinopathy-infantile spasms syndrome Aicardi syndrome, see there

corpus luteum REPRODUCTIVE MEDICINE A yellow secretory structure that forms from the ovarian follicle after ovulation; if the egg is fertilized, the CL ↑ in size, produces progesterone, persists for several months as a CL of pregnancy, which prepares the endometrium for implantation of a fertilized egg; if an egg is not fertilized, the CL degenerates and shrinks–CL of menstruation. See Luteal phase defect

corpus uteri Body of uterus; that part of the uterus north of the isthmus and medial to the fallopian tubes

corpuscular radiation PHYSICS Radioactive particles–electrons, protons, alpha particles, deuterons, pi-mesons, and others, which vary in terms of mass, charge, and energy proportional to speed. Cf Electromagnetic radiation

168 **correction** Making right, righting OPHTHALMOLOGY The use of specific lenses to improve–correct–vision. See Contact lenses, Myopia

correctional medicine The field of medical care which addresses the health needs of prisoners serving sentences in correctional facilities

correlation STATISTICS The degree to which an event, factor, phenomenon, or variable is associated with, related to, or can be predicted from another; the degree to which a linear relationship exists between variables, measured by a correlation coefficient. See Cervical biopsy-cytology correlation, Clinical correlation, Correlation coefficient, Intertemporal correlation, Pearson correlation, Rank correlation

corridor procedure HEART SURGERY Giraudon procedure A procedure for A Fib in which the atrium is divided into 3 compartments–right atrium, corridor from sinus node to A-V node, left atrium; post-op, the corridor is free of A Fib, sinus rhythm is preserved; atrial synchrony is lost. Cf Maze procedure

Corrigan's pulse CARDIOLOGY A clinical sign of aortic regurgitation 2° to ↑ pulse pressure, which is characterized by a bounding carotid pulse and a rapid downstroke. See Aortic regurgitation

corroborating physician TERMINAL CARE A physician who performs a 2nd examination verifying brain death. See Appropriate period of observation, Attending physician, Brain death

corrosive injury Caustic injury, corrosion of tissue PUBLIC HEALTH An injury of mucocutaneous surfaces–eg, eyes, esophagus, skin, with tissue destruction due to direct contact with a strong acid–coagulation necrosis, or with a strong base–liquefactive necrosis EPIDEMIOLOGY ±26,000/yr–US, 1988

corset ORTHOPEDICS A device that encircles and supports the trunk or vertebral column, which may be used for trauma or for congenital deformities of the spine. See Milwaukee brace

cortical contusion NEUROLOGY A hemorrhagic softening of superficial gray & white matter, caused by forceful contact with adjacent structures. See Coup, Countercoup. Cf Extradural hemorrhage, Subdural hemorrhage

cortical dysarthria NEUROLOGY A form of motor aphasia–a deficit in language output or speech production, characterized by intact comprehension and ability to write ETIOLOGY Lesion of motor speech center–aka Broca's area in brain. See Aphasia, Motor aphasia

cortical oscillations NEUROLOGY Cortical rhythms that range from 4-7 Hz–theta waves during sleep, to 14-60 Hz waves while aroused; COs are generated and synchronized by layer 5 pyramidal neurons of the neocortex. See Chaos

cortical screw ORTHOPEDICS A type of orthopedic hardware used to provide fixation by itself or in conjunction with other devices; CSs have fine threads along the shaft and are designed to anchor in cortical bone. See Screw. Cf Cancellous screw

corticosteroid CLINICAL PHARMACOLOGY Any of the steroids elaborated by the adrenal cortex–excluding the sex hormones of adrenal origin in response to the release of ACTH by the pituitary gland, to any of their synthetic equivalents or to angiotensin II; corticosteroids are used to manage arthropathies, inflammation, CA

corticotropin ACTH, see there

corticotropin-releasing hormone See CRF

cortisol Hydrocortisone A major hormone produced by the adrenal cortex, which is the primary glucocorticoid secreted by the adrenal gland in response to ACTH stimulation or stress; cortisol has anti-inflammatory activity, and is involved in gluconeogenesis, glycogen storage in the liver, immune regulation, mediation of physiologic stress responses, Ca^{2+} absorption, secretion of gastric acid and pepsin, conversion of proteins to carbohydrates, and nutrient metabolism; it is secreted in a diurnal pattern–levels rise early morning, peak ± 8 am, and flatten in the evening, diurnal cycling is lost in Cushing syndrome; cortisol secretion is influenced by heat, cold, infection, trauma, excercise, obesity, intercurrent illness ↑ IN Adrenal CA, Cushing's disease, ectopic ACTH, ectopic CRH, hyperthyroidism, depression, alcoholism, substance abuse, anorexia, heavy smoking, CA, ulcers, DM, chronic pain, strokes, CVA, Parkinson's disease, MS, psoriasis, acne, eczema, stress, aging, Alzheimer's disease, AIDS, space adaptation syndrome ↓ IN Addison's disease, hypopituitarism, hypothyroidism. See Corticosteroid, Dexamethasone suppression test

cortisone A glucocorticoid derived from cortisol, the term cortisone may be used generically to refer to all synthetic glucocorticoids

Corvac™ tube LAB MEDICINE A proprietary 1-15 ml blood collection tube, that contains a silicone-based goo for separating serum

Corynebacterium A genus of small, nonmotile, gram-positive bacteria which come in a number of shapes from straight and curved rods to club shapes; most are facultative anaerobes with some similarities to mycobacteria and nocardiae

Corynebacterium diphtheriae The causative agent of diphtheria, which produces a potent exotoxin RESERVOIR Humans EPIDEMIOLOGY Airborne, infected fomites, infected skin; more common in winter, with crowding, hot, dry air INCIDENCE $5/10^5$–US; attack rate in minorities is 5- to 20-fold that of whites MICROBIOLOGY In culture, *C diphtheriae* are often arranged in aggregates fancifully likened to Chinese letters

Corynebacterium equi *Rhodococcus equi*, see there

Corynebacterium minutissimum MICROBIOLOGY The erythrasma agent, which is causes a common superficial skin infection CLINICAL Pruritic, scaling, red-brown macular patches; rarely, *C minutissimum* may disseminate, if the mucocutaneous surface is disrupted MANAGEMENT Erythromycin

coryza MEDTALK → VOX POPULI Runny nose

Cosgrove™ clamp SURGERY A vascular clamp used to temporarily occlude a blood vessel; indicated for use in pulmonary and GI procedures, minimally invasive and standard open cardiovascular procedures; can be used to clamp over indwelling catheters

cosmesis Cosmetic results. See Hair replacement

cosmetic surgery Esthetic surgery Plastic surgery designed to sculpt an Adonis or Venus from lumps of mortal clay; CS techniques include chemical peels, dermabrasion, facial sculpturing, fat injections, liposuction, silicon implants; nearly 6.6 million people had CS in 2002, per the Am Society of Plastic Surgeons–ASPS, a decline of 12% from 2001, due to the highest unemployment rate in 8 yrs; CS procedures remained stable with a 1% increase in 2002, according to ASPS statistics, with > 1.6 million people having procedures; non-surgical cosmetic procedures decreased 15% to 4.9 million in 2002; Botox® surged to the top cosmetic procedure, due to its 2002 FDA approval for cosmetic use; >1.1 million people chose to have Botox®, an ↑ of 31% over 2001; the top 5 surgical cosmetic procedures in 2002 were nose reshaping–354,327, liposuction–282,876, breast augmentation–236,888, eyelid surgery–230,672, facelift–117,831 TOP 5 NON-SURGICAL COSMETIC PROCEDURES, 2002 Botox® injection–1,123,510, chemical peel–920,340, microdermabrasion–900,912, laser hair removal–587,540, sclerotherapy–511,827 GENDER ♀ represent most Pts; > 5.6 million ♀ –85% and nearly 1 million ♂ –15% had cosmetic plastic surgery in 2002; the top 5 surgical cosmetic procedures for in 2002 were breast augmentation–236,888, liposuction–230,079, nose reshaping–209,123, eyelid surgery–186,522 and facelift–105,850; the top 5 non-surgical cosmetic procedures for ♀ were Botox® injection–991,114, chemical peel–771,542, microdermabrasion–771,314, sclerotherapy–495,610 and laser hair removal–484,787; the top 5 surgical cosmetic procedures for ♂ in 2002 were nose reshaping–145,204, liposuction–52,797, eyelid surgery–44,150, hair transplantation–26,501 and ear surgery–21,316; the top 5 non-surgical cosmetic procedures for ♂ were chemical peel–148,798, Botox® injection–132,396, microdermabrasion–129,598, laser hair removal–102,753 and collagen injection–41,193 AGE The 35–50 age group made up 45 % of all CS Pts with 2.9 million people choosing CS; liposuction was the number one cosmetic procedure for this age category with 141,186 patients and Botox® injection topped the non-surgical cosmetic procedures for this age group with 610,226 people; the 19–34 age group had 1.6 million and 24% of the cosmetic total in 2002. Breast augmentation was the number one surgical cosmetic procedure with 126,643 people, and microdermabrasion was the top non-surgical cosmetic procedure for this age group with 253,016 people; the 51-64 age group had 1.4 million people, representing 22 percent of all cosmetic surgery patients in 2002; eyelid surgery was the number one surgical cosmetic procedure with 104,859 people, and Botox® injection topped the non-surgical cosmetic procedures for this age group with 272,592 people; the 65 and over category made up 6 percent of the overall cosmetic plastic surgery population with 396,993 people in 2002. The number one surgical cosmetic procedure was eyelid surgery with 37,790 people and chemical peel was the top non-surgical cosmetic procedure for this age group with 76,163 people; the age category with the least patients was the 18 or younger group with 223,673 people–3% of all cosmetic surgery patients in 2002. Nose reshaping was the number one surgical cosmetic procedure and chemical peel was the top non-surgical cosmetic procedure with 51,734 people; ASPS 2002 statistics www.plasticsurgery.org/news_room/index.cfm. See Plastic surgery

cost MANAGED CARE Any input, both direct and indirect, required to produce an intervention See Adjusted average per capita cost, Administrative cost, Average cost, Average cost per claim, Capital cost, Consumable cost, Direct cost, Fixed cost, Health care cost, Incremental cost, Indirect cost, Intangible cost, Marginal cost, Sunk cost, Variable cost. Cf Price LAB MEDICINE The money expended by a provider to produce goods or services.

cost-benefit analysis Cost-benefit evaluation CLINICAL TRIALS A form of economic analysis from a social perspective, in which the costs of medical care are compared with the economic benefits of the care provided, with both the costs and benefits being expressed in monetary units; the benefits evaluated include projected ↓ in future health care costs and ↑ earning as a result of the intervention of interest. Cf Cost-effectiveness analysis

cost-effective MANAGED CARE *adjective* Referring to an intervention that is considered financially optimal if there is no other available intervention that offers a clinically appropriate benefit at a lower cost

cost-effectiveness analysis Cost-utility analysis CLINICAL TRIALS A form of economic analysis in which alternative interventions are compared in terms of the cost per unit of clinical effect–eg cost per life saved, per mm Hg of lowered BP, per yr of quality-adjusted life gained, etc HEALTH CARE POLICY Analysis related to the effectiveness of therapies or interventions and their associated costs. Cf Cost-benefit analysis

CoStasis® hemostat SURGERY An atraumatic, bioresorbable, liquid hemostat spray to control bleeding in hepatic, orthopedic, cardiothoracic, and general surgical procedures

co-stimulation IMMUNOLOGY The delivery of a 2nd signal from an antigen-presenting cell to a T cell, which rescues an activated T cell from anergy, allowing it to produce the lymphokines necessary for production of additional T cells

costochondritis Inflammation of rib cartilage, especially sternal CLINICAL Peristernal pain, tenderness

cotinine A urinary metabolite of nicotine used to monitor exposure to environmental tobacco smoke–ETS. See Environmental tobacco smoke

cotrimoxazole Bactrim®, Septra®, Sulfatrim®, Cotrim®, smx-tmp, tmp-smx INFECTIOUS DISEASE A combination of 2 antimicrobials–sulfamethoxazole & trimethoprim, used for bacterial and protozoal infections. See Trimethoprim-sulfamethoxazole

Cotswolds Staging System HEMATOLOGY The newest staging system for Hodgkin's disease, which adds the dimensions of bulky disease and presence of multiple sites of disease into the staging process, which is based on data provided by staging laparotomy and CT imaging. Cf Ann Arbor classification, Rye Staging System

cotton ball patch IMAGING-BONE A desciptor for the irregular, rounded, 'fluffy' patches of sclerotic bone seen on a plain skull film in the thickened diploë of advanced Paget's disease of bone; CBPs are associated with exuberant chaotic bone formation in a background of osteosclerosis, also be seen in metastases, hereditary hypophosphatasia and chondroblastomas LUNG A descriptor for the multiple rounded fluffy nodules seen on a CXR in pulmonary histoplasmosis

cotton wool patch Retinitis angiospastica OPHTHALMOLOGY A descriptor for the fluffy exudates and axoplasmic debris seen after microinfarcts in the retinal nerve fiber layer; CWPs are typical of grade III hypertensive retinopathy, but may also be seen in Pts with connective tissue disease, DM–nonproliferative retinopathy, immune suppression, infections–*Babesia microti*, CMV, *Pneumocystis carinii*, ischemia, pheochromocytoma, BMT

Cotton's fracture ORTHOPEDICS A trimalleolar fracture of the ankle involving both–medial and lateral—malleoli and the posterior articular margin of the distal tibia–posterior malleolus—with the latter fragment being dislocated posteriorly and/or superiorly

couch potato An Americanism for a sedentary person, usually ♂, whose predominant non-work activity consists in lying on a couch, watching TV. See Television intoxication 'syndrome'. Cf Vigorous exercise

cough *noun* A voluntary or involuntary explosive expulsion of air from the lungs *verb* To explosively expulse air from the lungs after previously halting same at the glottis. See Brassy cough, Whooping cough

cough suppressant MEDTALK A drug used to control a dry, annoying cough

Coumadin® Warfarin, see there

counseling PSYCHOLOGY An interaction between a professional or trained individual and a Pt, intended to help the latter solve difficulties in psychosocial adjustment; counselors may also advise, opine, and instruct, in order to direct another's judgement or conduct. See Biblical counseling, Pregnancy counseling, Prevention counseling, Psychotherapy, Telephone counseling. Cf Co-counseling

count LAB MEDICINE The enumeration of a thing. See Absolute eosinophil count, Bacterial count, CD4 count, Colony count, Collateral frame count, Differential count, Platelet count, Pollen count, Red cell count, Reticulocyte count, Sperm count, White cell count SURGERY A needle and instrument count. See Sponge count

counterdetailing CLINICAL PHARMACOLOGY Any effort by managed care organizations to control drug costs by educating prescribing physicians on less expensive equivalent or generic alternatives. Cf Detailing

counterfeit drug PHARMACOLOGY A formulation sold or marketed as if it were a particular proprietary substance produced by a particular manufacturer with specified ingredients, which it may or may not, in fact, contain. See Generic drug, Proprietary drug

counterirritant Any substance applied to the skin–eg, capsaicin-from chili peppers—which, by acting as an irritant on a painful zone, attenuates pain sensation. See Capsaicin

countersuggestion PSYCHIATRY A technique in psychotherapy of negativistic-'negative' Pts, in which the therapist deliberately suggests the opposite of what is intended. See Reverse psychology

coup injury NEUROLOGY A cerebral 'bruise' that occurs when the head is stationary and receives a blow; the bruise is directly below the site of the injury IMAGING CT; MRI. Cf Contrecoup

courtesy Professional courtesy, see there

courtesy interview GRADUATE EDUCATION An intervew of a 4th-yr medical student who has completed an audition rotation at, but is not affiliated with, a hospital or institution. See the Match

courtesy staff HOSPITAL PRACTICE A medical staff status for a physician or other health care provider who admits or sees relatively few Pts/month. See Adjunct staff, Medical staff

cover *verb* INFECTIOUS DISEASE To administer antibiotics as a prophylaxis to a person at ↑ risk for bacterial infection after surgery or open trauma MEDTALK To provide coverage for another physician–eg, when the latter is out of town. See Coverage

coverage CLINICAL MEDICINE The provision of medical services by one physician for another, who is usually board certified in the same specialty as the physician for whom he is 'covering'; in the US, a physician must ensure 'coverage' when absent from his practice, or he is legally liable for 'abandonment,' should a Pt need care during his absence MEDICAL MALPRACTICE The provision of liability insurance–eg, typical 'coverage' for medical malpractice insurance ranges from $1-3 million per incident and $3-5 million for complete liability coverage. See Nose coverage, Tail coverage MANAGED CARE The extent of insurance or benefits afforded by an insurance policy. See Carve-out coverage, Coordinated coverage, Creditable coverage, Extended coverage, Portable coverage, Primary coverage, Secondary coverage, Vision care coverage

covered HEALTH CARE *adjective* Referring to a procedure, test or other health care service to which a policy holder or health insurance beneficiary is entitled under the terms of an insurance policy or payment system, eg, Medicare. Cf Not covered

covered person HEALTH INSURANCE An insured person who is eligible for medical benefits or other services covered by a health policy

covered service Covered health care service MANAGED CARE 1. A health care service to which a policy holder is entitled under the terms of a contract 2. A service by a primary care provider in a managed care organization, which is not referred to a specialist 3. Any medically necessary service, drugs or supplies–eg medical transportation services, which are covered under, and which plan members are entitled to receive under, a group benefits agreement. Cf Noncovered service

covering physician MEDTALK A licensed doctor of medicine or osteopathy with an agreement to provide services to another provider's Pts when the provider is not available

COX-1 Cyclooxygenase-1 A constitutive enzyme that converts arachidonic acid into prostaglandins–eg, PGE$_2$, a key player in inflammation; it protects the GI tract, kidneys, and platelets; it is induced by injury

COX-2 Cyclooxygenase-2 An enzyme primarily of immune-mediating cells–eg, macrophages, PMNs, and synoviocytes, which converts arachidonic acid into PGs 2° to inflammation. See COX-2 inhibitor

COX-2 inhibitor Cyclooxygenase-2 inhibitor PAIN MANAGEMENT A class of analgesics with fewer side effects than those of conventional NSAIDs–which inhibit both cyclooxygenases–COX-1 and COX-2; COX-1 protects the gastric mucosa, preventing ulcers, bleeding, and other digestive tract problems. See COX-2, Prostaglandin

coxa valga ORTHOPEDICS A hip deformity in which the angle between the neck and the shaft of the femur is > 140°

coxa vara ORTHOPEDICS A hip deformity in which the femoral neck is too horizontal

Coxiella burnetii INFECTIOUS DISEASE The single species of genus *Coxiella*, family Rickettsiaceae, a short, rod-shaped bacterium; it is global in distribution, causes Q fever, spreads by aerosol, primarily infects cattle, sheep, goats, multiplies well in the placenta, and is shed during parturition. See Q fever

Coxsackie MICROBIOLOGY A family of picornaviruses, genus *Enterovirus* with 22 virotypes of Coxsackie A, 6 virotypes of Coxsackie B CLINICAL Herpangina-Coxsackie A, hand, foot, mouth disease, summer grippe, aseptic meningitis–A and B, epidemic pleurodynia, acute nonspecific pericarditis, myocarditis–group B

CP 1. Cerebral palsy, see there 2. Certified prosthetist, see there

CPAP Continuous positive airway pressure, see there

CPB Cardiopulmonary bypass. See Port-Access cardiopulmonary bypass

CPC 1. Chronic passive congestion. See Nutmeg liver 2. Clinicopathologic conference, or clinical pathology conference

CPK-MB Creatine phosphokinase, muscle band. See CK-MB

CPM/cpm 1. CCNU–Lomustine, Procarbazine, methotrexate 2. Central pontine myelinolysis, see there 3. Counts per minute. See cpm 4. Cycles per minute

CPR EMERGENCY MEDICINE Cardiopulmonary resuscitation Those activities–artificial breathing and external chest compression intended to maintain the heart pump, performed on a person to revive him/her from apparent death, when the heart and/or lungs are not functioning. See ABC sequence, CAB sequence, Cough CPR

CPT Current procedural terminology MANAGED CARE A systematic listing and coding of procedures/services performed by US physicians; a physician-related procedure identification system that serves as the basis for health care billing; CPT coding assigns a 5-digit code to each service or procedure provided by a physician. See Coding, HCFA Common Procedures Coding System, HCPCS level I, HCPCS physician

CPU Central processing unit COMPUTERS The hardware that comprises the 'thinking' part of a computer; the simple-minded often equate the CPU to all the electronic circuitry in the box containing the hard and floppy drives. See Computer

CR 1. Cardiorespiratory 2. Cathode ray 3. Chylomicron remnant 4. Clinical record 5. Closed reduction 6. Clot retraction 7. Colon resection 8. Complement receptor 9. Complete remission 10. Continuous release–pharmacology 11. Crown-rump length

crabs Pubic lice, see there

crack SUBSTANCE ABUSE A 'free-base' form of cocaine which, when smoked in a pipe, produce a brief, very intense 'high'. See Bingeing, Cocaine, Crack 'smile', Free base cocaine

crack baby An infant born to a crack-addicted mother, who is often premature, ↓ birth weight, and has birth defects, respiratory, and neurologic defects; CBs are 4 times more likely to be premature, more commonly suffer SIDS, and given the mothers' high M&M, are often in foster care. See Baby, Crack, Neonatal withdrawal syndrome

cracked pot sign MacEwen sign NEONATOLOGY A finding of late hydrocephalus, where ↑ intracranial pressure leads to palpable separation of cranial sutures; percussion of the skull evokes a 'jagged' sound, unlike the clear sound of normal noggin knocking, the CP is likened to a 'cracked vessel' CLINICAL Thin, shiny skin, prominent veins, high-pitched cry. See Hydrocephalus, Setting sun sign

cracker sign Dry cracker sign RHEUMATOLOGY Discomfort when eating dry foods–eg, salted crackers, a finding typical of xerostomia, seen in Sjögren syndrome

cradle cap NEONATOLOGY A popular term for thick, crusty, yellow scales on infants ± 6 months of age, which may be the presenting, or only, manifestation of seborrheic dermatitis; the lesions may involve other 'seborrheic' regions, ie ears, nasal alae, eyebrows, eyelids

cramp MEDTALK A painful, involuntary muscle spasm. See Heat cramp, Menstrual cramp, Pianist's cramp, Word-processor's cramp, Writer's cramp

cranial arteritis Giant cell–temporal arteritis, see there

cranial base surgery SURGERY Surgery performed in the most difficult region, given the anatomic complexity of the deep intra- and extracranial structures, and the wide range of diseases affecting boundary regions–eg, acoustic neuromas—which occurs between the ear and brain, pituitary tumors—between the roof of the nasal cavity and base of brain, and meningiomas

cranial electrotherapy stimulation NET, neuroelectric therapy, TCET, transcranial electrotherapy PSYCHIATRY A maneuver in which electrodes are placed on or near the ears to pass low level–less than that used in electroshock therapy–electricity through the brain; the person is awake and alert; CES can relieve anxiety for limited periods, though an unknown mechanism, and is an adjunct to anxiolytics and/or psychotherapy SIDE EFFECTS Headache, lightheadedness, skin irritation CONTRAINDICATIONS Hx of epilepsy or seizures. Cf Electroconvulsive therapy

cranial nerve rhizopathy NEUROLOGY Any condition–eg, hemifacial spasm, glosopharyngeal neuralgia, trigeminal neuralgia, that affects cranial nerves

cranial nerve I Olfactory nerve

cranial nerve II Optic nerve

cranial nerve III Oculomotor nerve

cranial nerve IV Trochlear nerve

cranial nerve V Trigeminal nerve

cranial nerve VI Abducent nerve

cranial nerve VII Facial nerve

cranial nerve VIII Vestibulocochlear nerve

cranial nerve IX Glossopharyngeal nerve

cranial nerve X Vagus nerve

cranial nerve XI Accessory nerve

cranial nerve XII Hypoglossal nerve

cranial vault OBSTETRICS The bones that form the movable part of the fetal skull–bones–2 frontal, 2 parietal, occipital, and mold themselves to the ♀ birth canal, allowing passage of a cephalic-presenting infant

craniectomy NEUROSURGERY The surgical excision of bone to access the brain. See Metal plate in the head

craniofacial dysostosis Crouzon's disease PEDIATRICS An AD condition characterized by cranial suture defects, widened skull, a high forehead, ocular hypertelorism, exophthalmos, beaked nose and maxillary hypoplasia

craniopathy MAINSTREAM MEDICINE A general term for any organic disease of the head

craniopharyngioma NEUROLOGY A benign 1° pituitary tumor that is often functionally active, secreting ↑ pituitary hormones–eg, hGH, causing gigantism–if prepubertal and acromegaly—in adults; they comprise < 5% of brain tumors of children CLINICAL Changed vision, headache, weight gain TREATMENT Surgery + RT. See Histologically benign

craniosynostosis Craniostosis PEDIATRICS Premature closure of one or more cranial bony sutures; sagittal suture CSO is more common in ♂; coronal suture CSO is often associated with inherited disease

craniotomy NEUROSURGERY An operation in which an opening is made in the skull, a portion of bone is removed to gain access to the brain, and the bone repositioned

crash COMPUTERS noun An abrupt computer malfunction CRITICAL CARE MEDICINE noun An abrupt decompensation of a Pt's clinical status. See Crashcart PUBLIC HEALTH noun An MVA involving another vehicle, other object(s), or pedestrian(s), which is usually accompanied by noise and structural damage, and may result in bodily injury to the vehicle's occupants and others; MVAs have 4 basic modes: frontal, side, or rear impact, and rollover; death is largely related to the g-force of deceleration, calculated as g = V² miles/hr/stopping distance–in feet x 30 See Crashworthiness, Seatbelt SUBSTANCE ABUSE noun The complex that follows the abrupt withdrawal of a substance of abuse. See Cold turkey verb To sleep off effects of drugs TRANSPLANTATION MEDICINE verb To abruptly decompensate, as in a Pt on a waiting list for organ transplantation, who suddenly becomes 'supercritical,' who is likely to die quickly if the needed organ is not transplanted 'stat'. See Transplantation VOX POPULI noun 1. An MVA, see there. See Underride crash 2. A socioeconomic collapse. See Nuclear collapse

crash cart CAC cart EMERGENCY MEDICINE A cart that is readily accessible to health care workers and strategically placed in sites in a hospital where Pts commonly 'crash', ie undergo acute cardiovascular decompensation, including recovery room, ER, ICUs. See Arrest, CPR

craving PSYCHOLOGY A strong desire to consume a particular substance–eg of abuse, or food; craving is a major factor in relapse and/or continued use after withdrawal from a substance of abuse and is both imprecisely defined and difficult to measure. Cf Addiction, '–holic'

cream cheese appearance GYNECOLOGY A descriptor for the colposcopic appearance of vaginal candidiasis, which has a thickened, whitish, cheesy or tree bark-like exudate that adheres in plaques to the mucosa

creatine kinase Creatine phosphokinase INTERNAL MEDICINE An 82 kD dimeric enzyme that catalyzes the reaction ATP + creatine = ADP + phosphocreatine, which exchanges high-energy phosphate and consumes energy; CK is concentrated in skeletal muscle, heart, brain; since CK levels peak 12-24 hrs after an MI and returns to normal by 48 hrs, specimen timing is critical REF RANGE 0-250 U/L. See Creatinine kinase isoenzymes

creatine kinase isoenzyme An isoenzyme, designated CK-MM–skeletal muscle, CK-MB-cardiac muscle and CK-BB-brain; CK is ↑ in serum when tissue suffers trauma or infarction. See CK-MB, Troponin

creatinine NEPHROLOGY The end product of creatine metabolism, which is excreted into the urine; creatinine can be used to diagnose and monitor renal failure; creatinine may be measured in amniotic fluid to determine gestational age–fetal maturity index; normal ranges vary according to the lab. See Creatinine clearance

creatinine clearance NEPHROLOGY The ratio of the rate of creatinine excretion in urine to its concentration in serum, a value that reflects the body's ability to excrete creatinine; it is used to diagnose and monitor renal function

credential verb To determine or verify titles, qualifications, documents, completion of required training, and continuing education, in those persons who function in a professional or official capacity–eg, ER physician, neurosurgeon, etc. Cf Credentials

credentials The sum of a person's earned documents. See Medicaid. Cf Credential

crepitation 'Crunching' of tissue caused by presence of gas, which may occur in LUNG DISEASE Spontaneous rupture of small pulmonary blebs–most common in young ♂, which causes mediastinal or apical emphysema of little clinical significance ORTHOPEDICS A spongy quality on palpation–feeling with the open hand of a fracture site–broken bones; crepitation suggests a fracture

crescent Quarter moon *adjective* Referring to a sharply circumscribed, smoothly curved radiolucent or light-colored mass in a dark background, which converges at both ends

CREST complex A pentad of clinical signs seen with mixed connective tissue disease: Calcinosis, Raynaud's phenomenon, Esophageal dysmotility with stricture, Sclerodactyly, Telangiectasia; CREST has a slightly better prognosis than other connective tissue disorders, but has late complications–eg, biliary cirrhosis and pulmonary HTN LAB Anticentromere antibodies are characteristic of CREST, but may be seen in progressive systemic sclerosis, older ♀, or with HLA-DR1

cretin Congenital hypothyroidism ENDOCRINOLOGY A person with defective thyroxine or thyroglobulin synthesis ETIOLOGY Goiter, iodine deficiency in the mother while pregnant; thyroid gland defects–aplasia, hypoplasia or dysgenesis CLINICAL Cold intolerance, serosal effusions, myxedema, ↓ metabolic rate, ↑ cholesterol, profound mental retardation–hypothyroid idiocy, ↓ growth. See Thyroid gland

Creutzfeldt-Jakob disease Spastic pseudoparalysis, subacute spongiform encephalopathy NEUROLOGY An invariably fatal, kuru-like infection which causes a form of presenile dementia primarily affecting middle-aged adults, caused by an aberrant protein, a prion CLINICAL 2-4 yrs after exposure → pyramidal tract disease with Babinski sign, nervousness, loss of facial expression; cerebellar dysfunction with ataxia, myoclonus, unsteady gait; basal ganglia involvement is manifest by rigidity, bradykinesia, intention tremor, dystonic postures, choreoathetosis, muscle spasms, imbalance, dementia PROGNOSIS No treatment, no cure; CJD progresses rapidly to coma and death in 3 to 12 months. See Bovine spongiform encephalopathy, Kuru, Prion, Scrapie

CREUTZFELD-JAKOB DISEASE AGENT–DISINFECTION

INACTIVATED by autoclaving for 1 hour at 132ºC and 20 psi, 5% hypochlorite, and 0.03% permanganate solutions. Instruments 0.1 N sodium hydroxide for 2 hours; 5% bleach–NaOCl for 4 hours; all surfaces should be scrubbed with iodine or phenol-based detergents

NOT INACTIVATED by boiling, 10% formalin, 70% alcohol, UV light

CRF Chronic renal failure

CRH Corticotropin-releasing hormone A potent neuropeptide that stimulates synthesis and secretion of proopiomelanocortin-derived peptides; it is markedly ↑ in pregnancy, Cushing syndrome due to ectopic CRH; its activity is turned off by a 37 kD binding protein. See Proopiomelanocortin

cri-du-chat syndrome Lejeune syndrome PEDIATRICS An embryopathy more common in ♀, due to loss of short arm of chromosome 5, occasionally due to a ring chromosome CLINICAL High-pitched feline mewing which often diminishes with age, low birth weight, mental and physical retardation, hypertelorism, hypotonia, microcephaly, micrognathia, epicanthal folds, a moon-like facies, low-set ears, congenital heart defects, short metacarpals and metatarsals, pes planus, and partial syndactyly PROGNOSIS Lifespan relatively normal

crib death Sudden infant death syndrome, see SIDS

cricothyroidotomy Coniotomy EMERGENCY MEDICINE An emergency procedure in which a passageway is created in the cricothyroid membrane–CTM by direct incision coupled with the insertion of a tracheostomy tube CONS ±Perichondritis with subsequent laryngeal stenosis; incision in CTM may cause permanent voice changes; cricothyroid arteries may cross the space and be a source of major bleeding

criminal abortion OBSTETRICS Deliberate and illegal termination of pregnancy. See Abortion, *Roe v Wade*. Cf Webster decision

'criminal' nerve of Grassi A branch of the right posterior vagus which passes to the left behind the esophagus, ending in the gastric cardia

Crinone® GYNECOLOGY A topical progesterone used to manage infertility and 2º amenorrhea. See Hormone replacement therapy

crisis CLINICAL MEDICINE 1. An abrupt–paroxysmal change in the course of a disease, usually for worse–eg, an acute exacerbation of adrenal insufficiency 2. An abrupt intensification of a symptom or other manifestation of a disease, a paroxysm. See Adrenal crisis, Aplastic crisis, Blast crisis, Healing crisis, Hemolytic crisis, Hypertensive crisis, Myasthenic crisis, Therapeutic crisis, Thyrotoxic crisis, Tumarkin crisis, Vaso-occlusive crisis PSYCHIATRY A state of psychologic disequilibrium; turning point in a person's life. See Adolescent crisis, Identity crisis, Legal crisis, Midlife crisis

crisis intervention PSYCHIATRY The counseling of a person suffering from a stressful life event–eg, AIDS, cancer, death, divorce, by providing mental and moral support. See Hotline

crispy critter MEDICAL SLANG A cruel and inappropriate term for a Pt with severe burns. See Medical slang

'crit A widely used short form for hematocrit

criteria MEDTALK 1. A set of rules for assessing or categorizing a thing 2. A listing of the signs and symptoms of a disease. See Exclusion criteria, Harvard criteria, Inclusion/exclusion criteria, Initial peer review criteria, Light's criteria, Pittsburgh criteria, Release criteria, Sequestration criteria, Transfusion criteria, Witherspoon criteria

critical access hospital Rural primary care hospital, see there

critical care HOSPITAL MEDICINE Medical care administered to seriously ill hospitalized Pts, which is best provided in the controlled environment of an ICU; CC may be provided in the OR, recovery room, and ER. See Code–blue, Crash cart, Critical illness/injury, ICU

critical illness neuromyopathy Acute polyneuropathy of the critically ill, critical illness polyneuromyopathy, critical illness polyneuropathy CRITICAL CARE MEDICINE A combination of neural damage and muscle degeneration in Pts requiring prolonged critical care; a predominantly motor axonal polyneuropathy of acute onset occurring in a background of multisystem organ failure and systemic inflammation, often due to gram-negative sepsis CLINICAL Flaccid tetraparesis, respiratory paralysis, hyporeflexia, muscle atrophy, distal sensory defects DIAGNOSIS EMG studies reveal axonopathic features–eg, fibrillations and sharp positive waves on concentric needle studies, ↓ amplitude of muscle or nerve action potentials after stimulation; in ½, conduction velocity is ↓ and distal latency times ↑; CIP is associated with ↑ mortality and rehabilitation problems. See Critical care unit

critical value Critical results LAB MEDICINE A lab result from a Pt that must be reported immediately to care provider, which may require urgent therapeutic action. See Decision levels

CRITICAL VALUES IN THE LABORATORY

CHEMISTRY

ANALYTE	SI UNITS	US UNITS
Calcium	< 1.65 mmol/L	< 6.6 mg/dl
	> 2.22 mmol/L	> 12.9 mg/dl
Glucose	< 2.60 mmol/L	< 46 mg/dl
	> 26.9 mmol/L	> 484 mg/dl

K+	< 2.8 mmol/L	< 2.8 mEq/L
	> 6.2 mmol/L	> 6.2 mEq/L
		> 8.0 mmol/L if hemolyzed
Na+	< 120 mmol/L	< 120 mEq/L
	> 158 mmol/L	> 158 mEq/L
CO_2–plasma	< 11 mmol/L	< 11 mMol
	> 40 mmol/L	> 40 mMol

HEMATOLOGY, eg blasts or sickle cells on peripheral smear, may indicate leukemia or sickle cell anemia

MICROBIOLOGY, eg positive gram stain or culture from blood, serosal fluids or CSF, acid-fast stain or positive mycobacterial culture results

TRANSFUSION MEDICINE Incompatible cross-match, positive VDRL

Crixivan® Indinavir, see there

CRNA Certified registered nurse anesthetist

Crohn's disease Granulomatous enteritis, regional enteritis, terminal ileitis GI DISEASE A chronic recurring inflammatory disease with periods of remission and exacerbation, located primarily in the distal small and proximal large intestines, which may occur anyplace in the GI tract between mouth and anus EPIDEMIOLOGY CD is most common at ages 15-25 CLINICAL Recurrent abdominal pain, fever, N&V, weight loss, diarrhea–± bloody, possibly also, reddish tender skin nodules, inflammation of joints, eyes, liver DIAGNOSIS Barium enema, colonoscopy, Bx confirmation COMPLICATIONS GI bleeding, fistulas, anal fissures, deep ulcers can puncture bowel wall, leading to peritonitis MANAGEMENT Anti-inflammatories, immune suppression–infliximab ↓ draining fistulas, corticosteroids, antibiotics, possibly, fish oil; if severe, surgery

cromolyn sodium Nasalcrom®, Opticrom® ENT An anti-inflammatory delivered by nebulizer to ↓ airway hyperresponsiveness, treat allergic rhinitis, and prevent antigen-, exercise-, cold air-, hyperventilation-, SO_2-provoked, early and late asthma. See Allergic rhinitis

cross-allergy IMMUNOLOGY Immune response to 2+ cross-reactive allergens–allergy to one allergen is assocated with allergies to other, similar antigens

crossbite Malocclusion of teeth DENTISTRY A form of malocclusion in which there is a reversal of the normal relationship of the mandibular and maxillary teeth, with lateral displacement of opposing teeth COMPLICATIONS Periodontitis, caries

cross contamination MEDICAL PRACTICE The passsage of pathogens indirectly from one Pt to another due to use of improper sterilization procedures, unclean instruments, or recycling of products

cross-cover noun A physician who 'covers' for another physician and provides Pt care when the other physician is not available, the other physician reciprocates verb To provide health care for another physician's Pt. See Cover, Coverage

cross eyes Convergent strabismus OPHTHALMOLOGY A process affecting 3% of children; once recognized, it should be treated immediately to allow maximum development of visual acuity, binocular function, cosmetic results

cross-facial nerve graft RECONSTRUCTIVE SURGERY The use of a segment of the facial nerve from the intact, nonparalyzed side–opposite a damaged facial nerve, to partially restore functional, unconscious, symmetrical facial movements

cross infection EPIDEMIOLOGY A converging pattern of infection in which person A is infected with microorganism A, and person B with microorganism B. See Ping-pong infection MEDTALK Nosocomial infection, see there

cross-locking screw ORTHOPEDICS A stabilizing screw screwed at a 90° angle into an intramedullary rod placed in a long bone with an unstable fracture

cross-over design CLINICAL RESEARCH A clinical trial design in which Pts receive, in sequence, the treatment–or the control, and then, after a specified time, are switched to the control–or treatment. See Crossover

cross-reaction IMMUNOLOGY A partial reaction or 'recognition' of an epitope by an antibody generated in response to another antigen

cross-training Multiskilling SPORTS MEDICINE 1. The regular participation in multiple sports–eg, basketball and long-distance running 2. The exercising of muscle groups or participation in a sport differing from than an athlete's primary sport. See Training

crossed syndrome NEUROLOGY A cranial nerve lesion opposite the side of a hemiplegia, which may occur in transient occlusion of the basilar artery

crossing over Recombination VOX POPULI A popular phrase for a change in a person's sexual preference, usually heterosexual → homosexual

crossmatch Serologic crossmatch TRANSFUSION MEDICINE An agglutination test that determines donor-recipient blood compatibility. See Electronic crossmatch, Immediate spin crossmatch, Major crossmatch, Minor crossmatch

CROSSMATCH

MAJOR CROSSMATCH Patient serum–which may contain antibodies is cross-reacted against the donor's red cells and

MINOR CROSSMATCH Patient RBCs are incubated with donor serum, this is of less clinical significance; it reveals donor antibodies against uncommon antigens–eg C^w, -Wr^a, -Li^a

crossmatch/transfusion ratio TRANSFUSION MEDICINE The ratio of units of packed RBCs that are cross-matched–in the blood bank for potential transfusion during surgical procedures, to the number of units actually transfused. See Type & screen

crossover noun CLINICAL TRIALS The switching of a Pt from one arm–eg therapeutic or placebo of a clinical trial to another; crossing-over of Pts may be either part of the original study protocol-crossover study or a decision motivated by ethics, as a Pt might be deteriorating in the less effective arm

crossover unit TRANSFUSION MEDICINE A unit of autologous blood product(s) made available for general use after the donor's surgery and potential need for the unit of blood has passed. See Autologous blood

croup Acute laryngotracheobronchitis, angina trachealis, laryngostasis PEDIATRICS A heterogeneous group of acute infections, which cause upper-airway obstruction in children EPIDEMIOLOGY Annual incidence–3/100 children < 6 yrs–US; 1.3% require hospitalization ETIOLOGY Parainfluenza type 1, less often type 2 and 3, RSV, influenza virus, rubeola, adenovirus, *Mycoplasma pneumoniae* CLINICAL Brassy, seal-like barking or 'croupy' cough accompanied by inspiratory stridor and hoarseness, 2° to intense edema, laryngeal mucus, subglottic stenosis, progressive or episodic dyspnea, tachypnea, cyanosis, sternal and intercostal retractions, dysphagia, low-grade fever, chills, recent URI, ↓ breath sounds, restlessness DIFFDX Angioneurotic edema, bacterial tracheitis, epiglottitis, foreign body aspiration, retropharyngeal abscess MANAGEMENT Dexamethasone, prednisolone, budesonide–nebulizer–effective for croup, mist tents, vaporizers, antibiotics, decongestants, cough suppressants, pain medication, fluids, nebulized racemic epinephrine; hospitalization if stridor at rest

Crouzon's disease Craniofacial dysostosis PEDIATRICS An AD condition characterized by cranial suture defects, widened skull, a high forehead, ocular hypertelorism, exophthalmos, beaked nose and maxillary hypoplasia

crowding DENTISTRY Too many teeth in too little space in a mouth SOCIAL MEDICINE Too many people in too little space in a city. See Gangs

crowing Cantus galli, laryngismus stridulus PEDIATRICS Noisy respiratory 'cawing', stridor and severe respiratory distress

CROWING, CAUSES OF

CONGENITAL LARYNGEAL STRIDOR Crowing may be the first sign of congenital epiglottic and supraglottic deformity or flabbiness–laryngomalacia and tracheomalacia with collapse and partial inspiratory airway obstruction, a condition more common in ♂; during the paroxysms, the children are hoarse, aphonic, dyspneic, have inspiratory muscle retractions and if prolonged, fail to thrive

DOUBLE AORTIC ARCH

OTHERS, eg branchial cleft cysts, chondromalacia, congenital goiter, croup, intraluminal webs, laryngeal masses, lymphangioma, macroglossia, mandibular hypoplasia, mucus retention cysts, Pierre-Robin syndrome, and thyroglossal duct remnants

CRP LAB MEDICINE C-reactive protein, see there

CRPS NEUROLOGY Complex regional pain syndrome, see there

cruciate ligament *ligamentum cruciatum genus* SPORTS ANATOMY Either of 2 major ligaments which form a cross in the knee joint, ensuring proper movement; the anterior CL is more susceptible to injury and rupture than the posterior, which is deeper in the joint. See Anterior cruciate ligament injury

crude birth rate The number of live births in a yr divided by population size

crude mortality rate EPIDEMIOLOGY The mortality rate from all causes of death for a population

cruise *verb* A popular term for the seeking, watching, or flirting with potential sexual partners. See Casual sex

cruise-associated diarrheal disorder EPIDEMIOLOGY Big boat bouts of bowl-bound bowels; outbreaks of CADD occurs in 1.4/1000 cruises or 2.3 outbreaks/10 million passenger-days; an etiologic agent is implicated in ⅔ of cases; half are bacterial–enterotoxigenic *E coli*, *Salmonella enteritidis*, *Shigella sonnei*, half viral–Norwalk-like agent; a vehicle of transmission is identified in ½ No-NO FOODS Scallops, eggs, food from caterers in onshore excursions. See Sanitation score

crunch A crepitance. See Mediastinal crunch

crush syndrome Traumatic rhabdomyolysis TRAUMATOLOGY A condition that results from prolonged and continuous external pressure on the limbs, which reflects the disintegration of muscle and influx of myolytic products into the circulation LAB ↑ K⁺, purines, phosphates, lactic acid, thromboplastin, creatine kinase, creatine, BUN, hemoglobinuria, myoglobinuria

crusted scabies DERMATOLOGY Severe scabies, seen in institutionalized persons–eg, with Down syndrome or either debilitated or immunosuppressed Pts CLINICAL Many mites—*Sarcoptes scabei* var *homini*, with a psoriasis-like pachydermia, variably accompanied by thickened nails, generalized hyperpigmentation, eosinophilia, pyoderma, lymphadenopathy. See Seven-year itch

crutch palsy NEUROLOGY Injury to the radial nerve at the axilla by direct pressure–eg, due to improper use of crutches or pressure caused by hanging the arm over the back of a chair.

cry for help PSYCHOLOGY A popular expression for verbalizations–eg telephone call to crisis intervention hotlines or actions—eg standing on outer ledge of tall building, notes left in conspicuous places, which indicate a state of extreme mental distress or anguish, and the potential for suicide. See Suicide

cryocautery The destruction of tissue by applying extreme cold–eg, with liquid nitrogen

cryoglobulin HEMATOLOGY An abnormal protein–eg, polymeric IgG3, detected by cooling serum to 32°C FOUND IN Myeloma, Waldenström's macroglobulinemia, rheumatoid arthritis, Sjögren syndrome, CLL, SLE. See Cryoglobulinemia

cryoglobulinemia Primary cryoglobulinemia HEMATOLOGY A condition caused by proteins that precipitate in vivo on cooling of acral parts, which are often associated with immune complex-related disease; cryoglobulinemia has been divided into 3 clinical forms CLINICAL Pain, cyanosis, arthralgias, vascular purpuras, cold intolerance, HTN, CHF LAB ↓ C4 and other complement proteins

CRYOGLOBULINEMIA

TYPE I Monoclonal cryoglobulinemia Underlying disease is often malignant; IgG (malignant myeloma), IgM macroglobulinemia or lymphoma/CLL, rarely others (eg IgA nephropathy), benign monoclonal gammopathy

TYPE II Poly-monoclonal cryoglobulinemia A complex of immunoglobulins, eg mixed IgM-IgG, G-G, A-G or other combinations that may be associated with lymphoreticular disease or connective tissue disease (rheumatoid arthritis, Sjögren syndrome, mixed essential cryoglobulinemia)

TYPE III Mixed polyclonal-polyclonal cryoglobulinemia Mixed IgG & IgM, ± IgA, due to rheumatoid arthritis, SLE, Sjögren syndrome, EBV, CMV, subacute bacterial infections, poststreptococcal, crescentic and membranoproliferative glomerulonephritides, DM, chronic hepatitis, biliary cirrhosis

cryoprecipitate 'Cryo' TRANSFUSION MEDICINE A product derived from a unit of whole blood, which has a volume of 15 ml and provides 80 units of factor VIII:C procoagulant–for hemophilia A, factor VIII:vWF–von Willebrand's disease, factor XIII, fibronectin, fibrinogen–for DIC, dysfibrinogenemia

cryopreservation The reversible freezing of tissues–eg, packed RBCs and pre- and post-fertilization products, and autologous BM, often with glycerol for autologous BMT REPRODUCTION MEDICINE A technique for freezing cells to preserve for later use. See Frozen cells, in vitro fertilization

cryosurgery AMBULATORY SURGERY A technique that uses liquid nitrogen to freeze and destroy malignancy–eg, CA of uterine cervix, prostate CA. Cf Cone biopsy, Radical prostatectomy CARDIOLOGY A technique used to modify AV node conduction in Pts with treatment-refractory AV node reentrant tachycardia See Cryotherapy

cryotherapy MAINSTREAM MEDICINE The use of cold to ↓ discomfort, limit progression of edema or interrupt muscle spasms, which acts as a counterirritant OPHTHALMOLOGY A safe, effective therapy for retinopathy of prematurity CONS ↓ visual acuity and smaller visual field

cryptococcal meningitis NEUROLOGY An opportunistic infection of the meninges and spinal cord by *Cryptococcus neoformans* AT-RISK Pts AIDS, lymphoma, DM CLINICAL Severe headache, confusion, photosensitivity, blurred vision, fever, speech difficulties MANAGEMENT Amphotericin B plus flucytosine, then consolidation with fluconazole or itraconazole PROGNOSIS Untreated, coma, death

cryptococcosis INFECTIOUS DISEASE An infection by *Cryptococcus neoformans*, the most life-threatening fungal pathogen of Pts with AIDS, due to its CNS affinity AT-RISK Pts AIDS, cancer, lymphoma, corticosteroids CLINICAL In non-AIDS Pts, cryptococcosisis may cause transient pneumonia; in immunocompromise, *C neoformans* disseminates to bone, skin, viscera, or brain, causing meningitis, meningoencephalitis MANAGEMENT Amphotericin B, flucytosine. See Cryptococcal meningitis, *Cryptococcus neoformans*

cryptogenic *adjective* Of unknown origin; idiopathic

cryptorchidism Undescended testicles A condition in which one or both testicles fail to move from the abdomen, where they develop before birth, into the scrotum; cryptorchidism uncorrected in early childhood is linked to azoospermia and ↑ testicular CA. Cf Anorchia

cryptosporidiosis Crypto, cryptosporidium enterocolitis An opportunistic infection caused by *Cryptosporidium parvum*, an animal parasite EPIDEMIOLOGY Oral-fecal transmission; it grows in the intestines and bile ducts especially immunocompromised and HIV-infected Pts CLINICAL Cryptosporidium enterocolitis causes watery diarrhea, abdominal pain, flatulence, fever; CDC considers chronic intestinal cryptosporidiosis ≥ 1 month an AIDS-defining condition MANAGEMENT No standard therapy PROPOSED THERAPY Paromomycin–humatin, azithromycin, letrazuril, concentrates of cow and chicken antibodies

cryptosporidium enterocolitis Cryptosporidiosis, see there

crystal DRUG SLANG A popular term for a crystallized form of methamphetamine; PCP; amphetamine; cocaine UROLOGY Kidney stone, see there VOX POPULI A formed structure, often composed of a single type of material, which has a characteristic appearance by LM. See Birefringent crystal, Calcium oxalate crystal, Charcot-Leyden crystal, Coffin lid crystal, Hemoglobin C crystal, Jackstraw crystal, Lead crystal, Parking lot crystal, Piezoelectric crystal, Reinke crystal, Rhomboid crystal, Space crystal, Uric acid crystal, Washington monument crystal

crystallization period PSYCHIATRY A period in a person's life–which usually occurs between ages of 35 and 45, during which a person takes serious account of himself, his accomplishments to date, lack of progress, mistakes made and experience. See Mid-life crisis

crystalloid solution TRANSFUSION MEDICINE A balanced isotonic solution–eg, Ringer's lactate or saline fluid solution, used for volume expansion. Cf Colloid solution

C&S Culture & sensitivity MICROBIOLOGY A set of tests performed on a clinical specimen, where isolation of a potentially pathogenic bacterium is followed by antibiotic susceptibility testing. See MIC

CsA Cyclosporin A, see there

CSF Cerebrospinal fluid, see there

CT Computed tomography, see there; also 1. Carpal tunnel 2. Chemotherapy 3. Chest tube 4. Cognitive therapy, see there 5. Connective tissue 6. Continue treatment 7. Crossmatch:transfusion 8. Cytotechnologist

CT myelography NEUROIMAGING A technique that combines myelography with CT imaging to evaluate the spinal cord, and related neural and bone structures. See Computed tomography, Magnetic resonance imaging

C/T ratio Cross-match/transfusion ratio, see there

CT scan Computed tomography, see there

CT venography Spiral CT venography IMAGING The use of a spiral CT device to evaluate blood flow in peripheral veins, and diagnose DVT, using 80% < contrast media than conventional venography. See CT

CTAB Clear to auscultation bilaterally, see there

CTD 1. Connective tissue disease, see there 2. Cumulative trauma disorder, see there

CTLM IMAGING Computed tomography laser mammography A diagnostic technique using lasers to produce a 3-D cross-sectional image of the breast without x-rays. See Mammography

CTZ Chemoreceptor trigger zone, see there

cubital *adjective* Relating to the elbow/forearm. See Cubital fossa

cubital fossa Antecubital fossa ANATOMY The fossa of the anterior elbow, which is bounded laterally and medially by the humeral origins of the flexor and extensor tendons of the forearm and superiorly by a virtual line connecting the humeral condyles

cue PSYCHOLOGY Any sensory stimulus that evokes a learned patterned response. See Conditioning

cul-de-sac French, blind pouch An anatomic 'blind alley,' seen in the conjunctiva, cecum, dura, pouch of Douglas, see there

culdocentesis CYTOLOGY The transvaginal passage of a needle to collect fluid and cells from the cul-de-sac of Douglas, aka rectouterine pouch–a space in the peritoneal cavity between the uterus and rectum; the fluid obtained may contain RBCs–in ruptured ectopic pregnancy, PMNs–in acute salpingitis, or malignant cells–in various cancers

cultural competence SOCIAL MEDICINE The ability to understand, appreciate, and interact with persons from cultures and/or belief systems other than one's own

culture-negative endocarditis CARDIOLOGY Endocarditis in which a causative microorganism are not identified

culture & sensitivity See C&S

culture shock SOCIAL MEDICINE Feelings of isolation, rejection and alienation experienced by a person or group when transplanted from a familiar to an unfamiliar culture–eg, from one country to another; disorientation and confusion when visiting or relocating to culture different from one's own

Culturette® A proprietary product used to screen for group A streptococci; it doesn't differentiate between sick and merely colonized carriers

cumulative trauma disorder Repetitive motion injury, repetitive stress disorder OCCUPATIONAL MEDICINE Any of a group of conditions characterized by repeated stress on muscles, bones, tendons, nerves, which have psychologic and/or physical ramifications–eg, avoidance personality disorder, carpal tunnel syndrome, etc. See Americans with Disabilities Act, *Saroka v. Dayton Hudson*

cunnilingus SEXOLOGY Erotic stimulation of the female external sex organs with the tongue/lips of a partner, as part of normal loveplay which may induce orgasm. See Fellatio

cupping PEDIATRIC IMAGING A widened, metaphyseal concavity caused by muscular and ligamentous pulling on soft bone, which may occur at the sternal ends of the ribs, the proximal tibia and humerus, and the distal radius and ulna; cupping was first described as radiologic evidence of repeated trauma to the growth plates of long bones and thus is suggestive of child abuse DIFFDX Achondrogenesis, cretinism, congenital syphilis, diastrophic dwarfism, hypervitaminosis A, homocystinuria, hypophosphatasia, infarction, infection, leukemia, metaphyseal dysostosis, phenylketonuria, rickets, scurvy, sickle cell anemia, thanatophoric dwarfism, thermal injury, trauma. See Child abuse

curare ANESTHESIOLOGY A neuromuscular-blocking alkaloid used as an adjuvant in surgical anesthesia for skeletal muscle relaxation and to prevent trauma in electroconvulsive therapy

curative *adjective* Relating to that which promotes recovery

curbside consultation Sidewalk consultation An informal and unofficial consultation obtained from a health professsional by either a layperson or a fellow health care professional

---

CURBSIDE CONSULTATION

LAYPERSON A layperson may 'corner' any physician, seeking an opinion about a medical condition, diagnostic modality, or therapeutic option; this form of consultation is particularly dangerous to the physician offering the opinion, as 1. the physician being cornered may not have expertise in the area–eg, a plastic surgeon being questioned about minutiae related to the complications of chemotherapy 2. The person may be asking for information about another person–eg, Aunt Gertrude with gallstones, in which case the information being exchanged with the consultant is confusing–for both the consultant and the surrogate consultee and/or becomes complete gibberish by the time that Aunt Gertrude recieves the 2nd-hand consultation, and 3. The consultant may be liable for a lawsuit for misinformation that a damaged party may allege was provided

PHYSICIAN A physician may ask a colleague in another specialty for the best method for managing a particular clinical problem NEJM 1995; 332:474c

176 **curbstone sign** UROLOGY Sharp intravesicular pain occurring as a Pt with urinary calculi descends a staircase or steps down from a curbstone, caused by calculi bouncing on the bladder trigone

cure *noun* MEDTALK Restoration to a usual state of health. See Natural cure *verb* MEDTALK To heal, restore to health

curettage Curettement GYNECOLOGY The removal of tissue, growths or other material from the wall of a cavity or other surface, with a curette. See Endometrial curettage, Vacuum curettage

curing MEDICAL PRACTICE The removing of all traces of a disease. See Cure

curing light DENTISTRY A special UV light used to attach brackets to the teeth

currant jelly sputum An endobronchial plug of blood, sputum, mucus, and debris, typical of untreated *Klebsiella pneumoniae* pneumonia

current The amount of electrical charge carried/unit time. See Alternating current, M current, Radiofrequency current

Current Procedural Terminology See CPT

curriculum GRADUATE EDUCATION A formal didactic program for teaching a particular subject which, in a residency program includes rotations, conferences, syllabi on assigned reading and areas of emphasis. See Residency, 2+2 curriculum

curriculum vitae CV, resume MEDICAL PRACTICE A formal listing of a person's professional education, objectives, work history, including location and dates of service at a particular hospital, health care facility, university, the role filled at the time of service, academic appointments, professional accomplishments, honors, board eligibility/certification–if relevant, list of publications–eg abstracts, articles, chapters, books, personal, professional references–individuals who can attest to the candidate's competence and integrity; personal data may include marital status, extracurricular interests. See CV weighing

Cushing syndrome Hypercortisolism A condition characterized by excess corticosteroids, due either to an hypersecretion of cortisol by a hyperfunctioning or neoplastic adrenal cortex, or due to exogenous corticosteroids ETIOLOGY Exogenous coticosteroid administration; less commonly, pituitary hyperplasia, adenomas, or cancer CLINICAL Amenorrhea, hirsutism, HTN, impotence, muscular wasting, skin atrophy, neuropsychiatric dysfunction, osteoporosis, truncal–central obesity, weight gain–water retention, moon face, weakness, fatigue, backache, headache, ↑ thirst, ↑ urination, DM, osteoporosis LAB Hypersecretion of cortisol, loss of the usual circadian rhythm of ACTH and cortisol, and loss of suppressibility of cortisol production by administration of dexamethasone; ↑ adrenal androgens suppresses pituitary output of LH and FSH and causes ↓ sperm production or ovulatory failure MANAGEMENT Surgical excision if tumor is ID'd. See Dexamethasone suppression test, Ectopic hormone syndrome. Cf Pseudo-Cushing syndrome

Cushing's ulcer An acute gastric ulcer linked to severe CNS disease in the form of head trauma or tumor

cushingoid appearance An appearance of Pts with Cushing syndrome, which consists of 'moon' faces, truncal obesity, abdominal striae, atrophied skin, and hyperpigmentation of skin folds

cusp A rigid, sharp point, especially on a leaf CARDIOLOGY One of the triangular segments of the pulmonic or aortic valve which opens and closes with the flow of blood

custody SOCIAL MEDICINE The legal control of a child or other person, usually one who resides with a custodial guardian or parent. See Guardian

cut-off ANESTHESIOLOGY The point at which elongation of the carbon chain of the 1-alkanol family of anesthetics results in a precipitous drop in the anesthetic potential of these agents–eg, at > 12 carbons in length, there is little anesthetic activity, beyond 14 carbons, none LAB MEDICINE 1. A time after which a specimen cannot be processed due to logistics or necessity for 'batching' specimens to ↓ labor in performing certain assays and 2. A critical value for an analyte which is ≥ 2 standard deviations above or below a mean or 'cut-off' and thus abnormal. See Decision level, Panic value

cut-off sign Colonic cut-off sign RADIOLOGY An abrupt 'amputation' of the colonic gas column occurring at the splenic flexure, characteristic of acute pancreatitis, also seen in ischemic colitis or thrombosis of the mesenteric vasculature

cut & paste *verb* MEDICAL SLANG To open a Pt, discover that there is no hope–due to massive trauma or CA, and immediately sew up. See Medical slang

cutaneous horn DERMATOLOGY A focal hyperkeratosis seen in solar, seborrheic, or inverted follicular keratoses, in marsupialized tricholemmal or epidermoid cysts, verruca vulgaris, and in SCC or sebaceous gland carcinoma

cutaneous T cell lymphoma Mycosis fungoides ONCOLOGY A heterogeneous group of relatively uncommon NHLs, which first appear on the skin EPIDEMIOLOGY ±1000 cases occur/yr, US, affecting ages 50-60; ♂:♀ ratio, 2:1; black:white, 2:1 CLINICAL Long–4-10 yrs Hx of eczematous rashes, which may evolve to parapsoriasis en plaque or poikiloderma atrophicans vasculare PATHOLOGY Epidermotropism, paracortical proliferation in lymph nodes, periarterial proliferation in spleen; aggregates of epithelioid cells, prominent vascular channels, BM involvement MANAGEMENT Nitrogen mustard, BCNU, PUVA therapy, RT RESPONSE RATE 80% if T1; 30–60% if T2; 20% if T3. See Redman syndrome, Sézary syndrome

cutdown EMERGENCY MEDICINE A small surgical created incision over a major blood vessel, usually a vein–hence, venous cutdown, to facilitate rapid and direct venous access

cutter Self-mutilator PSYCHIATRY A person, more often ♀, who causes self injury in the form of razor cuts, slashes with hot knives, and other mentally cathartic acts; there are said to be up to 2 million cutters in the US CLINICAL Cross-hatched scars, depression, distrust of others RISKS Scars, suicide, blood loss, infection–eg, HIV, death. See Self mutilation

cutting edge *adjective* Advanced–eg, cutting edge research. Cf State of the art

cutting teeth PEDIATRIC DENTISTRY The eruption of baby/deciduous teeth through the oral mucosa, classically associated with malaise and irritability–of the child *and* caregivers. See Deciduous teeth

CV 1. Cardiovascular 2. Cell volume 3. Central vein 4. Closing volume 5. Condensing vacuole 6. Coefficient of variation, see there 7. Conjugata vera 8. Coronavirus 9. Corpuscular volume 10. Curriculum vitae, see there

c-v wave CARDIOLOGY An abnormally prominent pulsation of the jugular vein in tricuspid valve insufficiency, variably accompanied by systolic liver pulsations, and a blowing systolic murmur in the 4th and 5th left parasternal spaces, which is more intense with inspiration

CVA Cerebrovascular accident, see there

CVC Central venous catheter. See Central line

CVD Cardiovascular disease, see there

CVID Common variable immune deficiency, see there

CVP Central venous pressure, see there

CVS 1. Cardiovascular system, see there 2. Chorionic villus sampling, see there 3. Clean voided specimen, see there

cw 1. Clinically warranted 2. Consistent with

CWP Coal workers' pneumoconiosis, see there

CXR Chest X-ray, see there

¹⁴C-xylose test A test for general malabsorption METHOD 1.0 g–5-10 µCi of 'hot' xylose is administered per os; ↑ breath radioactivity at 30 min indicates overgrowth in the small intestine by gram-negative anaerobes

CYA Cover your ass. See Defensive medicine

cyanide TOXICOLOGY A reactive ion–CN⁻ with a high affinity for metal ions–eg, iron of cytochrome oxidase; in vivo, cyanide usually exists as a salt–eg, hydrogen cyanide–HCN, KCN, etc LAB Mean cyanide concentration of victims who died in fatal fires in one study was 116 µmol/L, and 22 µmol/L in those who lived; serum $T_{1/2}$ for HCN is 1 hr; plasma lactate levels in those who died was 29 mmol/L; lactate levels correlate well with HCN levels and may serve as a surrogate marker MANAGEMENT Agents used to treat cyanide poisoning, include those that transform Hb to metHb, which avidly binds cyanide; this maneuver is limited by metHb's ferric state–Fe³⁺, which cannot bind O_2, and at > 30% exacerbates tissue hypoxia caused by the cyanides

cyanocobalamin Vitamin B_{12} A water soluble B vitamin, central to proper CNS function, and carbohydrate, protein and fat metabolism. See Vitamin B_{12}

cyanosis PHYSICAL EXAMINATION A bluish discoloration of skin and mucosae due to excessive concentration of reduced Hb in the blood. See Hypoxia

cyanotic heart disease See Congenital heart disease, Ischemic heart disease

cyberdoc A computer-based interactive program that fills some of the roles traditionally filled by a physician–eg, explaining to Pts the risks, potential benefits, and possible negative outcomes of various diagnostic or therapeutic interventions. Cf Computer-based diagnostic system

cybersickness Simulator sickness, cyber side effects A constellation of findings that may affect those subjected to 'total immersion' virtual reality–VR, in which external audio, visual, possibly–when using a 'virtual glove' also tactile sensation, are replaced by computer-generated information CLINICAL Autonomic Sx–eg, cold sweats, N&V, motion sickness occurring while 'plugged in' to VR devices, post-session residua in the form of altered perceptions and flashbacks. See Porn addiction.

cybersurgery SURGERY A surgical procedure in which the operative field is accessed and manipulated with a digital interface controlled at a distance, usually with joy sticks. See DaVinci

cycle of violence A repeating pattern of physical abuse which occurs in abusive relationships, victims of which may be treated in the ER for injuries sustained during the abuse. See Domestic violence

cyclic edema A transient condition affecting ♀ who work on their feet, possibly of psychogenic origin, as it occurs in the emotionally labile; CE is a diagnosis of exclusion, to be considered after ruling out angioneurotic edema, IgE-dependent or complement-mediated urticaria and other immune-related nosologies

cyclic neutropenia A disorder characterized by episodic severe neutropenia occurring in 3-4 wk cycles, accompanied by maturational arrest of myeloid precursors in BM CLINICAL Fever, oral ulceration, cervical lymphadenopathy, multiple acute local infections which are rarely fatal MANAGEMENT If symptomatic, antibiotics–eg, aminoglycosides, penicillin

cycling SPORTS MEDICINE A weight training term for taking multiple doses of steroids over a specified period of time, stopping for a time and starting again. See Anabolic steroids, Weight training

cyclodestructive procedure OPHTHALMOLOGY Laser or surgical ciliary body destruction to control intraocular pressure in glaucoma Pts unresponsive to conservative therapy COMPLICATIONS Phthisis. See Glaucoma

cyclooxygenase-2 inhibitor See COX-2 inhibitor

cyclophosphamide Cytoxan®, Neosar® ONCOLOGY An alkylating chemotherapeutic used for lymphomas, CAs, and for immunosuppression in nephrotic syndrome ADVERSE EFFECTS BM suppression, cystitis, alopecia, N&V CONTRAINDICATIONS Hypersensitivity, marked thrombocytopenia, leukopenia, hemorrhagic cystitis, lung toxicity related to previous therapy with an alkylating agent

cyclopia NEONATOLOGY The presence of a single, large centrally located eye, which may be accompanied by a single upper front incisor, and is due to a developmental defect related to cholesterol transport. See Fraser syndrome

cycloplegia OPHTHALMOLOGY Paralysis of the ciliary muscle, resulting in paralysis of accommodation

cycloserine INFECTIOUS DISEASE A broad-spectrum antibiotic used for UTIs and TB ADVERSE EFFECTS CNS Sx, allergy, skin rash, ↑ LFTs

Cyclospora cayetanensis PARASITOLOGY A *Cryptosporidium*-like coccidian protozoan, family Eimeriidae, which is implicated in episodic traveler's diarrhea; it infects the GI tract of immunocompetent and immunocompromised hosts–especially with AIDS. See Cyclosporiasis

cyclosporine IMMUNOLOGY A cyclic undecapeptide, which induces potent T-cell immunosuppression; cyclosporine mitigates GVHD, allograft rejection, ulcerative colitis, autoimmune disease, schistosomiasis; it may be of use in aplastic anemia, in combination with antilymphocyte globulin and methylprednisone for psoriasis ADVERSE EFFECTS Nephrotoxicity, HTN, neurotoxicity–eg, tremor, seizures, encephalopathy, headache, coma, hyperlipidemia, hyperkalemia, hypomagnesemia, HTN, anemia, anaphylaxis, nausea, paresthesias, ↑ EBV, lymphoma, pseudolymphoma, fluid retention, thromboses, hirsutism, liver toxicity LAB ↑ Creatinine, ↑ uric acid, ↑ BR, ↑ cholesterol. Cf Tacrolimus

cyclosporiasis INFECTIOUS DISEASE Infection with *Cyclospora* spp, in particular *C cayetanensis* VECTOR Poorly cleaned imported fruits & vegetables CLINICAL Protracted diarrhea described as profuse, malodorous, watery, and can cause dehydration and weight loss; the diarrhea may be associated with nonspecific complaints–eg, intermittent abdominal colic, N&V, low-grade fever, malaise, myalgias, anorexia, bloating, flatulence, and/or profound fatigue, clinically identical to *Cryptosporidium* and *Isospora* infections MANAGEMENT TS for 1 wk; AIDS Pts require longer therapy. See Traveler's diarrhea. Cf Cryptosporiasis

cyclothymic disorder PSYCHIATRY *'A chronic, fluctuating mood disturbance involving numerous periods of hypomanic symptoms and...depressive symptoms'*, which is unrelated to external circumstances or situations. See Bipolar illness, Depression, Hypomanic symptoms

cyclotron RADIATION THERAPY A device used to accelerate charged particles–eg, protons to higher energy levels, in which a magnetic field causes the particles to orbit, which are accelerated by an oscillating electric field. See Radiotherapy

cynophobia PSYCHIATRY A morbid fear of dogs, rabies. See Phobia

cyprianophobia PSYCHOLOGY Fear of prostitutes, STDs. See Phobia

cyst MEDTALK Any closed sac, cavity, or capsule, usually filled with fluid or semisolid material, and lined by epithelium often contains liquid or semisolid material SURGERY A usually benign closed sac or capsule lined by epithelium, usually filled with fluid or semisolid material. See Aneurysmal bone cyst, Baker's cyst, Benign ear cyst, Blue domed cyst, Branchial cleft cyst, Bronchogenic cyst, Chocolate cyst, Dermoid cyst, Gartner's duct cyst, Giant acute pancreatic pseudocyst, Horn pseudocyst, Meibomian cyst, Multilocular renal cyst, Nabothian cyst, Odontogenic cyst, Odontogenic keratocyst, Pilonidal cyst, Pseudocyst, Rathke's cleft cyst, Sebaceous cyst, Synovial cyst, Thyroglossal duct cyst, Twisted ovarian cyst

cystectomy

178 **cystectomy** Surgery 1. Excision of the bladder. See Radical cystectomy 2. Excision of a cyst

cystic acne Mutilating acne Dermatology A form of acne vulgaris which may affect teenagers, 2° to bacterial infection of clogged sebaceous ducts deep in the skin Management Antibiotics, isotretinoin Prognosis Untreated CA may result in scarring. See Acne vulgaris

cystic fibrosis Pediatrics An AR disease of infants, children and young adults caused by exocrine gland dysfunction, which is characterized by chronic pulmonary disease—due to viscid mucus production in the respiratory tract, pancreatic deficiency, ↑ electrolytes in the sweat, occasionally by biliary cirrhosis; ineffective immune defense against bacteria in the lungs; CF is most common in those of Celtic stock–1:2000 births, US; carriers 1:25 white, 1:250 blacks in which there is global exocrine gland dysfunction, resulting in excess chloride and sodium in secretions, causes thickening of mucus and ↓ clearance of secretions, in part related to defective cAMP-dependent phosphorylation of a chloride-selective channel of both epithelial cells and lymphocytes Clinical Progressive lung disease, pneumonia—*Pseudomonas aeruginosa* or *P cepacia*, exocrine pancreas insufficiency, ↓ growth, ↑ sweat electrolytes, especially chloride, meconium ileus, nasal polyposis hepatobiliary disease Clinical–lungs–RTIs, ↓ pulmonary function, cyanosis, hemoptysis, chronic sinusitis GI Tract fatty stool, fat and protein malabsorption, vitamin deficiency, rectal prolapse, esophageal varices Sex organs ♂ infertility due to blocked vas deferens; ↓ fertility in ♀ Endocrine system Abnormal glucose tolerance, type 2 DM Musculoskeletal Bone pain due to hypertrophic osteoarthropathy Signs hypoproteinemia, salty taste, hyponatremia Lab Sweat chloride > 60 mEq/L Management Recombinant DNase, physiotherapy, bronchodilators, antibiotics, anti-inflammatories, pancrease to ↓ malabsorption; aerosolized α₁-antitrypsin to counteract ↑ production of PMN elastase and aerosolized DNAse, which digests partially degraded PMNs that accumulate in the alveoli; CF is usually diagnosed in early childhood because of meconium ileus, FTT, chronic sinus and bronchopulmonary infections, sterility, deafness, intraocular damage. See Passive smoking

cysticercosis Parasitology A condition caused by encystment by larvae from taeniid tapeworms–*Taenia saginata*–cattle tapeworm or *T solium*–swine tapeworm in the CNS and muscle Clinical Seizures, ischemic cerebrovascular disease. Management Symptomatic–control seizures, CSF diversion, steroids. See Cerebral cysticercosis, *Taenia saginatum*, *Taenia solium*

cystinosis Molecular medicine Any of 3–types I, II, III–AR conditios characterized by impaired transport of cystine across lysosomal membranes; the accumulation of cystine in lysosomes results in crystal formation in various tissues, in particular the kidneys; early renal tubular involvement results in Fanconi syndrome with FTT, dehydration, renal tube acidosis in infancy; cystal-related loss of glomerular function leads to uremia and death by age 10 Clinical Growth retardation, photophobia, hypothyroidism, and in later survivors, visual impairment, corneal ulcerations, pancreatic insufficiency, distal myopathy, dysphagia, CNS involvement Management Cysteamine-β-mercaptoethylamine, or phosphocysteamine removes cystine crystals. See Nephropathic cystinosis

cystinuria Molecular medicine An AR inborn error of metabolism in which there is an ↑ in cystine excretion in urine, with formation of cystine bladder stones 3 forms Based on the amount of cystine excreted; in type I, there is an ↑ urine cystine, ornithine, arginine and lysine; in types II and III, cystine excretion is lower Genetics The variability in expression is attributed to allelism on the genes for each type. Cf Cystinosis

cystitis Urology Bacterial infection and inflammation of the bladder and urethra, which is most common in ♀, attributable in part to the short urethra and short distance between the urethral opening and the anus–ages 20 to 50; it is rare in ♂ with anatomically normal urinary tracts; older adults are at high risk for cystitis–incidence in elderly is up to 33%, due to incomplete emptying of bladder associated with BPH, prostatitis, urethral strictures, dehydration, bowel incontinence, ↓ mobility; most cystitis is caused by *E coli*; sexual intercourse ↑ risk of cystitis because bacteria can be introduced into the bladder via the urethra Risk factors Bladder or urethral obstruction with stasis of urine, insertion of instruments into urinary tract–catheterization or cystoscopy, pregnancy, DM, analgesic nephropathy or reflux nephropathy. See Acute cystitis, Honeymoon cystitis, Interstitial cystitis, Urinary tract infection

cystocele Dropped bladder, fallen bladder Herniation of the bladder into the vagina, which is often accompanied by stress incontinence; the frequency of cystoceles ↑ with age and multiparity

cystometrogram Urology A coordinated electromyographic evaluation of the bladder sphincter; cystometrography is used to determine bladder capacity, presence of voluntary or involuntary contractions of the detrusor muscle, and bladder compliance, and evaluate the integrity of the affector–sensory limb of the detrusor reflex arc

cystoscopy Urology The use of a cystoscope or urethroscope with a fiberoptic light source, to directly visualize the urethra and bladder, and detect lower urinary tract disease

cystourethrocele A herniation of the urinary bladder and urethra, accompanied by abnormal protrusion of adjacent tissues into the bladder or urethra

cystourethropexy Urogynecology A bladder neck suspension and sling procedure for managing ♀ incontinence

cytapheresis Transfusion medicine 1. The separation and collection of blood cells by hemapheresis. Cf Leukapharesis, Plateletpheresis 2. The collection of cells for therapeutic transfusion. See Apheresis

cytarabine AraC, cytosine arabinoside Oncology An antimetabolite analogue of deoxycytidine, used synergistically with antifolates, alkylating agents, and cis-platinum to treat AML Pharmacokinetics Poorly absorbed–±20% absorbed orally, better IV; T₁/₂, ±2.5 hrs; 90% metabolized to the inactive arabinosyl uracil Side effects N&V, myelosuppression; in high doses, cerebral dysfunction, ataxia

cytochrome P-450 Clinical pharmacology A superfamily of heme-containing enzymes–CYPs of the electron transport chain, which have multiple activities, including activation of molecular O_2 for incorporation into unactivated hydrocarbons, xenobiotic metabolism, bioremediation, and breakdown of drugs and toxins

CytoGam® Cytomegalovirus immune globulin, intravenous, CMV-IGIV Virology An Ig concentrate used as prophylaxis against CMV disease associated with kidney, lung, liver, pancreas, and heart transplants

cytokine Biological response modifier Any of a number of small 5–20 kD polypeptide signaling proteins of the immune system, which are produced by immune cells and have specific effects on cell-cell interaction, communication and behavior of other cells. See Biological response modifiers, Colony stimulating factor(s), Fibroblast growth factor, Interferons, Interleukins, Platelet-derived growth factor, Transforming growth factor β, Tumor necrosis factor

cytology 1. The formal discipline in which cells are studied and the changes seen correlated with the clinical findings in Pts 2. Cytologic examination, cytologic study The microscopic examination of body fluids for the detection of disease; in cytology, the most common specimen is the Pap smear, a normal component of a gynecologic examination and is the best means of detecting early, curable stages of cancer of the uterine cervix–formerly the most common cause of death in sexually active ♀ as well as viral, fungal and other infections of the ♀ genital tract; cytology specimens can be obtained from various fluids–urine, CSF, or sputum or discharges, specifically as a means of detecting abnormal or malignant cells. See Aspiration cytology, Automated cytology, Bile cytology, Brush cytology, Exfoliative cytology, Fine needle aspiration cytology,

Needle aspiration cytology, Ocular cytology, Pap smear, Screening, Touch cytology, Urine cytology

cytomegalic inclusion disease INFECTIOUS DISEASE A viral infection illness of newborns and immunocompromised adults CLINICAL In utero infection leads to spontaneous abortion, stillbirth, or congenital defects; opportunistic CID in immunocompromised hosts may cause fever, hepatosplenomegaly

cytomegalovirus pneumonia PULMONOLOGY A viral pneumonia characterized by fever, non-productive cough, SOB and association with PCP, especially in AIDS; hypoxia, if extreme, may be fatal. See *Pneumocystis carinii* pneumonia

cytomegalovirus retinitis OPHTHALMOLOGY CMV infection of the eye, a common opportunistic infection in AIDS, a complication of disseminated CMV infection CLINICAL Retinal inflammation, blindness MANAGEMENT Antivirals–eg, foscarnet, cidofovir, gancyclovir, systemically and as a ganciclovir implant PROGNOSIS Without treatment, retinal necrosis, vision loss

cytometry The measurement of various physical characteristics of cells. See Flow cytometry, Laser scanning cytometry

cytopenia A ↓ number of cells in the circulation–eg, thrombocytopenia, leukopenia

cytophilic *adjective* Referring to substances that are attracted to cells–eg, cytophilic antibodies

cytostatic *adjective* Referring to that which suppresses cell growth and multiplication *noun* An agent that suppresses cell growth and multiplication

cytotoxic drug ONCOLOGY An anticancer drug which acts by killing or preventing the cell division ADVERSE EFFECTS Damage to noncancerous tissues or organs with a high proportion of actively dividing cells–eg, BM, hair follicles, GI tract, thereby limiting the amount and frequency of drug administration. See Chemotherapy

cytotoxic T lymphocyte CTL, cytotoxic T cell IMMUNOLOGY A subset of T cells with a CD8 receptor on the surface that recognizes and lyses malignant or virally-infected self cells bearing self, ie 'haplotype restricted', class I MHC molecules. See Helper cells, Perforin, Suppressor cells

Cytovene® Ganciclovir VIROLOGY An agent approved for prophylaxis against CMV in at-risk Pts who receive heart, lung, pancreas, kidney, or liver transplants See AIDS

D Symbol for: 1. Daunorubicin 2. Dead, deceased 3. Dead–air space 4. Decimal 5. Delivered, delivery 6. Density 7. Dental 8. Deuterium 9. Dexamethasone 10. Dextrose 11. Diagnosis 12. Dietician 13. Diopter 14. Diplomate 15. Distal 16. Diverticulum 17. Doctor 18. Donor 19. Dopamine 20. Dorsal 21. Dose 22. Doxorubicin 23. Drug 24. Duration 25. Mean dose

d Symbol for: 1. Day–SI 2. deci–in SI units 3. Deoxyribose 4. dextro– 5. Dextrorotary 6. Diameter

D & C Dilation & curettage OBSTETRICS A diagnostic procedure in which the cervix is dilated to permit the cervical canal and endometrium to be scraped with a curette–curettage

D-dimer test LAB MEDICINE A test that detects FDPs using latex beads coated with monoclonal antibodies to the D-dimer of fibrinogen. See Deep vein thrombosis, Fibrin split products

D&E Dilation & extraction GYNECOLOGY A method used to terminate 2nd trimester pregnancies as an alternative to PG-induced abortion–PIA. Cf Prostaglandin-induced abortion

D need Deficiency need. See Basic need

D-Val Diazepam, see there

D&X Dilation & extraction. See Partial birth abortion

D₄ receptor NEUROLOGY A receptor present on GABAergic neurons in the cerebral cortex, hippocampus, thalamic reticular nucleus, globus pallidus, and substantia nigra–pars reticulata; D_4R is of clinical interest given its high affinity for clozapine, an atypical neuroleptic agent, which ↓ the positive and negative symptoms in acutely psychotic and treatment-resistant schizophrenics, without producing extrapyramidal signs; a recent report suggests that this effect may be due to D_4R-mediated GABA modulation through disinhibition of excitatory transmission in intrinsic cortical, thalamocortical, and extrapyramidal pathways

d4T Stavudine, Zerit AIDS An anti-HIV reverse transcriptase inhibitor/nucleo-

side analogue ADVERSE EFFECTS Neuropathies of hands and feet, stomach upset, pancreatitis, liver damage,. See AIDS. Cf AZT, DDC, ddC

dacarbazine ONCOLOGY An alkylating chemotherapeutic used for lymphoma and metastatic melanoma ADVERSE EFFECTS BM suppression, pain on administration, vomiting

daclizumab Dacliximab, Zenapax® TRANSPLANT MEDICINE A monoclonal antibody used with other agents to ↓ transplant-related GVHD and organ rejection in kidney and liver graft recipients, and Pts with treatment refractory psoriasis, uveitis, tropical spastic paraparesis and some leukemias. See Kidney transplant

dacryoadenitis Inflammation of lacrimal glands and tear ducts

dacryostenosis Stenosis of a tear duct

dactinomycin ONCOLOGY An antitumor antibiotic member of a class of drugs first isolated from *Streptomyces*, and used to treat Wilms' tumor, Ewing sarcoma, embryonal rhabdomyosarcoma, choriocarcinoma, seminoma, KS, lymphoma

dactylitis Inflammation of a digit–a finger or a toe

DAD Diffuse alveolar damage, see there

Dafilon® SURGERY A nonabsorbable polyamide surgical suture for skin closure; used in plastic, ophthalmic, and microsurgery procedures

dahlite Carbonated hydroxyapatite being developed as an 'instant bone' administered in a paste form, hardening within 12 hrs to provide 55 megapascals of compressive strength. See Bone paste

daily census See Census

daily value CLINICAL NUTRITION A recommendation for the quantity–expressed in percentage of a specific nutrient that a person should consume per day in the diet; the food packaging labels on foods in the US express DVs in a percentage corresponding to the total percentage of the daily requirements for a particular nutrient based on a hypothetical 2000-calorie diet. Cf Recommended daily allowance

dalfopristin See Synercid®

Dalkon shield An IUD produced by AH Robins that was withdrawn from the market in 1974. See Pelvic inflammatory disease. Cf Copper-7, Intrauterine device

DALM Dysplasia-associated lesion/mass GI DISEASE An acronym referring to ↑ incidence of colorectal CA with ulcerative colitis when the pathologist finds dysplasia and the endoscopist finds a tumor mass; DALM alone may be a sufficient criterion for performing a proctocolectomy

Dalmane® See Flurazepam

daltaparin Fragmin® CARDIOLOGY An LMW heparin used as an anticoagulant to prevent or treat DVT or thromboembolism after abdominal surgery. See Low-molecular heparin. Cf Lovenox®

damage control VOX POPULI Any activity that ameliorates the effects of prior poor performance, a blot on one's record, or foot-in-mouth syndrome

Damocles' syndrome PEDIATRIC ONCOLOGY A state of long-term uncertainty, stress and anxiety experienced by the Pts and families of children treated for leukemia, as only 50% are cured and up to 10% of successfully treated Pts suffer future–often lymphoproliferative–malignancies. See Leukemia

danaparoid Organan® CARDIOVASCULAR DISEASE An injectable, LMW heparin that ↓ risk of postop DVT and PE after hip replacement surgery; it is also used for heparin-induced thrombocytopenia. See Deep vein thrombosis, Heparin-induced thrombocytopenia, Pulmonary embolism

danazol ENDOCRINOLOGY A non-virilizing anabolic androgen used to manage HANE, fibrocystic breast disease, endometriosis, and possibly autoimmune hemolytic anemia and autoimmune thrombocytopenia, but may mediate immune thrombocytopenia ADVERSE EFFECTS Weight gain, edema, acne, hirsutism, deepening of voice, clitoromegaly, menometrorrhagia, ↓ HDL-C, ↑ liver enzymes, peliosis hepatis, benign intracranial HTN. See Hereditary angioneurotic edema

dander Desquamated epithelium, and a mixture of microorganisms, hair and sebum from domestic animals–dogs and cats, which evokes allergic reactions in atopic persons

danger space SURGICAL ANATOMY A popular term for a region posterior to the retropharyngeal space, which descends in the posterior mediastinum to the diaphragm

danger space infection A life-threatening infection of the 'danger' space, which occurs by extension from the retropharynx, causing dysphagia, dyspnea, fever, nuchal rigidity, esophageal regurgitation TREATMENT Immediate open drainage

dangle NURSING A popular term for the first movement a Pt is allowed, either after surgery under general anesthesia, or 'under local', where the recuperee allows his/her feet to dangle over the side of the bed

darbopoietin alpha Aranesp™ ONCOLOGY A recombinant hematopoietic factor used to treat anemia in Pts with chronic renal failure, and chemotherapy-related anemia in nonmyeloid CA. See Breast CA

darifenacin THERAPEUTICS An agent in clinical trials for treating irritable bowel syndrome and overactive bladder

dart-out PUBLIC HEALTH The running of a child into a street with motor vehicles–MVs; the child is undetected until directly in front of the MV, as he/she is hidden from view by parked vehicles and tall stationary objects

Darvocet® Propoxyphene PHARMACOLOGY A commonly perscribed analgesic which is little more effective than acetaminophen. See Acetaminophen

Darvon® Propoxyphene HCl

dash'n dine Dine'n dash DRUG MARKETING A popular term for a marketing ploy by drug reps which consists of going to a Chinese takeout and sitting with a detailer and maybe an invitee while food is being prepared. See Detailer, Drug rep. Cf Gas & go

dashboard fracture ORTHOPEDICS A shear fracture with a hip dislocation seen in a seated–usually the front seat passenger who is thrown forward, striking the knee(s) against the dashboard, transmitting axial force to a flexed and adducted femur, dislocating the hip with a tear in the posterior capsule or rupture of the posterior rim of the acetabulum by the femoral head

DAT Direct antiglobulin test, see Antiglobulin test

data Singular, datum Factual information in the form of measurements or statistics; data is often quantifiable in terms of reproducibility TYPES Binary–either/or data, categoric-descriptive data, quantitative–instrument-measurable data, and semiquantitative–based on a limited number of categories data; nonquantitative data–eg, transcripts or videotapes may be coded or translated into numbers to facilitate analysis CLINICAL RESEARCH Information collected by a researcher, which is often statistical or quantitative. See Baseline data data, Binary data, Categoric data, Cellular digital packet data, Chart, Contaminated data, Continuous data, Discretely sampled data, Fragile data, GenMoreData, GenRunData, Graph, Hard data, Health data, Health outcome data, Incidence-based data, Inconclusive data, Individual data, Mydata, Microarray data, Orphan data, Quantitative data, Raw data, Semiquantitative data, Smoker data, Soft data, Table, Tobacco data

Data Bank National Practitioner Data Bank, see there

data synthesis Meta-analysis, see there

data to decision EMERGENCY MEDICINE The time that elapses–±15 mins–between the time when a managing physician has enough data–EKG, clinical Hx, CK/CK-MB to be reasonably sure that an acute MI is occurring and when the decision is made to administer thrombolytic therapy. See Door to drug, Thrombolytic

database Register CLINICAL TRIALS Any repository–often computerized for observations and related information about a group of Pts–eg, adult males living in Göteborg or a disease–eg, HTN or an intervention–eg, antihypertensives or other events or characteristics INFORMATICS An aggregation of records or other data that is updatable; DBs manage and archive large amounts of information; a pool of information that can be accessed or retrieved by any of the parameters or 'fields' used for data entry–eg, Pt name, medical record number, date of admission, date of surgery, etc. See Administrative database, ADROIT database, Curated database, Factual database, Hazardous Substances & Health Effects Database, LOINC database, Medical error super-database, Relational database, SAGE database

date rape drug PUBLIC HEALTH A popular name for Rohypnol, which is 10-fold more potent than Valium as a sedative hypnotic; its notoriety derives from its alleged ability to ↓ inhibitions and defenses in ♀, helping the female's partner make unwanted sexual advances

Datril® Acetaminophen, see there

daunorubicin ONCOLOGY An antineoplastic antibiotic isolated from *Streptomyces* spp, effective against ALL and AML, and used for other CAs TOXICITY Dose-related BM suppression, mucositis, alopecia, severe local reaction with extravasation

da Vinci SURGERY A surgical robot for performing certain surgeries–eg, mitral valve repair and laparoscopic procedures–eg, cholecystectomy and gastric ulcer repair. See Laparoscopic surgery, Robotics, Surgical robot

dawn phenomenon DIABETOLOGY Early morning hyperglycemia not preceded by hypoglycemia, linked to ↑ insulin requirements in type 1 DM, possibly due to nocturnal pulses of hGH; DP also occurs in insulin-pump users receiving continuous subcutaneous insulin infusion, who are presumed to be euglycemic; DP is attributed to residual endogenous insulin or insulin-like molecules. Cf Somogyi phenomenon, Subcutaneous insulin resistance syndrome

day VOX POPULI A 24-hr period. See NDA day, Personal day, Unhealthful day, Unhealthy day

day care center PEDIATRICS A facility in which infants and pre-school children are supervised, and their needs attended to while their parents work; because of the childrens' proximity to each other, miniepidemics of GI and RTIs are common in DCCs. See Quality time

day care diarrhea PEDIATRICS An ad hoc term for a miniepidemic of gastroenteritis seen in children in day care PATHOGENS Rotavirus–attack rate 70-100%, cryptosporidium–55%, *G lamblia*–55%, *Shigella*–50%, *Campylobacter jejuni*–35%, *C difficile*–30%

days per thousand MANAGED CARE A standard measure of utilization that refers to an annualized use of hospital or other institutional care, which corresponds to the number of hospital days used/yr per 1000 covered lives

Dazamide® Acetazolamide, see there

D/C 1. Discharge 2. Discontinue

DCA 1. Directed coronary atherectomy, see there 2. Directional color angiography, see there

DCIS Ductal carcinoma in situ, see there

DDAVP® 1-deamino, 8-DD-arginine vasopressin, desmopressin acetate ENDOCRINOLOGY A long-acting antidiuretic vasopressin analog given intranasally, to manage diabetes insipidus and primary nocturnal enuresis. See ADH

ddC Dideoxycytidine, Hivid AIDS An anti-HIV nucleoside analogue ADVERSE EFFECTS Neuropathy of hands and feet, pancreatitis, and oral sores. See AIDS, Zalcitabine

DDDR CARDIOLOGY A formal mode designation–atrial and ventricular pacing, atrial and ventricular sensing, dual response and rate-adaptive, used for dual chamber pacemakers. See Dual-chamber pacemaker. Cf VVIR

ddI 2',3'-dideoxyinosine, Videx AIDS An anti-HIV nucleoside analogue ADVERSE EFFECTS Neuropathy of hands and feet, pancreatic damage, pancreatitis, diarrhea. See AIDS, Didanosine

D-dimer HEMATOLOGY A fibrin split product that can be used in a sensitive assay of plasmin activity, which is often ↑ in systemic consumptive coagulopathy, ie DIC. See Disseminated intravascular coagulation, Fibrin split products

DDT Dichloro-diphenyl-trichloroethane ENVIRONMENT A highly hepatotoxic and potentially neurotoxic insecticide that accumulates in fat; DDT is non-biodegradable and concentrates up the food chain. See Pesticide

DEA Drug Enforcement Administration, see there

dead VOX POPULI Not alive, deceased, goners, kaputt

dead fetus syndrome Macerated fetus syndrome OBSTETRICS The clinical complex due to intrauterine death of a gestational product, with retention of a fetus for > 48 hrs–missed labor, occurring after the 20th wk of pregnancy MANAGEMENT In the US, generally await spontaneous onset of labor for up to 2 wks, then induce labor barring fetal or maternal dystocia, with either PGE₂ in a vaginal or cervical gel, or oxytocin COMPLICATION DIC

dead space CLINICAL THERAPEUTICS That part of a syringe's tip and needle that contains medication that can't be administered; DS is very important in insulin therapy, and for medications where the syringe has < 0.5 mL capacity

dead wood ACADEMIA A popular term for a person–eg, a professor, or other tenured academic whose function has been served but who can't be removed from a position or payroll–eg, because of tenure–to free the position for a more qualified, innovative, or often younger person

dean ACADEMIA The top banana in a college, university, or medical school's hierarchy, who is responsible for all professorial faculty and academics. See Mentor, Professor

death MEDTALK The permanent and irreversible cessation of vital functions–eg, cerebral, cardiovascular and respiratory activities. See Angel of death, Brain death, Cause of death, Cell death, Certifying death, Dance of death, Doctor death, Good death, Harvard criteria, Heat-related death, Immediate cause of death, Leading cause of death, Little death, Man-made death, Manner of death, Mechanism of death, Medicolegal death, Mister death, Monday death, Natural death, Private human-caused death, Public human-caused death, Red death, Reproductive death, Sudden death, Sudden cardiac death, Sudden unexplained nocturnal death, Underlying cause of death, Unnatural death, Voodoo death. Cf Angel of death, Kevorkian–'Dr Death' REGULATORY DEATH Unless you meet the requirements of the Uniform Determination of Death Act passed by the US Congress, 1981, you ain't officially dead; a person *is* dead if there is 1. Irreversible cessation of circulatory and respiratory functions or 2. Irreversible cessation of all functions of brain, including the brain stem, a concept endorsed by the AMA and the Am Bar Assn Harvard.

death certificate HEALTH STATISTICS A document in which a certifying physician formally states, to the best of his/her knowledge, the immediate, intermediate, and underlying cause(s) of death. Cf Unnatural death

death, fear of PSYCHOLOGY Necrophobia, thanatophobia. See Phobia

death rate The number of deaths in a population divided by the average population

death rattle A sound characteristic of end-stage lung disease–eg, terminal lung CA or pulmonary edema, that occurs when the clearance of large airway secretions becomes nearly impossible; as air moves to and fro in the bronchi, the sound acquires a gurgling or rattling quality, often presaging death. Cf Medicine rattle

DeBakey VAD® CARDIOLOGY A miniaturized ventricular assist device which provides ↑ blood flow in Pts with CHF

debrancher deficiency Glycogen storage disease III, see there

debridement SURGERY The cleansing of wounds, by excising 'dirty' edges, producing fresh margins, while removing necrotic tissue and foreign debris. See Wound care

debt That which one owes to another; a financial obligation. See Medical student debt

debulking operation Cytoreductive surgery, debulking surgery SURGICAL ONCOLOGY A surgical procedure to decrease mass effect by removing a portion of a tumor or dead tissue; excision of large malignant tumor masses. See Downstage

Decadron® Dexamethasone A glucocorticosteroid used to ↓ cerebral edema

decay DENTISTRY Caries, see there MEDTALK Putrefaction, see there

decedent A dead person

deceleration Dip OBSTETRICS A periodic & transient slowing of the fetal heart rate in response to uterine contractions, ie stress. See Fetal heart monitor, Uniform deceleration, Variable deceleration

DECELERATION

UNIFORM DECELERATION The fetal heart response to uterine contractions; UCs are symmetrical, have a uniform temporal relation thereto and are divided into:

EARLY DECELERATION/TYPE I DIP Due to vagal stimulation elicited in the first stage of labor by fetal head compression

LATE DECELERATION/TYPE II DIP Due to uteroplacental insufficiency, potentially associated with a less favorable outcome; may signal early vasomotor lability

VARIABLE DECELERATION The fetal heart response is asynchronous with respect to uterine contractions; the curves on the fetal heart monitor are more angled and saw-toothed, and may be related to compromise in placental blood flow, eg umbilical cord compression and, like late decelerations, may signify parturition-related difficulties

deceleration injury EMERGENCY MEDICINE An MVA-related injury, in which a freely-mobile heart in the pericardial cavity is thrown forward, and either tears the fixed ligamentum arteriosum or, if the deceleration is extreme, ruptures the aorta, causing fatal hemopericardium. See Steering wheel 'syndrome'

decerebrate posture Extensor posturing NEUROLOGY A clinical state characterized by hyperextension of extremities, pronation of arms and hyperflexion of hands; DP in humans implies tentorial herniation, often associated with paralysis of the contralateral 3rd cranial nerve. Cf Decorticate posture

decidua GYNECOLOGY Endometrial tissue that has been structurally and functionally transformed by pregnancy; 'decidualization' begins at conception's inception, and is complete by the end of the first month

decidual cast The entire decidual components with hypersecretory endometrium, which is characterized by Arias-Stella changes. See Arias-Stella changes

decision analysis Clinical decision analysis DECISION MAKING 1. An approach to decision making under conditions of uncertainty, involving modeling of sequences or pathways of multiple possible strategies–eg, of diagnosis and treatment for a particular clinical problem to determine which is optimal; DA is based on available estimates–drawn from the literature or from experts of the probabilities that certain events and outcomes will occur and the values of the outcomes that would result from each strategy. See Decision tree 2. An analysis in which '...*a problem is stated, assumptions concerning probabilities and utilities are made, and a conclusion is reached based on the results. If the reader (*of the analysis) *agrees with the structure, assumptions, and probabilities of the analysis, then he or she must agree with its conclusion.*' See Algorithm, Critical pathway

decision level LAB MEDICINE An alternative to a reference value for reporting lab test results; when DLs are exceeded, a response by a managing clinician is required. Cf Panic values, Reference value

decision to drug EMERGENCY MEDICINE The time that elapses–±7 mins–between the time that the decision is made to administer thrombolytics and they are actually administered. See Door to drug, Thrombolytic therapy

decision tree DECISION-MAKING A schematic representation of the major steps taken in a clinical decision algorithm; a DT begins with the statement of a clinical problem that can be followed along branches, based on the presence or absence of certain objective features, and eventually arrive at a conclusion

decitabine ONCOLOGY An antimetabolic anticancer agent

declaration Dec page, face sheet HEALTH INSURANCE A part, often on the first page, of an insurance policy, which personalizes the policy by specifying certain information

decompensation MEDTALK An acute exacerbation or worsening of a clinical condition–eg schizophrenia, renal failure, liver failure, which had been held in check by compensatory mechanisms PSYCHIATRY The exacerbation of a mental condition–eg schizophrenia, that occurs when corrective mechanisms cannot maintain the individual at an optimal level of functioning; the deterioration of existing defenses, leading to an exacerbation of pathologic behavior. See Nervous breakdown

decompression MEDTALK The therapeutic reduction of pressure in a limited space–eg, in the cranial cavity caused by cerebral edema; pericardium 2° to effusion; an extremity due to an expanding hematoma over a fracture encased in a cast. See Microvascular decompression

decompression sickness EMERGENCY MEDICINE A condition caused by rapid whole body decompression, with intravascular 'boiling' of nitrogen, and resultant morbidity or mortality; DS occurs in scuba divers and high-altitude pilots or workers in high-pressure environments CLINICAL, ACUTE The 'bends' headache, N&V, vertigo, tinnitus, dyspnea, tachypnea, convulsions, shock, pain in joints and limbs–the bends, chest–the chokes and abdomen; nitrogen gas in brain causes 'boxcar' air bubbles in leptomeningeal vessels, which separates the blood 'column' and may cause death CLINICAL, CHRONIC Caisson disease Dysbaric–ischemic–osteonecrosis with medullary infarcts of the femoral and humoral heads and rarely, malignancy—malignant fibrous histiocytoma, arising in bone infarction MANAGEMENT Hyperbaric oxygen. See Hyperbaric oxygen therapy

deconditioned NEUROLOGY *adjective* Referring to a musculoskeletal group that had previously been trained for a particular activity–eg, pole vaulting, cross-country running, etc, which has been underutilized, or suffered prolonged disuse. See Conditioned

decongestant PHARMACOLOGY An agent that ↓ swelling or congestion–eg, nasal decongestants–eg pseudoephedrine, phenylpropanolamine, which constrict blood vessels, ↓ blood flow to nasal mucosa and sinuses and ↓ mucosal edema ADVERSE EFFECTS Insomnia, irritability; HTN, renal failure, arrhythmias, psychosis, strokes, seizures, rebound effect; used with caution in Pts with HTN, heart disease, seizure disorders, or hyperthyroidism, or in those

receiving MAOIs. See Nasal decongestant, Rebound effect, Rhinitis medicamentosa, Steam decongestant

decontamination PUBLIC HEALTH Use of physical or chemical means to remove, inactivate, or destroy bloodborne or other pathogens on a surface or item, to the point where they are no longer capable of transmitting infectious particles, and the surface or item is rendered safe for handling, use, or disposal. See Photochemical decontamination. Cf Disinfection

decorticate posture NEUROLOGY A posture caused by diffuse and severe cortical dysfunction, seen in a deep coma, where primitive reflex posturing prevails after the loss of higher cortical control; DP is characterized by fisted hands, arms flexed on the chest, extended legs, often in response to painful stimuli, which indicates midbrain dysfunction. Cf Decerebrate posture

decubitus ulcer Pressure ulcer, see there

deep hypothermia CARDIOVASCULAR SURGERY An induced ↓ of a Pt's core body temperature, to slow metabolism during major surgery. See Cardiopulmonary bypass

deep shave biopsy Saucerization biopsy DERMATOLOGY A biopsy with a broad rim of epidermis and dermis, performed by dermatologists for certain small–< 1 cm pigmented lesions–eg, lentigo senilis, seborrheic keratosis, and dysplastic nevus. See Melanoma

deep sleep SLEEP DISORDERS A popular US term for NREM stages 3 and 4 sleep. See Intermediary sleep, Light sleep

deep vein thrombosis INTERNAL MEDICINE A condition characterized by blood clots in veins, most often of the lower extremity, often giving rise to embolism and tissue necrosis; DVT occurs in ½ of total hips without prophylactic anticoagulation, 2-3% of which evolve to fatal PE; acute DVT occurs in 1:1000 of the general population; 92% are idiopathic, ±8% are due to isolated deficiencies of protein C, protein S, antithrombin III, plasminogen RISK FACTORS ↑ Age, immobilization, prior DVT, anesthesia, surgery, pregnancy, CA, hypercoagulability–↓ AT III, ↓ protein C, ↓ protein S, activated protein C resistance, antiphospholipid syndrome, polycythemia vera, erythrocytosis, tissue trauma–which activates coagulation, and ↑ coagulation factor XI CLINICAL Vague–SOB ±leg swelling, pain, edema, discoloration DIAGNOSIS Phlebography, impedence plethysmography, compression ultrasonography–real time B mode, Doppler flow velocity, MR venography, radionuclide venography, thermography, D-dimer assay MANAGEMENT Anticoagulation–eg, heparin or warfarin or thrombolytic therapy–eg, alteplase or streptokinase; inferior vena caval filters; thrombectomy. See Total hip replacement

deer mice *Peromyscus maniculatus* PUBLIC HEALTH The murine vector for Hantavirus. See Hantavirus

defect MEDTALK A malformation or abnormality. See Acquired platelet function defect, Atrial septal defect, Atrioventricular conduction defect, Birth defect, Developmental field defect, Enzyme defect, Epigenetic defect, Fibrous cortical defect, Filling defect, Homonymous field defect, Mass defect, Neural tube defect, Slot defect, Ventricular septal defect

defense MEDICAL MALPRACTICE Any legal argument offered by a defendant that would either preclude or mitigate recovery of damages for a wrong allegedly committed by a defendant. See Affirmative defense, Black rage defense, Character defense, Insanity defense PSYCHOLOGY A mechanism by which a person minimizes harm to his/her psyche or to control anxiety. See Defense mechanism

defense mechanism PSYCHOLOGY An unconscious intrapsychic process by which a person obtains relief from emotional conflict and anxiety EXAMPLES Compensation, conversion, denial, displacement, dissociation, idealization, identification, incorporation, introjection, projection, rationalization, reaction formation, regression, sublimation, substitution, symbolization, undoing. See Defense mechanism

184 **defense wound** FORENSIC PATHOLOGY A wound sustained when a victim places a hand, arm or other body part to prevent or minimize a blow or slashing by a sharp weapon

defensive medicine A style of Pt management defined as those '...*objective measures taken to document clinical judgement in case there is a lawsuit*–costing ± $7 billion/yr US...'; DM is designed to minimize lawsuits and includes such 'devices' as

---

DEFENSIVE MEDICINE

INFORMED CONSENT A document to indicate Ps understanding of the intended outcome and potential risks of a procedure

DOCUMENTATION Formal paperwork generated by a physician that justifies his reasoning for managing a Pt, which may be viewed as being 'unreasonably excessive'

MEDICAL WORKUP Over-ordering of diagnostic tests to rule out 'zebras'–unusual diseases that are not seriously considered as diagnoses, which may rarely be seen in similar circumstances–a form of highly prevalent CYA–cover your ass mentality

DM is virtually a standard of practice in the US; its financial impact is difficult to quantify, and is to ± ↑ the cost of US health care by 20-40% Note: The disadvantage of providing a list of potential complications–each of which may be extremely rare may overwhelm the Pt, causing him to forego a needed procedure, resulting in 'misinformed consent' This highly colloquial and vulgar abbreviation is commonly used at all levels of medical practice and training, and has appeared in at least one major medical journal; 'CYA', ie diagnostic 'overkill', has a mystical overtone, as the physician may be advised to 'CYA' to ward off the evil humors of litigation; DM is practiced by ± 84% of US physicians, in order to protect themselves from potential malpractice-related lawsuits AMN 25/5/92 p3 In the US, anything less than a perfect outcome is unacceptable to a consumer, for whom the threshold for litigation appears to ↓ as medical technology ↑, despite the known risks for certain procedures

---

deferment Delaying of an obligation. See Default, Medical student debt. Cf Forbearance

deferoxamine Deferoxamine mesylate HEMATOLOGY A trihydroxamic acid produced by *Streptomyces pilosus* that ↑ urinary iron excretion; it is the only FDA-approved iron-chelating agent for clinical use; early use of deferoxamine in thalassemia major ↓ transfusion-related iron overload and protects against DM, cardiac disease, early death. See Hemochromatosis

deferred diagnosis SURGICAL PATHOLOGY An intraoperative consultation in which tissues of interest to the surgeon are cut, stained, and examined by LM, but a final diagnosis delayed pending final–paraffin–sections. See Frozen section

defibrillation CARDIAC PACING The termination of unsynchronized quivering of the ventricular myocardium by discharging a high energy, asynchronous electrical stimulus to the myocardium; the defibrillatory discharge often restores normal rhythm. See Low-energy defibrillation

defibrillator CARDIOLOGY A device used to synchronize erratic electrical signaling through the heart, which prevents efficient ventricular pumping. See Crashcart, Pacemaker syndrome

deficiency GENETICS Loss of a segment of a chromosome. See Chromosome LAB MEDICINE An inadequacy in procedure, record-keeping, policy, or implementation thereof, that has been identified by a regulatory agency MEDTALK Any absolute or relative lack of an exogenous or endogenous substance in the body. See Aldolase A deficiency, Alpha₂-antiplasmin deficiency, Alpha-1 antitrypsin deficiency, Androgen deficiency, Apolipoprotein C-II deficiency, Apolipoprotein E deficiency, Arginase deficiency, Arginine deficiency, Biotin deficiency, Calcium deficiency, Carnitine deficiency, Carnosinase deficiency, Chromium deficiency, Condition level deficiency, Congenital antithrombin III deficiency, δ-sarcoglycan deficiency, Diphosphoglycerate mutase deficiency, Eosinophil peroxidase deficiency, Factor V deficiency, Factor VII deficiency, Factor X deficiency, Glucocerebrosidase deficiency, Glucose-6-phosphate dehydrogenase deficiency, Gonadotropin deficiency, Hageman factor deficiency, HDPRT deficiency, Hexokinase deficiency, Hexose phosphate isomerase deficiency, HMG-CoA synthase deficiency, Immunodeficiency, Immunoglobulin A deficiency, Immunoglobulin M deficiency, Iodine deficiency, Iron deficiency, Lactase deficiency, L-CHAD deficiency, Late-onset immune deficiency, LFA-1 deficiency, Lipoprotein lipase deficiency, Lysyl-protocollagen hydroxylase deficiency, Magnesium deficiency, Manganese deficiency, Medium chain acyl-coenzyme A dehydrogenase deficiency, Methemoglobin reductase deficiency, 5,10-methylenetetrahydrofolate reductase–MTHFR deficiency, MHC class II deficiency,

Myeloperoxidase deficiency, Neuraminidase deficiency with beta-galactosidase deficiency, Ornithine transcarbamylase deficiency, Plasminogen activator inhibitor-1 deficiency, Protein deficiency, Protein C deficiency, Protein S deficiency, Purine nucleoside phosphorylase deficiency, Secondary deficiency, Selenium deficiency, Severe combined immune deficiency, Sucrase-isomaltase deficiency, Testosterone 17β-dehydrogenase (NADP⁺) deficiency, Triosephosphate isomerase deficiency, Vitamin A deficiency, Vitamin C deficiency, Vitamin D deficiency, Vitamin E deficiency, Vitamin K deficiency, Zinc deficiency

deficiency disease CLINICAL NUTRITION A condition due to lack of an essential nutrient–eg, protein, minerals or vitamins METABOLIC DISEASES Inborn error of metabolism, see there

deficit A lack of a substance or factor of interest

definition IMAGING The clearly delineated limit of a thing visible by the eye. Cf Resolution

definitive diagnosis CLINICAL MEDICINE A diagnosis reached by an extensive workup, on which therapy is based. Cf Clinical suspicion

definitive radiation RADIATION ONCOLOGY Administration of radiation as a sole or 1° therapy, usually for CA, with curative intent; DR doses average 40-50 Gy/4-5000 rads. See Radiation therapy

definitive test NIHSPEAK A test that generates adequate data to determine the particular hazard of a substance without additional testing; a test upon which decisions regarding safety can be made

defloration Loss of virginity GYNECOLOGY Rupture of the hymen, which often occurs at the time of the first sexual intercourse, or during digital vaginal examination, masturbation, or with tampons. See Hymen

deformation Deformity NEONATOLOGY A change from the normal size or shape of a part that differentiates normally, but cannot develop fully due to in utero constraints–eg, compression, or oligohydramnios. See Defect, Dysmorphology

deformity Deformation VOX POPULI A popular term for a change in the normal size or shape of a part. See Acorn deformity, Boutonniere deformity, Christmas tree deformity, Dishface deformity, Fishmouth deformity, Gooseneck deformity, Gunstock deformity, Haglund's deformity, Hockey stick deformity, Hourglass deformity, Lobster claw deformity, Main-en-lorgnette deformity, Mitten hand deformity, Phyrigian cap deformity, Pigeon breast deformity, Popeye deformity, Spade deformity, Swan neck deformity, Windsock deformity

degeneration MEDTALK The deterioration or compromise in function of a part. See Age-related macular degeneration, Ballooning degeneration, Fatty degeneration, Feathery degeneration, Granulovacuolar degeneration, Hereditary degeneration, Liquefactive degeneration, Macular degeneration, Myxoid degeneration, Paraneoplastic cerebellar degeneration, Red degeneration, Spongy degeneration of infancy, Wallerian degeneration, Waxy degeneration

degenerative joint disease Osteoarthritis, see there

degloving injury EMERGENCY MEDICINE An avulsion-type injury in which the skin and subcutaneous tissue of the hand are torn off in a glove-like fashion, leaving the musculofascial plane intact TREATMENT Clean, debride, sew clean flaps, light compressive dressing, antibiotics–eg, cefazolin, hospitalize. Cf Wringer injury

degree ACADEMENTIA A document that indicates completion of a course of study

dehiscence SURGERY The pulling apart of apposed or sutured margins

dehydration fever NEONATOLOGY An ↑ temperature in a neonate due to inadequate fluid intake, most severe in high ambient temperatures or when the infant is overclothed. Cf Prickly heat

dehydration INTERNAL MEDICINE The loss of intracellular water that leads to cellular desiccation and ↑ plasma sodium concentration and osmolality, often due to GI tract–eg, vomiting, diarrhea CLINICAL Rapid ↓ weight loss of 10% is severe, ↑ thirst, dry mouth, weakness or lightheadedness, worse on standing, darkened or ↓ urine; severe dehydration can lead to changes in the body chemistry, kidney failure, ±life-threatening MANAGEMENT Fluid replacement, 5% dextrose. Cf Volume depletion

dehydroandrosterone A weak androgenic steroid produced in the adrenal cortex and secreted by ♂ and ♀; DHEA levels ↑ during childhood, rise rapidly after puberty, peak at age 20 and then ↓ ABNORMAL VALUES ↑ IN Adrenal tumors and hyperplasia, Cushing's disease, hirsutism, polycystic ovary syndrome, ↓ IN Addison's disease, hyperlipidemia, psychosis, psoriasis. See Unproven methods for cancer management

de-identification PATIENT PRIVACY The removal of identifying information–Pt name, medical record number, birthdate, social security number from medical records, to protect Pt privacy. See HIPAA, Patient privacy

déjà French, previously PSYCHIATRY A group of paramnesias in which there is a perception of being familiar with, or having previous–déjà–experiences which have not occurred, or complete absence–jamais–of memory for events known to have been experienced by the subject, each of which has been associated with neurotic depersonalization and temporal lobe epilepsy

Delaney Clause PUBLIC HEALTH An addition to the US Food, Drug & Cosmetics Act, prohibiting the use of food additives known to be carcinogenic in experimental animals. See Alar, Ames test, Food & Drug Administration, Risk assessment

delavirdine mesylate Rescriptor® AIDS A nonnucleoside reverse transcriptase inhibiting antiretroviral used with nucleoside RTIs ADVERSE EFFECTS Rash. See AIDS, Antiretroviral, HAART, Reverse transcriptase inhibitor. Cf Nevirapine, Nonnucleoside reverse transcriptase inhibitor, Protease inhibitor

delay in diagnosis MALPRACTICE A cause of action in which a plaintiff sues a physician because of alleged failure to diagnose a condition early enough to effect a cure or achieve maximum survival. See Lost opportunity doctrine

delayed reconstruction SURGERY Breast reconstruction not done at the same time as a mastectomy. See Bowel rest

delayed sleep phase SLEEP DISORDERS 1. A condition that occurs when the clock hour at which sleep normally occurs is moved back in time in a given, 24 hr sleep-wake cycle, causing a temporarily displaced–delayed occurrence of sleep in the 24 hr cycle 2. Chronic sleep schedule disturbance

delayed stroke NEUROLOGY A stroke that occurs months after complete carotid occlusion CAUSE Unknown INCIDENCE Uncertain

delayed union ORTHOPEDICS A delay in the healing of the ends of a fracture

delegation Delegation of authority ADMINISTRATION The transfer of authority and responsibility from a person of higher to a person of lower ranking

deletion syndrome CLINICAL GENETICS Any of number of hereditary conditions caused by a major loss of chromosome segments; all are rare, often have microcephaly, and an IQ < 50

delinquency PSYCHOLOGY The participation in any of a number of antisocial acts–eg, truancy, vandalism, sexual promiscuity, shoplifting, gang wars, homicide; delinquency most often occurs during adolescence, hence, juvenile delinquent

delirium NEUROLOGY An acute organic brain disorder caused by a defect in cognate functions with global impairment and a ↓ clarity of awareness of the environment, which may progress or regress ETIOLOGY May be multifactorial–eg, due to toxins; substance abuse; acute psychosis, medication–eg anticholinergics; anemia, brain lesions–eg 1° tumors or metastases;

chemotherapy–eg, MTX, corticosteroids, asparginase, vincristine; endocrinopathies–eg, hypoglycemia; fever; infection; metabolic derangement–eg, ↑ Ca²⁺, ↓ Na⁺ paraneoplastic syndromes CLINICAL Disturbance of sleep-wake cycle, with insomnia and/or daytime drowsiness, altered psychomotor activity, perceptual disturbances, and behavior changes–eg anger, anxiety, depression, fear, irritability, paranoia, withdrawal, most prominent at night DIAGNOSIS EEG–slowing of brain waves; cognitive capacity screening examination; mini-mental state test; trail-making B test. See Black patch delirium, Fatal excited delirium, Pseudodelirium. Cf Dementia

delirium tremens Complicated alcohol abstinence ALCOHOLISM An acute organic psychosis seen 3-10 days after abrupt alcohol withdrawal CLINICAL Confusion, sensory overload, hallucinations–eg snakes, bugs, tremor, seizures, autonomic hyperactivity, cardiovascular defects, diaphoresis, dehydration LAB ↓ K⁺, Mg²⁺ MANAGEMENT Hallucinations require hospitalization and haloperidol; abrupt alcohol withdrawal requires CNS depressants–eg benzodiazepines, phenobarbital; antipsychotics–eg clopromazine should not be used; anticonvulsants are not used in absence of seizure history. See Othello syndrome

delivery OBSTETRICS The passage of a fetus and placenta via the birth canal to the stage called life. See Breech delivery, Vaginal delivery, Vaginal delivery after cesarean section delivery, Vertex delivery PHARMACOLOGY The actual, constructive, or attempted transfer of any item regulated under a jurisdiction's controlled substance legislation. See Drug delivery THERAPEUTICS See Drug delivery

dellen OPHTHALMOLOGY Foci of stromal degeneration with reversible corneal attenuation, caused by a break in the tear film layer, due to a local elevation of the cornea–eg, pterygium, filtering blebs, suture granuloma, or limbal tumors

delta wave CARDIOLOGY An EKG finding in WPW syndrome, seen as a slow upstroke of the QRS wave in a background of short P-R intervals. See Wolff-Parkinson-White syndrome

delusion of grandeur PSYCHIATRY A popular term for what the Am Psychiatric Assn terms 'delusional disorder, grandiose subtype'; DOGs are characterized as '*delusions of inflated worth, power, knowledge, identity, or special relationship to a diety or famous person*'. See Delusional disorder

delusional disorder DSM IV–297.1 PSYCHIATRY A mental disorder characterized by the presence of one or more nonbizarre delusions that persist for more than one month in a person who has not ever had a symptom presentation meeting Criterion A for schizophrenia TYPES Erotomanic, grandiose, jealous, persecutory, somatic, mixed, unspecified. Cf Schizophrenia

Demadex® Torsemide NEPHROLOGY A loop diuretic as effective as furosemide, with longer duration of action. See Diuretic, Loop diuretic

demand model HEALTH CARE POLICY A simplistic model for determining the existence of 'units'–eg pathologists, radiologists, etc, in a system, based on supply dynamics and environmental factors. Cf HMO–extrapolation model, Needs model, Supply model

demand pacemaker Demand pulse generator, inhibited pacemaker CARDIAC PACING A pacemaker which, after sensing spontaneous depolarization, withholds its pacing stimulus. See Pacemaker

dementia Chronic brain failure, chronic brain syndrome, chronic organic brain syndrome, cortical and subcortical dementia, organic mental syndrome, presbyophrenia, senility NEUROLOGY A general term for a diffuse irreversible condition of slow onset seen in older Pts, due to dysfunction of cerebral hemispheres; it is an end stage of mental deterioration, and is characterized by a loss of cognitive capacity, leading to ↓ social &/or occupational activity PREVALENCE Age-linked–affects ± 10.5% of those 80-85; 12.6-47.2% ≥ 85 ETIOLOGY 47% Vascular-type–potentially treatable; 44% Alzheimer's type CLINICAL ↓ mental ability, memory loss, often with emotional disturbances, personality changes

IMAGING MRI, PET, SPECT DiffDx Alzheimer's disease, multi-infarct–vascular dementia, Pick's disease, diffuse Lewy disease, Wernicke-Korsakoff syndrome, frontal lobe-type dementia, progressive supranuclear palsy, progressive subcortical gliosis, corticobasilar degeneration; other causes of dementia include alcohol, schizophrenia, subdural hematoma, normal pressure hydrocephalus, vitamin B_{12} deficiency and repeated trauma–'punch-drunk' syndrome, torture victims. See Dialysis dementia, Lewy body dementia, Multiinfarct dementia, Pseudodementia, Vascular dementia

DEMENTIA, TREATABLE CAUSES OF

ENDOCRINE & METABOLIC DISEASE Thyroid or parathyroid disease, pituitary/adrenal dysfunction, hepatic encephalopathy, Wilson's disease, chronic renal failure

INFECTIONS Cryptococcal meningitis, neurosyphilis

INTRACRANIAL DISORDERS Hydrocephalic dementia, tumors

VASCULITIS SLE, periarteritis nodosa, temporal arteritis

VITAMIN DEFICIENCY Thiamin, nicotinic acid, folic acid, vitamin B_{12}

Others Drugs and/or toxins, heart failure, 'respiratory' encephalopathy, sarcoidosis, functional psychiatric disorders

Demerol® Meperidine, see there

demineralized bone matrix putty ORTHOPEDICS A product used in spinal, reconstructive, trauma and oral/maxillofacial surgical grafting procedures; it is easily shaped and fills bone cavities, promotes bone growth and may be combined with autogenous BM and bone while maintaining malleability. See Bone paste

demographics EPIDEMIOLOGY Objective characteristics of a population–eg, age, marital status, family size, racial origin, present or prior disease, religion, income, and education. See Patient demographics. Cf Confounding variable, Covariate, Epidemiology

demulcent PHARMACOLOGY A protective, often viscid preparation, used to alleviate irritation of mucous membranes or abraded skin; by providing a rapid cover, the effects of local mechanical, chemical or bacterial irritants are diminished, as is pain and spasms, and drying prevented. See Binding agent

demyelination Demyelinization NEUROLOGY A condition in which nerve trunks are not myelinated *ab initio*–eg, leukodystrophy or which suffer loss after myelinization has been completed–eg, myelinoclasia. See Leukodystrophy

dendritic keratitis OPHTHALMOLOGY Keratitis characterized by linear and arborescent ulcerations on the anterior corneal surface seen in the relatively mild keratitis caused by herpes simplex or herpes zoster. Cf Geographic ulcer

denervated NEUROLOGY Nervelessness; loss of neural connections. See Chemical denervation

denial PSYCHIATRY A primitive–ego defense–mechanism by which a person unconsciously negates the existence of a disease or other stress-producing reality in his environment, by disavowing thoughts, feelings, wishes, needs, or external reality factors that are consciously intolerable. See In denial

Denis classification ORTHOPEDICS Classification of compression fractures of vertebrae TYPES A–superior and inferior endplates; B–superior endplate; C–inferior endplate; D–anterior cortex buckling with intact inferior and superior endplates

denominator EPIDEMIOLOGY The lower part of a fraction used to calculate a rate or ratio; in a rate, the denominator is usually the population–or population experience, as in person-yrs, etc. at risk

dense deposit disease Type II membranoproliferative glomerulonephritis NEPHROLOGY A glomerulopathy in which electron-dense material–usually complement C3 is deposited in the glomerular capillary basement membrane, with ↓ serum C3, due to alternate complement pathway activation, in the face of normal C4; there is patchy mesangial proliferation in Bowman's

capsule, and accumulation of basement membrane material in the peritubular capillaries and arterioles PROGNOSIS Poor. See Tramtrack appearance

density The amount of a substance per unit volume IMAGING 1. The compactness in a scan which reflects the type of tissues seen in CT and MR scans 2. The amount of 'hard' or mineralized tissue in a plain film. See Bone mineral, Current density, Muscle fiber density, Spin density, Vapor density

density dependent 1. Effects whose intensity changes with ↑ population density 2. Effects whose intensity ↑ with ↑ population density

dental amalgam DENTISTRY A filling material that contains up to 50% mercury, silver and other metals. See Alternative dentistry, Fluoridation, Gutta percha, Mercury

dental dam Rubber dam DENTISTRY A thin sheet of rubber latex punctuated with small holes stretched around the crown of the teeth USES Isolate teeth from mucosal secretions during dental procedures; prevent aspiration during oral surgery; protect mucosa during oral sex

dental impaction DENTISTRY The interlocking of teeth, especially of wisdom teeth, in which the crown of a developing/emerging tooth is locked by another crown CLINICAL Pain, displacement of teeth, changed "bite" MANAGEMENT Analgesics, antibiotics, surgical removal

dental implant Osseointegrated implant DENTISTRY A prosthesis anchored in the maxilla or mandible, allowing subsequent placement of artificial teeth. See Cosmetic dentistry

dentatorubral & pallidoluysian atrophy An AD condition CAG repeat-disease characterized by selective destruction of cerebellar neurons CLINICAL Progressive ataxia, choreoathetosis, dystonia, seizures, myoclonus, dementia DiffDx Machado-Joseph disease, spinocerebellar ataxia type 1. See CAG repeat-disease, Trinucleotide repeat disease

dentistry The field of health care dedicated to managing diseases of the teeth and surrounding tissues. See Biogical dentistry, Forensic dentistry. Cf Alternative dentistry, Holistic dentistry, Natural dentistry, Psychic dentistry

dentition Teeth, natural or artificial. See Dental implants, Dentures, Postpermanent teeth

dentures Removable artificial teeth. See Bridge

Denver developmental screening test PSYCHOLOGY A screening test that assesses a child's neurodevelopmental maturation. See Psychological testing

Denver shunt SURGERY A peritoneovenous shunt that relieves malignant ascites and improves renal function COMPLICATIONS Altered coagulation parameters; spontaneous brain hemorrhage after shunt placement

deoxycorticosterone ENDOCRINOLOGY An adrenal steroid with potent mineralocorticoid properties, and minimal glucocorticosteroid properties ↑ IN Congenital adrenal hyperplasia, Cushing syndrome, primary aldosteronism, low renin HTN, adrenal CA ↓ IN Adrenal insufficiency or hypoplasia

11-deoxycortisol Compound S An adrenocortical steroid marker for adrenal 11-β-hydroxylase deficiency ↑ IN Congenital adrenal hyperplasia ↓ IN Adrenal insufficiency or hypoplasia. See Metyrapone test

deoxygenated blood CARDIOLOGY Blood with a low O_2 saturation relative to blood leaving the lungs

deoxypyridinoline BONE DISEASE A crosslink of type I collagen present in bone which is excreted unmetabolized in urine and is a specific marker of bone resorption, and ↑ in Pts with osteoporosis. See Osteoporosis, Pyrilinks-D®

deoxyribonucleic acid See DNA

Depakene® Valproic acid, see there

department head ADMINISTRATION The director of a service—ie a specialty field—eg, surgery, who is often appointed by the chief of the medical staff, whose function is to foster, facilitate, and participate in institutional activities, coordinate departmental activities, maintain the quality of medical care rendered by the department, arbitrate intradepartmental disputes, interact with administration, nursing, and other departments, and serve as a liaison with the community. See Chairperson, Chief of service

dependence PSYCHIATRY A CNS adaptation to the persistent presence of a sedative. See Substance dependence, Tobacco dependence. Cf Addiction SUBSTANCE ABUSE A psychological or physiologic need to use a substance—usually a narcotic on a chronic and repeated basis; dependence on the drug may become overwhelming, compelling the user to sacrifice quality of life for the drug. See Addict. Cf Addiction

dependency needs PSYCHIATRY Vital needs for mothering, love, affection, shelter, protection, security, food, warmth, which may indicate regression when they reappear openly in adults. See Regression

dependent *adjective* PSYCHIATRY See Dependent personality disorder *noun* MANAGED CARE A health plan's spouse and eligible child or other member who meets the applicable eligibility requirements of a Group Benefits Agreement and is enrolled in Plan under its provisions

dependent personality disorder PSYCHIATRY A condition of early adulthood onset, which is characterized by a '...*pervasive and excessive need to be taken care of* (by others) *that leads to submissive and clinging behavior and fears of separation*'

dependent variable EPIDEMIOLOGY An outcome variable/variables that reflects other, independent, variables in the relationship being studied. See Variable

depersonalization PSYCHIATRY A sense of unreality or strangeness vis-á-vis the environment and/or self; a personality disorder in which the Pt thinks that either he or those in his environment have been changed into other people or life-forms; depersonalization classically occurs in schizophrenia, but may also occur in hysteria, depression, drug-induced states, temporal lobe epilepsy, and fatigue. See Derealization, Neurosis. Cf Dehumanization

DEPERSONALIZATION DISORDER–

A Persistent or recurrent sensation of detachment from one's own body, as if in a dream

B During the depersonalization experience, the subject's reality testing remains intact

C The depersonalization results in significant distress or impairment of social, occupational, other function

D The experience does not occur exclusively during the course of another mental disorder

DSM-IV™, American Psychiatric Association, Washington, DC, 1994

depletion The loss or reduction of a thing normally present. See Ozone depletion, Volume depletion

depolarization CARDIAC PACING 1. The sudden change in electrical potential from negative to slightly positive which occurs during phase O of an action potential in an excitable cell membrane—in nerve and heart muscle 2. A rapid alteration of the resting potential in a large mass of tissue—eg, heart ventricles, which usually results in a contraction. See Action potential, Latency period, Resting potential

depolarizing agent ANESTHESIOLOGY An agent—eg, succinylcholine, that depolarizes motor end plates, to maintain airways, control alveolar ventilation, abolish motor reflexes and relax muscle for surgery. See Succinylcholine. Cf Nondepolarizing agent

DepoMorphine™ DepoDur™ PAIN MANAGEMENT A sustained-release DepoFoam™ formulation of morphine used for acute post-surgical pain

Depo-Provera Medroxyprogesterone, Provera® ENDOCRINOLOGY A synthetic progestin that is structurally an androgen; it is a highly effective contracep-

tive, given q 3 months; it is also used for ♀ precocious puberty, DUB, dysmenorrhea, endometriosis, threatened abortion, to suppress postpartum lactation SIDE EFFECTS Weight gain, dysmenorrhea, fatigue, vertigo, nervousness, headaches, abdominal pain. See Androcur; Cyproterone

deposition MEDICAL MALPRACTICE A testimony obtained under oath from a witness or expert outside of court, which can be subsequently used in a criminal or civil action METABOLIC DISEASES The accumulation of crystallizable or nonmetabolizable material in certain organs

depraved heart murder FORENSIC PSYCHIATRY The killing of a person by extreme atrocity, with malicious intent inferred by the nature of the act. See Manslaughter, Murder, Serial murder

Deprax Trazodone, see there

deprenyl Eldepryl, selegiline A selective irreversible inhibitor of type B monoamine oxidase that delays the onset of disability in Pts with early untreated Parkinson's disease—PD. See MPTP

depressant PHARMACOLOGY An agent that depresses neuromuscular activity EXAMPLES Benzodiazepine, chloral hydrate, chlordiazepoxide, chlorpromazine, ethchlorvynol, gamma hydroxybutyrate, methaqualone, phenobarbital, secobarbital

depression Dejection, low spirits PSYCHIATRY A spectrum of affective disorders characterized by attenuation of mood, accompanied by psychogenic pain, diminution of self-esteem, retardation of thought processes, psychomotor sluggishness, disturbances of sleep and appetite, and not uncommonly, suicidal ideation; depression can be triggered by stressful life events, associated with medical or mental disorders, or may be idiopathic CLINICAL Apathy, anorexia, lack of emotion—flat affect, social withdrawal, fatigue TYPES Major depression, dysthymia, bipolar disorder; depression may run in families. See Anaclitic depression, Bipolar disorder, Clinical depression, DART, Depressive disorders, Double depression, Endogenous depression, Inbreeding depression, Major depression, Masked depression, Postoperative depression, Postpartum depression, Reactive depression

DEPRESSION

ATYPICAL DEPRESSION A term retired from the DSM, which some clinicians NEJM 1991; 325: 633 use to refer to combinations of mood reactivity, including anhedonia, overeating, oversleeping, chronic poor self-esteem; those with AD are thought to have a better response to MAOIs

MAJOR DEPRESSIVE DISORDER-RECURRENT A condition defined as

A. 2 or more major depressive episodes—MDE, which is defined as ≥ 5 of the following present during the same 2-week period, and represent a change from previous functioning and at least one of the 5 is either 1. depressed mood or 2. loss of interest

  1. Depressed mood most of the day, nearly every day, as indicated either subjectively—self or by observation of others—eg, tearfulness or in children irritability

  2. Marked decreased interest or pleasure in all or most activities for most of the day, nearly every day for the defining period

  3. Significant–≥5%, unintentional weight loss or weight gain or loss of appetite

  4. Insomnia or hypersomnia nearly every day

  5. Psychomotor agitation or retardation nearly every day

  6. Fatigue or loss of energy nearly every day

  7. Feelings of worthlessness or excessive or inappropriate guilt nearly every day

  8. Decreased ability to concentrate or think nearly every day

  9. Recurrent thoughts of death, recurrent suicidal ideation and/or suicidal plans

B. The MDE is not better explained for by schizoaffective disorder, or is not superimposed on schizophrenia, schizophreniform disorder, delusional disorder, or psychotic disorder NOS

C. There has never been a manic episode or hypomanic episode

MELANCHOLIC DEPRESSION Endogenous depression Characterized by pervasive sadness, hopelessness, loss of interest in activities, and physical symptoms, eg weight loss, sleep problems; in MD, there may be an ↑ 'threshold' that requires little external input to initiate recurrence

REACTIVE DEPRESSION—an excess response to stressful life events

188 **depressive disorder** PSYCHIATRY Any of a number of conditions characterized by one or more depressive episodes–major DD, depressed mood–dysthymic disorder and adjustment disorder with depressed mood, and those that do not fit the criteria of other conditions–aka, DD not otherwise specified

Deprol® Valproic acid, see there

Deproxin Fluoxetine, see there

depth, fear of PSYCHOLOGY Bathophobia. See Phobia

Deptran Doxepin, see there

deQuervain's fracture Typical intercarpal dislocation fraction ORTHOPEDICS A fracture dislocation of the wrist in which the scaphoid is fractured and the proximal fragment, and lunate, is dislocated volarward

derailment NEUROLOGY Speech characterized by abrupt changes in a parallel direction ETIOLOGY Organic brain disease, schizophrenia

derealization PSYCHIATRY An altered and unreal perception of things and objects in space/time, which may be accompanied by depersonalization. See Hallucination

derivation set DECISION-MAKING A group of Pts with a clinical finding of interest–eg, chest pain, studied retrospectively, to identify facets of their disease that might help predict the intensity of observation–eg admission to ICU, appropriate diagnostic workup–lab tests, imaging, and need for therapy. See Validation set. Cf Cohort

DermaBond® DERMATOLOGY A proprietary–Men's Hair Now, Queens, NY–non-medical hair replacement system that glues individual hairs into a goo–Liquid Seal™, applied to the skin. See Hair replacement surgery

dermabrasion DERMATOLOGY A technique in which a dermatome or abrading device is used to remove the epidermis and superficial dermis, allowing regeneration of the epithelium to occur from underlying adnexal structures–eg pilosebaceous unit; dermabrasion is used to treat postacne scarring, scars caused by surgery, trauma, or varicella; it may be used to treat actinic keratosis, nevi, rhinophyma, seborrheic hyperplasia, seborrheic keratosis, solar elastosis, tattoos CONTRAINDICATION Radiodermatitis, as radiation damages adnexal structures and blood supply

Dermagran® WOUND CARE A hydrophilic dressing for removing dead and damaged tissue from wounds used for surgical incisions, skin ulcers, pressure sores, lacerations, cuts, abrasions, diabetic ulcers, partial-thickness and hypothermia burns

dermatitis herpetiformis Brocq-Duhring disease, dermatitis multiformis, Duhring's disease DERMATOLOGY A chronic idiopathic skin disorder characterized by groups of severely pruritic blisters and papules, often associated with gluten-sensitive enteropathy. See Bullous disease

dermatitis Inflammation of the skin. See Allergic contact dermatitis, Atopic dermatitis, Caterpillar dermatitis, 'Club Med,' Contact dermatitis, Diaper dermatitis, Estrogen dermatitis, Herpetic dermatitis, Hot tub dermatitis, Neurodermatitis, Seborrheic dermatitis, Stasis dermatitis

dermatoglyphics DERMATOLOGY 1. The formal study of the patterns of skin ridges on the fingers and toes, palms, and soles, or 2. The patterns of skin ridges on the fingers and toes, palms, and soles. See Simian crease, Triradius

dermatographism Dermographism DERMATOLOGY A wheal-and-flare reaction evoked by firmly stroking the skin

dermatology The specialty focused on diagnosing and treating diseases of the skin and cutaneous manifestations of systemic disease. See Immunodermatology, Sports dermatology

dermatomycosis A fungal skin infection

dermatomyositis A collagen vascular disease caused by complement-mediated microangiopathic muscle fiber destruction, leading to loss of capillaries, muscle ischemia and necrosis and perifascicular atrophy CLINICAL Patches of reddish or scaly rash on bridge of nose, sun-exposed areas of neck, chest, dorsal hands, periorbital edema RISK OF CANCER ↑ MANAGEMENT Azathioprine, cyclophosphamide, cyclosporine, MTX, and prednisone are generally unsuccessful; high-dose IV Ig may ↑ muscle strength and ↓ neuromuscular symptoms in refractory dermatomyositis. See Polymyositis, Scleroderma

dermatopathy Dermopathy Any disease of the skin. See Factitial dermatopathy, Onchocercal dermatopathy

dermatophyte MYCOLOGY A fungus—*Epidermophyton* spp, *Microsporum* spp, *Trichophyton* spp, that primarily causes superficial infections of skin, hair, and fingernails

dermatoplasty DERMATOLOGY Skin repair surgery

dermoid cyst PATHOLOGY A benign teratoma of the ovary characterized by mature tissues, commonly sebaceous material mixed with hair, teeth and skin adnexal structures. See Teratoma

derotation arthroplasty ORTHOPEDICS A type of arthroplasty used to realign deformed–overlapping, underlapping, hammertoes, claw toes, curled toes, etc. See Overlapping toe, Underlapping toe

Deroxat® Paroxetine, see there

DES Diethylstilbestrol GYNECOLOGY A synthetic estrogen more potent than natural estrogens; DES use during pregnancy was banned in 1972, as in utero exposure before the 18th gestational wk results in a 3-fold ↑ in congenital abnormalities–without impairing fertility or sexual function—and vaginal wall adenosis in 35-70% of exposed ♀ infants; 0.14% of cases progress to vaginal adenocarcinoma; other DES changes include obliteration of vaginal fornices, microglandular hyperplasia of the cervix–cervical ectropion, transverse ridging–appearance, and a 2.5-fold ↑ in 1° infertility TREATMENT Aggressive surgery–eg, vaginectomy, hysterectomy and lymphadenectomy. See Cockscomb cervix

descending paralysis NEUROLOGY The opposite of the paralysis seen in botulism and paralytic shellfish poisoning

descending perineum syndrome A condition characterized by inhibition of pelvic floor muscles, which partially overlaps Sx of anismus, which occurs in a background of constipation; DPS may be an end stage of anismus, resulting from denervation. See Anismus

Desconet Diazepam, see there

desensitization therapy Allergy desensitization, immunotherapy IMMUNOLOGY Stimulation of the immune system with gradual ↑ doses of the substances to which a person is allergic–to modify or stop allergic responses, ↓ IgE and its effect on the mast cells; DT is used for allergies to pollen, mites, cats, stinging insects–eg, bees, hornets, yellowjackets, wasps, velvet ants, fire ants

desflurane ANESTHESIOLOGY An inhalation anesthetic, with a rapid onset of action and elimination CONS Requires specialized, heated vaporizers, expensive, produces tachycardia at ↑ dose, ↑ intracranial pressure. See Inhalation anesthetic

desiccation MEDTALK → VOX POPULI Drying

design See Double-masked design, Enrichment design, Experimental design, Masked study design, Open design, Rational molecular design, Single-masked design, Structure-based design, Universal design, Zelen design

designated driver PUBLIC HEALTH A person at a social function who volunteers, or is 'volunteered' to chauffeur inebriated revellers chez elles at festivity's end. Cf Squash it

designer drug SUBSTANCE ABUSE A substance of abuse synthesized or manufactured to give the same subjective effects as well known illicit drugs, produced in a clandestine lab; it is generally difficult to locate the manufacturer to identify adverse effects. See Adam, 'Ice'

designer estrogen SERM, see there

desipramine Norpramin®, Pertofran THERAPEUTICS An imipramine-like tricyclic antidepressant. See Imipramine

desirable weight NUTRITION The optimal weight for a particular height. See Diet, Healthy weight. Cf Obesity, Overweight

Desirel Trazodone, see there

deskercise Exercises performed while seated at the workplace, suited for those whose jobs require them to sit for prolonged periods–eg, at a computer workstation; deskercises stretch and loosen the muscles of the upper body, including the back, neck, arms, fingers, shoulders

Desloneg Diazepam, see there

desmoid Aggressive fibromatosis, desmoid tumor, musculoaponeurotic fibromatosis A nonencapsulated mass on the anterior abdominal wall, most commonly in postpartum ♀, often 2° to trauma which, according to anecdotal reports, may regress with local progesterone injection

desquamation MEDTALK The sloughing of a mucocutaneous surface. See Potato chip desquamation

desquamative interstitial pneumonitis A nonspecific interstitial pulmonary reaction of adults; DIP is often idiopathic but may be associated with inhalation of inorganic particles RADIOLOGY Bilateral ground-glass opacifications TREATMENT Corticosteroids. See Diffuse alveolar damage

desulfatohirudin Hirudin, see there

Desyrel® Trazodone, see there

detachment PSYCHIATRY A behavior pattern characterized by general aloofness in personal interactions; may include intellectualization, denial, and superficiality

detailing MANAGED CARE An educational activity by sales representatives–'detailers'–eg, from pharmaceutical companies or manufacturers of medical devices, to provide details or scientific information on a product's potential uses, benefits, side/adverse effects. See Project house

detection See Cancer detection

determinant EPIDEMIOLOGY A factor, event, characteristic, or other definable entity that brings about change in a health condition, or in other defined characteristics IMMUNOLOGY Antibody determinants. See Antigenic determinant, Public idiotype determinant

detoxication therapy INTERNAL MEDICINE 1. The removal of a toxic excess of any agent–eg, a therapeutic agent, drug of abuse or toxic agent–eg, pesticide, heavy metal or venom by induction of vomiting, administration of activated charcoal, hemodialysis, peritoneal dialysis or use of metabolic interference–eg, treating methanol intoxication with ethanol overloading 2. Medically supervised withdrawal from a substance of abuse and treatment of Sx of withdrawal. See Activated charcoal, 'Cold turkey', Detoxification therapy, Heat stress detoxification

deuteranopia OPHTHALMOLOGY Red-green color blindness caused by a lack of middle wavelength–530 nm photopigment. See Color blindness

development gap The time period between the development of products and processes with commercial potential in academia and basic research, and their use by industry in the form of licensing agreements, spin-off companies, and products in the marketplace

developmental delay Developmental disability NEUROLOGY A disability that affects a person's development, such as, mental retardation, epilepsy, autism, cerebral palsy or similar disability PEDIATRICS A lag in reaching developmental milestones by the expected age TYPES Biological–eg, chromosomal defects or in utero infection, environmental–eg, maternal mental malady or marital malaise, or 'transactional'–ie an interplay between biological and environmental factors DIAGNOSIS Developmental screening tests to identify developmental red flags. See Developmental milestone, Developmental red flag

developmental disorder PSYCHIATRY An impairment in normal development of language, motor, cognitive and/or motor skills, generally recognized before age 18 which is expected to continue indefinitely and constitutes a substantial impairment ETIOLOGY Mental retardation, cerebral palsy, epilepsy, other neurologic conditions–eg, autism

developmental milestone PEDIATRICS Any of a series of activities, eg, raising the head, rolling over, walking or other significant points in a child's physical and/or mental development that may be used to assess maturation and detect developmental delays. See Developmental red flag

developmental 'red flag' PEDIATRICS An objective finding that indicates a delay in achieving developmental milestones; DRFs in assessing infants, toddlers, and preschoolers fall into 5 major areas:

DEVELOPMENTAL RED FLAG

GROSS MOTOR SKILLS, eg does not roll over–5 months, can't hop–4 years

FINE MOTOR SKILLS, eg doesn't hold rattle–4-5 months, can't copy a circle–4 years

LANGUAGE SKILLS, eg not babbling–5-6 months, doesn't understand prepositions–4 years

COGNITIVE SKILLS, eg does not search for dropped object–6-7 months, doesn't know colors or any letters–5 years

PSYCHOSOCIAL DEVELOPMENT, eg does not smile socially–5 months, in constant motion, resists discipline, does not play with other children–3-5 years NEJM 1994; 330:478cc. See Developmental milestone

developmental screening test PSYCHOLOGY A test or questionnaire used to evaluate a child's achievement of developmental milestones. See Psychological testing

deviancy VOX POPULI A major abnormality, usually understood to be mental. See Paraphilia, Sexual deviancy

deviation VOX POPULI A departure from a norm. See Septal deviation, Standard deviation

device MEDTALK An instrument designed to perform a particular task THERAPEUTICS An instrument, apparatus, or contrivance, including components, parts, and accessories, intended for use in the diagnosis, cure, mitigation, treatment, or prevention of disease in man or animals or to affect the structure or any function of the body of man or animals. See Medical device

devitalization DENTISTRY The destruction of the pulp in tooth root

dewdrop appearance Dewdrop on a rose petal appearance PEDIATRICS A fanciful description of the appearance of fresh vesicles of chickenpox, see there

DEXA scan Dual energy X-ray absorptiometry IMAGING An imaging system to assess bone mineral density; commonly used to screen perimenopausal and menopausal ♀ before beginning HRT, to evaluate Pts with 1° or 2° osteoporosis or metabolic diseases affecting the skeleton and monitor treatment and progression of osteoporosis. See Bone mineral density, Hormone replacement therapy, Osteoporosis

dexamethasone suppression test ENDOCRINOLOGY A clinical test that measures the ability of dexamethasone to suppress ACTH and cortisol secretion by the adrenal gland. See Cushing's disease

Dexedrine® PHARMACOLOGY An amphetamine used to manage narcolepsy, ADD and possibly AIDS-related fatigue. See AIDS

dexfenfluramine Redux® CLINICAL PHARMACOLOGY An appetite suppressing amphetamine analogue that acts centrally–ie CNS as an appetite suppressant to ↓ hunger sensation. See Obesity

dexmedetomidine ANESTHESIOLOGY A selective α_2-adrenoceptor agonist, which ↓ the dose of isoflurane required for anesthesia during abdominal hysterectomy. See Isoflurane

dexrazoxone Zinecard® CARDIOLOGY A bispiperazinedione used to protect against dose-dependent cardiotoxicity by doxorubicin-induced intracellular complexing with iron and formation of O_2 free radicals. See Adriamycin

dextran TRANSFUSION MEDICINE Dextran-40, dextran-70, dextran-1 A colloid-type volume expander consisting of a large glycogen-like molecules which may occasionally be used in surgical blood management by hemodilution; these substances have the desired properties of being viscid, and gelatinous, resulting in oncotic pressure to retain fluids in vessels; they are widely used as replacement fluids and volume expanders PROS ↓ Allogeneic transfusions, ↓ postoperative bleeding, ↓ blood viscosity CONS Interferes with platelet and RBC function, crossmatching; may cause anaphylaxis and peripheral edema. See Colloid solutions, Crystalloids, Hemodilution, Surgical blood management

dextro– Latin dexter, right side

dextrocardia A heart misplaced in the right chest, with the apex of the heart pointing right; the cardiac function is normal if there is concomitant situs inversus of the abdominal organs–mirror-image dextrocardia; if the dextrocardia is due to immotile cilia, sinusitis and bronchiectasis–Kartagener syndrome may also be present EKG Mirror-image electrical activity with P, QRS, and T waves in leads I, aVR, aVL that are the reverse of the normal; the right activity resembles that normally seen in the left side of the chest; if the abdominal organs are not reversed, dextrocardia may be associated with ventricular inversion, single-chamber ventricle, pulmonary valve stenosis and anomalies of venous return. See Kartagener's syndrome. Cf Dextroposition of the heart

dextromethorphan DXM, poor man's Ecstasy SUBSTANCE ABUSE A opiate in some cough syrups which, in megadoses required to get the desired mind-altering effect, has many of the effects of Ecstasy–eg euphoria, hallucinations ROUTE Oral PHARMACOLOGIC EFFECTS ↓ Cough-center response; mimics codeine's effects without analgesia USES Suppresses nonproductive cough TOXICITY Respiratory depression, coma, death. See Ecstasy, Roboing

dextroposition of the heart PEDIATRIC CARDIOLOGY A congenital anomaly in which an anatomically correct heart is displaced to the right in the thoracic cavity, due to a shift of left intrathoracic content, or ↓ in right thoracic cavity contents–eg, by right lung collapse. Cf Dextrocardia

DHEA Dehydroepiandrosterone, see there

DHFR Dihydrofolate reductase, see there

DHT Dihydrotestosterone, see there

di- Greek, two

diabesity A popular term for the common clinical association of type 2 DM–adult-onset DM and obesity, a subgroup of which has been termed 'syndrome X', see there

diabetes See Diabetes insipidus, Diabetes mellitus

diabetes insipidus ENDOCRINOLOGY A condition characterized by defective water homeostasis due to a defect in ADH secretion, resulting in inadequate reabsorption of water by the kidney tubules and excretion of a large volume of dilute urine TYPES Central–due to a defect in the neurohypophysis (posterior pituitary) and/or hypothalamus; peripheral/nephrogenic–defect in the renal tubule

diabetes mellitus ENDOCRINOLOGY A chronic condition which affects ±10% of the general population, characterized by ↑ serum glucose and a relative or absolute ↓ in pancreatic insulin production, or ↓ tissue responsiveness to insulin; if not properly controlled, the excess glucose damages blood vessels of the eyes, kidneys, nerves, heart TYPES Insulin dependent–type I and non-insulin dependent–type II diabetes SYMPTOMS type 1 DM is associated with ↑ urine output, thirst, fatigue, and weight loss (despite an ↑ appetite), N&V; type 2 DM is associated with, in addition, non-healing ulcers, oral and bladder infections, blurred vision, paresthesias in the hands and feet, and itching CARDIOVASCULAR MI, stoke EYES Retinal damage, blindness LEGS/FEET Nonhealing ulcers, cuts leading to gangrene and amputation KIDNEYS HTN, renal failure NEUROLOGY Paresthesias, neuropathy DIAGNOSIS Serum glucose above cut-off points after meals or when fasting; once therapy is begun, serum levels of glycosylated Hb are measured periodically to assess adequacy of glucose control MANAGEMENT Therapy reflects type of DM; metformin and triglitazone have equal and additive effects on glycemic control PROGNOSIS A function of stringency of glucose control and presence of complications. See ABCD Trial, Brittle diabetes, Bronze diabetes, Chemical diabetes, Gestational diabetes, Insulin-dependent diabetes, Metformin, MODY diabetes, Nephrogenic diabetes insipidus, Non-insulin-dependent diabetes mellitus, Pseudodiabetes, Secondary diabetes, Starvation diabetes, Troglitazone

DIABETES MELLITUS–TYPE 1 VS TYPE 2		
FINDING	**TYPE 1**	**TYPE 2**
% OF DIABETICS	10%	90%
AGE OF ONSET	Usually < 35	Usually > 40
WEIGHT	Not overweight	Overweight
SPEED OF ONSET	Often abrupt/acute	Asymptomatic, slower onset
CLINICAL FINDINGS	↑ Thirst, urine production appetite; rapid weight loss fatigue	Poorly healing cuts, paresthesias of hands/feet; recurring skin, oral, infections
LAB FINDINGS	Ketonuria	

www.diabetes-symptoms-resource.com/type-1-diabetes-mellitus.htm

diabetes, type 1 Type 1 diabetes mellitus, see there, aka insulin dependent diabetes mellitus, juvenile-onset diabetes mellitus

diabetes, type 2 Type 2 diabetes mellitus, see there, aka non-insulin dependent diabetes mellitus, adult-onset diabetes mellitus, insulin-resistant diabetes mellitus

diabetic angiopathy Vascular disease in Pts with long-term DM. See Diabetic macroangiopathy, Diabetic microangiopathy

diabetic embryopathy A complex of congenital anomalies that affect ± 10% of children born to diabetic mothers, which develops in the 5th-8th wk of gestation CLINICAL Sacral agenesis, holoprosencephaly, atrioventricular septal defects, tetralogy of Fallot, other cardiac, genitourinary, and GI defects. See Honeybee syndrome

diabetic foot A foot with a constellation of pathologic changes affecting the lower extremity in diabetics, often leading to amputation and/or death due to complications; the common initial lesion leading to amputation is a non-healing skin ulcer, induced by regional pressure, pathogenically linked to sensory neuropathy, ischemia, infection

diabetic ketoacidosis A hyperglycemia-induced clinical crisis most common in type 1 DM CLINICAL N&V, thirst, diaphoresis, hyperpnea, drowsiness, fever, prostration, coma, possibly death LAB ↑↑↑ Glucose, often > 33.6 mmol/L–US: > 600 mg/dL, ↑ ketone bodies, relative ↑ in protein, albumin, Ca^{2+}, BR, alk phos, AST, CK, anion gap, acidosis, dehydration, ↓ K⁺, Na⁺, phosphate MANAGEMENT Insulin, fluid and electrolyte replacement, treat initiating factors–eg, leukocytosis or hypothermia, avoid complications–eg, hypokalemia, late hypoglycemia

diabetic macroangiopathy Diabetic angiopathy A finding in long-standing DM, characterized by accumulation of lipids and blood clots within large blood vessel walls, obstructing blood flow

diabetic microangiopathy Microvascular disease Any clinical or pathological changes resulting from small vessel disease in PTs with DM MEASUREMENT Capillary HTN can be measured directly by microcannulation of nailfold capillaries with a glass micropipette PROGNOSIS Progression can be slowed by tight control of serum glucose levels, especially with long-term intense insulin therapy

diabetic nephropathy Diabetic kidney disease The constellation of renal changes attributed to DM–eg Armanni-Ebstein lesion, arterionephrosclerosis, arteriolonephrosclerosis, chronic interstitial nephritis, diabetic glomerulosclerosis, fatty changes in renal tubules, glomerulonephritis, Kimmelstiel-Wilson disease—focal and segmental glomerulosclerosis, nephrotic syndrome, papillary necrosis, and pyelonephritis; DN is the most common cause of ESRD in the West DIAGNOSIS Microalbuminuria MANAGEMENT Antihypertensives–eg, ACE inhibitors–eg, captopril, protect kidneys against further deterioration in type 1 DM, and a 50% ↓ risk in end points–death, dialysis, and transplantation; if renal failure is in an early stage, the Pts are good transplant candidates; ESRD requires dialysis and protein restriction. See End-stage renal disease

diabetic neuropathy Diabetic nerve damage NEUROLOGY A neuropathy that affects up to 50% of Pts with DM with slowing of nerve conduction and/or Sx of neuropathy–eg, distal, bilateral, usually symmetrical often sensory polyneuropathy with hyperesthesias of the hands and feet, focal and multifocal neuropathy, and trophic changes of extremities–eg, hair loss, thin skin, disorders of sweating, and sensation of cold FORMS OF DN Diabetic ophthalmoplegia, thoracoabdominal radiculopathy, acute mononeuropathy, mononeuropathy multiplex–a painful, asymmetrical neuropathy TREATMENT None

diabetic retinopathy OPHTHALMOLOGY A condition characterized by progressive deterioration of the retina with microaneurysms, hemorrhage, neovascularization; it is the major cause of visual impairment in the US < age 60; at 20 yrs, 40% of Pts with type 1 DM and 5% of those with type 2 DM have proliferative DR; clinically significant macular edema occurs in 10–15% of Pts with DM for > 15 yrs MANAGEMENT Glycemic control, photocoagulation, vitrectomy

diagnosis DECISION-MAKING The process of determining, through examination and analysis, the nature of a Pt's illness; the process of identifying a disease by signs and symptoms; the label for a particular condition. See Computer-assisted diagnosis, Differential diagnosis, Deferred diagnosis, Definitive diagnosis, Direct diagnosis, Electrodiagnosis, Indirect diagnosis, Leading diagnosis, Misdiagnosis, Primary diagnosis, Secondary diagnosis, Wastebasket diagnosis, Working diagnosis

diagnosis of exclusion DECISION-MAKING A disease or clinical nosology that is extremely rare, and often unresponsive to therapy, the diagnosis of which is seriously considered only when all other possible–potentially treatable conditions–eg 'growing pains' or idiopathic midline granuloma, have been completely excluded. Cf Working diagnosis

diagnosis-related group MANAGED CARE A prospective payment system used by Medicare and other insurers to classify illnesses according to diagnosis and treatment; DRGs are used to group all charges for hospital inpatient services into a single 'bundle' for payment purposes. See DRGs

diagnostic center MEDICAL PRACTICE A place that offers diagnostic services to the medical profession or general public. See Imaging center

diagnostic evaluation Workup MEDTALK An evaluation used to diagnose disease COMPONENTS Medical Hx, CXR or other images, collection of specimens from blood for lab analysis

diagnostic overkill CLINICAL DECISION-MAKING The use of excess or overlapping tests that provide virtually no additional diagnostic information–eg, an MRI of the brain when a prior CT identified an intracranial mass. Cf Defensive medicine

Diagnostic & Statistical Manual PSYCHIATRY The standard reference for diagnosing psychiatric disorders; published by the Am Psychiatric Assn, the DSM presents theoretical definitions of disorders; the DSM describes biological, psychological, and social aspects of various conditions. See DSM-IV

diagnostic-therapeutic pair CLINICAL MEDICINE The performing of a specific therapeutic intervention after an abnormality is found by a particular diagnostic procedure

diagnostic yield The likelihood that a test or procedure will provide the information needed to establish a diagnosis. See Clinical decision making

dialect SOCIOLOGY A sublanguage system spoken in a region or by a particular group of people. See Ebonics. Cf Jargon, Slang

dialysis NEPHROLOGY The separation of molecules in solution based on differences in size; in renal failure, dialysis is used to separate macromolecules from low-molecular-weight molecules, using a semipermeable membrane THERAPEUTICS The clearance of a drug by a hemodialysis unit. See Hemodialysis, Peritoneal dialysis

dialysis-associated peritonitis Dialysis ascites NEPHROLOGY Exudative peritonitis related to chronic dialysis, attributed to bacteria–especially pneumococci and staphylococci introduced during dialysis MANAGEMENT The LeVeen shunt may be effective in treating DA

dialysis dementia Dialysis encephalopathy NEPHROLOGY Aluminum toxicity affecting Pts with ESRD, who require long-term dialysis; aluminum is present in the phosphate binders in the dialysate solutions, and in the oral $Al(OH)_3$ required to control terminal renal failure, and accumulates in the brain CLINICAL DD is defined by ≥ 2 of following neurologic signs: stuttering, stammering, dysnomia, hypofluency, mutism, seizures–generalized tonic-clonic, focal, or multifocal, or motor disturbances–myoclonic jerks, motor apraxia or immobility, EEG–presence of bilateral synchronous spike and wave activity; other changes include speech defects, seizures, mental deterioration, bone pain, osteolysis–fractures, pseudofractures, microcytic anemia, porphyria cutanea tarda, and delta waves by EEG–similar to those of metabolic encephalopathy MANAGEMENT Chelation–deferoxamine; recognition of the association between aluminum and dialysis has led to the ↓ in frequency of DD

Diamond-Blackfan syndrome Congenital hypoplastic anemia HEMATOLOGY An AR, occasionally consanguineous condition associated with pure red cell aplasia of early infancy onset with anemia, pallor, failure to thrive; 30% have minor physical abnormalities–eg short stature, thumb deformities, ocular changes TREATMENT Transfusion, corticosteroids, BMT, growth factor therapy

Diamox® Acetazolamide, see there

Diana complex PSYCHIATRY A reversal of roles by ♀ affecting the appearance, dress, and mien of men. See Delilah syndrome, Wild woman phenotype

Diapam Diazepam, see there

diaper rash Diaper dermatitis PEDIATRICS A dermatopathy of infancy caused by prolonged contact with soiled diapers, soaps, or topical lotions, resulting in maceration of a skin that is scaly and erythematous with papulovesicules or bullae; DD may be extensive, but often spares crural folds; 2° bacterial and yeast–especially *Candida* spp, infections may complicate recalcitrant cases; chronic DD may undergo papular induration

diaphoresis MEDTALK Sweat, ↑ sweat

diaphragm GYNECOLOGY A barrier contraceptive consisting of a thin flexible rubber disk that covers the uterine cervix to prevent the entry of sperm during sexual intercourse

diaphragmatic hernia A common–1:2000 live births congenital defect in which a loop of bowel passes through the diaphragm, due to a defective closure of the pleuroperitoneal membrane–through the foramen of Bochdalek, resulting in a common cavity, with the abdominal organs–usually left-sided prolapsing into the chest cavity, which compromises respiration; permutations of DH may be discovered later in life–eg, parasternal hernias or membranous defects; less common sites of DH–through esophageal hiatus or Morgagni's foramen TREATMENT Surgical repair, ventilatory support and bicarbonates MORTALITY 40%

diarrhea An abnormal frequency of defecation accompanied by abnormal liquidity of the feces; a daily stool weight of > 200 g; acute diarrhea is < 4 wks; chronic diarrhea is > 4 wks in duration. See Antibiotic-associated diarrhea, Brainerd diarrhea, Chewing gum diarrhea, Chronic idiopathic diarrhea, Day care diarrhea, High-output diarrhea, Magnesium-induced diarrhea, Osmotic diarrhea, Overflow diarrhea, Toddler's diarrhea, Traveler's diarrhea, Trench diarrhea, Weaning diarrhea, White stool

diarrhea of undetermined origin An uncommon diagnosis of exclusion; most DUOs are due to laxative abuse, IBD—irritable bowel syndrome, anal sphincter dysfunction, and bacterial overgrowth; DUO may require an inpatient workup. See Diarrhea

diastole CARDIOLOGY The dilatory–relaxation phase of the cardiac cycle during which the heart's chambers fill with blood

diastolic collapse *y* descent CARDIAC PHYSIOLOGY A normal finding in the JVP which is produced when tricuspid valve opens and there is a rapid inflow of blood into the right ventricle. See Jugular venous pulse

diastolic heart failure CARDIOLOGY Heart failure with preserved left ventricular systolic function–LV ejection fraction of ≥ 50%, no segmental wall motion abnormalities, and no evidence of significant coronary, valvular, infiltrative, pericardial, or pulmonary disease; although DHF is less severe than typical systolic HF in terms of pathophysiologic characteristics, both have severely reduced exercise capacity, neuroendocrine activation and impaired quality of life. See Heart failure. Cf Systolic heart failure

diathermy SPORTS MEDICINE The use of high-frequency electromagnetic waves to ↑ temperature of deep tissues due to resistance to the passage of energy TYPES Microwave diathermy, shortwave diathermy. See Microwave diathermy, Shortwave diathermy SURGERY Cauterization, see there

Diatran Diazepam, see there

diatrizoate IMAGING A high-osmolality–up to 7 times plasma osmolality–normally ±300 mosmol/kg compound used as intravascular radiographic contrast media. Cf Low-osmolality contrast media

Diazebrum Chlordiazepoxide, see there

diazepam Valium® PHARMACOLOGY A class V benzodiazepine muscle relaxant, sedative, hypnotic, anxiolytic, anticonvulsant, sometimes for panic disorders ADVERSE EFFECTS Physical, psychological dependence. See Benzodiazepine, Little Yellow Pill

diaziquone ONCOLOGY An anticancer drug that crosses the blood-brain barrier

DIC Disseminated intravascular coagulation, see there

dicing TRAUMA SURGERY Multiple 0.5-1.0 cm, cube-like lacerations of the skin seen in MVA victims who strike shattered tempered glass car windows

dictionary A book, database document that contains multiple definitions, which may be on a particular subject, usually in alphabetial order

dicumerol A polycyclic aromatic compound from sweet clover that is an anticoagulant analogue of vitamin K, which uncouples oxidative phosphorylation. See Warfarin

didanosine 2',3'-dideoxyinosine, ddI, Videx AIDS A purine analogue that inhibits HIV-1's reverse transcriptase, which may be more effective than zidovudine and may be an anti-retroviral nucleoside of first choice– ddI is associated with an ↑ in CD4–T-helper cells, and ↓ p24 antigen–an indicator of HIV in the blood; ddI is better tolerated than zidovudine and causes less myelosuppression SIDE EFFECTS Peripheral neuropathy, pancreatitis LAB Hyperuricemia, ↑ aminotransferases. See AIDS, Zidovudine

dideoxynucleoside AIDS Any of a family of compounds eg, zidovudine, didanosine, and zalcitabine used to treat HIV-positive Pts. See ddC, Didanosine

Didronel® Etidronate, see there

DIEP flap procedure Deep inferior epigastric perforator flap procedure SURGERY A technique of post-mastectomy breast reconstruction, in which abdominal skin, fat, and vasculature–but not the rectus abdominalis muscle is rotated up to the breast region to act as 'filler' PROS ↓ Hospital time, improved healing, improved cosmetic result, no risk of anterior hernia due to a loss of rectus muscle; none of the risks of breast implants

diet NUTRITION Eating and drinking either sparingly or according to a prescribed regimen; diets are either for supplementation–ie weight gain or restriction–ie weight loss; in restrictive diets, the intent is to limit one or more dietary components–eg, gluten or oxalate, or to globally ↓ caloric intake

diet pill DRUG SLANG A euphemism for an amphetamine VOX POPULI An agent that either ↓ appetite or ↑ basal metabolic rate–eg, amphetamines–by prescription and OTC diet aids–eg phenylpropanolamine, ephedrine, caffeine; in high doses, DPs may cause marked agitation, HTN, seizures, rarely death due to cerebral hemorrhage. See Artificial sweeteners, Diet, Phenolpropanolamine, Starch blocker, Sugar blocker

dietary fiber NUTRITION Indigestible plant-derived residues composed predominantly of cellulose, hemicellulose, and cell wall polymers; ↑ DF is associated with ↓ colonic malignancy and tumor regression in premalignant familial adenomatous polyposis and diverticulosis; ↓ DF intake is linked to colorectal CA, diverticulitis, ↑ cholesterol, gallbladder disease, constipation, appendicitis. See Bran, Fiber, Healthy food, Oat bran, Pectin, Soluble fiber

dietary guidelines CARDIOLOGY A series of dietary recommendations from the Nutrition Committee of the Am Heart Assn, that promote cardiovascular health. See Caloric restriction, food pyramid, French paradox

DIETARY GUIDELINES-AMERICAN HEART ASSOCIATION

1. Fat should comprise < 30% of total calories
2. Saturated fat should comprise < 10% of total calories
3. Polyunsaturated fat consumption should be < 300 mg/day
4. Carbohydrates (especially complex type) should constitute $\frac{1}{2}$ of calories in diet
5. Protein constitutes the remainder, ie 100% − (1% + 4%)
6. Sodium should be < 3 g/day
7. Alcohol consumption should be ≤ 60 g (2 oz)/day
8. Calories should be sufficient to maintain the body weight
9. A wide variety of food should be consumed

dietary mineral Minerals required for optimal functioning of the body; dietary requirements for minerals range from molar to trace amounts/day; some–eg, nickel, tin, and vanadium, may be required by some plants or animals, but are not known to have a role in humans

diethylstilbestrol See DES

dietician Nutritionist A health professional with specialized training in diet and nutrition

Dieulafoy ulcer Dieulafoy's vascular malformation GI DISEASE A vascular malformation of the stomach which may present as a bleeding ulcer

differential Differential diagnosis, see there

differential count Diff, white blood cell differential count HEMATOLOGY The relative number of leukocytes–eg segmented and band forms of granulocytes, eosinophils, lymphocytes and monocytes in the peripheral circulation, expressed in percentages of the total WBC count. See White cell count

differential diagnosis 1. A list of conditions that may cause a particular clinical sign or symptom 2. The arrival at a diagnosis by means of comparing the similarities and differences in various clinical signs

differentiation ONCOLOGY The degree to which tumor cells resemble normal cells; differentiated cells grow more slowly than undifferentiated tumor cells. See Dedifferentiation

differentiation therapy ONCOLOGY A strategy used to treat malignancies with a block in the normal cell differentiation; the intent with DT is to drive malignant cells into a mature nonproliferating state of remission

difficult patient MEDTALK A Pt who is troublesome, tedious, or noncompliant with physician instructions. See Abandonment, Ama, Contributory negligence. Cf 'Good' Pt

diffuse alveolar damage DAD The histologic findings in ARDS, which is characterized by an acute onset of diffuse pulmonary infiltrates ETIOLOGY AIDS, air embolism, cardiopulmonary bypass, connective tissue disease–SLE, rheumatoid arthritis, scleroderma, dermatomyositis, drugs–therapeutic–eg, bleomycin, busulfan, cytoxan, MTX, nitrofurantoin or drugs of abuse, eosinophilic granuloma, heat injury, hemosiderosis, high altitude, iatrogenic–PEEP, infections–viruses–eg, herpes, CMV; protozoans–eg, toxoplasma, pneumocystis, molar pregnancy, noxious fumes–eg beryllium, cadmium, mercury, zinc, toxins–eg ammonia, kerosene, paraquat, phosgene, or gases–eg CO_2, NO_2, acute pancreatitis, shock, uremia. See Adult respiratory distress syndrome

diffuse axonal injury NEUROLOGY A form of post-traumatic brain damage which results in significant neurologic sequelae in survivors. See Retraction balls

diffuse hypertrophy–breast Virginal hypertrophy A cause of breast enlargement in periadolescence, characterized by symmetrical mammary enlargement, which may be significant; in most Pts there is rapid growth over several months which begins near menarche, after which there is no further ↑ in size

diffuse ischemic injury NEUROLOGY A pathologic change in the brains of stroke victims who had an ischemic episode of sufficient severity to damage neurons *and* the supporting vessels and glia, resulting in a change in the brain's water content and a lesion that is detectable by CT or MRI. See CVA, TIA. Cf Ischemic neuronal damage

diffuse large B-cell lymphoma ONCOLOGY A B-cell lymphoma that is the most common type–accounting for 30-40%–of NHL, which occurs in children and adults. See Lymphoma, Non-Hodgkin's lymphoma, WHO classification

diffuse lymphoma A lymphoma composed of sheets of cells without germinal centers, which often spill into adjacent adipose and other tissues; DLs are often more aggressive than follicular lymphomas. See Lymphoma

diffusely positive 'review of systems' MEDTALK A Pt who reports findings or complaints in each system of the body during physical exam

diffusing capacity PULMONARY MEDICINE A measure of a substance's efficiency in transversing a particular barrier, which in the lungs corresponds to the ability of gases in the alveolar space to enter the blood, and of the gases in the blood to enter the alveoli for removal from the body by exhalation. See Pulmonary function test

diffusion chamber THERAPEUTICS A biohybrid device in which cells of interest–pancreatic islet cells–are placed in chamber and implanted via trocar or incision in the peritoneum or subcutis; the DC is like the earlier perfusion chamber, and the currently preferred microsphere, a type of encapsulated cell therapy, which in some cases, achieves adequate glucose control. See Artificial pancreas, Biohybrid artificial pancreas

diffusion-weighted imaging IMAGING A type of MRI that provides metabolic and hemodynamic information of the brain shortly after a stroke. See Magnetic resonance imaging. Cf Perfusion-weighted imaging

Diflucan® Fluconazole, see there

DiGeorge syndrome Hypoplasia of thymus & parathyroids, 3rd & 4th pharyngeal pouch syndrome A disorder characterized by 1. low blood calcium levels–hypocalcemia due to underdevelopment–hypoplasia of the parathyroid glands needed to control calcium; 2. underdevelopment–hypoplasia of the thymus, an organ behind the sternum in which lymphocytes mature and multiply; and 3. defects involving the outflow tracts from the heart; most cases of DGS are due to a very small deletion–microdeletion in chromosome band 22q11.2; a small number of cases have defects in other chromosomes, notably 10p13; named for the US pediatric endocrinologist Angelo DiGeorge

digestive aid A substance–eg protein, enzyme–bromelain, pancreolipase, papain, betaine, lecithin, ox bile, which is said to help digest foods. See Diet

digital *adjective* Pertaining to computer-based technology and data

digital autoeroticism Masturbation, see there

digital certificate Digital ID INFORMATICS An official electronic identity document based on public/private key encryption and obtained through a certificate authority; includes user's name and registered serial number, user's public key and expiration date. See HIPAA. Cf Digital signature

digital imaging INFORMATICS The creation of computerized image files and transferring them to computers located elsewhere in a network or across the internet. See JPEG

digital literacy INFORMATICS The ability to understand computer-based information. See Literacy

digital mammography IMAGING The capture of mammographic images on a digital grid PROS ↑ resolution and clarity than conventional mammography; DM is of use as a screening technique, and allows faster, earlier, and more accurate detection of early breast abnormalities

digital radiography IMAGING A format for producing x-rays in which film used to produce conventional x-ray images is replaced with more sensitive sensitive electronics; DXRs produce images with ½ the amount of radiation, and the images are easier to store, copy and send, as they are digital ab initio

digital rectal examination The insertion of a gloved & lubricated index finger per rectum by an examiner to palpate the prostate–and detect ↑ firmness, and anorectal mucosa, given its efficacy as a screening tool for identifying lower rectal and prostatic lesions–BPH or CA

digital subtraction angiography RADIOLOGY A diagnostic technique that uses video equipment and computers to enhance images obtained with conventional angiography

digitalis CARDIOLOGY A cardiac glycoside first found in foxglove, *Digitalis purpurea*, now of historic interest; the synthetic derivatives, digoxin and digitoxin are the most popular of the cardiac glycosides

digitalis toxicity Digoxin toxicity CARDIOLOGY Clinical findings of digoxin overdose CLINICAL Loss of appetite, N&V, defects in color vision–reds and greens, or seeing halos around lights, psychotic changes, weakness, fatigue, or dizziness; new onset of arrhythmias typical of DT include frequent or multiform PVCs, ventricular tachycardia, atrial tachycardia with block, accelerated junctional rhythms, Wenckebach rhythms, and A Fib with slowed ventricular responses, hyperkalemia MANAGEMENT Discontinue digoxin, monitor arrhythmias with telemetry, correct acid-base, electrolyte, and volume abnormalities; treat medical conditions–eg, hypoxia, ischemia, and arrhythmias; arrhythmias are common and often respond to lidocaine and/or dilantin; in extreme emergencies, digoxin antibody fragments, which bind the active portion of the digoxin molecule, may be needed to ↓ DT

digitalization CARDIOLOGY Treatment with digitalis or related cardiac glycosides. See Digitalis

digitoxin CARDIOLOGY A cardiac glycoside used like digoxin, which binds more strongly to proteins, but for a similar pharmacologic effect, requires a 10-fold greater concentration

digoxin CARDIOLOGY The most widely used cardiac glycoside, used primarily for CHF, and less commonly atrial arrhythmias; digoxin is often the first drug used in new-onset A Fib, as it↑ parasympathetic tone, which explains digoxin's relative inefficacy in treating rapid A Fib, as rapid A Fib is often a manifestation of ↑ SNS tone, which commonly occurs in acute illness and is associated with ↑ catecholamines; CCBs–eg, verapamil, diltiazem may be more effective than digoxin for slowing of very rapid A Fib especially with an overactive SNS. See Digitalis toxicity

dihydrotestosterone Dehydrotestosterone PHYSIOLOGY A potent C19 androgen produced from testosterone, which is the predominant androgen in skin, prostate, seminal vesicles, epididymis ↑ IN Hirsutism, polycystic ovary syndrome, ↑ 5-alpha reductase activity ↓ IN Hypogonadism, ↓ 5-alpha reductase activity due to drug interactions

dilatation catheter CARDIOLOGY A device used when balloon dilating a stenosed coronary artery or bypass graft stenosis to improve myocardial perfusion. See Balloon angioplasty

dilatation & curettage D&C GYNECOLOGY A minor operation in which the cervix is expanded enough–dilated–to permit the cervical canal–endocervix and endometrium–to be scraped with a curette and the material submitted for pathologic evaluation

dilated cardiomyopathy CARDIOLOGY The most common cardiomyopathy in the US, which is usually idiopathic and characterized by ↑ ventricular size and impaired ventricular function ETIOLOGY Infection–eg coxsackievirus, CMV, HIV, diphtheria, trichinosis, inflammation–eg connective tissue disease, sarcoidosis, metabolic–eg hypothyroidism, thyrotoxicosis, DM, Cushing's disease, thiamine, selenium deficiency, or toxic–eg cocaine, antiretroviral agents, lead, cobalt, ethanol, phenothiazines insults; ±20% of DCs, there is a familial component PROGNOSIS Often progresses to CHF accompanied by mitral and tricuspid valve insufficiency COMPLICATIONS Conduction defects–eg atrial & ventricular tachyarrhythmias and fibrillation MANAGEMENT Supportive–eg rest, weight control, smoking cessation, ↓ physical activity during exacerbation; therapies that may be effective include ACE inhibitors, anticoagulation, digoxin, diuretics, implantable defibrillators, nitrates, potassium, magnesium repletion INVESTIGATIONAL MODALITIES Amiodarone, amlodipine, beta-blockers, dual-chamber pacing, felodipine, pimobendan, vesnarinone; recombinant hGH has been reported to ↑ myocardial mass and ↓ left ventricular chamber size, resulting in improved hemodynamics, myocardial energy metabolism, clinical status

dilation OPHTHALMOLOGY A process by which the pupil is temporarily enlarged with mydriatics to improve visualization of the fundus

dilation & curettage See Dilatation & curettage

dilator THERAPEUTICS A device used to stretch/enlarge an opening or tubular structure–eg, esophagus, to allow the passage of food. See Bougienage

Dilaudid® Hydromorphone, see there

dildo Consoler, vibrator SEXOLOGY An elongated cylindrical device used as a surrogate erect penis inserted vaginally or anally. See Sextoy

diltiazem Cardizem®, Tiazac® CARDIOLOGY A CCB used to the control atrial flutter or A Fib, angina, paroxysmal SVT, and HTN, which may prevent ↓ coronary artery diameter after heart transplantation ADVERSE EFFECTS Atrial flutter, tachyarrhythmias, slowing of AV conduction and sinus node automaticity, pruritus, sweating, constipation, N&V CONTRAINDICATIONS WPW syndrome, constipation, peripheral edema, left ventricular dysfunction. See Calcium channel blocker

dilute *verb* To weaken the potency of a thing by adding more vehicle or diluent. See Single molecular dilution

dilutional anemia A pseudoanemia mimic due to a relative ↑ of plasma, which usually occurs in a health care setting, which is due to the administration of excessive fluids; DA results in a relative–but not absolute ↓ in Hb concentration, RBC count, or Hct

dilutional coagulopathy A coagulopathy evoked by massive transfusion, resulting in factor V and VIII deficiency, and quantitative and/or qualitative platelet defects MANAGEMENT Administer 4-6 units of platelets and 1-2 units of plasma for each 15 units of RBCs

dilutional hyponatremia See SIADH

Dimetapp® PHARMACOLOGY A popular OTC antihistamine and nasal decongestant with brompheniramine & phenylpropanolamine

dimorphic anemia A dual population of RBCs in the peripheral blood smear with both microcytic hypochromic and normocytic macrocytic RBCs, a finding typical of iron deficiency and B_{12}/folic acid deficiency, and idiopathic acquired sideroblastic anemia with B_{12}/folic acid deficiency

dimple sign A clinical finding evoked by pinching a subcutaneous lesion, which results in a central dell/dimple, seen in well-circumscribed, often benign superficial dermal tumors–eg, dermatofibromas, inclusion cysts, lipomas, neurofibromas. Cf 'Tent' sign

dinging Mild traumatic brain injury, see there

dinner fork appearance ORTHOPEDICS A descriptor for the deformity seen in fractures of the distal radius with dorsal angulation–Colles fracture, which is common in osteoporotic post-menopausal ♀ who fall on outstretched hands, where the distal fractured ends of the radius and ulna–and hand––are deviated in a palmar direction

Diogenes syndrome Senile neglect GERIATRICS Dementia-related lassitude in which a subject allows his home and personal environment to deteriorate, and may collect objects of little value–eg, string, newspapers

dioxin TOXICOLOGY Any of a family of highly toxic chlorinated hydrocarbons CLINICAL In humans, intense chronic exposure causes weight loss, myalgias, insomnia, dyspnea, cold intolerance, irritability, peripheral neuropathy, hepatomegaly, hemorrhagic cystitis, chloracne, actinic elastosis, loss of libido, impotence LAB ↑ PT, ↑ lipid levels. See Agent Orange, Times Beach.

DIP 1. Desquamative interstitial pneumonia, see there 2. Distal interphalangeal–joint–anatomy

dip 1. Deceleration–obstetrics, see there 2. Morning dip, see there

diphenhydramine Diphenadril An anticholinergic and sedative antihistamine. See Imipramine

diphtheria Diphtheritis INFECTIOUS DISEASE An acute, potentially fatal infection, primarily of the upper respiratory tract, throat. See DTP–Diphtheria-Tetanus-Pertussis and DTaP–acellular Pertussis vaccines

diphtheria toxin INFECTIOUS DISEASE A 62 kD protein responsible for *C diphtheriae*'s cardiotoxic and neurotoxic effects, and mucosal damage. See *Corynebacterium diphtheriae*, Diphtheria

DIPJ Distal interphalangeal joint, see there

diploid *adjective* Referring to diploidy, see there *noun* Having 2 haploid sets of chromosomes, one from an egg, one from sperm

diploidy A DNA complement double the haploid number, *n*–ie, 2*n*. See Haploid. Cf Aneuploidy

diplomate A person with a degree of higher education, a diplomate GRADUATE EDUCATION A physician who is board-certified in a particular specialty and holds a diploma from a specialty board. See Boards, Board-certified, Board-eligible

diplopia Double vision OPHTHALMOLOGY A condition whereby a single object appears as 2

dipsophobia PSYCHOLOGY Fear of drinking. See Phobia. Cf Dipsomania

dipstick Reagent strip LAB MEDICINE A blotting paper impregnated with enzymes or chemicals sensitive to various parameters of clinical interest which, when dipped in urine, undergoes a color change, allowing a substance to be semiquantified–eg, BR, glucose and reducing substances, Hb, nitrates, ketones, pH, protein, specific gravity, urobilinogen. See Urinalysisis

dipyridamole-thallium scintigraphy CARDIOLOGY A diagnostic procedure that permits the assessment of myocardial perfusion in Pts who are unable to undergo exercise testing. Cf Dipyridamole-thallium SPECT

dipyridamole-thallium SPECT CARDIOLOGY A refinement of dipyridamole-thallium scintigraphy, which assesses myocardial perfusion without exercise testing. See SPECT. Cf Dipyridamole-thallium scintigraphy

direct access MANAGED CARE *adjective* Referring to a Pt's ability to be treated by a specific provider without gatekeeper approval. See Gatekeeper

direct antiglobulin test Direct Coombs test IMMUNOLOGY A test to detect immune hemolysis caused by binding of Ig and/or complement components to RBCs after sensitization to an antigen–eg Rh factor–on the RBC surface; a DAT helps differentiate autoimmune and 2° immune hemolytic anemia, which can be drug-induced or associated with underlying disease such as lymphoma. See Indirect antiglobulin test

direct billing MANAGED CARE The submission of bills for services rendered–eg lab work directly to the party–ie Pt or financially responsible third party–insurance company, for whom the service was performed, rather than to the physician who ordered the test

direct-to-consumer advertising DRUG INDUSTRY The use of mass media–eg, TV, magazines, newspapers, to publicly promote drugs, medical devices or other products which, by law, require a prescription, which targets consumers, with the intent of having a Pt request the product by name. See Advertising. Cf 'Yellow' professionalism

direct contract MANAGED CARE A direct contract between a network of providers–physicians, hospitals, etc, and a block of employers or business coalitions, bypassing an HMO or PPO intermediary, insurer or third-party payer. See Third-party payer

direct Coombs test See Antiglobulin test

direct current cardioversion CARDIOLOGY The use of 100-300 J–or higher electrical pulses to convert a cardiac arrhythmia into a normal pattern. Cf Chemical cardioversion

direct fluorescent antibody method IMMUNOLOGY A technique in which a molecule is detected directly using an antibody labeled or tagged with a fluorochrome–eg, FITC–fluorescein-isothiocyanate. Cf Indirect immunofluorescence

direct transmission EPIDEMIOLOGY The immediate transfer of an agent from a reservoir to a susceptible host by direct contact–skin, mouth, open wounds, touching, biting, kissing, sexual contact, or by droplet spread. See Droplet spread. Cf Indirect transmission

directed donation TRANSFUSION MEDICINE The donation of blood products for use by a designated recipient; pre-AIDS DDs were carried for 1. Donor-specific transfusions prior to renal transplantation 2. Platelet pheresis transfusions 3. Transfusions of rare blood types. See Dedicated donation. Cf Autologous donation, Intraoperative autologous donation

directional coronary atherectomy CARDIOLOGY A method of coronary revascularization, in which targeted atherosclerotic plaques are excised and removed. See Balloon coronary angioplasty, Coronary revascularization, Stent placement

directly observed therapy THERAPEUTICS A strategy for ensuring Pt compliance with therapy, where a health care worker or designee watches the Pt swallow each dose of prescribed drugs. See Patient compliance. Cf Directed observation

director A person responsible for an enterprise; one who directs. See Clinical director, Laboratory director, Medical director

dirithromycin Dynabac® INFECTIOUS DISEASE A macrolide antibiotic similar to erythromycin, clarithromycin, azithromycin

dirt, fear of PSYCHOLOGY Molysmophobia, molysomophobia, rhypophobia, rupophobia. See Phobia

disability OCCUPATIONAL MEDICINE An inability to work because of physical or mental impairment, which precludes performing expected roles or tasks DEGREE Partial–some types of labor can be performed; total–degree of impairment precludes any type of gainful employment; disability is affected by various factors, including age, education, economic and social environments SOCIAL MEDICINE Handicap A limitation in a person's mental or physical ability to function in terms of work, learning or other socially required or relevant activities, to the extent that the person might be regarded as having a need for certain benefits, compensation, exemptions, special training because of said limitations EXAMPLES Impaired hearing, mobility, speech, vision, infection with TB, HIV, or etc, malignancy, past Hx of alcohol or drug abuse, mental illness. See Ambulatory disability, Americans with Disabilities Act, Handicap, Learning disability, Reading disability, Reversible ischemic neurologic, Political correctness, Serious emotional or behavioral disability/disorder, Temporary partial disability, Temporary total disability. Cf Impairment

disability insurance MEDICAL PRACTICE An insurance policy that pays a person in the event of temporary or permanent disability. See Insurance

disaffiliation SOCIAL MEDICINE The loss or absence of social cohesion and contact with family and/or former friends and peers. See Homelessness, Mission, Runaway

disarticulation ORTHOPEDICS The amputation of an extremity at the joint. See Amputation

disaster PUBLIC HEALTH Any unanticipated event that requires urgent response, bringing people and/or property out of harm's way in order to minimize loss of life or destruction of property; disasters are described by certain parameters VOX POPULI A cataclysmic event in which there is a loss of multiple lives and/or major property damage. See Climatologic disaster, Geological disaster, Man-made disaster, Natural disaster, Tsunami

DISASTER CLASSIFICATIONS

NATURE, ie either 1. Natural, geophysical–eg earthquakes, volcanoes or weather-related–eg floods, hurricanes 2. Man-made–transportation-related, structural collapse, war, hazardous materials, explosions, fires

LOCATION Single site–eg explosion or multiple sites–eg hurricanes

PREDICTABILITY Regular–eg hurricane season or sporadic–eg toxic spill

ONSET Gradual–eg armed conflict or abrupt–eg accident

DURATION Brief–eg natural disaster or extended–eg armed conflict

FREQUENCY Often–eg flood, or rare–eg fire

disaster syndrome PSYCHOLOGY A response in survivors of major natural or man-made disasters. Cf Concentration camp syndrome, Survivor syndrome

DISASTER SYNDROME, CHRONOLOGIC STAGES

MINUTES TO HOURS Stunned apathy, disorientation

DAYS Inefficiency while helping other victims in worse condition than self, guilt at having survived

WEEKS Euphoria and enthusiasm at rebuilding and renewing activities, sense of fellowship with co-victims

MONTHS Resolution

disc Fibrocartilaginous material between spinal vertebrae which provides a cushion-like support against shock

discharge MANAGED CARE *verb* (pron. dis charj´) To formally terminate a person's care in, and releasing from, a hospital or health care facility. See Complex repetitive discharge. Cf Admit MEDTALK *noun* (pron. dis´ charj) A secretion or material eliminated from a wound or orifice. See Autogenic discharge, Nipple discharge, Prune juice discharge, Vaginal discharge *verb* To release a secretion or material from a wound or orifice

discharge summary A document prepared by the attending physician of a hospitalized Pt that summarizes the admitting diagnosis, diagnostic procedures performed, therapy received while hospitalized, clinical course during hospitalization, prognosis, and plan of action upon the Pt's discharge with stated time to followup

disciplinary order MEDICAL PRACTICE A disposition suspending or revoking licensure privileges or imposing civil penalties or ordering the restoration of money or ordering corrective action or medical or other professional treatment or monitoring, or censuring, or reprimanding a licensee

disciplinary procedure A sanction, or restriction of the right to practice medicine, imposed on a professional

discitis Diskitis, spondylodiskitis ORTHOPEDICS Intervertebral disk/disk space inflammation that may lead to disk erosion ETIOLOGY Discography, myelography, lumbar puncture, paravertebral injection, obstetric epidural anesthesia, chymopapain chemonucleolysis, often accompanied by infection

disclosure RESEARCH ETHICS A formal statement about a person's or institution's financial relationship with a company or other commercial enterprise, by means of employment, consultancy, or through ownership of stock, or other significant equity. See Conflict of interest, Nondisclosure, Self-disclosure

discogenic pain ORTHOPEDICS Pain related to damaged spinal disks. See Intradiscal electrothermal therapy

disconnection syndrome NEUROLOGY Hemispheric disconnection syndrome, see there PSYCHIATRY Dissociative syndrome, see there

discontinuous metastases Skip metastases, see there

discovery MALPRACTICE Events, documents and facts assembled between the time a lawsuit has been filed and before trial which can be used to defend or prosecute a malpractice case

discretely sampled data Data collected at intevals in time, as is digitalized data; analog data is sampled continuously because a value can be defined at any point in time. See Chanos. Cf Analog data, Continuous data

disease MEDTALK 1. A condition in which bodily functioning is interfered with or damaged, resulting in characteristic signs and symptoms 2. The loss of a state of wellness due to a either a failure in physiologic adaptation mechanisms or an overwhelming of the natural defenses by a noxious agent or pathogen. See Acute disease, Adjuvant disease, Addison's disease, 'Aguecheek's disease,' AIDS-defining disease, Air travel disease, Akureyri disease, Alcoholic liver disease, Aleutian mink disease, Alpha heavy chain disease, Alzheimer's disease, Anchor disease, Anterior horn disease, Asbestos airways disease, Atherosclerotic heart disease, Atypical GERD, Autoimmune disease, Bachelor's disease, Batten's disease, Behçet's disease, Best's disease, Bird handler's disease, Black cardiac disease, Black liver disease, Black lung disease, Blount's disease, Blue disease, Borna disease, Bornholm disease, Bowen's disease, Bread & butter disease, Brill-Zinsser disease, Brisket disease, Brown-Symmers disease, Bubble & hole disease, Bubble boy disease, Bulky disease, Bullous disease, Caffey disease, CAG disease, Caisson disease, Calcium pyrophosphate deposition disease, Canavan's disease, Caroli's disease, Castleman's disease, Cat scratch disease, Catabolic disease, Cave disease, Celiac disease, Central core disease, Cerebrovascular disease, Chaga's disease, Charcot-Marie-Tooth disease, Cheese-washer's disease, Christian-Weber disease, Christmas disease, Chronic disease, Chronic Lyme disease, Circling disease, Clinical disease, Coats' disease, Cold agglutinin disease, Collagen vascular disease, Communicable disease, Complex disease, Complicated disease, Congenital heart disease, Connective tissue disease, Constitutional disease, Controlled disease, Coronary artery disease, Creutzfeldt-Jakob disease, Crohn's disease, Crouzon's disease, Crumpled bone disease, Cushing's disease, Cytomegalic inclusion disease, Deficiency disease, Degenerative joint disease, Delta heavy chain disease, Dense deposit disease, Dent's disease, Dialysis-associated cystic disease, Disappearing bone disease, Disialotransferrin developmental disease, Dread disease, Dual diagnosis disease, Eales' disease, Ebola disease, End-stage renal disease, Endemic disease, Environmental disease, Environmental lung disease, Ethnic disease, Evaluable disease, Exanthematous viral disease, Extrapyramidal disease, Fabry's disease, Fahr disease, Fibrocystic disease, Fifth disease, Finger & toe disease, Fish-eye disease, Foot & mouth disease, 'Foot-in-mouth disease', Forestier's disease, Fourth disease, Fox den disease, Freiberg disease, Fulminant disease, Fyrn's disease, G protein disease, Gamma heavy chain disease, Gastroesophageal reflux disease, Gaucher type 1 disease, Genetic disease, Gestational trophoblastic disease, Glycogen storage disease, Goldstein's disease, Goodpasture's disease, Graft-versus-host disease, Graves' disease, Green urine disease, Gum disease, Haff disease, Hailey-Hailey disease, Hand-Schüller-Christian disease, Hard metal lung disease, Hartnup disease, Heart disease, Heat stress disease, Heavy chain disease, Hemolytic disease of newborn disease, Hemorrhagic disease of the newborn, Hirschsprung's disease, His disease, Hodgkin's disease, Hoffa's disease, Human adjuvant disease, Huntington's disease, Hyaline membrane disease, Hydatid cyst disease, Hydroxyapatite deposition disease, Hyperendemic disease, Hypertensive cardiovascular disease, I cell disease, Iceland disease, Idiopathic midline destructive disease, Immune complex disease, Inflammatory bowel disease, Interesting disease, Interstitial kidney disease, Interstitial lung disease, Ischemic disease, Isobaric counterdiffusion gas lesion disease, Japanese cerebrovascular disease, Jodbasedow disease, Katayama disease, Kawasaki's disease, Keshan disease, Kikuchi's disease, Kimmelstiel-Wilson disease, Kimura's disease, Kinky hair disease, Kissing disease, Kohler's disease, Kuf's disease, Kyasanur forest disease, Lafore's disease, Legg-Perthes disease, Legionnaire's disease, Leigh's disease, Letterer-Siwe disease, Lifestyle disease, Light chain disease, Light chain deposition disease, Limb-girdle disease, Lipid storage disease, Little's disease, Liver disease, Localized disease, Low motion disease, Lyme disease, Lysosomal storage disease, Machado-Joseph disease, Maple syrup urine disease, Marble bone disease, Marburg disease, Margarine disease, Maroteaux-Lamy disease, McArdle's disease, McCune-Albright disease, Measurable disease, Medullary cystic disease, Ménière's disease, Meningococcal invasive disease, Metabolic bone disease, Metastatic disease, Microvillus inclusion disease, Minamata disease, Minimum change disease, Mitochondrial disease, Mixed connective tissue disease, Molecular disease, Most litigated disease, Motor neuron disease, Moya-moya disease, Mseleni disease, Mu chain disease, Multiple concurrent disease, Multivessel disease, Nantucket disease, Neutral lipid storage disease, Newcastle disease, Non-alcoholic fatty liver disease, Norrie's disease, Notifiable disease, Oasthouse urine disease, Obesity-related disease, Obstructive airways disease, Oguchi's disease, 'Oid-oid disease', Orphan disease, Osgood-Schlatter disease, Paget's disease of bone, Paget's disease of breast, Paget's extramammary disease, Parkinson's disease, Pathologists' disease, Pelvic inflammatory disease, Pendular disease, Periodontal disease, Peripheral vascular disease, Peroxisomal disease, Pet-associated disease, Peyronie's disease, Phagocytic disease, Pick's disease, Pigeon breeders' disease, Pink disease, Plummer's disease, Polycystic kidney disease, Polycystic liver disease, Polycystic ovarian disease, Polygenic disease, Post-transfusion graft-versus-host disease, Preimplantation genetic disease, Proliferative breast disease, Pseudo-Hirschprung's disease, Pseudo-Whipple's disease, Pseudo-von Willebrand disease, Psychosomatic disease, Pulmonary veno-occlusive disease, Quincke's disease, Ragged red fiber disease, Reactive hemophagocytic disease, Receptor disease, Redmouth disease, Red pulp disease, Red urine disease, Re-emerging disease, Refsum's disease, Regional disease, Reportable occupational disease, Residual disease, Restrictive lung disease, Rheumatic heart disease, Rickettsial disease, Rippling muscle disease, Ritter's disease, Round heart disease, Runt disease, Saint disease, Salla disease, SC disease, Schindler's disease, Seever's disease, Self-limited disease, Seventh-day disease, Sex-linked disease, Sickle cell disease, Silicone-reactive disease, Silo-filler's disease, Single-gene disease, Sixth disease, Slavic-type Wilson's disease, Slim disease, Small airways disease, Small duct disease, Small vessel disease, SS disease, Stable disease, Stargardt's disease, Still's disease, Storage disease, Storage pool disease, Tea-drinker's disease, Thatched roof disease, Transfusion-associated graft-versus host disease, Trinucleotide repeat disease, Tsutsugamushi Scrub typhus disease, Undifferentiated connective tissue disease, Uremic heart disease, Vagrant's disease, Valvular heart disease, Vector-borne disease, Veno-occlusive disease, Virgin's disease, Von Hippel-Lindau disease, Von Recklinghausen's disease, Warm agglutinin disease, Water-borne disease, Whipple's disease, White's disease, White muscle disease, White pulp disease, Wilson's disease, Winter vomiting disease, Wolman's disease, Woolly hair disease, Woringer-Kolopp disease, X disease, X-linked lymphoproliferative disease, Yellow disease, Yellow fat disease, Yellow ovary disease, Yu-Cheng oil disease, Zollinger-Ellison disease

disease flare-up A transient ↑ in severity of the manifestations of a disease

disease-free survival ONCOLOGY The time that a person with a disease lives without known recurrence; DFS is major clinical parameter used to evaluate the efficacy of a particular therapy, which is usually measured in 'units' of 1 or 5 yrs. See Cure, Remission

disease-modifying antirheumatic drug DMARD RHEUMATOLOGY Any agent—eg, azathioprine, gold, cyclophosphamide, hydroxychloroquin, and MTX—which slows the rate of joint destruction in rheumatoid arthritis

disease registry PUBLIC HEALTH A surveillance system that collects and maintains structured records on the new cases of a specific disease or condition for a specified time period and population; a DR analyzes, and interprets data those with a common illness or adverse health condition

disease of regulation A condition, the Sx of which are attributed to a loss of the homeostatic mechanisms responsible for maintaining in balance a hormone—eg PTH or autonomic function—eg blood pressure

disease of the week 'syndrome' A hypochondriacal symptom complex described in medical students who, as they learn about a disease, discover that they, too, have all of the symptoms of the disorder being discussed

disenfranchised population SOCIAL MEDICINE A group of persons without a home or political voice, who live at the whims of a host EXAMPLES Homeless, refugees of war and natural disasters. See Homelessness, Refugee

disfiguring surgery A popular term for surgery that mutilates, especially if it affects the face, hands and arms; the term mutilating surgery generally refers to other body regions

disinterested *adjective* Unbiased, objective

disk syndrome Disc syndrome REHAB MEDICINE A general term for pressure-related radiculopathy Sx—eg low back pain, paresthesia, sensory loss, weakness, ↓ reflexes, possibly muscle wasting, 2° to defects—eg prolapse of the intervertebral disk

diskectomy ORTHOPEDICS Microsurgical laminotomy A technique that avoids vertebral exploration before disk surgery, reducing the risk of 'failed disk syndrome' as there is minimal excision of bone and epidural fat and minimal nerve root adhesion. See Laminectomy, Percutaneous diskectomy

dislocation ORTHOPEDICS The complete displacement of a joint surfaces. See Acromioclavicular dislocation, Congenital hip dislocation. Cf Subluxation

dismemberment The removal of limbs from a body, an expediency used by some serial killers to more efficiently store body parts for future use, consumption, or disposal. See Jeffrey Dahmer, Mad dog

Disopam Diazepam, see there

disopyramide Norpace® An antiarrhythmic

disorder An abnormality, alteration, or derangement. See Antisocial personality disorder, Anxiety disorder, Asperger disorder, Arousal disorder, Attention deficit disorder, Autistic disorder, Bipolar disorder, Body dysmorphic disorder, Borderline personality disorder, Central auditory processing disorder, Chromosome disorder, Compulsive personality disorder, Conversion disorder, Cruise-associated diarrheal disorder, Cumulative trauma disorder, Delusional disorder, Dependent personality disorder, Depersonalization disorder, Depressive disorder, Developmental disorder, Disease, Dissociative identity disorder, Dysthymic disorder, Eating disorder, EBV-associated lymphoproliferative disorder, Endometrial disorder, Expressive language disorder, Factitious disorder, Functional disorder, Gender identity disorder, Generalized anxiety disorder, Hearing disorder, Histrionic personality disorder, Identity disorder, Internet addiction disorder, Iodine deficiency disorder, Language disorder, Late luteal phase dysphoric disorder, Lymphoproliferative disorder, Major depressive disorder, Martha Stewart disorder, Mendelian disorder, Mental disorder, Motor speech disorder, Movement disorder, Multiple autoimmune disorder, Multiple personality disorder, Musculoskeletal disorder, Myeloproliferative disorder, Narcissistic personality disorder, Neurodegenerative disorder, Neurogenic communication disorder, Neurotic disorder, Nonmendelian disorder, Obsessive-compulsive disorder, Obsessive-compulsive personality disorder, Panethnic disorder, Panic disorder, Partial syndrome eating disorder, Passive-aggressive personality disorder, Post-transplantation lymphoproliferative disorder, Post-traumatic stress disorder, Premenstrual dysphoric disorder, Psychotic disorder, Reactive attachment disorder of infancy or early childhood, Reading disorder, S-100–positive T-cell lymphoproliferative disorder, Schizoid personality disorder, Seasonal affective disorder, Seizure disorder, Sexual pain disorder, Shared psychotic disorder, Silicone-reactive disorder, Single gene disorder, Sleep disorder, Sleep terror disorder, Smell disorder, Somatization disorder, Speech disorder, Swallowing disorder, Syndrome, Taste disorder, Thought disorder, Throat disorder, Thyroid disorder, Urea cycle disorder, Urologic disorder, Voice disorder, X-linked disorder

disorientation PSYCHIATRY Loss of awareness of the position of one's self in relation to space, time, or other persons; confusion. See Delirium, Dementia

dispense as written PHARMACOLOGY An order on a prescription commanding the pharmacist to provide the recipient with the prescription exactly as it was written. Cf Voluntary formulary permitted

displacement PSYCHIATRY An unconscious ego defense mechanism in which a person's normal emotions and reactions are repressed, changed, or transferred to more socially appropriate responses, often to allay anxiety. See Acting out

display INFORMATICS A monitor or viewing device. See Electroluminescent display, Field emission display, Flat-panel display, LCD, Plasma display, Vacuum fluorescent display

disposable NURSING *adjective* Referring to that which is discarded or disposed of *noun* An item used in health care-related Pt contact which is discarded after use—eg masks, gloves, gowns, needles, paper products, syringes, wipes. See Biohazardous waste

disposition CLINICAL PHARMACOLOGY The fate of a therapeutic agent after absorption, which corresponds to the sum of its distribution and elimination, including the alpha and beta portions of a declining serum dose

concentration vs time curve. See Distribution MANAGED CARE A Pt's destination after discharge from a hospital

disruption OBSTETRICS The *in utero* destruction of a previously formed normal fetal body part, caused either by 1. 'Amputation', the result of pressure-induced strangulation by an amniotic band or 2. Interruption of regional blood supply, which causes ischemia, necrosis, and sloughing of the dead part. See Dysmorphology

disruption sequence PEDIATRICS The constellation of malformations that occur when a normally developing fetus is subjected to a toxin–eg, thalidomide or pathogen–eg, rubella

dissecting aortic aneurysm CARDIOVASCULAR DISEASE An aneurysm of the aorta in which there is an internal split in the wall of the aorta, caused by either ASHD or cystic medial hyperplasia. See Aneurysm, Aortic aneurysm

dissection SURGERY The separation of tissues or of one tissue plane from another. See Aquadissection, Axillary dissection, Balloon dissection, Blunt dissection, Full axillary dissection, Low axillary dissection, Modified neck dissection, Neck dissection, Radical neck dissection, Selective neck dissection, Sharp dissection

dissector SURGERY A surgical instrument used to separate one tissue or tissue plane from another. See Endoscopy

disseminated gonococcal infection STD A form of *N gonorrhoeae* infection affecting ±3% of Pts; it is 4-fold more common in ♀, especially during pregnancy and perimenstrual period CLINICAL Classic triad–dermatitis, tenosynovitis, migratory polyarthralgia; DGI follows mucosal infection–eg, genitourinary tract, rectum, oropharynx CLINICAL FORMS 1. Arthritis-dermatitis syndrome–tenosynovitis with skin rash and fever, often positive blood culture, asymmetric polyarthralgia; $1/4$ have monoarthralgia SITES Knee, elbow, wrist, metacarpophalangeal, ankle joints 2. Suppurative/septic arthritis–±$1/3$ of Pts, synovial fluid PMNs at 50,000–200,000 cells/mm^3, maculopapular, pustular, vesicular or necrotic rash, often of torso, limbs, palms, soles; tenosynovitis in $2/3$ of Pts typically of hands and fingers; rarely, DGI causes pericarditis, endocarditis, meningitis, death. See *Neisseria gonorrhoeae*

disseminated intravascular coagulopathy HEMATOLOGY An acquired bleeding diathesis with a generally bad outcome in which the balance between coagulation and fibrinolysis tips toward the former; DIC is characterized by accelerated platelet consumption with coagulation factor depletion–↑ PT, ↑ PTT and stimulation of fibrinolysis–generation of fibrin split products ETIOLOGY Severe sepsis–30-65% of DIC is caused by infection, extensive burns, trauma, retained dead fetus, heat stroke, mismatched blood, metastatic CA, leukemia CLINICAL FORMS FAST DIC–acute, fulminant, uncompensated consumptive coagulopathy with severe bleeding, and ecchymosis due to abruptio placentae, septic abortion, amniotic fluid embolism, toxemia, malignancy, massive tissue injury–eg burns, surgery, trauma, infections, gram-negative sepsis, meningococcemia, RMSF, incompatible blood; 'fast' DIC requires replacement of deficient or consumed factors SLOW DIC Due to chronic compensated illness, with few overt signs of bleeding; clinical picture is painted by thrombosis, microcirculatory ischemia, end-organ infarction, due to acute promyelocytic leukemia, coagulation factor transfusion, CA of pancreas, prostate, lung, stomach, aortic aneurysm, cocaine, IMAOs, giant hemangioma–Kasabach-Merritt syndrome, liver disease, vasculitis, chronic and/or low-grade infections–eg, histoplasmosis, aspergillosis, malaria, obstetrics,–eg, hemolysis, eclampsia, dead fetus, infection, hypoxia, acidosis, respiratory distress; 'slow' DIC may respond to heparinization LAB ↑ PT, aPTT, FDPs, fibrinopeptide A; ↓ fibrinogen, prothrombin, platelets, factor V, factor VIII, antithrombin III, plasminogen, ↓ factors VII, IX, X, XI

disseminated tuberculosis Miliary tuberculosis INFECTIOUS DISEASE A chronic, contagious infection by *M tuberculosis*, which has spread from the lungs to other organs by blood or lymphatics, after inhaling droplets sprayed

into the air from a cough or sneeze by a person infected with *M tuberculosis*; DT is characterized by granulomas of involved organs and develops in Pts with immune defects TISSUES AFFECTED Pericardium, peritoneum, larynx, bronchus, cervical lymph nodes, bones, joints, genitourinary system, eye, stomach, meninges, skin RISK OF CONTRACTING TB Crowding, unsanitary living conditions, poor nutrition, Hispanics, Native Americans, African Americans, HIV infection, homelessness, elderly, infants, those with incomplete treatment of TB, leading to proliferation of drug-resistant strains. See Tuberculosis

dissemination MEDTALK The spread of a pernicious process–eg, CA, acute infection ONCOLOGY Metastasis, see there

disseminator EPIDEMIOLOGY A person who spreads an infection. See High disseminator, Typhoid Mary

dissociation CARDIOLOGY Electromagnetic dissociation, see there. See also Atrioventricular dissociation, Pulse-temperature dissociation PSYCHIATRY A mental response that diverts consciousness from painful or traumatic associations EXAMPLES Shock, numbing, paralysis, loss of speech or other sensory perception, or even loss of consciousness

dissociative identity disorder Multiple personality disorder The *"presence of 2 or more distinct identities or personality states...that recurrently take control of behavior."* DID is accompanied by an inability to recall important personal information that exceeds ordinary forgetfulness; there are ±20,000 DIDs in the US

dissociative syndrome Cerebral disconnection syndrome, disconnection syndrome, dissociative language syndrome, Geschwind syndrome, interhemispheric connection syndrome NEUROLOGY 1. Any condition caused by a disruption in the connection between the cerebral hemispheres 2. Any of a number of disorders of language caused by an interruption of association pathways; the anatomic basis for these lesions is uncertain

distal axonopathy Dying-back neuropathy, see there

distended MEDTALK Enlarged, bloated. Cf Nondistended

distinguished fellow ACADEMIA Any of a group of really–really, really, *really* smart people selected to form an intellectual cadre and opine, guide, and influence decision-making in a particular area

distorted thinking PSYCHOLOGY Any of a number of 'emotional traps' that prevent a person from addressing negative emotions FORMS OF DT All-or-nothing thinking, overgeneralization, mental filtering, personalizing blame. See Cognitive therapy, Dereitic thought, Distortion break, Humanistic psychology, Magical thinking, Thinking

distortion break PSYCHIATRY An 'emotional trap' that prevents a person from dealing with negative emotions. See Cognitive therapy, Humanistic psychology

distractibility PSYCHIATRY The inability to maintain attention; shifting from one area or topic to another with minimal provocation SIGNIFICANCE Sign of organic impairment, or a part of a functional disorder–eg, anxiety states, mania, or schizophrenia

distraction PAIN MANAGEMENT A pain relief method that takes attention away from the pain. See Attentional distraction

distress syndrome MEDTALK A nonspecific term for a condition that impacts on one or more organ systems EXAMPLES Respiratory distress syndrome, Inflammatory bowel disease

distribution CLINICAL MEDICINE The pattern of involvement of a tissue by a particular condition. See Batwing distribution, Fat distribution, Mocassin distribution, Stocking & glove distribution EPIDEMIOLOGY The frequency and pattern of health-related characteristics and events in a population PHARMACOLOGY The location–eg intravascular or extravascular of a therapeutic agent after absorption, which corresponds to the sum of its distribution and elimination; disposition includes both the alpha and beta portions of a declining serum dose concentration versus time curve. See Disposition, Elimination

disulfiram Antabuse® An antioxidant that interferes with alcohol metabolism, resulting in ↑ acetaldehyde concentrations; it is effective in treating alcoholism as it produces aversive symptoms if combined with alcohol

disuse atrophy A generic term encompassing the degenerative changes that tissues undergo when they are functioning at suboptimal levels; involvement of the musculoskeletal unit is characterized by atrophy of muscles, contraction of tendons and osteoporosis; diversion of GI tract flow results in a DA-type phenomenon known as diversion colitis

Ditropan® Oxybutynin UROLOGY An agent for treating overactive bladder with urge incontinence, urgency, frequency

diuresis NEPHROLOGY Excretion of urine, especially in excess. See Overdiuresis

diuretic PHYSIOLOGY *adjective* Referring to or evoking diuresis *noun* Water pill An agent that ↑ excretion of fluid from the renal tubules, which are most commonly used in CHF ADVERSE EFFECTS Hemodynamics–eg, reflex tachcardia, SNS activation with catecholamine release, activation of RAA system, renal function–eg, ↓ perfusion leading to ↑ BUN and creatinine, electrolytes–eg, ↓ potassium and/or magnesium. See Potassium-sparing diuretic

diversion colitis A condition characterized by inflammation in bypassed segments of the colorectum, which occur after surgical diversion of the fecal stream CLINICAL Asymptomatic or bloody discharge, colicky anorectal pain, tenesmus, purulent or hemorrhagic discharge; DC occurs in most diverted colons up to yrs after diversion ENDOSCOPY Friable, erythematous and granular mucosa mimicking Crohn's disease and ulcerative colitis TREATMENT Topical use of short-chain fatty acids

diversion status Bypass status HOSPITAL CARE A temporary status for a health care facility, where its administration informs its emergency medical services that the hospital is full. See Antidumping laws

diverticula Plural of diverticulum

diverticular disease GASTROENTEROLOGY The presence of multiple diverticula–prolapsed mucosa-lined intestine through the muscularis propria of the large intestine EPIDEMIOLOGY DD affects 5–10% of those in developed countries > age 45; 80% of those > age 85; 20% have Sx CLINICAL Asymptomatic; or pain, N&V, farts PREVENTION DD is linked to ↑ intraluminal pressure–IP↓ Stool bulk → ↑ GI transit time → ↑ IP → ↑ diverticulsosis; ↑ dietary fiber softens stools, ↓ IP relieves Sx. See Acute diverticulitis

diverticulitis SURGERY Inflammation of one or more diverticula which, when acute, may rupture causing peritonitis CLINICAL More common in the elderly; ± abrupt onset with tenderness in left hypogastrium, pain of variable severity may radiate to the back, rebound tenderness, fever, anorexia, constipation, GI tract discomfort, N&V LAB ↑ WBCs, left shift of myeloid series, ↑ ESR DIAGNOSIS Hx, barium enema, ultrasound, sigmoidoscopy, colonoscopy MANAGEMENT-MEDICAL Oral antibiotics, low-fiber foods; for severe diverticulitis with high fever and pain, hospitalization, IV antibiotics MANAGEMENT-SURGICAL Excision of bleeding diverticula or resection for persistent bowel obstruction or abscesses not responding to antibiotics PROGNOSIS Pain waxes, wanes, spontaneously remits

diverticulosis The presence of multiple diverticula in the colon which, if inflamed/infected, is termed diverticulitis CLINICAL Sx, *if* present, include abdominal colic, constipation, diarrhea, bloating DIAGNOSIS Colonoscopy, CT of abdomen MANAGEMENT High fiber diets may delay progression of diverticulosis or diverticulitis

dizygotic twins Fraternal twins Twins resulting from 2 separate fertilized eggs, liberated simultaneously from the ovaries, which develop in separate or partially fused chorion and placenta, and usually a separate amniotic sac. Cf Monozygotic twins

dizziness AUDIOLOGY Unsteadiness, imbalance, lightheadedness associated with balance disorders NEUROLOGY An imprecise term often used by Pts to describe various sensations–eg, rotation, nonrotatory swaying, weakness, swaying, faintness, lightheadedness, unsteadiness, presyncope, vertigo. Cf Dizzy spell, Dysequilibrium, Pseudovertigo, Vertigo

DJD Degenerative joint disease, see there

DKA Diabetic ketoacidosis, see there

DM Diabetes mellitus, also, doctor of medicine–MD is used in the US

DN Diabetic nephropathy, see there

DNA Deoxyribonucleic acid MOLECULAR BIOLOGY A double-stranded linear macromolecule which encodes an organism's genetic information

DNA amplification MOLECULAR DIAGNOSTICS Any method used to ↑ the copy number of a sequence of DNA. See Cycling probe technology, Gap LCR–gap ligase chain reaction, Gene amplification, NASBA–nucleic acid sequence-based amplification, PCR, SDA–strand-displacement amplification, TMA–transcription-mediated amplification

DNA analysis Any technique used to analyze genes and DNA. See Chromosome walking, DNA fingerprinting, Footprinting, In situ hybridization, Jeffries' probe, Jumping libraries, PCR, RFLP analysis, Southern blot hybridization

DNA hybridization MOLECULAR MEDICINE A technique for determining the presence of a target DNA in a sample of tissue or cells. See HLA analysis, Paternity testing, RFLP analysis

DNA ploidy analysis The determination of the number of single copies of a complete haploid–*n*–set of chromosomes present in a particular cell population, which is most efficiently performed by flow cytometry. See DNA index, Flow cytometry, Ploidy analysis, Proliferation index

DNA repair syndrome DNA instability syndrome Any of a heterogeneous group of diseases with damaged genomes, chromosomal 'instability', hypersensitivity to irradiation and mutagenic chemicals and an ↑ risk of suffering malignancy EXAMPLES Ataxia-telangiectasia, Bloom syndrome, dyskeratosis congenita, Fanconi's anemia, progeria, xeroderma pigmentosum

DNA sequence GENETICS The precise order of bases–A,T,G,C–in a segment of DNA, gene, chromosome, or an entire genome. See Base pair, Base sequence analysis, Chromosome, Gene, Genome

DNAR Do not attempt resuscitation. See DNR

DNI status Do not intubate status TERMINAL CARE An advance directive, in which a Pt or next-of-kin specifically requests that extraordinary measures–eg, intubation, are not taken to resuscitate a terminally ill Pt. See DNR

DNR Do not resuscitate END OF LIFE DECISIONS An order written in a Pt's chart that explicitly and unequivocally states that CPR–eg, intubation, pounding, defibrillation, should not be initiated if a Pt is found in cardiac arrest. See Medical futility. Cf Advance directives, Euthanasia, Living will

do not resuscitate See DNR

DOA 1. Date of admission 2. Dead on arrival 3. Drug of abuse, see there

DOB Date of birth

dobutamine CARDIOLOGY A β–agonist used to treat standard therapy-refractory acute CHF; in contrast to other beta agonists, e.g., isoproterenol, dobutamine is not associated with a ↑↑↑ in myocardial O_2 consumption–MOC, which precluded their use; dobutamine has only moderate vasodilating activity, ↑ cardiac output with little change in MOC. See Vascular-Ventricular coupling

DOC Date of confinement/delivery 2. Deoxycorticoid 3. Death from other causes

'doc-in-a-box' A deprecating sobriquet for a physician who provides primary health care, usually at an hourly rate, at a free-standing ambulatory care clinic

docetaxol Taxotere® ONCOLOGY A taxane that inhibits CA cell growth by blocking mitosis; it is a first-line therapy for metastatic and refractory breast and non-small cell lung CAs. See Taxol

doctor VOX POPULI A title for 1. A person with an advanced degree in a healing art, including medicine–MD, osteopathy–DO, podiatry–DPM, pharmacy–DPh, chiropractic–DCM, dentistry–DDS, or veterinary medicine–DVM *or* 2. A person with the highest university degree–PhD *or* 3. A pretender to advanced knowledge. See Abrams, Barefoot doctor, Fertility doctor, Pitch doctor, Spin doctor

doctor-nurse game The complex 'pas de deux' between physician and nurses. See Nurse practitioner, Physician extender

doctor-patient interaction The doctor-patient interplay comprises the social aspects of a confidential relationship shared by physician and Pts. See Bedside manner

doctor shopping PSYCHIATRY The visiting of multiple physicians, each time with a new symptom SUBSTANCE ABUSE The seeking of doctors who will prescribe opioids and opiates. See Drug-seeking behavior

doctrine A theory or posit widely accepted by leading authorities in a particular field. See Assumption-of-risk doctrine, Borrowed servant doctrine, Captain-of-the-ship doctrine, De minimus doctrine, Emergency doctrine, Feres doctrine, Humoral doctrine, Hypothesis, Lost-opportunity doctrine, Posit, Therapeutic privilege

document *noun* ONLINE Any aggregate of data, whether it is on paper or in an electronic format; a document may be handwritten or typed, have illustrations and graphics, and may be legally binding *verb* HEALTH CARE To formally record information, usually in a permanent, legally acceptable fashion, usually as written and/or signed notes, order forms, and others. See Documentation

documentation MEDICAL PRACTICE The creation of a formal record, especially those that record Pt-physician contacts with dates–and often times and 'document' various aspects of Pt management; in certain types of Pt care–eg, cosmetic surgery, photographs are used as documentation of such interactions. See Defensive medicine, Forensic documentation, Medical documentation, Photodocumentation

DOD 1. Date of death 2. Date of discharge 3. Dead/died of disease

DOE Dyspnea on exertion/exercise, see there

dofetilide Tikosyn® CARDIOLOGY A selective potassium channel blocking antiarrhythmic used in CHF and left ventricular dysfunction to convert A Fib, prevent its recurrence, and ↓ risk of future hospitalization; it has no effect on mortality CONTRAINDICATIONS Succinylcholine-type anesthesia, sick sinus syndrome, seizures ADVERSE EVENTS Torsade de pointes. See Atrial fibrillation

dog VETERINARY MEDICINE A relatively obtuse homeothermic quadriped long domesticated in Europe, and longer a culinary staple in Asia PROS Dogs may belong to a family unit, providing companionship, unqualified affection, etc CONS Dogs are vectors for multiple infections–anthrax, blastomycosis, *Bergeyella (Weeksella) zoobelcum*, *Brucella canis*, campylobacteriosis, *Capnocytophaga canimorsus*, *Capnocytophaga cynodegmi*, CDC group EF-4a, CDC EF-4b, CDC group NO-1, cheyletiellosis, coenurosis, cryptosporidiosis, cutaneous larva migrans, *Demodex folliculorum*, dermato-

phytosis, *Dipylidium caninum*, echinococcosis, *Francisella tularensis*, *Gastrospirillum hominis*, granulocytic ehrlichiosis, leptospirosis, Lyme disease, *Neisseria canis*, *Neisseria weaveri*, *Pasteurella multocida*, plague, rabies, RMSF, salmonellosis, scabies, *Staphylococcus intermedius*, *Strongyloides stercoralis*, trichinosis, visceral larva migrans, *Y enterocolitica*. See Detection dog, Mad dog, Service dog

dog bite PUBLIC HEALTH The clamping of skin and subjacent soft tissues between the upper and lower mandible of a canine, which may cause infections, acting as a disease vector or even death. See Dog

dog boning INTERVENTIONAL CARDIOLOGY A complication caused when an angioplasty balloon becomes excessively expanded at either end of a stent–likened to a dog bone. See Stent

dog-ear appearance PLASTIC SURGERY A one-sided mound of redundant tissue, which is seen after the repair of certain skin lesions and defects SURGERY Puckering at the end of a scar

doll's head sign NEUROLOGY A clinical sign for evaluating brainstem function in a comatose Pt; in a normal person, as the head is turned rapidly to one side, the eyes conjugately deviate in the direction opposite to the head's movement; loss of this reflex implies dysfunction of brainstem or oculomotor nerves; inferolateral deviation of the eyes in combination with pupillary dilation implies dysfunction of 3rd cranial nerve, possibly due to tentorial herniation

Dolophine® Methadone, see there

domain knowledge MEDICAL EDUCATION A format for acquiring knowledge based on rote memorization of large blocks of information–eg, the findings of a particular disease. Cf Socratic method

domain name INFORMATICS The unique "address" name that identifies an Internet site. See IP Number

domestic violence Battering PUBLIC HEALTH A pattern of psychological, economic, and sexual coercion of one partner in a relationship by the other, often punctuated by physical assaults, or credible threats of bodily harm; physical abuse by a 'significant other'–boy/girlfriend, lover, spouse at home RISK FACTORS Partner abuse of substances, alcohol; intermittent employment, unemployment; less than high school education; perpetrator: former spouse, ex-boyfriend. See Abusive behavior, Criminal victimization

dominance GENETICS The ability of a dominant gene to express itself in a phenotype, when the gene is paired with another (recessive) gene that would have expressed itself in a different way NEUROLOGY Cerebral dominance The tendency of one brain hemisphere to be more controlling than the other in mediating neural activity PSYCHIATRY A predisposition to play a controlling role when interacting with others SEXOLOGY See Domination

dominant GENETICS A phenotype expressed when a particular gene is present in a cell, regardless of whether the allelic set contains 2 different forms of expression; the allele with the masked phenotype is termed recessive DOMINANT DISORDERS Achondroplasia, familial hypercholesterolemia, Huntington's disease. See Filial generation, Homozygote, Trait. Cf Recessive

dominant hemisphere NEUROLOGY That portion of the brain involved in guiding activities requiring manual dexterity; for those who are right-handed, the left hemisphere is dominant

domination SEXOLOGY A paraphilia in which "A" performs verbal–eg, humiliation or physical–eg, spanking, whipping, chaining, etc–acts of sexual sadism, on consenting "B"–weird, just plain weird. See Sadomasochism

domino donation Transplantation of heart and lungs of cadaveric origin into Pt A, who has long standing lung disease–eg, cystic fibrosis, but who has a heart suitable for donation, transplanting A's heart into B. See Heart-lung transplantation, Multiorgan donation, UNOS

domoic acid An excitatory kainic acid analogue and neurotoxic glutamate agonist, which ↑ neuronal activity, causing food poisoning

Don Juan syndrome Male hypersexuality, satyrism PSYCHIATRY A ♂ paraphilia in which insecurity about masculinity and/or latent homosexuality is masked by myriad sexual liaisons with ♀. Cf Delilah syndrome, Nymphomania

donation See Directed donation, Egg donation, Eye donation, Multiorgan donation, Predonation, Preoperative autologous donation

donepezil Aricept® NEUROLOGY A cholinesterase inhibitor used to maintain cognition in Alzheimer's disease CONTRAINDICATIONS Succinylcholine-type anesthesia, sick sinus syndrome, seizures ADVERSE EVENTS N&V, insomnia, muscle cramps, anorexia, fatigue. See Alzheimer's disease

donor The giver of a tissue, an organ, blood or blood products; in the usual parlance, an altruistic person who contributes blood products, often regularly. See Anencephalic organ donor, Oxydonor, Universal donor

donor deferral TRANSFUSION MEDICINE The nonacceptance of a potential donor based on lifestyle criteria or prior exposures to pathogens. See Donor exclusion criteria

donor exclusion criteria TRANSFUSION MEDICINE Those criteria used by a blood bank to exclude a person's blood from a donor pool. See Blood shortage, Donor deferral

donormobile SPORTS MEDICINE A macabre sobriquet for an all-terrain vehicle–ATV, or motorcycle–"donorcycle", which has a high fatality rate among healthy young people, whose organs may donated to those on wait-lists for organ transplants

door-to-balloon CARDIOLOGY The time between the moment a Pt with a possible acute MI enters an ER and he/she undergoes balloon angioplasty. See Acute MI, PCTA

door-to-data EMERGENCY MEDICINE The time that elapses–±6 mins–between when a Pt with chest pain arrives at the hospital and there is sufficient data–EKG, clinical Hx, CK/CK-MB to be reasonably certain that an acute MI is occurring. See Door to drug

door-to-drug Time to thrombolysis CARDIOLOGY The time between the moment a Pt with an acute MI enters the ER and when the Pt receives a thrombolytic–eg tPA, streptokinase. See Acute MI, tPA

dopamine NEUROLOGY A catecholamine hormone and neurotransmitter essential to CNS activity; it is involved motor control, cognition, and reward; abnormal dopamine levels occur in Parkinson's disease, paranoia, memory and concentration defects; dopamine may modulate endorphin levels, altering perception of pain and pleasure; it restricts prolactin, affecting libido See Biogenic amine, Catecholamine, Indolamine, Serotonin

dopamine hypothesis PSYCHIATRY A theory that attempts to explain the pathogenesis of schizophrenia and other psychotic states as due to excess dopamine activity in various areas of the brain

dope SUBSTANCE ABUSE A generic and popular term for any recreational drug, almost invariably referring to a narcotic–eg, heroin, but also marijuana

doping SPORTS MEDICINE Popular for the use of drugs and other nonfood substances to improve performance. See Anabolic steroids, Blood doping, Weight training

doppelgänger PSYCHIATRY A delusion that a double of a person or place exists elsewhere; it is related to other defects in recognition and suggests organic disease in the nondominant parietal lobe. See Depersonalization disorder, Schizophrenia

Doppler echocardiography Doppler cardiac ultrasonography CARDIOLOGY A noninvasive imaging modality for assessing hemodynamics, based on the Doppler effect on ultrasound–US waves; DE estimates transvalvular gradients and calculates valve–aortic, mitral area; it has been used to evaluate congenital shunts, atrial septal defect, patent ductus arteriosus, left ventricular inflow–eg parachute deformity, cor triatriatum and outflow–eg congenital aortic valve stenosis, subaortic stenosis, hypertrophic obstructive cardiomyopathy obstruction, atrioventricular valve regurgitation and other lesions

Doppler mapping CARDIOLOGY The analysis of the spatial relationship of Doppler signals to a 2-D image of the heart, based on pulsed-wave Doppler echocardiography. See Doppler echocardiography

Doppler velocimetry OBSTETRICS A technique used to analyze blood flow waveforms, allowing accurate noninvasive measurement of volume and velocity of blood flow

Dor procedure Endoventricular patch plasty HEART SURGERY A procedure in which the dilated and ischemic left ventricle is ↓ in size and reconstructed OPERATIVE MORTALITY 1.5–3%; 23–38% ↑ in ejection fraction. Cf Acorn prosthesis, Batista operation, Dynamic cardiomyoplasty

DORA Directory of Rare Analyses CLINICAL CHEMISTY A reference book published by the Am Chemical Soc that catalogs rarely ordered clinical tests and provides details on the labs performing them CRITERIA FOR INCLUSION IN DORA Test of interest is not performed by > 2 labs

Doriden® Glutethimide, see there

Dorito™ syndrome POPULAR NUTRITION A sense of emptiness and dissatisfaction resulting from consumpton of junk foods–eg Doritos, Pringles, other, often salt-laden, snacks. See Couch potato, Junk food, Twinkie defense

dorsal root Posterior root, see there

dorsal root ganglion syndrome NEUROLOGY A condition characterized by severe pain without sensory loss in the arm, caused by a hyperflexion or hyperextension injury to the cervical region, exacerbated by flexion or extension of the head ETIOLOGY Injury and/or hemorrhage around the dorsal root ganglion

dorsalis pedis pulse PHYSICAL EXAMINATION A pulse palpable over the dorsal surface of the foot

dorsiflex *verb* To bend toward the head

dose A quantity of a substance, medication, or radiation that is administered or absorbed during a specific time period OCCUPATIONAL MEDICINE The total amount of a toxicologically relevant material reaching the target site over a specific time period; determination of dose requires information on the agent, its concentration, the form of the agent–eg aerosol, topical, ingested, length of exposure, and weight of person exposed. See MSDS RADIATION ONCOLOGY The concentration of energy divided by the tissue mass of energy deposited by radiation in the body, measured in grays–formerly, rads. See Effective dose, Gray THERAPEUTICS The amount of medicine taken. See Absorbed dose, Effective dose, Equivalent dose, Genetically significant dose, Imputed dose, Law of infinitesimal dose, Lethal dose, Loading dose, Maintenance dose, Maximum tolerable dose, Median lethal dose, Minimum lethal dose, Occupational dose, Organ dose, Overdose, Pharmacologic dose, Ramped dose, Reference dose, TCID

dose control trial A therapeutic trial that attempts to prove efficacy of an investigational agent by comparing 2 doses of the agent. See Placebo

dose dependence Concentration dependence THERAPEUTICS The change in certain pharmacokinetic parameters–clearance, half-life, bioavailable fraction–as a function of administered dose

dose rate THERAPEUTICS The amount of drug administered per unit time, without regard to frequency of intermittent dosing. See Bioavailability, Clearance

dose-response curve A graphic representation of the effects that various doses of an agent–eg, ionizing radiation or a chemotherapeutic agent, have on a given parameter–eg, cell viability, mutation frequency, DNA damage, tumor growth or metastasis or other behavior THERAPEUTICS A graphic representation of the effectiveness or toxicity of a drug vs the dose administered

dosimetry RADIATION PHYSICS The formal science that measures and calculates doses and format of radiation to be administered to a Pt with a disease requiring RT, in particular cancer. See Definitive radiation, EPR dosimetry, Radiation

dosing interval THERAPEUTICS The frequency of intermittent drug administration, based on the drug's half-life. See Slow-release drug

dot dystrophy OPHTHALMOLOGY A pattern of microcystic dystrophy seen in normal persons, characterized macroscopically by clusters of tiny, round, or comma-shaped, gray-white opacities in the pupillary zones, uni- or bilaterally, corresponding to minute cystoid spaces

DOT physical INTERNAL MEDICINE A physical examination & drug screen required by the Dept of Transportation for drivers carrying commercial loads on public roads

dothiepin A tricyclic antidepressant, see there

dotted-Q sign MEDTALK Wardspeak for a Q sign in which one or more flies rest on the Pt's tongue. See Q sign

double aorta A congenital defect of the aorta in which both of the dorsal embryonic aortae persist and encircle the trachea and esophagus; after birth, and consequent to the aorta's growth in a region with limited space, the ring produces a relative constriction, resulting in stridor and dysphagia. Cf Double-barreled aorta

double-barreled aorta An aorta with a second vascular lumen formed in the media of the aortic wall, which connects the proximal and distal intimal tears in an aorta with a dissecting aneurysm. Cf Double aorta

double blinded study Double-masked design CLINICAL RESEARCH A clinical trial in which neither the Pt nor the researcher know which arm–eg a gold standard therapy, an experimental, or placebo–of the study the Pt has been placed, preventing, theoretically, Pt or researcher bias. See Bias, Blinding, Triple blinding; Cf Anecdotal

double boarded *adjective* Referring to a physician who is board-certified by 2 separate specialty boards. Cf Dual career

double bolus approach CARDIOLOGY A regimen in which tPA is administered shortly after vascular occlusion in Pts in the early myocardial ischemia; DBA may improve coronary artery revascularization and may minimize the size of MIs

double contrast study IMAGING A technique used to ↑ visualization of the intestinal mucosa by administration of 2 different radiocontrast mediums. See Colonoscopy

double crush syndrome ORTHOPEDICS A type of peripheral nerve compression syndrome in which there is a 'central' compression that impacts on a nerve bundle–eg, at the thoracic or pelvic outlet, and a 2nd more peripheral compression–eg, at the carpal or tarsal tunnel; optimal therapy requires surgical release of both. See Carpal tunnel, Thoracic outlet syndrome

double depression POP PSYCHIATRY Major depression superimposed on chronic depression. See Depression, Major depression

double discordance Corrected transposition of pulmonary arteries & aorta PEDIATRIC CARDIOLOGY A congenital anomaly in which the great arteries are in a mirror-image of their normal location; the aorta arises from the anterior left heart and the pulmonary artery from the right posterior heart,

the blood from the morphologic right atrium reaches the pulmonary trunk by traversing a mitral valve and morphologic left ventricle; blood from the morphologic left atrium traverses the tricuspid valve and morphologic right ventricle reaching the aorta; the coronary arteries are similarly reversed, ie the right is anterior, the left, posterior; many of those with this complex have other cardiac malformations including conduction and ventricular septal defects, requiring surgical repair OPERATIVE MORTALITY 25% TEN-YR SURVIVAL 60%. See Transposition of great vessels

double effect MEDICAL ETHICS The use in terminally ill Pts of sedative and analgesic dosages high enough to relieve both pain and suffering–the 'good' effect, and hasten their demise–the 'bad' effect. See Euthanasia

double fracture A general term for two or more fractures in the same bone

double gloving The use of 2 gloves when performing medical interventions, in which there is contact with biohazardous materials; DG is a practice that minimizes the hazards inherent in contact with body fluids, as latex gloves are prone to perforation

double helix A structural motif of nucleic acids in which 2 complementary chains of DNA and/or RNA spiral around each other as paired nucleobases attached to a deoxyribose phosphate backbone

double-jointed *adjective* A popular term for extreme joint laxity

double innervation NEUROLOGY Referring to a nerve containing sympathetic and parasympathetic nerves. See Innervation

double lumen MEDTALK A duplication of a vessel or tubular structure

double pneumonia 1. An obsolete term for bilateral lobar pneumonia, a now rare condition 2. Mixed viral-bacterial pneumonia, where one follows another–eg, *Staphylococcus aureus*, infection, causing a 100-fold enhancement of infectivity and ↑ multiplication rate of influenza viruses, by producing a hemagglutinin-cleaving enzyme PROGNOSIS Mortality up to 42%

double reading The interpretation of any form of visual images–eg, from pathology–cytology, surgical pathology or radiology–CT, MRI, mammography by either a second 'pair of eyes'–ie pathologist or radiologist, or by the same physician at a different diagnostic session

double set-up examination OBSTETRICS A 2-team approach for a high-risk–eg, placenta previa, vaginal delivery, where the 1st team is prepared for a normal–ie, uneventful, vaginal delivery, while the 2nd team–which includes a gynecologist, anesthesiologist, sundry nurses, etc is on alert, ready to perform an immediate cesarean section, should the vaginal delivery 'go sour'. See Cesarean section

doubling time ONCOLOGY A parameter used to determine tumor aggressiveness, which serves to prognosticate, measure therapeutic success, and quantify tumor kinetics and growth rate. Cf Gompertzian growth curve

douching GYNECOLOGY The rinsing of the vagina and cervix with water or other solutions; as a contraceptive method, it is essentially useless; because the vagina has a normal acidic environment which is protective, frequent douching is ill-advised

doughnut sign IMAGING A descriptor for the radionuclide image of a large anterior wall MI, in which there is ↓ uptake of ^{99m}Tc stannous pyrophospate in the central infarcted zone; the zone is surrounded by a region of intermediate uptake; it is seen in larger infarcts and is associated with a poor prognosis. See Acute MI

Down syndrome Chromosome 21 syndrome, mongolism, trisomy 21 PEDIATRICS A chromosome disorder, first described by Langdon Down, due to an extra chromosome 21–trisomy 21 CLINICAL Mental retardation, characteristic facies, multiple malformations, major congenital heart defects, duodenal atresia, acute leukemia, ↑ risk of pneumonia, other infections PRENATAL SCREENING ↓

Maternal serum AFP levels, ↓ unconjugated etriol, ↑ chorionic gonadotropin. See Triple marker test

download ONLINE *noun* A file, regardless of the format and type of data, which is obtained from the internet or an intranet *verb* To transfer a data file, regardless of the format and type of information from the internet or intranet to one's computer system or PC. See Browser

downstream procedure MEDTALK Any therapy performed as a direct result of an abnormality identified in a diagnostic (upstream) procedure. See Diagnostic-therapeutic pair

downtime INSTRUMENTATION The amount of time a device is nonoperational, due to failure, malfunction, servicing needs, or shutdown. See Crash, Mean time between failure

doxacurium ANESTHESIOLOGY A long-acting muscle relaxing anesthetic. See Muscle relaxant, Nondepolarizing agent. Cf Depolarizing agent

doxazosin mesylate Cardura® GI DISEASE/UROLOGY A proton pump used for GERD, duodenal ulcers, hypersecretory conditions–eg, Zollinger-Ellison syndrome, BPH ADVERSE EVENTS Headache, asthenia, fever. See Benign prostatic hypertrophy, GERD

Doxepin® Zonalon® PHARMACOLOGY A tricyclic antidepressant, see there

Doxil® ONCOLOGY A liposome formulation of doxorubicin, used for KS, ovarian and other CAs. See AIDS, Doxorubicin, Kaposi sarcoma

doxycycline INFECTIOUS DISEASE A broad-spectrum antibiotic used for rickettsiosis–eg, Rocky mountain spotted fever, typhus fever, Q fever, rickettsialpox, tick fevers, RTIs from *Mycoplasma pneumoniae*, lymphogranuloma venereum, trachoma, inclusion conjunctivitis, etc, due to *Chlamydia trachomatis*, psittacosis–*C psittaci*, nongonococcal urethritis–*Ureaplasma urealyticum*, relapsing fever–*Borrelia recurrentis*, gram-negative microorganisms: Chancroid–*Haemophilus ducreyi*, plague–*Yersinia pestis*, tularemia–*Francisella tularensis*, cholera–*Vibrio cholerae*, *Campylobacter fetus*, *Brucella* spp, *Bartonella bacilliformis*, *Calymmatobacterium granulomatis*, malaria prophylaxis–*Plasmodium falciparum* in travelers to chloroquine and/or pyrimethamine-sulfadoxine resistant areas ADVERSE EFFECTS GI tract disturbances, anorexia, N&V, diarrhea, glossitis, dysphagia, enterocolitis, anogenital Candida overgrowth, photophobia

doxorubicin Adriamycin ONCOLOGY An anthracycline antibiotic used for leukemias, lymphomas, sarcomas, solid tumors ADVERSE EFFECTS BM suppression, alopecia, vomiting, stomatitis, dose-dependent cardiomyopathy. See Chemotherapy

DP Dorsalis pedis pulse, see there

DP/PT Dorsalis pedis, posterior tibial pulses

DPT vaccination An immunization that protects against diphtheria, pertussis–whooping cough, tetanus ADVERSE REACTIONS Anaphylactic shock within 24 hrs, encephalitis, residual seizure disorder, shock-like state, hypotonia, ↓ responsiveness by end of day 3. See also Vaccination, DTaP

drag queen Female impersonator, gynemimetic SEXOLOGY A ♂ with ♀ affect–often 'overplayed'; a ♂ homosexual and ♀ wannabe, with ♂ genitalia; DQs may take hormones to ↑ breasts, and thus are hormonally, but not surgically "enhanced." See Transsexuals, Transvestite. Cf Andromimetic

dragged disc phenomenon A finding seen in advanced retrolental fibrodysplasia–retinopathy of prematurity, in which a scar in the optic fundus 'drags' the disc and retinal vessels, displacing the macula; when severe, a 'V'-shaped or funnel-shaped detachment of the retina may occur. See Cat's-eye reflex

drainage SURGERY 1. The withdrawal of fluids from a body cavity or region 2. Discharge, see there. See Percutaneous transhepatic biliary

drape SURGERY *verb* To cover and mark off a field before performing a sterile procedure

draw LAB MEDICINE *noun* A popular term for a specimen of blood obtained by a phlebotomist and submitted to a lab for diagnostic purposes. Cf Short draw *verb* To obtain blood by venipuncture for diagnostic purposes. See Phlebotomy

drawer sign ORTHOPEDICS A clinical finding characterized by forward sliding–caused by disruption of the anterior cruciate ligament or backward sliding–posterior cruciate of the tibia

drawing station LAB MEDICINE A facility that collects specimens and sends them to a central lab, but does not perform tests of any kind. See CLIA 88, Laboratory. Cf Collecting station

dread disease A disease with a significant impact on lifestyle–eg, multiple sclerosis, longevity–eg AIDS, CA, which incurs high costs–eg, extensive burns, persistent vegetative state, and/or cause significant and permanent residual morbidity, ie loss of eyes or limbs

dream NEUROLOGY A series of images and thought processes that occur during sleep which, in the framework of psychoanalysis, are believed to have latent and manifest content; eyelid movement and REM sleep coincide during dreams; dreaming is more common during REM sleep. See Wet dream

dream therapy PSYCHIATRY The use of dreams as 'raw material' for analysis of a person's psyche. See Psychoanalysis

dressing SURGERY Any natural or synthetic material used to cover a wound. See Composite/foam dressing, Fabric, Hydrocolloid, Hydrogel, Pressure dressing, Polyurethane film

Dressler syndrome Post-myocardial infarction syndrome, see there

DREZ Dorsal root end zone NEUROLOGY A CNS pathway that may be surgically sectioned to relieve chronic pain–eg, anterolateral cord and trigeminal tract; DREZ lesions are produced by thermal coagulation or laser, and relieve 65-70% of intractable pain associated with brachial and lumbar plexus avulsion and spinal cord trauma

DRG Diagnosis-related group MANAGED CARE A unit of classifying Pts by diagnosis, average length of hospital stay, and therapy received; the result is used to determine how much money health care providers will be given to cover future procedures and services, primarily for inpatient care. See RBRVS, Utilization review. Cf CPT coding

drift ORTHOPEDICS See Pronator drift VIROLOGY Antigenic drift, see there

drink VOX POPULI An orally ingested liquid. See Carbohydrate drink, Isotonic drink, Protein drink, Sports drink, Standard drink

drip STD A popular term for the purulent penile discharge beginning 2-5 days after *N gonorrhoeae* infection

drive-by shooting PUBLIC HEALTH A phenomenon in which one or more persons–commonly members of street gangs, open fire à la Al Capone from moving vehicles, often in retaliation for an alleged wrong-doing by a rival gang

driveling Double talk, jargon agrammatism PSYCHIATRY Speech in which associations are tightly linked, syntax preserved, but meaning is lost CONDITIONS CAUSING Organic brain disease, schizophrenia

driving under the influence (of alcohol) DUI, driving while intoxicated, DWI PUBLIC HEALTH The act of operating a moving vehicle while under the influence of alcohol. See Alcohol, Moving vehicle accident

dronabinol Marinol® A synthetic Δ⁹THC-tetrahydrocannabinol, an active ingredient of marijuana, which may be used to manage glaucoma, pain,

204

N&V, anorexia, wasting linked to chemotherapy, AIDS ADVERSE EFFECT Drowsiness, confusion, poor coordination. See Marijuana, THC

drop attack NEUROLOGY An episodic and precipitous loss of motor function, where the victim is either standing or walking, and abruptly plummets, fully conscious to the floor, as the legs give way; idiopathic DAs are most common in older ♀, and attributed to age-related defects in reflexes; DAs may also occur in vertibrobasilar ischemia, acute labyrinthine vertigo, cataplexy, 'plateau waves'; DAs with loss of consciousness occur in syncope and seizures

dropfoot gait Equine gait, footdrop gait, steppage gait NEUROLOGY A gait in which the foot is lifted high enough for the sagging forefoot to clear the walking surface; DG is seen in paralysis of anterior tibial and peroneal muscles, and in lesions of anterior motor horn cells and cauda equina. See Gait

droplet spread EPIDEMIOLOGY The direct dissemination of a pathogen from a reservoir to a susceptible host's conjunctiva, nose or mouth–by spray with relatively large, short-ranged–±1 meter–aerosols produced by sneezing, coughing, or talking

dropout ADOLESCENT MEDICINE A person, often a high school student, who abandons his/her education before graduating CLINICAL TRIALS A person admitted to a clinical–phase III–trial for efficacy of a therapeutic agent or procedure, who abandons the role of 'guinea pig'

dropped bladder Cystocele, see there

dropping Lightening, see there

drowning PUBLIC HEALTH A mechanism of death that claims 7000 lives/yr–US, comprising 15% of non-MVA deaths; death is by asphyxia due to submersion, with aspiration of fluid; 90% of decedents were hypoxic. See Fresh water drowning, Salt water drowning, Wet drowning

DROWNING

FRESH WATER DROWNING Hypoosmolar water affects the surface tension of alveolar surfactant, causing an imbalance in the V/Q ratio with a collapse of some alveoli, resulting in both true (absolute) and relative intrapulmonary shunting; the V/Q abnormality is further compromised by pulmonary edema; the shifts of fluids and electrolytes in fresh water drowning result in hemodilution, hemolysis, circulatory overload, and hyponatremia

SALT WATER DROWNING Sea water aspiration results in fluid-filled but perfused alveoli, accompanied by a V/Q abnormality due to pulmonary edema; the shifts of fluids and electrolytes in salt water drowning result in hemoconcentration, CHF, and hypernatremia

drowsiness MEDTALK Semiconsciousness; grogginess, sleepiness

DRT Dead right there MEDTALK A macabre adjective referring to a Pt who has been clinical kaputt long enough to minimize the likelihood of resuscitation

drug NIHSPEAK Any chemical compound that may be used on or administered to humans to help diagnose, treat, cure, mitigate, or prevent disease or other abnormal conditions REGULATORY DEFINITION An article or substance that is 1. Recognized by the US Pharmacopoeia, National Formulary, or official Homeopathic Pharmacopoeia, or supplement to any of the above 2. Intended for use in the diagnosis, cure, mitigation, treatment or prevention of disease in man or animals 3. Intended to affect the structure or any function of the body of man or animals SUBSTANCE ABUSE Any medication; the word drug also carries a negative connotation–implying abuse, addiction, or illicit use. See Alternative drug, Antithyroid drug, Antituberculosis drug, Blockbuster drug, Brake drug, Butterfly drug, Category X drug, Cholesterol-lowering drug, Club drug, Club of Rome drug, Crude drug, Designer drug, Disease-modifying antirheumatic drug, Door-to-drug, Free drug, Gateway drug, Generic drug, Group C drug, Hard drug, Immunomodulatory drug, INAD drug, Investigational drug, Legend drug, Me too drug, Lifestyle drug, Narrow therapeutic index drug, Natural drug, New drug, Non-legend drug, Nonsteroidal antiinflammatory drug, Oligonucleotide drug, Orphan drug, Over-the-counter drug, Overseas mail-order drug, Performance enhancing drug, Pocket drug, Prescription drug, Probe drug, Prodrug, Pseudo-orphan drug, Psychoactive drug, Radioactive drug, Radiomimetic drug, Recreational drug, Second-line drug, Selective cytokine inhibitory drug, Soft drug, Treatment-investigational new drug, Wonder drug

drug abuse See Substance abuse

drug of abuse A self-administered agent with neurophamacologic activity, including illegal 'recreational' drugs–eg marijuana, cocaine, heroin, controlled prescription drugs, and 'pleasure poisons'–alcohol and nicotine See Substance abuse

drug allergy An immune response to a therapeutic. See Allergy

drug challenge The administration of a drug suspected of causing an adverse drug reaction, to assess culpability; DCs are used to detect aspirin sensitivity, and identify adverse reactions to contrast media and local anesthetics

drug delivery The method–eg, time-release pellets and route–eg, oral, parenteral by which a therapeutic is administered

Drug Enforcement Administration SUBSTANCE ABUSE The federal agency that enforces controlled substances laws and regulations. See Controlled drug substances

drug holiday PSYCHIATRY A period in which a prescribed drug is proscribed, to evaluate baseline behavior or ↓ the dose of psychoactive drugs, and possibly incidence of tardive dyskinesia. See Tardive dyskinesia

drug-induced hypoglycemia Hypoglycemia associated with drugs ENDOCRINOLOGY A state of ↓ serum glucose linked to use of various therapeutics–eg, sulfonylureas, oral hypoglycemics–eg chlorpropamide, tolbutamide, tolazamide, acetohexamide, alcohol, aspirin, phenformin, insulin

drug-induced immune hemolytic anemia HEMATOLOGY A hemolytic anemia linked to therapy with various drugs EXAMPLES Penicillins, cephalosporins, levodopa, methyldopa, mefenamic acid, quinidine, salicylic acid, sulfonamides, thiazide diuretics, chlorpromazine, INH, streptomycin, NSAIDs. See Autoimmune hemolytic anemia

drug-induced lupus erythematosus An SLE-like syndrome that may develop in Pts receiving certain drugs, especially procainamide, hydralazine, INH, phenytoin, mesantoin, D-penicillamine and ergot compounds. See Systemic lupus erythematosus

drug-induced psychosis NEUROLOGY A drug-induced psychiatric illness in which reality is impaired, typically linked to hallucinations or delusions, leading to communication or social problems

drug-induced tremor NEUROLOGY A tremor, typically worsened with activity and purposeful movement, unassociated with other Sx, caused by a drug or medication AGENTS Lithium carbonate, theophylline, Alupent, cyclosporine, and stimulants–eg, caffeine. See Tremor

drug interaction PSYCHIATRY The effects of 2 or more drugs when coadministered, resulting in altered effects of either drug taken alone; the drugs may have a potentiating or additive effect and serious side effects

drug of choice CLINICAL THERAPEUTICS A therapeutic that is regarded as being the best agent or first agent to use when treating a particular condition; DOCs usually have the lowest toxicity and the widest therapeutic range of the available agents. Cf Second-line drug

drug paraphernalia Controlled paraphernalia SUBSTANCE ABUSE As defined in a regulatory context, DP is a hypodermic syringe, needle, metal or plastic (snorting) tube, or other instrument or implement or combination adapted for the administration of controlled substances under circumstances which reasonably indicate an intention to use such instruments for the purposes of illegally administering any controlled dangerous substance. See Controlled substance

drug promotion PHARMACEUTICAL INDUSTRY Any activity–advertising, 'detailing,' 'freebies,' and sponsoring of conferences and symposia–by a drug company which is intended to ↑ the sales of its products

drug-related neuropathy Neuropathy secondary to drugs NEUROLOGY A condition characterized by a loss of sensation and/or movement due to an agent that damages peripheral nerve axons, blocking neural conduction; DRNs often begin as a distal polyneuropathy and progress proximally AGENTS Amiodarone, hydralazine, perhexiline, vincristine, cisplatin, metronidazole, nitrofurantoin, thalidomide, INH, dapsone, phenytoin, disulfiram

drug 'rep' PHARMACEUTICAL INDUSTRY A drug company employee who has regular contact with prescribing physicians; DRs provide details on the proper administration of new agents, and usually try to have the physician change prescribing practices in favor of the DR company's agents; some say that drug reps' only function is to provide "sponsored"/free meals. See Detailing

drug resistance The ability of bacteria and other microorganisms to withstand a drug to which they were once sensitive

drug sample Drug 'freebie' CLINICAL PHARMACOLOGY A unit of a prescription drug not intended to be sold, which is given in subprescription-sized dollops to promote the drug's sales

drug screening CLINICAL TOXICOLOGY A method for identifying the presence of one or more drugs of abuse, usually in urine, by 'stat' methods–eg, EMIT–enzyme-mediated immunologic technique, FPIA–fluorescence polarization immunoassay, TLC; DS is used in the ER, for preemployment testing, and identifies the most commonly abused substances–eg, cocaine, opiates, tranquilizers, sedatives–barbiturates, and hallucinogens–PCP, LSD, tetrahydrocannabinol

drug-seeking behavior MEDTALK Any activity–eg, visiting the ER with spurious complaints of pain, claiming allergy to other agents–especially, analgesics–with the same effect, which are not the sought agent; DSB is almost invariably focused on obtaining prescriptions to addictive controlled substances–eg, oxycodone

drug tampering CLINICAL PHARMACOLOGY The inappropriate or illegal alteration of a drug formulated under specified conditions; DT may occur either at the manufacturer or at the pharmacy before its distribution to the consumer. See Tylenol™ incident

drug tapering THERAPEUTICS The gradual discontinuation or reduction of a therapeutic dose, required by a Pt over a prolonged period of time, of a particular drug

drug tolerance PSYCHIATRY Repeated use of some substance or drug, often narcotics, so that ever larger doses are required to produce the same physiologic and/or psychologic effect obtained previously by a smaller dose.

drug utilization review HEALTH INSURANCE A study of drug prescriptions to evaluate appropriateness and cost-effectiveness of drug therapy

drunkenness The state of acute alcohol-induced inebriation, which is a factor in ½ of the 35,000 MVAs/yr–US; it plays a role in domestic violence, drownings, falls, fires, homelessness, homicides, suicides. See Sleep drunkenness

drusen OPHTHALMOLOGY Acellular and amorphous, yellow-white, occasionally confluent nodules composed of aggregated abnormal glycoproteins and glycolipids produced by, and adjacent to, the basal cells of the retinal pigment epithelium

dry drowning A phenomenon seen in 90% of resuscitated drowning victims, in whom aspirated water is minimal, presumably due to reflex laryngospasm with airway obstruction and asphyxia; a rapid return of spontaneous ventilation is typical, if CPR is begun early and period of anoxia brief. See Drowning. Cf Wet drowning

dry eye syndrome Conjunctivitis arida, keratitis sicca, keratoconjunctivitis sicca, xerophthalmia MEDTALK Dryness of eyes, often due to ↓ tear secretion CLINICAL Dry, greasy, thickened and focally denuded cornea, which may

progress to keratomalacia, corneal ulceration, necrosis, and 2° infections; DES may also occur in Sjögren syndrome, autoimmune disorders–SLE, rheumatoid arthritis, scleroderma, sarcoidosis, amyloidosis, hypothyroidism, vitamin A deficiency, and in periorbital lymphoproliferative disorders

dry gangrene A condition caused by chronic vascular occlusion that slowly progresses to severe tissue atrophy and mummification, often associated with peripheral vascular disease–eg, DM, ASHD. See Gangrene

dry labor OBSTETRICS Labor that follows spontaneous rupture of membranes with loss of amniotic fluid

dry mouth Xerostomia INTERNAL MEDICINE A manifestation of salivary gland dysfunction with ↓ secretion, resulting in discomfort accompanied by dry, atrophic and/or inflammed oropharyngeal mucosa; the salivary glands become inflamed and eventually undergo fibrosis; it is most common in older ♀; it is associated with salivary gland dysfunction ETIOLOGY Sjögren syndrome, autoimmune disorders–SLE, rheumatoid arthritis, scleroderma, systemic sclerosis, sarcoidosis, amyloidosis, hypothyroidism, RT ≥ 4000 cGy, drugs–eg, atropine, antihistamines, amphetamines, anticholinergics, antidepressants–Prozac, Paxil, Zoloft, Xanax, valium, antihypertensives–Lopressor, Vasotec, opioids, thallium poisoning, fever, dehydration, mechanical–eg calculus in salivary duct, etc–eg, vitamin A deficiency, HIV-1, *Candida* infections, CA, inadequate function of salivary glands–eg, parotid glands TREATMENT Pilocarpine

dry socket Alveolar osteitis ODONTOLOGY A complication in 1-2% of all tooth extractions, most commonly in molar teeth; the wounds consist of focal osteomyelitis, in which the clot in the socket disintegrates prematurely and becomes a nidus for oral bacteria CLINICAL Severe pain and foul odor without purulence; DS responds poorly to therapy; it must therefore be aggressively prevented, and local or systemic antibiotics at the time of extraction

dry tap True dry tap HEMATOLOGY A needle biopsy of BM, usually from the iliac crests, in which either blood without clot, or no material at all, is obtained ETIOLOGY Reticulum fibrosis, marrow necrosis or an extremely packed marrow, DTs occur in 5-10% of all BM biopsies; 60+% of DTs are malignant, most are lymphoproliferative disorders–eg hairy cell leukemia, AML, myelofibrosis, lymphoma, myeloma, metastatic CA; benign disease–eg iron deficiency anemia, marrow hypoplasia, hemosiderosis, granulomas, pernicious anemia, accounts for the remaining DTs

dryness See Vaginal dryness

DSM-III-R PSYCHIATRY Diagnostic & Statistical Manual of Mental Disorders–3rd Edition Revised; a classification system for mental illnesses developed by the American Psychiatric Association, currently in its 4th edition, DSM IV

DSM-IV PSYCHIATRY The current DSM, which standardizes criteria for diagnosing psychiatric nosologies. See DSM

DSM-IV-PC PSYCHIATRY A streamlined DSM-IV intended for primary care providers, which offers simple algorithms for diagnosing mental disorders. See DSM-IV

DSMB Data & Safety Monitoring Board CLINICAL RESEARCH A committee of independent clinical research experts who review data in ongoing clinical trials, ensuring that participants are not exposed to undue risk, and look for any differences in effectiveness between experimental and control groups. See Clinical research

DT vaccine VACCINOLOGY A diphtheria and tetanus vaccine that does not protect from pertussis; it is reserved for persons with adverse reactions to a DPT shot or a personal or family history of a seizure disorder or brain disease

DTaP Diphtheria-Tetanus-acellular Pertussis vaccine. See DTP vaccine

DTIC Dacarbazine, see there

DTP INFECTIOUS DISEASE Diphtheria-Tetanus-Pertussis vaccine. See DTP vaccine NEUROLOGY Distal tingling on percussion

DTP vaccine A vaccine against diphtheria, tetanus and pertussis, which contains a mixture of formaldehyde-inactivated diphtheria and tetanus toxoids, and a sterile suspension of killed *Bordetella pertussis* CONTRAINDICATIONS Acute febrile reaction; during suspected evolving neurologic illness; when a previous DTP resulted in an allergic reaction

DTR Deep tendon reflex, see there

DTs Delirium tremens Popular term for an acute organic psychosis seen 3-10 days after abrupt alcohol withdrawal CLINICAL Confusion, sensory overload, hallucinations–eg snakes, bugs, tremor, seizures, autonomic hyperactivity, cardiovascular defects, diaphoresis, dehydration LAB ↓ K^+, Mg^{2+} MANAGEMENT Hallucinations require hospitalization and haloperidol; abrupt alcohol withdrawal requires CNS depressants–eg benzodiazepines, phenobarbital; antipsychotics–eg, clopromazine should not be used. See Othello syndrome

dual-chamber pacing Physiologic pacing CARDIAC PACING Pacing in both the atria and the ventricles to artificially restore the natural contraction sequence of the heart. See Pacemaker

dual choice MANAGED CARE A health benefit offered by an employment group permitting those eligible of the group a voluntary choice of health plans, usually the employer's primary insurer and an HMO; an HMO-like health plan with an indemnity plan. See Point-of-service plan

dual diagnosis NEUROLOGY A diagnosis of an emotional disorder *and* a developmental delay, drug and alcohol use or a mental illness in the same person

DUB Dysfunctional uterine bleeding, see there

Duchenne muscular dystrophy An X-R disease caused by a deficiency of a muscle protein, dystrophin, which affects 1:3500 ♂, resulting in progressive muscular atrophy, wasting, and death by age 20, often related to respiratory–due to compromised diaphragm activity, or cardiac failure; calf and deltoid muscles display the typical finding of pseudohypertrophy. See Climbing up on oneself, Dystrophin, Pseudohypertrophy

ductal carcinoma in situ Intraductal carcinoma, DIN 3 SURGICAL ONCOLOGY A localized form of breast CA, in which malignant cells are confined to the duct wall; DCIS has a heterogeneous biologic behavior and morphology, and is detectable by mammography EPIDEMIOLOGY DCIS has ↑–2/10⁵–1973 to 14/10⁵–1992; it accounts for 12% of breast CAs–US SUBTYPES 1. Low nuclear grade without central necrosis–low to minimal risk of recurrence after wide excision 2. High nuclear grade *with* central necrosis–high recurrence despite RT and complete–conservative excision MANAGEMENT Lumpectomy, lumpectomy with RT, or by mastectomy

ductal carcinoma Infiltrating duct carcinoma, infiltrating carcinoma–not otherwise specified ONCOLOGY The major pathologic form of breast CA, which accounts for 50-75% of all invasive breast CAs PATHOLOGY To be defined as DC, 90% of tissue examined must have a ductal pattern; grossly, DC is indurated with a stellate pattern of extension; DC imparts an unripe pear sensation when cut with a fresh scalpel. See Axillary dissection, Estrogen receptors. Cf Lobular carcinoma, Medullary carcinoma

ductal intraepithelial neoplasia PATHOLOGY A range of preneoplasia that arises in the ductal epithelium of the breast; the DIN schema is intended to standardize the language used by various pathologists, restricting the term carcinoma to those neoplasms that have extended beyond the ductal basement membrane. Cf Breast cancer, Cervical intraepithelial neoplasia

ductal lavage CYTOLOGY The harvesting of cells from the mammary ducts in an attempt to identify CA at an early, possibly more treatable stage. See Breast cancer

due diligence INFORMATICS A legal term for efforts to intercept potential problems before they occur–eg, monitoring for fraudulent claims, ensuring privacy and security under HIPAA, via audit trails, user authentication and access controls. See HIPAA

dueling experts phenomenon Dueling PhDs syndrome EXPERT OPINIONS A popular term of uncertain utility for one-sided information which may be provided in court or a public forum, by partially informed scientists who, in absence of complete information about a condition, may appear to joust with opposing views

DUI Driving under the influence (of alcohol), see there

Duke Activity Status Index CARDIOLOGY A measure of a person's functional capacity based on a 12-item questionnaire that correlates with peak O_2 uptake during exercise testing

Duke's classification SURGICAL ONCOLOGY A system for staging colorectal carcinoma–which has been modified with time. See TNM classification

DUKE'S CLASSIFICATION

STAGE	DEPTH OF INVASION	5-YEAR SURVIVAL
A	Limited to mucosa	100.0%
B1	Muscularis propria, negative nodes	66.4%
B2	Penetrates muscularis propria, negative nodes	53.9%
C1	Limited to wall, positive nodes	42.8%
C2	Through wall, positive nodes	22.4%
(D)	Non-Duke's designation for distal metastases	3.0%

duly qualified MEDICAL PRACTICE *adjective* Referring to the satisfactory completion of a residency program approved by the Accreditation Council for Graduate Medical Education. See Accreditation

dumb down *verb* A popular term for simplifying language to a less sophisticated–ergo, 'dumb'–audience

dumping MEDICAL ETHICS Patient dumping The practice, often by private, for-profit hospitals, of transferring indigent, uninsured Pts to other, usually public, hospitals for economic reasons; Pt-transfer guidelines and laws are generally limited to cases of 'unstable' emergencies and ♀ in active labor. Cf 'Anti-dumping' laws SURGERY See Dumping syndrome

dumping syndrome GASTROENTEROLOGY A disease complex confusing to both those who read about it and to those who write about it, seen in about 20% of those subjected to gastric surgery, including resection, gastroenterostomy with total gastric vagotomy, and gastric bypass; most inculpated in 'dumping' are pyloric ablation and bypass CLINICAL Diaphoresis, nausea, dizziness, palpitations, colicky abdominal pain and diarrhea, due to rapid movement–dumping of gastric contents into the small intestine; these syndromes can be divided into LAB ↑ Glucose–worse with high carbohydrate meals, ↑ hematocrit, ↓ blood volume related to dehydration, ↓ K^+ MANAGEMENT, MEDICAL ↓ Carbohydrate intake, smaller meals, pectin–a dietary fiber, acarbose, anticholinergics, L-dopa, opiates TREATMENT, SURGICAL 2-5% are medical failures, and require surgical conversion to a Roux-en-Y

duodenal atresia EMBRYOLOGY A birth defect in which a segment of the duodenum is lumenless. See Atresia

duodenal 'sweep' GI DISEASE A term used by both radiologists and endoscopists, for the 1st and 2nd segments of the duodenum as it sweeps superiorly and posteriorly past the gastric antrum

duodenal ulcer An ulcer of the duodenum EPIDEMIOLOGY *H pylori* infection of stomach, NSAIDs and other anti-inflammatory agents, and cigarettes CLINICAL Pain, often post-prandial which may not correlate with presence or severity of ulcers DIAGNOSIS Barium swallow, endoscopy COMPLICATIONS Bleeding, perforation, gastric obstruction MANAGEMENT Antibiotics–eg, amoxicillin, metronidazole to eradicate *H pylori*, ↓ risk factors, prevent complications

duplicate testing LAB MEDICINE The inappropriate repeating of lab or other diagnostic evaluations–eg, CBC, U/A, CK-MB, BMP, more often than allowed by Medicare or third party payers

duplicative drug 'Me too' drug, see there

DUR Drug utilization review, see there

durable power of attorney Continuing power of attorney DEATH & HEALTH CARE An 'advance directive' document that allows Pts to appoint a surrogate decision-maker to implement preferences for continued life support in the event of incapacitation. See Advance directive, Euthanasia, Living will

DurAct® Bromfenac PAIN MEDICINE A prescription NSAID-analgesic similar in efficacy to ibuprofen that was 'pulled' from the market by its producer, Wyeth-Ayerst, when it was linked to fulminant liver failure, death or need for liver transplantation

Duragesic® Fentanyl transdermal, see there

dural graft matrix NEUROSURGERY A biosynthetic substrate indicated for repair of the dura mater in the cranial cavity and spinal cord

Duraquin® Quinidine, see there

DuraSite® OPHTHALMOLOGY A sustained-release system designed to permit the gradual release of a drug into the eye over a period of hrs, overcoming problems common with conventional drug delivery

duration of action THERAPEUTICS The length of time that a particular drug is effective

dust OCCUPATIONAL MEDICINE A suspension of solid particles in air. See Coal dust, Inhalant, Nonasbestos dust, Nuclear dust

dust mite House dust mite, see there

duty to warn AIDS A legal concept indicating that a health care provider who learns that an HIV-infected Pt is likely to transmit the virus to another identifiable person must take steps to warn that person

DV 1. Daily value, see there 2. Dorsoventral

DVD Digital versatile disk, digital video disk INFORMATICS An optical disk capable of storing ≥ 4 GB of audiovisual data. Cf CD

DVT Deep vein thrombosis, see there

dwarf megakaryocyte 1. Micromegakaryocyte, see there 2. A misnomer for a fragment of cytoplasm from a megakaryocyte

dwarfism Nanosomia Excessively short stature–eg, ≤ 152 cm/5 ft in ♂ and ≤ 145 cm/4'9" in ♀; 35% of dwarfism is familial, 25% is idiopathic, 10% is due to pituitary failure, 10% to hypothyroidism, 10% to congenital gonadal aplasia, and the rest, etc; proper classification of the more than 55 congenital conditions associated with dwarfism allows determination of the likelihood of conceiving a similarly afflicted child. See Bird-headed dwarfism, Pituitary dwarfism, Psychosocial dwarfism, Silver-Russell dwarfism, Thanatophoric dwarfism

DWI Driving while intoxicated, see there

Dx Diagnosis

dyad PSYCHIATRY A 2-person relationship–eg, a therapeutic relationship between doctor and Pt in psychotherapy. See Folie á dieux. Cf Shared delusion

dyazide A potassium-sparing diuretic. See Diuretics, Potassium-sparing diuretic

dying GENERAL MEDICINE A poorly understood phenomenon characterized by a gradual systemic shutdown, followed by an absence of criteria that define life; dying and death eventually occur in the elderly, even without identifiable disease. See Brain death, DNR, Hospice

dying-back neuropathy Distal axonopathy NEUROLOGY A pattern of neuropathy seen in 'toxic' damage to large diameter peripheral sensorimotor nerves, affecting the long axons–eg, lower extremities, before the short–eg, cranial nerves; the condition is a generic reaction to beri-beri, pontocerebellar atrophy, spastic paraplegia, and thallium intoxication

Dynabac® Dirithromycin INFECTIOUS DISEASE A broad-spectrum antibiotic used to manage exacerbation of chronic bronchitis by *H influenzae* and skin infection due to *S pyogenes*

dynamic cardiomyoplasty HEART SURGERY A technique for treating moderately severe–NYHA class III heart failure in which a skeletal muscle, often the latissimus dorsi, is transplanted from its usual insertions in the proximal humerus to the chest wall and wrapped around the heart, and from the fascia to the pericardium; the muscle is stimulated to convert fibers from slow to fast twitch, while synchronously pacing the muscle wrap to ventricular systole INDICATIONS For Pts who aren't heart transplant candidates and don't respond to medical therapy. See Heart failure. Cf Batista operation

dynamic graciloplasty SURGERY A procedure for treating intractable fecal incontinence where the gracilis muscle is transposed into the anus, with implantation of stimulating electrodes and a pulse generator

dynamic hip screw ORTHOPEDICS Orthopedic hardware designed to resist angular deformation, while permitting early fracture impaction, with shortening along the lag screw's axis; the DHS is designed to treat intertrochanteric fractures, but may be used for subtrochanteric fractures; it has a side plate attached to the distal femur with several cortical screws

dynamic MRI ORTHOPEDICS A format for MRI in which a joint is viewed in real time. See Magnetic resonance imaging

dynamic testing LAB MEDICINE A testing format in which 2+ samples of Pt blood or urine are obtained at a specified time interval. See Glucose tolerance test, Timed specimen, Xylose absorption test

dyphylline A theophylline-like bronchodilator and vasodilator. See Imipramine

dys- MEDTALK *prefix* Greek; abnormal, bad, impaired, mis-, not good, difficult, nonfunctional

dysacusis Dysacousia, dysacusia NEUROLOGY 1. A hearing impairment caused by a signal processing defect of the CNS, auditory nerve, or organ of Corti; dysacusis can't be improved by increasing the signal amplitude and thus can't be measured in decibels. Cf Hearing impairment 2. Pain or discomfort in the ear caused by loud sounds

dysarthria NEUROLOGY A group of speech disorders caused by disturbances in the strength or coordination of speech muscles due to damage to the brain or nerves; dysarthria may indicate ↑ posterior fossa pressure on the brainstem/medulla oblongata CLINICAL Difficulty in speaking or forming words. See Speech pathology

dysautonomia NEUROLOGY Any condition characterized by sympathetic or parasympathetic derangements; autonomic hypofunction or failure is most often caused by drugs and disease-associated polyneuropathies–eg, DM and amyloidosis, but may be idiopathic CLASSIFICATIONS Primary dysautonomia system, sanctioned by the American Academy of Neurology; Goldstein classification. See Familial dysautonomia aka Riley-Day syndrome

dyscrasia See Blood dyscrasia, Plasma cell dyscrasia, Plasma cell dyscrasia with polyneuropathy

dysentery GASTROENTEROLOGY A generic term for functional colitis ETIOLOGY Pathogens, chemical irritants CLINICAL Colicky pain, tenesmus, loose or frequent stools, accompanied by blood and/or mucus. See Amebic dysentery. Cf Inflammatory bowel disease

dysequilibrium NEUROLOGY A derangement in equilibrium, attributed to altered signaling to and/or from the visual, vestibular, proprioceptive sensory pathways CLINICAL Vertigo, ataxia of gait, speech. See Dizziness, Vertigo. Cf Equilibrium

dyserythropoiesis HEMATOLOGY Any defect of RBC production characterized by morphologic abnormalities of the nuclei and cytoplasm in the BM, which may be acquired–eg pernicious anemia, sideroblastic anemia, 2° to myeloproliferative disorders or erythroleukemia, or congenital–eg thalassemia, or congenital dyserythropoietic anemia. See Congenital dyserythropoietic anemia

dysfibrinogenemia A group of qualitative, usually AD, fibrinogen defects ranging in severity from innocuous to hemorrhagic diathesis; most are asymptomatic and detected by presurgical screens, given the abnormalities in coagulation parameters; these subjects suffer frequent spontaneous abortion, bleeding, poor wound healing, and thrombosis LAB Normal fibrinogen and clotting times; ↑ PT, ↑ thrombin time, ↑ reptilase time. See Fibrinogen

dysfluency NEUROLOGY A speech rhythm disorder–eg, stuttering, often characterized by the repetition of a sound, word, or phrase. See Speech disorder

dysfunction A defect in function of one or more tissues. See Axillary nerve dysfunction, Biomechanical dysfunction, Brachial nerve dysfunction, Chemosensory dysfunction, Common peroneal nerve dysfunction, Constitutional liver dysfunction, Distal median nerve dysfunction, Erectile dysfunction, Femoral nerve dysfunction, Medication-induced allograft dysfunction, Primary autonomic dysfunction, Radial nerve dysfunction, Sciatic nerve dysfunction, Sexual dysfunction, Sinus node dysfunction, Somatic dysfunction, Tibial nerve dysfunction, Ulnar nerve dysfunction

dysfunctional family PSYCHOLOGY A family with multiple 'internal'–eg sibling rivalries, parent-child– conflicts, domestic violence, mental illness, single parenthood, or 'external'–eg alcohol or drug abuse, extramarital affairs, gambling, unemployment—influences that affect the basic needs of the family unit

dysfunctional uterine bleeding GYNECOLOGY Excess menstrual hemorrhage of hormonal origin, related to 'breakthrough bleeding' or estrogen withdrawal, often occurring in anovulatory cycles ETIOLOGY No cause is found in 75% of cases, although adolescent DUB is attributed to immaturity of the hypothalamic-pituitary-ovarian axis; peri- and post-menopausal DUB often occurs in endometria that are deaf to the ovary's curtain call; DUB in ♀ elderly requires curettage to exclude malignancy. See Dysmenorrhea

dysgenic GENETICS Tending to promote survival or reproduction of less well-adapted individuals especially at the expense of well-adapted individuals of a population. See Adaptation, Population, Reproduction, Survival

dysgeusia NEUROLOGY A distortion or perversion of the sense of taste, impaired taste. See Taste

dyskinesia NEUROLOGY An alteration in muscle movement. See Biliary dyskinesia, Tardive dyskinesia

dyskinetic gait Athetotic gait, see there

dyslexia PEDIATRICS An inability or unexpected difficulty in learning to read despite adequate IQ, motivation and education CLINICAL Word-blindness, tendency to reverse letters and words DIAGNOSIS Texas Primary Reading Inventory, functional MRI. See Congenital word blindness, Developmental reading disorder, Learning disability, Primary reading disability, Specific reading disability, Texas Primary Reading Inventory, Word blindness. Cf Reading retardation

dyslipidemia METABOLIC DISEASE Any defect in lipoprotein metabolism–eg ↑ cholesterol, ↑ TGs, combined hyperlipidemia, and ↓ HDL-C; dyslipidemias may be 1°–ie nosologies *a sui generis* or 2° to other medical conditions–eg, DM or hypothyroidism

dysmenorrhea MEDTALK Menstrual pain, painful periods. See Dysfunctional uterine bleeding

dysmorphology NEONATOLOGY The systemic study of structural defects of prenatal onset, a complex field in which single or multiple primary malformations are idiopathic or related to chromosome defects–recurrence rate of 2-5%, drugs, chemicals, toxins or radiation. See Deformation, Disruption, Malformation, Multiple malformation syndrome, Sequence

dysmyelopoietic syndrome HEMATOLOGY A hematologic cancer/precancer–eg idiopathic sideroblastic anemia, refractory anemia with excess blasts, subacute and oligoblastic leukemia, CML, preleukemia–which has considerable overlap; most predictive of future behavior is the presence of excess blasts in BM and circulation, neutropenia, thrombocytopenia. See Dyserythropoiesis

dysosmia NEUROLOGY A defect in the sense of smell. See Smell

dyspareunia Painful sexual intercourse in ♀. See Coital migraine

dyspepsia GASTROENTEROLOGY 1. Formally, a compromised ability to digest food 2. Popularly defined as postprandial epigastric discomfort. See Nonulcer dyspepsia

dysphagia INTERNAL MEDICINE Difficulty or inability to swallow, a finding that may indicate a brainstem tumor. See Malignant dysphagia. Cf Deglutition

dysphasia Dysphrasia NEUROLOGY A speech impairment and/or inability to produce recognizable speech CLINICAL Defects in perception, sound discrimination, auditory memory, comprehension, word-finding, dysplexia ETIOLOGY Tumors of dominant cerebral hemisphere–ie, frontal, temporal, parietal lobes. See Aphrasia

dysphonia NEUROLOGY Speech impairment or difficulty, often due to vocal cord dysfunction. See Spasmodic dysphonia

dysphoria NEUROLOGY Unpleasant mood. See Gender dysphoria

dysplasia Defective growth PEDIATRICS 1. Altered growth of a tissue or structure EXAMPLES Anhidrotic ectodermal dysplasia–Christ-Siemans syndrome, Holt-Oram syndrome–atrial digital dysplasia, chondroectodermal dysplasia–Ellis-van Creveld syndrome, hidrotic ectodermal dysplasia–Clouston syndrome, metaphyseal dysplasia–Pyle's disease, oculoauriculovertebral dysplasia–Goldenhaar syndrome, oculodentodigital dysplasia–ODD syndrome, olfactogenital syndrome–Kallmann syndrome, and progressive diaphyseal dysplasia–Camurati-Engelmann syndrome. See Anhidrotic ectodermal dysplasia, Congenital hip dysplasia, Coronal dysplasia, Cortical dysplasia, Familial focal facial dermal dysplasia, Fibrodysplasia ossificans progressiva, Hypohidrotic dysplasia, Metaphyseal chondrodysplasia, OSMED dysplasia, Osteofibrous dysplasia, Otopalatodigital dysplasia, Radicular dentin dysplasia, Retinal dysplasia, Right ventricular dysplasia, Septo-optic dysplasia, Spondyloepiphyseal dysplasia, Valve dysplasia 2. Bronchopulmonary dysplasia, see there 3. Fibrous dysplasia, see there IMAGING A term for unusual shadows seen on a mammogram which suggest possible breast CA. See Breast cancer, Mammography. Cf Mammary dysplasia SURGICAL PATHOLOGY A premalignant lesion of epithelia, especially squamous epithelium of the uterine cervix, oral cavity, upper respiratory tract, penis, anus, etc; epithelial dysplasia may be induced by HPV, especially types 16, and is synonymous with intraepithelial neoplasia, the term preferred by pathologists. See Bowenoid dysplasia, Minimal cervical dysplasia. See Ductal intraepithelial neoplasia, Fibrocystic disease, Mammary dysplasia

dysplastic nevus DERMATOLOGY A premalignant skin lesion characterized by irregular, > 5 mm in diameter macules numbering from a few to hundreds with a central papule, variegated dark color and lenticular changes

dyspnea Breathlessness, shortness of breath, SOB PULMONARY MEDICINE Difficult painful breathing, SOB or respiratory distress; dyspnea is subjective, difficult to quantify, and may indicate serious disease of the heart, lungs, or airways. See Nocturnal dyspnea, Paroxysmal nocturnal dyspnea

dyspnea on exertion CARDIOLOGY Shortness of breath which occurs with effort, often a sign of heart failure or ischemia

dyspraxia 1. Clumsy child syndrome, see there 2. An extinct term for impaired or painful function of any organ of the body. See Speech dyspraxia NEUROLOGY A condition characterized by defective voluntary movement despite intact sensory and motor function TYPES Constructional, dressing, ideational, ideomotor, oromotor

dysthymic disorder Minor depression PSYCHIATRY A condition characterized by '…*a chronically depressed mood that occurs for most of the day, for more days than not, for at least 2 yrs*…(persons so afflicted) *describe their mood as sad or 'down in the dumps'*. Cf Major depressive episode

dystocia Difficult childbirth, difficult labor OBSTETRICS A period of nonprogression of labor ≥ 4 hrs after the cervix has dilated to 3 cm ETIOLOGY Abnormal presentation, too small a birth canal or uterine dysfunction

dystonia NEUROLOGY Involuntary, often acute movement and prolonged contraction of one or more muscles, resulting in twisting body motions, tremor, and abnormal posture ETIOLOGY Inherited, ↑ during voluntary movement, nervousness, and emotional stress; dystonias may affect the extrapyramidal system–despite lack of demonstrable lesions, task specific–eg, writer's cramp or associated with medications–especially antipsychotics or disease–eg, a form of lung cancer LOCATION Tongue, jaw, eyes, neck, trunk, extremities, occasionally the whole body MANAGEMENT Dopamine, sedatives–benzodiazepines–eg diazepam, anticholinergics–eg, trihexyphenidyl, benztropine–which cause dry mouth, mydriasis, urinary retention, and visual hallucinations; surgery. See Cranial dystonia, Dopa-responsive dystonia, Focal dystonia, Idiopathic torsion dystonia, Oromandibular dystonia, Segawa's dystonia, Torsion dystonia

dystonic tic Complex tic NEUROLOGY A distinct, coordinated pattern of sequential movements, which appear purposeful–eg, touching the nose, touching other people, smelling objects, jumping, copropraxia, and echopraxia, See Tic

dystrophic calcification The combination of fat necrosis and caseating necrosis, resulting in the focal deposition of hydroxyapatite crystals in previously damaged tissues–eg, heart valves, scars, foci of TB and atherosclerotic blood vessels–arising in mitochondria, calcification in hyperparathyroidism which develops in the basement membrane of the renal tubules; DC may occur without hypercalcemia or defects of calcium metabolism

dystrophy MEDTALK Partial atrophy of tissue or an organ attributed to ↓ nutrition. See Becker's muscular dystrophy, Benign pseudohypertrophic muscular dystrophy, Hyperplastic dystrophy, Lattice dystrophy, Map-dot dystrophy, Myotonic dystrophy, Neuroaxonal dystrophy, Reflex sympathetic dystrophy, Spider dystrophy, Third dystrophy, Twenty nail dystrophy. Cf Atrophy

dysuria Pain on urination, difficulty urinating UROLOGY A common Sx of bladder infection, which also occurs in prostate lesions, eg, BPH, accompanied by sensation of pressure and tenderness around the bladder, pelvis, and perineum, which may ↑ as the bladder fills and ↓ as it empties; ↓ bladder capacity; an urgent need to urinate; painful sexual intercourse; in ♂, hyperpain in the penis and scrotum. See Cystitis

E Symbol for: 1. Electricity 2. Electronic 3. Emergency 4. Energy 5 Enzyme 6. Eosinophil 7. Epinephrine 8. Erythrocyte 9. Esophagus 10. Estradiol 11. Ethanol 12. Etiology 13. Expire 14. Redox potential 15. Voltage

e Symbol for: 1. Electrical potential 2. Electron 3. Ethyl 4. Natural logarithm–2.7187818285

E-CABG Endoscopic coronary artery bypass graft CARDIOVASCULAR SURGERY Endoscopic CABG performed with optical and surgical instrumentation through tiny incisions in the intercostal spaces. See CABG

E-care INFORMATICS The automation–E = electronic–of all aspects of care delivery processes across administrative, clinical and departmental boundaries in the healthcare delivery system

E-code MANAGED CARE A special ICD-9-CM diagnosis code used on the HCFA 1500–the physician claims form, generally to report accidents, injuries or diseases which, for a short period of time, included secondhand smoke; E-codes can be used in conjunction with ICD-9CM codes. See ICD-9-CM

E coli O157:H7 See *Escherichia coli* O157:H7

e-commerce E business INFORMATICS The service, sales and collaborative business conducted over the Internet, either business ↔ consumer or business ↔ business. See E-care, E-health

e-disease management Online disease management, see there

E5 E5 murine antiendotoxin monoclonal antibody therapy INFECTIOUS DISEASES An IgM anti-endotoxin monoclonal antibody reactive against lipid A of a mutant strain–J5 of *E coli*. Cf HA-1A therapy

E-health INFORMATICS A philosophy that empowers–E = electronic–health care consumers by bringing information, products and services online

e-mail Electronic mail, email COMPUTERS Scripted communication and attachments sent over a local or wide-area network or Internet to a distant recipient. See Listserv®, Mail-list. Cf Snail mail, Voice mail

e-prescribing THERAPEUTICS The use of handheld electronic products to communicate with pharmacies and provide prescribing information

E/M services Evaluation & management services, see there

EAEC Enteroadherent *Escherichia coli*, see there

ear PHYSICAL EXAM The auditory apparatus, which is divided into the external ear–a conical tube that collects sound that vibrates the tympanic membrane–the outer barrier of the middle ear, which contains the ossicles–malleus, incus, and stapes, that mechanically amplify the sound transmitted at the oval window to the cochlea; the cochlea's neuroepithelial hair cells convert the mechanical signal into an electrical/neural signal that is identified by the brain as sounds, speech, music, etc. See Blue ear, Cauliflower ear, Inner ear, Lop ear, Malrotated ear, Middle ear, Mozart ear, Outer ear, Outstanding ear, Satyr ear, Swimmer's ear, Third ear

ear infection AUDIOLOGY The presence and growth of bacteria or viruses in the ear. See Middle ear otitis, Otitis

ear inflammation 1. Middle ear otitis 2. Otitis

ear tag Preauricular tag A common minor skin defect, consisting of a rudimentary tag of tissue, often with central cartilage, usually located just in front of the ear

ear trauma Acoustic trauma, see there

ear wax AUDIOLOGY A yellow secretion from glands in the outer ear–cerumen that keeps the skin of the ear dry and protected from infectionVOX POPULI → MEDTALK Wax blockage, see there

eardrum Tympanic membrane, see there

early abortion OBSTETRICS An abortion performed before the 12th wk of gestation. See Abortion

early diastolic murmur CARDIOLOGY A heart murmur that begins right after the second heart sound, which is typical of aortic regurgitation. See Aortic regurgitation

early disseminated Lyme disease Secondary Lyme disease, see there

early dumping syndrome see Dumping syndrome

early pregnancy termination OBSTETRICS An abortion induced early in pregnancy, ideally before the 3rd month; EPT can be carried out using RU 486–mifepristone, and antiprogestin, misoprostol, a PGE₁ analogue. Cf Late abortion, Partial birth abortion

early pregnancy OBSTETRICS First trimester of pregnancy

early puberty PEDIATRICS The development of signs of sexual maturity before age 8 in ♀ and before age 9 in ♂; some children have changes as early as age 3 or 4; in general there is no identifiable cause in ♀; half of ♂ have underlying medical problems–eg, hormone secreting tumors. See Puberty

early stage breast cancer Breast CA confined to the breast–ie, not spread locally or metastasized

early uterine evacuation Vacuum abortion, see there

Eastern equine encephalitis A rare, sporadic, and aggressive enzootic infection by a single-stranded RNA Togavirus that primarily affects birds VECTOR Ornithophilic mosquito, *Culiseta melanura*, largely confined to the Northeast US, especially Massachusetts; infection of horses and humans is an accidental 'dead-end', occurring when the virus is transmitted to other mosquitoes–eg, *Aedes vexans, Aedes sollicitans, Coquillitidia perturbans*; ± 5 human cases/yr in the US–CLINICAL. Meningismus, lethargy, stupor, high fever, spinal pleocytosis PROGNOSIS 30-70% mortality; severe neurologic sequelae. Cf St Louis equine encephalitis, Western equine encephalitis

Easy Rider microcatheter CARDIOLOGY An over-the-wire catheter used for neurovascular applications–eg management of AV malformations, brain aneurysms, head and neck tumors. See Embolyx™

eating disorder PSYCHIATRY Any of a group of conditions–eg, anorexia nervosa–AN, bulimia nervosa–BN, binge-eating disorder–BED and variants–characterized by a serious disturbance in eating–eg, a marked ↓ in intake or bingeing and distress or excessive concern about body shape or weight, which may have an adverse effect on health due to physiologic sequelae of altered nutrition or purging EPIDEMIOLOGY More common in industrialized societies, where 3% of young ♀ have ED; 5-15% of AN and 40% of BED occur in ♂; 50% of EDs are unrecognized CLINICAL Extremely low weight, hypotension, bradycardia, hypothermia, dry skin, hypercarotenemia, lanugo, acrocyanosis, breast atrophy, amenorrhea, delayed puberty, swollen salivary glands, abnormal dentition, perimolysis, abrasions on the dorsum of the hand, prolonged QT interval, reduced left ventricular mass LAB Normal in absence of emesis; ↓ K⁺ with ↑ bicarbonate is linked to frequent vomiting or diuretic use; non-anion gap acidosis is linked to laxative abuse; ↓ Na⁺ occurs in AN, reflecting ↑ water intake or SIADH; also hypoglycemia, leukopenia, neutropenia, anemia, thrombocytopenia and ↓ TSH; ↑ cortisol MANAGEMENT Medical treatment for complications of abnormal weight and purging, TPN for severe malnutrition; psychiatric treatment–individual, group or family therapy; psychodynamic psychotherapy; psychopharmacology is ± effective for most EDs except AN PROGNOSIS 80% recovery partially or completely. See Anorexia nervosa, Bulimia nervosa, Binge-eating disorder

ebastine ENT A nonsedating antihistamine for managing seasonal allergic rhinitis. See Allergic rhinitis

EBD Emotional or behavioral disorder

eburnation Bone sclerosis A marbled appearance of weight-bearing joints with complete cartilaginous erosion, leaving polished, sclerotic bone as the new articular surface. See Osteoporosis

EBV Epstein-Barr virus

EC 1. Electronic commerce, see there 2. Enzyme Commission, see there

ECASS CARDIOLOGY An international, double blinded, randomized trial–European Cooperative Acute Stroke Study which evaluated effects of thrombolytics in Pts with stroke. See Thrombolytic therapy, tPA

ECC Endocervical curettage, see there

eccentric *adjective* Not central, peripheral *noun* A person not regarded as 'normal' by his peers, friends, or family

eccentric contraction Negative contraction SPORTS MEDICINE Muscle contraction that occurs while the muscle is lengthening as it develops tension and contracts to control motion by an outside force. Cf Concentric contraction

eccentric training SPORTS MEDICINE The lengthening of a muscle tendon unit while active, resulting in a negative movement, required under conditions of rapid deceleration; eccentric forces are required to reverse the body's trajectory after a particular athletic move–eg, jumping and throwing

ecchymosis Internal bruising or bleeding

ECFMG Educational Commission for Foreign Medical Graduates GRADUATE EDUCATION An organization formed by the Am Hospital Assn, AMA, Am Bd of Medical Specialties, Assn of Am Medical Colleges, etc, to establish standards and evaluate qualifications of foreign medical school graduates. See Foreign medical graduate, International medical graduate

ECG 1. Electrocardiogram, see there, EKG 2. Echocardiogram, see there

echinococcal cyst PARASITOLOGY A cyst caused by *Echinococcus gran-*

ulosus LAYERS Outer or pericyst–composed of host-derived inflamed fibrous tissue; intermediate acellular layer–ectocyst; innermost endocyst from a developed infection. See *Echinococcus* spp

echinococcosis Hydatid cyst disease, see there

***Echinococcus* spp** A genus of small taeniid tapeworms with 2–5 segments in adults–found in the definitive carnivore hosts–dogs, foxes; echinococcal larvae breed within cysts which develop in the liver, brain or elsewhere in the intermediate hosts–sheep, pigs, rodents, horses, humans; exposure and current infection are evaluated by measuring antibodies. See Hydatid disease

echinocyte Burr cell, crenated cell HEMATOLOGY An RBC with 30+ crenations, bumps or spurs, which reflect damage to the *normal* cell membrane by various lytics–eg, saponin, bile salts, ionic detergents, lecithin; slow drying; aged blood; rarely, echinocytes reflect disease–eg, uremia or pyruvate kinase deficiency. See Red blood cells

echocardiography Doppler ultrasonography CARDIOLOGY A group of noninvasive 2-D imaging techniques that take advantage of the Doppler effect to provide information on the size, shape, and motion of cardiac structures, pressure differences in chambers, and blood flow through the heart and great vessels APPLICATIONS Assess pericardial effusion, endocarditis, cardiac hypertrophy, congenital, valvular disease–mitral valve stenosis or prolapse, aortic insufficiency, aortic stenosis, subaortic stenosis, tricuspid valve, ischemic disease. See Biplane transesophageal echocardiography, Contrast echocardiography, Interventional cardiology, Intracardiac echocardiography, M-mode echocardiography, Myocardial contrast echocardiography, Stress echocardiography, Transthoracic echocardiography, Two-dimensional echocardiography

EchoFlow CARDIOLOGY A diagnostic device based on Doppler technology, which measures blood flow and velocity in the circulation during and following vascular/cardiovascular procedure

echolalia NEUROLOGY The parroting by a Pt of another person's words or speech fragments

echopraxia NEUROLOGY The parroting by a Pt of another person's actions, which may be seen in catatonic schizophrenia. See Schizophrenia

eclampsia/pre-eclampsia (From Greek *eklampsis*, shining forth) Metabolic toxemia of pregnancy OBSTETRICS A condition which usually develops in late pregnancy or the immediate puerperium CLINICAL HTN, hemoconcentration, sodium retention with resultant edema LAB Albuminuria, proteinuria, hypoproteinemia, ↑ nitrogen/BUN; pre-eclampsia is most common in primigravidas, after the 24th gestational wk, but may occur as soon as trophoblastic tissue is present TREATMENT If mild, bed rest and sedation; if severe, antihypertensives–eg, vasodilators, α methyldopa; if convulsions, magnesium sulfate. See HELLP syndrome

Eclipse™ VASCULAR SURGERY A laser for performing transmyocardial revascularizationn–TMR and percutaneous transluminal myocardial revascularization–PTMR

ECM External cardiac massage

ECOG Eastern Cooperative Oncology Group

ecstasy Hug drug, love drug SUBSTANCE ABUSE An oral designer analogue of amphetamine, a 'schedule I' controlled substance which may be fatal due to heat exhaustion and dehydration, combination with methadone, LSD, opiates–eg, heroin or Fentanyl, or anesthetics–eg, Ketamine; it is a popular 'recreational' drug of abuse, especially in a dance-party–see Rave–setting; at moderate doses, it causes euphoria, sense of well-being, enhanced mental or emotional clarity; at higher doses, hallucinations, sensations of lightness, depression, paranoid thinking, violent behavior TOXICITY Serotonin neurotoxicity, sweating, dilated pupils, blurred vision, tachycardia, arrhythmias, fever, spasticity, hypotension, bronchospasm, acidosis, anorexia, N&V, HTN, faintness, chills, insomnia, convulsions, loss of voluntary muscle control, anxiety, or paranoia. See Designer drugs, 'Ice', Rave party. Cf Eve

ECT Electroconvulsive therapy, see there

ecthyma DERMATOLOGY A chronic subcutaneous infection by β-hemolytic streptococci that develops usually on the legs, in a background of a pruritic lesion–eg, insect bite, scabies, pediculosis; skin becomes crusted, weeping, and later scarred MANAGEMENT Warm compresses, antibacterial soap, H_2O_2, systemic penicillin. See Impetigo

ectopia 1. Ectopic beat, see there 2. Ectopic pregnancy, see there

ectopic ACTH syndrome A condition associated with production of ACTH by tissues other than the pituitary–eg, by small cell lung cancer CLINICAL Similar to pituitary-dependent Cushing syndrome MANAGEMENT Bilateral adrenalectomy if site of ectopic ACTH is unknown, excision of nonmalignant ACTH-producing tumors. See Cushing syndrome

ectopic beat Spontaneous ectopic depolarization

ectopic Cushing syndrome ENDOCRINOLOGY Cushing syndrome due to extra-pituitary production of ACTH ECTOPIC SITES Thymoma, medullary CA of thyroid, pheochromocytoma, islet cell tumors of pancreas, small cell CA of lung. See Cushing syndrome

ectopic hormone A hormone produced by a tissue or site not normally associated with its production; EH production is common in CA and may cause a paraneoplastic syndrome. See Ectopic hormone syndrome, Paraneoplastic syndrome

ectopic kidney NEPHROLOGY A kidney located outside of the normal position. See Horseshoe kidney

ectopic pregnancy Ectopic gestation OBSTETRICS The implantation of a fertilized outside of the uterus–eg, fallopian tube, ovary, peritoneum, and other tissues not designed to accommodate the vasculature required by a growing fetus CLINICAL Lower abdominal pain, vomiting, amenorrhea MANAGEMENT Surgery

ectropion OPHTHALMOLOGY The outward rolling of an anatomic margin, usually understood to be that of the eyelid

eczema DERMATOLOGY A generic term for a dermatopathy characterized by vesicle formation, papules and crusting overlying an erythematous rash, typically in areas of high concentration of sebaceous glands

eczematous *adjective* Referring to eczema, see there

edatrexate ONCOLOGY An antimetabolic anticancer agent in clinical trials

EDC/EDL Expected date of confinement/labor OBSTETRICS An estimate of the 'usual' duration of pregnancy

edema Fluid retention, water retention PHYSICAL EXAM An excess of fluid in cells or tissues due to disease or injury, quantified as 1+, 2+, 3+. See Angioneurotic edema, Brawny edema, Cerebral edema, Cyclic edema, Cytotoxic edema, Flash pulmonary edema, Hereditary angioneurotic edema, High-altitude cerebral edema, High-altitude pulmonary edema, Leukoedema, Macular edema, Malignant edema, Pedal edema, Pseudopapillaedema, Pulmonary edema

edentulous *adjective* Toothless

Edex™ Alprostadil UROLOGY An agent used to treat erectile dysfunction. See Erectile dysfunction

Edronax® Reboxetine mesylate PHARMACOLOGY A selective noradrenaline reuptake inhibitor–NARI for treating depression

EDTA Ethylenediaminetetraacetic acid, edetic acid A chelator that binds divalent–eg

arsenic, calcium, lead, magnesium, trivalent cations LAB MEDICINE EDTA is added to blood collection tubes to transport specimens for analysis in chemistry–eg CEA, lead, renin; hematology–it is the preferred anticoagulant for blood cell counts, coagulation studies, Hb electrophoresis, ESR; in the blood bank, it prevents hemolysis by inhibiting complement binding

educable PSYCHIATRY *adjective* Capable of achieving at least a 4th grade academic level, typical of mild mental retarded–IQ, 52-67. See Challenged, Mental retardation

education VOX POPULI The training of a person or system in performing a task or process. See Appropriate education, Character education, Graduate medical education, Health education, Neuromuscular reeducation, Patient education, Undergraduate education

Edwards syndrome Trisomy 18, see there

Edwardsiella tarda MICROBIOLOGY An enterobacter which infects post-traumatic wounds, linked to aquatic accidents CLINICAL Extraintestinal infections, liver abscesses, mild gastroenteritis

EEA stapler See End-to-end stapler

EEC group Enterovirulent *Escherichia coli* group, see there

EECP® Enhanced external counterpulsation, see there

EEG Electroencephalogram, electroencephalography

Efasedan Lorazepam, see there

efavirenz Sustiva™ AIDS A reverse transcriptase inhibitor used with other antivirals–eg, zidovudine, lamivudine for HIV-1 infection ADVERSE EFFECTS N&V, maculopapular rash, fatigue, headache, dizziness. See Non-nucleoside reverse transcriptase inhibitor

effacement OBSTETRICS The process that occurs during the last month of pregnancy, extending through the 1st stage of labor in which the cervix and endocervix become thinner and shorter; the canal is thus converted from a tube to a flaring funnel

effect modification EPIDEMIOLOGY An interaction among multiple possible cause-and-effect relationships, where the estimate of the effect of one factor on a disease process depends on other factors in the study

effective reproductive ratio EPIDEMIOLOGY The number of 2° cases or ♀ offspring produced in a host population not consisting entirely of susceptible individuals–microparasites or within which density dependent constraints limit parasite population growth–macroparasites. Cf Basic reproducion ratio

effectiveness The benefit–eg, to health outcomes, of using a technology for a particular problem under general or routine conditions–eg, by a physician in a community hospital or by a Pt at home. See Outcomes

effector An organ, such as a gland or muscle, that responds to a motor stimulation. See Allosteric effector

efferent *adjective* Conveying away from the center of an organ or structure

efferent neuron See Motor neuron

Effexor® Venlafaxine PSYCHIATRY A mood-enhancing SSRI used for major depression, general anxiety disorder CONTRAINDICATIONS Should not be combined with MAOIs. See Depression, Major depression, Monoamine oxidase inhibitor

efficacy An index of the potency of a drug or disease treatment; for a vaccine, efficacy is the percentage of persons who are protected by the vaccine

efficiency LAB MEDICINE The relative ability of a test to detect a disease, while maintaining the rate of false positive results to a minimum; the efficiency of a test is defined as the number of true positives and true negatives multiplied by one hundred, divided by the sum of true positives, true negatives, false positives and false negatives. Cf Four cell diagnostic matrix

efflux MEDTALK That which flows outward

effort thrombosis Paget-Schroetter syndrome A blood clot that forms within a vessel–eg axillary vein of a muscle group–which was subjected to strenuous exercise; ET may also occur in thoracic outlet syndrome, see there

effusion Accumulation of fluid in various spaces of the body, or the knee itself, which is a frequent byproduct of injury. See Ascites, Peritoneal effusion, Pleural effusion, Pseudochylous effusion, Subdural effusion.

EFM External fetal monitor, see there

EGD Esophagogastroduodenoscopy, see there

egg PARASITOLOGY A fertilized gamete which may give rise to an adult. See Chicken footprint egg REPRODUCTION MEDICINE A female reproductive cell, also called an oocyte or ovum

egg crate NURSING A popular term for a foam mattress with a chicken wire-like pattern of elevations and depressions, likened to that of an egg carton, most useful for Pts with recalcitrant decubitus ulcers

egg harvesting REPRODUCTION MEDICINE The obtention of human eggs from a ♀ donor; the normal format for EH consists of inserting a long needle through the vaginal wall and aspirating tissue from ovaries. See Artificial reproduction, Egg brokerage, Egg donation

EGJ Esophagogastric junction, see there

ego PSYCHIATRY A major division in the Freudian model of the psychic apparatus, the others being the id and superego; ego is the sum of some mental mechanisms–eg, perception and memory, specific defense mechanisms, and mediates the demands of primitive instinctual drives–the id, of internalized parental and social prohibitions–the superego, and reality–the compromises between these forces achieved by the ego tend to resolve intrapsychic conflict and serve an adaptive and executive function Vox POPULI Self-love, selfishness

ego dystonic PSYCHIATRY *adjective* Referring to the aspects of a person's behavior, thoughts, and attitudes viewed as repugnant or inconsistent with the total personality. Cf Ego-syntonic *noun* A tendency towards a particular behavior–eg, homosexuality, in a person who seeks to negate that behavior

ego identity PSYCHOLOGY The sense of connection or belonging between a person and a particular social–religious, or political group, the values of which a person shares; an EI is formed by early adulthood and is rooted in early developmental experiences; sexual orientation is a facet of EI. See Cult, Gang

egophony PULMONARY MEDICINE An extreme form of bronchophony, in which spoken words assume a nasal or bleating–Greek, *aigos*, goat–quality, which is most common when there is simultaneous lung consolidation and pleural fluid accumulation, but also heard over uncomplicated lobar pneumonia or pulmonary infarction. See Bronchophony

EHEC Enterohemorrhagic *Escherichia coli*, see there

EHL Electrohydraulic lithotripsy

ehrlichiosis INFECTIOUS DISEASE A rare tick-borne infection, caused by *Ehrlichia canis*, that usually affects dogs CLINICAL Fever, chills, rigors, malaise, nausea, myalgia, anorexia, encephalopathy, acute respiratory failure with infiltrates, acute renal failure with ↑ creatinine LAB ↓ platelets, ↑ transaminases TREATMENT Chloramphenicol, tetracycline

eicosanoid PHYSIOLOGY A 20-carbon cyclic fatty acid which with its arachidonic acid metabolites–eg, HETE, HPETE, leukotrienes, PGs, and thromboxanes, are site-specific, ↑ during shock and after injury, have diverse functions–eg, bronchoconstriction, bronchodilation, vasodilation, vasoconstriction. See Arachidonic acid, Bad eicosanoid, Good eicosanoid, Zone-favorable diet

eidetic image PSYCHIATRY Unusually vivid and apparently exact mental image–eg, memories, fantasies, dreams

EIEC Enteroinvasive *Escherichia coli*, see there

eight-hour rule ANESTHESIOLOGY The observation that general anesthesia is safely performed–ie, with a relatively low risk of gastric aspiration in a Pt who has not received anything by mouth for the past 8 hrs. See Gastric aspiration

ejaculate *noun* Seminal fluid, which has a 2 to 5 mL volume and contains semen, prostate secretions and various fluids. See Ejaculation

ejaculation UROLOGY The discharge of semen from the male urethra at the time of sexual climax or orgasm. See Nocturnal emission, Premature ejaculation, Retrograde ejaculation

ejaculatory incompetence 1. Impotence, see there 2. Retarded ejaculation, see there

ejection CARDIOLOGY See Ejection fraction, Ejection murmur EMERGENCY MEDICINE The throwing of a person from a vehicle

ejection click A cardiac sound in early systole, related to cardiac dilation or HTN in the great vessels–aorta and pulmonary artery; ECs may be so close to the first heart sound that they simulate a splitting thereof; aortic ECs are constant and best heard at the left lower sternal border, and occur with aortic dilation–aortic stenosis, Fallot's tetralogy, truncus arteriosus; pulmonary ECs occur with pulmonary stenosis, are best heard at the left midsternum and disappear with inspiration; a midsystolic EC heard at the apex preceding a late systolic murmur, suggests mitral valve prolapse

ejection fraction Left ventricular ejection fraction CARDIOLOGY The percentage of blood present in the left ventricle that is effectively pumped forward during systole to supply the peripheral circulation. See Congestive heart failure

ejection murmur CARDIOLOGY A diamond-shaped systolic murmur ending before the 2nd heart sound, produced by ejection of blood of aortic and pulmonary valve stenosis

EKG Electrocardiogram, see there

elaboration PSYCHIATRY An unconscious process of expansion and embellishment of detail, especially with reference to a symbol or representation in a dream

ELAD® Extracorporeal liver assist device, see there. See Artificial liver

elastic fiber A protein strand found in connective tissue which has contractile properties

elastic recoil PHYSIOLOGY The inherent resistance of a tissue to changes in shape, and the tendency of the tissue to revert to its original shape once deformed; a sensitive indicator of ER is the coefficient of retraction; ER is the effective pressure driving maximal expiratory air flow, and is ↑ after lung-reduction surgery for severe emphysema. See Coefficient of retraction, Elastance, Lung-reduction surgery. Cf Compliance

elastomer SURGERY A type of silicone used in the outer shell of a breast implant. See Silicone elastomer

elastomeric infusion device THERAPEUTICS A device used to administer therapeutics–eg, analgesics, antimicrobials, chemotherapy PROS Portable, single use CONS Low infusion pressure, expensive. See Infusion device, Minibag. Cf Central venous catheter

Elatrol Amitriptyline, see there

Elavil® Amitriptyline, nortriptyline, see there

elbow The synovial joint between the brachium and the antebrachium. See Cat's elbow, Coal workers' elbow, Golfer's elbow, Mouse elbow, Nursemaid's elbow, Student's elbow, Tennis elbow

elbow reflex Triceps reflex, see there

ELBW Extremely low birth weight infant, see there

Eldepryl® Deprenyl, see there

elder abuse Elderly abuse, geriatric abuse, senior abuse GERIATRICS Physical or psychological mis/maltreatment of an elderly person by a family member or other close associate, in the form of physical injury, restraint, financial exploitation, threats, ridicule, insult or humiliation, forced physical or social isolation, or change in living arrangements. See Domestic violence, Elder neglect

ELDER ABUSE*

FEATURES OF ELDER ABUSE

- **PHYSICAL VIOLENCE** Intent to cause bodily harm: Hitting, slapping, or striking with objects, resulting in bruises, abrasions, fractures, burns

- **EMOTIONAL/PSYCHOLOGICAL ABUSE** Intent to cause mental or emotional pain or injury: Verbal aggression, statements that humiliate or infantilize, insults, threats of abandonment or institutionalization

- **MATERIAL EXPLOITATION** -an 'optional' form of EA–misappropriation of money or property, theft of social security checks, changing person's last will and testament, and so on (NEJM 1995; 332:437RA)

*EA is an act of commission, defined in an arbitrary and somewhat nebulous fashion; in the current absence of a consensus definition; some experts prefer to use alternative terms, eg inadequate care of the elderly or mistreatment of elderly, which include acts of commission and omission, do not assign blame and at the same time are 'politically correct'

INDICATORS OF ELDER ABUSE

- **PHYSICAL INDICATORS** Alcohol or substance abuse, burns, bruises, contusions, decubital ulcers, dehydration, duplication of medication, fecal impaction, fecal incontinence, lacerations, malnutrition, poor hygiene, repeated hospital admissions, urine burns/excoriations

- **FAMILY/CAREGIVER INDICATORS** Elder not allowed to speak without 'supervision' by family member/care giver (FM/CG), obvious absence of supervision, aggressive behavior toward elder, indifference or anger toward elder, FM/CG accuses elder of deliberate incontinence, unwillingness of FM/CG to comply with services provided, previous history of elder or alcohol and/or substance abuse, or mental illness, or conflicting accounts of incidents by the family, supporters, and victims

- **BEHAVIORAL INDICATORS** Agitation, anger, anxiety, confusion, depression, disorientation, fear, hesitation to talk openly, implausible stories, isolation, non-responsiveness, withdrawal (AMN 6/1/92, p17)

elderly neglect The failure for a caregiver to meet the needs of a dependent elderly person, which may be intentional–eg, withholding of food, medications, failure to clean or bathe, or unintentional, resulting from genuine ignorance of–or physical inability to address–a particular need

elderly primigravida OBSTETRICS A ♀ who delivers her first child after age 35; these ♀ are often professionals who delay childbearing to pursue a career; there is no ↑ risk to the pregnancy per se, although medical risks due to ↑ age may cause ↑ pregnancy-induced complications

elective *adjective* Referring to that which is planned or undertaken by choice and without urgency, as in elective surgery, see there *noun* GRADUATE EDUCATION *noun* A short–±6 wk period of clinical contact undertaken by a 4th-yr medical student, during which he/she learns about a field related to a planned specialty. See Match, Rotation

elective abortion Therapeutic abortion OBSTETRICS A voluntary interruption of pregnancy before fetal viability, which is performed voluntarily at the request of the mother for reasons unrelated to concerns for maternal or fetal health or welfare; most abortions are elective; there is 1 EA per 3 live births in the US. See Abortion

elective surgery SURGERY Any operation that can be performed with advanced planning–eg, cholecystectomy, hernia repair, colonic resection, coronary artery bypass

Electra complex PSYCHIATRY The ♀ version of the Oedipus complex, in which a daughter perceives her mother as a rival and the father is the psychosexual source of nourishment. Cf Delilah syndrome, Diana complex, Oedipus complex

electrical alternans CARDIOLOGY Marked swings in the QRS complex amplitude, which occur every 2 to 3 beats, caused by 'circus movement' in the myocardium ETIOLOGY Cardiac tamponade, pericardial effusion, pneumopericardium, cardiac muscle dysfunction, and paroxysmal SVT; EA may presage sudden cardiac death, and may precede V Fib in Pts undergoing coronary angioplasty or those with Prinzmetal's angina, congenital prolonged QT syndrome, acute MI, catecholamine excess, electrolyte derangements; EA affecting the ST segment and T wave is common in Pts at risk for ventricular arrhythmics–VAs and may be a noninvasive marker of susceptibility to VAs

electrical galvanic stimulation An electrical therapeutic modality sending a current to the body at select voltages and frequencies to stimulate pain receptors, disperse edema, or neutralize muscle spasms.

electrical muscle stimulator MAINSTREAM MEDICINE A device that stimulates muscle contraction by electrical impulses; these devices are used in mainstream physical therapy to ↓ muscle spasms, prevent development of blood clots after surgery or CVAs, prevent disuse atrophy of muscle

electrical storm CARDIOLOGY A cardiac event defined as multiple recurrent episodes of ventricular fibrillation, or hemodynamically destabilizing ventricular tachycardia, with a very poor prognosis; ES is most common in older men with CAD, often in a background of acute MI or heart surgery MANAGEMENT Most antiarrhythmics are ineffective in ESs; amiodarone may be effective. See Ventricular arrhythmia PUBLIC HEALTH See Lightning

electrically-enhanced drug delivery Iontophoresis, see there

electrocardiogram ECG, EKG CARDIOLOGY A non-invasive test of the electrical activity of heart's conduction system, which is transformed into recordings on graph paper–an electrocardiograph; in an EKG, electrodes–leads are placed on 12 specific sites of the body: standard limb leads–I, II, III, augmented limb leads–aV_R, aV_L, and aV_F, and precordial or chest leads–V_1 to V_6; EKG tracings consist of 3 major components: the P wave, which indicates atrial depolarization, the QRS complex–ventricular depolarization, and the T wave–ventricular repolarization; the Holter monitor is a portable EKG recording device worn by an individual for continuous monitoring; the EKG is used to detect cardiac damage by evaluating alterations in the electrical conduction the heart, and can be performed at rest or during excercise–eg thallium stress test; the Holter monitor is a portable device worn by a Pt for continuous cardiac monitoring; the EKG is used to detect the presence and location of myocardial ischemia or infarction, cardiac hypertrophy, arrhythmias, conduction defects. See His bundle electrocardiography, Signal-averaged electrocardiography, Sleep electrocardiography

electrocautery probe SURGERY An electro-cautery cutting and coagulation device with a suction irrigator for use in general and plastic surgery procedures

electrocauterization Electrocoagulation, electrosurgery, fulguration SURGERY Cauterization by passage of high frequency current through electrically heated tissue; excision of abnormal or diseased tissue or control of bleeding in small blood vessels of the skin and in a surgical field with controlled electric current

electrocerebral silence Electrocerebral inactivity, flat electroencephalogram, isoelectric encephalogram NEUROLOGY An EEG without cerebral activity over 2 µv from symmetrically placed electrode pairs > 10 cm apart, and with interelectrode resistance between 100 and 10,000 ohms; if present for 30 mins in a clinically brain dead adult and if drug intoxication, hypothermia, and recent hypotension have been excluded, the diagnosis of cerebral death is supported. See Brain death

electrocochleography AUDIOLOGY A test for measuring sound-evoked cochlear potentials, which is part of the battery of auditory evoked potential tests, used to diagnose inner ear disease and for intraoperative monitoring of neurosurgery

electroconvulsive therapy Electroshock PSYCHIATRY A therapy for mental disorders that consists in iatrogenic induction of generalized tonic-clonic seizures which, if of adequate duration, has an antidepressant effect INDICATIONS Refractory depression, need for rapid clinical response—inanition/starvation, psychosis, suicidality, Hx of previous response to ECT, nonaffective psychotic disorders, manic episodes, bipolar disease—manic type, schizo-affective disorders, or catatonic schizophrenia; chronic disease requires longer therapy COMPLICATIONS 18%, usually minor; memory loss, ± permanent. See Major depressive episode, Vagus nerve stimulation. Cf Cranial electrotherapy stimulation, Lobotomy

electrode CARDIAC PACING A part of an electric conductor through which a current enters or leaves; uninsulated conductive part of a pacing lead or a unipolar implantable pulse generator's casing which makes electrical contact with tissue; electrodes are used to record the electric activity of contracting muscles; electromyographic data is collected by surface electrodes, fine wire and needle electrodes. See Ring electrode, SilverBullet™ electrode, Tip electrode

electrodesiccation SURGERY The drying of tissue by a high-frequency electric current applied with a needle-shaped electrode. See Dissection

electroencephalogram NEUROLOGY A graphic recording of minute electric currents produced by neuronal activity scalp electrodes, which is used to diagnose neurologic disorders and in neurophysiologic research

electroencephalography NEUROLOGY A technique in which the brain's electrical activity is measured by electrodes, and sent to a device that records impulses—brain waves on a continuous feed sheet of paper; EEG is used to evaluate seizure disorders, diminished consciousness, intracranial masses or vascular lesions. See Sleep encephalogram

electrogram CARDIOLOGY The recording of the cardiac waveforms as taken at an electrode in the heart; electrogram may be transmitted from an implanted pacemaker by telemetry

electrolarynx HEAD & NECK SURGERY A battery-operated instrument that makes a humming sound, used to enable speech in Pts with laryngectomies. See Artificial voice, Laryngectomy

electrolyte imbalance CRITICAL CARE A general term for a derangement of major electrolytes—Na+, K+, chloride; thus defined, EI is common; in practice, EIs are only of interest if they cause clinical disease

electrolyte panel LAB MEDICINE A battery of tests used to evaluate serum—or rarely, stool electrolytes—eg, Na+, K+, chloride, CO₂. See Panel

electrolyte/fluid balance panel LAB MEDICINE A battery of tests performed by a batch autoanalyzer, which detects common imbalances of electrolytes and fluids, measuring, Na+, K+, chloride, pH, PCO₂, CO₂ content, plasma and urine osmolality, BUN. See Basic metabolic panel, Organ panel

electromagnetic interference CARDIAC PACING Radiated or conducted electrical or magnetic energy that can interfere with/disrupt pulse generator function

electromechanical dissociation CARDIOLOGY A pattern of cardiac arrest for which there is often inadequate treatment, characterized by mechanical failure with adequate, albeit occasionally bizarre electrical activity, seen in ⅔ of sudden cardiac deaths MANAGEMENT Atropine might be effective; nonpharmacologic interventions—eg MAST suit, pericardiocentesis, fluid challenge, needle thoracostomy, do not improve survival. See Wandering pacemaker

electromyography EMG NEUROLOGY A technique that measures minute electrical discharges produced in skeletal muscle, at rest and during voluntary contraction; EMG is used to diagnose neuromuscular disease; the electrode for EMG is inserted percutaneously and the resulting electrical discharge or motor unit potential is recorded

electronic claims processing Electronic billing The electronic—ie, by modem—submission of a bill to third-party payers, for physician services rendered

electronic data interchange INFORMATICS A standard transmission format for 'secure' bidirectional flow of information between health care providers and reimbursement agencies/entities and other health care providers. See Telemedicine

electronic fetal monitoring OBSTETRICS The use of electronic devices during L&D to assess the baby's heartbeat and uterine contractions. See External fetal monitoring, Fetal monitoring, Internal fetal monitoring

electronic journal An 'on line' journal that allows immediate access to a researcher's findings

electronic medical record Computerized Pt record INFORMATICS An evolving computer-based document containing various forms of Pt data routed through a health care system. See Panel

Electronic Residency Application Service GRADUATE EDUCATION A service of the AAMC—Am Assn of Med Colleges which allows residency training candidates to transmit their applications via the Internet, using a common application form. See Match, Residency

electronic signature A digital signature which, assuming effective security procedures, can only be generated by the writ's owner; ESs are as valid as handwritten signatures

electronic syringe THERAPEUTICS An electronically controlled device used to administer therapeutics—eg analgesics, antimicrobials, chemotherapy. See Infusion device, Minibag. Cf Central venous catheter

electronystagmography OPHTHALMOLOGY A battery of neurologic and neuro-otologic examinations that record eye movements, used to separate vestibular and oculomotor deficits of the CNS, from deficits of the peripheral vestibular system

electrophoresis LAB METHODS A method of separating large molecules—eg, DNA fragments or proteins from a mixture of similar molecules, by passing an electric current through a medium containing the mixture; each molecule travels through the medium at a different rate, depending on its electrical charge and size; agarose and acrylamide gels are media commonly used electrophoretic media

electrophysiologic study CARDIAC PACING An invasive study of the electrical behavior of the heart to diagnose and study arrhythmias

electroporation therapy THERAPEUTICS A form of drug delivery that generates electrical pulses via an electrode placed in a tumor to enhance the ability of a chemotherapeutic–eg, bleomycin to enter tumor cells involves using electric fields to open pores in human cells to allow easier and more efficient entrance of beneficial pharmaceutical products or genes. See Drug delivery, Iontophoresis

electrosurgical loop excision LEEP, see there

elemental diet NUTRITION A basic diet composed of oligopeptides and amino acids, disaccharides or partially hydrolyzed starch, and minimal fat; EDs provide proton–hydrogen ion neutralization sufficient to maintain gastric pH above pH 3.5. See Diet

elementary hallucination NEUROLOGY An unformed hallucination

Elenium Chlordiazepoxide, see there

elephant ear appearance PEDIATRIC RADIOLOGY A descriptor for flaring of the iliac wings, flattening of the acetabular roofs, and ischial tapering; the iliac index, which is $\frac{1}{2}$ the sum of the iliac and acetabular angles, is characteristically ↓ in Down syndrome

elephantiasis Pachydermoid cutaneous induration elicited by chronic lymphatic blockage

ELEPHANTIASIS

LEGS, causing edema, chronic inflammation and eventually pachydermia, due to lymphatic plugging by microfilaria, eg *Wuchereria bancrofti, Brugia malayi, Onchocerca volvulus*

PENIS/SCROTUM, due to lymphogranuloma venereum; the scrotal lymphedema may extend cranially to the renal lymphatics and rupture into the renal pelvis, causing chyluria

ARM, often secondary to axillary lymph node dissection in a modified radical mastectomy

eligibility VOX POPULI Having a required qualification

eligibility period HEALTH INSURANCE The time following the eligibility date–usually 31 days–during which a member of a group may apply for insurance without evidence of insurability

eligible expense HEALTH INSURANCE An expense defined in a health plan as being eligible for coverage–eg, specified health services fees or "customary & reasonable charges"

elimination diet A diet in which a food or small groups of foods are eliminated in turn, in order to detect a food allergy/intolerance, which may be to milk, eggs, peanuts, and others; EDs are used, especially in children with atopy, for those suspected of having allergies to certain foods, and for detecting a ↓ in allergy-related Sx. See Lactose intolerance. Cf Desensitization

elimination disorder CHILD PSYCHIATRY A condition characterized by a lack of control over bladder–enuresis or bowel–encopresis, unrelated to a physical disorder

ELISA Enzyme-linked immunosorbent assay LAB MEDICINE A heterogeneous immunoenzymatic assay that approaches the sensitivity of RIA PROS Lower cost, simpler equipment, faster 'turn-around time', and none of the problems inherent in handling radioactive substances; ELISA may be used to measure any antigen and antibody. See Avidin-biotin method, EMIT, RIA, Sandwich method

ELITE CARDIOLOGY A clinical trial–Evaluation of Losartan in the Elderly in which Pts with NY Heart Association–NYHA class II–IV heart failure were randomized to receive either captopril or losartan. See Angiotensin II receptor antagonists

elite athlete SPORTS MEDICINE An athlete with potential for competing in the Olympics or as a professional athlete; EAs are at ↑ risk for injuries, given the amount of training, for psychological abuse by coaches and parents, and self abuse. See Female athlete triad, Overtraining

ELK Ears–nose and throat, lungs, kidneys An acronym for the organs involved in

Wegener's granulomatosis–WG; limited WG spares the kidneys and lacks signs of systemic vasculitis; generalized WG involves the kidneys and/or has signs of systemic vasculitis; disease exacerbation is best monitored by measuring titers of antineutrophil cytoplasmic antibodies by indirect immunofluorescence. See Wegener's granulomatosis

ellagic acid NUTRITION A phenol that inhibits nitrosamine-induced esophagus and lung CA SOURCE Fruits. See Fruits & vegetables

ellipse sign An oblong 'mass' seen in an upper GI radiocontrast study, which corresponds to simple–non-malignant pooling of contrast material in an ulcer base

ELM test PEDIATRICS A short outcomes-based test that measures an infant's expressive, receptive, and visual language abilities; in the ET, the parents are questioned about certain components of the infant's speech, and the infant is asked to carry out certain simple tasks; serial ELM testing can be used to monitor response to various interventions intended to improve communication. See Developmental milestone

ELOS Estimated length of stay. See Length of Stay

Emadine® Emedastine difumarate OPHTHALMOLOGY A topical agent for temporary relief of allergic conjunctivitis

emasculated hormone PHARMACOLOGY A drug designed to simulate a target hormone involved in a receptor-hormone ligand interaction EXAMPLES Partial agonists or antagonists of a target hormone–eg, propranolol, a β-adrenergic receptor antagonist, 'emasculated' adrenaline, cimetidine, a histamine H2-receptor antagonist and 'emasculated' histamine

embolic stroke NEUROLOGY A stroke caused by an embolus. See Transient ischemic attack, Stroke

embolism The presence of extraneous material in blood vessels. See Air embolism, Amniotic fluid embolism, Arterial gas embolism, Fat embolism, Gas embolism, Nitrogen embolism, Paradoxical embolism, Pulmonary embolism

embolization THERAPEUTICS The blocking of an artery by a clot or foreign material, to prevent blood flow to a tumor

embryo An early stages of a developing organism, which follows fertilization of an egg including implantation and very early pregnancy–ie, from conception to the 8th wk of pregnancy. See Preembryo. Cf Fetus

embryonal rest A nonmalignant nonneoplastic primitive–embryonal element–eg wolffian and müllerian duct elements, which is retained after birth

embryonic *adjective* Undeveloped, related to an embryo

embryopathy A general term for any constellation of malformations attributed to a particular exogenous agent. See AIDS embryopathy, Alcohol embryopathy, Diabetic embryopathy, Retinoic acid embryopathy, Varicella embryopathy, Vitamin A embryopathy

embryoscopy OBSTETRICS An imaging technique in which an ultrasound-guided small-bore needle with an endoscope is inserted through the abdominal wall into the uterus–without violating the amniotic cavity, to view a living embryo. Cf Fetoscopy

embryotoxicity TOXICOLOGY Adverse effects on the embryo due to a substance that enters the maternal system and crosses the placental barrier; the effects of the substance may be expressed as embryonic death or abnormal development of one or more body systems, and can be deleterious to maternal health. Cf Fetotoxicity

EMC 1. Electronic medical claim, see there 2. Endometrial curettage, see there

EMD Electromechanical dissociation, see there

emergency *adjective* Referring to an emergency *noun* An acute unexpected development or situation that endangers life or limb and requires immediate action

emergency care Medical or other health treatment, services, products or accomodations provided to an injured or ill person for a sudden onset of a medical condition of such nature that failure to render immediate care would reasonably result in deterioration of the injured person's medical condition

emergency contraception Morning after pill GYNECOLOGY The use of 2° contraception in the event of failure or suboptimal 1° contraception. See RU 486. Cf Back-up contraception

emergency doctrine A guiding principle that permits health care providers to perform potentially life-saving procedures under circumstances where it is impossible or impractical to obtain consent. See Informed consent. See Good Samaritan laws. Cf 'Rule of rescue'

emergency medical service A system, created by the US government in 1973 that provides emergency care intimately linked to a universal emergency telephone number, 911; the care is provided from vehicles, ie ambulance or helicopter, by certified and licensed personnel—eg, emergency medical technicians, in restricted geographic regions; EMS services include: emergency medical communications, transportation, disaster plans, consumer training programs. See Air ambulance, EMT-A, EMT-D, EMT-I. Cf First responder

Emergency Medical Treatment & Active Labor Act EMTALA MANAGED CARE A law passed as part of the Consolidated Omnibus Budget Reconciliation Act of 1986 that was enacted to protect against Pt dumping. See COBRA legislation, Dumping (patient)

emergency psychiatric commitment The temporary admission to a mental institution, of a person with an acute psychotic reaction, for a period of observation, not to exceed 7 working days. See Malpractice

emergency services EMERGENCY CARE '...*services ...necessary to prevent death or serious impairment of health and, because of the danger to life or health, require the use of the most accessible hospital available and equipped to furnish those services*'

emergency surgery SURGERY Surgery that cannot be delayed, for which there is no alternative therapy or surgeon, and a delay could result in death or permanent impairment of health EXAMPLES Open fracture of skull, some gunshot and stab wounds, extensive burns, urinary obstruction, intestinal obstruction, ruptured appendix, twisted ovarian cyst, ruptured fallopian tube. See Surgery

emergent *adjective* Referring to that which is becoming manifest. See Emergent disease

emerging infection PUBLIC HEALTH Any infection caused by recently recognized or 'feisty' bugs—eg, viruses—eg Ebola virus, new strains of influenza virus, hantavirus and bacteria—eg cholera, multidrug-resistant TB, Creutzfeldt-Jakob disease and bovine spongiform encephalopathy, human monocytic ehrlichiosis, human granulocytic ehrlichiosis, leptospirosis, microsporidiosis, Ebola hemorrhagic fever, cyclosporiasis. Cf Reemerging infections

emerging pathogen PUBLIC HEALTH Any pathogen that ↑ incidence of an epidemic outbreak EXAMPLES *Cryptosporidium*, E coli 0157:H7, Hantavirus, multidrug resistant pneumococci, vancomycin-resistant enterococci. See Emergent disease

emeritus staff HOSPITAL PRACTICE Thse physicians, dentists, osteopaths, or independent practitioners on a hospital staff, who have retired from active medical practice. See Medical staff

Emery-Dreifuss muscular dystrophy Scapulohumeral dystrophy MOLECULAR MEDICINE A form of muscular dystrophy characterized by contractions of elbow, Achilles tendon, and postcervical muscles in childhood, with slowly progressive wasting and weakness of humeroperoneal muscles; by adulthood, Pts with EDMD have conduction system disease, usually, heart block. See Dilated cardiomyopathy, Emerin, Lamin A/C

emesis Vomiting, puking. See Chemotherapy-induced emesis

emetic THERAPEUTICS Any agent that causes vomiting

emetophobia PSYCHOLOGY Fear of vomiting. See Phobia

EMG Electromyogram, electromyography

emission The release of a gas, fluid or solid. See Alpha emission Otoacoustic emission

EMLA Eutechtic mixture of local anesthetics PAIN MEDICINE A formulated local anesthetic–2.5% lidocaine, and eutectic mixture of 2.5% prilocaine, effective when applied in a thick coat covered with an occlusive plastic wrap

emmenagogue MEDTALK A drug or chemical that triggers menstruation. See Abortifacient

Emmett tenaculum GYNECOLOGY A device for grasping and manipulating the cervix, fundus, and intravaginal soft tissue during gynecologic procedures

emollient DERMATOLOGY A hydrating agent composed of fat or oil applied topically to soften skin, especially laminated keratin or hyperkeratotic scales–eg, of psoriasis

emotion PSYCHOLOGY A mood, affect or feeling of any kind–eg, anger, excitement, fear, grief, joy, hatred, love. See Negative emotion, Positive emotion, Toxic emotion

emotional abuse PEDIATRICS The infliction of '...*coercive, demeaning, or overly distant behavior by a parent or other caretaker that interferes with a child's normal social or psychological development...*' See Child abuse, Elderly abuse, Psychological abuse

emotional disorder Emotional disability PSYCHIATRY Behavior, emotional, and/or social impairment exhibited by a child or adolescent that consequently disrupts the child's or adolescent's academic and/or developmental progress, family, and/or interpersonal relationships

emotional distress Intentional infliction of emotional distress, Outrage tort MALPRACTICE *adjective* Referring to a civil action against a person who allegedly said or did something so completely outrageous or insulting to the plaintiff that he/she suffered emotional damage. See Damages, Malpractice, Medical malpractice. Cf Punitive damages

emotional outlet Any venue used to relieve psychologic stress–eg, strenuous exercise, vigorous sexual activity, video games, etc

emotional stability PSYCHOLOGY Consistency of mood and affect

Emotival Lorazepam, see there

emphysema INTERNAL MEDICINE Accumulation of air in tissue–eg, lungs, dermis; pulmonary emphysema is characterized by ↑ size of air spaces distal to terminal bronchioles, ↓ vital capacity and ↑ airway resistance, often with alveolar wall destruction and fibrosis; emphysema commonly is often associated with chronic bronchitis and COPD MECHANISMS Atrophy, bronchitis, bronchiolitis, scarring. See Centriacinar emphysema, Coal workers' pneumoconiosis, COPD, Distal acinar emphysema, Giant bullous emphysema, Panacinar emphysema, Senile emphysema. Cf Focal emphysema

empirical MEDTALK *adjective* Based on experience or observational information and not necessarily on proven scientific data

Empirynol™ GYNECOLOGY An antibacterial spermicide gel for preventing STDs–chlamydia, gonorrhea. See Spermicide

employee MEDICAL PRACTICE A person who works for another person, an organization or business. See Service employee. Cf Independent contractor

Employee Retirement & Income Security Act ERISA MANAGED CARE The federal law that governs all employee benefit plans–eg, pension plan, health plans

employer group Association of employers MANAGED CARE An entity with a current group benefits agreement in effect with a health plan to provide covered health care services to its employee-subscribers and eligible dependents. See Covered health service, Dependent, Group benefits agreement, Subscriber

empower *verb* To encourage or provide a person with the means or information to become involved in solving his/her own problems

empty nest syndrome PSYCHOLOGY A popular term for the understudied constellation of Sx described in middle-aged ♀ whose children have left home/the 'nest' for college/university, career, marriage CLINICAL Depression, loss of self-esteem, loneliness, as mom has lost her principle raison d'etre–ie, raising children, who no longer depend on her for their needs

empty sella syndrome Empty sella turcica, intrasellar arachnocele NEURORADIOLOGY The finding of a moderately enlarged sella turcica, due to a partial/complete absence of sellar diaphragm; it is most common in obese, middle-aged ♀; compression of hypophysis against the floor and posterior wall of the sella, by the extended suprasellar cisterns may be accompanied by pituitary hypofunction, although TSH, gonadotropin, and prolactin levels may be ↓ and/or accompanied by diabetes insipidus CLINICAL Vague headaches, systemic HTN, pseudotumor cerebri; if secondary, CSF rhinorrhea, but may be asymptomatic

empty vault RECTAL EXAM No poop in shute

empyema MEDTALK A collection of pus in a body cavity. See Pyothorax, Subdural empyemia

EMR Electronic medical record, see there

EMS Emergency medical service, see there

EMT EMERGENCY MEDICINE An abbreviation for any of a number of grades of Emergency Medical Technician, based on the number of hrs of formal training, and type of specially training

EMT-A Basic EMT EMERGENCY CARE An EMT with 81 to 140 hrs of training, standardized by the DOT; EMT-As are the backbone of the EMT workforce in the US; they know basic principles of Pt care, how to identify clinical signs central to Pt assessment and care and how to treat specific emergencies. See EMT. Cf First responder

EMT-P Paramedic EMT EMERGENCY CARE An EMT with advanced life support skills–eg, assesses organ systems, inserts IV catheters, administers emergency fluids/medications, noninvasive airway management, uses defibrillators, manages the emotionally disturbed, obstetric emergencies. See EMT

EMTALA Emergency Medical Treatment & Active Labor Act, see there

emulsion PHARMACOLOGY A suspension of droplets of one liquid in another–eg, oil, water. See Emulsifier

en bloc resection ONCOLOGY The resection of a large bulky tumor virtually without dissection SURGERY EBR is used in certain cancers to remove a primary lesion, the contiguous draining lymph nodes, and everything in between, as in a modified radical mastectomy

en plaque *adjective* Referring to a flattened lesion that is often whitish and fibrous in consistency, which is located on an organ's surface, as in 'en plaque' meningioma, 'en plaque' mesothelioma, and so on; because the term describes gross morphology, it may not correlate with histologic findings; when used alone, it is nonspecific and of little diagnostic utility

enabler SUBSTANCE ABUSE A family member, friend, co-worker, cleric or other person who, by being deeply concerned for the well-being of a substance abuser, facilitates the person's continued abuse by attempting to help the abuser–eg, shouldering responsibility, making excuses for him or her, in fact encourages continued alcoholic or substance-abusing behavior. See Co-dependency

Enafon Amitryptiline, see there

enalapril maleate Teczem®, Vasotec® CARDIOLOGY A long-acting IV ACE inhibitor/antihypertensive ADVERSE EFFECTS Fatigue orthostatic hypotension, diarrhea, N&V, dizziness, headache, cough, dyspnea. See ACE inhibitor, CONSENSUS II, SOLVD

Enbrel® Etanercept RHEUMATOLOGY A TNF inhibitor that targets the immune system, ↓ M&M of moderately to severely active rheumatoid arthritis in Pts with an inadequate response to one or more disease-modifying antirheumatic drugs–DMARDs; can be used in combination with MTX for nonresponders to MTX alone

encainide Enkaid® CARDIOLOGY An antiarrhythmic that was removed from the market as it causes arrhythmias and ↑ mortality. See CAST. Cf Flecainide

encapsulated cell therapy THERAPEUTICS The enclosing of cells capable of producing a desired substance, in a semipermeable nonimmunogenic sheath, to restore a lost function

encapsulated Localized ONCOLOGY *adjective* Confined to a specific area, surrounded by a thin layer of fibrous tissue; encapsulation generally refers to a tumor confined to a specific area, surrounded by a capsule. See Islet encapsulation

encephalitis Meningoencephalitis NEUROLOGY Inflammation of the brain, usually caused by a virus CLINICAL Findings of aseptic meningitis–headache, stiff neck and back, fever, N&V IMAGING CT without and with contrast; MRI after initial evaluation. See Acute disseminated encephalitis, California encephalitis, Eastern equine encephalitis, Flea bite encephalitis, Granulomatous amebic encephalitis, Herpes simplex encephalitis, Japanese encephalitis, La Crosse encephalitis, Lyme meningoencephalitis, Murray Valley encephalitis, Primary amebic meningoencephalitis, Rasmussen's encephalitis, St. Louis encephalitis, Venezuelan equine encephalitis, Viral encephalitis, Von Economo's encephalitis, Western equine encephalitis. Cf Encephalopathy

Encephalitozoon INFECTIOUS DISEASE A genus of microsporidia that most commonly infects the small intestine of immunocompromised Pts, in particular those with AIDS; in most Pts, the infection is confined to the small intestine and accompanied by diarrhea; in others, *Encephalitozoon* spp can cause disseminated disease & may be present in macrophages & epithelial cells. See AIDS cholangiopathy, Microsporidiasis

encephalomalacia NEUROLOGY Softening of brain tissue, usually due to ischemia or infarction

encephalomyelitis Infection of the brain and spinal cord. See Experimental allergic encephalomyelitis

encephalomyelopathy A condition affecting the brain *and* spinal cord. See Subacute necrotizing encephalomyelopathy

encephalopathy NEUROLOGY A metabolic, toxic, neoplastic, or degenerative disease of the brain. See Alcoholic encephalopathy, Bismuth encephalopathy, Bovine spongiform encephalopathy, Boxers' encephalopathy, Hepatic encephalopathy,

encopresis Fecal incontinence, see there. See Elimination disorder

encounter MANAGED CARE The process of a Pt coming to a provider, receiving services, and leaving. See also Visit, Window of service MEDTALK A meeting, an interaction. See Semi-sexual encounter, Sexual encounter

encounter data LAB MEDICINE Pt-related data–EKGs, CXRs, H&P–generated at the time he/she is seen in a health care setting

encounter group Sensitivity group PSYCHIATRY A psychotherapist-directed meeting of similarly out-of-their-minded people who attempt, en masse, to resolve shared conflicts or anxieties, by ↑ self-awareness and understanding of group dynamics

end-diastolic pressure CARDIAC PACING Fluid pressure in heart chamber at the end of diastole just before systole

end-to-end anastomosis stapler EEA stapler SURGICAL ONCOLOGY A proprietary–US Surgical–device used to reanastomose the rectosigmoid colon during surgical therapy

end-of-life CARDIAC PACING *noun* The point at which a pacemaker signals need for replacement, as its battery is nearing depletion MEDTALK *adjective* Referring to a final period–hrs, days, wks, months in a person's life in which it is medically obvious that death is imminent or a terminal moribund state cannot be prevented. See Hospice

---

END-OF-LIFE CARE–MAKING DECISIONS

INITIATE DISCUSSION

- Establish supportive doctor-Pt relationship
- Designate surrogate decision maker
- Identify Pt's general preferences

CLARIFY PROGNOSIS

- Keep message clear, avoid misunderstanding
- Acknowledge prognostic limitations

IDENTIFY END OF LIFE GOALS

- Determine if preferences have changed
- Identify individual priorities

DEVELOP TREATMENT PLAN

- Help Pt understand treatment options
- Discuss resuscitation
- Discuss palliative care (RB Balaban, Harvard U, in Qual Life Matters 9,10/00)

---

end point Outcome, see there CLINICAL TRIALS A measured outcome that marks the closure of a particular clinical question; EPs may be true–eg, reduction of mortality, prolongation of life, or surrogate–eg, lowering of BP, ↓ arrhythmias LAB MEDICINE Any final point in a reaction or process, at which time the results are analyzed or interpreted

end-stage renal disease End-organ disease, end-stage renal failure The decompensated stage of chronic renal failure, defined as renal insufficiency of a degree that requires dialysis or kidney transplantation for survival EPIDEMIOLOGY 30% of ESRD is linked to DM and HTN, the most common causes of ESRD; ESRD is seen in kidneys subjected to chronic dialysis COSTS ESRD programs for 200,000 Pts–0.08% of US population, cost $6 x 10⁹–0.8% of US health care budget COMPLICATIONS Infection, possibly due to impaired macrophage Fc-receptor function MANAGEMENT Kidney transplantation, see there PROGNOSIS 1-yr survival of graft, ±94%; half-life from living donor grafts, ±22 yrs, from cadaveric grafts, ± 14 yrs

endarterectomy CARDIOLOGY The removal of atherosclerotic plaques from the intima of an artery. See Atherosclerosis, Carotid endarterectomy

endemic *adjective* Referring to an infection or condition which doesn't widely fluctuate over time in a defined place, or which persists in a population without being reintroduced from outside

endemic disease EPIDEMIOLOGY The presence of a disease or infectious agent in a given geographic area or population group; ED also refers to the usual prevalence of a given disease in an area or group

Endep® Amitryptiline, nortriptyline, see there

Ender nail ORTHOPEDICS An unreamed intramedullary nail used for femoral shaft and intertrochanteric fractures; ENs allow early weight-bearing and can be placed with closed technique, which avoids damage to soft tissue and to the periosteal and muscular blood supply

endergonic *adjective* Denoting a chemical reaction that requires extended energy input in order to proceed.

Endo-Flo® system SURGERY A device for irrigation and hydrodissection by providing high-pressure pulsed flow of irrigation solution

endobronchial biopsy PULMONOLOGY A Bx obtained via an endoscope passed through the nose or mouth, to visualize the upper airway mucosa–oropharynx, trachea, bronchi, upper bronchioles, identify physical changes or lesions and obtain tissue; if a lesion is identified, an instrument–eg, alligator forceps, cup forceps, curette, is passed through the endoscope's central channel to obtain tissue PURPOSE Diagnose tumors, pulmonary fibrosis, infection–eg PCP, TB, inflammation–eg, sarcoidosis COMPLICATIONS Bleeding, bronchospasm. Cf Transbronchial–bronchoscopic biopsy

endocardial border enhancement IMAGING Any imaging procedure–eg, use of sonicated albumen, fluocarbonated agents that ↑ visualization of the endocardium

endocardial lead transvenous lead CARDIAC PACING A pacing lead that is passed transvenously and lodged in either the right atrium or right ventricle

endocarditis CARDIOLOGY Inflammation of the endocardium. See Culture-negative endocarditis, Non-bacterial endocarditis, Subacute bacteria endocarditis

ENDOcare nitinol stent UROLOGY A stent for temporary treatment of urinary obstruction where catheterization would have been indicated

endocervical brush GYNECOLOGY A device used to obtain cells from the endocervical canal, which replaces wooden spatulas, cotton swabs. See Pap smear

endocervical curettage GYNECOLOGY The scraping of the endocervical mucosa with a spoon-shaped curette, usually performed with a cervical Bx to identify changes in the transition zone. See Transition zone

endocervical speculum GYNECOLOGY An instrument used to visualize the entire transformation zone, see there

endocervicitis GYNECOLOGY Inflammation of the endocervix

endocrine cancer ONCOLOGY Any malignancy that arises in endocrine glands–eg, thyroid CA, adrenal CA, etc

endocrine kidney Subinfarct NEPHROLOGY A popular term for atrophy of renal tubules due to renal artery stenosis or healed infarcts which, absent interstitial fibrosis, imparts a low-power appearance of an endocrine gland

endocrine pancreas The part of the pancreas that reacts to signals from the vascular system by secreting hormones, including insulin–B cells, glucagon–A cells, somatostatin–D cells, pancreatic polypeptide–PP cells, gastrin, etc EMBRYOLOGY The EP may arise embryologically from the neuroendocrine system. See Endocrinologist, Endocrinology. Cf Exocrine pancreas

endocrinologist A physician trained in managing–diagnosing and treating–disorders of the endocrine system–thyroid, parathyroid, adrenal glands, hypophyseal and hypothalamic axes, pituitary, pineal body, ovaries, testes, pancreas MEAT & POTATOES DISEASES Poorly controlled DM, early menopause,

endocrinology

hyperthyroidism SALARY $132K + 11% bonus. See Endocrine pancreas, Endocrinology

endocrinology The subspecialty of internal medicine dedicated to studying and managing diseases of the endocrine system–eg, conditions affecting the thyroid, adrenal glands, hypophysis and hypothalamic axes, ovaries, testes, pancreas. See Endocrine pancreas, Endocrinologist

endodontist DENTISTRY A dentist specialized in root canals and treating disease and infection of tooth pulp SALARY $151K + 12% bonus

endogenous anxiety PSYCHIATRY Anxiety due to internal stressors, typical of an anxiety disorder, perhaps involving hormone or neurotransmitter dysfunction. Cf Exogenous anxiety

endogenous depression Melancholia PSYCHIATRY A form of depression that occurs either *de novo* or without external events severe enough to warrant the degree of depression CLINICAL Pervasive sadness, hopelessness, loss of interest in daily activities; physical Sx–weight loss, insomnia, reduced libido; in ED, there may be an ↑ 'threshold' to stressful life events that requires little external input to initiate recurrence. See Depression. Cf Reactive depression

endolymphatic hydrops Ménière's disease, see there

endolymphatic sac decompression ENT A reduction of endolymph volume by ↑ drainage or ↑ absorption without damaging the vestibular labyrinth; control of vertigo is seen in 65% of Pts, which diminishes to ≤ 50% in a 10 yr follow-up; hearing stabilization occurs in 55% EFFICACY Uncertain

endometrial aspiration Vacuum curettage, see there

endometrial biopsy GYNECOLOGY The sampling of endometrial tissue to evaluate abnormal menses, heavy menstruation or post-menopausal bleeding, infertility FINDINGS Endometrial CA, hyperplasia, uterine fibroids, endometrial polyps

endometrial cancer Corpus carcinoma, endometrial adenocarcinoma, endometrial carcinoma GYNECOLOGY A malignancy of the endometrial epithelium EPIDEMIOLOGY It is the most common CA in ♀ age 55 to 70, and 4th most common cancer in ♀ DIAGNOSIS Pelvic examination, pap smear, endometrial biopsy from a D&C PRECURSOR Adenomatous hyperplasia with atypia. Cf Endometrial hyperplasia, Endometrioid cancer

ENDOMETRIAL CANCER

STAGE I	Cancer confined to the uterus, but not the cervix
STAGE II	Cancer spread to the cervix
STAGE III	Cancer spread outside the uterus to the vagina and/or pelvic lymph nodes; confined to the pelvis
STAGE IV	Cancer spread to the bladder or rectal mucosa or metastasized

endometrial curettage GYNECOLOGY A scraping of the endometrium to obtain tissue for histologic evaluation. See D&C

endometrial hyperplasia Adenomatous hyperplasia of endometrium GYNECOLOGY A premalignant endometrial lesion of older ♀

ENDOMETRIAL HYPERPLASIA

HYPERPLASIA WITHOUT ATYPIA Glands are crowded w/o cytologic atypia; these have a < 2% progress to carcinoma

SIMPLE HYPERPLASIA Glands are not back-to-back

COMPLEX HYPERPLASIA Glands are back-to-back

HYPERPLASIA WITH ATYPIA Glands are crowded with cytologic atypia; ± 23% progress to carcinoma

endometriosis Endometrial implants GYNECOLOGY A condition affecting up to 50% of ♀, defined as the presence of functioning endometrial glands and stroma outside of uterine cavity, occurring, in descending order of fre-

quency, the ovaries, broad ligaments, rectovaginal septum, umbilical scars, intestine, lungs, etc CLINICAL Often accompanied by dysmenorrhea, cyclical pain, low back pain, thigh pain, hypermenorrhea, repeated miscarriages, infertility; bleeding per rectum or bladder; regional swelling with vicarious ectopic bleeding parallels menses EVALUATION Laparoscopy MANAGEMENT Surgery if anatomy is distorted; TAH-BSO is definitive therapy; laparoscopic resection or ablation of minimal lesions ↑ fecundity ♀

endometritis GYNECOLOGY Inflammation of the endometrium ETIOLOGY Complication of first TM abortion, IUD CLINICAL Pelvic or lower abdominal pain, rank, funky discharge if infected MANAGEMENT Antibiotics

endomyocardial biopsy CARDIOLOGY A Bx of the endocardium and subjacent myocardium, to detect inflammation, anthracycline cardiotoxicity, or tumors. See Dallas criteria

Endopearl™ ORTHOPEDICS A bioabsorbable device for fixation of soft tissue grafts in the femoral socket during anterior cruciate ligament–ACL reconstruction

endoprosthesis INTERVENTIONAL MEDICINE Any prosthetic device that is inserted or placed within a vessel or tubular structure. See Stenting

endorphin PHYSIOLOGY An endogenous opioid–eg, endorphin, leu-enkephalin, met-enkephalin, dynorphin; each binds to a cognate receptor; endorphins act as neurotransmitters and neuromodulators, they are concentrated in the brain and associated with analgesia. See Enkephalin, Neuropeptide, Neurotransmitter, POMC

Endosac™ ENDOSCOPY A polyurethane pouch used in endoscopic procedures for specimen collection

endoscope A semirigid or flexible device with a long firm coil that is inserted into the region of interest, which has a light source, an optical system for viewing mucosa, camera, and a channel that allows insertion of sampling devices–eg alligator forceps, cup forceps, or curette for obtaining biopsies or surgical instruments to perform simple–minor surgeries. See Needle endoscope, Sigmoidoscope, Stereoendoscope

endoscopic hemostasis SURGERY The use of an endoscope to identify and locate a source of internal or intraluminal hemorrhage–due to bleeding ulcers, vascular malformation or esophageal varices, and guide therapy

endoscopic ligation Treatment of esophageal varices–EVs by endoscopically ligating the veins with small O-shaped elastic rings. See Endoscopic sclerotherapy

endoscopic papillotomy Endoscopic sphincterotomy, endoscopic retrograde sphincterotomy GI DISEASE The use of an endoscope to dilate and treat defects of the ampulla of Vater–eg, impaction by stones or ascaris COMPLICATIONS 10%–eg, pancreatitis, bleeding. See ERCP. Cf Precut spincterotomy

endoscopic retrograde cholangiopancreatography Pancreatic ductography IMAGING A diagnostic procedure in which an endoscope is snaked to the duodenum, then inserted through the ampulla of Vater with injection of radiocontrast to examine the extrahepatic biliary tract and pancreatic duct system–pancreatic duct, hepatic duct, common bile duct, duodenal papilla, gallbladder; ERCP can be used to remove bile duct stones and obstructions INDICATIONS Suspected pancreatic CA–PC, in particular of the periampullary region and chronic pancreatitis FINDINGS In PC, the duct of Wirsprung is often stenosed or occluded; visualization success rate ranges to ±85%. See Rat tail tapering. Cf Percutaneous transhepatic cholangiography

endoscopic sclerotherapy The treatment of esophageal varices–EVs by endoscopic injection of a sclerosant–sodium tetradecyl sulfate into the veins; ES has been used to manage bleeding EVs; medical therapy may be more effective and have fewer complications, vis-à-vis control of active bleeding, prevent recurrences, and survival; ES may be equal or superior to other non-

medical management–eg, insertion of a portacaval shunt, or selective splenorenal shunt for survival and preservation of liver function COMPLICATIONS Rebleeding–± 50%, esophageal ulceration, strictures, perforations, rebleeding, fever–40%, pulmonary and renal effects, and death in 1-2%. See Endoscopic ligation, Esophageal varices, Nadolol with isosorbide mononitrate

endoscopic sympathectomy PAIN MANAGEMENT A minimally invasive technique of selective sympathectomy, in which only the rami communicanti of the T1-T4 ganglia are divided; ES may eventually replace open thoracic sympathectomy in treating palmar hyperhidrosis COMPLICATIONS Compensatory sweating

endoscopic ultrasonography ENDOSCOPY A technique in which an echoendoscope is used to identify masses below the resolution of conventional imaging modalities. See Ultrasonography

endoscopist A health professional who performs endoscopic procedures. See Nurse endoscopist

endoscopy ENDOSCOPIC SURGERY The use of an endoscope to view internal structure–eg, mucosa of GI tract, upper respiratory tract–eg, oropharynx, trachea, bronchi, upper bronchioles, etc, which is usually well-tolerated. See Bronchoscopy, Colonoscopy, Fetal endoscopy, Laparoscopy, Nasal endoscopy, Sigmoidoscopy, Upper GI endoscopy, Virtual endoscopy

Endosolv E DENTISTRY A resin used to fill root canals. See Root canal

endospeculum GYNECOLOGY A device for direct visualization of the cervical os and endometrium

endosteal *adjective* Within a bone

endotoxin Bacterial endotoxin, lipid A MICROBIOLOGY A heat-stable lipopolysaccharide on the outer coat of gram-negative bacteria–eg, those causing cholera, meningitis, pneumonia, plague, whooping cough, et al CLINICAL Leukopenia, thrombocytopenia, fever, chills, hemorrhagic shock, dec resistance to infection

endovascular stent-graft Stent graft HEART SURGERY A prosthetic intravascular graft placed transluminally as an alternative to invasive surgical replacement of an artery with an aneurysmal dilatation; in aneurysms of the thoracic aorta, surgical interposition of a synthetic graft carries a 50% mortality if performed as an emergency procedure, and a 'mere' 12% mortality when performed electively; ESGs are less invasive, less expensive, and have a lower risk than standard operative repair, and are used to treat aneurysms of the abdominal aorta, thoracic aorta, subclavian artery, AV fistula, femoral occlusive disease

endowed professorship Chair ACADEMIA A university or academic appointment supported by income from an endowment, usually awarded to a person who is already a fully-tenured professor. See Professor. Cf 'Chair'

endowment A whole bunch of money given by a person to an academic institution, generally to support research

endurance The ability to continue performing a given task over a prolonged period of time SPORTS MEDICINE *'The ability to perform repetitive submaximal contractions'*, often understood to occur after a long period of time. Cf Eccentric contraction

enema A fluid infused per rectum, generally to cleanse the colon. See Barium enema, Colonic irrigation, Herbal enema

energy The capacity to do work, measured in joules TYPES Potential/stored energy, kinetic/in motion energy. See Activation energy, Adaptation energy, Binding energy, Biomass energy, Bond dissociation energy, Department of Energy, Orgone energy

energy-return system ORTHOPEDICS A technology or system–eg, Nike's Shox running shoes, designed to provide rebound from input kinetic force.

engagement OBSTETRICS The descent of the widest part of the presenting part of the baby's head through the pelvic inlet, which requires that the cervix is completely dilated; it is the first movement made by the baby's head during L&D. See Labor & delivery

engineering HOSPITAL CARE A hospital department that oversees preventive maintenance and safety of equipment in the hospital, managing service and inspection records, safety manuals, alert/recall notices, mechanical and electrical supplies and clearing all new equipment brought into the hospital from outside the facility. See Biotechnical engineering, Genetic engineering, Human factors engineering, Personal protective engineering

engineering controls OCCUPATIONAL SAFETY An OSHA term for devices–eg, sharps disposal containers, self-sheathing needles, etc, intended to isolate or remove blood-borne hazards from the workplace

engraftment TRANSPLANTATION The process by which transplanted or transfused cells–eg, in BM–from an allogeneic donor grow and reproduce with a recipient. See Bone marrow transplantation

enhanced external counterpulsation CARDIOLOGY A nonsurgical treatment of angina pectoris and CAD which ↑ blood flow to the heart by compressing blood vessels in the lower extremities. See MUST-EECP

enlarged adenoids Adenoidal hypertrophy, see there

enlarged liver Hepatomegaly, see there

enlarged lymph nodes Lymphadenopathy, see there

enlarged spleen Splenomegaly, see there

enoxacin ANTIBIOTICS A broad-spectrum–gram-negative bacilli, staphylococci–fluoroquinolone with ↓ activity against streptococci and anaerobes. See Fluoroquinolone

enoxaparin Lovenox® VASCULAR DISEASE An LMW heparin anticoagulant used to prevent DVT, venous TE in Pts with acute medical disease, and to prevent ischemic complications of unstable angina and non-Q-wave MIs ADVERSE EVENTS Hemorrhage. See Deep vein thrombosis, Low-molecular-weight heparin

enrichment FOOD INDUSTRY The addition of vitamins or minerals to a food–eg, wheat, which may have been lost during processing. See White flour; Cf Whole grains

enrollee HEALTH INSURANCE An eligible person who is enrolled in a health plan or a member's qualifying dependent. See Beneficiary

enrollment CLINICAL TRIALS The placing of a subject in a clinical trial. See Federal open enrollment, Open enrollment, State open enrollment HEALTH INSURANCE A term that refers to the total number of enrollees in a health plan; may also be used to refer to the process of enrolling people in a health plan

ENT Ears, nose & throat; formally, otorhinolaryngology

entacapone Comtan® NEUROLOGY A COMT inhibitor which may improve motor performance in Parkinson Pts receiving levodopa/carbidopa. See Parkinson's disease

Entamoeba histolytica PARASITOLOGY A protozoan that normally resides in the large intestine and may, under abnormal conditions, become pathogenic, enter the mucosa, producing flask-like ulcers and amebic dysentery; it may seed to other organs–eg, lungs, brain etc. Cf *Entamoeba polecki*

enteral nutrition Enteral feeding CRITICAL CARE The provision of nutrients by catheter, NG tube or if needed for > 4 wks, by gastrostomy or jejunostomy, to a Pt who cannot ingest food–eg, due to an upper GI tract malignancy, but whose GI tract does not require complete 'rest' PROS Preserves GI architecture, prevents bacterial translocation from the colon, fewer complications than TPN, ↓ cost. See Cancer cachexia, Malnutrition

enteric adenovirus VIROLOGY A serotype–eg, type 40, 41–of adenovirus which produces gastroenteritis CLINICAL Diarrhea; keratoconjunctivitis and nasopharyngitis–typical of infection with other adenoviruses–do not occur MANAGEMENT Symptomatic with rehydration. See Gastroenteritis

enteritis Inflammation of the small intestine

enteroadherent *Escherichia coli* EAEC An *E coli* serotype implicated in rare forms of chronic diarrhea in infants with FTT EAEC SEROTYPES IN TRAVELER'S DIARRHEA O55, O111, O119, O125-128, O142

Enterobacteriaceae MICROBIOLOGY A family of gram-negative, rod-shaped facultative anaerobic bacteria, most of which are motile–peritrichous flagella, oxidase-negative and have relatively simple growth requirements; Enterobacteriaceae are primarily saprobes, are widely distributed in nature in plants and animals, and are important pathogens; they are part of the intestinal flora, and popularly termed gram-negative rods–GNRs; they cause ± $\frac{1}{2}$ of all nosocomial infections in the US, most commonly by *Escherichia*, *Enterobacter*, *Klebsiella*, *Proteus*, *Providentia*, and *Salmonella* spp; less pathogenic Enterobacteriaceae include *Citrobacter*, *Edwardsiella*, *Erwinia*, *Hafnia*, *Serratia*, *Shigella*, *Yersinia* spp. See *Citrobacter*, *Edwardsiella*, *Enterobacter*, *Erwinia*, *Escherichia*, *Hafnia*, *Klebsiella*, *Proteus*, *Providentia*, *Salmonella*, *Serratia*, *Shigella*, *Yersinia*

Enterobius vermicularis Pinworm The nematode that is the most common intestinal parasite of young children in developed nations and the US. See Scotch tape test

enterocele SURGERY A hernia that contains intestine

Enterococcus faecium A nosocomial pathogen resistant to most antibiotics–eg, penicillin, teicoplanin, aminoglycosides, glycopeptides; ID of *E faecium* in a clinical specimen requires Pt isolation with barrier precautions. See Vancomycin-resistant *Enterococcus faecium*

enterocolitis GASTROENTEROLOGY Simultaneous inflammation of the small and large intestine. See Necrotizing enterocolitis

Enterocytozoon bieneusi PARASITOLOGY The most common microsporidiosis of humans; in AIDS, *E bieneusi* infects the small intestinal mucosa, hepatobiliary tract, and is implicated in previously unexplained–ie pathogen-'negative' AIDS-related cholangitis; Pts with *E bieneusi* may be coinfected with *Giardia lamblia*. See AIDS-related cholangitis

enteroglucagon ENDOCRINOLOGY A proglucagon-derived peptide of the gut–eg, glucagon-like insulinotropic peptide–GLIP, and glucagon-like peptide–GLP-1 that ↑ after oral glucose and fat, and have regulatory pathways that differ from pancreatic glucagon

enterohemorrhagic *Escherichia coli* EHEC Any of the *E coli* serotypes–eg O29, O39, O145 that produces shiga-like toxins, causing bloody inflammatory diarrhea, evoking a HUS. See *Escherichia coli* O157:H7, Hemolytic uremic syndrome

enterohepatic recirculation Biliary recycling THERAPEUTICS The cycling of drugs and metabolites after excretion in the biliary system, which are reabsorbed in the intestine. Cf Absorption

enteropathic *Escherichia coli* EPEC An agent causing epidemic diarrhea TREATMENT Symptomatic

enteropathogenic human orphan virus ECHO virus

enteropathy Any condition that causes triggers structural or functional defects of the small intestine. See AIDS enteropathy, NSAID enteropathy, Protein-losing enteropathy, Radiation enteropathy

enterorrhaphy MEDTALK The surgical fixation of a floppy loop of small intestine

enterotoxic *Escherichia coli* ETEC A group of *E coli* serotypes–implicated are O6, O8, O15, O20, O25 etc, which, like *Vibrio cholerae* and *Yersinia* spp, cause secretory or 'traveler's' diarrhea, due to a 80 kD heat-labile toxin that 'locks' the adenylate cyclase into the 'on' position

enterotoxin INFECTIOUS DISEASE A toxin with a direct effect on the intestinal mucosa, eliciting net fluid secretion; the 'classic' enterotoxin is cholera toxin, which evokes intestinal fluid secretion, by activating adenylate cyclase. See Endotoxin, Exotoxin

enterovirulent *Escherichia coli* **group** See Enterotoxigenic *E coli*–ETEC, enteropathogenic *E coli*–EPEC, enterohemorrhagic *E coli* O157:H7–EHEC, Enteroinvasive *E coli*–EIEC

enterovirus A genus of picornavirus comprised of more than 100 closely related viruses–eg, coxsackievirus, echoviruses, polioviruses and others, which cause gastroenteritis and viral encephalopathy. See Virus

enterovirus 71 VIROLOGY A virulent epidemic enterovirus serotype identified in 1969 in California CLINICAL Asymptomatic or diarrhea, rashes, hand-foot-mouth disease–vesicular lesions of hands, feet and oral mucosa, herpangina, aseptic meningitis, encephalitis, myocarditis, acute flaccid paralysis, rhombencephalitis, Guillain-Barré syndrome, rapidly fatal pulmonary edema, hemorrhage

enterprise liability MALPRACTICE That liability that results when a person–eg, a physician, is part of a commercial network that distributes a product. See Liability

enthesopathy RHEUMATOLOGY An abnormality involving an attachment of a tendon or ligament to bone (enthesis), and a prominent finding in seronegative spondyloarthropathy–ankylosing spondylitis, Reiter syndrome, psoriatic arthritis IMAGING Edema, erosion and proliferation of adjacent bones, reactive sclerosis, syndesmophyte formation, bone eburnation are seen; the bones may have a frayed, irregular surface or a fluffy appearance TREATMENT NSAIDs

enthusiasm hypothesis DECISION-MAKING A posit that explains the tendency of physicians in a geographic region to become enamored with a particular type of therapy or diagnostic modality, as an effect of influential–ie, thought—leaders involved in regional medical politics, policies, and therapeutic modalities

entomophagy GLOBAL NUTRITION The dietary consumption of insects. See Forensic entomology

entraining A technique for teaching Pts to focus on an extraneous factor–eg, music, rather than on a normal focus of attention. See Biofeedback

entrance wound FORENSIC PATHOLOGY The first lesion that a bullet or other projectile causes when entering the body. See Gunshot wound. Cf Exit wound

entrapment neuropathy Entrapment syndrome NEUROLOGY Any of a group of neuromuscular disorders caused by anatomic restriction or compression, usually of a single peripheral sensorimotor nerve in a bony or fibrous canal–eg, carpal tunnel syndrome, thoracic outlet syndrome, ulnar neuropathy CLINICAL Pain, especially at night, paresthesias, painful tingling, muscle weakness which, if not relieved, results in atrophy of the innervated muscles EXAMPLES Carpal tunnel syndrome, obturator canal syndrome, tarsal tunnel syndrome. See Carpal tunnel syndrome, Peripheral neuropathy, Thoracic outlet syndrome

entropion OPHTHALMOLOGY Inversion of the eyelid or, generically, any part with a free margin

entry-to-practice requirements PROFESSIONAL PRACTICE The educational, competency, and experience-related qualifications, required in a particular jurisdiction–state or country before a person can practice a profession

enucleated *adjective* Referring to an eye that has been traumatically or surgically removed from the orbit. Cf Anucleated

enuresis PSYCHIATRY Nocturnal and daytime incontinence of urine VOX POPULI Bed-wetting. See Elimination disorder

environmental cancer A cancer caused by environmental carcinogens EXAMPLES Lung CA–tobacco, mesotheliomas–asbestos, cervical CA–HPV, EBV-related lymphoproliferative disorders, lymphoma–EBV, liver CA–thorotrast, HBV, HCV, aflatoxin B. See Cancer

environmental lung disease PULMONARY MEDICINE The lung changes caused by exposure to environmental toxins EXAMPLES Asthma–acidic aerosols, nitrogen dioxide, photochemicals, COPD and emphysema–acidic aerosols, cigarettes, oxidant gases, lung cancer–cigarettes, radon, silica, volatile organic chemicals, mesothelioma-asbestos, ↓ lung capacity–acidic aerosols, tobacco, nitrogen dioxide, photochemicals, silicosis–silica PATHOGENESIS Environmental toxins induce bronchoconstriction, CA, fibrosis, inflammation. See COPD

environmental tobacco smoke The smoke from burning tobacco products to which a person is unintentionally exposed, in particular in public places–ie restaurants, hospitals, government buildings, aircraft. See Passive smoking, Second-hand smoke, Smoking.

enzyme A protein that catalyzes most chemical reactions in biological systems without itself being destroyed or altered by the reaction; enzymes accelerate the rate of reactions by lowering transition state energy ENZYME COMMISSION GROUPS Oxidoreductase, transferase, hydrolase, lyase, isomerase, ligase.

enzyme defect A structural or functional defect in an enzyme needed to catalyze a normal biochemical reaction in the body. See Enzyme, Inborn error of metabolism

enzyme therapy THERAPEUTICS Any therapy, in which an enzyme present in adequate amounts under normal conditions, is supplemented with a related or identical enzyme to perform a specific task–eg, rapid lysis of blood clots in an evolving MI by streptokinase, tissue plasminogen activator or urokinase

EOG Electrooculogram, see there

EOM Extraocular movements, see there

eosinophilia Eosinophilic leukocytosis HEMATOLOGY An absolute eosinophil count of > 500/mm³

eosinophilic gastroenteropathy Eosinophilic gastroenteritis A rare idiopathic condition characterized by eosinophilic infiltration of the GI tract ETIOLOGY $\frac{1}{2}$ occur in a background of allergy and atopy; EG may be linked to autoimmune connective tissue disease CLINICAL Diarrhea, ↓ weight, abdominal pain, fever, rebound tenderness, mesenteritis

eosinophilic granuloma MEDTALK A benign clinical form of histiocytosis X, characterized by circumscribed cystic osseous lesions in children and adolescents composed of eosinophils and histiocytes. See Langerhans' cell histiocytosis

eosinophilic panniculitis DERMATOLOGY A heterogeneous condition in which the subcutaneous fat is infiltrated with eosinophils ETIOLOGY Gnathostomiasis, leukocytoclastic vasculitis, erythema nodosum, atopic dermatitis, contact dermatitis, eosinophilic cellulitis, injection granuloma,

arthropod bites, bacterial infections, toxocariosis, lymphomas, toxocariosis, & RAEB

eosinophilic pneumonia CLINICAL MEDICINE A nonspecific term for disease processes in which the lung parenchyma is inflamed and diffusely infiltrated with eosinophils and histiocytes, ± PMNs, lymphocytes, plasma cells, mast cells ETIOLOGY *Ancylostoma duodenale*, *Ascaris lumbricoides*, *Dirofilaria immitis*, *Fasciola hepatis*, *Necator americanus*, *Strongyloides stercoralis*, *Toxocara canis*, *Trichinella spiralis*, *Wuchereria malayi*, drugs; $\frac{1}{3}$ are idiopathic CLINICAL From mild and transient–Löffler syndrome with fever, cough, and wheezing to severe and progressive–chronic EP with weight loss, dyspnea; an acute form is accompanied by respiratory failure and recuperation CLINICAL FORMS Loeffler syndrome, tropical eosinophilia, drug-induced pulmonary eosinophilia, EP associated with asthma LAB Eosinophils in BAL, peripheral blood MANAGEMENT Treat cause if known–eg, stop offending drug, kill offensive parasite, default to corticosteroids if idiopathic

EPA 1. Eicosapentaenoic acid. See n-3 fatty acids–biochemistry 2. Epidermolysis bullosa acquisita

EPEC Enteropathic *Escherichia coli*, see there

ependymoma Ependymal tumor NEUROLOGY An indolent ependymal cell–brain tumor, that arises in the walls of the ventricles–lateral, 3rd, 4th and central canal of the spinal cord. See Myxopapillary ependymoma

ephebiatrics Adolescent medicine, see there

ephedrine THERAPEUTICS An adrenergic bronchodilator, diuretic, mydriatic, vasoconstrictor ROUTE Oral EFFECTS CNS stimulant USED FOR Bronchodilator for mild asthma

ephelis Freckle DERMATOLOGY The most common pigmented lesion of young light-skinned Caucasians, often of Celtic stock, consisting in light brown 1-10 mm macules which fade in winter, and are accentuated in summer

ePhysician THERAPEUTICS A general term for a handheld electronic prescribing product, see there

epi-sick MEDTALK The pale, green, nauseated, intense, tachycardiac appearance of a Pt who received aggressive epinephrine therapy for anaphylaxis or status asthmaticus, as the emergency resolves and the Pt waits for the exogenous catecholamines to leave the system

epicardial lead Myocardial lead CARDIOLOGY A pacing lead placed on the epicardium. See Lead

epicondylitis ORTHOPEDICS Inflammation of the elbow due to overuse

epidemic *adjective* Referring to an epidemic *noun* The occurrence of more cases of a disease or illness than expected in a given community or region or among a specific group of people over a particular period of time; a wave of infections in a region by an organism with a short generation time; epidemics are usually heralded by an exponential rise in number of cases in time and a decline as susceptible persons are exhausted. See Hidden epidemic, Media epidemic, Pseudoepidemic, Tobacco epidemic. Cf Endemic, Pandemic

epidemic period EPIDEMIOLOGY A timespan when the number of cases of a disease reported is greater than expected

epidemic pleurodynia An acute viral infection, most commonly by coxsackievirus–usually B1-6; A4, A6, A10; enteroviruses 1, 6, 9, 19; EP is a summer, early fall disease first described on the Danish island of Bornholm CLINICAL Paroxysmal crushing, 'vise-like' pain, in the chest of adults or upper abdomen of children, SOB; ± $\frac{1}{2}$ of Pts have multiple recurrences COMPLICATIONS Septic meningitis in 5%, orchitis

epidemic viral gastroenteritis Gastroenteritis presumed to be of viral etiology and epidemic in nature; EVG is most common in children, occurs in the winter months, and resolves in 1-3 days ETIOLOGY 1. Norwalk and Norwalk-like agents–eg Hawaii, Snow Mountain, Montgomery agents, most common 2. Caliciviruses 3. Astroviruses 4. 'Also-rans', including Cockle, Paramatta, Wollan, et al CLINICAL N&V, diarrhea, abdominal discomfort, anorexia, headache, malaise, low-grade fever; EVG rarely requires hospitalization; it may kill the elderly– 2 typical patterns: 1. Nonfebrile, confined to the GI tract; 2. Febrile and systemic TREATMENT Fluids

epidemiologic study A study that compares 2 groups of people who are alike except for one factor, such as exposure to a chemical or the presence of a health effect; the investigators try to determine if any factor is associated with the health effect

epidemiologic surveillance The ongoing, systematic collection, analysis, and interpretation of health data essential to planning, implementing, and evaluating public health practice, closely integrated with the timely dissemination of these data to those who need to know

epidemiology 1. The study of the distribution of disease and its impact upon a population, using such measures as incidence, prevalence, or mortality 2. The study of the occurrence and causes of health effects in human populations 3. The science of public health, which studies the frequency, distribution, and causes of diseases in a population–rather than in an individual, and examines the impact of social and physical factors in the environment on morbid conditions. See AIDS epidemiology, Analytical epidemiology, Cancer epidemiology, Clinical epidemiology, Developmental epidemiology, Intersecting epidemiology, Inverted epidemiology, Prospective epidemiology, Retrospective epidemiology

Epidemiology Intelligence Service A branch of the Centers for Disease Control & Prevention–CDC, founded in 1951 to 1. Train field epidemiologists 2. Assist the CDC in preventing and controlling communicable diseases and 3. Provide public health services to state and local heath departments, thereby improving national–US disease surveillance. See CDC

epidermal inclusion cyst Epidermal cyst, epidermoid cyst DERMATOLOGY A benign cystic space lined by squamous epithelium and filled with keratinaceous debris and sebaceous goo

epidermoid carcinoma Squamous cell carcinoma, see there

epidermoid cyst Epidermal inclusion cyst, see there

epidermolysis DERMATOLOGY The shearing of the epidermis from the dermis, generally with an accumulation of fluid and formation of blebs and bullae. See Epidermolysis bullosa

epidermolysis bullosa Epidermolysis bullosa, epidermolysis bullosa lethalis, Epidermolysis bullosa simplex, Weber-Cockayne syndrome PEDIATRICS A general term for an array of dermatopathies with manifestations ranging from minor blisters of mucocutaneous surfaces, to formation of large bullae which may appear following minor trauma, that later rupture, leaving scars and dysphagia

epididymitis UROLOGY Inflammation of the epididymis, which is the most common cause of scrotal or testicular pain in ♂ > age 18. See Blue balls, Epididymis

epidural *adjective* Referring to the space between the bone of vertebral column and the meninges of the brain or spinal cord *noun* An epidural injection administered in the epidural space of the vertebral column

epidural block Epidural anesthesia ANESTHESIOLOGY The most popular locoregional anesthesia used in obstetrics, administered as a single injection or intrathecal 'drip', inserted in the L2-3 or L3-4 interspaces PROS Little local anesthetic is used and the bearing-down reflex is not abolished

epidural hematoma Extradural hematoma NEUROLOGY A medical emergency in which blood accumulates in the epidural space due to blood leaking from the middle meningeal artery, resulting in compression of the brain/dura mater; sans evacuation, tentorial herniation ensues, followed by a consult with the grim reaper. See Subdural hematoma

epigastric *adjective* Referring to the body region between the costal margins and the subcostal plane

epiglottitis PEDIATRICS Inflammation of epiglottis and oropharyngeal region CLINICAL Abrupt high fever, dysphagia, drooling, muffled speech, cyanosis, stridor, inspiratory retractions, sniffing dog position, ±respiratory arrest IMAGING Thumbprint sign DIFFDX Angioneurotic edema, bacterial tracheitis, croup, foreign body aspiration, retropharyngeal abscess ETIOLOGY In pre-HIB vaccination era, *H influenzae*; newer culprits include group A streptococci, *S pneumoniae*, *Corynebacterium diphtheriae*, TB MANAGEMENT Stat intubation; antibiotics–eg, ceftriaxone, ampicillin. See Thumbprint sign

epilation The removal of hair by extraction, electrolysis, or chemical lysis. See Hirsutism

epilepsy NEUROLOGY Any syndrome characterized by paroxysmal, usually transient, defects in cerebral function which are manifest as episodic impairment of neurologic activity, loss of consciousness, abnormal motor activity, sensory defects and alterations in the autonomic nervous system IMAGING MRI, PET, SPECT. See Absence, Automatic epilepsy, Focal epilepsy, Gelastic epilepsy, Juvenile myoclonic epilepsy, Progressive myoclonus epilepsy, Seizure disorder

epileptogenic zone NEUROLOGY A region of the brain which is more susceptible to seizures

epinephrine PHYSIOLOGY A sympathomimetic catecholamine hormone synthesized in the adrenal medulla and released into the circulation in response to hypoglycemia and sympathetic nervous system–splanchnic nerve stimulation due to exercise and stress; it acts on α– and β-receptors, resulting in vasoconstriction or vasodilation, ↓ peripheral blood flow, ↑ heart rate, ↑ force of contractility, ↑ glycogenolysis, ↑ lipolysis; pharmacologic epinephrine is used as bronchodilator for acute asthma to ↑ BP and in acute MIs to improve myocardial and cerebral blood flow. See Fight-or-flight response, High-dose epinephrine

EpiPen® ALLERGY MEDICINE A proprietary EM Pharmaceuticals device used to autoinject–IM epinephrine in Pts with potentially fatal anaphylactic reactions. See Anaphylactic reaction

episiotomy OBSTETRICS An incision at the introitus performed at the end of the second stage of labor to facilitate delivery and to avoid jagged perineal tears

episode MEDTALK A period of time or duration of action or interaction. See Acute episode, Hypomanic episode, Major depressive episode, Mixed episode, Reflux episode

episode of care MANAGED CARE Healthcare services provided for a specific illness during a set time period

episodic memory NEUROLOGY A 'cognitive' form of memory based on personal experience. See Memory

epispadias Hyperspadias UROLOGY A congenital defect in the dorsal urethra, resulting in an opened upper meatal orifice. Cf. Hypospadias

epistaxis MEDTALK → VOX POPULI Nosebleed

epithelial tissue One of 4 basic tissue types which covers or lines all exposed body surfaces

epithelioid Epithelial-like *adjective* Referring to cells, especially histiocytes, whose morphology mimics large epithelial cells

epithelioid sarcoma A low-grade sarcoma of the upper extremity, ♂:♀ ratio, 2:1, of Pts age 10-35 TREATMENT Wide local excision or amputation

DIFFDx-BENIGN Fibromatosis, fibrous histiocytoma, nodular fasciitis, infectious granuloma, necrobiosis lipoidica, rheumatoid nodule DIFFDx-MALIGNANT Synovial sarcoma, fibrosarcoma, melanoma PROGNOSIS Recurrence is common; 45% metastasize–eg, to lung, regional lymph nodes, scalp

epitope IMMUNOLOGY Any site on a molecule–an antigenic determinant that can evoke antibody formation. See Idiotype, Immunogenicity

EpiTouch™ DERMATOLOGY A ruby laser for removing tattoos and pigmented lesions, and hair

Epivir® Heptodin®, Lamivudine VIROLOGY An antiviral used with other agents–eg, zidovudine–Retrovir®, to manage HIV and HBV infections TOXICITY Renal toxicity. See HAART, Lamivudine

epizootic EPIDEMIOLOGY An outbreak or epidemic of disease in an animal host populations

Epley maneuver Canalith repositioning procedure, modified liberatory maneuver NEUROLOGY A technique used to manage BPPV, which involves sequential movement of the head into 4 positions. See Benign paroxysmal positional vertigo

EPO Erythropoietin, see there

epoetin alfa Epogen® HEMATOLOGY A recombinant erythropoietin that stimulates production of RBCs and platelets, during chemotherapy; a hematopoeitic cytokine. See Colony-stimulating factor, Erythropoietin

epoietin beta Recombinant human erythropoietin. See erythropoietin

eponym MEDTALK A syndrome, lesion, surgical procedure or clinical sign that bears the name of the author who first described the entity, or less commonly, the name of the index Pt(s) in whom the lesion was first described

Epoxide Chlordiazepoxide, see there

eprosartan mesylate Teveten® THERAPEUTICS An angiotensin II receptor antagonist thiazide combination agent used for HTN ADVERSE EVENTS URI, rhinitis, pharyngitis, cough. See Hypertension

EPS Extrapyramidal side effects

ε–epsilon Symbol for: 1. Hemoglobin ε, an 'early' Hb chain that disappears by the 3rd month of fetal development 2. Immunoglobulin E heavy chain 3. Molar absorptivity

Epsom salts A bitter, water-soluble substance used internally as a purgative and as a mouthwash for toothaches

EPSP Excitatory postsynaptic potential PHYSIOLOGY A graded depolarization of a postsynaptic membrane in response to stimulation by a neurotransmitter; EPSPs can be summated but transmitted only over short distances; they can stimulate production of action potentials when a threshold level of depolarization is reached

Epstein-Barr virus Human herpesvirus-4 VIROLOGY A double-stranded DNA virus that causes infectious mononucleosis, and belongs to the 8-member herpesvirus family; immature EBV particles are 75-80 nm and appear in the cytoplasm and nucleus; mature infectious particles are 150-200 nm and are cytoplasmic; it is associated with aplastic anemia, Burkitt's lymphoma–usually African type, hairy leukoplakia, histiocytic sarcoma in renal transplants and immunocompromise; EBV facilitate lymphoproliferative disorders–eg, Hodgkin's disease–ID'd by PCR, Southern blot, and in situ hybridization, angiocentric immunoproliferative lesions, NHL, Izumi fever, immunoblastic lymphoma, thymic carcinoma, undifferentiated nasopharyngeal carcinoma–mainland China; several serologic markers are used to detect active EBV infection, and determine the stage of infection

EPSTEIN-BARR VIRUS

EARLY ANTIGEN AB Present early in infection, usually disappears.

VIRAL CAPSID AB-IgG ↑ After infection and remains positive; ↑ dramatically with re-infection.

VIRAL CAPSID AB-IgM ↑ With infection and disappears; ↑ with re-infection.

NUCLEAR AB –IgG ↑ With infection and remains positive.

NUCLEAR AB –IgM ↑ For 4-6 weeks; post-infection becomes negative; ↑ with re-infection

Epstein-Barr virus-associated hemophagocytic syndrome A severe multisystem disorder characterized by fever, hepatic dysfunction, pancytopenia, in a background of EBV and HIV infection MANAGEMENT Foscarnet, acyclovir, prednisone, vinblastine. See Epstein-Barr virus

eptifibatide Integrilin® CARDIOLOGY An antithrombotic GP IIb/IIIa receptor antagonist that blocks fibrinogen and von Willebrand factor from binding to the surface of activated platelets

Epworth Sleepiness Scale SLEEP DISORDERS A testing instrument used to indicated a person's risk of dozing in specific situations, as well as daytime sleepiness. See Sleep disorder

Equagesic® Meprobamate, see there

equal opportunity employer An employer or enterprise that does not discriminate against a job candidate, or subject him/her to adverse exclusionary criteria, based on race, sex, religion, or national origin. See Equal employment opportunity

Equal™ Aspartame, see there

Equanil® Meprobamate, see there

equilibrium A state of constancy in a system; a population might be in static equilibrium–no pasa nada–ie, no births or deaths, or in dynamic equilibrium–ie, same numbers of births and deaths; the state to which a system evolves–eg, sustained periodic oscillations. See Chemical equilibrium, Linkage equilibrium, Sedimentation equilibrium NEUROLOGY A state of balance in the body, where forces are appropriately offset by counterforces. Cf Dizziness, Equilibrium, Vertigo ORTHOPEDICS A state of biomechanical homeostasis that enables persons to know where their bodies are in the environment and to maintain a desired position. See Fixed point equilibrium

equilibrium radionuclide angio(cardio)graphy CARDIOLOGY A technique in which an objective signal–eg, the EKG, is used to 'gate' or physiologically mark the otherwise static imaging of the cardiac blood pool; ERNA is used to evaluate the right and left ventricles–hence the synonym equilibrium radionuclide ventriculography in terms of volumes, as well as systolic and diastolic function, and regional or global myocardial performance, which is altered in MI. Cf First-pass radionuclide ventriculography

equine gait Steppage gait, see there. See Gait

equinovalgus Talipes equinovalgus ORTHOPEDICS A foot deformity in which the heel is raised and everted from the body midline

equinovarus Talipes equinovarus ORTHOPEDICS A type of clubfoot which involves different muscle groups: tibialis anterior, tibialis posterior, extensor hallucis longus, gastrocnemius and lack of peroneal activation; the heel is turned toward the midline, the foot plantarflexed, the inner border of the foot is supinated, and the anterior foot is displaced medial to the vertical axis

equinus ORTHOPEDICS Fixation of the foot, or part of a foot, in a plantarflexed position it would assume if the distal end of the part is farther away from the tibia TYPES Talipes equinus–fixed position of plantarflexion of the ankle joint; forefoot equinus—the fixed position of plantar flexed attitude of the forefoot in relation to the rearfoot. See Clubfoot

226

equipoise MEDICAL ETHICS A state of uncertainty regarding the pros or cons of either therapeutic arm in a clinical trial

equivalence trial CLINICAL TRIALS A study–eg, COBALT, intended to circumvent the ethical dilemmas of comparing a new therapy to a placebo which is known to be less effective than a proven therapy, by running the new agent against a standard therapy. See Clinical trial, COBALT

ER Emergency room–ward, also 1. Ejection rate 2. Electrical resistance 3. Endoplasmic reticulum 4. Erythrocyte rosette 5. Estrogen receptor, see there 6. Evoked response 7. Exchange ratio 8. External rotation

ER+ Estrogen-receptor positive, see there

ER- Estrogen-receptor negative, see there

ER/trauma setting MEDTALK *'A chaotic environment characterized by an unpredictable number of Pts with unscheduled arrivals, unplanned situations requiring intervention, allocation or limited resources, need for immediate care as perceived by the Pt or others and unknown variables that include severity, urgency, diagnosis'* **NEVA Standards of emergency Practice**

Erb's palsy Erb-Duchenne paralysis NEONATOLOGY Paralysis of multiple upper arm and shoulder muscles–eg, biceps, brachialis, brachioradialis, deltoid, due to an abnormal position during a vaginal delivery, resulting in C5 and C6 injury in cervical plexus

erbium:YAG laser SURGERY A laser for skin resurfacing and tissue sculpting, for general surgical use, for incision, excision, coagulation and vaporization of soft tissues, including subcutaneous, striated and smooth muscle, cartilage meniscus, calculi, mucosa, lymph nodes and vessels, organs glands

ERCP Endoscopic retrograde cholangiopancreatography, see there

ERDA Energy Research & Development Administration

erectile dysfunction UROLOGY A consistent inability to sustain an erection/penile tumescence sufficient for sexual intercourse ETIOLOGY Medical–eg, severe heart disease, psychologic–lack of sexual desire, defects in ejaculation and orgasm, surgical; ED affects 20–30 million ♂–US; up to 23% of impotent ♂ have ↓ testosterone MANAGEMENT Vacuum pump system, penile injection therapy, penile implant surgery, microvascular bypass surgery. See Impotence, Penile implant. Cf Blue balls

ergogenic drug Performance-enhancing drug SPORTS MEDICINE A weight training-term for any performance-enhancing substance. See Anabolic steroids, Weight training

ergonomic standards OCCUPATIONAL MEDICINE A series of guidelines developed by OSHA–to address activities in the workplace with a high risk for injury

ergonomics OCCUPATION MEDICINE The formal study of work situations, which attempts to evaluate, and if necessary, reconfigure a workplace by taking into account the anatomic and psychological variables of those working in the environment. See Ergogenic engineering, Human factor

ergophobia PSYCHOLOGY Fear of work. See Phobia

ergotism A condition caused by an excess of ergot compounds, derived from *Claviceps purpura*, resulting in intense vasoconstriction and gangrene of extremities

Eridan Diazepam, see there

ERISA Employee Retirement Income Security Act of 1974 MANAGED CARE A legislative act for regulating employee benefit plans–eg, healthcare plans sponsored and/or insured by an employer; ERISA exempts from state regulations, those companies that provide self insurance, or which fund their own insurance plans

ERNA Equilibrium radionuclide angiocardiography, see there

erogenous zone SEXOLOGY A region, specifically on the genitalia, on the skin surface which, when stimulated, often causes a sexual partner to become aroused. See Tanzen

erosion A wearing away, ulceration. See Apple core erosion, Cervical erosion

erotic *adjective* Referring to eroticism, sexual love or to its imagery in daydream, fantasy, or dream, either autonomously or in response to a perceptual stimulus, and either alone or with one or more partners. See Sexual

eroticism Erotism SEXOLOGY Personal experience and expression of one's genital arousal and functioning as ♂ or ♀, alone or with a partner, vis-á-vis arousing ideation, imagery, and sensory input. See Pornography, Sexuality

erotomania SEXOLOGY Hypersexuality A morbid exaggeration of, or preoccupation with sexuoerotic imagery and activity. See Cherambault-Kandinsky syndrome, Don Juan syndrome, Nymphomania

ERPF Effective renal plasma flow, see there

ERRLA (pupils) Equal, round, responsive to light & accommodation

error An unintentional deviation from standard operating procedures or practice guidelines LAB MEDICINE An erroneous result from a Pt sample, the frequency of which reflects the lab's QC procedures and adherence to well-designed procedure manuals MEDTALK See Misadventure, Honest error, Human error PATIENT CARE The failure of a planned action to be completed as intended–error of execution or the use of the wrong plan to achieve an aim–error of planning STATISTICS see Type I error, Type II error VOX POPULI → MEDTALK Opportunity for improvement

eructation GI DISEASE Belching, burping; the liberation of gas in the upper GI tract via the esophagus

eruption A rash. See Polymorphous light eruption, Sea-bathers' eruption

erysipelas Cellulitis, St Anthony's fire INFECTIOUS DISEASE A superficial infection of the very old or very young, caused by β-hemolytic group A streptococci, group C streptococci, staphylococci, pneumococci CLINICAL Abrupt fever, malaise, vomiting, skin lesions with cellulitis, brawny induration, geographic discoloration, butterfly rash on face with minimal necrosis LAB ↑ ESR, ↑ WBCs MANAGEMENT Rest, hot packs, penicillin

erysipeloid INFECTIOUS DISEASE An infection by the microaerophilic gram-positive *Erysipelothrix rhusiopathiae*, which is almost exclusive to those who occupationally handle animal products CLINICAL Sharply-demarcated red maculopapular lesions of the hands, which may spontaneously heal COMPLICATIONS Arthritis, endocarditis

erythema A general term for reddening, especially of a mucocutaneous surface, rash

erythema infectiosum Fifth disease A childhood exanthema caused by the moderately contagious B19 parvovirus, so named as it was the 5th childhood disease typically accompanied by a rash DIFF DX The other nosologies often associated with rashes in childhood are rubella, measles, scarlet fever, and a mild, atypical variant of scarlet fever–Filatov-Dukes disease CLINICAL Low-grade fever, fatigue, "slapped cheeks rash"; 80% of adults have symmetrical arthritis which may become chronic with stiffness in the morning, redness and swelling; B19 can infect the fetus before birth. See B19

erythrasma INFECTIOUS DISEASE A patchy red-brown rash in the axilla and inguinal region, due to the presence of *Corynebacterium minutissimum* in the stratum corneum

erythroblastosis fetalis Hemolytic disease of the newborn, see there

erythrocyte RBC, red blood cell, discocyte HEMATOLOGY A mature, nonnucleated cell averaging 7–8 μm in diameter, which is round or ovoid on peripheral smear, contains Hb and has a zone of central pallor due to the cell's biconcavity. See Fetal erythrocyte

erythrocyte count See Red cell count

erythrocyte sedimentation rate 'sed' rate, ESR A test that measures the rate at which RBCs in venous blood settle to the bottom of a test tube, a nonspecific indicator of inflammation; ESR ↑ in collagen vascular disease, neoplasia, pregnancy, hyperproteinemia; ↓ in polycythemia, microcytosis, sickle cell anemia; it is used to monitor Pts with rheumatic diseases, as therapy may require frequent adjustment of doses. See C-reactive protein

erythroleukemia Di Guglielmo syndrome HEMATOLOGY An acute myelocytic leukemia–FAB classification, M6, which is usually acquired affecting the elderly, but rarely also AD CLINICAL Anemia, fever, hepatosplenomegaly, hemorrhagic diathesis LAB ↑ Erythroblasts in circulation, BM, ↑ gammaglobulins, false positive positive rheumatoid factor, ANA, Coombs antiglobulin GENETICS Aneuploidy–63%; cytogenetic defects, especially chromosomes 5 and 7 PROGNOSIS Poor. See French-American-British classification

erythrophagocytosis The ingestion of RBCs by PMNs or macrophages, which occurs commonly in paroxysmal cold hemoglobinuria, cold agglutinin disease, and incompatible blood transfusion; erythrophagocytosis may also occur in Heinz-body anemia, autoimmune hemolytic anemia, isoimmune hemolytic anemia, sickle cell anemia, and in viral and parasitic infections

erythroplakia ORAL PATHOLOGY An oral lesion characterized by a reddened patch with a velvety surface and epithelial changes ranging from mild dysplasia to CIS to frankly invasive CA DIFFDX Candidiasis, TB, histoplasmosis, denture irritation. Cf Leukoplakia

erythropoietin EPO PHYSIOLOGY A 46 kD glycoprotein colony-stimulating factor produced predominantly by cells adjacent to the proximal renal tubules in response to signals from an oxygen-sensitive substances in the kidneys–eg, heme ADVERSE EFFECTS Chest pain, swelling, tachycardia, headache, HTN; erythropoietin–EP binds to receptors in erythroid precursors that mature into RBCs; EP is ↑ by hypoxia or by ectopic production from tumors–eg, cerebellar hemangioblastoma, hepatoma, pheochromocytoma, uterine leiomyoma, and renal cell carcinoma; it may not be ↑ in anemic premature infants, and is ↓ in 2° anemia, chronic inflammation, P vera, and certain CAs and may be useful in myeloma-related anemia; EP therapy is indicated for HIV-related anemia, anemia of renal failure and prematurity; it ↑ number of units of autologous RBCs that may be donated before surgery, for ↑ number of units that may be phlebotomized in Pts with hemochromatosis and to ↑ units that may be drawn from a person with a rare blood type

escape beat CARDIOLOGY An automatic beat occurring after an interval longer than the dominant cycle length, ie a normal ventricular contraction, which occurs when the usual cardiac pacemaker–the sinoatrial node–defaults

escapee A popular term for older relatives of those at risk for Huntington's disease, who didn't develop the disease. See Huntington's disease

escharotomy SURGERY An incision into an encircling scar–eg, of a 3rd degree burn to an extremity, to lessen the pressure on neurovascular structures

Escherichia coli MICROBIOLOGY The type species of genus *Escherichia*, and part of the normal colonic flora; some *E coli* serotypes are associated with hemorrhagic colitis, dysenteric syndrome, and watery diarrhea. See *Escherichia* spp

Escherichia coli 0157:H7 MICROBIOLOGY A shiga-like verotoxin-producing serotype of *E coli* inculpated in outbreaks of hemorrhagic diarrhea, due to undercooked meat in 'fast-food' restaurants CLINICAL Colic, bloody diarrhea RADIOLOGY Submucosal edema, 'thumbprinting'

escitalopram oxalate Lexapro® NEUROLOGY A serotonin reuptake inhibitor-type antidepressant, an isomer of citalopram ADVERSE EVENTS N&V, insomnia, ejaculation disorder, diaphoresis, fatigue

Esclim® GYNECOLOGY An estrogen replacement system for managing vasomotor Sx of menopause, Sx of vulval/vaginal atrophy, and treat hypoestrogenism due to hypogonadism, castration, or primary ovarian failure. See Hormone replacement therapy

esmolol Brevibloc® CARDIOLOGY An ultrashort β-blocker used to treat A fib/flutter, unstable angina, HTN, SVT, PVCs, and for 2° protection of MI survivors CONTRAINDICATIONS Bronchospastic lung disease, left ventricular dysfunction. See β-blocker

esomeprazole Nexium® INTERNAL MEDICINE An agent for short-term relief of esophagitis pain ADVERSE EFFECTS Headache, diarrhea, abdominal pain. See Esophagitis, GERD

esophageal achalasia Cardiospasm GI DISEASE A condition characterized by a failure of the lower esophageal sphincter to relax, accompanied by defective contraction of the thoracic esophagus, resulting in functional obstruction and dysphagia. See GERD

esophageal cancer GI DISEASE A malignancy of the esophagus, most commonly, SCC and adenoCA EPIDEMIOLOGY ♂:♀, 3:1; age 55-70; ↑ in China, Japan, Scotland, Russia, Scandinavia; ±12,000 new cases/yr US; blacks have a 4-fold greater risk than whites–rate 1/10⁵ ♀, 4/10⁵ ♂ RISK FACTORS Alcohol, tobacco use, poor nutrition, Hx of achalasia, corrosive esophagitis, Barrett's esophagus, tylosis palmaris et plantaris CLINICAL Dysphagia, retrosternal discomfort and pain, pressure, burning, and a sensation of food being stuck, eventually, narrowing of esophagus becomes severe, with choking, vomiting, weight loss TYPES SCC, ≥ 70%; adenoCA in Barrett's esophagus–10-20% DIAGNOSIS Barium swallow to ID lesions arising in the esophageal mucosa, upper GI endoscopy/esophagoscopy with Bx and cytology brushings STAGING CT, MRI, ultrasound TREATMENT Cisplatin, 5-FU, RT–5000 cGy is better than RT alone–6400 cGy for local control of CA, metastasis, ↑ survival, but ↑ side effects PROGNOSIS Median survival, 10 months; 5-yr survival, 20–36%

ESOPHAGEAL CANCER–STAGES

I   CA in esophageal mucosa; no spread to nearby tissues, lymph nodes, or organs

II CA in all layers of esophagus and/or regional lymph nodes; no spread to other tissues

III CA has spread to tissues or lymph nodes near the esophagus; no metastasis

IV CA has metastasized

esophageal candidiasis Candida esophagitis INFECTIOUS DISEASE Inflammation by *Candida* spp RISK FACTORS Immunocompromise–eg, AIDS, heart, kidney, other transplants, leukemia, lymphoma, chemotherapy CLINICAL Dysphagia; fever if systemic. See Candidiasis

esophageal hernia See Paraesophageal hernia, Sliding esophageal hernia

esophageal perforation GI DISEASE A defect in the esophagus where the lumen communicates with the thoracic cavity

esophageal ring Esophagogastric ring A partially encircling intraluminal mass, or overhanging fold of mucosa that constricts the esophageal lumen, causing dysphagia TREATMENT Surgery is rarely indicated or successful. See Esophageal web

esophageal spasm GI DISEASE A dysmotility of the LES, characterized by retrosternal anginal pain or heartburn. See GERD, Lower esophageal sphincter

esophageal stricture GI DISEASE A narrowing of the esophageal lumen which may result from prior exposure to caustic agents–eg, bleach. See Caustic burn

esophageal ulcer GI DISEASE A hole in the esophageal mucosa, which may be due to acid reflux from the stomach, *H pylori* infection, NSAIDs, cigarette smoking CLINICAL Pain may not correlate with severity of ulceration DIAGNOSIS Barium swallow, endoscopy COMPLICATIONS Bleeding, perforation TREATMENT Antibiotics to eradicate *H pylori*, eliminating risk factors. See Caustic injury to esophagus, GERD

esophageal varices The presence of varices under the esophageal mucosa, which most commonly occurs in a background of advanced liver disease ETIOLOGY Portal HTN, schistosomiasis MANAGEMENT Acute hemorrhage of EVs is treated by balloon compression; rebleeding is common, and is preemptively managed with endoscopic sclerotherapy, which in turn is often complicated by rebleeding, stenosis, & esophageal ulceration; some data suggest that combined modality therapy with a β-blocking agent–nadolol, propranolol and an anti-hypertensive–isosorbide mononitrate is better than endoscopic sclerotherapy in treating EVs PROGNOSIS Good if unrelated to cirrhosis. See Endoscopic sclerotherapy, Nadolol with isosorbide mononitrate

esophageal web A 2-3 mm in thickness stricture composed of mucosa and submucosa only, which is located anywhere along the length of the esophageal lumen; upper esophageal webs occur in the upper 2-4 cm of the esophagus, are lined by squamous epithelium, often associated with the Plummer-Vinson syndrome, and after yrs may evolve into postcricoid carcinoma; webs in the body of the esophagus may be multiple, represent embryonal remnants and may be associated with esophageal reflux; the lower esophageal web is a thin membrane marking the squamocolumnar junction and is seen in ±10% of normal subjects; symptomatic subjects may suffer intermittent dysphagia and impaction of a bolus of food TREATMENT Intraluminal balloon dilatation. See Café coronary

esophagectomy SURGERY Resection of part of the esophagus, which may require that the tissue be replaced by a segment of large intestine INDICATIONS CA, corrosive injury

esophagitis GI DISEASE Inflammation of the esophagus, primarily affecting the mucosa. See Candidia esophagitis, CMV esophagitis, Herpes esophagitis

esophagogastroduodenoscopy ENDOSCOPY An endoscopic examination of the esophagus, stomach, and upper small intestine, usually a tad beyond the ampulla of Vater

esophagoscopy ENDOSCOPY Examination of the esophagus (only) using an endoscope

esophagram Barium swallow GI DISEASE A series of x-rays of the esophagus taken after the Pt drinks a barium solution, which coats and outlines the esophagus

esophagus See Esophageal etc

esoteric test LAB MEDICINE The analysis of 'rare' substances or molecules that are not performed in a routine clinical lab. See DORA

esoterica MEDTALK A synonym for 'oddballs'–unusual causes of common complaints. See Anecdotal, Fascunomia

esotropia Convergent strabismus OPHTHALMOLOGY Strabismus characterized by a convergence of the visual axes. See Strabismus

esprit de corps GRADUATE EDUCATION The degree of happiness of the 'campers' in a place

ESR Erythrocyte sedimentation rate, see there

ESRD End-stage renal disease, see there

ESSENCE CARDIOLOGY A clinical trial–Efficacy & Safety of Subcutaneous Enoxaparin in Non-Q-Wave Coronary Events–which compared the effectiveness of enoxaparin antithrombolytic therapy vs heparin in treating acute MI. See Low-dose heparin

essential cryoglobulinemic vasculitis INTERNAL MEDICINE Vasculitis with cryoglobulin immune deposits in small vessels–arteries, capillaries, venules, typically of the skin and glomeruli, accompanied by cryoglobulins in the serum. See Small vessel vasculitis, Systemic vasculitis

essential dietary component Essential nutrient NUTRITION A required dietary component, without which a deficiency state develops EDCs Water– 1-2 L/day, calories–2000 to 3500 kcal/d, carbohydrates, fat, protein, vitamins, minerals, fiber. See Essential amino acids, Essential fatty acids, Fiber, Trace minerals, Vitamins

essential hematuria A condition that causes symptomatic episodic gross hematuria, most common in 2-11 yr-old ♂, often in a background of chronic glomerulopathy, 'nil' disease, IgA nephropathy PROGNOSIS Spontaneous resolution

essential hypernatremia A disease complex characterized by ↑ secretion of ADH in response to volume contraction, but lack of response of ADH to hyperosmolarity; EH may occur in sodium retention, and in burn victims with central pontine myelinolysis, see there. Cf SIADH

essential hypertension Primary HTN, A condition comprising 90% of all cases of HTN; EH is associated with impaired endothelium-mediated vasodilation, which may play an important role in functional defects of resistance vessels. See Goldblatt kidney, Hypertension, Pheochromocytoma

essential thrombocythemia Essential thrombocytosis HEMATOLOGY A primary myeloproliferative disorder of older–age 55-75–adults, or less commonly of young ♀, with a platelet count is > 600 x 10⁹/L; ET has many clinical features of P vera; it affects the same age group, is accompanied by splenomegaly, has similar BM findings, intensity of leukocytosis DIAGNOSIS Polycythemia Vera Study Group criteria TREATMENT Hydroxyurea, a myelosuppressant, prevents recurrent thromboses. See Polycythema vera

---

ESSENTIAL THROMBOCYTHEMIA–DIAGNOSTIC CRITERIA

1. Platelets > 1 x 10⁹ /L (US: < 1000/mm³)

2. Hb < 2.05 mmol/L (US < 13.0 g/dL)

3. Iron in BM or if absent, little ↑ in Hb after 1 month of oral iron therapy

4. Absent marrow fibrosis by biopsy and

5. Absent Philadelphia chromosome

Polycythemia Vera Study Group

---

essential thrombocytopenia A condition of young adults with thrombotic complications occurring in < ½ of cases, associated with vasocclusive headaches, erythromelalgia TREATMENT Conservative, anegrelide if Sx

essential tremor NEUROLOGY A benign idiopathic disorder characterized by rhythmic, moderately rapid tremor of voluntary muscles–hands, arms, head, larynx, eyelids, voice, evoked by activity and exacerbated by purposeful movement; if ET occurs in more than one member of a family, it is termed a familial tremor; emotional stress may ↑ tremors. See Tremor

estazolam ProSom® PHARMACOLOGY A benzodiazepine used to manage insomnia

esterase An enzyme–EC class 3.1 that catalyzes the hydrolysis of esters. See Leukocyte esterase, Nonspecific esters

esthesioneuroblastoma Olfactory neuroblastoma ENT A tumor of the nasal cavity, retrobulbar region or middle cranial fossa CLINICAL Nonspecific, similar to other intranasal lesions, ie congestion, rhinorrhea DIFFDX Lymphoma, plasmacytoma, embryonal rhabdomyosarcoma PROGNOSIS 5-yr survival 50-60%; late recurrence is common

esthetic rehabilitation A branch of cosmetology that is dedicated to covering wounds, burns, and scars, with silicone-based prostheses and makeup, for victims of trauma, violence, crashes, and fires

estimate A popular term for an educated guess about a thing or process. See Cookie cutter estimate, Demand-based estimate, Objective probability estimate, Subjective probability estimate

estimated hematocrit LAB MEDICINE A guestimate of the Hct by 'eyeballing' a spun blood tube

estradiol E2 ENDOCRINOLOGY The most potent natural estrogen, produced chiefly by the ovary and in small amount by the testis; it is measured to evaluate postmenopausal status or suspected hypogonadism ↑ IN Ovarian tumors, pregnancy, adrenal feminizing tumors, precocious puberty, liver disease, gynecomastia ↓ IN OCs, ovarian failure, menopause, hypogonadism

estramustine ONCOLOGY A combination of estradiol and nitrogen mustard used in the palliative therapy of prostate CA. See Prostate CA

Estrasorb® GERIATRICS A topical cream–17ß estradiol for ERT in symptomatic menopausal ♀

estriol PHYSIOLOGY A steroid hormone, the synthesis of which depends on a sequence of biochemical reactions in the placenta or fetus in a functional compartment, the fetoplacental unit–FPU. See Triple marker

estrogen Any of the estrus-related steroids, which include estradiol, estriol, estrone SITE OF PRODUCTION Ovaries, adrenal cortex, adipose tissue, in the fetus, placenta; estrogen is responsible for the 2° ♀ sex characteristics, and, during the menstrual cycle, prepares the endometrium for implantation ↑ IN Ovarian tumors, adrenal feminizing tumors, some adrenal and testicular tumors, precocious puberty, gynecomastia ↓ IN OCs, ovarian failure. See Designer estrogen, Estradiol, Estriol, Estrone, Estrus, Hormonal replacement therapy, Oral contraceptive, Phytoestrogen, Progesterone

estrogen dermatitis A condition characterized by severe premenstrual exacerbations of papulovesicular eruptions, urticaria, eczema, generalized pruritus ETIOLOGY Sensitivity to estrogens DIAGNOSIS Intradermal tests for estrogen sensitivity TREATMENT Antiestrogen therapy with tamoxifen or elimination of estrogen

estrogen receptor A protein of a superfamily of nuclear receptors for small hydrophilic ligands–eg, steroid hormones, thyroid hormone, vitamin D, retinoids; the presence of ERs in breast CA generally is associated with a better prognosis, as they respond to hormonal manipulation. See Tamoxifen. ERs are transcription factors that are regulated allosterically by binding a ligand; extracellular estradiol diffuses across the cell membrane, binds to the ER, leading to its dimerization and binding of ER to the estrogen responsive element, the ER's specific DNA target

estrogen receptor assay ONCOLOGY The ER is a protein found in high concentrations in the cytoplasm of breast, uterus, hypothalamus, and anterior hypophysis cells; ER levels are measured to determine a breast CA's potential for response to hormonal manipulation–60% of breast CAs are 'estrogen positive'; $1/2$ of ER-positive Pts respond favorably to anti-estrogen–tamoxifen citrate therapy, in contrast to < $1/3$ of ER-negative Pts. See 'Flare' phenomenon. Cf Progesterone receptor assay

estrogen receptor negative ONCOLOGY Breast CA cells without a receptor to which estrogens can attach; this is associated with a poorer prognosis as the CA usually doesn't respond to antiestrogen therapy. See Estrogen receptor

estrogen receptor positive ONCOLOGY Breast CA cells with a receptor to which estrogens can attach; this is associated with an improved prognosis as the CA usually responds to antiestrogen therapy that blocks the receptors. See Estrogen receptor

estrogen replacement therapy Estrogen administered to postmenopausal ♀, often in the form of a vaginal cream, which ameliorates the effects of lost ovarian function, ↓ progression of osteoporosis, has a cardioprotective effect, ↓ hot flashes, urogenital symptoms–eg, vaginal dryness, burning, itching, dyspareunia, bleeding, skin changes, depression; ERT ↓ PAI-1 by 50%, which may explain its benefit in CAD CONS ↑ in endometrial CA, HTN, thromboembolism, gallbladder disease, breast neoplasia– ↑ use of ERT, ↑ risk of breast CA

estrogen-responsive tissue Any tissue influenced/altered by estrogens

estrone E1 An estrogen with very low biologic activity that is the major source of estrogen in children and postmenopausal ♀ ↑ IN pregnancy, postmenopausal, ↑ age, obesity, luteal phase of menstrual cycle ↓ IN hypogonadism. See Androstendione

Estrostep® GYNECOLOGY A member of a new class of Estrophasic™ OCs that release gradually increasing amounts of estrogen–ethinyl estradiol, and a constant low dose of progestin–norethindrone acetate. See Oral contraceptive

ESWT Extracorporeal shock wave lithotripsy, see there

Eta-1 Osteopontin, see there

etanercept MOLECULAR MEDICINE A genetically engineered molecule that blocks TNF activity, used to manage therapeutically refractory rheumatoid arthritis. See Rheumatoid arthritis

état lacunaire Lacunar state, see there

ETEC Enterotoxic *Escherichia coli*, see there

ethacrinic acid An ototoxic loop diuretic. See Diuretic, Loop diuretic

ethambutol Myambutol AIDS A combination therapy agent used to treat TB and MAC ADVERSE EFFECTS ↓/distorted vision, pain and swelling of joints, burning pain, N&V, headache, mental confusion CONTRAINDICATIONS Ethambutol should not be given to children too young to be monitored for changes in vision. See AIDS

ethanol See Alcohol

Ethernet INFORMATICS A high speed LAN hardware standard for communicating among PCs, which can link up to 1024 nodes in a bus network. See Local area network, Packet switching. Cf Fiber distributed data interface

ethics committee A multidisciplinary hospital body composed of a broad spectrum of personnel–eg, physicians, nurses, social workers, priests, and others, which addresses the moral and ethical issues within the hospital. See DNR, Institutional review board

ethmoiditis Ethmoid sinusitis ENT Inflammation of the ethmoid sinus. See Sinusitis

ethnic group SOCIAL MEDICINE A group whose members (1) have a sense of common origins; (2) claim a common and distinctive Hx and destiny; (3) possess one or more dimensions of collective cultural individuality; (4) have a sense of unique collective solidarity. See Cultural awareness, Race

ethnicity VOX POPULI Racial status–ie, African American, Asian, Caucasian, Hispanic

ethosuximide NEUROLOGY An anticonvulsant used for absences ADVERSE EFFECTS BM suppression, aplastic anemia. See Epilepsy, Seizures

ethotoin NEUROLOGY An anticonvulsant used for generalized tonic clonic seizures ADVERSE EFFECTS BM suppression, aplastic anemia

ethyl alcohol Ethanol, see there

ethyl ether TOXICOLOGY An agent used as a CNS depressant; induces general anesthesia–ie, analgesia, amnesia, loss of consciousness, inhibition of sensory and automatic reflexes, skeletal muscle relaxation

ethylene diamine tetraacetic acid EDTA, see there

ethylene glycol TOXICOLOGY A chemical used as an antifreeze, which is highly toxic–50-100 ml and may be fatal ethanol surrogate occasionally used as an inebrient by alcoholics EG INTOXICATION STAGES 1. CNS Sx, occurring within first 24 hrs 2. Cardiovascular Sx, up to 72 hrs in duration 3. Respiratory arrest and renal failure with anuria LAB Anion-gap metabolic acidosis, ↑ serum osmolality, osmolar gap, hypocalcemia DIAGNOSIS GLC, fluorometry, colorimetry TREATMENT Gastric lavage, emesis, charcoal and catharsis, calcium gluconate for hypocalcemia

ethylene oxide OCCUPATIONAL MEDICINE A gas used to sterilize medical supplies and other materials

etidronate disodium Didronel® METABOLISM An organic biphosphonate used to manage osteoporosis, Paget's disease of bone–osteitis deformans, heterotopic ossification, hypercalcemia of CA. See Coherence therapy, Osteoporosis

etiology 1. The study of the cause of a disease, including its origin and what pathogens, if any, are involved 2. Cause of a disease 3. The branch of philosophy which deals with factors of causation or the factors associated with the causation of disease or abnormal body states MEDTALK → VOX POPULI Cause

EtOH Ethanol, see there

etoposide ONCOLOGY A chemotherapeutic active against monocytic leukemia, CNS lymphoma, refractory small cell carcinoma, testicular CA, and KS, which may be co-administered with cyclophosphamide and doxorubicin ADVERSE EFFECTS BM toxicity, hair loss, nausea, incoordination, stomatitis, dyspnea, anorexia. See AIDS, ALL

Etrafon® Amitriptyline + perphenazine, see there

Etravil Amitryptiline, see there

etretinate PHARMACOLOGY A retinoic acid derivative that stimulates epithelial differentiation and inhibits the malignant transformation of skin and mucosa; it is a 2nd line agent for psoriasis in combination with PUVA, to ↓ dose of methoxsalen. See Psoriasis

EUA Examination under anesthesia, see there

euglobulin clot lysis time Fibrinolysis time HEMATOLOGY A test of in vivo fibrinolysis, in which diluted and acidified plasma is cooled, causing precipitation of fibrinogen, plasminogen, plasmin, plasminogen activator; ECLT measures the time from the formation of a blood clot to its dissolution; ECLT is ↑ in DIC; ↓ 1° fibrinolysis. See Clotting time

eunuch ENDOCRINOLOGY A ♂ with sexual characteristics and appearance of a person who failed to mature at puberty

eunuchoid *adjective* Like a eunuch, see there

euploidy GENETICS The presence, in a particular organism of a normal complement of chromosomes which, in humans, is $2n$, where n is the haploid number

Euroblood TRANSFUSION MEDICINE A unit of packed RBCs imported from Europe which, until 2002, comprised a significant minority of the units transfused; in the wake of concern about mad cow disease, the FDA implemented a pan-European deferral, ending the 30-year old program that provides 140,000 packed units/yr to the New York metropolitan area

eutechtic mixture of local anesthetics See EMLA

euthanasia MEDICAL ETHICS The induction of death, or painlessly putting to death, a Pt suffering from an incurable disease; deliberate administration of medications–eg narcotics or barbiturates to an terminally ill Pt at the Pt's own request, to end his/her life. See Advance directive, DNR, Initiative 119, Kevorkian, Physician-assisted suicide, Slow code, Social euthanasia

euthyroid goiter ENDOCRINOLOGY A goiter–enlarged thyroid–caused by a paucity of iodine in the diet, which triggers compensatory hyperplasia of functionally normal thyroid tissue. See Goiter

euthyroid sick syndrome LAB MEDICINE A "condition" in Pts who are critically ill with nonthyroid diseases that alter serum levels of thyroid hormones which, in absence of underlying nonthyroid illness, would be correctly interpreted as indicating a disease of the thyroid 'axis' LAB Peripheral ↓/inhibition of 5'-deiodinase, the deiodination enzyme, resulting in ↓ peripheral 5'-monodeiodination of thyroxine–T_4, reversed, free, and total T_3; TSH, TRH, and usually free thyroxine levels are normal. See Hyperthyroidism, Hypothyroidism

Eutonyl® Pargyline, an MAO inhibitor. See Pargyline

euvolemic *adjective* Referring to a normal blood volume

evaluable disease MEDTALK A condition that cannot be measured directly by tumor size but can be evaluated by other methods specific to a particular clinical trial

evaluate *verb* MEDTALK Assess, examine

evaluation CHILD PSYCHIATRY A process conducted by mental health professionals resulting in an opinion about a child's mental or emotional capacity, which may include treatment or placement recommendations. See Assessment MEDTALK The assessment of a person's past medical history and present status. See Cost effectiveness evaluation, Economic evaluation, PART evaluation, PRIME-MD evaluation, Psychiatric evaluation, Urodynamic evaluation

evaluation & management service E/M service MEDICAL PRACTICE Any diagnostic and therapeutic procedure that may be performed by a health care provider at a specific location. See History of present illness, General multi-system examination, Past, family and/or social history, Review of Systems, Single organ system examination. Cf Physician test, Procedure

event MEDTALK 1. Error, see there 2. Misadventure, see there. See Accelerated compensation event, Adverse event, Adverse drug event, Cardiac event, Life event, Mutually exclusive event, Qualifying event, Sentinel event, Signal event, Terminal event, Unusual life event STATISTICS One or more outcomes of a probability experiment

event-free survival Failure-free survival ONCOLOGY The timespan that follows therapy for a malignancy or other dread disease, during which there are no objective signs of recurrence

event model PATIENT RIGHTS A model for obtaining informed consent, in which the Pt's signature for a procedure is obtained by a particular person at a particular time and place, after presenting the information in a codified fashion on a consent form. See Informed consent; Cf Process model

eversion ORTHOPEDICS Movement on the frontal plane in which the plantar aspect of the foot is tilted away from the midline; the axis lies on the sagittal and transverse planes

everything, fear of PSYCHOLOGY Panphobia. See Phobia

evidence-based medicine DECISION-MAKING *'The use of scientific data to confirm that proposed diagnostic or therapeutic procedures are appropriate in light of their high probability of producing the best and most favorable outcome'*. See Meta-analysis

Evista® Raloxifene, see there

evoked response Evoked potential CARDIOLOGY An area under an R wave that depends on the rate of myocardial depolarization; ↓ in cumulative R-wave area mandates ↑ pacing rate NEUROLOGY Any stimulus-evoked electrical potential recorded by EEG, which varies according to intensity, modality, location, and level of consciousness; ER is detectable over the appropriate cortical receptive areas by EEG after stimulation of sense organs or peripheral nerves–eg, brainstem auditory, transcortical motor, pattern shift visual, and somatosensory ERs; ERs may be used to guide surgical removal of tumors growing around important nerves. See White noise

Ewart sign PHYSICAL EXAM A dullness with bronchial breathing and bronchophony over the angle of the left scapula, a classic sign of pericardial effusion. See Pericardial effusion

Ewing sarcoma Primitive neuroectodermal tumor, PNET ONCOLOGY A primitive neuroectodermal tumor, which primarily affects the midshaft of long bones, which is closely related–if not biologically identical to peripheral neuroepitheliomas CLINICAL Locoregional bone pain, or pathologic fractures TREATMENT Excision, RT; few respond to chemotherapy. See Peripheral neuroepithelioma

ex officio HOSPITAL PRACTICE Referring to service provided by a medical practitioner by virtue of an office or position held, which does not have voting rights; *ex officio* refers to privileges or services that are implied without being officially delineated

ex vivo therapy MOLECULAR MEDICINE A therapy in which cells or genes are modified out of the body and reintroduced into a person with a genetic disease. See Gene therapy

examination Physical examination MEDTALK The physical evaluation of a Pt and relevant organ systems, which may be problem-focused, detailed, or comprehensive. See Abdominal examination, Baseline examination, Cardiac examination, Chest examination, Constitutional evaluation, Digital rectal examination, Double setup examination, Evaluation & management services, FLEX examination, Mental examination, National Boards examination, Neck examination, Neurologic examination, Oral examination, Physical examination, Rectal examination, Review of systems, Self examination, Sexual assault nurse examination, Single-organ system examination, Specialty board examination, SPEX examination, Stool examination

examination under anesthesia ORTHOPEDICS A format for testing joint integrity and ROM with the Pt anesthetized PROS Examinations on awake Pts have poor interobserver/intraobserver reproducibility CONS Intensity of Sx can't be assessed. See Laxity test, Provocative test

exanthema subitum Roseola infantum, sixth disease PEDIATRICS A mild, self-limited disease caused by HHV-6, which affects children age 6 months to 2 yrs CLINICAL High fever for 3–5 days, discrete, rose pink macules and maculopapules that may be tender to touch and blanch on pressure, with periorbital edema–Berliner sign and palatal edema. See HHV-6

exanthematous viral disease MEDTALK Any viral illness associated with a rash–eg, measles, varicella, HS

Excedrin® PAIN MANAGEMENT An OTC salicylate used to manage severe headaches, including migraines

EXCEL CARDIOLOGY A clinical trial–Expanded Clinical Evaluation of Lovastatin of lipid-lowering on CAD-related M&M in Pts with moderate hypercholesterolemia. See Bad cholesterol, Hypercholesterolemia, Lipid-lowering therapy, Lovastatin, Statin

excessive daytime sleepiness SLEEP DISORDERS A subjective difficulty in maintaining an awake state, and an increase ease of falling asleep when the person is sedentary; EDS may be quantified with subjective rating scales of sleepiness

exchange plasmapheresis Plasmapheresis, see there

exchange transfusion NEONATOLOGY A therapeutic procedure for reducing immunotoxins in the neonate by administration of compatible packed RBCs INDICATIONS HDN–due to maternal IgG antibodies against fetal antigens; neonatal hyperbilirubinemia, due to RBC and bilirubin metabolic defects COMPLICATIONS Thrombocytopenia, iron deficiency, HBV–rare, HIV–very rare

excimer laser Cold laser CARDIOLOGY A laser used in coronary artery angioplasty that delivers pulsating UV light to excise atherosclerotic plaques in stenosed arteries. Cf Percutaneous transluminal coronary angioplasty OPHTHAMOLOGY ELs may be used to vaporize part of the surface layer of the cornea

excision MEDTALK Surgical removal

excisional biopsy A surgical procedure intended to completely remove–ie, excise a lesion submitted for pathological evaluation; in EBs, the nature of the lesion–ie benign vs malignant is often unknown at the time of operation, and thus the margin of normal tissue obtained is based on clinical judgement. See Biopsy. Cf Incisional biopsy

excitotoxicity NEUROLOGY Neuronal injury caused by excessive release of excitatory neurotransmitters–glutamate and aspartate causing damage to nerve and glial cells, which occurs in diverse neurologic diseases that may be acute–eg hypoglycemia, seizures, stroke, or trauma or chronic neurodegenerative disease–eg AIDS-dementia complex, amyotrophic lateral sclerosis, Huntington's disease, and possibly Alzheimer's disease

exclude MEDTALK *verb* To rule out (from diagnostic consideration)

exclusion colitis Diversion colitis, see there

exclusion criteria AIDS Donor exclusion criteria, see there

exclusive provider organization MANAGED CARE A managed care organization similar to a PPO–preferred provider organization in purpose and organization, which allows a Pt to go outside the network for care, but must pay the full cost of the services received; it is similar to an HMO in that primary care physicians act as gatekeepers to a network of other providers, an authorization system, etc. See Gatekeeper, HMO, Managed care organization, PPO

excretion THERAPEUTICS 1. The final elimination of a drug or other compound from the circulation–eg, via the kidneys in urine, biles into stool, saliva, sweat 2. A product that has been eliminated

execute *verb* COMPUTERS Do FORENSIC MEDICINE Do in

execution wound FORENSIC MEDICINE A type of gunshot wound intended to kill the victim, which carries with it the legal implication of premeditation–ie, 1st-degree manslaughter; EWs are created by weapons discharged at close range or in contact, specifically to vital sites, usually the head. See Murder one

executive committee HOSPITAL PRACTICE A hospital committee that administers the medical staff–MS, acting as an interface between the MS and the hospital's governing body–eg, board of trustees, and the general public; the EC consists of MS officers and is chaired by the chief of the medical staff. See Hospital committee, House staff

executive profile LAB MEDICINE A battery of lab analytes that may be measured annually in quasi-important people, ie 'executives', to detect any potentially morbid condition that may require intervention. See Organ panel

exenteration SURGICAL ONCOLOGY The excision of large blocks of tissue in a particular region–eg, pelvis, as required by extensive malignancy TYPES Anterior–en bloc removal of bladder, urethra, uterus and tissues lateral thereto, and upper 2/3 of the vagina; posterior–rarely indicated; total pelvic–used in Pts with stage IV cervical CA, with removal of bladder and rectum. See Commando operation, Hemipelvectomy, Heroic surgery, Hysterectomy

exercise PUBLIC HEALTH The rhythmic contraction of muscles against a force PROS ↓ risk of cholecystectomy, ↓ risk of CAD, CHD, CA–colorectal, breast, prostate, DM–improved insulin utilization, obesity, stroke, osteoporosis, stress, anxiety; ↑ sexual pleasure, strength, flexibility, stamina, psychological well-being, general health; improved reaction time, memory, moods, immune resistance, sleep, self-confidence, control of arthritis, weight, quality of life. See Aerobic exercise, Anaerobic exercise, Breathing exercise, Cardiovascular exercise, Codman's pendulum exercise, Hoshino exercise, Isometric exercise, Isotonic exercise, Pritikin exercise, Vigorous exercise

EXERCISE
MUSCLE
- **ISOMETRIC** Exercise against an unmoving resistance; isometric exercises consist of muscle contraction with a minimum of other body movements; isometric exercises build muscle strength and include weight-lifting or squeezing a tennis ball
- **ISOTONIC** Dynamic exercise Isotonic exercise consists of continuous and sustained movement of the arms and legs; isotonic exercises are beneficial to the cardiorespiratory systems and include running and bicycling

WHOLE BODY
- **LOW-IMPACT AEROBICS** Any type of aerobic exercise that promotes physical fitness, but does not stress musculoskeletal tissues, and joints; low-impact aerobic exercises include walking, swimming, bicycling
- **HIGH-IMPACT AEROBICS** Any type of aerobic exercise that promotes physical fitness, at the risk of stress to musculoskeletal tissues, and joints; high-impact aerobic exercises include aerobic dancing, basketball, running, volleyball

EXERCISE-KCAL CONSUMED/HOUR
Distance running (15 km/hour)	1000
Contact sports (wrestling, karate)	900
Bicycling (25 km/hour)	800
Swimming, freestyle	800
Basketball, volleyball	700
Jogging (9 km/hour)	600
Tennis	500
Coitus	450
Walking	400

exercise echocardiography CARDIOLOGY A technique in which an echocardiogram is performed near the time that the subject exercises on a treadmill, ramp, bicycle or other device. See Bruce protocol, Cornell protocol

exercise hypertension An ↑ in BP that occurs in most people at maximum/near-maximum workload, evoked by vigorous exercise. Cf Exercise hypotension

exercise hypotension A ↓ in systolic BP that occurs at maximum/near-maximal workload in 0.23% of ♂ and 1.45% of ♀ during vigorous exercise, unrelated to age. Cf Exercise hypertension

exercise-induced amenorrhea Exercise-associated amenorrhea Menstrual dysfunction in ♀ long-distance runners which may be accompanied by osteopenia, osteoporosis, hypoestrogenic amenorrhea. See Running

exercise-induced anaphylaxis A form of allergy manifest by a sensation of skin warmth, pruritis and 2° erythema, urticaria, hypotension, upper airway obstruction DIFFDX Cholinergic urticaria, anaphylaxis. See MK-571

exercise-induced asthma A condition in which intense physical exertion results in acute airway narrowing in persons with airway hyperreactivity CLINICAL Cough, wheezing, dyspnea, cough, chest tightness, hyperinflation, airflow limitation, hypoxia TREATMENT Cromolyn and β_2-agonist

exercise-induced bronchospasm SPORTS MEDICINE A post-exertional event defined as a ↓ of 15% of peak expiratory flow; EIB affects up 35% of athletes and 90% of asthmatics; others at risk for EIB are blacks and those living in urban poverty areas. See Free running test

exercise/movement therapy A health-enhancing system of exercise and movement–eg, aikido, dance therapy, t'ai chi, yoga. See Exercise

exercise pyramid POPULAR HEALTH A proposed schematic that recommends the amount and types of exercise to be performed for optimal health. See Exercise, Vigorous exercise. Cf Food pyramid

exercise test CARDIOLOGY Any clinical method used to evaluate a person's cardiovascular responsiveness–ie, tolerance–to exercise. See Treadmill exercise test

exergonic *adjective* Referring to chemical reactions that liberate energy

exertional myopathy SPORTS MEDICINE A condition that may affect any individual who exercises beyond his/her normal capacity ETIOLOGY Dehydration, HIV, alcohol and illicit drug use, prescription drugs MANAGEMENT Short-term, nada; long-term, regular exercise ↑ mitochondria in slow-twitch muscles, allowing them to process O_2 more efficiently for longer periods without damaging muscle cells

exfoliative dermatitis Pityriasis rubra DERMATOLOGY A condition characterized by scaling and erythema rashes, which may occur in a background of psoriasis or lichen planus

exhaustion A state of fatigue or physical consumption, pooped. See Secretory exhaustion

exhibitionism SEXOLOGY A paraphilia in which sexuoeroticism hinges on exposing the genitalia to a stranger; largely a male thing. Cf Voyeurism

exhibitionist An exhibitor exhibiting exhibitionism, see there

exit block CARDIAC PACING Failure of a pacemaker to capture a heartbeat when the stimulation threshold exceeds the pacemaker's output. See Pacemaker

exit-site infection INFECTIOUS DISEASE A catheter-related infection which occurs in central venous catheters CLINICAL Erythema, tenderness, induration of skin and subcutaneous tissue that extends > 2 cm from the skin exit site. See Central venous catheter

exit wound FORENSIC PATHOLOGY The "goodbye" lesion that a bullet or other projectile causes when leaving the body; EWs are often larger than the entrance wound, due to tumbling and deformation of the bullet. Cf Entrance wound, Execution wound

exogenous anxiety PSYCHIATRY Anxiety caused by external stressors, which is usually a healthy and normal reaction to the environment; when EA is excessive it may be due to an anxiety disorder. See Anxiety disorder. Cf Endogenous anxiety

exogenous depression Reactive depression, see there

exostectomy ORTHOPEDICS The surgical excision of an exostosis or other bony bump. See Exostosis, Subungual exostosis

exostosis cartilaginea Osteochondroma, see there

exotropia Divergent or external strabismus, wall eye OPHTHALMOLOGY A form of strabismus characterized by a permanent deviation in the visual axis of one eye away from the other, causing double vision or diplopia. See Dipolopia

expanded rubella syndrome NEONATOLOGY A complex of symptoms that may affect infants with the congenital rubella syndrome; in addition to the typical findings of rubella–congenital heart disease, corneal clouding, microcephaly, mental retardation, deafness, ERS Sx include hepatosplenomegaly, thrombocytopenic purpura, intrauterine growth retardation, interstitial pneumonia, myocarditis, metaphyseal bone lesions. See Congenital rubella syndrome, TORCH

expectancy The expected occurrence of a thing. See Active life expectancy

expectant *adjective* CLINICAL MEDICINE Referring to a stance of non-treatment adopted for an illness for which the benefit of therapy is uncertain, following the dictum, '…*primum non nocere*' VOX POPULI Pregnant

expectant patient MEDICAL ETHICS A Pt who is not expected to survive a particular illness with a meaningful life, even if all efforts were expended to save the Pt

experience DECISION-MAKING The set of skills acquired through the repeated performance of a particular activity or operation, which in turn, implies competence in performing the task of interest

experience hypothesis The largely verified posit that the outcomes in terms of survival and complications of certain surgical procedures–eg transplantation of heart, kidney, and liver is a function of frequency with which the procedure is performed, the experience of the surgical team, and the physicians caring for the Pt after the procedure

experiment CLINICAL RESEARCH A study in which a researcher has control over some of the study's conditions and over some aspects of the independent variables being studied. See Binomial experiment, Dachau hypothermia experiment, Found experiment, Jackpot experiment, Marker rescue experiment, Meselson-Stahl experiment, Minnesota experiment, Mount Everest experiment, Noble experiment, Origin-of-life experiment, PAJAMA experiment, the Plutonium experiment, Pulse-chase experiment, Quasi-experiment, Science Club experiment, Shotgun experiment, Study, Trial

experimental device CLINICAL RESEARCH A therapeutic device that has not been approved by the FDA for use in humans

experimental procedure Unproven procedure HEALTH INSURANCE Health care services, supplies, procedures, therapies, or devices that the health plan determines regarding coverage for a particular case to be either: (1) not proven by scientific evidence to be effective, or (2) not accepted by health care representatives as being effective

experimental treatment INTERNAL MEDICINE An unproven therapy which may or may not be superior to a current 'gold standard' therapy

EXPERIMENTAL TREATMENT–CRITERIA

NOT GENERALLY ACCEPTED by the medical community as effective and proven

NOT RECOGNIZED by professional medical organizations as conforming to accepted medical practice

NOT APPROVED by the FDA or other requisite government body

ARE IN CLINICAL TRIALS or need further study

ARE RARELY USED, NOVEL, OR UNKNOWN and lack authoritative evidence of safety and efficacy (AMN 18/9/95, p16)

expert laboratory A lab regarded as having a particular expertise–eg, in microbiology or molecular biology. See Esoteric testing, Reference laboratory

expert MEDTALK A cove (British slang term) with book (learned) or street smarts who is held to have special knowledge and capable of convincing a jury black is white, or vice versa; for every expert, there is an equal and opposite expert. See Physician expert. Cf Hired gun

expert system INFORMATICS A topic-specific software program designed to imitate human decision-making using detailed knowledge of a particular subject and rules for applying facts to a scenario; an AI application designed to assist in a particular decision-making process; the key component of an ESS is a knowledge base, which combines a database of facts, beliefs, and an algorithm based on heuristic logic–eg, expert systems of internal medicine, CADUCEUS, ILIAD, INTERNIST. See Artificial intelligence, Computer-based diagnostic system, Neural networking

expert witness FORENSIC MEDICINE A person qualified by education, training, experience, occupation, present position, degrees held, publications and professional organization membership that establishes authority as an expert to give opinions; EWs are permitted to offer opinions in court related to their area of expertise which would not be permitted a witness without such status. See Expert. Cf Hired gun, Whore

expiratory reserve volume LUNG PHYSIOLOGY The maximum volume of air that can be forcibly exhaled after a quiet expiration has been completed from the end-expiratory position. See Lung volumes

expire *verb* PHYSIOLOGY To exhale. See Exhalation VOX POPULI To die

explant TRANSPLANTATION MEDICINE An organ that requires removal due to rejection, after transplantation. See Transplantation

exploitation SOCIAL MEDICINE The taking unfair advantage of a situation or person(s). See Sexual exploitation

exploratory laparotomy SURGERY A 'look-see' operation usually of the peritoneal cavity, in which the surgeon examines all surfaces for lesions–eg, abscesses and tumor nodules; during EL, the operator may biopsy the tissue or obtain peritoneal washings from which a specimen for cytology is processed INDICATIONS Surgical staging of regional malignancy–eg, ovarian CA COMPLICATIONS Adhesion formation, especially if Pts receive RT. See Laparoscopy

explosion 1. The rapid expansion of a thing. See Charge explosion, Code explosion, Cultural explosion 2. Boom(!)

explosive OCCUPATIONAL SAFETY A chemical that causes a sudden, virtually instantaneous release of pressure, gas, and heat when subjected to sudden shock, altered pressure, or high temperature; explosives seen in the context of terrorism or violent acts include pipe bombs, hand-grenades, dynamite, detonators, C-4 plastic explosives PSYCHOLOGY See Explosive personality, Explosive syndrome

explosive personality PSYCHOLOGY A popular term for a personality trait characeized by abrupt 'detonations' of negative emotive forces–anger, snarling, snapping and/or violence, in response to seemingly mild external stimuli. Cf Explosive syndrome

explosive syndrome NEUROLOGY A complex characterized by episodic outbursts of verbal abuse and physical violence in response to minor provocation, unrelated to psychiatric disease, seen in organic brain disease, after trauma, and metabolic dysfunction–eg, hypoglycemia, Wilson's disease, uremia, hyperammonemia, androgen or 'roid rage–abuse, premenstrual syndrome, Cushing's disease; Pts are usually pleasant between outbursts and apologetic for the explosions TREATMENT Some Pts respond to propranolol. Cf Explosive personality

exposure EPIDEMIOLOGY A state of contact or close proximity to a chemical, pathogen, radioisotope or other other substance by swallowing, breathing, or direct contact–eg, on skin or eyes; exposure may be short term–acute or long term–chronic. See Acute exposure, Athlete exposure, Chronic exposure, Intermediate exposure, Occupational exposure to bloodborne pathogens, Perinatal substance exposure IMAGING An image, as an AP exposure of the chest MEDICAL LIABILITY A general term for the degree of malpractice risk borne by a health care provider while performing a particular medical service See Risk management.

exposure assessment NIHSPEAK The part of risk assessment which quantifies exposure to a substance or agent by humans or other target species.

exposure-prone *adjective* Referring to a risk of contact with an adverse effect–eg malpractice liability, hazardous substance–eg, cyanide in environment, or pathogen–eg, HIV or HBV in a body fluid

exposure therapy Systematic desensitization, see there

expressive language disorder NEUROLOGY A disorder resulting in ability below that expected in vocabulary, production of complex sentences, and word recall ETIOLOGY Idiopathic, brain damage, head trauma, malnutrition

exsanguinate *verb* To be bled dry

exsanguination TRAUMA SURGERY A condition that is "...the most extreme form of hemorrhage, with an initial blood loss of > 40% and ongoing bleeding which, if not surgically controlled, will lead to death." See Salvage surgery–trauma, Staged surgery

extended family SOCIAL MEDICINE A family unit related by blood or marriage that extends over 3+ generations, and may include 'collateral' relatives, spouses, and progeny. See Companionship, Most significant other; Cf Nuclear family, Single-parent family, Social isolation

extended radical mastectomy Extended thorough mastectomy, Urban mastectomy BREAST SURGERY Surgery which is similar to the Halsted procedure but even more draconian in that the lymph nodes of the internal mammary chain are also removed. See Radical mastectomy

extended reporting endorsement Tail coverage, see there

extended-reach needle GYNECOLOGY A needle for administering anesthetic regional blocks–paracervical, uterosacral, pudendal during labor and/or delivery, and minor surgery

extended-spectrum beta-lactamase Third generation cephalosporinase MICROBIOLOGY A beta-lactamase produced by gram-negative enteric bacteria, in particular *K pneumoniae* and *E coli*, which are resistant to third-generation cephalosporins. See Third-generation cephalosporin

extension The making larger of a thing IMAGING The broadening of a lesion–eg, a cancer or focus of infection as seen on an imaging technique ORTHOPEDICS A ↑ in the angle between parts of a joint; a straightening of a flexed limb TERMINAL CARE See Extension of life

extension of life TERMINAL CARE Any maneuver intended to ↓ morbidity of various conditions in the elderly, thereby ↑ lifespan. See Dying

extern GRADUATE EDUCATION A student in the 3rd or 4th yr of medical school–in North America, who attends ward rounds of a teaching hospital, and learns clinical medicine by example and quasi-active participation in Pt management. See Pimping. Cf Intern

external *adjective* Located on or toward a surface

external-beam radiation therapy External radiation RADIATION THERAPY The aiming of high-energy radiation at the center of a CA, which may also be used to manage choroidal neovascularization in age-related macular degeneration. See Radiation oncology

external cephalic version OBSTETRICS A procedure that externally rotates the fetus from a breech position to a vertex presentation. See Emergency C-section

external chest compression EMERGENCY MEDICINE A technique of basic life support, consisting of serial, rhythmic applications of pressure on the lower half of the sternum, which provides circulation to the heart, lungs, brain, and other organs with a general ↑ in intrathoracic pressure and/or direct compression of the heart

external fetal monitoring OBSTETRICS The use of 2 straps–one over the upper abdomen attached to a pressure gauge and record contractions, the 2nd measures the fetal heart rate. Cf Internal fetal monitoring

external fixation ORTHOPEDICS Open reduction, stabilization and use of external fixators to manage fracture bone fragments INDICATIONS Open frac-

ture with massive soft tissue damage; instant fixation in polytrauma; fractures with deficient bone stock, infection IF DEVICES Fracture fixation, radius, tibia, pelvis, bone lengthening, Ilizarov device COMPLICATIONS Pin loosening and/or infection. Cf Internal fixation

external hemorrhoids SURGERY Engorged veins under the anal skin. Cf Internal hemorrhoids

external pacemaker CARDIAC PACING A pulse generator intended to be worn outside the body, used for temporary pacing

external rotation Lateral rotation BIOMECHANICS The act of turning about an axis passing through the center of the leg; ER of the leg occurs with closed chain supination; the talus acts as an extension of the leg in frontal and transverse planes

external version OBSTETRICS An active intervention carried out over the abdominal wall in late pregnancy, consisting of gentle rotation of the fetus, with the intent of manually converting an undesirable breech position, to a deliverable cephalic position. See Cesarean section. Cf Internal version

externship Acting internship, see there

extinction PSYCHIATRY A facet of operant–classical conditioning, in which the conditioned response is weakened and eventually disappears by nonreinforcement. See Operant conditioning, Respondent conditioning, Sensory extinction

extraaxial *adjective* Outside the CNS

extracampic hallucination NEUROLOGY A hallucination that occurs outside of the normal sensory field

extracardiac murmur CARDIOLOGY A sound heard over the precordium, arising in structures other than the heart–eg, related to gas embolism, pericardial friction rubs

extrachorial placentation NEONATOLOGY A placenta in which there is a rim of placental tissue that extends beyond the vascular plate, accompanied by fibrin on the margin OUTCOME EP is always associated with placenta previa, and often with peripartum maternal bleeding

extracorporeal *adjective* Referring to circulation of blood that occurs outside the body–eg, via apheresis. See Apheresis

extracorporeal membrane oxygenation A form of artificial oxygenation of blood, in which a cannula in the jugular vein is connected to a small reservoir, into which the blood drains by gravity; the blood is then pumped from the reservoir through a membrane oxygenator and heat exchanger, and returned to the Pt via the right carotid artery cannula; ECMO is part of the surgical repair of congenital diaphragmatic hernia in infants, and is standard therapy for many full-term infants with poor lung function; it has had limited success in older children and adults

extracorporeal shock-wave lithotripsy ESWL, lithotripsy, shock-wave lithotripsy NEPHROLOGY A non-surgical, non-invasive method for dissolving renal–chenodeoxycholic and ursodeoxycholic acids, and biliary tract calculi; the Pts lie prone and partially immersed in a large bathtub-like vat; shock waves are generated extracorporally by high-energy underwater spark discharge focused on the Pt's ventral aspect by a reflector

extracranial/intracranial bypass NEUROLOGY A type of surgery that restores circulation to hypoxic brain tissue by rerouting a healthy artery in the scalp to the brain tissue affected by a blocked artery

extract *noun* A concentrate of a drug, cells, or a supernatant. See Adrenal extract, Cell-free extract, Fluid extract, Green extract, Plasmid extract *verb* PSYCHOLOGY Obtain

extraction DENTISTRY See Dental extraction GYNECOLOGY See Menstrual extraction

extradural hematoma Epidural hematoma NEUROLOGY Accumulation of blood in the epidural space, due to leakage of blood from the middle meningeal artery. Cf Subdural hematoma

extradural hemorrhage Epidural hemorrhage NEUROLOGY Hemorrhage into the extradural space, which is a potentially fatal result of head trauma; it may be accompanied by herniation of the uncal gyri. Cf Subdural hemorrhage. Cf Subdural hemorrhage

extramammary Paget's disease A lesion similar to Paget's disease of breast, less often linked to underlying malignancy; the condition arises from skin adnexae, visceral malignancy or rarely, de novo, in scrotum, perineum, labia majora; cells are more often mucin-positive than in Paget's disease of breast

extraneous tissue See Floater, surgical pathology

extranodal MEDTALK *adjective* Outside of a lymph node

extranodal lymphoma HEMATOLOGY A lymphoproliferation that arises in nonlymphoid tissues, which are often histologically diffuse and more aggressive than nodal lymphomas SITES Stomach, tonsils, skin, small intestine, salivary glands, breast. See Lymphoma, MALToma

extranodal recurrence ONCOLOGY The reappearance of the signs and symptoms of CA at extranodal or extrasplenic site after the—successful completion of the first course of radiotherapy and/or chemotherapy. See Relapse

extraordinary treatment Heroic treatment *'Treatment or care that does not offer a reasonable hope or benefit to the Pt, or which cannot be accomplished without excessive pain, expense, or other great burden'; ET is an ethical determination about rendering care depending upon the Pt's condition and prognosis.'*

extrapulmonary TB INFECTIOUS DISEASE Clinical TB outside the lungs—eg, lymph nodes, pleura, brain, kidneys, or bones

extrapyramidal disease NEUROLOGY Any condition affecting the basal ganglia and their interconnections; EDs include Huntington's disease, Parkinson's disease, Tourette syndrome, Wilson's disease, dystonias, hemiballism, myoclonias, tardive dyskinesia, tremors

extrapyramidal syndrome NEUROLOGY A condition characterized by a range of findings—eg, rigidity, tremors, drooling, shuffling gait–parkinsonism, akathisia–restlessness, dystonia–odd involuntary postures, akinesia–motor inactivity, and other neurologic disturbances ETIOLOGY Extrapyramidal dysfunction, often a reversible side effect of certain psychotropics–eg, phenothiazines. See Simpson-Angus Scale, Tardive dyskinesia

extrapyramidal system ANATOMY That part of the CNS that controls and coordinates postural, static, support and locomotor mechanisms COMPONENTS Corpus striatum, subthalamic nucleus, substantia nigra, red nucleus, and their interconnections with the reticular formation

extravasation MEDTALK The seepage of fluid—eg, plasma, from a mucocutaneous surface, from underlying capillaries

extreme fighting SPORTS MEDICINE A modern blood sport, promoted as a barbaric combination of boxing, kick boxing, wrestling, judo, karate, and other martial arts. See Blood sport. Cf Boxing, Toughman fighting, Ultimate fighting

extreme sport SPORTS MEDICINE Any sport or recreational activity that is dangerous and, if performed optimally, even by the highly skilled, risks loss of life or limb. See BASE jumping, Canyoning, High-sensation seeking trait, Novelty seeking behavior

extreme thrombocytosis LAB MEDICINE A platelet count of $> 10^6/mm^3$ ETIOLOGY Reactive, myeloproliferative disorders, idiopathic

EXTREME THROMBOCYTOSIS–PLATELETS > $10^6/MM^3$

REACTIVE–82.5%

Infection–31%

Postsplenectomy hyposplenism–19%

Malignancy–14%

Trauma

MYELOPROLIFERATIVE DISORDERS–13.6%

CML–42%

Primary thrombocythemia–29%

UNKNOWN–3.9%

extremely low birth weight infant NEONATOLOGY An infant weighing ≤ 1000g at birth, who is at high risk for neurobehavioral dysfunction and poor school performance. See Low birth weight, Limits of viability. Cf Very low birth weight

extremities examination PHYSICAL EXAM A general term for an extremity's physical features, including appearance

extrication EMERGENCY MEDICINE The process of removing a person from an entrapment, usually from a motor vehicle, often requiring the use of special tools. See Jaws of life

extrinsic pathway HEMATOLOGY The arm of coagulation activation which, like the intrinsic pathway, converges on the common pathway, ie factor X activation; the EP is activated by exposure of blood to tissue factor, circulating factor VII, Ca^{2+} and phospholipid. Cf Common pathway, Intrinsic pathway

extubation NURSING The removal of a previously inserted tube used for respiratory support or gastric feeding. See Mechanical ventilation. Cf Intubation, Terminal weaning

exudate INTERNAL MEDICINE A cell and protein-rich fluid that extravasates from the capillaries. See Hard exudate, Pleural exudate, Waxy exudate

eye bank TRANSPLANTATION A repository of corneas obtained from decedents

eye donation TRANSPLANTATION The harvesting of decedents' corneas with deposition in an eye bank

eye injury EMERGENCY MEDICINE Trauma impacting on the eye; 5% EIs are severe and include intraocular foreign bodies, ruptured globe, hyphema, orbital fractures. Cf Anterior segment

eye-mouth gap NUTRITION The discrepancy between the actual caloric intake and exercise, and that reported by a person—usually with an eating disorder

eye-opener SUBSTANCE ABUSE A popular term for the first drink of an alcoholic's day, which he believes helps steady himself, or 'treats' a hangover

eye rolling NEUROLOGY Rhythmic eye movements which accompany rotation of the head, seen in the Pelizaeus-Merzbacher form of leukodystrophy VOX POPULI Etc.

eyelid inflammation Blepharitis, see there

Eysenck Personality Inventory PSYCHOLOGY A short reliable device for evaluating 2 'dimensions' of personality–extroversion/introversion and mental stability; it is used when there is reason to believe that personality and individual differences might affect the dependent variables in a study

F Symbol for: 1. Degrees Fahrenheit 2. Factor 3. Fluorine 4. Fragment–of antibody 5. Phenylalanine

f Symbol for: 1. Breathing frequency–pulmonary function testing 2. femto-, SI abbreviation for 10^{-15} 3. Frequency

F wave CARDIOLOGY An atrial flutter wave on EKG, which appears as a 'sawtooth' pattern in leads II, III and aVF; less commonly, F waves are undulating, firing at a rate of 280-320/min, a rate often associated with a 2:1 block and alternating F waves merge with QRS or T waves NEUROLOGY F response An undulation of an EMG that corresponds to the time between application of a stimulus to the axon of the α motor neuron as it propagates andromically to the anterior horn of the spinal cord, and then returns orthodromically along the same axon. See Electromyography

FAB classification French-American-British classification of acute leukemia HEMATOLOGY A schema that divides acute leukemias into lymphoid–ALL or myeloid–AML cell lines; of childhood ALL, 70% are predominantly L1, 27% are L2, and 3% or less are L3 or Burkitt cell type, in adults with ALL, 30% are L1, 65% are L2, and 5% are L3

FAB CLASSIFICATION, ACUTE LEUKEMIAS

ACUTE LYMPHOCYTIC LEUKEMIA (ALL)

L1	Small monotonous lymphocytes
L2	Mixed L1- and L3-type lymphocytes
L3	Large homogeneous blast cells

ACUTE MYELOID LEUKEMIA (AML)

M1	Myeloblasts without maturation
M2	Myeloblasts with maturation (best AML prognosis)
M3	Hypergranular promyelocytic leukemia (faggot cells)
M3V	Variant, microgranular promyelocytic leukemia
M4	Myelomonocytic leukocytes
M5	Monocytic, subtype
	a. Poorly differentiated monocytic leukemia
	b. Well differentiated monocytic leukemia
M6	Erythroleukemia/DiGuglielmo syndrome
M7	Megakaryocytic leukemia Pleomorphic undifferentiated cells with cytoplasmic blebs; myelofibrosis or ↑ BM reticulin; positive for platelet peroxidase antifactor VIII

Fab therapy THERAPEUTICS The use of antibody fragments–Fab fractions against certain antigens–eg, digoxin, to treat dioxin overdose. See Fab fragment, Psoriasis

Fabry's disease Alpha-galactosidase deficiency, angiokeratoma corporis diffusum PEDIATRICS An X-linked lysosomal storage disease caused by a defect in trihexosylceramide α-galactosidase CLINICAL Chronic pain, angiokeratomas, hypohidrosis

FACC Fellow, American College of Cardiology

FACCP Fellow, American College of Chest Physicians

face Facies MEDTALK 1. The frontal portion of the head, which includes mouth, nose, eyes, forehead, cheeks, chin 2. The exposed surface of a structure. See Battered prize fighter face, Bird face, Chipmunk face, Cow face, Dishface face, Dogface face, Doll face, Elfin face, Fat face, Flat face, Fish face, Frog face, Gargoyle face, Hatchet face, Heart-shaped face, Hippocratic face, Mask-like face, Mitral face, Monkey face, Moon face, Orphan Annie face, Peter Pan face, Porcelain doll face, Sagging face, Shrewmouse face, Triangular face

face-to-face time MEDICAL PRACTICE The time that a health care provider interacts with a Pt. See Specialty

facelift Rhitidectomy COSMETIC SURGERY A procedure in which wrinkles and sagging soft tissues of face and neck are 'tightened', imparting a more youthful appearance COMPLICATIONS Asymmetry, hematoma, injury to nerves–facial, great auricular, scars, sloughing of skin; signs of aging may recur in one yr. See Weekend face lift, Nose job

face presentation OBSTETRICS An uncommon presenting position of the baby at the time of delivery, where the face is the presenting part; if the chin is extended, the head becomes caught in the symphysis pubis and can't be delivered, therefore mandating a C-section. See Cesarean section, Presentation

facesheet Any coversheet to a multipage document contains the relevant points covered in the document itself MANAGED CARE Declaration of health insurance

facet A small, smooth surface of a bone where articulation occurs. See Bilateral locked facet, Unilateral locked facet

facet joint Zygapophyseal joint ORTHOPEDICS The synovial joint between the articular processes of the vertebral bodies

facet syndrome ORTHOPEDICS A low back pain syndrome attributed to osteoarthritis of the interarticular vertebrae CLINICAL Low back pain that ↑ on extension, irradiates to the posterior thigh, and ends at the knee; x-ray and CT imaging reveal narrowing of disk space, osteophyte formation TREATMENT NSAIDs, intraarticular injections with anesthetics, low back fusion with osteodegeneration, endurance training

facial droop NEUROLOGY A unilateral sagging of the face, which usually indicates paralysis of facial muscles due to trauma, infection or tumor removal near or at the facial nerve. See Facial palsy

facial fracture See LeFort classification

facial nerve palsy Facial palsy, see there

facial palsy Bell's palsy, cranial mononeuropathy VII, facial mononeuropathy, facial nerve palsy, facial neuralgia NEUROLOGY Acute peripheral paralysis of the face due to a herpes simplex immune-mediated condition often characterized by severe pain in the trigeminal nerve EPIDEMIOLOGY Risk of FP ↑ with age; age 10 to 19, 2:1, ♀:♂; age 40, 3:2, ♂:♀; pregnant ♀ have 3.3 times ↑ risk than nonpregnant; DM = 4.5 times↑ risk of FP; 10% of Pts have positive family Hx of FP PATHOGENESIS FP is due to reactivation of the virus leading to replication of virus within the ganglion cells; the virus travels down the axons, inducing inflammation CLINICAL Abrupt onset, drooping mouth, unblinking eye, twisted nose, uneven smile, distorted expressions; paralysis hits maximum in 1 to 14 days; retroauricular pain, facial numbness, epiphora, parageusia, ↓ tearing, hyperacusis, hypoesthesia or dysesthesia of cranial nerves–CN V and IX, motor paresis of CN IX and X, papillitis of tongue DiffDx, UNILATERAL Tumors or masses, otitis media, sarcoid, Lyme disease, skull fracture, facial injury DiffDx, BILATERAL Guillain-Barré syndrome, Melkersson-Rosenthal syndrome, Möbius syndrome, motor neuron disease, myasthenia gravis ETIOLOGY Trauma, Bell's palsy, stroke, parotid tumors, intracranial tumors MANAGEMENT Microvascular and micro-neurosurgical tissue transfers allow restoration of functional, unconscious, symmetrical facial movements, acyclovir, steroids–efficacy is uncertain, artificial tears, neuromuscular retraining–eg, mirror/visual feedback, biofeedback or electromyography feedback PROGNOSIS 60 to 80% recover, especially if incomplete paralysis, and Pt is young

facies 1. The appearance of an individual's face, physiognomy. See Acromegalic facies, Congenital syphilic facies, Cretinoid facies, Gargoyle facies, Mongoloid facies, Myasthenic facies, Myopathic facies, Myxedematous facies, Peter Pan facies, Tabetic facies 2. An anatomic term for the aspect of a a body structure or parts of the head. See Facet

facilitate MANAGED CARE To remove mental and/or physical obstacles in a pathway; Cf Delegate PHYSIOLOGY To lower the threshold for neural activity, making nerves more susceptible to depolarization. See Social facilitation VOX POPULI Help, assist

facility VOX POPULI A place where an activity occurs. See Approved health care facility, Correctional facility, Health care facility, Intermediate care facility, Plan skilled nursing facility, Residential facility, Screening facility, Skilled nursing facility, Treatment facility, Waste encapsulation storage facility

facioscapulohumeral muscular dystrophy Landouzy-Dejerine disease NEUROLOGY A benign AD type of muscular dystrophy characterized by marked atrophy of the muscles of the upper limb girdle and face, resulting in the so-called myopathic face. See Muscular dystrophy, Myopathic face

FACS 1. Fellow, Am College of Surgeons 2. Fluorescence-activated cell sorter, see there

factitious *adjective* Referring to symptoms driven by an unconscious, compelling need to assume a 'sick role', usually in absence of an external incentive. See Munchausen disease

factitious disease Factitious disorder, illness PSYCHIATRY Any of a number of self-produced lesions or biochemical changes seen in persons with neuroses to gratify various self-motivated needs–eg, sympathy and narcotics; these conditions share the same raison d'etre, differing only in the site of injury and agent used to produce the lesions. See Munchausen syndrome, Psychosomatic disorder, Self-mutilation

factitious hypoglycemia Surreptitious ingestion of hypoglycemics–eg, sulfonylureas or insulin, often by ♀ age 30-40, in health professions DIAGNOSIS Measure oral hypoglycemics, serum glucose, insulin antibodies, C-fragment of insulin is present in the serum of normal subjects at a ratio of 5-15:1; it is ↓ in insulin injecters. See Hypoglycemia, Munchausen disease

factitious thyrotoxicosis Factitious hyperthyroidism Falsely elevated thyroid hormones due to intake of exogenous thyroid hormones LAB ↑ total T_4, T_3 resin uptake, free T_4; ↓¹³¹I–RAIU by the thyroid. Cf Thyrotoxicosis

factor A substance or activity that produces a result or outcome CLINICAL RESEARCH 1. In analysis of variance, an independent variable–ie, a variable presumed to cause or influence another variable 2. In factor analysis, a cluster of related variables distinguishable from a larger set of variables 3. A number by which another number is multiplied, as in the statement: real estate values ↑ by a factor of 3–ie, tripled MEDTALK A molecule or substance known to exist in a system, which is poorly characterized when the system is first described; with time, the molecules may be identified and/or

sequenced, such that the original 'factor' designation may fall into disfavor and retain historic interest. See Act of smoking factor, Age/sex factor, Angiogenic factor, Antitermination factor, ARF factor, Atrial systolic factor, Autocrine motility factor, Basal factor, Bioaccumulation factor, Bioconcentration factor, Brain-derived neutrophic factor, Causal factor, Chemotactic factor, Co-carcinogenic K factor, Colonization factor, Colony-stimulating factor, Contact inhibiting factor, Conversion factor, Cord factor, DAF factor, Dilution factor, Elongation factor, Environmental factor, Epidermal growth factor, Faith factor, Father factor, Fibroblast growth factor, Fuzz factor, GAGA factor, GATA-1 factor, Glial cell-derived neurotrophic factor, Growth factor, Growth hormone factor, GM-CSF factor, Hageman factor, Hanukkah factor, Harmonic factor, Hassle factor, Heparin-binding fibroblast growth factor, Heparin cofactor II, Hepatocyte growth factor, HGF/SF factor, Horizontal scaling factor, Host factor, Human factor, IGF-I, IGF-II, Initiation factor, Integration host factor, Intrinsic factor, IRF-1 factor, Keratinocyte growth factor, Leukemia inhibitory factor, Leukocyte inhibitory factor, Macrophage activating factor, Macrophage chemotactic factor, Macrophage/monocyte inhibitory factor, Magic factor, Male factor, Mast cell growth factor, Mesoderm-inducing factor, Methuselah factor, Migration inhibition factor, Mitochondrial transcription factor, Monocyte colony inhibitory factor, Motivational factor, Motor neuron growth factor, MPF factor, Myocardial depressant factor, Nerve growth factor, Neurotrophic factor, Nuclear roundness factor, Osteoclast-activating factor, Ovarian factor, Partitioning factor, Partner risk factor, Peptide supply factor, Personal risk factor, Platelet factor, Platelet-activating factor, Platelet-derived growth factor, Prognostic factor, R factor, Release factor, Replication licensing factor, Resistance transfer factor, Rheumatoid factor, Rho factor, Risk factor, S factor, Safety factor, Satiation factor, Scatter factor, Serum spreading factor, Shrinkage factor, Sigma factor, Smooth muscle cell-derived growth factor, Sociodemographic factor, Soft risk factor, Somalia factor, Stem cell factor, Stringent factor, Stromal-cell-derived factor 1, Sun protection factor, Suppressive factor, Testis-determining factor, Thymic factor, Thymic humoral factor, Tissue factor, Transcription factor, Transforming growth factor, Trend factor, Trigger factor, Trypanosome lytic factor, Tumor necrosis factor, Uncertainty factor, Upstream binding factor, Uterine factor, V factor, Vascular endothelial growth factor, Vascular integrity factor, Vascular permeability factor, Von Willebrand factor, Windchill factor, X factor, Yates correction factor

factor I Fibrinogen, see there

factor II Prothrombin, see there

factor III Tissue factor, see there, aka thromboplastin

factor V Proacelerin HEMATOLOGY A coagulation factor synthesized in the liver which, when activated, is a cofactor with activated factor X–factor Xa–in the formation of thrombokinase; factor V is ↓ in liver disease. See Factor V deficiency

factor V deficiency Parahemophilia A condition characterized by mild bleeding, petechial hemorrhage, or menorrhagia that is either congenital–due to the AR defect in the gene for factor V, or acquired–due to IgA or IgG antibodies to factor V LAB ↑ PTT, ↑ PT TREATMENT Fresh plasma

factor V Leiden HEMATOLOGY A variant of factor V present in 3%-8% of Caucasians associated with a ↑ risk of DVT. See LETS, Hereditary thrombophilia

factor V Leiden mutation A HEMATOLOGY A mutation of factor V gene resulting in resistance to activated C protein; a common inherited cause of thrombosis

factor VII A coagulation factor that forms a complex with thromboplastin and calcium to activate factor X

factor VII deficiency MOLECULAR MEDICINE A coagulopathy which may be either inherited–AR due to a mutation resulting in a defect in factor VII, or acquired either due to a vitamin K deficiency, appearing in the neonatal period or due to an excess of factor VII in some Pts with thromboembolism CLINICAL Epistaxis, mucosal hemorrhage. See Factor X

factor VIII Anti-hemophilic factor HEMATOLOGY A heterotrimeric coagulation factor that forms a complex with factor IX, platelets, and calcium, thereby activating factor X; factor VIII is present in cryoprecipitated plasma, and is also used to treat hemophilia A; recombinant factor VIII–rF VIII therapy is used for hemophilia A

factor VIII:C HEMATOLOGY The LMW component of factor VIII, which has coagulant activity, and is deficient in classic hemophilia. See Hemophilia

factor VIII inhibitor IgG alloantibodies to factor VIIIc that inhibits coagulation, which occur in up to 50% of Pts with severe hemophilia A; factor VIII antibodies may arise spontaneously in inflammatory diseases–eg, SLE, rheumatoid arthritis, and ulcerative colitis, or with ↑ age, during post-partum period, or in drug reactions MECHANISM Factor VIII inhibitor inactivates factor VIII by steric prevention of interaction of factor VIII with vWF, phospholipids, activated factors IX and X, and thrombin, and by proteolysis. See Bethesda units, Factor VIII

factor IX Christmas factor, coagulation factor IX HEMATOLOGY An intrinsic pathway factor which, when activated–IXa, combines with Factor VIII and a phospholipid to activate Factor X in the common pathway. See Common pathway, Factor IX deficiency

factor IX deficiency Christmas disease, hemophilia B HEMATOLOGY An X-R disorder due to a deficiency in Factor IX resulting in a bleeding diathesis EPIDEMIOLOGY Incidence–1/40,000 or 15-20% of hemophilias; severe disease is usually of neonatal onset CLINICAL, SEVERE DISEASE Factor IX activity is < 1% of normal; neonatal onset of bleeding, hematomas after injections or circumcision, hemarthroses and deep tissue hemorrhage, spontaneous bleeding COMMON SITES Hemarthrosis of elbows, knees, ankles; pain, swelling, limited ROM of affected joint; muscle hematomas–pain, swelling, may cause muscle atrophy; mucosa–mouth, teeth, nose (epistaxis), GI tract; peripheral nerve lesions–femoral–iliopsoas muscle, sciatic–buttock; tibial–calf; perineal–anterior leg compartment; median & ulnar–forearm flexors; high-risk hemorrhages–intracranial, intraspinal, retropharyngeal, retroperitoneal; hematuria LAB ↑ PTT, ↓ factor IX, normal PT, bleeding time, thrombin time, platelets MANAGEMENT, SUPPORTIVE–avoid trauma, anticoagulants–ASA, add pad to crib/playpen; pressure compresses on bleeding sites; immunize against HBV REPLACEMENT Factor IX Concentrate; FFP. See Factor IX, Thrombin

factor X Coagulation factor X, prothrombase, prothrombinase, Stuart Prower factor HEMATOLOGY A vitamin K-dependent enzyme which is the key protein in all–intrinsic, extrinsic, common coagulation pathways, converting prothrombin to thrombin. See Factor X deficiency

factor XI Plasma thromboplastin antecedent, PTA HEMATOLOGY A 140–160 kD vitamin K-dependent serine proteinase of the intrinsic pathway of coagulation encoded on chromosome 4q35, which is activated in vitro by contact with a thrombogenic surface–eg, glass or thrombin and activates factor IX–IXa ROLE IN HEMOSTASIS Balances procoagulant action–fibrinogenesis and fibrinolysis–protects fibrin LAB ↑ PTT–corrects when Pt's serum is mixed with equal volumes of normal plasma, normal PT MANAGEMENT Fresh frozen plasma. See Deep vein thrombosis, Fibrin, Intrinsic pathway, Thrombin

factor XII Hageman factor A coagulation factor which, activated–XIIa, is a serine proteinase that activates factors VII–VIIa and XI–XIa, plasminogen and plasma prokallikrein

factor XIII Coagulation factor XIII, fibrin stabilizing factor, FSF A coagulation factor that stabilizes formed clots; there is no established reference range; factor XIII deficiency is characterized by an ↑ time for clot formation; factor XIII deficiency is detected by mixing studies, where coagulation factors are mixed with normal plasma and the Pt's plasma to determine which combination normalizes the clot formation; the factor deficiency is deduced from this exercise

factor A C3-complement, see there

factor B Complement C3 proactivator, a protein of the alternate complement pathway

factor D Protein activator of factor B of the alternate pathway of complement activation

factor H 1. A complement glycoprotein that inactivates C3b–alternate pathway 2. Biotin, see there 3. Vitamin B_{12} precursor or analogue

factor R 1. Release factor–any proteins that release a polypeptide chain from the ribosome 2. Folic acid, see there

faculty 1. A group of similarly trained or educated individuals. See Clinical faculty 2. A power or ability

faculty practice plan ACADEMIA An organized group of physicians and other health care professionals that treat Pts referred to an academic medical center

fad diet POPULAR NUTRITION Any of a number of weight-reduction diets that either eliminate one or more of the essential food groups, or recommend consumption of one type of food in excess at the expense of other foods; FDs rarely follow modern principles for losing weight. See Diet

Fagerstrom Tolerance Questionnaire ADDICTION DISORDERS An instrument for assessing tobacco dependence, which evaluates, among other factors, depth of inhalation, time from awakening to day's first cigarette, smoking when bedridden with illness, and difficulty in smoking cessation. See Smoking, Tobacco

fail-safe DECISION-MAKING *adjective* Referring to a mechanism or device incorporated in a system that ensures safety, should the system fail to function properly. See Clinical pathway

failed code EMERGENCY MEDICINE CPR in which all mechanical and pharmacologic armamentarium is used in an attempt to resuscitate a Pt who, despite the best efforts by the managing team, dies. See Cardiopulmonary resuscitation, Code. Cf Slow code

failure VOX POPULI A nonfunctioning state. See Acute renal failure, Acute skin failure, Atrial systolic failure, Autonomic failure, Backward failure, Baroreflex failure, Chronic renal failure, Decompensated low-output backward failure, Defibrillation failure, Failure to thrive, Forward failure, Fulminant hepatic failure, Graft failure, High-output heart failure, Induction failure, Intrinsic renal failure, Left ventricular failure, Low-output heart failure, Mean time between failure, Medical failure, Multisystem organ failure, Postpartum renal failure, Premature ovarian failure, Right ventricular failure, Zidovudine failure

failure of induction ONCOLOGY The nonachievement of remission in a Pt in whom it is otherwise expected. See Induction, Remission

failure to diagnose MALPRACTICE *adjective* Referring to a scenario in which a treatable diagnosis was not diagnosed at all, precluding therapy or resulting in death

failure to progress OBSTETRICS Cervical changes of < 1 cm/hr for 2 consecutive hrs; active management of labor does not ↓ the rate of C-section but may ↓ length of labor and ↑ Pt satisfaction in nulliparas

failure to thrive PEDIATRICS The inability of a child to gain weight or loss of weight without discernible cause ETIOLOGY-ENVIRONMENTAL DEPRIVATION More common; children have poor appetites, are apathetic and withdrawn; these findings are typical of abused children, offspring of schizophrenics or in children with physical deformities or 2° problems causing the parents to subconsciously reject them ORGANIC DISEASE Cerebral lesions, chromosome defects, chronic infection or inflammation, cystic fibrosis, eclampsia, endocrinopathy, congenital heart disease, idiopathic hypercalcemia, malabsorption, CA, renal failure or renal tube defects, TORCH complex. See Child abuse. Cf Infanticide

fall PUBLIC HEALTH A precipitous drop from a height, or from a higher position, which may be accompanied by injuries EPIDEMIOLOGY 30% of those > 65 yrs old fall/yr; 10-15% suffer injuries–eg, hip Fx–1% and other sites–5%, and soft tissue injuries–5%; it is the 6th leading cause of death in the elderly RISK FACTORS Postural hypotension, use of sedatives, use of 4+ prescription medicines, impaired arm or leg movement, strength, balance, or gait; fall survivors suffer from functional decline in ADL and a ↑ risk of institutionalization; fall risk in the elderly can be ↓ with exercise and endurance, flexibility, dynamic balance, and resistance training, behavior modification, adjustment of medications

fallen arch ORTHOPEDICS A popular term for a flattening of the longitudinal and transverse tendinous arches of the foot

Fallot's tetralogy CARDIOLOGY A group of 4 congenital heart defects–VSD, pulmonary valve or infundibular stenosis, a dextropositioned aorta which straddles the interventricular septum and right ventricular hypertrophy. See Fallot's pentalogy, Fallot's trilogy, Pink tetralogy of Fallot, Ventricular septal defect

false aneurysm A blood-filled pseudo-vascular space that parallels the native vessel lumen, which may act as a vascular channel and carry blood, if there is a site of exit and re-entry; the 'vessel' wall is formed by reactive connective tissue; FAs are secondary to trauma, transmural rupture of the vessel wall, or dehiscence of vessels

false knot False umbilical cord knot, varicosity NEONATOLOGY A prominent, often tortuous aggregation of dilated ectatic vessels with ↓ in covering by Wharton's jelly. Cf True knot

false labor OBSTETRICS Parturition that results from disordered uterine action where regular painful contractions are not accompanied by effacement or dilation of the cervix that may either cease or be followed by onset of true labor. See Labor

false memory Recovered memory, repressed memory PSYCHIATRY A series of suggestions and cues that cause a person to believe an event occurred, which in fact did not MECHANISM OF FM Source amnesia. See Memory, Source amnesia

false negative DIAGNOSTIC MEDICINE A test result–eg, Pap smear, mammography, from a Pt who has a particular disease, which is negative or does not detect the presence of an analyte that is usually abnormal in the disease of interest. See Four-cell diagnostic matrix, Cf False positive NIHSPEAK An active substance or result incorrectly identified as negative by an assay or test

false negative error Type II error STATISTICS An error which occurs when the statistical analysis of a trial detects no difference in outcomes between a treatment group and a control group when in fact a true difference exists

false negative Pap smear False negative smear GYNECOLOGY A Pap smear from a woman with a premalignant or malignant lesion that was classified incorrectly as containing no abnormal cells during routine Pap smears. See Cervical cancer

false negative rate NIHSPEAK The proportion of all positive–active–substances falsely identified as negative

false pelvis Greater pelvis OBSTETRICS The functional pelvic region which encompasses the area above the pelvic inlet. See True pelvis

false positive LAB MEDICINE A test result from a Pt who does *not* have a particular disease, which is positive or detects an analyte that is usually normal. See Four cell diagnostic matrix, Cf False negative

false positive error Type I error STATISTICS An error that occurs when the statistical analysis of a trial detects a difference in outcomes between a treatment group and a control group when in fact there is no difference

false positive HIV testing MEDTALK Antibodies cross-reactive to HIV-1 core antigen are rare, but may occur in normal subjects or those with autoimmune disease, cutaneous T-cell lymphoma and multiple sclerosis; low-level false positive ELISA occurs in alcoholic hepatopathy, pregnancy, non-HIV infected IV drug abusers and in hemodialysis. Cf AIDS, HIV

false positive rate STATISTICS The proportion of all negative–inactive substances that are incorrectly identified as positive; FPR is an indicator of test performance

false teeth Dentures, see there

falsification of clinical credentials MEDTALK The misrepresentation of one's professional background in terms of education–internship, residency and/or specialty board certification

Falstaff snore A loud snore of the sleeping obese, which may occur in the sleep apnea syndrome

FAM 5-FU, adriamycin/doxorubicin, mitomycin C ONCOLOGY A chemotherapeutic regimen used with varying degrees of failure for advanced gastric CA. See Stomach cancer

famcyclovir Famvir® VIROLOGY An oral antiviral used to manage acute herpes zoster–shingles and recurrent genital herpes METABOLIC ROUTE Metabolized in kidney SIDE EFFECTS Headache, nausea

familial adenomatous polyposis Familial polyposis An AD condition affecting ±50,000–US, characterized by progressive development of hundreds of adenomatous colorectal polyps; progression to cancer MOLECULAR PATHOLOGY *APC* gene on chromosome 5q21 is mutated in FAP, which may also be mutated in sporadic colorectal tumors DIAGNOSIS Allele-specific expression assay, in vitro synthesized protein assay TREATMENT Sulindac

familial combined hyperlipidemia METABOLIC DISEASE A common–1:300 AD disorder with ↑ TGs and/or cholesterol LAB ↑ apoB, ↑ LDL-C, ↑ VLDL-C, mild ↓ HDL-C, apoA1 CLINICAL CAD, first MI as early as age 40, overweight, HTN MANAGEMENT Diet–weight loss, exercise, lipid-lowering drugs; smoking is forbidden as it exacerbates ASHD

familial dysautonomia Riley-Day syndrome PEDIATRIC NEUROLOGY An AR condition in Jews, which affects peripheral sensorimotor autonomic and CNS neurons CLINICAL FTT, episodic vomiting, URIs, autonomic dysfunction–skin blotching, lacrimation, temperature dysregulation, diaphoresis, HTN and postural hypotension, early death

familial dysbetalipoproteinemia Broad beta lipoproteinemia, hyperlipoproteinemia type III METABOLIC DISEASE An uncommon 1:1,000 to 10,000 AD condition with a defective apoE–the apoE$_{2/2}$ phenotype, poor catabolism of beta-migrating remnants, and ↑ production of TG-rich lipoproteins–eg, VLDL CLINICAL Palmoplantar tuberoeruptive xanthomas, ASHD < age 50, peripheral vascular and CAD, which may be accompanied by hyperthyroidism LAB ↑ TGs, ↑ cholesterol, floating beta lipoproteins MANAGEMENT Diet–weight loss, exercise, drugs–eg bile acid-binding resins and nicotinic acid; in post-menopausal ♀, low-dose estrogens

familial focal facial dermal dysplasia An AD condition characterized by hairless lesions with fingerprint-like puckering of the skin, especially at the temples, due to alternating bands of dermal and epidermal atrophy

familial hypercholesterolemia METABOLIC DISEASE A common–1:500 congenital AD defect in the LDL receptor gene, resulting in dysfunctional or absent receptors CLINICAL Early CAD in ♂, first MI by age 40–♀ may be asymptomatic throughout life, tendinous xanthomas, corneal arcus, xanthelasma; homozygotes have LDL-C > 600 mg/dL, tuberous xanthomas and fatal CAD in adolescence LAB ↑ LDL-C–300-500 mg/dL-20% of cholesterol in this range is due to FH MANAGEMENT Smoking cessation, diet, exercise, drugs–bile-acid binding resins–eg, cholestipol, cholestyramine, nicotinic acid, ↓ cholesterol and ↓ saturated fat diet, liver transplant may provide LDL receptors

familial hypertriglyceridemia METABOLIC DISEASE A common–1:200 AD disorder, due to ↑ in production of TGs, cholesterol, cholic acid, with ↑ VLDL and TG transport CLINICAL ↑ TGs, obesity, alcohol consumption, drug therapy–β-adrenergics, diuretics, estrogens, corticosteroids MANAGEMENT Diet, exercise, lipid-lowering drugs–eg, clofibrate, gemfibrozil

familial hypertrophic cardiomyopathy An uncommon disease that is a common form of obstructive cardiomyopathy CLINICAL Angina, arrhythmia, dyspnea, syncope, possibly, sudden death by young adulthood DIAGNOSIS EKG–asymmetric hypertrophy of the septum–usually of left side, systolic anterior movement of mitral valve, and midsystolic closure of aortic valve

familial intraosseous fibrous swelling of the jaws Cherubism, see there

familial juvenile nephrophthisis 1. An AR form of chronic interstitial nephritis which may accompanied by retinal dysplasia, hepatic fibrosis, skeletal defects CLINICAL Urine concentration defect with polyuria and polydipsia, possibly due to a primary tubular defect, often sodium wasting, persistent hypokalemia, metabolic acidosis, possibly with growth retardation 2. Medullary cystic disease, see there

familial malignancy The development of a malignancy in 2+ or more blood-related members of a cohort–eg, hepatocellular carcinoma. See Family cancer syndrome

familial Mediterranean fever Familial paroxysmal polyserositis An AR disease affecting eastern Mediterranean rim Jews, Armenian and Sephardic–the latter comprise ½ of cases–and Arabs, Greeks, Turks, and other 'rim' inhabitants, causing episodic serosal–especially peritoneal–inflammation, more common in ♂ CLINICAL Onset by adolescence–recurring peritonitis, arthritis, pleuritis, diffuse abdominal pain due to serositis, episodic fever, amyloidosis, muscle 'guarding', leukocytosis, malabsorption that resolve within 24 hrs PROGNOSIS Good in absence of renal amyloidosis, which causes terminal nephropathy, an event prevented by colchicine. See *MEFV*

familial tremor NEUROLOGY A benign possibly AD disorder with a familial tendency characterized by tremors that worsen with activity and purposeful movement, especially in older persons, affecting 1:15,000. Cf Essential tremor

family 1. A group of related organisms, proteins, or chemicals. See Superfamily 2. A unit of related individuals. See Cancer family, Dysfunctional family, Extended family, Hernandez family, Immediate family, Jukes family, Multiproblem family, Nerve growth factor family, Nuclear family, Single-parent family GENETICS A category in the biological nomenclature of livings things which falls between an order and above a genus

family cancer syndrome See Hereditary cancer syndromes

family cluster EPIDEMIOLOGY A grouping of disorders found in ≥ 2 members of a family

family history A summary of diseases present in immediate blood relatives–eg cardiovascular disease, DM, malignancy, cancer, etc, which may be linked to hereditable DNA mutations. Cf Social history

family planning SOCIAL MEDICINE The constellation of activities, in particular, fertility control–timed contraception and planned conception, undertaken by a heterosexual couple of childbearing yrs to achieve the desired birth spacing and family size. See Fertility, *Griswold v. Connecticut*

family practice Family medicine, family & community medicine A specialty of medical practice in which the physician is involved in most aspects of the health care of a family unit, but is limited in provision of obstetric and surgical services; in the US, board certification in FP requires 4 yrs of residency training. Cf Internal medicine

family therapy Family psychotherapy PSYCHIATRY A treatment model that involves interaction with family members and family interactions as well as with the individual. Cf False memory

family violence PUBLIC HEALTH A global term for domestic violence–eg, mental and physical abuse of those in the nuclear and extended family–eg, child, elder, spousal abuse. See Domestic violence, Spousal abuse

famotidine Pepcid™ THERAPEUTICS An agent used to treat peptic ulcer disease by blocking binding of histamine to H_2 receptors, resulting in ↓ intracellular concentration of cAMP and acid secretion by gastric parietal cells ADVERSE REACTIONS Diarrhea, headache, drowsiness, fatigue, muscle pain, constipation, rarely, also mental confusion, agranulocytosis, gynecomastia, impotence, allergic reactions, myalgias, tachycardia, arrhythmias, interstitial nephritis, etc. See Histamine receptor antagonist

Fanconi's anemia Fanconi's pancytopenia NEONATOLOGY A rare and often fatal inherited disease in which the BM fails to produce RBCs, WBCs, platelets, or a combination thereof, which may with time evolve toward a myelodysplastic syndrome or leukemia.

fantasy PSYCHIATRY An imagined sequence of events or mental images–eg, daydreams, that serves to express unconscious conflicts, to gratify unconscious wishes, or to prepare for anticipated future events. See Fetishism, Paraphilia SEXOLOGY A series of mental images connected by a story line or dramatic plot that may be translated into actuality. Coital fantasy, Copulation fantasy, Masturbation fantasy, Sexual fantasy

FAP 1. Familial adenomatous polyposis 2. Familial amyloid polyneuropathy 3. Fibrillating action potential 4. Flurouracil, Adriamycin, cisplatin 5. *fos*-associated protein

FAQs ONLINE A list on a website that answers basic–Frequently Asked Questions–that might be asked by a first-time visitor to the site

Fareston® Toremifene citrate ONCOLOGY An antiestrogen used for postmenopausal Pts with estrogen-receptor-positive or receptor-unknown metastatic breast CA. See Breast cancer

farmer's lung Farm worker's lung IMMUNOLOGY An IgG1-mediated form of extrinsic allergic alveolitis or hypersensitivity pneumonitis that occurs in non-atopic individuals, who after repeated exposure to organic dust and fungi–eg, *Aspergillus* spp, become allergic to thermophilic actinomycotic organisms; 90% of Pts have antibodies to moldy hay, an ideal growth medium for the fungi–*Microspora vulgaris*, *Thermoactinomyces vulgaris*, and *Micropolyspora faeni*–implicated in FL CLINICAL Attacks of several days duration from May to October–the growing season in the Northern Hemisphere, causing rales, cyanosis, fever, dry cough, rhonchi, dyspnea, beginning 4-8 hrs after exposure to stored corn, barley, and tobacco; with time, weight loss PULMONARY FUNCTION TESTS ↓ Functional lung volumes, impaired gas exchange RADIOLOGY Normal or diffuse interstitial reticular pattern, occasionally with fine nodular shadows TREATMENT Corticosteroids COMPLICATIONS Pulmonary HTN, right ventricular hypertrophy, failure. See *Faeni rectivirgula*. Cf Animal House fever, Silo filler's lung

FARS Fatal accident reporting system

farsightedness Hyperopia, see there

FAS Fetal alcohol syndrome, see there

fasciectomy ORTHOPEDICS The removal of fascial tissue, generally to reduce pressure in a compartment in the body or in an extremity

fasciitis ORTHOPEDICS Inflammation of a fascia. See Eosinophilic faciitis, Necrotizing fasciitis, Nodular fasciitis

fascinoma Great case MEDTALK An obscure lesion or condition that evokes considerable interest, especially from other doctors in the process rather than to the human concern for the Pt afflicted with the problem. See Interesting disease

fasciodesis ORTHOPEDICS The surgical anchoring of a fascia to another or to a tendon

fasciotomy ORTHOPEDICS The incision through a fascial plane, usually to relieve neurovasular pressure in an underlying muscle compartment

fashion victim ORTHOPEDICS A person with pathology 2° to dressing á haute couture–eg, high heels, which ↑ DJD of the feet–ie, person victimized by fashion VOX POPULI A popular term for a person who feels to compelled to dress á haute couture–ie, a victim of fashion. See Degenerative joint disease

fast NUTRITION A period of abstinence from solid foods or fluids as well, which may be required of a Pt before general surgery or before obtaining a specimen of blood for certain analytes. See 72-hr fast, Vegetable juice fast, Water fast VOX POPULI Vroom (!)

fast CT imaging A format of CT used in cardiac imaging, in which a scan requires less than a fraction of a second, allowing the creation of multislice images, which can be repeated at frequent intervals for a specified period; fast CT provides information about cardiac anatomy, pulmonary and coronary arteries, myocardial perfusion, and microcirculation. Cf Spiral CT

fast food NUTRITION Prepared food from a restaurant that specializes in providing a full 'meal,' often consisting of a permutation of hamburger or chicken, French fries, pommes frites (chips), and a soft drink or a milkshake, in < 2 mins; the medical community is laying much of the blame for the obesity in the US on the purveyors of fast foods; a diet consisting solely of FF overloads the body with protein, fat, calories, salt, and highly saturated vegetable–eg, palm, coconut oils and is low in vitamins, minerals, and fibers. See Cafeteria diet, Empty calories, 'Junk food,' Nibbling diet

fast track CARDIOLOGY *adjective* Referring to prioritized Pt management, specifically FT thrombolytic therapy in Pts undergong acute MI. See Myocardial infarction

fastin DRUG SLANG A street term for amphetamine, and any other neuropharmacologic agent used by dieters

fasting glucose Fasting blood sugar, fasting plasma glucose ENDOCRINOLOGY Glucose obtained from a Pt who has had nothing–except water by mouth for 8+ hrs; FG is used in evaluating Pts for possible DM REF RANGE 65-115 mg/dL non-diabetic; 110-140 mg/dL, possible GTT warranted; > 140 DM. See Diabetes mellitus, Glucose tolerance test

fasting specimen LAB MEDICINE A blood specimen drawn from a Pt who has not eaten for 12 hrs; fasting is an absolute requirement for a limited number of tests–eg, GTT; prolonged fasting causes a marked–240% ↑ in bilirubin, ↑ plasma TGs, glycerol and free fatty acids, and a marked–± 50% ↓ in glucose. See Glucose tolerance test

fat Any of a class of neutral organic compounds formed by a molecule of glycerol linked to 3 fatty acids–a glycerol ester; fats are water-insoluble, ether soluble, solid at ≤ 20°C, combustible, energy-rich–9.3 kcal/g. See Animal fat, Baby fat, Fatty acids, Fish oil, Monounsaturated fat, Olive oil, Polyunsaturated fat, Saturated fatty acid, Tropical fat, Tropical oil, Unsaturated fat

fat balance A state of equilibrium at which the amount of fatty acids in the circulation drives fat oxidation; at the FB point, all fat that is consumed is metabolized, and weight is maintained at a status quo. See Body mass index, Obesity

fat embolism An embolus containing fat, an event that follows long bone fractures, and less commonly, hepatic trauma; embolic fat 'metastasizes' to the lungs, causing dyspnea, shock, to the brain causing coma, to the kidneys causing lipiduria. See Embolism

fat embolism syndrome Emboli composed of fat are common, relatively innocuous and may occur in alcoholism, BM biopsy, cardiopulmonary bypass, compression injury, DM, lymphangiography, pancreatitis, sickle cell anemia, corticosteroid therapy; contrarily, the FES is neither common nor trivial; clinically significant FE may be endogenous or exogenous in origin; most are due to major fractures, especially of long bones, and trauma to parenchymal organs–eg, the liver–most deaths in the immediate post-trauma period have significant fat embolism, burns, blast injury, severe infections, especially α-toxin-producing *Clostridium* spp CLINICAL Hypoxia–50% of femoral shaft fractures have ↓ arterial PO$_2$ within the first few days, acute onset of dyspnea, tachypnea, cyanosis, tachycardia with sudden onset of right-sided cardiac failure, showers of petechiae, thrombocytopenia, cerebral embolism–with changes in personality, confusion, drowsiness, weakness, agitation, spasticity, defects of the visual field, and rarely, extreme pyrexia DIAGNOSIS It had been reported that fat droplets in a BAL was indicative of fat embolism, a finding which in one small–34 group of Pts proved to have a low specificity of 26.5%; > 3% oil red O positive macrophages in the BAL are often found in trauma Pts, and may indicate FES or silent FE TREATMENT No therapy is effective. See Embolism

fat metabolizer See Invalid claims of efficacy

fat necrosis Liquefactive necrosis initiated by trauma and effected by lipolytic enzymes. See Calcium soap, Necrosis

fat pad atrophy ORTHOPEDICS Subcutaneous atrophy, which primarily affects the elderly, characterized by soft flattened heel pads, which may allow palpation of calcaneal tubercles; nonradiating pain 'maxes' at the center of weight-bearing part of the heel pad MANAGEMENT Viscoelastic heel pad, *not* surgery. See Heel pad

fat spurt PEDIATRICS A transient ↑ in subcutaneous fat in preadolescence. See Growth spurt

fatality PUBLIC HEALTH Death; a death, usually understood to be accidental in nature. See Occupational fatality

father SOCIAL MEDICINE The biological and/or rearing male figure in a family unit. See Secondary father

father 'factor' PSYCHOLOGY A popular term for the ill-defined constellation of components that a father contributes to a person's personality development and psychologic maturation. See Two-parent advantage

fatigue MEDTALK A state of mental and/or physical exhaustion, which is often understood to be unusual for the affected person. See Chronic fatigue, Driver fatigue, Nightshift fatigue, Taste fatigue

fatigue fracture Insufficiency fracture, stress fracture SPORTS MEDICINE A stress fracture affecting the feet of formerly fit foot soldiers, caused by repeated, relatively 'trivial' trauma to normal bone, resulting in local bone resorption. Cf Flat feet

fatigue syndrome Chronic fatigue syndrome, see there. Cf Burnout

fatty acid BIOCHEMISTRY A straight-chain monocarboxylic acid, which can be either saturated–ie, has no double bonds or unsaturated, which is, in turn, either monounsaturated–having a single double bond, or polyunsaturated–having more than one double bond. See Cholesterol-raising fatty acid, n-3 fatty acid, Polyunsaturated fatty acid, Unsaturated fatty acid

fatty food attack INTERNAL MEDICINE A popular term for severe transient colicky abdominal pain that occurs in response to ingestion of fried or fat-laden foods, which is considered a common clinical sign of cholelithiasis. See Gallbladder disease

fatty liver of pregnancy Acute fatty liver of pregnancy A rare idiopathic complication of pregnancy, most often affecting primiparas with ♂ infants or twins, ± associated with pre-eclampsia CLINICAL Onset after 35th wk–occasionally by the 13th wk–caveat medicus, possibly progressing to fulminant hepatic failure with jaundice, encephalopathy, DIC, death MANAGEMENT Terminate pregnancy PROGNOSIS ±20% death. See MTP complex

fatty metamorphosis of viscera White liver disease An AR rapidly fatal condition characterized by massive hepatic–and corporal-steatosis, progressive hypotonicity, lethargy, coagulopathy and jaundice LAB ↑ TGs, ↑ chylomicrons, ↑ HDL-C, hypoglycemia, ↓ Ca^{2+}

Favoxil Fluvoxamine, see there

favus A disfiguring scalp dermatophytosis caused by *Trichophyton violaceum* and *Microsporum gypseum*, with destruction of hair follicles and alopecia

FBI sign Fat-blood interface ORTHOPEDICS A semi-lunar soft tissue effusion, seen in post-traumatic lipohemarthrosis, using a 'horizontal' beam X-ray; the FS is most commonly seen in the knee and in the shoulder, because of ↑ marrow fat and blood in the joint space

FBS Fasting blood sugar. See Fasting glucose

FD&C Yellow No. 5 Tartrazine A ubiquitous dye used in foods and drugs that cross-reacts with aspirin, exacerbates asthma, and may cause life-threatening anaphylactic reactions, asthma, urticaria. See Artificial dye, Food additives, Food dye

FDA Food & Drug Administration PUBLIC HEALTH The agency charged with determining the safety and efficacy of drugs and therapeutic devices before marketing and assuring that certain labeling specifications and advertising standards are met while marketing the product. See Investigational new drug, Phase I, II, III studies

FDA classification of devices A system of stratifying devices used for various health care needs based on the potential for causing morbidity

---

FDA CLASSIFICATION, MEDICAL DEVICES

I Devices used by and easily accessible to the public; regarded as having minimal potential for incorrect use when used in health care

II Must undergo an approval process that includes special controls, eg performance standards and general controls required of all devices

III In addition to the above, the device or product must meet a rigorous premarketing approval standard, which may delay the release of the product by months or years

---

FDPs Fibrin degradation products, see there

fear conditioning A conditioned response induced by linking an intense noxious stimulus to an unrelated stimulus–eg, auditory stimulus. See Fight-or-Flight response

febrile seizure Fever-induced seizure PEDIATRICS A generalized tonic-clonic–grand mal seizure seen in infants to toddlers after rapidly rising fevers lasting from seconds to minutes; most are idiopathic; FSs may be more common in families DIFFDX Various intoxications, meningitis, encephalitis, roseola, or infection with HHV6, *Shigella*

fecal impaction MEDTALK The presence of an indurated bolus of feces in the rectum, which is difficult to pass and which may, with time, cause ulceration of the rectal mucosa. See Fecalith

fecalith An indurated bolus of partially dried fecal material

fecaloid *adjective* Feces-like

feces Doo-doo, poop, stool GASTROENTEROLOGY The semisolid material–composed of undigested food residue, bacteria, secretions–which is discharged from the anal sphincter by defecation

federal medical privacy rules HEALTH POLICY A series of federal mandates which provide the rules for use or disclosure of protected health information, for varous purposes, including for research. See HIPAA. Cf De-identification

Federal Register NIH GRANTSPEAK An official federal daily that notifies the public of any changes of federal regulations, rulings, legal notices, proclamations, and documents generated by the executive branch of the government

fee for service MEDTALK *adjective* Referring to the traditional form of reimbursement for health care services, where a fee is paid to a provider, according to the service performed, by a Pt or a conventional indemnity insurer, after a service is rendered. See Fee schedule. Cf Capitation

feeding VOX POPULI The providing of nutrients, usually to a person or animal incapable of obtaining those nutrients by itself. See Sham feeding

feeding gastrostomy SURGERY A procedure in which an opening is created in the anterior wall of the stomach to allow suction decompression and improved respiratory function by eliminating the need for a nasogastric feeding tube. See Nasogastric tube

FEF$_{25\%-75\%}$ PULMONARY MEDICINE A clinical test that measures the forced expiratory flow from 25% to 75% of vital capacity, which was once thought to be one of the most sensitive, simplest, and least expensive methods for evaluating small airways disease

FEIBA Factor VIII inhibitor–see there, or factor VIII bypassing activity. See Prothrombin complex concentrate

felbamate Felbatol® NEUROLOGY A second-line antiepileptic which ↓ seizures in Lennox-Gastaut syndrome, a severe form of epilepsy of childhood onset that is poorly controlled with other agents

fellow A physician who has completed medical school, internship and a residency, and is in fellowship, see there

fellowship GRADUATE EDUCATION A post-residency training period of 1–2 yrs in a subspecialty–eg, hand surgery, which allows a specialized physician to develop a particular expertise that may have a related subspecialty board; fellowship time is often used to prepare for specialty boards examinations. See Training

felodipine Plendil® CARDIOLOGY A beta-blocking antihypertensive

felon Paronychia, whitlow, run-around A purulent infection in the tight fascial plane adjacent to the terminal intraphalangeal joint of the fingers or toes, due to an open wound; as the inflammatory mass expands within the confined space, the vascular supply is compromised, predisposing the site to osteomyelitis, pulp necrosis and sloughing of tissue; the pain is very intense and seemingly disproportionate with the scant amount of swelling and erythema clinically evident TREATMENT Drainage by incision directly over the site of maximum swelling; the term has also been applied to a localized painful herpetic skin infection 'seeded' in an open abrasion by contact exposure

female androgenic alopecia COSMETIC SURGERY Hair loss in ♀, in which there is a usually preservation of a rim of frontal hair with central hair loss. See Hair replacement

female athlete triad SPORTS MEDICINE A 'condition' seen in some ♀ elite athletes–especially in those who overtrain, which results in M&M—characterized by disordered eating, menstrual dysfunction, osteoporosis. See Elite athlete, Gymnastics, Overtraining

female condom Vaginal pouch An externally placed contraceptive device, which offers some protection against pregnancy and STDs. See Contraceptives. Cf Condom

female impersonator Vox POPULI Drag queen, see there

female pseudohermaphroditism ENDOCRINOLOGY A type of intersex, in which the ovaries and müllerian derivatives are normally developed, and the anatomical ambisexuality is limited to the external genitalia–ie, 2 ovaries and a 46,XX genotype; FP is most often caused by congenital adrenal hyperplasia. See Intersex disorder

feminist therapy PSYCHIATRY Psychotherapy that incorporates feminist/women's rights–I am woman, HEAR ME ROAR–philosophy into therapeutic goals; FT attempts to empower ♀ and battle societal barriers to self-actualization

feminization ENDOCRINOLOGY The development of ♀ secondary sex characteristics in a genotypic ♂, with regression of body hair and change to a ♀ body contour. See Testicular feminization SOCIAL MEDICINE The shift in the demographics of a particular profession or type of work. See Occupational feminization

femoral hernia SURGERY A hernia in which a loop of intestine droops into the femoral canal. See Hernia

femoral pulse PHYSICAL EXAM A pulse palpated in the inguinal region which, if absent, indicates severe regional ASHD or ↓ blood flow

FemPatch® GYNECOLOGY A low-dose transdermal estrogen–17 beta-estradiol system for menopausal Sx. See Hormone replacement therapy

Femprox® Alprostadil A topical hormone analogue in clinical trials for treating ♀ sexual dysfunction. See Sexual dysfunction

Fen-Phen CARDIOLOGY A once-popular combination of 2 anorectic agents, fenfluramine and phentermine; the theoretical advantage of using 2 different agents is that the dose of each agent could be lowered ADVERSE EFFECTS The fen-phen combination is associated with valvular heart disease and fatal pulmonary HTN; it was withdrawn from the market. See Fenfluramine

fenestration Laminotomy, see there

fenfluramine Pondimin®, Redux® INTERNAL MEDICINE An adrenergic appetite suppressing–anorectic amphetamine congener, which, unlike amphetamine, depresses the CNS; fenfluramine ↑ serotonin levels and imparts a sensation of fullness from a lower intake of food and was prescribed as a short-term adjunct in obesity; it was withdrawn from the market following its association with heart valve abnormalities and other cardiac complications ADVERSE EFFECTS Pulmonary HTN, attributed either to serotonic vasoconstriction or altered depolarization of pulmonary vascular smooth muscle membrane. See Fen/Phen

fenofibrate Tricor®, Lipanthyl® An agent used to manage super-high serum TG levels uncontrolled by diet alone or in Pts at risk for pancreatitis

fentanyl Duragesic®, Sublimaze® PAIN MANAGEMENT An opioid analgesic used for severe chronic pain–it is 100-fold more potent than morphine–eg, CA, AIDS-related terminal pain ROUTE Injected, topical ADVERSE REACTIONS Hypoventilation, N&V, constipation, dry mouth, somnolence, asthenia, diaphoresis. See Opioid anesthetic, Patient-contolled analgesia. Cf Muscle relaxant

ferritin HEMATOLOGY A major iron storage protein present in small amounts in serum; in healthy adults, serum ferritin reflects the body's iron store, and can be used to detect iron deficiency REF RANGE ♂ 33-236 ng/mL; ♀, < 40 yrs 12-263 ng/mL; > 40 yrs 11-122 ng/mL; ↓ FERRITIN Hypochromic anemia, microcytic anemia. iron deficiency anemia↑ FERRITIN Iron overload, hemochromatosis, inflammation, malignancy

fertility 1. The ratio of live births/yr in a population to the number of women of child-bearing age 2. The ability to conceive; conceivable 3. The ability to produce living progeny. Cf Subfertility

fertility drug Fertility pill GYNECOLOGY An agent that either replaces sex hormones in ♀ whose production is insufficient for succesful gestation, or induces ovulation–eg, clomiphene citrate, which stimulates the hypothalamus, which in turn stimulates the pituitary to release LH and FSH. See Clomiphene citrate, Human menopausal gonadotropin

fertility-focused intercourse REPRODUCTIVE MEDICINE Coitus in which the impregnation is the primary goal; various maneuvers of FFI maximize the 'yield' of this activity, including coitus during the critical fertilization window, prior to ovulation, limiting the sorties to one/day, use of boxer shorts by ♂, and maintaining ♀ legs in the 'patas arriba' position. See Fertilization window

fertilization Fusion of ovum & spermatozoon REPRODUCTION MEDICINE The penetration of an egg by sperm, resulting in combined genetic material that develops into an embryo

fertilization window REPRODUCTIVE MEDICINE The time period in the normal menstrual cycle during which fertilization is most likely to occur–day 6 up to ovulation; the FW is a datum of use in both fertility-focused intercourse, and in postovulatory contraception. See Fertility-focused intercourse. Cf Rhythm method

festinating gait Parkinsonian gait NEUROLOGY Gait characterized by flexed trunk, hips and knees, in which the steps get progressively shorter and faster; FG is a clinical finding typical of Parkinson's disease. See Parkinsonism

fetal acoustic stimulator OBSTETRICS A device applied to the abdominal wall of woman in late pregnancy, which sends a sound wave to shock the baby, and markedly ↑ the fetal heart if the baby is healthy. Cf Internal fetal monitoring

fetal alcohol syndrome Alcohol embryopathy A condition due to in utero exposure to alcohol, resulting in a menagerie of birth defects; alcohol-related birth defects–ARBDs; when multiple ARBDs are present, the term fetal alcohol syndrome is used CLINICAL Prenatal and postnatal growth retardation, facial dysmorphia–short palpebral fissure, epicanthal folds, short up-turned nose, thin upper lip, long smooth or poorly-developed philtrum, short nose, micrognathia, maxillary hypoplasia, prematurity, perinatal asphyxia, mild to profound developmental delays, mental dysfunction, microcephaly, cranial defects, atrial septal defect, muscular hypotonia, bone anomalies–vertebral malformation and spina bifida with joint contraction. See Alcoholism, Embryopathy

fetal circulation EMBRYOLOGY Prenatal circulation which bypasses the lung and right heart, and is returned to the systemic circulation at the aorta via a patent ductus arteriosus, which usually closes at or shortly after birth, after which the blood flows to the lungs

fetal diabetic syndrome A complex seen in the children born to diabetic mothers, who have a 3- to 5-fold ↑ in various congenital and acquired anomalies, including visceromegaly, ↑ body fat, respiratory distress, and hyaline membrane disease, cardiomegaly, ventricular septal hypertrophy, skeletal anomalies, hypoplastic left colon syndrome–aganglionosis-like presentation, hypocalcemia, immaturity

fetal distress syndrome OBSTETRICS Intrauterine fetal hypoxia caused by prematurity or antepartum maternal infection, DM, eclampsia, HDN, hemorrhage, and others CLINICAL Tachycardia >160/min or bradycardia <100/min, which carries a worse prognosis, and ↓ pH

fetal heart rate OBSTETRICS A rate which, in the non-stressed fetus, reflects cardioaccelerator and cardiodecelerator reflexes; analysis of the FHR requires evaluation of a baseline FHR between uterine contractions or periodic changes in the FHR and non-periodic, short-term fluctuations in the FHR. See Deceleration

fetal hemoglobin Hemoglobin F An immature Hb composed of 2 α and 2 γ chains that usually disappears in the neonatal period. See Hereditary persistence of fetal hemoglobin

fetal hydantoin syndrome A congenital complex caused by in utero exposure to anticonvulsants, affecting the infants of pregnant ♀ being treated with these agents CLINICAL Growth retardation, microcephaly, midfacial hypo- or dysplasia, hypertelorism, short nose, broad depressed nasal bridge, cleft lip and palate, onychodigital dysplasia, cardiac malformations, mental retardation, rarely neuroblastoma. See Fetal trimethadione syndrome

fetal lung maturity OBSTETRICS A parameter that determines the likelihood a neonate will develop RDS; infants delivered at 40 ± 2 wks have 0% incidence of RDS; at 36 wks 0-2%, at 34 wks 8-34%—depending on birthweight

fetal monitoring OBSTETRICS A general term which can refer to any maneuver used to evaluate the fetus' status during pregnancy–eg, measurement of heartbeat and visual examination of the amniotic sac; however, as used, FM usually refers to the use of electronic devices during L&D to assess the baby's heartbeat and uterine contraction. See External fetal monitoring, Internal fetal monitoring

FETAL MONITORING*

INDIRECT FM–Electrodes are placed on abdominal skin over uterus, which detects fetal heart beats and uterine contractions

DIRECT FM involves the placing of an electrode directly on the fetus while it is in the uterus, which is done after the membranes around the fetus have ruptured

*Formerly limited to high risk pregnances, FM is a common procedure that been linked to a ↓ in labor-related death and disability; a consequence of FM is ↑ possibility of the obstetrician performing a C- section

fetal mortality rate The ratio of fetal deaths to the sum of the births–the live births + the fetal deaths in that year; in the US, the FMR dropped from 19.2 /1,000 births–1950 to 9.2/1,000 births–1980

fetal movement Kicking OBSTETRICS The constellation of activity by the fetus in the uterus which, in healthy infants, averages 10/hr

fetal posture NEONATOLOGY The position assumed by the fetus in utero, which may be adopted ex utero, by those with severe mental retardation, or by a normal person in a state of extreme mental stress

fetal tobacco syndrome NEONATOLOGY A malformation complex affecting infants born to ♀ smoking ≥ 1 pack of cigarettes/day during pregnancy; FTS neonates are 200 g lighter, infant mortality is ↑ by 40%; smoking is indirectly responsible for 4000 excess infant deaths, in the form of ↑ spontaneous abortion, fetal wastage, perinatal mortality, SIDS; children with FTS have impaired cognitive and emotional development. See Fetal alcohol syndrome, Passive smoking, Smoking

fetal varicella syndrome NEONATOLOGY An embryopathy affecting ± 2% of fetuses of mothers infected with HVZ in the first 20 wks of pregnancy–CLINICAL Scarring along dermatomes, muscular and osseous hypoplasia, eye defects–cataracts, microphthalmia, chorioretinitis, neurologic defects–mental retardation, microcephaly, sphincter dysfunction

fetal wastage A loss of a gestational product, either voluntary or involuntary, that occurs between the 20th wk of pregnancy and the 28th day of life, a value known for epidemiological purposes, as 'total pregnancy wastage'

fetation Pregnancy. See Superfetation

fetish SEXOLOGY A device–eg women's undergarments, bra, shoes, or other wearing apparel that is the object of sexual arousal, which may, in extreme cases, replace the need for a sexual partner for sexual arousal or orgasm

fetishism A paraphilia–sexual deviation that involves the use of nonliving objects–fetishes for sexual arousal; as defined by the DSM-IV, fetishism occurs over a period of ≥ 6 months, is distressful to the subject, and is not limited to those articles of female clothing used in cross-dressing, known as transvestic fetishism, or devices–eg vibrators, designed for tactile genital stimulation. See Fetish, Sexual deviation

fetomaternal hemorrhage Fetomaternal transfusion The passage of blood from the placenta–via the umbilical cord into the mother at the time of delivery; FH has considerable importance in Rh-negative ♀ who deliver Rh-positive babies, against whom the mother may form antibodies, possibly resulting in 'rejection' of the fetus in subsequent pregnancies, with–in extreme cases—hydrops fetalis

fetoplacental unit OBSTETRICS A functional compartment responsible for synthesizing hormones that maintain pregnancy. See Estriol

fetoscopy OBSTETRICS An imaging technique in which a ultrasound-guided, 3 mm in diameter needle with an endoscope is inserted through the abdominal wall into the uterus to view a living fetus; fetoscopy can be performed in the 2nd and 3rd trimesters of pregnancy and carries a significant risk of fetal wastage, related to rupture of fetal membranes. Cf Embryoscopy

fetotoxicity TOXICOLOGY Adverse fetal effects due to toxins entering the maternal circulation and crossing the placental barrier, which may compromise maternal health and appear as fetal malformations, altered growth and in utero death. Cf Embryotoxicity

fetus OBSTETRIS 1. The unborn child developing in the uterus–after the embryonic stage, circa age 7 to 8 wks to birth 2. The product of conception from the time of implantation until delivery; if the delivered or expelled fetus is viable, it is designated an infant. See Harlequin fetus, Nonviable fetus. Cf Embryo

FEV Forced expiratory volume PULMONARY MEDICINE The maximal amount of air that can be exhaled in a period of time, usually one–FEV_1 or less commonly, three–FEV_3 seconds; FEV_1 is usually ↓–and thus is a major parameter measured in obstructive airways disease, a generic term that encompasses both asthma and COPD

FEV_1/FVC PULMONARY MEDICINE The ratio of forced expiratory volume in one second to forced vital capacity

Fevarin Fluvoxamine, see there

fever INTERNAL MEDICINE A body temperature of ≥ 37.2°C/99.0°F in early morning or ≥ 37.8°C/100.0°F in the evening that is a complex and coordinated adaptive response, which is part of the reaction to an immune challenge; this response is stereotyped, and largely independent of a causitive agent; as with other integrated responses–eg, regulation of energy metabolism, BP and volume, and reproduction, fever depends on humoral cues and is orchestrated by the hypothalamus, which coordinates autonomic, behavioral, endocrine, and metabolic responses– when corporal temperature is raised by endogenous pyrogen, T cell production ↑ 20-fold; endogenous pyrogen also shifts iron–needed by bacteria away from plasma; hyperthermia–up to 40°C may trigger CA regression. See 'Animal House fever' Argentine hemorrhagic fever, Blackwater fever, Blue fever, Breakbone fever, Bolivian hemorrhagic fever, Bullis fever, Cabin fever, Cat-bite fever, Coley's toxin, Colorado tick fever, Congo-Crimean hemorrhagic fever, Cotton fever, Dehydration fever, Dengue hemorrhagic fever, Epidemic louse-borne typhus fever, Familial hibernian fever, Familial Mediterranean fever, Filarial fever, Fort Bragg fever, Haverhill fever, Hay fever, Hectic-septic fever, Hypothalamic fever, Izumi fever, Korean hemorrhagic fever, Lassa fever, Malta fever, Mediterranean fever, Metal fume fever, Omsk hemorrhagic fever, O'nyong-nyong fever, Polymer fume fever, Pontiac fever, Potamic fever, Q fever, Quartan fever, Quintan fever, Rat-bite fever, Relapsing fever, Rheumatic fever, Rift Valley fever, Rocky Mountain spotted fever, Sandfly fever, Scarlet fever, Seven-day fever, Simian hemorrhagic fever, Spring fever, Swamp fever, Tertian fever, Thirteen day fever, Three day fever, Trench fever, Valley fever, West Nile fever, Yellow fever

fever blister Any of the minute, often multiple, perioral vesicles filled with clear fluid, caused by HSV. See Canker sores

fever-induced seizure Febrile seizure, see there

fever of unknown origin INFECTIOUS DISEASE A febrile state with temperature of ≥ 37ºC of 2 or more wks in duration, for which a cause cannot be identified despite thorough physical examination and aggressive and relevant lab work-up ETIOLOGY Infectious in 30-40%, collagen vascular in 15-20%; in adults, 20-30% of rest are due to CA, which comprises 10% of the rest in children; rare causes of FUO include sarcoidosis and colitis; hereditary FUOs are rare and appear in Fabry's disease, familial Mediterranian fever, type 1 hyperlipidemia, cyclic neutropenia

fever phobia A popular term for the response of parents to childhood fever, which may result in inappropriate overmanagement with antipyretics. See Fever

fever therapy Controlled hyperthermia ONCOLOGY Induced hyperthermia, which enhances immune function, related to the release of a wide variety of pyrogenic—and nonpyrogenic—cytokines; FT may be used to enhance tumor cell lysis, and is most successful when combined with chemotherapy and radiotherapy. Cf BCG therapy, Fever

FFP Fresh frozen plasma, see there

FGR Fetal growth restriction. See Intrauterine growth restriction

FHR Fetal heart rate, see there

fiber NUTRITION Bulk, roughage The indigestible part of fruits and vegetables, which may have a cancer-preventive effect. See Bran, Crude fiber, Dietary fiber, Oat bran, Soluble fiber

fiberoptic bronchoscopy See Bronchoscopy, Bronchial brushings, Bronchial washings

fibric acid analogue CARDIOLOGY A cholesterol-lowering agent—eg, gemfibrozil, used to manage hyperlipidemia by enhancing lipoprotein lipase activity and inhibiting hepatic synthesis of VLDL EFFECTS ↓ TGs, ↓ LDL-C, ↑ HDL-C. See Cholesterol

fibrillation CARDIOLOGY Unsynchronized random and continuously changing electrical activity in the myocardium, causing inefficient pumping of blood. See Atrial fibrillation, Ventricular fibrillation

fibrin-fibrinogen degradation products Fibrin 'splits' HEMATOLOGY Polypeptide fragments of fibrin degradation, which are generated by unchecked 1º fibrinolysis. See Disseminated intravascular coagulation

FIBRIN DEGRADATION PRODUCTS

INTRAVASCULAR LESIONS Associated with DIC, DVT, and PTE—ie, pathological fibrinolysis

EXTRAVASCULAR CONDITIONS eg, hematomas, neoplasia, sepsis, allograft rejection, glomerular disease, severe liver disease and obstetric complications, eg abruptio placentae, eclampsia, retained dead fetus

fibrin glue Fibrin sealant SURGERY A liquid commercial product composed of purified fibrinogen and thrombin used to seal operative wounds, by partially re-enacting the final stage of the coagulation cascade, in which fibrinogen is converted to fibrin in the presence of thrombin, factor XIII, fibronectin and calcium ions; FS are mixed at the 'table,' and used to seal complex surgical wounds—eg, anorectal fistulas

fibrin sheath NURSING A tubular scar that is a potential complication of long-term catheterization, in which the catheter becomes encased in a fibrotic sheath, which may harbor bacteria and make it difficult to withdraw blood from the line. See Central line

fibrinogen Coagulation factor I, factor I HEMATOLOGY A soluble 340 kD plasma glycoprotein required for normal platelet function and wound healing; it is converted into fibrin in the common pathway of coagulation, and provides physical scaffolding for permanent hemostatic plugs, which is orchestrated under thrombin's baton; fibrinogen is an 'acute phase reactant,' which may be markedly ↑ in various types of nonspecific stimuli—eg, inflammation, hemostatic stress, pregnancy, autoimmune diseases; it is ↑ in hyperfibrinogenemia; ↓ in afibrinogenemia

fibrinogen Detroit A variant fibrinogen first described in Detroit, a city well-known for acquired bleeding disorders. See Dysfibrinogenemia

fibrinoid necrosis 'Smudgy' eosinophilic fibrin-like deposits, of degenerated collagen or ground substance, in arterial walls of Pts with malignant HTN and periarteritis nodosa; FN may also occur in the Arthus reaction, acute rheumatic fever, SBE, near peptic ulcers, rheumatoid arthritis, immune complex disease, HBV, malignancy, complement C2 deficiency, Henoch-Schönlein purpura, SLE and other collagen vascular diseases. See Fibrosis, Necrosis

fibroadenoma SURGERY A firm, round, and almost invariably benign tumor that occurs in the breast of younger–age 20-45 ♀, which is composed of a dense stromal tissue, within which are ribbons of compressed glands VOX POPULI Lump in the breast. Cf Breast CA

fibrocystic disease Chronic cystic mastitis, fibrocystic disease SURGICAL PATHOLOGY A common benign disease of the ♀ breast, first seen circa age 40, which presents as diffuse rubbery induration, punctuated by dilated ducts and variably-sized cysts MAJOR PATTERNS Cysts plus fibrosis; sclerosing adenosis; epithelial hyperplasia, which ↑ CA risk. See Ductal intraepithelial neoplasia

FIBROCYSTIC DISEASE

NO RISK FOR FUTURE CANCER: FCD with adenosis–sclerosing or florid, apocrine metaplasia, cysts–macro- or microscopic, ductal ectasia, fibroadenoma, fibrosis, hyperplasia

MINIMAL RISK FOR FUTURE CANCER: with glandular 'crowding', ie 2-4 epithelial cells in depth, mastitis, especially periductal, squamous metaplasia

1.5-2-FOLD ↑ RISK OF FUTURE CANCER: with hyperplasia–moderate or florid, solid or papillary or in a papilloma with a fibrovascular core

2-5-FOLD ↑ RISK OF FUTURE CANCER: with atypical hyperplasia–borderline lesion, either ductal or lobular

8-10-FOLD ≠ RISK OF FUTURE INVASIVE CANCER Carcinoma-in-situ

fibrodysplasia ossificans progressiva FOP, Generalized myositis ossificans, myositis ossificans progressiva, stone man An idiopathic or AD condition of irregular penetration and pre-pubertal onset, in which connective/interstitial tissues undergo extensive fibrosis and heterotopic ossification of ligaments, tendons, muscle, fascia, aponeuroses, skin, first seen in late childhood as firm masses CLINICAL Microdactly, focal, transient, ±painful ossifying tumors in neck, back and extremities, with bony replacement of fasciae, ligaments and fibrotendinous tissue; also baldness, deafness, mental retardation, fever; tragically, Pts may find jobs in circus 'freak' shows; usually glossal, diaphragmatic, laryngeal, and perineal muscles are spared and Pts die of respiratory infections as intercostal muscles become petrified DIFFDX Osseous metaplasia, myositis ossificans, extraskeletal osteosarcoma. Cf Parana hard skin syndrome, Stiff man syndrome

fibroid VOX POPULI A benign tumor composed of fibrous and muscular tissue; fibroma. See Leiomyoma

fibromyalgia syndrome Fibrositis, tension myalgia PSYCHIATRY A condition characterized by muscular pain, fatigue, sleep disorders, anxiety, depression, headaches, IBS–possibly linked to anxiety and panic disorders MANAGEMENT Exercise, benzodiazepines, SSRIs, TCAs. See Chronic fatigue syndrome

fibromyositis Tension myalgia, see there

fibrosing alveolitis PULMONOLOGY A progressive chronic inflammatory disease of adults–45 to 65, which later develops into diffuse interstitial fibrosis and honeycomb lung PREVALENCE 4/10,000; accounts for ±15% of honeycomb lung disease CLINICAL Dry cough, dyspnea, clubbed fingers; FA progresses to respiratory failure and may produce cor pulmonale PFTs Restrictive changes COMPLICATIONS ↑ lung CA, often peripheral adenoCA

fibrosis MEDTALK A proliferation of fibroblasts and fibrous tissue. See Bridging fibrosis, Hepatic fibrosis, Idiopathic interstitial fibrosis of lung, Pipestem fibrosis, Radiation fibrosis, Stellate fibrosis, Systemic idiopathic fibrosis

Fibrostat™ BURNS A topical for ↓ hypertrophic scarring after serious burns or surgery. See Burns

fibrous coagulum UROLOGY An aggregate of soft fibrous tissue in which mineral fragments are embedded following extracorporeal shock wave lithotripsy; further lithotripsy is contraindicated in the face of FC, as it inhibits stone fragmentation and passage

FICSIT Fraility & Injuries: Cooperative Studies of Intervention Techniques, *pron* 'fix-it' GERIATRICS A series of randomized placebo-controlled trials that assessed various interventions, in ↓ falls and frailty in elderly Pts. See Geriatrics, Gerontology

fictive imagery PSYCHOLOGY The representation of mental images acquired in dreams, confabulation, or fantasy. See Imagery

field cancerization Field carcinogenesis MOLECULAR ONCOLOGY The constellation of locoregional changes triggered by long-term exposure of a field of tissue to a carcinogen; FC may induce CA, CIS or dysplasia, which can be recognized histologically; the remaining 'field,' despite adequate resection, is grossly normal but more susceptible to future CA; FC occurs in any embryologic tissue–eg, colorectal, breast ducts, bladder, bronchial, laryngeal epithelia, and aerodigestive tract in tobacco exposure. See 'Cancerization'

fifth vital sign INTERNAL MEDICINE A popular term for a "new" vital sign in a basic workup, identification and location of pain; the other, true, vital signs are temperature, blood pressure, pulse, respiratory rate

55 rule CARDIOLOGY A clinical guideline used in Pts with aortic regurgitation to gauge the timing of surgery, which should be performed when the ejection fraction falls below 55%, or before the left ventricle can no longer contract to the end-systolic dimension is > 55 mm or less. See Aortic regurgitation

fight bite A jagged laceration on dorsal hand, often over the knuckles, seen when Hooligan A's fist strikes Ne'er-do-well B's teeth, causing abrasions, deep lacerations, puncture wounds, and an inoculum of mixed oral flora–eg, staphylococci–50% of which produce penicillinase, β-hemolytic streptococci, *Eikenella corrodens*; sans treatment, FBs may become complicated, resulting in clenched fist syndrome. Se Clenched fist syndrome

fight-or-flight response Flight-or-fight response, general adaptation syndrome, stress response PHYSIOLOGY A constellation of physiologic responses to fear or perceived stress imminent danger or anticipated pain, which triggers full-scale CNS activation and release of 'stressors' by adrenal medulla–eg epinephrine and norepinephrine and cortex–eg corticosteroids, mineralocorticoids, kidneys–renin, pancreas–insulin CLINICAL Tachycardia, ↑ blood flow to muscle, ↑ BP, muscle tone, ↑ O₂ consumption, sweating, ↑ respiratory rate, tremor, pallor, ↑ inotropism, vasoconstriction, mydriasis, bronchodilation, hyperglycemia. Cf Relaxation response

FIGO staging system ONCOLOGY A system delineated by the International Federation of Gynecologists & Obstetricians for staging a particular gynecologic cancer. See Ovarian CA

FIGO STAGING, OVARIAN CARCINOMA

I Malignancy of one (Ia) or both (Ib) ovaries, without ascites; 5-year survival: 60%

II Malignancy of one (IIa) or both (IIb) ovaries, with pelvic extension and ascites; 5-year survival: 40%

III Malignancy involves one/both ovaries, intraperitoneal metastases outside pelvis and/or positive retroperitoneal lymph nodes; 5-year survival: 5%

IV Involvement of one/both ovaries with metastases and histologically confirmed extension to pleural cavity or liver; 5-year survival, 3%

figure 8 distribution GYNECOLOGY A descriptor for the gross appearance of advanced lichen sclerosis et atrophicus in the labia majora and minora, where the labia have resorbed, leaving a residual patch of whitish skin around the vagina and anus

file INFORMATICS *noun* A basic unit of storage on a computer; a collection of data that can be stored, accessed, and transferred as a single unit; a file may contain text, calculations, graphics, or software routines. See Client-user options file, Corrupted file, RSV file *verb* To store information.

filgrastim Recombinant granulocyte colony-stimulating factor IMMUNOLOGY A colony-stimulating factor that stimulates production of blood cells, especially platelets, during chemotherapy. See Granulocyte colony-stimulating factor

filling DENTISTRY A material–eg amalgam, cement, porcelain or synthetics, used to restore lost tooth structure. See Amalgam, Caries

filling defect IMAGING A displacement of radiocontrast by a bulky lesion–eg a polyp of the stomach or colon, which often corresponds to a space-occupying mass in a hollow organ. See Defect

film badge RADIATION SAFETY A dosimeter worn by radiation workers to monitor radiation exposure; FBs contain a piece of film that is darkened by radiation, and may contain filters which shield parts of the film from certain types of radiation. See Dosimetry

filter IMAGING A layer of absorbing material, usually a metal–eg, Al, Cu, Pb, Sn that increases the ratio of hard X-rays to soft X-rays, the latter of which are of greater diagnostic value, given their ability to penetrate the imaged tissues. See Radiation MEDTALK A device used to separate one material from another. See Absolute filter, Adhesion filter, Blood filter, Inferior vena cava filter, Microaggregate filter, Red-free filter, Standard filter, Water filter

financial triage MANAGED CARE A popular term for evaluating a Pt's ability to pay for hospitalization or anticipated medical services, before rendering the services. See Dumping. Cf Triage

finasteride Proscar®, Propecia™ HAIR A competitive inhibitor of 5α-reductase, the enzyme responsible for converting testosterone to DHT or dihydrotestosterone; it is used to ↓ BPH Sx–↓ obstructive Sx, ↓ prostate volume, ↑ urinary flow–reportedly less effective than Hytrin; it binds with 5α-reductase type 2, inhibiting production of DHT; its uses include reversal of male-pattern baldness, hirsutism, acne ADVERSE EFFECTS ↓ Libido, impotence, ejaculatory defects. See Androgen ablation therapy

finder's fee Remuneration paid to a party that finds a client of a particular type. Cf Fee splitting

finding MEDTALK A clinical feature or datum of interest. See Incidental finding

fine crackle Fine rale, crepitation PULMONOLOGY A type of discontinuous, interrupted 'explosive' lung sound, which is less loud than coarse crackles and higher pitched. Cf Coarse crackles

fine needle aspiration DIAGNOSTICS A method of in which a thin or "skinny"–18- to 23-gauge needle is used to suck in cells or tissue bits for diagnoses; the sites selected for FNAs are often guided by radiologists with fluoroscopy, CT, MRI

fine needle aspiration biopsy Aspiration biopsy SURGICAL PATHOLOGY The removal of minute tissue fragments by a needle and syringe under suction to prepare a smear for cytologic examination or to make a cell block; a biopsy obtained by FNA from various body sites. Cf Needle biopsy

finger agnosia NEUROLOGY Inability to recognize ones own fingers

finger-in-glove appearance RADIOLOGY 'Gloved finger' shadow of Simon A descriptor for mucoid plugs within a bronchiectatic segment of a fibrotically thickened bronchus or bronchiole, as seen on a plain film of the chest

finger-to-nose test NEUROLOGY A test of voluntary motor function in which the person being tested is asked to slowly touch his nose with an extended index finger; the FTNT is used to evaluate coordination, and is altered in the face of cerebellar defects. See Heel-knee test

fingerbreadth PHYSICAL EXAM A clinical ruler for evaluating certain landmarks–eg the size of the liver based on the fingerbreadths from right costal margin

fingers, toes, penis, nose MEDTALK Anatomic doggerel for those acral parts which, per time-honored clinical dicta, should not be managed with vasoconstrictors, because they contain end-arterioles

fingerstick specimen Capillary specimen LAB MEDICINE A blood sample obtained by piercing a finger using spring-loaded device with a chisel or with a blade type lancet; once punctured, the fingertip is squeezed so that blood drops can be collected into a capillary tube or onto a glass slide; FSs can be used for micromethods–eg, to measure cholesterol, glucose, and other analytes

fire ant MEDICAL ENTOMOLOGY A nonwinged hymenopteran arthropod which is omnivorous, attacking livestock, crops, electric insulation; the FA injects a venom, causing intense burning and pruritus, due to necrotoxin or solenamine; reactions range from a wheal-and-flare response, a sterile pustule within 24 hrs to anaphylaxis-related death

firearm PUBLIC HEALTH A weapon–eg, pistol, rifle, shotgun, that uses an explosive charge to propel a projectile–eg bullet or shot STATISTICS 330,000 deaths 1980-1989–US; of 35,000 gun-related deaths–US, 1989, 52% were suicides, 42% homicides; from 1960 to 1980, there was a 100% ↑ in homicide rate, and 150% ↑ in homicide by firearm; from 1933 to 1982, rate of suicides by firearms ↑ 139%. See Ballistics, Black Talon bullet, Drive-by shooting, Shotgun

fire PUBLIC HEALTH A conflagration. See Residential fire, Superfire

fire retardant PUBLIC HEALTH A chemical used to resist combustion, which may contain polybrominated biphenyls and antimony oxide

fireman's hats CARDIOLOGY A popular term for the appearance of ST segment elevations on the EKG of a Pt with acute MI

first-degree block See Heart block

first heart sound S1 CARDIOLOGY A heart sound that occurs in systole, primarily produced by AV valve closure

first-pass elimination First-pass effect, first-pass metabolism, presystemic elimination PHARMACOLOGY The rate at which circulating drugs are metabolized as they traverse the liver, before they reach the systemic circulation

first-pass radionuclide angiocardiography First-pass radionuclide ventriculography INTERVENTIONAL CARDIOLOGY A technique used to obtain images of the heart during transit of a radiopharmaceutical–eg, ^{99m}Tc pertechnate through the central circulation; the high-velocity components of ^{99m}Tc pertechnate passage through the heart are recorded and analyzed quantitatively; it is regarded by some workers as the method of choice for evaluating right ventricular function and ventricular ejection fraction, right and left ventricular volumes, systolic and diastolic function, and regional or global myocardial performance. Cf Equilibrium radionuclide angiocardiography, MUGA

first responder First response personnel EMERGENCY MEDICINE A person employed in the public sector–EMT, fire fighter, police, volunteer EMS–whose duties include provision of immediate medical care in the event of an emergency; FRs have basic emergency care equipment, O_2 and mask combinations, tools for extrication–eg Excaliber, Jaws of life, and defibrillators. See ACLS, CPR. Cf EMT, Good Samaritan laws

first stage reconstruction SURGERY A procedure when a temporary expander or a permanent adjustable implant is inserted or an autologous flap procedure is performed

first trimester pregnancy First three months of pregnancy OBSTETRICS. See Pregnancy

fiscal intermediary Part A Contractor MEDICARE A private company that has a contract with Medicare to pay part A and some part B bills. See Medicare, Part A

FISH Fluorescent in situ hybridization MOLECULAR MEDICINE A hybrid of 3 technologies: cytogenetics, fluorescence microscopy, and DNA hybridization, which is used to determine cell ploidy and detect chromosome segments by evaluating interphase–non-dividing nuclei; in FISH, fluoresceinated chromosome probes are used for cytologic analysis and cytogenetic studies, and to detect intratumoral heterogeneity. See Chromosomal paint box

fish face A physiognomy characterized by antimongolic palpebral fissures, colobomata, fish-like mouth, total deafness, malformed ears, fancifully likened to an aquatic poikilothermic vertebrate SEEN IN Mandibulofacial dysostosis, Treacher-Collins syndrome, Franceschetti syndrome

fish tapeworm *Diphyllobothrium latum* A tapeworm that parasitizes freshwater fish of temperate zones in the Northern hemisphere CLINICAL Infection is usually limited to one worm, which causes CNS and GI symptoms, abdominal discomfort, weakness, loss of weight, malnutrition, and megaloblastic anemia TREATMENT Niclosamide

fish vertebrae RADIOLOGY A descriptor for biconcave, fish-like vertebrae, caused by infarction and central bone collapse due to thrombosis of the vertebral arteries, a finding typical of sickle cell anemia, which often occurs before the 2nd decade; peripheral perforating metaphyseal arteries are relatively spared, explaining lesser involvement of the anterior and posterior faces of the vertebrum DIFFDx FV may also be seen in hereditary spherocytosis, homocystinuria, Gaucher's disease, osteoporosis, osteopenia, renal osteodystrophy or hyperparathyroidism, osteomalacia, thalassemia major, blood dyscrasia

fishmouth incision AMBULATORY SURGERY A wide horizontal incision made on the tip of the finger to drain for a subungual abscess

fishmouthing CRITICAL CARE MEDICINE A sign characterized by lower jaw depression with inspiration, an ominous sign of ↑ medullary damage, with apneic potential in the face of autonomic respiratory failure

fishy odor A piscine odor described in various conditions–eg, vaginosis, caused by a newly described *Mobiluncus* genus, *Gardnerella vaginalis,* excretion of trimethylaminuriae–due to large doses of L-carnitine, 'rotting' fish, di-*N*-butylamine, diethylamine, stools infected by *Vibrio cholera,* which have a 'rice-water' appearance; a rancid fish odor is described in tyrosinemia, seen in hereditary tyrosinosis/tyrosinemia type I or liver failure. See Odors

fissure DERMATOLOGY A groove, cleft, or sulcus, which may or may not be normal. See Anal fissure, Slanted palpebral fissure NEUROLOGY A groove or narrow cleft that separates 2 parts, such as the cerebral hemispheres of the brain

fistula SURGERY An abnormal communication or conduit between one internal organ and another or with the skin surface. See Gastrointestinal fistula, Inner ear fistula, Urinary fistula

fitness HEALTH The ability or capacity to perform a particular task. See Aerobic fitness, Cardioivascular fitness, Physical fitness

fitness gadget POPULAR HEALTH A device, the use of which converts flab to fab, providing the meek and weak with a chic sleek unique physique EXAMPLES Rowing machines, treadmills, proprietary devices–eg, Bowflex, Dynabands, Nordic Track, Stairmaster, Thigh Master

five-part differential LAB MEDICINE A standard automated differential count of WBCs generated by an automated hematology analyzer–AHA from the peripheral circulation, which divides WBCs into neutrophils–PMNs, eosinophils, basophils, lymphocytes, monocytes. See Diff. Cf Three-part differential

five-year survival EPIDEMIOLOGY The timespan that a person survives with a particular dread disease, in particular CA; 5YS facilitates standardization of survival statistics. See Cancer-free survival

fix DRUG SLANG *noun* Street slang for a person's usual 'dose' of a drug of abuse *verb* To inject a drug of abuse PHLEBOTOMY *verb* A popular term for the manual positioning of a vein before performing venipuncture for drawing blood or placing an IV 'line'

fixation PHYSICAL EXAMINATION The immobility of a relatively well-circumscribed mass in soft tissue, as is typical of breast CA. See Postural fixation PSYCHIATRY The arrest of psychosocial development; fixation may be considered pathologic, if intense

fixed drug reaction DERMATOLOGY An idiopathic skin eruption more common in blacks, which often recurs at the same place, every time a particular drug or a related congener is administered; FDRs may also occur with chemically unrelated drugs or disappear with repeated administration of the same drug CLINICAL Sharply circumscribed edematous red-brown or purplish plaque that may be surmounted by a bulla, most often located on the extremities, the hands, and glans penis which with time, becomes lichenified, scaly, ±accompanied by hypermelanosis AGENTS CAUSING FDR phenazone, barbiturates, sulfonamides, quinine, tetracycline, oxyphenbutazone, chlordiazepoxide, food dyes, toothpaste, mothballs

fixed vocal cord Fixed cord SURGICAL ONCOLOGY Replacement of a vocal cord with infiltrating glottic CA with ↓ phonation. See Larynx, Vocal cord

flabby heart CARDIOLOGY An atonic, dilated, fat-laden heart typical of *Corynebacterium diphtheriae*-induced myocarditis

flaccid paralysis NEUROLOGY Paralysis characterized by complete loss of muscle tone and tendon reflexes. Cf Spastic paralysis

flag LAB MEDICINE A determined value on a diagnostic test at or above which a certain action is taken; eg fasting glucose > 7.8 mmol/L–US: > 140 mg/dL, is a flag for notifying the attending physician, who might not otherwise suspect DM. See Decision level, Panic value, Red flag

Flagyl® Metronidazole INFECTIOUS DISEASE A prescription formulation for treating bacterial vaginosis. See Vaginal candidiasis

flail chest TRAUMATOLOGY Paradoxic movement of the anterior chest wall during inspiration, as the free-floating 'flailed' part of the wall moves inward as the chest expands, in response to the negative intrathoracic pressure; FC is caused by multiple anterior rib fractures, often due to MVAs or aggressive CPR

flake Cocaine, see there

Flake maneuver EMERGENCY MEDICINE A type of Heimlich maneuver, in which the victim lies head down–eg, on stairs and pumps his own diaphragm. Cf Cough CPR, Heimlich maneuver

flank pain CLINICAL MEDICINE Pain in the side DIFFDX Adrenal tumor, hydronephrosis, polycystic kidney, pyelonephritis, renal tumor, renal cyst

flap PLASTIC SURGERY A pedicle of tissue used to cover a defect, usually of the skin. See Frechet flap, TRAM flap

flapping tremor Asterixis HEPATOLOGY An involuntary jerking tremor of wide amplitude elicited upon dorsiflexion of the pronated wrist and spreading of extended fingers; in full-blown FTs, there is abrupt flexion of the fingers at the metacarpophalangeal joint and flexion of the wrist, occurring asynchronously with each other every few secs, due to exaggerated reflexes; bilateral FT is quasi-pathognomonic for metabolic, often alcohol-related, hepatic encephalopathy seen in end-stage–post-fibrotic cirrhosis due to ↑ blood ammonia NEONATOLOGY Coarse bilateral tremors, accompanied by limb rigidity, hyperreflexia, resistance to flexion and extension, described in infants born to heroin-addicted mothers who undergo 'withdrawal' at birth

flare MEDTALK A period during which disease Sx reappear or exacerbate RHEUMATOLOGY An acute exacerbation of the Sx and disease activity of SLE LAB ↑ anti-double-stranded DNA antibodies, plasma C3a, serum complex of complement 5b-9 and plasma Bb and ↓ serum complement C3 and C4 UROLOGY 1. An abrupt ↑ in alk phos–ALP that occurs within 28 days after orchiectomy for prostate CA, followed by a decline to baseline 2. A worsening of clinical disease seen in the early stages of hormonal manipulation for metastatic prostate CA

flare-up MEDTALK An acute worsening of a condition

flashback PSYCHOLOGY A non-drug-related repetition of frightening experiences or images, which may affect ex-soldiers, as is well-described in veterans of the Vietnam conflict SUBSTANCE ABUSE Hallucinogen persisting perception disorder An involuntary recurrence of some aspect of a hallucinatory experience or perceptual distortion often with negative overtones and accompanied by fear and anxiety; flashbacks are an adverse effect classically associated with psychedelic drugs–eg, LSD and PCP, which occur days to wks after the last dose; flashbacks are common in heavy users and disappear with time. See LSD, PCP

FLASHBACK-HALLUCINOGEN PERSISTING PERCEPTION DISORDER

A The re-experiencing, after discontinuating use of a hallucinogen, of 1+ perceptual symptoms experienced while intoxicated with the hallucinogen, eg geometric hallucinations, flashes of colors, macropsia, micropsia, etc

B Symptoms in A cause significant distress or impairment of social, occupational, or other important function

C Symptoms are not due to a general medical condition, or otherwise accounted for by another mental disorder

*DSM-IV™ American Psychiatric Association, Washington, DC 1994

flasher PSYCHIATRY A person, usually a man who derives sexuoerotic stimulation from 'flashing'–ie, opening a coat, under which his doodads flap freely to the open air. See Bakerloo syndrome

flat affect PSYCHIATRY A marked attenuation of emotional range, a sign often associated with major depression. See Major depression, Schizophrenia

flat face PEDIATRICS Any facial dysmorphia characterized by attenuated malar prominences and broadened facies ETIOLOGY Achondroplasia, Apert syndrome, arteriohepatic dysplasia, camptomelic dysplasia, chondrodysplasia punctata–Conradi-Hünermann type, Down syndrome, Escobar syndrome, Kniest dysplasia, Larsen syndrome, lethal multiple pterygium syndrome, Marshall syndrome, partial 10q syndrome, rhizomelic chondrodysplasia, Stickler syndrome, trisomy 20p syndrome, XXXXX syndrome, XXXXY syndrome, Zellweger syndrome. Cf Dishface deformity, Fat face, Moon face

flat foot Pes planus ORTHOPEDICS A common complaint, which affects many age groups; true FF are uncommon; often the parent will perceive flattening of the foot when a child first ambulates; laxity of the ligaments may result in collapse of the foot with valgus on the hindfoot, and eversion or pronation of the forefoot; a valgus deformity of > 10% requires therapy; often a shoe will suffice as therapy

flat line NEUROLOGY A popular term for a complete lack of cerebral activity as measured by EEG, a finding equated with 'brain death'. See Harvard criteria, Persistent vegetative state VOX POPULI → MEDTALK Complete asystole

flat liner A popular term for a brain dead Pt. See Flat line, Harvard criteria

flat waist sign IMAGING A loss of concavity of the left cardiac border, seen on a plain AP chest film, which corresponds to a slight anterior, right and oblique rotation of the heart, seen in left lower lobe collapse of the lung

flatulence GI DISEASE Excess passage of noxious volatiles per anus; borborygmi are associated with legumes, nonabsorbable carbohydrates–eg, fruits, vegetables, lactose, wheat, cryptococcal infection, and iron and vitamin E deficiencies; in most subjects, CH_4 production is low, but ↑ dramatically in colorectal cancer, reflecting a change in flora WORLD RECORD FARTING Bernard Clemmens, London, UK, sustained a fart for a record 2 mins, 42 secs MANAGEMENT *A fabis abstinentis*; gastric gas may respond to simethicone, intestinal gas to activated charcoal. See Legumes

flatus MEDTALK *noun* Fart

Flavimonas oryzihabitans *Pseudomonas oryzihabitans*, CDC group Ve-2 BACTERIOLOGY A bacterium of moist hospital environments–eg, respiratory equipment–not part of normal flora INFECTIONS Catheter infections, septicemia, peritonitis due to continuous ambulatory peritoneal dialysis, etc

Flaviviridae VIROLOGY A large group of small viruses that have their entire life cycle in cytoplasm, without an intermediate DNA form EXAMPLES Dengue, Omsk hemorrhagic, St Louis encephalitis, West Nile, yellow fever viruses

flavonoid Bioflavonoid NUTRITION Any biologically-active polyphenol found in fruits, especially in the pulp, vegetables, tea, red wine, which are potent antioxidants and platelet inhibitors

FLB Funny looking beat MEDTALK Slang for indeterminate or chaotic aberrancies on an EKG that are not readily recognized

flea A wingless, 1-4 mm blood-sucking member of order Siphonaptera VECTOR FOR Bubonic plague, rickettsiosis FLEAS OF INTEREST Human flea–*Pulex irritans*, oriental rat flea–*Xenopsylla cheopis*, water flea–Cocepod

flea-bitten kidney A descriptor for the petechial hemorrhages and microinfarctions seen on the renal cortical surface, typical of malignant HTN, caused by thromboses in the arcuate and interlobular arteries DIFFDX SLE, polyarteritis nodosa, leukemia, lymphoma

flecainide Tambocor® CARDIOLOGY An antiarrhythmic that fell into disfavor after the CAST trials demonstrated a 3.5-fold ↑ in arrhythmia-induced death in treated Pts. See CAST, Encainide. It also may relieve neuropathic pain, the burning, stabbing, or stinging pain that may arise from damage to nerves caused by some types of cancer or cancer treatment

flesh Skin and/or muscle. See Fish flesh, Proud flesh

flesh-eating bacteria A variant of *Streptococcus* group A, which causes toxic shock-like syndrome. See Toxic shock-like syndrome

FLEX Federal licensing exam GRADUATE EDUCATION An examination required of physicians before they can practice medicine in the US, which consists of a 3 day multiple-choice assessment of knowlege in 'basic' and 'clinical' sciences

flexible bronchoscopy PULMONOLOGY Examination of the airways using a flexible bronchoscope, often performed at the bedside of critically ill Pts who may be too unstable to move to the OR or bronchoscopy suite; FB is used to visualize distal airways; generally, conscious sedation is sufficient for the procedure. Cf Rigid bronchoscopy

flexibility exercise An exercise intended to elongate soft tissues to prepare for the rigors of sport

flexibility SPORTS MEDICINE The ability of a muscle or extremity to relax and yield to stretch and stress forces; the ROM of a joint, affected by muscles, tendons, ligaments, bones, and periarticular structures FACTORS INFLUENCING Age, sex– ♀ are generally more flexible–and prior level and type of activity

flight of ideas PSYCHIATRY A virtually continuous flow of accelerated speech in which a person abruptly changes from one to another topic, usually based on understandable associations; in extreme FOI, the speech is disorganized or incoherent. See Bipolar disorder

flight-or-fight response Fight-or-flight response, see there

flip flap PLASTIC SURGERY A popular single-stage procedure for hypospadias repair; an incision is made in the glans and penile shaft; another incision releases the prepuce, and a flip-flap of skin and soft tissue is molded to the distal urethra

flipped LDH CARDIOLOGY An inversion of the ratio of LD isoenzymes LD_1 and LD_2; LD_1 is a tetramer of 4 H–heart subunits, and is the predominant cardiac LD isoenzyme; it migrates more rapidly at pH of 8.6 than LD_5, normally the LD_1 peak is less than that of the LD_2, a ratio that is inverted–flipped in 80% of MIs within the first 48 hrs DIFFDX LD flips also occur in renal infarcts, hemolysis, hypothyroidism, and gastric CA

flipping PATIENT CARE The conversion of an 'attached' Pt–already being seen by a particular physician or group–to another doctor for future followup after being seen in an ER or other neutral health care setting, where one of the ER staff suggested that the Pt switch physicians. See Attached patient

FLK Funny-looking kid PEDIATRICS A popular descriptor for nonspecific facial dysmorphias that are typically accompanied by growth and/or mental retardation

FLM S/A Fetal lung maturity surfactant/albumin ratio assay NEONATOLOGY An assay that uses fluorescence polarization to determine the relative concentrations of surfactant and albumin, and assess the risk of respiratory distress syndrome. See L/S ratio

float HEALTH CARE STAFFING A skilled responsible person–eg, house staff officer, resident physician, or supervisory nurse, who 'floats' about an institution addressing needs, assuring continuity of care, allowing staff to take breaks, and relieve personnel. See Night float

floater FORENSIC PATHOLOGY A popular term for a body that rises due to bacterial putrefaction and gas production, often accompanied by a nauseating stench; putrefaction is more rapid in fresh, stagnant water, slower in salt water; it may not occur in very cold water OPHTHALMOLOGY Muscae volitantes Any of the proteinaceous aggregates in the vitreous humor of the eye, which correspond to degenerative debris

Flomax® Tamsulosin UROLOGY An alpha blocker used to manage BPH, which can be used with antihypertensives, to manage hyperlipidemia ADVERSE EFFECTS Headache, infection, asthenia, dizziness. See Benign prostatic hypertrophy

Flonase™ Fluticasone propionate, see there

flooding Forced exposure, implosion PSYCHIATRY A behavior therapy for phobias and other problems linked to maladaptive anxiety, in which triggers are presented in intense forms, either in imagination or in real life; the presentations are continued until the stimuli no longer produce disabling anxiety; the hope is that by 'overloading'–ie flooding the person's psyche with the dread event or object, anxiety is exhausted and the Pt learns to cope with largely irrational fears. See Aversion therapy, Behavioral therapy, Encounter group therapy, Imaging aversion therapy, Systematic desensitization

floor PATIENT CARE A ward or place in a hospital or medical center where Pts are housed, usually for non-emergent or non-urgent care

floppy infant Floppy infant syndrome NEONATOLOGY A neonate with poor muscle tone and/or response to stimulation of extremities caused by a heterogeneous group of neuromuscular and musculoskeletal disorders

FLOPPY INFANT CAUSES

BONE DISEASE Osteogenesis imperfecta, rickets

CNS Atonic diplegia, cerebellar ataxia, cerebral lipidosis, chromosome defects, kernicterus, Lowe's oculocerebronal syndrome, Prader-Willi syndrome, Zellweger cerebrohepatocellular syndrome

MUSCLE DISEASE Central core disease, glycogen storage disease type IIa–Pompe's disease, mitochondrial myopathies, muscular dystrophy, myotonic dystrophy, nemaline myopathy

NEUROMUSCULAR JUNCTION DISEASE Botulism, myasthenia gravis

PERIPHERAL NERVE DISEASE Amyotonia congenita, anterior horn cell disease, congenital sensory neuropathy, familial dysautonomia, Guillain-Barré syndrome, polyneuritis

SPINAL CORD DISEASE Poliomyelitis, spinal cord trauma and tumors, transverse myelopathy, Werdnig-Hoffmann disease

NON-NEUROMUSCULAR DISEASE Endocrinopathies, metabolic disease, vitamin deficiency

flora The bacteria, fungi, and other microbes that normally inhabit a space in the environment or in/on the body–eg intestinal flora, oral flora, etc. See Upper respiratory tract

flosequinan CARDIOLOGY A quinolone that improves heart mechanics; it was withdrawn from the market when it was linked to ↑ mortality. See FACET, REFLECT

flossing DENTAL HYGIENE The use of synthetic thread to dislodge plaque, food, and microbes from the lateral borders of teeth; flossing ↓ halitosis, caries, periodontal disease. See Alternative dentistry, Caries

flow cytometry LAB MEDICINE Analysis of biological material by detecting the light-absorbing or fluorescing properties of cells or subcellular fractions such as chromosomes that have been labeled with monoclonal antibodies raised against various antigens, tagged with fluorochrome markers and passed in a narrow stream through a laser beam; the cells can be separated with automated sorting devices by size, intensity and type of fluorescence, and DNA ploidy analyzed. See Cell sorting, Cf Image analysis, Laser scanning cytometry

flow-limiting lesion CARDIOLOGY An atheromatous lesion large enough–usually 70+% ↓ of blood flow—to cause clinical Sx–eg, pain. See Atherosclerosis

flow murmur Hemic murmur CARDIOLOGY A murmur caused by blood turbulence due to ↑ flow, which may be linked to severe anemia

flow-volume tracing PULMONOLOGY A graphic representation of flow of air through the airways, which is altered in Pts with various forms of COPD–eg, asthma, bronchitis, and emphysema–see figure. See Chronic obstructive pulmonary disease

floxuridine ONCOLOGY An antimetabolic fluoropyrimidine, a metabolite of 5-FU. See 5-Fluorouracil

flu Influenza, see there. See China flu, Hong Kong flu

fluconazole Diflucan® INFECTIOUS DISEASE An antifungal used as a first-line therapy for mucocutaneous and systemic candidiasis, cryptococcal meningitis, and prophylaxis for BM recipients ADVERSE EFFECTS Hepatotoxicity, anaphylaxis, gastric pain, rashes especially in immunocompromised Pts.

flucytosine 5-FC, Ancobon INFECTIOUS DISEASE An anti-fungal given with amphotericin B to ↓ emergence of resistant strains ADVERSE EFFECTS Liver damage. See AIDS

fludrocortisone ENDOCRINOLOGY A synthetic steroid used to replace steroids normally produced by the adrenal gland, when the gland is structurally or functionally compromised

fluffy infiltrate IMAGING A descriptor for patchy perihilar parenchymal infiltrates on a plain CXR, which correspond to alveolar lesions of advanced pulmonary sarcoidosis

fluid A liquid or gas which conforms to the shape of its container. See Body fluid, Cerebrospinal fluid, Pericardial fluid

fluid imbalance METABOLISM A relative ↑ or ↓ in intracellular or extracellular H_2O. See Hypervolemia, Hypovolemia

fluid level IMAGING An interface between fluid and air, which is always horizontal; FLs in the small intestine indicate obstruction

fluid overload Hypervolemia, plethora MEDTALK A systemic excess of fluids. Cf Volume depletion

fluid resuscitation CRITICAL CARE MEDICINE The infusion of isotonic IV fluids to a hypotensive Pt with trauma; aggressive FR may disrupt thrombi, ↑ bleeding, and ↓ survival

fluid retention Edema, see there

fluke Any of a large family of trematodes that infect man as a definitive or accidental host–eg, genuses *Heterophyes, Metagonimus, Fasciola, Opisthorchis, Paragonimus, Schistosoma, Clonorchis*

flunisolide Aerobid® An inhalant corticosteroid used to manage asthma

flunitrazepam Rohypnol PHARMACOLOGY A benzodiazepine which has acquired notoriety by its slang name, the date rape drug, see there

fluorescein angiography OPHTHALMOLOGY A technique used to diagnose chorioretinal disease, based on the enhancement of anatomic and vascular details in the retina after IV injection of fluorescein, a dye; FA is used to evaluate retinal disease–it delineates abnormal areas, and demonstrates defects–eg, choroidal neovascularization, proliferatrive diabetic retinopathy, and light toxicity, and can be used to plan laser therapy for vascular lesions of the retina. See Retinopathy

fluorescence in situ hybridization. See FISH

fluorescent treponemal antibody absorption FTA-ABS A highly sensitive–±100% and sensitive–96+% serologic test for diagnosing congenital, secondary, tertiary syphilis and neurosyphilis, which is used when the RPR screening test is positive. See Biological false positive, RPR test, VDRL test

fluoride poisoning Fluoride intoxication TOXICOLOGY An acute excess of fluoride, which may be fatal–accidental/suicidal, given its affinity for calcium; it is present in some rodenticides, insecticides, fertilizers, industrial and anesthetics CLINICAL, INHALED Cough, choking, chills, fever CLINICAL, INGESTED N&V, salivation, paresthesias, diarrhea, abdominal pain CLINICAL–CONTACT HF is similar to HCl, and causes severe skin burns; acute intoxication may be due to an excess in the water supply. See Fluorosis

fluorography Fluororadiography, photofluorography IMAGING The photography of fluoroscopic images. See Electronic fluorography

fluoroquinolone ANTIBIOTICS A quinolone with an added flourine on the parent quinolone structure. See Quinolone

fluoroscopy IMAGING An x-ray imaging technique used to evaluate moving pulmonary and cardiac structures, and help in needle localization of masses being biopsied CONS Fluoroscopy exposed Pts to more radiation than a standard film; small lesions can be overlooked, there is no permanent record. Cf Computed tomography

fluorosis Chronic fluoride poisoning TOXICOLOGY A chronic low-level intoxication that occurs where drinking water has fluoride > 2 ppm CLINICAL Weight loss, brittle bones, anemia, weakness, ill health, stiffness of joints, mottled enamel and chalky white discolored teeth with a normal resistance to caries; fluorosis is common, given flouride's availability in mouth rinses, toothpastes, misuse of fluoride treatments. See Fluoride, Fluoride poisoning, Fluoride treatment, Fluorine

fluorouracil 5-FU, Efudex® ONCOLOGY An antimetabolite that interferes with DNA synthesis, and deprives DNA of functional thymidine; 5-FU is used for bladder CA, for terminal CA and, topically, for actinic keratosis SIDE EFFECTS BM toxicity, mucosal inflammation

fluoxetine Prozac® NEUROPHARMACOLOGY A selective inhibitor of serotonin reuptake used for clinical depression and other psychiatric disorders ADVERSE EFFECTS Anxiety, nervousness, tremor, insomnia, diarrhea, nausea, anorexia, undesired weight loss, sexual dysfunction CONTRAINDICATIONS MAOI therapy. See Serotonin-selective reuptake inhibitor

fluoxymesterone ENDOCRINOLOGY A 17α-alkylated derivative of testosterone used for androgen replacement therapy PROS Oral CONS Hepatotoxicity–cholestasis, peliosis hepatis, liver tumors. See Androgen replacement therapy

flupentixol PHARMACOLOGY A dopamine receptor antagonist with antidepressant activity at low doses, and neuroleptic effects at high doses, which may be of use in treating cocaine abuse and dependence

fluphenazine Permitil®, Prolixin® PHARMACOLOGY A tranquilizer, antipsychotic, neuroleptic

flurazepam Dalmane PHARMACOLOGY A benzodiazepine sedative and hypnotic; it is rapidly metabolized to desalkylflurazepam

flush method PEDIATRICS A method for measuring bp in a restless infant, which is based on color changes in the skin

flutamide ENDOCRINOLOGY A nonsteroidal antiandrogen which, unlike steroidal antiandrogens, lacks androgenic, estrogenic, antiestrogenic, progestational, adrenocortical, and antigonadotropic activity; flutamide has been used for metastatic androgen-responsive prostate CA and, when combined with LHRH analogues, achieves a relatively complete androgen blockade; flutamide has been used to treat acne and hirsutism ADVERSE EFFECTS GI tract–diarrhea, nausea, dizziness, gynecomastia, hepatic dysfunction or toxicity. Cf Nilutamide

fluticasone propionate Flovent®, Rotadisk® ALLERGY MEDICINE An aqueous nasal spray indicated for treating seasonal and perennial allergic rhinitis in children age 4 and older. See Asthma

flutter CARDIOLOGY A family of cardiac tachyarrhythmias characterized by rapid regular atrial–250-350/min or ventricular–200/min rhythms. See Atrial flutter, Ventricular flutter. Cf Fibrillation

FLUTTER TYPES

ATRIAL FLUTTER occurs at 200-350 beats/min (with a 2:1 block, so that the ventricle fires at ± 150 beats/min); AF results from a circus pathway, occurs in atrial dilatation, primary myocardial disease, or rheumatic heart disease and responds poorly to antiarrhythmics

VENTRICULAR FLUTTER is characterized by a continuous and regular depolarization rate of greater than 200 beats/min, and demonstrates high-amplitude zigzag pattern on the EKG, without clear definition of the QRS and T waves, a pattern that may revert spontaneously to a normal sinus rhythm or progress to ventricular fibrillation

fluvastatin Lescol® CARDIOLOGY An HMG-CoA reductase inhibitor/statin that improves lipid profile of Pts with hypercholesterolemia and mixed dyslipidemia EFFECT Slowing coronary ASHD in CAD by ↓ total-C, LDL-C, ↑ HDL-C, ↓ TGs ADVERSE EVENTS Rash. See LDL-C, Hypercholesterolemia, Statin

fluvoxamine NEUROPHARMACOLOGY An antidepressant that blocks serotonin reuptake at the synapse, approved for managing OCD. See Obsessive-compulsive disorder, Serotonin-selective reuptake inhibitor

FMF 1. Familial Mediterranean fever 2. Fetal movement felt 3. Forced mid-expratory flow 4. Free molecular flow

FMG 1. Foreign medical graduate, see there 2. Frequency modulation generator

FMS Fibromyalgia syndrome, see there

FNA Fine needle aspiration, see there

foam dressing WOUND CARE A highly absorbent dressing, which allows less frequent changing of dressings and ↓ maceration of surrounding tissues INDICATIONS Heavily exudating wounds, especially after debridement or desloughing, when drainage peaks; deep cavity wounds; weeping ulcers. See Dressing

foam stability index Foam stability test, shake test OBSTETRICS A semi-quantitative bedside test for determining fetal lung maturity, based on the stability of bubbles when amniotic fluid is shaken in test tubes filled with various concentrations–42% to 58% of ethanol. See Surfactant

FOBT Fecal occult blood testing, see there. See Occult bleeding

focal epilepsy NEUROLOGY A seizure disorder arising from a specific population of neurons, often in a background of a tumor or scar from recent injury; FE may be triggered through a kindling reaction. See Seizures

foci Plural of focus

focus A center, often of a disseminated disease–ie, cancer, infection

fogging RADIOLOGY Haziness of diagnostic X-ray films due to aging of unexposed film, or leakage of radiation or light before developing a films

folate deficiency Folic acid deficiency HEMATOLOGY A condition caused by a decrease in dietary folic acid, resulting in megaloblastic anermia, GI tract complaints–eg, glossitis, stomatitis, malabsorption, infertility, neural tube defects, and possibly also psychiatric complaints and peripheral neuropathy. See Megaloblastic anemia

folded lung syndrome Shrinking pleuritis with rounded atelectasis A condition associated with asbestosis and pleural plaques RADIOLOGY See Comet tail sign PATHOLOGY Chronic fibrosing pleuritis, inflammation; asbestos fibers are absent from pleural plaques and effusions, but may be found in the subpleural lymphatic plexus

Foley catheter NEPHROLOGY A pliable urinary catheter passed through the urethra and anchored in the bladder neck by an inflatable balloon; FCs can be left in place for prolonged periods

folic acid Folate A family of water-soluble B vitamins not synthesized by mammals, which are required for normal hematopoiesis; it is used to treat megaloblastic anemia, and folate deficiency, prevent cervical and other CAs, neural tube defects; ↓ in alcoholism, malabsorption, anticonvulsants. See Megaloblastic anemia

folie à deux Induced psychotic disorder, shared psychotic disorder PSYCHIATRY A condition in which 2 closely related people share a delusional system–eg, a mother who believes a son who believes that he is Jesus Christ

follicle ANATOMY 1. A sac, outpouching, cavity. See Hair follicle 2. Lymphoid follicle. See Lollipop follicle, Target follicle REPRODUCTION MEDICINE Ovarian follicle A structure in the ovaries which contains a developing egg. See Graafian follicle

follicular lymphoma

follicular lymphoma Follicle center lymphoma A heterogeneous group of NHLs arising in follicular center cells, which comprises 50% of all NHLs in adults–US, especially in older adults, rare < age 20 and in African Americans; FLs are usually confined to lymph nodes; affected cells resemble normal lymphoid follicles, and carry a translocation between chromosomes 14 and 18 TYPES Predominantly small cleaved cells–ie, large cells comprise < 20%; > 50% large cells; mixed, small cleaved and large cells DIAGNOSIS LM, Ig gene rearrangements, cytogenetics, Southern blotting, PCR. See Lymphoma, Translocation, WHO classification

follicular phase Proliferative phase, see there

follow *verb* To maintain clinical surveillance on a Pt

follow-up *noun* The constellation of future activities–eg return visits, imaging modalities etc, by a Pt after hospitalization or therapy, intended to help in return to a desired state of health *verb* To ensure that a follow-up has occurred. See Passive followup

fomite EPIDEMIOLOGY An inanimate object–sheets, clothing, in an environment that may harbor pathogens and thus be a passive vector for infection. See Vector

Fontan's operation Fontan's procedure HEART SURGERY A technique for correcting tricuspid valve atresia, which results in hypoplastic left heart syndrome COMPLICATION Protein-losing enteropathy. See Hypoplastic left ventricle syndrome. See Computed tomography

fontanel Fontanelle, soft spot ANATOMY A membrane-covered region on the skull of a fetus/infant, which corresponds to the convergence of suture lines, where ossification has not yet occurred

food NUTRITION An ingestable substance which serves to maintain the corporeal status quo; nutrient; comestible. See Bioengineered food, Chinese restaurant, Ciguatera poisoning, Designer food, Diet, Dietary fiber, Enriched food, Fast food, Fats, Fish, Food groups, Food pyramid, Fortified food, Four food groups, Frankenfood, Functional food, Health food, Healthy food, Hot food, Junk food, Live food, Low-fat snack food, Medical food, No food, Nonstandard food, Organic food, Scombroid poisoning, Soft food, Soy food, Spicy food, Standard food, Sticky food, Succotash, Sushi, Unhealthful food, Yang food, Yin food

food allergy ALLERGY MEDICINE A condition, the incidence of which–0.3-7.5%–is obscured by controversial data and differing disease definitions; food-induced reactions of immediate-hypersensitivity type are common and include anaphylaxis, angioedema, urticaria; food-induced reactions of delayed hypersensitivity type or those mediated by antigen-antibody complex formation are uncommon and include gluten-sensitive enteropathy CLINICAL Edema and pruritus of oropharyngeal mucosa, followed by various responses in the GI tract as the offending, and ultimately offensive, content traverses the system–eg, vomiting, colic, abdominal distension, flatulence, diarrhea, less commonly, occult blood loss, malabsorption, protein-losing enteropathy, functional GI obstruction, and eosinophilic gastroenteritis DIAGNOSIS Elimination/challenge diet, rotation diet, in vivo–intradermal, multi-test, sublingual testing, and in vitro–IgE, IgG, IgG_4, RAST, cytotoxic, histamine release testing. Cf Food intolerance

food group NUTRITION A family of foods in the diet. See Balanced diet, Essential dietary component, Food pyramid, Four food groups, Mineral, Vitamin

FOOD GROUPS

CARBOHYDRATES Bread, cereal, rice, oats, pastas

CITRUS FRUITS Grapefruits, lemons, melon, oranges, papaya, strawberries, tomatoes

DAIRY PRODUCTS Cheese, milk, yoghurt

FATS Butter, margarine, fish or vegetable oil, animal fat

GREEN/YELLOW VEGETABLES Brussel sprouts, cabbage, carrots, celery, green beans, kale, spinach

HIGH PROTEIN FOODS Eggs, fish, legumes, meat, nuts, poultry

OTHER FRUITS & VEGETABLES Apples, bananas, grapes, pineapples; beets, potatoes

YELLOW VEGETABLES Carrots, corn, cauliflower

food hypersensitivity See Food allergy, food intolerance

food intolerance NUTRITION Food sensitivity An adverse reaction to specific foods, seen in ±10% of the population, which are often chronic and may cause severe illness; FI is not synonymous with food allergies, which are predictable, often severe, involve IgE and release of histamine from mast cells. Cf Food allergy

food poisoning PUBLIC HEALTH A popular term for clinical intoxication by food contaminated with various pathogens, usually understood to mean bacteria ETIOLOGY Milk & dairy products, mayonnaise, eggs, parsley, exposure to turtles CLINICAL Diarrhea, N&V, chills, ↑ temperature, headache, colic, abdominal pain, confusion, seizures MANAGEMENT Fluids, antipyretics PRECAUTIONS Religious handwashing after Pt care. Cf Food allergy, Food intolerance

food pyramid NUTRITION A schematic guideline for ratios of food types that should be consumed daily in a healthy diet; at the base of the pyramid are carbohydrates, followed by fruits and vegetables, dairy products, protein foods; at the pyramid's peak are 'discouraged' foods, to be eaten sparingly–eg, fats, oils, refined sugars. Cf Fast food, Four food groups, Exercise pyramid, Five food groups, Mediterranean pyramid

food services HOSPITAL SERVICES A 24/7 department in a hospital that provides for the nutritional needs of inpatients–eg, those needing special diets, preparing meals and transporting them to the floor and, through the cafeteria, the hospital staff and visitors. See Nutrition

foodborne *adjective* Referring to that which is carried by food, either by pathogens: viruses–HAV, bacteria–eg salmonellosis, parasites–eg anisakiasis, toxins–eg botulinum, aflatoxin B_1 or chemicals–eg organophosphates, lead. See Foodborne pathogen. Cf Fomite

foodborne parasite PUBLIC HEALTH A parasite acquired from food–eg, *Giardia lamblia, Entamoeba histolytica, Cryptosporidium parvum, Cyclospora cayetanensis, Anisakis* spp and related worms, *Diphyllobothrium* spp, *Nanophyetus* spp, *Eustrongylides* spp, *Acanthamoeba* and free-living amoebae, *Ascaris lumbricoides, Trichuris trichiura*

foodborne pathogen PUBLIC HEALTH A pathogen–especially bacteria, for which the 'vector' is itself a food. See Airline food

FOOSH MEDTALK Fall Onto Outstretched Hand, a mechanism of injury

foot ANATOMY The distal part of the lower extremity on which a person stands and uses to walk COMPONENTS Tarsal bones, metatarsal bones, phalanges, muscles, tendons, nerves, blood vessels, soft tissues. See Athlete's foot, Claw foot, Club foot, Diabetic foot, Green foot, Immersion foot, March foot, Rocker bottom foot, Skew foot, Trench foot, Tropical immersion foot

foot-drop NEUROLOGY A manifestation of peripheral neuropathy, characterized by an inability to actively dorsiflex and evert the foot at the ankle; the foot drops when the Pt lifts the foot off the ground in the swing phase of ambulation–thus requiring a high stepping gait; it is associated with paresthesiae of feet, loss of vibratory and position sense, spasticity, exaggerated tendon reflexes in legs ETIOLOGY Diabetic mononeuropathy, Charcot-Marie-Tooth syndrome, severe vitamin B_{12} deficiency, residual poliomyelitis, old injuries to lateral popliteal–LP or sciatic nerves, cauda equina tumor, massive intervertebral disc prolapse at L_5S_1 level and, formerly in India, LP nerve paralysis due to leprosy. See Wrist drop

footballer's migraine SPORTS MEDICINE A migraine evoked by abrupt jarring of the head, as occurs in a tackled football player. See Migraine

footling breech presentation NEONATOLOGY A breech presentation in which one leg flops out of the birth canal

foramen magnum ANATOMY The large opening at the lower part of the occipital bone and outlet through which the medulla and spinal cord pass from the skull to the vertebral column

foraminal *adjective* Referring to a foramen

force feedback CYBERSURGERY The physical sensation of resistance, a property being integrated into virtual reality surgical instruments, based on tissue-mechanics data. See Telemedicine, Telesurgery

forced expiratory volume FEV PULMONARY MEDICINE The maximal amount of air that can be exhaled in a period of time, usually 1–FEV_1, or less commonly, 3–FEV_3 seconds; FEV_1 is usually reduced–and thus is a major parameter measured in obstructive airways disease, a generic term that encompasses asthma and COPD. See Pulmonary function test

forced vital capacity FVC PULMONARY MEDICINE The volume of air exhaled with maximum effort and speed after a full inspiration; FVC is usually ↓ and thus is a major parameter measured in obstructive airways disease, a term that encompasses both asthma and COPD

forceps OB/GYN A 2-part surgical instrument that articulates–hinges at the center—which is placed around the neonatal head to extract an infant in an operative vaginal delivery COMPLICATIONS Subdural or cerebral hemorrhage, facial nerve injury, brachial plexus injury, mechanical ventilation. See BiCOAG bipolar forceps, Biopsy forceps, Bissinger detachable bipolar coagulation forceps, Cold cup forceps, Mosquito forceps, Mousetooth forceps. Cf Vacuum extraction

forefoot valgus ORTHOPEDICS A fixed structural defect in which the plantar aspect of the forefoot is everted on the frontal plane relative to the plantar aspect of the rearfoot; the calcaneum is vertical, the mid tarsal joints are locked and fully pronated

forefoot varus Metatarsus adductus ORTHOPEDICS A fixed frontal plane deformity seen when the forefoot plane is everted to the rearfoot–ie, the 5th metatarsal head is more dorsal than the 1st metatarsal head; FV can be compensated by abnormal rearfoot pronation

foreign body A microscopic or macroscopic object introduced into the human economy at the time of an invasive procedure–ie iatrogenic, by accident, or by intent

FOREIGN BODY TYPES

IATROGENIC, eg sutures, sponges, instruments left during surgery, metals and plastics that replace or enhance failing or non-functioning body parts, eg artificial joints, limbs and pacemakers

ACCIDENTAL/UNINTENTIONAL, eg from abrasions and open wounds in various accidents, or in gun shot wounds, which may elicit foreign body-type granuloma formation, or

INTENTIONAL, eg introduced in the context of sexual deviancy, for inflicting pleasure or pain, commonly, in the anorectum or vagina, including an array of 'jeux d'amour', eg vibrators, bottles, light bulbs, eggs, and others

foreign medical graduate A physician who graduated from a medical school outside of the US, Canada or Puerto Rico. See International medical graduate, USFMG

forensic autopsy A postmortem examination of a body performed with the intent of determining the cause and manner of a death in question; a complete FA may require evaluation of evidence attached to the body and/or found at the scene, and reconstruction of the scene itself. Cf Hospital autopsy

forensic documentation EMERGENCY MEDICINE The creation of permanent records of an event or state, that are of sufficient credibility to serve as evidence should it be so needed in court. See Chain of custody

forensic drug screening LAB MEDICINE A format for drug screening which requires specific specimen handling protocols from the time of collection to the point of analysis and final reporting. See Drug screening. Cf Therapeutic drug monitoring

forensic nursing The application of forensic aspects of health care combined with the biopsychosocial education of the registered nurse in investigation and treatment of trauma, and death of victims and perpetrators of violence, criminal activity, and traumatic accidents within the clinical or community institution. See Sexual assault nurse examiner

forensics VOX POPULI A popular term for the investigative armamentarium–eg, documents and other writings, used to link an actor with an act, usually understood to be of criminal nature including but not limited to homicide and violent death

forequarter amputation SURGERY A major surgical procedure in which the upper extremity and a variable portion of the supporting shoulder girdle is amputated, to treat either advanced malignancy–eg, malignant melanoma, or for 1° CA of soft or bony tissues–eg, chondrosarcoma or osteosarcoma. See Heroic surgery, Mutilating surgery

foreskin UROLOGY Prepuce; the fleshy mucocutaneous tissue that covers an uncircumcised penis. See Circumcision, Penis

forgoing of treatment MEDTALK Any nonintervention–eg, withholding and/or withdrawing one or more of the following: CPR, intubation, mechanical ventilation, vasopressors, supplemental O_2, PEEP, blood transfusions, diagnostic and therapeutic procedures–eg, dialysis, and surgical intervention, antibiotics, antiarrhythmics, nutrition, fluids. See Advance directive, Against medical advice, Assisted suicide, DNR, Double effect, Slow code. Cf Watchful waiting

formaldehyde $H_2C=O$, methanal OCCUPATIONAL SAFETY A highly toxic, flammable gas that is highly irritating to the respiratory and conjunctival mucosa at concentrations > 2 ppm; formaldehyde is soluble in water and forms methylene bridges between denatured proteins. See Formalin

formalin PATHOLOGY A 37% solution of formaldehyde gas in water, that reacts with the amine groups of proteins and DNA, which acts to disinfect, and when buffered, to denature–'fix' tissues for histologic examination. See Formaldehyde

forme fruste An aborted, attenuated, or atypical expression of a clinical entity or pathologic condition

formication PSYCHIATRY The tactile hallucination or illusion that insects are crawling on the body or under the skin. Cf Fornication

formulary HOSPITAL CARE A list of preferred pharmaceuticals to be used by a managed care plan's network physicians, chosen based on the drugs' efficacy, safety, and cost-effectiveness; the list varies from one organization to another; in a healthcare system, providers are expected to use the listed products. See Hospital formulary, National Formulary, Open formulary, Stepped formulary

fornication Coitus. Cf Formication

Fort Bragg fever Anicteric leptospirosis A condition caused by *Leptospira autumnalis*, which is more common in children, and characterized by an abrupt 'toxic' state, with fever, shaking chills, headache, N&V and severe myalgias–especially of the legs, lethargy, dehydration, photophobia, orbital pain, generalized lymphadenopathy, and hepatosplenomegaly. See Leptospirosis

fortification NUTRITION The addition of a required dietary component to a food that doesn't usually have it. See Fortified food, Supplementation

fortification phenomenon 'Maginot line' phenomenon NEUROLOGY An angulated figure near the point of fixation, which gradually spreads and assumes a lateral convex shape with an angulated scintillating edge, leaving a varying amount of scotomata in its wake. See Classical migraine

forward failure CARDIOLOGY A concept that refers to ↓ cardiac output due to ↓ renal blood flow, and altered glomerular filtration with retention of salt and water, causing a secondary ↑ in blood volume. Cf Backward failure

Fosamax® Alendronate, see there

foscarnet AIDS An antiviral for treating CMV colitis, hepatitis, pneumonia, retinitis; acyclovir resistant HSV, nonresponders to gancyclovir; foscarnet inhibits other herpes viruses and HIV; being studied for treatment of KS ADVERSE EFFECTS Renal damage, meatus rash, which may disappear even if foscarnet is discontinued. See AIDS. Cf ddC, ddI, Zidovudine

fosphenytoin Cerebyx® NEUROLOGY A prodrug of phenytoin, which is pharmacologically equivalent to, but causes less phlebitis, soft tissue damage. See Phenytoin, Seizure. Cf Status spilepticus

Founier's gangrene A fulminant subcutaneous bacterial infection of the genital and anorectal regions, characterized by necrosis–cellulitis, fasciitis, myositis rapid progression, severe systemic toxicity, lack of suppuration; most Pts are ♂, ± age 50 ETIOLOGY Periurethritis with urine extravastion, post-surgery, postinstrumentation, septic injection in dorsal vein of penis UNDERLYING DISEASES DM, IVDA, CA CLINICAL Systemic toxicity, urinary retention, abdominal discomfort, regional necrosis, fever, leukocytosis TREATMENT Debridement, broad-spectrum antibiotics CONTROVERSIAL THERAPIES High-dose prednisone, hyperbaric oxygen

four Bs GRADUATE EDUCATION 1. Bells, Bowels, Bladders, Big Shots! A popular term for 0500 to 0700 when organ systems wake, require service and documentation, pre-rounds of junior physicians and rounds with attending physicians and an entourage of personnel 2. Breakfast, Beer, Benadryl™, & Bed That which is sought after an all-night-shift, to restore the physical and mental self

four Fs Fat, female, flatulent, forty A mnemonic with clinical currency as factors classically associated with cholelithiasis and acute cholecystitis

four food groups Basic four foods groups NUTRITION A series of nutritional guidelines promulgated by the USDA, which divided foods into 4 groups–meat; milk/dairy; fruits and vegetables; cereals and grains; in a balanced diet, each food group would represent $\frac{1}{4}$ of the foods consumed/day. See Food groups, Food pyramid, Balanced diet. Cf Five food groups

four Hs Hypoxemia, hypoglycemia, hypovolemia, high bladder MEDTALK The 4 main causes of unexplained restlessness, agitation, or combativeness in Pts with altered mental status

four-month pill See Seasonale

4S CARDIOLOGY A clinical trial–Scandinavian Simvastatin Survival Study–of the effect of simvastatin therapy on the M&M in Pts with hypercholesterolemia and CAD. See Lipid-lowering therapy, Simvastatin. Cf MAAS, PLAC I, PLAC II, WOSCOPS A randomized, double blind, placebo controlled, multicenter study

fourth generation cephalosporin Cefepime, Maxipime® INFECTIOUS DISEASE A cephalosporin active against a broader range of bacteria than 3rd generation cephalosporins. See Cephalosporin

fourth heart sound Atrial gallop CARDIOLOGY A heart sound which occurs late in the presystolic filling phase of the cardiac cycle and is caused by forceful atrial ejection; the 4th HS is best heard over the apex with the Pt in supine or left semilateral position. Cf Third heart sound

fourth therapy ONCOLOGY A popular term for the next generation of CA therapies BACKGROUND CA is traditionally treated with one plus of the 3 tumor-ablative modalities, ie surgery, RT, chemotherapy; biological response modifiers–IFNs, interleukins, monoclonal antibodies, CSFs, TNF–the '4th therapy', are based on the immune system–eg, immunomodulation, antiproliferative, tumorilytic activities. See Biological response modifiers

fourth ventricle syndrome NEUROLOGY A constellation of symptoms that arise from expansile lesions–neoplastic or inflammatory or infarction in the floor of the 4th ventricle, which involve cranial nerves V-VII

FPG Fasting plasma glucose, see there

fraction A part of a whole. See Ejection fraction, Plasma protein fraction, Purified protein fraction, S phase fraction, Volume fraction

fractionation RADIATION ONCOLOGY The parceling of a dose of radiation over time. See Accelerated fractionation, Hyperfractionation, Radiation therpy

fracture A breaking of bone or other hard tissue. See Avulsion fracture, Axial-load teardrop fracture, Basal skull fracture, Bennett fracture, Blow-out fracture, Boxer's fracture, Bucket handle fracture, Bumper fracture, Cavalry fracture, Chauffeur's fracture, Chip fracture, Chisel fracture, Clay shoveler's fracture, Closed fracture, Colles' fracture, Comminuted fracture, Complete fracture, Compound fracture, Compression fracture, Dashboard fracture, DeQuervain's fracture, Direct fracture, Dislocation fracture, Don Juan fracture, Double fracture, Dupuytren fracture, Essex-Lopresti fracture, Fatigue fracture, Greenstick fracture, Hangman's fracture, Hip fracture, Hairline fracture, Impacted fracture, Incomplete fracture, Insufficiency fracture, Jefferson fracture, Jones fracture, LeFort fracture, Linear fracture, Linear skull fracture, Maisonneuve ankle fracture, March fracture, Microfracture fracture, Monteggia fracture, Nightstick fracture, Nonunion fracture, Oblique fracture, Odontoid fracture, Open fracture, Open book fracture, Pathologic fracture, Piedmont fracture, Ping pong fracture, Ring fracture, Rolando fracture, Skull fracture, Stellate fracture, Stress fracture, Teardrop fracture, Telescoping fracture, Torus fracture, Tripod fracture, Wagon wheel fracture, Wagstaffe fracture, Wedge fracture

fracture threshold BONE PHYSIOLOGY A theoretical cancellous bone density below which osteoporosis-related fractures occur. See Osteoporosis IMAGING A benchmark–below which fractures are more common–determined by bone densitometry

fragile X syndrome MOLECULAR ONCOLOGY A condition that is the most common–1:1500 cause of inherited mental deficiency in ♂–30% of ♀ carriers are also mentally deficient CLINICAL Moderate mental retardation, neuropsychiatric disorders–hypotonic or hyperactive state, autism, large forehead, macroorchidism, enlarged chin, jaw, ears. Cf CAG repeat disease

fragmentation SLEEP DISORDERS The interruption of any stage of sleep due to appearance of another stage or waking, leading to disrupted NREMS-REMS cycles; SF usually refers to an interruption of REMS by movement arousals or stage 2 activity; SF connotes repetitive interruptions of sleep by arousals and awakenings

fragmented speech NEUROLOGY Speech characterized by consecutive unlinked phrases. See Flight of ideas

frailty VOX POPULI A state of delicacy or weakness which, which encompasses age-related fragility, in particular osteoporosis. See FICSIT, Osteoporosis

Frank-Starling law Frank-Starling mechanism A law that states that the energy liberated with each cardiac contraction is a function of the length of the muscle fibers in the ventricular wall; as preload ↑, so does end-diastolic pressure, which ↑ force of ventricular contraction; in CHF, the FSL results in marked inefficiency in cardiac pumping

Frankfort line Frankfort horizontal line, Reid's base line RECONSTRUCTIVE SURGERY An imaginary line that projects from the median line of the occipital bone and upper rim of the external auditory canal–the auricular point, to the lower rim of the orbit–the infraorbital point; the FHL divides the head into upper and lower halves from gnathion to trichion, and is used for craniometric studies–as it approximates the base of the skull and may be used as a point of reference in otoplasty. See LeFort fracture

fraternal twins Dizygotic twins Twins resulting from 2 separate fertilized eggs liberated simultaneously from the ovaries that develop in separate or partially fused chorionic sacs; 70-80% of twins are dizygotic. Cf Identical twins

fraud MANAGED CARE The intentional misrepresentation or deception resulting in payment(s) for services not rendered or payment above that normally paid. See Medicaid fraud, Medicare fraud PATIENT CARE Dishonest practice; breach of confidence. See AIDS fraud, Health fraud. Cf Misrepresentation

freak out SUBSTANCE ABUSE A verb, popularized in the US in the '60s–to experience nightmarish hallucinations including by LSD or a similar drug. See 'Bad trip', Flashback

freckles Ephilides Brown macules, often exacerbated on sun-exposed zones of the skin surface, which disappear during the winter, and most commonly affecting the fair-skinned, especially of Celtic stock. See Macule. Cf Nevus

free association PSYCHOANALYSIS Spontaneous, uncensored verbalization by a Pt of whatever comes to mind. See Word association test

free base cocaine SUBSTANCE ABUSE An aqueous form of cocaine that allows it to be injected IV or smoked, producing a more intense 'high'–and more intense addiction. See Cocaine, Crack cocaine

free drug Unbound drug PHARMACOLOGY An active drug or other compound that is not bound to a carrier protein–eg, albumin or alpha-1-acid glycoprotein

free flap RECONSTRUCTIVE SURGERY An autologous tissue flap with anastomosed blood vessels

free-floating anxiety PSYCHIATRY Severe, generalized, persistent anxiety not specifically ascribed to a particular object or event and often a precursor of panic

free PSA UROLOGY PSA in the circulation that is unbound to its usual carrier molecules, the protease inhibitors; free PSA is used to distinguish prostate CA from BPH, etc; free PSA levels are measured in Pts with a total PSA level between 4 and 10 ng/mL and nonsuspicious digital rectal examinations; ↑ free PSA indicates a ↓ risk of CA. See Prostate-specific antigen

free radical PHYSIOLOGY Any of a family of highly reactive molecules containing an unpaired electron in the outer orbital–eg, the excited variants of O_2; FRs cause random damage to structural proteins, enzymes, macromolecules, DNA, playing major roles in inflammation, hyperoxidation, postischemic tissue damage, infarction, possibly also CA and tissue damage in transplants. See Antioxidants, Free radical scavenger, Free radical theory VOX POPULI Freed radical A paroled political polemicist

free radical scavenger Free radical inactivator Any compound that reacts with free radicals in a biological system, ↓ free radical-induced damage, and protects against the indirect effects of free radicals produced by ionizing radiation, etc EXAMPLES Ceruloplasmin, cysteine, glutathione, SOD, transferrin, vitamins A, C, E, D-penicillamine. See Antioxidant, Antioxidant therapy, Free radical, Free radical theory, Superoxide dismutase

free-standing MANAGED CARE *adjective* Referring to a physically and, often, financially discrete entity–eg, a surgical center, that is separate from, but may be affiliated with, a hospital; FS facilities may provide ambulatory surgery, emergency or primary care. See Walk-in clinic

free thyroxine index FT₄I, T7 assay, T12 assay ENDOCRINOLOGY A lab value for T₃ uptake combined with total T₄; FTI is a clinical parameter measured by RIA, used to evaluate thyroid function, calculated by T₄ x %T₃RU–resin uptake; the FTI is ↑ in hyperthyroidism and factitious hyperthyroidism and ↓ in hypothyroidism; it is falsely ↑ in heparin therapy and falsely ↓ in phenytoin and valproic acid therapy, and in the euthyroid sick syndrome. See Thyroxine

freezing spell NEUROLOGY Episodic immobility–'off' component of the on-off phenomenon–seen in advanced Parkinson's disease COMPONENTS Hesitation on gait initiation–start-hesitation, upon stopping–stop hesitation or when walking in crowded places. See On-off phenomenon

French sizes EMERGENCY MEDICINE The system for sizes of tubes used in endotracheal intubation

frenectomy Excision of the frenulum of the tongue, see there

frequency STATISTICS The number of times that a particular periodic event occurs in a unit time. See Collision frequency, Cumulative frequency, Larmor frequency, Observed frequency, Order frequency, Pulse repetition, Recombination frequency, Relative frequency

frequent flyer HOSPITAL PRACTICE A popular term for a Pt who is regularly admitted to a particular ER or health care facility, for various reasons

fresh frozen plasma THERAPEUTICS A blood component, separated from whole blood which provides 80-120 mg of fibrinogen and ≥ 80 units each of factors VIII and XIII INDICATIONS PT > 16 secs, acute liver decompensation, massive hemorrhage, transfusions where > 10 units of packed RBCs cause coagulation factor depletion; FFP is administered pre-operatively in Pts with coagulopathies–eg hemophilia B, in therapeutic apheresis, DIC, ITP, TTP, HDN

fresh water drowning A type of drowning in which hypo-osmolar water compromises the surface tension of alveolar surfactant, causing an imbalance in the ventilation-perfusion–V/Q ratio, with a collapse of some alveoli, and both true–absolute, and relative–intrapulmonary shunting. See Drowning. Cf Salt water drowning

friction rub CARDIOLOGY A scratchy triphasic–occasionally, biphasic or monophasic sound extending over the entire precordium, best heard along the left midsternum with the Pt leaning forward, which changes in quality with inspiration and positional changes; FR is pathognomonic for pericarditis and must be differentiated from to-and-fro or machinery-like murmurs and 'crunching' sounds of emphysema; the 3 phases of the triphasic rub are due to pericardial-epicardial contact during ventricular systole, diastole, and atrial systole

FRIEDA GRADUATE EDUCATION An annual database produced by the AMA, which supplants the 'Green Book' as *the* source of information on residency programs. See Green Book, Residency

Friedreich's ataxia NEUROLOGY An AR neurodegenerative disorder, which is the most common–prevalence in European stock is ±1:50,000—cause of hereditary ataxia CLINICAL Ataxia of all limbs, cerebellar dysarthria, absent reflexes in legs, sensory loss, pyramidal signs; onset up to age 25; skeletal deformities and cardiomyopathy are found in most Pts, as are impaired glucose tolerance and DM. See Ataxia

Friend sign CARDIOVASCULAR SURGERY A sign consisting of a pulsatile rising and falling of the abdomen in Pts with medium-to-large abdominal aortic aneurysms. See Abdominal aortic aneurysm

fringe *adjective* Referring to health care methods and philosophies that lie between therapies that may represent viable alternatives to mainstream medical treatment–eg acupuncture, chiropractic, homeopathy and others, and those that are purely speculative, questionable, or frankly unethical–eg laetrile, psychic surgery. Cf Alternative, Integrative, Mainstream Note: In the present work, the author uses the terms fringe medicine, fringe nutrition, fringe oncology, fringe pharmacology, and others to refer to this 'gray area' of alternative health

frog leg position PEDIATRICS A descriptor for a position that may occur: 1. As an incorrect sleeping position in infants with an 'out-toeing' deformity of the leg, which may evolve into a Charlie Chaplin-like gait, prevented by sewing together the legs of the child's pajamas; 2. In infants with fulminant scurvy, where tenderness and irritability cause the children to assume the least painful position, resulting in a pseudoparalysis with the hips and knees semiflexed and the feet rotated externally, often accompanied by edematous swelling of the femoral shafts and occasionally, palpable subperiosteal hemorrhage; 3. In children with CHF

front load THERAPEUTICS *verb* To 'accelerate' administration of a particular agent, which either requires immediate high levels to maximize its clinical effects–eg, tPA, or to rapidly achieve drug equilibrium

frontal bossing Bossing, see there

frontotemporal dementia NEUROLOGY A form of dementia that affects speech and personality, while stimulating visual perception; FD has been linked to chromosome 17. See FTDP-17, Prion disease

frost Uremic frost, see there

frostbite Tissue damage or destruction induced by temperatures below 0°C, which is divided into superficial–frostnip and deep forms; in deep frostbite, subcutaneous tissue, muscle, and bone are involved CLINICAL Numbness, prickling, itching, if severe paresthesia, stiffness, bullae formation, necrosis, gangrene TREATMENT-IMMEDIATE Rewarm in water 40-42°C/104°F-107.6°F, *never warmer* TREATMENT-POST EMERGENT Debride blister, topical aloe vera gel, tetanus prophylaxis, analgesia, NSAIDs, penicillin, hydrotherapy, physical therapy

frozen pelvis SURGICAL ONCOLOGY A term for significant involvement of the pelvic floor by malignancy, usually carcinoma, in which there is massive extension of tumor from the bladder, ♀ genital tract, and sigmoid colon; adequate resection of a FP is virtually impossible; chemotherapy and RT are palliative at best CLINICAL FP Sx are related to compression and stenosis of pelvic floor organs, formation of fistulas which may cause sepsis–a common terminal event, difficulty in defecation, dyspareunia. See All-American operation

frozen section Cryostat section, 'frozen,' quick section SURGICAL PATHOLOGY A rapid diagnostic procedure performed on tissue obtained intra-operatively, which is frozen in a synthetic material–eg, OCT–Miles Labs, sectioned with a cryostat, stained with H&E and viewed with a LM, allowing a rapid diagnosis of a pathologic tissue; a FS provides a surgeon with information necessary to guide therapy, and to determine the extent of further surgery at the time of surgical procedure

FROZEN SECTION—INDICATIONS

1. Differentiate between benign and malignant
2. Determine type of malignancy–eg, lymphoma v. carcinoma
3. Evaluate tissue margins for involvement by malignancy, eg basal cell carcinomas
4. Determine adequacy of tissue for further studies after the Pt is closed
5. Determine type of tissue, eg differentiate lymphoid tissue from parathyroid gland

frozen shoulder ORTHOPEDICS A shoulder with incapacitating pain due to bursitis and inflammation, due to 1° or 2° osteoarthritis, rheumatoid arthritis, cuff tear arthropathy and the clinically similar 'Milwaukee shoulder', postmastectomy, avascular necrosis, and calcific tendonitis, and tearing of the rotator cuff muscles

fructosamine DIABETOLOGY Fructose with an bound amine group, the serum levels of which reflect the degree of glycation of serum proteins; fructosamine levels reflect glycemic control over a period of 2-3 wks. See Glycosylated hemoglobin

fructose intolerance syndrome Fructose intolerance The AR deficiency of fructose-1-phosphate aldolase; Pts are asymptomatic until exposed to fructose–or sucrose CLINICAL Hepatomegaly, jaundice, edema; with time, cirrhosis TREATMENT Dietary, avoid foods with fructose

fruit & vegetable PREVENTIVE MEDICINE A unit of food that is widely regarded as healthy, given its content of potassium, antioxidants, carotenoids, ellagic acid, flavonoids, and other as-yet unidentified substances. See Food pyramid

FSH 1. Facioscapulohumeral 2. Follicle-stimulating hormone A 30 kD glycoprotein gonadotropin secreted by the anterior pituitary which binds membrane receptors, activates adenyl cyclase and ↑ intracellular cAMP; FSH stimulates spermatogenesis, ovulation and estrogen secretion in the ovary; FSH is ↑ in primary gonadal failure, testicular or ovarian agenesis, Klinefelter syndrome, FSH-secreting tumors; it is ↓ in anorexia nervosa, hypogonadotropic hypogonadism, panhypopituitarism, malignancy of the ovaries, testes, adrenal glands

FSP Fibrin/fibrinogen split products. See Fibrin degradation products

FT 1. Family therapy, see there 2. Fatigue time 3. Ferritin 4. Free thyroxine

FTA-ABS Fluorescent treponemal antibody-absorption, see there

FTI Free thyroxine index, see there

FTT Failure to thrive, see there

5-FU Fluorouracil GYNECOLOGY A topical cream applied to the cervix and vagina for managing cervical CA ADVERSE EFFECTS Burning, inflammation

fugue state NEUROLOGY A state in which the Pt denies memory of activities for a period of hrs to wks; to external appearances these activities were either completely normal or the Pt disappeared and traveled extensively; most are functional; short fugues rarely occur in temporal lobe epilepsy. Cf Jamais vu PSYCHIATRY A state of personality dissociation characterized by amnesia and possibly physical flight from the customary environment or field of conflict. See Dissociation, Multiple personality, Schizophrenia

fulcrum test ORTHOPEDICS A 'provocative' joint laxity test used to diagnose shoulder instability. See Provocative test, Shoulder instability. Cf Laxity test

fulguration SURGERY Destroying tissue with an electric current; cauterization. See Electrocauterization

full axillary dissection SURGICAL ONCOLOGY A procedure for treating breast CA, consisting of en bloc resection of the axillary content–ie, level I, II, and III lymph nodes, from the latissimus dorsi laterally, to the subclavius–Halsted's ligament medially, with clearing of axillary vein superiorly with either preservation or conservation of the pectoralis minor. Cf Low axillary dissection

full-body CT scan IMAGING A self-explanatory procedure which, in absence of valid clinical criteria, is probably useless. See X-ray

full range of motion A strutural or functional compromise in the ability of a joint to move around an axis. See Range of motion

full-service laboratory LAB MEDICINE A lab with ≥ 3 of the following areas: chemistry, hematology, microbiology, blood banking, anatomic pathology–surgical pathology and/or cytopathology. See Specialized laboratory

fulminant MEDTALK *adjective* Abrupt, intense, critical

fulminant hepatic failure GI DISEASE An acute and/or severe decompensation of hepatic function, defined as '…*onset of hepatic encephalopathy within 2 months after diagnosis of liver disease*', which may be linked to brain edema

fume OCCUPATIONAL MEDICINE A solid suspension resulting from condensation of the products of combustion. See Inhalant VOX POPULI *verb* To be in the midst of a mental mini-meltdown

function VOX POPULI The activity of a system. See Cognitive function, Core function, Hotel function, Sexual function

functional disorder A condition believed to be psychogenic in nature–eg, chronic abdominal pain, low back pain, chronic fatigue, psychogenic chest pain, chronic headaches

functional endoscopic sinus surgery Functional endonasal endoscopic sinus surgery ENT A procedure that removes diseased nasal cavity and paranasal sinus tissue and restores mucociliary clearance Applications Chronic and/or recurrent sinusitis in Pts who fail conventional medical therapy, suppurative sinusitis, severe asthma

functional exercise Rehab medicine The '…*restoration of strength and agility through dynamic exercise*'; in FE and functional rehab, several muscle groups are exercised simultaneously, the set of exercises is usually sport specific, and there is little need for special equipment–eg, paddleball, skip rope, basketball, badminton–for knee rehabilitation, cycling, rowing, swimming

functional illiteracy Social medicine The inability to read and write enough to effectively function in an office or business. Cf Complete illiteracy

functional instability Orthopedics A joint instability that exists when neuromuscular deficits lead to repeated episodes of instability, which may occur with/without mechanical instability; FI is associated with impairments in postural control, joint position sense, joint conduction velocity, strength. See Instability. Cf Mechanical instability

functional MRI Fast MRI Imaging A brain imaging technique that measures ↑ blood flow–BF which, like PET, relies on changes in BF and oxygenation due to brain activity; aerobic metabolism in some neurons creates a local ↑ in deoxyHb, which triggers ↑ BF, maintaining O_2 in the immediate vicinity, followed by regional ↑ in BF Neuroscience MRI fitted with special hardware to accelerate imaging speeds and advanced software that turns the static images into movies; FM allows real-time evaluation of cerebral activity as the mind thinks, listens, dreams, and imagines. See MRI ↑

functional reach Clinical anatomy The distance between the relaxed shoulder pivot point and the tip of the outstretched fingers

functional residual capacity Lung medicine The volume of air that remains in the lungs after a normal–ie not forced–expiration Components Expiratory reserve volume, residual volume See Lung volumes. Cf Total lung capacity, Vital capacity

functional test Clinical medicine A test in which the activity of a particular organ or gland is measured based on the activity evoked by the hormone, factor, or other product

fundoplication Surgery The operative revision of the esophagogastric junction to prevent gastroesophageal reflux, GERD and recurring lung aspiration of gastric content. See Gastroesophageal reflux disease, Nissen procedure

fundoscopy Fundoscopic examination Physical exam The examination of the ocular fundus to assess retinal manifestations of hypertension, DM, retinal detachment, melanoma, and sundry ilk

fundus Anatomy The part of a hollow organ that is farthest away from the organ's opening; the bladder, gallbladder, stomach, uterus, eye, and middle ear cavity all have a fundus Ophthalmology The interior of the eyeball–ie, retina, optic disc, macula, which is seen during a fundoscopic eye examination

fungal *adjective* Pertaining to fungi, mycotic

fungal body infection Tinea corporis, see there

fungal foot infection Tinea pedis, see there

fungal inguinal infection Tinea cruris, see there

fungal nail infection Onychomycosis, see there

fungal scalp infection Tinea capitis, see there

fungoides Bacteria that mimic true fungi, both morphologically and clinically–eg, *Actinomadura, Actinomyces, Nocardia, Streptomyces* spp

fungus ball Aspergilloma, mycetoma A tumor-like mass of fungi, classically from the saprobic form of *Aspergillus* spp, which colonizes a preexisting pulmonary cavity Radiology A solid rounded mass within a cavity, rimmed by an 'air density' crescent; surgical excision of large lesions carries a 5-10% intraoperative mortality rate and a 25-35% complication rate, but sans surgery, potentially fatal hemoptysis occurs in 50-83% DiffDx-radiology Abcesses, ankylosing spondylitis, congenital lung cysts, cystic bronchiectasis, emphysematous bullae, cavitary histoplasmosis, neoplasia, radiation fibrosis, sarcoidosis, and AIDS

funnel chest Pectus excavatum A congenital, often isolated, skeletal anomaly associated with upper airway obstruction or segmental bronchomalacia; surgical correction is not clearly beneficial; some cases may resolve spontaneously

funny bone Vox populi A popular term for the median epicondyle of the humerus; the ulnar nerve crosses over the region; blows thereto cause a painful/'funny' tingling

funny looking kid see FLK

FUO Fever of unknown origin, see there

furosemide A loop diuretic. See Diuretic, Loop diuretic

furuncle Boil Infectious disease A pus-laden staphylococcal skin infection characterized by reddening, pain, swelling and central necrosis, which may require antibiotics and excision

fusiform *adjective* Elongated, spindled, string-like

fusion Medtalk The joining of ≥ 2 distinct entities. See Binaural fusion Orthopedics Operative joining of 2 bones or a single bone with a pseudarthrosis. See Pseudarthrosis, Spinal fusion Psychoanalysis The joining of instincts and objects Sports medicine The trendy combination of 2 or more types of exercise–eg, martial arts, swimming and free weights, theoretically to improve fitness. See Exercise

fusion beat Cardiology The superimposing of an ectopic beat on a paced natural sinoatrial node beat; an FB is usually narrower than a paced impulse and has various morphologies that reflect the relative contributions of the impulse to the venticular depolarization. See Pseudofusion beat

futile medical intervention Futility care Terminal care A medical intervention that does not lead to improvement in the Pt's prognosis, comfort, well-being, or general state of health. See Medical futility, Terminal illness

FVC Forced vital capacity, see there

Fx Fracture, see there

G Symbol for: 1. Conductance 2. Ganglion 3. Gastrin 4. Giga-, SI for 10⁹ 5. Globulin 6. Glucose 7. Glycine 8. Glycogen 9. Gravida 10. Guanidine 11. Guanine 12. Guanosine 13. Gynecology

g Symbol for: 1. Gas 2. Genome, see there 3. Gram, see there 4. Gravity, see there

G-CSF Granulocyte colony-stimulating factor MOLECULAR THERAPEUTICS A biological response modifier, the recombinant DNA form of which is filgrastim–Neupogen® ACTIVITY G-CSF stimulates granulocyte production in BM suppressed by chemotherapy, RT, serving to ↓ infections in Pts with malignancy; G-CSF can induce terminal maturation of myeloid leukemia cell lines and suppress self-renewal, and is thus useful in Pts with AIDS or leukemic relapse after allogeneic BMT ADVERSE EFFECTS Nausea, bone pain, rash. See Biological response modifiers, Colony-stimulating factors. Cf GM-CSF

G protein disease Any condition that blocks G protein activity at the cell membrane–eg, pertussis and cholera, preventing signal transduction.

G6PD deficiency Glucose-6-phosphate dehydrogenase deficiency, see there

GABA γ-aminobutyric acid An amino acid that is the major inhibitory neurotransmitter in the vertebrate gray matter; it is excitatory in the hippocampus; GABAergic neurons are classified according to the direction of the cell processes, and signal transmitted/received. See Disinhibition, Stiff man syndrome

gabapentin Neurontin® NEUROLOGY An antiepileptic structural analogue of GABA, which binds to a poorly characterized brain receptor INDICATIONS Drug-refractory complex partial seizures, generalized seizures, with/without 2° generalization ADVERSE EFFECTS Vertigo, somnolence, headaches; it has little interaction with other agents. See Convulsion

gabexate GI DISEASE A small synthetic non-antigenic protease inhibitor used to prevent pancreatic damage related to ERCP; IV gabexate is associated with ↓ liver enzymes, pancreatic pain, acute pancreatitis. See ERCP

GAD General(ized) anxiety disorder, see there

gadolinium A rare element–AW, 157.25 used as a contrast medium for MRI of the CNS, to enhance visualization of neoplasms, parenchymal, and congenital lesions, infections, and post-operative 'failed back' syndromes

gag reflex NEUROLOGY A reflex evoked by direct contact with the posterior oral cavity at the fauces; an intact GR indicates that the cranial nerves 10 and 11 are functioning properly

gag rule MEDICAL ETHICS A rule arising in the US Supreme Court's decision in *Rust* v. *Sullivan* which prohibits physicians employed by the Title X projects from fully counseling ♀ who are unintentionally pregnant, thereby preventing physicians from offering nondirective advice on prenatal, infant, foster care, adoption, and pregnancy termination. See *Rust* v. *Sullivan*, Title X projects

gain VOX POPULI An increase in a thing. See Primary gain, Secondary gain

gain-of-function mutation GENETICS Any mutation that results in a new activity–eg, activation of a proto-oncogene, Cf Loss-of-function mutation

gait Manner and style of walking. See Ataxic gait, Athetotic gait, Broad-based gait, Cerebellar gait, Charlie Chaplin gait, Choreoathetotic gait, Dromedory gait, Dropfoot gait, Drunken–staggering gait, Dystonic gait, Festinating gait, Gait of sensory ataxia, Hemiplegic gait, Hysterical gait, Myopathic gait, Penguin gait Senile gait, Shuffling gait, Spastic–paraplegic gait; Steppage–equine gait, Toppling gait, Trendelenburg Waddling–myopathic gait

gait analysis REHAB MEDICINE Evaluation of the gait of Pts with a neurologic or orthopedic condition affecting the motor control system–eg, brain injury, spinal cord injury, cerebral palsy, stroke, multiple sclerosis, musculoskeletal actuator systems, post polio, peripheral nerve injuries, orthopedic trauma/injuries or joint degeneration and amputation; an evaluation process that attempts to understand various aspects of gait–eg, gait mechanics, effects of disease, and how to manage gait disorders. See Gait

gait assistive device REHAB MEDICINE Any of a number of tools that facilitate ambulation–eg, crutches, cane, walker

gait control NEUROLOGY The electromechanics of walking, a '…*dazzlingly complex process which has an intrinsic focus on planning, execution and adaptation of movements by the CNS*' See Gait

galactography IMAGING A technique in which contrast medium is injected into spontaneously discharging ducts to delineate intraluminal abnormalities; the technique is of uncertain value

galactorrhea MEDTALK The spontaneous or excessive discharge of milk from breast, usually understood to be outside of the context of nursing a baby. See Breast milk

galactosemia NEONATOLOGY An AR condition characterized by defective metabolism of galactose, due to a lack of galactosyl-1-phosphate uridyltransferase, with accumulation of galactose-1-phosphate, detected by prenatal screening CLINICAL Hepatosplenomegaly, cirrhosis, mental retardation, cataracts LAB Albuminuria, aminoaciduria, galactosuria, ↑ in galactose, galactose-1-phosphate, galactitol MANAGEMENT Eliminate galactose from diet PROGNOSIS Fatal, if undetected

galactosialidosis GM1 gangliosidosis AR condition due to a defective gene on chromosome 10, resulting in neuroaminidase and β-galactosidase deficiencies CLINICAL Neonatal onset with mental and physical retardation, seizures, visual defects, deafness, gargoyle facies, corneal clouding, and a cherry red spot of the macula. See Cherry red spots

Galeazzi's fracture Dupuytren's fracture ORTHOPEDICS A fracture of the radial shaft above the wrist accompanied by dislocation of the distal end of the ulna. See Fracture. Cf Monteggia fracture

gallbladder attack A popular term for an episode of cholecystitis-related pain. See Acute cholecystitis

gallbladder biopsy SURGERY The sampling of gallbladder tissue to determine whether a condition arising therein is benign or malignant. See Cholecystitis

gallbladder disease SURGERY A popular term for any condition associated with dysfunctional bile ducts, including cholecystitis, cholelithiasis or gallstones, and cancer

gallbladder radionuclide scan HIDA scan, see there

gallbladder series Cholecystography, see there

gallium nitrate METABOLISM A hypocalcemia-inducer used to manage calcium levels in malignant hypercalcemia due to bone metastases. See Hypercalcemia

gallium scan NUCLEAR MEDICINE A radioscintillation imaging method in which ^{67}gallium citrate–$T_{1/2}$, 25 days, is injected IV, and images of its pattern of distribution are obtained; GS is used to stage lymphoma, lung CA, hepatoma, melanoma, metastases–to bone, brain, lung, head & neck, GI and genitourinary neoplasia

Gallivardin's phenomenon CARDIOLOGY A systolic ejection murmur radiating to the neck and reappearing over the cardiac apex, mimicking mitral regurgitation. See Ejection murmur

gallop CARDIOLOGY Cardiac auscultatory phenomena cvharacterized by a tripling or quadrupling of heart sounds, likened to a horse's canter; gallops may be the first sign of cardiac disease, but are often unrecognized, misinterpreted or ignored; gallops occur in diastole, and separated by the phase in which they occur; ventricular–S3 or protodiastolic gallop follows normal 1st and 2nd heart sounds, occurs in early diastole coinciding with rapid ventricular filling, and causes high-pitched vibrations of the ventricular wall as the blood is abruptly stopped; it connotes serious heart disease or decompensation and is associated with coronary, hypertensive, rheumatic, and congenital cardiac disease; it may be normal in young adults; once diagnosed, the average ventricular 'galloper' survives 4-5 yrs; the atrial–S4 gallop occurs during presystole or atrial systole, and is typical of left ventricular hypertrophy or ischemia; if ventricular failure accompanies ventricular hypertrophy, an S3 gallop may also be heard; the S4 gallop may occur in absence of cardiac decompensation or in 1° myocardial disease, coronary artery disease, HTN, and severe valvular stenosis, accompanied by an ↑ P-Q interval; if the P-R interval is prolonged or the heart rate sufficiently rapid S3 and S4 merge resulting in a 'summation' gallop. See S$_3$ gallop, Summation gallop

gallstone GASTROENTEROLOGY A concrement in the gallbladder or the cystic duct EPIDEMIOLOGY 10% of adults have gallstones–GS; ↑ with age; ♀:♂ = 2:1; highest in Scandinavia, Chile, Native Americans; ↑ risk with childbearing, ERT, OCs, obesity, rapid weight loss TYPES Cholesterol, bilirubin, calcium salts; cholesterol GSs constitute 75% of total in Western nations; up to 80% of the volume is cholesterol; non-cholesterol GSs are either black or brown GSs CLINICAL Biliary colic, recurrent upper-quadrant pain; fatty food intolerance, while suggestive, is nonspecific; GSs may be associated with acute cholecystitis which causes severe abdominal pain, N&V, fever, leukocytosis DIAGNOSIS Ultrasonography, cholescintigraphy, cholecystography MANAGEMENT Laparoscopic cholecystectomy, percutaneous dissolution of gallstones by MTBE–methyl-*tert*-butyl ether, via a percutaneous transcutaneous catheter, shock-wave lithotripsy. See Black gallstone, Brown gallstone, Soluble fiber

gallstone disease SURGERY The constellation of Sx associated with gallstones. See Cholecystitis, Gallstone

gallstone(s) in bile duct Cholelithaisis, see there

gallstone pancreatitis Inflammation of the pancreas due to common bile duct obstruction, with acute cholecystitis

GALOP syndrome NEUROLOGY A cerebellar disorder characterized by the acronym, GALOP: Gait disorder–ataxia, Autoantibodies–IgM against central myelin antigen; Late age of Onset; Polyneuropathy, sensory > motor, symmetric MANAGEMENT Cyclophosphamide, human Ig

GALT Gut-associated lymphoid tissue The GI immune system, which is present in the mucosa and submucosa of the GI tract, but especially prominent in the oropharynx–tonsils, subjacent to the mucosa–Peyer's patches and appendix

galvanic skin response biofeedback Electrodermal skin response biofeedback A type of biofeedback therapy in which subtle changes in the autonomic system are monitored by the Pt with a device that measures changes in skin conductivity caused by a minimal sweating; GSRB may be useful in treating asthma, desensitization of phobic disorders, psychotherapy, stuttering. See Biofeedback device, Biofeedback training, Electromyograph feedback

gambling VOX POPULI An activity in which a person wagers against another person–eg, friend, acquaintance, bookmaker or 'bookie', or organization–eg, casino, horse race track, internet company engaged in said activity, either legal or illegal, on the likelihood of a particular outcome, either in a game of chance, or sports event or other activity for which the outcome is not known in advance. See Compulsive gambling

gamekeeper's thumb An avulsion fracture at the ulnar base of the proximal phalanx of the thumb, linked to the ulnar collateral ligament of the metacarpophalangeal joint; GT occurs in 'week-end warriors', ie football or skiing injuries–as in a twisted ski pole, often due to falls FEATURES Valgus instability, treated by a cast in adduction; if point instability is marked, open repair of ligament

Games SPORTS MEDICINE A sports event, usually understood to be an elite competition among athletes of multiple nations, states, colleges or other autonomous units. See Paralympics, Special Olympics, World Medical Games, X games

gamete GENETICS A mature ♀ or ♂ reproductive cell–sperm or ovum/egg with a haploid set of chromosomes–23 for humans. See Chromosome, Haploid, Macrogamete, Microgamete, Nullisomic gamete

gamete intrafallopian transfer See GIFT

gamma γ Symbol for: 1. Heavy chain of immunoglobulin G–IgG 2. Hemoglobin monomeric chain 3. Photon 4. The 3rd carbon in an aliphatic organic molecule GENETICS A value calculated by the ratio between synonymous DNA mutations, which don't result in a different amino acid being translated from a codon, and nonsynonymous mutations–which result in a different amino acid being encoded IMAGING A measure of contrast 1. Film–The slope of the density vs. exposure curve 2. Electronic display terminology–The slope of the brightness distribution curve; a large gamma indicates a steep slope and high contrast

gamma-aminobutyric acid See GABA

gamma camera NUCLEAR MEDICINE A device that evaluates the distribution of a radionuclide in the body post-injection

gamma-glutamyl transferase Gamma-glutamyl transpeptidase LAB MEDICINE An enzyme that catalyzes the transfer of a γ-glutamyl group from glutathione or γ-glutamyl peptide to another peptide or amino acid; GGT is located on the cell membrane and microsomal fractions and is involved in amino acid transport across cell membranes; GGT is highest in the liver–GGT is the best single screening assay for detecting latent or chronic liver disease–eg, CA; in hepatobiliary diseases, a 10-fold + ↑ in GGT is seen in liver cancer, hepatic metastases, and PBC; a 4-fold + ↑ in GGT is seen in chronic active hepatitis, intrahepatic cholestasis, alcoholic hepatitis, extrahepatic biliary obstruction, and inactive cirrhosis

gamma heavy chain disease Franklin's disease A disorder of older ♂, ranging from fulminant–ie, death in wks to prolonged–lasting 20 yrs; death is the norm in the 1st yr, often due to infection CLINICAL Presents as a lymphoproliferation with fever, fatigue, anemia, angioimmunoblastic lymphadenopathy, hepatosplenomegaly, uvular and palatal edema, eosinophilic infiltrates, leukopenia, autoimmune disease, TB, lymphoma LAB ↑ IgG₁; most excrete < 1 g/day of paraprotein, rarely up to 20g/day TREATMENT Cyclophosphamide, vincristine, prednisone

Gammaherpesvirinae A subfamily of viruses that includes lymphotrophic viruses: EBV, Saimirivirus, Kaposi sarcoma-associated herpesvirus, HHV-8

gamma-hydroxy-butyrate SUBSTANCE ABUSE An agent used as an anesthetic adjunct PHARMACOLOGIC EFFECTS CNS depressant TOXICITY Severe respiratory depression, seizures, N&V, amnesia, vertigo, hypnagogic effect, coma. See Club drug. Cf Ecstasy

gamma knife RADIATION ONCOLOGY An RT device that provides 201 stereotactically focused beams of γ radiation from ⁶⁰Co that destroys intracranial targets; the GK and linear accelerators can be adapted to irradiate localized regions of the brain INDICATIONS AV malformations, meningiomas, acoustic neuromas, pituitary adenomas, craniopharyngiomas, 1° and 2° tumors

gammopathy An abnormal ↑ in immunoglobulin production; monoclonal gammopathies–MGs are usually malignant EXAMPLES Myeloma, Waldenström's disease, CLL, heavy-chain disease, but may also be benign, appearing in amyloidosis and MG of undetermined significance; polyclonal gammopathies are usually benign and appear in inflammatory conditions–eg, angioimmunoblastic lymphadenopathy, cirrhosis, leishmaniasis, rheumatoid arthritis, SLE, TB. Note: Polyclonal gammopathies may occur as epiphenomena in lymphomas, Hodgkin's disease, metastatic adenoCA. See Biclonal gammopathy, Monoclonal gammopathy of undetermined significance

ganciclovir Cytovene, DHPG VIROLOGY An IV protease inhibiting antiviral used for CMV-induced retinitis, gastroenteritis, and hepatitis and other infections, eliminating CMV from blood, urine and respiratory secretions INDICATIONS Immunosuppression 2° to AIDS, BMT, chemotherapy ADVERSE EVENTS Severe neutropenia/sepsis, changed mental status, thrombocytopenia. See Cytomegalovirus. Cf Acyclovir

ganciclovir implant AIDS An ocular implant that delivers ganciclovir directly into the vitreous cavity in CMV retinitis. See Cytomegalovirus retinitis

gang bang SEXOLOGY, STDs A series of coital acts by multiple ♂ with a single ♀; the current, unverified, record is held by a ♀ named Houston who allegedly brought multiple ♂ partners to orgasm 620 times in a 10 hour period; per Cicero, *O Tempora, O Mores*

ganglion cyst Tendon cyst SURGERY A common soft tissue mass of the hand that is 1. Not a ganglion, ie it is not neural in origin and 2. Not a cyst; GCs correspond to mucoid degeneration of tendinous tissue and are often located on the wrist in middle-aged ♀, and cause carpal tunnel syndrome, see there

ganglionectomy NEUROSURGERY Excision of a sensory or autonomic ganglion–eg, dorsal root ganglionectomy, usually to manage treatment refractory pain–eg, from occipital neuralgia SURGERY Excision of a ganglion cyst, see there

ganglioneuroma NEUROLOGY A benign, embryologically differentiated tumor of the sympathetic nervous system, associated with well-differentiated neurofibromatous elements. Cf Neuroblastoma

gangrene PATHOLOGY Tissue death, often due to loss of adequate blood supply, which is most common in the distal lower extremities or internal organs–eg, the large intestine; the gangrene type is a function of the environment or host

GANGRENE TYPES

DRY GANGRENE A condition caused by chronic occlusion that slowly progresses to severe tissue atrophy and mummification, often associated with peripheral vascular disease, eg DM, atherosclerosis

GAS GANGRENE A condition most common in open or dirty wounds infected by gas-producing gram-positive anaerobes, eg *Clostridium perfringens, C histolytica, C septicum, C novyi*, and *C fallax* which release histolytic enzymes, eg collagenase, fibrinolysin, hyaluronidase, and lecithinase

WET GANGRENE A condition caused by relatively acute vascular occlusion, eg burns, freezing, crush injuries, and thromboembolism, resulting in liquefactive necrosis, causing bleb and bullae formation with violaceous discoloration

ganja SUBSTANCE ABUSE 1. A bitter plant chewed for its stimulatory effects, indigenous to eastern Africa 2. Marijuana, see there, especially from Jamaica

GANT Gastrointestinal autonomic nerve tumor. See there

Gantrisin® Sulfisoxazole, see there

gap VOX POPULI A separation between 2 or more objects or processes. See Anion gap, AC-BC gap, Developmental gap, Gender gap, Osmolar gap, Utilization gap

GAPO syndrome An AR condition characterized by growth retardation, alopecia, pseudoanodontia–teeth are present but unerupted, and optic atrophy, often in a background of parental consanguity, as well as hydrocephalus, high palate, low-set ears

garden hose appearance The appearance of tubular lumens with extensive transmural fibrosis and stenosis, which may be seen in the 1. Small intestine, usually affecting the terminal ileum in advanced Crohn's disease and 2. Esophagus in well-developed progressive systemic sclerosis

garden variety MEDTALK *adjective* Referring to lesions or diseases that are both common and/or have relatively routine clinical, radiologic, or pathological findings. Cf Meat & potatoes disease

Gardner syndrome GENETICS An AD condition characterized by multiple–100s to 1000s of premalignant polyps of the large intestine; carcinoma may develop in the teens in the Pts

gargle *noun* A liquid preparation of H_2O_2 or other substance used as an intraoral wash–eg, for sore throat, which is not intended to be ingested *verb* To tilt an opened and fluid, often mouthwash-filled mouth heavenwards, allowing air to emanate from the trachea, resulting in a gurgling sound

gargoyle face The characteristic facies of gargoylism, formally, mucopolysaccharidoses–MPS; the classic gargoyle face is seen in MPS type I-H/Hurler syndrome and MPS type IV/Morquio syndrome and characterized by thickening and coarsening of facial features due to subcutaneous deposition of MPSs, most commonly seen after the first yr of age; the head is large and dolichocephalic, with frontal bossing and prominent sagittal and metopic sutures, with mid-face hypoplasia, depressed nasal bridge, flared nares, and a prominent lower ⅓ of face, thickened facies, widely spaced teeth and attenuated dental enamel, gingival hyperplasia

garment nevus Giant melanocytic nevus DERMATOLOGY A congenital pigmented nevus covering a large area, thus also designated as 'stocking', 'cap' or 'coat sleeve' nevi; GNs may be ≥ 20 cm in greatest dimension with satellite lesions, deeply pigmented with moderate hair growth, require multiple excisions; ± 12% undergo malignant degeneration; head & neck involvement may be associated with epilepsy, mental retardation, leptomeningeal melanoma; draining lymph nodes are often pigmented, corresponding to benign nevus cell aggregates. See Nevus

gas A volatilized liquid. See Biogas, Compressed gas, Flammable gas, Greenhouse gas, Mustard, Natural gas, Oxidant gas, Phosgene gas OCCUPATIONAL MEDICINE A gas phase contaminant. See Inhalant

gas-bloat syndrome GI DISEASE The inability to vomit and/or 'degas' after gastric fundoplication for reflux esophagitis, a complication of the Nissen repair of a hiatal hernia, a procedure that corrects 96% of esophageal reflux; bloating is attributed to vagal injury. See Nissen repair

gas-producing 'syndromes' Excess gas in the GI tract is due to aerophagia or ↑ gas production by intestinal bacteria, which may be facilitated by a deficiency of pancreatic enzymes; gastric gas is accompanied by bloating, pain, eruction, and flatulence; intestinal gas is often accompanied by abdominal distension, flatulence, and hypo- or hypermotility. See Flatulence

GASA Growth adjusted sonographic age OBSTETRICS A sonographic estimation of fetal age based on 2 determinations of biparietal diameter, one at 26 wks, and one at 30-33 wks. Cf Biophysical profile

gastralgia MEDTALK Stomachache

gastrectomy SURGERY The partial or, rarely, complete surgical removal of the stomach INDICATIONS Stomach cancer COMPLICATIONS Dumping syndromes. See Dumping, Vagotomy

gastric analysis A procedure in which a tube is inserted in the stomach to obtain gastric fluid, which is then analyzed for pH, volume, electrolytes, mucin, pepsin, gastrin, intrinsic factor. See BAO, MAO

gastric aspiration CRITICAL CARE The flow of gastric content into the upper respiratory tract due to a ↓ antireflux reflex. See Aspiration pneumonia, Eight-hr rule, GERD

gastric atrophy An end-stage result of prolonged chronic gastritis, in which the gastric mucosal glands become attenuated and eventually lose their ability to secrete gastric juices; GA may be followed by stomach cancer. See *Helicobacter pylori*, Stomach cancer

gastric balloon Garren's balloon A doughnut-shaped inflatable polyurethane cylinder designed to ↓ the available stomach volume, used in morbid obesity to ↓ gnawing hunger; when placed for prolonged periods, it induces hyperplasia of the G or gastrin-producing cells or pressure ulcers, See Diet, Ileal bypass operation, Morbid obesity

gastric bypass See Stomach stapling

gastric cancer Stomach cancer, see there

gastric 'cannonball' RADIOLOGY Any smoothly contoured, large and non-ulcerated filling defect, seen in radiocontrast studies of the stomach ETIOLOGY Metastatic hepatoma, but also seen with hematomas, multiple submucosal leiomyomas, lymphomas, neurofibromas, metastatic intraperitoneal CA, and other lesions that deform the gastric mucosa. Cf 'Golfball' metastases

gastric inhibitory polypeptide GIP, see there

gastric intubation Gavage, oral intubation Insertion of a nasogastric tube into the stomach to administer or withdraw substances from the stomach

gastric juice A fluid containing water, electrolytes, HCl, mucin, pepsin, gastrin and intrinsic factor–necessary to absorb vitamin B_{12}. See Gastric analysis

gastric lavage Gastric washing INTERNAL MEDICINE A procedure in which a nasogastric tube is passed into the stomach, and fluid obtained

gastric lymphoma ONCOLOGY A diffuse lymphoma composed of monotonous mature or atypical lymphocytes, or of large lymphocytes ETIOLOGY GL, especially MALT lymphoma, is linked to *H pylori* infection; other causes of GL are speculative CLINICAL Epigastric pain, N&V, tarry stool, ↓ appetite, abdominal mass PROGNOSIS 5/10-yr survivals–57%/46%, respectively. NOTE Histologic type, clinical stage, mode of therapy have little prognostic value. See Lymphoma, MALT lymphoma

gastric outlet obstruction GASTROENTEROLOGY A manifestation of gastric dysmotility; the rate of gastric emptying is controlled by duodenal receptors for fat or acid ETIOLOGY Ulcers, benign or malignant tumors, inflammation–cholecystitis, acute pancreatitis or Crohn's disease, caustic

strictures, pyloric stenosis CLINICAL Vomiting–often daily, intermittent epigastric pain; GOO may occur in neonates associated with antral hyperplasia, which has been attributed to therapy with PGE_1, used to maintain patency of the ductus arteriosus in the face of congenital heart disease

gastric stapling Stomach stapling, see there

gastric stump GASTROENTEROLOGY That part of the stomach that remains after partial resection

gastric ulcer A hole in gastric mucosa due to gastric secretions, related to H pylori in the mucosa, NSAIDs, cigarette smoking etc; the pain of a GU may not correlate with the severity of ulceration DIAGNOSIS Barium x-ray, endoscopy COMPLICATIONS Bleeding, perforation, gastric obstruction

gastrinoma An often multicentric tumor of the pancreatic islet delta cells that arises spontaneously or is associated with MEN-1; rarely the tumors may be very small–eg, 2-6 mm and located in the duodenum and/or are associated with Zollinger-Ellison syndrome; it is usually accompanied by gastrin hyperproduction MANAGEMENT Resection

gastritis MEDTALK Inflammation of the stomach mucosa. See Atrophic gastritis, Giant hypertrophic gastritis

gastroduodenal anastomosis SURGERY The end-to-end joining of the upper stomach to the duodenum after resecting the lower stomach

gastroenteritis Gastrointestinal inflammation INTERNAL MEDICINE Acute inflammation of the upper GI tract ETIOLOGY Viruses–rotavirus, enteric adenovirus, Norwalk virus, Norwalk-like viruses, calicivirus, astrovirus, bacteria–*Salmonella*, *Shigella*, occasionally toxins, food poisoning, stress and other factors CLINICAL Diarrhea, N&V, anorexia. See Epidemic viral gastroenteritis, Traveler's diarrhea, Viral gastroenteritis

gastroenterologist An internist subspecialized in the study and management of disease of the GI tract, from esophagus to anus MEAT & POTATOES DISEASES Ulcers, colon polyps SALARY $220K. See Gastroenterology. Cf Hepatologist

gastroenterology The subspecialty of internal medicine dedicated to the study of the GI tract, from esophagus to the anus. See Gastroenterologist, Cf Hepatology

gastroenteropathy A condition affecting the stomach and intestines. See Eosinophilic gastroenteropathy

gastroesophageal reflux GASTROENTEROLOGY A backflow of stomach content into the lower esophagus, causing heartburn, esophageal scarring and stricture, which requires stretching/bougienage–dilating of the esophagus

gastrointestinal *adjective* Referring to the stomach and intestine

gastrointestinal bleeding Any hemorrhage into the GI tract lumen, from esophagus–eg, from ruptured esophageal varices, to anus–eg from hemorrhoids

gastrointestinal endoscopy ENDOSCOPY A diagnostic procedure in which a flexible fiberoptic endoscope is passed into the esophagus, stomach, and upper small intestine–depending on the level at which lesions are anticipated INDICATIONS Dyspepsia, persistent N&V, dysphagia–odynophagia, persistent chest pain, caustic ingestion, small intestinal biopsy, follow-up of operations for obesity, surveys for malignancy, sclerotherapy for esophageal varices. See Endoscopy

gastrointestinal eosinophilia GI DISEASE An ↑ in GI tract eosinophils ETIOLOGY Idiopathic, parasites–anisakiasis–*Anisakis* spp, *Angiostrongylus costaricensis*, *Pseudoterranova* spp, food or drug allergies, IBD, systemic disease–eg polyarteritis nodosa, Churg-Strauss disease, acute radiation change. See Eosinophilia

gastrointestinal stromal tumor GIST SURGICAL PATHOLOGY A non-mucosal GI tumor most common in the stomach CLINICAL Benign–leiomyoma or malignant–leiomyosarcoma, determined histologically by ↑ mitotic activity and bizarre cells, findings seen in aggressive lesions. See Leiomyoma, Leiomyosarcoma

gastrojejunostomy GI SURGERY A procedure in which the duodenum is excised or bypassed and the stomach is end-to-end anastomosed to the jejunum

gastropathy INTERNAL MEDICINE Any disease of the stomach. See Stomach

gastroplasty SURGERY Partial excision of the stomach to manage morbid obesity. See Gastric bubble, Gastric stapling, Morbid obesity, Obesity

gastroscope GI DISEASE A thin, flexible endoscope inserted by the mouth through the esophagus to examine the stomach and, if warranted, obtain biopsy material

gastroscopy Gastric endoscopy GI DISEASE An internal examination of the stomach using a gastroscope passed through the mouth and esophagus

gastrostomy GI DISEASE A surgical opening into the stomach which can be used for feeding usually via a gastrostomy tube. See PEG tube

gated blood pool scanning CARDIOLOGY A radionuclide technique for calculating various hemodynamic parameters–eg, cardiac output, right and left ventricular ejection fractions, stroke volume ratio–which allows quantification of valvular regurgitation at rest and during exercise, and defects in regional ventricular wall movement

gatekeeper MANAGED CARE 1. A person, organization, or legislation that selectively limits access to a service; in health care, primary-care physicians–eg family practitioners, general practitioners, internists, pediatricians and PROs and utilization review committees, respectively, function as direct or indirect gatekeepers 2. Care coordinator A physician who manages a Pt's healthcare services, coordinates referrals and helps control healthcare costs by screening out unnecessary services; many health plans insist on a gatekeeper's prior approval for special services or the claim will not be covered

gatekeeping An activity that '...*has come to imply the medically limited and bureaucratic function of opening or closing the gate to high-cost medical services. This simplistic view ... is controversial, both because of its menial connotation and because of the implication that the physician is the agent of the third-party payers, not the patient.* The more politically correct way to define gatekeeping is '... *matching patients' needs and preferences with the judicious use of medical services.*'

gateway drug SUBSTANCE ABUSE Any drug or addictive subtance–eg, nicotine and alcohol, that may be abused, and is allegedly linked to subsequent abuse of illicit 'soft' drugs–eg, marijuana and/or 'hard' drugs–eg, cocaine and heroin. See Glue sniffing, Hard drug, 'White-out'

Gatorade® SPORTS MEDICINE A proprietary sports drink that replaces fluids, carbohydrates, electrolytes consumed during intense exercise. See Sports drink

Gaucher disease, type 1 An AR noncerebral juvenile form of GD most common in Ashkenazi Jews caused by a defect in glucocerebrosidase, leading to glucocerebroside accumulation in the spleen, liver, lymph nodes CLINICAL Splenomegaly, anemia, thrombocytopenia, ↑ skin pigmentation, yellow fatty spot on white of eye–pinguecula, severe bone involvement can lead to pain and collapse of hips, shoulders, vertebral column

gavage feeding NEONATOLOGY Nasogastric feeding of Pts–eg, premature infants with weak sucking reflexes or nasogastric hyperalimentation

gay *adjective* Referring to either ♂ and ♀ homosexual orientation, as in gay rights activism *noun* Homosexual; vernacular for ♂ homosexuals and ♂ homosexuality. See Homosexuality. Cf Closet

gay bowel syndrome An array of infectious and non-infectious GI symptoms described in homosexual ♂ before AIDS CLINICAL Proctalgia 80%, changed bowel habits 50%, condyloma acuminata 52%, cramping diarrhea, bloating, flatulence, nausea & vomiting, adenomatous polyps, fissures, fistulas, hemorrhoids, perirectal abscess, shigellosis, proctitis, rectal ulcers, giardiasis, STDs–eg HSV, syphilis, gonorrhea, *C trachomatis*; 'gay bowel' organisms include 'gay CLOs', ie *Campylobacter*-like organisms–eg, *C fennelliae* and *C cinaedi*, HPV–associated with anal carcinoma and dysplasia, CMV, HAV, HBV, parasites–eg, *Entamoeba histolytica, E coli, Endolimax nana, Enterobius vermicularis, Strongyloides stercoralis, Iodamoeba beutschlii* and bacteria–*N meningitis, Haemophilus ducreyi, Escherichia coli, Salmonella* spp; other findings include rectal dyspareunia, pruritis ani, anal incontinence, trauma–eg secondary to 'fisting' which, like insertion of variably-sized/shaped foreign objects, may cause colorectal perforation and abscess formation TREATMENT Acute proctitis–penicillin, probenicid, doxycycline, changing to specific agents if organism identified. See Anal intercourse, Foreign bodies

GB Gigabyte INFORMATICS 1,000 megabytes/Mb of data

GBM 1. Glioblastoma multiforme, see there 2. Glomerular basement membrane

GBM antibody test See Antiglomerular basement membrane antibody

GBS Group B streptococcus, see there

GC 1. Galactocerebroside, see there 2. Ganglion cell 3. Gas chromatography, see there 4. Germinal center 5. Glucocorticosteroid 6. Gonococcal, gonococcus-*Neisseria gonorrhoeae* 7. Guanine, cytosine

GC-MS Gas chromatography-mass spectroscopy. See there

GE reflux Gastroesophageal reflux, see there

geezer *noun* MEDTALK American slang for an offensive and/or dull-witted old person, especially a ♂ in hospitals, geezer is a highly derogatory term for an elderly, cantankerous, often poorly-educated ♂ Pt *verb* DRUG SLANG A regionally popular verb, to inject a drug

gegenhalten NEUROLOGY Involuntary resistance to passive movement, that ↑ with velocity of movement and continues through the full arc of motion; gegenhalten may be due to diffuse forebrain dysfunction, as occurs in anterior cerebral artery occlusion–'arteriosclerotic parkinsonism' or pseudoparkinsonism, and may be accompanied by the grasp reflex. Cf Clasp-knife phenomenon, Cogwheel rigidity

gel penetration 'Bleed' The passing of microdroplets of silicone through the semipermable gel envelope of a breast implant. See Breast implant, Silicone

gelastic epilepsy NEUROLOGY A form of complex partial seizure due to temporal lobe discharges, characterized by inappropriate laughter as a sign of automatisms

gelling RHEUMATOLOGY Stiffness after rest, which is typical of rheumatic diseases–eg, in juvenile rheumatoid arthritis–Still's disease, variably accompanied by polyarthritis and guarding of joints against activity

gem MALPRACTICE A popular term for a latent problem in Pt management that stems from poor communication, faulty documentation, and other factors, including human factors, such as Pt noncompliance, which sets the stage for events leading to suboptimal care or medical errors. See Malpractice

gemcitabine Gemzar™ ONCOLOGY An IV antimetabolic antineoplastic used with cisplatin for inoperable non-small cell lung CA, and as a single agent for inoperable pancreas CA SIDE EFFECTS ↓ WBCs and platelets; ↑ liver enzymes; 10% discontinued medication due to flu-like Sx, N&V, rash. See Cisplatin. Cf 5-FU

gemeprost OBSTETRICS A PG analogue administered vaginally, used with mifepristone–RU 486 as an abortifacient; it is safe but expensive, and requires specific conditions for storage and transportation. See Abortion, Misoprostol, Prostaglandin, RU 486, Sulprostone

gemfibrozil Lopid® CARDIOVASCULAR DISEASE A drug that inhibits VLDL synthesis, which ↓ M&M due to CAD, nonfatal MIs and strokes, by ↑ HDL-C and ↓ TGs. See Cholesterol-lowering drugs, Lovastatin, VA-HIT

Gen-Probe CT assay STD Noninvasive, highly sensitive test for detecting *C trachomatis* in urine or swab samples. See *Chlamydia trachomatis*

GenBank MOLECULAR BIOLOGY An electronic repository of publicly available DNA sequences, which is maintained by the NIH. Cf Swiss-Prot

gender Sex; one's personal, social, and legal status as ♂ or ♀, based on body and behavior, not on genital and/or erotic criteria. See Gender-identity/role

gender-based illness Vive la difference? Any condition that affects one sex more than the other and/or at an earlier age. See Female-based medicine

GENDER DIFFERENCES IN DISEASE & HEALTH

AUTOIMMUNE DISEASE ↑ in ♀, eg fibromyalgia, SLE, rheumatoid arthritis, multiple sclerosis

♀ **ADVANTAGES** ↑ Igs, ergo ↑ resistance to viral infections; ♀ regain from anesthesia faster; ♀ develop CAD ±10 years later

♂ **ADVANTAGES** ↑ Bone mass at peak maturity; ↓ susceptible to tobacco exposure, TMJ syndrome, chronic fatigue syndrome, interstitial cystitis, and UTIs

OTHERS Organ transplants are more successful in same-gender donation; there are ♂:♀ differences in drug metabolism; ♀ have faster heart rates and may have higher LDL-C (AMN 15/4/96, p16)

gender bias Any difference in the amount of screening activities, diagnostic work-up, or aggressiveness of therapy, based on a person's sex

gender identity Core gender identity *The inner conviction that one is male, female, ambivalent, or neutral.*'GI is a major personality trait, that develops in the first 2 yrs of life, and is 'fixed' by the 3rd yr. Cf Transsexual

gender identity disorder Transsexualism A clinical condition in which a person has a persistent desire to be of the opposite phenotypic sex–cross-gender identification, and experiences discomfort about his/her assigned sex; this desire may take the form of simple 'cross-dressing', or may be of such intensity to compel the person to seek sexual reassignment. See Sexual reassignment

gender role SEXOLOGY The private experience of gender role–GR which is, in turn, the public manifestation of gender identity–GI–a person's individuality as ♂, ♀, or ambivalent, especially re self-awareness, behavior, sexuoerotic arousal & response

gender-neutral *adjective* Referring to anything, toys and other products, activities and services that can accommodate the needs of either sex

gender-specific medicine INTERNAL MEDICINE A recently developed 'twiglet' of internal medicine that formally studies the relationship between gender and disease. See Women's health

gender transposition SEXOLOGY The switching or crossing-over of attributes, expectancies, or stereotypes, of gender-identity/role from ♀ → ♂, or ♀ → ♂. See Gender crosscoding, Gender dysphoria, Sex change

gene GENETICS CLASSIC MENDELIAN DEFINITION A unit of inheritance carrying a single trait and recognized by its ability to mutate and undergo recombination–this definition is widely recognized as primitive CURRENT

DEFINITION A segment of DNA nucleotides, comprised of 70 to 30,000 bp including introns, that encodes a sequence of mRNA, capable of giving rise to a functional producte–eg, enzyme, hormone, receptor–polypeptide; genes may be structural, and form cell components, or functional, and have a regulatory role; a biological unit of heredity which is self-reproducing and located at a definite position on a particular chromosome; genes are working subunits of DNA; each of the body's 50,000 to 100,000 genes contains the code for a specific product, commonly for making a specific protein–eg an enzyme; the functional and physical unit of heredity passed from parent to offspring. See Cellular oncogene, Crime gene, Expressed gene, Functional gene, Housekeeping gene, Lethal gene, Mutable gene, Mutator gene, Pseudogene, Reporter gene, Structural gene, Suppressor gene, Tumor suppressor. Cf Chromosome

gene amplification GENETICS A process by which specific DNA sequences are replicated disproportionately greater than their representation in the parent source; GS of cellular oncogenes occurs in malignancy, where the copy number is a crude benchmark of tumor aggressiveness. See PCR

gene chip MOLECULAR DIAGNOSTICS A postage stamp-sized plastic wafer coated with oligonucleotides that allows identification of genetic disease and viral and bacterial infections

gene dose GENETICS The number of copies of a particular gene present in a genome. See Gene, Genome

gene expression GENETICS The process by which a gene's coded information is translated into the structures present and operating in the cell. See Gene, mRNA, Protein, RNA, Transcription, Translation, tRNA

gene library Genomic library GENETICS A molecular 'database' created when the mRNA extracted from a given tissue is reverse transcribed into cDNA (complementary DNA) segments, which in toto comprise the sequences of genes expressed in the tissue of interest. See Library MOLECULAR BIOLOGY A random collection of DNA fragments–typically representing the entire genome of an organism that have been inserted into a cloning vector

gene map GENETICS 1. A linear designation of mutant sites in a gene, based upon the various frequencies of interallelic or intragenic recombination 2. The DNA sequence of a gene annotated with sites of regulatory elements, introns, exons and mutations.. See DNA sequence, Exon, Gene, Intron, Mutation, Recombination, Recombination frequency

gene product The biochemical material–either RNA or amino acid sequences–polypeptides or proteins expressed by a gene; the amount of GP indicates how active a gene is; abnormal amounts can be correlated with disease-causing alleles. See Gene, Gene expression, Protein, RNA

gene testing GENETICS The evaluation of blood, other body fluids or tissue for biochemical, chromosomal, or genetic markers of genetic disease. See Molecular diagnostics

gene therapy MOLECULAR MEDICINE Treatment of disease by replacing, altering or supplementing the genetic structure of either germline–reproductive or somatic–nonreproductive cells a structure that is absent or abnormal and responsible for disease; any of a group of techniques in molecular biology, in which a gene of interest is manipulated, either by mutational inactivation–eg, the 'knock-out mouse', or by replacement, if it causes a particular disease; GT encompasses any therapy that specifically targets the core defect in inherited diseases, either by affecting somatic cells or germ line cells which are usually inserted into the host's genome; strategies for GT include 1. Introduction of a recombinant retrovirus with the missing gene, the promoter, and the gene regulator sequence in the 'package', and 2. Implantation of the colonies of cells producing the missing factor(s)–eg, α₁-antitrypsin deficiency with the missing enzyme introduced into 'carrier' fibroblasts

GENE THERAPY STRATEGIES

ANTIBODY GENES Interfere with cancer-related protein activity in tumor cells

ANTISENSE Block synthesis of proteins encoded by a defective gene in the host

CHEMOPROTECTION Add proteins to cells that protect them from the toxic effect of chemotherapy

IMMUNOTHERAPY Enhance host defense against cancer

ONCOGENE DOWNREGULATION Turn off genes involved in uncontrolled growth and metastases of tumor cells

SUICIDE GENE/PRO-DRUG THERAPY Insert proteins that metabolize normal drugs and ↑ their toxicity to proliferating–ie tumor cells

TUMOR SUPPRESSOR GENES Replace defective/deficient cancer-inhibiting genes

general anesthesia ANESTHESIOLOGY The administration of pharmacologic agents, via parenteral or inhalation routes, to establish a controlled state of unconsciousness, accompanied by a complete loss of protective reflexes–eg, inability to independently and continuously maintain an airway and regular breathing pattern and respond purposefully to physical stimulation or verbal commands and/or physical stimulation. Cf Conscious sedation, Local anesthesia

general health screen A battery of serum assays used to assess a person's basic state of health–eg, albumin, alk phos, ALT, AST, BUN/creatinine, calcium, total BR, cholesterol, glucose, K+, LDH, total protein, Na+, TGs, uric acid. See Executive profile, Organ panel. Cf General health panel

general paresis NEUROLOGY A symptom of late tertiary syphilis 10+ yrs after the initial infection, caused by chronic meningoencephalitis resulting in progressive dementia and generalized paralysis. See Syphilis

general practitioner A physician who practices 'general medicine', often an older physician who did not specialize in any field of medicine after graduation from medical school. Cf Family practitioner, Generalist

generalist A physician who sees the Pt as a whole 'unit', ie not as an 'organ system' specialist; generalists include family practitioners, internists, pediatricians. See Family practitioner, Holistic approach, Internist. Cf General practitioner

generalized anxiety disorder PSYCHIATRY A situation-independent syndrome characterized by unrealistic or excessive anxiety and worry about life circumstances, often in a background of depression CLINICAL Motor tension, autonomic hyperactivity, vigilance and scanning, episodes of severe anxiety

generalized cortical hyperostosis van Buchem's disease, hyperostosis corticalis generalisata, sclerosteosis An AR osteoporosis complex of early onset with bony overgrowth of 1. BM, resulting in chronic cytopenias and 2. Cranial foramina, causing stenosis and facial nerve paresthesia, loss of vision, deafness LAB ↑ Alk phos

generalized tonic-clonic seizure Generalized seizure, grand mal seizure, tonic-clonic seizure NEUROLOGY A seizure of the entire body, characterized by muscle rigidity, violent rhythmic contractions, and loss of consciousness, with abnormal electric activity in neurons; GCTS may occur in anyone at any age, as a single episode or as a repeated, chronic condition–epilepsy, and are caused by abnormal electrical activity at multiple locations in the brain, or over most of the brain, which may be accompanied by changes in mental status–alertness, awareness and/or focal neurologic Sx ETIOLOGY Idiopathic, congenital defects and perinatal, metabolic defects, DM complications, electrolyte imbalances, kidney failure, uremia, nutritional deficiencies, PKU, alcohol or drugs, brain injury, tumors and space occupying lesions–eg, hematomas

generic drug Generic equivalent PHARMACOLOGY A drug that is no longer under patent protection, which may be produced by any manufacturer who follows good manufacturing protocols. See Drug monograph, 'Me too' drug

genesis The beginning of a process

genetic code GENETICS A sequence of nucleotides, coded in triplets/codons along the mRNA, that determines the sequence of amino acids in a protein; a gene's DNA sequence can be used to predict an mRNA sequence; the 'words' and 'language' that govern the way in which genetic information–DNA is 'written' in the genome and translated into the proteins that perform the genes' activities. See Amino acid, Code, Codon, DNA, Gene, DNA sequence, Nucleotide, Protein, RNA,

genetic counseling CLINICAL GENETICS The education of an individual, couple, or family known or suspected of being at risk for a genetic disease. See Genetic counselor

genetic disease A generic term for any–inherited condition caused by a defective gene–eg, an 'inborn error of metabolism'

genetic effect GENETICS The result of exposure to substances–eg, radiation that cause damage to the genes of germinal cells–ie, sperm or egg. See Teratogenicity

genetic engineering Biological engineering, genetic modification, recombinant DNA technology MOLECULAR BIOLOGY The manipulation of a living genome by introducing or eliminating specific genes through recombinant DNA techniques, which may result in a new capability–eg production of different substances or new functions, gene repair or replacement

genetic medicine Molecular medicine, see there

genetic mother A ♀ who for various physical reasons–eg, anatomic defects or antibody production, cannot herself carry a fertilized egg to full-term; under the usual scenario, the GM's egg is fertilized in vitro, and implanted into a surrogate–gestational–mother, who carries the product to term. See Gestational mother

genetic parent Biological parent SOCIAL MEDICINE A person who conceived a child, typically the biological parent of an adopted child

genetic predisposition MOLECULAR MEDICINE The tendency to suffer from certain genetic diseases–eg, Huntington's disease, or inherit certain skills–eg, musical talent

genetic screening MOLECULAR DIAGNOSTICS The screening of a person's serum for molecular markers that indicate an ↑ susceptibility to inherited or acquired diseases with a genetic component GS CANDIDATES Cystic fibrosis, PKU, sickle cell disease, Tay-Sachs disease

genetic testing MOLECULAR PATHOLOGY The analysis of human DNA, RNA, chromosomes, proteins and certain metabolites in order to detect heritable disease-related genotypes, mutations, phenotypes or karyotypes for clinical purposes; the analysis of a human DNA or chromosomes to detect a genetic alteration that may indicate an ↑ risk for developing a specific disorder caused by a gene mutation. See DNA chip

genetically engineered *adjective* Recombinant, see there

geneticist A scientist whose work hinges on genetics research or management of Pts with genetic disorders SALARY $81K + 12% bonus. See Clinical geneticist, Clinical biochemical geneticist, Clinical cytogeneticist, Clinical molecular geneticist, PhD medical geneticist

genetics GENETICS The study of the patterns of inheritance of specific traits, and how qualities or traits are transmitted from parents to offspring. See Behavioral genetics, Cancer genetics, Classic cytogenetics, Heredity, Inheritance, Medical genetics, Molecular cytogenetics, Pharmacogenetics, Reverse genetics, Trait, Variance

Geneva Convention Declaration of Geneva GLOBAL VILLAGE A standard established in 1864 regarding the conduct of the military towards medical personnel, and obligations of medical personnel during acts of war. See Helsinki Declaration, Nuremburg Code of Ethics, Unethical medical research. Cf Geneva Protocol

Geneva Protocol GLOBAL VILLAGE A document prepared by the League of Nations during the 1925 Geneva Conference intended to ban chemical weapons. See Chemical Weapons, Ypres. Cf Geneva Convention

genital *adjective* 1. Referring to the external and/or internal ♂ and/or ♀ sex organs and/or genitals 2. Pertaining or referring to reproduction 3. Referring to the key components of sexuality–penis and vagina–in the psychoanalytic construct of the mental universe. See Eroticism, Sex, Sexuality

genital culture Genital exudate culture MICROBIOLOGY The placing of swabs from external ♂ or ♀ genital surfaces in a growth medium, under conditions intended to optimize proliferation. See Culture

genital herpes An infection transmitted by contact with genital mucosa by oral, vaginal, or other mucosal surfaces; HS-2 enters the mucosa through microscopic tears, travels to nerve roots near the spinal cord and remains there; herpetic flareups are linked to immune suppression through stress, disease, or drugs and result from HSV traveling down the nerve fibers to the site of the original infection; once it reaches the skin, the classic redness and blisters occur. See Herpes simplex

genital mutilation The destruction or removal of a portion or the entire external genitalia, which may occur in the context of a crime of passion or as part of a cultural rite. See Bobbittize, Cutter, Female circumcision, Self-mutilation

genital phase Psychosexual development, see there

genital prolapse The prolapse of internal organs through a weak pelvic floor–eg, uterine prolapse, cystourethrocele, enterocele, rectocele CLINICAL Pelvic pressure, urinary incontinence, rectal discomfort, related to irritation or ulceration of exteriorized mucosa MANAGEMENT Pessary, topical estrogens, exercises to strengthen pelvic floor; hysterectomy. See Rectocele

genital wart An asymptomatic verrucous tumor induced by HPV, found primarily on the orogenital mucosa, which can be transmitted to an infant during childbirth; GWs are the most common STD in the US EFFECTS GWs, depending on HPV type, ↑ risk of cervical CA. See Cervical cancer, Condyloma acuminatum, HPV, Pap smear

genitourinary Urogenital *adjective* Referring to the genitourinary/urogenital system

genitourinary system UROLOGY The body system that includes the organs of reproduction and elimination of waste products in urine

genius PSYCHOLOGY *adjective* Referring to a marked superiority in intellectual prowess *noun* A highly intelligent person, whose IQ is > 140 and/or in the top 1% of those subjected to IQ testing

genome Genetic structure GENETICS All the genetic information in an organism's chromosomes and mitochondria; its size is given in base pairs. See Base pair, Chromosome, Mitochondrial genome, Nuclear genome

genome project 1. The Human Genome Project, see there 2. A general term for a coordinated research initiative for mapping and sequencing the genome of any organism

genomic *adjective* Referring to the complete chromosomal genetic complement

genomic medicine Molecular medicine The application of molecular methods to clinical medicine; MM encompasses diagnostic methods–eg, Southern, Western, and Northern blot hybridizations, DNA signal amplification by PCR, and use of agents produced by recombinant DNA tech-

niques–eg, biological response modifiers and vaccines, and insertion and manipulation of the genome itself

genomics The science of interpreting genes; the study of an organism's genome using information systems, databases and computerized research tools. See the Human Genome Project

genotophobia PSYCHIATRY A pathologic fear of old people and aging. See Elder abuse. Cf Agism

genotype GENETICS The entire genetic makeup of an organism, the type species of a genus, defined by the complement of allelic forms of each gene or genetic markers present in an organism's genome. See Gene, Genetic marker, Nucleus, Phenotype

Gensini score CARDIOLOGY A scoring system for evaluating collateral circulation of the coronary arteries

gentamicin Garamycin® INFECTIOUS DISEASE A broad-spectrum aminoglycoside antibiotic obtained from *Micromonospora purpurea* ADVERSE EFFECTS Ototoxicity, nephrotoxicity THERAPEUTIC RANGE Peak 5-10 mg/L; trough < 2 mg/L TOXIC RANGE Peak > 10 mg/L; trough > 2 mg/L. See Aminoglycosides

gentian violet PODIATRY A topical antifungal used to manage dermatomycosis. See Gentian

genu recurvatum ORTHOPEDICS Hyperextension of the knee, linked to paralysis of either the hamstrings or quadriceps. Cf Genu Valgum

genu valgum Knock knees ORTHOPEDICS A frontal plane deformity at the knee in which the distal tibia is directed away from the midline/median sagittal plane; GV is usually associated with coxa vara–the knees are together and ankles apart and the Pt has an awkward gait, where the knees rubbing together ↑ side-to-side movement of the pelvis and trunk. See Coxa vara

genu varum Bowleg ORTHOPEDICS A frontal plane deformity of the knee in which the distal tibia is directed towards the midline/median sagittal plane; GV is usually associated with coxa valga–the ankles are together and knees apart, and the Pt stands with feet together, knees separated and the tibias angled downward and inward. See Coxa valga

genus TAXONOMY A taxonomic grouping of one or several species which is subordinate to a family, tribe or subtribe and superior to a species. Cf Family, Species, Trait, Tribe

geographic exclusion TRANSFUSION MEDICINE A ban on collecting or using blood donated by persons from certain countries–eg, Africa except for Arab nations, which have a low incidence of AIDS

geographic pattern A general descriptor for lesions in which large areas of one color, histologic pattern, or radiologic density with variably scalloped borders sharply interface with another color, pattern or density, fancifully likened to national boundaries and/or coastlines IMAGING Broad areas of patchy destruction, seen in such diverse conditions as Gaucher disease, histiocytosis X, osteolytic tumors–eg, metastatic bronchogenic carcinoma and osteosarcoma

geographic practice cost index MANAGED CARE A scale defined by Medicare based on the relative costs of practicing medicine in specific geographic location. See RBRVS, Work adjuster

geographic tongue Benign migratory glossitis, glossitis areata migrans ORAL DISEASE A condition characterized by idiopathic inflammation, possibly due to emotional stress; more common in children and adolescents CLINICAL Abrupt onset, burning, irritation, or asymptomatic; denuded filiform papillae TREATMENT Rarely successful, ie empirical–eg, vitamins, antibiotics, psychotherapy DIFFDX Lingual syphilis with dense white patches

geographic ulcer OPHTHALMOLOGY A sharply demarcated corneal ulcer with scalloped margins, arising in dendritic keratitis, caused by HSV; ± ½ respond to idoxuridine therapy

geologic disaster PUBLIC HEALTH A generic term for a natural disaster due to geological disturbances, often caused by shifts in tectonic plates and seismic activity EXAMPLES Earthquakes, tsunami, volcanic eruptions, avalanches. Cf Climatological disaster

geomedicine Global medicine, see there

geophagy Clay-eating The consumption of dirt–eg, mud or clay, a former practice in many cultures, regionally extant in the southern US

GERD Gastroesophageal reflux disease, heartburn, reflux gastroenteritis GI DISEASE A constellation of findings caused by the chronic backflow of gastric acid into the esophagus; affects 20-40 million, US; 80% also have a hiatal hernia CLINICAL Heartburn, dyspepsia, regurgitation, aspiration, coughing DIAGNOSIS Esophagoscopy, barium swallow, Bernstein test ENDOSCOPY 90% GERD Pts have endoscopic inflammation at EG junction DIFFDX Angina/AMI MANAGEMENT Antacids, lifestyle modification, antisecretory prescription drugs–proton pump inhibitors, H₂ blockers, rarely, surgery PROGNOSIS GERD can lead to scarring and stricture of the esophagus, and require dilating; 10% develop Barrett's esophagus which ↑ the risk of adenoCA of esophagus; 80% of GERDs also have hiatal hernia. See Proton pump inhibitors, H2 blockers

geriatric intervention A maneuver intended to improve evaluation, ergonomics–ie, physical plant, discharge, rehabilitation, implementation of changes in health for older Pts. Cf Elder abuse

geriatrics The specialty dedicated to the care and management of the elderly; life expectancy–LE in 1900–US, 45 yrs; LE in 2002, 77 in ♂; 81 in ♀. Cf Gerontology, Todeserwartung

gene therapy Germline therapy A format for gene therapy which would prevent passage of parental disease to children by genomic manipulation. See Gene therapy. Cf Eugenics

German measles Rubella, see there

germ-fighting toy PEDIATRICS A toy treated with Microban, which permanently bonds microbicidal pellets to plastic or fiber surfaces, stopping mold, fungi, bacteria–eg, staphylococcus, streptococcus, E coli, Salmonella. See Microban

germline mutation Hereditary mutation MOLECULAR MEDICINE An inherited mutation transmitted via the germline from parent to progeny

gerontology GERIATRICS The systematic study of aging and age-related phenomena; senescence is attributed to 1. Accumulation of degradation products, coupled with a cell's ↑ inability to metabolize the products and/or 2. Activation of longevity-determining or aging genes, that may be intimately linked to certain oncogenes–eg, c-fos, which evokes uncontrolled cell proliferation. See Garbage can hypothesis

gerontophobia PSYCHOLOGY Fear of old people/of growing old. See Phobia

Geschlechtsverkehr Sexual traffic, Sexual activity

gestagen ENDOCRINOLOGY 1. A molecule with gestational activity–eg, medroxyprogesterone acetate. See Progestin 2. A synthetic progesterone, aka progestin; progestogen

gestalt psychology PSYCHIATRY A school of psychology that emphasizes a total perceptual configuration and interrelationships of its components

gestation Pregnancy REPRODUCTION MEDICINE The period from fertilization to delivery/birth. See Multiple gestation

gestational diabetes mellitus Glucose intolerance first detected during pregnancy ASSOCIATIONS ↑ Maternal and fetal perinatal complications, tendency to develop glucose intolerance in absence of pregnancy 5-10 yrs later INCIDENCE Up to 5% of pregnancies COMPLICATIONS–MATERNAL Preterm labor, HTN, polyhydramnios, C-section for macrosomia, ↑ subsequent DM COMPLICATIONS–FETAL Macrosomia, shoulder dystocia, perinatal mortality–2 to 5%, congenital malformation–cardiac, CNS. See Fetal diabetes 'syndrome', Glucose tolerance curve

gestational hypertension OBSTETRICS The development of high BP/HTN without other symptoms of preeclampsia or eclampsia after 20 wks of gestation in a previously normotensive ♀. See Eclampsia, Hypertension, Preeclampsia

gestational mother SURROGATE MOTHER A ♀ who carries a fertilized embryo to completion of pregnancy. See Assisted reproduction, In vitro fertilization, Surrogacy. Cf Genetic mother, Gestational carrier

gestational thrombocytopenia Immune thrombocytopenic purpura first seen in pregnancy, which carries the risk of neonatal thrombocytopenia and intracranial hemorrhage MANAGEMENT Conservative if ITP–a not uncommon event in ♀ of child-bearing yrs, first appears in pregnancy; no circulating antiplatelet antibodies are detected

gestational trophoblastic disease GYNECOLOGY A general term for the proliferative processes arising in trophoblastic tissue in ♀ of child-bearing age TYPES Gestational trophoblastic tumor/neoplasia, molar pregnancy, choriocarcinoma. See Choriocarcinoma, Mole, Trophoblastic tumor

get high DRUG SLANG verb To smoke marijuana or use any other mood elevating, often illicit, agent

GFR Glomerular filtration rate, see there

GGT Gammaglutamyltransferase, see there

GH Growth hormone, see there

GHB 1. Gamma-hydroxybutyrate, γ-hydroxy-butyrate. See GABA 2. Glycosylated hemoglobin, see there

GHb Glycosylated hemoglobin, see there

GHF-1 Growth hormone factor-1, see there

ghost villi OBSTETRICS Rounded, pale eosinophilic masses seen by LM that are surrounded by inflammatory cells and correspond to chorionic villi in a placental infarction; ghost villi may also be seen with retained products of conception–missed abortion

GHRH Growth hormone regulatory hormone

GI adjective Gastrointestinal

GI series See Upper GI series

giant cell Any markedly enlarged cell seen in benign or malignant lesions; although GCs are highly nonspecific, their presence in the proper setting supports the diagnosis of certain diseases; the epithelioid GCs of Langerhans and Touton are associated with infections and other 'benign' processes–eg, sarcoidosis, have abundant cytoplasm and a rim or clutch of enlarged histiocyte-like nuclei; GCs in tumors are less inhibited by rules of cytologic etiquette and are anointed with adjectival modifiers–eg, bizarre, monster, osteoclastoma-like, Reed-Sternberg; the cell may be markedly enlarged and mitotically active

giant cell arteritis Cranial arteritis, Horton's disease, temporal arteritis NEUROLOGY A condition that affects > age 50, incidence 18:100,000/yr, characterized by arterial vasculitis; it can lead to blindness and/or stroke CLINICAL New-onset headache, scalp tenderness, jaw claudication, polymyalgia rheumatica DIAGNOSIS Arterial Bx MANAGEMENT High dose corticosteroids. See Vasculitis

giant cell carcinoma A highly malignant epithelial tumor with fulminant clinical course, bizarre histologic appearance and poor prognosis; it is most common in 1. Lung An aggressive, poorly differentiated carcinoma that arises peripherally, grows rapidly and is often too large for adequate therapy by the time of clinical presentation; 2. Thyroid An anaplastic carcinoma arising in a pre-existing well-differentiated thyroid carcinoma CLINICAL Aggressive infiltration of neck structures with 100% mortality; most die in 6 months

giant cell hepatitis Giant cell transformation of liver NEONATOLOGY A nonspecific reaction of the newborn liver to ↑ conjugated hyperbilirubinemia; conjugated BR is ↑ in infants with intra- or extrahepatic biliary atresia, erythroblastosis fetalis, TORCH–toxoplasmosis, rubella, CMV, HS, and other in utero and neonatal infections–eg, coxsackie, hepatitis, *E coli*, syphilis, metabolic defects–eg, α_1-antitrypsin deficiency, cystic fibrosis, hereditary fructose intolerance, galactosemia, parenteral nutrition and tyrosinosis, choledocal cysts, idiopathic, congenital hepatic fibrosis, Byler's disease, Lucy-Driscoll disease, Niemann-Pick disease, trisomy 18, and Zellweger's disease. Cf Syncytial giant cell hepatitis

giant cell reaction Any reparative tissue reaction with multinucleated epithelioid histiocytes, that may be due to exogenous material–eg, sutures, or endogenous material–eg, the contents of a ruptured epidermal inclusion cyst, chalazion, or fat

giant cell tumor of bone A lesion that comprises 5% of all osseous tumors, which is most common in the weight-bearing epiphysis of ♀ > age 20, which affects the distal femur, proximal tibia, distal radius; 60% recur with curettage; 10% metastasize, an event far more common in deep–to the fascial planes–tumors DIFFDX A GCT-like histopathologic lesion may be seen in the brown tumor of hyperparathyroidism, chondroblastoma, giant cell reparative granuloma, benign and malignant fibrous histiocytoma, nonossifying and chondromyxoid fibroma, unicameral bone cyst, aneurysmal bone cyst, fibrous dysplasia, osteoblastoma, osteosarcoma, osteoid osteoma, eosinophilic granuloma

giant cell tumor of tendon sheath SOFT TISSUE TUMORS A potentially recurring lesion of the acral flexor tendon sheath which affects young Pts, is slightly more common in ♂, often with a long Hx, most common in the knee and ankle; GTTS is a variant of fibrous histiocytoma; it reaches a maximum of 3 mm, and may erode into bone. See Malignant giant cell tumor of tendon sheath

giant condyloma acuminatum Büschke-Loewenstein tumor, see there

giant hemangioma of Kasabach-Merritt Hemangioma-thrombocytopenia syndrome An AD condition characterized by thrombocytopenia, 'giant'–± 5 cm in diameter hemangiomas associated with DIC, microangiopathic hemolytic anemia TREATMENT Mild cases respond to corticosteroids or spontaneously regress in 5 yrs

giant hypertrophic gastritis A condition characterized by rugal folds of the stomach and late development of parietal cell autoantibodies and gastric atrophy; GHG is associated with Menetrier's disease–gastric mucosal hypertrophy, ↓ acid secretion, protein loss in the stomach, edema, weight loss, abdominal pain, nausea, and hypertrophic hypersecretory gastropathy; it is morphologically similar to Menetrier's disease, but with ↑ acid secretion, more common in ♂, age 30-50 TREATMENT Some success with anticholinergics, cimetidine, vagotomy, and pyloroplasty

giant peristaltic contraction GI DISEASE A prolonged, high-amplitude forward propagating contraction that represents a heightened peristaltic contraction; GPCs may be induced by cholera toxin, are present in Pts with irritable bowel syndrome, and identical in nature to the migrating action potential complex. Cf Local reflex peristalsis

giant pigmented hairy nevus Giant hairy nevus, giant pigmented nevus DERMATOLOGY A premalignant congenital melanocytic nevus, which measures > 5 cm in greatest dimension, and has a 'garment-like' distribution-bathing trunk, cap, coat sleeve, stocking, often with scattered satellite lesions; up to 12% develop melanoma; leptomeningeal involvement may be accompanied by epilepsy and mental retardation

giant platelet A platelet that is larger than usual, a finding typical of Bernard-Soulier syndrome–BSS; up to 10% of normal platelets are 'giant', ie 'squashed' and oversized; when > 20% of platelets are giant, other conditions must be considered in addition to BSS–eg, ITP, lympho- and myeloproliferative disorders, reticulocytosis, DIC, SLE, gray platelet syndrome, May-Hegglin anomaly, Montreal platelet syndrome, TTP

giant platelet syndrome Bernard-Soulier syndrome, see there

giant T wave inversion CARDIOLOGY An EKG finding characterized as an inversion of the normal T wave, seen in the mid- to left precordial leads, and suggestive of apical hypertrophic cardiomyopathy

giantism Gigantism, see there

giardiasis INFECTIOUS DISEASE Intestinal infection with *Giardia lamblia* EPIDEMIOLOGY Acquired by ingesting water or food contaminated by *G lamblia* cysts or by fecal–oral; post cyst ingestion; liberated trophozoites live non-invasively in small intestine and attach to the brush border and may cause local damage with ↓ of brush-border enzymes CLINICAL Acute, intermittent or chronic diarrhea unresponsive to conservative management; foul smelling, greasy stools; occasionally fever and vomiting; otherwise, malaise, flatulence, bloating, nausea, abdominal pain, weight loss LAB ↓ Intestinal trophozoites are associated with ↑ *G lamblia*-specific secretory IgA; cysts on LM of fresh stool specimens; trophozoites in duodenal aspirates or liquid stool; *G lamblia* antigens in feces TREATMENT Rehydration, metronidazole. See *Giardia lamblia*

gibbous *adjective* Humped, protuberant

gibbus An anterior angular deformity of the lower back, due to hypoplasia or 'wedging' of one or more lower thoracic or upper lumbar vertebrae, resulting in beaked projections on the infero-anterior aspects and hypoplasia of the upper portions of vertebral bodies, seen in mucopolysaccharidosis, type I-H Hurler syndrome, TB–Pott's disease, or trauma; the gibbus or 'buffalo hump' seen in Cushing's disease and syndrome, is due to accumulation of soft tissue 2° to prolonged endogenous or exogenous corticosteroids, and is located in the cervicothoracic region

GIFT Gamete intrafallopian transfer REPRODUCTION MEDICINE Removal of eggs from the ovary, joining with sperm, and laparoscopically placing same in the fallopian tube via small abdominal incisions. See Assisted reproduction

gigabyte INFORMATICS 1000 *or* 1024 megabytes. See Byte, Kilobyte, Megabyte

Gilbert syndrome Constitutional liver dysfunction, low-grade chronic hyperbilirubinemia An inherited defect in bilirubin metabolism CLINICAL Jaundice, weakness, fatigue, nausea, abdominal pain. Cf Criggler-Najjar disease

Gilex Doxepin, see there

Gilles de la Tourette syndrome See Tourette syndrome

gingivectomy ORAL SURGERY The surgical excision of the gingiva. See Periodontal disease

gingivitis ORAL SURGERY Inflammation of gingiva. See Acute necrotizing ulcerative gingivitis, Pregnancy gingivitis

gingivostomatitis Inflammation of the gums & oral cavity

gingko *Gingko biloba* MAINSTREAM MEDICINE Gingkolides are effective in cerebrovascular insufficiency, which causes lacunar defects of memory, migraines, strokes, vertigo TOXICITY Gingko may cause hypersensitivity; it should not be used in pregnancy, or in Pts with coagulation defects

268 **ginseng** PHARMACOGNOSY An herb used as a herbal remedy, as an anxiolytic and antidepressant PHYSIOLOGIC EFFECTS ↑ testosterone, corticosteroids, gluconeogenesis, CNS activity, HTN, ↑ pulse and BP, GI motility, hematopoiesis; ↓ cholesterol TOXICITY Ginseng should not be used in Pts with asthma, arrhythmias, HTN, or post-menopausal bleeding. See Unproven methods for cancer management

ginseng abuse syndrome A complex caused by excess daily ingestion—3+ g/day—of ginseng CLINICAL Diarrhea, nervousness, insomnia, rash, ↑ motor activity. See Ginseng

GIP 1. Gastric inhibitory polypeptide Gastrin inhibitory principle A 43-residue insulinotropic polypeptide hormone of the incretin family produced by the pancreas, that is secreted in response to oral–but not parenteral administration of nutrients–eg, glucose and lipids. See Incretin 2. General insertional protein 3. Giant cell interstitial pneumonia, see there

GIPP Gonadotropin-independent–ie nonpulsatile precocious puberty A condition caused by a lack of suppression in response to LHRHa with non-cyclic steroidogenesis CLINICAL Ranges from testicular immaturity to maturity. See Precocious puberty

G-I/R Gender-identity/role, see there

GISG AIDS A clinical trial–Ganciclovir Implant Study Group which evaluated the use of a sustained-release ganciclovir implant in Pts with CMV retinitis. See CMV retinitis

GISSI CARDIOLOGY A clinical trial–Gruppo Italiano per lo Studio della Streptochinasi nell'Infarcto miocardico–that evaluated the effect of various intervention on M&M in Pts with acute MI; GISSI defined the timespan between the onset of Sx and time when streptokinase was still effective. See Acute myocardial infarction

GIST Gastrointestinal stromal tumor, see there

giving up-given up complex PSYCHOLOGY A state of chronic depression and despair, which some workers suggest ↑ a person's susceptibility to various conditions–eg, ↓ immune responses and cancer. See Psychoneuroimmunology

glanders Infection by *Pseudomonas mallei*, a gram-negative aerobic bacillus, which affects large domestic animals, most commonly horses, and is rare in the US; human disease occurs in the form of acute, often fatal, septicemia, chronic mucocutaneous disease or pulmonary infection CLINICAL From cellulitis to necrosis and granulomas with draining ulcers, pleuritis, necrotizing lobar or bronchopneumonia, nasal septal necrosis, fever, chills, malaise, headaches, pustular rash, lymphadenopathy, splenomegaly; chronic disease is associated with hepatosplenomegaly, pulmonary abscess, granulomas TREATMENT Sulfadiazine, tetracycline, chloramphenicol, aminoglycosides. See Farcy

glandular differentiation The appearance in a carcinoma of glands and gland-like elements which, in an adenoCA, indicates some maturation, which generally is associated with a better prognosis

glandular fever Infectious mononucleosis, see there

Glanzmann's disease Glanzmann's thrombasthenia HEMATOLOGY An AR inborn error of metabolism characterized by coagulation defects CLINICAL Epistaxis, bruising, excess bleeding LAB ↑ Bleeding time, defective clot retraction, ↓ platelet aggregation with ADP, collagen, thrombin MANAGEMENT Platelet transfusions; avoid antiplatelet agents–eg, aspirin, NSAIDs

Glasgow coma scale CRITICAL CARE A method for evaluating the severity of CNS involvement in head injury, which measures 3 parameters–maximum score of 15 for normal cerebral function, 0 for brain death. Cf Harvard criteria, Pittsburgh score

GLASGOW COMA SCALE
BEST MOTOR RESPONSE, ie subject obeys commands 6, if none 0
BEST VERBAL RESPONSE, ie oriented 5, if no response 0
EYE OPENING, if spontaneous 4, if none 0; the GCS is less useful in

glass eye Artificial eye, see there

glaucoma OPHTHALMOLOGY A disorder characterized by ↑ intraocular pressure, which damages the optic nerve causing vision loss and eventually blindness TYPES Open-angle glaucoma–more common and of adult-onset; acute angle-closure glaucoma; in glaucoma, the fluid drains too slowly out of the anterior chamber of the eye, resulting in fluid buildup and damage to the optic nerve and eye and loss of vision; open-angle glaucoma is so named because the anterior angle of the eye stays open CLINICAL Initially, asymptomatic; later pain and narrowed visual field and blindness DIAGNOSIS Air puff test or other tests that measure intraocular pressure MANAGEMENT–surgery Surgery can also help fluid escape from the eye and thereby reduce the pressure, but is reserved for Pts whose pressure cannot be controlled with eyedrops, pills, or laser surgery HIGH-RISK GROUPS Family Hx of glaucoma; anyone > age 60; blacks > age 40; glaucoma is 5 times more likely to occur in blacks than whites and about 4 times more likely to cause blindness in blacks

Gleason grading system ONCOLOGY The most widely used system for stratifying the histologic features of prostate CA, delineated by Gleason, dividing the lesions into 5 groups of ↓ glandular differentiation and worsening prognosis

Gleason score ONCOLOGY A value derived from the Gleason grading system which is the sum of the 2 most predominant histologic patterns seen in prostate CA

Gleevec Imatinib mesylate ONCOLOGY An anticancer monoclonal antibody that blocks signaling of defective proteins, which may be effective against CML, brain tumors, GI and lung CAs

Gliadel® NEUROSURGERY A chemotherapy–polifeprosan + carmustine–saturated wafer implanted during surgery to ↓ local recurrence of glioblastoma SIDE EFFECTS Seizures, pain, brain edema. See Glioblastoma multiforme

glial tumor NEUROLOGY A 1° CNS tumor arising in neurons and neural processes. See Brain tumor

glimepiride Amaryl® ENDOCRINOLOGY An oral hypoglycemic sulfonurea used in type 2 DM, if hyperglycemia does not respond to diet and exercise SIDE EFFECTS Vertigo, asthenia, nausea, hypoglycemia

glioblastoma Astrocytoma, grade III-IV, glioblastoma multiforme, malignant astrocytoma NEUROSURGERY A highly aggressive 1° brain tumor–BT arising in glia, comprising 30% of the 5000 new BTs/yr, US AT RISK OCCUPATIONS Anatomists, pathologists, dentists, ophthalmologists TREATMENT Poor response to surgery, chemotherapy, RT. See Gliadel® wafer

glioma NEUROLOGY A tumor of the brain and spinal cord arising from glial/support cells. See Brainstem glioma, Glioblastoma, Mixed glioma, Oligodendroglioma, Optic glioma, Pseudoglioma

gliopathy NEUROPATHOLOGY Any condition primarily affecting the glia, usually understood to be benign. See Dying back gliopathy

gliosarcoma NEUROPATHOLOGY A glioblastoma with sarcomatous elements PROGNOSIS Poor, highly aggressive; most Pts die within 1 yr

gliosis Astrocytosis, astrogliosis NEUROPATHOLOGY An abnormal ↑ in astrocytes 2° to destruction of nearby neurons, typically due to ↓ glucose or O_2

GLIP Glucagon-like insulinotropic peptide, see there

glipizide Glucotrol® THERAPEUTICS A sulfonylurea used to control insulin in type 2 DM. See Diabetes mellitus

global aphasia Total aphasia, see there

global medicine Geomedicine *'The branch of medicine concerned with the influence of environmental, climactic, and topographic conditions on health and prevalence of disease in different parts of the world.'* See Disaster

global surgery package GRADUATE EDUCATION A billing format in post-graduate medical education and residency training, which bundles all pre-, post-, and operative services performed by a teaching surgeon into a single unit. See Teaching surgeon

globalization GLOBAL VILLAGE The dissolution of national borders in all aspects of human endeavor. See G7, Global medicine, Third world

globoid cell leukodystrophy Krabbe's disease NEUROLOGY An AR defect in sphingolipid metabolism due to galatocerebroside β-galactosidase deficiency, resulting in in utero demyelinization, and death in early infancy CLINICAL Spastic paralysis, seizures, pyrexia, vomiting, cortical blindness, deafness, dysphagia, pseudobulbar palsy, quadriplegia, mental deterioration MANAGEMENT CNS disease may be reversed by allogeneic hematopoietic stem cell transplantation. Cf Leukodystrophy

globoside dysfunction syndrome A group of enzymopathies characterized by defective addition of sugar residues to ceramide–eg, Fabry's disease and lactosyl ceramidosis

globulin Globular protein LAB MEDICINE An older term which now correspond to immunoglobulins, enzymes and other proteins. See Antithymocyte globulin, Core-lipopolysaccharide immune globulin, Human milk fat globulin, Hyaline globulin, Hyperimmune globulin, Immunoglobulin A, Immunoglobulin D, Immunoglobulin E, Immunoglobulin G, Immunoglobulin heavy chain, Immunoglobulin light chain, Immunoglobulin M, Intravenous immune globulin, Macroglobulin, Mammoglobin, Microglobulin, Monoclonal immunoglobulin, Polyspecific anti-human globulin, Pyroglobulin, Rh immune globulin, Sex hormone binding globulin, Specific immune globulin, Thyroxin-binding globulin

globus hystericus Globus, globus pharyngeus PSYCHIATRY A subjective sensation of compression or a lump–bolus in the throat, considered a symptom of neurosis. See Conversion disorder, Factitious disorders

glomerular filtrate Glomerular ultrafiltrate NEPHROLOGY Fluid filtered through the glomerular capillaries into the glomerular capsule of the renal tubules

glomerular filtration rate NEPHROLOGY The volume of filtrate produced/min by both kidneys, which is a principle determinant of sodium excretion by the kidneys; when GFR falls, sodium is reabsorbed by the renal tubules, activating the RAA system. See Glomerular filtrate, RAA system

glomerular syndrome NEPHROLOGY A condition characterized by a defect in glomerular function. See Glomerulonephritis

glomerulonephritis NEPHROLOGY Inflammation of the renal glomeruli, associated with fluid retention, edema, HTN, proteinuria. See Acute glomerulonephritis, Capsular 'drops', Casts, Chronic glomerulonephritis, Heymann glomerulonephritis, Membranoproliferative glomerulonephritis type 1 and 2, Membranous glomerulonephritis, Mesangial proliferative glomerulonephritis, Crescents, Deciduous tree in winter appearance, Dense deposits, Fingerprints, Fusion of foot processes, Heymann glomerulonephritis, Humps, Poststreptococcal glomerulonephritis, 'Shunt' nephritis, Spike, Tramtrack pattern

glomerulosclerosis NEPHROLOGY A degenerative lesion of the renal glomeruli caused by ASHD or DM. See Kimmelstiel-Wilson disease

glomus tumor Glomangioma A neoplasm that is most common in age 20-40 which arises in the neuromyoarterial glomus, an AV shunt CLINICAL When located in the 'usual' subungual site, the abundant innervation makes the tumor exquisitely painful; when located elsewhere–eg, middle ear, stomach, it is painless. Cf Hemangiopericytoma

glossary Dictionary, lexicon VOX POPULI A whole bunch of words/phrases arranged alphabetically and defined, sans pronunciation–eg, present work

glossitis ENT Inflammation of the tongue. See Geographic tongue, Median rhomboid glossitis

glossolalia PSYCHIATRY Gibberish, 'speaking in tongues'

glossy skin Neuritic atrophoderma A descriptor for the smooth skin characteristic of denervation atrophy

glove INFECTIOUS CONTROL A disposible sheath composed of latex or plastic used to cover an examiner's hands during many phases of Pt management to minimize the risk of transmitting pathogens either from the Pt to the health care provider or, for Pts in reverse isolation, from the provider to the Pt. See Hand washing, Personal protection garment

glove & stocking anesthesia NEUROLOGY A characteristic pattern of ↓ sensation affecting the hands and feet, seen in leprosy

GLP Glucagon-like insulinotropic peptide, see there

glucagon ENDOCRINOLOGY A 29-residue polypeptide hormone, produced by pancreatic islet α cells that opposes insulin, activates hepatic phosphorylase, ↓ gastric motility, secretion and muscle mass, promotes glycogenolysis, ↑ serum glucose, ↑ ketogenesis and liver incorporation of amino acids and urinary excretion of Na+ and K+ ↑ IN Neonates, glucagonoma, DM ↓ IN Some Pts with DM, hypoglycemia

glucagon-like insulinotropic peptide A natural fragment of glucagon-like peptide-1–GLP-1, which itself is a fragment of proglucagon; GLIP is antidiabetogenic and ↓ post-prandial release of insulin, glucagon, and somatostatin, ↓ insulin requirements

glucagonoma syndrome A symptom complex associated with glucagonoma, a tumor of post-menopausal ♀ who often have DM, blistering dermatitis–necrolytic migratory erythema, epidermal necrosis, subcorneal pustules, suppurative folliculitis, confluent parakeratosis, epidermal hyperplasia, and prominent papillary dermal hyperplasia, weight loss, normochromic anemia, ileus, constipation or diarrhea, glossitis, angular cheilitis, venous thrombosis, hypoproteinemia, neuropsychiatric changes

Gluco-Pro™ ENDOCRINOLOGY A flavored nutritional supplement for Pts with hyperglycemia and DM. See Glucola

glucocerebrosidase Glucosylceramidase An enzyme that catalyzes cleavage of glucose from glucocerebrosides to form acylsphingosine, a cerabmide; its absence is linked to Gaucher's disease

glucocorticoid METABOLISM A steroid hormone that primarily affects carbohydrate metabolism and, to a lesser extent, fats and proteins EXAMPLES Cortisol–hydrocortisone, the major human glucocorticoid, cortisone; glucocorticoids are produced naturally in the adrenal cortex, less in the gonads, can be synthesized; they have anti-inflammatory and immunosuppressive effects

glucometer Blood glucose meter, glucose meter LAB MEDICINE A dedicated point-of-care device that quantifies serum glucose at the bedside in < 5 minutes. See Phlebtech, Point-of-care testing

gluconeogenesis The formation of glucose from noncarbohydrate molecules–eg, amino acids, lactic acid

Glucophage® Metformin, see there

270 **glucose** BIOCHEMISTRY The hexose sugar that is the main source of energy in mammals. See Random glucose

glucose clearance A value corresponding to the rate of glucose uptake divided by plasma glucose concentration

glucose monitoring LAB MEDICINE The periodic evaluation of any analyte abnormal in Pts with DM, to assess short and long-term control with antiglycemic agents. See Glucose, Glycated hemoglobin

glucose test See 1. Glucose, see there 2. Glucose tolerance test, see there

glucose tolerance test A standardized test that measures the body's response to an oral–challenge dose of glucose, which is primarily used to confirm the diagnosis of DM, and may be performed in individuals with symptoms–eg ↑ fasting glucose level, hypertriglycerdemia, neuropathy, impotence, glycosuria that are suggestive of DM; in pregnancy GTT is used to diagnose gestational DM

glucose-6-phosphate dehydrogenase deficiency Moth ball syndrome A common X-R condition caused by the congenital deficiency of glucose-6-phosphate dehydrogenase, resulting in accumulation of glucose-6-phosphate with hemolytic crises when exposed to naphthalene in moth balls, sulfonamides or sulfones, primaquine, nalidixic acid, fava beans, etc CLINICAL Acute exposure by ingestion causes N&V, diarrhea, hematuria, anemia, jaundice, oliguria, possibly renal shut-down

glucosuria see Glycosuria

Glucotrol® Glipizide, see there

GlucoWatch™ DIABETOLOGY A painless, bloodless, automatic, wristwatch device worn for up to 12–hr monitoring of serum glucose, which incorporates an alarm triggered when hypoglycemic or hyperglycemic levels are detected. See Glucose monitoring

glue VOX POPULI A substance used to join one material to another. See Fibrin glue, Krazy glue, Superglue

glue ear Secretory otitis media, see there

glue-sniffing Solvent abuse SUBSTANCE ABUSE A potentially fatal form of substance abuse, more often practiced by young ♂ adolescents, in which model glue is placed in a paper or plastic bag and deeply sniffed to obtain the maximum desired effect–a combination of euphoria and CNS depression caused by toluene, an organic solvent EPIDEMIOLOGY ♂:♀ ratio is 5:1–UK, peak abuse 13–15; solvents implicated include toluene, trichloroethylene, trichloroethane, tetrachloroethylene; toluene vapors causes irritation of mucosae, lacrimation, arrhythmias, N&V, mental confusion, hypoxia and hypercarbia, hallucinations, chemical pneumonitis, respiratory arrest, sudden death; direct contact may cause erythema, defatting dermatitis, paresthesias, conjunctivitis, keratitis. See Gateway drugs. Cf 'White-out'

glut pronounced as rut, slut VOX POPULI An excess of a service or skilled labor in a particular area. See Physician glut

Glutasorb™ CRITICAL CARE An elemental diet for critically ill tube-fed Pts who need a high glutamine content. See TPN

glutathione γ-Glutamyl-cysteinyl-glycine A ubiquitous antioxidant tripeptide involved in CNS metabolism, which serves as a coenzyme for some enzymes of oxidation-reduction systems, transmembrane amino acid transport, maintaining RBC integrity, and prevention of H_2O_2 accumulation in RBCs. See Antioxidant, Antioxidant therapy, Free radical, Free radical scavenger

glutathione peroxidase deficiency MOLECULAR MEDICINE An AR inborn error of metabolism caused by a defect in glutathione peroxidase, characterized by HDN, compensated hemolytic anemia, neonatal hyperbilirubinemia. See Glutathione

glutathione reductase deficiency MOLECULAR MEDICINE An AR inborn error of metabolism characterized by hemolytic anemia due to a defect of glutathione reductase. See Glutathione

gluteal *adjective* Pertaining to the buttocks, formed by the gluteal muscles–gluteus maximus, medius, and minimus

gluteal flap RECONSTRUCTIVE SURGERY A free flap of tissue from the buttock, which may be used in breast reconstruction

gluten enteropathy Celiac sprue, gluten-sensitive enteropathy A condition characterized by malabsorption of nutrients through the small intestine due to an immune reaction to gluten, a protein in wheat or related grains and other foods CLINICAL Diarrhea, weight loss, which may be associated with dermatitis herpetiformis DIAGNOSIS Bx of involved small bowel MANAGEMENT Gluten-free diet, see there

gluten-free diet A diet sans gluten–wheat, rye, oats, barley, beans, cabbage, cucumbers, dried peas, plums, prunes, turnips, as well as beer (DUDE!!!), instant coffee, malted milk, Postum ALLOWED GRAINS Corn, rice, GF wheat, dairy products, seafoods, poultry; a GFD is supplemented with vitamins, minerals, and digestive aids–eg, enzymes–eg, bromelain, pancrelipase, papain, betaine, lecithin, and ox bile. See Celiac disease, Diet, Digestive aid, Gluten

glutes prounounced like shoots SPORTS MEDICINE A popular term for gluteal muscles; often, "nice GLUTES" which is safer than saying, "nice butt"

glyburide Micronase DIABETOLOGY A sulfonurea used to ↓ blood glucose in Pts with type 2 DM ADVERSE EFFECTS Jaundice, liver function abnormalities, ↑ transaminases, nausea, bloating, heartburn. See Diabetes, Sulfonurea

glycation BIOCHEMISTRY The binding of glucose of proteins, in particular to HbA_{1c}

glycemic index NUTRITION A benchmark of a food's ability to trigger ↑ insulin production; refined foods have a high GI, which ↑ the long-term risk of type 2 DM. See Diabetes, The Zone

glycine An inhibitory neurotransmitter that acts at the level of the spinal cord, the receptors for which are distributed in a heterogeneous pattern

glycogen Animal starch A polysaccharide of glucose-produced primarily in the liver and skeletal muscle, which is analogous to plant starch, but contains more highly branched chains of glucose subunits

glycogen branching enzyme deficiency Glycogen storage disease IV, see there

glycogen cardiomyopathy Glycogen storage disease, type IIb, see there

glycogen debrancher deficiency Glycogen storage disease III, see there

glycogen disease of muscle Glycogen storage disease VII, see there

glycogen loading Carbohydrate loading, see there

glycogen storage disease Glycogenosis Any of a group of 12 inherited AR defects in the ability to store and/or retrieve glucose from intracellular depots, resulting in accumulation of glycogen in liver, muscle, heart, kidney, and other tissues enzyme defects, and hepatosplenomegaly, cardiomegaly, mental retardation–eg, dancing eyes syndrome–GSD VIII

GLYCOGEN STORAGE DISEASE

TYPE	DEFICIENT ENZYME
0	Hepatic glycogen synthetase
I	Glucose-6-phosphatase
II	Lysosomal acid maltase alpha-1,4 glucosidase
III	Amylo-1,6 glucosidase ('debrancher' disease)
IV	Amylo-1,4-1,6-trans-glucosidase–'brancher' disease

V   Myophosphorylase

VI   Hepatic phosphorylase

VII   Phosphofructokinase

VIII   Inactive hepatic phosphorylase

glycogen storage disease 0 Liver glycogen synthase deficiency METABOLIC DISEASE An AR metabolic disorder caused by a deficit of liver glycogen synthetase CLINICAL Morning fatigue, disorientation, convulsions, ketotic hypoglycemia when fasting which disappears on eating MANAGEMENT Protein-rich diet, night-time feedings of infants with uncooked corn starch

glycogen storage disease Ia Glucose-6-phosphatase deficiency, hepatorenal glycogenosis, von Gierke disease METABOLIC DISEASE A rare AR disorder of glycogen storage, due to a defect in glucose-6-phosphatase, resulting in glycogen accumulation primarily in liver and kidney CLINICAL Hypoglycemia, lipidemia, xanthoma formation, ↑ uric acid, ↑ lactic acid, liver adenomas, which may become malignant, hepatomegaly, bleeding diathesis, vasoconstrictive pulmonary HTN, convulsions, failure to thrive, lordosis

glycogen storage disease Ib Glucose-6-phosphate transport defect METABOLIC DISEASE An AR metabolic disorder due to mutation in the microsomal transport system for glucose-6-phosphatase in hepatocytes CLINICAL Growth failure, hypoglycemia, lactic acidosis, oral and anal lesions, chronic IBD, neutropenia, recurrent infections, impaired neutrophil chemotaxis and metabolism TREATMENT GM-CSF may reverse Sx

glycogen storage disease Ic METABOLIC DISEASE A rare AR metabolic disorder caused by mutations in glucose-6-phosphate translocase gene CLINICAL Impaired carbohydrate tolerance, glycosuria, neutropenia, neutrophil dysfunction, hypoglycemia, hepatomegaly, fasting intolerance, ↑ lactic acid, recurrent infections, stomatitis, IBD

glycogen storage disease II Acid maltase deficiency, alpha-1,4–glucosidase deficiency, Pompe disease METABOLIC DISEASE An AR lysosomal storage disorder caused by a defect of α-1,4-glucosidase; the classic infantile form, GSD II, is characterized by cardiomyopathy, muscular hypotonia, occasionally glossomegaly and, in the juvenile and adult forms, skeletal muscle involvement CLINICAL Marked cardiomegaly; death by 1st yr LAB ↑ CK MANAGEMENT High-protein, low carbohydrate diet

glycogen storage disease IIb Glycogen cardiomyopathy, pseudoglycogenosis type II METABOLIC DISEASE An X-linked lysosomal glycogen storage disorder without acid maltase deficiency CLINICAL Proximal muscle weakness, hypertrophic cardiomyopathy, mental retardation, ± hepatomegaly

glycogen storage disease III Amylo-1,6-glucosidase deficiency, Cori disease, debrancher deficiency, debrancher disease, Forbes disease, glycogen debrancher deficiency, limit dextrinosis METABOLIC DISEASE An AR disorder of adult onset primarily affecting Jews of north African descent caused by a defect of amylo-1,6-glucosidase or debrancher enzyme, which removes side chains from stored glycogen, resulting in an inability to degrade glycogen closer than 4 units beyond a branch point; such 'closely-shaved' glycogen is known as limit dextrin CLINICAL Massive hepatomegaly and biventricular cardiac hypertrophy, slowly progressive weakness, muscle wasting, protuberant abdomen, hypoglycemia in childhood, hypoglycemic response to epinephrine and glucagon EMG Myopathic changes, abnormal electrical irritation LAB Hypoglycemia, ↑ ALT, AST, LDH, alk phos activity until puberty at which time hepatomegaly may regress

glycogen storage disease IV Andersen disease, brancher deficiency, brancher disease, amylopectinosis, glycogen branching enzyme deficiency An AR metabolic disorder, caused by an absence of α-1,4-glucan:α-1,4-glucan 6-glucosyltransferase, the 'brancher' enzyme, resulting in muscle weakness, especially of the tongue, hepatosplenomegaly, hepatic fibrosis, followed by cirrhosis, ascites, early death with normal mental development CLINICAL Growth failure, hypoglycemia, lactic acidosis, oral and anal lesions, IBD, neutropenia, ↓ neu-

trophil chemotaxis and metabolism TREATMENT Liver transplantation ameliorates the pancellular enzyme deficiencies by mechanism of microchimerism; GSD IV responds to liver transplantation with ↓ deposition of amylopectin in liver and ↓ neuromuscular or cardiac morbidity due to extrahepatic amylopectin deposits. Cf Debranching enzyme

glycogen storage disease V McCardle disease, muscle glycogen phosphorylase deficiency, myophosphorylase deficiency METABOLIC DISEASE An AR disorder caused by muscle phosphorylase deficiency CLINICAL Painful muscle cramps when exercising, second wind phenomenon, intermittent myoglobinuria. See Second wind phenomenon

glycogen storage disease VI Hers disease, phosphorylase deficient glycogen storage disease of liver METABOLIC DISEASE A clinically benign AR metabolic disorder caused by a deficit of glycogen phosphorylase CLINICAL Mild to moderate hypoglycemia, mild ketosis, fatigue, growth retardation, ↑ hepatomegaly

glycogen storage disease VII Glycogen disease of muscle, muscle phosphofructokinase deficiency, Tarui disease METABOLIC DISEASE An AR metabolic disorder caused by a deficit of phosphofructokinase CLINICAL Muscle cramps with exertion, myoglobinuria, hemolysis, ↑ CK

glycogen storage disease VIII Hepatic phosphorylase kinase deficiency METABOLIC DISEASE An X-linked metabolic disorder, caused by a deficiency of liver phosphorylase kinase; GSD VIII is the mildest glycogen storage disease CLINICAL Hepatomegaly, growth retardation, ↑ ALT, ↑ AST, ↑ cholesterol, ↑ TGs, fasting hyperketosis; with age, clinical and biochemical abnormalities disappear; most adults are asymptomatic

glycogenosis Glycogen storage disease, see there

glycohemoglobin Glycosylated hemoglobin, see there

glycoside PHARMACOLOGY A molecule formed from the condensation of either a furanose or a pyranose with another molecule as an acetal, nitrogen glycoside, or phosphate ester glycoside; cardiac glycosides include digitoxin, digoxin, ouabain

glycosuria Glucosuria The spillage of excess glucose into the urine, an event common in uncontrolled DM and in 'renal' glycosuria. See Renal glyosuria

glycosylated hemoglobin A_1 Glycated hemoglobin A_1, glycohemoglobin A_1, glycosylated hemoglobin, hemoglobin A_1 The sum of negatively charged, posttranslationally glycosylated Hb A_1, which includes Hb A_{1a}, A_{1b}, and A_{1c}; GHA_1 is a reliable marker for controlled DM; the higher the HbA_1, the poorer the control, and possibly, the less compliant the Pt in taking medication; Hb A_{1c} is the most abundant GHA_1 measured clinically REF RANGE > 12% poor glucose control; < 8% good glucose control. See Glycosylated hemoglobin A_{1c}

glycosylated hemoglobin A_{1c} Glycated hemoglobin A_{1c}, glycohemoglobin A_{1c}, hemoglobin A_{1c} The most abundant and measured GHA_1 REF RANGE > 10% poor glucose control; < 7-8% good glucose control. See Advanced glycosylation endproducts, Fast hemoglobin, Glycosylated hemoglobin A_1

GM-1 gangliosidosis A lysosomal disorder caused by β-galactosidase deficiency with accumulation of GM1 ganglioside, a monosialoganglioside of gray and white matter

GM-2 gangliosidosis A group of lysosomal disorders caused by hexosaminidase–A and/or B deficiency with accumulation of GM2 ganglioside

GM-1 GANGLIOSIDOSIS

TYPE 1–infantile form, characterized by severe mental retardation and neurological, somatic, and osseous defects, accumulation of β-galactoside and death by age 2

TYPE 2–juvenile and adult form–milder than type 1, with death by age 10

GM-2 GANGLIOSIDOSIS

Type I Tay-Sachs disease, infantile amaurotic idiocy An AR–carrier frequency, Ashkenazi Jews, 1:25 condition due to hexosaminidase A deficiency, resulting in a 100-fold ↑ in GM-2 ganglioside in the brain CLINICAL Onset in early infancy with generalized hypotonia, apathy, and poor head control, 'cherry-red spots' in the optic macula

Type II Sandhoff's disease An AR condition caused by a complete deficiency of hexosaminidase A and B, with the CNS changes of Tay-Sachs disease, and visceral involvement; 100 to 200-fold ↑ in GM-2 gangliosides and a 50-100-fold ↑ in GA2, the asialo derivative of GM-2 in the brain, liver, spleen and kidney; type II is not more common in eastern European Jews

TYPE III Juvenile GM-2 gangliosidosis An AR condition, onset, age 2 to 6, with ataxia, psychomotor retardation, death by age 5-15; it is not associated with organomegaly and, like type I, more common in Ashkenazi Jews

GM-CSF Granulocyte macrophage-colony stimulating factor, sargramostim, Leukine, Prokine A hematopoietic growth factor and immune modulator produced by granulocytes, macrophages, monocytes, lymphocytes, fibroblasts, endothelial cells; rGM-CSF–sargramostim may be used to ↑WBCs in AIDS, or stimulate hematopoiesis after high dose chemotherapy in autologous BMT; it may be used as an immune 'tonic' in CA and AIDS Pts, anemia, ↑ survival of BMTs, ↓ infections in congenital neutropenia ADVERSE EFFECTS Bone pain, rash, fever. See Biological response modifiers, Sargramostim. Cf G-CSF

GME Graduate medical education, see there

GMP 1. Guanosine monophosphate, see there 2. Good manufacturing practice, see there

GMS stain Gomori-Grocott methenamine silver stain A chromic acid, sodium bisulfite stain used in histology and cytology to identify fungi and *Pneumocystis carinii*; the slide is placed in a hot bath for penetration and stains fungi black with sharp margins, and a cleared center. See Sealed envelope appearance

GN 1. Glomerulonephritis, see there 2. Gram-negative, see there

GNR Gram-negative rods INFECTIOUS DISEASE Bacilli that don't absorb gram stain–ie, are pink; most clinically important GNRs are coliforms: Enterobacteriaceae–eg, *Escherichia, Klebsiella, Proteus, Pseudomonas, Salmonella, Sbigella*, growth of GNRs usually implies fecal contamination, as in peritonitis due to appendicitis, ruptured diverticulosis, gunshot wound

GnRH Gonadotropin-releasing hormone, see there. See LHRH

GnRH stimulation test UROLOGY A test used to evaluate adequacy of testosterone production REF RANGE At 30–45 mins, serum LH levels ↑ 3 to 6 fold; FSH levels ↑ 20 to 50% ABNORMAL VALUES ↑ LH and FSH in 1° testicular failure; normal to ↓ in hypothalamic or pituitary disease. See Testosterone

go down INFORMATICS *verb* To interrupt the flow of information in a computer–eg, to or from the central processing unit, which usually refers to larger computer systems; when the information flow is restored, it is said to be 'back up'. See Computer VOX POPULI *verb* 1. To engage in sexual activity 2. To engage in oral sexueoerotic activity

go live INFORMATICS *verb* To initiate the flow of data in a computer system, often for the first time the system is up and running; when a system goes live after 'going down' or off line, one often says that the system is 'back up'. See Go down

go sour Go South MEDTALK A popular verb for an abrupt and/or unanticipated deterioration of a Pt's clinical status

goalie finger SPORTS MEDICINE An injury that results when the goalkeeper in soccer is wearing a ring and suffers partial or complete avulsion of the finger when the ring is caught by one of the hooks used to anchor the goal's net

Göbell-Stöckel procedure GYNECOLOGY An abdominal fascia lata sling operation designed to relieve urinary stress incontinence by ↑ intraurethral pressure to > intravesical pressure at rest and under stress. See Urinary continence

God complex MEDTALK A personality flaw common among physicians, especially surgeons, who perceive themselves as omniscient–ie, God-like, thus treat others as mere mortals

goiter belt ENDOCRINOLOGY A popular term for an inland region of the US, encompassing the Great Lakes, Midwest, and intermountain regions, where goiter was once common, as diet was based primarily on foods grown on iodine-depleted soil, which triggered the development of goiter. See Colloid nodular goiter

goiter *guttur*, Latin, throat ENDOCRINOLOGY A nonneoplastic thyroid enlargement of any cause, which may be euthyroid, hypothyroid, or hyperthyroid, endemic or sporadic, simple–colloid or multinodular; goiters are most often due to ↑ pituitary secretion of TSH, stimulated by ↓ levels of circulating thyroid hormone; congenital goiter occurs in the rare Pendred syndrome–accompanied by deafness or with in utero exposure to antithyroid drugs or iodides; acquired goiter is idiopathic or may be due to goitrogens–eg, lithium carbonate, amiodarone; endemic goiter is subdivided into: 1. A nervous system syndrome, with ataxia, spasticity, deaf-mutism and mental retardation 2. A myxedematous syndrome characterized by poor growth, mental and sexual development and myxedema; in nodular goiters the lack of available iodine induces hyperplasia with excess colloid being stored in nodular, enlarged follicles; goiters are common in Graves disease in hyperactive middle-aged ♀ and are multinodular and 'hot', displaying hyperactivity on a gallium-67 scan GOITROGENIC FOODS & MEDICATIONS Large amounts of iodine are in seaweed, expectorants–eg SSKI, Lugol's solution–for cough, asthma, COPD, amiodarone–Cardorone, an iodine-rich medication used for arrhythmias. See Toxic multinodular goiter

gold compound RHEUMATOLOGY Any 2nd-line gold-based anti-rheumatic–eg, aurothioglucose, gold sodium thiomalate, auranofin to treat rheumatoid arthritis and other arthropathies SIDE EFFECTS GI tract–eg diarrhea, abdominal pain, N&V, partially relieved by cromolyn sodium, renal–eg, nephrotic syndrome, proteinuria, rash, blood dyscrasias, hepatitis. See Gold lung

gold standard Criterion standard The best or most successful diagnostic or therapeutic modality for a condition, against which new tests or results and protocols are compared. See Standard of practice, Practice guidelines

goldbricking Benign self-mutilation OCCUPATIONAL MEDICINE A condition in which a benign skin disorder–eg, an occupational dermatosis, is exacerbated by self-induced excoriation prior to a workman's compensation review, to ensure that pension benefits will continue, rather than be told to return to work

golden hour CRITICAL CARE The timespan after a trauma victim's arrival at a health care facility, during which the physician has an opportunity to have a positive impact on the Pt's outcome, ie to order appropriate diagnostic tests and initiate Pt management, either surgical or continued observation. Cf Golden period

golfball liver TOXICOLOGY Acute hepatitis caused by toxins sprayed on greens, affecting golfers who lick their balls to clean them–goofy golfers. See Acute hepatitis

golfball metastases IMAGING A pattern of pulmonary metastases typical of sarcoma, clear cell carcinoma, seminoma and melanoma, seen on CXR, where the nodules are between the 'cannonball' metastases of renal cell carcinoma, and 'miliary'/millet-seed sized metastases of thyroid, lung breast CAs. See Cannonball, Coin lesion

golfer's elbow Medial epicondylitis SPORTS MEDICINE An injury characterized by pain and tenderness of medial humeral epicondyle at origin of flexor tendons—caused by too much golfing off TREATMENT Rest, corticosteroid injection if severe. Cf Tennis elbow

golfer's wrist SPORTS MEDICINE Pain and tenderness of the wrist or palm on the side of the hand opposite the thumb caused by small stress fracture(s) of hook of hamate PATHOGENESIS Repeated swinging in golf, tennis or baseball DIAGNOSIS MRI, x-rays be negative MANAGEMENT Excision of hook of hamate. See Hamate

golfing injury PHYSICAL THERAPY Any of a number of injuries that are typical of golfing–eg, DeQuervain syndrome, hook-of-hamate fracture, golfer's cramp, golfer's elbow, golfer's wrist, rotator cuff injury, back injuries, herniated disk

gomer MEDTALK A derogatory acronym–Get Out of My Emergency Room–for any cantankerous Pt, used by some medical students and interns on the wards of some US teaching hospitals. Cf Gork

gonadal dysgenesis A condition characterized by underdeveloped or imperfectly formed gonads; the prototypic GD is Turner syndrome–45, X0, which occurs in 1:2-7000 ♀ births CLINICAL Short stature, webbed neck, cubitus valgus, micrognathia with high arched palate, epicanthal folds, lymphedema of hands and feet, aortic coarctation, renal malformation, osteoporosis, DM, widely spaced nipples, sexual infantilism. See Intersex syndromes

gonadectomy UROLOGY Uni- or bilateral excision of a gonad, either an ovary or a testis; castration. See Castration, Oophorectomy

gonadoblastoma GYNECOLOGY Dysgenetic gonadoma A mixed germ cell-sex cord-stromal tumor which is most commonly seen in sexually abnormal persons, especially with gonadal dysgenesis–GD with a Y chromosome–ie, XY GD and XO-XY mosaicism; gonadoblastomas carry a 25% risk of neoplasia. See Sex-cord tumor UROLOGY A mixed germ cell-sex cord-stromal tumor which almost invariably arises in a background of a gonadal disorder, either pure or mixed gonadal dysgenesis, male pseudohermaphroditism or, less commonly, in a phenotypically or karyotypically normal male. See Sex cord-stromal tumor

gonadotropin deficiency Hypogonadotropic hypogonadism ENDOCRINOLOGY An X-R condition characterized by cryptorchidism, gonadotropin deficiency, glycerol kinase deficiency

gonadotropin Gonadotrophin ENDOCRINOLOGY A hormone that regulates ♂ and ♀ reproduction EXAMPLES LH and FSH, produced by the anterior pituitary, stimulate the ovaries or testicles to secrete progesterone, testosterone, estrogen; chorionic gonadotropin is produced by the placenta and drives secretion of progesterone and estrogen, which are critical for maintaining the placenta during gestation. See FSH, hCG, LH

gonadotropin-releasing hormone analogue A polypeptide analogue of gonadotropin-releasing hormone–GnRH; some are more potent than GnRH in stimulating gonadotropin release EXAMPLES Leuprolide, histrelin, administered in a pulsatile fashion to restore lost GnRH, normalize pituitary-gonadal function, as in congenital GnRH deficiencies–eg, hypogonadotropic hypogonadism with Kallmann syndrome or without anosmia, or in acquired GnRH deficiency secondary to RT of the CNS, pituitary tumors, or panhypopituitarism. See Gonadotropin-releasing hormone

goniometry SPORTS MEDICINE The assessment of joint flexibility, using a large protractor to measure the extreme points in a joint's range of motion. Cf Flexometry

gonococcal arthritis INFECTIOUS DISEASE The most common cause of inflammatory monoarthritis in young adults, caused by *N gonorrhoeae* CLINICAL Asymmetric polyarthralgia of knee, elbow, wrist, metacarpophalangeal, ankle joints MANAGEMENT Penicillin. See Gonorrhea, *Neisseria gonorrhoeae*

gonococcal bacteremia Gonococcemia STD *N gonorrhoeae* in the peripheral circulation following acute local infection CLINICAL Fever, chills, malaise, pain in tendons and joints, which may be mono- or polyarticular, rash. See Tenosynovitis-dermatitis syndrome

gonococcal pharyngitis INFECTIOUS DISEASE An STD acquired through oral sex with an infected partner CLINICAL Usually asymptomatic; Sx occur with disseminated gonococcemia; untreated gonorrhea may cause orchitis, prostatitis in ♂ or PID in ♀ RISK FACTORS Recent exposure to gonorrhea, multiple partners, ♂ homosexuality. See Gonorrhea

gonorrhea STD An STD caused by *Neisseria gonorrhoeae*, which commonly affects the genitourinary tract in the form of PID, salpingitis, and urethral involvement; hematogenous spread may result in arthritis, hepatitis, and myocarditis; the gold standard for detecting *N gonorrhoeae* is culturing the organism in the microbiology lab EPIDEMIOLOGY 150 cases/100,000 population in 1995, most prevalent in young adults, especially with multiple partners; of infected ♀, 25-40% are co-infected with other bacteria–eg, chlamydia CLINICAL > Half of ♀ with gonorrhea are asymptomic; Sx include burning or urinary frequency, yellowish vaginal discharge, redness and swelling of genitals, vaginal burning or itching; untreated gonorrhea can lead to severe pelvic infections SPECIMEN Swab from infected site–eg, vaginal, cervical, throat, anal, or urethral; DNA probes are the current diagnostic method of choice, and are performed from a swab. See *Neisseria gonorrhoeae*

good child syndrome PEDIATRIC PSYCHIATRY A condition described in children with one or more siblings who are "bad" by societal standards–eg, drug abuse, oppositional defiant disorder, sexually promiscuous

'good' cholesterol A popular term for HDL-cholesterol, see there. Cf 'Bad' cholesterol

good death VOX POPULI Any death that others view as a comforting and 'smooth' transition from a living to nonliving state. See End of life decisions

good habit Healthy habit CLINICAL MEDICINE A behavior that is beneficial to one's physical or mental health, often linked to a high level of discipline and self-control EXAMPLES Regular exercise, consumption of alcohol in moderation–if at all, a properly balanced diet, monogamy, etc. Cf Bad habit

good intern case GRADUATE EDUCATION A routine Pt management case of some complexity, tediousness, or disagreeable aspects of no interest to the senior physician that can be palmed off on an underling to do the interview, examination, and scut for its putative educational value. See Intern

good mood PSYCHOLOGY A mental state characterized by positive mood parameters–eg ↑ energy, alertness, sense of well-being, friendliness. Cf Bad mood

good patient METDTALK A Pt who 1. Provides reliable information to the physician, 2. Follows the prescribed regimen, drug therapy, or recommended change in lifestyle, if appropriate for the Pt's condition, and 3. Reliably returns for 'check-up' visits at appropriate intervals. Cf Difficult Pt

Good Samaritan laws FORENSIC MEDICINE Legislation tailored to each jurisdiction, for health care professionals and citizens who provide emergency medical care in 'good faith', and act to help a person needing medical attention. See Herd mentality

Goodpasture syndrome Goodpasture's disease, lung purpura-glomerulonephritis, pneumorenal syndrome An idiopathic autoimmune disorder most common in young white ♂, characterized by alveolar hemorrhage, lung infiltrates, renal failure due to rapidly-progressive glomerulonephritis CLINICAL Hemoptysis, SOB, URIs, fatigue, chest pain, iron-deficiency anemia, acute renal failure PROGNOSIS Poor; average survival, 15 wks TREATMENT Plasmapheresis, immunotherapy

goose-flesh Cutis anserina Diffuse 1-3 mm bumps, fancifully likened to a goose's skin post feather plucking INFECTIOUS DISEASE Diffuse, finely papular rash of scarlet fever overlying an erythematous base SUBSTANCE

ABUSE Horripilation, an early symptom of heroin withdrawal or 'cold turkey', which reflects marked sympathetic discharge, seen ±8 hrs after last dose, accompanied by yawning, lacrimation, mydriasis, insomnia, hyperactivity of GI tract, diarrhea, tachycardia, systolic HTN

gooseneck deformity CARDIOLOGY The convoluted twisting of a short tubular structure, likened to a goose's neck, TYPES 1. 'Gooseneck' describes the distorted left ventricular outflow tract seen by selective left ventriculography in Pts with partial AV canal defects–ostium primum and common AV valve, where the failure of the endocardial cushions to fuse produces an abnormally low AV valve with an abnormally high anterior position of the aortic valve 2. A nodularity palpable in Mönckeburg's arteriosclerosis, caused by dystrophic calcified rings within the media of medium to small blood vessels, like flexible 'gooseneck' lamps

gopher GRADUATE EDUCATION A denigrating term for a medical student who is treated as a scut monkey and told to get things for the department through which he or she is rotating at the time. See Rotater, Scut monkey

Gore-Tex® SURGERY A versatile proprietary synthetic composed of expanded polytetrafluoroethylene, which is biologically similar to fascia and used for abdominal and thoracic wall, pediatric and diaphragmatic reconstructions, rectal, vaginal and urethral suspension, hernia repair, fascial defects

gorging NUTRITION The consumption of 1 or 2 'mega-meals'/day, linked to ↑ total cholesterol, LDL-C, apolipoprotein B. See Bingeing. Cf Nibbling

gork God Only Really Knows MEDTALK A deprecating acronym for a Pt with a difficult diagnosis. Cf Gomer

Gorlin Syndrome Basal cell nevus syndrome, see there

goserelin acetate Zoladex® ONCOLOGY A LHRH/GRH analogue used to suppress pituitary secretion of gonadotropins INDICATIONS Hormone-responsive prostate and breast CAs, uterine fibroids, endometriosis ADVERSE EFFECTS Hypersensitivity, rash, hypotension, HTN. See Androgen ablation therapy

GOT Glutamic oxaloacetic transaminase. See Aspartate amino transferase, AST

gothic arch appearance Steeple sign ENT Narrowing of the glottis and subglottis due to edema and soft tissue inflammation, seen in a lateral neck film of young children with croup–eg, due to parainfluenza, other viruses and bacteria, an appearance fancifully likened to that of a gothic arch

Gottron sign DERMATOLOGY Scaly, patchy erythema over the knuckles in Pts with dermatomyositis. See Dermatomyositis, Polymyositis

gourmand syndrome NEUROLOGY A rare, benign, eating disorder that may follow a stroke, with residual damage of the right frontal lobe, characterized by an obsessive focusing on eating, thinking, talking, and writing about fine foods

gout ORTHOPEDICS A condition characterized by ↑ uric acid in the blood and joints, resulting in recurring painful arthritis, renal function and deposits of uric acid in the urinary tract, linked to 2 X-linked enzymes RISK FACTORS Hereditary defects of HG-PRT–hypoxanthine-guanine phosphoribosyl transferase, obesity, weight gain, alcohol intake, HTN, renal dysfunction, drugs LAB ↑ uric acid, often ≥ 410 μmol/L–US: 7.0 mg/ml, a level found in a significant minority of ♂–90% of gout occurs in ♂ DIAGNOSIS Crystals in joints, body fluids, tissues; family Hx of gout is present in ± ½ of cases; ½ present with the classic 'podagra'–1ˢᵗ metatarsophalangeal joint; up to 90% suffer podagra at some time during their disease ACUTE TREATMENT Colchicine, NSAIDs, corticosteroids INTERVAL Diet CHRONIC Allopurinol, sulfinpyrazone, salicylates. See Congenital gout, PseudoGout, Saturnine gout

gouty arthritis RHEUMATOLOGY A chronic arthropathy characterized by uric acid crystal deposits in the joint which, over time, result in joint erosion. See Acute gouty arthritis, Gout

gown GRADUATE EDUCATION The black sari worn during graduation from an educational experience INFECTIOUS CONTROL A cloth, paper, or synthetic garment which covers the body to a greater–from the wrists and neck northward to the knees or below, southward–surgical gowns, or lesser–from upper arms and neck to mid-thigh for Pt gowns–degree. See Personal protection garment

GP General practitioner MEDICAL PRACTICE A physician in general practice

gp120 HIV A 120 kD glycoprotein on the surface of HIV-1, which binds to cells–T cells, macrophages at the CD4 receptor. See AIDS, AIDS vaccine, HIV

GPi pallidotomy NEUROSURGERY A surgery intended to ↓ Parkinsonism, using localized microprobes. See Parkinson's disease

grade CARDIOLOGY The degree of inclination of an exercise treadmill, measured in percentage. See Murmur grade ONCOLOGY The degree of differentiation of the cells in a malignant tumor; in general, the lower the grade of tumor, the slower its growth, and the better the prognosis. See Broders grade, Scharff-Bloom-Richardson grade

grade 1 injury ORTHOPEDICS A mild injury in which ligament, tendon or other musculoskeletal tissue may have been stretch or contused, but not torn

grade 2 injury ORTHOPEDICS A moderate injury when tissue has been partially, but not totally, torn which appreciable limitation in function of the injured tissue

grade 3 injury ORTHOPEDICS A severe injury in which tissue has been significantly, and in some cases, totally torn which causes a virtual loss of function of the injured tissue

graduate medical education Any type of formal, usually hospital-sponsored or hospital-based training and education, that follows graduation from a medical school, including internship, residency, or fellowship. Cf Continuing medical education–CME, Fifth pathway

graft IMMUNOLOGY Any tissue taken from one part of the body of the same or different person and used to replace diseased or injured tissue in another part of the body. See Allograft, Bone graft, Bypass graft, Corneal graft, Coronary artery bypass graft, Dermagraft, Endovascular stent graft, Endoscopic coronary artery bypass graft, Fetal brain graft, Hair graft, Hemi-homograft, Hemobahn endovascular graft, Irradiated chondral graft, Isograft, Micrograft, Minigraft, Skin graft, Split thickness graft, Standard hair graft, Strip graft, Test graft, Tissue graft, White graft

graft failure CLINICAL IMMUNOLOGY Loss of function in a transplanted organ or tissue. See Graft

graft rejection Rejection CLINICAL IMMUNOLOGY The constellation of defenses mounted by the immune system of the recipient of an allograft–eg kidney, liver, pancreas, etc, which compromise the continued viability of grafted tissue. See Graft

graft survival rate IMMUNOLOGY The percentage of Pts with functioning grafts–eg, for 1, 2, or 5 yrs. See Graft rejection

graft-versus-host disease CLINICAL IMMUNOLOGY A reaction of donated BM against a Pt's own tissue, which is a major cause of M&M in allograft BMTs; GVHD is less significant in transplanted kidneys, heart, liver, and skin; viable donor T cells react immunologically against host CLINICAL Fever, morbiliform rash–central erythematous maculopapular eruption that may spread to the extremities with bulla formation, anorexia, N&V, severe watery or bloody diarrhea, lymphadenopathy, infections, hepatosplenomegaly, ↑ LFTs, jaundice, hemolytic anemia PREVENTION Irradiation of donated blood may prevent active leukocytes from rejecting recipient tissues PROPHYLAXIS Cyclosporin, MTX, tacrolimus TREATMENT ½ of Pts who develop post-BMT GVHD respond to high-dose steroids. See Bone marrow transplantation, Rapamycin, Tacrolimus, Transfusion-associated graft-versus-host disease

graft-versus-tumor IMMUNOLOGY An immune response to a graft recipient's tumor cells by a donor's transplanted immune cells in the BM or peripheral blood. See Graft-versus-host disease

grafting of skin See Skin graft

Graftpatch™ SURGERY A product comprised of layers of collagen derived from porcine small intestine, used in minimally invasive and general surgical procedures–eg, bladder and hernia repair to reinforce soft tissue. See Gore-Tex®

Graham Steel murmur CARDIOLOGY An early diastolic murmur heard over Erb's Point heart murmur typical of pulmonary regurgitation/insufficiency, 2° to severe pulmonary HTN linked to mitral valve stenosis

–gram VOX POPULI The root form for something produced in a graphic format. See Angiogram, Nettergram, Scan

gram-negative bacteremia INFECTIOUS DISEASE Bacteremia due to organisms–eg, Enterobacteriaceae–eg, *E coli*—that stain negatively by the gram stain. See Sepsis

gram-negative meningitis INFECTIOUS DISEASE An acute meningitis caused by coliform bacteria–eg *E coli*, *Enterobacter aerogenes*, *P aeruginosa*, *Proteus morgagni*, *K pneumoniae*; GNM starts elsewhere the body and spreads to the brain or spinal cord via the bloodstream RISK FACTORS Recent neurosurgery, open head trauma, CSF shunts, UTIs in children CLINICAL Fever, severe headache, N&V, stiff neck, photophobia, change in mental status LAB Bacteria, PMNs in CSF MANAGEMENT Antibiotics COMPLICATIONS Endotoxic shock, brain abscess. See Meningitis

gram-positive bacteremia INFECTIOUS DISEASE Bacteremia due to organisms–eg, *S pneumoniae*—that stain positively by the gram stain. See Sepsis

gram stain BACTERIOLOGY A stain formulated by a great Dane, HCJ Gram, for identifying broad groups of bacteria; GS may be performed on specimens from skin, tissue, urethral discharge–for *N gonorrhoeae*, endocervix, joint fluid, pericardial or pleural fluid, sputum, stool

grand mal seizure Generalized tonic-clonic seizure NEUROLOGY A seizure disorder arising between infancy and early adulthood, with attacks triggered by fever or unidentified environmental cues–eg, psychologic and emotional stress CLINICAL Prolonged tonic-clonic seizures with the risk of intraictal cerebral hypoxia; sequelae include intellectual impairment, behavioral changes, or rarely ataxia and spasticity. See Epilepsy, Psychomotor epilepsy. Cf Petit mal epilepsy

grand rounds ACADEMICS The formal presentation of a specific aspect of clinical medicine, with discussion of pathogenesis, symptomatology, therapy, usually of a specific disease. See Rounds. Cf Clinicopathologic conference

GRANDDAD syndrome PEDIATRICS An AD condition characterized by Growth Retardation, Aged facial appearance, Normal Development, Decreased subcutaneous fat, Autosomal Dominant inheritance

grandfather clause Any policy or rule that exempts a group of individuals, organizations, or drugs, from meeting new standards or regulations; eg when a new subspecialty board in internal medicine is created, the physicians practicing in that area may be 'grandfathered' into the subspecialty and not required to meet residency or other educational requirements

grandfather device A device exempted from meeting newer standards–eg, pre-1976 breast implants were not required to be regulated by the FDA. See Breast implant

grandiosity PSYCHIATRY An exaggerated belief or claims of one's importance or identity, manifest by delusions of wealth, power, or fame. See Manic episode, Bipolar disorder

granisetron Kytril® ONCOLOGY A 5-HT$_3$–serotonin receptor antagonist antiemetic used for Pts receiving chemotherapy–eg, cisplatin, cyclophosphamide ADVERSE EFFECTS Headache–35%, constipation–30%, asthenia, somnolence, diarrhea

grant RESEARCH A sum of money or financial support provided to a worker or researcher for performing innovative or novel work. See Block grant, Matching grant, Medicaid block grant, Modular grant, Noncompeting grant, Research grant, Research project grant, Subgrant

granular cast NEPHROLOGY A finding in urinalysis characterized by molded forms from renal tubules punctuated by multiple granules; fine GCs arise from degenerating renal tubular cells; coarse GCs may correspond to a degenerating cell in the renal tubule. See Cast

granular cell tumor Granular cell myoblastoma A small–< 3 cm painless subepithelial tumor of the tongue, skin, breast, other epithelia and muscle, which primarily affects older blacks with a ♂:♀ ratio of 2:1 DIFFDX Lesions with large cells and granular cytoplasm–eg, oncocytoma, metastatic Hürthle cell tumor, rhabdomyosarcoma, hybernoma, xanthogranuloma; < 2% become malignant

granular dystrophy An AD variant of early-onset corneal stroma dystrophy with central 'bread-crumb'-like opacities, episodic irritation, and photophobia

granulation tissue A post-inflammatory reaction characterized by edema, chronic inflammation–lymphocytes, macrophages, plasma cells, and proliferating endothelial cells and blood vessels

granulocyte colony-stimulating factor See G-CSF

granulocytic leukocytosis Neutrophilia, see there

granulocytopenia Agranulocytosis, see there

granuloma 1. A nodular aggregate of epithelioid histiocytes or macrophages surrounded by lymphocytes, often accompanied by multinucleated epithelioid cells, and scattered CD4 T cells in the center, which function in antigen recognition, surrounded by a rim of collagen, rare CD8 T cells and proliferating fibroblasts DIFFDX Chronic infection, sarcoidosis, and drug toxicity; if necrotic, TB is suspected; eosinophils favors allergy 2. A nonspecific term which may be used by nonpathologists for any rounded mass or bump. See Central giant cell granuloma, Ceroid granuloma, Doughnut granuloma, Eosinophilic granuloma, Epithelioid granuloma, Fish tank granuloma, Giant cell granuloma, Giant cell reparative granuloma, Juvenile xanthogranuloma, Lethal midline granuloma, Lymphogranuloma venereum, Midline granuloma, Naked granuloma, Peripheral giant cell granuloma, Pregnancy granuloma, Pulse granuloma, Pyogenic granuloma, Swimming pool granuloma

granuloma inguinale Donovanosis STD An indolent, skin ulcer caused by *Calymmatobacterium granulomatosis*, transmitted by sexual contact MANAGEMENT Streptomycin, chloramphenicol, tetracycline, ampicillin, gentamicin. See Sexually-transmitted diseases

granulomatosis Any condition characterized by multiple nodules or granulomas. See Bronchocentric granulomatosis, Lymphomatoid granulomatosis, Necrotizing sarcoid granulomatosis, Wegener's granulomatosis

granulomatous angiitis of CNS Isolated or primary angiitis of the CNS A condition most common in middle-aged persons, characterized by angiitis–giant cell, lymphocytic, or necrotizing in leptomeningeal, or less often, in parenchymal vessels; diagnosis of GACNS by histologic criteria, ♂:♀ is 2:1; when angiography is used, the ♂:♀ is 1:2

granulomatous arteritis Large-vessel encephalitis NEUROLOGY A herpes zoster-induced condition characterized by a stroke that develops wks or months after zoster of contralateral trigeminal nerve distribution, caused by bland, or less commonly, hemorrhagic infarction, due to large vessel arteritis; most Pts are > age 60; mean time of onset is 7 wks after development of herpes zoster; TIAs and mental Sx are common; up to 25% die See Herpes zoster. Cf Small vessel arteritis

granulomatous prostatitis UROLOGY A heterogeneous condition characterized by prostate inflammation ETIOLOGY Infection, nonspecific, postsurgical, and rare 2° systemic causes CLINICAL Low-grade fever, dysuria, frequency. See Prostate

granulomatous thyroiditis De Quervain's thyroiditis, giant cell thyroiditis, pseudotuberculous thyroiditis, subacute thyroiditis ENDOCRINOLOGY Thyroid inflammation possibly of viral origin, more common in ♀, age 30–50 CLINICAL Sore throat, dysphagia, marked tenderness on palpation, often accompanied by fever, malaise, pressure Sx, mild hypothyroidism. See Thyroiditis

granulomatous uveitis Posterior uveitis OPHTHALMOLOGY Ocular inflammation characterized by impaired vision, watering of the eyes, and photophobia. Cf Anterior uveitis

granulosa cell tumor GYNECOLOGY A sex cord tumor of older ♀ that comprises 1-2% of ovarian neoplasms, which is 'driven' by unopposed estrogen stimulation; the endometrium of Pts with GCTs often has cystic hyperplasia and 5-25% of Pts also have a concomitant well-differentiated endometrial CA

grass DRUG SLANG Marijuana, see there

GRATEFUL MED MEDICAL INFORMATICS User-friendly software that facilitates literature searches and accessing data from the National Library of Medicine's database, MEDLARS; MEDLARS' most popular database is MEDLINE

Graves' disease Basedow disease ENDOCRINOLOGY The most common cause of hyperthyroidism, due to excess thyroid stimulation by autoantibodies with diffuse toxic goiter CLINICAL Hyperthyroidism, exophthalmos, dermatopathy–painless, reddish lumpy rash on anterior leg, familial tendency, tachycardia, ↑ metabolic rate RISK FACTORS Stress, smoking, RT to neck, medications–eg IL-2 and IFN-alpha, and viral infection DIAGNOSIS Thyroid scan–diffuse ↑ uptake, ↑ TSI–Thyroid Stimulating Ig level. See Exophthalmic goiter, Primary hyperthyroidism

Graves' ophthalmopathy ENDOCRINOLOGY A potentially disfiguring, sight-threatening condition that occurs in 25-50% of Pts with Graves' disease CLINICAL Gritty sensation in the eyes, diploplia, blurred vision, ↑ lacrimation, exophthalmos, extraocular muscle dysfunction, periorbital edema, lid edema, retraction, conjunctival chemosis and injection, exposure-type keratitis, photophobia. See Graves' disease

gravid *adjective* Pregnant

gravidity OBSTETRICS The state of being, or having been, pregnant. Cf Gravity

gray Gy RADIATION PHYSICS The SI unit for radiation, based on actual radiation absorption, as measured by a thermoluminescent dosimeter placed within a Pt or a phantom; 1 Gy is equal to 1 joule/kg of absorber–100 rads

gray hepatization A phase of lobar pneumonia, typically occurring at the 1st wk of infection, at which time the lobe is covered with fibrin; the cut surface is gray, dry, and granular; the alveoli are filled with fibrinous exudate composed of degenerating PMNs and inflammation is interconnected through the pores of Kohn. Cf Gray infiltration, Red hepatization

the Gray Journal A popular term for certain US journals that may be requested in a medical library by color; the classic 'GJ' is the gray-green *Annals of Internal Medicine*. Cf Green journal

gray literature SCIENCE COMMUNICATION A popular term for a report or manuscript circulated or published by unconventional routes; GL may not receive the attention accorded communications submitted to peer-reviewed journals or other forums. See Throwaway journal. Cf Peer-reviewed journal

gray-out CARDIOLOGY A descriptor for faintness due to vasodepressor illness/vasovagal syncope that may be seen in cardiac disease–eg, in hypertrophic cardiomyopathy

gray patch ringworm MYCOLOGY Fungal folliculitis in children, caused by *Microsporum canis* and *M audouinii*, which progresses from an erythematous papule at the hair follicles, extends peripherally, and forms annular lesions

gray scale IMAGING See Ultrasonography INFORMATICS See Bit depth

gray top tube LAB MEDICINE A blood collection tube containing powdered sodium fluoride and/or potassium oxalate, which inhibits glycolysis; 'gray tops' are used for GTT–as glycolysis in the RBCs would cause false-low glucose, for measuring lactate–transported on ice, lactate tolerance. See Glucose tolerance test

Gray's Gray's Anatomy MEDTALK The prototypic textbook of anatomy, first written and published in the UK in 1858, in the US in 1859

great case Fascinoma MEDTALK A popular term for a person or lesion that is instructive or interesting to the health care managing team which is often aggressive, or difficult to diagnose and may have a poor outcome

great imitator Great imposter A nonspecific term for any condition that is difficult to diagnose due to its polymorphous presentation; the classic 'great imitator' is syphilis, which has a broad clinical palette, especially affecting the skin and CNS; more recent lesser 'great imitators include HIV, hypothyroidism, tertiary Lyme disease, pheochromocytoma. Cf Internists' tumor RHEUMATOLOGY Systemic lupus erythematosus, see there

the Green Book GRADUATE EDUCATION Directory of Graduate Medical Education Programs A catalog produced by the AMA each yr that provides information on accredited internships and residency training programs in North American teaching hospitals. See FRIEDA, Residency PUBLIC HEALTH The schedule of adult immunizations published by the Am Coll of Physicians. Cf Red book

green bottle fly An opportunistic insect pathogen, which causes a myasis/maggot infestation, pupating in open sores or purulent discharges

the Green Journal LIBRARY MEDICINE A popular term for certain journals in medical libraries; the classic 'GJ' is the *Am J of Medicine*–British racing green

green stool syndrome Transient neonatal passage of thick, bile-stained stools, possibly linked to bacterial infection that alters urobilin metabolism, and responds within 24 hrs to penicillin

green tea POPULAR HEALTH A beverage prepared from leaves of an eastern Asian evergreen shrub, *Camellia sinensis*. See Tea. Cf Caffeine, Coffee, Maté

green top tube LAB MEDICINE A heparinized blood collection tube with a green stopper used to obtain blood for ammonia, carboxyHb, O_2 saturation, cholinesterase, Hb, methemoglobin, pH, histocompatibility testing–eg, HLA-A, B, HLA-D–mixed lymphocyte culture, NBT–nitrotetrazolium blue–assay, phagocytosis, T- and B-cell studies and surface markers

green urine disease A 'condition' characterized by ↑ verdohemoglobin in the urine, which fluoresces with UV light, a sign of fulminant *Pseudomonas* *aeruginosa* infection; GU may be accompanied by poor response to aminoglycosides

Greenfield filter Filter umbrella CLINICAL MEDICINE A vena caval filter for preventing PE, DVT. See Inferior vena caval filter

greens NUTRITION Green vegetable–eg, leafy vegetables–eg cabbage, lettuce, spinach, fruit vegetables–eg, green peppers, etc. See Leafy greens. Cf Fruit vegetable DRESS CODE Hospital uniform; scrubs, see there. Cf Whites

greenstick fracture ORTHOPEDICS A partial, angled fracture that causes bone bowing with rupture of the periosteum on the convex side of the bone, ie the opposite cortex is intact, without the fracture line traversing the bone; GFs are more common in rickets

Greer's goo PEDIATRICS A topical formulation of hydrocortisone, nystatin, and zinc oxide, used for diaper rash

grepafloxacin Raxar® A broad-spectrum quinolone for acute bacterial exacerbation of chronic bronchitis caused by *S pneumoniae*, *H influenzae*, *Moraxella catarrhalis*; community-acquired pneumonia, *Mycoplasma pneumoniae*; uncomplicated gonorrhea caused by *N gonorrhoeae*; nongonococcal cervicitis and urethritis due to *Chlamydia trachomatis*. See Fluoroquinolone

Grey Turner sign CLINICAL MEDICINE Local discoloration of the inguinal skin occasionally seen in acute hemorrhagic pancreatitis

GRID Gay-related immune deficiency An acronym used early in the AIDS epidemic, as the first cases affected homosexual/gay ♂. See AIDS

grieving Mourning, see there

grimace NEUROLOGY A humorless facial 'mask' typically seen in Pts with catatonia. See Amimia

Grinfeld cannula HEART SURGERY A triple-lumen cannula used during heart surgery which enables cardioplegia, arterial return and aortic clamping with one device

griseofulvin MICROBIOLOGY An oral antifungal with an affinity for skin, used to treat dermatophytoses–eg, *Epidermophyton* spp, *Microsporum* spp, *Trichophyton* spp ADVERSE EFFECTS Headache, N&V, diarrhea, photosensitivity, fever, rash, dysfunction of hepatic, CNS, hematopoietic systems; it is teratogenic and carcinogenic in rodents. Cf Amphotericin B

grommet See Tympanostomy tube

groove sign STD A finding characterized as linear fibrotic depressions parallel to the inguinal ligament, bordered above and below by enlarged and matted lymph nodes and covered by adherent, erythematous skin, seen in 10-20% of lymphogranuloma venereum–*Chlamydia trachomatis*

grooved tongue Scrotal tongue. See there

gross anatomy The branch of anatomy concerned with structures of the body that can be studied without a microscope

gross evidence FORENSIC MEDICINE Physical evidence that can link a perpetrator or victim to a particular location or crime EXAMPLES Garments, footwear, bed linen, rope, bullets, knives, glass, paint chips

gross negligence MALPRACTICE The reckless provision of health care that is clearly below the standards of accepted medical practice, either without regard for the potential consequences, or with wilful and wanton disregard for the rights and/or well-being of those for whom the duty is being performed; grossly negligent health care is best described as 'sloppy'. See Malpractice, Negligence

ground glass appearance A descriptor for a homogenous translucency in radiology, microbiology, or histology, which may obscure cellular detail

OPHTHALMOLOGY GG opacification of cornea occurs in children with mucopolysaccharidosis–MPS type I-H or Hurler syndrome, with global MPS deposition in the ocular structures, photophobia, retinal degeneration, attenuation of vessels, optic atrophy, loss of vision, hydrocephalus RADIOLOGY, ABDOMEN A GGA with haziness and erasure of organ silhouettes, is typical of massive ascites, and related to an ↑ water density and peritonitis RADIOLOGY, BONE The GGA corresponds to a relatively uniform loss of osseous density which may be accompanied by intravascular calcification and subcutaneous ossification, seen in bones affected by fibrous dysplasia, scurvy–located at the epiphyses, osteomalacia, agnogenic myeloid metaplasia, and osteoporosis RADIOLOGY, CHEST The GGA corresponds to an alveolar, acinar or amorphous pattern seen in bronchiolitis obliterans and obstructive pneumonia–BOOP

ground itch PARASITOLOGY A hypersensitivity reaction to hookworms occurring in a sensitized person CLINICAL Erythema, inflammation, blistering, and intense pruritus at the site of larval penetration–feet AGENTS *Necator americanus*, *Ancylostoma duodenale*

group HEALTH INSURANCE The coverage of a number of individuals under one contract; the most common "group" is employees of the same employer. See Employer group MEDICARE Two or more physicians, non-physician practitioners, or other health care providers/suppliers who form a practice together–as authorized by state law–and bill Medicare as a unit. See Independent group, Multispecialty group VOX POPULI A collection of related things. See Alkyl group, Ambulatory patient group, Andover Working group, Blood group, Carboxyl group, Clinical Context Object Workgroup, Clinical cooperative group, Control group, Cooperative group, DRG group, Encounter group, Ethnic group, Experiential group, Exposed group, Five food group, Food group, Four food group, HACEK group, Health Information & Application Working group, High risk group, Hyperchromatic crowded group, Incompatibility group, Initial review group, Jackson Hole group, JPEG group, MedisGroup, MLS group, New drug studies group, Newsgroup, P blood group, Resource utilization group, Scientific review group, Sensitivity group, Self-help group, Social group, Support group, Treatment group

group A streptococcus INFECTIOUS DISEASE A group of bacteria that are normally skin and saprobes of the skin and oral cavity; when pathogenic, GAS spreads by direct person-to-person contact CONDITIONS Strep throat, impetigo, scarlet fever, rheumatic fever, postpartum fever, wound infections, pneumonia; invasive GAS cause necrotizing fasciitis, toxic shock syndrome

group B streptococcus *Streptococcus agalactiae* A streptococcus classified into 7 capsular serotypes, which is the leading cause of sepsis and meningitis in neonates; GBS affects 1.8/1000 infants aged ≤ 90 days, and causes morbidity in ≥ 50,000 pregnant ♀/yr–US; it is a major pathogen of nonpregnant adults, affecting 2.4-4.4/10⁵

group B streptococcal septicemia of newborn PEDIATRICS A severe neonatal infection which causes ±3/4 of sepsis in the newborn RISK FACTORS Mother infected with GBS, premature rupture of membranes, congenital defects in infant CLINICAL Labile temperature and/or pulse, poor feeding, grunting, tachypnea, apnea, cyanosis, lethargy, shock LAB Cultures–blood, CSF, urine, films for effusions, ABGs MANAGEMENT IV antibiotics–ampicillin, gentamicin, fluids; O₂ therapy, assisted ventilation, ECMO COMPLICATIONS DIC, respiratory arrest, meningitis

group benefits agreement MANAGED CARE A written agreement between a health plan and an employer group which provides coverage for covered health care services to be provided to plan members

group dynamics PSYCHIATRY The interactions and interrelations among members of a therapy group and between members and therapist; effective use of GD is essential in group treatment

group practice MEDICAL PRACTICE The practice of medicine as a team, with physicians and administrators being 'players' who share various overhead costs and expertise. See Partnership. Cf Solo practice

group theory PSYCHOLOGY A theory that explains human behavior by studying the regular interactions of social groups that have a degree of association and interdependence TYPES OF SOCIAL GROUPS Formal, informal, informal allied, institutionalized allied

group therapy Group psychotherapy PSYCHIATRY The regular treatment of a small–2 to 20–group of people with a similar mental issue; GT techniques are as varied as the methods used in psychoanalysis, and tend to be therapist-specific. See Psychotherapy

growing pains PEDIATRICS Benign idiopathic leg pains in growing children, usually in the legs; experts disagree on whether this condition actually exists; some workers regard it as a type of overuse syndrome; if it does exist, it is a diagnosis of exclusion DIFFDX Trauma, infection, avascular necrosis of bone, tumors, collagen vascular disease, Lyme disease, psychosomatic disease, congenital, and developmental defects CLINICAL GP complaints are bilateral, intermittent, most often occur after going to bed, and are linked to edema of muscles encased within 'tight' fascia; they often follow a day of strenuous exercise, although a temporal or activity-relatable pattern may not be detected; GPs may be accompanied by restlessness and disappears with the cessation of growth TREATMENT Local heat, massage, rest, NSAIDs and, if severe, quinine sulfate. See Fat spurt, Growth spurt

growth factor Any natural cytokine that facilitates cell division and proliferation, which are produced by normal cells during embryonic development, tissue growth, and wound healing EXAMPLES EGF, erythropoietin, fibroblast growth factor, IGF-I, IGF-II, nerve growth factor, PDGF, relaxin, somatomedins A and B, TGF-α. See Cytokine, Transforming growth factor

growth failure 1. Dysplasia–Pediatrics, see there 2. Failure to thrive, see there

growth hormone Somatotropin A 21.5 kD growth-promoting protein hormone secreted in a pulsatile fashion by the anterior pituitary in response to the hypothalamic regulatory hormones, GHRH and somatostatin; GH influences protein, carbohydrate, fat metabolism. See Recombinant human growth hormone

growth hormone deficiency Hypopituitarism ENDOCRINOLOGY A condition which affects 1:4000 children; δ : φ, 3-4:1 ETIOLOGY 70% of GHD is idiopathic and attributed to a prenatal insult, possibly due to hypothalamic dysfunction, given that GHD children secrete hGH after stimulation with GH-releasing hormone; GHD is associated with midline CNS and facial defects–eg, cleft lip and cleft palate; hypopituitarism may be linked to hypotelorism or a single giant upper central incisor; other causes include septooptic dysplasia, craniopharyngioma, intrasellar or suprasellar tumor CLINICAL Infants with intrauterine hypopituitarism may present at birth with hypoglycemic seizures, prolonged jaundice, and, if δ, micropenis and undescended testes; linear growth rates as slow as 3 cm/yr; 10% with early onset disease have hypoglycemic seizures; 20% have chemical hypoglycemia; GHD children are proportional for age, have a prominant calvarium, and are often overweight for height, with prominant subcutaneous deposits of abdominal fat; they may have delayed puberty; electrolyte imbalances, diabetes insipidus, hypothyroidism are rare in Pts with idiopathic hypopituitarism; skeletal maturation is usually delayed; most have heights for bone age < third percentile SCREENING Assessment of bone age, x-rays of sella turcica, measure somatomedin C which, if normal virtually excludes GHD; pituitary function testing–provocative tests: maximal GH <7 ng/ml indicates impaired GH secretion; > 10 ng/ml excludes GHD MANAGEMENT GH replacement typically ↑ growth rate to 8-10 cm in first yr of therapy; treatment failure mandates workup for hypothyroidism or GH antibodies. See Septooptic dysplasia

growth hormone insensitivity syndrome Laron dwarfism, pituitary dwarfism II An AR condition characterized by severe growth retardation, delayed bone age, occasionally blue sclerae, and physical manifestations of GH deficiency, due to a defect in the GH receptor

growth hormone suppression test ENDOCRINOLOGY A clinical maneuver used to detect autonomy of hGH secretion. See Growth hormone

growth spurt PEDIATRICS A period of rapid growth in middle adolescence; φ ↑ ±8 cm/yr ± age 12; δ ↑ ±10 cm/yr ± age 14; GS is orderly, affecting acral parts–ie, hands and feet grow before proximal regions, partly explaining adolescent clumsiness

grübelsucht PSYCHIATRY German for hair-splitting, a feature of obsessive-compulsive disorders See Obsessive-compulsive disorder, Obsessive-compulsive personality disorder

grunting NEONATOLOGY A deep-pitched gutteral rumble which may be heard by infants suffering from respiratory distress. See Respiratory distress syndrome

GTT Glucose tolerance test, see there

guaiac-negative PHYSICAL EXAMINATION *adjective* Referring to stool that is negative for occult blood, see there

guaiac-positive PHYSICAL EXAMINATION *adjective* Referring to stool that is positive for occult blood, see there

guaifenesin Monafed® COMMON COLD An OTC agent for relief of cold/cough. See Common cold

Guanarito virus VIROLOGY An emerging viral pathogen of the Lassa and Arenavirus family. See Emerging pathogen

guardian Guardian ad litem, law guardian SOCIAL MEDICINE A person authorized under applicable state or local law to give permission on behalf of a child to general medical care; a person, usually an attorney who is appointed by the court to represent the best interests of a child or other person of "diminished capacity". Guardianship. Cf Emancipated minor

Guardian™ catheter INTERVENTIONAL CARDIOLOGY A balloon catheter that uses high- and low-inflation pressures in angioplasty, coronary dilatation, rotational atherectomy, coronary stenting. See Balloon angioplasty

guarding PHYSICAL EXAM Involuntary muscle spasm elicited by fulminant acute peritonitis, leaving the anterior abdominal muscles in a state of tonic contraction, imparting a board-like consistency to the rectus abdominis muscles, distinguished from voluntary guarding, seen in 'ticklish' Pts. See Abdominal guarding

Gubex Diazepam, see there

Guglielmi Detachable Coil® NEUROSURGERY A soft platinum microcoil insert delivered to disease site via catheter for treating intracranial aneurysms, vascular lesions, AV malformations, fistulas, tumors. See Berry aneurysm

guidance counselor CHILD PSYCHOLOGY A school worker trained to screen, evaluate and advise students on career and academic matters

guide VOX POPULI 1. A written document which orients a novice about the particular use of a thing 2. A device used to ensure the proper movement or placement of a device, which is not, per se, integral to the device's operation

guideline MEDTALK A series of recommendations by a body of experts in a particular discipline. See Cancer screening guidelines, Cardiac profile guidelines, Gatekeeper guidelines, Harvard guidelines, Transfusion guidelines

Guillain-Barré syndrome Acute inflammatory demyelinating polyradiculopathy NEUROLOGY A peripheral neuropathy of abrupt onset which follows a precipitating event–eg, a viral infection–eg, with HIV, HBV, EBV, or *Camplyobacter jejuni*, vaccination, bee stings, sarcoidosis, leukemia, lymphoma CLINICAL Rapidly ascending weakness of all 4 extremities with involvement of respiratory, facial, and bulbar muscles, with sparing of sphincter muscles MANAGEMENT Supportive, meticulous pulmonary hygiene, physical therapy, aggressive management of ulcers PROGNOSIS Generally good with few sequelae in those who respond to therapy. See Polyneuropathy

guillotine amputation A sharply defined loss or autoamputation of part or entire segments of extremities, ranging from digits to major amputations, seen in the congenital aglossia-adactylia syndrome

guilt PSYCHIATRY Emotion resulting from doing what one perceives of as wrong, thereby violating superego precepts; results in feelings of worthlessness and at times the need for punishment. See Shame

Gulf War syndrome MILITARY MEDICINE A condition described in veterans of the Gulf War CLINICAL Fever, headache, ↓ short-term memory, loss of vision, SOB, coughing, diarrhea, skin changes, bleeding gums, loss of hair and teeth, numbness, tingling, aching joints, fatigue

GULF WAR SYNDROMES

IMPAIRED COGNITION SYNDROME Sx include distractibility, memory loss, depression, insomnia, fatigue, slurring of speech, mental confusion, migraine-like headaches

CONFUSION-ATAXIA SYNDROME Defects in thinking, balance, reasoning ability, confusion, disorientation, carrying a diagnosis of posttraumatic stress diorder, depression, liver disease, and sexual impotence

ARTHROMYONEUROPATHY SYNDROME Generalized joint & muscle pain, ↑ difficulty in lifting heavy objects, fatigue, tingling/numbness of hands, feet, arms, legs AMN 20/1/97, p18, from 15/1/97 JAMA

gull wing pattern Seagull appearance A descriptor for a gull wing-like pattern with a broad, flat and gently curved V shape IMAGING A double curved shadow seen on a plain lateral film of the entire pelvis in either fracture-dislocation of the posterior acetabular rim or dislocation of the femoral head; the 2 wings are contributed by the intact and fractured acetabulum

gum TOBACCO CONTROL See Nicotine gum

gum disease DENTISTRY Gingival disease, often in the form of gingivitis and bone loss 2° to toxins produced by bacteria in plaque accumulating along the gum line CLINICAL Early–painless bleeding; pain appears with advanced GD as bone loss around the teeth leads to gum pocket formation; bacterial infection of gum pockets causes swelling, pain, and further bone destruction; advanced GD can cause loss of otherwise healthy teeth

gumboil Parulis ORAL PATHOLOGY A dental abscess, converted into a cyst with chronic drainage through a fistulous tract, seen by a plain film of the jaw

gumma STD Lentil-to-cherry-sized masses characteristic of tertiary syphilis, which represent a hypersensitivity response to treponemal products–the organism itself is rarely found. See Syphilis

gums VOX POPULI Gingiva

gun *noun* DRUG SLANG A regionally popular street term for a syringeVOX POPULI A device used to discharge a projectile in a targeted direction. See Gene gun, Heavy guns *verb* DRUG SLANG To inject an illicit drug–regionally popular

gunstock deformity ORTHOPEDICS A reversal of the arm's carrying angle, resulting from 1. Inadequate correction of the medial angulation of the distal fragment of a supracondylar fracture of the humerus or 2. stimulation of the lateral condylar epiphysis from the fracture itself–even when the fracture has not been displaced

gurney NURSING A wheeled cot or stretcher for transporting Pts in emergency care, in surgery or used to restrain a prisoner before a lethal injection

gustatory agnosia NEUROLOGY The inability to recognize tastes, a common finding in parietal lobe tumors. See Agnosia

Gustilo classification SURGERY A system for classifying severity of open tibial fractures, which divides them into grades I, II, and III, subdividing the last group into A, B, or C, depending on the vascular status and the degree of soft-tissue destruction, which forms of the basis for deciding whether to reconstruct or amputate

GUSTO CARDIOLOGY A series of clinical trials that have examined a series of strategies to reduce the M&M of acute MI; the GUSTOs include: Global Utilization of Streptokinase & tPA for Occluded coronary arteries trial–GUSTO I; Global Use of Strategies to Open Occluded arteries: recombinant hirudin versus heparin as adjuncts to thrombolysis and aspirin–GUSTO II; primary coronary angioplasty with tPA–GUSTO IIb; efficacy of alteplase and reteplase–GUSTO III; platelet glycoprotein IIb/IIIa receptor inhibitor–abciximab and reteplase v. reteplase alone–GUSTO V; Strategies for Patency Enhancement in Emergency Department–GUSTO-SPEED. See Abciximab, Acute coronary syndrome, Acute myocardial infarction, Heparin, Hirudin, Reteplase, Streptokinase, tPA

gut-associated lymphoid tissue GALT. See there

gut feeling Intuition, visceral sensation

GVH disease Graft-versus-host disease, see there

GVHD Graft-versus-host disease, see there

Gy Gray, see there

GYN is short for gynecology–or a gynecologist

gynandroid Female pseudohermaphrodite A person who merges traditional ♂ and ♀ stereotypes

gynecologic cancer GYNECOLOGY Any malignancy of the ♀ reproductive tract, including cervix, endometrium, fallopian tubes, ovaries, uterus, vagina and, for some the breast

gynecologic oncologist ONCOLOGY A physician specialized in managing cancer of ♀ reproductive organs

gynecologist A physician specialized in health maintenance and management of non-pregnant ♀, and in treating diseases of ♀ reproductive organs; gynecologists generally train simultaneously in obstetrics SALARY $193,000. See Obstetrician

gynecology The specialty of health care that addresses the physiology and pathology of nonpregnant ♀. See Alternative gynecology, Urogynecology. Cf Obstetrics

gynecomastia Benign enlargement of ♂ breast that most commonly affects boys and adolescents, often regresses at puberty; it is usually due to a proliferation of glandular component; may occur in Klinefelter syndrome or after malnutrition AGE-RELATED PEAKS 1. Perinatal, in 60-90% of ♂ at birth, due to transplacental passage of estrogens 2. Pubertal, up to 70%, depending on stringency of definition 3. Involutional, age 50-80% TREATMENT Clomiphene, tamoxifen, testolactone, surgery. See Adolescent gynecomastia

GYNECOMASTIA ETIOLOGY

ENDOCRINOPATHY Orchitis, hypogonadism–androgen deficiency, androgen resistance, tumors/hyperplasia of adrenal gland, testes, eg Leydig cell tumor–↑ hCG, lung CA, Klinefelter syndrome–↑ risk of breast CA and thyroid hyperplasia

DRUGS α-methyldopa, amphetamine, androgens, benzodiazepines, cimetidine, chemotherapeutics, digitalis, INH, marijuana, penicillamine, phenothiazine, reserpine, spironolactone, tricyclic antidepressants

OTHER CONDITIONS Starvation diet–mechanism: testicular, hepatic hypofunction or on resuming feeding, hemodialysis, liver disease–cirrhosis, hepatomas, hemochromatosis, due to ↓ hepatic metabolism of estrogens, mycosis fungoides, myotonic dystrophy with spastic paraplegia, leprosy

gynephobia PSYCHOLOGY Fear of ♀. See Phobia. Cf Misogyny

H 1. H antigen, see there 2. *Haemophilus* 3. Halothane 4. Heavy chain 4. Hemagglutinating 5. Hemin 6. Heparin 7. Histamine 8. Histidine 9. *Histoplasma* 10. Hospital 11. Hounsfield unit 12. Hour 13. Human 14. Hydrogen 15. Hypermetropia 16. Hyperopia 17. Hyperplasia 18. Hypodermic

h Symbol for: 1. hecto-10^2 2. Height 3. Hour

H antigen TRANSFUSION MEDICINE The trisaccharide stem chain of the ABO blood group, located on RBC membranes; the enzyme, α-L-fucosyl-transferase, is encoded by the H gene on chromosome 19q, and produces the trisaccharide stem; the RBC membrane component is destroyed by the so-called soluble substance. See ABO, Blood groups, H gene

H blocker See Histamine receptor antagonists

H&E Hematoxylin & eosin, also 1. Hemorrhage & exudate 2. Heredity & environment

H flu **meningitis** See *Haemophilus influenzae* meningitis

H flu **vaccination** See HIB vaccination

H gene TRANSFUSION MEDICINE A gene that encodes the enzyme, α-L-fucosyl-transferase, which produces an oligosaccharide, the H antigen; homozygous deficiency of H gene results in a Bombay–or hh phenotype individual who, lacking the above transferase, cannot attach the fucose to the galactose on the precursor H substance. See Bombay phenotype, H antigen, Secretor, *Ulex europa*

H&H Hemoglobin & hematocrit. See 10/30 rule, Transfusion trigger

H&P History & physical examination, see there

H pylori *Helicobacter pylori*, see there

H- or U-shaped vertebra PEDIATRIC RADIOLOGY A vertebra with a flattened central body with mineralization of its posterior aspects, which may be seen in sickle cell anemia and in thanatophoric dwarfism

H-type fistula PEDIATRICS A type of esophageal atresia and tracheo-esophageal fistula in which the esophagus communicates with the stomach in the usual fashion and has a small fistulous tract to the trachea, causing recurrent aspiration pneumonia TREATMENT Surgical closure, nutritional support

HAART Highly active antiretroviral therapy, triple combination therapy AIDS The concurrent administration of 2 nucleoside reverse transcriptase inhibitors–eg, AZT and 3TC, and a protease inhibitor–eg, Indinavir in newly diagnosed HIV infection EFFECTS ↓ HIV levels in plasma, ↓ opportunistic infections, ↓ mortality, ↑ circulating T cells. See AIDS, HIV

habit VOX POPULI A practice routinely or regularly performed by a person. See Bad habit, Good habit, Oral parafunctional habit

habitual abortion Recurrent abortion OBSTETRICS The loss of ≥ 3 consecutive pregnancies, or ≥ 3 spontaneous abortions with no intervening pregnancies FACTORS Stress, nutritional status; occur in ±1:200 ♀. See Abortion

habituation PSYCHOLOGY An adaptive response characterized by a ↓ reactivity to a repeated stimulus–eg, a substance of abuse or repeated electrical stimulation of a nerve

habitus A general body type; an asthenic habitus is seen in tall thin subjects whose organs are said to hang low in the body; the hyperthenic subject is plethoric and his organs are higher in the body and the hyposthenic habitus lies between the two. See Marfanoid habitus

HACEK Acronym for bacteria that cause infective endocarditis–*Haemophilus, Actinobacillus, Cardiobacterium, Eikenella, Kingella* spp. See Infective endocarditis

Hachinski ischemic score NEUROLOGY A system for quantifying features seen in multi-infarct dementia. See Multi-infarct dementia

hadephobia PSYCHOLOGY Fear of hell. See Phobia

Haemonetics® Cell Saver 5, plasma separator SURGERY A device that separates plasma from blood, returning Pt RBCs to the body, ↓ need for transfusions. See Blood salvage

Haemophilus MICROBIOLOGY A genus of nonmotile gram-negative rods that require blood for growth and cause RTIs, meningitis, and STDs

Haemophilus influenzae **meningitis** NEUROLOGY A bacterial meningitis most common in unimmunized children from 1 month to 4 yrs; Hib ↓ HIM after a URI CLINICAL May be abrupt, spread from the nasopharynx to the circulation, then meninges; absent therapy, rapidly fatal RISK FACTORS Hx of otitis media, sinusitis, pharyngitis or other URI, Hx of *H influenzae* infection in family; native American Indian and Eskimos > 3 times that of the average population; day-care

Haemophilus influenzae **type B vaccination** See HIB vaccination

Hageman factor Factor XII, see there

Haglund's deformity Pump bump PODIATRY An exostosis over posterolateral aspect of the calcaneus caused by recurrent friction, ± bursitis CAUSE Multifactorial–eg, abnormal mechanics leading to excess motion of calcaneus, congenital deformity of calcaneus, irritation from high heeled shoes–'pumps' MANAGEMENT Open-heeled/flat shoes, moleskin padding of calcaneus, cryomassage, orthotics to control excess motion; surgical excision

hair DERMATOLOGY A threadlike epidermal appendage consisting of keratinized dead cells that extrudes from a dividing basal layer. Related terms are Anagen hair, Bamboo hair, Bayonet hair, Bundled hair, Catagen hair, Corkscrew hair, Green hair, Paintbrush hair, Pubic hair, Ringed hair, See-through hair, Telogen hair, Terminal hair, Vellus hair, Whisker hair

HAIR-AN Hyperandrogenic insulin-resistant acanthosis nigricans GYNECOLOGY A type of hyperthecosis affecting 5% of hirsute ♀, often with polycystic ovaries CLINICAL Insulin-resistance–Pts may have clinical diabetes; hyperandrogenesis–↑ testosterone, ↑ androstenedione; early, premenarcheal hyperandrogenism; acanthosis nigricans, possibly an epiphenomenon TREATMENT Bilateral oophorectomy, OCs, corrects hyperandrogenism, but not hyperinsulinemia

hair analysis The use of scalp hair as primary analytic specimen MAINSTREAM MEDICINE HA is of limited usefulness in trace element analysis SUBSTANCE ABUSE The use of samples of hair to detect chronic drug abuse METHODS RIA, EIA, GC/MS DRUGS DETECTED Amphetamines, cocaine, heroin LEVELS OF DETECTION 10 pg or 10 ng/mg of hair TOXICOLOGY The most reliable use of HA is to detect chronic heavy metal poisoning–eg, arsenic, lead, mercury

hair-on-end appearance Crewcut appearance IMAGING A radiologic pattern seen as calcified spicules perpendicular to bone surface, corresponding to a periosteal reaction to disturbed bone repair with neoosteogenesis of the outer cranial table, marked calvarial thickening, external displacement and thinning of the inner table; classically seen in children/adolescents with thalassemia DIFFDX Periostitis of long bones in congenital syphilis–syphilitic periosteitis of tibia, giving rise to Saber shins, see there, congenital hemolytic anemias–pyruvate kinase deficiency, hereditary elliptocytosis and spherocytosis, sickle cell anemia, metastatic neuroblastoma, iron-deficiency anemia, cyanotic–right-to-left shunt congenital heart disease, osteomyelitis, polycythemia vera, thyroid acropathy, hemangiomas

hair loss in men Male-pattern baldness, see there, aka androgenic alopecia

hair loss in women Androgenic alopecia, see there

hair removal Epilation COSMETIC DERMATOLOGY The use of lasers or other instruments to remove unwanted facial and body hair by 'zapping' the hair follicle to slow or stop hair growth. See Cosmetic laser surgery

hair replacement Hair restoration COSMETIC SURGERY A general term for any procedure in which encroaching baldness is surgically reversed by grafting, tightening, or flap rotations, using standard grafts, minigrafts, and micrografts, scalp flaps, scalp-lifting procedures, scalp reduction, and tissue expanders. See Male pattern baldness, Peter/Paul pilferage

hairbrush appearance A radiologic descriptor for elongated periosteal bony spicules projecting from the femoral metaphyses in achondrogenesis, similar to a focal hair-on-end reaction

hairline fracture Capillary fracture ORTHOPEDICS A fracture without separation of bone fragments. See Fracture

hairy *adjective* VOX POPULI Hirsute; replete with hair, pellose

hairy cell leukemia Leukemic reticuloendotheliosis, tricholeukocytic leukemia A low-grade B-cell leukemia comprising 2% of adult leukemia, commonly affecting ♂-♂:♀ ratio, 4:1, age 50-55, leading to progressive pancytopenia CLINICAL Insidious onset, weight loss, bruising, abdominal fullness due to splenomegaly, pancytopenia with normocytic, normochromic anemia, rarely aplastic anemia due to infiltration of BM, spleen, lymph nodes; 10% have platelet counts of < 20 X 10⁹/L–US: < 20,000/mm³, 20% have thrombocytosis; 1-80% of nucleated RBCs are hairy LAB ↑ Acid phosphatase, especially isoenzyme 5–which is ↑ in bone metastases and in Gaucher's disease MANAGEMENT Purine analogues 2'-deoxycoformycin–pentostatin and 2-chlorodeoxyadenosine–cladribine, a purine nucleoside used in low-grade lymphoproliferations–eg, CLL and NHL is more effective than IFN-α CAUSE OF DEATH Infections, gram-negative bacteria, atypical mycobacteria, fungi. See Dry tap

hairy leukoplakia ORAL PATHOLOGY An EBV-associated condition seen in HIV infection, characterized by a condyloma-like tongue mass–75% have HPV, 95% have EBV in epithelial cell nuclei preceding clinical AIDS. See AIDS

hairy polyp Teratoid tumor, dermoid tumor of nasopharynx A benign congenital malformation of young, often ♀ children arising from ecto- and mesodermal totipotent cells in the nasopharynx, oropharynx and tonsils DIFFDX Intranasal glioma, rhabdomyosarcoma, meningoencephalocele, Rathke pouch cyst, pharyngeal hypophysis, craniopharyngioma

hairy tongue Benign elongation of tongue's filiform papillae and ↓ desquamation, a nidus for microorganisms, often with halitosis ETIOLOGY Idiopathic, cigarette, alcohol abuse, RT to head & neck, dehydration, systemic illness, antibiotic therapy, causing ↓ flow of saliva, with unrestrained fungal growth

halberd bone Battle ax bone PEDIATRIC RADIOLOGY Marked flaring of iliac crests, typical of metatropic dwarfism, a rare metaphyseal dysplasia with bulbous joints and multiple, markedly deformed and shortened bones

Halcion® Triazolam PHARMACOLOGY A hypnotic with serious side effects–eg, paranoia and severe anxiety

Haldol® Haloperidol, see there

half moon sign BONE RADIOLOGY A semilunar shadow that may disappear in posterior dislocation of the humerus with the glenoid capsule as, normally, the medial humeral head overlaps the glenoid fossa

half-life $T_{1/2}$ The amount of time required for a substance to be reduced to one-half of its previous level by degradation and/or decay–radioactive half-life, by catabolism–biological half-life, or by elimination from a system–eg, half-life in serum HEMATOLOGY The time that cells stay in the circulation–eg, RBCs 120 days–which ↑ after splenectomy, platelets–4-6 days, eosinophils–3-7 hrs, PMNs–7 hrs IMMUNOLOGY The time an Ig stays in the circulation: 20-25 days for IgG, 6 days for IgA, 5 days for IgM, 2-8 days for IgD, 1-5 days for IgE THERAPEUTICS The time that a therapeutic agent remains in the circulation, which reflects its rate of metabolism and elimination of parent drug and metabolites in the urine and stool. See Effective half-life

HALF LIFE IN HOURS

DRUG	ADULT	CHILDREN
Digoxin	6–51	11–50
Gentamycin	2–3	
Lithium	8–35	
Phenobarbital	50–150	40–70
Phenytoin	18–30	12–22
Procainamide	2–4	
Quinidine	4–7	
Theophylline	3–8	1–8
Tobramycin	2–3	
Valproic acid	8–15	
Advance/Lab Feb 1995, p19		

halfway house PSYCHIATRY A specialized sheltered environment for persons undergoing rehabilitation for mental illness or addiction disorders, including drug and alcohol abuse, who do not require in-patient hospitalization, but need an intermediate degree of care before returning to independent community living; HHs are often staffed by their own 'graduates' or professionals who provide guidance and if necessary, treatment. See Co-counseling. Cf Homelessness, Shelter

halitosis Bad breath An offensive oral odor caused by either oral pathology–eg, poor dental hygiene with bacterial growth in plaques, acute or chronic gingivitis, or fungal overgrowth, GI pathology–eg, food entrapment in Zenker's diverticulum. See Body odor, Odors

Hallervorden-Spatz syndrome Hallervorden-Spatz disease NEUROLOGY An AR condition first seen in childhood or adolescence CLINICAL Progressive neurologic degeneration with defects in muscle tone–rigidity, choreoathetosis, torsion spasm–dystonia, parkinsonism, cerebellar ataxia, speech impairment, mental deterioration, generalized rigidity IMAGING Cerebral atrophy by CT DIFFDX Dementia with extrapyramidal motor defects, Wilson's disease MANAGEMENT None PROGNOSIS Poor; death within 10 yrs of onset

Hallpike maneuver NEUROLOGY A test used to evaluate vertigo–eg, benign paroxysmal positional vertigo, by observing nystagmus induced by positional changes

hallucination NEUROLOGY A complex sensory perception that occurs without external stimulation, characterized by false or distorted perception of objects or events–eg, sights, sounds, tastes, smells, or sensations of touch, often accompanied by a powerful sense of reality. See Command, Functional, Hypnogenic, Hypnopompic, Olfactory hallucination. Cf Illusion, Schizophrenia

hallucinogen Psychedelic agent, psychomimetic PHARMACOLOGY A chemical agent that induces hallucinations AGENTS Cannabis, DMT, Dronabinol, Ketamine, LSD, Phencyclidine, Psilocybin. See Hallucination

hallux abductus PODIATRY A fixed angulation of the hallux directed away from the body midline

hallux valgus PODIATRY A condition characterized by lateral deviation of the metatarsal head, which is angled dorsomedially, ↑ metatarsal angle, and abduction and valgus rotation of great toe CLINICAL Pressure, pain over 1st metatarsal head, ± shooting pains at rest MANAGEMENT Orthotics to control pronation, accommodative padding, wide shoes, shoe modification, cryomassage, NSAIDs, extra-articular steroid injections

halo appearance *adjective* Referring to a doughnut-shaped light density within and surrounded by a rounded darker dense zones BONE PATHOLOGY A rimming of osteocytic lacunae, seen by LM in X-linked hypophosphatemia, due to delayed maturation CARDIAC IMAGING A halo of radiolucency seen on a plain AP chest film in pneumopericardium GI IMAGING The 'halo' may be seen either in duodenal ulcers, where the radiocontrast settles in a fixed location, typically occurring in an acute ulcer within a mound of radiolucent edema surrounding the crater; alternately, 'halo' refers to a saccular collection of radiocontrast surrounded by radiolucency, seen in the rare intraluminal duodenal diverticulum, causing a windsock deformity GYNECOLOGIC IMAGING The 'halo' or wall sign is a smooth-contoured delineation of one tissue density from another, seen in a mature cystic teratoma or dermoid on a plain abdominal film, enhanced by the fibrous capsule investing these benign tumors HEPATIC IMAGING An avascular radiolucency surrounded by radiodensity, which corresponds to ↑ vascularity, seen by hepatic angiography in hemangiomas OBSTETRIC IMAGING Spaulding's halo sign A semispherical shadow seen on a plain film of the maternal abdomen, corresponding to the skull of a dead infant that is covered by edematous skin and soft tissue over the skull RENAL IMAGING Hypernephroma halo A radiolucent rim surrounding perinephric fat caused by a diffuse renal mass, first described in renal cell carcinoma or hypernephroma that may be seen in abscesses, hematomas, disseminated malignancy and pancreatitis

halo device ORTHOPEDICS A device used to manage cervical spine injuries to minimize neurological damage, requiring long-term immobilization; in the halo device, pins are inserted on the outer skull for skeletal traction, using a 2-3 kg weight for upper cervical injuries and 10-15 kg for lower cervical injuries

halo effect The beneficial effect of a physician or other health care provider on a Pt during a medical encounter, regardless of the therapy or procedure provided. See Hawthorne effect, Placebo effect, Physician invincibility syndrome

haloperidol Haldol® PSYCHIATRY A dopamine-blocking antipsychotic for treatment-refractory schizophrenia; it ↓ hallucinations, delusions, thought disturbances, ameliorates withdrawal and apathy, used to control Sx, and to prevent relapse; it may be used for Tourette syndrome ADVERSE EFFECTS

Sedation, weight gain, extrapyramidal signs–which may be may be immediate–eg, parkinsonism, akathisia and acute dystonia, or late with long-term therapy–eg, tardive dyskinesia, 'rabbit' syndrome, orthostatic hypotension, cardiovascular, GI effects, potentially fatal neuroleptic malignant syndrome. See Neuroleptic malignant syndrome, Positive & Negative Symptom Scale, 'Rabbit syndrome,' Schizophrenia. Cf Clozapine

halothane ANESTHESIOLOGY A potent inhalation anesthetic, which is most commonly used in the US in children. See Inhalation anesthetic. Cf Isoflurane

halothane hepatitis A potentially fatal hepatitis evoked by halothane, a carbon tetrachloride-related halocarbon used as a maintenance anesthetic–thiopental is often used for induction

HALT-C VIROLOGY A clinical management trial–hepatitis C antiviral long-term treatment to prevent cirrhosis

HAM HTLV-I-associated myelopathy. See Tropical spastic paraparesis

HAM syndrome Hypoparathyroidism, Addison's disease and mucocutaneous candidiasis. See MEDAC syndrome

HAMA Human anti-mouse antibody, see there

hamartoma A tumor-like, non-neoplastic disordered proliferation of mature tissues that are native to a site of origin–eg, exostoses, nevi and soft tissue hamartomas; although most hamartomas are benign, some histologic subtypes–eg, neuromuscular hamartoma, may proliferate aggressively. See Mesenchymal cystic hamartoma, Sclerosing epithelial hamartoma, Sclerosing metanephric hamartoma

hamartosis Any congenital disease complex with hamartomas, often associated with neoplasms–eg, dyskeratosis congenita, incontinenti pigmenti, linear nevus sebaceous syndrome, von Recklinghausen's neurofibromatosis, tuberous sclerosis, xeroderma pigmentosa and Gardner, Goltz, Klippel-Trenaunay-Weber, Mafucci, neurocutaneous/McCune-Albright, melanosis, multiple lentigines/LEOPARD, multiple neuroma, Sturge-Weber and von Hippel-Lindau syndromes

Hamman sign PHYSICAL EXAM A crunching sound synchronous with heartbeat, which may be heard in Pts with mediastinal emphysema, pneumopericardium

hammer toe Hallux valgus PODIATRY A flexion deformity of the proximal interphalangeal joint–PIP of lesser toes, due to an imbalance of the intrinsic foot muscles; HT may occur when longer toes are pressed back into line with other toes most commonly from tight shoes, affecting 2nd to 4th; curling can cause a painful corn on dorsum of PIP or tenderness on plantar aspect of affected toe CONSERVATIVE MANAGEMENT Make deeper toe box, slitting shoe over affected toe, strapping toe into extension, or manipulating the toe to maintain mobility SURGICAL TREATMENT Keller arthroplasty to remove exostosis, proximal phalangectomy

Hampton sign RADIOLOGY A wedge-shaped lung infarction seen on a CXR in PTE. Cf Westermark sign

hamstring injury SPORTS MEDICINE A muscle injury of biceps femoris, seen in sprinters and runners, when a contracted muscle meets a lengthening force, overpowering intrinsic muscle resiliency MANAGEMENT RICE, NSAIDs, gradual ↑ of pain-free activity–eg, isometric, isotonic, and isokinetic stages. See RICE

HANC™ network OUTPATIENT CARE A computerized system, allowing Pts to conduct EKG, BP, pulse, temperature and pulse oximetry testing at home, transmitted by modem to a central nursing station for 24/7 Pt support. See Chemscope/HANC HealthTech

hand See Alien hand, Bayonet hand, Clawhand, Machinist's hand, Main-en-trident, Rosebud hand, Spade hand, Windmill hand

hand-arm vibration syndrome White finger syndrome OCCUPATIONAL MEDICINE A Raynaud-like complex due to cold-induced vasospasm resulting from prolonged use of vibrating hand-held tools AT RISK Pts Assembly line workers, grinders, mechanics, jack-hammer operators DIAGNOSIS Plethysmography to ID changes in digital blood flow

hand-assisted laparoscopic surgery SURGERY A type of surgery which combines open surgery and minimal invasion/rapid recovery of laparoscopy APPLICATIONS Minimally invasive kidney donation; repair of renal, ureteral, and bladder defects; partial nephrectomy, gastrectomy, splenectomy, nephrectomy, sigmoidectomy. See Laparoscopic surgery

hand-eye coordination Eye-hand coordination SURGERY Oculomanual synchronization, required by surgeons, especially for laparoscopic surgery. See Laparoscopic surgery, Paradoxical movement

hand-foot syndrome Sickle cell dactylitis HEMATOLOGY A 'crisis' in sickle cell anemia caused by sludging of RBCs in vessels and characterized by symmetric infarction of the small bones of the hand and foot, periosteal neoosteogenesis, pain and swelling that may occur as early as age 18 months, often resolving spontaneously within 1-4 wks. See Sickle cell anemia

hand-foot-flat face syndrome An AD condition characterized by a flattened physiognomy, flexion and extension deformities of the hands and feet, mental and growth retardation

hand-foot-genital syndrome Hand-foot-uterus syndrome An AD condition characterized by minor digital anomalies of hands and feet and uterine duplication extending to cervix and vagina, with ↑ stillbirth and perinatal death in fetuses of afflicted mothers; genital abnormalities in affected ♂ include hypospadias with chordee

hand, foot & mouth syndrome NEONATOLOGY An infection spread oral-fecally to infants due to Coxsackievirus A16, as well as by A5, A9, A10, B2, B7 and enterovirus, resulting in vesicular stomatitis and exanthema, with summer-fall cycles CLINICAL After a 4-6 day prodrome of malaise, low-grade fever, anorexia, respiratory Sx, 0.5 cm intraoral maculopapular then vesicular ulcers–sore throat and refusal to eat, vesicular lesions of hands, buttocks and feet, resolving in a wk MANAGEMENT Symptomatic

hand grip strength NEUROLOGY A measure of muscle strength, evaluated with a Jamar dynamometer, often ↓ in older folks

hand-washing HOSPITAL CARE Digitocarpal ablution, usually understood to follow Pt contact, long recognized as the easiest way to prevent transmission of pathogens from one Pt to another

Hand-Schüller-Christian disease A form of histiocytosis X, characterized by a childhood onset of osteolytic lesions of skull and sella turcica, loss of teeth, chronic draining ears. See Langerhans' cell histiocytosis

handheld electronic prescribing THERAPEUTICS A format for prescribing drugs in which the prescribing physician keys in Pt demographics and a script, which is then sent via the internet to the pharmacy

handicap SOCIAL MEDICINE Any of a broad range of physical and mental disabilities which substantially limit a person's major life abilities and opportunities. See Americans with Disabilities Act, Disability

handicapped health care provider A health care worker or physician with physical limitations, usually of sufficient duration to have learned to compensate for them. Cf 'Impaired' health care provider, 'Incapacitated' health care provider

HANE Hereditary angioneurotic edema, see there

HANES Health And Nutrition Examination Survey A series of dietary surveys first carried out in 1971 by the NIH–US; HANES I determined that Americans consumed suboptimal levels of iron, calcium and vitamins A and C; HANES III is under the auspices of the National Center for Health Statistics and will determine the weights of Americans

hang a shingle MEDICAL PRACTICE An Americanism for the opening of a private office by a professional–eg, physician, lawyer. See Private practice

hangman's fracture ORTHOPEDICS A bilateral avulsion fracture which extends through the pedicles of C2 at the pars interarticularis, without injuring the odontoid process, causing acute central cervical spinal cord damage with axial dislocation in the C3 vertebral body; HF occurs by extension/distraction–judicial hanging or extension/axial loading–eg, hitting a windshield in an MVA; neurologic injury is rare in hospitalized cases; forward subluxation of C2 on C3 may be present; ± 80% of HFs occur in MVAs, the rest from falls and diving–resulting in marked distraction at fracture site and at C2-C3 disc; immobilization is by collar in Pts with extension/distraction injury; collar or tongs with extension/axial load injuries; bracing with a halo is usually the most effective treatment; a collar should be used only if there is < 2 mm fracture displacement. See C2 fracture. Cf Capital punishment

hangover headache SUBSTANCE ABUSE Intense cephalgia and malaise on waking, after a night of binge drinking or with prolonged use of benzodiazepines; the HH is often accompanied by mental dulling, hyperacusia, mild incoordination, tremor and nausea, due to the toxic effects of alcohol and its metabolites and dehydration. See Eyeopener

hantavirus EPIDEMIOLOGY A zoonotic virus of the family Bunyaviridae; it is spherical, ±100 nm in diameter, has a single RNA strand in 3 segments surrounded by a lipid envelope, destroyed by lipid solvents–eg, alcohol, disinfectants, bleach VECTOR Deer mouse, caused "4 Corners" outbreak; other HVs cause pulmonary Sx–Black Creek Canal virus, Florida VECTOR Cotton rat; Bayou virus, Louisiana VECTOR Rice rat; NY-1 virus, New York VECTOR White-footed mouse

hantavirus pulmonary syndrome An often fatal RTI caused by a hantavirus; the first cluster occurred in the Four Corners region of Southwestern US EPIDEMIOLOGY Mean age 32, 61% ♀, 72% Native American CASE DEFINITION Unexplained bilateral interstitial infiltrates on chest film, arterial O_2 saturation < 90% PRODROME Fever, myalgia-100%, cough or SOB-76%, GI Sx, headache, tachycardia, tachypnea, hypotension LAB Leukocytosis–± 26 x 10^9/L, often with myeloid precursors, ↑ Hct, thrombocytopenia of ±64 x 10^9/L, ↑ PT, ↑ aPTT, ↑ LD, ↓ serum protein, proteinuria DIAGNOSIS Immunohistochemistry, Western blot, ELISA, IgM capture ELISA; massive acute pulmonary edema is common at death

HAPC Hospital-acquired penetration contact, see there

HAPE High-altitude pulmonary edema

haploid GENETICS *adjective* Referring to: 1. A normal chromosome complement–expressed as **n** 2. A cell with only one copy of each chromosome type–ie, half the number of chromosomes present in other cells; gametes–ie, sperm and ova are haploid. See Chromosome, Diploid, Polypoid, Prokaryote

'happy' A popular adjective referring to a zealous inclination to perform a particular task or activity, eg, trigger-happy.

happy letter A letter sent by a physician, hospital, lab, or other health care provider to a Pt regarding test results–eg, 'pap' test ('happy pap' letter) or battery of tests which indicates no abnormalities requiring therapy. Cf Recall letter

Happylite™ PSYCHOLOGY A 10,000-lux that simulates natural daylight, administered in brief 'zaps' to symptomatic Pts with SAD. See Seasonal affective disorder

haptic interface ROBOTICS A user interface that allows a computer or robotics-driven device to interpret the sensation of touch; HIs are used in surgical rehearsal systems to imitate the body's reaction to touch, and incorporated into minimally invasive instruments to give surgeons a better "feel" for the tissue without invasive access

haptic sense Sense of touch

haptoglobin A protein in the circulation that migrates in the α_2 portion of serum subjected to electrophoresis, a so-called acute phase reactant that ↑ in serum in acute inflammation or infection, stress, or necrosis ROLE Bind Hb released from RBCs undergoing hemolysis, preventing the accumulation of Hb in plasma; after iron has been removed, the haptoglobin-bound Hb is eliminated by the reticuloendothelial system.

harassment See Sexual harassment

hard clot An insoluble, usually intravascular thrombus in which the fibrin monomers are terminally cross-linked by coagulation factor XIII with calcium ions

hard data Tangible data–eg, serum levels of an analyte, or number of people in a defined region with a certain disease. See Data. Cf Soft data

hard drug Any intensely addictive substance of abuse that may compel its user to commit crimes to obtain the drug–eg, crack cocaine, heroin. See Gateway drug

hard liquor A popular term for beverages with a high–often > 30% by volume–ie, 60 proof alcohol content–eg, gin, rum, vodka, whiskey; HLs are preferred by alcoholics as a steady state of low-level inebriation is easier to maintain. See Standard drink

hard medical services MANAGED CARE Tangible medical services–eg, diagnostic and therapeutic procedures. Cf Soft medical services

hard-on VOX POPULI Erection, penile tumescence. See Nocturnal penile tumescence

hardware INFORMATICS The electronic and mechanical, non-software components of a computer–eg, central processing unit, monitor, disk drives, keyboard, 'peripherals'–eg, printer, modem, scanners ORTHOPEDICS Any device–eg, nut, bolt, pin, plate, screw, and wire placed in bone to stabilize a fracture or to anchor a prosthetic joint. See Bone glue

hard water Water with a high content of calcium and magnesium salts–eg, carbonates and sulfates, which may interfere with certain lab tests. See Mineral water, Spring water; Cf Soft water

hard x-rays RADIATION PHYSICS Short wavelength, high-frequency and highly penetrating megavolt range–eg, produced by ^{60}Cobalt–X-rays used in RT or generated by nuclear 'incidents'. Cf Soft X-rays

hardening of the arteries Atherosclerosis, see there

harlequin nail A fingernail in a person who abruptly stopped smoking; the proximal zone of new growth is pink while the outgrowing nail formed during nicotine exposure–the 'nicotine sign' is a dirty yellow-brown

harlequin skin reaction PEDIATRICS A benign erythema that longitudinally 'sections' the prone infant into a pale upper and a vividly flushed lower half–most prominent from forehead to pubis, related to autonomic vascular lability; HSR is rare and more common in premature or LBW neonates; like erythema neonatorum, it is a transient vasomotor phenomenon of no clinical significance

harmonic scalpel SURGERY An ultrasound-powered cutting tool that cuts and seals tissue simultaneously. See Ultrasonography

harmonic ultrasound imaging IMAGING A method designed to improve ultrasonographic detection of contrast agents in blood-containing cavities and vascularized tissue. See Contrast echocardiography. Cf Sonicated albumin, Transient response imaging

harp player's thumb PERFORMING ARTS MEDICINE A type of nerve compression caused by irritation of the ulnar and radial digital nerves of the thumb by harp strings CLINICAL Numbness, paresthesias, painful hypo- and hypersensitivity, which may lead to perineural fibrosis and formation of a painful nodule TREATMENT Rest, change strumming and plucking patterns; surgery is not indicated

Harrell v Total Health Care MALPRACTICE A landmark case decided in Missouri in 1989 that ruled that an HMO owes a duty to the Pt to conduct a reasonable investigation of a physician's credentials and reputation in the community. See Credentialing

HART CARDIOLOGY A clinical trial–Heparin-Aspirin Reperfusion Trial that compared adjuncts to tPA thrombolysis in Pts with acute MI. See Acute myocardial infarction, Heparin, tPA

Hartmann procedure SURGERY An operation for an obstructing unresectable rectosigmoid CA in which a proximal colostomy is created in the left lower quadrant and the distal rectal stump is sewn/stapled shut, forming a Hartmann pouch. See Colorectal cancer

Hartnup disorder Hartnup's disease MOLECULAR MEDICINE An AR inborn defect in tryptophan metabolism caused by a defect in intestinal brush-border amino acid transport CLINICAL Photosensitivity, ataxia, emotional lability, aminoaciduria

Harvard criteria A series of 4 parameters delineated by the the Harvard Medical School ad hoc committee for irreversible coma

HARVARD CRITERIA FOR BRAIN DEATH
- Unreceptivity and unresponsiveness
- No movement or breathing
- No reflexes
- Flat electroencephalogram (confirmatory)

IN ADDITION, THE FOLLOWING MUST BE PRESENT
- Body temperature ≥32° C
- Absence of CNS depressants

Harvard-Hsiao study HEALTH CARE FINANCING The study that resulted in a relative value scale based on resources necessary to generate medical services. See Health Care Financing Administration, RBRVS

Harvest Moon phenomenon SOCIAL MEDICINE An ↓ in mortality in a person or population that occurs before a symbolically meaningful event or occasion

Hashimoto's thyroiditis Autoimmune thyroiditis, Hashimoto's autoimmune thyroiditis, Hashimoto's disaease, lymphadenoid goiter, struma lymphomatosa A form of thyroiditis most common in ♀ age 30-50, presents as a diffuse firm thyroid enlargement and/or with a familial Hx of goiter, hypothyroidism, Graves' disease or antithyroid antibodies; other diseases with autoimmune substrate associated with HT include chronic active hepatitis, DM, megaloblastic anemia, PBC, rheumatoid arthritis, Sjögren syndrome CLINICAL Goiter, often little else, but, if extreme, esophageal or tracheal compression LAB Antimicrosomal antibodies; while euthyroid state is maintained, ↑ RAIU, ↓ TSH, normal T_3, normal T_4; with nonresponse to TSH, ↓ RAIU, ↓ T_4, ↑ T_3 which reflects maximum stimulation MANAGEMENT Hormone replacement PROGNOSIS Excellent COMPLICATIONS Rarely, lymphoma

hashish Marijuana, see there

hashitoxicosis Hashimoto's thyroiditis–chronic autoimmune thyroiditis presenting as hyperthyroidism

HASI CARDIOLOGY A clinical study–Hirulog Angioplasty Study Investigators–that compared treatment with hirulog to heparin in Pts undergoing coronary angioplasty for unstable or post-infarction angina. See Acute myocardial infarction, Heparin, Hirulog, Unstable angina

hassle factor MANAGED CARE Any time-consuming and/or paperwork-ridden maneuver required of physicians, pharmacologists and other health care professionals before a 3rd party payer–eg, Medicaid, Medicaid, insurance company—approves payment for a medical therapy, service or drug

hatband sign A unilateral sweat stain on a hat, seen in Horner syndrome, classically associated with lung CA

hatchet face The characteristic physiognomy of advanced myotonic dystrophy; the face is drawn and lugubrious, with hollowing of muscles around temples and jaws; eyes are 'hooded,' lower lip droops, and global weakness of facial muscles, causes sagging of lower face, accompanied by marked wasting of the neck muscles, especially the flexors, which imparts a 'swan-neck' appearance; HF may be seen in amyotrophic lateral sclerosis, Curschmann-Batten-Steiner syndrome

HAV Hepatitis A virus, see there

hawkinsinuria An AD form of tyrosinuria, named after the index family, which presents in infancy with severe metabolic acidosis, ketosis, FTT, transient tyrosinemia, ↑ excretion of *p*-hydroxyphenylpyruvic and *p*-hydroxyphenylacetic acids as well as unusual tyrosine metabolites, one of which is hawkinsin MANAGEMENT Restrict phenylalanine and tyrosine in diet–resolves spontaneously with age without mental retardation or hepatopathy. See Failure to thrive

Hawthorne effect PSYCHOLOGY A beneficial effect that health care providers have on workers in most settings when an interest is shown in the workers' well-being. See Halo effect, Placebo effect, Placebo response. Cf Nocebo

hay fever A popular term for a seasonal allergic rhinitis caused by pollen and characterized by itching and tearing of eyes, swelling of nasal mucosa, attacks of sneezing, often asthma. See Allergic rhinitis

hazard OCCUPATIONAL HEALTH An adverse health or ecologic effect; a source of risk if an exposure pathway exists, and if exposures have possible adverse consequences. See Health hazard, Hospitalization hazard, No apparent public health hazard, Physical hazard

Hazardous Substances & Health Effects Database OCCUPATIONAL HEALTH The database developed by ATSDR to manage data collection, retrieval, analysis, and utilization through the sophisticated technologies provided by computerization; HazDat allows ATSDR to locate information on the release of hazardous substances into the environment, and to ascertain the effects of hazardous substances on health with improved uniformity, efficiency, and precision

hazardous waste OCCUPATIONAL HEALTH Any unwanted waste product that poses a hazard or potential hazard to human health, which may be generated by a manufacturing process–eg, radioactive gas cylinders, chemicals, pesticides, acids, and liquid or by a health care facility, including regulated biohazardous waste. See Regulated waste, Toxic dump

Hb Hemogolobin, see there

HBIg Hepatitis B immunoglobulin A 3-5 ml preparation of antibodies against HBV, derived from donor pools and administered at the time of presumed exposure to HBV. See HBV

HBLV Human B-lymphotropic virus, see Herpesvirus-6

HBsAg Hepatitis B surface antigen, see there

HCFA Health Care Financing Administration, pronounced HICK-fah MANAGED CARE The preferred term is now Centers for Medicare & Medicaid Services–CMS, an agency of the US Dept of HHS that administers Medicare, the federal part of Medicaid and oversees Medicare's health financing; HCFA establishes standards for medical providers that require compliance to meet certification requirements. See CMS, JCAHO, Medicaid, Medicare

HCFA 1500 MANAGED CARE The official standard form used by physicians and other providers when submitting bills/claims for reimbursement to Medicare or Medicaid for health services; it is also used by private insurers and managed care plans; HCFA 1500 contains Pt demographics, diagnostic codes, CPT/HCPCS codes, diagnosis codes, units. See Compliance, HCFA, Medicare, Red tape. Cf UB92

hCG Human choriogonadotropic hormone A dimeric glycoprotein hormone synthesized and secreted by the placenta composed of a 92 amino acid α subunit similar to pituitary hormones–TSH, FSH and LH, and a unique 145 amino acid β-subunit linked to lactose and hexosamine that is produced by the syncytiotrophoblast; the β subunit is an early marker for pregnancy; in the fetal liver and kidney, and certain tumors; hCG is released in a pulsatile fashion in normal pregnancy, peaks at the 10th gestational wk with serum levels of 50-140 000 IU/L, later falling to 10-50 000 IU/L²; hCG may be low to undetectable in ectopic pregnancy or ↑↑ ↑ in multiple gestation, polyhydramnios, eclampsia, erythroblastosis fetalis or trophoblastic disease–eg, hydatidiform mole, choriocarcinoma and placental site trophoblastic tumor; hCG levels may be ↑↑↑ in tumors producing either the β subunit or both α & β subunits, measured by ELISA or RIA

hCG-like activity Any functional cross-reactivity of the α subunit of TSH, LH, and FSH with the α subunit of hCG, such that ↑ TSH/LH/FSH mimics ↑ hCG if measured by RIA

HCO₃ Bicarbonate, see there

HCT Hematocrit, see there

HCV 1. Hepatitis C virus, see there 2. Human coronavirus. See Coronavirus

HCW Health care worker, see there

HDL-cholesterol Good cholesterol Cholesterol bound to high-density lipoprotein; ↑ HDL-C and HDL-C:total cholesterol ratio are linked to ↑ longevity and ↓ morbidity and death from MI, CAD, cholesterol-related morbidity ; ↑ HDL-C commonly affects ♀ with HTN or obesity, and may be related to environmental causes–eg, alcohol consumption, or therapy with H₂-blockers or estrogens. See HDL/LDL ratio. Cf LDL–'bad' cholesterol

HDL/LDL ratio The ratio of cholesterol carried by high-density lipoprotein to that carried by low-density lipoprotein. See HDL cholesterol, LDL-cholesterol

HDN Hemolytic disease of the newborn, see there

HE 1. Hemoglobin electrophoresis, see there 2. Hereditary elliptocytosis, see there 3. Human enteric

head banger tumor PEDIATRICS A lesion that may develop in children who bang their heads as part of a 'routine' for falling asleep, which consists of a mass of organizing fibrous tissue covered by hyperpigmented skin

head drop sign PEDIATRIC NEUROLOGY A finding evoked by raising an infant's trunk at the shoulders; a head that drops backward suggests nuchal limpness typical of paralytic and non-paralytic poliomyelitis

head gear SPORTS MEDICINE A device worn on the head to minimize impact. See Football helmet, Helmet

headhunter A popular term for a person–or employment agency who recruits physicians, upper echelon executives or other professionals, matching potential employees with employers

head lice

head lice *Pediculosis capitis* PUBLIC HEALTH A louse transmitted in crowded conditions–eg, day care centers, homeless shelters TREATMENT Topical insecticides–permethrin, synergized pymethrin, malathion. See Crabs

head & neck cancer ONCOLOGY A CA of head & neck–brain, ocular, salivary gland, skin, thyroid gland, and nonepithelial neoplasms–eg, lymphomas and sarcomas; in practice, the term refers to SCC of H&N mucosae–oral cavity, oropharynx, hypopharynx and larynx, or one that excludes CA of skin, CNS, eye, thyroid, lymph nodes STATISTICS ±43K new cases–21K oral, 12,500 laryngeal, 9K pharyngeal/yr in the US, 1992 with 11,600 deaths, most common in ♂ > age 50; worldwide ± 500,000 new cases/yr ETIOLOGY Tobacco–smoked, smokeless, alcohol, woodworking, textile fiber exposure, nickel refining; dietary factors–eg, carotenoids, ↑ fruit and vegetable consumption may be protective; nasopharyngeal CA is associated with EBV infection, especially in southern Asia CLINICAL Signs and Sx are site-specific–eg, in oral cavity leukoplakia, erythroplakia; elsewhere, nasal obstruction, epistaxis, otalgia, serous otitis media, unilateral sore throat, cranial neuropathy MANAGEMENT–LARYNX, HYPOPHARYNX Stage III, IV, total laryngectomy, or induction chemotherapy–cisplatin, 5-FU, followed by RT, the latter of which allows voice preservation in ±40% of Pts OROPHARYNX Platinum-based chemotherapy + RT PROGNOSIS Metastatic or recurrent disease is usually incurable

head of department Chief of service The physician or scientist in charge of a department of a hospital or medical school

head-tilt/chin-lift maneuver EMERGENCY MEDICINE A maneuver used in rescue breathing, which is required for performing mouth-to-mouth and mouth-to-nose resuscitation. See Rescue breathing. Cf Jaw-thrust maneuver

HEAD-TILT/CHIN-LIFT MANEUVER

HEAD TILT One hand is placed over victim's forehead and firm, backward pressure is applied with palm to tilt the head back

CHIN LIFT Fingers of other hand are placed under bony portion of the lower jaw near the chin to bring the chin forward and the teeth almost to occlusion, which supports the jaw and helps tilt the head back; the fingers must not press deeply in the soft tissues under the chin, as this might obstruct the airway; for mouth-to-nose breathing, ↑ pressure is applied by the chin hand to further close the mouth; dentures are removed only if necessary

headache PAIN MANAGEMENT A painful sensation localized in the cranium. Related terms are Alarm clock headache, Benign headache, Benign exertional headache, Cluster headache, Food headache, Hangover headache, Hotdog headache, Ice cream headache, Migraine headache, Orgasmic headache, Postoperative headache, Roller coaster headache, Tension headache, Weight-lifter's headache

HeADDFIRST NEUROLOGY A clinical trial–Hemicraniectomy And Durotomy for Deterioration From Infarction Relating Swelling Trial

headlight effect Flashbulb effect IMAGING A hyperintense bright region seen in brains of Pts with early strokes when examined by diffusion-weighted imaging–a type of MRI, which corresponds to slowed movement of water through intra- and extracellular compartments. See Diffusion-weighted imaging, Magnetic resonance imaging, Stroke

HEADSS PSYCHOLOGY An interviewing technique that addresses the major psychological stressors–Home, Education, Activities, Drugs, Sex, Suicide–for adolescents, in a format that encourages development of rapport with the adolescent client, beginning with the least stressful–home–and progressing to areas of ↑ stress

healer MAINSTREAM MEDICINE A romantic synonym for physician. See Traditional healing

healing by first intention Primary union

healing VOX POPULI The process of returning to a previous state of health; the term is often used by alternative medical practitioners

Healon® OPHTHALMOLOGY A viscoelastic material used in capsulorrhexis and foldable intraocular lens implant procedures. See Intraocular lens

Healos™ ORTHOPEDICS A fully resorbable, sponge-like matrix of collagen mineralized with resorbable hydroxyapatite to promote bone healing, as an alternative to autologous bone graft; Healos is a bioactive scaffold that mimics bone's biochemical and architectural properties. See Bone graft

health EPIDEMIOLOGY A state of complete physical, mental, and social well-being and not merely the absence of disease or infirmity PUBLIC HEALTH Per WHO's Ottawa charter, '...a concept for everyday life...emphasizing social–peace, shelter, education, food, income, a stable ecosystem, sustainable resources, social justice & equity *and* personal–physiologic–resources; the WHO also defined health as '...*a state of complete physical, mental, and social well-being and not merely the absence of disease or infirmity'*. See African holistic health, E-health, Baseline health, Health promotion, Ottawa Charter, Russian health, Sexual health, Telehealth

health balance PUBLIC HEALTH A dynamic equilibrium between a person and the environment. See Health, Health promotion

health care access SOCIAL MEDICINE The ability of a person to receive health care services, which is a function of 1. Availability of personnel and supplies and 2. Ability to pay for those services

health care facility MEDTALK Any place where people receive health care–eg, hospital or clinic

Health Care Financing Administration See CMS/HCFA

health care provider A person who provides any form of health care–eg, physician, nurse, dentist, mental health worker, birth control counselor, STD manager. See Provider

health care proxy END-OF-LIFE A power of attorney for health-care decision-making in which a person designates another to make medical decisions in the event that he/she becomes too incapacitated to make such decisions. See Advance medical directive, Living will

health care rationing The limitation of access to or the equitable distribution of medical services, through various gatekeeper controls. See Gatekeeper. Cf Coby Howard, Oregon plan, Rule of Rescue, 'Squeaky wheel'

health care reform The 'Clinton Plan', proposed by President(s) Bill & Hillary Clinton in 1993, to revamp and address inequities in US health care. See Managed care

health care service 1. A business entity that provides inpatient or outpatient testing or treatment of human disease or dysfunction; dispensing of drugs or medical devices for treating human disease or dysfunction 2. A procedure performed on a person for diagnosing or treating a disease

health consultation A response to a specific question or request for information pertaining to a hazardous substance or facility, including waste sites; HCs often contain a time-critical element. Cf Health assessment

health data EPIDEMIOLOGY Information related to health conditions, reproductive outcomes, causes of death, and quality of life

health education PUBLIC HEALTH Activities that promote health and provide information and training about hazards in the environment that would ↓ exposure, illness, or disease.

health food POPULAR HEALTH A food defined by the lay public as a food with little or no preservatives, little processing, enrichment or refinement, often grown without pesticides. See Food preservatives, 'Go' food, Healthy food, Organic food. Cf Diet, Junk food, 'No' food

health fraud A general term for those practices in which either 1. Health services are promised and/or paid for, but not provided at an appropriate standard of professionalism or skill. See Quackery or 2. Practices in which health care is provided–or allegedly provided, but reimbursement claims to Medicare, Medicaid or other 3ʳᵈ-party payer are fraudulent. See Medicare fraud

health hazard OCCUPATIONAL SAFETY Any agent or activity posing a potential hazard to health. Cf Physical hazard

health history HEALTH INSURANCE A form used by underwriters to assist in evaluating groups or persons to determine whether they are acceptable risks to insure.

health information system EPIDEMIOLOGY A combination of health statistics from various sources, used to derive information about health status, health care, provision and use of services and health impact INFORMATICS Hospital information system A system that provides information management features that hospitals need for daily business FEATURES Pt tracking, billing and administrative programs; may include clinical features

health information trustee HEALTH INFORMATION A party designated by the Bennett-Leahy bill that would be allowed access to a person's personal and privileged–private medical information. See Anne Sexton, Bennett-Leahy bill, Confidentiality, Doctor-patient relationship, HIPAA, Medical privacy, Therapist-patient privacy

Health Insurance Portability Accountability Act See HIPAA

health insurance Accident & health insurance, accident & sickness insurance, medical expense insurance, sickness Insurance HEALTH FINANCING Insurance that provides lump sum or periodic payments in case of losses occasioned by bodily injury, sickness or disease, and medical expense

health intervention HEALTH CARE An activity undertaken to prevent, improve, or stabilize a medical condition

health literacy HEALTH CARE A measure of a person's ability to understand health-related information and make informed decisions about that information; HL includes interpreting prescriptions and following self care instructions. See Literacy

health maintenance organization See HMO

health outcome HEALTH CARE INDUSTRY An outcome or result of a medical condition that directly affects the length or quality of a person's life

health outcome data Data derived from databases at local, state, and national levels, and from data collected by private health care organizations and professional institutions TYPES OF HOD M&M, birth statistics, medical records, tumor and disease registries, surveillance data, previous health studies

health plan Health benefits package MANAGED CARE Any plan or organized format for delivering health care services–eg, HMO, indemnity, medigap, preferred provider organization, point-of-service plans

health professional affiliate HOSPITAL PRACTICE A person other than a practitioner whose Pt care activities require independent health care judgment within his/her area of professional competence, and provide specified Pt care services; HPAs include clinical psychologists, dental auxiliaries, nurse clinicians/practitioners, physician assistants, therapists–respiratory and physical, nurse anesthetists. See Allied health professional

health profile Health panel, health screen A battery of tests designed to identify one or more morbid conditions in a person who is asymptomatic at the time of the screening

health promotion Health advocacy PUBLIC HEALTH Any activity that seeks to improve a person's or population's health by providing information about and ↑ awareness of 'at risk' behaviors associated with various conditions, and ↓ those behaviors. See Health, Ottawa charter

health-related quality of life measure Functional status measure, health status measure, quality of life measure SOCIAL MEDICINE A patient outcome measure that extends beyond traditional measures of M&M, including dimensions such as physiology, function, social activity, cognition, emotion, sleep/rest, energy/vitality, health perception, general life satisfaction; some are also known as health status, functional status, or quality of life measures. See Quality of life

health services MANAGED CARE The benefits covered under a health contract

health statistics review Evaluation of information and relevant health outcome data for a population–eg, reports of injury, disease, or death in the community; databases may be local, state, or national, from private health care providers and organizations, M&M, tumor and disease registries, birth statistics, surveillance data

health surveillance Medical monitoring The periodic screening of a defined population for a specific disease or for biological markers of disease for which the population is, or may be, at significantly ↑ risk. See Screening

health yuppie An often egocentric 'yuppie'–young, upwardly-mobile professional–who is a physician, dentist, or other highly paid health care worker

healthy diet CLINICAL NUTRITION Any diet based on sound nutritional principles; the HD philosophy is often coupled with the belief that organic and/or unprocessed foods–ie, produced without pesticides and chemical preservatives, are superior to adulterated foods

HEALTHY DIETS, FEATURES OF

- High consumption of fruits & vegetables
- Low consumption of red meat & fatty foods
- Raw foods & whole grains are preferred to processed or refined foods
- Protein primarily from fish, dairy products, nuts
- Consumption of salt, pepper, sugar, coffee & other caffeinated beverages, and alcohol, is discouraged

healthy food Healing food, Health-promoting food NUTRITION Any food believed to be 'good for you', especially if high in fiber, natural vitamins, fructose, etc; HFs may ↓ cholesterol, ↓ ASHD, stroke, help control glucose, halt progression of osteoporosis, ↓ infections, CA EXAMPLES Apples, beans, carrots, cranberry juice, fish, garlic, ginger, nuts, oats, olive oil, soy foods, tea, yogurt. See Fiber, Fruit and vegetables. Cf Health food, Junk food, Unhealthy foods

healthy habit Good habit, see there

Healthy People 2000 PUBLIC HEALTH A series of health goals promulgated by the US Public Health Services that provides insight into the state of American health; the priorities include health promotion–ie, targeting risk behaviors and health protection, which includes environmental or regulatory measures that confer broad-based protection, as well as preventative services and surveillance systems, addressing numerous specific issues. See Mortality, YPLL

healthy weight NUTRITION The ideal weight for a particular height. See Diet, Obesity. Cf Desirable weight

heaped-up appearance GI DISEASE An endoscopic morphology seen in malignant gastric ulcers where the margins of the lesion are irregular, with overhanging borders, and a 'dirty', necrotic base. Cf Punched-out appearance. See Meniscus sign of Carmen

hearing AUDIOLOGY A series of events in which sound waves in air are converted to electrical signals that are sent as nerve impulses to the brain where they are interpreted

hearing aid AUDIOLOGY A battery-powered electro-acoustic device that brings amplified sound to the ear to improve hearing, generally worn in the ear. See Hearing loss

hearing disorder AUDIOLOGY Any disruption in the normal hearing process, where sound waves are not converted to electrical signals or nerve impulses are not transmitted to the brain for interpretation

hearing loss An ↓ ability to hear, which may be genetic or acquired, pre- or postlingual, and syndromic or nonsyndromic. See Noise-induced hearing loss, Occupational hearing loss, Prelingual hearing loss, Postlingual hearing loss, Sensorineural hearing loss

hearing loss of aging Presbycusis AUDIOLOGY The progressive loss of high pitch auditory discrimination seen in advanced age, affecting ±50% of those > age 75. See Hearing loss

HEARING LOSS TYPES

CONDUCTIVE HL caused by damage to any mechanical components of the ear, eg accumulation of cerumen, disruption of tympanic membrane, fusion of one or more middle ear ossicles; conductive HLs may be amenable to surgery

SENSORINEURAL HL due to a defect in the neural pathways, which may be at the level of the cochlea, auditory nerve, or in the cerebral cortex ETIOLOGY Infections–especially intrauterine, eg rubella, drug toxicity, eg aminoglycosides, and tumors, eg acoustic neuroma

MIXED HL due to a combination of mechanical and neural defects ETIOLOGY Down syndrome, cystic fibrosis, cerebral palsy

hearing sparing procedure ENT A procedure intended to spare the vestibular labyrinth, and maintain the auditory apparatus at status quo ante EXAMPLES Endolymphatic sac decompression, labyrinthine fistula procedures; procedures sparing cochlea & destroying labyrinth–vestibular neurectomy, chemical vestibular ablation

hearing test Hearing function test, pure tone audiometry AUDIOLOGY Any test that measures hearing or quantifies hearing loss; sound is perceived by intensity–loudness and by tone; both are measured in HTs, as is the ability to hear sound through air–air conduction and bone–bone conduction. See Hearing

heart See Abiomed implantable heart, Athlete's heart, Baby Fae heart, Bleeding heart, Crisscross heart, Dextroposition heart, Depraved heart, Egg-shaped heart, Flabby heart, Flask-shaped heart, Holiday heart, Left heart, Mongolian heart, Old soldier's heart, Penn State heart, Right heart, Sabot heart, Second heart, Stone heart, Swinging heart, Water bottle heart

HEART CARDIOLOGY A clinical trial–Healing & Early Afterload Reducing Therapy

heart attack VOX POPULI Acute MI–myocardial infarction. See Myocardial infarction

heart block CARDIOLOGY An arrhythmia in which an impulse is not conducted normally from the atria to the ventricles due a damaged cardiac conduction system TYPES 1st, 2nd, 3rd degree; various HBs can occur in the bundle of His or bundle branches

HEART BLOCK

FIRST-DEGREE Prolonged interval ≥ 200 ms/0.2 sec between atrial and ventricular depolarizations

SECOND-DEGREE A block of some, but not all, impulses traveling from the atria to the ventricles. It can take 2 forms

• Mobitz Type I or Wenckebach phenomenon–successive prolongation of the P-R interval until one P-wave does not elicit a QRS response

• Mobitz Type II block–occurs when occasional P-waves are blocked from ventricles

THIRD-DEGREE Complete Heart Block Condition in which all impulses from the atria are blocked; usually an ectopic focus in the ventricles takes over

heartburn VOX POPULI A popular term for burning retrosternal, often postprandial discomfort due to reflux of gastric contents into the esophagus, associated with dysfunction of the lower esophageal sphincter; heartburn may be idiopathic or associated with Barrett's esophagus, duodenal ulcers, reflux, scleroderma and Zollinger-Ellison syndrome. See GERD

heart disease Cardiac disorder, cardiovascular disease CARDIOLOGY Any disease that affects the heart–eg, alcoholic cardiomyopathy, angina, arrhythmias, ASHD, cardiogenic shock, cardiomyopathies–dilated, hypertrophic, idiopathic, ischemic, peripartum, CHF, CAD, HTN, mitral regurgitation, mitral stenosis, mitral valve prolapse, etc. See Atherosclerotic heart disease

heart failure Cardiac failure, cardiac insufficiency, congestive heart failure CARDIOLOGY The loss of the heart's ability to effectively pump blood, which may affect the right, left, or both sides of the heart; with the loss of pumping action on the right side, blood may back up into other areas of the body–eg, the liver, GI tract, extremities; the heart may be unable to pump blood efficiently to the lungs. See Congestive heart failure, Diastolic heart failure

heart laws OCCUPATIONAL MEDICINE Legislature that distinguishes cardiovascular injury and its consequences due to work-related physical exertion and/or mental stress, from the natural progression of underlying cardiovascular disease

heart-lung machine Extracorporeal circulation, pump-oxygenator HEART SURGERY A device with a pump and a blood oxygenator, used during open heart–eg, cardiothoracic and cardiovascular surgery; the HLM does the work of the heart–pump blood *and* lungs–oxygenating the blood, by routing the blood through the machine before returning it to the arterial circulation

heart-lung transplantation Heart-and-lung transplantation THORACIC SURGERY The surgical removal of the heart and lung block in a Pt in whom both are failing; HLT is performed at specialized centers OUTCOME Adequate ventilation despite loss of innervation and ↑ tidal volume INDICATIONS PRIMARY LUNG DISEASE–chronic disturbance in gas exchange and alveolar mechanics and 2° ↑ in pulmonary vascular resistance PULMONARY VASCULAR DISEASE–1° high-resistance circulatory disorder–HLT is beneficial in absence of tracheobronchial and alveolar defects. See Heart transplantation. Cf Domino donor transplantation, Lung transplantation, UNOS

HeartMate® HEART SURGERY A battery-operated left ventricular-assist system–LVAS implanted between the heart and aorta which pumps blood, a bridge for Pts awaiting heart transplant, allowing Pt outpatient mobility. See Heart transplant, Jarvik 7

heart murmur See Murmur

heart rate The number of heart beats/min; the rate at which the heart pumps blood; normal range, 50 to 90 beats/min

heart rate variability test Swing test CARDIOLOGY A test that measures swings in heart rate over 24 hrs via a portable EKG, identifying high-risk CHF Pts; if HRV is high, prognosis is good; if low, treatment is adjusted to prevent early demise. See Chaos theory

heart scan See MUGA

heart sound CARDIOLOGY Any normal or abnormal sound heard by auscultation of the heart. See Fourth heart sound, Murmur, Third heart sound

heart transplantation Cardiac transplantation CARDIAC SURGERY The surgical replacement of a defective heart with another, usually donated by an MVA victim; ± 2000 Pts/month–US need HTs; ± 100/month are performed CAUSE OF DEATH Infection, acute or chronic rejection, embolism, pancreatitis, peptic ulcer, etc Actuarial survival 74% at 6.5 yrs; mortality ↑ with ↑ age, > in ♂, more common in Pts without prior dilated cardiomyopathy REDUCTION OF REJECTION Steroids, immunosuppressants, daclizumab, an IgG1 MAb which blocks high-affinity IL-2 receptors. See Daclizumab

heart tumor A rare tumor of the heart; 1° HTs are usually benign—myxoma > rhabdomyoma > osteoclastoma; malignant HTs: angiosarcoma, rhabdomyosarcoma; 2° HTs are usually malignant and implant in the pericardium from lung or hilar lymphoma or melanoma. See Atrial myxoma

heart valve prosthesis HEART SURGERY A natural—eg, porcine or synthetic valve used to replace a damaged—stenosed or 'insufficient' cardiac valve; ±50,000 are performed/yr—US. See Shiley valve,

HeartWatch® cardiac-event recorder CARDIOLOGY A dual-electrode, cardiac-monitoring device worn on the wrist and activated for recording transient cardiac events—eg, arrhythmias, palpitations, dizziness and SOB, as Sx occur for later evaluation of EKGs. See Holter monitoring

heat index PUBLIC HEALTH A measure of the effect of combined elements—heat & humidity—on the body. See Heat-related death

heat intolerance SPORTS MEDICINE A condition caused by the thermal challenges of exercise, resulting in various responses from cramps and exhaustion to heat syncope, stroke, death. See Heat-related death, Heat wave

heat lamp Infrared lamp, see there

heat-related death FORENSIC MEDICINE A death with a core body temperature ≥ 40.6°C/105°F with no other reasonable explanation of death AT-RISK GROUPS Elderly, those living alone, alcoholics. See Heat wave

heat stress disease CRITICAL CARE A group of conditions due to over-exposure to or overexertion in excess environmental temperatures

HEAT STRESS-FORMS IN INCREASING SEVERITY

HEAT CRAMPS Non-emergent and treated by salt replacement

HEAT EXHAUSTION More serious, treated with fluid and salt replacement

HEAT STROKE Most commonly affecting extremes of ages, especially the elderly, accompanied by convulsions, delusions, coma and treated by cooling the body and replacement of fluids and salts

Note: The body's reaction to heat is a function of controllable—use of anticholinergics, phenothiazines, alcohol, heavy exercise, clothing, obesity, direct exposure and acclimatization and uncontrollable factors—high ambient temperatures or humidity, lack of air circulation, underlying fever, old age or infancy, ectodermal dysplasia

heat stroke PUBLIC HEALTH A condition characterized by rapidly rising internal body temperature that overwhelms the physiologic mechanisms releasing heat, resulting in death if not appropriately managed

heat wave PUBLIC HEALTH A climatologic event in which the air temperature is > 42°C or 103°F for a period of 5 days or more

heavy chain disease Any of a family of monoclonal gammopathies or paraproteinemias characterized by excess production of an immunoglobulin Fc fragment, detectable in the serum and/or urine and accompanied by lymphoproliferative disease. See Hyperimmunoglobulin D, E, M syndromes

HEAVY CHAIN DISEASE

ALPHA HEAVY CHAIN DISEASE Seligmann's disease The most common heavy chain disease or paraproteinemia, in which there is an excess production of an incomplete IgA1 molecule—partial heavy chain and no light chain, which affects Sephardic Jews, Arabs and Mediterranean rim inhabitants CLINICAL Onset in childhood or adolescence as either a lymphoproliferative disorder confined to the respiratory tract or an enteric form—see IPSID with severe diarrhea, malabsorption, steatorrhea, weight loss, hepatic dysfunction, hypocalcemia, lymphadenopathy, marked mononuclear infiltration which may eventuate into lymphoma—see Mediterranean lymphoma; AHCD may remit spontaneously, respond to antibiotic therapy, or if clearly monoclonal, may require combination chemotherapy, potentially causing death by ages 20-30 LABORATORY ↑ Alkaline phosphatase, ↓ Ca²⁺ TREATMENT Antibiotics, or if monoclonal, chemotherapy

GAMMA HEAVY CHAIN DISEASE A disorder of older ♂, ranging from fulminant, ie death in weeks to prolonged, lasting 20 years, most die in the first year, often due to infection CLINICAL GHCD presents as a lymphoproliferation with fever, fatigue, anemia, angioimmunoblastic lymphadenopathy, hepatosplenomegaly, uvular and palatal edema, eosinophilic infiltrates, leukopenia, associated with autoimmune disease, tuberculosis and lymphoma LABORATORY ↑ IgG1; most cases excrete less than 1 g/day of paraprotein, rarely up to 20 g/day TREATMENT Cyclophosphamide, vincristine, prednisone

MU CHAIN DISEASE A rare paraproteinemia that affects the middle-aged to elderly, most of whom have or slowly progress to CLL Clinical Lymphadenopathy, hepatosplenomegaly and BM infiltration by vacuolated plasma cells, often accompanied by ↑ kappa chain production TREATMENT As with CLL

heavy guns *adjective* Formidable, serious ONCOLOGY Referring to the use of aggressive chemotherapy in advanced malignancy, which is hit with HGs—eg, multiple agents and therapeutic modalities. See Induction

heavy metal TOXICOLOGY A metal that is 5 times heavier than water and often toxic, a feature linked to the HM's tight cationic binding to circulating proteins EXAMPLES Antimony, arsenic, bismuth, cadmium, copper, lead, mercury; cells respond to HMs by ↑ transcription of various genes—eg, the eukaryotic heat shock system, metallothioneins, prokaryotic mercury resistance gene and iron uptake systems CLINICAL General, fine tremor, speech slurring, stomatitis, drooling, cataracts, neurasthenia TREATMENT EDTA chelation. See Mad Hatter, Minamata disease, Queen of Poisons, Pink disease

hebephilia Ephebophilia SEXOLOGY Sexuoeroticism that hinges on imagery of or activity with an adolescents. Cf Gerontophilia, Nepiophilia, Paraphilic adolescentilism, Pedophilia

Heberden's node PHYSICAL EXAM A rounded lentiform exostosis seen on terminal phalanges in Pts with osteoarthritis

hectic fever Spiking fever INFECTIOUS DISEASE A highly nonspecific term for either a fever characterized by a daily spike in temperature, or one in which the peak and trough temperatures differ by 1.4°C; the term may be so clinically meaningless as to be completely abandoned. See Fever

Hectorol™ ENDOCRINOLOGY An agent for managing 2° hyperparathyroidism linked to ESRD, which may used in predialysis Pts

HEDIS Health Plan Employer Data & Information Set MANAGED CARE An initiative by the National Committee on Quality Assurance to develop, collect, standardize, and report measures of health plan performances. See National Committee for Quality Assurance, Satisfaction survey

hedonism Pleasure-seeking behavior. Cf Anhedonia

heel-knee test Heel-to-knee-to-shin test NEUROLOGY A test of voluntary motor coordination in which a person is asked to slowly touch the knee with the heel of the opposite leg, which is altered in cerebellar dysfunction. See Finger-nose test

heel pad ORTHOPEDICS—FOOT Calcaneal fat pad A flat encapsulated disk of fat that lies below the plantar surface of the calcaneus, which acts as a cushion for ambulation; normally < 21 mm; enlarged in the obese, with steroids, or in acromegaly ORTHOTICS A popular term for cushion placed inside athletic shoes for support to minimize the effects of high-impact sports

heel pad syndrome Fat pad impingement syndrome PODIATRY Pain under the weight-bearing part of calcaneus, due to one major or multiple minor trauamatic episodes, which disrupt the fibrous septae and fat pad MANAGEMENT Heel cup, donut padding, insoles, rest, NSAIDs, cryomassage

heel strike Heel contact The beginning of stance phase, at the point of heel strike there is zero reaction. Immediately after contact there is an ↑ in ground reaction, known as heel strike transient, which pre-empts the major ↑ in ground reaction forces which complete the first of the double peaks in normal stance phase.

Hegsted score A formula for determining a diet's relative lipid composition—a high value indicates ↑ saturated fatty acids—FAs and cholesterol and ↓ polyunsaturated FAs

290 **heights, fear of** PSYCHOLOGY Altophobia. See Phobia

Heimlich maneuver PUBLIC HEALTH A technique for removing a bolus of food stuck in the oropharynx potentially causing acute asphyxia. See Cafe coronary, Flake maneuver

Heinrichs–Carpenter Quality of Life Scale PSYCHIATRY A testing device–range of possible scores, 0–126 used to evaluate social functioning & behavior in Pts with schizophrenia–lower scores represent poorer mental health. See Schizophrenia. Cf Quality of life

helical scanning Spiral computed tomography, see there

Helicobacter pylori *Campylobacter pylori* GI DISEASE A curved gram-negative microaerophilic bacillus which causes most cases of gastritis; it is present in 10-50% of healthy young persons and up to 60% of those ≥ age 60 EPIDEMIOLOGY *H pylori* may be spread by houseflies–hand washing may ↓ transmission, by intrafamilial contact, or domestic cats Note: 90% of Pts with intestinal-type gastric CA have been infected by *H pylori*, which may be a co-factor for gastric CA, acting either to stimulate production of cellular mutagens or induce ↑ proliferation after cell damage; it has been implicated in antibiotic-responsive duodenal ulcers, and linked to other gastric lesions, benign and malignant DIAGNOSIS Gastric Bx with special–eg, Diff-Quik stain, serologic tests–90-95% accuracy, urea breath test DISEASE COURSE *H pylori* infection → chronic superficial gastritis, peptic ulcer disease, lymphoproliferative disease, ± → gastric lymphoma, chronic atrophic gastritis, ± → gastric adenoCA MANAGEMENT Proton-pump inhibitors–eg, omeprazole, lansoprazole, esomeprazole; antibiotics–amoxicillin, clarithromycin, metronidazole; ranitidine; bismuth subcitrate. See Duodenal ulcer, Gastric lymphoma, Urea breath test

heliophobia PSYCHOLOGY Fear of the sun. See Phobia

heliotrope rash DERMATOLOGY The classic rash seen on the face of Pts with dermatomyositis, so named for its purplish hue, likened to that of the fragrant herb, *Heliotropium peruvianum*

heliox CLINICAL THERAPEUTICS A mixture of O_2 and helium used to treat obstructive lung disease. See COPD

hell, fear of PSYCHOLOGY Hadephobia, stygiophobia, stigiophobia. See Phobia

Heller-Dor procedure SURGERY A laparoscopic procedure for treating esophageal achalasia by ↓ LES resistance. See Lower esophageal sphincter

HELLP syndrome OBSTETRICS A condition linked to eclampsia or severe pre-eclampsia, characterized by the acronym: Hemolysis, Elevated Liver function tests, Low Platelets, which may transiently worsen after delivery; other Sx include BP ≥ 160 systolic and/or 110 diastolic, urinary volume ≤ 400 mL/24 hrs, nonspecific cerebral abnormalities, pulmonary edema, cyanosis LAB Proteinuria ≥ 5 g/24 hrs, schistocytes–RBCs sheared and fragmented by intravascular fibrin deposition, ↑ LFTs–ALA, AST, ALT, BR, ↑ PT; aPTT, normal fibrinogen levels. See MTP complex

helmet PUBLIC HEALTH A personal protective device of hardened plastic worn on the head to ↓ severity of injuries in the event of an accident. See Pro cap helmet

helper T cell Helper T lymphocyte, CD4+ T cell IMMUNOLOGY A subset of T lymphocytes with the antigen determinant CD4, which are presented with a foreign antigen in the context of both a self MHC class II antigen and IL-1; once immune recognition or response occurs, HTCs produce various cytokines–eg, IFN-γ, IL-2 and osteoclast-activating factor, which are critical in hematopoietic differentiation, collagen synthesis, and antibody formation; HTCs play a central role in normal and pathologic immune responses by secreting various cytokines–eg, IL-2, IFN-γ, and TNF are involved in cell-mediated immunity TYPES Type 1–produces IL-2 and IFN-γ, ↑ in infectious granulomatous processes; type 2–produces IL-4 and IL-5; ↑ in atopic disorders, lepromatous leprosy, visceral leishmaniasis and as a clonal expansion in hypereosinophilic syndrome. See CD4, Flow cytometry, Suppressor cells, T cells

helper:suppressor ratio IMMUNOLOGY The ratio of CD4 helper T cells–lymphocytes to CD8 suppressor T cells, which normally is 1.5-2.0 See Flow cytometry

hem/onc A popular term for either 1. The combined specialty of hematology & oncology, see there or 2. A physician specialized in hematology & oncology. See Hematologist/oncologist

hemangioma SURGERY A tumor composed of clustered blood vessels, seen in the skin and elsewhere. See Capillary hemangioma, Congenital hemangioma, Postnatal hemangioma, Sclerosing hemangioma, Strawberry hemangioma, Swiss cheese hemangioma

hemapheresis Apheresis TRANSFUSION MEDICINE Removal of whole blood from a Pt or donor, followed by separation into components, some of which are removed; the rest is returned to the Pt INDICATIONS See Cytapheresis, Hemodialysis, Leukapheresis, Plasmapheresis

HEMAPHERESIS, THERAPEUTIC INDICATIONS
- Leukocytes in hyperleukemic leukostasis with > 100 x 10⁹/L blasts
- Platelets in thrombocytosis with > 1000 x 10⁹/L platelets, if symptomatic
- Defective RBCs, replacing them with normal RBCs, as in sickle cell anemia with crisis
- Igs causing hyperviscosity syndrome in macroglobulinemia/multiple myeloma
- Autoantibody production in myasthenia gravis, Goodpasture syndrome, SLE, factor VIII antibodies and
- Lipoproteins in Pts with homozygous familial hypercholesterolemia

hemaphobia PSYCHOLOGY Fear of blood. See Phobia

hemarthrosis ORTHOPEDICS Blood in a joint, which may be caused by trauma or linked to a coagulopathy–eg, hemophilia

Hemaseel APR™ SURGERY A hemostatic and sealing adhesive and tissue glue comprised of fibrinogen, thrombin, aprotinin INDICATIONS Adjunct to hemostasis–replicates natural coagulation–in cardiopulmonary bypass, splenic repair, colostomies. See Fibrin glue

HemAssist TRANSFUSION MEDICINE A proprietary–Baxter Intl–non-FDA-approved artificial blood, consisting of polyethylene glycol conjugated to cross-linked Hb from outdated blood. See Artificial blood, Blood substitutes

hematemesis Spitting up blood MEDTALK The passage of blood from the mouth, most often due to ruptured esophageal varices

hematocele UROLOGY Pooled blood in the scrotum. Cf Spermatocele

hematochezia Passage of bright red or maroon stool, usually due to blood; upper GI tract hemorrhage resulting in hematochezia implies a blood loss of ≥ 1000 ml, often accompanied by hypovolemia, hypotension, and tachycardia; lower GI or rectal hemorrhage may be guaiac-positive with ±25 ml–ie, blood is brighter red when more caudal

hematocrit Hct LAB MEDICINE The ratio of RBCs to total blood volume in a centrifuged sample of blood, expressed as a percentage. See Hemoglobin, Mean corpuscular volume

hematogenous *adjective* Originating in the blood or spread via the blood-stream

hematologic malignancy Hematologic cancer HEMATOLOGY Any CA of blood-forming tissues, BM, or lymph nodes–eg, leukemia and lymphoma

hematologist A physician specialized in disease of blood cells; most hematologists are also board certified in oncology MEAT & POTATOES DISEASES Anemia, leukemia, coagulopathies. See Hematology, Hem/onc, Oncologist

hematologist/oncologist A physician trained in hematology and oncology TRAINING 3 yr of internal medicine + 2 yrs of specialized hematology and oncology MEAN SALARY $182K

hematology The subspecialty of internal medicine that diagnoses and treats diseases of the blood and its cell components. See Hematologist, Hematology/oncology

hematology/oncology The combined specialty of hematology and oncology

hematoma A tumor-like mass produced by coagulated blood in a cavity. See Cerebral hematoma, Epidural hematoma

hematopoietic *adjective* Blood-forming

hematopoietic stem cell transplantation HEMATOLOGY A therapy in which defective hematopoietic cells are replaced with normal BM cells after chemotherapy and/or RT INDICATIONS AML, breast CA, CML, germ cell tumors, lymphoma, myelodysplastic syndrome, myeloma, neuroblastoma, ovarian CA, Wilms' tumor; nonmalignant lesions–congenital metabolic diseases, SCID

hematoporphyrin derivative ONCOLOGY An agent used in photodynamic therapy that is absorbed by tumor cells which is activated when exposed to light, killing tumor cells. See Photodynamic therapy

hematotoxic *adjective* Referring to substances or activities with adverse effects on coagulation, hemorrhage or hemolysis, leading to circulatory collapse. Cf Neurotoxic

hematuria MEDTALK Blood or RBCs in urine, which is either grossly visible or microscopic; the finding of hematuria requires consideration of the entire urinary tract DIAGNOSIS IVP, cystoscopy, urine cytology MANAGEMENT Directed toward underlying cause. See Benign persistent hematuria, Microscopic hematuria

heme An iron-containing red pigment which, with a protein, globin, forms hemoglobin

heme-negative MEDTALK *adjective* Referring to stool that does not appear to contain the heme commonly associated with colorectal CA; heme-negativity makes CA less likely. See Occult blood. Cf Heme-positive

heme-positive MEDTALK *adjective* referring to stool that contains heme, often associated with colorectal CA. See Occult blood. Cf Heme-negative

hemi- *prefix*, Latin, half

hemi-homograft HEART SURGERY A new valve replacement surgical technique in which only half of the valve is replaced with a cadaveric valve; only the diseased half is removed; the healthy half remains. See Porcine valve

hemianopia Hemianospia NEUROLOGY Loss of one half of the field of vision

hemiballism NEUROLOGY A hyperkinetic movement disorder characterized by violent flailing movements of the arm and leg contralateral to a lesion–hemorrhage or infarction, or rarely tumor of the subthalamic nucleus; hemiballism follows recovery from a stroke-induced hemiparesis and hemisensory defect CLINICAL Motor Sx may have a rotatory component in the shoulder or hip, accompanied by choreiform movements–repeated flexion and extension of hands and feet; movements usually disappear during sleep MANAGEMENT Neuroleptics–eg, chlorpromazine, haloperidol, reserpine, antipsychotics, or other agents that interfere with dopamine transmission–eg, reserpine

hemihypertrophy A unilateral ↑ in paired organs or tissues of the body ETIOLOGY Idiopathic, possibly related to neural, vascular, lymphatic, endocrine or chromosomal abnormalities APPEARS IN Curtius, Klippel-Trenaunay, Silver-Russel syndromes and hepatoblastoma

hemilaminectomy ORTHOPEDICS Unilateral excision of the vertebral lamina with removal of variable parts of the adjacent facet , a procedure used to treat a herniation of an intervertebral disk. See Laminectomy

hemilaryngectomy SURGICAL ONCOLOGY The partial excision of the larynx for invasive cancer, which is a voice-conserving procedure that consists of dividing the thyroid cartilage in the midline and resecting in continuity the thyroid cartilage with the corresponding true and false vocal cords and ventricle. See Laryngectomy

hemiparesis NEUROLOGY Weakness of one side of the body

hemipelvectomy SURGICAL ONCOLOGY A form of 'heroic' surgery in which an entire lower extremity including the hemipelvis with disarticulation of the sacroiliac joint and symphysis pubis, and removal of all major muscles to the lower extremity except the iliopsoas; hemipelvectomy is required for large malignant soft tissue tumors of the buttock and anterior or lateral proximal thigh. See Forequarter amputation, Hindquarter amputation

hemiplegia NEUROLOGY Complete paralysis of one side of the body. See Alternating hemiplegia

hemispheric disconnection syndrome NEUROLOGY A functional dissociation between the right and left cerebral hemispheres, such that one hand does not visually or tactually recognize that which is presented to the ipsilateral hemisphere CLINICAL Tactile agnosia, ideomotor dyspraxia, agraphia of left hand and constructional apraxia of right hand

hemivertebra ORTHOPEDICS A congenital vertebral body defect arising from a simple–ie, nondysplastic, nonmetabolic embryonal defect, in which the anterior half of a vertebral body is absent and adjacent vertebrae expand to fill the void, accompanied by preservation of interspaces. See Vertebra. Cf Butterfly vertebra

hemlock TOXICOLOGY Any of the poisonous herbs of the carrot family, especially *Conium maculatum*, but also *Tsuga* spp, which contain an alkaloid, conine, that evokes CNS hyperactivity, followed by medullary depression and respiratory failure TREATMENT Activated charcoal

hemo- *prefix*, Latin, pertaining to blood

Hemobahn endovascular prosthesis CARDIOLOGY A self-expanding nitinol stent coated with an ultra-thin PTFE–polytetrafluoroethylene. See Stent-graft

Hemoccult sensitivity LAB MEDICINE A change in sensitivity to the Hemoccult test for occult blood in stool. See Heme positive

hemochromatosis METABOLIC DISEASE An AR excess of GI iron absorption with progressive iron loading in parenchymal organs, the gene for which is linked to the HLA locus on the short arm of chromosome 6 EPIDEMIOLOGY The gene occurs in 10% of the white population, most are unaffected heterozygotes; 5/1000 of US whites are homozygotes CLINICAL Fatigue, cold intolerance, edema, cardiac arrhythmias, dyspnea, hepatosplenomegaly, ascites, spider telangiectasia, hyperpigmentation, palpitations, joint pain, ↓ ROM, impotence, and amenorrhea related to hypothalamic, cardiac, hepatic and pancreatic dysfunction; cardiomegaly 300-fold ↑, cirrhosis 13-fold ↑–once cirrhosis occurs, there is a 200-fold ↑ in liver cancer, gray-bronze skin pigmentation and a 7-fold ↑ in DM–'bronze diabetes', arthropathy and hypogonadism–impotence, testicular atrophy ASSOCIATED PATHOLOGY DM, hypopituitarism, hypogonadism, hypoparathyroidism, certain infections–*E coli, K pneumoniae, Listeria monocytogenes, Mucor* spp, *Rhizopus arrhizus, Salmonella enteritidis* serotype typhimurium, *V vulnificus, Y enterocolitica* LAB Transferrin saturation–TS > 62% warrants workup, ↑ iron, ↑ transferrin saturation, ↑ serum ferritin, ↑ transaminases, hyperglycemia; ↓ TSH, ↓ gonadotropins DIAGNOSIS Screen for iron, TS > 60% ≥ 50% ♀ ≥ 60% ♂ ; ferritin; liver biopsy. See HFE. Cf Juvenile hemochromatosis, Neonatal hemochromatosis

hemochromatosis triad Hepatomegaly, DM, skin pigmentation, seen in hemochromatosis, caused by the deposition of iron pigment in the liver, pancreas, heart, pituitary, glands–adrenal, parathyroid, thyroid, and joints. See Hemochromatosis

hemodialysis NEPHROLOGY A therapeutic procedure for removing low-molecular-weight toxins by allowing the blood to flow past a semipermeable membrane where the toxins diffuse away from the blood down a concentration gradient, either via an external AV shunt, or a surgically-placed AV fistula; hemodialysis is used in renal failure to ↓ BUN, creatinine, hyperkalemia and correct metabolic acidosis; prolonged dialysis results in a poor quality of life–anemia, infections, myalgia, peripheral neuropathy, cerebral edema, acute MI, aluminum toxicity, ergo renal transplants are always preferred to long-term hemodialysis COMPLICATIONS Pyogenic reactions due to gram-negative endotoxemia, with chills, fever, hypotension, nausea and myalgia, ↓ NK cell activity, ↓ serum calcitriol. See Dialysis dementia. Cf Hemapheresis

hemodilution CARDIOVASCULAR SURGERY A technique used during cardiopulmonary bypass to ↓ risk of intraoperative thrombosis diluting the blood with crystalloids, ↓ hematocrit to ≤ 0.25-0.30 TRANSFUSION MEDICINE The removal of blood from a Pt at induction of anesthesia while replacing the volume with crystalloids–eg, saline or lactated Ringer's solution, or with colloids–eg, albumin, hydroxyethyl starch, dextrans, or purified protein fractions, or a combination of the 2. See Albumin, Dextrans, Hydroxyethyl starch, Intraoperative hemodilution, Purified protein fractions, Surgical blood management

hemodynamic dysfunction CARDIOLOGY Disturbance or impairment of the circulatory system's ability to circulate blood ETIOLOGY Ischemic heart disease, valvular heart disease, HTN, etc

hemodynamic instability CLINICAL MEDICINE A state requiring pharmacologic or mechanical support to maintain a normal blood pressure or adequate cardiac output

hemodynamic monitoring CLINICAL MEDICINE A general term for the ongoing evaluation of hemodynamics

hemodynamic response CARDIOLOGY Response of the circulatory system to stimuli such as exercise, emotional stress, etc

hemodynamic success HEPATOLOGY A term referring to a post-TIPS ↓ of the portosystemic gradient below a threshold. See TIPS

hemodynamics CARDIOLOGY 1. The formal study of blood circulation 2. The status of blood flow in the circulation, the sum result of cardiac output and resistances–eg, vascular resistance to flow

hemoglobin Hb PHYSIOLOGY A tetrameric 64 kD protein that is the major constituent of RBCs, which transports O_2, and buffers CO_2 produced by respiration; Hb transports O_2 and CO_2 and which comprises 99% of the protein weight of RBCs; it is composed of 2 α chains, each 141 amino acids in length, encoded from the zeta chain gene on chromosome 16 and 2 β chains, each 144 amino acids in length, encoded from the contiguous eta, $G\gamma$, $A\gamma$ and delta chain genes on chromosome 11 FORMS OF HB HbF is formed in the fetus and is the major Hb until birth; at birth up to 30% of the hemoglobin is HbA; most adult Hb is HbA with small amounts of HbF and HbA_2; Hb defects are inherited and termed hemoglobinopathies. See Carboxyhemoglobin, Chemically modified hemoglobin, Fetal hemoglobin, Reduced hemoglobin

hemoglobin A1$_c$ Glycosylated hemoglobin, see there

hemoglobin C disease HEMATOLOGY An AR hemoglobinopathy most commonly affecting blacks, characterized by intermittent abdominal pain, arthralgias, headaches, splenomegaly, cholecystitis, gallstones LAB Mild–33% Hct anemia, 2–6% reticulocyte count; 90% of RBCs are targetoid, often with HbC crystals; ↑ iron turnover, ↓ RBC survival. Cf Sickle cell anemia

hemoglobin SC disease HEMATOLOGY A sickling disorder resulting from inheritance of a HbS gene from one parent and a HbC gene from the other; RBCs contain ± equal amounts of each Hb; HbA is absent; HbF may be ↑ CLINICAL Similar to, but less severe than, sickle cell anemia; by preadolescence, episodic skeletal or abdominal pain, moderate splenomegaly; unique to SC disease is an ↑ risk of retinal disease–proliferative retinopathy, retinal vascular disease, aseptic necrosis of femoral head, acute chest syndrome after fat embolism due to bone infarction. See Sickle cell anemia

hemoglobinopathy HEMATOLOGY A defect in either α or β hemoglobin, which may be quantitative or qualitative, congenital or–rarely —acquired; while the more common Hb defects–eg, HbS, HbC and thalassemias, cause a characteristic clinical picture, *'rare hemoglobin variants are variously ignored, misunderstood, misdiagnosed, feared, shunned or rejected...'* and are not accompanied by clinical disease. See Hemoglobin C disease, Hemoglobin SC disease, Sickle cell anemia, Thalassemia

HEMOGLOBINOPATHIES–MAJOR BIOCHEMICAL FORMS

Sickle Cell	Hgb S
Sickle/C disease	Hgb S, Hgb C
Hemoglobin C Disease	Hgb C
Thalassemia major	Hgb F
Thalassemia minor	Hgb A2

CLINICAL PRESENTATIONS OF HEMOGLOBINOPATHY

SICKLING PHENOTYPE, eg HbS, HbSC, HbS-Thalassemia

THALASSEMIC PHENOTYPE, eg Constant Spring, HbE, Lepore, Kenya, Vicksburg, Indianapolis

↑ **OXYGEN AFFINITY PHENOTYPE**, eg Bristol, Bucuresti/Louisville, Caribbean, Etobicoke, Hammersmith, Moscva, Okaloosa, Peterborough, Seattle, Torino

↓ **OXYGEN AFFINITY PHENOTYPE**, eg Altdorf, Istanbul, Baylor, Belfast, Boras, Buenos Aires, Cranston, Duarte, Djelfa, Freiburg, Geneva, Hopkins II, Koln, Lyon, Niteroi, Nottingham, Pasadena, Sabine, Santa Ana, St Louis, Shepherds Bush, Tak, Tours, Toyoake, Tübingen, Zürich

β hemoglobinopathy 1. Sickle cell anemia, see there 2. β-thalassemia, see there

hemoglobinuria HEMATOLOGY The presence of Hb in the urine which, if of sufficient quantity, colors urine, the intensity of which directly correlates with the quantity of Hb. See Paroxysmal cold hemoglobinuria, Paroxysmal nocturnal hemoglobinuria

hemolysis Destruction or lysis of RBCs

HEMOLYSIS

INTRACORPUSCULAR HEMOLYSIS

• Membrane defects, eg hereditary elliptocytosis, spherocytosis, stomatocytosis and paroxysmal nocturnal hemoglobinuria

• Metabolic defects, eg G6PD, pyruvate kinase deficiency

• Abnormal Hbs see Hemoglobin

EXTRACORPUSCULAR HEMOLYSIS

1° immune reactions, eg autoimmune hemolytic anemia

2° immune reactions, due to

• Infections, eg *Bartonella*, *Clostridia*, malaria, sepsis

• Neoplasia, eg lymphoma, leukemias

• Drug reactions due to the 'Innocent bystander' phenomenon (drug-antibody complex activates complement, causing intravascular hemolysis, eg quinidine), hapten-mediated —a protein-bound drug attaches to the red cell membrane, eliciting an immune response when the hapten-protein complex is recognized as foreign, evoking an immune response, eg penicillin acting as a hapten

• Induction of autoimmunity by RBC antigen alterations, eg Rh antigen

PHYSICAL, eg thermal, concentrated glycerol due to inadequate washing of frozen blood, bladder irrigation, cardiac valves

EXTRAVASCULAR Less severe, IgG-mediated and does not activate complement, eg Rh, Kell, Duffy LABORATORY ↓ haptoglobin, ↓ $T_{1/2}$ of circulating RBCs, ↑ indirect BR as liver capacity to conjugate BR–ergo direct BR is overwhelmed by massive hemolysis, ↑ LDH, Hb in blood and urine, hemosiderinuria, MetHb and metalbumin, ↑ urobilinogen in urine and feces, ↑ in acid phosphatase, K⁺, and prostatic acid phosphatase Clin Chem 1992; 38:575; peripheral smears demonstrate anisocytosis, polychromatophilia, nucleated RBCs, basophilic stippling; immune hemolysis is suggested by spherocytes NEJM 2000; 342:722cpc

INTRAVASCULAR More severe, IgM-mediated and requires complement activation, eg ABO blood groups **LABORATORY** ↑ free Hb Note: Clinically significant hemolysis is usually detected by hemagglutination, less commonly by hemolysis per se, which detects anti-P, $-P_1$, $-PP_1P^k$, $-Jk^a$, $-Le^a$, occasionally also anti-Le^b and -Vel

hemolytic anemia HEMATOLOGY Anemia which occurs when more RBCs are lysed than are produced; HA may be immune-mediated or non-immune, due to intrinsic or extrinsic RBC defects. See Nonimmune hemolytic anemia IMMUNOLOGY Immune hemolysis is alloimmune, autoimmune, and drug-induced, and may be intravascular or extravascular MICROBIOLOGY Hemolysis is characteristic of certain strains of streptococci and is divided into α- and β-hemolysis; γ 'hemolysis' is a complete misnomer. See Gamma-hemolysis, Innocent bystander hemolysis

hemolytic crisis HEMATOLOGY A 'crisis' common in sickle cell anemia, which consists of rapidly evolving anemia, leukocytosis, jaundice and fever. See Sickle cell anemia, Vaso-occlusive crisis

hemolytic disease of the newborn Alloimmune anemia of newborns, erythroblastosis fetalis NEONATOLOGY, PEDIATRIC HEMATOLOGY Hemolysis due to incompatibility of fetal antigens with maternal immune system, caused by production of maternal IgG antibodies in response to fetal RBCs that enter the maternal circulation; if the IgG response and sharing of circulations–as occurs in low-grade fetomaternal hemorrhage, is intense, erythroblastosis fetalis occurs CLINICAL See Hydrops fetalis LAB If hemolysis is intense, the excess unconjugated/indirect BR overloads infant's liver; because of blood-brain barrier immaturity, BR deposits in basal ganglia of the brain, causing cell death, and kernicterus. See Alloimmune anemia of newborn, Kell blood group, Kernicterus. Cf Hemorrhagic disease of the newborn

hemolytic transfusion reaction TRANSFUSION MEDICINE A therapy-related event mediated by 2 different mechanisms: 1. Intravascular hemolysis mediated by complement-fixing antibodies, 2. Extravascular hemolysis mediated by noncomplement-fixing antibodies CLINICAL Fever, chills, pain at infusion site, intense back pain, hypotension, sense of impending doom, chest tightness, acute dyspnea, brochospasm, anaphylaxis, hyperbilirubinemia, hemoglobinuria, DIC with fibrinolysis MANAGEMENT Stop transfusion, treat shock–vasopressors, IV fluids, cortiocosteroids, maintain high fluid throughput, monitor anemia, transfuse with compatible blood

hemolytic-uremic syndrome INTERNAL MEDICINE A condition often accompanied by a prodrome of bloody diarrhea, more common in summer and microangiopathic hemolytic anemia, thrombocytopenia and platelet defects, not uncommon in infants < age 2 LAB Impaired aggregation, depleted platelet serotonin, ADP, β-thromboglobulin. See TTP-HUS

HEMOLYTIC-UREMIC SYNDROME ETIOLOGY

- Prototypic or 'classic' form
- Post-infectious, eg *Shigella dysenteriae*-1, *S pneumoniae*, *Salmonella typhi*, occasionally viruses
- Hereditary forms: AD or AR associated with HTN
- Immune-mediated forms
- Associated with other diseases, eg HTN, connective tissue disease, immunosuppression, RT to kidneys
- Related to pregnancy and oral contraceptives

hemophilia HEMATOLOGY Any of 3 inherited coagulopathies due to deficiency of coagulation factor VIII–hemophilia A, factor IX–hemophilia B, factor XI–Rosenthal's disease or hemophilia C. Cf Thalassemia

hemophilia A Classical hemophilia, factor VIII deficiency hemophilia HEMATOLOGY An X-R coagulopathy due to a marked ↓ of factor VIII

PHYSIOLOGY Factor VIII circulates as a noncovalent complex with von Willebrand factor, which once cleaved by thrombin or by factor Xa enables factor VIII to bind to phospholipid surfaces of damaged cells and adherent activated platelets CLINICAL Hemophilia A is heterogenous as moderate–1—4% normal; or mild–5—25% normal factor VIII deficiency occurs; classic findings–joint hemorrhage of knee > elbow >etc, muscle hemorrhage, bruising, prolonged and potentially fatal post-traumatic or post-op hemorrhage LAB Normal platelets, normal prothrombin time, ↑ aPTT, ↓ factor VIII TREATMENT Recombinant factor VIII. See AIDSgate, Factor VIII

hemophilia B Christmas disease, factor IX hemophilia HEMATOLOGY An X-R form of hemophilia caused by a defect in factor IX; it is clinically similar to hemophilia A, and often first manifest by hemarthrosis with joint destruction; these Pts are often infected with blood-borne viruses–eg, HBV, HCV, HIV, parvoviruses; with time, IgG antibodies may develop against factor IX. See Hemophilia. Cf Thalassemia

hemophilia & HIV AIDS Hemophilia is intimately linked to AIDS–heat-treated factor VIII was not a blood product standard until 1985. See AIDS, AIDSgate, HIV

***Hemophilus influenzae* vaccine** see Hib

hemoptysis The passage of blood by mouth

HemoQuant GI DISEASE A home test for detecting occult blood in stool; samples are mailed to a lab and results provided in ± 1 wk. See Occult blood

hemorrhage MEDTALK *noun* Bleeding, which may be pooled or active *verb* To bleed. See Cerebral hemorrhage, Fetomaternal hemorrhage, Intracerebral hemorrhage, Intracranial hemorrhage, Lobar intracerebral hemorrhage, Splinter hemorrhage, Subarachnoid hemorrhage, Subdural hemorrhage

hemorrhagic diathesis MEDTALK Bleeding–like crazy, like stuck pig

hemorrhagic disease of the newborn A neonatal condition caused by vitamin K deficiency, the combined result of a lack of unbound maternal vitamin K, immaturity of the fetal liver and lack of vitamin K-producing bacteria in the infant colon CLINICAL Abrupt early postpartum onset with spontaneous nasogastric or intracranial hemorrhage, affecting up to 1/1000 neonates and carrying a 5-30% mortality, if untreated; the condition may be more common in breast-fed infants and is more severe and of earlier onset in infants of mothers receiving anticonvulsives–antagonistic to warfarin during pregnancy LAB ↑ PT due to extrinsic factor depletion, ↑ clotting time, ↓ liver-dependent coagulation factors. Cf Hemolytic disease of the newborn

hemorrhagic familial nephritis Alport syndrome, see there

hemorrhagic stroke NEUROLOGY An ischemic stroke in which blood enters necrotic brain tissue, which may not be accompanied by a worsening clinical status RISKS FOR HS Hemophilia, thrombocytopenia, sickle cell anemia, DIC, anticoagulants, HTN. See Stroke. Cf Intracerebral hemorrhage, Ischemic stroke

hemorrhoidectomy Hemorrhoid surgery SURGERY Excision of hemorrhoids INDICATIONS Refractory itching, pain, clots, bleeding. See Hemorrhoids

hemorrhoids Lump in the rectum, piles SURGERY Engorged veins under the rectal mucosa, associated with constipation, straining while squatting, pregnancy, prolonged sitting, anal infection. See External hemorrhoids, Internal hemorrhoids

hemosiderosis An iron overload syndrome arbitrarily differentiated from hemochromatosis by the reversible nature of the iron accumulation in the reticuloendothelial system. See Hemochromatosis

HemoSleeve™ ENT Bipolar electrosurgical sheath used with microdebriders to obtain hemostasis during ENT procedures

hemostasis INTERNAL MEDICINE Any natural or interventional stopping of blood flow. See Injection sclerotherapy

hemostats SURGERY A hand-held surgical instrument with flattened opposing surfaces used to occlude blood vessels for hemostasis

hemostatic plug A plug of platelets and proteins that forms at the site of tissue trauma or ruptured vessels. See Coagulation

hemothorax EMERGENCY MEDICINE The accumulation of blood in the pleural cavity, most often due to trauma, but also due to CA, surgery or infarction; a large hemothorax may cause respiratory failure. Cf Pneumothorax

Hemovac drain A blood recovery device used during surgery. See Blood salvage

hemp Marijuana, see there

HEMPAS Hereditary erythrocytic multinuclearity with positive acidified serum lysis, Congenital dyserythropoietic anemia, type II An AR condition characterized by an IgM autoantibody against the RBC's i antigen–'anti-HEMPAS', an antigen is present in $\frac{1}{3}$ of normal sera CLINICAL Mild or subclinical congenital dyserythropoietic anemia LAB Normocytic aniso-poikilocytosis and multi-nucleated erythroblasts; ↑ agglutinability of serum with anti-i antibodies TREATMENT None; splenectomy is partially beneficial. See CDA

Hennebert's phenomenon NEUROLOGY A sign associated with inner ear fistula characterized by vertigo and ↑ CNS pressure caused by nose blowing or lifting. See Inner ear fistula, Tulio's phenomenon

Henoch-Schönlein purpura Anaphylactoid purpura, vascular purpura INTERNAL MEDICINE An acquired form of small vessel vasculitis with IgA-dominant immune deposits affecting small vessels–arteries, capillaries, venules, typically of the skin, gut, and glomeruli, associated with arthralgia and/or arthritis; HSP is most common in younger Pts CLINICAL Red maculopapules on legs and buttocks, glomerulonephritis, abdominal pain ± infarction, diarrhea, fever, arthritis LAB IgA deposits in basement membrane of skin and glomerulus TREATMENT NSAIDs, aspirin, steroids–eg, prednisone, especially in Pts with major abdominal pain or kidney disease PROGNOSIS Usually self limited. See Small vessel vasculitis, Systemic vasculitis

HEPA respirator High-efficiency particulate air filter respirator A device with a replaceable filter that reduces particulate matter more effectively than simple isolation masks

heparin HEMATOLOGY A sulfated glycosaminoglycan anticoagulant that inhibits activated factors IXa, Xa, XIa, XIIa and thrombin, ↓ local antithrombin-III, promoting its inactivation by neutrophil elastase; interaction of heparin with endothelial cells results in displacement of platelet factor 4, which inactivates heparin INDICATIONS Thromboembolism, CAD, post acute MI, PTE MONITORING Titrate heparin so that aPTT is 1.5-2.0-fold normal SIDE EFFECTS Hemorrhage, thrombocytopenia, osteoporosis, skin necrosis, alopecia, hypersensitivity, hypoaldosteronism. See Low-molecular weight heparin

heparin-associated antiplatelet antibodies LAB MEDICINE Antibodies that develop in 2-5% of Pts receiving heparin, regardless of route of administration; HAAbs are associated with ↑ in risk of thrombotic events

heparin-induced thrombocytopenia Acquired thrombocytopenia affecting some heparin-treated Pts–HTPs, defined as a ↓ platelet count during or shortly after heparin exposure; HIT is a markedly prothrombotic disorder seen in Pts who are at a high baseline risk for venous thrombosis DIFFDX Antiphospholipid antibody syndrome, DIC, Trousseau syndrome–migratory thrombophlebitis & malignancy, TTP/HUS, thrombocytopenia and 'showers' of microthrombi CLINICAL HIT occurs 6-10 days after heparin exposure DIAGNOSIS Aggregation studies using platelets from Pts known to be sensitive to antibody; quantification of serotonin release from platelets radiolabeled with serotonin in assay using Pt serum

heparin lock INTENSIVE CARE A cylindrical multiport device with a rubberized access port which is incorporated into a venous access catheter and designed to facilitate IV administration of medications. See Access device, Heparin. Cf Saline lock

heparinize *verb* To manage a Pt long-term with heparin, see there

hepatectomy SURGERY Segmental resection of the liver INDICATIONS Cancer, parasites, major trauma–eg, MVAs

hepatic clearance THERAPEUTICS The hypothetical calculation of the volume of distribution in liters of unmetabolized drug cleared through the liver in 1 min–L/min. See Clearance

hepatic encephalopathy Hepatic coma NEUROLOGY A complication of hepatic failure due to ↑ ammonia in circulation, which may occur in decompensated cirrhosis or portocaval anastomosis; major finding in HE is cerebral edema, which ↑ morbidity; the Mayo Clinic devised a CT-based system–Brain Edema Severity Score for classifying HE MANAGEMENT It is uncertain whether treatment of cerebral edema ↑ survival in HE

HEPATIC ENCEPHALOPATHY CLINICAL FORMS

ACUTE HE Marked cognitive impairment and ↓ consciousness

SUBACUTE HE Subtle; spontaneous or due to surgical portal-caval shunt

CHRONIC PROGRESSIVE HE Associated with dysarthria, ataxia, intention tremor, and choreoathetosis, and dementia

hepatic function panel Hepatic panel LAB MEDICINE A standard battery of liver-related analytes used to evaluate the liver's baseline status

hepatic metabolism THERAPEUTICS The constellation of chemical alterations to drugs or metabolites that occur in the liver, carried out by microsomal enzyme systems, which catalyze glucuronide conjugation, drug oxidation, reduction and hydrolysis. See Metabolism

hepatic porphyria METABOLIC DISORDERS A general term which encompasses acute hepatic porphyrias, porphyria cutanea tarda and hepato-erythropoietic porphyria. See Porphyria

hepatic vein obstruction Budd-Chiari syndrome HEPATOLOGY A condition characterized by clotting of blood in the hepatic vein linked to the use of OCs

hepatitis HEPATOLOGY Liver inflammation ETIOLOGY-INFECTIOUS HAV, HBV, hepatitis non-A, non-B–HCV, HDV, HEV, CMV, coxsackievirus, herpesvirus, EBV, measles, mumps, rubella, rubeola, bacteria, parasites, fungi ETIOLOGY-NONINFECTIOUS Alcohol, drugs, chemicals and toxins, hyperthermia, radiation CLINICAL Anorexia, N&V, malaise, jaundice, myalgia, arthralgia, photophobia, bleeding diathesis LAB ↑ Transaminases–ALA, AST, GGT, BR, Igs, ↓ vitamin K-dependent coagulation factors, ergo ↑ prothrombin time; viral hepatitis is diagnosed by serology, measuring viral antigen(s) or antibodies formed against the antigens; non-viral hepatitides are diagnosed by history and exclusion of virus MANAGEMENT Acute hepatitis, no treatment–steroids, IFN-α are *not* recommended; chronic hepatitis–corticosteroids, IFN-α *may* prolong survival and improve outcomes. See Acute hepatitis, Biochemical hepatitis, Chronic hepatitis, Giant cell hepatitis, Halothane hepatitis, Hepatitis A, Hepatitis B, Hepatitis C, Hepatitis D, Hepatitis E, Hepatitis F, Hepatitis GB, Hepatitis non-A–G, Lupoid hepatitis, Neonatal hepatitis, Non-A, non-B hepatitis,

hepatitis A Acute viral hepatitis A HEPATOLOGY An acute, rarely fatal infection by a picornavirus; globally, it is a common cause of morbidity; it is the mildest of viral hepatitis; ± 30% of the US population has been exposed EPIDEMIOLOGY Contaminated food–eg, shellfish, and other foods prepared by HAV carriers CLINICAL Fever, nonspecific GI malaise–eg, nausea, hepatosplenomegaly, jaundice, pruritus CLINICAL Duration 4-6 wks, followed by recovery; antibody formation protects against repeat infection LAB ↑ Liver enzymes–transaminases, dark urine due to ↑ BR; ↑ anti-HAV IgM and IgG antibodies; IgM–acute-phase antibody is detectable within days of onset and usually disappears in 6 wks; the IgG antibody is usually detected 5-6 wks post onset and is positive for life AT-RISK GROUPS/FACTORS Oral-fecal transmission–travelers, military personnel, institutionalized persons, day care center inmates, children & adolescents, Native Americans, raw shellfish, 'high-risk' sex VACCINE Formalin-inactivated hepatitis A vaccine is well tolerated; a single dose is effective against clinical HAV

hepatitis A vaccination A vaccination for those in high-risk settings–frequent world travel, sexually active with multiple partners, gay guys, illicit drug use, day care centers, certain health care setting, sewage exposure VACCINES HAVRIX, VAQTA DOSING 2 doses are recommended for adults, 3 for children < age 18 for prolonged protection. See Hepatitis A

hepatitis B HEPATOLOGY Liver inflammation by HBV, a small, highly contagious DNA virus of the Hepadnaviridae family EPIDEMIOLOGY 60-110 day incubation, via exchange of body fluids–eg, blood, semen, etc CLINICAL Sx from malaise to death; acute HBV infection is similar to HAV, with headache, anorexia, nausea, ± vomiting, low fever, jaundice, itching, RUQ tenderness LAB ↑ transaminases, ↑ BR, mild hemolysis, normal/↓ hematocrit & Hb concentrations, mild hemolysis, ↓ PMNs, relative lymphocytosis–total WBC ± 12,000/mm STAGES OF HBV INFECTION Acute, convalescent, chronic, determined by various viral antigens & antibodies; early hepatitis B is characterized by sublobular involvement of all cells, later stages by scattered antigen-positive hepatocytes SEROLOGY See Hepatitis panel

hepatitis B antigen Hepatitis B surface antigen, see there

hepatitis B serology Hepatitis B serological markers LAB MEDICINE A generic term referring to hepatitis B antigens and antibodies to these antigens

---

HEPATITIS B SEROLOGY

CORE ANTIGEN The HBc particle that contains double-stranded DNA and DNA polymerase, and is associated with the HBe antigen; HBc is not directly detected by currently-used assays; its presence indicates persistently replicating HBV virus

CORE ANTIBODY A long-term serologic marker for HBV, with 2 antibodies

- **IgM HBcAb** A marker of acute infection, which rises early–within 2-4 weeks of HBV infection and slowly disappears; ↓ levels of IgM HBcAb indicate resolving infection; IgM HBcAb is the best serologic marker for acute HBV infection

- **IgG HBcAb** A 'convalescent' antibody that indicates prior HBV infection; it rises 4-6 months after infection and persists for life, especially in those with active liver disease; partially protective anti-HBc antibody levels can be induced by recombinant vaccination, but are short-lived

e ANTIGEN An antigen that rises and falls parallel to HBsAg, and derives from the proteolytic cleavage of the nucleocapsid; its presence implies a carrier state

e ANTIBODY Anti-HBe An antibody that rises as HBe falls, appearing in convalescent Pts, persisting for up to several years after resolution of hepatitis

SURFACE ANTIGEN HBsAg The first marker to appear after HBV infection, preceding clinical disease by weeks, peaking with the onset of symptoms and disappearing six months post-infection; as long as HBsAg is positive, the Pt is considered infectious and must follow prescribed sanitary procedures to avoid infecting others; if the hepatitis does not resolve, HBsAg persists and can be detected for many years or life.

ANTIBODY TO SURFACE ANTIGEN HBsAb, anti-HBs, antibody to surface antigen HBsAb begins to rise as the HBsAg falls; it is detectable 8-10 weeks post infection, is regarded as being protective against re-infection, and persists for life; HBsAb is formed after using the HBV vaccine, and is not present in the chronic phase of the disease.

hepatitis B surface antigen Australia antigen, HBsAg, Hepatitis B antigen VIROLOGY The protein coat of HBV; it is produced in excess during acute HBV infection, and correlates epidemiologically with an ↑ risk of liver CA and infectivity. See Dane particle, HBV

hepatitis B vaccine Engerix-B®, recombinant HBV vaccine HEPATOLOGY A vaccination that provides prolonged protection against hepatitis B AT-RISK GROUPS Healthcare workers, dentists, intimate and household contacts of Pts with chronic hepatitis B infection, ♂ homosexuals, those with multiple sexual partners, dialysis Pts, IV drug users, recipients of repeated transfusions

hepatitis C Non-A, non-B hepatitis INFECTIOUS DISEASE A transfusion-related hepatitis EPIDEMIOLOGY ± 4 million people–US have HC; 8000-10,000 die/yr; 1000 require liver transplants DIAGNOSIS EIA-2; supplemental RIBA-2 and/or RNA PCR testing; HCV genotyping for prognosis PROGNOSIS ⅓ of Pts with HCV progress to cirrhosis within 20 yrs LAB Screen for HCV with anti-HCV antibodies, ↓ the risk for seroconversion from 0.45%/unit transfused to 0.03% COMPLICATIONS Cryoglobulinemia, membranoproliferative glomerulonephritis, possibly due to glomerular deposits of immune complexes containing HCV, anti-HCV IgG, and IgM rheumatoid factors MANAGEMENT IFN-α results in histologic reversal and ALA response PROGNOSIS 20% with post-transfusion hepatitis develop chronic hepatitis, cirrhosis, or liver CA

hepatitis D Delta hepatitis HEPATOLOGY A severe form of hepatitis in which HDV infects a Pt with hepatitis B TREATMENT IFN-α normalizes liver enzymes, and causes HDV RNA to disappear from serum, with histologic improvement; relapse occurs shortly after stopping IFN therapy

hepatitis E A hepatitis caused by a single-stranded RNA virus; it has been implicated in major epidemics where sanitation is poor, drinking water contaminated, and population malnourished

hepatitis F An agent tentatively identified as a new, enteric virus, a finding that was not substantiated

hepatitis GB HEPATOLOGY Any of the single-stranded RNA viruses of family Flaviviridae, first isolated from a surgeon–initials GB; HGBV-C is genetically similar to HGBV-A and HGBV-B, and resembles HCV, and has been isolated from Pts with non-A-E hepatitis; HGBV-C infection is a risk for Pts on chronic hemodialysis, may be transmitted by transfusions, produces persistant infections, and is identified with reverse transcriptase-PCR

hepatitis non-A, non-B Non-A, non-B hepatitis, NANBH Any hepatitis unaccounted for by known hepatitis viruses. See Hepatitis C. Cf Hepatitis non-A-G

hepatitis non-A–G Any hepatitis not accounted for by major known hepatitis viruses–A, B, C, D, E, GB . Cf Hepatitis F, Hepatitis non-A, non-B

hepatitis panel LAB MEDICINE A standard battery of lab tests that are the most cost-effective in evaluating the clinical and immune status of a person with possible hepatitis. Cf Organ panels

hepatitis serology See Hepatitis B serology

hepatitis surface antigen Hepatitis B surface antigen, see there

hepatization PULMONOLOGY The transformation of a tissue that is normally fluffy to one with a liver-like consistency, classically described as acquired firmness of lungs, commonly seen in lobar pneumonia–LP, especially if caused by S pneumoniae types 1, 3, 7 and 2, as well as other streptococci, Klebsiella spp, staphylococci and gram-negative rods; hepatization occurs primarily in the elderly and infants

hepatobiliary scan IMAGING Scintigraphy used to identify bile duct obstruction, seen as a filling defect near the gallbladder. See Cholecystography, HIDA scan

hepatocerebral degeneration HEPATOLOGY Combined liver and brain degeneration. See Wilson's disease–aka hepatolenticular degeneration

hepatolenticular degeneration Wilson's disease, see there

hepatologist 'Liver guy' A gastroenterologist subspecialized in study of the liver. See Hepatology, Cf Gastroenterologist

hepatoma Hepatocellular carcinoma, see there

hepatomegaly Enlarged liver MEDTALK ETIOLOGY Acute hepatitis, alcoholic fatty infiltration, bile obstruction, CA. Cf Splenomegaly

hepatopathy Liver disease, see there

296

hepatopulmonary syndrome A condition in which hypoxemia due to intrapulmonary shunting and/or a V/Q mismatch develops in a Pt with liver cirrhosis; usually there is no apparent parenchymal lung disease, but Pts may have orthodeoxia, an unusual finding of ↑ hypoxemia with a change from the supine to the erect position; the pathogenesis of HPS is uncertain but may be due to an ↑ production of endogenous nitric oxide; shunting of HPS may respond to IV methylene blue

hepatorenal glycogenosis Glycogen storage disease–type Ia, see there

hepatorenal syndrome Bile nephrosis HEPATOLOGY A complication of liver failure characterized by acute renal dysfunction without renal pathology, due to ↓ perfusion, hypovolemia, and hyperaldosteronism, 2° to liver disease–eg, cirrhosis, acute fatty liver, hepatic failure, obstructive jaundice, sepsis, infectious hepatitis ETIOLOGY Unclear CLINICAL ↓/absent urine, jaundice, bloating, delirium, confusion, N&V PROGNOSIS Very poor; ⅓ relentlessly deteriorate. Cf Cirrhotic glomerulonephritis

hepatosplenomegaly MEDTALK Enlarged liver and spleen

Her's disease Glycogen storage disease, type VI. Cf His disease

HER-2 c-erbB-2, HER-2/neu ONCOLOGY An erbB family oncogene related to epidermal growth factor; HER-2 overexpression, especially breast CA, and stomach CAs, which have a ↓ survival and ↓ time to relapse. See Breast cancer

herald bleed SURGERY An episode of hemorrhage, often accompanied by abdominal pain, which may precede by hrs to wks a catastrophic hemorrhage; HBs are typical of arterial-enteric fistulas with false aneurysms, often at the site of the proximal anastomosis of a prosthetic graft of the abdominal aorta

herald patch DERMATOLOGY A lesion early pityriasis rosea consisting of a solitary oval, 1-10 cm annular macule with raised border and fine, adherent scales. See Pityriasis rosea

herald wave phenomenon EPIDEMIOLOGY The finding that the strains of influenza virus present in a population at the end of one season's epidemic are the same strains responsible for the next season's influenza syndromes

herb PHARMACOGNOSY A generic term for a plant/parts–roots, bark, stems, leaves, flowers, fruit, seeds, or others which can be used for its medicinal value. See Chinese herb, Herbal medicine, Pharmacognosy

herbal adjective Referring to herb(s) noun A book or treatise on herbs

Herbal Ecstasy SUBSTANCE ABUSE A proprietary form of ma huang which, given its content of ephedrine-like alkaloids, has been self-administered to produce euphoria; HE is composed of ephedrine–ma huang or pseudoephedrine and caffeine–koala nut, stimulants that simulate the effects of Ecstasy ROUTE Oral PHARMACOLOGY CNS stimulant, hallucinogenic properties, alertness, paranoia, sensory distortion ADVERSE REACTIONS HTN, seizures, acute MI, strokes, death. See Herbal medicine, Natural drug, Rave party. Cf Peyote

herbal medicine Botanical medicine, botanomedicine, herbalism, phytomedicine, phytotherapy, vegotherapy ALTERNATIVE HEALTH As usually defined in alternative medicine, the therapeutic use of extracts from flowers, fruits, roots, seeds, and stems, alone or as an adjunct to other forms of alternative health care or physical manipulation–eg, massages. See Alterative, Alternative medicine, Analgesic, Anthelmintic, Antibiotic, Antiseptic, Antispasmodic, Astringent, Botanical toxicity, Carminative, Cathertic, Decoction, Demulcent, Diaphoretic, Diuretic, Douche, Emetic, Emmenagogue, Enema, Ethnomedicine, Fluid extract, Green extract, Hepatic, Infusion, Laxative, Naturopathy, Nervine, Ointment, Poultice, Stomachic, Suppository, Syrup, Tincture

Herbert screw ORTHOPEDICS A specialty screw that is cannulated and threaded at both ends, used in fractures of small articular bones such as the carpals. See Screw

Herceptin® Trastuzumab MOLECULAR ONCOLOGY A humanized monoclonal antibody to the HER-2 growth factor receptor used to treat metastatic breast CA that overexpresses HER-2. See Breast cancer, Monoclonal antibody

hereditary adjective Transferred via genes from parent to child

hereditary amyloidosis METABOLIC DISEASE Any of the usually AD conditions caused by tissue accumulation of defective amyloid proteins in the brain, heart and kidneys. See Amyloidosis

hereditary angioneurotic edema HANE, hereditary angioedema, Quinke's disease IMMUNOLOGY An immune complex-induced condition caused by deficiency of C1q esterase inhibitor–C1q-INH–which normally prevents activation of a cascade of proteins leading to swelling, characterized by episodic consumption of activated C1, C4 and C2, triggered by physical stimuli–trauma, cold, vibration, histamine release, and menstruation; fewer attacks occur in the last 2 trimesters of pregnancy CLINICAL Recurrent attacks of acute swelling–non-pitting, non-pruritic, non-urticarial swelling that peaks at 12-18 hrs, lasts 1-4 days, affects extremities, lips, face, fingers, toes, knees, elbows, buttocks, GI mucosa, oropharynx, causing potentially fatal epiglottic and upper airway edema–33% mortality, often accompanied by abdominal pain with nausea and vomiting; edema also affects abdominal serosa and subserosa; edema of extremities is self-limited and requires no therapy 4 TYPES 2 are congenital; HANE Type I Common–85% with C1q-INH ↓ to 30% of normal levels; HANE Type II Variant form–gene product is present but dysfunctional; type II may be 1. Acquired, associated with lymphoproliferative disorders–eg, IgA myeloma, Waldenström's macroglobulinemia, CLL, and other B-cell proliferations or 2. Autoimmune, associated with IgG1 autoantibodies, due to uncontrolled activation of C1s DIAGNOSIS Hx of recurrent angioedema; very low C1 esterase inhibitor levels MANAGEMENT Prophylactic androgens, antihistamines, antifibrinolytics; vapor-heated–inactivate HBV and HIV C1 inhibitor concentrate may be effective in both acute management and in prophylaxis. See C1-INH

hereditary cancer syndrome Any of a group of often AD conditions characterized by tumors that are often site-specific, of early onset and multiple and/or bilateral; HCSs include nevoid basal cell carcinoma syndrome, dysplastic nevus–B-K mole syndrome, Cowden syndrome, familial polyposis, Gardner syndrome, Gorlin syndrome, Li-Fraumeni syndrome, MEN I, MEN II. See Beckwith Wiedemann syndrome, Cancer family, Dysplastic nevus sydrome, Familial malignancy, Hereditary neoplasmia, Hereditary preneoplasia, Li-Fraumeni syndrome

hereditary cerebellar ataxia AAA syndrome An AR condition characterized by cerebellar ataxia, degeneration of esophageal myenteric ganglion cells, and achalasia

hereditary cerebral hemorrhage with amyloidosis, Dutch type An AD form of cerebral amyloid angiopathy characterized by recurrent intracerebral hemorrhages, death by 6th decade. See Cerebral amyloid angiopathy

hereditary coproporphyria METABOLIC DISEASE An AD porphyria caused by a 50% ↓ in coproporphyrinogen oxidase activity CLINICAL Neurologic dysfunction, photosensitivity LAB ↑ Fecal protoporphyrin. See Porphyria

hereditary elliptocytosis Hereditary ovalocytosis HEMATOLOGY An AD condition, affecting ± 1:2500 of the general population; 15% of the RBCs are ovoid; hemolysis is uncommon, but severe in 5%

hereditary fructose intolerance Aldolase B deficiency, fructosemia, fructose-1-phosphate aldolase deficiency METABOLIC DISEASES An AR condition characterized by an aversion or intolerance for fructose-containing foods due to a defect in aldolase B CLINICAL Vomiting, hypoglycemia with weaning, and cirrhosis in childhood; in adults exposed to fructose later in life–GI bleeding, Fanconi syndrome, acute jaundice, proximal tubular acidosis, DIC. See Fructose

hereditary hemorrhagic telangiectasia Osler-Rendu-Weber syndrome, Rendu-Osler-Weber syndrome MOLECULAR HEMATOLOGY An AD condition characterized by telangiectases of mucocutaneous surfaces—tongue, nose, lips, hands, feet CLINICAL Episodic epistaxis in childhood, chronic GI hemorrhage, palmo-plantar, liver telangiectasias, and lung AV malformations; telangiectasias may be seen on spleen, brain and spinal cord; rupture of the thin-walled vessels may result in hemorrhage of varying severity TREATMENT Aminocaproic acid, an inhibitor of fibrinolysis. See Endoglin

hereditary motor & sensory neuropathy A group of conditions dignified by eponym and subdivided as follows.

HEREDITARY MOTOR & SENSORY NEUROPATHY

TYPE I Charcot-Marie-Tooth disease, hypertrophic form An AD—type 1a type 1b, or X-linked condition that is the most common form of HMSN, characterized by a slowly progressive disease of childhood onset with predominantly motor symptoms, pes cavus, calf atrophy, very slow motor impulse conduction with segmental demyelination and remyelination with 'onion-bulb' formation

TYPE II Roussy-Levy syndrome, or Charcot-Marie-Tooth disease, neuronal form An AD condition characterized by slowly progressive disease of adolescent onset with predominantly motor symptoms, clubfoot deformity, calf atrophy, mild reduction in impulse transmission and 'onion-bulb' formation

TYPE III Dejerine-Sottas disease A rare relentlessly progressive AD condition of infant onset with short stature, scoliosis, pes cavus, calf atrophy, very slow conduction of motor impulses with segmental demyelination and remyelination 'onion-bulb' formation. See Hereditary sensory neuropathy

hereditary mutation Germline mutation GENETICS A genetic change in germ cells—egg or sperm that is incorporated into the DNA of each cell of offspring

hereditary neoplasia ONCOLOGY A tumor or growth with a genetic component—eg, chemodectomas, medullary CA of thyroid, neurofibroma, pheochromocytoma, polyposis coli, retinoblastoma, trichoepithelioma; other tumors occur more frequently in certain families—eg, leukemia, melanoma. See Hereditary cancer syndromes

hereditary nephritis Alport syndrome, see there

hereditary neuralgic amyotrophy NEUROLOGY An AD peripheral neuropathy characterized by father to son transmission CLINICAL Either transient pain of abrupt onset occurring ± once/month or, more typically, pain and weakness for wks to months, with muscle atrophy

hereditary nonpolyposis colon cancer Lynch I syndrome ONCOLOGY A relatively distinct AD form of cancer that may account for 5-10% of all colorectal CA CLINICAL Family Hx of colorectal CA at relatively young age, primarily of proximal colon, tendency toward multiple primary tumors; HNPCC may be ssociated with extracolonic tumors—eg, of endometrium DIAGNOSIS Amsterdam criteria. See Amsterdam criteria, Colorectal cancer hMSH-2. Cf Familial adenomatous polyposis

hereditary persistence of fetal hemoglobin Hbα₂γ₂ An AD condition caused by a defect in the Hb 'switch' mechanism, where the usual transition from γ to β chain production does not occur; in HPFH, the number of cells with Hb F in the maternal circulation is greater than the baby's total blood volume, and therefore the condition's presence is suspected by mere calculations of maternal and fetal volumes; Type I—uniform distribution of HbF in all RBCs, as in Greek, Kenyan and Black American forms 2. Type II—heterogeneous distribution of HbF in RBCs, as in British and Swiss forms. See Class switching, Fetal hemoglobin

hereditary preneoplasia A general term for a neoplastic substrate related to a syndrome EXAMPLES Phakomatoses—eg, Cowden's disease, multiple exostoses, Peutz-Jegher syndrome, von Hippel-Lindau syndrome, neurofibromatosis, tuberous sclerosis, genodermatoses—eg, albinism, dyskeratosis congenita, epidermodysplasia verruciformis, polydysplastic epidermolysis bullosa, Werner syndrome, xeroderma pigmentosa, chromosome instability syndromes and immune deficiencies—eg, ataxia-telangiectasia, CVID, Wiskott-Aldrich syndrome, X-linked agammaglobulinemia and X-linked lymphoproliferative syndrome. See Cancer families, Common variable immunodeficiency. Cf One-hit, two-hit model

hereditary sensory neuropathy NEUROLOGY A disorder characterized by chronic pain, skin ulcers due to hypesthesia, dyskinesis, autonomic dysregulation, ↓ pain, touch-pressure, and temperature sense, affecting small nerves with degeneration of large myelinated nerves

hereditary spherocytosis Spherocytosis NEONATOLOGY An AD condition characterized by ↑ osmotic fragility of red cells, autohemolysis, splenomegaly and mild anemia partially corrected by splenectomy; HS is linked to a defect in a 28 kD RBC membrane protein MANAGEMENT Splenectomy. Cf Spherocytosis

hereditary thrombophilia HEMATOLOGY A hereditary ↑ tendency to form intravascular blood clots ETIOLOGY Factor V Leiden, prothrombin G20210A polymorphism, defects in homocysteine metabolism, protein C, protein S or antithrombin III. See LETS

heredity The transmission of characteristics from one generation to the next. See Progeny. Cf Congenital

Hering-Breuer reflex NEUROLOGY A reflex in which lung distension stimulates stretch receptors, which act to inhibit further distension

heritability GENETICS The likelihood of suffering from a hereditary disease when the defective gene is in a person's gene pool. See Additive genetic variance, Genetic variance, Phenotypic variance, Population

hermaphroditism Intersexuality A state characterized by the presence of both testicular tissue—ie, seminiferous tubules and ovarian tissue—ie, follicular structures in the same organ, yielding an 'ovotestis,' in which the tissues are arranged end-to-end and may be accompanied by a left-sided ovary and a right-sided testis; 60% of Pts are 46 XX, 12% 46 XY, the rest, mosaics CLINICAL Most true hermaphrodites have asymmetrical external genitalia—eg, labioscrotal folds; phenotypic ♂ may have gynecomastia, phenotypic ♀ may be amenorrheic or have successful gestation; 2.6% develop germ cell tumors. See Female hermaphroditism, Male hermaphroditism. Cf Pseudohermaphroditism

Hermes' wing Comb over COSMETIC SURGERY A popular term for a long frond of hair on one side of a bald man's scalp, which is combed over to cover the chromed dome. See Hare Krishna ponytail, Male pattern baldness

hernia SURGERY The protrusion of tissue or prolapse from a normal site. See Abdominal hernia, Diaphragmatic hernia, Direct hernia, Femoral hernia, Hiatal hernia, Incarcerated hernia, Incisional hernia, Indirect hernia, Inguinal hernia, Sliding hernia, Stangulated hernia, Umbilical hernia, Ventral hernia

hernia repair Herniorrhaphy SURGERY The surgical elimination of a hernia, by either tightening the tissue or shoring it with synthetics—eg, Gore-Tex®. See Hernia

Herniamesh SURGERY Polypropylene monofilament mesh and plug products to reinforce herniorrhaphy procedures, for femoral, crural, inguinocrural or large recurrent inguinal hernias. See Hernia

herniate *verb* To protrude through an abnormal body opening

herniated disk Herniated intervertebral disk, herniated nucleus pulposus, prolapsed intervertebral disk, slipped disk NEUROLOGY The herniation of an intervertebral disk, most commonly, lumbar; the term herniation in this context describes a spectrum of disk defects

HERNIATION DISK TYPES, used for MRI exams

BULGE—circumferential symmetric extension of the disk beyond interspace

herniated disk 'syndrome'

PROTRUSION–focal or asymmetric extension of the disk beyond interspace

EXTRUSION–more extreme extension of the disk beyond interspace Note: Bulges and protrusions on MRI examination are common findings in normal subjects, and appear to be coincidental findings–NEJM 1994; 331:69oa

herniated disk 'syndrome' An acquired condition most commonly affecting active middle-aged adults, classically occurring after a minor trauma or torsion stress of the vertebral column, punctuated by the sensation of a 'snap', which corresponds to a prolapse of the nucleus pulposus into the nerve roots or spinal cord, causing progressive and distal radiation of pain; the backache subsides as the sciatica 'syndrome' develops with its sensorimotor consequences

herniation A bulging of tissue through an opening in a membrane, muscle or bone. See Brain herniation, Cerebellar herniation, Disk herniation

HERO CARDIOLOGY A clinical trial–Hirulog® Early Reperfusion/Occlusion trial

heroic surgery SURGICAL ONCOLOGY An aggressive procedure–eg, radical mastectomy for malignancy, in which wide resection margins are obtained, as the tumor has spread beyond a resectable size; HS is performed to alleviate pain or improve the quality of life in terminally ill Pts. See All-American operation, Commando operation, Forequarter amputation. Cf Heroic therapy

heroic therapy MAINSTREAM MEDICINE Aggressive treatment of a dread disease regarded by a Pt's care-givers as incurable with standard therapies at usual and prudent doses of potentially toxic drugs; Cf Heroic surgery

heroin Diacetyl morphine SUBSTANCE ABUSE A semisynthetic narcotic PHARMACOLOGIC EFFECTS Blocks severe, constant pain, ↓ cough; respiratory sedative CARDIOVASCULAR *S aureus* endocarditis, especially right-sided and tricuspid valve LIVER Hepatitis–HAV, HBV, HCV, HDV, HEV LUNGS Pulmonary edema, due to direct heroin toxicity to capillaries or myocardium, hypoxic endothelial damage, CHF, central vasomotor effect–↑ protein in edema SKIN Track marks, circular scars with necrotic ulcers. See Brown heroin, Opium

herpangina INFECTIOUS DISEASE A coxsackievirus infection characterized by a prodrome with fever, sore throat, headache, followed by painful papules that ulcerate. See Coxsackievirus

herpes gestationis Duhring-Brocq disease OBSTETRICS A rare pruritic, polymorphic, subepidermal bullous dermatitis of pregnancy linked to↑ infant mortality; it may recur with subsequent pregnancies and be activated by OCs TREATMENT Corticosteroids

herpes gladiatorum Traumatic herpes SPORTS MEDICINE H simplex-1-induced lesions of the eyes and skin of the head, neck, trunk, or extremities, accompanied by lymphadenopathy, sore throat, fever, chills and headache, described in modern 'gladiators'–eg, wrestlers and rugby players. See Herpes simplex. Cf Tinea gladiatorum

herpes simplex 1. Herpesvirus-1, see there 2. Herpesvirus-2, see there

herpesvirus-1 Herpes simplex-1 A herpesvirus that typically affects the body above the waist, causing most oral herpes–cold sores and fever blisters, and is often accompanied by stomatitis, conjunctivitis, necrotizing meningoencephalitis, encephalitis; HHV-1 resides in sensory ganglia near the site of infection and multiples during episodic periods of impaired cellular immunity, resulting in disease recurrence; evoked by heat, cold, sunlight, UV radiation–ie, sunlight, hypersensitivity reactions, corticosteroids, and physical or emotional stress PORTAL OF ENTRY Fibroblast growth factor receptor. See Congenital herpes

herpesvirus-2 Genital herpes, herpes simplex-2 A herpesvirus that is most commonly transmitted sexually–only 5% of venereal herpes is caused by HHV-1, and rare under age 15 and primarily affects the body below the waist, causing venereal, vulvovaginal, and penile herpetic ulcers; like HS1, HS2 resides in sensory ganglia near the site of infection and multiplies during episodic periods of impaired cellular immunity, resulting in recurrence; this may be caused by heat, cold, sunlight, UV radiation–ie, sunlight, hypersensitivity reactions, corticosteroids, physical or emotional stress; HHV-2 is more prevalent in multiply married, ♀, urban African Americans with lesser educations PREVALENCE HSV-2 is detectable in ±20%, US HERPES IN PREGNANCY Early gestational infection may cause spontaneous abortion; products of late gestation infection suffer microcephaly, mental retardation, retinal dysplasia, hepatosplenomegaly; recurrence rate in subsequent pregnancies approaches 70%; HS-2 causes most neonatal herpes, infecting via the birth canal; the infants are healthy at birth and become symptomatic 1-4 wks post-partum; early neonatal herpes causes vesiculobullous lesions–absent in 20%, lethargy, irritability, hypotonia and loss of gag and sucking reflexes; late neonatal herpes may cause generalized seizures, jaundice, hypotension, DIC, acidosis, apnea, thrombocytopenia, shock DIAGNOSIS LM to identify cellular changes–eg, on a Pap smear, ELISA linked to an amplification culture system, complement fixation, culture in fetal fibroblasts, fluorescent antibody against membrane antigen; RFLP analysis allows epidemiologic tracking of sources of infection TREATMENT Antivirals–eg, cytosine arabinoside-5-iodo-2'-deoxyuridine, adenine arabinoside-Acyclovir, trifluorothymidine–Trifluidin shorten duration of attack

herpesvirus-3 Herpes varicella-zoster HHV-3 has 2 clinical forms: Acute HHV-3 infection–chickenpox and chronic HHV-3 infection–shingles. See Herpes zoster

herpesvirus-4 Epstein-Barr virus, see there

herpesvirus-5 Cytomegalovirus, see there

herpesvirus-6 Human B lymphotropic virus A lymphotropic virus that infects most Americans early in life, causing a fever and rash; although usually benign, HHV-6 may cause fulminant hepatitis and roseola–exanthema subitum with acute infectious mononucleosis-like Sx; it is a major cause of acute febrile illness in young children, and is associated with varied clinical manifestations and viremia; 10% of ER visits and hospitalization up to age 2, and 20% of visits age 6-12 months for febrile seizures are linked to HHV-6

herpesvirus-7 INFECTIOUS DISEASES A herpes virus linked to roseola infantum–exanthem subitum

herpesvirus-8 Kaposi sarcoma-related herpesvirus, see there

herpes zoster Human herpesvirus-3, shingles, zoster NEUROLOGY A peripheral manifestation of reactivated HHV3, which affects 300,000 people/yr–US primarily, elderly, immunocompromised Pts CLINICAL Severe sharp radicular pain, varicellar vesicular rash along dermatomes, itching dysesthesia; HZ in immunocompetent Pts causes large-vessel granulomatous arteritis, in immunocompromised Pts, small vessel encephalitis, myelitis, postherpetic neuralgia MANAGEMENT Analgesics–eg, acetaminophen, codeine; acyclovir or famciclovir ↓ new vesicles, ↓ acute pain. See Ramsay-Hunt syndrome, Varicella-zoster virus

herpetic lesion A vesicle filled with serous fluid and herpesvirus particles

herpetic whitlow An acute HS-induced paronychia seen in occupationally-exposed health care workers–eg, in neurosurgery units, appearing as periungual blisters with a honeycombed appearance, followed by purulence, with regional lymphadenopathy. See Herpes simplex, Whitlow

herpetophobia PSYCHOLOGY Fear of reptiles. See Phobia

HERS AIDS A study–HIV Epidemiology Research Study CARDIOLOGY A randomized, placebo controlled clinical trial–Heart & Estrogen/Progestin Replacement Study–which showed that in ♀ average age 67 with known/pre-existing CAD, mortality due to acute MIs, stroke, and hypercoagulation was ↑ with E/P therapy. See Study, Trial

hesitation wound FORENSIC MEDICINE One of multiple, usually superficial, incisions that are self-inflicted in an attempted suicide

hetero *prefix*, Latin, different

heterocyclic amine Any of a family of potential carcinogens present in grilled meat–eg, PhIP and AαC, compounds that volatalize. See Animal fat

heterogeneous *adjective* Dissimilar; possessing different or opposing characteristics in the same individual; composed of varied cell types

heterologous artificial insemination Semiadoption REPRODUCTION MEDICINE The insertion of sperm into the uterus sans intercourse, when the sperm from the husband or regular partner is of such poor quality that sperm from a donor is preferable, or there is a risk that the ♂ would pass a hereditary disease. See Artificial insemination. Cf Homologous artificial insemination

heterosexism PSYCHOLOGY The belief that heterosexual activities and institutions are better than those with a genderless or homosexual orientation. See Homophobia

heterosexual PSYCHOLOGY *adjective* Referring to heterosexuality straight; not gay *noun* A person who is heterosexually oriented, or prefers the opposite sex. See Heterophilia. Cf Homosexual

heterosexuality Heterosexualism A state of sexuality–pairing with a partner of the complementary genital morphology–directed to those of the opposite sex. See Heterosexualism. Cf Homosexuality

heuristic method DECISION MAKING A form of problem-solving based, not on scientific proof but rather on plausible, possible, or creative conclusions to questions that cannot be answered in the context of, or the 'logic' of which lies outside of, a currently accepted scientific paradigm; the heuristic process is of use as it may stimulate further research. See Medical mistake, Representative heurisic. Cf Aunt Millie approach, Stochastic method

hexachlorophene An antibacterial used as a pre-operative scrub solution and detergent

hexokinase deficiency MOLECULAR MEDICINE An AR inborn error of metabolism caused by a deficit of hexokinase, resulting in a defect in regulation by glucose-1,6-diphosphate and inorganic phosphate CLINICAL Early-onset hemolytic anemia

HFV High-frequency ventilation

Hg Mercury, see there

Hgb Hemoglobin, see there

hGH Human growth hormone. See Growth hormone

HGPRT See Hypoxanthine guanine phosphoribosyl transferase

HGSIL High-grade squamous intraepithelial lesion, see there

HHH syndrome An AR condition of early childhood to late adulthood onset characterized by hyperornithinemia, postprandial hyperammonemia and homocitrullinemia, caused by a defect in ornithine transport from the cytoplasm into the mitochondria–where the urea cycle occurs CLINICAL Chronic vomiting, repeated neurologic attacks after high-protein meals, resulting in seizures, FTT, lethargy, ataxia, choreoathetosis, acute episodic hyperammonemia, or coma, growth and mental retardation TREATMENT Protein restriction, dietary supplements with ornithine or arginine

HHHO syndrome Hypomentia, hypogonadotrophic hypogonadism, muscular hypotonicity & obesity PEDIATRICS A condition seen in Prader-Willi syndrome which is often accompanied by short stature, type 2 DM, acromicria–small hands and feet, micrognathia, strabismus, fish-like or Cupid's bow mouth, clinodactyly, absent auricular cartilage, hypoventilation with pulmonary HTN

HHRG Home health resource group MEDICARE A case-mix classification in which Pt characteristics and health status information are obtained from an OASIS assessment in conjunction with projected therapy use during a 60-day episode are used to determine Medicare reimbursement.

HHV 1 Human herpes virus-1. See Herpesvirus-1

HHV 2 Human herpes virus-2. See Herpesvirus-2

HHV 3 Herpes varicella-zoster: Acute HHV-3 infection. See Chickenpox, Shingles

HHV 4 Human herpes virus-4, aka Epstein-Barr virus, see there

HHV 5 Human herpes virus-5, aka Cytomegalovirus, see there

HHV 6 Human herpes virus-6. See Herpes virus-6

HHV 7 Human herpes virus-7. See Herpes virus-7

HHV 8 Human herpes virus-8. See Kaposi sarcoma-related herpesvirus

5-HIAA 5-Hydroxy-indoleacetic acid, see there

hiatal hernia GI DISEASE A herniation of the GE junction through the esophageal hiatus of the diaphragm into the thoracic cavity; HH affects up to 1% of the population, more common > age 50, 5% are symptomatic RISK FACTORS Smoking, obesity. See Fundoplication, Nissen repair

HIATAL HERNIA

SLIDING HIATAL HERNIA 90% of cases, characterized by axial and craniad displacement of esophagogastric–EG junction, which slides in and out of the chest depending on intrathoracic and intraabdominal pressures; the SHH is ensheathed in its own peritoneal sac TREATMENT Symptomatic cases are repaired by surgically returning the distal esophagus back to the peritoneal cavity with a valvoplasty

PARA-ESOPHAGEAL HIATAL HERNIA Less common and often accompanied by a sliding component; pure hiatal hernias are rare and associated with chronic hemorrhage and gastric volvulus, both indications for surgical repair

HIB vaccination Vaccination intended to prevent *H influenzae* type B infection, which causes meningitis and epiglottitis with airway obstruction DOSING 2, 4, 6 months of age, followed by a booster at 12-15 months of age ADVERSE REACTIONS Rare

hibernating myocardium CARDIOLOGY Regional dysfunction of myocardial tissue due to prolonged local hypoperfusion, which is completely reversible upon restoration of adequate blood flow; hibernation occurs in Pts with CAD and impairment of left ventricular function at rest. Cf 'Stunned' myocardium, Thallium imaging

hiccup Hiccough, singultation CLINICAL MEDICINE An abrupt inspiratory muscle contraction, followed within 35 msec by glottic closure; the hiccup center is in the spinal cord between C3 and C5; an afferent impulse is carried by the vagus and phrenic nerves and thoracic sympathetic chain; the efferent impulse is carried by the phrenic nerve with branches to the glottis and accessory respiratory muscles ETIOLOGY Idiopathic, psychogenic, abdominal disease–gastric distension, GI hemorrhage, bowel obstruction, esophagospasm, or inflammation including hepatitis, peritonitis, gastritis, enteritis, appendicitis, pancreatitis, abrupt temperature change, alcohol, inferior wall MI, irritation of tympanic membrane, metabolic derangements–azotemia, hyponatremia, uremia, diaphragmatic irritants, diseases of chest wall, lung, and heart–mediastinitis, tumors, aortic aneurysms, subphrenic abceses, pericarditis, foreign bodies, excess smoking, excitement or stress, toxins, drugs–general anesthesia, barbiturates, diazepam, α-methyldopa, tumors, pneumonia, herpes zoster, central and peripheral nervous system disease–encephalitis, tumors, meningitis, brainstem infarcts, phrenic nerve compression, cervical cord lesions; intractable hiccupping may result in inability to eat or sleep, arrhythmias or reflux esophagitis, or may be compatible with a normal life MANAGEMENT No therapy is consistently effective. Cf Burping, Flatulance, Sneezing

Hickman® catheter Chronic tunneled central venous catheter NURSING An indwelling silicone elastomer device that provides long-term IV access for administering total parenteral nutrition, hyperalimentation, blood products, drugs, high-dose chemotherapy. Cf Port-A-Cath

the hidden epidemic Secret epidemic INFECTIOUS DISEASE A popular term for ↑ STDs in the general population. See Sexually-transmitted disease

hide-bound skin A descriptor for the smooth shiny indurated skin associated with adherent fibrosis and sclerosis that 'glues' the subcutaneous tissue to underlying structures, classically described in scleroderma

hierba DRUG SLANG Spanish for marijuana, see there

HIFU High-intensity focused ultrasound SURGERY A method that focuses ultrasound to heat/ablate target tissue without injuring surrounding structures. See Sonablate 200™

HIG Human immunoglobulin, see there

high SUBSTANCE ABUSE A popular term for a state of pleasant and/or manic euphoria, which is often a desired end-point for users of narcotics, hallucinogens, or other potentially addicting substances of abuse. See Bad trip, 'Stoned'

high-altitude cerebral edema A syndrome attributed to vasogenic cerebral edema CLINICAL Headaches, nausea, disorientation, impaired cognitive function, death MANAGEMENT Transport Pt to a lower altitude. See Mountain sickness

high arch Pes cavus ORTHOPEDICS A foot characterized by a high antero-posterior arch, which is due to either orthopedic or neuromuscular defects

high blood pressure Hypertension, see there

high blood sugar Hyperglycemia

high cholesterol Hypercholesterolemia, see there

high-definition imaging ultrasonography IMAGING A diagnostic technique that improves imaging of breast tissue with ↑ density, which is relatively more common in premenopausal ♀, or in postmenopausal ♀ with estrogen replacement

high-density lipoprotein cholesterol See HDL-cholesterol

high-discretion procedure MEDICAL PRACTICE A surgical procedure–eg, CABG, stapedectomy, strabismus operation, total knee or hip replacement, that is expensive and performed on a discretionary basis, and more often if the client is well-insured. See Elective surgery

high disseminator EPIDEMIOLOGY A person who is a carrier of a highly virulent and easily transmissible pathogen. See Typhoid Mary. Cf Fabian Bridges

high-dose chemotherapy ONCOLOGY The administration of chemotherapeutics in excess of BM toxicity; given the risk of aplastic anemia, HDC requires autologous BMT and use of 'rescue' factors such as G-CSF, GM-CSF, and erythropoietin. See Bone marrow transplantation

high-dose epinephrine EMERGENCY MEDICINE A 'megadose' of epinephrine–15 mg vs 1 mg for standard therapy used as an initial treatment of prehospital cardiac arrest–PCA. See Epinephrine

high-fat diet A diet rich in fats, often saturated–animal or tropical oils—fats ADVERSE EFFECTS Arthritis, CA, vascular disease, DM, HTN, obesity, stroke. See Fat, Fatty acids, Saturated fat acis, Cf Low-fat diet

high-fiber diet High-residue diet, high-roughage diet NUTRITION A diet with ≥ 13–20 g/day of crude dietary fiber. Cf Low-fiber diet

high-grade lymphoma ONCOLOGY An aggressive lymphoma that responds poorly to chemotherapy and comprises 20% of lymphomas PROGNOSIS HGLs have a mean survival of < 1 yr without therapy. See Lymphoma

high-grade squamous intraepithelial lesion GYNECOLOGY A lesion defined by cytologic findings–cells occur singly or in syncytia-like sheets, ↑ nuclear:cytoplasmic ratio, nuclear hyperchromasia that translate into moderate-to-severe dysplasia–CIN 2 to 3/carcinoma in situ of the uterine cervix–a diagnosis made on histology of biopsied tissue, a precancerous lesion. Cf Low-grade squamous intraepithelial lesion

high-impact aerobics SPORTS MEDICINE Aerobic exercise that stresses musculoskeletal tissues, joints HIA EXERCISES Aerobic dancing, basketball, jogging, running. See Aerobic exercise, Exercise, Cf Low-impact aerobics

high-impact sport SPORTS MEDICINE An activity or sport charaterized by intense and/or frequent wear and trauma of weight-bearing joints–foot, knee and hip EXAMPLES HIS Baseball, basketball, football, handball, hockey, karate, racquetball, running, soccer, waterskiing; participation in HISs is discouraged after hip and knee arthroplasty. Cf Low-impact sport, Moderate impact sport, No-impact sport

high-mortality outlier HOSPITAL CARE A hospital identified by CMS/HCFA as having a mortality rate far in excess of what is considered 'normal' for the other (±5500) health care facilities in the US. See Health Care Financing Administration, Hospital mortality rate

high-osmolality contrast medium IMAGING An ionic compound–eg, diatrizoate, iothalamate which is used as intravascular radiographic contrast media; HOCMs have up to 7 times the osmolality of plasma. See Diatrizoate, Iothalamate. Cf Low-osmolality contrast media

high-output diarrhea High stool output diarrhea Diarrhea, usually caused by infectious gastroenteritis, in which the stool production is > 10 mL/kg/hr MANAGEMENT Rehydration–eg, with rice flour in water

high-output heart failure CARDIOLOGY CHF due to an ↑ in circulating blood volume without functional myocardial abnormalities, resulting in a hyperkinetic state; the cardiac pump activity is a function of preload–ventricular end-diastolic fiber tension, myocardial contractility, afterload and heart rate; HOHF may be a physiologic response to such 'insults' as anemia, cor pulmonale, exercise, fever, high humidity, systemic HTN, obesity, pregnancy, emotional stress and temperature extremes, or non-physiologic in Albright's disease–polyostotic fibrous dysplasia, carcinoid syndrome–serotonin-producing usually hepatic metastases, arteriovenous fistulas–trauma, Paget's disease of bone, hemangiomatosis, glomerulonephritis, hemodialysis, hepatic disease–alcohol-related thiamine deficiency ↓ peripheral arterial resistance, hyperkinetic heart syndrome, P vera, thyrotoxicosis–T_3 ↑ heart rate, cardiac sensitivity to epinephrine and peripheral vasodilation. Cf Low-output cardiac failure

high potassium VOX POPULI Hyperkalemia; often also, hyperpotassemia

high-power institution ACADEMIA A popular term for highly regarded and highly competitive hospitals, research institutes and universities that perform 'cutting edge'/'world class' science, and are highly selective in whom they train

high-resolution computed tomography IMAGING CT at slice–collimation scan interval widths of ≤ 4 mm, which is narrower than the usual 1-3 cm interval 'slices' obtained in conventional CT imaging. Cf Spiral computed tomography

high-risk *adjective* Referring to an ↑ risk of suffering from a particular condition INFECTIOUS DISEASE Referring to an ↑ risk for exposure to blood-borne pathogens, which occurs with blood bank technicians, dental professionals, dialysis unit staff, EMTs, ER staff, IV therapy teams, lab, and medical technologists, morticians, OR staff, pathologists, phlebotomists, surgeons, etc

high-risk behavior PUBLIC HEALTH A lifestyle activity that places a person at ↑ risk of suffering a particular condition. See Safe sex practices

high-risk & complex procedure Major procedure SURGERY A surgical procedure, defined in the context of post-graduate medical education and residency training, which requires the teaching surgeon's presence during all or most of the procedure. Cf Minor procedure

high-risk group EPIDEMIOLOGY A group of people in the community with a higher-than-expected risk for developing a particular disease, which may be defined on a measurable parameter–eg, an inherited genetic defect, physical attribute, lifestyle, habit, socioeconomic and/or educational feature, as well as environment

high-risk infant NEONATOLOGY An infant at ↑ risk of suffering comorbidity and potentially fatal complications due to fetal, maternal or placental anomalies or an otherwise compromised pregnancy. See High risk preganancy

high-risk pool HEALTH INSURANCE A group of persons who have been denied health insurance by insurers, because of a medical Hx that may include CA, heart disease, emphysema, etc, placing them at high risk for future claims and medical costs

high-risk pregnancy OBSTETRICS A pregnancy which is at risk for ↑ M&M due to fetal, maternal or placental defect. See High risk infant

---

HIGH-RISK PREGNANCY

FETAL HIGH-RISK FACTORS APGAR score of <4 at 1 min, birth weights of < 2500 g or > 4500 g, gestational age < 37 or > 42 weeks, fetal malformation, fetal-maternal blood group incompatibility and twinning

MATERNAL HIGH-RISK FACTORS Previous in utero or neonatal death, infection, true DM–gestational diabetes is less risky to the infant, premature rupture of membranes, maternal age < 16 and > 40, alcohol, drug or tobacco use, gestation beginning within 6 months of previous delivery, poor prenatal care, severe emotional stress, accidents or general anesthesia, use of teratogenic medication

PLACENTAL AND INTRAUTERINE HIGH-RISK FACTORS Placenta previa, short umbilical cord, single umbilical artery, abruptio placentae, oligohydramnios

---

high-risk sex Safe sex practices, see there

high sensitivity C-reactive protein LAB MEDICINE A method of measuring CRP, with higher sensitivity; HSCRP identifies Pts at risk for stroke, cardiovascular and peripheral vascular disease. See C-reactive protein

high tibial osteotomy ORTHOPEDIC SURGERY A procedure used for osteoarthritis in which a wedge of bone is excised from the tibial plate at the point of greatest contact with the femur; HTOs redistribute weight, and may ↓ cartilaginous wear

high ticket *adjective* Popular, referring to a high volume of sales or consumption. Cf Big ticket

high touch SOCIAL MEDICINE *adjective* Referring to physical and/or direct emotional interaction with Pts, distinct from modern medicine's trend toward technological and innovative aspects of healing; 'high-touch' fields include nursing, psychiatry, family practice. See Doctor-patient interaction

high-yield procedure A diagnostic procedure that often results in a definitive diagnosis. Cf Low-yield procedure

higher multiple OBSTETRICS Multigestation ≥ triplets: quadruplets, quintuplets, sextuplets, septuplets, octuplets, etc tuplets

highly complex test LAB MEDICINE A test requiring multiple and/or significant steps in preparation, processing, interpretation. See CLIA '88. Cf Moderately complex test, Waived test

highly toxic OCCUPATIONAL MEDICINE *adjective* Referring to a chemical that 1. Has a median lethal dose–LD_{50} of ≤ 50 mg/kg when administered orally to 200-300 g albino rats 2. Has an LD_{50} of ≤ 200 mg/kg when administered by continuous contact for 24 hrs on the shaved skin of 2.0-3.0 kg albino rabbits 3. Has an LD_{50} of ≤ 200 ppm of volume of gas or vapor, or ≤ 2 mg/L of mist or dust, when adminstered by continuous inhalation to 200-300 g albino rats

Hill sign CARDIOLOGY A peripheral sign of aortic regurgitation due to ↑ pulse pressure, defined as a popliteal systolic BP of ≥ 30 mm Hg greater than the BP of the brachial artery. See Aortic regurgitation

hilum overlay sign IMAGING A radiologic finding consisting of a well-circumscribed mass in the anterior or posterior mediastinum which may simulate an enlarged cardiac border and be misinterpreted as cardiomegaly ETIOLOGY Tumors, abscesses, TB

Hinchey staging system GI DISEASE A system used to stratify the complications of diverticular disease. See Acute diverticulitis, Complicated diverticulitis, Diverticular disease, Perforated diverticulitis

hindfoot ORTHOPEDICS The posterior part of the foot, comprised of the calcaneus and talus. See Midfoot

hip fracture ORTHOPEDIC SURGERY A femoral fracture which affects 1/6 white ♀–US during life EPIDEMIOLOGY 250,000/yr–US SPECIFICS Proximal femur; 90+% femoral neck, intertrochanteric; 5-10% are subtrochanteric RISK FACTORS Tall, thin ♀, osteoporosis, previous Fx or stroke, white, use of walking aids, alcohol consumption, poor health, sedentary lifestyle, Rx with benzodiazepines, anticonvulsants; HRT may protect ♀ < age 75, institutional residence, visual impairment, dementia DIAGNOSIS Hx, plain AP film, MRI, ^{99m}Tc bone scan. See Falls, Total hip replacement

hip joint replacement Total hip replacement, see there

hip-pointer contusion SPORTS MEDICINE A bruise of the iliac crest seen in contact sports, caused by a blow by the knee, elbow, football helmet or other blunt object, resulting in a subperiosteal hematoma and either avulsion of abdominal muscle or quadriceps or a fracture of the iliac crest

HIPAA Health Insurance Portability & Accountability Act of 1996, Kennedy-Kassenbaum Bill PRIVACY A bill enacted by Congress in 1996 which established a comprehensive and uniform federal standard for ensuring privacy of genetic information; HIPAA broadens the scope of existing fraud and abuse provisions, and ↑ penalties for fraud violations by health care providers. See Fraud, Rebundling, Unbundling, Upcoding

Hippocratic face A physiognomy characteristic of advanced, untreated, preterminal peritonitis, who Hippocrates described as having '...*hollow eyes, collapsed temples; the ears, cold, contracted and their lobes turned out; the skin about the forehead being rough, distended and parched; the color of the whole face being brown, black, livid or lead-colored...*'; these facial features are also described in celiac sprue. Cf Triangular face

Hippocratic Oath The ethical guide for physicians, delineated in the 5th century BC; although it has been attributed to Hippocrates, the father of medicine and his school, the 'Oath' is of uncertain origin. See Code of Hammurabi

---

HIPPOCRATIC OATH

'I swear by Apollo the physician, by Aesculapius, Hygeia and Panacea, and I take to witness all the gods, and all the goddesses, to keep according to my ability and judgement the following Oath:

To consider dear to as my parents him who taught me this art; to live in common with him and if necessary to share my goods with him; to look upon his children as my own brothers, to teach them this art if they so desire without fee or written promise; to impart to my sons and the sons of the master who taught me and the disciples who have enrolled themselves and have agreed to the rules of the profession, but to these alone, the precepts and instructions. I will prescribe regimen for the good of my patients according to my ability and judgment and abstain from whatever is deleterious and mischievous. I will give no deadly medicine to anyone if asked nor give advise which may cause his death. Nor will I give a woman a pessary to procure abortion. I will preserve the purity of my life and practice my art. I will not cut for stone, even for patients in whom the disease is manifest, but will leave this operation to be done by practitioners of this art. Into whatever house where I come, I will enter only for the benefit of the sick, keeping myself far from all intentional mischief and corruption, and especially from the pleasures of love with women or with men, be they free or slaves. All that come to my knowledge in the exercise of my profession or outside of my practice or in daily commerce with men, which ought not to be spoken abroad, I will keep secret and not divulge. If I keep this oath unviolated may I enjoy my life and practice my art, respected by all men and in all times, but should I trespass this oath, may the reverse be my lot'

hippus A spasm of the iris, resulting in exaggerated rhythmic contractions and dilatations of the pupil that are not consonant with accommodation and light. See Hiccup

hired gun FORENSIC MEDICINE A popular term for a physician, lawyer or other highly paid expert who is not a regular employee of a particular enterprise, whose services are paid only as long as necessary; the term is an analogy from the use of mercenaries to fight in a conflict. Cf Physician expert witness, Prostitute

Hirschsprung's disease Congenital aganglionic megacolon PEDIATRICS Absence of myenteric nerves in the lower colon, extending from the anus a variable distance northward; HD is 5 times more common in ♂ CLINICAL Failure to pass meconium or stool, abdominal distention, vomiting; in older children, HD is characterized by chronic constipation, abdominal distention, ↓ growth

hirsutism Excess body hair, which is divided into 1. Androgen-independent hirsutism—entire body is covered with vellous hair evenly distributed over androgen-dependent and -independent regions ETIOLOGY Congenital disease—eg, Cornelia de Lange and Seckel syndromes, drugs—eg, androgen analogues, anticonvulsants, corticosteroids, cyclosporine, minoxidil, phenytoin and progesterone analogues, metabolic disorders—eg, anorexia nervosa, porphyria cutanea tarda 2. Androgen-dependent hirsutism—↑ terminal hair over 'androgenic' regions of the face and upper chest

hirsutism profile ENDOCRINOLOGY A battery of tests performed on a ♀ to ID the cause of ↑ hair growth ANALYTES 3-alpha-androstanediol, androstenedione, DHEA-sulfate, 17-hydroxyprogesterone, total & free testosterone

hirudin HEMATOLOGY The most potent known inhibitor of thrombin, produced by the medicinal leech, *Hirudo medicinalis*

Hirudinea PARASITOLOGY A class of segmented annelids—ie, leeches, the most well-known of which is *Hirudo medicinalis*, that evolved from earthworms and are found in fresh water and soil in the subtropics and tropics GENERA *Haemadipsa*, medicinal leech, *Dinobdella*, *Haementeria*, *Helobdella*, *Hirudinaria*, *Hëmopis*, *Limnatis*, *Macrobdella*, *Poecilobdella*, *Pontobdella*; the leech is flat, has 2 suckers, the cranial sucker houses a mouth—the impression left by a leech bite has been fancifully likened to the Mercedes-Benz insignia; the caudal sucker is involved in crawling

Hirulog® Bivalirudin HEMATOLOGY A synthetic thrombin inhibitor, which may restore coronary blood flow after an acute MI. See HASI, TIMI-7

HIS Hospital information system, see there

His bundle electrocardiography CARDIOLOGY Use of a bipolar cardiac catheter electrode system for recording His bundle activity, studying physiology of recurrent arrhythmias, optimizing pacemaker implantation and differentiating true AV blocks from pseudo-AV block. See EKG

His disease His-Werner disease A louse-borne, rickettsial—*R quintana* infection LOCATIONS Mexico, N. Africa, E Europe, etc TRANSMISSION Direct contact—rubbing infected lousy feces into abraded skin or conjunctivae CLINICAL Abrupt onset of high fever, generalized myalgia, shin pain, vertigo, malaise; relapse is common. Cf Her's disease

Hispanic MULTICULTURE A person of Mexican, Puerto Rican, Cuban, Central or South American, or other Spanish culture or origin, regardless of race SOCIAL MEDICINE Any of 17 major Latino subcultures, concentrated in California, Texas, Chicago, Miam, NY, and elsewhere

histamine ALLERGY MEDICINE A bioactive amine/neurotransmitter produced by decarboxylation of histidine, stored in mast cells and basophils, and secreted by monocytes, neural, and endocrine cells; it is a potent mediator of immediate hypersensitivity reactions, and evokes a range of responses—bronchoconstriction, vasodilation, hypotension, tachycardia, flushing, headache, ↑ vascular permeability and secretion by nasal and bronchial mucous glands; it is responsible for Sx of hay fever, urticaria, angioedema, and bronchospasm in anaphylaxis. Cf Antihistamines

histamine receptor antagonists H₁ receptor antagonist, H₂ receptor antagonist THERAPEUTICS A family of agents that counter histamine activity, which are used to treat conditions linked to ↑ histamine release—eg, mast cell disease, basophilic leukemia ADVERSE EFFECTS Antiandrogenic—eg, gynecomastia and impotence. Cf *Helicobacter pylori*

histidinemia MOLECULAR MEDICINE A common—1:10,000—often asymptomatic AR condition characterized by a deficiency of histidase, the enzyme that converts histidine to urocanic acid in the liver and skin CLINICAL Variable which, when symptomatic, is of neonatal onset with impaired speech, growth and mental retardation LAB Some excess histidine is transaminated to imidazole, the urinary metabolite of which, imidazolepyruvic acid, is detectable by the ferric chloride test or by Phenistix, otherwise used to diagnose PKU

histiocytic lesion An aggregate of histiocytes, which may be benign, indeterminant, or malignant in proliferative potential; to reduce confusion, it has been suggested that histiocytoses be subdivided into different categories

---

HISTIOCYTIC LESIONS-BEHAVIOR CLASSIFICATION

BENIGN

- **FAMILIAL HISTIOCYTOSIS WITH EOSINOPHILIA** A chronic disease of infants with recurring bacterial infections, diarrhea, eczema, alopecia, associated with immunodeficiency

- **SINUS HISTIOCYTOSIS WITH MASSIVE LYMPHADENOPATHY** Rosai-Dorfman disease A disease most common in adolescent blacks with massive cervical lymphadenopathy as well as enlargement of extranodal (orbit, skin, bone, salivary gland, testis) lymphoid tissues

- **VIRUS-ASSOCIATED HEMOPHAGOCYTIC SYNDROME** A condition induced by viral infections, often accompanied by abnormal liver function tests, coagulation assays and pancytopenia *Pathology* Histiocyte hyperplasia, hemophagocytosis and replacement of native bone marrow elements

INTERMEDIATE

- **HISTIOCYTOSIS X**, aka Langerhans' cell histiocytosis, see there
- **REACTIVE HEMOPHAGOCYTIC SYNDROME**, see there

MALIGNANT

- **HISTIOCYTIC MEDULLARY RETICULOSIS**, see there
- **HISTIOCYTIC PROLIFERATIONS,** eg—acute monocytic leukemia (FAB M3), histiocytic lymphoma see there, malignant histiocytosis–see there

---

histiocytosis X Langerhans' cell histiocytosis, see there

histocompatibility testing IMMUNOLOGY A method of matching self antigens—HLA in transplant donor tissues with those of the recipient; the closer the match, the better the chance that the transplant will "take". See HLA typing

histocompatibility typing See HLA typing

Histofreezer® SURGERY A cryotherapy device for treating benign skin lesions–eg, warts–verruca vulgaris, verruca plantaris, HPV, acrochordon, molluscum contagiosum, seborrheic keratosis, verruca plana. See Cryotherapy

histologic examination The study of a tissue specimen by staining it and examining it by LM. See Light microscopy

histologically benign *adjective* Referring to a lesion that does not meet any criteria of malignancy–eg, marked cellular atypia, ↑ mitosis, disruption of basement membranes, metastasize

histopathology Anatomic pathology, tissue pathology The field that studies diseased tissue by microscopic examination of tissues

histoplasmosis MYCOLOGY Infection with *Histoplasma capsulatum*, 2° to inhalation of spore-laden dusts CLINICAL Asymptomatic to an acute respiratory illness that evolves to chronic cavitary lung infection–lesions may undergo calcification, to disseminated disease with low-grade fever, hepatosplenomegaly, lymphadenopathy and multiorgan involvement AT-RISK Elderly with COPD, immunocompromised hosts DIAGNOSIS Pt travel Hx, serology–CF, immunodiffusion, latex agglutination to identify antibodies MANAGEMENT Amphotericin B PROGNOSIS Poor despite therapy

historian A Pt who *provides*–in contrast to the usual dictionary definition, one who records–history; the term is used in positive or negative terms, as in a good/poor historian, when the Pt is able/unable to provide reliable information

historical control EPIDEMIOLOGY A control group that is chosen from a group of patients who were observed at some time in the past or for whom data are available through records; HCs are used for comparison with subjects being treated concurrently. See Control

history & physical examination MEDICAL PRACTICE A critical component of a Pt encounter in which information relevant to a present complaint is obtained, by asking questions about family and personal medical history and the organ systems examined in as great detail as necessary to manage the present conditon or evaluate–workup–the Pt. See Family history, Medical history, Natural history, Psychiatric history, Reproductive history, Sexual history, Social history

history of present illness MEDICAL PRACTICE A chronologic description of the development of the Pt's present illness, from the 1st sign and/or Sx or from the previous encounter to the present; HPI includes location, quality, severity, duration, timing, context, modifying factors, and associated signs and Sx. See Evaluation and management services

histrionic personality disorder Hysterical personality disorder PSYCHIATRY A state characterized by '...*pervasive and excessive emotionality and attention-seeking behavior*', which begins by early adulthood, and is present in various contexts; HPD is diagnosed by the finding of 5 or more of a list of criteria

---

HISTRIONIC PERSONALITY DISORDER ≥ 5 CRITERIA

1. Person is uncomfortable unless he/she is center of attention
2. Interactions with others may be sexually inappropriate or provocative
3. Volatile and/or shallow emotions
4. Use of physical appearance to draw attention to self
5. Impressionistic speech pattern
6. Theatricality, exaggerated emotions
7. Suggestible, ie easily influenced by others
8. Regards relationships as more intimate than they are

Diagnostic & Statistical Manual of Mental Disorders, 4th ed, Washington, DC, Am Psychiatric Assn, 1994

Histussin® D–hydrocodone bitartrate/pseudoephedrine HCl An antitussive-decongestant for symptomatic relief of cough accompanying upper respiratory tract congestion due to the common cold, influenza, bronchitis, sinusitis

HIT CARDIOVASCULAR DISEASE Either of 2 clinical trials: 1. HDL-cholesterol Intervention Trial–which evaluated the effect of gemfibrozil therapy on M&M in Pts with CAD; those with ↓ HDL-C often had ↑ TGs; conclusion Gemfibrozil ↓ mortality by 22%2. Hirudin for Improvement of Thrombolysis MEDTALK Heparin-induced thrombocytopenia

hit SUBSTANCE ABUSE A small dose of any illicit psychotropic drug

hit & run accident PUBLIC HEALTH An MVA in which the driver–often intoxicated at the time of the accident strikes a person–pedestrian, bicyclist or other vehicle, and leaves the scene of the accident, often without informing authorities or EMS; abandoning a scene is a criminal act that implies a lack of concern for the victim's safety and welfare. See Driving while intoxicated, Motor vehicle accident

the Hite Report SEXOLOGY A mid-1970s report on ♀ sexuality based on a questionnaire developed by Shere Hite. Cf Lesbianism

HITT Heparin-induced thrombocytopenia/thrombosis, see there

HIV-1 AIDS-related virus–ARV, Human immunodeficiency virus, human T-cell lymphotrophic virus, type III–HTLV-III, lymphadenopathy-associated virus–LAV The retrovirus intimately linked to AIDS SKIN IN HIV See HIV dermatopathy DIAGNOSIS PCR, ELISA to screen for anti-HIV IgM antibodies; ELISA-positive sera is subjected to Western immunoblot hybridization. See HIV tests DISINFECTION 750 ppm/1:10 dilution for 40 mins, formaldehyde–2% for 10 hrs; high laundry temps–90% ↓ in viable HIV–eg, 25 mins at 71°C or 10 min at 80°C may be effective; low temperature washing without bleach does not remove HIV– INFECTIONS HIV-positivity worsens responses to other infections; T-cell response to infections in HIV-positive subjects may trigger multiplication of dormant HIV; AIDS Pts have an ↑ susceptibility to disseminated vaccinia after immunization, neurosyphilis, TB, herpes and other infections that respond to standard therapy; several viruses may co-infect with HIV-1–eg, herpesvirus, HHV-6, papovavirus, adenovirus and HTLV-I LONG TERM SURVIVORS Nonprogressive HIV-1 infection, see there PRECAUTIONS See Universal blood & body fluid precautions SEROPREVALENCE IN AFRICA Sub-Saharan Africa ♂ 1:40 are infected; ♀ 1:40, regionally up to 1:5; North America ♂ 1:75; ♀ 1:700 SEROPREVALENCE, US 650,000-1,400,000–0.2-0.5% of the population is infected with HIV-1, ranging from 0.1% in rural regions with low 'risk' activities to 7.8% in urban populations; in low prevalence regions, HIV-1 positivity is more common in ♂; in high prevalence regions, the ♂ : ♀ ratio is 2.9:1; 20% of ♂ in such regions are HIV-positive; seropositivity is up to 9-fold greater in those refusing to be tested ARMY (US) PERSONNEL Seroconversion rate in soldiers is 0.29/1000 person-yrs CHILD-BEARING WOMEN US inner city 8.0/1000, urban–not inner-city and suburban 2.5/1000 suburban and rural 0.9/1000; 4.5-5.8/1000 in NY, New Jersey, Washington DC and Florida and 1.5/1000 in the US; rate of HIV transmission to the child is 30% ER PATIENTS–US, 1987 3% of all and 16% of 25-34 yr-old ER Pts are HIV positive, 80% of whom were unsuspected HEALTH CARE WORKERS–HCW HIV positivity in HCW reflects HIV positivity in the general population; in the US, < 100 HCW without other known risk behaviors have seroconverted; seroconversion after needle or mucosal exposure to HIV-infected Pts is ± 0.3%; dentists–see Acer cluster MANAGEMENT Multiagent antiretroviral therapy–eg, protease and reverse transcriptase inhibitors NEWBORNS Rural NY 0.16%+, NY City 1.25%+ PRISONERS Prevalence 2-7.6%, ♂; 2.5-14.7%, ♀ SEXUALLY ACTIVE ADULTS 5% of those with STD are HIV-1 positive, especially if Hx of syphilis or genital herpes TRANSMISSIBLE FLUIDS/TISSUES Blood, tissues, breast milk are recognized HIV 'vectors'; casual household contacts, feces, skin, tears and urine are not known to transmit HIV; saliva inhibits the ability of HIV to infect lymphocytes. See Bergalis TRANSMISSION Worldwide, most transmission is by heterosexual intercourse. See AIDS, HIV testing, Mosquito connection, Monkey connection, Western blot, Window period, Zagury, Zidovudine

HIV antibody A self antibody specifically directed against one or more proteins or antigens on the surface of HIV, which may be minimally protective against HIV

HIV-1 dermatopathy Skin disease occurs in 80% of HIV+ subjects—seborrheic dermatitis, herpes simplex, herpes zoster, KS, dermatophytosis, cellulitis, abscess formation, drug reactions etc

HIV drug resistance Antiretroviral drug resistance AIDS The resistance of a strain of HIV to an agent—eg, a reverse transcriptase inhibitor, which occurs in 5%-20% of those newly infected with HIV

HIV drug resistance testing Phenotype testing AIDS The testing of various agents—eg, reverse transcriptase inhibitors, for efficacy against a particular strain of HIV TYPES Genotypic tests¹, phenotypic test. Cf Genotype testing

HIV encephalopathyAIDS Encephalopathy

HIV exceptionalism PUBLIC HEALTH A stance in which health care policies favor the rights for privacy of HIV-infected individuals over concerns for public safety. See ACT-UP, AIDS advocacy, *Behringer* v *Medical Center of Princeton*

HIV-1 gene Any of the genes present in the human immunodeficiency virus. See AIDS

HIV GENES

art/trs **gene** *rev* gene, see there

env‡ **gene** Encodes viral coat proteins gp 120 and gp 41, which mediate CD4 binding and membrane fusion, controlled by tat and rev

gag‡ **gene** Encodes nucleocapsid core proteins including p24

nef† **gene** 3´ orf, B gene, ORF-2 gene Encodes a protein of unknown function found in infected Pts that may down-regulate viral expression—nef deletion results in a five-fold increase in viral DNA synthesis and replication

pol‡ **gene** Encodes reverse transcriptase, protease, integrase and ribonuclease

R† **gene** Encodes TAR—transcription activating response element, which has a nonspecific immunodeficiency effect

rev **gene** *art/trs* gene Encodes a 19 kD post-transcriptional protein regulator required for HIV replication, up-regulating HIV synthesis by transactivating anti-repression, ↑ levels of envelope RNA by regulating the *env* gene; inactivation of *rev* prevents viral replication—as measured by the successful production of the p24 glycoprotein by infected monocytes can be massively ↑ by adding cytokines to culture medium

tat† **gene** Encodes a potent 14 kD transcription activator that amplifies HIV replication, the inactivation prevents viral replication

vif† **gene** Facilitates infectivity of free HIV

vpr† **gene** Encodes a weak transcription activator

VPU† **GENE** Unique to HIV-1, encodes a 16 kD product which, when mutated, has a 5-10-fold Ø in replicative capacity and is critical for efficient budding of virions NEJM 1991; 324:308rv

†Gene encoding regulatory protein

‡Gene encoding structural protein

HIV genotype testing AIDS The testing of blood from HIV-infected individuals for HIV strains associated with certain patterns of resistance Cf HIV drug resistance testing

HIV & health care workers OCCUPATIONAL MEDICINE The average risk of HIV infection after percutaneous exposure to HIV-infected blood is 0.3% RISK FACTORS Deep wound; visible contamination of sharp of device; procedure involved a needle placed in 'source' artery or vein, source Pt died within 2 months of exposure due to AIDS; exposed HCWs are significantly—67% less likely to acquire HIV if they take AZT after exposure

HIV nephropathy AIDS-associated nephropathy Renal disease linked to HIV infection—eg, focal or global glomerulosclerosis with mesangial deposition of C3, IgM, and occasionally IgG, and HIV-associated IgA nephropathy CLINICAL Severe nephrotic syndrome of abrupt onset, ↓ albumin, rapidly progressive ESRD. See End-stage renal disease

HIV-PARSE AIDS A survey for HIV-infected individuals that encompasses use of health care services, disability, symptoms, and quality of life

HIV prophylaxis AIDS The administration of one—eg, zidovudine or more anti-HIV agents after contact with HIV-infected blood. See HAART, HIV

HIV reporting PUBLIC HEALTH The reporting of a person's HIV status to state health authorities. See AIDS, HIV surveillance, Notifiable disease

HIV RNA AIDS RNA of HIV origin, a serum marker of a Pt's 'HIV-ness,' now the standard by which Pt response to antiretrovirals is evaluated; HIV RNA levels correlate with CD4+ count, response to antiviral therapy, clinical stage and disease progression. See AIDS, CD4:CD8 ratio, HIV, HIV tests, Reverse transcriptase, Western blot

HIV seroconversion syndrome Acute HIV infection, see there

HIV-1 superinfection AIDS Simultaneous infection with 2 or more strains of HIV

HIV surveillance EPIDEMIOLOGY The identification and monitoring of HIV-infected persons through a regional or national database. See HIV reporting

HIV test Various tests have been used to detect HIV and production of antibodies thereto; some HTs shown below are no longer actively used, but are listed for completeness and context. See HIV, Immunoblot

HIV TESTS

CULTURE The direct culture of HIV in the appropriate cell—human lymphocytes

IgA ASSAY An immunoblot-type assay which allows the early diagnosis in infants of perinatal HIV infection

p24 ANTIGEN The measurement of HIV's p24 antigen by immunoassay, confirmation by neutralization; very low sensitivity

PCR Amplification of HIV nucleotide sequences by PCR, used to confirm indeterminate Western blot results

RNA TESTING

WESTERN BLOT An immune assay which detects specific antibodies to HIV antigens, including p24—often the first antibody to appear; low-risk individuals with a persisting indeterminant Western blot at 3 months may be regarded as negative and require no further followup

HIV vaccine AIDS As of mid-2005, there is no viable anti-HIV vaccine. See AIDS

HIV viral load AIDS A measure of the amount of HIV RNA in blood, expressed as number of copies/mL of plasma. See AIDS, HIV

HIV virion VIROLOGY A virion with a 100 nm in diameter lipid envelope surrounding a dense cylindrical nucleoid containing core proteins, reverse transcriptase and a genome with sequence similarity to non-cell-transforming lentiviridae—eg, visna and caprine encephalitis viruses; envelope proteins gp41, gp120; nucelocapsid proteins p24, p17, p9, p7

HIV wasting syndrome AIDS A clinical complex linked to HIV infection and characterized by marked weight loss, renal failure, and caused by poor nutrition and anorexia, endocrine dysfunction, and catabolic stress—eg, infection, uremia, dialysis; HWS is mediated by TNF, IL-1 and IFNs, resulting in ↓ lipoprotein lipase, ↓ fatty acid synthesis, and ↑ lipolysis in fat cells TREATMENT Recombinant hGH . See AIDS therapy

HIV-2 LAV-2, HTLV-IV, SIV/AGM An immunodeficiency virus identified in West Africans with abnormal reactions to HIV-1 and simian immunodeficiency virus—SIV CLINICAL Some are similar to AIDS; others are relatively 'benign' EPIDEMIOLOGY HIV-2 is largely confined to West Africa and has an AIDS epidemiology pattern II—heterosexual transmission

HIV-3 AIDS A putative 'third' human immunodeficiency virus described by workers peripheral to mainstream HIV research

hives Urticaria, see there

Hivid Zalcitabine, see there. See AIDS, ddC

hIVIG AIDS A passive immunotherapy for HIV infected Pts, which is a sterile concoction of concentrated antibodies from the blood from an HIV+ person with high levels of antibodies is given to an HIV+ person with low levels of antibodies ADVERSE EFFECTS Headaches, low-grade fever, allergic reaction, transient rashes. See AIDS

HLA Human leukocyte antigen complex IMMUNOLOGY A system of genes unique to a person and statistically shared with 1 in 4 siblings; HLA regions are within the major histocompatibility complex–MHC, on the short arm of chromosome 6, 32 centimorgans from the centromere, divided into 3 classes

HLA-B27-related arthropathies ORTHOPEDICS A group of joint diseases more common in persons with HLA-B27 antigen–eg, ankylosing spondylitis, juvenile rheumatoid arthritis, psoriatic arthritis, Reiter syndrome, *Salmonella*-related arthritis, *Yersinia*-related arthritis

HLA & disease IMMUNOLOGY The association of HLA haplotypes with specific diseases is determined by calculating 1. RELATIVE RISK Patients with a particular HLA antigen X, Number of control subjects without the antigen, divided by Pts without the antigen X, Control subjects with the antigen–but not the disease and 2. ABSOLUTE RISK Patients with the HLA antigen divided by the control subjects with the antigen X

DISEASES LINKED TO HLA SPECIFICITIES

Addison's disease–B8, DR3, Dw3; ankylosing spondylitis–B27; Behçet's disease–B5; Buerger's disease–B12; celiac disease–gluten-sensitive enteropathy-B8, Dw3; chronic active hepatitis–DR3; de Quervain thyroiditis–Bw35; dermatitis herpetiformis–Dw3; Goodpasture syndrome–DR2; mesangioproliferative glomerulonephritis–DR4; Graves' disease–B8 in whites, Bw35 and Dw12 in Japanese, Bw46 in Chinese; hemochromatosis–A3, B7, B14; IDDM–B8. B15, DRw3, DRw4; IgA nephropathy–B35; juvenile rheumatoid arthritis–B27, DR5, DR8; late onset adrenal hyperplasia with hirsutism–Aw33/B14; myasthenia gravis–A1, A3, B8, Dw3, DR3; psoriasis–Cw6; psoriatic arthritis–B27; Reiter syndrome–B27; rheumatoid arthritis–DR4; Salmonella arthritis–B27; Sjögren syndrome–DR3, B8; lupus erythematosus–DR2, DR3; Takayasu's disease–B52; Yersinia arthritis–B27

HLA match IMMUNOLOGY The number of HLA antigens at the HLA-A, B, and DR loci of donor and recipient that match, from a maximum of 6 to minimum of 0. Cf HLA mismatch

HLA mismatch The nonmatching of HLA antigen in recipients of renal and other allografts. See Transplantation. Cf HLA match

HMG CoA reductase inhibitor Any of a family of drugs, statins, that inhibits the activity of 3-hydroxy-3-methylglutaryl coenzyme A, which is involved in early cholesterol synthesis. See Atherosclerosis, Cholesterol

HMG-CoA synthase deficiency METABOLIC DISEASE A condition caused by a congenital defect in mitochondrial HMG-CoA synthase, resulting in a poor tolerance to fasting, which may evoke hypoketosis–since HMG-CoA is pivotal in the control of ketogenesis MANAGEMENT Prevent fasting. See HMG-CoA synthase

HMO Health maintenance organization MANAGED CARE 1. A comprehensive health care system that provides or ensures delivery of basic and supplementary health maintenance and treatment services including inpatient and ambulatory care, in a defined geographic region to a voluntarily enrolled group of persons for a premium; the HMO requires that its enrollees use the services of designated (participating) physicians, hospitals and other providers of medical care and receive reimbursement through a predetermined, fixed periodic prepayment by or on behalf of each individual or family unit regardless of the amount of services provided; use of physician outside the HMO nework requires preapproval PROS Reduced out-of-pocket costs–ie, no or minimal deductible, no paperwork–ie, insurance forms, and a minimal copayment for each office visit to cover the paperwork handled by the HMO HMO HEALTHCARE SERVICES Hospital, physician, often prescription

drug and vision care, home care, and other services HMO MODELS Staff model, group model, independent practice association, network model. See Gatekeeper 2. An organized group of physicians paid in advance for providing health care services to a voluntary population of individuals. Cf Group model, HMO network, Independent Practice Association model, Network model, Risk HMO, Staff model

HMO network MANAGED CARE An HMO that contracts with local hospitals to provide in-patient medical services, and with 2 or more independent groups of physicians to provide health services; the group is paid a set amount per HMO enrollee per month; in some, staff physicians may be HMO employees. See HMO

HMO penetration MANAGED CARE The proportion of Pts in a geographic region enrolled in an HMO. See HMO

hoarseness AUDIOLOGY An abnormally rough or harsh voice caused by vocal abuse and other disorders–eg, GERD, thyroid problems, or trauma to the larynx

Hodgkin's disease Hodgkin's lymphoma HEMATOLOGY A type of lymphoma most common in young adults, accounting for ± 0.7% of all new cancers in the US–±7500 cases; HD is clinically distinct as it often responds well to chemotherapy, and histologically distinct as its *sine qua non* diagnosis requires Reed-Sternberg cells CLINICAL Asymptomatic enlargement of lymph nodes, spleen, or other lymphoid tissue, ±fever, weight loss, fatigue, night sweats STAGING Ann Arbor classification TREATMENT RT, MOPP, mechlorethamine, vincristine, procarbazine, prednisone; 20% are treatment 'failures' and may respond to various cocktails–eg, ABVD–doxorubicin, bleomycin, vinblastine, dacarbazine, salvage therapy, third line therapy. See Ann Arbor classification, Reed-Sternberg cell

HODGKIN'S DISEASE STAGES

I CA present in one lymphoid region or in only one extranodal region or organ

II CA present in two or more lymphoid regions on same side of diaphragm, *or* cancer is found in only one extranodal region or organ and in surrounding lymph nodes; other lymph node areas on the same side of the diaphragm may also be involved

III CA present in lymphoid tissues on both sides of diaphragm, and may have spread to an area or organ near lymphoid regions and/or to spleen

IV CA has spread to extranodal organ/organs, on both sides of diaphragm

Hoffa's disease Infrapatellar fat pad syndrome ORTHOPEDICS Post-traumatic proliferation of fat, which may follow mild trauma in young people accompanied by quadriceps laxity

Hoffmann sign Digital reflex, Hoffmann reflex NEUROLOGY A clinical finding in hemiplegia in which the tips of the fingers are nipped and the thumb and finger undergo a reflex flexing. See Hemiplegia

Hogan Personality Inventory PSYCHOLOGY A psychologic test that assesses a person's customer orientation, ability to tolerate stress, work with supervisers, and get along with co-workers

HOHD syndrome Hair-onychodysplasia, hypohydrosis-deafness syndrome An AR condition characterized by baldness, microdontia, dysplastic toenails, hypohidrosis, palmo-plantar as well as knee-elbow hyperkeratosis that may be accompanied by sensorineural defects

holding bed Observation bed, see there

hole VOX POPULI Perforation

holiday heart A Fib or flutter that follows 1. Binge alcohol abuse or 2. loud noises–eg, firecrackers on July 4th, the US day of Independence

holism PSYCHIATRY An approach to the study of the individual in totality, rather than as an aggregate of separate physiologic, psychologic, and social characteristics

holistic approach A term used in alternative health for a philosophical approach to health care, in which the entire Pt is evaluated and treated. See Alternative medicine, Holistic medicine

holmium "pulsed" laser OmniPulse SURGERY A laser system, with related disposable fiberoptic components, used in soft-tissue applications in ENT surgery–tonsillectomy, uvula and soft palate surgery–uvulopalatopharyngoplasty or UPPP, as well as in sinus surgery and minimally invasive procedures–arthroscopy, spinal disk decompression, urology, lithotripsy, gynecology, peripheral artery laser angioplasty, and general surgery. See Laser

holmium laser resection of prostate UROLOGY A minimally invasive procedure using a holmium laser for treating BPH as an alternative to the more invasive transurethral resection of the prostate TURP. See Benign prostatic hypertrophy

Holt-Oram syndrome Heart-hand syndrome MOLECULAR CARDIOLOGY An AD disorder with structural defects of the heart and upper limbs CLINICAL Upper limb defects may be uni- or bilateral and involve structures of the embryonic radial ray causing aplasia, hypoplasia, fusion and anomalous development of the radial, carpal and thenar bones; defects include triphalangeal or absent thumbs, foreshortened arms and phocomelia; cardiac abnormalities are variably present and are either structural–eg, single or multiple atria, VSD, or functional–eg, bradycardia, various degrees of AV block; other defects in HOS include vertebral, anal, tracheoesophageal, and renal defects

Holter monitoring Ambulatory electrocardiographic monitoring CARDIOLOGY The monitoring of a Pt's EKG over 24 hr period while the Pt records symptoms and activities in a diary; a small portable EKG device is worn in a pouch around the neck; after completion, a correlation is made between the symptoms–or activities recorded and the EKG pattern that was obtained simultaneously. See Myocardial infarction

holy grail MEDTALK A literary way of expressing 'elusive goal'

Homans' sign ORTHOPEDICS Pain in back of calf or knee when the foot is dorsiflexed, which is typically associated with thrombosis in veins of calf. See Deep vein thrombosis

home access HIV testing Home HIV testing AIDS A format for allowing a person to directly submit blood specimens to ID the HIV in blood, obtaining the specimen in private. See AIDS, HIV

home/beeper call GRADUATE EDUCATION A call schedule in which the house officer–HO is on call outside of the hospital and can be reached by beeper/pager. See Call schedule, House officer

home birth Home delivery OBSTETRICS The birth of a child at parents' home PROS Cheaper, more comfortable CONS Complications and high-risk pregnancies have ↑ mortality. See Bimanual pelvic examination

home-bound HEALTH CARE *adjective* Referring to a person who is 'truly unable' to leave the home

home field advantage SPORTS MEDICINE A phenomenon in which a baseball–or other–team playing competitive sports has a slight winning edge over the visiting team, which is attributed to psychological factors, but may be linked to interruption of circadian rhythm or jet lag, especially in teams traveling west → east

home health care Home care MANAGED CARE Medical care that could be administered in a hospital or on an outpatient basis to Pts who are not sufficiently ambulatory to make frequent office or hospital visits EXAMPLES IV therapy given at Pt's residence, usually by a health care professional; HC ↓ need for hospitalization. Cf Ambulatory care, Inpatient care

home infusion therapy The IV administration of therapeutics–analgesics, antibiotics, chemotherapy, parenteral nutrition–outside of a formal health-care environment. See Hyperalimentation, Patient-controlled analgesics, TPN

home page ONLINE The first screen or main page of a website when entering the web; the main web page for a business, organization, person or the main page of a collection of web pages. See Browser, Internet, Web

HOME scale Home Observation for Measurement of Environment scale PEDIATRICS A measure of a child's at-home stimulation

home test Do-it-yourself testing LAB MEDICINE A diagnostic evaluation using a simple kit or test available 'over-the-counter' in a pharmacy, intended to be self-adminstered. See CLIA

home uterine activity monitoring OBSTETRICS A form of prenatal maternal care in which a ♀ in late pregnancy–24 to 36 wks–attaches an external tocodynamometer over the uterus twice a day for 1–2 hrs; the recording is transmitted by telephone to a central station. See Deceleration

homelessness SOCIAL MEDICINE A state of disenfranchisement, in which a person's lack a permanent residence, often living on the streets without protection from the environment and/or ready access to sanitation facilities. See 'Fourth World', Shelterization. Cf Refugee

homeopathic *adjective* MAINSTREAM MEDICINE Referring to a dose or serum level of a particular substance that is too low to be clinically effective

homeostasis PHYSIOLOGY The dynamic constancy of the internal environment; the self-regulating biologic processes that maintain an organism's equilibrium; the ability to maintain a constant state under various conditions of stress

homeric laughter Mirthless laughter NEUROLOGY Uncontrolled spasmodic laughter induced by mirthless stimuli, a symptom of organic brain disease that indicates a poor prognosis; HL may be seen in multiple sclerosis, pseudobulbar palsy, epilepsy, intracranial hemorrhage, frontal lobotomy, and kuru, which causes 'laughing death'

homicidal health care worker FORENSIC PSYCHOLOGY A rogue physician, nurse, physical therapist or other health care worker who deliberately causes the death of one or more Pts in his/her care, usually in absence of dread disease. See Shipman, Swango. Cf Kevorkian

homicide FORENSIC MEDICINE The killing of a person by another, regardless of intent. See Manslaughter, Murder. Cf Suicide

homicide by fright FORENSIC MEDICINE Death of a victim of a potentially violent criminal act in which there are no fatal injuries

homicide/suicide cluster FORENSIC PSYCHIATRY One or more homicides with the subsequent suicide of the perpetrator who typically is a ♂ married to, or living with, a ♀ in a relationship marked by physical abuse, who has a Hx of alcohol or substance abuse, and access to firearms. See Homicide, Suicide. Cf Postal worker syndrome

homocysteine PHYSIOLOGY An amino acid not found in proteins and a critical intermediate in cysteine and methionine metabolism; ASHD is linked to ↑ homocysteine–it promotes smooth muscle growth and inhibits endothelial cell growth, possibly by promoting cyclin gene expression. See Homocysteinuria

homocystinuria METABOLIC DISEASE An AR condition due to a defect of cystathionine β–synthase , characterized by ↑ (> 300 µmol/L) homocysteine in serum CLINICAL Overgrowth of long bones, mental retardation, osteoporosis, ectopia lentis, failure to thrive, sparse blond hair, genu valgum, thromboembolism, fatty liver; most die before age 30 of arterial and venous occlusive disease; lesser ↑ of homocysteine is seen in heterozygotes, in those with ↓ folic acid, vitamin B$_{12}$, in renal failure and after heart transplants LAB ↑ homocysteine in urine MANAGEMENT Pyridoxine or vitamin B$_6$

homogeneous *adjective* Composed of identical cell types; of the same type, having the same characteristics

homologous artificial insemination REPRODUCTION MEDICINE The insertion of sperm from the husband or regular partner into the uterus by means other than by intercourse. See Artificial insemination. Cf Heterologous artificial insemination

homophobia PSYCHOLOGY An irrationally negative attitude toward those with homosexual orientation, or toward becoming homosexual. See Closet, Gay-bashing, Heterosexism. Cf Gay, Homosexual, Phobia

homosexual SEXOLOGY *adjective* Characterized by same-sex contact; of, relating to, or having a sexual orientation as a person of the same sex *noun* A homosexual person: gay man, lesbian

homosexual panic SEXUALITY An acute severe attack of anxiety based on unconscious conflicts involving gender identity. See Circumstantial homosexuality. Cf 'Don Juan' syndrome

homosexuality SEXOLOGY Sexual attraction or erotosexual pairing with a partner of the same genital morphology. See AIDS, Circumstantial homosexuality, HIV, Latent homosexuality. Cf Lesbianism, Sexual deviancy

homovanillic acid HVA PHYSIOLOGY A major dopamine metabolite used to synthesize epinephrine and norepinephrine and excreted in urine; HVA is usually measured with the major catecholamines and other catecholamine metabolites; HVA is ↑ in neuroblastoma and ganglioneuroma REF RANGE 0-15 mg/24 hr urine. See Hypertension, Pheochromocytoma

honest error in judgement Honest mistake MEDICAL LIABILITY A non-malicious deviation in a provider's standard of practice. See Human error. Cf Negligence

honeycomb pattern A reticulated or net-like pattern with relative periodicity in a 2-D plane BONE RADIOLOGY An HP is seen in a plain skull film as patchy new bone fills in underlying osteoporosis circumscripta is typical of Paget's disease of bone PULMONOLOGY Accented interstitial markings A term for the coarsened pulmonary parenchyma seen on a plain chest film, in which there is partial alveolar wall destruction, incomplete replacement, and alveolar thickening by interstitial fibrosis–in contrast, emphysema demonstrates alveolar wall attenuation, a pattern typical of advanced interstitial pneumonia DIFF DX AdenoCA, asbestosis, berylliosis, chronic allergic alveolitis, chronic granulomatous infections, chronic interstitial pulmonary edema, collagen vascular disease–SLE, progressive systemic sclerosis, diffuse alveolar damage–in the organization phase, eosinophilic granuloma, hemosiderosis, interstitial fibrosis, interstitial pneumonia–linked to drugs, Langerhans' cell histiocytosis, lymphangiomatosis, lymphocytic interstitial pneumonia, pneumoconiosis, radiation, recurrent aspiration pneumonia, rheumatoid arthritis, end-stage sarcoidosis, scleroderma, tuberous sclerosis. Cf Chicken wire pattern

honeymoon cystitis INFECTIOUS DISEASE A form of nonspecific urethritis caused by local irritation due to prolonged, frequent, recent or first-time sexual activity in ♀ with a low baseline of sexual traffic; damage to the vesicovaginal wall may evoke cystitis, hematuria, dysuria, ↑ frequency and urgency; although often culture-negative, HC may respond to antibiotics; resolution may be hastened by a rest period

honeymoon period A timespan after diagnosing a disease before its impact is manifest, fancifully likened to the HP of early marriage, during which the husband and wife are most cordial and passionate with each other DIABETOLOGY A period of residual β cell function and insulin secretion in early-onset type 1 DM that follows stabilization of the Pt's hyperglycemic presentation; in the honeymoon period, most children require ½ of the cal-

culated insulin dose–ie, 0.5 U/kg or may even go into temporary 'remission', signaled by recurring hypoglycemia at the initial dose, a time period that is short-lived, usually < 1 yr DRUG SLANG Early stages of drug use before addiction or dependency develops NEPHROLOGY A period of optimism and euphoria accompanying the early stage of chronic dialysis by Pts in renal failure; the HP lasts from 6 wks to 6 months and is followed by a 'mourning period', see there PEDIATRIC SURGERY A postoperative time period in an infant with congenital diaphragmatic hernia characterized by stable vascular resistance, relative ease of ventilation and normalization of blood gases; this deteriorates within hrs to days and the infant develops pulmonary HTN and resultant right-to-left shunting through the ductus arteriosus and foramen ovale. See Extracorporeal membrane oxygenation

honk PEDIATRICS A widely-transmitted precordial whoop, described as a high-pitched, musical, late systolic murmur in some Pts with mitral valve prolapse–MVP, a sound attributed to resonation of the valve leaflets and chordae; non-honkers with MVP may be made to honk by having them stand or lean

honorarium A sum of money given to a person of importance to thank the person for her or his time and trouble

honorary co-authorship Spurious co-authorship A practice in which the name of senior researchers are added to research reports that are the work of others–eg, undergraduates and postdoctoral fellows working in their lab. See Authorship, Baltimore affair, CV-weighing, Darsee affair

honorary staff HOSPITAL PRACTICE A medical staff status accorded to a practitioner who is retired from active service to the hospital, or who has made noteworthy contributions to the hospital, or who, by reason of outstanding reputation or achievement, is otherwise honored. See Medical staff

hooch SUBSTANCE ABUSE 1. A street term for marijuana. See Marijuana 2. Moonshine, see there

hook & deliver technique GYNECOLOGY A method for removing a Norplant contraceptive. See Contraceptive. Cf IUD

hooked *adverb* Addicted

hookworm PARASITOLOGY A hematophagous nematode of family Ancylostomatidiae–eg, Old World hookworm–*Ancylostoma duodenale* and New World hookworm–*Necator americanus* that sensitizes the penetration site–eg, skin, causing 'ground itch', or lungs–eg, Loeffler syndrome as the worms wiggle through, causing eosinophilia and, due to bloodsucking, anemia LAB Rhabditidiform larvae may be confused with *Strongyloides stercoralis*; eggs may be confused with *Trichostrongylus* and *Meloidogyne* spp. See *Ancylostoma duodenale, Necator americanus*

HOPE CARDIOLOGY A randomized, multicenter, multinational trial–Heart Outcomes Prevention Evaluation–which found that ramipril/Altace®, an ACE inhibitor, in ↓ mortality in Pts with heart failure. See Calgary-HOPE

hopeless TERMINAL CARE Futile. See Medical futility

hopelessness PSYCHOLOGY Bleak expectations, usually about oneself or one's future. See Depression

hoplophobia PSYCHOLOGY Fear of firearms. See Phobia

hopped up DRUG SLANG A popular phrase for being influenced by drugs

hordeolum Sty, see there

horizontal fracture LeFort I fracture, see there

horizontal integration MANAGED CARE An informal network of providers of like service and/or with hospitals that form an integrated health delivery system . Cf Vertical integration

308

horizontal plane Transverse plane IMAGING A directional plane that divides the body, organ, or appendage into superior and inferior or proximal and distal

horizontal transmission EPIDEMIOLOGY The transmission of an infection from one to another person of the same generation in the same population. Cf Hereditary transmission, Vertical transmission

hormonal cyclicity ENDOCRINOLOGY The regularly recurring changes in serum levels of various hormones–eg, changes in pituitary and ovarian hormones in synchrony with the menstrual cycle

hormonalization ENDOCRINOLOGY Hormonal manipulation of ♂ or ♀ somatic or behavioral characteristics

hormone *adjective* Referring to a hormone, hormonal *noun* ENDOCRINOLOGY A chemical messenger produced in an endocrine gland and released into the circulation to effect a change in a specific target organ; hormones regulate the internal environment, effecting homeostatic control, regulate reproductive processes, and affect mood and behavior TYPES Steroid hormones–cortisol, estrogen, testosterone; nonsteroid hormones–choleckystokinin, epinephrine, dopamine, insulin, norepinephrine, serotonin, vasopressin. See ACTH, Beef growth hormone, Biological thyroid hormone, Candidate hormone, Designer hormone, Ectopic hormone, Emasculated hormone, Fat mobilizing hormone, FSH, Gonadotropin-releasing hormone, Growth hormone, HCG, LH, LH-RH, Melanocyte-stimulating hormone, Neurohormone, Pheromone, PTH, Somatotropin release-inhibiting hormone, Thyrotropin releasing hormone, TSH. Cf Neurotransmitter

hormone receptor test ENDOCRINOLOGY A test to quantify the hormone receptors, in a particular tissue–eg, the breast in breast CA. See Estrogen receptor, Progesterone receptor

hormone replacement therapy Hormonal replacement therapy GYNECOLOGY The administration of estrogen or progestins to ♀ to alleviate Sx of menopause–eg, urogenital atrophy and psychological symptoms, and slow or reverse osteoporosis and ↓ risk of ischemic heart disease and, more recently, Alzheimer's disease. See Estrogen replacement therapy. Cf Androgen replacement therapy

hormone resistance syndrome A condition caused by a reduced or absent end-organ responsiveness to a biologically active hormone, which may be due to a hormone receptor defect or a post-receptor defect

hormone therapy Endocrine therapy ENDOCRINOLOGY The use of hormones to manage deficiency states–eg, pharmacologic doses of steroids are antiinflammatory or immunosuppressive. See Ectopic hormones, Estrogen receptors GYNECOLOGY See Estrogen replacement therapy, Hormone replacement therapy ONCOLOGY A treatment modality in which hormonal effect is eliminated, blocked, or added in CAs known to respond to hormone receptor manipulation BREAST CANCER HT blocks receptors for estrogens and, to a lesser extent, progesterone, ↓ cancer aggressiveness LYMPHOPROLIFERATIVE DISORDERS Adjuvant therapy uses corticosteroids in both induction of remissions and maintenance doses PROSTATE CANCER *HT inhibits* gonadotropin in the pituitary with potent analogues–eg, buserelin, leuprolide acetate, blocking gonadotropin-releasing hormone, *or blocks* peripheral action of androgens at the cellular level–eg, flutamide, nilutamide

Horner syndrome Bernard-Horner syndrome, Horner's ptosis NEUROLOGY A clinical complex consisting of the sinking in of one eyeball, ipsilateral ptosis and miosis, anhidrosis and flushing of affected side of face ETIOLOGY Paralysis of cervical sympathetic nerves

horseshoe kidney A relatively common–1/700 live births–lesion–♂ : ♀ ratio, 2:1, seen alone or associated with trisomy 18, characterized by fusion of the renal poles ASSOCIATED ANOMALIES Urogenital anomalies–eg, polycystic kidneys, ureter reduplication, as well as GI, cardiac, and skeletal anomalies and a slight ↑ in transitional cell carcinoma of the renal pelvis CLINICAL From asymptomatic to advanced renal disease–eg, hydronephrosis, infec-

tion, kidney stones, which may be accompanied by chronic GI Sx and pain; the 'classic' Rovsing sign–abdominal pain localized to the kidneys accompanied by nausea on spinal retroflexion is rarely useful but, when present, is due to pressure of the isthmus of kidney on nerves and vessels and may occur in S-shaped or L-shaped kidneys

hospice MANAGED CARE An institution which provides comfort care and a combination of inpatient, outpatient, and home health services–pain relief, symptom management and support, for terminally ill Pts (and their families) with CA, AIDS and other dread diseases. See Comfort care

hospital A place where medical and surgical procedures are perfomed on inpatients. See America's Best Hospitals, American Biologics Hospital, Community hospital, For-profit hospital, Magnet hospital, Massachusetts General Hospital, Not-for-profit hospital, Public hospital, Plan hospital, Rural primary care hospital, 'Safety net' hospital, Single disease hospital, Virtual hospital

hospital-acquired penetration contact A break of skin or mucosal barriers by needle sticks, paper cuts, broken glass, resulting in contact infection by highly infectious, blood-transmitted, usually viral infections–eg, HIV-1, hepatitis B, or Jakob-Creutzfeldt agent; the frequency of HIV infection via this route is low, generally regarded as << 1/100 infected penetrations. See Sharps

hospital-acquired pneumonia Nosocomial pneumonia INFECTIOUS DISEASE Pulmonary infection acquired during a hospital stay which is often more severe than community-acquired pneumonia RISK FACTORS Immune compromise, alcoholism, elderly, aspiration due to intubation. Cf Community-acquired pneumonia

hospital affiliation HEALTH INSURANCE A contract whereby one or more hospitals agree to provide benefits to members of a specific health plan. See Affiliation

hospital-based physician A physician who provides 'clinical support' for Pt management, performing medical services within a hospital/health center EXAMPLES Radiologists, anesthesiologists, pathologists, ER physicians–acronym, 'RAPERs', physicians specialized in nuclear medicine, radiation oncology, rehab medicine, occupational medicine, public health, forensic pathology. See Hospitalist, RAPERs. Cf House physician, Medical specialties, Primary care, Surgical specialties

hospital bed 'syndrome' Cot sides 'syndrome' An event with medicolegal implications, seen in brain-damaged Pts–who often are, in addition, elderly with osteoporosis, who have fallen out of a hospital bed with the sides up–ie, from a greater height than had the sides been down PROGNOSIS Poor; coma, death are common

hospital chart A type of medical record that documents the course of a Pt from the moment of admission to a hospital to the time of discharge. See Chart war

hospital committee HOSPITAL PRACTICE Any committee in a particular hospital, often chaired by a physician, that addresses a particular aspect of Pt management, which may address issues related to financial or operational management of the facility. See Committee, Credentials committee, Executive committee, Patient care committee, Safety committee, Tissue review

hospital formulary THERAPEUTICS The drugs and therapeutics available in a particular hospital; agents included in the HF are at the discretion of the appropriate parties on the medical staff in conjunction with the hospital pharmacy department

hospital information system INFORMATICS The computer hardware and software that processes a hospital's data, including financial, Pt-related, and 'strategic' management data, Pt accounts, Pt tracking, payroll, reimbursements, taxes, statistics

hospital insurance Part A, see there

hospital mortality rate The death rate/1000 admissions at a particular hospital. See High mortality outlier

hospital utilization The usage rate of a particular health care facility; a group of statistics referring to a population's use of hospital services

hospital waste PUBLIC HEALTH Any waste, biologic and nonbiologic, that is discarded by a hospital and not intended for further use. See Infectious waste, Medical waste, Medical Waste Tracking Act of 1988

hospitalist HOSPITAL PRACTICE A physician specialized in inpatient medicine who acts as a Pt's primary doctor while the Pt is hospitalized, ensuring that tests are evaluated in a more timely manner than possible for private physicians PROS ↑ Management efficiency, ↓ hospital stay CONS ↓ Personalized physician-patient contact. See Hospital-based physician

hospitalization The period of confinement in a health care facility that begins with a Pt's admission and ends with discharge; hospitalization also refers to a group health insurance program that pays employees all the expenses incurred during a period of hospitalization, including hospital costs per se–part A, and part or all of the physicians' costs–part B. See Bed, Involuntary hospitalization, Part A, Part B, Per diem

hospitalization hazard Any risk to health linked to hospitalization–eg, medication prescribing errors, normal complications of anesthesia or surgery, and others. See July phenomenon

hospitalization insurance HEALTH INSURANCE Insurance that reimburses, within contractual limits, hospital and specific related expenses arising from hospitalization caused by injury or sickness. See Insurance. Cf Part A

host EPIDEMIOLOGY Any organism that can be infected by a pathogen under natural conditions. See Definitive host, Intermediate host, Paratenic host, Transport host IMMUNOLOGY Graft recipient. See Graft, Transplant INFORMATICS A networked computer that performs centralized functions–eg, providing access program or data files to computers in a network; a host may be self-contained or located on Internet; computer that acts as a source of information or capabilities for multiple terminals, peripherals and/or users. See Node, Network. Cf Server

host defense The protection an organism is afforded against infections TYPES 1. Nonimmunologic–eg, mucocutaneous or integumental barriers, cilia, microvilli; mechanical–eg, urinary outflow, vascular perfusion of tissues; native flora, which 'outcompete' pathogens; 2. Immunologic–eg, chemotaxis, phagocytosis, immunoglobulins, complement, T cell defense

host factor EPIDEMIOLOGY An intrinsic factor–age, race, sex, behaviors, etc. which influences an individual's exposure, susceptibility, or response to a causative agent

HOT CARDIOLOGY A trial–Hypertension Ongoing Trial designed to:; (1) determine the optimal target diastolic BP, and (2) assess whether low-dose aspirin ↓ cardiovascular M&M in Pts with HTN. See Antihypertensive, Felodipine, Hypertension, J curve phenomenon

hot *adjective* Inflamed

hot flashes Hot flush GYNECOLOGY A symptom afflicting 80-85% of middle-aged ♀, first occurring during the perimenopause, continuing with ↓ intensity for yrs, manifesting itself as transient waves of erythema and uncomfortable warmth beginning in the upper chest, face and neck, followed by fine sweating and chills; HFs are precipitated by emotional stress, meals and environmental cues, and are more intense if ovaries are surgically removed than if the decline of ovarian function is less abrupt ETIOLOGY Idiopathic, due to response of autonomic nervous system to ↓ estrogens; they are responsible for osteoporosis, atrophy of vaginal epithelium, leukorrhea and pruritus; relationship of ↓ estrogen to CAD is unclear THERAPY Although hormones–eg, estrogens in ♀ and androgens in ♂ ameliorate the symp-

toms, they are contraindicated in ♀ with breast CA, and in ♂ with prostate CA; megestrol acetate ↓ HFs by 85%–vs 20% with placebos

'hot' food See Spicy foods

hot nodule NUCLEAR MEDICINE A focal ↑ in radioisotope uptake on a ¹²³I scintillation scan in a solid organ–eg, liver or thyroid, seen by immunoscintigraphy; thyroid HNs often correspond to toxic nodular or multinodular goiter, as functional thyroid lesions suppress TSH synthesis; HNs are rarely malignant. See Scintigraphy. Cf Cold nodule, Warm nodule

hot potato speech SPEECH PATHOLOGY A fanciful term for a disorder of resonance in which the speech has a muffled quality, due to space-occupying lesions–eg, lymphoid masses, tumors of the vallecula between the epiglottis and the base of the tongue

hot scan see Hot nodule

hot spot EPIDEMIOLOGY A region with a marked ↑ of a particular disease–eg, HIV infection among disenfranchised inner-city inhabitants. See Designer antibodies

hot tub disease INFECTIOUS DISEASE A type of hypersensitivity pneumonitis related to recreational use of a hot tub caused by *Cladosporium* spp, and, aeroallergenic fungi of the class Deuteromycetes

hotdog headache NEUROLOGY Pulsating cephalgia with facial flushing, caused by sodium nitrite-preserved meat, especially hotdog–frankfurter meat; in addition to preserving meat, nitrites impart a red tinge that imparts a desirable–'marketable' appearance for displaying meats to shoppers

hotel function MEDICAL PRACTICE Any of the intangible facets of Pt satisfaction with a provider–ie, physician's office or hospital–contact which facet is unrelated to the actual quality of care received

hotline A telephone system often manned by volunteers and dedicated to answering questions about a particular disease or group of diseases–eg, AIDS 'hotline'; or designed to facilitate access to a particular service–eg, abused women 'hotline'. See National Disaster Medical hotline

Hottentot apron Excessive elongation of the labia minora seen in the Hottentot tribe of southern Africa, which when seen elsewhere has been attributed to masturbation

house VOX POPULI A residence, domicile, place where an activity occurs. See Almshouse, Clearinghouse, Halfway house, Phoenix House, Project house

house call A visit to a Pt by a physician providing some form of primary care or followup to prior therapy

house dust mite *Dermatophagoides farinae, D pteronyssoides* A mite that feeds on household detritus, which is often highly allergenic; exposure to HDMs can be measured by RAST

house officer A resident or residency

house physician A physician who 'covers' the medical needs of Pts in a hospital; an HP may not be board-certified, but usually has a number of yrs of formal training–ie, residency in clinical medicine in a teaching hospital, and works in a secondary care facility, usually carrying out the dictates of the Pt's 'private' physician. See Hospitalist

house staff MEDICAL PRACTICE The body of physicians and other health care providers who participate in Pt management within the hospital. See Hospital-based physician, House physician, Medical staff, RAPER

house-tree-person test PSYCHIATRY A test in which facets of a person's character and internal conflicts are inferred from the manner in which he/she draws a house, a tree, and a person

housebound syndrome PSYCHIATRY A type of panic disorder that most commonly occurs in ♀–ratio, 2-4:1, affecting 2-6% of ♀ ages 18–64; HBS is characterized by agoraphobia, difficulty in functioning in public, generalized anxiety, cardiovascular symptoms–eg, chest pain, tachycardia, arrhythmia, GI distress, headache, vertigo, syncope and paresthesiae. See Panic disorder

HP Health professional, see there

H&P History and physical examination

HPLC High-performance liquid chromatography LAB INSTRUMENTATION A highly sensitive analytic method in which analytes are placed at high pressure–500-1500 psi in a chromatography column to separate them, allowing highly specific identification. See Chromatography

HPRT Hypoxanthine-guanine phosphoribosyl transferase, see there

HPV Human papillomavirus INFECTIOUS DISEASE A DNA virus that causes warts on acral parts which, in those with multiple sexual partners, may be premalignant, especially in those infected with type 16—65 genotypes of HPV have been characterized–by a genotype is distinct if it has < 50% DNA sequence similarity or 'homology' with its closest relative; HPV has been identified by in situ hybridization in epithelial proliferations that are benign–eg, condyloma acuminatum, or malignant–eg, squamous cell carcinoma of the penis, anus, and uterine cervix, or of uncertain clinical behavior–eg, inverted papillomas of the nasopharynx. See Oncogenic HPV, Cf Cervical intraepithelial neoplasia

HPYLORI GASTROENTEROLOGY A clinical trial comparing the cost-effectiveness of 2 treatment regimens in eradicating *H pylori* in persons with dyspepsia

HR 1833 Partial Birth Abortion Ban Act of 1995 OBSTETRICS, MEDICAL ETHICS A congressional act banning abortions in which the physician 'partially vaginally delivers a living fetus before killing it and completing the delivery.'. See Partial birth abortion

HRQL Health-related quality of life. See Quality of life

HRT Hormone replacement therapy, see there

HS *hora somni*, bedtime

HSGYV ENT A mnemonic for mechanical maneuvers to resolve the sensation of a plugged ear: heat, steam, gum, yawn, Valsalva

HSIL High-grade squamous intraepithelial lesion, see there

HSV Herpes simplex virus. See Herpesvirus-1, Herpesvirus-2

HTLV Human T cell leukemia/lymphoma virus A family of single-stranded retroviruses, subfamily Oncoviridae, that produce a DNA copy from viral RNA using reverse transcriptase; HTLVs are capable of immortalizing and transforming T cells

HTLV-I VIROLOGY A retrovirus that immortalizes T cells; it is associated with 1. Adult T-cell lymphoma/leukemia in endemic regions of southern Japan–where up to 40% of leukemic Pts are HTLV-I-positive, and the Caribbean 2. Chronic progressive myelopathy–tropical spastic paralysis, a condition endemic to the Caribbean, tropical Africa, and South America and 3. HTLV-I-associated neuropathy; HTLV-I, like HIV-1, is blood-borne CARRIER PREVALENCE–US, EUROPE up to 0.10%; 1% of HTLV-I-infected subjects develop clinical disease, which may reflect host susceptibility and variant viral strains

HTLV-II A retrovirus that may transform CD4 T cells, and may rarely be linked to disease–eg, hairy cell leukemia

HTLV-III HIV-1, see there

HTLV-IV HIV-2, see there

http Hypertext transfer protocol INFORMATICS The most commonly used language protocol for transmitting information in Web sites; http on a web address–a URL indicates to Web browsers, "HTML spoken here". See Client, HTML, Server, Web browser, World Wide Web. Cf Gopher

Huckleberry Finn 'syndrome' Truancy syndrome A psychodynamic complex in which the obligations and responsibilities avoided as a child, eventuate into frequent job changes and absenteeism as an adult; the HFS may be a defense mechanism linked to parental rejection, low self-esteem and depression in an intelligent person. Cf Peter Pan syndrome

the hug drug Ecstasy, see there

Humalog® Pen ENDOCRINOLOGY A Humalog–insulin lispro injection rDNA origin, pen which allows a Pt to dial a specific dose in 1-unit increments up to 60 units.

human B lymphotropic virus Herpesvirus-6, see there

human chorionic gonadotropin See hCG

human chorionic sommatomammotropin Placental lactogen A single-chain 21 kD polypeptide hormone produced by the syncytiotrophoblast that is lactogenic, somatotrophic, luteotropic, and mammotropic

human factors engineering ERGONOMICS The body of knowledge about human abilities and characteristics that are relevant to design, and the application of this knowledge to the design of tools, machines, systems, and jobs for safe, comfortable, and effective human use. See Ergonomics

Human Genome Project A multinational collaborative effort to sequence–map all 80,000+ genes and 3×10^9 nucleotides of a haploid set of the human genome, begun in 1986 and completed in 2002 at a total cost of US 3×10^9. See Bioinformatics, Genomics, Junk DNA, Rough draft, Sequencers

human granulocytic ehrlichiosis INFECTIOUS DISEASES Infection by *Ehrlichia chaffeensis*, a rickettsia-like organism genetically similar to *E phagocytophila* and *E equi*, transmitted by the Lyme disease tick, which may be associated with Lyme disease CLINICAL High fever, chills, sweating, myalgia, nausea and vomiting, headache, shaking chills, coughing, severe pain–like being 'hit by a bus,' followed by septic state MANAGEMENT Doxycycline. Cf Human monocytic ehrlichiosis

human immune globulin A therapeutic prepared from a donor pool screened and negative for HIV-1 INDICATIONS FOR USE Primary immunodeficiencies–eg, X-linked agammaglobulinemia, SCID and combined variable immune deficiency, and in acute or chronic ITP. See Immunoglobulin

human immunodeficiency virus HIV-1, see there

human in vitro fertilization Any fertilization involving human sperm and ova that occurs outside the human body. See Assisted reproduction

human leukocyte antigens See HLA

human monocytic ehrlichiosis INFECTIOUS DISEASE An infection by *Ehrlichia chaffeensis* VECTOR Lone Star tick–*Amblyomma americanum*, possibly also *Dermacentor variabilis* RESERVOIR Deer, possibly dogs or others REGIONS US, Europe, Africa CLINICAL Fever, malaise, headache, myalgia, rigors, sweating, anorexia, arthralgia, N&V, pharyngitis, rash, cough, diarrhea; with progression of HME, fever, headache, myalgia, anorexia, and arthralgia become more intense LAB CSF pleocytosis, *E chaffeensis* inclusions in monocytes, spleen and other tissues, BM hyperplasia–58%, hypoplasia–17% DIAGNOSIS Romanovsky stains of circulating monocytes, EM, indirect immunofluorescence, PCR COMPLICATIONS Meningitis, interstitial pneumonia, GI and pulmonary hemorrhage, disseminated hemorrhage. Cf Human granulocytic ehrlichiosis

human papillomavirus See HPV

human placental lactogen Chorionic somatomammotropin, hPL, human placental lactogen, human chorionic somatomammotropin A hormone produced during implantation of a fertilized egg, and secreted by the placenta; hPL regulates and coordinates fetal growth and metabolism and maternal metabolism MATERNAL EFFECTS Relative insulin resistance, ↑ circulating free fatty acids; hPL may optimize metabolism of nutrients by the fetus in 1st half of pregnancy; in 2nd half, there is little correlation between hPL levels and fetal well-being; hPL stimulates milk production and breast enlargement, and is somatotopic and luteotropic; urine and plasma HPL levels reflect placental size and are higher in diabetic mothers REF RANGE Rises during gestation, plateauing at 37 wks at 10 mg/mL; ♂ <0.5 mg/mL; non-pregnant ♀ <0.5 mg/mL; wks of gestation 5-27 <4.6 mg/mL; 28-31 wks, 2.4-6.1 mg/mL; 32-35 wks, 3.7-7.7 mg/mL; 36 wks to term, 5-8.6 mg/mL; hPL is ↓ in postmaturity syndrome, retarded growth, toxemia of pregnancy, threatened abortion; it is ↑ DM, Rh isoimmunization, hydatiform mole, choriocarcinoma CRITICAL VALUE <4 μg/mL after 30 wks gestation

human shield FORENSIC MEDICINE A person used to protect a kidnapper, terrorist, or combatant from gunfire

human subject Participant DRUG APPROVAL A person who partipicates in a clinical trial of a new drug

humanistic medicine Patient-centered medicine SOCIAL MEDICINE A philosophy in medicine which encompasses positive mentoring skills, community service, compassion and sensitivity, collaboration, observation of professional ethics

humanitarian use designation See Orphan drug

humanized antibody MOLECULAR MEDICINE A recombinant DNA product which ↓ non-human monoclonal antibody immunogenicity, by transferring the hypervariable genes of rodent antibodies–which encode peptide segments capable of recognizing the desired epitope, into normal human genes. Cf Designer antibody

humanized milk NEONATOLOGY A modified formulation of cow milk, the fat ratio of which closely mimics human milk–40% casein, 60% whey; HM is indicated in infants < 2000 g. See Breast milk, Milk. Cf White beverages

Humate-P® Antihemophilic factor/von Willebrand factor complex HEMATOLOGY A plasma-derived product for treating Pts with hemophilia A and refractory von Willebrand's disease. See Hemophilia A, von Willebrand's disease

Humatrope™ Injectable rDNA somatropin ENDOCRINOLOGY A growth hormone for adult somatropin deficiency and short stature in girls with Turner syndrome. See Turner syndrome

humeral *adjective* Referring to the humerus. Cf Humoral

humidifier NURSING A device that puts moisture in air. Cf Dehumidifier

humidifier lung A transient condition related to mechanical ventilation of buildings–heating and cooling systems, possibly caused by water-borne amoebae, which develops 1-3 days after exposure CLINICAL Malaise, cough, chest tightness, dyspnea, weight loss; Sx appear 4-6 hrs after re-exposure to a workplace, often post-wkend, ergo the synonym of 'Monday sickness'; Sx resolve spontaneously, regardless of continued exposure RADIOLOGY Negative LAB Antibodies to *Negleria gruberi* and *Acanthamoeba* spp; other HL organisms include *Aspergillus fumigatus*, *Aureobasium pullulans*, *Micropolyspora faeni*, *Thermoactinomyces vulgaris* and water-borne organisms. See Sick building syndrome

humiliating bias DECISION MAKING A bias introduced by a previous error in which the diagnostician makes what in his or her mind was a grievous–humiliating error. See Bias, Medical mistake

humor 1. A fluid or gel-like substance. See Aqueous humor, Vitreous humor 2. Hardiharharness. See Laughter

humoral *adjective* Referring to the humoral–B cell lineage immune system. Cf Hormonal, Humeral

hump & dip sign IMAGING A small, focal, smooth contoured bulge in a breast sonogram, which is contiguous with a small sulcus, initially reported to be characteristic of fibroadenoma, but later found in carcinoma

hump sign IMAGING A distortion of the gas-fluid levels in a plain abdominal film caused by an intraluminal bolus of *Ascaris lumbricoides* within the intestine, an uncommon finding in massive intestinal involvement by roundworms

Humulin® Human insulin via recombinant DNA technology. See Insulin

hunchback ORTHOPEDICS A trivial name for angular kyphosis–AK which, in children, is either congenital, due to a lack of segmentation or lack of formation of one or more vertebral bodies TREATMENT Surgical fusion of vertebrae; acquired AK in children is idiopathic and often accompanied by a compensatory ↑ in lumbar lordosis; in adults, AK may be due to infections–eg, midthoracic TB or neoplastic–eg, myeloma or infiltration by an osteophilic CA–eg, breast or kidney

hungry bones syndrome Post-parathyroid surgery recalcification tetany METABOLISM Transient hypoparathyroidism due to resection of significant portions of a hyperactive parathyroid gland; HBs follow high pre-operative PTH–due to excess PTH secretion as occurs in parathyroid hyperplasia and adenoma, high alk phos and extensive bone demineralization, resulting in a rapid 'rebound' recalcification of bones after prolonged hypocalcemia; bone hunger is exacerbated by a pre-existing compromise in renal function RADIOLOGY Mottled bone hypodensity LAB ↑ Calcitriol, ↓↓↓ Ca²⁺–eg, 1.5 mmol/L–US: 6 mg/dl or lower, ↓ phosphorus, ↓ magnesium, reactive ↑ in PTH–if there is residual secretion; the danger lies in attributing Sx to acidosis, which may accompany and exacerbate the complex TREATMENT Vitamin $1,25(OH)D_2$

Hunt syndrome Ramsay Hunt syndrome, see there

hunter SPORTS MEDICINE A person who hunts game animals–eg, deer, elk, turkeys, with a firearm–eg, rifle, shotgun or other weapon–eg, compound bow, crossbow etc

Hunter syndrome Mucopolysaccharidosis type II MOLECULAR MEDICINE An X-R inborn error of metabolism caused by a deficit of sulfoiduronate or iduronosulfate sulfatase CLINICAL Type A–early onset is associated with a large skull, coarse facial features, profound mental retardation, spasticity, stiffness, aggressive behavior; type B is milder

Huntington's disease Huntington's chorea NEUROLOGY An AD degenerative disease of adult onset–ages 40-50 that leads inexorably to death CLINICAL Slowly progressive mood and personality changes, mental deterioration, loss of coordination, chorea, cognitive decline, chronic fatigue, apathy TREATMENT None. See 'Escapee', Trinucleotide repeat disease

huperzina A NEUROLOGY A semisynthetic from *Huperizia serrata* that may improve memory in Alzheimer's disease. See Tacrine

Hurler syndrome Hurler's disease, mucopolysaccharidosis IH METABOLIC DISEASE An AR condition caused by a defect in lysosomal α-L-iduronidase; Sx develop by end of first yr CLINICAL Gargoylism–coarse thick features, Breshnikov–prominent dark–eyebrows, cloudy corneas, progressive stiffness, mental retardation, heart and heart valve defects; death in early teens due to heart disease. See GL

hurricane GEOLOGICAL MEDICINE A natural disaster characterized by highly destructive winds with speeds of 210 km/hr–130 mph or more, heavy rains, and tidal surges, causing floods that once claimed most lives–reduced by early warning systems; hurricanes occurs in the tropical zones of North Atlantic or Pacific Oceans and are classified by severity on a scale of 1 to 5

HUS Hemolytic uremic syndrome, see there

husband-coached birth Bradley method OBSTETRICS A method of prepared birth in which the husband/significant other helps the wife learn to relax, and breathe to facilitate the birth

hustler SEXOLOGY A ♂ paid to service–nudge, nudge, wink, wink–♀ or other ♂

hustling MEDICAL PRACTICE The illegal soliciting of victims of accidents or dread disease, to provide them with services; after being hustled, the Pt's insurance company is usually billed for office visits and treatment. See Ambulance chaser

Hutchinson-Gilford syndrome An AD condition seen in infancy, characterized by poor growth, early onset ASHD, periarticular fibrosis, ↓ subcutaneous fat, dwarfism, small face, beaked nose, baldness, brownish discolored parchment-like skin, poor dentition, poor muscle development, early death. Cf Cockayne syndrome

H-V interval CARDIOLOGY A value measured by electrophysiologic studies of the heart equal to the time from the beginning of the H deflection to the earliest onset of ventricular depolarization recorded in *any* lead

Hx History

Hyalgan® ORTHOPECIS A hyaluronate used to treat osteoarthritic knee pain in Pts unresponsive to conservative therapy or simple analgesics; it is injected directly into the knee; relief lasts up to 6 months. See Osteoarthritis

hyaline arteriolosclerosis Arteriolar hyalinosis The accumulation of amorphous, eosinophilic material in the walls of small arteries and arterioles, a common degenerative change in the elderly, which is most often seen in the spleen and kidneys in Pts with accelerated/malignant HTN, and DM; HA may occur in the pancreas, adrenal glands, liver, GI tract, choroid, and retina; it is rare in the heart, lungs. See Atherosclerosis

hyaline membrane disease Respiratory distress syndrome of the newborn PEDIATRICS A morbid condition linked to up to 50% of neonatal deaths in the US–40,000/yr CLINICAL Atelectasis, hypoventilation, hypotensive shock, pulmonary vasoconstriction, alveolar hypoperfusion, shut-down of cell metabolism; 60% of HMD affects infants < 28 wks of age; 5% in infants > 37 wks PATHOGENESIS Surfactant deficiency, related to prematurity–insufficient phosphatidyl glycerol, intrapartum hypoxia, 'subacute' fetal distress, acidosis, family predisposition, α_1-antitrypsin deficiency, thyroxine, prolactin, cortisol, estrogen; HMD is more frequent in the 2nd twin delivered, twin-to-twin transfusion recipient infant, ♂ infants, children of diabetic mothers and in cesarean sections; a vicious cycle begins where ↓ surfactant results in atelectasis, ↓ ventilation–↑ pCO_2, ↓ pH, ↓ O_2 and hypoxia exacerbating the lack of surfactant, causing shock and more hypoxia CLINICAL Early onset of tachypnea, prominent grunting, intercostal retractions–air hunger, cyanosis; the infants may not respond to O_2; blood pressure and corporal temperature fall, asphyxia intervenes and causes death, or the symptoms peak at 3 days and the infant recovers DIFFDx Neonatal pneumonia, birth-related asphyxia, group B streptococcal sepsis, cyanotic heart disease TREATMENT Supportive–O_2, correction of acidosis, surfactant therapy

Hybrid Capture assay LAB MEDICINE A proprietary system used to detect and monitor viral–eg, *Chlamydia* spp, CMV, HBV, HPV infections. See HBV

hybrid revascularization procedure CARDIOLOGY The use of 2 or more methods for treating 2 or more severely atherosclerotic coronary arteries–eg, use of minimally invasive surgery for one vessel and balloon angioplasty for the other(s)

hybridization MOLECULAR BIOLOGY The formation of a complex of complementary nucleotides; hybridization allows determination of the relatedness or sequence 'homology' between 2 strands of nucleic acids, and precise ID of short–up to 20 kb segments of DNA–Southern blot or RNA–Northern blot

Hycamtin®/™ Topetecan ONCOLOGY A 2nd-line agent for small cell lung and ovarian CAs, or with ara-C–cytosine arabinoside, for myelodysplastic syndrome, chronic myelomonocytic leukemia, acute and chronic leukemias, and pediatric, breast, colorectal, and lung CAs

hyCURE™ WOUND CARE A hydrolyzed powder used as a wound filler and exudate absorber treat decubitus ulcers, stasis ulcers, superficial wounds, to control odor and promote healing. See Decubitus ulcer

hydatid cyst disease Echinococcosis, hydatid disease PARASITOLOGY A condition caused by cysts of tapeworm larvae, *Echinococcus granulosa*, *E multilocularis* DEFINITIVE HOSTS Canine carnivores–eg, coyote, wolves INTERMEDIATE HOSTS Sheep, cattle, pigs, especially in Southern hemisphere; it is an accidental 'tourist' in man; once tainted meat is ingested, ova hatch, penetrate the intestinal wall, migrate to liver, proliferate there or pass to the lung, kidney, heart, skeletal muscle, and CNS before reproducing; the unilocular hydatid cyst is most often located in the liver, but may develop in the lungs MANAGEMENT Abendazole, a benzimidazole-type scolicidal agent, percutaneous drainage of hepatic hydatid cysts

hydatid disease See Hydatid cyst disease

hydatid mole Hydatidiform mole, see there

hydatidiform mole Hydatid mole OBSTETRICS A pathologic product of pregnancy or trophoblastic tumor, which occurs in 1:2000 pregnancies in the US; it is a benign neoplasm or 'allograft' of trophoblastic tissue, characterized by edematous chorionic villi with proliferative trophoblastic tissue LAB Marked ↑ in hCG

HYDATIDIFORM MOLES

COMPLETE MOLE The chorionic villi are swollen, often accompanied by trophoblastic proliferation but not fetal tissue; CMs are 10-20 times more common in Southeast Asia, and demonstrate innumerable grape-like avascular chorionic villi, the usual fate is passage of the product; most CMs have a 46 XX genotype, of which both X chromosomes are of paternal origin; in 10-15% of the complete moles that are 46 XY, both the X and Y chromosomes are of paternal origin; CMs may be due to abnormal gametogenesis and fertilization, while in 46 XX moles, there is fertilization of an empty ovum with no effective genome by a haploid sperm that duplicates without cytokinesis; in the 46 XY complete moles, there may be fertilization of an empty ovum by 2 haploid sperm with subsequent fusion and replication

PARTIAL MOLE A mass characterized by a mixture of fetal tissue with normal edematous villi and/or hydropic degeneration; most partial moles are triploid–47 XXY > 47 XXX, > 47 XYY, rarely also trisomy 16, and may be the result of unsuccessful dual fertilization of a single ovum; the conceptus does not die, but remains as a proliferating 'tumor'

INVASIVE MOLE A trophoblastic proliferation that penetrates the myometrium, and may undergo malignant degeneration into a choriocarcinoma

hydergine NEUROLOGY A therapeutic cocktail used to alleviate some Sx of Alzheimer's disease, consisting of ergoloid mesylate–a 'metabolic enhancer,' hydergine's efficacy has not been well established. See Alzheimer's disease, Tacrine

hydralazine THERAPEUTICS A vasodilator which, with isosorbide dinitrate, is a 2nd line therapy in Pts with CHF for whom ACE inhibitors are contraindicated. See Congestive heart failure

hydramnios Polyhydramnios OBSTETRICS The presence or ≥ 2 liters of amniotic fluid; 20% of fetal malformations are accompanied by hydramnios, 20% are associated with anencephaly, 10% with multiparity, 5-10% with DM; other conditions linked to hydramnios include esophageal and duodenal atresia, hydrops fetalis, hydrocephalus, spina bifida, achondroplasia, toxemia, toxins–eg, lithium therapy; ± 50% have no fetal or maternal defects

hydrarthrosis ORTHOPEDICS Joint effusion. See Synovial fluid

Hydrasorb® WOUND CARE A foam wound dressing. See Dressing

hydrocarbon pneumonia TOXICOLOGY Pulmonary inflammation due to ingestion or inhaling of organic hydrocarbons–oils or solvents–eg, gasoline, kerosene, furniture polish, paint thinner, etc CLINICAL Edema, hemorrhage. See Pneumonia

hydrocele UROLOGY A fluid space in the spermatic cord–SC due to failed closure of the tract through which the testis descends from the abdomen into the scrotum; peritoneal fluid drains through the open tract from the abdomen into the scrotum where it becomes trapped, causing scrotal enlargement; most resolve shortly after birth; in older men, hydroceles may be caused by inflammation or trauma of the testicle or epididymis or by fluid or blood in the SC DIAGNOSIS Transillumination

hydrocephalus Water on the brain NEUROLOGY Distension of the cerebral ventricles due to an abnormal accumulation of CSF in the cerebral ventricles due to blockage of flow, ↑ production, or ↓ absorption; adult hydrocephlaus may be 1. ex vacuo, seen in severe cerebral atrophy, with loss of brain tissue, as in Pick's or Alzheimer's diseases, thus being hydrocephalus by default, as the production and absorption of cerebrospinal fluid are normal or 2. normal pressure, occurring 2° to trauma or infection with reflux into the ventricles. See Obstructive hydrocephalus

HYDROCEPHALUS ETIOLOGY

↓ **ABSORPTION**, eg blockage by congenital malformations, obliteration of the aqueduct, hemorrhage, infection, neoplasms and trauma

↑ **PRODUCTION**, an uncommon event; as the ventricle enlarges, the ependymal lining separates from the ventricles, permeability ↑, the brain becomes edematous and the gyri flatten; incomplete resolution results in chronic hydrocephalus; in infants with unclosed cranial sutures, the cranial circumference ↑, cerebral parenchyma is destroyed, gyri flatten, sulci are obliterated and the 'setting sun' sign appears, which is usually associated with a poor prognosis; treatment by ventriculo-peritoneal shunting is merely palliative

hydrochloric acid BIOCHEMISTRY The acid in gastric juice, which is linked to the pain of GERD, heartburn. See GERD

hydrocodone An opioid analgesic, antitussive, which also depresses respiration, used for cough and upper respiratory Sx of allergies or colds. See Opioid

hydrocolloid dressing WOUND CARE An occlusive and adhesive wafer dressing for moderate amounts of exudate

hydrocolonic ultrasonography Transabdominal ultrasonography with retrograde instillation of water–≤ 1500 ml to improve imaging

hydrocortisone Cortisol The principal corticosteroid secreted by the adrenal cortex, which has mineralocorticoid activity, and relieves Sx of certain hormone deficiencies, and is immunosuppressive

hydrogel WOUND CARE A polymer absorptive wound dressing. See Dressing

hydrogen peroxide enterocolitis Hydrogen peroxide enteritis An inflammatory response of the intestinal mucosa to varying concentrations of H_2O_2, which causes mucosal damage and contact injury ETIOLOGY H_2O_2 enemas for various clinical indications–eg, removal inspissated meconium, ID of rectovaginal fistulae, and treatment of constipation and fecal impaction; H_2O_2 may be used by radiologists to help eliminate gas from the intestine during diagnostic procedures ENDOSCOPY HPE may mimic acute ulcerative colitis, ischemic colitis, and pseudomembranous colitis PROGNOSIS Ranges from reversible and self-limited to toxic fulminant ulcerating colitis associated with perforation and death. See Colon therapy, Enteroclysis, Hydrogen peroxide therapy, Snow white sign

hydromorphone An opioid analgesic which, in excess, causes respiratory depression; blocks nociception to brain; miosis. See Opiate, Opioid, Pain management

hydronephrosis NEPHROLOGY Uni- or bilateral expansion of a renal pelvis and calyces, often due to obstructive uropathy which may be linked to HTN and result in permanent renal damage. See Bilateral hydronephrosis, Reflux nephropathy, Unilateral hydrosis

hydrophobic pocket theory ANESTHESIOLOGY A hypothesis on the mechanism of anesthesia, which holds that anesthetics disrupt oscillations of single electrons in hydrophobic pockets. Cf Lipid theory

hydrops fetalis Kernicterus, Rh incompatibility, Rh-induced hemolytic disease of newborn OBSTETRICS An accumulation of fluid in neonates, resulting in a 'puffy', plethoric or hydropic appearance that may be due to various etiologies CLINICAL Ascites, edema, ↓ protein or chronic intrauterine anemia, hepatosplenomegaly, cardiomegaly, extramedullary hematopoiesis, jaundice, pallor COD Heart failure. See Hemolytic disease of the newborn

HYDROPS FETALIS, CAUSES

IMMUNE Mother produces IgG antibodies against infant antigen(s), often an RBC antigen, most commonly, anti-RhD, which then passes into the fetal circulation, causing hemolysis

NON-IMMUNE Hydrops may result from various etiologies including

- **Fetal origin**, eg congenital heart disease (premature foramen ovale closure, large AV septal defect), hematologic (erythroblastosis fetalis, α-thalassemia due to hemoglobin Barts, chronic fetomaternal or twin-twin transfusion), infection (CMV, herpesvirus, rubella, sepsis, toxoplasma), pulmonary (cystic adenomatoid malformation, diaphragmatic hernia, with pulmonary hypoplasia, lymphangiectasia), renal (vein thrombosis, congenital nephrosis) and teratomas, skeletal malformations (achondroplasia, osteogenesis imperfecta, fetal neuroblastomatosis, storage disease, meconium peritonitis, idiopathic)

- **Placental** Chorangioma, umbilical or chorionic vein thrombosis

- **Maternal** DM, toxemia

hydrosalpinx GYNECOLOGY A fluid-filled, dilated fallopian tube/oviduct, which is closed at the fimbriated end, thereby precluding successful pregnancy; hydrosalpinx is typically an end-stage of gonorrhea, which follows resolution of the infection

hydroureter A urine distended ureter. See Hydronephrosis

hydroxyapatite deposition disease RHEUMATOLOGY A crystal-induced arthropathy CLINICAL Mono- or polyarticular periarthritis with joint erosions and destruction; HADD may be due to collagen vascular disease, renal failure and osteoarthritis

17-hydroxycorticosteroid Any of the steroid hormones synthesized in the adrenal gland by action of 17-hydroxylase–cortisol, cortisone, 11-deoxycortisol and tetrahydro derivatives; urine excretion of 17-OHCSs is a rough guide of functional status of the adrenal gland and rate of catabolism ↑ IN Pregnancy, Cushing's disease, obesity, pancreatitis ↓ IN Addison's disease, hypopituitarism REF RANGE < age 1, 1.4-2.8 μmol/24–US: < 1 mg/24; adult 8.2-27.6 μmol/24–US: 3-11 mg/24

5-hydroxyindole acetic acid 5-HIAA A biogenic amine found especially in the brain and other organs, which accounts for 1% of degraded tryptophan; 5-HIAA may be markedly–25-50-fold normal ↑ in carcinoid tumors, especially of the midgut–less commonly of lung and ovary–features of carcinoid syndrome–eg, bronchoconstriction, diarrhea, flushing, right-sided heart lesions, facial telangiectasia; 5-HIAA is also ↑ in Hartnup's disease, nontropical sprue, in some mental conditions DRUGS Acetaminophen, caffeine, L-dopa, glyceryl, guaiacolate, heparin, nicotine, phenothiazine, reserpine, salicylates; 5-HIAA is ↓ in major GI tract resection, renal insufficiency, PKU, phenothiazine therapy

hydroxyproline BIOCHEMISTRY A hydroxylated amino acid which is most concentrated in collagen ↑ IN URINE Indicates ↑ turnover of bone matrix, often caused by osteoporosis ↓ IN Malnutrition, ↑ by acromegaly, acute osteomyelitis, burns, congenital hyperphosphatasia, fibrous dysplasia, healing fractures, hyperparathyroidism, hyperthyroidism, Marfan syndrome,

hydroxyprolinemia

314 osteomalacia, Paget's disease of bone, parathyroid adenoma, rickets, bone tumors

hydroxyprolinemia MOLECULAR MEDICINE An AR condition caused by hydroxyproline oxidase deficiency CLINICAL Profound mental retardation LAB ↑ Hydroxyproline in serum, urine

hydroxyurea Droxia®, Hydrea® A non-alkylating, myelosuppressive chemotherapeutic of low toxicity, used to treat myeloproliferative disorders and hemoglobinopathies AIDS HU ↓ HIV viral loads if given as part of a protease-sparing triple combination with nucleoside analogues ddI–didanosine–Videx® and d4T–stavudine–Zerit®. See AIDS HEMATOLOGY Hydroxyurea induces ↑ HbF synthesis in sickle cell–SC anemia, which may comprise 25% of the total Hb–the remainder is HbS–seemingly enough to prevent formation of HbS polymers, the bête noire of SC disease; hydroxyurea may be used to ↑ fetal Hb production–by ↑ γ-globulin production and ↑ RBC survival and ↓ bilirubin and LD in Pts with SC anemia; used in Pts with SC disease ≥ age 18, to prevent/↓ painful crises and ↓ need for blood transfusions, for Pts who have had at ≥3 painful crises in previous yr. See Sickle cell anemia ONCOLOGY Hydroxyurea is used as a single agent to control blast transformation in CML, manage P vera, essential thrombocythemia and, with prednisone, treat idiopathic hypereosinophilic syndrome; long-term therapy may prevent thrombosis; hydroxyurea therapy–HT has also been used to manage melanoma and inoperable ovarian CA LONG-TERM EFFECTS Unknown; in Pts with P vera, leukemia is 3-fold higher in HT than in those treated with phlebotomy

hydroxyzine Vistaril, Rezine, Atarax A sedative and tranquilizer THERAPEUTIC RANGE 10-100 ng/mL

HYE Healthy-years equivalent, see there

hygiene MEDTALK The science of health and health maintenance VOX POPULI Cleanliness. See Natural hygiene, Poor sleep hygiene, Sleep hygiene

hygiene hypothesis ALLERGY MEDICINE The theory that a clean modern lifestyle alters the immune system, ↑ susceptibility to allergies. See Leipzig disparity

Hymenoptera MEDICAL ENTOLOGY An order of insects with 2 pairs of membranceous wings. See Bee, Fire ant, Wasp

hymenopterans MEDICAL ENTOLOGY Insects of the order Hymenoptera–eg, honeybee, yellow jacket, yellow hornet, white-faced hornet, polistes wasp and ants. See Immunotherapy, Killer bees

HYMENOPTERAN FAMILIES

APIDAE Honeybees–*Apis mellifera*–may react to pheromones released by a hivemates and swarm victim

BOMBIDAE Bumblebees are relatively docile and usually have laisez-faire attitude

FORMICIDAE Ants–*Solenopsis invicta*–en masse, may be fatal

VESPIDAE Wasps–*Chlorion ichneumonidae*, hornets–*Vespula maculata*, yellowjackets–*V maculiformis*–are less often assocated with allergic reactions than Apidae

Hypaque® Diatrizoate, see there

hyperactivity NEUROLOGY A state of generalized restlessness and excess movement; hyperactivity in children is common but, if extreme, reaches ADD; idiopathic hyperactivity in adults is often abnormal and may indicate a manic state. See Attention-deficit disorder, Bipolar disorder

hyperacusis ENT A marked ↑ in sensitivity to sounds

hyperaldosteronism Conn syndrome, primary aldosteronism, ENDOCRINOLOGY A condition caused by overproduction of aldosterone by an adrenal cortex tumor, resulting in ↓ K⁺, sodium retention, arrhythmias, alkalosis, muscle weakness, polydipsia, polyuria, HTN

hyperalimentation Total parenteral nutrition, see there

hyperammonemia METABOLISM A heterogeneous group of five largely AR inborn errors of metabolism; each has a defect in a urea cycle enzyme–arginase, argininosuccinase, argininosuccinic acid synthetase, carbamyl phosphate synthetase, ornithine transcarbamylase; all begin in late infancy or childhood, except arginase deficiency, which is neonatal CLINICAL Accumulation of urea precursors–eg, ammonia, glutamine causes progressive lethargy, hyperthermia, apnea, hyperammonemia DIAGNOSIS may be established in utero by restriction fragment–RFLP analysis TREATMENT Restrict dietary protein; activate alternate pathways of waste nitrogen excretion–eg, sodium benzoate or dietary supplementation with arginine

hyperbaric oxygen therapy The administration of O₂ in a chamber at > sea-level atmospheric pressure, which ↑ O₂ dissolved in the blood from 1.5g/dL to 6.0 g/dL and O₂ tension in tissues to nearly 400 mm Hg; this surfeit of O₂ has various biochemical, cellular, physiologic benefits EMERGENCY MEDICINE HOT is used in decompression sickness, in gas embolism, extreme blood loss anemia, and as adjunctive therapy for clostridial myonecrosis and gangrene, crush injury and compromised skin grafts and flaps, suturing of severed limbs, and in preventing osteoradionecrosis MAINSTREAM MEDICINE HOT is also used in smoke inhalation, cyanide intoxication, acute carbon monoxide poisoning, extreme blood loss anemia, traumatic ischemia, as in compartment syndrome(s) and crush injury, to enhance healing of recalcitrant or necrotic wounds, as an adjunct therapy for clostridial myonecrosis–associated with acute tissue hypoxia, compromised skin grafts and flaps, chronic osteomyelitis, actinomycosis. and to prevent osteoradionecrosis COMPLICATIONS Barotrauma–air embolism, pneumothorax, tympanic membrane damage, O₂ toxicity–CNS, pulmonary, reversible visual changes, fire or explosion and claustrophobia

hyperbilirubinemia HEPATOLOGY An ↑ of BR in the peripheral circulation; BR is either unconjugated, or conjugated to glucuronic acid, which ↑ BR's water solubility and facilitates its entry into the bile; hyperbilirubinemia in neonates is due to ↓ metabolism, clearance, and liver immaturity. See Jaundice

hypercalcemia Calcium excess ↑ Ca²⁺ in serum ETIOLOGY Hypercalcemia may be due to excess PTH–parathyroid tumors or hyperactivity, excess milk and/or vitamin D, ↑ bone turnover or aluminum intoxication, calcitriol, thyrotoxicosis, drugs–eg, corticosteroids, thiazides, and may be associated with granulomas–eg, berylliosis, sarcoidosis, silicon injection, TB, thyrotoxicosis CLINICAL N&V, constipation, arrythmias, ↑ QT interval, anorexia, CNS depression, fatigability, muscle weakness, renal tube defects, urinary frequency, dystrophic calcification–eg, cornea, kidneys, peptic ulcers, pancreatitis, joint pain, finger clubbing; Ca⁺⁺ > 2.95 mmol/L–US: 12 mg/dL is a medical emergency; prolonged hypercalcemia causes LAB ↑ Alk phos, cAMP in urine; if extreme–serum levels > 3.7–4.5 mmol/L or 15–18 mg/dL, coma, cardiac arrest. See Calcium; Cf Hypocalcemia, Long Q-T syndrome

HYPERCALCEMIA-DIFFERENTIAL DIAGNOSIS

Ca²⁺	PO₄	
↑	↑	Parathyroid carcinoma, adenoma or hyperfunction, tertiary hyperparathyroidism
↑	↓	Myeloma, hypervitaminosis D, bony metastases, sarcoidosis, milk-alkali syndrome
↓	↑	Malabsorption, chronic diarrhea
↓	↓	Status/post parathyroid ablation
N	N	Hypervitaminosis A, healing of fractures, adolescence

hypercalcemia of malignancy A clinical complex, 50% of which results from hypersecretion of parathyroid hormone-related protein–PTHRP, aka parathyroid hormone-related peptide; HCM may result from either direct replacement–eg, in lympho- and myeloproliferative malignancies–eg, leukemia, adult T cell lymphoma, Burkitt's lymphoma, lymphosarcoma, myeloma or solid tumors–eg, CA of breast, lung, pancreas, or due to factors; normal subjects have low–<2.0 pmol/L levels of PTHRP, while those with HCM may have plasma levels > 20.9 pmol/L; hypercalcemia may be moderate 10.8-12 mg/dL; or severe >12 mg/dL; HCM is uncommon–± 1% and occurs in Pts with more advanced disease, distant metastases, and poor prognoses DIAGNOSIS RIA TREATMENT HCM may respond to bleomycin

hypercapnia ↑ CO_2 in blood. See Permissive hypercapnia

hyperchromatic 1. Dark 2. Referring to ↑ chromatin

hypercoagulability 1. Disseminated intravascular coagulation, see there 2. Hypercoagulable state, see there

hypercoagulable state Hypercoagulability, thrombophilia HEMATOLOGY A condition in which there is an abnormal ↑ in clotting WORKUP PT, PTT, fibrinogen, anticardiolipin antibodies, APC resistance, factor V Leiden, prothrombin G20210A polymorphism, defects in homocysteine metabolism, protein C, protein S, AT III. See Coagulation, Hereditary thrombophilia, Hypercoagulable state profile

hyperechoic IMAGING *adjective* Referring to an abnormal ↑ in echoes by ultrasonography, due to a pathologic change in tissue density. See Ultrasound

hyperemesis Puke, puke, much, much

hyperemesis gravidarum OBSTETRICS A condition that respects no race, parity status or social class, characterized by vomiting in the 1st trimester–3.5/1000 pregnancies, of a severity that may induce renal failure and require hospitalization; if Pt stabilizes, there is no risk of toxemia during later gestation, nor ↑ risk of spontaneous abortion or deformities

hyperemia An excess of blood flow through a particular tissue. Cf Congestion

hypereosinophilic syndrome A heterogeneous group of conditions characterized by persistent peripheral eosinophilia with infiltration of eosinophils in BM, heart, and other organs, which may be a form of myeloproliferative disorder–possibly eosinophilic leukemia as 1. 25% have chromosomal abnormalities–eg, aneuploidy and ↓ vitamin B_{12} with persistent eosinophilia–> 1500 x 10^9/L for > 6 months; 2. Absence of 2° causes of eosinophilia, despite aggressive workup; 3. Sx due to organ involvement or dysfunction CLINICAL Age of onset, 20–50, generalized weakness, dyspnea, cough, malaise, myalgia, anorexia, angioedema, rash, low-grade fever, night sweats, weight loss, rhinitis, and Sx of Löffler's endocarditis; eosinophils cause dysfunction in heart, BM, liver, spleen, CNS, either focal–due to emboli–or diffuse, causing altered behavior and cognition, psychosis, ataxia, spasticity, peripheral symmetrical polyneuropathy, coma, GI tract–diarrhea, abdominal pain, malabsorption, lungs–interstitial inflammation with eosinophils, and skin–nonspecific rashes; once the heart is involved–eg, endocardial fibrosis, thrombosis and restrictive cardiomyopathy, death < 1 yr MANAGEMENT Possibly corticosteroids

hyperesthesia NEUROLOGY A state of excess sensitivity to touch or other sensory stimuli

hyperextension Overextension REHAB MEDICINE The extension of a limb or other body part beyond its normal range of movement

hyperfractionation RADIATION ONCOLOGY The delivery of radiation in small-dose fractions 2 to 3 times/day to minimize damage to normal tissue typical of smaller fractions. See Fractionation. Cf Accelerated fractionation

hyperglobulinemia IMMUNOLOGY An ↑ in serum Ig for any reason, often neoplastic, as seen in clonal expansions of B-plasma cells, as in myeloma, Waldenström's macroglobulinemia, and other lymphoproliferations–eg, leukemia, lymphoma; it may also occur in immune dysfunctions–eg, hyper-IgD, hyper-IgE, hyper-IgM syndromes. See Benign monoclonal gammopathy of undetermined significance, Myeloma, Waldenström's macroglobulinemia

hyperglycemia METABOLISM An abnormal ↑ in serum glucose, most commonly due to DM. See Diabetes mellitus, Glucose tolerance test, Hyperglycinemia

hyperglycinemia Isolated nonketotic hyperglycemia A group of AR conditions characterized by glycine accumulation 2° to a catabolic defect in glycine cleavage CLINICAL Presents shortly after birth with CNS depression, seizures, convulsions, FTT, mental retardation, followed by atonia and areflexia LAB ↑↑↑ Glycine in CSF, serum, urine, ↑ plasma osmolality, dehydration. Cf Hyperglycemia

hypergonadism Any condition characterized by an ↑ end-organ–testes, ovaries response, or ↑ sensitivity to sex hormones

hypergonadotropic hypogonadism Primary hypogonadism A condition due to a lack of target organ response to pituitary hormones–eg, FSH and/or LH. See Hypogonadism

hyperhidrosis Sweat, sweat, much, much

hyperhydration Overhydration, see there

Hypericum perforatum Popularly, St John's wort, see there

hyper-IgE syndrome Hyperimmunoglobulin E syndrome, see there

hyper-IgM syndrome Hyperimmunoglobulin M syndrome, see there

hyperimmunoglobulin E syndrome Hyper-IgE syndrome, Job syndrome IMMUNOLOGY An AR immunodeficiency characterized by chronic eczematoid lesions, recurring abscesses of skin, lungs, joints, recurring otitis media, sinusitis, life-threatening staphylococcal infections, chronic eczemoid lesions and recurring abscesses of lungs, skin, joints, eyes, ears by *Staph aureus, H influenzae, S pneumoniae*, group A streptococci, *Candida* spp LAB ↑ Eosinophils, IgE > 5000 IU/mL, ↓ antibody response to vaccines and MHC antigens, ↓ CD4:CD8–T cell helper:suppressor ratio, defects in PMN and monocyte chemotaxis, defects in PMN and monocyte chemotaxis, ↑ IgE PROGNOSIS Prolonged survival is possible

hyperimmunoglobulin M syndrome Hyper-IgM syndrome IMMUNOLOGY An X-R immunodeficiency CLINICAL Recurrent bacterial infections–eg, otitis media, pneumonia or opportunistic infections, recurrent neutropenia, lymphoid hyperplasia, neoplasia at an early age, and autoimmune phenomena LAB ↓/absent IgG, IgA, IgE; normal/↑ IgM, IgD

hyperimmune *adjective* Referring to an immune state characterized by an abundance of one or more immunoglobulins, due to repeated exposure to one or a limited palette of microorganisms

hyperimmune globulin An Ig concentrate with high titers of antibodies against an antigen of interest

hyperinfection syndrome Disseminated parasitosis in immunosuppressed, malignant, or malnourished hosts, caused by autoinfection with *Strongyloides stercoralis* CLINICAL Abrupt onset of high fever, abdominal pain, bloating, intestinal ulcerations, gram-negative sepsis and shock; intense transpulmonary nematodal migration results in dyspnea, cough, hemoptysis TREATMENT Thiabendazole PREVENTION Wear shoes, boil water

hyperinsulinemia Hyperinsulinism ↑ insulin–eg, due to ↑ production or secretion by pancreatic beta cells, or ↓ hepatic clearance

hyperkalemia Hyperpotassemia, potassium excess METABOLIC DISEASE A state characterized by ↑ serum K+ ETIOLOGY Oliguria, tissue injury, burns, K+ in IV solutions, metabolic acidosis, renal failure, K+-sparing diuretics–eg, spironolactone, antibiotics–eg, cephalosporins, isoniazid, penicillin, epinephrine, histamine CLINICAL Lethargy, nausea, diarrhea, colicky pain, oliguria, bradycardia, flaccid paralysis REF RANGE 3.8-5.0 mmol/L–US: 3.8-5.0 mEq/L

hyperkalemic periodic paralysis See Periodic paralysis

hyperkeratosis DERMATOLOGY An ↑ in superficial keratinized layers of certain epithelia, skin, and uterine cervix; hyperkeratosis usually represents a reaction to irritation, and generally overlies benign epithelium

hyperkinesis

hyperkinesis Move, move, much, much

hyperlearning INFORMATICS A process of acquiring information common in multiple work tasks and occurs in a multidimensional space formed by a matrix of digitally integrated information technologies with human and nonhuman components

hyperlipidemia An ↑ of circulating lipids–fatty acids, TGs, and cholesterol, often linked to ↑ lipoproteins–hyperlipoproteinemia and/or ↓ degradative enzymes–eg, lipoprotein lipase. See Acquired hyperlipidemia, Familial combined hyperlipidemia

HYPERLIPIDEMIA

PRIMARY Dyslipidemia intimately linked to CAD

SECONDARY
- **SECONDARY HYPERCHOLESTEROLEMIA**, seen in acute intermittent porphyria, cholestasis, hypothyroidism and pregnancy
- **SECONDARY HYPERTRIGLYCERIDEMIA**, seen in DM, acute alcohol intoxication, acute pancreatitis, gout, gram-negative sepsis, glycogen storage disease I, oral contraceptive use
- **COMBINED HYPERCHOLESTEROLEMIA & HYPERTRIGLYCERIDEMIA**, seen in nephrotic syndrome, chronic renal failure, corticosteroid therapy and immunosuppression

Note: 30-40% of the population is sensitive to dietary cholesterol and ↑ dietary cholesterol may ↑ cholesterol in the circulation

hyperlipoproteinemia An ↑ in lipoproteins in the circulation, which have been traditionally divided into 5 clinical forms, with variable hereditary components and different responses to dietary and pharmacologic intervention CLINICAL Xanthomas, pancreatitis, thromboembolism and in hyperlipoproteinemia, accelerated ASHD with acute MI at an early age

HYPERLIPOPROTEINEMIA

I FAMILIAL LIPOPROTEIN LIPASE DEFICIENCY An AR condition characterized by the inability to release TGs from chylomicrons, causing marked hypertriglyceridemia–> 20 g/L CLINICAL Childhood onset, diarrhea, pancreatitis–potentially fulminant, xanthomata, lipemia retinalis, hepatosplenomegaly, but no ↑ risk of atherosclerosis; this condition is similar to apoC-II deficiency LABORATORY TGs > 4.0 g/L–US: 400 mg/dl, cholesterol normal, ↑ chylomicrons, ↓ post heparin lipolytic activity; this lipoprotein pattern may be mimicked by SLE

II FAMILIAL HYPERCHOLESTEROLEMIA An AD condition characterized by tuberous xanthomas of tendons, accelerated CAD, early MIs and ischemic events, onset ages 20–50 MOLECULAR PATHOLOGY Absence or defect in LDL receptors due to mutant alleles Rbo, Rb and Rtio, causing inability to absorb cholesterol from LDL, complicated by ↑ hepatic production of LDL, due to loss of negative feedback LABORATORY ↑ Cholesterol–5–6 g/L, homozygotes; 3–4 g/L, heterozygotes, phospholipids, LDL, normal/↑ triglycerides

TYPE IIa ↑ LDL due to a defect in LDL receptor, or apoB-100, resulting in ↑ cholesterol, normal VLDL, normal TGs

TYPE IIb ↑ LDL, ↑ cholesterol, ↑ VLDL–↑ TGs

III REMNANT REMOVAL DISEASE Dys-β-lipoproteinemia, broad beta disease, fused beta band disease An AD condition characterized by decreased intermediate-density lipoprotein catabolism CLINICAL Obesity, accelerated atherosclerosis, thromboembolism, strokes, palmo-plantar xanthomas, measuring up to 10 cm, associated with DM, obesity, hypothyroidism LABORATORY ↑ TGs, cholesterol, phospholipids, LDL, pre-beta- and beta-lipoproteins, and an abnormal–diabetic glucose tolerance test; this pattern may be mimicked by myeloma

IV HYPERTRIGLYCERIDEMIA An AD condition which is the most common hyperlipoproteinemia CLINICAL Often asymptomatic, rarely peripheral and coronary vascular disease LABORATORY Mildly elevated TGs, VLDL, and pre-beta-lipoproteins TREATMENT Clofibrate, gemfibrozil; acquired causes of hypertriglyceridemia include DM, chronic uremia, dialysis, obesity, estrogens, alcohol, diuretics, glucocorticoids and β-adrenergics

V MIXED HYPERLIPOPROTEINEMIA A heterogeneous group of conditions characterized by eruptive xanthomata, pancreatitis, lipemia retinalis, ↑ VLDL and chylomicrons

hyperlucency IMAGING A region on a plain film with ↓ tissue density, allowing for ↑ transmission of X-rays

hyperlysinuria with hyperammonemia Periodic hyperlysinemia MOLECULAR MEDICINE An AR inborn error of metabolism caused by L-lysine dehydrogenase deficit CLINICAL Physical/mental retardation. Cf Lysinemia

hypermagnesemia Magnesium intoxication A state characterized by ↑ serum magnesium, which is associated with ESRD, eclampsia therapy with magnesium sulfate, adrenocortical insufficiency–Addison's disease, uncontrolled DM, leukemia, hypothyroidism, magnesium-based antacid and laxative therapies CLINICAL Lethargy, shallow respiration, ↓ BP, ↓ tendon reflexes, ↓ neuromuscular transmission, ↓ CNS; symptoms correspond to magnesium serum levels: nausea occurs at 2-2.5 mmol/L–4-5 mEq/L → sedation, ↓ tendon reflexes, muscle weakness; at 2.5-5 mmol/L–5-10 mEq/L, hypotension, bradycardia, and systemic vasodilatation; arreflexia, coma, respiratory paralysis appear above these levels. See Magnesium

hypermobile joint syndrome ORTHOPEDICS 1. Joint hypermobility syndrome A common benign childhood condition involving hypermobile joints that can move beyond the normal ROM CLINICAL Pain in knees, fingers, hips, elbows, ↑ tendency to dislocate, ↑ in scoliosis, which usually improves with age 2. Systemic joint laxity A generalized ↑ in joint mobility, which may be seen in various rheumatic conditions, associated with TMJ dysfunction

hypermobility Instability ORTHOPEDICS Any motion occurring in a joint in response to the reactive force of gravity at a time when that joint should be stable under such a load; hypermobility is often misused to describe extra movement as seen in a contortionist

hypernatremia ↑ Na⁺ in blood

hyperopia Farsightedness OPHTHALMOLOGY A refractive error in which the cornea and/or lens focus images behind the retina, resulting in ↓ visual acuity for near objects, correctable with glasses or contact lenses. Cf Myopia

hyperostosis A proliferation of bony matrix

hyperostosis corticalis generalisata Generalized cortical hyperostosis, see there, aka Van Buchem's disease

hyperostosis corticalis infantalis Infantile cortical hyperostosis, see there

hyperostosis frontalis interna Morgagni-Stewart-Morell syndrome A form of osteopetrosis more common in middle-aged ♀, associated with obesity, hirsutism, fatigue, hemiplegia and hemiparesis; HFI affects cranial bones, structurally compromising the hypophysis–causing dysmenorrhea, virilism, hirsutism, diabetes insipidus, and glucose intolerance, and cranial nerve foramina–causing vertigo, tinnitus, anosmia, and visual defects

hyperparathyroidism ENDOCRINOLOGY Parathyroid hypersecretion, which causes several relatively specific clinical complexes: myopathic syndrome, skeletal syndrome, urologic syndrome, CNS syndrome, in addition to a characteristic peptic ulcer, hypercalcemia and acute pancreatitis. See Hypercalcemia, Parathyroid gland, PTH, Secondary hyperparathyroidism

hyperphagia MEDTALK Characterized by an ↑ appetite

hyperphenylalaninemia A group of 8 different congenital enzymopathies characterized by an accumulation of phenylalanine and related metabolites–eg, phenylpyruvate, phenyllactate, phenylacetate, phenylacetylglutamine, a deficiency of T₄, T₃, melanin and other proteins; the most common or 'classic' phenylketonuria–PKU or hyperphenylalaninemia, type 1 is an AR condition characterized by a deficiency in phenylalanine hydroxylase, which if not recognized early–urine has a 'mousy' odor—and treated by restricting dietary phenylalanine, results in tremors, seizures, eczema, hyperactivity, and hypopigmentation and IQ of < 50; other hyperphenylalaninemia types–eg, types 2, 3 and 7, are relatively benign

hyperphilia SEXOLOGY Sexuoerotic excess. See Don Juan syndrome, Nymphomania

hyperphosphatasia Hyperostosis corticalis deformans juvenilis, juvenile Paget's disease An AR condition characterized by enlarged and defective bones, ↑ bone density and Fx CLINICAL Dwarfism, macrocephaly, blue sclerae LAB ↑ Alk phos, normal Ca^{2+}, PO_4

hyperphosphatemia NEPHROLOGY ↑ Phosphate > 1.5 mmol/L–US: 4.5 mg/dL in blood ETIOLOGY ↑ hGH, either physiologic with growth spurts or pathologic in gigantism and acromegaly, ↓ PTH, pseudohypoparathyroidism or in renal failure

hyperpituitarism ENDOCRINOLOGY Growth hormone hypersecretion which may be of early onset–resulting in gigantism or of adult onset, resulting in overgrowth of acral parts–acromegaly. See Acromegaly, Gigantism

hyperplasia An abnormal ↑ in number of cells in an organ or tissue, resulting in an increase in size

hyperpotassemia Potassium excess, hyperkalemia

hyperproteinemia An excess of protein in the circulation. See Hypergammaglobulinemia

hyperreflexia NEUROLOGY A state of excessive response to stimuli. See Autonomic hyperreflexia

hyperresonance A clang where a thump is expected–percussion

hypersecretion A state of excess secretion. See Rebound hypersecretion

hypersensitivity IMMUNOLOGY An abnormal immune response that may be immediate–due to antibodies of the IgE class, or delayed–due to cell-mediated immunity. See Cold hypersensitivity, Immediate hypersensitivity NEUROLOGY Exaggerated sensitivity An ↑ in a person's sensitivity to light, sound, smell, taste, touch, temperature, balance, and even emotional issues, which may be linked to anxiety and panic disorders

hypersensitivity pneumonitis A disorder caused by an exuberant pulmonary reaction to aerosolized immunogens, which with time may develop interstitial lung disease; the prototypic HP is farmer's lung CLNICAL Fever, malaise, cough, chest tightness, myalgias. See *Faeni rectivirgula*, Farmer's lung, Humidifier lung

HYPERSENSITIVITY PNEUMONITIS

SYNDROME	SUSPECTED ANTIGEN
Bagassosis	*Thermoactinomyces vulgaris*
Bird fancier's lung	Bird droppings
Pigeon handler's lung	*Histoplasma capsulatum*
Cheese worker's lung	*Penicillium* spp
Detergent worker's lung	*Bacillus subtilis* enzyme
Fish meal lung	Fish proteins
Farmer's lung	*Micropolyspora faeni, Thermoactinomyces vulgaris*
Humidifier lung	Thermophilic actinomyces
Malt worker's lung	*Aspergillus clavatus*
Maple bark disease	*Cryptostroma corticale*
Mushroom picker's lung	*Micropolyspora faeni*
Sequoiosis	*Graphium* spp, *Auerobasidium* spp
Suberosis (oak bark)	*Micropolyspora faeni*
Vineyard sprayer's lung	Copper
Wood dust pneumonitis	Oak and mahogany dust
Wood pulp worker's lung	Alternaria spp

hypersensitivity pneumonitis profile ALLERGY MEDICINE A battery of antigens and organisms to which a host may be allergic MEASURES Antibodies to *Aspergillus fumigatus, A niger, Micropolyspora faeni*, pigeon serum, *Saccharomonospora viridis, Thermoactinomyces vulgaris*

hypersensitivity syndrome A severe idiosyncratic reaction to certain drugs–eg, anticonvulsants, sulfonamides, allopurinol, which is characterized by rash–eg, exfoliative dermatitis and fever, and may be accompanied by arthralgias, carditis, hepatitis, lymphadenopathy LAB Atypical lymphocytes, eosinophilia, abnormal LFTs

hypersensitivity vasculitis Allergic vasculitis IMMUNOLOGY An immune dysfunction and systemic vasculitis characterized by small vessel involvement of the skin–accompanied by palpable purpura, leukocytoclastic inflammation, often also of the GI tract and kidney; HV may be associated with a drug reaction, bacterial infection, or underlying malignancy. See Vasculitis

hypersomnia SLEEP DISORDERS Excessive or prolonged sleep, which may be associated with difficulty in awakening, staying awake or sleep drunkenness EXAMPLES Sleep apnea, narcolepsy, nocturnal myoclonus, obstructive sleep apnea, isolated sleep paralysis, central sleep apnea, idiopathic hypersomnia, respiratory muscle weakness associated sleep disorder. See Narcolepsy, Sleep-apnea syndrome, Sleep disorders. Cf REM sleep

HYPERSOMNIA TYPES

PRIMARY HYPERSOMNIA

- **HYPERSOMNIA-BULIMIA SYNDROME OF KLEIN-LEVINE** Characterized by semiannual bouts of hyperphagia followed by a 2-5 day 'sleep-off', seen in young ♂
- **II. HYPERSOMNIA-SLEEP APNEA SYNDROME** A condition affecting obese and hypertensive middle-aged ♂, which is characterized by daytime grogginess and loud snoring; these Pts are at ↑ risk for AMI and CVAs

SECONDARY HYPERSOMNIA A symptom caused by focal CNS disease, eg brain tumors, especially those of the posterior hypophysis or diencephalon, encephalopathia lethargica and meningitis or systemic disease, eg hypothyroidism, trypanosomiasis

hypersomnolence Sleep, sleep, much, much

hypersplenism Pathologic enlargement of the spleen, often accompanied by pancytopenia, hyperplasia of BM precursors, and improvement upon splenectomy ETIOLOGY HYPERSPLENISM Blood stasis often 2° to cirrhosis, thrombosis and other vascular abnormalities, anemia, lymphoproliferative disorders–eg, lymphoma, leukemia–especially CML, infection–eg, infectious mononucleosis, kala azar, malaria, TB; infiltrations–eg, Gaucher disease, Niemann-Pick disease, connective tissue disease CLINICAL If isolated, asymptomatic; primary diseases may be accompanied by malaise, abdominal pain, fever, fullness, purpura, hematemesis, GI bleeding LAB ↓ RBC survival, reticulocytosis, 'left shift' of myeloid series. Cf Splenosis

hypertelorism ↑ separation of paired tissues–eg, breasts, eyes BREAST Rare–seen in Turner syndrome OCULAR A craniofacial defect seen in congenital syndromes–eg, Apert syndrome/acrocephalosyndactyl; Crouzon's disease/craniofacial dysostosis; Opitz' syndrome/hypertelorism-hypospadia/BBB syndrome; Taybi syndrome/otopalatodigital syndrome

hypertelorism-hypospadias syndrome PEDIATRICS An AD condition of neonatal onset, characterized by pulmonary aspiration at birth due to a laryngotracheoesophageal cleft, stridor, ocular hypertelorism, a broad nasal bridge, cleft lip and palate, cardiac defects, imperforate anus and hypospadias, and mental retardation. Cf G deletion syndrome

hypertension High blood pressure CARDIOVASCULAR DISEASE An abnormal ↑ systemic arterial pressure, corresponding to a systolic BP of > 160 mm Hg and/or diastolic BP of 95 mm Hg and graded according to intensity of ↑ diastolic BP; HTN affects ± 60 million in the US WORKUP Evaluation of HTN requires clinical Hx for Pt, family Hx, 2 BP determinations, funduscopy, ID of bruits in neck & abdominal aorta, evaluation of peripheral edema, peripheral pulses and residual neurologic defects in stroke victims, chest films to determine cardiac size and lab parameters to rule out causes of secondary HTN RISK FACTORS Race–blacks more common, ♂, family history of HTN, obesity, defects of lipid metabolism, DM, sedentary lifestyle, cigarette

hypertension assessment profile

smoking, electrolyte imbalance–eg, ↑ sodium, phosphorus, ↓ potassium, tin TREATMENT Diet–eg, sodium restriction, ↓ calories, alcohol and cigarettes–the weight gain accompanying smoking cessation tends to offset the minimal ↓ in BP, calcium supplements, lifestyle manipulation–eg, biofeedback, ↑ exercise; antihypertensives–eg, diuretics–benzothiadiazines, loop diuretics, potassium-sparing diuretics, sympatholytic agents–central and peripheral-acting α-adrenergics, β-adrenergics, mixed α- and β-blockers, direct vasodilators, ACE inhibitors–the preferred agent to use ab initio, dihydropiridine CCBs. See ACCT, ACE inhibitor, Borderline hypertension, Borderline isolated systolic hypertension, Calcium channel blocker, Drug-induced hypertension, Essential hypertension, Exercise hypertension, Familial dyslipemic hypertension, Gestational hypertension, Idiopathic intracranial hypertension, Isolated systolic hypertension, Malignant hypertension, MRC, Obetension, Paradoxic hypertension, Pill hypertension, Pregnancy-induced hypertension, Pseudohypertension, Pulmonary hypertension, Refractory hypertension, Renovascular hypertension, SHEP, STOP-Hypertension, TAIM, TOHP-1, TOMHSTyramine hypertension, White coat hypertension

HYPERTENSION

CLASS I–MILD Diastolic pressure 90-104 mm Hg

CLASS II–MODERATE Diastolic pressure 105-119 mm Hg

CLASS III–SEVERE Diastolic pressure > than 120 mm Hg

HYPERTENSION TYPES

ESSENTIAL HYPERTENSION Idiopathic HTN The major form comprising 90% of all HTN

MALIGNANT HYPERTENSION A sustained BP > 200/140 mm Hg, resulting in arteriolar necrosis, most marked in the brain, eg. cerebral hemorrhage, infarcts, and hypertensive encephalopathy, eyes, eg papilledema and hypertensive retinopathy and kidneys, eg acute renal failure and hypertensive nephropathy; if malignant HTN is uncorrected or therapy refractory, Pts may suffer a hypertensive crisis in which prolonged high BP causes left ventricular hypertrophy and CHF

PAROXYSMAL HYPERTENSION Transient or episodic waves of ↑ BP of any etiology, punctuated by periods of normotension, typical of pheochromocytoma

PORTAL HYPERTENSION ↑ portal vein pressure caused by a backflow of blood through splenic arteries, resulting in splenomegaly and collateral circulation, resulting in esophageal varices and/or hemorrhoids; PH may be intra- or extrahepatic, and is often due to cirrhosis, or rarely portal vein disease, venous thrombosis, tumors or abscesses

PULMONARY HYPERTENSION A condition defined as a 'wedge' systolic/diastolic pressure > 30/20 mm Hg–Normal: 18-25/12-16 mm Hg, often secondary to blood stasis in peripheral circulation, divided into passive, hyperkinetic, vasoocclusive, vasoconstrictive and secondary forms. See Pulmonary HTN

RENOVASCULAR HYPERTENSION see there

SECONDARY HYPERTENSION

- **AGING**
- **CARDIOVASCULAR** Open heart surgery, coarctation of aorta, ↑ cardiac output–anemia, thyrotoxicosis, aortic valve insufficiency
- **CEREBRAL** ↑ Intracranial pressure
- **ENDOCRINE** Mineralocorticoid excess, congenital adrenal hyperplasia, glucocorticoid excess, eg Cushing syndrome, hyperparathyroidism, acromegaly
- **GYNECOLOGIC** Pregnancy, oral contraceptives
- **NEOPLASIA** Renin-secreting tumors, pheochromocytoma
- **↓ PERIPHERAL VASCULAR RESISTANCE** AV shunts, Paget's disease of bone, beri-beri
- **RENAL DISEASE** Vascular, parenchymal

hypertension assessment profile A relatively complete array of serum tests used to evaluate HTN, and ID causes of 2° HTN TYPICAL HAP Plasma renin activity, urine albumin, albumin excretion rate, albumin/creatinine ratio, serum aldosterone, creatinine, urine potassium, urine sodium, fractionated metanephrines and catecholamines, VMA. Cf Hypertension panel, Organ panel

hypertension panel A battery of serum tests to evaluate a person with HTN, including the most common causes of 2° HTN: BUN/creatinine, chloride, CO_2 content, free urinary cortisol, potassium, sodium, thyroxine, VMA, urinalysis, bacterial colony count. Cf Hypertension assessment profile, Organ panel

hypertensive *adjective* Referring to HTN *noun* 1. An agent–eg, CCBs, diuretics, and others, used to treat HTN 2. A Pt with hypertension

hypertensive crisis A rare clinical event characterized by a severe and/or acutely ↑ diastolic BP > 120-130 mm Hg; an HC is a medical emergency if accompanied by rapid or progressive CNS–encephalopathy, infarction or hemorrhage, cardiovascular–myocardial ischemia, infarction, aortic dissection, pulmonary edema, and renal deterioration, eclampsia or microangiopathic hemolytic anemia ETIOLOGIC FACTORS Pre-existing chronic HTN; renovascular HTN; renal parenchymal disease; scleroderma and collagen vascular disease; drugs–sympathomimetics, tricyclic antidepressants, withdrawal from antihypertensives, recreational–eg, crack cocaine; spinal cord syndromes; pheochromocytoma CLINICAL Severe headache, transient blindness, vomiting, rapid deterioration of renal function COMPLICATIONS Acute end-organ damage–eg, myocardial ischemia/infarction, renal failure, aortic dissection, stage 3 or 4 hypertensive retinopathy TREATMENT Organ-targeted therapy with CCBs, Lobetalol, loop diuretics, nitroglycerin, nitroprusside PROGNOSIS Untreated 5-yr mortality is 100%

hypertensive heart disease Hypertensive cardiovascular disease CARDIOLOGY A condition characterized by ↑ BP which ↑ cardiac workload; over time, the left ventricle dilates, cardiac output ↓ and CHF ensues HHD ASSOCIATIONS Ischemic heart disease, ASHD, Obesity. See Hypertension

hypertensive intracerebral hemorrhage An intracerebral hemorrhage due to chronic uncontrolled HTN, associated with microaneurysms that leak into the brain, especially the basal ganglia, possibly resulting in a hematoma; blood acts as an irritant, evoking cerebral edema; bleeding into the ventricles or subarachnoid space results in meningeal irritation CLINICAL A function of the amount of necrosis and location; Sx are usually abrupt and may occur during activity, or be episodic, or slowly progressive. See Cerebrovascular disease

hypertensive retinopathy OPHTHALMOLOGY The retinal changes induced by HTN, which includes 'copper wire' and 'silver wire' changes of chronic HTN, or retinal and disc edema following abrupt ↑ in systemic BP–eg, malignant HTN

hyperthecosis A condition occurring in ♀, age 20-30, characterized by hyperplasia of the theca interna of a maturing ovarian follicle, often with hirsutism and amenorrhea. See Hirsutism

hyperthermia Hyperpyrexia MAINSTREAM MEDICINE A condition defined as a corporal temperature of ≥ 42°C; the body defends itself with peripheral vasodilation–↓ effective volume, resulting in ↑ pulse rate–a response to perceived blood loss, ↓ cardiac efficiency, hypoxia, ↑ permeability of cell membranes with ↑ potassium, followed by cardiac failure. See Malignant hyperthermia ONCOLOGY A type of treatment in which tissue is exposed to high temperatures to damage and kill CA cells, or ↑ CA cell sensitivity to RT and chemotherapy. See Induced hyperthermia, Malignant hyperthermia

hyperthrombocytosis See Essential thrombocythemia, Reactive thrombocytosis

hyperthyroidism Thyroid excess ENDOCRINOLOGY A state characterized by excess thyroid activity, due to ↑ secretion of thyroid hormones and ↓ response of hypothalamic–long and pituitary–short feedback loops ETIOLOGY Graves' disease, iatrogenic, toxic nodular goiter, thyroiditis, neonatal hyperthyroidism, exogenous iodide, factitious illness, malignancy struma ovarii CLINICAL ↑ O_2 consumption, ↑ basal metabolic rate, exophthalmos, nervousness, asthenia, weight loss LAB ↑ T_3 and/or T_4 MANAGEMENT Antithyroid drugs–methimazole, carbimazole, propylthiouracil, radioiodine, surgery. See Apathetic hyperthyroiditis, Factitious hyperthyroiditis, Subclinical hyperthyroiditis, Thyroid storm. Cf Hypothyroidism

hypertrichosis Hair, hair, much, much. See Congenital generalized hypertrichosis, Hirsutism

hypertrichosis lanugosa Malignant down Lanugo over large parts of the body, classically associated with internal CA–of breast, ovary, bladder, and other malignancies, or induced by some drugs; HL may also be inherited, more commonly AD than AR

hypertriglyceridemia An excess of TGs in the circulation. See Triglyceride

hypertrophic cardiomyopathy Idiopathic hypertrophic subaortic stenosis CARDIOLOGY A condition affecting 1:500 persons, causing an array of clinical and pathologic changes, either symmetric–concentric or asymmetric–eccentric hypertrophy, with disproportionate thickening beneath the mitral valve, seen without other cardiac disease; $\frac{1}{2}$ are congenital, with an AD pattern; in general in HC, ↑ obstruction is associated with ↓ venous return CLINICAL Younger Pts, ranging from asymptomatic to diastolic dysfunction, dyspnea, fatigue, chest pain and syncope, an ↑ incidence of severe obstruction, CHF, sudden death, simulating acute MI CLINICAL–OLDER PTS SOB, angina, syncope EKG ↑ QRS complexes, T-wave inversion, Q waves in inferior and left precordial leads MANAGEMENT Symptomatic–ie, relief of dyspnea or chest pain DRUGS β-adrenergics are effective short-term; CCBs–which ↑ diastolic ventricular filling may be effective long-term SURGERY Recalcitrant HC may require a transaortic ventricular septal myotomy-myectomy. See Cardiac myosin-binding protein C. Cf Dilated cardiomyopathy

hypertrophic pulmonary osteoarthropathy Hyperplastic osteoarthritis, hypertrophic osteoarthropathy PULMONOLOGY A condition characterized by distal expansion and/or periostitis of long bones, clubbing of fingers and toes, arthritis–related to cartilaginous erosion, and/or synovial proliferation, joint pain and, occasionally, autonomic dysfunction–eg, pallor, flushing, profuse diaphoresis; HPO is classically associated with lung disease–eg, abscesses, cancer, COPD, emphysema, chronic interstitial pneumonitis, but may rarely be idiopathic–as in pachydermoperiostosis. Cf Clubbing

hypertrophic pyloric stenosis PEDIATRICS A condition affecting neonates, primarily boys, in the 1st few wks after hatching PATHOGENESIS Defect in pyloric relaxation and pylorospasm CLINICAL Gastric outlet obstruction, projectile bile-free vomiting, metabolic alkalosis, dehydration

hypertrophic scar Keloid, see there

hypertyrosinemia Tyrosinemia, see there

hyperuricemia INTERNAL MEDICINE An accumulation of uric acid–a byproduct of normal metabolism, a side effect of cancer therapy. See Gout

hyperventilation PULMONOLOGY An ↑ in respiratory frequency or volume EFFECT ↓ CO_2, intracranial pressure PHYSIOLOGY pH-mediated cerebrovascular constriction; hypocarbia may restore cerebral auroregulation, alkalinze CSF, ↑ perfusion of ischemic brain tissue COMPLICATIONS Cerebral hypoxia, inverse steal, rebound intracranial HTN, myocardial ischemia

hyperventilation syndrome PSYCHIATRY 1. Panic disorder, see there 2. Tachy-dyspnea 'syndrome' A clinical complex affecting neurotics with anxiety attacks; the hyperventilation, in addition to the characteristic EEG changes–bilateral synchronous theta wave followed by delta activity with spike and slow-wave discharges causes respiratory alkalosis, tightness in the chest, dizziness without syncope, numbness of hands and feet and tetany TREATMENT Rebreathe air in a paper bag, as the ↑ CO_2 facilitates physiologic compensation

hyperviscosity syndrome LAB MEDICINE A clinical condition caused by an abnormal sluggishness of blood flow through peripheral vessels, especially with serum IgM levels > 3 g/dL, which may trigger oronasal bleeding, blurred vision, headache, dizziness, vertigo, ataxia, encephalopathy, or altered consciousness FUNDUSCOPIC EXAM Venous dilatation, "sausage formation" hemorrhages, exudates; serum viscosity correlates poorly with clinical findings among Pts, but correlates well in the same Pt. See Waldenström's macroglobulinemia

hypervitaminosis POPULAR NUTRITION A condition caused by the ingestion of vitamins in extreme excess of physiologic requirements, or pharmacologic doses; it most commonly is caused by excess consumption of fat-soluble vitamins–eg, vitamins A and D, as they accumulate in body fat; water-soluble vitamins B and C, are readily excreted. See Vitamins. Cf Pseudovitamins

hypervitaminosis A Vitamin A toxicity, see there

hypervitaminosis D Vitamin D toxicity, see there

hypesthesia Diminished sensitivity to tactile stimuli

hyphema OPTHALMOLOGY Hemorrhage in the eye's anterior chamber ETIOLOGY Blunt or perforating trauma to eyeball, iritis, cancer, vascular defects; hyphema may be associated with acute glaucoma. Cf Anterior segment

hyphephilia SEXOLOGY A fetish in which the sexuoerotism is associated with touching skin, hair, leather, fur, and fabric, especially if worn on or near erotically significant body parts. See Haptic

hypnagogic imagery Hypnagogic hallucination NEUROLOGY A pseudo-hallucination that occurs when a person is asleep SLEEP DISORDERS Vivid sensory images occurring at sleep onset but especially vivid with sleep-onset REMS periods; feature of narcoleptic REMS naps

hypnagogic startle Sleep start SLEEP DISORDERS A sudden body jerk, observed normally at sleep onset and resulting in momentary awakening

Hypnodorm Flunitrazepam, see there

hypnogenic hallucination NEUROLOGY A vivid visual and/or auditory hallucination associated with narcolepsy, which is potentially frightening, and occurs during sleep attacks, or at the onset of sleep at night

hypnosis PSYCHIATRY A technique involving relaxation and voluntarily ignoring conscious thought processes; hypnosis attempts to access the unconscious mind. See Highway hypnosis PSYCHOLOGY A technique that may be effective in behavior modification–eg, control of habits, relaxation, and biofeedback, in which a person learns to focus attention on thoughts or images unrelated to a particular stimulus–eg, cancer-related pain

hypnotherapy PSYCHOLOGY Hypnosis has some support in psychiatry and anesthesiology; the major effect is relaxation, and possibly control of habits; hypnotherapy has been used as an adjunct in controlling acute and chronic pain, addiction–alcohol, tobacco, and abuse substance disorders, and ↓ anesthetics. See Alternative medicine. Cf Hypnosis, Mesmerism

hypnotic *adjective* 1. Relating to hynosis 2. Inducing sleep 3. Referring to a trance-like state 4. Relating to a hypnotic agent *noun* An agent that induces hypnosis, trance state or sleep; a sedative or CNS depressant, of which benzodiazepines is a drug of choice for 'primary' insomnia; short-acting hypnotics–eg, triazolam and oxazolam are used to induce sleep; to maintain sleep throughout the night, long-acting hypnotics–eg, flurazepam, are used

hypnotic suggestion PSYCHIATRY The modification of unconscious thought through hypnosis, which may be useful for specific/simple phobias, but rarely for agoraphobia, social phobia, or anxiety and panic disorders. See Hypnosis

hypnotized, fear of PSYCHOLOGY Hypnophobia. See Phobia

hypo *prefix*, Latin, low, deficient, insufficient

hypoalbuminemia ↓ Albumin in circulation–normal 32-56 g/L, US: 3.2-5.6 g/dL; albumin has a central role in maintaining oncotic pressure, as it comprises $\frac{2}{3}$ of the protein in circulation; it is also a major repository for amino acids, and is a carrier protein for many organic and inorganic ligands

320 **hypoaldosteronism** A condition characterized by ↓ aldosterone secretion, often with ↓ renin release by kidney CLINICAL Most Pts are 50-70, have unexplained, chronic, asymptomatic hyperkalemia and moderate renal failure–creatinine ≥ 15 mL/min, muscle weakness, arrhythmias ASSOCIATED CONDITIONS DM–50%, SLE, myeloma, renal amyloidosis, sickle cell anemia, cirrhosis, AIDS MANAGEMENT ↓ K+ consumption

hypocalcemia Calcium deficiency A condition characterized by ↓ Ca²⁺–< 2.1 mmol/L or 8.5 mg/dL ETIOLOGY ↓ ingestion or malabsorption of Ca²⁺, magnesium, or vitamin D, absent/ineffective PTH, acute renal failure, cancer, vitamin D-resistant rickets CLINICAL Neurologic and neuromuscular Sx–eg, irritability, depression, psychosis, muscle and laryngeal spasms, convulsions, arrhythmias, GI cramping, malabsorption. See Calcium; Cf Hypercalcemia

hypocapnia ↓ Arterial CO₂

hypochlorite Bleach, ClO⁻ and salts

hypochondriac *adjective* 1. Referring to hypochondriasis–less commonly used than hypochondriacal 2. Referring to the hypochondriac region, *hypochondrium noun* An individual with hypochondriasis, see there

hypochondriasis Hypochondria PSYCHIATRY An exaggerated concern of diseases or medical disorders that can result in psychosomatic Sx; excessive concern for perceived bodily Sx despite assurance that such concern is unfounded. See Medical school syndrome

hypochromia Hypochromic red cell HEMATOLOGY A condition in which the MCH–mean corpuscular Hb and the MCHC–mean corpuscular Hb concentrations are less than normal, a finding in iron deficiency. See Microcytic anemia

hypodermic *adjective* Subcutaneous *noun* Short for 1. Hypodermic injection 2. Hypodermic needle

hypoechoic IMAGING *adjective* Relating to an abnormal ↓ in echoes by ultrasonography, due to a pathologic change in tissue density

hypofibrinogenemia ↓ Fibrinogen in circulation. See Congenital afibrinogenemia

hypogammaglobulinemia IMMUNOLOGY A gallimaufry of conditions characterized by ↓ production of proteins, usually Igs, which migrate in the gamma region of a protein electrophoretic gel; hypogammaglobulinemia may be congenital, as in Bruton's disease, or other B-cell defects or acquired, as in CLL, and accompanied by monoclonal gammopathies TREATMENT Human immune globulin. See Immunodeficiency, B cell

hypogenitalism ↓ External and/or internal reproductive plumbing, usually due to hypogonadism

hypogeusia A state of ↓ sensitivity to tastes

hypoglycemia ENDOCRINOLOGY A ↓ in blood glucose, often linked to systemic disease CLINICAL Headache, tremors, sweating, pallor, syncope, ↓ concentration, coma; fasting hypoglycemia occurs in endocrinopathies–eg, hypopituitarism, Addison's disease, adrenogenital syndrome, islet cell tumors, factitious insulin ingestion, non-pancreatic neoplasms–eg, retroperitoneal sarcoma producing ectopic insulin, hepatic disease, glycogen storage disease; postprandial hypoglycemia may be functional and idiopathic–associated with preclinical DM, gastrectomy, or drug-related–eg, sulfonylureas, oral hypoglycemics–eg, chlorpropamide, tolbutamide, tolazamide, acetohexamide, alcohol, aspirin, phenformin, insulin. See Drug-induced hypoglycemia, Glucose tolerance test, Reactive hypoglycemia

hypoglycemia associated with drugs Drug-induced hypoglycemia, see there

hypoglycemic *adjective* Referring to hypoglycemia, see there

hypogonadism ENDOCRINOLOGY 1. Inadequate gonadal function, as manifest by defects in gametogenesis, secretion of, and/or response to, gonadal hormones. Cf Androgen replacement therapy 2. A clinical condition with ↓ or absent phenotypic expression of a person's sexual genotype, which may be 1°, due to a lack of end organ response to FSH or LH produced normally by an intact pituitary gland–hypergonadotropic hypogonadism, or 2° to defective hypothalamic or pituitary hormonal activity–hypogonadotropic hypogonadism

HYPOGONADISM, ETIOLOGY

PRIMARY

♀ **HYPERGONADOTROPIC** Turner syndrome, XX Turner sydrome, XX pure gonadal dysgenesis, mixed gonadal dysgenesis, autoimmune ovarian disease

♂ **HYPERGONADOTROPIC** Congenital anorchia, rudimentary testes, germ cell hypoplasia–del Castillo syndrome, XY Turner phenotype–Noonan syndrome, Klinefelter syndrome and variants, XX males, XYY males

SECONDARY

♀ **HYPOGONADOTROPIC** Carpenter syndrome, hypopituitarism, Lawrence-Moon-Biedl, multiple lentigines syndrome, polycystic ovaries

♂ **HYPOGONADOTROPIC** Amyloidosis, Carpenter syndome, fertile eunuch syndrome, Fröhlich syndrome, Sheehan syndrome, Kallmann's disease, Laurence-Moon-Biedl disease, Lowe syndrome, Prader-Willi syndrome

hypogonadotropic hypogonadism Hypogonadotropic eunuchoidism, Kallman syndrome A rare condition with a highly variable hereditary pattern, characterized by secondary hypogonadism–↓ gonadotropin-releasing hormone due to hypothalamic or pituitary dysfunction with testicular failure, and anosmia–due to hypoplasia or aplasia of the olfactory bulbs and tracts, 2° to a defect in the migration of olfactory neurons, and neurons producing GRH–gonadotropin-releasing hormone; ↓ FSH and LH impairs sperm and androgen production CLINICAL Delayed puberty, micropenis, eunuchoid features, cryptorchidism, midline defects–eg, cleft lip and palate, unilateral renal agenesis, horseshoe kidney, nerve deafness and hearing loss, color blindness, skeletal abnormalities; synkinesia, spatial attention defects, spastic paraplegia, cerebellar dysfunction, horizontal nystagmus, pes cavus, mental retardation MANAGEMENT Androgens to induce anatomic maturation; gonadotropins or LH-releasing factor for spermatogenesis. See *Kalig-1*

hypohidrosis Reduced sweat production

HYPOHIDROSIS

I **INHERITED CONDITIONS**, eg hereditary anhydrotic ectodermal dysplasia, ichthyosis, or angiokeratoma corporis diffusum universale

II **ACQUIRED CONDITIONS**

COLLAGEN VASCULAR DISEASES—Sjögren syndrome, progressive systemic sclerosis

DERMATOPATHIES—Miliaria profunda, pemphigus vulgaris, psoriasis

DRUGS Anticholinergics, eg atropine, scopolamine; ganglionic blockers

ENDOCRINOPATHIES Diabetes insipidus, hypothyroidism, hypothalamic lesions

ENVIRONMENTAL STRESS Heat stroke and dehydration

PERIPHERAL NEUROPATHY Alcohol, amyloidosis, DM, Horner syndrome, leprosy

hypokalemia Hypopotassemia, low blood potassium METABOLISM ↓ K+ in serum ETIOLOGY Crash dieting, Cushing syndrome, DKA, dehydration, familial periodic paralysis, GI tract wasting, hyperaldosteronism, licorice–due to aldosterone-like effects of glycerrhizic acid, malnutrition, metabolic acidosis, nasogastic suctioning, starvation, stress–burns, surgery, trauma, vomiting, drugs–eg, aspirin, corticosteroids, potassium-wasting diuretics, estrogen, insulin, laxatives, lithium, sodium polystyrene sulfonate/Kaylexate™ CLINICAL Hypotension, weak, rapid, irregular pulse, muscle weakness, paresthesias, confusion EKG Flattened T wave REF RANGE 3.8-5.0 mmol/L–US: 3.8-5.0 mEq/L

hypokalemic periodic paralysis Familial periodic paralysis, see there

hypolipoproteinemia A group of conditions characterized by ↓ lipoproteins, either acquired–eg, 2° to malabsorption, anemia, hyperthyroidism, or congenital

hypomagnesemia Low blood magnesium, magnesium deficiency METABOLISM A serum magnesium ≤ 1.5 mg/dL, often manifest by muscular hyperirritability ETIOLOGY Alcohol abuse, burns, dehydration, DKA, diarrhea, ↑ Ca²⁺, ↑ aldosterone, ↓ K⁺, ↓ PTH, post-bowel resection, malabsorption, malnutrition, pancreatitis, renal insufficiency, therapy with amphotericin B, calcium gluconate, diuretics, insulin, neomycin CLINICAL Cramping, ↑ tendon reflexes, tremors. See Magnesium

hypomanic episode Hypomania PSYCHIATRY A discrete period–≥ 4 days of an abnormal and persistently elevated, expansive, or irritated mood, accompanied by 'ancillary' Sx–eg, ↑ self-esteem, non-delusional grandiosity, flight of ideas, ↑ distractibility, ↓ need for sleep, ↑ in goal-directed activities, or psychomotor agitation. See Bipolar disorder, Manic episode

hypomenorrhea A menstrual period of ↓ days or ↓ blood flow

hyponatremia ↓ Sodium. See Sodium

hypoparathyroidism ENDOCRINOLOGY A condition characterized by ↓ PTH and serum Ca²⁺ ETIOLOGY Congenital absence of the parathyroid glands, accidental excision or injury of parathyroid glands during thyroidectomy or other neck surgery; massive regional RT, magnesium deficiency RISK FACTORS Recent thyroid or neck surgery, family Hx of parathyroid disorder, or autoimmune disease–eg, Addison's disease CLINICAL Tetany. See PTH

hypoperfusion syndrome A 'prerenal' circulatory disorder 2° to 1. Occlusive renal arterial disease with stenosis and ischemia, resulting in ↑ renin, HTN and azotemia, affecting one or both kidneys, 2. ↓ Effective circulatory volume, due to ↑ vascular capacitance–eg, induced by sepsis, sequestration of fluids in compartments–eg, in the hepatorenal syndrome or ascites, or the inability to effectively transfer fluids from the venous compartment to the arterial compartment due to CHF or pericardial limitations due to constrictive pericarditis or effusions; the ↓ volume form of hypoperfusion is associated with oliguria, azotemia, ↓ sodium excretion and ↑ renin, but not HTN

hypophosphatemia Low blood phosphate ENDOCRINOLOGY Low serum phosphate which may be linked to management of DKA, starvation, etc. See Phosphate

hypophysectomy NEUROSURGERY A procedure in which the hypophysis is destroyed or excised, mandated by a mass lesion in the sella turca, which requires permanent hormonal support. See Neuroablative procedure

hypophysis Pituitary gland ANATOMY The pea-sized powerhouse at the base of the brain which orchestrates the endocrine symphony, saddled in the sella turcica and attached by a stalk to the hypothalamus–which controls pituitary function. Cf Hypothalamus

hypopigmented MEDTALK Abnormal skin color

hypopituitarism ENDOCRINOLOGY A condition characterized by ↓ secretion of one or more anterior pituitary or adenohypophyseal hormones ETIOLOGY Idiopathic, pituitary tumor, hypothalamic or infundibular cyst or tumor, infiltrative, vascular, etc, RT, trauma, postsurgery, encephalitis, hemochromatosis, stroke, dwarfism CLINICAL Sx and signs of pituitary hormone deficiency are similar to those of a 1° defect of the target organ–eg, corticotropin deficiency is associated with asthenia, headache, anorexia, N&V, abdominal pain, and altered mental status MANAGEMENT Prolonged absence of cortisol and thyroxine is fatal; hormone-replacement is a required 'life sentence'

hypoplasia of parathyroids Hypoplasia of thymus, DiGeorge syndrome, third and fourth pharyngeal pouch syndrome A condition characterized by 1. low blood

calcium levels–hypocalcemia due to underdevelopment–hypoplasia of the parathyroid glands which control calcium; 2. underdevelopment—hypoplasia of the thymus, an organ behind the breastbone in which lymphocytes mature and multiply; and 3. defects of the heart involving the outflow tracts from the heart

hypoplastic anemia See Aplastic anemia

hypoplastic left heart syndrome PEDIATRIC CARDIOLOGY A group of congenital often AR cardiac defects characterized by hypo- or agenesis of the left ventricle, aortic and mitral valves, an atrial right-to-left shunt; right-sided hypoplasia of tricuspid or pulmonary valves, pulmonary stenosis, or right ventricular hypoplasia; the ascending and transverse aorta is narrowed with a diaphragm-like aortic coarctation at the preductal aortic isthmus; postnatal life hinges on adequate blood supply–ie, is ductus dependent, unrestricted atrial shunting, and a balance between the pulmonary and systemic vascular resistances; surgery is difficult as there are extensive malformations MORTALITY High. See Baby Fae heart

hypopotassemia Hypokalemia, see there

hyposensitization Allergen immunotherapy, see there

hyposmia Diminished sensitivity to smell

hypospadias PEDIATRIC UROLOGY A congenital defect in the positioning of the urethral meatus on the penis or vagina TYPES Mild male hypospadias–opening is slightly displaced below the tip of the penis; severe male hypospadias, opening is in the female position at the base of the scrotum; the penis has an open gutter under the penis; the penis may be small. See Epispadia, Micropenis/microphallus

hyposthenuria UROLOGY The excretion of urine with a low specific gravity ETIOLOGY Inability to concentrate urine, ↑ in urinary volume, as seen in diabetes insipidus

hypotension ↓ BP. See Exercise hypotension, Orthostatic hypotension

hypothalamic dysfunction ENDOCRINOLOGY Any alteration in hypothalamic activity–eg, heart rate, ↑ breathing, ↑ blood flow, dilate pupils.

hypothalamic fever ↑ Temperature due to an altered hypothalamic thermostat, caused by hemorrhage, trauma or tumor

hypothalamic-pituitary axis ENDOCRINOLOGY Any feedback system that coordinates the activity of major peptide hormones; the hypothalamus synthesizes releasing hormones, which act on the pituitary, which evokes end-organ responses

---

HYPOTHALAMIC-PITUITARY–HP 'AXES'

- HP adrenocortical-ACTH-adrenal gland axis
- HP-FSH/LH-gonadal axis
- HP-TSH-thyroid gland axis
- HP-growth hormone-somatotroph axis
- HP-hypothalamic-lactotroph-breast axis

---

hypothalamic-pituitary-adrenal axis A tightly-linked, interdependent endocrine unit which, with the systemic sympathetic and adrenomedullary systems, comprises a major peripheral limb of the stress system, the main function of which is to maintain basal and stress-related homeostasis; the hypothalamus and pituitary form the central part of the HPA axis, and are active even at rest, responding to blood-borne or neurosensory signals–eg, cytokines–eg, IL-1, IL-6, TNF-α; at the highest level, CRH and noradrenergic neurons innervate and stimulate each other, which is controlled by an autoregulatory, ultrashort negative-feedback loop, in which CRH and noradrenergic collateral fibers inhibit presynaptic CRH and α₂-noradrenergic receptors. Cf Hypothalamic-pituitary axis

hypothalamo-hypophyseal portal system A vascular system that transports releasing and inhibiting hormones from the hypothalamus to the anterior pituitary

hypothermia CRITICAL CARE A ↓ in core/rectal body temperature ≤ 35°C–95°F, due to long-term occupational or recreational exposure to ↓ air or water temperatures EPIDEMIOLOGY Hypothermia causes 750 deaths/yr; racial differences in mortality 3.2–white vs 13.1–blacks deaths/10^6 ♂; 1.4 vs 4.1 deaths/10^6–♀ –US CLINICAL ↓ Respiratory rate, metabolic acidosis, ↓ pulse, ↓ blood pressure, ventricular fibrillation, hypo- or hyperglycemia, coagulopathy, hemoconcentration, pneumonia, renal failure, pancreatitis; when extreme and prolonged, drowsiness, delirium, coma, shivering, numbness, fatigue, poor coordination, slurred speech, impaired mentation, blue and/or puffy skin, irrational thinking RISK FACTORS Extremes of age–related to ↓ shivering mechanisms, less protective fat, ↓ mobility, ↓ metabolic rate, and chronic illness, alcohol use/abuse, use of neuroleptic agents, hypothyroidism, mental illness, starvation, dehydration, poverty, immobilizing illness, and young adults in winter sports TREATMENT External rewarming, best performed in a warm tub at 40-42°C–104-107.6°F; internal rewarming is recommended for those with severe hypothemia and includes extracorporeal blood warming using a femorofemoral bypass and/or repeated peritoneal dialysis with 2 L of warmed–43°C K⁺-free dialysate solution. See Accidental hypothermia NEUROLOGY The intentional cooling Pts with traumatic brain injury and 'salvageable' Glasgow coma scores–5 to 7 on admission to 33° for 24 hrs is associated with an improved survival–62% good outcomes in the hypothermia group vs 38% good outcomes in the normothermia group MECHANISM Hypothermia ↓ 2° brain injury by an unknown mechanism, possibly by ↓ brain metabolism, ↓ extracellular concentrations of excitatory neurotransmitters–eg, glutamate, ↓ post-traumatic inflammatory response, with ↓ cytokine release

hypothermic circulatory arrest Suspended animation CARDIOVASCULAR SURGERY A surgical procedure in which a heart-lung machine is used to cool the body during surgery, which ↓ BP and slows circulation to near standstill; HCA is critical in some neurosurgeries–eg, excision of aneurysms of the vertebrobasilar circulation. See Berry aneurysm

hypothesis EPIDEMIOLOGY A supposition, arrived at from observation or reflection, that leads to refutable predictions; a conjecture cast in a form that will allow it to be tested and refuted

hypothyroid adjective Referring to hypothyroidism, see there

hypothyroidism ENDOCRINOLOGY A condition characterized by underproduction of thyroid hormones CLINICAL Fatigue, hypersomnolence DIAGNOSIS Serum TSH. See Central hypothyroidism, Primary hyperthyroidism, Secondary hypothyroidism. Cf Hyperthyroidism

hypotonicity ↓ Muscle tone; limp muscles

hypoventilation MEDTALK A ↓ in depth/frequency of respiration. Cf Hyperventilation

hypovolemic shock CRITICAL CARE A rapid fall in BP due to a ±20% ↓ of blood volume ETIOLOGY GI tract and other internal or external hemorrhage, loss of blood volume and body fluids–eg, diarrhea, vomiting, intestinal blockage, inflammation, burns, etc. See Shock, Volume depletion

hypoxanthine-guanine phosphoribosyltransferase An enzyme encoded on chromosome X found in high concentrations in the brain, which is responsible for the transfer of phosphoribosyl groups

hypoxemia Hypoxia ↓ O_2 in the blood

hypoxia CARDIOLOGY A low O_2 concentration in arterial blood. See Cerebral hypoxia

hysterectomy GYNECOLOGY The 2ⁿᵈ most common–after C sections–operation performed in the US ± 590,000/yr, ±4/1000 ♀ /yr; annual cost ± $5 × 10^9; in developed nations, there is a 6-fold range in frequency of hysterectomy–highest in US, lowest in Sweden, Norway, UK INDICATIONS Uterine leiomyomas–30%, dysfunctional uterine bleeding–20%, endometriosis and adenomyosis–20%, genital prolapse–15%, chronic pelvic pain–10%, PID, endometrial hyperplasia, and other malignancy–eg, CIN, endometrial CA, and other indications–eg, massive postpartum hemorrhage, septic endometritis, hormone therapy where estrogen receptors are positive PROCEDURES Total abdominal, total vaginal, laparoscopic vaginal COMPLICATIONS Fever, infection, intra- and post-operative hemorrhage, urinary Sx, early ovary failure, retained ovary syndrome, constipation, fatigue, ↓ sexual interest and function, depression, psychiatric morbidity MORTALITY 7-20 deaths/10^5 hysterectomies for cancer, 3-4 deaths/10^5 for pregnancy-related indications, 0.6-1.1 deaths/10^5 hysterectomies for other indications SEXOLOGY ↑ Frequency, orgasms, sexual desire, ↓ dyspareunia posthysterectomy. See Abdominal hysterectomy, Complete hysterectomy, TAH-BSO, Total abdominal hysterectomy, Vaginal hysterectomy. Cf Laparoscopic vaginal hysterectomy

hysteresis pacing CARDIAC PACING A pacing parameter which usually allows a longer escape interval after a sensed event, giving the heart a greater opportunity to beat on its own. See Pacemaker

hysteria PSYCHIATRY A 16ᵗʰ century term for excessive emotional lability, anxiety etc. See Mass hysteria, Vapors

hysterical POP PSYCHOLOGY adjective Referring to a state of extreme agitation VOX POPULI Laugh, laugh, much, much; hilarious; jocular

hysterosalpingography Uterohysterography REPRODUCTIVE MEDICINE A nonsurgical method for evaluating uterotubal pathology, in which radiocontrast is instilled transcervically in the uterine cavity and fallopian tubes followed by fluoroscopic examination or obtention of a plain film, as a means of defining the shape and size of the uterine cavity and tubal patency. Cf Electronic fluorography, Salpingoscopy

hysteroscopy GYNECOLOGY Visualization of the uterine cavity using a hysteroscope to evaluate abnormal uterine bleeding, identify and resect lesions of the endometrial cavity–eg, uterine synechiae and septae, submucosal leiomyomas, endocervical and endometrial polyps, IUDs, and for endometrial ablation COMPLICATIONS Perforation, bleeding, infection

Hytrin® Terazosin, see there

Hyzaar® Losartan K⁺-hydrochlorothizide, see there

Hz Hertz, see there

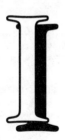

I Symbol for: 1. Electric current, expressed in amperes 2. Hypoxanthine 3. Illumination 4. Incisor 5. Index 6. Inosine 7. Intelligence 8. Intensity 9. Intermediate 10. Intern 11. Internal 12. Internist 13. Intestine 14. Invasive 15. Iodine 16. Isoleucine 18. Luminous intensity

I-cell disease Type II mucolipidosis An AR condition caused by a defect in protein trafficking and sorting, resulting in massive accumulation of intracellular and extracellular waste products, especially glycolipids CLINICAL Hurler-like/type I-H mucopolysaccharidosis disease with a gargoyle face, early onset of psychomotor retardation, joint contracture, hepatosplenomegaly and cardiac decompensation, with death in childhood LAB Normal urinary mucopolysaccharides, vacuolated lymphocytes, ↑ lysosomal enzymes in serum, CSF, urine

I&D Incision and drainage, see there

I-125 radioactive seed RADIATION ONCOLOGY A small radioactive seed implanted into the prostate under ultrasound guidance for treating prostate CA–brachytherapy. See Ultrasound-guided transperineal permanent palladium-103 implantation of prostate

IADL Instrumental activities of daily living, see there

IAP 1. Intermittent acute porphyria. See Acute intermittent porphyria 2. Intraabdominal pressure 3. Intraarterial pressure

-iasis Greek root form, for pathologic state–eg, cholelithiasis

iatrogenic *adjective* Referring to a physical or mental condition caused by a physician or health care provider–eg, iatrogenic disease, due to exposure to pathogens, toxins or injurious treatment or procedures

iatrogenic illness Any complication related to diagnosis and treatment of disease, regardless of whether the condition occurs as a known risk of a procedure or through errors of omission or commission

iatrophobia PSYCHOLOGY Fear of doctors. See Phobia

IBD Inflammatory bowel disease, see there

IBS Irritable bowel syndrome, see there

ibuprofen Advil®, Motrin®, Rufen® THERAPEUTICS An OTC oral NSAID tightly bound to albumin, excreted through the kidneys $T_{1/2}$ 2 hrs INDICATIONS Headaches, muscle or joint pain associated with anxiety disorders. See NSAIDs

IBW Ideal body weight, see there

Icarus complex PSYCHIATRY A constellation of mental conflicts, the degree of which reflects the imbalance between a person's desire for success, achievement, or material goods, and the ability to achieve those goals; the greater the gap between the idealized goal and reality, the greater the likelihood of failure. See Type A, 'Toxic core'

ICD 1. Implantable cardioverter-defibrillator, see there 2. International Classification of Disease, see there

ICD-9-CM International Classification of Disease, 9th edition, Clinical Modification A standardized classification of disease, injuries, and causes of death, by etiology and anatomic localization and codified into a 6-digit number, which allows clinicians, statisticians, politicians, health planners and others to speak a common language, both US and internationally. Cf SNOMED, SNOP

ice DRUG SLANG A street term for any cystallized abuse substance–eg, cocaine, crack cocaine, methamphetamine, smokable methamphetamine, PCP, MDM SUBSTANCE ABUSE The street name for a smokable, crystalline form of the psychostimulant, D-methamphetamine; IV ice induces an effect similar to IV methamphetamine, and is far more intense than that achieved orally COMPLICATIONS Pulmonary edema, dilated cardiomyopathy, acute MI, cardiogenic shock, death. See 'Designer' drug

ice climbing SPORTS MEDICINE An 'extreme sport' in which participants climb ice formations with pickaxes, often without ropes INJURY RISK Hypothermia, death. See Extreme sports, Novelty seeking behavior

ice cream headache A migraine triggered by oropharyngeal irritation due to cold foods, the ingestion of which may cause bifrontal headaches in Pts with migraines

iceberg sign Thoraco-abdominal sign IMAGING A sharply-demarcated radiopacity seen in plain films of the chest or abdomen that may be either sub- or supradiaphragmatic, corresponding to a paravertebral TB abscess, aneurysm, esophagogastric lesion or the azygous continuation of the inferior vena cava

ichthyosis Common ichthyosis, fish scale disease, ichthyosis vulgaris DERMATOLOGY A group of hereditary–1° or acquired–2° diseases characterized by dyskeratosis and non-inflammatory scaling of the skin. See Harlequin ichthyosis, X-linked ichthyosis

iconoclast SURGERY A surgical instrument used for blunt dissection, which may be used below the galea aponeurotica in preparation for scalp reduction-browlift in hair restoration. See Hair replacement

ICRT Intracoronary radiation therapy CARDIOLOGY An investigational therapy for preventing in-stent restenosis. See Beta-Cath™ system, START trial

ictal *adjective* Referring to the period of a seizure or seizure-like attack

icterus Jaundice, see there

ICU Intensive care unit MANAGED CARE A hospital unit, in which 20% of all Pts spend some time during hospitalization; criteria for admission and continued stay in the ICU–eg, chest pain and suspected MI, pulmonary edema and syncope

id PSYCHIATRY The unconscious source–per the freudian construct of mental energies, libido, unstructured desires and drives. See Ego, Superego. Cf Id reaction

ID number Identification number A numerical 'fingerprint'–eg, social security number, that a person acquires at some point in life which is used to identify personal data; Cf PIN

id reaction Dermatophytid reaction DERMATOLOGY A localized or generalized rash of sudden onset associated with, but located at a distance from, a site of cutaneous inflammation and/or infection; IRs occur on the hands and arms, as sterile papulovesicular pustules, classically associated with dermatophytosis–eg, tinea pedis, tinea capitis and may be associated with stasis dermatitis, eczema and contact dermatitis, which disappear after successful therapy; because the IR is similar to the original lesion, it is thought to represent an immune to circulating antigens. Cf Id, Isomorphic phenomenon

idarubicin 4-demethoxydaunorubicin ONCOLOGY An antitumor antibiotic

IDDM Insulin-dependent diabetes mellitus; now known as type 1 diabetes mellitus

IDEAL CARDIOLOGY A clinical trial–Incremental Decrease in Endpoints through Aggressive Lipid Lowering PSYCHOLOGY See Ego ideal

ideal body weight A person's optimum weight, defined by the Metropolitan Life Tables, see there

ideas of reference PSYCHIATRY Incorrect interpretation of casual incidents and external events as having direct reference to oneself, which may be suffficiently intense as to constitute delusions

ideation PSYCHIATRY The collective representation of thoughts and ideas, recalled from memory, or projected into the future, singly or combined. See Paranoid ideation, Suicidal ideation

ideational apraxia NEUROLOGY The inability to execute a sequence of movements in an orderly fashion; although individual components of the IA are executed, the entire act remains uncompleted–eg, a match may be removed from its cover but not struck

ideational dyspraxia NEUROLOGY A form of dyspraxia in which simple–but not complex or multistep–activities are executed in response to verbal commands. See Dyspraxia

idee fixe Fixed idea PSYCHIATRY An obsessive idea, delusion, or compulsion

identical twins Monozygotic twins Twins resulting from the division of a single fertilized egg, which usually share a common chorion and placenta, although usually each may have a separate amnion. Cf Fraternal twins

identification PSYCHIATRY An unconscious defense mechanism by which a person assimilates or copies another's activities, behavior, and reactions, a process that plays a major role personality and superego development. See Complementation, De-identification, Gender identity/role. Cf Imitation/role modeling

identity PSYCHIATRY A person's global role in life and perception of a sense of self. See Core identity, Gender identity SOCIAL MEDICINE A sense of individuality including one's distinct personality, talents, abilities, and flaws

identity crisis PSYCHOLOGY A conflict in a person's perceived role in society, which may be accompanied by a loss of the sense of self and historical continuity. See Crisis, Psychosocial development

identity disorder PSYCHIATRY Severe subjective distress caused by a child's inability to achieve an integrated sense of self

ideokinetic apraxia NEUROLOGY The dissociation of an idea and an act–eg, a Pt cannot whistle when commanded to do so, but may do so spontaneously

ideomotor apraxia NEUROLOGY The inability to demonstrate the use of simple objects in absence of motor weakness

ideomotor dyspraxia NEUROLOGY Inability to perform single motor tasks when requested to do so verbally or by visual imitation–eg, combing hair, waving goodbye. See Dyspraxia

IDET Intradiscal electrothermal therapy, see there

idiopathic *adjective* Unknown; of unknown origin, as in a symptom or syndrome that is apparently spontaneous

idiopathic aplastic anemia HEMATOLOGY A BM stem cell defect of unknown, presumed autoimmune, etiology, in which all 3 cell–erythropoietic, myeloid and megakaryocytic–cell lines are ↓. See Aplastic anemia, Granulocytopenia, Thromboycytopenia

idiopathic cardiomyopathy CARDIOLOGY A cardiomyopathy of unknown etiology which may represent end stage myocarditis RISK FACTORS Obesity, Hx of myocarditis, drugs, abuse substances–tobacco, alcohol, prior scarring CLINICAL Right, then left-sided heart failure. See Cardiomyopathy

idiopathic dilated cardiomyopathy CARDIOLOGY '...*primary myocardial disease of unknown cause characterized by left ventricular or biventricular dilatation* (sic) *and impaired myocardial contractility*'. See Actin, Dilated cardiomyopathy

idiopathic hydrocephalus Normal pressure hydrocephalus, see there

idiopathic hypercalciuria NEPHROLOGY Excess renal excretion of calcium, which ↑ risk of kidney stones, see there

idiopathic hypertrophic subaortic stenosis Hypertrophic cardiomyopathy, see there

idiopathic midline destructive disease Lethal midline granuloma of Stewart A diagnosis of exclusion–criteria, consist of a midline granuloma of unknown etiology which, unlike most granulomas of the midline, are secondary manifestations of a primary process–eg, vasculitis–eg, Wegener's granulomatosis, malignancy–eg, lymphoma, nasal carcinoma, infections–eg, destructive fungal infections in immunocompromised Pts TREATMENT RT. See Lethal midline granuloma, Midline granuloma

IDIOPATHIC MIDLINE DESTRUCTIVE DISEASE

1. Local destruction of upper respiratory tract

2. No progression to system disease–average follow-up, 7 years

3. Acute/chronic inflammation with variable necrosis, without vasculitis, atypia or malignancy

4. No evidence of infection or malignancy

idiopathic pulmonary hypertension Primary pulmonary hypertension, see there

idiopathic pulmonary fibrosis Idiopathic interstitial fibrosis of lung PULMONOLOGY An idiopathic condition characterized by scarring and fibrosis of alveolar septae more common in middle-aged men, possibly related to collagen vascular disease, with positive 'rheumatoid' serology CLINICAL Aggressive–rapid onset of dyspnea, orthopnea, hemoptysis, cyanosis, clubbing of fingers and toes, pulmonary HTN, bibasilar rales, non-productive cough, death in 3-6 yrs IMAGING Diffuse reticulonodular infiltrates. See Diffuse interstitial fibrosis, Lymphoid interstitial fibrosis, Usual interstitial fibrosis

idiopathic pulmonary hemosiderosis CLINICAL IMMUNOLOGY A rare possibly autoimmune condition affecting children < age 10 characterized by hemosiderin deposition in lungs CLINICAL Recurrent hemoptysis, anemia, weakness, clubbing, hepatosplenomegaly PROGNOSIS Poor, recurrent episodes; death often in 5 yrs due to cor pulmonale; it may be associated with celiac disease and improve with a gluten-free diet. See Hemosiderosis

idiopathic sideroblastic anemia Acquired refractory sideroblastic anemia, idiopathic acquired sideroblastic anemia HEMATOLOGY A pluripotent stem cell defect characterized by ineffective erythropoiesis with a slight ↓ in RBC survival, mild maturational impairment of all hematopoietic cell lines, coupled with defective iron metabolism and iron accumulation; ISA affects older Pts, often related to radio- or chemotherapy, 10-30% of whom later develop acute non-lymphocytic leukemia CLINICAL Pallor, fatigue, weakness, dyspnea and palpitations on exertion LAB 'Dimorphic' anemia with hypochromic-microcytic and macrocytic features; 40% of erythroblasts are ring sideroblasts, anisocytosis, basophilic stippling, reticulocytes with impaired heme synthesis, ↓ delta ALA synthetase and protease activity TREATMENT Pyridoxine is often administered, usually without therapeutic response PROGNOSIS ISA is usually characterized by chronic stable anemia, and less commonly by evolution to leukemia, or to BM failure. Cf Refractory anemia with excess blasts

idiopathic thrombocytopenic purpura Werlhof's disease HEMATOLOGY A diagnosis of exclusion characterized by thrombocytopenia without known exogenous etiologic factors, or conditions known to be associated with 'secondary' thrombocytopenia; in most Pts, ITP is attributed to immune-mediated accelerated platelet destruction; acute ITP is more common in children, causing a self-limited wave of ecchymotic hemorrhage 2° to viral infection or vaccination; chronic ITP is more common in adults and often is autoimmune with mucocutaneous, CNS, cardiac and renal hemorrhage–and potentially, infarction, bruisability, transient thrombocytopenia–with normal or ↑ megakaryocytes in BM; ⅔ of cases have IgG antiplatelet antibodies and hemolysis in splenic sinusoids CLINICAL ♀:♂ ratio 3:1, microangiopathic hemolytic anemia, fever, transient neurologic defects, renal failure, microthrombolic 'showers' to brain, heart, lungs, kidneys, adrenal glands, spleen and liver DIFFDX Microthrombi of TTP of young ♀, microangiopathic hemolytic anemia, neurologic defects, renal failure TREATMENT Efficacy of most therapies used in ITP is uncertain, especially given that spontaneous remission occurs, the small populations studied, and the wide difference in Pt populations; modalities include splenectomy–up to 75% markedly improve with splenectomy, prednisone, IV gamma globulin, immunoadsorption apheresis on staphylococcal protein A columns, plasmapheresis; low-dose corticosteroids or to pulsed high-dose dexamethasone, combination chemotherapy–eg, azathioprine, cochicine, cyclosporine, danazol, vincristine

ITP CLINICAL FEATURES	ACUTE	CHRONIC
Peak age	Children, age 2-6	Adults, age 20-40
Sex (♀:♂ ratio)	1:1	3:1
History, recent infection	Common	Rare
Onset	Abrupt	Insidious
Hemorrhagic bullae-oral	Present if severe	Rare
Eosinophilia, lymphocytosis	Common	Rare
Platelet count	< 20 x 10⁹/L	30-80 x 10⁹/L
Duration	2-6 weeks	Months, years
Spontaneous remission	Common, ± 80%	Rare, ± 20%

After TC Bithell, in GR Lee, et al Eds, Wintrobe's Clinical Hematology, 9th ed, Lea & Febiger, Philadelphia, 1993

idiosyncrasy THERAPEUTICS A Pt-specific constellation of reactions to a particular drug–eg, insomnia, tremor, weakness, dizziness, or cardiac arrhythmias, which may be seen in some Pts taking adrenergic amines

idiot savant Savant syndrome A mentally retarded person incapable of abstract reasoning, with a remarkably overdeveloped skill–eg, mathematical or calendar 'calculation', memory, mechanical ability or talent in music and the visual arts

IDL Intermediate-density lipoprotein, see there

idoxifene ONCOLOGY A selective estrogen receptor modulator–SERM in clinical trials for treating advanced breast CA and preventing osteoporosis. See Breast cancer, Osteoporosis, SERM

idoxuridine VIROLOGY A topical antiviral available in an ophthalmic solution for treating herpetic and other viral infections of the cornea

IEP Immunoelectrophoresis, see there

IF Intrinsic factor, see there

IFE Immunofixation electrophoresis, see there

IFN Interferon, see there

ifosfamide ONCOLOGY An alkylating antineoplastic agent approved under an IND protocol as a 3ʳᵈ-line therapy for germ-cell testicular malignancy; it is also used for cervical CA, soft tissue sarcomas SIDE EFFECTS Myelosuppression, encephalopathy, confusion, coma, hemorrhagic cystitis TREATMENT Hemorrhagic cystitis may respond to mesna, encephalopathy may respond to methylene blue

IgA nephropathy Berger's disease, IgA glomerulonephritis NEPHROLOGY Idiopathic IgA nephropathy is the most common form of primary glomerulonephritis in the world; primary IN is mediated by immune complexes and defined immunohistochemically by glomerular deposition of IgA; in 20-40% of Pts, it progresses to renal failure 5-25 yrs after diagnosis; IN may occur in HIV-infected Pts CLINICAL Macroscopic hematuria which may coincide with an URI and be accompanied by flank pain, a presentation typical of younger Pts; older Pts tend to be asymptomatic and IN is detected by sediment and proteinuria TREATMENT ACE-inhibitors, corticosteroids, n-3 polyunsaturated fatty acids provided in dietary fish oil may slow progression of renal failure, kidney transplantation PROGNOSIS More rapid progression occurs in ♂, older Pts, HTN, persistent proteinuria, baseline of impaired renal function, glomerulosclerosis or interstitial fibrosis at the time of initial evaluation

IGF-I Insulin-like growth factor I, somatomedin-C A polypeptide hormone structurally similar to proinsulin, synthesized in the liver and fibroblasts, giving fibroblasts a paracrine function; serum levels correlate with development of 2° sex characteristics in puberty ↑ IN Acromegaly, gigantism ↓ IN starvation, anorexia nervosa, African pygmies, ↓ somatomedins, Laron dwarfism, kwashiorkor, liver disease in children, GH deficiency, hypopituitarism

IGF-II Insulin-like growth factor II A protein homologue of IGF-I that may be produced by tumors and cause reactive hypoglycemia ↑ IN Acromegaly, hepatoma, leiomyosarcoma, mesothelioma, Wilms' tumor ↓ IN hGH deficiency, infants; IGF-II receptor is similar, if not identical to the mannose-6-phosphate receptor

IHSS Idiopathic hypertrophic subaortic stenosis, now known as hypertrophic cardiomyopathy, see there

Iktorivil Clonazepam, see there

IL Interleukin, see there, also 1. Ilium 2. Incisolingual–dentistry

IL372 GRADUATE EDUCATION A series of regulations from Medicaid and Medicare that prohibit attending physicians who are teaching medical students from billing for services that are not performed by the physician's own hand, regardless of how closely the medical students are being supervised. See Attending physician

ileal bypass operation A surgical procedure that may be indicated for extreme obesity or high serum cholesterol ADVERSE EFFECTS Jejuno-ileal bypass is a more extensive variation on the 'morbid obesity surgery' theme, resulting in significant weight loss, but complicated by steatorrhea, diarrhea, hepatic failure, cirrhosis, oxalate deposition, and concrement formation, bile stone formation, electrolyte imbalance–↓ Ca²⁺, K⁺, Mg²⁺, hypovitaminosis, psychogenic problems, polyarthropathy, hair loss, pancreatitis, colonic pseudo-obstruction, intussusception, pneumatosis cystoides intestinalis, and blind loop syndrome. See Gastric bubble, Jaw wiring, Morbid obesity

ileitis 1. Inflammation of the ileum. See Backwash ileitis 2. Crohn's disease, see there

ileostomy GI DISEASE An opening from the ileum to outside the body, which provides an exit for feces when the entire colon has been removed INDICATIONS Active ulcerative colitis with dysplasia or cancer, familial polyposis, Crohn's disease, injury–eg, shotgun blast to abdomen. See Colostomy

ileus GI DISEASE Impairment of the fecal flow of GI contents CLINICAL Sx reflect point of obstruction and type of ileus; paralytic ileus causes little pain and is first evident through abdominal distension and vomiting; post-operative paralytic ileus may manifest itself through ↑ nasogastric secretions or oliguria; mechanical ileus is associated with vomiting, abdominal colic, distension and constipation, which may be episodic with intermittent relief by production of voluminous, watery stools MANAGEMENT Stabilize, decompress, repair. See Meconium ileus, Paralytic ileus

ILEUS TYPES

ADYNAMIC Paralytic ileus, 2° to electrolyte derangements, mesenteric arterial vascular accidents, peritoneal irritation, surgery, trauma, paraneoplastic phenomena

MECHANICAL

• **OBSTRUCTIVE** Intraluminal tumors, intussusception, gallstones, bezoar, feces, foreign bodies

• **INTRINSIC LESIONS** Atresia, stenosis, strictures due to neoplasms, inflammation, chemicals, vascular lesions

• **EXTRINSIC LESIONS** Adhesive bands from C-sections, previous surgery, hernias, neoplasia, abscesses, volvulus, hematomas

iliotibial band syndrome Tensor fasciae latae syndrome, TFL syndrome SPORTS MEDICINE A common running injury, which is the most common cause of lateral knee pain in runners MUSCLES INVOLVED Gluteus maximus, tensor fasciae latae CLINICAL Lateral knee pain, a quasi-pathognomonic finding, often worse after running, especially after climbing hills or stairs; may be associated with a 'snapping hip', in which muscles crossing the outside of the hip can be felt to snap or click during walking or running ETIOLOGY Over-training, attributed to recurrent friction of the iliotibial band sliding over the lateral femoral epicondyle pulling on the lateral insertion near the knee bursa, which becomes inflamed MANAGEMENT Acute phase–↓ activity, ice massage, NSAIDs, phonophoresis, steroid injections for refractory swelling; rehab–progressive stretching, sprints, gradual ↑ to running distance. See Running

Ilizarov technique Transosseous osteosynthesis Any of a group of techniques for bone regeneration, in which a long bone, usually of the lower extremity is intentionally fractured and held in place for a short period or time, then slowly distracted with an external fixator APPLICATIONS Bone defects–eg, discrepancies in leg length, dwarfism, fracture nonunions, osteomyelitis, angular defects

illegal use of controlled dangerous substances SUBSTANCE ABUSE The use of controlled dangerous substances–eg, cocaine, heroin obtained illegally, or the use of controlled dangerous substances not obtained pursuant to a valid prescription or taken according the directions of a licensed health care practitioner

illegitimacy Nonmarital, out of wedlock SOCIAL MEDICINE The legal status of a child born to unwed parents

illicit drug Street drug, see there

illiteracy SOCIAL MEDICINE The inability to read and write. See Complete illiteracy, Functional illiteracy

illness The state of being unwell, a term used by regulatory agencies–eg, the FDA, which modifies 'illness' with certain adjectives, in order to allow Pts to receive experimental drugs that do not have FDA approval. See Life-threatening illness, Severely debilitating illness

illusion NEUROLOGY A misperception of a real external stimulus–eg, rustling of leaves interpreted as the sound of voices. See Hallucination VOX POPULI A false image. See Japanese illusion

illustration MEDICAL COMMUNICATION A photograph, print, drawing, plate, diagram, facsimile, map, table, or other representation or systematic arrangement of data designed to elucidate or decorate the contents of a publication. See National Library of Medicine

ILO classification International Labour Office Classification of Radiographs of Pneumoconiosis PULMONARY MEDICINE A classification of black lung disease. See Coal worker's pneumoconiosis

iloperidone Zomaril™ An antipsychotic for treating schizophrenia and related disorders

IM Intramuscular, also 1. *Index Medicus* 2. Infant mortality 3. Infectious mononucleosis 4. Intermetatarsal 5. Internal medicine 6. Intramedullary

image See Afterimage, Body image, Eidetic image, Fictive image, Glamor image, Perceptual image, Professional image

IMAGE CARDIOLOGY A clinical trial–International Multicenter Angina that evaluated medical management to reduce ischemia in Pts with stable angina. See Metoprolol, Nifedipine, Stable angina

image aversion therapy PSYCHOLOGY An aversion therapy in which a person attempting to modify a particular behavior imagines something bizarre or unsettling, associated with carrying out the undesired behavior–eg, eating a huge meal in front of a malnourished person from a developing nation. See Aversion therapy, Behavioral therapy, Encounter group therapy, Flooding, Systematic desensitization

ImageChecker™ IMAGING A hardware and software hybrid that reviews mammograms to identify microcalcifications, masses or tissue distortions which may represent CA; IC improves radiologic detection of CA from 80% to 88%. See Mammography

image-directed surgery SURGICAL ONCOLOGY Surgical excision of a nonpalpable malignancy diagnosed by image-directed biopsy or needle localization and excision

image management INFORMATICS The rearrangement and alteration of image files after storage

Imagent® AF0150 IMAGING An investigational perfluorochemical-based IV contrast to improve ↑ detection of lesions and defects of myocardial perfusion and cardiac function

imagery PSYCHOLOGY The evoking of a visual, audio, or other internalized mental image, that retains the 'flavor' and sensory qualities of an original external stimulus; a technique in which a person focuses on positive mental images. See Chemical imagery, Fictive imagery, Guided imagery, Hypnagogic imagery, Interactive guided imagery, Memory imagery, Perceptual imagery, Relaxing imagery

IMAGES NEUROLOGY A clinical trial–Intravenous Magnesium Efficacy in Stroke

imaging The term is used in 2 different areas of diagnostic medicine RADIOLOGY The production of non-invasive images of body regions using ionizing radiation–eg, CT or mammography, or electromagnetic radiation–eg, MRI or ultrasonography, with/without radiocontrast; the information obtained is then analyzed by a computer to produce a 2-D display TYPES OF INFORMATION PROVIDED Anatomic–CT, MRI, mammography, ultrasonography, metabolic–PET, SPECT-single photon emission CT or data on electrical activity–SQUID. See Brain imaging, Cardiac blood pool imaging, Contrast imaging, Diffusion-weighted imaging, Digital imaging, Document imaging, Doppler sonographic imaging, Echo planar imaging, Fast CT imaging, Fluorescence imaging, fMRI imaging,

4-D imaging, Freeze-fracture imaging, ¹H imaging, Harmonic ultrasound imaging, Magnetic resonance imaging, Molecular imaging, Multiband imaging, Multiple plane imaging, Myocardial perfusion imaging, Native™ tissue harmonic imaging, Neuroimaging imaging, ³²P imaging, Perfusion-weighted imaging, Real-time imaging, Sequential plane imaging, Somatostatin-receptor imaging, Spin-echo imaging, SQUID imaging, Structural imaging, Transient response imaging, Volume imaging

imaging center A facility with the equipment to produce various types of radiologic and electromagnetic images and a professional staff to interpret the images obtained

IMAO Inhibitor of monoamine oxidase, see MAOI

imatinib mesylate Gleevec, STI571 MOLECULAR ONCOLOGY A protein tyrosine kinase inhibitor that targets PDGF receptor, inhibits the fusion product of Philadelphia chromosome–BCR-ABL tyrosine kinase, and targets c-kit, a protein tyrosine kinase ADVERSE EVENTS Superficial edema, nausea, muscle cramps, weight gain, splenic rupture. See Platelet-derived growth factor

Imatron See Ultrafast CT

imbalance A loss of equilibrium or homeostasis. See Chemical imbalance

imciromab pentetate MyoScint CARDIOLOGY An indium-labelled monoclonal antibody that binds the heavy chain of cardiac myosin, which is exposed after loss of the myocyte membrane, as occurs in acute MI and is detected using immunoscintigraphy. See Monoclonal antibody

IMDD Idiopathic midline destructive disease, see there

IMG International medical graduate, see there

iminoglycinuria Familial iminoglycinuria MOLECULAR MEDICINE A benign AR condition characterized by defective tubular resorption and urinary spilling of proline, hydroxyproline and glycine; iminoglycinuria is a normal physiologic event that occurs in neonates, whose renal transport mechanisms are immature; iminoglycinuria may also occur in Fanconi syndrome and hyperprolinemia. See Hyperglyinuria

imipramine NEUROPHARMACOLOGY A dibenzazepine class tricyclic antidepressant used for depression, enuresis ADVERSE EFFECTS Insomnia, numbness, tingling, tremors, seizures, dry mouth, blurred vision

Imiprex Imipramine, see there

Imitrex® Sumatriptan succinate, see there

immature teratoma Malignant teratoma A teratoma that has mature elements from all 3 germ cell layers–ectodermal, mesodermal, and endodermal, and immature tissues, most commonly of neural origin; the neural tissues in ITs consist of primitive neurotubules, neuroepithelial rosettes, immature glial elements, immature ependyma

immediate action values Critical results, see there

immediate adjunctive coronary angioplasty CARDIOLOGY A procedure in which an acute MI-related artery is dilated after administering thrombolytics. See Coronary angioplasty, Thrombolytic therapy

immediate cause of death '…*the disease, injury, or complication that directly precedes death*', which is *the ultimate consequence of the underlying cause* of death. Cf Underlying cause of death

immediate family A person's spouse, children, siblings and parents, spouse's siblings and parents, and spouses of person's children

immediate family member MEDICAL PRACTICE A person's spouse, child, child's spouse, stepchild, stepchild's spouse, grandchild, grandchild's spouse, parent, stepparent, parent-in-law, or sibling. See Next-of-kin

immediate hypersensitivity Immediate hypersensitivity reaction IMMUNOLOGY An event unleashed by cross-linking of IgE antibodies on the surfaces of mast cells and basophils, after exposure to an allergen to which a person has been previously sensitized, resulting in release of histamine and related compounds

immediate post-op prosthesis ORTHOPEDICS A prosthesis fitted to a residual limb immediately after surgery, before the Pt wakes up, followed by ASAP physical therapy. See Prosthesis

immediate reconstruction SURGERY Cosmetic reconstruction of the breast at the same time as a mastectomy

immediately dangerous to life or health PUBLIC HEALTH The maximum environmental concentration of a contaminant from which one could escape within 30 mins, without suffering escape-impairing Sx or irreversible health effects

immobilized enzyme An enzyme fixed by physical or chemical means to a solid support–eg, a bead or gel to confine a reaction of interest to a particular site

immotile cilia syndrome Kartagener syndrome An uncommon–1:20 000 AR disease of childhood onset due to defective or afunctional cilia in the respiratory tract, resulting in chronic sinusitis, defective mucociliary transport and bronchial clearance, bronchiectasia, chronic otitis media and incapacitating headaches, related to immotility of ependymal cilia in the walls of cerebral ventricles; other defects associated with IC include cardiovascular, renal and ocular defects, and absence of frontal sinuses REPRODUCTION ♂ are infertile, ½ of ♀ are impregnatable, the other ½ sterile; ½ have Kartagener's triad, ie chronic sinusitis, bronchiectasis and situs inversus totalis, the last of which may be due to an in utero defect with contrary beating cilia. See Cilia

immune competence Competence IMMUNOLOGY The ability of the immune system to respond appropriately to an antigenic stimulation, and unleash an immune response 'cascade'. Cf Anergy, Antigenic competence

immune complex IMMUNOLOGY A molecular aggregate formed between antigens and antibodies. See Immune complex disease

immune complex disease IMMUNOLOGY Any of a number of conditions—caused by circulating antigen-antibody-immune complexes, which in the face of mild antigen excess, lodge in small vessels and filtering organs of the circulation CLINICAL Fever, enlarged and/or tender joints, splenic congestion, proteinuria due to glomerular IC deposition, eosinophilia, hypocomplementemia, lymphadenopathy, glomerulonephritis–HTN, oliguria, hematuria, edema, skin–purpura, urticaria, ulcers, carditis, hepatitis, myositis, necrotizing vasculitis. See Immune complex

IMMUNE COMPLEX DISEASE

ARTHUS REACTION Acute hemorrhagic necrosis that follows re-exposure to an antigen, which attracts PMNs, activates complement, binds ICs by the Fc receptor, causing phagocytosis, ↑ production of chemotactic factors, especially C5b67 and ↑ anaphylatoxins C3a and C5a, resulting in vasodilation

SERUM SICKNESS A reaction that is milder than the Arthus reaction, occurring 8-12 days after exposure to the antigen, at the time of 'equivalence' (antigen and antibody are in a 1:1 ratio), after injection of a foreign protein mixture, eg horse serum for antitoxin to tetanus

immune hemolytic anemia HEMATOLOGY Premature RBC destruction due to circulating antibodies, which may follow pregnancy, transfusion, or drug exposure. See Autoimmune hemolytic anemia

immune response IMMUNOLOGY The constellation of responses of the immune system to foreign antigenic stimuli

immune thrombocytopenic purpura Idiopathic thrombocytopenic purpura, see there

immunity

328 **immunity** 1. A state in which a host is not susceptible to infection or disease 2. The mechanisms by which this is achieved. Immunity is achieved by an individual through one of three routes: natural or innate immunity genetically inherited or acquired through maternal antibody, acquired immunity conferred after contact with a disease, and artificial immunity after a successful vaccination Also termed specific immunity, resistance or specific resistance, specific immunity is divided into cellular immunity, acting via the direct involvement of T cells and humoral immunity involving antibodies and B cells. See Acquired immunity, Active immunity, Adoptive immunity, Allograft immunity, Cell-mediated immunity, Herd immunity, Maternal immunity, Mucosal immunity, Natural immunity, Passive immunity, Sterilizing immunity, Superinfection immunity

immunization IMMUNOLOGY The process of inducing immunity by administering an antigen to allow the immune system to prevent infection or illness when it subsequently encounters the same pathogen. See Adult immunization, Alloimmunization, Anthrax immunization, Childhood immunization, Intracellular immunization, Passive immunization, Vaccination

immunoassay LAB MEDICINE Any assay that measures an antigen-antibody response, the sensitivity of which varies according to the method; an immunologic test that quantifies a particular antigen using a cognate antibody. See ELISA, RIA

immunocompetent *adjective* Capable of developing an immune response; possessing a normal immune system. Cf Immunoincompetent

immunocompromised *adjective* IMMUNOLOGY Having an impaired immune responsiveness caused by acquired or congenital immunodeficiency, infection—eg, HIV, TB, or chemotherapy

immunodeficiency IMMUNOLOGY Any partial or complete, congenital or acquired defect in the immune responsiveness. See Adenosine deaminase deficiency, Combined immunodeficiency, Combined variable immunodeficiency, Immunodeficiency disorder, Purine nucleoside phosphorylase deficiency, Severe combined immune deficiency, Radial immunodeficiency

IMMUNODEFICIENCIES-GENERAL TYPES

ACQUIRED IMMUNODEFICIENCY A generalized ↓ in the immune response to antigenic stimuli due to various conditions eg aging, AIDS, Alzheimer's disease, amyotrophic lateral sclerosis, burns, chemotherapy, coeliac disease, corticosteroids, depression and mental stress, IBD, leprosy, NSAIDs, radiation, sarcoidosis, sepsis, hematologic and lymphoproliferative disease—Hodgkin's disease, NHL, leukemia, myeloma, Waldenström's macroglobulinemia, aplastic and agranulocytic anemias, sickle cell disease, systemic disease—malnutrition, chronic diarrhea, fulminant mycosis, sepsis, terminal cancer, DM, uremia and nephrotic syndrome, splenectomy, surgery and trauma

CONGENITAL IMMUNODEFICIENCY A heterogeneous group of relatively rare diseases which may be accompanied by autoimmune disease, allergy, ↑ incidence of malignancy, GI abnormalities INCIDENCE IgA deficiency 1:500, agammabulinemia 1:50,000, severe combined immunodeficiency 1:100,000

immunodeficiency disorder Immunodeficiency syndrome CLINICAL IMMUNOLOGY Any condition characterized by a defect in immune system, resulting in an inability to produce an appropriate immune response. See AIDS, CVID, SCID

immunoglobulin A highly-specific molecule of the immune system, produced by mature B cells in response to an antigen STRUCTURE 2 identical light-L, 2 identical heavy–H chains; the L and H chains have constant and variable regions, the variable regions are critical for antigen recognition and binding; immunoglobulin production requires prior rearrangement of the variable, diversity and joining gene segments, that form part of a potential repertoire of 10^{10}-10^{12} antibodies that may be encoded in response to a molecule's surface binding site or epitope TYPES Idiotype–evoked by a particular epitope; isotype–Ig subtype–IgG, IgA, IgM, IgD, IgE that all normal persons have; allotype–a subtype shared by population groups–eg, with racial differ-

ences. See Hinge, Polyclonal immunoglobulin, Protein electrophoresis, Sporidin-G™ Bovine immunoglobulin

immunoglobulin A deficiency Selective immunoglobulin A deficiency An AD condition that is the most common–1:600 in US–primary immune system disease, characterized by ↓ IgA to < 50 mg/L–normal 760-3900 mg/L; US 76-390 mg/dl CLINICAL ↓ immune function, ↑ RTIs, ↑ GI infections, autoimmune disease–eg, SLE, RA, arthritis; fatal hemolysis with blood transfusions, due to production of natural anti-IgA antibodies; other features include intestinal lymphangiectasia, gluten-sensitive enteropathy, allergies, 7S IgM antibodies to food, especially to milk TREATMENT IV immune globulin may be effective in some Pts

immunoglobulin M deficiency syndrome IgM deficiency syndrome A rare condition, often without clinical consequences, that may have ↓ complement activation and ↑ susceptibility to RTIs, sepsis, tumors, associated with deficiency in 5-ectonucleotidase on the B-cell surface

immunology MEDTALK The study of the body's immune system. See Clinical lab immunology, Clinical & lab immunology, Neuroendocrinoimmunology, Neuroimmunology, Psychoneuroimmunology

immunomodulatory drug CLINICAL PHARMACOLOGY A therapeutic agent that suppresses the immune system, inhibiting lymphocyte functions, especially T and NK cells; IDs are used to treat IBD–ulcerative colitis, Crohn's disease EXAMPLES Azathioprine, mercaptopurine ADVERSE EFFECTS BM suppression, pancreatitis; ↑ lymphoma; cyclosporine ADVERSE EFFECTS Renal dysfunction; MTX ADVERSE EFFECTS BM suppression, liver toxicity

immunoscintigraphy IMAGING A procedure in which radioactively-labeled substances are administered and images obtained from body sites where the antibody localizes; the scintigraphic imaging of a CA after administrating a γ-emitting monoclonal antibody directed against a particular tumor-specific antigen

immunosuppressive therapy IMMUNOLOGY The use of immunosuppressants to suppress immune responsiveness–eg, to ↓ allograft rejection AGENTS Alkylating agents, toxic antibiotics, ionizing radiation TUBERCULOSIS Therapy that suppresses, or weakens, the immune system

immunotherapy Biological response modifier therapy, biological therapy, hyposensitization therapy ALLERGY MEDICINE 1. A therapy in which an allergen–eg, hymenopteran–bee, wasp venom, is administered in ↑ doses to those with potentially fatal hypersensitivity thereto; IT elicits production of blocking IgG antibodies, interferes with antigen-Fab part of Ig binding, prevents fixation of IgE–which causes anaphylaxis, down-regulates T-cell responses, inhibits inflammatory responses to allergens, and attenuates anaphylactic reactions; IT in Pts with seasonal ragweed-exacerbated asthma and allergic rhinitis evokes improvement of Sx that is not sustained with time. See Active immunotherapy, Adoptive immunotherapy, Allergen immunotherapy, Venom immunotherapy 2. A treatment to stimulate or restore a person's immune system's ability to fight infection and disease, or ameliorate the adverse effects of chemotherapy. See Biological response modifier ONCOLOGY A therapy that nonspecifically stimulates the immune system to destroy malignant cells; some success is reported with BCG IT–which may be effective in treating melanoma, AML, solid tumors; others include Coley's toxin and heat-killed formalin-treated *Corynebacterium parvum*, an immunopotentiator and immunomodulator that evokes reticuloendothelial hyperplasia, stimulation of macrophages and B cells and which may enhance T-cell function. See BCG, Coley's toxin, Malariotherapy. Cf Immunoaugmentive therapy

IMPACT CARDIOLOGY A clinical trial 1. Integrilin™ to Manage Platelet Aggregation to Prevent Coronary Thrombosis 2. International Mexiletine or Placebo Antiarrhythmic Coronary Trial PUBLIC HEALTH Initiatives to Mobilize for the Prevention and Control of Tobacco use A CDC-sponsored anti-tobacco initiative. See Smoking cessation, Tobacco control

IMPACT II CARDIOLOGY A clinical trial–Integrelin to Manage Platelet Aggregation to Combat Thrombosis–that evaluated the efficacy of integrilin as an adjunct to coronary intervention. See Acute myocardial infarction, Heparin, Hirulog, Unstable angina

impact boosting TOBACCO CONTROL *adjective* Referring to a tobacco-industry maneuver to ↑ the mental "jolt" received by smokers, caused by adding ammonia to cigarettes. See Smoking

impacted fracture ORTHOPEDICS A fracture in which parts of a bone "telescope" into bone across the fracture line. See Fracture

impacted tooth Dental impaction, unerupted tooth DENTISTRY An unerupted or partially erupted tooth positioned against another tooth, bone, or soft tissue so that complete eruption is unlikely; IT is most common in the 3rd molars COMPLICATIONS Misalignment of adjacent teeth, eventually causing the bite to shift; partial ITs can trap food, plaque, and debris in the surrounding soft tissue, leading to inflammation and tenderness of gums and halitosis, formally, pericoronitis

impaired health care provider A physician or health care worker whose ability to function in his usual role has been reduced or otherwise compromised by various internal and external forces. See Impairment. Cf 'Handicapped' health care provider, 'Incapacitated' health care provider

impaired memory Dementia, see there

impaired speech Dysarthria, see there

impaired taste Dysgeusia, see there

impairment MEDTALK An objective handicap, partial disability, loss of function, anatomic or functional defect, which may be temporary or permanent–persisting after appropriate therapy, without reasonable prospect of improvement, ranging from mild to severe, the latter of which precludes any form of gainful employment. See Disability, Handicap, Hereditary hearing impairment, Nonsyndromic hereditary hearing impairment, Specific language impairment, Syndromic hearing impairment. Cf Disability, Incompetence

Impedance CARDIAC PACING The total opposition that a circuit presents to alternating electrical current. Cf Resistance

impedance plethysmography CARDIOVASCULAR DISEASE A noninvasive method that measures changes in electrical resistance between 2 probes, which indicates changes in the volume of different regions of the body, as may be seen in obstruction to venous outflow. See Deep vein thrombosis

imperforate anus Anal atresia, anorectal atresia GI SURGERY A congenital obstruction of the anal opening which affects 1:5000 children; in some Pts there may be communication between the GI tract and vagina and bladder MANAGEMENT Surgical anastomosis

impertinent *adjective* Not pertinent; irrelevant

impetigo DERMATOLOGY A contagious streptococcal or staphylococcal infection that erodes the skin and dries to form a yellow-crusted sore; impetigo in children is associated with poor hygiene; it may follow RTIs in adults CLINICAL Pruritus, blisters, oozing, tightly adherent crusting, which tends to spread and form deep ulcers

impingement NEUROLOGY Compression of a nerve or blood vessel through a constricted space. See Nerve root impingement

impingement syndrome REHAB MEDICINE A condition caused by the limiting of space between bones and fascia, compromising blood flow and irritating nerves passing through the space EXAMPLE Carpal tunnel syndrome which affects middle-aged ♀; shoulder IS, where the space beneath the coraco-acromial arch for the supraspinatus and biceps tendons is ↓, resulting in a painful arc of movement and paresthesias, common in competitive swim-

mers MECHANISM Ischemia due to vascular stenosis or an osteophyte rubbing the acromium, which is common in throwing, serving, and other sports. See Carpal tunnel syndrome

implant A material that is not original to its location. See Baerveldt glaucoma implant, Bioimplant, Dental implant, Molteno implant, Penile implant, Silicone implant, Visual implant AUDIOLOGY See Cochlear implant BREAST SURGERY A silicone shell shaped in the form of a breast and filled with saline or silicone gel DENTISTRY A titanium device–eg, Brånemark system, Nobelpharma that is surgically placed in the mandible or maxilla, allowed to 'fuse' in 3 to 6 months to the bone–osseointegration, which serves as an anchor for attaching artificial teeth ONCOLOGY Tumor implant, see there SURGERY *noun* A device that is inserted to preserve or maintain a function–eg, a hip or knee prosthesis, or to preserve, enhance or alter a contour–eg, a breast or chin implant *verb* To surgically place such a device in its appropriate site. See Breast implant, Permanent adjustable implant

implant radiation therapy Brachytherapy, see there

implantable cardioverter-defibrillator CARDIOLOGY A device implanted subcutaneously–anterior chest wall, which provides automatic electrical impulses in Pts with severe–life-threatening or treatment refractory ventricular tachycardia or fibrillation. Cf Ablation therapy

implantable contact lens OPHTHALMOLOGY A refractive lens implanted over the natural lens of the eye for correcting myopia and farsightedness. See Myopia

implantable hearing aid ENT An electromagnetic or piezoelectric device used to correct hearing loss in as 'physiologic' a manner as possible; IHAs may improve sound fidelity in person with sensorineural hearing loss

implantable pulse generator CARDIAC PACING A pacemaker used for permanent pacing, which is placed inside a pocket under the skin; the leads are positioned in or on the heart

implantable pump THERAPEUTICS A device installed under the skin to administer a steady dose of drugs

implantation CARDIAC PACING The permanent placement of a pacemaker in a tissue pocket under the skin, with leads positioned in or on the heart

implantation bleeding OBSTETRICS A minimal hemorrhage seen at the time of implantation of the egg, attributed to angiogenesis

implantation recurrence SURGICAL ONCOLOGY A rare phenomenon at the site of a surgical resection of CA, where tumor cells, often epithelial, 'spill' from an operative field or are carried by needles or stapling devices, 'seeding' malignancy into previously uninvolved sites

implement MEDTALK Do

impotence Ejaculatory incompetence, erectile difficulty, erectile dysfunction, erectile failure, frigidity–female MEDTALK The inability to achieve or maintain a penile erection adequate for the successful completion of intercourse, terminating in ejaculation; penile erection is mediated by nitric oxide EPIDEMIOLOGY Prevalence of minimal, moderate, and complete impotence in the Massachusetts Male Aging Study was 52%; age is the most important factor; complete impotence ↑ from 5%–age 40 to 15%–age 70; for an erection to achieve a successful outcome, it requires 1. An intact CNS, ie without underlying medical or psychological disease, or central-acting drugs, intact sympathetic and parasympathetic circuitry, ie without spinal trauma or degenerative disease 2. An intact vascular supply to the penis and 3. An intact, anatomically correct penis; 25% of impotence may be psychologic or 'partner-specific', 25% has an organic component and 50% of impotence is organic in nature; in organic impotence, nocturnal penile tumescence is absent MANAGEMENT-SURGICAL Microvascular surgery to bypass occluded vessels–most effective in younger ♂, penile prosthesis MANAGEMENT–MEDICAL Combined therapy with phentolamine and papaverine–self-injected by the Pt, wielding an erection of 1 hr's duration is useful for arterial, neurologic, psychogenic impotence; other therapies–zinc, bromocriptine–Parlodel, isoxsuprine-Vasodilan, Voxsuprine, nitroglycerine, yohimbine–Yocon, Yohimex ETIOLOGY Smoking, CAD, HTN, DM, medications–hypoglycemic agents, vasodilators, cardiac drugs, antihypertensives, anger and depression; it is inversely correlated to dehydroepiandrosterone, HDL-C, and an index of dominant personality PRIMARY IMPOTENCE Complete absence of successful sexual coupling SECONDARY IMPOTENCE Priapism, penile plaques, Peyronie's disease; drugs linked to impotence: antihypertensives–eg, methyldopa, guanethidine, reserpine, clonidine, due to ↓ BP, antidepressants–eg, phenelzine, isocarboxazide, amitriptyline–causing altered moods and decreased libido, tranquilizers–eg, chlordiazepoxide and lorazepam, and the muscle-relaxing diazepam, cimetidine, which ↑ prolactin, and is associated with impotence and loss of libido. Cf Infertility, Orgasmic dysfunction

impotent MEDTALK *adjective* Referring to impotence

IMPRESS CARDIOLOGY A clinical trial–Inhibition of Metalloprotease by BMS-186716 in a Randomized Exercise and Symptoms Study

Impress™ Softpatch UroMed incontinence patch UROGYNECOLOGY A foam pad worn externally to help prevent urine leakage in ♀ with stress incontinenc. See Stress incontinence

Impril Imipramine, see there

imprimatur VOX POPULI Sanction; approval

imprinting MOLECULAR BIOLOGY The variable phenotypic expression of a gene, depending on paternal or maternal origin, a function of the methylation pattern; imprinted regions are more methylated and less transcriptionally active; the 'imprints' are erased and generated in early embryonic development of mammals EXAMPLES Insulin-like growth factor-2 and its receptor, fragile X syndrome, Prader-Willi syndrome, Angelmann/happy puppet syndrome, Wilms' tumor. See Allele, Gene, Genomic imprinting, Inheritance, Locus PSYCHOLOGY Developmental learning restricted to certain early, critical or sensitive time periods of life, which stops when definitive learning occurs or when a critical period has passed; it is irreversible and characteristic of the species of organism being imprinted. See Critical period

impulse CARDIAC PACING An electrical stimulus delivered by a pacemaker PSYCHIATRY A psychic striving; an instinctive urge

impulse control PSYCHOLOGY The degree to which a person can control the desire for immediate gratification or other; IC may be the single most important indicator of a person's future adaptation in terms of number of friends, school performance and future employment. See BarOn test, Emotional intelligence, Marshmallow test

impulse control disorder PSYCHIATRY A psychiatric disorder characterized by a failure to resist an impulse, drive, or temptation to perform an act that is harmful to the person or to others EXAMPLES Compulsive gambling, pyromania

IMRT Intensity-modulated radiation therapy RADIATION ONCOLOGY A format for delivering high-dose RT to regions–eg, nasopharynx, that are surrounded by radiation-sensitive areas; in IMRT, a broad radiation field is divided into hundreds of small pencil beams, the intensity of which is determined by computer optimization

ImuLyme™ *Borrelia burgdorferi* outer surface protein A INFECTIOUS DISEASE A vaccine to prevent or ↓ occurrence of Lyme disease. See Lyme disease

Imuran® Azathioprine, see there

IN Intraepithelial neoplasia, see these

in denial PSYCHIATRY To be in a state of denying the existence or effects of an ego defense mechanism. See Denial

in extremis At the point of death; circling the drain

In-Fast system UROGYNECOLOGY A bone screw system for incisionless, transvaginal cystourethropexy and sling procedures for ♀ urinary incontinence. See Incontinence

In-Flow catheter UROGYNECOLOGY A remote-controlled valved intraurethral valved catheter for treating ♀ urinary incontinence in atonic bladder. See Incontinence

in house *adjective* Within the confines of a campus, complex of buildings, health care center, hospital, or other institution, as in a test performed 'in house', ie not referred to an independent lab

in-line skating injury Rollerblade™ injury EMERGENCY MEDICINE An injury occurring in a person using in-line skates, most commonly Fx of the distal radius

in network MEDTALK *adjective* Referring to the inclusion of a provider in a health care network

in & out procedure CLINICAL MEDICINE A popular term for any diagnostic or therapeutic procedure–eg, bronchoscopic suctioning of a critically ill, intubated Pt, which is performed as quickly as possible, to minimize morbidity. Cf 'Quick and dirty,' 'Scoop and run'

in-service In-service training *adjective* Referring to any form of on-the-job training *noun* In-service training of an employee

in-shoe device BIOMECHANICS An instrument used to measure the interface between the plantar surface of the foot and the superior surface of the foot orthosis

in situ Latin, in the original place

in situ carcinoma See Carcinoma in situ

in situ hybridization A method for localizing a sequence of DNA, mRNA, or protein in a cell or tissue; the use of a DNA or RNA probe to detect a cDNA sequence in chromosome spreads or in interphase nuclei or an RNA sequence of cloned bacterial or cultured eukaryotic cells; ISH is used to map gene sequences to chromosomal sites and detect gene expression. See cDNA, Chromosome, CISH–Chromogenic in situ hybridization, DNA sequence, Eukaryote, FISH–Fluorescent in situ hybridization, GISH–Genomic in situ hybridization, Gene expression, Gene mapping, Hybridization, Immunoblotting, Interphase, Probe. Cf Southern blot

in-stent restenosis INTERVENTIONAL CARDIOLOGY Scar-induced reclosure of a previously stenosed coronary artery, a complication seen in ±20% of Pts undergoing stent placement for CAD. See Coronary artery disease, Stent. Cf Late stent thrombosis

In-Tac anchoring system UROGYNECOLOGY A technique for incisionless cystourethropexy procedure for managing ♀ incontinence, performed transvaginally with a bone anchor inserter stapler. See Cystourethropexy

in utero Occurring in the uterus or fetus before birth. Cf *in vitro*

in vitro Isolated from living organisms or systems, but artificially maintained; in a test tube, in glass; outside the body. Cf In vivo

in vitro fertilization REPRODUCTION MEDICINE A procedure that bypasses some causes of infertility, by removing eggs from the ovaries and fertilizing them ex vivo; the resulting early embryo is then transferred into the uterus via the cervix. See Assisted–'artificial' reproduction, Infertility, Surrogate motherhood, Test tube baby

in vivo Latin In a living organism or system. Cf In vitro

inability to conceive OBSTETRICS Infertile, see there VOX POPULI Inconceivable

inability to perform useful movements Dyspraxia, see there

inability to sleep Insomnia, see there

inability to speak Speech impairment, see there

inaccessible SURGERY *adjective* Unreachable; referring to a lesion that unmanageable by standard surgical techniques—eg, lesions deep in the brain or adjacent to vital structures—ie, not accessible. See Accessible

inactive arm Control arm, see there

inactive ingredient Additive, Excipient CLINICAL PHARMACOLOGY A substance regarded by the FDA as having no effect on a drug's absorption or metabolism, which is added for manufacturer expediency. Cf GRAS substances

inactive residual extremity syndome REHAB MEDICINE A pain complex in an amputee who, despite a well-fitted prosthesis and good surgical outcome, complains of pain, swelling, instability, and difficulty in wearing the prosthesis for extended time MANAGEMENT Ertl procedure. See Ertl procedure, Phantom limb syndrome

inactivity Sedentary activity INTERNAL MEDICINE An absence of physical activity and/or exercise, a predictor of obesity. See Couch potato. Physical activity, Vigorous exercise

inadequate nutrition Malnutrition, see there

inappropriate MEDTALK *adjective* A diagnostic or therapeutic procedure proven to be unnecessary for the efficient management of a particular Pt. See Appropriateness, Canadian plan, Practice guidelines NEUROLOGY *adjective* Referring to a response or behavior discordant with a particular situation

inappropriate affect PSYCHIATRY A display of emotion that does not reflect a Pt's reality

inappropriate care Care which, according to the RAND Corporation, is defined as '...*that for which the expected risks or negative effects significantly exceed the expected benefits for the average patient with a specific clinical scenario.*'

inborn error of metabolism Any of the expanding group, now in the hundreds, of inherited metabolic and biochemical disorders, that are divisible into those affecting 1. Small molecules—eg, simple sugars, amino or organic acids, that often have an acute onset in infancy/childhood; 2. Large molecules—eg, ↑ in 'storage diseases'—eg, mucopolysaccharidoses and glycogen storage diseases that affect older children

INBORN ERRORS OF METABOLISM CONSEQUENCES

LOSS OF CERTAIN MOLECULES—eg, albinism—defect of tyrosinase or Ehlers-Danlos disease—defect of lysyl-hydroxylase or others of a vast array of enzymes

ACCUMULATION OF NORMAL METABOLITES—eg, alkaptonuria—defect of homogentisic acid oxidase or galactosemia—defect of galactose-1-phosphate uridyl transferase

TRANSPORT DEFECTS—eg, cystinuria—dibasic amino acids or intestinal disaccharidase deficiency

DEFECTS IN ERYTHROCYTE METABOLISM—eg, glucose-6-phosphate dehydrogenase deficiency

PIGMENT DEFECTS—eg, acute intermittent porphyria

DEFECTS IN MINERAL METABOLISM—eg, Wilson's disease

VITAMIN DEFECTS—eg, vitamin D-dependent rickets

DEFECTS IN INTESTINAL ABSORPTION—eg, cystic fibrosis

OTHER DEFECTS OF UNKNOWN ORIGIN—eg, achondroplasia

incapable of making an informed decision CHOICE IN DYING The inability of an adult Pt, because of mental illness or retardation, or other mental or physical disorder that precludes communication or impairs judgement, which has been appropriately diagnosed and certified in writing by 2 or more health professionals, to make an informed decision about providing, withholding, or withdrawing a specific medical treatment because he cannot understand the nature, extent, or probable consequences of a proposed medical decision, or to make a rational evaluation of the risks and benefits of alternatives. See Informed consent, Informed decision

incapacitated health care provider A physician or health care worker who is physically or mentally impaired in the ability to provide Pt care. Cf 'Handicapped' health care provider, 'Impaired' health care provider

incapacity MEDTALK The inability to understand information presented, appreciate the consequences of acting—or not acting on that information, and to make a choice. See Incompetence

incarcerated hernia The herniation of a tissue, classically a loop of intestine, into a mesothelial sac, that cannot return to its original position without surgery. Cf Strangulated hernia

incardia valve system HEART SURGERY A system for performance of minimally invasive heart valve implantation surgical procedures. See Heart valve surgery

incentive arrangement MANAGED CARE A formal arrangement between a payer and a health care provider in a position to make therapeutic decisions or recommendations for potentially expensive Pt management. See Kickback

incentive spirometry PULMONOLOGY A form of spirometry which may prevent the complications—eg, atelectasis and infiltrates—of acute chest syndrome in sickle cell anemia. See Acute chest syndrome

INCERT SURGERY Bioabsorbable, implantable sponge designed to prevent postsurgical adhesion formation. See Adhesions

incest VOX POPULI Sexual intercourse among close kin—eg, brother/sister, parent/offspring, first cousins, based on genealogic or totemic descent, or by reason of marriage or adoption; incest is illegal in most societies. See Conguinity

incidence EPIDEMIOLOGY 1. The number of new cases—in the form of a count or rate of a disease or condition, often an infection diagnosed each yr—classically measured as an attack rate 2. The rate of occurrence of new cases of a disease or condition in a population at risk during a given period of time, usually 1 yr

incidence rate EPIDEMIOLOGY A measure of the frequency with which an event—eg, a new case of illness, occurs in a population over a period of time; denominator is the population at risk; numerator is the number of new cases occurring during a given time period. See Incidence

incident HEALTH INSURANCE An event or happening that causes unanticipated harm to a Pt HOSPITAL ADMINISTRATION Variance A marked negative deviation—eg, major substitution of medications or leaving a Pt unattended for a prolonged period of time, from the 'standard of care,' occurring in a health care facility. Cf Misadventure VOX POPULI An event. See Berkeley Lab incident, the Mouse incident, Multicasualty incident

incident to MANAGED CARE *adjective* Referring to non-physician services or supplies furnished as an integral but incidental, part of a physician's professional services in the course of diagnosing or treating an injury or illness, usually under a physician's direction. See Ancillary services

incidentaloma MEDTALK An incidentally discovered mass or lesion, detected by CT, MRI, or other imaging modality performed for an unrelated reason. See Pathologist's tumor. Cf Ulysses syndrome

incision & drainage SURGERY The incising of an abscess to allow pus to drain

incision SURGERY A cut in the body during surgery. See Fishmouth incision, Kerr incision, Kronig-Selheim incision, Lazy-S incision, Mercedes-Benz incision, Pfannenstiel incision

incisional biopsy A procedure to remove part of a lesion or mass for pathologic evaluation. See Biopsy. Cf Excisional biopsy

inclusion body myositis A type of idiopathic myositis that is not autoimmune and does not respond to immunosuppressive therapy, a clinical diagnosis of exclusion, confirmed by typical histologic features CLINICAL Slowly progressive disease of middle-aged ♂, beginning in legs, causing atrophy and weakness of quadriceps, sparing facial and oropharyngeal muscles EMG Abnormal electrical 'irritation', slowing of nerve conduction and ↑ wave amplitude

inclusion cyst DERMATOLOGY Epidermal inclusion cyst, see there GYNECOLOGY Gartner's duct cyst, see there

inclusion/exclusion criteria CLINICAL RESEARCH The medical or social reasons why a person may/may not qualify for participation in a clinical trial

incoherence Not understandable; disordered; without logical connection. See Schizophrenia

incompatibility group MOLECULAR BIOLOGY A number of different types of plasmid, often related to each other, that are unable to stably coexist in the same cell

incompatibility TRANSFUSION MEDICINE See Transfusion reaction. See ABO incompatibility, Rh incompatibility

incompetence Incompetency CARDIOLOGY See Chronotropic incompetence FORENSIC MEDICINE The inability of a physician or health care provider to perform his/her duties; a physician may be referred to as incompetent; the euphemistic/'politically correct' adjective 'impaired' is increasingly preferred. See Impairment, Incapacity GYNECOLOGY See Cervical incompetence MEDTALK The inability to perform a task or function, defined in terms of organ dysfunction; this use of incompetence is waning in popularity, and being replaced by insufficiency, as in cardiovascular, hepatic, renal, or other insufficiency or failure EXCEPTIONS Incompetence or competence of valves–eg, cardiac or ileocecal or of the cervical os. See Cervical incompetence, Chronotropic incompetence PSYCHIATRY The lack of capacity to legally consent or to contract–ie, the inability to appropriately exercise free will, as in Alzheimer's disease; incompetence in a legal framework requires a formal declaration that a person is incompetent to make his/her own decisions, and appointment of a surrogate decision-maker to be the person's 'advocate'. See Competency to stand criminal trial, Incompetent, Informed consent, Testamentary capacity

incompetent *adjective* 1. Referring to an inability to perform a task or function 2. Referring to a valve or other structure that is incapable of closing properly

incompetent cervix Cervical incompetence, see there

incomplete abortion OBSTETRICS Partial expulsion of fetus and placenta with pain and bleeding, which is potentially fatal for the mother. See Abortion

incomplete fracture ORTHOPEDICS A fracture without separation of bone fragments. See Fracture

incomplete penetrance *adjective* Referring to the presence of a gene that is not phenotypically expressed in all members of a family with the gene. See Penetrance

incomplete testicular feminization Androgen insensitivity syndrome, see there

incontinence UROLOGY The inability to control the flow of urine from the bladder. See Overflow incontinence, Paradoxic incontinence, Stool incontinence, Urge incontinence

incoordination MEDTALK Clumsiness

increased appetite Hyperphagia, polyphagia

increased body hair Hirsutism, see there

increased head circumference Macrocephaly, see there

increased intracranial pressure Intracranial hypertension, see there

increased rate of breathing Tachypnea, see there

increased salivation Sialorrhea, see there

increased sweating Diaphoresis, see there

increased tearing Hyperlacrimation

increment-decrement life table Population increment-decrement life table EPIDEMIOLOGY A life table that integrates opposing time-dependent trends in the size of a cohort; an IDLT is a device developed by demographers to analyze population mobility phenomena. See Life table

incremental cost CLINICAL DECISION-MAKING The additional financial resources, beyond the cost of usual care, that may be required when changing from one therapeutic option to another. Cf Incremental effect

incubation INFECTIOUS DISEASE The asymptomatic development of an infection LAB MEDICINE The maintenance of controlled environmental conditions to facilitate growth of microorganisms or cells in culture NEONATOLOGY The maintenance of an 'enhanced' environment to optimize growth of a premature or otherwise compromised infant

incubation period The time elapsed between infection and appearance of disease Sx. Cf Latent period EPIDEMIOLOGY A period of subclinical or inapparent pathologic changes after exposure, ending with the onset of Sx of an infection

incubator NEONATOLOGY A device that maintains 'enhanced' environmental conditions–↑ humidity, O₂, temperature, that optimize the growth of a premature or otherwise compromised infant

IND Investigational new drug THERAPEUTICS A status assigned by the FDA to a drug before allowing its use in humans, exempting it from premarketing approval requirements so that experimental clinical trials may be conducted. See Phase 1, 2, 3 studies, Sponsorship

InDDEx NEUROLOGY A clinical trial–Investigation into Delay to Diagnosis of Alzheimer's disease with Exelon™. See Alzheimer's disease

indemnity insurance MANAGED CARE A type of health insurance in which a Pt can choose the hospital and provider, and the insurer reimburses the Pt or provider for a set percentage of the cost, minus deductibles and co-payments

independent contractor MEDTALK A nonsalaried employee–eg, 'temp' nurse, 'per-diem' physician, consultant, etc, who provides a service at a rate higher than that of a salaried full-time employees, who is hired on an 'as-needed' basis; ICs don't receive health benefits, retirement contributions, paid vacation, or have part of their pay deducted for taxes. See Employee, Locum tenens, Per diem

independent investigator Independent research investigator NIHSPEAK A well-established scientist whose research accomplishments have resulted in the bestowal of "tenure", ie, long-term commitment of salary, personnel and research resources

independent practice association MANAGED CARE A group of providers, typically physicians, who organize into a corporation that contracts with HMOs; in IPAs, physicians retain their traditional practice autonomy while integrating themselves into self-directed groups that solve group problems and exert political influence. See HMO, Managed care, Risk pool. Cf Group model, Network model, Staff model

independent symposium Independent conference DRUG INDUSTRY A conference sponsored by one or more pharmaceutical companies which meets FDA guidelines regarding impartiality in data presentation. See Authorship, Ghost writing, Project house, Thought leader. Cf Sponsored symposium

Inderal® Propranolol, see there

Indermil™ adhesive SURGERY A tissue sealant for point-of-care wound closure as an alternative to tape, sutures, staples. See Fibrin glue

indeterminant colitis Inflammatory bowel disease, see there

index 1. A table of contents 2. A guiding principle 3. A formula expressing the relationship of one value, property, form, or ratio to another. See Ankle-brachial index, Atherogenic index, Barthel index, Bessman index, Biotechnology index, Bispectral index, Body mass index, Cardiac index, Case-mix index, Cephalic index, Clamp-derived insulin-sensitivity index, DNA index, Duke Activity Status index, England index, Family suffering index, Foam stability index, Free thyroxine index, Framingham Physical Activity index, Geographic practice cost index, Glycemic index, Greenhouse index, Heat index, Hemacytology index, Hemogram index, Hepatic iron index, HERP index, Hollingshead index, Icterus index, IgG index, IgM index, International sensitivity index, Insulin sensitivity index, Krimsky index, Labeling index, Lesquesne index, Life Events index, Locomotion index, Maturation index, Medicare Economic index, Mentzler index, Merck index, Misery index, Mitotic activity index, Mother index, National Death index, Nuclear contour index, Obstructive apnea-hypopnea index, Organism-specific antibody index, Pearl index, Phagocytic index, Pollution Standards index, Poverty index, Prognostic nutritional index, Proliferation index, Singh index, Psoriasis Area and Severity index, Quetelet index, Rate-adjusted mortality index, Shine & Lal index, Side Effects & Symptoms Distress index, Sleep Impairment index, Splenic index, Standard deviation index, Stress index, Therapeutic index, Thymidine labeling index, Transitional-dyspnea index, Uncitedness index, UV index

index case See Index patient, see there

index family EPIDEMIOLOGY A family with a case of a disease of interest, the genetic tree of which is examined both laterally and across generations to identify similar cases. See Extended family

index of inequality SOCIOLOGY A summed measure of differences in mortality according to the level of education or income status among persons of similar sex, race, and family status, based on mortality ratios; the IoI in the US between whites and blacks—based on education and economic status widened from 1960 to 2003

index patient Index case CLINICAL RESEARCH The first medically-identified Pt in a family or other group, with a particular condition, often an infection, which triggers a line of investigation . See Cohort, Proband

index of suspicion MEDTALK A phrase broadly used to indicate how seriously a particular disease is being entertained as a diagnosis; as an example, there is a high IOS that rapid and unexplained weight loss in an elderly Pt is due to pancreas CA, and a low IOS that it would be due to AIDS

Indiana pouch UROLOGY An ileocecal segment reconstructed to create a continent urinary pouch, sewn into the anterior abdominal wall in Pts post-radical cystectomy for bladder CA. See Bladder cancer, Urinary incontinence

indication INTERNAL MEDICINE A clinical Sx or circumstance indicating that the use of a particular intervention would be appropriate. Cf Contraindication

Indigo LaserOptic™ system UROLOGY A proprietary device for treating BPH that uses a diode interstitial laser to destroy enlarged prostate tissue by coagulation. See Benign prostatic hypertrophy

indinavir Crixivan®, MK-639 AIDS A potent HIV protease inhibitor used alone or with other antiretrovirals and nucleoside analogues for HIV Pts with ↓ CD4 cells ADVERSE EFFECTS ↑ LFTs, kidney stones, nausea, lipodystrophy–hyperglycemia, ↑ fats, ↑ waist, ↓ fat from face, arms, legs, ↑ BR. See AIDS, HIV, Protease inhibitor

indirect compensation MEDICAL PRACTICE The aggregate of benefits—eg, payment of health and malpractice insurance permiums, pension contribution, books and journal subscriptions, paid CME, paid vacation, car leasing, etc, given to an employee in addition to a base salary or wages. Cf Direct compensation

Indirect transmission EPIDEMIOLOGY A nonperson-to-person transmission of an agent—eg, parasite from a reservoir to a susceptible host by suspended air particles or by animate–vector or inanimatex–vehicle intermediaries

individual psychology PSYCHIATRY A system of psychiatric theory, research, and therapy that stresses compensation and overcompensation for inferiority feelings. See Complex

individualization PSYCHIATRY An individual's personal development, in which a partner—eg, lover, wife is viewed as a 'an individual sentient being.' See Animus, Jungian psychoanalysis

indolent MEDTALK *adjective* Referring to a condition which may linger longer, but often slowly progresses to a more advanced stage, as in an indolent CA VOX POPULI Slow growing

indolent myeloma Stable myeloma ONCOLOGY A slowly progressive myeloma with an average survival of 10 yrs, which contrasts with untreated–6-month survival and treated myeloma–3 to 5 yr survival. See Monoclonal gammopathy of undetermined significance, Myeloma

indomethamin Indocin® THERAPEUTICS A NSAID used for arthropathies EFFECTS Platelet and neutrophil function, cerebral, mesenteric, renal hemodynamics, renal tubular function, GI tract O_2 consumption. See NSAIDs

induced abortion Termination of pregnancy The voluntary termination of gestational products, a procedure performed by instruments—eg, dilatation and curettage, if performed in the first trimester or by saline infusion–saline abortion if performed later. See Abortion

induced psychotic disorder See Folie à deux

induced sputum INFECTIOUS DISEASE Sputum obtained by having the Pt inhale a saline–salt water mist, causing the Pt to cough deeply

induction OBSTETRICS Induction of labor, see there ONCOLOGY 1. Induction chemotherapy, see there 2. Induction of remission, see there

induction failure ONCOLOGY Complete nonresponsiveness to CA therapy, characterized by ↑ size of lesions or appearance of new lesions. Cf Complete remission, Minimal response, Partial remission

induction chemotherapy ONCOLOGY The use of chemotherapy as a primary treatment for Pts presenting with advanced CA for which no alternative treatment exists. See Salvage treatment

induction of labor The use of artifical maneuvers to hasten the onset of labor, including the use of prostaglandin E_2 gel, oxytocin infusion, and amniotomy; in post-term pregnancy, induction of labor results in a lower rate of cesarean section than serial antenatal monitoring

induction of remission ONCOLOGY The initial phase of a 3-step chemotherapeutic regimen used to treat acute leukemia; in ALL, prednisone and vincristine induce remission in 90% of children and 50% of adults, which ↑ to 80% by adding a third drug–eg, L-asparginase, daunorubicin or doxorubicin; in AML, therapeutic levels of these agents approach BM toxicity. See Consolidation chemotherapy, Maintenance chemotherapy

induration DERMATOLOGY A hardened bump on the skin or subcutaneous tissue MEDTALK Scleroma A term for hardened tissue. See Keloid TUBERCULOSIS A palpable welling at the site of injection after a Mantoux skin test; the reaction size is the diameter of the indurated area–excluding redness

industrial anthrax Anthrax transmitted to humans by contact with contaminated animal products. See Anthrax

industrial blindness OCCUPATIONAL MEDICINE Vision that is so poor that a worker can't pursue an occupation

indwelling catheter Any catheter, usually understood to be for the urinary bladder–eg, a 'Foley' left in place for a prolonged period of time

inevitable abortion OBSTETRICS The 'terminal' stage of threatened abortion where the cervix dilates and membranes rupture MANAGEMENT Curettage. See Abortion

infancy Babyhood The period of development between birth and very early childhood, up to age 1 (or 2)

infant A child between birth and age 1 (or 2). See High-risk infant, Premature infant, Very-low-birth-weight infant

infant botulism PEDIATRICS An acute, potentially fatal infection by spores from *Clostridium botulinum*, a spore-forming bug found in dust, honey, and elsewhere, affecting infants up to 10 months RISK FACTORS Unknown, breast feeding, honey in diet

infant mortality EPIDEMIOLOGY Death of a child < age 1 ETIOLOGY Congenital defects, short gestation, low birth weight, pneumonia, influenza, neonatal infections, complications of placenta, cord, membranes, intrauterine hypoxia, birth asphyxia, respiratory distress syndrome, accidents, SIDS. See Infant mortality rate. Cf Postneonatal mortality

infant of diabetic mother NEONATOLOGY An infant born to a ♀ with a high serum glucose during pregnancy FEATURES Larger than other infants with enlarged organs, pospartum hypoglycemia, given ↑ fetal production of insulin, ↑ risk of stillbirth. See Gestational diabetes

infant seat Child safety seat, see there

infanticide FORENSIC MEDICINE The active or semi-passive killing of a viable conceptus > 20 gestational wks, which breathes spontaneously. See Battered child syndrome, Child abuse. Cf Stillbirth

---

INFANTICIDE, DIAGNOSIS OF

'HARD' CRITERIA

- **COMPARISON OF GASTRIC FLUID COMPOSITION** with that of a toilet bowel-active drowning

- **PEURAL SURFACES WITH PETECHIAE** Seen in induced suffocation, most significant when coupled with hematomas and petechiae on the mouth and epiglottis; the lingual frenulum may be torn and the lips bruised, indicating active attempts to suffocate infant

- **LUNGS** Stillbirth lungs are not aerated and do not float

- **EDEMATOUS FOAM ON NOSTRILS** An indicator of active breathing

- **MECONIUM** Resuscitation of a true stillborn may push meconium into the perianal region, but extensive staining of the placenta and umbilical cord is due to antenatal stress

'SOFT' CRITERIA

- **DENIAL OF PREGNANCY** If the woman is obese or a dullard, she may not know she was pregnant

- **RIGOR MORTIS** A finding that is poorly appreciated in neonates

---

- **IMPRESSION OF THE BODY** in soil, blood, or fomites, requiring diligent and timely scene investigation

- **MACERATION OF SKIN** A finding typical of stillbirth

- **PUTREFACTION** Stillborns do not putrefy as they have sterile bowels

- **UMBILICAL CORD** A cut cord indicates active intervention-time undetermined; an intact cord is consistent with stillbirth

- **DETERMINATION OF AGE** Viability, most fetuses born before 18 wks of gestation die despite resuscitative efforts, age is determined by skeletal dating, antenatal studies corroborating fetal death, eg Spaulding sign of in utero death characterized by overlapping cranial bones

---

infantile agranulocytosis Infantile genetic agranulocytosis IMMUNOLOGY An AR condition of early infant onset, characterized by neutropenia–< 0.3 x 10^9/L–US: < 300/mm³, absolute monocytosis, eosinophilia, recurring pyogenic infections of skin and lungs TREATMENT None, universally fatal in childhood

infantile cortical hyperostosis Caffey disease An AD condition characterized by early onset hyperostosis and neo-osteogenesis, of facial and, less commonly, long bones, soft tissue swelling, hyperirritability, dysphagia, fever, pleuritis LAB ↑ ESR, alk phos

infantile digital fibromatosis Kissing tumor A tumor of infants of the terminal phalanges of the fingers and toes, which measures < 2 cm PROGNOSIS Benign, although 60% recur locally

infantile eczema 1. Atopic dermatitis, see there 2. Diaper dermatitis

infantile myofibromatosis A condition characterized by fibrosis affecting soft tissue, bone, and viscera of infants and young children; IM is usually solitary but may be multicentric; it rarely recurs, and may regress with time. See Fibrous tumor of childhood

infantile polyarteritis Kawasaki syndrome, see there

infarct PATHOLOGY Dead/necrotic tissue. See Acute myocardial infarct, Anemic infarct, Lacunar infarct, Myocardial infarct, Non-Q-wave infarct, Pseudoinfarct, Q wave infarct, Red infarct, Reperfusion-eligible acute myocardial infarct, Watershed infarct, White infarct. Cf Infarction

infarct-avid scintigraphy CARDIOLOGY A form of scintigraphy used to document acute MI, for Pts in whom EKG and marker evidence is inconclusive; in IAS, myocardial uptake of the isotope indicates heart disease DIFFDX healed MIs, myocarditis, pericarditis, left ventricular aneurysm, rib fractures, etc

infarction MEDTALK Dying of tissue, necrosis

Infasurf® Calfactant NEONATOLOGY A suspension administered intratracheally to prevent RDS in high-risk premature infants < 29 wks of gestational age who require endotracheal intubation. See Respiratory distress syndrome

infected abortion Septic abortion, see there

infection EPIDEMIOLOGY The presence of a pathogen in a host which may or may not be associated with clinical disease. See Acute HIV infection, Atypical mycobacterial infection, Breakthrough infection, Chronic symptomatic HIV infection, Close contact infection, Congenital CMV infection, Cross-infection, Danger space infection, Ear infection, Emerging infection, Exit-site infection, Fungal infection, HIV-1 superinfection, Hyperinfection, Latent clostridial infection, Mixed wound infection, Multiple infection, Multiplicity of infection, Mycobacterial infection, Nail infection, Nonprogressive HIV infection, Nosocomial infection, Occult infection, Opportunistic infection, Parasitic infection, Perivascular inflammation, Ping pong infection, Pocket infection, Reemerging infection, Respiratory tract infection, Satellite infection, Silent infection, Spillover infection, Sterile infection, Subclinical infection, Superinfection, Surgical site infection, Tap water infection, Tunnel infection, Urinary tract infection. Cf Disinfection

infection control MEDTALK A hospital department that identifes infections occurring in the facility, which is also responsible for instituting safeguards to prevent the transmission of infections

infectious *adjective* 1. Referring to an infection 2. Capable of spreading infection–eg, expulsion of infected aerosol by coughing or sneezing

infectious agent Pathogen, see there

infectious arthritis Septic arthritis, see there

infectious disease See Infection

infectious mononucleosis Glandular fever, kissing disease An acute EBV infection, most common in adolescents EPIDEMIOLOGY Spread by kissing; 4 to 8 wk incubation CLINICAL Fever, sore throat, malaise, fatigue, weakness, lymphadenopathy, hepatosplenomegaly LAB Monocytosis, atypical lymphocytes, high titers of sheep RBC agglutinins MANAGEMENT Vigorous contact sports should be avoided to prevent splenic rupture

infectious mononucleosis 'syndrome' Any peripheral monocytosis accompanied by Sx typical of classic EBV-induced infectious mononucleosis–eg, CMV, herpes virus, HHV-6, HIV-1, *Toxoplasma gondii*

infectious myringitis Bullous myringitis ENT Inflammation of the tympanic membrane intimately linked to otitis media. See Otitis media

infectious period The period during which an infected person can transmit a pathogen to a susceptible host

infective endocarditis Acute endocarditis; bacterial endocarditis; subacute endocarditis CARDIOLOGY An infection of the endocardium which may involve the valves and extend to the myocardium, often occurring in Pts with underlying heart disease SOURCES OF INFECTION Transient bacteremia, common during dental, upper respiratory, urologic, and lower GI diagnostic and surgical procedures; IE may result in vegetations on valves, endocardium and the vascular intima, which may become dislodged and send clots to the brain, lungs, kidneys, or spleen AGENTS *S viridans* causes ±50% of IE, staphylococcus and group D streptococcus, *Pseudomonas* spp, *Serratia* spp *Candida* spp CLINICAL Sx develop slowly over months–subacute or abruptly; fever, fatigue, malaise, headache, night sweats, splinter hemorrhages under fingernails, heart murmurs due to vegetations especially on the mitral valve, splenomegaly, anemia RISK FACTORS Congenital heart disease, ASD, PDA, etc; prior rheumatic heart disease; heart valve defects–eg, mitral insufficiency, prosthetic valves PROPHYLAXIS Antibiotics before surgery in in at-risk Pts. See Endocarditis

Infergen® Interferon alfacon-1 IMMUNOLOGY A recombinant IFN-alpha in clinical use for managing HCV infection Business Wire 7/10/97

inferior capsular shift ORTHOPEDICS The main procedure for managing multidirectional shoulder instability which directly addresses capsular laxity by reducing capsular volume by up to 57%. See Multidirectional shoulder instability. Cf Thermal capsular shift

inferior petrosal sinus sampling ENDOCRINOLOGY A test used to localize the source of excess ACTH production, based on the drainage routes of pituitary–hypophysis venous blood, which flows first into the cavernous sinuses and then into the inferior petrosal veins

inferior vena caval filter VASCULAR DISEASE A device implanted in leg veins at high risk of developing DVT and PTE INDICATIONS For Pts in whom anticoagulation is either contraindicated or has been tried and failed, or as prophylaxis for high-risk Pts–eg, those with cor pulmonale or after surgical pulmonary embolectomy. See Deep vein thrombosis, Greenfield filter, Pulmonary thromboembolism

inferior vena caval interruption VASCULAR DISEASE Any therapy that prevents the free flow of blood–and emboli from the lower half of the body via the inferior vena cava FORMATS IVC filters–most common, IVC ligation, external clips, both rarely used

inferiority complex PSYCHIATRY A popular term for a constellation of behaviors, including diffidence, timidity, etc, which may accompany a deep-rooted sense of inadequacy. Cf Superiority complex

infertility GYNECOLOGY The inability to produce children; the involuntary inability to conceive, which contrasts to sterility, the complete inability to reproduce; 8.5% of married couples–US are infertile– or inconceivable. See Assisted reproduction, Hemizona assay, In vitro fertilization, Involuntary infertility, Mechanical infertility, Sperm penetration assay

infiltrate *noun* INFECTIOUS DISEASE A collection of fluid and cells seen on a CXR in various form of lung inflamation. See Fluffy infiltrate MEDTALK A collection of fluid and cells seen on a plain film that corresponds to loco-regional extension of cancer

infiltrating *adjective* Referring to a tumor that penetrates the normal, surrounding tissue

infiltration anesthesia Infiltration anesthesia AMBULATORY SURGERY The injection of local anesthetics for the repair of traumatic lacerations, or for minor surgery

infiltrative *adjective* Referring to that which infiltrates or spreads, as in a cancer

infirmary MEDTALK A health clinic–eg, in a college or university–US or small hospital–UK. See Surgery. Cf Outpatient clinic

inflammation A response to injury which is characterized by pain, swelling, heat, redness, and/or loss of function. See Round cell inflammation

inflammation of heart muscle Myocarditis, see there

inflammation of kidney Nephritis, see there

inflammation of rectum Proctitis, see there

inflammatory bowel disease A general term for the symptoms associated with specific types of inflammation of the large intestine, including Crohn's disease–CD, ulcerative colitis–UC as well as idiopathic inflammatory bowel disease, a diagnosis of exclusion that clinically and pathologically mimics and/or overlaps CD and UC HEREDITY First-degree relatives of Pts with CD and UC have an 8- to 10-fold ↑ risk of developing the same form of colitis DIFFDX Ischemia, radiation, uremia, cytotoxic drugs, heavy metal intoxication; IBD Pts often have ≥ 2 of following: Visible abdominal distension, relief of pain upon defecation, looser and more frequent bowel movements when the pain occurs; ♂:♀ ratio, 2:1; more common in Jews; IBD is associated with mucocutaneous disease–pyoderma gangrenosa, erythema nodosum, oral ulcers, annular erythema, vascular thromboses, epidermolysis bullosa, ocular disease–uveitis, iridocyclitis, hepatopathy–CAH, cirrhosis, sclerosing cholangitis, arthropathy–ankylosing spondylitis. Cf Irritable bowel syndrome TREATMENT Corticosteroids–or corticotropin–favored by some clinicians, aminosalicylates–sulfasalazine and 5-aminosalicylic acid, immunomodulatory agents–azathioprine, mercaptopurine, cyclosporine, MTX, metronidazole, antibiotics and nutritional support–both more effective in Crohn's disease, resulting in ↓ Sx, ↓ inflammatory sequelae, improved nutritional status

INFLAMMATORY BOWEL DISEASE

CLINICAL	ULCERATIVE COLITIS	CROHN'S DISEASE
Rectal bleeding	Common	Rare
Abdominal mass	Rare	10-15%
Abdominal pain	Left-sided	Right-sided
Abnormal endoscopy	95%	< 50%

Perforation	12%	4%
Colon carcinoma	5-10%	Rare
Response to steroids	75%	25%
Surgical outcome	Excellent	Fair
Rectal involvement	> 95%	!0%
RADIOLOGY		
Ileal involvement	Rare	Usual
Cross-hatched ulcers	Rare	Occasional
Thumbprinting	Absent	Common
Fissuring	Absent	Common
Skip areas	Absent	Common
Strictures	Absent	Common
PATHOLOGY		
Distribution	Diffuse	Focal
	Superficial	Transmural
Mucosal atrophy	Regeneration	Marked Minimal
Hyperemia	Often marked	Minimal
Crypt abscess	Common	Rare
Cytoplasmic mucin	Decreased	Intact
Lymphoid aggregates	Rare	Common
Lymph nodes	Reactive hyperplasia	-----
Edema	Minimal	Marked
Granulomas	Absent	60%

inflammatory breast cancer ONCOLOGY Breast CA characterized by ↑ warmth, redness, swelling caused by cancer cells blocking skin lymphatics; skin has a pitted "peau d'orange" appearance. See Breast cancer

inflammatory carcinoma Erysipeloides CA characterized by invasion of dermal lymphatics by malignant cells, most often breast, less commonly, CA of lung, GI tract, uterus CLINICAL Diffuse swelling, erythema, pain, edema, simulating acute mastitis or cellulitis, potentially delaying diagnosis and therapy PROGNOSIS Poor

inflammatory polyp ENT 'Allergic polyp' A non-neoplastic, reactive, recurrent, often bilateral, 'tumor' formed in response to infection, allergy, mucoviscidosis GI DISEASE Postinflammatory polyposis, 'pseudopolyposis' Any of a number of often bizarre polyps and/or bridges of mucosa, which when single may mimic malignancy, and when multiple, mimic a polyposis syndrome; IPs may arise in a background of inflammatory bowel disease–Crohn's disease, ulcerative colitis, amebic colitis, bacterial dysentery, and chronic schistosomiasis, and are benign, nonspecific sequelae of inflammation of the intestinal mucosa CLINICAL IPs are rarely symptomatic, but pain, obstruction, and if large, intussusception may occur. Cf Inflammatory fibroid polyp

infliximab Remicade® MOLECULAR MEDICINE A monoclonal antibody that neutralizes TNF activity

influential physician Thought leader, see there

influenza INFECTIOUS DISEASE A viral infection which costs the health care system–US ± $12 billion/yr; it causes 10-40,000 excess deaths/yr, 80% in the elderly, most preventable by annual vaccination COMPLICATIONS Secondary bacterial infections–eg, otitis media, viral URIs, Reye syndrome, liver and CNS disease MANAGEMENT Oseltamivir, zanamivir, a selective neuraminidase inhibitor, ↓ symptoms in influenza A or B PREVENTION Annual vaccination. See Cold-adapted influenza vaccine, Zanamivir

influenza A INFECTIOUS DISEASE An avian virus, especially of ducks–which in China live near the pig reservoir and 'vector'; periodic mutations of the virus–13 hemagglutination and 9 neuraminidase subtypes cause 'flu' epidemics and pandemics. See Antigenic drift, Antigenic shift. Cf Influenza B, Influenza C

influenza B INFECTIOUS DISEASE An influenza virus which causes epidemics in 3-5 yr cycles. Cf Influenza A, Influenza C

influenza C INFECTIOUS DISEASE An influenza virus which causes nonepidemic, cold-like illness. Cf Influenza A, Influenza B

influenza syndrome A condition characterized by fever, headache, malaise, vertigo, ataxia, leukopenia, thrombocytopenia, muscular weakness, N&V, diarrhea, possibly renal failure, a side effect of rifampin therapy

influenza vaccine Flu vaccine A vaccine recommended for those at high risk for serious complications from influenza: > age 65; Pts with chronic diseases of heart, lung or kidneys, DM, immunosuppression, severe anemia, nursing home and other chronic-care residents, children and teenagers taking aspirin therapy–and who may thus be at risk for developing Reye syndrome after influenza infection, and those in close or frequent contact with anyone at high risk; persons with egg allergy should not receive IV. See Influenza, Vaccine

informant Historian MEDTALK A person who provides a medical history

information bias EPIDEMIOLOGY The bias that arises in a clinical study because of misclassification of the level of exposure to the agent or factor being assessed and/or misclassification of the disease itself; a type of bias that occurs when measurement of information–eg, exposure or disease—differs among study groups. See Bias

information overload An excess of information provided to a person in certain situations, resulting in either a wrong decision or the passing of misinformation 1. Patients must have sufficient information at their disposal to give an informed consent for–eg, an elective surgical procedure; this is problematic when a Pt is informed of very rare complications or side effects of a procedure or the anesthesiology. See Uninformed consent. Cf Informed consent

informed consent MEDTALK A voluntary, legally documented agreement by the Pt to allow performance of a specific diagnostic, therapeutic, or research procedure. See Emergency doctrine, Informed decision, Malpractice, Therapeutic privilege doctrine

informed decision Informed choice PATIENTS RIGHTS A decision by a Pt about a diagnostic or therapeutic procedure that is based on choice, which requires that the decision be voluntary and that the Pt has the capacity for choice, which rests on 3 elements: possession of a set of values and goals; ability to understand information and communicate decisions; ability to reason and deliberate. See Informed consent

infra- prefix, Latin, below, under

infracalcaneal adjective Below the calcaneus

infrapatellar adjective Below the patella

infrared capnography ANESTHESIOLOGY A method of capnography–analysis of CO_2 in exhaled air, which is useful for the diagnosis and management of esophageal and endobronchial intubation, airway obstruction, bronchospasm, hypermetabolic states, PTE, venous air embolism, and cardiogenic shock

infundibulopelvic adjective Referring to an infundibulum–a funnel-like structure and a pelvis, as seen in the renal calices. See Caliceal

InfusaSleeve II™ catheter DRUG DELIVERY A device that provides local drug delivery during angioplasty. See Angioplasty

infusion MAINSTREAM MEDICINE The administration of IV fluids. Parenteral nutrition. Cf Bolus THERAPEUTICS IV infusion The introduction of a fluid, including medications, into the circulation. See Autoinfusion, Continuous infusion, Hepatic arterial infusion, Intracarotid infusion, Intrahepatic infusion, Intraosseous infusion, Intraperitoneal infusion, Intraventricular infusion

infusion device THERAPEUTICS A device used to administer therapeutics–eg, analgesics, antimicrobials, blood products, chemotherapy, nutrients. See Elastomeric infusion device, Electronic syringe infusion device, Mechanical infusion pump, Minibag. Cf Central venous catheter

infusion pump A device designed to deliver drugs and/or 'biologicals', at low doses and at a constant or controllable rate; ↑ rates of delivery in such devices may be associated with local hemolysis, compromising the potential benefits of a calibrated delivery system. Cf Pancreas transplant

ingrown toenail Onychocryptosis PODIATRY The growth of the edge of a toenail–usually of the great toe–into the skin surrounding the nail, which may be accompanied by inflammation or pain ETIOLOGY congenitally curved nails, fungal infection, trauma RISK FACTORS Improperly fitted shoes, poorly trimmed toenails; may be more severe in Pts with DM MANAGEMENT Segmental chemical ablation with phenol; partial or total nail matricectomy

inguinal canal A cylindrical conduit in the abdominal wall through which a testis descends into the scrotum.

inguinal hernia SURGERY The prolapse of a loop of intestine into a patent inguinal canal

inguinal orchiectomy UROLOGY A procedure in which the testicle is removed via an inguinal incision

INH Isoniazid, see there

inhalant PUBLIC HEALTH A potentially hazardous particle, liquid or solid, which may be present in environmental air, and inhaled into small airways and alveoli. See Aerosol, Dust, Fiber, Fume, Gas, Mist, Vapor SUBSTANCE ABUSE A term used in 2 contexts: 1. Nitrates–vasodilators that allegedly prolong orgasm; 2. Solvents, which produce euphoria through neurotoxicity SA, NITRATES EFFECTS Vasodilation, ↑ blood flow to heart; may prolong orgasm USES Relieves anginal pain EXAMPLES Amyl nitrate, butyl nitrate, isobutyl nitrate, isosorbide dinitrate, nitroglycerin, nitrous oxide, isobutyl-nitrite SA, NEUROTOXINS EFFECTS Bronchodilation, CNS depressant, metabolic inhibitor EXAMPLES Freon, tetrahydrocarbons, benzol-based emollients and derivatives, toluene-based compounds, ethyl ether, volatile solvents, aerosols, adhesives, etc COMMONLY ABUSED PRODUCTS Air freshener, carburator cleaners, correction fluid, spray deodorant, gasoline, glue, hairspray, lighter fluid, markers, nail polish remover, octane booster, paint thinner, rubber cement, spot remover, spray shoe polish, spray paint–especially gold and silver

inhalation anesthetic ANESTHESIOLOGY An agent used to induce narcosis and absence of sensation EXAMPLES Halothane, isoflurane–a mainstay, desflurane, sevoflurane. See Desflurane, Halothane, Isoflurane, Nitrous oxide, Sevoflurane

inhalation anthrax Pulmonary anthrax, woolsorter's disease PULMONOLOGY Occupational anthrax caused by inhalation of *Brucella anthracis* spores, affecting those exposed to aerosols during early processing of goat or other infected animal hair CLINICAL Biphasic, early Sx are influenza-like, lasting 2-3 days, followed by presumed resolution, then severe respiratory distress, hypoxia, dyspnea, mediastinal LN involvement, subcutaneous edema of upper body, meningeal signs, shock, death in 24 hrs MORTALITY High TREATMENT Penicillin

Inhale™ THERAPEUTICS An investigational system that delivers therapeutics–eg, insulin, as an alternative to IV therapy. See Drug delivery VOX POPULI Breathe in

inhaler THERAPEUTICS A device that delivers a therapeutic agent. See Metered-dose inhaler, Nicotrol inhaler

inherited *adjective* Referring to a trait or characteristic that is intrinsic to one's genotype; in this context, the synonym hereditary is widely preferred

INHIBIT CARDIOLOGY A clinical trial–INtimal Hyperplasia Inhibition with Beta In-stent Trial

inhibited pacemaker Demand pacemaker, see there

inhibited sexual desire Hypoactive sexual desire, Sexual anhedonia, sexual apathy PSYCHOLOGY ↓ Sexual desire and interest manifest by failure to initiate or respond to a partner's initiation of sexual activity TYPES 1°–never had sexual desire/interest; 2°–had, no longer has, sexual interest; situational–interest in others, *not* current partner; general–no interest; nada. See Sexual anhedonia, Sexual aversion

inhibition PSYCHIATRY Behavior that reflects an unconscious defense against forbidden instinctive drives, which may interfere with or restrict specific activities. See Competitive inhibition, Disinhibition, Enzyme inhibition, Feedback inhibition, Multidrug-resistance inhibition, Outlaw inhibition, Postsynaptic inhibition, Presymptomatic inhibition, Reciprocal inhibition

inhibitor MEDTALK A substance that inhibits the activity of protein or downregulates a pathway. See ACE inhibitor, Aldose reductase inhibitor, Angiogenesis inhibitor, Bovine pancreatic trypsin inhibitor, Bowman Birk protease inhibitor, Cartilage-derived inhibitor, Cholinergic inhibitor, COMT inhibitor, Factor VIII inhibitor, FLAP inhibitor, HMG CoA reductase inhibitor, Lipoprotein-associated coagulation inhibitor, Matrix metalloproteinase inhibitor, Monoamine oxidase inhibitor, Neuraminidase inhibitor, Neuropeptide Y inhibitor, Nonnucleoside reverse transcriptase inhibitor, NOS inhibitor, Plasminogen activator inhibitor, Postsynaptic serotonin antagonist-presynaptic serotonin reuptake inhibitor, Protease inhibitor, Reverse transcriptase inhibitor, Selective serotonin reuptake inhibitor, Serotonin-norepinephrine reuptake inhibitor, SLPI inhibitor, SSRI inhibitor, Tissue factor pathway inhibitor, Topoisomerase inhibitor, Trojan Horse inhibitor, Tyrosine kinase inhibitor, Vasopeptidase inhibitor

initial phase INFECTIOUS DISEASE The first period of therapy for a condition–eg, for TB, the IP encompasses the 1st 8 wks of treatment, during which most of the tubercle bacilli are killed

INJECT CARDIOLOGY A clinical study–International Joint Efficacy Comparison of Thrombolytics–that compared the efficacy of IV streptokinase and rtPA–reteplase double bolus injection in managing acute MI See Acute myocardial infarction, Alteplase, Heparin, Myocardial infarction, Reteplase, Streptokinase, tPA. Cf COBALT, GUSTO I

injection Forced adminstration of a fluid usually understood to be by needle. See Bolus injection, Intracellular sperm injection, Intrathecal injection, Intravenous injection, Nipent® (pentostatin) injection, Therapeutic injection

injection sclerotherapy Endoscopic sclerotherapy GI DISEASE The intralesional administration of a sclerosant–eg, ethanolamine oleate, polidocanol with a vasoconstrictor–eg, epinephrine, to achieve hemostasis in Pts with bleeding ulcers, vascular malformation or esophageal varices. See Sclerosants. Cf Endoscopic ligation

injection technique OTOLARYNGOLOGY Phonosurgery–eg, vocal fold augmentation with alloplastics and bioimplants, or management of spasmodic dysphonia–coupled with transcutaneous or peroral botulinum toxin injection. See Phonosurgery

injections, fear of PSYCHOLOGY Trypanophobia. See Phobia

injury MEDTALK Trauma, wound, hurt. See Acceleration-deceleration injury, Acute spinal injury, Anterior cruciate ligament injury, Arachnic injury, Bite-mark injury, Blunt injury, Boot-induced anterior cruciate ligament injury, Brachial plexus injury, Chemical injury, Chemical eye injury, Closed fist injury, Cold injury, Corrosive injury, Deceleration injury, Degloving injury, Diffuse axonal injury, Diffuse ischemic injury, Golfing injury, Grade I injury, Grade II injury, Grade III injury, Hamstring injury, In-line skating injury, Lateral collateral ligament injury, Lye injury, Mass injury, Medial collateral ligament injury, Mild traumatic brain injury, Needle-stick injury, Overuse injury, Parachute-related injury, Patterned injury, Perversion injury, Phantom foot anterior cruciate ligament injury, Reperfusion injury, Repetitive motion injury, Reversible injury, SCIWORA, Sharp injury, Sliding injury, Spinal cord injury, Splash injury, Sports injury, Thoracic inlet injury, Transfusion-related acute lung injury, Trauma, Ventilator-induced lung injury, Violence-related injury, Weapons-related injury, Whiplash injury, Wound, Wringer injury PUBLIC HEALTH ±60 million people are injured, US/yr; total cost, ±$200 billion; direct costs account for 29%; in 1994, 151,000 US deaths were due to injuries, > ⅓ due to MVAs. See Burns, Drowning, Falls, Firearms, Hip fractures, Motor vehicle accidents, Poisoning

Injury Severity Scale EMERGENCY MEDICINE A numerical scoring system for calculating the probability of survival, and seriousness of wounds or trauma sustained; an ISS of > 20 implies a poor prognosis. See Abbreviated Injury Scale, Revised Trauma Score, Trauma Score

injury to ear 1. Acoustic trauma 2. Barotrauma, see there

ink blot test Rorschach test, see there

inkwell GI SURGERY A surgically constructed vagination-'intussusception' of a short sleeve of esophagus sewn into the stomach which, as intragastric pressure ↑, is compressed, forming a functional valve–eg, Nissen fundoplication. See Nissen procedure

inlay DENTISTRY A cemented restoration made to fit an internal/external preparation necessitated by loss of healthy tooth structure. See Cosmetic dentistry, Filling

inlet patch GI DISEASE A remnant of heterotopic but normal glandular epithelium, usually gastric, which may be found in the esophagus of otherwise asymptomatic subjects, which may be an embryonic rest. See Barrett's esophagus

inner ear decompression sickness A condition affecting deep sea divers due to gas bubbles CLINICAL Vertigo, tinnitus and hearing loss on ascent, particularly after rapid surfacing after deep, prolonged underwater dives TREATMENT Immediate recompression; delayed therapy results in permanent damage to the vestibular and auditory system

inner ear fistula ENT An abnormal communication between the perilymphatic space and middle ear or an intramembranous communication between the endolymphatic and perilymphatic spaces ETIOLOGY Barotrauma, penetrating trauma, surgery–eg, stapedectomy, cholesteatoma surgery, trauma following physical exertion, may rupture the labyrinth's limiting membranes CLINICAL Positional vertigo, motion intolerance, occasional disequilibrium, auditory Sx–eg, episodic incapacitating (Meniere's attack-like) vertigo, which may follow ↑ CSF pressure such as nose blowing or lifting, or vertigo after exposure to loud noises DIFFDX Meniere's disease, infection, acoustic neuroma, CNS lesions DIAGNOSIS Fistula test–positive pressure introduced into the suspect ear, either by rapid pressure on the tragus, compressing the external canal, or via a pneumatic otoscope, while observing the eyes; a positive fistula sign consists of conjugate contralateral slow deviation of eyes followed by 3 or 4 ipsilaterally directed beats of nystagmus; fiberoptic examination of the middle ear by myringotomy may help select Pts for surgical exploration, especially in Pts with a clinically suspicious Hx and a positive fistula sign TREATMENT Bed rest; head elevation; laxatives for ↑ intracranial pressure; surgical exploration if hearing loss worsens or vestibular Sx persist. See Hennebert's phenomenon, Tulio's phenomenon

inner ear infection Otitis interna, see there

innervated *adjective* Containing or characterized by nerves

innidation 1. Colonization, see there 2. Seeding, see there

innocent murmur CARDIOLOGY A cardiac murmur in a normal person–ie, without anatomic defects. See Murmur

Innohep® See Tinzaparin

InnoSense™ NEUROLOGY A battery-operated device used to manage urinary stress incontinence via electromyography feedback and electronically stimulated pelvic-floor muscle contraction. See Stress incontinence

innovation *noun* 1. Something new or different 2. The act of innovating, or introducing a new product, technology, or method

innovative medicine 1. Alternative medicine, see there 2. A popular term for cutting edge (mainstream) medicine

Inocor® Amrinone, see there

inoculum EPIDEMIOLOGY A gob of a pathogens to which a host is exposed at the time of transmission of an infection

Inocybe TOXICOLOGY A genus of North American poisonous mushrooms CLINICAL Abrupt onset 15 mins after ingestion with profuse sweating, nausea, abdominal colic, blurred vision, salivation, rhinorrhea, dizziness MANAGEMENT Atropine. See Poisonous mushroom

inoperable *adjective* Referring to a Pt's condition, especially, a disseminated CA, that won't benefit from surgery

INOS PEDIATRICS A clinical trial–Inhaled Nitric Oxide Study which evaluated the effect of NO therapy–NOT in hypoxic respiratory failure–HRF in neonates. See Respiratory distress syndrome

inosine prabonex Isoprinosin® AIDS An inosine-based salt that enhances immune function, possibly by ↑ IL-1 and/or IL-2 production, resulting in T-cell proliferation and ↑ NK cell activity; IO is used for viral infections–eg, HIV-1, flu, hepatitis, herpes simplex, and HPV. See AIDS therapy

inositol BIOCHEMISTRY A structure in phospholipids–eg, phosphoinositol; its main isomeric form is *myo*-inositol; it is present in breast milk, and may ↓ complications of prematurity and ↓ death due to lung disease, bronchopulmonary dysplasia, retinopathy of prematurity

inotropic agent Positive inotropic agent CARDIOLOGY A therapeutic agent that ↑ the strength and/or force of myocardial contraction; 2 major classes of positive inotropes: those that ↑ intracellular cAMP by stimulating the β-adrenergic receptor or inhibiting phosphodiesterase–eg, milrinone and those that ↑ intracellular sodium–eg, digoxin, vesnarinone, see there

Inoue balloon CARDIOLOGY A device for performing percutaneous mitral valve commisurotomy, a preferred method for treating mitral valve stenosis, as it is non-invasive and allows resumption of full-time employment within 1 wk vs 6-8 wks of recuperation required after an open-heart surgical procedure. See Balloon valvoplasty. Cf Percutaneous transluminal coronary angioplasty

inpatient care MANAGED CARE Services delivered to a Pt who needs physician care for > 24 hrs in a hospital

Inpatient Multidimensional Psychiatric Scale PSYCHIATRY A series of questions posed to Pts on admission to a mental institution, which enables immediate stratification into psychotic types, a term descriptive of attitude and behavior rather than clinical types, schizophrenia, manic-depressive, etc: hostile-paranoid, excited-hostile, excited-grandiose, intro-punitive, disorganized, retarded

inpatient service MANAGED CARE A service provided to a hospitalized Pt. Cf Outpatient service

INR International normalized ratio, see there

insane *adjective* Referring to unsoundness of mind; *non compos mentis*, dangerously fruity or nutty. See Insanity

insanity FORENSIC MEDICINE A legal and social term for a condition that renders the affected person unfit to enjoy liberty of action, because of the unreliability of his behavior with concomitant danger to himself and others; insanity denotes, by extension, a degree of mental illness that negates legal responsibility for one's actions. See Psychosis, Temporary insanity PSYCHIATRY A vague obsolete term for psychosis

insanity defense FORENSIC PSYCHIATRY A legal defense that a person cannot be convicted of a crime if he lacked criminal responsibility by reason of insanity–a term defined as a matter of law; the premise is that where there is no *mens rea* because of insanity, there is no criminal responsibility. See American Law Institute Formulation, Durham Rule, Irresistible impulse test, Long Island Rail Road massacre, M'Naughton Rule. Cf 'Black rage' defense, Television intoxication, 'Twinkie' defense

insecticide poisoning TOXICOLOGY Acute accidental or suicidal OD with insecticides, a common reason for rural ER visits TYPES Organochlorines–eg, DDT and chlordane–little used due to long-term persistence in environment, and cholinesterase inhibitors–either organophosphates–esters of phosphoric acid or thiophosphoric acid, carbamates–synthetic derivatives of carbamic acid CLINICAL Drooling, defecation, lacrimation, urinary incontinence, pupillary constriction, bradycardia, bronchoconstriction, muscle twitching, respiratory and/or cardiovascular depression, death Dx Hx, signs and Sx, acetylcholinesterase in RBCs and serum, which falls to ≤ 50% of normal; signs and Sx of organophosphate poisoning occur when acetylcholinesterase has fallen to < 20% of normal. See Organophosphate

insemination 1. The deposition of seminal fluid in the vagina, especially during intercourse 2. The fertilization of an ovum with semen. See Donor insemination, Heterologous insemination, Homologous insemination, Intrauterine insemination

insertable loop recorder CARDIOLOGY An implantable Pt-activated monitoring system that records subcutaneous EKGs in Pts with Sx suggestive of arrhythmia. See Holter monitoring

insertional tendonitis Enthesitis ORTHOPEDICS An inflammatory reaction in the tendon at its insertion, with regional tenderness IMAGING Calcification on the superior calcaneus at insertion

inside-out technique ORTHOPEDICS A method in shoulder surgery in which the arthroscope in the posterior access portal localizes the position of the anterior portal and a rod passed though the arthroscopic cannula to exit the anterior shoulder; a cannula is then passed retrograde over the rod and a portal established. See Outside-in technique

INSIGHT VASCULAR DISEASE A clinical trial–International Nifedipine once-daily Study–Intervention as a Goal in Hypertension Treatment

insomnia SLEEP DISORDERS The perceived or actual inability to sleep one's usual amount of time; a condition characterized by any combination of difficulty with falling asleep, staying asleep, intermittent wakefulness, and early-morning awakening; episodes may be transient, short-term–lasting 2 to 3 wks, or chronic TRIGGERS Illness, depression, anxiety, stress, poor sleep environment, caffeine, abuse of alcohol, heavy smoking, physical discomfort, daytime napping, medical conditions, poor sleep habits–eg, early bedtime, excessive time awake in bed EXAMPLES Psychophysiologic–learned insomnia, delayed sleep phase syndrome, hypnotic dependent sleep disorder, stimulant dependent sleep disorder. See Circadian rhythm, Conditioned insomnia, Familial fatal insomnia, Jet lag, Pseudoinsomnia, Rebound insomnia, REM sleep, Sleep disorder, Sleep-onset insomnia

INSOMNIA

CHRONOLOGIC CLASSIFICATION

- **TRANSIENT**–eg, 'jet lag'; does not require treatment

- **SHORT TERM** < 3 weeks in duration, due to travel to high altitudes, grieving loss of loved one, hospitalization, pain

- **LONG TERM** > 3 weeks in duration, eg related to medical, neurologic or psychiatric disorders or addiction

ETIOLOGY

- **PHARMACOLOGIC** Due to coffee, nicotine, alcohol

- **REBOUND (WITHDRAWAL)** Related to abrupt discontinuation of hypnotic drugs

- **DELAYED SLEEP PHASE** Due to shift work, chronic pain, sleep apnea and restless leg syndrome

inspissation Plugging of a tubular lumen–eg, bile ducts, intestine, with a thickened viscid material having a decreased fluid content

instability Lacking complete stability. See Hemodynamic instability ORTHOPEDICS The inability to maintain a reduced joint. See Functional instability, Mechanical instability, Multidirectional instability, Shoulder instability

instinct PSYCHIATRY Inborn drive An unreasoning response to an environmental cue, attributed to the Freudian id PRIMARY HUMAN INSTINCTS Self-preservation, sexuality; per some, aggression, ego instincts, heroism, social instincts are also primary instincts. See Death instinct, Id

institute VOX POPULI Institution An organized group of persons based in a particular place, who have a similar agenda. See American National Standards Institute, Cold Spring Harbor Institute, High-power institution, Howard Hughes Institute, Institution, Minority Institution, National Cancer Institute, National Institute on Aging, National Institute for Standards & Technology Institute, NIAAA, NIAMS, NICHD, NIDA, NIDCD, NIDDK, NIEHS, NIGMS, NIH, NIMH, NINDS, NINR, NIOSH, Salk Institute, Touch Research Institute, Whitehead Institute

institutional effect The biases in therapeutic decisions, diagnoses, or interpretation of data related to guidelines or 'cut-off' values for lab or other parameters promulgated in a health care facility or hospital. See Referral bias

institutional review board MEDICAL ETHICS A review body of physicians and lay persons established or designated by an entity–eg, a university hospital or academic health care facility, to protect the safety and welfare of human subjects participating in biomedical or behavioral research; the IRB debates and approves or rejects research projects performed at the institution. See Ethics committee, Helsinki Declaration, Nuremburg code

institutionalization syndrome SOCIAL MEDICINE The constellation of psychologic changes that occurs in persons who have been maintained for long periods of time in segregated communities–eg, mental institutions, state-run nursing homes CLINICAL Apathy, dependence, depersonalization, retreat from reality. See Homeless(ness), Shelterization. Cf Survivor syndrome

instructional simulation GRADUATION EDUCATION A simulation used to help students acquire knowledge and skills through surrogate experiences; simulation success hinges on the degree to which it resembles a real experience

instrumental activities of daily living A series of life functions necessary for maintaining a person's immediate environment–eg, obtaining food, cooking, laundering, housecleaning, managing one's medications, phone use; IADL measures a person's–elderly, mentally handicapped or terminally ill ability to live independently. See Dependency

insufficiency MEDTALK A lack of complete function. See Adrenal insufficiency, Chronic venous insufficiency, Haploinsufficiency, Pancreatic insufficiency, Renal insufficiency, Respiratory insufficiency

insufflation LAPAROSCOPY The process of injecting gas–room air, CO_2, or NO_2 via a machine into the abdominal cavity to create a tent to allow for the safe placement of trocars, visualization of the operative field, and manipulation of instruments used during the procedure. See Laparoscopic surgery

insulin PHYSIOLOGY A disulfide-linked polypeptide hormone produced by the beta cells of the pancreatic islets, which controls serum glucose and anabolism of carbohydrates, fat, protein. See Biphasic insulin, PEPCK, Proinsulin, rDNA insulin

340

insulin antibody Any anti-insulin autoantibody that develops in diabetics receiving insulin. See Insulin

insulin pump Insulin infusion device ENDOCRINOLOGY A device used for timed insulin delivery in DM BASIC COMPONENTS Insulin reservoir, a pump, a power source, electronic controls, a glucose sensor; IPs are either portable or implantable. Cf Biohybrid artificial pancreas, Islet cell transplantation

insulin receptor A heterodimeric membrane receptor composed of α and β chains, which has tyrosine kinase activity after binding insulin; IR deficiency is a rare cause of DM and may be due to a gene rearrangement, causing a deletion in the tyrosine kinase domain, a point mutation with a loss of the ATP binding site or other genetic defect

insulin receptor antibody An antibody found in Pts with insulin resistance with acanthosis nigricans, type 2 DM, non-endocrine autoimmune disease. Cf Insulin antibody

insulin resistance ENDOCRINOLOGY A suboptimal hypoglycemic response to physiologic insulin, which may be due to local allergy to insulin, acute exacerbation of chronic leukemia, DKA; IR–measured by oral GTT, and hyperinsulinemia are risk factors for type 2 DM; IR may also be due to ↑ circulating glucagon or a relative insulin receptor deficiency, where plasma insulin is high relative to glucose; in uremia, IR may be due to nonspecific receptor antagonism, as insulin binds poorly to receptors and hypocalcemia with ↑ proinsulin secretion. See Subcutaneous insulin-resistance syndrome. Cf Syndrome X

INSULIN RESISTANCE

TYPE A 'PRE-RECEPTOR', due to defects in synthesis and secretion of insulin, seen in the HAIR-IN syndrome, with hyperandrogenism (to a degree suggesting hyperthecosis, ovarian neoplasia, or polycystic ovaries), which is of pubertal onset, with ↑ levels of endogenous insulin, overt DM and resistance to exogenous insulin, related to anti-insulin antibodies

TYPE B 'RECEPTOR', related to ↓ insulin binding to receptors, ↓ number of receptors, or to insulin resistance, which may be associated with acanthosis nigricans, autoimmune disease, and overt DM due to circulating anti-receptor antibodies

TYPE C 'POST-RECEPTOR', related to defects arising after the insulin-receptor complex has formed, eg a defect in the cascade that follows receptor-ligand interaction

insulin sensitivity The systemic responsiveness to glucose, which can be measured by 1. The insulin sensitivity index–measures the ability of endogenous insulin to ↓ glucose in extracellular fluids by inhibiting glucose release from the liver and stimulating the peripheral consumption of glucose, and 2. The glucose-clamp technique, which measures the effect of changes in insulin concentration on glucose clearance–glucose uptake rate divided by plasma glucose concentration per unit of body surface area

insulin shock ENDOCRINOLOGY 1. A rare clinical event in which excess insulin is administered, causing profound hypoglycemia to levels below that required for normal brain function, causing anxiety, delirium, convulsions, coma, death 2. Hypoglycemia, see there

insulin spike ENDOCRINOLOGY An abrupt ↑ in insulin levels 2° to carbohydrates, the peak of which is said to be sharpest after consumption of simple carbohydrates and refined flours

insulinogenic index The ratio of insulin to glucose, a measure of the effect of meals on prediabetics

insulinoma Insuloma ONCOLOGY A pancreatic islet cell tumor, which is characterized by overproduction of insulin, resulting in hypoglycemia of sufficient severity to trigger a coma; insulinomas are typical of MEN I. See MEN I

insult MEDTALK noun Any stressful stimulus which, under normal circumstances, does not affect the host organism, but which may result in morbidity, when it occurs in a background of preexisting compromising conditions

insurance VOX POPULI A contractual relationship when one party–an insurance company or underwriter, in consideration of a fixed sum–a premium, agrees to pay on behalf another–an insured, or policyholder for covered losses, up to the limits purchased, caused by designated contingencies listed in the policy. See Adoption insurance, Cancer insurance, Catastrophic health insurance, Co-insurance, Comprehensive major medical insurance, Disability insurance, Group insurance, Hospitalization insurance, Indemnity insurance, Major medical insurance, Medical expense insurance, Medicare supplement insurance, National health insurance, Nationalized health insurance, Noncancellable insurance, Personal insurance, Reinsurance, Self-insurance, Workers compensation insurance

insurance carrier Insurer INSURANCE A commercial enterprise licensed in a state to sell insurance; carrier is widely preferred term for companies offering medical malpractice coverage

insurance only billing MEDICARE A general term for the practice by a physician of waiving of that part of a bill–the Pt's copayment–which is not covered by insurance. See Professional courtesy

insurer of last resort An insurance plan that accepts 'uninsurable' persons who have expensive and/or chronic diseases, and cannot obtain coverage at market rates. See Blues

InSync™ CARDIOLOGY An investigational cardiac resynchronization device that stimulates both sides of the heart and all chambers in Pts with CHF. See Pacemaker

int cib Between meals

int noct During the night

Intacs™ OPHTHALMOLOGY A device inserted into the periphery of the cornea, reshaping it to correct myopia of −1.0 to −3.0 diopters. See LASIK

INTACT CARDIOLOGY A clinical trial–International Nifedipine Trial on Antiatherosclerotic Therapy Study–that evaluated the effect of antihypertensive therapy with nifedipine, a CCB, in preventing heart disease. See Calcium channel blocker, Coronary artery disease, Nifedipine

intact dilation & evacuation Partial birth abortion, see there

Integra™ DERMATOLOGY An acellular artificial skin used to cover severe burns and wounds. See Artificial skin, Burns

integrated delivery system Integrated provider MEDICAL PRACTICE A coordinated health care system formed by physician groups and hospitals which ↑ efficiency and ↓ redundancy in providing health care; IDSs coordinate delivery of a broad range of health services–eg, hospital and ambulatory care services for members by contracting with several provider sites and health plans

integration INFORMATICS The successful interfacing of disparate platforms, versions of software, and devices into a coherent functioning information system PSYCHIATRY The incorporation of new and old data, experience, and emotional capacities into the personality; also refers to the organization and amalgamation of functions at various levels of psychosexual development VOX POPULI The incorporation of multiple units into one; assimilation. See Horizontal integration, Osteointegration, Seamless integration, Vertical integration

integrative medicine The 'new medicine' A term for the incorporation of alternative therapies into mainstream medical practice. See Alternative medicine, Integrative technique; Cf Mainstream medicine, Osteopathy

INTEGRITI CARDIOLOGY A clinical trial–Integrilin® and Tenecteplase–TNKase™ in Acute Myocardial Infarction

integumentary *adjective* Referring to a covering or skin; cutaneous

intellectual processing power NEUROLOGY A measure of the human brain's computational capacity, estimated to range from 10^{12} to 10^{16} synapse operations/sec

intellectualization PSYCHIATRY The use of reasoning in response to confrontation with unconscious conflicts and accompanying stressful emotions

intelligence quotient A ratio that compares a person's cognitive skills with that of the general population, usually calculated as the mental age divided by the chronologic age, multiplied by 100

INTELLIGENCE QUOTIENT TESTS

PRESCHOOL Bayley Scale of Infant development, Stanford-Binet Intelligence Scale, McCarthy Scales of Children's Abilities, Kaufman Assessment Battery for Children

SCHOOL AGE Wechsler scales, Stanford-Binet Intelligence Scale, Kaufman Assessment Battery for Children

ADULT Wechsler scales, Stanford-Binet Intelligence Scale

ADULT		
	20-35	Severe mental retardation
	36-51	Moderate mental retardation
	52-67	Mild mental retardation
	68-83	Borderline mental retardation
	90-110	Average
	>140	Gifted–'genius'

---

intensification chemotherapy Intensification ONCOLOGY Chemotherapy that is more aggressive than the standard therapy; IC improves survival in children with ALL and may be used in AML. See Adjuvant chemotherapy, Consolidation chemotherapy, Induction chemotherapy, Maintenance chemotherapy, Salvage chemotherapy. Cf Adjuvant chemotherapy

intensity-modulated radiation therapy See IMRT

intensity of care MEDICAL PRACTICE A measure of the number, technical complexity, or attendant risk of services provided. See Overtreatment, Undertreatment

Intensity of Sexual Desire & Symptoms Scale "Horniness scale" PSYCHOLOGY An instrument that assesses 3 types of behavior: sexual interest/desire, sexual activity, sexual fantasies

---

INTENSITY OF SEXUAL DESIRE & SYMPTOMS SCALE

SEXUAL INTEREST/DESIRE Nature, intensity, and frequency of sexual thoughts and desires in the previous month

SEXUAL ACTIVITY Number of times masturbate or engages in overt sexual acts/month; number and nature of incidents of abnormal sexual behavior (if any)

SEXUAL FANTASIES Pt is asked re erotic fantasies, and described the object of the fantasies, the intensity of same and frequency/month

All 3 behaviors are rated on an 8-point scale

---

intensive care unit See ICU

intensive monitoring INTENSIVE CARE The continuous monitoring of Pt vital signs, with electronic hookups to the nursing station; IM encompasses real time measurement of BP and ABGs via arterial lines, pulse oximetry, continuous cardiac monitoring, respiration, pulse, wedge pressure, measuring I/O–input and output

intensive nursing INTENSIVE CARE The nursing part of intensive monitoring–IM, with additional IM of therapy. See Intensive monitoring

intensivist Intensive care specialist MEDTALK A hospital-based critical care physician who works primarily in an ICU. See Hospitalist

intention tremor Kinetic tremor, Volitional tremor, Voluntary tremor NEUROLOGY A tremor that occurs during voluntary precision movement caused by disorders of the cerebellum and connections thereto/therefrom. See Tremor

inter- *prefix*, Latin, between

Inter Fix™ ORTHOPEDICS An implantable threaded spinal fusion "cage" used for vertebral fusion in degenerative disk disease; the device is first packed with a previously harvested autologous bone graft and implanted between 2 vertebrae; spinal bone grows and fuses with the bone in the device to form a permanent fusion. See Spinal fusion

interaction VOX POPULI The reciprocal activities of 2 or more entities in a shared environment. See Adhesive interaction, Drug interaction, Hydrophobic interaction, Integrin-mediated adhesive interaction, Paracrine interaction, Social interaction, Statistical interaction, VA interaction

interception of pregnancy Menstrual extraction, see there

intercostal retractions PHYSICAL EXAM The furrowed indentation of the intercostal spaces at inhalation, which is particularly prominent in Pts with difficulty in breathing. See Air hunger

intercourse SEXOLOGY Sexual activity between 2 people–generally understood to be penovaginal or penoanal interaction; an obsolete usage defines intercourse as a nonsexual connection, interaction or conversation between people. See Anal, Coital fantasy, Coitus, Copulation, Fertility-focused intercourse, Mount

interest See Conflict of interest, Liberty interest, Significant beneficial interest, Significant financial interest

interesting disease A clinical or pathologic nosology that is rare, difficult to diagnose, challenging to treat, poorly understood pathogenically, or any combination of the above

interface INFORMATICS 1. The electronic connection where 2 parts of a system are joined–eg, software program meets a hardware component, or where hardware meets an input device 2. Software that joins 2 different information systems. See Application program interface, Bidirectional interface, Command line interface, Fiber distributed data interface, GUI interface, Haptic interface, Messaging API interface, Parallel interface, Serial interface

interference LAB MEDICINE The effect that unmeasured components in a specimen or system have on the accuracy of a component being measured. See Electromagnetic interference, Matrix, Matrix interference, Matrix effect, Nerve interference

interference screw ORTHOPEDICS A specialty screw occasionally used to repair the anterior cruciate ligament. See Screw

interferon CELL BIOLOGY A family of immune regulatory proteins–immunomodulators–produced by T cells, fibroblasts, and other cells in response to double-stranded DNA, viruses, mitogens, antigens, or lectins; IFNs ↑ the bactericidal, viricidal and tumoricidal activities of macrophages TYPES α–20 subtypes, IFN-β–2 subtypes, both produced by macrophages, IFN-γ, IFN-omega, IFN-tau ACTIONS 1. Antiviral, causing those cells playing host to certain viruses–eg, rhinovirus, HPV, and retrovirus to produce proteins that interfere with intracellular viral replication 2. Antiproliferative, acting by unknown mechanisms, possibly ↓ translation of certain proteins, slowing cell cycling 3. Immunomodulatory, stimulating certain immune effects–T-cell activation, maturation of pre-NK cells, and ↑ phagocytosis and cytotoxicity by macrophages ADVERSE EFFECTS Flu-like symptoms, GI tract–N&V, anorexia, diarrhea, dysgeusia, xerostomia, neurologic–confusion, somnolence, poor concentration, seizures, transient aphasia, hallucinations, paranoia, psychoses, cardiopulmonary–tachycardia, dyspnea, orthostatic hypotension, cyanosis, hepatorenal–↑ transaminases, ↑ BUN, proteinuria, hematologic–neutropenia, thrombocytopenia Sx. See Biological response modifier, MAF, MIF

interferon-alpha IFN-α IMMUNOLOGY A family of 20 leukocyte-derived immunomodulating glycoproteins with antiproliferative and antiviral activity INDICATIONS Recombinant IFN-α is used to treat hairy cell leukemia, CML,

KS, HPV-related proliferations—eg, condyloma acuminatum, respiratory papillomatosis, pulmonary hemangioendotheliosis, chronic hepatitis—eg, HCV, as well as renal CA ±vinblastine, NHL, and to inhibit angiogenesis SIDE EFFECTS Flu-like Sx, malaise, fatigue, headache, anxiety, vertigo, depression, ↑ aminotransferase, BM suppression—granulocyte toxicity, SVT with hypotension or HTN, potentially fatal CHF, exacerbation of multiple sclerosis; ½ of Pts, especially with cirrhosis and hypersplenism don't tolerate therapy because of fatigue, profound thrombocytopenia and neutropenia. See Angiogenesis inhibitors

interferon beta Fibroblast interferon IFN-β A 20 kD anti-viral protein with 30% 'homology' with IFN alpha, encoded on chromosome 9, produced by fibroblasts in response to viruses or polyribonucleotides

interferon gamma IFN-γ A 21-25 kD glycoprotein lymphokine encoded on chromosome 12q and produced by activated T and NK cells; IFN-γ is antiviral, regulates class II MHC antigen expression, Fc receptors and immunoglobulin production and class switching, activates monocyte cytotoxicity and enhances NK cell activity; IFN-γ is ↓ in IgA deficiency, lymphoma, CLL, infections—eg, CMV, EBV, rubella, lepromatous leprosy, TB, SLE, rheumatoid arthritis, sickle cell anemia, post-transplantation INDICATIONS Recombinant IFN-γ is used to treat condylomata acuminata, CLL, Hodgkin's disease, mycosis fungoides, rheumatoid arthritis, and possibly leprosy, TB, nontuberculous mycobacterial infections, toxoplasmosis, chronic granulomatous disease—to prevent infections; IFN-γ suppresses collagen synthesis by fibroblasts, ↓ the size of keloids, causing a local ↑ in inflammatory cells and mucin production, and may be of use in controlling abnormal fibrosing conditions SIDE EFFECTS Acute renal failure, rash, headache, chills

Intergel™ Lubricoat® SURGERY A ferric hyaluronate gel intended to prevent or reduce postsurgical adhesion formation. See Adhesions

interictal NEUROLOGY *adjective* Referring to an interval between seizures or convulsions

interim analysis CLINICAL TRIALS Analysis of early data accumulated in a clinical trial, before all of the Pts have been enrolled, to detect trends that might warrant modification of the protocol, change in type or format of data being collected or, if extreme, trial termination. See CONSENSUS II

interleukin IMMUNOLOGY Any of a family of cytokines produced by lymphocytes, monocytes, and other cells, which induce growth and differentiation of lymphoid cells and primitive hematopoietic stem cells;. See Biological response modifier

IL-1 Leukocyte-activating factor An 11 kD cytokine produced by a wide range of cells; IL-1 elicits the acute phase response, acts on the CNS as a pyrogen, stimulates fibroblast, B- and T-cell proliferation and differentiation, ↑ lymphokine, collagenase and PG production, activity of NK cells against tumor targets, stimulates myocytolysis—via PGs, elicits hormone release from the pituitary, evokes the release of PMNs from the BM, and PMN degranulation, ↑ oxidase activity and hexose monophosphate shunt activity; in synovial cells, IL-1 stimulates proliferation and production of collagen, PG, and plasminogen activator IL-1 PRODUCTION Stimulated by various agents—eg, calcium ionophores, IFN-α, IFN-γ, lipopolysaccharides, muramyl dipeptide, aluminum hydroxide, phorbol myristate acetate, staphylococci, silica, and others; inhibited by corticosteroids, PGE2—via the cyclooxygenase pathway, suppressor T cells, and cyclosporine—which specifically inhibits T cell-induced IL-1 production IL-1 & DISEASE IL-1 may have a role in 1. Type 1 DM, ↑ IL-1 production by macrophages in the early β-cell destructive lesions of type 1 DM 2. ASHD—n-3 fish oils ↓ circulating IL-1 3. Rheumatoid arthritis—IL-1 acts with substance P

IL-1α THERAPEUTIC USES IL-1α is reportedly effective in accelerating recovery of platelet counts after high-dose carboplatin and may be useful in thrombocytopenia induced by chemotherapy

IL-2 Aldesleukin, IL-2, T-cell growth factor IMMUNOLOGY An immunomodulating biological response modifier produced at low baseline levels by CD4 T cells; after antigen presentation by antigen-presenting cells in presence of IL-1, T cell production of IL-2 and IL-2 receptor—IL-2Rβ on membrane ↑, peaks at 6 hrs and falls to baseline levels; IL-2 is also produced by medullary thymocytes and a subset of large granular lymphocytes—NK cell activators; IL-2 up-regulates the immune system, causing lymphokine-activated killer—LAK cells to lyse tumor cells, see IL-2/LAK cells THERAPEUTICS Recombinant IL-2 has been used for various conditions, including metastatic renal cell CA ADVERSE EFFECTS IL-2 therapy requires wks of intensive care due to toxic effects, including a 'capillary leakage syndrome', malaise, gastritis, GI symptoms, anemia, thrombocytopenia, rigors, chills, fever, hypotension, azotemia, jaundice, hyperbilirubinemia, rash—erythroderma globalis, confusion, ascites, fluid retention, pruritus, agitation, respiratory insufficiency, cardiac failure, irreversible demyelinization and hypothyroidism. See IL-2/LAK cells

IL-2/LAK cells Interleukin-2/Lymphokine-activated killer cells NK cells that have been activated by co-incubating in IL-2 and injected into cancer Pts in a therapeutic modality known as adoptive immunotherapy, which may evoke temporary tumor regression in non-Hodgkin's lymphoma, melanoma, colorectal CA, and renal cell CA, and induce regression of hepatic and pulmonary metastases ADVERSE EFFECTS Pulmonary edema, CHF

IL-6 IFN-β-2 A glycosylated 22-27 kD cytokine that mediates host response to injury and infection and plays a role in growth and differentiation of B cells, T cells, myelomas, hepatocytes, hematopoietic stem cells and nerve cells; IL-6 is critical for development of IgA-related immunity; IL-6 is a major mediator of the acute phase response—APR, stimulating hepatic production of APR proteins

interlocking OBSTETRICS A rare complication of vaginal delivery of twins; the 1st twin presents in breech and descends 'locking' his head above the head of the 2nd twin in vertex presentation; if 2nd baby is already in the pelvis, loss of 1st baby is almost inevitable and interventional decapitation of 1st infant may be necessary to salvage the 2nd infant PREVENTION C-section. Cf Selective termination

intermediate care facility MANAGED CARE A state-licensed facility that provides nursing care to those who don't require the degree of care which a hospital or skilled nursing facility provides. Cf Hospice, Nursing home, Shelter, Tertiary care center

intermediate density lipoprotein PHYSIOLOGY A plasma lipoprotein with a density of 1.006-1.019 g/dl, formed by lipoprotein lipase hydrolysis of VLDL partially depleted of TGs COMPOSITION 15% protein, predominantly apoB and apoE, 25% phospholipid, 30% cholesterol, 30% TGs ROLE Transports cholesterol from intestine to liver; once in the liver, is delipidated by hepatic lipoprotein lipase to form a low-density protein

intermediate lymphocytic lymphoma Mantle-zone lymphoma HEMATOLOGY An indolent B-cell lymphoma of older adults, related to small cell follicular lymphoma, composed of cells similar to those of WD lymphocytic lymphoma. See Lymphoma

intermediate resistance INFECTIOUS DISEASE Partial resistance of a bacterium to a particular antibiotic that falls short of resistance

intermediate syndrome TOXICOLOGY A condition caused by organophosphorus insecticides, characterized by chronic distal motor polyneuropathy, possibly due to a neuromuscular junction defect CLINICAL 5% to 10% of those exposed develop paralysis of cranial motor nerves, proximal limb, cervical flexor and respiratory muscles; onset, 1-4 days after a cholinergic phase

intermittent catherization UROLOGY Periodic catheterization to remove residual urine CONS Discomfort, ↑ UTIs, urethral trauma. See Bladder ultrasonography, Postvoid residual urine volume

intermittent claudication VASCULAR DISEASE A condition caused by ASHD-related muscle ischemia CLINICAL Asymptomatic at rest; with minimal movement, intense pain, weakness, limping

intermittent regimen THERAPEUTICS A treatment schedule in which the Pt takes each prescribed medication 2 or 3 times wkly at appropriate dosage

intermixed stroma-rich neuroblastoma Ganglioneuroblastoma, see there

intern GRADUATE EDUCATION A term of art equivalent to 'apprentice', used in North America for a graduate of a medical, osteopathic or dental school serving a first yr–'internship' of graduate clinical training, in a teaching hospital. See GME. Cf CME, Extern, Fellow, Internist

internal *adjective* Toward the center, away from the body surface

internal fetal monitoring OBSTETRICS The use of 2 electronic catheters inserted through the vagina and cervix; one is attached to the baby's scalp and measures fetal heart rate; the 2nd is placed between the fetus and the wall of the uterus and measures the rate and intensity of uterine contractions. Cf External fetal monitoring

internal fixation Intraosseous fixation ORTHOPEDICS Open reduction and stabilization of fractured bone fragments by direct binding to each other with IF devices–eg, screws, plates, wires, pins, intramedullary rods, nails, spinal fixation devices. Cf External fixation

internal hemorrhoids SURGERY Engorged veins under the rectal mucosa. Cf External hemorrhoids

internal radiation therapy Brachytherapy, see there

internal rotation Medial rotation The act of turning about an axis passing through the center of the leg, which occurs with closed chain pronation; the talus acts as an extension of the leg in the frontal and transverse planes. Cf External rotation

internal version OBSTETRICS A transvaginal procedure that converts a difficult fetal presentation to a vaginally deliverable situation by hand-rotating the fetus in utero. Cf External version

international medical graduate Foreign medical graduate A physician who graduated from a non-North American–ie, Canada or US medical school. See Off-shore medical school, USFMG

International Normalized Ratio HEMATOLOGY A method of reporting prothrombin time–PT results for Pts receiving oral anticoagulant therapy; the INR is defined by the formula, $PT_{Patient}/PT_{MNPT}^{ISI}$ that uses the international sensitivity index to adjust PT results from thromboplastin of different sources and facilitates monitoring anticoagulant therapy when using more than one lab and comparing a Pt's PT results with published standards for treatment. See Anticoagulation, International sensitivity index, Prothrombin time

international sensitivity index HEMATOLOGY A measure of thromboplastin-activating capacity, or the sensitivity of a thromboplastin reagent to anticoagulation. See International normalized ratio

international unit Any arbitrarily defined and internationally sanctioned unit of measurement for a naturally-occurring substance–eg, hormone, enzyme or vitamin. See SI THERAPEUTICS Drug concentrationin–µg/mL X 10^3/MW of drug–g/mole

International Workshop classification for chronic lymphocytic leukemia A clinical staging system that blends features of Rai and Binet classifications of CLL

INTERNATIONAL WORKSHOP CLASSIFICATION, CLL
STAGE A No anemia or thrombocytopenia and < 3 involved lymphoid regions
STAGE B No anemia or thrombocytopenia and > 3 involved lymphoid regions
STAGE C Anemia and/or thrombocytopenia, regardless of amount of involved lymphoid tissue

internet addiction disorder Internet misuse PSYCHOLOGY A maladaptive pattern of Internet use, leading to clinically significant impairment or distress

INTERNET ADDICTION DISORDER–MANIFEST BY 3 OR MORE OF THE FOLLOWING, OCCURRING AT ANY TIME IN THE SAME 12-MONTH PERIOD
1. **TOLERANCE** defined by either of the following: A. Need for markedly ↑ amounts of time on the Internet to achieve satisfaction B. Markedly diminished effect with continued use of the same amount of time on Internet
2. **WITHDRAWAL** manifest by either of the following:
   A. A characteristic withdrawal syndrome
   1. Following cessation of–or reduction in—Internet use that has been heavy and prolonged
   2. Two or more of the following, developing within several days to a month after Criterion 1:
      a. Psychomotor agitation
      b. Anxiety
      c. Obsessive rumination regarding events presently occurring on the Internet
      d. Fantasies or dreams about the Internet
      e. Involuntary typing movements of the fingers
   3. The symptoms in Criterion 2 cause distress or impairment in social, occupational, or another important area of functioning
   B. Use of Internet or a similar on-line service is engaged to relieve or avoid withdrawal symptoms
3. **INTERNET** Accessed more often or for longer period of time than was intended
4. **DESIRE TO CONTROL USE** Persistent desire or unsuccessful efforts to cut down or control Internet use
5. **TIME SPENT** A great deal of time is spent in activities related to Internet use, eg buying Internet books, trying out new web browsers, researching Internet vendors, organizing files of downloaded material
6. **NORMAL ACTIVITIES IGNORED** Important social, occupational, or recreational activities are given up or reduced because of Internet use
7. **USE CONTINUES DESPITE RECOGNITION OF ILL EFFECTS** Use continues despite knowledge of having a persistent or recurrent physical, social, occupational, or psychological problem that is likely to have been caused or exacerbated by Internet use–sleep deprivation, marital difficulties, lateness for early morning appointments, neglect of occupational duties, or feelings of abandonment in significant others Internet Addiction Support Group psy-doc@netcom.com

internet pharmacy ONLINE A website that offers prescription drugs from the comfort of home CONS The IP or prescribing physician may not be qualified or licensed to prescribe drugs in all states. See Operation Cure-All, VIPPS

internet service provider Internet provider INFORMATICS A company–eg, AOL, Earthlink, that provides a gateway to the Internet and the Web, via an electronic connection–either directly or through a modem, as well as software and services, for which the user pays a monthly fee

internist MEDTALK A practitioner of general medicine is certified by the Am Board of Internal Medicine–ABIM, who has had 3 yrs of formal training in internal medicine. Cf Family practitioner, Intern

the internist's tumor MEDTALK A popular term for renal cell CA, known for its protean manifestations and ability to mimic–aka the great mimic–other malignancies or benign conditions. See the Great imitator

internship MEDTALK A period–often 1 yr following the close of a period of formal education during which a health professional gains practical experience. See Acting internship

interpersonal psychotherapy PSYCHIATRY A semistructured treatment in which the Pt is educated about depression and depressive Sx, and the

Pt's relation to the environment, especially social functioning; unlike traditional psychotherapy, IP focuses on the present tense and not on underlying personality structures. Cf Family therapy, Group therapy

INTERSALT VASCULAR DISEASE A collaborative epidemiologic study–International Study of Electrolye Excretion and Blood Pressure–that evaluated the relationship of salt intake and BP. See Hypertension

intersecting epidemics The linking of 2 epidemics, in such way that one predisposes to the second epidemic–eg, the crack cocaine epidemic and AIDS, as sexual work may be performed to obtain the crack. See AIDS, Crack cocaine

intersex syndrome Intersex disorder A group of clinical complexes that occur in subjects with ambiguous genitalia. See Female pseudohermaphroditism, Male pseudohermaphroditism, Mixed gonadal dysgenesis, Pure gonadal dysgenesis, True pseudohermaphroditism

---

INTERSEX SYNDROMES

FEMALE PSEUDOHERMAPHRODITISM Ovaries are present, but the infant is masculinized by in utero androgen exposure during fetal development–maternal ingestion or the result of congenital adrenal hyperplasia with virilization

GONADAL DYSGENESIS Underdeveloped or imperfectly formed gonads; the prototypic gonadal dysgenesis is Turner syndrome 45, X0, seen in 1/2-7000 female births CLINICAL Short stature, webbed neck, cubitus valgus, micrognathia with high arched palate, epicanthal folds, lymphedema of the hands and feet, aortic coarctation, renal malformation, osteoporosis, diabetes, widely-spaced nipples, sexual infantilism PATHOLOGY Ovaries are small and thin–'streak' ovaries; a variant, mixed gonadal dysgenesis is characterized by a mosaic phenotype 45,X/46,XY, and a streak ovary on one side and a testis or germ cell tumor on the other side, accompanied by intense virilization

MALE PSEUDOHERMAPHRODITISM Testicles are present and cryptorchid, but testosterone production is inadequate–due to ↓ LH or hCG receptors on the Leydig cells; Pts are raised as ♀–Morris syndrome and may have defects of the CNS, eg defective gonadotropin response, primary gonadal defects, eg idiopathic, defective pregnanediol–3-β 17-α, 17,20 des and 17 β synthesis, regression of müllerian tubes, Leydig cell agenesis, androgen insensitivity, ↑ susceptibility to breast cancer, Sertoli adenoma, germinoma in situ, seminoma, Leydig cell tumor

TRUE HERMAPHRODITISM Gonads contain both ovarian and testicular tissue, genotypically either 46, XX or 46, XY; 75% are raised as ♂; the testicular tissue is dysgenic, doesn't produce sperm and may undergo malignant degeneration–requiring prophylactic removal; ovarian function in those raised as ♀ may be adequate to produce term pregnancy

---

Interstim® UROLOGY An implantable device that electronically stimulates sacral nerves to help control bladder function in Pts with urgency-frequency ↓ Sx and urine retention sans mechanical obstruction. See Urge incontinence

interstitial cystitis UROLOGY Idiopathic bladder wall inflammation, primarily affecting ♀ CLINICAL Pain, pressure, tenderness around bladder, pelvis, perineum; urgency; ↓ bladder capacity; painful sexual intercourse; in ♂, discomfort, pain in penis, scrotum DIAGNOSIS Hx, Sx, cystoscopy, biopsy; it is a Dx of exclusion MANAGEMENT Symptomatic therapy COMPLICATIONS Scarring, stiffening, ulceration, bleeding

interstitial disease MEDTALK 1. Interstitial cystitis, see there 2. Interstitial lung disease, see there 3. Interstitial nephritis, see there

interstitial keratitis OPHTHALMOLOGY Inflammation of the deep corneal stroma, often with neovascularization ETIOLOGY Syphilis, leprosy, TB CLINICAL Impaired vision, pain, excessive watering, photophobia

interstitial laser coagulation UROLOGY A type of therapy for BPH; a standard cystoscope is introduced into the urethra delivering a fiberoptic probe into the prostate; low-power laser energy delivered through the probe, heats and destroys a controlled amount of prostate, which is resorbed by the body over time. See Benign prostatic hypertrophy. Cf Transurethral resection of prostate or TURP

interstitial lung disease Diffuse interstitial pulmonary fibrosis PULMONOLOGY A group of disorders characterized by scarring of deep lung tissue, leading to SOB and loss of functional alveoli, limiting O_2 exchange; ILD is more common in smokers ETIOLOGY Inorganic and organic dusts, gases, fumes, vapors, medications, radiation, and certain lung infections, hypersensitivity pneumonitis, coal worker's pneumoconiosis, silicosis, byssinosis, idiopathic

Interstitial nephritis Tubulointerstitial nephritis NEPHROLOGY Inflammation of the space in and around the renal tubules; IN may be a transient drug response or chronic and progressive; it may accompany analgesic nephropathy and acute interstitial allergic nephritis DRUGS LINKED TO IN Antibiotics–penicillin, ampicillin, methicillin, sulfonamides, etc; NSAIDs, furosemide, thiazide diuretics CLINICAL IN ↓ renal function–from mild dysfunction to acute renal failure–ARF; ±½ of cases exhibit ↓ urine output and other signs of ARF–eg, failure to concentrate urine or regulate acid/base levels–resulting in metabolic acidosis, progression to ESRD. See Renal failure

interstitial therapy Brachytherapy, see there

intertrigo DERMATOLOGY Superficial dermatitis due to prolonged moisture and bacteria in skin fold areas, often in overweight people SITES Inner thighs, armpits, under breasts. See Morbid obesiity

interval VOX POPULI A period of time or space between 2 discrete periods or objects. See AH interval, AV interval, Class interval, Confidence interval, Conference interval, Dosing interval, Escape interval, Lucid interval, PA interval, QT interval, V-A interval

interval cancer ONCOLOGY A cancer that develops in the intervals between routine screening for a particular cancer–eg, prostate CA, breast CA, etc

intervention PUBLIC HEALTH A device or procedure capable of ↓ injuries. See Administrative intervention, Behavioral intervention, Crisis intervention, Health intervention SURGERY An operation. See Routine intervention, Motivational intervention, Percutaneous intervention, Pharmacist intervention, Remedial intervention

interventional cardiology CARDIOLOGY The subspecialty of cardiology dedicated to the diagnosis, medical and mechanical therapy, pre- and post-procedure management of adult patients with acute and chronic forms of cardiovascular disease amenable to catheter-based therapy IC TECHNIQUES Balloon angioplasty, intracoronary stent deployment, rotational atherectomy, extraction atherectomy, directional coronary atherectomy, balloon valvuloplasty, peripheral angioplasty. See Invasive cardiology. Cf Echocardiography, Noninterventional therapy

interventional neuroradiology A subspecialty of neuroradiology in which minimally invasive therapy can be effected by advancing various devices within a blood vessel to a point of a previously identified lesion–eg, an intracranial aneurysm. Cf Clipping, Trapping

interventional radiology IMAGING A subspecialty of radiology that provides DIAGNOSTIC INFORMATION–eg, CT-guided 'skinny' needle biopsies and dye injection for analysis of various lumina and tracts–eg, arteriography, cholangiography, antegrade pyelography or THERAPEUTIC OPTIONS–eg, percutaneous nephrostomy or biliary drainage

interventricular *adjective* Between 2 ventricles–eg, of the brain or heart

intestinal 'angina' Chronic intermittent occlusion of intestinal arteries, analogous to angina pectoris, causing sporadic claudication of the vascular supply to the intestine; ingestion of food causes IA as digestion ↑ blood flow through the gut, which is supplied by the celiac axis, superior and inferior mesenteric arteries; postprandial abdominal pain implies major atherosclerotic narrowing of > one vessel given the rich anastomotic network among these vessels

intestinal bacterial overgrowth Afferent loop syndrome, gastrojejunal loop obstruction, stagnant loop syndrome INFECTIOUS DISEASE A condition characterized by excess growth of opportunistic bacteria, which occurs when normal gut flora is eradicated with antibiotics CLINICAL Necrotizing enterocolitis, pseudomembranous colitis. See Pseudomembranous colitis

intestinal cancer Colorectal cancer, see there

intestinal malrotation GI DISEASE The malrotation of the intestines in utero, occurring to a greater or lesser degree than normal. See Ileus, Megacolon

intestinal obstruction Bowel obstruction SURGERY Obstruction of the intestine due to either mechanical causes–eg, volvulus, fecal impaction or nonmechanical causes–eg, paralytic ileus, see there

intestinal perforation Gastrointestinal perforation SURGERY The loss of integrity of the bowel wall which may be due to trauma–eg, shotgun blast to abdomen or ischemic breakdown of intestinal wall. See Fecal peritonitis

intestinal polyp Colorectal polyp. See Polyp

intimate distance PSYCHOLOGY A zone of space ranging from intimate contact to less than an arm's length in which people operate under specific social situations. See Proxemics. Cf Personal distance, Public distance, Social distance

intimate part SEXOLOGY Any primary genital area–groin, inner thigh, buttock or breast. See Boundary violation

InTIME CARDIOLOGY A clinical trial–Intravenous nPA for Treatment of Infarcting Myocardium Early–comparing efficacy of a weight-adjusted single bolus of nPA/lanoteplase to tPA–administered by infusion in restoring blood flow to the heart in Pts within 6 hrs of Sx onset. See Lanoteplase

intoeing ORTHOPEDICS A group of conditions in which a child walks/stands with the toes pointing inward, due to a malalignment of the hip and toes; it is often more noticeable in toddlers and usually improves with age CLINICAL Recurrent tripping and falling, but usually asymptomatic MANAGEMENT In most children, treatment is not necessary; some may benefit from a night splint

intolerance MEDTALK 1. Extreme discomfort with a particular environmental condition. See Cold intolerance, Heat intolerance, 2. Inability to metabolize or excrete a particular substance. See Food intolerance, Hereditary fructose intolerance, Lactose intolerance, Milk intolerance, Orthostatic intolerance

intoxication 1. A pathologic state induced by an exogenous or,less commonly, endogenous toxic substance 2. Drunkenness, inebriation TOXICOLOGY Too much of a bad or, less commonly, a good thing. See Arsenic intoxication, Chromium intoxication, Iodine intoxication, Scombroid intoxication, Selenium intoxication, Toxicology, Vitamin A intoxication, Vitamin C intoxication, Vitamin D intoxication, Vitamin E intoxication, Vitamin K intoxication, Water intoxication, Zinc intoxication

intra- *prefix*, Latin, within, inside

intra-abdominal abscess Peritoneal abscess SURGERY An abscess in the peritoneal cavity caused by a panoply of conditions–eg, ruptured appendix or diverticula, intestinal parasitosis–*Entamoeba histolytica*, etc

intra-aortic balloon pulsation CARDIOLOGY A therapeutic modality used to treat unstable angina, which consists of counterpulsation to ↑ the diastolic pressure gradient for coronary arterial blood flow, by simultaneously ↓ the left ventricular end-diastolic BP and ↑ intra-aortic diastolic pressure CONS Vascular complications are more common in ♀, Pts with DM or HTN

intra-aortic balloon pump INTERVENTIONAL CARDIOLOGY A device that changes hemodynamic loading conditions in the heart by ↑ mean and peak diastolic BP, and ↓ peak systolic BP; these changes result in ↓ left ventricular end-diastolic pressure, ↑ stroke volume, ↑ cardiac output; the IABP is used primarily for cardiogenic shock after an acute MI, or in severe mitral insufficiency of abrupt onset; the IABP ↑ coronary blood flow, myocardial perfusion, ↑ collateral coronary circulation, improves subendocardial oxygenation, ↓ myocardial O_2 consumption CONTRAINDICATIONS Acute or significant aortic regurgitation, as IABP exacerbates regurgitation from the aorta into the left ventricle and ↑ left ventricular diastolic filling pressure; IABP is

used as a ventricular assist device, pre-, intra- or postoperatively when medical therapy is insufficient by itself

intra-articular steroid injection Cortisone injection, steroid injection ORTHOPEDICS The direct injection of corticosteroid into a joint space, a conservative modality for managing degenerative joint disease, which may relieve pain for up to months

intra-axial *adjective* Within the CNS

intracardiac echocardiography CARDIOLOGY A minimally invasive imaging modality used to guide complex transseptal catheterization; IE is used to guide transseptal puncture for percutaneous transvenous mitral commisurotomy in mitral valve stenosis, as well as giant left atria, atrial septal aneurysms, or severe kyphoscoliosis

intracardiac electrophysiologic study CARDIOLOGY A study of the cardiac conduction system, performed by introducing multipolar catheter electrodes into the vascular system, and positioning them in various parts of the heart, allowing simultaneous recording of multiple leads; the catheters are used to record local electrical activity and stimulate the heart; IESs measure the AV, HA, and PA intervals, the last of which, the PA interval, has little clinical value

intracarotid *adjective* Within a carotid artery

intracarotid infusion THERAPEUTICS The introduction of fluids and drugs directly into the carotid artery, ergo the brain without passing through the liver

intracavitary *adjective* Within a space, often understood to mean the center of a necrotic lesion

intracavitary therapy Brachytherapy, see there

intracerebral hematoma A hematoma which develops in the cerebral parenchyma, most commonly caused by hypertension. See Epidural hematoma, Subdural hematoma

intracerebral hemorrhage Intracranial hemorrhage, thalamic hemorrhage NEUROLOGY Hemorrhage in the brain, often cause by hypertensive small vessel disease LOCATIONS Lobar or deep–affecting the thalamus, basal ganglia, pons, cerebellum ETIOLOGY Trauma, vascular defects–aneurysm, hemangioma, HTN, idiopathic PATHOPHYSIOLOGY Cerebral edema and hematoma result in ↑ intracranial pressure with destruction of brain tissue CLINICAL Sx are usually abrupt and occur without warning, often during activity, less commonly develop in a stepwise, episodic manner or progressive manner RISK FACTORS DIC, hemophilia, sickle cell anemia, leukemia, thrombocytopenia, aspirin, anticoagulants, thrombolytics, liver disease, cerebral amyloid, brain tumors, prematurity–associated with intra-ventricular hemorrhage, which may result in severe disability DIAGNOSIS Ultrasound, CT, MRI MANAGEMENT Surgical evacuation. See Hemorrhage. Cf Hemorrhagic stroke

intracoronary radiation therapy See ICRT

intracoronary stenting Coronary stenting CARDIOLOGY A treatment for vessel closure after percutaneous transluminal coronary angioplasty which ↓ the rate of restenosis, in which a cylindrical stent is placed within vessel(s) that are stenosed by atherosclerotic lesions COMPLICATIONS Thrombotic occlusion of stent, hemorrhage and peripheral vascular complications due to intensive anticoagulation

intracranial abscess NEUROLOGY A localized pus pocket that may be related to trauma, acute otitis interna, sinusitis IMAGING CT or if stable, MRI

intracranial hemorrhage Intracerebral hemorrhage NEUROLOGY Periventricular or cerebral hemorrhage, which may be subdural, parenchymal, intraventricular, subarachnoid; IH is more common in preterm infants ETIOLOGY Tentorial tears and skull fractures and birth-related trauma linked to operative delivery IMAGING CT in first 24 hrs, MRI after 24 hrs. Cf Periventricular leukomalacia

intractable hiccups NEUROLOGY Episodic, life-threatening hiccuping of > 1 month duration. See Hiccups

intractable pain Refractory pain PAIN MEDICINE Persistent pain which does not respond to at least 3 dosease of parenteral analgesics given over a 12-24 hr period; pain that does not respond to appropriate doses of opioid analgesics.

intradiscal electrothermal annuloplasty ORTHOPEDICS A minimally-invasive procedure used to manage severe low-back pain due to degenerative disk disease caused by damaged intervertebral disks where a wire is guided around the vertebral body, then heated to 90° for up to 15 mins–shrinking collagen fibers and destroying pain receptors RESULTS Improved functionality, sexual activity, QOL. See Low back pain

intraductal carcinoma of breast Ductal carcinoma in situ, see there

intraductal papilloma Ductal papilloma BREAST DISEASE A benign epithelial intraductal mass seen in the breast of pre- and perimenopausal ♀, arising in ductal epithelium CLINICAL Breast pain, nipple discharge, palpable mass . See Breast cancer

intraepithelial *adjective* Within the skin

intraepithelial neoplasia An in situ carcinoma confined to an epithelium that may superficially penetrate adnexal glands, measuring < either 3 mm or 5 mm depending on the criteria used; IN is adjectivally modified according to site of origin. See CIN

INTRAEPITHELIAL NEOPLASIA

AIN	Anal intraepithelial neoplasia
CIN	Cervical intraepithelial neoplasia, see there
DIN	Ductal intraepithelial neoplasia, see there
OIN	Oral intraepithelial neoplasia
PAIN	Perianal intraepithelial neoplasia
PIN	Prostatic (or rarely Penile) intraepithelial neoplasia
VAIN	Vaginal intraepithelial neoplasia
VIN	Vulvar intraepithelial neoplasia

intrahepatic infusion THERAPEUTICS Delivery of anticancer drugs directly into hepatic blood vessels

intralobar pulmonary sequestration Bronchopulmonary sequestration, see there

intraluminal brachytherapy RADIATION ONCOLOGY A form of locoregional RT ± endoluminal laser for treating non-small-cell lung CA

intramedullary *adjective* Within a medulla of an organ or structure–eg, the inner portion of the spinal cord, the medulla oblongata or BM

intramucosal carcinoma ONCOLOGY A high-grade/severe dysplasia–CIS of the GI mucosa, which extends to the lamina propria

intramural esophageal diverticulosis Pseudodiverticulosis GI DISEASE A condition characterized by multiple small diverticuli lined by squamous epithelium in the upper esophagus or occupying its length; apparently arising from mucous gland ducts. Cf Barrett's esophagus, Inlet patches

intramyocardial bridge Intramural coronary artery SURGICAL ANATOMY A band of myocardium that encases a segment of a large coronary artery–most often the left anterior descending branch; if the IB is compressed–as may occur in strenuous physical exertion, perfusion of a large region of myocardium may be compromised and cause sudden death due to V fib

intraocular pressure OPHTHALMOLOGY The pressure, in mm Hg, of the vitreous, which varies among individuals

intraoperative blood salvage Intraoperative autologous transfusion TRANSFUSION MEDICINE A procedure in which the blood shed or otherwise lost into an operative field is collected sterilely, washed, filtered and reinfused as a packed unit of RBCs; IAT is suited for 'bloody' heart–eg, aortic aneurysms or CABG or orthopedic surgery–eg, hip arthrodesis, ↓ Pt exposure to multiple donors and may be used with pre-deposit autologous donation, where packed red cells are collected from the Pt before surgery. See Surgical blood management. Cf Autologous transfusion

intraoperative hemodilution TRANSFUSION MEDICINE A form of perioperative autologous blood transfusion, in which 1-2 units of blood are withdrawn from the Pt before surgery and replaced with crystalloid; the blood is stored at room temperature in the OR, thereby eliminating need to formally collect, test, and store blood. Cf Autologous blood transfusion, Intraoperative blood salvage

intraoperative radiation therapy RADIATION ONCOLOGY RT directly in a tumor during surgery. See Radiation oncology

intraosseous infusion EMERGENCY MEDICINE A format for rapid infusion of fluids in Pts in shock and trauma, using a 15-18 gauge sternal or iliac BM—aspiration needle for intraosseous infusion in emergencies as an alternative to IV access COMPLICATIONS Placement failure, local cellulitis, abscesses, osteomyelitis . See Cutdown

intraperitoneal hyperthermic chemotherapy ONCOLOGY The administration of heated chemotherapeutics in solution circulated in the peritoneal cavity. See Chemotherapy

intraperitoneal infusion THERAPEUTICS A method of delivering fluids and drugs directly into the peritoneal cavity via a cannula

intrapsychic *adjective* Referring to that which occurs in the psyche or mind

IntraSite gel WOUND CARE A hydrogel for treating wounds that creates and maintains a moist environment. See Wound care

intrastromal corneal ring ICR® OPHTHALMOLOGY A ring comprised of 2 thin, transparent, half-rings made of a polymer inserted into the periphery of the cornea, reshaping it, to correct myopia

intrathecal *adjective* Within a sheath, specifically, the spinal canal–in the subarachnoid or subdural space

intrauterine growth retardation Fetal growth restriction NEONATOLOGY A generic term for any delay in achieving intrauterine developmental milestones, most commonly related to maternal drug, tobacco and alcohol abuse; IUGR affects high-risk infants with perinatal asphyxia, hypoglycemia, hypothermia, pulmonary hemorrhage, meconium aspiration, necrotizing enterocolitis, polycythemia and complications of infections, malformations and syndromes; IUGR fetuses have weight < 10th percentile for gestational age, abdominal circumference < 2.5th percentile TYPES Symmetric–body is proportionately small; asymmetric–head is disproportionately bigger than body, which implies undernourishment–growth of vital organs–heart, brain is at expense of liver, muscle and fat, often due to placental insufficiency; IUGR is the 2nd most common–after prematurity–cause of perinatal M&M; it affects ±5% of the general obstetric population. See Low birthweight, Small for gestational age

INTRAUTERINE GROWTH RESTRICTION
Placental insufficiency

- Unexplained elevated maternal alpha- fetoprotein level
- Idiopathic
- Preeclampsia
- Chronic maternal disease

- Cardiovascular disease
- Diabetes
- Hypertension

Abnormal placentation
- Abruptio placentae
- Placenta previa
- Infarction
- Circumvallate placenta
- Placenta accretia
- Hemangioma

Genetic disorders
- Family history
- Trisomy 13, 18 and 21
- Triploidy
- Turner's syndrome (some cases)
- Malformations

Immunologic
- Antiphospholipid syndrome

Infections
- Cytomegalovirus
- Rubella
- Herpes
- Toxoplasmosis

Metabolic
- Phenylketonuria

Other
- Poor maternal nutrition
- Substance abuse (smoking, alcohol, drugs)
- Multiple gestation
- Low socioeconomic status

intrauterine insemination Turkey baster insemination REPRODUCTION MEDICINE The direct introduction of sperm in the uterus, a maneuver used in unexplained or ♂-factor infertility. See Artificial reproductive technology

intrauterine pressure catheter OBSTETRICS A device inserted between the uterine wall and the fetus to measure the intensity of contractions in a ♀ who has received oxytocin. Cf Internal fetal monitoring

intrauterine transfusion The in utero administration of RBCs to a fetus with HDN. See Fetal paralysis

intravascular brachytherapy INTERVENTIONAL CARDIOLOGY The intracoronary administration of beta radiation to prevent restenosis of coronary arteries–which occurs in ±50% of Pts after balloon and other forms of angioplasty

intravascular lymphomatosis Angiotropic large cell lymphoma, malignant angioendotheliomatosis, neoplastic angioendotheliomatosis ONCOLOGY A lymphoma in which the lymphocytes home to vascular endothelium, possibly related to expression of specific lymphocyte-endothelium adhesion molecules PROGNOSIS Poor; most die, often *before* diagnosis

intravenous drug use Intravenous drug abuse The habitual IV injection of drugs of abuse EPIDEMIOLOGY In the US ± 2.5 million–population ± 235 million have used IVDs INFECTIONS Pyogenic–eg, endocarditis, pneumonia, sepsis COMMON AGENTS *S pneumoniae, H influenzae*, hepatitis–HBV, alcoholic, STD, TB

intravenous immune globulin A formulation of concentrated antibodies–aka immune globulins, predominantly IgG, prepared by pooling plasma from ±1000 donors, with a broad spectrum of activity against CMV, HAV, HBV, measles, rubella, tetanus, varicella zoster INDICATIONS LBW children with repeated infections, children with defects in humoral immune responses–eg, AIDS, X-linked agammaglobulinemia, CVID; IVIG may be of use in ITP, autoimmune phenomena–hemolysis, neutropenia, throm-

bocytopenia, Kawasaki's disease ADVERSE EFFECTS Pyrogenic, hypersensitivity, anaphylaxis, systemic–eg, headache, myalgia, fever, vasomotor and cardiovascular abnormalities, BP lability, tachycardia. Cf Human immune globulin

intravenous pyelogram IMAGING The 'hard copy' of intravenous pyelography. See IVP

intravenous pyelography IMAGING An imaging study of the transitional mucosa of the urinary tract after IV injection of a radiocontrast that concentrates in urine; an IVP outlines the renal pelvis, ureters, and bladder

intravesicular therapy ONCOLOGY Irrigation of urinary bladder with topical agents–eg, thiotepa, mitomycin C, BCG, to halt progression of superficial transitional cell CA; other agents include adriamycin, epodyl, 5-FU

intrinsic *adjective* Within or referring to an internal origin CARDIAC PACING Inherent; belonging to or originating from the heart itself–eg, an intrinsic or natural heart beat

intrinsic factor PHYSIOLOGY A 45 kD low-affinity vitamin B_{12}-binding glycoprotein secreted by gastric parietal cells, which closely parallels HCl secretion; IF secretion is stimulated by histamine, gastrin and methionine, and usually greatly exceeds that required for B_{12} absorption; IF is re-absorbed by specific receptors in the ileum; the high-affinity B_{12} binder, R protein, attaches to vitamin B_{12} in the acidic stomach, later releasing B_{12} to IF after cleavage by pancreatic enzymes; IF is ↓ in Pts with low gastric acid production or by agents which ↓ gastric acid secretion by blocking parietal cell receptors–eg, H_2-blockers, but is unaffected by agents that block H^+/K^+-ATPase-induced gastric acid secretion. See H_2-blockers, Vitamin B_{12}

intrinsic pathway HEMATOLOGY An arm of the coagulation cascade, initiated by negatively charged surfaces–contact factors–eg, sulfatide micelles, kaolin which bind factor XII and HMW kininogen–HMWK; HMWK binds prekallikrein and factor XI, activating the latter; XIa then activates IX which in turn activates factor X, initiating the common pathway. See Common pathway, Extrinsic pathway

intrinsicoid deflection EKG An abrupt deflection that falls between Q and R seen on unipolar precordial leads–eg, V_1 to V_6 in Pts with BBBs. See Bundle branch block, Deflection

introceptive *adjective* Referring to internal neural information. Cf Exteroceptive

introducer Introducer sheath INTERVENTIONAL CARDIOLOGY A short plastic tube placed within a vein or artery, through which various catheters are passed for local or central diagnostic or therapeutic interventions. See Catheterization

introitus MEDTALK The entrance to a cavity, commonly understood to be the vagina opening

Introl® UROGYNECOLOGY An intravaginal prosthesis that elevates the bladder neck in urinary stress incontinence. See Stress incontinence

intromission SEXOLOGY Insertion of one part–eg, the penis, into another–eg, the vagina

introspection PSYCHIATRY Self-observation; examination of one's feelings, especially through psychotherapy

introvert PSYCHIATRY A person who is introspective, self-conscious, often meticulous, a poor social mixer, who takes criticism too seriously. Cf Extrovert

intubation The placement of a tube for respiratory support or gastric feeding. See Gastric intubation. Cf Extubation

Intuitive™ system CYBERSURGERY A proprietary minimally invasive surgical system which provides 3D visualization and translates the surgeon's hand movements into precise microsurgical movements via 8 mm ports. See Cybersurgery

intussusception PEDIATRIC SURGERY The telescoping of one segment of the intestine into another, which compromises the involved segment, causing intestinal obstruction and strangulation, especially of the distal segment, inflammation, swelling, ↓ blood flow and, if prolonged, necrosis accompanied by bleeding, perforation, infection, dehydration, shock CLINICAL ♂:♀ ratio, 3:1; common from 5 months and 1 yr of age; onset is often abrupt with intense colicky intermittent abdominal pain that ↑ in intensity and duration, accompanied by vomiting and fever, weakness and eventually, shock; half of infants pass bloody mucus popularly termed "currant jelly" stool; early diagnosis is critical to salvage the bowel and the baby MANAGEMENT Surgery. Cf Volvulus

invasion ONCOLOGY The penetration of a basement membrane and extension into the stroma by a neoplasm which usually, but not invariably, implies a malignancy with metastatic potential. See Bulldozing invasion, Lymph node inclusions, Metastasis, Perineural invasion, Pseudoinvasion, Stabbing invasion, Vascular invasion

invasive *adjective* MEDTALK 1. Interventional Referring to therapy in which mucocutaneous barriers are violated–eg, invasive cardiology. Cf Non-invasive 2. Diffuse, infiltrating Referring to the spread of a lesion, usually understood to be CA, beyond confining or natural boundaries. Cf Encapsulated

invasive cardiology CARDIOLOGY The subspecialty of cardiology that focuses on diagnostic or therapeutic cardiovascular procedures–eg, coronary angiography, imaging and nonimaging total stress tests–thallium stress test, SPECT, gated blood pool studies, echocardiography, coronary artery revascularization procedures–stenting, percutaneous transluminal coronary angioplasty. See Interventional cardiology, Invasive hemodynamic monitoring. Cf Noninterventional cardiology

invasive hemodynamic monitoring CARDIOLOGY Any maneuver used to measure in vivo hemodynamics: arterial line(s), pulmonary artery catheter, central venous line, and cardiac output monitoring. See Interventional cardiology

invasive mole DERMATOLOGY Melanoma, see there GYNECOLOGY Choriocarcinoma, see there

invasive *Streptococcus* A Killer bug A variant of *Streptococcus* group A linked to necrotizing fasciitis and myositi

invasiveness INFECTIOUS DISEASES '...*the ability*–of a microorganism—*to enter and move through tissue*' ONCOLOGY See Metastasis

inventory PSYCHOMETRICS A questionnaire or test designed to measure aptitudes, interest or traits. See Hogean Personality Inventory, Miller Clinical Inventory, Multiaxial Inventory, MMPI, NEO Personality Inventory, Social Support Inventory, State-Trait Anxiety Inventory, Texas Primary Reading inventory, Tobacco Withdrawal Symptoms Inventory

inversion ORTHOPEDICS A frontal plane movement of the foot, where the plantar surface is tilted to face the midline of the body or the medial sagittal plane; the axis of motion lies on the sagittal and transverse planes; a fixed inverted position is referred to as a varus deformity

inverted *adjective* Referring to a foot, or part thereof, which is tilted in the frontal plane, so the plantar surface faces toward the midline of the body and away from the transverse plane, or another specified reference point

inverted nipple Innie GYNECOLOGY A nipple wannabe, the central part of which is introspective. See Nipple. Cf Retracted nipple

inverted papilloma ENT An inverted or exophytic tumor of nasal cavity and paranasal sinuses, which recurs in ± ⅔ of cases; malignancy develops in 2% SURGICAL PATHOLOGY A proliferation of thin epithelium overlying papillary fibrovascular fronds, found in either transitional epithelium–renal pelvis, ureters, bladder, urethra or stratified cuboidal epithelium–paranasal region, ¾ of Pts also had HPV types 6 and 11 and upper respiratory tract lesions

inverted S appearance IMAGING An appearance in central bronchogenic CA where the interlobar fissure is accompanied by atelectasis

inverted U sign IMAGING A radiologic appearance caused by a massively dilated–possibly extending to the diaphragm, redundant loop of sigmoid colon that twists on its mesenteric axis and may be seen in a volvulus. See Hot air balloon sign

inverted V sign IMAGING A radiologic appearance caused by lateral umbilical ligaments made prominent by gas percolated on either side of the ligaments, characteristic in a plain supine film of pneumoperitoneum, when the Pts are too sick for erect studies. See also Double wall sign

inverted Y field Inverted Y port RADIATION ONCOLOGY A large RT field for treating contiguous lymph nodes involved in Hodgkin's disease or other lymphomas, which covers para-aortic, splenic hilar, iliac, inguinal and femoral lymph nodes. See Abdominal bath, Mantle port, Radiotherapy

investigation MEDTALK Evaluation of or research on a particular event or phenomenon. See Cluster investigation, Community health investigation, Contact investigation, the Exposure investigation, Kumar investigation, Preclinical investigation, Study, Trial, Urban investigation

investigational new drug or device NIHSPEAK A drug or device permitted by FDA to be tested in humans, which is not yet regarded as safe and effective for a particular use in the general population and not yet licensed for marketing

Invirase® Saquinavir mesylate AIDS An antiviral used for advanced HIV infection which inhibits HIV protease by blocking cleavage of HIV precursors in lymphocytes and monocytes; it has an additive to synergistic effect when combined with nucleoside analogues–eg, zidovudine or ddC ADVERSE EVENTS Diarrhea, abdominal discomfort/pain, nausea, intraoral ulcers, asthenia, rash. See AIDS, HIV. Cf Nucleoside analogues

'invisible' profession Nursing, see there

involuntary hospitalization FORENSIC PSYCHIATRY A civil commitment in which a person is formally confined to a mental health institution, due to mental illness, incompetence, alcoholism, drug addiction, or other, as he/she is deemed dangerous to him/herself or others. See Commitment

involuntary infertility Sterility GYNECOLOGY The incapacity of a ♀ to bear a living child during the span of reproductive yrs; II may be due to anatomic reproductive tract abnormalities, malfunction of ovulation precluding conception, recurrent intrauterine loss of pregnancy, or specific diseases–eg, infections

involuntary oscillation of eyes Nystagmus, see there

involuntary slow & twisting movement Choreoathetosis

involuntary weight gain MEDTALK An unintentional ↑ in body mass ETIOLOGY Overeating, lack of exercise, poor eating habits–↑-carbohydrate, ↑-calorie diet, emotional factors–guilt, depression, and anxiety, social pressure, slowing of metabolism with age, drugs that ↑ fluid retention or abnormal pooling of fluids in tissue, smoking cessation, compulsive eating, alcohol consumption, hypothyroidism, drugs–eg, corticosteroids, cyproheptadine, lithium, OCs, tranquilizers, phenothiazines, TCAs. See Morbid obesity, Obesity

involution GYNECOLOGY See Uterine involvement MEDTALK A ↓ in organ size or functional capacity, generally understood to be age-related

iodine deficiency disorder Any condition–eg, cretinism and brain damage, goiter, or hypothyroidism, attributable to iodine deficiency and corrected by adequate dietary iodine. See Iodine

iodine intoxication The consequences of a toxic iodine overload–especially administered orally; IT is most marked in persons with low iodine at time of exposure to excess iodine CLINICAL Transient inhibition of thyroid hormone synthesis, ↑ TSH–thyrotropin, thyroid autoantibodies, exacerbation of goiter, failure to ↓ goiter size, thyrotoxicosis

Iodosorb WOUND CARE A wound cleansing gel with antibacterial properties containing cadexomer iodine. See Wound care

IoGold® RADIATION ONCOLOGY Radioactive seeds for treating prostate CA. See Brachytherapy, Prostate CA

IOL Intraocular lens, see there

Iontocaine™ ANESTHESIOLOGY A topical anesthetic–lidocaine w/ epinephrine administered by iontophoresis–in gel adhesive pads Iomed 6/3/97. See Iontophoresis, Lidocaine

iontophoresis THERAPEUTICS A type of transcutaneous drug delivery in which electric current is applied to the skin to enhance absorption of large polar or hydrophilic molecules and peptides–eg, insulin, and control therapeutic delivery. See Transcutaneous drug delivery. Cf Phonophoresis

IOP Intraocular pressure, see there

Iotrex™ RADIATION ONCOLOGY A radioisotope solution used for brachytherapy. See Brachytherapy

IP 1. Internet protocol, see there 2. Interphalangeal, see there 3. Interstitial pneumonia. See Interstitial lung disease

IPA Independent practice association. See IPO

IPG Implantable pulse generator, see there

ipecac MAINSTREAM MEDICINE A medicinal–*Cephalis ipecacuanha* that stimulates the CNS and stomach, evoking emesis; its sole purpose is to facilitate vomiting in a person who has ingested poison or overdosed on a drug or medication. Cf Ipecac abuse

ipecac abuse EATING DISORDERS The repeated use of ipecac syrup as a purgative by bulimics–interesting factoid–chanteuse extraordinaire Karen Carpenter suffered from an eating disorder and died from IA ADVERSE EFFECTS Cardiac arrest, arrhythmias, irreversible damage to myocardium, seizures, shock, hemorrhage, blackouts, HTN, respiratory complications, dehydration, electrolyte abnormalities, death. See Bulimia, Eating disorders

IPO Independent practice organization A legal entity in the US, in which physicians and/or dentists enter an arrangement to provide services through an entity–eg, a prepaid health plan, while maintaining their own private practices. See HMO, PPO

IPPV Intermittent positive pressure ventilation. See PEEP

ipratropium Br Atrovent® ENT A beta adrenergic aerosol bronchodilator for COPD, for symptomatic relief of rhinorrhea in allergic and nonallergic perennial rhinitis in children ADVERSE EFFECTS Cough, nervousness RELATIVE CONTRAINDICATIONS Narrow-angle glaucoma, BPH, bladder neck obstruction

IPSID Immunoproliferative small intestinal disease Mediterranean lymphoma, α heavy chain disease A heterogeneous group of conditions characterized by monoclonal ↑ in production of Ig–usually α heavy chain, without accompanying light chains, ie 'truncated' immunoglobulins CLINICAL Malabsorption, diarrhea, weight loss, abdominal pain–due to marked expansion of the proximal small intestine and mesenteric lymphoid tissue, clubbing of fingers, toes

TREATMENT Without antibiotic therapy–eg, tetracycline, IPSID may evolve to lymphoma or B cell immunoblastic sarcoma

IPTD Intrapericardial therapeutics & diagnostics INTERVENTIONAL CARDIOLOGY The percutaneous delivery of therapeutics in the pericardium to manage cardiac or pericardial disease or perform diagnostic procedures. See Cath lab

IQ test Intelligence quotient test PSYCHOLOGY A test that measures the complex interplay between attention, processing and retrieval of information, and execution of processes regarded as the domain of intelligence

IR 1. Immune response 2. Infrared, see there 3. Internal rotation–rehabilitation medicine 4. Inversion recovery–MRI

irbesartan Avapro® CARDIOLOGY An ACE inhibitor used to manage HTN LAB ↑ in BUN ADVERSE EFFECTS Diarrhea, dyspepsia, musculoskeletal fatigue, URIs CONTRAINDICATIONS Impaired renal function. See Hypertension. Cf Losartan

iridectomy EYE SURGERY Partial excision of the iris, see there

iris hook GYNECOLOGY A device for grasping and manipulating the cervix, fundus, and intravaginal soft tissue during gynecologic procedures

Irish's node ONCOLOGY Left anterior axillary lymph node, a site often involved by metastatic gastric CA. See Sentinel node, Sister Mary Joseph node

iron Iron-binding capacity, total iron-binding capacity, TIBC A metallic element–atomic number 26; atomic weight 55.8 essential to life, bound to Hb and responsible for O_2 transportation–total iron refers to the amount of iron actually present in serum; iron and TIBC are performed on autoanalyzers REF RANGE Iron, 40-180 µg/dL; TIBC, 250-390 µg/dL. See Total iron-binding capacity, Transferrin

iron deficiency A relative or absolute deficiency of iron which may be due to chelation in the GI tract, loss due to acute or chronic hemorrhage or dietary insufficiency Sources Meat, poultry, eggs, vegetables, cereals, especially if fortified with iron; per the National Academy of Sciences, RDA of iron is 15 mg/day for ♀, 10 mg/day for ♂

iron-deficiency anemia An anemia due to ↓ Hb production due to ↓ iron; in idiopathic IDA, 62% of Pts have lesions of the upper, lower, or both ends of the GI tract causing iron deficiency–ie, blood loss CLINICAL FTT, ↑ infections LAB Hypochromia, microcytosis. See Anemia, iron

irradiation 1. Radiation therapy, see there 2. Blood irradiation, see there

irrational *adjective* Unreasonable, illogical

irregular periods GYNECOLOGY A popular term for a wide variation in menstrual cycles–eg, ranging from 21 to 42 days or an even broader range ETIOLOGY Hormonal imbalance especially due to ↓ progesterone, crash dieting, thyroid disease, iron deficiency, platelet deficiency, emotional stuff, endometrial hyperplasia or CA, perimenarche, perimenopausal, stress

irregular sleep-wake syndrome SLEEP DISORDERS A 'condition' characterized by irregular periods of sleep and wakefulness seen in shift workers and those who frequently travel through time zones CLINICAL Insomnia, sleepiness, malaise DIFFDX, IRREGULAR SLEEP Sleep apnea, depression, restless leg syndrome, leg cramps, pain, COPD, asthma, CHF, nocturia–often linked to BPH or prostate CA. See Sleep disorder, Sleep studies

irresistible impulse test Control test, volition test FORENSIC PSYCHIATRY A test used by some states to define a person as insane under the law if, due to a mental disorder, the defendant was unable to control her action *or* conform her conduct to the law. See Insanity defense. Cf Durham rule, M'Naughten test, Substantial capacity test, Temporary insanity

irrigation WOUND CARE The cleansing of a space, wound or cavity with a fluid

350

irritability Inconsolability PEDIATRICS Overresponse by an infant to harmless stimuli; fussiness, whining, fretfulness despite attempts to comfort and console by caregiver; irritability may be a harbinger of an infection, metabolic disorder, head trauma, CA, and other medical conditions

irritable bladder Urge incontinence, see there

irritable bowel syndrome Irritable colon GI DISEASE A condition characterized by chronic abdominal pain, bloating, mucus in stools, irregular bowel habits, alternating diarrhea and constipation; IBS may accompany anxiety and panic disorders; Sx tend to wax and wane over yrs; the primary defect appears to be abnormal GI tract contractions—motility, which does not lead to any serious organ problems; it is a diagnosis of exclusion MANAGEMENT Symptomatic—high fiber diet, exercise, relaxation techniques, avoid caffeine, milk products, sweeteners, medications

irritant contact dermatitis OCCUPATIONAL DERMATOLOGY A form of CD that usually affects the hands and arms, caused by acids, alkalis; the intensity of skin response reflects intrinsic nature of the chemical, its concentration, and duration of contact CLINICAL Erythema, chapping, blistering, ulceration HOST FACTORS Skin occlusion, sweating, current or prior skin disease ENVIRONMENTAL FACTORS Temperature, humidity, friction, pressure, lacerations. See Contact dermatitis

irritative nystagmus Nystagmus in Meniere's disease which beats toward the affected ear

ischemia MEDTALK A blood flow rate through an organ that provides insufficient O_2 to maintain aerobic respiration in that organ. See Chronic mesenteric ischemia, Myocardial ischemia, Silent myocardial ischemia

ischemic cardiomyopathy CARDIOLOGY A disorder caused by myocardial hypoxia, which compromises the heart's ability to efficiently pump blood; IC may cause heart failure and is a complication of cardiac ischemia, especially affecting older ♂, a disparity that ↓ with advancing geezhood RISKS FACTORS Personal or family Hx of MI, angina, unstable angina, ASHD, or other CAD; HTN, smoking, DM, high fat diet, ↑ cholesterol, obesity, stress can all precipitate IC MANAGEMENT rGH improves hemodynamics, clinical function. See Growth hormone

ischemic cascade VASCULAR DISEASE A series of events lasting for hrs to several days after an initial ischemic event that results in extensive necrosis and tissue damage beyond the tissue zone first affected by the initial lack of blood flow

ischemic colitis GI DISEASE A condition characterized by intermittent abdominal colic, accompanied by nausea, tenesmus, fever, bloody diarrhea, due to ASHD of the mesenteric arteries which primarily impacts on the descending colon PROGNOSIS Relatively good, due to the high rate of turnover of glandular epithelium. See Intestinal angina

ischemic contracture Volkmann's contracture, see there

ischemic heart disease Ischemic disease CARDIOLOGY A condition characterized by impaired myocardial function due to ↓ blood flow through the coronary arteries 2° to CAD; IHD is the most common cause of death worldwide; its prevalence and incidence continues to ↑; the US death rate for CAD has ↓; the hospital discharge rates for acute MI has remained constant, in 1995, ±2.2 million Pts were discharged from US hospitals with IHD, of whom ±350,000 had new-onset angina. See Angina, Antiplatelet therapy, Balloon angioplasty, CABG, Stenting

ischemic penumbra NEUROLOGY An ischemic zone in the brain at risk of infarction, which lies in a 'no-man's land' between a zone of low blood flow—< 25 ml/100 mg brain tissue/min and a zone where brain tissue is undergoing necrosis—flow—< 8 ml/100 mg/min. Cf Hemorrhagic stroke

ischemic time TRANSPLANT SURGERY The time that an organ is outside the body when the heart is not beating or supplied with O_2 by the coronary arteries

iScribe THERAPEUTICS A handheld electronic prescribing doodad

ISDN Integrated services digital network INFORMATICS A digital telecommunications line with a data–text, graphics, video and audio–throughput of up to 128 kps. See Internet, Local area network, Modem

ISI International Sensitivity Index, see there

ISIS-4 CARDIOLOGY A clinical trial—International Study of Infarct Survival that compared medical therapies–eg, oral captopril, oral mononitrate, IV magnesium on M&M in Pts with suspected acute MI. See Acute myocardial infarction, Streptokinase

islet antibody Any autoantibody seen in type 1 DM, the clinical significance of which is uncertain. See Diabetes mellitus

islet cell tumor Islet of Langerhans tumor ENDOCRINOLOGY A benign or malignant tumor of the pancreatic islet cells which may produce one or more hormones causing distinct Sx ICT TYPES Insulinoma, glucagonoma, gastrinoma–Zollinger-Ellison syndrome, etc RISK FACTORS Family Hx of MEN-1. See MEN I, Zollinger-Ellison syndrome

ISO 9001 A series of standards that certifies commercial products and processes as meeting or surpassing certain rigorous physical and chemical standards of quality

isobaric *adjective* Having equal weight or pressure over time or space; referring to barometric pressure

isobaric counterdiffusion gas lesion disease PULMONOLOGY A condition seen in Pts treated with hyperbaric O_2, where gas bubbling into skin and subcutaneous tissue is caused by more rapid inward diffusion of helium from the ambient atmosphere than outward diffusion of nitrogen or neon from the capillaries to the atmosphere. See Scuba diving. Cf the Bends, Caisson's disease

isochronous *adjective* Simultaneous

isoechoic *adjective* Referring to echo similarity of 2 or more tissues as measured by ultrasonography, see there

isoflavone NUTRITION Any of a family—eg, genistein, daidzen, of phytoestrogens in peanuts, beans, soy beans, lentils, that are carcinoprotective and ↓ ASHD. Cf Genistein, Nutraceutical, Soy, Textured vegetable protein

isoflurane ANESTHESIOLOGY The most commonly used inhalation anesthetic, which maintains bp, heart rate, and cardiac output better than halothane. See Inhalation anesthetic. Cf Halothane

isolated angiitis of CNS Granulomatous angiitis of CNS, see there

isolated hepatic perfusion ONCOLOGY A procedure for administering high dose chemotherapy only to the liver

isolation INFECTIOUS DISEASE The segregation of a Pt, body fluids, and fomites to prevent transmission of an infection to other Pts or hospital personnel. See Biosafety levels, Disinfection, Immunoisolation, Precautions, Sterilization. Cf Reverse isolation, Reverse 'precautions'

isolation room NURSING A Pt room designed to minimize the spread of droplet nuclei expelled by Pt with TB and other contagious pathogens; IRs require negative-pressure ventilation. See Precautions, Universal precautions

isometric exercise SPORTS MEDICINE An exercise against an unmoving resistance, with muscle contraction and minimal other body movement; IEs build muscle strength EXAMPLES Weight-lifting, squeezing tennis balls, rowing, etc. See Exercise, Working out; Cf Isotonic exercise

isomorphic effect Köbner's phenomenon DERMATOLOGY The induction of skin changes at site 'B,' which follows minimal nonspecific trauma—heat or light when identical lesions are already present elsewhere–site A; IE is typical of psoriasis and seen in lichen planus, eczema, verruca. Cf Id reaction

isoniazid INH A first-line anti-TB drug used with other drugs to treat and prevent TB ADVERSE EFFECTS Liver damage, hepatitis, neuropathy. See Multi-drug-resistant tuberculosis, Tuberculosis

isoretinoin 13-*cis*-retinoic acid A proprietary vitamin A analogue used to treat refractory acne, leukoplakia, and induce long-term clinical and lab remissions in juvenile CML TERATOGENICITY It is a category X drug–ie, teratogenic, contraindicated in pregnancy–causes cardiac malformations–VSD, aortic arch and conotruncal defects, external ear deformity, cleft palate, micrognathia, CNS malformations. See Vitamin A analogues

isosorbide dinitrate CARDIOLOGY A venodilating nitrate used with hydralazine in CHF; ID ↓ cardiac filling pressure, ↑ exercise tolerance. See Nitrates

Isospora belli PARASITOLOGY A sporozoite parasite common in the tropics/subtropics of the Western Hemisphere/Southeast Asia; it may be underreported, as in uncompromised hosts, it mimics viral gastroenteritis; in AIDS, *I belli* may cause fulminant disease CLINICAL Watery diarrhea ± hemorrhage, colicky pain, weight loss, steatorrhea MANAGEMENT T-S

isosthenuria NEPHROLOGY The excretion of urine with the same osmolality as plasma

isotonic *adjective* 1. Referring to a uniform muscle tone 2. Referring to uniformity of osmotic pressure *noun* Isotonic drink, see there

isotonic drink SPORTS MEDICINE 1. A sports drink used to simply replace fluid and electrolytes lost during prolonged exercise. See Sports drink 2. A sports drink that replaces water and electrolytes *and* contains either fructose or glucose polymers allowing slow release of carbohydrates for replenishing reserves of energy consumed while exercising. See Sports drink

isotonic exercise Dynamic exercise An exercise that consists of continuous and sustained movement of the arms and legs; IEs–eg, running, bicycling for aerobic training that improves cardiorespiratory function. See Exercise. Cf Isometric exercise

isotretinoin Retinoic acid, 13-cis retinoic acid A retinoid used to treat acne and psoriasis and may ↓ tumor development SIDE EFFECTS Inflammation of skin and mucosa, hepatitis

ISS Injury severity score, see there

issue VOX POPULI A euphemism for a problem, as in "he has issues". See Core issue, Guaranteed standard issue

itch An irritation that mandates scratichization. See Jock itch, Seven-yr itch, Summer itch, Swimmer's itch

itching in groin MEDTALK Tinea cruris, aka jock itch

itching Pruritus DERMATOLOGY A tingling or irritation of skin that compels scratching, due to local or generalized causes TYPES Insect bites/stings; chemical irritants–poison ivy, stinging nettle, laundry detergents; environmental–drying, sunburn; hives; parasites–lice; infections–folliculitis, chickenpox; allergies; liver disease with jaundice; drug reactions DIAGNOSTIC WORKUP CBC and differential, ESR, SPEP, allergy tests, "rule out" tests MANAGEMENT Topical corticosteroids, antihistamines, tranquilizers

ITP Idiopathic, aka immune, thrombocytopenic purpura, see there

itraconazole Sporanox® A broad-spectrum antifungal used for onychomycosis, tinea corporis, tinea cruris, tinea capitis by *Trichophyton tonsurans* ; as well as for life-threatening systemic fungal infections–histoplasmosis, blas-

tomycosis, refractory aspergillosis, oropharyngeal and esophageal candidiasis, cryptococcal meningitis ADVERSE EFFECTS N&V, diarrhea, edema, fatigue, rash, headache, hypopotassemia

IU International Units, see there

IUD 1. Intrauterine death 2. Intrauterine device—a contraceptive device that prevents pregnancy primarily by mechanical disruption of the endometrium TYPES Coil, loop, triangle, or T-shaped, made of plastic or metal CO-MORBIDITY Actinomycosis affects 85% of ♀ with an IUD in place for ≥ 3 yrs; ± 20% of ectopic pregnancies occur in IUD users; pelvic infections are 3 to 7-fold ↑ in IUD users, often polymicrobial–eg, aerobic and anaerobic bacteria, mycoplasma, *Chlamydia* spp. See Copper-Seven, Dalkon Shield, Pearl index

IUGR Intrauterine growth retardation, see there

IUI Intrauterine insemination, see there

IUPC Intrauterine pressure catheter, see there

IV Intravenous, also 1. Interventricular 2. Intervertebral 3. Intravascular 4. Intraventricular–cardiology

IVAT CARDIOLOGY A clinical trial–Intermediate sized Vessel Atherectomy Trial

IVBAT Intravascular bronchiolar and alveolar tumor, interstitial vascular sarcoma A rare tumor of ♀ presenting as multifocal slowly-growing intrapulmonary nodules that mimic pulmonary metastases; first described as a variant of bronchoalveolar cell carcinoma, IVBAT is a neoplasm a sui generis caused by vascular proliferation CLINICAL Asymptomatic or minimal SOB, 40% of Pts are < 30 yrs old; 50% of Pts die of disease, 25% within 1st yr; it may be due to endotheliomatosis or identical to epithelioid hemangioendothelioma IMAGING Multiple bilateral pulmonary nodules < 2 cm in diameter

IVDA 1. Intravenous drug abuse. See Injecting drug use 2. Intravenous drug abuser. See Injecting drug user

IVF 1. In vitro fertilization, see there 2. Intravascular fluid

IVH Intraventricular hemorrhage, see there

IVIC syndrome An AD condition characterized by multiple congenital defects, eg, a defect in the radial 'ray'–an embryologic structure from which the radial bone and related musculoskeletal structures arise, strabismus, deafness, thrombocytopenia

IVIG Intravenous immunoglobulin, see there

IVP Intravenous pyelography IMAGING An imaging study of the transitional mucosa of kidneys, ureter, and bladder after IV injection of a radiocontrast which concentrates in the urine; an IVP outlines the renal pelvis, ureters, bladder

IVUS Intravascular ultrasound, see there

Ivy bleeding time HEMATOLOGY A quantitative coagulation assay based on a standardized skin wound, which measures platelet and vascular responses to injury; ↑ IN Bernard-Soulier disease, Glanzmann's thrombasthenia, platelet defects–eg, thrombocytopenia, storage pool disease, vascular defects–eg, Ehlers-Danlos disease, von Willebrand's disease

Ixodes scapularis Deer tick A tick with a 2-yr life cycle, and 3 feeding seasons; the cycle begins in spring with soil deposition of fertilized eggs; by summer, larvae emerge and imbibe a blood meal from small vertebrates–eg, white-footed mouse–*Peromyscus leucopus* which may be infected with *Borrelia burgdorferi*–maintaining the spirochete in the tick population. See *Borrelia burgdorferi*, Lyme disease

J 1. Flux 2. J chain–immunoglobulin 3. Joule 4. Juvenile DRUG SLANG Joint–a marijuana cigarette

J curve phenomenon EPIDEMIOLOGY A relationship between the risk factors for, and mortality from, a particular process, such that, as a risk factor–eg, alcohol consumption-AC, blood pressure-BP, total cholesterol-TC ↑, so does mortality; in diseases with a JCP, there is a critical point below which the AC, BP, TC, etc, is too low and there is opposite trend, and↓ of risk factors below a critical point is associated with ↑ risk of morbidity and death. Cf U curve

J pouch COLORECTAL SURGERY A reservoir formed from a J-shaped loop of terminal ileum where the loops are sectioned, forming a pouch and then anastomosed to create a continent anorectum, preserving anal sphincter function INDICATIONS After total proctocolectomy for familial polyposis coli or ulcerative colitis totalis. See S pouch

J-shaped sella A shallow, elongated or 'boot-shaped' sella turcica with an elongated anterior recess, extending below the anterior clinoid process, classically seen in Hurler's mucopolysaccharidosis, due to accumulation of dermatan and keratan sulfates or glycosaminoglycans; J-shaped sellas may also be seen in orodigitofacial syndrome and mannosidosis

J wave CARDIOLOGY A quasi-pathognomonic EKG change seen in ⅓ of Pts with hypothermia, as a positive 'hump' at the end of a QRS complex, which disappears on rewarming Pt, bradyarrhythmia, atrial flutter, A Fib EKG Prolonged P-R and S-T intervals and T-wave inversion

J-1 visa Exchange visitor visa A visa for those who are not citizens and wish to live in the US Cf Green card, H1-B visa

J1 waiver MEDTALK A waiver of the J1 visa requirement that a foreign physician return to his country after completing a residency or fellowship, if he agrees to accept a practice position in a state or federally-designated underserved area. See Green card

jackknife phenomenon Clasp knife phenomenon, see there

jackpot syndrome MEDICAL MALPRACTICE A popular term for the demand by a plaintiff for a larger settlement for a seemingly minor tort than is reasonable for the harm suffered

Jackson scale Dominance subscale, Jackson Personality Research Form E PSYCHOLOGY An instrument that measures the frequency at which a person tries to control the environment, influence others, and forcefully opine

Jacksonian seizure Partial seizure, see there

jail bars sign RADIOLOGY Dense osteosclerosis of ribs, seen on a plain AP CXR, resulting in horizontal bands fancifully likened to bars in a prison window; first described in agnogenic myeloid metaplasia; it may be seen in sickle cell anemia and osteopetrosis

Jakob-Creutzfeldt disease Creutzfeldt-Jakob disease, see there

Jamaican neuropathy A condition characterized by spasticity and other signs of corticospinal tract disease TYPES 1. Ataxic–more common in Nigeria, accompanied by sensory ataxia, numbing and burning of feet, deafness, visual defects with optic atrophy, central scotoma, spasticity, leg atrophy, footdrop, findings that may be due to subclinical malnutrition; 2. Tropical spastic paraparesis–subacute neuropathy with predominantly pyramidal tract disease which affects the posterior column, causing paresthesia, loss of sensation, bladder dysfunction and girdling lumbar pain; 80% of JN Pts have antibodies to HTLV-I

Jamaican vomiting sickness An intoxication by 'bush tea,' made from unripe fruit of the Jamaican ackee tree–*Blighia sapida* and *Crotalaria spectabilis*, caused by hypoglycin A metabolites CLINICAL Violent vomiting, prostration, drowsiness, convulsions, hypoglycemia MORTALITY High, often 24 hrs after ingestion

jamais vu French, never seen PSYCHIATRY A group of paramnesias with complete absence of memory for events known to have been experienced by the Pt; each is associated with depersonalization and temporal lobe epilepsy. Cf Deja vu

Jamar dynamometer NEUROLOGY A device used to measure muscle strength. See Hand grip strength

Jane Doe EMERGENCY MEDICINE A name assigned to a ♀ admitted to an ER without identifying documents. See John Doe FORENSIC MEDICINE A name for an unidentified decedent ♀. See End-of-life debate

Japanese encephalitis NEUROLOGY The most common epidemic viral encephalitis, it has an incidence of ± 0.5/1000; JE is often fatal or crippling, especially in Thailand CLINICAL Abrupt onset with fever, headache, meningeal irritation, convulsions, muscular rigidity, mask-like facies, coarse tremor, paresis, ↑ deep tendon reflexes VECTOR *Culex* mosquito

Japanese illusion NEUROLOGY A clinical test used to elicit right-left confusion in unilateral anesthesia

jargon SOCIOLOGY A specialized term, phrase, or acronym, that is either created for a particular purpose–eg, nutmeg liver or is a new use–eg, organ transplant for computers–for an extant term; language peculiar to a group or profession, medical, legal, etc. Cf Dialect, Slang

jaundice Icterus HEPATOLOGY A condition characterized by the deposition of excess–> 2 mg/dL, 34 µmol/L free or conjugated BR in peripheral circulation, and in skin, mucosa and sclerae; it is either physiologic–eg, due to hemolysis or pathologic–eg, seen in hepatitis or bile stasis. See Breast milk jaundice, Hyperbilirubinemia, Obstructive jaundice

JAUNDICE TYPES–unconjugated hyperbilirubinemia

PHYSIOLOGIC JAUNDICE Jaundice develops ≥ 72 hrs after birth; total BR rises to greater than 15 mg/dL; direct BR is < 15% of total BR; jaundice resolves within 1–2 weeks ETIOLOGY Sluggish glucuronyl transferase activity, ↑ of BR 'load,' ↓ plasma clearance of BR MANAGEMENT Phototherapy, exchange transfusion

PATHOLOGICAL JAUNDICE Jaundice develops in 72 hrs after birth; total BR peaks at ≤ 15 mg/dL; direct BR is > 15% of total BR; jaundice may require > 2 weeks to resolve ETIOLOGY HDN, hemolysis, extravascular loss of blood, ↑ enterohepatic circulation, breast feeding, defective BR metabolism, sepsis, metabolic disease MANAGEMENT Phototherapy, exchange transfusion

jaundice of the newborn Breast milk jaundice, see there

jaw-thrust maneuver EMERGENCY MEDICINE A jaw displacement maneuver used in 'rescue breathing,' required for mouth-to-mouth and mouth-to-nose resuscitation. See Rescue breathing. Cf Head-tilt/Chin-lift maneuver

jaw winking Marcus-Gunn phenomenon NEUROLOGY Elevation of a ptoptic eye by jaw movement, as seen in AD congenital ptosis, due to faulty innervation of levator palpebrae

jaw wiring OBESITY An extreme treatment of morbid obesity that utilizes the same methods and devices as those used for jaw fractures, allowing only the intake of liquids. See Gastric bubble, Ileal bypass surgery, Morbid obesity, Stomach stapling

JCAHO Joint Commission on Accreditation of Healthcare Organizations, see there

Jeep seat MILITARY MEDICINE A popular term for pilonidal cyst/sinus inflammation of congenital sacrococcygeal malformations with focal persistence of the neuroendocrine canal and ingrowth of hair into the cyst/sinus; JS follows repeated trauma to the 'seats' of military recruits who bounce on seats of shock absorber-challenged military vehicles/jeeps

Jefferson fracture Burst of the ring ORTHOPEDICS A burst fracture of C1 caused by an axial load to vertex of the head; because C1 fragments are displaced laterally, direct injury to the spinal cord is rare MANAGEMENT Nondisplaced–collar or SOMI; displaced < 7 mm–collar or halo; displaced > 7 mm–halo or obtain an MRI; if transverse atlantal ligament is disrupted, fusion may be needed. See C1 fracture

jejuno-ileal bypass See Ileal bypass surgery

Jekyll & Hyde syndrome A complex described in the elderly who cyclically improve with hospitalization–stabilization, rehydration, appropriate medication–and mentally and physically deteriorate at home

Jeliffe formula NEPHROLOGY An equation used to assess renal function by calculating creatinine clearance. See Creatinine clearance

Jello™ sign OBSTETRICS A popular term for the characteristic undulation of the scrotum that occurs with fetal limb movement, a 'soft' criterion for determining an infant's sex by ultrasonography; seen after the 22nd gestational wk, the movement is likened to the jiggle of a proprietary brand of instant gelatin

Jendrassik maneuver NEUROLOGY A method of enhancing the patellar reflex by briskly yanking at the leg

jeopardy call GRADUATE TRAINING A call schedule in which a house officer is placed on call on short notice, to cover a sick colleague or 'pitch in' with a heavy workload. See Call schedule

jerky leg syndrome Restless leg syndrome, see there

Jervell & Lange-Nielsen syndrome Long Q-T syndrome, see there

jet lag SLEEP DISORDERS An acute shift in the circadian rhythm, caused by travelling across multiple–≥ 3 time zones CLINICAL Altered mood, performance efficiency, temperature rhythms, rapid eye movement and slow-wave sleep. See Circadian rhythm, Insomnia, Shift work, Sleep disorders

jet lesions Zahn-Schmincke pockets CARDIOLOGY Vegetations that develop where a regurgitant 'jet' of turbulent blood flow strikes the endocardium, causing fibrosis and roughening, typical of anomalies of blood flow from a high-to-low pressure region–eg, aortic stenosis or coarctation, mitral stenosis, VSD, or PDA, as occurs in rheumatic fever or congenital heart disease; roughened JLs may give rise to small emboli and produce cerebral thromboembolism

jet ventilation THORACIC SURGERY A technique used during tracheal reconstructive surgery, in which a catheter is passed through the endotracheal tube into the distal main stem bronchus; a small tidal volume is delivered through the catheter at a high–60-150 'breaths'/min to maintain lung expansion, alveolar ventilation, oxygenation

JIC tube Just in case tube MEDTALK A popular term for an extra blood tube drawn from a Pt, "just in case" the physician later adds more tests to the lab requisition form

jitter IMAGING Low amplitude irregularities in echo location on an ultrasound display, attributed to electronic noise, mechanical disturbances, and other variables NEUROPHYSIOLOGY Muscle jitter The normal electric variability–'chaos'–measured by single-fiber EMG–in the interval between 2 action potentials of successive discharges of the same single muscle fiber in the same motor unit; jitter is characterized as instability in subcomponents of motor unit action potentials, and is due to the variation in the synaptic delay at the branch points in the distal axon and at the neuromuscular junction; like fiber density, jitter is ↑ in neuropathic conditions–motor neuron diseases–eg, myasthenia gravis, is accompanied by denervation and reinnervation, and attributed to inefficient transmission of impulses in recent neural collaterals, or due to blocking–abnormal neuromuscular transmission; it is normal or near-normal in myopathic disease. See Fiber density, Single-fiber electromyography PSYCHOLOGY Jitters, see there

jitters 'Butterflies' PSYCHOLOGY An episode of nervousness or anxiety that often precedes a public event; jitters is a type of performance anxiety which may affect actors in a stage production–stage fright or soloist musicians; it may respond to anxiolytics

job burnout OCCUPATIONAL MEDICINE End-stage work-related stress, in which an employee functions at a 'ground state'; at greatest risk for JB are those with low incomes, no college education, and single mothers. See Burnout. Cf Compassion fatigue

job lock HEALTH CARE A situation in which a person is in effect–but not in actual fact–forced to remain in a job, due to fear of losing health care coverage or because a future employer's health plan won't cover a medical circumstance–eg, a pre-existing condition, in the employee or dependents. See Pre-existing condition

Job syndrome Hyperimmunoglobulin E syndrome, see there

Jocasta complex PSYCHIATRY 1. The sexual desire, usually latent, that a mother has for a son or 2. The domineering and intense, but non-incestuous love that an affect-hungry mother has for an intelligent son, and an often absent or weak father figure. See Oedipus complex. Cf Folie á deux, Phaedra complex

'jock' VOX POPULI An Americanism for an athlete. See Jock itch

jock itch Tinea cruris SPORTS MEDICINE A popular term for fungal infection in a ♂ athlete who uses an athletic supporter or 'jock strap'. See 'Jock', Tinea cruris

Joffroy sign NEUROLOGY A loss of arithmetic skills as an early indicator of organic brain disease

jogger's nipple SPORTS MEDICINE Tenderness or frank pain of the nipples of long-distance runners of both sexes, caused by repeated rubbing of the runner's shirt against the protruding nipples which, after 20+ km, abrades the skin PREVENTION Band-Aids™ or plastic strips placed in an 'X' on the nipple before race

John Doe A generic name for a nameless ♂, used in clinical and forensic medicine, as a temporary identifier for persons without identification. See Jane Doe

Johnson v University of Chicago Hospitals MEDICAL ETHICS A case that involved the routing of a seriously injured infant to another hospital, when the U of Chicago's–UCH pediatric ICU was 'on bypass', as its

13-bed unit was filled to legal capacity; the infant subsequently died; the infant's mother sued UCH under EMTALA. See COBRA, Dumping

joint SUBSTANCE ABUSE A popular term for a cigarette made from dried marijuana, *Cannibas sativa* leaves, which is 'toked' to produce a 'high' and, if smoked in excess, 'get stoned'. See Hallucinogen, Marijuana, Substance abuse, THC receptor

Joint Commission on Accreditation of Healthcare Organizations HOSPITAL PRACTICE An independent, nonprofit organization sponsored by a number of medical associations that performs accreditation reviews primarily on hospitals, other institutional facilities, outpatient facilities and healthcare organizations, the purpose of which is to maintain a high standard of institutional care, by both establishing guidelines for the operation of hospitals and other–psychiatric, ambulatory and long-term healthcare facilities, 'policing' those facilities through surveys and periodic inspections

joint fluid analysis See Synovial fluid analysis

joint mobilization OSTEOPATHY The passive movement of joints over their entire ROM, to expand the ROM and eliminate restrictions. See Osteopathy

joint mouse ORTHOPEDICS A fanciful term for the free bodies in a synovial space, especially of the knee; JM are composed of fibrous tissue covered by cartilage and measure 0.5–1.5 cm in diameter, classically seen in degenerative joint disease DIFFDX JMs are a relatively nonspecific finding; they may also be seen in synovial osteochondromatosis, chondrometaplasia, neuropathic arthropathy, osteoarthritis dissecans, pigmented villonodular synovitis, gout

joint venture MEDICAL PRACTICE An arrangement in which physicians own the health care facility to which they refer Pts for services, but do not themselves practice medicine. See Kickback, Safe harbor

Jones fracture ORTHOPEDICS A fracture of the 5th metatarsal diaphysis. See Autoeponym

Jones' criteria CARDIOLOGY Criteria for diagnosing acute rheumatic fever, first proposed by TD Jones in 1944, later revised and updated by the Am Heart Assn; ARF is diagnosed with reasonable certainty with 2, or more, major criteria, or 1 major and 2 minor criteria, assuming previous evidence of group A streptococcal infection

---

JONES' CRITERIA–ACUTE RHEUMATIC FEVER

MAJOR MANIFESTATIONS

- Carditis
- Polyarthritis
- Chorea (Sydenham's chorea)
- Eythema marginatum
- Subcutaneous nodules

MINOR MANIFESTATIONS

CLINICAL FINDINGS

- Arthralgia
- Fever

LABORATORY FINDINGS

- ↑ Acute phase reactants
- Erythrocyte sedimentation rate
- C-reactive protein
- Prolonged PR interval

EVIDENCE OF PREVIOUS GROUP A STREPTOCOCCAL INFECTION

  + Throat culture or rapid streptococcal antigen test

  ↑ (or rising) streptococcal antibody titer

---

after JAMA 1992; 268:2069

Joubert syndrome NEONATOLOGY A condition characterized by episodic hyperpnea, abnormal eye movements, ataxia, and mental retardation linked to agenesis of the cerebellar vermis

journal club A form of graduate–and less commonly, continuing medical—education used by physicians during the residency training period, in which a small group of physicians convenes, discusses, analyzes, and reviews a limited number of articles from major medical journals, often on a weekly or monthly basis

judicial bypass FORENSIC MEDICINE A form of surrogacy in which a guardian's authority is circumvented and decision-making autonomy passed to the person for whom the guardian had been appointed or designated. See Christian Science, Emancipated minor

jughandle view RADIOLOGY A modified basal view of the skull used to visualize the zygomatic arches, of interest in evaluating midfacial fractures. See LeFort fracture

juice MEDTALK See Gastric juice POPULAR NUTRITION A liquid concentrate of a fruit or vegetable. See Cranberry juice SPORTS MEDICINE A regionally popular term for anabolic steroids. See Monkey juice

juicy baby A fanciful term for an infant who produces excess mucus in the early post-partum period; 'juiciness' is a soft criterion for esophageal atresia which, when accompanied by respiratory distress–cyanosis and tachypnea–implies concomitant tracheoesophageal fistula

July phenomenon GRADUATE EDUCATION A popular myth in North America that holds that the quality of medical care ↓ and mortality ↑ in teaching hospitals during July–the time when interns, fresh from medical school begin training. See Libby Zion, 405 Regulations. Cf DRGs, 'Quicker and sicker'

jump position PEDIATRICS The posture of a child who stands on his/her knees with hips flexed and ankles in equinus position, a characteristic stance in spastic paraplegia of cerebral palsy. See Cerebral palsy

jumper syndrome Vertical deceleration injury TRAUMA MEDICINE Blunt trauma that follows jumping or falling from heights, usually > 5 stories CLINICAL Injury severity score is 41–predicted survival, 50%; actual survival is less; all had multiple fractures–eg, 'ring fracture' of skull base, separating the rim of the foramen magnum from the rest of the base and compression fractures of vertebrae, both of which occur when the victim lands on the feet or buttocks; many also have coup and/or contrecoup injuries of the brain; over ½ arrive at the ER in shock; most have angiographic evidence of retroperitoneal hemorrhage. See Lover's heels

Jungian psychoanalysis PSYCHIATRY A form of psychoanalysis which guides a Pt to merge the personal unconscious with that of a 'collective unconscious'. See False memory. Cf Freudian analysis, Hypnosis, Psychotherapy

junior internship Acting internship, see there

junk food NUTRITION A popular term for any food low in essential nutrients and high in salt–eg, potato chips/crisps, pretzels, refined carbohydrates–eg, candy, soft drinks, or saturated fats–eg, cake, chocolates. See Cafeteria diet, Cafeteria model, 'Couch potato', 'Fast' food. Cf Fruits and vegetables

junkie POPULAR HEALTH A popular term for a person, usually an IV narcotic abusing addict, whose life is disorganized vis-á-vis family and societal structure, whose existence revolves around obtaining–often through theft, prostitution or other illicit means of another 'fix' of narcotic. See Cold turkey, Shooting galleries

juvenile *adjective* Between an infant and an adult

juvenile laryngeal papillomatosis A neoplasm in children caused by HPV types 6 and 11, which may also occur in adults in the upper respiratory tract–known as recurrent respiratory papillomatosis; JLP is analagous to condyloma acuminatum of the genital tract; it rarely undergoes malignant degeneration, although it may be accompanied by severe airway compromise, its major complication; JLP is so recalcitrant, hundreds of surgical resections may be required MANAGEMENT Although IFN-alpha significantly ↓ growth rate of JLP in early therapy, the effect is not sustained

juvenile melanoma Spitz nevus, see there. Cf Melanoma

juvenile myoclonic epilepsy NEUROLOGY A seizure disorder that comprises ± 4% of seizure disorders CLINICAL Normal IQ, onset in adolescence, affecting the flexor muscles of the head, neck and shoulders; the attacks tend to occur as clonic-tonic-clonic seizures upon awakening EEG 4-6 Hz multispike and wave pattern; 40% of relatives, especially ♀, have myoclonus MANAGEMENT Valproic acid. See Epilepsy

juvenile-onset diabetes mellitus Type 1 diabetes, see there

juvenile periodontitis DENTISTRY Early onset periodontitis of adolescents, ♂ : ♀ ratio, 3:1, characterized by an early loss of alveolar bone surrounding permanent teeth; 84% have underlying endocrinopathies, 12% systemic disease–eg, DM, neutropenia, Down syndrome, Ehlers-Danlos syndrome, hyperkeratosis palmaris et plantaris, larger histiocytosis X, hypophosphatasia ETIOLOGY *Actinobacillus actinomycecomitans* and others

juvenile polyp GI DISEASE A potentially premalignant GI tract polyp in children. See Polyp

juvenile rheumatoid arthritis Still's disease, systemic-onset juvenile chronic arthritis RHEUMATOLOGY An early-onset arthropathy characterized by high intermittent fever, a salmon-colored skin rash, lymphadenopathy, hepatosplenomegaly, pleuritis, pericarditis. See Rheumatoid arthritis

juvenile xanthogranuloma Juvenile lipogranuloma, nevoxanthoendothelioma A yellowish tumor of early childhood, involving the face, head, neck, extremities PROGNOSIS Spontaneous involution

juxtaglomerular cell tumor A benign tumor most common in those ages 10–19, which arises in the pericytes of the juxtaglomerular apparatus and causes extreme HTN DIFFDX Renal cell carcinoma, interstitial cell tumor, Wilms' tumor variants, leiomyoma MANAGEMENT Resection

juxtaovarian adnexal tumor GYNECOLOGY An often asymptomatic wolffian tumor, located in the broad ligament, which affects Pts ages 30–60 PROGNOSIS Often benign, rarely, JATs may recur or metastasize

JVD Jugular-venous distention, see there

K Symbol for: 1. Equilibrium constant 2. Karyotype 3. Ketamine 4. Kidney 5. Kilobyte 6. Kinetic energy 7. Lysine 8. Phylloquinone–vitamin K 9. Potassium DRUG SLANG A regionally popular street term for PCP

'K Mart' model A popular term for a workplace in which the loss of relatively low-paid personnel–eg, 'floor' nurses in a hospital, is compensated for by ↑ the salary of those who remain

K-wire fixation Kirschner wire fixation ORTHOPEDICS The use of Kirschner wire to fix fractured bone and soft tissues for traction

KAFO Knee-ankle-foot orthosis, see there

Kahn Test of Symbol Arrangement KTSA PSYCHOMETRY A 3-D projective instrument in which a subject manipulates plastic objects of different shapes and colors; the KTSA provides insight on psychosexual development, regression, maturity, etc

kainic acid NEUROPHYSIOLOGY An excitatory neurotoxin, which stimulates glutamate receptors. See Kainate receptors

Kallmann syndrome Hypogonadotropic hypogonadism A condition with a highly variable hereditary pattern, characterized by 2° hypogonadism CLINICAL Delayed puberty, micropenis, eunuchoid features, cryptorchidism, cleft lip and palate, unilateral renal agenesis, horseshoe kidney, nerve deafness and hearing loss, color blindness, skeletal defects; synkinesia, spatial attention defects, spastic paraplegia, cerebellar dysfunction, horizontal nystagmus, pes cavus, mental retardation MANAGEMENT Androgens; gonadotropins or LHRH for spermatogenesis. See Kalig-1

Kallmiren Buspirone HCl, see there

Kalma Alprazolam, see there

Kalmalin Lorazepam, see there

Kaltostat WOUND CARE An absorbent calcium alginate dressing for covering exudative wounds or rope packing. See Wound care

kanamycin MICROBIOLOGY A broad-spectrum aminoglycoside antibiotic isolated from *Streptomyces kanamyceticus*, which is effective against gram-negative rods, and some gram-positive organisms TOXICITY Ototoxicity, nephrotoxicity MOLECULAR BIOLOGY Kanamycin inhibits protein synthesis by binding to the 30s ribosomal subunit and preventing translocation

Kaneda anterior spinal/scoliosis system ORTHOPEDICS A spinal intervertebral body-fixation orthosis system consisting of 2 spinal staples–single- or 2-hole design, KASS blunt-tip, open and closed screws, standard Isola open and closed screws, and ³⁄₁₆-inch diameter Isola spinal rods. See Scoliosis

Kansas v Hendricks 65 USLW 4564 PUBLIC HEALTH A decision by the US Supreme Court upholding the constitutionality of civil commitment of sexual predators for treatment until they are safe to be released into society. See Chemical castration, Megan's law, Sexual predator

Kaposi sarcoma AIDS A once rare, indolent malignancy affecting older Italian or Jewish ♂; in the AIDS era, KS develops in homosexual ♂ infected with human herpesvirus 8; in those immunocompromised by transplantation, immunosuppression or lymphoproliferation; KS is characterized by a proliferation of lymphatic or vascular channels, driven by growth and regulatory factors–eg, IL-1-β, IL-6 and tat protein CLINICAL FORMS Classic–elderly European ♂; AIDS related, homosexual ♂ TREATMENT If lesion is single, surgical excision or 8–12 Gy RT–85% respond, or intralesional IFN-α; if multiple, chemotherapy–eg, doxorubicin, bleomycin, vincristine; multimodality therapy PREVENTION Antivirals–eg, ganciclovir, foscarnet, cidofovir, adefovir. See AIDS, Anti-LANA antibody, Human herpesvirus 8, Promontory sign

kappa-lambda test HEMATOLOGY An application of flow cytometry, consisting of comparitive enumeration of B lymphocytes expressing kappa vs lambda light chains; the KLT is used to detect clonal expansion in Pts with B cell lymphomas. See B cell lymphomas

Karnovsky scale LONG-TERM CARE A scale of objective criteria for measuring QOL in Pts with incapacitating diseases–eg, CA, AIDS. See Activities of daily living, Quality of Life

Kartagener syndrome Immotile cilia syndrome A trio of sinusitis, bronchitis and situs inversus–lateral reversal of all organs in the chest and abdomen–ie, heart and stomach on right, liver on left, etc–ie, opposite or inverted from their usual position

Kassebaum bill S 1477 A bill proposed by NL Kassebaum–R, Kan—that would expedite clinical investigation of new drugs, devices, and biologicals, and allow manufacturers to disseminate information on 'off-label' uses of drugs. See Biologicals, Off-label use

Kassebaum-Kennedy bill Health Insurance Portability & Accountability Act of 1996, see there

katagelophobia PSYCHOLOGY Fear of ridicule. See Phobia

Katz ADL Scale REHAB MEDICINE MEDICINE An interviewer- or Pt-based instrument, used to assess physical functions, specifically basic self-care PROS Simple, useful in rehab settings CONS Limited range of activities assessed; ratings subjective. See ADL scale

Kawasaki disease Mucocutaneous lymph node syndrome A disease of children < age 5 that often follows a 1-2 wk prodrome ETIOLOGY Uncertain; bugs implicated include *Proprionibacterium acnes* or retroviruses, none definitively CLINICAL Fever, cervical lymphadenopathy, palmoplantar and mucosal erythema and edema, conjunctivitis, gingivitis, maculoerythematous glove-and-sock rash over the hands and feet which becomes hard, swollen–edematous, and sloughs, aneurysms of small and medium-sized coronary, occasionally peripheral arteries, with arteritis LAB ↑ ESR, CRP, complement, Igs MANAGEMENT Igs in IV bolus, aspirin IV PROGNOSIS Possibly sudden death; 1-5% die of disease

---

KAWASAKI DISEASE*

1. Fever of > 5 days

2. Bilateral ocular conjunctival injection

3. One or more changes of oral mucosa, including erythema, fissuring and xerostomia, conjunctival edema, mucosal edema of upper respiratory tract, eg pharyngeal injection, dry, fissured lips and 'strawberry tongue'

4. One or more changes of extremities, including acral erythema or edema, periungual and/or generalized desquamation, polymorphous exanthematous rash, truncal and cervical lymphadenopathy

*CDC definition, requires 4+ of the above

kava kava TOXICOLOGY Kava kava's use as an herbal antidepressant has been linked to fulminant liver failure, accompanied by jaundice, fatigue, weight loss, concomitant renal failure and progressive encephalopathy, requiring a liver transplant. See Liver failure

Kaye catheter NEPHROLOGY A tamponade nephrostomy balloon catheter used for percutaneous nephrostolithotomy

Kb Kilobase, see there

KB Kilobyte, see there

Kearns-Sayre syndrome A mitochondrial disease characterized by chronic progressive external ophthalmoplegia–paralysis of ocular muscles and mitochondrial myopathy combined with retinal deterioration, heart disease, hearing loss, DM, renal disease See Mitochondrial disease

Kehr sign CLINICAL MEDICINE Severe pain in the left shoulder due to a ruptured spleen

Kelly plication GYNECOLOGY A vaginal procedure for urinary stress incontinence. See Urinary continence

keloid Hypertrophic scar DERMATOLOGY A thick, irregular and indurated skin scar of adults aged 15-45 that is 6-fold more common in dark-skinned persons and in ♀; keloids occur in Rubinstein-Taybi syndrome and are associated with infections, burns, trauma, insect bites MANAGEMENT Local steroid injections to relieve pruritus or ↓ size of early lesions; post-excisional recurrence is common

Kenalog® Triamcinolone, see there

Kennedy-Kassenbaum Bill HIPAA, see there

Kenny Self-Care Scale CLINICAL MEDICINE An interviewer-based instrument similar to the Barthel index, used to assess physical functions, specifically self-care and ambulation PROS Useful in rehab settings CONS Range of activities assessed doesn't detect small defects; subjective. See ADL scale

keratinaceous cyst Sebaceous cyst, see there

keratitis OPHTHALMOLOGY Corneal inflammation, caused by nonspecific irritants, or microorganisms. See Interstitial keratitis

kerato- prefix for 1. Cornea–as in keratitis, keratocornea 2. Horny tissue–as in keratin and keratosis

keratoacanthoma DERMATOLOGY A benign proliferation of squamous epithelium caused by infundibular hyperplasia and squamous metaplasia of sebaceous glands, which may histologically mimic WD (well differentiated) SCC. Cf Squamous cell carcinoma

keratoconjunctivitis INFECTIOUS DISEASE Acute inflammation of the anterior eye–cornea and conjunctiva, often accompanied by nasopharyngitis; both are features of the common cold and usually due to viruses. See Common cold

keratolytic *adjective* Referring to a keratolytic agent, *noun* Keratolytic agent A substance that promotes the softening and peeling of epidermis–eg, diluted topical salicylic acid to ↓ the thickness of laminated keratin or hyperkeratotic scales–eg, in psoriasis

keratoplasty Corneal transplant, see there

keratorefractive surgery Photoreactive keratotomy, see there. Cf Radial keratotomy

keratosis DERMATOLOGY A condition characterized by ↑ keratin production See Actinic keratosis, Seborrheic keratosis, Stucco keratosis

keratotomy A surgical incision in the cornea. See Radial keratotomy

keraunophobia PSYCHOLOGY Fear of thunder and lightning. See Phobia

kerion DERMATOLOGY A severe form of tinea capitis, in which well-circumscribed parts of the scalp are transformed into a painful, boggy, inflamed, confluent mass with loosened hair, purulent folliculitis, crusting, often with lymphadenopathy, due to a zoophilic superficial mycosis, *Trichophyton verrucosum* or *T mentagrophytes*; geophilic or anthropophilic fungal infections may abruptly become kerions

Kerley lines IMAGING Vague horizontal lines on AP chest CXRs, typical of mitral stenosis. See Mitral stenosis

kernicterus Bilirubin encephalopathy NEONATOLOGY The staining of parts of the infant brain, especially the basal ganglia and hippocampus by BR that has penetrated the blood-brain barrier which, in older children, is more impervious to bilirubin; kernicterus is classically linked to Rh HDN, when the immune system of a mother who does not have the RhD–less commonly C, c, E, e, or other antigen on her RBCs, comes in contact with the infant's RBCs and forms antibodies to them; this causes a brisk hemolysis and ↑ BR; serum levels of ≥ 20 mg/dL of BR pose a high risk for kernicterus, and represent a medical emergency; severe kernicterus is often fatal, and characterized by lethargy, poor feeding, hypertonicity, seizures and apnea; survivors have sequelae in the form of dental dysplasia, cerebral palsy, hearing loss CLINICAL, FULL TERM INFANTS Severe jaundice, lethargy, poor feeding, choreoathetoid cerebral palsy, mental retardation, sensorineural hearing loss, gaze paresis. See Hemolytic disease of the newborn, Jaundice

Kernig sign NEUROLOGY Pt lies on back and flexes thigh upward, then complete extension of leg is impossible, typical of meningitis, see there

Kerr incision OBSTETRICS A low transverse incision in C-sections, preferred as there is less blood lost during surgery, less risk of rupture in future pregnancies, less postoperative infections and it is easier to repair. See Cesarean section. Cf Kronic-Selheim incision

Kestenbaum sign OPHTHALMOLOGY ↓ number of arterioles crossing optic disk margins, a finding in optic atrophy. See Optic atrophy

Ketalar Ketamine, see there

ketamine SUBSTANCE ABUSE A surgical anesthetic withdrawn from the market due to disorientation and violent behavior that occurred when Pts regained consciousness; on the club scene, ketamine is in liquid form or a white powder snorted or smoked with marijuana or tobacco products PHARMACOLOGIC EFFECTS Dissociative anesthesia, CNS stimulant, hallucinations. visual distortion, a loss of senses, sense of time, and orientation for 30 mins to 2 hr, delirium, amnesia, impaired motor function, HTN, depression, recurrent flashbacks, potentially fatal respiratory problems

ketoacidosis Diabetic ketoacidosis, DKA A type of metabolic acidosis seen in poorly controlled DM, with 'starvation amidst plenty'; despite hyperglycemia, ↓ insulin makes the excess glucose unavailable to the cells, which rely on lipid metabolites–ketone bodies–for energy, due to incomplete lipid metabolism CLINICAL Systemic acidosis, ↓ cardiac contractility and vascular response to catecholamines with thready pulse, hypotension, poor organ perfusion and diabetic ketoacidosis may be the presenting sign in previously undiagnosed DM, accompanied by an acute abdomen and

ketoconazole

marked leukocytosis LAB Ketonuria, ketonemia, hyperglycemia, glycosuria, ↓ pH, ↓ bicarbonate, ↑ anion gap, hyperlipidemia. See Alcoholic ketoacidosis

ketoconazole Nizoral INFECTIOUS DISEASE An oral imidazole that inhibits fungal cytochrome P450-dependent steroid synthesis, which is effective against candidiasis, blastomycosis, coccidioidomycosis, histoplasmosis, chromomycosis, paracoccidioidomycsis ADVERSE EFFECTS Liver problems, ↓ production of hormones—eg, testosterone

ketogenic steroid Any of a group of corticosteroids and their metabolites which have a hydroxyl group and a short carbon chain at the 17 position of the steroid nucleus ↑ IN Adrenal hyperplasia with precocious puberty, Cushing syndrome, obesity, physiologic stress—burns, surgery, infection ↓ IN Addison's disease, ACTH deficiency, hypopituitarism

ketorolac tromethamine Acular® PAIN MANAGEMENT A short-acting NSAID and analgesic equivalent of morphine used to manage seasonal allergic conjunctivitis and postoperative inflammation after cataract extraction

ketosis An abnormal ↑ in serum concentration of ketone bodies that does not produce acidosis. Cf Ketoacidosis

17-ketosteroids 'Male' hormones primarily produced by the testes and adrenal cortex—androsterone, DHEA, epiandrosterone, etiocholanolone, 11-keto- and 11-β-hydroxyandrosterone, 11-keto- and 11-β-hydroxyetiocholanolone; serum levels reflect adrenocortical and testicular function ↑ IN Adrenal hyperplasia or tumors, testicular tumors, pregnancy, physical or mental stress, polycystic ovarian disease ↓ IN Adrenal hypofunction, Klinefelter syndrome, castration, hypothyroidism, anorexia nervosa

17-ketosteroids, total A screening tool; if 17-KSs are ↑, each individual steroid is analyzed, by the much more expensive fraction procedure REF RANGE 0–11 yrs, 0.1–3.0 mg/24 hrs; 11–14 yrs, 2–7 mg/24 hrs; ♂ 6–21 mg/24 hrs; ♀ 4–17 mg/24 hrs ↑ IN Adrenal hyperplasia, CA, adenoma, adrenogenital syndrome ↓ IN Addison's disease, panhypopituitarism, eunuchoidism

ketotifen fumarate An antihistamine that stabilizes mast cells, ↓ secretion of immune mediators

keyhole configuration Keyhole pattern GYNECOLOGY A descriptor for the gross appearance of advanced lichen sclerosis in the labia majora and minora—labia are resorbed, leaving a residual patch of whitish skin around the vagina and anus

keyhole surgery A popular term for endoscopic surgery

keyway SURGERY The 'female' part of a 2-part interlocking surgical device

Ki-1 lymphoma(s) A heterogeneous group of childhood lymphomas, the cells of which react against the monoclonal antibody Ki-1, which usually involves the skin, soft tissue, bone, GI tract, rarely CNS and pericardium DIFFDX Metastatic undifferentiated CA, amelanotic melanoma, malignant sinus histiocytosis; despite its 'ugly' histology, KLs may respond to chemotherapy with prolonged remission or complete cure. See Lymphoma

kick *noun* CARDIOLOGY See Atrial kick *verb* DRUG SLANG To get off a drug habit

kickback alert A warning by the Office of the Inspector General indicating that certain arrangements may violate federal law

kicking Fetal movement, see there

kidney See Cake kidney, Goldblatt kidney, Horseshoe kidney, Large white kidney, Polycystic kidney, Rat-bitten kidney

kidney cancer AdenoCA of renal cells, hypernephroma, kidney cancer, renal cell cancer, renal adeno CA ONCOLOGY A renal tubule CA EPIDEMIOLOGY 30,600 new, 12,000 deaths/yr–US, 1996 RISK FACTORS 1.6 ♂, 1.9 ♀, obesity in ♀–5.9 odds ratio in highest 5% of upper body mass; phenacetin analgesics, ↑ meat consumption, tea drinking–♀, petroleum exposure DIAGNOSIS Hx, exam, imaging, Bx MANAGEMENT Surgery only; RT, chemotherapy are ineffective; IL-2 and IFN alpha yield a 20% response and 20% 1 yr event-free survival

KIDNEY CANCER

STAGE I Tumor ≤ 7 cm

STAGE II Tumor ≥ 7 cm

STAGE III CA extended into major renal veins or spread to a single lymph node

STAGE IV Metastasis to lymph nodes or organs

kidney damage Kidney injury NEPHROLOGY A structural or functional compromise in renal function due to external–eg, athletic, occupational, or other trauma, resulting in bruising or hemorrhage, which can be profuse and life threatening ETIOLOGY Vascular disease–eg, arterial occlusion, renal vein thrombosis; toxins–eg, analgesic nephropathy, lead, solvents, fuels; waste products–eg, uric acid in gout; lysis of malignant cells in chemotherapy; immune response to medications, infection; glomerulonephritis or acute tubular necrosis. See Acute renal failure, Chronic renal failure

kidney disease Renal disease NEPHROLOGY See Acute kidney failure, acute nephritic syndrome, analgesic nephropathy, atheroembolic renal disease, chronic kidney failure, chronic nephritis, congenital nephrotic syndrome, end-stage renal disease, Goodpasture's syndrome, IgA nephropathy, IgM mesangial proliferative glomerulonephritis, interstitial nephritis, kidney cancer, kidney damage, kidney stones, lupus nephritis, membranoproliferative GN I, membranoproliferative GN II, membranous nephropathy, minimal change disease, necrotizing glomerulonephritis, nephroblastoma, nephrocalcinosis, nephrogenic diabetes insipidus, nephrotic syndrome, polycystic kidney disease, post-streptococcal GN, reflux nephropathy, renal artery embolism, renal artery stenosis, renal disorders, renal papillary necrosis, renal tubular acidosis type I, renal tubular acidosis type II, renal underperfusion, renal vein thrombosis

kidney failure Any of a number of conditions–eg, diabetic nephropathy, ESRD, lupus nephritis, characterized by ↓ renal function MANAGEMENT Dialysis; success is monitored by URR and Kt/V. See Acute renal failure, Chronic renal failure, End-stage renal disease

kidney infection Pyelonephritis, see there

kidney panel LAB MEDICINE A battery of tests for evaluating renal functional status ANALYTES Albumin, BUN/creatinine, chloride, CO_2 content, creatinine clearance, glucose, K^+, total protein, Na^+, 24-hr urinary creatinine, protein. See Organ panel

kidney stones Renal calculus, nephrolithiasis, kidney stone disease UROLOGY Small irregular solid structures, often composed of calcium, uric acid, phosphate, found in the urine or ureters, which are a major cause of morbidity; up to 12% of those in developed nations have KS CLINICAL Renal colic–the worst pain ever experienced, hematuria, ureteral or renal pelvic obstruction, which may lead to hydronephrosis or facilitate infection RISK FACTORS Familial tendency, prematurity, bowel disease, ileal bypass for obesity, renal tubule acidosis and nephrocalcinosis. See Gout, Oxaluriasis

KIDNEY STONE TYPES

CALCIUM 75-85% of stones; more common in men, composed of calcium oxalate, carbonate or phosphate, controlled by altering diet

URIC ACID More common in ♂; ± 50% of Pts with UA stones also have gout

CYSTIC ACID Formed in Pts with cystinuria

MAGNESIUM AMMONIUM PHOSPHATE Struvite stones More common in ♀, due to bacterial–eg *Proteus*—spp which produce specific enzymes UTIs; MAP stones can be very large, fill renal pelvis, develop a staghorn appearance, obstruct urinary tract, and cause kidney damage

kidney tumor 1. Kidney cancer, see there 2. Wilms' tumor, see there

killed vaccine A vaccine consisting of dead but antigenically active viruses or bacteria, which evokes production of protective antibodies without causing disease. Cf Live attenuated vaccine

killer VOX POPULI A person or thing that kills, in fact or figuratively. See Internet serial killer, Self-killer, Serial killer, Silent killer

Killip scale Killip classification CARDIOLOGY A system used to stratify the severity of left ventricular dysfunction and determine clinical status of Pts post MI

KILLIP CLASSIFICATION

CLASS 1 No rales, no 3rd heart sound

CLASS 2 Rales in < ½ lung field or presence of a 3rd heart sound

CLASS 3 Rales in > ½ lung field–pulmonary edema

CLASS 4 Cardiogenic shock–determined clinically

kilocalorie Big calorie, Calorie A unit of heat equal to 1000 calories–the heat required to raise the temperature of 1 kg of H_2O 1°C

kindling NEUROLOGY The tendency of some regions of the brain to react to repeated low-level electrical stimulation by progressively boosting electrical discharges, thereby lowering seizure thresholds. See Cocaine, Seizures

kindred VOX POPULI An extended family, see there

kinesiology BIOMECHANICS The science of body movements especially vis-á-vis therapy

kinky hair disease Menke syndrome, trichopoliodystrophy METABOLIC DISEASE An X-R condition due to defective copper metabolism with copper accumulation in fibroblasts, kidneys, GI mucosa, and a relative deficiency of copper in other tissues–eg, brain and liver, accompanied by defective copper enzymes–eg, lysyl oxidase, tyrosinase CLINICAL FTT, seizures, hypothermia, ↑ infections, kinky hair–pili torti due to defective disulfide bond formation, seborrheic dermatitis, mental and growth retardation, myoclonic seizures, scurvy-like radiologic changes of long bones PROGNOSIS Death in early infancy with progressive neurologic and vascular degeneration. Cf Giant axonal neuropathy, Uncombable hair syndrome, Woolly hair disease

kinship VOX POPULI Relationship by marriage or, specifically, a blood ties

Kirby-Bauer test Agar diffusion test, see there

Kirschner wire K wire ORTHOPEDICS An unthreaded segment of extruded wire which is drilled into bone like a drill bit, either temporary or permanent, and alone or with cannulated screws to reduce and stabilize fractures; K wires can be placed between bones, or used as an intramedullary device to bridge fractures of small tubular bones. See Wire

'kiss of death' surgery A surgical procedure which, under certain clinical conditions, is contraindicated and associated with a high mortality–eg, abdominal surgery during acute pancreatitis is considered a KOD procedure

'kiss of death' test CRITICAL CARE A sobriquet for any clinical or lab test which, when positive, infers a poor prognosis and may presage a Pt's demise

KISS OF DEATH TEST

Delta osmolality > 40 mosm/mL

Lactate levels > 10 mEq/L

'LDH_6' An abnormal LDH band that migrates cathodic to the usual LDH isoenzymes–85% mortality

Multifocal atrial tachycardia–43% mortality

'kissing balloon' coronary angioplasty CARDIOLOGY A technique used to treat stenoses of the coronary arteries at bifurcations, which normally have a difficult access, by placing 2 stents

kissing, hazards of MEDTALK 1. Infectious mononucleosis, see there 2. Allergic reactions due to transmission of allergens to the kissee, which occurs in ±5% of the general population, resulting localized itching, swelling, urticaria, and possibly wheezing

Kleine-Levin syndrome PSYCHIATRY Periodic episodes of hypersomnia; first seen in adolescence, usually in boys, and accompanied by bulimia

kleptomania PSYCHOLOGY Compulsive stealing, usually of objects which may have symbolic significance

Klinefelter syndrome XXY gonadal dysgenesis An inherited condition caused by an extra complete–most commonly–or partial X chromosome; it is the most common cause of 1° ♂ hypogonadism CLINICAL Often asymptomatic; may first come to medical attention for infertility, complaints of inadequate androgenization, small testes, ↓ production of testosterone, and infertility; other features of KS: cryptorchidism, subnormal intelligence, bone defects, long-limbed, long-trunked, relatively tall, slim build, little facial hair growth; one-third develop gynecomastia MANAGEMENT Secondary sex characteristics are usually poorly developed and do not respond well to testosterone

Klippel-Feil sequence/syndrome The combination of short neck, low hairline at nape of the neck and limited movement of head, due to a defect in the early development of the spinal column in the neck–the cervical vertebrae

Klippel-Trenaunay-Weber syndrome Angio-osteohypertrophy A usually sporadic, occasionally AD condition characterized by hypertrophy of extremities, port-wine stains, and venous malformations, syndactyly and polydactyly. See Syndactyly

Klonopin Clonazepam, see there

Klopoxid Chlordiazepoxide, see there

Kluver-Bucy syndrome PSYCHIATRY A syndrome following bilateral temporal lobe removal characterized by ↓ recognition of people, loss of fear, rage reactions, hypersexuality, excess oral behavior, memory defects, overreaction to visual stimuli

knee bursitis Inflammation of one or all 3 major bursae of the knee, which may also be infected MANAGEMENT RICE–Rest, ice, cold, elevation; if infected, antibiotics, aspiration, surgery

knee injury See Anterior cruciate ligament, Lateral collateral ligament, Meniscus injury, Medial collateral ligament, Patellar dislocation, Posterior cruciate ligament, Runner's knee

knee jerk Knee-jerk reaction, knee reflex, patellar reflex NEUROLOGY A reflex tested by tapping just below the bent knee on the patellar tendon, causing the quadriceps muscle to contract and bring the lower leg forward

kneeing SPORTS MEDICINE A minor penalty which occurs when an ice player uses a knee to hit his opponent in the leg, thigh, lower body. See Sports medicine

knob sign IMAGING A ↓ or absent aortic 'knob' on a plain AP CXR, a finding typical of 2° atelectasis of the left lower lobe of lung

knock knees Genu valgum ORTHOPEDICS An internal deviation of the knee joint and outward angulation of the lower legs –the knees touch, the ankles are separated; causes, incidence, and risk factors; some degree of KK is present in *all* children from 2-6 yrs of age; most autocorrect with time; the degree of KK is determined by measuring the distance between the medial malleoli–ankles; KK may be seen in Ellis-van Creveld syndrome, rickets, or a complication of epiphysiodesis

knot A popular term for any mass or perceived mass OBSTETRICS Umbilical cord knot False knots of the umbilical cord–UC result from twisting and meandering of the umbilical vein and are of no clinical significance; the umbilical arteries are relatively linear; true knots require that the UC be long enough to permit the infant to pass through a loop of cord; since the cord is composed of erectile tissue, irreversible cord knotting is rare POPULAR HEALTH 1. A group of muscles in spasm or 2. A knot in the stomach, a manifestation of anxiety, which does not per se represent a palpable mass

knotless anchor suture SURGERY A device that simplifies Bankart repair by avoiding the technically difficult arthroscopic knot-tying. See Bankart repair

Knowles pin ORTHOPEDICS Orthopedic hardware once commonly used for intertrochanteric fractures

K$_{nuc}$ value The phylogenetic 'distance' between one or more related organisms based on sequence data

Koala vascular clamp SURGERY A reusable, ring-handled, radiopaque, stainless steel instrument for clamping delicate vessels during surgery. See Clamp

Koch's postulates A series of 4 conditions that must be met to establish an infectious agent as the cause of a particular disease or condition. See Molecular Koch's postulates

KOCH'S POSTULATES

1. The agent must be present in all cases of the disease
2. The agent must be isolated from someone with the disease and grown in pure culture
3. Inoculation into a susceptible organism of the agent–from a pure culture—must produce the disease
4. The agent must be recovered from the infected–inoculated organism and grown again in culture

Kock pouch Kock internal reservoir UROLOGY A segment of ileum used to create a continent urinary pouch, which may be sewn into the anterior abdominal wall in Pts with a radical cystectomy for bladder cancer. See Bladder cancer

Koebner's phenomenon Isomorphic reaction The appearance of new lesions of a skin disease–in particular psoriasis, but also lichen planus or eczema at a site of trauma. See Id reaction, Psoriasis

Kogoj's abscess DERMATOLOGY A spongiform aggregate of PMNs located in the parakeratotic stratum corneum of psoriasis. Seew Psoriasis

Kohn v Delaware Valley HMO MALPRACTICE A landmark 1991 case in Pennsylvania that ruled that the theory of apparent–ostensible agency may be invoked to hold an HMO liable for physician negligence. See Apparent agency

Korotkov sounds CARDIOLOGY Low frequency vibrations–< 200 mHz arising in vascular walls and heard distal to cuff compression of a peripheral artery; KSs are subdivided into an initial transient murmur–opening tap and compression murmur–rumble. Cf Doppler ultrasonography

Korsakoff psychosis Korsakoff syndrome NEUROLOGY A condition characterized by impaired memory and cognitive skills–eg, problem solving, learning, fabrication of experiences to bridge brain damage-related memory gaps. See Alcoholic encephalopathy, Wernicke's encephalopathy

Kotex Sanitary napkin, see there

Krazyglue™ SURGERY A superadhesive that may be used as a surgical adhesive

Kremer excimer laser OPHTHALMOLOGY A laser for LASIK treatment of myopia ± astigmatism. See LASIK

Kronic-Selheim incision OBSTETRICS A low vertical incision for C-sections, which may be used instead of the widely preferred Kerr incision when the fetus is large, or there is a lack of intrauterine space. See Cesarean section. Cf Kerr incision

KS 1. Kaposi syndrome, see there 2. Ketosteroid, see there 2. Klinefelter syndrome, see there

Kt/V NEPHROLOGY A formula for measuring dialysis adequacy, where K = dialyzer clearance–volume of fluid completely cleared of urea in a single treatment–in mL/min; t = time; V = volume of water in a Pt's body. Cf URR

KUB Kidneys, ureters, bladder UROLOGY A plain–ie, without radiocontrast–AP film of the abdomen, a crude method for detecting kidney stones; KUB films may be used as a 'scout' examination before the definitive IVP

Kuf's disease Lipofuscinosis METABOLIC DISEASE A cerebroretinal degenerative disorder 2° to lipid storage in the cerebral gray matter

Kussmaul breathing Air hunger CLINICAL MEDICINE Rapid, deep respiration 2° to stimulation of the respiratory center of the brain triggered by ↓ pH, normal during exercise, and common in Pts with severe metabolic acidosis–eg, DKA. See Metabolic acidosis, Diabetes

Kussmaul sign H&PE A paradoxic ↑ in jugular venous distension when Pt inhale. Cf Cardiac tamponade

kyphoscoliosis Scoliosis A combination of kyphosis and scoliosis–lateral curving of the spine

kyphosis Humpback, spinal kyphosis ORTHOPEDICS Angular curvature of the spine, with a posterior convexity, usually situated in the thoracic region involving a variable number of vertebrae ETIOLOGY TB, osteoarthrosis, rheumatoid arthritis, poor posture

Kytril™ Granisetron, see there

L Symbol for: 1. Labor, see there 2. Laboratory 3. Lateral 4. Left 5. Lens 6. Leucine 7. Levorotary 8. Library 9. License 10. Licensed—eg, to practice medicine 11. Ligament 12. Light chain immunoglobulin, see there 13. Lingual—dentistry 14. Liter—abbreviation used in US 15. Lumbar 16. Lung

l Symbol for: 1. Length 2. Ligament 3. Liter—abbreviation as used in the SI, International System

Ls of dermatology, the 5' A group of skin lesions associated with dense patchy infiltrates of upper dermal lymphocytes, including SLE, lymphoma, pseudolymphoma of Spiegler-Fendt, polymorphous light eruption of plaque type, Jessner's lymphocytic infiltration DiffDx Actinic reticulosis, angioimmunoblastic lymphadenopathy, arthropod bites, lymphomatoid papulosis, phenytoin-induced drug eruption

L&D Labor & Delivery An obstetrics unit in North American hospitals

LA 1. Latex agglutination 2. Left arm 3. Left atrium 4. Left axilla 5. Linguoaxial—dentistry 6. Linoleic acid 7. Local anesthetic 8. Long-acting 9. Lower arm 10. Lymphadenopathy

Laband syndrome An AD condition characterized by gingival fibromatosis, whittling of terminal phalanges, floppy ears, defective or absent fingernails, joint hypermobility, ± profound mental retardation

label PHARMACOLOGY A display of written, printed or graphic matter upon a container or article; all information placed on the container must, in the case of medications, also be placed on the product's outside container or wrapper

labeling CLINICAL PHARMACOLOGY The affixing or attaching of labels and other written, printed or graphic matter on an article or any of its containers or wrappers, or accompanying such article

labile *adjective* PSYCHOLOGY Moody, volatile

labile diabetes Brittle diabetes, see there

lability NEUROLOGY Neurologic instability, as in an evolving CVA/TIA or spinal cord injury. See Cerebrovascular accident, Spinal cord injury, Transient ischemic attack

labioscrotum UROGYNECOLOGY An external genital structure in Pts with sexual ambiguity, which kinda looks like labia, but not really and kinda looks a scrotum, but no cigar there either. See Hermaphroditism

labor Parturition OBSTETRICS The physiologic process that results in the expulsion of a conceptus and placenta from the uterus via the cervix and vagina. See Back labor, False labor, Prechaotic labor, Precipitate labor, Prolonged labor VOX POPULI Work. See Child labor

labor market A place where labor is exchanged for wages; an LM is defined by geography, education and technical expertise, occupation, licensure or certification requirements, and job experience

laboratory LAB MEDICINE A facility for the biological, microbiological, serologic, chemical, hematology, immunohematologic, biophysical, cytologic, pathologic, or other examination of materials derived from the human body for the purpose of providing information for diagnosing, perventing, or treating disease or impairment of, or assessment of human. See Accredited lab, Behavioral care lab, Bio-analytical lab, Core lab, Cardiac cath lab, Expert lab, Family Research lab, Full service lab, MediaLab lab, Meth lab, Mobile lab, Physician office lab, Reference lab, Special-function lab, Specialized lab, Stat lab. Cf CLIA 88, Collecting station

lab error Any error in results or result reporting that can be attributed to a clinical lab or its workers

lab test An analysis of one or more substances from a subject—ie, a Pt's specimen in a site—lab dedicated to assuring accurate and timely results; LTs are performed to: 1. Detect—screen for or diagnose a disease, or to exclude its presence 2. Determine the severity of a disease 3. Monitor the progress of a disease, its response to therapy and prognosis 4. Monitor drug toxicity

labs A popular 'short form' for laboratory work or other studies of analytes performed in a clinical laboratory

labyrinthine concussion ENT Trauma implicated in peripheral vestibular damage, often associated with variable hearing loss due to blunt head trauma or barotrauma; vestibular and auditory complaints are usually transient; spontaneous resolution is the norm; persistent Sx are attributed to 2° endolymphatic hydrops; labyrinthine hemorrhage may trigger inflammation, leading to fibrosis and ossification, ergo audiovestibular dysfunction. See Concussion

labyrinthine fistula procedure ENT Procedures—Cody procedure, round window shunt, cochleosacculotomy, of historical interest, intended to provide controlled release of pressure from the endolymphatic space to the perilymphatic space without damaging the vestibular labyrinth; results were disappointing and inconsistent

labyrinthine hydrops ENT An excess in labyrinthine fluid, resulting in pressure in the ears, hearing loss, vertigo, imbalance

labyrinthitis ENT Viral or bacterial infection or other inflammation of the inner ear ETIOLOGY Otitis media, URI, allergy, cholesteatoma, ototoxic drugs CLINICAL Loss of balance, vertigo, temporary hearing loss

laceration Shearing of a mucocutaneous or other surface, often with visible briding of connective tissue. See Cerebral laceration

Lachman test SPORTS MEDICINE A clinical maneuver used to determine the effects of anterior shear loads applied to the knee at 30° flexion; the LT is preferred to the anterior drawer test for evaluating the integrity of the anterior cruciate ligament. See Anterior drawer test

LACI CARDIOLOGY 1. A clinical trail—Laser Angioplasty for Critical Ischemia 2. Lipoprotein-associated coagulation inhibitor, aka Tissue factor pathway inhibitor, see there

lactamase beta lactamase A bacterial enzyme capable of breaking down the beta lactam ring of certain antibiotics, including penicillins, inactivating them

lactate dehydrogenase CARDIOLOGY An oxidoreductase present in the cytoplasm of all cells that catalyze LACTATE + NAD^+ $\leftrightarrow$ PYRUVATE + NADH + H^+, the equilibrium of which favors lactate + NAD^+ at neutral pH; LD_1 is classically $\uparrow$ in acute MI, peaking by post-infarct day 4, and associated with a flip in normal ratio of LD_1 and LD_2. See Flipped LD, LD_6. Cf CK-MB, Troponin I, Troponin T

lactate EMERGENCY MEDICINE A salt or ester of lactic acid; the time required to normalize lactate is a useful prognostic tool in trauma victims. Cf 'Kiss of death' test

lactation 1. Milk secretion–the production and secretion of milk by the mammary glands 2. The time during which a ♀ provides breast milk to an infant. See Breast feeding, Weaning

lactic acidosis METABOLISM Metabolic acidosis due to $\uparrow$ lactic acid resulting from tissue hypoxia or $\downarrow$ conversion of lactate to pyruvate ETIOLOGY Exercise, endogenous or exogenous metabolic defects LAB $\downarrow$ Bicarbonate, $\downarrow$ pH, $\uparrow$ anion gap, $\uparrow$ PO_4 TREATMENT Dichloro-acetate treatment of Pts with severe LA results in statistically significant but clinically unimportant changes in the pH and arterial blood lactate and does not alter the hemodyanimcs or survival

Lactobacillus acidophilus A bacterium in yogurt, and in the colon of infants; consumption of L acidophilus may be therapeutic as it prevents overgrowth of opportunistic and/or potentially pathogenic bacteria and fungi. See Yogurt

lactose intolerance β-D-galactosidase deficiency, lactase deficiency INTERNAL MEDICINE An acquired, or AR condition characterized by an inability to digest lactose–often due to lactase deficiency, which may $\uparrow$ with age; LI is more common in African Americans, Native Americans, Mediterraneans, Asians, due to deficiency of lactase on intestinal brush borders ETIOLOGY GI disease–celiac sprue, viral or bacterial gastroenteritis, post-gastroduodenal surgery CLINICAL Abdominal bloating, nausea, cramps, flatulence, inability to metabolize disaccharides, resulting in osmotic diuresis, diarrhea, acidic stools MANAGEMENT Lactose restriction in diet unless pretreated with lactase; cultured milk products–eg, yogurt, buttermilk may be well tolerated. See Lactose, Lactose tolerance test

lactose tolerance test A test for lactase deficiency, based on oral loading of lactose. See Lactose, Lactose intolerance. Cf Glucose tolerance test

lacunar amnesia NEUROLOGY Amnesia only for certain events. Cf Selective memory

lacunar infarct Small vessel diseae NEUROLOGY Any of multiple small cerebral infarcts in the corona radiata, internal capsule, striatum, thalamus, basis pontis, cerebellum, occasionally preceded by transient Sx, due to occlusion or stenosis of small penetrating branches of the middle and posterior cerebral and median branches of the basilar arteries; resolution of infarcts is characterized by residual 1-3 mm cavities or lacunae, characteristic of longstanding HTN CLINICAL Pure motor hemiparesis, pure sensory stroke, ataxic hemiparesis, sensorimotor stroke. See Lacunar state, Multi-infarct dementia. Cf Hemorrhagic stroke

lacunar state NEUROLOGY A condition characterized by multiple minute infarcts—lacunes in the basal ganglia, which may be seen in severe HTN; when numerous, LS causes dementia or a lacunar 'syndrome' CLINICAL Loss of recent memory, altered time-space orientation, paranoia, headache, vertigo, giddiness, convulsions; the focal nature of the infarcts explains various neurologic defects–eg, homolateral cerebellar ataxia, isolated hemiplegia, pure segmental sensory stroke and dysarthria and clumsy hand syndrome

lacy appearance BONE RADIOLOGY A descriptor for the coarsely trabeculated shadows in long bones in 2º syphilis, which correspond to gumma and syphilitic periostitis; an LA is also seen in early hyperparathyroidism or chronic osteomyelitis; when marked, bone resorption imparts a 'scooped-out' appearance, trabeculation and radiolucency of the medullary cavity. Cf Ground-glass appearance, Veiled appearance

LAD Left anterior descending coronary artery

ladder diagram Laddergram CARDIOLOGY A ladder-like graph used to illustrate propagation of a cardiac impulse and analyze complex arrhythmias. See Electrophysiology. Cf Vectorgram

ladder pattern LAB MEDICINE Light chain ladder A phenomenon that may be seen on routine high-resolution agarose gel electrophoresis–AGE and immunofixation of urine specimens; LCLs are electrophoretically more homogeneous, are usually kappa–κ light chains–less commonly lambda–λ light chains, and widely distributed over the γ-globulin zone; LCLs appear in Pts with lymphoproliferative diseases–eg, lymphoma, myeloma, monoclonal gammopathy of undetermined significance, some infections, and inflammatory conditions PEDIATRICS A sign on the abdomen in children with an obstructed lower GI tract, seen as parallel loops of distended small intestine, causing a 'stepped' pattern

LADME THERAPEUTICS An acronym for processes that affect serum drug levels after administration–Liberation, Absorption, Distribution, Metabolism, Excretion. See Metabolism

laetrile Amygdalin QUACKERY A cyanide-rich bitter almond, apricot or peach pit extract, claimed to be effective in treating cancer. See Manner cocktail, Tijuana, Unproven methods of cancer management

Lafora's disease An AR neurometabolic disorder characterized by progressive myoclonus epilepsy and mental deterioration beginning in adolescence, ending in dementia MANAGEMENT Seizure control PROGNOSIS Poor; death within 10 yrs of onset

lag phase EMERGENCY MEDICINE The period between when a person is exposed to a toxic inhalant–eg, cadmium fumes, dimethyl sulfate, methyl bromide, ozone, nitrogen oxides, phosgene, phosphorus compounds and others and development of pulmonary edema–up to 12 hrs

lag screw ORTHOPEDICS Any screw used to achieve interfragmental compression; LSs do not protect fractures from bending, rotation or axial loading forces, and other devices should be used to provide these functions. See Screw

lalling stage NEUROLOGY The stage of linguistic development that follows babbling–± 6 months after birth and consists of the interspersing of babbling with pauses, inflections, and intonations from what the infant hears. Cf Babbling stage, Cooing stage

Lamaze technique ALTERNATIVE OBSTETRICS A program of instruction that orients first-time, less commonly, experienced mothers to uncomplicated vaginal delivery, with participation of the father or partner. See Alternative gynecology, Doula, Midwife, Natural childbirth

LAMB syndrome Carney's complex An AD clinical triad of young adult onset, which consists of spotty mucocutaneous Lentigenes, cutaneous and cardiac–Atrial Myxomas, and Blue nevi, endocrine hyperactivity, multifocal myxoid fibroadenomas of breast, adrenocortical hyperplasia and calcifying Sertoli cell tumors of testes. Cf NAME syndrome

lambda λ Symbol for: 1. Light chain–immunoglobulin 2. Wavelength

Lambda Plus® laser SURGICAL ONCOLOGY A dye laser used for light activation in photodynamic therapy for lung and esophageal CAs. See Photodynamic therapy, Photofrin®–porfimer sodium

Lambert-Eaton syndrome Carcinomatous myasthenia, carcinomatous myopathy, myasthenic syndrome NEUROLOGY A neuromuscular disorder characterized by chronic progressive muscular weakness usually of legs, aching, fatigability, autonomic dysfunction–dry mouth, impotence, constipation, blurred vision, dyshidrosis, absent deep tendon reflex; LES is often paraneoplastic, and commonly linked to primary lung carcinoma, 'classically' small cell type, but also SCC and adeno CA MANAGEMENT 3,4-Diaminopyridine may be enhance acetylcholine release

lambswool WOUND CARE A cushioning material used to the risk of decubitus ulcers. See Pressure ulcer

lamellar ichthyosis Collodion baby, nonbullous congenital ichthyosiform erythroderma DERMATOLOGY An AR condition characterized by a brittle shell-like skin surface in infancy, which gives rise to coarse scaling of the skin surface, especially on flexures of extremities. See Ichthyosis

Lamictal® Lamotrigine NEUROLOGY An adjunct agent for treating generalized seizures in Lennox-Gastaut syndrome. See Lennox-Gastaut syndrome

laminectomy ORTHOPEDICS A procedure for managing intervertebral disk herniation; the 'classic' laminectomy entails bilateral removal of the lamina of a vertebral body adjacent to a diseased disk as well as varying portions of both articular facets. See Cervical laminectomy, Diskectomy, Hemilaminectomy, Laminotomy

laminotomy Fenestration ORTHOPEDICS A procedure for treating herniation of an intervertebral disk, consisting of removal of a portion of the superior and inferior aspects of the lamina adjacent to the diseased disk. See Laminectomy

Lamisil Terbinafine DERMATOLOGY An antifungal for tinea pedis, tinea cruris, tinea corporis due to *Epidermophyton floccosum*, *Trichophyton mentagrophytes*, *T rubrum*, fluconazole-resistant oral candidiasis. See Fluconazole

lamivudine® Epivir, 3TC AIDS An anti-HIV nucleoside analogue with activity against HBV ADVERSE EFFECTS Headache, nausea, ↓ WBCs, rare alopecia. See AIDS, HIV

lamotrigine Lamictal An antiepileptic that blocks Na+ channels and causes presynaptic inhibition of the excitatory amino acids glutamate and aspartate, which may benefit peripheral neuropathy ADVERSE EFFECTS Somnolence, exanthema, vomiting, laryngitis, worsened convulsions

LAMP CARDIOLOGY A clinical trial–Locally Advanced Multimodality Protocol

Lamprene® Clofazimine, see there

Lamra Diazepam, see there

LAN Local area network INFORMATICS A computer network confined to a limited area, usually of the same floor or building. See Ethernet, Cf Internet, WAN

Lance Armstrong tumor Seminoma, see there, testicular cancer

lancet 1. A small pointed, two-sided knife. See Spring-loaded lancet 2. A major peer-reviewed medical journal in England

landmark case LAW & MEDICINE A civil or, far less commonly, criminal action that has had an impact on a particular area of medicine.

landmark walking MEDTALK The practice, by those with ataxia, motion sickness, or too much "sauce" on board, of walking from one supporting structure–wall, desk, etc–to another, without which the gait would be lurching, festinating, or semi-controlled falling

land mine GLOBAL VILLAGE An antipersonnel device buried in a road, field, and countryside, intended to, when stepped on, explode and kill/maim the enemy du jour

Langerhans' cell histiocytosis Histiocytosis X, Langerhans cell granulomatosis, multifocal eosinophilic granuloma HEMATOLOGY An autonomous proliferation of a lymphoreticular cell, the Langerhans cell, which stains positively with antibodies to ATPase, S-100 and CD1a; LC aggregates are accompanied by eosinophils, foamy cells, neutrophils, fibrosis; histiocytosis X is divided into 3 clinical forms

---

LANGERHANS CELL HISTIOCYTOSIS–HISTIOCYTOSIS X

SOLITARY BONE INVOLVEMENT Eosinophilic granuloma A lesion of younger Pts; may affect any bone–spares hands & feet, most commonly, the cranial vault, jaw, humerus, rib & femur. Radiology Mimics Ewing sarcoma. Treatment Curettage. Prognosis Excellent

MULTIPLE BONE INVOLVEMENT Polyostotic eosinophilic granuloma, Hand-Schüller-Christian disease A lesion that variably affects the skin, accompanied by proptosis, diabetes insipidus, or chronic otitis media or combination thereof, marked by a prolonged course with waxing and waning symptoms Prognosis Relatively good

MULTIPLE ORGAN INVOLVEMENT Letterer-Siwe disease A lesion that affects bone, lung and skin, which while histologically indistinct, but more aggressive than the other forms Prognosis Poor if < 18 months at time of diagnosis, hemorrhagic skin lesions, hepatomegaly, anemia, thrombocytopenia, BM involvement

---

language disorder SPEECH PATHOLOGY Any defect in verbal communication and the ability to use or understand the symbol system for interpersonal communication. See Dyslexia

language mismatch SPEECH PATHOLOGY The degree of discordance between a person's reading skills and the material being read. See Word agnosia.

lanoteplase Novel plasminogen activator CARDIOLOGY A thrombolytic for restoring coronary blood flow in Pts with an acute MI. See Acute myocardial infarction

Lanoxin® Digoxin, see there

lansoprazole Prevacid® THERAPEUTICS A proton pump–H+/K+-ATPase inhibitor similar to omeprazole which, with antibiotics–eg, amoxicillin, clarithromycin, and/or metronidazole, is used to eradicate *H pylori* in Pts with peptic ulcers, erosive esophagitis ADVERSE EFFECTS Diarrhea CONTRAINDICATIONS CAD, MI, ASHD

Lantiseptic® DERMATOLOGY A proprietary formulation of lanolin, used to clean and protect skin, and manage decubitus ulcers. See Pressure ulcer

Lantron® Amitriptyline, see there

Lantus Insulin glargine. See rDNA insulin

Lanza scale GASTROENTEROLOGY A system that scores the severity of NSAID-induced GI tract ulcers–eg, grade 2 < 5 erosions, grade 4 > 10 erosions. See ASTRONAUT, OMNIUM

LAP 1. Laparoscopy 2. Laparotomy 3. Left atrial pressure

lap chole Popular slang for a laparoscopic cholecystectomy, see there

lap traveler PUBLIC HEALTH A child seated on an adult's lap in a car who is unrestrained in an child safety seat. See Child safety seat, MVA

laparoscope SURGERY A flexible lighted endoscope inserted via a small incision in the anterior abdominal wall for assessing defects identified by various imaging modalities. See Laparoscopy

laparoscopic-assisted resection SURGERY A procedure in which one or more components of the operation is performed with a laparoscope–eg, anastomosis, lymphadenectomy. See Laparoscopic surgery

laparoscopic cholecystectomy Lap chole SURGERY The laparoscopic removal of a gallbladder though a small abdominal wall incision. See Laparoscopy

laparoscopic colon resection Laparoscope-assisted colectomy, laparoscopic colectomy SURGERY An 'advanced' laparoscopic procedure, in which a fiberoptic laparoscope is used to resect a segment of the colon; LCRs use a combination laparoscope/video camera, which allows a small incision and major, but minimally invasive surgery. See Laparoscopic surgery, Port-site metastasis

laparoscopic morcellator SURGERY A device that crushes tissue for laparoscopic removal of large hardened masses, supracervical hysterectomy, and laparoscopic myomectomy. See Laparoscopic hysterectomy

laparoscopic staging SURGICAL ONCOLOGY The use of a fiberoptic laparoscope to determine the extent of involvement by a malignant tumor–eg, gastric, esophagus, pancreas, liver. See Staging

laparoscopic surgery The use of a fiberoptic laparoscope and specialized instruments to diagnose and/or treat 'surgical' disease. See Laparoscopic staging, Paradoxical movement

laparoscopic vaginal hysterectomy GYNECOLOGY A hysterectomy performed by a laparoscope passed through the vaginal vault. See Hysterectomy

laparoscopist Laparoscopic surgeon

laparoscopy Peritoneoscopy PROCEDURAL MEDICINE A procedure in which a laparoscope is used for various procedures–eg, lyse adhesions, tubal sterilization, remove foreign bodies, tissue and organs, stage cancer, fulgurate–'zap'–endometriosis implants, biopsy the liver, ovary, etc, diagnose abdominal pain, visualize the liver, evaluate pelvic masses–eg, lymphoma, and obtain biopsies. See Laparotomy

laparotomy OBSTETRICS Open pelvic surgery, a surgical format which has waned in popularity–WIP, but may be used for sterilization See Minilaparotomy tubal sterilization SURGERY Open surgical exploration of the abdomen, which has WIP–laparoscopy has equal diagnostic yields with ↓ complications INDICATIONS Pronounced cholangitis, sepsis, perforation, hemorrhage. See Exploratory laparotomy. Cf Laparoscopy

lapse HEALTH INSURANCE Termination of coverage due to nonpayment within a specified time period

lapsus linguae PSYCHOLOGY A slip of the tongue, attributed to unconscious thoughts

large cell lymphoma ONCOLOGY A lymphoma that accounts for 30% of NHLs in children SUBTYPES Large cell lymphoma and immunoblastic lymphoma–both arising in B cells and anaplastic large cell lymphoma, which typically develops from T cells LOCATION Neck, chest–mediastinum, thymus, throat, or abdomen, often spreading to skin, rarely to BM or brain. See Lymphoma, WHO classification

large cell undifferentiated carcinoma of lung Large cell carcinoma An aggressive pleomorphic carcinoma which corresponds to a PD SCC, adenoCA or small cell carcinoma, which may be associated with marked eosinophilia or leukocytosis CLINICAL 50% metastasize to the brain; if > 40% of cells are 'giant', it is designated giant cell carcinoma and has a very poor prognosis–average survival of < 1 yr. See Squamous cell carcinoma. Cf Small cell carcinoma

large for gestational age NEONATOLOGY *adjective* Referring to a fetus weighing ≥ 4.0 kg, 9lbs or ≥ 90 percentile for weight

large head Macrocephaly, see there

large simple trial CLINICAL TRIALS A prospective, randomized controlled trial that uses large numbers of Pts, broad inclusion criteria, multiple study sites, minimal data requirements, electronic registries PURPOSE Detect small treatment effects, gain effectiveness data, improve external validity. See Trial

large vessel vasculitis INTERNAL MEDICINE Vasculitis of the aorta and its major tributaries, which supply the extremities, head, neck EXAMPLES Giant cell/temporal arteritis, Takayasu's arteritis. See Systemic vasculitis

Lariam® Mefloquine, see there

Laron dwarfism GENETICS An AR condition characterized by target resistance to the action of hGH, due to a defect in the hGH receptor CLINICAL Consanguinity, severe growth retardation LAB ↑ IGF-I

Laroxyl® Amitriptyline, see there

larva currens A form of larva migrans due to an infestation by ♀ larvae of *Strongyloides stercoralis*, which penetrate the skin, causing intense, transient pruritus, urticaria, and accompanied by systemic symptoms–eg, bronchitis, abdominal pain, diarrhea, constipation, N&V, anorexia, weight loss. See Hyperinfection syndrome

larva migrans PARASITOLOGY An infestation by non-human nematodes that penetrate and migrate through the skin, but which cannot complete their life cycle in humans

LARVA MIGRANS

CUTANEOUS LM Due to dog & cat hookworm, *Ancylostoma braziliense*, the larvae of which migrate a few mms/day, producing pruritic serpiginous tracks in the stratum germinatum that are visible below the skin

VISCERAL LM Due to dog–*Toxocara canis* or cat parasites–T cati, the life cycles of which in their usual hosts resemble that of Ascaris lumbricoides in humans; in VLM, embryonated eggs are accidentally ingested and hatch in the intestine, penetrating the mucosa, aimlessly wander through the circulation, passing to liver and/or lung vessels and may be fatal if they involve the myocardium or CNS; in the eye, the resulting retinal granuloma mimics retinoblastoma

laryngeal cancer Cancer of voice box, larynx cancer, throat cancer HEAD & NECK SURGERY An epithelial CA–usually SCC, that arises in the vocal cords in Pts > age 55, linked to smoking/tobacco use CLINICAL Painless hoarseness, dysphagia DIAGNOSIS Laryngoscopy MANAGEMENT RT, surgery; chemotherapy is rarely useful. See Oral cancer

LARYNGEAL CANCER

STAGE I The cancer is confined to site of initial lesion; the definition of stages I & II depends on where the cancer started

• Supraglottis CA is only in one area of the supraglottis; the vocal cords can move normally

• Glottis CA is only in the vocal cords; the vocal cords move normally

• Subglottis CA is confined to the subglottis

STAGE II CA is confined to the larynx and has not spread to lymph nodes in the area or to other parts of the body

• Supraglottis The cancer is in > one area of the supraglottis; the vocal cords can move normally.

• Glottis CA has spread to supraglottis and/or subglottis; the vocal cords may/may not be able to move normally

• Subglottis The cancer has spread to the vocal cords, which may/may not be able to move normally

STAGE III CA is confined to the larynx, but vocal cords are fixed or CA has spread to perilaryngeal tissues; or CA has spread to one lymph node on the same side of the neck as original tumor, and measures ≤ 3 cms

STAGE IV CA has spread to tissues around the larynx, eg, pharynx or perilaryngeal tissues; regional lymph nodes in the area may/may not contain cancer; or, the cancer has spread to > one lymph node on the same side of the neck as the cancer, to lymph nodes on one or both sides of the neck, or to any lymph node measures > 6 cms; or metastasized

laryngeal electromyography A technique analogous to conventional EMG used to differentiate neurogenic from myogenic disorders, localize a neurogenic lesion–ie, upper vs lower neurons, and differentiate organic from functional defects; LEMG is used to evaluate vocal fold dysfunction, tremor, myoclonus, pyramidal and extrapyramidal disorders, and primary muscular disorders, delineating specific disorders–eg, superior vs recurrent laryngeal nerve involvement, provide prognostic information. See Electromyography

laryngeal-mask airway Anesthesiology A device which rests in the hypopharynx and has an inflatable cuff that surrounds the glottic opening; its proximal end is similar to standard endotracheal tubes, and is used to deliver anesthetic gases for spontaneous or controlled ventilation. See Endotracheal tube

laryngeal microsurgery ENT Any phonosurgery in which microsurgical techniques are used to manage benign vocal fold pathology–eg, edema, nodules, polyps, sulci, tumors. See Phonosurgery

laryngeal nerve transplant ENT The surgical repair of recurrent laryngeal nerve–RLN-damage, which may unilateral or bilateral, temporary or permanent and occurs in 0.2% of cases of thyroid surgery for non-malignant conditions and 5% of cases treated for thyroid CA

laryngeal nodule Singer's node, see there

laryngeal papillomatosis Juvenile papillomatosis, warty growths on vocal cords ENT A condition characterized by the presence of multiple, often recurring warts on the vocal cords, most commonly seen in young children, often linked to HPV infection via the birth canal

laryngeal paralysis ENT Loss of function of one or both vocal folds. See Recurrent laryngeal nerve

laryngectomy Surgical oncology The surgical removal of part or all the larynx for invasive cancer Types Hemilaryngectomy, supraglottic, total. See Hemilaryngectomy, Subtotal laryngectomy, Supraglottic laryngectomy, Total laryngectomy

laryngismus stridulus Pseudocroup Pediatrics A childhood cough with a whoop due to spasmodic laryngeal closure with crowing inspiration, a sign of tetany Clinical Cyanosis, apnea, paresthesia, tingling of hands and feet; LS is common at night, often waking Pt Lab ↓ Ca^{2+}, normal magnesium. See Latent tetany syndrome. Cf Whooping cough

laryngitis ENT Inflammation of the larynx, resulting in a hoarse voice and/or complete voice loss 2° to vocal cord irritations. See Reflux laryngitis

laryngomalacia A soft floppy larynx, due to softening of the cartilage

laryngoscopy ENT Examination of the larynx, either with a mirror–indirect laryngoscopy or with a laryngoscope–direct laryngoscopy; a diagnostic and/or therapeutic procedure in which a laryngoscope is passed through the larynx to detect lesions, strictures or foreign bodies. See Larynx

laryngotracheal reconstruction ENT Any of the operative techniques used to enlarge and stabilize the upper airway

Lasègue sign Neurology A clinical sign used to distinguish hip from low back pain. See Low back pain

laser Light amplification by stimulated emission of radiation, Physics A device that provides a focused beam of light with enough energy to cut or coagulate tissue Surgery A tool that focuses light into an intense, narrow beam to cut or destroy tissue Applications Microsurgery, photodynamic therapy, diagnostics Types of lasers Argon-405 nm wavelength, CO_2-630 nm, tunable dye-1600 nm, YAG-neodymium-yttrium-aluminum-garnet-10,600 nm. See Argon laser, Excimer laser, Holmium pulsed laser, Nd:YAG laser, PTP laser, Roller ball technique, Selective photothermolysis, Yellow krypton laser

LASERS IN MEDICINE

Type/General uses	Specialty/Wavelength/Power or energy
Alexandrite	DR, ENT, O, U/700-800 nm//≤ 1 J/pulse
Argon/Coagulation	DN, DR, ENT, O, U/450-515 nm//≤ 6 W
CO_2/Cutting, vaporization	C, DN, DR, ENT, GS, GY, NS, OS, U//10.6 μm/≤ 100 W
Diode	GS, U/800 nm/≤ 50W
Er:YAG	DN, O/2.94 μm/≤ 1 J//pulse
Excimer	C, 0/193 nm/≤ 0.6 J//pulse
Holmium YAG	C, O, OS/2.10 μm//≤ 60 W
Nd:YAG/Volume coagulation	DN, DR, 0/532, 1064 nm//≤ 0.5 J/pulse
Nd:YAG, CW	C, DR, ENT, GI, GS, NS, OS, U/532, 1064, 1044 nm//≤ 125 W
Pulsed dye	DR, O, U/504-620 nm//≤ 2 J/pulse
Ruby	DR/694nm//≤ 2 J/pulse

C–Cardiology, DN–Dentistry, DR–Dermatology, ENT, GI, GS–General surgery, GY–Gynecology NS–Neurology, O–Ophthalmology, OS–Orthopedic surgery, PS–Plastic surgery, PU–Pulmonology, U–Urology

laser bronchoscopy Pulmonology The deployment of a laser through a bronchoscope to relieve obstruction by a neoplasm of a central airway Action Cutting, coagulation, vaporization, necrosis. See Bronchoscopy, Laser

laser dentistry Dentistry Any use of lasers in dentistry–eg, zapping caries, cosmetic dentistry. See Cosmetic dentistry

laser hair transplantation Laser-assisted hair transplantation Cosmetic surgery A format for hair replacement which combines pulsed CO_2 laser and slit grafting, micrografting and minigrafting. See Selective photothermolysis. Cf Laser-assisted hair transplantation

laser myringotomy ENT A procedure for treating serous otitis media by lasering a hole in the tympanic membrane–myringotomy without inserting ventilation tubes; the opening allows drainage and endoscopic evaluation of the middle ear; it remains open for wks, followed by spontaneous and complete healing of the tympanic membrane; LM is done in a physician's office (surgery for the Brits) under local anesthesia on Pts > age 6. See Myringotomy

laser resurfacing Cosmetic dermatology The use of CO_2 lasers and, more recently, lower temperature erbium lasers, to bloodlessly eliminate wrinkles and sun damage by vaporizing the epidermis, stimulating the dermis to produce collagen and rejuvenate the skin. Cf Chemical peel

laser surgery Surgery Treatment through revision, destruction, incision, or other structural alteration of human tissue using laser technology. See Selective photothermolysis

laser thrombolysis Vascular medicine A developing technology in which a fiberoptic catheter with a 570 nm laser is passed through an artery–eg, a coronary artery, and blood clots vaporized using ultrashort bursts of energy

Lasette® Lab medicine A laser finger perforator for capillary sampling of glucose and other chemistries, which provides less-painful monitoring for Pt and eliminates accidental needle sticks. See Needle stick

LASIK Laser-assisted in situ keratomileusis Ophthalmology A type of laser ocular surgery in which the cornea is flapped open and the cornea shaved to correct either near– or farsightedness Statistics 750,000 Americans were LASIK'd in 2000; 90+% of treated Pts achieve better than 20/20 vision Criteria < –12—15 diopters for nearsightedness; +4–+6 for farsightedness Complications Ocular blowout 2° to corneal perforation, flap wrinkling/misalignment, epithelial ingrowth into stroma resulting in vision defects–glare, halos, diminished contrast, ↓ light passing to the pupil, lifting of the cornea with irregular astigmatism. See Custom LASIK, Intraocular lenses. Cf INTACS, Photorefractive keratotomy, Radial keratotomy

last menstrual period Gynecology The most recent time that a ♀ notes menstruation, a datum recorded in a chart during a routine gynecologic visit. See Menstruation

last Podiatry A form–combination, inflare, narrow heel, outflare, straight–over which a shoe is constructed

LAT Ambulatory surgery A cocktail of topical anesthetics–lidocaine, adrenaline, tetracaine–used for lacerations of vascularized regions of the head and face; LAT is as effective as TAC, but is not associated with seizures and/or death. Cf TAC

latanoprost Xalatan™ OPHTHALMOLOGY An agent used to ↓ intraocular pressure in Pts with 1° and 2° glaucoma. See Glaucoma

latchkey child SOCIAL MEDICINE A child who arrives home after school, lets him/herself into the house–with a 'latchkey', and is unattended until the parents' arrive from work. Cf 'Supermom'

LATE CARDIOLOGY A clinical trial–Late Assessment of Thrombolytic Efficacy that compared the effect of late–6 to 24 h after onset of Sx–tPA therapy in Pts with acute MI. See Acute myocardial infarction, tPA

late abortion An abortion performed after the 12th wk of gestation. See Abortion

late dumping syndrome see Dumping syndrome

late luteal phase dysphoric disorder Premenstrual syndrome, see there, aka premenstrual dysphoric disorder

late-onset immune deficiency An idiopathic condition associated with gastric CA, atrophic gastritis, pernicious anemia, autoimmunity, and malabsorption, variably accompanied by lactose intolerance, small intestinal atrophy, thymoma and agnogenic myeloid metaplasia with Ig defects and the Prasad syndrome

late phase reaction CLINICAL IMMUNOLOGY A delayed or 2° response in asthmatics to an antigenic challenge, in which PMNs release histamine, stimulating mast cell and basophil degranulation, in turn evoking bronchial hyperreactivity TREATMENT β-adrenergic aerosols

late potential CARDIOLOGY A high-frequency, low-amplitude signal at the tail of a QRS complex, which is attributed to fragmented and delayed electrical conduction through the borders of a myocardial scar; delayed conduction of LPs allows reentry of electrical impulses and ↑ susceptibility to ventricular arrhythmias

late stent thrombosis CARDIOLOGY The scar-induced closure of a previously stenosed coronary artery, a complication in ±20% of Pts who have undergone stenting for CAD. See Stent. Cf In-stent restenosis

late systolic murmur CARDIOLOGY A diamond-shaped murmur late in systole, often accompanied by mid or late systolic click. See Mitral valve prolapse

late-term abortion Post-viability abortion MEDICAL ETHICS Any abortion performed after the fetus would be viable if delivered to a nonspecialized health center. See Partial birth abortion

latency period EPIDEMIOLOGY A period of subclinical or inapparent pathologic changes following exposure to a noxious agent, ending with the onset of Sx of disease. Cf Incubation period PSYCHOLOGY See Psychosexual development VIROLOGY 1. A period in which a virus–eg, EBV, HSV, HIV present in the body is undetectable or asymptomatic; viral LPs are attributed to a lack in host factors critical for expressing early viral gene products; during latency the virus absconds itself in certain cells–eg, EBV in epithelial cells and B lymphocytes; activation of specific cellular–host transcription factors in response to extracellular stimuli may induce the expression of viral regulatory proteins. leading to a burst of lytic viral replication 2. The period that follows the 1° infection–chickenpox in Pts infected with varicella-zoster virus. See Herpes zoster

latent *adjective* Dormant. Cf Active

latent clostridial infection INFECTIOUS DISEASE A reinfection by dormant *C perfringens* spores retained in a site of previous–up to 10 yrs ago–trauma. See *Clostridium* spp

latent content PSYCHIATRY The unconscious significance of thoughts or actions, especially in dreams or fantasies; LC in dreams is expressed in distorted, disguised, condensed, and symbolic form

latent homosexuality Unconsciously repressed homosexuality. Cf Circumstantial homosexuality, Closet, Homosexuality

latent tetany syndrome METABOLISM A state of neuromuscular hyperexcitability, defined by physical exam, electromyography CLINICAL Trousseau and Chvostek signs of hypocalcemia, bowel irritability, laryngismus stridulus, anxiety, asthenia, migraines, mitral valve prolapse. Cf Hypocalcemia

latent tuberculosis INFECTIOUS DISEASE Infection with *M tuberculosis* that has been contained by the host's immune system and thus does not infect others DIAGNOSIS Tuberculin skin test; release of IFN-γ in blood after PPD stimulation. See Tuberculosis

lateral *adjective* Away from the midline, referring to the side; situated on either side of the cardinal sagittal plane

lateral collateral ligament injury ORTHOPEDICS A sports injury due to medial pressure on the knee-joint, or varus stress; LCL tears in children may also cause epiphyseal fractures in the epiphyseal plate of the femur or tibia

lateral decubitus position ORTHOPEDICS One of 2 positions–the other is the beach chair position—for placing Pts undergoing shoulder arthroscopy. See Position. Cf Beach chair position

lateral epicondylitis Tennis elbow, see there

lateral film IMAGING A radiologic image taken from the side of the body

lateral rotation External rotation, see there

latex A lactescent gel of molecular homogeneity, obtained from plants and composed of microglobules of natural rubber; latex may be airborne, and is present in latex gloves, dental rubber dams, condoms, barium enema catheters, other medical devices, and tires/tyres LAB MEDICINE Latex-like particles–eg, neoprene, polyvinylchloride, polystyrene, and synthetic 'rubbers'; latexes are inert vehicles that may be used to carry antibodies or antigens in latex agglutination immunoassays; or rubber latex-like plastic monomer used to manufacture minute plastic beads of polystyrene

latex agglutination test Latex fixation test LAB MEDICINE An assay that uses visible agglutination as an end-point, to detect a reaction between derivatized particles and an analyte

latex allergy ALLERGY MEDICINE An IgE-mediated sensitivity to latex proteins CLINICAL Anaphylaxis, angioedema, asthma, conjunctivitis, contact urticaria, rhinitis, following sensitization to latex allergens; LA is common, affects ±7% of US population, ≥ 15% of health care workers. See Latex

latex glove allergy OCCUPATIONAL MEDICINE A type of latex allergy with a prevalence rate of ±9%, clinically characterized by contact urticaria, affecting HCWs who use latex gloves, dental rubber dams, condoms, barium enema catheters, other medical devices

latissimus dorsi flap SURGERY A tissue flap consisting of skin, tissue, blood vessels, and the latissimus dorsi

Latrodectus A genus of venomous spiders–Family Theridae–eg, black widow spider, *Latrodectus mactans*, endemic to the US, with a characteristic orange-red hourglass-shaped marking on the abdomen; the venom is a non-hemolytic neurotoxin CLINICAL Latency of hours, followed by severe myalgia, myospasm, truncal rigidity, N&V, diaphoresis, shock; most cases resolve spontaneously, small children are at ↑ risk of M&M TREATMENT Establish airway, supportive care

α-latrotoxin A neurotoxin produced by the black widow spider–*Lactrodectus mactans*, which causes synaptic vesicle exocytosis and the release of neurotransmitters from presynaptic nerve terminals, acting with synaptotagmin, a synaptic vesicle-specific membrane protein. See Black widow spider, Synaptotagmin

LATS protector An antibody in 90% of Pts with Graves' disease which, in vitro prevents inactivation of LATS; the LATS-P assay is a sensitive marker for Graves' disease, but is too cumbersome to use as a diagnostic tool. See Long-acting thyroid stimulator

lattice dystrophy OPHTHALMOLOGY A type of corneal dystrophy characterized by local deposition of amyloid in an irregular stroma, betwixt focally eroded epithelium and Bowman's membrane

laudanosine ANESTHESIOLOGY A metabolite of atracurium, an intermediate-acting nondepolarizing muscle relaxing anesthetic; laudanosine is epileptogenic, crosses the blood-brain barrier, and may accumulate in the elderly, and in renal failure. See Muscle relaxant. Nondepolarizing agent. Cf Depolarizing agent

laughter therapy Healing humor PSYCHOTHERAPY The use of humor to cope with major life trauma–stress, personal loss, and disappointment PHYSIOLOGY Laughter evokes endorphin release, ↓ BP, ↑ oxygenation of blood, IgA production

laundry list A popular term for a long list of Sx, diseases, or etiologies that share something in common–eg, differential diagnosis of acute abdomen

LAUP Laser-assisted uvulopalatoplasty A surgical alternative to UPPP–uvulopalatopharyngoplasty for treating obstructive sleep apnea and other sleep disorders, in which throat and palate tissues are removed to open the airway. See Sleep apnea

lavage MEDTALK The washing out of a body cavity–pleural, peritoneal, pericardial cavity or hollow organ to obtain fluids for diagnostic cytology, to detect hemorrhage in blunt trauma, or remove toxins–eg, gastric lavage in overdose. See Bronchoalveolar lavage, Gastric lavage, Tracheobronchial lavage

law A uniform principle or constant. See Boyle's law, Charles' law, Farr's law of epidemics, Fick's law of diffusion, Frank-Starling law, Gown's law, Gresham's law, Hardy-Weinberg law, Harvard law, Heart law, Leborgne's law, Moore's law, Murphy's law of genetics, Natural law, Natural sexual law, Ohm's law, Periodic law, Roemer's law, Sutton's law, Talion law GOVERNMENT A legislative act that compels compliance. See AIDS disclosure law, Annie's law, Anti-dumping law, Antikickback law, Antisubstitution law, Antitrust law, Baby Doe law, Bad baby law, Chinese Law on Maternal & Infant Health Care law, Company doctor law, Due process law, Good Samaritan law, Heart law, Megan's law, Preemptive tobacco control law, Prompt payment law, Roemer's law, Seat belt law, Son of Sam law, Stark law, Sunset law

lawsuit LAW & MEDICINE An action in which one party, a plaintiff, seeks a remedy for a wrong allegedly committed by another–eg, defendant. See Class action lawsuit, Parasitic lawsuit, *Qui tam* lawsuit

laxative Purgative PHARMACOLOGY Any agent used to encourage the onward march of the waste stream. See Bulk-forming laxative, Stimulant laxative

laxative abuse GI TRACT A phenomenon often accompanied by factitious diarrhea found in ± 4% of new Pts seen by gastroenterologists and up to 20% of those evaluated in a tertiary referral center CLINICAL Finger clubbing, skin hyperpigmentation, colonic inflammation, steatorrhea, osteomalacia, protein-losing enteropathy, nephropathy, melanosis coli–sigmoidoscopy, ahaustral right colon–barium enema LAB ↓ K⁺, ↑ uric acid, kidney stones; ↓ stool osmolality–eg, < 250 mOsm/kg, ↑ stool sulfate, phosphate

laxity test ORTHOPEDICS Any of a number of joint tests which measure the looseness–of articulating bones–eg, of the humeral head in the glenohumeral joint. See Load & shift test, Posterior/anterior drawer, Shoulder instability, Sulcus sign. Cf Provocative test

lay midwife Community midwife, independent midwife OBSTETRICS A midwife who may have had little formal training or recognized professional education in midwifery, who learned by accompanying doctors or midwives attending home births; LMs became active in the counterculture movement of the 1970s and are the main attendants at home births. See Midwife

lazy bladder That which occurs in children with a Hx of infrequent–1 to 2 times/day voiding and overflow incontinence; because of the bladder decompensation from chronic overdistension, the bladder does not properly empty and the children are prone to UTIs

lazy eye Suppression amblyopia OPHTHALMOLOGY Subnormal visual acuity in the non-dominant eye despite appropriate correction of refractive errors, due to an early visual defect/amblyopia–eg, strabismus, farsightedness, myopia, astigmatism, or cataract; with time, the stronger eye dominates and retains good vision; neural connections between the brain and nondominant eye fail to develop normally, and the brain eventually ignores visual information from that eye; LE vision thus lacks depth perception. See Strabismus

lazy leukocyte syndrome An idiopathic condition due to defective PMN chemotaxis after appropriate stimuli–eg, endotoxin or epinephrine and inefficient egress of PMNs from the BM CLINICAL ↑ Pyogenic infections–eg, gingivitis, abscess formation, pneumonia, neutropenia PROGNOSIS Uncertain

lazy pituitary ENDOCRINOLOGY Transient functional hypopituitarism seen in children with growth retardation and delayed puberty, often termed constitutional growth delay; persons with LP often fall below the third percentile as children and, while reaching a normal height as adults, do so only at age 20 or more. See Growth hormone

Lazy-S incision COSMETIC SURGERY An S-shaped incision on the posterior vertex, which has been advocated by Stough for scalp reduction in hair replacement. See Hair replacement

LBBB Left bundle-branch block

LBW Low birth weight, see there

LC₅₀ Median lethal concentration, see there

LCAT Lecithin:cholesterol acyl transferase, see there

LCD Liquid crystal display COMPUTERS A flat panel display used for laptops, PDAs/handhelds, etc. See Laptop computer

LCHAD deficiency Long-chain 3-hydroxyacyl-CoA dehydrogenase deficiency METABOLIC DISEASE A condition characterized by defective fatty acid oxidation due to a defect in the 3rd enzyme in the β-oxidation pathway in mitochondria CLINICAL Recurrent episodes of nonketotic hypoglycemia, hypoglycemia, a Reye syndrome-like encephalopathy of acute onset, fatty degeneration of the liver, sudden death in infancy, skeletal myopathy, lactic acidosis, retinal pigmentary changes, hypertrophic cardiomyopathy of varying intensity, and sensory-motor polyneuropathy, chronic severe anemia, recurrent sepsis, delayed CNS myelination; LCHAD deficiency is linked to maternal HELLP and acute fatty liver of pregnancy and attributed to an E474Q mutation in the LCHAD gene in itochondria. See MTP complex

LCIS Lobular carcinoma in situ, see there

LCL injury Lateral collateral ligament injury, see there

LCM Lymphocytic choriomeningitis, see there

LD₅₀ Lethal dose₅₀, median lethal dose The radiation dose that is lethal to 50% of those exposed–2.5-6.0 Gy, assuming that optimal care—ie transfusions, trauma and nursing care, antibiotics and nutrition—is available

LDL Low-density lipoprotein Beta-lipoprotein A 3000 kD plasma lipoprotein–normal serum concentration 60-155 mg/dL–which has a central role in

LDL cholesterol

transporting cholesterol from the intestine to the liver ↑ IN Nephrotic syndrome, obstructive jaundice, uncontrolled DM, type II hypercholesterolemia; ↓ IN Familial LDL deficiency. See Cholesterol, Small LDL. Cf HDL

LDL cholesterol Bad cholesterol Cholesterol carried by LDL which, when elevated, is a major risk factor for ASHD. See HDL/LDL ratio. Cf HDL-cholesterol

LE 1. Lateral epicondyle 2. Left extremity 3. Left eye 4. Leukocyte esterase 5. Lupus erythematosus. See Systemic lupus erythematosus

lead TOXICOLOGY *pronounced* Lead, as in dead A heavy metal that paints a broad clinical palette EPIDEMIOLOGY Inorganic lead sources–gasoline, old paints, burning car batteries, 'moonshine' liquor distilled in tubing soldered with lead, foods, beverages LAB RBCs with coarse basophilic stippling, anemia, reticulocytosis, erythroid hyperplasia, autofluorescence of RBCs and erythroid precursors CLINICAL–CHRONIC POISONING Neuromuscular disease with wrist drop and encephalopathy–convulsions, mania, delirium, paresis, paralysis, abdominal colic, nausea, constipation, weight loss, fatigue, headache, tremor, myalgia, loss of concentration, Fanconi syndrome, aminoaciduria, glycosuria, fructosuria, phosphaturia, protoporphyria-like symptoms, HTN NONTOXIC RANGE Serum < 10 µg/dL; urine < 100 µg/24 hr TOXIC RANGE > 25 µg/dL MANAGEMENT Chelation–eg, dimercaprol, calcium EDTA, D-penicillamine, succimer. See Lead crystal, Lead line, Lead poisoning, Port Pirie Cohort Study, Saturnine gout, Succimer

lead *pronounced* Leed, as in deed CARDIAC PACING Catheter, wire The 'deployment' part of a pacemaker, which has 3 components–the wire(s) which carry the electrical signals and pulses to and from the heart, a connector pin and stimulating/sensing electrode(s). See Active fixation lead, Atrial lead, Coaxial lead, Endocardial lead, Inferior lead, Low-threshold lead, Myocardial lead, Permanent lead, Silicone lead, Sprint™ tachyarrhythmia lead, Steroid eluting lead, Temporary lead, Ventricular lead CARDIOLOGY Any specific site for placing electrodes on an EKG STANDARD 12-LEAD EKG 3 bipolar limb leads–I, II, III, 3 augmented unipolar limb leads–aV$_R$, aV$_L$, aV$_F$, 6 precordial leads–V$_1$-V$_6$ PHARMACEUTICAL INDUSTRY A candidate unknown substance with properties that suggest to industry experts therapeutic properties

lead pipe rigidity NEUROLOGY A 'smooth' rigidity in flexion and extension that continues through the entire range of a stretched muscle, seen in atherosclerotic parkinsonism; LPR is identical to gegenhalten. Cf Garden hose

lead-time bias *Pronounced* Leed A bias introduced into a long-term study of the efficacy of a particular therapeutic maneuver–eg, RT or chemotherapy for malignancy; if the disease is diagnosed early–due to a newer or more sensitive diagnostic procedure or technique, the maneuver is viewed as being effective, when in fact the Pt survives 'longer' because his disease was diagnosed earlier. See Will Rogers effect

leader 1. One who leads. See Thought leader 2. That which is further or more advanced than others. See Loss leader

leading cause of death EPIDEMIOLOGY The most common cause of death in a particular region or population group

leak MEDTALK An oozing of blood or other fluid–eg, CSF–from a place where it was previously contained. See Cerebrospinal fluid leak, Warning leak

leaky gut syndrome Intestinal permeability syndrome A GI tract dysfunction caused by ↑ permeability of the intestinal wall, which allows absorption of toxic material–bacteria, fungi, parasites, etc; LGS may be linked to allergy and various autoimmune conditions

lean body weight THERAPEUTICS A person's body weight minus fat, which can be roughly calculated by measuring height, weight, girth and the person's sex. See Body-mass index, Ideal body mass

learned helplessness GERIATRIC MEDICINE A state of overdependency discordant with the degree of physical and mental disability seen in nursing home Pts PSYCHIATRY A condition in which a person attempts to establish and maintain contact with another by adopting a helpless, powerless stance

learned profession *pronounced* LERN-ed, as in burn-bed Profession, see there

learned treatise INFORMATICS A standard text–eg, *Sabiston's Textbook of Surgery* or other written authoritative source–eg, *Dorland's Medical Dictionary* which may be used as an 'expert' in a court of law

learning disability PSYCHIATRY A suboptimal ability to read–dyslexia, write–dysgraphia, perform mathematical operations–dyscalculia, or other cognitive skills in a child of presumed normal intelligence. See ADD, Dyslexia, Minimal brain dysfunction

learning disorder CHILD PSYCHIATRY A chronic condition that interferes with development, integration and/or demonstration of verbal and/or non-verbal abilities

Leber hereditary optic neuropathy NEUROLOGY A mitochondrial disease characterized by neurologic abnormalities, infantile encephalopathy, and transient or permanent blindness due to optic nerve damage. See Mitochondrial disease

lecithin-cholesterol acyl transferase deficiency LCAT deficiency An AR condition of adult onset characterized by corneal opacification, proteinuria, renal insufficiency, HTN, premature ASHD, hemolytic anemia, obstructive jaundice, liver failure LAB ↑ Ratio of free cholesterol to cholesteryl esters, ↑ phospholipids, TGs, lipoprotein X. See Fish eye disease

lecithin/sphingomyelin ratio L/S ratio OBSTETRICS The ratio of lecithin–phosphatidyl choline to sphingomyelin, a 'bench' parameter used to determine lung maturity and predict an infant's ability to survive without developing respiratory distress. See Biophysical profile, Lung profile, Respiratory distress syndrome

lecture GRADUATE EDUCATION A didactic monologue by an expert scattering pearls of wisdom *ad lecturnum* before disciples

lecturer A person who is primarily–if not entirely—involved in the teaching activities of an academic center, who is not expected to perform research or Pt management; in general, lectureships are non-tenured positions

leech *noun* A segmented annelid of fresh water or soil in the tropics and subtropics; classic medicinal leech is *Hirudo medicinalis*, others include *Poecilobdella, Dinobdella, Limnatis, Haemadipsa, Macrobdella* MEDICAL USES Remove excess blood from operative field; stimulate capillary ingrowth in reimplanted, traumatically amputated extremities and in plastic surgery; extract hirudin, a potent anticoagulant, and undelineated substances in leech saliva that inhibit tumor spread. See Hirudin, *Limnatis nilotica verb* To treat with a leech, to let blood

LEEP Loop extra/electrosurgical/electrical excision procedure GYNECOLOGY Partial excision of a uterine cervix with dysplasia or CIN, using a specially designed wire loop under local anesthesia; LEEP loops deliver high-frequency, low-voltage, alternating electric current, minimizing thermal damage, while preserving hemostasis. See CIN, Dysplasia. Cf Cone biopsy PULMONOLOGY Left end-expiratory pressure

LeFort fracture ORTHOPEDICS A bilateral fracture of the maxilla, which is divided into 3 types, defined by R LeFort in 1901

LEFORT FRACTURES

I Dentoalveolar dysjunction Fracture lines are transverse through the pyriform aperature above the alveolar ridge and pass posteriorly to the pterygoid region; the diagnosis is suggested by lip lacerations, clinical malocclusion, mobility of the fractures bone when the examiner moves the incisor teeth

II Pyramidal fracture Superior fracture lines are transverse through the nasal bone and/or maxillary articulation; diagnosis is suggested by free mobility of anterior maxilla

III Craniofacial dysjunction Central third of face is separated from base of skull; diagnosis is suggested by major facial edema, ecchymosis, and facile mobility of middle third of face by examiner; LeFort III is the most severe midfacial fracture, and may require open reduction and internal fixation Sabiston, Ed, Textbook of Surgery, 14th ed, WB Saunders, 1991

left axis deviation CARDIOLOGY Any shift in the pattern of EKG leads; when seen with a counterclockwise loop abnormality in the frontal plane of a vectorcardiogram, LAD is typical of an ostium primum type atrioventricular canal defect

left bundle branch block CARDIOLOGY A condition in which ventricular contraction is not completely synchronized due to a block in conduction of an electrical impulse to the ventricles; in LBBB, right ventricular endocardial activation begins before, and is often completed before, initiation of left ventricular endocardial activation; *benign LBBB is rare*; preexisting LBBB in absence of clinical evidence of heart disease is rare; newly acquired LBBB has a 10-fold ↑ in mortality CLINICAL Generally asymptomatic, or absent or diminished 1st heart sound–HS and reversed splitting of 2nd HS EKG Lead V$_1$–QS or rS; lead V$_6$–late intrinsicoid deflection, monophasic R; lead I–monophasic R MANAGEMENT Pacemaker, if syncope or significant arrhythmias FOLLOWUP Telemetry PRN. See Bundle branch block. Cf Right bundle branch block

left heart catheterization INTERVENTIONAL CARDIOLOGY The introduction of a catheter into the aorta, left ventricle and left atrium with cannulation of coronary arteries and bypass grafts. Cf Right heart catheterization

left lower quadrant PHYSICAL EXAM The region of the body that contains the left ovary and adnexae and rectosigmoid colon

left shift HEMATOLOGY An ↑ in a peripheral blood smear of immature granulocytes with ↓ nuclear segmentation, ie 'band' forms, due to ↑ production of myeloid cells in the BM, caused by acute infection PHYSIOLOGY An ↑ in Hb's affinity for O$_2$–as represented by the oxygen dissociation curve, where the P$_{50}$ is ↓ and shifted to the left, as occurs with ↑ pH–Bohr effect or ↓ temperature. Cf Right shift

left-sided 'appendicitis' EMERGENCY MEDICINE A popular term for the clinical findings in acute diverticulitis with impending rupture, which is a mirror image mimic of left-sided appendicitis, accompanied by pain, rebound tenderness, leukocytosis

left upper quadrant PHYSICAL EXAM The region of the body containing the stomach, spleen and tail of pancreas

left ventricular assist device CARDIOLOGY A mechanical device to ↑ force and volume of blood flowing through the heart. Cf CABG, Jarvik-7

left ventricular failure Left heart failure CARDIOLOGY CHF due to insufficient output or ↑ filling pressure, resulting in pulmonary vascular congestion CLINICAL DOE, SOB, exercise intolerance, orthopnea, paroxysmal nocturnal dyspnea, rest dyspnea, chronic cough, nocturnal urination, pulmonary congestion–rales, pleural effusion, wheezing, S3 gallop RISK FACTORS Smoking, obesity, alcohol, high fat, high salt intake PRECIPITATING FACTORS HTN, heart valve disease, congenital heart diseases, cardiomyopathy, myxoma or other heart tumor, ↑ physical activity, hypervolemia, ↑ salt intake, fever or complicated infections, anemia, arrhythmias; hyperthyroidism, renal disease, drugs affecting heart contractility–eg, beta blockers, CCBs, ACE inhibitors, digoxin, diuretics. Cf Right ventricular failure

left ventricular hypertrophy CARDIOLOGY Enlargement of the left ventricle often linked to the prolonged hemodynamic stress of CHF, characterized by myocardial cell hypertrophy, ↑ left ventricular wall thickness, ↓ ventricular compliance, ↑ ventricular stiffness and, in elderly, ↑ vascular resistance. See Congestive heart failure, Remodeling

leg See Bird leg, Champagne bottle leg, Tennis leg

leg length discrepancy Limb length discrepancy ORTHOPEDICS A difference in leg lengths, clinically significant at > 3 cm, affecting heart rate, muscle activity and O$_2$ consumption COMPENSATION STRATEGIES Steppage, circumduction, vaulting, hip hiking. See Gait analysis

legal VOX POPULI Licit, lawful, allowed by law

legal crisis MEDTALK Any event with a potentially adverse impact on a particular physician, health care organization or service–eg, a civil suit, indictment, or administrative action, and result in subpoenas, search warrants, and/or employee interviews

legally authorized representative Surrogate decision maker PATIENT RIGHTS A person authorized by statute or court appointment to make decisions for another

Legg-Perthes disease Avascular necrosis of femoral head, femur, hip, osteochondrosis of femoral head, coxa plana, hip click syndrome PEDIATRICS A condition characterized by necrosis of the epiphyseal center of the femoral head, caused by disrupted vascular supply, more often seen in boys–♂:♀ ratio 4:1, ages 4–8, rare in African Americans

Legionnaire's disease *Legionella* pneumonia INFECTIOUS DISEASE A lung infection caused by *Legionella pneumophila*, which is either sporadic or epidemic with a mortality of up to 15%; depending on the population, *Legionella* spp cause 1–27% of CAPs; ♂:♀ ratio, 3:1 EPIDEMIOLOGY Most *L pneumophila* infections are associated with buildings and ventilation systems; most epidemics occur during the summer with 0.5 to 5% attack rates; the only documented mode of spead is by air RISK FACTORS Cigarettes, alcohol, renal transplant, ↑ age CLINICAL 2-10 day incubation, followed by an abrupt malaise, headache, myalgia, a dry, initially non-productive, later productive, cough, hemoptysis; fever to 40ºC, rigors often associated with bradycardia; less commonly, nausea, diarrhea, delirium, septicemia, abscesses, acute myocarditis, pericarditis, rhabdomyolysis DIFFDX Sporadic LD mimics myoplasma pneumonia, Q fever, tularemia, plague, psittacosis, influenza and other viral pneumonias CXR Patchy, segmental, or lobar, often bilateral, alveolar infiltrates, often progressing to nodule LAB ↓ Na$^+$, ↓ PO$_4^-$, ↑ liver enzymes, proteinuria, microscopic hematuria, relative leukocytosis, ↑ ESR, hyponatremia, ↑ ALT, ↑ AST, ↑ BR, azotemia COMPLICATIONS Empyema, shock, DIC, renal failure, neurologic sequelae, peripheral neuropathy MANAGEMENT Erythromycin, T-S, penicillin PROGNOSIS 15-20% death w/o therapy; up to 50% death in immunocompromised Pts POOR PROGNOSTIC FEATURES Tachycardia, tachypnea, WBCs ≥ 14,000/mm³, ↑ BUN, ↑ creatinine, hyponatremia, hypoxia, leukopenia, bilateral infiltrates on CXR PREVENTION Chlorination; UV irradiation of water supplies. See Community-acquired pneumonia, *Legionella pneumophila*. Cf Pontiac disease

Leigh's disease Subacute necrotizing encephalopathy A mitochondrial disease of infants CLINICAL Progressive loss of motor and verbal skills PROGNOSIS Usually fatal by age 2. See Mitochondrial disease

leiomyoma Fibroma, plural, leiomyomata or, incorrect, but increasingly popular, leiomyomas A benign, well-circumscribed smooth muscle tumor most common in the uterus and stomach. See Fibroid, Fibroma, Intestinal leiomyoma

leiomyosarcoma SURGERY A malignant smooth muscle tumor which may occur anywhere in the body but is most common in the uterus and GI tract MANAGEMENT Complete surgical excision, if possible

Leipzig disparity ALLERGY MEDICINE The observation that children in the highly polluted ex-East German city of Leipzig had one-half the incidence of allergies as children living in the cleaner West German cities. See Hygiene hypothesis

Lembrol Diazepam, see there

lemon sign OBSTETRICS An ultrasound finding seen when a major neural tube defect accompanies Arnold-Chiari malformation, with herniation of cerebellar tonsils and midbrain structures into the foramen magnum, causing ventriculomegaly due to compression of outflow from the 3rd and 4th ventricles; downward traction of the brain causes a reduction in the anterior calvarium, resulting in a triangular-shaped head in the biparietal diameter, fancifully likened to a lemon

Lenal Temazepam, see there

length The linear distance between 2 points. See Collateral length, Pulse length, Stride length

length of stay MANAGED CARE The total number of days–measured in multiples of a 24-hr day that a Pt occupies a hospital bed. See Admission, 23-hr admission

Lennert's lymphoma Lymphoepithelioid lymphoma, malignant lymphoma with high content of epithelioid cells A diffuse mixed cell NHL with abundant benign epithelioid histiocytes; affects older Pts; LLs are nodal, tend to involve skin and bone, and have a relatively good prognosis CLINICAL Generalized lymphadenopathy, hepatosplenomegaly, 'B' symptoms–fever, night sweats, weight loss, often in stage III or IV. See Lymphoma, Working classification. Cf Progressive transformation of germinal centers

Lennox-Gastaut syndrome Lennox syndrome NEUROLOGY A severe encephalopathic form of epilepsy that constitutes 5% of all childhood epilepsies CLINICAL Early onset of multiple types–eg, absence, atonic, and others of seizures, slow spike-wave EEG pattern, cerebral atrophy, and often progressive mental retardation; seizures are poorly controlled even with multiple anticonvulsants PROGNOSIS For cognitive development, poor; felbamate may ↓ Sx. See Felbamate

lens opacity Cataract, see there

lentigo maligna Hutchinson's melanotic freckle, lentigo maligna melanoma A lesion comprising 10% of melanomas, affects those > age 60; flat, indolent, arises in premalignant freckle, located on sun-exposed aging skin; begins as an unevenly pigmented macule with an irregular border, slowly extends peripherally ETIOLOGY Solar overexposure, $\frac{1}{3}$ progress to melanoma in 10-15 yrs PROGNOSIS > 95% 5-yr survival. See Lentigo, Melanoma

Lentivirinae A subfamily of retroviruses which includes HIV, bovine immunodeficiency virus, feline lymphotropic virus–FTLV, simian lymphotropic virus–STLV, etc. See HIV, Retroviruses, STLV. Cf Prions

leonine facies Leontiasis A deeply furrowed 'lumpy' face with prominent superciliary arches, classically seen in lepromatous leprosy DIFFDX Hyperimmunoglobulin E syndrome, chronic granulomatous disease, van Buchem's disease, leontiasis ossium–idiopathic leonine facies; features due to overgrowth of bones, as in Paget's disease or McCune-Albright syndrome, polyostotic fibrous dysplasia or soft tissue, as in hypothyroidism with myxedema of periorbital tissues, epidermoid carcinoma, Sézary syndrome, which is characterized by generalized exfoliative dermatitis, edema, erythema, pachydermia, palmoplantar keratoderma

LEOPARD syndrome Multiple lentigines syndrome An AD condition with thousands of 1-5 mm darkly pigmented macules on the skin, not mucosae CLINICAL characterized by the acronym LEOPARD–Lentigines, EKG disturbances, Ocular hypertelorism, Pulmonary stenosis, Abnormalities of genitalia–gonadal or ovarian hypoplasia, Retarded growth, neural Deafness

lepirudin Refludan® HEMATOLOGY A recombinant hirudin anticoagulant used to manage heparin-induced thrombocytopenia type II, thromboembolic disease and prevent complications in Pts with HIT MECHANISM Inhibits thrombin activity. See Heparin-induced thrombocytopenia

leprechaunism Donohue syndrome, Donohue-Uchida syndrome, dysendocrinism An AR polydysmorphic complex with parental consanguinity, which is more common in ♀–↑ ♂ fetal wastage in utero and characterized by a coarse

gnome-like face with a saddle nose, broad mouth, large, low-set ears, hirsutism, cutis laxa, atrophy of subcutaneous fat, dwarfism, extreme wasting, mental retardation, dysphagia, enlarged nipples, breasts, clitoris, penis, kidneys, pancreatic islets and ovaries–with premature follicular maturation, hepatic nodules, insulin receptor dysfunction, early death. Cf Williams syndrome

leptin ENDOCRINOLOGY An *obese* gene-encoded adipocyte-derived hormone with hypothalamic and CNS receptors, which induces a sensaton of fullness. See Obesity. Cf *agouti yellow*, *diabetes*, Ghrelin, *obese*, Orexin, *tubby*

leptocyte *leptos*, Greek, thin A wafer-thin RBC with marginated Hb seen in thalassemia and obstructive liver disease, in iron-deficiency anemia, and in chronic inflammation

leptomeningeal metastases ONCOLOGY CA that has spread to the pia and arachnoid membranes of the brain and spinal cord

leptomeningitis NEUROLOGY Inflammation of the pia and arachnoid membranes of the brain and spinal cord

leptospirosis Icterohemorrhagic fever, Weil's disease TROPICAL DISEASE Infection with *Leptospira* spp–eg, *Leptospira canicola* serotype *canicola*, which may follow ingesting water contaminated by infected livestock, rodent urine CLINICAL Bleeding, jaundice, pulmonary hemorrhage, pulmonary edema

LES 1. Lambert-Eaton–myasthenic syndrome 2. Lower esophageal sphincter, see there

lesbian *adjective* Referring to female homosexuality *noun* A female homosexual

Lesch-Nyhan syndrome An X-R condition caused by a deficiency of 24 kD hypoxanthine-guanine phosphoribosyl transferase, with accumulation of uric acid crystals in renal pelvis and bladder, pseudo-gouty arthritis, erosive changes of fingertips, compulsive self-mutilation, choreoathetosis, mental retardation LAB ↑ Oxypurines, hypoxanthine, xanthine in CSF due to purine overload

Lescol® Fluvastatin, see there

lesion MEDTALK 1. A wounded or damaged area; an anatomic or functional tissue defect; an area of abnormal tissue change 2. A nebulous nonspecific term used by a physician when discussing a lump or bump with a Pt. See Mass

lesionectomy NEUROSURGERY A popular term for a stereotactic resection of poorly-circumscribed intra-axial–a region often considered inoperable–brain masses–eg, vascular malformations and glial tumors, identified by MRI that may be associated with epileptiform seizure activity. Cf Lumpectomy

Lesquesne Index ORTHOPEDICS A measure of pain and dysfunction of the knee in Pts with osteoarthritis. See Osteoarthritis

let-down reflex OBSTETRICS A physiologic postpartum response evoked by sucking–or negative mechanical pressure on the ♀ nipple or by psychologic stimuli, causing the release–'let-down' of breast milk in a nursing mother. See Breast milk

lethal midline granuloma A condition confusing to those who read about it and often to those who write about it; LMG is best considered a clinical syndrome rather than a specific histologic lesion; LMG is a destructive lesion of the upper respiratory tract–nose, nasopharynx, palate and midface; it is either idiopathic or 2° to lymphoma, malignant histiocytosis or Wegener's granulomatosis. See Idiopathic midline destructive disease, Malignant histiocytosis, Midline granuloma, Wegener's granulomatosis

lethargy NEUROLOGY A level of consciousness characterized by ↓ interaction with persons or objects in the environment; sluggishness, abnormal drowsiness, stupor

lethologia PSYCHIATRY Temporary inability to remember a proper noun or name

letrozone Femara™ ONCOLOGY An agent for treating advanced breast CA in postmenopausal ♀. See Breast cancer

LETS HEMATOLOGY A case-control study–Leiden Thrombophilia Study which evaluated relative and absolute risks of a first episode of DVT in the general population relative to known risks. See Factor V Leiden, Hereditary thrombophilia

letter See Dean's letter, Dear Doctor letter, Happy letter

letter of pre-determination SURGERY A letter written to an insurance company explaining the procedure the surgeon intends to perform

letter of recommendation GRADUATE EDUCATION A letter sent on the behalf of a candidate for a residency, job or other position which is filled with glorious blah blah blah about the candidate

Letterer-Siwe disease A form of Langerhans' cell histiocytosis or histiocytosis X, with an onset in infancy of fever, eczematoid rash, lymphadenopathy, hepatosplenomegaly, anemia, death. See Histiocytosis, Langerhans' cell histiocytosis

leukocyte adherence test IMMUNOLOGY A test for cell-mediated antitumor immunity and related serum blocking factors PRINCIPLE WBCs from a CA, when mixed in vitro with antigenic extracts of tumors of the same histologic type, become less adherent than normal to glass surfaces; serum from a Pt with an identical tumor blocks the reaction of her own WBCs. See Cell-mediated immunity

leucovorin rescue ONCOLOGY A strategy–administration of a formulation of vitamin B for preventing 'collateral damage' to normal cells when using MTX, an antimetabolite, to treat lymphomas and other malignancies, which may cause BM suppression and GI Sx. See Methotrexate

leukapheresis THERAPEUTICS The removal of blood to collect specific blood cells; plasma is returned to the body. See Apheresis TRANSFUSION MEDICINE The separation of leukocytes from whole blood with/without platelets with continuous or intermittent return of the RBCs and platelet- and leukocyte-poor plasma to a donor. See Cytapheresis, Hemapheresis, Plateletpheresis

LEUKAPHERESIS TYPES

'HARVESTING' LEUKAPHERESIS WBCs are donated from healthy subjects to immunocompromised and leukopenic Pts, which may be effective in short-term therapy of acute infections; long-term, Pts become either immunized against the donor antigens or infected with pathogens; 10¹⁰ granulocytes are needed for adequate 'coverage' against infections; to maximize the harvest, the donor receives corticosteroids

REDUCTION LEUKAPHERESIS WBCs that compromise normal circulation are removed to temporarily relieve Sx of hyperleukemia–> 100 x 109/L; in a typical procedure, ± 6 liters of whole blood are processed to remove 5-10 x 109 WBCs

leukemia HEMATOLOGY An uncommon–incidence, US 3.5/10⁵/yr– malignant clonal expansion of myeloid or lymphoid cells characterized by an ↑ in circulating WBCs; leukemias may be an incidental finding when evaluating an unrelated clinical problem, or when the expansion compromises BM production of one or more cell lines causing anemia, thrombocytopenia, granulocytopenia; leukemias are divided by chronology–acute or chronic, by cell lineage–lymphoid, myeloid/myelocytic, monocytic or megakaryocytic and divided by stage of maturation or cell size CLINICAL BM infiltration by leukemia, resulting in anemia, thrombocytopenia, granulocytopenia, immune paralysis, ↓ B cells and CD4–helper T cells, ↑ CD8–suppressor T cells, infiltration and leukostasis, cranial nerve palsies, meningitis, lymphadenopathy, hepatosplenomegaly, testicular and cutaneous involvement, metabolic derangements–eg, ↑ Ca²⁺, K+, LD, ammonia, weight loss,

less commonly, autoimmune hemolytic anemia, pallor and arthralgia DIAGNOSIS Hx, physical exam, peripheral smear MANAGEMENT Chemotherapy, RT, BMT. See Accelerated leukemia, Acute leukemia, Acute lymphocytic leukemia, Acute myelocytic leukemia, Acute promyelocytic leukemia, Adult T-cell leukemia-lymphoma, Aleukemic leukemia, Biphenotypic leukemia, Central nervous system leukemia, Chemotherapy-induced leukemia, Chronic leukemia, Chronic lymphocytic leukemia, Chronic myelocytic leukemia, Chronic myelomonocytic leukemia, Congenital leukemia, Erythroleukemia, FAB classification, Hairy cell leukemia, Hand mirror cell leukemia, Herald state of leukemia, Mast cell leukemia, Megakaryoblastic leukemia, Multilineage leukemia, Plasma cell leukemia, Preleukemia, Prolymphocytic leukemia, Promyelocytic leukemia, Smoldering myeloid leukemia

LEUKEMIA ACUTE V. CHRONIC

ACUTE LEUKEMIA More common in children, 80% are ALL, often before age 10, peak at ages 3–7 in whites, ♂:♀ ratio, 1.3:1 CELL TYPES Early pre-B cell 67%; pre-B cell 18%; B cell 1%; T cell 14%; 50-85% are cALLA positive–common acute lymphocytic leukemia antigen, CD10; 5% have Philadelphia chromosome CLINICAL ALL is more abrupt than AML, with petechial hemorrhage, bone and abdominal pain, headache and vomiting due to ↑ intracranial pressure, lymphadenopathy, splenomegaly, hepatomegaly LAB 70% have low-grade lymphocytosis–< 20 x 10⁹ when diagnosed EVALUATION Acute leukemia immunophenotypic profile SPECIMEN EDTA–lavender top tube *and* sodium heparin–green top tube, peripheral blood smears METHOD OF ANALYSIS Flow cytometry, immunofluorecesence MARKERS MEASURED CD1, -2, -3, -4, -5, -7, -8, -10, -19. -20, -21, -33, -34, -56, megakaryocytic markers, HLA D/DR, kappa, lambda, TdT MANAGEMENT Protocols vary according to standard- or high-risk clinical features, and may include BMT

CHRONIC LEUKEMIA More common in adults/older children, often myelocytic; CML is Philadelphia chromosome positive; may occur < age 5 with myelomonocytosis, anemia, thrombocytopenia, lymphadenopathy; WBC count < 50 x 109, ≠ HbF, ≠ muraminidase; adult CML comprises 20% of all leukemias Clinical Gradual onset of fatigability, anorexia, splenomegaly; lymphadenopathy is uncommon Lab > 25 x 109/L leukemic cells in blood–often an absolute lymphocytosis of > 15 x 1010/L, < 10% blasts in BM, myeloid:erythroid ratio is 10-30:1, 90% of cases have low-to-absent leukocyte alkaline phosphatase and rarely also, ≠ vitamin B12 and B12-binding capacity Evaluation Chronic leukemia immunophenotype profile Specimen EDTA–lavender top tube and sodium heparin–green top tube, peripheral blood smears Method of analysis Flow cytometry, immunofluorescence Markers measured CD3, -5, -19, -20, -21, kappa, lambda Management see Chemotherapy, Induction Prognosis see Remission

LEUKEMIA–PROGNOSTIC FEATURES

ACUTE LYMPHOCYTIC LEUKEMIA

Good Age 2-10, CD10 positivity, hyperdiploid karyotype

Poor Age < 2; >10, B-cell phenotype, especially L2 phenotype by FAB classification, presence of chromosome translocations, CNS involvement, mediastinal masses, high initial WBC count

ACUTE MYELOCYTIC LEUKEMIA

Good Younger, presence of Auer rods, rapid therapeutic response

POOR Older, prior malignancy or therapy, multiple chromosome defects

leukemic coagulopathy A complication of leukemia, characterized by hemorrhagic diathesis due to 1. Abnormalities related to the leukemia per se–eg, BM infiltration, dysmorphic megakaryocytes, ↓ platelet lifespan, qualitative platelet defects–↓ platelet aggregation with ATP and collagen stimulation or defects in ADP release and DIC due to sepsis or transfusion reactions and coagulopathy due to leukemic therapy 2. BM toxicity due to combination chemotherapy that may ↓ fibrinogen, factors IX, XI, plasminogen, antithrombin III, ↑ factor V, uric acid nephropathy–due to massive cytolysis of malignant cells, vitamin K deficiency–treatment of infections due to granulocytopenia, ↓ vitamin K-producing intestinal bacteria and heparin anticoagulation

leukemogenesis A process in which successive transformational events enhance the ability of hematopoietic progenitor cells to proliferate, differentiate, and survive. See Autonomous proliferation

leukemogenic *adjective* Causing leukemia, an acute or chronic disease characterized by an abnormal number of leukocytes or the presence of abnormal leukocytes

leukemoid reaction

leukemoid reaction Pseudoleukemia HEMATOLOGY 1. Left shift, see there 2. Benign abnormal polyclonal proliferation of leukocytes, defined as > 25 x 10⁹/L; LRs reflect a normal BM response to trauma, stress, metabolic disease, drugs, inflammation, connective tissue disease, or malignancy, 2° to secretion of CSF, often associated with immaturity of other cell lines LAB ↑ Leukocyte alkaline phosphatase which is ↓ or absent in leukemia, 'left shift' of myeloid series–↑ bands, metamyelocytes, myelocytes, plasma cells, plasmacytoid lymphocytes, toxic granulation, Döhle inclusion bodies, vacuolization–which implies intracellular bacterial phagocytosis

Leukeran® Chlorambucil, see there

Leukine® Sargramostim ONCOLOGY A formulation of GM-CSF used for cell recovery in Pts with ↓ cell counts due to chemotherapy with Immunex. See GM-CSF

leukocyte adhesion deficiency syndrome LFA-1 immunodeficiency syndrome A co-dominant or AR, often consanguineous, immune deficiency due to a defect in lymphocyte function-associated antigen–LFA-1, which facilitates cytolytic T-cell mediated killing and helper T-cell response CLINICAL Inflammation, delayed separation of umbilical cord, recurrent pyogenic infections, pneumonia and poor wound healing due to poor cell adherence, chemotaxis, ↓ respiratory burst. See CD11, CD18, Cell adhesion molecule, Leu-CAM, LFA-1

leukocyte count see White cell count

leukocyte-poor blood Leukocyte-poor packed red blood cells TRANSFUSION MEDICINE A unit of packed RBCs intended for transfusion in which ≥70% of the WBCs are removed–by inverted centrifugation, manual or automated saline washing of whole or packed RBCs, or with microaggregate filters–with ≤ 20% loss of RBCs. See Leukocyte reduction, Packed RBCs

leukocyte reduction Leukocyte depletion, leukodepletion TRANSFUSION MEDICINE Any technique that ↓ WBCs in transfused blood products–eg, RBCs, platelets; LR to < 5 x 10⁸ virtually eliminates nonhemolytic (immunologic) transfusion reactions. See Blood filters, Transfusion effect

leukocytoclastic vasculitis Cutaneous necrotizing vasculitis A form of vasculitis with fragmentation of neutrophil nuclei, immune complex deposition–direct immunofluorescence demonstrates IgG, IgM and complement deposition, that elicits neutrophilic 'suicide' and deposition of abundant nuclear 'dust', necrotic debris and fibrin, most common in small post-capillary venules; the condition may be local–eg, cutaneous or systemic. See Vasculitis

leukocytosis ↑ in WBCs–WBC count > 11 x 10⁹/L–US: 11,000/mm³, benign or malignant. See Reactive leukocytosis, WBC. Cf Leukemia, Leukopenia

LEUKOCYTOSIS

PHYSIOLOGIC Follows nonspecific immune stimulation, eg intense exercise; it may be idiopathic or hereditary, neonatal, induced by heat or solar irradiation, diurnal, ↑ in afternoon, related to stress, eg pain, nausea, vomiting, anxiety, womanhood–↑ during ovulation and near term, ↑ during labor, ether anesthesia, ↑ adrenalin, convulsions, paroxysmal tachycardia, pain, nausea, vomiting, anoxia, exercise, convulsions

PATHOLOGIC May be due to infections, often bacterial, inflammation, severe burns, postoperative, MI, strangulated hernias, intestinal obstruction, gouty attacks, acute glomerulonephritis, serum sickness, rheumatic fever, immune disorders and connective tissue diseases, metabolism–ketoacidosis, uremia, eclampsia, heavy metals–lead, mercury, petrochemicals–benzene, turpentine, drugs–phenacetin, digitalis, black widow spider venom, endotoxin or toxoid injection, Jarisch-Herxheimer reaction, hemorrhage–often into cranial cavity, serosal surfaces–pleural pericardium and peritoneum or acute hemolysis, malignancy–GI tract or hematopoietic, and Cushing syndrome

leukodystrophy NEUROLOGY A heterogeneous group of disorders of cerebral myelin or its metabolism–eg, Krabbe's disease, metachromatic leukodystrophy, 'sphingolipidoses', that share certain pathological features CLINICAL

'White matter disease', ie predominantly motor, dominated by progressive paralysis and ataxia rather than dementia. See Globoid cell leukodystrophy

leukoedema Edema of the oral mucosa of undetermined significance that may clinically mimic early leukoplakia, an often premalignant condition, characterized by acanthosis, intracellular edema and superficial parakeratosis

leukoerythroblastic reaction An ↑ in the peripheral blood of immature RBCs, ie normoblasts, and immature granulocytes, metamyelocytes and bands, which may be associated with metastatic cancer, hematopoietic malignancy, hemolytic anemia, Gaucher's disease, polytrauma, BM infiltration by various processes, including infection–eg, fungal, viral, TB, sarcoidosis, histiocytosis, hypoxia; ⅓ of Pts with LER have no known underlying disease. Cf Leukemoid reaction

leukopenia HEMATOLOGY A ↓ total WBC count–< 0.4 x 10¹²/L–US: < 0.4 x 10⁶/μL that results from BM depression due viral–eg, hepatitis, infectious mononucleosis, rubella or bacterial–eg, *Salmonella* spp infection or exposure to myelotoxic agents–eg, benzene, chemotherapy. See Benign glandular leukopenia, Differential white cell count

leukoplakia A potentially precancerous white patch or plaque on a mucosa characterized by epithelial hyperplasia and hyperkeratosis, often caused by chronic irritation; leukoplakia–LP affects the mucosa of oral cavity, upper respiratory tract, vulva, uterine cervix, renal pelvis, urinary bladder; in each site, the significance differs ENT Smoker's keratosis A white plaque or patch on the oral mucosa. See Hairy leukoplakia OB/GYN A white plaque or patch on the vaginal mucosa, seen without magnification or acetic acid, and often elevated from surrounding surfaces with a sharp border and Lugol's non-staining HISTOLOGY Hyperkeratosis, possibly epithelial proliferation. See Speckled leukoplakia

leukorrhea GYNECOLOGY A nonspecific whitish malodorous vaginal discharge accompanied by dyspareunia and intense pruritus, which may be caused by infection–eg, *Candida albicans*, *Gardnerella vaginalis*, *T vaginalis*, *N gonorrhoeae*, foreign bodies, estrogen depletion, neoplasms, and as a postpartum phenomenon

leukotriene receptor antagonist PHARMACOLOGY Any of a family of agents used to treat asthma by interfering with the binding of leukotriene D₄

leuprolide Viadur® An analogue of gonadotropin-releasing hormone, which has been used to manage PMS, and block ovarian or testicular hormone production in Pts with hormonally active CA. See PMS

leuvectin ONCOLOGY An agent for treating metastatic melanoma, renal cell carcinoma, and sarcoma

levamisole ONCOLOGY A broad-spectrum anthelmintic that stimulates immune function, enhancing hypersensitivity; it has been used with 5-FU to ↑ Pt survival in advanced colorectal and other CAs ADVERSE EFFECTS N&V, diarrhea, rash, flu-like Sx. See Unproven methods for cancer management PARASITOLOGY An L- form of tetramisole, which is most commonly used anthelmintic for roundworm, hookworm, *Strongyloides* infections

levan DENTISTRY A fructose homopolymer linked by β-2,6 bonds, formed by the partial digestion of sucrose by *Bacillus* and *Streptococcus* spp, which is a component of dental plaque representing the first biochemical event in cariogenesis. See Caries, Plaque. Cf Periodontal disease

Levanxol Temazepam, see there

Levaquin® Levofloxacin, see there

Levate Amitriptyline, see there

LeVeen shunt LONG-TERM CARE A hydraulic device with a specialized valve that opens with a low pressure gradient–3 cm H₂O, used to drain ascitic fluid from the peritoneal cavity Pros Improves systemic hemodynamics, liver function, nutrition, survival, ↓ renin and aldosterone Cons Coagulopathy, infections, CHF, pulmonary edema, variceal bleeding, vena cava obstruction with superior vena cava syndrome Indications Management of ascites with/without cirrhosis, hepatorenal syndrome Contraindications Rapidly progressive liver failure, encephalopathy, organic–ie, not mechanical renal failure, heart failure, present/past variceal bleeding, coagulopathy, severe infections, especially peritonitis

level A relative position; reference point; a specified amount of a thing

level of care The intensity of medical care being provided by the physician or health care facility

LEVEL OF CARE

Primary Coordinated, comprehensive and personal care, available on both a first-contact and continuous basis; it incorporates the tasks of medical diagnosis and treatment, psychological assessment and management, personal support, communication of information about illness, prevention and health maintenance; primary care is that provided by the family physician, general practitioner and by physicians in the emergency room

Secondary That medical care available in the community hospital, comprising the bulk of in-patient medical care provided in the US; secondary care centers are equipped to provide all but the most specialized of care, surgery and diagnostic modalities

Tertiary Highly specialized medical care for Pts who are usually referred from 2° care centers, which consists in subspecialty expertise in 1. Surgery—Organ transplantation, pediatric cardiovascular surgery, stereotactic neurosurgery, and others 2. Internal medicine—Genetics, hepatology, adolescent psychiatry and others 3. Diagnostic modalities—PET–positron emission tomography and SQUID–superconducting quantum interface device scanning, color Doppler electrocardiography, electron microscopy, gene rearrangement and molecular analysis and 4. Therapeutic modalities—Experimental protocols for treating advanced and/or potentially fatal disease–eg AIDS, cancer and inborn errors of metabolism

level 1 test LAB MEDICINE Any simple test that is loosely regulated under the provisions of CLIA '88. See CLIA '88. Cf Waived tests

level 1 trauma center EMERGENCY CARE A hospital equiped to handle any level of severity of trauma, and has a trauma surgeon on-site 24/7 and an OR ready at all times for trauma cases. See Trauma center

level 2 study CLINICAL RESEARCH A retrospective study which employs preexisting data–eg, hospital charts to test a hypothesis or project possible cause-and-effect relationships. Cf Prospective study

levetiracetam NEUROLOGY An antiepileptic adjunct for adults with partial-onset seizures with/without secondary generalization. See Seizures

LevLite® GYNECOLOGY A low-dose OC containing levonorgestrel and ethinyl estradiol. See Oral contraceptive

levo- Latin, laevus, left side. Cf Dextro

levobupivacaine Chirocaine™ ANESTHESIOLOGY A long-acting local anesthetic used for pain during childbirth, surgery. See Chiral chemistry

levocardia Reversal of all abdominal and thoracic organs–situs inversus except the heart, which is still in its usual location on the left; levocardia is always associated with congenital heart disease, including transposition of great vessels

levodopa NEUROPHARMACOLOGY The L- isomer of dopa, which is converted in vivo into dopamine, and used as a first-line anti-Parkinsonian agent THERAPEUTIC RANGE 0.2–0.3 µg/mL

levofloxacin Levaquin® INFECTIOUS DISEASE An advanced-generation, broad-spectrum fluoroquinolone with improved activity against streptococci and anaerobes INDICATIONS Community-acquired URIs, acute maxillary sinusitis, bacterial exacerbation of chronic bronchitis, skin, UTIs ADVERSE EVENTS Restlessness, N&V, diarrhea, dizziness, insomnia. See Fluoroquinolone

levonorgestrel Norplant®, see there

levosimendan Simdax® CARDIOLOGY A calcium-sensitizing agent which ↑ heart pump efficiency w/o ↑ O₂ consumption or causing arrhythmias

Levoxyl® Levothyroxine ENDOCRINOLOGY Thyroid hormone for treating hypothyroidism. See Hypothyroidism

Lexapro® Escitalopram oxalate, see there

Lexotan® Bromazepam, see there

Lexxel® VASCULAR DISEASE An extended-release combination of 2 ACE inhibitors—enalapril and felodipine for managing treatment-resistant HTN. See Enalapril, Felodipine, Hypertension

Leydig cell hyperplasia ENDOCRINOLOGY Autonomous hyperplasia and hyperfunction of Leydig cells ETIOLOGY Activating mutation of LH receptor gene CLINICAL Precocious puberty MANAGEMENT Orchidectomy

LFT Liver function tests, see there

LGA Large for gestational age, see there

LGL syndrome Lown-Ganong-Levine syndrome, see there

LGSIL Low-grade squamous intraepithelial lesion, see there. Cf High-grade squamous intraepithelial lesion

LGV Lymphogranuloma venereum, see there

LH Luteinizing hormone, see there

Lhermitte sign NEUROLOGY Flexion of the neck evokes abrupt spasms over the body, a finding typical of multiple sclerosis as well as cervical spine disease

LHON Leber's hereditary optic neuropathy, see there

LHRH Luteinizing hormone-releasing hormone, GnRH, gonadotropin-releasing hormone, LRH, LRF ENDOCRINOLOGY A decapeptide synthesized by hypothalamic neurons which controls production of ♂ and ♀ sex hormones, and stimulates FSH and LH release in response to CNS stimulation. See Hypothalamus

Li-Fraumeni syndrome SBLA syndrome An AD condition with ↑ risk of multiple malignancies—eg, sarcomas, carcinomas of adrenal cortex, breast, larynx and lung, brain tumors, leukemia and lymphomas at any time from infancy to adulthood, due to a defect in the p53 tumor suppressor gene

liability MALPRACTICE All character of obligation, amenability, and responsibility for an act before the law. See Corporate liability, Current liability, Limits of liability, Malpractice Product liability, Professional liability, Strict liability

liarozole fumarate ONCOLOGY A drug that stimulates retinoic acid production within tumors, which may have some effect in metastatic prostate CA, KS. See Retinoic acid

Libby Zion GRADUATE EDUCATION A young ♀ who died after admission to the ER of a NYC hospital in 1984; her death was attributed to inadequate care provided by overworked and undersupervised medical house officers. See 405 Regulations

liberation THERAPEUTICS The release and dissolution of a drug from its dosage form

libido plural, libidos, sex drive PSYCHIATRY Sexual drive, urge or desire; the psychic and emotional energy associated with instinctive biologic drives, generally equated to sexual drive; normal libido requires adequate testosterone and dopamine–which inhibits libido-attenuating prolactin. See Anorgasmia

Libnum

Libnum Chlordiazepoxide, see there

Librium® Chlordiazepoxide, see there

lice, fear of PSYCHOLOGY Pediculophobia. See Phobia

license Authorization by a governmental or other regulatory agency that allows a person, group of persons, or enterprise to carry out a particular activity; the certificate itself. See Revoked license, Unrestricted license

licensee MEDICAL PRACTICE Any person licensed or authorized to engage in the health care profession regulated by a state medical board

licensing authority MEDICAL PRACTICE Any professional or occupational licensing board charged with granting, suspending, or revoking licensure or certification privileges

licensure by endorsement ONLINE MEDICINE The acceptance by a state board of medicine of a license to practice medicine granted by another state with similar standards. See Telemedicine

licensure The public or governmental regulation of health or other professions for voluntary private-sector programs that attest to the competency of an individual health care practitioner. See License. Cf Certification

lichen planus DERMATOLOGY A pruritic dermatopathy of flexor surfaces–especially wrist, orogenital mucosa, characterized by papular thickening of upper dermis, a shiny violaceous surface MANAGEMENT Corticosteroids

lichen sclerosis GYNECOLOGY A pruritic mucosal lesion which is more common in older, often post-menopausal white ♀, that may have a vague genetic component in a background of autoimmunity; the cutaneous lesions consist of flat-topped coalescing white macules; LS evolves towards malignancy in 5% of ♂, but it is not premalignant in ♀

lichen simplex chronicus DERMATOLOGY A type of atopic dermatitis characterized by chronic pruritus, resulting in thick leathery hyperpigmented skin, exacerbated by emotional stress and associated with psoriasis, anxiety, depression, and mental retardation. See Atopic dermatitis

lichenification Induration of the skin with hyperpigmentation, acanthosis, hyperkeratosis, often linked to chronic pruritus

lidocaine AMBULATORY SURGERY An anesthetic used for topical and dental anesthesia, cardiac arrhythmias

Lidoderm® NEUROLOGY A transdermal lidocaine patch for postherpetic neuralgia. See Herpetic neuralgia

lie detector PSYCHOLOGY A device that detects chest and abdominal movement during respiration, heart rate, BP, and galvanic skin conductance due to sweating. See Polygraph test

life 1. A constellation of vital phenomena–organization, irritability, movement, growth, reproduction, adaptation. See Quality of life, Sex life, Sexual life 2. The duration of a product or material in its as-manufactured form. See Shelf life 3. The duration of a substance in a system. See Biological half-life

LIFE CARDIOLOGY A clinical trial–Losartan Intervention for Endpoint Reduction in Hypertension comparing the effects of 2 antihypertensives on cardiovascular M&M. See Antihypertensive, Atenolol, Hypertension, Losartan

life-enhancing therapy CLINICAL MEDICINE Pt management–eg, use of antiretrovirals, narcotics for CA pain, or palliative surgery, intended to improve quality of life without an expectation that life will be significantly extended. See Life-extending therapy

life event Any major change in person's circumstances–eg, divorce, death of spouse, loss of job, etc, that affects interpersonal relationships, work-related, leisure or recreational activities; LEs can be usual, ie not unexpected and therefore not evoking stress, or unusual–ie, unexpected and commonly associated with stress. See Unusual life event

life expectancy Longevity, period life expectancy EPIDEMIOLOGY The average length of life of persons in a population; the average number of yrs of life remaining for a population of persons, all of age x, and all subject for the remainder of their lives to the observed age-specific death rates corresponding to a current life table. See Life table

life-extending diet Anti-aging diet NUTRITION A general term for any dietary maneuvers believed to slow the aging process. See Diet, Melatonin, Zone-favorable diet

life support measures INTENSIVE CARE The care provided to a profoundly obtunded or moribund Pt, usually in an ICU to maintain a stable and/or 'compensated' clinical state, which requires 24-hr monitoring. See Advance directive, DNR orders

life table PUBLIC HEALTH A table that presents the results of a clinical study in which subjects enter and leave the trial at different times; each subject has a well-defined point of entry–onset of treatment and end point–relapse, death or other; all subjects may be evaluated at determined intervals with respect to the expected survival of an idealized person, based on actuarial analysis of census data and mortality rates

lifeguard lung PUBLIC HEALTH A popular term for granulomatous pneumonitis, caused primarily by *Pseudomonas* spp–which affects lifeguards who watch over swimmers at municipal indoor pools with water sprays, waterfalls, and water slides

LifePak® 12 EMERGENCY MEDICINE A defibrillator and monitoring system designed for first responders treating out-of-hospital cardiac arrests. See First responder

LifeSite® NEPHROLOGY An implantable venous access device which may ↓ occlusion, infection, and mechanical breakdown associated with other hemodialysis access devices. See Hemodialysis

Lifeskin DERMATOLOGY A proprietary artificial skin produced from cultured grafts of self keratinocytes, used in Pts needing self 'skin'–eg, to cover burns. See Artificial skin

lifespan Longevity EPIDEMIOLOGY The genetically endowed limit to life for a person, if free of exogenous risk factors. See Average lifespan, Life expectancy

lifestyle disease PUBLIC HEALTH Any condition–eg, obesity, HTN, cardiovascular disease, sports injuries, and some cancers–attributable, in part, to lifestyle choices, including diet, smoking, exercise, etc. See Lifestyle drug

lifestyle drug POPULAR HEALTH A prescription agent that allows its user to perform an activity 'on demand' or without consequences, ameliorate an imprudent binge, or modify effects of aging

lifetime maximum HEALTH INSURANCE The maximum benefits available to a member during his/her lifetime, which all benefits furnished are subject to this maximum unless stated as unlimited

lifetime risk EPIDEMIOLOGY The likelihood of suffering a particular condition during the lifespan of an average person, X

ligase chain reaction Ligation amplification reaction MOLECULAR BIOLOGY A DNA amplification technique for detecting minimal amounts of a known DNA sequence, similar in principle to PCR. See PCR

ligate *verb* To tie, close with a suture

ligation SURGERY The process of tying off blood vessels to block blood flow to a part of the body or to a tumor. See Band ligation

ligature SURGERY 1. A material–silk, gut, wire, etc used to ligate 2. A tissue plus the ligating material

light Electromagnetic radiation, usually understood to be in the range of visible light–ie, 390 to 770 nm. See Curing light, Ultraviolet light

light blue code Slow code, see there

light bulb appearance ORTHOPEDICS A morphology seen on a plain film of a posterior dislocation of the shoulder, which is almost invariably accompanied by a fracture of the medial part of an internally rotated humeral head

light chain nephropathy NEPHROLOGY A condition characterized by deposition of monoclonal light chains and electron-dense material in tubular and glomerular basement membranes CLINICAL Anorexia, nausea, weight loss, anemia, ↑ creatinine, ESRD; LCN may accompany myeloma

light chain deposition disease Bence-Jones myeloma, light chain disease ONCOLOGY A paraproteinemia linked to renal amyloidosis, which occurs in 20% of Pts with immunoproliferative disorders; both are characterized by deposits of fibrillar and nonfibrillar monoclonal light chains in various tissues CLINICAL Rapidly progressive renal failure due to renal tubular blockage by Bence-Jones proteins, nephrotic syndrome, heart failure, arrhythmias, liver disease, neurologic disease; similar clinical findings are associated with heavy chain deposition

light for date PEDIATRICS *adjective* Referring to a child who is underweight for his/her age or height, a finding suggesting malnutrition or child abuse or neglect. See Child abuse

light-headed PSYCHIATRY *adjective* Experiencing faintness

light sleep stage SLEEP DISORDERS NREMS stage 1–and sometimes stage 2. See Deep sleep stage

light therapy 1. The use of certain segments–in particular, the visible range of the electromagnetic spectrum as a therapeutic modality; LT may act via the hypothalamus, which releases neurotransmitters and releasing factors, after receiving impulses from retina FORMATS Full spectrum light–eg, sunlight, bright light–2 to 10,000 lux, UV light, colored light, hemoirradiation MAINSTREAM MEDICINE LT may be effective in seasonal affective disorder–SAD, and shiftwork-related sleep disorders. See Bright light therapy 2. Bright light therapy, see there 3. Heliotherapy, see there

Light's criteria Criteria which help differentiate pleural exudates and transudates

LIGHT'S CRITERIA FOR EXUDATES–ANY OF FOLLOWING
PLEURAL FLUID/SERUM PROTEIN RATIO > 0.5
PLEURAL FLUID/SERUM LDH RATIO > 0.6
PLEURAL FLUID LDH > ⅔ the normal upper limit for serum

lightening Dropping OBSTETRICS The descent of the fetus deeper into the pelvis, resulting in a ↓ in the fundal height, an event that occurs from days to wks–usually 2 wks before labor and is caused by physiologic changes of the uterus; lightening is characterized by an abrupt sensation of 'lightness' or relief from the weightiness felt by the mother which corresponds to the drop of the baby's head into the pelvis. Cf Lightning

lightning PUBLIC HEALTH A meteorologic phenomenon caused by a massive discharge of electricity. Cf Lightening

lightning and thunder, fear of PSYCHOLOGY Brontophobia or karaunophobia. See Phobia

lightning mark FORENSIC PATHOLOGY An arborescent charring of skin, caused by high voltage electricity–lightning and hydroelectric generators, which may be seen in victims of electrocution

lightning pains NEUROLOGY Sudden, sharp painful crises that are idiopathic or elicited by cold and stressants, which may require opiates for relief; LPs are a classic symptom of tabes dorsalis, and usually accompanied by progressive degeneration of the posterior and lateral columns of the spinal cord CLINICAL Loss of reflexes, vibration and position sense, ataxia, urinary incontinence, impotence DIFFDX Post-herpetic neuralgia, glossopharyngeal and trigeminal neuralgia–tic douloureux, triggered by peri- or intraoral stimulation, occurring as clusters of stabbing pain followed by refractory pauses, exacerbated in the spring and fall

likehood ratio DECISION MAKING The frequency that a test result is positive in subjects with–as opposed to without–a disease of interest; LR is a method for evaluating the informational properties of a nonbinary test SCREENING The relative odds that a given test result is expected in a Pt with–vs without–a disease of interest. Cf ROC curve

Likert scale A subjective scoring system that allows a person being surveyed to quantify likes and preferences on a 5-point scale, with 1 being the least important, relevant, interesting, most ho-hum, or other, and 5 being most excellent, yeehah important, etc

LIMA-Lift™ SURGERY A spreading system for accessing and visualizing the internal mammary artery during cardiac revascularization procedures. See CABG

limb The arm or leg. See Parasitic limb

limb-girdle disease Limb-girdle muscular dystrophy NEUROLOGY Any of a group of often AR muscular dystrophies that first involve either the shoulder–Erb type or the pelvic–Leyden-Möbius type girdle CLINICAL Limb-girdle 'syndromes' are of later onset than Duchenne's dystrophy, with marked muscular atrophy of the affected region accompanied by relentless deterioration, later involving the extremities DIAGNOSIS EMG, muscle biopsy DIFFDX LGD also occurs in acid maltase deficiency, Becker, Duchenne and Emery-Dreifuss dystrophies, endocrine myopathies, lipid storage disease, late onset–nemaline, central-core myopathies, polymyositis, sarcoidosis. See Duchenne's dystrophy, Muscular dystrophy

limb perfusion ONCOLOGY The delivering of chemotherapy directly to an arm or leg, in which the flow of blood through the limb is temporarily stopped with a tourniquet, and anticancer drugs are injected IV

limb salvage ORTHOPEDICS The returning of a limb to a state of reasonable functionality after severe trauma that might otherwise result in amputation. See Amputation

limbic system NEUROLOGY The 'peripheral' component of the CNS, which is linked to the autonomic nervous system and carries out the nonmotor and non-sensory aspects of cerebral function, including control of emotion, eating, drinking, sexual activity, other behaviors, olfaction COMPONENTS Olfactory system, hippocampus, dentate nucleus, cingulate gyri, amygdalus, septum, fornicate gyrus, parts of the thalamus and hypothalamus and connections with the septum, hypothalamus, and medial mesencephalic tegmentum; the LS controls autonomic activity–eg, changes in BP and respiration

Limbitrol® A proprietary combination of amitriptyline and chlordiazepoxide

limerance The state of being love-stricken. See Infatuation

LIMIT-2 CARDIOLOGY A clinical trial–Leicester Intravenous Magnesium Intervention Trial-2— that evaluated the effect of IV magnesium sulfate on survivors of acute MI. See Acute myocardial infarction, Magnesium sulfate. Cf ISIS-4

limit A ceiling, boundary, endpoint, maximum. See Abbe limit, Age limit, Catastrophic limit, Ceiling limit, Class limit, Confidence limit, Control limit, Flammable limits in air, Hayflick limit, Kerley limit, Lowest explosive limit, Nyquist limit, Permissible exposure limit, Product limit, Short-term exposure limit, Speed limit, Time limit on certain defenses, Upper explosive limit, Upper flammable limit

limitation A restriction or ceiling. See Salary limitation, Site limitation

Limitrol™ POPULAR NUTRITION A source of dietary fiber derived from fenugreek, a Middle Eastern vegetable, said to ↓ glucose absorption in DM and ↓ cholesterol in Pts with hypercholesterolemia. See Dietary fiber

limits of liability HEALTH INSURANCE The maximum amount an insurer will pay out under the terms of a policy

limit of viability PEDIATRICS The minimum weight or age of a fetus that is compatible with extrauterine life; the LoV has been lowered by improved obstetric and neonatal interventions and aggressive resuscitation

LIMITS OF VIABILITY				
WEEKS OF GESTATION	**22**	**23**	**24**	**25**
Average birthweight (g)	490	592	695	761
Number stillbirths	7/29	4/40	0/34	0/39
Number 6-month survival/total	0/29	6/40	19/34	31/39
Days of hospital stay	NA	120±23	107±37	76±23
Survival w/o cerebral defects	0%	12%	21%	69%

lincomycin Lincocin® An older generation antibiotic produced from *Streptomyces lincolnensis*, which has been used for gram-positive cocci

lindane TOXICOLOGY The γ-isomer of benzene hexachloride; a carcinogenic, lipid-soluble insecticide used topically to control lice and scabies, and occasionally for suicide; 20-30 g produce serious toxicity or death CLINICAL Onset is similar to DDT poisoning–tremors, ataxia, violent tonic-clonic convulsions, pulmonary edema and vascular collapse of neurogenic origin; massive hepatic necrosis ensues, accompanied by hyaline degeneration of renal tubules, aplastic anemia

line CLINICAL MEDICINE A popular term for an IV catheter DRUG SLANG An elongated trail of relatively pure pulverized cocaine, usually snorted through a straw TELEMEDICINE A telecommunication wire or other conduit. See Copper line, Digital subscriber line, Hotline, Medline, Leased line,T-1 line, T-3 line VOX POPULI A 2-D streak or mark or strip. See Bottom line, E-1 line, Fine line, Frankfort (horizontal) line, Gatekeeper guideline, Harvard guideline, Iliopectineal line, Intertrochanteric line, Lead line, Marionette line, Midclavicular line, Ohngren's line, Regression line, Skinner's line, Squamocolumnar Z line

lines of Blaschko DERMATOLOGY Alterating stripes of affected and unaffected skin, which are seen in certain skin diseases. Cf Dermatomes

line-of-sight transmission EPIDEMIOLOGY A mode of 'transmission' of a pseudoepidemic, in which those afflicted are within view of others similarly affected. See Mass sociogenic illness, Pseudoepidemic

linear dose-response THERAPEUTICS A consistent ↑ in biologic response as ↑ quantities of a test substance are administered

linear fracture ORTHOPEDICS A fracture that extends over part or the entire length of a bone. See Hairline fracture

linear regression STATISTICS A statistical method defined by the formula $y = a + bx$, which is used to 'fit' straight lines to scattered data points of paired values Xi, Yi, where the values of Y–the ordinate or vertical line are observations of a variable–eg, systolic BP and the values of X–the abscissa or horizontal line ↑ in a relatively nonrandom fashion–eg, age

linearity STATISTICS A straight line relation between 2 quantities, where when a value 'X' is ↑ or ↓, 'Y' is proportionately ↑ or ↓; linearity assumes that the relation between X and Y–abscissa and ordinate can be summarized in a straight line, known as a least-squares regression method

linezolid INFECTIOUS DISEASE An oxazolidinone antimicrobial in trials for gram-positive bacteria, possibly for managing VRE. See Vancomycin-resistant enterococcus

lingual *adjective* Referring to the tongue

link A connection between 2 or more things. See Body-mind link, Hypertext link, Missing link, Oncolink, Physician office-to-laboratory link

linkage analysis GENETICS A gene-hunting technique that traces patterns of heredity in large, high-risk families, in an attempt to locate a disease-causing gene mutation by identifying traits co-inherited with it; the formal study of the association between the inheritance of a condition in a family and a particular chromosomal locus; LA is based on certain ground rules of genetics. See Lod score

lip stripping DERMATOLOGY Excision and advancement of oral mucosa, a technique used in plastic surgery when the vermilion border becomes indistinct, due to squamous metaplasia or labial hyperkeratosis, a potentially preneoplastic lesion of light-skinned sun-exposed elderly Pts. Cf Chemical peel, Mohs surgery

Lipanthyl® Fenofibrate, see there

lipectomy COSMETIC SURGERY Removal of fat. See Abdominal lipectomy, Liposuction

lipemia retinalis INTERNAL MEDICINE Ocular fundus whitening due to extreme hypertriglyceridemia–serum TGs of 45.2 mmol/L–4000+ mg/dL, US, associated with DM

LIPID CARDIOLOGY A clinical trial–Long-Term Intervention with Pravastatin in Ischemic Disease–which evaluated preventing cardiovascular M&M with pravastatin in Pts with CAD and a range of initial cholesterol levels

lipid-associated sialic acid ONCOLOGY A glycoconjugate of sialic acid associated with the lipid bilayer of the cell membrane, which may be secreted by malignant–breast, colon, lung CAs, leukemias, lymphomas, Hodgkin's disease, melanomas and benign lesions REF RANGE 50-150 mg/L

lipid hypothesis CARDIOLOGY A widely accepted postulate that hyperlipidemia–eg, cholesterol, and to a lesser degree, other lipids in the circulation is responsible for CAD, the major cause of death in the US, levels which, when altered by dietary or pharmacologic manipulation, ↓ risk of ASHD-related morbidity. Cf Lipid theory

lipid profile LAB MEDICINE A battery of tests performed on a multichannel chemical analyzer–total cholesterol, LDL-C, HDL-C and TGs, which is used to stratify the risk of suffering ASHD-related M&M

Lipid Research Clinics Prevalence Study CARDIOLOGY A long-term–average 10.1 yrs—study of 2541 men, age 40–69, 17% of whom had CAD at base line, which evaluated the relationship between HDL-C and LDL-C and total and CAD-related mortality. See Cholesterol

lipid storage diseases A group of rare conditions–eg, Fabry's disease, Niemann-Pick disease, and sea-blue histiocytosis syndrome, which are often fatal in early childhood, usually due to a catabolic defect of lipid metabolism and characterized by the accumulation of lipids in one or more organs. See Pseudo-Gaucher's disease, Sphingolipidosis

lipidation See Protein lipidation

Lipiodol® ONCOLOGY A proprietary iodized oily agent that selectively remains in tumors for prolonged periods of time, and enhances the efficacy of certain chemotherapeutics

Lipitor® Atorvastatin CARDIOLOGY A statin which ↑ HDL-C, ↓ LDL-C, ↓ total cholesterol, ↓ TGs. See Statin

lipo(a) See Lipoprotein (a)

lipochrome Any natural, fat-soluble pigment–eg, lipofuscin, carotenes, lycopenes

lipodystrophy syndrome AIDS A clinical state characterized by redistribution of fat, increased waist size, and loss of fatty tissue from the face, arms and legs, caused by protease inhibitors used to manage HIV-1-infected Pts CLINICAL Redistribution of fat ADVERSE EFFECTS N&V, maculopapular rash, headache, fatigue, headache, vertigo. See AIDS, Nonnucleoside reverse transcriptase inhibitor

lipofuscin A pigmented lipid degradation product thought to derive from peroxidative destruction of mitochondrial polyunsaturated lipid membrane or the mitochondrion itself; the malonaldehyde produced by mitochondrial peroxide damage may block DNA template, activity contributing to heart failure; lipofuscin accumulates with age in the heart, muscle, liver, nerve, and in lysosomes

lipoid pneumonia Golden pneumonia A pneumonitis caused by exogenous oils that percolate into the lung after intranasal instillation of mineral oil, oral ingestion of cod liver, castor or other oils, or due to a congenital defect in oropharyngeal diaphragm–eg, cleft palate or intense gag reflex. See Mineral oil. Cf Aspiration pneumonia.

lipoid proteinosis An AR condition of childhood onset characterized by coalescent aggregates of lipid and mucopolysaccharides, resulting in numerous yellowish plaques, papules, nodules, and induration of skin–pachydermia, eyelids, oropharynx, and larynx with hoarseness, hyperkeratosis of the knees and elbows, hyalinization of blood vessels; calcification of hippocampal gyri, while rare, is pathognomonic and causes the associated convulsions

lipoleiomyoma A uterine leiomyoma-like neoplasm of obese, postmenopausal ♀ with cholecystitis, that may cause vague abdominal pain, backache, vaginal discharge, or hemorrhage

lipoma Adipose tumor DERMATOLOGY A benign fatty tumor which may occur in virtually any site in the body. See Myelolipoma, Myolipoma, Spindle cell lipoma

lipoma arborescens ORTHOPEDICS An arborescent intraarticular lipoma, usually seen in the knee associated with degenerative joint disease, rheumatoid arthritis, or joint trauma

lipomatosis dolorosa Dercum's disease A perimenopausal condition characterized by multiple circumscribed masses of adipose tissue accompanied by local pain at the sites of accumulation CLINICAL Neuroasthenia, headache, depression, ecchymoses and cardiovascular decompensation due to cardiac overload TREATMENT Weight loss. Cf Weber-Christian disease

lipoperoxide PATHOPHYSIOLOGY A peroxide–an O-O containing free radicals–which has peroxidized the carbon atoms close to the double bonds in an unsaturated fatty acid. See Peroxide

lipoplex MOLECULAR MEDICINE A complex of plasmids mixed with lipids that spontaneously forms a vesicle containing therapeutic genes of interest

lipopolysaccharide-binding protein A trace plasma protein that binds to the lipid A moiety of bacterial lipopolysaccharide–LPS or to endotoxin–a glycolipid present in the outer membrane of all gram-negative bacteria

lipoprotein(a) Lp(a) PHYSIOLOGY A lipoprotein with a range of serum levels, a lipid content similar to LDL, which binds to the LDL receptor with lesser affinity than LDL; Lp(a)'s role in ASHD is controversial; it is ↑ in Pts at risk for CAD REF RANGE 0-30 mg/dL

lipoprotein lipase deficiency An AR condition characterized by lack of lipoprotein lipase, resulting in massive hypertriglyceridemia of neonatal onset and recurrent episodes of pancreatitis CLINICAL Fatty food intolerance, eruptive xanthomas, hepatosplenomegaly that regresses with dietary control

lipoprotein X An abnormal lipoprotein composed of 65% lecithin, 30% cholesterol and 5% protein–apoC and albumin, seen in lecithin-cholesterol acyltransferase deficiency and in obstructive biliary disease, and associated with cholestatic jaundice. See Lecithin-cholesterol acyl-transferase deficiency

liposhaving COSMETIC SURGERY A technique used to layer off subcutaneous fat, which may be better than liposuction for facial procedures. Cf Liposuction

liposome Lipid vesicle DRUG DELIVERY A synthetic, uniform bilayer lipid membrane-bound vesicle formed by emulsification of cell membranes in dilute salt solutions, which is used to deliver toxic drugs–eg, amphotericin B, doxorubicin, pentavalent antimony–tagging them with an organ-specific antibody. See Liposome-encapsulated amphotericin B, Stealth liposome

liposuction Suction-assisted lipectomy COSMETIC SURGERY A technique for removing focal fat deposits. See Body sculpting, Cosmetic surgery, Tumescent liposuction. Cf Liposhaving

Lipoxide Chlordiazepoxide, see there

lipstick sign PSYCHOLOGY A clinical sign indicating resolution of transient depression, manifest by application of lipstick or use of a hairbrush, indicating a sense of future, self-perception, self-image and esteem, and a willingness to leave behind the sense of illness. Cf Suitcase sign.

liquid diet CLINICAL NUTRITION A very low-calorie diet that fulfils the daily fluid requirements and places little functional demand on the GI tract; LDs have little fiber and do not provide adequate protein or calories–circa 1000 kcal/day. See Bland diet, Diet

Liquid Embolic System THERAPEUTICS A fluid used for therapeutic embolization, administered via a pre-positioned microcatheter under continuous fluoroscopic monitoring

LiquiVent® Perflubron, see there

liquor MEDTALK A fluid, usually aqueous containing a medicinal. See Herbal liquor VOX POPULI A beverage with a high–generally > 20%–concentration of ethanol. See Hard liquor

LIS 1. Laboratory information system, see there 2. Left intercostal space

lisinopril Prinivil® CARDIOLOGY An ACE inhibitor used to manage HTN ADVERSE EFFECTS Hypotension, neutropenia, anaphylactoid reactions. See GISSI-3

lisofylline INFECTIOUS DISEASE An investigational antibiotic for treating serious infections in CA Pts immunocompromised by therapy–eg, induction chemotherapy for AML, high-dose RT and/or chemotherapy followed by BMT

lissencephaly NEUROLOGY A brain malformation that is isolated or associated with other syndromes–eg, Miller-Dieker syndrome, Norman-Roberts syndrome, characterized by absence–agyria or incomplete development–pachygyria of cerebral gyri CLINICAL Microcephaly, seizures, profound mental retardation, feeding difficulties, growth retardation, impaired motor activity

listeriosis INFECTIOUS DISEASE Infection by *Listeria* spp, most by *L monocytogenes* EPIDEMIOLOGY Outbreaks linked to contaminated milk products and cheese MORTALITY 20-40% CLINICAL $\frac{1}{3}$ of cases of listeriosis occur in pregnant women, causing transplacental infection with abortion, stillbirth, and premature delivery; infants may present with septicemia, diarrhea, vomiting, cardiorespiratory distress, meningoencephalitis; immunocompromised adults may suffer meningoencephalitis, endocarditis, disseminated granulomatosis, lymphadenitis, peritonitis, cholecystitis TREATMENT Ampicillin, gentamicin, tobramycin, erythromycin, tetracycline, doxycycline, T-S

literacy INFORMATICS The ability to read and write. See Digital literacy MEDICAL COMMUNICATION The ability to read and understand written text, instructions and medically relevant materials. See health literacy, Science literacy

the literature MEDICAL COMMUNICATION A shifting body of published information of varying validity and timeliness; a popular 'short form' for information written in peer-reviewed journals about a particular subject. See Literature review SEXOLOGY See Gay literature

lithium carbonate $LiCO_3$ PHARMACOLOGY An alkali used to treat bipolar I disorder, which blocks neurotransmission at the 'second messenger' phosphoinositide-mediated cholinergic neurons in the hippocampus, inhibiting release and uptake of norepinephrine at nerve endings by inhibiting receptor-mediated synthesis of cAMP NEUROPHARMACOLOGIC EFFECTS-ANTIMANIC Blocks development of dopamine receptor supersensitivity, $\uparrow$ GABA function, $\uparrow$ acetylcholine function ANTIDEPRESSANT $\uparrow$ 5-HT function, $\downarrow$ β-adrenoceptor stimulation of adenylate cyclase, $\downarrow$ α_2-adrenoceptor function THYMOLEPTIC $\downarrow$ Neurotransmitter-coupled adenylate cyclase activity and cAMP formation, $\downarrow$ receptor-G protein coupling, $\downarrow$ phosphoinositide metabolism, alters kinetics of alkali cations–Na^+, K^+, Ca^{2+}, Mg^{2+} SIDE EFFECTS Hyperirritability, hyperpyrexia, stupor, coma, gastroenteritis, cardiovascular disease–eg, arrhythmia, hypotension, $\downarrow$ ST wave, T inversion, osteoporosis TERATOGENESIS Cardiac malformations in 10% of infants born to lithium-treated ♀ TOXICITY Overdose causes death in $1/4$ of Pts TREATMENT K+-sparing drugs

LITHIUM TOXICITY

< 1.5 MMOL/L Nausea, tremor, mild polyuria

1.5–2.5 MMOL/L Diarrhea, vomiting, polyuria, coarse tremor, muscle fasciculation, ataxia, weakness, sedation

2.5–4.0 MMOL/L Muscle hypertonia, choreiform movement, $\uparrow$ deep muscle reflexes, seizures, focal neurologic signs, impaired consciousness, confusion, stupor

> 4.0 MMOL/L Coma, death

lithotripsy Shock-wave lithotripsy A nonsurgical, noninvasive method for fractionating renal, and recently, bile tract calculi

lithotripter UROLOGY A device used during cystoscopy to crush and remove bladder, ureteral, and renal stones

liticaphobia PSYCHOLOGY Fear of lawsuits. See Phobia

litogen FORENSIC MEDICINE A drug used during pregnancy that is not teratogenic or known to be teratogenic, but which nevertheless triggers lawsuits. See *Daubert v Merrell Dow*, *Frye* decision

litter MILITARY MEDICINE A mobile bed for transporting wounded military personnel VOX POPULI Trash strewn in a public or open place

Little League elbow Pitcher's elbow SPORTS MEDICINE A form of medial epicondylitis, manifest as apophyseal tenderness with ulnar nerve irritation; if severe, the LLE may require surgery with possible lifelong arthralgia, due to injury of the physeal cells under the articular cartilage; permanent damage results from repeated and excess axial loading, which compresses the cells against the osseous matrix; late sequelae, osteochondritis dissecans

Little League shoulder SPORTS MEDICINE A popular term for a fracture of the proximal humerus through the epiphyseal growth plate–the weakest point of growing long bones, the result of a shoulder articulation chronically insulted by throwing a baseball

Little's disease Bilateral congenital spastic diplegia A type of cerebral palsy with agenesis of the lower motor neurons in the legs, which affects $\pm$ 300,000, US; congenital cerebrovascular malformations and immune-related kernicterus account for some cases CLINICAL Spasticity, muscle weakness, mental retardation, and ocular defects, due to neonatal hypoxia, mechanical trauma and prematurity DIFFDx Brain tumors, 'floppy infant' syndromes, leukodystrophy, muscular dystrophy. See Cerebral palsy, Floppy infant

live attenuated vaccine A vaccine that induces an immune response, which more closely resembles that of a natural infection, than that elicited by killed vaccines, as the organisms contained therein actively reproduce until held in check by the recipient's own antibodies, thus often conferring life-long immunity EXAMPLES Measles, mumps, polio, rubella. See Killed vaccine

live birth REPRODUCTION MEDICINE The "Complete expulsion or extraction from its mother of a product of conception...which, after such separation, breathes or shows *any* other evidence of life such as the beating of the heart, pulsation of the umbilical cord, or definite movement of voluntary muscles, whether or not the umbilical cord has been cut or the placenta is attached". See Birth, Low birthweight

livedo reticularis DERMATOLOGY A bluish mottling of the skin evoked by low temperatures and characterized by vasodilation. See Raynaud's phenomenon

liver abscess Bacterial liver abscess, pyogenic liver abscess HEPATOLOGY A circumscribed focus of infection in the liver ETIOLOGY Intraperitoneal seeding from appendicitis, diverticulitis, perforated bowel, blood-borne, ascending bile tract infection, or post-traumatic CAUSATIVE AGENTS *E coli*, *Proteus vulgaris*, *Enterobacter aerogenes*

liver adenoma Adenoma of liver, hepatic adenoma, hepatocellular adenoma A rare solitary rounded, well-circumscribed and encapsulated mass–10% are multiple of benign liver tissue, which occurs in ♀ age 30-40, and is pathogenically linked to the used of oral contraceptives; 70% are solitary; rarely 10 or more may be present, a condition termed liver cell adenomatosis MANAGEMENT Excision

liver cancer Hepatocellular carcinoma, hepatoma, liver cell cancer, liver cell carcinoma, ICD-0 code 8170, HEPATOLOGY A CA affecting the liver, of which 7,000 arise in the gallbladder, 5,000 in extrahepatic and intrahepatic biliary ducts, and 3,000 are primary hepatocellular carcinomas–HCCs or primary liver cancer EPIDEMIOLOGY Incidence $\uparrow$–$1.4/10^5$ from 1976-80 to $2.4/10^5$ from 1991-95, primarily in younger–age 40-60–Pts; blacks are 2-fold more common than whites RISK FACTORS HBV-ness carries a 7-fold $\uparrow$ risk; HCV-ness has a 4-fold $\uparrow$ risk; mycotoxin–eg, aflatoxin B1 in stored grains, drinking water; 3 yr risk for HCC is 12.5% in Pts with cirrhosis; 4% in Pts with chronic liver disease, others include hemo CLINICAL Abdominal pain, weight loss, weakness, anorexia, vomiting, and jaundice, hepatomegaly, ascites, splenomegaly, wasting, fever; paraneoplastic changes include $\uparrow$ cholesterol, $\uparrow$ RBCs and sex changes LAB AFP, PIVKA II–protein induced by the absence of vitamin K, biopsy, hepatitis serology, LFTs, CBC, coagulation profile IMAGING MRI, CT with angiography, CTs of head, chest, bone scan for metastastatic workup PROGNOSIS 5 yr postoperative recurrence is 80%, reflects adequacy of surgical margins, detected by AFP. See Alpha-fetoprotein, Liver cancer profile

liver cancer profile A battery of serum assays that assess liver cancer susceptibility, including markers often associated with liver cancer MARKERS AFP, CA 19-9, CEA, ferritin, HB_sAg, HB_cAb, HB_eAg, HB_cAg, HBC. See Liver cancer

liver dialysis THERAPEUTICS The use of hepatocytes either in cartridges, as in an bioartificial liver or via cross-perfusion with other organisms, either human or animal origin to provide metabolic support for a Pt in acute hepatic failure. See Artificial liver

liver disease Hepatopathy HEPATOLOGY Any disease or disorder that causes the liver to function improperly or cease functioning. See Alcoholic liver disease, Amebic liver abscess, Autoimmune hepatitis, Biliary atresia, Chronic active hepatitis, Chronic persistent hepatitis, Cirrhosis, Coccidioidomycosis, Hemochromatosis, Hepatitis A, Hepatitis B, Hepatitis C, Hepatitis D, Hepatitis GB, Hepatocellular carcinoma, Liver abscess, Liver cancer, Primary biliary cirrhosis, Pyogenic liver abscess, Reye's syndrome, Sclerosing cholangitis, Wilson's disease

liver failure CLINICAL MEDICINE Liver insufficiency that results in death, requires a liver transplant, or is characterized by recovery after encephalopathy, or while awaiting a transplant; also defined as a condition with ≥ 3 of following: albumin < 3.5 g/dL; prolonged prothrombin time–PT; jaundice; ascites; PT is expressed as a ratio vs value from pooled normal plasma, is considered prolonged when the values are > the upper limit of normal–ie, 0.84-1.18; jaundice is defined as a BR concentration ≥ 2X upper normal range; ascites is ID'd by ultrasonography EPIDEMIOLOGY 27,000 die/yr–US ETIOLOGY Viral hepatitis; drugs–eg, valproic acid, INH, halothane, acetaminophen, mushroom, phosphorous, aspirin, etc; alcohol; idiopathic; myocarditis, heart surgery, cardiomyopathy, Budd-Chiari syndrome; metabolic disorders–eg, galactosemia, tyrosinemia, iron storage, mitochondrial disease, fatty acid oxidation CLINICAL Jaundice, fatigue, weight loss, if extreme, renal failure, hepatic encephalopathy COMPLICATIONS Cerebral edema, infection, renal failure, bleeding TREATMENT Symptomatic support; liver transplantation; possibly in the future, bioartificial liver

liver function tests Clinical parlance for a battery of serum analytes that reflect the liver's metabolic reserve capacity

LIVER FUNCTION TESTS

TEST ABILITY TO EXCRETE

- Endogenous substances, eg bilirubin, bile acids, ammonia
- Exogenous substances, eg drugs, dyes, galactose

TEST ABILITY TO METABOLIZE

- Conjugation and synthesis of proteins

ELEVATED SUBSTANCES

- Liver disease, inflammation or necrosis–↑ transferases, other enzymes, vitamin B_{12}, iron, ferritin
- Biliary tract obstruction–bilirubin, cholesterol, enzymes and lipoprotein-X

Note: Nonfunctional biochemical and immunologic markers of hepatic disease include those for hepatitides–HAV, HBV, HCV, HDV, HEV, and HIV, autoimmune diseases–anti-mitochondrial antibodies, PBC, malignancy–alpha-fetoprotein, relatively specific for liver cancer and metabolic diseases–ceruloplasmin in Wilson's disease, transferrin in hemochromatosis

liver metastasis ONCOLOGY Cancer that has spread from the original–1°–tumor to the liver PRIMARY CANCER Colorectal CA, neuroblastoma, pancreatic CA, Hodgkin's lymphoma. See Metastasis

liver panel LAB MEDICINE A battery of tests used to evaluate the liver's functional status and its ability to produce proteins and metabolize toxins, and detect inflammation, measuring transferases–AST, AST, γ-glutamyl transferase, alkaline phosphatase, total BR, conjugated BR, total protein, albumin, PT. See Liver function tests, Organ panel, Panel

liver rounds A popular term for a regular social function in which there is beer and/or other inebriants–hence, 'liver', and comestibles. Cf Grand rounds

liver-spleen scan NUCLEAR MEDICINE A scintigraphic imaging technique in which a radiopharmaceutical–eg, ^{99m}Tc, ^{198}Au is injected IV to visualize metabolically-active, isotope-labeled, potentially malignant masses of ≥ 2 cm in the liver and spleen INDICATIONS Determine liver and spleen size, ID dysfunction, detect focal lesions. See Radionuclide imaging

liver spots Age spots, lentigos, senile lentigines, sun-induced skin changes DERMATOLOGY A nonspecific lay term for red-brown skin lesions associated with aging–eg, pigmented seborrheic keratosis and lentigo senilis. See Lentigines

liver transplant Hepatic transplant TRANSPLANT SURGERY A procedure that replaces a cancer conquered, metabolically defeated, or substance subjugated liver with one no longer required by its owner, many of whom donate same after an MVA DISEASES REQUIRING TRANSPLANT Cirrhosis, chronic hepatitis, primary sclerosing cholangitis, PBC, HBV, hepatitis D, α_1-antitrypsin deficiency, LDL-receptor deficiency, cancer; in high-risk candidates for orthotopic LT, auxiliary heterotopic LT, leaving the Pt's own liver in place, may be compatible with up to 12 yrs survival EARLY COMPLICATIONS Hypothermia, hyperglycemia LATE COMPLICATIONS Infection–eg, CMV, gram-

positive bacteremia, renal insufficiency, due to cyclosporine nephrotoxicity, HTN, hypokalemia, metabolic alkalosis, fever; 1 yr survival, 83%; 2 yr survival 70%

living will RIGHT TO DIE An advance medical directive in which a mentally-competent adult formally expresses his/her preferences regarding medical treatment, in the event of future incapacitation or incompetence to make medical decisions. See Advance directive, DNR, Health care proxy. Cf Durable powers of attorney, Euthanasia

livor mortis Postmortem lividity A blanchable discoloration of the skin–pink, early, purple, due to the settling of blood in the most dependent tissues; with time, the discoloration becomes fixed. See Rigor mortis

LLE Left lower extremity

LM Light microscopy, also 1. Laboratory manager 2. Lateral meniscus 3. Left main coronary artery, see there 4. National Library of Medicine 5. Licentiate in medicine 6. Low molecular 7. Lower motor 8. Lumen, an SI–International System unit of luminous flux

LMP Last menstrual period, see there

LN Lymph node, also 1. Background noise level 2. Laser nephelometry 3. Lesch-Nyhan syndrome 4. Licensed nurse 5. Lot number

load A measurable quantity of a thing. See Patient load DRUG SLANG 25 bags of heroin MEDTALK The content of a substance or material. See Afterload, Front load, Standard load

load & shift test ORTHOPEDICS A joint laxity test used to diagnose shoulder instability. See Laxity test, Shoulder instability. Cf Provocative test

loaded DRUG SLANG A street term for a state of drug or alcohol-induced intoxication/euphoria

loading PHYSIOLOGY The adding of a substance to a system or to the body. See Bicarbonate loading, Carbohydrate loading, Phosphate loading, Soda loading

loading dose Initial dose PHARMACOLOGY A first dose of a drug administered in excess of the maintenance dose, administered to rapidly achieve therapeutic drug levels. Cf Maintenance dose

loaned servant doctrine See Captain of the ship doctrine

lobar intracerebral hemorrhage Intraparenchymal hemorrhage An intracranial hemorrhage of the frontal, parietal, temporal, or occipital cortex which represents $1/3$ of cases of 1°, nontraumatic intracerebral hemorrhage INCIDENCE 8.4/10⁵/yr PATHOGENESIS LIH is more commonly linked to cerebral amyloid angiopathy–CAA than to hypertensive small vessel disease DIAGNOSIS CAA is more common in those with the apolipoprotein E ε2 and ε4 than with the normal variant, ε3. See Cerebral amyloid angiopathy, Intracerebral hemorrhage. Cf Subarachnoid hemorrhage

lobar pneumonia PULMONOLOGY Pneumonia which affects part of one or more lobes of the lungs, characterized by virtually homogeneous consolidation; those with ↑ susceptibility in those with DM ETIOLOGY *Streptococcus pneumoniae* with bacterial pores of Kohn; ♂ : ♀ 3:1. See Community-acquired pneumonia

lobe 1. A subdivision of an organ–eg, liver–2 lobes; lung–2 lobes left, 3 lobes right; breast–15-20 lobes; brain–2 lobes, divided by fissures, connective tissue or other natural boundaries. See Temporal lobe 2. A rounded projecting portion, such as the lobe of the ear

lobectomy SURGERY The surgical excision of an entire lobe of the lung

lobucavir VIROLOGY A nucleoside analogue broad-spectrum antiviral for HBV, HS-1, HS-2, herpes zoster. See Antiviral, Nucleoside analogue

lobular carcinoma in situ Atypical lobular hyperplasia, LCIS ONCOLOGY A precancerous epithelial lesion of the female breast; $\frac{1}{3}$ of those with LCIS develop invasive CA; 8% of LCIS do not form a discrete mass. Cf Ductal carcinoma in situ

lobular carcinoma ONCOLOGY A major morphologic form of breast carcinoma which, like ductal breast carcinoma, arises in the terminal duct/lobular unit; the division is of morphologic, but no clinical, significance. Cf Ductal carcinoma, Medullary carcinoma

local *adjective* In the area of a lesion; confined to a specific area *noun* Regionally popular for a local anesthetic, as in, 'under local'

local anesthetic PAIN CONTROL An agent administered at the site of pain or anticipated pain–eg, before dental work or surgical incision TYPES Esters–hydrolyzed by pseudocholinesterase and amides–metabolized in the liver and excreted in urine; a vasoconstrictor, usually epinephrine, diluted to 1:200,000 may added to limit absorption, and enable use of more anesthetic. Cf General anesthesia

local recurrence ONCOLOGY The reappearance of the signs and Sx of CA at a site that was previously treated and responded to therapy. See Relapse

local therapy Local treatment ONCOLOGY Treatment that affects cells in a tumor and adjacent region. Cf Systemic therapy

localization procedure IMAGING Any imaging modality intended to identify the location of particular lesion–eg, a breast or lung mass, or foreign body–eg, a bullet, shrapnel, pet rocks (no kidding–been reported!)

localized *adjective* Restricted to a site of origin or primary infection without evidence of spread or extension

localized disease MEDTALK Any condition, generally understood as malignant, which is confined to a tissue or organ. Cf Regional disease

locally advanced cancer ONCOLOGY Cancer that has spread to nearby tissues or lymph nodes, but not metastasized

loci Plural of locus, see there

Locilex™ Cytolex™, MSI-78, pexiganan acetate INFECTIOUS DISEASE A topical antibiotic for treating infected diabetic foot ulcers; Locilex is the first in a class of antibiotic/anti-infective magainins, derived from the skin secretions of the African clawed frog, which kill bacteria by attacking cell membranes. See Magainin. NOTE: It was deemed not approvable.

lock VOX POPULI A device that prevents a particular action or activity. See Bloxham air lock, Heparin lock, Job lock, Vapor lock

locked-in syndrome NEUROLOGY Flaccid tetraplegia with facial paresis and complete incapacity of expression–ie, anarthric and aphonic; LIS is due to damage or dysfunction of descending motor pathways or peripheral nerves, 2° to bilateral destruction of the basis pontis or medulla and sparing of tegmentum, caused by infarcts or central pontine myelinolysis; LIS Pts are conscious and alert and only capable of communicating by moving their eyes–voluntary eye movement and eyelids–blinking

locked knee Trick knee A knee limited in its range of movement by a loose joint body or 'mouse', a meniscus–usually medial–tear, or patellofemoral derangement

locked lung Paradoxic bronchospasm, refractory status asthmaticus A state due to excess use of nebulized isoproterenol, resulting in complete loss of response to epinephrine, aminophyllin, corticosteroids and intermittent positive pressure, which requires weaning from isoproterenol

lockjaw Tetanus, trismus Spasm of the masseter muscles with stiffness of the jaw caused by tetanospasmin, a neurotoxin produced by *Clostridium tetani*, which causes an unrestrained muscle firing, and sustained muscular contraction which, if severe, causes dysphagia or acute respiratory insufficiency by causing prolonged diaphragm contraction

locum tenens A term for either 1. A person or professional filling another's place for a defined temporary period of time, usually for vacation coverage or 2. The position being covered; LT physicians or 'locums' usually have the same qualifications as the person for whom they are covering (he said petulantly)

locus minoris resistentiae MEDTALK Point of least resistance

Lodine® RHEUMATOLOGY A formulation of etodolac for managing osteoarthritis, rheumatoid arthritis. See Osteoarthritis, Rheumatoid arthritis

lofepramine A tricyclic antidepressant, see there

log LAB MEDICINE *noun* A book in which certain types of Pt-related information is recorded, usually by hand. See Master log RESEARCH Logbook A journal or ledger containing data, and dates when the data was collected; logs are used to record scheduling information, document Pt-physician encounters, to record quality control data in the lab and may be used as a legal document. Cf Notebook

log in INFORMATICS *noun* The account name used to access a computer system; an approved account on a computer that must be created by a system manager before the system can be accessed. Cf Password *verb* To enter a computer system, or to type in a password and access a computer network or system LAB MEDICINE *verb* To enter a Pt's demographic data and test(s) requested into an infomation system, which constitutes formal acceptance of a laboratory specimen for processing. Cf Log on

log on COMPUTERS *verb* To tap into a database or access the Internet. Cf Log in

log-rolling EMERGENCY MEDICINE A popular term for the moving of a Pt–as if rolling a log–without allowing lateral movement of the head and neck, usually to a board or other flat rigid device

logic DECISION-MAKING The sum total of education and experience that is integrated into a physician's medical decision-making processes. See Aunt Millie approach, Bayesian logic, Heuristic logic, Markov process, ROC analysis, Stochastic process

logizomechanophobia Fear of computers. See Phobia

loin The lower back from below the ribs to the pelvis

loin pain-hematuria syndrome A benign, rare condition characterized by recurring episodes of uni– or bilateral loin pain–which may be severe enough to require denervation or autotransplantation accompanied by hematuria of unknown origin; it is more common in ♀ and may be related to hormonal factors–eg, OC use, or due to defects in renal vasculature CLINICAL Episodic gross hematuria, pelvic pain, mild HTN that resolves with D/C of OCs

Lolita syndrome Adolescentalism, ephobophilia SEXOLOGY A paraphilia deviancy in which adolescents are the focus of sexuoerotic fantasies and/or acts. See Chronophilia

lollipop tree appearance A cholangiographic pattern consisting of multiple, variably-sized cystic spaces that freely communicate in and along the intrahepatic biliary ducts in Caroli's disease, congenital hepatic fibrosis. See Caroli's disease

Lombard voice-reflex test A clinical maneuver used to determine whether a person is feigning–ie, 'faking' unilateral deafness, in which one side is exposed to masking noise; a person alleging hearing loss will ↑ intensity of his/her voice above the masking sound, whereas a person with true hearing loss will not ↑ voice intensity

lomefloxacin Maxaquin® INFECTIOUS DISEASE A broad-spectrum quinolone for treating uncomplicated and complicated UTIs, prophylaxis of bacterial infection in posttransurethral surgery and transrectal prostate biopsy, for acute bacterial exacerbation of chronic bronchitis gram-negative bacilli, staphylococci. See Fluoroquinolone

Lomotrim® Clotrimazole, see there

lomustine CCNU ONCOLOGY An alkylating nitrosurea derivative used to treat lymphomas and other CAs ADVERSE EFFECTS BM suppression

Lonazep Clonazepam, see there

lone atrial fibrillation CARDIOLOGY A Fib in a Pt < age 60, in absence an underlying pathology–eg, DM, CHD, HTN. See Atrial fibrillation

Lone Star tick Amblyomma americanum A 3-host–wild animal, domestic animal, hard tick native to southern US, Central and South America, which is a vector of RMSF and occasionally Lyme disease. See Lyme disease, Rocky Mountain spotted fever

loner PSYCHIATRY A single young man estranged from society and family, who suffers from psychogenic pain, and tends to live 'on the edge', vacillating between aggression and depression; loners often have unrealistic goals, but are unable to work towards those goals PROGNOSIS Guarded; progressive depression, poor functionality, suicidal tendencies

longboarding SPORTS MEDICINE A street sport in which the practitioner is balanced atop a 130 cm–52 inch brakeless skateboard and careens down long asphalt paved roads at speeds of up to 40 kph–25 mph RISKS Road rash, getting squished

long call GRADUATE EDUCATION A call schedule in which the house officer–HO admits Pts throughout a 24-hr period. See Call schedule, Nightfloat

long feedback loop PHYSIOLOGY A self-adjusting circuit in the 'central' endocrine system, where the hypothalamic hormones are schematically represented as an 'axis' consisting of 2 circuits, the short hypothalamic-adenohypophysis–or pituitary loop and the long adenohypophysis–or pituitary-end organ loop. See Loops

long leg syndrome Short leg syndrome A condition caused by inequality of leg length, which causes mechanical disturbances of gait and posture CLINICAL Back and knee pain, due to compensatory pelvic tilt, lumbar scoliosis, backache and rheumatologic symptoms; the longer leg is held in flexion with excess lateral strain, causing premature degenerative changes and valgus deformity due to a collapse of the lateral compartment

long Q-T syndrome Long Q-T interval syndrome CARDIOLOGY A clinical complex, most common in otherwise healthy young ♀, evoked by physical–eg, exercise or mental–eg, fright stress, resulting in episodic syncope or, if stimulus is extreme, sudden death related to ↑ autonomic tone–eg, while exercising on a hot day, accompanied by sudden onset of ventricular arrhythmia; once diagnosed, all blood relatives should have an EKG, as a prolonged Q-T interval is associated with ↑ malignant ventricular arrhythmia–eg, 'torsades de points'. See NKX2-5

long-term care HEALTH INSURANCE A continuum of maintenance, custodial, and health services for the chronically ill or disabled, which may be provided on an inpatient–rehabilitation facility, nursing home, mental hospital, outpatient, or at home

long-term memory Anterograde memory, long-term potentiation, remote memory NEUROLOGY Memory in which information is stored in a permanent or semipermanent fashion. See Memory. Cf Short-term (immediate) memory

longevity The condition of having a long life, or having lived a long life; the average life expectancy of adults continues to spiral upward; the upper limit of average human life expectancy may range from 85 to 100. See Lifespan

longitudinal study Diachronic study STATISTICS A study that follows the same persons over time, evaluating the effects of one or more variables on a processtime EXAMPLES Cohort studies, case-control studies. Cf Cross-sectional study, Horizontal study

look-back program PUBLIC HEALTH A study of persons exposed to a disease, from whom specimens–eg, sera, epidemiologic, demographic or other data is examined retrospectively to determine whether they are currently infected

loop diuretic NEPHROLOGY Any of a family of potent–'high-ceiling'–diuretics that act on Henle's loop, evoking excretion of 20-25% of filtered Na⁺ INDICATIONS Pulmonary edema, CHF, renal failure EXAMPLES Furosemide, ethacrinic acid, bumetanide, demadex

loop extrasurgical excision procedure LEEP, see there

loop-o-gram UROLOGY A radiocontrast study in which a catheter is introduced into an ileal conduit to evaluate the presence of residual/recurrent transitional cell CA; ileal conduits are usually created surgically as a urinary diversion in Pts who have had significant resections of the urinary bladder–eg, for bladder cancer

loosening of associations PSYCHIATRY Disordered thinking in which ideas shift from one subject to another in an oblique or unrelated manner, without the speaker being aware of same; when severe, speech may be incoherent. Cf Flight of ideas

loperamide An antidiarrheal drug

Lopid® Gemfibrozil, see there

Lopressor® Metoprolol, see there

Loprox® Ciclopirox INFECTIOUS DISEASE A topical for tinea pedis, tinea cruris, tinea corporis due to Trichophyton rubrum, T mentagrophytes, or Epidermophyton floccosum, and topical treatment of seborrheic dermatitis of scalp. See Seborrheic dermatitis

Lorabid® Loracarbef, see there

loracarbef Lorabid® THERAPEUTICS A synthetic β-lactam antibiotic of the carbecepham family, most effective against gram-positive cocci–eg, S pneumoniae, H influenzae, Branhamella catarrhalis, and Streptococcus pyogenes

loratidine Claritin®, see there

lorazepam NEUROPHARMACOLOGY A benzodiazepine anxiolytic, antidepressant, sedative, hypnotic

lordosis ORTHOPEDICS Spinal lordosis, swayback The forward curvature of the lumbar spine, causing the normal concavity of the lower lumbar region of the back ETIOLOGY↑ abdominal contents–eg, pregnancy or extreme obesity, poor posture, flexion contracture of hip, rickets, TB of spine.

loricrin DERMATOLOGY An insoluble glycine/serine-rich transglutaminase substrate protein cross-linked to the cornified cell envelope of the upper dermis

LOS Length of stay, see there

losartan Cozaar® PHARMACOLOGY A sartan which blocks angiotensin II–A-II from binding to AT₁, ↓ proteinuria and filtration fraction, inhibits secretion of aldosterone by the adrenal cortex, and ↑ uric acid excretion; in Pts with CHF, losartan ↓ afterload, ↑ cardiac output–effects shared by ACE inhibitors—and prevent left ventricular hypertrophy SIDE EFFECTS Vertigo. See Angiotensin II receptor antagonist, ELITE

losartan-hydrochlorothiazide Hyzaar® An antihypertensive with an angiotensin II receptor antagonist and a diuretic

loss VOX POPULI A diminution, attenuation of a process or activity. See Acute blood loss, Conductive hearing loss, Fractional allelic loss, Hard loss of stability, Hearing loss, Noise-induced hearing loss, Nitrogen loss, Occupational hearing loss, Postlingual loss, Prelingual loss, Specific ionization loss, Stop loss

lossless compression TELEMEDICINE A data compression format that allows both efficient transmission and accurate decompression/expansion at the receiving end. See Data compression, Telemedicine; Cf Codec, Lossy compression, Modem, T-1

loss of appetite MEDTALK Anorexia, see there

loss of smell MEDTALK Anosmia

loss of taste MEDTALK Ageusia

lossy compression TELEMEDICINE A format for data transmission which compresses, transmits and decompresses at the receiving end with a 'tolerable' loss of imaged data. See Data compression, Telemedicine; Cf Codec, Lossless compression, Modem, T-1

lost-opportunity doctrine MEDICAL LIABILITY A legal doctrine which holds that a missed chance to diagnose, and therefore treat, a particular condition can be negligent–ie, wrongful, even if it does not result in death

lot A batch of a manufactured product–eg, chemicals, drugs, reagents, or specimen tubes, produced or packaged from one production run and simultaneously subjected to quality control testing

Lotronex An agent used to manage irritable bowel syndrome pulled from the market 6 months after release as constipation occurred in 25% of Pts, and potentially fatal ischemic colitis. See Irritable bowel syndrome

lottery See Malpractice lottery

Lou Gehrig disease Amyotrophic lateral sclerosis, see there

louse A flat wingless parasitic insect

OF LICE & MEN

BITING LICE, Order Mallophaga, which rarely affect humans

SUCKING LICE, Order Anoplua, family Pediculidae, which are global in distribution, and serve as either

• Disease vectors, eg Borrelia recurrentis–Bhermisi turcatae, B parkeri or

• Themselves cause disease—Pediculus humanis capitis, head lice, Pediculus humanis corporis, body lice, Phthirus pubis, crabs, pubic lice

louse-borne typhus Classic typhus, epidemic, typhus, European typhus, jail fever A severe acute disease with prolonged high fever up to 40°C/104° F, intractable headache and a pink-to-red raised rash, caused by Rickettsia prowazekii PROGNOSIS Mortality ↑ with age; over half of untreated persons > age 50 die

Lovan® Fluoxetine, see there

lovastatin Mevacor® CARDIOLOGY A lipid-lowering agent used to manage hypercholesterolemia and other dyslipidemias including primary mixed and primary moderate hyperlipidemias, diabetic dyslipidemia and hyperlipidemia of nephrotic syndrome in healthy volunteers, ↓ LDL-C by 35%, apoB by 25%, VLDL-C and IDL-C by 30-40%, ↑ HDL-C by 7% LAB ↓ Total cholesterol, ↓ LDL-C, ↓ TGs, ↑ HDL-C. See Statin. Cf Cholesterol-lowering drugs, Gemfibrozil, HMG CoA reductase inhibitors

the 'love drug' see Ecstasy

love handles Popular for the bilateral overhangs of fat and soft tissue on the anterolateral flank common in older ♂

love-hate relationship Ambivalence PSYCHIATRY A clinical complex characterized by Freudian impulses; love-hate is normal for children passing through the 'anal-sadistic' phase of development, in which there is often simultaneous love and 'murderous' hatred toward the same object–person. Cf Passive-aggression

Lovejoy staging NEUROLOGY A system for staging severity of neurologic impairment in Reye syndrome

LOVEJOY'S CLASSIFICATION–COMA STAGES IN REYE SYNDROME

I Vomiting, lethargy, sleepiness

II Disorientation, delirium, combativeness, ↑ ventilation, ↑ reflexes, normal response to noxious stimuli

III Obtunded, comatose, decorticate positioning; intact cranial nerves and oculomotor reflexes, inappropriate response to noxious stimuli

IV Deepening coma, decerebrate rigidity, fixed dilated pupils, caloric stimulation causes dysconjugate eye movement, oculovestibular reflexes lost

V Flaccid paralysis, deep tendon and respiratory reflexes absent, seizures

Lovenox® Enoxaparin CARDIOLOGY An LMW heparin antithrombolytic for managing unstable angina and non-Q-wave MI, and preventing DVT after-hip or knee replacement, and high-risk abdominal surgery. See Deep vein thrombosis

Lovett scale PHYSICAL THERAPY A 6-point–0, no evidence of contractility to 5, normal, unimparied ROM, measure of musculoskeletal strength based contractility and ROM

Lovium Diazepam, see there

low axillary dissection SURGICAL ONCOLOGY A procedure used to treat breast CA, which consists of the en bloc resection of level I axillary lymph nodes, from the latissimus dorsi muscle–lateral, to the lateral border of the pectoralis minor muscle–medial, with clearing of the axillary vein–superior. Cf Full axillary dissection

low back pain Discomfort of the lower lumbar region, in the US, the 2nd most common cause–after the common cold–for seeking medical care; LBP affects ± 31 million (US) at any given time and costs $8 x 10⁹/yr DIAGNOSIS MRI of lumbosacral spine

low birth weight NEONATOLOGY Referring to an infant weight from 1500 g to 2500 g at birth; LBW is a risk factor a sui generis for M&M in early infancy, defined as < 2500 g at birth; moderate LBW–1500-2500 g; very low birth weight infants < 1500 g account for 50% of neonatal deaths SURVIVAL 85-95% > 1250 g; 65-75% > 800 g, 2% if < 600 g. See Appropriate for gestational age, Extremely low birth weight, Intrauterine growth retardation, Prematurity, Small for gestational age

VERY LOW BIRTHWEIGHT CHILDREN–OUTCOMES

Birthweight	≤ 750 g	> X <	≥ 1.5 kg
Sample number	68	65	61
MPC score*	87	93	100
Mental retardation–IQ < 70	21%	8%	2%
Poor cognitive function	22%	9%	2%
Poor academic skills	27%	9%	2%
Poor gross motor function	27%	9%	0%
Poor adaptive function	25%	14%	2%
Cerebral palsy	9%	6%	0%
Severe visual disability	25%	5%	2%
Hearing disability	24%	13%	3%
< Normal Wt/Ht/HS	22/25/35%	11/5/14%	0/0/2%%

*Mental Processing Composite score

low blood VOX POPULI → MEDTALK Anemia, see there

low blood pressure VOX POPULI → MEDTALK Hypotension

low blood sugar VOX POPULI → MEDTALK Hypoglycemia, see there

low-cost physician MEDICAL PRACTICE A popular term for a generalist or primary care physician. Cf High-cost physician

low-density lipoprotein See LDL

low-dose aspirin VASCULAR DISEASE A minimal dose of aspirin administered daily to a person known to be at risk for coronary artery occlusion

low-energy defibrillation CARDIOLOGY A resuscitation format in which a biphasic waveform of low–150 joules–energy is administered in cardiac arrest. See Defibrillation

low-fat diet A diet low in fats, especially saturated fats, which has a positive effect on arthritis, CA, ASHD, DM, HTN, obesity, and strokes. See Diet, Low-fat snack; Cf Animal fat, High-fat diet

low-flow oxygen therapy Administration of low–0.5-2.0 L/min of supplemental O_2 may ↓ severity of hypercapnia in Pts with neuromuscular disease

low-grade B-cell lymphoma of mucosa-associated lymphoid tissue type MALToma, see there

low-grade lymphoma Indolent lymphoma Any relatively indolent lymphoma classified by the Working Formulation, with a prolonged survival of 5 to 7.5 yrs, often with minimal therapy EXAMPLES Small lymphocytic–plasmacytoid lymphoma, follicular small cleaved cell lymphoma, follicular mixed small cleaved & large cell lymphoma. See Lymphoma, REAL classification, Working Formulation

low-grade squamous intraepithelial lesion GYNECOLOGIC CYTOLOGY A lesion of the uterine cervix which is characterized by cells occurring singly or in sheets, nuclear abnormalities in cells with mature cytoplasm, bi- or multinucleation, well-defined optically clear perinuclear halo, distinct cell borders and others that translate into either HPV infection or mild dysplasia–CIN 1 of uterine cervix–a diagnosis made on biopsied tissue. Cf High-grade squamous intraepithelial lesion

low-impact sport SPORTS MEDICINE Any physical sport with minimal wear and trauma to weight-bearing joints, especially of the foot, knee, hip EXAMPLES Bicycling, bowling, golfing, sailing, swimming, scuba diving; LIS participation is encouraged in those engaging in physical activities after hip and knee arthroplasty. Cf High-impact sport, Moderate impact sport, No-impact sport

low-level waste Low-level radioactive waste A specific form of man-made radioactive waste for which there is reasonable assurance that public exposure–should it occur, presents only a fraction of the current dose limits. See Plutonium, Radioactive waste. Cf High-level waste

low magnesium Hypomagnesemia, see there

low-molecular weight heparin Enoxaparin/Lovenox®, dalteparin, fraxiparin PHARMACOLOGY A heparin with advantages over unfractionated heparin, which blocks thrombosis earlier in the coagulation cascade than conventional heparin by inhibiting factor Xa; it less likely to cause thrombocytopenia or thrombotic thrombocytopenia syndrome INDICATIONS DVT. See Deep vein thrombosis

low motion disease Any condition associated with a sedentary lifestyle, ↓ physical activity–eg, arthritis, osteoporosis, weakness. See Egoscue method, Motion starvation

low muscle tone VOX POPULI → MEDTALK Hypotonia

low-osmolality contrast medium RADIOLOGY Any of the radiologic contrast media first used in Europe which may have fewer side effects after IV injections; LOCMs are more expensive and may cause hypercoagulability

low-output gastrointestinal fistula An external–ie, communicates with skin–GI fistula that produces < 200 ml of fluid, originating from the distal small intestine and large intestine. Cf High-output fistula

low-output heart failure Forward failure syndrome, low cardiac output syndrome A clinical condition in which the cardiac output falls below tissue needs for O_2 LAB ↑ Vascular resistance and O_2 consumption, lactic acidosis, ↓ cardiac index, O_2 saturation TREATMENT Digitalis, vasopressors, dopamine, dobutamine, vasodilatation PROGNOSIS Poor if unresponsive to drugs

low phosphate Hypophosphatemia, see there

low potassium VOX POPULI Hypokalemia; increasingly, hypopotassemia

low-protein diet CLINICAL NUTRITION A diet that provides < 1.5 g/kg/day of protein during growth periods, or less in adults; adults in renal failure should receive no < 0.6 g/kg/day of protein, to avoid a negative nitrogen balance; LPDs are indicated for Pts with renal failure, as reduction of protein ↓ anorexia, N&V, and if begun early, may slow disease progression. See Diet

low-quality protein CLINICAL NUTRITION A protein usually of plant origin that lacks one or more essential amino acids–eg, corn–low in lysine, or beans–low in tryptophan. See Liquid diet; Cf Succotash

low sugar VOX POPULI → MEDTALK Hypoglycemia

low T_3 syndrome Euthyroid sick syndrome, see there

low T_4 syndrome A 'laboratory' disease seen in Pts undergoing regular hemodialysis or continuous ambulatory peritoneal dialysis, in which 40% have serum T_4–thyroxine < than 5 µg/dl but who have normal free T_4 levels

low-yield procedure A diagnostic procedure that rarely results in a definitive diagnosis; Cf High-yield procedure

Lowe syndrome Oculocerebrorenal syndrome, see there

lower cervical spine injury ORTHOPEDICS Any major injury affecting the lower cervical spine. See Bilateral locked facet, compression fracture, spinous process fracture, teardrop fracture, traumatic disk ruptur, eunilateral locked facet

lower esophageal sphincter A 3-5 cm in length zone of ↑ pressure at the junction of the distal esophagus with the gastric cardia, located at the hiatus, which forms a physical barrier in preventing gastric reflux; substances that ↑ the LES tone include gastrin, acetylcholine, serotonin, $PGF_{2\alpha}$, motilin, substance P, histamine, pancreatic polypeptide

lower GI series Barium enema, see there

lower rate CARDIOLOGY A programmed default rate at which atrial tracking dual-chamber–DDD and VDD–pacemakers pace the heart in absence of intrinsic heart beats. See Pacemaker

lowest-observed-adverse-effect level TOXICOLOGY The lowest concentration of a chemical in a study, or group of studies, that produces statistically or biologically significant ↑ in frequency or severity of adverse effects between the exposed population and an appropriate control. See Ames test. Cf No-observed-adverse-effect level

Lown-Ganong-Levine syndrome CARDIOLOGY A cardiopathy defined by EKG–atrial tachycardia, short PR and normal QRS

loxapine Loxitane® A tricyclic antidepressant

Loxitane® Loxapine, see there

Loxosceles reclusa See Brown spider

lozenge Troche DRUG DELIVERY A sweetened disk-shaped pill containing a topical therapeutic–eg, an astringent, antiseptic, or analgesic which is dissolved in the mouth for optimal efficacy

Lozepam Lorazepam, see there

LPS Lipopolysaccharide

LQT1, **LQT2**, **LQT3** MOLECULAR MEDICINE Three genes, located on chromosomes 11, 7, and 3 respectively, that have been linked to the long QT syndrome. See Long QT syndrome

LQTS Long QT interval syndrome, see there

LRTI Lower respiratory tract infection

LS & A Lichen sclerosis et atrophicus. See Lichen sclerosis

L/S ratio OBSTETRICS The ratio of lecithin–phosphatidyl choline to sphingomyelin, a 'bench' parameter used to determine infant lung maturity and predict the infant's ability to survive without developing respiratory distress. See Biophysical profile, Lung profile, Respiratory distress syndrome

LSD D-Lysergic acid diethylamide A synthetic indole amine with hallucinogenic activity derived from ergot alkaloids, which produces mood elevations, sensory distortion, panic attacks, flashbacks ROUTE Oral DOSE LSD is 3000– to 5000-fold more potent that mescaline; in adults, 100-150 µg is enough for a 'trip' CLINICAL-PSYCHOMIMETIC EFFECTS Spatial and temporal distortion–hallucinogenic, illusions, animation, hyperacusis and background amplification, distortion of body image, sensory hallucinations with hearing of smells and sights, smelling of images and sounds, seeing smells and sounds, etc SYMPATHETIC & PARASYMPATHETIC EFFECTS Dilated pupils, ↑ heart rate, ↑ temperature, ↑ salivation, ↑ lacrimation, ↑ sweating, N&V, loss of appetite, sleeplessness, dry mouth, tremors DIAGNOSIS Physical exam, laboratory methods: GC-MS–quantitative, HPLC, RIA–qualitative, TLC ADVERSE EFFECTS Bad trips with fear of insanity, depersonalization, panic attacks, flashbacks, which may occur 5-10 x/day, up to 18 months after last use of LSD. See Designer drugs, Hallucinogens, 'Ice'

LSF Lisofylline, see there

LSIL Low-grade squamous intraepithelial lesion, see there

LT 1. Laboratory test 2. Lead time, see there 3. Leukotriene, see there 4. Levothyroxine 5. Locum tenens, see there 6. Lymphotoxin

LTD 1. Laron-type dwarfism 2. Leukotriene D 3. Long-term depression, see there 4. Long-term disability

LTK™ system OPHTHALMOLOGY A holmium laser-based system for non-invasive thermal keratoplasty–reshaping of the cornea based on the collagen-shrinking Sand process. See Keratoplasty

LTS 1. Latent tetany syndrome, see there 2. Low-threshold spike–neurology

lubricant MEDTALK An oily or slippery substance to reduce friction SEXOLOGY An endogenous–eg, vaginal secretions, or exogenous–eg, K-Y jelly, substance that facilitates coitus

Lucey-Driscoll Syndrome Transient familial neonatal hyperbilirubinemia NEONATOLOGY A potentially fatal AR disorder characterized by severe hyperbilirubinemia present at birth, which accumulates in the brain. Cf Gilbert syndrome

lucid interval NEUROLOGY A period preceding the loss of consciousness and coma in Pts with subdural and epidural hematomas and intracranial edema. Cf Window period

Ludiomil® Maprotiline, see there

Ludwig's angina Cellulitis of neck, neck abscess, neck infection ENT Severe cellulitis of the neck–submaxillary, sublingual and subspaces due to infection of the oral cavity CLINICAL Dysphagia, glottal edema, fever, tachypnea, ↑ WBCs

LUE Left upper extremity

lues Syphilis, see there

LUF syndrome Luteinized unruptured follicle syndrome A condition characterized by the development of a dominant follicle without disruption and release of the ovum, an abnormality diagnosed by ultrasonography or laparoscopy; LUD is a rare sporadic cause of infertility

Lufyllin® Dyphylline, see there

Lugol solution Acetic acid solution GYNECOLOGY A solution of iodine and potassium iodide in acetic acid, which colors the normal uterine cervix a homogeneous brown, seen by colposcopy. See Acetowhite lesion, Schiller test

lumbar puncture Spinal tap NEUROLOGY A diagnostic procedure in which a very long needle is inserted into the subarachnoid space between the 3rd and 4th lumbar vertebrae in order to obtain CSF; an LP is used to measure intracranial pressure, which may be ↑ 2° to hemorrhage, tumors, or edema, measure CSF chemistries–eg, glucose, proteins, diagnose inflammation of the CNS, especially infections–eg, meningitis, and stroke, spinal cord tumors and metastases to the CNS, or inject a dye into the spine before myelography COMPLICATIONS Uncommon; meningitis, bleeding into spinal canal; if intracranial pressure is ↑, removal of CSF from spinal canal may cause fatal herniation of cerebellar tonsils. See Cerebrospinal fluid

Luminal® Phenobarbital, see there

lumone A hormone of the GI lumen. See Incretin

lump in rectum VOX POPULI Hemorrhoid

lump in breast VOX POPULI → MEDTALK 1. Cancer 2. Fibroadenoma

lumpectomy Local wide excision, segmental mastectomy, segmental resection, tumorectomy, tylectomy, wedge excision, wide excision SURGICAL ONCOLOGY A partial mastectomy for early breast CA, in which the tumor mass and a margin of grossly uninvolved tissue is excised. Cf Excisional biopsy, Lesionectomy, Radical mastectomy

lumping Reductionism CLINICAL DECISION-MAKING The practice of aggregating diseases or pathologic nosologies with variably distinct features under a common term. Cf 'Splitting'

lumpy jaw INFECTIOUS DISEASE A jaw characterized by painful, 'wood-hard' fibrotic induration of the parotid and submandibular regions, arising in a background of dental disease–caries, periodontitis, or extractions, due to cervicofacial actinomyces, the most common form of actinomyces infection; LJ in humans is caused by *A israeli* and others; other findings in cervicofacial actinomycosis include trismus, multiple draining sinus tracts bearing the classic yellow-white sulfur granules, fever, leukocytosis, extension to facial soft tissue, bone and, if untreated, the CNS TREATMENT Penicillin. See Actinomycosis

LUNA Laparoscopic uterine nerve ablation GYNECOLOGY A procedure for treating pelvic pain consisting in disrupting the uterine nerves. See Pelvic inflammatory disease

Lunelle® GYNECOLOGY A contraceptive injected IM once monthly containing estradiol cypionate and medroxyprogesterone acetate. See Contraception

lung cancer ONCOLOGY An epithelial malignancy of the lungs, which accounts for 30% of all cancer deaths in the US, most of which are directly attributable to cigarette use CLINICAL Persistent cough, hemoptysis, chest pain, weight loss, nonresponsive pneumonia TYPES Non-small cell cancer–eg, squamous cell carcinoma, adenocarcinoma, bronchoalveolar; small cell (undifferentiated) carcinoma EPIDEMIOLOGY US, ±170,000 new cases, 1999; ±150,000 die/yr due to LC SITES OF METASTASIS Brain, bone, liver, adrenal glands ETIOLOGY Most lung CA–83% is directly linked to cigarette smoking; the risk is higher with ↑ number of cigarettes smoked/day and earlier age of smoking; up to 3,000 lung CAs are attributable to second-hand smoke; other factors include high levels of pollution, radiation, and asbestos; cooks and chemists have an ↑ risk DIAGNOSIS Cytology of sputum, biopsy by bronchoscopy, needle localization, or surgical excision THERAPY Surgery for *resectable* non-small cell CA, based on the size of the primary tumor and extent of lymph node involvement and metastases; chemotherapy and/or RT for small cell CA PROGNOSIS Once diagnosed, average Pt survives 1–2 yrs; 5-10% survive 5 yrs after diagnosis; Pts with small cell–undifferentiated carcinoma have a slightly better prognosis than those with squamous cell and bronchoalveolar carcinomas, *if it responds to chemotherapy*

LUNG CARCINOMA, CLASSIFICATION

% TOTAL & TYPE	5-YEAR SURVIVAL BY STAGE		
	I	II	III
38% SQUAMOUS CELL–EPIDERMOID	38%	16%	9%
23% ADENOCARCINOMA	32%	7%	3%
Papillary adenocarcinoma			
Alveolar cell			
Bronchiolar carcinoma			
Mucinous adenocarcinoma			
Adenosquamous carcinoma			
29% SMALL CELL/UNDIFFERENTIATED	0%	0%	0%
Oat cell carcinoma			
Intermediate cell type			
Combined oat cell type			
9% UNDIFFERENTIATED LARGE CELL	30%	6%	5%
Rare GIANT CELL CARCINOMA	Rapidly fatal		

lung collapse Pneumothorax, see there. See Atelectasis

lung compliance See Compliance

lung disease Pulmonary disease PULMONOLOGY Any condition causing or indicating impaired lung function TYPES OF LD Obstructive lung disease–↓ in air flow caused by a narrowing or blockage of airways–eg, asthma, emphysema, chronic bronchitis; restrictive lung disease–↓ amount of air inhaled, due to ↓ in elasticity or amount of lung tissue; infections. See Asbestosis, Aspergillosis, Atelectasis, Eosinophilic pneumonia, Lung cancer, Metastatic lung cancer, Necrotizing pneumonia, Pleural effusion, Pneumoconiosis, *Pneumocystis carinii* pneumonia, Pneumonia, Pneumothorax, Pulmonary actinomycosis, Pulmonary alveolar proteinosis, Pulmonary anthrax, Pulmonary arteriovenous malformation, Pulmonary edema, Pulmonary embolus, Pulmonary hypertension, Pulmonary venoocclusive disease, Rheumatoid lung disease, Tuberculosis

lung metastases ONCOLOGY Cancer that has spread from the original–primary tumor to the lung; primary lung cancer spreads to the brain, bone, BM, liver. See Metastasis

lung-reduction surgery Lung volume reduction surgery SURGERY The excision of poorly functioning peripheral areas of emphysematous lung to improve functional air-exchange and function by ↑ airway conductance, conductance to lung volume ratio, elastic recoil; LRS may be used in Pts with COPD to alleviate SOB and improve exercise tolerance; the best indicator of clinical response to LVRS is lung resistance; Pts with a marked ↑ in inspiratory resistance have predominantly intrinsic airway disease and don't benefit from LVRS. See COPD, Elastic recoil, Emphysema, Transitional-dyspnea index

lung squeeze SPORTS MEDICINE An unsual form of barotrauma which may occur in deep–>100 fsw (feet seawater) breathhold diving CLINICAL SOB, dyspnea, hemoptysis after surfacing from a deep dive, hypoxia IMAGING Pulmonary edema MANAGEMENT O_2, respiratory support

lung transplant SURGERY Transplant of a lung allograft into a Pt with failing lungs; 90 US centers perform LT; 35 centers perform ≥ 10/yr MEAN WAIT TIME 18 months INDICATIONS COPD–eg, emphysema due to α_1-AT deficiency, cystic fibrosis–CF, bronchiectasis, emphysema, Eisenmenger syndrome, end-stage heart disease IMMUNOSUPPRESSION Cyclosporine, tacrolimus, azathioprine, mycophenolate mofetil, prednisone PROGNOSIS 1/3/5 yrs–71%/55%/43% Cost $240,000, $47,000 annual maintenance. See Cyclosporin, Domino-donor transplantation, Tacrolimus, UNOS

lung volumes PHYSIOLOGY A group of air 'compartments' into which the lung may be functionally divided

LUNG VOLUMES

EXPIRATORY RESERVE CAPACITY–ERV The maximum volume of air that can be voluntarily exhaled

FUNCTIONAL RESIDUAL CAPACITY (FRV) Volume left in the lungs at the end of a normal breath which is not normally part of the subdivisions

INSPIRATORY CAPACITY–IC The maximum volume that can be inhaled

INSPIRATORY RESERVE CAPACITY–IRC The maximum volume that can be inhaled above the tidal volume

TIDAL VOLUME–V_T The normal to-and-fro respiratory exchange of 500 cc; vital capacity is the maximum amount of exhalable air; after a full inspiration, which added to the residual volume, is the total lung capacity

TOTAL LUNG CAPACITY–TLC The entire volume of the lung, circa 5 liters

VITAL CAPACITY–VC The maximum volume that can be inhaled and exhaled

lung water VOX POPULI Pulmonary edema, see there

lupoid *adjective* Relating to a lupus erythematosus-like condition

lupoid hepatitis Autoimmune chronic active hepatitis, autoimmune hepatitis HEPATOLOGY An autoimmune hepatitis common in young ♀, who often produce anti-nuclear, anti-smooth muscle and antimitochondrial antibodies, termed 'lupoid' for the presence of an LE cell phenomenon, which occurs in merely 15%; LH is characterized by chronic hepatitis, a low risk of CA and response to corticosteroids. See Systemic lupus erythematosus

Lupron® ONCOLOGY A depot leuprolide, a GnRH–gonadotropin releasing hormone, for palliating advanced prostate CA. See Prostate CA

lupus anticoagulant Lupus inhibitor LAB MEDICINE Any IgG or IgM class antibody that arise spontaneously in Pts with SLE; while LAs produce in vitro interference with phospholipid-dependent coagulation–eg, activated partial thromboplastin time–aPTT and kaolin clotting time assays in specimens from Pts with LE, they *do not* produce in vivo coagulopathy without other platelet or coagulation defects or drug-induced antibodies; LAs also occur in Pts with HIV, DVT, and other conditions LAB ↑ aPTT; LAs interfere with derivative assays for factors VIII, IX, XI, and XII. See Systemic lupus erythematosus. Cf Anticardiolipin antibodies

lupus carditis CARDIOLOGY A heart affected by SLE; lupus endocarditis occurs in ±15% of Pts, is due to valve disease and causes hemodynamic alterations in the form of regurgitation; lupus myocardiopathy occurs in 10% of SLE and is manifest by arrhythmias and CHF, linked to HTN. See Systemic lupus erythematosus

lupus erythematosus Disseminated lupus erythematosus, lupus, systemic lupus erythematosus RHEUMATOLOGY An idiopathic multisystem collagen vascular (connective tissue) disorder EPIDEMIOLOGY Affects ±40/10⁵–North America, Europe, blacks/Hispanics > whites, ♀:♂ ratio = 3:1; 80% onset during childbearing yrs CLINICAL Vasculitis, serositis, synovitis, cerebral, renal, skin involvement CAUSE OF DEATH Sepsis 40%, CVS 20%, CNS 10%, 10% nephritis with possible renal failure, others 20%. See ANA-negative systemic lupus erythematosus, Anticardiolipin antibody, Antinuclear antibody, Butterfly rash

SYSTEMIC LUPUS ERYTHEMATOSUS–1982 REVISED CRITERIA

1. MALAR RASH Fixed erythema, in particular over the malar eminences

2. DISCOID RASH Raised erythematous patches with adherent hyperkeratotic scaling; atrophic scarring in some old lesions

3. PHOTOSENSITIVITY Unusual skin rashes in response to sunlight

4. ORAL ULCERS Oral or nasopharyngeal ulcers

5. ARTHRITIS Nonerosive arthritis of 2+ joints, accompanied by tenderness, swelling, or effusions

6. SEROSITIS
  a. Pleuritis OR
  b. Pericarditis

7. RENAL DISEASE
  a. Persistent proteinuria > 0.5 g/day OR
  b. Cellular casts

8. NEUROLOGIC DISORDER
  a. Seizures without substance use or medical disease OR
  b. Psychosis in absence of substance use or medical disease

9. HEMATOLOGIC DISORDER
  a. Hemolytic anemia OR
  b. Leukopenia OR
  c. Lymphocytopenia
  d. Thrombocytopenia

10. IMMUNOLOGIC DISORDER
  a. Positive LE cell prep–this test was abandoned in the mid-1990s
  b. Anti-DNA antibody
  c. Anti-Sm antibody
  d. False positive serological test for syphilis

11 ANTINUCLEAR ANTIBODY Abnormal titers of ANA in absence of drugs known to be associated with drug-induced lupus erythematosus

lupus nephritis Lupus glomerular disease, lupus nephropathy RHEUMATOLOGY Any nephropathy seen in systemic lupus erythematosus PATHOGENESIS Classes II-V are attributed to deposition of DNA-anti-DNA immune complexes containing Igs, complement, cryoglobulins MANAGEMENT Prednisone, cyclophosphamide

LUPUS NEPHRITIS–WHO CLASSIFICATION

CLASS I Normal–rarely recognized

CLASS II Mesangial lupus GN–10% of Pts Minimal clinical disease, mild hematuria, mild proteinuria LM Granular mesangial deposition of Ig and complement

CLASS III Focal proliferative glomerulonephritis–± V_3 of Pts Moderate clinical manifestations, recurrent hematuria, moderate proteinuria, possible progression to renal failure LM Focal swelling and proliferation of endothelial and mesangial cells, neutrophil infiltration, fibrinoid debris

CLASS IV Diffuse proliferative GN–45-50% Overtly symptomatic, microscopic or gross hematuria, proteinuria ± nephrotic syndrome, ± hypertension, often ↓ GFR LM Global glomerular involvement, proliferation of endothelial and mesangial, and sometimes epithelial cells

CLASS V Membranous GN–10% Overtly symptomatic, microscopic or gross hematuria, severe proteinuria with nephrotic syndrome, HTN, ↓ GFR LM Thickening of capillary walls similar to idiopathic membranous GN

GFR Glomerular filtration rate GN Glomerulonephritis LM Light microscopy WHO classification

Note: Plasmapheresis does not improve the clinical outcome in LN NEJM 1992; 326:1373aa

lupus profile LAB MEDICINE A battery of serum assays used to diagnose SLE or evaluate an SLE Pt's clinical status MARKERS Anti-double-stranded DNA antibody, ANAs, rheumatoid factor, anti-cardiolipin antibody–IgG, IgM, lupus anticoagulant. See Systemic lupus erythematosus

lupus psychosis NEUROLOGY A condition seen in 12% to 52% of Pts with SLE, broadly defined to include delusional and severe affective disease CLINICAL Frank psychosis, atypical psychosis, schizophreniform disorder, and severe affective disorders–eg, major depression and anxiety or panic disorders; some authors argue that major depression, anxiety, or panic may be caused by corticosteroids or may simply represent an appropriate reaction to the illness rather than genuine LP; LP, if it exists, is attributed to an antibody to P protein, a polypeptide in ribosomal phosphoproteins. See Systemic lupus erythematosus

lurking variable CLINICAL TRIALS A variable in a study population that is undetected/unsuspected and may alter interpretation of causal relationships. See Observational study

lust murder Erotophonophilia SEXOLOGY A paraphilia in which sexuoeroticism hinges on staging and murdering an unsuspecting sexual partner; the erotophonophile's orgasm coincides with the expiration of the partner. Cf Autassassinophilia

Lustra® Hydroquinone DERMATOLOGY A topical agent used to manage UV-induced skin discoloration and hyperpigmentation due to trauma, pregnancy, OCs, HRT. See Tanning

Lustral® Sertraline, see there

luteal phase defect GYNECOLOGY A deficiency–seen in 3 to 5% of infertile women and cause of ± 1/3 of recurrent early spontaneous abortions–in the amount of progesterone produced—or the length of time produced, which translates into an inability of the endometrium to support pregnancy. See Corpus luteum

luteinizing hormone Interstitial cell stimulating hormone A glyocoprotein hormone/gonadotropin secreted by the anterior pituitary which, in ♀, is secreted cyclically, as is FSH; the cyclic release of LH–and FSH–is responsible for ovulation and transforms the ovarian/graafian follicle into a corpus luteum, which produces progesterone and estrogens; episodic LH release is linked to milk production; in ♂, continuous LH release stimulates testosterone release from the testes; LH and FSH act in concert to stimulate and maintain spermatogenesis; LH and FSH are measured for infertility in ♀, and testicular dysfunction in ♂. Cf FSH

luteinizing hormone-releasing hormone agonist LH-RH agonist ENDOCRINOLOGY Any substance that simulates LH-RH which, with time, result in a ↓ sex hormone secretion. See H-RH

luteoma GYNECOLOGY A non-neoplastic yellow-gray mass seen in pregnancy which measures up to 20 cm in diameter; luteomas represent hyperplasia of luteinized cells, and involute spontaneously postpartum

lutetium texaphrin ONCOLOGY A photosensitizing anticancer agent used in photodynamic therapy which, when activated by light, kills cancer cells. See Photodynamic therapy

Luvox® Fluvoxamine maleate PHARMACOLOGY An antidepressant used for obsessive compulsive disorder. See Antidepressant

luxation ORTHOPEDICS The complete dislocation of a joint. Cf Atlantoaxial subluxation, Subluxation

luxury perfusion syndrome NEUROLOGY A cerebrovascular state in which there is ↑ blood flow to the brain but ↓ O_2 uptake by cerebral tissue, resulting from acute lactic–metabolic acidosis accompanied by cerebral edema; LP is nonspecific and may occur in strokes, trauma, tumors, alcoholism, sickle cell anemia, DKA, meningoencephalitis. Cf Carotid steal syndrome, Robin Hood syndrome

luxury room HOTEL SERVICES An expensive hospital room for Pts wishing amenities not found in 'economy-class' rooms, who also receive preferential treatment by nursing staff. See Hotel services

LVEF Left ventricular ejection fraction. See Ejection fraction

LVH Left ventricular hypertrophy, see there

LWBS Left Without Being Seen MEDTALK *adjective* Referring to a Pt who left a healthcare facility without examination or treatment. See Against medical advice

LWSD Left without see a doctor Referring to an impatient would-be inpatient who leaves the ER without physician contact. See Against medical advice

lye injury TOXICOLOGY A condition often caused by ingestion of bleach, suicidal in adults; accidental in children; 5% develop SCC 20-40 yrs after LI CLINICAL, ACUTE Tachycardia, retrosternal pain, copious frothy mucus, hematemesis, sloughing of esophageal mucosa, dysphagia followed by fibrosing stricture IMAGING Barium 'swallows' demonstrate pencil-thin esophageal lumen MANAGEMENT Bougienage is a therapeutic mainstay; given the potential for perforation, it is best begun ≥ 5 days post-insult; some prefer large-bore nasogastric tubes to bougienage, as the strictures formed are of a large and useful diameter; pharmacologic doses of corticosteroids may minimize fibrosis

Lyell syndrome Erythema multiforme, see there

Lyme arthritis CLINICAL IMMUNOLOGY An antibiotic-resistant disorder affecting ±10% of Pts with Lyme disease, which typically affects one knee for months to yrs. See Lyme disease

Lyme disease Lyme borreliosis, Primary lyme disease INFECTIOUS DISEASE An infection by *Borrelia burgdorferi*, possibly mediated by IL-1 EPIDEMIOLOGY 8000 cases were reported in 1993–US, making it the most common zoonosis in the US; *B burgdorferi* has been identified in Northern Europe, Australia VECTORS Deer tick–*Ixodes dammini*, Eastern USA, up to 60% of which carry the spirochete, white-footed mouse tick–*I pacificus*, Western US, ±1% carry the organism, wood tick–*I ricinus*, Europe, Lone Star tick–*Amblyomma americanum*, and rarely deerflies and horseflies HOST Deer mice, field mice LAB Nonspecific findings include ↑ ESR, IgM cryoglobulins, ↓ C3 and C4, ↑ IgG and IgM antibody titers to *B burgdorferi*; definitive diagnosis requires identification of IgG antibodies to *B recurrentis* by the 'Western'–immunoblot PROGNOSIS 60% of untreated subjects develop recurring arthritis–chronic Lyme arthritis lasting up to yrs after infection SEROLOGY may be positive in Pts who are also infected with *Ehrlichia* spp, which may be due to a co-infection with the same tick bite; PCR for human granulocytic ehrlichiosis is required to confirm the latter infection TREATMENT 1 month of doxycycline or amoxicillin or 2 wks of IV ceftriaxone or penicillin VACCINE OspA vaccine. See Chronic Lyme disease

LYME DISEASE STAGES

STAGE I Erythema chronicum migrans Rash stage, associated with wood tick bites and confined to Northern Europe until 1970 when the first US cases were described, presenting as a solitary reddish papule and plaque with centrifugal expansion–up to 20 cm, peripheral induration and central clearing, persisting for months to years; potentially pruritic with IgM and C3 deposition in vessels

STAGE II Cardiovascular–myocarditis, pericarditis, transient atrioventricular block, ventricular dysfunction; neurologic–Bell's palsy, meningoencephalitis, optic atrophy, polyneuritis symptoms

STAGE III Migratory polyarthritis Lyme disease may be accompanied by headache, stiff neck, fever, and malaise that is subsequently manifest as migratory polyarthritis, intermittent oligoarthritis, chronic arthritis of the knees, chronic meningoencephalitis, cranial or peripheral neuropathy, migratory musculoskeletal pains, cardiac abnormalities

Lyme embryopathy A complex of congenital malformations described in infants born to ♀ who had Lyme disease while pregnant CLINICAL Syndactyly, cortical blindness, intrauterine fetal death, prematurity and neonatal rash

Lyme encephalopathy See Neuroborreliosis

Lyme meningoencephalitis NEUROLOGY Some workers believe the neuropsychological changes may be less severe than previously reported PSYCHIATRY Anxiety, insomnia, panic attacks, hallucinations, déjà vu, obsessions, anorexia. See Neuroborreliosis

LYMErix™ PUBLIC HEALTH A Lyme vaccine comprised of lipidated recombinant OspA–outer surface protein A of *Borrelia burgdorferi*.

lymph follicle hypertrophy Lymphadenopathy, see there

lymph gland infection/inflammation Lymphadenitis, see there

lymph node necrosis A nonspecific finding that can be divided into 1. Focal necrosis, usually benign, seen in infection by bacteria, cat-scratch disease, EBV, fungemia, LGV, toxoplasmosis, TB, tularemia, trauma, vascular compromise, post-vaccination lymphadenitis or autoimmunity–eg, SLE, mucocutaneous lymph node syndrome or Kawasaki's disease, necrotizing lymphadenitis or Kikuchi's disease and 2. Global necrosis, 80% of which is associated with lymphoma

lymphadenectomy Lymph node dissection SURGERY A procedure in which the lymph nodes are removed and evaluated for cancer. See Axillary tail

lymphadenitis Lymph gland infection Inflammation of lymph nodes

lymphadenopathy Enlarged lymph nodes, follicular hypertrophy, swollen lymph glands HEMATOLOGY Enlargement of lymph nodes of any etiology; the differential diagnostic considerations are multiple and divided into reactive patterns; benign lymphadenopathy is characterized by 1. Variability of the follicle–germinal center size; 2. Lack of capsular or fat invasion; 3. Mitotic activity confined to the germinal center; 4. Cortical localization and inhomogeneous distribution of the follicles. See Angiography lymphadenopathy, Benign lymphadenopathy, Dermatopathic lymphadenopathy, Phenytoin lymphadenopathy, Shotty lymphadenopathy

lymphangiectasis Dilatation of multiple lymph vessels. Cf Telangiectasia

lymphangiography IMAGING A radiologic exam of the lymphatic circulation by injecting radiocontrast or dye at the feet, to examine the legs, inguinal and iliac regions and retroperitoneum as high as the thoracic duct–or the hands, to visualize axillary and supraclavicular lymphatic circulation and lymph nodes

lymphangioma A benign, often multiloculated lesion characterized by a localized proliferation of dilated lymphatics lined by benign endothelial cells CLINICAL May cause acute abdominal cysts, laryngeal masses, intraabdominal masses MANAGEMENT Resection PROGNOSIS Excellent

lymphangitis Inflammation of the lymphatic vessels. See Sclerosing lymphangitis

lymphatic pump Thoracic pump, rib raising OSTEOPATHY An osteopathic technique in which the rib cage and thoracic spine are manipulated with the intent of improving the lymphatic circulation of the chest. See Manual lymphatic drainage massage, Massage therapy, Osteopathy

lymphedema Lymphatic obstruction HEMATOLOGY A condition in which interstitial fluid–lymph accumulates–eg, in the arm, legs, scrotum–due to interference with lymphatic drainage–eg, obstruction of lymphatics, lymph node disease or surgical removal of lymph nodes in the face of cancer, with expansion of the interstitial fluid compartment

LYMPHEDEMA, ETIOLOGY

CONGENITAL, eg Milroy's disease, due to underdeveloped lymphatic channels

ACQUIRED, eg through microfilarial infection, malignancy or by modalities for treating cancer, eg lymphadenectomy or regional lymphoid irradiation

IDIOPATHIC–affects young ♀ with unremitting swelling in one+ extremities

lymphoblastic lymphoma ONCOLOGY A lymphoma composed of diffuse monomorphous sheets of large cells, which is most common in children and adolescents; it arises in the mediastinum and, untreated, is highly aggressive, causing death within a yr of presentation; it derives from thymic T-cells and develops in the anterior mediastinum, neck or tonsils; ♂ : ♀, 2:1;

it spreads rapidly to other lymph nodes, BM, brain, lungs, heart. See Lymphoma, WHO classification

lymphocytic choriomeningitis An aseptic meningitis of low morbidity caused by an arenavirus, transmitted through rodent excreta, affecting adults in winter, when rodents move indoors CLINICAL Biphasic fever with flu Sx, followed by meningitis with fever, headache, lymphocytosis in CSF, often with leukopenia and thrombocytopenia–which may be a hyperimmune response DIFFDX Infectious mononucleosis, enterovirus, HZV **R**

lymphocytic interstitial pneumonia A diffuse pulmonary disease of insidious onset that is most common in middle-aged ♀, which may be accompanied by Sjögren's disease, hyper- or hypo-gammaglobulinemia CLINICAL Progressive SOB, cough IMAGING Reticulonodular infiltrates on a plain CXR, often accompanied by Kerley 'B' lines. See Interstial pneumonia

lymphocytic lymphoma Non-Hodgkin's lymphoma, see there

lymphocytic thyroiditis Atypical subacute thyroiditis, hyperthyroiditis, lymphocytic thyroiditis with spontaneously resolving hyperthyroidism, painless thyroiditis, slient thyroiditis, subacute lymphocytic thyroiditis ENDOCRINOLOGY A condition in which the thyroid is infiltrated with lymphocytes CLINICAL Hyperthyroidism for ≤ 3 months or less RISK FACTORS ♀, > age 30, current pregnancy. See Thyroiditis

lymphocytosis Lymphocytic leukocytosis HEMATOLOGY An absolute lymphocyte count of > 4000/mm³ in adults and > 8,000/mm³ in children

lymphocytotoxicity assay A complement-mediated assay commonly used in HLA typing laboratories, which tests for cytotoxic antibodies in the serum of a potential recipient that may react with the lymphocytes of a potential donor. See Human leukocyte antigen, Mixed lymphocyte culture

lymphoepithelioma Undifferentiated nasopharyngeal carcinoma A CA with a marked lymphocytic component TYPES–SCHMINCKE TYPE Diffuse mixing of epithelial and lymphoid elements, making the lesion difficult to distinguish from lymphoma, since the epithelial cells may mimic lymphoblasts or the lacunar cell variants of Reed-Sternberg cells REGAUD TYPE Cohesive carcinoma cell aggregates that are large enough to make the epithelial nature of the tumor obvious DIAGNOSIS Immunoperoxidase stains for cytokeratin

lymphogranuloma inguinale Lymphogranuloma venereum, see there

lymphogranuloma venereum LGV, lymphogranuloma inguinale, lymphopathia venereum An STD caused by one of 3 immunotypes–L₁, L₂ and L₃ of *Chlamydia trachomatis*; LGV is uncommon in USA, regionally endemic in Asia, Africa and South America CLINICAL Papuloulcer that heals at the inoculation site, then matted and painful peri-regional–inguinal and perirectal–lymphadenopathy, described as 'kissing' lesions with a 'groove' sign, sloughing of skin, purulent drainage, hemorrhagic proctocolitis, malaise, fever, headache, aseptic meningitis, anorexia, myalgia, arthralgia, hepatitis, conjunctivitis, erythema nodosum LAB Antibody assays–eg, immunofluorescence, counterimmunoelectrophoresis, complement fixation titers > 1:32 TREATMENT Tetracycline, excision LATE COMPLICATIONS Urethral, rectal strictures, lymphedema, rectovaginal fistulas. See Bubo

lymphoid *adjective* Referring to lymphocytes or related tissue(s)

lymphoid granular complex Lymphogranular complex IMMUNOLOGY A microbursa of the large intestine and local recipient site for antigens destined for future immune recognition; LGCs are an integral part of GI-associated lymphoid tissue and can be seen by LM. See GALT, Inflammatory bowel disease

lymphokine-activated killer cells LAK cells, see there

lymphoma ONCOLOGY A malignant neoplasm of B or T lymphocytes, arising from a monoclonal, ie derived from a single progenitor cell, proliferation of lymphocytes; the proliferative process is considered lymphomatous in the appropriate clinical setting, given that not all monoclonal expansions are

malignant CLINICAL Painless swelling of lymph nodes in neck, underarm, groin PROGNOSIS Favorable in follicular lymphomas, especially cleaved, mixed and large non-cleaved cell types; some diffuse lymphomas–eg, small lymphocytic, cleaved cell, Burkitt's, non-cleaved cell and convoluted cell types, have OK-ish prognoses; Ls with poor prognosis include diffuse plasmacytoid lymphocye, mixed cell, mixed small noncleaved cell and large noncleaved cell types; advanced age, anemia and high mitotic activity are associated with a poor prognosis. See Angiotropic lymphoma, B-cell lymphoma, Biclonal lymphoma, Burkitt's lymphoma, Composite lymphoma, Cutaneous cell lymphoma, Diffuse lymphoma, Diffuse large cell lymphoma, Diffuse mixed (small & large) cell lymphoma, Diffuse small cleaved cell lymphoma, Discordant lymphoma, Extranodal lymphoma, Follicular lymphoma, Gastric lymphoma, High-grade lymphoma, Histiocytic lymphoma, Hodgkin's disease, Intermediate lymphocytic lymphoma, Ki-1 lymphoma, Large cell lymphoma with filopedia, Lennert's lymphoma, Low-grade lymphoma, Lymphoblastic lymphoma, MALToma, Mantle zone lymphoma, Mediterranean lymphoma, Monoclonality, Monocytoid B-cell lymphoma, Non-Hodgkin's lymphoma, Pediatric lymphoma, Pinkus lymphoma, Pleomorphic non-Burkitt's lymphoma, Polylobulated lymphoma, Prelymphoma, Primary lymphoma of brain, Pseudolymphoma, Pseudopseudolymphoma, REAL classification, Reversible lymphoma, Serous lymphoma, Signet ring cell lymphoma, Small lymphocytic lymphoma, Small non-cleaved cell lymphoma, T-cell lymphoma lymphoma, Testicular lymphoma, Working classification. Cf Leukemia

lymphomatoid granulomatosis A lymphoproliferative disorder which presents in middle-aged subjects with well-circumscribed bilateral nodules seen on CXR; some cases occur in immunosuppressed renal transplant recipients and in Pts with Sjögren syndrome CLINICAL 80% have extrapulmonary involvement–eg, skin, CNS, kidneys, liver, spleen, adrenal glands, heart, GI tract, etc PROGNOSIS 64% mortality; median survival of 14 months, death is due to pulmonary destruction accompanied by sepsis; severe T-cell impairment may explain the tendency for malignant degeneration. Cf Lymphoid interstitial pneumonia

lymphomatoid papulosis A benign recurring papular eruption of the skin which histologically resembles lymphoma; although most lesions are indolent, gene rearrangement studies demonstrate a T cell clonal expansion; 10-20% evolve toward T-cell lymphoma

lymphomatosis ONCOLOGY Extensive lymphomatous involvement of various body regions. See Intravascular lymphomatosis, Neurolymphomatosis

lymphopenia HEMATOLOGY A ↓ in peripheral lymphocytes–normal, 1,000-4,000/mm³ in adults–or 8,000/mm³ in children, due to developing immune system ETIOLOGY Bacterial sepsis, active TB, uremia, NHLs, Hodgkin's disease, autoimmune disorders, immune deficiency syndromes, chemotherapeutics, glucocorticoids, malnutrition

lymphopoietic *adjective* Referring to formation of lymph or lymphocytes

lymphoproliferative disorder ONCOLOGY Any clonal expansion of lymphatic system cells–eg, clearly malignant conditions–eg, leukemia, lymphoma, Hodgkin's disease and relatively indolent lesions–eg, lymphomatoid granulomatosis and lymphomatoid papulosis

lymphostatic verrucosis Lymphedematous keratoderma, mossy foot A feature of chronic lymphedema of the leg characterized by a lawn of velvety hyperkeratotic filiform projections

Lynx over-the-wire catheter CARDIOLOGY A line of cardiovascular balloon catheters using Focus technology, with high burst pressure, low balloon profile, ↓ catheter shaft size. See Balloon angioplasty

Lyon Heart Study A study which evaluated the effects of changes in diet on Sx and progression of ASHD. See Ornish regimen

lyonization GENETICS A normal genetic event that consists of inactivation of all X chromosomes–portions of the 'inactivated' chromosome may remain functional if in excess of one

lyophilization Freeze drying A method for preserving foods or biologicals, where a substance–eg, coagulation factor VIII, is 'snap-frozen' in liquid nitrogen—70°C and placed in a high vacuum to remove the water vapor as it sublimes; once the water is removed, the substance is brought to room temperature and stored. Cf Quick-freeze technique

lyophilized liposomal delivery system THERAPEUTICS An oral delivery system that allows a drug to pass through the GI tract to the small intestine without degrading the active agent; once in the small intestine, the liposome attaches itself to the intestinal wall allowing delivery of the therapeutic agent. See Liposome

lysergic acid diethylamide LSD, see there

lysinemia A heterogeneous group of diseases with ↑ lysine or its metabolites in the blood, including hyperlysinemia, types I and II, saccharopurinuria, hydroxylysinuria, pipecolic acidemia and α-ketoadipic aciduria

lysis Destruction of cells with release of contents. See Antibody-mediated lysis, Cytolysis, Follicle lysis, Hemolysis, NK-mediated lysis

lysosomal storage diseases A heterogeneous group of diseases with specific lysosomal enzyme defects. Cf Inborn errors of metabolism

lytes Short form for electrolytes, including Na^+, K^+, Cl^-, and CO_3^-, and possibly also Mg^{2+}, Ca^{2+}, and PO_4^-. See Electrolytes

-lytic *suffix* Referring to lysis; characterized by lysis

M Symbol for: 1. Male 2. Malignant 3. Man 4. Married 5. Mass 6. Maximum 7. Mega- 10^6, per SI–International System 8. Melphalan 9. Memory 10. Mercaptopurine 11. Mesial–dentistry 12. Methionine 13. Midline 14. Minimum 15. Minute–time 16. Mitochondria 17. Mitomycin 18. Molar concentration 19. Molar–dentistry 20. Mole–chemistry 21. Monoclonal 22. Monocyte 23. Movement–neurology 24. Multipara 25. Muscle/musculus 26. Mycoplasma 27. Myopia 28. Myosin

m Symbol for: 1. Mass 2. Median–statistics 3. Mean of a sample 4. meta–organic chemistry, benzene ring position 5. Messenger–RNA 6. Meter 7. milli—10^3, per SI-International System 8. Molal concentration

M component Monoclonal immunoglobulin A narrow peak or 'spike' seen on SPEP which is presumptive evidence of a monoclonal proliferation of mature B cells producing IgG, IgA or IgM WHERE SEEN Myeloma, Waldenström's macroglobulinemia, heavy chain disease, lichen myxedematosus–a rare disease of proliferating fibroblasts. See Serum protein electrophoresis

M current NEUROPHYSIOLOGY A time- and voltage-dependent potassium current that persists at slightly depolarized membrane potentials

M-mode echocardiography Unidimensional echocardiography CARDIOLOGY Echocardiography based on one-dimensional–'ice-pick' analysis of the heart in motion; MME was the first application of ultrasound in cardiology and continues to be used

M2 protocol ONCOLOGY A protocol for treating myeloma AGENTS vincristine, BCNU–carmustine, cyclophosphamide, melphalan, prednisone

M-tropic strain AIDS A strain of HIV-1 that infects macrophages and activated T cells, the entry of which is facilitated by CCR5, a coreceptor for chemokines, RANTES, MIP 1α, MIP 1β; binding of these chemokines to CCR5 blocks the entry of M strains. Cf T-tropin strain

M-VAC Methotrexate, vinblastine, Adriamycin–doxorubicin & cis-platinum A chemotherapeutic regimen for bladder CA

M-Zole® pack GYNECOLOGY An antibiotic for vaginal candidiasis and vulvar itching and irritation

MAAS CARDIOLOGY A clinical trial–Multicentre Anti-Atheroma Study–of the effect of simvastatin on coronary atheromas in Pts with hypercholes-

terolemia and CAD. See Atheroma, Hypercholesterolemia, Lipid-lowering therapy, Simvastatin. Cf PLAC I, PLAC II, 4S, WOSCOPS

MAB Monoclonal antibody, Monoclonal antibodies

MAC 1. Mammalian artificial chromosome 2. Maximum allowable change, see there 3. Maximum allowable concentration 4. Membrane attack complex, see there 5. Minimum alveolar concentration 6. Monitored anesthesia care, see there 8. *Mycobacterium avium* complex

MAC regimen GYNECOLOGIC ONCOLOGY A regimen for managing choriocarcinoma AGENTS MTX, dactinomycin, chlorambucil; MR 'failures' may respond to other chemotherapy–eg, bleomycin, cisplatin, hydroxyurea, vinblastine

maceration OBSTETRICS The sloughing of wads of immature skin from a fetus that died in vivo and wasn't immediately evacuated from the uterus WOUND CARE Generic nastiness of a wound or ulcer which has been wet way too long

Machado-Joseph disease NEUROLOGY An AD trinucleotide repeat disease characterized by progressive spinocerebellar ataxia, choreoathetosis, dystonia, seizures, myoclonus, dementia. See Trinucleotide repeat disease

machine See Heart-lung machine, Suicide machine

machinery murmur CARDIOLOGY A continuous harsh, rasping or rumbling cardiac murmur characteristic of patent ductus arteriosus that begins shortly after the first sound, reaches a maximum at the end of systole and wanes in late diastole; the MM is best heard in the 2nd left intercostal space, transmits to the chest and neck and may be felt by the Pts as a palpable thrill or 'buzz'; it may also be heard in VSD, rarely in pulmonary HTN and in penetrating soft-tissue injury with formation of an AV fistula

machinist's hand RHEUMATOLOGY A descriptor for the changes in the hands of Pts with dermatomyositis CLINICAL Periungual erythema, linear erythematous discoloration around the nailbeds, accompanied by scaling, hyper- and hypopigmentation, and eventually brauny induration with darkened or 'dirty' horizontal line across the lateral and palmar aspects of the fingers–from whence the name–and pink to violaceous scaling overjoint flexures–eg, the knuckles, elbows, and knees. See Polymyositis-dermatomyositis

MACOP-B ONCOLOGY A '3rd-generation' combination chemotherapy regimen consisting of MTX with leucovorin rescue, doxorubicin, cyclophosphamide, vincristine, prednisone, bleomycin. See CHOP

macro- *prefix*, Greek, large, long. Cf Micro-

macroamylasemia A condition characterized by persistent 6-to-8–fold ↑ of macroamylase that occurs in 1% of older subjects, which is linked to a false ↑ in amylase levels, often accompanied by abdominal pain, associated with alcoholism, pancreatitis, malignancy, DM, cholelithiasis, autoimmune disease.

macrobiotic diet ALTERNATIVE NUTRITION A diet of whole grains, vegetables–eg, barley, millet, oats, rice, wheat, comprising 50% of dietary intake–DI, vegetables–freshly picked in season, 20-30% DI, soups–eg, vegetables, seaweed, grains, 5-10% DI, oils, juices, nuts, seeds–eg, sunflower, herbs, pulses–beans, lentils, peas, and seaweed, 5-10% DI and enough animal foods–eg, 'white meat' fish, 5-15% DI to prevent malnutrition. See Kushi, Macrobiotic Shiatsu™, Raw food diet, Zen macrobiotic diet. Cf Unproven methods for cancer management

macrocephaly ↑ head circumference, megalocephaly PEDIATRICS An abnormally big head/brain; in children, macrocephaly is defined as an occipitofrontal circumference of > 3 SD ≥ the mean; in adults, macrocephaly is any brain weighing > 1800 g, due to expansion of any subdural component–eg, cerebral tissue, liquid, blood, tumor or storage disease DIFFDX–NON-HYDROCEPHALIC CAUSES Benign familial form–sibs also have large heads, achondroplasia, Banayan syndrome, cerebral gigantism–with macrosomia or Sotos syndrome, cutis marmorata telangiectatica congenita, fragile X syndrome, Klippel-Trenauny-Weber syndrome, mucopolysaccharidosis, neurofibromatosis, Weaver syndrome

macrocytic anemia HEMATOLOGY Anemia characterized by macrocytic RBCs, usually due to a discordance between maturation of the cytoplasm and nucleus, typical of vitamin B$_{12}$ or folate deficiency

macrocytosis HEMATOLOGY The presence of abundant RBCs with an MCV of > 105 fL and a diameter of $\geq$ 8.5 µm; macrocytosis is normal in newborns and infants, but uncommon in adults, and is typical of vitamin B$_{12}$ or folate deficiency. See Megaloblastic anemia. Cf Microcystosis

macroglobulin Any large serum protein, usually $\geq$ 400 kD–eg, IgM–900 kD, α_2-macroglobulin–820 kD; macroglobulins are detected by sharp peaks on a simple zone electrophoresis, usually in the γ-region

macroglossia Enlargement of the tongue due to accumulation of various substances, edema, ectopic tissue, tumors, etc DIFFDX Amyloidosis, Beckwith-Wiedemann syndrome, congenital hypo-thyroidism, cystic hygroma, Down syndrome, ectopic thyroid, glycogen storage disease, type II–Pompe's disease, hemangioma–of tongue, Hurler syndrome, intestinal duplication, lymphangioma, mannosidosis, neurofibromatosis, rhabdomyoma, Sandhoff's disease

macrolide antibiotic INFECTIOUS DISEASE A broad-spectrum antibiotic–eg, erythromycin, produced by *Streptomyces* spp, that contains a lactone ring and inhibits protein synthesis in target bacteria. See Antibiotic resistance

macronutrient NUTRITION An essential dietary component—proteins, fats, carbohydrates consumed in large amounts. See Diet. Cf Micronutrient, Nonnutritive dietary component

macroparasite INFECTIOUS DISEASE A parasite–eg, helminths, arthropods–which does not multiply in its definitive host, cycling instead through transmission stages–eggs and larvae–which pass into the external environment; immune responses evoked by macroparasites are transient and depend on the parasite load. Cf Microparasite

macropenis Macrophallus UROLOGY The largest medically verified penis was 13.5 inches long and 6.25 inches around, documented in the early 20th century by Dr. RL Dickinson. Cf Micropenis

macrosomia A larger than normal body typical of infants of mothers with gestational diabetes. See Gestational diabetes

macular *adjective* Related to 1. A macule 2. The macula

macular degeneration Senile macular degeneration OPHTHALMOLOGY Deterioration of the retinal macula, resulting in $\downarrow$ central vision and destruction of the retinal nerve, as fluid leaks into the choroid and forms scar tissue RISK FACTORS $\uparrow$ after age 50; affects 35% of general population by age 75, family Hx, cigarette smoking, White DIAGNOSIS Funduscopy. Cf Retinopathy

macular edema OPHTHALMOLOGY Waxing and waning retinal thickening due to the accumulation of fluid, seen in Pts with diabetic retinopathy, seen with a binocular slit lamp or stereoscopic fundus photography; while ME has no visual consequences, it is often accompanied by hard exudates, which are associated with loss of vision. See Diabetic retinopathy, Hard exudate

macule MEDTALK A small flat colored spot. See Coal dust macule

maculopapular rash DERMATOLOGY Any rash characterized by minibumps overlying macules, which may be caused by drug allergy, West Nile agent, Rocky Mountain spotted fever, atypical measles, meningococcemia, *S pneumoniae*, see there

maculopapule DERMATOLOGY A papule that arises in a macule

mad cow disease Bovine spongiform encephalopathy, see there

mad dog FORENSIC PSYCHIATRY A popular term for a person so depraved as to be likened to a rabid dog. See Dahmer, Jeffrey, Depraved heart murder, Serial killer

MADD Mothers Against Drunk Drivers PUBLIC HEALTH An organization that advocates stricter legislation against DUI and underage drinking, and provides support services for victims of DUI collisions. See DUI

MADIT CARDIOLOGY A clinical trial–Multicenter Automatic Defibrillator Implantation Trial that evaluated the effects of implanted defibrillators–IDs in Pts with CAD at high risk of ventricular arrhythmia

magenstraße CLINICAL ANATOMY The pliable, linear rugal folds of the gastric mucosa that follow the lesser curvature, bounded externally by the gastrohepatic ligament, the site of most spontaneous gastric rupture, due to the lesser curvature's lower distensibility

maggot MEDICAL ENTOMOLOGY Larvae–a worm-like feeding state of flies–order Diptera–eg, green–*Phaenicia sericata*, black–*Phormia regina* bottle flies. Cf Leeches, Roach

MAGIC CARDIOLOGY A clinical trial–Magnesium in Coronaries

magic bullet IMMUNOLOGY Paul Ehrlich's term for an ideal therapeutic, which would target a highly specific site and treat ONLY that site; in modern immunology, MB refers to any agent with the specificity of an antibody and lethal potential of a toxin. See Orthozyme CD5plus

magic mushrooms See Peyote

magical thinking PSYCHOLOGY Dereitic thinking, similar to a normal stage of childhood development, in which thoughts, words or actions assume a magical power, and are able to prevent or cause events to happen without a physical action occurring; a conviction that thinking equates with doing, accompanied by an unrealistic understanding of cause and effect EXAMPLES Dreams in children, in primitive peoples, and in Pts under various conditions

maginot line pattern Fortification phenomenon NEUROLOGY A fanciful descriptor for the jagged, slightly off-center scintillating lines characteristic of visual aura that may precede visual migraines

magnesium METABOLISM An alkaline earth–atomic number 12; atomic weight 24.3–intracellular cation and essential mineral required for bone and tooth formation; Mg^{2+} is bound to ATP, plays a pivotal role in neuromuscular activity, nerve conduction and signaling, muscle contraction; Mg^{2+} is a cofactor in enzymatic reactions in carbohydrate, protein, and nucleic acid metabolism, and is linked to the electrochemical properties of living systems $\uparrow$ IN Renal failure, Addison's disease, magnesium infusion $\downarrow$ IN DM, neuromuscular disease, $\uparrow$ Ca^{2+}, $\uparrow$ K$^+$, malabsorption syndromes, hyperparathyroidism. See Hypermagnesemia, Hypomagnesemia, Minerals. Cf Manganese

magnesium ammonium phosphate stone Struvite stone A type of stone more common in $\female$, linked to UTIs–eg, by *Proteus* spp, which produce specific enzymes; MAPSs can become very large, fill the renal pelvis, develop a staghorn appearance, obstruct the urinary tract and cause kidney damage

magnesium deficiency Hypomagnesemia, low magnesium A clinical situation due to inadequate intake or impaired intestinal absorption of magnesium, often associated with $\downarrow$ Ca^{2+}, and $\downarrow$ K$^+$ CLINICAL Irritability of nervous system with tetany–spasms of hands & feet, muscular twitching, cramps, laryngeal spasm

magnesium-induced diarrhea GI DISEASE Chronic diarrhea induced by ingestion of excess magnesium, present in antacids–Maalox, Mylanta or diet DIAGNOSIS Fecal quantification

magnesium sulfate CARDIOLOGY An agent which, administered IV, may slow the ventricular response to rapid AF, and act in tandem with pharmacotherapy to ↓ the heart rate attributed to intracellular magnesium depletion seen in ethanol abuse, renal failure, in hypokalemia, hyponatremia, or in Pts receiving digoxin or diuretics. See Atrial fibrillation

magnet hospital HOSPITAL CARE A hospital that features flat organization structure, unit-based decision making, investment in education; MH nurses have less burnout, Pts have better care, ↓ time in hospital and in ICU

magnet reaction NEUROLOGY A reflex in which light finger pressure on a toe pad causes a slow reflex contraction of the lower extremity, which seems to follow the examiner's hand, as if drawn by a magnet; the MR is seen in animals that have had the cerebellum removed, or in human infants, whose cerebellums are immature

magnetic media MANAGED CARE A general term for tapes, floppies, superfloppies which are used to archive data

magnetic resonance The absorption or emission of electromagnetic energy by nuclei in a static magnetic field after excitation by a suitable resonance frequency magnetic field; the peak resonance frequency is proportional to the magnetic fiel, and given by the Larmor equation. See Magnetic resonance imaging

magnetic resonance angiography Peripheral MR angiography IMAGING An imaging technique in which a contrast dye is injected into a blood vessel and magnetic resonance is used to create an image of the flowing blood through the vessel; often used to detect stenosis of the cerebral arteries. See Contrast enhanced method–MRI, Phase contrast method–MRI, Time-of-flight method

magnetic resonance imaging IMAGING A noninvasive technique for imaging anatomic structures which involves placing a person in a strong magnetic field and then, using magnetic gradients and brief radiofrequency pulses, determining the resonance characteristics at each point in the area being studied; the creation of images by magnetic resonance–MR, which is a function of the distribution of hydrogen nuclei–protons in the body; the MR image is a computerized interpretation of the physical interaction of unpaired protons with electromagnetic radiation in the presence of a magnetic field; although both have multiplanar capabilities, MRI often trumps CT: no ionizing radiation; contrast between normal and pathologic tissue is greater; confounding bone artifacts do not occur; rapidly moving components appear dark, therefore blood flow through large vessels can be analyzed as a type of natural contrast; the images derive from analysis of the amplitudes and frequencies of the weak signals produced by MR, allowing deduction of the sample's chemical composition, with protons providing the best images. See Burst MRI, Diffusion-weighted imaging, Dynamic imaging, Perfusion-weighted imaging, Short-bore MRI

magnetic stereotaxis NEUROSURGERY A surgical procedure in which a magnetic probe towing a catheter is inserted through a small hole in the brain, and positioned by superconducting magnets under NMR guidance; compared to conventional stereotactic surgery, MS↓ operating time and ↓ brain destruction. Cf Stereotaxis

magnetic stimulation NEUROLOGY A noninvasive method for stimulating the brain and nerves, with a high-current magnetic pulse passed through a coil of wire

magnetoencephalograph NEUROBIOLOGY A device used to study electromagnetic impulses in the cerebral cortex. See Binding problem

magnetoencephalography Biomagnetism NEUROIMAGING An imaging technique used to evaluate subtle fluctuations in the electrical magnetic field produced by the body; MEG measures the brain's electrical activity by evaluating the magnetic field associated with neuronal activity. Cf Electroencephalography, PET scanning

magnification IMAGING Amplification of an image. See Biologic magnification, Microscopy

mail-order medicine A derogatory phrase for basing therapeutic decisions solely on the results of tests–eg, radioallergosorbent test–RAST sent to a reference lab. See RAST

main d'accoucheur French, hand of birthing coach MEDTALK A descriptor for the cramped and coned posture of the hand and fingers seen with carpal tetany in acute hyperventilation syndrome

main-en-lorgnette deformity lorgnette, French, opera glass RHEUMATOLOGY A classic descriptor for a hand afflicted by arthritis mutilans, which is accompanied by extensive compressive erosion, collapse of the proximal phalanges with 'telescoping' of bones on themselves and overriding of fingers by metacarpophalangeal dislocation, resulting in destroyed, contracted hands; although typical of advanced rheumatoid arthritis, MEL may occur in psoriatic arthritis, erosive osteoarthritis, chronic infection, DM, leprosy

mainline DRUG SLANG verb To inject a drug

mainstream medicine Allopathic medicine, allopathy, conventional medicine, modern medicine, orthodox medicine, traditional medicine, Western medicine CLINICAL MEDICINE The approach to health care practiced in developed nations, based on scientific data for diagnosing and treating disease; MM assumes that all physiologic and pathological phenomena can be explained in concrete terms; tools of MM include nonhuman model systems, blinded studies, statistical analysis, to ensure reproducible results. Cf Alternative medicine

mainstreaming Inclusion PEDIATRICS The education of a student with disabilities in a regular classroom in a neighborhood school with the support so the student can participate fully SOCIAL MEDICINE The placing of learning- or otherwise impaired children in the same environment as other–normal children, while supplementing learning with various educational maneuvers

maintenance ANESTHESIOLOGY The maintaining of a Pt undergoing general anesthesia in an unconscious state for the duration of surgery by injecting small boluses or by infusing an injectable drug, or by administering halothane, isoflurane, sevoflurane. Cf Induction

maintenance chemotherapy ONCOLOGY The administration of a chemotherapeutic regimen on an ongoing basis after managing the CA in a traditional manner–eg, with surgery, RT, or conventional chemotherapy ADVERSE EFFECTS Neutropenia, pancytopenia, N&V, myelodysplastic syndrome, peripheral neuropathy. Cf Adjuvant chemotherapy, Consolidation chemotherapy, Induction chemotherapy, Intensification chemotherapy, Salvage chemotherapy. Cf Adjuvant chemotherapy

maintenance dose PHARMACOLOGY A dose of a drug administered after achieving stable levels in the body, to maintain a therapeutic status quo. Cf Loading dose

maintenance therapy ONCOLOGY Chemotherapy to ↓ risk of relapse in Pts with CA or leukemia in remission. See Remission

Mainz pouch UROLOGY An ileocecal segment constructed as a continent urinary pouch, sewn to the anterior abdominal wall in Pts undergoing a radical cystectomy for bladder cancer. See Bladder cancer, Ileostomy

MAIS complex Mycobacterium avium-intracellulare-scrofulaceum INFECTIOUS DISEASE Mycobacterial spp indistinguishable from each other vis-á-vis surface lipids and antigens, pigment production, biochemical reactions, antibiotic susceptibilities, often in the same clinical company INCIDENCE Uncommon; affects 5-8% of Pts with AIDS, especially at CD4T cells < 0.1 x 10^9–US: 100/mm³ CLINICAL Persistent fever, night sweats, chronic diarrhea, abdominal pain, extrahepatic obstruction, anemia TREATMENT Ciproflozacin, clofazimine, ethambutol, rifampicin, amikacin, rifabutin, clarithromycin, azithromycin

major cancer ONCOLOGY A CA which, if untreated, results in death of its host; most malignancies are 'major' and once developed must be eradicated aggressively and Pt followed for indications of recurrence. Cf Minor cancer

major depression Unipolar depression PSYCHIATRY A form of depression with Sx that interfere with the ability to work, sleep, eat, and enjoy once pleasurable activities CLINICAL Feelings of guilt, hopelessness, persistent thoughts of death or suicide, difficulties in concentration, memory, decision-making capacity, behavior–changes in sleep patterns, appetite, weight; physical well-being; persistent Sx–eg, headaches or digestive disorders, that don't respond to treatment; disabling episodic MD can occur several times in a lifetime See Depression

major depressive disorder PSYCHIATRY A chronic, relapsing illness affecting 3–6% of the population at a given time LIFETIME RISK 10–15%; it is linked to a high–10% to 20% rate of suicide, and high morbidity when compared with other medical illness STATISTICS, INTL, LOW Taiwan 1.5%, Korea 3%, Puerto Rico 4.3%, US 5% HIGH Lebanon 19%, France 16.4%, New Zealand 12% OTHER FINDINGS Positive dexamethasone test, sleep changes–eg, ↓ REM latency DIFFDX AIDS, acute intermittent porphyria, amphetamine withdrawal, CA, endocrine disease–eg, Addison's disease, Cushing's disease, hypothyroidism, infectious mononucleosis, influenza, malnutrition, multiple sclerosis, drugs–eg, alpha-methyldopa, benzodiazepines, cimetidine, clonidine, corticosteroids, INH, OCs, propranolol, reserpine, thiazide diuretics

---

MAJOR DEPRESSIVE DISORDER, 5 OR MORE CRITERIA

- ↓ APPETITE or loss of weight
- ↓ CONCENTRATION
- DYSPHORIC MOOD Sad, anxious, irritable
- FATIGUE or decreased energy
- GUILT or excessive self blame
- ↓ INTEREST in pleasurable activities
- PSYCHOMOTOR RETARDATION OR AGITATION
- SLEEP DISTURBANCES
- SUICIDAL IDEATION or suicidal attempt AMN 16/9/96, p17

---

major depressive episode PSYCHIATRY A condition defined as '…*a period of at least 2 wks, during which there is either depressed mood or the loss of interest or pleasure in nearly all activities…*(and) …*experience at least 4 additional symptoms* (including) …*changes in appetite or weight, sleep, and psychomotor activity; decreased energy; feelings of worthlessness or guilt; difficulty thinking, concentrating, or making decisions; or recurrent thoughts of death, or suicidal ideation, plans, or attempts.'*. See Major depression

major life activity SOCIAL MEDICINE Any activity that constitutes economic, intellectual, and functional self-sufficiency–eg, ability to maintain a job, learning, mobility, self-direction. See Activities of daily living

major medical Major medical expense(s) insurance MANAGED CARE A health insurance policy or 'rider'–extension to an insurance policy–that finances medical expenses incurred in injuries, catastrophic or prolonged illness, providing benefit payments above the base paid by the insurance company

major procedure High-risk & complex procedure, see there

major surgery 1. A surgical operation within or upon the contents of the abdominal–or pelvic, cranial, or thoracic cavities, or 2. A procedure which, given the locality, condition of Pt, level of difficulty, or length of time to perform, constitutes a hazard to life or function of an organ or tissue. Cf Minor surgery

malabsorption GI DISEASE A group of Sx–eg, gas, bloating, abdominal pain, diarrhea due to the inability to properly absorb nutrients ETIOLOGY Cystic fibrosis–number one cause in US, abetalipoproteinemia, acrodermatitis enteropathica, biliary atresia, bovine lactalbumin intolerance–cow's milk protein, celiac disease–gluten-induced-enteropathy, sprue, juvenile pernicious anemia, lactose intolerance, parasites–*Diphyllobothrium latum*, *Giardia lamblia*, *Necator americanus*, *Strongyloides stercoralis*, soy milk protein intolerance, vitamin B_{12} malabsorption

malaise INTERNAL MEDICINE A vague feeling of general discomfort, sensed as something "just ain't right". See Fatigue

malakoplakia *malakos*, Greek, soft plaque A lesion characterized by soft, yellow, elevated and friable 3-4 cm mucosal plaques of the GU tract–bladder, renal pelvis, ureter, uterus, broad ligament, endometrium, testes, epididymis, prostate and rarely, retroperitoneum, colon, stomach, appendix, lymph nodes, lungs, bone, skin; more common in immunosuppressed transplant recipients TREATMENT Long-term antibiotics, ascorbic acid, cholinergics; if recalcitrant, excision

malaoxon TOXICOLOGY An anticholinesterase organophosphate used against insects–aphids, spiders, mites, houseflies, and other creepy crawlies. See Intermediate syndrome, Organophosphate pesticide

Malassezia furfur MYCOLOGY A lipophilic yeast associated with tinea vesicolor

malathion PUBLIC HEALTH A topical anticholinesterase-type organophosphate pesticide, used for insects–aphids, spiders, mites, head lice; houseflies, and other creepy crawlies. See Intermediate syndrome, Organophosphate pesticide

maldigestion 1. Indigestion, see there 2. Malabsorption, see there 3. Steatorrhea, see there

male Dude, guy, feller, man, chap, bloke

male bonding PSYCHOLOGY The formation of a close nonsexual relationship between 2 or more men; guy stuff. Cf Bonding

male chromosome complement 46, XY

male factor REPRODUCTION MEDICINE Any cause of infertility due to deficiencies in sperm quantity, function, or motility–ability to move that make it difficult for a sperm to fertilize an egg under normal conditions

male hermaphroditism Male pseudohermaphroditism ENDOCRINOLOGY Any of the congenital syndromes of sexual ambiguity occurring in genetically correct 46,XY, phenotypically incorrect male ETIOLOGY Induced in the fetus by a lack of hormonal masculinization secondary to either the quantity or type of ♂ hormone available, or to insensitivity of the tissues to ♂ hormone. Cf Male pseudohermaprhroditism

male menopause Andropause, male climacteric ENDOCRINOLOGY A popular term for the ♂ equivalent of menopause which is attributed to blunting of hypothalamic-pituitary feedback loop, sensitivity to androgenic hormones, ↓ Leydig cell mass, ↓ serum testosterone, ↓ testicular function, ↓ in mental and physical components of arousal, ↓ ejaculatory force, ↓ semen volume and viscosity, ↓ urethral and prostatic contractions, ↑ sexual refractory periods, which occur with ↑ age; other features of MM include↑ body fat, ↓ bone and muscle mass, energy, virility, fertility CLINICAL Hot flashes, weakness, exhaustion, insomnia, irritability, impotence. See Midlife crisis. Cf Menopause. Cf Androgen replacement therapy

male pattern baldness Androgenetic alopecia, androgenic hair loss, male-pattern hair loss DERMATOLOGY A popular term for androgenetic alopecia; MPB is typical of aging ♂; hair that once proudly waved o'er the crown, creeps south, eastward and westward, ne'er more to be seen RACIAL DIFFERENCES White:black, 4:1; white:Asian, 3:1. See Hair replacement

394 **male pseudohermaphroditism** ENDOCRINOLOGY A type of intersex in which the testes are present and cryptorchid–ie, 46,XY–male genotype and female phenotype; '...*a heterogeneous condition in which the gonads are exclusively testes, but the genital ducts and/or external genitalia are incompletely masculinized*' CLINICAL The phallus is either very small or there is severe hypospadias; the testes may not have descended in the scrotum, resulting in ambiguity on examination; MP may be caused by defects during sexual differentiation; testosterone production is inadequate–due to ↓ LH or hCG receptors on Leydig cells; Pts are raised as ♀–Morris syndrome and may have CNS defects–eg, defective gonadotropin response, primary gonadal defects–eg, idiopathic, defective pregnanediol–3-β 17-α, 17,20-des-17 β synthesis, regression of müllerian tubes, Leydig cell agenesis, androgen insensitivity, ↑ risk of breast CA, Sertoli adenoma, germinoma in situ, seminoma, Leydig cell tumor. See Intersex disorder. Cf Male hermaphroditism

malemission Failure to ejaculate during intercourse. Cf Nocturnal emission

malformation NEONATOLOGY An isolated birth defect caused by abnormal growth of an organ, which, if surgical correction is possible, usually has a good prognosis. See Arnold-Chiari malformation, Arteriovenous malformation, Birth defect, Cerebral cavernous malformation, Cleft palate, Cystic adenomatoid malformation, Dandy-Walker malformation, Dysmorphology, Spina bifida, Teratogenesis

malignancy A cancer capable of metastasizing. See Congenital malignancy, Conjugal malignancy, Occult primary malignancy, Occupational malignancy, Post-trauma malignancy, Premalignancy, Secondary malignancy

malignancy of unknown origin A cancer, the origin of which cannot be determined even by full autopsy examination; many such lesions are undifferentiated PROGNOSIS Six month survival

malignant *adjective* MEDTALK Tending to harm, kill, maim, pernicious, life-threatening, resistant to treatment, severe–as in malignant HTN, threatening to life, virulent. Cf Benign ONCOLOGY Cancerous; a tendency to invade and destroy nearby tissue and spread to other regions; referring to a neoplasm that invades surrounding tissue, which generally is unencapsulated, grows by invasion, with vascular and lymphatic metastases. See Cancer, Carcinogens, Congenital malignancy, Multiple primary malignancy syndrome, Metastasis, Occult primary malignancy, Occupational malignancy, Oncogenes, Post-trauma malignancy, Premalignancy, Secondary malignancy

malignant ascites Excess peritoneal fluid evoked by malignancy, which causes subdiaphragmatic lymphatic obstruction–eg, of the thoracic duct and ↑ intraperitoneal fluid production ETIOLOGY Ovarian, breast, gastric, pancreatic, hepatic, colorectal CA, lymphoma, mesothelioma CLINICAL Abdominal distension, weight gain, indigestion, dyspnea, orthopnea, tachypnea, intestinal obstruction with N&V IMAGING Ground glass appearance, central positioning of small bowel loops, and obscured psoas sign on plain abdominal films; MA is confirmed with ultrasonography, CT, and MRI DIFFDX TB, CHF, nephrotic syndrome, hepatic disease, bacterial peritonitis, malnutrition–due to hypoalbuminemia MANAGEMENT Drainage of fluid; intraperitoneal chemotherapy, antibiotics, radioactive colloids, biological response modifiers–eg, IL-2, α-IFN, external-beam RT, and surgical placement of LeVeen or Denver shunts PROGNOSIS Poor

malignant blue nevus A rare skin lesion that usually involves the scalp, presents as a multinodular plaque > 2.0 cm, a Hx of progressive enlargement, often associated with a cellular blue nevus; MBN's behavior is unpredictable. See Blue nevus, Melanoma

malignant dysphagia GI TRACT Dysphagia due to irreversible strictures–eg, due to lye injury or 1° esophageal CA. See Esophageal cancer

malignant fibrous histiocytoma A pleomorphic mesenchymal malignancy of older adults, which affects deep soft tissues–involving muscle 60% or fascia 20% of the lower–50% and upper–20% extremities, retroperitoneum 15% and abdominal cavity; MFH metastasizes to the lung 80%, lymph nodes 30%, liver, bone DIFFDX Sarcomas, pleomorphic or primitive carcinoma, bizarre melanoma PROGNOSIS 40-65% of tumors recur; 25-50% metastasize

malignant histiocytosis HEMATOLOGY A term that has been used for 3 different conditions 1. Malignant lymphoma. See Lymphoma 2. 'Regressing atypical histiocytosis' An indolent pre-histiocytic lymphoma accompanied by chromosome defects CLINICAL Vague heterogenous clinical picture; peak onset in the 3rd decade, commonly with extranodal involvement of the GI tract, skin, BM PROGNOSIS Good 3. Malignant histiocytosis of Robb-Smith A rapidly fatal disease associated with aggressive proliferation of atypical histiocytes and precursors in lymph nodes, splenic red pulp, BM, skin, GI tract, kidneys, adrenal glands and lungs; although idiopathic, MH is associated with ALL and AML, post-renal transplantation immunosuppressive therapy and EBV viremia, lethal midline granuloma CLINICAL ♂:♀ ratio, 2-3:1; at any age; fever, weakness, weight loss, diaphoresis, chest and back pain, rash, lymphadenopathy, hepatosplenomegaly, subcutaneous tumor nodules, pancytopenia, ↑ BR followed by jaundice, it may cause rapid deterioration DIFFDX AML-FAB M5, hairy cell leukemia, 'histiocytic', and Hodgkin's lymphomas, melanoma, anaplastic or 'large cell' carcinoma, virus-associated hemophagocytic syndrome, infectious mononucleosis, sinus histiocytosis with massive lymphadenopathy, familial hemophagocytic reticulosis, Langerhans' cell histiocytosis; may cause lethal midline granuloma TREATMENT Multidrug regimen–eg, vincristine, cyclophosphamide, doxorubicin, prednisone. See Histiocytosis

malignant hypertension Accelerated HTN, arteriolar nephrosclerosis CARDIOLOGY A condition characterized by severe HTN, retinopathy, renal insufficiency, fibrinoid necrosis of renal arterioles, rapidly progressive and fatal disease RISK GROUPS ♂, African Americans, hypertensives CLINICAL Headache, blurred vision, N&V, lethargy COMPLICATIONS Stroke, MI, blindness, renal failure. See Hypertension

malignant hyperthermia syndrome Malignant hyperpyrexia An AD condition–of variable penetration in which the individual when subjected to the anesthetic–halothane, diethyl-ether, cyclopropane, enflurane and certain psychotropics, develops a potentially fatal–up to 70% mortality in acute episodes, occurring in 1/15,000 administrations of anesthesia in children, 1/50-100,000 adults; ½ of cases had not been previously sensitive to anesthesia; anesthetized Pts may develop high fever and muscle rigidity with rhabdomyolysis, release of myoglobin, renal damage and acute renal failure CLINICAL Tachycardia, tachypnea, cyanosis, labile BP, muscle rigidity, rapid and marked hyperpyrexia, acidotic, hyperkalemic, possibly DIC, renal failure; similar reactions may be evoked in these subjects by warm weather, exercise, emotional stress or without known environmental cues and are initiated by muscular hypermetabolism, due to an idiopathic ↑ in sarcoplasmic calcium occurring under general anesthesia DIAGNOSIS Muscle contraction test with halothane or caffeine challenge TREATMENT Hypothermia, hydration, sodium bicarbonate infusion, mechanical hyperventilation, diuretics to ↑ urine flow, dantrolene–an agent which blocks excitation-contraction coupling between T tubules and sarcoplasmic reticulum

malignant melanoma Melanoma, see there

malignant mesothelioma Mesothelioma, see there

malignant mixed mesodermal tumor see Mixed mesodermal tumor

malignant mole DERMATOLOGY Melanoma, see there GYNECOLOGY Choriocarcinoma, see there

malignant narcissism PSYCHIATRY A range of psychopathic personality disorders characterized by the coexistence of marked narcissistic and antisocial traits. See Sociopath. Cf Serial killer

malignant nephrosclerosis Renal disease associated with 'malignant' or accelerated phase HTN, which may arise de novo, but usually occurs in a background of benign essential HTN; MN comprises 5% of HTN, and occurs in young African American ♂. See Malignant hypertension

malignant neuroleptic syndrome A complex that affects < 1% of Pts exposed to neuroleptics–phenothiazine, butyrophenones CLINICAL Onset 1-3 days after beginning antidepressant therapy, which causes hyperthermia, autonomic instability, muscle rigidity, myoglobinuria

malignant osteoporosis Osteolysis 2° to infiltration by CA, often accompanied by pain; 80% of bone metastases arise from CA of breast, kidney, lung, prostate, thyroid; most weaken the bony trabeculae, with the exception of prostate CA, which is often osteoblastic; 70% of bone metastases involve the axial skeleton; the remainder affect the proximal appendicular skeleton LAB ↑ Acid phosphatase, if extensive, hypercalcemia TREATMENT Remove 1° carcinoma, RT–effective in most cases and fixation of unstable bone. See Hypercalcemia of malignancy

malignant otitis externa ENT Otitis externa accompanied by osteomyelitis and bone erosion. See Otitis externa

malignant plasmacytoma Myeloma, see there

malignant teratoma Immature teratoma, see there

malignant transformation ONCOLOGY The constellation of changes in the growth properties of cells in culture evoked by various agents–eg, radiation, toxins, and viruses that result in development of tumors

malingering OCCUPATIONAL MEDICINE The willful production of symptoms for specific external incentives, or the fraudulant simulation of illness or exaggeration of the Sx of a minor illness or injury, usually to avoid work or school. See Factitious disease(s). Cf Munchausen syndrome

malleolar torsion REHAB MEDICINE A condition in which the knee is extended and the subtalar joint in neutral, the angular relationship between the transverse bisection of the malleoli and frontal plane of the knee describes MT

mallet finger ORTHOPEDICS A flexion deformity of the terminal phalanx of a finger caused by striking the dorsal surface of the finger tip against a hard surface; the MF is due to a partial or complete rupture of extensor tendon of the terminal phalanx, or fracture and/or avulsion of a bony fragment of the extensor insertion at the dorsal base of the distal phalanx, which occurs in a background of closed hyperflexion-type blunt trauma

mallet toe A flexion deformity of the distal interphalangeal joint of the lesser toes, affecting 1 toe or 2 adjacent toes; MT is less common than hammer toe–a flexion deformity of the proximal interphalangeal joint–and becomes symptomatic in adolescence or adulthood with development of a painful 'corn' at the tip of the toe. See Hammer toe

Mallophaga An order of biting lice that may affect humans

Mallory-Weiss tear Mallory-Weiss lesion EMERGENCY MEDICINE A laceration of the esophagogastric junction, which accounts for 5–15% of upper GI bleeding; the classic scenario is vomiting, retching, or violent coughing in an alcoholic Pt MANAGEMENT Bleeding stops spontaneously in 80–90% of Pts; most require symptomatic relief; active management strategies include bipolar electrocoagulation by endoscopy, injection therapy, transcatheter embolization, and intra-arterial vasopressin infusion. Cf Esophageal varices

malnutrition The result of prolonged nutritional defects–eg, ↓ proteins, minerals, vitamins, or calories ETIOLOGY–EXOGENOUS Poverty, alcohol, mental disorders–eg, severe depression, infection–eg, TB, malignancy, or nosocomial due to TPN ENDOGENOUS Metabolic defects–congenital or acquired, malabsorption CLINICAL Weakness, apathy, anorexia, diarrhea, skin pigmentation and/or ulceration LAB ↓ Folic acid, iron, magnesium, bile synthesis, disaccharidase activity, protein, vitamin B_{12}. See Kwashiorkor, Marasmus

malocclusion of teeth Crossbite, crowded teeth, misaligned teeth, overbite ORTHODONTICS A form of malocclusion in which there is a reversal of the normal relationship of the mandibular and maxillary teeth, with lateral displacement of opposing teeth–ie, a poor bite; and difficulty in mastication normally, mandibular teeth lie inside maxillary teeth and the outside mandibular cusps or incisal edges meet the central portion of opposing maxillary teeth COMPLICATIONS Periodontal disease, caries. See Crossbite, Crowded teeth, Misaligned teeth, Overbite

malpractice MODERN MEDICINE Failure to provide professional services with the skill usually exhibited by responsible and careful members of a profession, resulting in injury, loss, or damage to the party contracting those services; misconduct or unreasonable lack of skill in the performance of an act by a professional. See Chiropractic malpractice, Medical malpractice

malpractice stress syndrome MEDTALK A popular term for the mental and financial consequences of being sued for health care provider-related activities. See Going bare

MALT lymphoma Mucosa-associated lymphoid tissue lymphoma ONCOLOGY A lymphoma that arises in the lymphatic tissue of the GI tract, which affects the thyroid, salivary glands, lungs and stomach, often preceded by chronic lymphadenitis. See Lymphoma, WHO classification

Maltese Cross appearance A microscopic pattern likened to a Maltese cross, which may correspond to granules of talc or cholesterol crystals JOINT Maltese crosses occur in arthroscopic fluid associated with post-trauma URINALYSIS 'Maltese crosses' are anisotropic or birefringent cholesterol-rich fat droplets, associated with finely granular renal casts, which have a cruciform appearance by polarized light and are found both within and outside of the cells in the urine sediment of Pts with nephrotic syndrome, eclampsia, renal toxicity, fat embolism, after crush injury and in Fabry's disease–due to aggregates of glycosphingolipids

maltreatment SOCIAL MEDICINE Any of a number of types of unreasonable interactions with another adult. See Child maltreatment, Cf Child abuse

malunion ORTHOPEDICS The partial and incorrectly angled joining of large fracture fragments

MAMA Mid-arm muscle area, see there

mammary duct ectasia GYNECOLOGY The dilation of lactiferous ducts which, if blocked, form cysts; if the cysts rupture, they become inflammed, resulting in plasma cell mastitis. See Plasma cell mastitis

mammary dysplasia SURGICAL PATHOLOGY A nonspecific term for various benign microscopic nosologies in the breast–eg, periductular fibrosis, ductal dilatation, apocrine metaplasia and others. See Fibrocystic disease of breast

mammary souffle CARDIOLOGY An innocent systolic or continuous cardiac murmur of presumed arterial origin heard during late pregnancy or in the early post-partum period that is differentiated from pathologic lesions as it is unaffected by the Valsalva maneuver. See Murmur, Valsalva maneuver

mammogram IMAGING A diagnostic x-ray of the breast. See Mammography.

mammographic abnormality A lesion of the breast or breast tissue identified by mammography, see there

mammography IMAGING The radiologic examination of the breast using a dedicated device designed specifically for imaging the breast; mammography is the single best noninvasive screening procedure for detecting breast CA; mammography–MG yields a false negative rate of 5-15% and a false positive of 10% RADIOLOGY Finely-stippled microcalcifications–white pinpoint-sized dots; suspicious–eg, poorly-circumscribed–geographic densities; other findings–as classified by Wolf are thought to be less reliable; radiation dose during a MG is 25-35 kV/0.025-.035 rem–2.5-3.5 $\times$ 10^{-4} Sievert; the Am Cancer Society and the NCI–US recommend breast self examination–BSE after age 20, a baseline MG between ages 35-45, and annual or biennial MG thereafter–frequency of mammography is a function of the person's relative risk factors for breast CA–first-degree relative with breast CA, Caucasian, later pregnancy; 24% of biopsies of non-palpable breast masses with calcification–> 15 calcifications or calcifications in a linear or branching fashion have ductal or lobular CA GUIDELINES NATIONAL CANCER ADVISORY BOARD High risk ♀–annual MG & BSE < age 40; average risk ♀–annual MG & BSE 40–49; low risk ♀–annual MG & BSE > age 50. See Breast cancer, Cancer screening, Scintimammography. Cf Lumpectomy

Mammography Quality Standards Act IMAGING A regulation requiring mammography clinics to mail Pts–in addition to those sent to the Pt's physician–clear, easily understood, written reports of mammography within 30 days of the examination or sooner if results are suspicious. See Mammography

mammoplasty SURGERY Any form of surgery to the breast. See Augmentation mammoplasty, Breast implants, Mastopexy, Reduction mammoplasty

man VOX POPULI A male human. See Hole-in-the-stomach man, Ice man, Marlboro Man, Reference man, Renaissance man, Visible Man

man-bling NEUROLOGY Quasi-random vocalizations in infants that precede language acquisition

man-made death ETHICS Death that occurs by acts of man, which may be divided into civilian forms–MVAs, smoking, AIDS, accidents, eg, Bhopal and Chernobyl, possibly also, induced abortions, and non-civilian forms, related to war or totalitarianism. See Man-made disaster. Cf Natural disaster

man-made disaster Technological disaster PUBLIC HEALTH An event in which a significant number of people are injured or die as a result of human devices or activities, unrelated to conflicts, and attributed to operator error–eg, Exxon *Valdez* and/or equipment failure–eg, NASA's *Challenger*, *Columbia*, Chernobyl Cf Natural disaster

manage MEDTALK *verb* To be responsible for a Pt's clinical care; to diagnose and treat

managed care Managed health care HEALTH CARE A system that provides for the coordination of health services encompassing early intervention to control price, volume, delivery site, and intensity of health services provided, to maximize the health of the insured, and maximize the value of health benefits; a philosophy in which the goal is a system that delivers quality, cost effective health care by monitoring and recommending utilization of services, and controlling cost of services. See Managed care organization, Opt-out managed care

managed care organization HEALTH INSURANCE A health care delivery system consisting of affiliated and/or owned hospitals, physicians and others which provide a wide range of coordinated health services; an umbrella term for health plans that provide health care in return for a predetermined monthly fee and coordinate care through a defined network of physicians and hospitals EXAMPLES HMO, POS. See HMO, Point of service plan, PPOs

managed service organization Physician equity group MEDICAL PRACTICE A type of managed care organization in which physicians have a greater degree of control than is typical for a traditional MCO. See Managed care organization

management MEDTALK The control of an activity or enterprise by direction or persuasion. See Care management, Case management, Computer-assisted therapeutics management, Demand management, Disease management, Hierarchical storage management, Image management, Knowledge management, Multispecialty management, Nurse case management, Online disease management, Outcomes management, Population health management, Process management, Risk management, Solid waste management, Stress management, Supply chain management, Surgical blood management, Utilization management, Workflow management

management *service* organization MANAGED CARE An integrated health care organization that emphasizes efficient management of primary care networks or medical group practices. Cf Management *services* organization

management *services* organization Physician practice management company MEDICAL PRACTICE An organization contracted by a health care provider/supplier to furnish administrative, clerical, and claims processing functions of the provider/supplier's practice. Cf Management *service* organization

management test Any test specifically intended to guide the therapy of a disease or pathologic entity–eg, DM, HTN, known to be present in a Pt and which is controlled by titrating therapy according to results of the test. Cf Diagnostic test, Screening test

Manchester operation GYNECOLOGY A surgical procedure designed for 2° or 3° uterine prolapse with cystourethrocele, ideal for elderly Pts GOAL Change angle of uterus in pelvis. See Uterine prolapse

mandated benefit MANAGED CARE A benefit that a health plan is required by law to provide EXAMPLES In vitro fertilization, defined days of inpatient mental health or substance abuse treatment, special-condition treatments. See Benefit, ERISA

mandated choice TRANSPLANTATION A mechanism suggested by the AMA's Council on Ethical and Judicial Affairs to ↑ number of cadaveric organs available for transplantation; in MC, a person would be required to choose–to donate or not while registering for a drivers' license, filing income tax form or performing other tasks mandated by the state. See Cadaveric organ transplantation, Presumed consent

mandatory assignment MANAGED CARE A format for reimbursing health care services that requires physicians and other providers to accept Medicare reimbursement as payment in full; under MA, balance billing is not allowed. See Balance billing

mandatory reporting The obligatory reporting of a particular condition to local or state health authorities, as required for communicable disease and substance abuse INFECTIOUS DISEASE State boards of health maintain records and collect data resulting from MR of communicable or other diseases that represent a hazard to the public. See Reportable disease PSYCHIATRY Physicians must report abuse or suspected abuse of children, spouses, or the elderly, and are granted prosecutorial immunity if they report their suspicions in good faith. See Abuse

mandibular advancement device SLEEP DISORDERS A wire and plastic device inserted in the mouth of Pts with obstructive sleep apnea before sleeping, which moves the mandible forward 4–6 mm, to ↓ sleep apnea and improve sleep quality. See Sleep apnea

mandibulectomy SURGICAL ONCOLOGY The resection of part of the mandible for CA of the oral cavity, especially of the floor of the mouth; mandibulectomies can be either marginal, in which only the bone, teeth, and adjacent soft tissues are resected and the mandible's continuity is maintained, or segmental, where a complete segment of the mandible is removed. See Head & neck surgery

Manegan Trazodone, see there

Manerix® Moclobemide, see there

maneuver MEDTALK A method or technique for performing a task. See Abdominal thrust maneuver, Doll's head maneuver, Epley maneuver, Flake maneuver, Hallpike maneuver, Head-tilt/chin-lift maneuver, Heimlich maneuver, Jaw thrust maneuver, Jendrassik maneuver, Lichtenstein maneuver, Semont maneuver, Triple airway maneuver, Valsalva maneuver

manganese deficiency A rare mineral deficiency state due to a lack of dietary intake of manganese CLINICAL MD during gestation is linked to skeletal defects—shortened limbs, chondrodystrophy, incoordination, ataxia, tremors and hyperreflexia. See Manganese

manganese poisoning Acute or chronic intoxication due to manganese excess ETIOLOGY Industrial exposure to manganese-laden fumes and dusts in mining, steel foundries, welding, battery manufacture CLINICAL Acute–pneumonitis; chronic–psychotic or schizophrenia-like episodes, parkinsonism, movement disorders MANAGEMENT O$_2$ administration, supportive. See Manganese

mania PSYCHIATRY A hyperkinetic psychiatric reaction or Sx complex seen in bipolar disorder and other affective disorders, temporarily or regularly affecting up to 1% of the population CLINICAL Inappropriate elation or irritability, exaggerated gaiety, severe insomnia, grandiose thoughts, and a sense of invincibility, ↑ talking speed and/or volume, disconnected and racing thoughts, hypersexuality, ↑ energy, poor judgment, poor concentration, hyperactivity, inappropriate social behavior. See Hypomania

manic-depressive illness Bipolar I disorder, see there

manic episode PSYCHIATRY A period characterized by a persistently elevated, expansive, or irritable mood, with ↑ energy, ↓ sleep, distractibility, impaired judgement, grandiosity, flights of ideas, and so on, most often affecting Pts < age 25; MEs are seen in those with primary–idiopathic affective illness or bipolar I disorder, in which Pts vacillate between hypermania and abject depression

MANIC EPISODE-CRITERIA

A A distinct period of abnormally or persistently elevated, expansive, or irritable mood of ≥ 1 week or less if hospitalization is required

B During the period, ≥ 3 of following symptoms

1. Grandiosity or inflated self-esteem

2. ↓ Need for sleep

3. ↑ Talkativeness

4. Flight of ideas, or impression that thoughts are 'racing'

5. Distractibility

6. ↑ Goal-oriented activity–socially, work- or school-related or psychomotor agitation

7. Involvement in activities with potentially dire consequences, eg buying sprees, sexual indiscretions, inappropriate business transactions

C Symptoms do NOT meet criteria of a mixed episode

D The mood disturbance may markedly impair occupational or social function

E Symptoms are unrelated to the direct physiological effects of a substance–of abuse, medication, or other therapy or to a general medical condition–eg hyperthyroidism

DSM-IV™, American Psychiatric Association, Washington, DC, 1994

manic phase of bipolar disorder Manic phase of manic-depressive disorder PSYCHIATRY A phase characterized by elation, hyperactivity, over-involvement in activities, inflated self-esteem, distractability, ↓ sleep, lasting from days to months. See Bipolar disorder

manifest content PSYCHIATRY The remembered content of a dream or fantasy, as contrasted with latent content, which is concealed and distorted

manipulation The mechanical changing of a body position as a therapy.

manipulator SURGERY A device used to mechanically lock something down or hold it in place; a device used as an extension of a surgeon's hand, often in the context of a laparoscopic procedure, to perform a particular task that may be at the limits of the operative field. See DaVinci system, Endoscopy

manner of death FORENSIC MEDICINE The fashion or circumstances that result in death, which are designated natural or unnatural. See Natural death, Unnatural death. Cf Cause of death

mannosidosis Alpha-D-mannosidase deficiency METABOLIC DISEASE An AR condition caused by a defect in lysosomal alpha B mannosidase CLINICAL Macrocephaly, thickened calvaria, coarse face, macroglossia, wide-spaced teeth, prognathism, deafness

manometry The recording of pressure in an inelastic liquid or gas; manometry is the standard for the diagnosis of motor disorders of the body of the esophagus and lower sphincter, and allows intraluminal evaluation of pressure based on perstaltic performance, contraction wave configuration, and the sphincter's basal pressure and relaxation

manpower shortage A dearth of persons with a particular skill which, in a free market economy driven by 'supply-and-demand', may result in ↑ salaries and difficulty in obtaining their services. Cf Physician 'glut'

manslaughter FORENSIC MEDICINE The unlawful, unjustifiable, and/or inexcusable, killing of one human by another, under circumstances lacking premeditation, deliberation, and express or implied malice. See Serial killer. Cf Murder

MANSLAUGHTER

VOLUNTARY That which is committed voluntarily in a heat of passion

INVOLUNTARY That which occurs when a person commits an unlawful act that is not felonious or tending to cause great bodily harm, or when a person is committing a lawful act without due caution or requisite skill–eg a surgeon performing an operation while intoxicated, and inadvertently kills another

mantle port Mantle field RADIATION ONCOLOGY A radiotherapy field that covers axillary, mediastinal, hilar, cervical, supra- and infraclavicular lymph nodes, used to treat multiple contiguous lymphoid regions involved by Hodgkin's lymphoma. See Abdominal bath, Inverted 'Y' field, Radiotherapy

mantle-zone lymphoma Mantle cell lymphoma A low-to-intermediate grade B cell lymphoma by the Working Formulation, which accounts for 1-8% of NHLs, and is most common in ♂ ≥ age 55; it characterized by a proliferation of small lymphocytes in the mantle zone, surrounding benign germinal centers; MZLs are often aggressive and involve the spleen, liver, BM, etc CLINICAL Massive splenomegaly, generalized lymphadenopathy, ⅓ present with 'type B' symptoms–eg, fever, weight loss, night sweats MEDIAN SURVIVAL 31 months; 40% achieve complete remission with chemotherapy. See Lymphoma

Mantoux test Mantoux tuberculin skin test An intracutaneous test used to diagnose TB based on hypersensitivity to tuberculin, a concentrate of TB antigen, the standard preparation of which is PPD–purified protein derivative. See PPD, Tuberculin test, Tuberculosis

manual MEDTALK A written guide, handbook or text. See Plan manual, Policy manual, Procedure manual, Reference manual, Safety manual

manual expression Manual encoding PSYCHOLOGY The ability to express ideas in meaningful gestures; most Pts with difficulties in ME also have difficulties in demonstrating concrete activities and concepts to others

manuscript JOURNALISM A document, in particular, a rough draft of a paper/document intended for publication in a journal, book, other printed, or electronic form. See Authorship Cf Peer review

many things, fear of PSYCHOLOGY Polyphobia. See Phobia

MAO 1. Maximum acid output A measure of the maximum secretory capacity of gastric hydrogen ions, defined as the sum of 4 15-minute acid outputs NORMAL 5-60 mmol of titratable acid/hr, after either pentagastrin or histamine stimulation; MAO is ↑↑↑ in Zollinger-Ellison syndrome, and in ↑ gastrin production. See also BAO, PAO 2. Monoamine oxidase, see there

MAOI Monoamine oxidase inhibitor A family of therapeutic drugs–eg, phenelzine, that are used to treat atypical depression or when tricyclic antidepressants fail; MAOIs have traditionally been relegated to a 2° role in treating depression, given their tendency towards inducing hypertensive crises when MAOI-treated Pts ingest tyramine-containing products; MAOIs may be most effective when depression is accompanied by anxiety

MAP

398

MAP 1. Mean airway pressure 2. Mean arterial pressure 3. Medication administration record, see there 4. Medical assistance program 5. Melphalan, adriamycin, prednisone-oncology 6. Modular assembly prosthesis-orthopedics 7. Monoclonal antibody production 8. Monophasic action potential-neurology 9. Muscle action potential-neurology

map-dot dystrophy OPHTHALMOLOGY Bilateral, symmetric intraepithelial, grayish cystic corneal opacities, thought to be aggregates of basement membrane material which wax and wane; the opacities have no pathologic significance, but impart a foreign-body sensation

MAPCath CARDIOLOGY A sterile, disposable plastic stylet with a magnetically-activated position–MAP sensor in its tip, and comes with a Tuohy-Borst adapter and catheter adapter for assembly with various CV catheters

maple syrup urine disease Branched chain ketoaciduria NEONATOLOGY A rare AR inborn error of metabolism due to ↓ branched-chain α-keto acid dehydrogenase complex activity, resulting in defective amino acid metabolism, characterized by urine that smells like maple syrup FREQUENCY General population 1:200,000; in Pennsylvania Mennonites of German descent 1:176; the defect in oxidative decarboxylation of branched chain amino acids–BCAA, valine, leucine and isoleucine results in accumulation of BCAA CLINICAL Neonatal onset, ↓ Moro reflex, dyspnea, spasticity, opisthotonos, mental and growth retardation, severe hypotonia, feeding difficulties, hypoglycemia, convulsions, decorticate rigidity LAB ↑ BCAA, ↓ threonine, serine, alanine in urine and serum, a positive dinitro-phenylhydrazine test for α keto amino acids, which form insoluble hydrazines TREATMENT Dietary ↓ of BCAA, plus dietary overload–20-fold excess of thiamine PROGNOSIS Mortality was formerly 100%, often due to intercurrent infection; with BCAA-free infant formulas, the survival is virtually 100% and mental retardation completely preventable; since acute decompensation by BCAA and BCKA is due to a breakdown of endogenous proteins resulting in metabolic acidosis, ketosis, anorexia, emesis and potentially fatal encephalopathy, Pts may respond to parenteral solutions of BCAA-free amino acids

maprotiline Ludiomil® NEUROPHARMACOLOGY A tetracyclic antidepressant similar to tricyclic agents INDICATIONS Major depression, bipolar disorder. See Tricyclic antidepressant

MAR Medication administration record, see there

marasmus A state of severe malnutrition, due to ↓ ingestion of protein and calories, due to poor diet, feeding habits, parent-child relationship or metabolic defects CLINICAL FTT, weight loss, emaciation, ↓ skin turgor, skin and muscle atrophy; children are wizened, listless, abdominal distension and edema, 'starvation stools', hypotonia, hypothermia, ↓ metabolism. Cf Kwashiorkor

marathon SPORTS MEDICINE A foot race measuring 26.2 miles/42.2 km; muscle changes in marathon runners: local muscle tenderness, pain, swelling, inflammation. See Ultramarathon

marble bone disease Albers-Schönberg disease, malignant osteopetrosis An AR form of early onset osteopetrosis with FTT, bone fragility, multiple Fx, osteomyelitis and other infections, proptosis, blindness, deafness and hydrocephalus due to bony overgrowth of cranial foramina; replacement of BM evokes extramedullary hematopoiesis in liver and spleen, hepatosplenomegaly LAB ↑ acid and alk phosphatases, ↓ Ca²⁺, pancytopenia, defective T cell functions. See Osteopetrosis

marble brain disease An AR condition due to carbonic anhydrase II deficiency CLINICAL Mental and growth retardation, facial dysmorphia, dysodontogenesis, cerebral calcification with a 'veined' pattern, osteopetrosis, renal tubular acidosis, restrictive lung disease due to rib deformity LAB Metabolic acidosis, hyperchloremia, alkaline urine. Cf État marbre

marbling FORENSIC PATHOLOGY 'Venous patterning' A mosaic of discoloration due to prominent subdermal vessels on the skin of a body in early decomposition/decay SPORTS MEDICINE Skin marbling Mottling of skin seen in divers with pulmonary barotrauma and air embolism. See Air embolism, Pulmonary barotrauma

Marcaine® Bupivacaine, see there

MARCATOR CARDIOLOGY A clinical trial–Multicenter American Research Trial with Cilazapril After Angioplasty to Prevent Transluminal Coronary Obstruction & Restenosis that evaluated ACE inhibition in preventing restenosis after PCTA in Pts with CAD. See Coronary artery disease, Percutaneous transluminal angioplasty

march fracture Insufficiency fracture ORTHOPEDICS A metatarsal stress fracture seen in military recruits unaccustomed to the repeated, otherwise trivial trauma to the feet associated with long marches while carrying heavy equipment

march hemoglobinuria Exertional rhabdomyolysis An episodic hemoglobinuria due to hemolysis caused by repeated mechanical injury to RBCs that travel through small vessels overlying the bones of hands and feet in long-distance marching–soldiers, marathon running, calisthenics, karate LAB Myoglobinuria, proteinuria, ↑ BUN, ↑ enzymes–creatinine phosphokinase, ↑ lactic acid

Marcus Gunn phenomenon Marcus Gunn pupillary sign, swinging flashlight NEUROLOGY A change in the pupillary reflex, caused by unilateral optic nerve or retinal disease; a light is shined, first on one, then the other eye with the opposite eye covered; the affected eye demonstrates a slight contraction or dilation of the pupil

Marfan syndrome An AD connective tissue disease with a prevalence of 1:10, 000; 25% occurs without parental defects, implying de novo mutation; MS expression varies by family affected CLINICAL Ocular–ectopia lentis, myopia, cardiovascular–dissection of ascending aorta, mitral valve prolapse, aortic valve regurgitation, skeletal–scoliosis, arachnodactyly, abnormally long fingers and extremities, pneumothorax, thickened skull, etc TREATMENT β-Adrenergic blockade–eg, with propranolol reportedly slows pace of aortic dilation and may ↓ aorta-related complications in some Pts with Marfan syndrome. See Fibrillin, FBN1

marfanoid habitus A leptosomic body type that is tall and thin with long hands; marfanoid features may be familial in nature or pathological as occurs in homocystinuria and MEN type IIb, mimicking some of the changes of Marfan syndrome, but not accompanied by luxation of lens, funnel chest, dissecting aneurysm of aorta

Margrilan Fluoxetine, see there

marijuana Cannabis sativa, C indica MJ HERBAL MEDICINE MJ is listed in ancient pharmacopeias of China, and used for pain, insomnia, nervous complaints MAINSTREAM MEDICINE MJ has been evaluated as an appetite stimulant, and to control asthma, glaucoma, seizures, and nausea due to chemotherapy. See Herbal medicine, THC SUBSTANCE ABUSE A substance derived from the hemp plant Cannabis sativa, the leaves of which are smoked, producing a hallucinogenic effect due to the neurochemical Δ⁹-tetrahydrocannabinol–THC, which has a cognate THC receptor in the brain IMMUNE SYSTEM THC blocks monocyte maturation NERVOUS SYSTEM Impaired motor skills, defective eye tracking and perception; THC receptors are most abundant in the hippocampus, where memory is consolidated, explaining MJ's detrimental effect on memory and least abundant in the brainstem, explaining why death by overdose is unknown with chronic marijuana abuse; heavy use is associated with residual neuropsychological effects, as evidenced by ↑ perseverations on card-sorting, and ↓ learning of lists RESPIRATORY TRACT MJ is inhaled or 'toked' in a fashion that differs from that of tobacco; in order to maximize THC absorption and elicit the desired 'high', the subject prolongs inhalation, markedly ↑ carbon monoxide and tar, and thus is possibly more detrimental than tobacco smoke THERAPEUTIC USES MJ is an analgesic, but unusable as such, due to the inseparable hallucinogenic effect; it is of use for 1. Control of N&V in terminal CA–2 antiemetic cannabinoids are commercially available, nabilone–Cesamet, a synthetic derivative of MJ and dronabinol–Marinol the principle psychoactive substance in MJ; both are 2nd-line therapies, given their psychotomimetic effects and side effects–drowsiness, dizziness, vertigo, loss of ability to concentrate and mood swings and 2. Control of intraocular pressure in open-angle glaucoma, administered orally, in topical drops or smoked; MJ may evoke anxiety or panic attacks ROUTE Inhaled, oral PHARMACOLOGIC EFFECTS Hallucinations, euphoria, relaxed inhibitions, ↑ appetite, disorientation, ↑ pulse rate, reddening of conjunctiva TOXICOLOGY THC and metabolites are detectable in urine 1 hr post-pot puffery, later if used as a garnee–ie, pot in pan. See Amotivational syndrome, Joint, Medical marijuana, Substance abuse, THC receptor, Toke

marimastat ONCOLOGY An angiogenesis inhibitor and matrix metalloproteinase inhibitor in clinical trials as an ancillary therapy for advanced malignancy–ovary, pancreas, small cell of lung, stomach, glioblastoma. See Angiogenesis inhibitors

marital adjustment scale PSYCHOLOGY A 15-item inventory in which individuals rate various aspects of marital life, allowing the subjects to rate the satisfactoriness of their marriage

mark See Birthmark, Bite mark, Bookmark, Cervix mark, Chair's mark, Class mark, Earmark, Hourglass mark, Pockmark, Stork mark, Stretch mark

marker INTERNAL MEDICINE A diagnostic indicator for presence of a disease. See Biomarker, Bone marker, Breast marker, Cardiac marker, Cell marker, Cell surface marker, Gene marker, Genetic marker, Kinesic marker, Microsatellite marker, Pan B-cell marker, Pan T-cell marker, Selectable marker, Selected marker, Surface marker, Surrogate marker, Tumor marker

marker of choice A lab parameter used to evaluate disease response to therapy and monitor for recurrence; MOCs include RNA for progression of HIV infection and CEA for colorectal cancer

market share VOX POPULI The percentage of total market potential for a particular service or goods that a provider of those services or goods has captured. See Market

marketing MEDICAL PRACTICE A constellation of activities–advertising, giving of labeled trinkets–pens, mugs, etc and other forms of public dissemination of information, to maximize advertiser's marketshare for a particular service or goods. See Advertising, Cigarette marketing, Detailing, Yellow professionalism

Markov process THEORETICAL MEDICINE A stochastic process in which the conditional probability distribution for a system's state at any given instant is unaffected by that system's previous state

markovian analysis DECISION-MAKING The use of the Markov process to project future events in a system with multiple hypothetical components. See Receiver operator characteristic

Marlex® SURGERY A proprietary mesh used to reinforce tissues during herniorrhaphy, to repair and prevent recurrence. Cf Gore-Tex®

Maroteaux-Lamy disease Mucopolysaccharidosis VI, arylsulfatase B deficiency METABOLIC DISEASE An AR condition caused by arylsulfatase deficiency CLINICAL Coarse facies, corneal clouding, progressive dysostosis multiplex, hepatomegaly MANAGEMENT BM transplant. See Gargoyle.

Marquis of Queensbury rules SPORTS MEDICINE Boxing rules promulgated in 1866, that require gloves in all rounds, a 10-second count poist-knockdown, and matching fighters by weight. Cf Broughton rules

marriage bonus PSYCHOLOGY A popular term for health and longevity benefits linked to being married; the 'bonus' is attributed to better diet, ↓ bad habits, ↓ risk-taking, improved medical care, and companionship in married persons. See Companionship, Extended family, Most significant other, Nuclear family. Cf Social isolation

MARSA Methicillin-aminoglycoside resistant *Staphylococcus aureus*, see there

Marseille Classification GI DISEASE A classification of pancreatitis that divides it into acute pancreatits, which may be a single episode or recurrent, and chronic pancreatitis. See Acute pancreatitis, Chronic pancreatitis

Marshall-Marchetti procedure Marshall-Marchetti-Krantz procedure GYNECOLOGY An operation for relieving urinary stress incontinence by suspending periurethral vaginal tissue from the conjoined tendon and making the bladder neck an intra-abdominal organ. See Urinary incontinence

marshmallow test PSYCHOLOGY A test of impulse control, in which a young child is given a marshmallow by an examiner, and promised a 2nd one, if he doesn't eat the first marshmallow until the examiner returns–in a period of 15 minutes; ±15% pass. See BarOn test, Emotional intelligence

marsupialization SURGERY A procedure in which a cyst–eg, a Bartholin cyst abscess, or a cyst-like space needs to be opened–a lá kangaroo pouch, incised and drained, followed by suturing the incised wall to remain an opened position, so healing occurs by granulation tissue formation from the base of the pouch

Martha Stewart disease VOX POPULI A 'condition' described by D Lypchuk, in which a victim is immersed in planning parties, decorating, and other irrelevant post-modern activities

martial arts *Budo* HEALTH ENHANCEMENT Any physical and mental discipline based on self-defense, intended to achieve self-awareness and expression in movement; some Asian MAs evolved from ritual dances, or exercises intended to relieve tension caused by hrs of meditation. See Aikido, Chi kung–qi gong, T'ai chi ch'uan

masculinization ENDOCRINOLOGY Virilization, see there PSYCHOLOGY A developmental differentiation and/or assimilation of masculine features and characteristics

mask AUDIOLOGY *verb* To diminish or attenuate a sound with another INFECTIOUS CONTROL *noun* A usually disposable personal protection device that covers the nose to prevent transmission of potentially infected aerosols from a Pt to a health care worker–HCW or from an HCW to a Pt. See Isolation, Personal protection garment, Reverse isolation PSYCHOLOGY Barrier A concealment of wishes or needs, often as a ego defense

mask-like face Mask A hypomimic, expressionless physiognomy or complete lack of facial affect, a finding characteristic of Parkinson's disease, which may be seen in depression, facioscapulohumeral-type muscular dystrophy, infantile botulism, Möbius' syndrome, myotonic dystrophy, Prader-Willi disease, Wilson's disease. See Parkinson's disease, Parkinsonism

mask of pregnancy Melasma, see there

mask phenomenon Post-emetic purpura that follows prolonged vomiting–'retching', appearing as evanescent punctate macules on the face/upper neck, thought to be due to abruptly ↑ intrathoracic pressure; similar lesions occur in violent coughing, Valsalva maneuver, or strangulation, and accompanied by conjunctival petechiae

masked CLINICAL RESEARCH Blinded, see there

masked depression PSYCHOLOGY A form of depression that occurs in adolescents, who deal with depression through denial, somatization–headaches, abdominal or other pain, or 'acting out'–truancy, substance abuse, multiple accidents

masochism PSYCHIATRY A paraphilia/sexual deviancy in which there is a need–or preference for humiliation, physical abuse, or other form of suffering in order to achieve sexual arousal or orgasm. Cf Paraphilia, Sadism PSYCHOLOGY Moral masochism A pattern of behavior in which a person tolerates abuse and exploitation by others, possibly linked to unresolved childhood conflicts and a low self-esteem. See Self-esteem

mass A cohesive aggregate of often similar components, composition, cells or molecules; the amount of matter contained in a body. See Biomass, Critical mass PHYSICAL EXAM An aggregate of tissue with a different consistency than is normal for a particular body region. See Breast mass

mass disaster PUBLIC HEALTH A man-made or natural disaster in which so many persons are injured that local emergency medical services may be overwhelmed or destroyed. See Man-made disaster, Natural disaster, National Disaster Medical System

mass effect NEUROLOGY The result of ↑ intracranial pressure of any cause–brain tumor, blockage of CSF egress or accumulation of CSF in cranial cavity–which, in the nondistensible cranial cavity, acts like a mass

mass hysteria PSYCHIATRY The 'transmission' of anxiety among a group of individuals

mass immunizer VACCINOLOGY An enterprise that vaccinates a large population group with a particular vaccine–eg, pneumococcal, influenza

MASS phenotype GENETICS An acronym–Mitral valve, Aorta, Skin, Skeletal–for a heterogeneous complex seen in Pts/families with mitral valve prolapse with thoracic cage deformity and spontaneous pneumothorax, with defects in the skin–in the extracellular matrix, striae atrophicae and skeleton–joint hypermobility, long limbs; involvement of the aorta is integral to the phenotype, progressive dilatation and dissection of the aorta does not cause significantly ↑ morbidity. See Marfan syndrome, Mitral valve prolapse

mass shooting PUBLIC HEALTH Multiple discharging of firearm(s) onto a group of unarmed victims. Cf Genocide, Serial killing

mass sociogenic illness Epidemic hysteria The occurrence of a group of nonspecific physical symptoms for which no organic cause can be determined, which is transmitted among members of a group by 'line of sight'; MSI is more common in ♀, and characterized by headache, nausea, weakness, dizziness, hyperventilation, and fainting, a general lack of Sx in those sharing the same physical environment, but in a different timeframe

mass tort litigation Mass injury claim CIVIL LITIGATION A class of civil actions in which multiple plaintiffs are injured in a similar fashion by a defective product, hazardous substance, or disaster. See Asbestos, Breast implant, Class-action, Dalkon shield

Massachusetts General Hospital HEALTH CARE The major teaching hospital for Harvard Medical School, widely regarded as one of the best health care centers in the world

massage parlor SEXOLOGY An establishment that advertises nonsexual manipulation and massage services, which may be provided by 'sex workers' who, for an appropriate fee, provide sexual services. See Sexual service. Cf Massage therapy

massage therapy ALTERNATIVE MEDICINE Any of a number of techniques in which the body surface and musculoskeletal system are therapeutically stroked, kneaded, pounded, and yanked; MT has a time-honored history in medicine that stretches back to ancient Greece; massages are intended to relax the body–and mind, mobilize stiff joints, ↑ flow of blood and lymph, ↓ muscular tension and chronic pain, ↓ swelling and inflammation and ↓ tension and stress; MT is believed to integrate the mind and body, improve skin tone, ↑ energy flow through the nervous system, wastes through the GI tract, and enhance body systems. See Contemporary Western massage, Shiatsu, Swedish massage, Swedish/Esalen massage, Traditional European massage. Cf Acupressure, Massage parlor, Reflexology

massive transfusion TRANSFUSION MEDICINE The infusion, in a 24-hr period, of a blood volume that approaches or exceeds the recipient's calculated blood volume

MAST EMERGENCY MEDICINE 1. Military antishock trousers A pressure device designed to provide life support in Pts with external or internal subdiaphragmatic hemorrhage, which stabilizes the lower extremities and pelvis addresses hemorrhagic-traumatic shock by producing hemostasis and ↑ systemic vascular pressure. Cf Pressure pants 2. Military Assistance to Safety & Traffic A program in which the US military contributes helicopters and medical assistance to low-population density areas Note: Major MAST users are high-risk infants in rural settings SUBSTANCE ABUSE Michigan Alcoholism Screening Test A screening tool used to identify alcoholics. See Alcoholism

mast cell leukemia ONCOLOGY An aggressive neoplasm seen in 15% of Pts with malignant systemic mastocytosis CLINICAL Fever, anorexia, weight loss, fatigue, abdominal colic, diarrhea, pruritus, bone pain, duodenal ulcer, hepatosplenomegaly, lymphadenopathy PROGNOSIS Poor; survival 6 months. See Mastocytoma

mastectomy SURGERY The removal of the breast or as much of breast tissue as possible. See Bilateral mastectomy, Extended radical mastectomy, Modified radical mastectomy, Outpatient mastectomy, Partial mastectomy, Prophylactic mastectomy, Radical mastectomy, Simple mastectomy, Subcutaneous mastectomy

master-servant doctrine See Respondeat superior

mastitis Breast inflammation, often bacterial, often staphylococcal, often during breast-feeding MANAGEMENT Warm wet compresses, oral antibiotics. See Breast abscess, Plasma cell mastitis

mastocytoma PEDIATRICS Urticaria pigmentosa, see there

mastocytosis A heterogeneous group of uncommon, poorly understood lesions characterized by ↑ mast cells in one or more tissues or organs, especially skin; mastocytosis may be classified according to extent and behavior. See Mast cell leukemia. Urticaria pigmentosa

---

MASTOCYTOSIS

LOCALIZED MASTOCYTOSIS

- Focal: Single skin lesion: mast cell 'nevus'
- Generalized: Urticaria pigmentosa

SYSTEMIC MASTOCYTOSIS

- Indolent
- Progressive
- Malignant

MAST CELL LEUKEMIA

MAST CELL SARCOMA

PATHOGENESIS Some cases may be reactive to ↑ soluble mast cell growth factor–*kit*-ligand, possibly due to an ↑ in proteolytic processing, a cytokine that causes mast cell accumulation, melanocyte proliferation and ↑ melanin production NEJM 1993; 328:1302na REACTIVE MASTOCYTOSIS A focal ↑ in mast cells due to immediate or delayed hypersensitivity reactions, which may also occur in lymph nodes draining benign or malignant lesions, eg chronic liver or renal disease, leukemia, lymphoproliferative disorders or Hodgkin's disease; benign mast cell diseases include localized mastocytosis, which may be cutaneous or extracutaneous and urticaria pigmentosa SYSTEMIC MASTOCYTOSIS A potentially aggressive condition characterized by mast cell proliferation in the skin, liver, lymph nodes, BM, GI tract CLINICAL Histamine hyperproduction with flushing, vertigo, palpitations, pruritus, colic, dyspnea, nausea; Sx range from mild and intermittent, to severe, disabling, and progressive '**MALIGNANT**' SYSTEMIC MASTOCYTOSIS A form of mast cell disease that is fatal within 2 years of conversion to an aggressive form

---

mastodynia MEDTALK → VOX POPULI Breast pain

mastoid obliteration operation ENT An operation to eradicate disease when present and obliterate a mastoid or fenestration cavity

mastoid surgery AUDIOLOGY A procedure to remove infected tissue from the mastoid bone. See Mastoidectomy

mastoidectomy HEAD&NECK SURGERY Partial excision of part of the mastoid cells or mastoid process. See Modified radical mastoidectomy, Radical mastoidectomy

mastoiditis ENT Inflammation of the mastoid, often 2° to middle ear infection–acute otitis media. See Mastoid

mastopathy Any disease of the breast. See Diabetic mastopathy

mastopexy BREAST SURGERY The surgical lifting of the breast with removal primarily of the skin; a tightening procedure for breasts sagging from the viscissitudes of aging, atrophy, lactation, pregnancy, gravity COMPLICATIONS Hematomas, infection, breast asymmetry, altered sensation, loss of function in nipple areolar region, scars. See Mammoplasty

masturbation Automanipulation, digital autoeroticism, secret vice SEXOLOGY Sexuoerotic stimulation of one's own–or another's–external genitalia, by manual contact or means other than sexual intercourse. See Coitus. Cf Social vice

MAT Multifocal atrial trachycardia, see there

match The fitting together of one or more things. See HLA match, Language mismatch, Logic mismatch, Mismatch. Cf Crossmatch

the Match GRADUATE EDUCATION A system used in North America by which both teaching hospitals rank their preferences for candidates to fill their first yr post-graduate training positions, and graduating medical students rank their preferences for those–internship yr positions; some graduate programs allow certain graduate training positions to be filled 'outside of the match'

MATCH–OTHER NRMP & NON-NRMP MATCHING PROGRAMS

ARMED FORCES MATCH Similar in format to the NRMP; is the route by medical school graduates with military obligations are paired to their residency slots

CANADIAN MATCH Similar in format to the NRMP, begun in 1970

COUPLES MATCH The NRMP allows any 2 people to be matched with residency programs in the same region

OSTEOPATHIC MATCH Similar in format to the NRMP, run by the National Matching Service; requires a 1-year osteopathic internship

SHARED SCHEDULE MATCH The NRMP allows 2 people to share the duties and responsibilities of 1 residency position; the SSM requires double the time in residency and may be chosen by those with family or research responsibilities

SPECIALTY MATCHES Dermatology, neurology, neurosurgery, ophthalmology, otorhinolaryngology–ENT, and urology each have their own matches

match test RESPIRATORY THERAPY A screening test of expiratory capacity and expiratory airway obstruction, in which lighted match is held at 10-15 cm from the opened mouth; failure to blow out the match after 6 tries–a finding typical of advanced emphysema–indicates a maximum breathing capacity < 40 L/min, and maximum mid-expiratory flow rate of < 0.6 L/sec

matching CLINICAL TRIALS A process by which a study group and comparison group are 'equalized' with respect to factors viewed as extraneous to the study's central questions, that might interfere with interpreting data on study completion. See Case-control study. Cf Match

matching grant ACADEMIA Non-peer-reviewed funding in which a commercial enterprise, foundation, or philanthropy, federal government, contributes a sum of money that 'matches' a financial contribution made by an institution, university or hospital. See 'Approved but not funded'

material VOX POPULI Stuff; lots of stuff; related stuff. See Barium contrast material, Biomaterial, Cross-reacting material, Flammable material, Healos bone-grafting material, Interphotoreceptor material, Licensed material, Open cell material, Pressure-generating material, Pyrophoric material, Radioactive material, Reactive material, Water-reactive material, Z-band material

Material Safety Data Sheet MSDS OCCUPATIONAL SAFETY A document containing information and instructions on hazardous materials present in the workplace; MSDSs contain details about hazards and risks relevant to the substance, requirements for its safe handling, and actions to be taken in the event of fire, spill, or overexposure. See Hazardous materials

maternal age effect The adverse impact that ↑ age has on obstetric events: ↑ complication rate, ↑ fetal defects–possibly due to an unknown effect of aging on the uterus and eggs; ↑ frequency of non-disjunctional events in Down syndrome and other aneuploidies, as well as Prader-Willi syndrome, in which there is uniparental disomy of chromosome 15. See Elderly primigravida. Cf Paternal age

MATERNAL AGE & CHROMOSOME DEFECTS		
AGE	TRISOMY 21	OTHERS
<20	1/1900	1/526
25	1/1200	1/476
30	1/885	1/384
35	1/365	1/178
40	1/109	1/63
45	1/32	1/18
49	1/12	1/7

maternal deprivation syndrome Non-organic failure to survive NEONATOLOGY A condition affecting socioeconomically disadvantaged infants and toddlers FACTORS Absent father or support group–family, friends, etc; low maternal education or socioeconomic status, mental illness, teen or unwanted pregnancy. See Failure to thrive

maternal-fetal conflict MEDICAL ETHICS A medicolegal dilemma that arises when a mother wishes to carry out an activity–eg, drinking alcohol or working at a job with an occupational exposure to high levels of lead, that is potentially harmful to her fetus. Cf Emancipated minor

maternal immunity Passive immunity Immunity in a neonate provided by IgG antibodies from the mother passing across the placenta to the fetus; MI is immunoprotective for up to 6 months

maternal-infant transmission EPIDEMIOLOGY Transmission of a pathogen from mom to infant. See Vertical transmission. Cf Horizontal transmission

maternal inheritance GENETICS An inheritance pattern displayed by mitochondrial genes that are propagated from one generation to the next through mom; the mitochondria of the zygote come almost entirely from the ovum. See Gene, Inheritance, Mitochondria, Zygote

maternal mortality rate EPIDEMIOLOGY The number of pregnancy-related deaths/100,000 ♀ of reproductive age; the number of maternal deaths related to childbearing divided by number of live births–or number of live births + fetal deaths/yr. Cf Maternal mortality ratio

maternal mortality ratio EPIDEMIOLOGY The number of pregnancy-related deaths/100,000 live births. Cf Maternal mortality rate

maternity home OBSTETRICS Birth center, see there SOCIAL MEDICINE A residence for pregnant ♀

mathematics disorder Math disability PSYCHOLOGY A condition characterized by difficulties in performing simple mathematic equations–eg, counting, adding. See Math anxiety

matrix LAB MEDICINE The principal constituents of a material of interest; for blood chemistries, the matrix includes serum, specific proteins, and synthetic material. See Decision matrix, Four cell diagnostic matrix, Job exposure matrix, Interference, Matrix effect, Matrix interference, Scoring matrix

matrix point NEUROLOGY A type of trigger point that is a primary site of local pain, or referred pain located at a distance–eg, in the shoulders or neck muscles, from the primary pain–eg, the head. See Bonnie Prudden myotherapy, Trigger point; Cf Satellite point

matroclinous inheritance GENETICS Inheritance in which all progeny have the mother's phenotype. See Inheritance, Phenotype, Progeny

matter ANATOMY Stuff that occupies cavities. See Gray matter, White matter

maturity onset diabetes mellitus of the young MOLECULAR ENDOCRINOLOGY A mild form of type 2 DM, with disease onset before age 25; MODY is linked to a DM susceptibility gene possibly on 20q13 and a defect of the glucokinase gene on chromosome 7. See Type 2 DM

Maveral Fluvoxamine, see there

Mavik® Trandolapril, see there

max An enzyme that interacts with myc, the protein product of *myc*, by means of a leucine zipper motif, instructing cells to mature, divide, or undergo autodestruction or apoptosis

Maxalt® Rizatriptan benzoate NEUROLOGY An agent used for the acute management of migraine. See Migraine

Maxamine® Epileukin ONCOLOGY A phagocyte H2 receptor agonist administered with IL-2 and IFN-alpha to manage AML, melanoma, myeloma, renal cell CA, prostate CA, HCV infection. See IL-2, IFN-alpha

Maxaquin® Lomefloxacin, see there

maxide A potassium-sparing diuretic that inhibits aldosterone–resulting in ↓ sodium retention. See Diuretics, Potassium-sparing diuretic

maxillofacial injury TRAUMA SURGERY Any injury of the upper jaw bone and face. See Le Fort fracture

maxillofacial surgeon A surgeon who deals with the diagnosis and treatment of diseases, injuries, and deformities of the mouth and supporting structures

maximum androgen ablation A form of hormonal therapy for prostate CA using combined therapy–eg, gonadotropin-releasing hormone agonist plus flutamide, to ↓ effects of gonadal and adrenal androgens

maximum containment facility PUBLIC HEALTH A 'level 3 to 4' research facility that is equipped to, and experienced in, handling exotic, dangerous, and potentially life-threatening pathogens–eg, Ebola, Lassa fever viruses. See Biosafety, Ebola virus

maximum tolerable dose The highest dose of a substance that can be given without causing serious weight loss or other signs of toxicity. See Maximum contaminant level

maximum voluntary ventilation Maximum beathing capacity A nonspecific clinical benchmark of the integrated functionality of the airways, lung tissue, thoracic cage, and respiratory muscles. See Pulmonary function tests

Maxipime® Cefepime, see there

May-Hegglin anomaly HEMATOLOGY A rare AD condition characterized by the triad of thrombocytopenia, giant platelets, and Döhle-body-like inclusions in WBC cytoplasm CLINICAL Recurrent epistaxis, gingival bleeding, easy bruising, menorrhagia, excess bleeding postsurgery MANAGEMENT Most Pts are asymptomatic; corticosteroids, splenectomy are useless; platelet transfusions may be needed. See Hereditary thrombocytopenia

Mayo Clinic syndrome A popular term for the treatment of a Pt as organ or organ system by a specialist, named after the Mayo Clinic, a 'center of excellence' dedicated to the diagnosis and treatment of difficult diseases. See Center of Excellence. Cf Marcus Welby syndrome

mayoism Mayo-ism Any term coined by health care workers at the Mayo Clinic. See Vasculopath

maze procedure Cox procedure HEART SURGERY A procedure used to manage A Fib. See Atrial fibrillation. Cf Corridor procedure

mazindol Teronac PHARMACOLOGY An agent that inhibits cocaine binding to dopamine transporters in the brain. See Cocaine

MB Megabyte, see there

MB fraction CK-MB, see there

mBACOD ONCOLOGY A chemotherapy protocol–MTX, bleomycin, doxorubicin–Adriamycin, cyclophosphamide, vincristine–Oncovorin, dexamethasone; with folinic acid and GM-CSF to stimulate WBC production ADVERSE EFFECTS–MAY OVERLAP SX OF AIDS N&V, hair loss, leukopenia; B-symptoms–eg, fever, night sweats, weight loss of ≥ 10%; enlarged and asymmetrical lymph nodes. See AIDS, BACOD. Cf CHOP

MBC Minimum bactericidal concentration, see there

MBD Minimal brain dysfunction, see there

MBEST A type of MRI consisting of a heavily-weighted T2 sequence, based on the echo-planar technique, used for ultra-high-speed imaging of the brain, which allow rapid screening, functional imaging and analysis of the CSF fluid and blood flow patterns and rapid 'shooting' of restless Pts. See BEST, Magnetic resonance imaging

MBP Mean blood pressure

MCAD deficiency Medium-chain acyl-CoA dehydrogenase deficiency, see there

McArdle's disease Glycogen storage disease V METABOLIC DISEASE An AR condition caused by muscle phosphorylase deficiency CLINICAL Exercise intolerance, premature fatigue, myalgia, and cramping

MCAT Medical college admission test, pronounced, EM-cat A preadmission exam administered by the Psychological Corp., required in the US before entrance to medical school. See Graduate medical education, Medical student

McBurney's incision Gridiron incision SURGERY A muscle splitting abdominal incision descibed in 1894 by McBurney, used for appendectomy; it parallels the external oblique , 2.5 to 5 cm from the right ilium's anterior superior spine, through the external oblique to the internal oblique and transversalis muscles. See Appendicitis

McBurney's point PHYSICAL EXAM The point on the right lower flank, midway between the anterosuperior iliac spine and belly button, that corresponds to the appendix; finger pressure on MP elicits a major pain response–*Dang!, that HURTS*–from Pts with acute appendicitis. See Appendicitis

McCollough effect A phenomenon observed by persons who have worked for a prolonged period with a computer monitor that displays green lettering on a darkened background, who find that white paper acquires a pink hue. See Video display unit

McCune-Albright disease Albright syndrome, osteitis fibrosa cystica, polyostotic fibrous dysplasia An AD condition attributed to altered regulation of cAMP CLINICAL Precocious puberty, polyostotic–cystic fibrous dysplasia–spontaneous fractures at young age, café-au-lait spots of skin, ovarian cysts, endocrinopathy–eg, hyperthyroidism, hypophosphatemia and cyclical–4-6 wk-fluctuations of plasma estrogen; afflicted young girls have ↓ gonadotropins, ↓ response to LH-RH LAB ↑ Testosterone, ↑ alk phos TREATMENT Aromatase inhibitor testolactone. See Luteinizing hormone-releasing factor

McDougal clamp UROLOGY A clamp used for dissection of the dorsal vein complex during radical retropubic prostatectomy procedures. See Prostatectomy

McDougall diet NUTRITION A strictly vegetarian diet with ↓ fat, sugar, salt, ↑ fiber, ↑ complex carbohydrates, low cholesterol. See Diet, Ornish regimen, Pritikin diet, Vegan

mcg Microgram, see there; the preferred written abbreviation is µg

McGill Pain Questionnaire NEUROLOGY A 2-part instrument used to evaluate subjective components of pain

MCGN Minimal-change glomerulonephritis, see there

MCH 1. Maternal and Child Health 2. Mean corpuscular hemoglobin A measurement of the Hb per individual erythrocyte REF RANGE 26-34 pg/red cell–SI

MCHC Mean corpuscular/cell hemoglobin concentration A value derived on automated cell counters from measured parameters REF RANGE 31-36 g/dl

MCL Median collateral ligament, see there

MCL injury Median collateral ligament injury, see there

MCLN Mucocutaneous lymph node syndrome. See Kawasaki's disease

McMurray sign ORTHOPEDICS A clinical maneuver used to identify the presence of a torn meniscus. See Meniscus

MCO Managed care organization, see there

MCT 1. Mean circulation time 2. Medullary carcinoma of the thyroid

MCTD Mixed connective tissue disease, see there

MD 1. Macular degeneration, see there 2. Medical department 3. Medical discharge 4. Medical doctor–*medicinae doctor*, see there 5. Megadalton 6. Mitral disease 7. Moderately differentiated–refers to cancer differentiation 8. Multinomial distribution–statistics 9. Muscular dystrophy, see there 10. Myotonic dystrophy, see there

MD/PhD ACADEMIA A person with a degree in medicine and a doctorate of philosophy. See Triple threat

MDA 1. Mento-dextra anterior–obstetrics 2. Methydopamine 3. 3,4-Methylenedioxymethamphetamine. See Ecstasy

MDC CARDIOLOGY A clinical trial–Metoprolol in Dilated Cardiomyopathy that evaluated the effect on M&M in Pts with idiopathic dilated cardiomyopathy treated with metoprolol, a beta blocker. See Heart failure, Metoprolol

MDD Major depressive disorder, see there

MDILog™ monitor THERAPEUTICS A portable, electronic device that monitors and logs metered dose inhaler–MDI use, with warning tones to alert Pt of improper usage. See Compliance

MDM 1. Methylenedioxymethamphetamine. See Ecstasy 2. Mid-diastolic murmur. See Murmur

MDMA 3,4 methylenedioxy-methamphetamine. See Ecstasy

MDPIT CARDIOLOGY A clinical trial–Multicenter Diltiazem Post Infarction Trial in which Pts were randomized 3-15 days post MI to receive diltiazem or a placebo. See Diltiazem, Myocardial infarction

MDR FDA See MDR system NUTRITION Minimum daily requirements, see there ONCOLOGY Multidrug resistance Simultaneous cross-resistance to multiple chemotherapeutic agents–eg, antitumor antibiotics–eg, daunorubicin, vinca alkaloids and epidophyllotoxins; mutants resistant to the effect of drug X occur with a frequency of $10^{-5}–10^{-8}$

MDR system Medical Device Reporting system FDA A program requiring manufacturers and importers of medical devices to inform the FDA should they receive information suggesting that one of their devices may have caused/contributed to death/serious injury. See Medical device

MDS 1. Maternal deprivation syndrome, see there 2. Myelodysplastic syndrome, see there

'me too' drug PHARMACOLOGY A popular term for a generic drug with an identical formulation and stated indications as a previously FDA-approved agent; MTDs are usually approved 'automatically' based on their virtual identity with other formulations. See Seeding trial, Switch campaign

MEA Multiple endocrine adenomatosis. See Multiple endocrine neoplasia

mean STATISTICS The sum of values divided by number of values. See Arithmetric mean, Geometric mean, Weighted mean

measles Hard measles, rubeola, 10-day measles A highly contagious viral infection primarily of children characterized by high fever, lethargy, cough, conjunctivitis, coryza, and a maculopapular rash; measles is prevented by vaccination; it is airborne and can spread quickly, especially in school DIAGNOSIS Clinical Hx, epidemiologic evidence–ie, other children have same complaints, direct examination of multinucleated giant cells from scrapings of buccal mucosa by LM or fluorescence microscopy LAB Lymphopenia; ↑ IgM anti-measles antibody See Atypical measles, Hard measles, Subacute sclerosing panencephalitis

measurable disease ONCOLOGY A condition characterized by a lesion or tumor that can be measured in size, information which is used to evaluate response to treatment. See Occult disease

measure MEDTALK A scale by which a thing can be quantified. See Binary outcome measure, Health-related quality of care measure, Outcomes measure, Performance measure

measurement MANAGED CARE See Measure MEDTALK The International System–SI officially sanctions the use of certain prefixes for SI units. See SI

MEASUREMENT

googa	100^{100}	deci	10^{-1}
exa	10^{18}	centi	10^{-2}
peta	10^{15}	milli	10^{-3}
tera	10^{12}	micro	10^{-6}
giga	10^{9}	nano	10^{-9}
mega	10^{6}	pico	10^{-12}
kilo	10^{3}	femto	10^{-15}
hecto	10^{2}	atto	10^{-18}
deka	10^{1}	zepto	10^{-21}

Note: 10^{12} corresponds to a British billion

10^9, the American billion, to a British milliard

meatal stenosis Urethral meatal stenosis UROLOGY Meatal closure, seen only in ♂, virtually exclusive to uncircumcised penises, due to irritation of the urethral opening at the penile tip, leading to tissue overgrowth and scarring; back pressure causes painful urination and occasionally bleeding at the end of urination. See Circumcision

mebendazole Vermox® PARASITOLOGY A systemic anthelmintic that interferes with glucose metabolism, used to manage ascariasis–roundworms, enterobiasis, oxyuriasis–pinworms, trichuriasis–whipworms, uncinariasis–hookworm ADVERSE EFFECTS Fever, rash, itching, sore throat, asthenia. Cf Anthelmintic

mecamylamine SUBSTANCE ABUSE A ganglion-blocker approved for HTN, possibly smoking cessation, to be used with nicotine replacement therapy–eg, gum or patches; it may ↓ tics and rage reactions in children with Tourette's syndrome. See Nicotine replacement therapy

mechanical infertility GYNECOLOGY Infertility attributable to a blockage in passage–eg, status post tubal ligation—of one or both participants in fertilization. See Ectopic pregnancy. Cf Transcervical balloon tuboplasty

mechanical infusion pump THERAPEUTICS A device for controlled administration of therapeutics–eg, analgesics, antimicrobials, chemotherapy. See Infusion device, Minibag. Cf Central venous catheter

mechanical instability ORTHOPEDICS Instability where injury to a joint results in pathological laxity. See Instability. Cf Functional instability

mechanical restraint Physical restraint A device used on a person to restrict free movement–eg, seatbelt, straitjacket–camisole, vest, or physical confinement INDICATIONS Unsteadiness, wandering, disruptive behavior, often 2° to psychiatric conditions and/or dementia; Pts may also require pharmacologic restraints. See Restraint, Pharmacologic restraint

mechanical ventilation Assisted mechanical ventilation Mechanically assisted respiration in which inspiration is driven at a preset frequency or triggered by the Pt INDICATIONS PO_2 < 60 mm Hg, despite non-interventional efforts–mask, bronchodilators, diuretics, and physical therapy, chest wall restriction–kyphoscoliosis, thoracoplasty, COPD, CNS and brainstem disease–central apnea, 1° alveolar HTN, tumors, vascular malformation, degenerative disease–Shy-Drager disease, spinocerebellar degeneration, neuromuscular disease–amyotrophic lateral sclerosis, multiple sclerosis, muscular dystrophy, myopathy, phrenic nerve damage, poliomyelitis, spinal cord disease–cervical trauma, quadriplegia, syringomyelia

mechanism MEDTALK The manner by which a process occurs; the arrangement or association of the elements or parts of a thing in relation to the effect generated

mechanism of death FORENSIC MEDICINE '…the process that causes one or more vital organs or organ systems to fail when a fatal disease, injury, abnormality, or chemical insult occurs; it is the functional–physiologic or structural change that makes independent life no longer possible after a lethal event has occurred EXAMPLES Hemorrhage, hypovolemic shock, acidosis, alkalosis, asystole, V fib, respiratory depression and paralysis, cardiac tamponade, sepsis with profound bacterial toxemia, etc; MODs are divided into terminal events and nonspecific physiological–functional derangements. Cf Immediate cause of death, Underlying cause of death

mechanotherapy Any method–eg, massage, or device, used to mechanically stimulate a part, usually on the outer human chassis. See Bodywork, Massage therapy

mechlorethamine ONCOLOGY An alkylating nitrogen mustard chemotherapeutic used for lymphomas ADVERSE EFFECTS GI Sx, BM suppression, skin vesiculation at injection site. See MOPP regimen

Meckel's diverticulum An outpouching in the small intestine, which corresponds to the omphalomesenteric or vitelline duct remnant located along the antimesenteric border; MD may contain gastric tissue and be associated with peptic ulcer disease–ulceration, perforation, or pancreatic tissue, or be linked to small bowel obstruction

meclofenamate Meclomen® THERAPEUTICS An oral anthranilic-type NSAID excreted via the kidneys and GI tract. See NSAIDs

meconium A dark green viscid glue-like material found in the intestine of all neonates; it is the first stool passed by the newborns and is passed in the first 24-48 hrs of life; prenatal passage of meconium may indicate fetal distress

meconium aspiration syndrome NEONATOLOGY A symptom complex caused by the aspiration of meconium at the time of delivery, especially if post-term CLINICAL Low Apgar scores, tachypnea, dyspnea, cyanosis, which either resolves in the first 3 days of life, or if aspirated meconium was significant, patchy infiltrates on CXR, with atelectasis, emphysema, rales RISK FACTORS Post-term, difficult delivery, fetal distress, intrauterine and/or peripartum hypoxia, triggering increased GI peristalsis. Cf Apgar

meconium ileus Meconium plug syndrome OBSTETRICS A condition characterized by obstruction of the neonatal intestine by meconium that may be confined to the ileus, a finding characteristic of cystic fibrosis CLINICAL Nonpassage of stool in the first 2 days of life, vomiting, abdominal distension COMPLICATIONS Volvulus, intestinal infarction

meconium-stained amniotic fluid Amniotic fluid with a greenish discoloration, which may indicate fetal distress. See Fetal distress

Mectizan® Ivermectin, see there

MED 1. Mental or emotional disability/disorder 2. Microendoscope diskectomy

MEDAC syndrome Multiple endocrine deficiency, Addison's disease, candidiasis. See APECED–autoimmune polyendocrinopathy-candidiasis-ectodermal dystrophy

Medea complex PSYCHIATRY Murderous hatred by a mother for her child or children, arising from a desire for revenge on her husband

MEDENOX CARDIOLOGY A clinical trial–Medical patients with ENOXaparin

media epidemic A popular term for a flurry of interest displayed by the news media–newspapers, television, radio for an important medical event–eg, a new therapy, 'breakthrough', or disease, which follows a report in a major medical journal. See Embargo arrangement, Ingelfinger rule

media malpractice The misuse of the mass media to slander a person or group, provide an unbalanced viewpoint, or otherwise misinform the public. See Embargo rule

medial *adjective* Toward the center, middle, median of the body. Cf Lateral

medial collateral ligament injury MCL injury ORTHOPEDICS An injury to the collateral tibial ligament, which results in medial instability of the knee

MEDIAL COLLATERAL LIGAMENT INJURY

FIRST DEGREE MCL is stretch over its length; knee joint tightness is unchanged

SECOND DEGREE A partial tear of the medial collateral ligament from the bone attachment or anywhere else along its length

THIRD DEGREE A complete tear of the medial collateral ligament from the bone attachment or anywhere else along its length

medial tibial stress syndrome SPORTS MEDICINE A condition characterized by dull, aching, diffuse pain along the posteromedial shin, which may be linked to stress fractures of the tibia. Cf Chronic exertional compartment syndrome

median nerve dysfunction Carpal tunnel syndrome, see there

median rhomboid glossitis A condition now considered a manifestation of chronic infection by *C albicans*, facilitated by hyperglycemia as in DM CLINICAL Rhomboid or diamond-shaped red plaque or patch on the dorsum of the tongue anterior to circumvallate papillae TREATMENT Nystatin or amphotericin B may cause regression

mediastinal crunch MEDTALK A substernal crepitance, often synchronous with the heartbeat, caused by percolation of air into the mediastinum–emphysema–eg, due to tracheal perforation or due to expansion of gas in rapid ascent in scuba divers. See Mediastinal emphysema

mediastinal emphysema Pneumomediastinum SURGERY Air in mediastinal soft tissues which may be linked to perforation of the trachea. See Mediastinal crunch, Mediastinum

mediastinal tumor SURGERY Any benign or malignant mediastinal mass–eg, epithelial lesions, lymphomas, thymoma, pseudotumors. See Mediastinum

mediastinitis SURGERY Inflammation of the mediastinum. See Mediastinoscopy, Mediastinum

MEDIASTINITIS

ACUTE MEDIASTINITIS A fulminant infectious process with a high M&M, in which organisms rapidly spread through areolar planes of mediastinum ETIOLOGY Traumatic esophageal perforation, foreign bodies, suture line leakage, post-emetic rupture CLINICAL Chest pain, dysphagia, respiratory distress, cervical-upper thoracic subcutaneous crepitus CXR May be normal early, followed by evidence of mediastinal and subcutaneous air MANAGEMENT Aggressive and early with antibiotics, fluid resuscitation, chest tubes for pneumothorax or effusions, thoracotomy and surgical repair

CHRONIC MEDIASTINITIS A relatively indolent process characterized by chronic–'round cell' inflammation and fibrosis *Etiology* Most often the result of a granulomatous process, eg TB, histoplasmosis, but may be idiopathic, as in sclerosing mediastinitis *Clinical* Often silent if no esophageal obstruction occurs *CXR* May be normal or mediastinum widened *Management* Thoracotomy to confirm diagnosis or relieve obstruction

mediastinoscopy SURGERY A procedure in which an endoscope is inserted in the mediastinum, and regional structures—lungs and lymph nodes are visualized to detect neoplasms or other lesions needing evaluation or therapy; mediastinoscopy may be part of a 'staging procedure', where a Pt has a known malignancy, and metastases are identified to determine further management INDICATIONS Widened mediastinum of unknown cause, cancer staging, confirmation of TB or sarcoidosis, diagnosis of mediastinal fibrosis TYPES Cervical mediastinoscopy–for right paratrachial and subcarinal LNs; anterior mediastinoscopy–for left mediastinum, especially in presence of left upper lobe lesions; anterior mediastinotomy–on either side COMPLICATIONS < 3%; hemorrhage, pneumothorax, vocal cord paralysis, esophageal perforation. See Endoscopy, Mediastinotomy, Thoracotomy

mediastinotomy PULMONOLOGY A minimally invasive surgical procedure used to gain direct access to mediastinal structures for visualization and biopsy INDICATIONS Lesions or suspected lesions of the mediastinum which are inaccessible by mediastinoscopy INDICATIONS Idiopathic mediastinal widening, cancer staging, confirmation of TB or sarcoidosis COMPLICATIONS < 3%; hemorrhage, pneumothorax, vocal cord paralysis, esophageal perforation. See Mediastinoscopy, Thoracotomy

Medicaid MEDICAL PRACTICE A federally-funded, state-operated and administered program authorized by Title XIX of the Social Security Act of 1965, which provides medical assistance to low-income groups–eg, elderly, blind, disabled, single-parent families, unemployed under age 65. See HMO

Medicaid buy-in MANAGED CARE A proposed system to allow those who are not eligible for Medicaid coverage to enroll by paying premiums on a sliding scale

Medicaid fraud The fraudulent billing of Medicaid by physicians or other health care providers, especially international medical graduates and psychiatrists. See Medicaid

Medicaid mill A for-profit enterprise that provides health care, usually ambulatory, where few medical services are available–eg, inner city and rural communities. See Family 'ganging', 'Ping-ponging'

Medicaidization HEALTH FINANCING The shift of costs for medical care of financially disadvantaged Pts to the public sector. Cf 'Dumping', 'Safety net' hospital, 'Skimming'

medical *adjective* Referring to medicine *noun* See Major medical

medical abortion OBSTETRICS An elective nonoperative abortion effected in the 1st trimester by abortifacients. See Abortion

medical abuse MEDICAL ETHICS The use of excessive life support or resuscitative measures for a terminally ill Pt. See End-of-life debate. Cf Abuse

medical castration ENDOCRINOLOGY The use of drugs to suppress the function of the ovaries or testicles. See Castration

medical center MANAGED CARE A health care organization defined by **STRUCTURE**–a physical plant–eg, a hospital and buildings in which health care, research, staff support, and ancillary services are provided and **FUNCTION**–medical services that may be more complex than that provided by a traditional community hospital

medical condition A disease, illness, or injury MEDICAL PRACTICE Any condition–eg, physiologic, mental, or psychologic conditions or disorders–eg, orthopedic, visual, speech, or hearing impairments, cerebral palsy, epilepsy, muscular dystrophy, multiple sclerosis, CA, CAD, DM, mental retardation, emotional or mental illness, specific learning disabilities, HIV disease, TB, drug addiction, alcoholism

medical device Any article or health care product intended for use in the diagnosis of disease or other condition or for use in the care, treatment, or prevention of disease that does not achieve any of its primary intended purposes by chemical action or by being metabolized EXAMPLES Diagnostic test kits, crutches, electrodes, pacemakers, catheters, intraocular lens. See Class I, II, III devices, Significant risk device, Silicone implant. Cf Nonsignificant risk

medical direction MEDICAL PRACTICE The direct oversight of procedures performed by others–eg, by resident physicians, CRNAs, physician assistants See Physician extender. Cf Medical supervision

medical directive END-OF-LIFE DECISIONS A specific and comprehensive advance care document–being developed for health care at the end of life. See Advance directive, Durable power of attorney, Living will

MEDICAL DIRECTIVE–OPTIMAL COMPONENTS

INTRODUCTION Provides an explanation of the document's purpose

PARADIGMATIC SCENARIOS Provides examples that help the individual understand various illness circumstances and evaluate the types of life-sustaining interventions that might be employed; the PSs would–in theory–help the individual designate his/her preferences with respect to specific treatments

PROXY DECISION-MAKER Section provides details on who would make the decisions in the event of the individual becoming mentally incompetent

ORGAN DONATION Yes/no, what, to whom, for what

PERSONAL STATEMENT The individual's 'wrap-up'

medical documentation MANAGED CARE A term relating to a Pt care record or medical record, which may be written and include EKG, tracings, X-rays, videotapes and other media; typically, operative notes, progress notes, physician orders, physician certification, physical therapy notes, ER records, other notes and/or written documents are included

medical ethics The moral construct focused on the medical issues of individual Pts and medical practitioners. See Baby Doe, Brophy, Conran, Jefferson, Kevorkian, Quinlan, *Roe v Wade*, Webster decision

medical examiner FORENSIC MEDICINE A medical doctor–MD or DO–who performs postmortem examinations on decedents; MEs are public officials appointed by a particular jurisdiction–usually a state whose chief responsibility is to investigate and provide official interpretation regarding the manner and possible cause(s) of unexplained deaths. See Forensic pathology. Cf Coroner

medical 'failure' A Pt who does not respond to a non-interventional modality used for a non-malignant but potentially pernicious condition, and who, due to this failure, may benefit from surgery

medical fetal therapy Fetal therapy OBSTETRICS Any therapy in which a pharmacologic agent is administered to a ♀ or her fetus in order to avoid or alleviate fetal disease EXAMPLES Use of blood products or other agents that prevent neural tube defects, endocrine or metabolic disorders–eg, congenital adrenal hyperplasia, thyroid disease

medical food NUTRITION A food formulated for enteric administration under physician supervision and intended for the specific dietary management of a disease or condition for which distinctive nutritional requirements are established by medical evaluation. See Diet, Total parenteral alimentation

406 **medical fragility** SOCIAL MEDICINE A state of medical lability and ↑ risk of clinical deterioration due to an underlying disease–eg, as in crack cocaine babies. See Lability

medical futility Futile resuscitation, futility BIOMEDICAL ETHICS A subjective term that encompasses a range of probabilities that a Pt will benefit from efforts designed to improve his life and survive to discharge from a health care facility MEDTALK The lack of efficacy of a particular maneuver in ↓ M&M. See Advance directive, DNR, Futility. See DNR orders. Cf Euthanasia

medical history CLINICAL MEDICINE The part of a Pt's life Hx important for determining the risk factors for, diagnosing, and treating a disorder–eg, history of exposure, Sx, occupational, exposure to causative agents linked to a particular condition, infection or cancer VOX POPULI → MEDTALK Anamnesis

medical informatics COMPUTERS The use of computers and information technology in support of Pt care, teaching, biomedical research. See Artificial intelligence, Computers, Expert system, Electronic journal, Electronic publishing, MEDLINE, Online database

medical mall BUSINESS OF MEDICINE A free-standing unit that provides a "one-stop shopping" format for delivering health care MM EXAMPLES Cardiac evaluation, imaging, outpatient surgery, rehabilitation therapy, pulmonary clinics, etc. See One-stop shopping

medical malpractice LAW & MEDICINE Negligent conduct or unreasonable lack of skill in the performance of a medical task, on the part of a physician or a party–eg, a health care facility in which that act or task occurs STATISTICS, US In a 5-yr period, 48% of surgeons and surgical specialists, 34% of obstetricians-anesthesiologists and 15% of other physicians had malpractice claims; 85% of all payments were made on behalf of 3% of policy holders. See Abandonment, Assault and battery, Blood shield laws, Borrowed servant doctrine, Causation, Compensatory damages, Confidentiality, Consent, Contributory neglect/negligence, Countersuit, Damages, Defensive medicine, 'Difficult Pt', DNR, Emergency doctrine law, Emergency psychiatric committment, Emotional distress, Expert witness, Frivolous lawsuit, Liability, Good Samaritan laws, Informed consent, Jehovah's Witness, Medical record, Misdiagnosis, Negligence, Patient-physician relationship, Punitive damages, Quinlan case, Referral and consultation, *Res ipsa loquitur*, *Respondeat superior*, Standard of care, Statute of limitations, Therapeutic privilege doctrine, Tort, Wrongful birth

medical massage Rehabilitation massage, see there

Medical Mercy Mission Inc PRO BONO PUBLICO A nonprofit organization founded that sponsors 30 missions each yr to treat indigent Pts in the US and internationally

medical movie POPULAR MEDIA A film–eg, Coma, Hospital, MASH, Patch Addams, etc, that fictionalizes a theme based on some aspect of medical care or care gone awry. See ER

medical necessity MEDICAL PRACTICE A general term referring to diagnostic procedures needed to detect a potentially treatable condition, or therapeutic modalities required by a particular condition

medical oncologist ONCOLOGY An oncologist who diagnoses and treats cancer with chemotherapy, hormones, biologicals, or immunologic agents; the MO becomes a cancer Pt's de facto primary care giver, and coordinates treatment provided by other specialists. See Oncologist, Radiation oncologist. Cf Surgical oncologist

medical paternalism MEDICAL ETHICS A philosophy that certain health decisions–eg, whether to undergo heroic surgery, appropriateness of care in terminally ill Pts, are best left in the hands of those providing health care. See *Arato v Avedon*. Cf Informed consent

medical privacy Clinical privacy LAW & MEDICINE The right of a Pt not to have personal health-related or other information open to public view;

freedom from public dissemination of personal medical records. See Anne Sexton, Bennett-Leahy bill, Health information trustee, Nondisclosure ceiling, Therapist-patient privacy

medical record The documents pertaining to a Pt's medical history, diagnoses and therapies, and status when last seen by health care providers. See Documentation, Hospital chart, Problem-oriented medical record, SOAP

medical risk CLINICAL DECISION-MAKING The risk of an adverse outcome when performing a diagnostic or therapeutic procedure. See Dread risk, Risk, Unknown risk

medical school GRADUATE EDUCATION A period of formal education–4 yrs in the US, prerequisite 4 yrs of undergraduate education, generally divided into 2 yrs of basic sciences–eg, anatomy, physiology, pathology, pharmacology, etc, and 2 yrs of clinical sciences in which students have courses that combine lectures with hands-on learning experience in a hospital setting in OB, surgery, psychiatry, etc. See International medical graduate, Offshore medical school

medical school syndrome Medical student syndrome, see there

medical service HOSPITAL PRACTICE A health care-related service, which includes orthodontics and psychoanalysis. See Service. Cf Surgical service

medical services organization MANAGED CARE A for-profit entity that offers medical practice management services to physicians and hospitals

medical Spanish MEDTALK Spanish language as applied to the provision of health care and communication. See Hispanic, Spanish

medical specialty Any specialty that provides non-interventional Pt management, ie with drugs, or with minimum intervention–eg, balloon catheterization EXAMPLES Internal medicine–allergy and immunology, cardiology, gastroenterology, hematology/oncology, neurology, infectious and pulmonary diseases, dermatology, pediatrics, psychiatry, occupational medicine, preventive medicine, aerospace medicine. Cf Hospital-based physicians, 'Surgical' specialty

medical student syndrome Medical school syndrome PSYCHOLOGY A collection of psychosomatic Sx that "appear" in a medical student learning about a particular disease in medical school. See Hypochondriasis

medical staff Staff HOSPITAL PRACTICE The organized body of licensed physicians and other health care providers, who are permitted by law and by a hospital–through admitting privileges–to provide medical care, within that hospital or health care facility. See Admitting privileges, House staff, Staff courtesy, Staff privilegesSummary suspension

medical student abuse A common practice, in which medical students are psychologically 'abused' by superiors—interns, residents, fellows, attending physicians who badger, belittle and have them perform menial, degrading tasks. See Pimping, Scut work

medical student debt The amount of financial obligations or monies owed with (accruing) interest to various parties by a medical student

medical supervision MEDICAL PRACTICE The indirect oversight of multiple procedures performed by others–eg, resident physicians, CRNAs, physician assistants See Physician extender. Cf Medical direction

medical surveillance EPIDEMIOLOGY The monitoring of individuals who may have been exposed to a pathogen or noxious thing in the environment, or detection of the early symptoms of disease

medical team The group of physicians and health care workers who are responsible for a Pt's medical needs

medical tourism GLOBAL VILLAGE The combination of medical care–usually of a type–eg, pediatric neurosurgery or experimental cancer therapy, that is not available to a Pt near his/her home–with pleasure–eg, visit to a theme park, Disneyworld, MGM studios

medical underwriting MANAGED CARE The process of determining the medical needs of an individual or group before providing coverage. See Health insurance

medical waste PUBLIC HEALTH Any waste–regardless of whether it is potentially infectious–generated as a result of Pt diagnosis and treatment. See Hospital waste, Infectious waste, Medical Waste Tracking Act of 1988

medicalization SOCIAL MEDICINE A term for the erroneous tendency by society–often perpetuated by health professionals–to view effects of socioeconomic disadvantage as purely medical issues

medically necessary MANAGED CARE *adjective* Referring to a covered service or treatment that is absolutely necessary to protect and enhance the health status of a Pt, and could adversely affect the Pt's condition if omitted, in accordance with accepted standards of medical practice. See Futility

---

MEDICALLY NECESSARY, CRITERIA

A. Appropriate for the Sx and diagnosis or treatment of a condition, illness or injury

B. Provided for the diagnosis or the direct care and treatment of the condition, illness or injury

C. In accordance with the standards of good medical practice in the service area

D. Not primarily for the convenience of a plan member or a plan provider

E. The most appropriate level or type of service or supply which can safely be provided to the plan member

---

medically underserved area A region that has a relative or absolute deficiency of health care resources–eg, hospital beds, equipment and/or medical personnel

medically uninsured A person or group that has/have no health insurance. See Underinsured

medicamentosa rhinitis ENT Noninfectious nonallergic rhinitis caused by rebound vasodilation linked to overuse of vasoconstricting intranasal sprays

Medicare MANAGED CARE A federal program run by HCFA/CMS that provides hospital and medical insurance protection for a significant minority in the US COMPONENTS-PART A Compulsory hospitalization insurance, financed by contributions from employers, employees, and participants; PART B Voluntary supplementary medical insurance, which covers outpatient services, financed in part by monthly premiums paid by enrollees and by the federal government. See HCFA, HMO, Part A, Part B. Cf Medicaid, Socialized medicine

Medicare fraud Medifraud MEDICAL PRACTICE Any unlawful act which results in the inappropriate billing of Medicare for services by a health care provider–eg, physicians, hospitals and affiliated providers. See Medicare

Medicare Identification Number MEDICARE A number–eg, UPIN, OSCAR number, NSC number–that uniquely identifies a health care provider and is used on billing forms submitted to Medicare. See NPI, NSC, OSCAR, UPIN

Medicare risk contract MANAGED CARE An HMO-like format for delivering care under Medicare in which a Pt/client pays a flat fee to a Medicare risk contractor, which is then responsible for delivering health care; a person covered under an MRC receives only listed services provided by listed providers. See Medicare

medication Drug THERAEUTICS Any chemical substance, which may be natural or synthetic, that has a medical or pharmacologic affect on the body. See Co-medication, Herbal remedy

medication administration record HOSPITAL PRACTICE A computer-generated schedule for administering medications to a Pt for a defined period of time, including physician's orders and time to adminster the agents

medication error MALPRACTICE An error in the type of medication administered or dosage. See Adverse effect, Error

Medication Event Monitoring System CLINICAL THERAPEUTICS A system with an electronic umbilical cord that records the dates and times of bottle cap openings as a means of ensuring adherence to therapeutic regimens . See Patient compliance

medication nurse EMERGENCY MEDICINE A nurse responsible for administering medications–epinephrine, procainamide, bretilium, etc–during CPR. See Cardiopulmonary resuscitation NURSING A nurse responsible in some hospitals for administering routine scheduled medications

medicine MEDTALK A discipline devoted to understanding and treating disease, often referring to physical and chemical mechanisms. Related terms are Addiction medicine, Aerospace medicine, Behavioral medicine, Botanical medicine, Boutique medicine, Cardiovascular medicine, Chinese herbal medicine, Community medicine, Complementary medicine, Cookbook medicine, Correctional medicine, Critical care medicine, Defensive medicine, Electromedicine, Emergency medicine, Energy medicine, Environmental medicine, Ethnic traditional Chinese medicine, Evidence-based medicine, Evolutionary medicine, Field medicine, Folk medicine, Fringe medicine, Gender-specific medicine, Global medicine, Herbal medicine, Humanistic medicine, Integrative medicine, Japanese medicine, Legal medicine, Low yield medicine, Mail-order medicine, Mainstream medicine, Mickey Mouse medicine, Military medicine, Mind/body medicine, Molecular medicine, Mountain medicine, Nuclear medicine, Occupational medicine, Organized medicine, Orthomolecular medicine, Palliative medicine, Patient-oriented medicine, Pain medicine, Performing arts medicine, Preclinical medicine, Preventive medicine, Psychosomatic medicine, Rehabilitation medicine, Schüssler's biochemical system of medicine, Social medicine, Socialized medicine, Space medicine, Sports medicine, Telemedicine, Traditional medicine, Traditional Chinese medicine, Transfusion medicine, Translational medicine, Travel medicine, Tropical medicine, Vibrational medicine, Wilderness medicine THERAPEUTICS A drug or therapeutic agent See Black medicine, Natural medicine, Outdated medicine, Patent medicine, Pink medicine,

medicolegal death FORENSIC MEDICINE Any death that requires official–medical examiner or coroner investigation EXAMPLES Unexpected or violent death, death of an infant or child, unidentified or prominent person, anyone not under a doctor's care; death occurring < 24 hrs after hospital admission; death involving negligence or malfunctioning equipment; death of anyone in police custody

Medicus Nursing Classification Score LONG-TERM CARE A global measure of Pt acuity that reflects the amount of nursing care required for a Pt

Medifil® WOUND CARE Highly absorbent granules, pads and paste to promote healing, remove exudates, ↓ edema and inflammation. See Wound care

medigap MANAGED CARE A generic term for employer-sponsored, individually purchased health insurance that supplements monies reimbursed by Medicare for medical services

Medipramine Imipramine, see there

MedisGroups Medical illness severity grouping system A system for classifying Pts admitted to a hospital, based on disease severity. See Kiss of death test. Cf DRGs, High mortality outlier

meditation FRINGE HEALTH A technique in which a person empties the mind of extraneous thought, with the intent of elevating the mind to a different level, and transcend–hence the term transcendental meditation–mundane concerns PHYSIOLOGIC EFFECTS ↓ heart and breathing rates, BP, brain activity. See Auto-suggestion therapy, Body scan meditation, Bubble of light technique, Color meditation, Mantra, Meditation sickness, Microcosmic orbit meditation, Yoga

Mediterranean diet NUTRITION A diet that differs by country, characterized by ↑ consumption of olive oil, complex carbohydrates, vegetables, ↓ red meat. See Diet, Mediterranean diet pyramid. Cf Affluent diet

Mediterranean fever Familial Mediterranean fever see there

Mediterranean food pyramid NUTRITION A schematic representation of recommendations on the amounts of specific food groups that should be consumed daily. See Diet. Cf Exercise pyramid, Food pyramid, Mediterranean diet

Mediterranean lymphoma Immunoproliferative small intestinal disease, see there, aka IPSID

medium IMAGING A substance used to enhance imaging of a particular structure. See Contrast medium, High-osmolality contrast medium, Low-osmolality contrast medium, INFORMATICS A material on which data is stored. See Magnetic medium, Multimedia, Output medium.

medium chain acyl-coenzyme A dehydrogenase deficiency An AR disease of fatty acid oxidation seen in the first 2 yrs of life as either sudden unexplained death at home or, if in the hospital, as Reye syndrome, due to an inability to break down fat to ketones and energy CLINICAL Intolerance to fasting, episodic vomiting, lethargy, coma, seizure, sudden death LAB Hypoketotic hypoglycemia, medium-chain dicarboxylic aciduria DIAGNOSIS Mutation analysis of paraffin-embedded blocks of postmortem tissue

medium-sized vessel vasculitis INTERNAL MEDICINE Vasculitis affecting main visceral arteries and branches EXAMPLES Polyarteritis nodosa, Kawasaki's disease, primary granulomatous CNS vasculitis. See Systemic vasculitis

MEDLARS MEDical Literature & Analysis Retrieval System MEDICAL INFORMATICS The computerized bibliographic retrieval system from the 40+ databases of Index Medicus which provides access to biomedical literature, managed by the NLM in Washington, DC. See National Library of Medicine

MEDLINE MEDICAL INFORMATICS National Library of Medicine's bibliographic database for scientific publication, and computer version of the printed Index Medicus. See National Library of Medicine

MedModeler IMAGING A device that creates 3D anatomical models by software conversion of CT and MRI data. See CT, Magnetic resonance imaging

Medoff sliding plate ORTHOPEDICS A plate designed for stabilizing and fixing proximal and supracondylar femoral fractures

medomalacuphobia PSYCHOLOGY Fear of losing an erection. See Phobia

medroxyprogesterone acetate Provera® GYNECOLOGY A contraceptive injected every 3 months, whichs inhibits gonadotropin secretion which, in turn, prevents follicular maturation and ovulation, and causes endometrial thinning ADVERSE EFFECTS Menstrual irregularities, weight changes, headache, nervousness, abdominal pain/discomfort, dizziness, asthenia–weakness or fatigue. See Contraceptive

meds A popular term for physician-prescribed medications

Medtalk Medspeak Medicine's working parlance with a core vocabulary of abbreviations, jargon, acronyms and neologisms based on the classics–Greek & Latin, video games, movies, the Internet, television, and other sources

Medtronic CARDIOLOGY A major manufacturer of pacemakers, defibrillators, and other cardiac devices

Meduck monitor ANESTHESIOLOGY A device that measures the depth of anesthesia via the sympathetic nervous system using noninvasive sensors

medullary carcinoma of breast A type of ductal carcinoma affecting ♀ < age 50, said to be more common in Japanese ♀ PROGNOSIS 84% 10-yr survival vs ductal CA, 63% 10-yr survival

medullary carcinoma of thyroid Atypical carcinoma, compact small cell carcinoma, solid carcinoma, of thyroid A tumor comprising 3-10% of thyroid CAs,

arising in the C–parafollicular–cells of ultimobranchial cleft–neural crest origin CLINICAL Tumors metastasize to cervical lymph nodes, mediastinum, lungs, liver, bone, adrenal glands; ↑ calcitonin causes hypercalcemia, HTN, paraneoplastic syndromes–eg, Cushing syndrome and neuromas; MCTs may produce ACTH-like substance, biogenic amines, CEA, corticotropin-releasing factor, NGF, prolactin-releasing hormone, PGs, melanin-stimulating hormone, histaminase, β-endorphin, 5-hydroxytryptamine, serotonin, somatostatin, thyroglobulin RISK FACTORS RT to head or neck, family Hx of multiple endocrine neoplasia, Hx of pheochromocytoma, mucosal neuromas, hyperparathyroidism TREATMENT Total thyroidectomy PROGNOSIS 70-80% 5-yr survival; sporadic MCT has a 50% 10-yr mortality–10- and 20-yr survival is 63% and 44% respectively; in contrast, papillary thyroid carcinoma has 95% 5- and 10-yr survival. See Anaplastic carcinoma of thyroid, Follicular tumor of thyroid, Papillary carcinoma of thyroid, Thyroid lymphoma

---

MEDULLARY CARCINOMA OF THYROID–CLINICAL FORMS

SPORADIC MCT 80-90% of cases, mean age of onset 45, presenting as a solitary 'cold'–by thyroid scan–nodule variably accompanied by intractable diarrhea and Cushing's syndrome

FAMILIAL MCT 10-20% of cases–in one study with ≈ 23.5 years follow-up 11% were familial, mean age of onset 35, presenting as a multifocal and bilateral mass, accompanied by C-cell hyperplasia of residual thyroid tissue; familial MCT is often AD and associated with multiple endocrine adenomatosis, usually type II–which has a germline abnormality on chromosome 10, an earlier age of onset, and is often bilateral, or occasionally type III–IIb, less common, but more aggressive and rarely a non-aggressive form of MCT that is not associated with other neural or endocrine lesions

---

medullary cystic disease Familial juvenile nephrophthisis, Senior-Loken syndrome NEPHROLOGY A group of AD and, less commonly AR, renal diseases characterized by renal cysts located at the corticomedullary junction in a background of scarring; MCD presents as functional tubular defects and Fanconi syndrome associated with azotemia, uremia and high-output renal failure 3-5 yrs after onset CLINICAL Retarded growth, renal osteodystrophy, tapetoretinal degeneration LAB Salt wasting, ↓ Na⁺, acidosis with hyperchloremia

medullary sponge kidney Polycystic kidneys, see there

medulloblastoma Primitive neuroectodermal tumor NEUROLOGY A malignant brain tumor that comprises ±¼ of intracranial tumors in children and adolescents; it is most common at the midline of the cerebellum, and on the roof of 4th ventricle CLINICAL Presents with N&V, headache, ataxia, papilledema, nystagmus, irritability, lethargy, cranial nerve palsy, dizziness, altered vision PROGNOSIS 2 yr-survival after relapse is 46% in those treated with salvage chemotherapy or RT, 0% in untreated Pts. See Brain tumor. Cf Glioblastoma

medulloepithelioma A highly malignant undifferentiated primitive neuroepithelial tumor of children that may contain bone, cartilage, skeletal muscle; it tends to metastasize extracranially; MEs of the eye arise in the ciliary epithelium, and are termed diktyomas

medullomyoblastoma NEUROLOGY Malignant teratoma A rare tumor of the cerebellar vermis of young children PROGNOSIS ≥ 50% death

Medusa head appearance A fanciful descriptor for a structure with undulating or serpentine lines radiating from a central mass HEPATOLOGY Caput medusae, see there

meet'em, treat'em, & street'em MEDTALK An aphorism for efficient health care, which entails a rapid problem-oriented workup, management and discharge. See Quicker & sicker. Cf Million-dollar workup

MEFR Maximum expiratory flow rate

mega- Greek, big, abnormally large

Megace® Megestrol acetate, see there

megacephaly Big-headedness; macrocephaly is more widely used

megacolon A massively distended colon with ↓ activity, due to defective innervation—congenital, intraluminal overgrowth of microorganisms or of psychogenic origin CLINICAL Intestinal obstruction, constipation, vomiting, abdominal distension, poor weight gain, retarded growth MANAGEMENT Temporary colostomy for bowel rest, followed by resection of affected bowel segment PROGNOSIS Sx eliminated in ± 90% of Pts with surgery; outcomes better with early intervention

MEGACOLON

CONGENITAL MEGACOLON Congenital aganglionosis, Hirschsprung's disease A disease affecting 1:5000 live births, with a sibling risk of 1% for girls and 5% for boys; Hirschsprung's disease–HD is ten-fold more common in Down syndrome; other anomalies in HD include hydrocephalus, VSD, cryptorchism, diverticulosis of the urinary bladder, renal cysts and agenesis, polyposis coli, Laurence-Moon-Biedl syndrome TREATMENT Resection of aganglionic colon

ACQUIRED MEGACOLON A condition related to narcotics or disruption of ganglionic innervation—eg idiopathic hypomotility, neuropathies–parkinsonism, multiple sclerosis, myotonic dystrophy, diabetic neuropathy, Chagas' disease, smooth muscle disorders—amyloidosis and progressive systemic sclerosis and metabolic disease—hypokalemia, lead poisoning, porphyria, pheochromocytoma, hypothyroidism and may be due to intraluminal overgrowth of microorganisms in Crohn's disease and ulcerative colitis–toxic megacolon–characterized by mucosal necrosis, transmural inflammation and systemic 'toxicity' associated with high fever, tachycardia, leukocytosis and diarrhea; in psychogenic megacolon, no radiologic or pathologic defects are present–the condition may be related to a 'fixation' in Freud's anal retentive stage of psychosexual development, with constipation of later onset than in HD, possibly 2° to abuse of anthracine laxatives

megadosing SPORTS MEDICINE A weight training term for ingesting massive amounts of steroids, by injection or pill. See Anabolic steroids, Weight training

megalencephaly NEONATOLOGY Big-brained

megaloblastic anemia Pernicious anemia HEMATOLOGY A type of anemia characterized by enlarged RBCs, ↓ WBCs, ↓ platelets ETIOLOGY MA is usually due to a deficiency of folic acid and/or vitamin B$_{12}$ or due to leukemia, myelofibrosis, myeloma, hereditary disorders, drugs affecting nucleic acid metabolism—eg, chemotherapeutics–MTX. See Anemia

megaloblastic madness NEUROLOGY The neurologic changes of vitamin B$_{12}$ deficiency—eg, altered personality, dementia, spastic weakness, ataxia, due to demyelination of lateral and posterior columns of spinal cord. See Subacute combined degeneration

megalocephaly Big-headedness; macrocephaly is in wider use

megalophobia PSYCHOLOGY Fear of large things. See Phobia

-megaly Bigness. See Hepatomegaly, Splenomegaly

Megan's law PUBLIC HEALTH A popular term for legislation aimed at registration and community notification of the movements of convicted sexual assailants. See child molestation.

megarectum A feces-filled rectum in which intrinsic electromechanical activity that would otherwise stimulate the external anal sphincter and puborectalis muscle has ceased, most common in the elderly, resulting in constipation. See Megacolon

megaureter A large ureter of any etiology, divided by some urologists into reflux megaureter, obstructed ureter and non-reflux or idiopathic megaureter

megavitamin therapy The administration of excess or 'hyper-doses' of water-soluble vitamins, either physician-guided–eg, to treat neuropathies, or self-prescribed by health-food advocates. See Decavitamin, Orthomolecular medicine, Vitamin

MEGAVITAMINS, ADVERSE EFFECTS

THIAMIN CNS hyperresponsiveness–convulsions, Parkinson's disease–thiamin antagonizes L-dopa, sensory neuropathy–destruction of dorsal axon roots

NIACIN/NICOTINIC ACID & NIACINAMIDE/NICOTINAMIDE Exacerbation of asthma–histamine release, cardiac disease–arrhythmias, GI symptoms, eg nausea, vomiting, diarrhea, anorexia, DM–hyperglycemia, gout–↑ uric acid, liver disease–enzyme leakage, hepatocellular injury, portal fibrosis or massive necrosis, cholestatic jaundice, peptic ulcer disease–histamine release, ↑ acidity, skin disease

VITAMIN B$_6$ Paresthesia, headaches, asthenia, irritability

VITAMIN C ↑ Iron absorption, possibly iron overload, evoking diarrhea, renal calculus formation and possibly inhibiting the bacteriolytic activity of neutrophils, G6PD deficiency–↑ red cell lysis, megaloblastic anemia–↓ vitamin B$_{12}$ absorption, nephrolithiasis–oxaluria Diagn Clin Testing 1990; 28:27

megestrol acetate Megace® ENDOCRINOLOGY A synthetic progestin/progestational antiandrogen that blocks estrogenic effects, androgen production and androgen-mediated activity, ↓ serum gonadotropin and testosterone ONCOLOGY A progestin used to manage weight loss, anorexia in Pts with AIDs, breast, endometrial, and rarely prostate CA ADVERSE EFFECTS Dysmenorrhea, chest pain, depression, edema in hands, feet, brown spots on skin, hirsutism, breast tenderness, HTN, venous thrombosis

meibomian cyst Chalazian, tarsal cyst OPHTHALMOLOGY Inflammation of the sebaceous gland of the eyelid

Meige syndrome NEUROLOGY A movement disorder characterized by blepharospasm, oromandibular dystonia, sometimes spasmodic dysphonia, See Blepharospasm

melan- Prefix, dark, black

melancholia melan, Greek, black; chole, bile PSYCHIATRY Psychotic depression similar/identical to the depression of bipolar disease, characterized by severe depression, loss of interest in life activities, early morning awakening with intensification of Sx, marked ↑/↓ functionality, anorexia, weight loss, inappropriate sense of guilt. See Involutional melancholia. Cf Melancholy

melancholy PSYCHIATRY Depression that follows external events–eg, mourning loss of a loved one. Cf Melancholia

melanin PHYSIOLOGY A dark natural body pigment found in the epidermis or skin adnexal structures. See Albinism, DOPA, Melanoma

melanocyte-stimulating hormone PHYSIOLOGY A group of polypeptide hormones derived from prepro-opiomelanocorticotropin secreted in the middle lobe of the hypophysis in response to MRH–an oxytocin-related releasing hormone, inhibited by MRIH–a tripeptide release inhibitor, both of which are secreted by the hypothalamus; MSH's function in humans is unknown

melanocytic nevus The most common nevus, characterized by melanocytes in the dermis or epidermis

MELANOCYTIC NEVI

COMPOUND NEVUS Nevus cells and nests at the dermal-epidermal interface–the 'junction' and dermis

INTRADERMAL NEVUS Nevus cells and nests confined to dermis

JUNCTIONAL NEVUS Nevus cells and nests confined to dermal-epidermal interface

SPINDLE & EPITHELIOID CELL NEVUS OF SPITZ A compound nevus with elongated and/or epithelioid nevus cells and nests present at the dermal-epidermal interface–the 'junction' and in the dermis; it is histologically bizarre but benign

melanoma Malignant melanoma DERMATOLOGY A tumor which comprises 1-3% of all new cancers–18,000/yr, causes 6500 deaths/yr–US, most age 30–50; the incidence of melanoma is ↑ at ± 7%/yr, and now affects 9/10⁵, primarily head & neck in ♂ and 12/10⁵, primarily the legs in ♀; melanoma is rare, but more aggressive in chidren RISK FACTORS Giant congenital melanocytic nevus, dysplastic nevus, xeroderma pigmentosum, immunodeficiency, moles with persistent pigment changes–especially > age 15, large or irregularly pigmented lesions, familial moles, congenital moles; white–12-fold greater risk than blacks, previous melanoma, melanoma in 1ˢᵗ-degree relative, immunosuppression, photosensitivity, ↑ sun exposure; ocular melanomas may *not* ↑ melanoma risk S<small>ITE OF METASTASIS</small> Liver, lung, intestine, pancreas, adrenal, heart, kidney, brain, spleen, thyroid. See Acral lentiginous melanoma, Amelanotic melanoma, Congenital melanoma, Dysplastic nevus syndrome, Lentigo maligna melanoma, Nodular melanoma, Ocular melanoma, Premalignant melanoma, Pseudomelanoma, Radial growth phase melanoma, Superficial spreading melanoma, Thin melanoma, Vertical growth phase melanoma, Vertical growth phase melanoma

M<small>ELANOMA</small>

S<small>TAGE</small> I Confined to the epidermis and/or the upper dermis, and measures ≤ 1.5 mm thick

S<small>TAGE</small> II 1.5 mm to 4 mm thick, spread to lower dermis, but not beyond or to adjacent lymph nodes

S<small>TAGE</small> III Any of the following: (1) > 4 mm thick; (2) Spread beyond the skin; (3) Satellite lesions within 2 cms of the original tumor; or (4) Spread to nearby lymph nodes or satellite lesions between original and regional lymph nodes

S<small>TAGE</small> IV Metastases to other organs or to lymph nodes far from the original lesion

melanosis coli Brown bowel syndrome A benign condition characterized by segmental or global darkening of the colonic mucosa coupled with chronic constipation, due to prolonged abuse of cascara/sagrada-type laxatives C<small>LINICAL</small> Abdominal pain, pancreatitis, biliary atresia, cirrhosis, peptic ulcer disease, mucoviscidosis. See Laxative

melanotic whitlow Melanoma that presents under or adjacent to the fingernail, which may extensively involve the finger. See Whitlow

MELAS Mitochondrial Encephalomyopathy with Lactic Acidosis & Stroke-like episodes N<small>EUROLOGY</small> A childhood condition, associated with intermittent vomiting, proximal limb weakness, and recurrent cerebral insults resulting in hemiparesis, hemianopia or cortical blindness. Cf MERRF

melasma Chloasma, mask of pregnancy O<small>BSTETRICS</small> Darkening of face and neck with 'blotchy' coalescent hyperpigmented macules seen in pregnancy or with OC use, due to oxidation of tyrosine to melanin, often regressing with delivery; a similar mask may appear in ♂ without abnormal hormone levels, and in Pts treated with phenytoin

melena Stool or vomit blackened by heme pigments

melioidosis Pseudoglanders A tropical infection primarily of rats by *Pseudomonas pseudomallei*, an aerobic gram-negative bacillus found in wells and stagnant waters; epizootic infection occurs in sheep, goats, pigs; human infection is water-borne, transmitted via the skin or by inhalation, most commonly in Southeast Asia; C<small>LINICAL</small> Asymptomatic to fulminant sepsis with multiple abscesses in liver, spleen, lungs, resolving as granulomas—mortality without antibiotics, 90%; with antibiotics 50%, high fever, chills, tachypnea, myalgia; chronic form has a ± 10% mortality and is characterized by an intermittent, TB-like pneumonia, lung cavitation, chronic drainage T<small>REATMENT</small> 3<small>rd</small>-generation cephalosporins, ceftazadime, also tetracycline, chloramphenicol, aminoglycosides

Melipramine Imipramine, see there

melituria The spilling of sugar into the urine—eg, fructose, glucose, maltose, pentose

melorheostosis Lèri's disease An idiopathic defect of long bone growth, characterized by cortical thickening of long–tubular bones which, when stripped of muscle, simulates a candle with wax dribbled down a side C<small>LINICAL</small> Pain, limitation of movement, contraction and/or fusion of joint spaces, usually of one limb

melphalan Alkeran® O<small>NCOLOGY</small> An oral alkylating anticancer nitrogen mustard agent used to treat ovarian CA and myeloproliferative disease A<small>DVERSE EFFECTS</small> Myelosuppression, mutagenesis

member H<small>EALTH INSURANCE</small> A person covered under a health plan—enrollee or eligible dependent. See Group member, Nonaffiliated member, Plan member

membrane A very thin layer of tissue covering a surface, lining a body cavity, or dividing a space or organ

membrane potential C<small>ARDIOLOGY</small> The voltage difference between the inside and outside of resting excitable–neurons and muscle, not-yet depolarized cells

membranes O<small>BSTETRICS</small> A popular term for the amniotic sac, see there

membranoproliferative glomerulonephritis Mesangiocapillary glomerulonephritis–GN Chronic GN which accounts for 5-10% of all GN, caused by immune deposition in glomeruli; it often progresses to chronic renal failure

membranoproliferative glomerulonephritis type 1 N<small>EPHROLOGY</small> The more common form of MG, characterized by subendothelial deposits of antibodies in the glomerular membrane, primarily under age 50 C<small>LINICAL</small> Acute nephrotic syndrome, ↓ urine output–↓ GFR; edema–protein loss; HTN–sodium retention, ↑ renin production by kidney L<small>AB</small> WBCs in urine, accumulation of nitrogenous wastes–BUN and creatinine–azotemia due to poor renal function. See Nephrotic syndrome

membranoproliferative glomerulonephritis type 2 N<small>EPHROLOGY</small> A condition that is far less common than MPGN 1, but similar clinically; in type 2, the glomerular membrane is more permeable to protein and blood cells. See Nephrotic syndrome

membranous glomerulonephritis N<small>EPHROLOGY</small> A common idiopathic nephropathy usually affecting those > age 40, caused by immune complex deposition in the capillary wall of the glomerular basement membrane; it is the most common cause of nephrotic syndrome C<small>LINICAL</small> Asymptomatic proteinuria or nephrotic syndrome; possibly, edema, weight gain, hypertension, nocturia, anorexia, GFR is near normal U<small>RINALYSIS</small> Oval fat bodies, hyaline, granular, fatty casts R<small>ISK FACTORS</small> HBV, malaria, malignant solid tumors, NHL, SLE, syphilis, etc, drug or substance exposure–gold, mercury, penicillamine, trimethadione, skin-lightening creams T<small>REATMENT</small> Glucocorticoids and/or immunsuppressants may be used in some Pts. See Glomerulonephritis

meme N<small>EUROLOGY</small> A block of mental cues associated with experience, which relies on the brain's 'pattern-evolving machinery' P<small>SYCHOLOGY</small> A thought construct that endows a person with certainty about his/her fate. See Mediational unit

memory N<small>EUROLOGY</small> The persistence of the effects of learning and experiences on an organism's behavior, a process attributed to molecular transformation in incoming neuronal branches–dendritic trees. See Emotional memory, Episodic memory, Long-term memory, Immediate memory, Procedural memory, Recent memory, Repressed memory, Semantic memory, Short-term memory, True memory, Visual memory, Working memory

memory impairment screen N<small>EUROLOGY</small> A 4-minute card-based screening test to differentiate between Pts with normal forgetfulness and those with memory defects typical of early Alzheimer's disease and other memory disorders. See Minimental test

MEN Multiple endocrine neoplasia, Multiple endocrine adenomatosis A group of AD, often overlapping diseases characterized by hyperplasia or neoplasia of > one endocrine gland, many of the APUD system, see there

MEN–M<small>ULTIPLE ENDOCRINE NEOPLASIA</small>

MEN TYPE 1–I A complex characterized by pituitary adenoma or hyperplasia, adrenal adenoma or hyperplasia, parathyroid adenoma or hyperplasia, acromegaly, pancreatic islet cell tumors–insulinoma, carcinoid, Zollinger-Ellison syndrome, WDHA syndrome, ↑ gastrin secretion, peptic ulcer, Note: Nonhereditary factors may influence the expression of MEN I as the condition may be unequally expressed in identical twins; the defective gene is located on chromosome 11q13 Lab Invest 1995; 72:494

MEN TYPE 2A–I/IIA A complex characterized by medullary carcinoma of the thyroid–bilateral and present in nearly 100% of cases producing calcitonin, histaminase, prostaglandins, ACTH, potentially causing Cushing syndrome, pheochromocytoma and parathyroid adenoma or hyperplasia DIAGNOSIS Conventional diagnostic modalities–eg radiologic imaging, histopathology, and laboratory values–↑ plasma calcitonin, ↑ urinary catecholamines, and catabolites, for MEN-2A identifies relevant tumors, but not carriers MOLECULAR PATHOLOGY The *MEN-2A* gene mutation identified by linkage analysis in the *RET* proto-oncogene, which maps to chromosome 10q11.2, specifically at exon 10 or 11, is highly specific–no false positives and highly sensitive–no false negatives, allowing early diagnosis of MEN-2A. See *RET* gene PROGNOSIS Excellent

MEN TYPE 2B–III/IIb Mucosal neuroma syndrome A complex characterized by bilateral medullary carcinoma of the thyroid–producing calcitonin, histaminase, prostaglandins, ACTH and potentially, Cushing syndrome, pheochromocytoma, parathyroid adenoma or hyperplasia, mucosal–gut neurofibromas, associated with thickened lips, submucosal oral nodules, intestinal ganglioneurofibromatosis, infantile intestinal dysfunction and cranial nerve hyperplasia and connective tissue disease–Marfanoid habitus, scoliosis, kyphosis, pectus excavatum PROGNOSIS 20% mortality

menarche OB/GYN The time that menstruation begins; the first menstrual discharge. Cf Menopause

mendelian disorder CLINICAL GENETICS A popular term for any genetic disease which follows simple mendelian patterns of inheritance–eg, autosomal recessive disorders, such as cystic fibrosis. Cf Nonmendelian disorder

Ménière's disease Endolymphatic hydrops; hydrops NEUROLOGY An idiopathic inner ear disorder that can affect both hearing and balance, causing vertigo, profound hearing loss in one or more ears, tinnitus, and a sensation of fullness in the ear ETIOLOGY INITIAL HEARING LOSS Head trauma (acoustic and physical), viral labyrinthitis (mumps, influenza), mastoiditis, meningitis, diphtheria, measles, idiopathic MANAGEMENT Dehydrating agents–eg, furosemide, improve the vestibular response of the affected ear and form the basis of the lasix test; labyrinthectomy

meningeal carcinomatosis Leptomeningeal carcinomatosis, carcinomatous leptomeningitis ONCOLOGY The spread of cancer to the meninges, most common in small cell lung cancer–SCLC, seen in 26% of Pts with SCLC

meningeal sign NEUROLOGY Any clinical sign that indicates meningeal irritation; of the 3 signs–Brudzinski's signs–chin to chest evokes hip flexion, Kerdnig's sign–resistance to knee extension evokes pain in hamstrings, and nuchal rigidity, the last is most useful; none are reliable

meningeal metastases ONCOLOGY Cancer that has spread from the primary site to the meninges of the brain and/or spinal cord

meningioma NEUROSURGERY A tumor of meninges and meningeal cells, most common in middle-aged ♀ CLINICAL Often asymptomatic, usually slow-growing masses, attached to dura, usually where arachnoid villi are prominent; Sx are related to tumor growth and compression PROGNOSIS 5-yr survival 70%. See Malignant meningioma

meningismus A constellation of signs and Sx–eg, headache, neck stiffness suggestive of meningitis, characterized by meningeal irritation without objective findings; it is more common in young Pts with systemic infections–eg, the 'flu', pneumonia

meningitis NEUROLOGY A condition characterized by potentially fatal meningeal inflammation PATHOGENESI Infants, *Streptococcus agalactiae, E coli, Klebsiella-Enterobacter, Pseudomonas, Serratia, Listeria monocytogenes, Citrobacter diversus*; children, *H influenzae, S pneumoniae, Neisseria meningitidis* WORKUP CT, MRI, EEG–rarely performed, CBC & CSF–↑ WBCs/PMNs, ↑ CSF protein MANAGEMENT Neonates ampicillin ±ceftriaxone ±gentamicin; older children, ceftriaxone, cefotaxime. See Aseptic meningitis, Bacterial meningitis, Cryptococcal meningitis, Eosinophilic meningitis, Gram negative meningitis, *H influenzae* meningitis, Lymphocytic choriomeningitis, Meningococcal meningitis, Mollaret's meningitis, Neoplastic meningitis, Pneumococcal meningitis, Pseudomeningitis, Purulent meningitis, Staphylococcal meningitis, Tuberculous meningitis

MENINGITIS, CLINICAL FINDINGS BY AGE

NEONATES Irritation, respiratory distress, altered sleep patterns, vomiting, lethargy, labile temperature, shock bulging fontanelles

PRE-TODDLERS Irritation, altered sleep patterns, vomiting, lethargy, fever, shock, nuchal rigidity, coma, shock

TODDLERS Headache, stiff neck, photophobia, myalgia, lethargy, fever, shock, nuchal rigidity, coma, shock

meningitis belt A popular term for a region of sub-Saharan Africa where epidemics of group A meningococcal infection occur in cycles of ± 10 yrs

meningocele NEONATOLOGY A neural tube defect with protrusion of the spinal meninges and part of the spinal cord through a bony defect in the vertebral column, caused by a failure of closure during embryonic period at the base of the neural tube. See Neural tube defect, Spina bifida

meningococcal meningitis NEUROLOGY Meningeal infection by *N meningococcus*, primarily of children, typically of rapid onset after a URI EPIDEMIOLOGY Often in winter or spring, possibly with local epidemics at boarding schools or military bases RISK FACTORS Exposure to another person with MM, recent URI. See Meningitis

meningococcal septicemia Meningococcemia INFECTIOUS DISEASE An infection arising in the respiratory tract linked to aggressive *N meningococcus* strains EPIDEMIOLOGY Transmitted by droplet to family members or close exposures; more in winter/early spring, especially < age 5. See *Neisseria meningococcus*

meningoencephalitis NEUROLOGY Inflammation of the brain and its covering. See Primary acquired meningoencephalitis, West Nile meningoencephalitis

meningomyelocele NEUROLOGY Protrusion of the spinal meninges and cord through a defect in the vertebral column; the bone defect is termed spina bifida, see there

meniscus sign Air crescent sign RADIOLOGY A semilunar radiolucency peripheral to a pulmonary mass described as typical of echinococcal infection/hydatid cyst disease; most common cause in the US for the MS is a fungus 'ball', usually due to *Aspergillus fumigatus*; it may be seen in lung abscesses, tumors, hematomas, granulomatous infections and Rasmussen's aneurysm

meniscus sign of Carman RADIOLOGY A large semi-lunar hypodense zone that may be seen in ulcerated gastric adenoCAs, where there is a flattened polypoid mass with a broad central ulceration; the gastric mucosa adjacent to the polyp forms a smooth inner margin. See Whalebone in a corset. Cf Quarter moon sign

meniscus tear Meniscal tear SPORTS MEDICINE A laceration of the meniscus most commonly caused by twisting or hyperflexion of the knee joint. See Meniscus

Menkes syndrome Menkes kinky hair syndrome METABOLIC DISEASE An X-R condition linked to ↓ serum copper CLINICAL Progressive mental deterioration, kinked or twisted brittle hair, skeletal defects, death in infancy

menometrorrhagia GYNECOLOGY Excessive uterine bleeding during menstruation accompanied by irregularity

menopause Change of life, climacteric, 'time of life' GYNECOLOGY The cessation of menstrual activity due to failure to form ovarian follicles, which normally occurs age 45–50 CLINICAL Menstrual irregularity, vasomotor instability, 'hot flashes', irritability or psychosis, ↑ weight, painful breasts, dyspareunia, ↑/↓ libido, atrophy of urogenital epithelium and skin, ASHD, MI, strokes and osteoporosis–which can be lessened by HRT. See Estrogen replacement

menorrhagia

therapy, Hot flashes, Male menopause, Premature ovarian failure, Premature menopause. Cf Menarche

MENOPAUSE—"...WHAT A DRAG IT IS GETTING OLD." JAGGER, RICHARDS

BLADDER	Cystourethritis, frequency/urgency, stress incontinence
BREASTS	↓ Size, softer consistency, sagging
CARDIOVASCULAR	Angina, ASHD, CAD
ENDOCRINE	Hot flashes
MUCOCUTANEOUS	Atrophy, dryness, pruritus, facial hirsutism, dry mouth
NEUROLOGIC	Psychological, sleep disturbances
PELVIC FLOOR	Uterovaginal prolapse
SKELETON	Osteoporosis, fractures, low back pain
VAGINA	Bloody discharge, dyspareunia, vaginitis
VOCAL CORDS	Deepened voice
VULVA	Atrophy, dystrophy, pruritus

menorrhagia GYNECOLOGY Excessive uterine bleeding at the regular menstrual times lasting longer than usual

menorrhea 1. Normal menses 2. Abnormal excessive menses

menses 1. The period during which menstruation occurs 2. The endometrium per se

menstrual cramps Spasmodic dysmenorrhea GYNECOLOGY Painful cramps, spasms, lower abdominal discomfort, generally occurring on the first day of the menstrual period; the pain may extend to the low back, thighs, pelvis, and be accompanied by N&V, dizziness, fainting; MCs tend to lessen in severity after age 30. Cf Frozen section

menstrual cup GYNECOLOGY A reusable, cup-shaped device placed inside the vagina, held in place by suction and used to collect menstrual flow. See Menstruation

menstrual extraction GYNECOLOGY Menstrual regulation, minisuction The use of a plastic syringe or cannula to extract the menses in lieu of the natural, albeit prolonged and messy period. See Menstruation OBSTETRICS Interception of pregnancy, preemptive abortion Suction extraction of the intrauterine content to prevent a possible pregnancy

menstrual flow Menstrual discharge, period GYNECOLOGY The vaginal bleeding that occurs monthly from menarche to menopause, which lasts for 2–7 days of a discharge averaging 60–70 mL. See Menstruation

menstrual migraine NEUROLOGY A migraine that waxes/wanes in intensity with menstruation MANAGEMENT Percutaneous estradiol may control the headaches in ♀ with dysmenorrhea or ovarian dysfunction. See Classical migraine

menstrual sponge GYNECOLOGY A semidisposable natural sponge placed in the vagina during menses, used by some ♀ instead of manufactured pads or tampons. See Tampon. Cf Contraceptive sponge

menstruation Menorrhea The discharge of blood and tissue or the process itself from the uterus at the end of a menstrual cycle, occurring at ±4 wk intervals, ±2 wks after ovulation. See Vicarious menstruation. Cf Amenorrhea

mental decline NEUROLOGY The loss of mental abilities with advancing age, a part of normal aging; the sharpest decline in mental abilities occur in mathematics ability, the least in spatial orientation for ♂, and in inductive reasoning for ♀. See Sundowning

mental disorder Mental illness PSYCHOLOGY '...a clinically significant behavioral or psychological syndrome or pattern that occurs in an individual...associated with present distress–eg, a painful Sx or disability–ie, impairment in one or more important areas of functioning or with significantly increased risk of suffering death, pain, disability, or an important loss of freedom...this syndrome or pattern must not be merely an expectable and culturally sanctioned response to a particular event, for example, the death of a loved one'

mental examination An examination of a person's mental state which evaluates affect, affective, alert and oriented x 3–to person, place and time. Cf Neurologic examination

mental filtering PSYCHOLOGY The selective evaluation of a complex situation with both positive and negative elements; positive MF occurs when a person ignores or downplays negative aspects of a situation or criticism, is typical of manic reactions, and indicates a loss of reality sense; negative MF prevents coping with internal conflicts and emotions. See Cognitive therapy, Distorted thinking

mental health professional MEDTALK A person who by education and experience is professionally qualified to provide counseling interventions designed to facilitate individual achievement of human development goals and remediate mental, emotional, or behavioral disorders, and associated distresses which interfere with mental health and development. See Psychiatrist, Psychologist. Cf Mental health provider

mental health (service) provider MEDTALK A person, partnership or professional corporation comprised of appropriately licensed persons–eg, certified substance abuse counselors, clinical psychologists, clinical social workers, licensed substance abuse treatment practitioner, licensed practical nurse, marriage and family therapist, mental health professional, physician, professional counselor, psychologists, registered nurse, school psychologist, or social worker . Cf Mental health professional

mental illness Mental disorder, see there

mental retardation NEUROLOGY Below-average general intellectual function with associated deficits in adaptive behavior that occurs before age 18; intellectual activity that is significantly below average for a population, and associated with impaired social function. Cf Psychological testing

mental status See Mini-mental test

mental stress PSYCHOLOGY A general term encompassing mental arousal and/or emotional stress; MS can be evoked by a number of mental tasks–eg, mental arithmetic, public speaking, mirror trace, type A structured interviews; MS evokes pathophysiologic responses–eg, myocardial ischemia measurable by radionuclide ventriculography. Cf Physiological stress

Mentalium Diazepam, see there

mentally defective SEXUAL OFFENSES adjective Referring to a person whose mental defect renders him/her temporarily or permanently incapable of appraising the nature of his/her own conduct. See Rape

mentally disabled See Cognitively impaired

mentally incapacitated FORENSIC PSYCHOLOGY adjective Referring to a person rendered temporarily incapable of appraising or controlling his/her conduct due to the influence of a narcotic, anesthetic or other substance administered to that person without the person's consent, or due to any other act committed upon that person without the person's consent. See Date-rape

Mentax® Butenafine DERMATOLOGY An antifungal for tinea corporis, tinea cruris, interdigital tinea pedis, onychomycosis. See Dermatophytosis

menthol An aromatic oil from peppermint–Mentha piperita or produced synthetically, used topically for arthritic pain, hemorrhoids, itching, sore muscles and orally in cough medicines and throat lozenges TOXICITY Pure menthol oil is toxic–5 mL may be fatal. See Botanical toxicity, Herbal medicine, Poisonous plants; Cf Peppermint, Wintergreen

mentoplasty RECONSTRUCTIVE SURGERY Surgery of the chin to correct defects/deformities or alter mandibular contour–eg, remove fat, implant silicone prosthesis. See Cosmetic surgery

mentor GRADUATE EDUCATION A professional and a role model who gives attention and feedback to a junior colleague

menu INFORMATICS A display on a computer monitor of the function it is currently capable of performing LAB MEDICINE Test menu, see there

meperidine Demerol PAIN MANAGEMENT An IV agonist opioid analgesic, with an active metabolite, normeperidine ROUTE Parenteral, oral PHARMACOLOGIC EFFECTS Powerful blocker of pain messages to the brain CONTRAINDICATIONS Not coadministered with MAOI antidepressants–may cause dizziness, lightheadedness, nausea. See Opioids

mephenytoin Mesantoin® A drug used to control seizures, convulsions, epilepsy; it is metabolized in liver to pharmacologically active normephenytoin, excreted in urine. See Therapeutic drug monitoring

meprobamate PHARMACOLOGY A potentially addicting agent with anticonvulsant, anxiolytic, muscle relaxing, sedative, hypnotic, tranquilizing properties that blocks spinal interneurons, depressing brainstem functions, coma, death; may cause hypotension METABOLISM Liver; excreted in urine, stool. See Therapeutic drug monitoring

Mepron® Atovaquone INFECTIOUS DISEASE An antiprotozoal agent used to treat and prevent PCP in Pts who are allergic to trimethoprim-sulfamethoxazole. See AIDS, *Pneumocystis carinii* pneumonia

mEq Milliequivalent, see there

purinethol 6-mercaptopurine ONCOLOGY An antimetabolic chemotherapeutic that targets rapidly dividing cells–eg, in ALL, AML ADVERSE EFFECTS Myelosuppression, anorexia, N&V, jaundice

MERCATOR CARDIOLOGY A clinical trial–Multicenter European Research Trial with Cilazapril After Angioplasty to Prevent Transluminal Coronary Obstruction & Restenosis that evaluated ACE inhibition with Cilizapril in preventing restenosis after PCTA in Pts with CAD. See Coronary artery disease, Percutaneous transluminal angioplasty

Mercator catheter CARDIOLOGY A high-density array catheter used in the right atrium to map and diagnosis complex arrhythmias and assess the effectiveness of ablation treatment. See Ablation therapy

Mercedes-Benz sign IMAGING The finding of gallstones on plain films of the abdomen; gallstones contain gas which fills the spaces left by cholesterol crystals and appears as stellate–triradiate translucent areas, fancifully likened to an automobile logo

Mercurochrome A proprietary merbromin which is nearly useless as a topical antibacterial

mercy killing MEDICAL ETHICS The termination of a person's life as a humane act. See Euthanasia

merger MANAGED CARE The integration of ≥ 2 hospitals/health care facilities to form a single unit. See Virtual merger

Meridia® Sibutramine OBESITY An anti-obesity agent used to facilitate and maintain weight loss, coupled to a low-calorie diet CONTRAINDICATIONS Hx of stroke, CAD, CHF, uncontrolled HTN. See Obesity. Cf Xenical

merit compensation MEDICAL PRACTICE A salary based on the quality and cost-effectiveness of the work being performed

Merkel cell carcinoma Cutaneous neuroendocrine carcinoma A highly malignant skin tumor, usually head & neck, most common in the elderly PROGNOSIS Poor, 3-yr survival 68% ♀, 36% ♂ TREATMENT Wide excision, prophylactic LN dissection, RT, chemotherapy

Merlit® Lorazepam, see there

MERRF Myoclonus Epilepsy with Ragged Red Fibers NEUROLOGY A mitochondrial myopathy characterized by myoclonus, epilepsy and ataxia, maternal inheritance. See Ragged red fibers. Cf MELAS

mesalamine Asacol® THERAPEUTICS A delayed-release antiinflammatory salicylate developed for delivery to synovial tissue in arthritics; it is used maintain remission in Crohn's disease, and treat ulcerative colitis. See Inflammatory bowel disease, Ulcerative colitis

mescaline SUBSTANCE ABUSE A hallucinogenic psychotropic alkaloid, derived from the peyote cactus–*Lophophora williamsii*; it is similar to indole alkaloids–eg, psilocin, bufotenin, ibogaine, and LSD. See Hallucinogen

mesenchymoma A tumor that contain 2 or more mesenchymal elements in addition to fibroblasts; the term has specific significance in each affected organ CARTILAGE Vascular hamartoma, cartilaginous hamartoma A benign chest wall tumor of infancy LIVER Hepatic mesenchymoma is a large, aggressive embryonal or 'primitive' sarcoma of the pediatric liver with a median survival of < 1 yr, characterized by necrosis, hemorrhage, cystic degeneration, atypical fibroblasts, and exuberant mitotic activity, entrapping hyperplastic bile duct-like structures MUSCLE Ecto-mesenchymoma An embryonal rhabdomyosarcoma with ganglionic differentiation. See Triton tumor

mesenteric artery syndrome Mesenteric artery ischemia Any condition linked to occlusion of one or more mesenteric arteries; seen in older subjects 2° to ASHD, occasionally linked to OCs possibly due to vasospasm CLINICAL, EARLY Nonspecific GI complaints, abdominal pain LATE Abdominal distension, shock, peritonitis TOO LATE 45% mortality

mesenteric venous thrombosis VASCULAR DISEASE A blood clot occluding the mesenteric vein, a major vein located in the tissue that connects the intestine to the posterior abdominal wall

MeSH Medical Subject Headings INFORMATICS The vocabulary of about 16,000 terms used for MEDLINE and other MEDLARS databases. See MEDLINE, MEDLARS

mesna ONCOLOGY A uroprotective agent that protects the urogenital tract from toxicity of ifosfamide and cyclophosphamide ADVERSE EFFECTS Rash, itching, diarrhea, N&V. See Cyclophosphamide, Ifosfamide

Mesocestoides A genus of non-human tapeworms that may rarely afflict humans who eat poorly-cooked encysted muscle

mesoridazine besylate An antipsychotic

mesothelioma SURGERY A neoplasm of serosal surfaces–pleura, peritoneum, pericardium, tunica vaginalis, scrotum, seen in 5-10% of those occupationally exposed to asbestos, with a latency period of 20-40 yrs; incidence of mesothelioma ↑ exponentially if the subject is also a smoker; up to 10% of heavily asbestos-exposed workers die of mesothelioma. See Asbestos, Cystic mesothelioma

MESS Mangled Extremity Severity Score TRAUMATOLOGY A scoring system used to evaluate the need for amputation in an extremity; a MESS of ≥ 7 indicates a need to amputate ALTERING FACTORS Loss of posterior tibial nerve causes trophic ulceration, ischemia > 6 hrs doubles score. See Amputation

---

MESS SCORING POINTS
Skeletal & soft tissue injury

- Low energy–stabs, simple Fx, civilian gun shot wounds
- Medium energy–open or multiple Fx, dislocation
- High-energy–close range shotgun, military gunshot wounds, crush injury
- Very high energy–as in line just above, with gross contamination or tissue avulsion

Limb ischemia

- Pulse reduced, normal perfusion

Mestinon®

- Pulseless, paresthesia, ↓ capillary filling
- Cool, paralyzed, numb

Shock
- Systolic BP ≥ 90 mm Hg
- Transient hypotension
- Permanent hypotension

Age
- < Age 30
- Age 30–50
- > Age 50

Mestinon® Pyridostigmine, see there

meta-analysis Data synthesis, quantitative overview DATA ANALYSIS A systematic method that uses statistical techniques for combining results from different studies to obtain a quantitative estimate of the overall effect of a particular intervention or variable on a defined outcome; MA produces a stronger conclusion than can be provided by any individual study. See Cochran Collaboration, Cumulative meta-analysis

metabolic acidosis PHYSIOLOGY A condition in which there is a ↓ pH due to either an ↑ in acids or loss of bicarbonate LAB pH < 7.35, HCO_3^- < 22 mEq/L, and in 'compensating' cases $PaCO_2$ < 35 mm Hg ETIOLOGY Bicarbonate–HCO_3^- depletion due to renal disease, diarrhea, fistulas of GI tract; ↑ production of organic acids due to liver disease, endocrinopathies–eg, DM, hypoxia, shock, drug overdose; ↓ excretion of acids due to renal disease CLINICAL Rapid, deep breathing, fruity breath, headache, lethargy, N&V, coma. See Metabolic alkalosis, Respiratory acidosis, Respiratory alkalosis

metabolic alkalosis PHYSIOLOGY A condition in which there is a ↑ pH due to either an ↓ in acids or excess bicarbonate LAB pH > 7.42, HCO_3^- > 26 mEq/L, $PaCO_2$ > 45 mm Hg ETIOLOGY Loss of acids due to hyperemesis, gastric suction, loss of K^+ due to ↑ renal excretion–eg, diuretic therapy, steroid use, excess–eg, overuse of antacids CLINICAL Slow shallow breathing, irritability, confusion. See Metabolic acidosis, Respiratory acidosis, Respiratory alkalosis

metabolic bone disease Any defect in bone absorption or deposition that alters the PTH/calcium-phosphate/vitamin D axis, often with ↑ bone fragility ETIOLOGY Fibrous dysplasia, Langerhans' cell histiocytosis/histiocytosis X, acromegaly, corticosteroid therapy, heparin, hyperparathyroidism, hyperthyroidism, rickets, immobilization syndrome, bone metastases, metabolic disease, congenital–Ehlers-Danlos syndrome, homocystinuria, hypophosphatasia, Marfan syndrome, osteogenesis imperfecta, osteoporosis, Paget's disease of bone–osteitis deformans DIAGNOSTIC WORKUP Measure Ca^{2+}, PO_4^-, PTH, other hormones, bone biopsy, tetracycline test

metabolic neuropathy NEUROLOGY A neuropathy attributable to a nutritional or other deficiency–eg, of vitamin B_{12} or folate. See Neuropathy

metabolic panel LAB MEDICINE An abbreviated battery of analytes–BUN, chloride, CO_2, glucose, potassium, sodium–to screen for metabolic defects. See Panel

METABOLIC PANELS

BASIC MP G0096 CO_2–bicarbonate, chloride–blood, creatinine–blood, glucose–quantitative, potassium, sodium, urea nitrogen–quantitative

COMPREHENSIVE MP Albumin, alkaline phosphatase, bilirubin–total, calcium–total, carbon dioxide–bicarbonate, chloride–blood, creatinine–blood, glucose–quantitative, potassium–serum, protein–total, sodium–serum, transferase–AST, urea nitrogen–quantitative

metabolic screen Metabolic profile NEONATOLOGY A battery of tests on a neonate that identifies the most common treatable metabolic diseases–eg, cystic fibrosis, galactosemia, homocystinuria, hypothyroidism, maple syrup urine disease, PKU. See Neonatal panel/profile, Neonatal screen

metabolic syndrome VASCULAR DISEASE A clustering of medical conditions–ASHD, type 2 DM, HTN, hyperlipidemia, abdominal obesity, hypertriglyceridemia, insulin resistance, often linked to claudication

metabolic toxemia of pregnancy Eclampsia, see there

metabolizer A person who metabolizes. See Poor metabolizer

metacarpectomy Surgical excision of the wrist

metachromatic leukodystrophy NEUROLOGY An AR lysosomal storage disease caused by arylsulfatase. A deficiency, characterized by sulfated sphingolipid accumulation, ↓ myelin CLINICAL FORMS Infantile form–most severe; adult form–least severe; juvenile form–intermediate CLINICAL Onset by age 2, death by age 5 with upper and lower motor neuron disease, ↓ nerve conduction, spasms, ataxia, oculomotor paralysis, bulbar palsy, blindness, deafness, dementia. See Arylsulfatase A, Leukodystrophy

metachronous *adjective* Referring to conditions that appear at times differing from an event of interest–ie, not synchronous

metal fume fever OCCUPATIONAL MEDICINE An acute self-limited influenza-like illness due to occupational exposure to metal–zinc, copper, magnesium–dust or fumes, often inhaled as a sulfide ore AT RISK OCCUPATIONS Mining, welding, brass foundry work, due to metal fume exposure CLINICAL Several hrs after exposure, cough, dry throat, tightness in chest, chills, fever, then sweating, weakness, nausea, then recovery; other Sx: myalgia, upper respiratory tract irritation, leukocytosis. Cf 'Monday death'

metanephrine Methoxyepinephrine LAB MEDICINE Any exogenous or endogenous catecholamine metabolite excreted in the urine; metnephrines ↑ in neuroblastoma, pheochomocytoma, metastases, stress, sepsis

metaproterenol sulfate PULMONOLOGY A bioequivalent of Alupent for treating asthma and bronchospasm. See Asthma

metastasectomy ONCOLOGY The surgical excision of one or more foci of metastatic malignancy NOTE Because lymphoproliferative neoplasms–eg, lymphomas are ab initio multifocal, metastasectomy is used for carcinoma, soft tissue sarcoma, melanoma and other solid cancers

metastasis Plural, metastases ONCOLOGY The distal spread of a malignancy, either by penetration of a blood or lymphatic vessel or by spread along a serosal membrane with later development into a 2nd focus of malignancy. See Blow-out metastasis, Cannonball metastasis, Micrometastasis, Skip metastasis

metastasize *verb* To spread to another part of the body, usually through blood vessels, lymph channels, or spinal fluid; metastastatic or 2° foci of CA are morphologically similar to the 1°/original CA, or may be more primitive

metastatic cancer Secondary cancer, secondary tumor Cancer that has spread from a primary site elsewhere

metastatic cancer to brain Metastatic brain tumor, secondary brain tumor A brain tumor arising from metastatic CA–commonly, 1° lung or breast CA. See Brain tumor

metastatic cancer to lung Metastatic lung cancer ONCOLOGY CA that has spread to the lung EXAMPLES Breast CA, NHL, osteosarcoma, neuroblastoma, Wilms' tumor. See Lung cancer

metastatic disease panel LAB MEDICINE A battery of tests–albumin, alk phos, AST, Ca^{2+}, CEA, LDH–intended to detect metastatic or recurrent CA

metastatic pleural tumor ONCOLOGY A primary tumor that metastasizes to the pleural cavity, especially breast CA. Cf Mesothelioma

metatarsalgia ORTHOPEDICS Pain of the forefoot at the plantar aspect of the metatarsal head DIFFDx Neuroma, intermetatarsal bursitis DEFECTS CAUSING EXCESS LOCAL PRESSURE Depressed metatarsal, elongated metatarsal with abnormal metatarsal parabola, ↓ plantar fat pad, accessory sesamoids, over-pronation with hypermobile first ray MANAGEMENT ↓ pressure under affected metatarsal head, metatarsal pads to unload metatarsal, arch supports or orthotics, Spenco insoles to ↓ shear forces and pad, thick, curved mid-sole–rocker-bottom–shoes to ↓ forces under the metatarsals

metatarsectomy ORTHOPEDICS Excision of the metatarsus, see there

metatarsus adductus Forefoot varus ORTHOPEDICS A foot deformity characterized by a sharp inward angle of the front half of the foot; flexible deformity; the foot can be straightened and poses little risk for the infant; most cases resolve voluntarily; the rest only need simple exercises

metatarsus equinus ORTHOPEDICS A rare condition characterized by plantar flexion at midtarsal joints. Cf Forefoot equinus

metatarsus varus 1. A forefoot deformity in which the medial foot is lifted upward and outward, such that the victim is walking on the outer edge of the foot 2. Clubfoot, see there 3. Metatarsus adductus, see there

metered-dose inhaler PHARMACOLOGY A device used to deliver a spec-ified number of doses of a therapeutic inhalant–eg, β-agonist for asthma

metformin Glucophage® DIABETOLOGY A biguanide antihyperglycemic and antidiabetic used for type 2 DM, alone or with sulfonurea; metformin sensitizes cells to insulin, ↓ serum glucose and insulin, ↓ insulin resistance, ↑ glucose utilization, ↓ TGs, ↓ weight; in some Pts, it suppresses appetite. See Diabetes mellitus. Cf Troglitazone

methadone Dolophine®, Methenex® SUBSTANCE ABUSE A synthetic, rela-tively long-acting oral opioid analgesic USES Severe pain; narcotic detoxifi-cation, controlled maintenance of addiction EFFECTS ↓ intestinal motility; respiratory depression; analgesic; antitussive OVERDOSE EFFECTS Respiratory depression, stupor, coma, loss of short-term memory. See Heroin, Opioids

methamphetamine Speed, crank, crystal, Desoxyn SUBSTANCE ABUSE A sympathomimetic methylated derivative of amphetamine; it is more potent in its CNS stimulatory effect and, by extension more likely to be abused; methamphetamine abuse has ↑ to epidemic proportions ABUSER PROFILE White, age 20-35, high school education FORMS IV 62%, 'snorting' 18%, oral 13%, smoking 7% USED FOR Narcolepsy, ADD, obesity FETAL EFFECTS Prematurity, LBW. See Adam, Designer drugs, Ecstasy, Eve, 'Ice,' Tweaking

methanol TOXICOLOGY A polar alcohol used as an industrial solvent mis-cible with water, ethanol, ether, petroleum derivatives–eg, gasoline, in canned fuel and in antifreeze, where it may be abused as an inebriant by indigent alcoholics; methanol is metabolized to formaldehyde and formate, causing significant metabolic acidosis and optic nerve damage and blind-ness TOXIC RANGE 60-250 ml; as little as 15 ml has been fatal MANAGEMENT Overload Pt with ethanol–which competes with methanol for sites on alco-hol dehydrogenase, reducing methanol metabolites and toxicity

methaqualone Quaalude, 'ludes' SUBSTANCE ABUSE An addictive Schedule II hypnotic-sedative quinazolone with barbiturate-like effects; methaqualone was a popular drug of abuse and linked to physical or psy-chologic dependence CLINICAL Delirium, headache, nausea, pyramidal signs, convulsions, renal and cardiac failure; rarely, aplastic anemia ROUTE Oral, injected, sniffed, smoked USED FOR Anxiety, tension MANAGEMENT Hemoperfusion. See Controlled drug substance

methemoglobinemia HEMATOLOGY A condition characterized by excess–> 3%–methemoglobin in the circulation CLINICAL Cyanosis, headache, dyspnea, fatigue, drowsiness, ataxia, stupor. See Acquired methemo-globinemia, Methemoglobin

methicillin-resistant *Staphylococcus aureus* Methicillin-aminoglycoside resistant *Staphylococcus aureus*, MRSA An organism with multiple antibiotic resist-ances–eg, aminoglycosides, chloramphenicol, clindamycin, erythromycin, rifampin, tetracycline, streptomycin, cephalosporin; some strains of MRSA have ↓ sensitivity to antiseptics TREATMENT Vancomycin. See Antibiotic resistance

methionine malabsorption syndrome Oasthouse disease An AR condi-tion characterized by albinism, hyperpnea, convulsions and mental retar-dation after methionine loading, with ↑ α-hydroxybutyric acid, which causes the oasthouse odor

method MEDTALK The manner in which a particular thing is performed. See Cold turkey method, Comparative method, Confidence profile method, Confirmation method, Contrast-enhanced method, Delphi method, Designated comparison method, Diary method, Direct fluorescent antibody method, Empirical method, Heuristic method, Immunoperoxidase method, Micro method, Natural experiment method, Parametric method, Pisano method, Reference method, Rhythm method, Sandwich method, Shotgun method, Socratic method, Street intercept method, Substitute method, Time sampling method, Unproven method for cancer management, Valid method, Validated method, Westergren method

method of choice A diagnostic or therapeutic maneuver preferred in prac-tice, based on ease of performance or efficiency

methotrexate Folex® PFS, Rheumatrex® ONCOLOGY A widely used but toxic antimetabolic chemotherapeutic which may be used alone to treat certain malignancies–eg, choriocarcinoma, and with other agents for lymphopro-liferations–ALL, ANLL, Hodgkin's disease, non-Hodgkin's, Burkitt's and histi-ocytic lymphomas, mycosis fungoides, myeloma, head & neck, ovarian and small cell carcinomas, osteosarcoma, medulloblastoma; nonmalignant con-ditions–eg, recalcitrant psoriasis, rheumatoid arthritis; combined with cyclosporine and prednisone, MTX ↓ the incidence of acute GVHD from 23% to 9%, without affecting disease-free survival; MTX is a potent folic acid antagonist–antifolate and competes with dihydrofolate–the natural sub-strate, for binding sites on dihydrofolate reductase–DHFR, blocking produc-tion of tetrahydrofolate–folate is the vitamin co-factor in methyl group transport for purine and thymidilic acid synthesis and DNA synthesis. See Leucovorin rescue, MDR gene TOXICITY N&V, anorexia, stomatitis, CNS changes, hypersensitivity, liver damage, ocular irritation, dose-limiting myelotoxicity may appear 4-7 days after beginning therapy, nephrotoxicity–crystalliza-tion within renal tubules LAB Transient ↑ of LFTs

methoxamine CARDIOLOGY A vasopressor used for hypotension and PAT; it improves performance in left ventricular dysfunction, which may be linked to exercise-induced vasodilation of airway vessels

methoxsalen Oxsoralen® DERMATOLOGY A photosensitizing psoralen, given systemically or topically, after which the skin is exposed to UV light. See Psoriasis, PUVA

methsuximide Celontin® NEUROLOGY An antiepileptic used for absence seizures/petit mal epilepsy. See Therapeutic drug monitoring

methyl bromide TOXICOLOGY An insecticide and rodenticide, which is a volatile fumigant 3-fold denser than air and absorbed through skin, produc-ing narcosis, pulmonary edema, renal tubule damage, jacksonian convul-sions, CNS depression, peripheral neuropathy; permanent neurologic seque-lae may follow prolonged exposure

3,4-methylenedioxymethamphetamine MDMA, see Ecstasy

5,10-methylenetetrahydrofolate reductase deficiency MTHFR reductase An AR condition that is more common in ♀, characterized by devel-opmental delay, motor, gait and typical EEG changes and early death LAB ↑ Homocysteine in serum and urine, ↓ methionine

methylmalonic acid Isosuccinic acid An intermediate in fatty acid metabo-lism, which is ↑ in vitamin B_{12} and/or folate deficiency

methylmercury Dimethyl mercury TOXICOLOGY An inorganic mercury industrial pollutant; it is concentrated up the food chain, teratogenic and causes severe CNS defects in children whose mothers consumed MM-contaminated seafood while pregnant. See Mercury, Minamata disease

methylphenidate Ritalin® NEUROPHARMACOLOGY An amphetamine derived CNS stimulant used to control ADD and hyperactivity in children ADVERSE EFFECTS ↓ Appetite, 10-15% of children have major weight loss; insomnia–most suffer sleep delay, abdominal pain, headaches, dry mouth, dizziness, depression, tachycardia; link with ↓ growth is uncertain USED FOR Hyperactivity; narcolepsy; ADD. See Amphetamine

methyltestosterone ENDOCRINOLOGY A 17α-alkylated derivative of testosterone used for androgen replacement therapy PROS Oral CONS Hepatotoxicity–cholestasis, peliosis hepatis, liver tumors. See Androgen replacement therapy

metoprolol succinate Toprol-XL® CARDIOLOGY A β₁-cardioselective beta-blocker used for angina, HTN, A Fib, PVCs, heart failure, and to protect against recurrent MI ADVERSE EFFECTS Headache, fatigue, dizziness, dry cough, angioedema CONTRAINDICATIONS Sinus bradycardia, cardiogenic shock, heart block > 1st degree, bronchospastic lung disease, left ventricular dysfunction. See ACE inhibitor, Beta blocker, MDC, TIMI-2B

metrics MANAGED CARE A popular term for standards by which the quality of a product, service, or outcome of a particular form of Pt management is evaluated. See TQM

metrifonate NEUROLOGY A cholinesterase inhibitor which is anthelmintic and used to manage dementia. See Alzheimer's disease

Metrix® defibrillator CARDIOLOGY An implantable atrial defibrillator for restoring normal sinus rhythm in Pts with A Fib. See Atrial fibrillation, Implantable defibrillator

metrizamide IMAGING An iodinated radiocontrast used for myelography and for enhancing CT images

metrizoate IMAGING A water-soluble, tri-iodobenzene-based radiocontrast used for angiography and urography

metronidazole Noritate® DERMATOLOGY An antibiotic used for managing dermatitides and erythema of rosacea. See Rosacea INFECTIOUS DISEASE An antibiotic used to treat bacterial, fungal, and parasitic infections; it is the most effective agent for *H pylori* infections

Metropolitan Life table A correlative table generated by the Metropolitan Life Insurance Co comparing the weight of subjects to minimums of mortality in actuarial data, ie. an obesity/mortality ratio. Cf Obesity

metrorrhagia Uterine bleeding at irregular intervals

metyrapone A diagnostic tool for evaluating the hypothalamus-pituitary-adrenal 'axis', especially, abnormal pituitary activity

Meuse fever His disease. See Werner-His disease

Mevacor® Lovastatin, see there

Meval Diazepam, see there

mevalonic acid A molecular precursor of many endogenous molecules–eg, cholesterol, coenzyme Q, carotenoids, dolichol, steroids, and terpenes

mevalonic aciduria A disease of infant onset characterized by FTT, retarded development, hepatosplenomegaly, anemia, cataracts, dysmorphias, ↑ in urinary cholesterol and nonsterol isoprene precursors including mevalonic acid

mexiletine Mexitil® CARDIOLOGY An antiarrhythmic that may have some currency in treating HIV-related neuropathy ADVERSE EFFECTS GI distress and dizziness

MFD 1. Minimum dose 2. Minimum focusing distance 3. Monostotic fibrous dysplasia. See Fibrous dysplasia

MFO Mixed function oxidase, see there

MFS Medicare fee schedule

MG 1. Marcus Gunn pupil, see there 2. Membranous glomerulopathy, see there 3. Myasthenia gravis, see there 4. Myoglobin, see there

mg Milligram, see there

MGN Membranous glomerulonephritis, see there

MGUS Monoclonal gammopathy of unknown significance, see there

μg Microgram, 1×10^{-6} gram.

MH 1. Malignant histiocytosis, see there 2. Malignant hyperthermia 3. Master of Hygiene 4. Medical history 5. Menstrual history 6. Mental health

MHC Major histocompatibility complex A small region of the genome that is highly conserved in vertebrate evolution, which encodes 3 classes of polymorphic molecules known as the immune recognition unit; the MHC is located on the short arm of chromosome 6 in humans; the products of the MHC gene complex are membrane-bound receptors for antigens and peptides which, when bound, are displayed to T cells; if the bound peptides are recognized by the T cells, an immune response against those peptidesis initiated

MHC class II deficiency MOLECULAR MEDICINE A genetically heterogeneous AR immunodeficiency disease in which MHC II molecules are absent, with altered gene regulation due to defects in several transactivating genes that regulate expression of MHC class II genes. See RFXAP

MHC restriction IMMUNOLOGY The ability of T cells to recognize antigens when associated with the organism's own MHC haplotype, providing a dual recognition system critical to T-cell function. See Antigen-presenting cell, MHC

MHz Megahertz COMPUTERS A measurement of clock speed for computers, related to the efficiency of the circuit design and the microprocessor. Cf MIPS

MI Myocardial infarction, see there, also 1. Medical inspection 2. Mental illness 3. Metabolic index 4. Migration inhibition 5. Mitotic index 6. Mitral insufficiency 7. Motility index–GI tract

Miacalcin® Calcitonin-salmon, see there

MIB Medical information bureau, see there

MIBB™ system SURGICAL PATHOLOGY A minimally invasive breast biopsy system for obtaining multiple Bx samples with one insertion, used with the ABBI™ and Sonopsy™ systems. See Breast biopsy

mibefradil Posicor® CARDIOLOGY A CCB that vasodilates without reflex tachycardia, used to manage HTN and angina pectoris; it was pulled from the market as it inhibits liver metabolism of certain drugs–eg, antihistamine, Histamal, and statin cholesterol-lowering agents. See Calcium channel blocker

MIC 1. Methyl isocyanate 2. Minimum inhibitory concentration, see there

Micardis® Telmisartan, see there

mice 1. Mouse, see there 2. See Joint mice

Mickey Finn 1. A mixture of whiskey with chloral hydrate, the latter of which causes hypotension, pinpoint pupils, cardiac arrhythmia and, in high doses, gastric irritability with perforation, hepatotoxicity, nephrotoxicity 2. Croton oil and alcohol DRUG SLANG A street term for a depressant, see there

miconazole A broad-spectrum antifungal, used topically for cutaneous candidiasis, IV for systemic mycosis–candidiasis, coccidioidomycosis, paracoccidioidomycosis, intrathecally for cryptococcal meningitis ADVERSE EFFECTS Pruritus if topical, N&V, fever if systemic

MICRO-HOPE CARDIOLOGY A HOPE substudy–Microalbuminuria, Cardiovascular & Renal Outcomes–Heart Outcomes Prevention Evaluation–that found that ramipril prevents macro- and microvascular complications of hypertensive Pts with DM. See HOPE

microabscess A focal aggregate of PMNs seen in skin disease–eg, mycosis fungoides–Pautrier's microabscess, psoriasis–Munro's microabscess, bullous pemphigoid–papillary microabscess, or elsewhere–eg, in perivascular tissues of lung in Wegener's granulomatosis. See Munro's abscess

microaggregate filter Micropore filter TRANSFUSION MEDICINE A 2nd-generation blood filter with a pore size of 20-40 μm which removes 75-90% of WBCs, used to transfuse packed RBCs. See Blood filter, Irradiation, Leukocyte reduction

microalbuminuria NEPHROLOGY The excretion of 30-300 mg albumin/day; ↑ albumin excretion predicts hemodynamic and morphologic changes of diabetic nephropathy; when ↑, is an early indicator of renal failure the risk of microalbuminuria in DM ↑ when Hb$_{A1}$ value rises > 10% . See Diabetic nephropathy

microampere CARDIOLOGY A unit of electric current—one millionth of an ampere–for measuring very small electric currents; most pacemakers draw 10–30 μAmp continuously from a battery

microaneurysm A teensy-weensy aneurysmal dilation of small arteries, arterioles, or capillaries 'classically' seen in the retina of Pts with long-standing DM, often associated with edema and hemorrhage; MAs occur elsewhere in DM, and in other conditions–eg, thrombotic thrombocytopenic purpura. See Cherry-red spot

microangiopathy VASCULAR DISEASE Any defect of very small blood vessels, usually capillaries, which is most common in DM. See Diabetic microangiopathy

Microban® PUBLIC HEALTH A proprietary process that bonds microbicidal pellets to plastics or fibers–eg, surgical drapes, mattresses, pillow covers, children's toys

microbe A teensy-weensy organism–eg, bacterium, fungus, protozoan; bug

microcalcification IMAGING An aggregate of precipitated calcium salts, which by LM appears as purplish crystalline debris; MCs *usually* lack clinical significance, and occur in various tissues–eg, blood vessel walls in ASHD, or in necrosis; MCs appear on mammography as scattered or aggregated white dots, and are a 'soft' diagnostic criterion suggestive of breast CA

microcephaly NEONATOLOGY An abnormally small head, often associated with developmental delay and mental retardation; any brain or head that is ≥ 3 standard deviations below the mean for a person's age, sex, height, weight, race. Cf Macrocephaly

microcirculation The circulation of blood at the terminal arterioles, capillaries, small venules

microcolon PEDIATRIC IMAGING The inactive portion of the large intestine below a site of small or large intestinal atresia, which demonstrates a pencil-thin column of radiological contrast. See Colon

microcytic anemia An anemia in which the RBCs are < 6.5 μm in diameter, < 80 fl, due to iron deficiency, thalassemias, lead poisoning

microcytosis HEMATOLOGY RBCs with an MCV of < 65 fL and a diameter of ≤ 6 μm; microcytosis is typical of iron-deficiency anemia and tha-

lassemia; RBCs may be low in iron. See Iron-deficiency anemia, Thalassemia. Cf Macrocystosis

microdeletion syndrome MOLECULAR MEDICINE A clinical condition caused by a loss of a teensy-weensy part of a chromosome–eg, Angelman syndrome, DiGeorge syndrome, Prader-Willi syndrome, Williams syndrome

microencapsulated THERAPEUTICS *adjective* Surrounded by a thin layer of biodegradable substance–eg, a microsphere, as a means of protecting a drug or vaccine antigen from rapid breakdown, or of enhancing antigenic absorption and immune response thereto

microendoscope SURGERY An endoscope with a diameter of < 3.0 mm. See Endoscopy

MicroFlow OPHTHALMOLOGY A phacoemulsification needle used in ocular surgery.

microfocal cancer UROLOGY A focus of prostate CA measuring ≤ 1 mm, which is *clinically microfocal*, if < 1 mm on a needle biopsy, or *pathologically microfocal* if < 1 mm on a radical prostatectomy

microfracture BONE-OLOGY A linear defect in articular cartilage that extends beyond the articular calcified cartilage into adjacent tissue, accompanied by a reparative reaction. Cf Microcrack

microhematocrit A hematocrit performed on blood in a capillary tube

microinvasive carcinoma A superficially invasive epithelial malignancy; if you must ask, it usually isn't UTERINE CERVIX Stage Ia carcinoma A squamous cell carcinoma–SCC that penetrates < 5 mm from the base of the epithelium or < 7 mm in horizontal spread; anything larger is Stage Ib; cervical MC has > 95% 5-yr survival; lymph nodes are involved in ±1% of MCs VULVA A SCC measuring < 2 cm in diameter, and < 5 mm stromal invasion; 5% have lymph node metastases TREATMENT Vulvectomy, lymph node excision if involved. See Carcinoma in situ

microlaparoscope LAPAROSCOPY A 2 mm endoscope that allows a surgeon to perform most procedures–diagnose endometriosis, treat tumors, and repair hernias for which a larger–5 mm laparoscope was once used PROS Less pain, more rapid recovery CONS Greater skill required, cannot perform other laparoscopic procedures–eg, appendectomies, cholecystectomies. See Laparoscopy

Microlet Vaculance™ PHLEBOTOMY A lancet with 4 adjustable depths; indicated for use in obtaining blood glucose levels from areas other than fingertips–base of thumb, edge of palm, outer thigh. See Lancet

micromanage ADMINISTRATION A popular term for excess oversight of lower management by upper management

micrometastasis SURGICAL PATHOLOGY A metastatic CA identified only by histologic examination; micrometastases are detectable by neither physical exam nor imaging techniques

Micromonosporaceae A family of fungi–eg, *Micropolyspora faeni* and *Thermoactinomyces vulgaris*, which cause farmer's lung. See Farmer's lung

micronutrient CLINICAL NUTRITION A minor and necessary component of a balanced diet–eg, vitamins, minerals. See Diet. Cf Macronutrient, Minerals, Non-nutritive dietary component, Vitamins

microorganism A organism detected by microscopy–eg, viruses, bacteria, fungi and intracellular parasites–protozoans; bug

micropenis Microphallus, penile agenesis, wee wee-wee UROLOGY A small penis; a normal infantile penis measures 3.9 cm ± 0.8 when stretched; micropenis may be idiopathic, a common finding in obese infants and boys or may be related to ↓ hormonal axis activity, as in hypogonadotropic hypogonadism–Kallmann syndrome, Prader-Willi syndrome, septo-optic dysplasia, 1° hypogonadism or Klinefelter's disease, partial androgen insufficiency. See Macropenis

microphthalmia A congenital ↓ in eye size, with the ocular bulb measuring $\frac{1}{2}$ of the normal volume in extreme cases, due to an abnormal development of the optic vesicle in the optic cup, which may be 1. Congenital, as in encephalo-ophthalmic dysplasia, focal dermal hypoplasia, Hallermann-Streiff syndrome, incontinentia pigmenti, Lenz's microphthalmia syndrome, retinopathy of prematurity, trisomy 13-15 or 2. Infectious–eg, CMV, rubella, toxoplasmosis. See TORCH

microscopic esophagitis GASTROENTEROLOGY Chronic round cell esophageal inflammation, which may be the only physical change of GERD. See Gastroesophageal reflux disease

microscopic hematuria Microhematuria Hematuria that can only be detected by LM examination of the urine

microscopic metastasis ONCOLOGY A focus of metastasis which can only be identified by light microscopic evaluation

microscopic polyangiitis Hypersensitivity angiitis, microscopic periarteritis, microscopic polyarteritis INTERNAL MEDICINE The term recommended for necrotizing vasculitis with few or no immune deposits, affecting small vessels–ie, arterioles, capillaries, venules; MP is often accompanied by ANCA, necrotizing glomerulonephritis and pulmonary capillaritis. See Small vessel vasculitis, Systemic vasculitis

microsleep Nodding SLEEP DISORDERS A brief episode of sleep lasting a few secs, which occurs in night-shift fatigue; it is associated with excess daytime sleepiness and automatic behavior. See Circadian rhythm, Libby Zion, Night-shift fatigue, Sleep disorders

microsomia Corporal smallness; somatic undersize

Microsporidia PARASITOLOGY A phylum of ubiquitous unicellular obligate intracellular protozoans–eg, *Enterocytozoon*, which infect insects and vertebrates, and humans. See *Encephalitozoon* species

microsporidiosis INFECTIOUS DISEASE Infection by *Microsporidia*, which usually infects immunocompromised hosts, especially AIDS Pts CLINICAL–GI TRACT Chronic, watery diarrhea, ↓ weight, abdominal pain N&V; CLINICAL––SYSTEMIC Sx of cholecystitis, renal failure, RTIs, red eyes, photosensitivity, asymptomatic URIs MANAGEMENT Albendazole

Microsporum A genus of fungi causing tinea capitis, tinea corpus, ringworm, and other dermatophytoses. See Tinea

microsurgery SURGERY A surgical procedure performed with the aid of a low-power–7x to 15x microscope, using special equipment, surgical thread, clamps, scalpels, to repair severed blood vessels or nerves or other structures. See Free flap microsurgery, Laryngeal microsurgery

microsurgical diskectomy Microsurgical laminectomy NEUROSURGERY, ORTHOPEDICS A procedure for treating a herniated intervertebral disk, in which the vertebral body is not explored before surgery; ML ↓ the risk of the feared 'failed disk syndrome', as there is minimal excision of bone and epidural fat and minimal nerve root adhesion. See Failed disk syndrome, Laminectomy. Cf Percutaneous diskectomy

microtrauma ORTHOPEDICS Small, usually unnoticed injuries caused by repetitive overuse. See Overuse syndrome

microvascular decompression Corridor procedure NEUROSURGERY A procedure for cranial rhizopathies, to ↓ pressure on nerves compressed by vessels in a confined space–eg, in the cranial cavity INDICATIONS Hemifacial spasm, glossopharyngeal neuralgia, trigeminal neuralgia

microvascular disease See Diabetic microangiopathy

microvascular free toe transfer HAND SURGERY The transfer of the great toe to a hand in which the thumb was lost due to trauma

microwave THERAPEUTICS Deep heat therapy The administration of electromagnetic waves that pass between electrodes placed on the Pt's skin, creating heat that ↑ blood flow and relieves muscle and joint pain

microwave catheter ablation CARDIOLOGY A procedure for correcting atrial flutter/fibrillation and restoring normal sinus rhythm by creating a linear lesion across the isthmus between the inferior vena cava–IVC and the tricuspid annulus, interrupting the abnormal electrical circuit causing the arrhythmia. See Atrial fibrillation

microwave diathermy SPORTS MEDICINE A form of diathermy that delivers shorter waves of higher frequency electromagnetic waves than deliverable by shortwave diathermy. See Diathermy. Cf Shortwave diathermy

microwave prostatectomy The reduction of prostate volume by 'zapping' the tissue with microwaves. See TURP

microwave radar Sublethal exposure has been associated with headaches, insomnia, irritability, photophobia, and diastolic HTN; after acute exposure, there is a sensation of warmth, and ↑ CK

micturition Urination; the process of voiding urine

MICU Mobile intensive care unit EMERGENCY MEDICINE A vehicle, usually a specially-designed minivan or truck with the capacity for providing emergency care and life support to the severely injured or ill at the scene of an accident or natural disaster and transporting the Pts to a medical facility where treatment may continue. See Air ambulance

MID Multi-infarct dementia, see there

mid-arm muscle area A derived value for estimating lean body mass as a function of skeletal muscle; ≥ 30% below a standardized value–52–55 for ♂; 31–35 for ♀–Health & Nutritional Examination Surveys–HANES data indicates a depletion of lean body mass, ie malnutrition.; Cf Triceps skin fold

midcycle pain Mittelschmertz, ovulation pain GYNECOLOGY Unilateral pain that occurs at the time of ovulation, attributed to release of an egg from the ovary, accompanied by peritonitis. See Dysmenorrhea

middiastolic murmur CARDIOLOGY A heart sound which begins after the AV valves have opened in diastole, typical of mitral stenosis

midlevel provider MANAGED CARE A health care provider–eg, nurse practitioner, physician assistant, etc, whose activities are directed and/or dictated by a supervising physician on whom liability for those activities rests

midlife crisis PSYCHOLOGY A popular term for a mental 'crisis' occurring in a person's 'middle' yrs–from late-30s to early 50s who reflects on life events to date vis-á-vis accomplishments and possible future accomplishments before turning in the 'feedbag'

midline granuloma HEAD & NECK Destruction and necrosis of tissues of midline lower neck and upper mediastinal structures ETIOLOGY Idiopathic midline destructive disease; malignant midline reticulosis–a pleomorphic, often T-cell, lymphoma; Wegener's granulomatosis; infections–eg, bacterial–*Actinomyces, Brucella* spp, *Mycobacterium leprae, M tuberculosis, Klebsiella rhinoscleromatis, Treponema pallidum*, fungal–blastomycosis, candidiasis, coccidioidomycosis, histoplasmosis, phycomycosis, rhinosporidiosis, parasitic–leishmaniasis and myiasis; also sarcoidosis, relapsing polychondritis, necrotizing sialometaplasia CLINICAL Progressive, ulceration of nasopharyngeal and midline facial tissues; pathogenesis is unclear; MG may represent a poor immune response to various infections in an immune- or otherwise compromised Pt–eg, a diabetic or malignant–SCC, rhabdomyosarcoma, lymphomatoid granulomatosis, lymphoma. See Idiopathic midline destructive disease, Lethal midline granuloma, Wegener's granulomatosis

midrange EPIDEMIOLOGY The halfway point or midpoint in a set of observations; for most data, MR is calculated as the sum of the smallest observation and the largest observation, divided by 2; for age data, one is added to the numerator; a midrange is usually calculated as an intermediate step in determining other measures STATISTICS The mean of the highest and lowest values. (Max + Min)/2 . See Range

mid-systolic click syndrome Mitral valve prolapse syndrome, see there

midwife OBSTETRICS A trained person, often an advanced practice registered nurse, who assists in childbirth or, in many situations, is the primary provider of obstetric care SALARY $66K + 9% bonus. See Alternative birthing center, Alternative gynecology, Certified nurse midwife, Doula, Granny midwife, Lay midwife, Natural childbirth; Cf Lamaze technique

MIDAS CARDIOLOGY A clinical trial–Multicenter Isradipine Diuretic Atherosclerosis Study which evaluated effects of antihypertensives–eg, isradipine, a CCB in hypertensive Pts. See Calcium channel blockers

midazolam Versed® PHARMACOLOGY A short-acting anxiolytic, hypnotic anticonvulsant INDICATIONS IV sedation, general anesthesia, therapeutically refractory status epilepticus, muscle relaxing ADVERSE EFFECTS Ataxia, incoordination, mental and psychomotor impairment, confusion, dysarthria, retrograde amnesia, paradoxical effects; rarely, N&V, venous complications. See Phenytoin, Seizure, Status epilepticus

MIDCAB Minimally invasive direct coronary artery bypass CARDIOLOGY Minimally invasive heart surgery in which coronary bypass–usually with a single graft is performed on a blood-filled beating heart without a heart-lung machine via a small incision in the chest wall. See Minimally invasive cardiac surgery, OPCAB. Cf CABG, Transmyocardial revascularization

middle-aged *adjective* Referring to a person between age 45 and 65, used in taking a history. Cf Elderly, Older

middle ear infection Otitis media ENT A condition characterized by inflammation, fluid overproduction–which may rupture the tympanic membrane, providing a portal of entry for bacteria and viruses, purulence, bleeding; MEI is more common in children as their eustachian tubes are shorter, narrower, and more horizontal than adults RISK FACTORS Infections, sinusitis, allergy-induced eustachian tube blockage or adenoidal enlargement, recent illness–lowered resistance, crowding, poor hygiene, heredity, high altitude, cold climate, bottle feeding–with pooling of fluid at the eustachian tube

middle lobe syndrome Chronic atelectasis and collapse of the middle lobe and/or lingula of the right lung due to extrinsic compression of the right middle bronchus by thymoma, hilar lymphadenopathy–eg, TB, sarcoidosis, or lymphoma, tumors or obstruction–due to intraluminal tumors or foreign bodies; compression results in chronic pneumonitis, bronchial obstruction, bronchiectasis and ↓ lung capacity; primary disease findings include calcified hilar lymph nodes, granulomas and erosion into the bronchopulmonary apparatus

middle molecule toxins A group of small, ie 3-3.5 kD molecules–eg, uric acid, guanidinos, metabolic end products–eg, phenols, that are not removed during dialysis of Pts with chronic renal failure and linked to the peripheral neuropathy and pericarditis common in uremia

middle stage HIV disease Chronic symptomatic HIV infection, see there

mifepristone RU 486 OBSTETRICS An antiprogestin used as a morning-after contraceptive, with misoprostol, a PGE₁ analogue, for early pregnancy termination. Cf Late abortion

migraine Hemicrania, sick headache NEUROLOGY An intense idiopathic, episodic, uni/bilateral, pulsating–vascular headache, often exacerbated by physical activity, linked to dilation of branches of the carotid artery CLINICAL 'Classic' migraines are most common in ♀ age 30-49, and in lower income households, and appear to have a hereditary component; migraines first appear before puberty and remit at menopause; they may be accompanied or preceded by N&V, photophobia, other visual phenomena–eg, hemianopia, scotomas, fortification phenomenon, phonophobia TREATMENT Analgesics–eg, aspirin, acetaminophen, propoxyphen, codeine, NSAIDs–eg, naproxen, ibuprofen, ketorolac, 5-HT agonists–eg, ergotamine, sumatryptan, dopamine antagonist–eg, chlorpromazine, metoclopramide PREVENTION Avoid precipitating factors; if conservative measures fail and the attacks are > 1/wk, pharmacologic prophylaxis is indicated, which may be 5-HT-influencing–eg, amitriptyline, methysergide, β-adrenergic antagonist–eg, propranolol, metoprolol, calcium channel blocker–eg, nifedipine, verapamil, NSAIDs–eg, ketoprofen, mefenamic acid, aspirin, sumatriptan. See Aura, Basilar migraine, Classical migraine, Common migraine, Footballer's headache, Menstrual migraine, Mixed tension, Ophthalmoplegic migraine, Retinal migraine

migraine equivalent NEUROLOGY A rare presentation of a migraine, in which the Pt suffers from the neurologic and somatic epiphenomena of a migraine attack, but is spared the headache. See Migraine

Migranal® Dihydroergotamine mesylate NEUROLOGY An antimigraine agent. See Migraine

migrating testis Elevator testicle A testicle that is mobile within the inguinal canal, which may migrate into the abdominal cavity; such cases require surveillance, as they are at ↑ risk for testicular torsion

migratory thromboembolism Trousseau sign Thrombophlebitis seen transiently in multiple sites, a finding seen in up to 10% of Pts with CA, classically in mucin-secreting GI adenoCA, but also in breast, lung, ovarian, prostate CA, and others

MiKasome® Amikacin liposome INFECTIOUS DISEASE An agent in clinical trials for treating complicated UTIs, acute infections in cystic fibrosis, nosocomial pneumonia

Miketorin Amitryptiline, see there

mild traumatic brain injury Dinging, mild traumatic brain injury SPORTS MEDICINE An '… *immediate and transient impairment of neural function such as alteration of consciousness, disturbance of vision, equilibrium, and other similar symptoms.*' features common to MTBI are '…*limited or absent loss of consciousness, limited post-traumatic amnesia, and an initial Glasgow Coma Scale of ≥ 13 of 15.*'

miliaria Prickly heat DERMATOLOGY A condition characterized by multiple vesicles associated with ↑ heat, moisture, occlusion CLINICAL Pruritus, hypohidrosis may cause irritability and insomnia. Cf Diaper dermatitis

miliary *adjective* Referring to a disseminated process comprised of innumerable millet-seed sized lesions, classically seen in miliary TB in the pre-antibiotic period–the 'millet seeds' correspond to granulomas, also seen in disseminated histoplasmosis and CMV pneumonitis

miliary tuberculosis TUBERCULOSIS TB that has disseminated throughout the body, especially the lungs; the minute, millet seed-sized lesions correspond to granulomas. See Tuberculosis

milium Whitehead DERMATOLOGY One of multiple, small subepithelial keratin cysts arising in eccrine sweat ducts, often on the face, present from infancy onwards, especially in young ♀ after sunbathing

milk A whitish fluid derived from the mammaries or simulates same in color or constitution. See Breast milk, Certified milk, Humanized milk, Raw milk, Unpasteurized milk, Witch's milk

420 **milk-alkali syndrome** METABOLIC DISEASE A condition caused by hypercalcemia due to excess consumption of dairy products, or overuse of calcium-containing-> 5g/day antacids–eg, $CaCO_3$ or alkalis–eg, sodium bicarbonate for treating peptic ulcer CLINICAL Lethargy, constipation, renal decompensation LAB Hypercalcemia, severe compensated metabolic alkalosis, normo- to hyperphosphatemia; metabolic derangements may cause renal failure through nephrocalcinosis, inability to compensate for alkalosis and dehydration

milk intolerance Lactose intolerance, see there. See Lactose tolerance test

'milk' leg Phlegmasia alba dolens OBSTETRICS Extensive deep vein or iliofemoral thrombosis due to stasis of uterine blood, accompanied by painful swelling and pallor of the entire leg, a condition once common in parturition–hence, 'milk leg', more often seen after abdominal or pelvic surgery; with progression, veins become thrombosed, blood can't return to heart, leg becomes cool, painful, cyanotic–phlegmasia cerulea dolens–painful blue leg. See Deep vein thrombosis

milk let-down see Let-down reflex

'milking' HEART SURGERY The gentle squeezing of a blood vessel, often a distal vein or the extremity itself, 'milking' it proximally to extract a non-adherent thrombus

Milkman syndrome Generalized osteomalacia A 'radiologist's disease' more common in middle-aged ♀, characterized by alternating symmetrical radiolucent bands or pseudofractures corresponding to resorption and mineralization near arteries, a condition that responds to vitamin D therapy

Millennium™ system DENTISTRY An erbium laser system that uses high-speed water particles to cut hard and soft tissues, as an alternative to a conventional dental drill and surgical scalpel. See Dentistry

Miller Behavioral Style Scale PSYCHOLOGY A 32-item questionnaire that evaluates a person's means of coping with stress, based on whether the individual wants more or less information in the presence of ↑ mental stress

milli- Latin, one one-thousandth–1/1000–of a thing

million-dollar workup MEDTALK An extensive, expensive, intensive, sometimes painful, diagnostic evaluation initiated by a particular Pt complaint–eg, worst headache of my life, Pt request–eg, comes with x-rays in hand from another hospital, wanting a 2nd opinion, physician fear of lawsuit, or attempt to achieve a definitive diagnosis of a fascinoma

millivolt CARDIOLOGY A thousandth of a volt, a unit of measure for low voltage levels which forms the basis for measuring intrinsic P and R waves

Millon Clinical Multiaxial Inventory-II PSYCHOLOGY A psychometric instrument used to place a person in one of 20 categories, assessing acute clinical disorders, and chronic–normal and abnormal personality characteristics and personality disorders in the DSM. See DSM-IV

Mills' test SPORTS MEDICINE A clinical maneuver for diagnosing lateral epichondylitis–tennis elbow

Milontin® Phensuximide, see there

milrinone Primacor® CARDIOLOGY A positive inotropic phosphodiesterase inhibitor that provides short-term benefit; it may have a deleterious effects on chronic heart failure. See PROMISE trial

Miltown® Meprobamate, see there

Milwaukee Brace Moe brace ORTHOPEDICS A whole body brace that extends into neck and is used in the conservative therapy of scoliosis

Milwaukee shoulder ORTHOPEDICS A painful, destructive, bilateral upper girdle dysfunction of elderly ♀, consisting of capsular calcification, joint effusions–↑ collagenase without inflammation, erosion of rotator cuff tendons and glenohumeral joint degeneration on the dominant side; the condition is worse at night or after heavy usage with accumulation of basic calcium phosphate crystals in inflamed joints

MIM number A numerical assignment for inherited diseases, genes and functional segments of DNA listed in the comprehensive catalog, *Mendelian Inheritance in Man*, begun and maintained by Victor McKusick. See OMIM

mimic *noun* VOX POPULI That which simulates another thing. See Malignant mimic

mind Psyche PSYCHOLOGY The consciousness that originates in the brain, and is evident in emotion, imagination, memory, perception, thought, and volition

mind/body medicine 1. Alternative medicine, see there 2. Behavioral medicine. See Behavioral medicine, Biopsychosocial model, Body-mind technique, Evocative breath therapy, Flight-or-fight response, Movement & exercise technique, Psychoneuroimmunology

mineral NUTRITION A popular term for a nonvitamin nutrient needed to maintain health

DIETARY MINERALS

MAJOR MINERALS—IN BONE Calcium, phosphate, magnesium

MAJOR MINERALS—IN ELECTROLYTES Sodium, potassium, chloride

MINOR MINERALS—IN METALLOPROTEINS Iron, copper, manganese, iodine, cobalt, molybdenum, selenium, chromium, fluoride, zinc

TRACE MINERALS Nickel, silicon, vanadium, tin

mineral oil NUTRITION A mixture of liquid petroleum-derived hydrocarbons–specific gravity, 0.818-0.96, which was formerly used as a vehicle for pharmaceuticals or as a GI tract lubricant. See Lipoid pneumonia

mineralocorticoid Any of the steroid hormones–the most important is aldosterone–that regulate water and electrolyte equilibrium, acting specifically on renal tubule

miniantibody IMMUNOLOGY A small antibody with 'nonessential' oligopeptides, which ↓ its size by ⅔ but has the same binding affinity as a garden variety antibody; MAs have properties desirable in tumor immunology and other clinical applications

minifilm RADIOLOGY A 10 cm^2 film used for producing chest X-rays that requires 2.0 mrad–in contrast to ≥ 9.2 mrad—for the 'standard' 35 x 42.5 cm film

minilap Minilaparotomy An abbreviated laparotomy used to obtain cells, document intraperitoneal hemorrhage, obtain fluids for assessing bile amylase, bacteria, or fecal material by peritoneal lavage OBSTETRICS Minilaps are used for sterilization See Laparotomy tubal sterilization

mini-med school Any outreach program developed by a medical school, which provides a 4-8 wk classroom education on multiple topics of public interest, in particular, discoveries in health sciences

Mini-Mental Status Exam MMSE of Folstein PSYCHOMETRIC TESTING A screening mental status tests; a perfect score on the Folstein is 30; a score < 17 corresponds to probable dementia.

mini-mental test NEUROLOGY A brief clinical test of mental status, where each correct answer in a series of questions is given one point–total score 30

MINI-MENTAL TEST

ORIENTATION IN TIME: Year, season, month, date, day–total 5 points–pts

ORIENTATION IN SPACE Country, state, county, town, place, hospital ward–5 pts

COGNITION Serial 7s–x 5 or spell world backwards–5 pts

SHORT RECALL Name 3 objects–total 3 pts

MEMORY Rename 3 above objects–3 pts

FOLLOW A THREE-PART COMMAND Take a paper, fold it, put it on the floor–3 pts

COMMON OBJECT RECOGNITION Name 2 familiar objects–2 pts

RECOGNITION OF COMMON PHRASE 'No ifs, ands, or buts'–1 pt

READ AND OBEY 'Close your eyes'–1 pt

WRITE SIMPLE SENTENCE–1 pt

COPY DRAWING Intersecting pentagons–1 pt

A change in mental status and a score > 27 points is most often associated with affective depression; depressed Pts with cognitive impairment have scores of ± 20, those with true dementia often have scores of < 10 **J Psych Res** 1975; 12:189

mini stroke Transient ischemic attack, see there

minimal brain dysfunction Hyperactive child, hyperkinetic child syndrome, hyperkinetic syndrome, minimal brain damage, minimal cerebral dysfunction NEUROLOGY A term used in the 1960s for children with learning problems of implied neurological basis. See ADD

minimal breast cancer ONCOLOGY An invasive breast CA measuring ≤ 5 mm. See Breast cancer

minimal change disease Idiopathic nephrotic syndrome of childhood, lipoid nephrosis, minimal change nephrotic syndrome, nil disease NEPHROLOGY A cause of nephrotic syndrome, so named because, by LM, the glomeruli appear normal; by EM, characteristic glomerular changes are seen–eg, fusing of the epithelial layer; MCD is most common in children ETIOLOGY Unknown; risk is ↑ in those with immune disorders, recent immunizations, bee stings PROGNOSIS MCD is not linked to oliguria or progressive renal failure. Cf Renal failure

minimal detectable concentration LAB MEDICINE The lowest concentration at which an analyte can be reliably detected

minimal response ONCOLOGY A measurement of response to therapy for malignancy. Cf Complete remission, Induction failure, Partial remission

minimally invasive direct coronary artery bypass surgery See MIDCAB

minimally invasive cardiac surgery INTERVENTIONAL CARDIOLOGY Any of a number of techniques–MIDCAB, off-pump coronary-artery bypass, minimally invasive valve surgery, port-access coronary surgery, and port-access valve surgery–increasingly being used to ↓ the discomfort, morbidity, and costs associated with conventional cardiac surgery. See Interventional cardiology, MIDCAB, Off-pump coronary-artery bypass, Port-access coronary surgery. Cf CABG

minimally invasive surgery Laparoscopic surgery, see there. See Laparoscopic cholecystectomy

minimum bactericidal concentration LAB MEDICINE The lowest concentration of an antibiotic that is bactericidal to ≥ 99.9% of an original inoculum. See Persistence phenomenon, Paradoxic effect, and Tolerance

minimum daily requirements NUTRITION The amount of a particular macronutrient–essential fats, proteins, carbohydrates, or micronutrients–vitamins, minerals, needed by a person/day in absence of special metabolic needs

minimum effective concentration THERAPEUTICS The minimum serum drug concentration necessary to produce a desired pharmacologic effect in most Pts. See Minimum bactericidal concentration

minimum inhibitory concentration LAB MEDICINE The minimum antibiotic concentration needed to inhibit bacterial growth from a clinical isolate–eg, a bloodborne infection, which is a form of antimicrobial susceptibility testing. See Minimum bactericidal concentration

minimum risk level TOXICOLOGY An estimate of daily human exposure to a concentration of a chemical which has a minimal appreciable risk of cancerous effects over a specified time period

minipill GYNECOLOGY An OC that contains a very low dose of progestin, a low-dose synthetic progesterone norethindrone, and minimal estrogen; MP ↓ sperm penetration of cervical mucus, interferes with luteinization and implantation and ↓ gonadotropin secretion; it may be less effective than combined OCs and IS associated with dysmenorrhea. See Oral contraceptive. Cf Norplant, RU 486

Minipress® Prazosin, see there

ministroke Transient ischemic attack, see there

minisuction Menstrual extraction, see there

Minneapolis plan SUBSTANCE ABUSE A popular term for inpatient treatment of alcoholism. See Alcoholics Anonymous

Minnesota Multiphasic Personality Inventory MMPI PSYCHOLOGY A widely used true-false test for evaluating a person's psychological and personality 'profile'. See Psychological testing

Minnesota tube EMERGENCY MEDICINE A device used to control bleeding from esophageal varices at the gastric junction. See Esophageal varices. Cf Sengstaken-Blakemore tube

minor A person under 18 yrs of age. See Emancipated minor

minor cancer A CA which, even if untreated, is not clinically aggressive EXAMPLES BCC, CIN 1. Cf Indolent malignancy, Major cancer, Pathologist's cancer, Watchful waiting

minor depression Dysthymic disorder, see there

minor salivary gland Any of the multiple small unnamed salivary glands in the oral cavity and palate

minor surgery Minor procedure SURGERY Any surgical procedure that can be performed in a brief period of time–usually < 1 hr under local anesthesia, does not–under normal circumstances–constitute a major hazard to life or function of organs or body parts; MS does not generally require hospitalization and may be performed electively, usually by a general–board-certified surgeon in a secondary-care hospital setting. See Minor procedure. Cf High-risk & complex procedure, Major surgery

minority institution NIHSPEAK An institution of higher learning that draws most of its enrollment from minorities. See HBCU, Traditionally or Predominantly African American Institution

minoxidil Rogaine® HAIROLOGY An agent first used as vasodilator and potassium channel opener in Pts with severe HTN, now used for androgenetic alopecia, alopecia areata, congenital hypotrichosis, loose anagen syndrome ADVERSE EFFECTS Sodium & water retention, pericardial effusion, hypertrichosis, hirsutism. See Hair, Hair replacement. Cf Terminal hair

MINT CANDY CLINICAL ROTATIONS A mnemonic of uncertain utility, which may be used by interns in training to recall broad categories of differential diagnoses for a particular symptom or clinical finding–Metabolic, Infection, Neoplastic, Trauma, Collagen vascular disease, Allergies 'N'ything else, Drugs, Youth–congenital disease.

minute volume & respiration rate CARDIAC PACING Variations in inhalation and expiration to meet metabolic demand, which cause changes in lung volume and intra-pleural pressure; an ↑ in minute volume or respiration rate correlate to a need for ↑ pacing rate

minutes *pronounced* min-nutts MEETINGOLOGY The record that summarizes the proceedings of a meeting and its important points

miosis Contraction of the pupil. Cf Mydriasis

Mipralin Imipramine, see there

MIRA RHEUMATOLOGY A clinical trial–Minocycline In Rheumatoid Arthritis

MIRACL CARDIOLOGY A clinical trial–Myocardial Ischemia Reduction with Aggressive Cholesterol Lowering

MIRACLE CARDIOLOGY A clinical trial–Multicenter InSync™ Randomized Clinical Evaluation–a study of resynchronization therapy in Pts with end-stage heart failure

MIRAGE NEUROLOGY A study–Multi-Institutional Research in Alzheimer Genetic Epidemiology

Miraluma™ Tc⁹⁹ᵐ sestamibi IMAGING A radionuclide used in planar imaging as a 2ⁿᵈ-line post-mammography diagnostic tool for evaluating breast lesions in Pts with an abnormal mammogram or a palpable breast mass. See Mammography

Mirapex® Pramipexole, see there

Mircette™ GYNECOLOGY An OC that contains desogestrel/ethinyl estradiol and ethinyl estradiol. See Oral contraceptive

mirror INFORMATICS *noun* An archiving device written to more than one hard disk simultaneously, so that if one disk fails, the computer continues to function without loss of data *verb* To maintain an exact copy of a file or database. See FTP, Web PARANORMAL See Mind mirror

mirror image biopsy A biopsy of the contralateral breast when lobular carcinoma-in-situ–CIS is found in one breast, a procedure used to exclude invasive carcinoma. See Breast self-examination, Mammography. Cf Lumpectomy

mirror movement NEUROLOGY The involuntary copying by one extremity of a movement occurring in another

mirror test ENT A crude clinical test used to obtain material from the upper respiratory tract, in which the Pt coughs on a mirror held in place in the upper larynx, which is sprayed with bronchial 'juices'

mirtazapine Remeron® NEUROPHARMACOLOGY A presynaptic-postsynaptic noradrenergic-serotoninergic reuptake antagonist, used as an antidepressant; mirtazapine blocks 2 specific serotonin receptors–5-HT2 and 5-HT3; but is not an SRI

MIS 1. Management information system, see there 2. Medical information system. See Hospital information system, Laboratory information system 3. Minimally-invasive surgery, see there 4. Müllerian(-duct) inhibiting substance, see there

misadventure LEGALESE An unintended result of an action, as in an occupation-related 'homicide by misadventure'. See Mistake. Cf Miscall

miscall Misdiagnosis, see there

miscarriage OBSTETRICS A popular term for the inadvertent loss of a pregnancy before the fetus is viable; spontaneous abortion. See Spontaneous abortion

misdiagnosis Miscall FORENSIC MEDICINE The incorrect diagnosis of a morbid condition; a diagnostic error–eg, misdiagnosing a benign tumor as malignant. See Misdiagnosis. Cf Misadventure. See Overcall, Undercall

misery index GLOBAL VILLAGE A term for 'how bad things are' in a country; the lower the MI–eg, Denmark = MI of 1, the more content the populace. See Third world

misinformed consent See Information overload. Cf Informed consent

mismatch PULMONOLOGY A marked difference between ventilation–V and perfusion–Q on V/Q scan, typical of PTE. See Pulmonary thromboembolism, V/Q scan

misophobia PSYCHOLOGY Fear of being contaminated with dirt of germs. See Phobia

misoprostol OBSTETRICS A synthetic PGE₁ analogue administered vaginally with mifepristone–RU 486 as an abortifacient, also used to manage peptic and duodenal ulcers. See Abortion, Gemeprost, Mifepristone, Sulprostone

misrepresentation MALPRACTICE Representing falsely or imperfectly; lying. Cf Fraud

MISS Modified Injury Severity Scale URGENT CARE A method for quantifying pediatric multi-trauma (excluding burns) injuries; MISS is the sum of the squares of the 3 most injured body regions, after the AMA's abbreviated injury scale–AIS, substituting the Glasgow coma scale for neurologic evaluation. See Glasgow coma scale

missed abortion The retention of a fetus known to be dead for ≥ 4 wks MANAGEMENT Expectant–spontaneous delivery occurs usually by the 6ᵗʰ postmortem wk. See Abortion

mission MANAGED CARE Why a hospital does what it does. See Mission statement SOCIAL MEDICINE A facility with 100–150 beds, which provides nightly accommodations and meals for single–usually homeless persons–usually ♂. See Homelessness, Runaway, Shelter

mission statement MANAGED CARE A formal 'writ' by an organization, indicating its raison d'etre. See TQM

mistake DECISION-MAKING 1. An act, omission or error in judgement by a health care provider that has/may have serious consequences for a Pt and that would be judged to be wrong by knowledgeable peers 2. An incorrect act, decision, or statement that is knowledge-based, judgemental, heuristic, or based on subconscious bias. See Medical mistake. Cf Misdiagnosis, 'Overcall', 'Undercall'

Mister British title for a male surgeon

Mr Green HOSPITAL CARE An announcement made over a hospital's paging system notifying personnel that a previous fire warning is in an all clear status. See Code. Cf Mister Red

Mr Red Code Red HOSPITAL CARE An announcement made over a hospital's paging system notifying personnel of a fire in the building. See Code. Cf Mister Green

mistreatment of elderly 1. Elderly abuse, see there 2. Elderly neglect

mitgehen NEUROLOGY The allowing by a Pt to have a limb moved in a particular position, regardless of given instructions. Cf Mitmachen

MITI-1 CARDIOLOGY A multicenter trial–Myocardial Infarction Triage & Intervention that evaluated the effect on M&M of preadmission thrombolytics by paramedics in Pts with suspected acute MI. See Acute myocardial infarction, Thrombolytic therapy, tPA

MITI-2 CARDIOLOGY A randomized, multicenter trial–Myocardial Infarction Triage & Intervention that evaluated the efficacy of prehospital vs. hospital-initiated thrombolytic therapy with alteplase in Pts with suspected acute MI. See Acute myocardial infarction, Thrombolytic therapy, tPA

mitmachen German, to do with NEUROLOGY A finding in catatonic Pts, who despite instructions, allow an extremity to be placed in any position without resistance to light pressure, and return the body part to the original resting position, once the examiner releases the extremity. Cf Mitgehen

mitochondrial disease Any clinically heterogeneous multisystem disease characterized by defects of brain–mitochondrial encephalopathies and/or muscle–mitochondrial myopathies due to alterations in the protein complexes of the electron transport chain of oxidative phosphorylation; MDs include Alper syndrome, Leber's hereditary optic neuropathy, Lowe syndrome, Luft syndrome, Menke's kinky hair syndrome, Zellweger syndrome, MELAS, MERRF, mitochondrial myopathy, rhizomelic chondrodysplasia punctata, and stroke-like episodes. See MELAS, MERRF

MITOCHONDRIAL DISEASES

GROUP 1 PROGRESSIVE EXTERNAL OPHTHALMOPLEGIAS
- Kearns-Sayre disease Ophthalmoplegia plus syndrome
- Ocular myopathy
- Leber's hereditary optic neuropathy–due to a point mutation

GROUP 2 MITOCHONDRIAL ENCEPHALOMYOPATHIES
- Mitochondrial encephalomyopathy with lactic acidosis and stroke-like episodes
- Myoclonus epilepsy with ragged red fibers
- Leigh syndrome

GROUP 3 UNDEFINED MITOCHONDRIAL ENCEPHALOMYOPATHIES, eg congenital lactic acidosis

GROUP 4 MITOCHONDRIAL MYOPATHIES
- Luft syndrome
- Enzyme defects, eg ATPase, cytochrome oxidase

mitochondrial inheritance The inheritance of a trait encoded in the mitochondrial genome, always of maternal origin

mitomycin ONCOLOGY An antineoplastic antibiotic produced from *Streptomyces caespitosus*, which acts as an alkylating agent and cross-links DNA, thereby inhibiting DNA synthesis INDICATIONS Leukemia

mitotane Lysodren® ONCOLOGY A cytotoxic agent that controls the endocrine Sx in Pts with adrenocortical CA and ACTH producing pituitary tumors that cause Cushing's disease ADVERSE EFFECTS Anorexia, N&V, neurologic disturbances–ataxia, speech defects, somnolence, vertigo, lethargy, liver enzyme changes–↑ ALA, ALT, alkaline phosphatase, GGT, ↑ corticosteroid-binding globulin, ↑ sex hormone-binding globulin, ↑ cholesterol, ↓ uric acid, ↓ thyroid hormones–normal TSH. See Adrenocortical carcinoma

mitotic activity ONCOLOGY The degree to which a cell population proliferates, often an indicator of tumor aggressiveness; MA is measured by frequency of cell division; it is semiquantified by counting mitotic figures/high-power field, or flow cytometry. See Flow cytometry, High power fields

mitotic activity index CYTOMETRY The number of mitotic figures/10 high power fields

mitoxantrone Novantrone® ONCOLOGY A synthetic antitumor antibiotic INDICATIONS Combined with steroids, for treating hormone-refractory prostate CA, breast CA, AML and other lymphoproliferative malignancy and multiple sclerosis PHARMACOKINETICS Similar to doxorubicin SIDE EFFECTS Dose-limiting granulocytopenia, thrombocytopenia, N&V

Mitraflex WOUND CARE A self-adhesive wound dressing for oozing or exudating wounds. See Wound care

mitral regurgitation Chronic mitral valve regurgitation, mitral insufficiency CARDIOLOGY Backflow of blood from the left ventricle to the left atrium due to a defective mitral valve, which ↓ forward flow of blood and ↑ work by heart to pump more blood to compensate for inefficiency; acute MR may be due to valve dysfunction or injury post MI or infectious endocarditis, which may cause rupture of the valve, papillary muscle, or chordae tendineae–which anchor the valve cusps, resulting in valve leaflet prolapse into the atrium, leaving an opening for the backflow of blood; chronic MR is prolonged and progressive, and often associated with mitral valve prolapse, characterized by weakening and ballooning ETIOLOGY Rheumatic heart disease–thickening, rigidity, retraction of mitral valve leaflets; ASHD, HTN, left ventricular enlargement, connective tissue disorders–eg, Marfan syndrome, congenital defects, endocarditis, heart tumors, late syphilis, untreated acute MR. See Mitral valve

mitral ring calcification Mitral annular calcification CARDIOLOGY A noninflammatory chronic degenerative process characterized by a massive deposition of calcium in the fibrous support structure of the base and leaflet of the mitral valve; primarily a disease of ♀ and elderly, MRC affects younger

Pts with Marfan syndrome or conditions linked to metastatic calcification–eg, obesity CLINICAL Hemodynamic problems–left ventricular hypertrophy, systemic HTN, A fib, arrhythmias, heart block, CHF, stroke MANAGEMENT Replace valve

mitral stenosis Mitral valve obstruction CARDIOLOGY A sequela of rheumatic heart disease, primarily affecting ♀ and more common in developing nations CLINICAL Left-sided heart failure–DOE, orthopnea, paroxysmal nocturnal dyspnea; less commonly, hemoptysis, hoarseness, signs of right-sided heart failure; Sx may be triggered by A Fib, pregnancy or other stress–eg, RTI, endocarditis or other heart disease EXAMINATION Diastolic rumble after opening snap; S_1 is usually load, because the mitral valve remains open by a transient gradient until closed by the systolic force; pulmonary HTN is indicated by a loud P_2, right ventricular lift, ↑ neck veins, ascites, edema WORKUP Doppler echocardiography–↓ valve diameter, severity of stenosis MANAGEMENT Medical therapy in asymptomatic Pts–antibiotics for endocarditis; mild Sx–diuretics; with A Fib, digoxin, beta-blocker, or CCB; moderate Sx–balloon valvotomy if calcification not excessive, otherwise, valve repair or replacement. See Mitral valve

mitral valve prolapse syndrome Barlow syndrome, billowing mitral valve, floppy mitral valve, myxomatous mitral valve, prolapsing mitral leaflet syndrome, systolic click-murmur syndrome CARDIOLOGY A common, heterogeneous condition which affects up to 7% of young ♀, in which the mitral valve prolapses into the left atrium CLINICAL MVPS is usually asymptomatic, it may be associated with fatigue and/or palpitations; it may cause sudden death by arrhythmia or rupture of cordae tendinae, leaflet thickening and leaflet redundancy; untreated Pts are at ↑ risk for infectious and hemodynamic complications of MVP EKG Inverted T waves in II, III, aV_F leads, prolonged Q wave DIAGNOSIS Clinical, echocardiography. See MASS phenotype

Mitran Chlordiazepoxide, see there

mitriptyline Etrafon® A tricyclic antidepressant, see there

mivacurium ANESTHESIOLOGY A short-acting nondepolarizing muscle relaxing anesthetic, a congener of atracurium. See Atracurium, Muscle relaxant, Nondepolarizing agent. Cf Depolarizing agent

mix MANAGED CARE A term referring to certain types of ratios vis-á-vis beneficiaries or service types rendered in a managed care setting. See Case mix, Patient mix, Payer mix, Service mix

mixed connective tissue disease A connective tissue disease with features of SLE, dermatomyositis, rheumatoid arthritis CLINICAL Pleuritis, Raynaud's phenomenon, sclerodactyly, good response to steroids LAB MCTD has a unique speckled nucleolar pattern due to presence of a specific circulating antibodies to ribonucleoprotein; no antibodies to double-stranded DNA and Sm antigen. See Antinuclear antibodies, 'Chinese menu' diseases, Overlap syndrome

mixed cryoglobulinemia Mixed polyclonal-polyclonal cryoglobulinemia NEPHROLOGY A form of cryoglobulinemia characterized by a IgG & IgM, ± IgA cryoglobulins, evoked by to rheumatoid arthritis, SLE, Sjögren syndrome, EBV, CMV, subacute bacterial infections, poststreptococcal, crescentic and membranoproliferative glomerulonephritides, DM, chronic hepatitis, biliary cirrhosis. See Cryoglobulinemia

mixed episode PSYCHIATRY A period of ≥ 1 wk in which criteria for manic *and* depressive episodes are met on nearly every day, with abrupt mood swings from sadness, irritability, euphoria CLINICAL Agitation, insomnia, appetite dysregulation, psychotic features and suicidal ideation; MEs are of sufficient severity to cause impairment of social and occupational function and/or require hospitalization. See Bipolar disorder

mixed message PSYCHOLOGY A form of approach-avoidance conflict, in which a person projects both a desire *and* lack of enthusiasm for a particular task, goal, or relationship, leaving the MM's recipient as perplexed as the responsible party is confused

mixed tumor Pleomorphic adenoma A usually benign salivary gland tumor which may also be seen in breast and pancreas; MT comprises 60% of parotid gland tumors; it is 10 times more common than in the submandibular gland, often in younger ♀. See Collision tumor. Cf Composite tumor

mixed wound infection A skin infection with many different organisms including aerobes, anaerobes, occasionally saprobic fungi

MJ DRUG SLANG A popular term for marijuana, see there

MKSAP Medical Knowledge Self-Assessment Program GRADUATE EDUCATION A 'do-it-yourself' continuing medical education program sponsored by the Am Coll Physicians, in written–book or electronic–CD-ROM format

mL Milliliter

ml Milliliter

MLC 1. Minimum lethal concentration. See Minimum bactericidal concentration 2. Mixed lymphocyte culture, see there

MLD 1. Median lethal dose 2. Metachromatic leukodystrophy, see there 3. Minimum lethal dose, see there

MLEL Malignant lymphoepithelial lesion, see there

MLO Mycoplasma-like organism(s)

MLR Myocardial laser revascularization, see there

MLS group INFECTIOUS DISEASE A family–macrolides, lincosamide, and streptogramins–of antibiotics with similar properties and patterns of antibiotic resistance

MLT 1. Maximum lethal time 2. Median lethal time 3. Medical laboratory technician, see there

mm Hg Millimeters of mercury CLINICAL MEDICINE A measure of blood pressure, referring to the height to which the pressure in the blood vessels push a column of mercury. See Blood pressure, Hypertension

MMM Myeloid metaplasia with myelofibrosis, see there

MMPI CHILD PSYCHIATRY A personality assessment tool widely used in making psychologic evaluations, which is normally given at age 16 and older. Personality testing

MMR vaccine Live measles-mumps-rubella vaccine A trivalent vaccine containing an aqueous suspension of live attenuated strains of measles, mumps, and rubella viruses grown in chick or duck embryo cells. See Killed vaccine, Live attenuated vaccine

MMWR Morbidity & Mortality Weekly Report EPIDEMIOLOGY A news bulletin published by the CDC, which provides epidemiologic data–eg, statistics on the incidence of AIDS, rabies, rubella, STDs and other communicable diseases, causes of mortality–eg, homicide/suicide, divided by region, sex, age, race. See Centers for Disease Control & Prevention

M'Naughten test FORENSIC PSYCHIATRY The most popular test for defining a person as insane under the law, which will find a defendant insane if she had a diseased mind that caused a defect of reason, such that when she acted, she either didn't know the act was wrong *or* didn't understand the nature and quality of her actions. See Insanity defense. Cf Durham rule, Irresistible impulse test, Substantial capacity test

mnemonic Any artifice–eg, rhyme, formula, acronym, used to jog the memory

Moban® Molindone, see there

mobile laboratory Portable lab LAB MEDICINE A 'supercart' containing various instruments that provide Pt's test results near the bedside, and incorporate an extensive test menu. See Point-of-care testing, Turnaround time

mobile medical service AMBULATORY CARE A diagnostic–eg, mammography, cardiac catheterization, bedside lab tests or therapeutic–eg, lithotripsy medical service performed at the bedside

MobilExcimer® system OPHTHALMOLOGY A traveling eye laser system for places lacking permanent facilities for photorefractive keratectomy, which uses the VISX excimer laser system. See Photorefractive keratectomy

mobility testing Motion palpation OSTEOPATHY A technique of classic osteopathy, in which the examiner evaluates each spinal segment for proper mobility in all planes of motion, and in relationship to above and below vertebrae. See Classic osteopathy, Osteopathy

mobilization therapy REHAB MEDICINE A group of treatments including traction, massage, manipulation, which may help control pain and ↑ joint and muscle motion

Mobiluncus A genus of gram-positive, curved bacteria with spinning motility, implicated in bacterial vaginosis

Mobin-Uddin filter See Inferior vena caval filter

Möbius sign ENDOCRINOLOGY A clinical sign characterized by impaired ocular convergence–accommodation, typical of Grave's disease. See Grave's disease

mobs, fear of PSYCHOLOGY Ochlophobia. See Phobia

moccasin distribution A pattern of pedal involvement in chronic tinea pedis due to *Trichophyton rubrum*, most common in adult ♂, with Cushing's disease or lymphoproliferative disorders–implicating defective cell-mediated immunity CLINICAL Asymptomatic, finely scaling diffuse erythematous lesion

moclobemide Aurorix®, Manerix® NEUROLOGY A mono-amine oxidase inhibitor used to manage depression

modafinil Provigil® NEUROLOGY A nonamphetamine antinarcoleptic used as a 1st-line treatment of excess daytime sleepiness in narcolepsy. See Narcolepsy

mode MEDTALK The way in which a thing occurs. See Asynchronous transfer mode, Pacing mode, Syntaxic mode STATISTICS The most frequent number or observation in a data set

mode of dying Mechanism of death, see there

model A conceptual representation of a thing or concept. See Acucare model, Age-structured model, Animal model, Biopsychosocial model, Brownian rachet model, Civil defense model, Coalescence model, Compartment model, Component object model, Conceptual model, Conflagration model, Coronary Heart Disease Policy model, Danger model, David Eddy cervical cancer model, Demand model, Deterministic model, Discrete time model, Disney model, Effector inhibition model, Emergency Medical Services model, Event model, Extrapolation model, Five factor model, Fixed effects model, Failure rate model, Frailty model, Framework model, Group model, Hebbian model, HMO model, Hobson model, *Homo economicus* model, Independent Practice Association model, K Mart model, Kirk model, Linear model, Mathematical model, Mouse model, MPM–mortalities probability model, Needs model, Open access model, Partnership model, Point-of-service model, Prediction model, Prevalence model, Process model, Pyramid model, Radial unit model, Remodeling model, Risk adjustment model, RITARD model, Scissors grip model, SEIR model, Self-nonself model, Sinclair swine model, Sliding filament model, Staff model, Supply model, Three-tiered model, Two-tiered model

modeling PSYCHOLOGY A normal process of personality development, in which a child learns appropriate social and cognitive behaviors by imitating a significant other who is socially accepted; these behaviors are positively reinforced and eventually integrated into the child's personality profile. See Face modeling

modem Modulator/demodulator ONLINE An acoustic coupling device that converts a computer's digital signals to analog signals, allowing transmission of data through standard telephone lines. See Cable modem, Codec

moderate-impact sport SPORTS MEDICINE A recreational activity or sport in which there is relatively intense wear and trauma of weight-bearing joints, especially of feet, knees, hips. Cf High-impact sport, Moderate-impact sport, No-impact sport

modern medicine Mainstream medicine, see there

modified community rating MANAGED CARE A method of determining rates for medical services based on data from a given geographic area. See Service mix

modified fee-for-service MANAGED CARE A situation where reimbursement is made based on the actual fees subject to maximums for each procedure

modified neck dissection SURGERY A subtotal resection of the neck region, usually for CA of the floor of the mouth; most MNDs preserve the spinal accessory nerve, internal jugular vein, and sternocleidomastoid muscle. See Commando operation. Cf Radical neck dissection

modified radical mastectomy Total mastectomy with axillary node dissection BREAST SURGERY A procedure for localized breast CA, which excises the breast, an ellipse of skin, usually with the nipple and axillary lymph nodes, leaving underlying chest muscle intact. See Lumpectomy

modified radical mastoidectomy ENT An operation to eradicate disease of the middle ear cavity and mastoid process, in which the mastoid and epitympanic spaces are converted into an easily accessible common cavity by removing the posterior and superior external canal walls. Cf Radical mastoidectomy

modulator A substance that regulates or changes the activity of another. See Neuromodulator, Selective estrogen receptor modulator

MODY Maturity onset diabetes mellitus of young ENDOCRINOLOGY A mild form of type 2 DM, with onset before age 25. See Type 2 DM

Mogadon® Nitrazepam, see there

mogul sign Third mogul sign RADIOLOGY A sharply demarcated–suggestive of serosal covering–ie, extrapulmonary-margin seen in the left side of a plain CXR; the 3rd mogul is an abnormal protuberance located below the left mainstem bronchus and pulmonary artery corresponding to an enlarged or herniated left atrial appendage seen in rheumatic heart disease, pericardial defects, chordae tendineae or papillary muscle defects, left atrial tumors and cardiomyopathy; the left ventricle may be elevated in tetralogy of Fallot, Ebstein anomaly, or in a left-sided ascending aorta in corrected transposition of great vessels

Mohs surgery SURGICAL ONCOLOGY A therapy for broad-based, shallow BCC or SCCs, especially for lesions that are 1-2 cm, recurring or recurrence-prone sites–nose, eyes, ears, aggressive histologic subtypes–eg, morphea-like BCC. See Basal cell carcinoma, Squamous cell carcinoma

moist rales PHYSICAL EXAM Rales with gurgling sounds due to fluid in the tracheobronchial tree. See Rales

molar *adjective* Referring to the number of moles of solute/L of solution *noun* DENTISTRY Any of the posterior, multicusped teeth adapted to grinding Latin mola, millstone. See Peg molar

molar pregnancy See Hydatidiform mole, Trophoblastic disease

mold VOX POPULI A form that provides shape for a gel or substance–eg, a resin set in a particular shape

molding OBSTETRICS The changes that the fetus's skull–which has a rigid face and base and a mobile cranial vault–undergoes to slide through the birth canal

mole DERMATOLOGY A nonspecific lay term for any pigmented lesion, benign or malignant. See Melanoma, Nevus OBSTETRICS Hydatidiform mole, see there

molecular biology A major discipline that marked its birth with Watson and Crick's seminal report, *'General Implications of the Structure of Deoxyribonucleic Acid'*, which elucidated the double helical nature of DNA; MB seeks to understand the mechanisms controlling gene expression, in physiologic and disease states, and provide the tools necessary for treating genetic diseases. See DNA, Human Genome Project

molecular disease Any condition caused by a defect in a gene encoding one or a few proteins–eg, sickle cell anemia, Gaucher disease. See Inborn errors of metabolism

molecular epidemiology MOLECULAR MEDICINE An evolving field that combines the tools of standard epidemiology–case studies, questionnaires and monitoring of exposure to external factors with the tools of molecular biology–eg, restriction endonucleases, Southern blot hybridization, PCR and other amplification techniques and agarose gel electrophoresis. See PAH-adducts

molecular genetic analysis A nonspecific term for the detection or diagnosis of a disease by analyzing DNA or RNA transcripts, rather than by indirect methods that detect transcriptional products, ie proteins, enzymes and other 'processed' molecules; eg lipids and carbohydrates

molecular genetic test Cytogenetic test, molecular cytogenetic test MOLECULAR MEDICINE The analysis of human DNA, RNA, and chromosomes to detect heritable or acquired disease-related genotypes, mutations, phenotypes, or karyotypes for clinical purposes, which include predicting risk of disease, identifying carriers, and establishing prenatal or clinical diagnoses or prognoses. See Applied genomics

molecular imaging IMAGING Imaging technology based on the molecular nature of biologic changes caused by a particular disease. See PET, PET/CT. Cf Structural imaging; an emerging field in which the tools of molecular & cell biology are being married to state-of-art technology for noninvasive imaging. See PET scanning

molecular medicine Genomic medicine The application of molecular methods to clinical medicine; MM encompasses 'diagnostic' methods–eg, Southern–Western and Northern blot hybridizations, DNA signal amplification by PCR, and use of agents produced by recombinant DNA techniques–eg, biological response modifiers and vaccines, and insertion and manipulation of the genome per se

molecular mimicry IMMUNOLOGY A mechanism that may explain some forms of autoimmune disease, where the immune system attacks self antigens that are structurally similar to nonself antigens

molecular prognostication Molecular prognosis The use of molecular biology and immunopathology to prognosticate malignant, premalignant, or other disease processes

molecular wellness testing PREVENTIVE MEDICINE Genetic analysis as a means of risk assessment in various populations in which illness may not be present. See BRCA1, BRCA2, Wellness testing

Molipaxin Trazodone, see there

Mollaret's meningitis NEUROLOGY A term of waning popularity for recurring and self-limited episodes of aseptic meningitis; MM is of abrupt onset, associated with fever, and pleocytosis of mononuclear cells ETIOLOGY Uncertain; at least some cases are due to HSV CLINICAL Malaise, headache, meningeal signs. See Meningitis

molluscicide PUBLIC HEALTH A chemical which kill snails or mollusks

molysmophobia PSYCHOLOGY Fear of dirt. See Phobia

mom VOX POPULI Mother, mum, mammy, Mutti. See Mother, Supermom

mometasone furoate Asmanex®, Nasonex®, Twisthaler® THERAPEUTICS An inhaled steroid used for seasonal allergic rhinitis and asthma

Monafed® Guaifenesin, see there

MONARCS IMMUNOLOGY A clinical trial–Monoclonal Anti-TNF: a Randomized Controlled Sepsis trial which evaluated the efficacy of afelimomab in managing sepsis and septic shock, selecting the Pts using Il-6, a cytokine marker for severity of inflammation; treated Pts had a 36% mortality; untreated Pts, 33%. See Afelimomab, Sepsis

Monday morning sickness 1. Exertional rhabdomolysis, see there 2. Humidifier lung, see there 3. A condition that may affect workers in plants producing calcium carbimide, an ethanol inhibitor; a person with ethanol in the system may suffer headache, nausea, vertigo

mongolian heart A complex of cardiac anomalies affecting 40-60% of those with trisomy 21–eg, VSD and atrioventricular canal–endocardial cushion defects, less often, tetralogy of Fallot, secundum ASD, PDA; aortic regurgitation and mitral valve prolapse may also occur. See Atrial septal defect, Patent ductus arteriosus, Ventricular septal defect

mongolian spot Mongolian spot DERMATOLOGY A large slate-gray–due to the Tyndall effect–macule with variable margins, usually over the presacral regions, posterior thighs, legs, back, shoulders, most common in blacks and Asians, less in whites, often fading with age–persisting in 4% of Japanese adolescents CLINICAL SIGNIFICANCE None

MONICA CARDIOLOGY A WHO initiative–Multinational Monitoring of Trends & Determinants of Cardiovascular Disease–which evaluated the effects of various factors on mortality in Pts MIs

monilethrix Beaded hair disease An AD condition characterized by short brittle, beaded hair that may evolve to alopecia, accompanied by cataract formation

monilial *adjective* Relating to *Candida* spp

moniliasis of oral cavity Oral thrush, see there

monitor CLINICAL MEDICINE A device which assesses the status of a particular parameter. See Apnea monitor, Holter monitor MANAGED CARE Any parameter regularly and consistently used to evaluate quality of care LAB MEDICINE A component of an instrument that detects physical or chemical fluctuations in electromagnetic radiation

monitored anesthesia care ANESTHESIOLOGY A philosophy for administering local anesthesia, which ↑ Pt comfort and safety, through use of formal anesthesiology services–eg, an anesthesiologist or a certified nurse/registered nurse anesthetist

monitoring MEDTALK The ongoing measurement of a process or substance of interest. See Ambulatory monitoring, Bedside glucose monitoring, Biological monitoring, Blood glucose monitoring, Electronic fetal monitoring, External fetal monitoring, Fetal monitoring, Glucose monitoring, Hemodynamic monitoring, Home uterine activity monitoring, Internal fetal monitoring, Masked monitoring, Ovarian monitoring, Self blood glucose monitoring, Therapeutic monitoring, Uterine monitoring

mono Infectious mononucleosis, see there

mono- *prefix*, Latin, one

monoamine PHARMACOLOGY A class of hormones or neurotransmitters–eg, catecholamines–dopamine, epinephrine, norepinephrine and indoleamines–serotonin, melatonin, which have 1 amine. See Catecholamine

monoamine oxidase inhibitor PHARMACOLOGY An inhibitor of monoamine oxidase, which inactivates monoamines EXAMPLES Brofaromine, isocarboxazid, moclobemide, pargyline, phenelzine, selegeline, tranylcypromine; herbal MAOIs–eg, hypericum, yohimbe ADVERSE INTERACTIONS Cheeses and wines with high concentrations of tyramine–metabolized by monoamine oxidase, may trigger epinephrine and norepinephrine release, hypertensive crises, stroke; MAOIs should not be combined with SRIs

monoamniotic *adjective* Occurring in or referring to a single amniotic cavity, as in monoamniotic twins

monochorionic *adjective* Referring to a single chorionic sac, which occurs in monozygotic twins

monoclonal antibody MAb, MoAb DIAGNOSTICS MAbs are used in diagnostics by radioactively-labelling MAbs to target malignant cells, detect metastases, and screen body fluids for microorganisms or measure levels of circulating hormones. MAbs are used in pathology to differentiate tumor subtypes with batteries of MAbs against intermediate filaments or membrane antigens. See Hybridoma IMMUNOLOGY A highly specific antibody formed by a clone of B lymphocytes, either naturally–eg, in cold hemagglutinin disease, plasma cell dyscrasia, or produced synthetically by fusing an immortal cell–mouse myeloma to a cell producing an antibody against a desired antigen. See Monoclonal immunoglobulin ONCOLOGY MAbs are viewed as a therapy for CA, as they directly inhibit the growth of certain tumors, can be chemically bound to toxins that are lethal to malignant cells, stimulate the complement system–a nonspecific arm of the immune system, which may destroy malignant cells, can be used to purge the BM of malignant cells, form the basis for CA vaccines, and for drug delivery systems ADVERSE EFFECTS Allergic reactions, fevers, chills, hypotension, liver and kidney problems. See Biological response modifier, Clone bank, HAMA. Cf Humanized antibody

monoclonal gammopathy A condition characterized by the clonal proliferation of immunoglobulin-producing B cells

MONOCLONAL GAMMOPATHY TYPES

MALIGNANT

Multiple myeloma

Variants of multiple myeoloma

- Solitary plasmacytoma of bone
- Extramedullary plasmacytoma
- Plasma cell leukemia
- Non-secretory myeloma

Lymphoproliferative disorders

- Waldenström's macroglobulinemia
- Lymphoma

Heavy chain disease (γ, α, μ)

Amyloidosis

OF UNDETERMINED SIGNIFICANCE

Benign (IgG, IgA, IgM, IgD, or light chains)

Associated neoplasms that rarely produce monoclonal proteins

Biclonal gammopathies

monoclonal gammopathy of undetermined significance Benign monoclonal gammopathy, MGUS HEMATOLOGY A condition defined by lab criteria CRITERIA Normal Hb, serum albumin < 20 g/L–US: < 2 g/dl, an M-component, no Bence-Jones proteinuria, < 5% plasma cells in the BM, no osteolytic lesions; MGUS differs from myeloma in that the plasma cells in the

BM are < 20%, and the ESR and monoclonal component are lower PROGNOSIS 20-40% of MGUS progress to malignant monoclonality MANAGEMENT Plasma exchange may be improve the neuropathy associated with IgA or IgG MGUS

monoclonal immunoglobulin M spike, M spot, myeloma component, myeloma protein A protein produced by clonally expanded immunoglobulin-producing cells seen in various malignancies–eg, myeloma, Waldenström's disease, CLL, and other lymphoproliferative disorders; MI production may occur in other malignancies–eg, bladder, cervix, liver CAs, angiosarcoma, KS NONMALIGNANT MONOCLONAL CONDITIONS Infections–eg, *Acanthocheilonema perstans, Endolimax nana, Entamoeba hartmanni,* filariasis, schistosomiasis, septicemia, syphilis, trichuriasis, TB, viral hepatitis, hematologic disorders–eg, anemia, autoimmune hemolytic anemia, hereditary spherocytosis, thalassemia, autoimmune disease–eg, glomerulonephritis, pemphigus vulgaris, scleroderma, liver disease, kidney disease, amyloidosis, acute porphyria, chronic salpingitis, systemic capillary leak syndrome, Gaucher's disease, uterine fibromas, cerebrovascular disease, ASHD. See Hybridoma

monocyte HEMATOLOGY A phagocytic WBC that arises in BM from a common progenitor, CFU-GM; 'daughter' monocytes circulate in the blood, forming resident and transient populations in various sites; resident monocytes–histiocytes include Kupffer cells–liver, Langerhans cells–dermis, microglial cells–brain, pleural, peritoneal, alveolar macrophages and osteoclasts; monocytes normally constitute 2%–8% of peripheral WMCs, measure 12-25 µm, have a reniform nucleus with lacy chromatin, an N:C ratio of 4:1 to 2:1, and gray blue cytoplasm containing lysosomal enzymes–eg, acid phos, arginase, cathepsins, collagenases, deoxyribonuclease, lipases, glycosidases, plasminogen activator and others, and surface receptors–eg, FcIgG and C3R; monocytes are less efficient in phagocytosis than PMNs, but have a critical role in antigen processing. See CFU-GM, White blood cell

monocytoid B-cell lymphoma HEMATOLOGY A low-grade B-cell lymphoma affecting lymph nodes and extranodal sites is the malignant counterpart–ie, a clonal expansion of monocytoid cells—in reactive lymphoid hyperplasias arising in a background of toxoplasmosis and other benign lesions. See Lymphoma

monocytosis A relative or absolute ↑ in number of monocytes, which may be benign and reactive, premalignant or malignant–NHL, Hodgkin's disease. See Reactive monocytosis

monofilament NEUROLOGY A nylon bristle passed over the sole of the foot to detect loss of sensation in a diabetic foot and circulatory defects. See Diabetic foot

monogamy The state of having one spouse or mate

monolaurin Glylorin™ DERMATOLOGY An orphan drug that may be used to manage congenital primary ichthyosis, nonbullous congenital ichthyosiform erythroderma, ichthyosis vulgaris, possibly seborrheic dermatitis. See Ichthyosis vulgaris

monomer A single unit of a multiunit molecule, which are joined to form dimers, trimers, polymers; hydrolysis of polymers yields monomers. Cf Polymer

mononeuropathy Mononeuritis NEUROLOGY A peripheral neuropathy characterized by loss of movement or sensation in a single nerve group ETIOLOGY Damage/destruction of an isolated nerve/nerve group, which may be systemic or, more commonly, due to direct trauma, prolonged pressure, or compression by adjacent structures

mononuclear-phagocytic system Macrophagic system, monocytic-phagocytic system, reticulendothelial system, reticulohistiocytic system The constellation of cells and nonspecific immune responses by spleen-based monocyte and macrophage cells and progeny

mononucleosis See Infectious mononucleosis

monophasic basal body temperature pattern REPRODUCTION MEDICINE A pattern of temperature in a ♀ that remains relatively constant throughout the entire menstrual cycle, which may correspond to an anovulatory cycle. See Basal Body temperature

monophasic oral contraceptive Monophasic pill GYNECOLOGY An OC in which all pills have the same hormones at the same concentration. See Oral contraceptive

monophasic pill Monophasic oral contraceptive, see there

monophobia PSYCHIATRY 1. A phobic state evoked by specific–ie, only one, mono–animals or situations, characterized by circumscribed anxiety, childhood onset, a 1:1 ♂:♀ ratio, a normal physiologic and mental substrate, and response to therapy. See Phobia. Cf Agoraphobia, PAD syndrome 2. Fear of solitude or being alone.

monoplegia NEUROLOGY Paralysis of a single extremity or part thereof

monosaccharide Simple sugar A monomer of a more complex carbohydrate EXAMPLES Glucose, fructose, galactose. Cf Disaccharide, Polysaccharide

monosodium glutamate allergy Chinese restaurant syndrome, see there

monosomy GENETICS A type of aneuploidy defined by the formula $2n - 1$, where n is the haploid number–eg, Turner syndrome or monosomy X. See Aneuploidy

Monospot test LAB MEDICINE A proprietary test that detects heterophil antibodies, typical of EBV infection; the MT is abnormal in infectious mononucleosis, chronic EBV infection, occasionally in chronic fatigue syndrome, Burkitt's lymphoma, and hepatitis

Monostrut® valve CARDIOLOGY A free-floating, hingeless, metallic, mechanical valve for replacing aortic or mitral valves. See Prosthetic heart valve

monotherapy Single drug therapy or treatment THERAPEUTICS The management of a particular condition–eg, HTN, with a single agent. See Hypertension. Cf Polypharmacy

monounsaturated fat A saturated fatty acid–ie, an alkyl chain fatty acid with one ethylenic–double bond between the carbons in the fatty acid chain. See Fatty acid, Saturated fatty acid; Cf Polyunsaturated fatty acid, Unsaturated fatty acid

monovalent vaccine A vaccine containing one antigen. See Vaccine

monozygotic twins Identical twins Twins resulting from the division of a single fertilized egg, which usually share a common chorion and placenta; usually each has a separate amnion. Cf Fraternal twins

monster *adjective* Relating to anything–eg, a cell or entire organism that is strange, bizarre, freakish, or extremely unusual *noun* MEDTALK A popular term of waning popularity for a person with severe external malformations; freak. See Acardiac monster

Monteggia fracture Parry fracture ORTHOPEDICS An angled forearm fracture at the junction of the proximal and middle third of the ulna with an anterior dislocation of the radial head. See Fracture. Cf Galeazzi fracture

montelukast Singulair® INTERNAL MEDICINE A leukotriene receptor antagonist used to manage asthma. See Asthma

Montevideo unit OBSTETRICS A graphic portrayal of uterine activity that corresponds to the product of the uterine contractions/10 mins multiplied by the intensity of the contractions–the average intrauterine pressure peaks of all contractions in the same 10 min span. See Deceleration

Monurol™

Monurol® Fosfomycin tromethamine INFECTIOUS DISEASE A single dose antibiotic for treating uncomplicated UTIs. See Urinary tract infection

mood PSYCHIATRY A pervasive and sustained emotion that, in the extreme, markedly colors one's perception of the world EXAMPLES Depression, elation, anger. See Affect, Bad mood, Emotion, Good mood

moon face A rounded face with a double chin, prominent flushed cheeks, and fat deposits in the temporal fossa and cheeks, which is typical of Cushing's disease/syndrome and cri-du-chat–5p-syndrome; the name also refers to a broad, round face with bright-red cheeks in well-developed, broad-chested children with pulmonic stenosis and a right-to-left shunt

moonlighting PHYSICIAN INCOME An Americanism, for working at a 2nd job after regular working hrs–ie, 'by moonlight'. See Libby Zion, Medical school debt, 405 Regulations

moonshine TOXICOLOGY Illicitly distilled whiskey. See Lead poisoning, Saturnine gout

MOPP Mechlorethamine–nitrogen mustard, vincristine–Oncovorin, Procarbazine, Prednisone ONCOLOGY A 4-drug chemotherapy regimen for stage III Hodgkin's disease ADVERSE EFFECTS Sterility, N&V, immunosuppression, ↑ infections, 2° myelosuppression, myelodysplasia, ANLL. See Chemotherapy, Hodgkin's disease

moral masochism PSYCHOLOGY The need by a person to seek verbal abuse or castigation from another through extreme passiveness, subservience to the demands of others, or provocation of negative reactions in others; MM is attributed to unresolved conflicts in childhood. See Masochism

morbid fear Phobia, see there

morbid obesity Superobesity BARIATIRCS A condition defined as 45 kg > ideal body weight, 2 times > ideal/standard weight or, for children, a triceps skin fold > 95th percentile of all children; despite significant weight loss following jejuno-ileal bypass, the procedure is complicated by steatorrhea, hepatic failure, cirrhosis, oxalate deposition, bile stone formation, electrolyte imbalance–↓ Ca²⁺, Mg²⁺, K⁺, hypovitaminosis, psychologic problems, polyarthropathy, hair loss, pancreatitis, colonic pseudoobstruction, intussusception, pneumatosis cystoides intestinalis, blind loop syndrome. See Gastric balloon, Obesity, Pickwick syndrome

morbidity Disease, illness MEDTALK Any departure, subjective or objective, from a state of physiological or psychologic well-being

morbidity rate EPIDEMIOLOGY The number of cases of a particular disease in a unit of population

morbilliform rash *morbilliform* Latin, measles-like An exanthema commonly due to echovirus 9, consisting of fine, discrete maculopapules on the head and neck, rarely elsewhere DIFFDx Rubella, meningococcal petechiae-Waterhouse-Friderichsen syndrome, Kawasaki's disease; the rash and characteristic low-grade fever usually resolve in a wk. See Rash

MORE OSTEOPOROSIS A clinical trial–Multiple Outcomes of Raloxifene Evaluation–that evaluated the effect of raloxifene on ↓ fracture risk in ♀, and the secondary impact of raloxifene therapy on ♀ cancer. See Osteoporosis, Raloxifene

morgue HOSPITALS The place where decedents or cadavers are stored before being sent either to a funeral home for burial services or to the state medical examiner's office, if the cause of the death is uncertain or questionable

moricizine CARDIOLOGY A phenothiazine derivative once used to manage life-threatening ventricular arrhythmias

morning-after pill Emergency contraception, interception pill GYNECOLOGY A high-dose estrogen given in the early post-ovulatory period to prevent implantation of a potentially fertilized egg after unprotected intercourse. See Contraception, DES, Norplant, Pearl index, RU 486

morning dip PULMONARY MEDICINE A popular term for a measurable ↓ in expiratory air flow rates that commonly occurs between 2 and 4 AM, resulting in sleep loss due to coughing and breathlessness

morning sickness Nausea gravidarum OBSTETRICS Pregnancy-related nausea often accompanied by vomiting upon awakening, possibly related to hunger; MS occurs in ½ of pregnancies in the first 2-12 wks of gestation and, if severe, causes dehydration and acidosis. See *Daubert v. Merrell Dow Pharmaceuticals*

morphea A condition characterized by subcutaneous sclerosis, divided by some authors into 1. A generalized form: Scleroderma and 2. A localized form, subdivided into (a) Circumscribed morphea–characterized by one or more round-to-oval firm reddish plaques measuring up to several centimeters in diameter with a yellow-white center and a lilac telangiectatic border; (b) Linear morphea or linear scleroderma; (c) Frontoparietal lesions–en coup de sabre with or without hemiatrophy of the face

Morphelan® PAIN MANAGEMENT A proprietary formulation of morphine. See Morphine

MorphiDex® PAIN MANAGEMENT A morphine with dextromethorphan, an NMDA receptor antagonist, for relieving severe CA pain PROS Faster onset of action, lasts 8 hrs. See Morphine

morphine PAIN MANAGEMENT An opium alkaloid with potent analgesic effect that owes its narcotic properties to its particular aromatic ring structure ROUTE Parenteral, oral EFFECTS Blocks pain signals to brain and spinal cord USES Relieves severe acute and chronic pain; facilitates induction of anesthesia CLINICAL Euphoria, respiratory depression, drowsiness, N&V, ↓ GI motility, risk of addiction, miosis. See Controlled drug substances, Designer drugs, Heroin, Substance abuse

Morquio syndrome Mucopolysaccharidosis IV METABOLIC DISEASE An AR mucopolysaccharide storage disease characterized by 6-sulfo-N-acetylhexosaminide sulfatase deficiency and excretion of keratosulfate in urine CLINICAL Coarse facies, short stature, and skeletal, joint defects; onset after first yr, life expectancy 20+ yrs; may be mentally normal. See Mucopolysaccharidosis

mort d'amour Coition death, going–not coming CARDIOLOGY Death due to coitally-induced cardiac overload, which may occur in anyone with underlying cardiac disease, especially with HTN, arrhythmias, or cerebral aneurysms. See Coital coronary

mortality Death rate EPIDEMIOLOGY A health statistic that corresponds to the total number of deaths per unit time in a population divided by the population's number, ergo deaths/1000 population. See Infant mortality, Neonatal mortality, Operative mortality, Post-neonatal mortality, Proportionate mortality

MORTALITY–DATA OF INTEREST

LEADING CAUSES OF MORTALITY–US Cardiovascular–ASHD and aneurysm disease 39%, CA 22%, CVAs 7.6%, accidents 4.6%, pneumonia or influenza 3%, lung disease 3%, DM-related 1.8%, suicide 1.4%, cirrhosis 1.3%, nephritis 1.0%, homicide 1.0%, etc to 100%

MORTALITY RATE IN VIRAL INFECTIONS Rabies 99%, HIV 50+%, Ebola 20-80%, HBV 3-5%, polio ± 0.1%

MORTALITY < AGE 19 Fatal injuries–MVAs 47%–33% occupants, 8% pedestrians, homicide 13%–usually firearms, suicide 9.6%–?:/, 4:1, drowning 9%–most common in those < age 4, 90% in residential pools, fire/burns 7%–most < age 4, black:white ratio, 3:1

mortality deceleration POPULATION BIOLOGY A term referring to the unexplained observation that death rates slow after a certain age–80. See Programmed cell death

mortality rate EPIDEMIOLOGY The frequency of death in a defined population during a specified time interval; the per capita death rate in a population; the reciprocal of the population life expectancy. See Fetal mortality rate–MR, Infant MR, Maternal MR, Neonatal MR, Postneonatal MR

Morton's neuroma ORTHOPEDICS A type of perineural fibrosis, which is not a true neuroma CLINICAL Sharp, burning pain, commonly between the 3rd and 4th metatarsal heads, worse with pressure, better with rest MANAGEMENT Conservative–metatarsal pads in insole, loose footwear–ie, tight high heels, ski boots, bike shoes ganz verboten; blocks with a local anesthetic and steroids; surgery if conservative measures fail; if Sx are limited to one web space, surgical outcome is excellent in up to 90%

Morton's test A maneuver used to identify metatarsal pain, which is positive when lateral pressure across the forefoot elicits pain

MOS-HIV AIDS A 45-item medical outcomes survey used for evaluating the effects of therapy on AIDS Pts

mosaic *adjective* A patchwork of one sharply-defined 'jig-saw'-shaped pattern imposed upon another of different color, tissue pattern or radiologic density *noun* GENETICS An individual with 2 or more genotypically or karyotypically distinct cell lines, arising from a single zygote by somatic mutation, crossing-over, or nondisjunction during mitotic division. See Chimera, Freemartin OB/GYN A vascular change of interconnecting vessels resulting in a cobblestone or honeycomb surface appearance by colposcopy, the mosaic pattern is often associated with CIN and mandates biopsy.

mosaic artifact DERMATOLOGY A mimic of fungal infection of mucocutaneous tissues when dystrophic epithelial tissue is stained with KOH; the 'mosaic' consists of an irregular band surrounding epithelial cells which is less translucent than the squamous cells

mosaic pattern CERVIX A colposcopic defect at the transformation zone of the cervix–atypical when the cervix is covered by 3% acetic acid; the fields of sharply demarcated 'mosaic' are separated by reddish–vascularized borders; MP may signify epithelial proliferation ranging from mild dysplasia to CIS SKIN 1. A haphazard arrangement of minute arthrospores seen ensheathing hairs in tinea capitis caused by *Microsporum*, which differs from tinea capitis caused by *Trichophyton*, as the arthrospores are larger and appear in parallel chains outside or within the hair shafts 2. A pattern of skin involvement in which there are alternating stripes–lines of Blaschko of affected and unaffected skin; diseases with a MP are often linked to the X chromosome–eg, Conradi-Hünnermann syndrome, focal dermal hypoplasia, incontinentia pigmenti

mosaic wart A large flat verrucoid plaque composed of confluent contiguous plantar warts, caused by HPV-2

mosquito MEDICAL ENTOMOLOGY An arthropod of the dipteran family Culicidae, the ♀ of which is a bloodsucker; eggs are laid on water–insecticides are sprayed on stagnant water for mosquito control–where larvae feed on debris or occasionally other living organisms MOSQUITO GENERA OF MEDICAL IMPORTANCE *Aedes, Anopheles, Culex, Stegomyia*, vectors for blood-borne parasites–eg, *Brugia malayi, Wuchereria bancrofti, Plasmodium* spp, *Trypanosoma* spp, and viruses–eg, alphaviridae, flaviviridae, togaviridae, that cause California, eastern equine, Venezuelan and western equine encephalitides, O'nyong-nyong, dengue fever, Rift valley fever, yellow fever SURGERY Mosquito forceps A small hemostatic forceps. See Forceps

most-significant other SOCIAL MEDICINE The person in another person's universe upon whom he/she most heavily depends for moral and physical support during crisis and/or stress EXAMPLES Parent, spouse, child, lover, sibling, or friend. See Bonding, Companionship, Extended family, Marriage bonus, Mothering, Negative most significant other, Positive most significant other; Cf Anaclytic depression, Holtism, Nuclear family, Social isolation

moth ball syndrome Popular for a congenital deficiency of glucose-6-phosphate dehydrogenase, which causes hemolysis after naphthalene–moth ball exposure See Glucose-6-phosphate dehydrogenase

moth-eaten pattern SKIN An oozing and encrusted reddish skin erosion with a sharp margin and raised borders, characteristic of Paget's disease of breast; extramammary Paget's disease is similar but often accompanied by pruritus; moth-eaten skin also appears in Bowen's disease and in pagetoid spread of melanoma SKULL Patchy variegated defects with a ground-glass center seen by a plain skull film in the diplöe in hyperparathyroidism; MP also refers to the patchy, skull lesions of myeloma; the bone defects of myeloma are usually sharply demarcated and termed 'punched-out'

mother GENETICS The progenitor of a cell or cell line. Cf Daughter cell VOX POPULI The female parent. See Birth mother, Genetic mother, Gestational mother, Surrogate mother, Working mother

mother-to-child transmission Vertical transmission, see there

mother index OBSTETRICS A measure of the status of mothers vis-a-vis access to health care, used of contraception and family planning, literacy, participation in government

mother-in-law, fear of PSYCHOLOGY Pentheraphobia. See Phobia

motherhood VOX POPULI The state of being a mother, which implies giving care to offspring borne of or "belonging" to the mother. See Bonding, Surrogate motherhood

mothering PEDIATRICS The constellation of physical and emotional interactions–eg, holding, cuddling, rocking, and others, that a caregiver has with an infant; mothering is nearly as important to an infant's survival as physical care. See Bonding; Cf Anaclytic depression; Holtism, Social isolation

motion MEDTALK Movement. See Closed chain motion, Foot motion, Quality of motion, Range of motion, Triplane motion

motion sickness PHYSIOLOGY Vertigo, often accompanied by nausea which occurs in those susceptible to vehicular motion; MS is attributed to vestibular system activation, which may be related to the central triggering zone TRAVEL MEDICINE A condition triggered by vestibular or by visual stimulation, through linear and/or angular acceleration of the head CLINICAL Autonomic Sx–hypersalivation, frequent swallowing; ↓ gastric motility, appetite–food may evoke nausea, then vomiting, vertigo, cold sweats, headache, malaise, pallor, ↓ digestion; hyperventilation ergo hypocapnia, pooling of blood in legs, predisposing victim to postural hypotension TREATMENT H₁ type antihistamines–scopolamine, dimenhydrinate, which ↓ vestibular nucleus activity; visual-vestibular conflict may trigger MS. See Visual-vestibular conflict, Weightlessness

Motival Nortryptyline, see there

motivation VOX POPULI The drive to perform a task. See Neuromuscular motivation, Positive motivation

motivational interview SUBSTANCE ABUSE A nonconfrontational counseling technique that may be used by a primary care giver to evaluate a substance–alcohol, illicit drug–abuser's receptiveness to treatment

motor aphasia Broca's aphasia NEUROLOGY A deficit in language output or speech production, ranging in severity from the mildest, cortical dysarthria, characterized by intact comprehension and ability to write due to a lesion of the motor speech center–aka Broca's area in the brain

motor neuron disease NEUROLOGY Any of a group of conditions characterized by progressive degeneration and dysfunction of the motor neuron or anterior horn cell–eg, amyotrophic lateral sclerosis–ALS; the terms are difficult to differentiate, and thus used interchangeably; the classic or sporadic form has an incidence rate of 2/10⁵, characterized by upper limb weakness, atrophy, focal neurologic signs See Amyotrophic lateral sclerosis

motor speech disorder AUDIOLOGY A disorder due to an inability to accurately produce speech sound or phonemes due to muscle weakness/incoordination or difficulty in voluntary muscle movement. See Speech pathology

430 **motor tic** NEUROLOGY A nonpurposeful repeated muscle spasm; simple MTs–SMTs are abrupt, brief, single, isolated EXAMPLES Eye blink, shoulder shrug, repetitive kicking of legs, head jerk, eye dart, nose twitch. See Torsion dystonia. Cf Akathitic movement, Hand caressing, Hyperekplexia, Ritual, Sterotypy

motor vehicle accident PUBLIC HEALTH A morbid condition that kills 45,000/yr–US; 60% are < age 35; MVAs account for 500,000 hospitalizations and most 20,000 spinal cord injuries, at a cost of $75 billion/yr

Motrin® Ibuprofen, see there

MOTT Mycobacteria other than *M tuberculosis* An acronym for non-TB mycobacteria–eg, *M avium-intracellulare* complex, *M chelonei*, *M kansasii*, *M malmoense*, *M xenopi*; the ratio of MOTT:*M tuberculosis* reflects the population; it is ↓ in the indigent. See Runyon classification, Tuberculosis

mottled enamel DENTISTRY Enamel punctuated by patches of white and/or brownish discoloration; ME may be due to infection and/or antibiotic therapy or fluorosis

mount Coitus SEXOLOGY The penovaginal part of sexual intercourse; in humans, either sex may mount; in most other animals, the ♀ presents, the ♂ mounts. See Coitus, Copulation, Intercourse

mountain biking SPORTS MEDICINE A sport in which participants use specialized bicycles to navigate rough, steep trails covered with unforgiving rocks INJURY RISK Concussions, fractures, death. See Extreme sport, Novelty seeking behavior

mountain fever Colorado tick fever, see there

mountain medicine A subspecialty of wilderness medicine that addresses medical/technical aspects of high-altitude travel/mountain sports AREAS OF INTEREST Mountain rescue and resuscitation, high-altititude physiology and pathophysiology, 'mountain diseases'–eg, frostbite, hypobaric hypoxia, mountain sickness, pulmonary edema, solar radiation. See Mountain sickness, Wilderness medicine

mountain sickness Acute MS, altitude anoxia, altitude sickness, high altitude cerebral edema, high altitude pulmonary edema WILDERNESS MEDICINE A syndrome affecting those living for various periods at high altitudes, which is diagnosed if a person has 3+ major symptoms: anorexia, dyspnea, fatigue, headache, insomnia, which respond to dexamethasone; MS is divided into acute and chronic forms. See High-altitude pulmonary edema, Höhendiurese

Mountain staging system ONCOLOGY A system for staging lung cancer per CF Mountain

mountain tick fever Colorado tick fever, see there

Mournier-Kuhn syndrome Tracheobronchomegaly A condition characterized by bronchiectasis, cough, recurrent infections

mourning Grief PSYCHIATRY Reaction to a loss of a loved object–significant or important person, object, role, status, or anything considered part of one's life, which is a process of emotional detachment from that object, freeing the subject to find other interests

mouse COMPUTERS A device with control buttons, used to manipulate files represented by icons, to access data and execute commands from pull-down menus. Cf Trackball RHEUMATOLOGY See Joint mouse

mouse elbow OCCUPATIONAL MEDICINE A repetitive-strain injury produced by the constant lifting and shifting of a computer mouse

mousetooth Mousetooth forceps SURGERY A forceps with one or two fine points at the tip of each blade, fitting into hollows between the points on the opposite blade; forceps with small teeth at the tip of each blade. See Forceps

moustache Pitchfork, Whale's tail INTERVENTIONAL CARDIOLOGY A popular term for the distal bifurcation of the left anterior descending coronary artery. See Collateral circulation

mouth cancer VOX POPULI Oral cancer, see there

mouth ulcer Oral ulcer, stomatitis, see there

movement VOX POPULI 1. The act of moving; motion. See Ballistic movement, Circus movement, Closed chain movement, Extraocular movement, Fetal movement, Intraocular movement, Line movement, Paradoxical movement 2. Bowel movement, defecation 3. A group of similarly minded individuals with an agenda. See Antiabortion movement, Health food movement, Outcomes movement, Popular health movement, Right-to-die movement

movement time SLEEP DISORDERS The term in sleep record scoring to denote when EEG and EOG tracings are obscured for > 15 secs due to movement, which may be combined with awake time

MOXCON CARDIOLOGY A study–Moxonidine Congestive Heart Failure trial–evaluating the efficacy of central sympathetic inhibition on heart failure. See Moxonidine, SIRA

moxonidine CARDIOLOGY An antihypertensive of the selective imidazoline receptor agonist family which may be useful in managing CHF. See MOXCON, SIRA

moya-moya disease *moya-moya*, Japanese, hazy, smoky NEUROLOGY An idiopathic condition named for the angiographic appearance of prominent collateral vessels of the basal ganglia, that accompany narrowed and distorted cerebral arteries with thin collateral vessels, which arise in a wispy, netlike fashion from the circle of Willis, with progressive occlusion CLINICAL Pts are in a 'fog', due to ischemia in young Pts, or subarachnoid hemorrhage in older Pts; most occur in Japanese as recurrent strokes in otherwise healthy ♀ children/adolescents, familial tendency, after a febrile illness; abrupt onset of hemiparesis, transient aphasia, convulsions, spontaneous resolution

6-MP 6-Mercaptopurine. See Purinethol

MP Metatarsophalangeal, see there

MPEG Moving Pictures Experts Group INFORMATICS The standard compression ratio for data transmission and pixels for video files. See Pixel

MPGN Membranoproliferative glomerulonephritis, see there

MPH Master's Degree in Public Health

MPJ Metatarsophalangeal joint

MPM II Mortality Probabilities Models II INTENSIVE CARE A '3rd-generation' system for estimating hospital mortality in adult ICU Pts based on logistic regression modeling techniques. See Prognostic scoring systems

MPO Myeloperoxidase, see there

MPS 1. Meters/sec 2. Mononuclear phagocytic system, see there 3. Mucopolysaccharidosis, see there 4. Myeloma progression score

MPTP 1-Methyl-4-phenyl-1,2,3,6-tetrahydropyridine, analogs MTMP, PEPAP NEUROLOGY A potent neurotoxin–which has an effect much like Meperidine or Demerol—that acts on neuromelanin, producing parkinsonism CLINICAL Bradykinesia, muscular rigidity, resting tremor, defects of posture and gait PHARMACOLOGIC EFFECTS MPTP initially blocks the brain's responses to pain, then metabolizes to MPP+, causing an irreversible syndrome resembling Parkinson's disease. See BDNF, Designer drugs, Parkinson's disease

MR 1. Magnetic resonance, see there 2. Manual removal 3. Maximum response 4. Medical record 5. Mental retardation, see there 6. Metabolic rate, see there 7. Mortality rate 8. Mortality ratio 9. Muscle relaxant

MR angiography Magnetic resonance angiography, MRI angiography IMAGING A noninvasive method for evaluating the blood vessels of Pts with severe peripheral vascular disease or acute renal failure PROS Less time and cost than conventional catheter angiography without puncture; it is painless, no known complications, much more rapid than cerebral contrast angiography–CCA and allows construction of images in any plane; MRA is useful for imaging of ASHD and dissecting aneurysms of neck and intracranial aneurysms. See Magnetic resonance imaging

MR IMAGES NEUROLOGY A clinical study–Magnetic Resonance in Intravenous Magnesium Efficacy in Stroke. See Stroke

MRA Magnetic resonance angiography, see MR angiography

MRC 1. CARDIOLOGY A clinical trial–Medical Research Council Trial of Treatment of Hypertension in Older Adults that evaluated the efficacy of pharmacologic therapy in older hypertensives. See Antihypertensive, Hypertension 2. Medical Research Council, see there

MRFIT Multiple Risk Factor Intervention Trial CARDIOLOGY A long-term prospective study designed to analyze the effects of modifying the risk factors for heart disease

MRI Magnetic Resonance Imaging, see there

MRI coronary angiography MRCA IMAGING A form of MR angiography used to ID clinically important coronary artery stenoses; with standard MRI spin-echo and gradient-echo techniques, coronary arterial visualization is suboptimal and limited to the proximal segments, apparently related to respiratory and cardiac movements; in MRCA, the image is obtained while holding the breath, reducing the respiratory 'noise'; cardiac movement is minimized by obtaining the image during mid-diastole–a point of the cardiac cycle with relative diastasis; the temporal resolution is minimized by using k-space segmentation

MRO Medical review official, see there

MRSA Methicillin-resistant *Staphylococcus aureus*. See MARSA

MS 1. Mass spectroscopy 2. Master of Science 3 Master of Surgery 4. Medical student 5. Mental status 6. Mitral stenosis, see there 7. Morphine sulfate 8. Multiple sclerosis, see there 9. Musculoskeletal

MSBOS Maximum surgical blood order schedule TRANSFUSION MEDICINE A list of common elective surgeries with the maximum number of units of blood to be cross-matched preoperatively. See Cross-match/transfusion ratio

MSDS Material Safety Data Sheets, see there

MSG Monosodium glutamate FOOD INDUSTRY A flavor-enhancing amino acid used in processed, packaged, and fast foods; it is an excitatory neurotransmitter and neurotoxin, and often added to Chinese food OTHER SOURCES WITH UP TO 40% MSG Autolyzed yeast, calcium caseinate, hydrolyzed protein, sodium caseinate SENSITIVITY SYMPTOMS Headaches, palpitations, skin flushing, tightness of chest. See Chinese restaurant syndrome, Domoic acid

MSOF Multisystem organ failure, see there

MSUD Maple sugar urine disease, see there

MT 1. Massage therapist 2. Medical technologist, see there 3. Metatarsal

MTBE Methyl-tert-butyl-ether SURGERY An aliphatic ether that rapidly dissolves cholesterol stones in vivo, introduced under local anesthesia via a percutaneous transhepatic cholecystectomy catheter, as a non-invasive method for treating gallstones; after injection, the dissolved 'slurry' is drained from the bladder COMPLICATIONS N&V, bile leakage, drowsiness, transient ↑ of liver enzymes, anorexia, hypotension; MTBE stone dissolution is effective if the Pts are selected for cholesterol stones, improved and accelerated by adding trans-

cutaneous ultrasound. See Lithotripsy ENVIRONMENT An octane-boosting additive used in reformulated gasoline that helps the gas burn cleaner

MTF Metabolize to freedom MEDTALK Worked-up sufficiently to exclude diagnoses other than alcohol intoxication, disposition/discharge planned when Pt achieves safe functional level

MTP-PE PHARMACOLOGY A synthetic lipophilic analog of the bacterial cell wall component, muramyl dipeptide, that is 'bundled' in fat, in a liposomal vesicle and delivered into the pulmonary vasculature, theoretically triggering the tumorilytic machinery of the pulmonary macrophages; MTP-PE has been evaluated for managing metastatic osteosarcoma

MTX Methotrexate, see there

mu μ Symbol for: 1. Linear attenuation coefficient–statistics 2. Mean–statistics 3. Micro—10^{-6} 4. Heavy chain of IgM

MUAC Mid-upper arm circumference. See Triceps skin fold

mucin clot test Rope's test, string test RHEUMATOLOGY A bedside test for evaluating synovial fluid–SF viscosity or 'quality' of the SF mucin clot, which reflects hyaluronidate polymerization; the further a drop of SF falls before separating–'stringing effect', the greater the viscosity, the more normal it is. Cf Spinnbarkeit, String test

mucinous *adjective* Referring to the presence of mucin

mucocele ENT Mucus retention cyst, see there

mucocutaneous lymph node disease See Kawasaki's disease

mucolipidosis Any of 4 AR lysosomal storage diseases characterized by accumulation of glycolipids in interstitial tissues, which have clinical features similar to the mucopolysaccharidoses

mucopolysaccharidosis A heterogeneous group of diseases each caused by a specific enzyme deficiency, resulting in an accumulation of substrate mucopolysaccharides–glycosaminoglycans–eg, dermatan sulfate, heparan sulfate, keratan sulfates CLINICAL Childhood onset of Sx–eg, developmental delay, mental retardation, short stature, skeletal anomalies–dysostosis multiplex, coarse facial features, hepatosplenomegaly. See Gargoyle face

mucopolysaccharidosis IH Hurler's disease An AR condition characterized by lysosomal alpha-L-iduronidase deficiency, with accumulation of mucopolysaccharides in the heart and other tissues CLINICAL Infants develop signs by the first yr of life, with characteristic facies–coarse thick features, prominent dark eyebrows, cloudy corneas, progressive stiffness, mental retardation, death in early teens, often from associated heart disease

mucopolysaccharidosis IS Scheie's disease An AR mucopolysaccharidosis which, like Hurler syndrome, is characterized by a defect of α-L-iduronidase and excretion of ↑ dermatan sulfate in urine CLINICAL MPS-IS is the mildest form mucopolysaccharidosis; onset at age 4–5, including inguinal hernia, broad mouth with full lips, early onset corneal clouding, early onset of stiff joints, claw hands and deformed feet, aortic valve defects

mucopolysaccharidosis II Hunter syndrome, iduronate 2-sulfatase deficiency METABOLIC DISORDERS An X-R–♀ carriers, ♂ with disease mucopolysaccharidosis due to a defect in iduronosulfate sulfatase, resulting in tissue accumulation of chondroitin sulfate B and heparan sulfate CLINICAL Early onset form, type A appears by age 2–large skull, coarse facies, profound mental retardation, spasticity, joint stiffness, aggressive behavior; late onset type B causes much milder Sx. See Mucopolysaccharidosis

mucopolysaccharidosis III Sanfilippo syndrome An AR condition characterized by α-N-acetyl-a-D-glucosaminidase deficiency, ↑ heparan sulfate excretion in urine CLINICAL Relatively late onset, coarse facies, slow mental development progressing to severe mental retardation, stiff joints, gait disturbances, speech defects, behavioral problems, survival into the twenties or later

mucopolysaccharidosis IV Beta-galactosidase deficiency, Morquio syndrome METABOLIC DISEASE An AR mucopolysaccharidosis characterized by 6-sulfo-N-acetylhexosaminide sulfatase deficiency and excretion of keratan sulfate in urine CLINICAL coarse facies, short stature, skeletal and joint defects; onset after 1st yr, life expectancy 20+ yrs; may be mentally normal. See Mucopolysaccharidosis; Beta-galactosidase

mucopolysaccharidosis V Obsolete for mucopolysaccharidosis IS, see there, aka Scheie syndrome

mucopolysaccharidosis VII An AR lysosomal storage disease of mice and men caused by β-glucuronidase deficiency, with accumulation of glycosaminoglycans–GAGs in brain and other tissues, resulting in progressive mental deterioration and death TREATMENT Transplanted neural progenitor cells into the cerebral ventricles of mice corrects lysosomal storage of GAGs in neurons and glia. See Beta-glucuronidase

mucopurulent cervicitis GYNECOLOGY Cervical inflammation due to STD bacterial–*N gonorrhoeae* and/or *C trachomatis*–infection. See Cervicitis. Cf Pelvic inflammatory disease

mucormycosis An opportunistic infection by fungi of the order Mucorales, that are found in decaying organic matter which grow in tissues as hyphae, thus being defined as molds; most grow within 2-5 days in usual culture media; most common pathogenic mucormycoses are *Rhizopus* spp and *Rhizomucor* spp; others include *Absidia* spp, *Apophysomyces* spp, *Cunninghamella* spp, *Mucor* spp, *Saksenaea* spp CLINICAL Fungi enter the respiratory tract and gain a foothold in the nasopharynx, usually in hosts immunocompromised by AIDS, corticosteroids, DM, malnutrition, terminal CA, transplantation CLINICAL FORMS Rhinocerebral, pulmonary, cutaneous, GI, cerebral, etc; once host defense is circumvented, hyphae are vasculocentric, explaining the commonly associated necrosis and thrombosis TREATMENT Amphotericin B, azoles–ketoconazole, etc, surgical debridement

mucosa-associated lymphoid tissue lymphoma See MALT lyphoma

mucosal lacerations of cardioesophageal junction Mallory-Weiss tear, see there

mucositis Inflammation of a mucosa which, in the GI tract, appears as oral ulcers. See Stomatitis

mucous *adjective* Referring to either mucosa–eg, mucous membrane, or to mucus–eg, mucous secretion

mucoviscidosis Cystic fibrosis, see there

mucus A clear viscid fluid produced by various mucosae–eg, nose, mouth, throat, vagina, containing mucopolysaccharides, enzymes, IgA, and other proteins, desquamated epithelial cells, inorganic salts in fluid. See Cervical mucus

mucus retention cyst Mucocele, mucus cyst ENT A semitransparent blister consisting of clear fluid trapped beneath the mucus membrane–eg, on the oral surface of the lips, attributed to sucking of the mucus membranes. Cf Cold sore

muddy complexion A characteristic patchy hyperpigmentation, seen on the face in malnourished children, accompanied by pallor, lassitude, hypochromic anemia, delayed epiphyseal development, delayed puberty, defects in dentition, anorexia, ↑ susceptibility to infection. See 'Crazy pavement' appearance

muddy lung A descriptor for the lungs of the those who drown or nearly drown in stagnant water; MLs are characterized by crystalline material, for-

eign body type giant cell and granulomatous reaction with carnification, massive fibrosis of pulmonary parenchyma, abundant diatoms

MUGA Multiple gated acquisition scan CARDIOLOGY A technique used to evaluate the heart's ability to respond to physical stress–ie, its functional status INDICATIONS Evaluate the functional impact of CAD, CHF, cardiomyopathies, cardiotoxic drugs. See Equilibrium radionuclide angiocardiography

Muir-Torre syndrome A condition characterized by multiple internal CAs–colon, bladder, lung, sebaceous proliferation, and keratoacanthomas

mule DRUG SLANG A street term for a carrier of illicit drugs. See Body packer syndrome

müllerian mixed tumor see Mixed mesodermal tumor

multiaxial system PSYCHIATRY A mental disorder classification schema used with the DSM-IV, which provides a more comprehensive evaluation of the whole person; it is best used for treatment planning and prognosis because it reflects the interrelated complexities of the various biological, psychological, and social aspects of a person's condition; all mental disorders are included under Axis I and II. See Axis I, Axis II, Axis III, Axis IV, Axis V

multicasualty incident PUBLIC HEALTH A 'minor' man-made–technological or natural disaster which can be managed by local emergency medical services. See National Disaster Medical System. Cf Mass disaster

multicenter *adjective* Referring to that which occurs in many hospitals, as in a randomized multicenter study

multicystic kidney see Polycystic kidney

multidirectional shoulder instability ORTHOPEDICS Symptomatic glenohumeral instability in >1 direction CLINICAL Vague Sx EVALUATION History, PE, radiographs, laxity tests, provocative tests TREATMENT Nonoperative; activity modification. See Shoulder instability

multidisciplinary approach A term referring to the philosophy of converging multiple specialties and/or technologies to establish a diagnosis or effect a therapy

multidrug resistance inhibition ONCOLOGY A management philosophy intended to ↓ cancer cell resistance to chemotherapeutics. See MDR protein

multidrug-resistant bacteria MICROBIOLOGY Bacteria that have acquired antibiotic-resistant genes. See Multidrug resistance

multifactorial *adjective* Referring to the influence of multiple factors in the etiology of a particular disease–eg, DM, HTN–attributable to genetic and environmental components

multifocal Multicentric *adjective* Having many points of origin, as in a multifocal cancer with a primary tumor and scattered satellites of cancer in surrounding tissues

multifocal atrial tachycardia Atrial tachycardia CARDIOLOGY A rapid cardiac arrhythmia caused by stimuli to the heart from multiple locations within the atria, and characterized by irregularity, variable 'P' waves and–in adults–a poor prognosis; in MAT, multiple atrial foci "fire," leading to tachycardia, often 100 to 180 beats/min, ↑ cardiac workload, but ↓ filling time, thus ↓ cardiac function; MAT is seen in 0.05-0.32% of the EKGs interpreted in general hospitals and is more common in the acutely ill–burns, sepsis, CHF, lung cancer, respiratory failure–COPD, PTE, CAD, elderly, recent surgery, hypoxia, theophylline or digitalis overdose, DM, bacterial infections, etc TREATMENT Magnesium, potassium, CCBs–eg, verapamil, β-adrenergic blockers–eg, metoprolol PROGNOSIS 43% of Pts with MAT died during the hospital stay in which the arrhythmia was documented, but death was usually related to the underlying disease

multifocality Multicentricity The simultaneous presence of a lesion–usually understood to be either an infection or a malignancy, in multiple sites

multifocal neuropathy Mononeuritis multiplex, Mononeuropathy multiplex NEUROLOGY A peripheral neuropathy with simultaneous or sequential damage of 2 or more nerve groups or regions; with time, damage is less multifocal and more symmetric, resembling polyneuropathy ETIOLOGY Idiopathic, polyarteritis nodosa, connective tissue disease–eg, rheumatoid arthritis, SLE, Sjögren syndrome, Wegener's granulomatosis, hypersensitivity vasculitis, leprosy, sarcoidosis, amyloidosis, diabetic neuropathy, hypereosinophilia, cryoglobulinemia. Cf Polyneuropathy

multigravida OBSTETRICS A ♀ who has been pregnant 2 or more times. Cf Multipara

multi-infarct dementia NEUROLOGY A condition characterized by global cognitive impairment due to ASHD-induced disease; MID is more common in ♀ and associated with DM, HTN, smoking, amyloidosis CLINICAL Gait and motor defects, defects of language, mood, abstract thinking, apraxia, agnosia, urinary incontinence DIFFDX Repeating 'mini-infarcts' of HTN mimic the gradual deterioration typical of the more common Alzheimer's disease, which lacks prominent motor and reflex changes. See Alzheimer's disease

Multikine™ CLINICAL IMMUNOLOGY An immunomodulating combination of natural cytokines, which may be useful in managing head & neck and prostate CAs and HIV infection. See Cytokine

MultiLight system COSMETIC SURGERY PhotoDerm® technology for long-term epilation and noninvasive treatment of benign vascular and pigmented lesions. See Hair removal

multilineage leukemia A leukemia that displays myeloid and other–eg, megakaryocytic and/or erythroid–clonal expansions, a finding associated with monosomy 7. Cf Biclonality, Composite tumor

multimodal therapy Multimodal treatment ONCOLOGY The combination of ≥ 2 therapeutic modalities–eg, RT, chemotherapy, and/or surgery, to treat a disease–eg, cancer. See Chemotherapy, Radiotherapy. Cf Heroic surgery

multiorgan transplantation SURGERY The simultaneous transplantation of 2+ organs, performed either because the organs being transplanted form an integral structural or functional unit, or because the transplanted organs are replacing failed/failing organs. See Domino liver transplant

multipara OBSTETRICS A woman who has completed 2 or more pregancies resulting in the birth of live infants, or to the point of viability. Cf Multigravida

multiple birth REPRODUCTION MEDICINE A pregnancy resulting in birth of ≥ 2 infants. See Clomifen, Higher multiple

multiple chemical sensitivity/sensitivities syndrome OCCUPATIONAL MEDICINE An acquired workplace disorder characterized by recurrent Sx referable to multiple organ systems, in response to demonstrable exposure to chemically unrelated compounds at doses far below those established in the general population to cause harmful effects; MCSS may be caused by indoor air pollution CLINICAL Respiratory problems, headaches, fatigue, flu-like symptoms, GI difficulties, skin disease, cardiovascular disease, muscle and joint pain, irritability and depression, eye, ear, nose, throat problems, confusion, short-term memory loss, disorientation. See Clinical ecology, Environmental disease. See Chemical imbalance

multiple cholesterol emboli syndrome A condition of subacute onset caused by showers of emboli occurring days to wks–or months–after coronary artery catheterization or aortic surgery, causing ischemia of affected organs

multiple endocrine neoplasia See MEN

multiple epiphyseal dysplasia Fairbank disease PEDIATRICS A mild AD form of dwarfism CLINICAL Pain in hips, knees or ankles in later childhood, due to developmental hip defects; height is slightly ↓; arm, leg, finger or toe lengths may be markedly ↓; movement may be restricted. Cf Dwarfism

multiple hamartoma syndrome Cowden's disease An AD condition characterized by ↑ susceptibility to ectodermal, mesodermal, endodermal mucocutaneous hamartomas, as well as malignancies–eg, papillary or follicular thyroid CA, breast CA, osteosarcoma. Cf Chromosomal breakage syndromes, Li-Fraumani syndrome

multiple lentigines syndrome LEOPARD syndrome DERMATOLOGY An AD condition with variable expression, characterized by thousands of 1-5 mm darkly pigmented macules on the skin often of the trunk and neck, but not on mucosae CLINICAL Sx = mnemonic acronym LEOPARD–Lentigines, EKG disturbances, Ocular hypertelorism, Pulmonary stenosis, Abnormalities of genitalia–gonadal or ovarian hypogonadism, cryptorchidism, delayed puberty, retarded growth, nerve deafness, prominent ears. See Lentigo

multiple malformation syndromes PEDIATRICS A group of disorders defined by developmental anomalies of 2 or more systems, possibly related to chromosomal damage, teratogens and environmental influences–eg, Cornelia de Lange syndrome, Prader-Willi syndrome, Rubinstein-Taybi syndrome, Williams' syndrome. See Dysmorphology, Sequence

multiple personality disorder Split personality PSYCHIATRY The 'presence of two or more distinct identities or personality states...that recurrently take control of behavior.' See Dissociate, Fugue, Personality, Schizophrenia

multiple plane imaging MRI A type of sequential plane imaging used with selective excitation that doesn't affect adjacent planes–which are imaged while waiting for relaxation of the first plane toward equilibrium, resulting in ↓ imaging time. See Magnetic resonance imaging

multiple primary malignancy syndrome ONCOLOGY The rare– < 1% of Pts–finding of 3+ primary malignancies, a 'syndrome' defined by Werthamer–MAXIMUM KNOWN MPMs IN 1 PT 6

MULTIPLE PRIMARY CANCERS–WERTHAMER CRITERIA

1. The malignancies must be primary in different organs
2. Paired-organ–breast, kidney malignancies–synchronous or metachronous are considered to be a single primary
3. Multiple malignant tumors originating in the same organ are viewed as a single primary
4. The lower intestine and uterus–with adnexae, are each considered single organs
5. The malignant nature of the lesions must be confirmed histologically
6. The lesion should be histologically proven to be non-metastatic–this may be impossible

Additional criterion–The malignancy should not have been induced by chemo- or RT

multiple sclerosis NEUROLOGY An idiopathic demyelinating disease in which infiltrating lymphocytes–primarily T cells and macrophages—phagocytose myelin EPIDEMIOLOGY Onset in younger, often ♀ adults, affecting 1:2500 (US), commonly associated with HLA-A3, B7, Dw2 haplotypes; MS is ↑ in a south-to-north gradient in the Northern hemisphere–ie, more common in colder climates CLINICAL Waxing/waning or slowly progressive neurologic changes with paresthesias, gait and visual defects, muscular weakness, absent abdominal reflexes, hyperactive tendon reflexes, cerebellar ataxia, retrobulbar neuritis, loss of proprioceptive sense, spastic weakness of legs, vertigo; the 'classic' Charcot's triad–dysarthria, nystagmus, intention tremor is rare DIAGNOSIS Clinical Hx, multiple brain defects by CT, MRI–

without and with contrast, oligoclonal ↑ IgG in CSF–present in 90% of Pts, 'evoked potentials' by EEG in visual cortex and brainstem; all tests are non-specific and must be correlated with clinical findings; definitive diagnosis requires a brain biopsy PROGNOSIS Worse if ♂, > age 40 at presentation, rapidly progressive, with motor dysfunction–eg, with cerebellar or corticospinal activity MANAGEMENT None consistently effective. See Cop-1, Faroe Islands, Holstein cow pattern, Oligoclonal bands, Shadow plaques

multiple-system atrophy NEUROLOGY A heterogenous group of neurologic syndromes– corticobasal ganglionic degeneration, olivopontine cerebellar atrophy, progressive supranuclear palsy, Shy-Drager syndrome, striatonigral degeneration–which share features caused by lesions of interrelated groups of neurons

multiproblem family SOCIAL MEDICINE A family with an ↑ potential for child and spousal abuse, due to multiple stressors–eg, single parent, drug abuse, etc. See Child abuse

multishot suicide NOT a cry for help FORENSIC MEDICINE A firearms suicide that was not fatal on the first round; MSs represent 1.7% of all firearm suicides

multispatula cervical sampler GYNECOLOGY A cervical sampling device with unique head modification to ensure adequate specimen collection to detect CA. See Pap smear

multispecialty management A philosophical and practical approach to Pts with complex diseases–eg, breast CA that requires the integrated activities of multiple health care providers–eg, radiologist, medical oncologist, surgical oncologist, radiation oncologist, surgical pathologist, RN, social worker, and others to obtain optimal results

multistage carcinogenesis A general term referring to the development of cancer through multiple steps of oncogene activation and tumor suppressor inactivation. See One-hit, two-hit model, p53, Tumor

Multistix® see Dipsticks

multisystem organ failure Multiorgan failure, multiple organ dysfunction syndrome CRITICAL CARE A 'physiologic' shut-down of multiple body systems in the face of critical injury or uncontrolled sepsis

Multitak suture ORTHOPEDICS Soft tissue anchor/bone fixation fastener for soft tissue to bone suture fixation, delivered to a predrilled hole in the bone with an introducer, after which sutures can be used to secure soft tissue to the bone INDICATIONS Shoulder rotator cuff tear repairs, ankle and foot instability repair and reconstruction, ulnar and radial collateral ligament reconstruction, knee and elbow ligament and tendon repairs

multitasking INFORMATICS The ability of an operating system to simultaneously run multiple programs–eg, graphics, spreadsheets, word processing or formats of software–eg, coding, compiling, testing

multivessel disease CARDIOLOGY CAD involving 2 or more coronary arteries. See Coronary artery disease

multivitamin An often self-prescribed OTC diet supplement containing lipid-soluble vitamins–A, D, E, K; water-soluble vitamins–thiamin–vitamin B_1, riboflavin–vitamin B_2, vitamin B_6, vitamin B_{12}, vitamin C, folic acid, niacin, pantothenic acid, biotin; minerals–eg, calcium, phosphorus, iron, iodine, magnesium, manganese, copper, zinc. See Decavitamin, Neural tube defects; Cf Megavitamin therapy

mumps PEDIATRICS Acute infection by a paramyxovirus, common in young children CLINICAL Fever, malaise, parotiditis COMPLICATIONS Aseptic meningoencephalitis, pancreatitis, orchitis. See Iodine mumps, Mumps vaccine, Nutritional mumps

mumps vaccine PEDIATRICS A live attenuated virus vaccine that confers lifelong protection against mumps INDICATIONS All children without immune compromise, anaphylactic reactions to eggs or other contraindications ADVERSE REACTIONS Rare–orchitis and parotitis; most people born before 1957 are naturally immune. See Mumps

Munchausen-by-proxy syndrome Polle syndrome PSYCHIATRY, CHILD ABUSE A form of child abuse in which children, often < age 5-6, are victims of factitious illnesses, either fabricated by–eg, reports of fever and seizure disorders, or induced by parents or guardians–eg, induction of apnea, bleeding, diarrhea, fever, seizures, vomiting, administration of laxatives, withholding antibiotics, friction-induced 'rashes'; MBPS may satisfy parents' aberrant psychologic needs. See Capper

Munchausen syndrome Münchhausen syndrome PSYCHIATRY A pseudo-disease seen in subjects who create bizarre lesions or fabricate Sx, to enjoy the perceived benefits of hospitalization STATISTICS ♀:♂ ratio, 2:1; 74% develop the condition by age 24, and on average are diagnosed as having MS by age 32. See Munchausen-by-proxy syndrome

munchies SUBSTANCE ABUSE A popular term for the craving for salt-rich and/or high-carbohydrate 'junk food,' associated with use of marijuana, amphetamines, and other recreational drugs. See Junk food

mupirocin DERMATOLOGY A topical bactericide that inhibits bacterial protein synthesis. See Methicillin-resistant *Staphylococcus aureus*

murder one DRUG SLANG A regionally popular term for heroin and cocaine FORENSIC MEDICINE Murder in the first degree

murmur Heart murmur CARDIOLOGY An auscultatory sound of cardiac or vascular origin, usually caused by an abnormal flow of blood in the heart due to structural defects of the valves or septum; murmurs may be benign or pathological. See Austin Flint murmur, Cardiopulmonary murmur, Cooing murmur, Graham Steel murmur, Innocent murmur, Late systolic murmur, Machinery murmur, Middiastolic murmur, Millwheel murmur, Musical murmur, Pistol shot murmur, Regurgitant murmur, Roger's murmur

murmur grade CARDIOLOGY A standard system for indicating the intensity of cardiac sounds, based on a scale of 1 to 6, by stethoscope. See Gallop, Murmur

MURMUR GRADES

1 Barely audible-requires concentration

2 Soft, easily heard

3 Loud w/o thrill

4 Loud with thrill

5 Loud with minimal contact between stethoscope and chest

6 Loud with no contact between stethoscope and chest

Murphy's kidney punch Murphy's test A clinical test in which the examiner makes jabbing thrusts under the Pt's 12th rib, evoking pain and/or tenderness in Pts with renal inflammation or infection

Murphy's sign CLINICAL MEDICINE Pain on inspiration when the examiner's fingers are placed under the right costal margin next to the rectus abdominalis, a finding associated with acute cholecystitis. See Cholecystitis

muscarine TOXICOLOGY A small alkaloid in some poison mushrooms; it is structurally similar to acetylcholine, and puts the parasympathetic nervous system in 'overdrive' ANTIDOTE Atropine. See Mushrooms

muscle antibody Smooth muscle antibodies, see there

muscle contraction headache Tension headache, see there

muscle dysmorphia SPORTS MEDICINE A mental disorder seen in bodybuilders who, although very muscular and physically fit, still see the proverbial "90-pound weakling" when viewing themselves in the mirror MANAGEMENT Antidepressants. Cf Anorexia nervosa

muscle relaxant ANESTHESIOLOGY An agent used in anesthesiology to facilitate airway management, control alveolar ventilation, abolish motor reflexes, and provide the muscle relaxation. **DEPOLARIZING AGENTS**, eg succinylcholine, cause a prolonged depolarization of the motor end plate. **NONDEPOLARIZING AGENTS**, eg pancuronium, are competitive inhibitors of acetylcholine at the motor end plate See Depolarizing agent, Nondepolarizing agent

muscular dystrophy Inherited myopathy NEUROLOGY Any of the primary degenerative myopathies, characterized by selective atrophy and weakness of voluntary muscles, pseudohypertrophy–see Champagne bottle legs, progressive deterioration, and early death; usually X-linked, often recessive , also AD and AR, affecting 1:3500 ♂ children, the most common form of which, Duchenne's muscular dystrophy, is fatal by age 20

MUSCULAR DYSTROPHY, SHORT CLASSIFICATION

NAME/HEREDITY	LOCATION	ONSET/DEATH
Becker/ X-R	Pelvifemoral	15-20/50-60
Duchenne/ X-R	Pelvifemoral	20-50/ ±20
Erb/AR	Scapula	20-30
Landouzy-Dejerine/AD	Facioscapulohumeral	15-30/benign
Leyden-Mobius/AR	Pelvic girdle	20-50/ ±20

AD Autosomal dominant; AR Autosomal recessive; X-R X-linked recessive

musculoskeletal disorder OCCUPATIONAL MEDICINE Job-related injuries and disorders of the muscles, nerves, tendons, ligaments, joints, cartilage, spinal disks EXAMPLES Carpal tunnel, rotator cuff, De Quervain's disease, trigger finger, tarsal tunnel, sciatica, epicondylitis, tendinitis, Raynaud's phenomenon, carpet layers' knee, herniated spinal disk. See Occupational medicine

mushroom *adjective* A descriptor for a thermonuclear explosion-like terminal expansion of a cylindrical structure. See Mushroom appearance, Mushroom lesion, Mushroom pattern *noun* DRUG SLANG A street term for psilocybin or psilocin TOXICOLOGY 50 of 2000 species of mushrooms are poisonous; the major toxin is amanitine, a selective RNA polymerase II inhibitor present in *Amanita* and *Galerina* species; *Amanita phalloides* causes most mushroom deaths MORTALITY 40-90% CLINICAL STAGE 1 Abrupt onset–6-24 hrs after ingestion, accompanied by abdominal pain, N&V, diarrhea, major fluid and electrolyte imbalances STAGE 2 Apparent resolution, with asymptomatic renal and hepatic deterioration STAGE 3 occurs by days 3-4, characterized by hepatorenal collapse, cardiomyopathy, DIC, convulsions, coma, death. See Poison mushroom, Power mushroom, Shiitake mushroom

mushroom of foam FORENSIC PATHOLOGY A frothy nasolabial 'spume' seen in drowning victims, which appears as pressure is applied to the chest–in either resuscitation or by removing clothing. Cf 'Shaving cream' appearance

mushroom grower's lung Mushroom picker's disease PULMONARY MEDICINE A hypersensitivity pneumonitis seen in mushroom processors exposed to *Trichophyton vulgaris*, *Micropolyspora faeni* and other related mushrooms. See Hypersensitivity pneumonitis

mushroom poisoning TOXICOLOGY Intoxication with substances derived from mushrooms, which are divided into those of rapid and delayed onset

MUSHROOM INTOXICATION
RAPID-ONSET SYMPTOMS

GI SYMPTOMS, due to unknown toxins; onset within 1-3 hrs of ingestion, most commonly due to *Chlorophyllum molybdites*–aka *Lepiota morgani*, characterized by severe vomiting and diarrhea

ALCOHOL SENSITIZATION, due to coprine, which results in an Antabuse-like reaction, which affects the ingestant for up to 5 days after mushroom intake, most commonly due to *Coprinus atramentarium*, the symptoms include face flushing, tingling of fingers, and a metallic taste in mouth

PSL SYNDROME A symptom complex occurring 15 mins to 1 hr after ingesting the mushroom and characterized by perspiration, salivation, and lacrimation, caused by muscarine, a cholinergic toxin, which is accompanied by vomiting, diarrhea, and ↑ urination; most commonly due to *Inocybe* spp, *Clitocybe* spp, and others; antidote atropine

HALLUCINOGENIC EFFECT Toxins psilocybin and psilocin; symptoms begin 15 mins to 3 hrs after ingesting *Psilocybe* spp and others, and include delusions, euphoria, anxiety, distortion of the time-space continuum, and seizures, most commonly in children

INTOXICATION/DELIRIUM Toxins ibotenic acid/muscimol; symptoms begin 30 mins to 2 hrs after ingesting *Amanita muscaria* and others, and include apparent intoxication, incoordination, hyperactivity, muscle spasms, collapse, anxiety, altered vision, vomiting, and a coma-like state

DELAYED-ONSET SYMPTOMS

GI/HEADACHE/LIVER DAMAGE Toxin monomethylhydrazine–MMH; symptoms begin within 6-12 hrs of ingestion, and include bloating, nausea, vomiting, diarrhea, abdominal pain, muscle cramps, and vertigo; MMH is present in *Gyromitra esculenta*

GI/LIVER DAMAGE Toxin amatoxin; symptoms occur in three stages: 12-24 hrs with nausea and severe abdominal pain, followed by a short period in which the symptoms appear to remit, followed by a resumption of symptoms, and fatal–in up to 50% of cases liver failure in 4-7 days; amatoxins are present in *Amanita phylloides*, and other *Amanita* spp

KIDNEY/LIVER DAMAGE Toxin orellanine; symptoms are delayed, beginning between 36 hrs and 14 days, and include extreme thirst, ↑ urinary frequency, lumbar pain, chills or fever, diarrhea or constipation

music therapy Music-facilitated psychoeducational strategy PSYCHOLOGY The use of music as an interventional modality. See Guided Imagery and Music, New Age music, Noxious music, Sedative music, Sensory therapy, Stimulative music. Cf Art therapy, Color therapy, Dance therapy, Play therapy, Recreational therapy

musician's wart PERFORMING ARTS MEDICINE A callus, hyperkeratosis, or comedonic lesion where a musical instrument contacts/rests on the body, anointed with descriptions–eg,, 'guitar nipple', 'cello scrotum', 'fiddler's neck', etc, variably accompanied by erythema, papule formation. See Performing arts medicine, Singer's nodule

Musset sign CARDIOLOGY A sign of aortic regurgitation, and pulsating aortic aneurysms due to ↑ pulse pressure, characterized by head bobbing that coincides with the pulse. See Aortic regurgitation

MUST CARDIOLOGY TRIALS 1. Medication Use Studies 2. Multicenter stent study 3. Multicenter stents ticlopidine–trial study MILITARY MEDICINE Medical Unit, Self-contained, Transportable An acronym for a species of medical equipment used by the US military in field situations. Cf DEPMEDS

MUSTT CARDIOLOGY 1. A clinical study–Multicenter Unstable Tachycardia Trial 2. A clinical study–Multicenter Unsustained Tachycardia Trial. See Calgary-MUSTT

mutilating arthritis Arthritis mutilans RHEUMATOLOGY A generic term for end-stage joint destruction accompanied by osteolysis, seen advanced psoriatic arthritis, also in rheumatoid arthritis, mixed connective tissue disease. See Psoriatic arthritis, Rheumatoid arthritis

mutilating surgery SURGERY A 'heroic' intervention entailing massive excision of tissue, often from a broadly invasive malignancy, to remove tumor and/or metastases, regardless of 'cost' vis-á-vis deterioration of quality of life, potential infection and other co-morbidity

mutilation Disfigurement; a major reduction or alteration of a limb or tissue, which may be intentional or accidental. See Genital mutilation, Self-mutilation

mutism NEUROLOGY Inability or refusal to speak

MVA Motor vehicular/vehicle accident, see there

436 **MVP** 1. Mitral valve prolapse 2. Mitral valvoplasty

Mx Metastasis, metastases

myasthenia gravis NEUROLOGY An autoimmune disease characterized by weakness and muscle fatigue EPIDEMIOLOGY 5-12.5/10⁵–± 25,000 active cases, peak incidence ♀ 2nd–3rd decade, ♂ 6-7th decade CLINICAL Ptosis and diplopia–ocular MG, weakness, skeletal muscle fatigability, generalized in most Pts, often proximal GRADES Based on severity, ranging from focal disease–grade I to life-threatening crisis with impaired respiration–grade IV DIAGNOSIS Anticholinesterase–Tensilon test, repetitive nerve stimulation, anti-acetylcholine receptor assay, single-fiber EMG ASSOCIATED CONDITIONS Thymoma, thymic hyperplasia, autoimmune diseases–eg, thyroiditis, Graves' disease, rheumatoid arthritis, SLE TREATMENT Anticholinergics–eg, pyridostigmine; thymectomy, immunosuppressants–corticosteroids, azathioprine, cyclosporine, immunotherapy–IV immunoglobulin, plasma exchange

myasthenic crisis Any of a number of clinical complexes characterized by an acute exacerbation of myasthenia gravis symptoms, which are divided into

MYASTHENIC CRISIS

MYASTHENIC CRISIS An acute ↑ in requirement for anticholinesterase therapy or refractoriness to same, diagnosed by a Tensilon test, with transient ↓ of symptoms

CHOLINERGIC CRISIS An acute ↓ in the need for anticholinesterase medication, resulting in 'overmedication' with the customary doses; the Tensilon test exacerbates this form of myasthenic crisis; cholinergic crises may be either

- **MUSCARINIC CRISIS** Abdominal pain, diarrhea, nausea, vomiting, lacrimation, blurred vision, bronchial hypersecretion due to parasympathetic hyperresponse

- **NICOTINIC CRISIS** Muscle weakness, fasciculations, cramping and dysphagia, due to overdepolarization at the neuromuscular junction. See Tensilon test

myasthenic reaction NEUROLOGY A progressive ↓ in muscle strength with repeated contraction, typical of myasthenia gravis

myasthenic syndrome Lambert-Eaton syndrome, see there

Mycelex® Clotrimazole, see there

mycetoma MYCOLOGY A slow, relentless, ulcerating fungal–true fungi infection which, when neglected, may result in osteomyelitis, in a background of impaired host defense, most common in the feet of young ♂–5:1 who work in the tropics/subtropics, which may be seen in the thigh and shoulders; clinical disease is rare unless accompanied by bacterial infection. See Pulmonary aspergilloma

mycobacteriology lab LAB MEDICINE A reference lab that deals specifically with *M tuberculosis* and other mycobacteria

Mycobacterium A genus of obligate aerobic bacteria, family Mycobacteriaceae, order Actinomycetales; all are capable of producing the typical chronic inflammation, Langhans' giant cells and caseating necrosis, and are indistinguishable by acid-fast staining–due to the high lipid concentration in the outer cell wall; skin is the usual portal of entry for non-TB mycobacteria. See Acid-fast stain, Atypical mycobacteria, Buruli ulcer, Langhans' giant cells, MOTT, Nontuberculous mycobacteria, Prosector's wart, Runyon classification, Scrofula, Tuberculosis

***Mycobacterium avium* complex disease** AIDS, INFECTIOUS DISEASE An opportunistic mycobacterial infection and serious, life-threatening complication of HIV infection CLINICAL Fever, night sweats, weight loss, diarrhea, ↓ survival in immunocompromised Pts MANAGEMENT Rifabutin, ethambutol, clarithromycin, azithromycin

Mycobacterium bovis A mycobacterium that causes a TB-like infection in cows; before pasteurization was common, *M bovis* spread to humans via contaminated milk

Mycobacterium kansasii INFECTIOUS DISEASE A mycobacterium that may involve any body site, but primarily lungs; dissemination is rare and associated with immunocompromise or immuosuppression MANAGEMENT INH, ethambutol; antibiotic resistance is uncommon

Mycobacterium leprae INFECTIOUS DISEASE The mycobacterium that causes leprosy. See Leprosy

Mycobacterium marinum INFECTIOUS DISEASE A mycobacterium that lives in fresh or salt water, causing chronic ulcerating granulomatous lesions. See Swimming pool granuloma

Mycobacterium tuberculosis Tubercle bacillus INFECTIOUS DISEASE The mycobacterium that causes TB. See Tuberculosis

Mycobacterium xenopi INFECTIOUS DISEASE A rare atypical mycobacterium associated with clinical disease CLINICAL Mucocutaneous abscesses, often with cavitation; pulmonary and disseminated disease are rare MANAGEMENT High-dose INH, streptomycin, kanamycin, cycloserine

Mycobutin® Rifabutin, see there

mycophenolate mofetil IMMUNOLOGY A mycophenolic ester derivative that may be better than azathioprine as an immunosuppressive purine analogue; it inhibits purine metabolism in T and B cells, and lymphocyte proliferation in vitro; MMF is used to prevent or manage GVHD, rejection of heart, kidney, lung and other allograft transplants, autoimmune disorders ADVERSE EFFECTS Leukopenia, anemia, diarrhea, emesis. See Heart transplant, Kidney transplant, Lung transplant

Mycoplasma INFECTIOUS DISEASE A pathogen that causes 'walking pneumonia'–resolving in 4-6 wks, and genitourinary infections. *M hominis* may cause PID, septicemia, urogenital infection

***Mycoplasma fermentans* incognitus strain** A strain of *Mycoplasma* linked to AIDS MANAGEMENT Tetracycline, doxycycline, chloramphenicol, clindamycin, lincomycin, ciprofloxacin

mycoplasma pneumonia Primary atypical pneumonia; walking pneumonia INFECTIOUS DISEASE A primary atypical pneumonia by *M pneumoniae*, which causes miniepidemics that spread by aerosol or close contact, causing up to ¾ of a 'closed population'–military 'boot' camps, boarding schools, colleges–pneumonias, affecting ages 5-20 CLINICAL Incubation ± 3 wks; insidious onset of fever, malaise, headache, myalgia, low-grade fever, cough, chest pain, and respiratory disease from asymptomatic to rhinitis, pharyngitis, tracheobronchitis, pneumonia LAB Leukocytosis, relative ↑ in PMNs, cold hemagglutinins, biologic false positive VDRL COMPLICATIONS Erythema multiforme, Raynaud's phenomenon, cold agglutinin hemolysis; less commonly, neurologic, cardiovascular, musculoskeletal defects TREATMENT Tetracycline, erythromycin VACCINE Under development

mycosis A fungal infection; the most common mycosis in the US is candidiasis, caused by *C albicans*, commonly vaginal in adult ♀, oral in children, and esophageal in AIDS and immunocompromise

mycosis fungoides HEMATOLOGY A rare–0.3/10⁵/yr, malignant lymphoproliferation of paracortical T cells, usually helper, less often, suppressor subtypes that is 2-fold more common in older blacks CLINICAL Skin involvement precedes Sx by up to 2 yrs; the leukemic phase–Sezary syndrome–occurs in 80% and is accompanied by fever, weight loss, lymphadenopathy, hepatosplenomegaly, lung involvement, eosinophilia, lymphocytosis, peripheral neuropathy, and periarteritis nodosa. Stages of ↑ aggression: Erythema stage, plaque stage, tumor stage, d'emblee stage MANAGEMENT Early RT and chemotherapy don't alter clinical disease

Mycostatin® Nystatin INFECTIOUS DISEASE An antifungal used for managing candidiasis

mycotic aneurysm An intravascular inflammatory response seen in 3-15% of Pts with infective endocarditis, which may arise from contiguous infected sites, but more commonly are of hematogenous spread, potentially resulting in thrombosis or rupture of arterial walls weakened by an inflamed vasa vasorum or by impaction-necrosis; because the vessels are a poor culture substrate for bacteria–eg, *S aureus*, smaller MAs may resolve spontaneously; those ≥ 1-2 cm require excision if surgically accessible; of greatest concern are the cerebral aneurysms; sites of symptomatic MAs include sinus of Valsalva 25%, visceral arteries 24%, extremities 22%, brain 15%. Cf Berry aneurysm

myelitis NEUROLOGY Spinal cord inflammation, often caused by VZV CLINICAL Paraparesis, sensory defects, sphincter impairment IMAGING T_2-weighted MRI reveals hyperintense lesions, ± fusiform swelling of spinal cord LAB CSF slight pleocytosis, ↑ protein PROGNOSIS Excellent in immunocompetent Pts; immunocompromised Pts may need acyclovir

myeloablative therapy ONCOLOGY The use of RT and/or high-dose chemotherapy to eliminate tumor load–eg, of myeloma in BM, followed by restitution of native BM with either allogeneic BMT or autologous–self BMT, which is reinfused after purging it of plasma cells, using monoclonal antibodies

myelodysplasia HEMATOLOGY Abnormal BM precursor cells, which may lead to CML. See Myelodysplastic syndrome NEUROLOGY A generic term for various developmental defects of the spinal cord and nerve roots–eg, myelomeningocele, sacral agenesis, spinal dysraphism and caudal regression syndrome; $\frac{1}{3}$ of infants with myelodysplasia develop external urethral sphincter dysfunction, often in the first 3 yrs of life, $\frac{1}{2}$ of which are permanent

myelodysplastic syndrome Preleukemia, smoldering leukemia HEMATOLOGY A preleukemia-like state characterized by compromised BM function; myelodysplasia is associated with monosomy 7 in PMNs and monocytes, and evokes PMN dysfunction. See Preleukemia

myelofibrosis Agnogenic myelofibrosis, agnogenic myeloid metaplasia, marrow fibrosis, myelofibrosis with myeloid metaplasia, idiopathic myelofibrosis, myeloid metaplasia, myelosclerosis, primary myelofibrosis HEMATOLOGY A chronic progressive condition—panmyelosis characterized by variable fibrosis of BM, massive splenomegaly 2° to extramedullary hematopoiesis and leukoerythroblastic anemia with dysmorphic RBCs, circulating normoblasts, immature WBCs, atypical platelets CLINICAL Often > age 50; insidious weight loss, anemia, abdominal discomfort due to splenomegaly, often hepatomegaly; 80% have nonspecific chromosome defects BONE MARROW Megakaryocytic hyperplasia, ↑ reticulin fibers, giant cells and immature forms, virus-like particles, granulocytic and erythrocytic hyperplasia, chromosomal abnormalities–eg, of 21q which may correlate with reverse transcriptase activity LAB Thrombocytosis–1-14 × 10^{12}/L with giant and bizarre forms, hypochromic and microcytic anemia or erythrocytosis, elliptocytosis, Howell-Jolly bodies, target cells, teardrop cells; leukocytosis–15-40 × 10^{12}/L with ↑ 'bands', ie left shift of granulocytes, juvenile metamyelocytes, ± eosinophilia, basophilia, splenic atrophy, ↑ leukocyte alk phos, platelet acid phosphatase, uric acid, vitamin B_{12}, low-grade DIC DIAGNOSIS BM Bx MANAGEMENT AMM has no specific therapy; packed RBCs for anemia, androgens may ↓ the transfusion requirements in some Pts, but are poorly tolerated in ♀; recombinant erythropoietin PROGNOSIS Survival ± 5 yrs, often → acute leukemia. See Idiopathic onychomycosis, Pseudonym syndrome

myelogenous *adjective* Produced by or originating in the bone marrow. Cf Myeloid

myelography IMAGING A technique for examining the spinal canal, in particular the lumbar region, by percutaneous injection of a water-soluble radiocontrast into the subarachnoid space by lumbar puncture INDICATIONS Evaluation of spinal cord defects due to compression of the spinal cord and/or its roots by a herniated disk, degenerative articular lesions, trauma, tumors or other masses of the spinal cord or meninges or metastases CONTRAINDICATIONS Multiple sclerosis, ↑ intracranial pressure, untreated coagulopathies, allergies COMPLICATIONS Meningitis, infection, hemorrhage, N&V, cerebral herniation, seizures, radiocontrast allergies. See CT myelography

myeloid *adjective* 1. Referring to the BM 2. Referring to granulocytes and maturation thereof, aka myelogenous 3. Referring to the spinal cord–rarely used in practice

myeloma Malignant plasmacytoma, multiple myeloma, multiple plasmacytoma of bone, myelomatosis, plasma cell myeloma HEMATOLOGY A neoplastic proliferation of plasma cells in BM and extramedullary sites; if circumscribed, plasmacytoma EPIDEMIOLOGY MM comprises 10% of hematopoietic malignancies, causing 10,000 deaths/yr–US; ↑ with age; black:white 2:1 CLINICAL As the myeloma cells proliferate in BM, they cause osteolysis, resulting in pathologic fractures and industrial strength pain, anorexia, nausea, thirst, fatigue, muscle weakness, restlessness, confusion; BM replacement by neoplastic plasma cells results in ↑ infections and anemia due to displacement of normal WBCs, and erythroid series; excess Igs in circulation plug up renal tubules LAB Monoclonal Igs, often light chains, in serum, urine; ↑ Ca^{2+}; normal PO_4, normal alk phos COMPLICATIONS Painful pathologic fractures, anemia, hypercalcemia, renal failure, recurrent bacterial infections TREATMENT 1° chemotherapy: Melphalan and prednisone; VAD–vincristine, Adriamycin, dexamethasone in new cases; IFN-α has been used for maintenance; myeloablative therapy, allogeneic BMT from HLA-matched siblings. See Clear cell, myeloma, Indolent myeloma, Myeloma kidney, Solitary myeloma

myeloma, Ig D type A condition that comprises 1-2% of all myelomas, most common in older ♂ CLINICAL Lymphadenopathy, hepatosplenomegaly, accompanied in 45% of cases by extra-osseous dissemination–which is present in 15% of IgA and IgG myelomas, hyperviscosity, severe anemia, azotemia, marked osteolysis, hypercalcemia and, commonly bizarre plasmacytes and plasmablasts; the M component is not markedly elevated

myeloma kidney The combination of structural and functional renal defects seen in ±40% of myelomas–eg, intraluminal eosinophilic 'blocked pipe' casts composed of PAS—positive homogeneous material and light chains–Bence-Jones proteins within flattened distal tubules and collecting ducts–pressure atrophy, spilling over of proteinaceous material, eliciting chronic interstitial nephritis, occasionally causing glomerulonephritis; the mesangial widening may mimic diabetic nephropathy; functional abnormalities cause renal failure in 20%, due to hypercalcemia and renal calcinosis, heavy Bence-Jones proteinuria–causing tubular damage, hyperuricemia–↑ tumor DNA turnover, proteinuria, amyloidosis, and chronic pyelonephritis, acquired Fanconi syndrome, defects in acidification and concentration, acute and chronic renal failure LAB A peak may be seen in γ-globulin region–usually of urine electrophoresis or may appear between the $α_2$ and β regions

myeloma profile Multiple myeloma profile A panel of lab tests used to evaluate Pts with possible myeloma SUBSTANCES MEASURED IgG, IgA, IgM, beta-2-microglobulin

myelomeningocele NEONATOLOGY A common birth defect characterized by nonclosure of the vertebral and spinal canal, resulting in protrusion of spinal cord and covering meninges from the dorsal surface; spina bifida is a more generic term which includes any defect characterized by incomplete spinal closure; myelomeningocele accounts for ±75% of all spina bifida, affecting ±1/800 infants; most of the rest are spina bifida occulta–the vertebral arches don't close, the spinal cord and meninges are simply covered by skin, and meningoceles–in which the meninges protrude through the vertebral defect while the spinal cord remains in place ETIOLOGY Unknown; folic acid deficiency may play a part in neural tube defects; it has a familial tendency, possibly triggered by a viral infection or environmental factors–eg, radiation; protrusion of cord and meninges damages the spinal cord and nerve roots, compromising function at or below the defect; most defects are lower lumbar or sacral, as these regions are the last to close in the early embryo CLINICAL Partial or complete paralysis, loss of sensation, bladder or bowel control; the exposed cord is at risk for infection. See Hydrocephalus, Hip dislocation, Spina bifida, Syringomyelia

myelopathy A spinal cord disorder. See Vacuolar myelopathy

myeloperoxidase deficiency MOLECULAR MEDICINE A common–1:500-2000 AR condition characterized by neutrophil dysfunction, resulting in a prolonged respiratory burst due to defective post-translational processing of an abnormal precursor protein CLINICAL Usually asymptomatic–*Candida* infection may occur in Pts with concomitant DM TREATMENT Unnecessary

myelophthisic anemia HEMATOLOGY A form of anemia due to a significant loss of BM or compromise of BM architecture with depressed cell production; most commonly due to space-occupying lesions–eg, metastatic infiltration of the BM from 1° cancers of lung, breast, etc, MA is characterized by ↓ in all 3 blood cell lines and appearance of immature RBCs and WBCs in peripheral circulation, due to irritation. See Space-occupying lesion

myelopoiesis HEMATOLOGY The production of myeloid cells in the BM; more generically, the production of any cell line–erythroid, megakaryotic, myelocytic in the BM

myeloproliferative disorder A hematopoietic stem cell disease. Cf Lymphoproliferative disorder

MYELOPROLIFERATIVE DISORDER

ACUTE MPD Acute myelogenous leukemia–myeloblastic, promyelocytic, myelomonocytic, monocytic, erythroid, megakaryocytic, eosinophilic, basophilic, acute biphenotypic–with myeloid and lymphoid markers, leukemia, and acute leukemia with lymphoid markers evolving from a prior clonal hemopathy

SUBACUTE MPD Oligoblastic–smoldering leukemia, refractory anemia with excess blasts–see RAEB, myelomonocytic leukemia

CHRONIC MPD Polycythemia vera, agnogenic myeloid metaplasia, primary thrombocythemia, chronic myelogenous leukemia–Philadelphia chromosome positive or negative, chronic monocytic leukemia, chronic neutrophilic leukemia

myelosclerosis with myeloid metaplasia Agnogenic myeloid metaplasia, see there

myelosuppression ONCOLOGY The suppression–usually an undesired side effect–of normal BM activity, often the result of RT, chemotherapy, or various toxins. Cf Myelosuppressive therapy

myelosuppressive therapy ONCOLOGY Any therapy intended to inhibit or frankly wipe out normal cell proliferation in the BM. Cf Myelosuppression

myocardial blush CARDIOLOGY The angiographic opacification of the myocardial microvasculature or vessels, the size of which can be planimetered so that the total area of the blush can be determined

myocardial bridging Coronary bridge, intramural coronary artery, intramyocardial tunneling, mural coronary artery, myocardial squeezing, myocardial tunneling CARDIOLOGY An anomaly in which a major coronary artery–CA leaves the epicardium, follows an intramyocardial course and subsequently returns to the surface; the CA is thus 'bridged' by a band of overlying myocardium; seen in 27% of autopsies, MB is a normal anatomic variant of the CAs; however, constriction of a bridged CA occurs in asystole and may cause myocardial ischemia or sudden death

myocardial contrast echocardiography CARDIOLOGY A technique in which microbubbles of albumin containing a high molecular weight gas are injected IV to evaluate the coronary arterial blood flow. See Contrast echocardiography, Sonicated albumin

myocardial contusion CARDIOLOGY A bruise of the myocardium caused by blunt trauma to the chest

myocardial hibernation Hibernation CARDIOLOGY Persistent left ventricular dysfunction when myocardial perfusion is chronically reduced but still enough to maintain tissue viability; MH is due to prolonged–months, yrs ischemia; ventricular dysfunction persists until blood flow is restored. Cf Myocardial stunning

myocardial infarction Acute necrosis of myocardial tissue; in the early post-MI period, there may be a need to rely on 'soft' data, especially if troponin I or CK-MB have yet to ↑, or there is a loss of sensation to the pain characteristic of MI, as occurs in circa 10% Pts with DM; older ♀ may have normal levels of CK after an MI RISK FACTORS FOR MI ASHD, ↑ cholesterol, HTN, smoking, DM, low selenium, etc LAB Cardiac enzymes, 'flipped' LD, troponins increase to normal size. PATHOLOGY Chronology of myocardial changes FATAL COMPLICATIONS OF MI Shock, arrhythmias, rupture of ventricular aneurysms or papillary muscle, acute CHF, mural thromboembolism RISKS ↑ risk with ↑ TGs, ↑ small LDL particle diameter, ↓ HDL-C

myocardial infarction scan NUCLEAR CARDIOLOGY A noninvasive method for determining the location and extent of an MI, based on the sequestration–pooling of a radiopharmaceutical–eg, ^{99m}Tc-Sn-pyrophosphate, which results in a 'hot spot' in the infarcted region of the heart, when an image is captured with a gamma camera

myocardial ischemia Hypoxia of myocardium, characterized by ↑ TNF-β, local production of superoxide anions, ↓ coronary vasodilation, and myocardial necrosis; if administered at the time of ischemia, recombinant TNF ↓ circulating superoxide anions, maintains endothelial-dependent coronary relaxation, ↓ myocardial injury mediated by endogenous TNF

myocardial laser revascularization HEART SURGERY A procedure in which a laser is used to drill channels in the left ventricle for improved perfusion and oxygenation of heart muscle, treating angina pectoris and myocardial ischemia. Cf Batista procedure

myocardial perfusion imaging A technique in which the regional distribution of blood throughout the myocardium, is determined by injecting a radiopharmaceutical–eg, ^{201}Tl. or ^{99m}Tc and capturing images with a gamma–scintillation camera; by determining blood and oxygen distribution, MPI informs on the hemodynamics and functional effect of coronary arterial stenosis

myocardial stunning Stunning CARDIOLOGY Prolonged nonpermanent postischemic systolic and/or diastolic ventricular dysfunction without myocardial necrosis; MS may follow successful thrombolytic therapy in evolving MI, PCTA, during exercise, and after relief of ischemia caused by vasospasm. Cf Myocardial hibernation

myocarditis CARDIOLOGY Myocardial inflammation ETIOLOGY Idiopathic, viral, toxic COMPLICATIONS Dilated cardiomyopathy. See Acute myocarditis, Borderline myocarditis, Fulminant myocarditis, Giant cell myocarditis, Lymphocytic myocarditis

myoclonus Lightning movement NEUROLOGY A rapid involuntary non-rhythmic spasm that can occur spontaneously at rest, in response to sensory stimulation, or with voluntary movements; myoclonias are *symptoms* and not, per se, diseases *a sui generis* MANAGEMENT Clonazepam, valproic acid. See Baltic myoclonus, Posthypoxic ischemic myoclonus, Sleep-related myoclonus

MYOCLONUS TYPES

ESSENTIAL MYOCLONUS Idiopathic/non-progressive, eg restless legs syndrome

PHYSIOLOGIC MYOCLONUS Associated with sleep jerks and hiccups

EPILEPTIC MYOCLONUS Associated with epilepsy and

SYMPTOMATIC MYOCLONUS Associated with encephalopathy, spinocerebellar degeneration, metabolic, toxic, or viral encephalopathy or trauma

myofacial pain syndrome INTERNAL MEDICINE 1. TMJ syndrome, see there 2. A referred pain pattern characteristic for an individual muscle MANAGEMENT Injection with local anesthetics, steroids, antiinflammatories. See Myofascial trigger point

myofascial syndrome NEUROLOGY A painful condition characterized by local or referred pain evoked at multiple trigger points, accompanied by pain, stiffness, weakness, ↓ ROM CLINICAL Pain everywhere–head, neck, chest, joints, pelvis, back, sciatica MANAGEMENT 3-step drug ladder

myofascial trigger point INTERNAL MEDICINE A self-sustaining hyperirritative focus that may occur in any skeletal muscle after strain produced by acute or chronic overload; MTPs produce a referred pain pattern characteristic for that individual muscle; each pattern becomes part of a single myofascial pain syndrome. See Myofascial pain syndrome

myofibroma Leiomyoma, see there

myofibromatosis See Infantile myofibromatosis

myogenic *adjective* Arising in muscle cells; used to describe self-excitation by cardiac and smooth muscle

myoglobinuria INTERNAL MEDICINE The loss of myoglobin in urine

MYOGLOBINURIA, ETIOLOGY

CONNECTIVE TISSUE DISEASE Polymyositis, dermatomyositis

HEREDITARY Carnitine palmityl transferase deficiency, glycogen storage disease type V–phosphorylase deficiency, McArdle's disease, myopathies associated with glycogen or lipid storage, paroxysmal familial myoglobinuria, malignant hyperthermia, periodic paralysis

INFECTION

Bacteria–eg, Legionnaire's disease

Parasites–eg, trichinosis, toxoplasmosis

Viral–eg, EBV, influenza, herpes

TOXINS Alcohol–alcoholic myopathy, carbon monoxide, phencyclidine–PCP, 'angel dust', ethylene glycol, insect venom, some diuretics, Haff disease–historic interest

TRAUMA Crush injury, excess/prolonged physical exertion–eg, marathon running, severe muscle injury, crush injury

ETCETERA Hyperthermia, infarction due to vascular occlusion, seizures

The diagnosis is based on clinical suspicion–especially with bright red urine and ↑ CK

myomectomy GYNECOLOGY Excision of a uterine leiomyoma–popularly, fibroma without hysterectomy. See Leiomyoma

myopathic gait ORTHOPEDICS A waddling duckoid or penguinoid sashay characterized by exaggerated lateral trunk swaying, hip raising SEEN IN Muscle dystrophy, spinal muscle atrophy, febrile polyneuropathy. See Gait.

myopathy ORTHOPEDICS Any disease of skeletal muscle. See Central core myopathy, Centronuclear myopathy, Colchicine myopathy, Exertional myopathy, Fetal alcohol myopathy, Mitochondrial myopathy, Myotubular myoathy, Nemaline myopathy, Spheroid body myopathy, Vacuolar myopathy, X-linked centronuclear (myotubular) myopathy, Zebra body myopathy. Cf Cardiomyopathy

myophosphorylase deficiency McArdle's disease, see there, aka glycogen storage disease V, see there

myopia Nearsightedness OPHTHALMOLOGY An error of refraction and accommodation in which the cornea and/or lens focuses an image in front of the retina, resulting in ↓ visual acuity for far objects; most myopia is correctable with glasses or contact lenses. Cf Hyperopia

myopotential CARDIOLOGY An electric signal arising in skeletal muscle which may be sensed by a pacemaker and falsely interpreted as a depolarization See Oversensing

myositis Myitis Muscle inflammation. See Dermatomyositis, Inclusion body myositis

myositis ossificans Bone formed within muscle; localized MO is 2° to trauma resulting from a blow or muscle tearing; generalized MO is AD, often accompanied by aplasia of the thumb, great toe, or rarely other digits, in which the first 'tumor' occurs in the paravertebral or cervical region, followed by multiple ossifying tumors, forming calcifying bridges across muscles and joints resulting in massive rigidity and the Pt is turned into 'stone'. See Zoning phenomenon

myotonia NEUROLOGY Delayed muscle relaxation after voluntary contraction–action myotonia or mechanical stimulation–percussion myotonia TYPES Chloride channel-related disorders–eg, myotonia congenita, Thomsen type; protein kinase-related disorders–eg, myotonic dystrophy; sodium channel-related disorders–eg, hyperkalemic periodic paralysis; idiopathic

myotonia congenita Thomsen's disease NEUROLOGY An AD or AR condition characterized by deficiency of true cholinesterase CLINICAL Inability of muscles to relax quickly, resulting in myotonia, gagging, dysphagia, stiff movement that improves with repetition; Pts are often muscular. See Myotonia

myotonic dystrophy MOLECULAR MEDICINE An AD condition affecting ± 1:8000, age of onset, age 20-25, ↓ IQ, which causes distal myopathy, preferentially affecting certain muscles–eg, levator palpebrae, facial, masseter, sternocleidomastoid, forearm, hand and pretibial muscles, resulting in diffuse muscular weakness and atrophy beginning in early adulthood causing the characteristic 'hatchet face'; other changes include lenticular opacities, endocrinopathies–testicular atrophy with androgen insufficiency, ovarian dysfunction which rarely interferes with fertility, DM, hypothyroidism, mild cerebral cortical atrophy, frontoparietal baldness, cardiac and smooth muscle–GI, especially esophageal motility defects, respiratory dysfunction and hyperostosis frontalis interna; death usually by age 50. See Myotonia, Trinucleotide repeat disease

Myotrophin® Mecasermin, rhIGF-1 NEUROLOGY A recombinant orphan drug for treating ALS. See Amyotrophic lateral sclerosis

myotube See Early myotube, Late myotube

myotubular myopathy Centronuclear myopathy A myopathy with various patterns of inheritance; all have centrally-located nuclei in muscle fibers, surrounded by cytoplasmic material with features of maturing myotubules, accompanied by atrophy of type I and hypertrophy of type II muscle fibers; the X-linked form results in neonatal death due to respiratory muscle insufficiency; the AD form is benign. See Myopathy

myringoplasty ENT An operation in which the reconstructive procedure is limited to repair of a tympanic membrane perforation

myringotomy ENT A circumferential incision in the inferior quadrants of the pars tensa of tympanic membrane for draining middle ear effusions, a procedure used in treating chronic otitis media; myringotomy with tympanostomy offers more disease-free time and improved hearing, at a 'cost' of otorrhea and persistent tympanic membrane perforation

myrmecophobia PSYCHOLOGY Fear of ants. See Phobia

mysophilia SEXOLOGY A paraphilia in which sexuoerotism hinges on smelling, chewing or using sweaty or soiled clothing or articles of menstrual hygiene. Cf Coprophilia, Urophilia

myxedema ENDOCRINOLOGY Severe hypothyroidism characterized by yellowish discoloration, nonpitting edema, especially facial, accompanied by periorbital puffiness, puffy lips, and tongue, hoarse voice and sluggish movement; myxedema elicits several reactions *a sui generis*

MYXEDEMA

MYXEDEMA COMA A complication of severe hypothyroidism, characterized by a loss of consciousness, which may be iatrogenic–sedatives in hypothyroidism are very slowly metabolized, may be due to infections or cold exposure; or rarely, occurring spontaneously MORTALITY 20-50%

myxoid

MYXEDEMA MADNESS A condition most common in the elderly, characterized by impaired hearing and memory, acalculia, somnolence, psychologic withdrawal, and paranoia

MYXEDEMA MEGACOLON Pseudo-obstruction due to ↓ GI motility

MYXEDEMA WIT Confabulation or use of humorous non-sequiturs by a hypothyroid Pt in order to draw the interviewer's attention away from the Pt's impaired memory

myxoid *adjective* 1. Mucin-like 2. 'Loose' pale-to-lightly basophilic stroma, as stained by H&E; the few cells present include fibroblasts and rare chronic inflammatory cells; myxoid stroma occurs in nodular fasciitis, intramuscular myxoma, ganglion cyst, chordoma, neurofibroma, carcinomas, as well as spindle cell lipoma and lipoblastoma and in myxoid variants of sarcomas, which generally have a better prognosis EXAMPLES Rhabdomyosarcoma, chondrosarcoma, MFH, liposarcoma.

N Shorthand for 1. Amino 2. Asparagine 3. *Neisseria* 4. Nerve 5. Neuraminidase 6. Neutron number 7. Nifedipine 8. *Nocardia* 9. Nitrogen 10. Normal solution–equivalents/L 11. Nuclear 12. Nucleoside 13. Nucleus 14. Number 15. Nystatin 16. Population size 17. Radiance

n Shorthand for: 1. Haploid number–genetics. 2. nano- SI– abbreviation for 10^{-9} 3. Neutron. 4. Refractive index

n-3 fatty acid n-3 polyunsaturated fatty acid, omega-3 fatty acid A family of long-chain polyunsaturated fatty acids, primarily eicosapentaenoic–C20:5 and docosahexanenoic acid–C22:6; ↑ dietary NFAs are cardioprotective and have a positive impact on inflammation, interfering with production of inflammatory mediators–eg, leukotrienes, PAF, IL-1, TNF. See Fish; Cf Olive oil, Tropical oil

N of 1 trial CLINICAL TRIALS a clinical trial in which a single Pt is the total population for the trial–eg, a single case study; an N of 1 trial in which random allocation is used to determine the order in which an experimental and a control intervention are given to a Pt is an N of 1 RCT. See Trial

NA 1. Needle aspiration 2. Neuraminidase activity 3. Neutralizing antibody 4. Nicotinic acid 5. Noradrenaline–norepinephrine is preferred in the US 6. Nucleic acid

NABCO National Alliance of Breast Cancer Organizations ONCOLOGY A coalition of ±370 US organizations that offer services for diagnosing and treating breast CA and providing support ☎ 800.719.9154 www.nabco.org

Nabi-HB™ IMMUNOLOGY An IM formulation of human HBV immune globulin used to manage acute exposure to blood containing HBsAg, perinatal exposure of infants born to HBsAg-positive mothers, sexual exposure to HBsAg-positive persons, and household exposure to persons with acuteHBV infection. See Hepatitis B

Nabothian cyst GYNECOLOGY A cyst-like space consisting of entrapped mucus from secreting columnar villae under the developing squamous epithelium

nadolol Corgard® CARDIOLOGY A nonselective β-blocker used to treat HTN, angina. See Beta-blocker

nadolol with isosorbide mononitrate A combination of β-blocker–nadolol and an anti-hypertensive–isosorbide mononitrate that may be better than endoscopic sclerotherapy in treating esophageal varices

nafamostat mesylate A synthetic serine protease inhibitor once used in Japan to treat acute pancreatitis; NM and its metabolites also reversibly inhibit amiloride-sensitive sodium conductance of renal cortical collecting ducts, thus impairing potassium excretion

nafarelin A gonadotropin-releasing hormone–GnRH analogue used to treat estrogen-driven conditions–eg, endometriosis or uterine leiomyomas ADVERSE EFFECT Osteopenia–prevented by co-administration of PTH in ♀ receiving long-term therapy with GnRH analogues

nafcillin A semisynthetic penicillinase-resistant penicillin

Naffziger's test NEUROLOGY A maneuver used to identify compression of nerve roots in the cervical spine, based on exacerbation of pain and paresthesias of the hand and fingers, when the examiner exerts pressure on the scalenus anterior muscles in Pts with scalenus anterior syndrome

Nagahara phaco chopper OPHTHALMOLOGY A lens nucleus emulsification device and technique developed by Dr. KB Nagahara. See Cataract

nail ORTHOPEDICS A cylindrical metal device constructed of stainless steel used to hold 2 or more pieces of fractured bone in place

nail-patella syndrome Fong disease, hereditary osteo-onychodysplasia, HOOD syndrome, osteo-onychodysplasia, Turner-Kieser syndrome NEPHROLOGY An AD condition affecting structures of both mesodermal and ectodermal origin with partial-to-complete absence of thumb and great toenails, flexion contractions of multiple joints, defective or absent patellae, lordosis, clinodactyly and campylodactyly, conical iliac horns, scapular thickening, radial head subluxation, interfering with full ROM–pronation and supination, renal abnormalities–mesangial proliferation, thickened glomerular basement membrane, collagen deposition with proteinuria, microscopic hematuria, glomerulonephritis, pyelonephritis, slowly progressive renal failure, and ocular disease–clover leaf pigmentation of iris, cataracts, microphakia, microcornea, keratoconus, ptosis MOLECULAR PATHOLOGY Linked genetically to the ABO blood, on chromosome 9q34

nail unit Any part of the nail–eg, plate, matrix, bed, proximal and lateral folds, hyponychium. See Nail disorder

nailfold capillary microscopy Widefield capillary microscopy A technique used to evaluate finger nailfold capillaries, using a wide-angle stereomicroscope or ophthalmoscope; enlargement, ↑ tortuosity, or ↓ number of capillaries occur in connective tissue diseases–eg, Raynaud's phenomenon, scleroderma–to evaluate extent of visceral involvement, MCTD, dermatomyositis, SLE

naked virus A non-enveloped virus–eg, a reovirus

nalbuphine Nubain® A synthetic narcotic chemically related to naloxone and similar in action to, but less addictive than, morphine, it has a similar analgesic intensity as morphine, but has a ceiling on the intensity of the respiratory depression it causes when in excess ADVERSE EFFECTS Sweating, N&V, vertigo, dry mouth, headache

NALC N-acetyl L-cysteine MICROBIOLOGY A mucolytic agent used to collect sputa destined for TB culture that liquefies the mucus by breaking disulfide bonds

Nalfon®

Nalfon® Fenoprofen, see there

nalidixic acid ANTIBIOTICS An early generation quinolone with a limited antimicrobial palate, primarily used for UTIs. See Quinolone

nalorphine A morphine derivative which reversses the activity of morphine and other narcotics. See Opioids

naloxone Narcan® An IV morphine and endorphin antagonist-type opioid analgesic. See Opioids, Reversal agent

naltrexone SUBSTANCE ABUSE An opioid antagonist used to manage opiate, alcohol and other abuse substances; naltrexone binds to endorphin receptors, preventing endophin receptor binding, reducing the craving for abuse substances. See Cocaine. Cf Antabuse

name GRADUATE EDUCATION A popular term for the degree to which a person, mentor or professor is recognized in a particular field. See Reputation PHARMACEUTICAL INDUSTRY A designation, official or otherwise for a chemical or other substance with therapeutic potential. See Blank canvas name

named reporting PUBLIC HEALTH The reporting of infected persons by name to public health departments, a standard practice for surveillance of certain infectious diseases–TB, syphilis, gonorrhea, that pose a public health threat. Cf Anonymous testing

NANB See Non-A, non-B hepatitis

nandrolone decanoate ENDOCRINOLOGY A long-acting parenteral 17β-hydroxyl ester of 19-nortestosterone with weak androgenic activity, used as an androgen–eg, for refractory anemia. Cf Androgen replacement therapy

nanism Dwarfism, politically correctly termed short stature

nanomedicine MODERN MEDICINE A developing field in which nanoscale–ie, teeny-weeny–sensors would detect internal signals–eg, glucose levels, and respond by releasing insulin or other biomolecule

Nantucket disease A blood-borne infection by *Babesia microti*, an intertriginous cyst-forming parasite, named after an island off Massachusetts, occurring along the entire eastern seaboard of the US RESERVOIR White-footed mouse VECTOR *Ixodes dammini*–'Lyme disease' tick; in Europe, it is most common in splenectomized Pts and often fatal; in the US, it is rarely fatal; splenectomized subjects comprise $\frac{1}{3}$ of cases CLINICAL 1-3 wk incubation, malaise, fatigue, anorexia, shaking chills, fever, headache, myalgias, mental depression, emotional lability TREATMENT Clindamycin, quinacrine

nap SLEEP DISORDERS An episode of sleep outside of the principal sleep period. See Driver fatigue, Sleep deprivation, Sleep disorders

NAPc2 Nematode Anticoagulant Protein c2 HEMATOLOGY An inhibitor of the enzymatic complex factor VIIa/tissue factor that initiates coagulation, possibly of use in the future for managing DVT. See Deep vein thrombosis

naphthalene TOXICOLOGY A crystal formed from 2 benzene rings, used for mothballs and insecticide TOXICITY Headache, N&V, hematuria; if severe or prolonged exposure, cataracts, convulsions, hepatocellular necrosis and marked hemolysis, especially in Pts with glucose-6-phosphate dehydrogenase deficiency

naphthol TOXICOLOGY A white crystalline phenol derivative intermediate in the synthesis of multiple compounds including pharmaceuticals TOXICITY Abdominal pain, glomerulonephritis, convulsions, circulatory collapse, skin pigmentation

napkin See Sanitary napkin

napkin ring lesion Apple core lesion RADIOLOGY A pattern of GI constriction caused by mucosal erosion and stenosis with 'shouldering' of the margins, corresponding to an exophytic encircling mass in the colonic lumen, usually corresponding to a large adenocarcinoma, more in the left colon; the mass may obstruct fecal flow, inducing pencil-thin stools, and be partially mimicked by concentric amebomas in *Entamoeba histolytica* granulomas, by exuberant submucosal fibrosis in Crohn's disease, in the stenosing fibrotic stages of diverticulosis coli, in an SCC arising in a dermoid cyst of the ovary with invasion of the muscularis of the rectosigmoid colon and in small intestinal adenoCA. See Squamous cell carcinoma

naproxen Naprosyn® THERAPEUTICS An oral NSAID excreted through the kidney, breast milk, which crosses the placental barrier. See NSAIDs

naratriptan Amerge® NEUROLOGY An agent used for acute treatment of migraine. See Migraine

'narc' American slang for narcotics enforcement agent. Cf Narks

Narcan® Naloxone, see there

narcissistic personality disorder Autophilia, narcism, narcissism, self-centeredness, self-love PSYCHIATRY A condition characterized by '...*a pervasive pattern of grandiosity (in fantasy or behavior), need for admiration, and lack of empathy that begins in early adulthood...*'; ±1% of the general population, and 2-16% of the clinical population has NPD. See Autoeroticism, Autosexual

---

NARCISSISTIC PERSONALITY DISORDER > 5 of following criteria

1. Requires excessive admiration
2. Grandiose sense of self-importance; believes self to be superior
3. Preoccupied with fantasies of unlimited success, power, brilliance
4. Believes that he/she is special and should have only the best
5. Has sense of entitlement, ie deserves special favors or treatment
6. Exploits interpersonal relations, ie takes advantage of others
7. Lacks empathy and concern for others
8. Is envious of others or believes them to be envious of him/her
9. Displays arrogance

Modified from °Diagnostic & Statistical Manual of Mental Disorders, 4th ed, Am Psychiatric Asso, 1994

---

narcolepsy Sleeping sickness SLEEP DISORDERS A condition characterized by uncontrollable, recurrent, brief episodes of sleep associated with excess daytime sleepiness, cataplexy, sleep paralysis, hypnagogic hallucinations, and often disturbed REM sleep; narcolepsy affects 125,000 US–prevalence 40/10⁵; it is defined as a daytime mean sleep latency of < 5 mins, with verification of REM in 2 of 5 daytime nap periods; narcoleptics may have amnesia for the 'absences,' have fallen asleep while driving or while at work and prefer shift work as 'drowsiness' is more socially acceptable; it may also be accompanied by sleep paralysis PREVALENCE 1:600–Japan, 1:2400–US, 1:500, 000–Israel, ♂:♀ ratio 1:1, onset age 15– 35; 98-100% of narcoleptics have HLA-DR2 and/or HLA-DQw1 CLINICAL Narcolepsy tetrad; if accompanied by cataplexy, Pt feels a sense of absolute urgency for sleep in often inappropriate situations–while standing, eating, carrying on conversations, accompanied by blurred vision, diplopia, ptosis MANAGEMENT Pre-planned 'catnaps' throughout day, analeptics–ie, long-term stimulants–eg, methylphenidate, dextroamphetamine or tricyclic antidepressants that inhibit re-uptake of norepinephrine and serotonin; MAOIs may eventually cause tardive dyskinesia. See Insomnia, Narcolepsy tetrad, Sleep apnea syndrome, Sleep disorders

narcosis PSYCHIATRY Drug-induced stupor of varying depth. See Nitrogen narcosis

narcosynthesis PSYCHIATRY Psychotherapy under partial anesthesia, induced by barbiturates, first used to treat acute mental disorders in a combat setting. See Post-traumatic stress disorder

narcotic SUBSTANCE ABUSE A substance causing euphoria and analgesia at the desired abuse levels and physical dependence and CNS depression, stupor, coma and death in excess. See Opiates

NARCOTIC TYPES

NATURAL Products extracted from the poppy plant, yielding morphine and heroin, or the coca plant, yielding cocaine and crack

SEMI-SYNTHETIC Products with opiate activity, eg meperidine and methadone or synthetics, see MPTP; under the umbrella term of narcotic, alkaloids, eg LSD, mescaline, barbiturates, alcohol, marijuana, cocaine, hallucinogens and stimulants, eg antidepressants

COMPLETELY SYNTHETIC Products created by synthesis alone, eg fentanyl

Nardil® Phenelzine, see there

Narol Buspirone HCl, see there

Naropin® Ropivacaine NEUROLOGY A long-acting anesthetic for obstetrics, locoregional anesthesia, post-op pain, surgical anesthesia, acute pain management

NARP Neurogenic muscle weakness, ataxia, & retinitis pigmentosa A mitochondrial disease characterized by ↓ muscle strength and coordination, regional brain degeneration, retinal deterioration. See Mitochondrial disease

narrow QRS complex tachycardia A relatively rapid–usually > 100 beats/min cardiac rhythm with a QRS duration of ≤ 100 msec which, while rarely fatal, may be symptomatic MECHANISM Reentry causes > 90% of NQCRs; other mechanisms include triggered activity, and automaticity TREATMENT IV adenosine

narrow therapeutic index drug CLINICAL PHARMACOLOGY Any pharmaceutical which has a < 2-fold difference between the minimum toxic concentration and minimum effective concentration in blood. See Therapeutic index

NAS 1. Neonatal abstinence syndrome. See Neonatal withdrawal syndrome 2. No abnormalities seen

NASA syndrome Not a surgical abdomen SURGERY A popular term for an acute abdomen examined by a surgeon, and deemed not to require surgical intervention and is thus 'punted back' to the nonsurgical attending physician for a diagnostic workup

nasal airway resistance ENT The state of the nasal passages during breathing, which reflects the degree of nasal obstruction EVALUATION Simultaneous measurement of transnasal pressure and airway resistance

nasal cannula CRITICAL CARE An O₂ delivery device loosely attached to the head with 2 prongs inserted in the nose; the FiO₂ delivered by an NC is 24–35%

nasal compliance ENT The change in nasal volume per unit pressure, which may be determined by the transition from nasal to oronasal breathing; NC is measured by acoustic rhinometry, which demonstrates that anterior NC is > posterior NC and vasoconstrictors ↓ midnasal compliance. See Acoustic rhinometry. Cf Pulmonary compliance

nasal congestion ENT Difficulty in nasal breathing, due to an ↑ vascular thickness of nasal mucosa . See Nasal stuffiness

nasal cycle RESPIRATORY PHYSIOLOGY Alternating congestion and decongestion of the nasal airway that occurs in 70% of adults, controlled by the autonomic nervous system, and which may be affected by circadian changes in hormone levels, temperature, humidity, posture, and emotion; in the face of a unilateral fixed obstruction, the congestion phase of the side opposite the obstruction may be interpreted as an abnormality of the normal side or 'paradoxic nasal obstruction'

nasal decongestant An oral or topically sprayed agent that ↓ swollen nasal mucosa, and facilitates breathing; NDs often cause a rebound effect, in which the Sx worsen when the ND is discontinued, due to tissue dependence on the drug

nasal endoscopy Rhinolaryngoscopy, rhinopharyngoscopy, rhinoscopy The use of a flexible fiberoptic endoscope to evaluate upper airways–nasal passages, nasopharynx, oropharynx, and larynx, a procedure usually carried out by ENTs or allergists INDICATIONS Idiopathic upper airway disease that is chronic, recurring, or persists despite adequate therapy; other indications include epiglottitis, laryngeal trauma, and evaluation of stridor which in children may be due to foreign objects, and in adults to tumors FINDINGS Nose–nasal polyps, vascular defects, inflammation; upper pharynx–ulcers, lymphoid hyperplasia, cysts; lower pharynx–lymphoid hyperplasia, cysts, vocal cord trauma COMPLICATIONS NE is a low-risk procedure; epistaxis, bronchospasm, laryngospasm, cardiac arrhythmias–due to vasovagal stimulation, may rarely occur

nasal flaring INTERNAL MEDICINE An ↑ in nostril size with breathing, a classic sign of severe asthma. See Asthma

nasal fossa See Nasal cavity

nasal obstruction ENT Difficulty in nasal breathing, due to any obstacle to breathing–eg, a thickened mucosa, polyps, a foreign body, deviated septum. See Nasal stuffiness

nasal packing ENT The filling of the nasal cavities with adaptic gauze impregnated with polysporin ointment, used in treating nasal fractures, reconstructive surgery, after septorhinoplasty and in posterior nosebleeds

nasal polyp ENT Any of the small, sac-like growths consisting of inflamed nasal mucosa, which can arise in clusters or individually, near the ethmoid sinuses, expanding into the open areas of the nasal cavity, possibly obstructing the airway and blocking drainage from the sinuses; sinusitis may arise in fluid accumulating in blocked sinuses CLINICAL Obstruction, mandatory mouth breathing due to chronic nasal obstruction, runny nose ETIOLOGY Asthma, allergic rhinitis/hay fever, vasomotor rhinitis, certain drugs, chronic sinusitis, cystic fibrosis

nasal stuffiness ENT A sensation of difficulty in nasal breathing, ± associated with ↑ nasal airway resistance. See Nasal congestion, Nasal obstruction

nasal tumor Juvenile angiofibroma, see there

nasal turbinate Nasal concha, see there

Nasalcrom® Cromolyn ENT An OTC agent for managing allergic rhinitis. See Allergic rhinitis

NASCET CARDIOLOGY A clinical trial–North American Symptomatic Carotid Endarterectomy Trial designed to compare the outcomes of surgery plus medical therapy with medical therapy alone in Pts at risk for TIAs and CVAs. See Stroke, Transient ischemic attack

NASCIS NEUROLOGY A multicenter trial–National Acute Spinal Cord Injury Study, which evaluated strategies for reducing the sequeale of SCI

nasopharyngeal carcinoma Nasopharyngeal cancer A rare malignancy, which is endemic to regions of southern China and Southeast Asia; persons with serologic markers for EBV–IgA antibodies against EBV capsid antigen and/or neutralizing antibodies against EBV-specific DNase have a 3-fold cumulative risk of nasopharyngeal cancer if one marker was positive and a 35-fold risk if both markers are positive. See Epstein-Barr virus

NASOPHARYNGEAL CANCER

STAGE I Lesion confined to nasopharynx

STAGE II

A Lesion extends to oropharynx and/or nasal fossa

B Lesion extends to nearby lymph nodes or to parapharyngeal region

STAGE III Lesion spread to lymph nodes on both sides of the neck or to nearby bones or sinuses

STAGE IV

A Spread beyond nasopharynx to other areas in head, and possibly to nearby lymph nodes

B Spread beyond nasopharynx to other areas in head and to lymph nodes above clavicle or are ≥ 6 cm

C Lesion has metastasized

nasopharyngitis INFECTIOUS DISEASE Acute nasopharyngeal inflammation, often with keratoconjunctivitis; both occur in common cold and are attributed to viruses. see Common cold

natal sex A baby's birth sex

natal tooth A deciduous tooth present in ± 1:2000 neonates, often located near the central mandibular incisors, with minimal gingival attachment; NT may presage early eruption of remaining deciduous teeth CLINICAL Loose NT annoy nursing infant; implanted NT annoy nursing mother COMPLICATION Amputation of tongue tip–Riga-Fede disease by the NT during delivery

National Bill of patients' rights MANAGED CARE A series of provisions that ensure that all Pts have certain rights when they are sick in a health care facility. Cf Principles for Consumer Protection, Putting Patients First

National Boards Examination A standardized exam administered in the US and Canada in lieu of state medical examinations, to assess medical knowlege of a candidate physician applying for a state's medical license. See ECFMG, FLEX

National Cancer Institute An institute which, like the others in the National Institutes of Health, is under the US Public Health Service; the NCI has a focused interest on cancer research, conducts its own research, and guides peer-review for funding of mainstream experimental protocols for treating CA. See Cancer screening. Cf International Association of Cancer Victors & Friends, National Health Federation, Office of Alternative Medicine, Unproven methods for cancer management

National Center for Complementary & Alternative Medicine An upgrade for the Office of Alternative Medicine, which operates under the NIH

national certification LAB MEDICINE A voluntary form of regulation that affirms that a person has the knowledge and skill to perform essential tasks in a given field, in the lab or in nursing; NC is granted by nongovernmental agencies or associations with predetermined qualifications. See Certification; Cf Registration

National Committee for Quality Assurance MEDICAL PRACTICE A private, not-for-profit organization which has become the leading accreditor of managed care plans; in site visits, NCQA reviewers evaluate a managed care plan in terms of quality management, physicians' credentials, members' rights and responsibilities, preventive health services, utilization management, medical records. See HEDIS, Managed care. Cf JCAHO

National Council Against Health Fraud An anti-quackery group. See Health fraud, Quackery

NATL COUNCIL AGAINST HEALTH FRAUD–MISSION

CONDUCT STUDIES on the claims made for health care products and services

EDUCATE the public, professionals, legislators, business people, organizations, and agencies about untruths and deceptions

PROVIDE A COMMUNICATION CENTER for individuals and organizations concerned about health misinformation, fraud, and quackery

SUPPORT SOUND CONSUMER HEALTH LAWS and oppose legislation that undermines consumer rights

ENCOURAGE LEGAL ACTIONS AGAINST LAW VIOLATORS

National Death Index EPIDEMIOLOGY A central mortality database compiled from data submitted by state vital statistics offices to the Natl Center for Health Statistics. See Mortality rate

National Domestic Violence Hotline PUBLIC HEALTH A tollfree hotline–1.800.799.SAFE which offers 24/7/365 counseling to US victims of DV–♂ & ♀; it also connects callers with related community services and shelters. See Violence Against Women Act

National Drug Code HEALTH INSURANCE A system for identifying drugs. See Branding

National Formulary One of 2–the other is the US Pharmacopeia–official compendia recognized by Federal Pure Food & Drug Act of 1906; the NF's approved therapeutics are described and defined with respect to source, chemistry, physical properties, tests for identitification and purity, dosage range, and class of use

national health care MANAGED CARE The financing and delivery of health care by a government. See Canadian plan, Clinton Plan, Managed health care

National Health Federation FRINGE MEDICINE An organization that exerts political pressure to secure 'health freedom' and 'freedom of medical choice' on behalf of alternative medicine practitioners, their families, and 'health food' consumers. See Alternative medicine; Cf National Council Against Health Care Fraud, Quackery, Unproven methods for cancer management

National Health Service BRITISH MEDICINE The 50+-yr-old UK government agency that controls the British form of socialized medicine. See Fund holding, Total purchasing

National Heart, Lung, & Blood Institute GOVERNMENT An NIH agency charged with conducting and coordinating research in its focused areas of interest. See NIH

National Information Infrastructure INFORMATICS A type of Internet from the NIST, which includes fiberoptics, videography, telecommunications, cable, satellites; healthcare emphases are on computer-based Pt records, secure data access, telemedicine

National Institute for Occupational Safety & Health GOVERNMENT The research arm and 'scientific conscience' of the US federal health & safety programs relating to the workforce. See NIOSH

NATL INSTITUTE FOR OCCUPATIONAL SAFETY & HEALTH–RESPONSIBILITIES

DEVELOPMENT OF 'CRITERIA DOCUMENTS' that recommend exposure limits to hazardous substances

TRAINING AND EDUCATION of occupational health professionals

DEVELOPMENT OF EXPOSURE MEASUREMENT AND SAMPLING METHODS and

PERFORMANCE OF INDUSTRY-WIDE STUDIES to evaluate the health effects of low level long-term exposure to potentially hazardous substances or processes

Exposure to environmental toxins in the workplace is usually measured in parts–1 to 5000 or more, depending upon the substance per million–ppm of exposure/8 hrs–Publications Div, NIOSH, 4676 Columbia Pkwy, Cincinnati, Ohio 45226

National Institutes of Health NIH, see there

National Institute of Standards & Technology Formerly, National Bureau of Standards GOVERNMENT A branch of the Commerce Dept's technology administration which maintains 1° reference standards and develops reference methods & materials. See Advanced Technology Program

National Library of Medicine MEDICAL INFORMATICS The world's leading repository of medical information, part of the NIH, located on its campus in Bethesda, MD. See GRATEFUL MED, *Index Medicus*, Literature search, MEDLINE

National Polyp Study GI DISEASE A 10-year study that found that colonoscopy was more effective than barium enema in detecting precancerous adenomas, reported in June 2000, NEJM.

National Practitioner Data Bank A database established by the Congress to facilitate professional peer review and restrict incompetent physicians' and dentists' ability to move from state to state, and elude discovery of previous substandard performance or unprofessional conduct

National Provider Identifier MEDICARE A unique 8 character ID assigned by the National Provider System to providers/suppliers who bill for services or goods. See Medicare Identification Number, NSC, OSCAR, UPIN

National Surgical Adjuvant Breast Project See NSABP

National Toxicology Program ENVIRONMENT A program that conducts toxicologic tests on substances frequently found at the EPA's National Priorities List sites, which have the greatest potential for human exposure

nationalization Expropriation GOVERNMENT The process whereby a government may transfer ownership of privately-held units to itself; the government then assumes either partial or complete responsibility for managing and controlling the expropriated enterprise

Native™ tissue harmonic imaging IMAGING A format for high-frequency, high-resolution ultrasonographic imaging without contrast material, using Microson™ technology. See Ultrasonography

natural antibody Normal antibody An antibody in the circulation, without prior exposure to its cognate antigen

natural carcinogen A substance normally present in foods which is carcinogenic when tested by mutagenic assays in rodents or bacteria. See Ames' test, Toxicity testing

natural childbirth ALTERNATIVE OBSTETRICS A vaginal delivery in which the mother is more actively involved in the parturitional mechanics–than in an 'unnatural' birth. See Alternative gynecology, Bonding, Breast milk, Doula, Lamaze method, Naturopathic obstetrics

natural death '*A death…caused solely by disease and/or the aging process…*'. Cf Unnatural death

natural disaster A disaster caused by forces of nature TYPES Climatologic–eg, hurricanes, tornados; geologic–eg, earthquake, volcanos. See Climatologic disaster, Geologic disaster. Cf Man-made disaster, Technological disaster

natural family planning Biological birth control Any FP that does not rely on artificial agents–eg, OCs, 'morning-after' pill, spermicidal foam, RU-486 or devices–eg, condoms, diaphragms, IUDs to prevent conception METHODS Rhythm–calendar method, coitus interruptus, prolonged breast feeding. See Breast feeding, Coitus interruptus, Contraception, Pearl index, Periodic abstinence, Rhythm method

natural father Biological father, see there

natural food movement Health food movement, see there

natural gas A combustible gas from underground petroleum deposits used for cooking, heating COMPONENTS Short-chain hydrocarbons–eg, methane, ethane, propane, butane, CO_2, N_2, H_2S TOXICITY At high ambient levels, volatile hydrocarbons induce hypoxia by replacing alveolar gas, crossing the alveolar-capillary barrier, causing CNS depression. Cf Flatulence

natural healing Alternative healing ALTERNATIVE HEALTH Any healing technique that may be rooted in supernaturalist methods. See Absent healing, Acupuncture, Acupressure, Alexander technique, Applied kinesiology, Ayurvedic medicine, Bioenergetics, Cayce therapies, Cranial osteopathy, Hydrotherapy, Iridology, Jin Shin Acutouch, Jin shin jyutsu, Jungian psychology, Laying on of hands, Lomilomi massage, Macrobiotics, Nature cure, Naturopathy, Neo-Shamanism, Orgonomy, Past life/lives therapy, Polarity therapy, Psychic healing, Radiesthesia, Radionics, Rebirthing therapy, Reiki, Rolfing®, Shamanism, T'ai chi ch'uan, Therapeutic touch, Touch for health, Yoga, Vibrational medicine

natural history of disease EPIDEMIOLOGY The timeline of a morbid condition from onset–inception to resolution; the course of a particular disease if it is not treated or manipulated in any way

natural medicine 1. Naturopathy, see there 2. Any form of health care–eg, diet, exercise, herbs, hydrotherapy, which allegedly enhances the body's natural healing powers

natural mother Biological mother, see there

natural pacemaker CARDIAC PACING That cell or a group of cells within the heart which initiates each contraction. Normally, the S-A node performs the function; but in arrhythmias, cells almost anywhere in the heart can assume the role of the dominant pacemaker

natural parent Biological parent, see there

natural psychotherapy ALTERNATIVE PSYCHIATRY Client-oriented psychotherapy that accepts a person's flaws/foibles, and helps maximize emotional/spiritual potential. See Blockade point, Humanistic psychology, Reflective communication

natural short sleeper SLEEP DISORDERS A person who functions normally with less than what is considered to be a normal-length sleep or rest for his/her age group, without known adverse effects. Cf Insominia, Sleep disorder

naturally fluoridated water system PUBLIC HEALTH A public water system that produces water that has fluoride from natural sources at levels that provide maximum dental benefits. See Public water

nature cure Natural care Any method of self-healing, based on fasting, diet, rest, hydrotherapy. See Fasting, Hydrotherapy

naturopath A person who practices naturopathy, a drugless system of therapy using physical forces–eg, heat, water, light, air and massage

naturopathic obstetrics ALTERNATIVE OBSTETRICS A 'cross-platform' speciality that emphasizes 'natural' prenatal care, 'natural' childbirth, non-use of anesthesia, noninvasive interventions in birthing, and postnatal care, in the form of massage and herbal infusions for mother and infant. See Alternative gynecology, Bonding, Breast milk, Doula, Lamaze method, Natural childbirth

naturopathy Natural medicine, natural therapeutics, naturopathic medicine, naturopathic therapy, naturology ALTERNATIVE MEDICINE A healing philosophy that attributes disease to a violation of natural laws, and uses the forces of nature as therapeutic modalities. See Alternative medicine; Homeopathy

NATUROPATHY'S SIX PRINCIPLES
FIRST, DO NO HARM–*primum non nocere*
PREVENT RATHER THAN CURE
NATURE HAS INNATE HEALING POWERS–*vis medicatrix naturae*
HOLISTIC APPROACH–the 'whole person' is treated
TREAT CAUSE OF DISEASE, not symptoms–*tolle causum*
TEACH PREVENTION

nausea GI DISEASE The urge to vomit CAUSES Systemic illness–eg, influenza, medications–especially chemotherapy, pain, inner ear disease

nausea gravidarum Morning sickness, see there

navel MEDTALK Umbilicus, see there VOX POPULI Belly-button

Navigator CARDIOLOGY An external hand-held locating instrument that emits low-level, high-frequency magnetic fields, detected by the sensor in the tip of a MAPCath catheter or guide wire

NB Abbreviation for 1. Needle biopsy 2. Nerve block 3. Newborn 4.. No bowel movement 5. Normal bowel movement 6. Nota bene, Latin, please note

NBE Non-bacterial endocarditis, see there

NBG Code CARDIAC PACING Five letter code for identifying pulse generators and pacing modes which supersedes the ICHD code. See Pacemaker

NBS Newborn screening, see there

NCCP Noncardiac chest pain, see there

NCEP National Cholesterol Education Program

NCI National Cancer Institute, see there

NCQA National Committee on Quality Assurance, see there

ND Symbol for 1. Natural death 2. Neonatal death 3. Neurologic deficit 4. New drug 5. Node dissection 6. No disease 7. Nondetectable 8. Nondisabling 9. Normal delivery 10. Normal development–pediatrics 11. Normal distribution 12. Not detectable 13. Not determined 14. Not diagnosed

Nd-YAG laser Neodymium-yttrium aluminum garnet laser SURGERY A photocoagulation unit used to control acute and chronic GI hemorrhage–eg, endoscopically guided Nd-YAG laser may be used to control esophageal varices, vascular ectasias, angiodysplasia, radiation-induced telangiectasia, watermelon stomach, telangiectasia of Osler-Weber-Rendu, BPH, palliation of CA and benign and malignant obstructive biliary tract lesions. See Lasers

NDI National Death Index, see there

near extremis Near death. See Circling the drain

'near miss' sudden infant death syndrome A prolonged, usually nocturnal apneic period in children, in whom the non-fatal outcome of apnea is attributed to continuous monitoring; 'near-miss' SIDS children may have enlarged adenoids or nasopharyngitis which responds to adenoidectomy. See SIDS

near poor Working poor SOCIAL MEDICINE A segment of the US population with earnings only enough for daily needs, who are not qualified for US government assistance programs; NP seldom have medical insurance and when ill, are a major burden to the health care system. See Underinsured. Cf Engel's phenomenon, 'Fourth World', Homelessness

nearsightedness Myopia, see there

Nebcin® Tobramycin, see there

nebulized lidocaine PULMONOLOGY A formulation of lidocaine, which ↓ steroid dependence in severe asthma

nebulizer MEDTALK A device that converts aqueous solutions containing medications into fine–±10 μm in diameter droplets, which is small enough to pass directly to the alveoli

NebuPent® Pentamidine, see there

NEC Symbol for 1. Necrotizing enterocolitis, see there 2. Nonesterified cholesterol

Necator americanus MEDTALK Hookworm, see there

necessary See Medically necessary

necessary fallibility DECISION-MAKING A term related to the chaos theory that explains the unpredictability of events results in nonlinear, non-coupled, nondeterministic systems. See Inevitable human fallibility

necessary stopping distance PUBLIC HEALTH The distance required to bring a vehicle with maximum-rated load at maximum tramming speed to a stop, plus the distance that the vehicle travels during the operator's 2-second reaction time. See MVA

necessity See Medical necessity

neck MEDTALK Any constricted part of an elongated structure–eg, neck of an organ–eg, uterine cervix. See Bottleneck, Bull neck, Frog neck, Genetic bottleneck, Gooseneck, Microscopist's neck, Swan neck, Webbed neck, Wryneck

neck-cracker neuropathy NEUROLOGY A clinical condition that may occur in those who habitually crack their cervical spine, resulting in a spinal accessory nerve neuropathy, which may lead to atrophy of the trapezius, slowed conduction of the spinal accessory nerve, and winging of the scapula

neck dissection SURGERY The excision of lymph nodes and other tissues grossly (lal) (macroscopically) involved by CA in the neck for the staging of cancer. See Commando operation, Radical neck dissection

neck examination PHYSICAL EXAM Evaluation of the neck, by observation, palpation, auscultation

neck face syndrome A transient complex characterized by oropharyngeal spasms, dysarthria, tachycardia, HTN, after beginning chlorpromazine therapy. See Chlorpromazine

neck hold FORENSIC MEDICINE A restraint used to subdue overactive, unruly, violent or inebriated subjects to prevent them from harming themselves and others. See Restraint

neck tongue syndrome A condition characterized by sharp pain and tingling of the upper neck and/or occiput on sudden neck rotation, with numbness of the ipsilateral half of the tongue; the NTS is attributed to stretching the C2 ventral ramus, which contains proprioceptive fibers from the lingual nerve to the hypoglossal nerve and 2nd cervical root

necrobiosis lipoidica DERMATOLOGY An inflammatory condition seen in 50-80% of DM, most often on the legs; in 10%, NL precedes onset of DM. See Non-insulin-dependent diabetes mellitus

necrolytic migratory erythema DERMATOLOGY A plaque dermatopathy seen on the lower trunk, buttocks, perineum, and thighs of Pts with a glucagonoma of the pancreas CLINICAL Erythematous, scaling, ± bullous, erosive lesions accompanied by weight loss, anemia, stomatitis LAB ↑ Glucagon. See Glucagonoma

necrophilia PSYCHIATRY A paraphilia in which sexuoerotism requires a corpse, which may involve engaging in sexual activity with same; the fantasies, sexual urges, or behaviors cause clinically significant distress or impairment in social, occupational, or other important areas of functioning; getting a stiffie with a stiff

necropsy Postmortem examination. See Autopsy

necrosectomy SURGERY Excision of necrotic tissue; generally, debridement, see there

necrosis PATHOLOGY Cell or tissue death due to disease, trauma, hypoxia, radiation, acute infection, etc; the constellation of changes that accompany and follow irreversible cell injury in living organisms. See Acute tubular necrosis, Aseptic necrosis, Bridging necrosis, Coagulation necrosis, Colliquative necrosis, Contraction band necrosis, Cystic medial necrosis, Fat necrosis, Fibrinoid necrosis, Liquefaction necrosis, Lymph node necrosis, Osteonecrosis, Osteoradionecrosis, Papillary necrosis, Piecemeal necrosis, PORN

necrospermia UROLOGY The absence of live sperm in seminal fluid. See Sterility

necrotic *adjective* Dead, referring to death of cells and tissues. See Necrosis

necrotic tumor ONCOLOGY A tumor with one or more areas of necrosis, often related to tumor growth beyond the reach of the tumor's vascular supply. See Metastasis

necrotizing alveolar capillaritis Pulmonary capillaritis A histologic finding of uncertain pathogenesis characterized by fibrinoid necrosis of alveolar septae, which is associated with collagen vascular disease and pulmonary hemorrhage

necrotizing arachnidism TOXICOLOGY Insect bites causing necrosis requiring surgical debridement–eg, bites of black widow and brown recluse spiders. See Black widow spider, Brown recluse spider

necrotizing colitis Pseudomembranous colitis, see there

necrotizing enterocolitis NEONATOLOGY A disease of premature infants, affecting the terminal ileum 3-10 days after birth, representing 2% of neonatal ICU admissions and 10% of admissions of premature infant, or low birth weight neonates, causing significant M&M CLINICAL Banal to fulminant with abdominal distension, vomiting, hematochezia, intestinal gangrene, perforation, sepsis, shock; survival 80%; NEC is rare in breast-fed children, who may be protected by IgA in maternal milk; oral IgA-IgG solution in LBW infants may be protective. Cf Pigbel

necrotizing fasciitis Necrotizing subcutaneous infection INFECTIOUS DISEASE A rapidly progressive bacterial infection that spreads along fascial planes which, absent effective therapy–eg, debridement, results in skin breakdown with bleb and bulla formation, small vessel thrombosis and 2° necrosis, leading to subcutaneous anesthesia AGENTS Streptococci, gram-negative and mixed bacteria

necrotizing lymphadenitis Kikuchi's disease, see there

necrotizing pneumonia PULMONOLOGY 1. Aspiration pneumonia, see there 2. Pneumonia with significant tissue necrosis, often seen in a background of aspiration or severe acute bacterial pneumonia

necrotizing sialometaplasia ORAL PATHOLOGY A benign self-limited reactive salivary gland inflammation, which may clinically and histologically mimic malignancy–eg, mucoepidermoid carcinoma; most occur in minor salivary glands–SG; it may occur in SGs of the upper aerodigestive tract, or larynx DIFFDX Mucoepidermoid carcinoma, SCC

NED No evidence of disease

need VOX POPULI The requirement for an activity, function or thing. See Basic need, Dependency need, Self-actualization need, Urgent need

needle MEDTALK An elongated device with a narrow central bore for injection and withdrawing fluids. See Biopsy needle, Butterfly needle, Colposuspension needle, Extended reach needle, Safe Step™ blood-collection needle

needle aspiration cytology A diagnostic preparation of cells–eg, smears and/or a 'cell block', see there, obtained from a clinically or radiologically identified mass, using a 'skinny' needle to spread the material on a glass

needle biopsy Fine-needle aspiration A diagnostic preparation similar to aspiration cytology, but uses a larger bore–eg, 19-gauge, needle to obtain architecturally intact tissue, yielding a higher diagnostic success rate than with cytology alone. Cf Needle aspiration cytology

needle exchange program Syringe exchange program PUBLIC HEALTH Any program intended to slow the spread of AIDS among IV drug users, in which a governmental or charitable agency exchanges sterile needles for dirty, potentially HIV-contaminated needles used by IVDAs when 'shooting' heroin or other substances. See Intravenous drug use, Safe injection room

needle-free blood testing LAB MEDICINE The use of a low-power–800 mJ, 300 μsec laser–eg, erbium-YAG to vaporize a small hole through the epidermis, allowing obtention of a minute amount of capillary blood for performing simple lab tests–eg, serum glucose, sodium etc

needle scope SURGERY A popular term for a thin-bore–< 3 mm in diameter laparoscope, used to stage and/or treat conditions amenable to laparoscopy. See Laparoscopic surgery

needlestick injury INFECTION CONTROL The unintentional exposure of a health care worker to a needle used in direct Pt management. See Hospital-acquired penetration contacts, Sharps

needs model HEALTH POLICY A model for assessing future needs for a particular product or supplier of a service–eg, surgeons–based on projected needs for the product or supplier in a hypothetically 'perfect' world. Cf Demand model, Extrapolation model, Supply model

nefazodone Serzone® PSYCHIATRY An anxiolytic serotonin antagonist reuptake inhibitor ADVERSE EFFECTS Sedation, vertigo, weakness

negative *noun* LAB MEDICINE A lab result that is normal; failure to show a positive result for the specific disease or condition for which the test is being done. See False negative, Hemoccult negative, True negative VOX POPULI Absence or nonexistence of a thing

negative acute phase reactant Any of the molecules produced in ↓ amounts during the acute phase reaction–eg, albumin, alpha-fetoprotein, α_2-HS glycoprotein, transferrin, transthyretin/negative acute phase protein. See Acute phase protein

negative autopsy PATHOLOGY An autopsy for which the gross and histologic findings do not provide an adequate explanation for the cause of death. See Autopsy

negative axillary lymph node ONCOLOGY A cancer-free axillary lymph node–often referring to Pts with breast CA. See Axillary tail

negative emotion Any adverse emotion–eg, anger, envy, cynicism, sarcasm, etc. Cf Positive emotion

negative end-expiratory pressure RESPIRATORY CARE A format for mechanical ventilation in which negative pressure–ie, suction is applied during ventilation. Cf PEEP

negative inotrope Negative inotropic agent CARDIOLOGY Any agent–eg, β–blockers–eg, metoprolol or CCBs–eg, diltiazem, verapamil, used to manage CHF due to pure diastolic dysfunction, as they ↓ cardiac contractility

negative most-significant-other PSYCHOLOGY The most-significant-other–eg, spouse, parent, child who has a negative influence on a person through a deprecating attitude, negative comments, and general kvetching; a person's link to a NMSO is often a relationship 'required' by blood, marital affiliation, or employment. See Most-significant-other. Cf Positive most-significant-other

negative pressure INFECTION CONTROL Referring to a ventilation system designed so that air flows from the corridors into an isolation room, ensuring that contaminated air doesn't pass from the isolation room to other parts of the facility

negative symptom Deficit symptom PSYCHIATRY Any Sx involving loss of normal mental function, seen in schizophrenia, depression, and other mental disorders EXAMPLES Blunting of or ↓ range of affect, loss of will, pleasure, fluency, and content of speech, range of emotion, sense of purpose, social drives, poverty of speech, loss of interests. See Positive and Negative Symptom Scale, Schizophrenia. Cf Positive symptom

negative triad SURGERY A popular term for any breast lump judged to be 1. clinically negative by the surgeon; 2. cytologically negative by fine-needle aspiration; 3. negative by mammography

neglect NEUROLOGY Neglect syndrome The inability to perform a particular motor activity, often residual to a CVA. See Neglect patient PATIENT CARE The conscious ignoring–by a physician or other provider–of a clinical finding that might, in another setting and/or in another Pt, trigger further evaluation or therapy. See Benign neglect. Cf Negligence PSYCHOLOGY '...*the failure of a caretaker to provide basic shelter, supervision, medical care, or support.*', neglect of children or elders, a form of maltreatment, which may be linked to poverty. See Child abuse, Elder abuse, Elder neglect, Self-neglect, Willful patient neglect

neglect patient NEUROLOGY A Pt with left-sided paralysis caused by damage to the right brain, where sensory signals for touch, joint position, proprioception, visceral sense, and pain are processed; because the damage is unilateral, there is no input from the left brain, and the Pt may believe that he can move the left side of body; when accompanied by left-sided brain damage, neglect does not occur

negligence MEDICAL MALPRACTICE The failure or alleged failure on the part of a physician or other health care provider to exercise ordinary, reasonable, usual, or expected care, prudence, or skill–that would usually and customarily be exercised by other reputable physicians treating similar Pts–in performing a legally recognized duty, resulting in forseeable harm, injury or loss to another; negligence may be an act of omission–ie, unintentional, or commission–ie, intentional, characterized by inattention, recklessness, inadvertence, thoughtlessness, or wantonness. See Adverse event, Comparative negligence, Contributory negligence, Gross negligence, Malpractice, Wanton negligence, Willful negligence. Cf Recklessness

NEGLIGENCE, REQUIRED ELEMENTS

DUTY A recognized relationship between Pt and physician

BREACH Failure of a medical practitioner to practice in accordance with standard of care

PROXIMATE CAUSE The plaintiff must show that injury is reasonably connected to physician's action

DAMAGES Plaintiff must show that alleged loss or damage has a quantifiable value such that a monetary payment can be made APLM 1997; 121:252

negligent nondisclosure MALPRACTICE, ETHICS A situation in which a physician discovers that the diagnosis on which the Pt's management strategy is incorrect, but chooses not to tell the Pt or alter the management

negligent referral MEDICAL MALPRACTICE The referral by a health care provider of a Pt to another physician or provider for management outside of the referring physician's expertise; the referral is deemed negligent when the referring physician knew or should have known that the physician to whom the case was referred would manage the Pt in a substandard manner. See Referral

negotiated payment schedule Negotiated fee schedule, see there

negotiated safety STD A term for a stance in which HIV-negative sexual–in particular homosexual–partners abandon the use of condom

Neisseria gonorrhoeae Gonococcus STD The gram-positive coccus that causes gonorrhea, the most common STD in the US

Neisseria meningitidis Meningococcus INFECTIOUS DISEASE A gram-positive coccus which is part of the normal nasopharyngeal flora; it is one of the most common causes of meningitis in the US

nelfinavir Viracept® AIDS A potent HIV protease inhibitor, used with reverse transcriptase inhibitors–eg, Retrovir–AZT, zidovudine and Epivir–3TC, lamivudine; it acts at late stages in the viral life cycle SIDE EFFECTS Diarrhea–20%, ↑ LFTs, lipodystrophy. See AIDS, Antiretroviral, HIV, Protease inhibitors. Cf Reverse transcriptase inhibitor

nemaline myopathy *nemaline*, Greek, rod-shaped A benign AD muscular dystrophy affecting 'floppy infants', characterized by non-progressive muscular weakness, ↓ deep tendon reflexes and hypotonicity, causing skeletal abnormalities, a typical facies–oval face, micrognathia, malocclusion, and a high arched palate, kyphoscoliosis, dislocation of hips and pes cavus; NM is compatible with a normal life span; 'nemaline' refers to the ultrastructural finding of rod-like Z-band material in both type I and type II myocytes. Cf Central core myopathy, Floppy infant syndrome

nematode Roundworm, see there

Nembutal® Pentobarbital, see there

neo- Prefix, new

neoadjuvant chemotherapy Induction chemotherapy, neoadjuvant therapy, preemptive chemotherapy, primary chemotherapy ONCOLOGY 1. The use of chemotherapy on solid–ie, non-hematopoietic neoplasms–eg, osteosarcomas, anal cancer, head&neck cancer; NC is used in Pts with clinically resectable but locally advanced disease that is not amenable to complete eradication surgery or RT; agents used in NC include cisplatin, 5-FU, vinblastine, and the M-VAC 'cocktail.' See Neoadjuvant therapy. Cf Induction therapy

neoadjuvant therapy 1. Neoadjuvant chemotherapy, see there 2. Treatment given before a primary cancer treatment–eg, chemotherapy, RT, or hormone therapy

Neocontrol® UROGYNECOLOGY A nonsurgical treatment for urinary incontinence–stress, urge, or mixed–related to pelvic-floor muscle weakness in ♀, based on ExMI™–extracorporeal magnetic innervation technology, directing magnetic pulses toward pelvic floor muscles, evoking contractions with each pulse, analogous to Kegel exercises. See Urinary incontinence

neologism NEUROLOGY/PSYCHIATRY A word created by a Pt with a mental disorder or dementia, which includes new usages for standard words and ad hoc substitutes for names forgotten by a Pt; neologisms are created by Pts with schizophrenia and organic mental disorders

neomycin INFECTIOUS DISEASE A broad-spectrum aminoglycoside antibiotic obtained from *Streptomyces fradiae*, which is generally effective against gram-negative rods and some gram-positive cocci

neonatal *adjective* Referring to the newborn period, by convention, the first 4 wks after birth

neonatal alloimmune thrombocytopenia Fetal alloimmune thrombocytopenia, see there

Neonatal Behavioral Assessment Scale Brazelton An instrument that measures various infant characteristics–eg, temperament, social behavior, orienting responses to stimuli, responses to disturbing stimuli, state of arousal, and motor skills; unlike earlier neonatal testing devices, the 'Brazelton' recognizes the uniqueness of each child

neonatal conjunctivitis Ophthalmia neonatorum Inflammation of the newborn conjunctiva acquired during passage through the birth canal, usually bacterial–eg, gonococcal or chlamydial, but also herpetic infection PROPHYLAXIS Formerly, silver nitrate, now, antibiotic drops

neonatal hemochromatosis METABOLIC DISEASE A fulminant disease characterized by massive hepatic iron loading and perinatal liver failure, accompanied by myocardial and pancreatic acinar iron overload MANAGEMENT Liver transplant–rarely successful. Cf Hemochromatosis, Juvenile hemochromatosis

neonatal 'hepatitis' A generic term for diseases of the newborn hepatic parenchyma, commonly associated with ↑ conjugated hyperbilirubinemia DIAGNOSIS Requires ≥ 3 of following: fatty changes, cholestasis, bile duct proliferation, fibrosis, pseudoacini, cirrhosis ETIOLOGY Infection–syphilis, listeriosis, HBV, rubella, CMV, echovirus, adenovirus, toxoplasmosis, metabolic disease–α1-antitrypsin deficiency, cystic fibrosis, Wilson's disease, galactosemia, fructosuria, tyrosinemia, mechanical–choledochal cysts, intrahepatic ductal atresia hypoplasia, familial intrahepatic cholestasis, HDN, etc. See Hemolytic disease of the newborn. Cf Giant cell hepatitis

Neonatal Infant Pain Scale NEONATOLOGY An assessment tool for measuring pain in preterm and full-term neonates, used to monitor a neonate before, during and after a painful procedure–eg, a venipuncture; NIPS assesses facial expression, strength of cry, breathing patterns, activity of arms and legs, state of arousal

neonatal intensive care unit NICU pronounced, Nick-you A ward in a 3° care center that provides intensive medical care for newborns REQUIREMENTS Trained medical personnel at all levels, 24/7 access to appropriate specialists, monitoring devices, alarm systems for continuous assessment of vital functions, equipment for resuscitation and respiratory therapy, drugs and full lab coverage. See Low birth weight

neonatal macrosomia A big baby; birthweight > 4000 g–8.8 lbs

neonatal mortality rate EPIDEMIOLOGY A ratio expressing the number of deaths in infants from birth up to, but not including 28 days of age divided by the number of live births reported during the same time period, usually per 1,000 live births

neonatal panel/profile PEDIATRICS A battery of low-cost lab tests performed on the serum and urine of newborns to identify a treatable disease or state of physiologic decompensation; the typical NP/P measures albumin, total bilirubin, blood group–ABO and Rh, BUN, calcium, electrolytes–Na⁺, K⁺, Cl⁻, CO_2, glucose. Cf Neonatal screen

neonatal screen PEDIATRICS A low-cost test performed on newborn infants to identify potentially treatable diseases; most NSs can be performed on minimal amounts of blood or urine. Cf Neonatal panel/profile

neonatal sepsis Sepsis of newborn, septicemia of newborn PEDIATRICS A severe systemic infection of the newborn caused primarily by group B streptococcus, a bacterium found in the GI and GU tracts, which causes ±³/₄ of neonatal sepsis RISK FACTORS Mom with group B strep infection with transplacental dissemination or by direct innoculation as the baby slides through the birth canal, PROM, prematurity, immune defects, some congenital defects

neonatal thrombocytopenia A state that is the most common cause of purpura during the neonatal period, defined by platelets < 150,000/µL

NEONATAL THROMBOCYTOPENIA–CAUSES
PLATELET CONSUMPTION
• IMMUNE-MEDIATED Autoimmune, alloimmune–ITP or SLE in mother, platelet antibodies from mother
• NONIMMUNE Thrombin generation, infections
↓ PLATELET PRODUCTION Amegakaryocytic thrombocytopenia, aplastic anemia, myeloinfiltrative processes
OTHERS Drugs, hypoxia, inherited metabolic disorders, hypersplenism, associated with giant hemangiomas–Kasabach-Merritt syndrome

neonatal withdrawal syndrome Neonatal abstinence syndrome NEUROLOGY A condition affecting infants of mothers who chronically used CNS-active substances during pregnancy, for which the infant developed an in utero tolerance and who, on delivery, undergoes withdrawal AGENTS Opioids–eg, heroin, methadone, meperidine, codeine, pentazocine, propoxyphene, crack cocaine, alcohol, clomipramine, sedative-hypnotics–eg, barbiturates, meprobamate, benzodiazepines CLINICAL Wakefulness, irritability, seizures, tremulousness, lability of temperature, tachypnea, hyperacusis, hyperreflexia, hypertonicity, diarrhea, sweating, respiratory distress and apnea, rhinorrhea, autonomic dysfunction, respiratory alkalosis, lacrimation, yawning, sneezing; Sx appear 12 h to 1 wk after birth TREATMENT

Swaddling–firmly wrapping in blankets, ↓ external stimuli, tincture of opium. See Crack babies

neonate An infant in the first 4 wks of life, newborn

neonatologist A pediatrician specialized in managing neonates/newborns MEAT & POTATOES ACTIVITY Management of newborns with respiratory disease, infections, birth defects; hemodynamically unstable, premature, critically ill, or post-C-section babies SALARY $146K + 13% bonus

neonatology Neonatal medicine The hospital-based subspecialty of pediatrics dedicated to managing infants in the early postnatal period; the art and science of caring medically for the newborn

neoplasia ONCOLOGY Abnormal and uncontrolled cell growth. See Anal intraepithelial neoplasia, Cervical intraepithelial neoplasia, Ductal intraepithelial neoplasia, Hereditary neoplasia, Hereditary preneoplasia, Papillary neoplasia, Prostatic intraepithelial neoplasia, Vulvar intraepithelial neoplasia

neoplasm ONCOLOGY 'An abnormal mass of tissue, the growth of which exceeds and is uncoordinated with that of normal tissue and persists in the same excessive manner after cessation of the stimuli evoking the change'; an autonomous proliferation of cells, benign or malignant. See Cancer, Doubling time, Intraductal papillary-mucinous neoplasm of pancreas, Metastases, Papillary & solid neoplasm of pancreas

NEOPLASM CLASSIFICATIONS
BEHAVIOR Benign, borderline or malignant
DEGREE OF DIFFERENTIATION Well-differentiated, ie the neoplastic cell simulates its parent or progenitor cell or poorly-differentiated, ie the neoplastic cell is bizarre and "ugly', as defined by pathologic criteria'
EMBRYOLOGIC ORIGIN Epithelial–eg adenocarcinoma, squamous cell carcinoma, lymphoproliferative–eg leukemia, lymphoma, mesenchymal–eg sarcoma, histiocytosis, neural crest–eg carcinoid tumor, some small cell carcinomas, etc
GROSS APPEARANCE Well-circumscribed or infiltrative; benign neoplasms, usually slow-growing, well-circumscribed, often with a fibrous capsule, and are symptomatic only if they compromise a confined space, eg massive meningioma of the cranial cavity, or encirclement of vital blood vessels; malignancies are often aggressive with ↑ mitotic activity, bizarre cells, necrosis and invasion of adjacent structures and have metastatic potential

neoplasm panel LAB MEDICINE A battery of tests considered to be the most cost-efficient means of delineating a malignancy of unknown origin–eg, measurement of acid phosphatase, alkaline phosphatase, AFP, CEA, hCG, and LDH

neoprene ORTHOPEDICS A synthetic rubber produced by polymerization of chloroprene PROS Durable, compressible, impermeable CONS Allergic contact dermatitis, prickly heat

Neoral® Cyclosporine for microemulsion IMMUNOLOGY An immunosuppressant for reducing organ rejection in kidney, liver, heart transplants; to manage severe, active rheumatoid arthritis unresponsive to MTX; to manage adult, nonimmunocompromised Pts with severe, recalcitrant, plaque psoriasis who are unresponsive to one or more systemic therapies–ie, PUVA, retinoids, or MTX--or in Pts for whom either systemic therapies are contraindicated or cannot be tolerated. See Immunosuppression

Neosar® Cyclophosphamide, see there

neostigmine PHARMACOLOGY An anticholinesterase used for the symptomatic therapy of myasthenia gravis and, in anesthesiology, to reverse the effects of depolarizing agents. See Myasthenia gravis, Reversal agent

neostigmine test NEUROLOGY A clinical test that consists of the administration of neostigmine methylsulfate, a muscarinic, that exacerbates the Sx of myasthenia gravis–MG. See Muscarinic, Myasthenia gravis. Cf Tensilon–edrophonium test

Neothylline® Dyphilline, see there

Neotrend®

Neotrend® NEONATOLOGY A system for monitoring blood gases in premature newborns without blood withdrawal. See Arterial blood gases

neovascularization The formation of new blood vessels–ie, capillary ingrowth and endothelial proliferation in unusual sites, a finding typical of so-called 'angiogenic diseases,' which include angiogenesis in tumor growth, diabetic retinopathy, hemangiomas, arthritis, psoriasis

nephrectomy SURGERY Surgical removal of a kidney. See Partial nephrectomy, Radical nephrectomy, Simple nephrectomy

nephritic syndrome Glomerulonephritic syndrome NEPHROLOGY An obsolete nonspecific term for a renal lesion characterized by inflammation and necrosis of glomeruli. Cf Nephrosis, Nephrotic syndrome

nephritis NEPHROLOGY Inflammation of the kidney. See Acute interstitial nephritis, Glomerulonephritis, Interstitial nephritis, Lupus nephritis, Pyelonephritis, Shunt nephritis

nephroblastoma Wilms' tumor, see there

nephrocalcinosis NEPHROLOGY A condition, more common in premature infants, especially if receiving loop diuretics, characterized by calcium oxalate or calcium phosphate deposits in renal tubules and interstitium, resulting in ↓ renal function ETIOLOGY Hypercalcemia, ↑ Ca^{2+} excretion, medullary sponge kidney, renal cortical necrosis, renal tubular acidosis, TB CLINICAL Obstructive uropathy, possibly renal failure. See Kidney stones

nephrogenic diabetes insipidus Gypsy's curse NEPHROLOGY A form of diabetes insipidus characterized by profound hyposmotic polyuria and inability to concentrate urine despite high plasma concentrations of arginine-vasopressin ETIOLOGY Congenital–X-R–or acquired–lithium or democlocycline toxicity defect of collecting tubules, resulting in ↓ responsiveness to vasopressin CLINICAL Dehydration, hypernatremia, and possibly, CNS damage. See Aquaporins, Diabetes insipidus

nephrolithiasis Kidney stone(s), see there

nephrologist A physician subspecialized in non-surgical kidney disease, who is board-certified in internal medicine and nephrology MEAT & POTATOES DISEASES Renal failure, dialysis SALARY $175K +10% bonus. Cf Urologist

nephrology The medical subspecialty dedicated to the study and management of non-surgical diseases of the kidney. Cf Urology

nephropathic cystinosis An AR lysosomal storage disease characterized by early-onset renal tubular Fanconi's syndrome, progressive photophobia, and renal failure severe enough to require either hemodialysis or transplantation by age 10, caused by defective trans-lysosomal membrane transport of cystine, resulting in tissue deposition of cystine with corneal erosions, DM, neurologic deterioration CLINICAL Dehydration, acidosis, vomiting, electrolyte imbalance, hypophosphatemic rickets, FTT TREATMENT β-mercaptoethylamine–aminothiol cysteamine to deplete intracellular stores and dissolve tissue crystals, which improves growth and delays renal failure. See Salla disease

nephropathy Kidney disease NEPHROLOGY A general term for any disease that affects nephrons. See Analgesic nephropathy, Balkan nephropathy, Contrast-induced nephropathy, HIV nephropathy, IgA nephropathy, Light chain nephropathy, Reflux nephropathy

nephropexy NEPHROLOGY The surgical fixation of a loose/floppy kidney

nephroptosis NEPHROLOGY The downward displacement of a loose or floppy kidney. See Nephropexy

nephrosclerosis Arteriolonephrosclerosis Global fibrosis and atrophy of glomeruli, most common in atherosclerotic kidneys. See Malignant

nephrosis NEPHROLOGY 1. Nephrotic syndrome, see there. See Nil disease, Myeloma kidney. Cf Nephritis 2. A noninflammatory, nonneoplastic disease of kidney. See Bile nephrosis, Cholemic nephrosis

nephrotic range NEPHROLOGY Proteinuria of ≥ 2 g/m²/24 h, a finding classically associated with nephrotic syndrome

nephrotic syndrome Ellis type 2 nephrosis NEPHROLOGY A non-inflammatory derangement of glomerular function, characterized by ↑ glomerular leakage with loss of albumin and other macromolecules; NS is characterized by the triad of edema; proteinuria, > 3.5 g protein/1.73 m²/24 h; hypoalbuminemia, < 30 g/L ETIOLOGY Diabetic nephropathy, 1° glomerular disease; membranous glomerulonephritis may by linked to neoplasia, myeloma, amyloidosis, DM, infection, SLE, toxins–eg, colloidal gold, 'street' heroin, penicillamine, rarely, HIV, preeclampsia CLINICAL Various conditions may lead to NS; the Sx reflect the cause MANAGEMENT Loop diuretics–eg, furosemide, thiazides and potassium-sparing diuretics; heparin, anticoagulants to ↓ risk of thromboembolism; extreme measures include hyperoncotic albumin, plasma ultrafiltration, bilateral nephrectomy to avoid the complications of severe hypoproteinemia and hypovolemia; ex vivo adsorption of plasma on protein A Sephadex columns to ↓ proteinuria. See Diabetic nephropathy

Nerbet Buspirone, see there

Nerozen Diazepam, see there

nerve VOX POPULI A popular adjective for cheekiness, brass, "balls," as in she's got nerve. Cf Nerves

nerve biopsy NEUROLOGY A biopsy used to evaluate focal neuropathies or systemic disease ABNORMAL IN Amyloidosis, sarcoidosis, metabolic polyneuropathy, leprosy, demyelination, alcoholic neuropathy, Charcot-Marie-Tooth disease, peroneal nerve dysfunction, mononeuritis multiplex, other polyneuropathies SITES Ankle, wrist SPECIAL STUDIES Biochemical analysis, EM, histology, immunohistochemistry, molecular, virology

nerve block ANESTHESIOLOGY A neurolytic or nonneurolytic modality used to block pain sensation COMPLICATIONS Bleeding, infection, organ or vessel puncture. See Block, Neurolytic block, Nonneurolytic block

nerve conduction study NEUROLOGY A noninvasive method for assessing a nerve's ability to carry an impulse, which quantifies latency periods and conduction velocities; larger peripheral motor and sensory nerves are electrically stimulated at various intervals along a motor nerve. See Carpal tunnel syndrome, F wave, H-reflex, Jitter, Latency period

nerve gas See Nerve agent

nerve pain VOX POPULI → MEDTALK Neuralgia, see there

nerve regeneration PHYSIOLOGY The regrowth and reconnection of viable and functional neural connections damaged by transection or other trauma

nerve root impingement Nerve root irritation NEUROLOGY Pressure on the nerve roots caused by disc herniation or subluxation of the vertebrae or ribs which can cause involuntary muscle contraction, numbness, tingling and/or pain.

nerve-sparing prostatectomy UROLOGY Any prostatectomy intended to preserve sexual function

nerves VOX POPULI A popular term, especially among those of lower socioeconomic strata for anxiety, as in she's got nerves. Cf Nerve

nervous *adjective* 1. Neural–neurology 2. Excitable–psychology

nervous bowel syndrome Irritable bowel syndrome, see there

nervous breakdown PSYCHIATRY A dated term that encompasses anxiety, panic and depressive disorders PSYCHOLOGY A popular term for being mentally overwhelmed by circumstance and accumulated acute stressors

nervous exhaustion 1. Nervous breakdown, see there 2. Neurasthenia, see there

nesting OBSTETRICS Frenetic house cleaning by a ♀ in late pregnancy, which is more common with the first-born child, fancifully likened to birds building a nest and preparing to lay and incubate chicks

Netherton syndrome Atopic diathesis, trichorrhexis invaginata, trichorrhexis nodosa DERMATOLOGY A rare skin disorder occurring almost exclusively in ♀, characterized by scaling in a circular pattern–ichthyosis linearis circumflexa CLINICAL Hair shafts held inside root–trichorrhexis invaginata; hair fragility "bamboo hair", allergies such as asthma, or food allergies that cause skin eruptions PATHOGENESIS Excess cornification of epidermis. See Bamboo hair

netiquette INFORMATICS Internet etiquette. See Internet

netizen ONLINE Any person who communicates on the Internet and belongs to the global electronic village; an Internet citizen, who uses networked resources, which connotes civic responsibility and participation. See Digerati, Internet

NETT PULMONOLOGY A clinical trial–National Emphysema Treatment Trial–examining the long-term effects of lung reduction surgery on function, M&M in emphysema. See Emphysema

Nettergram MEDICAL EDUCATION A color illustration rendered by Frank H Netter, MD–1906-1991 that summarizes in a lucid and graphic form, the clinical and pathologic findings of a disease state or aspect of physiology. See Cartoon

network COMPUTERS An interconnecting system of computer terminals and processors that interact with each other and/or share software programs and applications and 'peripherals,' eg printers, scanners, archiving and other devices, linked either by cable or wireless technology SOCIOLOGY *noun* A group of similarly-minded individuals that interacts socially or professionally. See Old boy network *verb* To interact socially

network model HMO MANAGED CARE An HMO that contracts with 2 or more independent physician group practices to provide health services. Cf Group model, Independent practice association, Staff model

Neupogen® Filgrastim ONCOLOGY An agent used to ↓ neutrophil recovery time and duration of fever after chemotherapy for AML. See Acute myelogenous leukemia

Neuprex® INFECTIOUS DISEASE An orphan drug for treating life-threatening infections–eg, meningococcemia–aka meningococcal septicemia, gram-negative bacteria, and *N meningitidis* infection in Pts with partial hepatectomy. See Meningococcemia

neural prosthesis NEUROLOGY Any electronic and/or mechanical device that connects with the nervous system and supplements or replaces functions lost by disease or injury. See Nerve regeneration

neural tube defect NEUROLOGY Any of the congenital developmental defects of the CNS characterized by defective closure of the neural tube at one or more segments; NTDs range from asymptomatic to extreme–anencephaly with absent cranial vault and most or all of the cerebral hemispheres and spina bifida cystica attributable to multifactorial events and noxious environmental agents; NTDs occur in 1:1000-5000 live births, ♂:♀ ratio 2-3:1, with regional differences–eg, higher in Ireland–2-7% recurrence rate CLINICAL Cinercephaly, cephalocele, spina bifida, and myelodysplasia; failure to close neural tube at 4th-5th fetal wk LAB ↑ α-fetoprotein, detected in antenatal screening of maternal serum or amniotic fluid PREVENTION Multivitamins in early pregnancy ↓ risk of NTDs

neuralgia Nerve pain NEUROLOGY Pain along a nerve, more common in the elderly ETIOLOGY Injury, irritation, infection–herpes, or idiopathic CLINICAL Burning pain, worsened by movement or contact with affected areas; trigeminal neuralgia and postherpetic neuralgia are most common. See Glossopharyngeal neuralgia, Neuropathy

neuraminidase inhibitor INFECTIOUS DISEASE Any antiviral that inhibits neuraminidase, an enzyme essential for replication of influenza and other viruses. See Influenza

neurapraxia NEUROLOGY Partial or complete conduction block over a segment of a nerve fiber, with temporary paralysis. See Cervical cord neurapraxia, Chronic recurrent root neurapraxia

neurasthenia PSYCHOLOGY Effort syndrome A nonspecific finding, often associated with depression or anxiety disorders, characterized by fatigue, and inability to function ACCOMPANIMENTS Autonomic changes–eg, tachycardia, sighing, blushing, dysdiaphoresis; Pts may believe neurasthenia is organic, not psychological

neuritis Inflammation of a nerve–neuropathy is also widely used. See Optic neuritis, Retrobulbar neuritis.

neuroablative procedure PAIN MANAGEMENT A general term for any neurosurgical procedure used to relieve pain–especially CA-related refractory to lesser forms of therapy–eg, analgesics, opiates, and either nonneurolytic or neurolytic nerve blocks. Cf Analgesic ladder, Nerve block, Neurolytic block, Nonneurolytic block

neuroarthropathy Neuropathic joint, see there; aka Charcot's joint

neuroblastoma ONCOLOGY A highly malignant childhood tumor of adrenal or related tissue in the nervous system, derived from neural crest, which is composed of undifferentiated neuroblasts; it is the 2nd most common pediatric neoplasm after leukemia and lymphoproliferative disorders CLINICAL Median age of onset < age 2; stage I survival is 80-90%, stage IV, 15%–stage IV-S is an exception, and has a better prognosis; higher stage neuroblastomas metastasize to bone, lymph nodes, liver LAB Catecholamines and metabolites are ↑ by TLC, GLC, HPLC, with a 5-fold ↑ in VMA and HVA PROGNOSIS Better with stage I, II, or IVa disease, younger age–< 1 yr, ↑ TRK–protein product of proto-oncogene *TRK*. See Esthesioneuroblastoma, Ganglioneuroblastoma, Homovanillic acid, Intraventricular neuroblastoma, Nerve growth factor receptor, 'One-hit, two-hit' model, TRK, Vanillylmandelic acid

neuroblastoma, IV-S syndrome PEDIATRIC ONCOLOGY A type of neuroblastoma (S for special) comprising 10-20% of all neuroblastomas, in which the primary tumor may be small, confined to the adrenal gland, but has widespread disease with massive involvement of the liver and skin; bone may be involved but osteolysis is not present; despite metastases, the tumor regresses spontaneously by maturation sequence neuroblastoma, the most immature lesion composed of neuroblasts, to ganglioneuroblastoma, finally to ganglioneuroma–the most mature of the sequence, composed of ganglion cells. See One-hit/two-hit theory

neuroborreliosis Lyme encephalopathy, Lyme meningoencephalitis NEUROLOGY CNS response to *Borrelia burgdorferi*; it is more common in adults in the 5th decade INCUBATION 7 wks, adults, 4 wks, children CLINICAL Radiculoneuritis-type pain–severe, burning, migratory, worse at night, hyperesthetic; other findings include cranial nerve palsy, meningoencephalitis, cerebral vasculitis, meningitis, peripheral neuropathy, encephalomyelitis PSYCHIATRY Anxiety, insomnia, panic attacks, hallucinations, déjà vu sensation, obsessions, anorexia

neurocardiogenic syncope Vasovagal syncope A syncope with a psychogenic substrate, and predisposition to bradycardia, hypotension, peripheral vasodilation, syncope CLINICAL Abrupt loss of vascular tone, nausea, diaphoresis, pallor TREATMENT Beta blockers–eg, metoprolol, theophylline, disopyramide

neurocirculatory asthenia 1. Post traumatic stress disorder, see there 2. Panic disorder, see there

NeuroCol™ sponge NEUROSURGERY A collagen sponge that absorbs over 40 times its weight, used to maintain exposed brain tissue moist and protected during neurosurgery

neurocutaneous syndrome Phakomatosis Any multisystem disease characterized by involvement of the brain, skin, eyes, etc EXAMPLES Neurofibromatosis type I–von Recklinghausen disease, tuberous sclerosis, von Hippel-Lindau disease–all AD, Sturge-Weber syndrome–no known hereditary pattern, and ataxia-telangiectasia. See Tuberous sclerosis

neurodegenerative disorder NEUROLOGY A chronic progressive neuropathy characterized by selective and generally symmetrical loss of neurons in motor, sensory, or cognitive systems TYPES BY AREA CEREBRAL CORTEX–Alzheimer's disease, Pick's disease, Lewy body dementia BASAL GANGLIA–Huntington's disease, Parkinson's disease BRAINSTEM & CEREBELLUM–dentatorubropallidoluysian atrophy, Freidreich's ataxia, multiple system atrophy, types 1, 2, 3, 6, 7 spinocerebellar ataxia MOTOR–amyotrophic lateral sclerosis, familial spastic paraparesis, spinal muscular atrophy, spinal & bulbar muscular atrophy. See COPD

neurodermatitis Atopic dermatitis, see there

neuroendocrine tumor NEUROPATHOLOGY A neoplasm with a characteristic morphology, often composed of clusters and trabecular sheets of round 'blue cells', granular chromatin, and an attenuated rim of poorly demarcated cytoplasm Examples Carcinoids, small–'oat' cell carcinomas, medullary carcinoma of thyroid, Merkel cell tumor, cutaneous neuroendocrine CA, pancreatic islet cell tumors, pheochromocytoma. See Neurosecretory granules

neurofibrillary tangle NEUROPATHOLOGY A characteristic histologic finding in the perikaryon of large cortical neurons of Pts with Alzheimer's disease, the number of NFTs loosely correlates with the severity of dementia DIFFDX Down's syndrome, subacute sclerosing panencephalitis, lead encephalopathy, tuberous sclerosis, Hallervorden-Spatz disease, lipofuscinosis, post-encephalitic parkinsonism, and normal adults–confined to the hippocampus. See Alzheimer's disease, Amyloid plaque, Granulovacuolar degeneration, Paired helical filaments, Senile plaques

neurofibromatosis, type 1 Von Recklinghausen disease An AD condition affecting ± 100,000–US, due to a mutation of a gene on chromosome 17

NEUROFIBROMATOSIS, TYPE 1 DIAGNOSIS–2+ OF FOLLOWING

6 or more 'cafe-au-lait' macules > 5 mm in greatest diameter in a prepubertal, or > 15 mm in greatest diameter in post-pubertal Pts

2 or more histologically-confirmed neurofibromas–or one plexiform neurofibroma

Freckling in the axillary or inguinal regions

Optic glioma

2 Lisch nodules–iris hamartomas, the most common feature of NF1 in adults

Distinct bone lesions, eg sphenoid dysplasia, cortical thinning of long bones, pseudoarthrosis

The person has a 1st-degree relative with NF1

neurofibromatosis, type 2 Central neurofibromatosis, bilateral acoustic neurofibromatosis An AD condition less common than NF1, which affects several thousand Pts in the US; in 95%, the gene defect is located on chromosome 22q11.21-q13.1, 1st seen in late adolescence; NF2 is diagnosed in the presence of either bilateral 8th nerve masses by CT or MRI, or when a Pt known to have a 1st-degree relative with NF2 presents with either a unilateral 8th nerve mass or 2+ 'neural crest' tumors–eg, neurofibroma, meningioma, glioma, spinal

neurofibromatosis, schwannoma, or juvenile posterior subcapsular lenticular opacity

neurogenic *adjective* Originating in nervous tissue; relating to nerves; forming nervous tissue

neurogenic bladder UROLOGY A urinary bladder with loss or impairment of voluntary control of micturition. The two types are **SPASTIC**, due to lesions of the spinal cord, accompanied by urgency, ↑ frequency, ↓ functional capacity, spastic contractions, and poor voluntary control; or **FLACCID**, due to segmental lesions at S2 to S4, interfering with voluntary and reflex control, ↓ of sensation of bladder fullness, causing 'overflow' incontinence, when the bladder contains 2+ liters

neurogenic communication disorder NEUROLOGY The inability to exchange information with others because of hearing, speech, and/or language problems due to nervous system impairment

neurogenic shock Spinal shock, see there

neuroimaging IMAGING 1. Any imaging technique–eg, PET scans, functional MRI, used to evaluate functional aspects of neural activity 2. Images obtained from the head which detect any abnormal mass, but which do not identify a specific type of tumor

neuroleptic malignant syndrome NEUROLOGY A disorder seen in those receiving antipsychotics–eg, haloperidol, major tranquilizers, and other agents–eg, phenothiazines, reserpine, butyrophenone, an effect attributed to dopamine blockade in the basal ganglia and hypothalamus; NMS may also be associated with anesthesia, affecting ± 1:50,000 Pts exposed to inhalation anesthesia, most commonly in young ♂, transiently weakened by exhaustion, dehydration CLINICAL Fever ≥ 41℃, extrapyramidal Sx–eg, rigidity, involuntary movements, facial dyskinesia, skeletal muscle hypertonicity, loss of consciousness, autonomic lability–pallor, sweating, tachycardia, arrhythmia, transient HTN which, if severe, may cause renal failure MORTALITY 20-30%, often between days 3-30, usually from renal failure TREATMENT Bromocriptine or dantrolene may shorten clinical disease

neuroleptic PSYCHIATRY An agent used to treat psychotic illnesses–eg, obsessive-compulsive disorder

neurologic examination A battery of clinical tests that evaluates a person's physiologic function and mental status, as well as the presence of any structural–organic lesions that may cause changes in neurologic function. Cf Psychiatric examination

NEUROLOGIC EXAMINATION-DATA OBTAINED

HISTORY of the Pt's chief complaint

MENTAL STATUS Coherency, memory, judgement, comprehension, and other cognitive processes

CRANIAL NERVES

1. Olfactory–1st cranial nerve Ability to identify common odors

2. Optic nerve–2nd cranial nerve Visual acuity, visual fields, pupil size and response to light, optic examination

3. Ocular muscles–3rd, 4th, 6th cranial nerves

4. Trigeminal nerve–5th cranial nerve

5. Facial nerve–7th cranial nerve

6. Auditory nerve–Rinne & Weber tests, 8th cranial nerve

7. Glossopharyngeal–9th cranial nerve, vagal nerve–10th cranial nerve

8. Accessory nerve–11th cranial nerve

9. Hypoglossal nerve–12th cranial nerve

MOTOR FUNCTION

Upper extremities: Finger-to-nose test, extension-flexion of elbow, power of grip, reflexes–eg biceps, brachioradialis, sensation–eg vibration, pain

Lower extremities: Heel-to-knee test, extension-flexion of knee, reflexes–eg patellar, plantar, sensation–eg vibration, pain

neurologic impairment NEUROLOGY Any damage to, or deficiency of, the nervous system

neurologic orthostatic hypotension Shy-Drager syndrome, see there

neurologic stability CRITICAL CARE A clinical state characterized by lack of change in mental status or level of consciousness; control of seizures; absence of new neurologic defects–eg, aphasia, ataxia, dysarthria, paresis, paralysis, visual field loss, or blindness. See Stroke, Transient ischemic attack

neurologist A physician specialized in the non-surgical diseases of the brain and nervous system, who is board-certified/eligible in internal medicine *and* neurology MEAT & POTATOES DISEASES Strokes, unusual headaches, seizures SALARY $166K + 12%. Cf Neurosurgeon

neurology The medical subspecialty dedicated to the study and management of non-surgical diseases of the brain and nervous system. See Pediatric neurology. Cf Neurosurgery

neurolymphomatosis A rare lymphoma with a predilection for peripheral and cranial nerves, lumbar and brachial plexi and nerve roots, meninges, and vessels in the brain; $\frac{1}{2}$ of cases are accompanied by systemic lymphoma, which may be clinically silent; NLM may present as a painful chronic sensorimotor neuropathy with asymmetric distribution and prominent bulbar involvement; it is associated with HIV and HTLV-I infections

neurolytic block PAIN MANAGEMENT A form of anesthesia in which a neurodestructive agent–eg, phenol and alcohol is injected at or near a nerve causing extreme pain; NBs are useful in cancer pain syndromes that are visceral or involve the torso, but rarely for pain of extremities, as it would cause paralysis. See Nerve block, Nonneurolytic block

neurolytic celiac plexus block ANESTHESIOLOGY The injection of 50% alcohol into the celiac plexus to control severe pain in pancreas CA. See Pancreatic cancer

neuromuscular blockade NEUROLOGY The partial or complete inhibition of motor activity at a neuromuscular junction ETIOLOGY 1. Reduction of post-synaptic receptors–eg, myasthenia gravis; 2. Defective acetylcholine release from storage vesicles–eg, botulism, myasthenia, or Eaton-Lambert syndrome; 3. Competition for binding sites, either pharmacologic blockade–eg, neostigmine, edrophonium, or toxins–eg, organophosphate insecticides. See Neuromuscular junction

neuromuscular reeducation REHAB MEDICINE The use of any manipulation-based therapeutic modality–eg, biofeedback training, intended to help a Pt recuperate functional activity, after trauma or a CVA. See Biofeedback training

neuronal nitric oxide synthase See nNOS

neuronal plasticity NEUROPHYSIOLOGY 1. The ability of neurons to stabilize or alter synapses 2. The malleability of cortical representations of sensory and motor innervation, which has a range of 10-14 mm in the somatosensory cortex in animal models that have undergone long-term loss of sensory nerves in the forelimbs. See Long-term potentiation, Synaptic plasticity

Neurontin® Gabapentin NEUROLOGY An agent used to manage postherpetic neuralgia and an adjunct for managing partial seizures in Pts > age 12 ADVERSE EVENTS Vertigo, somnolence, peripheral edema, asthenia, diarrhea, ataxia, fatigue, nystagmus. See Herpetic neuralgia, Seizures

neuro-oncologist A physician–neurologist, oncologist or neurosurgeon, subspecialized in treating Pts with brain tumors and/or their consequences on the nervous system

neuropathic joint Neuropathic arthropathy RHEUMATOLOGY The mechanical failure of a joint due to impaired sensory input; NA is a destructive and productive arthropathy with a loss of pain and nociceptive sensation that may be due to the cumulative effect of trauma and joint laxity RADIOLOGY Joint effusion, fragmentation of the articular surface and eburnation of bony surfaces, eventually complete joint disorganization ETIOLOGY DM, syringomyelia, tabes dorsalis; rarely, congenital indifference to pain, amyloidosis, meningomyelocele and other 'exotica'

neuropathy NEUROLOGY A disorder of peripheral nerves, which may be congenital–eg, hereditary sensory radicular neuropathy or hypertrophic interstitial neuropathy, traumatic–entrapment–eg, carpal tunnel syndrome, metabolic–eg, due to amyloid or DM, toxic–eg, tobacco or alcohol-related amblyopia, cis-platinum, vincristine, or infectious–eg, herpetic. See Alcoholic neuropathy, Autonomic neuropathy, Colchicine neuropathy, Diabetic neuropathy, Drug-related neuropathy, Dying back neuropathy, Entrapment neuropathy, Familial amyloidotic polyneuropathy, Giant axonal neuropathy, Hereditary motor & sensory neuropathy, Hereditary neuropathy with liability to pressure palsies, Hereditary sensory neuropathy, Jamaican neuropathy, Leber hereditary optic neuropathy, Localized hypertrophic neuropathy, Motor neuropathy, Neck crack neuropathy, Nutcracker neuropathy, Optic neuropathy, Retrobulbar neuropathy, Subacute myelo-optic neuropathy, Thalidomide neuropathy. Cf Nerve dysfunction

Neuropax Chlordiazepoxide, see there

neuropeptide Any of a family of LMW–< 5 kD intracellular peptides that transmit information in the CNS, GI tract, etc NEUROPEPTIDES ACTH, angiotensin II, bombesin, bradykinin, calcitonin gene-related products, carnosine, cholecystokinin, corticotropin-releasing factor, dynorphins, β-endorphin, leu-enkephalin, met-enkephalin, gastric inhibitory polypeptide, gastrin, glucagon, GH, growth hormone releasing factor, insulin, LHRH, α-melanocyte-stimulating hormone, melanotropin-inhibiting factor, motilin, neurotensin, oxytocin, prolactin, secretin, somatostatin-14 and -28, substance P, TRF, thyrotropin, VIP, vasopressin. See Endorphin, Hormone, Neurotransmitter, Synapsin

neuroprobe REHAB MEDICINE An electric stimulation device that may be used to identify tender areas or trigger points, and locate points for placing the electrodes of a TENS–transcutaneous electric nerve stimulation unit, or may itself be used to deliver electrical impulses to treat pain. Cf Shortwave diathermy

neuroprotective agent NEUROLOGY Any agent or drug that protects the brain from secondary injury caused by stroke. See Stroke

neuropsychiatry NEURAL SCIENCE The discipline dedicated to understanding the neurobiologic basis of behavior and behavioral development, encompassing neurology, psychiatry, psychology, neurogenetics

neuroradiologist A radiologist specialized in using various imaging techniques to diagnose diseases of the nervous system

neuroradiology The subspecialty of radiology that uses various imaging techniques–CT, MRI, PET to diagnose diseases of the nervous system. See Interventional neuroradiology

neuroscientist A researcher, often with an advanced degree–MD, MS, PhD–who investigates neural and brain-related phenomena

Neurosedin Diazepam, see there

Neurosine Buspirone, see there

neurosis PSYCHOLOGY An older term for a disorder characterized by excess anxiety and avoidance behaviors NEUROSES Anxiety disorder, dissociative disorder, mood disorder, personality disorder, bipolar I disorder, depression, histrionic personality disorder, obsessive-compulsive behavior, phobias. See Neurotic disorder, Semi-starvation neurosis, Sunday neurosis

neurosurgeon A surgeon specialized in managing diseases of the brain, spine and peripheral nerves MEAT & POTATOES DISEASES Brain tumors, spinal cord disease SALARY $245K + 15% bonus. See Critical care neurosurgeon

neurosurgery The surgical subspecialty dedicated to the management of diseases of the brain and peripheral nervous system

neurosyphilis NEUROLOGY Any of the rare neural changes of 3° syphilis CLINICAL Lightning-like pain, ataxia, optic nerve degeneration → blindness, urinary incontinence, loss of position sense, Charcot's joints, personality changes, aphasia, paralysis, seizures. See STS-RPR, VDRL. See Charcot's joints, Windswept cortex, Tabes dorsalis. Cf Quaternary syphilis

neurotic *adjective* Referring to neurosis, see there *noun* A person with a neurotic disorder; any person with a mental disorder other than a psychotic disorder

neurotic disorder PSYCHIATRY A mental disorder in which the predominant disturbance is a distressing symptom or group of Sx which one considers unacceptable and alien to one's personality without a marked loss of reality testing; behavior does not actively violate gross social norms although it may be quite disabling; the disturbance is relatively enduring or recurrent without treatment and is not limited to a mild transitory reaction to stress; there is no demonstrable organic etiology

neurotoxic *adjective* Referring to substances or activities with an adverse effect on the nervous system, leading to ↑ excitability, numbness, muscle cramps, paresthesias, systemic failure or cardiac arrest. Cf Hematoxic

neurotoxin TOXICOLOGY A substance that is toxic or destroys nerve tissue—eg, exotoxins, present in plants and animals and either block conduction of the nerve impulse or synaptic transmission, by binding to the voltage-gated Na+ channel protein or ↑ neuronal activity. See Conotoxin, *Conus*, Puffer fish, Red tide, Shiga neurotoxin

neurotransmitter Neurosynaptic transmitter PHYSIOLOGY Any of a number of small neuroregulating molecules—eg, catecholamines and acetylcholine, which are synthesized in the presynaptic terminals of neurons, stored in vesicles, and cause rapid and transient depolarization near their point of release in the synaptic cleft, where it stimulates production of either excitatory or inhibitory postsynaptic potentials; neurotransmission at synapses or neuromuscular junctions is due to binding of a neurotransmitter to its cognate receptor. See Amino acid neurotransmitter, Neuropeptide, Neuroregulator. Cf Hormone

neurotrauma NEUROLOGY Any trauma which impacts on the brain and spinal cord

Neurotrend™ CRITICAL CARE A system designed for continuous monitoring of ABGs to detect cerebral ischemia and/or hypoxia in Pts suffering from closed-head trauma and during surgery. See Arterial blood gases

neurotrophic factor A generic term for any of a family of substances with roles in maintenance and survival of neurons—eg, secretory proteins, nerve growth factors—see there, brain-derived growth factor, neurotrophin-3

Neutra-Caine™ ANESTHESIA A mixture of local anesthetics—lidocaine, mepivacaine, bupivacaine—eg, at time of injection to ↓ stinging pain at injection site. See Lidocaine

neutral lipid storage disease An AR condition characterized by congenital ichthyosis, hepatosplenomegaly, myopathy, and diffuse accumulation of TGs, vacuolated granulocytes

neutral position ORTHOPEDICS The position of a joint where the bones that make up the joint are placed in the optimal position for maximal movement

Neutralase™ Heparinase I CARDIOLOGY An agent used after cardiac bypass surgery to reverse heparin's anticoagulant effects. See Heparin

neutralization plate ORTHOPEDICS A plate designed to protect fracture surfaces from normal bending, rotation and axial loading force, often used with lag screws. See Plate

NeuTrexin® Trimetrexate glucuronate, see there

neutropenia HEMATOLOGY An absolute ↓ in neutrophils—normal, 2,500-7,000/mm³; serious infections occur if < 500/mm³ See Congenital neutropenia, Cyclic neutropenia. Cf Neutrophilia

neutropenic colitis with aplastic anemia Mulholland syndrome A clinical condition of young adults, which consists of right-sided colonic necrosis, profound agranulocytosis, aplastic anemia, fever, watery diarrhea, generalized abdominal pain without inflammation, accompanied by transmural bacterial infiltration, often following antibiotic therapy TREATMENT Non-interventional—bowel rest, ICU, fluids. See Pseudomembranous colitis

neutrophil A phagocytic WBC, normally constituting 60-70% of circulating WBCs. See Band, Dysplastic neutrophil, Hypersegmented neutrophil, Polymorphonuclear neutrophil

neutrophilia Granulocytic leukocytosis HEMATOLOGY An absolute neutrophil count of > 8,000/mm³, which may be physiologic or pathologic. See Hereditary neutrophilia, Physiologic neutrophilia. Cf Neutropenia

nevirapine Viramune® AIDS An antiretroviral non-nucleoside reverse transcriptase inhibitor, combined with nucleoside RTIs to manage HIV infection ADVERSE EFFECTS Rash. See Nonnucleoside reverse transcriptase inhibitor, TIBO. Cf Antiretroviral agent, Delavirdine

nevus Plural, nevi DERMATOLOGY An often congenital, usually benign, circumscribed pigmented—tan, brown, or flesh-colored, spot—eg, a mole, on the skin and/or mucosae, due to either an ↑ or ↓ of melanin; nevi are considered hamartomas and contain spindle-shaped melanocytes. See Atypical nevus, Becker's nevus, Blue nevus, Congenital nevus, Dysplastic nevus, Familial dysplastic nevus, Garment nevus, Giant (pigmented) hairy nevus, Halo nevus, IVLEN nevus, Malignant blue nevus, Melanocytic nevus, Port-wine nevus, Speckled lentiginous nevus, Spitz nevus, Spotted pigmented nevus, White sponge nevus. Cf Lymph node inclusions

nevus anemicus White nevus A sharply circumscribed, pale macule on the trunk, neck, or limbs, attributed to a functional defect in the superficial dermal vessels, as these are not affected by vasodilators; NA contrasts to achromic or depigmented nevi—eg, Ito's nevus, due to a focal loss of melanin pigment

nevus of Ota A bluish or gray-brown macular lesion of the periorbital skin innervated by the 1st and 2nd branches of trigeminal nerve TREATMENT Selective photothermolysis with a Q-switched ruby laser, surgical removal, skin grafting, dermabrasion, cryotherapy

New Age "…a metaphor for the expression of a transformative, creative spirit…for being in the world in a manner that opens us to the presence of God…in the midst of our ordinariness…". See Alternative medicine, Healer

new drug CLINICAL PHARMACOLOGY Any agent intended for use in man, the composition of which is not generally recognized, among experts qualified by scientific training and experience to evaluate the safety and effectiveness of drugs, as safe and effective for use under the conditions prescribed, recommended, or suggested in the labeling. See Investigational new drug

new drug application See NDA

New Jersey vesicular stomatitis virus A virus that replicates in infected host cell cytoplasm CLINICAL 60% of those exposed develop disease after a 1-2 day incubation—fever, chills, malaise, myalgias, N&V, pharyngitis; rarely, oral vesicles PROGNOSIS Spontaneous resolution in 1 wk

new physician MEDTALK A physician during the first 7 yrs of practice. Cf Good laboratory practice

New York Heart Association classification A functional classification of cardiac failure, used to stratify Pts according to severity of disease and the need for—and type of—therapeutic intervention

New York Heart Association–classification

I Asymptomatic heart disease

II Comfortable at rest; symptomatic with normal activity

III Comfortable at rest; symptomatic with < normal activity

IV Symptomatic at rest—Criteria Committee, NYHA, Inc: Diseases of Heart & Blood Vessels, 6th ed, Little Brown, Boston 1964

newborn jaundice Breast milk jaundice Physiologic jaundice of the newborn, see there, aka breast milk jaundice

newborn screening NEONATOLOGY The analysis of a neonate's blood for metabolic or other disorders to prevent mental retardation, disability or death

news embargo Embargo arrangement, see there

news release JOURNALISM A document of public record in which an institution or organization formally announces matters of interest to media 'consumers'

newsgroup INFORMATICS A USENET discussion group. See USENET. Cf Listserve

Nexacryl™ OPHTHALMOLOGY A tissue adhesive for treating corneal lacerations and perforations See Corneal laceration

Nexium® Esomeprazole, the Purple pill INTERNAL MEDICINE An agent used to manage diagnostically confirmed erosive esophagitis

next-of-kin LAW & MEDICINE A term "…with two meanings 1. nearest blood relations according to the law of consanguity and 2. those entitled to take under statutory distribution of intestate's estates…(which) may include a relationship existing by marriage, and embrace persons, who …bear no relation of kinship at all'

NF 1. National Formulary, see there 2. Neurofibromatosis, see there 3. Neurofilament, see there 4. Normal flow 5. Normal formula 6. Nursing facility

NGU Non-gonococcal urethritis

NHANES III Third National Health & Nutrition Examination Survey PUBLIC HEALTH A population-based survey conducted by the National Center for Health Statistics, designed to assess the health and nutritional status of the noninstitutionalized Americans

NHL Non-Hodgkin's lymphoma, see there

nibbling NUTRITION The consumption of multiple–up to 17–'mini-meals' per day, as opposed to the usual 3 meals/day. Cf Bingeing, Gorging

nibbling diet NUTRITION The consumption of food 'á petit pas,' between the 'normal' 3 meal-a-day regimen followed by most individuals, and either binging–gorging or snacking diets. See Binge, Junk food

Nibblit™ SURGERY A laparoscopic device used for aquadissection, sharp and blunt dissection for lysis of adhesions, biopsy, specimen retrieval. See Dissection

NIC PHARMACOLOGY Lorazepam, see there

NicCheck® TOBACCO CONTROL A proprietary semiquantitative urine dipstick test for nicotine. See Passive smoking

NICE CARDIOLOGY A clinical trial–National investigators collaborating on enoxaparin–Lovenox® CHILD PSYCHIATRY An information service that assists parents, educators, caregivers and others in ensuring that all children and youth with disabilities have a better opportunity to reach their fullest potential

niche IMAGING A recess which may correspond to an ulcer in the wall of a hollow organ which tends to retain contrast medium

nick-you Neonatal intensive care unit, see there

nickel & dime lesion A fanciful descriptor for the annular, waxing and waning maculopapular lesions seen at the mucocutaneous borders at the mouth and nasolabial folds in untreated 2° syphilis, more common in darker Pts; these lesions may coincide with anogenital condylomata lata

nickel toxicity A condition caused by nickel excess, which may be due to nickel-based jewelry, resulting in contact dermatitis, or occupational, resulting in liver necrosis and pulmonary congestion; long-term nickel exposure increases the risk for nasopharyngeal and lung cancer. See Minerals, Nickel

niclosamide PARASITOLOGY A salicylanilide that kills snails that harbor *Schistosoma*; internally it is an effective anthelmintic See Schistosomiasis

NICODARD National Information Center for Orphan Drugs & Rare Diseases. See NORD, Orphan disease, Orphan drug

Nicorette® See Nicotine gum

nicotine SUBSTANCE ABUSE A colorless pyridine alkaloid in tobacco ROUTES Inhalation, skin absorption, ingestion, either accidental or suicidal CLINICAL Transient CNS stimulation followed by depression or paralysis, nausea, hypersalivation, abdominal pain, vomiting, diarrhea, cold sweats, headache, vertigo, confusion, incoordination, ↓ pulse rate, dyspnea with possible respiratory paralysis and intense vagal stimulation, which may cause cardiac arrest; death occurs 1-4 hrs after ingesting a fatal adult dose–> 60 mg TREATMENT Emesis, gastric lavage, atropine–*Nicotiana tabacum* stimulates cholinergic receptors. See Cigarette, Coniotine, Nicotine gum, Passive smoking, Smokeless tobacco, Smoking, Tobacco

nicotine analogue NEUROPHARMACOLOGY Any of a family of compounds that structurally mimic nicotine–eg, ABT-418, Abbott labs, which may ↑ short-term and spatial memory, and reaction times in clinical testing

nicotine gum Nicotine polacrilex A masticant that slowly releases nicotine, ameliorating the effects of tobacco withdrawal and the intensity of relapse factors–eg, weight gain

nicotine nasal spray Nicotrol™ TOBACCO A tool for smoking cessation, in which nicotine in solution is spritzed–0.5 mg nicotine/spritz in each nostril, up to 40 doses/day ADVERSE EFFECTS Runny nose, nasal and/or throat irritation, watery eyes, sneezing RESULTS Cessation is 2-fold greater than with placebos. See Nicotine

nicotine patch Nicotine transdermal delivery system SUBSTANCE ABUSE A device used in smoking cessation SIDE EFFECTS Transient burning, itching–50%, erythema–14%; contact hypersensitivity–2.4%. See Nicotine replacement therapy

nicotine replacement therapy SUBSTANCE ABUSE The use of any of a number of formulations of nicotine–gum, sprays or patches to attenuate the Sx of nicotine withdrawal. See Nicotine gum, Nicotine nasal spray, Nicotine patch, Nicotrol inhaler, Smoking cessation

nicotine withdrawal syndrome Tobacco withdrawal syndrome SUBSTANCE ABUSE A condition of mild irritability, headache, and psychological craving associated with the abrupt cessation of nicotine exposure in a nicotine-dependent person. See Nicotine, Smoking, Tobacco

nicotinic acid deficiency Pellagra, see there

Nicotrol® inhaler Nicorette® inhaler TOBACCO CONTROL A nicotine replacement device for smoking cessation. See Nicotine replacement therapy

nictameral MEDTALK *adjective* Referring to a bimodal, ***night-day*** circadian rhythm. See Circadian rhythm

nidation EMBRYOLOGY Implantation of a blastocyst into the endometrium. See Blastocyst

NIDDM Non-insulin-dependent diabetes mellitus. See Type 2 diabetes mellitus

Niemann-Pick disease METABOLISM A rapidly progressive AR storage disease with progressive neurologic damage due to sphingomyelinase deficiency; NPD type A is AR, more common in Ashkenazi Jews and is characterized by accumulation of excess sphingomyelin and cholesterol in the brain, BM, liver, spleen CLINICAL Early infancy onset with feeding problems and delayed or regressing motor development, progressive vision and hearing loss; late onset NPD is characterized by movement disorders and seizures. See Niemann-Pick cell

nifedipine Procardia® CARDIOLOGY A dihydropiridine CCB vasodilator used for angina, as a short-acting antihypertensive–↓ BP, resulting in a ↓ left ventricular volume and myocardial mass, ↑ ejection fraction SIDE EFFECTS Tachycardia, headache, peripheral edema, cerebral ischemia, stroke, severe hypotension, AMI, conduction defects, fetal distress, death. See Calcium channel blockers, Hypertension, TIBBS

night blindness 1. Vitamin A deficiency, see there 2. Retinitis pigmentosa, see there 3. Nyctalopia Defective vision in ↓ illumination, often implying defective rod function with delayed dark adaptation and perceptual threshold; it is either congenital and stationary with myopia and degeneration of the disc–eg, retinitis pigmentosa, hereditary optic atrophy or progressive and acquired with retinal, choroidal or vitriretinal degeneration–eg, cataract, glaucoma, optic atrophy, retinal degeneration and, the 'classic' cause of nyctalopia, vitamin A deficiency

nightfloat GRADUATE EDUCATION A house officer–HO on a dedicated night schedule who takes admissions for an on-call team after a certain hr–eg, midnight; the nightfloat's admission are transferred to another team in the morning. See Call schedule. See Circadian rhythm, Libby Zion, Sleep disorders

nightmare Anxiety dream, dream anxiety attack PSYCHIATRY An anxiety-provoking dream occurring during REM sleep, accompanied by autonomic nervous system hyperactivity ONSET Begins in childhood usually before age 10, more common in girls, often seen in normal childhood unless they interfere with sleep, development or psychosocial development; nightmares in adulthood are often associated with outside stressors or coincide with another mental disorder; nightmares usually occur during REM sleep and include unpleasant or frightening dreams; they are most common in the early morning, and may follow frightening movies/TV shows or emotional situations, but may be associated with psychological disturbances or severe stress, especially in adults TREATMENT None. Cf Sleep terror disorder

night-owl insomnia Delayed sleep phase syndrome, see there

night-shift fatigue SLEEP DISORDERS Tiredness during the 'graveyard' shift– between 11 pm and 7 am, characterized by disruptions in memory, loss of attention, brief sleep episodes–microsleeps, ↓ work quality. See Circadian rhythm, Libby Zion, Sleep disorders

nightstick fracture FORENSIC MEDICINE A solitary ulnar fracture occurring when the arm is raised to parry the blow of a nightstick, used by police, security guards, etc

nightsweats Nocturnal, often drenching, diaphoresis, a finding classically associated with terminal Hodgkin's disease and TB, but also seen in trypanosomiasis, giant cell–temporal arteritis, and following prolonged intense exercise

NIH Stroke Scale NEUROLOGY A somewhat cumbersome system for stratifying stroke victims who are candidates for thrombolytics PARAMETERS MEASURED Level of consciousness, orientation, ability to obey simple commands, ability to visually trace an object, visual field, facial palsy, motor strength of all 4 extremities, coordination, language, speech clarity. See Cincinnati Prehospital Stroke Scale, Stroke, Thrombolytics

nihilistic delusion PSYCHIATRY The delusion of the nonexistence of the self, part of one's self, or of some object in external reality

NIHL Noise-induced hearing loss, see there

nil disease Minimum change disease, see there

nilutamide RU 23908 ONCOLOGY A nonsteroidal antiandrogen that binds irreversibly to androgen receptors INDICATIONS Metastatic androgen-responsive prostate CA; when combined with LHRH analogues, achieves a relatively complete androgen blockade, preventing fatal flare-up reactions seen in early treatment of metastatic prostate CA; it may be superior to flutamide. See Disease flare-up. Cf Flutamide

Nimbex Cisatracurium ANESTHESIOLOGY A neuromuscular blocker used during tracheal intubation, for skeletal muscular relaxation during surgery or mechanical ventilation

NINDS NEUROLOGY A multicenter, double blinded, randomized trial–National Institute of Neurologic Disorders and Stroke which evaluated the effects of tPA therapy in Pts with stroke. See Thrombolytic therapy, tPA

911 EMERGENCY MEDICINE A telephone number that in most of the US provides the public with rapid access to communication centers linked to mobile EMS units via a radio network, which are in turn linked to an on-line supervising physician at a hospital GLOBAL VILLAGE September 11, 2001, the day that three 747 jets were used to suicide bomb 2 key US buildings–the World Trade Center, NYC and the Pentagon in Washington, DC

ninhydrin Triketohydrindene CLINICAL TOXICOLOGY An oxidizing reagent that reacts with amino acids and proteins, used to screen urine specimens for presence of, and semiquantify, α-amino acids, as it changes colors

NINOS PEDIATRICS A clinical trial–Neonatal Inhaled Nitric Oxide Study which evaluated the effect of NO therapy–NOT in hypoxic respiratory failure–HRF in neonates. See Respiratory distress syndrome

Nintendo® surgery Laparoscopic surgery, telepresence surgery A popular term for surgical procedures, performed at a distance often with a joystick from a highly restricted operating field, using endoscopy and laparoscopy with applied video, miniaturization, computer-based technologies

NIOSH National Institute for Occupational Safety & Health, see there

NIOSH RECOMMENDATIONS FOR SAFETY & HEALTH STANDARDS

AGENT	NIOSH REL*/OSHA PEL†	HEALTH EFFECTS
Acetylene	≤2500 ppm/2500 ppm	Asphyxia
Alkanes-octane	75 ppm (≤385 15 min)/500 ppm	Skin, CNS
Ammonia	50 ppm/50 ppm	Respiratory and ocular irritant
Benzene	0.1 ppm, TWA†† (< 1 ppm 15 min)/	
	1 ppm, TWA (<5 ppm 15 min)	Leukemia
Carbon tetrachloride	2 ppm ceiling/20 ppm (100 ppm 30 min)	Liver cancer
Chlorine	0.5 ppm (15 min)/1 ppm (15 min)	Respiratory and ocular irritant
Ethylene oxide	0.1 TWA (5 ppm 10 min)/1 ppm TWA	Mutations, leukemia, mesothelioma
Formaldehyde	0.16 ppm (1 ppm 15 min)/	
	1 ppm, TWA (2 ppm 15 min)	Nasal cancer
Hydrogen cyanide	4.7 ppm (10 min ceiling)/10 ppm	Blood, thyroid, respiratory effects
Hydrogen sulfide	10 ppm/20 ppm (50 ppm 10min)	Irritation-CNS, respiratory effects
Methylene chloride	As low as possible/500 ppm (2000 ppm 5 min)	Cancer
Nitric oxide	2 ppm/2ppm	Dental erosion, nasal, lung irritation
Nitroglycerin	0.1 mg/m³ (20 min)/2 mg/m³	Circulatory effects
Phenol	5.2 ppm (15.6 15 min)/5 ppm	Skin, eye, CNS, liver, kidney effects
Sulfuric acid	1 mg/m³/ 1 mg/m³ avoid skin & eye contact	Pulmonary irritation
Toluene	100 ppm (200 ppm 10 min)/	
	200 ppm (500 ppm 10 min)	CNS depression
Xylene	100 ppm (200 ppm 10 min)/100 ppm	CNS depression, lung irritation

*Relative exposure limit

†Permissible exposure limit

††8 hr-time weighted average; all time values refer to 'ceiling' levels

Complete NIOSH table is available from Pubs Dissemination, Div Standards Development & Tech Transfer, NIOSH, 4676 Colombia Pkwy, 45226 ☎ 1.513.533.8287

niosome A microsphere–0.1–2 μm in diameter delivery device, at the center of which is an aqueous cavity enveloped by one or more regularly arranged layers of non-ionic amphilic lipids in lamellar phase

nip & tuck A popular term for cosmetic surgery, see there

Nipent® Pentostatin, see there

nipple discharge Breast discharge BREAST DISEASE Serous or serosanguinous fluid emanating from a nipple, most common in peri- and postmenopausal ♀, due to various lesions–eg, intraductal papilloma, nipple adenoma, ductal ectasia, Paget's disease of breast, advanced ductal carcinoma

nisoldipine core coat Sular® CARDIOLOGY A long-acting CCB with a slow onset of action INDICATIONS HTN, angina EFFECT ↑ exercise duration, ↑ time to ST segment depression, ↑ time to angina onset. See Calcium channel blockers

Nissen fundoplication NEONATOLOGY A laparoscopic procedure for treating reflux esophagitis and GERD. See Gastroesophageal reflux disease, Nissen fundoplication

nit MEDICAL ENTOMOLOGY An empty louse egg shell, deposited by *Pediculus capitis*, *P corporis*, *Phthirus pubis*–'crabs'. See Louse

nitric oxide A multifaceted bioregulatory agent and environmental pollutant, capable of causing genotoxicity ENVIRONMENT A gas byproduct of high temperature combustion–eg, internal combustion engines which, on exposure to light, results in NO_2 formation, an irritating air pollutant, and major greenhouse gas PHYSIOLOGY Endothelium-derived relaxing factor A neurotransmitter released when glutamate binds to the NMDA receptor, which is critical in regulating vascular tone. See Nitric oxide synthase

nitrogen embolism An air embolism-like event, which causes the 'bends' in divers who surface too rapidly; nitrogen 'boils' in vessels or tissues, causing joint and abdominal pain; if it affects the brain, it may be fatal. See the Bends, Caisson's disease, Embolism, Nitrogen narcosis

nitrogen mustard ONCOLOGY Any of a family of alkylating agents used primarily to treat NHLs, Hodgkin's disease, ALL, mycosis fungoides; NMs enter cells via the choline transport system; tumor cells may develop mustard resistance through enhanced repair of alkylated DNA or thiol-mediated mustard inactivation TOXICITY GI tract–N&V, myelosuppression with pancytopenia, alopecia, tissue injury, diarrhea, diaphoresis

nitrogen narcosis Rapture of the deep SPORTS MEDICINE A neurologic response to ↑ nitrogen gas dissolved in blood, resulting in euphoria, apathy, loss of judgement; NN occurs in scuba divers, and is most common at depths below 20 m PREVENTION Use helium-oxygen gas. See the Bends, Caisson disease

nitroglycerin CARDIOLOGY Glycerol trinitrate An organic nitrate that is a short-acting agent for treating anginal pain and CHF SIDE EFFECTS Headache, tachycardia, nausea, hypotension; other organic nitrates–eg, ethylene nitrate, trinitrotoluene–TNT, are used to produce explosives. See Monday death

Nitropress® Nitroprusside sodium, see there

nitrosamines CLINICAL TOXICOLOGY A class of complex organic nitrogen molecules, formed in the stomach by a reaction between nitrites and the amine groups of certain proteins, or ingested preformed in beer and certain drugs, or absorbed from cigarette smoke–the nitrosamine levels in smokers is up to 8-fold that of nonsmokers; the reaction between nitrites and amines can be inhibited with antioxidants–eg, vitamins C and E, etc; nitrosamines are implicated in CA of stomach, esophagus, nasopharynx, urinary bladder, etc. Cf Nitrates, Nitrites, Nitrogen, Sodium nitrate

nitrosourea ONCOLOGY Any of a group of lipid-soluble alkylating anticancer drugs–carmustine–BCNU, lomustine–CCNU, which cross the blood-brain barrier, and are used to manage brain tumors

nitrous oxide NO_2 ANESTHESIOLOGY The most commonly used inhalation anesthetic PHARMACOLOGIC EFFECTS Analgesia, delirium, loss of motor control, elation, excitement USED FOR Pre-operative, short minor surgery

nitrovasodilator THERAPEUTICS An agent–eg, hydralazine, nitroglycerin

Nivalen Diazepam, see there

nizatidine Axid® THERAPEUTICS An H_2-receptor antagonist used to treat peptic ulcer disease ADVERSE REACTIONS Diarrhea, headache, drowsiness, fatigue, muscle pain, constipation, confusion, agranulocytosis, gynecomastia, impotence, allergic reactions, tachycardia, arrhythmias, interstitial nephritis, etc. See Histamine receptor antagonists

Nizoral® Ketoconazole, see there

NMD Neuromuscular disease, see there

NMDA antagonist NEUROLOGY Any of a number of agents that protect against brain damage in neurologic disorders–eg, stroke; PCP–phencyclidine and ketamine are effective, but have psychotomimetic effects or damage neurons in the cerebral cortex; diazepam and barbiturates act at the GABA receptor channel complex, and may prevent the unwanted side effects of NMDA antagonists

NMP22® test UROLOGY A urine assay for identifying Pts at risk for recurrent bladder CA; Pts with high NMP22–nuclear matrix protein levels 10 days post-surgery tend to have recurrent CA within 3 months. See Bladder cancer

NMR 1. Neonatal mortality risk, see there 2. Nuclear magnetic resonance. See Magnetic resonance imaging

NNRTI Non-nucleoside reverse transcriptase inhibitor, see there

no choice plan MANAGED CARE A health care plan paid for by an employer, in which the employee has no choice of a health plan–ie, employer picks, employees join

no code orders See Do not resuscitate–DNR

no-fault insurance An insurance scheme under which a victim is compensated for an injury without an individual or organization being assigned responsibility, regardless of the cause of the injury

'no' food HEALTH A popular term for those foods and substances–eg, alcohol, junk food, refined sugar, tobacco, and others that should be curtailed or deleted from the diet. See Fast food, Junk food, Nutritional therapy. Cf 'Go' foods

'NO' FOODS

Alcohol–allowed in moderation	Drugs of abuse
Artificial sweeteners	'Fast' foods, eg pizza, hamburgers,
Bleached flours	hot dogs
Cakes, pastries, pies	'Junk' foods, eg pretzels, potato chips
Candy, ice cream, chocolate	Refined sugar
Carbonated soft drinks	Salt
Coffee and other caffeinated beverages	Tobacco

no-impact sport SPORTS MEDICINE A physical activity or sport in which there is minimal wear or trauma to weight-bearing joints EXAMPLES Bicycling, sailing, scuba diving, swimming. Cf High-impact sport, Low-impact sport, Moderate-impact sport

No-Man's land HAND SURGERY A fanciful term for the fibrous sheath of the flexor tendons of the hand, specifically in the zone from the distal palmar crease to the proximal interphalangeal joint. See Rule of threes

no masses No lumps BREAST EXAMINATION Written documentation that indicates that the breast was examined by a physician or other health care provider and no discrete firm regions or areas were identified

no meningeal sign NEUROLOGY Written documentation that indicates that a Pt was evaluated and found negative for clinical signs of meningeal irritation

"no nit" policy PUBLIC HEALTH A stance taken by many US public schools excluding children from attendence if *Pediculus capitis* eggs–nits are identified in the hair, allowing them to return to school only when they have been found negative for eggs. See Head lice (*Pediculus capitis*)

no observable effect ENVIRONMENT *adjective* Referring to a lack of observable changes associated with a potentially dangerous or toxic chemical. See Bounty hunter provision, No significant risk, Proposition 65

no observed adverse effect level TOXICOLOGY The concentration of a chemical in a study, or group of studies, that produces no statistically or biologically significant ↑ in frequency or severity of adverse effects between an exposed population and an appropriate control. Cf Lowest-observed-adverse-effect level

no-reflow phenomenon CARDIOLOGY The finding that restoration of antegrade flow in an occluded epicardial artery in acute MI may not result in recovery of microvascular tissue perfusion; Pts with NRP have lower 1-month ejection fractions, larger end-diastolic volumes, and more persistent heart failure; Pts at high risk with significant zones of no reflow after thrombolytics might need emergency coronary angiography, and benefit from early and aggressive ACE inhibition to prevent LV remodeling. See Ventricular remodeling

no-show A Pt who does not present for an appointment

Noan Diazepam, see there

Nobel Prize An award given to a person who has provided seminal thought and work in a particular area of human endeavor and has been recognized for that work by the Nobel Prize committee of the Swedish Academy; NPs are awarded in Physiology or Medicine, Chemistry, Physics, Peace, Economics

Nobelist Nobel laureate A person whose seminal thought and work in a particular area of human endeavor has been recognized by the Nobel Prize committee of the Swedish Academy. See Matthew effect, Nobel Prize

nocardiosis INFECTIOUS DISEASE Infection with *Nocardia* spp, in particular, *N asteroides*, which usually presents as a suppurative abscessing lesion of skin and lungs; dissemination to the brain is usually fatal

nocebo CLINICAL TRIALS A negative placebo effect that may occur when Pts in a clinical trial of a drug therapy recognize–or think they recognize that they are getting a placebo–ie, not receiving therapy, and fare worse due to the effect of negative suggestibility. Cf Placebo

nociceptor Pain receptor NEUROLOGY Any of a class of periarticular and mucocutaneous sense organs and neural receptors–eg, reflex loops for reception and response to pain; located primarily in the skin or viscera, nociceptors respond to chemical, mechanical, or other stimuli

nocturnal bruxism PSYCHIATRY Clenching of jaw or grinding of teeth while sleeping

nocturnal dyspnea SLEEP DISORDERS Respiratory distress, minimal during the day, which is disturbing in sleep

nocturnal emission Night visitor, polluting dream, sex dream, wet dream Semen seeping while sleeping; NE occurs during REM sleep and may be accompanied by erotic dream(s)

nocturnal enuresis MEDTALK Bed-wetting, see there

nocturnal myoclonus 1. Restless legs, see there; a condition of periodic lower-leg movements during sleep with associated daytime sleepiness, or complaints of insomnia. See Sleep disorder 2. Sleep-related myoclonus, see there

nocturnal penile tumescence SEXOLOGY The spontaneous erection of the penis during sleep occurring from birth to advanced old age, typically, 3 episodes/night, for a total of 2-3 hrs (!!!); NPT occurs during REM sleep and is accompanied by erotosexual dreams. See Postage stamp test

nocturnal sleep PHYSIOLOGY Nighttime, or major, sleep period dictated by the circadian rhythm of sleep and wakefulness; the conventional time for sleeping

node CARDIOLOGY An intrinsic pacemaker of the heart, composed of neural tissue. See Atrioventricular node, Sinoatrial node DECISION-MAKING Any point in a decision tree where choices occur or results assigned. See Chance node, Decision node, Outcome node PATHOLOGY 1. A circumscribed tissue mass. See Heberden's node, Singer's node 2. Lymph node, see there. See Axillary node, Irish's node, Potato node, Sentinel node, Virchow's node

node-negative ONCOLOGY *adjective* Referring to a cancer that has not spread to regional lymph nodes. See Axillary tail

nodovenous shunt SURGERY Surgical decompression to reduce future lymphedema due to lymphatic blockage

nodular *adjective* Referring to a nodule; bumpy, lumpy

nodular melanoma A melanoma that comprises 15% of melanomas; NM is similar clinically to superficial spreading melanoma; 50% average 5-yr survival. See Melanoma

nodule A small node, bump, swelling, protuberance. See Apple jelly nodule, Cold nodule, Hot nodule, Pseudorheumatoid nodule, Renal nodule, Rheumatoid nodule, Satellite nodule, Sister Mary Joseph nodule, Solitary thyroid nodule, Surfer's nodule, Tobacco nodule, Typhoid nodule, Warm nodule

NOGA map INTERVENTIONAL CARDIOLOGY A 3-D electromechanical image of the left ventricle, using an ultra-low magnetic-field energy source and a sensor-tipped catheter to locate the catheter position; NMs allow separation of normal and infarcted or ischemic myocardium

noise ELECTRONICS Random variation in signals of the electromagnetic spectrum that carries no useful information from the source. See Cymatics, Music therapy, Sound therapy TECHNOLOGY Poisson noise Fluctuation in the number of information carriers–photons, electrons, which appears as 'snow' in a cathode ray tube, a function of the statistical variation of the rays received by the detector and number of electrons produced by the photomultiplier. See White noise. Cf Chaos, Pink noise

noise-induced hearing loss Temporary or permanent hearing loss caused either by a single exposure to very loud sound(s) or by repeated exposure to louder sounds over an extended period. See Hearing loss

noise pollution OCCUPATIONAL HEALTH Noise and sounds in the workplace and environment that are annoying or excessive to the point of causing lost productivity PREVENTION Active noise control. See Cymatics, Sound therapy, Toning

noisy breathing PULMONARY MEDICINE Breathing in which there are random fluctuations in rhythm. See Ataxic breathing

NOK Next-of-kin, see there

Nolvadex® Tamoxifen citrate, see there

noma Gangrenous stomatitis, cancrum oris ENT An acute necrotizing, polymicrobial infection of orofacial tissues seen in malnourished children, which rapidly erodes to deep tissue, exposing bone and teeth MICROBIOLOGY Anaerobic fusospirochetes–eg, *Borrelia vincenti* and *Fusobacterium* *nucleatum*, rarely, *Bacteroides melaninogenicus* and filiform gram-negative bacteria TREATMENT High dose IV penicillin; correct dehydration and malnutrition. See Herpes stomatitis

nomenclature Any system for assigning names to a particular structure. See Binomial nomenclature, Classification, SNOMED, SNOP, Taxonomy ALTERNATIVE MEDICINE The names used in alternative health care often overlap with those of mainstream medicine and other fields and may confuse practitioners of both types of medicine, as well as Pts

NOMENCLATURE–ALTERNATIVE MEDICAL TERM–SOURCES OF CONFUSION

DIFFERENT USES FOR SAME TERM Eg, colonic irrigation is used in mainstream medicine for flushing the colon in preparation for emergency surgery, and in alternative health as a synonym for colon therapy, the practice of performing multiple enemas to flush out putative toxins; similarly, herbologists use the same names for medicinal plants as used by horticulturists, which may or may not refer to the same plants; Example: geranium for ornamental use and for medicinal use

DIFFERENT TERMS FOR THE SAME ENTITY Eg, homeopaths use a latinized term, *Natrum muriaticum*, for table salt–sodium chloride; similarly, some herbs are known by the trivial name, eg, rue, and blood root, while the homeopathic remedies based on these same plants take the Latin name, *Ruta*, and *Sanguinaria*, respectively

Nomina Anatomica ANATOMY The former official standard for anatomic terminology, sanctioned by the Intl Congress of Anatomists; the NA was replaced by the updated Terminologia Anatomica produced by the Federative Committee on Anatomical Terminology. See Terminologia Anatomica

nominal CARDIAC PACING *adjective* Referring to a set of parameters which are usually values that adequately pace most patients; a pacemaker can be programmed to nominal parameters *noun* The value of a setting at which a device will operate optimally under normal conditions. See Cardiac pacing

nomogram An alignment chart in which there are ≥ 3 abscissas, each measuring a specific parameter and mathematically or empirically defined as having a relationship to the other two; the abscissas are placed parallel to each other on an open 2-D field; a straight line drawn when 2 parameters are known allows determination of the value of a 3rd–unknown parameter

non-accidental trauma A euphemism for child abuse, see there

non-A, non-B hepatitis A group of hepatitides which are the major cause of transfusion-related hepatitis INCIDENCE 7/10⁵/yr–US RISK FACTORS 42% IV drug abuse, 40% unknown risk factors, 6% sexual contact, 6% blood transfusion, 3% household contact, 2% health care professional; of the ±150,000 new cases/yr–US, 30-50% become chronic carriers–20% of these develop cirrhosis; NANBH may also be enteric–parenteral; NANBH is most commonly due to HCV and enteric NANBH to HEV MANAGEMENT IFN-α2b results in significant histologic reversal and serum ALA response PROGNOSIS No ↑ in overall mortality with chronic NANBH, but ↑ liver-related deaths. See Hepatitis

non-articular rheumatism Fibromyalgia, see there

non-bacterial endocarditis Marantic endocarditis Endocarditis with 1-5 mm sterile vegetations on both faces of the valve leaflets composed of fibrin and clots with possible rupture of the papillary muscles, a condition clinically indistinguishable from infectious endocarditis characterized by petechiae, fever, murmurs and emboli ETIOLOGY Debilitating diseases–eg, terminal malignancy, protracted malnutrition, collagen vascular disease, chronic sepsis, renal failure

non-bacterial prostatitis UROLOGY Inflammation of the prostate of abrupt onset often caused *Chlamydia* spp or *Ureaplasma* spp CLINICAL Associated with or follows UTIs, urethritis, or epididymitis, more common in ♂ age 20 to 35, who have unsafe sex and/or multiple sexual partners. See Prostatitis

non-cardiac chest pain INTERNAL MEDICINE Chest pain that simulates cardiac nosologies, but is unrelated to cardiovascular disease; 50% of Pts with NCCP have known reflex and may have postprandial or noctural Sx. See Gastroesophageal reflux disease

NON-CARDIAC CHEST PAIN SHAMELESSLY TAKEN, VIRTUALLY VERBATIM FROM WWW.VNH.ORG/GMO/ CLINICALSECTION/08CHESTPAIN.HTML, FROM DEPT NAVY, BUREAU OF MED & SURG; INTERNALLY PEER REVIEWED

SOURCES OF CHEST PAIN The heart, great vessels, pericardium; GI tract; lungs & pleura; chest wall

HOW TO MINIMIZE YOUR RISKS OF MANAGING ACUTE CHEST PAIN Identification of ischemic chest pain requires a high index of suspicion; when the diagnosis of acute MI is overlooked and Pts are sent home–the mortality during the next 72 hrs is about 25%–how did you spell that phrase again, "…out-of-court settlement."–vs ± 6% for Pts with infarction who are hospitalized; being liberal in admissions for evaluation of CAD; incidence of acute MI in Pts hospitalized with acute chest pain is between 25 and 30%; despite conservative admission rates, clinicians misdiagnose ±5-10% of Pts with acute MI–ie, you're in good company if you screw up

HISTORY The Hx rules decision making; elements of the Hx important in discriminating cardiac from noncardiac chest pain are quality, severity, duration and frequency; knowledge of exacerbating features and maneuvers that ameliorate the discomfort are helpful; cardiac risk factors should not overly influence clinical thinking; the presence of risk factors simply implies that a person is more likely to develop overt signs of ASHD in the future, but are not exclusionary criteria

PAIN CHARACTER Chest pain due to coronary ischemia is classically a dull heavy pressure–but Pts have been pretty colorful in use of adjectives to describe this pain; they may have classic pain, DON'T expect a classic description of anginal pain; the pain may be confined to the chest or accompanied by aching in one or both arms, more often the left; neck or mandibular pain or aching confined to the shoulder, wrist, elbow, or forearm may manifest solely or with typical chest pressure; small zones of pain are generally not of myocardial origin; radiation of pain to the digits, brief zaplets of pain or discomfort that persists for days are not due to myocardial ischemia; effort or emotional stress commonly provokes angina; angina may occur at rest if perfusion is compromised; pain subsides within 1 to 5 mins if the triggering activity is discontinued; nitroglycerin hastens this relief

EKG The 12-lead ECG has limited value in excluding the presence of CAD; excluding the Dx of angina pectoris or acute MI because of a normal ECG is as great an error as inferring a diagnosis of CAD from the incorrect interpretation of nonspecific electrocardiographic abnormalities

COCAINE Cocaine causes ↓ coronary blood flow due to vasoconstriction; rhabdomyolysis, a complication of cocaine use, provides another mechanism for the chest pain; all chest painers should be questioned about cocaine use and, when appropriate, have a urine drug screen

GI TRACT Pain from the GI tract, especially the esophagus, may give rise to angina-like chest discomfort; GERD is the most common esophageal cause of noncardiac chest pain; it is described as a burning sensation or squeezing pain located in the retrosternal area between the xyphoid and suprasternal notch; listen for clues about association of Sx with meals, posture, and relief by belching or antacids; medical management involves dietary modifications, smoking cessation, and histamine type 2 (H2) antagonists or antacids; GI referral is warranted when these interventions are unsuccessful in alleviating Sx; the pain of peptic ulcer disease may also occur high in the epigastrium or lower chest; relationship to meals and relative nonresponse to nitroglycerine helps distinguish this pain from angina pectoris

ESOPHAGUS SPASM Diffuse esophageal spasm is a neuromuscular disorder characterized by chest pain and difficulty in swallowing; NOTE Nitroglycerin promptly relieves esophageal spasm causing confusion in the diagnosis; vigorous disordered contractions in the body of the esophagus are induced by ingestion of cold liquids or normal swallowing during a meal; anxiety and stress are also common precipitating factors; there is usually no exertional component but ↑ abdominal pressure from lifting, sit-ups, or running can cause reflux; diagnosis rests on history and verification of esophageal spasm by manometric studies

PULMONARY ORIGIN Pain of pulmonary origin characteristically has a distinct pleuritic quality varying with the respiratory cycle; intercostal nerves supply sensory afferents to the costal parietal pleura; inflammation arising from this region is appreciated in the adjacent chest wall; referred pain originating in the diaphragm is appreciated in the ipsilateral shoulder; differentiating features of pulmonic from musculoskeletal pain are the more intense nature of pleuritic pain and the worsening of musculoskeletal pain by extension, abduction, or adduction of the arm and shoulder; pain centered around involved muscle groups may also distinguish musculoskeletal from pleuritic chest pain; (a) Spontaneous pneumothorax tends to occur in young adult males producing sharp pleuritic chest discomfort and dyspnea; (b) Pulmonary embolus may produce pleuritic pain, however, dyspnea, and tachypnea are most frequent. Inciting factors for pulmonary embolus include the post-operative period after long recumbent or inactive periods and following trauma where the same immobility may result in venous stasis and thrombosis.

CHEST WALL Tietze's syndrome or costochondritis is a self-limiting discomfort. Its quality is sharp or burning and is exacerbated by mechanical activity of the chest wall, specifically respiration; the second or third costal cartilages on either side are the most common area of involvement, but any of the costochondral articulations can be involved; NSAIDs or aspirin may offer temporary relief but reassurance tends to be as useful.

ETC Rarely, no etiology is found on standard evaluation of chest pain from the cardiology or GI consultation; one should then rule out panic disorder, visceral hypersensitivity in irritable bowel syndrome, and other exotica

non-coital sex SEXOLOGY Any form of safe sexual activity that does not include vaginal penetration–eg, masturbation, frottage. See Safe sex practices

noncommitted operation CARDIAC PACING An operation in an A-V sequential–DVI pacemaker whereby intrinsic ventricular activity sensed during the A-V interval can inhibit delivery of a ventricular output pulse. See Committed operation

noncompete clause MEDICAL PRACTICE A clause in a contract in which the provider of a specific service, commonly understood to be physicians in private practice, agrees not to practice medicine–ie, compete–in the same geographic region–the size of which is defined by the contract

noncompliance MEDTALK The disregarding of a prescribed treatment plan; the degree to which Pt does not adhere to a prescribed diet or treatment, and whether the Pt returns for re-examination, followup or treatment. See Bad patient, Directly observed therapy. Cf Compliance

noncontrollable life-style specialty A specialty that cannot be scheduled in terms of work hrs EXAMPLES General surgery, obstetrics, primary care–eg, family practice, internal medicine, pediatrics. Cf Controllable life-style specialty

noncoverage *adjective* Referring to a lack of insurance benefits, usually used in the context of limited access to 'covered' medical care. Cf Coverage

nondepolarizing agent ANESTHESIOLOGY An agent–eg, pancuronium, used to facilitate airway management, control alveolar ventilation, abolish motor reflexes, and provide muscle relaxation for surgery. See Pancuronium, Pipercuronium, Vecuronium. Cf Depolarizing agent

nondisclosure MALPRACTICE Negligent nondisclosure, see there RESEARCH ETHICS The withholding of information about financial interests–stocks, consultancy fees, and other arrangements–that a researcher might have in the outcome of a clinical trial of a therapeutic device or agent

nondistended MEDTALK Not enlarged, of normal size. Cf Distended

nonduplication of benefits HEALTH INSURANCE A provision in some health insurance policies specifying that benefits will not be paid for amounts reimbursed by others

nonedematous *adjective* Not swollen; without edema

nonepileptic seizure Pseudoseizure, psychogenic seizure NEUROLOGY A condition characterized by episodic changes in behavior, accompanied by somatosensory or other seizure-like events, unrelated to electrical activity in the brain ASSOCIATIONS Pts with NS may also have mood disorders, bipolar disorder, dysthymic disorder, separation anxiety, school refusal DIFFDX Psychogenic disorders–eg, somatiform disorders, anxiety disorders, dissociative disorders, psychotic disorders, impulse control disorders, attention-deficit disorders, factitious disorders, cardiovascular disease–eg, syncope, arrhythmias, TIAs, breath-holding spells, migraines, movement disorders–eg, tremors, dyskinesis, parasomnia and sleep-related disorders, GI disease–eg,

cyclic vomiting syndrome, episodic nausea, others–eg, daydreams, malingering. See Psychogenic seizure

nonesterified fatty acids See Free fatty acids

nonfocal NEUROLOGY *adjective* Referring to a seizure which is not localized; generalized

non-gonococcal bacterial arthritis Septic arthritis, see there

non-gonococcal urethritis An STD defined as the presence of abundant PMNs in urine; NGU is more common in heterosexuals than gonococcal urethritis CLINICAL Dysuria, pyuria and Sx similar to gonorrhea ETIOLOGY *C trachomatis, Ureaplasma urealytica, Mycoplasma genitalium*; NGU is often culture negative

non-hemolytic transfusion reaction TRANSFUSION MEDICINE Immune reactivity to homologous WBCs in a previously sensitized blood product recipient, which occurs in 0.5–5.0% of all transfusions, and in up to 50% of Pts with β° thalassemia PREVENTION NTRs are minimized by using leukocyte-depleted blood products. See Leukocyte reduction

non-Hodgkins' lymphoma A lymphoid cancer that is not Hodgkins disease: NHL ALL, B cell lymphoma, Burkitt's lymphoma, diffuse cell lymphoma, follicular lymphoma, immunoblastic large cell lymphoma, lymphoblastic lymphoma, mantle cell lymphoma, mycosis fungoides, post-transplantation lymphoproliferative disorder, small non-cleaved cell lymphoma, T-cell lymphoma EPIDEMIOLOGY Incidence $\uparrow$ $8.5/10^5$ in 1973$\rightarrow$15/10^5 in 1990 (21/10^5, in 2000, Canada $\male$), ±53,900 new cases/yr, US in 2002; 60% of all lymphomas are NHLs, of which 55% are diffuse, 45% are nodular RISK FACTORS AIDS, primary immunodeficiency, immunosuppression, transplants, exposure to pesticides, hair dyes, smoking, alcohol, in older $\female$–linked to $\uparrow$ consumption of meat/animal fats– MANAGEMENT Chemotherapy, especially CHOP COMPLICATIONS Spinal cord compression occurs in up to 10%, which is often aggressive, and may respond to high-dose RT. See Lymphoma, REAL classification, Working Formulation

NON-HODGKIN'S LYMPHOMA

STAGE I CA present in only one lymph node area or in only one extranodal regional/organ

STAGE II CA present in ≥ 2 lymph node areas on same side of diaphragm; CA present in only one extranodal area or organ outside the lymph nodes and in the lymph nodes around it. Other lymph node areas on the same side of the diaphragm may also have cancer

STAGE III CA present in lymphoid tissue on both sides of the diaphragm; tumor may also have spread to an area or organ near the lymph node areas and/or to the spleen

STAGE IV CA has spread to > one organ or organs outside lymphoid tissue; CA has spread to only one organ outside the lymph system, but lymph nodes distant from that organ are involved

nonhypertension syndrome Small cuff syndrome, see there

nonimmune hemolytic anemia HEMATOLOGY A general term for hemolysis caused by various 1. chemicals–eg, antimalarials including quinolones, sulfonamides, nitrofurantoin, phenazopyridine, and others *or* 2. physical agents–eg, metals including arsenic, chromium, copper, lead, nickel, platinum, snake venom, hypotonic solutions. See Hemolytic anemia

noninflammatory arthritis Osteoarthritis, see there

noninflammatory joint disease See Osteoarthritis

noninnovative drug 'Me too' drug, see there

non-insulin-dependent diabetes mellitus Type 2 diabetes mellitus, see there

non-invasive Non-interventional *adjective* Referring to therapy in which the integrity of mucocutaneous barriers is not violated. Cf Invasive

non-invasive positive pressure ventilation PULMONOLOGY The delivery of assisted mechanical ventilation using a lightweight machine by means of a tight-fitting mask over the nose or nose and mouth. See Positive end pressure

non-ionizing radiation Electromagnetic radiation, the photons of which lack the energy required to ionize atoms or induce ion formation; NIR includes sound, ultraviolet, visible, and infrared light, microwaves, and radiowaves. Cf Ionizing radiation

non-ketotic hyperglycemic coma NKHHC; nonketotic hyperglycemic coma ENDOCRINOLOGY A complication of type 2 DM that results in high serum glucose levels without ketone production CLINICAL ↓ consciousness, extreme dehydration, ↑ glucose, no DKA; NHC is most common in previously undiagnosed or poorly controlled diabetics, precipitated by infection or drugs that impair glucose tolerance or ↑ fluid loss; normally the kidneys compensate for high serum glucose by spilling glucose; when dehydration occurs, kidneys conserve fluid and glucose levels ↑ further, creating a vicious cycle of ↑ dehydration and ↑ glucose RISK FACTORS Elderly, underlying renal failure, CHF; recent discontinuation of insulin or oral hypoglycemics, precipitating events–eg, infection, stroke, recent surgery or other trauma. See Diabetes mellitus, Diabetic ketoacidosis

nonlinear *adjective* Not linear, random

nonmaleficence MEDICAL ETHICS A central guiding principle of the ethical practice of medicine, first expressed by Hippocrates, and translated into Latin as *primum non nocere*, first do no harm

nonmelanoma skin cancer 1. Basal cell carcinoma, see there 2. Squamous cell cancer, see there 3. Skin adnexal carcinoma 4. Cutaneous lymphoma

nonmendelian disorder GENETICS Any complex genetic disease–eg, HTN, DM, ASHD, which does not follow a simple mendelian patterns of inheritance. Cf Mendelian disorder

nonmetastatic ONCOLOGY *adjective* Referring to a cancer that has not spread beyond the primary site to other sites in the body. See Metastasis

nonmodifiable risk factor MEDTALK Any risk factor–eg, heredity, for a particular condition–eg, breast CA, which cannot be modified

nonmotile sperm UROLOGY Adynamic sperm

nonneurolytic block ANESTHESIOLOGY A form of anesthesia in which a local anesthetic agent is injected locoregionally, occasionally in combination with corticosteroids; NBs include injections into trigger points, stellate ganglion, or other sympathetic chain sites. See Nerve block, Neurolytic block

nonnucleoside reverse transcriptase inhibitor AIDS Any of the antiretroviral–ie, anti-HIV agents–eg, delavirdine and nevirapine which inhibit viral nonnucleoside reverse transcriptase and are combined with nucleoside RTIs to manage HIV infection. See AIDS, Antiretroviral, Reverse transcriptase inhibitor. Cf Delavirdine, Nevirapine, Protease inhibitor

nonnutritive dietary component NUTRITION Any critical component of a well-balanced diet that has an 'ancillary' function; NNDCs include water, fiber–soluble and insoluble, and a host of molecules present in plant and animal foods with as-yet unknown functions. Cf Micronutrients

nonpalpable PHYSICAL EXAM *adjective* Referring to that which cannot be felt. Cf Palpable

nonparous Nulliparous, see there

nonparticipation The nonacceptance by a physician of the fees paid by Medicaid, or less commonly by Medicare. See Medicaid. Cf Participation

nonpitting edema PHYSICAL EXAM The lack of indentation when fingertip pressure is applied to the skin, which classically occurs in hypothyroidism–in which it is termed myxedema, but also occurs in rosacea, hand-foot syndrome, surrounding the eschar in cutaneous anthrax, etc. Cf Pitting edema

nonprogressive HIV-1 infection AIDS Chronic HIV infection; ± 5% of HIV-infected persons may survive for prolonged periods of time–eg, 10-15+ yrs without developing AIDS, a feat variously attributed to a feisty immune response to HIV, preservation of lymphoid tissue, and attenuated HIV-1

nonproprietary *adjective* Generic, see there

non-Q-wave infarction CARDIOLOGY A subendocardial MI in which the EKG demonstrates persistent abnormal ST-segment depression in all but the aVR lead–which has ST-segment elevation–often accompanied by T-wave changes. See Myocardial infarction, Thallium imaging

nonrandomized trial Nonrandomized control trial CLINICAL TRIALS A study in which Pts are assigned to an arm–intervention, nonintervention–in a nonrandom fashion. Cf Randomized trial

non-rapid eye movement sleep Non-REMS sleep. See Sleep stages

nonsense syndrome Ganser syndrome PSYCHIATRY A condition in which a person gives 'astonishingly' incorrect answers to simple questions; although considered to be factitious in nature, 'nonsense Sx' may occur in hysteria, schizophrenia, or transiently in normal subjects under stress or when fatigued. Cf Factitious 'diseases'

nonsequitur NEUROLOGY A totally unrelated response to a question ETIOLOGY Acute and chronic organic brain disorders, schizophrenia

nonsexual genital assault A kick in or aimed at the groin and external genitalia

non-small cell carcinoma of lung A differentiated lung CA–squamous cell carcinoma, adenocarcinoma, large cell carcinoma MANAGEMENT Preoperative chemotherapy–cisplatin, ifosfamide, mitomycin–and RT ↑ median survival in advanced stage NSCC–26 vs 8 months. Cf Small cell carcinoma of lung

NON-SMALL CELL LUNG CANCER

STAGE I CA confined to lung

STAGE II CA spread to nearby lymph nodes

STAGE III CA spread to structures near lung; mediastinal lymph nodes: or to lymph nodes on the opposite side of the chest or neck

IIIA Resectable

IIIB Nonresectable

STAGE IV Metastasis

nonsmoking tobacco see Smokeless tobacco

nonspecific back pain See Low back pain

nonsteroidal anti-inflammatory drug Any of a family of weak organic acids that 1. Inhibit prostaglandin biosynthesis, by inhibiting cyclooxygenase, and to a lesser degree lipooxygenase and 2. Interfere with membrane-bound reactions–eg, NADPH oxidase in neutrophils, phospholipase C in monocytes and G protein-regulated processes USES Rheumatoid arthritis, gouty arthritis, ankylosing spondylitis, osteoarthritis, serosal inflammation, Bartter syndrome, and other inflammatory conditions CHEMICAL CLASSES Carboxylic acid: acetylated–eg, aspirin–figure or nonacetylated–eg, sodium salicylate; acetic acid analogues–eg, indomethacin, tolmetin, sulindac; propionic acid analogues–eg, ibuprofen, naproxen; fenamic acid analogues–eg, mefenamic acid, enolic acid analogues–eg, oxyphenbutazone, phenylbutazone; nonacidic compounds–eg, proquazone SIDE EFFECTS Rash, pruritus, edema, vertigo, drowsiness, tinnitus, aseptic meningitis, N&V, gastric ulcers, potentially fatal GI hemorrhage, jaundice, Stevens-Johnson syndrome, Henoch-Schönlein syndrome, aplastic anemia, acute renal failure. See NSAID enteropathy

nonstress test OBSTETRICS An indirect non-invasive monitor of the well-being of a fetus, where the frequency of fetal movement, degree of heart rate acceleration and beat-to-beat variation of the heart rate are monitored to determine the 'health' of the placental vasculature. See Deceleration, Montevideo units. Cf Fetal heart monitoring, Stress test

nontender MEDTALK Relatively insensitive to pressure by palpation. Cf Tender

nontherapeutic research NIHSPEAK Research unlikely to produce a diagnostic, preventive, or therapeutic benefit to current subjects, which might benefit Pts with a similar condition in the future

nontoxic *adjective* Referring to that which does not cause damage or harm

nontoxic goiter See Goiter

non-traditional cancer therapy Unproven method for cancer management, see there

nontropical sprue Gluten enteropathy, see there aka celiac sprue

nontuberculous mycobacterium Any mycobacteria that does not cause TB, which is not usually spread from person to person–eg, *M avium-intracellulare* complex, *M kansasii*, *M marinum*, *M ulcerans*, and others, which may respond to IFN-γ therapy

non-ulcer dyspepsia GI DISEASE A condition characterized by ulcer Sx in absence of ulceration; 30-60% of dyspeptics have no demonstrable gross lesions CLINICAL Sx range from a classic duodenal ulcer–epigastric burning 1-3 hrs after meals, relieved by food or alkali to functional indigestion–bloating, belching, fullness, and nausea, not relieved by antacids and worsened by meals; fat intolerance is common ENDOSCOPY The duodenal mucosa demonstrates edema, erythema, petechial hemorrhage, erosions. See GERD. Cf Dumping syndrome

nonunion fracture ORTHOPEDICS A fracture unhealed after 9 months. See Pseudoarthrosis

nonverbal communication 'Body language', see there

nonviable fetus OBSTETRICS An expelled or delivered fetus which, although living, cannot possibly survive to the point of sustaining life independently, even with support of the best available medical therapy. See Viable Infant

Noonan syndrome Turner-like syndrome NEONATOLOGY A group of specific abnormalities affecting both males and females, both sporadic in appearance but also reflecting a hereditary component, possibly AD CLINICAL Webbing of neck, pectus excavatum, facial defects–low-set or abnormally shaped ears, ocular ptosis, hypertelorism, epicanthal folds, micrognathia, mild mental retardation, short stature, variable hearing loss, delayed puberty, undescended testicles, small penis, congenital heart disease–often, pulmonary stenosis. Cf Turner syndrome

noradrenaline Norepinephrine, see there

norastemizole ALLERGY MEDICINE A nonsedating antihistamine for allergic rhinitis, which is an active metabolite of astemizole–Hismanal® without its cardiac side effects. See Antihistamine

Norco™ PAIN MANAGEMENT An analgesic formulation of acetaminophen and hydrocodone. See Acetaminophen, Hydrocodone

NORD National Organization for Rare Disorders A private non-profit organization that 1. Acts as a clearing house of information for orphan diseases–see there and 2. Facilitates communication among governmental agencies and the research community–eg, by locating Pts for clinical trials WEBSITE www.rarediseases.org/. See NICODARD, Orphan disease, Orphan drug/product

nordiazepam NEUROPHARMACOLOGY A long-acting sedative/hypnotic. See Benzodiazepine

NORDIL CARDIOLOGY A clinical trial–Nordic Diltiazem Study which evaluated the effect of antihypertensives in Pts with HTN. See Calcium channel blockers

Norditropin® Somatropin ENDOCRINOLOGY A recombinant DNA product used for long-term treatment of children with growth failure due to inadequate endogenous pituitary GH. See Growth hormone

nordoxepin See Doxepin

norfloxacin ANTIBIOTICS A broad-spectrum–gram-negative bacilli, staphylococci–fluoroquinolone with limited activity against streptococci and anaerobes. See Fluoroquinolone

Norfranil Imipramine, see there

Norian SRS SRS injectable cement ORTHOPEDICS A skeletal repair system consisting of injectable cancellous bone and replacement material to provide structural support to fractures, ↓ risk of loss of anatomic positioning while healing. See Bone paste

Noritate® Metronidazole, see there

Norline Nortryptyline, see there

normal curve EPIDEMIOLOGY A bell-shaped curve that results when a normal distribution is graphed. See Normal distribution

normal for age MEDTALK *adjective* Referring to a physical characteristic or developmental milestone which is typical for an average person of similar age. See Developmental delay, Red flag

normal pressure hydrocephalus Adult hydrocephalus, communicating hydrocephalus, idiopathic hydrocephalus NEUROLOGY A form of gradual onset progressive hydrocephalus, which accounts for ±5% of all dementias ETIOLOGY Idiopathic, obstruction to CSF flow; CSF is produced normally but not reabsorbed; the brain's ventricles enlarge to accommodate the ↑ volume of CSF; CSF pressure remains normal; it is accompanied by brain atrophy due to compression by the fluid-filled ventricles RISK FACTORS Conditions causing obstruction of CSF–eg, closed head injury, neurosurgery with craniotomy, meningitis, subarachnoid hemorrhage. See Hydrocephalus

normal saline Physiologic saline solution, see there

normal sleep SLEEP MEDICINE A quality and length of sleep that is normal for a particular person DIVISIONS 2–5% stage 1, 45–55% stage 2, 13–23% stage 3–short-wave, 20–25% REM sleep. See REM sleep, Sleep apnea

normal volunteer NIHSPEAK A person used to study normal physiology and behavior or who does not have a condition being studied in a particular protocol, used to compare with subjects who do have the condition

Normaton Buspirone, see there

Normide Chlordiazepoxide, see there

Normiflo® SURGERY An LMW heparin indicated for preventing DVT in Pts undergoing hip or knee replacement. See Low-molecular weight heparin

Normison Temazepam, see there

normoactive *adjective* Referring to bowel sounds typical of a person without GI tract disease

Normodyne® Labetalol, see there

normophilia Heterosexuality SEXOLOGY Sexuoerotic activity that conforms to the dictates of custom, religion, or law. Cf Paraphilia

normotensive CARDIOLOGY *adjective* Referring to a blood pressure that is normal–117/72 mm Hg; daytime, 122/77; nighttime 106/6, based on met-analysis. See Hypertension, Hypotension

Norpace® Disopyramide, see there

Norplant® Levonorgestrel GYNECOLOGY An implantable contraceptive that prevents pregnancy by inhibiting ovulation and thickening the cervical mucus ADVERSE EFFECTS Dysmenorrhea, headache, nervousness, nausea, vertigo, ↑ size of ovaries and fallopian tubes, dermatitis, acne, weight gain, breast tenderness, hirsutism. See Contraceptive

Norpramin® Desipramine, see there

Norpress Nortryptyline, see there

Norrie's disease OPHTHALMOLOGY A bilateral XR condition of early onset, characterized by deafness, mental retardation, cataracts, and pseudoglioma

North American operation SURGICAL ONCOLOGY Radical surgery of a 'frozen pelvis', consisting of radical en bloc resection of the uterus and urinary bladder. See 'Frozen pelvis'. Cf 'All-American' and 'South American' operations

Northern blotting MOLECULAR BIOLOGY A technique used to detect the presence of a specific mRNA sequence. See Blotting, Hybridization, Probe, RNA, Southern blotting

Nortriptyline NEUROPHARMACOLOGY A tricyclic antidepressant. See Tricyclic antidepressants

Norvasc® Amlodepine besylate, see there

Norvir® Ritonavir, see there

Norwalk agent(s) Montgomery County virus, Norwalk virus VIROLOGY Any of a group of 27 nm parvoviruses–single-stranded DNA viruses of the Caliciviridae family, which cause 'winter vomiting disease', first described in Norwalk, Ohio, incriminated in up to 40% of nonbacterial epidemic gastroenteritis in the US and a frequent cause of traveler's diarrhea EPIDEMIOLOGY NAs are transmitted in an oral-fecal fashion–exposure to recreational swimming water, ingestion of raw shellfish, cake-frosting, stored water on cruise ships, etc CLINICAL After a 1-2 day incubation, it may present explosively, often accompanied by N&V, abdominal cramping, anorexia, malaise, myalgia; NA is difficult to identify–it doesn't grow in cell culture; there are no animal models PROGNOSIS Spontaneous resolution without sequelae. See Gastroenteritis

Norwalk-like virus VIROLOGY Any of a group of viruses with biologic, clinical, and immunologic findings similar to those of the Norwalk agent(s). see Gastroenteritis, Hawaii agent, Norwalk agent(s), Otofuke virus, Snow Mountain virus

NOS inhibitor THERAPEUTICS An agent that ↑ systemic vascular resistance and mean arterial pressure–MAP by inhibiting nitric oxide synthase activity. See Nitric oxide synthase

nose ANATOMY The double-barrelled structure at the center of the face, which is a conduit of air for non-mouth breathers, and a support for eyeglasses. See Cocaine nose, Internal nose, Potato nose, Rabbit nose, Saddle nose, Sculptured nose, Stinky nose, Tapir nose, WC Fields nose DRUG SLANG A regionally popular street term for cocaine

nose candy DRUG SLANG A street term for cocaine. See Cocaine, Snorting

nose coverage Prior acts coverage MALPRACTICE INSURANCE A supplement to a claims-made malpractice insurance policy that may be purchased from a new carrier when a physician changes carriers and had claims-made coverage with a previous carrier; nose coverage covers incidents that occurred before the beginning of the new insurance relationship but for which no claim has been made; in NC, the insured's liability is assumed by a prior malpractice insurance carrier, usually until the physician has a new policy. Cf Tail coverage

nose hose MEDTALK Nasogastric tube, see there

nose job Rhinoplasty, see there

nosebleed ENT Hemorrhage from the highly vascularized nasal mucosa ETIOLOGY Trauma, spontaneous with dried nasal mucosa, common in dry climates in winter, infection, trauma, rhinitis, HTN, alcohol abuse, inherited coagulopathies and therapy with anticoagulants–eg, coumadin, warfarin, aspirin, or other anti-inflammatory agents HERBAL MEDICINE Yarrow, see there VOX POPULI → MEDTALK Epistaxis

nosocomial *adjective* Relating to a hospital, commonly referring to an infection acquired while interned in a hospital. See Iatrogenic. Cf Community acquired

nosocomial AIDS AIDS HIV infection/AIDS caused during the course of providing some form of health care

NOSOCOMIAL AIDS, CAUSES

BLOOD TRANSFUSIONS A unit may be in HIV 'window' period–1:500,000 risk

CONTAMINATED PRODUCTS Factor VIII concentrates that were not heat-treated were administered to hemophiliacs in France, Japan, and elsewhere, causing thousands to become infected

CONTAMINATED DEVICES Re-use of needles in disadvantaged regions–in Russia, 2 outbreaks of HIV-1 seroconversion were reported, infecting 58 and 23 children respectively

nosocomial anemia Anemia of investigation, see there

nosocomial infection Cross infection EPIDEMIOLOGY An infection that begins ≥ 3 days after admission to a hospital or other healthcare facility; the microorganisms that cause nosocomial infection are a function of 1. The underlying disease: burns are associated with *Pseudomonas aeruginosa*, leukemia with enterobacteriaceae due to indwelling vascular accesses, DM with anaerobes, GNRs, and post-operative wounds with *S aureus* and 2. The organ system involved, often facilitated by an indwelling catheter, tracheostomy or other device; NIs affect 2-4 million Pts/yr–US at a cost of ± $4.5 × 10⁹ SITES Urinary 39%, lower respiratory 18%, surgical wound 17%, blood 7.5%, other 19%

nosocomial pneumonia An infection of lungs–bronchoalveolar unit–in a Pt who has been hospitalized ≥ 48 hrs, and directly attributable to pathogens acquired during the hospital visit ETIOLOGY *Pseudomonas* spp, *S aureus*, *Legionella* spp MANAGEMENT 3rd generation cephalosporins, aminoglycosides–eg, gentamicin. Cf Community-acquired pneumonia, Nosocomial infection, Reverse isolation

nosology INFORMATICS The branch of medical science that deals with orderly relationships among or classifications of disease. See Composite clinical data dictionary

not covered HEALTH CARE *adjective* Referring to a procedure, test or other health service to which a policy holder or insurance beneficiary is not entitled under the terms of the policy or payment system–eg, Medicare. Cf Covered

not palpable PHYSICAL EXAM *adjective* Referring to that which cannot be touched or felt, usually in the context of bedside examination of the breast or internal organs

notching RADIOLOGY Small grooves on the anterior aspect of ribs seen on a plain CXR of children with post-ductal–ductus arteriosus coarctation of the aorta, due to 'tracks' from the pressure of collateral vessels on the ribs

notebook INFORMATICS A small laptop computer, see there RESEARCH A book in which research data is recorded or otherwise documented. See Raw data. Cf Logbook

notes Hand-written or typed information, which documents a particular task. See OR notes, Progress notes

notice FORENSIC MEDICINE Information on a pending court action–eg, a hearing MANAGED CARE See Advance beneficiary notice

notice of privacy practices See HIPAA

notifiable disease PUBLIC HEALTH A disease, usually an infection, the occurrence of which must, by law, be reported to a local health officer or government authority, given the local, state, or federal interest in maintaining the disease under surveillance in order to control and prevent its dissemination. See Reportable occupational diseases

Novacor® An implantable left ventricular assist device, which provides self-regulating circulatory support for end-stage heart disease, serving as a mechanical bridge while Pt awaits a heart transplant

Novaldex® Tamoxifen, see there

Novantrone® Mitoxantrone, see there

Novazam Diazepam, see there

novelty diet See Diet, Fad diet

novelty-seeking behavior PSYCHOLOGY A behavioral pattern which may be typical of persons who engage in high-risk and extreme sports or who abuse drugs. See Extreme sports, High-sensation seeking trait

Novidorm Triazolam, see there

Novocain® Procaine, see there

Novolorazem Lorazepam, see there

NovoPen® 3 ENDOCRINOLOGY An insulin delivery system for injecting 2 to 70 units of various insulin formulations. See Diabetes mellitus

Novopoxide Chlordiazepoxide, see there

Novopramine Imipramine, see there

NovoSeven® HEMATOLOGY A recombinant coagulation factor VIIa indicated for treating bleeding episodes in hemophilia A or B Pts with antibodies to coagulation factors VIII or IX

Novotriptyn Amitriptyline, see there

Novus fusion cage ORTHOPEDICS A device used for spinal fusion, which is a threaded, hollow cylinder filled with bone graft and implanted into disk material between vertebrae. See Spinal fusion

noxious music NEUROLOGY Strident music capable of inducing seizures or other adverse responses. See Aura, Noise, Noise pollution. Cf Music therapy, Sedative music, Stimulative music

noxious thing FORENSIC MEDICINE A "…substance unlawfully administered to another, or taken by oneself with a deliberate intent to cause ill effects or death. (which) may be a poison or any substance capable of producing injury…". See Toxin

nPA Novel plasminogen activator CARDIOLOGY A thrombytic that dissolves clots and restores coronary blood flow in Pts with an acute MI. See InTIME trials, Lanoteplase, tPA

NPC Nasopharyngeal carcinoma, see there

NPI National Provider Identifier, see there

NPO Nothing by mouth

NREMS Non-rapid eye movement sleep

NREMS period SLEEP DISORDERS The NREMS portion of the REMS cycle; such a period consists primarily of sleep stages 3/4 early in the night and sleep stage 2 later. See Sleep cycle; Sleep stages

NSABP National Surgical Adjuvant Breast Project ONCOLOGY A series of ongoing multicenter clinical trials evaluating the effects of various therapies, including RT, surgery and chemotherapy–eg, tamoxifen and 5-FU, in treating advanced breast or colorectal CAs

NSAID Nonsteroidal anti-inflammatory drug, see there

NSAID enteropathy An enteropathy induced by NSAIDs CLINICAL Intestinal inflammation, occult blood loss, protein-losing enteropathy, iron-deficiency due in part to nonspecific small intestinal ulceration–in particular of the jejunum and ileum, with hemorrhage and perforation, intestinal strictures, ileal stenoses. See Nonsteroidal anti-inflammatory drug

NSCLC Non-small cell lung cancer, see there

NSILA Non-suppressible insulin-like activity An action displayed by acid-dissociable 7.5 kD serum complex with activity of IGF-I, IGF-II, somatomedins A and C; most NSILA is associated with a HMW protein–NSILP, which has significant homology with IgG's Fc fragment. See Insulin-like growth factor

NSR Normal sinus rhythm, see there

NST Nonstress test, see there

NSTEMI Non-ST elevation myocardial infarction

NSU Nonspecific urethritis, see there

NTD Neural tube defect, see there

nth admission A popular term for multiple hospital admissions–more than the 6th admission, and measured as an 'nth', of an ordinal number–eg, seventh, eighth, ninth, tenth, eleventh, etc. See 'Still' syndrome

NTHI Native tissue harmonic imaging IMAGING An ultrasound technique for high-frequency, high-resolution difficult-to-image Pts without contrast; NTHI provides clear images of structures as small as nerve bundles in the median nerve of the wrist

ν Greek nu; symbol for 1. Degrees of freedom 2. Frequency, as expressed in hertz 3. Neutrino

Nubain® Nalbuphine, see tthere

nuchal translucency sign OBSTETRICS A thickening of the nuchal fold seen by ultrasonography in the first trimester fetus, once said to be specific for Down syndrome. See Down syndrome

nuclear cardiology IMAGING The use of nuclear imaging techniques in the noninvasive study of cardiovascular disease–eg, myocardial perfusion imaging, planar imaging, SPECT–single-photon-emission computed tomography, infarction imaging. See First-pass myocardial infarction imaging, Myocardial perfusion imaging

nuclear:cytoplasmic ratio N:C ratio A crude parameter used in cytology and surgical pathology, where interphase–ie, nondividing, nuclei are proportionately larger than the surrounding cytoplasm; the N:C ratio is usually ↑ in malignant cells

nuclear family SOCIAL MEDICINE The core family unit, which typically consists of heterosexually-oriented ♂ and ♀ partners and their direct–usually unmarried genetic progeny. Cf Companionship, Extended family, Marriage bonus, Most significant other, Social isolation. Cf Single-parent family

nuclear magnetic resonance See Magnetic resonance imaging, NMR

NUCLEAR MAGNETIC RESONANCE

DIAGNOSTIC NM

- In vivo–eg injection of radiocontrast to detect an ↑/↓ in local 'signal'–uptake in suspected bone metastases
- In vitro–assays of clinical specimens, eg radio immunoassay–RIA for various hormones, eg human chorionic gonadotropin–β-HCG, insulin, and TSH

THERAPEUTIC NM Radiation oncology Administration of a 'hot' isotope to sterilize a particular area, and prevent malignant cells from growing & dividing

Note: RIAs are being replaced by ELISAs, which are easier to perform, the reagents are more easily stored, and do not have the problems inherent in using and disposing of radioactive waste; nuclear medicine also encompasses in vivo diagnostics in the form of scintillation counters to 'scan' various body regions for the presence of increased uptake of radionuclides, which when focal, implies primary neoplasia or metastasis

nuclear medicine A clinical discipline that uses radioisotopes to diagnose and treat disease

nuclear scanning Radionuclide imaging IMAGING Any diagnostic procedure–eg, bone scan, liver scan, thyroid scan–that use a radioisotope–eg, 99m-Technitium–^{99m}Tc or 123-iodine–^{123}I, linked to a molecule that selectively concentrates in a particular tissue; after administration, the compound's distribution in the body is evaluated using a scintillation camera; any region that is larger, brighter, or located in different sites than normal is regarded as diagnostic or at least suspicious of having a disease process. See Bone scan, Thyroid scan

nucleoside analogue MOLECULAR MEDICINE An artificial nucleoside which, when incorporated into viral DNA during replication, prevents production of new virus; NAs can inhibit DNA production in healthy cells; a molecule that structurally mimics a nucleoside. See AIDS, Zidovudine

nucleoside analog reverse transcriptase inhibitor NRTI, NARTI, "nuke" AIDS A class of drugs that includes AZT–Retrovir, ddI–Videx, ddC–HIVID, d4T–Zerit, 3TC–Epivir, abacavir–Ziagen; reverse transcriptase is the part of HIV that assembles viral DNA; NRTIs inhibit reverse transcriptase by acting as 'dummy' nucleosides. See AIDS

Nuctane Triazolam, see there

NUD Non-ulcer dyspepsia, see there

nuke VOX POPULI A popular term for 1. Any activity intended to render a population/location devoid of life–eg, by incineration, exposure to chemicals, and various forms of irradiation–eg, IR or UV light 2. Heating food or tissues with a microwave

null hypothesis STATISTICS A hypothesis that assumes that if there are no differences between 2 populations–or sets of data being compared, a statement of probabilities–P value can be made; the proposition, to be tested statistically, that the experimental intervention has "no effect," meaning that the treatment and control groups will not differ as a result of the intervention. The NH is a statistical assumption based on data which demonstrates an association of 2 events or factors in > 95% of cases. See Hypothesis testing. Cf Alternative hypothesis

nulligravida OBSTETRICS A ♀ who has not been pregnant

nullipara OBSTETRICS A ♀ who has not given birth to a viable child, or delivered an infant

number VOX POPULI A symbol of a value. See Accession number, Burst number, CLIA number, Common account number, Copy number, ID number, Linkage number, Magic number, Mass number, Medicare Indentification number, IP number, MIM number, Primary accession number, Reynolds' number, Secondary accession number, Threshold number, UPIN number, Winding number, Writhe number

number 'crunching' INFORMATICS A popular term for intense and complex mathematical computations–eg, 3-D rotation of complex molecules and images

number needed to treat DECISION-MAKING The minimum number of Pts to whom a particular intervention must be administered in a trial or controlled study to prevent a single target event. See Absolute risk reduction, Odds ratio, Relative risk reduction, Threshold NNT

numeracy Mathematical literacy NEUROLOGY The ability to understand mathematical concepts, perform calculations and interpret and use statistical information. Cf Acalculia

numerator EPIDEMIOLOGY The upper part of a fraction

nurse *noun* A person who has received the appropriate–determined by jurisdiction–education and training in the discipline of nursing; a person specially trained to provide services essential to or helpful in the promotion, maintenance, and restoration of health and well-being; a person skilled in nursing. See Advanced practice nurse, Charge nurse, Circulating nurse, Medication nurse, Registered nurse, Scrub nurse, Traveling nurse *verb* 1. To breast-feed 2. To care for an infirm individual

nurse endoscopist An advanced practice–ie, specialized nurse trained to perform endoscopic examinations–eg, of the lower GI tract. See Advanced practice nurse, Nurse practitioner, Physician extender

nurse midwife Certified nurse midwife, see there

nurse practitioner NURSING A registered nurse with an advanced nursing degree and/or training who can render medical services with a limited degree of autonomy under the aegis of a physician supervisor. See Physician extender

nursemaid's elbow Pulled elbow ORTHOPEDICS Subluxation of the radial head, caused by a longitudinal 'yank' on the forearm–by a nursemaid, etc, forcing the child's elbow into extension; the subluxation is reduced by firm supination at 90° and extension, followed by immobilization with a posterior splint or a sling. See Elbow

Nurses' Health Study CARDIOLOGY A large cohort study that evaluated the effect of exogenous HRT on the risk of cardiovascular disease. See Estrogen replacement therapy, Osteoporosis

nursing *adjective* 1. Breast-feeding, see there 2. The provision of nursing care. See Forensic nursing, Intensive nursing

nursing home MANAGED CARE A licensed facility which provides general long-term nursing care to those who are chronically ill or unable to handle their own necessary daily living needs; NHs are staffed by nurses, and have a physician on call. See Geriatrics, Home health care. Cf Hospice

nut NUTRITION A dry fruit with an edible kernel enclosed in a leathery or woody shell–eg, peanuts, almonds, walnuts, a vegetarian food staple. See Almond, Healthy foods, Walnut VOX POPULI A kook, a loony

nutate MRI *verb* To tip on an axis, specifically referring to that which occurs when hydrogen nuclei are subjected to magnetic fields in magnetic resonance

nutcracker esophagus Corkscrew esophagus, see there

nutcracker phenomenon Nutcracker syndrome A clinical finding in Hb SC disease, in which left-sided renal hemorrhage causes infarction of renal papillae, due to ↑ pressure as the left renal vein passes between the aorta and the superior mesenteric artery; ↑ pressure causes renal medullary hypoxia sufficient to sickle the RBCs

Nutrasweet™ Aspartame, see there

nutrient FOOD INDUSTRY A substance added to foods to ↑ vitamin, mineral and protein content NUTRITION A general term for proteins, carbohydrates, fats, vitamins and minerals, necessary for growth and maintenance of life. See Food additive, Macronutrient, Micronutrient

nutriphile VOX POPULI A formal term for a health food 'nut'

nutrition The study of the metabolic utilization of foods. See Applied nutrition, Malnutrition, Parenteral nutrition, Total parenteral nutrition

nutritional assessment ONCOLOGY The profiling of a Pt's current nutritional status and risk of malnutrition and cancer cachexia. See Cachexia, Malnutrition

nutritionist Dietitian, see there

Nutropin® Somatropin ENDOCRINOLOGY A recombinant growth hormone–GH for treating growth failure associated with Turner syndrome, GH inadequacy, and chronic renal insufficiency before kidney transplantation. See Growth hormone

NuvaRing® GYNECOLOGY An intravaginal contraceptive ring that releases low doses of progestins for 3 wks, followed by a ring-free wk. See Contraception

Nuvera™ GYNECOLOGY An oral 17-beta estradiol/norethindrone acetate for managing postmenopausal Sx. See Menopause

nyctophobia PSYCHOLOGY Fear of the dark or night. See Phobia

nymphomania Female hypersexuality PSYCHIATRY A popular term for a ♀ psychosexual disorder characterized by ↑↑↑ in sexual activity and desire, viewed in the psychoanalytic context as a reponse to an inferiority complex and/or a need for affection; the compulsive condition in a ♀ of recurrent sexual intercourse with different ♂ partners, promiscuously and without falling in love, but not as a paid prostitute or call girl. Cf Don Juan syndrome, Satyriasis, Sexual addiction, Sexual compulsivity

Nyotran™ INFECTIOUS DISEASE A lipid-based, IV formulation of nystatin for treating *Aspergillus* and *Candida* infections in Pts with immune suppression due to HIV infection, antirejection therapy for organ transplants, or adverse effects of chemotherapy. See Nystatin

nystagmus OPHTHALMOLOGY Rapid involuntary oscillary eye movements DIRECTIONS Horizontal, vertical, rotary. See Caloric nystagmus, Irritative nystagmus, Paralytic nystagmus, Recovery nystagmus, Opticokinetic nystagmus, Seesaw nystagmus

nystatin Mycostatin INFECTIOUS DISEASE An antifungal obtained from *Streptomyces noursei*, which is effective against oral, GI, and urogenital candidiasis ADVERSE EFFECTS Dysgeusia, N&V, diarrhea, stomach pain.

central groove and minimal cytoplasm **EM** Neurosecretory granules, containing hormones–ACTH, ADH, bombesin, calcitonin, CRF, estrogen, FSH, hGH, histaminase, HPL, LH, MSH, PTH, renin, serotonin **CLINICAL** ±3-month survival without therapy, often with cerebellar degeneration–80% have brain metastases **TREATMENT** Non-surgical; ±85% respond to combination chemotherapy–CCNU, cyclophosphamide, doxorubicin, vincristine, etoposide–VP-16 **PROGNOSIS** Often fatal in 2 yrs, 50% of those with limited metastases respond to therapy and 15-20% survive 2 yrs; RT may control bone pain, spinal cord compression, superior vena cava syndrome and bronchial obstruction. See Small cell carcinoma

OATS™ Osteochondral autograft transfer system **ORTHOPEDICS** A device for harvesting and transfer of autogenous bone plugs for repairing articular cartilage defects in the knee and ankle

OB Symbol for 1. Obstetrics, obstetrician, see there 2. OB protein

obese *adjective* Characterized by obesity, see there; excessively fat

obesity **ENDOCRINOLOGY** A state of excess body fat, which is regarded as a premorbid addiction disorder, defined as 20% above a person's standard weight; the ideal body weight is 21 kg/m^2 **EPIDEMIOLOGY** 59% of Americans are clinically obese, according to a 1995 report by the Institute of Medicine, there has been a 54% ↑ in obesity and a 98% ↑ in superobesity in children 6-9 yrs of age; an obese child is often an obese adult; the patterns may be established by 3 months of age and linked to ↓ energy expenditure in infants of obese mothers; diet-resistant obesity is characterized by an inability to lose weight despite ↓ caloric intake and ↑ exercise; a certain percentage of diet-resistant obesity is related to underreporting of actual caloric consumption and/or overreporting of physical activity, not due to low energy expenditure **ETIOLOGY, 2° OBESITY** Endocrine-hypothyroidism, Cushing syndrome, hypogonadism–Fröhlich syndrome, polycystic ovaries, pseudohypoparathyroism **PATHOGENESIS** ↑ Lipid deposit in fat cells, ↓ mobilization of lipids from adipocytes, and ↓ lipid utilization; obesity mimics lab findings of type 2 DM–insulin resistance, ↑ glucose, ↑ cholesterol, ↑ TGs, ↓ HDL-C and norepinephrine and depressed sympathetic and parasympathetic nervous systems **CO-MORBID CONDITIONS** See Obesity-related disease **MANAGEMENT** Diet–balanced hypocaloric or individualized, exercise, behavior modification, hypnosis, bariatric surgery, OTC appetite suppressants, prescription agents–eg, orlistat. See Abdominal obesity, Adipsin, Adult obesity, Body mass index, Central obesity, Childhood obesity, Diet, Eye-mouth gap, Gastric 'balloon', Ideal weight, Morbid obesity, Orlistat, Secondary obesity, Superobesity, Upper body fat obesity

OBESITY, CLASSIFICATIONS OF

AGE OF ONSET, eg juvenile, mature, in pregnancy or other

ANATOMIC

- **ANDROID OBESITY** Central obesity, 'beer-gut' obesity More common in ♂, more central/truncal in distribution; carries an ↑ risk for DM

- **GYNECOID OBESITY** More common in ♀; fat is distributed in the lower abdomen and legs and is less associated with ASHD

PRIMARY/SECONDARY

- **PRIMARY** A component of Allström, Blount, Cohen, Carpenter, Laurence-Moon-Biedl, Prader-Willi, and other eponymic syndromes

- **SECONDARY** Acquired obesity comprises the bulk of obesity

PSYCHOLOGICAL

TYPE OF TISSUE CHANGE, eg hyperplastic or hyperplastic-hypertrophic

obesity-hypoventilation syndrome Pickwick syndrome, see there

obesity-related disease **CLINICAL NUTRITION** Any condition linked in part to obesity–eg, cardiovascular disease, gallbladder disease–cholecystitis, cholelithiasis, gout, adverse lipid profile, ↑ post-operative complications–poor wound healing, insulin resistance, HTN, osteoarthritis, sleep apnea, abnormal GI transit, colorectal CA, strokes, PE, poor wound healing, atelectasis, hepatic steatosis/fibrosis, psychologic disorders. See Obesity

O Symbol for 1. Objective 2. Occiput 3. Oncovin 4. Ophthalmology 5. Oral 6. Orbit 7. Organ 8. Orotidine 9. Orthopedic 10. Os 11. Ovary 12. Ovulation 13. Oxygen

o Symbol for 1. Organ 2. Ortho

O sign **MEDTALK** The open-mouthed slack-jawed appearance of the narcotized, unconscious, stuporose, and dead-drunk. Cf Q sign

O/T Oral temperature

OA Symbol for 1. Occipital artery 2. Occiput anterior 3. Osteoarthritis

oak leaf spot A descriptor for the size and shape of the pigmented cutaneous macules seen in Pts with von Recklinghausen's disease and tuberous sclerosis

OASIS **CARDIOLOGY** A clinical trial–Organization to Assess Strategies for Ischemic Syndromes. See Heparin, Hirudin, OASIS-2 **MANAGED CARE** A system developed by the Centers for Medicare & Medicaid Services, CMS, formerly HCFA, as part of the required home care assessment for reimbursing health care providers; OASIS combines 20 data elements to measure case-mix across 3 domains–clinical severity, functional statuts and utilization factors. See HHRG

Oasis™ **CARDIOVASCULAR DISEASE** A catheter system for treating occluded hemodialysis grafts; a low-pressure water jet is used to break up a clot and remove clot fragments. See Deep vein thrombosis

oasthouse urine disease **METABOLIC DISEASE** A rare AR disorder characterized by ↑ α-hydroxybutyric acid in urine and stools, due to defective intestinal absorption of methionine and other amino acids; the GI bacteria ferment the excess methionine into α-hydroxybutyric, α-ketobutyric, and α-aminobutyric acids which are absorbed and excreted imparting an oasthouse-ish odor, **CLINICAL** White hair, failure to thrive, seizures, hypotonia, edema, mental retardation; α-hydroxybutyric acid may also rarely appear in the urine of Pts with PKU

OAT **CARDIOLOGY** A clinical trial–Open Artery Trial

oat cell carcinoma **SURGICAL PATHOLOGY** A type of small cell lung CA, characterized by a dense hyperchromatic oval nucleus, often with a vague

468   **Obesity Task Force classification** BARIATRIC MEDICINE A weight classification based on the body-mass index–BMI, per intl consensus. See Obesity

WEIGHT TYPE–OTF CLASSIFICATION

UNDERWEIGHT	BMI ≤ 18.5
NORMAL	BMI 18.5–24.9
OVERWEIGHT	BMI 25–29.9
OBESE	BMI 25–29.9
CLASS I	BMI 30–34.9
CLASS II	BMI 35–39.9
CLASS III	BMI ≥ 40

obetension A popular term for the relatively common clinical association of obesity with HTN. Cf Diabesity

OB/GYN A common abbreviation for obstetrics and gynecology

OBJECT UROLOGY A clinical trial–Overactive Bladder: Judging Effective Control and Treatment

objective *adjective* Referring to the perception of external events or phenomena in an impartial, impersonal, and unbiased fashion *noun* VOX POPULI A goal; the reason for doing a thing. See Treatment objective

obligate *adjective* Necessarily; without alternative

obligate carrier CLINICAL GENETICS The ♂ or ♀ parent of a child with an AR disorder who, by definition, is a carrier of the defective gene is question; similarly, mothers of ♂ children with an X-R condition are carriers of the defective gene. See Autosomal recessive

OBS Organic brain syndrome, see there

observation MANAGED CARE A nonemergent status for a Pt admitted to a hospital who does not meet acute care critieria and if any of the following apply: (1) stabilization and discharge expected < 24 h; (2) treatment is required for > 6 h; (3) clinical Dx is unclear and can be determined < 24 h. See Observation bed MEDTALK The act of observing; a state of vigilant nonintervention; in observation, a person's condition is closely monitored, but treatment does not begin until Sx appear or change. See Discharge, Post-op, Watchful waiting

observation bed Holding bed HOSPITAL CARE A status recognized by third-party payers–eg Medicare, health insurance companies and others, in which a Pt is admitted to the hospital for a period of 23 hrs and 59 mins–or more, depending on the 3rd party with either a specific 'rule/out' diagnostic consideration–eg, appendicitis, angina, MI, or pneumonia; observation may also refer to a known Pt status, in which a previously diagnosed condition is managed under observation–eg, dehydration, anemia, etc; a popular term for a bed occupied by a person in an outpatient observation status. See Observation services

observation services MANAGED CARE Those services furnished on a hospital's premises–eg, use of a bed and periodic monitoring by nursing or other staff, which are reasonable and necessary to evaluate an outpatient's condition and determine the need for a possible admission as an inpatient; Medicaid defines duration of OS as those lasting ≤ 48 hrs. See 24 hr rule

obsession PSYCHIATRY *'Recurrent & persistent thoughts, impulses, or images* (that are perceived) ... *as intrusive and inappropriate and cause marked anxiety or distress'*

OBSESSIONS ARE

1. Recurrent & persistent thoughts, impulses, or images–TII, that are perceived as intrusive and inappropriate and cause marked anxiety or distress

2. Or that are not simply excessive reponses to genuine real-life problems

3. Active attempts are made to suppress or neutralize the TIIs by some thought or action

4. The person recognizes that the TIIs are products of his/her own mind

COMPULSIONS ARE

1. Repetitive behaviors–eg handwashing, double-checking, mental acts–praying, repeating words silently that a person feels compelled to perform in response to an obsession, or in accord with strictly applied rules

2. Behaviors or mental acts aimed at preventing or reducing distress or preventing some dreaded event or situation, which are not realistically connected with what they are intended to neutralize or prevent, or behaviors that are clearly excessive DSM-IV, 1994

obsessive-compulsive disorder Obsessive-compulsive neurosis, OCD PSYCHIATRY A disabling anxiety disorder characterized by repetitive patterns of intrusive and persistent thoughts–obsessions and behaviors–compulsions that are senseless–eg, rituals, distressing, and extremely difficult to overcome; OCD affects 1-2% of Americans, has a neurophysiopathologic component and may respond to TCAs MANAGEMENT Clomipramine for trichotillomania and other forms of OCD. Cf Obsessive-compulsive personality disorder

obsessive-compulsive personality disorder PSYCHIATRY A condition characterized by '*A pervasive pattern of preoccupation with orderliness, perfectionism, and mental and interpersonal control at the expense of flexibility, openness, and efficiency...*'DSM-IV, 1994. See Pack rat. Cf Obsessive-compulsive disorder

obstetric hypercoagulability profile Obstetric hypercoagulability panel A battery of tests for a pregnant ♀ who may be at risk–eg, repeated abortions for coagulopathies, which supplements an obstetric screening profile, measuring in addition, proteins C and S, anti-thrombin III, lupus anticoagulant, fibrinogen, plasminogen

obstetric profile Obstetric screening panel LAB MEDICINE A standard–CPT-4 code 80055–panel of assays used to evaluate a pregnant ♀'s baseline health status; the OP is used for ♀ not known to have, or to be at risk for, conditions that might complicate L&D. See TORCH panel. Cf Biophysical profile

obstetrician Baby Doc A physician who practices obstetrics, see there MEAT & POTATOES WORK Pulling babies from babes SALARY $165K + 15% bonus

obstetrics The art and science of managing pregnancy, labor and post-delivery of the mother. See High-risk obstetrics, Naturopathic obstetrics

obstruction A blocking or clogging of a tube or vessel. See Ball-valve obstruction, Biliary obstruction, Bladder outlet obstruction, Bowel obstruction, Bypass obstruction, Nasal obstruction, UPJ obstruction

obstruction series IMAGING A series of x-rays obtained from a Pt with suspected obstruction or functional intestinal paralysis

obstructive airways disease Any lung disease–asthma, COPD with airway obstruction, hyperresponsiveness MANAGEMENT Inhaled corticosteroids, maintenance therapy with a β₂-agonist–eg, terbutaline, which ↓ mortality, airway obstruction, hyperresponsiveness. See COPD

obstructive hydrocephalus Noncommunicating hydrocephalus PEDIATRICS Hydrocephalus due to interference with the flow of CSF, resulting in enlarged ventricles; OH may be due to congenital aqueductal stenosis or atresia–eg, Dandy-Walker syndrome, a complication of intracranial infection, violent birth trauma or inherited in an X-R fashion. Cf Communicating hydrocephalus

obstructive jaundice HEPATOLOGY Jaundice due to obstruction of the flow of bile through the cystic duct into the ampulla of Vater and thence into the small intestine. See Jaundice

obstructive sleep apnea Obstructive sleep apnea syndrome SLEEP DISORDERS A clinical complex due to the pathophysiologic response to anatomic defects of the nasopharynx, characterized by loud snoring, nocturnal oxyHb desaturation, disrupted sleep, hundreds of apneic episodes during sleep, resulting in upper airway closure for ≥ 10 secs CLINICAL Daytime drowsiness, especially in obese middle-aged ♂, cardiovascular Sx–eg, apnea-induced arrhythmia, bradycardia, ↑ ventricular ectopic activity CONTRIBUTING FACTORS Alcohol or sedatives before sleep, obesity, nasal obstruction, adenoid/tonsillar hyperplasia, macroglossia, retrognathia, acromegaly, hypothyroidism, HTN, pulmonary HTN MANAGEMENT Individualized–eg, surgery; uvulopalatopharyngoplasty is successful in 50%. Cf Snoring. Cf Narcolepsy, Sleep disorders

obstructive uropathy Chronic bilateral obstructive uropathy, chronic urethral obstruction UROLOGY A condition caused by urine blockage, resulting in ↑ pressure in renal pelvis and ureters which, with time, leads to HTN, renal failure ETIOLOGY Common in ♂–BPH, less common in ♀–due to cystocele; other causes include tumors of the prostate, bladder, ureter CLINICAL Urinary frequency, spasms, incontinence, later, UTIs and renal failure. See Acute bilateral obstructive uropathy, Acute unilateral obstructive uropathy

obtunded NEUROLOGY *adjective* Mentally dulled; "out of it". See Comatose

obturator sign Obturator test INTERNAL MEDICINE A clinical finding in a Pt lying on his back with the right hip and knee flexed and rotated internally and externally; pain in the right lower quadrant is due to irritation of the medial internal obturator muscle, and is 'classically' associated with appendicitis, but may also occur in right pelvic abscesses IMAGING A unilateral ↑ in obturator muscle bulk, seen as a soft tissue bulge on the inner pelvis with medial displacement of the normal fat line, is characteristic of infectious arthritis, but may be seen in traumatic hemorrhage

occipital *adjective* Referring to the occipital region and structures related thereto or contained therein

occipital horn Any of the broad, calcified protrusions exostoses that extend caudally from the base of the skull, seen in X-linked type IX Ehlers-Danlos syndrome, which is related to defective copper metabolism, resulting in 2° lysyl oxidase deficiency and by extension, collagen defects

occiput presentation Vertex presentation, see there

occiput The back of the head

occlusion MEDTALK 1. The complete closure of a vessel with gas, liquid or solid 2. Obstruction 3. Closure of the upper and lower molars. See Acute vascular occlusion, Aortic occlusion, Central retinal artery occlusion, Malocclusion

occlusive dressing WOUND CARE A dressing that seals a wound to preventing contact with air or moisture

occult *adjective* Not obvious, hidden, of unknown cause *noun* PARANORMAL dee-dee-dee-dee-dee-dee-dee

occult bleeding See Occult blood

occult blood Grossly inapparent blood, usually understood to mean hematochezia, often an early indicator of colorectal CA DIFFDX Amebiasis, heavy metal poisoning, acute GI ischemia, celiac sprue; OB may, with time, cause iron deficiency anemia. See Occult blood testing UROLOGY A rarely used synonym for microhematuria

occult cancer 1. Occult primary malignancy, see there 2. Clinically inapparent malignancy, pathologist's tumor A lesion defined as malignant by histologic criteria, which is an incidental discovery that occurs either when a Pt is being evaluated for other reasons, or at the time of autopsy, as has been well-documented in CA of the prostate and thyroid

occult infection An infection first recognized by 2° manifestations–eg, ↑ PMNs in the circulation or FUO, often caused by a bacterial infection in an obscure site–eg, a subphrenic or other intraabdominal region. See Fever of unknown origin

occult primary malignancy Occult cancer, unknown primary A malignancy of unknown 1° site or origin that is symptomless, which first manifests itself as metastases or secondary–paraneoplastic phenomena, and usually has a poor prognosis; OPMs are problematic as appropriate therapy requires that the primary malignancy be eradicated, and many remain obscure despite aggressive diagnostic work-up; certain malignancies metastasize to certain sites with greater than expected frequency; in OPMs affecting the brain, the primary arises in the lungs in up to 85% TREATMENT Up to 30% of Pts with metastases from an occult primary adenoCA may respond to chemotherapy–mitomycin C, adriamycin, vincristine; poor response is more common in ♂ and in Pts with liver and/or infradiaphragmatic metastases

OCCULT PRIMARY MALIGNANCIES

BONE Breast, bronchus, prostate, thyroid, kidney

CNS Breast, bronchus, kidney, colon

HEAD & NECK Oropharynx, nasopharynx–most are squamous cell carcinoma; also adenocarcinoma, melanoma, rhabdomyosarcoma, oat cell, salivary gland, thyroid carcinomas

LIVER CA of stomach, colon, breast, pancreas, or bronchus

LUNG Breast, colon, kidney, melanoma, sarcoma, stomach, testis, thyroid

LYMPH NODES

• Cervix	Naso– and oropharynx, thyroid, larynx, lymphoma
• Supraclavicular	Bronchi, breast, stomach, esophagus, pancreas, colorectal, lymphoma
• Axillary	Breast, melanoma, lymphoma
• Inguinal	Urogenital tract, anus, melanoma, lymphoma

OVARY Stomach, colon

SEROSAL SURFACES Bronchi, breast, ovary, lymphoma

SKIN Melanoma, breast, bronchus, stomach, kidney

occupant protection device PUBLIC HEALTH Any protective device–seat belt, air bag, child safety seat, booster seat–which ↓ death and/or injury in MVAs. See MVA

occupational asthma CLINICAL IMMUNOLOGY A clinical complex that causes predominantly pulmonary Sx in previously healthy persons exposed to a noxious fumes or gases in the workplace; OA affect ± 3% of Americans, many of whom function adequately, despite Sx. See Hypersensitivity pneumonitis, Monday morning sickness, Sick building syndrome

occupational bronchitis Industrial bronchitis PULMONOLOGY Large airway inflammation affecting those occupationally exposed to cotton, flax, hemp dust, or coal ETIOLOGY Dusts, fumes, strong acids, other chemicals; exacerbated by smoking ASSOCIATED CONDITIONS Chronic bronchitis, coal worker's pneumoconiosis, silicosis, asbestosis. See Occupational pneumonitis

occupational dose RADIATION SAFETY A dose of ionizing radiation received by a person at work, where assigned duties involve exposure to ionizing radiation and radioactive materials

occupational exposure to bloodborne pathogens An event occurring in a healthcare setting, formally defined by OSHA as '…*any reasonably anticipated skin, eye, mucous membrane or parenteral contact with blood or other potentially infectious materials that may result from the performance of an employee's duties…*'

occupational feminization SOCIOLOGY A process in which ♀ account for an ↑ proportion of those employed in a traditionally ♂ occupation. See Gender gap, Glass ceiling

occupational hearing loss AUDIOLOGY A ↓ in auditory discrimination linked to prolonged exposure to excess noise in the workplace caused by various machines–eg, airline industry, automotive repair, construction, entertainment, in an occupational setting. See Hearing loss, Noise-induced hearing loss

occupational malignancy Any work-related malignancy linked to extragenetic factors

occupational medicine

OCCUPATIONAL MALIGNANCY

TYPE	INDUSTRY	AGENT
Hemangiosarcoma, liver	Vinylchloride polymerization	Vinyl chloride workers
	Industry vintners	Arsenical pesticides
Mesothelioma	Asbestos producers/consumers	Asbestos
Nasopharyngeal CA	Cabinet makers, woodworkers	Hardwood dusts
	Boot and shoemakers	Unknown
	Radium workers, dial painters	Radium
	Nickel smelting & refining	Nickel
Laryngeal CA	Asbestos producers and consumers	Asbestos
Bone CA	Radium workers, dial painters	Radium
Scrotal CA	Metal work, lathe operation	Mineral/cutting oil
Bladder CA	Rubber and dye workers	Benzidine, α- & β-naphthylamine, auramine, magenta, 4-aminobiphenyl, 4-nitrophenyl
Kidney, urogenital CA	Coke oven workers	Coke oven emissions
ALL	Rubber industry	Unknown
	Radiologists	Ionizing radiation
AML	Petrochemical industry	Benzene
	Radiologists	Ionizing radiation
Erythroleukemia	Petrochemical industry	Benzene

MMWR 1984, 33 in JAMA 1984; 251:1054

occupational medicine The medical specialty concerned with disease or dysfunction arising in work-related injuries and/or exposure to noxious agents or stimuli; common occupational hazards–eg, asbestos–resulting in pleural plaquing or tumors, noise–hearing loss, solvents, welding fumes, fiberglass–causing upper respiratory irritation and bronchitis and asthma, musculoskeletal dysfunction due to repetitive trauma, heavy metal intoxication, silicosis, toxic hepatitis, dysfunctional mental reactions to the workplace. See NIOSH, OSHA, Postal worker syndrome, Sick building syndrome

occupational therapist A person trained to help people manage daily activities of living–dressing, cooking, etc, and other activities that promote recovery and regaining vocational skills SALARY $51K + 4% bonus. See ADL

occurrence policy MALPRACTICE A type of medical malpractice–professional liability insurance which covers a physician–or other health care professional–for claims arising from medical incidents that occurred during the coverage period of the policy. Cf Claims-made policy

OCD Obsessive-compulsive disorder, see there

ochronosis Alkaptonuria, see there

OCP Oral contraceptive pill, the pill

OCT Oxytocin stress test, see there

OCTAVE CARDIOLOGY A clinical trial–Omapatrilat Cardiovascular Treatment Assessment Versus Enalapril

Octopus® HEART SURGERY An anastomosis site restraining device designed to immobilize a specific area during minimally invasive "beating heart" CABG. See Beating heart surgery, Percutaneous coronary angioplasty

OctreoScan® IMAGING A radiopharmaceutical that binds to specific receptors of small tumors, for early detection and treatment of cancer. See Scanning

octreotide acetate Sandostatin ENDOCRINOLOGY A depot somatostatin analogue with high GH affinity, resulting in a ↓ serum GH and amelioration of Sx in most Pts with acromegaly INDICATIONS (1) Reduction of growth hormone and IGF-1–insulin growth factor in acromegaly; (2) suppression of severe diarrhea and flushing associated with malignant carcinoid syndrome; (3) treatment of profuse watery diarrhea associated with VIPoma–vasoactive intestinal peptide tumor–and diarrhea and flushing associated with certain types of tumors; used to control of chemotherapy-induced diarrhea and

flushing in Pts with metastatic carcinoid tumors and VIPomas, and for treating acromegaly; now also in clinical trials as a potential treatment of severe chemotherapy-induced diarrhea in Pts with colorectal CA

octyl-cyanoacrylate SURGERY A tissue adhesive used to close skin wounds, which sloughs with healing. Cf Sutures

ocular *adjective* 1. Pertaining to the eyeball 2. Ophthalmic *noun* Either of the 2 eyepieces on a binocular LM. Cf Objective

ocular flow analyzer™ INTERNAL MEDICINE A device for measuring intraocular pressure and calculating blood flow of veins through the eye for diagnosing early glaucoma. See Glaucoma

ocular melanoma Melanoma of choroid ONCOLOGY A melanoma linked to sun exposure affecting Pts with light skin and blue eyes CLINICAL Asymptomatic early; eventually causes retinal detachment and vision distortion. See Melanoma

ocular photodynamic therapy OPHTHALMOLOGY A process for managing age-related macular degeneration in which vertoporfin is administered, concentrates in the eye, and the macula is zapped with a laser. See Age-related macular degeneration

ocular plethysmography Oculoplethysmography A noninvasive method for indirectly measuring blood flow through the ophthalmic artery to the brain; in OP, ocular pressure is measured through suction cups placed on the eyes

ocular ultrasonography Orbital ultrasound IMAGING A diagnostic technique that uses ultrasound to produce 2-D images of the eye and surrounding tissues; because MRI and CT images are superior, OU is limited to office practice to screen suspected orbital and periorbital masses

oculocerebrorenal syndrome Lowe syndrome An X-R disorder that maps to chromosome Xq24-26 CLINICAL Congenital cataract, corneal ulceration, hydrophthalmia, glaucoma, mental retardation, renal tubular dysfunction-Fanconi syndrome, aminoaciduria, vitamin D-resistant rickets, areflexia, hypotonia, idiopathic joint swelling LAB Proteinuria, ↑ muscle enzymes, $α_2$-globulin, HDL-C, metabolic acidosis MANAGEMENT Alkalinize urine; supplemental potassium, phosphate, calcium, carnitine

OD 1. Occupational disease 2. Optical density 3. Overdose, see there 3. Right eye–*oculus dexter*

ODD syndrome Oculodental dysplasia An AD condition characterized by hypertelorism, microphthalmia, myopia, hypoplastic teeth, syndactyly, camptodactyly, and visceral malformation, without mental retardation

odds ratio EPIDEMIOLOGY Cross-product ratio, exposure odds ratio overdispersion A measure of association in a case-control study which quantifies the relationship between an exposure and health outcome from a comparative study. See Absolute risk reduction, Number needed to treat, Relative risk

odontoid ORTHOPEDICS *adjective* 1. Tooth-shaped 2. Referring to the odontoid process, see there

odontoid fracture ORTHOPEDICS A fracture caused by 'sudden forward and backward movement of the head with respect to the trunk', with a shearing of the dens from the body of C2, with forward movement by the transverse ligament, with backward movement by the anterior arch of C1; flexion is the most common mechanism of injury; extension injuries result in posterior displacement of the dens. See C2 fracture, Odontoid

odontoma A tumor of odontogenic origin–ie, which arises from the dental epithelium. See Complex odontoma, Compound odontoma

odor Numerous, relatively uncommon conditions may be associated with typical odors of the urine and/or breath, often in Pts with inborn errors of metabolism

ODORS OF BIOMEDICAL INTEREST

Acetone–Russet apples	Chloroform, ethanol, isopropanol, ketoacidosis, lacquer
Acrid–pear-like melaninogenica	Chloraldehyde, paraldehyde; *Bacteroides melaninogenicus*, Prevotella
Ammonia	Renal failure, uremia, N-ethyl morpholine
Bitter almonds	Cyanide
Burned chocolate	*Proteus* spp
Cabbage	Methionine (see also Hops)
Camphorous	1,8-cineole
Carrots	Circutoxin
Coal gas	Carbon monoxide
Disinfectants	Phenol, creosote
Eggs, rotten	H$_2$S, mercaptans, disulfuram (Antabuse®)
Ether-like	Ethylene chloride
Fecaloid, putrid	*Clostridium* spp
Fishy	Vaginal infection by *Gardnerella vaginalis*, 'rice-water' choleric stools; di-N-butylamine, diethylamine, hepatic failure
Floral (sweet, fruity)	DM, acetone, ethyl- and isobutyl- acetate, phenyl methylethyl carbinol
Fruity/alcohol	Amyl nitrate, ethanol, isopropanol
Garlic	Arsenic, phosphorous, selenium, tellurium, thallium, malathion, parathion, DMSO
Grape juice	*Pseudomonas* spp
Halitosis	Oral infections
Hop-like	Oasthouse disease, α-hydroxybutyric acid
Maple syrup	Maple syrup urine disease
Mint	Menthone
Mothballs	Camphor-products
Mousy/musty	Phenylketonuria
Musty basement	*Streptomyces* spp, *Nocardia* spp
Musty (fish, raw liver)	Hepatic failure, zinc phosphide, pentadecanolacetone
Odorless urine	Acute tubular necrosis
Peanuts	RH-787 (Vacor, see Vacor diabetes)
Pungent	Ethylchlorvynol; formic acid
Putrid	Dimethyldisulfide
Rancid fish	Tyrosinemia
Rotting fish	Trimethylaminuria
Shoe polish	Nitrobenzene
Sour and pungent	Ethyl acrylate, 2-methyl-5-ethyl pyridine, propionic acid, 2,4-pentanedione
Sweaty feet	Glutaric acidemia, type II, isovaleric acidemia
Sweet and musty	Isobutylacrylate
and rancid	2,6 butanol
and sharp	Methylethylketone
Swimming pool	Hawkinsinuria
Tomcat urine	β-Methylcrotonylglycinuria
Violets	Turpentine
Wintergreen	Methylsalicylate

odynophagia Dysphagia, see there

odynophobia PSYCHOLOGY Fear of pain, algophobia

oedipal phase PSYCHIATRY A psychoanalytical stage that occurs between ages 4 and 6, which partially overlaps components of the phallic stage and represents a time of inevitable conflict between a child and parents

Oedipus complex PSYCHIATRY Normal attachment of a child to the parent of the opposite sex, accompanied by envious and aggressive feelings toward a same-sex parent; the OC is a constellation of consequences–per Freud–resulting from the sublimation of a boy's psychosexual desire for his mother, likened to Oedipus of Greek mythology, who killed his father and married his mother. See Jocasta complex

oesophagus British spelling for esophagus, see there

off campus *adjective* Relating to the location of an off-site affiliate of a large medical center complex. Cf 'On campus'

off-gassing Release of gaseous chemicals from a solid. Cf Flatulence

off-label use Unlabeled indication, unlabeled use PHARMACOLOGY The use of a drug–eg, tretinoin, an analog of vitamin A or medical device–eg, injectable collagen, to treat a condition for which it has not received approval by a regulatory agency–eg, the FDA; OLU is common in chemotherapy of difficult-to-treat cancers, for which there is no agreement on standardized therapy

off-ladder ACADEMIC MEDICINE *adjective* Referring to a non-tenure track position–eg, lecturer, assistant, associate or activity–eg, raising a family that hinders a person from climbing a career 'ladder'. See Glass ceiling, Sticky floor phenomenon

off period NEUROLOGY A period in Pts with Parkison's disease in which the antiparkinsonian medication does not ameleriorate Sx. See Parkinson's disease. Cf On period

off-shore medical school A medical school in the Caribbean–'off-shore' operated as a for-profit venture. See International medical graduate, USFMG

office The suite of rooms where a physician receives and treats Pts, and otherwise practices medicine, known in the UK as 'surgery' MEDTALK An agency or section of an official or governmental body. See European Patent Office

office D&C See Pipelle biopsy

office hypertension See White coat HTN

Office of Alternative Medicine A section of the NIH established in 1992 by the US Congress, the purpose of which is to investigate the claims of efficacy for various forms of alternative therapy, and their possible health benefits; it was replaced in 1998 by the National Center for Complementary and Alternative Medicine

office practice BUSINESS OF MEDICINE The facility or facilities at which a practitioner, on an ongoing basis, provides or supervises provision of health services to consumers. See Practitioner, Private practice

official compendium CLINICAL PHARMACOLOGY The US Pharmacopoeia, National Formulary or supplement thereto

official written order CLINICAL PHARMACOLOGY An order written on a form provided for that purpose by the DEA. See Drug Enforcement Administration

offspring Progeny, see there

ofloxacin Floxin®, Levaquin® ANTIBIOTICS A broad-spectrum (fluoro)quinolone used for GNR INDICATIONS Bacterial skin infections, lower respiratory, urinary, prostate infections, STDs, PID, otitis externa, chronic suppurative otitis media, and acute otitis media in Pts with tympanostomy tubes. See Fluoroquinolone

Oguchi's disease An AR form of night blindness characterized by retinal involvement, and virtual absence of rods

17-OH-corticosteroid See 17-Ketogenic steroids, urine

OHTS VASCULAR DISEASE A clinical trial–Ocular Hypertension Treatment Study

ointment MAINSTREAM MEDICINE A medicated formulation with an oil base; in contrast, creams are water-soluble

old boy network "In" group, the 'Mafia' Any 'inside track' to publishing articles in 'premier journals', access to grant monies–eg, from the NIH, hospital staff privileges in prestigious institutions–eg, Massachusetts General Hospital, and other seemingly unfair advantages linked to friendship. See Glass ceiling principle

old man sleeping after dinner NEUROLOGY A fanciful descriptor for the image evoked by full-scale parasympathetic nervous system activation, which may be accompanied by bradycardia, bronchoconstriction resulting in noisy respiration, snoring, meiosis, ↑ salivation/drooling

the old man's friend TERMINAL CARE A popular term for pneumonia in elderly Pts with strokes or other debilitating illness, which allows them the dignity of a quiet death, often while asleep. See Euthanasia, Pneumonia

old soldier's heart Post traumatic stress disorder, see there

Old Sparky A generic, ghoulishly facetious name for any electric chair used in capital punishment. See Capital punishment. Cf Hangman's fracture

oldest old GERIATRICS The population above age 85. See Elderly. Cf Elderly primigravida

olestra Sucrose polyester, Olean® A proprietary synthetic–no-calorie fat, approved by the FDA–for use in savory snack foods–eg, tortilla chips, potato chips, and crackers; SIDE EFFECTS GI discomfort including cramps, diarrhea; it binds to vitamins A, D, E, and K, and carotenoids. See Obesity

olfactory agnosia NEUROLOGY An inability to recognize smells, a common finding in parietal lobe tumors. See Agnosia

olfactory hallucination Phantosmia The misperception of a nasty odor that is either 1. Extrinsic–ie, of non-self origin, which is mildly annoying but not a pervasive problem for the Pt or 2. Intrinsic–ie, perceived to emanate from the Pt's own sweat, flatus or halitosis and which may prove disconcerting to the Pt. See Odors, Olfactory reference syndrome

olfactory reference syndrome PSYCHIATRY A condition in which a person complains of unpleasant odors emanating from the skin or from a specific body orifice, unrelated to any physiologic process. See Halitosis, Odors, Olfactory hallucination

oligo- A Greek root for few, soupçon, tad, wee, scanty, not a whole bunch

oligoanalgesia Underuse of analgesics in the face of valid indications–eg, intense bone pain of terminal CA–for its use. See Patient-controlled anesthesia

oligohydramnios NEONATOLOGY A relative paucity of amniotic fluid– ≤ 500 mL at term–seen when a fetus swallows more than usual ETIOLOGY Placental insufficiency, donor twin, urinary tract malformation, amount of fluid

oligomeganephronia A condition that may be a subtype of renal hypoplasia, characterized by ↓ nephrons with hypertrophy of remaining nephrons; kidneys are often small, glomeruli are enlarged, as are the tubules, which may become cystic; it is often first identified in early childhood as a cause of polyuria, polydipsia, and growth failure; it is accompanied by a defect in urinary concentration, sodium reabsorption, and acid excretion, often resulting in metabolic acidosis and slow progression to chronic renal failure

oligomenorrhea GYNECOLOGY Scant menstruation; less than usual menses; menstrual periodicity of 38 to 90 days

oligonephronic hypoplasia NEPHROLOGY A bilateral condition of neonatal onset, characterized by small kidneys and ↓ glomeruli CLINICAL Vomiting, dehydration, polydipsia, polyuria, renal failure, FUO, persistent hyposthenuria LAB Proteinuria, ↓ specific gravity–1.006-1.010, microscopic hematuria DIFFDX Fanconi syndrome, hypoplasia with renal dysplasia, bilateral simple hypoplasia, and secondarily contracted kidneys MANAGEMENT Hydration, titrate Na⁺, Ca²⁺, bicarbonate; phosphate-binding antacids, vitamin D PROGNOSIS If no response, renal failure is inevitable, requiring dialysis or transplantation

Oligon™ catheter NURSING A large-bore, infection-resistant, Foley-type catheter treated with Oligon, an antimicrobial which ↓ urinary catheter-related infections. See Urinary tract infection

oligospermia UROLOGY Abnormally low number of sperm in the semen, resulting in subfertility. See Azoospermia

oliguria NEPHROLOGY Excretion of < 400 mL of urine/day

Ollier's disease Osteochondrodysplasia

-ology *suffix*, Latin, study of

Olux™ ViaFoam™ Clobetasol mousse DERMATOLOGY A foam formulation of a potency corticosteroid, clobetasol propionate, for topical treatment of severe psoriasis and other steroid-responsive scalp dermatoses. See Psoriasis

Olympics SPORTS MEDICINE An international competition among (traditionally) nonprofessional athletes trained in a particular summer or winter sport, which is held every 4 yrs in a selected city. See Paralympics, Special Olympics, World Medical Games

-oma *suffix*, Latin, tumor, benign or malignant

omapatrilat CARDIOLOGY A vasopeptidase inhibitor that inhibits neural endopeptidase and ACE, used to manage HTN. See Isolated systolic hypertension, OPERA, Vasopeptidase inhibitor

OMD 1. Oculomuscular dystrophy 2. Organic mental disorder

omega sign CLINICAL MEDICINE A sign in melancholia in which the Pts have a furrowed brow due to sustained contraction of the corrugator muscle, often accompanied by Veraguth's folds, an upward, inward peaking of the upper eyelids, a finding likened to the Greek letter omega

omega-3 fatty acids n-3 fatty acids, see there

omentectomy Supracolic total omentectomy SURGICAL ONCOLOGY The complete excision of the omentum accessed by an extended midline incision INDICATIONS Surgical therapy of extensive peritoneal malignancy–eg, ovarian CA COMPLICATIONS Adhesion formation, especially if Pts receive RT. See Ovarian carcinoma

omeprazole Prilosec® THERAPEUTICS A proton pump inhibitor used to manage GERD; heartburn; erosive esophagitis; maintenance of healed erosive esophagitis; short-term management of active duodenal ulcer and active benign gastric ulcer; and for certain hypersecretory conditions; combined with clarithromycin to eradicate *H pylori* associated with duodenal ulcers. See GERD, *Helicobacter pylori*. Cf H2 blockers

-omics A neologistic root popular in some English-speaking regions, and in the US, which ties economics to a particular politician–eg, Reagonomics, Clintonomics

Omiderm WOUND CARE An occlusive, nonadhesive dressing for burns and other wounds. See Wound dressing

OMIM Online Mendelian Inheritance in Man ONLINE GENETICS The electronic–Web site-www.ncbi.nlm.mih.gov/omim version of Mendelian Inheritance in Man, a curated database See MIM catalog

Ommaya reservoir ONCOLOGY A reservoired device implanted under the scalp with a catheter to a ventricle, used to deliver chemotherapy to the CSF

OmniCath® INTERVENTIONAL CARDIOLOGY A catheter used to remove obstructing atherosclerotic plaques from blood vessels. See Endarterectomy

Omnicef® Cefdinir INFECTIOUS DISEASE A cephalosporin used for community-acquired pneumonia, exacerbation of chronic bronchitis, bacterial sinusitis, pharyngitis and tonsillitis related to strep throat, uncomplicated skin infections; in children, for acute bacterial otitis media. See Cephalosporin

Omniferon™ IMMUNOLOGY A purified IFN derived from human WBCs for treating viral and other immune disorders–eg, HBV, HCV, HIV/AIDS, herpes, CML, multiple sclerosis. See Interferon

OMNIUM GI DISEASE A clinical trial–Omeprazole vs Misoprostol for NSAID-Induced Ulcer Management–that compared the efficacy, tolerability, and safety of omeprazole and misoprostol in Pts requiring long-term NSAIDs. See Misoprostol, Omeprazole

omphalitis NEONATOLOGY Umbilical inflammation

omphalocele NEONATOLOGY A congenital periumbilical defect in which loops of small intestine prolapse into a sac covered by peritoneum and amnion

Omphalotus TOXICOLOGY A genus of poisonous mushrooms; intoxication may occur in casual mycophiles, who cook the wrong catch CLINICAL Abrupt onset after ingestion with profuse sweating, nausea, abdominal colic, blurred vision, salivation, rhinorrhea, dizziness. See Poisonous mushroom

OMS 1. Organic mental–brain syndrome, see there 2. Oral & maxillofacial surgery 3. Oral morphine sulfate 4. Otomandibular syndrome

on VOX POPULI adverb Taking, receiving, as in, he/she is on coumadin, on drugs, etc

on-campus *adjective* Referring to an on-site site of a medical complex with multiple buildings. Cf 'Off campus'

on-call HOSPITAL PRACTICE *adjective* Referring to a status in which a physician can be reached and arrive at the hospital within 30 mins of being paged

on-off phenomenon NEUROLOGY An ↑ refractoriness to L-dopa's ability to control the smooth skeletal muscle movement in Parkinson's disease, where periods of excess abnormal movements–'on,' alternate with periods of prolonged immobility or freezing–'off'; 'on-off' also refers to the waxing and waning of parkinsonism itself. See Parkinson's disease

on period NEUROLOGY A period in Pts with Parkinson's disease in which the antiparkinsonian medication ameleriorates Sx. See Parkinson's Disease. Cf Off period

onco- *prefix*, Latin, pertaining to a tumor or malignancy

OncoChek™ ONCOLOGY An immunoassay intended to detect breast, lung, colon, ovary, prostate, and other CAs, by measuring protein degradation products. See Tumor marker

oncocytoma A usually benign tumor characterized by aggregates of bland pink cells LOCATIONS Bronchial, lacrimal, salivary, parathyroid, and thyroid glands–Hürthle cell adenoma, anterior pituitary, kidney; rarely, malignant oncocytomas occur in salivary glands, nasal cavity, paranasal sinuses, mediastinum and thyroid and kidneys DIFFDX Oncocytic CA of pancreas, oncocytic carcinoid of lungs, clinically similar to 'garden variety' carcinoid, oncocytosis of salivary gland, an age-related hyperplasia

oncofetal antigen ONCOLOGY Any antigen expressed in the fetus during embryogenesis, which isn't produced in major quantities in adults. See α-fetoprotein, Carcinoembryonic antigen

oncogenic HPV A human papillomavirus–HPV genotype, especially types 16, 18, but also types 31, 33, and 51, which is pathogenically linked to intraepithelial neoplasia–eg, uterine cervix, termed CIN. See CIN, HPV

oncogenic virus Cancer virus, tumor virus ONCOLOGY A DNA virus or RNA virus capable of causing malignant transformation of cells, inducing a neoplasia in its host or causally linked to human tumors. Oncogenic viruses include Retroviruses, eg leukemia viruses of cats, cattle, chickens; Herpesviruses, eg EBV-induced Burkitt's lymphoma, Asiatic nasopharyngeal carcinoma; DNA virus, eg HPV

oncogenicity The capacity to induce tumors

OncoLink An online service operated by U Penn, which has a vast array of information on cancer and therapies, and various cancer-related forums, links and postings WEBSITE www.oncolink.com

oncologist A physician specialized in cancer management MEAT & POTATOES DISEASES Colorectal lung, breast, prostate CAs, lymphoma SALARY $157K + 13% bonus. See Gynecologic oncologist, Medical oncologist, Radiation oncologist, Surgical oncologist

oncology The study of cancer. See Gynecologic oncology, Hematology/oncology, Medical oncology, Psychooncology, Radiation oncology, Surgical oncology

oncology nurse NURSING A nurse specialized in treating and caring for people with cancer SALARY $53K + 2% bonus. See Oncology

Onconase® P30 Protein ONCOLOGY A ribonuclease-like protein derived from frog–*Rana pipiens*–embryos, said to be effective against some CAs–eg, mesothelioma. See Unproven methods for cancer management

oncoprotein A protein–eg, v-*ras*, encoded by an oncogene–eg, c-*ras*. See *bcr-abl*

oncornavirus 1. Oncogenic RNA virus 2. Retrovirus, see there

oncotic pressure Colloid osmotic pressure PHYSIOLOGY The colloid pressure in solutions produced by proteins; in plasma, OP counterbalances the egress of fluid from capillaries due to hydrostatic pressure

oncotope TUMOR BIOLOGY An epitope on a gene product, which is unique to a particular tumor, which stimulates a B- and T-cell response during progressive tumor growth in a syngeneic or autochthonous host

OnCyte™ system NEONATOLOGY A portable, point-of-care rapid CD11b assay system to detect sepsis in newborns. See Sepsis

ondansetron ONCOLOGY A selective serotonin S_3 receptor antagonist used to ameliorate chemotherapy-induced N&V. See Chemotherapy-induced emesis

one & one-half syndrome Fischer syndrome NEUROLOGY A unilateral pontine lesion involving the medial longitudinal fasciculus and the pontine paramedian reticular formation, resulting in ipsilateral gaze palsy and internuclear ophthalmoplegia on contralateral gaze; the only remaining horizontal movement is abduction of the contralateral eye; the eyes are straight or exodeviated ETIOLOGY Focal lesions of brain stem–eg, multiple sclerosis, 1° or 2° tumors–eg, glioma, hemorrhage or infarction, AV malformations, basilar artery aneurysm

one bone-two bone sign OBSTETRICS A simple method for differentiating the upper arm or thigh–each of which has 1 bone from the forearm or lower legs–2 bones in evaluation of the fetus by ultrasonography

one-eyed vertebra RADIOLOGY A descriptor for the unilateral–'one-eyed' destruction of a lumbar vertebral pedicle, fancifully likened to a one-eyed jack in playing cards, where the 'nose' is contributed by the spinous process; an OEV is a rare finding on a plain AP film in carcinoma metastatic to a vertebral body, 'classically' of breast origin

one-hour glucose challenge ENDOCRINOLOGY A 'quick test' for evaluating Pts with borderline ↑–110-140 mg/dL of fasting serum glucose REF RANGE < 130 mg/dL non-diabetic; > 140 mg/dL suggest gestational DM. See Glucose tolerance test

one-lung anesthesia One-lung general anesthesia ANESTHESIOLOGY A technique used in thoracic surgery, in which a double-ballooned, double-lumened tube is passed into the trachea and the airflow to mainstem bronchi is blocked by either inflating a balloon or with a mechanical endobronchial 'blocker'; OLA is usually performed on the dependent–'down' lung, while the airflow to the 'up' lung is blocked to improve the ease of surgery ABSOLUTE INDICATIONS Isolation of airflow–eg, in bronchopleural fistula, unilateral abscess or hemorrhage COMPLICATIONS Trauma resulting in edema, misplacement of tubes causing asphyxia, perforation resulting in bronchopleural fistula, V/Q mismatch resulting in ↓ O_2, ↑ CO_2

474 **$1 million/$3 million coverage** MEDICAL MALPRACTICE A standard liability limit for physicians, which provides for a settlement of $1 million in damages per lawsuit, and a total of $3 million for all lawsuits

One-Shot HEART SURGERY A device for automatic anastomosis of vessels–eg, coronary arteries in < 2 mins, used with Mini-CABG instruments, which places 12 vascular clips for a complete closure. See Coronary arterial bypass graft

one-stop shopping strategy Delivery system strategy BUSINESS & MEDICINE A strategy by a health care system to retain virtually all diagnostic–lab and radiology work and therapeutic services–eg, ambulatory surgery stay–within its system. See Medical mall, Multispecialty management

onion skin appearance A pattern characterized by concentric laminations of differing radiologic or histologic densities

online INFORMATICS *adjective* Referring to a device connected–eg, to a network or the Internet, or which is ready to send or receive data. Cf Offline

online CME Continuing medical education obtained from various sources on the Internet. See Continuing medical education

online disease management e-disease management, internet disease management CYBERMEDICINE An evolving format for managing disease at a distance in which primary care providers are informed regularly by their Pts who self monitor their own weight, BP, cholesterol and other parameters

Online Mendelian Inheritance in Man See OMIM

online safety PUBLIC HEALTH A general term for freedom from cyberstalking and violent crime committed against those who communicate on the internet. Cf Cybersex

onset of action PHARMACOLOGY The length of time needed for a medicine to become effective. See Therapeutic drug monitoring

ONTARGET CARDIOLOGY A clinical trial–The Ongoing Telmisartan Alone and in Combination with Ramipril Global Endpoint Trial–designed to compare the benefits of telmasartan, an angiotensin II type 1 receptor blocker with ramipril, an ACE inhibitor, in managing Pts with HTN. See Renin-angiotensin system. Cf TRANSCEND

ontogeny The biological history of the development and growth of a single individual or organism. See Phylogeny

onychodystrophy Any defect or malformation of nails; usually nail dystrophy

onychogryposis DERMATOLOGY Thickened and curved finger and/or toenails

onychomycosis Dermatophytic onychomycosis, fungal nail infection, mycotic toenail, ringworm of nail, tinea unguium DERMATOLOGY A fungal infection of nails, which makes them white, opaque, thicker, brittle HIGH RISK GROUPS Older ♀–linked to estrogen deficiency which ↑ risk of infection, DM, small vessel vasculitis, artificial nails–acrylic or 'wraps'–the nail surface is usually abraded with an emery board, which itself may be a vector for fungi that really really like the moist, warm environment of the nails

onychoosteodysplasia Fong disease, nail-patella syndrome, Turner-Kieser syndrome An AD condition characterized by dysplastic/absent nails, absent/underdeveloped–hypoplastic kneecaps–patellae, iliac horns, elbow joint defects which interfere with full ROM–pronation and supination, and glomerulonephritis-like kidney disease, which may be progressive and lead to renal failure; gene locus is linked to the ABO blood group on chromosome 9q34

oocyst PUBLIC HEALTH The infectious stage of coccidian sporozoites–eg, *Cryptosporidium parvum* and others, which has a protective wall that facilitates survival in water and other environments

oocyte Egg HISTOLOGY The female reproductive or germ cell. Cf Sperm

oogenesis EMBRYOLOGY The process of ♀ gamete formation

oophorectomy Ovariectomy GYNECOLOGY Surgical excision of one or both ovaries. See Prophylactic oophorectomy

oophoritis GYNECOLOGY 1. Inflammation of the ovary 2. Pelvic inflammatory disease, see there

OP DRUG SLANG A street term for opium. See Opium MANAGED CARE Outpatient, see there

O&P Ova and parasites, see there

opacification MEDTALK Clouding; making opaque. See Lenticular opacification

opacity MEDTALK A clouded or opaque area. See Corneal opacity

Opal™ tissue ablation device ENDOSCOPY An endoscopic device with hollow retractable needle electrode, which allows insertion of 5-French probe for delivering electrolytes, irrigants, anesthetics and other fluids to a site, used with rigid endoscopes. See RF tissue ablation device

OPAT Outpatient parenteral antibiotic therapy

OPCAB Off-pump coronary artery bypass INTERVENTIONAL CARDIOLOGY A type of minimally invasive cardiac surgery in which cardiopulmonary bypass is eliminated and the heart continues to beat while the surgery is being performed as with MIDCAB, but in which a full median sternotomy is performed to expose the entire heart for multiple bypasses to more than 1 coronary artery. See MIDCAB, Minimally invasive cardiac surgery

OPD syndrome Otopalatodigital syndrome, see there

open VOX POPULI → MEDTALK Patent, pronounced PAY tent

open architecture Open system architecture INFORMATICS An information system design that uses a network manager–a centralized electronic 'clearinghouse' to retrieve data and send it back to the requester. See Computer, Hospital information system, LAN

open artery theory Open infarct-related artery theory CARDIOLOGY A theory that holds that the duration of coronary artery occlusion affects the extent of infarction

open biopsy A biopsy in which the lesion is excised under direct visual examination during an open surgical procedure. See Biopsy

open capsulotomy SURGERY A procedure in which the capsule is released around the implant

open chain ORTHOPEDICS *adjective* Referring to a series of musculoskeletal links that are free to move on each other but capable of combining to support the body; the chain is said to be open when the distal link is free to move and closed when it is fixed; in OC motion, the calcaneum supinates and pronates about the subtalar joint

open colectomy GI DISEASE Removal of the colon via a surgical incision made through the abdominal wall

open enrollment MANAGED CARE An eligibility stance by a state Blue Cross/Blue Shield, which accepts all persons, regardless of medical history, lifestyle, occupation, or potential disease risk factors. Cf Closed enrollment

open formulary MANAGED CARE A relatively unrestricted list of drug choices available through an HMO's drug plan

open fracture Compound fracture ORTHOPEDICS A fracture in which the skin is broken and the potential for infection significant. See Fracture. Cf Closed fracture

open heart surgery HEART SURGERY A popular term for any surgery in which the thoracic and pericardial cavities are opened and the heart is surgerized to access and repair, on an elective or emergent basis, conditions affecting the heart, great vessels, coronary arteries, or valves TYPES CABG, repair of ASD, VSD, valve replacement, etc ADVERSE OHS OUTCOMES Need for reoperation, acute MI, dialysis dependency, DM, CHF, a low ejection fraction. See Atherosclerosis, Coronary artery bypass graft; Heart-lung machine

open-label trial CLINICAL RESEARCH A trial in which doctors and participants know which therapy is being administered. See Blinding

open lung biopsy PULMONOLOGY A procedure in which the chest cavity is opened to allow visually directed biopsy of lung tissue INDICATIONS Diagnose bronchiolitis, chronic interstitial lung disease, lung CA, eosinophilic granuloma, honeycomb lung, lymphoma, pulmonary HTN, sarcoidosis PROS Greater diagnostic yield than transbronchial Bx CONS ↑ complications COMPLICATIONS Pneumothorax, hemorrhage, air embolism. Cf Cf Bronchoalveolar lavage, Transbronchial biopsy

open panel MANAGED CARE A managed care plan that contracts–directly or indirectly, with private physicians to deliver care in their offices EXAMPLES Direct contract HMO, IPA; OPs reimburse members for health services obtained from outside of its provider network. See Network model HMO. Cf Closed panel

open protocol system THERAPEUTICS An FDA-approved protocol that allows use of drugs or other therapeutics outside of a controlled trial–and before their FDA approval. See Compassionate IND protocol

open reduction ORTHOPEDICS The reduction of a fracture by direct visualization after surgical exposure. Cf Closed reduction

open-set speech recognition NEUROLOGY Understanding speech without visual clues–speech reading. See Speech

open spaces, fear of Agoraphobia. See Phobia

OPERA CARDIOLOGY A clinical trial–Omapatrilat/Vanlev™ in Persons with Enhanced Risk of Atherosclerotic Events–that compared omapatrilat with placebo in managing Pts with stage I isolated systolic HTN. See Isolated systolic hypertension, Omapatrilat, Vasopeptidase inhibitor GYNECOLOGY Outpatient endometrial resection/ablation A minimally invasive procedure for treating abnormal uterine bleeding as alternative to hysterectomy. See Hysterectomy

operable *adjective* Referring to a condition, usually malignant, that is amenable to operative intervention, with ↑ survival. Cf Inoperable

operant PSYCHOLOGY *adjective* Referring to response(s) contingent upon or influenced by their impact on the environment or a situation, and/or by the ability of response(s) to achieve reward or reinforcement

operating team SURGERY The participants–surgeons, nurses, etc–in a sterile surgical procedure performed under general–less commonly, local anesthesia

operation SURGERY A surgical procedure. See All-American operation, Arterial switch operation, Berry-picking operation, Billroth's I operation, Caldwell-Luc operation, Commando operation, Committed operation, Curse operation, Debulking operation, Fontan operation, Ileal bypass operation, Le Fort operation, Look & see operation, Manchester operation, Mastoid obliteration operation, Noncommitted operation, North American operation, Pomeroy operation, Potato chip operation, Richardson composite operation, Second-look operation, South American operation, Speech-preserving operation, Wertheim operation

operative mortality The percentage of Pts who die while hospitalized during or after a surgical procedure

operative report A document produced by a surgeon or other physician(s) who have participated in a surgical intervention, which contains a detailed account of the findings, the procedure used, the specimens removed, the preoperative and postoperative diagnoses, and names of the primary performing surgeon and any assistants

ophidiophobia Morbid fear of snakes. See Phobia

ophthalmia neonatorum Acute neonatal conjunctivitis, neonatal conjunctivitis OPHTHALMOLOGY Inflammation of the neonatal conjunctiva, often infection by *N gonorrhoeae* or *Chlamydia trachomatis*; ON causes blindness in up to 4000 newborns/yr in Africa TREATMENT Silver nitrate, erythromycin, other antibiotics; povidone-iodine may be more effective, less costly, toxic, and may reduce ocular HIV and herpes simplex

ophthalmoplegic migraine NEUROLOGY A migraine characterized by repeated headaches lasting up to 1 wk associated with paresis of one or more ocular nerves. Cf Classical migraine

opiate Any natural–eg, opium semi-synthetic–eg, morphine or synthetic–eg, fentanyl, usually alkaloid narcotic agent with opium-like activity. See Drug screening, Narcotic

opioid NEUROLOGY A pain-attenuating peptide that occurs naturally in the brain, which induces analgesia by mimicking endogenous opioids at opioid receptors in the brain. See Opioid-mediated analgesia system

OPIOIDS

AGONISTS The most potent opioid agonists are morphine, meperidine, methadone; other opioids include hydromorphine–Dilaudid®, codeine, oxycodone–Percodan®, propoxyphene–Darvon®

ANTAGONISTS Naloxone–Narcan®

MIXED AGONSTS-ANTAGONISTS Pentazocine–Talwin®

opioid anesthetic ANESTHESIOLOGY Any agent used in anesthesia to modulate sympathetic responses–HTN and tachycardia to endotracheal intubation and surgical manipulation; nonmorphine OAs are 10 to 1000-fold more potent than morphine; they maintain cardiac output and BP and ↓ heart rate, making them ideal for cardiac anesthesia, as they ↓ myocardial O_2 demand, but maintain the O_2 supply within the limits of compromised coronary anatomy CONS Profound respiratory depression, skeletal muscle rigidity, CNS neurotoxicity with injury in the cortical regions and limbic system. See Alfentanil, Fentanyl, Remifentanil, Sufentanil. Cf Muscle relaxant

opioid intoxication SUBSTANCE ABUSE A mental state characterized by somnolence, confusion, or unconsciousness caused by opium or its derivatives–heroin, morphine, semisynthetic opioid narcotics. See Endorphin, Narcotic, Opiate, Opioid

opioid-mediated analgesia system NEUROLOGY A neurohumoral system that modulates the transmission of pain; OMAS neurons are activated by noxious stimuli, or by input from the pain-transmission neurons they inhibit. See Opioid-mediated analgesia system

opioid peptide Endogenous opiate Any natural polypeptide neurotransmitters involved in perception of pain, response to stress, regulation of appetite, sleep, memory, learning

opioid withdrawal SUBSTANCE ABUSE An acute state caused by withdrawal of opioid narcotics from a person addicted to narcotics CLINICAL Sweating, shaking, headache, craving, vomiting, abdominal cramping, diarrhea, insomnia, confusion, agitation, other behavioral changes. See Neonatal withdrawal syndrome

opiophobia PSYCHOLOGY A fear that physicians have of prescribing needed pain medications. See Phobia

opisthotonus NEUROLOGY A type of spasm in which the head and heels arch backward in extreme hyperextension and the body forms a reverse bow; opisthotonus may be seen in scorpion stings, due to cholinergic hyperstimulation by venom

opium SUBSTANCE ABUSE A narcotic from *Papaver somniferum* PHARMACOLOGIC EFFECTS Inhibits peristalsis–may induce constipation; used to ↓ GI cramps, diarrhea OVERDOSE In excess, respiratory depression. See Heroin, Narcotic

opportunistic MEDTALK *adjective* Relating to a microorganism that is part of the normal nonpathogenic flora, which causes disease given the opportunity–eg, by a compromised host immune defense

opportunistic infection INFECTIOUS DISEASE An infection by a microorganism, usually bacterial, that is part of the normal flora–eg, of the skin, GI tract, GU tract, respiratory tract or elsewhere, which becomes pathogenic when the host's immune system is compromised by an unrelated disease–eg, AIDS, chemotherapy, or DM, resulting in OIs of lungs, brain, eyes, etc

opportunity for improvement Error, mistake, screwup A euphemism for a booboo of biblical/near-biblical magnitude that is indefensible but, because it is systemic, is the type of error that can be corrected in the future. See Medical error

oppositional defiant disorder PSYCHIATRY A condition characterised by negativistic, hostile, defiant behavior towards authority figure(s), usually without violating societal norms; ODD is more prevalent in ♂, affecting ±20% of school children, beginning at age 8; parents of ODD kids may be overly concerned with power and control which may trigger oppositional defiant behavior. See Sociopathic disorder

OpSite WOUND CARE A transparent polyurethane membrane that creates a moist wound environment by retaining wound exudate; it is gas and water vapor permeable, but bacteria and water impermeable. See Wound dressing

opsoclonus NEUROLOGY Rapid, irregular movements of eyes in all directions. See Paraneoplastic opsoclonus

opsoclonus-myoclonus Dancing eyes-dancing feet syndrome, see there

optic glioma NEUROPATHOLOGY An astrocytoma arising in the optic nerve, which is associated with neurofibromatosis, type 1 CLINICAL Invasion and destruction of optic nerve tissue, loss of vision

optic nerve swelling VOX POPULI Papilledema, see there

optic nerve 2nd cranial nerve ANATOMY A bundle of > 1 million nerve fibers that carries afferent/senosry fibers from the retinal gangliion cells, passing out of the orbit via the optic foramen (canal) to the optic chiasm, where part of the fibers cross to the opposite side, passing though the optic tract to the geniculate bodys, pretectum, and superior colliculus in the brain

optic neuritis OPHTHALMOLOGY Inflammation of optic nerve, which may cause abrupt, albeit temporary, loss of vision of the affected eye ETIOLOGY Idiopathic, possibly due to viral infection, autoimmune disease, multiple sclerosis, resulting in optic nerve swelling and destruction of the optic nerve's myelin sheath

Optical Tracking System OTS™ IMAGING A CT- or MRI-guided system for localizing and targeting lesions and visualizing regional anatomy; used with craniotomies, catheter placement, tumor resections, ENT procedures. See Needle localization

opticokinetic nystagmus Railroad nystagmus NEUROLOGY A normal bilateral optical response to objects moving slowly across a visual field. Cf 'Seesaw' nystagmus

OPTIMAAL CARDIOLOGY A clinical trial–Optimal Trial in Myocardial Infarction with Angiotensin II Antagonist Losartan. See Losartan

OPTIME-CHF CARDIOLOGY A clinical trial–Outcomes of a Prospective Trial of Intravenous Milrinone for Exacerbations of Chronic Heart Failure

Optimil® Methaqualone, see there

option VOX POPULI Choice; alternative. See Point-of-care option, Quick-fix option, Triple option

optional surgery SURGERY A surgical procedure performed at the Pt's discretion EXAMPLES Breast reconstruction, varicose vein stripping, asymptomatic cholecystectomy, wart removal. See Elective surgery

Optison® CARDIOLOGY A '2nd-generation' ultrasound contrast medium consisting of a solution of gas-filled microbubbles with a mean bubble size 4 μm, range 1–10 μm produced by sonicating a 5% albumin solution. See Sonicated albumin. Cf Albunex®

Optrin™ OPHTHALMOLOGY An agent–lutetium texaphyrin–Lu-Tex used in photodynamic therapy of eye diseases–eg, age-related macular degeneration. See Age-related macular degeneration

OPUS CARDIOLOGY A clinical trial–Orbofiban in Patients with Unstable Coronary Syndromes. See Orbofiban

OR 1. Odds ratio 2. Operating room, operating suite 3. INFORMATICS A word used in programming that expands the available logical choices, such that either of 2 conditions may be met for a sequence of data flow or for processing to be allowed

OR notes Operating room notes SURGERY A document, usually written by the surgeon, that delineates in detail the course of a surgical procedure. See Operative record

oral cancer Cancer of mouth, mouth cancer ENT A malignancy of the lips, tongue, floor of mouth, salivary glands, buccal mucosa, gingiva, palate; most OCs are squamous cell carcinomas linked to tobacco use and/or smoking, and tend to spread rapidly HIGH RISK FACTORS FOR OCs Alcohol abuse, poor dental and oral hygiene, chronic irritation–eg, rough teeth, dentures, etc; OCs may begin as leukoplakia or mouth ulcers; ♂ are hit twice as often as ♀. See Squamous cell carcinoma

oral candidiasis INFECTIOUS DISEASE A yeast infection of the adult oral mucosa, caused by *Candida albicans*, an opportunistic pathogen linked to immune compromise–eg, with AIDS, immunosuppression in transplants, chemotherapy, corticosteroids, DM, ↑ age, poor health, inherited immune defects, xerostomia CLINICAL Whitish plaques on oral mucosa which, if scraped away, leave a reddish base and pinpoint bleeding; OC may spread to the esophagus, producing candida esophagitis with dysphagia, and disseminate throughout the body–mortality of systemic candidiasis may reach 70%. See Oral thrush

oral contraceptive GYNECOLOGY A preparation of synthetic hormones intended to make a ♀ inconceivable by inhibiting ovulation OC FORMATS Sequential method, combined method. See Biphasic contraceptive, Contraceptives, Monophase contraceptive, Third-generation contraceptive, Triphasic contraceptive

ORAL CONTRACEPTIVES, CONTRAINDICATIONS

Age–over 35

Breast CA or other estrogen-dependent malignancy

Breast-feeding and < 6 weeks after delivery

Cardiovascular defects–acute MI, ASHD, CVA/TIA

Circulatory defects–varicose veins, phlebitis

Cystic fibrosis

Diabetes and long-term OC use

Hypertension

Liver disease–hepatitis, CA, neoplasms

Migraines

Obesity–BMI > 30

Pregnancy–current, suspected, or recently ended

Sickle-cell disease

Smoking–especially > 1 pack/day

oral thrush NEONATOLOGY Infection of the infant mouth with *Candida albicans*. See Oral candidiasis

oral rehydration therapy The administration of osmotically balanced solutions to a dehydrated Pt–commonly with severe diarrhea. Cf Ringer's lactate

oral sex STD Sexual activity in which buccolingual mucosa contacts a partner's genital or anal mucosa–eg, cunnilingus, fellatio. See High-risk sex, Safe sex

orange person syndrome A rare clinical condition caused by an overdose of rifampin, which colors the skin and body fluids a deep orange, accompanied by altered liver function–↑ BR, alk phos, and transaminase, pruritus. See Red man syndrome

OraQuick® LAB MEDICINE A technology produced by OraSure, Inc used to rapidly assess HIV-1 antibody status in oral fluids

OraTest® ENT A toluidine blue mouthrinse for detecting oral cancer, specifically SCC which accounts for 95% of oral CA. See Oral cancer, Toluidin blue

ORBIT CARDIOLOGY A clinical trial Oral Glycoprotein IIb/IIIa receptor Blockade to Inhibit Thrombosis

orbital cellulitis OPHTHALMOLOGY Acute infection of the tissues of the eye, with potentially serious complications ETIOLOGY Bacterial infection, usually an extension from the ethmoid or para-nasal sinuses, regional abscesses, trauma to eye, or foreign object AGENTS *H influenzae, S aureus, S pneumoniae*, beta hemolytic streptococci

orbital pseudotumor OPHTHALMOLOGY Idiopathic swelling of the orbital soft tissues of the eye, characterized by bulging eyes and swollen eyelids. See Pseudotumor

orchiectomy Orchidectomy UROLOGY The surgical removal of one or both testicles INDICATIONS CA–eg, seminoma or other germ cell tumor, hormonal ablation in Pts with prostate CA; hormonal deletion in habitual sex offenders. See Inguinal orchiectomy

orchiopexy UROLOGY The surgical fixation of an undescended testes in the scrotum. See Cryptorchidism

orchitis Orchiditis UROLOGY Testicular inflammation ETIOLOGY Infections–eg, mumps, brucellosis, STDs–gonorrhea or chlamydia, polyarteritis nodosa, in conjunction with infections of the prostate or epididymis

order MEDTALK Formal, usually written, instructions from a physician. See Emergency medical services do not resuscitate order, Official written order, Orders, Physician order, PRN order VOX POPULI A mandate. See Disciplinary order, Verbal gag order

order entry MANAGED CARE An automated process of entering charges into the computer billing system; OEs often refer to ancillary services such as lab work and radiology

orderly *adjective* Neat, organized and, when pathologically so, compulsive *noun* An assistant or other health care worker who takes orders from virtually everyone in the hospital. Cf Scut-monkey

orders MEDTALK A written 'command' from an attending physician that clearly and specifically delineates how a diagnostic or therapeutic intervention is to be carried out by responsible supporting staff. Cf Closed order book, Libby Zion

ordinary means MEDICAL ETHICS The measures that a person, as the 'steward' of his/her own life, is required to use to ensure health and self-preservation. See Reasonable person. Cf Extraordinary means

ordinary negligence standard MALPRACTICE A standard of Pt care, based on a deviation from 'reasonableness', rather than a standard of care accepted by one's peers. See Contributory negligence, Negligence

ORF Oral rehydration fluid

orf Ecthyma contagiosum A benign self-limited parapoxvirus infection acquired by handling infected sheep and goat skins and flesh which, in young animals, causes watery warty lesions of the cornea, mucous membranes and lips; in canines, orf-orf CLINICAL Hypertrophic bullae at inoculation site which, in immunocompromised Pts, may be very large AGENT Orf virus is similar to that causing milker's nodes

Orfidal Lorazepam, see there

organ bank TRANSPLANT MEDICINE A repository, usually shared by multiple hospitals for long-term storage of certain tissues destined for transplantation–eg, acellular bone fragments, BM, corneas. Cf UNOS

organ-based specialty MEDTALK Any specialty–or subspecialty of medicine or surgery that focuses on diagnosing and treating diseases of a particular organ or organ system–eg, brain–neurology, neuropathology, neuroradiology, neurosurgery, etc

organ brokerage Organ theft TRANSPLANTATION The sale of an organ–eg, a kidney by a living donor, or any commercial transaction in which an organ for transplantation is obtained through coercion

organ cluster transplantation SURGERY A procedure used in 1° malignancies of the bile tract, duodenum, or stomach with 2° liver involvement. Cf Heroic surgery

organ donor TRANSPLANTATION A person/cadaver that donates his/her organ(s) to a recipient

organ dose RADIATION THERAPY The amount of radiation delivered to a particular organ

organ-limited autopsy A postmortem examination restricted to one organ–or body region. See Autopsy

organ panel Lab diagnosis-related group Any of a group of diagnostic tests regarded as the most cost-effective, sensitive, and specific test for evaluating a particular diseased organ, organ system, or disease. See Anemia panel, Bone/joint panel, Cardiac injury panel, Cardiac risk evaluation panel, Collagen disease and arthritis panel, Collagen disease panel, Coma panel, Diabetic panel, Electrolyte/fluid balance panel, General health panel, Hepatitis–immunopathology panel, Hypertension panel, Kidney panel, Liver panel, Metastatic disease panel, Neoplasm panel, Pancreatic panel, Parathyroid panel, Pulmonary panel, Thyroid panel, TORCH panel

organ shortage TRANSPLANTATION The gap between the number of organs transplanted and number needed. See UNOS. Cf Organ brokerage

organ system management INTERNAL MEDICINE A philosophical approach to human disease, which assumes that a particular condition arises from an organ system, and treatment of the identified condition will resolve associated systemic manifestations; Cf Holistic approach

organic *adjective* ALTERNATIVE NUTRITION Relating to foods that are grown without pesticides or artificial growth enhancers, which are processed and preserved without chemicals CHEMISTRY Relating to carbon-based chemicals CLINICAL MEDICINE Relating to a disease process that can be objectively evaluated, as it is organ-based, in contrast to mental disorders, which are not organic

organic acidemia A clinical presentation of a metabolic disorder, often first seen in infants who present with poor feeding, vomiting, tachypnea, acidosis, hyperammonemia, ketosis, ketonuria, irritability, and convulsions or hypotonia and lethargy, findings that are otherwise suggestive of neonatal sepsis DISEASES WITH OA Isovaleric and propionic acidemias, maple syrup urine disease, medium chain acyl dehydrogenase deficiency, glutaric, methylmalonic, formiminoglutamic acidurias

478

organic brain syndrome Chronic organic brain syndrome, organic mental disorder NEUROLOGY Cerebral degeneration as cortical atrophy with 'simplification' of myelinated tracts, a process that is usually irreversible, often age-related, and associated with atherosclerosis. See Lacunar state, Multi-infarct dementia PSYCHIATRY An alteration in mental status, behavior, or mood attributed to organic disease–eg, alcoholic liver disease–hepatic encephalopathy, infections–AIDS dementia, renal failure–uremia, and others

organic mental disorder 1. Dementia, see there 2. Organic brain syndrome, see there

organic mood syndrome, manic type A persistently expansive or elevated mood, caused by nonpsychogenic conditions evidenced by Hx, physical exam, or labs, often affecting those > age 35 ETIOLOGY CNS infections–eg, viral and cryptococcal meningitides, neurosyphilis, trauma–eg, thalamotomy, right hemispherectomy, tumors–1° or metastatic, CVA, TIA, eponymic disorders–Klinefelter syndrome and Huntington's, Kleine-Levin, Parkinson's, Pick's and Wilson's diseases, carcinoid, idiopathic cerebral calcification, hyperbaric O_2, systemic disease, endocrine–eg, hyperthyroidism, hypothyroidism with starvation diet, puerperal and premenstrual psychoses, systemic infections–eg, Q fever, infectious mononucleosis, renal failure–eg, uremia, hemodialysis, drugs–eg, bromide, bromocriptine, cocaine, corticosteroids, H_2-blocking agents, isoniazid, L-dopa, phencyclidine, procainamide, procarbazine, thyroid drugs, hypovitaminosis–eg, ↓ vitamin B_{12} and niacin

organization VOX POPULI A body of persons with a common agenda. See Administrative services organization, Animal Liberation Front, Chain organization, HMO, Exclusive provider organization, Independent practice organization, International organization, International relief organization, JCAHO, Learning organization, MADD, Managed care organization, Managed service organization, Management service organization, Medical services organization, Multiple physician health system organization, NIH, Nongovernmental organization, NORD, Peer review organization, Physician organization, Physician Hospital organization, Physician practice organization, Preferred provider organization, Provider service organization, World Health Organization

organomegaly An ↑ in the size of a particular organ

organophosphate pesticide A phosphorus-rich organic compound–eg, parathion, that contain a halide which phosphorylates cholinesterase and irreversibly inhibits its activity MANAGEMENT Atropine, pralidoxime

orgasm Climax, coming PSYCHIATRY Sexual climax; the peak psychophysiologic response to sexual stimulation SEXOLOGY The highest point of sexual excitement, characterized by strong feelings of pleasure and marked normally by ejaculation of semen–♂ and vaginal contractions–♀. See Arousal, Dry orgasm, Nocturnal emission

orgasmic headache GYNECOLOGY An intense throbbing headache that begins with intercourse and peaks during orgasm, attributed to dilated intracranial blood vessels, the result of ↑ blood pressure and heart rate characteristic of an orgasm; the mechanism of OHs is probably the same as migraines. See The not tonight headache, Orgasm

Oriental flush complex A facial erythema seen in up to 80% of Orientals who drink alcohol, possibly due to an atypical isomer of alcohol dehydrogenase that causes rapid metabolism of ethanol and high acetaldehyde levels

orientation NEUROLOGY The state of being oriented; the knowledge of one's self, and present situation–eg, awareness of one's environment with reference to time, place, and interpersonal relationships VOX POPULI Proclivity, tendency; mien. See Sexual orientation

oriented to person, place & time See Orientation

ORIF Open reduction and internal fixation, see there

origin VOX POPULI Source. See Diarrhea of undetermined origin, Fever of unknown origin, Race/ethnic origin

Orion® laser system DERMATOLOGY A switched Nd:YAG laser configuration used to remove tattoo ink–eg, blue & black. See Tattoo

orlistat Xenical® An anti-obesity therapy which interferes with pancreatic lipase, and allows up to $\frac{1}{3}$ of ingested fat to pass undigested; orlistat therapy results in a modest–10% loss of weight ADVERSE EFFECTS Loose greasy stools. See Obesity. Cf Olestra

Ornish regimen NUTRITION A health enhancing program developed by Dean Ornish, MD RESULT Weight loss; 40% ↓ in total cholesterol, 60% ↓ in LDL-C. See Diet, 'Health food', Organic food

ORNISH REGIMEN
- **DIET** Low-fat, vegetarian diet of beans, bean curd, grains, fruits, vegetables
- **STRESS MANAGEMENT** Daily meditation and yoga
- **EXERCISE** 30 minutes/day
- **RESTRICTED** Alcohol, fat-free yogurt
- **ABSTINENCE** Meat, poultry, fish, egg yolks, caffeine, dairy products, tobacco; no fat or oil added to foods

ornithine transcarbamylase deficiency An X-D condition due to an absence of ornithine transcarbamylase, an X-linked mitochondrial enzyme expressed in hepatocytes and small intestinal cells–enterocytes CLINICAL Chronic hyperammonemia, episodic hyperirritability, vomiting, lethargy, protein avoidance, ataxia, coma, delayed growth and development, often mental deterioration caused by a mutation in ornithine transcarbamylase–OTC gene LAB ↓ Arginine, citrulline, urea; during acute hyperammonemic episodes, orotic aciduria is common PROGNOSIS Poor in ♂ who lapse into hyperammonemic coma, or recover with mental retardation and cerebral palsy. See Hyperammonemia

ornithinemia, type I HHH–hyperornithinemia, hyperammonemia, & homocitrullinuria syndrome, see there

ornithinemia, type II Gyrate atrophy–with ornithine aminotransferase deficiency, see there

ornithophobia PSYCHOLOGY Fear of birds. See Phobia

oromandibular dystonia NEUROLOGY A condition that affects the muscles of the jaw, lips, and tongue; the jaw is either open or shut, and may be accompanied by dysphagia and/or dysphonia

oropharyngeal cancer ENT A malignancy of the lips, tongue, floor of mouth, salivary glands, buccal mucosa, gingiva, palate, and throat; most are SCCs linked to tobacco use and/or smoking, and tend to spread rapidly HIGH RISK FACTORS Alcohol abuse, poor dental and oral hygiene, chronic irritation–eg, rough teeth, dentures, etc; OCs begin as leukoplakia or mouth ulcers; 2:1, ♂:♀. See Squamous cell carcinoma

OROPHARYNGEAL CANCER-STAGES

I Lesion is ≤ 2 cm and confined to the oropharynx

II Lesion is between 2 cm and 4 cm and confined to the oropharynx

III Lesion is ≥ 4 cm and may involve a single lymph node on same side of neck

IV Lesion has spread to the hard palate, tongue, or larynx, to nearby lymph nodes, or metastasized

orphan SOCIAL MEDICINE A person, usually a dependent child, whose parents have died or are presumed dead and thus must be cared for by another person or agency

orphan disease MEDTALK A disorder affecting < 200,000 people–US–ie, < 1/1000 people. See NICODARD, NORD

orphan drug/product Any drug, biological, medical device, or food of potential or actual use in treating 'orphan' diseases–diseases regarded by the pharmaceutical industry as too rare for developing commercially viable products. See Orphan disease. Cf Pseudoorphan drug

orphan patient PSYCHIATRY A Pt with primary hypochondriasis that has its psychodynamic origin in the unconscious gratifications of bodily Sx and physical suffering that begins when a Pt mistakenly assigns serious disease to normal bodily functions or to benign Sx of trivial illnesses or to the somatic Sx of emotional arousal. See Munchausen syndrome, Self-mutilation

ORS Oral rehydration solution

ORT 1. Operating room technician 2. Oral rehydration therapy, see there 3. Registered Occupational therapist

Orthocomp™ ORTHOPEDICS An injectable, nonresorbable, bone-bonding cement used in vertebroplasty for treating spinal compression fractures due to osteoporosis or carcinoma. See Vertebroplsaty

orthodontics DENTISTRY A specialty of dentistry involved in correcting dental and, less commonly, dentofacial defects

orthopaedics Orthopedics

orthopedic hardware See Under Pin, Plate, Screw, Wire

orthopedic shoes A term coined by the shoe industry, *not by the orthopedic community at large*; OSs may harm a normal child's foot as they are too stiff. See Orthosis

orthopedist Orthopedic surgeon

orthopnea CLINICAL MEDICINE Dyspnea in a Pt with moderate CHF, due to ↑ venous return from failing ventricles; such Pts breathe better when sitting straight or standing erect

orthopod Slang for orthopedist, orthopedic surgeon

orthosis plural, orthoses ORTHOPEDICS The straightening of a deformity; an external device—eg, a cast, brace, or splint used to stabilize, reinforce or immobilize an extremity, ↓ sensory input to an extremity, prevent stretch weakness, ↓ contractures, functionally assist weak muscles, protect a limb with pressure sores, provide a mechanical block to prevent undesired movement TYPES Orthoses are available for spine, hip, foot, knee. See Foot orthotics, Scapular reaction

orthostatic hypotension CLINICAL MEDICINE An abrupt ↓ in BP which occurs either when one stands up or when one remains in for a prolonged period of time in an erect position CLINICAL Dizziness, faintness, syncope, dim or tunnel vision—↓ cerebral blood flow; Sx typically improve with recumbency; OH occurs in normal healthy people who rise quickly from a chair, especially after a meal ETIOLOGY Autonomic failure, which may be 1°, or 2° to sympatholytic drugs, or irritation of the sympathetic nervous system, as in DM and late syphilis, where sympathetic vasoconstrictor fibers attempt to compensate for effects of gravity. See Autonomic failure

orthotics Arch supports, inner soles, inserts PODIATRY A device placed on the insole of shoes to either make them more comfortable and provide added support, or to modify a person's balance in the shoe. Cf Orthoptics, see Vision therapy

orthotopic transplantation Transplantation of a donor organ graft into the same site as that occupied by the original organ that failed OT ORGANS Heart, lungs, liver

Orthovisc ORTHOPEDICS An injectable hyaluronic acid for treating osteoarthritis of the knee; viscoelastic and lubricating properties are similar to normal synovial fluid. See Hyaluronic acid

Ortrip Nortryptyline, see there

OS Symbol for 1. Occupational safety 2. Oculus sinister—left eye 3. Opening snap 4. Operating system 5. Oral surgery 6. Order sheet 7. Osgood-Schlatter's disease 8. Osteogenic sarcoma 9. Osteosclerosis 10. Oxygen saturation

OSA Obstructive sleep apnea. See Sleep apnea

oscillations See Cortical oscillations

oseltamivir Tamiflu® INFECTIOUS DISEASE An antiviral used for acute influenza type A and B ADVERSE EVENTS N&V. See Influenza

Osgood-Schlatter disease Apophysitis of tibial tubercle, Osteochondrosis ORTHOPEDICS A condition seen in ♂ age 10 to 15, characterized by unilateral or bilateral post-exercise pain with a painful bump, at the anterior tibial tubercle ETIOLOGY Unknown, possibly microtrauma before complete maturity of the tibial tubercle CLINICAL Running, jumping, climbing stairs, traction cause discomfort, especially in active, athletic adolescents. See Osteochondrosis

OSHA Occupational Safety & Health Administration OCCUPATIONAL MEDICINE A federal agency that recommends health and safety procedures in, and promulgates standards for, the workplace. See Occupational medicine

Osler sign CLINICAL MEDICINE Painful erythematous swellings in skin and subcutaneous tissues of hands and feet, described as typical of endocarditis. See Endocarditis

Osler-Weber-Rendu syndrome Hereditary hemorrhagic telangiectasia, hereditary multiple aneurysmal telangiectasia GENETICS An AD vasculopathy, which may lead to massive episodic hemorrhage CLINICAL Affected children develop reddish telangiectasias on the lips, tongue, nasal mucosa, face and ears; vascular defects also occur in the throat, larynx, GI tract, liver, bladder, vagina, and brain, where it may result in hemorrhage, causing seizures or death; early signs include frequent nosebleeds in children; the characteristic telangiectasias on the tongue and lips may be delayed until puberty. See Vasculopathy

osmolality TOXICOLOGY A measure of the amount of osmotically effective solute/1000 g of solvent; serum osmolality is ↑ IN Ethanol, azotemia, dehydration, DM, hypercalcemia, hyperglycemia, hypernatremia, ethylene glycol, glycerine, INH, ketosis, mannitol therapy, methanol, pyelonephritis, renal tubular necrosis, diabetes insipidus, shock, sorbitol, uremia ↓ IN Overhydation, ↑ fluid intake, ↓ Na⁺, paraneoplastic syndrome, SIADH; urine osmolality is ↑ IN SIADH, liver disease, heart disease, dehydration ↓ IN Overhydration, diabetes insipidus, ↓ K⁺ REF RANGE Serum, 275-295 mOsm/Kg; urine 50–1400 mOsm/Kg CRITICAL (PANIC) VALUES ≤ 265 mOsm/Kg; ≥ 320 mOsm/Kg. See Delta osmolality, Effective osmolality

osmole PHYSIOLOGY A volume-regulating organic solute that may accumulate in high concentrations in a cell without adverse effects on the cell's structure or function. See Idiogenic osmole

osmotic diarrhea INTERNAL MEDICINE An ↑ in volume and frequency of fecal flow due to ingestion of a poorly absorbable solute—either a carbohydrate or divalent ion or hypertonic material, resulting in a fecal osmolality higher than plasma osmolality; OD may occur in antacid therapy, disaccharidase deficiency, magnesium sulfate ingestion and others, lactulose therapy, malabsorption of glucose-galactose, fructose or generalized malabsorption, or mannitol, and sorbitol ingestion. See Chewing gum diarrhea

osmotic fragility Red blood cell osmotic fragility test HEMATOLOGY The susceptibility of RBCs to osmotic lysis; in hypotonic solutions, RBCs behave as perfect osmometers, where the free water rapidly equilibrates, causing the RBCs to swell. See Hereditary spherocytosis

OspA vaccine INFECTIOUS DISEASE A vaccine containing recombinant outer-surface protein A of *B burgdorferi*, the Lyme disease agent. See *Borrelia burgdorferi*, Lyme disease

480 **osseointegrated implant** See Dental implant

ossicular chain reconstruction ENT A procedure for tympanoplasty–see there, using malleus strut, peg-top, and hydroxyapatite cap prostheses, and revision stapedectomy using stapedial tendon reconstruction

ossify *verb* To change into bone

Ossigel® ORTHOPEDICS An injectable bioengineered bone matrix that accelerates fracture healing. See Bone paste

ostectomy ORTHOPEDICS The excision of bone or part thereof

osteitis fibrosa cystica Albright syndrome, McCune-Albright disease, polyostotic fibrous dysplasia An AD condition possibly related to altered regulation of cAMP CLINICAL Precocious puberty, polyostotic–cystic fibrous dysplasia with spontaneous fractures at young age, cafe-au-lait spots of skin, ovarian cysts, endocrinopathy–eg, hyperthyroidism, hypophosphatemia and cyclical–4-6 wk–fluctuations of plasma estrogen; afflicted young girls have ↓ gonadotropins and ↓ response to LH-RH LAB ↑ Testosterone, ↑ alk phos TREATMENT Aromatase inhibitor testolactone. See Luteinizing hormone-releasing factor

osteoarthritis Degenerative arthritis, degenerative joint disease, hypertrophic osteoarthritis, noninflammatory arthritis, osteoarthrosis wear-and-tear arthritis ORTHOPEDICS The most common type of arthritis characterized by inflammation, degeneration and eventual loss of the cartilage of the joints of finger, hands, feet, spine, and large weight-bearing joints–eg, hips and knees, local pain, without systemic disease ETIOLOGY Unknown; aging, metabolic, genetic, chemical, mechanical factors may play a role. See Cervical osteoarthritis. Cf Rheumatoid arthritis

osteochondritis ORTHOPEDICS Inflammation of bone and joint surfaces–usually aseptic; note: A legacy of the German school of medicine was eponymic immortalization of osteochondritis in each joint

OSTEOCHONDRITIS EPONYMS

FREIBERG'S DISEASE–metatarsal head

HAGLUND'S DISEASE–calcaneus

KÖHLER'S DISEASE–tarsal-navicular bones

LEGG-CALVE-PERTHES DISEASE–femoral head

OSGOOD-SCHLATTER DISEASE–tibial tubercle

PANNER'S DISEASE–humeral head

SINDING-LARSEN-JOHANNSON DISEASE–patella

THIEMANN'S DISEASE–metacarpal and metatarsal bones

WEGNER'S DISEASE–osteochondritis with epiphyseal separation–congenital syphilis

osteochondrosis Osgood-Schlatter disease, see there

osteoclasis ORTHOPEDICS The intentional breaking of a bone–eg, one in which the fracture line is malformed

osteodystrophy ORTHOPEDICS A general term for defective development of bone. See Renal Osteodystrophy

osteogenesis imperfecta Brittle bone disease, fragilitas imperfecta ORTHOPEDICS A heterogenous group of AD conditions with variable penetration due to various defects–eg, deletions, frame shifts, point mutations, rearrangements, glycine substitution, exon skipping in the genes responsible for collagen production CLINICAL OIs vary in presentation and have thin bones, multiple fractures, blue sclera, opalescent teeth, deafness due to middle ear osteosclerosis, scoliosis, thin skin, visceral herniation, dentinogenesis imperfecta, vascular lesions

osteogenic *adjective* 1. Producing bone 2. Derived from bone-producing tissues

osteogenic stimulation ORTHOPEDICS Electrical stimulation to hasten bone repair, either invasive or noninvasive INDICATIONS Nonunion of long bone Fx, failed fusion of Fx, congenital pseudoarthroses, an adjunct to spinal fusion in Pts at ↑ risk of pseudoarthrosis due to prior failed spinal fusion . See Ultrasonic osteogenic stimulation

osteointegration The direct structural and functional connection between living bone and the surface of a load-bearing implant; osteointegrated implants have been used to treat edentulism, and for head and neck reconstruction to facilitate retention of auricular mandibular, maxillary, nasal, and orbital implants, and for bone-anchored hearing aids. See Dental implant

osteolytic *adjective* Causing bone breakdown

osteomalacia Nutritional rickets A condition characterized by softened bones due to poor mineralization occurring in a background of vitamin D deficiency CLINICAL Weak, deformed, and deformable bone, which in children may be manifest by craniotabes, bowlegs, and knock knees, rachitic rosary, ↓ ventilation, often accompanied by pneumonia, muscular weakness, ↓ appetite, hypocalcemia TREATMENT Vitamin D. See Osteogenic osteomalacia, Rickets

Osteomark® METABOLISM A proprietary urine test for measuring bone loss, quantifying cross-linked N-telopeptides–NTx, which are degradation products of bone resorption and are unique biochemical markers specific for the type I collagen–found in bone. See Hormone replacement therapy, N-telopeptides, Osteoporosis

osteomyelitis ORTHOPEDICS Bone infection, usually bacterial, which may be 2° to infection elsewhere in the body that spreads hematogenously to bone predisposed to infection by minor trauma in the region resulting in a blood clot; in children, long bones are affected; in adults, vertebrae and pelvis. See Cervical osteomyelitis

osteopath A physician trained in osteopathic medicine; the differences between doctors of medicine–MDs and doctors of osteopathy–DOs have been largely erased; the chief distinction between osteopaths and mainstream physicians is that osteopaths rely more on 'manipulation' of various body parts; otherwise, DOs prescribe drugs and may train in the same hospitals as MDs

osteopathy A school of medicine practiced predominantly in the US, and based on Dr Andrew Taylor Still's theory of healing, first delineated in 1874; osteopathic theory holds that a body in a state of wellness is correctly adjusted, and that disease represents a loss of coherency of structure and/or function, and the inability to mount a normal defense against infection, malignancy, inflammation, toxins and other inciting agents. See Articulatory techniques, Classical osteopathy, Cranial manipulation, CranioSacral Therapy™, Integrative medicine, Joint mobilization, Lymphatic pump, Muscle energy manipulation, Myofascial release, Positioning techniques, Rule of the artery, Somatic dysfunction, Strain-Counterstrain Therapy™, Visceral manipulation; Cf Chiropractic, Naturopathy

OSTEOPATHY, KEY PRINCIPLES

HOLISM The body is an integrated unit or balanced musculoskeletal system which holds the key to optimal physiologic function

STRUCTURE & FUNCTION ARE INTERRELATED An alteration in the body's structure leads to functional defects

HOMEOSTASIS The body has an intrinsic mechanisms to heal itself; osteopathic manipulation, exercise, and medication are intended to enhance the body's intrinsic healing ability; chronic conditions are believed to occur when the healing capacity is compromised

OTHER PRINCIPLES

• PREVENTION Central to the school of osteopathy is the teaching of lifestyle alteration, in particular through diet and exercise

• RULE OF THE ARTERY All illness responds to improved blood circulation, which provides essential nutrients and releases toxins

• SOMATIC DYSFUNCTION A premorbid state in which tissues are functioning in a suboptimal state, but structural defects are not yet present

osteopenia A ↓ in bone linked to estrogen deficiency seen in ♀ with hypogonadism due to hyperprolactinemia, ↑ exercise, anorexia nervosa, hypothalamic amenorrhea, or in Pts receiving gonadotropin-releasing hormone–GnRH analogues. See Post-renal transplant osteopenia warch out

osteopetrosis Albers-Schoenberg disease, marble bone disease PEDIATRIC ENDOCRINOLOGY A heterogeneous group of rare AD and cortical and trabecular osteosclerotic disorders, characterized by ↓ osteoclastic activity and ↓ bone resorption, with accumulation of sclerotic bone, compromising marrow space MANAGEMENT IFN-gamma-1b, BMT, high-dose calcitriol

osteophyte ORTHOPEDICS A bony bump

osteopoikilosis An AD condition characterized by multiple small foci of osteosclerosis in the spongiosa of the pelvis, metaphysis of long bones, tarsal and carpal bones, often associated with subcutaneous bony nodules

osteoporosis Brittle bones ORTHOPEDICS A condition characterized by ↓ bone mass and density–demineralization, bone fragility; it is the most common morbid condition of elderly ♀ STATISTICS Age-related osteoporosis causes > 10^5 fractures/yr–US–vertebrae 54%, hip 23%, distal forearm or Colles fracture, 17%; 25% of ♀ > 70 have evidence of vertebral fractures, as do 50% of ♀ > 80; 90% of femoral head fractures–FHFs occur in those > 70, US CLINICAL Pain, loss of height, other deformities and fractures MORBIDITY Osteoporosis-related fractures occur in 1.5 million/yr–US RISK FACTORS White, elderly, ♀ thin, immobilization, space travel–weightlessness, extreme exercise and/or amenorrhea, alcoholism, endocrinopathies–eg, acromegaly, Cushing's disease, hypogonadism, hyperthyroidism, hyperparathyroidism, smoking POSSIBLE RISK FACTORS Heredity, inadequate exercise, ↓ Ca^{2+} intake, exercise, small body frame, levels of serum and urinary Ca^{2+} and creatinine; Pts receiving physiologic L-thyroxine may have ↓ bone density DIAGNOSIS Bone densitometry, N-telopeptide measurement TREATMENT 1. Antiresorptive agents–eg, estrogen, calcium, calcitonin, biphosphonates and others eg anabolic steroids, calcitriol; K^+ bicarbonate improves Ca^{2+} and phosphorous balance by ↓ hydroxyproline excretion–a marker of bone resorption, and ↑ osteocalcin–a marker of bone formation 2. Bone stimulation regimens–eg, sodium fluoride, PTH, and growth factors–eg, IGF-I, IGF-II, transforming growth factor-β and 3. Ca^{2+} supplementation–1000 mg/day, may slow axial and appendicular bone loss in postmenopausal ♀ See Congenital cranial osteoporosis, Malignant osteoporosis

OSTEOPOROSIS–PRIMARY INVOLUTIONAL

TYPE I OR 'POSTMENOPAUSAL' OSTEOPOROSIS A relatively common condition with a 6:1 ♀:♂ ratio, which affects ages 50-75, characterized by ↓ estrogen, accelerated trabecular bone loss, 'crush' fractures associated with abnormal PTH secretion and age-related ↓ in response to vitamin D 1,25(OH$_2$D)$_2$]; 15-20 years after the onset of menopause, type I osteoporosis may reach a 'burned-out' phase with no further bone loss

TYPE II OR 'AGE-RELATED' OSTEOPOROSIS A less common condition with a 2:1 ♀:♂ ratio, which affects > age 70 and is characterized by trabecular and cortical bone loss–the elderly ♀ typically suffers a 35% loss of cortical bone and a 50% loss of trabecular bone, affecting vertebral bodies and flat bones, resulting in hip fractures and wedge-type vertebral fractures, due to 1. ↓ osteoblast function–↓ IGF-I, hGH, and local regulators 2. Marked ↓ in calcium absorption–to 50% of 'normal' with ↓ vitamin D 1,25(OH$_2$D)$_2$, possibly due to ↓ activity of renal 1-α hydroxylase and 3. Other factors, including ↓ clearance of parathyroid hormone's carboxyl–COOH terminals and ↑ calcium resorption; calcitonin's role in this form of osteoporosis is unclear

osteopuncture A technique for relief of bone pain in which needles are inserted into the periosteum in one of 120 areas, where the bone is relatively accessible to the skin surface; once the needle is inserted, it may be stimulated with low-voltage electricity for up to $\frac{1}{2}$ hr. Cf Acupuncture

osteoradionecrosis Degenerative pathology caused by radiation-induced cell injury, resulting in ↓ cell repair, ↓ vascularity, local hypoxia, necrosis, defective wound healing CLINICAL Edema, ulceration, bone necrosis, poor wound healing MANAGEMENT Hyperbaric O_2. See Hyperbaric oxygen therapy

osteosclerotic myeloma A myeloma characterized by sclerotic bone lesions and progressive demyelinating polyneuropathy. See Plasma cell dyscrasia with polyneuropathy

osteotome ORTHOPEDICS A chisel-like bone cutter

ostitis pubis ORTHOPEDICS Inflammation of the anterior pelvic joint ETIOLOGY Running, trauma MANAGEMENT NSAIDs, rest or activity modification, steroid injections, surgery

ostium Plural, ostia MEDTALK An opening, aperture

ostomy GI DISEASE A surgical opening through which waste material is discharged due to functional rectum or bladder. See Appliance, Colostomy, Ileostomy, Stomach-partitioning gastrojejunostomy, Urostomy

OT Symbol for 1. Occipitotransverse 2. Occupational therapist, therapy, see there 3. Olfactory threshold 4. Olfactory tubercle 5. Operating time 6. Oral temperature 7. Orotracheal 8. Oxygen therapy 9. Oxytocin

OTC 1. Ornithine transcarbamylase, see there 2. Over-the-counter drugs, see there

OTC stimulant A nonprescription–over-the-counter substance–eg, phenylpropanolamine, ephedrine before its banning, pseudoephedrine, caffeine, used as an anorexiant, amphetamine substitute, or stimulant ADVERSE EFFECTS Overdose, adverse drug interactions, toxicity–eg, HTN, tachyarrhythmia, seizures, hypertension

Othello syndrome Erotic jealousy, alcoholic paranoia PSYCHIATRY A delusion of spousal infidelity, a form of psychotic paranoia that is primary or more commonly, a symptom of organic 'psychopathies'–senile dementia, boxer encephalopathy, alcoholism

other See Most-significant other, Negative most-significant other

otitis externa Outer ear infection, swimmer's ear ENT Inflammation of the outer ear and auditory canal to the tympanic membrane, which most often affects adolescents, linked to recent exposure to water or mechanical ear trauma from scratching or foreign objects in the ear, hair spray, hair dyes, shampoos, and other chemicals which irritate the skin of the ear canal; OE may be associated with otitis media, URIs, common cold; moisture in the ear predisposes it to infections– fungal, *Pseudomonas* spp, *Proteus* spp

otitis interna ENT 1. Inflammation of the inner ear 2. Inner ear infection

otitis media ENT Middle ear inflammation, most common in young children, often due to infection CLINICAL Bulging, discolored tympanic membrane with ↓ motility, ear pain, irritability, difficulty in sleeping and eating, fever, vomiting DIFFDX Abscesses, furuncles, foreign bodies, mumps, external otitis, toothache RISK FACTORS UTIs, rhinitis, trisomy 21, other trisomies, cystic fibrosis, hypothyroidism, passive smoking, day care centers MANAGEMENT Amoxicillin; T-S for resistant bacteria. See Acute otitis media, Secretory otitis media

otitis media with effusion Secretory otitis media, see there

oto- *adjective* Referring to the ear

otolithic crisis of Tumarkin Drop attack A sudden unexplained fall without loss of consciousness or vertigo, attributed to abrupt change in otolithic input, resulting in an erroneous vertical gravity reference which, in turn, generates an inappropriate postural adjustment via the vestibulospinal pathway, resulting in a sudden fall; drop attacks occur in < 2% of Meniere's disease Pts. See Meniere's disease

otopalatodigital dysplasia Otopalatodigital syndrome, Taybi syndrome An X-linked condition completely expressed in ♂ and partially expressed in carrier ♀ CLINICAL Conduction-type deafness, distinct craniofacial deformity with prominent supraorbital ridges–frontal & occipital bossing, hypertelorism, a broad nasal root, flattened midface, small nose, mouth and jaw, partial anodontia, cleft palate, digital abnormalities—eg, short broad fingers—brachydactyly; bone dysplasia with dislocation of radial heads and/or hips, short trunk dwarfism, mental retardation

otoplasty A technique for correcting auricular deformities, often using the Mustard technique combined with conchal setback; symmetry is best achieved by using a standard point of reference—eg, Frankfort line

otorhinolaryngologist A physician trained in the medical and surgical management of disease of the ears, nose & throat MEAT & POTATOES DISEASES Deafness, nosebleeds, hoarseness, nasal congestion, tonsillar & adenoidal hypertrophy, sinus defects SALARY $176 + 4% bonus

otorrhea A fluid or purulent discharge from the ear through a perforation of the tympanic membrane or through a surgically placed ventilating tube

otosclerosis ENT Abnormal growth of bone in inner ear, blocking normal function of ear bones with gradual loss of hearing CLINICAL Slow, progressive; hearing may be better in noisy environments, tinnitus DIAGNOSIS Ear examination may exclude other causes of hearing loss; audiometry/audiology may determine extent of hearing loss; CT, MRI, PET of head

otoscope A device with a light and a magnifying glass used to examine the outer ear

ototoxic drug ENT A drug that can damage the hearing and vestibular/balance organs of the inner ear

ounce An English system unit of weight that is obsolete in civilized nations, but is still popular in the US; the apothecary ounce is equal to 31.1 g; the avoirdupois ounce is equal to 28.35 g

out-of-hospital transfusion The transfusion of blood products in settings other than hospitals or hospital-affiliated outpatient medical or surgical facilities—eg, a Pt's home, hemodialysis centers, physicians' offices, convalescent homes, surgery centers, community blood centers, and elsewhere. See Quality assurance, Quality improvement

out-of-network MANAGED CARE *adjective* Referring to a provider/service that is not part of an MCO's network of contracted providers/services. See Managed care organization

out-of-pocket costs MANAGED CARE Health care costs that a covered person must pay out of pocket—eg, coinsurance, deductibles, etc. See Copayment

outcome MEDTALK A general term for the results of an intervention or process. See Composite outcome, Health outcome, Placement outcome STATISTICS The result of a single trial of a probability experiment

outcomes MANAGED CARE Assessment of a treatment's efficacy by considering its success as a care solution as well as its cost, side effects, and risk

outcomes management MANAGED CARE A system for assessing and identifying preferred medical or surgical interventions or noninterventions that leads to a desired clinical outcome

outdated medicine 1. Pharmaceuticals that are older than the stamped expiration date 2. Less commonly, a style of medical practice that is no longer regarded as a standard of care

outer ear infection VOX POPULI → MEDTALK Otitis externa, see there

outgrow *verb* To change the relationship with a condition or structure by dint of ↑ age or size; while children outgrow clothing, and certain behaviors, they rarely outgrow diseases—eg, asthma

outie VOX POPULI An everted belly button—umbilical cord residuum incapable of collecting lint. See Inverted nipple. Cf Innie

outlier MANAGED CARE A Pt who falls outside of the norm—ie, who has an extremely long length of hospital stay or has incurred extraordinarily high costs. See Extreme outlier, High mortality outlier

outpatient *adjective* Referring to that which occurs—eg, therapy, outside of an institutional health care environment or hospital setting *noun* A Pt undergoing a diagnostic workup, evaluation, or treatment outside of a hospital. Cf Inpatient PSYCHIATRY Mental health treatment in the community at a local clinic or from private therapists

outpatient mastectomy Drive-through mastectomy A mastectomy in which no complication is anticipated and there is minimal or no hospital stay

outpatient services Hospital-based services MANAGED CARE Medical and other services provided, to a nonadmitted Pt, by a hospital or other qualified facility—eg, mental health clinic, rural health clinic, mobile X-ray unit, freestanding dialysis unit EXAMPLES Physical therapy, diagnostic X-ray, lab tests. See Ambulatory services

output CARDIAC PACING The electrical stimulus generated by a pulse generator and intended to trigger a depolarization in the chamber of the heart being paced. See Impulse MEDICINE A thing produced—eg, urinary output. See Basal acid output, Cardiac output, Maximum acid output, Peak acid, Standard output, Stimulated acid output SEXOLOGY See Put out

outside-in technique ORTHOPEDICS A surgical technique which precisely places instruments in arthroscopic surgery of the shoulder, by passing a needle through the anterior shoulder and visualizing where it enters the joint; once an acceptable position is found, an incision is made and a portal established. See Inside-out technique

outsource *verb* To assign specific work to a 3rd party for a specific length of time at an set price and service level MANAGED CARE To use outside labor to perform functions—billing and collections, accounting, janitorial services, ER coverage and others—which could be performed at a higher price and/or less efficiently by in-house personnel. See Manager

outstanding ear An auricular deformity characterized by an excess protrusion of the ear from the head, attributed to a defective or absent antihelical fold formation and often accompanied by a large conchal bowl TREATMENT Surgical reduction or correction of conchal bowl cartilage

ovalocytosis HEMATOLOGY A condition seen in up to 30% of ethnic groups in southeast Asia caused by a defective band 3 protein, in which there is ↑ affinity of RBC membrane band 3 to ankrin, resulting in a ↑ in RBC rigidity. See Elliptocytosis

ovarian ablation Ovarian suppression GYNECOLOGY Surgery, RT, or chemotherapy to block ovarian hormonal activity

ovarian cancer Cancer of ovary, ovarian carcinoma ONCOLOGY The 5th most common malignancy in ♀–US, 22,000 new cases/yr, and 13,300 deaths/yr RISK FACTORS ↑ Risk with nulliparity or first birth after age 35; ↓ risk with childbirth < age 25 or use of oral contraceptives; familial ovarian cancer accounts for 5% of ovarian cancers CLINICAL Generally vague Sx in early disease—the silent killer; advanced disease presents with abdominal fullness and early satiety due to ascites and omental tumor implants; OC is rarely confined to the pelvis; early diagnosis is fortuitous SCREENING Serum assays for markers of OC are useless DIAGNOSIS Physical examination—eg, pelvic exam, ultrasound, x-ray tests, CA-125 serum test, biopsy TYPES Epithelial tumors/carcinomas comprise 90% of ovarian CAs and are more common > age 40, often asymptomatic; non-epithelial tumors—eg, stroma cell and germ cell tumors are more common in younger ♀; OC use may ↓ the risk of ovarian CA; 5–10% of are familial; 3 hereditary patterns are identified: ovarian CA alone, ovarian & breast CAs, ovarian & colon CAs THERAPY—LIMITED DISEASE (stage I, II): TAH-BSO and omentectomy with examination of peritoneal surface THERAPY—ADVANCED DISEASE (stage III, IV) Debulking of peritoneal tumors, followed by platinum–eg, cisplatin–or the less toxic carboplatin, or taxol-containing regimens; 'compassionate' protocols that may improve survival include intraperitoneal chemotherapy and autologous BM transplantation

FIGO* STAGING, EPITHELIAL CANCER OF OVARY

STAGE I Tumor limited to ovary

A One ovary, no ascites, intact capsule

B Both ovaries, no ascites, intact capsule

C Both ovaries, malignant ascites (positive peritoneal washings), ruptured capsule (capsular involvement)

STAGE II Tumor extends beyond ovary into pelvis

A Pelvic extension to uterus or fallopian tubes

B Pelvic extension to other pelvic organs, eg bladder, rectum, vagina

C Pelvic extension + IC findings

STAGE III Extrapelvic extension or positive lymph nodes

A Microscopic seeding outside of pelvis

B Gross lesions ≤ 2 cm

C Gross lesions > 2 cm and positive lymph nodes

STAGE IV Distant/extraperitoneal organ involvement, eg liver, pleura

* Federation of Gynecology and Obstetrics

ovarian cyst Functional ovarian cyst; physiologic ovarian cyst GYNECOLOGY A round enclosed space filled with fluid or a semisolid material that develops on the ovary; most OCs are functional and involute spontaneously. Cf Polycystic ovaries

ovarian factor REPRODUCTION MEDICINE A cause of infertility due to problems with egg production by the ovaries

ovarian hypofunction GYNECOLOGY A general term for ↓ functional activity of the ovaries

ovarian monitoring REPRODUCTION MEDICINE The use of ultrasound and/or lab tests to monitor ovarian follicle development and hormone production

ovarian stimulation REPRODUCTION MEDICINE The use of drugs to stimulate the ovaries to develop follicles

ovarian vein 'syndrome' A clinical complex due to an enlarged and tortuous right ovarian vein with incompetence of the venous valves, typically seen in pregnancy, which is accompanied by hydronephrosis and pyelonephritis CLINICAL Intermittent right flank pain coinciding with menstruation, recurring UTI and exacerbation with progesterone TREATMENT Surgical

ovariectomy Oophorectomy The excision of one or more ovaries; ♀ castration

ovation falloposcopy GYNECOLOGY A diagnostic self-steering system with "unrolling balloon" and real-time visualization with a falloposcope, for atraumatic evaluation of proximal fallopian tube for diagnosing tubal occlusion

overcompensation PSYCHIATRY A conscious/subconscious process in which a real/imagined physical/psychologic deficit generates an exaggerated correction

overdose SUBSTANCE ABUSE Consumption of a therapeutic agent, drug, or narcotic, in excess of that required to produce the desired effects; overdoses are either accidental or suicidal and many agents have their own relatvely characteristic clinical, diagnostic, and therapeutic 'fingerprint'. See 'Designer' drugs, Heroin, 'Ice', Withdrawal

overflow incontinence Paradoxic incontinence Dribbling of urine due to chronic overdistension of the bladder–volume from 1000 to 3000 mL, normal, ±500 ml, with attenuation of the muscle; this may be confused–and therefore is paradoxic–with pure stress incontinence

overflow proteinuria Persistent proteinuria without glomerular disease, with excess production of filterable, low-molecular-weight proteins, exceeding resorptive capacities of the renal tubules–eg, ↑ lysozyme in myelomonocytic–M4 leukemia or Bence-Jones proteinuria; OP is usually asymptomatic if the protein loss is < 2.0 g/day

overgrowth syndrome A condition with multifocal excess growth due to pituitary overactivity with excess growth hormone; OS occurs before, acromegaly after puberty EXAMPLES Fragile X syndrome, Beckwith-Wiedemannn syndromes

overkill VOX POPULI An excess of anything

overlap syndrome GI DISEASE The presence of histologic features of Crohn's disease and ulcerative colitis in the same Bx, precluding a specific diagnosis of either; overlap occurs in ±15% of Bxs of these 2 conditions NEUROLOGY See Parkinsonism plus syndromes RHEUMATOLOGY A connective tissue disease with features of 2 or more rheumatologic disorders–eg, lupus erythematosus-mixed connective tissue disease overlap, rheumatoid arthritis-lupus overlap–1% of cases, scleroderma-polymyositis overlap–8-12% of cases and lupus erythematosus-polymyositis overlap– 5-10% of cases VASCULAR DISEASE A combination of systemic necrotizing vasculitis with features of polyarteritis nodosa and Churg-Strauss disease–allergic angiitis and granulomatosis, involving small and medium muscular arteries of the lung, causing HTN and hypersensitivity

overlapping toe ORTHOPEDICS A congenitally deformed toe which naturally lies over the plane formed by the other toes and, by pushing on the other toes, cause irritation MANAGEMENT Shoes; if symptomatic, derotation arthroplasty. See Derotation arthroplasty. Cf Underlapping toe

overlying PEDIATRICS The accidental rolling of a person on top of an infant or child sleeping in a parent's bed, which may cause death by suffocatio. See Co-sleeping

oversensing CARDIAC PACING Inhibition of a pacemaker to events other than those which the pacemaker was designed to sense–eg, myopotentials, electromagnetic interference, T-waves, crosstalk, etc. See Cardiac pacing

over-the-counter drug A therapeutic agent that does not require a prescription, which the FDA feels can be safely self-prescribed by non-physicians. Cf Prescription drug, Under-the-counter

overtraining SPORTS MEDICINE A general term for any practice of, or training for, a particular sport which is in excess of that necessary to participate in the sport, which ↑ the physical stress on specific parts of the musculoskeletal system. See Elite athlete, Female athlete triad, Reflex sympathetic dystrophy, Training

OVERTURE CARDIOLOGY A clinical trial–Omapatrilat Versus Enalapril Randomized Trial of Utility in Reducing Events

overuse HEALTH CARE The common use of a particular intervention even when the benefits of the intervention don't justify the potential harm or cost–eg, prescribing antibiotics for a probable viral URI. Cf Misuse, Underuse

overuse injury SPORTS MEDICINE A sports- or occupation-related injury that involve repetitive submaximal loading of a particular musculoskeletal unit, resulting in changes due to fatigue of tendons or inflammation of surrounding tissues; OIs include tennis elbow and golf elbow. See Overuse syndrome

overvalued procedure SURGERY Any procedure used to treat nonmalignant conditions–eg, cholecystectomy, that Congress has deemed 'too expensive' and has targeted for budget cuts from Medicare reimbursement. See Resource-based relative value scale

overweight A condition defined as ≥ 75th percentile of body-mass index [wt in kg/(height in m)]²; overweight adults are at ↑ risk for ASHD, arthritis, cardiovascular disease, DM, gallbladder disease, gout, HTN, and certain CAs; overweight adolescents are at ↑ risk for CAD, stroke, colorectal CA. See Morbid obesity, Obesity

ox eye Buphthalmos A progressively enlarged eye most commonly seen in children with congenital glaucoma; over time, the head of the optic nerve undergoes cupping and atrophy, resulting in blindness. Cf Bull's eye

oxacillin A semisynthetic penicillin

oxalates METABOLIC DISEASE A general term for oxalic acid salt or ester endproducts of metabolism excreted in urine which, if in extreme excess, accumulate as oxalate crystals in urine and kidneys; oxalates are ↑ in cirrhosis, IBD, DM, kidney stones, excess vitamin C, antifreeze ingestion, methoxyflurane anesthesia; they are ↓ in renal disease

oxaliplatin Eloxatin® ONCOLOGY A parenteral platinum used with 5-FU and leukovorin for advanced colorectal CA. See Cisplatin, 5-FU, Leukovorin

oxandrolone Oxandrin® ENDOCRINOLOGY An anabolic steroid used to reverse involuntary weight loss and depletion of cell mass, in Pts with AIDS, trauma, chronic infection, major burns or surgery, and other debilitating conditions ADVERSE EFFECTS Hepatotoxicity, cholestasis, liver cell necrosis, peliosis hepatis, hirsutism, gynecomastia, acne, male pattern baldness in ♀. See Anabolic steroid

oxazepam NEUROPHARMACOLOGY A benzodiazepine anxiolytic/sedative/hypnotic THERAPEUTIC RANGE 0.2–1.4 mg/L; 200-1400 μg/L TOXIC RANGE Not defined

oxidative phosphorylation The formation of ATP within mitochondria using energy derived from the electron transport chain

oxidative stress The presence of oxygen free radicals, which are generated by various stressants–eg, tobacco, alcohol; the primary antioxidant is glutathione; other antioxidants include vitamins A, C, and E; ↑ oxidative stress may be one of the key factors in the early pathogenesis of AIDS in which the production of TNF ↑ the production of free radicals in T cells. See Free radicals

Oxsoralen® Methoxsalen, see there

oxybutynin Ditropan® UROLOGY A spasmolytic anticholinergic GI and GU tract colic used to ↓ urge incontinence, overactive bladder, bladder Sx associated with neurogenic bladder–eg, frequency, urgency, dysuria, urinary leakage CONTRAINDICATIONS Urinary retention, gastric retention, narrow-angle glaucoma ADVERSE EFFECTS Dry mouth, constipation, somnolence, diarrhea, blurred vision, dry eyes, dizziness, rhinitis. See Urge incontinence

oxycodone Hillbilly heroin, poor man's heroin, redneck heroin PHARMACOLOGY An oral opioid analgesic, which may be coadministered with other analgesics PHARMACOLOGIC EFFECTS Morphine-like actions; blocks CNS responses to pain; CNS depressant; analgesic. See Opioids, OxyContin

OxyContin® SUBSTANCE ABUSE A potent formulation of oxycodone, a highly addictive analgesic linked 291 deaths in 2000. See Oxycodone. Cf Heroin

oxygen consumption VO₂ ABG The difference between the amount of O₂ delivered to tissues and the amount of O₂ returned to the right ventricle PHYSIOLOGY The measurement of O₂ uptake, this measurement is used routinely to calculate the metabolic cost of different activities

oxygen dissociation curve Oxygen saturation curve PHYSIOLOGY A curve that describes the relationship between Hb O₂ saturation and tension; defined by a sigmoid curve which reflects the interaction of the 4 Hb molecules involved in O₂ uptake, transport and release; a 'right shift' of the curve indicates ↓ Hb affinity for O₂, as occurs in ↓ pH–ie, acidosis, ↑ temperature, ↑ PCO₂, while a 'left shift' indicates ↑ O₂ affinity with ↑ pH, ↓ temperature, ↓ 2,3 DPG and ↓ PCO₂ RIGHT SHIFTS Acidosis, hyperthermia, alveolar hypoventilation, anemia LEFT SHIFTS Alkalosis, hypothermia, hyperventilation, carboxyhemoglobinemia, hypophosphatemia, ↑ fetal Hb. See 2,3 DPG

oxygen radical Reactive oxygen metabolite Any molecule with an unpaired electron in its outer orbital which leads to an unstable and/or short half-life, which is capable of reducing or oxidizing other molecules. See Oxygen free radicals

oxygen saturation sO₂ The O₂ concentration of blood expressed as a ratio of its total O₂-carrying capacity; the OS is a measure of the utilization of O₂ transport capacity; sO₂ is ↑ at high altitude, hypocapnia, hypothermia, ↑ cardiac output, O₂ therapy, PEEP–positive end-expiratory pressure ventilation, and respiratory alkalosis. sO₂ is ↓ with AV shunting, carbon monoxide poisoning, congenital cardiac defects, emphysema, hypercapnia, hypoventilation, hypoxia, respiratory acidosis REF RANGE 95-98% arterial; 60-80% venous

oxygen toxicity Tissue and molecular damage due to the effects of O₂ free radicals in cellular and extracellular micro-environments; OT occurs in older subjects, shock and inflammation. See Oxygen radical

oxyphenbutazone Oxalid® THERAPEUTICS An oral NSAID tightly bound to plasma protein METABOLISM It is biotransformed in the liver; metabolites are excreted in urine. See NSAIDs

oxytocin ENDOCRINOLOGY A nonapeptide hormone which stimulates contraction of uterine smooth muscles and promotes milk ejection in females. Cf ADH

oxytocin stress test Contraction stress test, oxytocin challenge test OBSTETRICS A test of fetal well-being based on the principle that a hypoxic feturs will show late decelerations in the fetal heart rate in response to uterine contractions. See Deceleration. Cf Fetal heart monitoring, Non-stress test

ozena Pue-nez, Stinknase ENT A mucopurulent nasal discharge in chronic atrophic rhinitis, seen in long-standing systemic disease–eg, iron-deficiency anemia or chronic local infection, described in southern Europe; the etiologic role of *Klebsiella ozaenae* is speculative as this organism is sensitive to broad-spectrum antibiotics, while ozena is refractory to antibiotic therapy

ozonation PUBLIC HEALTH The bubbling of ozone through water as a method of water purification, a process that had been proposed as an alternative to chlorination, which has been linked to an ↑ in cancer. Cf Chlorination

P Symbol for: 1. Para-obstetrics 2. Partial pressure 3. *Pasteurella* 4. Patient 5. Peak flow, see there 6. Pectoral 7. Penicillin 8. Percussion 9. Phosphate-inorganic 10. Phosphorus 11. Plasma 12. *Plasmodium* 13. Placebo 14. Polyneuropathy 15. Population–genetics, see there 16. Position 17. Post 18. Posterior 19. Postpartum 20. Prednisolone 21. Premolar 22. Pressure 23. Primipara 24. Probability–P value 25. Procarbazine 26. Progesterone 27. Proline 28. Protein 29. Proteinuria 30. *Proteus* 31. Proton 32. Pulse 33. Pupil 34. Pyranose

p Symbol for: 1. Atomic orbital with angular momentum 2. The more common of 2 different alternative–allelic versions of a gene 3. The frequency of less common allele is q 4. pico—SI unit for 10^{-12} 5. Protein–biochemistry–eg, p53 is 53 kD protein 6. Proton 7. Sample proportion–binomial distribution in statistics 8. Short arm of a chromosomepeta–SI unit for 10^{15}

p arm The short arm of a chromosome–from French, petit, small; all human chromosomes have 2 arms: p and q

'P' pulmonale CARDIOLOGY A sharply peaked P wave on EKG, which is a relatively nonspecific finding of COPD, most prominent during exacerbation of clinical disease. See Chronic obstructive pulmonary disease.

P wave CARDIAC PACING The electrocardiographic representation of atrial depolarization

PA 1. Panic attack 2. Pathology 3. Percussion & Auscultation 4. Pernicious anemia 5. Photoallergic 6. Physical activity 7. Physician's assistant 8. Plasminogen activator 9. Platelet adhesiveness 10. Polyarteritis 11. Posterior anterior 12. Posterior aorta 13. Pregnancy-associated 14. Profile analysis 15. Pseudoaneurysm 16. Psychological age 17. Pulmonary angiography 18. Pulmonary artery 19. Pulpoaxial

PA interval CARDIOLOGY The interval between the onset of the P wave in the surface tracing of the EKG, which slightly precedes the onset of high right atrial recording, and the low right atrial deflection, measured in the His lead; the PAI corresponds to the intra-atrial conduction time, but has little clinical value. Cf AH interval, HV interval

p24 antigen AIDS The 24 kD core antigen of HIV-1, which is linked to clinical AIDS; p24 is the earliest marker of HIV-1 infection, and detectable days to wks before seroconversion to anti-HIV-1 antibody production, detected by ELISA. See CD4:CD8 ratio, HIV, HIV RNA, Reverse transcriptase, Western blot

P_{50} The O_2 tension at which Hb is $\frac{1}{2}$ (50%) saturated, a value equal to 26 torr–mm Hg in normal RBCs; the P_{50} value is obtained from the midpoint of the O_2 dissociation curve–O_2DC, and does not reflect the shape of the curve; with $\uparrow O_2$ affinity, P_{50} $\downarrow$, resulting in a 'left shift' of O_2; $\uparrow$ P_{50} indicates $\downarrow O_2$ affinity for Hb. See Oxygen dissociation curve

p53 A 53 kD nuclear phosphoprotein encoded by the proto-oncogene *p53*, on chromosome 17p13; in its wild form, p53 inhibits cell growth control and transformation; it activates transcription of genes that suppress cell proliferation, acting as a tumor suppressor protein; if p53 is physically lost or functionally inactived, cells can grow without restraint. See Li-Fraumeni syndrome, p21, Tumor suppressor genes, *WAF1*

P_{CO_2} Partial pressure of CO_2 in blood, expressed in kilopascals

P_{O_2} Partial pressure of O_2 in blood, expressed in kilopascals

PAC 1. Papular acrodermatitis of childhood–Gianotti-Crosti syndrome 2. Political action committee 3. Premature atrial contraction, see there 4. Pulmonary artery catheterization

PAC-A-TACH A clinical trial–PACing in Atrial Fibrillation & Tachycardia

pacemaker CARDIOLOGY An electronic device which generates timed electric pulses that stimulate ventricular contractions. See Artificial pacemaker, Asynchronous pacemaker, Automatic antitachycardia pacemaker, Cardiac pacemaker, Circadian pacemaker, Demand pacemaker, Dual chamber pacemaker, External pacemaker, Programmable pacemaker, Triggered pacemaker, Wandering pacemaker

pacemaker-mediated tachycardia CARDIAC PACING Tachycardia seen in atrial tracking pacemakers, which begins with and is sustained by ventricular events conducted retrogradely to the atria; the pacemaker senses retrograde atrial depolarization, then delivers a stimulus to the ventricle, causing depolarization, that again is conducted retrogradely to the atria which, if repeated, causes tachycardia

pacemaker syndrome CARDIOLOGY A common–up to 20%–complication of pacemakers caused by adverse hemodynamic effects of ventricular pacing, due to AV asynchrony CLINICAL Vertigo, syncope, dyspnea, weakness, $\downarrow$ exercise tolerance, postural hypotension, palpable hepatic and jugular veins ETIOLOGY Alternating AV asynchrony in which the atrium contracts against closed valves, raising venous pressure or ventricle contracts before blood arrives, causing transient $\downarrow$ in cardiac output MANAGEMENT Dual chamber pacing pacemaker, antiarrhythmics

Pacerone® Amiodarone CARDIOLOGY An antiarrhythmic similar to Cordarone®. See Antiarrhythmic

Paceum Diazepam, see there

pachydermoperiostosis An AD condition characterized by induration of skin in natural folds, accentuation of creases of face and scalp, finger clubbing and periostosis of long bones; acquired pachydermoperiostosis is linked to bronchogenic carcinoma, termed hypertrophic–pulmonary osteoarthropathy, see there

pachydermy A nonspecific term for leathery subcutaneous induration due to an accumulation of inelastic connective tissue, as occurs in acromegaly or due to accumulation of protein-rich mucin, collagen and fibroblasts, as occurs in myxedema

pacing CARDIOLOGY The timing of a physiologic event. See Burst pacing, Demand pacing, DDDR pacing, Dual-chamber pacing, Overdrive pacing, Physiologic pacing, Ramp pacing, Rate responsive pacing, Safety pacing, Transvenous pacing, Underdrive pacing

pacing mode CARDIAC PACING The manner in which a cardiac pacemaker provides artificial rate and rhythm support during arrhythmias; PMs are ID'd by the NBG Code. See NBG code

Pacitran

Pacitran See Diazepam

pack rat PSYCHIATRY A person who hoards objects of little worth–eg, string, old newspapers, etc–and lacking sentimental value; the PR mentality is typical of OCD. Cf Obsessive-compulsive disorder

pack-years MEDTALK A crude indicator of a person's cumulative cigarette consumption, equal to the number of packs of cigarettes smoked/day, multiplied by yrs of consumption. See Smoking, Tobacco

package insert PHARMACOLOGY A synopsis of key physicochemical, pharmacologic, clinical efficacy, and clinical safety properties of a prescription drug, bundled therewith, intended to be highly readable and helpful to clinicians looking for specific information–eg, indications for use; forms of administration, side effects, etc; it is more cynically described as a hard-to-handle and difficult-to-read package 'stuffer' printed in Lilliputian type on Bible paper. See Advertising

packed red cells TRANSFUSION MEDICINE A concentrated unit of RBCs prepared from a unit of whole blood by removing plasma INDICATIONS Active bleeding, excess intraoperative blood loss, low 'pre-op' or 'post-op' Hcts, chronic anemias–eg, sickle cell anemia, thalassemia, chemotherapy, dialysis, blood exchange. See CPDA-1, Quad pack, Single unit transfusion, Storage lesions, Whole blood

packing ENT Nasal packing, see there

paclitaxel Paxene®, Taxol™ ONCOLOGY An antimitotic anticancer taxane used for KS, breast and ovarian CAs. See Breast CA, Ovarian CA

PACT CARDIOLOGY Any of various clinical trials–1. Philadelphia Association of Clinical Trials study; 2. Plasminogen activator Angioplasty Compatibility Trial; 3. Prehospital Application of Coronary Thrombolysis; 4. Prourokinase in acute coronary thrombosis

PAD 1. Panic/anxiety disorder. See Panic disorder 2. Peripheral arterial disease. See Peripheral vascular disease 3. Preoperative autologous donation. See Autologous donation 4. Public access defibrillator CARDIOLOGY A portable defibrillator for on-scene management of cardiac arrest victims in public locations–airports, planes, malls, stadiums, first-response vehicle. See Defibrillator

pad VOX POPULI 1. A fleshy mass, often subcutaneous skin. See Dancer's pad, Heel pad 2. A wad of absorbent material. See Loofah pad, Superstat hemostatic wound pad

PAD syndrome Phobic-anxiety-depersonalization syndrome PSYCHIATRY A debilitating phobic state–75% of Pts are ♀, which begins with a stressor, causing sudden intense anxiety or emotional shock, followed by depersonalization of varying duration, then by generalized and pervasive anxiety CLINICAL Pt can't work, socialize, and is under constant stress. See Phobia; Cf Agoraphobia, Monophobia

padded dash(board) syndrome A rare form of trauma, primarily affecting right front seat occupant–'passenger side' in many countries in an automobile accident when he/she is restrained by a lap-type safety belt; in an abrupt stop, the passenger's body is thrown forward at the waist and the hyperextended neck strikes the dashboard at the level of the thyroid cartilage CLINICAL Respiratory distress due to upper airway obstruction, due to a hematoma, fluid accumulation, aspiration of saliva, fluids, subcutaneous emphysema, fractures of cartilage, pain on swallowing

page noun MEDTALK A beep or 'detonation' of a pager/beeper verb To contact a person by voice over a public address system

Page syndrome A vasomotor dysfunction characterized by periodic blotchy flushing and sweating of face, upper chest, abdomen, ±associated with cold extremities, headache, tachycardia, HTN, possibly due to vascular compression of brain stem

Paget's disease of bone Osteitis deformans An idiopathic bone-remodeling disorder of older Northern European ♂ especially of lumbosacral spine, pelvis, skull, affecting ±10% of those > age 80 CLINICAL Often asymptomatic; or pain, deformity, pathologic fracture, neural compression IMAGING Moth-eaten destruction of trabecular bone, endosteal scalloping, cortical penetration, soft tissue masses LAB ↑ Serum alk phos COMPLICATIONS Heart failure due to AV shunting; osteosarcoma, or other sarcomas in 1-25% TREATMENT Biphosphonate, calcitonin, mitrimycin, surgery for decompressing critical cranial structures, or joint replacement. See Mosaic bone

Paget's disease of breast Paget's disease of nipple A 'weeping', eczematoid nipple lesion often associated with underlying intraductal breast CA–DCIS and, if accompanied by a palpable mass, invasive CA CLINICAL Itching, burning at nipple, ± oozing, bleeding DIFFDX Bowen's disease, melanoma PROGNOSIS Survival reflects aggressiveness of underlying ductal cell carcinoma, or less commonly, lobular carcinoma. See Ductal carcinoma in situ. Cf Extramammary Paget's disease, Intraepidermal carcinoma

PAH 1. Polycyclic aromatic hydrocarbon, see there 2. Pulmonary artery HTN

PAI Plasminogen activator inhibitor, see there

pain NEUROLOGY 'An unpleasant sensory and emotional experience associated with actual or potential tissue damage or described in terms of such damage'-per Intl Assn for Study of Pain; a sensation of discomfort, distress, or agony, due to stimulation of specialized nerve endings; a sensation of marked discomfort, either sharp and well-localized–conducted along A-delta fibers or dull and diffuse–conducted along C nerve fibers. See Acute pain, Acute low back pain, Ankle pain, Back pain, Breakthrough pain, Brief Pain Inventory, Central stroke pain, Chest pain, Chest wall pain, Chronic pain, Discogenic pain, Elbow pain, Gait control theory, Growing pain, Intractable pain, Knee pain, Lightning pain, Low back pain, Noncardiac chest pain, Patient controlled analgesia, Phantom limb pain, Substantial pain, Suprapubic pain

pain clinic PAIN MANAGEMENT An outpatient facility in which a person with chronic, poorly controlled or intolerable pain–eg, cancer pain, refractory low back pain–can go to obtain physical and pharmacologic therapy for pain relief; PCs are often supervised by a physician–usually an anesthesiologist, neurologist, psychiatrist–trained in pain management.

pain diary PAIN MANAGEMENT A log kept by a Pt with chronic pain, indicating when the pain is greatest during the day, and medication needs for relieving pain. See Pain maangement

pain disorder Somatiform pain disorder, see there

pain med-er SUBSTANCE ABUSE A popular term for a Pt who abuses pain medication. See Drug-seeking behavior

pain medicine INTERNAL MEDICINE A new subspecialty of neurology dedicated to the relief and/or control of pain PHARMACOLOGY A popular term for an analgesic

SURGICAL MANAGEMENT OF PAIN

- **SECTION OF PERIPHERAL NERVE PATHWAY** Neurectomy, splanchnicectomy, dorsal rhizotomy
- **SECTION OF CNS PATHWAY** Anterolateral cordotomy, trigeminal tractotomy, dorsal root entry zone lesioning
- **PROCEDURES TO ALTER AFFECTIVE RESPONSE TO PAIN** Cingulotomy, obotomy, thalamotomy
- **SECTION OF EFFERENT ARC OF VASOMOTOR REFLEX** Sympathectomy
- **SUPPRESSION BY NONABLATIVE TECHNIQUES** Stimulation of peripheral nerve, spinal deep brain, intraspinal drug infusion

pain modulation NEUROLOGY An ↑ or ↓ of the sensation of pain, possibly due to a 2° neural pathway. See Opioid-mediated analgesia system

PainBuster™ infusion kit ORTHOPEDICS A disposable device that provides continuous local anesthesia–bupivacaine to an intraoperative site for relief of pain postorthopedic surgery. See Bupivacaine

painful bruising syndrome PSYCHOLOGY An idiopathic psychosomatic trauma-induced condition seen in emotionally-labile ♀ arising on legs, face, trunk, characterized by recurring painful ecchymoses with a ladder-like morphology, accompanied by syncope, N&V, GI, intracranial bleeding

painful crisis Vaso-occlusive crisis, see there

painful fat syndrome An atypical, chronic, symmetrical swelling and tenderness of the legs, more common in adolescent ♀; lipoedema is painful on pressure, non-pitting and fancifully likened to pigskin. Cf Lipomatosis dolorosa

painful heel An idiopathic condition of older ♂ causing tenderness of the heel associated with focal edema; ½ have calcaneal spur formation PROGNOSIS Persistent pain or spontaneous resolution

painful red leg syndrome Erythromelia A condition characterized by ↑ sensitivity to exact skin temperatures > 32ºC, with focal vasodilation and a burning sensation TREATMENT Aspirin; 2º erythromelia may occur in HTN or polycythemia vera

painless thyroiditis 1. Silent thyroiditis, see there 2. Subacute lymphocytic thyroiditis, see there

paintball SPORTS MEDICINE A sport in which marble-sized gelatin capsules filled with a nontoxic dye are shot at speeds of 300 kph/200 mph ***WARNING: Paintball-related eye injuries are common and include cataracts, retinal tears/detachment and damage to lens and optic nerve; 50% of eye injuries result in permanent blindness in affected eye; protective masks are a must****; other players must KNOW you're in "time out" to clean goggles*

pairbond PSYCHOLOGY A strong and long-lasting closeness between 2 persons–eg, between parent and child or 2 lovers

paired helical filaments NEUROPATHOLOGY Paired structures that are core constituents of the neurofibrillary tangles of Alzheimer's disease and occur in Down syndrome, Hallervorden-Spatz disease, lead encephalopathy, lipofuscinosis, subacute sclerosing panencephalitis, and tuberous sclerosis. See Alzheimer's disease, Neurofibrillary tangles, Senile plaques

palatoplasty ENT Surgery performed on the hard and/or soft palate to correct a structural or functional defect, usually, a cleft palate, see there

palindromic rheumatism A form of monoarthritis that may precede rheumatoid arthritis CLINICAL Intermittent recurring–> 5 attacks in 2 yrs episodes of intense gout-like pain and joint inflammation, more common in middle-aged men, affecting the knee, wrist or dorsum of hand, accompanied by transient subcutaneous nodules; although the condition is distinct between Pts, each attack tends to follow the same pattern in the individual Pt LAB Nonspecific ↑ ESR and acute phase reactants

palivizumab Synagis® IMMUNOLOGY A humanized monoclonal antibody used as prophylaxis against RSV infection ADVERSE EFFECTS Fever pneumonia, and injection site reactions. See Respiratory syncytical virus

palladium-103 implantation UROLOGY The implantation of radioactive seeds–brachytherapy to manage prostate CA, generally coupled to the use of external beam RT. See Prostate cancer

palliate *verb* To treat a disease without expecting a cure

palliative RADIATION ONCOLOGY *adjective* Referring to treatment to relieve Sx of a disease but not to cure it, especially alleviating pain

palliative care MEDTALK Care provided to a Pt, optimizing quality of remaining life and offering support and guidance to the Pt and family

palliative surgery An operation performed on an incurable CA, which is justified to ↓ severity of Sx and improve the quality of life, relieve pain–cordectomy, hemorrhage–cystectomy for bleeding urinary bladder, obstruction–colostomy or gastroenterostomy, or infection–amputation of a necrotic and malodorous tumor-ridden breast or extremity. See 'Heroic' surgery, Mutilating surgery

palliative therapy Palliative treatment MEDTALK Any treatment of a terminally ill Pt intended to alleviate pain and suffering, without performing aggressive–'heroic' procedures; PT does not alter course of disease, but improves the quality of life before death through various modalities–eg, surgery and RT, to ↓ tumor masses compressing vital structures. See 'Band-aid' therapy, Karnovsky scale; 'Heroic' therapy

pallidotomy NEUROSURGERY Incision into or partial destruction of the globus pallidum, a difficult high-risk procedure useful in Parkinsonism unrefractory to L-dopa therapy COMPLICATIONS Visual defects, loss of speech, seizures, coma, death. See GPi pallidotomy

Pallister-Hall syndrome Hypothalamic hamartoblastoma An AD condition characterized by hypothalamic neuronal hamartoma, pituitary agenesis, hypogenitalism, dwarfism, facial dysmorphia, postaxial polydactyly, anorectal atresia, renal and pulmonary malformations

pallor MEDTALK Paleness

Palmaz-Schatz™ stent CARDIOLOGY A balloon-expandable coronary stent delivered via balloon angioplasty, which remains in the expanded artery to maintain patency and prevent restenosis. See Percutaneous transluminal angiography

Palmaz® Corinthian™ system A balloon-expandable biliary stent and delivery system for palliation of biliary obstructions caused by malignant biliary tract tumors. See Biliary tumors

palmomental reflex Palm-chin reflex NEUROLOGY A unilateral contraction of the mentalis and orbicularis oris triggered by brisk scratching of the ipsilateral palm

palpable PHYSICAL EXAM *adjective* Referring to that which can be felt. Cf Nonpalpable

palpation CLINICAL MEDICINE 1. A part of a physical examination, in which an examiner lighly presses with the hands over the surface to ID lumps, bumps, beating, buzzing, and burbling of the body below. See Percussion. Cf Palpitation 2. Touching, feeling. Cf Touchy-feely

palpebral fissure The horizontal slit-like opening for the eyes between the eyelids

palpitation CARDIOLOGY A generally unpleasant subjective sensation of strong and/or irregular heart pulsations, often accompanying ↑ physical exertion Sx & CLINICAL CORRELATES Flip-flopping sensation, the heart seems to stop, then start with pounding–s/o premature atrial or ventricular contractions; fluttering sensation–s/o atrial or ventricular arrhythmias; pounding in neck–s/o AV dissociation–atria are contracting against closed tricuspid or mitral valves, as in reentrant supraventicular arrhythmias, especially AV nodal tachycardia or ventricular premature depolarization, producing cannon A waves in jugular veins with neck pulsations which, if prominent, cause a bulging–'frog sign' ETIOLOGY Anxiety, panic disorders, catecholamine excess–eg, in cool-down period after exercise, postural changes–eg, standing abruptly after recumbent position–causing syncope/near syncope DIAGNOSIS Hix, physical exam, 12-lead EKG, Holter monitor MANAGEMENT Reassurance, beta-blockers, CCBs, radio-frequency ablation, modification of sinus node

PALS Pediatric advanced life support

palsy NEUROLOGY Complete paralysis, see there, of a particular body region or extremity, rendering the region incapable of voluntary motor activity. See Bell's palsy, Cerebral palsy, Crutch palsy, Erb's palsy, Facial palsy, Progressive supranuclear palsy, Pseudobulbar palsy, Saturday night palsy

PAM Primary amoebic meningoencephalitis

Pamelor® Nortriptyline, see there

PAMI CARDIOLOGY Clinical trials–Primary Angioplasty in Myocardial Infarction that compared M&M and recurrent in-hospital ischemia of primary coronary angioplasty v Rx w/ rt-PA thrombolysis; PA won. See Coronary angioplasty

pamidronate Aredia® METABOLISM A 2nd-generation biphosphonate used to manage osteoporosis and Paget's disease of bone. See Biphosphonate, Osteoporosis

Panadol® Acetaminophen, see there

Panaeolus TOXICOLOGY A mushroom with significant quantities of psilocybin, a hallucinogen CLINICAL 20–60 mushrooms may be ingested to obtain the desired hallucinogenic effect; small amounts of *Panaeolus* spp evoke euphoria, dizziness, weakness; larger amounts alter temporal and spatial perception and induce hallucinations MANAGEMENT Supportive. See Poisonous mushroom

panagglutination Polyagglutination, see there

Panalgesic® Salicylate, see there

Panalok ORTHOPEDICS A resorbable soft tissue anchor used in shoulder–eg, rotator cuff, as well as elbow, knee, and ankle repair

Panama red DRUG SLANG A regionally popular term for marijuana from–no rocket science, here–Panama

pANCA Perinuclear ANCA. See ANCA

pancake omentum A thick, indurated omentum caused by diffuse infiltration by CA, usually epithelial, commonly, of ovary, but also colon, stomach or pancreas; by CT, the omental 'cake' is a flat layer of tumor separating the small or large intestine from the anterior abdominal wall. Cf 'Policeman of the abdomen'

Pancoast syndrome ONCOLOGY A syndrome characterized by a malignant neoplasm–Pancoast's tumor–of the cervical region with destruction of the thorax inlet and involvement of the brachial plexus and cervical sympathetic nerves, accompanied by severe pain in the shoulder region radiating toward the axilla and scapula along the ulnar aspect of hand muscles, atrophy of hand and arm muscles, Bernard-Horner syndrome, and compression of blood vessels with edema; PS may be found in inflammatory pseudotumor and in mycotic aneurysms. See Pancoast tumor

Pancoast tumor Pulmonary sulcus tumor ONCOLOGY A non-small cell lung CA arising in the upper lung, which extends to adjacent tissues–eg, ribs, vertebrae, etc. See Non-small cell lung CA, Pancoast syndrome

pancolitis GI DISEASE Inflammation of the entire colon, which most commonly occurs in ulcerative colitis. See Ulcerative colitis

pancreas ultrasonography Pancreatic sonography IMAGING Ultrasonography in which the recording probe–transducer is passed over the pancreas to ID neoplasms, circumscribed inflammation, cysts, pseudocysts. See Ultrasonography

pancreatectomy SURGERY The partial or radical–Whipple procedure–surgical removal of the pancreas INDICATIONS Pancreatic CA. See Whipple procedure

pancreatic abscess SURGERY An aggregate of PMNs in the pancreas which is caused by inadequate drainage of a pancreatic pseudocyst, a complication associated with pancreatitis. See Pancreatic pseudocyst

pancreatic cancer Cancer of pancreas, pancreatic carcinoma ONCOLOGY A 'silent' cancer that rarely causes early Sx unless it blocks the common bile duct and bile cannot pass into the digestive system, causing jaundice EPIDEMIOLOGY 5th leading cause of CA death–US; 11th most common CA; 24,000 deaths/yr–US INCIDENCE $9/10^5$ 5-yr survival, ± 5% RISK FACTORS ♂, black, smoking, ↓ consumption of fruits, vegetables, ↑ meats, alcohol, chronic pancreatitis

PANCREATIC CANCER STAGES

I CA confined to the pancreas, or direct extension to duodenum, bile duct, peripancreatic tissues

II CA extends directly to the stomach, spleen, colon, or adjacent large vessels

III Any direct tumor size with positive lymph nodes

IV Any known metastasis, eg to liver or lungs

pancreatic cholera syndrome(s) WDHA syndrome, see there

pancreatic endocrine tumor A neoplasm of the endocrine pancreas–eg, APUDoma, islet cell tumor, nesidioblastoma, common in the body and tail, the sites of ↑ concentration of islets of Langerhans; PETs are seen in 0.5-1.0% of unselected autopsies, hormonally active tumors comprise $< 1:10^5$; PET names reflect the predominant hormone–eg, gastrinoma, glucagonoma, VIPoma, PPoma etc, although most PETs produce ≥ 1 hormone PROGNOSIS Most PETs are indolent with a 10-yr survival post-resection; aggressive PETs may need chemotherapy, streptozocin. See Endocrine pancreas, WDHA syndrome, Zollinger-Ellison syndrome

pancreatic insufficiency A relative or absolute lack of pancreatic enzymes, a finding typical of cystic fibrosis. See Cystic fibrosis

pancreatic islets Islets of Langerhans ANATOMY Clusters of cells in the pancreas that form the endocrine portion and secrete insulin and glucagon

pancreatic islet transplantation The transplantation of a Pt's own pancreatic islets–pancreatic islet autograft into the liver via intraportal injection postpancreatectomy, mandated by severe chronic pancreatitis 2° to small duct disease, pancreas divisum, etc PROS Nonimmunogenic

pancreatic juice GI DISEASE Fluid produced by the pancreas containing digestive enzymes

pancreatic panel LAB MEDICINE An abbreviated battery of assays used to ID pancreatitis: Amylase, lipase, calcium, glucose. Cf Organ panel

pancreatic pseudocyst GI DISEASE Any of a circumscribed collection of pancreatic secretions surrounded by non-epithelial cell lined fibrous walls of granulation tissue; pseudocysts develop in 10% of Pts with chronic pancreatitis; most are small and resolve spontaneously; others hemorrhage, rupture, become infected TREATMENT Resection, external or internal drainage

pancreatic sufficiency A term referring to Pts with cystic fibrosis–equally applicable to DM with adequate insulin production–with enough exocrine pancreatic function to allow normal digestion without enzyme supplements; PS occurs in milder disease, in Pts older when diagnosed, with lower sweat chloride, milder respiratory disease, normal growth and better prognosis. See Cystic fibrosis

pancreatic transplantation A procedure designed to halt the progression of diabetic neuropathy and achieve glycemic control; pancreatic transplantation requires that a large segment or the entire pancreas be harvested; donors are usually cadaveric. Cf Biohybrid artificial pancreas, Islet cell transplantation

pancreatitis GI DISEASE Inflammation of the pancreas, common in alcoholics; typical nonalcoholic Pt is an older ♀ ETIOLOGY Idiopathic, alcohol, bile tract disease, etc–eg, hyperlipidemia, DKA, ESRD, post-transplant, pregnancy, trauma, surgery, ERCP, EBV, HIV, mycoplasma, hepatitis, drugs–eg, azathioprine, thiazides, furosemide, estrogens, tetracycline, sulfonamides, metronidazole, erythromycin, salicylates, toxins–eg, scorpions, methanol, organophosphates CLINICAL N&V, agitation, restlessness, abdominal pain, low-grade fever, painful subcutaneous fat nodules on abdomen DiffDx Perforated viscus, cholecystitis, bowel obstruction, vascular occlusion, renal colic, acute MI, pneumonia, DKA Imaging CT may show pancreatic pseudocysts in acute pancreatitis Lab Amylase, lipase, transiently ↑ BR, ↑ glucose, ↑ WBCs, ↓ Ca²⁺ Management Supportive–eg, nada per os; NG aspiration for Sx relief; somatostatin. See Acute pancreatitis, Chronic pancreatitis, Gallstone pancreatitis, Pancreatic abscess

pancuronium Anesthesiology A long-acting nondepolarizing muscle relaxing anesthetic Mechanism Competitive inhibition of acetylcholine at motor end plate Pros Metabolized by plasma cholinesterase or nonenzymatic pathways, thus better in multiorgan failure Cons Evokes histamine release, causing hypotension, tachycardia. See Muscle relaxant. Nondepolarizing agent. Cf Depolarizing agent

pancytopenia Hematology A global ↓ of 'blood cells' with hypoplasia or aplasia of hematopoietic precursors in BM–eg, hypoplastic myelodysplasia Etiology Aplastic anemia–drug-related, especially in chemotherapy, RT, toxins, BM replacement by hematopoietic, lymphoproliferative and metastatic CA, storage diseases, osteopetrosis, myelofibrosis, hypersplenism–congestive splenomegaly, hematopoietic malignancy, storage diseases, sarcoidosis, malaria, kala-azar, infection–eg, fungemia, septicemia, TB, megaloblastic anemia. See Fanconi's disease

pandiculation Stretching & yawning; not important, but useful

panel Lab medicine A group or battery of tests. See Anemia panel, Bone/joint panel, Cardiac injury panel, Cardiac risk evaluation panel, Collagen disease and arthritis panel, Coma panel, Custom panel, Diabetes panel, Electrolyte panel, Electrolyte/fluid balance panel, General health panel, Hepatic function panel, Hepatitis panel, Hypertension panel, Kidney panel, Lipid panel, Liver panel, Metabolic disease panel, Metastatic disease panel, Neonatal panel, Neoplasm panel, Organ panel, Pancreatic panel, Parathyroid panel, Profile, Pulmonary panel, Sensitire HP™ panel, Test panel, Radiation hybrid panel, Testicular cancer panel, Thyroid panel, TORCH panel Managed care A group of physicians. See Closed panel, Open panel Research See Advisory panel.

panel physician Managed care A physician who works in a health care network, and unlike physicians in private practice, does not have his/her 'own' Pts. See Deselection

panencephalitis Simultaneous inflammation of the gray and white matter. See Subacute sclerosing panencephalitis

panhypopituitarism Pituitary dwarfism, see there

panic attack Psychiatry A period of extreme anxiety, intense fear, apprehension or discomfort, in which ≥ 4 specific Sx develop abruptly, peak in 10 mins Clinical Abrupt onset of tachycardia, SOB, dizziness, tingling, anxiousness, temporary paralysis, syncope; PAs are often recurrent, unpredictable, sudden, intense; during a PA, subjects avoid public and thus also suffer transient agoraphobia Management Anxiolytics, psychotherapy. See Panic disorder, Status panicus

panic disorder Atypical depression, polysystemic dysautonomia Psychiatry An idiopathic psychogenic complex affecting 1.5% of US, characterized by recurrent and unpredictable episodes–panic attacks of sudden, extreme apprehension, fear, autonomic nervous system hyperactivity Clinical Dyspnea, palpitations, chest pain, choking sensation, tachycardia, vertigo, loss of reality sense, paresthesias, hot/cold flashes, sweating, faintness, trembling, a fear of dying or of 'going crazy'; fear of an attack in public may cause functional agoraphobia; Pts with PD have an 18-fold ↑ risk of suicidal ideation than a mentally 'fit' population. See Agoraphobia. Cf Panic attack

panic value Alert value, critical value Lab medicine Lab results from a specimen that must be reported immediately to a clinician–ie, of such severity as to mandate urgent therapy. See Decision levels

PANIC VALUES

CHEMISTRY PANIC VALUES

ANALYTE	SI UNITS	US UNITS
Calcium	< 1.65 mmol/L	< 6.6 mg/dl
	> 2.22 mmol/L	> 12.9 mg/dl
Glucose	< 2.60 mmol/L	< 46 mg/dl
	> 26.9 mmol/L	> 484 mg/dl
K+	< 2.8 mmol/L	< 2.8 mEq/L
	> 6.2 mmol/L	> 6.2 mEq/L
	> 8.0 mmol/L if hemolyzed	
Na+	< 120 mmol/L	< 120 mEq/L
	> 158 mmol/L	> 158 mEq/L
CO₂ in plasma	< 11 mmol/L	< 11 mMol
	> 40 mmol/L	> 40 mMol

Hematology, eg blasts or sickle cells on a peripheral smear, possibly indicating leukemia or sickle cell anemia

Microbiology, eg positive gram stain or culture from blood, serosal fluids or CSF, acid-fast stain or positive mycobacterial culture results

Transfusion medicine Incompatible cross-match and positive serology for VDRL; the panic values differ in each lab and the route of the communication is at the discretion of the lab director

panlobular emphysema Pulmonology Emphysema that is more common in ♀, of relatively early onset–usually in the 4th decade, often associated with alpha-1-antitrypsin deficiency; bronchitis may develop in late-stage disease. See Chronic bronchitis, Emphysema

panniculitis Adipositis Medtalk Inflammation of the subcutaneous fat, which may extend to the connective tissue septae. See Eosinophilic panniculitis, Popsickle panniculitis

pannus Rheumatology A reticulated membrane of granulation–reactive fibrovascular tissue typical of the chronic proliferodestructive phase of rheumatoid arthritis; immune complexes form at synovial membranes, evoking a nonspecific response, resulting in global destruction of chondrosseous tissue

panphobia Psychiatry A morbid fear of everything

PANSS Positive & Negative Symptom Scale, see there

pansystolic murmur Cardiology A heart murmur that occurs over the entire systole, from the 1st to the 2nd heart sounds; PMs are associated with blood flow between 2 chambers when there is a fairly constant pressure gradient throughout systole, typically seen in AV valve regurgitation and VSDs

PAO Peak acid output, see there

PaO₂/FiO₂ An index of arterial oxygenation efficiency that corresponds to ratio of partial pressure of arterial O₂ to the fraction of inspired O₂

PAP 1. Pap smear 2. Peak airway pressure 3. Peroxidase-antiperoxidase technique, see there 4. Positive airway pressure 5. Primary atypical pneumonia, see Interstitial pneumonia 6. Pulmonary alveolar proteinosis 7. Pulmonary arterial pressure

Pap smear Papanicolaou test, pap test Gynecology A test in which cells from the uterine cervix and endocervix are sampled, spread on a glass slide, stained, and interpreted by a cytotechnologist or pathologist based on the Bethesda System Abnormal values The PS is the 'gold standard' method for early detection of HPV, herpes, trichomonad infections, CIN/dysplasia, cervical CA Accuracy PS interpretation is a subjective art based on experience of the screening cytotechnologist or pathologist; in good labs, error rate–false negative is 5%–15%. See Bethesda System. Cf Papanicolaou classification

490 **Papile grading system** NEONATOLOGY A system used to stratify peri- and intraventricular hemorrhage into 4 grades of increasing severity, based on findings by cranial ultrasonography

papillary carcinoma ONCOLOGY Occult sclerosing carcinoma, papillary adenocarcinoma, papillary cystadenocarcinoma An indolent CA, primarily of the thyroid, which is linked to radiation exposure or RT

papillary necrosis Papillary necrosis of renal papillae NEPHROLOGY A complication of acute pyelonephritis, consisting of uni- or bilateral lesions of one or more pyramids, with a white-gray discoloration of the pyramid tips, in a background of renal failure CLINICAL Renal colic, pyuria ETIOLOGY DM, sickle cell anemia, analgesics–aspirin, phenacetin, cyclophosphamide, ischemia, post-obstruction pyelonephritis

papilledema OPTHALMOLOGY Swelling of the optic nerve and optic disk, indicating ↑ intracranial pressure on the optic nerve, often linked to brain tumors

papillomavirus VIROLOGY A group of viruses that cause noncancerous warty tumors on mucocutaneous surfaces–skin, larynx, uterine cervix. See Human papilloma virus

papovavirus A family of small icosahedral double-stranded DNA viruses–eg, SV40 and polyomavirus, which induce benign and malignant neoplasms

para-aortic/hepatic field Para-aortic/hepatic port RADIATION ONCOLOGY A field of RT that encompasses the splenic hilar and para-aortic lymph node chains and the entire right lobe of the liver–through a 50% transmission lead block, often joined across another 'gap' by a separate pelvic field. See Radiotherapy

paracentesis Abdominal tap, 'belly' tap, peritoneal fluid analysis CRITICAL CARE A minimally invasive procedure for differentiating a 'surgical' abdomen–ie, requiring surgery, from a 'non-surgical' abdomen INDICATIONS Blunt abdominal trauma, especially with concomitant substance abuse, acute pancreatitis, post-operative peritonitis or peritonitis in children with a 2nd disease; diagnostic paracentesis is performed to determine the cause of ↑ intraabdominal fluids, which is most often due to cirrhosis, but is also caused by carcinoma, inflammation–peritonitis, pancreatitis, ruptured diverticulitis, and abdominal trauma with rupture of organs or blood vessels; the protein level is low in transudates–eg, ascitic fluid and high in exudates–eg, inflammation and malignancy. Cf Thoracentesis

NORMAL VALUES-PARACENTESIS

Cells	
RBCs	None
WBCs	< 300/mm³
Enzymes	
Alk phos	50-250 U/L
Amylase	140-400U/L
Glucose	70-100 g/dL
Protein	< 4 0 g/dL
Volume	Minimal, usually ± 20 ml

paracervical block Paracervical anesthesia OBSTETRICS Locoregional anesthesia used in the 1st stage of labor, with injection of a local short-acting anesthetic in lateral paracervical region; ½ of infants experience post-anesthetic bradycardia. See Block

parachute reaction Anterior propping reaction PEDIATRICS Protective abduction of arms, extension of elbows and wrists and spreading of fingers, a normal defense reflex, elicited when an infant is held in ventral suspension and is tilted abruptly forward toward the floor, seen in 7–9th months of age, a response that is asymmetrical in infants with hemiparesis and an early sign of cerebral palsy

parachute valve complex Shone's complex A heart malformation tetrad comprised of a 'parachute' mitral valve–chordae tendinae of both leaflets of the mitral valve inserted into the left ventricular papillary muscle causing obstruction of the blood flow, supravalvular stenosis, subvalvular aortic stenosis and variably aortic coarctation

paradigm An example, hypothesis, model, or pattern; a widely accepted explanation for a group of biomedical or other phenomena that become accepted as data accumulate to corroborate aspects of the paradigm's explanation or theory, as occurred in the 'central dogma' of molecular biology. See Central dogma, Paradigm shift

PARADIGM ENDOCRINOLOGY A clinical trial–Pramlintide for Amylin Replacement Adjunct for Diabetes in Glycemic Management

paradigm shift A decay or collapse in a paradigm that occurs when new data accumulate, and either partially invalidate the previously-accepted theory–paradigm, or are completely at odds with the paradigm. See Central dogma

paradox VOX POPULI A thing that appears illogical or counterintuitive to that which is known to be correct. See Anion paradox, Asher's paradox paradox, C value paradox, Calcium paradox, French paradox, Glucose paradox, Grandfather paradox, Oxygen paradox, Sherman paradox

paradoxic embolus An embolus that passes through right-to-left cardiac shunts in congenital heart disease in the direction opposite–ie, 'paradoxically', to the blood flow; PE arise from thrombotic material on septic or other vegetations from the right side of heart–or from peripheral veins and pass through a patent foramen ovale into the systemic circulation, circumventing the filtering effect of the pulmonary vessels, potentially causing cerebral abscess CLINICAL Meningeal irritation–stiff neck, drowsiness, fever, headache, focal signs–eg, aphasia, hemiplegia, jacksonian convulsions, intracranial pressure, coma. See Embolism

paradoxic incontinence Overflow incontinence UROLOGY Constant or intermittent dribbling of urine due to chronic overdistension of the bladder–volume, 1000–3000 ml, normal, ± 500 ml, with attenuation of the muscle; this may be confused–and therefore is paradoxic with pure stress incontinence

paradoxic movement NEONATOLOGY A misnomer for the respiratory movement of newborns, whose breathing is entirely diaphragmatic; with inspiration, the anterior thorax draws inward and abdomen protrudes; in neonates, this is normal

paradoxic pulse Pulsus paradoxicus CARDIOLOGY A ↓ of ≥ 10 mm Hg in systolic BP on inspiration, a finding which, like the Kussmaul sign–an abnormal ↑ instead of a ↓ in jugular venous pressure with inspiration, suggests cardiac tamponade–impaired diastolic filling of the heart due to ↑ intrapericardiac pressure CLINICAL Air hunger, cyanosis, visible distension of neck veins

paraffin bath Wax bath SPORTS MEDICINE The immersion of a hand or foot either by dipping several times in a paraffin/mineral oil solution heated to 126°C for 20-30 mins, or wrapping after dipping in towels to maintain the temperature; PBs are used for arthritic complaints, and were once used to treat burns

paraganglioma ONCOLOGY A head & neck neural crest tumor more common in ♀; 2-9% of PGs in carotid body, vagal body and jugulo-tympanic region are malignant; 25% of laryngeal PGs are malignant. See Zellballen

PARAGON CARDIOLOGY A trial–Platelet IIb/IIIa Antagonism for Reduction of Acute Coronary Syndrome Events in a Global Organization Network

paragrammatism PSYCHOLOGY The garbling of speech syntax and omission or confusion of particles of inflection

parainfluenza INFECTIOUS DISEASE A virus that causes URIs–up to 50% of croup and 10–15% of bronchiolitis, bronchitis, pneumonias in toddlers

CLINICAL Rhinorrhea, cold-like Sx RISK FACTORS Preschool children; by school age most children have been exposed to parainfluenza virus; most adults have antibodies against parainfluenza. Cf Influenza

Paralium Diazepam, see there

parallel play PSYCHOLOGY Play typical of very young children, in which the child engages in independent play, without interacting with other children; the PP stage is normal until the child is toilet-trained, after which associative and interactive play develops

parallel tracking THERAPEUTICS A mechanism by which promising therapeutics are made available–early in drug development without interfering with research–to Pts ineligible to participate in clinical trials because of geographic or entry criteria; clinical trials proceed on one 'track'; the drug used for treatment–outside of trials, proceeds on a separate or 'parallel track'. See Clinical equipoise, Compassionate IND, IND

Paralympics SPORTS MEDICINE An international sports competition in which participants are elite athletes with physical or visual impairments, representing 4 intl federations for the blind, paraplegics and quadriplegics, amputees, people with cerebral palsy. Cf Special Olympics

paralysis NEUROLOGY The loss of voluntary movement–motor/ muscle function due to injury or disease of the nervous system which may be partial–palsy or total, such as in botulism. See Ascending tick paralysis, Fetal paralysis, Flaccid paralysis, Laryngeal paralysis, Periodic paralysis, Sleep paralysis, Spastic paralysis, Tick paralysis, Tourniquet paralysis, Unilateral vocal cord paralysis, Vocal cord paralysis

paralytic ileus GI DISEASE Functional 'obstruction' of intestinal flow, often following abdominal surgery, as well as electrolyte defects–eg, hypokalemia, drugs–eg, phenothiazine, narcotics, gram-negative sepsis, catecholamines, diabetic ketoacidosis, mesenteric vascular disease, porphyria, retroperitoneal hemorrhage, spinal and pelvic fractures. See Gastroparesis, Intestinal obstruction, Volvulus

paramedic A health professional certified to perform advanced life support procedures–eg, intubation, defibrillation and administration of drugs under a physician's direction; paramedics provide urgent care from an emergency vehicle or air service; in contrast, EMTs can only perform basic life-support. Cf EMT, Physicians' assistant, Physician extender

parameter CARDIAC PACING A term quantifying an operational element determining pacemaker behavior–eg, rate, pulse width, A-V interval, refractory period, etc

Paramyxoviridae VIROLOGY A family of RNA viruses including respiratory pathogens–eg, croup, measles, mumps, parainfluenza, RSV

paraneoplastic syndrome ONCOLOGY A co-morbid condition due to the indirect–remote or 'biologic' effects of malignancy, which may be the first sign of a neoplasm or its recurrence; PSs occur in > 15% of CAs, are caused by hormones, growth factors, biological response modifiers, and other as-yet unidentified factors, and may regress with treatment of the primary tumor. See Ectopic hormone

PARANEOPLASTIC SYNDROMES

GI TRACT, eg anorexia, vomiting, protein-losing enteropathy, liver disease

HEMATOLOGIC, eg leukemoid reaction, reactive eosinophilia, peripheral 'cytoses or 'cytopenias, hemolysis, DIC, thromboembolism, thrombophlebitis migrans

HORMONAL EFFECTS

METABOLIC DISEASE, eg lactic acidosis, hypertrophic pulmonary osteoarthropathy, hyperamylasemia, hyperlipidemia

NEUROMUSCULAR, eg peripheral neuropathy, myopathy, CNS, spinal cord degeneration, inflammation

RENAL, eg nephrotic syndrome, uric acid nephropathy

SKIN, eg bullous mucocutaneous lesions, acquired ichthyosis, acanthosis nigricans, dermatomyositis

OTHERS, eg callus formation, hypertension, and amyloidosis

paranoia PSYCHIATRY 1. An evolving or fixed persecutory delusional state; the term paranoia is not used in DSM-IV; paranoid delusions are an integral component of the paranoid personality disorder and paranoid subtype of schizophrenia 2. Paranoid personality disorder, see there 3. Delusional disorder, see there VOX POPULI An insidious pattern of unfounded thoughts and fears, often based on misinterpretation of actual events; Pts with paranoia may have highly developed delusions of persecution and/or of grandeur

paranoid personality disorder DSM 301.0 PSYCHIATRY A pattern of pervasive distrust and suspiciousness of others such that their motives are interpreted as malevolent; PPD begins by early adulthood and is present in various contexts

paranoid schizophrenia DSM 295.30 PSYCHIATRY A type of schizophrenia associated with feelings of being persecuted or plotted against. Affected individuals may have grandiose delusions associated with protecting themselves from the perceived plot. See 287 in DSM-IV. See Paranoid personality disorder, Schizophrenia

paraphilia Sexual deviancy PSYCHIATRY A mental disorder characterized by '...recurrent, intense sexually arousing fantasies, sexual urges, or behaviors generally involving, 1. nonhuman objects; 2. suffering or humiliation of oneself or one's partner; or 3. children or other non-consenting persons, that occur over a period of at ≥ 6 months...(causing) significant distress or impairment in social, occupational, or other important areas of functioning; sexual excitement to the point of erection and/or orgasm, when the object of excitement is considered abnormal in the context of the practitioner's societal norms FORMAL PARAPHILIAS, PER AM PSYCHIATRIC ASSN Exhibitionism, fetishism, frotteurism, pedophilia, sexual masochism, sexual sadism, transvestic fetishism, voyeurism, paraphilia–not otherwise specified, a 'wastepaper basket' category MANAGEMENT Psychotherapy, antidepressants, progestins, antiandrogens, surgical castration, triptorelin. See Coprophilia, Exhibitionism, Fetishism, Frotteurism, Masochism, Multiplexed paraphilia, Necrophilia, Pedophilia, Sadism, Sexual asphyxia, Sexual deviancy, Telephone scatologia, Transvestism, Urolagnia, Voyeurism, Zoophilia

paraphimosis UROLOGY The inability to pull the retracted foreskin–in an uncircumcised ♂–over the head of the penis ETIOLOGY Inflammation and stenosis of foreskin, due to infection and poor personal hygiene; it may follow direct trauma to the area, resulting in swelling

paraphrasia NEUROLOGY Use of words that approximate those intended ETIOLOGY Organic brain disease, schizophrenia PSYCHOLOGY Unintended garbling of speech while consciously attempting to speak

paraplegia NEUROLOGY Complete paralysis, paresthesias, loss of sensation and control of voluntary movement in legs, post spinal cord injury or disease

Paraquat lung PULMONOLOGY A lung disease caused by the weed killer Paraquat/dipyridylium, used by the ATF–Dept of Alcohol, Tobacco, Firearms to destroy marijuana plants; drug dealers may harvest the tainted marijuana and sell it which, smoked, may cause pulmonary fibrosis and ARDS; Paraquat may cause lung damage by inhalation, GI absorption or through intact skin; ingestion rapidly causes death due to respiratory failure due to proliferative alveolitis with ↓ O_2 exchange across alveoli, hemorrhage, bronchiolitis and hyaline membrane formation; paraquat also damages the kidneys, liver, mouth, esophagus. See Pulmonary fibrosis

parasellar syndrome Tolosa-Hunt syndrome NEUROLOGY A condition characterized by episodic orbital pain, with inflammation of the cavernous sinus ETIOLOGY Neoplasia–pituitary adenoma, craniopharyngioma, meningioma, neurofibroma, sarcoma, chondroma, giant cell tumor, myeloma, lymphoma, metastases; inflammation–herpes zoster, arachnoiditis, giant cell arteritis, Wegener's granulomatosis, syphilis, aspergillosis CLINICAL Diplopia, headaches, orbital pain MANAGEMENT Excise or decompress, chemotherapy, corticosteroids, antibiotics

parasite INFECTIOUS DISEASE 1. A disease-causing organism 2. An organism with an obligatory dependence on a host, to the host's detriment. See Microparasite, Opportunistic parasite

parasite infection INFECTIOUS DISEASE An infection by a parasite caused, in developed nations, by exposure to uncooked fish–sushi, linked to anisakiasis or uncooked meat–pork, linked to trichinosis, recent travel to a developing nation–eg, malaria, immune defects–eg, AIDS, malignancy with ↑ susceptibility to infection

parasomnia SLEEP DISORDERS A dyssomnic event occurring during sleep, or induced or exacerbated by sleep–eg, sleepwalking. See Sleep disorder

parasympathetic *adjective* Pertaining to the subdivision of the autonomic nervous system concerned with activities that, in general, inhibit or oppose the physiological effects of the sympathetic nervous system

parathion TOXICOLOGY An acetyl-cholinesterase-inhibiting organophosphate insecticide and neurotoxin; acute intoxication is characterized by nicotinic and muscarinic effects which, when severe, may cause rapid respiratory failure; chronic intoxication may cause demyelinating neuropathy and axonal degeneration TREATMENT Atropine. See Organophosphate poisoning

parathormone See PTH

parathyroid adenoma A benign tumor of the parathyroid, which may be functionally active.

parathyroid gland A small gland located at the posterior aspect of the thyroid gland, which regulates calcium by secreting PTH; is critical to calcium and phosphorus metabolism, and magnesium balance and in normal bone mineralization. See PTH

parathyroid squeeze test ENDOCRINOLOGY A clinical test in which the neck is gently squeezed on the side thought to harbor a parathyroid adenoma, which will respond by ↑ serum PTH

paravertebral facet joint denervation PAIN MANAGEMENT The destruction of a lumbar paravertebral facet joint nerve using a neurolytic INDICATIONS Facet joint arthropathy–eg, spondylysis or post-laminectomy syndrome–resulting in therapeutically recalcitrant low back pain. See Laminectomy

parent VOX POPULI A person who has produced one or more offspring from a sexual union. See Adoptive parent, Fertile adoptive parent, Foster parent, Genetic parent, Grandparent, Psychological parent, Real parent, Step-parent, Surrogate parent, Unwed parent

parenteral *adjective* Referring to a non-topical route of administration; by injection is parenteral

parenteral nutrition IV feeding, parenteral alimentation The administration of nutrients parenterally, usually IV. See Total parenteral nutrition. Cf Forced feeding

parenting The activities carried out by a parent–eg, supplying physical sustenance, emotional support instilling moral values, etc. See Bonding; Father 'factor', Motherhood. Cf Anaclitic depression, Child abuse

paresis NEUROLOGY Incomplete paralysis, weakness; partial paralysis of voluntary and involuntary muscles. See General paresis, Quadriparesis

paresthesia NEUROLOGY An abnormal tactile sensation, described as burning, pricking, tickling, tingling, or creeping, which indicates nerve irritation

pargyline Eutonyl PHARMACOLOGY An MAOI with antidepressant activity; it is less effective than TCAs; it requires abstinence from foods with tyramine; interacts metabolically with other drugs; its use is limited to Pts unresponsive to TCAs and tetracyclic antidepressants and Pts with atypical depression. See Monoamine oxidase inhibitor, Tricyclic antidepressants

Parinaud syndrome ENDOCRINOLOGY An often unilateral eye problem accompanied by lymphadenopathy, which may be linked to infections–eg, tularemia and cat-scratch disease. See Cat-scratch disease, Tularemia

PARIS VASCULAR DISEASE A clinical trial–Peripheral Artery Radiation Investigational Study

parite An intra-arterial fibrous plaque in ASHD, which may display dystrophic calcification

PARK CARDIOLOGY A clinical trial–Post-Angioplasty Restenosis Ketanserin that evaluated ketanserin in preventing restenosis post PCTA in Pts with CAD. See Coronary artery disease, Percutaneous transluminal angioplasty OPHTHALMOLOGY See Photorefractive astigmatic keratectomy, see there

Parkinson's disease Paralysis agitans NEUROLOGY A condition characterized by tremors and rigidity followed by inhibition of voluntary movement, a shuffling gait due to neuronal degeneration in the basal ganglia–substantia nigra and corpus striatum, focal ↓ of dopamine production and eventually severe mental deterioration CLINICAL Older subjects with static tremor, plastic rigidity of trunk and extremities, bradykinesia, masklike facies, progressive stoop, slow monotonous voice, slow shuffling gait, often accompanied by autonomic dysfunction–hypersalivation, sweating, rapid, coarse tremor, pill-rolling movements, cogwheel rigidity, drooling, akinesia MANAGEMENT L-Dopa is effective in early PD, as it is transported to the brain and converted into dopamine, but with time, loses efficacy, clozapine; 'brain-graft' surgery was disappointing; fetal nerve graft–mesencephalic dopamine neurons from 8-9 wk fetuses–may be more effective; deprenyl, an MAOI that blocks conversion of MPTP to MPP+; PD may be prevented by 3+ cups of coffee/day. See BDNF, Clozapine, Fetal-brain tissue grafting, MPTP, Unified Parkinson's Disease Rating Scale

Parkinson's tetrad NEUROLOGY 4 classic findings in Parkinson's disease: Resting tremor–eg, 'pill-rolling' of fingers, possibly extremities, head, trunk; rigidity–'cogwheeling' effect; bradykinesia–gait slow to start, shuffling, festinating, ↓ arm swing; postural instability

parkinsonism plus syndrome A Parkinson's disease-like complex accompanied by other neurologic changes–eg, impaired eye movement, orthostatic hypotension, cerebellar ataxia, dementia; PPSs include olivopontocerebellar degeneration with ataxia, parkinsonism-amyotrophic lateral sclerosis overlap, parkinsonism-dementia–normopressure hydrocephalus, gait disturbance and urine incontinence, progressive supranuclear palsy with ophthalmoplegia, Shy-Drager syndrome with orthostatic hypotension, striatal degeneration; 2° parkinsonism may be caused by stroke, encephalitis, meningitis; drugs–tranquilizers–eg, haloperidol, metoclopramide, phenothiazine, narcotics, anesthesics, toxins, carbon monoxide poisoning, recreational use of MPTP or other drugs

Parnate® Tranylcypromine, see there

paronychia Nail infection DERMATOLOGY A superficial infection of the skin around nails, most commonly caused by staphylococci or fungi, often due to local injury–eg, biting off or picking a hangnail, or manipulating, trimming, or pushing back the cuticle; fungal paronychia is often associated with DM. See Onychomycosis.

paronychial *adjective* Referring to paronychia, see there

parosmia AUDIOLOGY Any disorder or perversion of the sense of smell, especially a perception of nonexistent odors

parotidectomy SURGERY The surgical removal of a parotid gland

parotitis ENT Inflammation of the parotid glands, a finding characteristic of mumps. See Mumps

paroxetine Paxil® PSYCHIATRY An SRI antidepressant that blocks serotonin reuptake at the synapse more effectively than norepinephrine INDICATIONS Depression, panic disorder, OCD, social anxiety disorder. See Serotonin-selective reuptake inhibitor

paroxysm MEDTALK A spasm or convulsion; a constellation of findings that signal a manifestation of disease, as in fever and shaking chills with malaria

paroxysmal atrial tachycardia Paroxysmal supraventricular tachycardia, see there

paroxysmal cold hemoglobinuria HEMATOLOGY A disorder that is: (1) Rarely 'paroxysmal' clinically; (2) Not always precipitated by the cold; and (3) Not always associated with hemoglobinuria; PCH comprises 2-5% of autoimmune hemolytic anemias, and is caused by IgG–Donath-Landsteiner antibodies that react at < 15°C and are directed against the ubiquitous P antigen on red cells; PCH may be transient, and 2° to viral exanthemas of childhood CLINICAL After exposure to the cold, the Pt may experience myalgia, abdominal cramping, headaches, hemoglobinuria, Raynaud phenomenon, cold urticaria, jaundice LAB Positive direct Coombs test–using anti-C3 antiserum, anemia, hemoglobinuria, ↓ haptoglobin, ↑ LD, ↑ BR; the antibody is a non-agglutinating IgG that binds to RBCs at cold temperatures and, when warmed to 37°C, evokes complement-mediated hemolyses; antibody elutes from the red cells in vitro, and complement remains fixed, and thus is a 'biphasic' hemolysin; anti-P reacts with normal neutrophil antigens except for p and P^k PREVENTION Keep warm MANAGEMENT If chronic, corticosteroids, immunosuppressants

paroxysmal nocturnal dyspnea SLEEP DISORDERS Dyspnea several hrs after a Pt with severe and decompensated CHF falls asleep, due to fluid accumulation in lungs, usually relieved by a short period of sitting up; PND may be sudden and awaken the sleeper

paroxysmal nocturnal hemoglobinuria An acquired hemolytic disease, due to proliferation of an abnormal clone(s) of myeloid stem cells, the progeny of which are susceptible to complement-mediated membrane damage and hemolysis–CMH CLINICAL Thromboses, ↑ infections; PNH may evolve into aplastic or sideroblastic anemia, myelofibrosis, AML LAB Leukopenia, thrombocytopenia, dimorphic RBC population, iron-deficiency, ↓ leukocyte alk phos, ↓ RBC acetylcholinesterase, altered properdin–alternate pathway of complement lysis, hemoglobinuria, hemosiderinuria, positive Ham test, positive sucrose lysis test; negative direct Coombs' test, ↑ susceptibility of RBCs to CMH

paroxysmal supraventricular tachycardia Supraventricular tachycardia CARDIOLOGY Tachycardia triggered sporadically in the myocardium above the ventricles; PSVT is most common in younger subjects with normal hearts RISK FACTORS Smoking, caffeine, stress, alcohol abuse, overactive thyroid, excess thyroid hormone intake, drugs–eg, digitalis; PSVT can be a form of a re-entry tachycardia, resembling Wolff-Parkinson-White syndrome

Parsol® DERMATOLOGY A sunscreen that protects against UV A. See Ultraviolet A

Parstelin Tranylcypromine (Product development discontinued)

part A MANAGED CARE A component of the Medicare reimbursement system, consisting of compulsory hospital insurance financed by contributions from employers, employees and participants; Part A pays costs of hospitalization, but not physician fees. See Medicare, Part B, TEFRA

part B MANAGED CARE A component of the Medicare reimbursement system–US that provides supplementary payments for medical services and supplies that are not covered under part A; part B is voluntary, covers physician

fees and individual provider services; it is financed partly by monthly premiums paid by enrollees and partly by the federal government. See Medicare, Part A, Participation, TEFRA. Cf Medicaid

partial birth abortion Intact dilation & extraction, intact dilatation & evacuation, intact D&E, D&X MEDICAL ETHICS A late-term abortion in which the fetus is partially vaginally delivered alive before the skull is collapsed, killing the fetus before complete delivery

partial complex seizure Complex seizure NEUROLOGY A brief, temporary alteration in brain function caused by abnormal electrical activity in the neurons of a discrete area of the brain, especially temporal lobe, characterized by changed alertness or awareness, and behavioral or emotional Sx and temporary loss of memory; PCSs may occur at any age, as a single episode or as a seizure disorder; mental Sx may accompany ↓ consciousness ETIOLOGY Local hypoxia, trauma, brain tumor or other discrete lesions. See Temporal lobe seizure

partial disability HEALTH INSURANCE A condition in which, as a result of injury or sickness, the insured can perform some, but not all of the duties of his occupation.

partial draw tube HEMATOLOGY A blood collection tube with sodium citrate anticoagulant that is smaller–2.7 mL than the standard–4.5 mL purple top tube. See APTT. Cf CTAD tube

partial remission Partial response ONCOLOGY An incomplete response to therapy for CA; for lymphomas, PR is defined as a ↓ by ≥ 50% of the longest perpendicular diameter of all measurable lesions. Cf Complete remission, Minimal response

partial syndrome eating disorder PSYCHIATRY A condition of primarily ♀ adolescents, with features of anorexia nervosa

partial thromboplastin time Activated partial thromboplastin time, aPTT A test that evaluates the coagulation cascade by measuring the time required for citrated recalcified plasma to form a clot after the addition of partial thromboplastin; PTT is prolonged in deficiencies of factors V, VIII, IX, X, XI, XII, HMW kininogen, lupus anticoagulants–inhibitors, heparin, prekallikrein, prothrombin, fibrinogen, Passovoy factor NORMAL Standard PTT 68-82 sec; aPTT 32-46 sec

partial twinning A group of rare congenital anomalies of cloacally-derived structures–eg, focal doubling of the GI tract at Meckel's diverticulum, extending to anus, doubling of bladder, vagina, penis, sacrum, or lumbar vertebrae

participation MANAGED CARE An agreement between Medicare and health care providers–ie, hospitals, physicians, to 'accept assignment'–ie, Medicare's fees, as payment in full, for any health care services rendered; also, acceptance of an insurance or health plan's established fee as the maximum amount collectable for services rendered. See Participating provider

partner notification PUBLIC HEALTH Any formal and systematic means of informing the sexual partner(s) of a person with an STD, that the person being tested is infected with an organism–eg, HIV, *N gonorrhoeae*, *T pallidum*–of interest to public health officials. See Negotiated safety

partner risk factor STD Any factor that ↑ a sexual partner's risk of transmitting STD, especially HIV PRFs Sex with a prostitute, or someone with AIDS, or a STD, receptive anal intercourse, injection of drugs of abuse. See High risk behavior, Negotiated safety, Partner notification program. Cf Safe sex practices

parturition OBSTETRICS Childbirth

parvus et tardus CARDIOLOGY A small arterial pulse with a delayed systolic peak ± associated with an anacrotic 'shoulder'0 on the upstroke of the carotid pulse, a pattern seen in older Pts with severe aortic stenosis

Parzam Diazepam, see there

PAS 1. Periodic acid-Schiff–stain, see there 2. Para-aminosalicylic acid 3. Pulmonary artery stenosis

PASE CARDIOLOGY A clinical trial–Pacemaker Selection in the Elderly, comparing the effects of ventricular–single chamber pacing with dual-chamber pacing. See Dual chamber pacemaker, Ventricular pacemaker

passive-aggressive personality disorder PSYCHIATRY A personality disorder in which the Pt expresses personal conflicts through retroflexed anger in as covert obstructionism, procrastination, stubbornness, inefficiency; ±1% of the population exhibits passive-aggressive or passive-dependent behavior DEFENSES Turning against oneself–a form of sado-masochism, denial, rationalization, hypochondriasis PROGNOSIS Poor. Cf 'Anal-retentive'

PASSIVE-AGGRESSIVE PERSONALITY DISORDER–DIAGNOSTIC CRITERIA

A A pervasive pattern of negativistic attitudes and passive resistance to demands for adequate performance, beginning by early adulthood and present in various contexts, indicated by at least four of the following

1. Passive resistance to routine social or occupational obligations
2. Complains of being misunderstood or underappreciated
3. Complains of personal misfortune
4. Sullenness or belligerence (argumentative)
5. Highly critical of authority
6. Resents or is envious of those perceived as being more fortunate
7. Alternates between hostile defiance and contrition

B Not accounted for by dysthymic disorder or occurs exclusively during major depressive episodes

---

passive immunity IMMUNOLOGY Immunity conferred by an antibody produced in another host and acquired naturally by an infant from its mother or artificially by administration of an antibody-containing preparation–antiserum or immune globulin

passive learning EDUCATION The acquisition of knowledge without active effort–eg, by listening to audiocassettes, exposure in the working environment. See Spoon-feeding

passive smoking PUBLIC HEALTH Involuntary 'smoking' by non-smokers who breathe ambient air containing carcinogenic inhalants from an 'active' cigarette smoker; PS ↑ platelet activity, accelerates ASHD, ↑ tissue damage in ischemia or acute MI; PS ↓ both cardiac delivery of O_2 to the heart and myocardial ability to use O_2 to produce ATP, resulting in ↓ exercise capacity in passive smokers, and ↑ risk of fatal and nonfatal cardiac events; exposure to 3 hrs/day of PS is associated with an ↑ in cervical CA; in children, neonates, and fetuses, PS is linked to ↓ pulmonary function, bronchitis, pneumonia, otitis media and middle ear effusions, asthma, lower birth and adult weight and height, SIDS, poor lung–and physical development, and ↑ perinatal mortality–due to placental vascular disease–eg, placenta previa and abruptio placentae, breast CA; ♀ exposed to PS before age 12 had an odds ratio of 4.5; such children are more likely to become smokers and are at ↑ risk for developing CA in a dose-related manner, in all sites 50% higher than expected, and up to two-fold ↑ in NHL, ALL, and Wilms' tumors; PS by children with cystic fibrosis adversely affects growth and health, resulting in ↑ hospital admissions and poor performance in pulmonary function tests. See Conicotine, Environmental tobacco smoke

past, family and/or social history MEDICAL PRACTICE A historical evaluation of a Pt, which consists of a review of 3 areas, which is required for CPT-related documentation of a physician's evaluation and management services

---

HISTORY

PAST HISTORY The Pt's past experiences with illnesses, operations, injuries, and treatments

FAMILY HISTORY A review of medical events in the Pt's family, including disease which may be hereditary, or place the Pt at risk

SOCIAL HISTORY An age-appropriate review of past and current activities

---

pasta loading Carbohydrate loading, see there

paste DRUG SLANG A regionally popular term for crack cocaine

Pastia sign INFECTIOUS DISEASE Hemorrhagic transverse lines at bend of elbow, wrist, or inguinal region, which persists after desquamation, caused by scarlet fever. See Scarlet fever

PAT 1. Paroxysmal atrial tachycardia. See Supraventricular tachycardia 2. Patient 3. Pre-admission testing, see there 4. Pregnancy at term

Patanol® Olopatadine OPHTHALMOLOGY An agent used to relieve eye itching in allergic conjunctivitis

Patau syndrome Trisomy 13 syndrome NEONATOLOGY A syndrome characterized by multiple malformations–scalp defects, hemangiomas of face and nape of neck, cleft lip/palate, malformations of heart and GI tract, flexed fingers with extra digits, profound mental retardation; most die in early infancy

patch INFORMATICS An occasionally inelegant software or hardware "workaround" to solve a problem in data flow in an information system THERAPEUTICS A delivery system in which an agent of interest–eg, nicotine, testosterone is impregnated in a disposable material and placed on the skin for passive absorption. See Lidocaine patch, Nicotine patch, St. John's® transdermal patch, RapiSeal patch VOX POPULI A gob or wad of a thing. See Cottonball patch, Cotton wool patch, Herald patch, Inlet patch, Leaky patch, Peyer's patch, Shagreen patch

patch test Allergy skin test, contact dermatitis skin test, patch skin test IMMUNOLOGY An epicutaneous test of contact-type–delayed hypersensitivity, which consists of applying a patch with a low dose of an allergen–antigen to an unexposed area of the skin, usually the back, and observing the site 1-2 days later; the most common sensitizing haptens in North America are poison ivy–*Toxicodendron radicans*, nickel, chromate, paraphenylenediamine–a dye constituent, ethylenediamine–a solvent and emulsifier, local anesthetics–eg, benzocaine, rubber, neomycin, and others; PT materials have been standardized and are available commercially, either as individual allergens, or as batteries of allergens, including those for specific occupations–eg, hairdressers, printers, and others; incorrect PT results are common in the form of false-positives, due to too high concentration of allergens in the patches, misinterpretation of irritant reactions, and generalized erythema of the skin testing site; false-negative results are linked to technical errors and failure to simulate the 'real-world' situation in which the person is exposed to the allergen

patellar dislocation ORTHOPEDICS A subluxation, usually lateral, of the patella, due to a sudden change in direction while running and the knee is under stress; may follow injury, accompanied by pain and inability to walk. See GLC7

patellofemoral pain syndrome SPORTS MEDICINE An often bilateral condition of insidious onset seen in young ♀ athletes CLINICAL Diffuse knee pain exacerbated by stair descent, squatting and prolonged sitting, patellar crepitus, knee joint stiffness, ↓ ROM. See Moviegoer sign

patellofemoral stress syndrome Runner's knee, see there

patent *adjective pronounced*, pay-tent Open, unobstructed, referring to a duct, lumen, or vessel *noun pronounced*, pah-tent A document that grants an inventor in terms of a determined number of years, the exclusive right to make use of and sell his invention

patent ductus arteriosus PDA CARDIOLOGY An acyanotic congenital heart disease–CHD characterized by postpartum persistence of a ductus arteriosus–a fetal vessel that connects the pulmonary artery–PA to the aorta in utero, bypassing the unexpanded lungs–which normally closes soon after birth; PDA accounts for 10% of cases of CHD; $1/3$ of Pts with nontreated PDAs die of heart failure, pulmonary HTN or endoarteritis by age 40 RISK PDA is more common in pregnancies with persistent perinatal hypoxia, maternal rubella, prematurity and high altitude HEART SOUNDS Normal 1st HS, machinery murmur in 2nd left anterior IC space after 1st HS, peaking at beginning of 2nd; it declines during diastole EKG Asymptomatic if small; large PDAs with major L→R shunting have left atrial and LV hypertrophy CLINICAL SOB, DOE, palpitations, Rt heart failure, paradoxic embolism, recurrent pneumonia, small shunts are asymptomatic, ↑ risk of infective endocarditis and septic pulmonary embolism; moderate shunts appear in later childhood with fatigue; large shunts may cause LV failure DIAGNOSIS Catheterization to localize ASDs IMAGING CXR-lung plethora, proximal PA dilatation, prominent ascending aorta; pulmonary HTN with RV hypertrophy; 2D echocardiography demonstrates PDA, Doppler studies demonstrate continuous flow through pulmonary trunk; catheterization and angiography quantify severity of shunt and vascular resistance MANAGEMENT Ligation; once severe pulmonary vascular obstructive disease develops, repair is contraindicated. See Congenital heart disease

patent foramen ovale PFO CARDIOLOGY An opening between the left and right atria which allows blood to bypass the lungs in utero; the FO normally closes shortly after birth, but remains open in up to 20%; a PFO is, in absence of other cardiac defects, is of no consequence

pathogen Popularly, bug Any disease-producing microorganism. See Blood-borne pathogen, Emerging pathogen, Food-borne pathogen, Intracellular pathogen, Waterborne pathogen

pathognomonic MEDTALK *adjective* Referring to a distinctive sign, Sx, or characteristic of a disorder on which a diagnosis is made

pathologic fracture ORTHOPEDICS A fracture linked to marked attenuation of bone–eg, due to metastases. See Fracture

pathologist A physician trained in evaluating tissues–surgical pathology, cells–cytology and/or lab medicine-clinical pathology, to render a diagnosis AVERAGE SALARY $190 K. See Cytologist/cytopathologist, Dermatopathologist, Forensic pathologist, Hematopathologist, Histopathologist, Immunohematopathologist, Immunopathologist, Neuropathologist, Speech pathologist

pathologist's tumor Occult cancer, definition 2.

pathologist's disease A lesion defined as malignant by histologic criteria, identified as an incidental finding, and in practice regarded as a non-aggressive or not requiring treatment–eg, pancreatic endocrine tumor or well-differentiated adenoCA of prostate in an elderly male

pathologize A neologism for diagnosing a normal condition as pathologic, based on the assumption that it 'should' cause disease–eg, adopted children are assumed to develop psychologic dysfunction with a higher frequency than their nonadopted counterparts

pathology 1. The medical science and specialty dedicated to the study and diagnosis of disease processes, based on analysis of objective parameters–eg, gross examination of tissues, microscopy, chemical and immune-mediated assays, cultures of microorganisms, etc. See Anatomic pathology, Anatomic/clinical pathology, Chemical pathology, Clinical pathology, Comparative pathology, Digital pathology, Immunopathology, Neuropathology, Speech pathology, Stereopathology, Surgical pathology, Telepathology 2. A term used in working medical parlance for a pathologic lesion

pathology consultation A request for a pathologist's opinion on the nature of an anatomic change–cytopathology, tissue pathology, or chemical abnormality–clinical pathology

pathway The route by which a thing occurs. See Critical pathway, Fifth pathway

patient MEDTALK A person receiving health care. See Crossover patient, Difficult patient, Expectant patient, Good patient, Index patient, Negative patient, Noncompliant patient, Orphan patient, Outpatient, Private patient, Problem patient, Professional patient, Qualified patient, Qualifying patient, Service patient, Standardized patient, Violent/combative patient, Wandering patient

patient autonomy MEDICAL ETHICS The right of a Pt to have his/her carefully considered choices for health care carried out in a fashion that is consonant with his or her personal philosophy; PA also assumes that, in absence of explicit instructions to the contrary, the most aggressive efforts should be made to resuscitate a Pt *in extremis*. See Code. Cf Medical paternalism, Slow code

patient 'Bill of Rights' Patient Bill of Rights & Responsibilities A statement developed by the Am Hospital Assn that delineates the treatment that a person has a right to expect when in a hospital, and the behavior expected of that person with respect to his own therapy, follow-up, and conduct while there

patient compliance PHARMACOLOGY The degree of adherence of a Pt to a prescribed diet or treatment, and whether the Pt returns for re-examination, follow-up or treatment. See Directly observed therapy, Good Pt. Cf Bad Pt, Noncompliance

patient confidentiality MEDICAL PRACTICE A Pt's right to privacy and freedom from public dissemination of information that the Pt regards as being of a personal nature. See HIPAA, Medical privacy

patient-controlled analgesia PAIN MANAGEMENT A method for self-administration of narcotic-analgesics via a programmable pump; PCA is used for pain of terminal CA, postsurgery, angina pectoris, L&D AGENTS Fentanil, meperidine, morphine, sufentanil; PCA is delivered IV, subcutaneously or epidurally. See Pain

patient education Health and wellness information provided to a Pt in electronic, video or print forms

patient empowerment The providing of information regarding therapeutic options so that a Pt can actively participate in the decision on whether to undergo a diagnostic or therapeutic procedure, or pursue alternatives. See Patient Bill of Rights

patient informatics INFORMATICS Pt-oriented information, generally divided into consumer informatics and Pt education

patient-initiated lab testing Direct testing Direct access of Pts to medical testing, who either perform the tests–eg, pregnancy kits, occult blood in stool, or blood in urine, or request the tests directly from the lab

patient load GRADUATE EDUCATION A popular term for the number of Pts for which a physician in training is responsible during a shift. See Residency. Cf Turf–verb

patient medical record See Patient record

patient mix The demographics of a Pt population served by a hospital or other health care facility; the PM may be classified according to disease severity or socioeconomic parameters. See Case-mix index

patient-physician relationship MEDTALK A formal relationship that exists between the physician and the Pt, often equated to medical 'duties' that the physician must perform in a professionally acceptable manner. See Doctor-Pt interaction. Cf Abandonment

patient population MEDTALK The demographics and other particulars of a population being serviced–eg, ethnicity, socioeconomic status–eg, unsophisticated vs urbane, population density–ie, rural vs urban. See Call schedule

patient record Patient medical record A document relevant to Pt management–eg, medical Hx, chief complaint(s), diagnostic and therapeutic procedures performed and current status, usually arranged in chronologic order. See SOAP

patient release criteria AMBULATORY SURGERY An objective parameter used to determine whether a Pt can be released from a health care facility after minimally invasive surgery PRC Post-procedural voiding, hemodynamic stability, neurologically intact, oriented, and ambulation with minimal assistance

496

Patient Right to Know Act MEDTALK A US federal act that bars managed care plans from restricting physicians' communications–so-called gag clauses with Pts–that could influence them, and inappropriately encourage them to seek less expensive therapeutic options

Patient Self-Determination Act An act that requiring health professionals reimbursed by Medicare/Medicaid to inform Pts of their legal rights to refuse treatment and prepare advance directives. See Advance directive, DNR, Durable power of attorney, Living will, Right-to-die movement

patient tracking HEALTH CARE The process of following a Pt and her changing medical status, lab studies, imaging, and other diagnostic and therapeutic interventions, especially when that Pt is away from an institution, being managed by another provider

patient viewpoint standard FORENSIC MEDICINE A standard of disclosure of information used in wording informed consent documents, based on what a reasonable person in the Pt's position would want to know in similar circumstances. See *Arato v Avedon*, Cf 'Reasonable physician' standard

Patrick's test A clinical maneuver performed on a supine Pt, in which the hip and knee are flexed and the external malleolus is placed on the patella of the opposite leg; pain in the hip evoked by pressure on the knee is presumptive evidence of sacroiliac disease

PAVE CARDIOLOGY A clinical trial–Post AV Node Ablation Evaluation

Pavlik harness PEDIATRICS ORTHOPEDICS The gold standard device used in infants < 6 months for managing hip dysplasia FUNCTIONS The PV promotes hip flexion and abduction by prohibiting hip extension or adduction, while allowing some kicking and movement. See Abduction therapy, Developmental dysplasia of the hip, Spica cast

Paxate Diazepam, see there

Paxene® See Paclitaxel

Paxil® Paroxetine, see there

Paxium Chlordiazepoxide, see there

Paxon Buspirone, see there

Paxtibi® Nortriptyline, see there

payer mix MEDICAL PRACTICE The type–eg, Medicaid, Medicare, indeminity insurance, managed care–of monies received by a medical practice. Cf Patient mix, Service mix

payment VOX POPULI A wad of cash given for a service rendered or product received. See Bonus payment, Bundled payment, Pass-through payment, Prospective payment

payor MEDICAL PRACTICE A company or an agency that purchases health services

Ps the three Ps ENDOCRINOLOGY The association of pituitary adenoma, pancreatic neoplasia and parathyroid adenoma in multiple endocrine neoplasm, type I. See MEN 1 STD Permissiveness, promiscuity, ' pill'–3 factors associated with ↑ rates of STDs in the 1960s

PBC 1. Peripheral blood cells 2. Primary biliary cirrhosis, see there

PBS 1. Phosphate-buffered saline 2. Phosphate-buffered sodium

PC 1. Packed cells–RBCs 2. Personal computer, see PC 3. Plasma cell 4. Politically correct 5. Present complaint 6. Professional corporation

PC₁₅ A provocation test for evaluating bronchial responsiveness to histamine, which measures the concentration of histamine required to produce a 15% ↓ in the FEV_1–forced expiratory volume in one second 2° to bronchoconstriction

PCA 1. Parietal cell antibody, see there 2. Patient-controlled analgesia, see there 3. Percutaneous carotid arteriogram 4. Peripheral circulatory assist 5. Posterior cerebral artery

PCH Paroxysmal cold hemoglobinuria, see there

PCKD Polycystic kidney disease, see there

PCL 1. Plasma cell leukemia 2. Posterior cruciate ligament, see there

PCL injury Posterior cruciate ligament injury. See Posterior cruciate ligament

PCO 1. Patient complains of 2. Polycystic ovaries, see there

PCP 1. Phencyclidine, see there 2. *Pneumocystis carinii* pneumonia, see there 3. Primary care physician/provider, see there

PCR Polymerase chain reaction, see there

PCTA Percutaneous transluminal coronary angioplasty, see there

PDA 1. Patent ductus arteriosus, see there 2. Personal desk accessory, personal digital assistant

PDMS Polydimethylsiloxane PRODUCT LIABILITY A silicone used for the elastomeric capsule and gel filler of breast implants; silicone elastomers are constructed of PDMS high MW–500,000 to 700,000 kD polymers treated with fumed silica particles and highly cross-linked to ↑ consistency of final product. See Breast implant

PDQ Physician Data Query INFORMATICS An online database for state-of-the-art treatment, directory, protocol information available to primary care physicians

PDR Physicians Desk Reference A book published annually that lists all ± 2500 US therapeutics requiring a physician prescription

---

PDR 7 color-coded sections

WHITE Manufacturers' index, containing the company addresses and list of products

PINK Product name index, an alphabetical listing of the drugs by brand name

BLUE Product classification, where drugs are subdivided into therapeutic classes

YELLOW Generic and chemical name index

MULTICOLORED Photographs of the most commonly prescribed tablets and capsules

WHITE Product information, a reprint of the manufacturers' product inserts and

GREEN Diagnostic product information, a list of manufacturers of diagnostic tests used in office practice and the hospital; Cf Over-the-counter drugs

---

pea soup stool A descriptor for the khaki-green, slimy stools typically in the 3ʳᵈ wk of typhoid fever, at which point the Pts are in a 'toxic state' and at greatest risk for intestinal perforation and hemorrhage; similar stools occur in enteropathic *E coli* infections of infants. See Stool.

PEACE CARDIOLOGY A clinical trial–Prevention of Events with Angiotensin Converting Enzyme Inhibitor Therapy

peak expiratory flow rate Maximum expiratory flow, peak flow The greatest rate of airflow that can be obtained during forced exhalation, which follows a diurnal pattern of fluctuation and can be used clinically to evaluate airway tone; PEF is ↓–and airway tone ↑ in various forms of asthma; it can be plotted during the day and used to investigate occupational asthma and detect exacerbation before the onset of Sx. See Pulmonary function tests

peak level THERAPEUTICS The highest serum level of free or unbound drug in a Pt based on a dosing schedule, which is usually measured ± ½ hr after an oral dose of a drug. See Therapeutic drug monitoring. Cf Trough level

PeakLog™ **spirometric device** RESPIRATORY THERAPY A hand-held spirometer that logs use and provides visual and audible alerts to maintain Pt compliance with a respiratory regimen. See Respiratory therapy

peanut allergy IMMUNOLOGY A common cause of anaphylactic reactions which, unlike some allergies, is rarely outgrown; PA is the most common cause of food allergy in the US, and a leading cause of food-induced anaphylaxis and death after accidental exposure

PEARL™ Physiologic Endometrial Ablation/Resection Loop A method for resection and ablation procedures using isotonic irrigation–eg, normal saline, eliminating complications associated with non-isotonic irrigation systems which may occur when the tissues absorb irrigation solutions

Pearl Index OBSTETRICS A formula that allows comparison of the efficacy of contraceptive methods, calculated as the pregnancy rate in population divided by 100 yrs of exposure. See Breast feeding, Coitus interruptus, Condoms, Morning-after pill, Contraception, Natural family planning, Norplant, Rhythm method, RU 486

PEARL INDEX–pregnancies/100 years of use

PHYSIOLOGIC 15-30/100 years: Coitus interruptus, natural family planning (rhythm or safe period), eg calendar method, evaluation of cervical mucosa or temperature, breast feeding

CHEMICAL 15-20/100 years: Contraceptive sponges

BARRIER 2-20/100 years: Intrauterine devices, condoms

HORMONAL 1-3/100 years

SURGICAL << 1/100 years: Ligation of fallopian tubes, vas deferens

pearly penile papule UROLOGY A localized nodular anatomic variant of the penile corona, first seen at ages 20-50 as a verrucous lesion TREATMENT Unnecessary

Pearson syndrome A mitochondrial disease characterized by early onset BM dysfunction leading to pancytopenia & pancreatic failure, which may progress to Kearns-Sayre syndrome. See Mitochondrial disease

peau d'orange appearance A descriptive term for any bosselated, rugose surface, usually of the skin with deep, pin-point dimpling, likened by French authors to an orange's skin; the 'classic' peau d'orange change occurs in the skin overlying breast CA OPHTHALMOLOGY A funduscopic finding corresponding to diffuse, rugose hyperpigmentation of the retinal epithelium in angioid streaks SKIN Peau d'orange-like changes of the skin are classically seen in advanced ductal CA of breast or the highly aggressive inflammatory CA, in which there is subcutaneous 'puckering' accompanied by dermal edema, desmoplastic induration, superficial bossellation, erythema, local tenderness ±ulceration

PEBB Percutaneous excisional breast biopsy, see there

pectoral girdle The part of the skeleton that supports the upper extremities

pectus carinatum Pigeon-breast, see there

pectus excavatum Funnel chest, see there

pedal edema MEDTALK The accumulation of fluids in the feet most prominently on the dorsum; PE is characteristic of CHF. See Congestive heart failure

pederasty PSYCHIATRY Homosexual anal intercourse between men–active partners and boys–passive partners. See Ephebophilia, NAMBLA, Pedophilia

pediatric AIDS AIDS acquired HIV perinatally or by 'vertical'–maternal-infant transmission; children with PAIDS may become symptomatic–lymphoid interstitial pneumonia, encephalopathy, recurrent bacterial infection, *Candida* esophagitis in the 1st yr of life; the 5-yr mortality is ±50%. See AIDS

pediatric arthritis Juvenile arthritis Arthritis affecting children. See Still's disease

pediatric lymphoma An uncommon lymphoma, often extranodal and diffuse, that responds well to chemotherapy. See Lymphoma

pediatric patient Child, see there

pediatric trauma score EMERGENCY MEDICINE A triage tool with no advantage over the easily learned Revised Trauma Score–see there; the PTS measures weight, airway, systolic pressure, CNS status, open wound and skeletal trauma

pedicle flap Pedicle PLASTIC SURGERY A peninsula of donor tissue, muscle, and blood vessels transferred to reconstruct a recipient site

pedicle screw fixation system ORTHOPEDIC SURGERY A multi-component device constructed from stainless or titanium-based steel, consisting of solid, grooved, or slotted plates of rods that are longitudinally interconnected and anchored to adjacent vertebrae using bolts, hooks, or screws INDICATIONS Stabilize and fuse the spine in degenerative spondylolisthesis, fusion after spinal fusion, lumbar fractures, and in surgically repaired spinal pseudoarthroses. See Omni litigation

pediculosis Infestation with lice. See Louse

Pediculus capitis Head louse PUBLIC HEALTH A hematophagous louse; human infestation is generally asymptomatic, but may rarely cause severe pruritus TREATMENT Topical insecticides–permethrin, synergized pymethrin, malathion

pedigree ACADEMIA A popular term for a person's educational 'lineage', ie, where he or she studied; a physician with a pedigree has studied at 'premiere' undergraduate institution, medical school, and 'center of excellence' teaching hospital GENETICS An ancestral chart of the blood relatives and mates of an index Pt with a disease or a particular hereditary trait; a family health history diagrammed with a standard set of symbols to indicate the individuals in the family, their relationships to one another, and members with a particular disease

pedophile FORENSIC PSYCHIATRY A person with pedophilia; there are an estimated 500,000 pedophiles in the world. See Child prostitution, Megan's law, Pedophilia

pedophilia PSYCHIATRY A paraphilia involving heterosexual or homosexual activity or intercourse between adults with children, especially prepubertal. See Child abuse, Ephebophilia, Incest, Nepiophilia, Pederasty, Sex tourism. Cf Gerontophilia

peduncle A stalk, often referring to a sessile GI polyp

pee VOX POPULI Micturate, urinate

PEEP Positive end-expiratory pressure A therapeutic modality consisting in the active–interventional maintenance of a slightly positive pressure in the tracheobronchial tree during assisted pulmonary ventilation, such that alveoli are not allowed to completely collapse between breaths; PEEP is of greatest use in ARDS and generated by attaching an airflow threshold resistance device to the expiratory port of the non-rebreathing valve of a manual or mechanical ventilator, allowing a ↓ of airway pressure to a plateau level CONVENTIONAL PEEP Pressure is maintained at 5-20 cm H_2O; it is indicated where an inhaled oxygen fraction at 0.6 cannot maintain the PaO_2 above 60 Torr HIGH PEEP Pressure is maintained at 20-50 cm H_2O; it is used for marked hypoxia, as may occur in severe pulmonary edema

Peeping Tom 'syndrome' Voyeurism, see there

peer review The objective evaluation of the quality of a physician's or a scientist's performance by colleagues MEDTALK The evaluation of a practitioner's professional performance, including identification of opportunities for improving the quality, necessity, and appropriateness–suitability of care; peer review organizations–PROs in the US contract with the CMS, formerly HCFA. See Peer-reviewed journal, Peer review organization

peer review organization Professional review organization, quaility improvement organization MANAGED CARE An independent or sponsored group of physicians or other appropriate peers–eg, allied health professionals who conduct pre-admission, continued stay, services reviews, for Medicare Pts by

Medicare approved hospitals or physicians; PROs also review activities and records of particular health provider, institutions or groups. See Peer review

peer-reviewed journal Refereed journal ACADEMIA A professional journal that only publishes articles subjected to a rigorous peer validity review process. Cf Throwaway journal

PEFR Peak expiratory flow rate

PEG 1. Percutaneous endoscopic gastrostomy GI DISEASE An enteric feeding tube used to purge the colon presurgery 2. Physician equity group. See Managed service organization 3. Polyethylene glycol THERAPEUTICS An inert long-chain synthetic molecule that may be attached to various proteins making them invisible to the immune system; multiple PEGs have been attached to ADA–adenosine deaminase, one of the enzymes responsible for SCID, allowing long-term survival of the molecule within the body; PEG has been attached to Hb, and has potential as a transport vehicle for artificial blood

PEG-conjugated IL-2 Polyethylene glycol-conjugated interleukin-2 A formulation of IL-2 with a 10-fold $\uparrow$ in $T_{1/2}$ without loss of efficacy that may be administered parenterally; IL-2 promotes the growth and differentiation of CD4–helper/inducer, CD8–suppressor/cytotoxic T, NK, and lymphokine-activated cells; PC-IL-2 reported to be beneficial in CA, HIV infection, and in common variable immune deficiency

pejorative MEDTALK Bad...*real* bad

PEL 1. Permissible exposure level 2. Permissible exposure limit, see there

PELA Peripheral excimer laser angioplasty CARDIOLOGY Use of a nonthermal excimer laser for minimally invasive treatment of total occlusions in leg arteries that have not responded to medical therapy and/or cannot undergo bypass surgery. See Angioplasty

peliosis hepatis A liver enlarged by multiple cavernous blood-filled cysts due to use of OCs and androgenic steroids; PH may be associated with CA and TB; a distinct form–bacillary PH, occurs in AIDS. See Bacillary angiomatosis

pellagra Niacin deficiency, nicotinic acid deficiency NUTRITION A deficiency of niacin–nicotinic acid, a B vitamin, due to alcoholism or malnutrition CLINICAL Anorexia, glossitis, GI disorders, headaches, insomnia, bilateral symmetrical scaly rashes, inflammation of skin, vagina, rectum, mouth, polyneuropathy, confusion, depression, pseudodementia, psychotic Sx. See Niacin

Pelligrini-Steida disease Köhler's disease, post-traumatic pararticular osteoma ORTHOPEDICS A post-traumatic condition characterized by calcification and ossification in a hematoma due to partial avulsion of the medial cruciate ligament

pelvic field Pelvic port RADIATION ONCOLOGY An RT field that encompasses iliac, inguinal, femoral node chains, with lead–Pb shields for the rectum and bladder, iliac and upper femoral BM. See Radiotherapy

Pelvic Flex™ UROGYNECOLOGY An exercise program for treating some forms of ♀ urinary incontinence. See Stress incontinence

pelvic girdle The part of the skeleton to which the lower extremities are attached. Cf Shoulder girdle

pelvic inflammatory disease PID, salpingo-oophoritis GYNECOLOGY An imprecise term for intense pain due to direct extension of a lower genital tract infection–often sexually-transmitted along mucosae, first causing asymptomatic endometritis, then acute salpingitis, $\uparrow$ Sx as it spreads into fallopian tubes engorged with pus–pyosalpinx, pus in the peritoneum; PID is accompanied by leukostasis, fever, chills, N&V, extreme tenderness of uterine cervix and adnexae EPIDEMIOLOGY PID is 3–4-fold more common in IUD users and in those who douche 3+ times/month; ectopic pregnancy is 7–10-

fold more common in PID; 500,000 cases of PID are reported/yr–US ETIOLOGY $\frac{1}{2}$ are due to *N gonorrhoeae*, less commonly, *Chlamydia trachomatis*; 15% of Pts with gonococcal cervicitis develop PID CLINICAL Severe pain, peritonitis, low-grade fever COMPLICATIONS Fallopian tube scarring, in $\frac{1}{4}$, infertility EPIDEMIOLOGY Primarily disadvantaged urban ♀, affecting ±1 million ♀–US, cost ±$4 billion TREATMENT Cefoxitin, doxocycline, clindamycin, ofloxacin

pelvic inlet *apertura pelvis superior*, pelvic brim ANATOMY The upper opening of the minor pelvis, bounded by the crest and pecten of pubic bones, ilia arcuate lines, and anterior sacrum. Cf Pelvic outlet

pelvic ultrasonography Obstetric ultrasonography OBSTETRICS A form of sonography in which 'real time' images are analyzed to provide information on the fetus or status of ♀ pelvis organs and structures. Cf Biophysical profile, Radiologic pelvimetry

PELVIC ULTRASONOGRAPHY indications

OBSTETRICS Early diagnosis of pregnancy, twin–or 'higher multiple', or ectopic pregnancy; identifying placental abnormalties that may compromise pregnancy–eg placenta previa, abuptio placentae, differentiation between pregnancy and placental tumors–moles and choriocarcinomas, determine fetal age, fetus' position, placental position prior to amniocentesis, monitor the rate of fetal growth

GYNECOLOGY Identify the presence of and differentiate among lesions–cysts, abscesses, or benign or malignant tumors of the uterus, ovaries, and pelvic organs and tissues

pelvimetry IMAGING A radiologic study in which the diameters of the osseous birth canal are compared with that of the infant's head to determine whether the pelvis is of sufficient diameter to allow a normal vaginal delivery. See Pelvic ultrasonography, X-ray pelvimetry

pemoline Cytert® PHARMACOLOGY A psychostimulant used to manage ADD ADVERSE EFFECTS Toxic hepatitis-monitoring of LFTs may be inadequate to prevent liver failure, transient tics

pemphigoid *adjective* Referring to pemphigus *noun* A condition that simulates bullous pemphigoid. See Bullous pemphigoid, Juvenile pemphigoid

pemphigus DERMATOLOGY An autoimmune disease–pemphigus foliaceus, pemphigus vulgaris–characterized by acantholysis with bulla formation. See Anti-epidermal antibody, Paraneoplastic pemphigus, Pemphigus vulgaris

pemphigus vulgaris DERMATOLOGY An autoimmune disorder characterized by acantholysis and blistering of mucocutaneous surface primarily in older people ETIOLOGY Immune response to penicillamine and captopril; > 95% of Pts have specific HLA antigens; ±50% of Pts begin with blisters in the mouth, followed by skin

pen Penicillin, see there

penbutolol Levatol® An antihypertensive β-blocker. See β-blocker

penciclovir INFECTIOUS DISEASE An antiviral cream used for oral herpes. See Acyclovir, Herpes

'penciling' Mortar & pestle sign RHEUMATOLOGY A descriptor for marked resorption of distal phalanges–acroosteolysis, a radiologic finding typical of scleroderma DIFFDX Severe burns with soft tissue contracture, 'black toe' disease–ainhum, hyperparathyroidism, neural leprosy, SLE, neuropathies–eg, DM, tabes dorsalis, progeria, psoriatic arthritis, PVC exposure, Reiter syndrome, rheumatoid arthritis, sarcoidosis. Cf Spade deformity

Pendred syndrome ENDOCRINOLOGY Hereditary association of congenital deafness and goiter due to a defect in thyroid hormone production. See Goiter. CF Hypothyroidism.

pendulous abdomen Beer gut A loose, fat-filled abdominal wall that hangs over the belt, accompanied by weak abdominal wall muscles

penem A β-lactam broad-spectrum antibiotic that acts on cell walls, lysing bacteria in resting phase; penicillins kill bacteria in growth phase by inhibiting cell wall synthesis

penetrance Penetration The disruption of a surface, as in penetrating–eg, gunshot wounds, hospital-acquired penetration contact due to infected 'sharps', or forcible penetration in rape

penetrating trauma URGENT CARE An injury sustained as a result of either 1. Sharp force, which includes injuries from cutting or piercing instruments or objects and nonvenomous bites of pets or humans or 2. Firearm injuries from projectiles Cf Blunt trauma

penetration MANAGED CARE See HMO penetration SEXOLOGY Entrance of the penis into the vagina or anus. See Sexual penetration

penetration phobia Aninsertia SEXOLOGY A morbid fear that prevents the vagina, anus, or mouth from being penetrated, in particular by the penis

penguin gait NEUROLOGY A fanciful descriptor for the waddling gait of muscular dystrophy in whom there is marked exaggeration of the lumbar lordosis, a rolling of the hips from side to side in the stance phase of each forward step–in order to shift the weight of the body, exaggerated lateral tilting and rotation of the pelvis to compensate for the weakened gluteal muscles accompanied by overuse of the trunk and upper extremities during ambulation. See Muscular dystrophy

penicillamine METABOLIC DISEASE A drug that removes copper, lead, mercury, and other metals from the body, which inhibit angiogenesis in brain tumors

penicillin INFECTIOUS DISEASE An antibiotic that inhibits crosslinking of peptidoglycan chains in bacterial cell walls; bacteria growing in penicillin synthesize weak cell walls, causing them to burst due to the high osmotic pressure. See Ampicillin

penicillinase-producing *Neisseria gonorrhoeae* Any strain of *N gonorrhoeae* with penicillinase-producing plasmids; PPNG is common in non-Caucasian illicit drug abusers, prostitutes and their sexual partners MANAGEMENT Ceftriaxone. See Methicillin-resistant *Staphylococcus aureus*

penile anesthesia SEXOLOGY ↓ Penile erotism or sexual sensation; numbness. See Sexual anhedonia

penile cancer Cancer of penis UROLOGY An SCC usually seen in noncircumcised ♂ with poor penile hygiene or Hx of STD, genital herpes. See Circumcision, Phimosis

penile prosthesis An FDA Class 3 medical device composed of silicone polymers designed to allow penile erection. See Erection

PENILE PROSTHESIS TYPES

SEMIRIGID A sausage surrogate PROS << prone to infection and complications, eg mechanical failure CONS Equipment is always on standby, and may be difficult to conceal

INFLATABLE PI has a fluid reservoir that is surgically placed under the abdominal musculature, a pump in the scrotum, and inflatable cylinders in the penis; when 'yellow alert' goes to 'red alert', the Pt pumps up, transferring the fluid–usually saline into the cylinders; following appropriate use, the cylinders are emptied by means of a deflate button; of 500 reoperations for inflatable PIs, 64% required therapy for mechanical failure, 19% for surgical complications, and 10% for infections JAMA 1992; 267:2578MN&P

INTERMEDIATE A flexible cylinder; no moving parts; rises to the occasion by repeated compression of tip; 'at ease' achieved by bending device in half THE MAN'S HEALTH BOOK, M OPPENHEIM, PRENTICE-HALL, 1994

PIs have been available since 1973; ± 30 000 ♂ receive penile implants/year; the FDA receives ± 5000 complaints/year about PIs; most recipients are in their 50s and 60s, often the result of DM

penile reattachment The surgical reattachment of a penis severed by intentional or unintentional trauma, or unscheduled surgery; recovery of sexual function requires microsurgical techniques to sew vessels and nerves ≤ 1 mm. See 'Bobbittize'

penile wart Genital wart. See Condyloma, Condyloma acuminata

penis envy PSYCHIATRY The unconscious desire by ♀ to have a penis which, per psychoanalysts, corresponds to an unresolved castration complex. Cf Oedipus complex

Pennsaid™ Diclofenac PAIN MANAGEMENT A topical NSAID for osteoarthritic pain. See NSAID

penoclitoris Clitoropenis NEONATOLOGY A sex organ defect consisting of a structure with features of a small and deformed penis lacking a urethra, or an enlarged clitoris

Penrose drain SURGERY A pliable rubber tube drain used to promote surgical wound drainage

PENS Percutaneous electrical nerve stimulation PAIN MANAGEMENT A method for treating low back pain, consisting of insertion of thin needles into the soft tissue and/or muscles of the back to electrically stimulate peripheral sensory nerves. See TENS

pentachlorophenol TOXICOLOGY A toxic pesticide and wood preservative CLINICAL Hyperpyrexia, night sweats, diaphoresis, tachycarida, tachypnea, generalized weakness, N&V, abdominal pain, anorexia, headache, intense thirst, pain in extremities, coma, death MANAGEMENT Supportive–↓ body temperature, hydration, treat metabolic acidosis

pentafluoropropionyl derivative TOXICOLOGY Any cocaine metabolite–eg, benzoylecgonine, ecgonine methyl ester, measured in the urine by GC-MS, after extraction on a solid phase extraction column

pentagastrin test GI DISEASE A test in which gastric acid secretion–GAS is stimulated with pentagastrin, a synthetic analogue of gastrin; GAS ↑ 10 mins post-injection, peaks at 20-30 mins, and lasts for < 2 hrs; the peaks in GAS are similar to those evoked by histamine–once used for the same purpose, with fewer adverse effects; PT is used to exclude achlorhydria in Pts with gastric ulcers

PENTALYSE CARDIOLOGY A trial–Synthetic Pentasaccharide as an Adjunct to fibrinolysis in ST-elevation Acute myocardial infarction

pentamidine isoethionate Nebupent, Pentam AIDS A 2ⁿᵈ-line agent used to treat or prevent PCP–↓ episodes by 65% in AIDS Pts unresponsive to T-S ADVERSE EFFECTS Arrhythmia, azotemia, hypotension, sterile abscesses at injection site, pancreatitis, DM, dose-related, potentially fatal hypoglycemia, kidney disease, coughing, lung collapse. See AIDS, *Pneumocystis carinii* pneumonia

pentazocine PAIN MEDICINE An IV mixed opioid sedative and analgesic–Talwin and naloxone ROUTE Oral, injected. See Opioids

pentetic acid ONCOLOGY An EDTA that protects tissues from chemotherapy-induced toxicity. See Leukovorin rescue

pentobarbital A short-acting barbiturate. See Barbiturate

pentosan polysulfate ONCOLOGY An angiogenesis inhibitor that may block effects of growth factors on endothelial cells and protect the GI tract against RT UROLOGY A drug used to relieve bladder pain due to interstitial cystitis. See Angiogenesis inhibitor

pentostatin Nipent® ONCOLOGY An antimetabolic anticancer agent

pentoxifylline Trental® VASCULAR DISEASE An agent used to prevent blood clots, especially in the elderly; it may also slow weight loss in CA ; may ↓ TNF ADVERSE EFFECTS Headache, tremor, dizziness, indigestion, N&V. See AIDS

PEP 1. Post-exposure prophylaxis, see there 2. Protein electrophoresis, see there

Pepcid® Famotidine, see there

PEPI CARDIOLOGY A trial–Postmenopausal Estrogen/Progestin Interventions Trial evaluating the effect of combined hormonal–♀ –therapy on cholesterol levels and major CAD. See Estrogen replacement therapy, Hormone replacement therapy, Menopause, Osteoporosis

pepper syndrome PEDIATRIC ONCOLOGY Massive metastatic liver involvement by neuroblastoma resulting in 'peppering' of the parenchyma by innumerable small dark aggregates of tumor cells. See Neuroblastoma

peppermint stick candy bone A descriptor for the radiologic appearance of some cases of osteopetrosis, in which the defect in osteoclastic activity is intermittent, giving rise to radiodense and radiolucent bands, most often in the bony pelvis, paralleling the iliac crest

peptic ulcer disease See Duodenal ulcer, Gastric ulcer, GERD

per capita rate A rate proportional to the number of persons in a population

per diem Latin, by day HOSPITAL CARE *adjective* Referring to the hospital practice of charging 'daily' rates, where the expenses incurred on a daily basis are averaged over the entire hospital census *noun* A temporary employee–eg, a nurse, who receives a higher hourly salary, but does not get the benefits received by salaried employee–eg, vacation and pension

per member, per month See PMPM

per os By mouth, orally

per re nata As needed, *prn*

perception PSYCHOLOGY Mental processes by which intellectual, sensory, and emotional data are organized logically or meaningfully

perception hearing AUDIOLOGY The process of knowing or being aware of information via the ear. See Hearing

Percocet® Oxycodone, see there

Percodan® Oxycodone, see there

PercScope™ ORTHOPEDICS A reusable percutaneous diskectomy system for viewing, flushing and aspirating the surgical field during endoscopic spinal procedures. See Diskectomy

PercuSurge CARDIOLOGY A system consisting of a balloon-tipped angioplasty guide wire and an aspiration device for removing plaque and debris loosened during revascularization for blocked saphenous vein grafts. See Balloon angioplasty

percutaneous balloon valvuloplasty CARDIOLOGY A procedure for treating mitral stenosis with immediate improvement in hemodynamics and stenosis-related Sx. See NY Heart Association classification

percutaneous diskectomy NEUROSURGERY A form of diskectomy for managing lumbar disk prolapse OUTCOMES 67% relief of leg pain vs 98% pain relief for standard laminotomy-diskectomy; PD is confined to situations in which disk herniation is contained and neural compression is not produced by a free fragment of bone. Cf Microsurgical diskectomy

percutaneous endoscopic gastrostomy See PEG

percutaneous excisional breast biopsy SURGERY A minimally invasive method of removing suspicious breast tissue with minimal trauma to surrounding tissue. See Breast cancer

percutaneous intervention CARDIOLOGY An intravascular procedure performed without a large operative field TYPES Diagnostic catheterization, cardiac revascularization, angioplasty, stent placement

percutaneous myocardial revascularization Percutaneous transluminal myocardial revascularization, see there

percutaneous pin ORTHOPEDICS Orthopedic hardware commonly used there to treat humeral neck fractures, which has a self-threading screw tip. See Pin

percutaneous transhepatic biliary drainage GI DISEASE A type of percutaneous transhepatic cholangiophy that provides nonsurgical decompression of obstructed bile ducts PROS ↓ mortality, hospital stay INDICATIONS Symptomatic relief of biliary duct obstruction due to advanced malignancy; preop in jaundiced Pts to improve liver function and prevent bleeding. See Percutaneous transhepatic cholangiography. Cf Endoscopic retrograde cholangiopancreatography

percutaneous transhepatic cholangiography GI DISEASE A technique used to evaluate the gall bladder and intra- and extahepatic bile ducts; diagnostic success rate is 95% for dilated bilie ducts, 75% when ducts are normal or sclerosed CONS No anatomic information CONTRAINDICATIONS Radiocontrast dye allergies, ↑ bleeding time COMPLICATIONS Allergic reactions, bile peritonitis due to spillage post injection, < 5%–eg, sepsis, hemorrhage due to puncture of blood vessel, death. See Percutaneous transhepatic biliary drainage. Cf ERCP-endoscopic retrograde cholangiopancreatography

percutaneous transhepatic portal venography Portovenography IMAGING A technique in which radiocontrast is injected into the portal vein, allowing its visualization and that of the veins–splenic, inferior and superior mesenteric, gastroduodenal, and coronary that empty into it

percutaneous transluminal cerebral angioplasty Balloon angioplasty to ↓ the risk of strokes in atherosclerosis of cerebral arteries

percutaneous transluminal coronary angioplasty PCTA INTERVENTIONAL CARDIOLOGY A procedure in which an angioplasty balloon is inserted percutaneously into the arteries advanced to a stenosis, and inflated, reopening the lumen INDICATIONS Single and multivessel CAD, stable angina on exertion, unstable angina, acute MI–'primary' PCTA, stenosed renal arteries, arteries with fibromuscular hyperplasia, postthrombolytic therapy SUCCESS RATE 90%; re-stenosis in 30%; success is lower with stenoses that are chronic, long, eccentric, angulated, calcified, at branching, or with intraluminal thrombi, unstable angina, ↑ age, ♀. See Balloon angioplasty, Coronary artery bypass surgery, Excimer laser therapy. Cf Balloon valvoplasty

percutaneous transluminal myocardial revascularization Percutaneous myocardial revascularization, percutaneous transluminal endomyocardial revascularization CARDIOLOGY A catheter-based minimally invasive procedure for treating CAD by creating channels in O_2-starved areas of myocardium; the procedure can be performed without general anesthesia, for Pts who are poor risks for invasive surgery. See Transmyocardial revascularization. Cf Balloon angioplasty, CABG

percutaneous transluminal renal angioplasty CARDIOLOGY Balloon dilation of a renal artery for ASHD occlusion, linked to HTN. See Stent

percutaneous transvenous mitral commisurotomy PTMC CARDIOLOGY A minimally invasive procedure used for mitral valve stenosis; PTMC success is ↑ when coupled with intracardiac echocardiography to guide transseptal catheterization

percutaneous umbilical blood sampling PUBS Cordocentesis NEONATOLOGY An ultrasound-guided needle aspiration of umbilical cord blood, to ID fetal disease–eg, hemoglobinopathies, hemophilia, autoimmune thrombocytopenia, von Willebrand disease, alloimmunization–Rh disease, Kell and other RBC antigens, alloimmune thrombocytopenia, metabolic disorders, fetal infection–eg, B19, CMV, rubella, toxoplasmosis, varicella, fetal karyotyping, and for fetal therapy–eg, RBC and platelet transfusion COMPLICATIONS Cord hematoma, bradycardia, fetal wastage ±2.7%

PerDUCER™ CARDIOLOGY A device with a shielded needle for percutaneous pericardial access, ↓ risk of inadvertent puncture or injury to surrounding vessels and tissues

perfectionism PSYCHIATRY A personality trait of many physicians, consisting of obsessiveness, overwork, checking compulsions, and other behaviors regarding Pt management, and ↓ ability to enjoy family, friends, and basic human needs. See Anal. Cf Obsessive-compulsive disorder.

Perflex™ SURGERY A stent and delivery system to maintain patency of the biliary tree as a palliative treatment of pancreas CA

perflubron LiquiVent® NEONATOLOGY A highly oxygenated perfluorocarbon liquid intrapulmonary ventilating agent, used in neonates with severe RDS and adults with acute hypoxemic respiratory failure, which may prevent O_2 toxicity, ↓ barotrauma due to improved dynamic compliance, ↑ oxygenation. See Liquid ventilation, Respiratory distress syndrome

perforated diverticulitis GI DISEASE Diverticulitis complicated by rupture of a peridiverticular abscess into the peritoneal cavity resulting in purulent peritonitis. See Acute diverticulitis, Complicated diverticulitis, Hinchey staging system

perforating abscess An abscess in which hydrolytic enzymes–presumably from PMNs–continue to digest tissue on the surface, causing locoregional necrosis

perforating collagenosis See Reactive perforating collagenosis

perforation MEDTALK An abnormal transmural defect in a hollow organ. See Intestinal perforation

performance anxiety Performing anxiety, stage fright PSYCHOLOGY A 'flight-or-fight' reaction in an anxious person carrying out an activity in public–eg, entertaining, public speaking–or in front of others, as in sexual activity, for fear of poor performance CLINICAL Tachycardia, ↑ BP, ↑ respiration, ↑ muscle tone. See Anticipatory anxiety, Flight-or-fight response, Performing arts medicine

performance-enhancing drug Ergogenic drug SPORTS MEDICINE An agent–eg, amphetamines, androstendione, erythropoietin, hGH, testosterone, known or thought to improve performance in a particular activity. See Anabolic-androgenic steroids, 'Stacking'. Cf Blood doping, Carbohydrate loading

performance profile Quality report card, scorecard MANAGED CARE An aggregate of data on medical and surgical specialists used to monitor quality and drive referrals to specialty care. See Scorekeeping; Cf Gatekeeping

performance test PSYCHOLOGY Any non-language and nonverbal IQ test based on the manipulation of concrete objects–eg, blocks, pictures, and printed mazes, rather than on relationships rooted in linguistics and symbols. See IQ tests. Cf Psychologic test

performing arts medicine A developing subspeciality of occupational medicine that formally addresses the medical complaints of those who toot, tickle, trill, or tap, playing musical instruments, warbling, or dancing COMMON PROBLEMS Those of a specific muscle-tendon unit, ranging in severity from mild pain to complete incapacitation, related to a combination of relatively repetitive movements of a limited number of muscles, and awkward position required to hold the instrument and/or weight of instrument, overuse 'syndromes', nerve impingement, facial dystonia MANAGEMENT Rest; β-adrenergics for performance anxiety

perfusion Bathing an organ or tissue with a fluid. See Arterial perfusion, Hyperthermic perfusion, Isolated hepatic perfusion, Limb perfusion, Myocardial perfusion ONCOLOGY A technique used for a melanoma of an arm or leg; circulation to and from the limb is stopped with a tourniquet; chemotherapy is put directly into the circulation to ↑ regional drug dose TRANSPLANTATION The intravascular irrigation of an isolated organ with blood, plasma or physiologic substance, to either studying its metabolism or physiology under 'normal' conditions or for maintaining the organ as 'fresh' as possible, while transporting donated organs to recipients. See Slush preparation

perfusion scan Perfusion scintigraphy CARDIOLOGY A radionuclide technique for assessing myocardial or pulmonary blood flow, measured by IV injection of ^{99m}Tc microaggregated albumin, or thallium, TC-MIBI. See Myocardial infarction. Cf Ventilation scan

perfusion shunt THERAPEUTICS A biohybrid device in which cells of interest–eg, pancreatic islet cells are placed in capillary fibers, housed in a plastic chamber, and anastomosed to a blood vessel. See Biohybrid organ

perfusion studies See Perfusion scan

perfusion-weighted imaging IMAGING An MRI that provides metabolic and hemodynamic data of the brain in the first few hrs post-stroke, based on the movement of perfused contrast material. See Magnetic resonance imaging. Cf Diffusion-weighted imaging

periapical abscess Tooth abscess DENTISTRY A complication of caries, linked to trauma to enamel, allowing bacteria to infect pulp, and extend to the tooth root and bone, with necrosis, gum swelling, toothache, and periodontal disease. See Caries, Periodontal disease

pericardial adhesion A fibrous pericardial adhesion that follows heart surgery or other condition causing pericarditis. See Adhesion

pericardial effusion MEDTALK An abnormal collection of fluid in the pericardium

pericardial 'knock' CARDIOLOGY A loud 3rd heart sound occurring when the ventricular filling abruptly stops at the end of the early diastolic pressure dip–ie, at the end of the rapidly filling phase of the ventricles; classically associated with severe constrictive pericarditis, or with penetrating trauma to the pericardium, the PK has a relatively high pitch, often ↑ in intensity with inspiration and coincides with the nadir of the 'y' descent of the jugular venous pulse, resembling a premature 3rd sound. Cf Gallop

pericardiocentesis CARDIOLOGY Insertion of a long needle in the pericardial sac to obtain fluids and cells for analysis of inflammation or tumor cells INDICATIONS Controversial; experts believe pericardiocentesis should be limited to emergencies, and advocate an open–surgical approach, it is used to detect bacterial or tuberculous pericarditis or CA COMPLICATIONS Laceration of coronary arteries, lungs, or liver, arrhythmia due to needle irritation, vasovagal arrest, pneumothorax, infection

PERICARDIOCENTESIS--Normal values

Appearance	Clear to straw-colored
Cells	
RBCs	None
WBCs	< 300/mm³
Glucose	70-100 g/dL
Protein	< 40 g/L
Volume	Minimal, usually ±20 mL

pericarditis CARDIOLOGY Inflammation of the pericardium ETIOLOGY Infection–eg, polio, influenza, rubella, adenovirus, coxsackieviruses, TB, rheumatic fever, injury or trauma to chest, esophagus, heart; systemic diseases such as CA, kidney failure, leukemia, AIDS or AIDS related disorders, autoimmune disorders, acute MI, myocarditis, RT to chest, immunosuppressants CLINICAL Pain due to rubbing of pericardium against heart, pericardial effusion TYPES Bacterial pericarditis, constrictive pericarditis, post-MI pericarditis DIFFDX Restrictive cardiomyopathy. See Bacterial pericarditis, Constrictive pericarditis, Dressler syndrome

perichondritis ORTHOPEDICS Inflammation of the soft tissue surrounding cartilage, especially the ribs

perilymph fistula AUDIOLOGY Leakage of perilymph to the middle ear ETIOLOGY Idiopathic or associated with head trauma, physical exertion, or barotrauma. See Perilymph

perimenopausal *adjective* Referring to a period of a ♀'s life–age 45 to 55-ish–in which menstrual periods become irregular; perimenopause is immediately before, during and after menopause. See Menopause

perimetry OPHTHALMOLOGY A test in which a topographic 'map' is created of the visual field, to diagnose and evaluate diseases of optic nerve, retina, and neuroophthalmic Goldman perimetry; uses both stationary–static light sources of increasing intensity and moving light sources to delineate the visual field, defects of which can be either central or peripheral

perinatal *adjective* Referring to the period around birth, often between the 28th wk of pregnancy and the end of the first wk of life

perinatal substance exposure NEONATOLOGY The contact by a near-term fetus with a substances of abuse–SOA due to maternal ingestion; PSE is associated with neonatal and obstetric complications–LBW, prematurity, abruptio placentae, fetal distress, stillbirth, brain infarcts, congenital defects, neurobehavioral dysfunction

perineal pearl NEONATOLOGY A cyst filled with viscid green-white mucoid material, seen in the anterior perianal region of a neonate, which may extend to the scrotum, a finding pathognomonic imperforate anus

perineal prostatectomy UROLOGY Surgery to remove the prostate via an incision made between the scrotum and anus. See Prostatectomtry

perineural invasion SURGICAL PATHOLOGY Extension of epithelial cells around nerves which, while typical of malignancy, may be seen in sclerosing adenosis–breast, and is not per se an indication of malignancy

period VOX POPULI A discrete time frame. See Accumulation period, Blanking period, Collection period, Critical period, Crystallization period, Eligibility period, Golden period, Grant budget period, Honeymoon period, Incubation period, Infectious period, Initial eligibility period, Last menstrual period, Latency period, NREMS period, Off period, Open enrollment period, Postoperative period, Pre-ejection period, Pre-patent period, Probationary period, Project period, Refractory period, REMS period, Sleep stage period, Sleep-onset REMS period, Total sleep period, Waiting period, Window period

periodic fever syndrome IMMUNOLOGY A heterogeneous group of inherited diseases of uncertain pathogenesis, in which the diagnosis is based on clinical features rather than on specific tests. See Familial Mediterranean fever, Hyper-IgD syndrome

periodic paralysis NEUROLOGY Any of a group of conditions characterized by centrifugal 'attacks' of paralyzing, focal or systemic weakness of hrs to days in duration, accompanied by a loss of deep tendon reflexes, refractoriness of muscle fibers to electrical stimulation, profound changes in potassium levels, variable cardiac arrhythmias and complete recuperation between attacks; rest following vigorous exercise may evoke an attack in a group of muscle fibers without changing the serum K+ levels

PERIODIC PARALYSIS

HYPOKALEMIC PERIODIC PARALYSIS Periodic paralysis I An AD condition of late onset that is more intense in ♂ and occurs following strenuous exercise or carbohydrate meals, affecting the extremities, respiratory and cardiac muscle, potentially causing ventricular tachycardia and premature ventricular contractions TREATMENT KCl, acetazolamide; the severe-

ly afflicted may develop persistent weakness and dystrophic changes in muscle DIFFDX Carnitine palmityl transferase deficiency, glycogen storage disease, type V, all other forms of periodic paralysis

HYPERKALEMIC PERIODIC PARALYSIS Periodic paralysis II An AD variant of muscular dystrophy caused by a defective gene on chromosome 17, which encodes the α subunit of a sodium channel in muscle cell membranes, closely linked to the growth hormone gene GH1 CLINICAL Early onset, most intense in males in whom paralytic attacks follow strenuous exercise, affecting the legs and eyelids; hyperkalemia may be prevented by acetazolamide; with time, severely afflicted subjects develop persistent weakness and dystrophic changes in muscle

NORMOKALEMIC PERIODIC PARALYSIS Periodic paralysis III 1. Primary or hereditary A condition with attacks of childhood onset that may disappear by middle age; exposure to cold may provoke attacks and over time, result in vacuolar myopathy; the attacks may be provoked by high-carbohydrate, high-sodium diets during periods of excitement and may respond to oral potassium 2. Secondary or acquired A condition associated with thyrotoxicosis, hypokalemia or K+ wasting by the kidneys or GI tract or due to accidental ingestion of absorbable barium salts that block K+ channels, reducing the egress of K+ from the muscles, evoking systemic hypokalemia or hyperkalemia, which may be associated with renal or adrenal insufficiency

periodontal disease DENTISTRY Any disease of the periodontium–eg, chronic gingivitis, extension of infection into periodontal ligaments and alveolar bone destruction; PD is the most common cause of loss of teeth in adults, the result of combined bacterial infection and impaired host response; 300 different bacterial spp occur in healthy mouths; most are gram-positive–eg, actinomyces and streptococci; in gingivitis, oral flora changes, streptococci ↓, actinomyces ↑, *Fusobacterium nucleatum*, *Lactobacillus*, *Veillonella*, *Treponema* spp, *Actinobacillus actinomycetemcomitans*, *Bacteroides gingivalis* DIFFDX Hypophosphatasia, Langerhans' cell histiocytosis–histiocytosis X, leukemia, vitamin C and/or vitamin D deficiencies

periodontitis Gum disease, pyorrhea gum disease DENTISTRY A condition caused by progression of gingivitis, with inflammation and infection of tooth ligaments and bones supporting teeth. See Juvenile periodontitis

PerioGlas® DENTISTRY A bone grafting material used in ridge augmentation procedures and filling extraction sockets to prevent bone loss and aid regeneration of jaw. See Bone graft

perioperative normothermia SURGERY A maneuver that attempts to counter adverse immune effects of perioperative hypothermia, due to anesthetic-induced impairment of thermoregulation, exposure to cold, and altered heat production in tissues, ↓ chemotaxis, phagocytosis, and oxidative killing of bacteria by PMNs, and ↓ resistance to infection in certain bacteria–eg, *E coli*, *S aureus*

perioral dermatitis Clown lips DERMATOLOGY Inflammation of the skin–itching, redness, around the mouth, most often caused by licking of the lips with the tongue, possibly due to fluoride and other substances

periorbital cellulitis OPHTHALMOLOGY Acute infection of tissue around the eye–eg, retrobulbar fat pads, not generally accompanied by protrusion or limited movement; untreated PC may progress to orbital cellulitis. See Orbital cellulitis

periorbital edema Puffy eyes MEDTALK Accumulation of edema around the orbit ETIOLOGY Blowout Fx of the orbit, Graves' disease, Hashimoto/chronic lymphocytic thyroiditis, nephrotic syndrome, orbital sinusitis, rhinosinusitis, rosacea, SLE, sporotrichosis

Periostat® DENTISTRY A formulation of doxycycline for periodontitis as an adjunct to scaling and root planing procedures ACTIVE INGREDIENT Doxycycline inhibits collagen breakdown that forms pockets between teeth and gums. See Periodontal disease

periostitis Inflammation of the outer layer of a bone

peripartum cardiomyopathy CARDIOLOGY A rare but often fatal–50-85% mortality–form of dilated cardiomyopathy, defined as heart failure–HF occurring in the last month of pregnancy or within 5 months of delivery in absence of an identifiable cause of HF, and without demonstrable preexisting or concurrent heart disease; findings include cardiothoracic ratio > 0.55, a left ventricular ejection fraction < 50%, and a diastolic dimension > 95th percentile for age and body-surface area or recent pregnancy where heart muscle is weakened and cannot pump blood efficiently; ↓ heart function affects lungs, liver, and other systems RISK FACTORS Obesity, personal or family Hx of heart disease–eg, myocarditis, use of certain drugs, smoking, alcoholism, multiple pregnancies, race, malnourishment, muscle degeneration, fibrosis

peripheral blood CARDIOLOGY Blood circulating in the system/body

peripheral excimer laser angioplasty See PELA

peripheral giant cell granuloma Peripheral reparative giant cell granuloma ORAL PATHOLOGY A sessile or pedunculated gingival or alveolar growth of the young–age 5-15 mandible, ♀:♂ ratio, 2:1, possibly induced by trauma–eg, tooth extration RADIOLOGY Superficial erosion, peripheral cuffing of bone TREATMENT Curettage, but not–as was occasionally practiced–extraction of the teeth

peripheral MR angiography See MR angiography

peripheral neuropathy Peripheral neuritis NEUROLOGY A functional defect of nerves outside the spinal cord, due to damage of peripheral nerves, axons or myelin sheath CLINICAL Numbness, weakness, burning pain–especially nocturnal, loss of reflexes; Sx depend on whether the 1° disorder affects sensory or motor nerve fibers or both; damage to sensory fibers results in changes in sensation ranging from perception of abnormal sensation, to pain, to ↓ sensation or lack of sensation in the area; sensory changes usually begin in the feet or hands and progress centrally with axon degeneration; motor fiber damage ↓ movement or control of movement of the area supplied by the nerve; loss of neural function causes trophic changes in muscle, bone, skin, hair, nails, etc; structural changes caused by lack of nervous stimulation, not using the affected area, immobility, lack of weight bearing, muscle weakness and atrophy; recurrent injury to the area may cause infection or structural damage–ulcer, poor healing, loss of tissue mass, scarring, deformity TYPES Single nerve/nerve group–mononeuropathy, multiple nerves–polyneuropathy ETIOLOGY Demyelination Idiopathic, pressure injury caused by direct injury, or compression by peripheral nerve or other tumors, abnormal bone growth, cysts, casts, splints, braces, crutches, etc, systemic causes–connective tissue disease, vasculitis, hereditary, metabolic or chemical disorders, Charcot-Marie-Tooth disease, Friedreich's ataxia, sniffing glue, nitrous oxide, industrial agents--eg, solvents, heavy metals–lead, arsenic, mercury, etc, infections and inflammation–AIDS, botulism, Colorado tick fever, diphtheria, Guillain-Barre syndrome, HIV infection, leprosy, periarteritis nodosa, polyarteritis, rheumatoid arthritis, sarcoidosis, syphilis, systemic lupus erythematosus, neuropathy secondary to drugs, hypoxia, prolonged exposure to cold, damage to sensory nerves of hands and feet, causing tingling or ↓ sense of touch in the hands and feet, DM–diabetic neuropathy, dietary deficiencies–especially vitamin B, alcoholism–alcoholic neuropathy, uremia, systemic effects of CA, myeloma, lung CA, lymphoma, leukemia. See Autonomic nervous system, Entrapment

peripheral pulses PHYSICAL EXAM Pulses palpable at the periphery–eg, radial, dorsal pedal, which signal vascular compromise–especially in the legs

peripheral stem cell transplantation Peripheral stem cell support ONCOLOGY A method of replacing hematopoietic cells–HCs destroyed by chemotherapy; stem cells in circulating blood are removed before treatment, then readministered treatment to help BM recovery Transplantation may be autologous–a person's own HCs, allogeneic–HCs donated by another, or syn-

geneic–HCs from an identical twin. See Bone marrow

peripheral stem cells HEMATOLOGY Stem cells in peripheral circulation. See Stem cells

peripheral vascular disease Atherosclerosis of extremities VASCULAR DISEASE Any vasculopathy, generally of the major peripheral arterial branches of the aorta with ASHD, which supply the legs and feet, with resulting ↓ in blood flow DIAGNOSIS Angiography MANAGEMENT Angioplasty, peripheral bypass operation, amputation

peripheral vestibular disorder NEUROLOGY A hallucination of movement, either subjective or objective HISTORY Duration of an attack–eg, hrs v. days, frequency daily v. monthly, effect of head movement–eg, better, no effect, induced change position or posture, associated aural Sx–eg, hearing loss and tinnitus, concomitant or prior ear disease and/or ear surgery

peripheral vision OPHTHALMOLOGY Side vision; ability to see objects and movement outside the direct line of vision

peritoneal dialysis NEPHROLOGY A therapeutic modality used to clear toxic metabolites from Pts with terminal renal failure, using the peritoneum to filter waste products. See Middle molecules

PERITONEAL DIALYSIS TYPES

INTERMITTENT PERITONEAL DIALYSIS A modality requiring up to 8 hrs/session–only practical for home therapy

CONTINUOUS AMBULATORY PERITONEAL DIALYSIS A treatment modality in which the Pt exchanges 1.5-3.0 L of sterile dialysate containing hypertonic glucose, 3-5 times/day, requiring 30-40 mins/session, a therapy ideal for diabetics in renal failure who have poor venous access, as insulin may be delivered in dialysate SIDE EFFECTS CAPD results in hyperlipidemia and obesity due to the high glucose of dialysate, and sclerosing peritonitis; long-term failure may be due to peritoneal infections–eg, candidiasis and phaeohypomycosis–*Fusarium* spp

peritoneal effusion Fluid in the peritoneum that may be benign or malignant; in ♀, malignant ascites is often linked to ovarian, gastric, pancreas, endometrial CA; in ♂, malignant ascites is linked to pleural mesothelioma, gastric, colon and pancreas CA

peritoneal perfusion THERAPEUTICS A method of delivering fluids and drugs directly to tumors in the peritoneal cavity by direct instillation. See Peritoneal dialysis

peritonealization SURGERY The covering of the stump of a partially excised organ or tissue after intraabdominal surgery with peritoneum

peritonitis Inflammation of the peritoneum, often due to infection–especially bacteria, but also post-traumatic, injury and bleeding, or disease–eg, SLE. See Bile peritonitis, Dialysis-associated peritonitis, Idiopathic sclerosing peritonitis, Sclerosing metanephric peritonitis, Spontaneous bacterial peritonitis, Vernix caseosa peritonitis

peritonsillar abscess Quinsy ENT Advanced anaerobic infection that begins as aerobic pharyngitis or Vincent's angina CLINICAL Marked pharyngeal pain, dysphagia, low-grade fever, inflammation and medial displacement of the tonsil; usually unilateral, bilateral lesions may cause pharyngeal obstruction TREATMENT Penicillin, broad-spectrum antibiotics active against *Fusobacterium necrophorum*

periventricular leukomalacia NEONATOLOGY The presence of lucencies in the periventricular white matter, affecting extremely premature infants, often in a background of subependymal hemorrhage PREVENTION Vitamin E, ethamsylate may ↓ hemorrhage. Cf Intracranial hemorrhage

504 **PERLA** Pupils equal, reactive to light and accommodation A clinical acronym for normal oculomotor functions

Perlman syndrome An AR condition characterized by fetal gigantism, renal hamartomas, nephroblastomatosis, ± Wilms' tumor, unusual facies

Perma-Flow® graft VASCULAR SURGERY A synthetic blood vessel used in CABG for Pts lacking adequate donor vessels PROS Alternative to harvested vessels, ↓ operative time and trauma; it is used in beating-heart surgery. See CABG

Perma-Seal® graft NEPHROLOGY A dialysis access graft that minimizes blood loss for early graft utilization, negating need for temporary catheters. See Peritoneal dialysis

permanent adjustable implant SURGERY A tissue surrogate expanded to a specific volume as a permanent implant

permanent lead CARDIOLOGY A pacing lead that is implanted for long term use. See Lead. Cf Temporary lead

permanent teeth Adult teeth DENTISTRY The 32 teeth of the 2nd dentition. Cf Baby teeth

permethrin Acticin® INFECTIOUS DISEASE A topical insecticide used for scabies and head lice. See Head lice, Scabies

permission CLINICAL TRIALS Agreement of parent(s) or guardian to the participation of a child or ward in research MEDICAL COMMUNICATION A formal writ from an author and/or publisher allowing re-use of previously published information

permissive hypercapnia CRITICAL CARE An approach to management of acute respiratory failure in which the tidal volume–V_T is lower–5-8 mL/kg than that conventionally used–10-15 mL/kg, the arterial Pco_2 is allowed to rise above the 'normal' of 40 mm Hg, and no attempts are made to compensate for subsequent changes in blood pH–respiratory acidosis, the deleterious effects of which may have been overestimated

Permitil® Fluphenazine, see there

pernicious adjective Bad/not real good

pernicious anemia Addison's anemia, megaloblastic anemia HEMATOLOGY Anemia caused by ↓ vitamin B_{12}, due to a lack of intrinsic factor which facilitates vitamin B_{12} absorption CLINICAL Glossitis, neurologic signs, achlorhydria, gastric atrophy LAB Antibodies against intrinsic factor, parietal cells. See Anemia

peroxisomal disease Any of a heterogeneous group of diseases in which peroxisomes are either lacking or markedly ↓, resulting in metabolic defects in all major biosynthetic peroxisomal pathways and failure to synthesize lipids or oxidize long-chain fatty acids EXAMPLES Zellweger's cerebrohepatorenal syndrome, rhizomelic chondrodysplasia punctata, neonatal adrenoleukodystrophy, infantile Refsum disease, hyperpipecolic acidemia

perphenazine Etrafon, Trilafon® PHARMACOLOGY A phenothiazine and tricyclic antidepressant combination used to manage anxiety, depression, agitation, severe N&V, hiccups, pain ADVERSE EFFECTS Extrapyramidal symptoms–involuntary movement–opisthotonus, trismus, torticollis, retrocollis, aching, numbness of limbs, motor restlessness, oculogyric crisis, hyperreflexia, dystonia, sedative effects, jaundice, agranulocytosis, leukopenia, hemolytic anemia, thrombocytopenic purpura, pancytopenia. See Mitriptyline

PERRLA Pupils equal, round & reactive to light & accommodation. See PERLA

Persantine® CARDIOLOGY A formulation of dipyridamole used as an alternative to exercise in thallium myocardial perfusion imaging for evaluating CAD in Pts who cannot exercise adequately. See Exercise stress test

perseveration NEUROLOGY The repeating of the same verbal or motor response to varied stimuli ETIOLOGY Organic brain disease, schizophrenia

persistent anxiety PSYCHIATRY A popular term for chronic anxiety, variously attributed to a serotonin imbalance; PA has a familial tendency and may have hereditary factors SYMPTOMS Worry, irritability, insomnia, disturbed sleep, loss of concentration, tachycardia, tremor, hot flashes, nausea, diarrhea, upset stomach, SOB, headaches, myalgia MANAGEMENT Buspirone

persistent generalized lymphadenopathy 1. AIDS lymphadenopathy 2. AIDS

persistent vegetative state CHOICE IN DYING A condition caused by injury, disease or illness in which a Pt has suffered a loss of consciousness, with no behavioral evidence of awareness of self or surroundings in a learned manner, other than reflex activity of muscles and nerves for low level conditioned response, and from which to a reasonable degree of medical probability, there can be no recovery; PVS is characterized by a prolonged loss of upper cortical function that may follow acute–eg, infections, toxins, trauma, or vascular events, or chronic–eg, degenerative events; in PVS, Pt is bed-ridden, nutritional support is completely passive, either parenteral or by NG tube; PVS Pts do not require respiratory support or circulatory assistance for survival and are in a state of chronic wakefulness which may be accompanied by spontaneous eye opening, grunts or screams, brief smiles, sporadic movement of facial muscles and limbs; while the eyes blink upon stimulation, they do not do so in response to visual threats; some Pts chew or clamp their teeth; urinary and fecal incontinence is universal; recovery occurs within the 1st month–if at all, recovery is rare beyond the 3rd month. See Advanced directives, DNR, Harvard criteria, Living will, Quinlan. Cf Procurement

PERSISTENT VEGETATIVE STATE–CRITERIA

1. No evidence of awareness of environment; inability to interact with others

2. No evidence of sustained, reproducible, purposeful, or voluntary behavioral responses to visual, tactile, auditory, or noxious stimuli

3. No evidence of language comprehension or expression

4. Intermittent wakefulness manifested by the presence of sleep-wake cycles

5. Sufficiently preserved hypothalamic and brain-stem autonomic functions to permit survival with medical and nursing care

6. Bowel and bladder incontinence

7. Variably preserved cranial nerve reflexes (pupillary, oculocephalic, corneal, vestibulo-ocular, gag) and spinal reflexes

persister INFECTIOUS DISEASE A person in whom a pathogen–eg, HPV, persists, who has an unknown risk of suffering its long-term consequences–eg, CIN or invasive SCC.

Personal Attributes Questionnaire PSYCHOLOGY An instrument consisting of 24 trait descriptions, each arranged on a bipolar scale of masculinity and femininity, which evaluates a person's perception of sex role stereotypes. See Personal Attributes Questionnaire

personal physician A physician who assumes responsibility–or who is held responsible–for a particular Pt's care. See Family practitioner, General practitioner, Internist. Cf Private patient

personal protective equipment Personal protective gear INFECTION CONTROL Specialized–ie, not 'standard issue' clothing or equipment worn by an employee for protection against a hazard, in particular blood-borne pathogens APPROPRIATE PPE Gloves, gown, lab coats, face shields or masks, eye protection, mouthpieces, resuscitation bags, pocket masks, or other ventilation devices. See Double gloving, Gloves, Handwashing, Mask, Shield, Spacesuit

personal risk factor A person's risk factors for STD, in particular HIV, which include engaging in sexual activity with more than one partner, use of IV drugs, receiving money or drugs in exchange for sex, having been previously treated for STD, use of drugs or alcohol during sexual episodes–which are associated with nonuse of condoms. See Sexual work

personality PSYCHIATRY The distinctive attributes of a person or characteristic manner of thinking, feeling, behaving; the ingrained pattern of behavior that each person evolves, consciously and unconsciously, lifestyle, way of being in adapting to the environment TRAITS/'SUPERFACTORS' OF PERSONALITY 1. Extraversion–eg, positive emotionality 2. Neuroticism–eg, negative emotionality 3. Conscientiousness–eg, constraint 4. Agreeableness–eg, aggression 5. Openness–eg, absorption. See Borderline personality, Cancer-prone personality, Explosive personality, Multiple personality, Obsessive-compulsive personality, Type A personality, Type B personality

personality disorder PSYCHIATRY Any condition characterized by individual traits that reflect ingrained, inflexible, and maladaptive patterns of behavior that cause discomfort and impair ability to function TYPES Antisocial, avoidant, borderline, compulsive, dependent, histrionic, narcissistic, paranoid, passive-aggressive, schizoid CLINICAL People with PDs in general do not take responsibility for their own lives and feelings; they tend to blame others; they have inadequate coping mechanisms for stress, difficulties in interpersonal relationships

personality test PSYCHOLOGY Any psychological test–eg, Rorschach ink-blot test or multiple choice California Psychological Inventory–designed to objectively measure certain facets of a person's personality and possibly predict the ability to function in the workplace; some experts believe that PTs are of little use and poorly predictive of future behavior; CF IQ test

personnel Personnel department HOSPITAL ADMINISTRATION The department responsible for employees, their grievances, benefits, and other work-related issues VOX POPULI People employed by a particular enterprise. See Allied health personnel

perspective DECISION-MAKING The vantage for cost-effectiveness analysis, which affects the types of costs included in a study

Perthes' test Tourniquet test VASCULAR SURGERY A test that evaluates deep leg vein patency and competency–ie, whether they close–of the leg veins' valves, in the face of varicosities of the superficial veins

pertinent diagnosis MANAGED CARE Any diagnosis relevant to a Pt that forms the basis for managing and billing for services rendered. See Diagnosis

Pertofran Desipramine, see there

pertussis Whooping cough PEDIATRICS An acute contagious and potentially epidemic bacterial infection caused by *Bordetella pertussis*–less commonly, *B bronchoseptica* and *B parapertussis*, which affects children < age 5; it causes 600,000 deaths/yr in the world CLINICAL Paroxysmal cough, post-tussive emesis, cyanosis, apnea, whoop COMPLICATIONS Pneumonia, atelectasis DIAGNOSIS Culture, direct fluorescent antibody, lymphocytosis MANAGEMENT Supportive; infants may need hospitalization if coughing is severe TREATMENT Erythromycin, T-S VACCINE Whole cell–DPT vaccine

pervasive developmental disorder PSYCHIATRY A condition characterized by extreme distortions or delays in the development of socialization and communication

perversion injury SEXOLOGY Trauma induced by deviant sexual activities, most common in the anus and rectum or lower urogenital tract, consisting of torn tissues and infection, caused by various devices ranging from high-pressure hoses per rectum-'pumping', to various objects–sex 'toys' or body parts-'fisting', intended to stimulate or enhance sexual arousal. See Sex toy, Sexual asphyxia

pes cavus High arch ORTHOPEDICS A foot with a high longitudinal–toe to heel–arch ETIOLOGY Neuromuscular diseases CLINICAL Changed muscle tone, pain, especially when stress is placed on the arch, significant disability

pes planus Flat foot, flat feet, see there

Pesaro classification A classification for Pts with β-thalassemia, based on adequacy of iron chelation, and presence/absence of hepatomegaly, and portal fibrosis. See Thalassemia

pessary GYNECOLOGY A rounded 'donut' placed in the upper vagina as a nonsurgical means of providing support for a prolapsed uterus. See Uterine prolapse

pesticide TOXICOLOGY An annihilator of ambient arachnids, antagonistic arthropods, abominable animacules or pugnacious plants–eg, fumigants, fungicides, herbicides, insecticides; most are toxic and potentially fatal, with high arsenical or organophosphate content, and store in adipose tissue, given their lipid solubility TYPES Organochlorines–eg, DDT, chlordane, mirex, organophosphates–eg, parathion, diazinon, carbamates–eg, Aldicarb, carbaryl, carbofuran, metals–eg, copper, tributyl-tin oxide, pyrethroids–eg, permethrin, cypermethrin, etc–eg, 2,4-D, atrazine, paraquat. See Intermediate syndrome, Organophosphate pesticide

pesticide bomb PUBLIC HEALTH A disposable device containing chlorpyriphos used to eliminate cockroaches, termites, fleas. See Cockroaches

PET 1. Pancreatic endocrine tumor, see there NEPHROLOGY 1. Peritoneal equilibration test, which is used to assess peritoneal transport characteristics in CAPD Pts 2. Photoemission tube, see there

PET PET scan, positron emission transaxial tomography IMAGING A non-invasive imaging modality in which emission computed axial tomography is used to detect positron-emitting isotopes–radionuclides that reflect biochemical and pathologic defects in tissues and evaluate blood flow and metabolism in the cerebral cortex, heart, whole body scanning; PET scans may be used to evaluate AIDS-related neuropathology–response to AZT by local ↑ of glucose metabolism, gliosis, differentiating among Alzheimer's, multi-infarct, and other forms of dementia, Huntington's disease, tardive dyskinesia, epilepsy for localization of seizure focus, making surgical therapy viable, malignancy–gliomas, residual tumor, pituitary adenomas, Parkinson's disease–↓ dopamine, psychiatric disease–depression, schizophrenia, and analysis of radiopharmaceuticals; PET scanning may be used in cardiology to evaluate coronary arteriosclerosis, differentiate between benign and malignant tumors, stage CA, detect CA recurrence and metastases, monitor response to therapy and target biopsy sites, assess myoardial viability, regional myocardial blood flow, and ischemia, using $^{15}CO_2$; after an AMI, there is an severely attenuated vasodilator response in the resistance vessels in both the infarcted myocardium and in the myocardium perfused by normal vessels

pet-associated disease Any condition, often infectious or inflammatory linked to prolonged exposure to pets/domestic animals. See Cats, Cat-scratch disease, Cutaneous larva migrans, Dogs, Dog-bites, Fish-tank granuloma

PET-ASSOCIATED DISEASES

Animals act as vectors for various microorganisms, eg dogs–rabies, cats–toxoplasmosis, parrots–psittacosis

Animals may attack the owner–popularly, 'turning', an event that is relatively common in animals not bred for domestication, eg coyotes, lions, pythons, weasels

Given the often unusual clinical presentations that characterize pet-associated illness, a detailed history is imperative to establish a diagnosis Am Fam Prac 1990; 41:931

pet therapy Pet-facilitated therapy PSYCHOLOGY The use of a domestic pet as an adjunct to psychotherapy and for persons with a marginal role in society–eg, children in foster care, the elderly, in nursing homes, mentally retarded, physically handicapped, inmates of correctional facilities, for those in mental or physical isolation, or with a low self-esteem. See Companionship, Most significant other; Cf Social isolation

petechia Plural, petechiae DERMATOLOGY A pinpoint, unraised, round red spot under the skin caused by RBCs leaking from capillaries

506   **Peter Pan & Wendy complex** PSYCHIATRY A marital dyad composed of a narcissistic and/or unfaithful husband–Peter Pan who devotes considerable time to studies, sports or extramarital liaisons, and a chronically depressed long-suffering wife–Wendy. See Wendy dilemma

Peter Pan syndrome Any condition named for the 'boy who would not grow up' ENDOCRINOLOGY Physical immaturity due to a hypothalamic defect with underdeveloped 2° sexual characteristics occurring in ♂ children with microphalus and ↓ height

petit mal Absence NEUROLOGY A spell characterized by lapse of attention and awareness, with loss of recall, without convulsions or loss consciousness. Cf Grand mal seizure; psychomotor epilepsy

petrified man syndrome Fibrodysplasia ossificans progressiva, see there

Peutz-Jeghers syndrome SURGERY An AD condition characterized by brownish perioral and oral macules developed in infancy, accompanied by premalignant intestinal polyps causing abdominal cramping, intussusception, chronic bleeding, anemia due to chronic blood loss

peyote Trivial name for mushrooms of genus *Psilocybe*, which contain psychotropics, psilocybin, psilocin SUBSTANCE ABUSE The flowering heads–mescal buttons–are hallucinogenic CLINICAL Minutes after ingestion, euphoria, hallucinations, tachycardia, mydriasis, rarely fever, seizures. See Mescaline

Peyronie's disease Curvature of penis, penile fibromatosis UROLOGY A condition characterized by unilateral fibrotic nodules within the fascial sheath of one or both corpora cavernosa, leading to penile shaft curvature and painful erection. Cf Erectile dysfunction

Pfannenstiel incision SURGERY A transverse curved abdominal incision with downward convexity, above the symphysis pubis; the PI passes through the skin, superficial fascia, and aponeurosis, and exposes the pyramidalis and rectus muscles, which are separated at the midline, after which the peitoneum is opened vertically USED FOR Abdominal hysterectomy

Pfiesteria TOXICOLOGY A dinoflagellate–alga of rivers & estuaries of southeastern US associated with red & brown tides; it produces a potent–1000-fold more powerful than cyanide–neurotoxin; it is lethal to aquatic life, ranging from eelgrass and other underwater vegetation to vertebrates–eg, menhaden and manatees; in humans, the toxin causes confusion, wasting, crippling, oozing red sores

PG 1. Postgraduate 2. Prostaglandin, see there 3. Proteoglycan

PGA™ suture SURGERY An absorbable coated synthetic suture used in gynecology, orthopedics, general surgery. See Suture

PGY-1 Postgraduate yr 1 The first–second yr is PGY-2 and so on–yr of post-graduate medical education, which usually corresponds to an internship or a residency, in a formal training program in a teaching institution. See Graduate medical education, Residency

pH of urine LAB MEDICINE Urine pH–UP has a wider range than does blood; the kidneys and lungs maintain the pH of the circulation near neutral; UP is ↑ by alkaline excess, diet–↑ fruits & vegetables, drugs–acetazolamide, aldosterone, amphotericin B, cortisone, dichlophenamide, certain diuretics, methazolamide, potassium bicarbonate, potassium citrate, potassium carbonate, potassium gluconate, prolactin, sodium bicarbonate; UP is ↓ by achlorhydria, alkaptonuria, DM, COPD, diet–↑ proteins, cranberries, drugs–eg, ascorbic acid, ammonium chloride, certain anesthetics, diazoxide, hippuric acid, methenamine mandelate, methanol intoxication, PKU, respiratory acidosis, renal TB, sepsis, starvation RANGE 4.6-8.0; normal 6.0

pH study 12 to 24 hr pH study GI DISEASE A procedure that places a probe in the distal esophagus, to evaluate acid reflux and the need for medical or surgical therapy INDICATIONS Suspected or evaluate therapy for GERD or nocturnal pulmonary aspiration of gastric contents. See GERD

phacoemulsification OPHTHALMOLOGY The therapeutic dissolution or emulsification of a cataract by ultrasound with subsequent removal. See Cataract

PHADE CARDIOLOGY A clinical trial–Pneumatic HeartMate® Assist as Destination Evaluation

Phaedra complex PSYCHIATRY The libidinous desire of a stepmother for a stepson; since the 2 are not genetically related, a sexual liaison is not legally incestuous. Cf Electra complex, incest, Jocastra complex, Oedipus complex

phaeohyphomycosis MYCOLOGY An infection caused by dematiaceous hyphae with yeast forms in tissue. Cf Chromoblastomycosis

phag *prefix*, Latin, referring to eating or swallowing

phagocytic disorder Phagocytic disease IMMUNOLOGY Any rare qualitative disorder of phagocytic WBCs, which may be familial and accompanied by generalized metabolic dysfunction; the prototypic PD is chronic granulomatous disease. See Chronic granulomatous disease. Cf Histiocytic diseases

PHAGOCYTIC DISORDERS

STRUCTURAL & FUNCTIONAL DEFECTS
- Alder-Reilly anomaly
- Chédiak-Higashi anomaly
- Familial vacuolization of leukocytes (Jordan's anomaly)
- May-Hegglin anomaly
- Pelger-Huët anomaly
- Pseudo- (or acquired) Pelger-Huët anomaly
- Hereditary giant neutrophilia
- Hereditary hypersegmentation of neutrophilic nuclei

FUNCTIONAL DEFECTS–WITHOUT STRUCTURAL CHANGES
- Chronic granulomatous disease
- Myeloperoxidase deficiency
- CD11/CD18 adhesive protein deficiency
- Other enzyme defects

phakomatoses *phakos*, Greek, lens Neurocutaneous syndromes A group of inherited conditions–many are AD–that cause disordered growth of ectodermal tissues, with distinctive skin lesions and tumors and/or defects of the nervous system and/or retina

PHAKOMATOSES

ATAXIA-TELANGIECTASIA An AR disorder characterized by cerebellar ataxia, oculomotor apraxia, telangiectasias of bulbar conjunctiva, skin of ears, and skin folds–appearing by age 3–and sinopulmonary infections; telangiectasias later extend to the butterfly region of the face; most Pts die in adolescence

BASAL CELL NEVUS SYNDROME See there

NEVUS SEBACEOUS OF JADASSOHN An occasionally AD clinical condition characterized by a congenital solitary lesion most often present in the scalp which, when large, may be associated with internal derangements eg intracranial masses, seizures, mental retardation, skeletal abnormalities, pigmentary changes, ocular lesions and renal hamartomas; 10% of the skin lesions develop into basal cell carcinoma

STURGE-WEBER DISEASE Encephalotrigeminal angiomatosis An occasionally AD condition characterized by congenital capillary hemangiomas of the head & neck, following normal developmental milestones, mental retardation may ensue, caused in part by the sluggish flow of blood through the pial vessels and venous hemangiomas in the leptomeninges and frontoparietal cortex with ipsilateral port-wine nevi, 'tram-track' radiopacities on the skull caused by calcification of the cerebral cortex

TUBEROUS SCLEROSIS Bourneville-Pringle disease An AD disorder–50% arise de novo CLINICAL Convulsions, seizures, mental retardation, skin lesions–adenoma sebaceum, sebaceous gland atrophy, angiofibromas, dermal fibrosis with dilated capillaries, shagreen patches, cardiac rhabdomyomas, pulmonary fibrosis, bronchiolar hematomas, bilateral tubular adenomas of kidneys, pancreatic cysts, angiomyolipomas, myxedematous glossitis, spina bifida

VON HIPPEL-LINDAU DISEASE An AD condition with retinal hemangioblastoma, ↓ erythropoietin production, cerebellar hemangioblastoma CLINICAL Ataxia, headache, papilledema, angiomas of the liver, kidney, renal adenomas, papillary cystadenomas of epididymis, pancreatic cysts, adrenal pheochromocytomas; V_4 develop renal cell cancer

VON RECKLINGHAUSEN DISEASE A relatively common–1/3500 AD condition CLINICAL Neurofibromas, cafe-au-lait spots of skin, scoliosis, gliosis, glioblastoma multiforme, ependymoma, meningioma and schwannoma, 5-10% sarcomatous degeneration, spina bifida and glaucoma. See Neurofibromatosis

Note: Neurofibromatosis, tuberous sclerosis, and von Hippel disease constitute the 'classic' phakomatoses

phalangitis Inflammation of a finger

Phalen sign Phalen's maneuver, Phalen's test NEUROLOGY Paresthesias over the median nerve evoked by maximum passive flexion of the wrist for 1 min, a finding in entrapment neuropathy. See Carpal tunnel syndrome

phallic PSYCHIATRY *adjective* Referring to the period from about $2\frac{1}{2}$ to 6 yrs, during which sexual interest, curiosity, and pleasurable experience center about the penis in boys, and in girls, to a lesser extent, the clitoris

phalloplasty UROLOGY The construction or reconstruction of a penis as needed by a birth defect, ♀-to- ♂ transsexualism, and accidental or surgical amputation of penis

phallotoxin TOXICOLOGY A peptide in poisonous mushrooms–eg, *Amanita phylloides, A verna, A virosa, Galerina autumnalis, Cenocybe filaris,* etc which, with other cyclopeptides, causes most of the 100 deaths/yr–US due to poisonous mushrooms CLINICAL STAGE 1 Abrupt onset of abdominal pain, N&V, cramping, diarrhea with blood and mucus STAGE 2 Apparent recovery with ↑ liver enzymes STAGE 3 1-3 days postingestion–hepatic, cardiac, renal failure, coagulopathies, seizures, coma, death TREATMENT None. See Mushrooms

phantom RADIOLOGY A mass or dummy that approximates tissues in its physical properties that may be used to calibrate or determine the dose of radiation applied to a tissue. Cf Ballistic jelly

phantom foot anterior cruciate ligament injury SPORTS MEDICINE A mechanism of knee injury that affects downhill skiers, in which the tail end of the downhill ski acts like an unnatural foot pointing backwards, thereby facilitating the twisting and tearing of ligaments Cf Boot-induced anterior cruciate ligament injury

phantom limb syndrome Chronic intense pain localized to the site of an amputated or denervated limb; 60-70% of amputees have a PL sensation; 10-15% have PLS; the pain often reflects the amount of pre-amputation pain, and is often refractory to excision of amputation neuroma, rubbing, electrical stimulation, peripheral nerve or spinal blocks, narcotics, sympathectomy

pharmaceutical care MANAGED CARE An evolving concept, in which a pharmacist provides services beyond that which can be reasonably provided by physicians–eg, reviewing a drug's intended use, counseling Pts on compliance and optimal use

pharmaceutical sales representative Detailer, Drug rep DRUG INDUSTRY A drug company employee who regularly visits physicians and office practices, providing information on the company's products–usually putting a negative 'spin' on competitors' products. See Detailing

pharmacist Chemist–British PHARMACOLOGY A person qualified by a graduate degree in pharmacy, and licensed by a state to prepare, dispense, sell and control certain drugs TITLE RPh–registered pharmacist; a person holding a license in a particular jurisdiction to practice pharmacy. See Pharmacy, Practice of pharmacy. Cf Pharmacologist

pharmacognosy 1. Herbal medicine, see there 2. The scientific approach to and formal study of the effects and uses of medicinal plants. See Herbal medicine, Pharmacognosist; Cf Traditional herbalism

pharmacologic *adjective* ENDOCRINOLOGY Referring to administration of hormones or other substances, naturally present in the circulation–in doses far in excess of normal levels in the body. Cf Homeopathic, Physiologic, Supraphysiologic

pharmacologic dose PHARMACOLOGY A supraphysiologic dose of a substance–eg, mineralocorticoids, normally present in the body, to produce a 'pharmacologic' effect

pharmacologic phlebotomy CRITICAL CARE The use of morphine in pulmonary edema to ↓ the fluid load in pulmonary vessels by pooling blood in capacitance vessels

pharmacologic purging HEMATOLOGY The use of a cytotoxic agent to destroy the BM of a Pt with a malignant or nonmalignant lesion which requires hematopoietic stem cell transplantation

pharmacologic restraint A pharmacologic agent–eg, anxiolytics, hypnotics, neuroleptics, sedatives used to control or restrain an inmate or Pt in a sheltered environment–eg, mental institution or nursing home. Cf Physical restraint

pharmacologic stress imaging Thallium stress test, see there

pharmacologic study A study to assess the potential harmful or other effects of drugs

pharmacologic testing The administration of a pharmacological to either ↑ or↓ a physiologic response. See Catecholamines, Provocative testing

pharmacology The study of the science and clinical application of medications; the study of drugs, their sources, their nature and properties. See Clinical pharmacology, Cosmetic pharmacology, Recombinant pharmacology

pharmacopeia An official authoritative listing of drugs. See National Formulary

pharmacy PHARMACOLOGY An establishment or institution in which the practice of pharmacy is conducted; drugs, medicines or medicinal chemicals are dispensed or offered for sale, or a sign is displayed bearing the word or words 'pharmacist,' 'pharmacy,' 'apothecary,' 'drugstore,' 'druggist,' 'medicine store,' 'drug sundries,' 'prescriptions filled,' or similar words intended to indicate that the practice of pharmacy is being conducted. See Internet pharmacy, Pharmacist, Polypharmacy

Pharmadine Diazepam, see there

Pharnax Alprazolam, see there

pharyngeal *adjective* Referring to the pharynx or throat

pharyngeal facelift Palatopharyngoplasty ENT A 'tuck & tighten' surgical procedure for the soft palate and pharynx, which eliminates redundant mucosa in order to ↓ the noise level in snorers. See Obstructive sleep apnea syndrome, Snoring

pharyngitis Sore throat ENT Inflammation of the oropharyngeal mucosa, most often viral infection–95% or by bacteria–5% ETIOLOGY ACUTE Group A streptococcus, aka strep throat ETIOLOGY CHRONIC Caused by a continuing infection of the sinuses, lungs, or mouth; or by irritation from smoking, heavily polluted air, alcohol, or by swallowing substances that scald, corrode or irritate throat. See Strep throat, Throat culture

phase MEDTALK A step in a process or cycle. See Acceleration phase, Burned-out phase, Chronic phase, Conceptive phase, Delayed sleep phase, Initial phase, Lag phase, Oepidal phase, Plateau phase, Prevascular phase, Proliferative phase, Recovery phase, Resting phase, Secretory phase, Shock phase, Stance phase, Swing phase, Take-off phase

phase 1 clinical trial Phase 1 study. See Phase study

phase 1 study See Phase study

phase 1 vaccine trial

phase 1 vaccine trial IMMUNOLOGY A closely monitored study trial of a new vaccine conducted in a small number of healthy volunteers, to determine safety in humans, metabolism, pharmacologic actions, and side effects at various doses. See Phase study

phase 2 clinical trial Phase 2 study. See Phase study

phase 2 study See Phase study

phase 2 vaccine trial IMMUNOLOGY A closely monitored clinical study of a new vaccine to ID common short-term side effects and risks associated with the vaccine and to collect data on immunogenicity

phase 3 clinical trial Phase 3 study. See Phase study

phase 3 study See Phase study

phase 3 vaccine trial IMMUNOLOGY A large controlled study to determine the ability of a new vaccine to produce a desired clinical effect on the risk of a given infection, disease, or other condition at an optimum dose and schedule. See Phase study

phase 4 study See Phase study

phase 4 trial Phase 4 study. See Phase study

phase advance SLEEP DISORDERS A backward shift in the 24 hr sleep-wake cycle—eg, an 11 pm to 7 am sleep phase shift to 8 pm to 4 am. Cf Phase delay

phase delay SLEEP DISORDERS A forward shift in the 24 hr sleep-wake cycle—eg, 8 pm to 4 am sleep phase shift to 11 pm to 7 am. Cf Phase advance

phase study PHARMACOLOGY A series of clinical trials assessing the safety and efficacy of an 'investigational new drug'—IND being 'sponsored' by a drug company hoping to bring it to the marketplace. See Compassionate investigational new drug, IND, NDA, Premarket approval process, Treatment IND

PHASES—CLINICAL TRIALS

PHASE 1 EARLY CLINICAL STAGE Phase I studies are designed to examine the safety of a new medication and understand how it will work in humans by gathering extensive data on how it is absorbed, distributed, metabolized and eliminated from the human body; other data include assessment of how quickly therapeutic concentrations is achieved, how long the drug remains in the body, and what, if any, the effect drug metabolite by-products may have; with step-by-step increases in dose, the optimal dosage is determined where minimum side effects are coupled with maximum therapeutic effect, termed the toxic-therapeutic window; a trial to determine the best way to give a new treatment and what doses can be safely given; phase 1's involve 20-80 subjects and generate data on toxicity and maximum safe dose, to later allow a properly controlled trial; FDA's review at this point ensures that subjects are not exposed to unreasonable risks; phase I studies generally enroll only healthy persons to evaluate how a new drug behaves in humans, but may enroll Pts with the disease that the new drug seeks to treat; candidates are enrolled in a study only after a review of their Hx and physical confirms eligibility and informed consent for treatment is given; the number of Pts enrolled in a phase I trial will vary depending on the step-wise progression established for achieving optimal dosing as well as prior clinical experience with similar compounds and approaches; followup time periods may range from just a few days to 6+ months; further trials may continue only if phase 1 results indicate that the new therapy is reasonably safe in humans, and the FDA approves further work; the ultimate goal of phase 1 trials is to obtain sufficient information about the drug's pharmacokinetics and pharmacological effects to permit the design of well-controlled, sufficiently valid phase 2 studies; other examples of phase 1 studies include studies of drug metabolism, structure-activity relationships, and mechanisms of actions in humans, as well as studies in which investigational drugs are used as research tools to explore biological phenomena or disease processes Glossaries, nih.gov-6/99

PHASE 2 LATER CLINICAL STAGE Phase 2 studies are designed to evaluate the short-term therapeutic effect of a new drug in Pts who suffer from the target disease, and confirm the safety established in phase I trials; phase 2 studies are sometimes placebo-controlled, often double-blinded, enroll a larger number of Pts than in phase 1 and Pt followup may be for longer periods; phase 2 studies are tailored to specific treatment indications for which the company plans to seek broader approval; where compelling scientific evidence is present-

ed, the FDA expedites review of a company's application for market clearance; expedited review of phase 2 clinical data, and clearance of that early application, can obviate requirements for phase 3 trials; phase 2's involve several hundred Pts and generate enough data to 1. at least suggest—if not prove that the drug actually works—efficacy and 2. demonstrate the most common side effects Glossaries, nih.gov-6/99; phase 2 trials are sometimes divided into phase 2a pilot trials and phase 2b well-controlled trials

PHASE 3 FINAL CLINICAL STAGE Phase 3 trials are designed to demonstrate the potential advantages of the new therapy over other therapies already on the market; safety and efficacy of the new therapy are studied over a longer period of time and in many more Pts enrolled into the study with less restrictive eligibility criteria; phase 3 studies are intended to help scientists identify rarer side effects of treatment and prepare for a broader application of the product; phase 3 trials enroll 1,000-3,000 Pts to verify efficacy and monitor adverse reactions during longer-term use—sometimes divided into Phase 3a trials, conducted before regulatory submission and Phase 3b trials, conducted after regulatory submission, but before approval

PHASE 4 POST-FDA APPROVAL/POST-MARKETING Phase 4 studies involve many thousands of Pts and compare its efficacy with a gold standard; some agents have been withdrawn from the market because they increase the mortality rate in treated Pts; concurrent with marketing approval, FDA may seek agreement from the sponsor to conduct certain postmarketing—phase 4 studies to delineate additional information about the drug's risks, benefits, and optimal use, and could include, but would not be limited to, studying different doses or schedules of administration than were used in phase 2 studies, use of the drug in other Pt populations or other stages of the disease, or use of the drug over a longer period of time 21 CFR §312.85

Note: Occasionally, an agent's benefit is so obvious, eg zidovudine—AZT that the need for phase 3 studies, a stage immediately preceding an official NDA, may be obviated; NDA—new drug applications were often rejected because the data from the phase 2 and 3 trials revealed study design flaws, forcing the sponsor to repeat work, which has been largely eliminated by the '1987 rewrite' of the IND status

PHC Primary health care, see there

PhD Doctorate in Philosophy GRADUATE EDUCATION An advanced academic degree, requiring 3 to 6 yrs after basic college/univerity; a PhD can be obtained in a wide range of disciplines—eg, sociology, psychology, anthropology, mathematics, etc

phencyclidine PCP, 'angel dust' SUBSTANCE ABUSE A recreational hallucinogen with major side effects—neurologic dysfunction, with schizophrenia-like behavior, analgesia, dysarthria, nystagmus, ataxia, seizures, delirium, coma, GI Sx, ↑ BP, temperature, ↓ pulmonary function

Phenergan® Promethazine, see there

phenethyl isothiocyanate CLINICAL NUTRITION An organosulfur compound derived from *Alium* spp—garlic, onions—which inhibits nitrosamine-induced esophagus and lung CAs, and prevents DNA adduct formation in animals. See Garlic

phenmetrazine A schedule II controlled drug and sympathomimetic amine/CNS stimulant PHARMACOLOGIC EFFECTS Stimulates satiety center in hypothalamic and limbic regions, inducing CNS stimulation and anorexia. See Fastin, Therapeutic drug monitoring

phenobarbital NEUROLOGY A long-acting barbiturate used as a hypnotic, sedative, hepatic enzyme inducer, anticonvulsant, given as a monotherapy for partial seizures, 2° generalized seizures; also used to treat diarrhea and to ↑ the antitumor effect of other therapies. See Seizures, Therapeutic drug

phenol NUTRITION Phenolics A simple cyclic compound with a hydroxyl group on an aromatic ring—eg, tyrosine; phenols are concentrated in fruits—grapes/raisins, garlic, onions, green tea, and may protect against cardiovascular disease, CA, possibly viruses TOXICOLOGY Carbolic acid, hydroxybenzene, phenyl hydrate A toxic crystalline compound, with a hydroxyl group on a benzene ring; phenol was once used as a topical anesthetic, antiseptic, and antipruritic

phenomenon VOX POPULI An observable fact or event that can be described scientifically. Related terms are Aging phenomenon, Aha! phenomenon, 'Alice in Wonderland,' Alien limb phenomenon, Anesthesia cutoff phenomenon,

Anniversary phenomenon, Blowback phenomenon, Booster phenomenon, Clasp-knife phenomenon, Cogwheel phenomenon, Dawn phenomenon, Dragged disc phenomenon, Engel's phenomenon, Flare phenomenon, Fleck phenomenon, Fortification phenomenon, Glass ceiling phenomenon, Harvest moon phenomenon, Hennebert's phenomenon, Herald wave phenomenon, Hunting phenomenon, J curve phenomenon, Jet phenomenon, July phenomenon, Koebner's phenomenon, Mask phenomenon, Mismatch phenomenon, No reflow phenomenon, Nutcracker phenomenon, On-off phenomenon, Pass through phenomenon, R-on-T phenomenon, Ratchet phenomenon, Raynaud's phenomenon, Re-entry phenomenon, Satellite phenomenon, Second disease phenomenon, Second-wind phenomenon, Sticky floor phenomenon, T-on-P phenomenon, Tulio's phenomenon, U curve phenomenon, Uninvolved bystander phenomenon, Vacuum phenomenon, Vanishing cancer phenomenon, Variation phenomenon, Walk-through phenomenon, Waterfall phenomenon, Wavefront phenomenon, West-to-east phenomenon, Will Rogers phenomenon, Zoning phenomenon

phenothiazine A class of antipsychotics and tranquilizers–eg, chlorpromazine, compazine, prochlorperazine, thioridazine, thorazine, possibly used as an antiemetic

phenotype GENETICS 1. Any observable or identifiable structural or functional characteristic of an organism 2. The sum of the structural/physical and functional–biochemical, and physiologic characteristics of an organism, defined by genetics, modified by the environment. See Bombay phenotype, Mutation, Null phenotype, Para-Bombay phenotype, Swarmer cell phenotype, Trait. Cf Genotype

phenotype testing HIV drug resistence testing, see there

phentolamine test A pharmacologic test for pheochromocytoma, in which an α-adrenergic blocker, phentolamine, is administered IV; a significant drop in bp, confirmed by quantifying catecholamines or their metabolites in the urine. See Catecholamines, Pharmacological testing

Phenurone® Phenacemide, see there

phenylbutazone THERAPEUTICS An NSAID tightly bound to plasma proteins. See NSAIDs

phenylketonuria Phenylpyruvic oligophrenia, PKU PEDIATRICS An AR metabolic condition caused by defective phenylalanine 4-hydroxylase, with an inability to convert phenylalanine to tyrosine CLINICAL Mental retardation, microcephaly, light skin and hair, EEG abnormalities LAB ↑ phenylalanine in blood, ↑ excretion of phenylalanine phenylpyruvate, hydroxyphenylacetate metabolites in urine REFERENCE VALUE < 4 mg/mL–blood, negative dipstick DIFFDX Phenylalanine is also ↑ in LBW infants, galactosemia PROGNOSIS If untreated, PKU causes mental retardation

phenylzine Nardil® PHARMACOLOGY An antidepressant

phenytoin Dilantin® PHARMACOLOGY An antiepileptic and anticonvulsant widely used as a monotherapy for partial seizures–eg, 2° generalized seizures. See Imipramine, Seizures, Therapeutic drug monitoring

phenytoin-induced gingival overgrowth Phenytoin-induced gingival hyperplasia Hyperplasia of gums induced by direct stimulation of gingival fibroblasts, resulting in ↑ collagen synthesis; PIGO is seen in 10-30% of phenytoin-treated Pts. Cf Phenytoin lymphadenopathy

phenytoin lymphadenopathy A phenytoin-induced condition characterized by generalized lymphadenopathy, rashes, fever. See Pseudopseudolymphoma

pheochromocytoma 10% tumor ENDOCRINOLOGY A benign paraganglioma of adrenal medulla; 10% are associated with systemic disease–eg, von Recklinghausen's disease, von Hippel-Lindau syndrome, Sturge-Weber disease, MEN IIa and IIb; pheochromocytomas may produce ACTH, calcitonin, VIP CLINICAL The pheochromocytoma triad–headaches, sweating attacks and tachycardia in a hypertensive Pt has a 94% specificity and 91% sensitivity for pheochromocytoma–absence of all triad Sx in a hypertensive Pt excludes

pheochromocytoma; other Sx include anxiety, tremor, pallor, N&V, fatigue, chest or abdominal pain, weight loss LAB ↑ VMA, ↑ metanephrine, free catecholamines, MHPG, dopamine, HVA DIAGNOSIS SPECIFICITY OF IMAGING Abdominal ultrasonography 100%; abdominal CT 100%; abdominal MRI 97%; MIBG–metaiodobenzylguanidine scintigraphy 97% SPECIFICITY OF LAB URINE Epinephrine 98%; norepinephrine 95%; VMA 91% SPECIFICITY OF LAB PLASMA Norepinephrine 97%; chromogranin A 95%; epinephrine 91%

pheresis See Cytapheresis, Plasmapheresis, Plateletpheresis

Philadelphia chromosome A small acrocentric chromosome from the distal long–q arm of chromosome 22, transferred to chromosome 9q[t(9;22)(q34;q11)] in 95% of CML; PC is present in 3 to 5% of childhood ALL–for whom prognosis is poor, and 25% of adults ORIGIN OF PC Pluripotent stem cell, which generates myeloid, erythroid, megakaryocytic and lymphoid lines MOLECULAR PATHOLOGY Reciprocal translocation, with juxtaposition of the c-*abl* gene on chromosome 9 with a gene of unknown function, with a *bcr*–breakpoint cluster region on chromosome 22; the resulting hybrid *abl/bcr* gene encodes P210$^{bcr/abl}$, a phosphoprotein unique to CML that resembles v-*abl*, as it has disregulated protein-tyrosine kinase activity MANAGEMENT High-dose chemotherapy, BMT. See P210$^{bcr/abl}$

-phile MEDTALK Root for a person who loves (something)

-philia MEDTALK Root for the love of something. See Normophilia, Paraphilia

-philiac MEDTALK Root referring to a person who loves a thing

-philic MEDTALK Root referring to the love of a thing

phimosis UROLOGY A foreskin constriction that prevents its retraction over the glans penis. See Circumcision

phlebitis Venous thrombosis, thrombophlebitis SURGERY Venous inflammation. See Thrombophlebitis

phleborheography Plethysmography, see there

phlebotomy Venesection LAB MEDICINE 1. The obtention by venipuncture of blood for a diagnosis 2. The surgical opening of a vein to withdraw blood–eg, to ↓ blood volume, as in hemochromatosis . See Pharmacologic phlebotomy, Therapeutic phlebotomy. Cf Venipuncture

phlegmasia cerulea dolens SURGERY A severe form of DVTs, usually the upper leg, of abrupt onset, accompanied by swelling and loss of normal color in the leg below the blockage.

PHN Postherpetic neuralgia, see there

PHO Physician-hospital organization, see there

phobia Greek, πηοβοσ, fear PSYCHIATRY An irrational fear or an objectively unfounded 'morbid' dread of an element in the environment or particular activity, of such intensity as to evoke anxiety, panic, and adverse physiologic effects, and compel its victim to avoid contact therewith at virtually any social cost; phobias may result from displacing an internal conflict to an external object symbolically related to the conflict COMMON PHOBIAS Achluophobia–darkness, agora–open spaces, ailuro–cats, algo–pain, andro–♂, auto–solitude, batho–depths, claustro–closed spaces, cyno–dogs, demo–crowds, erythro–blushing; gyno–♀, hypno–sleep, myso–dirt/germs, pan–everything, pedo–children, xeno–strangers. See Agoraphobia, Cancer phobia, Displacement, Fever phobia, Homophobia, Monophobia, PAD syndrome, School phobia, Simple phobia, Social phobia

phobic disorder PSYCHIATRY A condition that causes extreme and irrational anxiety in particular situations, objects or activities. See Phobia

phocomelia TERATOLOGY A congenital malformation characterized by attachment of a hand to the shoulder or foot to the pelvis, imparting a seal

Phoma spp

flipper appearance, classically associated with exposure of a developing fetus to thalidomide. See Thalidomide

Phoma spp MYCOLOGY A genus of aeroallergenic fungi of the family Deuteromycetes, which may induce a specific form of hypersensitivity pneumonitis known as shower-curtain disease, see there

phone counseling service PUBLIC HEALTH A 24-hr service, in which trained nurses provide Pts with health-related information. Cf Hotline

phone sex SEXOLOGY Telephone conversations in which references to sexual acts are made by one party to the other for erotic stimulation. Cf Telephone scatologia

phonocardiography Echophonocardiography CARDIOLOGY A non-invasive technique that amplifies faint, low frequency sounds of blood flowing through the heart and great vessels, displaying them graphically; PCG encompasses carotid, apex, and venous pulse tracings APPLICATIONS Diagnose heart valve defects, ventricular hypertrophy, left heart failure; PCG is usually performed in synchrony with EKG, and M-mode echocadiography or Doppler echocardiography, to match changes heard by PCG with the point at which they occur in the heart beat

phonologic disorder Speech disorder SPEECH PATHOLOGY A disorder characterized by failure to use speech sounds appropriate for the person's age and dialect; PDs are more common in boys–10% < age 8, 0.5% > age 17 ETIOLOGY Genetics, low socioeconomic status, large family

phonosurgery ENT A procedure on the vocal cords and adjacent tissue intended to improve the voice's timbre, tone and quality ASSESSMENT TECHNIQUES Acoustic and aerodynamic measurements, laryngeal stroboscopy, manual compression, and intraoperative monitoring of voice SURGICAL TECHNIQUES Injection, laryngeal microsurgery, laryngeal framework modification–eg, medialization thyroplasty, laser techniques. See Injection technique, Laryngeal microsurgery. Cf Uvulopalatopharyngoplasty

phonotrauma Any abuse or misuse of the vocal folds, most common in those with professional voices, which gives rises to various lesions–eg, polyps, nodules, degenerative polyps, cysts, varices, papillomas, and other benign conditions. See Singers' nodes

phosgene gas Carbonyl chloride, COCl₂ TOXICOLOGY A gas produced when an organic material burns with chlorine or chloride–eg, chlorinated hydrocarbons, plastics, and other materials; once in the alveoli, phosgene is hydrolyzed and forms HCl, compromising the lungs' diffusing capacity and evokes edema

phosphate loading SPORTS MEDICINE Ingestion of phosphate–phosphorus to ↑ O_2 delivery by ↑ production of 2,3-DPG which, in RBCs, right shifting the O_2 dissociation curve, facilitating unloading of O_2 in tissues, improving oxygenation and, in theory, athletic performance. Cf Bicarbonate loading, Carbohydrate loading, 2,3-Diphosphoglyceride, Performance-enhancing drug

phospholipid antibody See Anticardiolipin antibody

phosphorus binder THERAPEUTICS Any oral calcium–acetate or aluminum salt-based agent used in Pts with renal failure to bind both dietary and endogenous phosphorus; PBs ↑ GI excretion of phosphorus in Pts with chronic renal failure

photoactivator Any ingested substance that enhances reactivity to light–eg, PUVA, tetracycline, psoralens–eg, celery, parsley. See PUVA

photochemotherapy Extracorporeal photochemotherapy, photopheresis, phototherapy ONCOLOGY A therapy in which a photosensitizer–eg, a hematoporphyrin derivative, is administered, post light exposure; PCT is used to eradicate, or ↓ size of superficial transitional cell carcinomas of the urinary bladder, systemic sclerosis and other autoimmune diseases

photocoagulation OPHTHALMOLOGY The use of argon, or less commonly, xenon, lasers to focally burn the retina to ↓ neovascularization, microaneurysms, macular edema in Pts with various retinopathies OUTCOMES 50% ↓ in severe visual loss, compared to nontreated eyes. See Diabetic retinopathy. Cf Vitrectomy

photodamage Photoaging DERMATOLOGY The structural and functional deterioration of sun-exposed skin, resulting in wrinkling, roughness, altered texture, discoloration, acral lentigines, mottled hyperpigmentation, ↓ epidermal thickness, basophilic degeneration of dermis, ↓ collagen, ↓ dermal vessels, epithelial atypia, dysplasia MANAGEMENT Photoaging is largely reversed by tretinoin

PhotoDerm® PL DERMATOLOGY An electro-optic device for noninvasive treatment of benign pigmented lesions–eg, age/liver spots, freckles, birthmarks, melasma, hyperpigmentation, tattoos using intense pulsed light

photodocumentation The use of photography to record and 'document' various aspects of Pt management.

photodynamic therapy ONCOLOGY The injection of a site-specific photosensitizer in the body which accumulates in rapidly dividing cells, activated by certain wavelengths of light to treat and diagnose CA–eg, in situ transitional cell CA of bladder and other proliferative disorders. See Phototherapy OPHTHALMOLOGY A nonthermal process for ↑ local reactive O_2 species which mediate local cellular, vascular, and immunologic injury; RT is used for outpatient treatment of macular degeneration–characterized by choroidal neovascularization of retina. See Age-related macular degeneration

photoelectric effect RADIOLOGY That fractional ↓ in beam intensity of ionizing radiation due to photoelectric effect in a medium through which it passes. See Attenuation coefficient

Photofrin Porfimer sodium ONCOLOGY A light-activated drug used in photodynamic therapy for esophageal CA, Barrett's esophagus, microinvasive non-small cell lung CA–NSCLC, palliation of obstructing NSCLC, where surgery and RT are not indicated. See Photodynamic therapy

photography See Photodocumentation

photomutilation DERMATOLOGY Destruction of the skin and possibly acral parts by an abnormal response to UV radiation. See Congenital erythropoietic porphyria, Porphyria

Photon Radiosurgery System RADIATION ONCOLOGY A system for treating malignant intracranial and possibly other tumors, by delivering high-dose x-rays via a needle-like probe directly to a target

photophobia Painful oversensitivity to light

photoprotection PUBLIC HEALTH The use of a physical–eg, hat, protective clothing, zinc oxide, titanium oxide or chemical barrier–eg, PABA esters, digalloyl trioleate, cinnamic acid, methyl anthranilate, to ↓ UV light exposure. See Sunblock, Sunscreen

photoreceptor NEUROLOGY A sensory nerve ending that responds to stimulation by light

photorefractive keratectomy OPHTHALMOLOGY A refractive surgery that corrects myopia by changing corneal conformation RESULTS 78% vision improvement; 3-7% complications–eg, painful healing of cornea, glare. See Refractive surgery. Cf LASIK, Radial keratectomy

photosensitivity An abnormal sensitivity of skin to light

photosensitizer ONCOLOGY A substance that sensitizes an organism, cell, or tissue to light; an agent used in photodynamic therapy which, when absorbed by CA cells and exposed to light, is activated, killing cancer cells. See Photodynamic therapy

phototherapy NEONATOLOGY Bright light therapy, see there ONCOLOGY Photochemotherapy, photodynamic therapy A therapy in which various conditions–eg, colorectal CA, cutaneous T-cell lymphoma, head & neck CA, KS, psoriasis, skin CA are treated by light after previous administration of agents–eg, hemoporphyrin that become active after exposure to light

PHS 1. Personal health survey 2. Pooled human serum 3. Public Health Service, see there 4. Pulmonary hemorrhagic syndrome, see there

physiatrist A physician specializing in physical medicine and rehabilitation, who restore optimal function to people with injuries to the muscles, bones, tissues, and nervous system–eg, stroke victims. Cf Physical therapist

physical *adjective* 1. Referring to the body 2. Referring to the laws of physics and the universe *noun* Popular for physical examination, see there

physical abuse PEDIATRICS "*…Inflicting bodily injury through excessive force or forcing a child to engage in physically harmful activity, such as excessive exercise*, PA ↑ with poverty. See Child abuse, Spousal abuse

physical activity Athletic, recreational or occupational activities that require physical skills and utilize strength, power, endurance, speed, flexibility, range of motion or agility; PA is a behavioral parameter used to evaluate a Pt's cardiovascular 'reserve'. See MET

physical deconditioning MEDTALK The deterioration of heart and skeletal muscle, related to a sedentary lifestyle, debilitating disease, or prolonged bed rest CLINICAL ↓ lean body mass, maximum O_2 uptake, exercise-induced cardiac output, stroke volume, impaired vasodilation, exercise intolerance, especially in CHF

physical dependence SUBSTANCE ABUSE A physiologic state of neuroadaptation to a specific opioid, characterized by a withdrawal syndrome if the drug is stopped or ↓ abruptly, or if an antagonist administered; PD may be relieved partly or completely by readministration of the substance. See Addiction, Tolerance

physical examination MEDTALK A systematic investigation by a physician or other health care provider of the body to identify evidence of disease

physical fitness PUBLIC HEALTH A state of physical well-being and higher-than-average tolerance to ↑ cardiovascular activity; PF is defined by exercise test tolerance to a standard treadmill protocol, which requires a cardiovascular 'reserve'; the degree to which a person meets or exceeds expected working capacity according to body weight; there is '*…a graded, inverse association between physical fitness and mortality from cardiovascular causes …independent of age and conventional coronary risk factors.*' See Exercise, Obesity

physical inactivity A sedentary state. Cf Physical activity

physical rehabilitation See Physical therapy

physical status classification A classifying of physical condition by the Am Soc of Anesthesiologists that stratifies Pts undergoing surgery into categories of relative risk of suffering complications during surgery or in the immediate post-operative period

PHYSICAL STATUS CLASSIFICATION

CLASS 1 No organic, physiologic, biochemical, or psychiatric disturbance; the pathologic process for which the operation is to be performed is localized, and does not entail a systemic disturbance, eg inguinal hernia repair in a robust ♂

CLASS 2 Mild to moderate disturbance caused either by the condition being treated surgically, or by a physiopathologic derangement, eg mild cardiac disease, mild DM, chronic bronchitis, essential hypertension

CLASS 3 Severe systemic disease or derangement of any cause, which may defy classification, eg severe cardiac disease, angina or status post-MI, severe diabetes with vascular complications, moderate to severe pulmonary compromise

CLASS 4 Severe systemic disease that is already life-threatening, which may not be corrected by surgery, eg organic heart disease with signs of severe cardiac insufficiency, advanced pulmonary, hepatic, renal or endocrine insufficiency

CLASS 5 A moribund patient with little chance of survival who is submitted to an operation in desperation, eg ruptured aortic aneurysm, major cerebral trauma with rapidly ↑ intracranial pressure

EMERGENCY OPERATION E A designation for any of the above classes when the operation 'goes sour', eg an incarcerated hernia with strangulation would be a class 1E

physical therapy Physiatry, physical rehabilitation REHAB MEDICINE Physiatry, physiotherapy The use of mechanical–eg, massage, manipulation, exercise, movement, hydrotherapy, traction and electromagnetic–eg, heat/cold, light, ultrasound, to manage Pts recuperating from sports injuries, MVAs, or who have musculoskeletal disease, ↓ joint mobility. See Massage therapy, Rehabilitation medicine

physically helpless SEXUAL OFFENSES *adjective* Referring to a person who is unconscious, asleep or for any other reason physically or verbally unable to communicate unwillingness to perform an act. See Date rape pill

physician MEDTALK A person trained, qualified, and licensed to practice medicine, osteopathy or dentistry. See Admitting physician, Attending physician, Chest physician, Complementary physician, Corroborating physician, Covering physician, Doctor of osteopathy, Exempt physician, Family physician, Fellow, High-cost physician, Hospital-based physician, Intern, House physician, Low cost physician, Medical doctor, Panel physician, Personal physician, Primary care physician, Rent-a-doc, Resident, Sentinel physician, Supervising physician, Surgeon, Teaching physician, Triple threat physician, Virtual physician

physician assistant PA MEDTALK A person qualified to perform various physician-supervised medical tasks–eg, taking Pt Hx, performing physical exams and autopsies EDUCATION 2 yrs beyond college or university. See Pathologists' assistant, Physician extender

physician-assisted suicide/euthanasia Assisted suicide MEDICAL ETHICS A passive intervention by a physician to help a person end life, including prescribing medications–eg, narcotics or barbiturates or counseling an ill Pt to use an overdose to end life; general situations in which a physician is called to assist in a final 'role call'; general concerns triggering requests for PAS&E include loss of control, dignity, being a burden on others. See Active euthanasia, Euthanasia, Initiative 119, Kevorkian, Slippery Slope

physician autonomy The physicians' right to determine his life events, without uninvited intervention:

physician bashing Verbal or written persecution of physicians, who may be viewed by some as greedy, opportunistic, callous, unprincipled, overpaid entrepeneurs (ouch!) Cf Gay-bashing

physician expert witness A principal actor in the drama of malpractice litigation, defined by guidelines from the Council of Medical Specialty Societies. See Expert, Hired gun

physician extender A popular term for a trained health professional who provides quasi-autonomous health care under a particular physician's license EXAMPLES Physician assistant, nurse practitioner, etc. See Physician assistant, Nurse, Nurse practitioner

physician-hospital organization MANAGED CARE A corporation formed by a hospital and its medical staff to contract with MCOs. See Managed care

physician income Physician salary The amount of direct compensation a physician receives as a result of services rendered; US PI ranges from < $100 K in primary care to > $500 K for heart surgery, cosmetic surgery and LASIK, see there

physician invulnerability syndrome PSYCHOLOGY A self-maintained delusion by a physician that he/she is not susceptible to the same diseases as Pts

512 **physician oversupply** Physician glut MEDTALK An excess of physicians in a particular geographic region or specialty. Cf Manpower shortage, Physician shortage area

physician-patient relationship MEDICAL MALPRACTICE A formal or inferred relationship between a physician and a Pt, which is established once the physician assumes or undertakes the medical care or treatment of a Pt; the establishment of a PPR is 'automatic' in certain situations–eg, in the physician's private office, but in others–eg, physical exam as a health screening procedure, it cannot be assumed to exist. See Abandonment, Doctor-Pt interaction

physician profiling MANAGED CARE A method of cost containment that focuses on the patterns of health care provided by a single physician or group, instead of on specific clinical decisions; the resulting profile is then compared to other norms based on practice–ie, other physicians' profiles or to standards of practice–practice guidelines. See DRGs, Shrinkage factor

physician referral A physician's recommendation to a Pt to consult another physician for a 2nd opinion. Cf Self-referral

physician shopping VOX POPULI The visiting by a Pt of multiple physicians, generally due to dissatisfaction with one or more aspects of care by previous physician(s)

physician shortage area A rural and/or low income region–eg, inner city with few practicing physicians–eg, < 1 physician/5000. See Manpower shortage. Cf Brain drain, Physician oversupply

physician-sponsored organization MANAGED CARE A managed care organization that is clinical–ie, physician and hospital oriented, thus differing from traditional managed care–HMOs, POSs, PPOs; PSOs are utilization driven and based on discounting strategies and treatment denials. Cf HMO, Point-of-service, PPO

physician stress MEDICAL PRACTICE The mental stress of medical practice–eg, responsibility for life and health, long working hrs, examination of unclothed Pts–with the potential for boundary violations, threat of malpractice litigation, ready access to psychoactive and addicting drugs, financial concerns, and need to keep current in one's scope of practice

physician test MANAGED CARE A noninvasive diagnostic procedure performed by a physician–eg, cardiac stress test, EKG monitoring, etc–eg, audiometry. Cf Evaluation and management service, Procedure

Physician's Desk Reference PDR, see there

Physicians Online INFORMATICS The largest medical online service; its use is limited to physicians, providing authenticated doctors with secure discussion groups, email, medical literature searches–eg, Medline, Pubmed, Clinical Pharmacology, Drug Interactions and other resources to locate relevant information quickly, online CME, access to honoraria, and a medical marketplace. See Internet, Medline, National Library of Medicine

Physio ring HEART SURGERY An ellipsoid device that provides annuloplasty support in Pts requiring repair for mitral valve regurgitation

physiologic *adjective* 1. Referring to administration of hormones or other substances normally present in the circulation, in doses similar to levels normally produced by the body. Cf Homeopathic, Pharmacologic, Supraphysiologic 2. Referring to physiology

Physiologic Endometrial Ablation/Resection Loop See PEARL

physiologic jaundice of newborn Neonatal jaundice NEONATOLOGY Yellowing of skin during the newborn period, caused by ↑ BR level in the blood, due to immaturity of liver function plus destruction of RBCs; seen between days 2 and 5, clears by 2 wks RISK FACTORS, JAUNDICE Prematurity DIFFDX Nonphysiologic, prolonged, or pathologic jaundice in newborn, biliary atresia, ABO and/or Rh incompatibility, galactosemia, cephalhematoma, polycythemia, G6PD deficiency, neonatal sepsis, congenital infection–CMV, toxoplasmosis, syphilis, herpes, rubella, late pregnancy use of sulfa drugs by mother, Crigler-Najjar syndrome, spherocytosis, cystic fibrosis, breast-milk jaundice, pyruvate kinase deficiency, thalassemia, Gilbert's syndrome, congenital hypothyroidism, Lucey-Driscol syndrome, Gaucher's disease, Niemann-Pick disease

physiologic neutrophilia Reactive leukocytosis A transient ↑ in neutrophils linked to vigorous exercise, pregnancy, epinephrine release; occurs in the newborn

physiologic pacing CARDIAC PACING Artificial pacing that maintains the heart's normal contraction sequence–ie, A-V synchrony, with resulting hemodynamic benefits

physiologic saline solution FluidCare solution, Normal saline A salt solution in water with electrolytic properties similar to those of a body fluid

physiology See Applied physiology, Cardiac electrophysiology, Clinical neurophysiology

physostigmine A reversible antiacetylcholinesterase, used in Pts with Sx of an anticholinergic crisis NEUROLOGY Memory drug Physostigmine may improve working memory by ↑ efficiency and ↓ effort. See Working memory

pianist's cramp A dystonia affecting muscles of the hand and sometimes forearm, when playing a piano–or another keyboard instrument–eg, harpsichord OTHER CRAMPS Writer's cramp, typist's cramp, musician's cramp, golfer's cramp

piano playing NEUROLOGY A fanciful descriptor for finger movements linked to the loss of position sensation, in which the Pt seeks to discover finger position in space by periodic movement; PP occurs in Dejerine-Sottas syndrome; PP also refers to intermittent flexion and extension of the hands in tardive dyskinesia, a complication of long-term use of antipsychotics–eg, phenothiazines, butyrophenones

pica PSYCHIATRY An eating disorder consisting of the craving and eating of unusual foods or other substances with no known nutritional value, 'classically' associated with iron-deficiency anemia OTHER INGESTANTS Paint, laundry starch, ice–pagophagia, newspapers; pica also occurs in zinc, copper and vitamin deficiencies ASSOCIATIONS Various medical conditions, pregnancy, mental disorders. See Geophagy

PICC Peripherally-inserted central catheter CRITICAL CARE An IV catheter inserted in the superior vena cava for long-term infusion of bolus or continuous delivery of therapeutics or TPN–drugs, fluids, nutrients, chemotherapy. See Catheter

Pick's disease Aphasia-agnosia-apraxia syndrome, Arnold Pick disease, Circumscribed brain atrophy, Lobar sclerosis NEUROLOGY A form of dementia characterized by a slowly progressive deterioration of social skills and changes in personality leading to impairment of intellect, memory, and language. See Presenile dementia. Cf Alzheimer's disease

picket fence fever A descriptor for a saw-tooth pattern of high temperature 'spikes', a finding typical of pyogenic hepatic abscesses, often accompanied by chills, sweating, N&V, anorexia, pain

pickoma DERMATOLOGY A popular term for a nodular reactive 'tumor' that develops in a Pt who has repeatedly squeezed and picked at an otherwise innocuous bump

Pickwick syndrome Cardiopulmonary obesity syndrome, sleep apnea/obesity hypoventilation syndrome A complication of extreme obesity, with marked cardiovascular compromise, ↓ tidal and expiratory reserve volumes, alveolar hypoventilation, hypoxia, cyanosis and hypercapnia, if severe and prolonged, dyspnea, polycythemia, cardiac hypertrophy, pulmonary HTN, pulmonary edema, CHF, extreme somnolence; O_2 therapy may be fatal, as it removes the chemoreceptor drive needed for respiratory movement. See Morbid obesity, Sleep apnea syndrome

Picornaviridae VIROLOGY A large family of RNA viral pathogens EXAMPLES Aphthovirus–foot & mouth disease; rhinovirus–common cold agents; enterovirus–eg, polioviruses, coxsackieviruses

picric acid 2,4,6 Trinitrophenol OCCUPATIONAL MEDICINE A strong–pK 1.0 acid once used as a dye, an antiseptic, and fixative; when dry, it is explosive, and used in manufacturing explosives and rocket fuels; occupational exposure results in a yellowish skin

picture frame appearance A descriptor for vertebral involvement by Paget's disease of bone–an isolated vertebral body is enlarged, centrally osteoporotic, and surrounded by a rim of osteosclerotic cortex, which involves the anterior and posterior margins, and vertebral end-plates; despite ↑ radiologic density, bone is fracture-prone

picture puzzle appearance Jigsaw puzzle appearance–cells, contour, model, tumor, see there

PID Pelvic inflammatory disease, see there

PID shuffle MEDTALK Regional slang for the shuffling gait and flexed posture of a ♀ with significant pain due to PID. See Chandelier sign, Pelvic inflammatory disease

PIE Pulmonary infiltration with eosinophilia, see there

PIE syndrome Pulmonary infiltrates with eosinophilia A condition characterized by intense, nonspecific Sx accompanied by chronic relapsing fever, cough and SOB, associated with chronic eosinophilic pneumonia

PIE SYNDROME

SIMPLE PIE Löffler syndrome Transient pulmonary infiltrates fever, dyspnea, eosinophilia of peripheral blood

TROPICAL EOSINOPHILIA Associated with microfilarial parasites–eg, *Ascaris lumbricoides* and *Toxocara canis*

SECONDARY CHRONIC PULMONARY EOSINOPHILIA Associated with allergic bronchopneumonia related to aspergillosis, bronchocentric granulomatosis, allergic angiitis and granulomatosis–Churg-Strauss syndrome, drugs–nitrofurantoin, sulfonamide, and infections–parasitic, fungal, and bacterial

Piedmont fracture ORTHOPEDICS An isolated fracture at the distal ⅓ of the radius

Pierre Robin syndrome Pierre Robin sequence PEDIATRICS A group of oral defects–very small lower jaw with a posterior-placed tongue that causes choking and dyspnea, especially while asleep; other findings include a high arched palate or cleft palate. See Cleft palate

piezoelectric crystal CARDIAC PACING A crystal that produces electrical signals when subjected to mechanical deformation; in pacing, PC sensors are used to detect motion, changes in pressure, etc, in certain rate-responsive pacemakers

PIF Prolactin-inhibiting factor, see there

pig VOX POPULI A food animal and occasional vector for human pathogens: BACTERIA *Bacillus anthracis*–anthrax, *Brucella suis*, *Clostridium botulinum*–botulism, *C perfringens*–pigbel, *Flavobacterium* group IIb-like bacteria, Leptospirosis, *Pasteurella aerogenes*, *Pasteurella multocida*, *Salmonella cholerae-suis*–salmonellosis, *Streptococcus dysgalactiae* (group L), *Streptococcus milleri*, *Streptococcus suis* type 2 (group R), *Yersinia enterocolitica*, *Y pseudotuberculosis* PARASITES *Ascaris suum*, cryptosporidiosis, *Entamoeba polecki*, *Erysipelothrix rhusiopathiae*, *Fasciolopsis buski*, sarcocystosis, scabies, *Taenia solium*, *Trichinella spiralis* VIRUSES Influenza, rabies, swine influenzae, swine vesicular disease. Cf Guinea pig

pigeon breast deformity Chicken breast deformity Anterior displacement of the sternum, adjacent cartilage, and anterior rib cage due to abnormal pulling by respiratory muscles on soft bone, enlargement of costochondral junctions and flattening of thorax, a finding typical of advanced vitamin D-induced rickets; PBD is an asymptomatic, asymmetric, deep depression of the costal cartilage on each side of the sternum; it is most apparent below the nipple level, involves the 4th to 7-8th costal cartilages, and is the most common protrusion deformity of the sternum–pectus carinatum. Cf Pouterpigeon

pigeon toe In-toeing, talipes varus ORTHOPEDICS A foot deformity of childhood, which is characterized by medial rotation of the forefoot–metatarsus varus and medial tibial or femoral torsion–the deformities require surgery

piggyback device CRITICAL CARE A device used to optimize the IV delivery of fluids and drugs being infused at different rates; in PBDs, the reservoir and the valve controlling the rate of delivery are separate, while the delivery port–eg, an IV access line is shared

pigment DERMATOLOGY A substance that imparts color to tissue–eg, skin, eyes, hair. See Accessory pigment, Bile pigment, Malarial pigment, Tattoo, Tyndall

pigment dispersion syndrome OPHTHALMOLOGY An AD condition affecting ♂ age 20 to 40; 50% develop high-pressure glaucoma, leading to blindness CLINICAL Subtle ocular pain, halos around lights, blurred vision MANAGEMENT Miotics

pigmented neuroectodermal tumor of infancy A tumor of neural crest origin affecting infants < 6 months of age, seen in the anterior maxilla, oral cavity and skull, less commonly in the mediastinum, thigh, forearm, epididymis CLINICAL Locally aggressive, 15% recur; few metastasize LAB ↑ VMA in urine. See Pseudonym syndrome

pigmented villonodular synovitis ORTHOPEDICS A lesion of knee–less commonly, ankle, hip, shoulder of young adults MANAGEMENT Excision, re-excision DIFFDX Fibrosarcoma, synovial sarcoma, incontinentia pigmenti

pigskin appearance Moroccan leather appearance DERMATOLOGY A descriptor for taut, shiny, non-pitting and painful skin of Pts with lipemia, usually on legs. See Painful fat syndrome

pigtail catheter A drainage catheter with side holes, used for draining clear non-viscid or coagulable collections of bile, urine or pancreatic fluids; the 'pigtail' is inefficient in draining abscesses from solid organs, but may be used for perihepatic abscesses

pilar sheath acanthoma An uncommon skin appendage tumor, of probable hair follicle origin; average age 66 yrs; ♂:♀ ratio 2:1 TREATMENT Excision

piles VOX POPULI Hemorrhoids, see there. See Sentinel pile

pili torti A hair shaft defect, seen in ash-blondes, where hair is grooved and flattened at varying intervals and twisted on its axis; the defect is usually recognized by age 2-3, the hair having a 'spangled' appearance; it is either an AR condition *a sui generis* or part of the X-R Menke's kinky hair syndrome, associated with mental retardation. Cf Ringed hair, Wooly hair

the Pill A popular term for any oral contraceptive

pill MEDTALK A medication formulated in tablet form, intended to be taken orally. See Little yellow pill, Minipill, Pressure pill, Sugar pill, Water pill

pill esophagitis Esophageal mucosal injury caused by oral medication–eg, aspirin, NSAIDs, anticholinergics, iron-preparations, KCl, tetracycline and quinidine CLINICAL Prolonged 'cancer-like' Sx–eg, retrosternal pain, progressive stricture, hemorrhage, perforation

514 **pill-rolling** NEUROLOGY *adjective* Referring to a tremor characterized by a circular movements of the tips of the index finger and thumb, typical of Parkinson's disease

pill whore SUBSTANCE ABUSE A popular term for a ♀ who exchanges goods–eg, oxycodone pills, for services–eg, sexual favors. See Raspberry, Oxycodone

Pillo™ Pro WOUND CARE A highly absorbent, polymer dressing used to control excessive wound drainage after liposuction, and other cosmetic and dermatologic procedures. See Wound care

pillow MEDTALK A functional 'unit' used to assess the severity of orthopnea in Pts with CHF, which refers to the number of pillows a Pt needs to sleep comfortably. See Congestive heart failure

pilocarpine THERAPEUTICS An alkaloid with mild β-adrenergic activity which, topically, ↓ intraocular pressure in glaucoma; it stimulates glands, resulting in diaphoresis, salivation, lacrimation it may be used to treat post-radiation or opioid-induced xerostomia, gastric and pancreatic secretion

pilonidal cyst DERMATOLOGY A type of abscess found in the cleft between the buttocks in adolescents after prolonged sitting

pilot study Pilot project MEDTALK A preliminary study designed to evaluate the effect of a particular change in policy. See CHIP

PIMI CARDIOLOGY A multicenter study–Psychophysiological Interventions in Myocardial Ischemia, which concluded that mental stress ↑ mortality in Pt with CAD. See Type A

pimp *verb* MEDTALK To interrogate another mercilessly regarding knowledge of a subject, deriving personal benefit from another's study without doing so oneself

pimping ACADEMIA See Pimp. Cf Pumping

pimple Zit DERMATOLOGY A superficial abscess, generally located on a 'sebaceous' region of the body, often associated with acne. See Acne, Blackhead, Cystic acne

PIN 1. Penile intraepithelial neoplasia 2. Personal identification number A number chosen by a person to verify ID; PINs are used for personal banking, voicemail retrieval, etc. 3. Prostatic intraepithelial neoplasia, see there

pin ORTHOPEDICS An internal fixation device used to join fractured bone. See Knowles pin, Percutaneous pin, Steinmann pin

pincer nail An idiopathic excess tranverse curvature of the nail bed associated with intense pain and loss of soft tissue at the fingertips. See Nail

pincers mechanism SPORTS MEDICINE The mechanical substrate on which cervical cord neurapraxia occurs where, in the presence of cervical canal stenosis, there is either cervical spine hyperextension, causing the posterior lower face of a superior vertebral body and the anterior upper face of the lamina of a lower vertebra to come together *or*, in flexion, the lamina of the superior vertebra and the posterior superior aspect of the subjacent vertebral body come together; approximation causes a sudden ↓ in the canal's AP diameter, compressing the spinal cord and causing temporary neurologic dysfunction and cord deformation. See Cervical cord neurapraxia

pinched nerve syndrome Carpal tunnel syndrome, see there

pindolol CARDIOLOGY A now-generic β-blocker used to treat HTN. ADVERSE EFFECTS Edema, insominia. See Beta-blocker

'ping pong' infection A descriptor for the epidemiology of *Trichomonas vaginalis* infection, an STD, where person A is treated during sexual partner B's incubation period for same; B then becomes symptomatic after A has responded to antibiotics, resulting in an infection that 'bounces' back and forth from A/B to untreated B/A, in ping-pong ball-like manner; β-hemolytic streptococci may also have a 'ping-pong ball' pattern of infection

pinguecula GERIATRICS Yellow nodules at the nasal side of cornea and conjunctivae, of little clinical significance unless they grow over the cornea and affect the vision

the Pink Sheet A specialized weekly report that provides business and US federal regulatory information on prescription pharmaceuticals and therapeutic biotechnology

pinkeye Conjunctivitis OPHTHALMOLOGY Acute contagious conjunctivitis by *Haemophilus aegyptius* or *H ducreyi*; 'pinkeye' has been obfuscated by the lay public, which may use the term for any condition in which the eyes are pink–eg, bilateral bacterial or viral conjunctivitis, 'misuse' of eyes–ie, prolonged exposure to smoky rooms, alcoholism, dissipated lifestyle, severe iritis, closed angle glaucoma, etc. See Red eye

pink puffer A descriptor for a Pt with COPD and severe emphysema, who have a pink complexion and dyspnea; PPs have ↑ residual lung capacity and volume, ↓ elastic recoil, ↓ expiratory flow rate and diffusing capacity and a ventilatory/perfusion–V/Q mismatch 2° to emphysema-related destruction of blood vessels CLINICAL SOB, hyperventilation ABGs Usually near normal due to compensatory hyperventilation; arterial pO_2 is in the mid-70s, pCO_2 is low to normal; PPs have ↑ tidal volume and retraction of accessory respiratory muscles. See Chronic obstructive pulmonary disease. Cf Blue bloater

pink up MEDTALK A popular verb referring to a normalization of (pink) skin color from a bluish tinge typical of cyanosis

Pinkus tumor Fibroepithelioma A polypoid variant of BCC, commonly located on the back. See Basal cell carcinoma

pinprick NEUROLOGY A sharply focused stimulation of the skin, often by a needle, used to evaluate the sense of touch

Pinsaun Amitriptyline, see there

pinup procedure GYNECOLOGY Surgery designed to relieve urinary stress incontinence by placing the urethrovesical angle in the abdomen, changing the focus of pressure applied through the abdomen during the Valsalva maneuver. See Urinary incontinence

pinworm *Enterobius vermicularis*, seatworm; threadworm MICROBIOLOGY A small–visible to the naked eye—whitish worm found in North America; adult pinworms live in the colon, laying eggs outside the anus during the night and spread directly from contaminated clothing, articles, hands, or indirectly by aerosol, from which the eggs are inhaled or swallowed; eggs hatch in the small intestine and travel to the colon where they mature

pinworm infection Enterobiasis, oxyuriasis; pinworm infection; INFECTIOUS DISEASE Infection with *E vermicularis*, a highly contagious intestinal parasite of children, especially in urban areas, crowded settings–eg, day-care centers, schools CLINICAL Perianal pricking and itching, restless sleep, abdominal discomfort DIAGNOSIS Freshly passed stool uncontaminated with urine, water, dirt, disinfectants; Scotch tape test MANAGEMENT Mebendazole, pyrantel pamoate, albendazole. See Pinworm

pinworm tape test Scotch™ tape test, see there

pioglitazone Actos® ENDOCRINOLOGY An oral antidiabetic that ↓ insulin resistance, used to manage type 2 DM; it inhibits hepatic gluconeogenesis, and improves glycemic control while ↓ circulating insulin ADVERSE EFFECTS Edema, ↑ URIs, headaches, myalgias, sinusitis . See Diabetes mellitus

PIOPED CARDIOLOGY A clinical trial–Prospective Investigation of Pulmonary Embolism Diagnosis which evaluated the diagnostic value of clinical suspicion and VQ scan in the diagnosing PTE

PIP 1. Phosphatidylinositol phosphate 2. Postinflammatory polyposis, see Inflammatory polyps 3. Proximal interphalangeal 4. Psychotic InPt Profile

PIP Proximal interphalangeal joint

Pipelle™ GYNECOLOGY A disposible device consisting of a plastic core and a drinking-straw-like sheath, used to obtain endometrial biopsies through gentle suction; the resulting tissue has a linguini-oid appearance

piperacillin THERAPEUTICS A semisynthetic broad-spectrum penicillin for parenteral use ADVERSE EFFECTS Thrombophlebitis, pain, erythema, induration at the injection site, ecchymosis, DVT, hematomas; GI–diarrhea, loose stools, N&V, ↑ liver enzymes–LD, SGOT, SGPT, ↑ BR, cholestatic hepatitis, rarely pseudomembranous colitis

piperacillin-tazobactam INFECTIOUS DISEASE Two penicillin derivatives used for infections in CA Pts

pipercuronium Arduan® ANESTHESIOLOGY A long-acting nondepolarizing muscle relaxing anesthetic derived from pancuronium. See Muscle relaxant. Nondepolarizing agent. Cf Depolarizing agent

pipestem calcification A fanciful term for the tubular mineralization of arteries typical of extensive ASHD, which may cause pseudohypertention, as affected vessels are relatively nondistensible

pipestem fibrosis A descriptor for the histologic appearance–portal vein fibrosis with hyaline thickening and tortuosity of vessels of portal spaces in liver involvement by *Schistosoma mansoni* and *S mekongi*; schistosomal hepatopathy mimics cirrhosis with hepatosplenomegaly, portal HTN, 2° HTN, and variceal bleeding; native architecture is preserved

pipestem ureter A thickened and fixed, aperistaltic ureter that course in a stiff pencil- or pipestem-like fashion from the kidney to bladder in advanced TB of the urinary tract

PIPJ Proximal interphalangeal joint

PIPP Premature Infant Pain Profile, see there

piroxicam Feldene® THERAPEUTICS An NSAID tightly bound to plasma proteins $T_{1/2}$ 30 hrs; hepatic biotransformation; metabolites excreted in kidneys. See NSAIDs

PISA Primary idiopathic sideroblastic anemia, see there

pistol PUBLIC HEALTH A handheld firearm that holds up to 14 rounds of ammunition. See Ballistics. Cf Assault weapon, Shotgun

pistol shot pulse A popular term for a loud, cracking sound heard by the stethoscope over an artery in which there is distension followed by an abrupt collapse, as classically occurs in large arteries in aortic regurgitation. See Water hammer pulse

pit MEDTALK A dell that remains transiently in edematous skin and subcutaneous tissue after firm fingertip pressure OBSTETRICS Pitocin, see there

pit recovery time MEDTALK The time needed for subcutaneous skin to 'fill' after firm fingertip pressure, used to differentiate peripheral edema in CHF–prolonged, from tissue swelling due to hypoalbuminemia–relatively short. See Congestive heart failure, Pitting edema

pitch workers' cancer OCCUPATIONAL MEDICINE SCC of the head & neck, scrotum, etc, linked to prolonged exposure to 3,4-benzpyrene-rich pitch, shale oil, and tars; PWC differs from chimney sweeps' cancer only in occupation involved

Pitocin OBSTETRICS A proprietary oxytocin, see there

pitting edema PHYSICAL EXAM A term used to describe the indentation caused when fingertip pressure is applied to the skin, forcing fluids into the underlying tissue; PE occurs when there is an ↑ amount of low protein fluid

in the interstitial space, associated with disorders caused by high capillary filtration–DVT, chronic venous insufficiency, or venous obstruction, or hypoalbuminemia; pitting is a subjective assessment using the grading scale of 1+ for mild and up to 4+ for deep pitting. Cf Nonpitting edema

Pittsburgh brainstem score CRITICAL CARE A scale used to determine the clinical status of a victim of brain trauma which measures carinal, corneal, 'doll's eye', eyelash and ice water caloric reflexes. Cf Glasgow scale, Mini-mental test

Pittsburgh pneumonia agent *Legionella micdadei* An often intracellular bacteria, and frequent contaminant of water supplies, which may cause bronchopneumonia with consolidation and a fibrinopurulent exudate, fever, pleuritic pain and cough, most often affects immunosuppressed children. See Legionnaire's disease

pituitary apoplexy Acute life-threatening hemorrhagic infarction of the anterior pituitary or adenohypophysis, often associated with an infarcted pituitary adenoma or other tumor; PA may occur spontaneously, or due to an 'obstetric' hemorrhage–eg, Sheehan syndrome, regional RT, ↑ intracranial pressure, or systemic anticoagulation CLINICAL The PA 'syndrome' may be transient or permanent; it is characterized by sudden headache, loss of vision, ophthalmoplegia, and if severe, shock LAB ↓ Growth hormone, ↓ gonadotropins, ↓ ACTH, ↓ TSH–with hypothyroidism, abnormal prolactin secretion, rarely, diabetes insipidus TREATMENT Hormone replacement, bromocriptine

pituitary insufficiency ENDOCRINOLOGY A condition in which the pituitary fails to produce enough of one or more hormones. See Pituitary apoplexy, Pituitary infarction

pityriasis rosea DERMATOLOGY A skin condition, most common in young adults, with a 3:2 ♀:♂ ratio, seen in the fall and spring, lasting 4 to 8 wks, manifest as a single larger patch called a herald patch followed several days later by more rash. See GLC7

PIVOT UROLOGY A clinical study–Prostate cancer Intervention versus Observation Trial

pixel resolution TELEMEDICINE The sharpness of a computerized image, based on pixel concentration, which determines display resolution

PIXI® densitometer OSTEOPOROSIS A device for rapid estimate of bone mineral density in g/cm^2 of the forearm and heel. See Bone densitometry

PiZZ The most common variant allele in α_1-antitrypsin deficiency; the common *normal* allele is PiMM. See α_1-antitrypsin deficiency

PKU Phenylketonuria, see there

PLAC I/II CARDIOLOGY A clinical trial–Pravastatin Limitation of Atherosclerosis in Coronary Arteries studying the effects of pravastatin on cardiac M&M in Pts with CAD and hypercholesterolemia. See Lipid-lowering therapy, Pravastatin. Cf MAAS, 4S, WOSCOPS

place of service Point of service, see there

placebo MEDTALK An inactive material, in the form of a capsule, pill, or tablet, which is visually identical, and administered by the same route as a drug being tested; a chemically inert substance given in the guise of medicine for its psychologically suggestive effect; used in controlled clinical trials to determine whether improvement and side effects may reflect imagination or anticipation rather than the drug's power. See Dose control trial, Equivalence trial, Putative placebo trial. Cf Nocebo

placebo effect Placebo response PSYCHIATRY The effect that an inactive or inert substance has on a clinical condition. See Biofeedback, 'Halo' effect, Hawthorne effect, Placebo, Placebo response; Cf 'Nocebo'

placenta abruptio Abruptio placentae, see there

placenta membranacea

placenta membranacea Membranaceous placenta OBSTETRICS A very thin placenta–most of the gestational sac is covered with functional chorionic villi; a distinct membranous sac is not present OUTCOME PM is always associated with placenta previa and often with peripartum maternal bleeding

placenta previa OBSTETRICS A condition in which the placenta implants in the lower uterus and obstructs the birth canal ETIOLOGY Scarred endometrium, a large placenta, abnormal placentation INCIDENCE ±1 in 200 births; 1 in 20 with multiparas, doubled in multiparas RISK FACTORS Multiparity, multiple pregnancy, prior C-section if scar is low and close to the cervix region

placental clock OBSTETRICS A hypothetical 'timer' that may be intrinsic to the placenta, and control length of gestation

placental dysfunction Placental insufficiency OBSTETRICS An abnormal slowing of fetal growth in pregnancy ETIOLOGY Defects of placental membranes–disruption or leaking, mixing of fetal and maternal blood–eg, with Rh incompatibility, umbilical cord defects, abnormal implantation site, multiparity

placental failure OBSTETRICS A ↓ in placental progesterone production, causing pregnancy loss. Cf Placental insufficiency

placental site trophoblastic tumor GYNECOLOGIC PATHOLOGY A rare uterine neoplasm of gestational tissue of reproductive age ♀, presenting as a 'missed' abortion, which ranges from microscopic to massive and from 'timid' to highly aggressive that actively secretes human placental lactogen PROGNOSIS 10% mortality

placental transfusion See Fetomaternal hemorrhage

placentation OBSTETRICS The formation of the placenta in the uterus; the process by which a placenta grows and develops in the uterus. See Extrachorial placentation

placentophagy ETHNOMEDICINE The practice of eating the placenta post delivery. Cf Cannibalism, Pica

Placidox Diazepam, see there

Placidyl® Ethchlorvynol PHARMACOLOGY A sedative hypnotic, CNS depressant and muscle relaxant, used for insomnia

plague INFECTIOUS DISEASE An epidemic infection by *Yersinia pestis* spread to humans by fleas bitten by infected rodents–bubonic or septicemic plague or by inhalation of highly virulent encapsulated *Y pestis* when in close quarters with infected Pts–1° pneumonic plague CLINICAL FORMS Bubonic–90%, septicemic, pneumonic and, as a complication of any of these, meningitis CLINICAL Fever, chills, prostration, headache, N&V, diarrhea TREATMENT Streptomycin, tetracycline, chloramphenicol. See Bubonic plague, Fifth plague of Egypt

plaintiff FORENSIC MEDICINE A person–eg, a Pt, who initiates a civil action to obtain restitution for injury caused by a negligent/allegedly negligent act

plan VOX POPULI An organized manner of doing a thing. See Bare-bones health plan, Basic plan, Budget plan, Cafeteria benefits plan, Canadian plan, Certified health plan, Clinton Plan, Competitive medical plan, Discounted fee-for-service plan, Faculty practice plan, Financial management plan, Health plan, Indemnity plan, Individual family services plan, Keogh plan, Minneapolis plan, Nonpoint source management plan, Oregon health plan, Other plan, Pepper Commission plan, Phoenix plan, Point-of-service plan, Section 125 plan, Self-funded plan, Service plan, Statewide plan, Statutory Health Insurance plan, Strategic plan

plan of care MEDICARE A formal treatment plan for a home health Medicare Pt delineated by the treating physician based on knowledge of the Pt's medical condition. See HHRG

plan providers MANAGED CARE The physicians, hospitals, skilled nursing facilities, home health agencies, pharmacies, ambulance companies, labs, imaging facilities, durable medical equipment suppliers, and other licensed health care entities or professionals which or who provide covered health care services to plan members through an agreement with the plan

plantar fasciitis Heel spur syndrome ORTHOPEDICS The most common cause of inferior heel pain, usually of the medial aspect of the plantar fascia as it attaches to the inferior medical calcaneal tuberosity; the pain is usually worse in the morning and persists as a dull, toothache-like pain, exacerbated by ↑ activity, lasting up to 6-12 months; the medial insertion of the plantar fascia on the calcaneus may be tender; extension of the great toe can cause Sx; cavus feet or pronation on gait may be evident on exam MANAGEMENT Cross-friction ice massage, arch exercises, stretches, heel cups, NSAIDs, arch pads/orthotics, night splints, physical therapy

plantar ischemia test A test used to evaluate the adequacy of peripheral circulation of the leg. See Trendelenberg test

plantar wart Verruca pedis PODIATRY A virally-induced bump on the bottom of the foot, linked to the immune system, which either responds to minimal therapy or not. See Abracadabra therapy, Podophyllin

Planum® Temazepam, see there

plaque CARDIOLOGY An early lesion of ASHD found in persons of any age in larger vessels DENTISTRY A soft sticky substance on teeth composed of bacteria and saliva; an indurated gob of polysaccharides and bacteria–eg, *Lactobacillus acidophilus, Streptococcus mutans.* See Periodontitis DERMATOLOGY A flat, solid, elevated ≥ 1.0 cm in diameter skin nodule formed either by extension or coalescence of papules of lichen amyloidosis, lichen simplex chronicus, lichen planus, psoriasis; a 'plaque' stage occurs in certain skin tumors–eg, the 2nd stages of KS, mycosis fungoides NEUROLOGY 'Shadow plaques' Multiple, irregularly shaped, and sharply demarcated lesions–focal demyelination in the gray and white matter in the brain of Pts with MS. Asymmetric unit membrane plaque, Atherosclerotic plaque, Complicated plaque, Fibrous plaque, Multiple sclerosis, Parietal pleural plaque, Senile plaque, Shadow plaque, Soldier's plaque

Plaquenil® Hydroxychloroquine sulfate, see there

-plasia *suffix*, Latin, growth, formation

-plasm *suffix*, Latin, growth, formation

plasma LAB MEDICINE A clear yellowish extracellular fluid that comprises 50-55% of the blood volume; it is 92% liquid, 7% protein, < 1% inorganic salts, gases, hormones, sugars, lipids; fibrinogen- and coagulation factor-depleted plasma is 'serum' TRANSFUSION MEDICINE See Fresh frozen plasma

plasma cell dyscrasia A lymphoproliferative disorder characterized by monoclonal proliferation of plasma cells, with clinical behavior ranging from innocuous extramedullary plasmacytomas to premalignant solitary plasmacytoma of bone to myeloma. Cf Monoclonal gammopathy of undetermined significance, Myeloma

plasma cell leukemia A neoplastic ↑ in circulating PCs PATHOLOGIC CRITERIA Nuclear vacuolization, monotonous sheets of cells, > 20% WBCs must be PCs or the absolute number of PCs in the peripheral blood must be > 2 × 10⁹/L–US: > 2000/mm³ CLINICAL Most PCLs are advanced when diagnosed, with massive tissue infiltration and BM replacement at the time of diagnosis, and a poor prognosis DIFFDX Reactive plasmacytosis, which may be associated with agranulocytosis, burns, chronic granulomatous disease, collagen vascular disease, exanthematous lesions, hypersensitivity reactions, nonmalignant liver disease, nonmyelomatous malignancy, sarcoidosis, syphilis, subacute bacterial infection, streptococcal sepsis, typhoid, viral infection–especially infectious mononucleosis

plasma protein fraction THERAPEUTICS A blood product containing ≥ 83% albumin, which expands fluid volume without risk of hepatitis or HIV; PPF–or albumin replaces large losses of colloid–eg, hypovolemic shock, burns, retroperitoneal surgery. See Albumin

plasma renin activity A test for renovascular HTN in which peripheral venous blood obtained by catheterization from the renal vein is measured for the ability of the renin to convert angiotensinogen to angiotensin I ↑ IN Addison's disease, renal HTN, 2° hypoaldosoteronism ↓ IN Hyporeninemic hypoaldosteronism. See Hypertension, Renin

plasmacytoma HEMATOLOGY A tumor composed of malignant plasma cells. See Plasma cell

plasmacytosis HEMATOLOGY An abnormal ↑ in circulating or tissue plasma cells DiffDx NEOPLASTIC–Myeloma, macroglobulinemia, gamma heavy chain disease, CLL, carcinoma–breast, lung, etc INFECTIOUS–BACTERIAL RMSF, streptococcus, syphilis, TB PARASITIC Malaria, trichinosis, visceral larva migrans VIRAL CMV, EBV–infectious mononucleosis, rubella, rubeola, varicella encephalitis, hepatitis ALLERGIC-DRUGS Penicillin, sulfa compounds, antitoxins–eg, diphtheria ETC–Chemotherapy, immune responses, transfusions, trauma

Plasmanate® A proprietary plasma protein fraction that is 90% albumin by volume and used as a volume expander. See Albumin, Colloid solutions. Cf Crystalloid solutions

plasmapheresis Plasma exchange THERAPEUTICS External shunting of plasma from the peripheral circulation, to remove an undesired substance–eg, toxins, medications–overdose, antibodies, or other 'noxins' or, less commonly, to obtain plasma for donation; pheresed fluid is resuspended in an appropriate fluid–eg, albumin or albumin in saline, then readministered to the Pt; the removal of one plasma volume–± 2500 ml effects a 65% ↓ in toxin or autoantibody in the circulation–2 volume exchanges ↓ toxin by another 20% INDICATIONS Hyperviscosity syndrome, myasthenia gravis, Eaton-Lambert syndrome, Goodpasture syndrome, post-transfusion purpura, acute Guillain-Barré syndrome, and to ↓ circulating toxins–eg, paraquat, methylparathion, mushrooms–*Amanita phylloides*. See Hemapheresis

plasmin A proteolytic enzyme formed from plasminogen that lyses blood clots; plasmin exists in free and bound–fibrin-adsorbed forms; the former is destroyed as it is formed by antiplasmins, the latter acts as a serine endopeptidase to solubilize fibrin clots; it hydrolyzes lysine and arginine bonds in certain proteins–eg, fibrinogen, coagulation factors V and VII. See tPA

plasminogen An 88 kD single-stranded plasmin proenzyme present in the circulation which is converted to plasmin by cleavage of the Arg-Val bond; it is synthesized in the liver, produced or stored in eosinophils and forms complexes with fibrinogen and fibrin; during coagulation, large amounts of plasminogen are integrated in the fibrin clot ↑ IN DVT, infection, inflammation, CA, MI, OCs, pregnancy, physical stress, surgery, trauma ↓ IN Cirrhosis, other liver disease, DIC, fibrinolysis, hyaline membrane disease–RDS of newborn, renal disease, post-surgery–eg, CABG, thrombosis NORMAL 2.5-5.2 U/mL–US: 20 mg/dL

Plastazote® ORTHOPEDICS A cross-linked polyethylene block foam used to create orthopedic prosthetics. See Prosthetics

plaster cast lung A popular term for a lung with extensive consolidation, typical of *K pneumoniae* pneumonia; each lung weighs up to 1500 g and is covered with fibrinopurulent exudate and copious slimy, mucoid pus

plastic surgeon A surgeon specialized in reconstruction or cosmetic enhancement of various body regions, most commonly the face–nose, chin, and cheeks, breasts and buttocks; PSs remove fat deposits through liposuc-

tion; PSs reduce scarring or disfigurement caused by accidents, birth defects, or prior surgery

-plasty SURGERY A procedure in which the shape of a tissue is altered or enhanced. See Abdominoplasty, Canalplasty, Mammaplasty, Rhinoplasty, Vaginoplasty

plate atelectasis RADIOLOGY Segmental atelectasis characterized by linear shadows of ↑ density at the lung bases that are horizontal, measure 1-3 mm in thickness and a few cm, typically seen after abdominal surgery or pulmonary infarction

plate ORTHOPEDICS Flattened orthopedic hardware with multiple holes for screws to stabilize fractures. See Blade plate, Buttress plate, Compression plate, Neutralization plate, Pressure plate Reconstruction plate

plate rounds INFECTIOUS DISEASE An exercise in which the physician(s) caring for Pts with infections passes through the microbiology lab on a regular basis–'rounds' to visually examine culture plates for growth of bacteria. See Susceptibility testing

plateau development PEDIATRICS A form of disease progression that occurs in infants who reach normal developmental milestones in the 1st few months or years of life, slows to a 'plateau' and then begins inexorable deterioration until death in early childhood; PD is typical of children with AIDS, Tay-Sachs disease, and certain 'floppy infant' syndromes. See Floppy infant

plateau phase MICROBIOLOGY A phase in the growth cycle of bacteria in culture, in which the nutrients are sufficient to sustain growth and the cells dying equal the number being produced de novo SEXOLOGY The 2nd of 4 sexual phases delineated by Masters and Johnson. Cf Excitement phase, Orgasmic phase, Resolution phase

plateauing SPORTS MEDICINE A weight training term for the point above which an anabolic drug becomes ineffective in increasing muscle mass. See Anabolic steroids, Weight training

platelet adhesion HEMATOLOGY The attachment of platelets to nonplatelet surfaces, which occurs after trauma when platelets contact exposed collagen fibers of the subendothelium of blood vessels ↓ IN Afibrinogenemia, anemia, azotemia, Bernard-Soulier disease, Chédiak-Higashi syndrome, cirrhosis and other liver disease, congenital heart disease, DIC, fibrinolysis, Glanzmann's thrombasthenia, glycogen storage disease, hyaline membrane disease–RDS of newborn, myeloma, plasma cell dyscrasia, platelet release defects, renal disease, post-surgery–eg, CABG, uremia, von Willebrand's disease, Waldenström's macroglobulinemia. Cf Platelet adhesion, Platelet aggregations studies

platelet antigen TRANSFUSION MEDICINE An antigen on the surface of platelets that may evoke production of antibodies, causing neonatal alloimmune thrombocytopenia and post-transfusion purpura, most commonly as a reaction to the P1^{A1} antigen in P1^{A1} antigen-negative recipients; other PAs causing purpura include P1^{A2}, HLA-A2 and Bak^a

platelet concentrate TRANSFUSION MEDICINE A product prepared from a single donor, which transiently ↑ platelet count by 5-10 x 10^9/L/M^2 body surface area, if thrombocytopenia is *not* due to ↑ destruction ADVERSE REACTIONS Febrile transfusion reactions are related to bioreactive substances–IL-1β and IL-6 in supernatant, which ↑ with ↑ storage and ↑ number of WBCs in stored unit. See Platelet transfusion

platelet count Enumeration of circulating platelets SPECIMEN Blood, EDTA–lavender top-tube REF RANGE 150-450 x 10^9/L–US: 150-400 000/μL–mm^3; 'panic' values: < 20 x 10^9/L–US: < 20 000/μL; > 1000 x 10^9/L–US: 1,000,000/μL ↑ IN Myeloproliferative disorders and/or polycythemia vera, and less commonly with infections, blood loss, and splenectomy; rare causes of ↑ PC include anemia–hemolytic, iron-deficiency, sickle cell, cirrhosis, collagen vascular disease, cryoglobulinemia, drugs–

epinephrine, OCs, exercise, hemorrhage, hypoxia, ITP, post-partum, pregnancy, rheumatoid arthritis, TB ↓ IN < 0.02 **x** 10^{12}/L–US: 20 000/mm³–µL are associated with ↑ bleeding tendency; platelets are ↓ in malignacies of bone, GI tract, brain, leukemia, kidney or liver disease, aplastic anemia, DIC, ITP, SLE, drugs associated with ↓ PC: aspirin, chemotherapeutic agents, chloromycetin, phenylbutazone, quinidine, thiazide diuretics, tolbutamide. See Platelet concentrate

platelet transfusion The administration of platelets to ↑ platelet concentration in the circulation. See Platelet antigens

PLATELET TRANSFUSION GUIDELINES

Platelet count–PC < 20 x 10^9/L–US: < 20 000/mm³

PC < 40 x 10^9/L in active hemorrhage

PC < 50 x 10^9/L in neonates, or Pts with documented coagulopathies, recurrent fever, severe infections or receiving drugs that cause platelet dysfunction

PC < 100 x 10^9/L before 'bloody' surgery, eg cardiopulmonary bypass or < 48 hrs after surgery

Bleeding time > twice upper limit of normal

For transfusing one unit of random-donor platelets/10 kg body weight/24 hrs

plateletpheresis Platelet apheresis TRANSFUSION MEDICINE The centrifugal separation of platelets from whole blood, with continuous or intermittent return of the RBCs and platelet-poor plasma to the donor; plateletpheresis is a form of exchange transfusion used for Pts with extremely high–> 1-1.5 **x** 10^9/L platelet counts, which is linked to severe thrombotic and hemorrhagic phenomena; to prevent a rebound ↑ of platelets after the procedure, plateletpheresis must be followed by cytotoxic therapy. See Therapeutic apheresis. Cf Cytapheresis, Leukapheresis

platinum TOXICOLOGY A metallic element–atomic number 78, atomic weight 195.1, used in catalytic converters for cars, various industries, dental restorations, chemotherapy–eg, cisplatin and carboplatin

Plavix® Clopidogrel bisulfate CARDIOLOGY An antiplatelet agent used to ↓ risk of stroke, acute MI, or vascular death in Pts with known ASHD. See CAPRIE

play or pay MANAGED CARE A universal health care coverage plan proposed by Presidents Bill & Hillary Clinton where an employer would either provide his workers with a basic health benefits package–'play', or pay into a government-managed insurance pool. See Clinton plan HOSPITAL STAFFING A plan in California in which physicians choose either to provide coverage to a department–play, or contribute to a fund for the coverage–pay

pleasure principle PSYCHIATRY The psychoanalytic concept that people instinctually seek to avoid pain and discomfort and strive for gratification and pleasure. Cf Reality principle

-plegia *prefix*, Latin, paralysis

pleomorphic adenoma See Mixed tumor

pleomorphic *adjective* Referring to a variable appearance or morphology

Plesiomonas shigelloides *Aeromonas shigelloides, Pseudomonas shigelloides* BACTERIOLOGY A bacterium of brackish or salt water, which may cause gastroenteritis due to exposure to contaminated water or seafood

PLESS UROLOGY A clinical trial–Proscar® Long-term Efficacy and Safety Study–confirming the benefits of Proscar–finasteride in treating BPH. See Benign prostatic hypertrophy

Pletal® Cilostazol, see there

plethoric *adjective* Fluid-filled, edematous

plethysmography A technique that detects changes in the volume of an organ, limb or the body, by measuring the flow of blood through its veins. See Arterial plethysmography, Impedance plethysmography, Ocular plethysmography

PLETHYSMOGRAPHY

WHOLE BODY PLETHYSMOGRAPHY measures the volume of gas in the lungs, including that which is trapped in poorly communicating air spaces, of particular use in COPD and emphysema

IMPEDANCE PLETHYSMOGRAPHY Venous impedance plethysmography A technique used to diagnose acute venous obstruction or vascular insufficiency of an extremity by measuring the change in limb volume with each arterial pulse and during cuff occlusion of the venous flow from the limb, the manipulation of which allows evaluation of either the arterial or venous flow

pleural biopsy A 'blind' percutaneous biopsy of the pleura, often performed in tandem with thoracentesis to determine the cause of pleural effusions, which may be due to bacterial or TB infection or malignancy–eg, adenocarcinomas and mesotheliomas CONTRAINDICATIONS Low platelet count, especially < 20,000/mm³, and low fluid volume

pleural effusion Fluid in the chest PULMONOLOGY An abnormal accumulation of fluid in the pleural space ETIOLOGY PEs may be benign or malignant and either a transudate–with low concentrations of proteins, due to CHF, hepatic hydrothorax, nephrotic syndrome, peritoneal dialysis, and others, or an exudate CLINICAL Dyspnea, pleural pain, ↓ breath sounds, dull to percussion, ↓ tactile fremitus CXR Blunting of costophrenic angle MANAGEMENT Thoracentesis. See Pleural exudate

pleural exudate PULMONARY MEDICINE An abnormal accumulation of protein-rich fluid in the pleural space ETIOLOGY Infection–bacterial, TB, viral, chylothorax, neoplasm, PTE with pulmonary infarction, GI disease, collagen vascular disease–eg, SLE, asbestosis, pancreatitis, traumatic tap, postcardiotomy, neoplasm MANAGEMENT Thoracentesis. See Pleural effusion

pleural needle biopsy A biopsy with a needle under local anesthesia to obtain a small sample of pleural tissue for LM DIAGNOSES CA–metastatic or 1°–eg, mesothelioma; infection–TB, fungal, viral, bacterial, collagen vascular disease COMPLICATIONS Pneumothorax, internal bleeding

pleural rub PHYSICAL EXAM A friction sound heard during breathing in Pts with peuritis, due to the rubbing of the visceral and parietal pleurae

pleural tag IMAGING A radiographic 'knob' seen in the lung periphery in ¼ of Pts with bronchoalveolar CA. See Lung cancer

pleuritis Pleurisy PULMONARY MEDICINE Pleural inflammation ETIOLOGY RTIs, TB, rheumatoid disease, lung neoplasms CLINICAL Pain over the chest wall at the site of inflammation, which ↑ with breathing, coughing, chest movement; the pleural surfaces, roughened by inflammation, rub together with each breath, causing a rough grating sound–"friction rub"–heard with a stethoscope; fluid accumulation separates parietal and visceral pleurae, ↓ chest pain; large effusions compromise respiration and may cause coughing, SOB, tachypnea, cyanosis, retractions. See Tuberculous pleuritis

pleurodesis THORACIC SURGERY A procedure in which the visceral and parietal pleura are deliberately fused, either by inducing chemical–eg, talc-inflammation or mechanically–eg, by stripping the pleura, to treat recurrent pneumothorax

pleurodynia Pleural pain. See Epidemic pleurodynia

pleuropneumonia-like organism *Mycoplasma pneumoniae*, see there

PLEVA Pityriasis lichenoides et varioliformis acuta DERMATOLOGY An idiopathic papulovesicular disease of acute onset more common in the young ♂, characterized by successive waves of lesions on trunk and extremities TREATMENT PUVA, corticosteroids, antibiotics, MTX

plexogenic pulmonary arteriopathy PULMONARY DISEASE A vasculopathy seen in pulmonary HTN, characterized by medial hypertrophy of muscular pulmonary arteries, fibrinoid necrosis ETIOLOGY Anorexiants–eg, fenfluramine; possibly phentermine. See Fen-phen

Plexonal® Scopolamine, see there

plexus A network of interlaced nerves or vessels. See Brachial plexus, Cervical plexus, Hemorrhoidal plexus, Lumbar plexus, Myenteric plexus, Sacral plexus, Submucous plexus

Plidan Diazepam, see there

PLLA screw Poly-L-lactic acid cortical screw ORTHOPEDICS A biodegradable fixation device. See Screw

ploidy analysis LAB MEDICINE A flow cytometry technique that evaluates a cell's chromosome content—a parameter of aggressiveness in CA; in general, diploidy—ie, the presence of 2 haploid sets of chromosomes, is a normal or near-normal state; in contrast, anaplastic and aggressive tumors are more often aneuploid or hyperdiploid; PA is used to prognosticate CA of bone—osteosarcoma, bladder, breast, colon, endometrium, lymphoma, ovary. See Flow cytometry, S-phase analysis

plucked chicken appearance PEDIATRICS An appearance classically described in children with Hutchinson-Gilford progeria, characterized by alopecia, midfacial cyanosis, atrophy of subcutaneous fat, sculptured nose with a 'beaked' tip, a disproportionately large head, prominent eyes and scalp veins, nail and dental dystrophy, micrognathia, xeroderma, pyriform thorax, a horse-riding stance, thin limbs, stiff joints and osteoporosis

plug & play VOX POPULI Referring to any device that interfaces with other devices in a system and simply needs to be plugged into an electric outlet and connected to the other units in the system to be fully operational

plugginess COSMETIC SURGERY A popular term for the somewhat 'unnatural' appearance of larger hair grafts which, while dense, are separated by hairless 'valleys'. See See-through hair

Plum® pump NURSING A line of drug-delivery systems for computer-regulated infusion from bags, bottles, syringes or vials. See Drug delivery

Plummer-Vinson syndrome Paterson-Kelly syndrome; sideropenic dysphagia METABOLIC DISEASE A possibly autoimmune condition affecting middle aged ♀, characterized by iron-deficiency anemia, oropharyngeal mucosal atrophy, koilonychia—spooned nails, angular cheilitis, pallor, glossitis, dysphagia due to a fibrous esophageal web, associated with oropharyngeal SCC. See Esophageal web

plump hilus sign Fleischer sign IMAGING An ↑ prominence of the hilar vasculature on a plain CXR on the side affected by acute PTE

pluripotent stem cell HEMATOLOGY The 'mother of all cells'—the progenitor of all hematopoietic cells—eg, platelets, RBCs, neutrophils, macrophages, lymphocytes. See Stem celll

PMA Progressive muscular atrophy

PMI Point of maximum impulse, see there

PML 1. Progressive multifocal leukoencephalopathy, see there 2. Promyelocytic leukemia, see there

PMN Polymorphonuclear leukocyte, see there

PMPA Tenofovir AIDS An anti-HIV nucleotide analogue. See AIDS

PMR 1. Percutaneous myocardial revascularization, see there 2. Perinatal mortality rate 3. Polymyalgia rheumatica 4. Proportionate mortality ratio, see there

PMS Premenstrual syndrome(s), Premenstrual dysphoric disorder A disorder characterized by affective, behavioral and somatic Sx during the luteal–2nd phase of menstrual cycle, resolve with onset of menstruation, weakly linked to ↓ estrogen and progesterone from luteal peaks CLINICAL PMS affects 10-30% of menstruating ♀; it is characterized by several days of mental or physical incapacitation of varying intensity, insomnia, headache, emotional lability—anxiety, depression, irritability, loss of concentration, poor judgement, mood swings and tendency towards violence evoked by environmental cues, acne, breast enlargement, fullness or tenderness, abdominal bloating with edema, craving for salty, sweet or 'junk' food TREATMENT Fluoxetine–Prozac may relieve tension, irritability, dysphoria, leuprolide. See Leuprolide

PMT 1. Pacemaker-mediated tachycardia, see there 2. Premenstrual tension. See PMS

PND Paroxysmal nocturnal dyspnea, see there

pneumatic antishock garment MAST garment, military antishock trousers EMERGENCY MEDICINE A garment-like device placed around a Pt's legs and abdomen—2 separate chambers and inflated to provide emergency treatment of shock until more definitive therapy—eg, volume replacement, transfusion, or surgery can be performed INDICATIONS Shock with BP < 80 mm Hg, or 80–90 mm Hg if evidence of hypoperfusion; splinting of pelvic and femoral fractures; tamponade of suspected intra-abdominal hemorrhage—eg, from a ruptured or leaking abdominal aortic aneurysm; as an adjunct to CPR in Pts with cardiac arrest

pneumatic larynx HEAD & NECK SURGERY A device that uses air to produce sound to help a laryngectomee talk

pneumatic tube system HOSPITAL ARCHITECTURE A system for transporting specimens and drugs in a hospital, which ↓ bottlenecks of inefficiency that occur with conventional–human transport by whooshing tubes of stuff and paperwork between patients and stations

pneumococcal meningitis NEUROLOGY Meningitis caused by *S pneumoniae*, the most common meningitis pathogen in adults, and 2nd most common in children > age 6, which typically has an abrupt onset RISK FACTORS Recurrent meningitis, meningitis with CSF leak, type 2 DM, head trauma, alcohol use, recent RTI, infective endocarditis, or recent ear infection. See Meningitis

Pneumocystis carinii Pneumocystosis INFECTIOUS DISEASE An opportunistic 'bug' causing pneumonia–PCP in immunocompromised hosts–eg, with AIDS, leukemia, lymphoma, organ transplants, corticosteroid therapy, cytotoxic drugs, the elderly DIAGNOSIS GMS staining of tissues; *P carinii* also causes otic infection and choroiditis PROPHYLAXIS In advanced HIV infection, aerosolized pentamidine, T-S, and high-dose dapsone are equally effective; the latter 2 are superior in Pts with < 100 CD4+/mm³ TREATMENT T-S, manipulation of inflammatory response, and immune enhancement pentamidine, atovaquone, trimetrexate glucuronate

pneumomediastinum Mediastinal emphysema CRITICAL CARE The presence of air in the mediastinum, either post-traumatic or induced during mediastinoscopy; the air may percolate into the thorax, resulting in pneumothorax, or into the pericardium, causing pneumopericardium. See Emphysema

pneumonectomy SURGERY The excision of an entire lung

pneumonia A viral or bacterial infection of lungs–bronchoalveolar unit characterized by inflammation and the oozing of exudate from the bronchial or bronchiolar mucosa and alveoli CLINICAL ↑ Sputum, dyspnea, fever/chills, night sweats, pleural pain, ↓ breath sounds, dull percussion, wheezing, ↑ tactile fremitus CXR Lobar consolidation, patchy infiltrates DIFFDX Aspiration–gastric, foreign body, atelectasis, congenital malformations–eg, pulmonary sequestration, CHF, COPD, tumors, infarction, collagen vascular disease MANAGEMENT Antibiotics. *See* Aspiration pneumonia, Atypical pneumonia, BOOP pneumonia, Bronchopneumonia, Chemical pneumonia, CMV pneumonia, Community-acquired pneumonia, Cytomegalovirus pneumonia, Diffuse interstitial pneumonia, Double pneumonia, Eosinophilic pneumonia, Giant cell pneumonia, Giant interstitial pneumonia, Hydrocarbon pneumonia, Lipoid pneumonia, Lobar pneumonia,

pneumonitis

Lobular pneumonia, Lymphocytic interstitial pneumonia, Necrotizing pneumonia, Nonspecific interstitial pneumonia, Nosocomial pneumonia, Post-operative pneumonia, Spherical pneumonia, Usual interstitial pneumonia, Ventilator pneumonia, Viral pneumonia, Walking pneumonia

pneumonitis PULMONOLOGY Inflammation of lung tissue. See Chemical pneumonitis, Desquamative interstitial pneumonia, Hypersensitivity pneumonitis, Radiation pneumonitis, Reflux pneumonitis, Rheumatoid pneumonitis

Pneumopent Aerosolized pentamidine, see there

pneumothorax Lung collapse PULMOLOGY The presence of air in the pleural space, which may be 1°–seen in tall, thin, young ♂, characterized by subpleural apical blebs, 2°–asthma, COPD, PCP, trauma, TB, iatrogenic–due to thoracentesis, subclavian line placement, PEEP, bronchoscopy CLINICAL Pleural pain, dyspnea, ↓ breath sounds, percussion hyperresonance, ↓ tactile fremitus MANAGEMENT Small blebs may heal spontaneously, larger pneumothoraces require chest tube drainage, pleurodesis. See Spontaneous pneumothorax

PNH Paroxysmal nocturnal hemoglobinuria, see there

PNS 1. Parasympathetic nervous system, see there 2. Peripheral nervous system, see there

PO 1. Parieto-occipital 2. Per os 3. Period of onset 4. Perioperative 5. Peroxidase 6. Posterior 7. Postoperative 8. Preoptic 9. Pulmonary valve opening

POC 1. Products of conception, see there 2. Postoperative care 3. Procarbazine, Oncovin–vincristine, CCNU–lomustine

pocket DENTISTRY A pathologically altered or enlarged gingival sulcus, so defined when the distance from the gingival margin is ≥ 3 mm; in a healthy periodontium, the gingiva is snug around the teeth, and the gingival crevice near zero; with inflammation, ↑ in bulk of gingival tissue around the teeth results in ↑ depth of tissue around the teeth which, if confined to the gingiva, is termed gingival pseudopocket; if it extends into the periodontium, it is termed periodontal pocket

pocket infection INFECTIOUS DISEASE A catheter-related infection occurring in an implanted subcutaneous central venous catheter with signs of inflammation or purulent exudate in the subcutaneous pocket occupied by the reservoir. See Central venous catheter

pocket simulation Pocket muscle stimulation CARDIAC PACING Unwanted stimulation of skeletal muscle around an implanted pacemaker by current flow at the pacemaker's indifferent electrode–ie, at the external metal casing of unipolar pacemakers

pocketing DENTISTRY The formation of a pocket, see there

pockets See Deepest pockets

pockmark Pock mark A deep, sharply circumscriped 'icepick' scar seen in Pts recovered from smallpox, now of historic interest; shallower pockmarks may be seen in chickenpox

POCs Products of conception, see there

POCT Point of care testing, see there

POD Postoperative day

podagra RHEUMATOLOGY Painful gouty toe. See Gout

podiatrist Chiropodist, podologist A person trained in podiatry, who, in the US, has graduated from a 4-yr education program in podiatry after graduating college or university; podiatrists are examined and licensed by a state's medical board, carry a title of Doctor of Podiatric Medicine–DPM, and diagnose and treat diseases of the feet by medicine or surgery. See Podiatry

podiatry Chiropody, podology The field of health care dedicated to understanding the anatomy, mechanics, and pathology of the foot, and diagnosis and treatment of its diseases. See Podiatris

POEM CARDIOLOGY A clinical trial–Patency, Outcomes and Economics of MIDCAB–comparing minimally invasive direct coronary artery bypass to traditional CABG vis-à-vis efficacy, Pt outcomes, recovery time, hospital costs. Cf MIDCAB

POEMS syndrome Crow-Fucase syndrome A multisystem disease characterized by an acronym, POEMS, derived from Polyneuropathy–distal symmetric progressive muscle weakness, paresthesias, ↓ nerve conduction velocity; Organomegaly–hepatosplenomegaly, lymphadenopathy; Endocrinopathy–hirsutism; Monoclonal gammopathy–myeloma and focal osteosclerosis

point-of-care testing LAB MEDICINE The analysis of clinical specimens as close as possible to the Pt, including bedside, ward–unit, or 'stat' regional response labs that service specified areas–eg, the ER or ICU

point VOX POPULI A small place. See Alarm point, Blockade point, Breakpoint point, Cell cycle restriction point, Checkpoint point, Critical control point, Dilution end point, Distal point, Dose point, Eye reference point, Fixed point, Flashpoint, Isobetic point, Joint point, Limit point, Loo point, Matrix point, McBurney's point, Murphy's point, Myofascial trigger point, Pressure point, Reorder point, Saddle point, Saddle node point, Satellite point, *Shu* point, Trigger point

point of service Place of service MANAGED CARE The 'where' a medical service is provided–eg, clinic, hospital inpatient, hospital outpatient, nursing facility, home, clinic, etc

poison DRUG SLANG A regional street term for heroin; fentanyl TOXICOLOGY A toxic substance that adversely affects the metabolism of a cell, tissue or entire organism, evoking biochemical and histologic changes, and possibly evoke irreversible cell damage and/or death MANAGEMENT–DIALYSIS Sedative-hypnotics–chloral hydrate, ethanol, ethylene glycol, methanol, barbiturates, meprobamate, analgesics–acetaminophen, aspirin, phenacetin, amphetamines, heavy metals–arsenic, lead, mercury, metallic salts–eg, of calcium or lithium, halides, alkaloids–quinine, strychnine, anilines, carbon tetrachloride, ergotamine, INH, nitrofurantoin, phenytoin, theophylline NONDIALYSIS POISONS Amitriptyline, anticholinergics, antidepressants, atropine, benzodiazepines, digitalis, hallucinogens, heroin, methaqualone, phenelzine, phenothiazines, propoxyphene

POISON POTENCY–MLD–MINIMUM LETHAL DOSE

AGENT	MLD, MOLE/KG
Botulinum toxin A	3.3×10^{-17}
Tetanus toxin	1.0×10^{-15}
Diphtheria toxin	4.2×10^{-12}
Agent Orange	3.1×10^{-9}
Curare	7.2×10^{-7}
Strychnine	1.5×10^{-6}
Cyanide	2.0×10^{-4}

poison control center TOXICOLOGY A nonprofit facility, often affiliated with a university or hospital, that provides emergency toxicology assessments by telephone, and treatment recommendations, primarily to parents of children who swallowed a household product, but also to physicians and hospitals. See Hotline

poison ivy 1. A plant that is highly allergenic due to urushiol, which is also found in mango, japanese lacquer tree, cashews; common urushiol-rich plants: poison ivy–*Toxicodendron radicans*, eastern US, poison oak–*T diversilobium* or *Rhus diversiloba*, western US; poison sumac–*T vernix*, southern US. See Urushiol 2. A popular term for any allergic reaction or dermatopathy caused by *T radicans*

poison oak A type of hypersensitivity dermatitis caused by *Rhus diversiloba*–western poison oak, and *R quercifolia*–eastern poison oak. See Poison ivy

poisoning VOX POPULI Intoxication with a substance or chemical; M&M linked to a poison. See Beryllium poisoning, Cadmium poisoning, Carbon monoxide poisoning, Ciguatera poisoning, Fluoride poisoning, Food poisoning, Insecticide poisoning, Lead poisoning, Manganese poisoning, Mushroom poisoning, Toxic oil poisoning

poisonous mushroom TOXICOLOGY A mushroom capable of causing toxic reactions or death CLINICAL Abdominal cramps, N&V. See *Amanita phalloides, Chlorophyllum molybdites, Copinus atramentarius, Gyromitra esculenta, Inocybe* spp, *Muscaria* spp, *Psilocybe* spp

poisonous plant PHARMACOGNOSY A plant capable evoking a toxic response and/or death. See Botanical toxicity, Herbal medicine

Poisson distribution STATISTICS The distribution that arises when parasites are distributed randomly among hosts. See Distribution

polar body biopsy REPRODUCTIVE MEDICINE The removal of the polar body, the genetic residua from the oocyte after the 1st meiotic division, which is metabolically inactive, and a mirror image of the oocyte, which can be removed with impunity. See Assisted reproduction, Blastocyst biopsy, Cleavage stage biopsy

polarization CARDIAC PACING The condition of an electrode in which its electrical potential differs from an equilibrium potential–ie, no current flow

Polarus® ORTHOPEDICS A rod and accessories for internal fixation of fractures of the proximal humerus to the distal third of the shaft

policy HEALTH INSURANCE A contractual agreement between an insurer and insured, which sets forth the rights and obligations of both parties to the agreement. See Any willing provider policy, Claims made policy, Commercial policy, Major hospitalization policy, Occurrence policy, Trolley car policy, Wrap-around policy VOX POPULI A series of rules or guidelines promulgated by an authorizing body. See Federal policy, General policy, Guidelines, HISP policy, Local Medicare review policy, No nit policy, Mexico City policy, PHS policy, Zero tolerance policy

policy manual LAB MEDICINE A formal document that reflects the philosophy and goals of a lab, hospital or other entity, which is approved and dated by a director and indicates a course of action or standard by which a worker will act in various situations

polio See Poliomyelitis

poliomyelitis NEUROLOGY A condition characterized by the selective destruction of anterior horn cells in the spinal cord and/or brain stem, ± leading to muscle weakness, paralysis and respiratory paralysis ETIOLOGY Viral, in particular poliovirus, but also coxsackie A7, enterovirus 71, and others CLINICAL If extreme, asymmetric flaccid paralysis LAB CSF pleocytosis, ↑ lymphocytes. Cf Aseptic meningitis

politically correct Politically sensitive *adjective* Referring to language reflecting awareness and sensitivity to another person's physical, mental, cultural, or other disadvantages or deviations from a norm; a person is not mentally retarded, but rather mentally challenged; a person is not obese but rather has an eating disorder, etc

POLITICALLY CORRECT–A MICROGLOSSARY

FORMER TERM	PC TERM
AMERICAN INDIAN	Native American
BLACK	African American
DEMENTED	Disoriented, severely confused
HANDICAPPED	Disadvantaged
HOMOPHOBIC	Heterosexually biased
HOUSEBOUND	Domestic
(AMERICAN) INDIAN	Native American
MENTALLY RETARDED	Mentally disabled or challenged
OBESE	Large, ample, right-sized
ORIENTAL	Asian
PHYSICALLY HANDICAPPED	Physically disadvantaged
POORLY EDUCATED	Educationally disadvantaged
RACIST	Culturally insensitive
STUPID	Educationally challenged

POLITICALLY CORRECT *AD ABSURDUM*–a microglossary

PC*AA* TERM	TRANSLATION
COLORFUL	Flaky, fruity
DETAIL ORIENTED	Anal-retentive or, if extreme, obsessive compulsive
ECCENTRIC	Nuts, weird
ENTHUSIASTIC & HOPEFUL	Insufferably arrogant
FOLLICLY CHALLENGED	Bald
KNOWLEDGE DEFICIENT	Ignorant
OBTUNDED	Stupid
SEXUAL ARTS SPECIALIST	Prostitute, hooker
SEXUAL ARTS AFICIONADO	Slut, sleaze
VERTICALLY CHALLENGED	Short
VERTICALLY ENHANCED	Tall
VISUALLY CHALLENGED	Myopic

POLL Physician Office-to-lab link LAB MEDICINE A system consisting of soft and hardware that links a physician's office computer to a diagnostic lab, so the results are transferred to the office when available in the lab

pollen The male gametophyte of flowering plants, a major cause of seasonal allergies; ragweed–and related henchmen–eg, feverfew, has garnered the greatest revulsion among the allergically challenged

pollen count ALLERGY MEDICINE An estimate of the number of pollen particles in a standard volume of air

Polly BIOTECHNOLOGY A Poll Dorset sheep cloned from sheep skin cells, which has a human gene in each cell. See Dolly

polyarteritis nodosa Periarteritis nodosa INTERNAL MEDICINE A connective tissue disease, characterized by necrotizing vasculitis of small to medium-sized arteries, affecting multiple organ systems–GI tract, kidneys, liver, muscle, skin; in > 50%, vasculitis affects peripheral nerves as mononeuritis multiplex or symmetric sensorimotor polyneuropathy CLINICAL Fever, malaise, myalgia LAB ↑ WBCs, ↑ ESR, 30-50% have HBsAg or HBsAb MANAGEMENT Immunosuppression–eg, prednisone, cyclophosphamide yields ± 90% 5-yr survival

polyclonal immunoglobulin Polyclonal antibody IMMUNOLOGY Any of a bouquet of Igs produced by multiple, usually non-malignant clones of cells summoned to arms by an immune challenge, which may evoke multiple clonal expansions, each responding to a different epitope on an antigen or group of antigens. See Epitope, Idiotype. See Polyclonal antiserum. Cf Monoclonal antibody

polyclonal antiserum IMMUNOLOGY A Gemische of antibodies with distinct epitope reactivities, produced in response to a broad antigenic stimulus, which may be harvested from a person exposed to a particular pathogen and administered to another whose response to the pathogen is inadequate. See Immune globulin

polycystic kidney Polycystic renal disease NEPHROLOGY An inherited disease characterized by the development of innumerable cysts in the kidneys filled with fluid–urine that replace much of the mass of the kidneys and reduce kidney function leading to kidney failure. See Autosomal dominant polycystic kidney disease, Multilocular cyst of the kidney

polycystic liver disease

Polycystic disorders of kidneys

Renal dysplasia Relatively common, often acquired, presents in infancy; unilateral or bilateral and segmental, focally irregular cystic kidneys, related to mesenchymal immaturity, accompanied by obstruction

Infantile polycystic kidneys Uncommon, AR, seen in infants with massively enlarged kidneys and aberrant bile ducts

Adult polycystic kidneys Common (1-2:1000, US), AD, located to a gene on chromosome 16, affects adults, large bumpy kidneys, cysts in liver, lung, pancreas, berry aneurysms in brain

Medullary sponge kidneys Common, bilateral, uncertain pattern of heredity, affects adults, inability to concentrate urine, hypercalcemia, nephrolithiasis, pyelonephritis, distal renal tubular acidosis and cystic dilation of collecting ducts; renal function and life span may be normal

Uremic medullary sponge kidney Rare, inherited, first seen in young adults; bilateral corticomedullary junction cysts and functional tubular defects, Fanconi syndrome and uremia

polycystic liver disease A condition characterized by multiple variably-sized cysts lined by cuboidal epithelium; liver disease is often obscured by accompanying adult polycystic kidney disease; 40% of affected livers also have von Meyenburg's complexes

polycystic ovarian disease Polycystic ovaries, sclerocystic ovary disease, Stein-Leventhal syndrome **Gynecology** An idiopathic condition affecting 3.5-7.0% of ♀, and most common cause of familial hirsutism **Clinical** Obesity, hirsutism, galactorrhea, 2° amenorrhea following dysmenorrhea, acne vulgaris, and an ↑ risk of endometrial carcinoma due to unopposed estrogenic stimulation; PCO may be associated with CNS trauma or injury in childhood **Diagnosis** Palpation, ultrasonography **Lab** ↑ LH or LH/FSH ratio, ↑ prolactin, ↑ response to nafarelin–a GnRH-gonadotropin-releasing hormone agonist that causes a ♂ pattern response, suggesting that PCO has defective regulation of 17-hydroxylase and C-17,20-lyase; sterility, menstrual abnormalities and hyperandrogenism ± associated with valproate therapy for epilepsy. See HAIR-AN syndrome

polycythemia Any ↑ RBC mass. See Relative polycythemia, Secondary polycythemia

Polycythemia types

Relative RBC mass is above normal but not pathologic. See Relative polycythemia

Secondary to various physiopathologic mechanisms, usually hypoxia or ↑ erythropoietin secretion

Neoplastic, ie polycythemia vera, see there

polycythemia vera Hyperglobulinemia, polycythemia rubra vera, primary polycythemia **Hem/Onc** A chronic myeloproliferation due to the expansion of abnormal pluripotent stem cell with ↑ erythropoietin-independent erythropoiesis and megakaryopoiesis **Clinical** More common in ♂, rarely < age 40, hyperviscosity, ↑ risk of stroke or acute MI, more common in Jews, indolent but may progress to AML **Lab** ↑ Leukocyte alk phos, ↑ platelets, basophils, ↑ vitamin B_{12}, vitamin B_{12} binding capacity–transcobalamin I and III, ↓ erythropoietin and stainable iron in BM, myelofibrosis, extramedullary hematopoiesis; in PV, erythropoietin is ↓ to ±2.1 U/L **Prognosis** 15-20% resolve in a so-called 'spent' phase with marrow fibrosis; 40% die of thrombosis and hemorrhage; others are at ↑ risk of myeloproliferative disease; 5-15% of Pts develop acute leukemia or myeloid metaplasia, or less commonly, AML FAB-M6 **Diagnosis** PV requires that either all of 'A' criteria are present or 2 'A' criteria and 2 'B' criteria are present **Management** Simple phlebotomy yields a 14-yr survival; hydroxyurea may be used for long-term treatment

Polycythemia vera

A Clinical

A1 ↑ RBC mass–♂ > 36 ml/kg, ♀ > 32 mg/kg

A2 Arterial O_2 saturation > 92%–near normal

A3 Splenomegaly–present in 75%

B Laboratory

B1 Thrombocytosis > 400 x 10⁹/L–most cases

B2 Leukocytosis > 12 x 10⁹/L, without fever or infection

B3 ↑ Alk phos

B4 ↑ Vitamin B_{12} > 666 pmol/L–US > 900 pg/ml

polydactyly Bunches of fingers

polydipsia Excessive thirst

polydrug therapy See Polypharmacy

polyene antibiotic A broad-spectrum antifungal produced by *Streptomyces* spp, eg amphotericin B, nystatin, pimaricin; toxicity may be ameliorated by newer drug delivery systems–eg, liposomes, lipid dispersions, colloidal suspensions

polyglandular autoimmune syndrome Either of 2, often overlapping endocrinopathies, characterized by gonadal failure, possibly 2° to hypothalamic defects with vitiligo and autoimmune adrenal insufficiency–80% have autoantibodies. Types: Type I APECED An AR condition of late childhood onset with hypoparathyroidism, mucocutaneous candidiasis, alopecia, pernicious anemia, malabsorption, chronic hepatitis; Type II Schmidt syndrome An AR condition of adult onset with Addison's disease, and autoimmune (Hashimoto's) thyroiditis and/or IDDM

polygraph Lie detector A device designed to detect deception by evaluating physiologic responses to various spoken questions, measuring and recording changes in electrical and mechanical impulses in various parameters–eg, bp, respiratory rate, galvanic skin reflex

polyhydramnios Hydramnios, see there

PolyMem® the Pink Dressing®, space-age dressing **Wound care** A dressing containing a cleanser, moisturizer and refined cornstarch, for treating diabetic ulcers PR Newswire 5/12/97. See Wound care

polymerase chain reaction **Molecular biology** A molecular technique that uses DNA polymerases from high-temperature bacteria–known as extremophiles to rapidly amplify–ie, ↑ the number of copies of–a sequence of DNA in a sample; starting from minimal amounts–<< 1 µg–as little as one copy of a sequence of DNA, PCR exponentially amplifies a target DNA sequence, which has been inserted between 2 oligonucleotide primers through multiple amplification cycles **Application** Prenatal Dx of hereditary disease–sickle cell anemia, PKU, cystic fibrosis; ID gene rearrangements in lymphoproliferative disorders, determine fetal sex, Lyme disease, TB, *Chlamydia trachomatis*, ID viruses–HIV, CMV, HPV, HBV, delineate viral link to cancer–HTLV-1, HPV, bacteria, parasites, pathogenic mechanisms–DM, pemphigus vulgaris, myasthenia gravis, oncogene-induced cancer **Sensitivity** In detecting leukemia in BM–Bx has a 65-75% sensitivity, Southern blot analysis of gene rearrangement, 98-99% sensitivity, PCR, 99.999%. See Allele-specific PCR, AP-PCR, DNA amplification, Fluorophore-enhanced repetitive sequence-based-PCR, Inverse PCR, Jumping PCR, Multiplex PCR, Nested PCR, Reverse-transcriptase PCR, Semi-nested PCR, Touchdown PCR. Cf Ligase chain reaction

polymicrobial *adjective* Referring to many different spp of 'bugs' in a specimen which, in lab lingo, refers to a menagerie of bacteria

polymorphic **Vox populi** Having ≥ 2 appearances or forms

polymorphic light eruption Polymorphous light eruption **Dermatology** An abnormal skin reaction to sunburn range UVB–290-320 nm light; PLE is more common in young adults 4-24 hrs after exposure to light as papular, papulovesicular, plaque or diffuse erythematous lesions; the classic PMLE lesion is a plaque in which patchy lymphocytic infiltrates mimic the lesions of early SLE **Management** Antimalarial drugs–eg, chloroquine

polymyalgia rheumatica Polymyalgia INTERNAL MEDICINE A condition characterized by an abrupt onset of myalgia and arthralgia of the neck and proximal 'girdle' muscles, most prominent in the morning and after rest; systemic Sx are vague–aching of $\geq$ 30 mins, affecting 2+ major joints, in descending order, shoulder, hips and thighs, or neck/torso PT PROFILE > age 70; ♀:♂, 2:1 CLINICAL Pain exacerbated by movement, fatigue, malaise, weight loss, depression, low-grade fever LAB ↑ ESR, anemia of chronic disease, mild ↑ LFTs, ↑ WBCs, ↑ Igs, ↑ acute phase reactants MANAGEMENT NSAIDs, aspirin; if brutal, corticosteroids; PR may be associated with other inflammatory conditions, giant cell/temporal arteritis, connective tissue disorders, CA

polymyositis RHEUMATOLOGY An inflammatory myopathy of adults that may be acute, subacute, or chronic; predominant feature is symmetrical proximal muscle weakness of insidious onset; polymyositis is accompanied by EMG changes, muscle necrosis, ↑ creatinine phosphokinase, skin lesions, myalgias, tenderness, later atrophy and fibrosis; the criteria for defining polymyositis–and dermatomyositis are those of Bohan and Peter TREATMENT Corticosteroids; if refractory, MTX, RT. See Polymyositis-dermatomyositis

polymyositis/dermatomyositis RHEUMATOLOGY An 'overlap' syndrome where polymyositis and dermatomyositis share multiple features–proximal distribution of muscle weakness, typical skin changes, 'machinist's hands', chronic 'round cell' inflammation, IgM rheumatoid factor, myopathic changes, spontaneous electrical discharges by EMG, response to corticosteroids. See Machinist's hands, Polymyositis

---

POLYMYOSITIS-DERMATOMYOSITIS—DEFINING CRITERIA

1. Symmetric proximal/limb-girdle muscle weakness of insidious onset
2. Typical skeletal pathology–eg, presence of 'skip' areas of involvement
3. ↑ Serum levels of skeletal muscle enzymes–ALA, aldolase, AST, CK-MM, LDH
4. A characteristic EMG triad of brief small polyphasic motor unit potentials, fibrillation potentials, positive waves, insertion irritability, and normal conduction velocity

DERMATOMYOSITIS IS DEFINED BY A FINAL CRITERION

5. Lilac/heliotrope discoloration of eyelids, periorbital edema; characteristic scaling erythematous rash over dorsal aspect of hands–Gottron sign, involvement of elbows, knees, medial malleoli and upper body

---

polyneuropathy Chronic inflammatory polyneuropathy, critical illness polyneuropathy, familial amyloidotic polyneuropathy, multiple neuropathy, peripheral neuropathy, polyneuritis, sensorimotor polyneuropathy NEUROLOGY The simultaneous inflammation of the motor and/or sensory components of 2 or more peripheral nerves. See Mononeuropathy

polyp An elevated 'tumor' mass, which is usually epithelial, and often neoplastic; polyps are common in the colon, ♀ genital tract, nasopharynx, stomach See Bladder polyp, Cervical polyp, Colon polyp, Hairy polyp, Inflammatory polyp, Inflammatory fibroid polyp, Juvenile polyp, Pseudopolyp, Retention polyp

---

POLYPS

COLON Colonic polyps are usually epithelial, and are acquired or hereditary

ACQUIRED POLYPS Adenomatous (tubular or villous) in morphology, ↑ frequency with age; although often asymptomatic, larger polyps are often announced by bleeding, or changed bowel habits; if really large, APs may form a leading 'front' of an intussusception; distinction between adenomatous polyps ('tight' round glands) and villous adenomas (finger-like fronds of elongated glands) has little practical importance–both have malignant potential; periodic colonoscopy and polypectomy yields a 3-fold ↓ in subsequent cancer; hyperplastic polyps are also acquired but are non-neoplastic

HEREDITARY POLYPS are epithelial and may overlap with each other

- **FAMILIAL ADENOMATOUS POLYPOSIS** (FAP) A premalignant, AD MIM 175100 condition presenting in early adulthood with 100s to 1000s of colonic polyps, related to a loss of the normal repression of DNA synthesis in the entire colonic epithelium; adenocarcinoma occurs in 70–100% of Pts, prevented by prophylactic colectomy
- **GARDNER SYNDROME** A rare AD MIM 175100 condition with premalignant polyps of the entire GI tract, which is identical to FAP, but has, in addition, extraintestinal tumors; most Pts

develop colon carcinoma; other neoplasms in GS Pts include bile duct carcinoma, osteomas of the mandible, skull, and long bones, soft tissue tumors (fibromas, lipomas), sebaceous cysts, and rarely, thyroid and adrenal gland cancers

- **TURCOTT SYNDROME** A rare AR MIM 276300 condition associated with brain tumors, eg medulloblastoma, glioblastoma
- **OTHER COLON POLYPS** Hamartomas, hyperplastic polyps, juvenile and retention polyps–little neoplastic potential
- **TURCOTT SYNDROME** A non-hereditary condition characterized by diffuse GI polyposis, accompanied by alopecia, nail atrophy, cutaneous hyperpigmentation, weight loss, protein-losing enteropathy, electrolyte imbalance and malnutrition
- **PEUTZ-JEGHERS SYNDROME** An AD MIM 175200 condition with hamartomas of the entire GI tract, predominantly of the small intestine, focal Paneth cell hyperplasia, melanin spots in buccal mucosa, lips, and digits, intussusception and bleeding; colonic adenocarcinomas, when seen in PJS, arise in adenomatous and not in hamartomatous polyps; PJS may be associated with Sertoli cell tumor with annular tubules, see SCTAT

FEMALE UROGENITAL TRACT Endometrial and endocervical polyps are circumscribed foci of cystic glandular hyperplasia of the mucosa and may cause abnormal bleeding; carcinoma arising in such polyps is rare; when smooth muscle is also present, they are designated as adenomatous polyps DiffDx Polypoid smooth muscle tumors, benign and malignant. See Müllerian mixed tumor

NASOPHARYNX Nasal polyps Inflammatory ('allergic') polyps of the nasal cavity are not neoplastic, but rather reactive to inflammation or allergy; unlike true polyps, nasal polyps display edema and chronic inflammation (eosinophils, plasma cells, and lymphocytes), are bilateral, recurrent, and intranasal

SKIN Squamous polyps and fibroepithelial polyps or 'skin tags' are benign prolapses of upper dermis onto the skin surface, which have no neoplastic potential

STOMACH POLYP Gastric polyp It is often (incorrectly) assumed that colon polyps are analogous to gastric polyps; hyperplastic polyps (type I and II polyps by Japanese authors) comprise 75% of all gastric polyps; they are neoplastic, but are usually benign

---

polyphagy Pathologic overeating. See Bulimia

polypharmacy THERAPEUTICS The use of mutiple drugs to treat one or a limited number of conditions; it is most common in elderly Pts. See Therapeutic drug monitoring

polypoid tumor ONCOLOGY Any tumor benign or malignant with a polyp-like appearance. See Polyp

polyposis PATHOLOGY Numerous polyps on a mucosa. See Familial adenomatous polyposis, Multiple lymphomatous polyposis

polyserositis Inflammation of multiple serosal surfaces as in familial Mediterranean fever, or connective tissue disease

polysomnography Cardiopulmonary sleep study, sleep apnea study SLEEP DISORDERS The continuous and simultaneous recording of multiple physiological variables during sleep–ie, EEG, EOG, EMG–the 3 basic stage scoring parameters, EKG, respiratory air flow, respiratory excursions, lower limb movement, and other electrophysiologic variables, and electrooculography; a polysomnograph uses noninvasive sensors for nasal airflow–thermocouple, oral airflow–end-tidal CO_2 gauge, tracheal sounds–microphone, thoracic and abdominal respiratory effort–inductance plethysmography, oxyhemoglobin–finger-pulse oximeter INDICATIONS Monitor defects in respiratory control–sleep apnea disorders, COPD, restrictive ventilatory disorders

polyurethane film WOUND CARE A transparent polymer-based wound dressing with variable vapor permeability

polyurethane lead CARDIAC PACING A thermoplastic polymer used as an insulating material–eg, on pacing leads

polyuria NEPHROLOGY Excessive urination due to ↑ production

polyvalent pneumococcal vaccine A vaccine against *S pneumoniae* with antigens against 23 of the most common pneumococcal serotype INDICATIONS Populations at risk for pneumococcal infections–eg, elderly, Pts with lung, cardiac, renal disorders, immunocompromised Pts

524 **polyvalent vaccine** IMMUNOLOGY A vaccine containing antigens from multiple serotypes of a pathogen, to induce immune responses against the broadest range of viruses or bacteria–eg, DTP, diphtheria, tetanus, pertussis

polyvinyl chloride PVC TOXICOLOGY A toxin that causes interstitial lung disease due to *bis*(2)-ethyhexylphthalate–DEHP, an agent for 'plasticizing' vinyl chloride polymers, with narcotic effects, and causes acroosteolysis, hepatitis, soft-tissue changes, Raynaud phenomena, hepatic hemangiosarcoma–with as little as 250 ppm, brain tumors, poorly differentiated large cell CAs and adenoCA of lungs. Cf PCB, Plasticizer, Toxic dump

Pomin Diazepam, see there

Pompe disease Glycogen storage disease, type II, see there

pompholyx Acute vesicular palmoplantar eczema A condition common in warm weather, characterized by intense pruritus, possibly psychogenic, related to ↑ autonomic nervous system activity, with crops of palmoplantar vesicles and bullae, which may evolve into eczema. See Factitious dermatitis

ponalrestat See Aldose reductase inhibitor

Pondimin® Fenfluramin, see there

Pontiac fever Pontiac disease A epidemic infection by *Legionella pneumophila* serogroup 6–and other *Legionella* spp, without pneumonia, described in Pontiac, Michigan in 1968 CLINICAL 24-48 hr incubation, fever, headaches, myalgia, cough, ±diarrhea, neurologic signs; resolution in 1 wk. Cf Legionnaire's disease

Pontius syndrome Limbic psychotic trigger reaction FORENSIC PSYCHIATRY A transient neuropsychiatric state in which a previously normal person, often a social loner, 'snaps' and commits unpremeditated and remorseless acts of extreme violence–eg, homicide and mass murder CLINICAL Episode may be preceded by hallucinations or déjà vu, and accompanied by nausea, tachycardia, incontinence, delusions of grandeur

pool MEDTALK The totality of a substance, material or resource in a 'universe'–eg, metabolic pool, donor pool, gene pool. See Gene pool, High-risk pool, Reinsurance pool, Risk pool, Storage pool, Whirlpool, Zero work pool

poor metabolizer PHARMACOLOGY A person who metabolizes a probe drug–the rate of which is related to the metabolizing cytochrome P-450 enzyme–slower than others; a person can be a PM of one probe drug, and an extensive metabolizer of another. See Probe drug. Cf Extensive metabolizer

poorly differentiated ONCOLOGY *adjective* Referring to a malignancy in which the malignant cells bear minimal resemblance to the cell from which they arose. Cf Well-differentiated

POP *noun* ORTHOPEDICS Plaster of Paris, gypsum *verb* DRUG SLANG Pop To subcutaneously inject heroin or other substance of abuse

popcorn calcification Popcorn densities BONE A descriptor for clusters of small scalloped radiolucencies with sclerotic margins seen in the epiphysis and metaphysis of actively growing knees and ankles of children with osteogenesis imperfecta; PC may be due to fragmentation and disordered maturation of the epiphysis with an irregular or defective growth plate, resulting in severe growth retardation LUNG A descriptor for the puffed and lobulated appearance typical of a well-circumscribed calcified solitary hamartoma, seen on a plain AP CXR; multiple 'popcorn' nodularities are suggestive of pulmonary histoplasmosis and may be seen on a plain CXR

pope ACADEMIA A popular term for a physician or researcher regarded by his peers as the ultimate expert–the 'pope' in a particular subject. See Name

popliteal cyst Baker's cyst, see there

popliteal pterygium syndrome An AD condition of neonatal onset and variable penetration and clinical expression CLINICAL A fibrous cord extends from the heel to the ischial tuberosity, limiting leg movement, accompanied by syndactyly, bone malformation, club feet, cleft lip and palate, cryptorchidism, and absence of labia majora

poppers DRUG SLANG A regional street term for amyl nitrate or isobutyl nitrite

population CLINICAL RESEARCH Universe A group of persons to be described or about which one wishes to generalize, assuming that the group is representative of an entire population. See Control population, Patient population GLOBAL VILLAGE The aggregate of persons in a specified area. See Zero population growth

population health management Community-based healthcare MEDTALK The coordination of care delivery across a population to improve clinical and financial outcomes, through disease management, case management and demand management

porcine skin dressing Pig skin dressing WOUND CARE A temporary skin graft that may be used as an occlusive dressing for burns, donor sites of a homograft and decubiti and superficial ulcers

porfiromycin ONCOLOGY An anticancer antibiotic

pork barrel funding Earmarking RESEARCH The practice by Congress of attaching costs for 'pet projects' to certain government spending packages

pork tapeworm *Taenia solium*, see there

PORN Progressive outer retinal necrosis OPHTHALMOLOGY A rare form of necrotizing herpetic retinopathy, most common in immunocompromised Pts–eg, with AIDS; cause unknown, probably VZV CLINICAL Vision loss, floaters, constricted peripheral visual field; funduscopy demonstrates perivenular lucency, followed by a cracked mud appearance of the retina PROGNOSIS Poor; retinal detachment and vision loss the norm. See Retinal detachment

pornography Sexually explicit erotic writings and images. See Child pornography, Erotic, Erotography, Fanny Hill, Obscenity, Potter Stewart standard, Sex work

poroma See Eccrine poroma, Malignant eccrine poroma

porphyria cutanea tarda METABOLIC DISEASE An AD condition caused by uroporphyrinogen decarboxylase deficiency CLINICAL Liver dysfunction, photosensitivity, hyperpigmentation, scleroderma-like skin changes LAB ↑ Uroporphyrin in urine

porphyria METABOLIC DISEASE Any of a family of inborn errors of porphyrin metabolism–most commonly acute intermittent porphyria CLINICAL Confusion, nausea, acute abdominal pain, extreme sensitivity to sunlight resulting in skin lesions; it is acutely exacerbated by alcohol and medications LAB ↑ urinary excretion and circulating levels of porphyrins or precursors–eg, porphobilinogen, δ-aminolevulinic acid. See Acute intermittent porphyria, Congenital erythropoietic porphyria, Erythropoietic porphyria, Porphyria cutanea tarda, Variegate porphyria

port RADIATION ONCOLOGY Field, portal The site on the skin where the radiation enters the body. See Mantle port, Radiation oncology, Subcutaneous port, Venous port

Port-A-Cath NURSING A proprietary indwelling device that provides long-term IV access for administering TPN, blood products, drugs, high-dose chemotherapy. See Total parenteral nutrition. Cf Hickman catheter

port-access cardiopulmonary bypass HEART SURGERY A method for cardiopulmonary bypass surgery using Heartport Port-Access System instruments to surgerize through a single 3- to 4-inch incision in the chest wall. See PSL Group

port-access coronary surgery INTERVENTIONAL CARDIOLOGY Minimally invasive surgery in which a system of catheters is introduced from peripheral blood vessels in the groin and arm, to establish cardiopulmonary bypass and isolate the heart by inflating a balloon in the ascending aorta before a cardiac intervention. See Minimally invasive cardiac surgery

port-access surgery HEART SURGERY MIDCAB—minimally invasive direct coronary artery bypass surgery in which the heart-lung machine used in major cardiovascular procedures is replaced with femoral artery and vein catheterization. Cf CABG, MIDCAB

port site metastasis SURGICAL ONCOLOGY The development of a recurrent tumor at the site of laparoscopic resection of CA, in particular colorectal. See Laparoscopic surgery

port-wine nevus Nevus flammeus, flat hemangioma, port-wine mark or stain A common congenital neurovascular malformation, appearing as deep redpurple macular lesions, corresponding to cutaneous angioma(s), often located in the ophthalmic branch of the trigeminal nerve; when located on the meninges, PWN may be confined to the occipitoparietal pial vessels, where sluggish blood flow predisposes to hypoxia of underlying cortex CLINICAL 5% of Pts with PWNs have convulsions, mental retardation, hemiparesis or hemianopsia contralateral to lesions; PWNs may occur in the normal population—eg, Mikhail Gorbachev, or be part of various syndromes—eg, Klippel-Trenaunay, Beckwith-Wiedemann, Cobb, Rubenstein-Taybi, trisomy 13 syndromes MANAGEMENT Flashlamp-pulsed tunable argon dye laser, most effective if administered < age 7; may require more therapy for facial lesions

port-wine urine A descriptor for the transparent, red urine seen in myoglobinuria due to muscle trauma or intense, prolonged, and/or violent exercise—eg, marathon-running and karate; in contrast, hemoglobinuria with RBC casts is turbid red

portable coverage MANAGED CARE A type of health insurance policy that can be continued with the same level and format of coverage when the insured changes employers. See Clinton Plan, Job lock

portacaval shunt SURGERY A procedure in which the portal vein is anastomosed with the inferior vena cava, diverting blood away from the portal venous system, reducing portal HTN—which causes the feared and potentially fatal complication of cirrhosis, exsanguination from esophageal varices

portal *noun* ORTHOPEDICS A small—eg, ±1 cm incision over a joint to provide access for arthroscopy RADIATION ONCOLOGY See Port *adjective* ANATOMY Referring to the portal vein

portal venography Percutaneous transhepatic portal venography, see there

Porterfield catheter CARDIOLOGY A multipurpose, deflectable electrophysiologic catheter for pacing and recording. See Electrophysiology

POS Point of service, see there

POSCH GI DISEASE A clinical trial—Program on the Surgical Control of the Hyperlipidemias—of the effect of partial ileal bypass on M&M due to CAD in Pts with hypercholesterolemia. See Hypercholesterolemia, Ileal bypass surgery

Posicor® Mibefradil, see there

position MEDTALK A stance or placement. See Beach chair position, Calcaneal neutral position, Dorsal lithotomy position, Figure of four position, Fixed structural position, Frog leg position, Jump position, Lateral decubitus position, Leapfrog position, Recovery position, Sims position, Sniffing dog position, Statue of Liberty position, Stress position, Thorburn's position, Tip-toe position, Waters' position

positioning OSTEOPATHY A method for determining, in a painful joint or muscle, the point of minimum pain; PT relaxes and relieves the pressure and strain on a musculoskeletal group and ↓ spasms. See Osteopathy PSYCHOLOGY Body language—eg, use of arms or crossed legs to face the other person and indicate to a potential sexual partner, interest in closing the physical space between 'preditor' and 'prey'. See Preening behavior

positive See False positive, True positive

positive axillary lymph node ONCOLOGY Lymph nodes in the axilla to which cancer has spread, a poor prognostic indicator in breast CA

positive end-expiratory pressure See PEEP

positive feedback See Feedback

positive inotropic agent CARDIOLOGY An agent—eg, digoxin, gitalin, lanatoside, beta-adrenoceptor agonists that ↑ force and velocity of myocardial contractility

positive most-significant-other PSYCHOLOGY An MSO—spouse, lover, close friend, parent, child, who provides positive mental and/or material support. See Most-significant-other. Cf Negative most-significant-other

positive motivation Positive incentive PSYCHOLOGY A technique in which a person is rewarded for accomplishing a certain task or changing a particular behavior—non-performance of the task is not punished, but rather ignored

Positive & Negative Symptom Scale PANSS PSYCHIATRY A testing device—range of scores, 30–210 used to evaluate the clinical state of a Pt with schizophrenia—high scores indicate worse Sx. See Negative Sx, Positive Sx, Schizophrenia

positive predictive value STATISTICS The number of true positives divided by the sum of true positives—TP and false positives—FP, a value representing the proportion of subjects with a positive test result who actually have the disease, aka 'efficiency' of a test. Cf Negative predictive value, ROC—receiver operating characteristic

positive symptom PSYCHIATRY A symptom due to mental distortion, typical of schizophrenia—eg, perceptual distortions—hallucinations, inferential thinking—delusions, disorganized thinking, agitation Sx are "positive" because the behavior adds to what is considered normal. See Positive & Negative Symptom Scale, Schizophrenia. Cf Negative Sx

positron emission tomography See PET scan

POST NEUROLOGY A clinical trial—1. Posterior Stroke Trial 2. Potassium-channel Opening Stroke Trial

postanginal sepsis A rare condition most common in adolescents and young adults, caused by a peripharyngeal abscess 2° to tonsillitis, pharyngitis, or dental procedures; the infection spreads by direct extension, or into the lymphatic or venous channels, causing thrombophlebitis of the internal jugular vein, septic emboli and metastatic abscesses, often to lung CLINICAL Local, abscess-related Sx—oral and facial edema, hoarseness, dysphagia; lung involvement manifest by high fever, rigors, cough, pleuritic chest pain, hemoptysis, dyspnea MICROBIOLOGY *Fusobacterium necrophorum*, peptostreptococci, bacteroides, *Eikenella corrodens*, *S aureus* etc LAB PMNs ≥ 30,000/mm³, ↑ LFTs, ↓ platelets, ↑ urinary sediment MANAGEMENT High-dose IV antibiotics, covering for anaerobes

postconcussive syndrome SPORTS MEDICINE A constellation of Sx that follow traumatic brain injury, more common in athletes CLINICAL Altered consciousness, anxiety, dizziness, limited post-traumatic amnesia, ability to concentrate, fatigability, headaches, sleep disturbances, disturbed vision, equilibrium, and other Sx, and an initial Glasgow Coma Scale of ≥ 13 of 15. See Concussion

postdatism Post-term pregnancy, see there

post exposure prophylaxis PUBLIC HEALTH The administration of a vaccine and Ig after exposure to a potentially fatal pathogen—eg, rabies . See Rabies vaccine

postexposure zidovudine OCCUPATIONAL MEDICINE Administration of zidovudine—1000 mg/day ±4 wks after percutaneous exposure to HIV; PEZ is associated with a lower—67% rate of HIV infection

postgastrectomy syndrome Dumping syndrome(s) GI DISEASE A condition seen in ± 20% of those subjected to gastric surgery–eg, resection, gastroenterostomy with total gastric vagotomy and gastric bypass, especially, pyloric ablation and bypass CLINICAL Diaphoresis, palpitations, abdominal colic, diarrhea, due to rapid movement or dumping of gastric contents into the small intestine; PGSs occur shortly after a meal–early dumping or several hrs later–late dumping. See Intestinal overgrowth

POSTGASTRECTOMY SYNDROMES

EARLY DUMPING SYNDROME A condition affecting 5-10% of Pts with sub-total gastrectomies, due to the release of vasoactive substances–eg, serotonin, bradykinin, glucagon CLINICAL Onset 20-30 mins after meals with early satiety, upper GI discomfort and vasomotor phenomena–flushing, diaphoresis, palpitations, tachycardia, hypotension, resolving in one hr, weakness, N & V, diarrhea, cramping and borborygmi, flatulence, aerophagia, anemia; when prolonged malabsorption, steatorrhea, weight loss and osteomalacia LAB ↑ Glucose–worse Sx with high carbohydrate meals, ↑ Hct, ↓ blood volume, related to dehydration, ↓ serum K+

LATE DUMPING SYNDROME Less common; more polymorphous clinically; most Sx are due to reactive postcibal hypoglycemia, as the rapid entry of glucose releases GIP–gastroactive intestinal polypeptide, inhibiting the hyperglycemic response to glucagon; spontaneous remission may occur 3-12 months after surgery TREATMENT, MEDICAL Smaller meals, ↓ carbohydrate intake, pectin–a dietary fiber, acarbose, anticholinergics, L-dopa and opiates TREATMENT, SURGICAL 2-5% are medical failures, requiring surgical conversion to a Roux-en-Y

Note: Other postgastrectomy syndromes include the small capacity, afferent and efferent loop syndromes, bile gastritis, anemia, postvagotomy diarrhea and metabolic bone disease

postgraduate year-1 See PGY-1

posthemorrhagic anemia HEMATOLOGY The normochromic normocytic anemia that follows a significant blood loss due to hemorrhaging

postherpetic neuralgia Shingles NEUROLOGY A painful, residuum of nerve injury post-shingles/herpes zoster, due to a reactivation of HZV. See Lidoderm®

posthypnotic suggestion HYPNOSIS The interjection of an 'alien' thought into the mind of a Pt undergoing hypnosis, which would compel him to commit an act that he would not normally commit. PS: Commonly occurs in grade B movies. See Hypnosis, Hypnotherapy

posthypoxic ischemic myoclonus NEUROLOGY Myoclonus that follows transient cerebral hypoxia–eg, post cardiopulmonary arrest CLINICAL Preserved cognitive function; defective voluntary movements of facial, vocal, and other muscles MANAGEMENT Clonazepam, and/or valproic acid

postlingual adjective Referring to a condition that develops after the onset of speech

postlingual hearing loss Postlingual deafness AUDIOLOGY Hearing loss that follows the onset of speech; it is less severe but more stable and more common than prelingual hearing loss; PHL affects 10% of the general population by age 60, 50% by age 80; most PHL is multifactorial. See Hearing loss. Cf Prelingual hearing loss

postmenopausal Change of life GYNECOLOGY adjective Referring to the time in ♀ when menstrual periods stop for ≥ 1 yr

post-mortem adjective After death noun A popular term for an autopsy, see there

postmyocardial infarction syndrome Dressler syndrome, postinfarction syndrome, post-cardiac injury syndrome, postcardiotomy pericarditis CARDIOLOGY A post-MI pericarditis that develops from 2 days to 11 wks after an acute MI in up to 4% of Pts, open heart surgery, stab wounds to the heart, or blunt chest trauma CLINICAL Severe malaise, fever, fibrinous pericarditis often with a friction rub, chest pain, and pleuritis; more aggressive use of aspirin and less aggressive anticoagulation has resulted in a ↓ in frequency of PMIS. Cf Postpericardiotomy syndrome

postnasal drip ENT The sensation that mucus, secretions, or inflammatory products are passing from the nasopharynx into the oropharynx; PNDs usually occur in a background of chronic sinusitis

postnatal NEONATOLOGY adjective After birth Pertaining to events occurring after birth, usually within 1 year of childbirth

postnecrotic cirrhosis Obsolete term, see Cirrhosis

postneonatal mortality PUBLIC HEALTH A standard indicator of health, defined as the number of infant deaths occurring between 28 days and 11 months of life. Cf Infant mortality

postoperative adjective After an operation or surgical procedure

postoperative headache A postoperative complication that occurs shortly after recovery from general anesthesia, which may be due to caffeine withdrawal

postoperative psychosis A symptom complex said to follow surgery, especially in Pts requiring general anesthesia

postoperative reinfusion TRANSFUSION MEDICINE The collection and subsequent reinfusion of blood lost in the operative field during and after surgery, after the blood has been anticoagulated, filtered, and washed INDICATIONS Significant–≥ 250 mL blood losses, the operative field was not contaminated and Pts do not have systemic infections or malignancy. See Surgical blood management

postorgasm blackout SEXOLOGY A loss of consciousness after an orgasm, more common in ♀, attributed to hypotension. See Orgasm

postpartum depression Postpartum 'blues' GYNECOLOGY A prolonged period of depression and flattened affect which begins within a few wks of delivery and may last for months; a stress reaction in ♀ after delivery, characterized by depression, fatigue, irritability, insomnia which, if extreme, may result in infanticide. See Baby blues

postpartum hypopituitarism Sheehan syndrome, see there

postpartum renal failure Postpartum hemolytic uremic syndrome An idiopathic condition with a poor prognosis, characterized by renal failure, microangiopathic hemolytic anemia, thrombocytopenia, DIC, beginning from days to 10 wks after a normal pregnancy and delivery; PRF may be preceded by HTN, proteinuria or preeclampsia CLINICAL Vomiting, diarrhea, flu-like illness may precede the oliguric or anuric phases of acute renal failure accompanied by hemolysis and coagulopathy MANAGEMENT No therapy is consistently effective; early diagnosis, control of HTN and early dialysis may be preventive PROGNOSIS Complete recuperation of renal function occurs in only 10%. See TTP-HUS

postpartum thyrotoxicosis ENDOCRINOLOGY Thyroid gland hyperactivity temporally linked to delivery–eg, de novo or recurrent Graves' hyperthyroidism, characterized by high radioiodine uptake and painless thyroiditis with hyperthyroidism–with low radioiodine uptake; PT is mild, transient, and may result from unmasking of associated autoimmune phenomena MANAGEMENT If needed, propranolol

postperfusion lung Pump lung A condition seen shortly after heart surgery CLINICAL Fever, dyspnea, cyanosis, hypotension, pulmonary edema, caused by anoxia, traumatic hemolysis of RBCs due to shearing against pump hardware, turbulence, possibly anaphylactic reaction to various materials–proteins and other allergens in tubing, CHF, acute renal tubular necrosis, UTIs TREATMENT Antibiotics, corticosteroids PROGNOSIS Relatively guarded

postperfusion syndrome A condition seen in 2% of Pts who undergo cardiac surgery, 3-7 wks after cardiopulmonary bypass, resembles infectious mononucleosis or hepatitis, and is attributed to viruses transfused with the blood during surgery CLINICAL Fever, splenomegaly, lymphadenopathy, maculopapular rash, anemia, atypical lymphocytes PROGNOSIS Benign; resolves spontaneously without therapy. See Post-perfusion lung. Cf Post-resuscitation syndrome

postpericardiotomy syndrome Postcommissurotomy syndrome A condition of acute onset characterized by fever, pericarditis, pleuritis that develops ≥ 2 wks after cardiac surgery, in which the pericardium has been 'violated' by wide incision and manipulation CLINICAL Malaise, fever, pericarditis, friction rub, chest pain, pleuritis MANAGEMENT Aspirin, NSAIDs, if unresponsive, corticosteroids PROGNOSIS PPS is usually self-limited, but often recurs and may be disabling. Cf Postmyocardial infarction syndrome

postpermanent dentition The rare appearance of supernumerary teeth after loss of permanent teeth; most teeth that appear after extraction of permanent teeth are due to eruption of previously impacted teeth

postpill amenorrhea GYNECOLOGY Failure to resume menstruation 3 months after discontinuing OCs; amenorrhea of > 6 months occurs in 0.2%; in 15% is accompanied by galactorrhea

postpolio syndrome NEUROLOGY A progressive late-onset condition occurring yrs after acute poliomyelitis, which affects previously involved muscles CLINICAL Fatigue, muscle weakness, fasciculations, atrophy, dyspnea; PPS is often benign and may reach a plateau phase. See Type grouping

postprandial Postcibal *adjective* After a meal, eating

postpump pancreatitis CRITICAL CARE A complication of cardiopulmonary bypass, characterized by hyperamylasemia, and locoregional inflammation, ±associated with preoperative renal insufficiency, post-op hypotension, peri-operative administration of $CaCl_3$

postpump syndrome CARDIOLOGY A complication of cardiopulmonary bypass characterized by multiorgan dysfunction in the early post-operative period, systemic inflammation with ↑ capillary permeability, interstitial edema, leukocytosis, fever, renal dysfunction, hemolysis, vasoconstriction and possibly ↑ susceptibility to infection. Cf Postperfusion syndrome

postremission therapy ONCOLOGY Chemotherapy to kill CA cells surviving remission-inducing therapy. See Remission

post-renal transplant osteopenia Renal transplant osteonecrosis A phenomenon characterized by ↓ mineral density of the vertebrae, ↓ PTH, phosphorus, and alk phos, and ↑ calcitriol, attributed to the toxic effects of corticosteroids. See Kidney transplantation

postresuscitation syndrome EMERGENCY MEDICINE A condition seen in 'arrested' Pts in whom CPR is delayed, characterized by protracted ↓ in cardiac output, despite normal BP, due to a combination of pump failure, microthromboembolism–2° to intravascular obstruction, causing ↑ systemic vascular resistence, DIC, vasospasm–multi-organ failure, lung insufficiency, and cerebral ischemia triggering arrhythmias, renal shutdown, pulmonary edema. Cf Post-perfusion syndrome

postsplenectomy syndrome HEMATOLOGY A constellation of findings that follow splenectomy, which are most marked in the erythroid series, as the spleen is responsible for 'pitting' and 'culling' effete or defective RBCs CLINICAL Thrombocytosis, hemolysis, ↑ risk of severe infections, especially to encapsulated bacteria–eg, *S pneumonia* LAB ↑ Lifespan of RBCs, codocytes or target cells, schistocytes, Howell-Jolly bodies-nuclear chromatin remnants, thrombocytosis

poststreptococcal glomerulonephritis Post-infectious glomerulonephritis, poststreptococcal GN, proliferative glomerulonephritis NEPHROLOGY A now uncommon form of glomerulonephritis, caused infection of remote sites–eg, the skin or pharynx, with a group A hemolytic streptococcus RISK FACTORS Recent Hx sore throat, strep throat, streptococcal skin infections–eg, impetigo, other streptococcal infections

postsynaptic serotonin antagonist-presynaptic serotonin reuptake inhibitor PSYCHIATRY A class of antidepressants–eg, nefazodone that antagonizes serotoninergic activity

postterm pregnancy Post-datism OBSTETRICS A gestation correctly dated by Naegele's rule and ≥ 42 wks in duration; post-mature infants have ↑ M&M: 1. they are bigger and 2. the placenta has 'planned obsolescence', and undergoes fibrosis and infarction after 40-42 wks CLINICAL Absent lanugo, attenuated vernix caseosa, long finger- and toenails, abundant scalp hair, pale, parchment-like or desquamating skin and ↑ alertness. Cf Prematurity

posttransfusion graft-versus-host disease A condition similar or identical to post-operative erythroderma, seen in immunocompetent blood recipients; PT-GVHD may result from engraftment of donor T cells in transfused blood products, which mount an immune attack against host tissues; PT-GVHD occurs when the donor is homozygous, and the recipient heterozygous for certain HLA antigens CLINICAL High fever, dermatitis, severe diarrhea, liver dysfunction, pancytopenia DIAGNOSIS Analysis of RFLPs and/or DNA probes is either uninformative and/or cumbersome; PCR amplification of polymorphic micro-satellite markers followed by gel electrophoresis can be used to identify Pts at risk for PT-GVHD DIFFDX Drug reactions, toxic shock syndrome, viral infections

posttransfusion purpura TRANSFUSION MEDICINE A nonimmune response to transfusion of packed RBCs and FFP, attributed to passive transfer of antibodies from donor plasma, resulting in complement activation and lung injury CLINICAL Bilateral pulmonary edema, hypoxia, tachycardia, fever and hypotension occurring within 6 hrs of transfusion MANAGEMENT Ventilatory and hemodynamic support PROGNOSIS Complete recovery in 48 h. See Transfusion

posttransplantation lymphoproliferative disorder A complication of 1-10% of organ transplant recipients, which may be poly- or monoclonal; most affected cells contain EBV, usually latent; post-transplantation Pts at risk for PTLD can be identified by detecting expression–as small mRNA of the EBER-1 gene; most cases that occur in solid organ recipients are of host origin

posttrauma malignancy ONCOLOGY A malignancy, usually a sarcoma arising in a site of previous injury–eg, shrapnel, bone infarction, or others; PTMs tend to be aggressive and have a poor prognosis

posttraumatic stress disorder PSYCHOLOGY A psychologic disorder linked to the mental stress of intense trauma or armed conflict; PTSD is defined as one or more of the following: Sx related to re-experiencing a traumatic event, Sx related to avoiding stimuli associated with the trauma, numbing of general responsiveness, or Sx related to ↑ arousal with long-term psychologic 'scars' ETIOLOGY Combat, rape, child abuse, witnessing a violent event, or any serious medical or psychological trauma CLINICAL Nightmares, inability to concentrate, and intrusive thoughts about the traumatic event, numbing, irritability, guilt–for having survived when others died, recurrent nightmares, flashbacks to traumatic scene, overreactions to loud noises, dissociation, anxiety or panic attacks, depression, or anger; PTSD is associated with ↑ alcoholism and may arise in a background of child abuse, PTSD is similar to the 'Vietnam syndrome'; the 'shell shock' form of PTSD occurs in less than 1% of the general population, 15-35% of Vietnam veterans, 30-50% of those exposed to natural disasters and up to 80% of those exposed to man-made disasters–eg, Bhopal. Cf Battle fatigue

posttussive syncope CLINICAL MEDICINE Syncope during a cough, or shortly after a paroxysm of coughing, from which consciousness is regained without sequelae; PSs are more common in ♂, presumably because they can generate greater intrathoracic pressures and thus ↓ cardiac output more

postvoid residual urine volume UROLOGY The amount of urine remaining in the bladder after void completion EVALUATION Bladder ultrasound, intermittent catheterization. See Bladder ultrasound, Intermittent catheterization

postal worker syndrome FORENSIC PSYCHIATRY A popular term for a condition allegedly more common in postal workers, who are said to 'snap' and shoot their colleagues in a bloody rampage

posterior cord syndrome Posterior column syndrome NEUROLOGY A condition due to the loss of vibration and position sense below a lesion of the posterior spinal cord; the PCS is accompanied by a positive Romberg sign, tingling in the affected regions, sensory ataxia, hypotonia, and preserved pain and temperature sense; because PC lesions interrupt the central projections of the dorsal root ganglia cells, they may mimic tabes dorsalis. See Tabes dorsalis

posterior cruciate ligament injury PCL injury ORTHOPEDICS A partial or complete tear, dislocation, or stretch of the PCL from the bone attachment to the knee, or anywhere else along its length; it is usually injured by hyperextension, or a direct blow to the flexed knee. See Posterior cruciate ligament

posterior/anterior drawer test ORTHOPEDICS A joint laxity test used to diagnose shoulder instability. See Laxity test, Shoulder instability. Cf Provocative test

postural fixation NEUROLOGY The loss of anticipatory and compensatory righting reflexes, resulting in the inability to make appropriate postural adjustments to involuntary movements of the trunk. limbs and head; PF is lost early in supranuclear ophthalmoplegia and late in Parkinson's disease, and is attributed to defects in the globus pallidus

postural hypotension Orthostatic hypotension, see there

posture MEDTALK A position of the body. See Decerebrate posture, Decorticate posture, Fetal posture

posturing NEUROLOGY The positioning of the body and limbs. See Decerebrate posturing, Decorticate posturing, Postural fixation PSYCHOLOGY 1. The adoption of a rationalized mental stance 2. The making of gestures

pot Marijuana, see there

potable CLINICAL NUTRITION *adjective* Drinkable *noun* A drinkable fluid

potassium K⁺, Kalium PHYSIOLOGY An alkaline metallic element–atomic number 19; atomic weight 39.09; it is the principal intracellular cation–positive ion and is critical for synthesis of new molecules, transfer of energy, muscle contraction, neural transmission, and maintaining BP; K⁺ in the circulation has a narrow range; it is ↓ in crash dieting, Cushing syndrome, DKA, dehydration, hyperaldosteronism, licorice–due to aldosterone-like effects of glycerrhizic acid, malnutrition, metabolic acidosis, nasogastric suctioning, starvation, stress–burns, surgery, trauma, vomiting, and drugs–eg, aspirin, corticosteroids, potassium-wasting diuretics, estrogen, insulin, laxatives, lithium, sodium polystyrene sulfonate–Kaylexate; it is ↑ with anuria or oliguria, tissue injury/necrosis, burns, potassium in IV solutions, metabolic acidosis, renal insufficiency or failure, and therapy with K⁺-sparing diuretics–eg, spironolactone, antibiotics–eg, cephalosporins, isoniazid, penicillin, epinephrine, histamine POTASSIUM-RICH FOODS Bananas, cereals, legumes, potatoes, prunes, raisins

potassium-sparing diuretic MEDTALK A diuretic that effects mild diuresis, while conserving K⁺ and Mg²⁺; PSDs may be more effective in maintaining the electrolyte balance than cation–K⁺, magnesium supplementation; PSDs include amiloride, triamterene, diazide, maxide, spironolactone See Diuretics. Cf Loop diuretic

potato node RADIOLOGY A descriptor for the enlarged nodular hilar and mediastinal lymph node(s) typical of sarcoidosis

potato nose A rare AD deformity characterized by a bulbous proboscis and developmental visual field defect. Cf 'WC Fields' nose

potato tumor A popular term for a carotid body paraganglioma–CBP arising at the angle of the jaw forming a massive tumor in young adults. See Zellballen

potency MEDTALK A measure of a substance's relative biologic or chemical activity and biologic or biochemical effects ONCOLOGY See Carcinogenic potency PHARMACOLOGY A term for the relative strength of a therapeutic agent TOXICOLOGY A term for the relative strength of a toxin TRANSFUSION MEDICINE The degree of 'antigenicity' or the intensity of agglutination that may be elicited by different alloantigens

potency ratio PHARMACOLOGY The ratio of the effect of a member of a particular family of drugs–eg, corticosteroids or analgesics, to that of a benchmark agent

potential VOX POPULI 1. The difference in electric charge between 2 points in a circuit, expressed in volts or mV. See Action potential, Evoked potential, Inhibitory post-synaptic potential, Late potential, Membrane potential, Spike potential, Ventricular late potential, Zeta potential 2. The inherent capacity to occur. See Biological hazard potential, Biopotential, Biotic potential, Chemical potential, Health potential, Maximum life-span potential

potentiation See Long-term potentiation

pothole sign MEDTALK A clinical sign of acute appendicitis, referring to the severe pain, evoked by every bump in the road–potholes felt by the victim on the drive to the hospital. See Appendicitis

Pott's puffy tumor A fluctuant swelling over frontal bones with osteomyelitis, accompanied by a subperiosteal–pericranial abscess, often 2° to chronic frontal sinusitis; the causative organism in children is often hematogenous and in adults due to direct traumatic extension ORGANISMS *S aureus*, β-hemolytic streptococci, anaerobes DIAGNOSIS Clinical finding; 'hot' lesion by ⁹⁹ᵐTc scanning

Potter syndrome Potter phenotype NEONATOLOGY A complex of findings caused by prenatal renal failure and oligohydramnios; without the cushioning amniotic fluid, produced by the kidneys, the uterus presses directly on infant's face–see Potter's face–and limbs, which are held in abnormal positions or contractures; oligohydramnios also cause hypoplasia of lungs–which don't function properly at birth PATHOGENESIS Intrauterine renal defects–eg, bilateral renal agenesis–and ↓ amniotic fluid production; no kidneys, no amniotic fluid.

Potts' procedure PEDIATRIC SURGERY A palliative operation for congenital pulmonary artery stenosis and tricuspid atresia, consisting of creation of an anastomosis between descending aorta and the left pulmonary artery

pouch VOX POPULI A bag or sac of tissue. See Hartmann pouch, Indiana pouch, J pouch, Koch pouch, Mainz pouch, Marsupial pouch, Pharyngeal pouch, S pouch

pouchitis GI DISEASE A popular term for acute inflammation of intestinal mucosa seen in an ileal reservoir that may extend transmurally, as a late complication of restorative proctocolectomy, possibly due to obstruction and stercoral ulceration DiffDx IBD, DiffDx with idiopathic IBD. See Inflammatory bowel disease

povidine-iodine A topical solution used for neonatal conjunctivitis–ophthalmia neonatorum; in contrast to silver nitrate, erythromycin, etc., PI is less costly, toxic, and more efficient, has a broad antibacterial spectrum, and antiviral–anti-HIV, herpes simplex activity

powder DRUG SLANG A pulverized abuse substance–eg, heroin, amphetamine, cocaine VOX POPULI A pulverized material. See Antler velvet powder, Dover's powder, Fluticasone propionate inhalation powder, Inheritance powder, James Fever powder, Talcum powder

powder-burn spots Mulberry spots A pattern of endometriosis consisting of multiple tiny puckered foci of hemorrhage, surrounded by minute stellate scars and varying fibrosis

powder tattoo FORENSIC PATHOLOGY A geographic appearance on the skin caused by a gun fired at close range, in which the still-burning gunpowder embeds in the skin and cannot be wiped away; the PT helps determine whether clothing was worn over an entrance wound. Cf 'Stippling'

power CLINICAL RESEARCH The probability of detecting a treatment effect of a given magnitude when a treatment effect of at least that magnitude truly exists VOX POPULI Energy. See Crypto-nuclear power, Durable power of attorney, Guild power, Intellectual processing power, Mass stopping power, Null hypothesis, Police power, Pyramid power, Scanning power, Social power

power take-off lesion OCCUPATIONAL MEDICINE Avulsion of the loose skin of the scrotum and penis, caused by moving parts from factory or farm equipment that may engage a trouser leg and twist upward; the skin may be torn from the glans penis—which is spared—and extend to the coronal sulcus

PowerHeart® CARDIOLOGY An AECD designed to monitor cardiac rhythm and to automatically defibrillate when life-threatening arrhthymias are detected. See Automatic external cardioverter-defibrillator

PowerVision ultrasound system IMAGING High frame-rate, fully-digital ultrasound system allows for superior image quality, especially of moving organs—as a fetal heart beating or Pts that are incapable of holding their breath during examination

pox INFECTION DISEASE See Chickenpox, Smallpox

PP 1. Pancreatic polypeptide 2. Paradoxical pulse 3. Partial pressure 4. Periportal 5. Permanent partial—dentistry 6. Plasma protein 7. Plasmapheresis 8. Polypeptide 9. Postpartum 10. Posterior parietal—cortex, brain 11. Posterior pituitary 12. Postprandial 13. Preferred provider 14. Protein phosphatase 15. Proximal phalanx 16. Psychological profile 17. Pulse pressure

PP-oma A pancreatic polypeptide-producing islet cell tumor—40% are malignant—and characterized by hypercalcemia, hyperglycemia, hypomagnesemia, muscle weakness. Cf Gastrinoma, Islet cell tumors, Pancreatic endocrine tumors, WDHA syndrome, Zollinger-Ellison syndrome

PPA 1. Palpation, Percussion & Ausculation 2. Pittsburgh pneumonia agent 3. Postpartum amenorrhea 4. Price per accession 5. Pure pulmonary atresia

PPD 1. Packs per day 2. Personal protective device 3. Postpartum depression, see there 4. Progressive perceptive deafness 5. Purified protein derivative—of tuberculin—the antigenic material used to detect exposure to TB and response to *M tuberculosis*. See BCG

PPD conversion INFECTIOUS DISEASE The conversion from a state of nonreactivity to reactivity to immune challenge by PPD, presumptive evidence of active TB unless proven otherwise. See Tuberculosis. Cf Anergy

PPD skin test PPD test TUBERCULOSIS A test for detecting exposure to *M tuberculosis*. See PPD

PPF Plasma protein fraction, see there

PPHN Persistent pulmonary hypertension of the newborn

PPLO Pleuropneumonia-like organism See *Mycoplasma pneumoniae*

ppm 1. Parts per million 2. Pulse per minute

PPNG Penicillinase-producing *Neisseria gonorrhoeae*, see there

PPO MANAGED CARE Preferred provider organization, see there INFECTIOUS DISEASE Pleuropneumonia-like organism, see there

PPS 1. Post-polio syndrome, see there 2. Postpartum sterilization 3. Postperfusion syndrome, see there 4. Postpump syndrome, see there 5. Prospective payment system, see there

PPV Positive-pressure ventilation

PR 1. Partial remission/response 2. Peer review, see there 3. Per rectum 4. Peripheral resistance 5. Progesterone receptor, see there

PRA 1. Plasma renin activity 2. Progesterone receptor assay, see there

practical nurse NURSING A trained and licensed person who provides custodial type care—eg, help in walking, bathing, feeding, etc PNs do *not* administer medication or perform other medically related services

practice MEDTALK *noun* Surgery–British The place where a physician practices medicine in a privately managed setting. See Better practice, Family practice, General practice, Group practice, Independent practice, Integrated group practice, Malpractice, Office practice, Reduced-risk practice, Solo practice SPORTS MEDICINE To train at a particular activity. See Spring practice VOX POPULI A habit, manner of performing something. See Good laboratory practice, Good manufacturing practice, Malpractice, Office practice, Reduced-risk practice, Solo practice, Spring practice *verb* To perform the art and science of medicine

practice guidelines MEDICAL PRACTICE A set of recommendations for Pt management that identifies a specific or range of range of management strategies. See Peer review organization, Practice standards. Cf 'Cookbook' medicine

practice standard MEDICAL PRACTICE An accepted principle for Pt management; practice variation related to Pt or physician-specific factors is not expected. See Practice guidelines, Practice parameters

practitioner BUSINESS OF MEDICINE Any individual certified or licensed by any health regulatory board of a particular jurisdiction. See Fringe practitioner, General practitioner, Health care provider, Nurse practitioner MEDTALK A duly licensed medical or osteopathic physician, dentist, podiatrist REGULATORY DEFINITION A physician, dentist, licensed nurse practitioner, licensed physician assistant, pharmacist, certified optometrist, veterinarian, scientific investigator, or other person licensed, registered, or otherwise permitted to distribute, dispense, prescribe and administer, or conduct research with respect to, a controlled substance in the course of professional practice in a particular jurisdiction

Prader-Willi syndrome NEONATOLOGY A complex developmental and neurobehavioral disease with an AD pattern of inheritance, occuring in $1:10^4$ births—US CLINICAL Dwarfism with small hands and feet, hypotonia, mental retardation, hyperphagia, obesity, DM, hypogonadism ± cryptorchidism. See Angelman syndrome, Oculocutaneous albinism type II

Pragmarel Trazodone, see there

PRAISE CARDIOLOGY A clinical trial—Prospective Randomized Amlodipine Survival Evaluation Trial—that evaluated the efficacy of amlodipine therapy in Pts with CHF. See Amlodipine, Heart failure

pramipexole Mirapex® NEUROLOGY A dopamine antagonist used to manage Parkinson's disease ADVERSE EFFECTS Constipation, dry mouth, vision defects, insomnia

pramlintide ENDOCRINOLOGY An amylin analogue of possible use in ↑ glucose control in Pts with DM. See Diabetes mellitus

Prandin® Repaglinide, see there

pranlukast Singulair® ALLERGY MEDICINE An antileukotriene that ↓ early and late response inflammation and bronchoconstriction of asthma EFFECT Improved FEV_1, peak expiratory flow rate, Sx—eg, night awakening, ↓ use of β_2 agonists MECHANISM LTD_4 receptor antagonism; pranlukast is used for asthma prophylaxis; it ↓ bronchial hyperresponsiveness in chronic asthma ADVERSE EFFECTS Zafirkulast ↑ in PT in Pts on warfarin. See Antileukotrienes, Asthma

530 **PRAVA-II** CARDIOLOGY A trial comparing pravastatin and placebo in ↓ fatal and nonfatal strokes; pravastatin therapy had a 24% ↓ in nonfatal strokes, but no effect on fatal strokes. See Pravastatin, Stroke

pravistatin Pravichol® CARDIOLOGY A cholesterol-lowering agent that inhibits HMG-CoA–3-hydroxy-3-methyl coenzyme A reductase, the rate-limiting enzyme of cholesterol synthesis, and ↓ risk of M&M in Pts with CAD, prior MI or at risk for TIA and strokes. See CARE, Cholesterol, LIPID

praziquantel Biltracide® The antiparasitic of choice for schistosomiasis–*S haematobium*, *S japonicum*; flukes or Trematoda–eg, *Clonorchis sinensis*, *Opisthorchis viverrini*, *Paragonimus buski*, *Heterophyes heterophyes*, *Metagonimus yokogawai*; tapeworms–*Hymenolepsis nana*, *Taenia solium* SIDE EFFECTS Drowsiness, headache, nausea, backache, abdominal discomfort

prazosin Minipress® An alpha blocker used as an antihypertensive

PRB Pharmaceutical Resources Branch

pre- *prefix*, Latin, before

preadmission testing LAB MEDICINE A battery of tests performed before a person is admitted to the hospital for elective surgery–eg, cataract extraction or cholecystectomy, which establish baseline values

preauthorization MANAGED CARE The requirement by an HMO that a costly surgery, specialist referral or non emergency health care services be approved by the insurer before it is allowed. See HMO

precancer Premalignant, preneoplasia Any of a broad group of conditions with a malignant predisposition; epithelial precancers may be 1. Glandular–eg, adenomatous hyperplasia–endometrium and adenomatous polyps–colon, stomach, which evolve towards adenocarcinoma of their respective organs or 2. Squamous–eg, dysplasia of the uterine cervix or other urogenital mucosae; premalignant lesions of mesenchymal origin include prelymphoma and 'presarcoma'–an ad hoc coinage, the latter of which may be due to predisposing factors–eg, osteosarcoma may arise in Paget's disease of bone, radiation, hereditary multiple exostoses, polyostotic fibrous dysplasia, enchondroma, Maffucci's enchondromatosis; osteosarcomas may be induced experimentally by various trauma–chemical–eg, turpentine, mechanical– eg, local pressure, indwelling foreign bodies, ischemia–eg, vessel clamping. Cf Preneoplastic state. See Fragile X syndrome, Hereditary neoplasms, Premalignancy

PRECANCER

CHROMOSOME BREAKAGE SYNDROMES Bloom syndrome, Fanconi syndrome

GENODERMATOSES Albinism, dyskeratosis congenita,epidermodysplasia verruciformis, polydysplastic epidermolysis bullosa, Werner syndrome, xeroderma pigmentosa

HAMARTOMATOUS SYNDROMES Multiple exostoses, neurofibromatosis, Peutz-Jegher syndrome, tuberous sclerosis, von Hippel-Lindau syndrome

IMMUNODEFICIENCY SYNDROMES Ataxia-telangiectasia, Wiskott-Aldrich syndrome, X-linked agammaglobulinemia

precancerous changes of cervix GYNECOLOGY Cervical dysplasia, aka CIN, see there

precancerous polyps ONCOLOGY A polyp that arises int the mucosa, which has an ↑ risk of cancer

precautions INFECTIOUS DISEASE The constellation of activities intended to minimize exposure to an infectious agent; precautions imply that the isolation of an infected Pt is optional, but not mandatory. See AIDS precautions, Barrier precautions, Universal precautions. Cf Isolation; Safe sex practices, Reverse precautions

PRECEDENT CARDIOLOGY A trial–Prospective Randomized Ectopy Evaluation on Dobutamine on Natrecor®–nesiritide Therapy

preceptorship GRADUATE EDUCATION A period of hands-on training under a physician or surgeon skilled in a technique–eg, placement of a stent in a coronary artery, or laparoscopic surgery. See Laparoscopic surgery, Paradoxical movement

precertification Preadmission certification review, precert MANAGED CARE The obtaining of authorization from a managed health plan for routine inpatient hospital admissions or outpatient therapy. See Capitation, Virtual capitation, 'Withhold'

precipitate labor OBSTETRICS Parturition in < 3 hrs in a primigravida. See Labor

precision LAB MEDICINE A measure of test or assay reproducibility–ie, capability of producing the same results when performed on the same specimen under the same conditions; data with high precision has a low standard deviation and a low coefficient of variation. Cf Accuracy

precision therapy RADIATION ONCOLOGY Stereotactic therapy for treating CA, directing multiple beams of low-dose radiation intersecting at the tumor site, and high-dose radiation to the exact site, while not harming surrounding healthy tissue. See Radiation oncology

preclinical medicine Basic sciences GRADUATE EDUCATION The 2-3 yr period of medical school, that precedes clinical instruction and training, which provides a core of basic knowledge required for success in clinical medicine–for the student's rotations though surgery, internal medicine, gynecology, pediatrics, and the other fields of clinical medicine; basic sciences include anatomy, physiology, biochemistry, molecular biology, pharmacology, microbiology, pathology

precocious puberty Premature puberty, pubertas precox ENDOCRINOLOGY The appearance of 2° sexual characteristics < age 8 in ♀–complete by age 9; < age 9 in ♂–complete by age 11; if caused by hypothalamic-pituitary axis activation, it is known as complete or true PP; if the precocity is 2° to ectopic production or autonomous secretion of end-organ hormones, it is designated incomplete PP . See Premature puberty, Pseudoprecocity. Cf Pubertal delay

PRECOCIOUS PUBERTY

TRUE OR COMPLETE PRECOCIOUS PUBERTY–affects both sexes

Idiopathic

Constitutional or familial

CNS disease, eg tumors (hypothalamic and pineal gliomas, craniopharyngiomas, germinomas, hamartomas of the tuber cinereum), encephalitis, abscesses, cysts, sarcoidosis, TB

McCune-Albright syndrome

Hypothyroidism

Virilizing syndromes, eg congenital adrenal hyperplasia

INCOMPLETE OR PARTIAL PRECOCIOUS PUBERTY

♂ Due to gonadotropin-secreting tumors, eg hepatoma, Leydig cell tumor, excessive androgen production or premature Leydig cell and germ cell maturation

♀ Due to ovarian follicle cysts or estrogen-producing neoplasms, eg granulosa cell tumor

preconscious PSYCHIATRY *adjective* Referring to thoughts that are not in immediate awareness but that can be recalled by conscious effort

Precose® Acarbose ENDOCRINOLOGY An agent used with metformin–Glucophage® or insulin to control postprandial glucose spikes in Pts with type 2 DM, or as monotherapy to ↓ blood glucose levels in type 2 DM unresponsive to diet alone, administered with a sulfonylurea– Glucotrol™–eg, when diet and a sulfonylurea don't adequately control glycemia. See Diabetes mellitus

precut sphincterotomy GI DISEASE An endoscopic sphincterotomy that requires endoscopic dissection of the papilla with various techniques–eg, needle-knife method, to access the bile duct. See ERCP

predetermination MANAGED CARE A requirement for prior approval from a health insurer before it will pay for a proposed treatment. See Red tape rationing

PREDICT-HD A clinical trial–Neurobiologic Predictors of Huntington Disease onset

predictive value P value DECISION-MAKING A value that predicts the likelihood that a result from a test reflects the presence or absence of a disease. Cf ROC curve

predisposition testing Susceptibility testing Analysis of testing of molecular markers to ID inherited mutations linked to CA–eg, inherited colorectal–MSH2, MSH1; breast–BRCA1, BRCA2; endocrine–RET, melanoma–p16; Li-Fraumeni syndrome–p53. See BRCA1, p16, p53, Tumor marker

predonation TRANSFUSION MEDICINE The depositing of one or more units of packed RBCs for autologous blood transfusion before an anticipated need–eg, elective surgery. See Autologous donation, Leapfrog method

preeclampsia OBSTETRICS A hypertensive disorder occurring in the 3rd trimester in ± 5% of pregnancies CLINICAL HTN, proteinuria, dependent edema, vasospasm, coagulation defects TREATMENT Low-dose aspirin ↓ preeclampsia in nulliparas, especially with systolic HTN; aspirin ↑ risk of abruptio placentae, but does not ↓ perinatal mortality. See Eclampsia

preejection period CARDIAC PACING The interval between electrical initiation of systole and first flow of blood through pulmonic or aortic valves. See Ejection fraction

preemptive analgesia Sensory blockade before incision ANESTHESIOLOGY A pain control strategy intended to counteract central sensitization–'windup' or neuroplasticity, a 'physiologic' state of ↑ sensitization of excitable spinal neurons, coupled to a ↓ threshold for peripheral afferent pain terminals; PA combines local anesthetics, neuraxial blockade, and inhibition of mediators of windup to prevent self-perpetuating pain. See NMDA receptor antagonist, Windup. Cf Multimodality pain control

preemptive chemotherapy Neoadjuvant chemotherapy ONCOLOGY Aggressive chemotherapy in a person with a previously diagnosed and treated CA, who is at high risk of tumor recurrence. Cf Chemoprevention

preexcitation syndrome Wolff-Parkinson-White syndrome, see there

preexisting condition HEALTH INSURANCE An injury, illness or medical condition–eg, cancer, DM, HTN, mental disorder that a person had before a (new) health insurance policy becomes effective TRAUMATOLOGY A chronic medical condition encoded as a 2° diagnosis at the time of discharge from a hospital

preferred provider organization MANAGED CARE A network of independent health care providers who provide medical services to a health plan's members or purchasers–eg, insurance companies, employers and other health care buyers at a discount; PPO members typically have autonomy over health care rather than needing to pass by a primary care (gatekeeper) physician like HMO members; use of in-network physicians is less expensive than using non-network providers. See Fee-for-service, HMO

pregnancy OBSTETRICS The state of gestation; the period of time from confirmation of implantation of a fertilized egg within the uterus–presumptive signs of pregnancy include missed menses or a positive pregnancy test [45 CFR 46.203(b)]–until the fetus has entirely left the uterus–ie, been delivered. See At-risk pregnancy, Cervical pregnancy, Clinical pregnancy, Crisis pregnancy, Ectopic pregnancy, Fatty liver of pregnancy, High-risk pregnancy, Mole pregnancy, Multifetal pregnancy, Postterm pregnancy, Pseudopregnancy, PUPPP, Selective termination of pregnancy, Sympathy pregnancy, Teenage pregnancy, Tubal pregnancy, Unwanted pregnancy

pregnancy granuloma Granuloma gravidarum, pregnancy tumor An exuberant, pyogenic granuloma-like inflammatory response of pregnant gums to an overhanging margin of tooth filling or crown with excess calcium buildup; PGs occurs in the 1st trimester in ±1% of pregnancies, grow until delivery and regress spontaneously without therapy, but may re-appear with subsequent pregnancies

pregnancy-induced hypertension A term that encompasses isolated–nonproteinuric HTN, pre-eclampsia or proteinuric HTN, eclampsia; PIH occurs in 5-15% of pregnancies, and is a major cause of obstetric and perinatal M&M MANAGEMENT Low-dose aspirin

pregnancy test Any test used to detect or confirm pregnancy; in early pregnancy, all PTs measure hCG, the developing placenta's principal hormone, which is detectable as early as 6 days after fertilization; in clinical laboratories, serum levels of hCG are quantified by ELISA or RIA; home testing kits are reliable, widely used, operate on the principle of agglutination, and have < 1% error rate REF RANGE Nonpregnant < 5 mIU/mL. See hCG, Estriol

preleukemia Myelodysplastic syndrome, smoldering leukemia Any of a heterogeneous group of clonal expansions of stem cells, characterized by dysmyelopoiesis–a complex of structural and functional defects–eg, abnormal cell morphology, ineffective hematopoiesis, chromosome defects–eg, aneuploidy and pseudodiploidy EXAMPLES Acquired idiopathic sideroblastic and nonsideroblastic anemias, pancytopenia with hypercellular marrow, paroxysmal nocturnal hemoglobinuria, which may all progress to ANLL

preliminary diagnosis Working diagnosis, see there

prelingual hearing loss Prelingual deafness AUDIOLOGY Hearing loss that precedes the onset of speech or learning spoken language; it is severe, stable, and less common than postlingual hearing loss; 50% of PHL is monogenic, 75% of PHL is AR; 20%, AD; 5%, X-R; the rest are caused by perinatal factors, infections in infancy, or trauma. See Hearing loss. Cf Postlingual hearing loss

preload CARDIOLOGY The amount of hemodynamic pressure upstream from the heart, which is ↑ in heart failure due to mitral or aortic valve regurgitation. Cf Afterload

Prelone® Prednisone, see there

prelymphoma HEMATOLOGY A condition characterized by monotonous aggregates of lymphocytes and a tendency to evolve to lymphoma–and thus warrants close clinical follow-up; prelymphomas include pseudolymphoma of the orbit and small intestine, lymphomatoid granulomatosis–an angiocentric lung disease with lymphoma-like extrapulmonary involvement, angioimmunoblastic lymphadenopathy, and lymphoid interstitial pneumonitis

premalignant Precancerous, see there

Premarin® GYNECOLOGY An estrogen conjugate obtained from pregnant horse mare urine commonly used as a hormone-replacement in postmenopausal ♀, to relieve the Sx of menopause–eg, hot flashes, osteoporosis, vaginal atrophy, etc ADVERSE EFFECTS Dizziness, lightheadedness, stomach upset, bloating or nausea, severe depression, calf pain, sudden severe headache, chest pain, SOB, lumps in breast, weakness or tingling in the arms or legs, dysmenorrhea

premarital profile Premarital screen LAB MEDICINE An abbreviated battery of tests–eg, automated reagin test for syphilis and rubella IgG antibody screen–required in some states for completing a marriage certificate

premarket approval MEDICAL DEVICES A scientific and regulatory review by the FDA to ensure the safety and effectiveness of a Class III device, before its approval for marketing. See Advisory panel, Medical device

premature atrial complex Atrial premature contraction, atrial premature depolarization CARDIOLOGY A premature heartbeat of atrial origin, and common cause of irregular pulse; PACs arise in normal hearts, but are more common in structurally abnormal hearts, and ↑ with age **EKG** Premature P wave, P-R interval of > 120 msec **MANAGEMENT** PACs are rarely serious and do not require therapy

premature contraction Ectopic heartbeat, see there

premature ejaculation Ejaculatio praecox **UROLOGY** A manifestation of sexual dysfunction, defined in the DSM-IV as *'Persistent or recurrent ejaculation with minimal sexual stimulation before, on, or shortly after penetration and before the person wishes it.*–Criterion A, which *'...causes marked distress or interpersonal difficulty'*–Criterion B, and is not substantially associated with the effects of a substance–ie, a drug or substance of abuse–Criterion C; in evaluating PE, factors weighed include age, novelty of sexual partner, recent sexual activity, etc. Cf Nocturnal emission, Onanism

premature infant Prematurity, premie; preterm infant **OBSTETRICS** An infant born before the 37th wk of gestation and after the 20th wk, who weighs 500–2500 g. See Very-low birth weight

premature ovarian failure Cessation of menses before age 40, often accompanied by ↑ serum gonadotropin **ETIOLOGY** Idiopathic, or 2° to ovarian receptor antibodies, viral infection, cytotoxic drugs, RT, etc

premature rupture of membranes Premature rupture of fetal membranes, PROM **OBSTETRICS** The leakage of amniotic fluid before the onset of labor; PROM occurs in 8% to 10% of term pregnancies, 15-20% of preterm pregnancies and associated with ↑ M&M **ETIOLOGY** Unknown, possibly due to bacterial and/or internal enzymes **COMPLICATIONS** Respiratory distress syndrome, fetal and neonatal infections–eg, congenital pneumonia or septicemia, fetal wastage, intraventricular hemorrhage **MANAGEMENT** Deliver baby within 36 hrs. See Amnion, Chorion

premature separation of placenta Abruptio placentae, see there

premature ventricular contraction Ventricular premature beat **CARDIOLOGY** A ventricular beat initiated by an ectopic focus, which occurs before the usual sinuatrial beat, characterized by premature, widened, bizarre QRS complexes, not preceded by a P wave; PVCs are extremely common, occur in ½ of normal adults monitored by EKG for 24 hrs, and are of no significance **ETIOLOGY** Anxiety, fever, stimulants, associated with drug toxicity–digitalis, quinidine, tricyclic antidepressants and an indication for discontinuing therapy **TREATMENT** Lidocaine, quinidine, propranolol, amiodarone

prematurity **OBSTETRICS** The state of being premature, which encompasses a constellation of clinical findings in an infant delivered before 37 wks–from the 1st day of the last menstrual period of gestation, who is also immature **CLINICAL** Anemia due to ↓ iron and vitamin E, bronchopulmonary dysplasia, intraventricular hemorrhage, especially in very LBW infants, permanent neurologic sequelae and retinopathy. See Low birth weight, Premature infant. Cf Post-term pregnancy

premed Premedical *adjective* Referring to preparing for a career in medicine *noun* A popular term for a student studying the core curriculum required for admission to medical school in the US

premenopausal *adjective* Referring to the period in a ♀ preceding menopause, aka 'change of life'. See Menopause

premenstrual dysphoric disorder The official term for PMS–premenstrual syndrome, see there

premenstrual syndrome PMS, Premenstrual tension, premenstrual dysphoric disorder **GYNECOLOGY** A cyclical disorder characterized by affective, behavioral, and somatic Sx that often occur during the luteal–2nd phase of the menstrual cycle, resolve with the onset of menses, and are weakly linked to the fall in estrogen and progesterone from luteal peaks **CLINICAL**

Premenstrual dysphoric disorder–PMS affects 10–30% of menstruating ♀ and is characterized by several days of mental or physical incapacitation of varying intensity, insomnia, headaches, emotional lability–anxiety, depression, irritability, loss of concentration, poor judgement, mood swings, violence evoked by environmental cues, acne, breast enlargement, tenderness, abdominal bloating, edema, craving for salty, sweet, or 'junk' foods **MANAGEMENT** Fluoxetine–Prozac® for tension, irritability, dysphoria, ovariectomy, anxiolytics are marginally better than placebos

PREMENSTRUAL DYSPHORIC DISORDER–RESEARCH CRITERIA

Five + of below symptoms in a cyclic fashion, at least one of which is 1-4

1. Depressed mood, self-deprecation, hopelessness
2. Anxiety, tension, feeling 'wired'
3. Emotional lability
4. Marked and/or persistent anger, irritability, or interpersonal conflict
5. Decreased interest in usual activities or relationships
6. Difficulty in concentrating
7. Lethargy
8. Change in appetite
9. Change in sleep habits
10. Subjective sense of loss of control
11. Physical symptoms, eg breast tenderness, headaches, arthralgia, myalgia, bloating, weight gain

Modified from DSM-IV

premie Premature infant, see there

Premier TOBACCO CONTROL A smokeless cigarette produced briefly by RJ Reynolds in 1987 which heated the tobacco to give the 'smoker' the desired nicotine 'high'; consumers didn't like the taste. See Eclipse

Premilene® SURGERY A polypropylene monofilament suture for use in cardiac, plastic, and vascular surgery and skin closure. See Suture

Prempro® GYNECOLOGY An OC with ↑ medroxyprogesterone acetate–Cycrin® and estrogen conjugates–Premarin®. See Oral contraceptives

prenatal diagnosis OBSTETRICS The examination of fetal cells taken from the amniotic fluid, the primitive placenta–chorion, or umbilical cord for biochemical, chromosomal, or gene defects

prenatal masculinization ENDOCRINOLOGY The masculinizing effect on the embryonic and fetal sexual anatomy and/or sexual pathways of the brain, induced by testosterone, estradiol, or other androgenizing hormones

prenatal test OBSTETRICS Any lab test or assay used to detect genetic and/or congenital fetal anomalies that would compromise the infant's well-being and quality of life to such a degree that the parents might prefer an abortion **EXAMPLES** α-fetoprotein levels in mother's serum or amniotic fluid, chromosome analysis, ultrasonography, chorionic villus biopsy–8th-10th gestational wks, risk of spontaneous abortion/miscarriage–1.0-1.5%, amniocentesis–16th gestational wk, risk of spontaneous abortion/miscarriage–0.5%, fetoscopy, embryoscopy **DISEASES DETECTED** α_1-antitrypsin activity, thalassemias, defects of sex and autosomal chromosomes, eg, trisomies, deletions, mosaicisms, fragile X syndrome, hemophilia, neural tube defects–anencephaly, spina bifida, polycystic kidney disease, Tay-Sachs disease. See Embroscopy, Fetoscopy. Cf Wrongful birth

preneoplastic state Precancer, see there

pre-op Pre-operative *adjective* Referring to the logistics of preparing for a surgical procedure/operation. See Pre-op time *noun* 1. A popular term for all forms of care–eg, medication provided to a Pt prior to a surgical intervention 2. The ward where Pts are held before surgery

pre-op time SURGERY The time required for readying the Pt and the OR for surgery ACTIVITIES Cleanup from prior case, pulling supplies for next case, checking equipment, suction, cauterization, camera, light bulbs etc, readying Pt for OR–eg, review of vital signs, Hx, physical examination, signed consent, preanesthesia.

pre-operative autologous donation Prestorage TRANSFUSION MEDICINE The deposition by a Pt of his/her own blood in a hospital's blood bank before elective surgery in which blood transfusion may be required INDICATIONS PAD is appropriate for procedures in which significant blood loss is typical–eg, total hip replacement. See Autologous blood, Surgical blood management

'prep' SURGERY The preparation of both the Pt and the operative site before the first incision, which consists of painting the skin surface with disinfectant, shaving hair, and inserting a catheter

preparticipation physical exam SPORTS MEDICINE A physical exam that may be required by law before a child or adolescent can participate in school-related sports; PPE are intended to detect any cardiovascular or musculoskeletal condition that may ↑ the risk of injury or death. See Station system

PREPIC CARDIOLOGY A French study–Prévention du Risque d'Embolie Pulmonaire par Interruption Cave that compared the effect of vena cava filter and LMW heparin in Pts at high risk of DVT. See Deep vein thrombosis, Pulmonary thromboembolism, Vena cava filter

prerenal azotemia Renal underperfusion NEPHROLOGY The most common form of acute renal failure, characterized by ↑ nitrogenous waste, due to ↓ blood flow to the kidney LAB ↑ nitrogenous wastes–eg, creatinine and urea, which act as poisons when they accumulate in the body, damaging tissues and compromising organ function RISK FACTORS ↓ blood volume–eg, dehydration, prolonged vomiting, diarrhea, bleeding, burns, etc; pump failure–eg, CHF, shock, kidney trauma or surgery, renal artery embolism, and other types of renal artery occlusion. See Renal failure

Presamine® Imipramine, see there

presbycusis Age-related hearing loss AUDIOLOGY A progressive loss of hearing 2° to age-related changes in the inner or middle ear, beginning with high-frequency sounds–eg, speech, which may have a genetic predisposition, as it tends to occur in families; it occurs ±25% > age 65 to 75 yrs old, 50% > age 75. See Hearing loss

presbyopia Old eyes OPHTHALMOLOGY The gradual loss of the lens's ability to accommodate–for near objects, due to ↓ elasticity, associated with aging; occurs in almost all people > age 45

prescription PHARMACOLOGY An order for drugs or medical supplies, written, signed or transmitted by word of mouth, telephone, or other means of communication to a pharmacist by a duly licensed physician, dentist, veterinarian or other practitioner, authorized by law to prescribe and administer such drugs or medical supplies. See Frontier prescription, Prescription

prescription drug Prescription medication PHARMACOLOGY An FDA-approved drug which must, by federal law or regulation, be dispensed only pursuant to a prescription–eg, finished dose form and active ingredients subject to the provisos of the Federal Food, Drug, Cosmetic Act. See Legend drug

presenile dementia 1. Alzheimer's disease, see there 2. Pick's disease, see there

present pron, PREE-sent verb intransitive CLINICAL MEDICINE To come to medical attention OBSTETRICS To appear–eg, a fetal part at the opening cervical os during labor verb transitive To formally provide information about a case or Pt

presentation CLINICAL MEDICINE The manner in which a Pt appears to his/her caregivers, generally at the time of initial examination. See Clinical

presentation OBSTETRICS The manner in which an infant presents in the birth canal. See Breech presentation, Brow presentation, Cephalic presentation, Face presentation, Footling breech presentation

preset CARDIAC PACING A parameter of a pacemaker that is programmed permanently when manufactured

pressure VOX POPULI A force or stress applied to a suface by a fluid or object, and measured in units of mass per unit area. See Blood pressure, Continuous positive airway pressure, Coronary perfusion pressure, End-diastolic pressure, End-systolic pressure, Intracranial pressure, Intraocular pressure, Intrauterine pressure, Negative pressure, Negative end-expiratory pressure, Oncotic pressure, Osmotic pressure, PEEP, Pulmonary-capillary pressure, Transpulmonary pressure, Wedge pressure

pressure dressing A misnomer for an occlusive, pressureless wound dressing that stabilizes and partially immobilizes a region of skin; PDs are used for burns

pressure pants EMERGENCY MEDICINE An intermittent pneumatic leg compression device used to ↓ deep and proximal vein thromboses, and prevent post-surgical venous thrombosis seen after general, prostatic, orthopedic, and neurosurgery. Cf MAST–Military anti-shock trousers

pressure point EMERGENCY CARE A point or zone on the body surface–femoral, brachial, or carotid region that can be compressed to control hemorrhage from the respective artery NEUROLOGY A zone highly sensitive to pain

pressure ulcer Bedsore A decubitus ulcer on dependent sites, usually lumbosacral, but also on heels, knees, or vertebrae, which is most common in the bed-ridden elderly, seen in up to ¼ of nursing home residents, and associated with an ↑ mortality RISK FACTORS Nonblanchable erythema, lymphopenia, immobility, dry skin, ↓ body weight, activity limited to bed or chair MANAGEMENT A 'cocktail' of recombinant PDGF, proteases, cell-adhesion molecules may induce healing of recalcitrant PUs

PRESSURE ULCER

STAGE I Nonresolving erythema with no break in skin

STAGE II Erythema with superficial disruption of skin, abrasions, vesiculation

STAGE III Full-thickness loss of skin with serosanguineous drainage

Stage IV Full-thickness loss of skin and invasion of deeper tissue

pressured speech Pressure of speech NEUROLOGY Speech that is rapid and voluminous, as if the speaker is under pressure PSYCHIATRY A virtually continuous flow of accelerated speech that may be difficult or impossible to interrupt by the listener; PS may be loud, emphatic, socially uninhibited, and continue even though no one is listening

PRESTO CARDIOLOGY A clinical trial–Prevention of Restenosis with Tranilast & its Outcomes–which will evaluate the effect of tranilast in preventing restenosis after revascularization procedures. See Isolated systolic hypertension, Omapatrilat, Vasopeptidase inhibitor

presumed consent TRANSPLANTATION The assumption that a particular action would have been approved by a person or party if permission had been sought. See Cadaveric organ transplantation, Mandated choice, Organ brokerage, Transplantation. Cf Informed consent

presymptomatic testing MOLECULAR MEDICINE The early testing of a Pt who may have–based on family Hx–a genetic disorder–eg, Huntington's disease or adult polycytic kidney disease, that manifests itself late in life. Cf Genotype

presynaptic-postsynaptic noradrenergic-serotoninergic reuptake antagonist PHARMACOLOGY A class of antidepressants–eg, mirtazapine that antagonizes noradrenergic, serotoninergic activity

534 **presyncopal dizziness** Vertigo preceding syncope, accompanied by a sense of impending blackout, and tunnel vision, all due to ↓ blood flow to the brain, typically worse when standing; PD may occur in healthy persons who rise too quickly from a chair, often after a meal, accompanied by brief disorientation. See Orthostatic hypotension.

presyncope NEUROLOGY An episode of near-fainting which may include lightheadedness, dizziness, severe weakness, blurred vision, which may precede a syncopal episode . See Syncope

presystolic murmur CARDIOLOGY A late diastolic murmur caused by atrial contraction against mitral stenosis or AV valve stenosis

preterm infant Premature infant, see there

pretty good privacy INFORMATICS *adjective* Referring to popular low-cost encryption tool for sending, receiving and storing secure email, including digital signatures. See Public key infrastructure

Prevacid® Lansoprazole INTERNAL MEDICINE A proton pump inhibitor used to manage–heal and relieve active duodenal ulcers, erosive esophagitis, long-term treatment of gastric hypersecretion–eg, Zollinger-Ellison syndrome, short-term treatment of GERD. See GERD

Prevacid® triple therapy INTERNAL MEDICINE A regimen for treating *H pylori* in Pts with duodenal ulcers, consisting of lansoprazole–Prevacid®, clarithromycin–Biaxin® and amoxicillin–Trimox®, packed in a daily Prevpac to aid Pt compliance. See GERD, *Helicobacter pylori*

prevalence EPIDEMIOLOGY 1. The number of Pts with a specific condition at a specified time divided by the total number of people in the population 2. The number or proportion of cases, events or conditions in a given population at a given time. See Disease prevalence, Period prevalence. Cf Incidence PHARMACOLOGY The proportion of positives among the agents tested. See Two-by-two table

prevalidation NIHSPEAK The process during which standardized test protocols are constructed for use in validation studies, and labs selected and determined to be competent to perform validation studies

prevascular phase ONCOLOGY A stable, relatively non-aggressive phase of tumor development that may persist for years, characterized by limited tumor growth; the PP may be 'switched' by an unknown mechanism to a 'vascular phase,' characterized by angiogenesis, rapid tumor growth, bleeding and metastatic potential. See Metastasis

Preven™ GYNECOLOGY An "Oh-my-God-that-color-couldn't-have-changed-he-wore-a-condom-what-will-I-do-now" kit, consisting of a Pt information booklet, a urine pregnancy test, and 4 emergency contraceptive pills, each with levonorgestrel and ethinyl estradiol INDICATIONS Prevent pregnancy after contraceptive failure or unprotected intercourse, within 72 hours of unprotected intercourse; the second two ECPs to be taken 12 hours later, if and only if the accompanying urine pregnancy test is negative, indicating an existing pregnancy. See Oral contraceptive, RU 486

PREVENT CARDIOLOGY Any of a number of clinical trials: 1. Prevention of Recurrent Venous Thromboembolism 2. Program in Ex Vivo Vein Graft Engineering via Transfection 3. Proliferation Reduction with Vascular Energy Trial 4. Prospective Randomized Evaluation of the Vascular Effects of Norvasc® Trial

prevention counseling AIDS Advising Pts on the risk of HIV infection and developing a plan to ↓ that risk for them and their partners

preventive medicine PUBLIC HEALTH The branch of medicine dedicated to preventing disease, injury, and disability, and promoting health; PM attempts to identify preventable diseases and risk factors thereof PREVENTIVE CARE Routine physical exams, surveillance screening–eg, mammography, immunizations–eg, measles, mumps, rubella, education on promoting safety–eg, use of bicycle helmets and ↓ high-risk–eg, smoking–behaviors. See Screening

Preveon® Adefovir dipivoxil, see there

Prevnar® PEDIATRICS A proprietary 7-valent conjugate vaccine for routine immunization against invasive pneumococcal disease in infants

Prevpac® GI DISEASE Packaging for Prevacid® triple therapy with Prevacid®, Biaxin®, Trimox® . See *Helicobacter pylori*

priapism UROLOGY A urologic emergency characterized by painful erection without sexual excitement or desire; 60% are idiopathic, the rest are due to various disorders–eg, leukemia, pelvic infection, pelvic CA, sickle cell anemia, abuse substances–alcohol, cocaine, marijuana, methaqualone, drugs–anticoagulants, antihypertensives, corticosteroids, neuroleptics, tolbutamide, papaverine, scorpion bites, penile or spinal cord trauma CLINICAL Painful, prolonged erection with a tense, congested corpora cavernosa PROGNOSIS Without decompression, interstitial edema and fibrosis of penile shaft ensue, causing impotence. See Bobbittize, Penile prosthesis, Peyronie's disease

price fixing PHYSICIAN PRACTICE The illegal practice by a group of health care providers of establishing a standard price for procedures, thus creating a monopoly on a particular market segment

PRICE principle SPORTS MEDICINE A rule for treating an acute joint injury, per a mnemonic, PRICE: *Protection* of injured body part with crutches, splints, or immobilizer, *Rest* of injured part, *Ice, Compression, Elevation* to ↓ swelling and facilitate early range of motion

prickly heat Miliaria, see there

Priftin® Rifapentine, see there

Prilosec® Omeprazole PHARMACOLOGY A proton pump inhibitor preferred to H_2 blockers for managing GERD, heartburn, erosive esophagitis, benign duodenal and gastric ulcers, and for certain hypersecretory conditions; combined with clarithromycin for eradicating *H pylori* associated with duodenal ulcers. See GERD. Cf H2 blockers

Prima® CARDIOLOGY A laser guide wire system for treating occluded coronary arteries by excimer laser ablation, indicated when mechanical guide wire and other treatment modalities have failed

primal scene PSYCHIATRY In psychoanalytic theory, the real or fancied observation by the infant of parental or other heterosexual intercourse

primary *adjective* First site or place of origin *noun* Primary cancer The site of origin of a CA, usually understood to be non-lymphoproliferative or non-myeloproliferative

primary acquired melanosis A condition of older adults, which is a unilateral, diffuse brown pigmentation of the conjunctiva; it may 1. Remain stationary or regress 2. Slowly enlarge, but remain benign or 3. Undergo malignant degeneration–17%–melanocytic atypia is a major predictor of malignancy

primary amebic meningoencephalitis An intracranial infection by free-living amoebae–eg, *Naegleria fowleri*, *N grubei*, *Acanthamoeba*, *Hartmannella*, *Entamoeba histolytica*, etc CLINICAL Acute, purulent meningoencephalitis–typically by *Naegleria* spp–in young healthy persons swimming in stagnant, artificial fresh water lakes; inflammation and hemorrhage is most intense along the olfactory tract, inculpating the cribriform plate as the portal of entry via the nose Prognosis Poor; the few Pts who survive are treated with parenteral amphotericin B, miconazole, rifampicin, and have major sequelae; PAM may also be subacute with a granulomatous reaction, a finding more common in immunocompromised hosts

primary amenorrhea GYNECOLOGY The absence of any menstrual flow by age 16, which affects 2.5% in US; median age for menarche is ± age 12.3; PA may occur with or without other signs of puberty ETIOLOGY Normal delay of menarche; drastic weight loss–poverty, fad dieting, anorexia nervosa, bulimia, etc; congenital defects of genital system, hypoglycemia, extreme obesity, gonadal dysgenesis, Turner/XO syndrome, hypogonadotropic hypogonadism, testicular feminization, true hermaphroditism, chronic illness, malnutrition, Crohn's disease, cystic fibrosis, cyanotic congenital heart disease, craniopharyngioma, ovarian tumors, adrenal tumors, hyperthyroidism, imperforate hymen, vaginal and/or cervical absence, adrenogenital syndrome, Prader-Willi syndrome, polycystic ovarian disease, congenital adrenal hypoplasia. See Amenorrhea

primary amyloid METABOLIC DISEASE An idiopathic condition in which insoluble protein fibers are deposited in tissues, impairing function CLINICAL Sx linked to organs affected–tongue, intestines, muscle, nerves, skin, ligaments, heart, liver, spleen, kidneys, resulting in cardiomyopathy, renal failure, carpal tunnel syndrome, malabsorption, GI reflux, etc; PA deposits result in loss of resilience, stiffening, ↓ absorption and diffusion of metabolites; 2° amyloidosis may follow infection or inflammation

primary autonomic dysfunction NEUROLOGY A heterogeneous group of autonomic nervous system dysfunctions characterized by hypoadrenergic postural hypotension with blunted vasomotor response to norepinephrine upon standing, ↓ sweating, heat intolerance, GI Sx, impotence, urinary and fecal incontinence DIFFDX Postural hypotension may also be due to cerebral and spinal cord lesions–eg, degeneration, infection, trauma, tumors, peripheral neuropathy–eg, alcohol, amyloidosis, DM, porphyria, toxins TYPES: TYPE I Characterized by low plasma levels of norepinephrine; TYPE II Shy-Drager disease, which is further characterized by parkinsonian-type cerebral degeneration

primary biliary cirrhosis A disease of ♀ age 30 to 65 CLINICAL Fatigue, pruritis, steatorrhea, hepatic osteodystrophy, renal tubular acidosis, 4-fold ↑ in liver and breast CAs; 80% of PBC have autoimmune or connective tissue disease–eg, autoimmune thyroiditis, scleroderma, rheumatoid arthritis, Sjögren–sicca syndrome LAB 20-50-fold ↑ alk phos, IgM antimitochondrial antibodies; anti-SS-A/Ro antibodies in 20% MANAGEMENT Liver transplant, colchicine–antifibrotic and anti-inflammatory, ursodiol–a non-toxic bile acid that may slow disease progression. See Ursodiol. Cf Cirrhosis

primary brain tumor NEUROLOGY A tumor that arises in the brain–eg, ependymoma, astrocytoma grade 3 or 4, glioblastoma multiforme, glioma, medulloblastoma, meningioma, neuroglioma, oligodendroglioma. See Metastatic brain tumor

primary care MEDTALK Basic–ie, 'low tech' health care services that are coordinated, comprehensive and personal; PC is available on both a first-contact and continuous basis, and incorporates medical diagnosis and treatment, mental assessment and management, personal support, provision of information about illness, its prevention and health maintenance. See General practitioner, Internist, Primary care physician. Cf Hospital-based medicine, Medical specialties, RAPERs, Secondary care, Surgical specialties, Tertiary care

primary care exception ACADEMIC MEDICINE An exception to the general rule that teaching physicians or surgeons cannot bill Medicare for E/M–evaluation & management health services provided by physicians in training–residents, fellows. See Teaching physician

primary care network MANAGED CARE A group of primary care physicians who contract among themselves and/or with health plans to care for a group of covered persons

primary care physician A physician who provides care to a Pt at the time of first–non-emergent contact, which usually occurs on an outPt basis; PCPs include internists (formerly general practitioners), family practitioners, or emergency room physicians. See Family practice, General practitioner, Internist

primary cause of death Underlying–proximate cause of death, see there

primary coronary angioplasty CARDIOLOGY An angioplastic procedure performed in Pts with evolving Q-wave MI, in which the infarct-related artery is dilated during the acute phase of MI; PCA was the treatment of choice for Pts with contraindications to thrombolytic therapy. See Coronary angioplasty, Percutaneous transluminal angioplasty, Stent

primary diagnosis Main diagnosis, principal diagnosis PSYCHIATRY The condition motivating admission of a patient to the hospital or for outpatient treatment

primary G cell hyperplasia GI DISEASE A cause of peptic ulcers clinically identical to Zollinger-Ellison syndrome, and caused by gastrinoma DIAGNOSIS ↑ Baseline gastrin; ↑↑↑ gastrin after meals; secretin injection ↓ gastrin. Cf Zollinger-Ellison syndrome

primary gain PSYCHIATRY An emotional gain directly from illness–eg, alleviation of anxiety in neurosis by 'converting' emotions into an organic disease, a defense mechanism–eg, hysterical dysphonia. Cf Secondary gain

primary hyperoxaluria Oxalosis An AR condition characterized by crystallization in tissues CLINICAL Calcium oxalate–CO is ↑↑↑ in urine, resulting in CO supersaturation crystalluria, stone formation, and deposition of CO in the renal parenchyma, leading to renal dysfunction and CO deposition in multiple organs–oxalosis TREATMENT Orthophosphate, pyridoxine PROGNOSIS With therapy, actuarial ESRD-free survival at 20 yrs is ± 75%; sans therapy, 80% have ESRD by the 3rd decade of life; renal transplants do not survive well; liver transplants may be used to correct the enzyme defect

primary hyperparathyroidism Parathyroid related hypercalcemia ENDOCRINOLOGY Parathyroid gland hyperactivity with excess PTH secretion because of hyperplasia or adenoma of 1 or more glands CLINICAL Calcium deposits may occur in bone, the GI tract, kidney, muscle, and CNS; it most commonly affects those between 30 and 50 yrs, and is more common in women. See Familial adenoma

primary lung cancer ONCOLOGY Lung cancer arising in lung tissue–eg, trachea, bronchial tree, parenchyma. See Bronchoalveolar carcinoma, Small cell carcinoma, Squamous cell carcinoma. Cf Metastatic lung cancer

primary lymphoma of brain Primary lymphoma of CNS ONCOLOGY A lymphoma arising in the lymphoid tissue of the CNS–brain, spinal cord, meninges, which is relatively more common in immunocompromised Pts, for whom EBV may play a role CLINICAL Personality changes, seizures, neurologic defects RISK FACTORS Immunosuppression from cancer, organ transplant, or autoimmune disorders

primary mental health care clinician Any health care provider who first provides mental health care to a particular person–eg, physician, physician assistant, psychiatrist, psychologist, nurse, social worker.

primary prevention PREVENTIVE MEDICINE Any intervention which ↓ risks of a particular disease–eg, ↓ cholesterol intake to ↓ risk of ASHD, or ↓ tobacco to ↓ risk of lung CA. See Healthy lifestyle, Prevention. Cf Secondary prevention PSYCHIATRY Measures to prevent a mental disorder–eg, nutrition, substitute parents, etc

primary process PSYCHIATRY In psychoanalysis, the unorganized mental activity of the unconscious, which is marked by the free discharge of energy and excitation without regard to demands of environment, reality, or logic. Cf Secondary process

primary response The response that the immune system displays when first exposed to an antigen. Cf Secondary response

primary sclerosing cholangitis Sclerosing cholangitis A chronic idiopathic progressive liver disease characterized by inflammation, necrosis, fibrosis and obliteration of intrahepatic and extrahepatic bile ducts, resulting in cholestasis, cirrhosis, portal HTN, liver failure, often associated with other autoimmune diseases–eg, Crohn's disease, ulcerative colitis, Addison's disease EPIDEMIOLOGY PSC is uncommon–1 to 6/10⁵, US; it is the 4th most common indication for liver transplant; 75% of Pts have IBD; 70% of Pts are ♂, average age 39 CLINICAL Often asymptomatic until end-stage liver disease; as albumin ↓ and BR ↑, pruritus, fatigue, jaundice, and weight loss dominate clinical picture, ± accompanied by fever, chills, night sweats, right upper quadrant pain IMAGING Multiple zones of narrowing and dilatation– "beading" of bile duct by ERCP and transhepatic cholangiography LAB ↑ Alk phos, ↑ aminotransferase, hypergammaglobulinemia, ↑ IgM MANAGEMENT Azacytidine may ↓ need for transfusion, iron chelation–early use of deferiprone, corticosteroids, penicillamine, MTX; ursodiol is useless PROGNOSIS Median survival after diagnosis, 12 yrs. See Primary biliary cirrhosis

primary syphilis INFECTIOUS DISEASE A common STD caused by a spirochete, *Treponema pallidum* transmitted through sexual contact–especially since toilet seats have been definitively excluded SOBERING FACTOID Risk of contracting syphilis from an infected sex partner after one episode of unprotected sex is 30% EPIDEMIOLOGY US, 100,000 new cases/yr–third most commonly reported STD, gonorrhea is first, herpes is numero dos; it is more common in urban areas, in the south, ages 15 to 25, and may be asymptomatic STAGES Primary, secondary, latent–hidden, benign late, tertiary CLINICAL Usually presents as a small painless ulcer–ie, a chancre, 10 days to 6 wks post exposure which, if on the penis, is easily diagnosed, but not on the labia, cervix, anal area, or mouth APPEARANCE Shallow, sharply defined borders, slightly raised edges, base of the ulcer is clean and free of debris, it is typically painless and indurated; untreated chancres heal spontaneously within 6 to 8 wks, leaving a thin, depressed atrophic scar, signaling the end of the primary stage; *T pallidum* continues to multiply in the body and, untreated, give rise to secondary syphilis. See Congenital syphilis, Secondary syphilis

primary tumor A neoplasm which, in clinical parlance, is regarded as malignant, arising in one site and capable of giving rise to metastatic or secondary tumors. See Metastasis. Cf Tumor of unknown origin

primidone Desoxyphenobarbital NEUROLOGY An anticonvulsant used as a monotherapy for partial seizures–eg, secondary generalized seizures. See Phenobarbital, Seizures

primigravida OBSTETRICS A ♀ who is pregnant for the first time. See Elderly primigravida

primipara OBSTETRICS A ♀ who has been pregnant with one fetus of ≥ 500 g and/or carried the pregnancy for ≥ 20 wks, regardless of whether the gestational product was born live or stillborn, and whether it was single or multiple

primitive *adjective* EMBROLOGY Undifferentiated; undeveloped; before development of 1° germ layers–ectoderm, endoderm, mesoderm PSYCHIATRY Pertaining to the early development of the personality

Primonil Imipramine, see there

Primsol® PEDIATRICS A formulation of Trimethoprim for managing acute otitis media in younger children . See Acute otitis media

primum non nocere First, do no harm MEDICAL ETHICS A guiding principle for physicians, that whatever the intervention or procedure, the Pt's well-being is the primary consideration. See Hippocratic oath

principle VOX POPULI A guiding rule or maxim. See Bateman's principle, Bolam principle, Ceiling principle, Dale's principle, Eggshell skull principle, Fortner principle, Handicap principle, Heuristic principle, Homeopathic principle, KISS principle, Mendelian principle, Pleasure principle, Polluter pays principle, PRICE principle, Reality principle

Prinivil® Lisinopril, see there

Prinox Alprazolam, see there

Prinzmetal's angina Coronary artery vasospasm, variant angina CARDIOLOGY A condition characterized by sudden and usually transient vasoconstriction of a coronary artery with ↓ myocardial O₂ CLINICAL Severe, crushing, chest pain at rest and ST segment deviation PRECIPITATING FACTORS Emotional stress, medications, street drugs–cocaine, cold exposure, stress, vasoconstrictors DIAGNOSIS Normal angiogram, ST segment deviation–usually ↑ without preceding ↑ in heart rate or BP MANAGEMENT Nitrates, nifedipine or diltiazem with aspirin; slow CCBs–eg, verapamil, may contribute to AV block. Cf Angina pectoris

PRIO study group AIDS A randomized French multicenter study–Prévention des Infections Opportunistes, which evaluated the effects of various interventions intended to ↓ the incidence of opportunistic infections in Pts with HIV. See Dapsone-pyrimethamine

prion Slow spongiform encephalopathy virus MOLECULAR MEDICINE An unconventional 33–35 kD sialoglycoprotein, the smallest known infective particle and implicated in diseases of man–Creutzfeldt-Jakob disease-CJD, fatal familial insomnia, Gerstmann-Straussler-Scheinker syndrome, kuru, and animals–scrapie of sheep and goats, bovine spongiform encephalopathy, transmissible mink encephalopathy, chronic wasting disease of captive mule, deer, elk. See Creutzfeldt-Jakob disease, Protein-only hypothesis

prior acts coverage Nose coverage, see there

prior authorization HEALTH INSURANCE A cost containment measure that provides full payment of health benefits only if the hospitalization or medical treatment has been approved in advance

prior probability DECISION MAKING The likelihood that something may occur or be associated with an event based on its prevalence in a particular situation. See Medical mistake, Representative heurisic

PRISM CARDIOLOGY A clinical trial–Platelet Receptor Inhibition in Ischemic Syndrome Management which compared the composite end point–Death, MI, and refractory ischemia at 48 hrs of Pts with unstable angina–who were already receiving aspirin, who were managed with either heparin–H or tirofiban–T– a specific, nonpeptide inhibitor of platelet gpIIb/IIIa receptor. See Tirofiban

PRISM score Pediatric Risk of Mortality score CRITICAL CARE A prognostic scoring system derived from 14 physiologic variables assessed during the first 24 hrs of care in an ICU for pediatric populations that derives from the PSI–physiology stability index. See Prognostic scoring systems

prisoner of war syndrome PSYCHOLOGY A condition characterized by 'withdrawal, apathy, and sometimes death as a reaction to capture, imprisonment, and hopelessness about reunion with one's love ones'. See Torture. Cf Concentration camp survivor syndrome

Pritikin diet NUTRITION A diet high–> 90% of calories in complex carbohydrates–eg, whole grains, fruits, and vegetables, ↓ in protein, fat, cholesterol, with severely restricted caffeine, salt, and sugar; the PD ↓ the risk of CAD and HTN, and may benefit arthritis, DM, gout, stroke, and other diseases of civilization. See Diet, Ornish regimen

privacy NIHSPEAK Control over the extent, timing, and circumstances of sharing oneself–physically, behaviorally, or intellectually with others

private human-caused death A death attributed to an individual act–homicide, suicide, accident, or which is beyond human control–eg, epidemics. Cf Public human-caused death

private patient A Pt whose care is entrusted to one physician, who usually has had a long-term relationship with the Pt, and who is often directly reimbursed for his services by a 'third party' payer or by the Pt. Cf Personal physician

private speech PSYCHOLOGY A verbalized, but internal monologue that accounts for 20–60% of the remarks made by a child < age 10. See Babbling

privilege to disclose AIDS The right of a health care provider who knows the identity of a partner at risk for HIV to inform him or her, when the infected client is unable or unwilling to inform. See HIPAA, Negotiated safety

privileged communication FORENSIC MEDICINE Communication between a psychotherapist or a physician and Pt which is protected from public disclosure, unless the therapist is legitimately concerned that criminal acts against others could be committed, and could be prevented by disclosure of said communication to proper authorities. See Privilege, *Tarasoff* v *Regents of the University of California*

privileges Permission granted by a hospital or other health care institution to a physician or other provider to render specific diagnostic or therapeutic services TYPES Admitting privileges–right to admit Pts; clinical privileges–right to treat. See Admitting privilege, Clinical privilege, Conversion privilege, Emergency privilege, Hospital privileges, Staff privilege, Temporary privilege, Therapeutic privilege

PRK Photorefractive keratectomy, see there

prn order CLINICAL PHARMACOLOGY Any physician-promulgated mandate or regimen–'doctor's orders' that allows use of a therapy or modality as needed–Latin *pro re nata*–by the Pt; in certain situations–eg, malignancy or other terminal illnesses, PRN orders, in particular those for controlling pain–schedule II through V controlled substances, are more liberal. See Patient-controlled analgesia

PRO Peer review organization, see there

PROACT NEUROLOGY A clinical trial–Prolyse in Acute Cerebral Thromboembolism

proarrhythmia CARDIOLOGY A drug-related arrhythmia, seen in 4–6% of Pts treated for V tach or V fib, especially in Pts treated with digitalis & diuretics, ↓ left ventricular function, longer pretreatment Q-T intervals PATHOGENESIS Altered reentry substrate and early after depolarization. See Arrhythmia

probability STATISTICS p value The likelihood that an event will occur by chance alone, and given a value between 0–impossible and 1–certain; the higher the p value, the more likely that 2 or more sets of overlapping variables occurred randomly–ie, the less the likelihood that the 2 events are associated; the lower the p value, the greater is the likelihood that the events are not random associations–counterintuitive, but think it out in a dark quiet room, you'll get it. See Conditional probability, Empirical probability, Gaussian probability, Personal probability, Prior probability, Theoretic probability VOX POPULI An expression of the likelihood that a specific event will occur

proband EPIDEMIOLOGY A Pt who triggers study of other members of the family to identify the possible genetic factors involved in a given disease, condition, or characteristic. Cf Index case

probation MEDTALK A period during which a health care provider whose practice of medicine was restricted by a licensing or certifying authority or by a hospital's medical staff due to questionable medical judgement, is evaluated to determine if he/she can be allowed an unrestricted license to practice medicine or retain medical staff privileges. See Restriction

probe SURGERY Explorer A long, thin, usually metal instrument with a blunt or bulbous tip which is used to poke around in cavities, fistulae, sinuses, and wounds

problem *adjective* Recalcitrant, refractory *noun* MEDICAL EDUCATION A didactic exercise, in which a Pt is presented as an 'unknown,' and clinical, lab and imaging data are evaluated as a means of improving diagnostic skills and teaching management principles MEDICAL PRACTICE A complaint or reason for seeking medical attention PSYCHIATRY A euphemism for mental illness VOX POPULI See Binding problem, File drawer problem, Moving target problem, Satisfaction problem, Sibling relational problem

problem-based learning MEDICAL EDUCATION An instruction strategy in which groups of students are presented with clinical problems without prior study or lectures. See Cooperative learning

problem drinker SUBSTANCE ABUSE A person who meets 2 of the 3 criteria in the last 12 months, for alcoholics. See Alcohol, Binge drinking. Cf Social drinker

---

PROBLEM DRINKER

5+ ALCOHOLIC DRINKS on any one occasion at least once/month–'heavy drinking'

ONE OR MORE ALCOHOL-RELATED SOCIAL CONSEQUENCES–eg, drunk-driving, public drunkenness arrests, alcohol-related criminal arrests, traffic or other accidents when drinking, confrontations about alcohol-related health problems by a medical practitioner, alcohol-related family or on-the-job problems

ONE OR MORE SYMPTOMS OF ALCOHOL DEPENDENCE–eg, having an alcoholic drink upon awakening–an 'eye-opener', shaking hands, awakening not remembering the events that occurred while drinking JAMA 1992; 268:1872oc

---

problem-oriented medical record A medical record in which each Pt's condition or complaint is formally addressed; a POMR may be organized by the acronym of SOAP–subjective criteria, objective criteria, assessment, plan. See Hospital record, Medical record, SOAP

problem patient MEDICAL PRACTICE A 'difficult' Pt, who may be garrulous, poorly compliant, belligerent, antagonistic. See Patient noncompliance

problem solving GRADUATE EDUCATION A strategy which confronts the student with a problem situation, and requires development and application of a plan to reach a solution

problem wound WOUND CARE An open wound–eg, diabetic foot infection, or leg ulcer due to arterial insufficiency, which is refractory to the usual medical or surgical therapy MANAGEMENT Hyperbaric O_2 may ↓ need for above knee amputations see Hyperbaric oxygen therapy

pro bono *Pro bono publico*, Latin, for the good of the public *adjective* Referring to a non-reimbursed service–health care, legal advice by an attorney to those who cannot afford to pay professional fees. See Learned profession, Profession

probucol Lorelco® CARDIOLOGY A hypocholesterolemic with antioxidant activity, that ↓ restenosis after balloon angioplasty. See Balloon angioplasty

procainamide CARDIOLOGY An antiarrhythmic that may contain N-acetylprocainamide–Procan®, Pronestyl®

procaine Novocain® AMBULATORY SURGERY An ester class local anesthetic of 15–45 min duration, which doubles with coadministration of epinephrine

procap helmet SPORTS MEDICINE A protective rubber pad attached to the outside of professional football helmets, used to ↓ head injuries

procarbazine Matulane® ONCOLOGY An alkylating therapeutic used as part of a multi-drug treatment for Hodgkin's disease ADVERSE EFFECTS BM suppression, N&V, neuropathy, confusion, swelling. See Hodgkins disease

Procardia® Nifedipine, see there

proceduralist A physician, usually a specialist or subspecialist who performs diagnostic or therapeutic procedures

procedure MEDTALK An 'invasive' service performed by a physician, which is arbitrarily divided into major–eg, general, orthopedic, cardiovascular, or other surgical procedures, ambulatory or outPt–eg, radial keratotomy procedures, and endoscopic procedures. See BAK procedure, Batista procedure, Booked procedure, Coronary revascularization procedure, Cough-inducing procedure, Cyclodestructive procedure, Diagnostic procedure, DIEP flap procedure, Disciplinary procedure, Dor procedure, Downstream procedure, Ertl procedure, Experimental procedure, Goebell-Stöckel procedure, Hartmann procedure, Hearing-sparing procedure, Heller-Dor procedure, High-discretion procedure, High-risk & complex procedure, High-yield procedure, In & out procedure, Indiana pouch procedure, Infection control procedure, Labyrinthine procedure, LEEP procedure, Life-prolonging procedure, Localization procedure, Low-yield procedure, LTPs procedure, Maintenance procedure, Manchester procedure, Marshall-Marchetti-Krantz procedure, Maze procedure, Minor procedure, Mumford procedure, Myocardial laser revascularization procedure, Neuroablative procedure, No/NA procedure, Overvalued procedure, Physician-performed microscopy procedure, Pinup procedure, Potts procedure, Regnauld procedure, Ross procedure, Sauvé-Kapandji procedure, Scalp-lifting procedure, Shirodkar procedure, Special firefighting procedure, Standard operating procedure, Upstream procedure, Ultra-rapid opiate detoxification procedure, Whipple procedure, Wrap-around procedure, Yes/no procedure, Yes/yes procedure. Cf Evaluation and management service, Physician test

proceptive phase SEXOLOGY The initial phase of reciprocal signaling and responding to attraction and solicitation, in a ritual of courtship before the acceptive/copulatory phase. See Acceptive phase, Conceptive phase

process VOX POPULI A manner of performing a procedure or activity. See Activated sludge process, Bioprocess, Cosmetic suturing process, Cyclodestructive process, Due process, Foot process, Graduation process, Hoffman quadrinity process, Identity process, Inner self-healing process, Markov process, Marshall-Marchetti-Krantz process, Mastoid process, Multirule process, Nonspecific anatomic process, Odontoid process, Primary process, Rating process, Sand process, Sauvé-Kapanji process, Secondary process,

process-oriented psychology Process psychology, see there

processing MANAGED CARE See Electronic claims processing NEUROLOGY See Natural language processing

prochlorperazine Emezine® ONCOLOGY An antiemetic equivalent to Compazine® used for chemotherapy-related nausea

Pro-Clude WOUND CARE A transparent film wound dressing. See Wound care

procoagulant *adjective* Favoring coagulation; referring to activation of coagulation factors *noun* Any agent–eg, thrombin, factor Xa, which clots blood

ProCrit® Epoetin alpha HEMATOLOGY Recombinant erythropoietin, used to treat anemias–eg, Hct 10–13 g/dL, anemia of chronic renal failure, perioperative transfusion, to ↓ blood transfusions in anemic Pts scheduled for elective noncardiac, nonvascular surgery, anemia linked to AZT-treatment of HIV Pts; anemia of prematurity SIDE EFFECTS Possibly HTN, hypersensitivity, thrombotic/vascular events. See Erythropoietin

proct- Greek, Anus, rectum

proctalgia 1. A pain in the rectum 2. Figuratively, A pain in the butt(ocks)

proctitis Rectal inflammation SURGERY Anal and rectal inflammation, linked to anal sex, high-risk sexual practices, homosexuality CLINICAL Tenderness, hemorrhage, ± discharge of mucus or pus ETIOLOGY STDs–gonorrhea, herpes, chlamydia, lymphogranuloma venereum, amebiasis; non-STD infections–rare–eg, in children, due to beta-hemolytic streptococcus; autoimmune proctitis–ulcerative colitis, Crohn's disease; physical agents–chemicals per rectum, drugs, RT. See Gay bowel disease, Pseudoinfectious proctitis, Streptococcal proctitis

proctocolectomy Surgical removal of the rectum and a variable length of colon INDICATIONS Prophylactic for Pts with extensive long-term ulcerative colitis

proctoscope Anoscope A rigid, 5-8 cm in length by 2 cm in diameter metal or plastic tube inserted per rectum to examine the anorectal mucosa

proctoscopy Anoscopy GI DISEASE Examination of the anorectal mucosa with a proctoscope inserted per rectum to detect neoplasms or other lesions which are biopsied and sent to pathology. See Colonoscopy, Endoscopy, Sigmoidoscopy. Cf Barium enema, Fecal occult blood test

proctosigmoidoscopy GI DISEASE Examination of the rectum and sigmoid colon with a sigmoidoscope. See Proctology

procurement TRANSPLANTATION The obtention of organs for transplantation; the best donor of multiple organs destined for transplantation is a recently brain-dead Pt with unimpaired circulation. See Slush preparation, Transplantion, UNOS

PROCUREMENT CRITERIA

PHYSICAL CRITERIA–eg, young age, previous excellent health before the trauma that left the donor in a persistent vegetative state, absence of substance abuse history

LEGAL CRITERIA–ie, appropriate permission for organ donation has been obtained from next-of-kin

LAB CRITERIA A battery of serologic tests–IgG, IgM ELISA–for HIV-1, HTLV-I, HAV, HBV, hepatitis C, RPR–*Treponema*

Prodep Fluoxetine, see there

Prodigy™ IMAGING A bone densitometer that provides rapid full-body scans. See Densitometry

prodrome MEDTALK A premonitory or early Sx of a disease or a disorder NEUROLOGY A premonitory Sx unrelated to a seizure Cf Aura

prodrug THERAPEUTICS A drug ingested in the inactive form which is transformed into an active form by in vivo metabolism

production SUBSTANCE ABUSE The manufacture, planting, cultivation, growing, or harvesting of a controlled substance or marijuana

products of conception OBSTETRICS The aggregate of tissues present in a fertilized gestation; in a pregnancy that has been terminated or aborted, chorionic villi and/or fetal tissue must be present in a specimen to make a definitive diagnosis of intrauterine–as opposed to ectopic pregnancy; decidualized and secretory endometrium may be found in either intrauterine and ectopic pregnancies

profession Learned profession, *pronounced* Lern-ed, as in burn-bed BIOMEDICAL ETHICS An occupation requiring intense preparation in a body of erudite knowledge–eg, law, medicine, which is applied in service to society, has a system of self-governance and in which success is measured by accomplishments in serving man and society and/or furtherance of knowledge in the field, rather than in personal gain. See Pro bono, Learned profession, Remedial profession, 'Yellow professionalism'

professional *adjective* Pertaining to a profession *noun* One who practices a profession, see there. See Behavioral health professional, Health care professional, Mental health professional

professional assistance program SUBSTANCE ABUSE A public or private program that offers services to help in rehabilitation of health care licensees suffering from chemical dependencies or other impairments. See Substance abuse

professional board A group of members of a particular profession who are responsible for maintaining specialty standards and certification of new members in the field of interest, through examinations and documentation of continued competency. See Board certification

professional boundary PROFESSIONAL ETHICS An ill-defined psychosocial 'frontier' maintained between a professional and a Pt or client. See Dual relationship, Sexual misconduct, Slippery slope

professional courtesy Professional discount MEDTALK The practice by a physician of waiving all, or a part, of the fee for services provided to a physician's office staff, other physicians and/or their families; PC has been extended to include the waiver of coinsurance obligations or other out-of-pocket expenses for physicians or their families. See Insurance only billing, Pro bono

professional image The constellation of tangible or visible representations and/or perceptions resulting from a person's conduct as a professional, linked to ethical behavior and competence

professional liability The obligation that a professional practitioner has to provide care or service that meets the standard of practice for his/her profession–ie, reponsibility; when a professional fails to provide the standard of practice, liability refers to the obligation to pay for damages incurred by negligent acts. See Liability, Malpractice

professional misconduct Behavior by a professional that implies an intentional compromise of ethical standards. See Impaired health care provider. Cf Fraud in Science

professional patient Fake Pt MEDTALK A person who feigns illness for various reasons; PPs include malingerers filling a psychologic need, drug entrepeneurs seeking prescriptions or drugs to be sold later, and those with Munchausen syndrome

professional voice PERFORMING ARTS MEDICINE A voice used–ie, singing, speaking, etc–to earn its owner's income TREATMENT Surgery–eg, endoscopic microflap techniques. See Performing arts medicine

professor ACADEMIA A member of the faculty at an institution of higher learning who has attained its highest possible academic rank, who possesses special knowledge in an occupation requiring special skills. See 'Chair'. Cf, Lecturer

professorial rounds GRADUATE EDUCATION A type of Pt rounds that forms part of the teaching activities at an academic institution in which the clinical, radiologic and pathologic data from one or a limited number of Pts are presented. See Socratic method. Cf Clinico-pathological conference, Grand rounds, Pimping, Rounds

profile LAB MEDICINE A panel of screening tests used to: 1. Establish a baseline of normalcy for either a certain population–eg, executive profile, or for a limited group of analytes–eg, lipid profile, and 2. Detect the presence of a particular category of disease–eg, inborn error of metabolism, or cardiovascular disease. See Alcohol profile, B-cell leukemia/lymphoma immunophenotyping profile, Biophysical profile, Bladder profile, Breast tumor profile, Criminal profile, Curvilinear profile, DNA profile, Executive profile, Fingerprint profile, Health profile, Hirsutism profile, *Histoplasma* antibody profile, Hybrid revascularization profile, Hypercoagulable state profile, Hypersensitivity pneumonitis profile, Hypergonadism profile, Hypertension assessment profile, Immunoprotein profile, Iron-deficiency profile, Lipid profile, Liver cancer profile, Lung profile, Lupus profile, Lymphocyte subset profile, Multiple myeloma profile, Multiple sclerosis profile, Natural killer cell leukemia/lymphoma immunophenotyping profile, Neonatal profile, Obstetric hypercoagulability profile, Obstetric profile, Organ panel, Ova & parasites profile, Perfomance profile, Pheochromocytoma profile, Physician profile, Premarital profile, Provider profile, Renal cancer profile, Respiratory antibody profile, Resume profile, Serum concentration-vs-time profile, Sickness Impact profile, Sjögren profile, Urethral pressure profile, Vasculitis profile, Viral profile MEDTALK A longitudinal or cross-sectional aggregation of health care data applied to any segment of a population being served or the individuals or groups providing the service and the statistics obtained therefrom; there are thus Pt, physician, and hospital profiles

Profile of Mood States PSYCHOLOGY A 65-item questionnaire that assesses a person's moods–eg, anger, anxiety, confusion, depression, fatigue, vigor

profiling MANAGED CARE The assessment of a health care provider's performance vis-á-vis utilization of services, preventive care, Pt satisfaction. See Provider profile

Profore four-layer system VASCULAR DISEASE A Charing Cross bandaging system, used to control venous HTN responsible for venous ulcers by providing effective pressure for a full wk. See Venous ulcer

progeria PEDIATRICS A condition characterized by rapid premature aging of childhood onset, in which morbidities usually seen in the elderly–eg, ASHD, CAD, cataracts, wrinkled skin, appear during puberty and death from 'old age' occurs by age 20 from heart disease or strokes CLINICAL Growth failure in 1st yr of life; affected children are small and thin with disproportionately large appearing heads, baldness, wizened narrow faces, and aged skin; congenital progerias–eg, Cockayne syndrome, Hutchinson-Gilford syndrome

Progesterone Plus™ GYNECOLOGY A progesterone-replacement product from soy and/or wild yam extract applied to skin with Derma-Trans™ roll-on system. See Progesterone

progesterone receptor A progesterone-binding protein complex found in the cytoplasm of certain cells in particular of the breast, which belongs to the nuclear receptor family. See Progesterone receptor assay. Cf Estrogen receptor

progesterone receptor assay ONCOLOGY A test that detects progesterone receptors in breast and endometrial CAs. See Progesterone receptor positivie. Cf Estrogen receptor

progesterone receptor negative ONCOLOGY A status of breast CA cells that lack a receptor to which progesterone attach, and thus don't respond to hormonal manipulation

progesterone receptor positive ONCOLOGY A status of breast CA cells that have a receptor to which progesterone can attach; up to 80% of Pts with PR+ breast and endometrial CAs respond–ie, have a better prognosis–to hormonal manipulation

progestin hermaphroditism ENDOCRINOLOGY Hermaphroditism induced in 46,XX gonadally ♀ fetus by synthetic progesterone and 19-nortestosterone derivatives, once used for threatened miscarriage with the hope of preserving gestation; in some, the hormone given to the mother masculinized the fetus' external genitalia

prognosis MEDTALK A projection of the likely, anticipated course and outcome of a particular condition in terms of M&M based on stage and Sx, and previous experience with Pts with the same disease, with or without treatment. See Molecular prognosis, Pseudoprognosis

prognostic estimate Probability estimate INTENSIVE CARE A value between 0 and 1 that measures a Pt's expected risk of dying or of having another defined outcome within a specified time period. See Objective probability estimate

prognostic factor MEDTALK Any factor–eg, Pt age, family Hx, lifestyle, stage of presentation, that is weighed in determining a prognosis. See Prognosis

Prognostic Nutritional Index Buzby Index ONCOLOGY An instrument used to define the nutritional status and predict clinical outcomes of Pts with CA, based on body weight, total body nitrogen and potassium, serum albumin and transferrin levels, triceps skin-fold thickness, and delayed skin hypersensitivity. See Cancer cachexia, Malnutrition

prognostic scoring system Any scoring to help predict outcome(s) and identify Pts and clinical situations in which the potential value of intensive care is low, while the burden of therapy is high, providing a numerical prediction of mortality. See APACHE II, MPS, PRISM, SAPS, TISS

prognosticate Prognose *verb* To project the outcome of a particular condition or state

programmable pacemaker CARDIAC PACING A pacemaker in which the pacing mode and/or parameters can be changed noninvasively with an external programmer ADJUSTABLE PARAMETERS Rate, pulse amplitude and width, sensitivity, mode of pacing–eg, asynchronous ventricular inhibited or ventricular triggered. See Pacemaker

540 **progress note** A brief summary of a hospitalized Pt's current clinical status, written sequentially in the chart, reflecting information provided by physical exam, lab tests, and imaging modalities. See Hospital chart

progressive angina Unstable angina, see there

progressive disease MEDTALK A chronic or dread disease–eg, CA, which ↑ with time in scope or severity. See Dread disease

progressive multifocal leukoencephalopathy NEUROLOGY A demyelinating CNS lesion, caused by a human papovavirus, usually JC virus, rarely also BK virus, affecting ±4% of AIDS Pts CLINICAL Dementia, spastic paralysis, blindness, opportunistic infections; PML occurs in immunocompromised hosts–eg, with leukemia, AIDS MANAGEMENT Cytarabine, antiretrovirals are useless

progressive myoclonus epilepsy NEUROLOGY A heterogeneous group of disorders that share clinical features, and thus generically termed PME syndrome CLINICAL Prominent sensitivity of myoclonus to all stimuli–eg, passive movement of a limb might evoke a generalized convulsions ETIOLOGY Unverricht-Lundborg's disease, Lafora body disease, neuronal ceroid lipofuscinosis–late infantile, juvenile, and adult types, sialidosis–types I/II, and MERRF–mitochondrial encephalopathy; rare causes of PME include Gaucher disease, GM_2 gangliosidosis, biotin-responsive encephalopathy, Hallervorden-Spatz disease, Ekbom syndrome, May-White syndrome

progressive purple mask of death MEDTALK A term for the deep facial cyanosis that extends to the earlobes, ear, neck, and chest, seen in a fading response to CPR

progressive spinal muscular atrophy Spinal muscular atrophy, see there

progressive stroke Stroke in evolution, see there

progressive supranuclear palsy Dementia-nuchal dystonia, Steel-Richardson-Olszewski syndrome NEUROLOGY An idiopathic condition characterized by progressive degeneration of basal ganglia, oculomotor nuclei, and periaqueductal gray matter CLINICAL Onset after age 50, akinetic rigid syndrome, mask-like facies, fixed stare–supranuclear gaze palsy, ophthalmoplegia, pseudobulbar palsy–with dysarthria and dysphagia, axial dystonia on extension, nystagmus, unsteady gait, balance, mild dementia, slow movements–bradykinesia, ↑ tendon reflexes PROGNOSIS Eventually bedbound, death in 5–10 yrs, often from URI

progressive systemic sclerosis Sclerodema An idiopathic connective tissue disorder, characterized by localized or generalized induration of skin due to ↑ collagen deposition in skin, GI tract, lungs, heart, muscle, kidneys; PSS may be confined to skin for long periods of time, but usually encroaches on the viscera, causing malabsorption, respiratory insufficiency before death ensues, usually from heart or renal failure CLINICAL Raynaud's phenomenon is the initiating event in most Pts; other features include arthralgias, arthritis, early–subcutaneous edema, late–skin induration, skin ulcers, pinched face, dyspnea due to lung fibrosis, cor pulmonale IMMUNOLOGY Autoantibodies–eg, anti-nuclear, anti-centromere, and anti-topoisomerase 1–Scl-70 antibodies LAB Normochromic, normocytic anemia, microangiopathic anemia, polyclonal gammopathy, ↑ ESR IMAGING Osteoporosis, interstitial markings in CXR, loss of esophageal and colonic peristalsis, irregular narrowing in renal arteriogram MANAGEMENT CCBs for vasodilation in Pts with Raynaud's phenomenon; penicillamine may slow the disease progression; preventive measures include avoiding tobacco and cold temperature, antibiotics and digitalis as needed PROGNOSIS 5-yr survival, ± 40%. See CREST complex, Raynaud's phenomenon

projectile vomiting PEDIATRIC NEUROLOGY Violent and 'explosive' vomiting without antecedent nausea, or vomiting at the peak of maximum inspiration without the usual rhythmic hyperactivity of the respiratory muscles associated with 'retching'–diaphragmatic spasms that precede vomiting; PV is associated with ↑ intracranial pressure, classically occurring in meningitis of young children

projection PSYCHIATRY A defense mechanism, operating unconsciously, in that which is emotionally unacceptable in the self is unconsciously rejected and attributed–projected–to others

projective identification PSYCHIATRY The projection of an emotion or personality trait–with which the person is uncomfortable–onto another person–eg, a child, as in the Munchausen-by-proxy syndrome. See Munchausen-by-proxy syndrome

projective test Projection test PSYCHOLOGY A psychologic tests in which a person is presented with unstructured external stimuli–eg, Rorschach inkblots, thematic apperception test, that are ambiguous and subject to subjective interpretation; analysis of responses to the situations or images in a PT provides–in theory, information on unconscious desires, personality traits, interpersonal dynamics. See Psychological testing, Rorschach test

prolactinoma ENDOCRINOLOGY A pituitary tumor more common in Pts < age 40 characterized by an excess secretion of prolactin–tumor size correlates with the prolactin level; smaller tumors, microprolactinomas, are more common in ♀; macroprolactinomas are more common in ♂ and associated with sexual dysfunction; women may become symptomatic while on OCs; menses may stop when OCs are discontinued; prolactinomas may grow during pregnancy and be diagnosed postdelivery. See Prolactin

prolapse The sinking or lowering of an organ or tissue. See Genital prolapse, Mitral valve prolapse, Rectal prolapse, Uterine prolapse

ProLease DRUG DELIVERY A technology that places injectable formulations in a sustained-release format. See Sustained release

Proleukin® Aldesleukin, IL-2 IMMUNOLOGY An agent used for metastatic kidney CA, melanoma, NHL, AML, TB, HIV infection, KS. See IL-2

pro-life Pro-choice, see there

proliferation Multiplication of cells or organisms. See Autonomous proliferation, Cell proliferation, Systemic immunoblastic proliferation

Prolixin® Fluphenazine

Prolixin™ shuffle NEUROLOGY A shuffling gait and bradykinesia of Pts high on high doses of a potent anti-psychotic, fluphenazine

prolonged labor OBSTETRICS Labor of > 24 hrs duration, which may be due to a prolonged latent phase–> 20 hrs in a primigravida or > 14 hrs in a multipara, or due to a 'protraction disorder' in which there is protracted cervical dilatation in the active phase of labor and protracted descent of the fetus. See Labor

PROloop™ SURGERY Electrosurgical device for cutting, coagulating, vaporizing tissues. See Cauterization

prolymphocyte HEMATOLOGY A large lymphocyte- 'wannabe' not normally found in the peripheral circulation which has a high–5:1 to 3:1–N:C ratio; the centrally placed, round, ovoid or slightly indented nucleus has minimal coarse chromatin clumps along the margin, indistinct parachromatin, a large, round bluish nucleolus; organelles are indistinct; cytoplasm is medium to deep homogeneous blue and greater in prolymphocytes than in lymphoblasts or mature lymphocytes. Cf Little blue cell, Lymphoblast

PROM Premature rupture of membranes, see there

ProMACE-cytaBOM ONCOLOGY A '3ʳᵈ-generation' chemotherapy regimen consisting of prednisone, doxorubicin, cyclophosphamide, etoposide, cytarabine, bleomycin, vincristine, MTX with lecovorin rescue. See CHOP

Prometrium® Micronized progesterone GYNECOLOGY A formulation of progesterone for treating 2° amenorrhea and preventing endometrial hyperplasia in postmenopausal ♀ who have not had hysterectomy and who are receiving conjugated estrogen CONTRAINDICATIONS Peanut allergy, severe liver disease, breast CA. See Endometrial hyperplasia

promiscuity SOCIAL MEDICINE The engaging in multiple and indiscriminate sexual relationships

PROMISE CARDIOLOGY A clinical trial–Prospective Randomized Milrinone Survival Evaluation–that evaluated the effect on M&M in Pts with severe CHF treated with milrinone. See Congestive heart failure, Heart failure, Milrinone

promotion DRUG INDUSTRY The act of furthering the sale of a product by advertising or through publicity. See Advertising, Detailing, Drug promotion ONCOLOGY Tumor promotion, see there VOX POPULI 1. The assignment of an employee to a higher job position or pay grade, which often entails ↑ responsibility and indicates ↑ employer confidence in the employee 2. The encouragement of an activity. See Health promotion

Promycin® Porfiromycin ONCOLOGY An orphan drug used to manage hypoxic CA cells, which are less susceptible to RT than other tumor cells. See Orphan drug, Radiation therapy

promyelocytic leukemia HEMATOLOGY A type of AML, a rapidly progressive CA characterized by an excess of promyelocytes in the blood and bone marrow. See Leukemia

pronator drift NEUROLOGY A drifting downward of the hand into pronation, a sign of weakness

prone MEDTALK *adjective* 1. Facing downward. See Position 2. Tending toward. See Exposure-prone

pronouncing death FORENSIC MEDICINE The act of monitoring–eg, with a stethoscope or an EKG a Pt's condition and determining the time of death. See Harvard criteria. Cf Certifying death

ProOsteon® ORTHOPEDICS A coralline hydroxyapatite, cancellous bone-like graft used to fill defects created by acute metaphyseal fractures, for rigid internal fixation when excising cysts and tumors in metaphysis or diaphysis of long bones. See Bone graft

propafenone Rythmol® CARDIOLOGY A beta-blocking antiarrhythmic for severe ventricular and supraventricular arrhythmias–A Fib, V Tach, conduction block, HTN COMPLICATIONS Ventricular proarrhythmia; conduction defects ADVERSE EFFECTS Neurologic–headache, dizziness, paresthesias, peripheral neuropathy, GI–eg, N&V, anorexia. See Antiarrhythmic, Beta-blockers

Propagest® Phenylpropanolamine, see there

Propam Diazepam, see there

Propecia® Finasteride ESTHETIC MEDICINE An agent for treating male pattern hair loss–androgenetic alopecia on the vertex and the anterior midscalp area CONTRAINDICATIONS ♀, children. See Male-pattern baldness

propentofylline NEUROLOGY An agent that may help improve mental performance in dementia. See Dementia

prophylactic MEDTALK *adjective* Preventive, protective *noun* A drug, vaccine, regimen, or device designed to prevent or protect against a given disorder VOX POPULI Condom, see there

prophylactic antibiotic therapy Administration of antimicrobials in absence of a known infection, a standard practice to ↓ risk of surgical wound infection COMMON SURGICAL WOUND PATHOGENS *Staphylococcus aureus*, *Bacteroides fragilis*, *Enterobacter cloacae*; most commonly used are 3rd-generation cephalosporins; PAT is best if administered 2 hrs presurgery

prophylactic cranial irradiation RADIATION ONCOLOGY RT to the head to ↓ risk of brain metastases. See Radiation therapy

prophylactic mastectomy SURGICAL ONCOLOGY Bilateral mastectomy in a ♀ at high risk–eg, with *BRCA1* or *BRCA2* mutations and breast CA in 1st-degree relatives, to ↓ CA risk. See Prophylactic oophorectomy, Radical mastectomy

prophylactic oophorectomy ONCOLOGY The excision of the ovaries in a ♀ with a mutation in *BRCA1* or *BRCA2*–associated with a 5–60% lifetime risk of ovarian CA–normal risk is 1.5%–undergoes prophylactic removal of ovaries. See Bilateral mastectomy, Prophylactic mastectomy

prophylactic surgery SURGICAL ONCOLOGY An excision of precancerous tissue–eg, mastectomy of a ♀ at high risk of developing breast CA–to minimize the risk of future malignancy

prophylaxis MEDTALK A medical maneuver intended to prevent disease. See Chemoprophylaxis, HIV prophylaxis, Malaria prophylaxis

propofol Diprivan® ANESTHESIOLOGY A versatile nonopioid inhalation sedative-hypnotic with rapid–< 45 secs onset, used to induce and maintain anesthesia, and possibly control refractory status epilepticus; it ↓ intracranial and intraocular pressure, cerebral blood flow, and O_2 consumption PROS Ideal for conscious sedation; useful for critically ill Pts; can be combined with other agents–eg, volatile anesthetics, NO_2, opiates or benzodiazepines; little post-op N&V ADVERSE EFFECTS Profound respiratory depression ± transient apnea; injection site pain, hypotension INDICATIONS Short procedures or outpatient surgery CONTRAINDICATIONS Hemodynamically unstable Pts. See Meyer-Overton rule. Cf Inhalation anesthetic, Opioid anesthetic

proportionate mortality EPIDEMIOLOGY The proportion of deaths in a specified population over a period of time attributable to different causes; each cause is expressed as a percentage of all deaths; the sum of causes must add to 100%. See Mortality

propoxyphene Darvon® An oral opioid analgesic, administered with other analgesics; it is metabolized to therapeutically active norpropoxyphene. See Opioids

propranolol Inderal® CARDIOLOGY A β-blocker used for A Fib, angina, HTN, PVCs, and 2° protect against acute MI MECHANISM Cell membrane stabilizing CONTRAINDICATIONS Bronchospasms, left ventricular dysfunction. See β-blocker

proprietary medicine PHARMACOLOGY A completely compounded nonprescription drug in its unbroken, original package, which is understood *not* to contain any controlled substance

proprioception NEUROLOGY The subconscious sensation of body and limb movement and position, obtained from non-visual sensory input from muscle spindles and joint capsules

proprioceptor neuromuscular facilitation stretching PNF stretching SPORTS MEDICINE A stretching technique that reeducates muscles by 'tricking' proprioceptors into allowing ↑ ROM, by ↓ tensing of muscles that normally occurs at preset limits to ROM. See Massage, Range of motion

proptosis MEDTALK Forward displacement of the eyeball. See Exophthalmos

Propulsid® Cisapride GI DISEASE An agent used for symptomatic treatment of nocturnal heartburn due to GERD removed from marketplace due to arrhythmias, death. See Proton pump inhibitor

Proscar® Finasteride, see there

ProSol™ NURSING A sulfite-free amino acid concentrate used for TPN INDICATIONS To offset nitrogen loss or negative nitrogen balance in Pts where: GI tract should not be used; protein absorption is impaired; metabolic requirements for protein are markedly ↑, as with extensive burns; also used to ↓ fluid intake in Pts requiring fluid restriction and TPN. See TPN

542 **prosopagnosia** NEUROLOGY Inability to recognize familiar faces unexplained by defective visual acuity or ↓ consciousness or alertness

Prosorba® column THERAPEUTICS A blood-filtration device used during apheresis–eg, for treating ITP; it may improve severe rheumatoid arthritis in Pts who failed or are intolerant to disease-modifying antirheumatic drugs–DMARDs. See Apheresis

prospective epidemiology A study over time of a cohort of persons who share a feature of clinical or other interest–eg, HTN, exposure to an environmental toxin, etc; the population is compared to a parallel population presumed not to be exposed to the same factor

prospective study EPIDEMIOLOGY A long-term predictive study designed to observe outcomes or events that occur after identifying a group of subjects to be studied; PSs collect data as the events being evaluated occur; PSs allow testing of hypothesized cause-and-effect relationships, and determine the effect of a therapy on disease progression. See Cohort analytic study. Cf Retrospective study

PROSPER CARDIOLOGY A clinical trial–Prospective Study of Pravastatin in the Elderly at Risk. See Pravastatin

prostaglandin-induced abortion GYNECOLOGY A method for terminating pregnancy in the 2nd trimester, in which PG is administered to induce uterine contractions, followed by cervical dilatation PROS Allows delivery of intact stillborn fetus CONS Vomiting, diarrhea, prolonged duration, discomfort, contraindicated in ♀ with prior uterine surgery. Cf D&E

Prostar® system VASCULAR SURGERY A suture-based percutaneous system for surgical closure of arterial access sites during minimally invasive procedures–eg, catheterization, angioplasty, stent placement, atherectomy

ProstaScint® Capromab pendetide, see there

ProstAsure® UROLOGY A computer-assisted test that discerns subtle changes in biochemical activity among groups of highly relevant biomarkers–eg, PSA, which is said to be 85% sensitive for prostate CA still confined within the prostatic capsule, thus leading to earlier diagnosis, treatment, cure. See Prostate cancer

prostate biopsy A 'needle' biopsy of the prostate of older ♂ with either clinical–eg, ↑ firmness of prostate on rectal examination, or lab–eg, ↑ PSA findings commonly associated with prostate CA APPROACHES Perineal, transurethral, transrectal. See Digital rectal examination, Gleason's grade, Ploidy analysis, Prostate cancer, Prostate-specific antigen

prostate cancer Prostatic adenocarcinoma ONCOLOGY A CA of older ♂, and 2nd leading cause of death of ♂ with 106,000 new cases and 30,000 deaths/yr, US; 35-50% of ♂ > 70 yrs of age have PC; it is more common and aggressive in African Americans DIAGNOSIS PSA, digital rectal examination, transrectal ultrasonography, needle biopsy STAGING Table MANAGEMENT Observation alone, RT, surgery, hormone, chemotherapy PROGNOSIS Uncertain; PC often remains occult; those with PC often die natural deaths; flow cytometry of tumor cells allows partial prediction of PCs most likely to progress. See PSA , Watchful waiting

PROSTATE CANCER STAGES

I Confined to prostate, not palpable during digital rectal examination, not visible by imaging, asymptomatic; usually found accidentally or detected by ↑ PSA; cancer cells may be found in only one or more areas of prostate

II Found by a needle Bx triggered by ↑ serum PSA or identified by DRE

III CA spread beyond the capsule but not to lymph nodes; seminal vesicles may be involved

IV CA has metastasized to lymph nodes or to organs and tissues far from the prostate–eg bone, liver, or lungs

prostate chips UROLOGY Crescent-shaped pieces of firm, rubbery prostatic tissue removed in a TURP–transurethral resection of prostate, for

BPH, which measure up to 1.5 cm in greatest dimension. See Benign prostatic hypertrophy

prostate-specific antigen See PSA

prostate ultrasonography IMAGING A noninvasive procedure in which images of the prostate and surrounding tissues are obtained by ultrasound; if a mass is ID'd by PU, and the PSA is ↑, the prostate is biopsied. See Prostate biopsy

prostate years VOX POPULI A popular term for age 60 or older, when ♂ begin to become symptomatic for urinary retention due to BPH or, less commonly, prostate CA. See Benign prostatic hypertrophy

prostatectomy UROLOGY The surgical removal of part or the entire–radical/total prostate and some surrounding tissue. See Microwave prostatectomy, Nerve-sparing prostatectomy, Perineal prostatectomy, Radical prostatectomy

prostatic *adjective* Relating to the prostate

prostatitis UROLOGY Inflammation of the prostate. See Granulomatous prostatitis

prostatitis syndrome UROLOGY Any inflammatory condition defined by the Intl Prostatitis Collaborative Network under the aegis of the NIH. See Prostate

PROSTATITIS SYNDROMES

ACUTE BACTERIAL PROSTATITIS Pts present with acute Sx of UTI–eg urinary frequency and dysuria, as well as malaise, fever, myalgia; bacteria and pus are usually present in the urine and due to uropathic bacteria–eg, E coli

CHRONIC BACTERIAL PROSTATITIS Pts have recurrent UTIs, usually due to the same organisms–eg, E coli , another GNR or an enterococcus; between symptomatic bouts, Pts continue to be carriers in cultures from the lower urinary tract

CHRONIC PROSTATITIS/CHRONIC PELVIC PAIN SYNDROME The most common of the prostatitis syndromes, CP/CPPS recognizes the limited understanding of the cause of this condition and implies that other organs beside the prostate may be involved and is a diagnosis of exclusion, after active urethritis, urogenital cancer, GU disease, functionally significant urethral and neurological disease have been completely excluded; it is divided into an inflammatory subtype with leukocytes in the urine and a noninflammatory subtype which has no manifestation of inflammation

ASYMPTOMATIC INFLAMMATORY PROSTATITIS A condition which is diagnosed in Pts who lack a Hx of GU complaints who have ↑ PSA or infertility

Prostatron® UROLOGY A device for treating BPH using microwave heat without injuring adjacent healthy tissue with TUMT®–transurethral microwave thermotherapy. See Benign prostatic hypertrophy

prosthesis plural, prostheses MEDTALK An artificial body part–eg, pseudobreast, artificial limb, etc. See Bioprosthesis, Heart valve prosthesis, Hemobahn endovascular prosthesis, Neural prosthesis, Seagull wing prosthesis

prosthetic disk nucleus device PDN® ORTHOPEDICS An implantable device for treating severe low-back pain from degenerative disk disease in Pts refractory to nonsurgical treatment, as an alternative to spinal fusion. See Low back pain

prosthetic speech ENT Monotonous speech produced by an artificial sound generator, required by Pts with laryngectomies for CA; electrical artificial larynges have a buzzing sound generator that competes with speech generated through the mouth, and thus is also 'mechanical' in nature. See Artificial sound generator

prosthetist ORTHOPEDICS A person who produces and/or fits a prosthetic device, usually understood to be a limb

prosthodontist DENTISTRY A dentist specialized in restoring teeth, replacing missing teeth and maintaining proper occlusion of teeth to restore appearance, comfort, and/or health

Prostin® Alprostadil, see there

prostitution STD Performance of sexual work–ie, sexual activity for hire EPIDEMIOLOGY There are 0.5–2 million prostitutes–US; enter the field ± age 14; arrests for prostitution/commercialized vice, 1992 ♀ 47,526; ♂ 24,401; 17% of ♂ have solicited prostitutes. See Child prostitution, Sexual work, Sexually transmitted diseases

ProStream™ DRUG DELIVERY A multiple sidehole infusion wire for delivering therapeutics to peripheral vessels. See Drug delivery

Prostsafe™ POPULAR HEALTH A concoction of zinc, ginseng, *Serenoa serrulata*, amino acids, bee pollen, vitamins A, B₆, and E, which allegedly controls prostate-related urinary frequency, and other Sx in ♂ > age 40. See *Serenoa serrulata*. Cf Invalid claims of efficacy

protanopia OPHTHALMOLOGY Red-green color blindness due to a lack of long wavelength–LW, 560 nm photopigment

protease inhibitor AIDS An antiretroviral agent that inhibits retroviral protease, which interferes with the ability of a virus to copy itself, used to manage HIV infection–eg, saquinavir, ritonavir, indinavir, nelfinavir

Protectaid® GYNECOLOGY A contraceptive sponge containing F-5 Gel*, a formula of 3 different low concentration spermicides–↓ tissue irritation, ↑ prophylactic effect; it is also kills microorganisms associated with STDs. See Contraceptive sponge

protected health information HEALTH INFORMATICS Any individually identifiable health information that is used or circulated by an entity that falls under the governance of HIPAA; the privacy regulations mandate safeguards for protected health information, and the responsibility for maintaining them also extends to third-party business partners. See HIPAA, Medical privacy, Patient confidentiality, Pretty good privacy, Privacy

protection See Absorptive eye protection, Neuroprotection, Personal protection, Photoprotection, Reflective protection, Self-protection

protein BIOCHEMISTRY A large molecule consisting of a long chain or sequence of amino acids with a general formula of $H_2N–CHR–COOH$–aka alpha amino acids, joined in a peptide likage; after water, proteins are the major cell constituent, and are critical for all biological structures–eg, organelles, mitochondria, enzymes and functions–eg, growth, development, immune function, motility TYPES Hormones, enzymes, antibodies

protein C deficiency A condition characterized by a deficiency of vitamin K dependent plasma protein C and protein S, both natural anticoagulants; PCD is either AD of variable penetration, or acquired, and due to DIC, warfarin therapy, hepatic disease and postoperatively

protein diet Liquid-protein diet, see there

protein electrophoresis See Serum protein electrophoresis

protein-losing enteropathy GASTROENTEROLOGY A condition characterized by excess transmucosal efflux of plasma proteins into the intestinal lumen, due to ↑ permeability 2° to mucosal cell damage, inflammatory ulceration, or leakage from lymphatic vessels 2° to obstruction ETIOLOGY Paraneoplastic syndromes, malignancy–gastric CA, leukemia, lymphoma, KS, carcinoid tumor, melanoma, nontropical sprue, ulcerative colitis, intestinal lymphangiectasia, CHF, constrictive pericarditis, superior vena cava thrombosis, Ménétrier's disease, eosinophilic gastroenteropathy, sarcoidosis, Whipple's disease, infections–giardiasis, bacterial overgrowth, viral enteritis, schistosomiasis, *C difficile*, pulmonary artery stenosis, SLE, amyloidosis CLINICAL Diarrhea, weight loss, peripheral edema IMAGING Thickened intestinal wall by CT LAB ↓ Albumin MANAGEMENT Treat underlying cause

protein malnutrition Kwashiorkor, see there

protein restriction CLINICAL NUTRITION A restriction of dietary protein from a 'normal' level–±1.3 g/kg/day, indicated in renal failure; extreme PR–very low protein diet, 0.28 g/kg/day does not significantly slow the progression of renal disease > moderate PR–low protein diet–eg, 0.58 g/kg/day. Cf Caloric restriction

protein S deficiency HEMATOLOGY An AD condition clinically and therapeutically similar to heterozygous protein C deficiency, characterized by pulmonary thrombosis, DVT, thrombophlebitis

protein-sparing NUTRITION *adjective* Referring to any maneuver that minimizes protein catabolism–eg, adding carbohydrates and fats to a low-protein diet. See Low-protein diet

proteinuria NEPHROLOGY The excretion of excessive (> 5 mg/dL) protein in the urine; normally, about 150 mg/day of protein is lost in the urine, ⅓ is albumin, ⅓ is Tamm-Horsfall glycoprotein; the rest is divided among actively secreted proteins–eg, retinol binding proteins, $β_2$-microglobulin, Ig light chains and lysozyme; in absence of disease, large proteins are retained due to their size, while the smaller proteins are actively resorbed; proteinuria is most often caused by kidney disease, due to glomerular defects, and defective renal tubular resorption, and most often detected by screening with reagent strip–dipstick. See Functional proteinuria, Overflow proteinuria

---

PROTEINURIA, SEVERITY

SEVERE ≥ 1.0 g/dL, due to glomerulonephritis, nephrotic syndrome, lupus nephritis, amyloidosis

MODERATE ≥ 0.2 g/dL, ≤ 1.0 g/dL, due to CHF, drugs, acute infections, myeloma, chemical toxins

MILD 0.05-0.2 g/dL, due to polycystic kidneys, pyelonephritis, renal tubular defects

---

PROTEINURIA, PATTERNS

GLOMERULAR PATTERN Due to a loss of fixed negative charge on the glomerular capillary wall, allowing albumin and other large (≥ 68 kD) molecules to leak into Bowman's space–eg, in glomerulonephritis and nephrotic syndrome LAB ↓ albumin, antithrombin, transferrin, prealbumin, $α_1$-acid glycoprotein, $α_1$-antitrypsin

HEMODYNAMIC PATTERN Due to rheostatic changes in the body, causing a loss of 20 to 68 kD molecules, seen in transient proteinuria, CHF, fever, seizures, excess exercise

OVERFLOW PATTERN Due to tissue/cell destruction that overwhelms renal capacity to excrete certain proteins–eg, Bence-Jones proteinuria and myoglobinuria

TUBULAR PATTERN Due to renal tubular dysfunction with loss of normally filtered low molecular weight (≤ 40 kD) molecules LAB ↓ $β_2$-microglobulin and lysozyme–eg, Fanconi syndrome, Wilson's disease, interstitial nephritis, antibiotic-induced injury and heavy metal intoxication

---

Proteus mirabilis MICROBIOLOGY A gram-negative pathogen linked to UTIs, wound infections HABITAT *P mirabilis* may be found in water, soil, feces

Proteus vulgaris MICROBIOLOGY A pathogen linked to UTIs and wound infections HABITAT *P mirabilis* is found in water, soil, and feces

Prothiaden Dothiepin, see there

prothrombin complex concentrate HEMATOLOGY Any commercial product–eg, FEIBA–Factor VIII inhibitor by-passing activity that contain nonactivated factors IX, in addition to II, XI and X; PCCs may be used to ameliorate the intensity and duration of joint and soft tissue bleeding in hemophiliacs who produce factor VIII inhibitors; PCCs are 50% effective in staunching hemorrhage. See Prothrombin

prothrombin time PT, protime HEMATOLOGY A coagulation test used to monitor oral anticoagulant therapy, in particular with warfarin, to maintaining the PT at 2-2.5 times > the normal control; if PT is < 2-fold normal, anticoagulation is inadequate, if > 2.5 times normal, anticoagulation is excessive; PT is ↓ in thrombophlebitis, PTE, MI, drugs–barbiturates, OCs, digitalis, diphenhydramine, diuretics, metaproterenol, vitamin K; PT is ↑ in afibrinogenemia, drugs–anticoagulants, antibiotics, chlorpromazine, chlordiazepoxide, methyldopa, reserpine, salicylates, sulfonamides, erythroblastosis fetalis, coagulation factor deficiencies REF RANGE 11.5–13.7 secs. Cf Partial thromboplastin time

protime Prothrombin time, see there

proto-oncogene MOLECULAR MEDICINE A normal gene with latent transforming potential, homologous to a retroviral oncogene EXAMPLES *c-erb, c-fos, c-jun, c-myb, c-myc, c-mos, c-raf, c-ras*; POs act in normal growth, differentiation, and induction and/or maintenance of CA. See *Abl, cyclin D1, RET/PTC*

protocol BARIATRICS See Fen/Phen protocol CARDIOLOGY See Cornell Protocol NEONATOLOGY See Philadelphia protocol NIHSPEAK 1. The formal design or plan of an experiment or research activity; specifically, the plan submitted to an IRB for review and to an agency for research support 2. A precise step-by-step description of a test, including the listing of all necessary reagents and all criteria and procedures for the evaluation of the test data ONCOLOGY Treatment protocol An organized and detailed description of a therapeutic method being used to manage a malignancy–eg, CHOP, MOPP. See M2 protocol RADIATION PHYSICS A written, detailed experimental design, reviewed and approved by the Radiation Safety Branch, for use of radioactive materials in excess of set limits or the use of volatile radioiodines THERAPEUTICS A formal outline of care; a treatment plan. See Compassionate investigational new drug protocal, Ponticelli protocol

proton pump inhibitor PHARMACOLOGY A compound–eg, Prilosec–omeprazole that is better than H2 blockers–eg, Zantac–ranitidine, Tagamet–cimetidine, for treating GERD. See Gastroesophageal reflux disease. Cf H2 blockers

Protropin® ENDOCRINOLOGY Recombinant growth hormone for short children

protuberant *adjective* Large, excessive, overhanging

proud flesh Exuberant granulation tissue seen in a poorly healed wound, characterized by florid, 'geographic' scarring on the skin surface, related to a defect of union of interrupted tissues by 'second intent' healing. See Keloid, Wound healing

PROVE IT CARDIOLOGY A clinical trial–Pravastatin or Atorvastatin Evaluation and Infection Therapy

PROVED CARDIOLOGY A clinical trial–Prospective Randomized Study of Ventricular Failure & Efficacy of Digoxin–which evaluated the effect of digoxin withdrawal in Pts with CHF. See Heart failure

Proventil® CRITICAL CARE An albuterol sulfate-based metered dose inhaler–MDI used to relieve asthma attacks. See Asthma

proverbs test NEUROLOGY A test of abstract thinking based on the ability to explain proverbs, a facet of intelligence that deteriorates in dementia. See Mini-mental test

Providentia MICROBIOLOGY A genus of Enterobateriaceae found in water, soil, and fecally contaminated materials; 5 SPECIES *P alcalifaciens, P heimbachae, P rettgeri, P rustigianii, P stuartii*

provider INFORMATICS See Internet provider MEDTALK Health care provider A person–eg, doctor, nurse, nurse practitioner, or institution–eg, hospital, clinic, or lab, that provides medical care. See Alternative site provider, Application service provider, Handicapped health care provider, Health care provider, Impaired health care provider, Incapacitated health care provider, Internet provider, Medical service provider, Mental health provider, Midlevel provider, Participating provider, Plan provider

provider-based MEDICAL PRACTICE *adjective* Referring to a medical practitioner's location, defined by HCFA–Health Care Financing Administration as any facility–eg, hospital or nursing home reimbursed by Medicare on a cost basis

provider number See UPIN

provider profile MANAGED CARE An examination of services provided, claims filed and benefits allocated by healthcare facilities, physicians and other providers to assess quality of care and cost management

Provigil® Modafinil, see there

Provir® SP-303 INTERNAL MEDICINE An agent used for diarrhea of unknown etiology and traveler's diarrhea.

provisional status HOSPITAL PRACTICE A medical staff status accorded to a practitioner during the first yr of service to the hospital. See Medical staff

provocation test MEDTALK 1. Any of a number of tests used to deliberately induce a suspected pathologic derangement–eg, provocation of ↑ intraocular pressure by ingestion of excess water 2. Neutralization, see there ORTHOPEDICS Any of a number of tests designed to reproduce the pain, instability or disability associated with a particular joint injury. See Apprehension test, Fulcrum test, Relocation test, Shoulder instability. Cf Laxity test

provost A high-level academic or administrator at a college or university

PROWESS INFECTIOUS DISEASE A clinical trial–Recombinant Human Activated Protein C [Zovant™] Worldwide Evaluation in Severe Sepsis

proximal carcinogen Ultimate carcinogen A chemical or physical agent that (hypothetically) initiates the first or induction step of a carcinogenic cascade

proximal renal tubular acidosis Renal tubular acidosis type II NEPHROLOGY A disorder caused by a partial defect in the secretion of hydrogen ions in the proximal renal tubule, with ↓ reabsorption of HCO_3 from the tubule into the circulation; the net loss of HCO_3 from the body results in metabolic acidosis, accompanied by a loss of glucose, amino acids, phosphate, Ca^{2+}, and potassium in the urine; because of the solute loss, water is also dragged out of the body; the metabolic acidosis results in loss of Ca^{2+} from bone, and ↑ serum Ca^{2+}; excess serum Ca^{2+} is excreted by the kidneys, causing a loss of total body Ca^{2+} resulting in osteomalacia or rickets, impaired growth, skeletal deformities, muscle weakness, kidney stones and nephrocalcinosis associated with the excessive excretion of Ca^{2+} and phosphate through the kidneys

proximate cause MALPRACTICE An element required to prove negligence; the plaintiff–Pt or Pt's estate must prove that the Pt's injury is reasonably connected to the physician's action, through either the 'but for' test or the 'substantial factor' test. See 'But for' test, Negligence, 'Substantial factor' test

proxy See Care proxy, Health care proxy

Prozac® Fluoxetine NEUROPHARMACOLOGY An oral selective serotonin reuptake inhibitor–SSRI used for depression, obsessive-compulsive disorder, bulimia nervosa ADVERSE EFFECTS 10-15% of Pts experience anxiety, nervousness, insomnia, weight loss, overstimulation, upset stomach, headache CONTRAINDICATIONS MAOI therapy. See SSRI–serotonin-selective reuptake inhibitor

Prozac-related suicide PSYCHIATRY A postulated association between use of the antidepressant, Prozac® and suicide. See Prozac®, Tricyclic antidepressant suicide

Prozepam Diazepam, see there

PrP A prion protein. See Prion

prudence See Reasonable prudence

prudent layperson standard EMERGENCY MEDICINE A standard for determining the need to visit the ER, which defines an emergency as a condition that a prudent lay person, 'who possesses an average knowledge of health and medicine.' expects, may result in 1. placing the Pt in serious jeopardy, 2. serious impairment of bodily function, or 3. serious dysfunction of any bodily organs

prune belly syndrome Eagle-Barrett syndrome PEDIATRICS A condition characterized by congenital lack of abdominal muscles, which imparts a rugose, prune-like appearance to the flaccid abdominal wall; 97% occur in ♂ and are accompanied by GU anomalies–eg, bilateral cryptorchidism, hypoplastic and dysplastic kidneys; affected ♀ have uterine defects; although considered an X-linked disease, no chromosome defect has been identified, and PBS may represent a 'sequence' initiated by in utero urethral obstruction, causing urinary tract anomalies–megaureters, megabladder, patent urachus or urachal cyst; other findings include Potter's facies, talipes, hip dislocation, musculoskeletal and cardiac defects TREATMENT Corsets, excision of redundant tissue PROGNOSIS Oligohydramnios may arise in utero, causing fetal pulmonary hypoplasia, 20% are stillborn, 50% die in infancy. See Oligohydramnios

prune juice discharge A descriptor for the dark brown vaginal discharge typical of a hydatidiform mole, likened to stewed prune–dried plum–juice

prune juice sputum A descriptor for the watery, dark brown, hemorrhagic sputum of advanced pneumococcal pneumonia. Cf Rusty sputum

pruritus Itching

pruritus gravidarum A condition characterized by intense itching, which occurs in 1:300 pregnancies, beginning in the 3rd trimester, first on the abdomen, later extending to the entire body surface TREATMENT Antihistamines. See Pregnancy-related conditions. Cf PUPPP

Pryleugan Imipramine, see there

PSA Prostate-specific antigen UROLOGY A 34 kD glycoprotein serine protease with significant homology with other neutral proteases; PSA is secreted exclusively by the prostate epithelium and is responsible for lysing the seminal coagulum; it is present in the serum and is ↑ in Pts with both benign–acute inflammation and BPH and malignant–CA prostate lesions; most serum PSA is bound to protease inhibitors, including α_1-antichymotrypsin–the form usually measured in the lab, α_1-macroglobulin, and in far lower amounts to α_1-antitrypsin; small amounts of PSA in the circulation are not bound–ie, free—to a carrier protein; prostate CA is present in 22% of those with PSA levels above 4.0 µg/L, 60% of those with levels above 10 µg/L

pseudoacanthosis nigricans A disease of obese, darkly pigmented adults, characterized by patchy hyperpigmentation TREATMENT Weight loss

pseudoachondroplasia Any of a heterogeneous group of often AD conditions; the most common is pseudoachondroplastic spondyloepiphyseal dysplasia CLINICAL Early onset with ↓ limb growth–irregular 'mushroomed' metaphyses, small, irregular and fragmented epiphyses, short bowed diaphyses, flattened vertebrae, lumbar lordosis, scoliosis, kyphosis, 'spatula' ribs, hypermobility of major and acral joints, short hands and feet, contractures of hips and knees, waddling gait, early onset osteoarthritis. Cf Achondroplasia

pseudoaddiction SUBSTANCE ABUSE A drug-seeking behavior that stimulates true addiction in Pts with pain who are receiving inadequate pain, medication. Cf Addiction, Drug-seeking

pseudoaldosteronism Pseudohyperaldosteronism, see there

pseudoallergy CLINICAL IMMUNOLOGY A nonimmunologic, anaphylaxis-like reaction of sudden onset, associated with food ingestion; PAs may be due to an anaphylactoid reaction, intolerance–eg, psychogenic response, metabolic defect–eg, enzymatic deficiency, tyramine reaction, or toxicity–eg, tetrodotoxin. See Cheese reaction, Food allergy, Food intolerance. Cf Allergy

pseudoaneurysm False aneurysm, see there

pseudoarthrosis ORTHOPEDICS The non-union of 2 fractured ends of a long bone, in which the bone is covered by fibrous tissue or fibrocartilage; in extreme cases, the false joint is surrounded by a bursal sac containing synovial fluid; congenital PAs may occur in von Recklinghausen's disease or osteogenesis imperfecta; acquired PA usually follows trauma or, less commonly, tumor-related osteolysis and fibrous dysplasia. Cf Arthritis

pseudoatrophy of brain An apparent ↓ in volume of cortical tissue, seen by CT, due to changes in CSF production and alterations in the blood-brain barrier, with 2° ↓ in cerebral interstitial fluid, due to mannitol therapy. Cf Brain atrophy

pseudoautosomal region A small terminal region of homologous DNA sequences shared by mammalian sex chromosomes, which pair and recombine during ♂ meiosis. Cf Autosomal region, X chromosome inactivation

pseudo-Bartter syndrome A condition attributed to furosemide therapy, characterized by hypokalemic-hypochloremic alkalosis, hyperactivity of renin-angiotensin-aldosterone system with ↑ aldosterone, normotension, pressor inactivity of angiotensin II, ↑ urinary PGE_2, ANP. Cf Bartter syndrome

pseudobulbar palsy Pseudobulbar paralysis, spastic bulbar paralysis A disease of middle age, characterized by bilateral spasticity of the facial and deglutitive muscles, resulting in dysarthria, dysphonia, dysphagia, drooling, facial weakness, hyperreflexia of extremities, and shuffling–parkinsonian gait; PP is remarkable for the variable psychiatric component, in which the Pts may have a flat affect–simulating apathy or severe depression, become enmeshed in trivialities, or have inappropriate responses to environmental cues, aka 'laughing sickness' for the characteristic pathologic–ie, inappropriate laughing–or crying ETIOLOGY Multifocal infarcts, often due to ASHD, but also to HTN, infections, trauma, degeneration TREATMENT Antibiotics PROGNOSIS Guarded. Cf Bulbar palsy

pseudocholinesterase A liver and plasma enzyme that rapidly metabolizes succinylcholine, a short-acting–5-10 mins–neuromuscular anesthetic, thereby controlling its duration of action; some people have congenital PChe variants, in which neuromuscular block is prolonged with usual doses of succinylcholine–and identified by determining the 'dibucaine' number; PChe activity may be ↓ in various acquired conditions–eg, liver disease–hepatitis, cirrhosis, metastasis, malnutrition, acute infections, anemia, MIs, CA, pregnancy, cytotoxic drugs, acetylcholinesterase inhibitors, dermatomyositis. See Dibucaine number

pseudochylous effusion A milky pleural effusion mimicking chylothorax, associated with ↑ lipids–cholesterol or lecithin-globulin complexes, seen in chronic pulmonary effusions, as in TB, rheumatoid arthritis or empyema. Cf Chylous effusion

pseudocoma A state mimicking acute unconsciousness with intact self-awareness, occurring in (1) Organic disease states–eg, 'locked-in syndrome' (2) Psychogenic unresponsiveness, due to catatonic states–eg, schizophrenia, severe depression, hysterical reactions or in frank malingering or (3) Near-death experiences, for which there is no acceptable scientific explanation. Cf Coma

pseudocryptorchidism Retractile testis A testis characterized by a hyperactive cremasteric reflex, which draws the testis into the inguinal canal; it is caused by cold temperature, fear, and genital manipulation, most common ± age 5 DIAGNOSIS Pseudocryptorchid testis can be pushed into the scrotum cryptorchid testis can't. Cf Cryptorchidism, Migrating testis

pseudo-Cushing syndrome A condition in which there are certain clinical–truncal obesity and purple striae or biochemical–↑ urinary cortisol levels that fall short of those seen in Cushing syndrome abnormalities that overlap with Cushing syndrome; PCSs occurs with major depressive disorder, in middle-aged ♀ with obesity–HTN and DM, and in alcoholics TREATMENT ↓ weight, psychotherapy. Cf Cushing syndrome

546   **pseudocyesis** PSYCHIATRY A Sx complex affecting ♀ with a strong and unfulfilled desire for children, resulting in amenorrhea, morning sickness, induration of breasts, ↑ abdominal girth. See Pseudopregnancy. Cf Sympathy 'pregnancy'

pseudocyst PARASITOLOGY A cyst-like 'mass' that corresponds to a macrophage laden with *Toxoplasma gondii* or other sporozoans which is most often seen in the brain, a finding typical of AIDS neuropathy PATHOLOGY A dilated space lined by neither epithelium nor mesothelium, classically seen in the pancreas as unilocular spaces lined by fibrous tissue, often following multiple bouts of acute pancreatitis, or in the ultrarare hereditary pancreatitis

pseudodelirium PSYCHIATRY A term proposed for transient delirium-like psychosis, seen in absence of a functional cause CLINICAL Inconsistent results on cognitive testing, history of psychiatric illness, depressive or manic behavior, systematized delusions, absence of fluctuations and nocturnal worsening, typical of delirium EEG Normal. Cf Delirium

pseudodementia NEUROLOGY Dementia-like Sx due to psychologic impairment–eg, depression or histrionic episode, characterized by cognitive impairment of short duration, with preserved attention and ability to concentrate, and a variable performance in tests with similar levels of difficulty; it is often transient, common in the elderly and may be linked to medications–anticholinergics, barbiturates, benzodiazepines, butyrophenones, corticosteroids, digitalis, IMAOs, TCAs or due to depression–with physical and emotional deprivation, accompanied by apathy, akinesia and anxiety; pseudodementia also occurs in normal pressure hydrocephalus, Creutzfeldt-Jakob, Huntington's, Parkinson's, Pick's, Wilson's diseases, and endocrinopathy. Cf Cerebral pseudoatrophy, Dementia

pseudodiabetes 1. Latent diabetes mellitus 2. A clinical condition characterized by defective carbohydrate metabolism 2° to chronic renal failure–uremia, with ↓ glucose tolerance–rapid post-prandial rise and delayed return of glucose to normal, mild baseline hyperglycemia and insulin 'resistance'. Cf Diabetes

pseudofailure CARDIOLOGY A finding on an EKG which simulates the failure of a cardiac pacemaker; in pseudofailure, the atrium is paced at a rate of ± 80 beats/min, and the ventricle responds normally to the atrial beats, which implies that the AV node and bundle branches successfully carry the intrinsic cardiac pulses, and thus do not need the pacemaker's help. See Pacemaker. Cf Heart failure

pseudofracture Looser zone A thin radiolucent line that mimics a true fracture and which is quasipathognomonic for osteomalacia in the adult, characterized by complex bony lesions typical of advanced renal failure. Cf Fracture

pseudofusion beat CARDIOLOGY A cardiac complex or an ECG waveform, which corresponds to the superimposition of an ineffective pacemaker spike on a spontaneous P or QRS complex originating from a single focus; the pacing stimulus is delivered after the chamber has already spontaneously depolarized; thus the stimulus does not cause contraction. Cf Fusion beat, Pseudopseudofusion beat

pseudogout Chondrocalcinosis RHEUMATOLOGY An arthropathy of the elderly ♀, characterized by deposits of calcium pyrophosphate dihydrate crystals in large joints–knees, shoulders, hips, due to either local overproduction of pyrophosphate or phosphatase deficiency. Cf Gout

pseudohermaphroditism ENDOCRINOLOGY A state in which a person has the gonadal tissue of one sex, but the wiring, plumbing, and/or chassis of the opposite sex. See Female pseudohermaphrodism, Male pseudohermaphroditism. Cf Hermaphroditism, Intersex, Virilization

PSEUDOHERMAPHRODITISM

FEMALE PSEUDOHERMAPHRODITISM A condition affecting a genotypic–46, XX female with ovaries, caused by a relative excess of androgen in utero, resulting in equivocal or masculinized genital duct derivatives, ie external genitalia and/or a male phenotype with genital ambiguity and/or virilization ETIOLOGY 1. Adrenogenital syndrome Defects of 21-hydroxylase, 11-β-hydroxylase or 3-β-hydroxysteroid dehydrogenase, or delta 5-4 isomerase deficiency, resulting in ↑ androgenic intermediates 2. Maternal ingestion of progestins or androgens and three maternal virilizing tumors–eg, luteoma of pregnancy

MALE PSEUDOHERMAPHRODITISM, A condition affecting a genotypic–46, XY male with testes, caused by a relative deficiency of androgen in utero, resulting in a phenotypic female with ambiguous genitalia ETIOLOGY 1. Gonadal defects Testicular regression syndrome, persistent müllerian duct origin, Leydig cell agenesis and defects in testosterone synthesis 2. End-organ defects Testicular feminization or androgen insensitivity syndrome, incomplete androgen insensitivity syndrome and 5-α reductase deficiency. See Testicular feminization

pseudo-Hirschsprung's disease Colonic inertia in children, of possible psychogenic origin, which lacks histologic evidence of defective myoenteric innervation. Cf Hirschsprung's disease

pseudohyperaldosteronism Liddle syndrome An AD aldosteronism-like condition characterized by severe HTN and spontaneous hypokalemia (K+ wasting) leading to hypokalemic alkalosis with ↓ aldosterone secretion TREATMENT KCl, amiloride and triamterene to prevent K+ wasting, sodium restriction; renal transplantation. Cf Hyperaldosteronism

pseudohyperinflation PULMONOLOGY A finding that simulates hyperinflation of the lungs, descibed as typical of ankylosing spondylitis, in which there is ↑ residual volume and functional residual capacity and ↓ vital capacity and total lung capacity. Cf Hyperinflation

pseudohyperkalemia LAB MEDICINE An in vitro phenomenon seen in megakaryocytic hyperplasia, thrombocytosis, leukocytosis, or myeloproliferative disease, where rapid clotting of blood releases potassium from RBCs LAB ↑ Serum K+, plasma K+ is normal. Cf Hyperkalemia

pseudohypertension Sphygmomanometric cuff pressure that is > actual BP, due to markedly calcified–'pipestem' brachial arteries; it is most common in the elderly with extensive ASHD DIAGNOSIS Osler's maneuver, a bedside method for assessing the palpability of the radial or brachial pulse distal to a point of presumed occlusion. Cf Hypertension, Small cuff syndrome, White coat HTN

pseudohypertrophy Any ↑ in organ size without ↑ in number or size of the organ's native cells; pseudohypertrophy is typical of Duchenne's muscular dystrophy, where the ↑ in girth of the extremities is due to infiltration of fat among bundles of atrophic muscle. Cf Hypertrophy

pseudohypoaldosteronism A heterogeneous group of salt wasting syndromes due to distal renal tubular insensitivity to mineralocorticoids–eg, aldosterone or due to a defect in the mineralocortioid receptor in the colonic mucosa, salivary gland, sweat glands, resulting in salt loss but normal adrenocortical and renal function, and a hyperactive renin-angiotensin system. Cf Hypoaldosteronism TYPES TYPE I Albright's hereditary osteodystrophy More common, and is almost invariably X-linked MIM 300100; AD MIM 103580 cases–designated type IA have been reported; there is inadequate cAMP response to PTH; TYPE II Due to inadequate end-organ response to ↑ cAMP levels; affected children are short and stocky with a round facies, brachydactyly, tetany, foci of bony demineralization, osteitis fibrosa

pseudohypohyperparathyroidism A condition characterized by congenital end-organ resistance to PTH–pseudohypoparathyroidism and osteitis fibrosa cystica, seen in hyperparathyroidism; the condition may be related to differences in the transduction of the bone remodeling response–ie, divergence of PTH secretion and expression of its receptor.

pseudohyponatremia Spuriously low sodium levels due to either 1. Intrinsic properties of a Pt sample–eg, hyperglycemia or hyperproteinemia–displacement of plasma water or due to the cationic nature of monoclonal proteins which bind sodium, hyperlipidemia or hyperviscosity or 2. Analytic factors, which occur when preparing samples for flame photometry or indirect potentiometry MANAGEMENT Both respond to vitamin 1,25–OH$_2$D$_3$.

pseudohypoparathyroidism ENDOCRINOLOGY A hypoparathyroid-like state–hypocalcemia, hyperphosphatemia due to end-organ resistance to PTH by bone and kidney, with a loss of renal tubule response to PTH, accompanied by ↑ PTH secretion and parathyroid gland–PG hyperplasia; despite excess PTH secretion in response to hypocalcemia by a normal or hyperplastic PG, the condition is associated with a 2° hypocalcemia-induced ↑ in parathyroid function–administration of pharmacologic doses of PTH normally results in ↑ urinary phosphate excretion and ↑ cAMP, but not in pseudohypoparathyroidism; the pattern of heredity is unclear, ♂:♀ ratio is 2:1; changes include a round face, dental dysplasia, dry coarse hair, mental retardation

pseudohypophosphatasia A condition characterized by clinical and radiologic features of vitamin D-resistant rickets and phosphoethanolaminuria, with ↑ pyridoxal 5'-phosphate–vitamin B₆ cofactor, caput membraneceum, osteopathy of the skull and long bones, failure to thrive, muscle hypotonicity. Cf Hypophosphatasia

pseudoincontinence The inability to retain urine, due to difficulty in reaching the toilet, advanced arthritis, or other physical handicaps, resulting in anger and frustration in the impaired. Cf Incontinence

pseudoinfarct A Q-wave inversion on EKG mimicking an acute MI, which may be seen in WPW syndrome, cardiac amyloidosis, and hypertrophic cardiomyopathy, due to elongation and partial stretching of nerve fibers. Cf Myocardial infact .

pseudoinfectious proctitis Noninfectious inflammation of the rectum due to anal-erotic activity–eg, trauma and erosion by inserted body parts, vibrators, bottles, eggs, and other objets d'art, allergic response to lubricants used in anal intercourse–cooking oil, suntan lotions, medicinal creams and reactions to toxins. See Gay bowel disease, eversion injuries, Sex toys, Sexual deviancy. Cf Proctitis

pseudoinsomnia Subjective difficulty in falling asleep, described in 10% of those who claim to suffer from insomnia, despite objective observations to the contrary. Cf Insomnia, Sleep disorders

pseudo-Kaposi sarcoma Kaposiform dermatitis, acroangiodermatitis A condition that clinically and histologically mimics Kaposi sarcoma–KS, seen in the AV hemangiomas or AV fistulas of Klippel-Trenaunay disease. Cf Kaposi sarcoma

pseudolymphoma A form of lymphoid hyperplasia, characterized by relatively monotonous infiltrates of lymphocytes, seen in breast, GI tract, lung, mediastinum, orbit, salivary gland, skin, soft tissue, thyroid, etc; pseudolymphomas occur in a younger population, have been associated with collagen diseases, SLE, Sjögren syndrome, phenytoin therapy, and may present in those with concomitant lymphoma or in those who later develop lymphoma. See Phenytoin lymphadenopathy. Cf Lymphoma

pseudomembrane A thin, adherent, gray-white exudative layer composed of necrotic epithelium and debris, fibrin, bacteria, PMNs, which overlies the mucosa of the–1 Colon, see Pseudomembranous colitis, and–2 Oropharynx, which extends from the tonsils to the contiguous soft and hard palates, and pharynx, the removal of which causes hemorrhage; pseudomembranes are classically seen in diphtheria–due to *Corynebacterium diphtheriae*, causing a bull-like neck; pseudomembranes of the oropharynx may also occur in shigellosis, staphylococcal infections, *Clostridium perfringens*, *C difficile*, less commonly in viral infections

pseudomembranous colitis Antibiotic-associated colitis, necrotizing colitis GI DISEASE An acute illness, with often severe diarrhea that follows antibiotic therapy with ampicillin, clindamycin, metronidazole, etc, which eliminate the Pt's native bacterial flora, resulting in superinfection by *Clostridium difficile*, causing most cases of PC; the condition may occur in compromised hosts or the elderly, in a background of colonic obstruction, leukemia, major

surgery, uremia, spinal injury, colon CA, burns, infections, shock, heavy metal poisoning, hemolytic-uremic syndrome, ischemia, Crohn's disease, shigellosis, necrotizing enterocolitis, Hirschsprung's disease CLINICAL Sx range from asymptomatic, to mild diarrhea and abdominal pain, to fulminant colitis with fever, ↑ WBC, vomiting, dehydration, perforation, peritonitis, shock

pseudometabolic acidosis A lab artifact seen when vacuum blood collection tubes–eg, Vacutainer™ are underfilled with blood, resulting in falsely ↓ bicarbonate value, falsely ↑ anion gap

PSEUDOMETABOLIC ACIDOSIS

Amount of blood in 10 ml tube	1 ML	3 ML	10 ML
Bicarbonate mmol/L, unvented tube	21.7	19.4	16.3
Bicarbonate mmol/L, vented tube	23.3	20.3	17.3
Anion gap mmol/L	16.7	17.5	19.2

Pseudomonas aeruginosa A normal soil inhabitant and human saprophyte that may contaminate various solutions in a hospital, causing opportunistic infection in weakened Pts CLINICAL Infective endocarditis in IVDAs, RTIs, UTIs, bacteremia, meningitis, 'malignant' external otitis TREATMENT Aminoglycosides–eg, gentamicin, amikacin, tobramycin, etc

pseudomyotonic reaction Myotonoid reaction, pseudomyotonia A myotonia-like reaction characterized by abnormally slow contraction and relaxation in response to mechanical or electrical stimulation of muscles and/or tendons; the PR reaction is typical of hypothyroidism and is attributed to a slowing of the reaccumulation of calcium ions in the endoplasmic reticulum and in the disengagement of thin actin and thick myosin filaments. Cf Myotonia

pseudomyxoma peritonei A rare condition characterized by poorly-circumscribed gelatinous masses filled with malignant mucin-secreting cells; 45% of PPs arise from the ovary, usually in a mucinous cystadenocarcinoma; 30% are due to mucin-producing CA of the appendix PROGNOSIS Mucinous ovarian cystadenocarcinomas of uncertain malignant potential–'borderline' tumors have a 10-yr survival of 95%; ovarian cystadenocarcinomas with pseudomyxoma have a 10-yr survival of 40%

pseudoobstruction Ogilvie syndrome, non-toxic megacolon, acute colonic pseudoobstruction GASTROENTEROLOGY Massive colonic dilation without mechanical obstruction, possibly due to a sympathetic nervous system defect, resulting in chronic peristaltic paralysis, affecting the cecum, right colon, distal small intestine, less commonly, the esophagus and stomach, resulting in nonabsorption of essential nutrients CLINICAL Initially painless abdominal distension with nausea, pain relieved by vomiting and diarrhea and intermittent Sx extending over yrs; the condition may be congenital–eg, hereditary hollow viscus myopathy, acquired–DM, hypothyroidism, collagen vascular diseases, myotonic dystrophy, parkinsonism, multiple sclerosis, amyloidosis, trauma, surgery, inflammation–pancreatitis, infections, radiation therapy, malignancy, cardiovascular–MI, neurologic, respiratory–pneumonia, metabolic–alcoholism, hypokalemia and other electrolyte imbalance, uremia, muscular dystrophy, familial dysautonomia–Riley-Day syndrome, porphyria, dysproteinemia, drug-related–phenothiazines, TCAs, ganglion blockers, clonidine, narcotics, anticholinergics MANAGEMENT Decompress intestine, correct electrolyte imbalance, cecostomy, or combined transplantation of portions of the small and large intestine and liver. See Paralytic ileus

pseudoparaproteinemia An ↑ of transferrin to ≥ 2-fold normal–2-4 g/L; US: 200-400 mg/dl, in response to severe iron-deficiency anemia; because transferrin exists as a single molecular species and migrates as a 'tight' band in the β region in serum electrophoresis, it mimics paraproteinemia

pseudopolyp An 'island' of preserved colonic mucosa, surrounded by an ulcerated 'sea' of hemorrhagic mucosa; pseudopolyps are typical of ulcerative colitis, but may be seen in IBD, bacterial dysentery, amebiasis due to *Entamoeba histolytica* and schistosomiasis. Cf Polyp

548

pseudoprecocity Isosexual pseudoprecocity, seen in ♀, which consists of signs of sexual maturation induced by functional ovarian tumors–eg, juvenile type of granulosa cell tumor, due to ↑ estrogens and/or androgens CLINICAL Development of breasts, pubic and axillary hair, stimulation and development of internal and external 2° sex organs, irregular uterine bleeding, whitish vaginal discharges, acceleration of somatic and skeletal growth, and occasionally clitoromegaly. Cf Precocity

pseudopseudofusion beat CARDIAC PACING A term related to dual chamber pacing that describes an atrial pacing artifact occurring in time with an intrinsic QT wave; the atrial output pulse has no effect on the intrinsic ventricular depolarization. Cf Pseudofusion beat

pseudopuberty The premature development of 2° sexual characteristics, while remaining inconceivable, a phenomenon due to ↑ androgen or estrogen secretion in the face of an immature hypothalamic-adenohypophysial-gonadal axis; pseudopuberty in ♂ infants with Leydig cell tumors is associated with growth of pubic hair and penile enlargement. Cf Puberty

pseudosac OBSTETRICS A small intrauterine sac, which may be seen by vaginal ultrasonography in ♀ with an ectopic pregnancy, which corresponds to decidualized reactive tissue. See Ectopic pregnancy

pseudosarcoma A tumor mimicking a sarcoma, the significance of which differs according to site of origin ORAL CAVITY Spindle cell carcinoma, a type of SCC. See Spindle cell cacinoma SOFT TISSUE 1. Pseudosarcomatous fasciitis Nodular fasciitis; 2. Pseudo-KS, see there UROGENITAL TRACT Inflammatory pseudotumor A small, sessile and/or friable 'tumor' with ↑ cellularity and ↑ mitotic activity, which bleeds easily and occurs at the site of recent surgery to the bladder, prostate or vagina, representing a florid inflammatory response to a local insult. See Inflammatory pseudotumor. Cf Sarcoma

pseudotolerance PHARMACOLOGY A diagnosis of exclusion defined as a need to ↑ dosage due to factors other than tolerance–eg, disease progression, new disease, ↑ physical activity, lack of compliance, change in medication, drug interaction, deviant behavior for psychosis. Cf Drug-seeking, Tolerance

pseudotumor Any circumscribed, nonneoplastic tumor-like mass–eg, gastric inflammatory fibroid polyps, a wad of helminths–eg, *Strongyloides* species–seen in Uganda, 'amyloidomas', endometriomas or other mass lesions. See Inflammatory pseudotumor. Cf Tumor OPHTHALMOLOGY Inflammatory pseudotumor of orbit An idiopathic proliferation of the lymphoid tissue surrounding the ocular orbit, which may be autoimmune in nature and related to orbital myositis CLINICAL Pain, exophthalmos, limitation of eye movement, lid erythema, edema, myositis, perineuritis, scleritis, dacryoadenitis; the lesion may be histologically impossible to differentiate from a true lymphoma, and may require molecular studies to determine clonality DIFFDx Dacryoadenitis, orbital myositis, vasculitis, sclerosing pseudotumors, lipogranuloma, epithelioid cell granuloma, xanthogranuloma. See Pseudolymphoma

pseudotumor cerebri Benign intracranial hypertension, NEUROLOGY A condition caused by ↑ intracranial pressure with normal CSF; PC is most common in young obese ♀ with dysmenorrhea of ovarian origin, and is diagnosed by 1. Presence of bilateral papilledema and objective evidence of ↑ intracranial pressure 2. Absence of focal neurological Sx or signs 3. Absence of an extracranial cause of papilledema 4. Normal CSF CLINICAL Vision defects–loss of acuity, diplopia, blind spots, headaches, N&V, vertigo, tinnitus ETIOLOGY Anemia, leukemia, hyper- or hypovitaminosis A, lead intoxication, nalidixic acid, poliomyelitis, Guillain-Barre disease, menarche, pregnancy, galactokinase deficiency, hypoxia, allergies, cerebral trauma, steroid therapy for cerebral edema, withdrawal of steroids, chronic hypocalcemia with hypoparathyroidism with 1° adrenal insufficiency, thyroid replacement therapy, endocrinopathies–Addison's or Cushing's diseases, OCs, tetracycline–in infants, intracranial venous occlusion, inflammation

pseudo-Turner syndrome Noonan syndrome An AD condition with a heterogeneous presentation that mimics some of the clinical findings of Turner syndrome; both are characterized by short stature, webbing of the neck, developmental delays, pectus carinatum or pectus excavatum, and cubitus valgus; PTS affects both sexes, and is further characterized by mild mental retardation, congenital heart disease–pulmonary valve stenosis, ASD, etc, a characteristic facies–hypertelorism, epicanthus, antimongoloid palpebral slant, ptosis, low-set ears, micrognathia and gonadal defects in ♂ including cryptorchidism, ↓ Leydig cell function, ↓ spermatogenesis. Cf Turner syndrome

pseudovitamin A substance that does not meet the accepted definition of a required human vitamin

pseudo-von Willebrand's disease Platelet-type von Willebrand disease–vWD An AD condition similar to type IIB vWD, with moderately severe Sx LAB ↑ Bleeding time, ↓ von Willebrand factor–vWF and factor VIII levels, ↑ ristocetin-induced platelet aggregation, absence of large vWF multimers and presence of the same multimers in platelets PATHOGENESIS Unknown; may involve platelet glycoprotein IB MANAGEMENT Cryoprecipitate. Cf Von Willebrand's disease

pseudo-Whipple's disease A Whipple's disease-like condition, which may occur in *Mycobacterium avium* complex in AIDS, also found in a CNS isolate from a young homosexual ♂ with AIDS-related complex, linked to *Corynebacterium equi*, agent of suppurative equine pneumonia. Cf Whipple's disease

pseudoxanthoma elasticum A progressive disorder of connective tissues of the skin, cardiovascular system, joints, eyes CLINICAL Early changes–lax, yellow, redundant 'plucked chicken skin' that coalesces into plaques, becoming thickened, grooved, leathery and inelastic, likened to 'Moroccan leather,' involving head, neck, trunk and upper legs, eyes, cardiovascular system–murmurs, HTN, CHF, intermittent claudication, angina, ↓ peripheral pulses, vascular occlusion, cerebral, visceral, GI hemorrhage

Psicofar Chlordiazepoxide, see there

Psilocybe TOXICOLOGY A genus of mushrooms that contains psilocybin, a hallucinogen CLINICAL 20–60 mushrooms may be ingested to obtain the desired hallucinogenic effect; small amounts of *Psilocybe* spp evoke euphoria, accompanied by dizziness, and weakness; larger amounts alter temporal and spatial perception, causing visual disturbances MANAGEMENT Supportive; effects are transient. See Poisonous mushrooms

psittacosis Bird fancier's lung, chlamydial pneumonia; ornithosis, parrot fever INFECTIOUS DISEASE An infection of birds by *Chlamydia psittaci* which may cause asymptomatic infection, an influenza-like disease or serious pneumonia in humans exposed to feathers, tissues or droppings from psittacine birds–parrots, parakeets, cockatoos, which may be sick or carriers of *C psittaci* CLINICAL Most Pts are asymptomatic; symptomatic cases have a 1-2 wk incubation, followed by chills, moderate to high fever, slow pulse, severe headache, myalgias, anorexia, N&V, arthralgia and mental clouding; pneumonic Sx are uncommon, with production of minimal mucoid sputum mixed with hemorrhage, and if severe, accompanied by hypoxia and cyanosis TREATMENT Tetracycline

PSL syndrome A symptom complex occurring 15 mins to one hr after ingesting certain poisonous mushrooms–eg, *Inocybe* spp, *Clitocybe* spp, etc CLINICAL Perspiration, salivation, lacrimation–PSL–caused by muscarine, a cholinergic toxin, vomiting, diarrhea, ↑ urination, pupillary constriction, bradycardia ANTIDOTE Atropine

PSO Progressive supranuclear ophthalmoplegia

psoas sign RADIOLOGY The loss of the sharp delineation of the psoas muscle border normally seen on a plain erect abdominal film SIGNIFICANCE Intra-abdominal or retroperitoneal pathology–eg, hemorrhage in trauma victims or florid acute inflammation in ruptured appendicitis

psoralen THERAPEUTICS A class of furocoumarins used to treat psoriasis and other skin conditions

psoriasis Plaque psoriasis DERMATOLOGY A common–± 3 million, US–chronic hyperproliferative and inflammatory skin disorder, characterized by erythematous papules that coalesce, forming plaques with sharply demarcated borders; removal of a 'virgin' yellow-white lesion results in pinpoint hemorrhage–Auspitz' sign; trauma may evoke lesions on new body sites–Koebner's phenomenon; lesions are prominent on scalp, knees, elbows, umbilicus, genitalia EXACERBATION Injury–solar, mechanical, infection–β-hemolytic streptococcus, HIV, drugs–ACE inhibitors, lithium, antimalarials, indomethacin RISK FACTORS Injury or irritation–cuts, burns, rash, insect bites, immunosuppression–eg, AIDS, chemotherapy for cancer, Pts with autoimmune disorders, certain medications, viral or bacterial infections, alcoholism, obesity, lack of sunlight, sunburn, stress, cold climate, friction on skin MANAGEMENT Symptomatic–emollients, keratolytics, topicals–anthralin, corticosteroids, vitamin D analogues, phototherapy–ie, UV light exposure–natural sunlight, artificial UVB light, photochemotherapy–methoxsalen + UVA light, PUVA therapy, oral agents–eg, cyclosporine, etretinate, MTX, calcipotriene

psoriatic arthritis RHEUMATOLOGY Joint inflammation associated with psoriasis, which is generally mild and involves few joints; in some Pts, the arthropathy is severe and affects the fingers and the vertebral column, where it mimicks ankylosing spondylitis. See Psoriasis

PSS Progressive systemic sclerosis, see there

PST Paroxysmal supraventricular tachycardia, see there

psyche Mind, see there

psychiatric *adjective* Pertaining to psychiatry, mental disorders

psychiatric evaluation The assessment of a person's mental, social, psychologic functionality. See DSM-IV-table multiaxial assessment, Personality testing, Psychiatric history, Psychiatric interview

psychiatric history A person's mental profile, which includes information about chief complaint, present illness, psychological adjustments made before onset of disease, individual and family Hx of psychiatric or mental disorders, and an early developmental Hx

psychiatric interview PSYCHIATRY The central vehicle for assessing a psychiatric Pt, during which there is a free exchange of information that forms the basis for therapy

psychiatrist A physician–MD or DO specialized in mental, emotional, or behavioral disorders, licensed to prescribe medication and provide verbal-based psychotherapy MEAT & POTATOES DISEASES Anxiety, depression, eating disorders SALARY $135K. See Addiction psychiatrist, Child & adolescent psychiatrist, Forensic psychiatrist, Psychoanalyst. Cf Psychologist

psychiatry The medical specialty concerned with physical and chemical interactions in the brain and how they affect mental and emotional processes; the study, treatment, and prevention of mental illness. See Consultation-liaison psychiatry, Forensic psychiatry, Geriatric psychiatry, Neuropsychiatry, Orthomolecular psychiatry, Orthopsychiatry

psychic *adjective* Referring to psychic phenomena; referring to the mind, its ideas and images

psychic dentistry Miracle in mouth QUACKERY The alleged healing of dental disease–eg, periodontal disease, caries, by noninterventional maneuvers–eg, faith healing, laying on of hands, psychic healing, etc. Cf Alternative dentistry

psychic energizer An antidepressant that ↑ mood, ↑ motivation, ↑ quality of life. See Psychoactive drugs. Cf Nervine

psychoactive drug SUBSTANCE ABUSE An agent that provides pleasure or ameliorates pain, and may cause physical dependence and tolerance, with a tendency to ↑ dose in order to achieve the same effect; use of non-prescribed psychoactive agents may be 'social' or casual, or evolve into frank addiction; in descending order of addictive potential cocaine/'crack', amphetamines, opiates, nicotine, alcohol, benzodiazepine, barbiturates, cannabis, hallucinogens, caffeine THERAPEUTICS An agent that improves the ability to function appropriately in clinical settings–eg, psych ward, nursing home TYPES Antidepressants–eg, desipramine, nortriptyline, antipsychotics–eg, haloperidol, thioridazine, benzodiazepines–eg, lorazepam, diazepam, and other hypnotics–eg, diphenhydramine

psychoanalysis PSYCHIATRY A branch of psychology concerned with issues of emotional conflict and repression attributed to formative childhood experiences; through analysis of free associations and interpretation of dreams, emotions and behavior are traced to the influence of repressed instinctual drives and defenses against them in the unconscious; psychoanalysis seeks to eliminate or diminish the undesirable effects of unconscious conflicts by making the analysand aware of their existence, origin, and inappropriate expression in current emotions and behavior. See Freudian analysis, Jungian psychoanalysis. Cf Psychotherapy

psychoanalyst PSYCHIATRY A person who diagnoses and treats emotional disorders by exploring a Pt's mental and emotional history and makeup; treatment is usually long-term. See Psychiatrist, Psychologist

psychobiology PSYCHIATRY A school of thought that views a person's biologic, psychologic, and social experiences as an integrated unit

psychodiagnostic test Inkblot test, Rorschach projective technique PSYCHOMETRIC TESTING A psychologic test based on the characteristic differences in perception and interpretation of pictorial form by normal persons and those suffering from hysteria, schizophrenia, etc, mental disorders

psychogenic dyspnea Anxiety-related dyspnea CLINICAL MEDICINE Dyspnea that occurs in a background of emotionally stress, characterized by irregular breathing and prominent deep sighs; severe PD may be marked by hyperventilation, light-headedness, tingling of hands and feet, tachycardia, T wave inversion, syncope. See Anxiety

psychogenic seizure Hysterical fit, nonepileptic seizure A seizure regarded as a conversion Sx, seen when a person cannot directly express mental stress CLINICAL Frequent seizures despite adequate antiepileptic medication, prolonged duration–> 5 mins, wild movements, pelvic thrusting, fluctuating intensity, resolution of Sx with distraction, nonphysiologic spread of Sx, crying, bilateral motor activity with preserved consciousness, lack of post-ictal confusion or lethargy PROGNOSIS Good, if ♀, of higher IQ, independent lifestyle, no prior psychotherapy, normal EEG; poor, if accompanied by seizures, Hx of psychiatric disorders, unemployed

psychogenic syndrome An anxiety-related conversion reaction caused by endogenous or exogenous stress EXAMPLES Hysterical reactions, psychogenic chest pain, psychogenic polydipsia, psychogenic purpura, psychophonasthenia and 'women who fall' syndrome, a conversion reaction to aggressive or erotic impulses. See Psychosomatic disorder; Cf Factitious disorders

psycholinguistics PSYCHOLOGY The study of factors affecting activities of communicating and understanding verbal information; the study of the manner in which language is acquired, stored, integrated and retrieved. See Kinesics, Language

psychological abuse Emotional abuse, mental abuse A form of mistreatment in which there is intent to cause mental or emotional pain or injury; PA includes verbal aggression, statements intended to humiliate or infantilize, insults, threats of abandonment or institutionalization; PA results in stress, social withdrawal, long-term or recalcitrant depression, anxiety

psychological adjustment

psychological adjustment PSYCHOLOGY The mental response of a person to a dread life situation–eg, AIDS, CA, imminent death STAGES Denial–this is not happening; anger–why me?; bargaining–if God lets me live, I'll join the Peace Corps; depression; acceptance

psychological autopsy PSYCHIATRY An autopsy that analyzes the cause(s) of death, examining the body *and* the circumstances–natural or unnatural that led to death; in the 'usual' death, a person suffers from a known set of morbid condition(s) and dies as a natural consequence of the terminal progression of those conditions; in 'unnatural' death–eg, homicide or suicide, determination of nosology is more difficult and requires analysis of circumstances preceding death; a PA focuses on the decedent's intentions relating to his own death, especially suicide

psychological distress The end result of factors–eg, psychogenic pain, internal conflicts, and external stress that prevent a person from self-actualization and connecting with 'significant others'. See Humanistic psychology

psychological well-being RESEARCH A nebulous legislative term intended to ensure that certain categories of lab animals, especially primates, don't 'go nuts' as a result of experimental design or conditions

psychologist PSYCHIATRY A nonmedical mental health care worker who has completed graduate education and training and is qualified to perform psychologic research, testing, or therapy SALARY $61K. See Child psychologist, Clinical psychologist, Neuropsychologist, Psychoanalyst, Psychotherapist, School psychologist; Cf Psychiatrist

psychology PSYCHIATRY The discipline concerned with behavioral, mental and emotional processes, especially vis-á-vis human behavior. See Analytical psychology, Archetypal psychology, Clinical psychology, Depth psychology, Developmental psychology, Ego psychology, Evolutionary psychology, Gestalt psychology, Humanistic psychology, Individual psychology, Parapsychology, Process psychology, Psychological, Psychotherapy, Reverse psychology, Spiritual psychology, Transpersonal psychology

psychometric test Any test used to quantify a particular aspect of a person's mental abilities or mindset–eg, aptitude, intelligence, mental abilities and personality. See IQ test, Personality testing, Psychological testing

psychomotor agitation PSYCHIATRY Physical and emotional overactivity in response to internal and/or external stimuli, as in hypomania

psychomotor retardation PSYCHIATRY A generalized slowing of physical reactions–eg, eye-blinking, common in depression

psychomotor seizure Complex partial seizure, psychomotor epilepsy, temporal lobe epilepsy NEUROLOGY A seizure disorder involving abnormal discharge of neurons of the temporal lobe with episodic changes in behavior accompanied by loss of consciousness, with retention of capacity to respond to environmental stimuli. See Paraphilia

psychomotor test NEUROLOGY Any test of the senses or perception that measures the speed and accuracy with which a person can carry out a particular task–eg, building with blocks or copying a design or picture. Cf Psychological test, Psychometric test

psychoneuroimmunology PNI The study of the effects of the mental and neurological status on the immune system.

psycho-organic 'syndrome' OCCUPATIONAL MEDICINE A petroleum solvent-induced neurologic dysfunction CLINICAL Fatigue, loss of memory and concentration, emotional lability, seen after 5-10 yrs of regular exposure to solvents–eg, styrene, toluene, xylene, which affects painters, and degreasers, plastic, and chemical workers

psychopathy Antisocial personality disorder, see there

Psychopax Diazepam, see there

psychosexual development PSYCHIATRY A series of stages from infancy to adulthood, relatively fixed in time, determined by the interaction between a person's biologic drives and the environment; with resolution of this interaction, a balanced, reality-oriented development takes place. See Tanner stage

psychosis PSYCHIATRY 1. Psychotic disorder, see there 2. A popular term for a wide range of major mental disorders of organic or emotional origin in which a person's ability to think, respond emotionally, remember, communicate, interpret reality, and behave appropriately is suffficiently impaired so as to interfere with the capacity to meet the ordinary demands of life CLINICAL Regressive behavior, social withdrawal, inappropriate mood, ↓ impulse control, abnormal mental content–eg, delusions, hallucinations, distortions of reality, loss of contact with environment and disintegration of personality. See Drug-induced psychosis, Lupus psychosis, Myxedematous psychosis, Postoperative psychosis, Psychotic disorder, Symbiotic psychosis

psychosocial development PSYCHIATRY Progressive interaction between a person and her environment through stages beginning in infancy, ending in adulthood, which loosely parallels psychosexual development. See Cognitive development

psychosocial dwarfism Abuse dwarfism, Kaspar Hauser syndrome, deprivation dwarfism SOCIAL MEDICINE Irreversible hyposomatism, largely attributed to abuse, neglect, and sensory deprivation. See Anaclytic depression, Genie, Social isolation, The Wild Child. Cf Companionship, Reversible hyposomatotropic dwarfism

psychosocial intervention PSYCHOLOGY A nonpharmacologic maneuver intended to alter a Pt's environment or reaction to lessen the impact of a mental disorder. See Attention-deficit-hyperactivity syndrome

psychosocial support A nontherapeutic intervention that helps a person cope with stressors at home or at work. See Companionship, Most significant other

psychosomatic disorder PSYCHIATRY A condition in which 1. A person has a biologic predisposition to a particular condition–genetic, trauma-related, etc; 2. The person has a vulnerable personality–ie, there is a type or degree of stress that the coping mechanism and ego structure cannot manage; 3. The person must experience a significant psychosocial stress in a susceptible personality area. See Factitious disorders, Psychogenic 'syndromes'

psychostimulant *adjective* Referring to agents that upregulate higher cortical activity *noun* An agent which upregulates higher cortical activity

psychosurgery NEUROSURGERY Neurosurgery intended to alleviate psychiatric Sx, by selectively destroying neural tracts; psychosurgery includes radioactive ^{90}Yt implants in the substantia innominata, cryoprobes, coagulation, proton beams and ultrasonic waves; psychosurgery is rarely performed, as it must be established that a Pt is unresponsive to all other therapies and that the condition is chronic–ie, > 3 yrs duration; significant improvement is reported in 60% of carefully selected Pts; in 3%, the Sx worsen after the procedure; the measurable IQ may ↑, given the ↑ ability to concentrate and memorize, while distraction is minimized COMPLICATIONS 1%; include infections, hemorrhage, seizures. Cf Psychic surgery

psychotherapy PSYCHIATRY The treating of mental–ie, emotional, behavioral, personality, and psychiatric disorders through verbal and non-verbal communication–eg, psychoanalysis with the Pt, rather by pharmacologic, surgical, or other physical intervention; the classic format of psychotherapy is based on the Freudian school of psychoanalysis, in which the focus is to bring repressed memories to the conscious mind; such therapies typically involve open discussion of emotional issues; psychotherapy differs from psychoanalysis in that it is more informal and interactive, less intense, and less concerned with repressed mental trauma; psychotherapy can be one-on-one with a therapist or in a group where other Pts participate TYPES Behavioral therapy, biofeedback, cognitive therapy, cognitive behavioral therapy, exposure and response prevention, eye movement desensitization and reprocessing, neurolinguistic programming, psychoanalysis, traumatic incident reduction, virtual reality exposure. See Biodynamic psychotherapy, Body-oriented psychotherapy, Hakomi body-oriented psychotherapy, Hypnotic psychotherapy, Interpersonal psychotherapy, Natural psychotherapy, Organismic psychotherapy, Psychiatry, Psychologic therapies, Psychoanalysis, Supportive psychotherapy

psychotic PSYCHIATRY*adjective* 1. Referring to a person suffering from a psychotic personality disorder 2. Referring to a behavioral 'impairment that grossly interferes with the capacity to meet the ordinary demands of life', which may be accompanied by the loss of ego boundaries and/or gross impairment of reality testing DSM-IV. See Psychotic disorder *noun* A person with psychosis, see there

psychotic disorder Psychosis PSYCHIATRY A broad class of mental disorders, classified in the DSM-IV under the umbrella of 'Schizophrenia and other psychotic disorders' EXAMPLES OF PD Schizophrenia, schizophreniform disorder, schizoaffective disorder, delusional disorder, brief psychotic disorder, shared psychotic disorder–eg, folie à deux, psychotic disorder due to a general medical condition, substance-induced psychotic disorder, psychotic disorder-not otherwise specified. See Psychotic

psychotic symptom PSYCHIATRY A Sx representing an acute mental decompensation–eg, delusions, hallucinations, disorganized speech or behavior, or catatonic behavior. See Pain

psychotomimetic *adjective* Mimicking psychosis or a psychotic disorder *noun* A drug–eg, LSD or mescaline, which may produce psychotic states

psychotropic drug Psychoactive drug PHARMACOLOGY A drug that affects brain activities associated with mental processes and behavior CATEGORIES Anti-psychotics; antidepressants; antianxiety drugs or anxiolytics; hypnotics. See Antidepressant, antipsychotic, Anxiolytic, Hypnotic

psyllium Plantago psyllium GI DISEASE A soluble dietary fiber that acts as a bulk laxative and cholesterol-lowering agent–ingestion protects against cholesterol gallstones. See Laxative, Soluble fiber

PT 1. Parathyroid 2. Patient 3. Pertussis toxin 4. Phototherapy 5. Physical therapy 6. Pneumothorax 7. Posterior tibialis pulse, see there 8. Proficiency testing, see there 9. Propylthiouracil 10. Prothrombin time, see there

PTCA Percutaneous transluminal coronary angioplasty, see there

PTE Pulmonary thromboembolism, see there

PTER Percutaneous transluminal endomyocardial revascularization, see there

pterygium OPHTHALMOLOGY A condition of older adults, characterized by a fleshy triangular fold of tissue that grows from the conjunctiva, encroaching on the cornea; it is clinically insignificant unless it affects the vision; it is usually on the nasal side, and may be bilateral RISK FACTORS Exposure to sun and UV light, dust, sand, wind

PTMR Percutaneous transluminal myocardial revascularization, see there

ptoptic *adjective* Fallen; drooping; descended from a normal position

ptosis Drooping eyelid NEUROLOGY A condition caused by weakened palpebral muscle, damage to nerves, or laxity of skin surrounding eye RISK FACTORS Aging, DM, stroke, Horner's syndrome, myasthenia gravis, brain tumor, or cancer, which can affect nerve or muscle response

PTP laser Potassium-titanyl-phosphate laser SURGERY A device attached to a flexible endoscope used to vaporize, coagulate, or cut. See Lasers

PTP Post-transfusion purpura

PTRA Percutaneous transluminal renal angioplasty, see there

PTSD Post-trauma stress disorder, see there

pubertal delay SEXOLOGY A delay in puberty beyond the normal upper age limit for its completion–ie, age 13 in ♀ and 15 in ♂; PD may reflect a timing error, or it may be associated with a permanent hormone defect in gonads, pituitary gland, or hypothalamus. See Premature puberty

puberty ADOLESCENT MEDICINE The period of hormone-induced transition to adolescence with development of 2° sex characteristics, ending in procreative maturity. See Early puberty, Precocious puberty, Premature puberty, Pseudopuberty, Pubertal delay

pubic lice Crabs DERMATOLOGY An ectoparasite that may reside in pubic hairs. See *Pthirus pubis*

public distance PSYCHOLOGY A space at the far end of the territory in which social humans operate. See Proxemics. Cf Intimate distance, Personal distance, Social distance

public health advisory A statement containing a finding that a release of hazardous substances poses a significant risk to human health recommending measures to be taken to ↓ exposure and eliminate or substantially mitigate the risk to human health

public health assessment The evaluation of data and information on the release of hazardous substances into the environment to assess current or future impact on public health, develop health advisories or other recommendations, and identify studies or actions needed to evaluate and mitigate or prevent human health effects; the document resulting from that evaluation

public health hazard A chemical or other substance known to be hazardous, based on the effects of long-term exposures thereto

public hospital A health care institution owned by a federal, state, or local government; PHs have had a significant role in caring for the sick in the US, but are regarded by some as anachronisms in an increasingly complex and costly health care environment with ↑ operating costs, economic recession, and governmental budgetary restraints

PUBS OBSTETRICS Percutaneous umbilical blood sampling, see there

PUD Peptic ulcer disease, see there

pudenda ANATOMY 1. The external female genitalia 2. Vulva, see there

pudendal block Pudendal anesthesia OBSTETRICS Locoregional anesthesia used in the 2nd stage of labor to relieve episiotomy-related pain, by transvaginally injecting local anesthetics into the pudendal nerve. See Block. Cf Epidural block

pufferfish *Fugu rubripes*, fugu TOXICOLOGY A raw fish delicacy; some tissues–intestine, liver, ovaries, skin, have a high concentration of tetrodotoxin, a sodium channel blocker and very potent toxin; it blocks the neuromuscular junction, causing numbness, motor weakness, ataxia, respiratory paralysis. See Tetrodotoxin

PUGH syndrome OPHTHALMOLOGY A condition defined by an acronym, PUGH: PseudoUveitis, Glaucoma, Hyphema with neovascularization of iris and occlusion of the central retinal vein

pugilistic stance Pugilistic attitude FORENSIC MEDICINE A 'defensive' position fancifully likened to that adopted by pugilists-boxers found in severely burned bodies, characterized by flexion of elbows, knees, hip, and neck, and clenching of hand into a fist; it is caused by high-temperatures in fire, resulting in muscle stiffening and shortening; it occurs even if the person was dead before the fire

pull the plug MEDICAL ETHICS *verb* A popular term for withdrawing nutritional support and ventilation from a Pt in a persistent vegetative state. See De-escalate, End-of-life debate, Persistent vegetative state

Pulmicort®

pulmonary actinomycosis INFECTIOUS DISEASE An infection by *Actinomyces israelii* or related actinomycetes, involving chest, mouth, and pelvis, often linked to poor dental hygiene and oral abscesses; in the thoracic cavity, it causes cavities in the lung and pleural effusion, and may spread through the chest wall via sinus tract formation

pulmonary alveolar proteinosis Alveolar proteinosis A rare disease most common in ages 30-50, ♂:♀ ratio 2.5-4:1; although idiopathic, ≥ 50% of Pts have been exposed to dusts, chemicals–eg, busulfan, infections–eg, nocardiosis, CMV, *Pneumocystis carinii*, toxins–eg, aluminum, antimony; PAP may be idiopathic, associated with immune compromise or thymic aplasia CLINICAL SOB, cough, fever, chest pain, weight loss, fatigue, finger clubbing CXR Symmetric bilateral 'bat wing'-like alveolar infiltrates, less commonly, unilateral patchy infiltrates MANAGEMENT Bronchoalveolar lavage–BAL with saline or heparin and acetylcysteine for removal of phospholipids, is required in ≥ ½ of Pts; without BAL, progressive dyspnea and deterioration of lung functions, ↑ mortality, and risk of superinfections, especially with *Nocardia* spp, which may be due to the enhanced growth of certain organisms 2° to ↑ phospholipids PROGNOSIS Spontaneous remission in some, respiratory failure others

pulmonary angiography Pulmonary arteriography IMAGING A technique in which radiocontrast is injected into the pulmonary arteries and its branches by percutaneous catheter placed in the internal jugular or the common femoral veins INDICATIONS Identification of PTE aneurysms, AV malformations, pulmonary artery stenosis

pulmonary anthrax Inhalation anthrax, see there

pulmonary arteriovenous fistula Pulmonary arteriovenous malformation PULMONOLOGY An abnormal arterioovenous communication in the lungs, resulting in right-to-left shunting of non oxygenated blood, a finding in ⅓ of hereditary hemorrhagic telangiectasia, aka Osler-Weber-Rendu disease CLINICAL May be asymptomatic; Sx may include SOB, hemoptysis, cyanosis, finger clubbing, and a heart murmur heard by stethoscope over the malformation MANAGEMENT Unnecessary if asymptomatic, or, surgical resection or embolization

pulmonary aspergilloma Aspergilloma; mycetoma PULMONOLOGY A cavitating lung infection by *Aspergillus fumigatus*, which grows in clumps into pre-existing pulmonary cavities–prior histoplasmosis, TB, sarcoidosis, abscess, cystic fibrosis, or lung CA IMAGING Typical cavity CLINICAL Cough, bloody sputum. See Aspergillosis

pulmonary aspergillosis–invasive type INFECTIOUS DISEASE An acute infection by *Aspergillus fumigatus*, which affects immunosuppressed or immunocompromised hosts; invasive infection can affect the eye as well as the heart, lungs, brain, kidneys; with severe disease; CNS, skin and other organs may become infected

pulmonary atresia PEDIATRIC CARDIOLOGY Stenosis or obstruction of the pulmonary heart valve, a congenital heart disease

pulmonary barotrauma SPORTS MEDICINE Life-threatening barotrauma seen in divers CLINICAL Mediastinal and subcutaneous emphysema produces changes in voice, dyspnea, dysphagia, supraclavicular crepitus, typical findings on chest and neck films; pneumothorax may cause abrupt pain, dyspnea SEQUELAE Arterial air embolism caused by passage of gas into pulmonary veins and thence to the systemic circulation; gas bubbles lodged in small arteries may occlude segments of the cerebral, coronary, and other systemic vascular beds, resulting in a stroke-like syndrome ranging from mild neurologic defects to collapse, unconsciousness, and death once the diver surfaces. See Barotrauma

pulmonary blastoma A biphasic CA of older adults CLINICAL Primitive mesenchymal stroma with a myxoid appearance, ± mature mesenchymal elements–eg, cartilage and malignant epithelial component reminiscent of fetal lung tissue or secretory endometrium PROGNOSIS Poor; most Pts die. Cf Pleuropulmonary blastoma

pulmonary burn Pulmonary parenchymal destruction caused by inhalation of irritating gases–eg, synthetic nitroso- compounds–eg, burning mattresses, plastics, generating PVCs and HCl, which generate toluene di-isothiocyanate; toxins and intense heat affect tracheobronchial tree, causing pulmonary edema, congestion, parenchymal hemorrhage, epithelial desquamation, lung necrosis

pulmonary capillaritis Necrotizing alveolar capillaritis, see there

pulmonary-capillary wedge pressure A key indicator of cardiac function and of intravascular and pulmonary venous volume; PCWP has been traditionally monitored in acutely ill Pts using a balloon-tipped flow-directed catheter wedged in an arteriole of pulmonary circulation, to determine pressure in the pulmonary capillaries and left atrium CONS Invasive; may cause serious complications; PCWP can also be determined by a noninvasive procedure, the pulse-amplitude ratio

pulmonary chemoreflex LUNG PHYSIOLOGY A triad of reactions to stimulation of the J–juxtacapillary receptors in the lungs, resulting in apnea, bradycardia, and hypotension. See J receptor

pulmonary compliance Pulmonary distensibility PULMONOLOGY The change in thoracic volume/unit pressure; the pressure required to ↑ lung volume; PC is ↓ in emphysema, congestion of pulmonary vessels, and interstitial pulmonary fibrosis. See Compliance, Pulmonary function tests. Cf Nasal compliance

pulmonary dirofilariasis A benign self-limited disease caused by lodging of microfilariae of *Dirofilaria immitis* in the pulmonary artery

pulmonary edema Lung water, water in the lung INTERNAL MEDICINE The exudation of protein-rich fluid due to heart failure–eg, left ventricular failure, aortic or mitral valve stenosis, post-MI, or high altitude MANAGEMENT Furosemide. See Congestive hear failure

pulmonary embolism Pulmonary embolus, pulmonary thromboembolism INTERNAL MEDICINE The migration of an embolus of material, often a blood clot, from elsewhere in circulation, lodging in the pulmonary arteries, occluding same and causing pulmonary infarction, right heart failure, and ↓ oxygenation; most–± 95% originate in the deep–popliteal, femoral, iliac leg veins; PE causes ± 4% of all US hospital deaths/yr–50,000, and is largely preventable CLINICAL Dyspnea, chest pain, tachycardia, tachypnea, low-grade fever, ± hemoptysis, syncope, ± signs of DVT–eg, right ventricular heave, right-sided S3, rales; PE is underdiagnosed as the classic signs of dyspnea–seen in 59%, chest pain–17%, and hemoptysis–3% are absent, or the Pt may be unable to communicate as he/she is comatose or sedated IMAGING–CXR Usually normal, rarely Hampton sign–wedge shaped infarct or Westermark sign–low blood in embolic lung zone; VQ SCAN Lung mismatch PULMONARY ANGIOGRAM Accurate but a high-risk procedure in these Pts ULTRASOUND DVT may be seen EKG Sinus tachycardia, nonspecific ST-T wave changes–pathognomonic SI, QIII, TIII ABG Respiratory alkalosis–↑ pH, ↓ CO_2 < 80 mm Hg RISK FACTORS Blood stasis, immobilization, CHF, surgery, trauma, venous endothelial damage, hypercoagulability–eg, pregnancy, post-partum, oral contraceptives, protein C deficiency, protein S deficiency, factor V–Leiden mutation, malignancy, severe burns PREVENTION Early post-surgical and postpartum ambulation, or exercise if Pt is bed-ridden, anticoagulation, inferior vena caval filters MANAGEMENT Heparinization; long-term anticoagulation with warfarin

pulmonary eosinophilic granuloma An idiopathic nonmalignant histiocytosis characterized by pulmonary hemorrhage, immune crescentic glomerulonephritis, and antineutrophil cytoplasmic antibodies ETIOLOGY Uncertain

pulmonary function test Any technique and maneuver–eg, spirometry, ventilation and perfusion scans, measurement of lung volumes, airway resistance, strength of respiratory efforts, efficiency of gas exchange, carbon monoxide diffusing capacity and arterial blood gases–that provide objective and quantifiable data of pulmonary function, and may be combined with endoscopy-bronchoscopy, mediastinoscopy and various forms of imaging–chest films, CT, MRI to evaluate morphologic abnormalities–eg, an abscess or tumor. See Lung volumes, Pulmonary panel, Spirometry

pulmonary hemorrhagic syndrome(s) Any of the nonneoplastic and noninfectious pulmonary pathologies that present with hemoptysis–eg, Goodpasture syndrome, idiopathic pulmonary hemosiderosis, hemorrhagic vasculitides, hypersensitivity angiitis and Wegener's granulomatosis. See Goodpasture syndrome, Wegener's granulomatosis

pulmonary hypertension Idiopathic pulmonary hypertension, primary pulmonary hypertension INTERNAL MEDICINE An idiopathic condition more common in ♀ age 20 to 40 characterized by ↑ blood pressure in pulmonary arteries in absence of other heart or lung disease; the major effect of PH is ↑ right ventricular load which, when prolonged, predisposes Pts to right ventricular failure, syncope, precordial pain, sudden death TYPES Idiopathic, 2° to Eisenmenger's complex, respiratory failure in cystic fibrosis and COPD, with inhibition of endothelium-dependent pulmonary arterial relaxation due to ↓ synthesis of nitric oxide or endothelium-derived growth factor PROGNOSIS No longer thought to be universally fatal TREATMENT High-dose CCBs may induce ↓ pulmonary artery pressure and pulmonary vascular resistance, which may be combined with warfarin

PULMONARY HYPERTENSION

PASSIVE PULMONARY HYPERTENSION, characterized by systemic congestion due to mitral stenosis, left ventricular failure, left atrial myxoma, anomalous drainage of the pulmonary circulation

HYPERKINETIC PULMONARY HYPERTENSION, with ↑ blood flow through lungs due to congenital heart defects

VASO-OCCLUSIVE PULMONARY HYPERTENSION, due to recurring vessel obstruction, seen in IV drug abuse and PTE, associated with hypoxia, alveolar hypoventilation (mitral stenosis, coarctation of aorta, Eisenmenger's complex, ventricular septal defect)

SECONDARY PULMONARY HYPERTENSION–comprises 10-20% of cases, treated by addressing the underlying disease, eg unilateral renal artery stenosis, coarctation of aorta, primary aldosteronism, pheochromocytoma

pulmonary nocardiosis Nocardiosis PULMONOLOGY A lung infection caused by a gram-positive fungus-like bacterium, *Nocardia asteroides*, acquired through inhalation, which cause pneumonia-like illness; it responds poorly to antibiotics, spreading to the brain and subcutaneous tissue RISK FACTORS Immunosuppression, long-term corticosteroid therapy

pulmonary panel A battery of cost-effective tests used to evaluate the functional reserve capacity of the lungs in Pts with a clinical diagnosis of obstructive or restrictive lung disease; the PPmeasures CO_2 content, $PaCO_2$, PaO_2, pH, O_2 saturation, a/A ratio. Cf Organ panels

pulmonary renal syndrome NEPHROLOGY An idiopathic condition characterized by pulmonary hemorrhage, immune crescent glomerulonephritis, and antineutrophil cytoplasmic antibodies; pulmonary-renal 'syndrome' may be defined as a heterogeneous group of multisystem diseases–eg, Goodpasture syndrome, Wegener's granulomatosus, collagen vascular disease–in particular SLE, polyarteritis nodosa, Henoch-Schönlein purpura, and various other conditions, which have prominent pulmonary and renal components and microangiopathic vasculitis CLINICAL Asymptomatic pulmonary infiltrates or pulmonary hemorrhage with episodic cough, hemoptysis, dyspnea and widespread alveolar infiltrates on CXR; renal involvement is characterized by microscopic hematuria, RBC casts, ↑ creatinine

pulmonary sequestration An uncommon–1:1000 adult lobectomy specimens congenital anomaly characterized by misplaced lung parenchyma, which lacks normal communication with the main tracheobronchial tree that may be intralobar or extralobar

PULMONARY SEQUESTRATION

	INTRALOBAR	EXTRALOBAR
Separate pleura	No	Yes
Location	Posterior basilar	Above or below diaphragm
Age of onset	50% > 20 years	60% < one year
Symptoms	Recurrent pneumonia	Respiratory distress
Laterality	60% left	90% left
♂:♀ ratio	1:1	4:1
Other defects	Uncommon	> 50%, eg diaphragmatic defects, tuberous sclerosis
Bronchial communication	Uncommon, small	None
Arterial supply	Systemic; single aorta	Systemic; multiple, small
Venous drainage	Inferior pulmonary vein	Systemic; azygous and hemiazygous vein

pulmonary stenosis Pulmonary valve stenosis CARDIOLOGY A narrowed pulmonary annulus, which constitutes ±11% of congenital heart disease–CHD in adults; 90% of PS is valvular, the rest is subvalvular or supravalvular CLINICAL If severe, DOE, fatigability, retrosternal pain, syncope with exertion HEART SOUNDS With moderate/severe disease, an RV impulse may be palpable at the left sternal border, and there may be a thrill at the 2nd IC space; the 1st HS is normal; the 2nd HS is widely split; a harsh crescendo-decrescendo murmur that ↑ with inspiration is heard along the left sternal border; an ejection click may precede the murmur, if the leaflets are pliable; with ↑ severity, the systolic murmur peaks later, the ejection click is earlier and superimposes on 1st HS CLINICAL Sx reflect severity of stenosis, RV function, tricuspid valve competence; if severe, DOE, fatigability, retrosternal pain, syncope with exertion EKG Rt axis deviation, RV hypertrophy IMAGING CXR–normal cardiac silhouette–if enlarged, indicates RV failure, tricuspid regurgitation; post-stenotic dilation of main pulmonary artery and ↓ pulmonary vascular markings; Doppler flow allows assessment of severity of obstruction; catheterization and angiography are unnecessary MANAGEMENT If asymptomatic, prophylactic antibiotics to prevent infective endocarditis; if severe, percutaneous balloon valvuloplasty; valve replacement for dysplastic valves. See Supravalvular pulmonary stenosis. Cf Pulmonary regurgitation, Aortic stenosis

pulmonary TB Pulmonary tuberculosis, see there

pulmonary thromboembolism Pulmonary embolism, see there

pulmonary toilet PULMONARY MEDICINE The use of a fiberoptic bronchoscope to clear inspissated secretions in Pts with atelectasis, which may be facilitated with saline lavage, or local instillation of N-acetylcysteine INDICATIONS Pts with lobar atelectasis unresponsive to conservative therapy, with gas-exchange abnormalities, or who are 'fragile' and require immediate clearing.

pulmonary tuberculosis INFECTIOUS DISEASE Infection by *Mycobacterium tuberculosis*, which occurs primarily–85% of cases, US–in the lungs EPIDEMIOLOGY TB is spread by aerosol from coughing or sneezing by infected Pt CLINICAL 1° infection is usually asymptomatic, 95% of Pts–US recuperate without further evidence of disease; PT develops in immunocompromised Pts, generally within wks after the primary infection or may lie dormant for yrs before causing disease AT RISK POPULATION Infants, elderly, and immunosuppressed–eg, chemotherapy, posttransplant Pts receiving immunosuppressants, or those with immunodeficiency–eg, AIDS; PT ranges from minimal to massive involvement, but progresses in all without therapy RISK FACTORS Frequent contact with Pts with TB, crowding, unsanitary living conditions, poor nutrition; TB is rising in the US, due to AIDS, homelessness, drug-resistant strains of TB, incomplete treatment of TB–eg, poor compliance. See Atypical mycobacterial infection, Disseminated tuberculosis

554 **pulmonary veno-occlusive disease** An idiopathic condition of children/young adults that causes progressive fibrous obliteration of veins, resulting in severe postcapillary pulmonary hypertension ETIOLOGY Uncertain, possibly linked to BM transplantation, chemotherapy–eg, carmustine, malignancy, viral infection CLINICAL Nodular zones of congestion, edema, hemorrhage, hemosiderosis RADIOLOGY CXR reveal prominent interstitial markings, Kerley B lines, pulmonary artery dilatation MANAGEMENT Temporary response may occur with vasodilators, nifedipine, α-adrenergic blocker, prazosin PROGNOSIS Poor, most die in 2 yrs

pulmonology The study of the lungs and respiratory function

Pulmozyme® Dornase alpha solution PULMONOLOGY A therapy for cystic fibrosis for FVC–forced expiratory volume < 40% to improve pulmonary function and ↓ RTIs requiring parenteral antibiotics. See Cystic fibrosis

pulp HEMATOLOGY See Red pulp, White pulp

pulp vitality test DENTISTRY Any technique–eg, application of temperature extremes, mechanical pressure, electrical stimulation, etc, to determine if tooth root tissues are intact and sensitive

Pulsar™ PROCEDURAL MEDICINE A CO_2 laser with pulsed and continuous energy used in skin resurfacing, dewrinkling, ENT, gynecology. See Laser

pulsatile VOX POPULI *adjective* Occurring discontinuously–ie, in bursts or pulses that may be either cyclical or regular or sporadic in frequency

pulsatile tinnitus ENT The perception of abnormal pulsing sounds in the ears or head, which are most often caused by conductive hearing loss; PHL may also be due to vascular abnormalities–eg, glomus tumor, aneurysms, carotid vaso-occlusive disease, AV malformation

pulsation See Intraaortic balloon pulsation

pulse CARDIOLOGY The rhythmic expansion of a blood vessel, which for certain large arteries can be evaluated clinically using the fingers or stethoscope; the 'ritual' of taking the Pt's pulse provides information about the heart rate, and a marked ↓ in the strength of the pulse suggests severe atherosclerosis, ↓ pumping activity by the heart, or vascular defects in the form of AV shunts or fistulas. See Bisferiens pulse, Corrigan's pulse, Dorsalis pedis pulse, Femoral pulse, Paradoxic pulse, Pistol shot pulse, Quincke's pulse, Radiofrequency pulse, Water hammer pulse. Cf Pulse diagnosis NUCLEAR MEDICINE 1. A brief exposure to a radioisotope, in order to label a substance and follow its path through a metabolic labyrinth 2. A discharge of electric current produced by radionuclides in an ionization chamber or scintillation counter

pulse generator Pacemaker CARDIAC PACING The part of the pacing system that produces periodic electrical pulse and contains the power supply and electronic circuit. See Pacemaker

pulse ox 1. Pulse oximeter, see there 2. Pulse oximetry, see there

pulse oximeter CRITICAL CARE A device that measures the difference in the light transmitted by oxygenated and deoxygenated Hb caused by their differences in light absorption within the circulation; the PO takes advantage of these differences and provides an indirect measurement of the arterial Hb O_2 saturation

pulse oximetry Oxygen saturation measurement, SaO CRITICAL CARE A method used to determine the O_2 saturation–SaO$_2$ and desaturation of blood in a continuous noninvasive fashion, through the noninvasive assessment of arterial Hb-bound O_2 saturation, based on the combined techniques of optical plethysmography and spectrophotometry; PO is accurate within 5% of a standard co-oximeter when the SaO$_2$ is in the 70-100% range; accuracy plummets below 70%; PO is used to detect O_2 desaturation accurately, inexpensively, quickly, safely, and is useful in endoscopy, recovery, intensive care, and for evaluating obstructive sleep apnea syndrome, for which the gold standard is the costly and onerous nocturnal polysomnography

pulse-temperature dissociation BACKGROUND The pulse rate ↑ 15-20 beats/min. for each degree ↑ in core body temperature > 39ºC; a lower than normal ↑ in pulse rate or relative bradycardia–ie, a PTD is not uncommon and occurs in burns, drug fever, hepatitis, intoxication–eg, trinitrotoluene, TNT, Legionnaires' disease, malaria–blackwater fever, acute MI, psittacosis, typhoid fever, yellow fever; relative tachycardia is far less common, but is typical of clinically silent PE, diphtheria and clostridial infections

pulse width Pulse duration CARDIAC PACING The duration of a pacing pulse in msecs

pulseless disease Takayasu arteritis, see there

pulselessness EMERGENCY MEDICINE Lack of blood flow in the circulation of the large arteries, detected by a rescuer by palpating the carotid pulse–in the groove between the trachea and the adjacent strap muscles for 5-10 secs. See External chest compression

PulseMaster Nd:YAG laser DENTISTRY A laser used for aphthous ulcers, soft-tissue procedures–eg, frenectomy, ↓ hyperplastic tissue, access gum-covered decay, treat gum disease, for cosmetic soft-tissue recontouring, decay removal in caries. See Y laser

PulsePak™ INFECTIOUS DISEASE A formulation of Sporanox® or itraconazole designed to ↑ Pt compliance with pulse-dosing.

pulses per minute CARDIAC PACING The unit used to express the frequency of events in a 60 sec period–eg, the pacemaker stimulation rate

PulStar™ NURSING A wrap used with PAS® pump units to provide intermittent limb compression to prevent DVT in bedridden Pts and those undergoing major surgical procedures. See Deep vein thrombosis

pulsus alternans CARDIOLOGY A pulse occurring at regular intervals which there is rhythmic attenuation of the pulse pressure heights–eg, every other beat; sustained PA is associated with severely depressed left ventricular function, accompanied by an altered blood flow in the aorta, left ventricular and systolic pressures, and often a 3rd ventricular sound PROGNOSIS Poor

pulsus paradoxus CARDIOLOGY A marked ↓ in pulse amplitude during normal quiet inspiration or a ↓ in the systolic pressure by > 10 mm Hg, a typical finding in cardiac tamponade, less common in constrictive pericarditis–quantifiable by a sphygmomanometer, superior vena cava obstruction, asthma, PE, shock, or after thoracotomy

pump THERAPEUTICS A device used to deliver a precise amount of medication at a specific rate. See Implantable pump, Implantable insulin pump, Infusion pump, Insulin pump VOX POPULI A device, structure or other artifice that causes a fluid or specific molecules to flow in a designated direction. See Breast pump, Colleague 3 infusion pump, Intra-aortic balloon pump, Lymphatic pump, Mechanical infusion pump, Proton pump

pump lung Adult respiratory distress syndrome, see there

punch biopsy SURGICAL PATHOLOGY A minor surgery performed in an outpatient setting or dermatology clinic, in which a hollow needle is used to obtain a 3 or 4 mm in diameter core–'punch' of skin, which is evaluated by LM. See Biopsy. Cf Shave biopsy, Skin biopsy

punch-drunk syndrome Boxer's encephalopathy, dementia pugilistica NEUROLOGY A syndrome affecting 10–20% of professional boxers, which is the cumulative result of recurrent brain damage and progressive communicating hydrocephalus due to extrapyramidal and cerebellar lesions that translate into dysarthria, ataxia, tremors, pyramidal lesions–causing mental deterioration and personality changes–eg, rage reaction and morbid jealousy–'Othello syndrome'. See Boxing. Cf Alzheimer's disease, Parkinson's disease, Torture, Vascular dementia

punched-out *adjective* Pertaining to rounded, well-circumscribed often multiple lesions seen in various sites GI DISEASE Punched-out lesions seen in the stomach by endoscopy usually correspond to benign gastric ulcers; they are well-demarcated with a sharply-defined wall and a smooth base. Cf Heaped-up IMAGING Rounded, sharply demarcated, cyst-like spaces without sclerotic margins, characteristic of myeloma of the diploë of the skull, causing sharply demarcated 'holes', due to osteoclast-activating factor secretion in plasma cells; punched-out bony defects also occur in well-circumscribed mutilating sarcoidosis of the small hand bones, chronic gouty arthritis as chondroosseous lesions that communicate with the urate 'crust' through cartilaginous defects, childhood hypophosphatasia, leukemic foci in skull, TB. See Myeloma.

punctation OB/GYN A zone of red dots representing stromal papillae and blood vessel loops reaching the surface epithelium, seen by colposcopy which, when found, mandates a biopsy, as the pattern may reflect blood vessel changes of neoplasia

puncture A rounded hole made with a pointed instrument. See Lumbar puncture, Osteopuncture

Punktyl Lorazepam, see there

pupillary reflex NEUROLOGY A part of a clinical exam in which the ability of the pupils to react to light is evaluated METHOD 1. Shining a beam of light in one eye and assessing its response, which under normal circumstances constricts 2. Shining a beam of light in one eye and assessing the response of the other eye, which normally constricts, this being known as the consensual light reflex 3. Observing the changes in diameter of the pupil as the subject observes near or distant objects

PUPPP OBSTETRICS Pruritic urticarial plaques and papules of pregnancy An erythematous papule and plaque-forming eruption seen late in the 3rd trimester in up to 75% of primigravidas, which does not recur in subsequent pregnancies TREATMENT Topical steroids. See Pruritus gravidarum

pure autonomic failure NEUROLOGY A sporadic, idiopathic cause of persistent orthostatic hypotensio in and other manifestations of autonomic failure, unaccompanied by other neurologic features

pure gonadal dysgenesis ENDOCRINOLOGY An intersex disorder characterized by 46,XX and normal external ♀ genitalia at birth; internally, bilateral streak gonads; Pts usually seek medical advice for pubescent amenorrhea; ♀ are sterile. See Intersex disorder

pure red cell aplasia HEMATOLOGY A type of anemia caused by selective depletion of erythroid cells

---

PURE RED CELL APLASIA TYPES

ACUTE Aplastic crisis A condition often preceded by viral gastroenteritis, pneumonitis, primary atypical pneumonia, mumps, viral hepatitis, pregnancy and drug toxicity CLINICAL Malaise, pallor and other symptoms of a chronic, compensated hemolyzing process TREATMENT The only effective modality is discontinuance of inculpated drug, if identified

CHRONIC Either 1.Congenital Diamond-Blackfan disease A condition due to a ↓ in erythrocyte stem cells with ↓ colony-forming units and burst-forming units and a poor response to erythropoietin TREATMENT Transfusions, corticosteroids or 2. Acquired 30-50% of chronic acquired PRCA is associated with thymoma, as well as rheumatoid arthritis, lupus erythematosus, chronic active hepatitis, hemolytic anemia and CLL

---

pure tone audiometry AUDIOLOGY A hearing test in which the person is exposed to a series of pure tones over a range of frequencies, the data from which are placed in graphic form–an audiogram. See AC-BC gap, Audiogram

pure white cell aplasia HEMATOLOGY Severe neutropenia, analogous to pure red cell aplasia; it is either associated with thymoma and hypogammaglobulinemia, and responds to plasmapheresis or is associated with other immune diseases eg Goodpasture's disease and responds to antithymocyte globulin or high-dose IV Ig

purge GI DISEASE 1. *verb* To evacuate the bowels 2. *noun* A megaevacua-

tion of the bowels 3. *noun* An agent that #1's, resulting in a #2

purging HEMATOLOGY The elimination of BM of a Pt with a malignant or nonmalignant lesion requiring hematopoietic stem cell transplantation; purging may be effected with cytotoxic chemotherapeutics, radiation, or by experimental immunologic methods PSYCHIATRY To eliminate by self-induced vomiting, a high volume of food eaten by a Pt with binge eating disorder. See Anorexia nervosa, Bulimia nervosa, Eating disorder, Pharmacologic purging

purging behavior PSYCHIATRY Emesis induced by ipecac, or use of laxatives, enemas, diuretics, anorexic drugs, caffeine, other stimulants DIFFDX IBD, DM, CA, thyroid disease. See Anorexia nervosa, Bulimia nervosa, Eating disorder

purified protein fraction TRANSFUSION MEDICINE A colloid-type volume expander consisting of 88% albumin and 12% globulins. See Albumin, Hemodilution, Surgical blood management

purine analogue PHARMACOLOGY A therapeutic that mimics the chemical structure of purine; in pathologic conditions, where there is ↑ production of DNA, PAs compete with normal purines–guanine and adenine and are incorporated in the DNA of mitotically active 'rogue' cells, eventually destroying them. See 6-MP, Nucleoside analogues

purine nucleoside phosphorylase deficiency An autosomal co-dominant condition caused by defective purine metabolism and accumulation of deoxyGTP, resulting in immune dysfunction by inhibiting ribonucleotide reductase, and blocking cell division, causing predominantly T-cell immune dysfunction CLINICAL Recurring opportunistic infections of lungs, skin, GU tract, autoimmune hemolytic anemia, BM hypoplasia LAB ↓ T cells, ↑ urine/serum uric acid, ↑ inosine, ↑ guanosine. Cf Adenosine deaminase deficiency

purity NIHSPEAK The relative absence of extraneous matter in a drug or vaccine that may or may not be harmful to the recipient or deleterious to the product

Purodigin® Digitoxin, see there

puromycin Puromycin aminonucleoside EXPERIMENTAL BIOLOGY An antibiotic that inhibits protein synthesis isolated from *Streptomyces alboniger*, used to induce experimental lipoid nephrosis in animals

the Purple Book GRADUATE EDUCATION The annually updated directory of transitional yr programs, which provides information in required and elective rotations, call schedule, etc. See Internship, Transitional yr program

purpura Visible hemorrhage into a mucocutaneous region. See Cocktail purpura, Fulminant neonatal purpura, Henoch-Schönlein purpura, Idiopathic thrombocytopenic purpura, Post-transfusion purpura

purpuric *adjective* Referring to purpura, see there

PURSUIT CARDIOLOGY A clinical trial–Platelet Glycoprotein IIa/IIIb in Unstable Angina: Receptor Suppression Using Integrilin Therapy–which evaluated eptifibatide in Pts with acute coronary syndromes. See Eptifibate

purulent 1. Containing or consisting of pus, exudate 2. Suppurative, pusy

purulent drainage WOUND CARE A drainage of material chock full of PMNs; pus-laden discharge

purulent meningitis INFECTIOUS DISEASE Acute inflammation of the meninges, which often extends to the brain and ventricles, causing ventriculitis ETIOLOGY Bacteria, fungi, rarely parasites CLINICAL Headache, stiff neck or back, N&V LAB CSF pleocytosis, ↑↑↑ PMNs MANAGEMENT Antibiotics PROGNOSIS A function of response to antibiotics. Cf Aseptic meningitis

push DRUG SLANG *verb* A street term, to sell drugs NURSING See IV push

push enteroscope/enteroscopy GI DISEASE A diagnostic instrument/treatment of jejunal angiodysplasia and small intestinal bleeding sites and lesions. See Angiodysplasia

556   **pusher** DRUG SLANG 1. A person who sells drugs, especially the 'heavies'—eg, heroin 2. A metal hanger or umbrella rod used to scrape residue in crack stems

pustule DERMATOLOGY A collection of pus, inflammatory debris and often dying/dead microorganisms. See Abscess, Malignant pustule, Spongiform pustule

put out SEXOLOGY *verb, active voice* To engage in penovaginal activity, usually at the female's discretion VOX POPULI *verb, passive voice* To be annoyed with someone, something

PUVA therapy Psoralen with UVA–light therapy DERMATOLOGY A therapy used for severe psoriasis, palmoplantar dermatosis—eg, palmoplantar pustulosis, atopic dermatitis, mycosis fungoides, vitiligo, 20-nail dystrophy, alopecia areata, vesicular eczema; PT causes irregular hyperpigmented macular lesions with ↑ melanocytes and epithelial atypia SIDE EFFECTS Nausea, burning, pruritus, with time, also, wrinkling, irregular hyperpigmentation, lentigines, premalignant keratoses, skin cancer. See Psoriasis

PV 1. Pemphigus vulgaris 2. Peripheral vascular/vein/vessel 3. Plasma volume 4. Polio vaccine 5. Poliovirus 6. Polycythemia vera, see there 7. Portal vein 8. Polyoma virus 9. Postvoiding 10. *Proteus vulgaris* 11. Pulmonary vein

PVC Premature ventricular contraction, see there

PVD Peripheral vascular disease, see there

PVS 1. Persistent vegetative state, see there 2. Pulmonary valve stenosis

pyelogram Intravenous pyelogram, see there

pyelography Intravenous pyelography, see IVP

pyelonephritis NEPHROLOGY Infection of the kidney and ureters TYPES Reflux nephropathy, acute uncomplicated pyelonephritis, complicated UTI or chronic pyelonephritis ETIOLOGY UTIs, cystitis, especially with backflow of urine from the bladder into the ureters or kidney pelvis—vesicoureteric reflux RISK FACTORS Hx of cystitis, renal papillary necrosis, kidney stones, vesicoureteric reflux or obstructive uropathy, Hx of chronic/recurrent UTIs, infection by virulent bacteria; elderly, immunosuppression—eg, AIDS, CA

Pylori-Chek GI DISEASE A method for detecting urease associated with *H pylori* infection in adults. See Helicobacter pylori

pyloric stenosis Congenital hypertrophic pyloric stenosis, hypertrophic pyloric stenosis GI DISEASE A narrowing of the gastric outlet into the duodenum due to thickening of pyloric muscle, which controls gastric flow to the duodenum; PS is more common in ♂; Sx appear shortly after birth.

pyloric string sign PEDIATRIC IMAGING Elongation and narrowing of the pylorus as seen in an upper GI radiocontrast series in a child with hypertrophic pyloric stenosis. Cf, Tit sign

pyloroplasty Pyloric stenosis repair, pyloromyotomy GENERAL SURGERY An elective procedure in which pyloric sphincter muscle near the serosa surface is cut longitudinally and resutured, relaxing muscle and widening the outlet to the duodenum, usually accompanied by vagotomy INDICATIONS Pts with gastric or peptic ulcer disease for whom conservative management has failed. See Peptic ulcer, Vagotomy PEDIATRIC SURGERY An incision on the serosa of the pylorus in an infant with pyloric stenosis, which is associated with projectile vomiting of *non*-bile-tinged goo; pyloroplasty is the only effective and efficient treatment for pyloric stenosis. See Pyloric stenosis

pylorospasm 1. Hypertrophic pyloric stenosis, see there 2. Irritable bowel syndrome

pyo- Greek, pus

pyoderma gangrenosum DERMATOLOGY A rare idiopathic condition characterized by skin ulceration; at least 50% of Pts have associated systemic disease, including infection, malignancy, vasculitis, collagen vascular diseases, DM, trauma CLINICAL An initial bite-like lesion, pain, arthralgias and malaise; the classic ulcers occur on the legs; a superficial variant, atypical PG, tends to occur on the hands; involvement of other organ systems manifests as sterile neutrophilic abscesses in the lungs, heart, CNS, GI tract, eyes, liver, spleen, bone, and lymph nodes ASSOCIATED CONDITIONS IBD–either ulcerative colitis or Crohn's disease, a symmetric polyarthritis, hematologic disorders–eg, leukemia or preleukemia, predominantly myelocytic, monoclonal gammopathies, especially IgA gammopathy; less common associations include arthritides–eg, psoriatic arthritis, osteoarthritis, or spondyloarthropathy; liver disease–eg, hepatitis and primary biliary cirrhosis; myeloma–especially IgA type and immunologic diseases–eg, SLE, Sjögren's DIFFDX Acute febrile neutrophilic dermatosis, aphthous stomatitis, atrophie blanche, Behçet disease, chancroid, Churg-Strauss syndrome–allergic granulomatosis, ecthyma, ecthyma gangrenosum, herpes simplex, hypersensitivity vasculitis–leukocytoclastic vasculitis, impetigo, insect bites, sporotrichosis, SCC, venous insufficiency, verrucous carcinoma, Wegener's granulomatosis MANAGEMENT Anti-inflammatories–eg, corticosteroids, immunosuppressants

pyogenic granuloma DERMATOLOGY A raised, red, highly vascularized skin bump seen in children, often at the site of trauma on the hands and arms or face, which bleeds easily, rarely exceeding 1 cm in diameter

pyosalpix GYNECOLOGY Pus in the fallopian tube, a finding classically associated with PID. See Pelvic inflammatory disease

pyramid POPULAR HEALTH A simplified schematic used to guide a person to optimize her diet and lifestyle choices. See Exercise pyramid, Food pyramid, Mediterranean food pyramid

pyramid diet NUTRITION A diet based on the USDA's food pyramid, promulgated as a guideline for the proportions of food groups to be eaten/day; carbohydrates are at the base of the pyramid; at the peak are the 'discouraged' foods–to be eaten sparingly–eg, fats, oils, refined sugars. See Mediterranean pyramid; Cf Five food groups, Four food groups

pyramid system GRADUATE EDUCATION A system used in prestigious US teaching hospitals, to limit the number of physicians graduating from highly-selective residency programs. See Residency. Cf 'Match'

pyrethrin PUBLIC HEALTH A topical insecticide used to manage head lice. See Head lice

pyridinium collagen crosslinks METABOLIC DISEASE A family of molecules that links collagen fibers to each other; the breakdown products are excreted in the urine with attached crosslinks including pyridinoline–PYD and deoxypyridinoline–DPYD; ↑ DPYD and PYD indicate bone matrix degradation and resorption, osteoporosis, primary hyperparathyroidism, Paget's disease. See Osteoporosis; Cf N-telopeptides

Pyridium® Butabarbital, see there

pyridostigmine Mestinon® PHARMACOLOGY A cholinergic used to manage Pts with myasthenia gravis, including crises ADVERSE EFFECTS ↑ salivation, fasciculation, abdominal cramps, diarrhea. See Reversal agent

pyridoxine NUTRITION A form of vitamin B_6 used with INH for TB to prevent peripheral neuropathy. See Vitamin B_6

Pyrilinks®-D assay OSTEOPOROSIS An immunoassay that measures deoxypyridinoline–Dpd crosslinks in urine, which are ↑ in osteoporosis. See Deoxypyridinoline

pyrimethamine PARASITOLOGY An anti-protozoal combined with sulfadiazine for managing toxoplasmosis ADVERSE EFFECTS BM damage, rashes; folic acid deficiency can cause a painful burning tongue, loss of taste, anemia. See Toxoplasmosis

pyrin Marenostrin, see there

pyrogen A fever-inducing substance

pyromania PSYCHIATRY A disorder of impulse control which is characterized by a pattern of fire setting for pleasure, gratification, or relief of tension. See Impulse control disorder

---

PYROMANIA–DIAGNOSTIC CRITERIA

DIAGNOSTIC CRITERIA

(1) The person has set fires deliberately and on purpose on more than 1 occasion (multiple episodes involved)

(2) The person feels a tension or affective arousal before setting the fire

(3) Fascination, interest, attraction and/or curiosity for fire making paraphernalia, fire fighting equipment or any fire-related topic

(4) Pleasure, gratification, or relief may be felt

(a) when setting fires

(b) while witnessing a fire

(c) when participating in the aftermath

EXCLUSION CRITERIA

The fire setting cannot be better explained by another disorder (mania, antisocial personality conduct disorder, other).

The fire setting is not done (1) for monetary gain–insurance etc (2) to express a sociopolit-ical ideology, (3) to conceal a criminal act, (4) as a conscious expression of anger or vengeance, (5) to improve one's living circumstances, (6) in response to a delusion or hallucination, and (7) as a result of impaired judgement due to delusion or intoxication.

---

pyrroloquinolone quinone PQQ, see there

pyruvate kinase deficiency HEMATOLOGY An AR condition affecting Nothern Europeans and the Amish; lack of PK in RBCs results in chronic hemolytic anemia and release of free Hb into the circulation; $\frac{1}{3}$ present in the neonatal period with jaundice requiring phototherapy; in older children, PKD ranges in severity from asymptomatic, to requiring blood transfusions, to jaundice and splenomegaly CLINICAL Mild to severe hemolysis, jaundice, anemia LAB ↑ 2,3-DPG, ↓ PK activity, ↓ ATP in RBCs MANAGEMENT Transfusions, splenectomy

PYtest® GI DISEASE A breath test for detecting gastric urease produced by *H pylori*. See *Helicobacter pylori*

pyuria UROLOGY The presence of abundant PMNs in the urine, usually due to bacterial URI. See Sterile pyuria

PZA Pyrazinamide, see there

PZI Protamine zinc insulin

Q angle Quadriceps angle ORTHOPEDICS The angle of alignment of the quadriceps; a Q angle of ≥ 20° is considered abnormal and creates a lateral stress on the patella, predisposing it to pathologic changes; contrarily, a normal Q angle does not preclude regional problems–eg, it may underestimate the lateral force on the knee where there is an imbalance between the vastus medialis and lateralis muscles

QMB Qualified Medicare beneficiary, see there

QNS Quantity not sufficient LAB MEDICINE An abbreviation indicating that the material–eg, blood, urine, submitted for analysis is insufficient for requested test

qod Latin, very other day

QoLITY CARDIOVASCULAR DISEASE A clinical trial–Quality of Life Trial hypertension

QRS complex R-wave CARDIAC PACING The deflections on an EKG tracing produced by a ventricular depolarization

QRS score A measure of the size of an MI, based on evaluation of a 12-lead EKG

QRST interval QT interval, see there

QT Interval CARDIAC PACING The interval between the beginning of a paced QRS complex and the peak of the subsequent T wave–evoked QT in response to metabolic needs in a normal EKG; QT shortening signals a need for an ↑ in pacing rate. See Electrocardiogram

Quaalude Methaqualone, see there

quack *adjective* Unproven, see there *noun* Charlatan, montebank A person who impersonates a physician. See Health fraud, Medicine show, Patent medicine, Quackery, Questionable doctor, Unproven methods for cancer management. Cf Alternative medicine

quackery Bogus therapy HEALTH FRAUD False representation of a substance, device or therapeutic system as being beneficial in treating a medical condition–eg, 'snake oil' remedies, diagnosing a disease, or maintaining a state of health; eliberate misrepresentation of the ability of a substance or device to prevent or treat disease. See AIDS fraud, AIDS quackery, Health fraud, Pseudovitamin, 'Snake oil' remedy, Unproven methods of cancer management. Cf Alternative medicine, Fraud

quad pack Quadruple pack TRANSFUSION MEDICINE A plastic blood collection bag with 3 attached 'peripheral' bags, allowing the sterile collection and separation of a unit–circa 500 mL–of whole blood into 4 sterile 125 ml aliquots which have a normal shelf life. See 'Cow method'

Quadramet™ Samarium Sm, Lexidronam™ ONCOLOGY A parenteral radio-pharmaceutical for pain relief in metastatic CA to the bone, seen by bone scans. See Bone scan

quadrant A quarter–eg, the liver is in the right upper quadrant of the abdomen. See Left lower quadrant, Left upper quadrant, Right lower quadrant, Right upper quadrant

quadrantectomy SURGERY The excision of a quadrant of tissue, usually from the breast, which includes both grossly identifiable malignancy and grossly normal soft-tissue margins. Cf Lumpectomy, Mastectomy

quadriceps angle Q angle, see there

quadriparesis NEUROLOGY Weakness of both arms and both legs, as seen in muscular dystrophy. See Muscular dystrophy, Paraplegia

quadriplegia NEUROLOGY Paralysis of both arms and both legs, as seen in a high spinal cord accident or stroke. See Spinal cord injury, Stroke

quads Quadriceps femoris, see there

quahog MEDTALK A derogatory regionalism for an obese, pregnant, unkempt, malodorous, socially/economically disadvantaged ♀. See Gomer

qualified patient TERMINAL CARE A Pt who has made an advance directive in accordance with prevailing legislation. See Advanced directive, Living will

quality-adjusted life yr QALY A unit of health care outcomes that adjusts gains or losses in yrs of life subsequent to a health care intervention by the quality of life during those yrs. See Disability-adjusted life yrs, Healthy-yr equivalent. Cf QALE

quality assurance MANAGED CARE The constellation of activities and programs intended to assure a high quality of care in a defined medical setting; the assessment of delivery of healthcare by managed care plans; the NCQA is a key agency in evaluating performance of managed care plans MECHANISMS OF QA Peer review, utilization review–identify and remedy deficiencies in quality. See National Committee for Quality Assurance, Peer review organization. Cf Quality control

quality control LAB MEDICINE The constellation of mechanisms used to determine accuracy, reliability and consistency of data, assays or tests, often in the context of a clinical lab. See Accredited lab, Multirule procedure . Cf Quality assurance, Total quality management

quality of life PSYCHIATRY The degree to which a person is able to function at a usual level of activity without, or with minimal, compromise of routine activities; QOL reflects overall enjoyment of life, sense of well-being, freedom from disease Sx, comfort, and ability to pursue daily activities. See Heinrichs-Carpenter Quality of Life Scale, Karnovsky scale, QALE, Q-TWiST, QWB scale, SF-12

quality of life therapy THERAPEUTICS A therapy for a non-life threatening condition–ie, not medically indicated; QOLTs treat balding–eg, Minoxidil, Propecia; impotence–eg, Caverject, Muse, Viagra; fertility–eg, Clomid. See Cosmeceutical, Impotence

quality of motion REHAB MEDICINE A term that encompasses range, direction, and symmetry of motion of a particular joint ASSESSMENT OF QOM

Appearance, feel and sound by passive and active examination, radiologic imaging, interarticulor changes by arthrography. See Range of motion

'quality' time Meaningful time SOCIAL MEDICINE The time spent in meaningful interaction with persons who are significant in a person's life. Cf Latchkey children

quantify *verb* To determine the quantity of; measure

Quantison™ IMAGING An ultrasound contrast agent for assessing CAD. Sonicated albumin

quantitative CT IMAGING A bone density scan used to assess the mineral content of trabecular bone of the lower vertebral column in Pts being evaluated for osteoporosis. See Osteoporosis

quantitative overview Meta-analysis, see there

quantity of motion The size or magnitude of motion available for active use, which is determined based on anatomic or geometric assessment of the assessment of the joint

Quantum™ PSV CRITICAL CARE A proprietary noninvasive pressure support ventilator for treating respiratory insufficiency. See Respiratory insufficiency

quarantine EPIDEMIOLOGY *noun* A period of isolation intended to control the spread of a contagious infection *verb* To restrict the freedom of movement in those with–or presumed to have been exposed to–a highly communicable disease, to prevent dissemination. See Notifiable disease, Proposition 64

quarter bag DRUG SLANG A street term for $25 worth of drugs

quarter moon sign RADIOLOGY A descriptor for pooled barium seen in an upper 'GI' series, a finding typical of benign gastric ulcers, which corresponds to an overhanging fold of mucosa surrounding the ulcer mouth. See Meniscus sign of Carman

quarter-strength formula NEONATOLOGY Highly-diluted or watered-down formula, given to infants with watery diarrhea. See Formula

Quartet™ RESPIRATORY CARE A system for diagnosing and managing obstructive sleep apnea MODES Continuous, bi-level pressure, automatic 'smart' CPAP modes. See Obstructive sleep apnea

quasi-courtship PSYCHOLOGY The use of sexual elements–eg, courtship behaviors, flirtation, or sexual acts themselves to achieve non-sexual goals

Quazepam A benzodiazepine, see there

Quebec classification EMERGENCY MEDICINE A classification used to stratify the severity of whiplash-related injury and need for therapy, based on the data from meta-analysis. See Whiplash

---

QUEBEC CLASSIFICATION–WHIPLASH-ASSOCIATED DISEASE

GRADE

0 No complaint about the neck; no physical signs

I Neck complaint–pain, stiffness, or tenderness only

II Neck complaint AND musculoskeletal signs[a]

III Neck complaint AND neurological signs[b]

IV Neck complaint AND fracture or dislocation

[a]Musculoskeletal signs include ↓ range of motion and point weakness

[b]Neurologic signs include ↓ or absent deep tendon reflexes, weakness, and sensory deficits

Symptoms that may occur in all grades include deafness, dizziness, tinnitus, headache, memory loss, dysphagia, TMJ pain Spine 1995; 20/8S.2S-68S

---

questionable doctor MEDTALK A physician who has been sanctioned for serious state and federal offenses and placed on a list by the Public Citizen's Health Research Group. See Impaired physician

questioning mania Folie du pourquoi PSYCHIATRY An obsessive-compulsive neurosis characterized by the need to ask questions. See Obsessive-compulsive disorder, Obsessive-compulsive personality disorder

quetiapine fumarate Seroquel® PHARMACOLOGY An agent used to manage psychotic disorders–eg schizophrenia

queue INFORMATICS A waiting line for a document VOX POPULI, UK A line. See Gender queue

Quibron® A proprietary OTC formulation of guaifenesin and theophylline INDICATIONS Expectorant, bronchodilator for common cold, allergies, hay fever, URIs

'quick and dirty' Crude but effective A popular adjective for a survey, lab procedure, or of test, using the tools at hand to answer an experimental question in a crude fashion. Cf Gold standard

QuickDraw™ venous cannula HEART SURGERY A device used for draining nonoxygenated blood from the venae cavae and/or right atrium during cardiopulmonary bypass. See CABG

quickening OBSTETRICS A subjective sensation experienced by the mother during early pregnancy, that occurs around the 16th gestational wk. See Bonding. Cf Lightening

QuickScreen™ SUBSTANCE ABUSE An OTC test for home testing of urine for drugs of abuse–cocaine, marijuana, opiates, amphetamine, and PCP; a positive result indicates the presence of one or more of these drugs. See Drug of abuse

QuickSilver guide wire CARDIOLOGY A hydrophilically coated guide wire used with the UltraLite flow-directed microcatheter when accessing low-flow or difficult anatomies

QuickVue® LAB MEDICINE A rapid pregnancy test designed for clinical lab use. See HCG

quiescence Inactive, resting, silent MEDTALK A period during which an infection is present but not active in a host EXAMPLES TB; chickenpox and recrudescence as herpes zoster–shingles. Cf Latency

QUIET CARDIOLOGY A clinical trial–the Quinapril Ischemic Event Trial–on ischemic events and progression of ASHD in CAD. See Coronary artery disease, Quinapril

'quiet zone' LUNG PATHOLOGY The terminal airways contribute little to the total airflow resistance–most resistance is contributed by bronchioles > 2 mm in diameter, thus although a disease process may begin in small airways, it may be clinically silent–ie, a 'quiet zone' until it affects the larger airways

Quietal Amitryptiline, see there

quinacrine CLASSIC CYTOGENETICS A fluorescent dye used to stain chromosomes, especially Y chromosome GYNECOLOGY TROPICAL MEDICINE An antiparasitic agent once used to manage giardiasis, helminths, malaria, protozoa, tapeworms

quinapril Accupril® An ACE inhibitor used to treat HTN which is used alone or in combination with a thiazide diuretic to manage hypertension, or as an adjunct with a diuretic or digitalis to treat CHF ADVERSE EFFECTS Vasodilation, tachycardia, HF, palpitations, chest pain, hypotension, MI, CVA, hypertensive crisis, angina pectoris, orthostatic hypotension, arrhythmias, shock, dry mouth, constipation, diarrhea, N&V, hepatitis, pancreatitis, hemorrhage; somnolence, vertigo, insomnia, sleep disturbances, paresthesias, depression, headache, dizziness, fatigue, agranulocytosis, BM depression, thrombocytopenia

Quincke's pulse Quincke sign CARDIOLOGY A sign of aortic regurgitation due to ↑ pulse pressure, characterized by alternating flushing and blanching of the skin, due to pulsating of subpapillary arteriolar and venular plexi; QP is seen in nailbeds, classically in Pts with aortic regurgitation–insufficiency, but may be seen in normal individuals. See Aortic regurgitation

quinidine Duraquin®, Quinaglute® CARDIOLOGY A class Ia antiarhythmic that ↓ myocardial excitability, conduction velocity, and contractility . See Therapeutic drug monitoring

quinolone Fluoroquinolone INFECTIOUS DISEASE Any of a family of broad-spectrum oral antibiotics–eg, ciprofloxacin, norfloxacin, and ofloxicin, which target bacterial DNA gyrase and concentration-dependent inhibitors of DNA synthesis; bacterial resistance to quinolones is rare and hinges on mutations of gyrase; quinolones are active against most aerobes, including bacteria resistant to other antibiotics; they are effective in GI, GU, prostatic, respiratory infections, STDs ADVERSE EFFECTS GI discomfort, vague CNS Sx

quinsy Peritonsillar abscess, see there

quinupristin See Synercid®

Quitaxon Doxepin, see there

Quixil® ORTHOPEDICS A fibrin sealant which ↓ bleeding during total knee and total hip replacement surgery. See Fibrin sealant

quorum HOSPITAL PRACTICE The mimumum number of members of a committee or decision-making body needed at a meeting to make decisions for that body. See Medical staff, Roberts' Rules of Order

QVAR® PULMONOLOGY An inhaled corticosteroid–beclomethasone dipropionate–administered by a metered dose inhaler to control asthma. See Asthma

QWB scale Quality of well-being scale A 30-symptom questionnaire used to rank the value that people place on alleviating certain Sx

glands of mammals EPIDEMIOLOGY Human rabies is rare 1 to 2 cases/year in US; internationally, 33,000 people die/yr from rabies in Asia, Africa CLINICAL 18-60 day incubation, followed by nonspecific Sx–eg, fever, headache, N&V, numbness at site of exposure, and early neurologic signs–anxiety, restlessness, depression; acute neurologic phase is characterized by agitation, confusion, delirium hydrophobia, laryngeal spasms, paralysis, complications VACCINATION VRG vaccine. See Negri bodies. Cf Pseudo-rabies

Rabson-Mendenhall syndrome A condition characterized by DM, insulin resistance, pineal body hypertrophy, adrenal cortical hyperplasia

RACA Rescue adjunctive coronary angioplasty, see there

Racal space suit Orange suit VIROLOGY A positive-pressurized, bright orange space suit with a battery-powered air supply used in fieldwork with extreme biohazards–Biosafety Level 4 organisms. Cf Chemturion space suit

raccoon eye Black eyes, raccoon eye appearance A descriptor for bilateral periorbital accumulations of blood or other substances, likened to the nocturnal North American omnivore, *Procyon lotor*; this Panda bear-like appearance is classically described in periorbital hematomas, often associated with anterior-posterior displacement-type automobile accidents or basilar fractures of the skull; less commonly, the descriptor is applied to the periorbital purpura due to skin infiltration in primary amyloidosis, as a spontaneous event, or after prolonged eye-strain

race SOCIAL MEDICINE Ethnic origin A subdivision of species which, while capable of genetic recombination, may nonetheless be divided in part based on biochemical, hematologic, immunologic, morphologic, serologic differences. See Equal opportunity SPORTS MEDICINE An athletic competition in which the fastest person wins

racial inequality Racial disparity SOCIAL MEDICINE, PUBLIC HEALTH A disparity in opportunity for socioeconomic advancement or access to goods and services based solely on race. See Women and health

rad PHYSICS An obsolete unit of ionizing radiation–X-rays, γ rays–corresponding to an energy absorption of 100 ergs/g of tissue. Cf Gray, Roentgen

Radepur Chlordiazepoxide, see there

radial growth phase melanoma Nontumorigenic melanoma DERMATOLOGY An intraepithelial melanoma not associated with metastasis, characterized by noninvasive radial enlargement EXAMPLES Melanoma in situ, lentigo maligna, dysplastic nevi. Cf Vertical growth phase melanoma

radial keratotomy OPHTHALMOLOGY A form of refractory surgery–a technique for correcting myopia by changing corneal conformation, causing it to flatten, ↓ myopia COMPLICATIONS Daily fluctuation in visual acuity, visual haziness, glare and halos, overcorrection, undercorrection, bacterial keratitis or endophthalmitis, corneal perforation, recurring corneal erosion and scarring, irregular astigmatism, cataract formation CONTRAINDICATIONS Corneal dystrophy, ocular surface disease, corneal scarring. See Laser Eye Surgery. Cf LASIK, Photoreactive keratectomy

radial nerve dysfunction Radial nerve neuropathy NEUROLOGY A peripheral neuropathy characterized by impaired movement and/or sensation of the back of the arm–triceps, forearm, or hand, due to radial nerve damage ETIOLOGY Radial nerve trauma or damage which may be a mononeuropathy, a finding typical of a local cause of nerve damage–direct trauma as occurs in crutch palsy, Saturday night palsy, fracture-related trauma, prolonged nerve pressure as occurs in odd positions while sleeping, compression resulting in hypoxia or entrapment as in carpal tunnel syndrome; systemic disorders may rarely cause isolated nerve damage–as in mononeuritis multiplex. See Carpal tunnel syndrome, Crutch palsy, Saturday night palsy

RADIANCE CARDIOLOGY A clinical trial–Randomized Assessment of Digoxin on Inhibitors of the Angiotensin Converting Enzyme that evaluated the role of

R Symbol for 1. Electrical resistance 2. Purine nucleoside 3. Radiation 4. Rate 5. Ratio 6. Record 7. Rectal 8. Recurrence 9. Relapse 10. Respiration 11. Rhodopsin 12. Rhythm 13. Ribose 14. *Rickettsia* 15. Right 16. Ring chromosome 17. Risk 18. Roentgen

r Symbol for: 1. Correlation coefficient 2. Drug resistance 3. Recombinant, see there 4. Reverse 5. Ribose 6. Ribosomal

R-on-T phenomenon CARDIOLOGY A premature ventricular depolarization so early in the cardiac cycle that it falls on the apex of the preceding T wave, possibly presaging ventricular tachycardia or fibrillation; 'R-on-T' indicates a need to ↑ antiarrhythmics–eg, lidocaine

r-tPA recombinant tissue Plasminogen activator

RA 1. Refractory anemia 2. Renin activity 3. Retinoic acid 4. Rheumatoid arthritis, see there 5. Right atrium 6. Risk assessment

RA cell Rheumatoid arthritis cell, see there

rabbit nose A descriptor for nose twitching and wrinkling by children with allergic rhinitis, which relieves pruritus or ↑ air passage; the characteristic upward rubbing of the nose–'allergic salute' may result in a groove formation at the tip of the nose

'rabbit' stool A descriptor for the small rounded, mucus-covered fecal 'pellets' produced in irritable bowel syndrome

rabbit syndrome NEUROLOGY A condition characterized by focal perioral tremors and nose twitching, which may be seen in parkinsonism as a late side effect of antipsychotic drug therapy; unlike tardive dyskinesia, the Sx of RS respond well to antiparkinsonian agents. See Tardive dyskinesia

rabeprazole Aciphex® GI DISEASE A proton pump inhibitor used to manage for heartburn, duodenal ulcers, gastric ulcers, GERD, hypersecretory conditions–eg, Zollinger-Ellison syndrome. See GERD

rabies A fatal infection rabiesvirus which follows 'injection' by an animal bite; the virion crosses the neuromuscular junction and infects nerves, spreading centripetally into the CNS and centrifugally into the salivary

radiate

digoxin in Pts receiving ACE inhibitors in CHF. See ACE inhibitor, Digoxin, Myocardial infarction

radiate *verb* To extend from a central point–eg, pain radiating from the neck to the elbow

radiation The combined processes of emission, transmission and absorption of highly energetic waves and particles on the electromagnetic spectrum treatment to kill cancer cells. See Acute radiation injury, Alpha radiation, Background radiation, Chemoradiation, Coherent radiation, Corpuscular radiation, Definitive radiation, Electromagnetic radiation, External radiation, Gamma radiation, Grenz radiation, Implant radiation, Internal radiation, Ionizing radiation, Non-ionizing radiation, Remnant radiation, Scattered radiation, Synchrotron radiation, Total body irradiation CLINICAL PRACTICE The direct, band-like extension of a sensation, in particular of pain, from a point of origin to another region of the body. Cf Referred pain ONCOLOGY The administration of ionizing radiation to kill malignant tumor cells. See Radiation fibrosis, Radiation therapy

RADIATION

ALPHA RADIATION 2 protons and 2 neutrons, eg plutonium, radon; α radiation travels 15 cm in air and is stopped by a piece of paper; proven role in soft tissue malignancy–see Radium Dial company, relationship with epithelial malignancy is uncertain; it is present in cigarette smoke and may have an additive effect to the known carcinogenic effect of tar; emitted by radium, thorium, uranium

BETA RADIATION Electrons, eg strontium-90, tritium–^{3}H; β radiation travels at the speed of light, is stopped by wood and thin metals and is carcinogenic to skin

GAMMA RADIATION Gamma photon A quantum of electromagnetic radiation of ≤ 1 nm, which is generated by unstable nuclei eg ^{60}Co; γ radiation is stopped by several feet of heavy concrete or 10-40 cm of lead and is linked to cancer, inducing mutations at the glycophorin A locus in survivors of atomic blasts; $183/10^5$ excess deaths in survivors of the Hiroshima and Nagasaki blasts, with a 13-fold $\uparrow$ in non-lymphocytic leukemia–peaking at 6 yrs post-blast, thyroid nodules and tumors–peaking at 15-20 years post-blast and multiple myeloma 6-fold $\uparrow$–peaking 30 yrs post-blast

radiation enteropathy Radiation enteritis, radiation enterocolitis ONCOLOGY Intestinal disease evoked by locoregional radiation, usually 2° to RT. See Radiation therapy

radiation fibrosis Radiation-induced fibrosis ONCOLOGY Scarring in a tissue due to ionizing radiation. See Radiation effects, Radiation therapy

radiation-induced cancer Radiogenic cancer ONCOLOGY CA induced by ionizing radiation–eg, ALL, thyroid cancer

radiation oncologist Radiation therapist A radiologist specialized in using radioactive substances and x-rays to treat tumors and CA; an oncologist who uses various formats of radiation to manage CA SALARY ± $200K. See Oncologist

radiation oncology ONCOLOGY The specialty of medicine that uses high-energy ionizing radiation to treat malignant neoplasms and certain nonmalignant conditions MODALITIES Teletherapy, brachytherapy, hyperthermia, stereotactic radiation. See Oncology, Surgical oncology

radiation pneumonitis Irradiation pneumonitis PULMONOLOGY A condition caused by exposure of lung tissue to radiation, a common complication of RT to mediastinal and thoracic tumors–eg, NHLs, Hodgkin's disease, breast and esophageal CAs; frequency of RP reflects dose and amount of tissue exposed to radiation INCIDENCE RP occurs in 65% of irradiated lung fields; clinical pneumonitis occurs in 6%, 2% die from RP. See Radiation fibrosis

radiation sensitizer ONCOLOGY Any agent which $\uparrow$ cell susceptibility to ionizing radiation EXPLANATION Most CA cells are hypoxic and thus more resistant to radiation; a hypoxic-cell RS–eg, etanidazole, misonidazole, nimorazole would, *theoretically*, improve radiation killing of CA cells–results have been disappointing. See Radiation therapy

radiation surgery Radiosurgery, stereotactic external beam irradiation ONCOLOGY An RT technique that delivers radiation directly to the tumor while sparing healthy tissue. See Radiation therapy

radiation syndrome Any condition induced by radiation EXAMPLES Radiation arthropathy, radiation carditis, radiation cytitis, radiation dermatitis, radiation enteropathy, radiation fibrosis, radiation hepatitis, radiation nephritis, radiation pneumonitis. See Radiation enteropathy, Radiation pneumonitis

radiation therapy Radiotherapy Administration of ionizing radiation to treat disease, usually malignant TYPES Local low energy radiation–brachytherapy or radioisotopes placed at or near the tumor or cancer cells–internal RT, implant radiation; high energy radiation delivered at a distance–teletherapy; most RT uses high-energy radiation from x-rays, neutrons etc to kill CA and shrink tumors delivered as external-beam radiation; systemic RT includes use of radiolabeled monoclonal antibodies that circulate in the body, binding target cells, effecting therapy. See Conformal radiotherapy, Intracoronary radiotherapy, Intraoperative radiotherapy, Plaque radiotherapy, Radiation oncology, Stereotactic radiotherapy

radical cystectomy UROLOGY The complete removal of the urinary bladder, pelvic lymph nodes, and adjacent organs with creation of an ileal conduit or internal reservoir for internal drainage of urine INDICATIONS Muscle-invasive bladder CA. Cf Bladder-conserving therapy, Ileal conduit

radical mastectomy Halsted radical mastectomy A mastectomy that removes the breast in toto, as well as skin, subcutaneous tissue, axillary lymph nodes, and muscle from the anterior thoracic wall. See Halsted radical mastectomy. Cf Modified radical mastectomy

radical mastoidectomy ENT An operation to eradicate disease of the middle ear cavity and mastoid process, in which the mastoid, epitympanic, and mesotympanic spaces are converted into an easily accessible common cavity by removing the posterior and superior external canal walls . Cf Modified radical mastoidectomy

radical neck dissection A common major operation for head & neck CA, most of which are SCC; in a 'radical neck'–RN, the neck is opened laterally, most of the sternocleidomastoid muscle is removed, as are cervical lymph nodes, jugular vein, spinal accessory nerve, submaxillary gland and most of the parotid gland; it may be combined with a partial resection of the mandible and tongue, depending on lesion topography. Cf Commando operation, 'Heroic' surgery, Mutilating surgery

radical nephrectomy SURGERY Excision of a kidney, the adrenal gland, adjacent lymph nodes, and other surrounding tissue

radical prostatectomy The complete surgical removal of the prostate–a widely used method for treating prostate CA TYPES Retropubic prostatectomy, perineal prostatectomy

radical skiing SPORTS MEDICINE An 'extreme sport' practiced on slopes of $\geq 50°$ angles and 15 + meter drops INJURIES Knee joints/ ligaments, death. See Extreme sport, Novelty-seeking behavior

radical surgery Surgery consisting of major excision or restructuring of a body region; RS is most often used aggressive or advanced cancer EXAMPLES 'Heroic' operations–eg, forequarter amputation, hindquarter amputation, hemipelvectomy, 'Commando' operation–radical neck. See Heroic surgery. Cf Palliative surgery

radiculitis Inflammation of a spinal nerve root

radiculopathy NEUROLOGY A spinal nerve root dysfunction characterized by back pain and nerve root irritation ETIOLOGY Nucleus pulposus herniates through a weakened part of the intervertebral disk. See Cervical radiculopathy, Herniated disk

radioactive *adjective* Referring to radioactivity or emission of radiation–eg, alpha or beta particles, or gamma rays from an atomic nucleus

radioactive drug NIHSPEAK Any substance defined as a drug in §201(b)(1) of the Federal Food, Drug & Cosmetic Act that exhibits spontaneous disintegration of unstable nuclei with the emission of nuclear particles or photons [21 CFR 310.3(n)]

radioactive iodine Radioiodine A radioisotope of iodine–eg, ^{131}I or ^{123}I, used for imaging or to treat CA. See Radioactive iodine uptake

radioactive iodine uptake RAIU, thyroid scan, thyroid scintigraphy NUCLEAR MEDICINE A method of assessing thyroid function, using radioactive iodine–eg, ^{131}I or ^{123}I; ↑ in hyperthyroidism, ectopic hormone production, iodine deficiency, and in response to thyroid hormone depletion; ↓ in hypothyroidism, after administration of exogenous thyroid hormone, in defects of hormone storage, after exposure to iodine overload NORMAL RANGE 5-25%. See T3, T4, TSH

radioactive isotope Radionuclide, see there

radioactive material RADIATION A substance that contains unstable–radioactive–atoms that give off radiation as they decay. See Radioactive decay

radioactive spill PUBLIC HEALTH The outpouring of a fluid from a container or receptacle containing low-level radiation or highly radioactive isotopes. See Chemical spill, Spill

radioactivity RADIATION The spontaneous transformation of an unstable atom, often resulting in the emission of radiation and a lower energy state. See Airborne radioactivity, Radioactive decay

radioallergosorbent test An allergy test on peripheral blood that assesses allergic sensitivity to specific substances. See RAST

radiocurable ONCOLOGY *adjective* Referring to a condition or lesion that regresses completely in response to ionizing radiation. See Radiation therapy

radio-frequency ablation ONCOLOGY The induction of thermal changes–coagulation necrosis in cells and tissue using a high-frequency alternating current, a modality used to "excise" tissue

radio-frequency catheter ablation CARDIOLOGY A technique that selectively destroys arrhythmogenic foci in the endocardium–and subjacent myocardium by the use of controlled heat production, avoiding general anesthesia, as pain is minimal, and skeletal muscle contraction does not occur

radio-frequency current CARDIOLOGY A level of electrical energy–31+ watts, delivered through a large-tip electrode to ablate accessory ventricular conduction pathways in WPW

radio-frequency energy ablation SURGERY Laser ablation of soft palate tissue of Pts with daytime sleepiness and snoring, who had a low respiratory disturbance index and O_2 saturation of ≥ 85% RESULTS ↓ Tissue volume, snoring, daytime sleepiness

radio-frequency thermal rhizotomy NEUROSURGERY Percutaneous heat ablation of a nerve or ganglion for medically intractable pain COMPLICATIONS Anesthesia dolorosa, keratitis. Cf Microvascular decompression

radiograph An x-ray; a film produced by X-ray

radiographic absorptiometry IMAGING A bone density scan used to assess the mineral content of bone of the hands in Pts being evaluated for osteoporosis. See Osteoporosis

radiography Roentgenography, radiographic imaging IMAGING The recording of an image of a region placed in a beam of radiation. See Angiography, Cholangiography, CT imaging, Fluoroscopy, GI series, IVP, Mammography, MRI, Venography, Xeroradiography

radioimmunoassay see RIA

radio-immunoguided surgery SURGERY A procedure in which radiolabeled substances are administered and tumors are excised based on the location of radioactive signals

radioimmunotherapy IMMUNOLOGY The use of radio-labeled–monoclonal antibodies that recognize specific antigens, and therefore concentrate in a particular tissue to treat disease TOXICITY Myelosuppression, nausea, infections, cardiopulmonary toxicity

radio-insensitive Radio-resistant *adjective* Unresponsive or insensitive to X-rays and other forms of radiation. Cf Radiosensitive

radioisotope stent CARDIOLOGY An intracoronary arterial stent loaded with a radioisotope that stops local immunoresponding cellular proliferation, thus inhibiting restenosis of stented arteries. See Stent

radio-labeled *adjective* Referring to a molecule or substance that has been linked to a radioactive substance, which can be detected in vivo. See Radionuclide scan or in vitro in a fluid

radiologic Radiological *adjective* Referring to radiology

radiologic pelvimetry X-ray pelvimetry A study in which radiography is used to assess the size of the infant and the birth canal to determine whether the bony pelvis is large enough to allow normal vaginal delivery. See Pelvic ultrasonography

radiologist A physician trained in the use of radioactive substances, x-rays, and other imaging techniques–eg, MRI, PET, SPECT, to reach a diagnosis INCOME $250 K to $1.5 million/yr. See Neuroradiologist

radiology Roentgenology The use of ionizing–eg, x-rays, and nonionizing–eg, ultrasound and MRI–radiation, to diagnose and treat disease. See Interventional radiology, Teleradiology

radiolucent *adjective* Referring to a material or tissue that allows the facile passage of x-rays–ie, has an air or near air density; radiolucent structures are black or near black on conventional x-rays. Cf Radiopaque

radiomimetic drug An immunosuppressant–eg, an alkylating agent, used to treat CA that affects nucleic acids–eg, DNA, mimicking ionizing radiation

radionuclide Radioactive isotope, radioisotope RADIATION PHYSICS A nuclide with an unstable neutron to proton ratio, which undergoes radioactive decay; an artificial or natural nuclide with an unstable nucleus, that decays spontaneously, emitting electrons–β-particles or protons–α-particles and γ-radiation, ultimately achieving nuclear stability; RNs are used as in vivo or in vitro labels, for RT, or as sources of energy

radionuclide angiocardiography See Equilibrium angiocardiography, First-pass angiocardiography

radionuclide scan Radiopharmaceutical study IMAGING Imaging of internal organs and tissues after injection or ingestion of a small amount of radioactive material which concentrates or localizes to a particular region or tissue, and is detected with a scanner PULMONOLOGY Ventilation-perfusion scan, see there

radionuclide venography Radioisotope thrombophlebography NUCLEAR MEDICINE A noninvasive method for detecting thromboembolism, in particular DVT of the lower extremity; in RV, a radiopharmaceutical, ^{99m}Tc is administered IV in the dorsal veins of the feet and images are captured using a gamma camera

radionuclide ventriculography IMAGING The use of radionuclides to study left ventricular changes during diastole INDICATIONS Hypertrophic cardiomyopathy, HTN, anthracycline-induced cardiomyopathy, CAD ; RV is the 'gold standard' for estimating right ventricular ejection fractions

564

radiopaque *adjective* Referring to a material or tissue that blocks passage of x-rays, and has a bone or near bone density; radiopaque structures are white or near white on conventional x-rays. Cf Radiolucent

radiopaque contrast Contrast dye IMAGING A nonradioactive material that stops or attenuates radiation passing through the body, creating an outline on film of the organ(s) being examined; RCAs can produce adverse reactions–some may be severe and possibly life-threatening

radiopharmaceutical Radioactive drug RADIOLOGY A drug, compound or other material labeled or tagged with a radioisotope, which is used to diagnose or treat cancer and manage pain of bone metastases

radioresistance RADIATION ONCOLOGY The relative resistance of cells and tissues to irreversible damage by radiation; RR is greatest in terminally differentiated tissues–eg, bone, cartilage, muscle, peripheral nerve, and in certain tumors–eg, glioma, melanoma, renal cell carcinoma, sarcoma. Cf Radiosensitivity

radiosensitivity The relative susceptibility of cells and tissues to irreversible damage by RT, which prevents mitosis or completion of normal metabolism; lymphoid, hematopoietic, and gonadal tissues are most susceptible to radiation damage; some CAs–eg, lymphoproliferative, gonadal malignancies–eg, seminoma and small cell carcinoma of lung 'melt away' with RT. Cf Radioresistance

radiosensitizer ONCOLOGY A drug that ↑ tumor cell sensitivity to radiation. See Radiosensitization

radiosurgery 1. Radiation surgery, see there 2. Stereotactic radiosurgery, see there

radium A radioactive metallic element–atomic number 88, atomic weight 226.025, which is an intermediate in the uranium decay series, produced when an α particle is emitted from ^{230}Th

RADIUS OBSTETRICS A study–Routine Antenatal Diagnostic Imaging with Ultrasound–involving 15,000 ♀ that evaluated fetal ultrasonography CONCLUSION No benefit

Radizepam Diazepam, see there

radon ^{222}Ra PUBLIC HEALTH A natural radioactive gaseous element, atomic number 86; atomic weight, 211.4 in the ^{238}U → ^{206}Pb decay chain; radon has a $T_{1/2}$ of 3.8 days, decays into 2 solid α particle-emitting daughters; radon exposure carries a relative risk of 12.7 for lung CA in non-smoking uranium miners and an ↑ risk of childhood CA, myelogenous leukemia, renal cell carcinoma, melanoma, prostate CA. See Radionics, Radium Dial Company

RAEB Refractory Anemia with Excess myeloBlasts, see there

rage A state of violent anger. See Black rage, 'Roid rage

ragweed *Ambrosia artemisiifolia* ALLERGY MEDICINE Any of the weedy composite herbs of family Compositae, the pollen from which is highly allergenic DIAGNOSIS Bronchial provocation testing, in which pollen is inhaled through a dosimeter MANAGEMENT Avoid pollen, stay indoors, cromolyn, antihistamines, sympathomimetics, theophylline, corticosteroids DRUG SLANG A street term for low quality marijuana or heroin

RAI staging system ONCOLOGY A system used to 'stage' Pts with CLL. See Chronic lymphocytic leukemia.

RAI STAGING SYSTEM

LOW

0 Lymphocytosis only

INTERMEDIATE

I Lymphocytosis + lymphadenopathy

II Lymphocytosis + splenomegaly with/without lymphadenopathy or hepatomegaly

HIGH

III Lymphocytosis + anemia, with or without organomegaly

IV Lymphocytosis + anemia + thrombocytopenia, with or without organomegaly

railroad track scar PLASTIC SURGERY A scar with cross-hatched stitch marks due to poor repair, excess tension on the skin or to a delay in suture removal, which should be ideally removed on the 3rd–5th days–except when the sutures overlie highly mobile sites; some body regions–eg, trunk, sternum and proximal extremities are more susceptible to cross-hatching

rale *râle*, French, to rattle CLINICAL MEDICINE An abnormal lung sound TYPES Sibilant–whistling; dry–crackling; wet–sloshy depending on the amount and density of fluid flowing back and forth in the air passages; rales may be discontinuous sounds or vibrations heard by auscultation in various lung disease–eg, bronchitis, pneumonia, atelectasis, pulmonary edema, heart failure, bronchiectasis, TB. See Dry rales, Moist rales

RALES CARDIOLOGY A series of trials–Randomized Aldactone Evaluation Study that demonstrated that the addition of Aldactone®–spironolactone to standard therapies–eg, ACE inhibitors, resulted in a 27% ↓ in total mortality in severe CHF Pts with systolic left ventricular dysfunction. See Spironolactone

raloxifene Evista® OSTEOPOROSIS An SERM–selective estrogen receptor modulator, that ↑ bone density–less extensively than estrogen, ↓ total cholesterol and LDL-C, ↓ risk of breast CA;. Cf Tamoxifen CONTRAINDICATIONS Pregnancy, nursing, active or prior venous thromboses PROS Lacks estrogenic effects on breast and uterus ADVERSE EFFECTS Hot flashes, leg cramps, DVT, PTE, retinal vein thrombosis. See Calcium channel blocker, MORE, Osteoporosis, STAR. Cf Biphosphonates

Ralozam Alprazolam, see there

RALS Remote Automated Laboratory System, see there

rambling NEUROLOGY Fragmented non-goal directed speech most often caused by acute organic brain disease. See Organic brain disease, Word salad

ramipil Altace®, see there

ramp pacing Ramp series CARDIAC PACING A form of burst pacing that ends tachycardia by delivering a series of pacing pulses, where intervals between pulses ↓ progressively at programmed decrements thereby ↑ pacing rate See Pacing

ramped dosing THERAPEUTICS The gradual introduction of a drug–eg, TCAs, generally at low dosages, to be increased at wkly intervals, until satisfactory therapeutic levels are achieved

Ramsay Hunt syndrome Hunt syndrome ENT A condition that occurs when herpes zoster affects the auditory nerves CLINICAL Intense ear and mastoid pain, facial nerve paralysis, hearing loss, vertigo, tinnitus, aguesia–loss of taste, dry mouth, dry eyes, herpetic lesions in the auditory canal, mouth, face, neck, and scalp MANAGEMENT Analgesics; pain may be refractory PROGNOSIS Residual loss of function is the norm. See Herpes zoster

random Occurring by chance alone–ie, not by design, pattern, plan, or selection CLINICAL TRIALS Referring to a formal chance process in which previous events have no bearing on future events. See Random allocation, Randomized trial

random glucose ENDOCRINOLOGY Serum glucose level obtained without considering the timing of meals. See Glucose tolerance test

random sample STATISTICS A population sample derived by selecting individuals, such that each has the same probability of selection; a group of subjects selected in such a way that each member of the population from which the sample is derived has an equal or known chance–probability–of being chosen for the sample. See Sampling

randomization Random assignment, random allocation, randomized allocation STATISTICS The selection of subjects or samples for each 'arm' of a study or experiment based on chance alone–ie, a theoretical coin toss, which is intended to minimize the influence of irrelevant details and selection bias, and produce statistically valid data. Cf Convenience sample

randomized clinical trial Randomized controlled trial RESEARCH A clinical trial–eg, of a therapeutic agent's efficacy, in which Pts are randomly assigned to different–treatment, placebo, or 'gold standard'–arms of a study. See Equipoise, Nocebo, Placebo, Randomization

range of affect NEUROLOGY The palette of a person's facial expression and emotiveness

range of motion PHYSICAL EXAM The range through which a joint can be moved, usually its range of flexion and extension, as determined by the type of joint, its articular surfaces, and that allowed by regional muscles, tendons, ligaments, joints and physiologic control of movement across the joint

ranitidine Zantac™ THERAPEUTICS An H_2-receptor antagonist used to treat peptic ulcer disease, gastric and duodenal ulcers, esophageal erosions, GERD, and conditions–eg, Zollinger-Ellison disease and systemic mastocytosis in which there is ↑ H_2 activity due to blocking of the binding of histamine to H_2 receptors, resulting in ↓ intracellular concentration of cAMP and acid secretion by gastric parietal cells ADVERSE REACTIONS Diarrhea, headache, fatigue, myalgia, constipation, confusion, agranulocytosis, gynecomastia, impotence, allergic reactions, tachycardia, arrhythmias, interstitial nephritis, etc. See H_2 blockers, Histamine receptor antagonists

rank list GRADUATE EDUCATION A list created at a residency program by its selection committee that places 1st-yr candidates in order of qualifications and desirability. See the Match

ranolazine CARDIOLOGY An anti-anginal metabolic modulator, which inhibits partial fatty acid oxidation, maintaining glucose oxidation without lactic acid buildup. See CARISA

rapamycin Sirolimus, see there

rape FORENSIC MEDICINE An unlawful, nonconsensual act of sexual intercourse carried out by force or other forms of duress. See Date rape, Prison rape, Rape-trauma syndrome, Rohypnol, Spousal rape, Statutory rape

rape trauma syndrome PSYCHOLOGY An acute stress reaction to a life-threatening situation in which sexual assault was attempted or successful. See Date rape, Rape, Sexual assault

rape kit FORENSIC MEDICINE A collection of receptacles–cups, envelopes, plastic bags, tubes, disposable items–cotton swabs, napkins, pipettes and tools–sterile comb for pubic hairs, sheets–used to obtain evidentiary specimens from a rape victim. See Date rape, PERK kit, Rape, SARN, SART

RAPER MEDTALK An acronymm for radiologists, anesthesiologists, pathologists, emergency room physicians, who are usually hospital-based specialists

RAPID CARDIOLOGY A clinical trial–Retaplase–rtPA v Alteplase–tPA Infusion in Acute Myocardial Infarction–that compared 2 thrombolytic regimens for reperfusing Pts with acute MI. See Acute myocardial infarction, tPA, rtPA

rapid eye movement sleep See REM sleep, Sleep stages

rapid sequence induction Rapid sequence induction of anesthesia with cricoid pressure ANESTHESIOLOGY A maneuver in which anesthesia is rapidly induced by thiopental, then succinylcholine, at the same time that a scrub nurse, anesthesia assistant or other applies manual pressure over the cricoid cartilage at the level of the C6 vertebra, which closes the esophagus until a cuffed endotracheal tube is established in the Pt's airway. See Gastric aspiration

rapid urease test CLO test, see there

rapidly progressive glomerulonephritis Crescentic glomerulonephritis, membranous glomerulonephritis, necrotizing glomerulonephritis NEPHROLOGY A type of kidney disease characterized by a rapid loss of renal function, with crescent-shaped deposits in at least 75% of glomeruli seen on kidney biopsies CLINICAL Acute nephritic syndrome or unexplained renal failure with progression to ESRD; RPGN occurs in ±1/10,000 people, especially between age 40 and 60 ASSOCIATED DISEASES Vasculitis, polyarteritis, abscess of internal organ(s), collagen vascular disease–eg, SLE, Henoch-Schönlein purpura, Goodpasture syndrome, IgA nephropathy, membranoproliferative GN, antiglomerular basement membrane antibody disease, Hx of malignancy or blood or lymphatic disorders, and exposure to hydrocarbon solvents. See End-stage renal disease, Glomerulonephritis

rapidly progressive glomerulonephritis with pulmonary hemorrhage Goodpasture syndrome, see there

rapids riding SPORTS MEDICINE An 'extreme sport' in which the participant rides white water rapids on a polystyrene float with a joystick INJURY RISK Concussions, drowning. See Extreme sports, Novelty seeking behavior

RapiSeal™ patch SURGERY A biodegradable collagen-based patch for creating a hemostatic seal on solid organs without stressing surrounding tissue. See Fibrin sealant

rapport PSYCHIATRY Harmonious accord and mutual responsiveness that contributes to a Pt's confidence in the therapist and willingness to work cooperatively

RAPPORT CARDIOLOGY A clinical trial–ReoPro® and primary PTCA organization and randomized trial

Rapunzel syndrome A fanciful term for the Sx of a massive trichobezoar caused by trichotillomania and trichophagia, seen mainly in mentally retarded or deranged ♀ CLINICAL Epigastric pain, bloating, N&V RADIOLOGY Mass lesion, usually gastric with 'strands' of hair extending into upper small intestine LAB Hypochromic microcytic anemia–hair is an iron chelator TREATMENT Surgery. See Bezoar

rare disease See Orphan disease

rare hemoglobinopathy HEMATOLOGY A hemoglobinopathy that affects a limited number of kindreds. See Hemoglobinopathy

RARS Refractory anemia with ringed sideroblasts. See Idiopathic sideroblastic anemia

rasburicase Elitek™ ONCOLOGY An agent used in CA Pts with ↑ uric acid due to tumor lysis

rash An exanthem or skin eruption. See Butterfly rash, Heliotrope rash, Herpetiform rash, Hot tub rash, Maculopapular rash, Morbilliform rash. Cf Petechia, Purpura

raspberry tongue A descriptor for the characteristic enanthema of scarlet fever, in which the tongue is bright red with edematous white papillae. Cf Strawberry tongue

RAST Radioallergosorbent test IMMUNOLOGY A 'solid phase' radioisotopic method for quantifying specific allergenic IgE antibodies in serum, which is similar to an agglutination test. See Mail-order medicine, Patch testing

rat INFECTIOUS DISEASE A rodent, genus *Rattus*, which is a vector and/or reservoir of disease–eg, Bunyaviridae, black plague, rat-bite fever VOX POPULI A dishonorable person. See Lab rat, Weasel

rat-bite fever INFECTIOUS DISEASE Either of 2 similar systemic bacterial infections caused by different gram-negative facultative anaerobes 1. Streptobacillary RBF–caused by infection with *Streptobacillus moniliformis*, transmitted in crowded urban environs by ingesting milk contaminated by rat feces; *S moniliformis* is present in the nasopharynx of up to 50%

of healthy lab and wild rodents CLINICAL 2-10 day incubation followed by an irregularly relapsing fever and asymmetric polyarthritis followed within 2-4 days by a maculopapular rash on extremities, palmoplantar surfaces; bite wound heals spontaneously; headache, N&V, myalgia, minimal regional lymphadenopathy, anemia, endocarditis, myocarditis, meningitis, pneumonia, and focal abscesses may occur; most cases resolve spontaneously within 2 wks, 13% of untreated cases are fatal DIAGNOSIS Streptobacillary RBF can only be diagnosed by blood culture; *S moniformis* is a slow grower with strict growth requirements, making it difficult for most labs to culture; no serologic test is available; biological false positives for syphilis occur MANAGEMENT Penicillin, tetracycline, streptomycin 2. Spirillary RBF caused by *Spirillum minus* which occurs worldwide, but is most common in Asia; it has a longer incubation period (1-3 weeks), rare arthralgia, and an inoculation wound which can reappear at the onset of Sx or persist with edema and ulceration MANAGEMENT Penicillin, tetracycline, streptomycin. See Sodoku

rat tail tapering IMAGING HEART A descriptor for a smooth progressive narrowing to a point of maximal stenosis of the left anterior descending coronary artery, seen by angiography. Cf Bridging LUNG A descriptor for the abrupt loss of normal bronchial arborization in bronchography due to intraluminal neoplasms–eg, bronchogenic carcinoma PANCREAS A descriptor for the abrupt narrowing of the pancreatic duct in ERCP, a finding in pancreas CA. See Endoscopic retrograde cholangiopancreatography

rate The number of events divided by the period of time over which they occur. See Average payment rate, Basal metabolic rate, Basic pacing rate, Baud rate, Case rate, Composite rate, Erythrocyte sedimentation rate, False negative rate, False positive rate, Fetal heart rate, Glomerular filtration rate, Graft survival rate, Growth rate, Heart rate, Instantaneous rate, Minute volume & respiration rate, Platelet production rate, Pulse rate, Respiratory rate, Secondary attack rate, Sedimentation rate, Slew rate, Success rate, Urinary flow rate, Upper rate EPIDEMIOLOGY An expression of the frequency with which an event–eg, disease or death–occurs in a defined population. See Adjusted community death rate, Age-adjusted death rate, Age/sex rate, Basic reproduction rate, Birth rate, Cause-fatality rate, Cause-specific mortality rate, Crude birth rate, Crude death rate, Crude mortality rate, Death rate, Fertility rate, Fetal mortality rate, Hospital mortality rate, Incidence rate, Infant mortality rate, Maternal mortality rate, Morbidity rate, Mortality rate, Neonatal mortality rate, Postneonatal mortality rate, Prevalence rate, Rate-specific mortality rate, Sex-specific mortality rate, Total fertility rate. Cf Ratio

rate-adjusted mortality index The expected in-hospital mortality rate, which is based on actual in-hospital rates for diagnoses, grouped by their diagnosis-related group–DRG code and adjusted for age, race, sex, the presence of co-morbidities and main surgical operative procedure(s). See DRG

rate of change The amount of change occurring in a interval divided by the timespan

rate response curve CARDIAC PACING A graphic representation of the relationship between the rate of response of the pacemaker–pacing rate, to detected activity levels or other physiologic parameter. See Rate responsive pacing

rate responsive pacing Rate adaptive pacing, Rate variable pacing CARDIAC PACING Artificial pacing in which pacemakers change pacing rate to meet metabolic demand for ↑ circulation, in response to changes in objective parameters, detected by sensors

rating VOX POPULI The comparison or a thing or process with another. See Community rating, Experience rating, Hazard risk rating, Modified community rating, RISCC rating, SPF rating, Therapeutic rating

ratio VOX POPULI The value obtained by dividing one quantity by another; a measure of association calculated by dividing one amount by another. See Abortion ratio, A/G ratio, A/P ratio, AT/GC ratio, Benefits-to-cost ratio, Branching ratio, Case-to-infection ratio, Case-fatality ratio, CD4/CD8 ratio, Cross-match/transfusion ratio, Death-to-case ratio, Duty ratio, Effective reproductive ratio, Exposure odds ratio, Flexor/extensor ratio, G:C ratio, HDL/LDL ratio, Helper:Suppressor ratio, HERT ratio, International Normalized ratio, Lecithin/sphingomyelin ratio, Likelihood ratio, Magnetization transfer ratio, Maternal mortality ratio, Medical loss ratio, Myeloid:Erythroid ratio, Nuclear:cytoplasmic ratio, Odds ratio, Packing ratio, Potency ratio, Prevalence ratio, Proportionate mortality ratio, Rate ratio, Signal-to-noise ratio, Reproduction ratio, Risk ratio, Sex ratio, Solubility ratio, Standardized mortality ratio, Thyroid hormone binding ratio, Urea reduction ratio, Waist-to-hip ratio

rationalization PSYCHOLOGY A defense mechanism in which a person attempts to justify or make consciously tolerable by plausible means, feelings, behavior, or motives that are otherwise intolerable; an explanation or justification for one's actions. See Ego defense mechanism, Projection

rationing MANAGED CARE The allocation or distribution of a scarce product, commodity or service. See Age-based rationing, Health care rationing, Oregon plan, Red-tape rationing

rattlesnake A pit viper of the genera *Crotalus* or *Sistrurus*, primarily of the southern Northern hemisphere, which has a characteristic tail that produces a rattling sound when its owner is disturbed

rave Rave party SUBSTANCE ABUSE A social activity of recent vintage consisting of an all-night dance session at a club or party, often accompanied by the ingestion of recreational doses–ie, not overdose levels–of the 'designer' drug of abuse, ecstasy–MDMA. See Ecstasy

RAVEL CARDIOLOGY A clinical trial–Randomized Study with the Sirolimus-Coated Bx Velocity Balloon-Expandable Stent in the Treatment of Patients with de Novo Native Coronary Artery Lesions CONCLUSION The treated stent effectively prevents neointimal proliferation, restenosis, and associated clinical events. See Sirolimus-eluting coronary stent

raw milk Unpasteurized milk PUBLIC HEALTH Milk that has not been pasteurized, which is associated with infection by *Campylobacter jejuni*, *Salmonella* spp–*S dublin, S enterica* Typhimurium DT104, *S typhimurium, S derby, Brucella* spp, *E coli, Listeria monocytogenes, M bovis, M tuberculosis, Corynebacterium pseudotuberculosis, Staphylococcus aureus, Streptococcus* spp, *Streptobacillus moniliformis, Yersinia enterocolitica*, brucellosis, hemorrhagic escherichiosis, Brainerd diarrhea; RM from other ruminants is equally susceptible to microbial infections–eg, toxoplasmosis–goat's milk, tick-borne encephalitis–sheep's milk. See Certified milk; Cf Milk, White beverages

Raxar® Grepafloxacin, see there

Ray TFC™ Ray threaded fusion cage ORTHOPEDICS A device used in posterior lumbar interbody fusion for stabilizing and maintaining height of disk space after discectomy. See Discectomy

Raynaud phenomenon Raynaud's disease CARDIOVASCULAR DISEASE A condition characterized by vasospasm of small vessels of the fingers and toes, resulting in skin discoloration ETIOLOGY Extreme temperatures–especially cold or hot or emotional events; initially, digits involved turn white because of ↓ blood supply, then blue because of prolonged hypoxia, then red, when the blood vessels reopen, causing a local flushing

Raysedan Chlordiazepoxidem, see there

RBBB Right bundle-branch block, see there

RBC Red blood cell, erythrocyte. See Nucleated RBC

RBE Relative biological effectiveness, see there

RBRVS Resource-based relative value scale MANAGED CARE A 'work unit' used to determine the value of various physicians' labor. See Medicare, Physician reimbursement

RCA 1. Red cell agglutination 2. Right coronary artery

RCS Read classification system, see there

RD 1. Radiation absorbed dose 2. Registered Dietician 3. Relative density 4. Renal disease 5. Respiratory disease 6. Retinal detachment 7. Ruptured disk

RD Recommended daily allowance, see there

RDS Respiratory distress syndrome, see there

RDT 1. Renal dialysis treatment 2. Retinal damage threshold

reabsorption PHYSIOLOGY The selective passage of certain substances–glucose, proteins, sodium–back into the circulation after it had been secreted into the renal tubules. See Countertransport

REACH CARDIOLOGY A clinical trial–Research on Endothelin Antagonism in Chronic Heart Failure

REACT CARDIOLOGY A community-based clinical trial–Rapid Early Action for Coronary Treatment–which compared the time it took a population educated on how to respond to a heart attack to one lacking such education. See Coronary artery disease, Quinapril

reaction VOX POPULI A response in a chemical, immunologic, physiologic, psychological, or other interaction. See Allergic reaction, Anaphylactic reaction, Anniversary reaction, Anxiety reaction, Conversion reaction, Cortical reaction, Cross-reaction, Downgrading reaction, Fixed drug reaction, Harlequin skin reaction, Hemagglutination inhibition reaction, Hemolytic transfusion reaction, Hysterical reaction, Late phase reaction, Leukemoid reaction, Leukoerythroblastic reaction, Leukotriene reaction, Magnet reaction, Mourning reaction, Myasthenic reaction, Nonhemolytic transfusion reaction, Parachute reaction, Pseudomyotonic reaction, Uphill reaction, Vasopermeability reaction, Wheal-and-flare reaction

reaction formation PSYCHIATRY An unconscious defense mechanism in which a person adopts affects, ideas, attitudes, and behaviors opposite of conscious or unconscious impulses

reactive airway disorder Asthma, see there

reactive arthritis Reiter syndrome, see there

reactive attachment disorder of infancy CHILD PSYCHOLOGY A disturbance of social interaction due to neglect of a child's basic emotional needs–eg, related to multiple care givers, preventing bonding; risk of RAD ↑ with isolation, poor parenting, teen parents, or a mentally retarded caregiver. See Day care center

reactive depression Exogenous depression PSYCHIATRY A situational depression triggered by a recent stressful external life event–eg, loss of a loved one, which is of greater duration than is regarded as normal. See Depression. Cf Endogenous depression

reactive hypoglycemia Plasma glucose measuring < 2.8 mmol/L–US: < 50 mg/dl with Sx of adrenergic neural activation–eg, weakness, palpitations, tremor, sweating and hunger, occurring after a meal or after oral glucose loading, caused by compensatory insulin hypersecretion

reactive leukocytosis Inflammatory leukocytosis HEMATOLOGY A non-neoplastic ↑ of peripheral WBC > 10,000/mm³, usually accompanied by fever, focal signs of infection, normal platelet count, normal Hb, paucity of immature WBCs TYPES OF RL Basophilia, eosinophilia, lymphocytosis, monocytosis, neutrophilia ETIOLOGY Pathogens, drugs, toxins. Cf Leukemoid reaction, Leukocytopenia

reactive marrow Reactive bone marrow A descriptor for a polyclonal BM response to a local or systemic 'insult', often inflammatory, which may be confined to one cell line, as in reactive granulocytosis, reactive mast cell hyperplasia, reactive thrombocytosis. See Bone marrow

reactive perforating collagenosis An AR condition of childhood onset characterized by recurring umbilicated papules due to an ill-defined collagen defect, which is triggered by trauma and exacerbated by cold weather

reactive thrombocytosis Reactive hyperthrombocytosis platelet count of ≥ 800 x 10⁹/L–US = ≥ 800 000/µl, seen in ± 1:200 of hospital Pts Associations Acute and chronic inflammation–eg, RA, SLE, malignancies, neonatal RDS, and after hemorrhage, surgery, trauma; in some cases of RHT–eg, burns, hypothermia, preeclampsia, low platelet numbers are present. Cf Essential thrombocytosis, Respiratory distress syndrome

reader ACADEMIC MEDICINE An academician in the UK who is an expert in a specific area of medicine. See Lecturer INFORMATICS A device that can access data from a machine-readable storage source–eg, a device that reads magnetic stripe cards. See OCR reader, Scanner

reading disability Reading disorder NEUROLOGY A difficulty in learning to read despite normal IQ and opportunity to learn with competent instruction, in absence of general health problems, emotional disturbances or sensory defects. See Dyslexia

reading disorder See Dyslexia, Reading disability

reading retardation PEDIATRICS Reading difficulties due to mental retardation or cultural deprivation. Cf Dyslexia, Reading disability

readmission MANAGED CARE The admission of a Pt to a health care facility for a condition–eg, stroke, MI, GI bleeding, hip fracture, cancer surgery, shortly after discharge. See nth admission. Cf Admission, Discharge

reagent LAB MEDICINE A substance or material used in a reaction to detect or measure substances of interest; reagents are integral and standardized participants in reactions or detection methods–eg, GLC, HPLC, GC-MS. See Lot-analyzed reagent, Maximum impurities reagent, Raichem® cholesterol rapid liquid reagent

reagent strip Dipstick, see there

REAL classification Revised European-American Lymphoma classification HEMATOLOGY A new system for classifying lymphomas, proposed by the Intl Lymphoma Study group, intended to classify lymphoma that are known clinical entities; unique to the REAL classification is recognition of arbitrary nature of the distinction between lymphocytic leukemias and certain lymphomas. See Lymphoma. Cf Working Formulation

real miss GYNECOLOGY A popular term for a pap smear of the cervix that was initially interpreted as negative which, upon re-examination, has obviously abnormal cells. See Pap smear, Undercall

real time *adjective* Referring to live action displayed on a monitor that is not subject to delay INFORMATICS Computer communications or processes that are so fast they seem instantaneous

'real-time' imaging Visualization of a dynamic process µsecs after occurring, which requires rapid information processing–ie, as the process occurs, as in 'B' mode ultrasound

real-time telemetry CARDIOLOGY A method that provides access to the measured performance, parameters, and detected signals of an implanted pacemaker–eg, lead impedance, battery voltage, marker channel, EKG. See Interrogation

reality testing PSYCHIATRY The ability to evaluate the external world objectively and differentiate it from the internal world. See Psychosis

REALM Rapid Estimate of Adult Literacy in Medicine MEDICAL COMMUNICATION An instrument used to rapidly assess a Pt's literacy skills and, presumably, ability to understand medical information. See Language mismatch

reappointment HOSPITAL PRACTICE The renewal of medical staff membership and privileges of a practitioner whose previous service on the medical staff has met the staff's standard of Pt care . See Appointment

rearfoot valgus Calcaneal valgus ORTHOPEDICS An everted rearfoot See Valgus

reason for visit AMBULATORY CARE A raison d'être of a medical encounter in an outpatient setting, equivalent to chief complaint(s) in Pts admitted to the hospital. See Chief complaint

reasonable & necessary MEDICARE *adjective* Referring to services that Medicare finds to be reasonable & necessary–ie, will reimburse–for managing a Pt. See Advance beneficiary notice, Medicare fraud

reasonable accommodations A standard of providing for a worker's or customer's needs, as mandated by the ADA, which requires that a business make appropriate changes in the environment to accommodate those with mental or physical disabilities as long as such changes do not create an undue–financial–burden to the employer. See Americans with Disabilities Act, Architectural barriers, Disability, Handicap

reasonable person standard Reasonable man standard LAW & MEDI-CINE A standard of behavior that is appropriate and expected for a mentally stable or 'reasonable' person under particular circumstances. See *Canterbury* v *Spence*, Contributory negligence, Negligence

reasonable physician standard Reasonable practitioner standard FORENSIC MEDICINE A standard of disclosure of information used in the wording of informed consent documents, based on customary practice or what a reasonable practitioner in the medical community would disclose under the same/similar circumstances. Cf *Arato* v *Avedon*, *Nathanson v. Kline*, 'Patient viewpoint' standard

reasonable prudence FORENSIC MEDICINE A standard of care which derives from a legal doctrine expounded upon by Judge Learned Hand in 1932 which has become a founding principle of medical malpractice law. See Negligence

reasoning MEDTALK The thought process behind an action. See Case-based reasoning, Symbolic reasoning

reattachment The reanastomosis of a thing detached. See Penile reattachment

rebeccamycin ONCOLOGY An anticancer antibiotic and topoisomerase I inhibitor which may be of use in managing CA of the gallbladder. See Daunomycin

Rebetol® Ribavirin, see there

Rebetron® VIROLOGY An antiviral that combines Rebetol®–ribavirin and Intron® A–recombinant IFN alfa-2b injection for treating chronic HCV in Pts with compensated liver disease. See Hepatitis C virus, Interferon

Rebif® Recombinant interferon beta-1a, see there

rebound *adjective* Referring to a reversed response when a stimulus is withdrawn. See REMS rebound PHYSICAL EXAMINATION A technique of clinical evaluation in which the forehand is pressed firmly on the abdomen and released. See Rebound tenderness

rebound effect The worsening of Sx when a drug–eg, a decongestant, is discontinued, attributed to tissue dependence on the agent

rebound hypersecretion GASTROENTEROLOGY A phenomenon that occurs when H_2-receptor antagonist therapy is discontinued and gastric acid secretion is transiently in excess. See Histamine receptor antagonists

rebound insomnia An ↑ in insomnia above a baseline, which may appear if long-term hypnotic therapy is abruptly stopped; the effect is greater with short-acting hypnotics. See Insomnia

rebound scurvy A vitamin C-dependency state that occurs in the fetus of a woman taking megadoses of vitamin C during pregnancy, caused by an ↑ production of vitamin C-metabolizing enzymes. See Vitamin C

rebound tenderness CLINICAL MEDICINE The presence of pain that is more intense when the examiner releases pressure when palpating the abdomen

recalcitrant *adjective* Poorly responsive to therapy

recall NEUROLOGY *noun–pronounced* ree CALL The process of bringing a memory into consciousness; the recollection of past facts, events, feelings; invoking of the memory of experiences or learned information *verb–pronounced* ricall To remember experiences or learned information. See Class recall, Immediate recall, Memory PUBLIC HEALTH *noun–pronounced* REE call, drug recall A public announcement by a manufacturer or producer of a particular product–eg, motor vehicles, toys, drugs, medical devices, foods, asking the purchaser of a particular 'lot,' or model to return the goods as they may have defects posing a health hazard

recall bias Memory bias PSYCHOLOGY A type of information bias that occurs when there are differences in how exposure groups or disease groups remember certain information. See Bias. Cf Interviewer bias

recent completed small stroke NEUROLOGY A stroke in which the signs and Sx of focal unilateral cerebral ischemia did not fully resolve within 24 hrs. See Stroke, Transient ischemic attack

recent memory NEUROLOGY Memory between short-term and long-term memory

recertification Recredentialing GRADUATE EDUCATION A process in which a professional is periodically re-evaluated–eg, every 10 yrs by an accrediting body to assure continued provision of safe, high-quality health care

recessive GENETICS A genetic trait that is not phenotypically expressed in a heterozygous or partially heterozygous cell, but rather only in a homozygous or hemizygous state. See Phenotype, Trait. Cf Dominant

RECIFE CARDIOLOGY A clinical trial–Reduction of Cholesterol in Ischemia & Function of Endothelium

recipient MANAGED CARE A person eligible for Medicaid benefits. See Beneficiary TRANSFUSION MEDICINE A Pt who receives a blood product. See Universal recipient TRANSPLANTATION A Pt who has received a donated tissue or organ

reciprocal inhibition PSYCHIATRY In behavior therapy, the hypothesis that if anxiety-provoking stimuli occur simultaneously with inhibition of anxiety–ie, relaxation, the link between the stimulus and the anxiety is weakened

reciprocity MEDICAL EDUCATION The recognition by one jurisdiction–eg, a state, of the validity of certificates and licenses–eg, to practice medicine issued by another jurisdiction

recognition PHARMACOLOGY See Drug recognition SUBSTANCE ABUSE See Drug recognition VOX POPULI The state or quality of being recognized or acknowledged. See Continuous speech recognition, Intelligent character recognition, Kin recognition, OCR, Open-set speech recognition, Speech recognition

recombinant *adjective* Referring to a structural rearrangement or 'shuffling' of genetic material that 1. Occurs normally during meiosis or 2. Is deliberately generated under controlled experimental conditions, as in recombinant DNA *noun* An organism with a combination of alleles from either parent due to crossing over or independent assortment of chromosomes during meiosis

recombinant DNA technology MOLECULAR BIOLOGY The chopping of DNA and moving the pieces, permitting direct examination of the human genome, and identifying the genetic components of various disorders; RDT is also used to develop diagnostic tests, drugs and biologicals for treating disease; the constellation of techniques that comprise 'genetic engineering', in which a gene producing a protein of interest from one organism is spliced into the genome of another organism–eg, a phage DNA integrated into a plasmid is inserted into a 'carrier' bacterium. See Genetic engineering, pBR322, PCR

recombinant human antihemophilic factor HEMATOLOGY A recombinant factor VIII for treating hemophilia A. See Hemophilia A

recombinant pharmacology MOLECULAR THERAPEUTICS The use of recombinant DNA techniques to produce DNA-derived biologicals **RP PRODUCTS** IFN-α, IFN-γ, interleukins, TPA, epidermal, fibroblast, platelet-derived growth factors, transforming growth factors, erythropoietin, GM-CSF, growth hormone, insulin, superoxide dismutase, TNF, factor VIII. See Biological response modifiers

recombinant protein MOLECULAR BIOLOGY A protein encoded by recombinant DNA or generated from a recombinant gene. See Recombinant pharmacology

recombinant toxin 'Magic bullet' A hybrid cytotoxic protein made by recombinant DNA technology, designed to selectively kill malignant cells. See Immunotoxin, Magic bullet

Recombivax HB® A recombinant hepatitis B vaccine. See Hepatitis B

recommended control range PUBLIC HEALTH A range–usually set by state regulation–within which adjusted fluoridated water systems should operate to maintain optimal fluoride levels. See Fluoridation

recommended daily allowance CLINICAL NUTRITION A guideline of essential nutrients recommended by the Food and Nutrition Board of the National Research Council, for daily ingestion in an idealized normal person engaged in average activities, in a temperate environment for optimal nutrition, in a 'standardized' 70 kg man; RDAs are ↑ for ↑ activity, body growth and size, pregnancy, lactation, and adverse environmental conditions; RDAs are designed for a state of wellness and are poorly applicable in sick, traumatized, or burned Pts. See Daily Value, Minerals, Vitamins

reconstruction plate ORTHOPEDICS A plate used to repair pelvic and calcaneal fractures and for posterior cervical spine fusion; RPs are malleable and can be trimmed to support fractures through complex bony surfaces. See Plate

reconstructive surgery Reconstruction SURGERY A procedure that attempts to restore a tissue as closely as possible to its original state. See Cosmetic surgery, Corrective surgery. Cf Mutilating surgery

record MEDTALK A permanent document in the form of a writing, files, photographs, etc. See Electronic medical record, Medical record, Medical administration record, Open record, Patient record, Problem-oriented medical record, Sealed record. Cf Chart

recorder EMERGENCY MEDICINE A nurse who records events–eg, intubation, medication administration, placement of endotracheal tube or IV line, etc–during CPR. See Cardiac event recorder, Cardiopulmonary resuscitation, Code Blue, Insertable loop recorder

recovered memory False memory, see there

recovery 1. Restoration of health and strength after an illness 2. The state a Pt is in after a therapeutic intervention; he/she is 'in recovery'. See Health recovery

recovery room Recovery unit A suite in a hospital next to the surgical suites, where recently operated Pts are monitored for a period of hrs to ensure recovery from anesthesia, physiologic stress of surgery, prevent post-surgical complications–eg, aspiration and suffocation, identify arrhythmias, hypotension, and other conditions and/or acute decompensation of pre-existing conditions

recreational drug SUBSTANCE ABUSE Any agent–most have significant psychotropic effects–used without medical indications or prescription in the context of social interactions–eg, parties. See Gateway drug, Ice, Rave

recreational therapy Play therapy *'Any free, voluntary and expressive activity…*(which may be)…*motor, sensory, or mental, vitalized by the expansive play spirit, sustained by deep-rooted pleasurable attitudes and evoked by whole emotional release.'* See Art therapy, Dance therapy, Laughter therapy, Music therapy

recredentialing Recertification, see there

recrudescence Reappearance of a disease after a quiescent stage; recurrence; relapse

recruitment MEDTALK The process of finding a suitable candidate for a position NEUROLOGY An ↑ in number of active motor units involved in a neuromuscular response, resulting from the temporal or spatial summation of a stimulus or from an ↑ intensity of a stimulus

rectal *adjective* Pertaining to the rectum *noun* A popular term for a digital rectal examination, see there

rectal abscess Anal abscess, anorectal abscess PROCTOLOGY An anorectal pus pocket due to an infection of an anal fissure, STD, or anal glands; deep RAs may be caused by intestinal disease–eg, Crohn's disease or diverticulitis; superficial abcesses may occur in infants with anal fissures MANAGEMENT Surgical drainage, antibiotics. See Anal fistula

rectal cancer See Colorectal cancer

rectal prolapse PROCTOLOGY The abnormal egress of the rectal mucosa through the anus, seen primarily in children < age 6. CLINICAL The prolapsed rectal tissue appears as a red 5–10 cm mass; mucosa is visible and may bleed ASSOCIATIONS Pinworms, whipworms, cystic fibrosis, celiac disease, constipation. Cf Uterine prolapse

rectal stenosis A narrowing of the rectosigmoid lumen, due to tumor or scarring

rectocele Proctocele SURGERY Herniation of part of the rectum into the vagina

recumbent MEDTALK Lying down

recurrence MEDTALK The return of Sx or of a tumor in the same site as the 1° tumor or in another location, after it had disappeared. See Extranodal recurrence, Implantation recurrence, Local recurrence, Marginal recurrence, Neurorecurrence, Regional recurrence, Transdiaphragmatic recurrence. Cf Remission

recurrent angina CARDIOLOGY Sharp precordial pain directly related to cardiac ischemia, seen in 3-5% of Pts with CABG, due either to progressive stenosis or occlusion of a CABG or progressive stenosis in an ungrafted artery. See Angina, CABG, Triple-bypass surgery

recurrent cancer ONCOLOGY A cancer that reappears in a site where it was eradicated or disappeared. Cf Remission, Residual cancer

recurrent cystitis UROLOGY Frequent/repeated episodes of cystitis with dysuria, urinary frequency, urgency, hematuria. See Urinary tract infection

recurrent seizure NEUROLOGY A repeated unprovoked seizure

recurrent UTI Recurrent urinary tract infection. See Urinary tract infection

recurrent vestibulopathy NEUROLOGY A condition characterized by Ménière's-like attacks of vertigo without audiologic Sx–eg, tinnitus, hearing loss, or sensation of fullness in ear; 15% of RV evolve toward Ménière's with addition of cochlear Sx MANAGEMENT Symptomatic control. See Ménière's disease

recuts SURGICAL PATHOLOGY Glass slides–GSs from paraffin-embedded tissue, obtained in addition to the original GSs, to either confirm the presence of a lesion identified on the first GSs or obtained for a second opinion, requested by the Pt or referring physician. Cf Levels

red bag *noun* INFECTION CONTROL A red plastic bag used for the disposal of 'nonsharp' and potentially infectious biohazardous waste by health care facilities *verb* To place biohazardous materials in a red bag. See Biohazardous waste

red blood cells Erythrocytes HEMATOLOGY The anucleated 'cells' which carry O_2 in the circulation to the peripheral tissue

Red Book INFECTIOUS DISEASE A publication of the Am Acad Pediatrics that contains recommendations and immunization schedules for all licensed vaccines, information on HBV control, hemophilus and measles, TB treatment, AIDS guidelines, recommendations, information on STD and infection control in day-care settings and hopitals. See Childhood immunizations PHARMACEUTICAL INDUSTRY Drug Topics Red Book A reference text used by pharmacies which provides the average wholesale price of drugs and other therapeutic agents

red cell count Erythrocyte count HEMATOLOGY The number of RBCs per volume of blood, measured in microliters–µL or cubic millimeters–mm³; at birth, the RCC is ↑, which is followed shortly by a ↓ that 'bottoms out' at ± 2 months of age, then slowly rises to adult levels; polycythemia is an ↑ in RBCs of any cause, which may be either neoplastic, as in polycythemia vera, or nonneoplastic–erythrocytosis or 2° polycythemia, due to various factors–high altitudes, smoking, cyanotic heart defects, and COPD; ↓ RBC count–anemia can, like polycythemia, be physiologic–as in marathon runners or pathologic REF RANGE ♂: 4.1-5.4 x 10^{12}/L–US: 4.1-5.4 x 10^6/µL; ♀: 3.8-5.2 x 10^{12}/L: 3.8-5.2 x 10^6/µL. RBCs, WBCs, platelets are counted by automated devices. See Anemia, Hematocrit, Red cell indices

red cell fragmentation syndrome A form of hemolytic anemia due to intravascular mechanical trauma with destruction of RBCs, related to cardiovascular defects and hemolytic anemia

red cell indices A group of values obtained from automated blood cell counters that provide information about size–ie, volume and concentration of Hb in RBCs; 3 abbreviations are found on the printouts of CBCs: MCH, MCHC, MCV

RED CELL INDICES

MCH Mean corpuscular hemoglobin A measurement of Hb/individual RBC **Ref range** 26-34 pg/red cell

MCHC Mean corpuscular hemoglobin concentration A value derived on automated–eg Coulter cell counters from measured parameters **Ref range** 31-36 g/dL

MCV Mean corpuscular volumes A calculated value for the average volume of peripheral RBCs **Ref range** 85-100 fL/cell

red cell osmotic fragility See Osmotic fragility test

red eye 1. Conjunctivitis evoked by allergens, bacteria, viruses, air-borne irritants, or linked to episcleritis, corneal ulcer–infectious or traumatic, uveitis, glaucoma–acute or chronic, cellulitis, etc. Cf Pink-eye 2. Blood-shot eye

'red flag' PEDIATRICS A sentinel finding which may indicate a more extensive disease process. See Developmental red flag SCIENCE JOURNALISM An article in a peer-reviewed scientific journal which, when published, reaches startling conclusions that are likely to evoke wide interest and therefore appear in the lay news media–eg, Washington Post, NY Times, and others, which may be reported less for their scientific value than for their sensationalist impact. See 'Media epidemic', Peer-reviewed journal SOCIAL MEDICINE A popular term for subtle physical findings and behavioral patterns may indicate domestic abuse. See Domestic violence, Wife-abuser

red-free filter GYNECOLOGY A green or blue filter used in colposcopy, which ↑ contrast between blood vessels and background, improving evaluation of superficial vessels. See Colposcopy

red herring MEDTALK An unusual clinical, radiologic or pathologic finding that should be ignored in the context of a Pt's disease. See Sutton's law. Cf 'Zebras'

red hot throat A nonspecific descriptor for an erythematous, acutely inflamed oropharynx seen in various infections–eg, *Streptococcus pyogenes*, *N gonorrheae*, *Corynebacterium diphtheriae*, *Bordetella pertussis*, *H influenzae*, as well as various viral infections

red ink HEALTH ADMINISTRATION A popular term for financial losses . Cf in the Black

red man syndrome HEMATOLOGY L'homme rouge A term referring to the skin changes of Sézary syndrome, characterized by generalized erythroderma, and may be accompanied by typical plaques or tumor nodules INFECTIOUS DISEASE A centripetal maculopapular erythema of abrupt onset accompanied by hypotension that may occur after rapid IV infusion of vancomycin for gram-negative septicemia in neutropenic Pts, resulting in histamine release, flushing or cardiac arrest; the rash may involve the head and neck–'red neck' syndrome or large areas of the body, and affect ♀–'red person' syndrome and resolves spontaneously in mins to hrs. Cf Red skin

red meat CLINICAL NUTRITION Carcass meat–skeletal muscle from grazing livestock–a major source of protein in the western hemisphere. Cf White meat

'red-out' A homogeneous red-orange color seen by GI endoscopy when the intestinal mucosa directly covers the endoscope's lens, allowing only the passage of light across the blood vessel-rich mucosa; red-out can be corrected by a puff of air to push the mucosa away from the visual field

red pulp disease An infiltrative process of the spleen that preferentially affects red pulp **Examples** CML, heavy chain disease, iron deficiency, hairy cell leukemia, malignant histiocytosis, rheumatoid arthritis. See White pulp disease

red sauce NUTRITION Any low-fat, low-calorie tomato-based sauce. Cf White sauce

red skin Reddened skin of any cause, which may be caused by intoxication by boric acid, carbon monoxide, cyanide, atropine, scopolamine. Cf Red man syndrome

red urine disease Any condition associated with red urine–eg, hematuria 2° to glomerulonephritides, bladder tumors, foreign bodies or calculi, infection, inflammation, thrombosis and other condition diagnosed by urinalysis. See Urinanalysis

redneck heroin See Oxycodone

redness VOX POPULI Erythema

re-do vascular surgery SURGERY Any repeated surgery performed in the renal, aorto-iliac, infrainguinal areas

Redomex Amitriptyline, see there

reduction COSMETIC SURGERY The surgical excision of redundant tissue and skin. See Reduction mammoplasty, scalp reduction OBSTETRICS See Multifetal pregnancy reduction, Selective reduction ORTHOPEDICS The positioning of displaced parts–eg, surgical or manipulative repositioning of dislocated bones in a joint. See Closed reduction, Open reduction.

reduction mammoplasty SURGERY A procedure intended to ↓ size of hypertrophied and/or ptotic breasts. See Mammoplasty. Cf Augmentation Mammoplasty

redundancy MEDTALK The duplication of time-sensitive or failure-prone equipment–eg, computer server, surgical supplies, etc–allowing the system or surgeon to switch to a backup device with minimal compromise in Pt management or critical systematic functions or activities. See Laparoscopic surgery

redundant test Redundant testing LAB MEDICINE A test that has already been performed on the same Pt in a brief time period. See Panel, Reflex testing

Redux® Dexfenfluramine, fenfluramin OBESITY An agent used to manage obesity, withdrawn from the market after it was linked to heart valve abnormalities and other heart complications. See Fen-Phen, Obesity

reemerging disease GLOBAL MEDICINE Any disorder, usually an infection–eg, cholera, malaria, TB, which was on the decline in the global population, reached a nadir and has now increased due to changes in the health status of a susceptible population. Cf Emerging disease

reemerging infection INFECTIOUS DISEASE An infection that has resurged in incidence, after having been brought under control through effective health care policy and improved living conditions EXAMPLES Malaria, diphtheria, TB, dengue. Cf Emerging infections

reentrant arrhythmia CARDIAC PACING An arrhythmia caused by a depolarization that spreads through one part of the heart, and then circles back to depolarize the part of the heart from which it began. See Arrhythmia

reentry phenomenon CARDIOLOGY The reexcitation of a region of the heart by a single electrical impulse, which may cause ectopic beats or tachyarrhythmia; RE is the common cause of paroxysmal atrial or supraventricular arrhythmia, which is coupled to premature ventricular depolarization

reentry tear VASCULAR SURGERY The most distal intimal tear in a dissecting aortic aneurysm, where the blood is presumed to return to the circulation. See Dissecting aneurysm

reference *adjective* Referring to a standard or norm *noun* MEDICAL COMMUNICATION *noun* 1. A note in an article or publication that refers the reader to another passage or source 2. An entry in a bibliography; a citation of previously published material, which includes author names, title of the article, journal, yr, volume and pages in which it was published

reference range LAB MEDICINE A set of values established as normal maximums or minimums for a given analyte. Cf Decision levels, Panic values

referral MANAGED CARE A formal process that authorizes an HMO member to get care from a specialist or hospital; most HMOs require Pts to get a referral from their primary care doctor before seeing a specialist MEDICAL PRACTICE 1. A Pt who has been sent–referred–for a 2nd opinion or therapy to a specialist or subspecialist with greater expertise, as the Pt has a disease or condition that the primary or referring physician cannot, or does not wish to, treat. See Second opinion 2. The sending of a Pt to another physician for ongoing management of a specific problem, with the expectation that the Pt will continue seeing the original physician for coordination of total care. See Negligent referral, Self-referral. Cf Consultation

referral center bias DEMOGRAPHICS A skewing in M&M statistics based on data generated from a referral and tertiary care center, which is rarely representative of the population that a hospital of a similar size might service. See 'Institutional effect,' Tertiary care center

refill *noun* A second allotment of a prescription agent obtained from a pharmacy, which is allowed by the original prescription *verb* PHARMACOLOGY To obtain more of a particular drug, after the initially prescribed amount of the agent has been used or administered. See Prescription

REFLECT CARDIOLOGY A clinical trial–Randomized Evaluation of Flosequinan on Exercise Tolerance that evaluated the efficacy of flosequinan, a quinolone

that acts via the inositol triphosphate pathway, in managing Pts with CHF. See Congestive heart failure, Flosequinan. Cf FACET

reflective communication Client-oriented psychotherapy in which the client is afforded the opportunity to examine behaviors and interactions with others, while the therapist acts as a verbal 'mirror,' often restating what the client has just said. See Humanistic psychology, Natural psychotherapy

reflex NEUROLOGY A rapid involuntary response to a mechanical or chemical stimulus. See Ankle reflex, Axon reflex, Babinski's reflex, Baroreflex, Cat's eye reflex, Consensual light reflex, Corneal reflex, Deep tendon reflex, Diving reflex, Doll's eye reflex, Gag reflex, Galant reflex, Gastrocnemius reflex, H reflex, Hering-Breuer reflex, J reflex, Let-down reflex, Moro reflex, Patellar reflex, Peristaltic reflex, Pulmonary chemoreflex, Pupillary reflex, Rooting reflex, Triceps reflex, Vestibulo-ocular reflex, Westphal-Piltz reflex

reflex arc NEUROLOGY The basic conduction pathway through the nervous system, consisting of a sensory neuron, an association neuron, and a motor neuron

reflex diagnostic testing Algorithmic testing, contingent testing, reflex testing, reflexive ordering, reflexive testing LAB MEDICINE The performance of a test on a Pt specimen only after a particular analyte is abnormal or outside of predetermined range. Cf Red flag

reflex sympathetic dystrophy Causalgia, complex regional pain syndrome, post-traumatic pain syndrome, reflex neurovascular dystrophy, shoulder-hand syndrome, Sudeck's bone atrophy NEUROLOGY Persistent pain of an extremity after prolonged autonomic nervous system stimulation ETIOLOGY AMI, cervical osteoporosis, CVAs, nerve injury, neurologic events, surgery, trauma CLINICAL Dysesthesia, pain, swelling of an extremity with trophic skin changes, hypertrichosis, osteoporosis MANAGEMENT Sympathetic block with local anesthetics SPORTS MEDICINE An exaggerated response of the sympathetic nervous system to minor trauma, especially if superimposed on healing injuries PATHOGENESIS Uncertain, possibly mental stress CLINICAL Severe, recurrent, chronic pain, affecting an entire extremity PROGNOSIS Inactivity and inadequate therapy result in muscle atrophy, demineralization, contractures. See Elite athlete, Female athlete triad

reflexology Ingham's method of compression massage FRINGE MEDICINE A massage therapy based on the belief that there are nerve endings–up to 72,000–for organs and body regions in the feet; reflexology consists of manual stimulation with the thumbs and fingers of cutaneo-organ reflex points on the feet as well as on the ears and hands; reflexologists believe that the body is divided into 10 distinct energy zones or energy channels that extend from the head to the feet, and that foot massage ↑ the flow of healing energy through any or all of these channels; this energy may be blocked by lumps of crystalline material which reflexologists claim to be able to identify during therapy. See Acupressure, Alternative medicine, Auricular reflexology, Bodywork, Hand massage, Vacuflex Reflexology System. Cf CranioSacral Therapy™, Integrative massage, Massage therapy

Refludan® Lepirudin HEMATOLOGY A recombinant hirudin for treating heparin-induced thrombocytopenia. See Heparin-induced thrombocytopenia

reflux MEDTALK The reversal of the normal flow of a fluid–eg, from the stomach into the esophagus. See GERD, Hepatojugular reflux. Cf Regurgitation, Vomiting PEDIATRICS Spit-up, see there

reflux episode GI TRACT An episode of esophageal pH of < 4.0 for ≥ 5 secs, a parameter used to define gastroesophageal reflux disease. See GERD

reflux esophagitis Gastroesophageal reflux disease, see there

reflux laryngitis A chemical inflammation/acid reflux caused by pooling of regurgitated gastric secretions on the laryngeal mucosa, characterized by "…hoarseness, persistent nonproductive cough, a sensation of pressure in

reflux nephropathy

the throat and a continual need to clear the throat. The classic Sx of reflux, such as heartburn and regurgitation are often minimal or absent" DIAGNOSIS ± Reflux, pH defects MANAGEMENT Omeprazole. See Acid-reflux disorders, GERD

reflux nephropathy Ureteral reflux, vesicoureteral reflux, vesicoureteric reflux NEPHROLOGY The retrograde flow of urine from the bladder into the ureters which occurs when the unidirectional valve-like flow between the ureters and bladder fails; bladder infection may cause pyelonephritis, and expose the kidney to high pressures, which over time damage the kidney and cause scarring ETIOLOGY Short or absent bladder wall tunnels, bladder infection, stones, bladder outlet obstruction, neurogenic bladder, abnormal ureters or number of ureters, following surgical reimplantation of ureters during kidney transplantation or due to ureteral trauma CLINICAL Repeated bladder infections ASSOCIATIONS Cystits, stones, bladder outlet obstruction, structural defects of ureters GRADING 5 grades; grades I and II don't require treatment RISK FACTORS Personal or family Hx of reflux, congenital urinary tract defects, and recurrent UTIs PROGNOSIS RN may lead to chronic renal failure, ESRD, nephrotic syndrome. See End-stage renal disease, Urinary tract infection

reflux pneumonitis Chemical inflammation of lung tissue caused by the pooling of regurgitated gastric secretions on respiratory mucosa and parenchyma. See Acid-reflux disorder, Aspiration pneumonia. Cf GERD

refractive error Ametropia, myopic shift OPHTHALMOLOGY The inability of images to focus properly on the retina, often corrected by glasses contact lenses, or refractive surgery. See Astigmatism, Farsightedness, Myopia, Presbyopia

refractive surgery OPHTHALMOLOGY Any technique that corrects myopia by changing the cornea's conformation TYPES Radial keratectomy–incisions are made at the periphery of the cornea, flattening and weakening it; photoreactive keratectomy–cornea is photoshaved with a laser in concentric circles, which imparts homogenous consistency. See PhotoRadial keratectomy

refractory *adjective* Intolerable, resistant to therapy, intractable, unresponsive, stubborn

refractory anemia with excess blasts HEMATOLOGY A myelodysplastic syndrome of older persons characterized by anemia or pancytopenia and BM hypercellularity CLINICAL Nonspecific–anemia of gradual onset, fatigue, weakness, exacerbation of underlying heart disease LAB Anisocytosis, megaloblastoid maturation of erythroid precursors, thrombocytopenia; 5-20% of the BM cells are blasts; < 5% of peripheral cells are blasts PROGNOSIS Guarded; average Pt survives < 1 yr; ±¼ undergo leukemic transformation, usually into ANLL TREATMENT Not all Pts require therapy; ⅓ die of bleeding or infectious complication linked to pancytopenia; intensive chemotherapy may substitute one cause of M&M for another; cytokines and growth factors may be effective. See Myelodysplastic syndrome, PISA

refractory cancer ONCOLOGY Cancer that is unresponsive to treatment. See Radioinsensitive

refractory hypertension A condition characterized by BP ≥140/90, or ≥160/90 if > 60 and absent features of 2° HTN, maximal dose of 2+ antihypertensives is being administered, and adequate time has passed to allow the usual antihypertensives to be effective DIFFDX Spurious–pseudoHTN, 'white coat' HTN, noncompliance with therapy, presence of exogenous substances that ↑ BP–alcohol, anabolic steroids, caffeine, chlorpromazine, cyclosporine, erythropoietin, MAOIs, nicotine, sympathomimetics, tricyclic antidepressants, cocaine, corticosteroids, NSAIDs, salt; obesity should be controlled

refractory osteomyelitis Osteomyelitis that persists for a prolonged period–eg, > 1 month, defined by the clinical context MANAGEMENT Surgical decompression, excision of necrotic bone, antibiotics–eg, penicillin; hyperbaric O_2 may be useful. See Hyperbaric oxygen therapy

refractory period CARDIAC PACING The time during which a pacemaker's sensing mechanism is nonresponsive–in full or in part to cardiac activity–eg, to a retrograde P-wave in a DDD pacemaker. See Pacemaker SEXUALITY A post-orgasm recovery period lasting from mins to hrs during which the penis is unerectable. See Erection

refractory septic shock Septic shock lasting > 1 hr, that does not respond to fluid resuscitation–IV therapy or pharmacologic–pressors. See Septic shock

Refsum's disease Phytanic acid oxidase deficiency An AR degenerative disorder caused by a defect in phytanic acid metabolism, resulting in fatty acid accumulation CLINICAL Chronic polyneuropathy, retinitis pigmentosa, cerebellar ataxia, EKG changes, nerve deafness, ichthyosis. See Phytanic acid

regimen MEDTALK A treatment plan that specifies dosage, schedule, duration of treatment. See CHAP, CHOP, CMF regimen, Daily regimen, Intermittent regimen, MAC regimen, Ornish regimen, STAMP I regimen, Treatment plan

regional cancer ONCOLOGY A cancer that has spread beyond the original–1° tumor to nearby lymph nodes and/or organs. Cf Carcinoma in situ

regional disease ONCOLOGY A CA that extends beyond the limits of the primary organ and spreads into adjacent tissues and locoregional lymph nodes. Cf Localized disease

regional perfusion The intravascular bathing of a specific area of the body–usually an arm or a leg with high dose chemotherapy, to manage CA confined to a region. See Perfusion

regional recurrence ONCOLOGY The appearance of the signs and Sx of malignancy at a site near–for lymphomas, on the same side of the diaphragm–CA that had been treated and responded to therapy. See Relapse

regionalization MANAGED CARE The subdivision of a broadly available service–eg, a blood bank, into quasi-autonomous regional centers, capable of making decisions and providing more cost-effective and/or faster service to hospitals and health care facilities, located the greatest distance from a 'centralized' service's hub

registered nurse HEALTH CARE A trained and licensed medical professional with a 4-yr nursing degree, who assists people in health care under the direction of a physician and is able to provide all levels of nursing care, including administration of medication. See Nurse. Cf Physician assistant

registration MEDICAL PRACTICE The least restrictive form of state regulation of persons with certain professional qualifications, which requires the filing of name, address and qualifications before practicing the field of expertise. Cf Certification LAB MEDICINE The documentation of Pt demographics and orders on the paperwork accompanying a specimen before it can be processed and analyzed

registry A system for collecting and maintaining, in a structured record, data on specific persons from a defined population, which allows preliminary analyses and reviews. See Disease Registry, Exposure registry, National Exposure Registry

Regnauld procedure ORTHOPEDICS A technique for managing metatarsalgia, by resecting the metatarsal heads, to restore correct metatarsal support INDICATIONS Severe forefoot deformities–eg, rheumatoid arthritis

Regranex® Becaplermin WOUND CARE An agent used to manage diabetic neuropathic leg ulcers that extend into the dermis or beyond and have an adequate blood supply. See Diabetic ulcer

REGRESS CARDIOLOGY A clinical trial–Regression Growth Evaluation Statin Study of the effect of pravastatin on progression or regression of CAD in symptomatic ♂ with hypercholesterolemia. See Lipid-lowering therapy, Pravastatin. Cf MAAS, PLAC I, PLAC II, 4S

regression Any return to an original state. See Atavistic regression, Generalized additive logistic regression, Hypnotic age regression, Least-squares regression, Linear regression, Past life regression, Psychoregression MEDTALK The subsiding of disease Sx or a return to a state of health ONCOLOGY A receding of CA PSYCHIATRY A partial, symbolic, conscious, or unconscious desire to return–regress to a state of dependency, as in an infantile pattern of reacting or thinking, which occurs in normal sleep, play, physical illness, and in various mental disorders

regressive behavior PSYCHOLOGY Thoughts or actions typical of early life stages–eg, infancy, childhood

regulatory agency An agency, organization or body of a federal, state or local government that creates, promulgates, and enforces rules concerning delivery of a service or product. See College of Am Pathologists, Environmental Protectional Agency, Joint Commission on Accreditation of Healthcare Organizations, OSHA–Occupational Safety and Health Administration

regurgitant murmur CARDIOLOGY A murmur caused by a leaking incompetent heart valve

regurgitation CARDIOLOGY The backflow of blood across an incompetent valve. See Aortic regurgitation, Mitral regurgitation. Cf GERD, Reflux

rehabilitation The return to function after illness or injury, often with the help of specialized medical professionals. See Esthetic rehabilitation

rehabilitation massage Medical massage PHYSICAL THERAPY Massage therapy for musculoskeletal rehabilitation and/or reduction in pain in persons who have lost a limb or suffered a CVA. See Massage therapy

rehabilitation medicine Physiatry, physiotherapy A field of therapeutics that bridges the gap between conventional and nonconventional medicine; rehabilitation physicians may adminsiter or prescribe mechanical–eg, massage, manipulation, exercise, movement, hydrotherapy, traction and electromagnetic–eg, heat and cold, light, and ultrasound modalities, psychologic and pharmacologic therapeutic modalities. See Massage therapy, Physical therapy. Cf Physiotherapy

rehydration solution A fluid used to manage severe bacterial–eg, *V cholerae*, *E coli*, or viral–eg, rotavirus diarrhea TYPES Glucose-based, which ↑ intestinal resorption of fluids and electrolytes; rice-syrup-based solution, which also ↓ stool output

reimbursement MANAGED CARE Payment by a 3rd party–eg, an insurance company, to a hospital, physician, or other health care provider for services rendered to an insured/beneficiary. See Prospecitve reimbursement, Third-party payer

reinfarction CARDIOLOGY Recurrence of clinical Sx of acute MI or new EKG changes of acute MI accompanied by a new ↑ of CK-MB. See Acute myocardial infarction

reinforcement PSYCHOLOGY Any activity, either a reward-positive reinforcement, or punishment-negative reinforcement, intended to strengthen or extinguish a response or behavior, making its occurrence more or less probable, intense, frequent; reinforcement is a process central to operant conditioning. See Contingency reinforcement

reinfusion TRANSFUSION MEDICINE The readministration of the Pt's own blood during the operation, which ↓ number of units needed for a particular surgery. See Postoperative reinfusion

Reinke space ENT/SURGICAL ANATOMY A potential space located between the internal elastic lamina of the thyroarytenoid–vocalis muscle and the external elastic lamina; RS is limited on the superior surface of the true vocal cord–TVC by superior linea arcuata, and extends from the vocal process of the arytenoid muscle to the anterior limit of the TVC; in the presence of lesions of the TVC–eg, phonotrauma, the RS fills with fluid

Reinsch test A qualitative test for identifying heavy–and toxic metals–antimony, arsenic, bismuth, mercury, selenium, tellurium, in a fluid of interest–eg, gastric juice or urine. See Arsenic intoxication, Mercury intoxication

reinsurance MANAGED CARE A type of protection purchased by HMOs or providers from insurance companies specializing in underwriting specific risks for a stipulated premium

Reiter syndrome Reactive arthritis ORTHOPEDICS A condition, more common in young ♂, especially with HLA-B27, characterized by arthritis, urethritis, conjunctivitis ETIOLOGY Unknown, may follow STD, or infection with *Chlamydia*, *Campylobacter*, *Salmonella*, *Yersinia* CLINICAL Urethritis days or wks after an infection, followed by a low-grade fever, conjunctivitis, asymmetric polyarticular arthritis over ensuing wks

rejected specimen LAB MEDICINE Any blood, urine or other specimen for which one or more of the tests ordered cannot be performed as the specimen does not meet laboratory acceptability critieria

rejection IMMUNOLOGY An immune reaction evoked by allografted organs; the prototypic rejection occurs in renal transplantation, which is subdivided into three clinicopathologic stages. See Cyclosporin A, Graft rejection, Graft-versus-host disease, Second set rejection, Tacrolimus, Transplant rejection

REJECTION TYPES

HYPERACUTE REJECTION Onset within minutes of anastomosis of blood supply, which is caused by circulating immune complexes; the kidneys are soft, cyanotic with stasis of blood in the glomerular capillaries, segmental thrombosis, necrosis, fibrin thrombi in glomerular tufts, interstitial hemorrhage, leukocytosis and sludging of PMNs and platelets, erythrocyte stasis, mesangial cell swelling, deposition of IgG, IgM, C3 in arterial walls

ACUTE REJECTION Onset 2-60 days after transplantation, with interstitial vascular endothelial cell swelling, interstitial accumulation of lymphocytes, plasma cells, immunoblasts, macrophages, neutrophils; tubular separation with edema/necrosis of tubular epithelium; swelling and vacuolization of the endothelial cells, vascular edema, bleeding and inflammation, renal tubular necrosis, sclerosed glomeruli, tubular 'thyroidization' Clinical ↓ Creatinine clearance, malaise, fever, HTN, oliguria

CHRONIC REJECTION Onset is late–often more than 60 days after transplantation, and frequently accompanied by acute changes superimposed, increased mesangial cells with myointimal proliferation and crescent formation; mesangioproliferative glomerulonephritis, and interstitial fibrosis; there is in general a poor response to corticosteroids

Relac Buspirone HCl, see there

Relanium Diazepam, see there

relapse Recrudescence The return of signs and Sx of a disorder after a period of improvement ONCOLOGY The reappearance of the signs and Sx of malignancy after successful completion of a first course of RT and/or chemotherapy. See Bone marrow transplantation, Chemotherapy, Cyclophosphamide, Extranodal recurrence, Local recurrence, Marginal recurrence, Regional recurrence, Terminal cancer, Transdiaphragmatic recurrence. Cf Remission TUBERCULOSIS The return of TB after a partial recovery from the disease

relapsing fever A tick-born bacterial infection seen primarily in the western US, characterized by multiple episodes of fever interspersed with disease free intervals EPIDEMIC BORRELIOSIS *B recurrentis*–louse-borne–*Pediculus humanis* and transmitted person-to-person CLINICAL History of recent outdoor camping, fevers with 'negative' blood cultures, Jarisch-Herxheimer-like hypotensive 'crises' after therapy with antibiotics, thrombocytopenia ENDEMIC BORRELIOSIS *B recurrentis*, *B hemisi*, *B turicatae*, *B parkeri* and others–transmitted by ticks–*Ornithodoros* spp, which inject borrelia during a blood meal CLINICAL Abrupt onset of high fever, rash, headache, photophobia, N&V, myalgias, arthralgias, abdominal and chest pain, hematuria, hematemesis, epistaxis, productive cough and minimal respiratory distress; after febrile wave passes, diaphoresis, weakness, hypotension, hypothermia; ±1 wk later, the cycle repeats itself; each successive cycle is less severe than previous one; late relapses typically involve the CNS–meningismus, peripheral neuritis, cranial nerve paralysis, seizures, coma; late disease may cause hepatitis, potentially fatal myocarditis TREATMENT Tetracycline, erythromycin, chloramphenicol

relapsing nodular nonsuppurative panniculitis Weber-Christian disease, see there

relapsing polychondritis DERMATOLOGY An uncommon condition characterized by inflammation and cartilaginous degeneration, beginning about age 40 CLINICAL Fever, vasculitis, arthropathy DIAGNOSIS 3+ of following Sx–in descending frequency: auricular chondritis with ear drooping, non-erosive arthritis, nasal chondritis with saddle nose deformity, upper respiratory obstruction, audiovestibular Sx, cardiovascular disease–eg, aortic valve insufficiency PROGNOSIS 74% 5–yr, 55% 10–yr survival

relationship VOX POPULI A state in which one thing or person is connected or related in some way to another. See Dual relationship, Dose-response relationship, Love-hate relationship, Physician-patient relationship, Volume-outcome relationship

relative bradycardia See Pulse-temperature dissociation

relative infertility Subfertility, see there

relative polycythemia A spurious ↑ in RBCs with a normal red cell mass, and a ↓ in blood volume; RP may occur in a background of mental stress, alcohol abuse, use of diuretics, or in acute nephritis; erythropoietin–EP levels are normal–± 7 U/L in contrast to P vera, in which EP is ↓–± 2 U/L or 2° polycythemia, in which the EP is ↑–± 120 U/L. Cf Polycythemia vera

relative tachycardia See Pulse-temperature dissociation

relative value scale Resource-based relative value scale, see there

relative value unit HEALTH INSURANCE A comparative financial unit that may sometimes be used instead of dollar amounts in a surgical schedule, this number is multiplied by a conversion factor to arrive at the surgical benefit to be paid. See Resource-based relative value scale

relaxant An agent that reduces tension. See Muscle relaxant

relaxation 1. The proactive act of not actively acting 2. Intentional inactivity, where a person performs active or passive exercises to ↓ mental and physical stress. See Longitudinal relaxation, Progressive relaxation, Relaxation training

relaxation training PSYCHOLOGY Any technique–eg, hypnosis, electromyograph feedback, relaxing imagery, etc, for relaxing mind or muscle. See Autogenic feedback training, Biofeedback training, Breathing exercises, Imagery & visualization, Meditation, Progressive relaxation, Systematic desensitization, Yoga

Rela® Carisoprodol & meprobamate

Relenza® Zanamivir INFECTIOUS DISEASE An neuraminidase inhibitor used to treat and prevent influenza A and B infection. See Influenza. Cf Amantadine, Rimantidine

reliability STATISTICS The extent to which the same test or procedure will yield the same result either over time or with different observers. See Interrater reliability, Test-retest reliability

Reliance® UROLOGY An intraurethral insert for managing ♀ stress incontinence. See Urgency

Reliberan Chlordiazepoxide, see there

religious diet Ethical diet NUTRITION A dietary regimen based on religious regulations, some of which may be extremely rigorous. See Kosher

religious faith Faith PSYCHOLOGY The belief in a 'higher' power or being, often linked to adherence to a particular religion or belonging to the congregation of a particular church

relocation test ORTHOPEDICS A 'provocative' joint laxity test used clinically to diagnose shoulder instability. See Provocative test, Shoulder instability. Cf Laxity test

rem Roentgen-equivalent in man RADIATION PHYSICS The traditional unit of equivalent dose/unit of absorbed radiation, ± equal to a rad or 0.01 Seivert–Sv, the unit that replaces the rem–1 Sv = 100 rem. See Gray, Rad, Sievert

REM behavior disorder SLEEP DISORDERS A type of psychosis linked to a lack of REM sleep and lack of dreaming. See Sleep disorder

REM sleep Rapid eye movement sleep, desynchronized sleep, paradoxical sleep NEUROLOGY A 5 to 20 min segment of a normal sleep cycle, characterized by irregular heart rate and respiration, BP, EEG similar to awake and alert state, involuntary or jerky muscle movement, and higher threshold for arousal; the usual high-amplitude slow brain waves seen by EEG are replaced by rapid eye movement and rapid, low-voltage irregular EEG activity. See Insomnia, Sleep disorders

REMATCH CARDIOLOGY Clinical trials–Randomized Evaluation of Mechanical Assistance Therapy as an alternative in Congestive Heart failure–related to use of a portable, electric left ventricular-assist system–LVAS–eg, HeartMate® or Novacor®, instead of drugs in CHF Pts who are not candidates for heart transplant. See Congestive heart failure

remedial profession OCCUPATIONAL MEDICINE Any specialty or area of allied health–eg, occupational, physical, recreational, or speech therapy, and rehabilitation psychology that provides assistance or therapy, usually in Pts with physical disabilities. See Rehabilitation medicine

remedial proficiency testing Off-schedule proficiency testing LAB MEDICINE A type of proficiency testing for a diagnostic or clinical lab, which may be temporarily mandated by a lab inspector–eg, from the CAP, JCAHO, or state, when the lab had a previously poor performance on a proficiency test, but must, by nature of its clientele, continue testing Pt specimens. See Proficiency testing

remedy A drug or agent used to treat a disease or ameliorate Sx. See Clinical remedy, Cold remedy, Constitutional remedy, Folk remedy, Herbal remedy

Remeron® Mirtazapine, see there

Remicade® Avakine™, cA2™, CenTNF™, infliximab IMMUNOLOGY An agent used to manage severe or mutilating rheumatoid arthritis, or severe or active–eg, fistulizing–Crohn's disease. See Rheumatoid arthritis

remifentanil Ultiva® ANESTHESIOLOGY An opioid anesthetic that is 250-fold more potent than morphine. See Opioid anesthetic. Cf Muscle relaxant

Reminyl® 1. Sabeluzole, see there. 2. Galanthamine

remission MEDTALK A period during which the signs and Sx of a disease disappear or diminish ONCOLOGY Regression of Sx or lesions in a malignancy, most commonly referring to the disappearance of a lympho- or myeloproliferative tumor by radio- or chemotherapy and amelioration of clinical Sx, which may be temporary, partial or complete. See Cure, Induction of remission, Leukemia management, Partial remission, Pathologic remission, Spontaneous regression of cancer. Cf Relapse

remission induction therapy ONCOLOGY Initial chemotherapy administered to a Pt with CA, intended to achieve remission

remnant-like particle cholesterol assay CARDIOLOGY A test for detecting remnant lipoproteins; Pts with ↑ RLP cholesterol are at ↑ risk for premature heart disease. See Hyperlipoproteinemia, type III

remnant radiation IMAGING X-rays that pass through an anatomic part and impact on film. See Radiation

remnant removal disease Hyperlipoproteinemia, type III, see there

remodeling CARDIOLOGY A process of structural deterioration of the left ventricle, which is transformed from its normal elliptical shape to a rounded or globoid shape; remodeling is typical of late stage CHF, and accompanied by 'functional' mitral regurgitation, ventricular dilatation, ventricular wall thinning ETIOLOGY Remodeling is due to a loss of myocardial tissue–eg, necrosis post acute MI, metabolic defects–eg, thyroid disease, toxins–eg, drugs or alcohol, inflammation–eg, viral myocarditis, or altered hemodynamics–eg, uncontrolled HTN or aortic stenosis EVALUATION Remodeling can be quantified by measuring the end-systolic and end-diastolic volumes

remote memory Long-term memory, see there

REMS SLEEP DISORDERS Rapid eye movements. See REM sleep

REMS latency SLEEP DISORDERS The lag between sleep onset–stage 2 sleep–and the first REM period minus any awake time. See REM sleep

REMS period SLEEP DISORDERS The REMS portion of a NREMS-REMS cycle; early in the night it may be as short as 30 secs, but in later cycles > 1 hr. See REM sleep, Sleep stage REM

Remune™ AIDS A therapeutic AIDS vaccine consisting of an inactivated virus to stimulate an immune response at a cellular level, used alone or with other antivirals to prevent or delay progression of HIV infection to AIDS. See AIDS, AIDS vaccine

remuneration MEDICARE Any compensation, directly or indirectly, overtly or covertly, in cash or in kind–which includes professional courtesy and forgiveness of debt. See Compensation arrangement

RENAAL NEPHROLOGY A clinical trial–Reduction in Endpoints in patients with Non-insulin-dependent diabetes mellitus with the Angiotensin II Antagonist Losartan. See Losartan

Renagel® Sevelamer, see there

RENAISSANCE RHEUMATOLOGY A clinical trial–Tandomized Enbrel®/etanercept North American Strategy to Study Antagonism of Cytokines. See Etanercept

renal ablation glomerulopathy NEPHROLOGY A nonimmune glomerular disease due to glomerular injury with loss of nephron mass and compensatory hypertrophy of remaining glomeruli. See Glomerular disease

renal agenesis NEONATOLOGY A rare disease characterized by bilateral RA, low-set floppy ears, a broad, flat nose and lung hypoplasia; infants with bilateral RA die shortly after birth; infants with unilateral RA have normal lungs and are asymptomatic in neonatal period

renal allograft See Renal transplantation

renal aplasia Nondevelopment of a kidney or renal tissue

renal artery stenosis 1. Acute renal arterial thrombosis 2. Arteriolonephrosclerosis, see there 3. Goldblatt kidney, see there

renal artery thrombosis Acute renal arterial thrombosis, renal artery occlusion NEPHROLOGY Abrupt occlusion of RA which, if complete, may cause permanent renal failure; loss of function of one kidney may be asymptomatic as 2nd kidney adequately filters blood–HTN is typical; without a 2nd functional kidney, RAT may cause acute kidney failure ETIOLOGY Acute RAT may follow abdominal injury, embolism, especially in Pts with heart disease–eg, A Fib, smoking, RA stenosis MANAGEMENT Surgery

renal biopsy Kidney biopsy A Bx guided by ultrasonography of a core of renal tissue to be examined by LM, immunofluorescence, EM INDICATIONS Nephrotic syndrome, idiopathic proteinuria, proteinuria with 'glomerular' hematuria, acute renal failure, lupus nephritis, rapidly progressive glomerulonephritis, transplant rejection, renal vasculitis COMPLICATIONS Microscopic hematuria–which occurs with most renal biopsies, and thus is regarded by some authors as normal, perineal hematoma, pain, worsened HTN, AV fistu-

la formation, renal laceration, puncture or laceration of aorta or arteries, pancreas, spleen, liver, GI tract, and death, which occurs in 1:3000 Pts

renal calculi Kidney stones, see there

renal cell carcinoma Adenocarcinoma of renal cells, hypernephroma, kidney cancer, renal cancer, renal cell cancer, renal adenocarcinoma ONCOLOGY A renal tubule CA EPIDEMIOLOGY 30,600 new, 12,000 deaths/yr–US, 1996 RISK FACTORS 1.6 ♂, 1.9 ♀, obesity in ♀–5.9 odds ratio in highest 5% of upper body mass, phenacetin analgesics, ↑ meat consumption, tea drinking–♀, petroleum exposure TYPES Clear cell–75%-80%, papillary–types 1, 2–10%-15%, chromophobe–5%, oncocytoma–5% DIAGNOSIS Hx, exam, imaging, Bx MANAGEMENT Surgery, RT, chemotherapy are ineffective; IL-2 and IFN alpha yield a 20% response, 20% 1 yr event-free survival

renal clearance THERAPEUTICS A hypothetical calculation of plasma volume in liters of unmetabolized drug cleared through the kidneys/unit time FACTORS INFLUENCING RC Renal blood flow, urinary pH, tubular resorption and secretion

renal colic NEPHROLOGY A colicky pain typical of Pts passing kidney stones

renal failure NEPHROLOGY See Acute renal failure, Chronic renal failure, End-stage renal failure, Postpartum renal failure

'renal' glycosuria A common–1:500 AR condition in which glycosuria occurs without hyperglycemia, unrelated to diet; subjects are asymptomatic, have a normal glucose tolerance test and utilization and storage of carbohydrates, but may become transiently ketotic in stress or pregnancy

renal infarction NEPHROLOGY Death of renal tissue, usually caused by renal artery sternosis

renal insufficiency A defect in renal ability to 'clear' waste products, a sign of inadequate glomerular filtration

renal stone Kidney stone, see there

renal transplant Transplantation of a kidney from a living donor or cadaver to a recipient with ESRD INDICATIONS–CHILDREN Congenital kidney/GU tract malformations–42%; focal segmental glomerulosclerosis-12% and others; 31% of children were ≤ age 5 yrs PROGNOSIS 1-yr survival of graft, ±94%; half-life from living donor grafts, ±22 yrs, from cadaveric grafts, ±14 yrs 3. See End-stage renal disease, Organ shortage

renal tubular acidosis NEPHROLOGY A condition characterized by functional defects in the distal renal tubules, with inability to form ammonia and exchange hydrogen ions; GFR is normal with persistent metabolic acidosis, hyperchloremia, ↓ urine excretion of acid; the acid pH results in hypercalcemia–kidney stones, excess calcium excretion, and bone demineralization LAB Acid urine, acidosis, ↓ bicarbonate excretion, ↓ ammonium clearance, ↓ K⁺ clearance, calcium loss

---

RENAL TUBULAR ACIDOSIS

TYPE I 'Classic' distal RTA, which is due to a selective defect–in secretion of H⁺–in distal tubule acidification, resulting in a defect in the pH gradient, causing hyperchloremia (with persistent bicarbonate excretion/↓ reabsorption of bicarbonate), hypokalemia and metabolic acidosis; the urinary pH is inappropriately high (> 6)

TYPE II Proximal RTA is accompanied by ↓ acidification of the proximal tubule; when the blood pH is ↓, tubular acidification occurs normally; when plasma bicarbonate normalizes, type II RTA wastes bicarbonate, causing metabolic acidosis, as well as hyperchloremia and hypokalemia, which may be accompanied by Fanconi syndrome with loss of glucose and amino acids, phosphate, calcium, potassium; these Pts are prone to osteopenia, rickets and kidney stones

'TYPE III' A designation once applied to infants with renal bicarbonate wasting, now considered a type I RTA subtype

TYPE IV RTA Generalized–nonselective distal RTA, due to aldosterone deficiency or antagonism; hyperchloremia, hyperkalemia, metabolic acidosis, salt wasting

576 **renal tubular necrosis** See Acute tubular necrosis

Renaquil Lorazepam, see there

renin PHYSIOLOGY An aspartyl proteinase with one substrate, angiotensinogen, which is secreted primarily by the granular cells of the juxtaglomerular apparatus, as well as the brain and endothelial cells, in response to ↓ renal perfusion pressure or ↓ kallikrein; renin cleaves angiotensinogen to yield angiotensin I–a decapeptide, the precursor of angiotensin II–an octapeptide, and angiotensin III–a heptapeptide; the latter 2 are potent vasoconstrictors, and stimulate thirst and ↑ aldosterone production ↑ IN Addison's disease, COPD, CRF hypersecretion, eclampsia–and preeclampsia, hyperthyroidism, cirrhosis, ↓ K+, ↓ salt diet, malignant HTN, pregnancy, renal failure, renovascular HTN, functional kidney tumors, drugs–eg, antihypertensives, diuretics, estrogens, OCs ↓ IN Cushing syndrome, DM, essential HTN, hypothyroidism, high salt diet, drugs–eg, antihypertensives, levodopa, propranol. See Hypertension, Plasma renin activity

Renografin® Diatrizoate, see there

renoprotection MEDTALK Any activity to minimize adverse effects–especially due to oxidative stress–on renal vasculature

renovascular hypertension Atherosclerotic renal artery disease, atherosclerotic renovascular disease, renal hypertension NEPHROLOGY Systemic HTN due to renal artery obstruction by ASHD, fibroplastic disease, aneurysms, embolism HIGH RISK ASHD, smoking, white, abdominal bruits, peripheral vascular disease, late–> age 60, onset of HTN CLINICAL Hemorrhage in cerebellum, pons, internal capsule, basal ganglia TREATMENT–MEDICAL Empirical, a function of disease severity and response to therapy–eg, diuretics, β blockers, vasodilators, ACE inhibitors TREATMENT-INTERVENTIONAL Percutaneous transluminal angioplasty of renal artery, 60-70% success in atherosclerosis; 90% success in fibromuscular hyperplasia. See Goldblatt kidney, Hypertension

Renovue™ Iothalamate, see there

'rent-a-doc' A deprecating noun and adjective for, or referring to, a physician in a locum tenens position. See Doc-in-a-box, Locum tenens, Per diem

reoperation SURGERY A 2nd surgical procedure in the same site for the same indications, which may be due to failed therapy–eg, persistent Sx from BPH or tumor recurrence–eg, ovarian CA. Cf Second look operation

ReoPro® Abciximab, see there

reovirus A family of non-enveloped, double-stranded RNA viruses MEMBERS Orthoreovirus, orbivirus, rotavirus DISEASE Generally asymptomatic, ±gastroenteritis, rhinopharyngitis, ±hepatitis, encephalitis, pneumonia, Colorado tick fever. See Rotavirus

repaglinide NovoNorm®, Prandin® ENDOCRINOLOGY An oral nonsulfonylurea hypoglycemic for Pts with type 2 DM, whose hyperglycemia does not respond to diet and exercise

repeatability NIHSPEAK The closeness of agreement between test results obtained within a single lab when the procedure is performed on the same substance under identical conditions in a given time period. See Reproducibility

reperfusion-eligible acute MI CARDIOLOGY An acute coronary syndrome with ST elevation which, in a Pt with chest pain, suggests acute coronary artery occlusion and benefit from aggressive therapy–thrombolytics or intervention aimed at restoring perfusion. See Acute coronary syndrome, Acute myocardial infarction

reperfusion injury CARDIOLOGY Myocardial injury caused by rapid flow of blood into areas previously rendered ischemic by coronary artery occlusion; RI is attributed to oxidative stress, which may cause arrhythmia, infarction, myocardial stunning. See Reperfusion TRAUMATOLOGY A component of crush syndrome, which occurs when blood flow is reestablished to an organ or tissue exposed to prolonged ischemia; renewed blood flow aggravates tissue damage either by causing additional injury or by unmasking injury sustained during the ischemic period; RI occurs in the heart, intestine, kidney, lung, and muscle, and is due to microvascular damage. See Calcium paradox, Oxygen paradox

reperfusion therapy CARDIOLOGY Any therapy–eg, thrombolytic therapy-tPA, stenting, or immediate percutaneous transluminal coronary angioplasty-IPCTA–intended to ensure continued blood flow–and oxygenation through a vascular bed acutely compromised by vasospasm or thrombosis, especially post acute MI. See tPA, Trials

repetition compulsion PSYCHOANALYSIS The impulse to reenact earlier emotional experiences, considered by Freud more fundamental than the pleasure principle. Cf Pleasure principle

repetitive motion injury Cumulative trauma disorder OCCUPATIONAL MEDICINE A work-related illness–eg, carpal tunnel syndrome caused by overuse of a particular musculoskeletal group to perform a task repeated hundreds to thousands of times/day; it is the fastest growing health problem in the US; it affects textile industry workers, meat-packers, keyboard operators, etc. See Alexander technique, Ortho-bionomy

replacement MEDTALK The plenishment or substitution of that which is lost or inadequate. See Estrogen replacement, Hair replacement, Total hip replacement, Total knee replacement

Replicare™ WOUND CARE A hydrocolloid paste for autolytic debridement, absorption of exudate, moist wound healing. See Wound care

replication study INTERNAL MEDICINE A clinical study that seeks to verify data from a prior study

Repliderm™ WOUND CARE A collagen-based dressing for treating pressure, stasis, diabetic and foot ulcers, 1st and 2nd-degree burns, surgical incisions, cuts, abrasions, partial-thickness wounds and for absorbing exudate. See Wound care

report card Performance report MANAGED CARE 1. An annual report on the quality and outcome of care by a particular health care provider or health care facility 2. Information on aspects of health care provided to the public to help consumers chose health plans or physicians; information on RCs include Pt outcomes, satisfaction, cost structures, etc of the plan's health care delivery. See *Consumer Guide to CABG*

report disclaimer MALPRACTICE A statement that may be placed on a report, eg of pap smears, to indicate that the test being performed has an undetermined but low error rate

reportable occupational disease Reportable event PUBLIC HEALTH An occupational or environmentally-related morbid condition that a local, state, or federal government wishes to maintain under surveillance; each state has occupational diseases of particular interest RODs Asbestosis, bronchitis and acute pulmonary edema due to fumes and vapors, byssinosis, caisson disease, coal worker's and other pneumoconioses, heavy metal, lead, pesticide and radiation poisoning, intoxication with acid, alkali, antimony, benzene, beryllium, cadmium, chlorinated hydrocarbons, chlorine, chromium, Freon™, hydrogen cyanide, manganese, mercury, petroleum products, other solvents and sulfur dioxide, pulmonary fibrosis, silicosis; non-occupational events that require reporting to central health authorities–eg, AIDS, child abuse, drug addiction, STDs, gunshot or stab wounds; for reportable infectious disease. See Notifiable disease

reporting VOX POPULI The act of providing information about a thing or event. See Case reporting, HIV reporting, Mandatory reporting, Named reporting, Unique provider reporting

Reposal Chlordiazepoxide, see there

Repose™ surgical system SURGERY A system for minimally invasive treatment of obstructive sleep apnea and snoring, in which the base of the tongue is stabilized, preventing it from collapsing and obstructing the airway. See Obstructive sleep apnea

repositioning LAPAROSCOPIC SURGERY The changing of a Pt's position during a procedure to improve access or visualization of the operative field, which may be linked to complications, as it changes anatomic planes of operation. See Laparoscopic surgery

repressed memory PSYCHOLOGY An event that occurred in a subject's past, the memory of which was actively repressed often because of the psychologically devastating impact of that memory–eg, childhood abuse, rape, molestation. See False memory, Source amnesia

repression PSYCHIATRY An unconscious defense mechanism, that blocks unacceptable ideas, fantasies, or impulses from consciousness or that keeps unconsciousness what never was conscious. Cf Suppression PSYCHOANALYSIS A mental block to acknowledging an uncomfortable memory or feeling

reprint An individually bound copy of an article in a journal or science communication

reproducibility LAB MEDICINE The degree of agreement among repeated measurements of a particular parameter, presented in terms of a standard deviation or coefficient of variation of the results in a set of measurements

reproduction GYNECOLOGY Conceiving; making babies. See Asexual reproduction, Assisted reproduction, Teratogenicity

reproductive death OBSTETRICS See Maternal mortality

reproductive history OBSTETRICS A set of 4 numbers that may be used to define a woman's obstetric Hx–eg, 4-3-2-1, would mean 4 term infants delivered, 3 preterm infants, 2 abortions, 1 child currently living

reproductive toxicity Any adverse effect attributable to exposure to a chemical, directed against the reproductive and/or related endocrine systems ADVERSE EFFECTS Altered sexual behavior, fertility, pregnancy outcomes, or modifications in other functions that depend on reproductive integrity of system. See Teratogenicity

Repronal® Renamed Repronex™, see there

Repronex® CenterWatch, menotropins PA, Repronal™ GYNECOLOGY A proprietary generic equivalent of Pergonal® for treating infertility. See Infertility

reptiles, fear of PSYCHOLOGY Herpetophobia. See Phobia

reputation GRADUATE EDUCATION The level of prestige of an institution or residency program. See High-power institution, Residency

Requip® Ropinirole PHARMACOLOGY A 2nd-generation dopamine agonist for treating Parkinson's disease. See Parkinson's disease

required surgery SURGERY A procedure that should be performed soon EXAMPLES Cataract, spinal fusion, sinus operaton, repair of heart or valve defect, cholecystectomy for symptomatic stone, excision of oversized fibroid of uterus

requisition form Lab slip LAB MEDICINE A form used to request specific lab tests–eg, chemistry, hematology, microbiology, cytology, HIV, pathology, etc. See Results

res ipsa loquitur The thing speaks for itself LAW & MEDICINE A legal doctrine under which a plaintiff's burden to prove negligence is minimal as the details of the incident are clear and understandable to a jury–eg, foreign objects left behind during surgery, eg towels. See Medical malpractice

Rescriptor® Delavirdine, see there

rescue EMERGENCY MEDICINE Any activity that brings a victim of disaster or accident to safety. Cf Disaster ONCOLOGY Rescue therapy. See Leucovorin rescue, Marker rescue

rescue adjunctive coronary angioplasty CARDIOLOGY A procedure in which an infarct-related artery is dilated immediately after thrombolytics have failed. See Coronary angioplasty

rescue breathing EMERGENCY MEDICINE Any of a number of life-saving maneuvers in which a rescuer–R¹ inflates the rescuee's–R² lungs by breathing into the R²'s airway access 'port'. See ABCs–of CPR, Head-tilt/Chin-lift maneuver

RESCUE BREATHING TYPES

MOUTH-TO-MOUTH Airway is opened by the head-tilt/jaw-lift maneuver, nose is pinched by R¹ who takes 2 deep breaths, seals his/her lips around R²'s mouth and gives 2 full breaths–1 to 1.5 seconds/breath, allows good chest expansion, average volume 800 mL.

MOUTH-TO-NOSE Used when there is major trauma to the face, trismus, or a tight mouth seal cannot be formed; airway is opened by the head-tilt/jaw-lift maneuver, mouth is closed by R¹ who takes 2 deep breaths, seals his/her lips around R²'s nose and gives 2 full breaths as above

MOUTH-TO-STOMA Used in Pts who have undergone laryngectomy; R¹ who takes 2 deep breaths, seals his/her lips around R²'s stoma and breathes as above

RESCUE CARDIOLOGY A clinical trial–Randomized Evaluation of Salvage Angioplasty with Combined Utilization of Endpoints that compared the effects of rescue coronary angioplasty with conservative therapy of occluded infarct-related arteries. See Coronary angioplasty, Rescue adjunctive coronary angioplasty

research Scientific inquiry to discover or verify facts, test hypotheses, and confirm theories. See Applied research, Bioprocess research, Clinical research, Effectiveness research, Ethnographic research, Extramural research, Health services research, Interactive research, Intramural research, Investigator-initiated research, Nontherapeutic research, Outcomes research, Survey, Targeted research, Twin research, Unethical medical research NIHSPEAK A systematic investigation and gathering and analysis of information designed to develop or contribute to generalizable knowledge. See Notebook Cf Fraud in science

resection Excision Surgical removal of a tumor or organ. See Abdominal-perineal resection, Colon resection, Endometrial resection, En bloc resection, Gastrocnemius resection, Human factors resection, Laparoscopic resection, Laparoscopic-assisted resection, Oncologic resection, Wedge resection

reserpine THERAPEUTICS A Rauwolfia–derived alkaloid that ↑ CNS 5-HT and catecholamine concentrations, used for HTN, mild anxiety

reservation of rights HEALTH INSURANCE A term referring to a situation arising when there is a question as to whether a medical service is covered; usually the insurer is obliged to defend a claim while a coverage issue between insurer and policyholder is being resolved

reserve PHYSIOLOGY A capacity or capability to be used in an emergency. See Cardiac reserve, Coronary vasodilator reserve, Resistance reserve

residency MEDICAL EDUCATION A period of formal graduate medical education that consists of on-the-job training of medical school graduates; completion of a residency program is required for board certification in a medical or surgical specialty. See Fellowship, GME, Internship, Resident. Cf CME, Extern, Intern

resident Registrar A physician who has completed medical school and internship, and is receiving training in a specialty. See Residency

residential facility NURSING A live-in institution for people of the same feather–eg, nursing home, correctional facility, homeless shelter

residential treatment PSYCHIATRY Health care provided at a live-in facility to a person with emotional disorders who requires continuous medication and/or supervision or relief from environmental stresses

578 **residual cancer** ONCOLOGY Any CA that remains after definitive management has been attempted. Cf Recurrent cancer

residual disease ONCOLOGY Malignant cells or neoplasia that remains after any form–chemotherapy, surgery, RT–of 1° treatment

resiniferatoxin NEUROLOGY An agent for treating urge incontinence in Pts with neurologic disease–eg, spinal cord injury and multiple sclerosis

resistance INFECTIOUS DISEASE The ability of a host to resist a pathogen; able to grow in the presence of a particular antibiotic. See Antibiotic resistance, Drug resistance, HIV drug resistance, Intermediate resistance MEDTALK The ability to function in a normal or near-normal fashion, in the face of a toxic environment. See Activated protein C resistance, Airway resistance, Cross-resistance, Hormone resistance, Insulin resistance, Multidrug resistance, Nasal airway resistance, Radioresistance, Variable resistance, Vasopressin resistance ONCOLOGY Failure of a cancer to regress after RT or chemotherapy PSYCHIATRY Conscious or unconscious psychologic defense against recall of repressed–unconscious thoughts

resistance training Weight training SPORTS MEDICINE Exercising muscle(s) against weight. See Exercise

resi-tern MEDICAL EDUCATION A junior resident–usually a 2nd yr resident who functions as both an intern–PGY-1 and resident–PGY-etc. See HMO

resolution CLINICAL MEDICINE The stage of a disease–often an infection, marked by subsidence of Sx

resolution phase SEXOLOGY The last of 4 sexual phases delineated by Masters & Johnson in the sexual cycle. Cf Excitement phase, Orgasmic phase, Plateau phase

RESOLVD CARDIOLOGY A clinical trial–Randomized Evaluation of Strategies for Left Ventricular Dysfunction of Pts with ↓ left ventricular function and clinical Sx of heart failure

RESOLVE REPRODUCTION MEDICINE A national, nonprofit consumer organization offering education, advocacy, and support to those experiencing infertility

resource The components of a system–eg, equipment, space, and labor, available to perform a task MANAGED CARE The source of financial support–eg, insurance, personal income

resource allocation MANAGED CARE The constellation of activities and decisions which form the basis for prioritizing health care needs

resource-based relative value scale MANAGED CARE A scale that ranks physician services by the labor required to deliver those services. See CPT codes, DRGs, Overrated procedures

resource utilization group HEALTH ADMINISTRATION Any of a number of groups into which a nursing home resident is categorized, based on functional status and anticipated use of services and resources. See Functional assessment

resources See Human resources

RespiGam™ NEONATOLOGY An IV immune globulin used to protect against RSV in high-risk children ≤ age 2 with bronchopulmonary dysplasia or prematurity. See Respiratory syncytial virus

respiration Breathing; to inhale and exhale; the exchange of gases between the external environment and an organism's cells. See Anaerobic respiration, Cellular respiration, Cheyne-Stokes respiration

respirator A device used to facilitate respiration. See BagEasy respirator, HEPA respirator

respiratory acidosis METABOLISM A condition caused by alveolar hypoventilation, tissue accumulation of CO_2, and ventilatory failure leading to ↑ $PaCO_2$–ie, hypercapnia; this leads, in turn to ↓ $HCO_3^-/PaCO_2$ and ↓ pH LAB pH < 7.35, HCO_3^- > 26 mEq/L–if 'compensating', $PaCO_2$ > 45 mm Hg ETIOLOGY Hypoventilation due to drugs, CNS depression, heart disease, lung disease–eg, COPD, neuromuscular disease–eg, ALS, diaphragm paralysis, major kyphoscoliosis, Guillain-Barré syndrome, myasthenia gravis, muscular dystrophy, obesity hypoventilation syndrome CLINICAL Cyanosis, diffuse wheezing, hyperinflation–ie, barrel chest, ↓ breath sounds, hyperresonance on percussion, prolonged expiration; depressed mental status due to ↑ CO_2, accompanied by asterixis, myoclonus, seizures, papilledema, dilated conjunctival or facial blood vessels. See Metabolic acidosis, Metabolic alkalosis, Respiratory alkalosis

respiratory alkalosis METABOLISM A condition characterized by ↑ pH due to excess CO_2 excretion ETIOLOGY Hyperventilation–eg, due to anxiety, pain, panic attacks, psychosis, CVA, fever, encephalitis, meningitis, tumor, trauma, hypoxemia–severe anemia, high altitude, right-to-left shunt, drugs–progesterone, methylxanthines, salicylates, catecholamines, nicotine, hyperthyroidism, pregnancy, ↑ ambient temperature, lung disease–pneumo/hemothorax, pneumonia, pulmonary edema, PE, aspiration, interstitial lung disease, etc–sepsis, liver failure, mechanical ventilation, heat exhaustion, recovery from metabolic acidosis DIFFDX Asthma, A Fib, flutter and tachycardia, heatstroke, metabolic acidosis or alkalosis, acute MI, and other causes of RA CLINICAL Hyperventilation, paresthesia, twitching–positive Chvostek and Trousseau signs, N&V, focal neurologic signs, depressed consciousness, or coma LAB pH > 7.42, HCO_3^- < 22 mEq/L–if compensating, $PaCO_2$ < 35 mmHg. See Metabolic acidosis, Metabolic alkalosis, Respiratory acidosis

respiratory distress syndrome CLINICAL MEDICINE Adult respiratory distress syndrome, see there NEONATOLOGY Respiratory distress syndrome of newborn, see there

respiratory distress syndrome of newborn Hyaline membrane disease, respiratory distress NEONATOLOGY A morbid process seen in up to 50% of neonatal deaths–40,000/yr CLINICAL Atelectasis, hypoventilation, hypotensive shock, pulmonary vasoconstriction, alveolar hypoperfusion, shut-down of cell metabolism; 60% of HMD occurs in infants, most are < 28 wks of age; 5% occur in infants > than 37 wks; HMD is more frequent in the 2nd twin delivered, twin-to-twin transfusion recipient infant, ♂ infants, children of diabetic mothers and in C-sections; a vicious cycle begins where ↓ surfactant results in atelectasis, ↓ ventilation–↑ pCO_2, ↓ pH, ↓ O2 and hypoxia exacerbating the lack of surfactant, causing shock and more hypoxia CLINICAL Early onset of tachypnea, prominent grunting, intercostal retractions–air hunger, cyanosis; the infants may not respond to O_2; blood pressure and corporal temperature fall, asphyxia intervenes and causes death, or the Sx peak at 3 days and the infant recovers DIFFDX Neonatal pneumonia, birth-related asphyxia, group B streptococcal sepsis, cyanotic heart disease TREATMENT Supportive–O_2, correct acidosis, surfactant

respiratory papillomatosis ONCOLOGY A condition characterized by proliferation of benign papillomas in the respiratory tract, primarily the larynx, which may extend into airways and lungs DIAGNOSIS Endoscopy MANAGEMENT Endoscopic excision, IFN alpha-n1, if refractory

respiratory syncytial virus VIROLOGY A 'pediatic' virus–Family Paramyxoviridae, genus, *Pneumovirus*, that causes RTIs with the greatest M&M in infants < age 2, especially if there is underlying cardiovascular or pulmonary disease EPIDEMIOLOGY In children, RSV causes 55,000 hospitalizations, 2000 deaths/yr MANAGEMENT High-dose RSV immune globulin VACCINE Under development

'respiratory syndrome' A relatively specific immune response to high-dose rifampin therapy, characterized by a flu-like complex, dyspnea and wheezing, leukopenia, thrombocytopenia; other hypersensitivity reactions caused by rifampin include flushing, fever, pruritus without rash, urticaria, eosinophilia, hemolysis, interstitial nephritis-induced renal failure

respiratory therapy MEDTALK Exercises/ treatments that help improve or restore lung function

respite care EXTENDED CARE Short-term care provided in a nursing home or hospice to a Pt needing long-term care, so that the normal care-givers can get a break (respite) from the 24/7 routine of caring for the Pt. See Hospice, Palliatrics

respondeat superior *Latin–let the master answer for the servant* MEDICAL MALPRACTICE A legal doctrine that holds an employer liable for an employee's wrongful or negligent act. See 'Captain of the ship', 'Deepest pockets', Malpractice

response The reaction to a stimulus. See Acute phase response, Adaptation response, Brainstem auditory evoked response, Clinical benefit response, Chronotropic response, Diving response, Evoked response, Fight or flight response, Frequency response, Galvanic skin response, Heat shock response, Hemodynamic response, Immune response, Intelligent voice response, Interactive voice response, Linear dose response, Minimal response, Placebo response, Pleotypic response, Sexual response, Triple response, Visual evoked response

responsibility VOX POPULI A duty to do/not do a thing. See Administrative responsibility, Clinical responsibility, Criminal responsibility, Sexual responsibility

responsiveness MEDTALK The ability to respond to a stimulus. See Airway responsiveness

rest VOX POPULI See Bowel rest

restenosis CARDIOLOGY The reocclusion of an artery or lumen after otherwise adequate therapy, which may occur in a mitral valve after replacement or valvoplasty, or in a coronary artery after CABG, PCTA, or stenting. See Atherosclerosis, CABG, In-stent restenosis

restless leg syndrome Nocturnal myoclonus SLEEP DISORDERS A clinical complex characterized by nocturnal cramping of the anterior calf, restlessness, a feeling of heaviness, aching, painful paresthesia and tingling in legs with uncontrolled twitching, relieved by movement; RLS is worse at night and with recumbency, and interferes with sleep; other features include stereotypic triple flexion movement of the legs that occur at regular intervals during sleep, myotonic movements while awake and asleep, insomnia, family Hx MANAGEMENT Home remedies–eg, hot baths, creams and cotton stockings may be as effective as clonidine, benzodiazepines, opiates, L-dopa, transcutaneous electrical nerve stimulation. See Sleep disorders. Cf Periodic limb movement disorder

Restocalm Chlordiazepoxide, see there

RESTORE CARDIOLOGY A multicenter trial–Randomized Efficacy Study of Tirofiban for Outcomes and REstenosis–to evaluate Aggrastat in Pts with unstable angina or acute MIs post-angioplasty. See Tirofiban

Restoril® Temazepam, see there

restraining garment FORENSIC PSYCHIATRY An article of clothing or other wearable-eg, straitjacket, harness, splints, gloves, etc, designed to mechanically restrain or prevent sleepwalking, self-injury, tooth-grinding, or other harm to Pts or others

restraint Control or prevention of an action NURSING Any device used to restrict the free movement of Pts with behavioral or physical problems, who may cause harm to themselves and others. See Mechanical restraint, Pharmacologic restraint

restricted area OCCUPATIONAL SAFETY An area, the access to which is strictly limited to essential personnel with the purpose of protecting individ-

uals against undue risk from exposure to various materials–eg, ionizing radiation, high-level pathogens, toxic chemicals and others

restriction HOSPITAL PRACTICE The narrowing or limiting of a health care provider's unrestricted practice of medicine by a licensing or certifying authority, due to activities determined to be illegal or at least of questionable medical judgement. Cf Revocation NUTRITION The limiting of ingestion of a substance. See Caloric restriction, Protein restriction VOX POPULI Any limiting of an activity. See Host-controlled restriction, Intrauterine growth restriction, MHC restriction

restrictive cardiomyopathy Infiltrative cardiomyopathy A group of myocardial disorders characterized by an inability of the ventricles to fill and pump blood efficiently–the least common cardiomyopathy in the US ETIOLOGY Myocardial and endomyocardial diseases CLINICAL If left and right atrial are chronic, systemic and pulmonary HTN→congestion and/or edema; Sx of right-sided congestion–↑ jugular venous pressure, dependent edema, liver congestion or ascites are common. Cf Hypertrophic cardiomyopathy

restrictive lung disease PULMONOLOGY A general term that encompasses the functional aspects of interstitial lung disease ETIOLOGY-ACUTE Infections–miliary TB, histoplasmosis, PCP, CMV, fungal; RT; pulmonary edema, inhalation-byssinosis; aspiration; Goodpasture's disease; lung CA; idiopathic-Hamman-Rich syndrome ETIOLOGY-CHRONIC Pneumoconiosis; RT; CA-lung, breast, stomach, pancreas; drugs-bleomycin, busulfan, hydralazine; systemic disease-sarcoidosis, collagen vascular disease, Langerhans cell histiocytosis; diffuse interstitial fibrosis CLINICAL SOB, tachypnea, concentric ↓ in lung volumes IMAGING Interstital and alveolar pattern FUNCTIONAL TESTS ↓ diffusing capacity. Cf Restrictive lung disease

resuscitation CRITICAL CARE The restoration of consciousness to a person who appears dead. See Active compression-decompression-cardiopulmonary resuscitation, Cardiopulmonary resuscitation, Fluid resuscitation, Mouth-to-snout resuscitation

resveratrol NUTRITION A phytochemical in grapes which may be partly responsible for the health benefit attributed to red wine. See Red wine

retained acid syndrome A condition occurring in 25% of those treated–onset 2nd day to 3rd wk of therapy–with all-*trans*-retinoic acid–eg, for acute promyelocytic leukemia CLINICAL Fever, dyspnea, pulmonary infiltrates, pleural effusions TREATMENT Corticosteroids; discontinuing all-*trans*-retinoic acid may be ineffective. See Acute promyelocytic leukemia

retained antrum syndrome A rare complication of inadequate resection of the distal antrum and pylorus during antrectomy and a Billroth II gastrojejunostomy. See Postgastrectomy syndromes

retained bullet FORENSIC MEDICINE A bullet that remains in the body after a gunshot, either because it wasn't found or because it lodged in an inaccessible site or would create vascular or neural problems if removed

retardation MEDTALK The slowing of a process or activity. See Intrauterine growth retardation, Mental retardation, Psychomotor retardation, Reading retardation

Retcol Chlordiazepoxide, see there

retention NEUROLOGY See Memory UROLOGY See Urinary retention.

retention polyp GI DISEASE A pedunculated mass arising in the GI tract mucosa, composed of dilated glands in an inflamed stroma; surface is often eroded, accompanied by inflammatory atypia; all are associated with polyposis syndromes. See Polyp

reteplase Retavase™ CARDIOLOGY A recombinant mutant of alteplase-tPA used to treat acute MI. See Acute myocardial infarction, GUSTO V

reticulocyte count The number of immature erythrocytes–reticulocytes in peripheral blood that have a basophilic reticulum–residual RNA when stained with supravital dyes–eg, methylene blue, brilliant cresyl blue; the RC is used to evaluate the rate of RBC production, and BM response to anemia ↓ IN ↓ adrenocortical and anterior pituitary activity, aplastic anemia, cirrhosis, megaloblastic anemia, exposure to radiation↑ IN Erythroblastosis fetalis, HbC, leukemia, after hemorrhage, pregnancy, thalassemia, during treatment of iron and megaloblastic anemia NORMAL RANGE Newborn < 6.5% of peripheral RBCs; children < 2%; adults < 1%

Retin-A Tretinoin DERMATOLOGY A topical vitamin A formulation used to manage wrinkling, age spots, acne

retinal *adjective* Referring to the retina *noun* 11-*cis*-retinal, vitamin A aldehyde A retinal pigment that absorbs visible light at 400-600 nm, resulting in an isomeric transition of the 11-*cis*-retinal moiety to a *trans*-retinal conformation causing a G protein-mediated depolarization event

retinal anlage tumor Pigmented neuroectodermal tumor of infancy, see there

retinal detachment OPHTHALMOLOGY Separation of the neurosensory–rods and cones–retina from the retinal pigment epithelium, accompanied by opening of the vestige of the cavity of the optic vesicle of the embryo; RD is a true ophthalmologic emergency that requires immediate therapy to prevent blindness; RD affects 18-30,000/yr–US ETIOLOGY 1. Retinal tear, which may be caused by vitreous traction, and results in vitreous fluid accumulation 2. Exudative detachment, in which the neurosensory retina is lifted away from the retinal pigment epithelium by fluids that accumulate in response to local tumors–eg, melanoma, local inflammation, or malignant HTN, idiopathic, trauma, aging, tumors, inflammation; in premature infants, RD is caused by retinopathy of prematurity CLINICAL Progressive loss of visual field, often accompanied by scotomas; bleeding from small arterioles vessels may cloud vitreus; if macula is involved, central vision is affected RISK FACTORS Myopia, family Hx of retinal detachment, Caucasian, ♂ TREATMENT A silicone oil may be used to treat complicated RTs uncorrectable with standard therapy. See Myopia

retinal migraine NEUROLOGY A migraine characterized by repeated attacks of monocular scotomata or blindness, lasting up one hr in duration. Cf Classical migraine

retinal vessel occlusion OPHTHALMOLOGY Blockage of the retinal–arterial or venous–vascular supply ETIOLOGY Blood clot, fat deposit, fragmented atherosclerotic plaque RISK FACTORS Glaucoma, HTN, DM, coagulation disorders, ASHD, hyperlipidemia CLINICAL Variable vision loss after retinal vein occlusion, hemorrhage; profound vision loss after retinal artery occlusion, stroke

retinitis Inflammation of the retina. See Cytomegalic retinitis, Uveitis

retinitis pigmentosa Night-blindness disease OPHTHALMOLOGY A heterogenous group of AR, less commonly, AD, and X-linked forms of retinal degeneration; RP affects 1:3500, and is characterized by nyctalopia and progressive centripetal loss of visual fields progressing to blindness by middle age

retinoblastoma Cancer of retina ONCOLOGY A CA, usually of children–affecting 1:15-30 000 infants–that arises in retinal cells, which has photoreceptor differentiation; 10-20% are hereditary, mostly AD; 70% are unilateral and arise *de novo*; bilateral tumors are associated with germ cell neoplasms and occasionally other tumors–eg, osteogenic sarcoma, Ewing sarcoma TREATMENT RT. See One-hit, two-hit model, Flexner-Wintersteiner rosettes

9-*cis*-retinoic acid AIDS A retinoid that may have some efficacy in topical management of KS ADVERSE EFFECTS Skin irritation, photosensitivity. See AIDSALRT 1057

(all-*trans*-)retinoic acid Tretinoin The natural form of vitamin A, which transports monosaccharides in glycoprotein synthesis, as occurs in the turnover of mucosal epithelia of the oral cavity, respiratory and urinary tracts. See Retinal, Retinoic acid embryopathy, Tretinoin, Vitamin A

retinoic acid embryopathy PERINATOLOGY A teratogenic complex induced by retinoic acid, a vitamin A analogue, resulting in a 26-fold ↑ of congenital defects–eg, microtia, anotia, cleft palate, cardiac–conotruncal and aortic arch, neural crest, craniofacial, thymic defects, hyperostoses, retinal and optic nerve abnormalities, CNS malformations, premature closure of the epiphyseal plates; an identical embryopathy occurs with 'megadose' ingestion of vitamin A

retinoid *noun* 1. Vitamin A or a vitamin A-like compound 2. Any of a class of drugs used to manage CA and other conditions–eg, hairy leukoplakia, molluscum contagiosum. See AIDS

retinoid X receptor One of 2 receptors for retinoids; RXR plays a key role in organ development, in particular of the skin. Cf Retinoic acid receptor

retinopathy A disorder that affects the retina. See Central serous retinopathy, Diabetic retinopathy, Hypertensive retinopathy, Preproliferative diabetic retinopathy, Retinitis, Retinopathy of prematurity

retinopathy of prematurity Retrolental fibroplasia NEONATOLOGY Inflammation of the retina of premature infants, characterized by retinal vascular proliferation, vitreous scarring with funneling of the retina, leading to extreme distortion of vision or complete blindness, retinal detachment RISK FACTORS Prematurity, O_2 excess, hypoxia, respiratory distress, infection, congenital heart disease. See Retinopathy

retirement syndrome PSYCHIATRY Acute or chronic maladjustment to a nonworking state–ie, retirement; the 'condition' is most common in those who had no activities other than the field of labor CLINICAL Irritability, apathy, asthenia, ↑ alcohol consumption, non-specific autonomic nervous system complaints

retort tube appearance GYNECOLOGY A descriptor for fallopian tubes affected by acute PID with hyperemia and fibrin deposition; the lumen is filled with pus and the fimbriae are sealed. See Pelvic inflammatory disease

retracted nipple GYNECOLOGY A nipple pulled inward by an underlying tumor or inflammation which, in breast CA, may be accompanied by a peau d'orange appearance of the surrounding skin; RNs may also be red, ulcerated or scaled. See Inflammatory breast cancer, Nipple, Peau d'orange appearance. Cf Inverted nipple

retraction INTERNAL MEDICINE A hollowing of tissue in the neck and intercostal spaces due to negative intrathoracic pressure, a classic sign of severe asthma. See Asthma SURGERY The drawing of tissues away from a particular site to open an operative field

retractor DENTISTRY An orthodontic device used to facilitate posterior movement of misaligned teeth SURGERY Any device used to hold wound margins, organs, or tissues away from an operative field. See Hohmann retractor, Scoville retractor

retrieval VOX POPULI The obtaining of a thing from a place stored or sequestered. See Antigen retrieval, Egg retrieval, Microwave antigen retrieval

retrobulbar neuritis NEUROLOGY Transient inflammation of the optic nerve, causing rapid loss of vision and ocular pain when the eyes move

retrocalcaneal bursitis PODIATRY Inflammation of the bursa at the calcaneus ETIOLOGY Pain with flexion and extension of foot, anterior to Achilles' tendon, above its insertion on the os calcis EXAM Positive 2-finger squeeze test

retroflexion Uterine retroflexion GYNECOLOGY The backward displacement of the uterus, a finding that occurs in the 1st trimester of ±10% of pregnancies; despite occasional reports of ↑ frequency of bleeding in early pregnancy, therapy is only indicated in rare cases of incarceration in the hollow of the sacrum

retrogasserian glycerol rhizotomy PAIN MANAGEMENT Injection of glycerol into the trigeminal cistern to manage intractable pain COMPLICATIONS Anesthesia dolorosa, keratitis. Cf Microvascular decompression, Pain medicine

retrograde amnesia NEUROLOGY Amnesia that extends to before the trauma or events that caused the loss of memory. See Amnesia. Cf Anterograde amnesia

retrograde conduction Ventriculoatrial conduction CARDIOLOGY The propagation of depolarization from the ventricles to the atria–ie, VA conduction. See Conduction

retrograde ejaculation UROLOGY A benign clinical condition characterized by entry of semen into the bladder rather than through the urethra during ejaculation ETIOLOGY Prostate or urethral surgery, DM, drugs–eg, antihypertensives and psychotropics. See Ejaculation

retrograde pyelography Retrograde urography IMAGING A method for visualizing the renal collecting system–ie, the pelvis and lower urinary tract, based on the excretion of radiocontrast after injection into the renal pelvis or ureter via a ureteral catheter INDICATIONS Suspected kidney disease–eg, pyelonephritis, kidney stones, or right upper quadrant colic

retrolental fibroplasia Retinopathy of prematurity, see there

retroperitoneal fibrosis Ormond's disease UROLOGY An idiopathic affliction of young adults–ages 30-45, ♂ : ♀ ratio, 2:1, characterized by retroperitoneal proliferation of fibrous tissue that encases the ureters, causing obstructive uropathy and possibly renal failure, which may evoke fibrosis elsewhere–eg, sclerosing cholangitis and mediastinitis, Riedel's thyroiditis, pseudotumor of orbit and generalized vasculitis CLINICAL Malaise, vomiting, backache, constipation, HTN RADIOLOGY Compression of intraabdominal structures LAB Oliguria, azotemia, proteinuria, ↑ ESR, anemia

retropharyngeal abscess ENT A disease of children < age 5, in which posterior throat tissue is susceptible to abscess formation, accompanied by high fever, severe sore throat, dysphagia and dyspnea, which may be life threatening. Cf Strep throat

retroposition GYNECOLOGY Retroflexion, see there

retropubic prostatectomy UROLOGY A surgical procedure in which the prostate is excised in toto through an incision made in the abdominal wall

retrospective study Case-control study EPIDEMIOLOGY An epidemiologic study that analyzes data collected before a certain point in time, to determine if past events are related to the present distribution of disease; a study–eg, case control study in which investigators select groups of Pts who have already been treated and analyze data from the events experienced by these Pts. Cf Prospective study

retrospectroscope The resolution of diagnostic dilemmas if viewed in hindsight

retrosternal thyroid Ectopic thyroid tissue located behind the sternum which, if benign, has no clinical significance. See Ectopic thyroid

retroversion of uterus Malposition of uterus, tipped uterus, uterus retroversion GYNECOLOGY A normal variant of pelvic anatomy–seen in 20% of ♀ especially at menopause–in which the body of the uterus is tipped toward the back rather than forward ETIOLOGY Laxity of pelvic ligaments; pregnancy or tumor-related uterine enlargement; salpingitis, prior PID, endometriosis

retrovirus VIROLOGY An RNA virus that encodes reverse transcriptase so that its RNA can be transcribed into DNA in a host cell; modified retroviruses are used as vectors to introduce genes–or portions thereof–of interest into eukaryotic cells. See Cloning vector, DNA, Eukaryote, Gene, HIV, HTLV, Reverse transcriptase, RNA, Rous sarcoma virus, Spumavirus, Transcription, Virus

Retrovir® Zidovudine, see there

Rett's disorder PEDIATRICS An X-R condition affecting ♀, characterized by mental retardation and delayed developmental milestones beginning from 6 months to 2 yrs the child's condition deteriorates.

return of spontaneous circulation CARDIOLOGY A palpable pulse which is present after clinically documented asystole. See Atrial fibrillation

revascularization VASCULAR DISEASE A surgical procedure in which compromised or stenosed blood vessels are bypassed to treat ischemia; revascularization is the treatment of choice for Pts subsequently found to have abnormal ventricular function and extensive CAD. See CABG

reversal agent ANESTHESIOLOGY Any drug used to reverse the effects of anesthetics, narcotics or potentially toxic agents EXAMPLES Antilirium, digibind, mestinon, narcan, neostigmine, protopam, pyridostigmin, romazicon, tensilon. See Induction agent

reverse '5' sign A finding on a plain AP film of an infant with hypoplastic left heart syndrome, in which the enlarged right atrium corresponds to the hip of the '5' and the superior vena cava corresponds to the shoulder of the '5'. See Baby Fae heart

reverse isolation INFECTIOUS CONTROL The placement of an immunocompromised person in a specialized room or unit that attempts to minimize exposure to pathogens. See Personal protection garment

reverse peristalsis Antiperistalsis A wave of contraction of GI tract smooth muscle which moves contents orally

reverse 'precautions' Infection control procedures–eg, sterilization and isolation, used to protect a Pt–rather than the care providers or other Pts who are immunocompromised, either as a congenital condition–eg, combined variable immunodeficiency syndrome or an acquired condition–eg, BM irradiation before BMT; RPs are required when the PMN count falls below 0.5 x 10^9–US: 500/mm^3. See Gnotobiotic environment. Cf Universal precautions

reverse psychology PSYCHOLOGY The application of an inverted suggestion that would compel a person in a 'contrary' state, to perform a desired task

reverse T$_3$ 3,3',5'-Triiodothyronine, rT$_3$ LAB MEDICINE A conversion product of T$_4$–thyroxine, the level of which reflects the rate of peripheral conversion of thyroid hormones of T$_4$ to T$_3$ NORMAL rT$_3$ LEVELS 0.15-0.77 mmol/L. See Euthyroid sick syndrome

reverse 3 sign IMAGING A term referring to 1. Broadening of the duodenal loop with a 'puckering' around the ampulla of Vater, classically associated with pancreatic adenocarcinoma at the head of the pancreas, seen in barium studies of the upper GI tract or 2. Indentation of the cecum, which may be seen on a plain abdominal film in acute appendicitis. Cf Figure 3 sign

reverse transmission INFECTIOUS DISEASES The transmission of an infection from a health care provider to a Pt; RT is rarely documented, and is a clinical rarity. See Acer cluster, High disseminator. Cf Reverse precautions, Typhoid Mary

reversible injury PATHOLOGY Ischemic changes–eg, of the myocardium, that can be reversed with timely return of normal circulation to the tissue of interest

reversible ischemic neurologic disability NEUROLOGY A TIA, defined as a focal ischemia-induced loss of neurologic function of abrupt onset, with a disability of > 24 hrs but < 3 wks in duration. Cf Multi-infarct dementia, Transient ischemic attack

reversion CARDIAC PACING The automatic suspension of pacemaker inhibition in the face of certain types of electrical activity. See Pacemaker

review

review MEDTALK The evaluation of a process by a person or party other than the person or party performing the task of interest. See Concurrent review, Drug price review, Drug utilization review, Health statistics review, Integrity review, Literature review, Peer review, Technical assistance review, Utilization review NIHSPEAK The concurrent oversight of research on a periodic basis by an institutional review board. See Clinical review, Grant compliance review

review of systems INTERNAL MEDICINE An organized and complete examination of a Pt's organ systems as part of the workup when the Pt is first seen by a physician; an ROS is an inventory of body systems obtained by verbal history, with the signs and/or Sx which the Pt is experiencing or had. See Evaluation and management services, Physical examination

reviparin VASCULAR DISEASE An LMW heparin for ↓ the risk of DVT in Pts requiring prolonged immobilization

Revised Trauma Score EMERGENCY MEDICINE A triage tool consisting of a numeric scoring system for calculating the probability of surviving an accident. See Injury Severity Score, Triage

revision MEDICAL COMMUNICATION A manuscript that has been submitted for publication in a peer-reviewed journal, which addresses the questions asked by the reviewers. See Manuscript NIHSPEAK Any change in a study's protocol which occurs between initial submission and CTEP approval and official filing VOX POPULI The alteration of a thing

Revive Ring/Vacuum Therapy Unit UROLOGY A proprietary system that provides a temporary mechanical solution–suction inflation + vascular constriction to manage impotence. See Impotence, Penile prosthesis. Cf Viagra

revocation HOSPITAL PRACTICE The suspension of a medical provider's professional license, often accompanied by suspension of the practitioner from a hospital's active medical staff. Cf Restriction

revolving door 'syndrome' PSYCHIATRY A cyclical pattern of short-term readmissions to the psychiatric units of health care centers by young adults with chronic psychiatric disorders. See Commitment, Homeless(ness) SOCIAL MEDICINE A general term for the lack of permanent jobs and homes typical of underskilled laborers or foreigners without permanent visas who emigrate to large cities and have a compromised quality of life

rewarming EMERGENCY MEDICINE The raising of the body temperature in a Pt with hypothermia. See Accidental hypothermia

Rey test NEUROLOGY Any test that evaluates verbal and nonverbal learning and memory. See Psychological testing

Reye syndrome NEUROLOGY A potentially fatal condition characterized by acute encephalopathy and fatty degeneration of liver, with a dose-dependent relationship to use of aspirin in children with viral infections–eg, influenza, varicella-zoster CLINICAL Vomiting, hepatic dysfunction, variable neurologic impairment–irritable and combative behavior, loss of consciousness, stupor, seizures, coma, death, often preceded by viral URIs or varicella LAB ↑ Transaminases, glutamine and ammonia in CSF, hypoglycemia, metabolic acidosis DIFFDX DRUG & CHEMICALS–Aspirin, other salicylates, acetaminophen, lead, valproic acid, methyl bromide, hydrocarbons, chlordane, ethanol, disulfiram INFECTIOUS–viral encephalitis, hepatitis METABOLIC–cystic fibrosis, carnitine deficiency, hereditary fructose intolerance, isovaleric acidemia, urea cycle disorders OTHERS–Pancreatic encephalopathy, acute encephalopathy with fatty metamorphosis of the liver due to cold agglutinin autoimmune hemolytic anemia, toxic encephalopathy TREATMENT None universally accepted. See Lovejoy's classification, Will Rogers phenomenon

Rezulin® Troglitazone, see there

RF 1. Radiofrequency, see there 2. Rate of flow 3. Renal failure 4. Respiratory failure 5. Rheumatic fever 6. Rheumatoid factor, see there

RFLP Restriction fragment length polymorphism MOLECULAR MEDICINE A local variation in the DNA sequence of an individual which may be detected by restriction endonucleases–which 'cut' the double-stranded DNA whenever they recognize a certain highly-specific oligonucleotide sequence or 'restriction' site. See DNA, DNA sequence, Linkage map, Marker, Mutation, Physical map, Polymorphism

rH Recombinant human

Rh antigen TRANSFUSION MEDICINE Any of a composite of multiple blood group antigens on RBC membranes–eg, Rh-C–c, C, CG, C*, Rh-D–D, weak D–formerly D^u, D^w, Rh-E–e, E, E*, Rh-G, Rh-LW, Rh-Ne^a and Rh$_{null}$. See Rh system

Rh immune globulin RhIg, Rh$_o$(D) immune globulin OBSTETRICS A sterile plasma-based preparation rich in anti-Rh antibodies used to prevent production of Rh$_o$(D) antibodies in Rh-negative mothers who may have an Rh-positive infant; its use virtually eliminates HDN due to maternal Rh-antibody production. See Hemolytic disease of the newborn

Rh incompatibility TRANSFUSION MEDICINE An incompatibility in Rh types between a donor and recipient, or between an fetus–who is Rh positive, and the mother–who is negative and produces antibodies against the fetus' RBCs, which pass to the placenta, and cause HDN. See Hemolytic disease of the newborn

Rh system TRANSFUSION MEDICINE A family of blood group antigens that can cause fatal transfuion reactions if improperly matched FREQUENCY OF RH ANTIGENS–WHITES Rh-e, 98%; Rh-D, 85%; Rh-c, 80%; Rh-C, 70%; Rh-E, 30%; Rh-D exposure is important in OB–an Rh-D-negative mother's anti-RH-D antibody is an IgG which may cross the placenta, and cause HDN–titer > 1:16 in the 8th month indicates maternal formation of alloantibodies, evidenced by stomatocytes in mom's blood; the Rh type is determined prenatally by PCR of amniotic cells PROPHYLAXIS 300 µg of anti-D–RhoGAM neutralizes 30 ml feto-maternal hemorrhage containing 15 ml of Rh+ RBCs. See Blood group, Hemolytic disease of newborn, Rh immune globulin

rhabdomyolysis Skeletal muscle destruction, with release of myoglobin in blood and urine ETIOLOGY Severe exertion–eg, marathons, calisthenics, muscle necrosis due to arterial occlusion, DVT, seizures, drug overdose–amphetamines, cocaine, heroin, PCP, trauma, shaking chills, heatstroke, alcohol–delirium tremens. See Crush injury, Exertional rhabdomyolysis, Myoglobinuria

rhaphy A surgical repair, eg herniorrhaphy

rheumatic fever RHEUMATOLOGY The late non-purulent sequelae of a URI by streptococcus group A DIAGNOSIS Major criteria–carditis, chorea, erythema marginatum, polyarthritis, subcutaneous nodules; minor criteria–arthralgia, fever, Hx of previous rheumatic fever, or evidence of cardiac involvement, lab parameters ↑ acute phase reactants, anti-streptolysin O titers, C-reactive protein, ESR, delineated by Jones and later modified. See 'Chinese menu disease,' Jones criteria

rheumatic heart disease CARDIOLOGY A condition characterized in the acute form by myocarditis–which may cause atrial arrhythmias, 'bread and butter'-type serofibrinous pericarditis, Aschoff bodies in subserosal fibroadipose tissue–which, when aggregated, form McCullum's plaques, valve lesions–fibrin deposits and WBC aggregates–ergo friable vegetations–verrucae along the line of valve closure

rheumatism An older term for a constellation of nonspecific complaints–eg, pain, stiffness, and limitation of movement vaguely attributable to derangements of joints and supporting connective tissue. See Arthritis, Desert rheumatism, Palindromic rheumatism

rheumatoid arthritis RHEUMATOLOGY A multisystem autoimmune disease characterized by chronic inflammation of multiple joints; RA is defined by the 1987 revised criteria, which requires that criteria 1-4 be present for > 6 wks LAB IgG autoantibodies, aka rheumatoid factors MANAGEMENT Etanercept ↓ disease activity in therapeutically refractive RA

RHEUMATOID ARTHRITIS–Revised criteria

1. Periarticular morning stiffness lasting ≥ one hour before maximum improvement
2. Soft tissue swelling ('arthritis') of ≥ 3 joints observed by a physician
3. Swelling ('arthritis') of proximal interphalangeal, metacarpophalangeal or wrist joints
4. Symmetric swelling ('arthritis')
5. Rheumatoid nodules
6. Presence of rheumatoid factor
7. Roentgenographic erosions and/or periarticular osteopenia

rheumatoid factor RHEUMATOLOGY Any of a group of often polyclonal IgM–rarely also, IgG or IgA antibodies directed against the Fc portion of denatured IgG; RFs are produced by synovial neutrophils in 80% of Pts with rheumatoid arthritis. See Jones criteria, Rheumatoid arthritis

rheumatoid nodule A mass in tendons, tendon sheaths, periarticular tissue, serous membranes–pleural, pericardium, meninges, cardiac valves, kidneys, lung parenchyma, skin, spleen, synovium, vessels, viscera, in 20% of Pts with rheumatoid arthritis DIFFDX Similar nodules occur in SLE, rheumatic fever. See Rheumatoid arthritis

rheumatoid pneumonitis RHEUMATOLOGY A form of diffuse interstitial pulmonary fibrosis seen in 2% of Pts with rheumatoid arthritis, accompanied by variable pulmonary compromise; RLD is rare in juvenile RA. See Rheumatoid arthritis

rheumatoid vasculitis A necrotizing vasculitis affecting a subset of Pts with rheumatoid arthritis CLINICAL Mononeuritis multiplex, GI involvement–infarction, ulceration, perforation, colitis, stricture formation, hemorrhage, skin infarction, constitutional Sx

rheumatologist An internist with a subspecialty interest in chronic diseases of the joints, connective tissues, and muscles; rheumatologists may also manage vasculopathies MEAT & POTATOES DISEASES Arthritis, gout, SLE

rheumatology A specialty of internal medicine dedicated to the diagnosis and management of musculoskeletal diseases, connective tissue–formerly, collagen vascular diseases, and vasculopathies. See Rheumatologist

RhIg See Rh immune globulin

rhinitis Inflammation of nasal mucosa. See Allergic rhinitis, Medicamentosa rhinitis, Seasonal allergic rhinitis, Vasomotor rhinitis

rhinitis medicamentosa ENT A complication of chronic topical nasal decongestant use, in which there is a progressive shortening of therapeutic efficacy, coupled with rebound rhinitis after treatment is discontinued. See Allergic rhinitis, Decongestant

rhinomanometry ENT A technique for evaluating nasal patency or obstruction, and airflow by simultaneously measuring transnasal pressure and airway resistance, and pressure-flow characteristics. TYPES–ANTERIOR A pressure tap is inserted into a sealed nostril. POSTERIOR A face mask is combined with a pressure tap in the oral cavity or oropharynx; used to measure net bilateral airflow resistance. UNILATERAL POSTERIOR One nostril is taped during the procedure. Cf Acoustic rhino(mano)metry, Nasal airway resistance

rhinophyma Bulbous nose, WC Fields nose ENT A nasal nosology not known in ♀, which is an end-stage acne rosacea, producing a bulbous proboscis. See Acne rosacea

rhinoplasty Nose job, nose reconstruction PLASTIC SURGERY A procedure in which the size and/or contour of the nose is revised. TYPES–ESTHETIC–which

constitutes the majority, and for which 25 000 are performed/year–US 70% in ♀; 11% < age 18. RECONSTRUCTIVE–for defects caused by trauma or surgery to the region, most often related to extensive locoregional malignancy

rhinorrhea MEDTALK → VOX POPULI Runny nose

rhinovirus VIROLOGY An RNA virus that may cause lower RTIs in immunocompromised Pts and exacerbations of bronchitis and asthma in children; rhinovirus is a small–30 nm, nonenveloped virus with a single-strand RNA genome in an icosahedral capsid; there are ±100 serotypes of RV; RVs belong to the Picornaviridae family, which includes genera Enterovirus–polioviruses, coxsackieviruses, echoviruses, enteroviruses, and Hepatovirus–HAV CLINICAL Common cold, accompanied by nasopharyngitis, croup, pneumonia; most cases are mild and self-limited EPIDEMIOLOGY Transmitted by aerosol or direct contact, with the nasal mucosa and conjunctiva. See Common cold

RHIT Registered Health Information Technician

rhizotomy PAIN MANAGEMENT The surgical disruption of a cranial or spinal nerve root, as a means of limiting pain. See Radiofrequency thermal rhizotomy

rhizoxin ONCOLOGY An anticancer drug of fungal origin of the vinca alkaloid family. See Vinca alcaloid

Rhodococcus equi *Corynebacterium equi* INFECTIOUS DISEASE A pleomorphic coccobacillary opportunistic pathogen that may infect immunocompromised hosts–eg, AIDS Pts CLINICAL Pneumonia, pleural effusion, empyema, abscesses of brain, skin, and elsewhere, osteomyelitis, lymphadenitis, endophthalmitis MANAGEMENT Erythromycin, rifampin. See AIDS

RhoGAM See Rh immune globulin

rhombencephalitis Brainstem encephalitis NEUROLOGY An infection of the brain corresponding to the rhombencephalon GRADE I–generalized myoclinic jerks with tremor and/or ataxia GRADE II–myoclonus with cranial nerve involvement–eg, ocular disturbance–nystagus, strabismus, or gaze paresis or bulbar palsy GRADE III–transient myoclonus followed by rapid onset of respiratory distress, cyanosis, poor peripheral perfusion, shock, coma, loss of doll's eye reflex, and apnea

rhonchus Sonorous rhonchus PULMONARY MEDICINE A type of continuous–> 250 msec, high-pitched, hissing lung sound, with a frequency of ≥ 400 Hz . Cf Wheeze

rhuMAb-E25 Monoclonal anti–IgE antibody, Omalizumab IMMUNOLOGY A recombinant humanized monoclonal antibody that complexes with free IgE and blocks IgE interaction with mast cells and basophils; rhuMAb-E25 may be used to manage Pts with severe allergic asthma

rhythm MEDTALK A periodic movement; an action which occurs at regular intervals. See Alpha rhythm, Biologic rhythm, Circadian rhythm, Kappa rhythm

rhythm method Calendar method, Vatican roulette OBSTETRICS A form of natural family planning, and the contraceptive method sanctioned by the Catholic Church, in which unprotected intercourse is allowed shortly after a menstrual period or before the onset of the next period; the RM is the least effective form of contraception, resulting in 20 pregnancies/100 ♀-yrs. See Breast feeding, Coitus interruptus, Contraception, Natural family planning, Pearl index

Rhythm™ catheter CARDIOLOGY An electrophysiologic catheter used for internal cardioversion/defibrillation of tachyarrhythmias. See Electrophysiology

RHYTHMS PSYCHOLOGY A 2 yr controlled effectiveness trial funded by Pfizer, which randomized 242 Pts receiving only anti-depressants to receive either usual care–control arm or written instructional material on depression and effective treatment

584 **rhytidectomy** Wrinkle removal PLASTIC SURGERY A procedure in which redundant, wrinkle-rich skin, most commonly, facial–from whence the term, facelift, is 'tightened' by excising excess tissue from bordering hair- and necklines. See Cosmetic surgery. Cf Tummy tuck

rib raising OSTEOPATHY Lymphatic pump, see there

rib-tip syndrome NEUROLOGY Intercostal neuralgia accompanied by sharp episodic pain at the costal margin, caused by hypermotility of the anterior end of the costal cartilage of–usually the 10th rib due to trauma

ribavirin Rebetol® VIROLOGY An oral ribavirin, a synthetic nucleoside analogue and broad-spectrum antiviral. See Rebetron

ribbon VOX POPULI A strip of satin in a single loop; ribbonmania began with the red ribbon, signifying AIDS awareness; it has continued ad absurdum with its wearers lending their maudlin support to virtually every disease, cause, and hackneyed do-gooder catchphrase in every known color

ribbon rib A descriptive term for ribs with marked costal hypoplasia or attenuation, a finding typical of trisomy 13-15 and trisomy 18 syndromes, and may be seen in neurofibromatosis, Gorham's angiomatosis, hyperparathyroidism, osteodysplasia, osteogenesis imperfecta, poliomyelitis, rheumatoid arthritis and scleroderma

riboflavin Vitamin B₂ NUTRITION A vitamin which combines with phosphate to form enzyme cofactors flavin mononucleotide–FMN and flavin adenine dinucleotide–FAD, both involved in oxidation-reduction reactions in many metabolic pathways, and in energy production in the mitochondrial respiratory chain. See B complex vitamins, Riboflavin deficiency, Vitamins

riboflavin deficiency A rare condition caused by a lack of riboflavin CLINICAL Hyperemia, erythema, and pain of oral mucosa, stomatitis, glossitis, angular cheilitis, seborrheic dermatitis LAB Normocytic and normochromic anemia, which may be accompanied by pure red cell aplasia in the BM. See Riboflavin

ribosomal RNA MOLECULAR BIOLOGY Any of a family of single-stranded nucleic acids ranging from 100 to 3000 bases in length, that assemble in heteromultimeric complexes with proteins, to form ribosomes, the 'docking stations' for mRNA and nascent polypeptide strands. See Ribosome, RNA. Cf mRNA, tRNA

RICE SPORTS MEDICINE An acronym–rest, ice, compression, elevation–for the first aid maneuvers of musculoskeletal and joint injuries. See Sports medicine

rice body ORTHOPEDICS Any of numerous elongated and indurated oval-to-rounded rice-like masses found in the joints of Pts with rheumatoid arthritis, SLE, septic arthritis, TB bursitis, synovial chondromatosis. Cf Joint 'mice'

rice water stool Clear, watery stools with a vaguely fishy odor and flecks of mucus, likened to water from boiled rice, an appearance classically seen in cholera; cholera stool is low in protein and isotonic with the plasma, and thus a 'secretory' diarrhea in which adenylate cyclase is locked in an 'on' position by enterotoxins produced by Vibrio cholerae and some strains of E coli; at early postmortem examination, the intestines are stiff and non-distensible and likened to 'iron rods, a finding due to the antemortem metabolic acidosis and ↓ K⁺. See Cholera cot

Richter syndrome The development of a clinically aggressive pleomorphic large–usually B cell–lymphoma in the background of CLL–which occurs in 3-10% of CL or Waldenström's macroglobulinemia CLINICAL Rapid onset of intractable fever, weight loss, lymphadenopathy, hepatosplenomegaly

ricin TOXICOLOGY A toxic vegetable poison from the castor bean plant–Ricinus communis which causes agglutination and fulminant hemolysis at very high dilutions–1/10⁶ CLINICAL Abdominal pain, nausea, cramps, convulsions, dehydration, hemolysis, cyanosis, renal failure–oliguria, hematuria, circulatory collapse. See Magic bullet

rickets Osteomalacia in children, renal osteodystrophy, renal rickets, vitamin D deficiency METABOLISM A rare condition caused by vitamin D deficiency, insufficiency or inefficiency, with interference of normal ossification in children, whose bones are poorly mineralized, accompanied by deficit of Ca²⁺ or PO₄ from the bone and supportive matrix; the parathyroid ↑ PTH secretion in response to ↓ serum Ca²⁺, resulting in ↑ loss of Ca²⁺ and PO₄, as it is reabsorbed from bone; bone cysts may develop in severe cases, especially during rapid growth when the body demands more Ca²⁺ and PO₄; rickets is usually seen in young children 6 to 24 months old and is rare in newborns ETIOLOGY ↓ vitamin D in diet, especially in strict vegetarians or those who are lactose intolerant, malabsorption syndromes–eg, fat malabsorption, steatorrhea, sprue, short bowel syndrome; dietary lack of Ca²⁺ and PO₄ is rare–both are present in milk and green vegetables; hereditary rickets is an X-linked vitamin D-resistant form of rickets caused when the kidney is unable to retain PO₄, renal tubular acidosis–which dissolves Ca²⁺ in the bones, leaving them soft and weak bones; rickets may occur in children with disorders of the liver or biliary system, when fats and vitamin D are poorly absorbed or when the vitamin D is not converted to its active form. See Hepatic rickets, Vitamin D, Vitamin D deficiency, Vitamin D-dependent rickets, X-linked hypophosphatemic rickets

rickettsial infection Rickettsial disease, rickettsiosis Any infection by Rickettsiae GROUPS 1. Typhus group–epidemic typhus, Brill-Zinsser disease, murine–endemic typhus, scrub typhus; 2. Spotted fever group–Rocky Mountain spotted fever, Eastern tick-borne rickettsiosis, rickettsialpox; 3. Q fever; 4. Trench fever

rickettsialpox Kew Gardens fever INFECTIOUS DISEASE An infection by Rickettsia akari, transmitted from rodents to humans by the hematophagous mouse mite, Liponyssoides sanguineus CLINICAL Typical eschar over the initial lesion consisting of a nodule followed by a vesicle at the mite bite site → fever, malaise, headache, backache, myalgia, conjunctivitis, sore throat, chest pain, cough, lymphadenopathy, maculopapules, papulovesicules TREATMENT Tetracycline

RID Radial immunodiffusion, see there

RID-HD NEUROLOGY A double-blind, placebo-controlled, multicenter dosing trial–Riluzole Dosing in Huntington Disease. See Riluzole

rider's bone A type of post-traumatic myositis ossificans of the upper femur of equestrians, near adductor muscles

Riedel's thyroiditis Fibrous thyroiditis, invasive thyroiditis, Riedel struma ENDOCRINOLOGY A rare condition of older adults and the elderly which presents clinically as an ill-defined thyroid enlargment often accompanied by severe SOB due to tracheal compression MANAGEMENT Steroids, surgical excision. Cf Hashimoto's thyroiditis

Rieger syndrome NEONATOLOGY An AR condition characterized by hypodontia, defect of anterior chamber of eye, myotonic dystrophy, anal stenosis

rifabutin Mycobutin® AIDS A semisynthetic rifamycin used to prevent MAC in AIDS Pts with < 75 T4 cell/mm³, or combined with other drugs to treat MAC; it ↓ anemia, fatigue, fever, hospitalization, Karnovsky performance score, prevents ↑ alk phos seen in advanced AIDS SIDE EFFECTS Kidney and liver damage, BM suppression, rash, fever, GI stress. See AIDS, Mycobacterium avium-intercellulare complex

rifalazil INFECTIOUS DISEASE A rifampin derivative effective against TB, which has a longer T₁/₂ and intracellular concentrations, allowing shorter treatment, fewer doses, better Pt compliance. See Rifampin

rifampin TUBERCULOSIS An anti-TB drug used in Pts with a positive skin test who have been exposed to INH-resistant TB ADVERSE EFFECTS Hepatitis, adverse drug interactions, gastritis. See AIDS. Cf Rifalazil

rifapentine Priftin® INFECTIOUS DISEASE An agent used for pulmonary TB and MAC with other agents–INH, pyrazinamide, ethambutol, streptomycin. See AIDS, Tuberculosis

Rift valley fever INFECTIOUS DISEASE A dengue-like viral disease spread by mosquitoes in floods, causing fatal enzootic hepatitis in ruminants–sheep, cattle and occasional human epidemics, by direct contact CLINICAL Abrupt onset with a biphasic fever curve, headaches, prostration, myalgias, anorexia, N&V, conjunctivitis, lymphadenopathy; death may result from hemorrhagic fever or encephalitis MORTALITY 5-20%. See Dengue

right bundle branch block CARDIOLOGY A condition in which the electrical impulse from the bundle of His to the ventricles is delayed or fails to conduct along the right bundle branch, resulting in right ventricular depolarization by cell-to-cell conduction spreading from the interventricular septum and left ventricle to the right ventricle–ie, slow and uncoordinated NATURAL HISTORY Surgically induced RBBB has few acute hemodynamic consequences and a generally benign course long term; rarely, progression to complete heart block and sudden death occur, especially if accompanied by major His-Purkinje system–eg, left anterior hemiblock, first-degree AV block–injury; tetralogy of Fallot repair with an RBBB and a markedly prolonged QRS duration >180 ms have an ↑ risk for ventricular arrhythmias and sudden death; familial RBBB may be benign or, if it occurs in Kearns-Sayre syndrome or Brugada syndrome, potentially fatal CLINICAL Children with RBBB may have Hx of congenital heart disease, heart surgery–eg, VSD, palpitations, ↓ energy/activity/exercise tolerance, dizziness, syncope, familial Hx of arrhythmias–eg, BBB, complete heart block, pacemaker/defibrillator, premature or sudden unexplained death, acute MI under age 45; persistently split 2nd HS EKG Lead V₁–late intrisicoid deflection, M-shaped QRS, wide R or occasionally qR; lead V₆–early intrinsicoid deflection, wide S; lead I–wide S MANAGEMENT Pacemaker, if syncope or significant arrhythmias FOLLOWUP Telemetry PRN; annual EKG. See Bundle branch block. Cf Left bundle branch block

right heart catheterization Pulmonary artery catheterization CARDIOLOGY A technique for direct measurement of cardiac function, consisting of the introduction of a catheter into the right atrium, right ventricle, pulmonary artery DATA Hemodynamic measurements, cardiac output, shunt determinations, blood sampling, oxygen arrival time INDICATIONS Valve disease, congenital heart disease, CHF, cor pulmonale, pulmonary HTN, intracardiac shunts. Cf Left heart catheterization

right lower quadrant PHYSICAL EXAM The region of the abdomen that contains the terminal ileum, appendix and cecum

right ovarian vein syndrome GYNECOLOGY Enlargement of the right ovarian vein at the expense of the left due primarily to valve incompetence, often linked to pregnancy CLINICAL Intermittent right flank and right-lower quadrant pain, often before menstruation, aggravated by UTIs and progesterone use, and linked to hydronephrosis and pyelonephritis MEDICAL MANAGEMENT Long-term antibiotics for UTIs SURGICAL MANAGEMENT Vein excision

right-sided heart failure Right heart failure CARDIOLOGY A disorder in which the right side of the heart loses its ability to pump blood efficiently, often a complication of other disorders ETIOLOGY Left-sided heart failure, COPD, emphysema, congenital heart disease, heart-valve disease, cardiomyopathy PRECIPITATING FACTORS ↑ Activity, ↑ fluids, ↑ salt intake, high fever or complicated infections, anemia, arrhythmias, hyperthyroidism, kidney disease, drugs that affect myocardial contractility–beta-blockers, CCBs, stopping some drugs–diuretics, digoxin, ACE inhibitors RISK FACTORS Smoking, obesity, alcoholism. See Heart failure

right-to-die MEDICAL ETHICS A philosophical stance essentially equivalent to a DNR order, to be honored outside of a hospital or health care setting. See Do not resuscitate

right to refuse treatment FORENSIC MEDICINE A doctrine that a person, even if involuntarily committed to a hospital, may not be forced to submit to any treatment against his will unless a life-and-death emergency exists

right upper quadrant PHYSICAL EXAM The abdominal region that contains the liver, duodenum and head of pancreas

right ventricular failure CARDIOLOGY CHF due to ↑ filling pressure, resulting in systemic–venous congestion CLINICAL ↑ Abdominal girth, pitting edema of lower extremities, ↑ JVP, liver congestion. See Congestive heart failure. Cf Left ventricular failure

right ventricular dysplasia Arrhythmogenic right ventricular dysplasia, right ventricular cardiomyopathy CARDIOLOGY An idiopathic form of right ventricular cardiomyopathy CLINICAL Strong familial tendency with variable infiltration or replacement of myocardium by fat and/or fibrous tissue FINDINGS 10% are asymptomatic, recurrent ventricular tachycardia with left bundle branch pattern–45%, CHF–25%, heart murmur–10%, sudden death–5% DIAGNOSIS T wave inversion in V₁-V₃ of the EKGl. See Apoptosis, Cardiomyopathy

right ventricular hypertrophy CARDIOLOGY An ↑ in myocardial mass which may be due to interventricular septal defects or ↑ blood flow–eg, hyperthyroidism

rigid bronchoscope PULMONOLOGY A device used for–dramatic pause–rigid bronchoscopy, which consists of a hollow stainless steel tube, through which instruments are passed; ventilation is usually needed due to tube size; instruments used with the RB include optical telescopes, cameras, biopsy tools, forceps, suction devices, needles for transbronchial biopsies and lasers NOTE For most applications, flexible bronchoscopy is preferred. See Rigid bronchoscopy. Cf Flexible bronchoscopy

rigid bronchoscopy PULMONOLOGY Examination of the airways using a rigid bronchoscope; for most applications, a flexible bronchoscope is preferred–but when you need a cadice rectifier, *you need a cadice rectifier*; RB is the method of choice for retrieval of foreign bodies, and is commonly used for (1) massive hemoptysis–allowing large-bore suction devices to clear blood and allow ventilation; (2) laser therapy–more precise targeting of beam; facilitates removal of chunks of zapped tissue; (3) cryotherapy; (4) endobronchial stenting; (5) pediatric cases–they're squirmy little boogers and have thin-walled airways CONS Need for general anesthesia–read, OR time, inability to visualize distal airways, although thanks to optical telescopes, RB can be used to visualize lobar orifices and segmental bronchi. See Rigid bronchoscope. Cf Flexible bronchoscopy

rigid hysteroscope GYNECOLOGY A small bore endoscope with improved imaging and narrow size which may negate the need for dilatation and anesthesia See Hysteroscopy

rigid spine syndrome(s) A heterogeneous group of early onset muscle dystrophies–eg, X-linked Emery-Dreifuss syndrome, in which muscular atrophy begins by early adolescence, often accompanied by multiple contractures of the spinal and other muscle groups

rigidity FORENSIC PATHOLOGY Rigor. See Rigor mortis NEUROLOGY Excessive muscle tone. See Extrapyramidal syndrome, Lead pipe rigidity, Parkinsonism PSYCHIATRY An unreasonable resistance to change VOX POPULI Stiffness

rigor mortis Post-mortem corporal rigidity, stiff stiffness PATHOLOGY Board-like contraction of skeletal muscles that first appears in jaw and other short muscles–eg, of hand, 2-4 hrs after death; RM later appears on the trunk and extremities, peaks at 24-48 hrs; it disappears in the same order as it developed. Cf Livor mortis

586 **Riley-Smith syndrome** An AD condition characterized by benign–ie, not associated with hydrocephalus or neurologic defects–macrocephaly, multiple subcutaneous hemangiomas, and pseudopapilledema

riluzole RILUTEK® NEUROLOGY A drug with some benefit to Parkinson's disease and amyotrophic lateral sclerosis. See Amyotrophic lateral sclerosis, Parkinsonism

rim sign IMAGING GI TRACT An opacification of the margin of a congenital choledochal cyst seen by a plain abdominal film GYNECOLOGY An attenuated annular radiopacity seen in the pelvis by infusion urography corresponding to cystic pelvic mass(es), usually of ovarian origin; a similar finding may occur in benign unilateral pelvic masses with a smooth serosal contour–eg, fibroma-thecomas, cystadenofibromas ORTHOPEDICS A finding on a plain film of a posterior dislocation of the shoulder, which is often accompanied by a fracture of the medial portion of the humeral head; in a positive RS, there is a widening–$\geq$ 6 mm between the anterior rim of the glenoid fossa and the medial margin of the humerus PEDIATRICS An annular, attenuated periadrenal radiopacity seen in neonatal hemorrhage in this region, seen by high-dose excretory urography UROLOGY A series of connected, overlapping physaliferous rims seen in the nephrogram phase of selective renal angiography of advanced hydronephrosis, where attenuated curved vascular tissue surrounds dilated calices of the renal pelvis

RIMA 1. Reversible inhibitor of monoamine oxidase 2. RMA Right internal mammary artery, see there

rimantadine Flumadine® An antiviral analogue of amantadine and ±75% effective in preventing influenza A during community epidemics; it is partially effective as therapy and post-exposure prophylaxis for influenza A, as drug-resistant strains rapidly appear. See Canyon region

RIND See Reversible ischemic neurological disability

ring block ANESTHESIOLOGY A form of anesthesia used in hair transplantation, in which local anesthetics are infiltrated around the inferior and lateral borders of the hair donor site

ring enhancement IMAGING A CT finding in the brain consisting of a radiolucent zone surrounded by a faint radiodense rim, in turn is surrounded by a 2nd radiolucent zone outside of the rim, where the rings correspond to regional edema, hypervascularity and hypercellularity with early ingrowth of fibroblasts; RE is a nonspecific finding typically linked to early cerebral abscesses DIFFDX Brain tumors–eg, cystic astrocytoma, or metastases with central necrosis; IV contrast material may be used to enhance faint radiodense 'rings'

ring fracture Vertical deceleration injury TRAUMA MEDICINE A type of blunt trauma due to jumping or falling from heights, usually > 5 stories CLINICAL Injury severity score is 41–predicted survival, 50%; actual survival is less; all had multiple fractures–eg, 'ring fracture' of the skull base, separating the rim of the foramen magnum from the rest of the base and compression fractures of vertebrae, both of which occur when the victim lands on his/her feet or buttocks. See Jumper syndrome

ring sign IMAGING Annular filling defect(s) in the renal calices by IVP, which correspond to necrotic sloughed renal papilla, seen in analgesic abuse, papillary necrosis, sickle cell anemia. See Analgesic abuse, Obstructive uropathy, Papillary necrosis, Sickle cell anemia

ringworm Tinea corporis, dermatophytid; dermatophytosis DERMATOLOGY A skin infection by mold-like fungi known as dermatophytes–eg, *Trichophyton rubrum, T mentagrophytes, Microsporium canis, M gypsum*, rarely also *Epidermophyton*; in children, *T canis* is the most common agent CLINICAL TYPES Tinea corporis; tinea capitis–scalp; tinea cruris–groin, aka jock itch;

tinea pedis–feet, aka athlete's foot DIFFDX Nonfungal dermatopathies–eg, erythema annulare, 'herald patch' of pityriasis rosea, atopic dermatitis, other dermatitides TREATMENT Most resolve without therapy; otherwise, miconazole, if severe, griseofulvin. See Black dot ringworm, Gray patch ringworm, Tinea capitis, Tinea corporis, Tinea cruris, Tinea pedis

Ringer's lactate solution A standardized sterile physiologic–ie, isotonic– 0.9% solution containing calcium chloride, KCl, NaCl, sodium lactate; it is used as a topical irrigant and as a crystalloid solution for restoring fluid volumes. See Physiologic saline

Rinne test AUDIOLOGY A test in which a tuning fork is used to determine whether hearing loss is conductive or sensorineural in nature. Cf Weber test

ripe cervix GYNECOLOGY A cervix ready for delivery, which is soft, anterior, dilated, and displays early effacement. See Effacement

Ripolin Chlordiazepoxide, see there

ripple effect EPIDEMIOLOGY See Signal event

rippling effect IMAGING A descriptor for the layered angiographic appearance of blood vessels in the cortical sulci peripheral to a cerebral abscess through which the blood flows in an undulating pattern; other cerebral lesions differ as they may be associated with gyral edema or with neovascularization

rippling muscle disease A usually AD, occasionally sporadic disorder of adolescent onset, that occurs when local muscle compression evokes myoedema, followed by longitudinal contraction, moving transversely across the muscle in a 10-20 muscle fascicle wave, fancifully likened to 'plucking a chromatic scale on a harp'; while resembling myotonia, RMD is electrically silent and clinically benign. See Myotonia

Risachief Chlordiazepoxide, see there

RISC CARDIOLOGY A clinical trial–Risk of Myocardial Infarction & Death by Treatment with Low Dose Aspirin & Intravenous Heparin in Men with Unstable Coronary Artery Disease that evaluated the risk of MI and death during therapy with low-dose aspirin and IV heparin in high-risk men. See Acute myocardial infarction, Low-dose aspirin

RISCC rating Cholesterol-saturated fat index NUTRITION The ratio of saturated fat and cholesterol to calories–RISCC, a parameter used to evaluate a diet's fat content, which is stratified into low- and high-fat diets. See Saturated fatty acid. Cf Unsaturated fat

risedronate Actonel® METABOLIC DISEASE A biphosphonate used to manage osteoporosis and prevent steroid-induced bone loss INDICATIONS Postmenopausal osteoporosis, corticosteroid-induced osteoporosis, Paget's disease of bone. See Osteoporosis, Paget's disease of bone

risk EPIDEMIOLOGY The chance or likelihood that an undesirable event or effect will occur, as a result of use or nonuse, incidence, or influence of a chemical, physical, or biologic agent, especially during a stated period; the probability of developing a given disease over a specified time period. See Minimal risk MANAGED CARE The chance or possibility of loss. See Risk sharing OCCUPATIONAL MEDICINE A value determined by the potential severity of the hazard and the frequency of exposure to the 'risky' substance or activity, usually understood to mean the probability of suffering from a particular disease RISK ASSESSMENT The probability that something will cause injury, combined with the potential severity of that injury. See Absolute risk, Acceptable risk, Assigned risk, Attributable risk, Cancer risk, Cardiac risk, Dread risk, Hazard risk rating, High risk/high impact, Incremental risk, Lifetime risk, No significant risk, Nonattributable risk, Thick conception of risk, Thin conception of risk, Unknown risk

RISKS OF DISEASE

INFECTION

HBV	1:63,000
HCV	1:103,000
HIV	1:493,000
HTLV I/II	1:641,000
HAV	1:1,000,000

OTHER MORBID CONDITIONS

MVA	1:6,700
Flood	1:450,000
Earthquake	1:600,000
Lightning	1:1,000,000

risk analysis HEALTH INSURANCE The process of determining what benefits to offer and premium to charge a particular group MANAGED CARE The combination of risk assessment and risk evaluation, performed at a particular point in time

risk assessment MANAGED CARE An activity that IDs risks and estimates their probability and the impact of their occurrence; RA is integral to system development as a means of estimating damage, loss, or harm that could result from a failure to develop individual system components. See Dose response assessment, Hazard identification, Risk characterization TOXICOLOGY The process by which new chemical substances are evaluated for their potential impact on human health, a process that entails determining its toxicity and number of people exposed to it. See Ames test, Morbidity, Toxicity testing

risk communication PUBLIC HEALTH Activities to ensure that messages and strategies designed to prevent exposure, adverse human health effects, and diminished quality of life are effectively communicated to the public

risk factor An aspect of personal behavior or lifestyle, an environmental exposure, or an inborn or inherited characteristic associated with an ↑ in disease or other health-related event or condition; a variable that affects the probability of a specified adverse event. See Risk. Cf Risk equivalent

risk management The constellation of activities–planning, organizing, directing, evaluation and implementation involved in ↓ the risks of injury to Pts and employees and ↓ property damage or loss in health care facilities MANAGED CARE A management approach designed to prevent or ↓ risks–eg, system development risks and mitigation their impact

risk stratification MEDICAL DECISION-MAKING The constellation of activities–eg, lab and clinical testing used to determine a person's risk for suffering a particular condition and need–or lack thereof–for preventive intervention

Risolid Chlordiazepoxide, see there

RIST Radioimmunosorbent test, see there

risus sardonicus A fixed 'sarcastic' grimace and anxious expression with drawing up of the eyebrows and corners of the mouth due to spasms of the masseter and other facial muscles, accompanied by rigidity of neck and trunk muscles and arching of back; the RS is seen in generalized tetanus, and caused by a neurotoxin produced by *Clostridium tetani*, a soil contaminant with a 7-10 day incubation PROGNOSIS Up to 90% mortality in the unvaccinated/susceptible–eg, narcotic addicts

RITA CARDIOLOGY A clinical trial–Randomized Intervention Treatment of Angina–comparing the outcome of PCTA vs CABG in Pts with angina. See Angina, Angioplasty, CABG, Percutaneous transluminal angioplasty

ritodrine Yutopar® OBSTETRICS An epinephrine-like β-adrenergic agonist that acts on β$_2$-adrenoceptors to relax smooth muscle in arterioles, bronchi, uterus. See Tocolytic

ritonavir Norvir® AIDS A potent HIV protease inhibitor which is combined with reverse transcriptase inhibitors, nucleoside analogues to manage AIDS ADVERSE EFFECTS GI tract complaints–nausea, diarrhea, perioral paresthesiae, ↑ liver function tests, lipodystrophy–↑ glucose, ↑ fat, ↑ abdominal girth, ↓ fatty tissue from face, arms, legs, ↑ TGs, paresthesias. See AIDS, Antiretroviral, Protease inhibitor; Cf Reverse transcriptase inhibitor

ritual PSYCHIATRY Repetitive complex movements, often a distorted or stereotyped elaboration of a daily routine, used to relieve anxiety, or seen in obsessive compulsive disorder. See Obsessive-compulsive disorder. Cf Motor tic

Rituxan® Rituximab ONCOLOGY A humanized mouse antibody used to treat relapsed or refractory low-grade or follicular B-cell NHL. See Humanized antibody

Rival Diazepam, see there

river surfing SPORTS MEDICINE An 'extreme sport' in which the participant surfs à la Hawaii on white water rapids INJURY RISK Concussions, death. See Extreme sports, Novelty seeking behavior

Rivotril® Clonazepam, see there

RK 1. Radial keratotomy, see there 2.Right kidney

RLC Residual lung capacity

RLE Right lower extremity

RLF 1. Retrolental fibroplasia 2. Right lateral femoral

RLP cholesterol Remnant-like particle cholesterol CARDIOLOGY A form of cholesterol that is ↑ in Pts who have had acute MIs and have otherwise normal LDL-C and HDL-C; remnant lipoproteins result from metabolism of VLDLs; RLP cholesterol occurs in Pts with type III dyslipidemia. See Dyslipidemia

RLS Restless leg syndrome, see there

RMT 1. Registered Massage Therapist 2. Renal mesenchymal tumor

RN Registered nurse, see there

RNA Ribonucleic acid MOLECULAR BIOLOGY A polymer of ribonucleic acids that functions in coding, storage, transfer and translation of genetic information. See Antisense RNA, Catalytic RNA, Chromosomal RNA, Heterogenous nuclear RNA, Pre-mRNA, Ribosomal RNA, Transfer RNA

Ro-Azepam Diazepam, see there

roach DRUG SLANG A street term for a marijuana cigarette or joint MEDICAL ENTOMOLOGY Cockroach, see there. Cf Entomophagy

roach clip DRUG SLANG A spring-loaded pincered device used to hold a marijuana cigarette or joint

road rash EMERGENCY MEDICINE Deep skin abrasions caused by falling on and scraping skin on asphalt, which may affect bike riders, skateboarders, MVA victims and others

road-test MEDTALK To assess discharge suitability criteria by testing arousal, level of self-sufficiency, cerebellar function–gait, ataxia, ambulation, ability to understand discharge instructions. See Discharge

robo-ing SUBSTANCE ABUSE A popular term for the ingesting of Robitussin, an OTC cough medicine, in megadoses, to achieve the 'high'– euphoria, hallucinations–evoked by the dextromethorphan, a weak opiate contained in Robitussin TOXICITY Respiratory depression, coma, death. See Dextromethorphan, Ecstasy, Rave

robotic system An integrated system of devices that automate production and manufacturing of goods and services SURGERY An AI-based surgical assistant system, which processes sensory input from haptic interfaces and/or allows surgeons to act with more accuracy than the unassisted human hand. See Artificial intelligence

robust STATISTICS *adjective* Referring to any method or procedure that is relatively insensitive to violations in the method's required assumptions or rules, or a method that makes few assumptions *ab initio*

ROC 1. Receiver operating characteristic, relative operator characteristic CLINICAL DECISION-MAKING A "...*global measure of the accuracy of a diagnostic system, independent of the cutoff point used to discriminate 'normal' from 'abnormal'*" JR Beck. See ROC analysis, ROC curve 2. Reduced oxygen concentration

ROC analysis CLINICAL DECISION-MAKING The analysis of the relationship between the true positive fraction of test results and the false positive fraction for a diagnostic procedure that can take on multiple values. See 4-cell decision matrix. Cf Likelihood ratio

Rocaltrol® Calcitriol ENDOCRINOLOGY An agent used to manage 2° hyperparathyroidism and resultant metabolic bone disease in Pts with renal failure who are not yet undergoing dialysis. See Hyperparathyroidism

Rocephin® Ceftriaxone INFECTIOUS DISEASE An antibiotic for acute bacterial otitis media in children. See Cephalsporine

Rochester study CARDIOVASCULAR SURGERY A single center clinical trial which compared the benefits of thrombolysis–by intraarterial recombinant urokinase with vascular surgery–eg, thrombectomy or bypass surgery in Pts with acute–< 7 days duration, limb-threatening ischemia of the legs. See Thrombolytic therapy. Cf STILE trial, TOPAS

rock climbing SPORTS MEDICINE An 'extreme sport' in which the participant climbs rock formations, with or without ropes INJURY RISK Fractures, abrasions, death. See Extreme sports

rocker bottom foot ORTHOPEDICS Congenital vertical talus, pes valgus A rigid flatfoot deformity caused by a malpositioned navicular bone at the neck of the talus; the ankle is in severe equinus and the forefoot in dorsiflexion–ie, rocker bottom-like, accompanied by contraction of the talonavicular, deltoid and calcaneal cuboidal ligaments, the peroneus brevis and triceps surae muscles; RBF are either isolated deformities or accompany trisomies 18 and 13 TREATMENT Early manipulation and plaster correction of the forefoot into plantar flexion, inversion and adduction

rocker bottom shadow IMAGING A horizontal, broad, curved soft tissue radiodensity that partially overlies the heart, extends into the hilum, corresponding to the thymus, seen in a plain AP CXR in neonatal pneumomediastinum, and derives its name from old wooden cradles. See Spinnaker sail sign

Rocky Mountain spotted fever Spotted fever INFECTIOUS DISEASE An exanthematous disease common in the eastern US from April to October even in large cities–eg, NYC AGENT *Rickettsia rickettsii* HOSTS Furry woodland creatures–rodents, et al VECTORS Wood–*Dermacentor andersoni* and dog–*D variabilis* ticks CLINICAL 1 wk incubation, followed by a discrete pale, blanchable centrifugal maculopapular rash, which may be very dark, hence the alias, 'black measles', persistent headache, fever, ±coughs, rales, myalgia, malaise, splenomegaly; N&V, abdominal pain; CNS Sx–delirium, stupor, ataxia, meningismus; myocarditis, EKG changes, thrombocytopenia, multiple coagulopathies, renal failure, shock TREATMENT Tetracycline, chloramphenicol MORTALITY 3-10%; ↑ in blacks, ↑ > age 40. See Rickettsial infection

Rocosgen Lorazepam, see there

rocuronium Zemuron® ANESTHESIOLOGY A nondepolarizing muscle relaxing anesthetic. See Muscle relaxant, Nondepolarizing agent. Cf Depolarizing agent

rod ORTHOPEDICS A metal fixation device used to stabilize fractures. See Harrington rod

rodent INFECTIOUS DISEASE A mammal of order Rodentia–eg, mice, rats, squirrels, gerbils, chipmunks, voles, moles, et al VECTORS FOR Argentine hemorrhagic fever, Bolivian hemorrhagic fever, endemic typhus, *Francisella tularensis*, Hantavirus pulmonary syndrome, *Helicobacter cinaedi*, hemorrhagic fever with renal syndrome, *Hymenolepsis diminuta*, *Hymenolepsis nana*, Lassa fever, leptospirosis, listeriosis, lymphocytic choriomeningitis, *Ornithonyssus bacoti*-induced dermatitis, plague, rabies, rickettsialpox, salmonellosis, *Spirillum minus*, *Streptobacillus moniliformis*, tick-borne relapsing fever, *Trichophyton mentagrophytes*, Venezuelan hemorrhagic fever, *Yersinia enterocolitica*

Roe v. Wade OBSTETRICS A 'landmark' case presented before the US Supreme Court in 1973, which was instrumental in legalizing abortion

rofecoxib Vioxx® PAIN MANAGEMENT An analgesic Cox-2 inhibitor. See Cox-2 inhibitor, Cyclooxygenase

Roger's murmur Bruit de Roger CARDIOLOGY A loud pansystolic murmur maximal at the left sternal border typical of a small ventricular septal defect. See Murmur

Rogerian therapy Client-centered therapy HUMANISTIC PSYCHOLOGY A form of psychoanalysis in which a therapist with an 'unconditioned positive regard' for the client attempts to ↓ negative aspects of overdependence on others, and ↑ self-reliance. See Humanistic psychology. Cf Psychoanalysis, Psychotherapy

rohypnol Date-rape drug, flunitrazepam DRUG ABUSE A benzodiazepine that is ± 10-fold more potent than Valium ROUTE OF ADMINISTRATION Oral, pulverized/snorted EFFECTS Slurred speech, impaired judgment, hence 'date-rape' drug, difficulty walking, sedation–blackouts of up to 24 hrs, amnesia, respiratory distress; may have paradoxic effects and trigger aggression; OD or death can occur if mixed with other drugs or alcohol

'roid rage SPORTS MEDICINE An acute psychotic response–uncontrolled outbursts of anger, frustration or combativeness–of unknown pathogenesis seen in those who abuse anabolic steroids, usually in body builders. See Anabolic steroids, Weight training.

Rokitansky syndrome Mayer-Rokitansky-Kuester syndrome A congenital defect characterized by impaired differentiation of Müllerian duct; the uterus is rudimentary and cordlike; the deep vagina is absent and the outer part is shallow; fallopian tubes may be defective; there may be other sporadic congenital anomalies; ovaries are normal and induce normal femininizing puberty, except for lack of menstruation due to defective uterus; individuals are phenotypically and psychosexually ♀

Rolando fracture ORTHOPEDICS A 3-part comminuted intraarticular fracture/dislocation at the base of the thumb/1st metacarpal, in which a large dorsal fragment of the metacarpal continues to articulate with trapezium. See Fracture. Cf Bennett fracture

role-playing PSYCHOLOGY The voluntary and conscious adoption of a particular role–eg, a child adopting role of parent under another person–eg, a phychotherapist; RP provides 1. Insight on how the person playing the role perceives other person(s) and 2. A window through which the person playing the role may see the other person's vantage. See Psychodrama

role socialization PROFESSIONALISM A process in which a person incorporates knowledge, skills, attitude and affective behavior associated with carrying out a particular role–eg, physician, nurse, technologist, etc. See Affective behaviors

ROLE SOCIALIZATION

STEP 1 Develop cognitive and psychomotor skills, and affective behaviors

STEP 2 Internalize behavior and values of profession

STEP 3 Individualize professional role

STEP 4 Incorporate professional role into other life roles

'roller ball surgery' GYNECOLOGY A popular term for laser excision of uterine leiomyomas, which may make successful pregnancy impossible by preventing normal uterine growth; with RBS, the tumors are 'shelled out' by rolling the tumors while cutting with the laser

RollerLOOP™ ENDOSCOPY Electrovaporization and resection electrode for endoscopic procedures. See RF tissue ablation

rollover AUDIOLOGY A popular term for the shape of the curve in speech discrimination–SD audiometry, related to a defect in retrocochlear cranial nerve VIII, in which SD improves as sound intensity ↑ until a maximum–PB_{max} is reached, after which SD worsens. See Hearing test

ROM GYNECOLOGY Rupture of membranes. See Premature rupture of membranes ORTHOPEDICS Range of motion The arc of a joint's movement, potentially limited by musculoskeletal defects

Romberg sign Romberg test NEUROLOGY A clinical test used to evaluate dysequilibrium; excess swaying implies a severe defect in postural sensation in the lower extremities, which is classically associated with tabes dorsalis; the RS differentiates central–cerebellar from peripheral ataxia, as in the former, there is no ↑ in the ataxic movements with the eyes closed

rongeur SURGERY A forceps-like instrument used to cut indurated tissue–eg, bone. See Intervertebral rongeur

rooming-in NEONATOLOGY The placing of a newborn in the same room as the mother in the early post-partum period, which may foster maternal-fetal bonding and facilitate breast-feeding. See Bonding, Rooting

root canal DENTISTRY A popular term for the complete removal of a tooth's pulp from the root canal and filling it with an inert material; RCs are performed when the decay is too deep for amalgam to provide adequate permanent therapy

rooting reflex Rooting NEONATOLOGY The instinctive searching for the mother's nipple by a neonate, which is often accompanied by grunting, opening of the infant's mouth and sucking; the RR is elicited by touching the baby's cheeks and by the smell of milk

ROP 1. Retinopathy of prematurity, see there 2. Right occipitoposterior

rope sign NEUROLOGY An acute angle between the chin and larynx, caused by weakness of the hyoid muscles, resulting in the posterior displacement of the hyoid bone, and narrowing of the hypopharyngeal passage

ropivacaine Naropin® ANESTHESIOLOGY A local anesthetic, which is less potent and less toxic than bupivacaine. Cf Bupivacaine

Rorschach test Ink blot test, Rorschach technique of projective assessment PSYCHOLOGY A personality test in which 10 ink blots are presented to an individual for an interpretation of what is seen in the 'picture'. See Psychological testing

rosacea Acne rosacea DERMATOLOGY An idiopathic skin disorder affecting light-skinned–of Celtic descent, middle-aged–♂ more severely, ♀ more commonly, characterized by chronic inflammation of cheeks, nose, chin, forehead, eyelids, often accompanied by erythema or acne-like eruptions, prominent subcutaneous blood vessels, tissue edema ASSOCIATIONS Migraines, other skin disorders–acne vulgaris, seborrhea, blepharitis, keratitis. Cf Acne

rosary bead appearance COLON IMAGING A descriptor for the exaggerated haustral contractions that may be seen in barium studies of IBS, often accompanied by constipation, pain, and hardened, dehydrated, and pelleted–rabbit-like stools. See Irritable bowel syndrome

ROSC Return of spontaneous circulation, see there

rose spots INFECTIOUS DISEASE Innumerable transient 1-5 mm in diameter blanchable reddish macules on the lower chest and upper

abdomen, caused by bacterial emboli in skin vessels accompanied by focal aggregates of macrophages; RSs classically occur in *Salmonella typhi*-induced typhoid fever in the 1st wk of disease, coinciding with the onset of splenomegaly. See Typhoid fever

Rosenkranz™ retractor SURGERY A retractor for pediatric open heart surgery

roseola Exanthem subitum PEDIATRICS An acute herpesvirus 6 infection of infants and young children–6 months to 2 yrs–characterized by a fever and skin rash with a spring and fall pattern of infection INCUBATION PERIOD 5–15 days; high fever 105°F–generally responds well to acetaminophen, which may persist for 3–5 days and be accompanied by convulsions; between the 2nd and 4th day, the fever falls dramatically and a transient–hrs to 1 day–rash appears on trunk and spreads to limbs, neck, face. See Herpesvirus-6

roseola infantum Exanthem subitum, see there

rosette blister DERMATOLOGY An annular arrangement of sausage-shaped bullae around an eschar on the trunk, genitalia and legs in IgA dermatosis, an eruption with pruritus, most common in infants TREATMENT Sulfapyridine, corticosteroids; most resolve spontaneously in 2-4 yrs

rosiglitazone Avandia® THERAPEUTICS A thiazolidinedione used in Pts with type 2 DM to ↑ the efficiency with which the body uses insulin ADVERSE EFFECTS Flu-like Sx, headache, weight gain CONTRAINDICATIONS CHF

Ross procedure Pulmonary autograft, total aortic root replacement HEART SURGERY A procedure used for aortic regurgitation, which preserves the entire–pulmonary autograft annulus, the leaflets, and the surrounding pulmonary artery wall as a single unit; the Pts' main coronary arteries are reimplanted into the side of the autograft and encompassing pulmonary artery wall as a single unit. See Aortic regurgitation

rotary gamma scalpel ONCOLOGY A scalpel that uses rotary gamma rays to destroy tumor cells in the brain while leaving healthy tissue relatively undamaged. See Gamma knife

rotating tourniquet CARDIOLOGY A modality for ↓ preload in acute cardiogenic pulmonary edema, in which the blood flow to the extremities is blocked by RTs; because preload is ↓ more precisely with nitroprusside, RTs are rarely used

rotation Movement around an axis GRADUATE EDUCATION A period of time during which a medical student, or a physician in an early period of his training works in a particular service. See Audition rotation, Clinical rotation, Extern, Intern OBSTETRICS The turning of a fetus around its long axis such that the presenting part changes. See External rotation, Internal rotation, Limb rotation

rotation diet CLINICAL NUTRITION A low-calorie diet in which the individual 'rotates' between extreme and less stringent dieting. See Diet, Low-calorie diet

rotational thrust CHIROPRACTIC One of the 2 most common techniques for chiropractic adjustment of the vertebral column. See Chiropractic

rotator cuff tendonitis Pitcher's shoulder, shoulder/rotator cuff impingement syndrome, swimmer's shoulder, tennis shoulder ORTHOPEDICS A microtrauma or overuse injury caused by stress on or tearing of the rotator cuff, which holds the humeral head in the scapular fossa ETIOLOGY Tearing and inflammation of the tendons occurring in sports requiring the arm to be moved over the head repeatedly as in tennis, pitching baseball, swimming, lifting weights over the head. See Rotator cuff

rotavirus infection VIROLOGY RI is usually mild, but may be severe in children ≤ 2 yrs due to intense vomiting MORBIDITY > 870,000 children < age 5 die of rotavirus infection in developing countries, in contrast to 75 to

150 in the US EPIDEMIOLOGY Transmitted by fecal-oral route CLINICAL Diarrhea of 2-12 days duration; rotavirus causes 30–60% of gastroenteritis in children, it is a major cause of epidemic and endemic gastroenteritis, including traveler's diarrhea; RI in immunocompromised Pts may disseminate and involve the liver and kidney MANAGEMENT Symptomatic, rehydration PREVENTION Vaccine. See Astrovirus, Rotavirus vaccine, VP2, VP4, VP6, VP7, West-to-East phenomenon. Cf Norwalk virus

rough sex Sexual activity in which one or both participants risk bodily harm or even death should one lose control. See Kinky sex, Paraphilia, Sexual asphyxia, Sexual deviancy

roughage NUTRITION Indigestible complex carbohydrates–eg, bran and cellulose of plant origin that form the bulk of stools; roughage absorbs water, acting as a laxative, sequesters bile acids and degradation products; ↑ dietary roughage is linked to ↓ risk of diverticulosis, colorectal CA, ↓ cholesterol. See Bran. Cf Slops

rouleaux HEMATOLOGY Stacks of RBCs in groups of 4–10 cells, likened to stacked coins; artifactual rouleaux occur in the thick area of any blood smear; true rouleaux formation occurs with ↑ protein, fibrinogen and globulins, ↑ ESR due to dextran and monoclonal gammopathies–myeloma, Waldenström's disease, cryoglobulinemia, sarcoidosis, cirrhosis. See LISS

round *verb* To visit those Pts for whom a physician is responsible. See Rounds

rounds MEDTALK Bedside visits by a physician–or other health professional, to evaluate treatment, assess current course and document the Pt's progress or recuperation. See Grounds rounds, Plate rounds, Professorial rounds, Protease inhibitor rounds, Round, SOAP. Cf Grand rounds

roundtable A conference or discussion involving several participants, often before an audience and open for questions. Cf Keynote address, Plenary session

routine intervention MEDTALK An intervention to monitor or maintain a clinical status quo–eg, daily lab tests, regular radiographic exams, routine determination of vital signs and weight, pulmonary hygiene, frequent turning, care of pressure sores

routine orders MEDTALK A battery of diagnostic and therapeutic procedures ordered by a physician when a Pt presents with a constellation of clinical findings. Cf Standing orders

roux-en-Y operation SURGERY A procedure with a Y-shaped anastomosis including small intestine; the distal resected end is implanted into an organ–eg, bile ducts–choledochojejunostomy and portoenterostomy, esophagus–esophagojejunostomy, and pancreas–pancreaticojejunostomy; the proximal end is implanted into the small intestine further 'downstream' to prevent reflux

Rovsing sign CLINICAL MEDICINE Pain in right abdomen at McBurney's point when pressure is exerted at the same level on left abdomen, a typical finding of appendicitis. See Appendicitis

Rowexetina Fluoxetine, see there

Roxicodone™ Oxycodone PAIN MANAGEMENT A formulation of oxycodone, an opioid analgesic indicated for severe pain. See Oxycodone

RPR test Rapid plasma reagin test, reagin test LAB MEDICINE A test for syphilis in which the Pt's serum is incubated with reagin, a lipoidal antigen containing cardiolipin–diphosphatidylglycerol, cholesterol, lecithin. See Biological false positivity

RPT Registered Physical Therapist

RQ Respiratory quotient. See Pulmonary function tests

RR 1. Recovery room 2. Recurrence rate 3. Rate regular, a measure of the heart rate 4. Relative risk 5. Respiratory rate

RSD Reflex sympathetic dystrophy, see there

RS₃PE syndrome Remitting seronegative symmetrical synovitis with pitting edema A subgroup of Pts with rheumatoid arthritis with sudden onset, seronegativity for rheumatoid factors MANAGEMENT Aspirin, NSAIDs PROGNOSIS Excellent. See Rheumatoid arthritis

RSR Regular sinus rhythm, see there

RSV 1. Respiratory syncytial virus, see there 2. Rous sarcoma virus, see there

RT 1. Radiation therapy 2. Radiologic technologist 3. Reaction time 4. Real time 5. Registered technician 6. Rehabilitation therapist 7. Rehabilitation therapy 8. Relaxation time 9. Renal transplantation 10. Respiratory therapist 11. Respiratory therapy 12. Resuscitation team 13. Room temperature

rT₃ Reverse T₃, see there

RTA Renal tubular acidosis, see there

RU 486 POPULATION CONTROL A synthetic 19-norsteroid abortifacient with potent antiprogestational and antiglucocorticoid properties SIDE EFFECTS Minimal pelvic pain, fatigue, nausea, headache; used for emergency postcoital contraception POTENTIAL USES FOR RU 486 Inhibition of ovulation, blocking of secretory endometrium, induction of labor, treatment of endometriosis and breast CA. See Contragestive. Cf Morning-after pill

rub See Friction rub

rubber dam Dental dam A sheet of latex rubber punctuated with holes that is stretched around the teeth during dental procedures to isolate the operative field from the oropharynx, and respiratory tract, and prevent pulmonary aspiration of fragments of teeth and amalgam

rubella German measles, Röteln, three-day measles, third disease INFECTIOUS DISEASE An acute, benign, but potentially teratogenic viral infection, most commonly affecting children CLINICAL, ACQUIRED Most common in children age 5-15; after a 2-3 wk incubation, an evanescent–3-5 day maculopapular rash begins on the face, neck and spreads caudally, accompanied by mild fever, malaise, sore throat, cervical lymphadenopathy and palatal enanthema–rose spots coalescing into a reddish blush in the fauces and an evanescent, rapidly extending maculopapular rash with innumerable, variably-sized lesions resolving by day 3 MANAGEMENT If mother is pregnant while acutely infected, abortion is advised, given frequency of congenital rubella syndrome. See Congenital rubella syndrome, Expanded rubella syndrome, TORCH. Cf Rubeola

rubella vaccine See MMR vaccine

rubeola Measles. see there. Cf Rubella

Rubin test GYNECOLOGY A test designed to detect obstruction of the fallopian tubes, which would prevent pregnancy; in the RT, CO₂ is blown into the cervix under carefully monitored pressure; if the tube is patent, the CO₂ passes into the peritoneal cavity and is absorbed; because the RT merely IDs obstruction, hysterosalpinogography is preferred by many. See Effacement

rubitecan ONCOLOGY A topoisomerase inhibitor in trials for treating ovarian, colorectal, gastric, liver, lung, breast, pancreas, prostate, cervical, head & neck CA, myelodysplastic syndrome, CML, melanoma, glioma, sarcoma

rubor Redness, a cardinal sign of inflammation

ruby laser DERMATOLOGY A high-speed laser used for hair removal and permanent hair reduction over large areas of the body. See Laser

RUE Right upper extremity, right arm

Rufen® Ibuprofen, see there

rugger jersey spine appearance A descriptor for a osteosclerosis most marked in vertebral end plates adjacent to joint spaces, fancifully likened to a 'rugger jersey'; RJSA is seen in 2° hyperparathyroidism—renal osteodystrophy, renal tubular disease—eg, Fanconi syndrome, renal tubular acidosis, vitamin-D-resistant rickets, chronic glomerulonephritis, DM, HTN

rule VOX POPULI A statement of the parameters usually associated with a particular condition or state. See Birthday rule, Chambon's rule, Cram-down rule, Discovery rule, Durham rule, Eight-hr rule, Federal Rules of Evidence, Federal Rules of Civil Procedure, Federal medical privacy rule, 55 rule, FFP rule, Frye rule, Gag rule, GU-AG rule, Ingelfinger rule, Haldane's rule, Locality rule, Loser pays rule, Marquis of Queensbury rule, M'Naghten rule, N-end rule, Normal rule, Prudent person rule, Safe harbor rule, Trapezoidal rule

rule of the artery OSTEOPATHY A tenet that holds that most, if not all, diseases that respond to osteopathic manipulation do so because of compromised blood circulation. See Osteopathy. Cf Rule of the nerve

rule of the forceps OBSTETRICS A simple rule that serves as a reminder of which handle of the forceps should be used on which side of the Pt

rule of the nerve CHIROPRACTIC A posit that holds that most, if not all, diseases that respond to chiropractic manipulation do so because of released impingement on the structures—nerves, but also blood and lymphatic vessels that pass through intervertebral foramina. Cf Rule of the artery

rule of nines CRITICAL CARE A method for rapidly assessing the extent of burns on the skin surface, which determines the amount of fluid required as replacement therapy; the head and arms each represent 9% of skin surface; anterior or posterior surface of legs represent 9% each; anterior and posterior truncal skin represent 18% each; inguinal area 1%

'rule of rescue' MEDTALK The perceived duty to save endangered life whenever possible. See Coby Howard, Health care rationing. Cf Good Samaritan laws, Oregon plan

rule-out CLINICAL DECISION-MAKING *verb* To eliminate as a serious diagnostic consideration, as in to 'rule out' the presence of acute MI; the term that has long had currency as a verb, but more recently has been popularized as a noun, as in a Pt–a 'rule out' whose medical condition is a diagnostic dilemma

'rum fits' SUBSTANCE ABUSE Generalized alcohol withdrawal-related convulsions that present as status epilepticus, seen in chronic alcoholics 12-48 hrs after a major decline in blood alcohol levels CLINICAL Hypocapnia and hypomagnesemia have a postulated but unproven role; $\frac{1}{3}$ of cases progress to delirium tremens. See Alcoholism. Delirium tremens. Cf 'Cold turkey'

runner DRUG SLANG A person who transports drugs, but may not sell them. See Body packer. Cf Dealer EMERGENCY MEDICINE A nurse or personnel responsible for obtaining supplies or medications needed during emergency CPR. See Cardiopulmonary resuscitation, Code blue

runner's 'high' SPORTS MEDICINE A state of euphoria experienced by those who run long distances, attributed to endorphins and enkephalins. See Acupuncture

runner's rump SPORTS MEDICINE A dermatopathy seen in long-distance runners, characterized by small ecchymoses, which occurs on the superior portion of the gluteal cleft; RR is attributed to the flapping of butt cheeks against each other with the running stride

runny nose VOX POPULI → MEDTALK Rhinorrhea

runs 1. Multiple of run, see there 2. Popular for diarrhea, usually expressed as 'the runs'

rupture MEDTALK A tearing or disruption of a membrane or flattened tissue that was subjected to pressure. See FASIAR rupture, Partial rupture, Premature rupture of membranes, Traumatic disk rupture

ruptured aortic aneurysm SURGERY The transmural disruption of the integrity of the aorta, which usually occurs in a background of aortic dissection CLINICAL Onset may be abrupt or gradual; pulsatile abdominal mass, early satiety, N&V, constipation, pain or claudication, waxing and waning abdominal paining which may remit spontaneously, rebound tenderness LAB ↓ RBCs, ↓ hematocrit DIAGNOSIS Hx, abdominal film, arteriography, ultrasound, CT MANAGEMENT Surgery is mandatory PROGNOSIS May be fatal if untreated. See Dissecting aneurysm

ruptured disk Herniated disk, see there

rural area DEMOGRAPHICS An area, defined by the US Census Bureau, as a non-urbanized area with < 2500 people EPIDEMIOLOGY An area outside of a metropolitan statistical area, defined by the Office of Management and Budget. See Paramedic

Rush rod ORTHOPEDICS An unreamed intramedullary rod with a chisel-like tip, and is commonly used for fibular shaft fractures, and occasionally in other tubular bones as well.

Russell-Silver syndrome Russell-Silver dwarfism PEDIATRICS An idiopathic and sporadic condition characterized by extremely short stature and body and facial asymmetry, excessive sweating, a small triangular face which makes the skull look large by comparison, inward curved fingers, cafe-au-lait spots. See Dwarfism

Russell traction ORTHOPEDICS A splintless type of balanced lower limb traction effected by holding the skin of the whole leg with adhesive plaster. See Traction

rusty lung Brown induration of the lungs due to accumulation of hemosiderin-laden macrophages. See Heart failure cells

rusty sputum A descriptor for the sputum produced in pneumonia caused by S pneumoniae and K pneumoniae; which is composed of bacteria, hemorrhage, mucus, sloughed necrotic lung tissue

RUTH CARDIOLOGY A multinational trial–Raloxifene Use for The Heart to determine the ability of raloxifene—Evista® to prevent acute MIs and cardiac deaths in postmenopausal ♀. See Osteoporosis

RVH 1. Renovascular HTN 2. Right ventricular hypertrophy, see there

RVU Relative value unit, see there

Rx Prescription, therapy

Rye Staging System A system for staging Hodgkin's disease delineated in 1966, which divided it into 4 distinct clinicopathological categories, based on data provided by the then popular method of lymphangiography. Cf Ann Arbor classification, Cotswolds classification

RYE CLASSIFICATION–Hodgkin's disease

LYMPHOCYTIC PREDOMINANT 10% of cases, more common in young adult ♂, usually early/low-stage disease without 'B' symptoms, 5-year survival: 90%

NODULAR SCLEROSING 40%; more common in young adult ♀, usually early and low-stage disease with or without 'B' symptoms, 5-year survival: 70%

MIXED CELLULARITY 40%; more common in young adult ♂, often stage II-III disease with or without 'B' symptoms, 5-year survival: 50%

LYMPHOCYTE-DEPLETED 10%; more common in older adult ?, usually stage III-IV disease with or without 'B' symptoms, 5-year survival, 30%

Rynatuss® Chlorpheniramine, see there

S 1. *Salmonella* 2. *Spirillum* 3. *Staphylococcus* 4. *Streptococcus*

S 1. Entropy 2. Mean dose/unit cumulated activity 3. Period of DNA synthesis in the cell cycle 4. Sacral 5. Saline 6. Secretin 7. Serine 8. Serum 9. Siemens–SI derived unit, electrical conductance 10. Slow muscle–physiology 11. Small 12. Sound–audiology 13. Standard 14. Standard deviation 15. Streptomycin 16. Subject–psychology 17. Subluxation–chiropractic 18. Substrate 19. Sulfamethoxasole 20. Sulfur 21. Superior 22. Svedberg unit 23. Synthesis–phase, cell cycle 24. System

s 1. *semi*, Latin, half 2. *sinistra*, Latin, left

s 1. Distance 2. Second-time 3. Secondary 4. Sedimentation coefficient–physical chemistry 5. Standard deviation

S₁ First heart sound, which corresponds to the closure of the mitral and tricuspid valves, which is a long, low, "lubb" that occurs during systole. See S₂, S₃, S₄

S₂ CARDIOLOGY The 2nd heart sound, which corresponds to the closure of semilunar–pulmonary and aortic valves, which is a short, higher pitched "dubb".

S₃ Third heart sound, a soft prolonged sound, which corresponds to vibrations in the wall of the ventricle caused by rapid filling of same; the S₃ is normal in young athletes, but is usually abnormal.

S₃ gallop Third heart sound CARDIOLOGY The heart sound that coincides with the onset of a low frequency diastolic wave generated by rapid filling; the SG is best heard at the apex of the left lateral decubitus position; the SG indicates ↓ ventricular compliance, and is characteristic of CHF; 'physiologic' SGs occur in children and young adults. See Congestive heart failure

S₄ 1. Fourth heart sound, which corresponds to atrial contraction during presystole; it is rarely heard in normal hearts.

S&M Sadism/masochism, bondage/discipline SEXOLOGY A paraphilia à deux, in which sexual partner A metes physical abuse/injury, to party B, for mutual eroticism. See Masochism, Paraphilia, Sadism

S phase analysis LAB MEDICINE Analysis of the S phase fraction or index–SPF of cells during the cell cycle, which corresponds to the rate of tumor cell proliferation; it is measured by flow cytometry–eg, in a 'work-up' for certain malignancies. See Ki-67, Mitotic activity

S phase fraction ONCOLOGY The proportion of cells in the S phase of the cell cycle, which reflects a tumor's rate of proliferation. See S phase analysis

S pouch Parks pouch SURGERY A reservoir that allows preservation of anal continence in ulcerative colitis after total colectomy. See J pouch

S/S Staples & sutures

S/TOP Selective tubal occlusion procedure GYNECOLOGY A procedure for non-surgical ♀ sterilization. See Tubal ligation

the 'S' word MEDTALK Slow, referring to a current workload or activity level. Cf the 'B' word, the 'Q' word

SA node Sinoatrial node, see there

SAARD Slow-acting antirheumatic drug, DIMARD–disease-modifying antirheumatic drug RHEUMATOLOGY An agent used to manage rheumatic complaints when first-line therapy–NSAIDs, fails to control disease; SAARDs have delayed onset of action and possible disease-modifying potential FIRST-LINE SAARDs Hydroxychloroquine, IM gold SECOND-LINE SAARDs D-penicillamine, MTX, azathioprine, due to toxicity. See Pyramid model

SAB Spontaneous abortion. See Abortion

Saber BT™ trocar SURGERY Single-use blunt-tip disposable trocar used for endoscopic surgery to establish a pathway for instrumentation and to permit maintenance of insufflation. See Endoscopic surgery

saber shin A descriptor for the thickened, anteriorly-bowed–with possible lateral bowing tibial cortex, caused by chronic periostitis which spares the epiphysis, first described in latent congenital syphilis, which may also be seen in advanced acquired syphilis

Sabolich socket system ORTHOPEDICS A socket system for improved comfort, stability, fit and control for Pts with prosthetic limbs

sabot heart Couer en sabot PEDIATRIC CARDIOLOGY A heart with a boot-like radiologic silhouette, in which the tip of the 'boot's toe' corresponds to an elevated cardiac apex, while the broad 'foot' portion corresponds to an ↑ prominence of the left cardiac border resulting from right cardiac hypertrophy; SH is typical of well-developed Fallot's tetralogy–infundibular pulmonary valve stenosis, VSD, right cardiac ventricular hypertrophy and a dextraposed/overriding aorta

sac MEDTALK A pouch or other enclosed structure. See Alveolar sac, Gestational sac, Pouch, Yolk sac

saccadic eye movement NEUROLOGY Rapid symmetrical jerking eye movements with constantly changing retinal foci from one point to another

saccharin NUTRITION A cyclic imine of 2-sulfobenzoic acid, which is 500 times sweeter than sugar, and used as an artificial sweetener. See Artificial sweeteners. Cf Aspartame, Sweet protein

Saccharomonospora viridis A fungus that causes hypersensitivity pneumonitis, detectable by RAST

saccharopinuria An AR condition characterized by abnormal lysine metabolism due to inactivity or absence of aminoadipic semialdehyde-glutamate reductase, EEG abnormalities CLINICAL Moderate mental retardation, spastic diplegia, short stature LAB ↑ Excretion of saccharopine, lysine, citrulline in urine

saccular aneurysm An eccentric aneurysm due to weakening of one side of a vessel wall

sacral *adjective* Referring to the sacrum

sacro-occipital technique OSTEOPATHY A type of craniosacral manipulation. See Cranial osteopathy, Osteopathy

SAD Seasonal affective disorder, see there

saddle back curve A descriptor for the febrile curve of trench fever, where 3-5 days of high temperature are followed by 3-5 days of low temperatures, occurring in up to 8 waves before resolution; a similar fever curve may also occur in bartonellosis, chinkungunya, Colorado tick fever. Cf Camelback curve, Pulse-temperature dissociation

saddle embolus A large thrombus lodged at an arterial bifurcation, where blood flows from a large-bore vessel to a smaller one; the 'classic' SE, which occurs at the bifurcation of the pulmonary arteries in fatal PTE, 2° to a centrally migrating venous embolus, is uncommon. See Pulmonary embolism

SADHAT CARDIOLOGY A clinical trial–Sertraline or Zoloft® Antidepressant Heart Attack Trial

sadomasochism S/M PSYCHIATRY A popular term for the coexistence of sadism and masochism either as 1 An integral part of sexuoeroticism. See Paraphilia, Sexual deviancy or 2 A component of a person's personality profile, where passive and submissive–masochistic attitudes coexist with aggressive and cruel–sadistic attitudes. See Passive-aggressive personality disorder

SADR 1. Severity-adjusted death rate, see there 2. Suspected adverse drug reaction, see there

safe VOX POPULI Any activity, or element–instrument in the environment for which the risks of its use and disposal are considered acceptable

'safe' blood TRANSFUSION MEDICINE Packed RBCs and blood products from persons with no known risks for exposure to transfusion-transmissible microorganisms–TTM. See Blood shield laws, Euroblood, Surrogate marker

SAFE CARDIOLOGY A clinical trial–Safety After Fifty Evaluation

safe cigarette TOBACCO CONTROL An oxymoron for a cigarette–fire-safe, low-tar, smokeless, said to ↓ risks–CA, emphysema, COPD, etc, associated with tobacco use. See Eclipse, Premier

safe harbor rule ANTITRUST LAW A federal guideline as to what constitutes antitrust activity, established by the FTC and Justice Dept, after specific legislation–which might be open to misinterpretation–is enacted. See Self-referral

safe sexual activity SEXOLOGY A group of guidelines on the relative safeness of various sexual activities, in terms of the potential for transmitting infection, in particular HIV-1; these guidelines are of use in advising Pts. See Paraphilia, Sex industry, Sexual deviancy, Sexually transmitted disease

SAFE SEX GUIDELINES

SAFE SEXUAL ACTIVITIES Mutual masturbation–male or female, social kissing–dry, taking a bath together, body massage, hugging, body-to-body contact–frottage to orgasm of any external body part, light sado-masochistic activities–without bruising or bleeding, licking dry healthy skin

RISKY SEXUAL ACTIVITIES–insufficient data is available–should be considered with caution French–wet kissing, vaginal or anal intercourse with a condom, fellatio without ejaculation, fellatio interruptus or into condom, cunnilingus, urine contact–'water sports', ie urination on sexual partner, stimulation to orgasm between the thighs or buttocks, digital-anal sex without a barrier

UNSAFE SEXUAL ACTIVITIES–risk increases with multiple partners Vaginal or anal intercourse without a condom, manual-anal intercourse–'fisting', nonprotected oral-anal contact–anilingus, 'rimming', nonprotected manual-vaginal contact, fellatio totalis–oral-semen contact, swallowing menstrual blood, swallowing vaginal fluids, blood contact, sharing of sex toys

safe weapon PUBLIC HEALTH An oxymoron for a next-generation device being developed for law enforcement as a means of incapacitating unruly persons or crowds without causing permanent injury

safety PUBLIC HEALTH The state of being secure or safe from injury, harm, or loss; a judgment of the acceptability of risk–a measure of the probability of an adverse outcome and its severity associated with using a technology in a given situation–eg, for a Pt with a particular health problem, by a clinician with certain training, or in a specified treatment setting. See Injury, Injury prevention, Negotiated safety, On-line safety

safety committee CLINICAL TRIALS A group of workers who function independently of the investigators of a trial, who are charged with ensuring the well-being of the trial's participants, in accordance with Helsinki Declaration principles. See Helsinki declaration.

safety net HEALTH ACCESS Any mechanism of health care provision for uninsured Pts by direct–ie, mandated–federal/state funding or indirect/tacit–write-offs vehicles

safety pacing–ventricular CARDIAC PACING In some A-V sequential–DVI and A-V universal–DDD pacemakers, following atrial pacing, the pacemaker is designed to trigger a ventricular pacing output if ventricular sensing occurs during the first portion–e.g., 110 ms of the programmed A-V interval. See Pacemaker

safety seat Car seat, child safety seat PUBLIC HEALTH A seat with a safety belt which is required by law for transporting infants in cars. See Air bag

SAFHS Sonic Accelerated Fracture Healing System ORTHOPEDICS A device that emits sound waves to the Fx site, accelerating healing

Safil® suture SURGERY An absorbable multi- or monofilament suture thread composed of synthetic polyglycolic acid polymer INDICATIONS Soft tissue approximation and/or ligation–eg, in ophthalmic surgery, but not for cardiovascular surgery, microsurgery or neural tissue. See Suture

SAGE GERIATRICS 1. A clinical study–Study Assessing Goals in the Elderly 2. A population-based dataset–Systematic Assessment of Geriatric Drug Use via Epidemiology–that contains data on nursing home Pts and combines information from the MDS–Minimum Data Set and the On-Line Survey & Certification Automated Record. See Geriatrics

sagging face COSMETIC SURGERY A change–eg, expressive furrows, moderate to severe wrinkles, nasal obstruction, jowling or neck laxity, brow ptosis, sagging skin and eyelids, caused by subtle degenerative changes of aging that impact on soft tissues and structural landmarks. See Aging, Aging face syndrome, Aging skin

sagging rope sign PEDIATRIC ORTHOPEDICS A thin opaque curved line, which in the AP–or posterior view extends laterally from the inferior border of the femoral neck–FN; the SRS is typical of Perthes' disease and is seen when ischemia and growth occur simultaneously; it implies epiphyseal growth plate damage DIFFDX Spondylo-epiphyseal dysplasia, achondroplasia, cretinism

SAHS Sleep apnea/hyponea syndrome. See Sleep apnea syndrome

sail sign IMAGING A sharply demarcated triangular radiopacity seen in the mediastinum of 10% of normal children, which corresponds to the thymus and disappears on inspiration. See Rocker-bottom shadow

sail vertebra A hook or wedge deformity of an upper lumbar vertebra linked to physical stress, which may be seen in early-onset hypothyroidism, achondroplasia, Morquio, Hurler, and Hurler-like syndromes, metatropic dwarfism, spondylometaphyseal dysplasia

St John's wort *Hypericum perforatum* HERBAL MEDICINE A perennial herb that contains flavonoids, glycosides, mucilage, tannins, volatile oil; it is antibacterial, anti-inflammatory, antimicrobial, astringent, expectorant, sedative; as an antidepressant, it is overrated, proving ineffective in double-blinded placebo-controlled trials TOXICOLOGY SJW may interfere with OCs, anti-HIV agents, transplant drugs, anti-depressants; it causes HTN, headaches, N&V; it may also interact with amphetamines, amino acids–tryptophan, tyrosine, antiasthmatic inhalants, beer, wine, chocolate, coffee, fava beans, cold and hay fever agents, narcotics, nasal decongestants, smoked or pickled foods. See Herbal medicine

salaam convulsion NEUROLOGY A spasm of early infancy onset that occurs 20-100 times/day; SCs either disappear by age 2 or evolve into grand mal seizures with mental retardation. See Seizure disorder

Salagen® ENT A pilocarpine used for dry mouth of Sjögren syndrome and RT for head & neck CA. See Pilocarpine

Saldino-Noonan syndrome Short rib-polydactyly syndrome NEONATOLOGY A lethal AR form of neonatal chondrodystrophy characterized by polydactyly, severe short limb dwarfism CLINICAL Hydropic appearance with narrow thorax, short flipperlike limbs, postaxial polydactyly, brachydactyly, prominent abdomen, cardiovascular defects–transposition of great vessels, hypoplastic lungs, anal atresia, genital defects

salicylate Salicylic acid PHARMACOLOGY The analgesic derivative of aspirin, used topically as a keratolytic. See Salicylism

salicylism TOXICOLOGY Acute Sx due to aspirin–salicylate overdose, causing a severe metabolic acidosis and a large anion gap; early, salicylates stimulate respiration by central mechanisms–$\downarrow$ PaCO$_2$ and plasma bicarbonate levels; salicylates interfere with mitochondrial metabolism; when mitochondria degenerate, there is a release of unmeasured organic acids, and $\uparrow$ anion gap CLINICAL Severe headache, N&V, ringing in ears, confusion, $\uparrow$ pulse, $\uparrow$ respiratory rate. See Salicylates

saline *adjective* NURSING Referring to salt or a saline solution *noun* NURSING A solution of salt and water. See Physiologic saline

saline abortion OBSTETRICS Voluntary termination of pregnancy during the 2nd trimester, effected by replacing 200 ml of amniotic fluid with 200 ml of 20% saline solution, which stimulates uterine contraction, followed by fetal delivery in 12-24 hrs. See Abortion

Salinetrode™ SURGERY A group of cutting loop and vaporization electrodes that allow use of isotonic saline solution for safer endoscopic surgery

saliva Spit The clear, semifluid secretion of the major and minor salivary glands, and mucus-secreting cells of the oral cavity; saliva keeps the oral cavity moist, lubricates food during mastication–which facilitates deglutition, and, via its enzyme content-alpha amylase, begins the process of digestion. See Salivary glands, Sputum

salivary duct stone Salivary gland stone, sialolithiasis ENT The presence of a crystallized concrement in one or more salivary glands; SGSs are most common in the submandibular/submaxillary gland, less in the parotid glands CLINICAL Swelling of affected gland, continuous pain, duct obstruction

salivary duct tumor ENT A salivary gland neoplasm, most of which arise in the epithelium; the most common SGT arises in the parotid gland, and is usually benign DIFFDX Swelling in response to local infection, stones, systemic disease–eg, cirrhosis, sarcoidosis, abdominal surgery, CA, infection. See Mixed tumor

salivary gland biopsy The removal of a small specimen of salivary gland to diagnose CA, Sjögren syndrome, etc RISKS Local bleeding, infection, facial nerve injury

salivary gland disorder Salivary gland disease ENT Any disorder that causes swelling or pain of the salivary glands. See Salivary duct stone, Salivary gland tumor, Sialadenitis

salivary gland infection Sialadenitis, see there

salivary insufficiency Dry mouth, see there, aka xerostomia

Salk vaccine Inactivated Polio Vaccine An inactivated vaccine used to prevent polio. See Immunization, Polio

Salla disease An AR condition seen in the Salla region of Finland CLINICAL Psychomotor retardation, coarse facies, clumsiness, slow speech, spasticity, ataxia LAB Pts excrete and store 10-30-fold normal quantities of N-acetyl-neuraminic acid–NANA, sialic acid

salmeterol A β$_2$-adrenoceptor agonist used to control asthma and prevent high-altitude pulmonary edema

***Salmonella* enteritis** Salmonellosis INFECTIOUS DISEASE Swelling of small intestinal mucosa by *Salmonella* spp, especially *S typhi*, due to ingestion of contaminated human or animal secretions or food, followed by a 8-48 hr incubation; acute illness lasts 1–2 wks; bacteria are shed in the feces for months to a yr or more ETIOLOGY Pet reptiles, consumption of poultry CLINICAL Fever with afternoon spikes, nonspecific abdominal pain, anorexia, weight loss, changed sensorium MANAGEMENT Generally symptomatic; if needed, chloramphenical, or multidrug antibiotic regimen PROGNOSIS Poor if perforation occurs

salpigectomy GYNECOLOGY Excision of a fallopian tube

salpigitis GYNECOLOGY 1. Inflammation of fallopian tube 2. Pelvic inflammatory disease, see there

salpingo-oophorectomy GYNECOLOGY Surgical excision of a fallopian tube and attached ovary

salpingo-oophoritis 1. Inflammation of the fallopian tube and ovary 2. Pelvic inflammatory disease, see there

salpingo-peritonitis 1. Inflammation of the fallopian tube and peritoneum 2. Pelvic inflammatory disease, see there

salpingoscopy GYNECOLOGY A method for evaluating uterotubal pathology, in which a fiberoptic endoscope–salpingoscope is inserted through the cervical canal and the uterus and fallopian tubes are visualized directly. Cf Electronic fluorography, Hysterosalpingography

SALT ONCOLOGY 1. Sequential aggressive local therapy 2. Skin-associated lymphoid tissue. See MALT

salt & pepper appearance An appearance in which minute hyperpigmented lesions or radiopaque 'granules' are mixed with equally-sized hypopigmented lesions or radiolucent granules

salt water drowning FORENSIC PATHOLOGY Drowning due to seawater aspiration, resulting in fluid-filled alveoli–pulmonary edema and a V/Q–ventilation/perfusion imbalance; the shifts of fluids and electrolytes in SWD result in hemoconcentration, CHF, hypernatremia. See Drowning. Cf Fresh water drowning

salutary Healthy, beneficial

salvage chemotherapy ONCOLOGY A treatment modality consisting of high-dose chemotherapy and BMT, for cancer Pts who have 'failed' one or more protocols. See Heroic therapy

salvage cryosurgery UROLOGY A type of prostate surgery for advanced prostate CA in Pts unresponsive to RT COMPLICATIONS Impotence, incontinence, obstructive urinary Sx. See Prostate cancer

salvage pathway ONCOLOGY A popular term for a metabolic pathway that uses substrates other than the usual biosynthetic intermediates for a product–eg, 1. The salvage of free purines from the hydrolysis of nucleotides for the generation of new nucleotides, or 2. The 2° pathway of nucleotide synthesis, in which thymidylate is produced by thymidine kinase from thymidine, rather than the usual path in which deoxyuridylate is methylated by thymidylate synthetase into thymidylate, a pathway utilized by Pts with megaloblastic anemia. See Leucovorin 'rescue'

Salvarsan MEDICAL HISTORY An antisyphilitic arsenical compound synthesized by Paul Ehrlich. See Magic bullet

SAM CARDIOLOGY Systolic anterior motion An anomalous movement of the mitral valve's anterior leaflet, which appears to strike the interventricular septum in early diastole; seen in Pts with idiopathic hypertrophic subaortic stenosis and obstruction

same-day surgery MANAGED CARE Any operation which, in absence of complications may be provided at a hospital on an outPt basis. See ASC surgical services

same-sex partner SOCIAL MEDICINE A domestic partner of the same genotypic sex. See Homosexual

sampling STATISTICS The obtaining of representative material from a population SURGERY A procedure that obtains a soupçon of material for pathologic evaluation, without a formal attempt at complete removal of a suspected or confirmed lesion. See Cluster sampling, Inferior petrosal sinus sampling

sampling error SURGICAL PATHOLOGY An error in a diagnostic work-up, in which insufficient, inadequate, or non-representative material is obtained for analysis, especially biopsy material, less commonly, cytologic samplings

sanction MANAGED CARE Any penalty, punitive or disciplinary action imposed on an institution by a regulatory or governmental agency, 3rd-party payer, or other MEDTALK Any penalty, punitive or disciplinary action against a physician imposed by an institution's medical staff

sanctuary site ONCOLOGY A region of the body–eg, CNS, testes–where leukemia cells are relatively protected from the cytolytic effects of systemic chemotherapy. See Remission

Sandimmune® Cyclosporin, see there

Sandoglobulin® A proprietary human immune globulin, see there

Sandostatin LAR® Octreotide, see there

sandpaper skin A descriptor for 1 The coarse, bumpy, cool, pale, hypotrichous skin characteristic of hypothyroidism 2 The skin surface in scarlet fever with indurated hair follicles; sandpaper mucosa refers to the bright bumpy, intraoral erythema of scarlet fever that desquamates with the resolution of infection

sandwich vertebrae IMAGING A fanciful term for the radiologic findings of ↑ density of vertebral end-plates with normal bodies and preservation of intervertebral spaces, classically seen in osteopetrosis. Cf Rugger jersey spine appearence

sane Normal, *compos mentis*, not nuts

SANE Sexual assault nurse examiner

Sanfilippo syndrome Alpha-N-acetylglucosaminidase deficiency, mucopolysaccharidosis type III A common AR Tay-Sachs-like disease of late infant onset CLINICAL Coarse facies, ↓ mental development progressing to severe retardation, stiff joints, gait disturbances, speech disturbances, behavioral problems,↑ startle reflex, early blindness, doll-like facies, mental and physical deterioration, cherry red spots on retina, macrocephaly; cornea is clear; survival is longer than with Tay-Sachs, often to age 20+. See Hurler syndrome, Mucopolysaccharidosis

SangCya™ ONCOLOGY A formulation of cyclosporine, an immunosuppressant for preventing rejection in transplants, rheumatoid arthritis, plaque psoriasis. See Cyclosporine

sanitarium HEALTH CARE A term of waning popularity for a health care facility that provides long-term inpatient care for a particular condition. See Saranac. Cf Hospice

sanitary *adjective* Healthy, healthful, conducive to health

sanitary napkin Kotex, sanitary pad GYNECOLOGY An absorbent pad worn externally to soak up menstrual flow

sanitation score EPIDEMIOLOGY A score generated by the Miami-based Vessel Sanitation Program–VSP, under the CDC, based on unannounced inspections of all cruise ships that travel to US ports from international areas. See Cruise-associated diarrheal disease

sanitize FORENSIC MEDICINE A euphemism for falsification by rewriting and reformatting various hospital records, in particular those related to peer review. See Malpractice

Santos syndrome Hirschsprung's disease with renal agenesis, polydactyly, hypertelorism, deafness, megacolon

SAPHfinder™ SURGERY A surgical balloon dissector for minimally invasive saphenous vein harvesting. See Varicose veins

SAPHO syndrome A syndrome characterized by synovitis, acne, pustulosis, hyperostosis, osteitis

sapphism Female homosexuality. See Lesbianism

SAPS II Simplified Acute Physiology Score INTENSIVE CARE A 'third-generation' system for estimating in-hospital mortality in adult ICU Pts, based on assessments of most severely affected values during the 1st 24 hrs in the ICU and subjecting the results to logistic regression modeling techniques. See APACHE III, MPM II, Prognostic scoring systems

saquinavir mesylate Fortovase, Invirase® AIDS An anti-HIV protease inhibitor ADVERSE EFFECTS Nausea, diarrhea, stomach upset, ↑ LFTs, lipodystrophy–characterized by ↑glucose, ↑ lipids, ↑ waist size, ↓ fat in face, arms, legs. See AIDS, AIDS wasting syndrome, Antiretroviral, HIV, Protease inhibitor; Cf Nucleoside analogues

SAR 1. Sarcoidosis 2. Scaffold attachment region. See Matrix attachment region 3. Sexual Attitude Reassessment 4. Standard admissions ratio

sarcoglycanopathy NEUROLOGY A genetic defect of α-, β-, γ, and δ-sarcoglycans–of the sarcoglycan complex; a mutation of a gene that encodes *any* sarcoglycan leads to destabilization of the entire complex and 2° deficiency of sarcoglycan proteins; some Pts with AR muscular dystrophy have underlying sarcoglycanopathies. See Dystrophin-associated proteins, Limb-muscle dystrophy. Cf Dystrophin

sarcoidosis IMMUNOLOGY An idiopathic multisystem disease characterized by nonnecrotizing–noncaseating granulomas EPIDEMIOLOGY Relatively more common in blacks, ♀, age 30 to 50, CLINICAL Cough, dyspnea, hemoptysis, endobronchial, upper respiratory tract and laryngeal 'bumpiness' due to mural granulomas, ↓ sputum; lungs are involved in ≥ 90%, followed by lymph nodes, skin, liver, eye, spleen, bone, joints, heart, muscle, CNS; lung function tests show restrictive, less commonly, obstructive disease; ocular disease occurs in ±25%–redness, photophobia, blurred vision; other findings: erythema nodosum, lymphadenopathy, hepato- and/or splenomegaly IMMUNOLOGY ↓ Cellular immunity–anergy; ↑ humoral immunity–polyclonal ↑ Igs, ↑ IgG in lung lavage fluid IMAGING 10% of Pts have a normal CXR; 15% have pulmonary infiltrates sans lymphadenopathy; 40% have hilar lymphadenopathy; 30%–50% have infiltrates + lymphadenopathy MANAGEMENT Prednisone DIAGNOSIS ↑ Angiotensin-converting enzyme CAUSE OF DEATH Heart, lungs. See Cardiac sarcoidosis, Scar sarcoidosis

sarcoidosis of nervous system Neurosarcoidosis NEUROLOGY A complication of sarcoidosis characterized by inflammation and abnormal amyloid deposits in nervous tissues, affecting ±5% of those with sarcoidosis CLINICAL Sudden, transient facial palsy affecting cranial nerve VII, variably affecting ocular, gustatory, olfactory and other cranial nerves, variably accompanied by muscle weakness and/or sensory losses with peripheral nerve involvement; SNS may cause permanent disability and death

sarcoma ONCOLOGY A CA of mesenchymal tissues–eg, bone–osteosarcoma, cartilage–chondrosarcoma, fat–liposarcoma, fibrous tissue–fibrosarcoma, smooth muscle–leiomyosarcoma, skeletal muscle–rhabdomyosarcoma, stroma–fibrosarcoma, and vessels–angiosarcoma, KS; certain tumors are of uncertain cell lineage, but designated sarcomas–eg, alveolar soft part sarcoma, epithelioid sarcoma, Ewing sarcoma, synovial sarcoma DIFFDX Pseudosarcomas, which are either 1. Mesenchymal and non-malignant soft tissue lesions–eg, fibrous histiocytoma and fibromatoses and 2. Non-mesenchymal and malignant, most commonly spindle cell squamous carcinoma of oral cavity, anaplastic carcinoma, melanoma PROGNOSIS Clinical behavior of sarcomas is determined by tumor size–eg, > 10 cm is worse, presence of necrosis, and histologic grading–based on mitotic activity and cellular pleomorphism; sarcomas in ♂ have a worse prognosis TREATMENT Wide excision; chemo- and RT are ineffective

sarcomatoid carcinoma An epithelial malignancy that can arise in any epithelial surface–eg, oral, anorectal, bladder, etc; the cells are spindled and simulate a sarcoma DIAGNOSIS Keratin ID's by immunohistochemistry

Sarcoptes scabiei Itch mite The globally distributed arthropod responsible for crusted scabies; its modus operandi is to burrow serpiginous subcutaneous tracks, laying eggs and evoking pruritus. See Crusted scabies, Scabies

SARECCO CARDIOLOGY A clinical trial–Stent or Angioplasty after Recanalization of Chronic Coronary Occlusions

Sargramostim® Recombinant GM-CSF IMMUNOLOGY A biological response modifier that accelerates myeloid recovery in Pts with lymphomas and ALL with BM suppression by chemotherapy and/or RT SIDE EFFECTS Rash, diarrhea, asthenia, malaise. See G-CSF, GM-CSF. Cf Pseudo-orphan drug

Sarin GB, isopropyl methylphosphonofluoridate MILITARY MEDICINE A nerve gas, chemically related to certain insecticides–eg, malathion, developed as a chemical weapon by the Germans in 1936 MECHANISM Sarin is an anticholinesterase that affects nerves, muscles, glands ROUTE Aerosol, skin contact; one drop may be fatal CLINICAL Pinpoint pupils, severe headache, drooling, N&V, convulsions, severe dyspnea, respiratory paralysis TREATMENT Atropine, PAM. See Chemical warfare. Cf Tabun

Saromet Diazepam, see there

Sarotex Amitriptyline, see there

SARS Severe acute respiratory syndrome VIROLOGY A potentially severe and fatal RTI caused by coronavirus. See Coronavirus

SARS WHO CASE DEFINITIONS
SUSPECTED SARS
- **HIGH FEVER** P> 38ºC
- **RESPIRATORY Sx** Cough, or SOB, or dyspnea
- **CONTACT** Close contact with a person previously diagnosed with SARS–having cared for, lived with, or had direct contact with bodily secretions of a person with SARS
PROBABLE SARS Pt meets criteria of a suspected case and there is radiologic evidence of infiltrates consistent with pneumonia or respiratory distress syndrome l

SART 1. Sexual assault response team, see there 2. Society for Assisted Reproductive Technology A voluntary organization dedicated to improving the quality and delivery of assisted reproduction 3. Standard acid reflux test, see there

sartan family See Angiotensin II receptor antagonist–eg, Losartan

satellite *adjective* Referring to lesions, masses, patterns or radiologic densities that surround a central point. See Minisatellite

satellite abscess A typical multifocal lesion of nocardiosis, in which there is slow, indolent extension of 'daughter' or satellite abscesses from a central purulent mass, each surrounded by an incomplete fibrotic layer

satellite phenomenon HEMATOLOGY 1. Platelet satellitism The rimming of PMNs by platelets in peripheral blood, a phenomenon that may cause spurious thrombocytopenia in automated platelet counters ETIOLOGY Cryoglobulinemia, hypergammaglobulinemia, immune complexes, use of EDTA anticoagulant during blood collection 2. The rimming of RBCs by erythroid siderophages in iron-deficiency anemia

satiation factor A substance postulated as having a pivotal role in controlling body weight, which may correspond to leptin

satigrel VASCULAR SURGERY An antiplatelet agent that inhibits smooth muscle cell proliferation, intimal thickening at anastomotic sites, and platelet accumulation at the sites of prosthetic vascular grafts. See Antiplatelet therapy. Cf Thrombolytic therapy

satumomab OncoScint ONCOLOGY A murine IgG-kappa monoclonal Fab fragment directed against TAG-72, a HMW tumor-associated glycoprotein expressed in some adenoCAs–83% of colorectal and 97% of ovarian CAs, detected by immunoscintigraphy. See Monoclonal antibody

saturated fatty acid NUTRITION An animal fats–eg, butter, margarine, meat and dairy fats are rich in SFAs–eg, stearic acid; ↑ consumption of SFAs is linked to CAD. See Fatty acid; Cf Unsaturated fatty acid

saturation OCCUPATIONAL MEDICINE A measure of the maximum amount of a particular task a person can perform. See Task saturation

Saturday night palsy SUBSTANCE ABUSE A group of transient neuromuscular defects affecting a person who falls in a stuporous state in an unnatural position, classically after an alcoholic 'binge', or overdose of sedatives. See Alcoholic neuropathy

saturnine gout Chronic lead intoxication TOXICOLOGY A gout-like complex linked to drinking 'moonshine' whiskey distilled in copper tubing joined with lead solder CLINICAL Interstitial nephritis, ↓ glomerular filtration rate, HTN. See Moonshine. Cf Pheasant hunter's toe

satyriasis Male hypersexuality. See Don Juan syndrome, Nymphomania, Sexual addiction

saucerization ORTHOPEDICS A flattened, disciform defect that parallels the shaft of long bones, which may be seen on a plain film, punctuated by microcalcifications; saucerized bone defects are typical of fibrosarcoma of bone SURGERY Saucerization biopsy, see there

saucerization biopsy Deep shave biopsy DERMATOLOGY A biopsy with a broad rim of epidermis and dermis, performed by dermatologists for certain small–< 1 cm pigmented lesions–eg, lentigo senilis, seborrheic keratosis, dysplastic nevus. See Melanoma

sausage link pattern OPHTHALMOLOGY A descriptor for the marked dilatation and tortuosity of retinal veins that are focally segmented at AV crossings, in the optic fundus in grade II hypertensive retinopathy, non-proliferative diabetic retinopathy, Waldenström's macroglobulinemia or hyperviscosity syndrome

sausage toe A fanciful descriptor for weenie-like toes with edematous tenosynovitis, occasionally seen in nonspecific arthritis, Reiter syndrome, psoriatic arthritis

Sauvé-Kapandji procedure ORTHOPEDICS Surgery at the distal radioulnar joint to restore pronation and supination

SAVE CARDIOLOGY A clinical trial–Survival and Ventricular Enlargement which evaluated antihypertensive therapy with captopril, an ACE inhibitor, on Pts with left ventricular dysfunction post MI. See ACE inhibitor, Acute myocardial infarction, Captopril

SAVED CARDIOLOGY An ongoing clinical trial–Saphenous Vein De Novo evaluating the efficacy of various interventions on Pts with prior CABGs. See Coronary angioplasty

SAVS INFECTIOUS DISEASE A clinical trial–Synercid® as an Alternative to Vancomycin in Staphylococcal infections

Savvy® GYNECOLOGY A contraceptive vaginal gel that ↓ transmission of STDs–eg, HIV, chlamydia, gonorrhea. See Contraceptive

saw ORTHOPEDICS Any of a number of manual or electric devices with a serrated cutting edge, used to cut bone and/or hard tissue. See Stryker saw

sawfish pattern CARDIOLOGY A descriptor for the jagged narrowing of the left anterior descending coronary artery by angiography, a finding typical of hypertrophic cardiomyopathy. Cf Bridging

sawtooth pattern DERMATOLOGY An appearance by LM corresponding to a jagged and thickened dermal-epidermal junction, classically described in lichen planus–RADIOLOGY A jagged radiocontrast column seen by barium studies of the colon in ischemic colitis, exudative enteropathy, cathartic colon, necrotizing enterocolitis 2° to congenital megacolon–Hirschsprung's disease, and rarely in diverticulosis, a pattern attributed to a combination of edema and erosion of the mucosa

SBE 1. Self breast examination 2. Subacute bacterial endocarditis

SBLA syndrome Li-Fraumeni syndrome, see there

SBP Spontaneous bacterial peritonitis, see there

SBT Symplastin bleeding time

SC disease Sickle-hemoglobin C disease A hemoglobinopathy affecting circa 1:800 US blacks, characterized by ↑ infections–eg, bacterial meningitis, *Salmonella* osteomyelitis due to a defect in the alternate–properdin complement pathway; other effects of SC disease include osteoporosis–resulting in formation of 'fishmouth vertebrae', nephropathy with ↓ renal concentration, acidification, ↑ glomerular filtration rate, retinopathy–in ± 75% of SCD vs 15% of Pts with sickle cell anemia FUNDOSCOPY 'Black sunburst' pattern, due to ↑ glycolysis in end-arterioles; 'seafoam' pattern of proliferative retinopathy LAB ↑ 2,3 DPG, ↑ factor VIII; ↓ osmotic fragility; reticulocytes comprise 5-25% of peripheral RBCs, which have a 'holly-leaf' or navicular shape. See Sickle cell anemia

SCAB ONCOLOGY Strepozotocin, CCNU, doxorubicin, bleomycin A 'salvage' chemotherapy regimen used for Pts with disease–eg, lymphoma relapse after RT or chemotherapy. See Salvage chemotherapy

scabies Mite infestation DERMATOLOGY A severe form of infestation with the itch mite, *Sarcoptes scabiei* var *hominis*, which most commonly affects children, and spreads by direct physical contact with infected individuals–ergo it is often familial or institutional; it is characterized by intense itching that develops ± 1 month after the first exposure to the mite, which infers that the manifestations are due to hypersensitivity CLINICAL Intensely pruritic linear eruption corresponding to the tracks of the burrowing beasts; the pruritus results in excoriation and secondary pyoderma, often in the head & neck with sparing of the palmoplantar regions; while the papules and 'tracks' of the burrowing mite are most prominent on the hands and arms, the 'scabies rash' is found on the armpits, waist, buttocks, inner thigh, and ankles, sites that where *S Scabiei* is rarely found DIAGNOSIS Established by dissecting organisms or eggs from mite tunnels, placing them in 20% KOH or mineral oil and examination by the LM TREATMENT Lindane lotion, benzene hexachloride. See Crusted scabies

scalded skin syndrome 1. Staphylococcal scalded skin syndrome, see there 2. Toxic epidermal necrolysis, see there

scale CLINICAL RESEARCH A group of related measures of a variable, which are arranged in some order of intensity or importance. See Abbreviated injury scale, Abnormal Involuntary Movement scale, ADL scale, Alzheimer's Disease Assessment scale, Baker scale, Barnes Akathisia scale, California relative value studies scale, Celsius scale, Center for Epidemiologic Studies scale, Cerebral performance category scale, Cincinnati Prehospital Stroke scale, Conflicts Tactics scale, Crohn's disease activity scale, Economy of scale, Epworth sleepiness scale, Family Environment scale, Framingham Disability scale, Glasgow coma scale, General Perceived Health scale, Goldman scale, Heinrichs–Carpenter Quality of Life scale, HOME scale, International Nuclear Event scale, Injury Severity scale, Inpatient Multidimensional Psychiatric scale, Instrumental ADL scale, Intensity of Sexual Desire & Symptoms scale, Jackson scale, Jenkins Activity scale, Karnovsky scale, Katz ADL scale, Kenny Self-Care scale, Killip scale, Lanza scale, Life event scale, Likert scale, Marital adjustment scale, Miller Behavioral Style scale, MISS scale, Modified Rankin scale, MRC scale, Neonatal Behavioral Assessment scale, NIH Stroke scale, Nominal scale, Ordinal scale, Overall Quality of Life scale, Paling scale, Positive & Negative Symptom scale, Prostate Symptomatology scale, Quality of Life scale, QWB-quality of well-being scale, RBRVS scale, Richter scale, Rosenberg scale, Safety-degree scale, Schneider scale, Scoville scale, Sexual Symptom Distress scale, Simpson-Angus scale, Specific Activity scale, Spielberger scale, Unified Parkinson's disease rating scale, Visual analogue scale, Wechsler Adult Intelligence Scale-Revised scale, Zung Depression scale

Scale for Assessment of Positive Symptoms PSYCHIATRY A measure of psychosis which rates severity of psychotic Sx on a scale from 0 to 176–the highest number represents more severe disease. See Psychosis

scaling PERIODONTICS The removal of dental plaque–an early lesion that predisposes to periodontitis and 'tartar' or calculus from the crown of a tooth and/or exposed root surfaces. See Periodontal disease

scalloping Scallop sign BONE RADIOLOGY A descriptor for a semilunar erosion at the ulnar aspect of the distal radius, seen in Pts with advanced rheumatoid arthritis, caused by spontaneous rupture of the digital extensors; the 'scallop' is often more prominent as it may be rimmed by an osteosclerotic margin CHEST RADIOLOGY A descriptor for the tethering of the visceral to parietal pleura, seen in pneumothorax which develops "over" prior lung disease GI RADIOLOGY A descriptor for the appearance of *Candida* esophagitis, where there are irregular serrations, seen by a barium swallow

scalp VOX POPULI The skin and subcutaneous tissue covering the neurocranium

scalpel SURGERY A surgical cutting knife, usually understood to have a curved and disposible blade. See Rotary gamma scalpel

SCAN Suspected child abuse or neglect, see there

scan IMAGING An image of internal body structures, often using radioactive materials, to diagnose, stage, and monitor disease TYPES Liver, bone, CT and MRI scans. See Abscess scan, Biliary scan, Bone density scan, Bone scan, Brain scan, CAT scan, Cold scan, Gastric emptying scan, Hepatobiliary scan, Liver-spleen scan, 111-Indium labeled leukocyte scan, Liver scan, Magnetic resonance imaging, Meckel's diverticulum scan, Myocardial infarction scan, Perfusion scan, PET scan, Radionuclide scan, Scintigraphy, Thyroid scan, Ventilation scan, Ventilation-perfusion scan, VIP-receptor scan, Virus scan *noun* A popular short form for the 'hard copy' of various imaging procedures–eg, scintiscan, CT scan

scanner IMAGING A popular term for a device used for various imaging procedures–eg, CT scanner INSTRUMENTATION A device that measures differences in chromatic or radioactive intensity on a 2-D matrix–eg, electropherogram or chromatogram, to quantify various substances

scanning speech NEUROLOGY Speech characterized by sliding and stretching of words, and slurring of phonation, which is associated with cerebellar defects, often accompanied by inappropriate rate, range, force, and direction of voluntary movements

598

scaphocephaly Boat-shaped skull PEDIATRICS A defect in infant skull shape caused by premature closure of the sagittal, parietotemporal, sphenotemporal, and sphenoparietal sutures, resulting in an anteroposterior elongation of the skull. See Craniostenosis

scaphoid abdomen PHYSICAL EXAM A hollowed anterior abdominal wall classically seen on physical exam of neonates with diaphragmatic hernia, especially through the pleuroperitoneal hiatus or foramen of Bochdalek at the periphery of the diaphragm near the attachments of the 10th and 11th ribs; SA also occurs in jejunal atresia and in infants with pyloric stenosis; SAs in adults may be linked to extreme weight loss–eg, terminal CA, AIDS, dementia, breavement. Cf Protuberant abdomen

scar VOX POPULI Fibrous tissue that fills in defects in skin and other tissues that remain after injury. See Icepick scar, Railroad track scar, Trapdoor scar, U-shaped scar

scar cancer A cancer often located in the pulmonary apices and associated with pre-existing scars, wounds or inflammation–eg, healed TB, infarcts, abscess cavities, bronchiectasis, and metallic foreign bodies–eg, bullets; 80% of SCs are adenoCAs, 15% are SCCs, the rest are of other histopathologic types. See Passenger/driver controversy

scarlatina Scarlet fever, see there

scarlatinoid Exanthem subitum, see there

scarlet fever Scarlatina INFECTIOUS DISEASE A reaction to pharyngitis by *Streptococcus* group A–which produces an erythrogenic toxin, consisting of an oral enanthema–'raspberry' tongue, 'strawberry' tongue, generalized blanching erythema–sparing the palmoplantar region and mouth with circumoral pallor and linear petechiae–Pastia's lines. See Strep throat

Scarlett O'Hara 'syndrome' PSYCHOLOGY A term referring either to 1. Pretentious eating habits when a person is in the public eye or 2. A peer-group accepted eating 'disorder', in which young ♀ socialites attend multiple social functions in a short period and self-induce emesis to control weight

scatologia The use of obscene language and phraseology with sexual overtones. See Telephone scatology. Cf Coprolalia

SCC 1. Small cell carcinoma, see there 2. Squamous cell carcinoma, see there

SCD 1. Sickle cell disease, see there 2. Subacute combined degeneration, see there 3. Sudden cardiac death, see there

scent The smell of a thing or person. See Pseudoscent

Scharff-Bloom-Richardson grading BREAST SURGERY A method for histologic grading of breast CA

Schatzki's ring Esophageal web, see there, aka lower esophageal ring

schedule MEDTALK A listing of events or services by dates or by cost. See Call schedule, Fee schedule, MSBOS schedule, Negotiated fee schedule

schedule I agent PHARMACOLOGY An agent defined by the DEA as one with a high potential for abuse, with no accepted medical use EXAMPLES Heroin, LSD, marijuana, methaqualone. See Controlled drug substances

schedule II agent PHARMACOLOGY An agent defined by the DEA as one with a high potential for abuse or which has a currently accepted medical use with severe restrictions on its use in the US; abuse of a schedule II agent may lead to severe psychologic or physical dependence EXAMPLES Morphine, cocaine, methadone, as well as methamphetamine, amphetamine, benzedrine, biphetamine, desoxyn, dexamyl, dexedrine, dextroamphetamine, eskatrol, fetamin, methylphenidate, obetrol, obotan, PCP, phenmetrazine, preludin, ritalin. See Controlled drug substances

schedule III agent PHARMACOLOGY An agent with an abuse potential < schedules I and II agents, which have accepted medical use in the US; abuse may lead to moderate physical dependence or high psychological dependence EXAMPLES Anabolic steroids, codeine, hydrocodone with aspirin or Tylenol, barbiturates. See Controlled drug substances

schedule IV agent PHARMACOLOGY An agent with a low abuse potential relative to schedule III agents; these agents have accepted therapeutic uses; abuse of these agents may lead to limited physical or psychological dependence EXAMPLES Darvon, Talwin, Equanil, Valium, Xanax. See Controlled drug substances

schedule V agent PHARMACOLOGY An agent with a low abuse potential relative to schedule IV agents, with an accepted therapeutic use; schedule V agents may lead to limited physical or psychologic dependence EXAMPLES Cough medicine with codeine. See Controlled drug substances

scheduled downtime INFORMATICS A planned, suspension of one or more computer functions to perform maintainance or enhance system operations. Cf Extended downtime

Scheie syndrome Alpha-L-iduronidase deficiency, mucopolysaccharidosis type IS An AR condition in which a deficit of L-iduronidase results in inefficient degradation and intracellular accumulation of mucopolysaccharides; involvement is less severe than MPS type IH–Hurler syndrome CLINICAL SS is manifest in early childhood as short, coarse features, a broad mouth with full lips, inguinal hernias, cataracts, hepatosplenomegaly, aortic incompetence, stiff joints, with development of claw hands and deformed feet, clouding of cornea, prolonged survival LAB ↑ dermatan sulfate in urine

Schiller's test GYNECOLOGY A clinical maneuver used to identify cervical abnormalities requiring biopsy and histologic evaluation; areas with ↓ glycogen are pale and, by LM, often demonstrate histologic changes–eg, hyperkeratosis, parakeratosis, neoplasia associated with premalignant–eg, HPV infection or malignancy–eg, CIN, or carcinoma. See Colposcopy

Schilling test Vitamin B_{12} absorption test A test that evaluates ability to absorb vitamin B_{12}, which is ↓ without intrinsic factor, a glycoprotein produced by the gastric parietal cells. See Vitamin B_{12} deficiency

Schindler's disease Alpha-N-acetylgalactosidase deficiency An AD condition characterized by severe psychomotor retardation with myoclonic seizures, decorticate posture, optic atrophy, blindness, marked long tract signs, and total loss of contact with the environment

Schirmer's test A maneuver for measuring tear production at baseline and in response to stimuli–eg, nasal irritants. See Dry eye

Schistosoma PARASITOLOGY A genus of elongated sexually dimorphic trematodes, the blood flukes, of phylum Platyhelminthes, Class Trematoda, of which there are 3 major human pathogens: *S hematobium, S japonicum,* and *S mansoni; Schistosoma* spp infect ±200 million worldwide, kill 800,000/yr; morbidity is due to exuberant tissue reaction to the eggs–they don't replicate in humans. See Circumoval body, Pipestem fibrosis

schistosome A blood fluke, in particular of genus *Schistosoma*

schistosomiasis Bilharziasis A parasitic infection by a trematode acquired from infested water, which can live in humans, producing liver, bladder, and GI disease

schizoaffective disorder 295.70 DSM-IV PSYCHIATRY A disorder primarily affecting ♀ with schizophrenic–disordered behavior, hallucinations, delusions, deteriorating function, and affective components; diagnosis is often provisional, due to uncertainty about predominant Sx ETIOLOGY Uncertain; genetic, biochemical, psychosocial factors play interconnected role RISK FACTORS Family Hx of schizophrenia or affective disorder; it is less common than schizophrenia or affective disorders. See Shizophrenia

schizoid personality disorder 301.20 DSM-IV PSYCHIATRY A mental disorder characterized by '…*a pervasive pattern of detachment from social relationships and a restricted range of expression of emotions in interpersonal settings.*' Persons with SPD may have difficulties relating to others, are often reclusive and, as older adults, may live with their parents; onset in early adulthood. See Loner. Cf Schizophrenia

schizophrenia PSYCHIATRY A heterogenous group of disorders characterized by progressive mental disturbances in thought, perception, affect, behavior, and communication that last longer than 6 months, which may be accompanied by psychotic Sx, bizarre behavior, or by negative–deficit Sx, including low levels of emotional arousal, mental activity, and social drive; schizophrenia is a diagnosis of exclusion–none of its clinical, biochemical, neuroradiologic, pathophysiologic, and psychologic features are sufficient to establish a definitive diagnosis DSM-IV CLINICAL Disordered thought processes and abnormal behavior–eg, hallucinations, delusions, social withdrawal, inappropriate or "blunted" affect, verbal incoherence, cognitive deficits, inappropriate or blunted affect TYPES Catatonic, paranoid, disorganized, undifferentiated, residual MANAGEMENT Antipsychotics are used to 1. ↓ hallucinations and delusions, and other thought disturbances and improve Sx of withdrawal and apathy 2. Control Sx through maintenance therapy, and 3. As long-term prophylaxis, neuroleptics include haloperidol and clozapine See Cultural schizophrenia, Disorganized schizophrenia, Fugue, Multiple personality, Paranoid schizophrenia. Cf Schizoid personality disorder

SCHIZOPHRENIA-diagnostic criteria[1]

A CHARACTERISTIC SX: 2 or more[2] of following, for a significant portion of 1+ months

Delusions

Disorganized (or catatonic) behavior

Disorganized speech (incoherence)

Hallucinations

Negative symptoms, eg flattening of affect, loss of volition

B SOCIAL/OCCUPATIONAL DYSFUNCTION

C DURATION 6+ months in duration with 1+ months of 'active' symptoms, defined by criteria A

EXCLUSIONS Symptoms are not better accounted for by

D OTHER MENTAL DISORDERS, eg Schizoaffective disorder or Mood disorder with psychotic features

E OTHER CONDITIONS, eg substances(medication, substance of abuse) and/or general medical conditions

[1]Modified from Diagnostic & Statistical Manual of Mental Disorders, 4th ed, Washington, DC, Am Psychiatric Assn, 1994

[2]Only one of criteria A is required if the delusions are bizarre, or hallucinations include 1 + 'voices in head'

schizophreniform disorder 295.40 DSM-IV PSYCHIATRY A schizophrenia-like condition that is identical to schizophrenia except: (1) the total duration of the illness–including prodromal, active and residual phases–is ≥ 1 month, but < 6 months and (2) impaired social and occupational functioning during some part of the illness is not required; 1/3 of Pts with SD recover completely; the rest progress to schizophrenia or schizoaffective disorder. See Schizophrenia

schizotypal personality disorder 301.22 DSM-IV PSYCHIATRY A schizophrenia-like condition characterized by defects in interpersonal relationships and disturbed thought patterns, appearance, behavior; Pts with SPD have bizarre speech, poor social skills, strained relationships with others; it is more common in relatives of schizophrenic. See Schizophrenia

Schneider scale SUBSTANCE ABUSE An instrument that assesses Sx that smokers experience during withdrawal, rating them according to how upsetting they are/were. See Tobacco Withdrawal Symptoms Inventory

school phobia CHILD PSYCHOLOGY Fear of going to school associated with anxiety about leaving home and family members

school refusal PSYCHIATRY An anxiety disorder affecting schoolchildren who, for various reasons, avoid attending school. See Psychogenic seizure

Schwabach's test AUDIOLOGY A hearing test comparing conduction of sound from tuning forks through bone–skull and through air, to determine whether hearing loss is due to defects in conduction of sound, or in sensorineural processing of sound. See Rinne test, Weber test

SCI Spinal cord injury, see there

sciatic nerve dysfunction Sciatic nerve neuropathy NEUROLOGY A mononeuropathy caused by sciatic nerve injury, which may be accompanied by sciatica, a descriptor for pain along the sciatic nerve ETIOLOGY Pelvic fracture, gunshot, trauma to butt or thigh, IM injections, prolonged sitting on butt, systemic disease with polyneuropathy–eg, diabetic neuropathy, polyarteritis nodosa, local pressure–tumor, abscess, bleeding, idiopathic, mechanical factors with regional ischemia, ruptured lumbar disk linked to sedentary lifestyle, improper lifting, old age, obesity CLINICAL ↓ ability to flex knee, ↓ movement of foot and toes, paresthesias in leg, sciatica DIAGNOSIS Physical exam, MRI, CT, nerve conduction velocity test, EMG, bone scan, myelography MANAGEMENT Conservative–eg, NSAIDs, heat, massage, muscle relaxants, exercise program; interventional–eg, diskectomy, laminectomy ± chymopapain, spinal fusion.

sciatica NEUROLOGY Lumbosacral pain that radiates down the posterior thigh and lateral leg into the foot, caused by compression of the sciatic nerve and lumbosacral nerve roots ETIOLOGY Injury, prolapse of intervertebral disk, tumors, sciatic nerve irritation or inflammation CLINICAL Hyporeflexia, paresthesias, ↓ muscle strength. Cf Sciatic nerve dysfunction

SCID Severe combined immune deficiency, see there

science VOX POPULI The formal and systematic study of natural phenomena. See Big science, Fraud in science, Little science, Junk science, Misconduct in science, Prediction science, Pseudoscience

science literacy A general term for the awareness a person or the public has of basic scientific facts, concepts, and theories

scientist VOX POPULI A person active in a field of science. See Cited scientist, Clinical lab scientist, Gentleman scientist, New scientist

scintigraphy IMAGING The creation of 2-D images of the distribution of a radionuclide in tissue after administration. See Brain scintigraphy, Dimercaptosuccinic acid scintigraphy, Dipyridamole-thallium scintigraphy, Immunoscintigraphy, Infarct-avid scintigraphy, Perfusion scintigraphy, Whole gut scintigraphy

scintimammography IMAGING A noninvasive technique in which a radioactive tracer is injected into a breast being screened for cancer. See Mammography

Scion set SURGERY Instruments–dissectors, graspers, optical retractors, scissors–used to harvest saphenous veins during CABG. See CABG

sciophobia PSYCHOLOGY Fear of shadows. See Phobia

scirrhous *skirrhos*, Greek, gypsum *adjective* Relating to a dense stroma seen in some CAs, which produce abundant connective tissue; scirrhous induration–SI or 'desmoplastic reaction' is typical of ductal CA of breast, often accompanied by microcalcifications; SI also occurs in pancreas adenoCA

scirrhous carcinoma A nonspecific term of waning popularity for the gross findings of 'classic' invasive ductal CA of breast, characterized by marked induration with 'chalky streaks' due to ↑ deposition of elastin fibers, often accompanied by microcalcifications detectable on mammography

scissors SURGERY An instrument with 2 opposing cutting blades. See Dissecting scissors

SCIWORA Spinal cord injury without radiologic abnormality NEUROSURGERY Serious spinal cord damage and disruption of tracts without a Fx, an event that is most common in children MECHANISMS Flexion, hyperextension, longitudinal distraction and ischemia causing complete severe partial cord lesions MANAGEMENT Regional stabilization, exploration

scleral icterus PHYSICAL EXAM Yellowing of the sclera, typical of liver disease, especially hepatitis. See Jaundice

scleredema PHYSICAL EXAM A diffuse, symmetrical idiopathic induration of skin which may follow staphylococcal skin infection, associated with DM, and resolve spontaneously

scleritis OPHTHALMOLOGY Inflammation of sclera, most common in older adults ETIOLOGY Idiopathic, or associated with rheumatoid arthritis, Wegener's granulomatosis, metabolic disorders, infection, chemical or physical injury

scleroderma Progressive systemic sclerosis DERMATOLOGY An idiopathic condition characterized by induration of skin due to ↑ collagen deposition CLINICAL Raynaud's phenomenon followed by arthralgia, arthritis, early–subcutaneous edema, late–skin induration, skin ulcers, pinched face, dyspnea due to lung fibrosis, cor pulmonale IMMUNOLOGY Autoantibodies–eg, antinuclear, anti-centromere, anti-topoisomerase 1–Scl-70 antibodies LAB Normochromic, normocytic anemia, microangiopathic anemia, polyclonal gammopathy, ↑ ESR IMAGING Osteoporosis, interstitial markings in CXR, ↓ esophageal and colonic peristalsis, irregular narrowing in renal arteriogram MANAGEMENT CCBs for vasodilation in Pts with Raynaud's phenomenon; penicillamine may slow disease progression; preventive measures include avoiding tobacco and cold, antibiotics and digitalis PRN PROGNOSIS 5-yr survival, ±40%

ScleroLaser™ VASCULAR DISEASE A laser for treating telangiectasia/veins of the leg instead of sclerotherapy

scleroma Induration, see there

ScleroPLUS™ DERMATOLOGY A laser for treating spider angiomata, port wine stains, and other vascular malformations

sclerosing cholangitis Primary sclerosing cholangitis, see there

sclerosing hemangioma LUNG PATHOLOGY A benign lesion of ♀ adults characterized by a circumscribed, slowly growing mass of polygonal cells thought to arise from type II pneumocytes with mesenchymal stroma SKIN PATHOLOGY Benign fibrous histiocytoma, see there; aka dermatofibroma, nodular fibrosis

sclerosing metanephric hamartoma A lesion of children < age 2, which may precede Wilms' tumor. Cf WAGR syndrome, Wilms' tumor

sclerosing peritonitis Extensive peritoneal fibrosis in response to asbestos, continuous ambulatory peritoneal dialysis, silicosis–IV drug abusers, in carcinoid syndrome or linked to beta-blockers CLINICAL Intestinal obstruction due to massive peritonial adhesions

sclerosis MEDTALK Induration of tissue. See Amyotrophic lateral sclerosis, Diffuse mesangial sclerosis, End plate sclerosis, Lichen sclerosis, Mesial temporal sclerosis, Multiple sclerosis, Nephrosclerosis sclerosis, Progressive systemic sclerosis, Systemic sclerosis

Sclerosol® CHEST DISEASE A proprietary sclerosing talc used to prevent recurrent malignant pleural effusions, administered by aerosol during thoracoscopy or open thoracotomy. See Sclerotherapy

sclerosteosis syndrome An AR condition of children that leads to deafness with bony overgrowth and occlusion of cranial foramina, accompanied by an asymmetrically enlarged mandible, syndactyly, onychodysplasia

sclerotherapy The use of a sclerosing agent to induce fibrous obliteration of pathologic blood vessels–eg, hemorrhoids or esophageal varices. See Endoscopic sclerotherapy, Injection sclerotherapy

scoliosis ORTHOPEDICS A lateral curvature of the spine typically seen in young ♀, often an incidental finding; severe symptomatic scoliosis may respond to surgery; scoliosis may occur alone or with kyphosis–kyphoscoliosis ETIOLOGY Congenital–eg, defect in vertebrae or fused ribs; acquired–eg, muscle paralysis due to polio, cerebral palsy, muscular dystrophy; idiopathic CLINICAL Fatigue, pain in spine after prolonged sitting or standing; as the spine curves laterally, a compensatory curve develops to maintain balance which may become more severe; severe scoliosis may cause respiratory problems

scombroid intoxication A histaminic reaction caused by ingesting spoiled fish of suborder *Scombroidea*–eg, saury, skipjack, maki-maki, dolphin, tuna, bonito, seerfish, butterfly kingfish, mackerel; these fish have free histamine in muscle which is decarboxylated when infected by *Proteus* spp; if infection is intense, oral antihistamines or activated charcoal may be used to reverse the histaminic effect. See Fish. Cf Ciguatera poisoning

'scoop and run' EMERGENCY MEDICINE A stance taken when a trauma victim's condition is of such severity that there is 1. Insufficient time for the usual format of medical stabilization and 2. The equipment and/or experts needed to save the victim's life are not present in the ambulatory field unit–eg, ambulance or helicopter. Cf Stay and play

'scope' *noun* The 'short form' for any complex optical system–eg, endoscope, microscope *verb* To perform endoscopy

scope of practice Scope of care PROFESSIONALISM The range of responsibility–eg, types of Pts or caseload and practice guidelines that determine the boundaries within which a physician, or other professional, practices

scopolamine THERAPEUTICS An atropine-like anticholinergic–or antimuscarinic, used in preanesthesia, where CNS depression is desirable, GI tract antispasmodic, to ↑ heart rate, and counteract vasodilation and low BP caused by choline esters METABOLISM GI tract absorption, ½ metabolized in liver, remainder in kidneys ADVERSE EFFECTS Dry mouth, tachycardia, palpitation, pupillary dilatation, blurring of vision, headache, dry hot-skin

scoptophilia Scopophilia SEXOLOGY A paraphilia in which sexuoeroticism hinges on watching others engaging in sexual activity. See Troilism

scopy MEDTALK The examination of a tissue or body region with an endoscope. See Bronchoscopy, Colonoscopy, Cystoscopy, Endoscopy, Hysteroscopy, Laser bronchoscopy, Mass spectroscopy, Rapid bronchoscopy, Salpingoscopy, Virtual colonoscopy

score *noun* HEALTH LEGISLATION Economic impact report A report by the Congressional Budget Office that evaluates the economic impact of legislation–eg, that of health care reform; the score obtained via projective prestidigitation may determine the fate of a legislative proposal. See Scorecard, Scorekeeper. Cf Report card VOX POPULI A numeric rating of a particular process. See APGAR score, Apopnea-hypopnea score, ASIA motor score, Borg score, Child-Pugh score, DeMeester score, Fagerstrom tolerance score, Family Environment score, Gleason score, Hachinski ischemic score, Hegsted's score, Lod score, Longitudinality score, Mayo risk score, Medicus modified score, Modified Bournemouth score, Nursing Classification score, Pediatric trauma score, Pittsburgh brainstem score, PRISM score, QRS score, Revised Trauma score, Sanitation score, SAPS II, Trauma score, Z score, Z score *verb* DRUG SLANG To purchase illicit drugs VOX POPULI To get a home run; to get lucky

Scorpaenidae TOXICOLOGY A family of venomous bony fish–eg, lionfish, and scorpion fish which contains 30 genera and 350 spp; they are considered to be the most dangerous of all fish in terms of number and severity of injuries to the groups–scuba divers, tropical fish fanciers–affected; the fish envenomate by erecting spines on their dorsal, anal, and pelvic fins that pierce the victim's flesh and release venom CLINICAL Intense local pain that peaks at 60-90 mins and persists for 8-12 hrs; other manifestations include erythema, ecchymosis, pallor, induration, swelling, edema, hypesthesia, anesthesia, paresthesia, N&V, regional lymphadenopathy, necrosis; the venom is a heat-labile protein that produces hypotension, vasodilation, muscular weakness, respiratory paralysis TREATMENT Immerse affected body part in hot water; antivenom is available

scorpion MEDICAL ENTOMOLOGY A nocturnal arachnid native to the hot, dry regions of the southwestern US and Mexico–the spp in Mexico are more toxic; scorpions most common in the US are *Centruroides gertschi* and *C sculpturatus*, which cause systemic disease through the cholinergic effects of their venom. See Scorpion sting

scorpion sting A toxic systemic response to scorpion venom CLINICAL SOB, opisthotonus, nasal and periorbital itching, dysphasia, drooling, gastric distension, diplopia, transient blindness, nystagmus, fecal & urinary incontinence, penile erection, HTN, arrhythmias, lasting up to 48 hrs MANAGEMENT Antivenom from poison control centers; also immobilization, ice water immersion, oxygen, ventilation; opiate analgesics may potentiate venom toxicity, should be avoided; atropine is used to combat parasympathetic effects. See Scorpion

Scotchtape™ test MICROBIOLOGY A lab method used to 1. Retrieve eggs from the perianal region in children infected with *Enterobius vermicularis*–pinworm and 2. Observe fungi in a fashion similar to their 'native' conformation in culture; a piece of transparent adhesive tape is touched to a colony of fungi, then adhered to a glass slide and stained with lactophenol blue

scotoma NEUROLOGY A vision defect PSYCHIATRY A figurative blind spot in a person's awareness

scout film IMAGING A preliminary film taken of a body region before a definitive imaging study–eg, an SF of the chest before a CT; 'scouts' serve to establish a baseline and are used before angiography, CT, MRI

Scoville retractor ORTHOPEDICS A double-hinged instrument for spinal surgery retraction

Scoville scale NUTRITION/MASOCHISM A system devised by WL Scoville for determining the relative 'spiciness' of hot peppers; a dried pepper is dissolved in alcohol, serially diluted in sugar water and given to a panel of tasters who sip increasingly diluted concentrations of peppers. See Capsaicin, Spicy food

(the) 'scramble' GRADUATE EDUCATION A popular term for the 'mad rush' to secure a residency slot by US medical students who did not obtain a position through the Match. See Match

scrape Abrasion, see there

screen PUBLIC HEALTH 1. Any systematic activity–eg, measuring BP, glucose or cholesterol, pap smear, or other activity, which attempts to identify a particular disease in persons in a particular population. See Drug screen, General health screen, Laxative screen, Memory Impairment screen, Metabolic screen, Neonatal screen, 2. A solar protection barrier. See Sunscreen

screening MEDTALK The evaluation of an asymptomatic person in a population, to detect an unsuspected disease process not known to exist at the time of evaluation; screening tests measure specific parameters–eg, bp–for HTN, sigmoidoscopy–colorectal CA, imaging–eg, mammography–breast CA or lab parameters–eg, cholesterol–CAD, guaiac-positive stools–colorectal CA or Pap smears of the uterine cervix–cervical CA; screening tests in general have high sensitivities and low specificities, which allows detection of most Pts with a morbid condition, while having the acceptable disadvantage of a high rate of false positivity. See Cancer screening, Colorectal screening, Developmental screening, Drug screening, Forensic drug screening, Genetic screening, Industry screening Microalbuminuria screening, Multiphasic screening, Newborn screening PSYCHIATRY An assessment or evaluation to determine the appropriate services for a client

SCREENING

BLOOD-PRESSURE Measured in normotensive persons-every 2 years, all age groups

BREAST EXAMINATION, ♀ By physician-every year > age 40; mammography-every 1-2 years, age 35+, every year > age 50

CERVICAL CYTOLOGY Pap smear every 1-3 years, starting at age of first intercourse

CHOLESTEROL Measured-every 5 years, but not in younger subjects

PROSTATE Rectal exam, ideally every year

screening colonoscopy GI DISEASE The use of flexible colonoscopy to detect malignant or premalignant colorectal lesions; SC is most cost effective ≥ age 50. See Colonoscopy

screening facility MEDICAL PRACTICE A location not licensed by a state dept of health, which offers services to the medical profession or to the public for one or more types of medical testing. See Screening, Screening test

screw *noun* ORTHOPEDICS A ubiquitous threaded internal fixation device, usually constructed of stainless steel, commonly used to fix bones in place, either alone or with a plate. See Bionx absorbable cannulated screw, Cancellous screw, Cannellated screw, Cortical screw, Cross-locking screw, Herbert screw, Interference screw, Lag screw VOX POPULI *verb* 1. Popular, 'to treat unfairly' 2. Vulgar; to have intercourse

screwup VOX POPULI Opportunity for improvement

SCRIP CARDIOLOGY A clinical trial–Stanford Coronary Risk Intervention Project that studied the effect of intensive multiple risk factor reduction on M&M and prevention of new atherosclerotic lesions in Pts with moderate CAD. See Coronary artery disease

SCRIPPS CARDIOLOGY A clinical trial–Scripps Coronary Radiation to Inhibit Proliferation Post Stenting

script PSYCHOLOGY *noun* A part of communication with a Pt that recognizes that different phrases may have the same meaning, but the order, choice of the words or the manner in which they are said can either stimulate or inhibit communication

scrotal exploration UROLOGY Scrotal palpation for evaluating infertility and presence of masses; SE may reveal testicular defects, obstruction, or congenital defects of vas deferens. See Infertility

scrotal mass UROLOGY A lump palpable in the scrotum, which may be benign–eg, hematocele, spermatocele, inflammation–eg, epididymitis, trauma, hernia, or malignant–eg, seminoma

scrotal tongue Grooved tongue Congenital tongue furrows seen in Melkerson-Rosenthal syndrome or Down syndrome; the furrows are attributed to sucking and mouth breathing; the finding has no pathologic significance and is only of interest, as food particles get stuck and are later colonized by oral bacteria, evoking halitosis TREATMENT Brush teeth, tongue

scrub *noun* The formal preoperative hand washing ritual that is a *sine qua non* prerequisite for performing an invasive procedure in a body cavity RULES OF SCRUBBING 1. Work from distal–fingertips to proximal 2. Complete scrubbing before rinsing 3. Let water drain off elbow. Cf Scrubs *verb* 1. To perform the holy ritual of the scrub *noun* 2. To cleanse vigorously as required in emergency treatment of MVAs in which wounds are 'dirty', deep, bloody, and studded with gravel, glass, debris, dirt, and sundry schmutz

scrub nurse A nurse–or technician who participates in a sterile surgical operation, prepares sterile supplies and passes them to the surgeon, assisting the surgeon during the procedure, accounting for needles, sharps, sponges and other supplies used during the operation, and teaching any new–and qualified–personnel details of operating room protocol. See Circulating nurse, Operating team, Physician assistant

scrub typhus Tsutsugamushi disease, mite-borne typhus, tropical typhus A mite-borne infection by *Rickettsia tsutsugamushi* EPIDEMIOLOGY ST occurs in Japan, India, Australia VECTOR Chigger–larval stage of a mite, *Leptotrombidium deliensis*, or *Trombicula pseudo-akamushi*, inhabitants of scrub vegetation that feed on host rodents CLINICAL 1-3 wk incubation with prodromal Sx of headache, malaise, anorexia; scarification of inoculation papule is followed by abrupt onset of high fever with pulse-temperature dissociation, headache, ocular pain, conjunctivitis, malaise, lymphadenopathy, and a dark crusted eschar or tache noire at the site of the chigger–mite larva bite, cardiac dysfunction with minor EKG changes–eg, T wave inversion, a pale pink, centrifugal maculopapular rash, lymphadenopathy, interstitial pneumonia DIAGNOSIS *Proteus* OX-K antigen seropositivity TREATMENT Tetracycline, chloramphenicol, ciprofloxin MORTALITY 10-30% if untreated

scrubs See Surgical scrubs

scuba Self-contained underwater breathing apparatus, SCUBA SPORTS MEDICINE A device consisting of one or more tanks of compressed breathable gases–'air' in optimized ratios, a regulator that can be adjusted to regulate the flow of air, and a meter to indicate the amount of 'air' remaining in the tanks. See–the Bends, Cf Caisson's disease

sculpting COSMETIC SURGERY The surgical reshaping of a tissue. See Deep tissue sculpting, Facial sculpting

scurvy Scorbutus, vitamin C deficiency NUTRITION A condition characterized by weakness, anemia, gingivitis, skin hemorrhage due to ↓ vitamin C in diet, which rarely occurs in developed nations, but may be seen in older, malnurished adults. See Rebound scurvy, Vitamin C

scut 'monkey' A popular, demeaning term for the medical student at the bottom rung of a Pt management team in a university-affilitated health care facility, who performs so-called 'scut' work. See Extern, Medical student abuse, Pimping, 'Scut work'

scut work Menial, non-Pt care-related activities passed to medical students–externs or interns, although they may be the duties of other healthcare workers. See Medical student abuse, Pimping, Scut 'monkey'

SD 1. Senile dementia 2. Septal defect 3. Skin dose 4. Spontaneous delivery 5. Standard deviation 6. Sudden death

SDA 1. Shoulder disarticulation 2. Steroid-dependent asthma

SDAT Senile dementia–Alzheimer type, see there

SEAL HEART SURGERY A clinical trial–Simple & Effective Arterial Closure Study which evaluated the safety and efficacy of the Duett™ vascular sealing device

seal PUBLIC HEALTH Any air- and/or water-tight closure on a product–blood components, medications intended to maintain a product sterilely until used by a consumer; products with broken seals should be discarded

seal finger A monoarticular infection of a puffy finger described in coastal Scandinavia and Canada, seen in individuals with wildlife and marine wildlife-related jobs, possibly caused by fastidious bacteria TREATMENT Tetracycline

sealant DENTISTRY A UV light-cured resin used to coat fissures in teeth and prevent cavities SURGERY A substance used to close gaps or vessels. See Fibrin sealant

seamless VOX POPULI *adjective* Referring to a smooth and seemingly uninterrupted transition from one task to another

search engine INFORMATICS A software tool–eg, www.google.com for finding information from various sources online, usually by typing a few keywords. See Keyword. Cf Web crawler

seasickness NEUROLOGY A type of motion sickness–type C, which occurs in susceptible persons subjected to the rock & roll of a ship at sea. See Motion sickness

seasonal affective disorder SAD, winter 'blues,' winter depression PSYCHIATRY A clinical form of depression with an onset in late fall and remission in spring. See Bright light therapy; Cf Melancholia, Melancholy

seasonal allergic rhinitis Allergic rhinitis in which Sx wax and wane as a function of environmental pollen. See Allergic rhinitis

Seasonale® GYNECOLOGY An OC designed to ↓ the number of menstrual cycles to 4/yr, each ±91 days in duration. See Contraception

seasonality EPIDEMIOLOGY Change in physiological status or in disease occurrence that conforms to a regular seasonal pattern

seat belt syndrome Contusion of anterior abdominal wall caused by lap seat belts, which may produce lumbar spine fractures with horizontal splitting of the vertebral body and posterior arch, trauma to bowel, vessels, spleen and liver; in the US, lap-type safety belts are found in the front seats of older cars, and in the middle back seats of current vehicles. Cf Dashboard fracture, Padded dash(board) syndrome

sebaceous cyst Epidermoid cyst, keratinaceous cyst, steatoma DERMATOLOGY An epidermal inclusion cyst containing greasy, pasty keratinaceous debris secreted by sebaceous glands. Cf Pilar cyst

seborrheic dermatitis Cradle cap; dandruff DERMATOLOGY An idiopathic dermatopathy characterized by greasy or dry white scales, variably accompanied by erythema SITES Scalp, face, nose, eyebrows, behind ears, external ear, skin over sternum and over skin folds RISK FACTORS Familial, stress, fatigue, temperature extremes, oily skin, infrequent shampoos or skin cleaning, acne, obesity, excess use of lotions with alcohol, AIDS, neurologic conditions–eg, Parkinson's disease, head injury, stroke

seborrheic keratosis Senile keratosis DERMATOLOGY A benign, often numerous lesion of the face and upper trunk of older adults, which appear as wart-like growths in various colors; SKs may become irritated and itch because of friction from clothes. See Keratosis

seclusion FORENSIC PSYCHIATRY A strategy for managing disturbed and violent Pts in psychiatric units, which consists of supervised confinement of a Pt to a room–ie, involuntary isolation, to protect others from harm

secobarbital NEUROPHARMACOLOGY An intermediate-acting sedative and hypnotic, replaced by benzodiazepines ADVERSE EFFECTS Prolonged drowsiness, paradoxic excitement, diffuse arthritis, myalgic, or neuralgic pain, hypersensitivity in Pts with upregulated immunity–eg, with asthma, angioedema, urticaria, drug interactions–eg, MAOIs, INH, ethanol; stimulation of hepatic microsomal enzymes may ↑ metabolism of drugs, endogenous steroids, and certain anesthetics PHARMACOLOGIC EFFECTS Like phenobarbital USED FOR Anxiolytic, relieves insomnia; pre-anesthesia

second cancer ONCOLOGY A 1° CA that is either a new CA in a different organ in a Pt with a Hx of a first CA or a new CA linked to prior cancer therapy. See Occult cancer

second-degree block See Heart block

second-degree burn See Burn

second disease phenomenon The tendency for a Pt with a particular condition to do less well clinically, or have a worse prognosis, when suffering from a 2nd disease. See Passenger phenomenon

second-generation oral contraceptive GYNECOLOGY A levonorgestrol-based OC, which has an ↑ risk of acute MI relative to third-generation OCs. See Oral contraceptive. Cf Third-generation oral contraceptive

second-hand smoke Passive smoking, see there

second heart sound CARDIOLOGY A heart sound that corresponds to closure of the semilunar–pulmonary and aortic valves, which is a short, higher pitched "dubb". See First heart sound, Fourth heart sound, Third heart sound

second impact syndrome SPORTS MEDICINE A catastrophic condition associated with boxing and other 'head-impact' sports, which occurs in 2 phases; a concussion or cerebral contusion 2° to blunt trauma to the head causes headaches, impaired cognition, incoordination, and ↓ speech and motor functions; if further trauma–ie, a 2nd impact occurs before the Sx resolve, a 2nd blow or impact, however minor, may cause coma or sudden death due to cerebral edema. See Boxing, Concussion, Ultimate fighting

second-line chemotherapy ONCOLOGY Any chemotherapeutic administered after the agent of choice fails

second-line drug Any therapeutic agent that is not the drug of choice, or the 1st normally used to treat a particular condition; in rheumatoid arthritis, 2nd-line agents are used when standard 'first-line' therapy–ie, anti-inflammatory agents and corticosteroids fail

'second-look' operation A 2nd surgical procedure in the same site as a previous operation, often abdominal, with the intent of continuing therapy that could not be completed for various reasons during the 1st operation GENERAL SURGERY An operation to re-examine questionably viable segments of small intestine 24-48 hrs after an initial massive resection for ischemia, with the hope that enough small intestine–at least 30% remains viable, thereby circumventing 'short bowel syndrome'. See Short bowel syndrome GYNECOLOGIC ONCOLOGY A laparotomy for advanced and presumably inoperable ovarian CA, which was later re-examined to determine whether RT and/or chemotherapy is successful in reducing tumor size to allow debulking surgery, or determine whether RT and/or chemotherapy may be discontinued or requires modification SURGICAL ONCOLOGY SLOs are performed in colons previously resected for adenoCA and monitored by CEA levels; an ↑ of CEA > 35% above the Pt's established baseline suggests metastatic recurrence; at operation, < ½ are resectable. See Debulking operation

second opinion MEDTALK Formal or informal advice sought from a 2nd health professional regarding the diagnosis and/or therapy proposed by a 1st provider. See Consultation

Second Skin® WOUND CARE A hydrogel dressing used for minor burns, blisters, insect bites or stings, to promote wound healing and protect against bacterial infection. See Wound care

second-stage reconstruction SURGERY Mammoplasty in which a temporary expander is replaced with an implant or the valve is removed on a permanent adjustable implant or any necessary flap revision is done. See Breast reconstruction

second surgical opinion MANAGED CARE A cost containment technique to help Pts and insurance companies determine whether a recommended procedure is necessary, or whether an alternative method of treatment can accomplish the same result. See Second opinion

second-wind phenomenon INTERNAL MEDICINE A substrate-dependent variation of exercise tolerance, in which previously fatiguing exercise can be performed with relative ease, after a period of rest RHEUMATOLOGY A surge of subjective 'energy' that follows rest, typical of the mid-afternoon fatigue of rheumatoid arthritis

secondary adjective 1. Not primary; generally that which follows another linked process 2. Metastatic, see there

secondary amenorrhea GYNECOLOGY A condition in which menstruation begins at the appropriate age but stops for 6+ months without usual causes–eg, pregnancy, lactation or menopause ETIOLOGY Excessive exercise,

body fat content < 15%, extreme obesity, emotional stress, busulfan, chlorambucil, cyclophosphamide, phenothiazines, oral and other contraceptives–eg, Norplant, Depo-Provera, D&C. See Amenorrhea

secondary aplastic anemia Acquired aplastic anemia HEMATOLOGY A failure of hematopoiesis in the BM caused by stem cell injury, resulting in a reduction in all types of blood cells: ↓ RBCs, ↓ WBCs, ↓ platelets–ie, pancytopenia ETIOLOGY Chemotherapy, immunosuppressants, RT, myelotoxins–eg, benzene or arsenic, drugs, pregnancy, congenital disorders, hepatitis, SLE, idiopathic CLINICAL Anemia–fatigue, weakness; neutropenia–↑ infection; thrombocytopenia–↑ mucocutaneous and internal organ hemorrhage; SAA may be acute or chronic and usually progresses. See Aplastic anemia

secondary cardiomyopathy Myocardial disease with a specific etiology or associated with systemic diseases; 2° cardiomyopathies may be linked to Adriamycin, alcohol, catecholamines–pheochromocytoma, cobalt intoxication, diphtheria, beriberi heart disease, and various chemicals and drugs. See Cardiomyopathy

secondary care center Secondary care facility A community–or 'voluntary' hospital equiped to provide all but the most specialized forms of health care, surgery, and diagnostic techniques. Cf Tertiary care center care

secondary deficiency NUTRITION A deficiency state due to either ↑ requirement for a substance–eg, iron in pregnancy, or ↓ availability or 'wastage' of the nutrient, as in proteinuria in nephrotic syndrome

secondary diabetes A form of DM due to mechanisms other than those causing 1° –type 1 DM and type 2 DM ETIOLOGY Pancreatitis, pancreas CA, pheochromocytoma, hemochromatosis, acromegaly or by drugs known to impair glucose metabolism–eg, corticosteroids

secondary female characteristic Any physical sign of change from a girl to a woman–eg, development of breasts. See Secondary sex characteristics

secondary gain PSYCHIATRY The external gain derived from an illness or Sx thereof–eg, ↑ attention and service, sympathy, social interaction, money, disability benefits, release from unpleasant responsibility. Cf Primary gain

secondary glaucoma OPHTHALMOLOGY Glaucoma in a background of other ocular disease–eg, uveitis and systemic disease, or after exposure to some drugs–eg, steroids. See Glaucoma

secondary hyperparathyroidism ENDOCRINOLOGY ↑ Parathyroid activity with overproduction of PTH due to parathyroid hyperplasia in response to an extraparathyroid defect that results in ↓ Ca²⁺ ETIOLOGY Rickets, vitamin D deficiency, chronic renal failure ORGANS AFFECTED Bone, GI tract, kidneys, CNS, muscle

secondary Lyme disease Early disseminated Lyme disease, stage 2 Lyme disease INFECTIOUS DISEASE A stage of Lyme disease that develops wks to months after the primary skin lesions disappear, characterized by neurologic or cardiac Sx; SLD spreads via the lymph or blood wks after the tick bite CLINICAL Lesions involve skin, CNS, and musculoskeletal systems; Sx occur in half of Pts and may be intermittent and disappear spontaneously; heart is involved in ±10%. See Lyme disease

secondary male characteristic Any physical sign of change from a boy to a man–eg, development of facial hair. See Secondary sex characteristics

secondary malignancy ONCOLOGY A cancer that arises in the background of another malignancy treated by RT or chemotherapy; SM is also defined as one caused by environmental toxins, physical agents, radiation EXAMPLES ANLL–eg, AML, acute promyelocytic leukemia, acute monocytic leukemia, erythroleukemia and myelodysplastic disorders–preleukemia LEUKEMOGENIC CHEMOTHERAPEUTICS Chlorambucil, melphalan and, combined, doxorubicin, cisplatin PHYSICAL AGENTS CAUSING SM–RADIATION–solar, X-rays to head & neck causes 2° BCCs; latency period to 2° neoplasm 3-65 yrs

secondary obesity

CHRONIC IRRITATION–Marjolin's ulcer causing skin cancer or chronic injury–eg, heat-induced SCC SCAR MALIGNANCY–eg, malignant fibrous histiocytoma arising in sites with metal objects or shrapnel or malignancy induced by mechanical trauma, a relation which in humans is anecdotal, ischemia and SCC near varicose veins. See Occult primary

secondary obesity Obesity due to endocrine or other systemic disease. See Morbid obesity, Obesity

SECONDARY OBESITY

CONGENITAL Alström-Hallgren syndrome, Carpenter syndrome, Cohen syndrome, Lawrence-Moon-Biedl syndrome, Prader-Willi syndrome, Vasquez syndrome

CNS DISEASE Defects of hypophyseal-hypothalamic axis, intracranial leukemia, and other lesions

ENDOCRINOPATHIES Hypothyroidism, insulinoma, Cushing syndrome, polycystic ovary disease, pituitary dwarfism

secondary peritonitis SURGERY Peritoneal inflammation due to entry of bacteria or enzymes ETIOLOGY GI tract perforation–eg, ruptured appendix with spilling of bacteria; chemical reaction from pancreatic enzymes or bile due to perforation; overdistension of GI tract with gas and fluid; necrotizing enterocolitis in premature infants. See Peritonitis

secondary polycythemia ↑ in RBC mass in response to a physiologic insult–eg, hypoxia due to pulmonary disease, cardiomyopathy, vena cava thrombosis, or in response to excess erythropoietin production in renal cell CA or cystic kidneys. See Polycythemia vera

secondary prevention PREVENTIVE MEDICINE Reduction of M&M in Pts with a known disease, treating it while it is asymptomatic–eg, optimizing glycemic control in DM. See Healthy lifestyle. Cf Primary prevention

secondary sex characteristic ENDOCRINOLOGY The constellation of changes in hair distribution, body configuration, and genital size in boys or girls at the time of puberty

SECONDARY SEX CHARACTERISTIC–♂

EXTERNAL GENITALIA Penis ↑ in length/diameter; scrotum becomes pigmented and rugose

INTERNAL GENITALIA Prostate, bulbourethral glands, and seminal vesicle enlarge, begin to secrete

BODY Shoulders broaden, muscle mass ↑

HAIR Hair, hair, everywhere; beard, back, chest, anus

MENTAL More aggressive, sexual interest awakens

SKIN Sebaceous glands ↑, 'Zit Follies' begin

VOICE Larynx, vocal cords ↑ in size and/or length; voice deepens

SECONDARY SEX CHARACTERISTIC–♀

EXTERNAL GENITALIA ↑ Size breast, vagina

INTERNAL GENITALIA ↑ Size uterus

BODY Shoulders are narrow, hips broad, thighs converge and arms diverge–broad carrying angle

HAIR More scalp hair, less body hair, ♀ escutcheon

VOICE Nada; voice unchanged

secondary syphilis STD The 2nd stage of syphilis that begins 1 wk to 6 months post initial infection, often manifest as a generalized skin rash simulating many other diseases; SS has been thus called the "great impostor" CLINICAL Scaly red-brown palmoplantar maculopapules lasting up to 6 wks, ± systemic Sx–fever, myalgias, arthralgias, moist genital warts or condylomata lata; SS is preceded by a genital ulcer and followed by a latency period, during which there are no overt signs of infection EPIDEMIOLOGY 1° and 2° syphilis are very contagious; after lesion clears, infection can become "latent" and lacks overt signs of infection HIGH RISKS Multiple and/or unknown sex partners, high-risk sex practices, urban areas, low socioeconomic status. See Benign late syphilis, Congenital syphilis, High-risk sexual activity, Latent syphilis, Primary syphilis, Tertiary syphilis

secondary systemic amyloid METABOLIC DISEASE Insoluble protein that is deposited in tissues and organs impairing function RISK FACTORS Myeloma, and conditions lasting 5+ yrs–eg, chronic arthritis, rheumatoid arthritis, SLE, TB, paraplegia, bronchiectasis, cystic fibrosis, osteomyelitis, recurrent pyogenic infection, abscess, decubitus ulcers, renal dialysis, Reiter syndrome, ankylosing spondylitis, Hodgkin's disease, Sjögren syndrome, hairy cell leukemia CLINICAL Sx reflect organ affected by deposits, compromising its functionality. See Amyloidosis

secretin injection test GI DISEASE A provocative test in which porcine–or other–secretin is injected IV and the gastrin levels are monitored; gastrin ↓ in antral G-cell hyperplasia and duodenal ulcers, ↑ in gastrinomas, Zollinger-Ellison syndrome

secretory otitis media Otitis media with effusion A noninfectious form of otitis media, characterized by effusions MANAGEMENT Antibiotics with decongestant-antihistamines may be ineffective

section OBSTETRICS See Cesarean section SURGICAL PATHOLOGY A slice of tissue, as prepared for histologic evaluation. See Frozen section, Gough section, Paraffin section, Permanent section, Poincaré section, Slab section, Thick section, Thin section

Sectral® Acebutolol, see there

secure *adjective* INFORMATICS Referring to a computer system in which data is protected from unauthorized access. See HIPAA

SED 1. Seriously emotionally disturbed, serious emotional disability/disorder 2. Spondyloepiphyseal dysplasia, see there

sed rate Erythrocyte sedimentation rate, see there

sedation The production/induction of a sedative state. See Conscious sedation, Terminal sedation

Sedatival Lorazepam, see there

sedative *adjective* Calming *noun* PHARMACOLOGY Any agent that acts on the CNS to attenuate responses to stimuli ACTIVITIES Anxiolytic, sedative, anticonvulsant ADVERSE EFFECTS Ataxia, loss of inhibitions, cardiac and respiratory depression, psychologic and physical dependence, tolerance EXAMPLES Amobarbital, butabarbital, chlordiazepoxide, diazepam, ethchlorvynol, flurazepam, meprobamate, methyprylon, nordiazepam, pentobarbital, trichlorethanol

sedimentation rate Erythrocyte sedimentation rate A test that detects and monitors inflammatory activity by recording the rate at which RBCs settle–sediment–in a test tube; the SR ↑ with inflammation, and usually performed by the Westergren method, see there

Sedizepan Lorazepam, see there

Seduxen Diazepam, see there

'see one, do one, teach one' MEDICAL EDUCATION A traditional format for acquiring medical skills, based on a 3-step process: visualize, perform, regurgitate

see-through phenomenon COSMETIC SURGERY The visual effect when hair is transplanted onto an exfoliated expanse in the chrome-domed; STH is more prominent in dark-haired coves and more common in Pts with severe alopecia, in whom donor sites themselves are threadbare. See Plugginess

seed RADIATION ONCOLOGY *noun* A cylindrical pellet containing radioactive material, used to deliver local RT. See Brachytherapy, 125I radioactive seed *verb* 1. To disseminate, as in the seeding of an infection or malignancy 2. To inoculate a culture plate with a clinical specimen; generally, *plate* is preferred

seed calculus Any of the innumerable minute, oval concrements that may form in a hydronephrotic renal pelvis in ureteropelvic obstruction

seesaw nystagmus Torsional-vertical ocular oscillation NEUROLOGY A rare oscillatory oculomotor phenomenon, in which one eye rises and intorts, while the other falls and extorts, seen in bitemporal hemianopia due to sellar or parasellar lesions ETIOLOGY Tumors of parasellar region and diencephalon, brainstem vascular lesions, multiple sclerosis, syringobulbia, trauma, Chiari malformation MANAGEMENT Surgical decompression. Cf 'Railroad' nystagmus

Seever's disease Calcaneal apophysitis ORTHOPEDICS A condition characterized by heel pain in adolescence, immediately before fusion of the calcaneal apophysis, due to overuse

Segard™ Afelimomab, see there

Segawa's dystonia NEUROLOGY A dopa-responsive dystonia of early–childhood/adolescent-onset with progressive difficulty in walking and, in some, spasticity CLINICAL Diurnal variability–relative mobility in morning, with disability later in day and after exercise. See Dopa-responsive dystonia. Cf Cerebral palsy

segment VOX POPULI A part of a larger region, structure, body. See Abominal segment, Bronchopulmonary segment, Cervical segment, Competitive segment, Diaphragmatic segment, Posterior segment

segmental dysfunction CHIROPRACTIC A focal misalignment–subluxation of vertebra, treated by spinal manipulations intended to free 'life forces' and allow the body to heal itself

segmental resection of breast 1. Partial mastectomy, see there 2. Lumpectomy, see there

Seinfeld syncope CARDIOVASCULAR DISEASE Syncope caused by hysterical laughter in a Pt with cerebrovascular disease, causing a reaction similar to that effected by a Valsalva maneuver, with ↓ O_2 flow to the brain

seizure NEUROLOGY A sudden convulsion, due to temporary disruption in electrical activity of the brain CLINICAL Uncontrollable body movements, sense of unusual smells or tastes, loss of consciousness

SEIZURES CLASSIFICATION
PARTIAL SEIZURES
Simple partial seizures–consciousness preserved
    Motor signs–jacksonian, adversive
    Somatosensory or special sensory symptoms
    Autonomic symptoms or signs
    Psychiatric symptoms
Complex partial seizures–consciousness impaired
    Simple partial seizure, followed by impaired consciousness
    Impaired consciousness at onset
Secondarily generalized seizures
    Simple partial seizure evolving to generalized tonic-clonic seizures
    Complex partial seizure evolving to generalized tonic-clonic seizures
    Simple partial seizure evolving to complex partial seizures, then to generalized tonic-clonic seizures
GENERALIZED-ONSET SEIZURES
Tonic-clonic seizure
Absence seizure
Atypical absence seizure
Myoclonic seizure
Tonic seizure
Atonic seizure
LOCALIZATION-RELATED/FOCAL SEIZURES
Idiopathic
    Benign focal epilepsy of childhood

Symptomatic
    Chronic progressive partial continuous epilepsy
    Temporal lobe epilepsy
    Extratemporal epilepsy
GENERALIZED SEIZURES
Idiopathic
    Benign neonatal convulsions
    Childhood absence epilepsy
    Juvenile myoclonic epilepsy
    Other generalized idiopathic epilepsy
Symptomatic
    West syndrome (infantile spasms)
    Early myoclonic encephalopathy
    Lennox-Gastaut syndrome
    Progressive myoclonic epilepsy
SPECIAL SYNDROMES
Febrile seizures
Drug-related seizures

seizure disorder Epilepsy NEUROLOGY A neurologic dysfunction characterized by a recurring tendency to suffer convulsions–seizures, either local–minor or generalized–major; Hughlings Jackson, 19th century British neurologist, believed epilepsy to be an intermittent derangement of nervous system due to a sudden, excessive, and disorderly discharge of cerebral neurons ETIOLOGY Usually, idiopathic; rarely metabolic disturbances, drug-related CLINICAL Presentation, severity, and duration of the seizure is a function of its type DIFFDX, SEIZURES Head injuries, tumors, lead poisoning, maldevelopment of brain, genetic, infections; half are idiopathic MANAGEMENT Anticonvulsants. See Absence seizure, Complex seizure, Complex partial seizure, Fever-induced seizure, Generalized tonic-clonic seizure, Grand mal seizure, Nonepileptic seizure, Psychogenic seizure, Recurrent seizure, Seizure, Serial seizure, Sexual seizure, Temporal lobe seizure

Seldane® An antihistamine-decongestant–terfenadine + pseudoephedrine-HCl COUNTRAINDICATIONS CAD, HTN, liver disease, therapy with erythromycin, itraconazole, ketoconazole, MAOIs; withdrawn from the market in 1998

selection VOX POPULI The chosing among a number of different options. See Adverse selection, Artificial selection, Group selection, Kin selection, Negative selection, Patient selection, Sex selection

selective amnesia PSYCHOLOGY Amnesia for certain events; as commonly used, SA refers to a deliberate inability to recall an event's details. See Amnesia. Cf Anterograde amnesia, Retrograde amnesia

selective estrogen receptor modulator Designer estrogen, SERM THERAPEUTICS A drug–eg, idoxifene, raloxifene–Evista®, which, like estrogen, prevents bone loss and ↓ cholesterol but, unlike estrogen, does not stimulate the endometrium and breast and thus does not ↑ risk of breast and endometrial CA. See Estrogen receptor, Raloxifene

selective imidazoline receptor agonist See SIRA

selective neck dissection SURGICAL ONCOLOGY A procedure for managing laryngeal CA, in which only the anatomic regions most likely to contain cancer-laden lymph nodes are removed, thus ↓ tissue loss, ↓ comorbidity. See Head and neck cancer. Cf Radical neck dissection

selective photothermolysis Selective thermophotolysis LASER SURGERY The bleaching of certain skin lesions–eg, port-wine nevi and tattoos–lesions with preferential light absorption via a short-pulsed CO_2 laser that delivers ultrashort 'zaps' of energy. See Laser, Laser surgery, Port-wine nevus

selective reduction OBSTETRICS Selective termination of pregnancy, see there

selective serotonin reuptake inhibitor Serotonin-selective-reuptake inhibitor, see there

selective termination of pregnancy Selective reduction OBSTETRICS Selective abortion of one or more products of a 'higher multiple' gestation for various indications–eg, chromosomal or physical abnormalities; STOP is a draconian measure to ↑ the odds of producing a viable baby from a multiple gestation by terminating one or more fetuses. See Induction, Interlocking, Oxytocin, Partial birth abortion

Selector® aspirator SURGERY An ultrasonic device that removes unwanted tissue from an operative field, leaving healthy tissue–eg, nerves and veins intact. See Aspiration

selegiline L-Deprenyl NEUROPHARMACOLOGY A selective monoamine oxidase type B inhibitor which, with L-dopa, may be useful for early symptomatic treatment of parkinsonism

selenium deficiency CARDIOLOGY An absence of selenium in the diet, with ↑ binding of hepatic nucleoproteins to DNA regulatory sequences, and activation of transcription in response to oxidative stress; SD has been implicated in colon polyp formation, endemic cardiomyopathy, Keshan disease–attributed to a defect in glutathione peroxidase activity, without which there is ↑ platelet aggregation due to impaired free radical salvage or in the catalytic function of type I deiodinizing thyroxine-activating enzyme. See Keshan disease

selenium intoxication TOXICOLOGY A rare condition caused by acute or chronic selenium excess; SI is linked to occupational exposure to selenium in manufacture of photocopiers, red lights, photoelectric cells CLINICAL Pallor, garlicky odor, metallic taste, GI complaints, nasal irritation, conjuctivitis, dermatitis, reversible balding, drowsiness, chest constriction, intestinal distress, weakness and slower mentation REF RANGE Negative; general population, 3–52 μg/g creatinine. Cf Selenium deficiency

self abuse PSYCHIATRY Intentional physical injury to oneself MANAGEMENT Psychotherapy, SSRIs. See Self-mutilation

self-actualization The process of self-discovery & personal growth

self-blood glucose monitoring Home blood glucose monitoring A process by which blood glucose is determined at home by pricking the finger, putting a drop of blood on a chemically treated test strip, and comparing the color changes to a chart

self-bougienage GI DISEASE A therapy in which a Pt auto-introduces a 44-46-F Maloney dilator tube to treat benign recurrent esophageal strictures. See Caustic injury, Esophageal web

self-esteem Self-worth PSYCHOLOGY The internalized sense of one's own worth

self-examination PUBLIC HEALTH Any format in which a person monitors self body regions for 'lumps or bumps'–breast SE or any changes in size, appearance, or color of 'spots'–skin SE; soft data suggest that SEs help detect breast and skin cancer–especially melanoma at earlier, more treatable, stages

self-fulfilling prophecy PSYCHIATRY A distorted prediction or statement about a person in a certain setting, which forms a substrate that ultimately leads the person to behave in the predicted manner

self help GYNECOLOGY Routine gynecologic care by a ♀ for herself–eg, breast self examination and pelvic examination, simple lab testing, fitting and inserting birth control devices, etc PSYCHIATRY Coping skills or recovery techniques practiced sans physician or therapist. See Self actualization

self-hypnosis PSYCHOLOGY The hypnosis of oneself–a modality for which a person can be trained INDICATIONS Control of habits, behaviors, pain, smoking, weight; SH can be used to evoke repressed memories, experiences–Freud's 'abreaction'. See Hypnosis

self-insured Self fund HEALTH INSURANCE adjective Referring to the practice of carrying an individual health insurance policy for oneself; self insurance is usually more expensive than group insurance

self-limited adjective Related to a condition, especially infection or inflammation, that resolves without therapy

self-medication PSYCHIATRY The consumption of a substance, without physician imput, to compensate for any medical or psychological condition. See Over-the-counter drug

self-mutilation NEUROLOGY The physical abuse of a person's body by its owner, often over a psychogenic substrate; SM is also an integral component of certain hereditary conditions–eg, Lesch-Nyhan syndrome, Cornelia de Lange syndrome, attributed to neurotransmitter imbalances PSYCHIATRY Auto-destruction, perpetuated in various guises by ± 1:1500 population; most SM is psychogenic in origin, and may be related to physical confinement, deprivation, depression, often related to childhood experiences; SM may be a form of self-cleansing to cauterize the 'pain of living'

self-neglect PSYCHOLOGY The relative lack of self-care, especially common in the elderly, who live alone and cannot provide for themselves and/or maintain physical and/or mental health

self-protection STD The obligation one has to oneself to prevent the transmission of an infection–eg, HIV; SP assumes that the other partner may not disclose HIV status, and if one engages in consensual sexual activity, caveat emptor. See Contributory negligence, Negotiated safety, Sexual responsibility

self-referral Physician self-referral HEALTH INDUSTRY The referral by a physician to a health facility–eg, imaging center at which he/she has a financial interest, but no professional responsibility. See Fee-splitting, Joint venture, Kickback, 'Safe harbor' rules. Cf Physician referral, Referral

semantic memory NEUROLOGY A 'cognitive' form of memory linked to acquisition and use of factual knowledge. See Memory

semen analysis LAB MEDICINE A procedure for evaluating possible ♂ causes of infertility; sperm are ↓ in some forms of infertility or post-vasectomy REF RANGE Count 20–200 x 10⁹; volume 1.5-6.0 mL; morphology > 70% mature and without defects; > 60% motility; pH 7.5–8.5; liquefaction > 70%. See Post-vasectomy semen analysis, Sperm, Vasectomy

semen quality UROLOGY The measurable parameters of semen–eg, sperm concentration, total sperm count per ejaculate, % of motile sperm, number of abnormal and immature sperm

semicomatose NEUROLOGY adjective Relating to a state in which a Pt is unresponsive unless shaken, not stirred. See Comatose

semipermeable membrane NEPHROLOGY A dialysis membrane with a pore size that permits passage of solvent and some solute molecules. See Cell membrane

semistarvation neurosis PSYCHIATRY A pre-anorexia nervosa affecting ♀ who, as children were described as 'good girls' PROFILE Perfectionist tendencies, effects of long-term caloric restriction–eg, fatigue, weakness, apathy, passivity, withdrawal, physical regression. See Anorexia nervosa, Binge-purge syndrome. Cf Starvation

seminoma Lance Armstrong tumor UROLOGY A testicular germ-cell neoplasm and most common tumor of ♂ < age 40 CLINICAL Mass, ± pain MANAGEMENT Excision, RT PROGNOSIS ± 90% 5-yr survival if treated early. See Anaplastic seminoma, Germ cell tumor, Spermatocytic seminoma

Semont maneuver Liberatory maneuver AUDIOLOGY A maneuver used to manage benign paroxysmal positional vertigo–BPPV; Pt is rapidly moved from lying on one side to the other. See Positional vertigo

semustine ONCOLOGY An alkylating anticancer nitrosourea. See Alkylating agent

SEN-V INFECTIOUS DISEASE A new hepatitis virus, named after the person so afflicted CLINICAL CLASSIC Classic hepatitis Sx with negative tests for other hepatitis antigens. See Hepatitis

Sengstaken-Blakemore tube EMERGENCY MEDICINE A 3-lumen rubber tube used to quench bleeding in esophageal varices; central lumen allows communication of mouth with stomach; 2nd tube communicates with a rounded 'bladder' that is inflated to anchor the device in the upper stomach; 3rd tube communicates with a long inflatable balloon that places gentle pressure on the esophageal lumen, which usually stops the bleeding. See Esophageal varices. Cf Minnesota tube

senile cardiac amyloid Cardiac amyloidosis CARDIOLOGY A disorder caused by myocardial deposition of amyloid protein with compromised cardiac function, causing restrictive cardiomyopathy or dilated cardiomyopathy, which may be associated with conduction defects; SCA is confined to the heart, usually occurring in older adults. See Amyloidosis

senile cerebral amyloid angiopathy Cerebral amyloidosis NEUROLOGY An idiopathic condition characterized by deposits of amyloid protein in cerebral arterial walls with ↑ risk of lobar intracerebral hemorrhage in the elderly. See Amyloidosis

senile dementia The cognitive decline and progressive neurodegeneration typical of older adults; ischemia of periventricular white matter may cause a significant proportion of SD, which disconnects the relatively intact cerebral cortex, resulting in true subcortical dementia or Binswanger type dementia. Cf Alzheimer's disease, Lacunar state, Multi-infarct dementia, Pseudodementia

senility GERIATRICS A state of advanced physical and mental deterioration associated with advanced age. See Dementia, Geriatrics, Senile dementia

senior clerkship Acting internship, see there

seniority VOX POPULI The length of time spent in a department, company or other organization, which usually forms the basis for ↑ privileges, salary, responsibility

Senning procedure HEART SURGERY An operation for Pts with transposition of great vessels, which switches blood flow at the atrial level so systemic venous return passes into the left atrium, left ventricle, and pulmonary artery; the right ventricle assumes the role of a systemic ventricle. Cf Mustard procedure

sensation MAINSTREAM MEDICINE The conscious recognition of a physical–audio, chemical, electrical, mechanical, visual stimulation which excites a sense organ. See Epicritic sensation

Sensaval Nortriptyline, see there

sense NEUROLOGY The ability to perceive a stimulus. See Haptic sense

sensi Antibacterial sensitivity test, see there

Sensibal Nortriptyline, see there

sensing threshold CARDIAC PACING The minimum atrial or ventricular intracardiac signal amplitude in mV required to inhibit or trigger a demand pacemaker

sensitivity CARDIAC PACING The degree to which a pacemaker responds to electrical activity in the heart. See Sensing Threshold LAB MEDICINE PID rate, positivity in disease rate, true positive rate The degree to which a test or clinical assay is capable of confirming–or supporting–the diagnosis of disease X–ie, the analyte is appropriately abnormal in a person with disease. See Analytical sensitivity, Predictive value, Two-by-two table. Cf Specificity NEUROLOGY

The degree to which one can sense a stimulus with a sense organ. See Contact sensitivity, Functional sensitivity, Insulin sensitivity, Multiple chemical sensitivity, Subsensitivity

607

sensitization IMMUNOLOGY The process in which a person acquires the ability to react to an antigen, usually of nonself origin. See Secondary–immune response. Cf Primary–immune response

Sensival Nortriptyline, see there

sensorimotor polyneuropathy NEUROLOGY A peripheral neuropathy characterized by ↓ movement or sensation, due to nerve damage; SP is a systemic process that damages nerves, with loss of myelin, which slows neural conduction or damages neurons, especially the axon, blocking conduction at the point of damage–pressure on nerves, inflammation, ↓ blood flow, connective tissue disorders, idiopathic, alcoholic neuropathy, diabetic neuropathy, chronic inflammatory neuropathy, Guillain-Barre syndrome, and drug-induced neuropathy. See Neuropathy

sensorineural hearing loss AUDIOLOGY Hearing loss caused by damage to the sensory cells and/or nerve fibers of the inner ear. See Hearing loss

sensory aphasia Wernicke's aphasia NEUROLOGY Aphasia comprised of 2 elements: (1) Impaired speech comprehension due largely to an inability to differentiate spoken and written phonemes–word elements–due to either involvement of the auditory association areas or separation from the 1° auditory complex; (2) Fluently articulated but paraphasic speech, which confirms a major role by the auditory region in regulating language; Pts with SA are voluble, gesticulate, and unaware of the incoherency of their speech; words are nonsubstantive, malformed, inappropriate–paraphasia. See Aphasia, Motor aphasia

sensory nerve transposition NEUROSURGERY A procedure in which sensation lost in a critical area is partially restored by transposing sensory nerve fibers into a region of interest

sensory tic NEUROLOGY A tic triggered by a general or local dysphoric somatic sensation of pressure, tickle, warmth, cold, pain, etc, seen in ±40% of Pts with Tourette syndrome. See Tourette syndrome

sentinel event HEALTH POLICY A term used by the JCAHO for a 'headliner' event that may cause an unexpected or unanticipated outcome or death, and trigger an investigation of a hospital's policies

sentinel loop IMAGING A dilated loop of jejunum seen in LUQ in upper GI radiocontrast studies and a typical finding of acute pancreatitis. Cf Colonic cut-off sign

sentinel node Sentinel lymph node, signal node of Virchow An isolated, enlarged often left-sided supraclavicular lymph node, classically associated with metastatic gastric CA which, when found, indicates that the malignancy is non-resectable. Cf Mary Joseph nodule

sentinel pile The swelling at the lower end of a chronic anal fissure, palpable as an anal mass, which may be the first or most prominent manifestation of a fissure, hence, a 'sentinel'

SEPA® Soft enhancer of percutaneous absorption THERAPEUTICS A technology that enhances transdermal drug delivery. See Transcutaneous therapy

SEPA/alprostadil A gel formulation of alprostadil or synthetic PGE with the SEPA transdermal-delivery of erectile dysfunction. See Erectile dysfunction

separation anxiety PEDIATRICS A normal developmental stage–between 6-8 months and 10-14 months–during which an infant experiences apprehension, uncertainty, discomfort when faced with anticipated or actual separation from a 1° care giver, mother or parent-surrogate; SA wanes by age 2, when toddlers begin to trust that their parents will return; SA reappears

Sepracoat™

whenever one is in an unfamiliar situations–eg, hospitals, illness, pain, or unfamiliar people

Sepracoat® SURGERY A liquid hyaluronic acid used to minimize postoperative adhesions by moistening exposed tissues during surgery. See Adhesion, Hyaluronic acid

sepsis INFECTIOUS DISEASE Sepsis is defined by clinical parameters as 'SIRS–systemic inflammatory response syndrome plus a documented–ie, 'culture-positive' infection', and is part of a continuum of an inflammatory response to infection that evolves toward septic shock CLINICAL Tachypnea, tachycardia, hyperthermia, hypothermia MANAGEMENT Ibuprofen ↓ prostacyclin, thromboxane, ↓ tachycardia, fever, O₂ consumption, lactic acidosis; NSAIDs do *not* prevent shock, ARDS, or improve survival. See Postanginal sepsis, Septic shock, Severe sepsis, SIRS

sepsis syndrome A constellation of signs, Sx, and systemic responses caused by a wide range of microorganisms that may eventuate into septic shock; SS is a systemic response to infection

---

SEPSIS SYNDROME, DEFINING PARAMETERS

- **TEMPERATURE** Hypothermia < 35°C–96°F or hyperthermia > 39°C–101°F
- **TACHYCARDIA** > 90 beats/minute
- **TACHYPNEA** > 20 breaths/minute
- **SITE OF INFECTION** Clinically evident focus of infection or positive blood cultures
- **ORGAN DYSFUNCTION** 1+ end organs with either dysfunction or inadequate perfusion or cerebral dysfunction
- **METABOLIC DERANGEMENT** Hypoxia–PaO₂ < 75 mm Hg, ↑ plasma lactate/unexplained metabolic acidosis
- **FLUID IMBALANCE** Oliguria–< 30 mL/hr
- **WBC COUNTS** < 2.0 x 10⁹/L; < 12.0 x 10⁹/L–US: < 2000/mm³; > 12 000/mm³

Note: The confusing semantics of the terms sepsis, sepsis/septic syndrome, and septic shock are unlikely to be resolved in the forseeable future; the terms sepsis and sepsis syndrome are essentially interchangeable and would in part overlap with septicemia–the early components of a pernicious infectious cascade that has spilled into the circulation; the term septic shock is used when the process becomes virtually irreversible

---

septal defect See Atrial septal defect, Ventricular septal defect

Septata intestinalis Encephalitozoon intestinalis INFECTIOUS DISEASE A microsporidium, seen in macrophages, fibroblasts, endothelial cells in the lamina propria of the large intestine in AIDS Pts, associated with chronic diarrhea and systemic disease. See Microsporidium

septic abortion Infected abortion OBSTETRICS An abortion caused by infection and complicated by fever, endometritis, parametritis BACTERIA Native vaginal flora, *Clostridium perfringens*, aerobic and anaerobic streptococci, gram-negative bacilli. See Abortion

septic arthritis Bacterial arthritis, infectious arthritis, non-gonococcal bacterial arthritis RHEUMATOLOGY A non-gonococcal infection of a joint, which develops when bacteria spread to a joint CLINICAL Rapid onset with joint swelling, intense pain, low-grade fever RISK FACTORS Concurrent bacterial infection, chronic illness, immunosuppression, rheumatoid arthritis, IVDA, recent joint trauma, or recent joint arthroscopy or surgery; it is not uncommon in children < age 3, affecting primarily the hip, in adults, knee involvement is more common AGENTS IN CHILDREN Group B streptococcus, *H influenzae*

septic shock Bacteremic shock INFECTIOUS DISEASE A condition identical to the sepsis syndrome with an added component of hypotension–systolic BP < 90 mm Hg or loss in the baseline systolic BP of > 40 mm Hg; SS has been formally defined by clinical parameters as 'Sepsis-induced hypotension despite fluid resuscitation plus hypoperfusion abnormalities…', which include lactic acidosis, oliguria, or an acute alteration in mental status; SS is end-stage of a continuum of a biologic inflammatory response to infection ETIOLOGY Gram-negative bacilli–eg, E coli, Klebsiella spp, gram-positive cocci–eg, Staphylococcus spp; viruses, fungi rarely cause SS EPIDEMIOLOGY SS

kills 100k to 175k people/yr–US, 13ᵗʰ major cause of death in US RISK FACTORS Underlying disease–DM; hematologic CA; GI, GU, hepatobiliary disease, recent infection, prolonged antibiotics, recent surgery CLINICAL Either reflect response to infection–ie, tachycardia, tachypnea, changes in temperature, leukocytosis, or the organ system(s) involved–eg, cardiovascular, hematologic, hepatic, renal, pulmonary; SS is severe if there is hypoperfusion, lactic acidosis, hypotension, or altered mental status MEDIATORS Bacterial endotoxin evokes a vehement response by the complement system, kinin system, and plasma phospholipases and release of various cytokines–eg, TNF-a, IL-1, IL-6, β-endophins, PAF, PGs, leukotrienes. See Sepsis, Sepsis syndrome, Severe sepsis, SIRS

septic syndrome Sepsis syndrome, see there

septicemia INFECTIOUS DISEASE A rapidly progressive, life-threatening infection characterized by bacteremia which may be 2° to local infection of respiratory, GI, or GU tracts, associated with osteomyelitis, meningitis, or infection of other tissues; it may rapidly lead to septic shock, death CLINICAL Onset heralded by spiking fever, chills, tachypnea, tachycardia, toxic appearance, sense of impending doom; Sx rapidly progress to shock–hypothermia, hypotension, changed mental status, clotting defects–eg, petechiae, ecchymosis; if caused by *N meningococcus*, shock, adrenal collapse, DIC. See Sepsis syndrome

septo-optic dysplasia sequence De Morsier syndrome NEONATOLOGY An idiopathic form of midface-CNS hypoplasia characterized by incomplete early morphogenesis of the anterior midline brain, causing hypothalamic defects, hypoplasia of optic chiasmata, absent septum pellucidum CLINICAL Pendular nystagmus, visual impairment, 2° hypopituitarism, sexual precocity, aberrant retinal vessels TREATMENT Growth hormone replacement

septoplasty Reconstruction of nasal septum ENT Any operation that alters the shape–eg, deviated nasal septum, often performed with rhinoplasty. See Rhinoplasty

Septra See TMP/SMX

septuplets OBSTETRICS *Seven* babies born at the same time to the same mother. See Higher multiple pregnancies

sequelae CLINICAL MEDICINE The consequences of a particular condition or therapeutic intervention

sequence PEDIATRICS Anomalad An array of multiple congenital anomalies resulting from an early single 1° defect of morphogenesis that unleashes a 'cascade' of 2° and 3° defects; a sequence is also defined as a set of clinicopathologic consequences of the aberrant formation of one or more early embryologic structures. See Dysmorphology

---

SEQUENCE TYPES

MALFORMATION Incorrect formation of tissues

DEFORMATION Abnormal forces acting on normal tissues

DISRUPTION Breakdown of normal tissue

Note: The Pierre-Robin sequence is caused by 1° mandibular hypoplasia, which results in a tongue that is too small for the oral cavity and which drops back–glossoptosis, blocking closure of the posterior palatal shelf, resulting in a high arched U-shaped cleft palate Examples of sequences include athyroidotic hypothyroidism sequence, DiGeorge sequence, early urethral obstruction sequence, bladder exstrophy sequence, cloacal extrophy sequence, holoprosencephaly sequence, jugular lymphatic obstruction sequence, Kartagener syndrome/sequence, Klippel-Feil sequence, laterality sequence, meningomyelocele, anencephaly, iniencephaly sequence, occult spinal dysraphism sequence, oligohydramnios sequence, Rokitansky sequence, septo-optic dysplasia–de Morsier sequence, sirenomelia sequence

---

sequential compression device SURGERY A pneumatic device used to prevent DVT from the legs and arms. See Deep vein thrombosis

sequential pulse cardioversion/defibrillation CARDIAC PACING Therapy that terminates a ventricular tachycardia and/or V fib episode by delivering 2 high voltage pulses one after another across 2 different pathways through the cardiac tissue. See Ventricular fibrillation

sequential study See Consecutive sample

sequestration MEDTALK 1. The development of a sequestrum. See Bronchopulmonary sequestration, Carbon sequestration, Pseudosequestration, Pulmonary sequestration 2. The removal or isolation of a chemical, molecule, cell, or tissue from general access–eg, binding of certain proteins–eg, profilin, thymosin β4, Gc protein to G-actin to prevent polymerization. See Carbon sequestration

sequestration complex PULMONOLOGY The separation of an aberrant pulmonary lobe from the pulmonary parenchyma by pleura with a vascular supply arising directly from the aorta; the venous drainage is by the azygous or hemiazygous veins; SC is often associated with diaphragmatic hernias and GI malformations; lung tissue may be normal or inflamed. See Pulmonary sequestration. Cf Folded lung, Scimitar syndrome, Trapped lung

sequestration crisis HEMATOLOGY A condition affecting sickle cell anemia Pts with splenomegaly, characterized by major sequestration of sickled RBCs in the spleen, leading to hypersplenism, hypovolemia, occasionally shock. See Sickle cell anemia

sequestrum ORTHOPEDICS A plug of necrotic bone separated from viable bone. See Button sequestrum, Kissing sequestrum

sequoiosis PULMONARY MEDICINE A hypersensitivity pneumonitis of redwood loggers caused by exposure to *Pullularia pullulans* spores. See Hypersensitivity pneumonitis

Serax® Oxazepam, see there

Seren Chlordiazepoxide, see there

Serentil® Mesoridazine, see there

Serepax Oxazepam, see there

Seresta Oxazepam, see there

Seretide™ Salmeterol/fluticasone propionate, see there

serial films IMAGING X-rays taken of the same side that has the putative lesion, over a period of days to wks or even yrs to evaluate progression or regressiveness of the lesion

serial induction OBSTETRICS The repeated dosing of oxytocin to induce labor. See Induction, Oxytocin

serial killer FORENSIC PSYCHIATRY A person who commits serial murders PROTOTYPIC SK White ♂ age 30; 97% are ♂; 80% are sociopaths. See Dahmer, Depraved heart murder, Ice Man. Cf Megan's law, Son of Sam law

serial murder FORENSIC PSYCHIATRY A series of homicides in which a single person–or a small group selects victims based on a common characteristic or, less commonly, at random. See Depraved heart murder, Murder, Serial killer

serial seizure NEUROLOGY Two or more seizures over a relatively brief–mins to hrs–period, from which the Pt regains consciousness between seizures. See Seizure, Status epilepticus

serial sevens test NEUROLOGY A test of subtracting 7 from 100, from 93, from 86, etc, to test consciousness, used when administering general anesthesia pre-surgery. See Mini-mental test

series Clinical series, series of consecutive cases CLINICAL RESEARCH An uncontrolled study–prospective or retrospective of a series–succession of consecutive Pts who receive a particular intervention and are followed to observe outcomes. See Case series, Time series IMAGING A set of images taken in a sequence. See Cardiac series, Lower GI series, Obstruction series

serious emotional or behavioral disorder PSYCHIATRY Emotional and/or social impairment in a child or adolescent that disrupts academic and/or developmental progress, family and/or interpersonal relationships, and impairs functioning for at least one yr, or with impairment of short duration and ↑ severity

serious reportable event PATIENT CARE A health care event resulting in death, serious injury or other significant harm TYPES Surgical; product/device-related; Pt protection; Pt care; environmental; criminal EXAMPLES Surgery on wrong body part, on wrong Pt, wrong procedure, retention of foreign object in Pt post surgery or other procedure. See Medical error

Serlect® Sertindole, see there

SERM Selective estrogen receptor modulator, see there

Sermonil Imipramine, see there

SEROCO AIDS A prospective French multicenter epidemiologic study, which assessed the natural history of HIV infection and outcome factors

seroconversion IMMUNOLOGY The development of antibodies detectable in the serum, after exposure to a particular organism or antigen, in a person who was previously immunologically 'naive' for–ie, previously unexposed to a particular antigen; seroconversion may indicate current infection–and transmissibility of a pathogen–eg, HIV-1–seroconversion to p24 and/or p41 antibody production or HBV–seroconversion to surface antibody-HBsAb or e antibody–HBeAb production. See Seropositive. Cf Seronegative

serologic test LAB MEDICINE A test that measures components–eg, antibodies, complement, and reactions–eg, complement fixation, agglutination, precipitation, etc, that reflect immune status, especially antibody titers. See Seroconversion

serologic test for syphilis Any assay for detecting antibodies to *Treponema pallidum* antigens. See RPR test, Syphilis, TPI, VDRL

serology The study of antigen-antibody reactions–eg, past or present infection. See Hepatitis B serology, HIV serology

seronegative IMMUNOLOGY The lack of antibodies or other immune markers in serum that would indicate exposure to a particular organism or antigen; in an immunocompetent person, exposure to an antigen results in a seroconversion and the subject is said to then be seropositive. Cf Seropositive

Seronil Fluoxetine, see there

seropositive IMMUNOLOGY The presence of antibodies or other immune markers in serum, that indicate prior exposure to a particular organism or antigen. Cf Seronegative

seroprevalence IMMUNOLOGY The proportion of a population that is seropositive–ie, has been exposed to a particular pathogen or immunogen; the seropositivity of a population is calculated as the number of individuals who produce a particular antibody divided by the total population

Seroquel® Quetiapine, see there

seroreversion IMMUNOLOGY A change from seropositivity–production of antibodies to a particular antigen or pathogen, to seronegativity–nonproduction of antibodies. See Seronegativity

serosanguineous *adjective* Referring to a body fluid that is both sticky and bloody

serositis Inflammation of the serous tissues of the body, which line the lungs–pleura, heart–pericardium, and abdomen–peritoneum

serotonin-norepinephrine reuptake inhibitor CLINICAL PHARMACOLOGY Any of a class of antidepressants–eg, venlafaxine, that block the reuptake of serotonin and norepinephrine

610 **serotonin-selective reuptake inhibitor** Selective serotonin reuptake inhibitor, serotonin reuptake inhibitor PHARMACOLOGY A class of antidepressants that slows reabsorption of serotonin by neurons; allowing it to remain in the synapse longer; SRIs should be started with ramped dosing EXAMPLES Buspirone, fluoxetine, fluvoxamine, mirtazapine, nefazodone, paroxetine, sertraline, trazodone, venlaxfazine INDICATIONS Dysthymia or minor depression, atypical depression, anxiety SIDE EFFECTS Insomnia, agitation, headache, nausea, diarrhea, adverse interactions with other agents, especially fluoxetine CONTRAINDICATIONS MAOIs. See Antidepressants, Obsessive-compulsive disorder

serotonin syndrome PSYCHIATRY A potentially life threatening condition caused by combined therapy with MAOIs and selective serotonin reuptake inhibitor–SSRI, antidepressants–Prozac, Zoloft, Paxil, Luvox, Wellbutrin–bupropion, tryptophan or tricyclic antidepressants, which abruptly ↑ serotonin resulting in neuromuscular, autonomic, and behavioral changes due to ↑ CNS serotonin activity CLINICAL Confusion, diaphoresis, ataxia, excitement, hyperthermia, hyperreflexia, myoclonus, hypotension, tremor COMPLICATIONS DIC, rhabdomyolisis, cardiovascular compromise MANAGEMENT D/C offending drug, support vital functions, dantrolene–muscle relaxant, Periactin–cyproheptadine–antihistamine, serotonin antagonist. See MAOI, SSRI

serous drainage MEDTALK A 'weeping' of protein-rich serous fluid from a body cavity

serous otitis ENT Noninfectious inflammation of the ear which may occur when there is a collection of sterile fluid in the ear

Seroxat Paroxetine, see there

serpiginous tract A descriptor for a twisted, vermiform radiolucency surrounded by a sclerotic rim, seen in long bones in pyogenic osteomyelitis–caused by streptococci, staphylococci, *Brucella* spp or infarction accompanied by intramedullary calcification

serpiginous ulcer OPHTHALMOLOGY A necrotizing corneal ulcer that occurs 1–2 days after inoculation with *S pneumoniae*, characterized by active ulceration at the leading edge and healing at the trailing border; it is often accompanied by hypopyon and dacrocystitis TREATMENT Cefazolin, penicillin G

Serratia marcescens MICROBIOLOGY The type-species of the gram-negative *Serratia*, widely present in the environment, and occasional cause of hospital-acquired infections ASSSOCIATIONS Contaminated fluids, equipment, cleaning solutions, hands, ↓ nurse-to-Pt ratio. See *Serratia*

sertraline Atruline, Lustral, Zoloft® PSYCHIATRY A serotonin reuptake inhibiting–SRI antidepressant that blocks reuptake of serotonin at the synapse more potently than that of norepinephrine INDICATIONS Panic disorder, major depression, dysphoria, OCD, PTSD CONTRAINDICATIONS MAOI therapy ADVERSE REACTIONS Nausea, insomnia, somnolence, diarrhea, 1° ejaculatory delay; in children, hyperkinesia, twitching, fever, malaise, purpura, ↓ weight, emotional lability, ↓ urine concentration. See Obsessive-compulsive disorder, Serotonin-selective reuptake inhibitor

serum 1. The fluid component of blood from which the coagulation factors have been removed. See Fetal bovine serum. Cf Plasma 2. A protein-rich fluid that contains a high concentration of antibodies to a particular antigen of interest; convalescent sera–from an individual who has recuperated from a particular infection–eg, scarlet fever may be of use in treating an individual who is suffering from the same infection. See Acute phase serum, Antilymphocyte serum, Convalescent serum

serum concentration THERAPEUTICS The amount of a drug or other compound in the circulation, both bound to proteins and unbound, the latter of which generally corresponds to the theraepeutically active fraction

serum lipid Any major lipid in the circulation–total cholesterol, HDL, LDL, TGs. See Cholesterol, Triglyceride

serum protein electrophoresis A method for determining protein 'homeostasis'; serum proteins are divided into prealbumin/albumin, α_1 and α_2, β, and γ zones; regions of the SPEP are ↑ or ↓ in certain conditions, or variable with gene polymorphisms

SERUM PROTEIN ELECTROPHORESIS

FRACTION	CLINICAL CONDITIONS
NORMAL–6.5-8.0 g/dL	
↑	Dehydration, vomiting, diarrhea, myeloma, respiratory distress syndrome
↓	Burns, cancer–GI tract, liver disease, malnutrition, malabsorption, renal failure, ulcerative colitis
↓ Prealbumin	↓ Functional hepatic mass, inflammation, malnutrition
ALBUMIN–4.0-6.0 g/dL	
↑	Dehydration, exercise
↓	Burns, cancer–GI tract, CHF, eclampsia, ↑ extracellular volume, inflammation, liver disease, malignancy, malnutrition, malabsorption, nephrotic syndrome, renal failure, SLE, ulcerative colitis
α_1–0.1-0.5 g/dL	
↑ antitrypsin	Inflammation, hepatocellular injury, malignancy, pregnancy, necrosis
↓ antitrypsin	Deficient allele causing emphysema
± antitrypsin	Polymorphism of $\alpha_1$antitrypsin
α_2–0.4-1.0 g/dL	
↑ macroglobulin	Biliary cirrhosis, burns, malignancy, inflammation, acute MI, nephrotic syndrome, rheumatic fever, selective proteinuria in age extremes
↑ haptoglobulin	Inflammation
↓ haptoglobulin	Hemolysis, hepatosplenic sequestration, liver disease
TOTAL β–0.3-1.2 g/dL	
β_1	
↑ transferrin	Iron deficiency, estrogens
↓ transferrin	Malnutrition, burns, inflammation
↓ lipoprotein	Hypercholesterolemia
±	Polymorphisms of transferrin
β_2	
↑ C3	Chronic inflammation, bile obstruction
↑ IgA	Malignancy, infection of mucosal surfaces, rheumatoid arthritis, ethanol, cirrhosis
↓ C3	Complement activation
± C3	Polymorphism of C3
γ–0.4-1.0 g/dL	
↑	Collagen vascular disease, inflammation, monoclonal antibody production, polyclonal stimulation, SLE
↓	Hypogammaglobulinemia, nephrotic syndrome

serum sickness IMMUNOLOGY An immune response to re-exposure to an antigen to which an organism had been previously sensitized CLINICAL Urticaria, fever, lymphadenopathy, ±arthritis, glomerulonephritis. See Immune complex disease, Zone of equivalence

servant See *Respondeat superior*

service 1. A group of physicians in a particular discipline in a hospital 2. A group of Pts for which a particular physician is providing care HOSPITAL PRACTICE A group of medical practitioners who have clinical privileges in a specific area of medicine, surgery, obstetrics, pediatrics, or other. See Ancillary service, ASC surgical service, Basic health service, Bulletin board service, Bundled service, Carve-out service, Clinical personal health service, Clinical preventive service, Clinical trials monitoring service, Covered service, Customer service, Directory service, Domestic service, Electronic Residency Application service, Emergency medical service, Epicell service, Enabling service, Environmental service, Epidemiology Intelligence service, Evaluation & management service, Fee for service, Fertility service, Free service, Hard medical service, Health care service, Hotel service, Inpatient service, Medical service, Medical transportation service, Mobile medical service, Modified fee for service, Observation service, Outpatient service, Phone counseling service, Preventive health service, Protective service, Public health service, Related service, Respite service, Social service, Soft medical service,

Support service, Surgical service, Transition service, Type I service, Unrelated service, Wraparound service

service area MANAGED CARE The area, allowed by state agencies or by a certification of authority, in which a health plan can provide services. See Health plan

service coordination Case management, see there

service dog A dog trained to care for a person with disabilities, most commonly those with visual impairment, but also those with ambulatory disabilities who are confined to wheelchairs. See Adaptive equipment, Americans with Disabilities Act, Physical barriers

service patient Public Pt SOCIAL MEDICINE A Pt who is the ward of a health facility, often by default; SPs lack insurance and have a condition requiring long-term care. See Homeless(ness), 'Safety net' hospitals. Cf Private Pt

Servium Chlordiazepoxide, see there

Servizepam Diazepam, see there

Serzone® Nefazodone, see there

sesamoidectomy ORTHOPEDICS Removal of a sesamoid bone, see there

sesamoiditis Inflammation of a sesamoid bone of the foot, linked to overuse injury, associated with sprinting or driving out of a stance in football PREDISPOSING FACTORS Forefoot valgus, rigid pes cavus, multipartite sesamoid CLINICAL Pain and swelling on ball of foot; standing on toes, running, walking MANAGEMENT GOAL Unload area with aperture padding, crutches, custom orthotics, or steroid injections

sessile MEDTALK *adjective* Referring to a broad base of attachment

sestamibi ⁹⁹ᵐTc or technicium-99m sestamibi IMAGING A myocardial perfusion agent for assessing an 'at-risk' myocardium–for acute MI, used with tomography to determine the final size of an acute MI

setback ENT An operation for cleft palate VOX POPULI A reversal in an improving trend, which may occur in a person with a physical disability, a phobic disorder, or with a chronic condition requiring long-term therapy. See Physical setback

setpoint hypothesis ENDOCRINOLOGY A deterministic theory that explains the interplay of appetite and other factors–eg, fats and carbohydrates in weight control; per the SPH, the brain is constantly adjusting the metabolic rate and manipulating behavior to maintain a target weight; although the SP changes with age, it reflects a fixed genetic program; while diet and exercise can shift the SP, the target is immutable. See Obesity; Cf Glycogen hypothesis, Settling point hypothesis

setting sun sign PEDIATRICS Inferior ocular deviation, typical of severe infantile hydrocephalus–occasionally seen in subdural hematomas; it is often accompanied by the 'cracked-pot' sign, prominant scalp veins, thinned and shiny skin, a high-pitched cry, and optic nerve atrophy, due to nerve and chiasm compression. See Hydrocephalus

setting VOX POPULI A place where something occurs. See ER/trauma setting

sevelamer Renagel® NEPHROLOGY A chelator used to ↓ phosphorus in Pts with ESRD and in chronic renal failure Pts not requiring dialysis. See Chelation, End-stage renal disease

Seven Countries study CARDIOLOGY A prospective 10-yr epidemiologic study which examined the relation of dietary fats in middle-aged ♂ with CAD, and linked CHD to ↑ saturated fats SCS TRACKED RISK FACTORS Diet, BP, weight, cigarette smoking, exercise habits

70s, the rules of A mnemonic for CNS tumors

70s, RULE OF

GENERAL 70% are primary CNS neoplasms, 70% of primary CNS neoplasms are glial, 70% of primary glial tumors are astrocytomas; 70% of astrocytomas are high grade

CHILDREN 70% of tumors arise in the posterior fossa, 70% of those occurring before age 2 are medulloblastomas; 70% of supratentorial tumors are craniopharyngiomas

ADULTS 70% are in the hemispheres, 70% of those in the pineal region are germinomas, 70% of those in the pituitary gland are adenomas, of which 70% are chromophobe adenomas

72-hour fast ENDOCRINOLOGY A test performed in a hospital to ID causes of hypoglycemia. See Fast, Hypoglycemia

severe combined immunodeficiency A heterogeneous X-linked or less commonly, AR condition, more common in blacks, onset in early infancy, characterized by dysfunctional T and B cells CLINICAL Morbiliform rash, hyperpigmentation, recurring infections–*Candida, Pneumocystis carinii,* CMV, EBV, HBV, varicella, FTT, early death TREATMENT BM transplant; gene therapy–enzyme replacement. See Adenosine deaminase deficiency, 'Bubble boy', Gnotobiotic, Purine nucleoside phosphorylase, SCID mice

SEVERE COMBINED IMMUNODEFICIENCY

DEFINING CRITERIA Coma, respiratory distress–pulmonary edema, hypoglycemia, circulatory collapse–clinical shock, repeated convulsions, severe anemia–< 5g/dL and > 10 000 parasites mm³, acidosis–plasma bicarbonate < 15 mmol/L, hemoglobinuria, renal failure, spontaneous bleeding

SUPPORTING CRITERIA Jaundice, prostration, hyperpyrexia, impaired consciousness, hyperparasitemia–> 500 000 parasites mm³ NEJM 1995; 332:1399ᴏᴀ, 1441ᴇᴅ

severe sepsis A condition defined clinically as '*Sepsis associated with organ dysfunction, hypotension, or hypoperfusion abnormalities*' (which include) ...*lactic acidosis, oliguria, or an acute alteration in mental status*; SS is part of a continuum of a biologic inflammatory response to infection that evolve toward septic shock. See Sepsis, Septic shock, SIRS

severely debilitating illness Any condition in which there is major irreversible morbidity–eg, AIDS, Alzheimer's disease, blindness, diabetic nephropathy, neurologic degeneration, Parkinson's disease, terminal CA, etc. See Dread disease, Illness, Life-threatening illness

severity-adjusted death rate A calculated rate of mortality, based on the severity of a morbid condition, a datum used to determine appropriate reimbursement from Medicare. See Case-mix index, DRGs, Medicare, Mortality

sevoflurane ANESTHESIOLOGY An inhalation anesthetic PROS Pleasant odor–used in children CONS Unstable in soda lime, used in anesthesia circuits to absorb CO₂; may break down to toxic products. See Inhalation anesthetic, Compound A

sex SEXOLOGY Personal and reproductive status as ♂, ♀, etc, generally based on external genitalia VOX POPULI Sexual activity, genital interaction. See Anonymous sex, Non-coital sex, Oral sex, Phone sex, Rough sex, Solitary sex

sex change Sexual reassignment, see there

sex derivative SEXOLOGY A characteristic of ♂ / ♀ differences that are 2° or subsidiary to sex-irreducible, 1° differences, and largely influenced by sex hormones

sex differential The ♂ to ♀ difference in M&M; in general, when all causes of death are considered, the mortality rate is lower, the likelihood of survival greater, and life expectancy longer in ♀. See Men, Sex-specific mortality rate, Women. Cf Sex ratio

sex drive Libido, see there

sex industry The commercial enterprises related to sale or purchase of sex-related services, ranging from individual 'workers' in prostitution to the pornographic end of the entertainment industry. See AIDS, Condoms, HIV, Safe sexual activities, Sexual work, Sexual deviancy, Sexually-transmitted diseases

612   **sex life** A popular term referring to a person's sexual relationships and level of sexual activity–ie, whether a person is "getting 'it'", or "doing 'it'". Cf Sexual life

sex linked X-linked, see there

sex-linked disease Inherited chromosome X-linked conditions, which are carried by the mother and expressed by the son–eg, Fabry's disease, pyruvate kinase deficiency, G6PD deficiency, Xg^a blood group, factor VIII, factor IX deficiencies. See Testis-determining factor, X-linked

sex-reversed person A person with 1. Pseudohermaphroditism, see there 2. Sexual reassignment, see there

sex role SEXOLOGY A traditional or stereotypical pattern of behavior and thought related to the sex organs and procreation, which is regarded as typical of, or especially suited to, either one sex or the other. See Genderidentity/role

sex-specific mortality rate EPIDEMIOLOGY A mortality rate for either ♂ or ♀

sex toy SEXOLOGY Any device used during sexual activity to enhance pleasure EXAMPLES Chains, dildos, special condoms, edible undergarments, whip PER CICERO O tempora! O mores!

sexist behavior PSYCHOLOGY Actions or language that discriminates based on a person's sex. See Sexual harassment

sexology SEXOLOGY The formal study of the differentiation and dimorphism of sex and of erotic/sexual pairbonding of partners. See Sexosophy

sextant biopsy UROLOGY A series of 6 biopsies from the right and left upper, middle and lower lobes of the prostate, which ↑ diagnostic yield, and likelihood of detecting prostate CA. See Prostate cancer

sexual *adjective* Referring to sex or the stimulation, responsiveness, and function of sex organs alone or with partner(s). See Erotic

sexual abuse PEDIATRICS '…*inappropriate exposure of a child to sexual acts or materials, the passive use of children as sexual stimuli for adults, and actual sexual contact between children and older people. Sexually abused children, in addition to their depressive and aggressive Sx, have an increased frequency of anxiety disorders and problems with sex role and sexual functioning.*'; SA is also defined as '… *sexual contact with a child that occurs as a result of force or in a relationship where it is exploitative because of an age difference or caretaking responsibility,*' the activities displayed by the offender range from exhibitionism to intercourse. See Child abuser, Child sexual abuse, Domestic violence

sexual activity SOCIOLOGY A general term for sexuoerotic interactions with oneself and/or others. See Masturbation

sexual addiction Sex compulsion SEXOLOGY Compulsive and ritualized sexuoerotic hyperactivity, generally under specific sexuoerotic conditions and stimuli. See Sexaholics Anonymous

sexual aggressor SEXOLOGY A person who comes on *real* strong in social situations (if you know what I mean) and is after you know what

sexual anhedonia Inhibited sexual desire SEXOLOGY Sexual dysfunction with a loss of libido; SA may be more common in ♂, and is characterized by normal erection and ejaculation without pleasure during orgasm; SA may be psychogenic–eg, situational to the partner or due to drug abuse–eg, cocaine TREATMENT Detoxification, psychotherapy. See Inhibited sexual desire, Sexual dysfunction

sexual arousal Horny/horniness, randy/randiness PHYSIOLOGY A state of sexual 'yellow alert' which has a mental component–↑ cortical responsiveness to sensory stimulation, and physical component–↑ penile sensitivity, neural response to stimuli, and sense of impending orgasm

sexual asphyxia Erotic self-strangulation FORENSIC MEDICINE, SEXOLOGY The intentional restriction of airflow, often by an adolescent ♂, during 'rough sex', during a sexual ritual or during cross-dressing sexuoeroticism; the passing of a ligature around the neck transiently ↓ O_2 to the brain to ↑ sexual excitement; if the person does not loosen the ligature after orgasm, sexual asphyxia occurs, death may occur

sexual assault FORENSIC MEDICINE '…*any sexual act performed by one person on another without permission…components: use or threat of force, inability of victim to give proper consent or both.*' INCIDENCE 80/100,000/yr; accounts for 7% of violent crimes reported in US– it is vastly underreported due to '…*humiliation, feelings of guilt, fear of retribution, lack of knowledge of legal rights, disillusionment with criminal justice system*' MORBIDITY Up to 5% of SA victims sustain major nongenital injuries; 1% have severe genital injuries requiring surgery; 0.1% sustain injuries that are ultimately fatal INJURIES 11 anatomic sites may be affected by SA, in particular, posterior fourchette; videocolposcopy ↑ rate of detection 3-fold, allows permanent forensic documentation. See Rape, Rape trauma syndrome

sexual assault response team A team of health care professionals–eg, ER physician, forensic nurse, social worker, specifically trained in responding to victims of sexual assault–rape; SARTs combine '…*medicine, law enforcement, victim advocacy to ensure that rape survivors receive comprehensive medical attention, evidentiary examinations, emotional support, and referral information*'. See Sexual assault

sexual behavior A person's sexual practices–ie, whether he/she engages in heterosexual or homosexual activity. See Sex life, Sexual life

sexual body signal Any unspoken signal to a prospective copulatory partner either sexual availability or desire. See Flirting

sexual contact SEXUAL OFFENSES The touching of a victim's, the defendant's, or any other person's intimate parts, or the intentional touching of the clothing covering the immediate area of the victim's, the defendant's or any other person's intimate parts, if that intentional touching can be reasonably construed as being for the purpose of sexual arousal or gratification. See Sexual assault

sexual desire Horny, horniness PSYCHOLOGY A subjective awareness of desire for sexual satisfaction, irrespective of sexual activity. See Sexual function

sexual development See Tanner staging

sexual deviancy Paraphilia PSYCHIATRY Sexual excitement to the point of erection and/or orgasm, when the object of that excitement is considered abnormal in the context of the practitioner's learned societal norms TYPES Exhibitionism, fetishism, frotteurism, pedophilia, sexual masochism, sexual sadism, transvestic fetishism, voyeurism, paraphilia, not otherwise specified, an informal 'wastepaper basket' category. See Child abuse, Paraphilia

sexual differentiation See Hermaphroditism, hirsutism, Müllerian ducts, Precocious puberty, Pseudoprecocious puberty, Tanner staging, Testis-determining factor, Virilization, Wolffian ducts, XXX, XXY, XXXY, XYY syndromes, Y Chromosome

sexual drive SEXOLOGY One of 2 primal drives–the other is the aggressive drive–per Freud's dual-instinct theory; the subjective desire for erotosexual activity. See Libido

sexual dysfunction PSYCHIATRY A term that encompasses disturbances in sexual desire, and psychophysiologic changes in the sexual response cycle, which may be accompanied by marked distress and interpersonal difficulty. See Sexual anhedonia, Sexual response cycle

sexual fantasy PSYCHOLOGY Private mental imagery associated with explicitly erotic feelings, accompanied by physiologic response to sexual arousal. See Sexual desire

sexual favor Any sexual act occurring in an employee-employer relationship, exchanged for privileged treatment in a workplace, ↑ salary, career advancement. See Sexual bribery, Sexual harassment

sexual feelings A constellation of psychological sentiments that constitute desire for sexual satisfaction or release of sexual tension

sexual harassment SEXOLOGY Socially inappropriate behavior defined by the Equal Employment Opportunity Commission–EEOC, as '...*unwelcome sexual advances, requests for sexual favors, and other conduct of a sexual nature*', or '...*creation of a hostile atmosphere or abuse of a position of power in a relationship through sexual behavior or language*.' Sexual favor

sexual history PSYCHIATRY A critical component of a person's mental history–including sexual development, orientation, sexual attitudes and behaviors, sexual conflicts or dysfunction. See Sex life, Sexual life

sexual incompatibility SEXOLOGY The mental substrate on which sexual dysfunction rests, where the 'chemistry' between 2 sexual partners is no longer, or never was, there

sexual intercourse, fear of PSYCHOLOGY Coitophobia. See Phobia, Sexual anhedonia

sexual interplay VOX POPULI Doing "it"; messing around

sexual life '*The directions and manifestations of the sexual drive that contribute to a person's life-style, sometimes confined to genital activity and sometimes referring to all of the manifestations of libidinal energy in the subject's personality and relationships*'. Cf Sex life

sexual maturity rating Tanner staging, see there

sexual misconduct PROFESSIONAL ETHICS Any behavior that violates a health professional's ethics through sexual contact of physician and his/her Pt. See Professional boundaries

SEXUAL MISCONDUCT–AMA COUNCIL OF ETHICAL & JUDICIAL AFFAIRS

1. Predatory physicians who systematically attempt to seduce Pts
2. Those who claim to use sex for therapeutic purposes
3. Abuse of the physical examination, in particular when it is not indicated according to his/her standard of care
4. Situations in which the physician asks for a date on first visit or on seeing the Pt in an ER
5. Situations in which a long-term professional relationship evolves into an infatuation
6. Raping or fondling a Pt while either awake or under anesthesia
7. Sexual harassment in which a physician makes erotic or suggestive remarks to a Pt

Note: The Medical Council of New Zealand has proposed a classification of sexual misconduct that is of use in determining the severity of the infraction, and possibly the degree or severity of the sanction levied against the perpetrator; the MCNZ divides SM into sexual impropriety, sexual transgression, and the most severe form, sexual violation JAMA1995; 273:1445; in an analysis of six such sex studies, up to 12% of ♂, and 4% of ♀ physicians have admitted to SM; 20% involved same sex dyads

sexual orientation A person's potential for responding with sexual arousal to persons of the opposite sex–ie, heterosexual, same sex–ie, homosexual, or both–ie, bisexual

sexual pain disorder SEXOLOGY A condition–eg, dyspareunia, vaginismus–more common in ♀, in which sexual intercourse and intimacy evoke discomfort and pain. See Inhibited sexual desire

sexual penetration SEXOLOGY Sexual intercourse, cunnilingus, fellatio, anal intercourse, or any other intrusion, however slight, of any part of a person's body or of any object into the genital or anal openings of the victim's, defendant's, or any other person's body; emission of semen is not required. See Sexual assault

sexual perjury PSYCHOLOGY The 'faking' of an orgasm by a ♀; ♂ can't fake orgasms

sexual reassignment Transsexual conversion, sex change, sex conversion The constellation of surgical and medical therapies intended to physically change a person from one sexual phenotype to the other; most SRs are in a *he → she* direction–the ♂:♀ ratio of SRs is 3 to 7:1; ♀→♂ reassignment is less common and more difficult; in a study of 40 ♀ transsexuals, ± ½ had endocrine dysfunction–eg, polycystic ovarian disease, gonadal dysgenesis. *Male-to-female RA* Bilateral orchiectomy, penectomy, vaginoplasty, breast augmentation, homone therapy AMN 17/2/92, p26. *Female-to-male RA* Partially successful; requires mastectomy, hysterectomy, androgenic hormone therapy

sexual response GYNECOLOGY A biochemical and physiological response to sexual stimulation that occurs in ♂ and ♀ after puberty PHASES Excitement, plateau, orgasm, resolution

sexual response cycle PHYSIOLOGY A term that encompasses the phases of a sexual act from prearousal to denouement; the SRC is divided into 4 phases. Cf Sexual dysfunction

SEXUAL RESPONSE CYCLE

DESIRE Consists of fantasies about sexual activity, desire to engage therein

EXCITEMENT Subjective component of sexual excitement is accompanied by physiologic changes, in ♂, penile tumescence and erection, in ♀ pelvic vasocongestion, vaginal lubrication, and swelling of external genitalia

ORGASM Peaking of sexual pleasure with release of sexual tension, rhythmic contraction of perineal muscles in ♂, ejaculation, in ♀, contractions of outer ⅓ of vagina; in both, anal sphincter contracts rhythmically

RESOLUTION Denouement Muscle relaxation, and a sense of well-being; during resolution, ♂ are unerectable for a variable period of time; ♀ remain on 'red alert'

sexual responsibility The ethical obligations that a person who knows he/she has an STD–especially HIV–has to others with whom he/she engages in penetrative sexual activity. See Negotiated safety, Sexual ethics. Cf Fabian Bridges, Self-protection

sexual Symptom Distress Scale PSYCHOLOGY TESTING A 5-item testing instrument used to evaluate distress with sexual dysfunction. See Overall Quality of Life Scale

sexual traffic Geschlechsverkehr Sexual activity, specifically, penovaginal interaction. German term is *Geschlechsverkehr*.

sexual violation A form of sexual misconduct defined as physician-patient sexual relations, regardless of who initiated the relationship, which includes genital intercourse, oral sexual contact, anal intercourse, mutual masturbation. See Professional boundary, Sexual misconduct–note

sexual work Sexual traffic–anal, oral, vaginal–performed or allowed by ♂ or ♀ in exchange for drugs or money. See Sex industry

sexuality SEXOLOGY 1. The human sexual response, which is a function of external cues for heterosexual or homosexual orientation, and ability to produce and respond to gonadotropin-releasing hormone; the personal experience and expression of one's status as male or female, especially vis-á-vis genitalia, pair-bondedness, reproduction; the stimulation, responsiveness, functions of the sex organs, alone or with one or more partners. See Ambiguous sexuality, Bisexuality, Eroticism, Heterosexuality, Homosexuality, Hypersexuality, Transsexuality 2. Sexual activity, see there

sexually-acquired reactive arthritis RHEUMATOLOGY A polyarthritic complex of uncertain clinical importance reported in HLA-B27-positive

Pts and thought to be triggered by sexually transmitted *Chlamydia trachomatis*

sexually aggressive *adjective* Relating to potentially violent behavior focused on gratification of sexual drives, regardless of the desire for participation on the part of the partner. See Sexually dangerous

sexually dangerous *adjective* Relating to a constellation of known sexual behaviors and previously committed acts—eg, rape and sodomy by a person—almost invariably ♂—a sexual predator—with potential for repeating these acts and causing significant physical injury. See Community notification, Megan's law, Rape, Sexual assault

sexually transmitted disease Venereal disease Any infection or tumor acquired by direct genital and orogenital contact; most common STD agents in US: *Chlamydia trachomatis* 3-5 million; HSV-2, 2-3 million; *N gonorrhoeae*, 720,000/25,300 penicillin-resistant; *T' pallidum*/2° syphilis, 40,117 STD AGENTS-BACTERIAL *C trachomatis, Mycoplasma hominis, N gonorrhoeae,* streptococcal spp, *Ureaplasma urealyticum* STD AGENTS-VIRAL EBV, HBV, HIV-1, HIV-2, HPV, HSV OTHERS *Trichomonas vaginalis*, pubic lice, scabies SEXUALLY-TRANSMITTED NEOPLASMS Intraepithelial neoplasia—IN, SCC of cervix, penis, and anus often arise in HPV infection; HPV 16 and 33 are implicated in penile IN; HPV 16 and 18 in CIN; HPV 6, 11 and 42 in condyloma acuminata

SEXUALLY TRANSMITTED DISEASE

AGENT/DISEASE	INCIDENCE	PREVALENCE
Chlamydia	3 million	2 million
Trichomoniasis	5 million	N/A
HPV	5.5 million	20 million
PID	1,000,000	
Gonorrhea	650,000	N/A
Herpes	1 million	45 million
Syphilis	70,000	N/A
Cong syphilis	3,400	
AIDS	80,000	
Hepatitis B	120,000	417,000
Chancroid	3,500	

N/A Not available; new STD agents HPV–1976, HTLV-I & *Mobiluncus* spp–1980, *Mycoplasma hominis*–1981, HTLV-II–1982, HIV-1–1983, HIV-2–1986, HHV-8–1995 AMN 3/2/87, p13 www.intelihealth.com/IH/ihtIH/WSIHW000/8799/9154/360682.html?d=dmtContent

Sezary syndrome Mycosis fungoides, see there

SF-12 Short Form Health Survey CLINICAL CARE A generic dual—ie, mental and physical health—scale measure of quality of life

SF-36 CLINICAL STUDIES 36-item Medical Outcomes Study Short-Form General Health Survey An instrument used to assess multidimensional health-related QOL, which measures 8 health related parameters: physical function, social function, physical role, emotional role, mental health, energy, pain, general health perceptions; each parameter is scored from 0 to 100 MANAGED CARE 36-Item Short-Form Functional and Perceived Health Status Survey A questionnaire which measures health status; the SF-36 also includes a list of 18 self-reported chronic conditions

SF-EMG Single fiber electromyography, see there

SGA Small for gestational age, see there. See Low birth-weight

shadow GRADUATE EDUCATION A medical student or intern who closely follows a clinician, resident, mentor, etc while doing rounds IMAGING A subtle change on a film suggesting a possible lesion. See Gloved finger shadow, Ring shadow, Rocker bottom shadow

shaggy heart sign IMAGING A descriptor for the 'ragged' cardiac contour seen in a plain CXR in some Pts with whooping cough—*Bordetella pertussis*

during the paroxysmal stage; the shagginess is due to densities that obscure the cardiac borders—and correspond to peribronchial thickening and infiltration of the basal triangle extending laterally from the hilum to the flattened and lowered diaphragm

shake & bake Sx INFECTIOUS DISEASE A popular term for amphotericin B's adverse effects: shaking chills and fever. See Amphotericin B SUBSTANCE ABUSE A popular term for intense muscle spasms, craving, N&V, fever and other Sx typical of withdrawal from opiates. See Cold turkey

shake test Foam stability index, see there

shaken baby syndrome Whiplash shaken baby syndrome FORENSIC MEDICINE A severe form of child abuse that may be either fatal, or leave its victim with permanent neurologic sequelae; severe shaking of an infant—who has virtually no neck muscle tone, may cause bilateral subdural hematomas—due to laceration of the veins bridging the dura mater and cerebral cortex. See Battered child syndrome, Bucket handle fracture. Cf Infanticide

sham feeding A method for assessing completeness of vagotomy; food is smelled, seen and chewed, but not swallowed. See Maximum acid output, Vagotomy

shaping PSYCHIATRY Reinforcement or responses in a Pt's repertoire that increasingly approximate a sought-after behavior

shared-decision program DECISION-MAKING A video-based system that provides Pts with information relevant to various therapeutic options for a particular disease—eg, BPH, low back pain, mild HTN, breast CA, etc, in an interactive touch-screen format. See *Arado* v *Avedon*, Empowerment, Informed consent, Paternalism

shared psychotic disorder The term preferred by the Am Psychiatric Assn for the widely used "folie à deux," see there

SHARP CARDIOLOGY A clinical trial—Subcutaneous Heparin & Angioplasty Restenosis Trial that evaluated effects of high dose unfractionated heparin in preventing restenosis after PCTA in Pts with CAD. See Coronary artery disease, Heparin, Percutaneous transluminal angioplasty

sharp dissection SURGERY The separation of tissues in a surgical plane using a scalpel or other sharp instrument. See Dissection. Cf Blunt dissection

sharp injury A cut, stab, or incision with skin penetration, which reflects patterns or characteristics consistent with the wounding object. See Needle stick injury. Cf Blunt injury

'sharps' INFECTION CONTROL Sharp objects—eg, needles, syringes with needles, scalpels, blades, disposable scissors, suture equipment, stylets, and trocars, broken test tubes, and glass that may contain human blood, fluids and tissues with pathogens. See Biohazardous waste, Contaminated sharps, Needle-stick injuries, Regulated waste

shave biopsy DERMATOLOGY A superficial skin biopsy, in which tissue is excised tangentially to the surface; SBs are used for raised lesions—eg, seborrheic keratosis, and include a flat part of the upper dermis. Cf Punch biopsy, Skin biopsy

SHBG Sex hormone binding globulin, see there

shed mediastinal blood salvage HEART SURGERY The reinfusion of blood lost into the mediastinal field of operation during coronary revascularization surgery; SMBS ↓ allogeneic blood needs by 50%. See Autologous transfusion, Intraoperative blood salvage

Sheehan syndrome Postpartum hypopituitarism GYNECOLOGY A condition that follows postpartum uterine hemorrhage severe enough to cause circulatory collapse, resulting in pituitary necrosis and hypopituitarism

sheep OCCUPATIONAL MEDICINE A cloven-hoofed barnyard beast raised for wool and meat SHEEP-RELATED INFECTIONS *Actinobacillus* spp, anthrax, brucellosis, campylobacteriosis, *Chlamydia trachomatis,* cryptosporidiosis, European tick-borne encephalitis, *Francisella tularensis,* giardiasis, leptospirosis, orf, Q-fever, rabies, salmonellosis, *Yersinia enterocolitica.* See Dolly

shelf life A food industry term, which, in health care, refers to the length of time that a blood product, therapeutic or other product may be stored under appropriate conditions before it must be discarded by law

shelf sign IMAGING A flattened horizontal mass often accompanied by mucosal irregularity that may be seen by barium enema in colorectal CA

shell shock Post-traumatic stress disorder, see there

sheltered environment An environment that provides protection and custodial care to those who cannot, for various reasons, fend for themselves EXAMPLES Nursing homes, institutions for mentally challenged, 'safe houses' for abused ♀, halfway houses for rehabilitating drug addicts. See Disenfranchised population

shelterization SOCIAL MEDICINE An adaptive response by those living in shelters for the homeless in the US, characterized by ↑ passive behavior and ↓ personal hygiene. See Homelessness, Institutionalization 'syndrome'

Shenton's line IMAGING A curved line, drawn on a plain film of a horizontal pelvis, that follows the lower margin of the femoral neck and follows the top of the obturator foramen; SL helps describe the axial relationship of a normal hip joint. Cf Shenton's line

SHEP CARDIOLOGY A clinical trial–Systolic Hypertension in the Elderly Program–that evaluated efficacy of antihypertensives–with diuretics or β-blockers on M&M and stroke in Pts with isolated systolic HTN. See Antihypertensive, Hypertension, Stroke

Sherlock™ anchor ORTHOPEDICS A tin-coated, threaded anchor for fixing absorbable braided polyester surgical suture to bone, used in scapholunate and ulnar collateral ligament reconstruction, and endoscopic brow lifts

Shewanella putrefaciens BACTERIOLOGY A bacterium of environment and foods–ie, not part of normal human flora CLINICAL Found in mixed cultures, respiratory tract, urine, feces, pleural fluid; implicated in cellulitis, otitis media, septicemia

shiatsu Shiatsu therapy ALTERNATIVE MEDICINE A type of Japanese acupressure/deep massage. See Acupressure, Massage

shield INFECTIOUS CONTROL A protective device that covers the face, which has a plasticine barrier through which the user sees. See Personal protection garment MEDTALK That which protects, a protective barrier. See Dalkon shield, Human shield

shift differential 'Shift diff' NURSING An hourly premium for a worker–eg, a skilled nurse, who works an 'undesirable'–eg, evening, night, or weekend–shift. See Sleep-wake shift

shift syndrome GERIATRICS A mental decline that may occur in an elderly person, who has been shifted from his/her domicile to a nursing home or long-term care facility. Cf Shift work maladaptation syndrome

shift work maladaptation syndrome OCCUPATIONAL MEDICINE A long-term inability to adapt to shift work, resulting from the stress of desynchronized circadian systems, sleep deprivation, domestic and social dysharmony, health problems, ↓ work performance. See Circadian rhythm. Cf Shift work

Shigella MICROBIOLOGY A genus of gram-negative bacilli of the family Enterobacteriaceae, and major cause of bacterial dysentery

Shigella dysenteriae *Shigella* group A MICROBIOLOGY The least commonly isolated and most virulent *Shigella* serotype

Shigella enteritis *Shigella* gastroenteritis, shigellosis INFECTIOUS DISEASE Inflammation of small intestinal mucosa by *Shigella*–*the* most communicable bacterial diarrhea, which is associated with poor sanitation and water supplies, contaminated food, crowded living conditions, fly-infested environments; it affects travelers to developing countries and those living on reservations, refugee camps, institutions INCUBATION 1-3 days CLINICAL Neurologic Sx can develop in children–febrile seizures, headache, lethargy, confusion, nuchal rigidity

Shigella sonnei *Shigella* group D MICROBIOLOGY The most commonly isolated, least virulent *Shigella* serotype

shigellosis *Shigella* enteritis, see there

Shiley heart valve Björk-Shiley 60 degree Converso-Concave prosthetic heart valve CARDIOVASCULAR SURGERY An artificial heart valve manufactured until 1985, when it was withdrawn from the market for valve failure due to strut fractures. See Product liability

shin splint 1. Innocuous pain over the antero-lateral tibial bone, relieved by rest, leg elevation, cold temperatures 2. Medial tibial stress syndrome, see there.

shingles Varicella-zoster NEUROLOGY An acute reactivation of a prior and latent infection by herpes zoster–HS-2, resulting in vesicular skin eruptions, usually distributed along the zone of the skin innervated by a nerve that supplies sensation; shingles pain is, in part, related to inflammation of the sensory nerve

shivering The involuntary contraction of skeletal muscle due to exposure to cold or fright, or which is temporally related to the onset of fever

SHOCK CARDIOLOGY A clinical trial–Should We Emergently Revascularize Occluded Coronaries for Cardiogenic shock comparing the efficacy of emergency revascularization and initial medical stabilization in managing Pts with MI complicated by cardiogenic shock. See Cardiogenic shock

shock A condition characterized by signs and Sx due to a cardiac output below that required to fill the arteries with blood of sufficient pressure to adequately perfuse organs and tissue CLINICAL Hypotension, poor peripheral perfusion, hyperventilation, tachycardia, oliguria, cyanosis, mental clouding, a sense of great anxiety and foreboding, confusion and, sometimes, combativeness CAUSES Trauma with major multiorgan system injury, septicemia, fluid loss–blood or intractible diarrhea, burns, high voltage electric current, abrupt loss of myocardial activity CLASSIFICATION Based on related mechanisms of cardiac dysfunction–pump failure, ↓ volume–loss of blood or extracellular fluid or changes in arterial resistance or venous capacity PATHOGENESIS Shock activates sympathetic nervous system via the carotic and aortic baroreceptors, ↑ catecholamines, vasoconstriction of 'non-essential' organs–intestine, kidneys, skin to maintain blood flow to vital organs–heart and brain; with time, hypotension becomes irreversible. See Anaphylactic shock, Bacteremic shock, Calcium shock, Cardiogenic shock, Cardiopulmonary obstructive shock, Culture shock, Heat shock, Hypotension, Hypovolemic shock, Insulin shock, Psychologic shock, Refractory septic shock, Septic shock, Spinal shock

shock lung Post-traumatic respiratory insufficiency, traumatic 'wet lung' CRITICAL CARE A condition in which changed pulmonary compliance and oxygenating capacity causes an ARDS-like picture with defective aeration due to multiple factors–eg, aspiration of gastric contents, atelectasis, cerebral injury–affecting respiratory rate, interstitial edema, microembolism, O_2 toxicity, sepsis, fulminant meningococcemia. See Adult respiratory distress syndrome

shock therapy Electroconvulsive therapy, see there

shoot up DRUG SLANG A street verb, to inject a drug, often an abuse substance

shooter MEDTALK *noun* An IV drug user; a person who inject illicit drugs IV. See Shooting gallery

616 **shooting gallery** SUBSTANCE ABUSE A place–eg, an abandoned building in an economically-depressed urban area–ie, a ghetto, where IV drug users congregate, purchase, inject–'shoot' heroin, cocaine, oxycodone or other drug. See Crack house, Needle exchange programs, 'Pocket shot', 'Skin popping'

shop See Sweatshop, Workshop

short bowel syndrome Small intestinal insufficiency SURGERY A complication of major small intestine resections–eg, for ischemia or inflammation, resulting in complex nutritional imbalances and/or malnutrition; massive small intestinal resection may be required in 1. Multiple congenital atresias or stenoses of the neonatal intestine; if loss is significant, diarrhea and malabsorption appear shortly after birth; barium studies reveal a malrotated colon and markedly shortened small bowel; if infant survives first few months, intestinal function improves; 2. Massive resection of gangrenous small intestine due to mesenteric arterial occlusion, traumatic interruption, volvulus, or Crohn's disease; in children, necrotizing enterocolitis is main cause of SBS; excess small intestinal resection results in inadequate absorption of nutrients, minerals and vitamins, causing hypovitaminosis, malnutrition, hypochromic and megaloblastic anemia, diarrhea, electrolyte imbalances, ↑ oxalates–derived from bile salt detergents that pass into the circulation, which may crystallize in renal tubules, causing renal failure, lactic acidosis, osteopenia, steatorrhea; up to 70% of the absorptive surface may be lost and tolerated; if the Pt survives surgery, residual tissue undergoes adaptive hyperplasia of absorptive villi, ↑ absorptive cells and ↑ in small intestinal caliber; after stabilization and temporary parenteral nutritional support, oral feeding must be initiated ASAP to stimulate adaption TREATMENT High 'quality'–ie, essential amino acids, low protein diet, middle chain TGs, vitamins, minerals. See 'Second look' operation

short call GRADUATE EDUCATION A call schedule in which a house officer–HO takes Pt admissions from the time he/she arrives at the hospital until a certain hr. See Call schedule

'short' draw LAB MEDICINE A tube of blood drawn for analysis of chemistries or cells that has less than the recommended volume

short PR syndrome Lown-Ganong Levine syndrome A condition comprising a short PR interval, normal QRS, tachycardia, attributed to atrio-His bypass

short rib-polydactyly syndrome A heterogeneous group of AR conditions characterized by short stature, horizontal ribs, and perinatal death

shortage See Baby shortage, Blood shortage, Manpower shortage

shortwave diathermy SPORTS MEDICINE Diathermy that delivers shortwave frequency electromagnetic waves; SD is used for chronic inflammation–eg, bursitis, neuritis, osteoarthrosis, rheumatoid arthritis, strains, tendinitis, etc CONTRAINDICATIONS ASHD, hemorrhage, metal implants, infections, malignancy, pacemakers, phlebitis, pregnancy, wet dressings. See Diathermy. Cf Microwave diathermy

shot Injection, see there

shot therapy Immunotherapy for desensitizing a person to allergens

shotgun FORENSIC PATHOLOGY A pump-action or single shot firearm, which fires projectiles of packed shot of various diameters; shotguns are popular among hunters, and an efficient 'execution' weapon when a hit-person must address a moving target; homicide and suicide by shotgun is common in the US, but exceedingly rare in civilized countries, given the strict gun control laws. See Ballistics, Cookie cutter wounds, Drive-by shootings, Execution wounds, Gauge, Sawed-off shotgun. Cf Assault weapon, Pistol

'shotgun approach' A diagnostic philosophy in which every conceivable parameter is measured, especially in a Pt with an obscure disease, to detect rare conditions that may cause a particular Sx. See Defensive medicine. Cf Screening

shotgun technique MALPRACTICE A strategy by a plaintiff in a civil action involving multiple potential defendants, where all possible parties are named, in a shotgun-like manner

shotty lymphadenopathy The clustering of multiple small, contiguous and indurated lymph nodes, palpable in the inguinal, cervical and other regions in children with viral infections, which, if palpated in adults, would suggest syphilis or metastatic CA

'shoulder' OBSTETRICS A descriptor for the gently-sloped acceleration rhythm seen on a paper printout of the fetal heart monitor that either precedes or follows a typical deceleration, in contrast to the usual 'acceleration'–a short-term ↑ in heart rate above baseline occurring in response to fetal movement. See Deceleration, Fetal heart monitor

shoulder bursitis Inflammation of either of the shoulder's major bursae, caused by trauma or overuse MANAGEMENT Rest, ice, NSAIDs INFECTIOUS BURSITIS Antibiotics, aspiration, surgery

shoulder complex CLINICAL ANATOMY The joints of the shoulder–acromioclavicular, sternoclavicular, glenohumeral, scapulothoracic joints, which rely on associated muscles–eg, those attached to the scapula, to provide dynamic stability. See Frozen shoulder, Little League shoulder, Milwaukee shoulder, Pathologists' shoulder, Rotator cuff, Scapular reaction

shoulder dystocia OBSTETRICS An obstetrical emergency that occurs when the anterior shoulder of the fetus becomes lodged behind the superior symphysis pubis, preventing further delivery; SD is not always preventable, and is usually not recognized until after the head has been delivered, and gentle downward traction of the fetal head fails to accomplish delivery INCIDENCE Up to 2% of deliveries–probably underreported RISK FACTORS Maternal obesity >250 lbs–9 X risk, excess weight gain during pregnancy; DM–5-16.7% vs 1.7% in controls; postterm pregnancy–fetal growth does not stop at 40 wks; inertia of macrosomia; abnormal labor with prolonged deceleration, prolonged 2nd stage of labor; operative vaginal delivery–forceps, vacuum extraction MANAGEMENT Mobilize anesthesia and pediatric support ASAP; gentle attempt at downward traction on head and moderate fundal pressure; large episiotomy; suprapubic pressure at midline; attempt to rotate shoulder; fracture the clavicle; cleidotomy; symphysiotomy–rarely done in US; McRobert's manuever–flexing legs on abdomen to ↑ diameter of pelvic outlet; Wood's screw manuever–apply pressure to ventral surface of posterior shoulder; deliver posterior shoulder. See Dystocia

shoulder-girdle syndrome NEUROLOGY A condition evoked by upregulation of the sympathetic nervous system ETIOLOGY Contusions, fractures, neurovascular injuries CLINICAL Seating pain, vasomotor lability, edema, osteoporosis

shoulder-hand syndrome NEUROLOGY A condition characterized by shoulder pain, swelling, stiffness, vasomotor Sx of arm and hand, skin edema/induration, in Pts > age 50 after an acute MI, or less commonly, a CVA or head trauma; SHS is attributed to reflex sympathetic stimulation; some Pts later develop adhesive capsulitis, sclerodactyly, and ↓ ROM with regional demineralization. See Rotator cuff

shoulder instability ORTHOPEDICS The weakening of the glenohumeral joint by subluxation or dislocation. See Multidirectional shoulder instability

shoulder pointer 'syndrome' SPORTS MEDICINE A 'sprain' of the acromioclavicular joint, associated with contact sports–eg, football, wrestling, karate, hockey, and non-contact sports due to dislocations or strain–eg, swimming, gymnastics without joint instability

shoulder presentation OBSTETRICS A rare presenting position during delivery, which occurs when the baby lies in transverse position, across the uterus; SP is easily diagnosed and virtually impossible to delivery vaginally. See Cesarean section, Presentation

Shprintzen syndrome VCF syndrome, velo-cardio-facial syndrome A congenital anomaly with cleft palate, heart defect, abnormal face, learning defects, short stature, microcephaly, mental retardation, ear anomalies, slender hands/digits, inguinal hernia

shrink Vox POPULI *noun* A psychiatrist

shrinkage OSTEOPOROSIS The loss of height with age, a typical finding in osteoporosis of vertebral column. See Osteoporosis UROLOGY Popular for a marked ↓ in penile length after exposure to cold water

'shrinking field' technique RADIATION ONCOLOGY A method used in RT for treating a large mediastinal lymphoma, in which the treatment field is ↓ or shrunk as the tumor responds or 'melts'. See Mantle port

shrinking lung A popular term for elevation of the diaphragm due to pleural adhesions, plate-like atelectases and chronic fibrosis, a finding seen in SLE; the radiologic and pathologic findings may not translate into clinical disease. Cf Pulmonary sequestration

shuffling gait NEUROLOGY A gait in which the foot is moving forward at the time of initial contact, with the foot either flat or at heel strike, or during midswing ETIOLOGY Basal ganglia degeneration with extrapyramidal effects in parkinsonism. See Parkinsonism

shunt The diversion of the flow of a fluid from its normal pathway to another, which may be accidental, as in a traumatic AV aneurysm, or by design–eg, portocaval shunt or ventriculoperitoneal shunt. See Arteriovenous shunt, Nodovenous shunt, Denver shunt, Distal splenorenal shunt, LeVeen shunt, Perfusion shunt, Portacaval shunt, Transjugular intrahepatic portosystemic shunt PEDIATRIC CARDIOLOGY Bypassing the pulmonary circulation–shunting is a normal physiologic process in utero; it becomes abnormal after birth TYPES 1. Those in which already oxygenated blood in the left heart passes back into the right heart–left-to-right shunt and 2. Those which partially bypass the lungs, with venous blood directly entering the systemic circulation–right-to-left shunt

SHUNT-ACYANOTIC, CYANOTIC

L→R SHUNT Acyanotic shunt Right and left sides of the heart communicate by an ASD or VSD and PDA; the blood flows from the region of highest–left heart to lowest–right heart and systemic circulation pressure, as occurs in VSDs and corrected transposition of great arteries; since the blood does not bypass the pulmonary circulation, it is well oxygenated CLINICAL The plethora of blood causes pulmonary congestion and HTN that becomes significant when the pulmonary blood flow is 1.5-2.0-fold greater than the systemic flow with diastolic overloading and cardiac dilatation which, without correction, results in cardiac failure; a late complication is bacterial pneumonia related to stasis within the pulmonary circulation; L→R shunts may be created surgically–eg, Blalock procedure

R→L SHUNT Cyanotic shunt Variable degree of pulmonary circulation bypass accompanied by obstruction of blood flow into the pulmonary circulation R→L SHUNTS Fallot's tetralogy–VSD, pulmonary valve stenosis, overriding or dextroposed aorta and 2° right ventricular hypertrophy, transposition of great vessels, tricuspid valve atresia, truncus arteriosus and total anomalous return of pulmonary veins; pulmonary blood flow is less than in L→R shunts CLINICAL Cyanosis with limited exercise tolerance, neurologic damage and compensatory polycythemia; as children, these Pts are often very sick and by adolescence may suffer acquired coagulopathies

shunt nephritis Immune complex glomerulonephritis seen in ± 4% of infants with infected ventriculoatrial, ventriculojugular or ventriculoperitoneal shunts CLINICAL Weight loss, lethargy, fever, lymphadenopathy, hepatosplenomegaly, HTN, arthralgia, nephrotic syndrome, often delayed recognition of infection LAB Anemia, microhematuria, proteinuria, azotemia; ↓ C3, rheumatoid factors, cryoglobulins; common bacteria–*S epidermidis, S albus* TREATMENT Antibiotics, high-dose prednisone PROGNOSIS $\frac{1}{2}$ resolve with therapy, $\frac{1}{4}$ have persistent urine abnormalities; $\frac{1}{4}$ die of the neurologic defects necessitating ventricular shunt

shunting LUNG PHYSIOLOGY The bypassing of alveoli by blood circulating through the lungs ETIOLOGY Atelectasis, portal hypertension, small airway obstruction, smoke inhalation injury DIAGNOSIS Hypoxemia that only par-

tially improves with high inspired O_2 concentrations–eg, arterial O_2 tension/$PaO_2 < 200$ mm Hg for inspired O_2 concentration $FIO_2 = 100\%$

Shwachman syndrome A condition characterized by pancreatic atrophy, neutropenia, growth failure, metaphyseal defects, hepatitis, anemia, thrombocytopenia, myocardial lesions–myofiber fragmentation, round cell inflammation, and fibrosis

Shy-Drager syndrome Neurologic orthostatic hypotension, Shy-McGee-Drager syndrome NEUROLOGY A rare idiopathic degenerative disorder affecting primarily older ♂, characterized by progressive damage to the autonomic nervous system CLINICAL Tremor, rigidity, slowed movement, neurologic defects. Cf Parkinsonism

SI LAB MEDICINE Systeme International d'unites, Intl System of units The international system for standardization of units of measurement; the SI is a refinement and extension of the metric system 1. Sacroiliac 2. Saline injection 3. Saturation index 4. School inventory–psychological test 5. Sex inventory–psychological evaluation 6. Solubility index

SI-SYSTEME INTERNATIONAL D'UNITES Common US to SI conversions		
	CF	Final units
Acetone	172.2	mmol/L
Albumin	10.0	mmol/L
Bilirubin	17.10	µmol/L
Calcium, mg/dL	0.2495	mmol/L
Cholesterol	0.0258	mmol/L
Creatinine	88.40	µmol/L
Glucose	0.0555	mmol/L
Hemoglobin	10.0	g/L
Iron	0.1791	µmol/L
LDL	0.2586	mmol/L
HDL	0.2586	mmol/L
Magnesium, mEq/L	0.4114	mmol/L
Phosphate	0.3229	mmol/L
Potassium, mEq/L	1.0	mmol/L
Protein	10.0	g/L
Sodium, mEq/L	1.0	mmol/L
Trioiodothyronine–T_3	0.1536	nmol/L
Triglycerides	0.0113	mmol/L
Urea nitrogen, mg/dL	0.3570	mmol/L

CF = Conversion factor

SIADH Syndrome of inappropriate antidiuretic hormone secretion, dilutional hyponatremia A complex characterized by ↑ vasopressin–ADH secretion despite low plasma osmolarity, water retention, dilutional hyponatremia ETIOLOGY Addison's disease, ACTH deficiency, AIDS, hypopituitarism, ectopic hormone production in CA–small cell, bronchogenic, pancreas, uterine, bladder, prostate, lymphoproliferative disorders, mesothelioma, thymoma, CNS disease–trauma, infection, chromophobe adenoma, metastases, lung disease–TB, pneumonia, PEEP ventilation, porphyria, drugs–eg, chlorpropamide, vincristine, etc LAB Hypervolemia, hypouricemia, ↓ creatinine, hyponatremia, natriuresis–urinary sodium > 20 mEq/L with ↓ BUN, no Sx of volume depletion, ↓ maximum urinary dilution, ↑ ADH, normal renal and adrenal function MANAGEMENT Corticosteroids to suppress ADH secretion

sialadenitis Salivary gland inflammation. ENT An infection of the salivary glands caused by viruses–eg, mumps–mumps parotiditis, or bacteria–linked to obstruction–as in salivary duct stones or to poor oral hygiene. See Myoepithelial sialadenitis

sialidosis type I Cherry-red spot myoclonus syndrome, see there

618 **sialidosis type II** An AR condition characterized by coarse facies and dysostosis multiplex, divided into a congenital form with ascites, hydrops fetalis, gargoyle-like facies, visceromegaly, mental retardation, myoclonus, tonic-clonic seizures, cherry-red spots, hearing loss, deficiency of alpha-N-acetyl-neuraminidase, usually with a partial defect of β-galactosidase; infantile and juvenile forms are less severe

sialography IMAGING A method for visualizing salivary glands and ducts by instilling radiocontrast INDICATIONS Sialadenitis, cysts, stones–sialolithiasis, dry mouth

sialolith Salivary duct stone, see there

sialorrhea MEDTALK Increased salivation

SIAM CARDIOLOGY A clinical trial–Streptokinase in Acute Myocardial Infarction which evaluated the effect of PCTA up to 48 h after IV streptokinase. See Percutaneous coronary angioplasty, Streptokinase

'Siamese' twins Conjoined equal twins A joined gestational product due to a failure in division of the yolk sac or due to delayed monovular separation; STs occur in ±1:200,000 term deliveries; most are joined at the chest–thoracopagus PROGNOSIS Depends on adequacy of surgical separation

sibilant PHYSICAL EXAM *adjective* Whistling, hissing

sibling A brother, sister or littermate

sibling rivalry PSYCHOLOGY The intense, emotional competition among siblings–brothers and/or sisters that pits one against the other to obtain parental affection, approval, attention, and love. See Cain complex. Cf Oy child, Sibling relational problem

sibrafiban Xubix™ CARDIOLOGY A 'superaspirin' GP IIb/IIIa receptor antagonist used to prevent post coronary-angioplasty coagulation

sibutramine Meridia® OBESITY A central, appetite-suppressing agent that blocks reuptake of serotonin and norepinephrine, and, to a lesser extent dopamine; it provides a sensation of fullness from less food, and ↑ metabolism INDICATIONS Obese Pts with an initial BMI of ≥ 30 kg/m² or ≥ 27 kg/m² with other factors–eg, HTN, DM, dyslipidemia CONS Sibutramine is a Schedule IV Controlled Substance ADVERSE EFFECTS Headache, dry mouth, anorexia, constipation CONTRAINDICATIONS Concurrent MAOIs, other central appetite suppressants, anorexia nervosa. See Obesity

sicca complex Symptoms related to generalized mucosal drying, affecting EYES–xeroconjunctivitis due to ↓ tears, thick, 'ropy' secretions on the inner canthus, foreign body–tired, itchy, sandy sensation, ↓ visual acuity MOUTH–xerostomia with ↓ salivation due to lymphocyte infiltration and duct obstruction, with soreness, adhesion of food to mucosa, 'cracker' sign, angular cheilitis, lingual fissuring, acceleration of caries. See Dry eye, Dry mouth

sick building syndrome Tight building syndrome PUBLIC HEALTH A condition defined by the WHO, as excess work-related irritation of mucocutaneous surfaces and other Sx–eg, headache, fatigue, difficulty concentrating, reported by workers in modern office buildings. See Building biology, Environmental disease

SICK BUILDING SYNDROME–clinical features

HYPERSENSITIVITY Hypersensitivity pneumonitis and allergic alveolitis in response to various microorganisms eg water-borne ameba, known as 'humidifier lung'

ALLERGIES Allergic rhinitis and asthma, due to dust mites

INFECTIONS Mini-epidemics, eg Legionnaire's disease, Pontiac fever, by low-level airborne pathogens that thrive in stagnate water and are disseminated through poorly-maintained air conditioning systems

MUCOCUTANEOUS IRRITATION Skin eruptions, due to fiberglass, mineral wool or other particles; contact lens wearers may suffer corneal abrasions

MUCOSAL IRRITATION Dry throat, cough, tightness in chest, sinus congestion and sneezing–formerly due to tobacco smoke, which is increasingly banned in buildings, solvents and cleaning materials, eg chlorine, reactions to photochemicals or other toxins, eg in laser printers due to the styrene-butadiene toners and ozone production by photocopiers

PSEUDOEPIDEMICS Due to 'mass hysteria'

sick Santa 'syndrome' OCCUPATIONAL MEDICINE Any condition sidelining seasonal Santas, who suffer infected aerosols, frostbite–eg, Salvation Army Santas, repressed anger, neuromuscular Sx–prolonged periods in uncomfortable positions–eg, holding children on knees

sick sinus syndrome Bradycardia-tachycardia syndrome, sinus node dysfunction CARDIOLOGY A diffuse cardiac conduction system defect characterized by a pathologically slow or erratic rate of sinus depolarization due to impaired automaticity; SSS accounts for up to 50% of pacemaker implantations INTRINSIC CAUSES Scarring, degeneration, damage to conduction system–eg, aging, MI; ischemia; infiltrative disease–eg, amyloidosis, hemochromatosis, sarcoidosis; collagen vascular disease–SLE, RA, scleroderma; myotonic muscular dystrophy; heart surgery–valve replacement, transplantation; infections-Chagas' disease, endocarditis EXTRINSIC CAUSES–autonomic syndromes–eg, carotid hypersensitivity, neurocardiac syncope, vagal stimulation; drugs–CCBs, beta-blockers, digoxin, clonidine, sympatholytics, antiarrhythmics; hypothyroidism; hypothermia; neurologic defects; ↑/↓ K⁺ CLINICAL Common in elderly; if severe, associated with vertigo, palpitations, exercise intolerance, syncope, cerebral dysfunction, persistent sinus bradycardia–30–60/min PLUS SVT, hence the trivial name, 'brady-tachy' syndrome TREATMENT Verapamil, diltiazem, pacemaker PROGNOSIS Mortality at 1, 3, and 10 yr is 5%, 9%, 25% respectively; unaffected by pacemakers

sickle cell anemia Sickle cell disease HEMATOLOGY An AR hemoglobinopathy affecting 0.15% of US blacks, caused by a point mutation on the β Hb gene, resulting in a defective Hb function, causing RBCs to 'sickle' with ↓ O₂ CLINICAL Hemolytic anemia, various sickling syndromes MANAGEMENT Intense pain which may respond to dexamethasone; hydroxyurea or butyrate See Sickle cell, Sickle cell disease

sickle cell disease Any syndrome associated with sickle cells–eg, sickle cell anemia–hemoglobin genotype S/S, sickle cell-HbC disease, sickle cell-β-thalassemia MANAGEMENT Allogeneic BMTP may be curative in young Pts with symptomatic SCD. See Sickle cell anemia

sickle cell trait Heterozygosity for HbS seen in 8% of US blacks, up to 30% of African populations CLINICAL No anemia, normal growth, development, lifespan, slight ↑ in sudden unexplained death; hematuria, splenic infarction. See Sickle cell disease

sickle chest syndrome A complication of sickle cell anemia due intravascular 'sludging' of RBCs IN terminal pulmonary arteries CLINICAL Chest pain, dyspnea, fever PROGNOSIS Guarded

sickness Disease. See Air sickness, Altitude decompression sickness, Cybersickness, Decompression sickness, English sweating sickness, Ghost sickness, Green tobacco sickness, Inner ear decompression sickness, Jamaican vomiting sickness, Meditation sickness, Monday morning sickness, Morning sickness, Motion sickness, Mountain sickness, Seasickness, Serum sickness, Space sickness

Sickness Impact Profile MEDTALK An instrument used to evaluate perceived health status–quality of life and changes in functional status in Pts being treated for a potentially fatal condition. Cf Karnovsky scale, Oswestry low back pain disability questionnaire, QWB scale

SICU Surgical intensive care unit. See ICU

SIDA Syndrome d'immunodeficience acquise, French for AIDS, see there

Side Branch Occlusion system SURGERY A system designed for minimally invasive treatment of peripheral vascular disease–PVD, requiring a small incision at the groin and below the knee for accessing the saphenous vein. See Peripheral vascular disease

side effects THERAPEUTICS Any result of a drug or therapy that occurs in addition to the intended effect, regardless of whether it is beneficial or undesirable EXAMPLE CHEMOTHERAPY–Fatigue, N&V, anemia, hair loss, mouth sores. See Adverse effect, Anticipatory side effects, Indication

side-impact air bag PUBLIC HEALTH A safety device in cars that protects the head, thorax, shoulders in side-impact crashes, which cause ± 30% of MVA deaths. See Air bag, Seat belts

Sidenar Lorazepam, see there

Sideril Trazodone, see there

siderosis 1. Occupational lung disease due to inhalation of iron dusts 2. Hemosiderosis, see there 3. Localized deposition of iron in the body. See Transfusion-related siderosis

SIDS Crib death NEONATOLOGY Sudden infant death syndrome Definition per 2nd Intl Conf on SIDS, WHO *'Sudden and unexpected death of an infant* (generally, from 2 wks to 6 months of age, while sleeping) *who was well or almost well before death which remains unexplained after an adequate autopsy'* SIDS BY ETHNICITY–US Asian 0.5/1000; white 1.3/1000; black 2.9/1000; Native American 5.9/1000 RISK FACTORS SIDS is ↑ in premature ♂ infants < 6 months old, lower socioeconomics, prior SIDS death with same mother, children of narcotic–heroin, methadone, cocaine–users, smokers, single mothers DIFFDX Involuntary smothering by exhausted mother who 'co-sleeps' with infant PATHOGENESIS Unknown–theories abound PROPHYLAXIS Am Acad Pediatrics recommends placing infants on backs to sleep. Cf Child abuse

SIESTA CARDIOLOGY A clinical trial–Snooze-Induced Excitation of Sympathetic Triggered Activity

sievert Sv RADIATION PHYSICS The SI unit of biologically effective dose–equivalent dose of ionizing radiation that produces the same biological effect–of ionizing radiaction–dose equivalence; 1.0 Sv = 1.0 joule/kilogram or 100 rem. See Gray, Rad; Roentgen

SIG IMMUNOLOGY Specific immune globulin, see there ONLINE Special interest group, see there

SIG E CAPS PSYCHIATRY A mnemonic–Sleep, Interest, Guilt, Energy, Concentration, Appetite, Psychomotor retardation, Suicidal ideation for the signs of major depression. See Major depression. Cf Antidepressants

sigh Augmented breath PULMONOLOGY A breath occurring at regular–predictable intervals, possibly in response to the slow ↓ in lung compliance seen when constant ventilation is maintained; sighs tend to be followed by smaller than normal breaths . Cf Noisy breathing

sigmoidoscope GASTROENTEROLOGY A narrow-bore, lighted device for examining the sigmoid colon mucosa, which can be rigid or flexible, and equipped to obtain biopsies from regions of interest. See Endoscopy

sigmoidoscopy GASTROENTEROLOGY A test in which a flexible sigmoidoscope is inserted per rectum and colonic mucosa visualized to detect lesions which, if present, are biopsied to determine whether they are benign or malignant See Endoscopy, Proctoscopy. Cf Barium enema, Fecal occult blood test

sign MEDTALK A defect or abnormality which is associated with a particular disorder. Related terms are Accordion sign, Air meniscus sign, Amputation sign, Angel wing sign, Aortic nipple sign, Apical cap sign, Applesauce sign, Arrowhead sign, Asterisk sign, Bald sac sign, Banana sign, Battle sign, Beak sign, Bird's sign, Bird beak sign, Blade of grass sign, Blue tongue sign, Bowler hat sign, Bozzolo sign, Branham sign, Broken ring sign, Broken straw sign, Brudzinski sign, Candy cane sign, Carman's meniscus sign, Cat's paw sign, Celery stick sign, Chaddock sign, Chadwick sign, 'Chandelier sign,' Chinese lantern sign, Chvostek sign, Clenched fist sign, Cluster of grapes sign, Cobrahead sign, Collar sign, Colon cut-off sign, Comet tail sign, Cracked-pot sign, Cracker sign, Curbstone sign, Cullen sign, Cut-off sign, Dagger sign, Deciduous tree in winter sign, Delta sign, DeMusset sign, Dimple sign, Doll's eye sign, Donohue-Fauver sign, Double bubble sign, Double condom sign, Double duct sign, Double wall sign, Dough sign, Doughnut sign, Drawer sign, Drooping lily sign, E sign, Einstein sign, Ellipse sign, Ewart sign, Eyelash sign, FBI sign, Fifth vital sign, Figure 3 sign, Fishhook sign, Flag sign, Flat waist sign, Football sign, Friend sign, Frog sign, Grey Turner sign, Groove sign, H bomb sign, Half moon sign, Hampton sign, Head drop sign, Heel pad sign, Hegar sign, Hill sign, Hilum overlay sign, Hoffman sign, Homan sign, Hot air balloon sign, Hump sign, Hump & dip sign, Iceberg sign, Inverted comma sign, Inverted mushroom & stem sign, Inverted 3 sign, Inverted U sign, Inverted umbrella sign, Jail bars sign, Jello™ sign, Jet sign, Joffrey sign, Kussmaul sign, Kehr sign, Kernig sign, Kestenbaum sign, Key sign, Keyhole sign, Lachman sign, Lasègue sign, Lemon sign, Lhermitte sign, Lipstick sign, Luftsichel sign, Marcus Gunn sign, McMurray sign, Meningeal sign, Meniscus sign, Mickey Mouse sign, Milk rejection sign, Möbius sign, Mogul sign, Monocle sign, Moro sign, Moviegoer sign, Mulder sign, Musset sign, Napoleon's hat sign, Nikolsky's sign, Notch sign, Nuchal translucency sign, Numb chin sign, Number 3 sign, Ober sign, Obturator sign, Okra sign, Omega sign, One bone-two bone sign, One sign, two sign, three sign, Overhanging ledge sign, Osler sign, Padlock sign, Pencil-in-cup sign, Peninsula sign, Phalen sign, Playboy Bunny sign, Plump hilus sign, Psoas sign, Puddle sign, Pyloric string sign, Pyriform sign, Quarter moon sign, Reverse 5 sign, Reverse 3 sign, Rigid loop sign, Rim sign, Ring sign, Romberg sign, Rope sign, Rovsing sign, Sagging rope sign, Sail sign, Sanctuary sign, Scarf sign, Scotty dog sign, Setting sun sign, Shaggy heart sign, Shelf sign, Signet ring sign, Snowman sign, Snow white sign, Square root sign, Step-off sign, String sign, String of beads sign, Sulcus sign, Terry Thomas sign, Thorn sign, Thumb & little finger sign, Thumb sign, Thymic wave sign, Tinnel sign, Tinted spectacles sign, Tit sign, Tooth sign, Toothpaste sign, Track sign, Trident sign, Twinkling star sign, Vacuum sign, Vital sign, 'Waiter accepting a tip' sign, Wall sign, Water lily sign, Westermark sign, Westphal sign, Whalebone in a corset sign, White line sign, Winterbottom sign, Wrist sign

sign language AUDIOLOGY A formal language of nonverbal communication based on hand shapes, facial expressions, and movements. See Americans with Disabilities Act

Signa™ system IMAGING An open design MRI which allows surgery while the Pt is being scanned, providing real-time, precision visualization. See Magnetic resonance imaging, Real-time imaging

signal MEDTALK A measurable or recognizable indicator of an event or process. See Magnetic resonance signal, Sexual body signal

signal averaged electrocardiography CARDIOLOGY A technique that amplifies late potentials–high-frequency, low-amplitude signals at the end of the QRS complex, attributed to fragmented and delayed conduction through the borders of a myocardial scar

signal event PUBLIC HEALTH An event, usually man-made, in which there is a tremendous 'signal,' often in the form of a 'disaster,' eg Love Canal or the Libby Zion case, which engenders multiple 'ripple' effects of legislation and 'landmark' legal cases, due in part to a popular outcry against prevailing policies that were inadequate or incapable of addressing the event. See Libby Zion, Love canal, Sentinel event

signal-to-noise ratio MRI The ratio obtained from the relative contributions of detected true signal to that of random superimposed signals–'noise'. See Magnetic resonance imaging

signature MEDTALK Any highly specific pattern that defines a substance of interest–eg, chemical, protein, DNA, as being unique; the term is often used interchangeably with 'fingerprint' VOX POPULI See Digital signature, Electronic signature

signet ring cell A usually malignant cell containing copious clear cytoplasm that flattens a hyperchromatic nucleus to one side, having an appearance fancifully likened to a signet ring; CAs composed predominantly of SRCs often carry a worse prognosis; the 'classic'–and most common SRC occurs in the stomach, but is well described in CA of the breast, colon, gallbladder, lung, nasal cavity, prostate, thyroid–medullary carcinoma, urinary bladder, malignant SRCs may also be seen in non-epithelial malignancies–eg, mesothelioma, rhabdomyosarcoma, balloon cell melanoma, oligodendroglioma, myxoid angioblastomatosis, myxoid liposarcoma, lymphoma, and is a morphology typical of normal fat cells and oligodendrogliocytes

significance CLINICAL MEDICINE A finding to be weighed in establishing a diagnosis, or influencing management of, a clinical state, which may be expressed as a finding *of significance* STATISTICS A measure of deviation of data from a statistical mean, defined by a probability–p value, where a p of 0.05 indicates a 5% possibility or 1 chance in 20 that a dataset differs from a mean and 19 chances that it will not. See Clinical significance, Statistical significance

significant other Most significant other, see there

SIL Squamous intraepithelial lesion, see there

SILENT CARDIOLOGY A clinical trial–Sonotherapy for In-Lesion Elimination of Neointimal Tissue

silent angina Silent ischemia, see there

silent infection An infection lacking significant clinical signs of disease; SIs may be recognized only in retrospect–eg, by a $\geq$ 4-fold $\uparrow$ in antibody titers

silent killer Silent lesion MEDTALK Popular for a condition that may progress to very advanced stages before manifesting itself clinically

silent myocardial ischemia Silent ischemia CARDIOLOGY Objective–eg, EKG–ST-segment depression, perfusion defects, radionuclide angiography or echocardiography of myocardial ischemia without associated Sx–ie, crushing precordial anginal pain PROGNOSIS SMI is an independent predictor of poor clinical outcome in Pts with CADSee Angina, Myocardial infarction, Total ischemic burden

SILENT MYOCARDIAL ISCHEMIA

TRUE SILENT ISCHEMIA The nociceptive pathways have a marked $\downarrow$ in sensitivity to pain, as occurs in DM, present in 10-20% of Pts with both CAD and DM resulting in significant coronary artery spasms

PSEUDOSILENT ISCHEMIA The Pt either

- Denies pain, recognizing both its import, and that myocardial ischemia would have an immediate impact on lifestyle or
- Recognizes the pain but attributes it to something else, ie heartburn

silent organ SURGERY A popular term for the pancreas, and the relative paucity of clinical signs when it has disease, especially, in tumors of the body and tail. See Whipple's operation

silent thyroiditis Atypical subacute thyroiditis, hyperthyroiditis, lymphocytic thyroiditis, subacute lymphocytic thyroiditis ENDOCRINOLOGY Self-limited hyperthyroidism that resolves in 2-5 months; thyroid is infiltrated with lymphocytes CLINICAL Hyperthyroidism for $\leq$ 3 months or less RISK FACTORS $\female$, > age 30, current pregnancy MANAGEMENT If severe, tranquilizers, β-blockers, prednisone. See Hyperthyroidism, Thyroiditis

silhouette symmetry A criterion for differentiating benign lesions–eg, Spitz nevi, compound nevi, which are usually symmetrical, from malignant pigmented lesions–eg, melanomas, which are often asymmetrical. See Melanoma

silicon A nonmetallic element–atomic number 14, atomic weight 28.086 present in nature as silica and silicates; silicon is integral to semiconductors and solar batteries, and essential for normal growth and skeletal development in rats and chickens; a silicon deficiency state is not known in man

silicone A polymer composed of a repeating unit $-R_2Si-O-$ in which $-R$ is a simple alkyl group–a hydrocarbon; silicones are produced in various forms–eg, adhesives, sponges, solid blocks, gels, and widely used in medicine, as they are stable, repel water and inert MEDICAL DEVICES Silicone is used for hydrocephalic shunts, pacemakers, implantable drug-delivery pumps, dialysis and chemotherapy ports, ostomy systems, tracheal and feeding tubes, central venous catheters, myringotomy tubes, cochlear implants, intraocular lenses, intra-aortic balloons, angioplasty devices, cardiac valves, vascular ports, various types of sheeting, and small-joint orthopedic devices 3 forms of silicone are used to fabricate implants: polymer–relatively hard; significant 'bleeding' is rare; elastomer–pliable; some silicone 'bleeding' occurs; gel–'bleeding' is common PLASTIC SURGERY Various formulations of silicone have been used in cosmetic surgery; one formerly popular silicone, polydimethylsiloxane, was enclosed in plastic bags of various sizes and shapes and implanted subcutaneously to impart cosmetically acceptable contours to soft tissues, most commonly in $\female$ for breast augmentation, and in $\male$ for chin augmentation; the complications of such implants in trained hands are minimal and confined to rupture of bags and/or fibrosis. See Breast implants, Human adjuvant disease, Mammoplasty

silicone arthroplasty ORTHOPEDIC SURGERY The implantation of prosthetic joints constructed of silicone to replace joints damaged by arthritis, avascular necrosis, trauma; silicone is also used as spacers in certain procedures–eg, Keller's bunionectomy LOCAL COMPLICATIONS Breakage/breakdown, dendritic synovitis or osteolysis, infection, loosening of joints from anchoring bone, multinucleated giant cell reaction, erosion through soft tissues and skin SYSTEMIC COMPLICATIONS Acute pneumonitis, delirium, fever, granulomatous hepatitis, ITP, lymphadenopathy, malaise, progressive systemic sclerosis, renal and respiratory failure, scleroderma, SLE. See Silicone

silicone implant An FDA class 3 medical device composed primarily of silicone or silicone gel–eg, gel and saline-filled breast implants, and gel-filled chin prostheses, testicular implants, Angelchik reflux valves, penile implants. See Breast implant, Human adjuvant disease, Medical device, Silicone

silicone-reactive disorder Siliconosis PRODUCT LIABILITY A proposed autoimmune condition said to be associated with leakage of silicone into surrounding breast and/or systemic penetration thereof CLINICAL Fatigue, myalgia, arthralgia, synovitis, serositis, sicca syndrome–dry eyes and mouth, dermatitis, alopecia, lymphadenopathy, GI Sx, fever, night sweats, and cognitive dysfunction DIAGNOSIS Clinical features, Hx, microscopy, inductively coupled plasma emission spectroscopy. Cf Silicone testing

siliconoma A circumscribed tissue response to extravasated silicone, accompanied by a foreign body-type giant cell reaction, histiocytes, brightly refractile silicone crystals

silicosis Classical silicosis, silicoproteinosis OCCUPATIONAL MEDICINE A form of pneumoconiosis caused by inhalation of silicates/silica dust from industries that process quartz and flint–potteries, foundries, sand pits, construction sites, etc CLINICAL Emphysema, $\downarrow$ respiratory function, pulmonary fibrosis, $\uparrow$ susceptibility to TB RADIOLOGY Bilateral symmetrical interstitial fibrosis, hilar lymphadenopathy with 'eggshell' calcification RISK FACTORS Mining, stone cutting, quarrying, blasting, road and building construction, foundry workers, abrasives manufacture, and other exposure to silicates; symptomatic disease requires 10+ yrs of exposure. See Acute silicosis, Pneumonoconiosis. Cf Asbestosis

SILIG Squamous intraepithelial lesion of indeterminant grade, see there

silk SURGERY A silkworm–*Bombyx mori* protein-based absorbable suture material, favored by many surgeons due to its superior handling characteristics; with time, silk loses strength and thus is not used for prosthetics–eg, Teflon vascular grafts or prosthetic heart valves, which require permanent sutures. Cf Catgut

silo-filler disease Silo-filler's lung OCCUPATIONAL MEDICINE A toxic gas-induced pneumonitis and bronchiolitis caused by inhalation of nitrogen oxides in freshly filled grain silos, often coupled to asphyxia CLINICAL Cough, light headedness, dyspnea, cyanosis, hemoptysis, choking FORMS OF PRESENTATION 1. Collapse and sudden death; 2. Acute alveolar damage with pulmonary edema; 3. Early reversible bronchiolitis obliterans and 4. Late irreversible bronchiolitis obliterans DIFFDX Mycotoxicosis, an allergic reaction TREATMENT High dose corticosteroids may prevent the bronchiolitis obliterans common in the heavily exposed survivors. See 'Animal House' fever, Farmer's lung

silver TOXICOLOGY A metallic element–atomic number, 47; atomic weight, 107.9, which may be ingested as silver nitrate, an antiseptic for conjunctivitis in newborns, and skin infections in Pts with extensive burns, or due to industrial exposure CLINICAL–ACUTE Pain and burning in mouth, vomiting, diarrhea, collapse, coma, death CLINICAL–CHRONIC Blackened mucocutaneous surfaces

silver wire appearance OPHTHALMOLOGY A descriptor for the funduscopic appearance of grade IV arteriosclerotic retinopathy, in which the arterial wall becomes so completely opaque that the blood column is not seen and light is completely reflected, yielding a white 'line,' likened to an SW, regardless of whether the lumen is occluded; arterial patency is determined by fluorescein angiography. Cf Copper wire appearance

Simantha GRADUATE EDUCATION A computerized humanoid simulator used to train medical personnel in catheterization, reponse to simulated medical emergencies, etc

Simasedan Diazepam, see there

Simdax® Levosimendan, see there

simian crease Simian fold NEONATOLOGY A dermatoglyphic pattern of a single deep transpalmar crease formed by fusion of proximal and distal palmar creases, classically seen in trisomy 21, but also in trisomies 13 and 9, fetal trimethadione syndrome. See Dermatoglyphics. Cf Triradius

Simon filter A vena caval filter for preventing PE, delivered via the subclavian vein

simple fracture Closed fracture, see there

simple goiter ENDOCRINOLOGY A goiter caused by ↓ thyroid hormone, resulting in compensatory hypertrophy TYPES Endemic–aka colloid goiter, sporadic

simple mastectomy Total mastectomy A mastectomy that removes the breast, but not muscle or lymph nodes; SM may be adequate therapy for in situ or intraductal CA which rarely spreads to lymph nodes. See Lumpectomy, Mastectomy, Prophylactic mastectomy. Cf Radical mastectomy

simple phobia Specific phobia PSYCHIATRY A phobia characterized by persistent irrational fears of a specific object, activity, or situation EXAMPLES Acrophobia, arachnophobia, agoraphobia, brontophobia, claustrophobia, emetophobia. See Phobia

simple suction probe SURGERY A cannula for end-of-case cleanup

simple test MEDTALK A test that is supersimple to perform–"idiot-proof", often in an outpatient setting or ambulatory service EXAMPLES Urine dipstick tests, EKG, X rays. See Waived test. Cf Complex test

simple traction ORTHOPEDICS Minimalist opposing forces, used to separate fracture fragments, usually followed by application of a plaster cast, which is used for simple fractures of long bone shafts. See Traction

Simplified Acute Physiology Score See SAPS II

Simpson-Angus Scale PHARMACOLOGY A testing instrument–range of scores, 0–40, ↑ scores indicate ↑ severity, for evaluating drug-related extrapyramidal syndromes. See Extrapyramidal syndrome

Sims position OBSTETRICS A position to facilitate a vaginal exam; the Pt lies on the side with the under arm behind the back, thighs flexed, the upper one more than the lower. Cf Lithotomy position

simulation MEDTALK The controlled representation of real world phenomena, used when real world experiences are either unavailable or undesirable; simulations are based on observing other system functions, or by assessing a hypothetical system created from existing data. See Casualty simulation, Instructional simulation, Monte Carlo simulation, Pocket simulation

simulator RADIATION ONCOLOGY A radiation generator operating in the diagnostic x-ray range, which can orient a radiation beam toward a Pt with parameters imitating that proposed for RT, affording direct x-ray fluoroscopic visualization and imaging of the treatment area. See Radiation oncology simulation, Simulation, Therapeutic radiology treatment planning, Treatment Port

Simulect® Basiliximab, see there

simultanagnosia NEUROLOGY Inability to comprehend > 1 element of a scene at the same time or integrate the parts into a whole

simultaneous pulse cardioversion Simultaneous pulse defibrillation CARDIAC PACING Therapy that terminates V tach and/or V fib by simultaneously delivering 2 high voltage pulses along 2 different pathways through the myocardium

simvastatin Zocor® THERAPEUTICS A lipid-lowering HMG-CoA reductase inhibitor used in hypercholesterolemia, to improve lipid profiles and ↓ mortality from CAD, DM, TIAs, stroke ADVERSE EFFECTS Constipation, flatulence, dyspepsia, abdominal pain, myalgia, muscle weakness. See Cholesterol-lowering drugs, Gemfibrozil, HMG-CoA reductase inhibitors, Hypercholesterolemia, Statin

'sin' tax A popular term for any tax levied on 'pleasure poisons'–eg, alcohol, tobacco. See Alcohol, Smoking

sine wave Sinusoidal waveform A waveform of periodic oscillations–eg, in alternating current in which the amplitude of each point in the wave is proportional to the sine of the time from a start point CARDIOLOGY An EKG finding described in severe hyperkalemia where the 'P' wave disappears and the QRS complex and 'T' wave merge in an oscillating pattern

Sinemet® NEUROLOGY An anti-parkinsonian containing levodopa, carbidopa

Sinequan® Doxepin, see there

Sinestron Lorazepam, see there

singer's node Laryngeal polyp A non-inflamed stromal reaction on the anterior vocal cord, caused by phonotrauma, seen in voice abuse–eg, singers, disk jockeys, often work-related. Cf Noise-induced hearing loss

single breath diffusing capacity Dt_CO₂ PHYSIOLOGY A screening test for identifying severe lung impairment PROS Noninvasive, amenable to rapid screening, requires little cooperation/effort by Pt, well-standardized. See Pulmonary function test

single breath imaging See Spiral computed tomography

single donor transfusion TRANSFUSION MEDICINE A form of 'directed donation' consisting of the administration of a particular blood component–eg, plasma, platelets, from the same donor to a specific recipient with a special need–eg, formation of antibodies to a 'public' antigen. See Directed donation. Cf Single unit donation

single drug therapy Monotherapy, see there

single fiber electromyography SF-EMG NEUROLOGY A test used to evaluate the activity of single fibers of selected muscles–eg, common extensor of fingers, displayed as an action potential on an cathode-ray oscilloscope;

SF-EMG may detect early defects in neuromuscular junction diseases–eg, myasthenia gravis, Eaton-Lambert syndrome, botulism, when a neurologic exam is still normal. See Electromyography, Fiber density, Jitter

single-gene disorder GENETICS A hereditary disorder caused by a mutant allele of a single gene–eg, cystic fibrosis, Duchenne muscular dystrophy, hemophilia, Huntington's disease, retinoblastoma, sickle cell disease; SGDs typically describe classic simple Mendelian patterns of inheritance–autosomal dominant, autosomal recessive, X-linked. See Gene, Mutation. Cf Polygenic disorder

single organ system examination MEDTALK A physical exam that focuses on one organ system–eg, cardiovascular, and evaluates other systems as pertains to the system of interest. See Evaluation and management services

single-parent family SOCIAL MEDICINE A family unit with a mother or father and unmarried children. See Father 'factor', Latchkey children, Quality time, Supermom. Cf Extended family, Nuclear family, Two parent advantage

single-payer system HEALTH REFORM Social medicine, in which all medical services are paid by a single reimbursement agency. See Canadian plan, Clinton Plan, Managed care, Socialized medicine

single pulse cardioversion/defibrillation CARDIAC PACING Therapy that terminates V tach and/or V fib by delivering a single high voltage pulse across one pathway through the myocardium

single unit transfusion TRANSFUSION MEDICINE A controversial therapeutic agent given to a Pts if a single unit of packed RBCs suffices as therapy, the transfusion probably was unnecessary. See Transfusion 'trigger'. Cf Single donor transfusion

single vessel disease Univessel disease CARDIOLOGY Coronary atherosclerosis that is very advanced in one artery

SingleBAR™ SURGERY An electrovaporization and resection electrode for use during endoscopic procedures. See RF tissue ablation. Cf Singles bar

singles bar SOCIAL MEDICINE A tavern that is a meat/meet market for unattached or allegedly unattached adults, usually understood to be heterosexually oriented. Cf Gay bar

singleton MEDTALK One baby. Cf Triplet, Twin

Singulair® Montelukast, see there

sink testing LAB MEDICINE The illegal practice of providing false test results on clinical specimens–eg, vials of blood, urine specimens, that were deliberately discarded–ie, down the sink, without actually testing them

sinoatrial exit block CARDIOLOGY Sinus-node dysfunction, aka sick sinus syndrome characterized by a normal P-wave axis EKG Progressive shortening of PP interval until one P wave fails to conduct–2nd degree, type I–or sinus pause is an exact multiple of the baseline PP interval–2nd degree, type II. See Sick sinus syndrome

sinoatrial node SA node, sinuatrial node, sinus node, heart's pacemaker ANATOMY A knot of specialized, spontaneously depolarizing cells located at the posterior wall of the upper right atrium, in the sulcus terminalis at the junction between the superior vena cava and right atrium, the heart's natural pacemaker NORMAL RHYTHM 70/min; conducts impulses via 3 Purkinje fiber tracts: (1) anterior internodal tract of Bachman; (2) middle internodal tract of Wenckebach; (3) posterior internodal tract of Thorel to AV node–normal rhythm, 45/min–in right posterior portion of interatrial septum; it is continuous with bundle of His–normal rhythm, 35/min BLOOD SUPPLY Rt coronary–65%, circumflex–25%, both in remainder FLOW OF IMPULSE From the SN through right atrium to AV node in low septal right atrium. See Conduction system. Cf Atrioventricular node

Sinonna® Butabarbital, see there

Sinquan Doxepin, see there

sinus arrest Sinus pause CARDIOLOGY A form of sinus-node dysfunction–sick sinus syndrome characterized by a normal P-wave axis EKG Every P wave is followed QRS complexes and accompanied by pauses of > 3 sec without atrial activity . See Sick sinus syndrome

sinus bradycardia CARDIOLOGY Sinus-node dysfunction–sick sinus syndrome characterized by a normal P-wave axis of < 60 beats/min EKG Every P wave is followed QRS complexes , or at < 50 beats/min. See Sick sinus syndrome. Cf Sinus tachycardia

sinus histiocytosis with massive lymphadenopathy Rosai-Dorfman disease HEMATOLOGY condition most common in young blacks CLINICAL Massive bilateral cervical lymphadenopathy, fever, leukocytosis, ↑ ESR, polyclonal hyperimmunoglobulinemia PROGNOSIS Uncertain; most resolve spontaneously; some are aggressive and ultimately fatal. Cf Sinus histiocytosis

sinus node Sinoatrial node, see there, SA node

sinus node dysfunction CARDIOLOGY A disturbance, impairment or defect in the behavior of the sinoatrial/SA node. See Sick sinus syndrome

sinus tachycardia CARDIOLOGY A heart rate triggered by the sinoatrial node at > 90 beats/min, usually in response to exogenous factors–eg, pain, fever, thyroid hormone, stress, hypoxia, stimulants–eg, caffeine, cocaine, amphetamines; ST may indicate heart failure, valve disease or other disease. Cf Sinus bradycardia

SinuSeal™ ENT A resorbable nasal packing for after sinus surgery PROS No need to remove postoperative packing, which may be painful and cause rebleeding . See Nasal packing

sinusitis ENT Infection and/or inflammation of paranasal sinuses, often caused by blocked drainage of fluid or purulent material, linked to swelling of nasal mucosa ETIOLOGY Allergic rhinitis, viral respiratory infections, deviated nasal septum, other obstruction; swimming or immersion of head in water may allow water and bacteria to enter the sinus, causing irritation and infection, fluid-related pressure pain. See Allergic fungal sinusitis, Sinus

Sipam Diazepam, see there

Sipramine Imipramine, see there

sips/chips NURSING Popular for the first ingestibles allowed either after a surgical procedure under general anesthesia, or before and after surgery 'under local,' where the recuperee is allowed to sip water and other clear liquids or chew on ice chips

SIRA Selective imidazoline receptor agonist CARDIOLOGY A class of drugs which ↓ sympathetic nervous system overactivity, linked to HTN and other cardiovascular risk factors, by selective stimulation of imidazoline receptors in the brain. See MOXCON, moxonidine

sirolimus Rapamycin, Rapamune® IMMUNOLOGY A macrolide immunosuppressant structurally similar to tacrolimus; it suppresses B- and T-cell proliferation, lymphokine synthesis, T-cell response to IL-2. See TOR. Cf Cyclosporine, Tacrolimus

sirolimus-eluting coronary stent CARDIOLOGY A coronary artery stent graft which has sirolimus, an imunosuppressive and antimitotic agent incorporated into its matrix, which significantly ↓ restenosis of stents. See RAVEL

SIRS 1. Subcutaneous insulin-resistance syndrome, see there 2. Systemic inflammatory response syndrome, see there

SISI test Short increment sensitivity index test AUDIOLOGY A hearing test used to determine differential sensitivity to changes in loudness. See Rinne test, Schwabach test, Weber test

sister 1. Nurse–UK, especially a head nurse 2. A ♀ sibling

Sister Mary Joseph nodule SURGICAL ONCOLOGY A nonulcerating periumbilical tumor metastatic from an adeno CA of the stomach, colon, ovary, or pancreas, which may be the first mass detected in gastric CA; it usually indicates a poor prognosis

Sisyphus 'syndrome' PSYCHIATRY A mindset typical of a stress-driven type 'A' person, who obtains no gratification from accomplishing the difficult goals he or she places upon himself or herself. See 'Anal-retentive', 'Toxic core', Type A personality

site visit MANAGED CARE A formal visit to a hospital or heath care facility, by representatives from an accrediting organization–eg, JCAHO, HCFA, now CMS, to assess the quality of care provided in the institution, as reflected by the facility's adherence to guidelines for providing such care. See CMS, JCAHO

SiteSelect™ SURGERY A system for removing a single tissue sample for faster diagnosis of suspicious breast lesions. See Breast biopsy

SITEtrac™ ORTHOPEDICS A nondisposable system for minimally invasive nerve root decompression COMPONENTS Tubular dilators, retractors, high-resolution endoscope. See Nerve compression

sitophobia Cibophobia, sitiophobia GI DISEASE Fear of eating due to the unpleasant Sx–eg, N&V and abdominal pain that follow eating, a feature of chemotherapy-induced anorexia; it may occur in Crohn's disease and chronic mesenteric artery insufficiency; unlike anorexia, there is an appetite, but it is curtailed by the anticipated emesis and nausea. See Phobia

16 Personality Factor Questionnaire PSYCHOLOGY An instrument used to evaluate adult personality ASSESSES Assertiveness, emotional maturity, self-sufficiency, shrewdness, anxiety, tension, rigidity, neuroticism. See Psychometric testing

sixth disease Exanthem subitum, pseudorubella, roseola, roseola infantum PEDIATRICS A herpesvirus 6 infection of infants/young children characterized by abrupt onset of high fever lasting several days, followed by a fine red rash. Cf Fifth disease, Seven-day disease.

size VOX POPULI An object's "bigness". See Effect size, French size, Sample size

size acceptance *adjective* Relating to a philosophy in which a person's ↑ girth/berth is accepted. See Obesity, Politically correct

size discrimination Sizeism, weight discrimination VOX POPULI Discrimination based largely–no pun intended–on a person's size. See Size acceptance

Sjögren syndrome RHEUMATOLOGY An autoimmune disorder more common in older ♀, associated with rheumatoid arthritis, SLE, scleroderma, polymyositis, and other connective tissue diseases CLINICAL Dry eyes, dry mouth LAB SS-A, SS-B antibodies MANAGEMENT Symptomatic, maintain hydration of mouth, eyes, skin

skeletal survey Metastatic series IMAGING The radiologic examination of the entire skeleton INDICATIONS Suspected child abuse, detection of bone metastases from 1° CA–eg, prostate, breast CA. See Bone scan

skeletal traction Pin traction ORTHOPEDICS Traction first achieved with tongs, followed by wire–eg, a K wire or pin–eg, Steinmann pin, placement in a bone–eg, tibia, femur and weights suspended therefrom to maintain proper alignment of the Fx. See Traction PHYSICAL THERAPY A technique that may relieve pain linked to certain neck disorders–eg, muscle spasm, nerve root compression, osteoarthritis, cervical spondylosis, myofascial syndrome, facet joint dysfunction; cervical traction–CT applies a stretch to muscles, ligaments, and tissue components of the cervical spine, providing relief by promoting separation of the intervertebral joint space, which contains the disc and may reduce bulging or impingement of structures in the foramen; it is not indicated for conditions of instability–eg, whiplash injuries; CT is most commonly used when the Pt is in the supine position–lying on the back with knees bent at a 45° with the neck placed at 20°-30° of flexion–forward tilt; traction in this position helps stretch the posterior neck muscles and facilitate intervertebral separation, relieving pressure that may be pinching nerves, promoting muscle relaxation and intervertebral separation

skeletonize SURGERY A popular term for the stripping of serosal tissues from certain structures–eg, the round ligament, fallopian tube, to facilitate clamping or excision and minimize bleeding

Skelid® Tiludronate, see there

skew foot S-shaped foot, serpentine foot, skewfoot deformity, Z foot deformity, zed foot ORTHOPEDICS A foot deformity consists of marked metatarsus adductus of forefoot and plantarflexion, hindfoot valgus, and lateral displacement of midfoot in abduction with lateral displacement of navicular on talar head, prominence of the talar head in medial arch with possible flattening of medial arch and presence of thickened callus over the head of the plantarflexed talus, tendoachilles contracture of variable degree, and deformity may be rigid or supple IMAGING Lateral displacement of navicular on head of talus, metatarsal adduction; in the normal foot on AP radiograph, a line drawn through long axis of first metatarsal and a line drawn through long axis of talus should be parallel or divergent laterally, widening of the talocalcaneal angle on AP radiograph usually greater than 35°, and increased lateral talocalcaneal angle with plantarflexion of talus MANAGEMENT-NONOPERATIVE Serial casting is mainstay of nonoperative care, similar to that for metatarsus adductus but heel is held in neutral to slight varus, successful nonoperative treatment improves appearance of foot and reasonably corrects radiographic features; cast treatment for SD is longer than that needed for metatarsus adductus; serial casting of ± 8–10 wks and static casting of 2–4 wks is usual OPERATIVE MANAGEMENT Usually not indicated until early childhood

skewed EPIDEMIOLOGY *adjective* Referring to an asymmetrical distribution of a population or of data

'skid row' stereotype Skid row syndrome SOCIAL MEDICINE A sociomedical entity affecting chronic alcoholics, primarily ♂. See Alcoholism, Homelessness

skier's tear ORTHOPEDICS Rupture of a skier's ulnar collateral ligament–UCL, seen when a thumb gets caught on the ski pole strap when falling, which prevents pulling the hand away from the pole; when the skier falls, the pole, especially poles with a saber-type grip, is pushed away from the palm, carrying the thumb with it, causing the UCL to snap. See Skiing

skill VOX POPULI Adeptness of performance. See Cognitive skill, Social skill

skilled nursing care NURSING Daily nursing and rehabilitative care performed by or under the supervision of skilled professional or technical personnel SNC Administering medication, diagnostics, minor surgery

'skimping' MANAGED CARE The delaying or denial of services to members of a prepaid or 'capped' health plan, to control costs–because the monies received by the health plan remain constant, providing 'extra' services is more costly to the plan. See Skimming, Capitation

skin VOX POPULI *adjective* Cutaneous *noun* ANATOMY Cutis The outer integument of the body which consists of epidermis and dermis, the latter of which rests on subcutaneous tissue. See Aging skin, Artificial skin, Blue skin, Cigarette-paper skin, Composite cultured skin, Diamond skin, Elephant skin, Glossy skin,

Harlequin skin, Hide-bound skin, Leopard skin, Lizard skin, Moleskin, Moroccan leather skin, Paper money skin, Red skin, Sandpaper skin, Second Skin®, Spray-on skin, Swiss cheese skin, Washerwoman skin

skin abscess Boil, subcutaneous abscess DERMATOLOGY A gob of pus and infected debris in the skin caused by localized, usually bacterial–especially *Staphylococcus aureus*, infection; SAs may follow minor injury, or follow folliculitis, furuncles, carbuncles, occur anywhere, and affect any age group

skin cancer ONCOLOGY A cutaneous malignancy NONMELANOMA SCs BCC, SCC–up to 10^6 new cases/yr, ±2700 deaths MELANOMA ±34,000 new cases/yr, very aggressive–±7000 deaths BIOLOGY 1° SC may be indolent, as in BCC, or serious, as in melanoma–depth of invasion is a critical prognostic parameter; metastatic SCs–eg, from breast, GI tract, have a poor prognosis. See Basal cell carcinoma, Melanoma, Squamous cell carcinoma.

skin graft Autologous, donated, or surrogate skin removed from one site to cover surfaces on another region with 3rd-degree burns or traumatic tissue loss. See Split-thickness graft. Cf Artificial skin, 'Spray-on' skin

skin 'popping' SUBSTANCE ABUSE Intradermal injection of narcotics. See Fight bite, Shooting gallery. Cf 'Pocket shot'

skin resurfacing COSMETIC SURGERY A procedure in which the epidermis is partially removed, chemically or by laser, to remove wrinkles, spots, scars and tighten skin. See Chemical peel

skin test IMMUNOLOGY A test of immune response to a substance by placing it on or under the skin. See PPD, Prick test

skin test conversion IMMUNOLOGY A change in skin test reaction status–often PPD reactivity–from negative to positive between screening intervals. See Anergy, PPD, Seroconversion, Tuberculosis

skin type Sun-reactive skin type DERMATOLOGY A classification of a person's reactivity to solar/UV light, based on a 45–60 min exposure after winter or minimal previous sun exposure

SkinLaser™ COSMETIC SURGERY A skin resurfacing system that focuses laser energy on the dermis without significant damaging–ie, burning or cutting the epidermis; it may be of use in revising stretch marks and in facial rejuvenation. See Cosmetic surgery

Skinlight™ COSMETIC SURGERY An erbium–YAG laser for skin resurfacing with minimal thermal damage. See Cosmetic surgery

Skinner's line IMAGING A horizontal line, drawn on a plain film of a horizontal pelvis, between the top of the greater trochanter of the femur to the uppermost part of the obturator foramen; SL helps describe the axial relationship of a normal hip joint. Cf Shenton's line

skinnerian PSYCHOLOGY *adjective* Relating to theories delineated by BF Skinner, vis-á-vis conditioned responses to highly controlled environments. See Behavioral intervention, Skinner; Cf Deterministic, Pavlovian

'skinny' needle–biopsy A 22-gauge needle used for percutaneous, often radiologically-guided biopsies or aspiration cytology specimens obtained from masses of the breast, lung, and sites of difficult access. See Interventional radiology, Fine needle aspiration biopsy

skip metastases Discontinuous metastases ONCOLOGY The metastatic spread of a CA in which contiguous regions are skipped, while distant foci of the CA are present, a finding with a poor prognosis. See Metastasis

skitching SPORTS MEDICINE An 'extreme sport' in which an inline skater launches him/herself from the bumper of a speeding car INJURY RISK Concussion, death. See Extreme sports, Novelty seeking behavior

skull fracture ORTHOPEDICS A fracture of one or more cranial bones, caused by MVAs, falls, assault, sports, occupational accidents and other forms of blunt trauma CLINICAL Headache, bleeding from wounds and orifices, loss of consciousness, confusion, seizures, restless, drowsiness, visual defects, slurred speech, N&V, stiff neck. See Fracture

sky surfing SPORTS MEDICINE An 'extreme sport' in which the participant skydives with a graphite snowboard, which allows performance of various gyrations while free falling. See Extreme sports, Novelty seeking behavior

SLAC wrist ORTHOPEDICS A wrist affected by scapholunate advanced collapse–SLAC, a common form of wrist osteoarthritis IMAGING Loss of joint space, periarticular sclerosis, loose fragments, scapholunate disassociation and rotation of scaphoid and, with time, carpal collapse. See Wrist

slang SOCIOLOGY A specialized lexicon of words that are exclusive or replace other words in function, and tend to have a short life cycle. Cf Dialect, Jargon

slanted palpebral fissure A facial anomaly in which the PFs slope upward, as viewed from outside inward; SPFs may be seen in congenital disorders–eg, Aarskog, Apert, Coffin-Lowry, Cohen, Conradi-Hünermann, DiGeorge, femoral hypoplasia-unusual facies, 5p-, Jarcho-Levin, Miller, Miller-Diker, Nager, Opitz-Frias, partial trisomy 10q, Pfeiffer, rhizomelic chondrodysplasia punctata, Rubenstein-Taybi, Säthre-Chotzen, Sotos, Treacher-Collins, trisomy 9 mosaic, trisomy 9p, trisomy 20p, XXXXX, XXXXY syndromes

slapped cheek appearance A circumscribed, intense facial erythema of sudden onset, followed by an maculopapular rash of the entire body, characteristic of '5th disease'–erythema infectiosum, due to parvovirus B19. See Fifth disease, B19

slave Bottom SEXOLOGY Vernacular for the masochistic partner in a sadomasochistic dyad, whose role is of subservience and obedience.

SLE Systemic lupus erythematosus, see there

sleep SLEEP DISORDERS Rest resulting from a natural suspension of voluntary bodily functions and consciousness. See Deep sleep, Delta sleep, Good habit, Light sleep, Microsleep, Non-REM sleep, Nocturnal sleep, Normal sleep, REM sleep, Twilight sleep. Cf Bad habit, Poor sleeping hygiene

sleep apnea A temporary cessation of breathing during sleep, usually lasting several secs; SA may be fatal if the person is asleep and stops breathing multiple times for ≥ 10 secs

sleep apnea syndrome Ondine's curse A condition defined by frequent episodes of sleep apnea, hypopnea, and Sx of functional respiratory impairment; it is potentially life-threatening, and associated with daytime hypersomnolence, MVAs, and cardiovascular M&M in the form of HTN, stroke, acute MI; it is more common in the obese and in heavy snorers; 2% of middle-aged ♀ and 4% of ♂ meet SAS criteria. See Narcolepsy, REM sleep

sleep architecture SLEEP DISORDERS The NREMS/REMS stage and cycle infrastructure of sleep understood from the vantage of the quantitative relationship of these components to each other. See Sleep structure

sleep attack Narcoleptic attack SLEEP DISORDERS A sine qua non feature of narcolepsy, consisting of an episode of irresistible sleep, which may occur at any time, but most commonly during boring or monotonous activities–eg, during lectures, driving a car, or watching TV, but also during emotional and physical exertion–eg, coitus, phone conversations

sleep deprivation SLEEP DISORDERS A prolonged period without the usual amount of sleep. See Driver fatigue, Poor sleeping hygiene, Sleep disorders, Sleep-onset insomnia

sleep disorders NEUROLOGY Any disruptive pattern of sleep CATEGORIES (1) Difficulty with falling or staying asleep–insomnias; (2) falling asleep at inappropriate times; (3) excessive total sleep time; (4) abnormal behaviors associated with sleep. See Insomnia, Pseudoinsomnia, Sleep apnea syndrome. Cf Shift work

---

SLEEP DISORDERS

INSOMNIAS Difficulty in initiating and maintaining sleep

HYPERSOMNIAS Disorder of excessive somnolence–eg, narcolepsy, sleep deprivation and obstructive sleep apnea

DISORDERS OF SLEEP-WAKE SCHEDULE–eg jet lag and shift work

DYSFUNCTIONS ASSOCIATED WITH SLEEP, SLEEP STAGES OR PARTIAL AROUSAL–eg night terrors and enuresis

---

sleep drunkenness SLEEP DISORDERS A form of hypersomnia in which full alertness is not achieved for a prolonged period of time after awakening; SD is characterized by automatic behaviors, disorientation, drowsiness, unsteadiness. See Narcolepsy, Sleep deprivation, Sleep disorders

sleep hygiene SLEEP DISORDERS The maintenance of habits conducive to sound sleep and rest; for good SH, naps should be limited, exercise, proper nutrition, caffeine, alcohol, nicotine, worrying, etc, minimized before sleeping

sleep latency SLEEP DISORDERS The time period from lights out/bedtime to sleep onset

sleep medicine MEDTALK The sub(subspecialty) of internal medicine that studies and manages poor sleep hygiene VOX POPULI A medicine with a sedating or calming effect, intended to promote sleep

sleep onset association disorder PEDIATRICS A condition in which sleep onset occurs most quickly when an object–eg, a blanket, or condition–stroking mother's hair is met. See Sleep disorder

sleep paralysis NEUROLOGY Flaccid paralysis at sleep onset or on waking in otherwise healthy persons; SP is characterized by an inability to 'kickstart' the voluntary muscles, resulting in a transient 'locked-in' syndrome; SP is associated with narcolepsy, Pickwick syndrome, sleep apnea MANAGEMENT Most SP is isolated and self-limited; if frequent and uncontrollable, clonipramine or desipramine, to prevent. See Narcolepsy, Pickwick syndrome, Sleep apnea

sleep-related myoclonus Nocturnal myoclonus SLEEP DISORDERS Muscle contractions in the form of 'jerks' or twitches, primarily of the flexors in the lower extremities, which have a characteristic frequency of 20-40 secs. See Restless legs

sleep stage 1 NREMS stage 1 SLEEP DISORDERS A stage of NREM sleep that immediately follows the awake state ACCOUNTS FOR 4-5% of total sleep time

sleep stage 2 NREMS stage 2 SLEEP DISORDERS A stage of NREM sleep ACCOUNTS FOR 45-55% of total sleep time

sleep stage 3 NREMS stage 3 SLEEP DISORDERS A stage of NREM sleep ACCOUNTS FOR 4-6% of total sleep time; with stage 4, it constitutes "deep" NREM sleep; often lumped with stage 4 into NREMS stage 3/4 because of lack of documented physiologic differences; appears usually only in 1st third of sleep period

sleep stage 4 NREMS stage 4 SLEEP DISORDERS The last period of sleep ACCOUNTS FOR 12-15% of total sleep time, SS4 is similar to NREMS stage 3, except that high-voltage, slow EEG waves cover $\geq$ 50% of the record; somnambulism, sleep terror, and sleep-related enuresis generally start in stage 4 or during arousals from SS4

sleep terror disorder Night terror, pavor nocturnus SLEEP DISORDERS An abrupt awakening from sleep with behavior consistent with terror–panic,

sweating, tachycardia, confusion, and poor recall for the event; STD occurs during stage 3/4 (deep) sleep, and is associated with emotional stress or conflict, especially in preadolescent boys and, in adults, is associated with alcohol use. See Sleep disorder. Cf Nightmare

sleepiness Drowsiness, somnolence SLEEP DISORDERS Difficulty in maintaining the wakeful state so that the person falls asleep if not actively kept aroused; sleepiness is *not* simply physical tiredness or listlessness. See Excessive daytime sleepiness

sleeping sickness 1. African trypanosomiasis, see there 2. Narcolepsy, see there

sleepwalking Somnambulism PSYCHIATRY A sleep disorder characterized by walking or other activity while seemingly asleep ETIOLOGY, CHILDREN Fatigue, sleep loss, anxiety ADULTS Mental disorders, drug reactions, abuse substances, alcohol, medical conditions–eg, partial complex seizures, elderly organic brain syndrome, REM behavior disorders; the activity may include sitting up and appearing awake, while actually asleep, arising and walking around, or complex activities–eg, moving furniture, going to bathroom, dressing and undressing, and other activities, including driving a car; the episode can be very brief–a few secs or mins or last for 30+ mins; sleep walkers may be confused or disoriented after awakening; injuries caused by such things as tripping and loss of balance are common for sleep walkers; SW is most common in children aged 6 to 12 yrs old and may run in families

slice IMAGING A popular term for a collimation scan interval in CT or equivalent in MRI

slipped disk Herniated disk, see there

'slippery slope' MEDICAL ETHICS An ethical continuum or 'slope,' the impact of which has been incompletely explored, and which itself raises moral questions that are even more on the ethical 'edge' than the original issue

slit-lamp microscope OPHTHALMOLOGY A low-power microscope fitted with a specialized illuminating system for generating collimated light, used to examine the anterior eye–conjunctiva, sclera, cornea, iris, lens, anterior chamber, and to visualize ocular fluids APPLICATION ID dendritic keratitis, foreign bodies on cornea, iris tumors, etc

slot defect COSMETIC SURGERY A complication of scalp reduction as part of hair replacement therapy for male pattern baldness, in which there is a slot or zone of hair growing the direction opposite from the scalp reduction scar MANAGEMENT Frechet flap. See Frechet flap, Hair replacement

sloughed off MEDTALK *adjectice* Desquamated

'slow code' Hollywood code, light blue code, partial code, show code MEDICAL ETHICS A response to an emergency CPR–a 'code' in a hospital in which the speed typically associated with emergency resuscitation is not exercised and full therapeutic 'firepower' not utilized. See Advance directives, DNR, Euthanasia, Hippocratic Oath, Kevorkian, Living will. Cf Code, Failed code

slow pathway CARDIOLOGY An anomalous conduction pathway in the heart which has no known functions; the SP provides the antegrade limb of a reentry circuit in atrioventricular nodal reentrant tachycardia–AVNRT–the fast pathway provides the retrograde limb–the 'slow-fast' or common form of AVNRT; selective catheter ablation of the atrial end of the SP eliminates AVNRT with little risk of AV block

slow virus VIROLOGY A virus that may cause fatal infectious encephalitides after prolonged latency TYPES Conventional and unconventional viruses–the latter now termed prions; conventional SVs include measles–a paramyxovirus causing subacute sclerosing panencephalitis, rubella–rare progressive rubella panencephalitis, papovavirus–causes progressive multifocal leukoencephalopathy; SVs are inactivated by bleach, ethanol iodine, autoclaving. Cf Prions

626 **SLUDGE syndrome** A symptom complex occurring 15–60 mins after ingesting certain mushrooms–eg, *Inocybe* spp, *Clitocybe* spp, etc CLINICAL ↑ Salivation, Lacrimation, Urination, Defecation, Gastritis, Emesis–whence the acronym; SS is caused by muscarine, an anticholinergic, and accompanied by constricted pupils, bradycardia MANAGEMENT Atropine

SMA 1. Smooth muscle antibody 2. Spinal muscle atrophy, see there 3. Superior mesenteric artery, see there

smack DRUG SLANG A popular street term for heroin. See Heroin

small airways A term for membranaceous bronchioles–noncartilaginous conducting airways with a fibromuscular wall and respiratory bronchioles–airways in which the fibromuscular wall is partially alveolated. See Small airways disease

small airways disease A condition in which airway obstruction is attributed to ↓ luminal dimension; SAD is largely confined to the small airways or bronchioles–< 2 mm in diameter, initiated by inhaled irritants and is most common in smokers; it is accompanied by inflammation, hypersecretion, and small airway changes, including fibrosis, ulceration, metaplasia, smooth muscle proliferation. Cf COPD

small capacity syndrome A post-gastrectomy syndrome characterized by ↓ functional volume in the stomach, leading to early satiation, possibly resulting in major weight loss, malnutrition. See Dumping

small cell ONCOLOGY A 9-14 μm in diameter cell with a faint or indistinct rim of cytoplasm, and an oval-to-elongated nucleus with relatively dense chromatin; SCs are 'classically' neuroendocrine–confirmed by presence of cytokeratin and secretory granules; when a vague 'crease' in the nucleus is also present, the descriptor 'oat cell' may be used; SCs occur in small cell carcinomas (duh!), common in lungs, but also occur in bladder, breast, cervix, endometrium, nasopharynx. See Small cell carcinoma

small cell carcinoma Small cell undifferentiated carcinoma, undifferentiated carcinoma A highly aggressive malignancy, usually of lung, which arises in proximal bronchi and spreads early to hilar and mediastinal lymph nodes EPIDEMIOLOGY SCC comprises 10% to 20% of lung CA, more common in ♂ > age 65; 85+% are smokers CLINICAL Cough, hemoptysis, weight loss TREATMENT Combination chemotherapy–eg, etoposide, cisplatin, cyclophosphamide, doxorubicin, vincristine with chest RT PROGNOSIS Without treatment, 3 month survival

small cuff syndrome Nonhypertension syndrome A form of pseudohypertension in which the cuff used to measure the BP is too small and both systolic and diastolic pressures are interpreted as 'hypertensive'. Cf Pseudohypertension, White coat hypertension

small for gestational age Intrauterine growth retardation NEONATOLOGY *adjective* Referring to an infant whose gestational age and weight gain are < expected for age. See Low birthweight

small intestine biopsy Small bowel biopsy, small intestinal mucosa biopsy A biopsy of the duodenal or upper jejunal mucosa, performed in an ambulatory or outpatient setting with an upper GI endoscope, often as part of a workup for nonmalignant disease, in particular diarrhea. See Biopsy

small lymphocytic lymphoma Well differentiated lymphoma ONCOLOGY A type of NHL involving mature appearing B-cells, more common in elderly Pts LOCATION Spreads in lymph nodes later involving spleen, liver, BM SUBTYPES Lymphoplasmacytic lymphoma, B-cell chronic lymphocytic leukemia, B-cell prolymphocytic leukemia; SLL is closely linked to CLL. See Chronic lymphocytic leukemia, Lymphoma, WHO classification

small meal syndrome A clinical feature of chronic mesenteric ischemia in which dull or cramping periumbilical abdominal pain develops 10-30 mins after eating; the pain gradually ↑ in severity, plateaus then subsides

within 1-3 hrs; initially the pain only occurs with large meals and Pts may obtain some relief by assuming a prone position or squatting, but with ↑ severity of ischemia, becomes unbearable; it may be accompanied by bloating, flatulence, constipation, and diarrhea, and ultimately forces its victims to eat minimalist rations–resulting in significant weight loss

small non-cleaved cell lymphoma A high-grade B-cell lymphoma that is subdivided into Burkitt's lymphoma and pleomorphic–non-Burkitt's lymphoma. See Lymphoma

small vessel disease NEUROLOGY Cerebrovascular disease due to stenoses in small arteries of the brain. See Ministroke

small vessel vasculitis INTERNAL MEDICINE Vasculitis affecting vessels smaller than arteries–eg, arterioles, venules, and capillaries CLINICAL Palpable purpura, nodules, ulceration, urticaria; 30-50% involve GI tract and are accompanied by fever, neuritis, glomerulonephritis EXAMPLES Churg-Strauss syndrome, cutaneous leukocytoclastic angiitis, essential cryoglobulinemic vasculitis, Henoch-Schönlein purpura, microscopic polyangiitis, Wegener's granulomatosis, etc. See Systemic vasculitis

smallpox Variola VIROLOGY A DNA virus, genus *Orthopoxvirus*, transmissible by aerosol and implanted on the oropharyngeal or respiratory mucosa CLINICAL After a 2-wk incubation, high fever, malaise, prostration, headache, backache, abdominal pain, delirium, oral and upper body rash, that became vesicular then pustular. See Maalin

SMART Any of a number of trials: CARDIOLOGY 1. Second Manifestations of Arterial Disease 2. Serum Markers Acute Myocardial Infarction and Rapid Treatment 3. Study of Microstent's Ability to Limit Restenosis Trial 4. Study of Medicine versus Angioplasty Reperfusion Trial IMMUNOLOGY Study of Monoclonal Antibody Radioimmunotherapy

smart defibrillator Automatic external defibrillator, see there

smart gun PUBLIC HEALTH A firearm that can only be used after a PIN–personal identification number–is keyed in to release the trigger. See Safe weapon

smart loop SURGERY A device for simultaneously cutting and coagulating tissue

'smart' terminal INFORMATICS A computer terminal with its own CPU, which allows input and output of data, thus being capable of free-standing operation. Cf 'Dumb' terminal

smarts VOX POPULI Intelligence.

SmartTack™ ORTHOPEDICS A self-reinforced, resorbable fixation device for repair of ruptures of the ulnar collateral ligament of thumb. See Skier's thumb

SMART™ biliary stent SURGERY Nitinol stent for treating biliary obstruction; SMART is an acronym for Shape Memory Alloy Recoverable Technology. See Biliary stent

SMD 1. Senile macular degeneration, see there 2. Spondylometaphyseal dysplasia

smear TUBERCULOSIS A specimen gobbed on a glass slide, stained, washed in an acid solution, and examined by LM to detect AFB in a specimen. See Acid-fast bacilli, Wet mount

smegma UROLOGY A cheesy secretion–yuck! of certain sebaceous glands, expecially under the prepuce and clitoris, which can harbor bacteria, mandating regular cleaning. See Circumcision

smell *noun* Popularly, an odor or scent *verb* To perceive odor or scent by stimuli affecting the olfactory nerves. See Olfaction

smell disorder NEUROLOGY The inability to perceive odors; it may be temporary–eg, cold or swelling or blockage of nasal passages, or permanent–eg, damage to olfactory system–eg, brain injury, tumor, disease, chronic rhinitis

SMILE CARDIOLOGY A trial–Survival of Myocardial Infarction Long-Term Evaluation that evaluated the Zofenopril, an ACE inhibitor on M&M in Pts with acute MI. See ACE inhibitor, Acute myocardial infarction, Zofenopril

Smith-Lemli-Opitz syndrome NEONATOLOGY A rare AR condition characterized by multiorgan birth defects, with microcephaly, hypotonia, dysmorphic facies–short nose with anteverted nares, ptosis of eyelids, micrognathia, poly- and/or syndactyly, ♂ genital disorders–cryptorchidism, hypospadias, endocrine defects, cataracts, cardiac and renal malformations, major mental retardation, FTT, high infant mortality LAB ↓ Cholesterol in < 5th percentile, ↑–2000 ✗ normal cholesterol precursor 7-dehydrocholesterol–detected by GC which may be incorporated into cell membranes, interfering with proper functioning

smog Photochemical smog ENVIRONMENT An acronym of smoke and fog caused by combustion of often synthetic materials with production of noxious volatile byproducts; smog mortality ↑ substantially when smoke and SO_2 levels exceed 750 µg/m³ CLINICAL Dyspnea, acute exacerbation of COPD; smog triggers asthmatiform Sx–wheezing and markedly ↑ respiratory effort or, if intense, marked cyanosis–PaO_2 < 50 mm Hg; $PaCO_2$ 50-100 mm Hg, CHF, cor pulmonale. See Air pollution, Bhopal, Clean Air Act, Particular air pollution, Pollution, Yokohama asthma

'smoke' IMAGING A haziness occasionally seen by transesophageal echocardiography in the left atrium, a sign of blood stasis, fancifully likened to smoke, which corresponds to the spontaneous presence of contrast; 'smoke' is associated with ↑ thromboembolism. Cf Atrial systolic failure, Moya-moya disease VOX POPULI Fumes produced by a lit cigarette and its slave. See Sidestream cigarette smoke

smoke-free workplace LABOR LAW A workplace where use of cigarettes and other tobacco smoke products–cigars, pipes, is not allowed indoors

smoke inhalation TOXICOLOGY A cause of death in fire victims linked to toxic gases, especially, CO and hydrogen cyanide–HCN. See Cyanide

smokeless tobacco SUBSTANCE ABUSE Any chewed–chew or snorted–snuff tobacco product, that is regionally popular among athletes, the indigent, Native Americans and, tragically, children

smoker A person who smokes tobacco, almost always understood to be cigarettes RATIO OF ♂ : ♀ SMOKERS Philippines $^{64}/_{19}$, China $^{61}/_{7}$, Saudi Arabia $^{53}/_{2}$, Russia $^{50}/_{12}$, Argentina $^{45}/_{30}$, US $^{28}/_{23}$ PHYSICIAN SMOKERS Canada, UK, US, < 10% of physicians smoke; China, 67.5% of ♂ physicians smoke. See Former smoker, Nonsmoker

smoker data The number of smokers internationally has ↑ by 75% in past 20 yrs; 10^9 smokers consume ± 10^{12} cigarettes/yr PER CAPITA CONSUMPTION US 54/yr–1900, 4500/yr–1968, 3200/yr–1988; CHINA 400/yr–1953, 1900/yr–1988 tobacco-industry 'loses' ±2.5 million smokers/yr, 2.1 million to health-related 'attrition'–ie, quitting, 400,000 to smoking-related deaths PEAK AGE OF STARTING 16 EDUCATION 8.3% of physicians smoke; <age 30, 4.5% smoke; 18% of college-educated, 34% of high school drop-outs RACE 34% black, 28% white, 27% hispanic–US

smoker's palate ENT A condition caused by smoking: palatal inflammation, minor salivary gland edema, hyperkeratosis in response to tobacco

smoking PUBLIC HEALTH An addiction causing the most preventable form of malignancy, lung CA ECONOMIC COSTS Direct: ± $16 ✗ 10^9/yr, hospitalization; indirect: ± $35 ✗ 10^9, lost productivity, earnings, disability, prematurity POPULATION COST OF SMOKING $200 per capita; 1 cigarette ↓ life span by 7 mins; smoking-related fires kill 1500/yr, injure 4000/yr–US; smokers suffer

excess mortality of ±400,000/yr; in 1960, 70% of Americans smoked; in 2001, 27%. See Passive smoking

smoking cessation PUBLIC HEALTH Temporary or permanent halting of habitual cigarette smoking; withdrawal therapies–eg, hypnosis, psychotherapy, group counseling, exposing smokers to Pts with terminal lung CA and nicotine chewing gum are often ineffective. See CASS, COPD, Emphysema, Nicotine replacement therapy, Passive smoking, Smokeless tobacco

'smoking gun' DECISION-MAKING adjective A metaphor for a definitive confirmation of a cause-and-effect relationship

smoldering leukemia See Myelodysplastic syndrome, Preleukemia

smoldering myeloid leukemia Subacute myeloid leukemia An indolent myeloma that doesn't evoke significant anemia CLINICAL Infections, hemorrhage, occasional blast transformation; SML may 'smolder' for 20 yrs, causing osteolysis, hypercalcemia, renal insufficiency DIFFDX Smoldering leukemic states–eg, preleukemia, RAEB–refractory anemia with excess blasts, subacute myeloid leukemia; most SMLs arise from a malignant clone, accompanied by refractory cytopenias, cell dysfunction PROGNOSIS Survival ≥ 5 yrs. See Myelodysplastic syndrome, Preleukemia

smooth muscle antibody Anti-smooth muscle antibody IMMUNOLOGY Serum IgM and/or IgG autoantibodies associated with liver disease–eg, chronic viral PBC, infectious mononucleosis, asthma. See Primary biliary cirrhosis

SMR 1. Sexual maturity rating. See Tanner stages 2. Severely mentally retarded 3. Standardized mortality ratio, see there

snake Trivial name for limbless members of suborder Ophidia, of medical interest, primarily because snakes have poisonous venom containing hemotoxins and/or neurotoxins; venomous North American snakes belong to either the viper family Crotilidae–rattlesnake, copperhead, water moccasin, or to Elapidae, coral snakes, related to cobras and kraits–India, Southeast Asia, and mambas, brown, black, tiger snakes of Australia; others–eg, constrictors encircle prey and asphyxiate; pet snakes are associated with infections: *Aeromonas hydrophila*, *Edwardsiella tarda*, *E coli*, mesocestoidiasis, *Morganella morganii*, *Mycobacterium ulcerans*, *Ophionyssus natricis* infestation, pentastosomiasis, *Proteus vulgaris*, *Providencia* spp, Q fever, salmonellosis, sparganosis. See Sea snake VOX POPULI A derogatory term for a ne'er-do-well–eg, lawyer, ex-spouse, etc

snakebite TOXICOLOGY A bite from a snake that may be nonpoisonous or poisonous–which may cause envenomation and, if severe, be fatal EPIDEMIOLOGY ± 7000 persons are bitten by poisonous snakes/yr

snakes, fear of PSYCHOLOGY Ophidiophobia. See Phobia

snap CARDIOLOGY noun A click or other sharp sound corresponding to an abnormal mitral valve opening–opening snap, or closing–closing snap PSYCHIATRY verb To suffer abrupt decompensation in social or interpersonal coping mechanisms. See Postal worker syndrome

snare SURGERY A looped device used to snag a polyp or other small and/or pedunculated structures from sites of difficult access

Sneddon syndrome A condition characterized by fixed patchy livido reticularis, cerebral ischemia LAB Antiphospholipid antibodies, leukocytoclastic vasculitis

sneeze Sternutation An abrupt and involuntary explosive expulsion of air from the lungs through opened glottis into the nose and mouth, in response to various irritants WORLD RECORD SNEEZING On 1/1/81, a girl in the UK began sneezing and continued for 977 consecutive days, sneezing ±1 million times in the first 365 days

sneezing gas Any inhaled substance that evokes sternutation

sniffing dog position PEDIATRICS A popular term for the posture classically adopted by a child with epiglottitis; the child sits with the neck hyperextended and chin protruded. See Epiglottitis

SNOMED Systemized Nomenclature of Medicine & Veterinary HEALTH INFORMATICS A computerized electronic vocabulary system for medical databases, which may become the standard vocabulary for speech recognition systems and computer-based Pt records; it is the vocabulary of choice for reporting terminology. Cf HL7, LOINC, Read classification system, Read codes

SNOMED International HEALTH INFORMATICS A comprehensive multiaxial nomenclature system created to index the entire medical vocabulary, including signs and Sx, diagnoses, procedures, allowing the integration of all the medical information in an electronic medical record to be integrated into a single data structure

snore Snoring SLEEP DISORDERS A harsh buzzing noise in a sleeper, produced primarily with inspiration during sleep due to vibration of soft palate and pillars of oropharyngeal inlet; snoring ↑ with age; it affects 60% of ♂, 40% of ♀; many snorers have incomplete obstruction of upper airway, and may develop obstructive sleep apnea; it is associated with ↑ risk of HTN, coronary ischemia, CVAs; alcoholism, arthritis, asthma, daytime drowsiness, depression, DM, insomnia, obesity WORLD RECORD A Swede who saws wood at 93 dB MANAGEMENT Isolated snoring needs no treatment; it may be ↓ with a nasal dilator, or surgery to tighten redundant soft palate. See Obstructive sleep anpea syndrome. Cf Sleep disorders, Uvulopalatopharyngoplasty

'snorting' SUBSTANCE ABUSE A popular method for consuming cocaine and opiates–one nostril is held closed, the other inhales pulverized cocaine. See Cocaine, Crack

snow DRUG SLANG A street term for a pulverized substance of abuse which can be snorted, classically, cocaine, but also heroin, amphetamine, oxycodone, etc

snowblindness SPORTS MEDICINE The transient loss of vision caused by UV-light-induced corneal damage CLINICAL Tearing, conjunctival hyperemia, corneal clouding, superficial punctate keratitis TREATMENT Topical anesthetics, homatropine drops, gentamicin, systemic analgesics

snowboarding SPORTS MEDICINE A winter sport in which the rider 'surfs' down a snow-covered mountain with both feet locked onto a flat waxed board while the upper body is freely mobile INJURIES Distal radial Fx occur in 1:4000 SB days, represent 25% of SB injuries–half require surgery, 17% of SB injuries are shoulder Fx; spinal injuries are 3X more common in SB than in skiing. See Intradiscal electrothermal therapy

snowman sign PEDIATRIC CARDIOLOGY A rounded, figure-of-eight-like cardiac contour seen on a plain AP CXR of infants with total anomalous drainage of the pulmonary veins accompanied by common pulmonary vein dilatation and supracardiac drainage; the unique contour is produced by the dilated left ventricle and superior vena cava and left innominate vein; a similar appearance may be caused by a prominent thymic shadow

snowstorm pattern RADIOLOGY A ground-glass appearance in plain CXRs of Pts with severe fat embolism, which corresponds to diffuse pulmonary infiltrates. See Fat embolism

snow white sign GI DISEASE An endoscopic finding seen in hydrogen peroxide–H_2O_2 enteritis, which consists of pure white GI mucosa with frothy bubbles extending over multiple mucosal crests and valleys, analogous to moguls on a ski slope; it was caused by use of H_2O_2 for cleaning endoscopes between use. Cf Hydrogen peroxide enteritis

snuff SUBSTANCE ABUSE A smokeless tobacco consumed by snorting; snuff may be more dangerous than smoking 1 pack of cigarettes/day, and have 2-fold more carcinogens. See Smokeless tobacco, smoking

snuffles PEDIATRICS Noisy breathing through a partially obstructed nasopharynx, accompanied by profuse, mucopurulent and hemorrhagic nasal discharge containing viable *Treponema pallidum*; snuffles may be seen in infants with early congenital syphilis manifest as syphilitic rhinitis; snuffles have also been described with trisomy 21–Down syndrome. See Congenital syphilis

Snugs™ WOUND CARE Tapeless wound care products: leg wrap, arm wrap, foot/plantar wrap, foot glove, mastectomy wrap, abdominal wrap, and hood–for head wounds. See Wound care

SOAP PATIENT RECORDS A standard format for physician charting of Pt exams on a problem-based Pt record; SOAP combines patient complaints and physician determinations. See Hospital chart, Medical record

---

SOAP

SUBJECTIVE DATA–supplied by the Pt or family
OBJECTIVE DATA–physical examination and laboratory data
ASSESSMENT–a summary of significant–if any new data, physician conclusions
PLAN–intended diagnostic or therapeutic action

---

soap bubble An adjective referring to a dilated, smooth-contoured cyst-like or ballooned, occasionally loculated space(s). See Physaliferous BONE RADIOLOGY An expansile, often eccentric, vaguely trabeculated space with a thin, sclerotic, sharply defined margin, which is characteristic of an aneurysmal bone cyst; other bone lesions with SB-like expansions include giant cell tumor, osteosarcoma, solitary bone cyst, non-ossifying fibroma, fibrous dysplasia, metastatic CA, chondromyxoid fibroma, 'combined' stage of Paget's disease of bone, expansile mandibular lesions of cherubism, metastatic renal cell CA; a radiologic finding of multiple 'punched-out' lytic lesions of bone–eg, in myeloma, angiosarcoma, ameloblastoma are described as SB-like GI RADIOLOGY A descriptor for the physaliferous air spaces seen in an abdominal plain film in gas-producing bacterial abscesses NEUROLOGY A descriptor for the LM appearance of *Cryptococcus neoformans* when stained with India ink, which is most prominent in the Virchow-Robin space

SOB Shortness of breath. See Dyspnea

social adaptation PSYCHIATRY The ability to live and express oneself according to social restrictions and cultural demands

social breakdown syndrome PSYCHIATRY A psychiatric Sx complex resulting from suboptimal treatment conditions and inadequate facilities–ie, not integral to the primary illness ETIOLOGY Social labeling, learning the role of chronically sick Pts, atrophy of work and social skills, identification with the sick. See Rehabilitation, Social labeling

social contract MEDICAL PRACTICE The implied understanding between physician and Pt that the former provides the best possible care in a truthful and timely fashion, in exchange for the latter's trust. See Doctor-patient relationship

'social disease' A euphemism for a sexually-transmitted disease, see there

social distance PSYCHOLOGY A zone of space in which most social interactions occur; SDs may be 1. Close–2.5 m–12-25 feet, which corresponds to informal situations, in which one–or more persons are 'in control', as in a teacher talking to students in a classroom, or a manager addressing subordinate and 2. Far–> 8m or >25 feet, which corresponds to 'formal' distances, such as in lectures, political rallies, etc. See Proxemics. Cf Intimate distance, Personal distance, Public distance

social drinker A person who consumes alcoholic beverages in moderation–ie, ≤ 2 'standard drinks'/day, often in a socially acceptable situation. See Alcohol. Cf Binge drinker, Problem drinker

social good MANAGED CARE Any benefit to the general public–eg, teaching and charity care, provided by physicians and health care workers

social history A summary of life-style practices–eg, diet, exercise, sexual orientation and level of sexual activity, occupation, and habits–eg, smoking, abuse of alcohol or other substance, which may have a direct or indirect effect on a person's health. See Psychiatric history. Cf Family history

social inequality in health SOCIAL MEDICINE Any disparity in the risk and incidence of disease and access to health care, based on socioeconomic factors

social isolation PSYCHOLOGY The virtual absence of interaction with others, outside of that required to perform basic life functions–eg, food shopping, transportation, work, and entertainment. See Anaclytic depression, Nuclear family. Cf Companionship, Extended family, Marriage bonus, Most significant other, Pet therapy, Social interaction

social maladjustment PSYCHIATRY An extreme difficulty in dealing appropriately with other people

social medicine A field of medicine that studies the impact of the collective behavior of organized society on individuals belonging to various, often disadvantaged, subgroups within the society. See Engel's phenomenon, Homelessness, Latchkey children, Supermom. Cf Socialized medicine

social phobia Social anxiety disorder PSYCHIATRY '*A marked & persistent fear of social and performance situations in which embarassment may occur...*(and)...*take the form of a situationally bound or ... predisposed panic attack*, while social anxiety is normal in children, in adults this fear is excessive or unreasonable; social or performance situations are avoided, or endured with dread; SP may begin in adolescence and be due to parental overprotectiveness or limited social opportunity MANAGEMENT Psychoanalysis, cognitive behavioral therapy, group therapy, support group; some respond to pharmacology–eg, paroxetine. See Panic attack, Phobia

social power SOCIAL MEDICINE The influence–ie, power that a person has in society and among peers, attributable to expertise, information, or emulation by others. See Decision-making

Social Security Administration GOVERNMENT A federal agency that administers social security and disability benefits. See Medicaid, Medicare

social skill PSYCHOLOGY A level of interpersonal savvy, which often determines future social adjustment and success EXAMPLES Friendliness; positive involvement in group activities; respect for rights, property, and ideas of others; self control; positive self image; appropriateness; obedience of group rules; conforming to social norms.Cf Social stress

social stress PSYCHOLOGY Threatened psychosocial stress which may be acute–eg, forgetting important event, or chronic–eg, chronic alcoholism

Social Support Inventory PSYCHOLOGY A 2-part instrument designed to evaluate a person's network of psychosocial support

social worker A professional, usually with a bachelor's or master's degree in social work, who provides counsel, aid, and other services to persons with domestic issues

socialized medicine A health care system in which 1. The entire population's health care needs are met without charge or at a nominal fee and 2. The organization and provision of all medical services are under direct governmental control. Cf Social medicine

sociopath PSYCHIATRY A popular term for a person with antisocial personality disorder, see there

Socratic method EDUCATION A teaching philosophy that differs from the traditional format as instruction is in the form of problem-solving and testing of hypotheses. See Layer cake education, Spoon feeding

SOD 1. Sphincter of Oddi dysfunction. See Biliary dyskinesia 2. Superoxide dismutase, see there

SODA *Salmonella* Outbreak Detection Algorithm EPIDEMIOLOGY A CDC program that tracks the ±50,000 clinical isolates of *Salmonella* serotyped each yr by state health departments–US

soda loading SPORTS MEDICINE The ingestion of multiple carbonated beverages 2-3 hrs before an athletic event; an ↑ in serum bicarbonate may delay the onset of lactic acidosis

sodium Na⁺ An alkaline metallic element–atomic number, 11; atomic weight 22.99, which is the main extracellular cation; it is a critical electrolyte in body fluid homeostasis, neuromuscular conduction, and enzyme activity REF RANGE 135-145 mmol/L–US: 135-145 mEq/L; urine: 40-220 mEq/24 h PANIC–CRITICAL VALUES Serum < 120 mmol/L; > 160 mmol/L–US: 135-145 mEq/L; urine 43–260 mmol/24 hr

SODIUM

SERUM SODIUM

↓ Burns, drugs–ethacrynic acid, thiazides, mannitol, salt-wasting diuretics–eg, furosemide, prolonged IV therapy, low-sodium diet, nasogastric suctioning, salt-wasting renal disease, tissue injury, SIADH, vomiting

↑ Adrenal hyperfunction, CHF, dehydration, hepatic failure, high-sodium diet, drugs–some antibiotics, corticosteroids, cough medicines, laxatives

URINE SODIUM

↓ Adrenal hyperfunction, CHF, hepatic failure, low-sodium diet, renal failure

↑ Adrenal hypofunction, anterior pituitary–adenohypophysis hypofunction, essential HTN, high-sodium diet

sodium azide NaN₃ MICROBIOLOGY A toxic salt added–concentration, 0.01%, to a transport medium of lab specimens–eg, urine for culturing bacteria, which prevents oxidative phosphorylation and bacterial overgrowth

sodium nitrite FOOD INDUSTRY A preservative and flavor-enhancer added to processed meats–bologna, salami, ham, and hot dogs. See Food additives, Nitrates, Nitrites, Nitrosamines

sodium oxybate Gamma-hydroxybutyrate, Xyrem® NEUROLOGY A neuropharmacologic used to manage narcolepsy and cataplexy, which causes abrupt loss of muscle control. See Date rape drug

sodomy SEXOLOGY 1. Anal intercourse, see there, buggery 2. A term variously applied to zoophilia, to orogenital or anogenital contact between humans, especially ♂. Cf Paraphilia

SofDraw™ syringe NURSING A syringe designed to collect body fluids–eg, for amniocentesis or cyst aspiration with a safety needle that protects against accidental needle pricks See Sharps

Sofsilk SURGERY Nonabsorbable silk suture for soft tissue approximation and/or ligation–eg, cardiovascular, ophthalmic, microsurgery, neural tissue

soft *adjective* 1. That which lacks statistical significance–ie, a statistical 'p' value > 0.05, as in 'soft' data or 'soft' risk factors. Cf Fragile data 2. That which is socially regarded as relatively innocuous, as in 'soft' drugs–eg, nicotine or alcohol where dependence is often considered psychological or 3. That which is not based on objective data, as in the 'soft' sciences of psychology or sociology

soft diet NUTRITION A diet of soft foods–eg, bananas, Jello™, oatmeal, soups, yogurt, that precludes hard foods–eg, carrots or 'chewy'–eg, broccoli, jelly beans, steak. See Diet, TMJ syndrome; Cf Bland diet

soft drug SUBSTANCE ABUSE Any illicit drug popularly perceived to have a relatively low abuse potential; for some, marijuana is viewed as an SD, heroin a a hard drug, and cocaine in the middle. Cf Hard drug

soft food NUTRITION Foods that require a minimal amount of chewing–eg, bananas, Jello™, oatmeal, soups, yogurt. See Bland food

'soft' medical services MANAGED CARE A general term for relatively intangible medical services, which encompass prevention initiatives–eg, anti-smoking programs, screening for CA, HTN, and ↑ cholesterol, domestic violence, teenage pregnancy, women's health, children's health. Cf 'Hard' medical services

'soft' risk factor Any factor that slightly ↑ risk of suffering a morbid process, which does not reach statistical significance

soft tissue A generic term for muscle, fat, fibrous tissue, blood vessels, or other supporting tissue matrix

soft tissue sarcoma ONCOLOGY A sarcoma that arises in muscle, fat, fibrous tissue, blood vessels, or other supporting tissues. See Sarcoma

SOFT TISSUE SARCOMA STAGING

I   A Tumor cells are morphologically similar to normal cells; tumor measures ≤ 5 cm

   B Ditto; tumor ≥ 5 cm

II A Tumor cells are mildly atypical; tumor measures ≤ 5 cm

   B Ditto; tumor ≥ 5 cm

IIIA Tumor cells are moderately or severely atypical/bizarre; tumor measures ≤ 5 cm

   B Ditto; tumor ≥ 5 cm

IVA Any cellular atypia; any sized tumor with lymph node involvement

   B Any cellular atypia; any sized tumor with metastasis

software INFORMATICS A sequence of programmed instructions used to operate a computer and perform specific tasks EXAMPLES Assemblers, compilers, programs, programming languages, routines, translators, and documentation; PC software includes word processors, databases, spreadsheets, graphics, desktop publishers. See Computer, Speech recognition software, Tunneling software. Cf Hardware

soiling Encopresis PEDIATRICS The involuntary passage of stools causing soiling of clothes by a child > age 4. See Encopresis, Fecal incontinence

Solanax Alprazolam, see there

solar *adjective* 1. Pertaining to the sun 2. Pertaining to the solar plexus

solar elastosis Degeneration of subdermal elastic tissue by prolonged sun exposure, causing wrinkling 'sailor skin'–aka, farmer's skin, golfer's skin, which predisposes skin to malignancy–eg, BCC, SCC, melanoma. See Actinic–solar keratosis

Solarase® DERMATOLOGY A topical gel for treating actinic keratosis. See Actinic keratosis

soldier's heart Post traumatic stress disorder, see there

sole *adjective* Only *noun* The bottom of the foot

Solenopsis saevissima Fire ant MEDICAL ENTOMOLOGY An imported–South America–black *S saevissima richteri* or red *S saevissima invicta* fire ant, which may evoke anaphylaxis, painful pustules, death

solid cancer Solid tumor ONCOLOGY A malignancy that forms a discrete tumor mass–eg, CA of brain, breast, prostate, colorectum, kidney; sarcoma; melanoma, in contrast to lymphoproliferative malignancies–eg, leukemia, which may diffusely infiltrate a tissue without forming a mass. See Lymphoproliferative disorder

solid organ transplant IMMUNOLOGY A transplanted solid organ–eg, heart, liver, kidney, as contrasted to 'liquid' transplanted tissues–eg, BM, pancreatic islets. See Transplant, Transplantation

solid tumor of childhood A nonlymphoproliferative–ie, not leukemia, malignancy in children–eg, Ewing sarcoma, glioma, medulloblastoma, neuroblastoma, osteosarcoma, retinoblastoma, rhabdomyosarcomas–alveolar and embryonal types, Wilms' tumor. See Small round cell tumor of childhood

Solis Diazepam, see there

solitary congenital nodular calcification DERMATOLOGY A condition affecting the extremities or head & neck region, in which the epidermis is acanthotic and hyperkeratotic, with calcified subcutaneous bumps of unknown significance

solitary myeloma Solitary plasmacytoma HEMATOLOGY A single focus of neoplastic myeloma cells, which comprises 3-5% of monoclonal gammopathies. See Monoclonal gammopathy, Myeloma

solitary rectal ulcer syndrome A disease of young adults, more common in ♀ with irregular bowel habits ETIOLOGY Idiopathic, linked to laxative abuse, anal intercourse CLINICAL Hematochezia, ± accompanied by anal or abdominal pain, passage of mucus per rectum, excess straining on defecation; ulcers ±2 cm in diameter, close to anal verge TREATMENT Stool softeners; when accompanied by rectal prolapse, rectopexy may be required

solitary sex A sexual act by a single person, usually private, often understood to mean masturbation. Cf Consensual sex

solitary thyroid nodule A discrete enlargement of an otherwise normal thyroid gland EPIDEMIOLOGY ♀:♂ 4:1; most are incidental findings during autopsy or surgical exploration for other reasons DIFFDx Colloid–adenomatous nodule–42-77% of STNs, follicular adenoma–15-40%, CA–8-17% DIAGNOSIS FNA biopsy, radionuclide scanning–^{131}I, ^{99m}Tc, ultrasonography MANAGEMENT Malignant and indeterminant lesions are treated by surgery

Solium Chlordiazepoxide, see there

solo practice Medical practice by a single physician–a solo practioner, usually understood to mean a nonspecialist. See Private practice; Cf Group practice

Solus® CARDIOLOGY An ASIR, single-chamber, rate-modulated pulse generator. See Pacemaker

SOLVD CARDIOLOGY A series of clinical trials–Studies of Left Ventricular Dysfunction that evaluated the effect of antihypertensives–eg, with enalapril, an ACE inhibitor, on M&M in Pts with CHF. See ACE inhibitor, Congestive heart failure Consensus II, Enalapril

SOLVD-Prevention Study CARDIOLOGY A clinical trial–Studies of Left Ventricular Dysfunction that evaluated the effect of enalapril, an ACE inhibitor on heart failure in Pts with ↓ left ventricular ejection fraction. See Congestive heart failure, Enalapril

SOLVD-Treatment Study CARDIOLOGY A clinical trial–Studies of Left Ventricular Dysfunction that evaluated the effect of enalapril, an ACE inhibitor on M&M in Pts with severe CHF and ↓ left ventricular ejection fraction. See Congestive heart failure, Enalapril

solve rate FORENSIC MEDICINE The percentage of homicides in which the killer is identified with reasonable certainty. See Homicide, Manslaughter, Serial killer

solvent abuse SUBSTANCE ABUSE The recreational inhalation of chemical solvents in model glue, paint thinner, 'white-out', nail polish remover, etc for psychotropic effects EPIDEMIOLOGY Adolescents–up to 15% have experimented with solvents, especially from lower socioeconomic strata with less parental supervision; adolescents with Hx of SA are at ↑ risk for IV drug abuse ACTIVE INGREDIENTS Acetone, benzene, carbon tetrachloride, naphtha, toluene, xylene CLINICAL At low concentrations, solvents induce euphoria; at higher levels, tinnitus, diplopia, hallucinations, confusion ensue; further

↑ in serum levels are accompanied by incoordination, ataxia, slurred speech, hyperreflexia, nystagmus, unconsciousness; long term SA may cause hematopoietic–eg, aplastic anemia and liver disease–eg, hepatocellular necrosis, liver CA, renal tubule acidosis, pulmonary, and CNS dysfunction–which may cause cerebellar degeneration, cognitive impairment, dementia, distal sensory polyneuropathy, etc. See Gateway drugs, Soft drugs, White-out

SOM 1. School of medicine 2. Secretory otitis media 3. Somnolent, Latin *somnus* 4. Superior oblique muscle

somatic *adjective* Pertaining to 1. The body 2. Not the viscera

somatic dysfunction OSTEOPATHY A defect in structure and/or function, which can be diagnosed by identifying tenderness, asymmetry, restricted motion, and tissue texture changes. See Osteopathy

somatic pain NEUROLOGY Pain arising in nerve endings of muscles, skin, bones; it is highly localizable–the "trademark" indicator of SP is the ability to localize it with "pin point" or fingerpoint precision; Pts describe SP as aching, gnawing EXAMPLES Bone Fx, wounds, large bruises, bone metastases. See Pain, Pain management

somatization PSYCHOLOGY A multifactorial tendency to experience and report somatic Sx with no pathophysiologic cause

somatization disorder PSYCHIATRY A chronic disorder that begins before age 30, more in ♀, characterized by multiple and recurrent Sx and complaints of various body systems, for which the Pt regularly seeks medical attention for unexplained somatic complaints; Sx suggesting organ dysfunction(s) but not supported by lab data or clinical findings, are often vague, don't appear to be voluntary and may fulfill a psychological need CLINICAL Sx are severe enough to lead to visits to health professionals, require medication, or interfere with living; Sx begin or worsen after stress–eg, loss of job, close relative, or friend; Pts believe themselves to be sickly, have pseudoneuralgias–visual defects, dysphagia, voice loss, urinary retention, convulsions, seizures, GI Sx–colicky pain, N&V, dysmenorrhea, ↓ libido, pain–low back, genitalia, joints, cardiopulmonary Sx–eg, dyspnea, palpitations, chest pain; SD results from psychological, behavioral, interpersonal, and social factors, making treatment complex and multimodal. Cf Factitious diseases, Hypochondriasis

somatostatinoma Delta cell tumor A somatostatin-producing tumor of pancreatic islet delta cells, characterized by ↓ gastric acid secretion and metastasis to liver and bone; it is associated with DM, cholelithiasis, steatorrhea, and hypochlorhydria; it is most common in the head and tail of the pancreas, but may rarely arise in the duodenum

somatostatinoma syndrome ONCOLOGY A paraneoplastic syndrome caused by ectopic somatostatin secretion CLINICAL Vomiting, hypochlorhydria, abdominal pain, diarrhea, malabsorption, weight loss, cholelithiasis, ↓ glucose control, causing 2° DM LAB ↑ somatostatin–an inhibitory hormone, evoking reactive ↓ insulin, ↓ glucagon, ↓ gastrin, steatorrhea, anemia

somatostatinoma triad A trilogy of Sx–gallstones, DM, diarrhea, typical of Pts with somatostatinoma

somatotropin Growth hormone, see there

somatotropin release-inhibiting hormone A hormone produced and released from the anterior hypothalamus in response to various stimulants–eg, high circulating glucocorticosteroids, which inhibits the release of growth hormone and TSH

Soma® Carisoprodol and meprobamate, see there

Somese® Triazolam, see there

somesthetics SEX THERAPY A treatment involving skin sensation as in touch, pressure–massage, hot/cold, wet/day, sensuous body-contact grooming

somnambulism Sleepwalking, see there

Somniton Triazolam, see there

Somnium See Lorazepam

somnolence MEDTALK → VOX POPULI Sleepiness, see there

somnophilia Sleeping princess syndrome PSYCHOLOGY A predatory paraphilia in which sexuoeroticism hinges on intruding and awakening a sleeping stranger with erotic caresses–eg, oral sex, without force or violence. Cf Rape

somnophilic *adjective* Referring to somnophilia, see there

Somnoplasty℠ system SLEEP DISORDERS A system based on the use of radiofrequency to ablate obstructive tissue in the upper airway INDICATIONS Habitual snoring, chronic nasal obstruction, obstructive sleep apnea See Sleep apnea

Somogyi effect Rebound hyperglycemia A phenomenon described in diabetics in whom hyperglycemia is a counter-regulatory overcompensation to nocturnal hypoglycemia. See Dawn phenomenon, Glucose tolerance test, Subcutaneous insulin-resistance syndrome

Sonablate® UROLOGY A device that uses high-intensity focused ultrasound–HIFU technology, combining focused ultrasound beams and imaging, to heat and destroy targeted tissue without affecting surrounding healthy tissue INDICATIONS Treatment of BPH. See Benign prostatic hypertrophy

Sonacon Diazepam, see there

Sonata® Zaleplon NEUROLOGY A hypnotic GABAA–gamma-aminobutyric acid type A benzodiazepine receptor agonist for insomnia ADVERSE EFFECTS Headache, dizziness, somnolence. See Insomnia

Songar Triazolam, see there

sonic fracture healing system ORTHOPEDICS A device that emits sound waves to a Fx site to accelerate healing

sonicated albumin CARDIOLOGY An echocardiography contrast agent–eg, Albunex®, Optison® administered in an IV bolus to enhance visualization of the left ventricule and endocardium. See Albunex®, Optison®

sonogram Ultrasonogram. See Ultrasonography

Sonopsy™ system SURGERY A system that compresses and stabilizes the breast for optimizing biopsies and 3-D ultrasound imaging. See Breast biopsy

sonorous rhonchus Rhonchus, see there

SonoRx® Simethicone-coated cellulose suspension IMAGING A gas-shadowing reduction agent used to enhance delineation of upper abdominal anatomy with ultrasound imaging

Sonotron PAIN MANAGEMENT A non-FDA approved modulated radiofrequency device for noninvasive treatment of pain and edema. See Pain management

Sontac® IMAGING An ultrasound-conducting gel pad. See Ultrasonography

SOP 1. Standard operating plan/procedure, see there 2. Surgical outpatient

sorbitol A polyhydroxyl alcohol or polyol synthesized from glucose by aldose reductase in neural tissue, produced in excess in DM; sorbitol may be further metabolized to fructose, which together cause ↑ osmotic pressure, intracellular edema, Schwann cell swelling, anoxia and nerve demyelination; it has been implicated in diabetic neuropathy; it has been used as an artificial sweetener. Cf Advanced glycosylation endproducts

sore An external skin lesion. See Canker sore, Cold sore, Oriental sore

Sorsby syndrome An AD condition characterized by bilateral macular coloboma and apical dystrophy of the hands and feet, which may be accompanied by hearing less, uterine anomalies, a single kidney

'sort of' test LAB MEDICINE A popular term for a test with a low level of sensitivity that 'sort of' (possibly; with broad caveats) suggests a particular diagnostic entity. Cf Confirmatory test

SOS® Segmented orthopedic system ORTHOPEDICS A modular proximal and distal femur implant system indicated for limb salvage in Pts with bone loss due to CA, trauma or failed implants

sotalol Betapace®, Sotacor® CARDIOLOGY A class II beta blocker that ↑ myocardial action potential duration, ↑ cardiac refractory period, lengthens QT interval; sotalol is used for conduction block, A Fib, torsade de pointes, angina, HTN, suppress PVCs, ventricular tachyarrhythmias ADVERSE EFFECTS Proarrhythmia due to ↑ β blockade; it may aggravate pre-existing arrhythmias–eg, sinus bradycardia, provoke new arrhythmias–eg, torsade de points–±2.5%, exacerbate CHF, hypotension, bronchospasm, fatigue, dizziness, dyspnea, headache. See β-blocker, SWORD

Sotos syndrome Cerebral gigantism An AD condition with rapid early somatic growth of neonatal onset, that is not accompanied by endocrinopathy, accelerated bone maturation, or precocious puberty CLINICAL Macrosomia, where growth is accelerated for the first 4-5 yrs of life, followed by a normal growth pattern, enlarged acral parts–hands and feet, thickened subcutaneous tissues, macrocephaly, mild mental retardation and perceptual defects; EEG abnormalities, dilated ventricles, dolichocephaly, antimongolic slanting of palpebral fissures, macrognathy, hypertelorism, convulsions, poor coordination, ↑ risk for CA–eg, Wilms' tumor, liver, parotid, ovarian CAs

souffle A soft blowing sound due to the flow of blood through various vessels, as heard by auscultation. See Mammary souffle

sound *noun* PHYSICS Electromechanical energy sensed by the auditory apparatus. See Bowel sound, Dive bomber sound, First heart sound, Fourth heart sound, Korotkov sound, Second heart sound, Third heart sound SURGERY A simple device used to assess the length of a nonpregnant uterus before performing an endometrial Bx. See Probe

sound generator Artificial sound generator ENT A device used to produced an artificial voice, as required by Pts with laryngectomies for cancer

sound vocalization NEUROLOGY The ability to produce sounds with the voice. See Voice

sounds of Korotkoff Korotkoff sounds, see there

SoundScan™ OSTEOPOROSIS An ultrasound used to assess bone density and thickness of cortical bone. See Bone densitometry

soup-to-nuts *adjective* Americanism referring to an organization or system that provides a full range of products or services

source VOX POPULI The point from which a thing originates, arises, emanates, springs forth, hell, you get the picture. See Carbon source, Knowledge source

source amnesia PSYCHOLOGY The inability to recall the origin of the memory of a given event. See Jamais vu, False memory, Repressed memory. Cf Déjà vu

'South American' operation SURGICAL ONCOLOGY A radical operation for 'frozen' pelvis, which consists of en bloc resection of the uterus and rectum. See Frozen pelvis. Cf 'All-American' and 'North American' operations

Southern blot hybridization Southern blotting MOLECULAR BIOLOGY A method delineated by EM Southern for detecting and manipulating specific DNA sequences previously separated by gel electrophoresis. See Autoradiography, Base sequence, Complementary sequence, DNA, Electrophoresis, Probe. Cf Northern blot, Western blot

sp gr Specific gravity, see there

space VOX POPULI A limited or confined area. See Confined space, Cyberspace, Danger space, Dead space, Deep fascial space, Fuzzy space, File space, Intercostal space, Interdigital space, Paraglottic space, Preepiglottic space, Reinke space, Sample space

space adaptation syndrome Space sickness The constellation of effects of space travel on human physiology, especially, motion sickness CLINICAL Vertigo, N&V, GI dysmotility, malaise, diaphoresis, sialorrhea, yawning, anorexia, hyperventilation–resulting in hypocapnia with vasodilation of lower extremities and pooling of blood, causing postural hypotension and syncope; prolonged space flight is associated with osteoporosis, disuse muscle atrophy, growth of 5+ cm in space, related to low gravity, and may be punctuated by inconvenience–eg, waste management systems fail on most space flights. See Motion sickness

space medicine A branch of aerospace medicine that evaluates the changes imposed on man by travel through and beyond the earth's atmosphere and flight in space ISSUES Weightlessness, space adaptation syndrome, cardiovascular deconditioning, radiation exposure, isolation, ↓ RBC mass, bone mineral loss. See Space adaptation syndrome

space-occupying lesion HEMATOLOGY A lesion that replaces BM, alters its architecture and/or depresses marrow production, resulting in myelophthisic anemia; most SOLs are due to metastases from 1° CAs of lung, breast, prostate, thyroid, adrenals, but may be due to myeloma, leukemia, lymphoma, osteosclerosis, myelofibrosis; all 3 blood cell lines are affected; immature RBCs and WBCs appear in peripheral circulation, due to irritation phenomena. See Myelofibrosis. Myelophthisic anemia

space sickness Space adaptation syndrome, see there

space suit OCCUPATIONAL MEDICINE A popular term for a fluid-resistant disposable garment that covers the arms and legs, worn with gloves–doubled, shoe and head coverings, and face mask and/or shield. See Personal protective garment

spaces, confined, fear of PSYCHOLOGY Claustrophobia. See Phobia

spaces, open, fear of PSYCHOLOGY Agoraphobia. See Phobia

spade deformity CARDIOLOGY A finding by contrast ventricular angiography, likened to a playing card spade, in focal concentric left ventricular hypertrophy confined to the ventricular apex, seen at end-diastole in the right anterior oblique ventriculogram and in 2-D echocardiography; the deformity may be accompanied by 'giant' negative T waves on EKG

spade field Spade port RADIATION ONCOLOGY A field of RT that encompasses the splenic pedicle, para-aortic and common iliac lymph nodes; double thickness lead shields are placed over gonads and pelvic structures–bladder, rectum, uterus. See Radiation therapy

SPAF VASCULAR DISEASE A clinical trial–Stroke Prevention in Atrial Fibrillation–comparing low-dose aspirin vs anticoagulation with warfarin in preventing stroke and systemic embolism–the 1° events in 150 Pts with A Fib. See Atrial fibrillation, Low-dose aspirin. Cf AFASAK, BAATAF

SPAF-2 CARDIOLOGY A randomized, multicenter study that compared warfarin and aspirin for preventing stroke in Pts with nonrheumatic A Fib–NAF after a TIA or minor stroke–and optimal oral anticoagulation in Pts with NAF and recent cerebral ischemia. See Atrial fibrillation, Low-dose warfarin. Cf BAATAF, EAFT, SPAF

spam ONLINE Unsolicited, commercial bulk–junk E-mail. See Mail list, USENET

Spanish ETHNICITY The language spoken by 40 million US inhabitants; Hispanics have surpassed African Americans as the largest minority in the US Primary care providers, especially those practicing in large urban areas, are increasingly obliged to become passably fluent in Spanish to be able to best care for their Pts. See Hispanic, Medical Spanish

spanking PEDIATRICS Corporal punishment, usually of children, in which the buttocks, are pummeled, swatted, or otherwise struck. See Corporal punishment SEXOLOGY Slapping, usually of the buttocks as a part of sexuoerotic activity. See Sadomasochism

SPARCL NEUROLOGY A clinical trial–Stroke Prevention by Aggressive Reduction in Cholesterol Levels

sparfloxacin Zagam® INFECTIOUS DISEASE An advanced, broad-spectrum–gram-negative bacilli, staphylococci fluoroquinolone with improved activity against streptococci and anaerobes, TB, MAC INDICATIONS Community-acquired pneumonia, acute bacterial exacerbations of chronic bronchitis ADVERSE EFFECTS Photosensitivity, GI upset. See Fluoroquinolone

spasm NEUROLOGY An abrupt, violent involuntary contraction of a muscle or group of muscles. See Blepharospasm spasm, Bronchospasm, Carpopedal spasm, Coronary artery spasm, Esophageal spasm, Vascular spasm

spasmodic dysphonia Laryngeal dystonia, spastic dysphonia NEUROLOGY A voice disorder characterized by spasmodic contraction of laryngeal muscles, which chokes off words as uttered, resulting in strained and strangled speech with breaks in rhythm; SD may be accompanied by other dystonias–eg, blepharospasm, oromandibular dystonia, torticollis, writers' cramp MANAGEMENT Sectioning of recurrent laryngeal nerve may be complicated by late failure; botulinum toxin injection into laryngeal muscles may be preferred

spasmodic torticollis Wry neck, see there, aka cervical dystonia

spasmus nutans NEUROLOGY A self-limiting acquired nystagmus characterized by a triad: horizontal or pendular nystagmus, head nodding, head tilting; SN may last months to yrs and be associated with optic gliomas and other brain tumors

spastic colon Irritable bowel syndrome, see there

spastic diplegia A feature of cerebral palsy, which affects both legs, often unequally, characterized by hip flexion and internal rotation, due to the overactivity of the iliopsoas, rectus femoris, hip adductors; knee extension, due to overactivity of hamstrings, especially medially, equinus deformity of foot with rearfoot eversion, due to overactivity of triceps surae and peronei. See Cerebral palsy

spastic paralysis NEUROLOGY Paralysis characterized by spasms and ↑ tendon reflexes of the muscle(s) in the paralyzed region, due to upper motor neuron disease. See Spasm. Cf Flaccid paralysis

spasticity NEUROLOGY A velocity-dependent ↑ in tonic stretch reflexes–involuntary muscle contraction, most common in Pts with spinal cord lesions above the conus medullarisis, developing months after spinal cord injury MANAGEMENT Baclofen, which potentiates GABA's inhibitory effect on reflexes

spatial agnosia NEUROLOGY Inability to recognize spatial relations; disordered spatial orientation

spatial resolution IMAGING Resolution in lateral direction of–ie, perpendicular to–ultrasound propagation in pulse-echo ultrasonography; SR reflects frequency of the ultrasonic pulse. Cf Axial resolution

spear tackling SPORTS MEDICINE A tackle in football in which a player's entire body is launched, spear-like against an opponent, resulting in sig-

nificant axial loading on the cervical spine, as the head is the 'spear point' of contact ADVERSE EFFECTS Transient neurologic episodes, slight torticollis, ↓ cervical ROM; if condition is chronic, atrophy or ↓ cervical paravertebral muscle bulk; permanent paralysis. See Spear tackler spine

spearing SPORTS MEDICINE–ICE HOCKEY A penalty which occurs when a player illegally jabs, or attempts to jab, his stick blade into another player's body. See Sports medicine

special need SOCIAL MEDICINE A condition or characteristic that makes a child difficult to place by adoption services; SNs may be unrelated to the health or temperament of the child

special test Specialized test LAB MEDICINE Any test not performed routinely in a diagnostic lab; STs may be sent to reference labs or performed 'in-house' on an 'as needed' basis. Cf Routine test

specialist MEDTALK A person with a recognized expertise in something; in the usual medical context, a specialist is board-eligible or board-certified in a recognized area of medicine–eg, pathology, psychiatry, radiology, who has undergone a formal residency training program of ≥ 3 yrs, and is entitled to sit for the closure–'specialty board' examination in that field. See Addiction specialist, Adolescent medicine specialist, cLinical nurse, Expert, Grants management specialist, Lab support specialist, State fluoridation specialist, Transfusion medicine specialist. Cf Primary care, Subspecialty

specialization GRADUATE EDUCATION A period of acquisition of special skills or specialty knowledge, usually in a formal educational setting, which is closed with an examination by a specialty board and a certificate of special expertise–a process known as board certification–awarded. See Board certification, Subspecialty

specialized laboratory A type of reference lab dedicated to a particular type of 'esoteric' testing–eg, allergy, coagulation, drugs–especially drugs of abuse, endocrinology, genetics, paternity, virology, etc. Cf Reference laboratory

specialoid MEDICAL PRACTICE A popular term for a physician trained in a specialty for which there are more specialists than Pts with the special problems the specialty was created to address

specialty MEDTALK A specialized area of medical or surgical practice. See Controllable life-style specialty, Medical specialty, Noncontrollable life-style specialty, Organ-based specialty, Subspecialty, Surgical specialty, Unrecognized specialty. Cf General practice

specialty board GRADUATE EDUCATION An organization that certifies, through standardized examinations, that a person has the knowledge to practice a chosen specialty. See Board certification, Peer review, Residency. Cf State board

specialty network Single specialty network MANAGED CARE A loosely cohesive group of physicians specialized in one area of medicine–eg, cardiology, oncology, Ob/Gyn, ophthalmology, radiology, etc, who form a network to attempt to capture a segment of a Pt population covered by HMO contracts

species-ism POPULAR SCIENCE A neologism coined by animal rights activists, who view use of animals in research as a crime. See Animal rights activism, Animal Welfare Act

Specific Activity Scale CARDIOLOGY A 4-point–1–best, 4–worst instrument used to assess disease-specific cardiovascular functional status

specific gravity LAB MEDICINE A measure of the solutes in a fluid; SG is detected in the urine, based on the response of a polymeric acid to surrounding ions. See Dipstick. Cf Nitrites NEPHROLOGY The SG reflects the kidney's ability to concentrate urine: if a random urine specimen has an SG > 1.023, the kidney's ability to concentrate is assumed to be normal ; SG ↑ in SIADH, uncontrolled DM, proteinuria, eclampsia and obstructive uropathy and ↓ in renal tubular damage, chronic renal insufficiency, diabetes insipidus, malignant HTN OCCUPATIONAL SAFETY A physical parameter of a liquid that indicates how heavy it is in relationship to air–≤ 1.0 = lighter than water; ≥ 1.0 = heavier than water, a datum of interest to OSHA, which requires listing of SGs in its Materials Safety Data Sheets

specific phobia PSYCHOLOGY A persistent, irrational fear of an object, activity or situation that compels a person to avoid it, evoking distress and functional impairment, disproportionate to the actual threat of a feared object or situation EXAMPLES Animals, insects, heights, lightning, flying, most common in children; SPs are not generally associated with mental disorders. See Phobia

specificity EPIDEMIOLOGY The proportion of persons without disease who are correctly identified by a screening test or case definition as not having disease IMMUNOLOGY The avidity of an antibody for an antigen PHYSIOLOGY The degree of a ligand's affinity for a receptor

specificity spillover syndrome ENDOCRINOLOGY A condition caused by low-affinity binding of a peptide hormone to a receptor other than its own, resulting in low levels of hormone signal transduction–stimulation

specimen A small sample of something–cells, organ, organism, plasma, tissue, *whatever*, that represents a whole, from which a diagnosis is rendered or other determination of said object's nature is made. See Fasting specimen, Fingerstick specimen, Frozen specimen, Rejected specimen

specimen radiograph An x-ray of tissue–usually from the breast, obtained from surgery, which helps identify lesions seen by mammography. See Breast biopsy, Mammography, Microcalcifications

SPECT Single photon emission computed tomography IMAGING A non-invasive technique for constructing cross-sectional images of radiotracer distribution, to evaluate CNS disease–acute ischemic episodes, epileptic foci, vascular and Alzheimer's dementia, myocardial perfusion, and detect subtle changes in bone metabolism. See Dipyridamole-thallium SPECT. Cf Computed tomography, Magnetic resonance imaging

spectacles OPTHALMOLOGY Glasses, gafas, occhiali, lunettes VOX POPULI A scene (spectacle)

Spectra Doxepin, see there

spectrophotometry LAB MEDICINE The use of a spectrophotometer to evaluate a fluid for a molecule of interest, based on the molecule's ability to alter the transmission of light at specific wavelengths. See Fluorescence spectrophotometry

speculum SURGERY An instrument used to widen a body opening or surgical field to ↑ visibility

speech The production of definite vocal sounds to form words to express thoughts and ideas. See Cricumstantial speech, Cued speech, Esophageal speech, Fragmented speech, Geek speech, Hot potato speech, Motor speech, Open set speech, Pressured speech, Private speech, Posthetic speech, Scanning speech, Tangential speech

speech disorder Articulation deficiency, dysfluency, speech disturbance, voice disorder AUDIOLOGY Any of a group of defects in speech involving abnormal pitch, loudness, or quality of sound produced by the larynx NEUROLOGY A disorder of impaired or ineffective verbal communication not attributed to faulty innervation of speech muscles or organs of articulation–eg, language and learning disabilities ETIOLOGY Neural damage, paralysis, structural defects, hysteria, mental retardation, hearing impairment, ADD, learning disabilities, autism, schizophrenia, cerebral palsy, cleft palate, vocal cord injury, disorders of palate, cri-du-chat syndrome, Tourette syndrome See Amimia, Dyslexia. Cf Agraphia, Aphasia, Apraxia

speech dyspraxia AUDIOLOGY The partial inability to consistently pronounce words by a person with normal tone and coordination of speech muscles. See Speech pathology

speech pathology A field of allied health care that evaluates abnormalities of language, speech, and voice, which may be developmental or acquired

speech-preserving operation Larynx preservation ENT A type of conservative operation for CA of the larynx, which allows a Pt to retain speaking voice. See Sound generator

speech processor AUDIOLOGY A part of a cochlear implant that converts speech sound into electrical impulses to stimulate the auditory nerve, allowing a person to understand sound and speech

speech recognition Speech understanding, voice recognition INFORMATICS The ability of a computer, through software, to accept spoken words as dictation or follow voice commands

speech therapist Speech pathologist, speech/language therapist A health professional trained to evaluate and treat voice, speech, language, or swallowing disorders–eg, hearing impairment, that affect communication. See Speech pathology

speech threshold audiometry AUDIOLOGY A type of hearing test in which a person's ability to discriminate recorded speech, most in the form of 'spondee' words, is transmitted through earphones. See Spondee word. Cf Pure tone audiometry

SPEED CARDIOLOGY A clinical trial–Strategies for Patency Enhancement in the Emergency Department

speed DRUG SLANG A street synonym for amphetamine, which may also refer to methamphetamine and crack

spell *noun* A period during which a person is in a particular state–eg, spell of hospitalization–hospital stay is preferred, spell–bout or period of sickness. See Blue spell, Dizzy spell MEDTALK A '... *sudden onset of a symptom(s) that is recurrent, self-limited, stereotypic...*' TYPES Endocrine–eg, hypoglycemia, thyrotoxicosis, carbohydrate intolerance; cardiovascular–essential HTN, angina, pulmonary edema, psychologic–eg, panic and anxiety disorders; hyperventilation; pharmacologic–eg, MAOI therapy, cheese, illicit drugs; neurologic–seizure disorders, migraine; etc–eg, mastocytosis, carcinoid, P vera, POEM syndrome CLINICAL Facial flushing attributed to vasodilation, accompanied by one of various spell phenotypes–eg, pheochromocytoma, carcinoid syndrome, or mast cell disease, manifest as diaphoresis, numbness, SOB, headaches, chest tightness, ↑ BP, etc PSYCHIATRY A trance-like state in which a person allegedly communicates with dead persons, or various–non-mineral, non-grain spirits, usually in a culture-specific context, most common among African-Americans and/or those from the southern US; thé importance lies in it being confused with a psychotic episode. See Culture-bound syndrome

SPEP Serum protein electrophoresis, see there

sperm analysis See Semen analysis

sperm bank REPRODUCTION MEDICINE A registered tissue bank that collects, stores, tests, and sells frozen sperm to be used for artificial insemination. See Artificial insemination

sperm banking REPRODUCTION MEDICINE Freezing of sperm before CA treatment for future fathering/fertility furthering

sperm count UROLOGY A measure of the concentration of sperm in semen NORMAL ±100 million/mL. See Post-vasectomy sperm count, Semen analysis

spermatocele UROLOGY A cyst-like mass within the scrotum that contains fluid and dead sperm

spermatogenesis PHYSIOLOGY The production of male sex gametes; the maturational changes that transform spermatids into spermatozoa. See Sperm

spermatorrhea Wet dream SEXOLOGY Ejaculation while asleep, often during an erotic dream; semen seeping while sleeping

spermicide Spermaticide A contraceptive agent with a high failure rate–11.9 pregnancies/100 woman-yrs FORMULATIONS Foams, creams, sponges; most contain a surfactant nonoxynol 9, an agent that ↓ risk of STDs–it is bactericidal and viricidal; spermicides are not associated with teratogenesis or trisomies, although some 'soft' data suggest possible limb reduction defects. See Contraceptives, IUDs, Litogens, Pearl index, RU 486

SPEX Special Licensing Examination MEDTALK An examination administered to ±1500 physicians/yr who seek relicensure and/or want to practice in a different state yrs after initial licensure

SPF 1. S-phase fraction, see there 2. Standard perfusion fluid 3. Sun protection factor A scale for rating sunscreens; sunscreens with an SPF of 15 or higher provide the best protection from the sun's harmful rays. See Malignant melanoma, Sunblock, Sunscreen, Tanning salons, Ultraviolet radiation

sphenoethmoidectomy ENT The removal of the ethmoid labyrinth in Pts with chronic hyperplastic rhinosinusitis–CHR

spherocytosis Congenital spherocytic anemia, congenital spherocytosis, hereditary spherocytosis, spherocytic anemia HEMATOLOGY A rare–1:5000 AD condition characterized by chronic hemolytic anemia with ↑ osmotic fragility and autohemolysis of globose RBC due to various defects in RBC membrane proteins CLINICAL Infants may be jaundiced; other Sx may be seen in older Pts fatigue, weakness, SOB, anemia, intermittent jaundice, splenomegaly, gallstones, leg ulcers MANAGEMENT Splenectomy. Cf Elliptocytosis

sphingolipidoses A group of inborn errors of sphingolipid metabolism in which lysosphingolipids accumulate, inhibiting protein kinase C activity in signal transduction, cell differentiation and in tumor promotion. See Ceramide lactoside lipidosis, Fabry's disease, Fucosidosis, Gaucher disease, Gangliosidosis, Globoid cell dystrophy, Krabbe's disease, Metachromatic leukodystrophy, Niemann-Pick disease, Sanhoff's disease, Tay-Sachs disease

sphingomyelin lipidosis Niemann-Pick disease, see there

sphygmomanometer Blood pressure cuff CARDIOLOGY A device used to measure arterial BP, which consists of an inflatable cuff, usually with a Velcro™ closure, a rubber inflating bulb, and a gauge to measure systolic and diastolic pressures. See Hypertension, Small cuff syndrome, White coat HTN

SPI-77 ONCOLOGY A liposomal formulation of cisplatin used to manage advanced CA unresponsive to other treatment. See Cisplatin

SPICA cast ORTHOPEDIC SURGERY A body cast that fits over both legs, encasing the lower body from the nipple line down

SPICE AIDS Ongoing clinical trials involving use of a formulation of saquinavir, with other nucleoside analogues and/or treatment regimens in HIV/AIDS Pts. See AIDS, Saquinavir CARDIOLOGY A clinical trial–Study of Protease Inhibitor Combination in Europe

spicy food NUTRITION Any comestible marinated in and/or which contains chili peppers, mustard with horseradish, curry or other spices that evoke a desired intraoral sensation that crosses pain with pleasure; SFs may elicit an autonomic nervous system response–eg, diaphoresis, arrhythmias. See Capsaicin, Scoville units, Seder syncope, Sushi syncope

spider DERMATOLOGY See Spider angioma ENTOMOLOGY A chelicerate arthropod of the class Arachnida, which has 8 legs, a cephalothorax, a smooth, round abdomen, and equipment for spinning webs; 2 spiders are of medical importance in the US: *Lactrodectus mactans*, the black widow spider and *Loxosceles reclusus*, the North American brown recluse spider. See Black widow spider, Brown recluse spider

spider angioma DERMATOLOGY A superficial spider-like cluster of capillaries composed of a central 'feeder' vessel and multiple minute tortuous and dilated radiating vessels with a peripheral erythema; while classically due to ↑ circulating estrogens–eg, in pregnancy and alcoholic cirrhosis, SAs may occur in chronic hepatic congestion 2° to constrictive pericarditis and may be a normal birthmark in children. See White spider

'spider' dystrophy Macroreticular dystrophy A branching arachnoid pigmentary pattern seen in an AR form of retinal pigment dystrophy of the epithelium, appearing as bilateral, symmetrical lesions that do not affect vision

spiders, fear of PSYCHOLOGY Arachnophobia. See Phobia

Spielberger scale Anger scale, hostility scale, Spielberger Anger Expression Scale PSYCHOLOGY A device that measures a Pt's levels of hostility–anger suppression and expression

spike *noun* DRUG SLANG A needle for injecting a drug *verb* To inject a drug ENDOCRINOLOGY See Insulin spike INFECTIOUS DISEASE A popular term for a sharply defined febrile peak NEUROLOGY A sharply defined depolarization on EEG VIROLOGY A projection on the surface of the virus seen by EM, corresponding to hemagglutinin or neuraminidase on the coat of a virus *verb* CLINICAL MEDICINE To place vascular access tubing into the appropriate port in an infusion bag–eg, containing Ringer's solution INFECTIOUS DISEASE To develop a sharp febrile peak

spike & dome contour CARDIOLOGY A descriptor for the carotid arterial pulse curve seen in hypertrophic cardiomyopathy–HC, which rises abruptly and falls during midsystole–spike, later rising a 2nd time at a slower rate during late systole–dome; the jugular pulse in HC is characterized by a prominent a–atrial wave. See Hypertrophic cardiomyopathy, Spade deformity

spike potential NEUROLOGY A single rapid, 100–200 msec voltage 'transient' that occurs spontaneously in electromyography in smooth muscle cells, when they have resting membrane potentials above the SP threshold, as occurs in the lower esophageal sphincter. See M spike

spike & wave pattern NEUROLOGY An EEG pattern seen in an absence, formerly petit mal epilepsy, occurring as symmetric and synchronous, ≥ 3 discharges/sec with an abrupt beginning and end. See Seizure

spiking fever Hectic fever INFECTIOUS DISEASE A highly nonspecific term for either a fever characterized by a daily spike in temperature, or one in which the peak and trough temperatures differ by 1.4°C; the term may be so clinically meaningless as to completely abandon it. See Fever

spillover infection An infection that occurs when a relatively confined infection–eg, meningitis, pneumonia, deep-seated abscesses of liver or kidneys 'breaks into' the vascular space–due to enzymes released by PMNs, or other artifice–resulting in bacteremia

spina bifida NEUROLOGY A group of birth defects caused by a defective fusion of the vertebral arch and one or more contiguous vertebrae, leaving a variably sized opening in the spinal canal; in SB, the spinal meninges and subcutaneous tissue may herniate through the defect, but remain covered by skin and subcutaneous tissue; protrusion of the spinal cord and meninges damages the spinal cord and nerve roots, compromising function at or below the defect CLINICAL Related to anatomic level of defect, usually in the lower lumbar or sacral region–partial or complete paralysis of legs, partial or complete lack of sensation, loss of bladder or bowel control; exposed spinal cord is susceptible to infection–meningitis CONCOMITANT CONGENITAL DEFECTS Hydrocephalus–in 90% of children with myelomeningocele, syringomyelia, hip dislocation LAB ↑ AFP; occult SB affects up to 10% of adults seen in 3° care centers. See Alpha-fetoprotein, Multivitamins, Myelomeningocele

spina bifida occulta NEUROLOGY Spina bifida not manifest clinically. See Spina bifida

SPINAF CARDIOLOGY A clinical trial–Stroke Prevention in Nonrheumatic Atrial Fibrillation that evaluated the effect of low-dose warfarin anticoagulation on risk of stroke in Pts with nonrheumatic A Fib. See Atrial fibrillation, Low-dose warfarin. Cf BAATAF, EAFT, SPAF

636 **spinal adjustment** CHIROPRACTIC The main type of treatment provided by chiropractors; the most common SA used in chiropractic is a high-velocity, low-force recoil thrust and rotational thrust. See Chiropractic, Recoil thrust, Rotational thrust, Spinal misalignment

spinal & bulbar muscular atrophy Kennedy's disease NEUROLOGY A rare, circa 1:50,000 ♂, X-R neuromuscular condition CLINICAL Adult onset of proximal muscle weakness, atrophy fasciculations, androgen insensitivity–gynecomastia, ↓ fertility, testicular atrophy; ♀ carriers are asymptomatic. See CAG repeat disease

spinal block ANESTHESIOLOGY Locoregional anesthesia that is rapid and complete at very low doses NEUROLOGY Obstruction in the flow of CSF through the subarachnoid space

spinal cord abscess NEUROLOGY A gob of pus, usually in the epidural space, causing spinal cord compression ETIOLOGY *Staphylococcus* spp spread through bone and meninges; rarely also, fungal, TB RISK FACTORS Skin abscesses, bacteremia, back injury or trauma, lumbar puncture, back surgery, any of which may cause osteomyelitis

spinal cord injury Spinal cord compression, spinal cord trauma NEUROLOGY Traumatic damage to the spinal cord resulting in a significant or complete loss of voluntary control of extremities or autonomic nervous system at and below the level of injury EPIDEMIOLOGY 10,000 to 15,000 new cases/yr; 200,000 currently with chronic SCI–US; MVAs cause 45%, falls 22%, acts of violence 16%, sports–eg, diving into shallow water 13%, etc–industrial accidents, gunshot wounds, assault; ♂:♀ ratio 4:1; mean age = 31; 59% occur at age 16-30 CLINICAL SC compression due to fluid or blood in sub- and epidural spaces, paralysis, loss of bladder, bowel control COMPLICATIONS-2ND YR UTI 59%, spasticity 38%, chills and fever 19%, pressure sore 16%, autonomic hyperreflexia, contractures COMPLICATIONS-≥ 30 YRS Pressure sores 17%, muscle and joint pain 16%, GI or cardiovascular problems, UTIs or other infections MANAGEMENT Depending on the severity; surgical if acute or unstable; physical therapy for chronic SCI. See SCIWORA

spinal cord resuscitation ORTHOPEDICS Any maneuver used to ↓ 2° neurologic sequelae of acute spinal cord trauma, due to hypoxia, edema, and aberrant cell membrane potential MANAGEMENT Maintain BP, respiration; ↓ spinal deformity ASAP to relieve cord deformation; stabilize injured segment of cervical spine; IV steroids–bolus, infusion of methylprednisolone, which may suppress cell membrane breakdown by inhibiting lipid peroxidation and hydrolysis at injury site, which ↓ vasoreactive products of arachidonic acid metabolism and ↑ blood flow at site of injury; use of cerebrogangliosides–eg, Sygen–GM-1, to facilitate neurologic recovery. See Spinal cord injury

spinal cord tumor Spinal tumor NEUROLOGY A neoplasm of the spinal cord–intramedullary, meninges–extramedullary, intradural, between meninges and vertebrae–extradural or overlapping; most are extradural TYPES 1°–arising in the spine, or 2°, usually metastatic–eg, CA of lung, breast, prostate, kidney, etc CLINICAL Sx by compressing the spinal cord or nerve roots, invading normal tissue or by ischemia due to intravascular obstruction. See Spinal shock

spinal fusion Spondylosyndesis ORTHOPEDICS A procedure in which multiple vertebrae are operatively fused, usually with diskectomy or laminectomy; while SF ↑ spinal stability, virtually eliminates pain, ROM is lost

spinal muscle atrophy Werdnig-Hoffmann disease NEUROLOGY A group of conditions that cause progressive muscle degeneration and weakness; SMA is the 2nd most common fatal AR disorder after cystic fibrosis, affecting 1:6000-20,000 newborns/yr–US; SMAs selectively affect the α motor neuron, and are characterized by anterior horn cell degeneration, accompanied by muscle weakness, atrophy, progressive symmetrical limb and trunk paralysis; in type I, the most severe form, infants have weak muscles, feeding and breathing problems, death usually occurs before age 3; other types affect increasingly older Pts

SPINAL MUSCLE ATROPHY–TYPES

I Acute/infantile onset form of Werdig-Hoffmann

II Intermediate/childhood onset form

III Juvenile onset form of Kugelberg-Welander

IV Adult onset form

spinal nerve ANATOMY Any one of the 31 pairs of nerves that arise from the spinal cord

spinal shock CRITICAL CARE Neurogenic shock A clinical complex caused by trauma to the vertebral column and spinal cord, resulting in a transient–3-6 wk in duration–loss of reflex activity due to functional or anatomic interruption of the corticospinal tracts; SS is seen immediately after complete injury at the T6 level or above, and is accompanied by arreflexia, loss of sensation, flaccid paralysis below the level of the lesion, flaccid bladder with urine retention, and lax anal sphincter OBSTETRICS An idiopathic postpartum vasomotor collapse that follows spinal anesthesia, 2° to various stressants of delivery–eg, acute blood loss, electrolytic imbalance, adrenocortical insufficiency, pre-eclampsia, anesthetics, amniotic fluid embolism

SPINAL SHOCK

ARREFLEXIA–complete 'failure' below lesion–eg, tetraplegia, paraplegia, overflow incontinence, paralytic ileus, gastric atony and depression of cremasteric reflex; arreflexia is followed several wks later by

HYPERREFLEXIA–exaggeration of reflexes with flexor spasms and autonomic dysreflexia, bladder distention, diaphoresis, HTN, and bradycardia; certain reflexes–eg, anal 'wink', bulbocavernosus and cremasteric reflexes, full penile erection, reflex leg withdrawal and Babinski sign, are retained after complete spinal cord transection since these reflexes don't require higher levels of control

spinal tap Lumbar puncture, see there

spinal vasculature steal syndrome A clinical complex characterized by an AV malformation of the spinal cord that causes spinal cord compression, which may respond to ligation of the 'offending' artery

spindle and/or epithelioid cell nevus Spitz nevus, see there

spindle cell carcinoma Carcinosarcoma, pseudosarcoma An aggressive and poorly differentiated carcinoma composed of sweeping fascicles of elongated epithelial cells of transitional, squamous, undifferentiated or rarely glandular origin that mimics a sarcoma clinically and pathologically; SCCs are most common in the oral cavity–♂:♀ ratio, 10:1, and a variant of squamous cell carcinoma; it also occurs in the larynx, upper respiratory and upper GI tracts, thyroid gland, breast, rarely in the female genital tract; ¼ of Pts with SCC have had regional RT TREATMENT Surgery

spine VOX POPULI Vertebral column, see there. See Kissing spine, Spear tackler's spine

spinnbarkeit German, stretchability GYNECOLOGY The 'stretchability' of cervical mucus, or the length that strands of cervical mucus reach before breaking–≥ 6 cm, a reaction that parallels 'ferning' reaction, peaking on the 14th day–ovulation of the menstrual cycle. See Ferning. Cf String test

spinocerebellar degeneration Friedreich's ataxia, see there

spiral computed tomography Helical scanning IMAGING CT imaging based on 'slip-ring' technology, in which a large image volume is acquired by continuous rotation of the detector. See Computed tomography, Cf High-resolution computed tomography

spirit PHARMACOLOGY A solution containing a volatile substance, usually alcohol PSYCHOLOGY Per Alexander Lowen '...*the life force within an organism manifested in the self-expression of the individual.*'

spiritual training MEDICAL EDUCATION The teaching of spiritual and religious dimensions to medical students, based on the belief that religiously committed Pts have a more rapid and effective recovery from illness. See Faith, Religion

spirometry The measurement of the movement of air in and out of the lungs during various breathing maneuvers, which is the most important pulmonary function test. See Incentive spirometry, Pulmonary function test

spironolactone A potassium-sparing diuretic, which inhibits aldosterone, resulting in less sodium retention. See Diuretics, Potassium-sparing diuretic

spit-up Reflux PEDIATRICS The regurgitation of stomach contents into the esophagus by peristalsis, a small amount of which is normal in infants; most lands on mommy's shirt. Cf GERD

Spitz nevus Epithelioid and/or spindle-cell–nevomelanocytic nevus A benign compound nevus on the face of children and young adults CLINICAL Average 8 mm in diameter and present in various guises–eg, a smooth elastic, pink-tan papule that flattens with external pressure, lightly pigmented papule, nevus or nevi arising in a large congenital nevomelanocytic nevus, in clusters, etc TREATMENT Excision

splash injury NURSING An accidental spraying of body fluids onto exposed body regions. Cf Needle-stick–injury

splashback FORENSIC PATHOLOGY Protrusion of tissue from a bullet entrance wound, resulting from the kinetic energy imparted by the bullet; although the tissue surrounding the entrance wound closes over behind the bullet, the gases expanding after the bullet find the path of least resistance for escape, leaving via the entrance wound, pushing out fragments of subcutaneous tissue. Cf Contact ring, Execution wound

splenectomy SURGERY Surgical removal of a spleen INDICATIONS 1° splenic tumors, hereditary spherocytosis, 1° hypersplenism, chronic ITP, splenic vein thrombosis with esophageal varices, splenic abscess, splenic trauma, autoimmune hemolytic anemia, HbH disease, hereditary elliptocytosis, Hodgkin's disease–for staging, Felty syndrome, myeloid metaplasia, chronic malaria, TB EFFECTS ↑ WBCs, ↑ platelets, possibly compromised immune system, post-splenectomy sepsis–frequency, ± 1%. See Functional splenectomy, Post-splenectomy, Spleen

splenic flexure syndrome Payr syndrome GI DISEASE A functional accompaniment of IBD, caused by swallowed gas that passes only as far as the transverse colon–due to kinking of the transverse and splenic flexure CLINICAL Left upper quadrant pain, abdominal distension, discomfort, constipation MANAGEMENT Autotherapy–defecation, flatulation. See Inflammatory bowel disease

splenomegaly Enlarged spleen Enlargement of spleen for any reason, which is usually a manifestation of underlying disease; the only specific finding in splenomegaly is dragging sensation in the upper right quadrant; megalic spleens may reach 4.0+ kg–eg, in agnogenic myeloid metaplasia

SPLENOMEGALY

CONGESTION Cirrhosis, CHF, thrombosis of portal or splenic veins

INFECTION

- **BACTERIA** Brucellosis, infective carditis agents, syphilis, TB, typhoid fever
- **FUNGI** Histoplasmosis
- **PARASITES** Echinococcosis, leishmaniasis, malaria, schistosomiasis, toxoplasmosis, trypanosomiasis
- **VIRUSES** CMV, EBV

INFLAMMATORY/IMMUNE-RELATED Rheumatoid arthritis, SLE

HEMATOPOIETIC DISEASE/LYMPHOID FUNCTION

- **MALIGNANT** Leukemias, eg ALL, CLL, myeloproliferative disorders–eg agnogenic myeloid metaplasia, CML, multiple myeloma, polycythemia vera; lymphomas–Hodgkin's disease, NHL
- **NONMALIGNANT** Hemolytic anemia, histiocytosis, ITP

STORAGE DISEASES Gaucher's disease, mucopolysaccharidosis, Niemann-Pick disease

ETC Amyloidosis, cysts, hypersplenism, metastases, primary tumors

splenosis The autotransplantation of splenic tissue to atypical sites after open splenic trauma–eg, MVAs, gunshot, or stab wounds; splenic pulp implants appear as red-blue nodules on the peritoneum, omentum, and mesentery, which are morphologically similar to multifocal pelvic endometriosis. Cf Hypersplenism

splinter hemorrhage A small linear subungual hemorrhage which is red when fresh and brown when aged, located at the distal $\frac{1}{3}$ of the nailbed, classically associated with mitral stenosis

split personality A popular synonym for schizophrenia or multiple personality disorder. See Multiple personality disorder, Schizophrenia

split thickness graft SURGERY A 'thin' skin graft–0.25-0.35 mm in thickness that includes epidermis and minimal dermis. See Skin graft. Cf Artificial skin, Spray-on skin

'splitting' ACADEMIA Hair-splitting The division of a morbid condition, lesion, or other entity into smaller subtypes–eg, dividing Laurence-Moon-Biedl-Bardet syndrome into Laurence-Moon and Bardet-Biedl syndromes. Cf 'Lumping' DRUG SLANG A street term for rolling marijuana and cocaine

spondee word AUDIOLOGY A word with 2 syllables–disyllabic, that is pronounced with equal emphasis the 1st *and* 2nd syllables–eg, baseball, toothbrush; SWs are used to test for auditory acuity in Pts with suspected hearing loss, and evaluate baseline hearing

spondylitis ORTHOPEDICS Inflammation of one or more vertebrae. See Ankylosing spondylitis

spondyloepiphyseal dysplasia A heterologous group of AD or AR conditions characterized by short trunk and short limbed dwarfism, due to a defect in ossification of the epiphyseal growth plates of the vertebral bodies and proximal centers of the long bones, affecting the shoulder and pelvic girdles

spondylolisthesis ORTHOPEDICS The forward slippage of a lumbar vertebra on the vertebra inferior thereto, which usually involves lumbar vertebra, most often at the level between the 5th lumbar vertebra and the 1st sacral vertebra; ranges from mild to severe ETIOLOGY Congenital defect in 5th lumbar vertebra, stress fractures, traumatic fractures, bone diseases; it may be associated with and produce lordosis CLINICAL Low back pain and pain in thighs and buttocks, stiffness, muscle tightness, tenderness in slipped area, paresthesias, sciatica due to nerve root compression

spondylosis ORTHOPEDICS Abnormal vertebral fixation or immobility . See Cervical spondylosis

sponge bath NURSING A bath performed on a Pt with prescribed bed rest. See Bath

sponge kidney Polycystic kidney, see there

sponge-worthy POPULAR CULTURE *adjective* The '…quality one…must possess before a ♀ … decides to sacrifice…her personal user-friendly, but discontinued contraceptive sponges for sex.'

spongialization Abrasion arthroplasty ORTHOPEDIC SURGERY Excision of patellar cartilage and subchondral bone, leaving well-vascularized cancellous bone exposed. See Autologous chondrocyte transplantation

spongy degeneration of infancy Canavan-Van Bogaert-Bertrand disease An AR condition characterized by diffuse vacuolization of the deep cerebral cortex, predominantly affecting the white matter, but also the gray matter, caused by hydropic degeneration of the glia and myelin CLINICAL Onset in early infancy with hypotonia and poor head control, hyperextension of legs and flexion of arms, optic atrophy and blindness, severe mental retardation, rigidity, hyperreflexia, seizures and progressive macrocephaly; death occurs by 18 months of age. Cf Leukodystrophy

sponsored symposium A collection of papers published as a separate issue or as a special section–eg, an appendix in a regular issue of a peer-reviewed medical or scientific journal; symposia are usually underwritten by a drug company with a marketing 'agenda'. See Seeding trial

spontaneous abortion Miscarriage, natural abortion OBSTETRICS A pregnancy ending in spontaneous loss of the embryo or fetus before 20 wks of gestation. See Abortion

spontaneous bacterial peritonitis Spontaneous peritonitis CRITICAL CARE A severe acute infection of the peritoneum that accompanies end-stage liver disease and ascites AGENTS *E coli, Klebsiella* spp, *S pneumoniae, Enterococcus faecalis* CLINICAL Abdominal pain, ascites, chills, encephalopathy, fever, rebound tenderness LAB Ascitic fluid has > 500–often 10,000+ PMNs/mm³, protein > 1.0 g/dL, monomicrobials; 40% are culture-negative RISK FACTORS Cirrhosis, nephrotic syndrome, peptic ulcer disease, appendicitis, diverticulitis TREATMENT 3rd-generation cephalosporins–eg, cefotaxime, + IV albumin MORTALITY 30-40%, less if treated early, worse if accompanied by signs of poor liver function–eg, upper GI bleeding, BR > 8 mg/dL, serum albumin < 2.5 g/dL, hepatic encephalopathy, hepatorenal syndrome. See Peritonitis

spontaneous pneumothorax EMERGENCY MEDICINE A condition affecting ± 17,000/yr–US, which may be idiopathic or 2° to underlying pulmonary disease–eg, COPD, most commonly occurring in previously healthy tall, thin ♂, age 20-40, 2° to rupture of subpleural blebs AT-RISK Smokers, family Hx of SP, asthma, emphysema, pneumothorax, rapid ascent to high altitude, histiocytosis CLINICAL Abrupt chest pain, with SOB proportionte to the size of the pneumothorax; tension pneumothorax, while rare, may compromise the circulation by a ball-valve mechanism TREATMENT Suction followed by water seal drainage PROGNOSIS 30% recur on the same side, a tendency that may be ↓ by intrapleural tetracycline; 10% occur de novo on the opposite side

spontaneous regression of cancer ONCOLOGY The partial or complete disappearance of a histologically-confirmed malignancy in absence of treatment or with treatment deemed inadequate to sufficiently alter its natural course. See Melanoma

spontaneous tumor ONCOLOGY A neoplasm that arises in a control animal/person that is not exposed to a known carcinogen or tumor-promoting factor–eg, ionizing radiation, HPV

spontaneously MEDTALK Without treatment

spooge Goo MEDTALK A fluid discharged from the body–often onto a freshly-changed bed. See Body fluid

spoon nail Koilonychia A rare acquired nail dystrophy seen in bronchiectasia in which the nail bed is atrophic and the nail has a 'hollowed' appearance or central concave. See Clubbing

spooning NEUROLOGY A descriptor for a hand deformity in Pts with chorea, in which the outstretched hands, slightly flexed wrists and metacarpophalangeal joints are slightly hyperextended, displaying a distinct concavity

sporadic EPIDEMIOLOGY *adjective* Referring to a noninherited condition that arises rarely in a particular population GENETICS *adjective* Referring to an event attributable to changes in somatic DNA, which may be inherited, but unpredictably

sporadic fatal insomnia NEUROLOGY A prion disease manifest by insomnia, marked tearing of eyes, delusional ideation, dementia, death. See Prion disease

Sporonox® Itraconazole, see there

sporotrichosis INFECTIOUS DISEASE A chronic skin infection by *Sporothrix schenckii* which spreads systemically via the lymphatics, especially in immunocompromised Pts EPIDEMIOLOGY Follows skin inoculation related to handling plants–rosebushes, briars, or mulch-rich soil, affecting farmers, horticulturists, etc CLINICAL Localized as a small painless red lump at the innoculation site which develops into an ulcer, often on the hands and forearms; as the lesion spreads into the lymphatics, it causes linear tracks, seen on the arms SYSTEMIC SPOROTRICHOSIS SOB, osteomyelitis, arthritis, meningitis TREATMENT Supersaturated potassium iodide, itraconazole

sport VOX POPULI A competetive physical activity, played according to specific rules, and which requires practice to attain a level of expertise. See Blood sport, Contact sport, Extreme sport, High-impact sport, Low-impact sport, Moderate-impact sport, No-impact sport, Sports medicine, Water sport

SPORT CARDIOLOGY A clinical trial–Stent Implantation Post Rotational Atherectomy Trial

sports 'anemia' SPORTS MEDICINE A red cell mass that is mildly anemic, typical of 'endurance' athletes. See Blood doping

sports dermatology The study of dermatopathies affecting athletes, which may be traumatic, environmental, infectious, exacerbation of pre-existing disease. See Herpes gladiatorum, Jock itch

sports drink Performance drink SPORTS MEDICINE A thirst-quenching beverage used in sports-related activities, which may boost energy and/or help build muscle mass; water, sugar, salt, potassium are common to all SDs. See Hydrotherapy, Water

---

SPORTS DRINK

TRUE ISOTONIC DRINKS Replace fluid and electrolytes lost during lengthy exercise

CARBOHYDRATE DRINKS Contain glucose polymers and are intended to replenish energy reserves during and after exercise

PROTEIN DRINKS Amino acid drinks Commonly made of whey, a bovine milk product and are used to help recuperate fatigued or overly stressed muscles NY Times 7/12/94; C6

---

sports injury A injury sustained practicing or competing in a sport SITES Thigh, foot, knee, lower leg, ankle, hip, finger TYPES Contusion, strain, sprain, heat exhaustion, lacerations, etc SPORTS WITH MOST Martial arts–judo, tae kwon do, wrestling, gymnastics, power lifting, track & field, soccer, etc

sports massage SPORTS MEDICINE A Western massage that addresses specific needs of athletes COMPONENTS Swedish massage, cross-fiber friction massage, deep compression massage, trigger point therapy TIMING During training, before or after events, to enhance performance, or promote healing postinjury. See Massage, Traditional European massage; Cf Deep tissue massage, Manual lymph drainage massage, Neuromuscular massage, Swedish/Esalen massage

sports medicine MEDTALK A health subspecialty usually practiced by orthopedic surgeons or by rehabilitation medicine physicians, involved in care of those who spring, sprint, splash, smash, whack, whoosh, bang, bash, bat, bounce, bogey or bop, for play or pay. See Anabolic steroids, Boxing, Exercise, Exercise-associated amenorrhea, Running, Sports dermatology, Sports injury. Cf Performing arts medicine

spot GYNECOLOGY See Spotting VOX POPULI A small lesion, usually on the skin. See Age spot, Bald spot, Black spot, Blind spot, Cafe-au-lait spot, Horder spot, Hot spot, Liver spot, Milk spot, Mongolian spot, Oak leaf spot, Powder burn spot, Rose spot, Strawberry spot, White spot

spot test LAB MEDICINE The testing of blood or urine by droppering it on a filter paper impregnated with enzymes or other substances which, in the presence of a–blood or urine constituent of interest would result in a specific color change. See Dipstick, Thin-layer chromatography

spotted fever Rocky Mountain spotted fever, see there

spotted leg 'syndrome' A condition characterized by patchy subdermal atrophy caused by diabetic vasculitis, commonly below the knee in older diabetics; affected skin is smooth, shiny and hyperpigmented due to hemosiderin deposition, ↑ melanin

spotting OBSTETRICS Any episode of minor hemorrhage, which may be associated with pregnancy or the menstrual period. See Leopard spotting, Vaginal spotting

spousal benefits SOCIAL MEDICINE Benefits, including health and life insurance, provided to a spouse–ie, husband or wife–of an employee; in socially advanced nations and in the US, SBs may be extended to unmarried–including same sex–partners

spousal rape FORENSIC MEDICINE Rape by a husband or common law partner, a violent crime and a component of battered wife syndrome. See Assault, Date rape, Domestic violence, Rape

spousal support SOCIAL MEDICINE The mental and psychological help that a spouse provides

spouse A legal marriage partner as defined by state law

sprain ORTHOPEDICS A ligament injury characterized a rupture of fibers without disruption of the ligament itself. See Lateral collateral sprain, Medial collateral sprain, Syndesmosis sprain

'spray-on' skin A polymer for covering superficial 2nd-degree burn wounds; SOS is no longer used as it is contraindicated in deep 2nd and 3rd degree burns–where it would have been most useful–due to ↑ superficial infections and non-adherence. See Artificial skin, Split-thickness graft

spread INFECTIOUS DISEASE The dissemination of an infection either systemically in the same Pt, or to others. See Droplet spread

Sprengel's deformity A condition characterized by a faulty descent of the scapula from its fetal position next to C4, accompanied by a simultaneous scapular rotation, resulting in a horizontal axillary margin. See Fracture

spring VOX POPULI 1. A source of natural water. See Natural spring 2. A coiled metal wire used to provide tension to mechanical devices 3. A season between winter and summer

spring fever VOX POPULI A constellation of mental changes–eg, brighter mood, positive attitude, joie de vivre, that accompany longer, sunnier days in spring. See Heliotherapy. Cf Bright light therapy, Seasonal affective disorder

spring-loaded lancet PHLEBOTOMY A mechanical device used to draw a capillary blood for microchemical analysis, which 'shoots' a sharp point/blade into the site from which the blood is drawn. See Sharps

spring practice SPORTS MEDICINE An off-season football training period associated with an ↑ risk of injury. See Sports medicine

SPRINT CARDIOLOGY A clinical trial–Secondary Prevention Reinfarction Israel Nifedipine Trial that evaluated the effect of antihypertensive therapy with nifedipine, a short-acting CCB, in early management of Pts with possible acute MI. See Calcium channel blockers, Hypertension

SPROM Spontaneous premature rupture of membranes. See PROM

sprue syndrome 1. Celiac–nontropical sprue, see there 2. Tropical sprue, see there

SPS3 NEUROLOGY A clinical trial–Stroke Prevention of Small Subcortical Strokes

spun blood LAB MEDICINE Blood that has been centrifuged, from which serum is obtained for analysis

spur ORTHOPEDICS A bony projection often arising in a calcified tendon. See Calcaneal spur

spurt VOX POPULI A surge or abrupt ↑ in the size or speed of a thing. See Fat spurt, Growth spurt

sputum Semiliquid diagnostic 'goo' obtained from deep coughs from the lungs, bronchi, trachea, which is collected sterilely and examined by cytology and/or cultured COMPLICATIONS Laceration of coronary arteries, or liver due to puncture, arrhythmias caused by needle irritation, vasovagal arrest, pneumothorax, infection. See Brick-red sputum, Currant jelly sputum, Induced sputum, Prune juice sputum, Rusty sputum

SPUTUM–DIAGNOSTIC UTILITY

CYTOLOGIC EXAMINATION Specimen is smeared on a glass slide, stained with one of several dyes, and examined by LM; the only cells seen in normal sputa are those of tracheobronchial tree and lungs

CULTURE & SENSITIVITY The specimen is swabbed on a culture plate in the microbiology laboratory to detect the growth of potentially harmful bacteria or fungi

squamocolumnar Z line ENDOSCOPY A zig-zag border seen by upper GI endoscopy which corresponds to the abrupt transition from stratified squamous or esophageal epithelium to columnar–gastric epithelium; the Z line's appearance is due to the change from the multicelled layer of the esophagus, to the single cell layer of gastric mucosa in intimate contact with blood vessels. See Barrett's esophagus

squamous cell carcinoma Squamous cell cancer ONCOLOGY A malignant epithelial neoplasm that arises in squamous cells of skin, hollow organs, respiratory and GI mucosa, either de novo or in actinic keratosis, or other lesions; SCCs have a high cure rate if treated early; 90+% occur on skin regularly exposed to sunlight or other UV radiation; SCC is more common in Pts with a genetic predisposition, especially those with light skin–↓ melanin in skin and eyes, exposure to chemicals, radiation, arsenic–present in some herbicides; SCC is more aggressive than BCC and carries a slight risk of metastasis if untreated for a long period of time CLINICAL Asymptomatic early, pain if ulcerated . See Actinic keratosis, Basal cell carcinoma, Skin cancer

squamous intraepithelial lesion CYTOLOGY A term that encompasses a spectrum of noninvasive cervical epithelial abnormalities–eg, flat condyloma, dysplasia/carcinoma in situ, CIN; the Bethesda Classification divides SILs into low-grade and high-grade lesions; the former encompass cellular changes associated with HPV cytopathic effect–koilocytotic atypia and mild dysplasia; high-grade lesions encompass moderate dysplasia, severe dysplasia, carcinoma in situ/CIN2–3. See CIN, High-grade squamous intraepithelial lesion, Low-grade squamous intraepithelial lesion, Pap smear

squamous intraepithelial lesion, indeterminant grade SIL-IG, ungraded dysplasia CYTOLOGY A finding on a pap smear in which squamous cells that qualify as an intraepithelial lesion–ie, neoplasia, that defy placement in the high-grade–HSIL, moderate-to-severe dysplasia–ie, CIN 2–3 or low-grade–LSIL, HPV effect or low-grade dysplasia–ie, CIN 1. See The Bethesda System, CIN, Pap smear, Squamous intraepithelial lesion. Cf ASCUS, High-grade squamous intraepithelial lesion–HSIL, Low-grade squamous intraepithelial lesion–LSIL

square root sign CARDIOLOGY A pressure contour recorded by cardiac catheterization, which consists of an elevation of the right ventricular diastolic pressure with early filling and a subsequent plateau, a finding suggestive of chronic constrictive pericarditis

square-shaped wrist test ORTHOPEDICS A test per the *Am J Physical Med & Rehab* for diagnosing carpal tunnel syndrome. See Carpal tunnel syndrome, Thenar weakness test

square wave CARDIOLOGY An abnormal BP response to the Valsalva maneuver–VM, seen in left ventricular failure; BP ↑ at the onset of the VM, remains elevated until VM is released, after which the BP abruptly drops to baseline without 'overshooting' the baseline pressure, accompanied by little change in the pulse pressure or tachycardia

square wave jerk NEUROLOGY A brief, intermittent and horizontal ocular oscillation, arising from defects in the primary gaze position, seen in cerebellar disease

'squaring' RADIOLOGY A descriptor for a sharp angulation of the lower patella, in which its normally voluptuous curve is converted to a virtually teutonic right angle, a resulting from multiple hemarthroses in hemophilia

squash drinking syndrome PEDIATRICS A condition in toddlers characterized by poor appetite, poor weight gain, loose stools, linked to intake of high-energy foods

'Squash It!' PUBLIC HEALTH A phrase integrated into street culture, which verbalizes an increasingly popular concept–ie, that it is socially acceptable to back down from violence or a fight and defuse a volatile situation. Cf Drive-by shooting, Saturday night special

squat jump 'syndrome' SPORTS MEDICINE A transient clinical complex caused by intense and violent exercise, first related to 'squat-jumping,' a calisthenic, resulting in myoglobinuria, swelling of quadriceps, proteinuria, hematuria, hemoglobinuria

squatting PEDIATRICS A position adopted by a child with cyanotic congenital heart disease, classically seen in Fallot's tetralogy, linked to acute hypoxia; squatting relieves exertion-induced dyspnea by ↓ right-to-left shunt and ↑ systemic vascular resistance and pulmonary blood flow. See Shunt

squeeze technique COSMETIC SURGERY The kneading of an established breast implant to improve flexibility CON ↑ 'bleeding' of silicone into surrounding tissues UROLOGY A method of controlling premature ejaculation, generally performed by the partner, in which the penis is squeezed at the tip, quelching the urge to splurge, the need to seed. See Premature ejaculation

squint OPHTHALMOLOGY See Strabismus VOX POPULI The closing of orbicular muscles to examine a distant object

SRI Serotonin Re-Uptake Inhibitor, see there, aka SSRI

SRUS Solitary rectal ulcer syndrome, see there

SS Social Services

S/S Signs & Sx

SS disease Sickle cell anemia due to homozygous HbS gene. See Sickle cell anemia, Sickle cell anemia

SSBOs Standard surgical blood orders TRANSFUSION MEDICINE A list of common surgical procedures and number of units that are normally typed and screened preoperatively for each, to anticipate potential blood loss EXAMPLES 2 units for repair of fracture, 4 for total joint arthroplasty. Cf MSBOS

SSPE Subacute sclerosing panencephalitis, see there

SSRI Selective serotonin re-uptake inhibitor, see there, aka SRI

SSSS Staphylococcus scalded skin syndrome, see there

4S CARDIOLOGY A clinical trial–Scandinavian Simvastatin Survival Study which enrolled 4444 persons, and evaluated M&M with a cholesterol-lowering agent–simvastatin, an HMG-CoA reductase inhibitor on mortality, risk of CHD, CA and other outcome data in hypercholesterolemic Pts with CAD. See Myocardial infarction, Warfarin

the 6 Ss A mnemonic for 6 risk factors for nasopharyngeal CA: smoking, spirits/alcohol, sepsis, sunlight, syphilis, spices–and EBV

St Louis encephalitis INFECTIOUS DISEASE The most common cause of epidemic viral encephalitis in the US; < 1% are clinically apparent CLINICAL Fever, headache, aseptic meningitis, encephalitis EPIDEMIOLOGY SLE is transmitted in passerine birds–eg, sparrows–*Passer domesticus* are the main vertebrate amplifying host; there are regional differences on the mosquito vector

stabilize CRITICAL CARE *verb* To intervene in a critically ill Pt in such a way as to minimize the possibility of acute decompensation of vital functions. See Scoop and run, Triage. Cf Stay and play

stable angina CARDIOLOGY Chest pain that may extend regionally due to ↓ myocardial blood flow ETIOLOGY CAD with stenosis, ↑ blood flow to heart–exercise, heavy meals, stress; other causes of angina include coronary artery spasm–Prinzmetal's angina, heart valve disease, heart failure, arrhythmias RISK FACTORS ♂ sex, cigarette smoking, ↑ LDL-C, ↓ HDL-C, HTN, DM, family Hx of CAD < age 55, sedentary lifestyle, obesity. See Coronary artery disease

stable disease ONCOLOGY A state in which a CA has been treated and clinical signs of disease do not appear to ↑ or ↓ in extent or severity. See Remission

staccato speech NEUROLOGY Clipped, abrupt speech which may occur in multiple sclerosis

Stachybotrys atra MYCOLOGY A mold found in straw and hay, which causes mycosis and fatal GI hemorrhage in cattle and horses; humans may be infected through occupational exposure or after floods, where moist carpets, paper and other fibers provide optimal growth conditions for *S atra*, a toxic mold; exposed infants may develop life-threatening pulmonary hemorrhage

Stachybotrys chatarum See Black mold

stacking SPORTS MEDICINE The illicit self-administration of various 'cocktails' of oral and injectable anabolic steroids by athletes, often during weight training–by body builders, resulting in serum levels up to 100–fold therapeutic levels. See Anabolic-androgenic steroids, 'Roid rage, Weight training

staff MEDTALK *noun* A group of physicians and health care professionals in a specific practice setting, or affiliated to a particular hospital or medical school, generally who provide Pt care. See Active staff, Adjunct staff, Consulting staff, Courtesy staff, Emeritus staff, Honorary staff, House staff, Medical staff *verb* To provide the personnel necessary for a service or operation to function in its intended capacity

staff privileges Admitting privileges The rights that a health professional has as a member of a hospital's medical staff, which includes hospitalization of private Pts, participation in committees, and in decisions relevant to the hospital's future. See Medical staff. Cf Hospital-based physician, RAPERs

stage ONCOLOGY *noun* The extent of a cancer, especially whether it has spread or metastasized *verb* To determine the extent of tissue involvement by a cancer, which is used to guide future therapy and determine prognosis PEDIATRICS *noun* A level of development. See Alarm stage, Babbling stage, Cooing stage, Deep sleep stage, Delta sleep stage, Lalling stage

stage fright Performance anxiety, see there

stage migration Will Rogers phenomenon, see there

staging ONCOLOGY Evaluation of a Pt to determine the severity and extent of a disease to guide therapy, a process required as each stage has a relatively standard treatment. See B Sx, Cotswolds staging, Dukes classification, FIGO staging, Laparoscopic staging, Lovejoy staging, TNM classification, Whitmore-Jewett staging

staging laparotomy ONCOLOGY A procedure in which a particular body region is surgically examined to assess the extent of disease with the purpose of determining the stage or extension of a cancer. See Stage

stagnant loop A segment of intestine overgrown with bacteria, see there

stagnant loop syndrome Blind loop syndrome, see there

stainless steel endoprosthesis Stent, see there

staircase ventilation A form of PEEP, used in early cardiac arrest, where the lungs are prevented from fully collapsing–exhalation, in order to 'recruit' collapsed or fluid-filled alveoli, thereby ↑ arterial PO₂; since a prolonged ↑ in intrathoracic pressure can both stop a weakly beating heart and, by distending the stomach, predispose to regurgitation, PEEP is 'stepped-down' as quickly as possible. See PEEP, Step-down therapy

staircase vertebra A broad, flattened vertebral body with prominent articular facets and spinal processes, seen on a plain AP film in spondylometaphyseal dysplasia, a condition with short trunk, short stature, and bone defects

stalk *verb* PUBLIC HEALTH To actively pursue, harass, or threaten a person who is an unwilling recipient of the stalker's advances

stalking PUBLIC HEALTH Harassing or threatening behavior that a person engages in repeatedly, such as following a person, appearing at a person's home or place of business, making harassing phone calls, leaving written messages or objects, or vandalizing a person's property. See Date rape, Domestic violence

STAMP I regimen ONCOLOGY Combination chemotherapy–cyclophosphamide, cisplatin, carmustine; 10% on regimen suffer life-threatening pneumonitis. Cf Intracranial hemorrhage

stance A body position. See Pugilistic stance

standard MEDTALK A benchmark for measuring and comparing similar or analogous activities or persons. See Air Quality standard, Community standard, Capacity standard, Double standard, Engineering standard, Ergonomic standard, Food standard, Gold standard, Internal standard, Medicare volume performance standard, Ordinary negligence standard, Patient viewpoint standard, Performance standard, Practice standard standard, Prudent layperson standard, Reasonable person standard, Reasonable physician standard, Small parts standard, Zero error standard

standard acid reflux test GI DISEASE An assay that evaluates esophageal pH for reflux esophagitis, using standard stress maneuvers. Cf BAO, GERD, MAO

standard of care FORENSIC MEDICINE '…a normative standard of effective medical treatment, whether or not it is provided to a particular community.'; the SOC corresponds to a level of competence in performing medical tasks accepted as reasonable, and reflective of a skilled and diligent health care provider. See Ethics, Malpractice, Practice guidelines. Cf Reasonable person standards

STANDARD OF CARE

ACADEMIC SOC Every possible diagnosis would be ruled in or out simultaneously, often through use of parallel testing

ECONOMIC SOC Costs of diagnostic or therapeutic interventions are a major consideration

IDEALIZED SOC Physician would have unlimited time to spend with a Pt to establish a warm personal relationship and unlimited resources to carry out a diagnosis and therapy

MANAGED CARE SOC A type of economic SOC, in which minimizing the cost of each Pt ↑ profits; here, a Pt is reduced to a 'unit' on which profit is made, or monies lost

MEDICOLEGAL SOC All hinges on limiting exposure to medical liability

PERSONAL SOC The individual physician draws from his/her education, training, and experience, and incorporates an ethical and humanistic code of professional conduct to do what is best for the Pt JAMA 1996; 275:1296

PRACTICAL SOC The level of care that can be provided by the resources at hand, or based on the access–not economic

standard of care test LAB MEDICINE A test used to diagnose or monitor a disease, regardless of whether the test–eg, immunohistochemistry to ID certain antigens in tissue–is FDA-approved. See Accommodation test

standard deviation Square root of the variance STATISTICS The most widely used measure of the dispersion of a set of values about a mean, which is equal to the positive square root of the variance, where a graphic representation of the data points is described by a curve with Gaussian distribution–GD–ie, bell-shaped. See Gaussian curve

standard drink SUBSTANCE ABUSE A 'unit' of alcohol consumption equal to 44 ml–a 'shot'–of 40% alcohol by volume, aka 'hard' liquor–gin, rum, vodka, whiskey, *or* 150 ml of 12% alcohol by volume–eg, wine, *or* 360 ml of 5% alcohol by volume–eg, beer. See Alcoholic type I, Alcoholic type II, Alcoholism

standard operating procedure MEDTALK A technique, method or therapy performed 'by the book,' using a standard protocol meeting internally or externally defined criteria; a formal, written procedure that describes how specific lab operations are to be performed. See Procedures manual

standard treatment guideline Critical pathway, see there

standard(ized) mortality ratio EPIDEMIOLOGY The ratio of the mortality of population A–with risk factor X to population B, without risk factor X; the ratio of observed to expected deaths, which is used as an estimate of relative risk

standardized patient Teaching patient, see there

standby MEDTALK *adjective* Referring to the immediate availability of a certain specialist–anesthesiologist, surgeon, who can be deployed in a medical emergency. Cf Concurrent

standing order MEDTALK Instructions for Pt management that are to be followed–usually by the nursing staff on a regular and consistent basis, unless instructed to the contrary–eg, SOs for medications or changing of wound dressings. Cf Standing order

stanozolol Androstanozole, stanozol ENDOCRINOLOGY An anabolic 17α-alkylated testosterone derivative used for primary osteoporosis

Stapam Lorazepam, see there

stapedectomy ENT A middle ear operation in which the footplate of the stapes is partially or completely removed with an argon laser and replaced with a ˜ prosthesis INDICATIONS Conduction-type hearing loss; the procedure attempts to ↓ conductive deafness to < 10 decibels COMPLICATIONS Cochlear deafness–< 1%, prolonged vertigo, facial nerve injury

staphylococcal meningitis NEUROLOGY A *S aureus* meningeal infection, secondary to a regionally invasive diagnostic or surgical procedure, or due to a hematogenous spread from an infection elsewhere in the body RISK FACTORS Previous meningitis due to a CSF shunt, infective endocarditis, prior brain abscess. See Meningitis

staphylococcal protein A immunoadsorption Pheresis for salvage treatment of refractory ITP in Pts unresponsive to splenectomy. See Idiopathic thrombocytopenic purpura, Refractory ITP

staphylococcal scalded skin syndrome Ritter disease INFECTIOUS DISEASE A potentially serious vesiculobullous dermatopathy affecting infants in hospital nurseries; SSSS resembles a 2ⁿᵈ-degree burn, and is characterized by erythema, then exfoliation ETIOLOGY Drugs–eg, sulfonamide, phenylbutazone, salicylate, penicillin, barbiturates, or chemicals–eg, acrylonitrile exposure, and has a mortality rate of 30-40% CLINICAL Prodrome of malaise, fever, irritability, generalized erythema and skin tenderness, followed by midepidermal bullae formation, variably accompanied by painful movement of involved areas, anorexia, diarrhea, N&V; SSSS is due to exfoliatin, a protein that loosens the skin's 'cement,' with blister formation and sloughing of epidermis; if it occurs over large body regions it can be lethal, similar to burns MANAGEMENT IV antibiotics, skin protection to prevent dehydration. Cf Stevens-Johnson syndrome, Toxic epidermal necrolysis, Toxic shock syndrome

staphylococcal skin infection Carbunculosis DERMATOLOGY A local staph infection involving deep subcutaneous fascia, consisting of multiple confluent furuncles that form a mass with multiple drainage points, especially on the neck and back; SSIs may affect family members RISK FACTORS Hygiene, friction from clothing or shaving, DM, immunosuppression, dermatitis, pernicious anemia, other systemic disorders

Staphylococcus aureus *Staphylococcus pyogenes* MICROBIOLOGY The most common pathogenic staphylococcus, which is often part of the normal human microflora, and linked to opportunistic infections PREDISPOSING FACTORS Nonspecific immune defects–Wiskott-Aldrich syndrome, chronic granulomatous disease, hypogammaglobulinemia, folliculitis; skin injury–burns, surgery; presence of foreign bodies–eg, sutures, prosthetic devices; systemic disease–eg, CA, alcoholism, heart disease, viral infection; antibiotic therapy CLINICAL Folliculitis, bronchopneumonia

***Staphylococcus aureus* food poisoning** INFECTIOUS DISEASE A state of abrupt onset–4 to 6 hrs–caused by ingesting food contaminated by a *S aureus* toxin, usually linked to a food handler and carrier SUSPECT FOODS Desserts, salads, baked goods, custards, mayonnaise, cream-rich desserts, served or stored at room temperature. See Carrier

Staphylococcus epidermidis MICROBIOLOGY A coagulase-negative staphylococcus that comprises up to 80% of clinical isolates INFECTIONS BY *S EPIDERMIDIS* Infective endocarditis, IV catheter infections, bacteremia, CSF shunt infections, UTIs, osteomyelitis, vascular graft infections, prosthetic joint infections

staple SURGERY A wire-like fastening device composed of steel-based alloys, used to close operative wounds, especially of skin, which minimizes infection by not introducing a foreign body that would connect external and internal body regions. See Surgical closure

STAR ONCOLOGY A 5-yr clinical study–Study of Tamoxifen And Raloxifene–to determine whether raloxifene helps prevent breast CA in ♀ and whether it has any benefits over tamoxifen–Nolvadex. See MORE, Raloxifene, Tamoxifen

'starch blocker' NUTRITION A crude amylase inhibitor said to allow weight-gainless gluttony, a claim torpedoed by well-designed studies. Cf Sugar blocker

starch granuloma Talc granuloma, see there

Stargardt's disease A hereditary condition characterized by progressive retinal degeneration, due to a defect on chromosome 1; SD is similar to age-related macular degeneration. See Macular degeneration

Stark law Physician self-referral law, 42 USC 1395nn MEDICARE A law that prohibits a physician from making a referral to an entity with which she or her immediate family has a financial relationship if the referral is for the furnishing of designated health services, unless the financial relationship fits into an exception set forth in the statute or impending regulations

STARRT™ OBSTETRICS A falloposcopy system, Selective Tubal Assessment to Refine Reproductive Therapy designed for visualizing fallopian tubes to diagnose proximal tubal occlusion. See Falloscopy

STARS CARDIOLOGY A clinical trial–St Thomas Atheroma Regression Study at the St Thomas Hospital, London, which evaluated the effect of a high-fiber diet and/or choleserol-lowering drugs on ASHD. See Acute myocardial infarction, LDL pattern A, LDL pattern B NEUROLOGY A clinical trial–Standard Treatment with Activase to Reverse Stroke

START AIDS A clinical trial–Selection of Thymidine Analogue Regimen Therapy CARDIOLOGY Any of a number of clinical trials 1. Saruplase and Taprostene Acute Reocclusion Trial 2. St Thomas' Atherosclerosis Regression Trial 3. Study of Thrombolytic therapy with Additional Response following Taprostene 4. Stent versus Angioplasty Restenosis Trial 5. Stent vs Directional Coronary Atherectomy Randomized Trial 6. Stents And Radiation Therapy Evaluated the safety and efficacy of the Beta-Cath™ system, an intracoronary RT–ICRT device, in treating in-stent restenosis. See Stent

starvation NUTRITION A condition resulting from prolonged global deprivation of food, which occurs in abnormal environmental conditions–eg, during war or famine, or in normal society through willful neglect of others–eg, children, the handicapped or elderly by parents, family, care-givers or guardians, or by self-neglect in the elderly, mentally feeble, anorectics, or those who, irrespective of means, choose to live in apparent poverty; without food and water, the body loses 4-5% of its total weight/day and few survive > 10 days; when water is provided, a starving person may survive up to 60 days CLINICAL Hypovitaminoses, malnutrition, ↓ subcutaneous fat with thin, dry and hyperpigmented skin stretched over bone prominences, atrophy of organs, marked attenuation of the GI tract, with an enlarged stone-laden gallbladder. See Fasting, Minnesota experiment NEUROLOGY A paucity of neurologic activity. See Motion starvation, sensory deprivation

starvation diabetes Transient glucose intolerance accompanied by glycosuria seen when a person ingests carbohydrates after prolonged starvation, an effect attributed to suboptimal glycogen synthesis and storage. See Diabetes mellitus, Somogyi effect

starvation diet Very low calorie diet NUTRITION A fad diet that provides 300-700 kcal/day, which must be supplemented with high quality protein; given the risk of death through intractable cardiac arrhythmias SIDE EFFECTS Orthostatic hypotension due to loss of sodium, ↓ norepinephrine secretion, fatigue, hypothermia, cold intolerance, xeroderma, hair loss, dysmenorrhea. See Crash diet, Fad diet, Diet

starvation stools Watery green feces that develop when a person is maintained on a clear-liquid starvation-type diet. See Starvation, Starvation diet

stash DRUG SLANG *noun* A place where illicit drugs are hidden

stasis A block in flow, usually of the peripheral circulation. See Venous stasis

stasis dermatitis DERMATOLOGY Skin changes that follow blood stasis due to varicose veins, CHF, etc with swelling of lower extremities, especially feet and ankles; due to extravasation of fluid into adjacent tissue, which interferes with regional nutrition and disposal of intracellular metabolites CLINICAL Skin pigmented, inflamed, open ulcers that heal slowly, early skin atrophy, followed by thickening due to itching

STAT Stat! CLINICAL MEDICINE *adverb* Fast, quickly, immediately, schnell, vite LAB MEDICINE *noun* A specimen, often from the ICU or ER, which is given priority in a lab to measure analytes with immediate potential impact on Pt management; 'stats' include serum glucose, hematocrit, WBC count, certain enzymes, PT, PTT, BUN, creatinine. Cf 'Stats' MEDTALK Jocularly, Some Time After Tomorrow, the likelihood of having an event occur in the desired timeframe NEUROLOGY A clinical trial–Stroke Treatment with Ancrod Trial

stat lab LAB MEDICINE A free-standing lab capable of performing an abbreviated battery of tests, and serving as a receiving 'node' for a larger lab to which the SL is ultimately responsible; SLs are intended to ↓ Pt length of hospital stay, ↓ turnaround time, ↑ Pt convenience, ↑ physician–client satisfaction, ↓ costs

state MEDTALK A condition. See Acute confusional state, Character state, Fugue state, Ground state, Herald state of leukemia, Hypercoagulable state, Lacunar state, Persistent vegetative state, Preneoplastic state, Silent carrier state, Thromboembolic state, Trance state, Vegetative state

state of the art *adjective* Referring to a method or device, at the 'cutting edge' of a particular field

state board State board of medicine A state agency that oversees the activities of the physicians and health care professionals licensed in that state, assuring that a high standard of practice by the physicians and others is maintained, and that the use of controlled drug substances is appropriate and without impropriety. Cf Specialty board, State boards

state boards Examinations administered by a US state board of medical examiners to license a physician in a particular state; these examinations play an ever-decreasing role in state medical licensure, as these bodies now rely on standardized national examinations for assessing a physician's knowledge of medicine. See FLEX exam. Cf State board

state regulatory agency A state body responsible for establishing professional standards, and for certifying professionals or organizations through appropriate documentation

statement VOX POPULI A verbal or written assertion of a thing. See Explanation of benefits statement, Mission statement, Modified summary statement

static resistance SPORTS MEDICINE A constant resistance occurring through a range of motion, when an isotomic contraction is used to move a load. Cf Variable resistance

–statin A popular term for a HMG-CoA reductase inhibitor, which ↓ cholesterol–eg, pravastatin, lovastatin, etc, which ↓ progression of ASHD/CAD, even in absence of ↑ cholesterol. See Atorvastatin, Cervastatin, Fluvastatin, Lovastatin, Pravastatin, Simvastatin

station OBSTETRICS The position or level of descent of the presenting part in the pelvis in a vaginal delivery; full engagement of the presenting part at the iliac spines is considered station 'zero'

statistic A value that defines a specified measure of a population or dataset–eg, mean, percentage, rate of reaction, standard deviation, coefficient of variation, or other datum; a number that quantifies a characteristic of data in a sample. See Durban-Watson statistic, F statistic, G statistic, Kappa statistic, Test statistic, V statistic, Z statistic. Cf Statistics

statistical significance Significance STATISTICS A statement of the probability that an observation represents a true causal relationship and not a chance occurrence; the probability that an event or difference occurred as the result of an intervention–eg, a vaccine, rather than by chance alone; this probability is determined by using statistical tests to evaluate collected data. See Significance

statistics STATISTICS 1. A collection of datapoints or numerical values that can be categorized and subject to analysis; statistics are the raw material on which conclusions about cause-and-effect relationships are based 2. The field that formally studies cause-and-effect relationships; the systematic collection, classification, and mathematical compilation of data vis-á-vis amount, range, frequency, or prevalence; those methods for planning experiments, obtaining data, and organizing, summarizing, presenting, analyzing, interpreting, and drawing conclusions. See Actuarial statistics, Coefficient of variation, Cusum statistics, Descriptive statistics, Health statistics, Mean, Standard deviation, t test

'stats' A popular term for statistics. Cf 'Stat'

Statue of Liberty position ORTHOPEDICS The mandatory position for the hand–when standing after reconstructive and corrective hand surgery or after traumatic injury

status MEDTALK A condition or state. See Code status, Diversion status, DNI status, ECOG performance status, Mental status, Provisional status, Serostatus, Socioeconomic status

status asthmaticus PULMONOLOGY A condition characterized by ↓ response in asthmatics to drugs for which they had previously been sensitive; alternatively, the failure to respond to 3 therapeutic interventions with adrenergic bronchodilators in the ER; Pts in SA are invariably hypoxic and require hospital admission for monitoring of arterial blood gases and pH MANAGEMENT If hypercapneic, rehydration, O₂, aminophylline, methylprednisone, high-dose IM triamcinolone. See Asthma

status epilepticus NEUROLOGY 1. Per the Intl League Against Epilepsy–a seizure that persists for a sufficient length of time or is repeated frequently enough that recovery between attacks does not occur 2. Seizures that persist for 20 to 30 mins, ± a time sufficient to cause injury to CNS neurons 3. Operational definition–either continuous seizures for 5+ mins or 2 or more discrete seizures without complete recovery of consciousness ETIOLOGY-ACUTE Metabolic defects–eg, electrolyte imbalances, renal failure, sepsis, CNS infections, strokes, head trauma, drug toxicity, hypoxia ETIOLOGY-CHRONIC Preexisting epilepsy where SE is due to breakthrough seizures or discontinuation of antiepileptics; chronic alcohol abuse; or tumors or stroke CLINICAL Initially, Pts are unresponsive and have obvious tonic, clonic, or tonic-clonic movements of the extremities; with time, the clinical findings become more subtle, and require EEG confirmation MANAGEMENT Airway control, monitor vitals–temperature, pulse oximetry, monitor cardiac function, measure glucose, administer thiamine and glucose, begin anticonvulsants MANAGEMENT-ANTICONVULSANTS Benzodiazepines–eg, lorazepam, et al, if no response–INR → phenytoin or fosphenytoin, INR → repeat phenytoin or fosphenytoin, INR → phenobarbital, INR → repeat phenobarbital, INR → anesthesia with midazolam or profonol, INR, inter MORTALITY ± 20%. See Seizure. Cf Serial seizures

status/post *adjective* MEDTALK Clinical shorthand referring to a state that follows an intervention–eg, S/P CABG, S/P cholecystectomy S/P mastectomy, S/P orchiectomy, or condition–eg, S/P acute MI

statute of limitations MALPRACTICE A doctrine that allows a plaintiff 2 to 3 yrs–depending upon the state in the US, from the time of the alleged malpractice or negligence–by a physician or hospital–to file a lawsuit. See Emancipated minor, Malpractice

statutory rape SEXOLOGY Sexual intercourse with someone who, based on age–usually < 18, cannot legally consent to intercourse. See Assault, Date rape, Rape, Spousal rape

stavudine d4T, Zerit® AIDS An antiretroviral nucleoside analogue used in multiagent management of AIDS ADVERSE EFFECTS Lactic acidosis, severe hepatomegaly with steatosis, severe peripheral neuropathy. See AIDS

stay MEDTALK A period of hospital ingression. See Average length of stay, Length of stay

stay & play EMERGENCY MEDICINE A stance in which a person's injuries are of such severity as to require some of treatment and/or stabilization on scene before transporting the Pt to a trauma center. See Stabilization. Cf Scoop and run

STD 1. Sexually-transmitted disease, see there 2. Standard test dose

STD litigation FORENSIC MEDICINE A civil lawsuit against a sexual partner who transmitted an STD to the plaintiff

steakhouse syndrome A condition caused by plugging of the lower esophagogastric sphincter with a large, poorly chewed bolus of food, usually meat, accompanied by intense epigastric pain that resolves spontaneously if the food passes into the stomach PREDISPOSING FACTORS Alcohol excess, edentulousness. Cf Cafe coronary, Sushi syncope

steal syndrome Steal, vascular steal syndrome Any Sx complex seen when there are extensive anastomoses between 2 vascular beds, and the arterial supply to one is stenosed or occluded, resulting in diversion of blood to the other vascular bed. See Coronary steal, Reverse cerebral steal. Cf 'Robin hood' syndrome

stealth hardware ORTHOPEDICS Orthopedic hardware constructed of radiolucent resorbable polycarbonates

stealth virus INFECTIOUS DISEASE A popular term for a cytopathic virus–eg, herpesvirus, that mutates and infects host cells without evoking the usual inflammatory response. See Prion

steatorrhea GASTROENTEROLOGY The passage of fatty stools ETIOLOGY Pancreatic insufficiency, malabsorption, ↑ fat in stools–eg, olestra, ingestion of substances–eg, orlistat–Xenical® that ↓ fat absorption

Stedicor® Azimilide, see there

steer MEDTALK *verb* To direct a Pt to a particular facility for various reasons–eg, types of benefits provided by the Pt's health care policy, and ability to pay VOX POPULI *noun* Moo

steering PATIENT CARE The directing by a physician or health care worker of an 'unattached' Pt to a particular physician or group for followup after being seen in an emergency room or other neutral health care setting. See Unattached patient. Cf Flipping

steering wheel 'syndrome' EMERGENCY MEDICINE A blunt chest injury to the driver in a 'head-on' MVA, which may cause acute MI-like signs and Sx; the heart bears the brunt of the injury with contusion of the anterior epicardium and myocardium, compression of the heart between the sternum and vertebral column, and an abrupt ↑ in intrathoracic pressure, potentially rupturing cardiac structures–eg, ventricular septum, chordae tendinieae or free wall. Cf Air bags, Dashboard fracture, Padded dash(board) syndrome, Seatbelt injury, Whiplash

Stein-Leventhal syndrome PCOD, Polycystic ovarian disease, polycystic ovaries, sclerocystic ovarian disease GYNECOLOGY A condition that develops shortly after puberty, characterized by irregular menses, scanty or absent menses, multiple small incompletely developed ovarian follicles–cysts–polycystic ovaries, mild hirsutism, infertility; many Pts with SLS also have insulin resistance DM; PCOs are 2 to 5-fold larger than normal ovaries, and have a white, thick, fibrous outer layer CLINICAL Dysmenorrhea, premature menopause, obesity, hirsutism, virilization, first-degree relatives with SLS; conception is possible with proper surgical or medicinal treatments; these pregnancies are normally uneventful. See Infertility

Steiner speculum SURGERY A speculum used in deep vaginal surgery. See Speculum

Stelazine® Trifluoperazine, see there

stellate fracture ORTHOPEDICS A bone fracture–eg, of the skull, patella, or elsewhere in which the fracture lines radiate centrifugally. See Fracture

stem cell therapy Cell therapy MOLECULAR MEDICINE A technology in which a person's own cells–eg, neuronal stem cells are triggered to revert to their primitive embryonic form, then redifferentiate into mature cells of various organs

stem cell transplantation Bone marrow transplantation The administration of primitive hematopoeitic cells to a Pt whose native RBC, WBCs, platelets were either destroyed by chemotherapy or are functially defective, to assist BM recovery

STEM CELL TRANSPLANTATION 5-YEAR SURVIVAL
LEUKEMIA

CML chronic phase 60-85%; blast crisis 10-20%

AML 1ST remission 50-70%; 2ND+ remissions 20-50%

ALL 30-50%

MYELOMA 10-25%

LYMPHOMA

First therapy after standard regimens 40-70%

Advanced disease 15-30%

MYELODYSPLASIA 20-60%

APLASTIC ANEMIA 60-90%

FANCONI'S ANEMIA 50-70%

SICKLE CELL ANEMIA 50-90%

THALASSEMIA 60-95%

Stemgen® Ancestim HEMATOLOGY An early growth factor with orphan drug status used with Neupogen®–filgrastim–rG-CSF to manage CA Pts receiving stem cell transplants, augmenting the effects of other growth factors to mobilize stem cells into blood for collection and transplantation. See Filgrastim

Stenger test NEUROLOGY A functional test for used to differentiate feigned from true unilateral hearing loss–subthreshold intensity sounds are presented to either ear in succession

stenosis An abnormal ↓ in the diameter of a lumen, in particular of an artery or heart valve–eg, in ASHD with plaque buildup on the inner wall of an artery. See Aortic stenosis, Carotid stenosis, Cervical stenosis, Fishmouth stenosis, Hypertrophic pyloric stenosis, Meatal stenosis, Mitral stenosis, Occlusion, Pulmonary stenosis, Restenosis, Supravalvular aortic stenosis, Subvalvular pulmonary stenosis, Supravalvular pulmonary stenosis. Cf Regurgitation

stent CARDIOLOGY Intracoronary stent An expansile tube positioned in a blood vessel, especially a stenosed coronary artery, to ↑ its diameter, ergo blood flow to the myocardium; stenting of obstructed CABGs results in superior outcomes, larger gain in luminal diameter, ↓ major cardiac events. See Coronary artery bypass graft, Coronary stent, Endovascular stent, Endovascular stent-graft, Harrel Y stent, Horizon temporary stent, INR stent. Cf Balloon angioplasty MEDTALK A synthetic tube placed in a tubular structure and intended to maintain that structure's open state. See Biliary stent

stent-graft Endovascular stent graft, see there

step-down therapy CARDIOLOGY A staged reduction in the doses and agents used to manage HTN; the 'steps' begin with thiazide diuretics, β blockers or ACE inhibitors for Pts with mild HTN, diastolic BP < 100 mm Hg, to hydralazine, prazosine, minoxidil, guanethidine or furosemide for severe HTN, diastolic BP > 120 mm Hg. Cf 'Staircase' ventilation

stepladder pattern Stepladder configuration GI IMAGING A pattern seen on a barium enema of the ascending colon, in which rigid fibrosis causes transverse linear fissures intersecting with gullies of contrast material in deep longitudinal ulcers, a finding in in Crohn's disease, when radiocontrast 'spills' past the constricted and narrowed terminal ileum, imparting a 'railroad track' appearance on gross examination. See Garden hose appearance, String sign RHEUMATOLOGY A descriptor for the consecutive subluxations in the cervical spine in rheumatoid arthritis, which may be accompanied by intervertebral disk space narrowing

step-off sign IMAGING A step-like central defect seen on multiple adjacent vertebral bodies, typical of sickle cell anemia, attributed to blood stasis and ischemia, which slows the growth of the vertebra's cartilaginous growth plate; because the peripheral portion of the vertebral growth plate has a different blood supply, the vertebral body develops in a lop-sided–step-like fashion; the SOS has been described in Gaucher disease, and is attributed to recuperation of peripheral growth after the vertebral body has collapsed

steppage gait Steppage BIOMECHANICS A simple swing phase gait modification, consisting of exaggerated knee and hip flexion to lift the foot higher than usual for ↑ ground clearance; the SG compensates for a plantar flexed ankle, where the dorsiflexors are inadequate to control foot drop. See Foot drop, Gait

stepped formulary PHARMACOLOGY A hospital formulary in which clinicians are encouraged to prescribe the least expensive equivalents–eg, NSAIDs, as first-line therapy, reflexing to more expensive agents *after* failure of first-line therapy

stercoral ulcer A colonic ulcer that develops in elderly or mentally retarded Pts with intractable constipation due to pressure of impacted fecal material and sluggish mesenteric circulation; with time, infected fistulas form under the ulcer. See Solitary rectal ulcer syndrome

stercoroma A tumor-like plug of feces in the rectum

StereoEndoscope LAPAROSCOPY A device used in minimally invasive surgery with a StereoScope camera head and 3-D visualization platform. See Laparoscopic surgery

stereotactic biopsy PATHOLOGY A biopsy–most commonly from breast, obtained via imaging–eg, CT, ultrasonography-based guidance TYPES Stereotactic needle aspiration–provides cytology specimen; stereotactic core–cutting needle biopsy–provides one or more cores of tissue. Cf Guided wire open biopsy

stereotactic neurosurgery NEUROSURGERY A procedure in which needles or electrodes are passed through brain tissue into designated trouble spots PROS Therapy is performed on inaccessible tissues deep in the brain CONS Healthy or essential neural tissue is damaged by the transit of the needles to the therapeutic site. Cf Magnetic stereotaxis

stereotactic radiosurgery Stereotactic radiotherapy RADIATION ONCOLOGY A nonsurgical RT technique that delivers numerous narrow, precisely aimed, highly focused beams of ionizing radiation that converge at a specific point–the tumor; SR uses heavy charged particles–protons or helium ions, photons–γ radiation or a linear-accelerator to treat intracranial lesions inaccessible to conventional neurosurgery. See Gamma knife

stereotype NEUROLOGY Stereotypy, see there VOX POPULI A preconceived and oversimplified idea of the characteristics which make up a person. See Sexual stereotype, Skid row stereotype

stereotypic movement disorder NEUROLOGY A disorder characterized by repetitive, seemingly driven and non-functional motor behavior lasting 4+ wks; movements ↑ with stress, frustration, boredom; ↑ boys ETIOLOGY Unknown

stereotypy NEUROLOGY A non-goal-directed automatic and/or persistent mechanical repetition of speech or motor activity; a series of repetitive complex movements that simulate motor tics, seen in Pts with hyperactivity, mental retardation, schizophrenia, psychosis.Cf Motor tic

Steri-Strip WOUND CARE A Band-Aid wannabe that provides wound security through ↑ tensile strength, used for nonsuture closure of wounds

sterile *adjective* Referring to 1. An inability to produce children; inconceivable 2. A product–gauze, surgical instruments–that has been packaged so as to have no organisms at the time it is opened

sterile field SURGERY A 'clean' environment that surrounds an incision, and relatively free of microorganisms, in particular bacteria; the SF is inhabited by the surgeon(s), scrub nurses, and occasionally, physicians in training. See Dirty wound

sterile 'infection' Any infection–eg, of lung, liver, pelvic, adnexae, appendix, bile tract, bone, chest, CNS, ♀ genitalia, respiratory, urinary tract, etc, in which the culture conditions–eg, high O₂ during specimen transportation, etc, preclude in vitro growth of anaerobic bacteria, even though they cause clinical disease

sterile pyuria MEDTALK Abundant PMNs in culture-negative urine; SP may occur in urethritis, urogenital TB, anaerobic bacteriuria

sterility 1. The absence of viable contaminating microorganisms; aseptic state, asepsis 2. Involuntary infertility–the inability to procreate, conceive or induce conception

sterilization REPRODUCTION MEDICINE The process of rendering an organism or person inconceivable, irreproducible, or infertile, through tubal ligation, vasectomy, or orchiectomy. See Involuntary sterilization, Thermic sterilization, Surgical sterilization, Voluntary sterilization

sternotomy SURGERY The midline vertical splitting of the sternum before thoracic surgery–eg, for CABG, heart valve replacement, etc, in thoracic cavity

steroid *adjective* Pertaining to steroid hormones *noun* 1. A cholesterol-derived lipid that is the parent compound for steroid hormones of the adrenal gland and gonads 2. Steroid hormone A hormone produced from modified cholesterol EXAMPLES Hormones from testis, ovary, adrenal cortex, etc USES Relief of swelling, inflammation. See Anabolic steroid, Corticosteroid, Glucocorticosteroid, Ketogenic steroid. Cf Peptide hormone 3. Any compound–eg, bile acids, cardiac glycosides, vitamin D precursors, that is struturally similar–ie, has a cyclopentaphenanthrene core–to steroid hormones

steroid injection Intraarticular steroid injection, see there

steroid therapy THERAPEUTICS Treatment with corticosteroids to ↓ swelling, pain, and other Sx of inflammation. See Steroid

sterologist Infertility specialist GYNECOLOGY A physician specialized in treating couples with infertility. See Infertility

stethoscope MEDICAL PRACTICE An instrument with a Y-shaped flexible tube that connects at one end to a bell-shaped device fitted with a piece of hardened plastic that amplifies sound and, at the other, to 2 ear pieces for listening to various sounds from the heart, lungs, GI tract, etc POPULAR MEDIA That really cool thingie that TV docs carry around to impress people. See Sphygmomanometer, White coat

Stevens-Johnson syndrome DERMATOLOGY A variant of erythema multiforme characterized by purpuric macules or atypical targetoid lesions with focal confluence and epithelial detachment covering < 10% of body surface; ½ are reactions to drugs–eg, sulfonamides, anticonvulsants, allopurinol, etc; 10-30% are accompanied by systemic Sx CLINICAL Fever, erosive stomatitis, lesions of anogenital mucosa, conjunctiva–keratitis, corneal erosions, RTIs; < 5% are fatal TREATMENT Delete offending drug

Stewart-Treves syndrome An angiosarcoma arising in chronic lymphedema, especially of the arm in ♀ after mastectomy and lymphadenectomy; it may occur in extremities with congenital, idiopathic or traumatic lymphedema

STICH CARDIOLOGY A clinical trial–Surgical Treatment for IntraCerebral Hemorrhage

stick NURSING *verb* To perform a venipuncture

stick-man syndrome A type of congenital muscular dystrophy characterized by toe walking in early childhood, contractures, weakness, muscle atrophy and extremely thin–ie, stick-like limbs. See Muscular dystrophy

sticky floor phenomenon ACADEMENTIA A term for a ♂/♀ disparity in career advancement in medicine; despite similar preparation of candidates for academic careers vis-á-vis board certification, advanced degrees, research during fellowship training, ♀ had fewer resources, office/lab space, grant support, and 'protected' reseach time. Cf Glass ceiling phenomenon

sticky food DENTISTRY Any food–eg, 'gummy bears,' jelly beans, chewing gum, chewy chocolates–proscribed post procedure–eg, placing of temporary crown–SFs tend to pull out anything not permanently anchored in the gums

stiff baby syndrome Hereditary stiff-man syndrome An AD complex of variable penetration, characterized by ↑ startle reflex and virtually continuous motor activity by EMG CLINICAL Choking, vomiting, dysphagia, which may improve with age. Cf Floppy infant syndrome, Stiff man syndrome

stiff heart syndrome CARDIOLOGY A nonspecific term for ventricular pump failure due to restrictive heart disease CLINICAL Chest pain, exertional dyspnea, ↑ venous pressure, extra-diastolic murmurs, hepatomegaly, ascites, edema ETIOLOGY Idiopathic, or related to amyloidosis, constrictive pericarditis due to radiation, mycosis, trauma, TB, hemochromatosis, myocardiopathy. Cf 'Stone' heart

stiff person syndrome Stiff-man syndrome NEUROLOGY A rare GABAergic autoimmune motor dysfunction with a 2:1 ♂:♀ ratio CLINICAL Stiffness of axial and appendicular muscles with intermittent superimposed painful muscle spasms precipitated by emotional or physical stress, low back pain, hyperlordosis, motor and gait defects, diaphoresis, tachycardia ETIOLOGY Probably autoimmune, given presence of anti-GAD65 antibodies against glutamic acid decarboxylase in 60% of Pts and pancreatic islet cells; remaining 40% have other autoantibodies; some CA-associated SPS have autoantibodies against a 128 kD synaptic protein; associated with epilepsy, type 1 DM and other organ-specific autoimmune disorders–eg, myasthenia gravis, thyroiditis, adrenalitis DIAGNOSIS Simultaneous video-electroencephalographic surface EMG demonstrates continuous motor unit activity in affected muscles at rest, abnormal activity of small gamma motor neurons TREATMENT Benzodiazepines, cortisol if adrenocortical dysfunction, plasma exchange, IVIG–well-tolerated, effective, expensive. See Anti-GAD65 antibodies

stiffie VOX POPULI–UK Erection, penile tumescence

stigma A sign, mark, feature, indicator of something, which generally has a negative connotation

STILE trial CARDIOVASCULAR SURGERY A clinical trial–Surgery versus Thrombolysis for Ischemia of the Lower Extremity, comparing thrombolysis–by intraarterial recombinant urokinase or tPA with vascular surgery–eg, thrombectomy or bypass surgery in Pts with arterial obstruction of legs. See Thrombolytic therapy, Tissue plasminogen activator. Cf Rochester study, TOPAS

stillbirth Death in utero OBSTETRIC Fetal death before complete extraction or expulsion of a product of conception, irrespective of duration of pregnancy REPRODUCTION MEDICINE A fetus or infant delivered without signs of life after 20 wks or more of gestation

Still's disease Juvenile rheumatoid arthritis RHEUMATOLOGY A persistent arthritis that presents with systemic Sx–eg, high intermittent fever, salmon-colored rash, lymphadenopathy, hepatosplenomegaly, pleuritis, pericarditis

stimulant *adjective* Relating to anything that ↑ activity, especially of the nervous system *noun* PHARMACOLOGY Any substance that evokes ↑ activity–eg, a CNS stimulant, cardiovascular stimulant, and others. See Amphetamine, Dextroamphetamine, Ephedrine, Herbal ecstasy, MDMA, Methamphetamine, Methcatinon, Methylphenidate, OTC stimulant, Phenmetrazine, Sexual stimulant

stimulant laxative Any laxative–eg, phenophthalein, sennosides, that promotes stool evacuation by ↑ peristalsis. See Laxative; Cf Bulk-forming laxative, Stool softener

stimulated acid output GI DISEASE A stimulatory test evaluating adequacy of vagotomy in Pts requiring a vagotomy for gastric ulcer disease. See BAO, MAO, PAO

stimulated cycle REPRODUCTION MEDICINE An ART cycle in which a ♀ receives drugs to stimulate the ovaries to produce more follicles. See Assisted reproduction

stimulation MEDTALK The evoking of a particular activity. See Costimulation, Fetal acoustic stimulation, Magnetic stimulation, Neural stimulation, Osteogenic stimulation, Ovarian stimulation, Pocket stimulation, Ultrasonic osteogenic stimulation, Vagus nerve stimulation

stimulation threshold CARDIAC PACING The minimum electrical stimulus needed to consistently elicit a cardiac depolarization, expressed in terms of amplitude–volts, milliamps; pulse width–millisecs; energy–micro joules

stimulative music MUSIC THERAPY Assertive or buoyant music that may prompt the body into joining the rhythm, evoking hand clapping, dancing, etc. Cf Sedative music

stimulus-response theory PSYCHOLOGY The theory that human responses hinge on external reward and punishment. See Operant conditioning

sting MEDTALK The injury caused by an injected venom from a plant or animal. See Hymenopteran sting, Scorpion sting, Wasp sting

stinger SPORTS MEDICINE A popular term for an injury to the brachial plexus due to abnormal stretching

stinging insect ENTOMOLOGY A hymenopteran that stings TYPES Aphids–eg, honeybees, bumblebees; vespids–eg, hornets, yellow jackets; wasps; stinging apparatus consists of a venom-filled sac attached to a barbed stinger; the multiple barbs on the honeybee's stinger cause it to detach, resulting in the honeybee's death; the vespid stinger has fewer barbs, allowing it to sting many times. See Hymenopteran hypersensitivity, Venom

stinky nose ENT A popular name for the unpleasant odor of atrophic rhinitis; the mucosa is attenuated and infected. See Atrophic rhinitis

stippling A punctate appearance or, in radiology, white granularity in a radiolucent background, similar to the 'salt-and-pepper' BONE RADIOLOGY Punctate calcifications in epiphyseal ossification centers, which may occur in congenital calcific chondrodystrophy, cretinism, ischemic necrosis–osteochondrosis, multiple epiphyseal dysplasia, pituitary gigantism, sclerotic osteopetrosis or osteopoikilosis COLONOSCOPY A pattern of fine granularity of the mucosa, seen in early ulcerative colitis ESOPHAGOSCOPY A pattern in esophagitides due to corrosive agents, reflux, infections–eg, candidiasis, RT RENAL IMAGING A pattern seen in papillary transitional cell carcinomas with incomplete filling of the pelvi-caliceal system due to tumoral replacement in the intravenous pyelogram

stirrups The footholds in a lithotomy table

stitch SPORTS MEDICINE Popular for a stabbing pain, often at the lower border of the ribcage. See Charley horse SURGERY A popular term for a single suture, as in a wound needing 14 stitches. See Baseball stitch VOX POPULI A measure of a laceration's severity/bragging rights with others–eg, it required 5 stitches–yawn, 20 stitches–wow!, 135 stitches–da-yammm!

stochastic *adjective* Referring to a random process; a process determined by a random distribution of probabilities; referring to a behavior not governed by known equations and initial conditions, thus unpredictable at any past or future time. See Chaos. Cf Deterministic

Stockholm syndrome PSYCHIATRY A syndrome in which hostages identify and sympathize with their captors on whom they depend for survival

stocking & glove distribution NEUROLOGY A pattern of peripheral nerve disease characterized by a relatively sharply demarcated loss of pain, touch, temperature, position and vibration sensation, accompanied by weakness, muscular atrophy, and loss of tendon reflexes–eg, the 'stocking' pattern of distal diabetic polyneuropathy, characterized by waxing and waning paresthesias that worsen at night

Stokes-Adams syndrome CARDIOLOGY A constellation of Sx due to complete heart block with a pulse of 30 to 45 bpm CLINICAL Inadequate blood to brain—vertigo, fainting, convulsions MANAGEMENT Permanent pacemaker. See Heart block

stoma Ostomy SURGERY A surgically created opening of a hollow viscus organ to the outside of the body. See Colostomy, Urostomy

stomachache VOX POPULI Gastralgia

stomach cancer Gastric cancer, cancer of stomach ONCOLOGY A stomach malignancy, usually 1° adenoCA EPIDEMIOLOGY Common in Japan, Finland, Iceland, Chile CLINICAL Often vague—eg, anorexia, weight loss DIAGNOSIS Endoscopy, biopsy RISK FACTORS Family Hx of gastric CA, ABO group A, pernicious anemia, chronic atrophic gastritis, adenomatous gastric polyp, intestinal metaplasia, *H pylori*, prior gastric surgery, consumption of salt cured and smoked foods, low vitamin C intake MANAGEMENT Gastrectomy with lymphadenectomy PROGNOSIS Usually poor; maximum 2 yr survival. See Gastric lymphoma, Signet ring carcinoma

stomach-partitioning gastrojejunostomy SURGERY A bypass procedure for Pts with unresectable gastric CA OUTCOMES Improved quality of life, improved survival. Cf Whipple procedure

'stomach virus' A popular term for gastroenteritis of presumed viral origin often accompanied by diarrhea, fever, N&V, abdominal pain ETIOLOGY Enteroviruses, rotavirus, possibly astroviruses

stomatitis Acute necrotizing ulcerative gingivitis; stomatitis, see there, aka trench mouth. See Herpetic stomatitis

stone Concrement MEDTALK An indurated material, generally composed of crystallized minerals, les commonly, organic materials—eg, bile, cholesterol. See Bladder stone, Gallstone, Kidney stone, Magnesium ammonium phosphate stone, Salivary duct stone

stone heart CARDIOLOGY A popular term for irreversible ischemia-induced cardiac rigor mortis, in which the heart undergoes global spastic contraction in systole; anoxia rapidly depletes glycogen and ATP, causing death; a heart of stone is exceptionally rare, clinically linked to severe heart disease—and romantically linked to the heartless. See Rigor mortis. Cf Stiff heart 'syndrome'

stone heart syndrome Fatal postoperative ischemic contracture of the left ventricle CARDIOLOGY A post-operative Sx complex, characterized by marked ventricular hypertrophy and contraction band necrosis

'stoned' 'Wasted' SUBSTANCE ABUSE Popular for a state of quasi-stupor induced by psychoactive substances of abuse—eg, heroin, marijuana, alcohol; to be pleasantly under the influence of drugs. See 'High'. Cf 'Bad trip'

stones, bones and groans A mnemonic for the clinical triad of hyperparathyroidism, where prolonged ↑ of PTH and end-organ response thereto result in disseminated Ca^{2+} stones, osteoporosis—bones, GI Sx—eg, N&V, anorexia, weight loss, peptic ulcers—groans

stool Feces *noun* Solid waste discharged in a bowel movement; fetid intestinal evacuate. See Black stool, Currant jelly stool, Milk stool, Pea soup stool, Rabbit stool, Rice water stool, Starvation stool

stool culture The placing of a sample of stool in an appropriate culture medium to detect growth of a pathogen—eg, rotavirus, for which specimen is frozen ASAP, or *Yersinia enterocolitica*—specimen stays at room temperature. See Ova and parasites

stool lipids GI DISEASE The presence of significant lipids in the stool is abnormal, and indicates malabsorption and, if voluminous, is malodorous, greasy, and nasty. See Malabsorption

stool softener A laxative—eg, docusate—that softens stool by adding fluid. See Laxative. Cf Bulk-forming laxative, Stimulant laxative

STOP S/TOP, selective tubal occlusion procedure GYNECOLOGY A system for non-surgical ♀ sterilization

STOP-Hypertension CARDIOLOGY A trial—Swedish Trial in Old Patients with Hypertension evaluating M&M on hypertensives age 70-84 receiving antihypertensives. See Antihypertensive, Calcium channel blocker, Hypertension

STOPIT OSTEOPOROSIS A trial—Steroid-Induced Osteoporosis Intervention Trial by Merck, evaluating the safety and efficacy of 2 doses of oral alendronate in treating glucocorticoid-induced bone loss

stopping power RADIATION ONCOLOGY The ability of a material to stop ionizing radiation; alpha paticles are stopped by a piece of paper, gamma radiation by thick lead shielding RADIOLOGY The density of a tissue reflected in an image's whiteness; white tissues—eg, bone have the greatest SP

storage disease HEMATOLOGY A condition, often designated 'inborn errors of metabolism,' in which a defective or functionally absent enzyme causes organ dysfunction through accumulation of precursor substances derived from the metabolism of glycogen, amino acids, often within lysosomes; each has a relatively distinct pattern of organ involvement—eg, in glycogen storage disease, the excess substances compromise the liver, skeletal muscle and cardiac muscle. See Brancher disease, Debrancher disease

storage lesion TRANSFUSION MEDICINE The constellation of changes occurring in a unit of packed red cells during storage. See Red cell preservatives

STORAGE LESIONS

↑ **AMMONIUM** to 470 µmol/L–US: 800 µg/dL

↑ **FREE HB** in plasma from 82 to 6580 mg/L–US: 8.2 to 658 mg/dL

↑ **K+** from 4.2 to 78.5 mmol/L–US: 4.2 to 78.5 mEq/L

↓ **ATP** from 100% to 45%

↓ **2,3 DPG** to < 10% of original levels–replenished within 24 hours of transfusion

↓ **LABILE PROTEINS**, eg complement, fibronectin and coagulation factors ↓ to negligible

↓ **Na+** from 169 to 111 mmol/L–US: 169 to 111 mEq/L

↓ **pH** from 7.6 to 6.7

Adverse physiologic effects of stored blood is negligible in the absence of a previous compromise of the Pt's–recipient's status

storage pool diseases METABOLIC DISEASE A group of platelet disorders caused by defects of platelet granules, divided into 1. α granule storage pool disease. See Gray platelet syndrome and 2. Dense granule deficiency or delta-storage pool disease CLINICAL Moderate bleeding LAB ↑ Bleeding time, ↓ platelet ADP content, ↓ serotonin levels TREATMENT Hemostasis with platelet transfusions, cryoprecipitate, desmopressin acetate–DDAVP, which releases von Willebrand factor from storage sites

stork mark Stork bite NEONATOLOGY 1. Birthmark, see there 2. A common capillary nevus of the newborn, consisting of one or more flat pink macules with irregular borders on the forehead, eyelids, tip of nose, upper lip or hairline on back of neck; SBs often darken with crying, fade with pressure, clear spontaneously in months and disappear by 18 months

stovepipe colon IMAGING A descriptor for a colon with long-standing ulcerative colitis, which has been converted into a shortened tubular structure of small caliber, with virtual loss of flexure, and facile reflux into the ileum. See Ulcerative colitis

strabismus Crossed eyes; exotropia, walleye OPHTHALMOLOGY Nonparallel position or movement of eyes, due to ↓ muscle coordination between eyes, with loss of stereoscopic vision and inability to focus simultaneously on a single point ETIOLOGY Extraocular muscle defects, neurotoxins, blindness, mechanical defects, unilateral vision obstruction in childhood, various brain disorders or systemic diseases, amblyopia, paralytic shellfish poisoning, botulism, hemangioma near eye, Guillain-Barré syndrome, Apert syndrome, Noonan syndrome, Prader-Willi syndrome, trisomy 18, congenital rubella, incontinentia pigmenti, cerebral palsy, Laurence-Moon-Biedl syndrome, pseudohyperparathyroidism. See Farsightedness, Myopia

straddle lesion EMERGENCY MEDICINE A complex perineal injury, usually caused by a fall with a point of impact between the legs–eg, falling on a fence, with major trauma to posterior urethra, extravasation of blood and urine through Buck's fascia, often extending into the scrotum, perineum, central tensor and anterior abdominal wall under Scarpa's fascia

straight Heterosexual, see there

straight back syndrome Straight back and flat chest syndrome A variant position due to ↓ of normal thoracic kyphosis, which ↓ AP chest diameter, ↑ prominence of pulmonary artery and right hilum, displacing heart to left, simulating cardiomegaly ASSOCIATIONS ASD, scoliosis; it may cause mild pulmonary vein obstruction and dilation, evoking a harsh late systolic ejection murmur; it is asymptomatic but may be misinterpreted by the examiner, causing an otherwise healthy person to become a cardiac 'cripple'

Straight-In™ system UROLOGY A system for treating urinary stress incontinence via laparoscopic bladder neck fixation and/or transvaginal repair of vaginal prolapse Influence 11/11/97. See Stress incontinence

strain *noun* AIDS An HIV isolate from a person or group of persons given its own unique identifier, or strain name–eg, MN, LAI ORTHOPEDICS An overuse injury *verb* To injure by overuse; to wear out or stress beyond normal limits; straining may be associated with tissue microtearing OPHTHALMOLOGY Overuse of eyes, resulting in transient discomfort VOX POPULI *verb* To filter; remove particles from a fluid

straitjacket Camisole A physical restraint device consisting of a canvas jacket with overly long sleeves topped with leather straps that buckle in the jacket's back, preventing a violent person from hurting him/herself or others. See Physical restraint

strangers, fear of Xenophobia. See Phobia

strangulated hernia The prolapse of a loop of intestine into a hernia sac with vascular compromise and, if unresolved or trapped for a prolonged period, infarction of the entire prolapsed loop. Cf Incarcerated hernia

strangulation FORENSIC MEDICINE Transient, or more commonly, permanent occlusion of the tracheal lumen; 3500 suicidal strangulations occur/yr–US, 1983; homicidal strangulations represent 5-10% of criminally violent deaths in urban populations. Cf Sexual asphyxia OBSTETRICS See Cord strangulation

STRANGULATION–MANNER, MECHANISM, SETTING

MANNER OF DEATH Hanging–usually suicidal, ligature, manual, postural

MECHANISM OF DEATH

- Mechanical constriction of neck structures–primary mechanism in suicidal stangulation; it is unknown whether arterial occlusion, venous occlusion, or asphyxia causes most deaths in 'mechanical' strangulation

- Injury to spinal cord and brainstem–due to drop, the intended cause of death in judicial hangings

- Cardiac arrest, possibly facilitated by pressure on the carotid sinus, and pericarotid sympathetic and parasympathetic networks

SETTING Suicidal, homicidal, accidental, judicial–no longer performed in developed nations, despite its alleged value as a crime deterrent and for pre-TV entertainment

Ann Emerg Med 1994; 13:179

strap muscles SURGICAL ANATOMY A popular term for the infrahyoid group of intrinsic laryngeal muscles; SMs are small, flat muscles below the hyoid bone that connect with hyoid bone and thyroid cartilage; SMs are antagonistic to the elastic suspensory ligaments and laryngeal elevators. Cf Laryngeal complex

strawberry cervix GYNECOLOGY A descriptor for the quasi-pathognomonic colposcopic appearance of subepithelial punctate petechiae of the cervix infected by *Trichomonas vaginalis*, variably accompanied by 'double hairpin' capillaries.

strawberry gallbladder A descriptor for the appearance of the gallbladder mucosa with cholesterosis, where aggregates of cholesterol/lipid-laden histiocytes overlie a reddish or green mucosa, which may be associated with a giant cell reaction

strawberry tongue A characteristic enanthema of the tongue, characterized by hypertrophy of the fungiform papillae, accompanied by changes of the filiform papillae in a bright red background; an ST is classically seen in scarlet fever, but also occurs in Kawasaki's disease–mucocutaneous lymph node syndrome, toxic shock syndrome, and in prodromal yellow fever

street drug Illicit drug, see there

street smarts VOX POPULI Worldly wisdom and wariness in human interactions. Cf Social smarts

the streets SOCIAL MEDICINE A lifestyle place, usually understood to be shared by those of a lower socioeconomic status in urban areas

strength NEUROLOGY The amount of force that a person can exert. See Back extensor strength, Ego strength, Hand grip strength PSYCHOLOGY The ability to withstand mental stress

'strep throat' Streptococcal pharyngitis INFECTIOUS DISEASE A term for any infectious erythema of the oropharynx; although the name implies a bacterial origin and treatment is usually antibiotics, it is commonly due to viruses–eg, EBV and CMV, and *less* commonly due to streptococci; other agents causing ST include diphtheria, tularemia, toxoplasmosis, brucellosis, salmonellosis, TB; true ST has a 2-4 day incubation, dysphagia, headache, malaise, fever, anorexia, N&V, abdominal pain EPIDEMIOLOGY Spread by droplet, direct contact PHYSICAL EXAM Extreme hyperemia covered by punctate or confluent yellow-gray exudate with edema, lymphoid hyperplasia; ±30 million cases/yr. See Beta-hemolytic streptococcus

strephosymbolia NEUROLOGY The reversing of letters and words in reading and writing, which may occur in learning disability. See Dyslexia

streptococcal pharyngitis Strep throat, see there

streptococcal proctitis Perianal streptococcal cellulitis INFECTIOUS DISEASE Anorectal inflammation due to streptococci, which occurs in children following 'strep throat'/pharyngitis due to autoinoculation CLINICAL Perianal erythema, itching, pain during defecation, fever. See Strep throat

streptococcal sex syndrome A rare recurrent erythroderma due to postcoital *Streptococcus agalactiae* bacteremia, with poor lymphatic drainage in ♀ who have had perineal RT or lymph node excision for malignancy

Streptococcus agalactiae A streptococcus normally found in the GI tract, which may cause UTIs, subacute bacterial endocarditis. See Group A *Streptococcus*, Group B *Streptococcus*

Streptococcus pneumoniae MICROBIOLOGY A pathogenic streptococcus with 90 serotypes associated with pneumonia, bacteremia, meningitis TRANSMISSION Person to person INCIDENCE Before 2000, *S pneumoniae* infections caused 100K-135K hospitalizations for pneumonia, 6 million cases of otitis media, and 60K cases of invasive disease–including 3300 cases of meningitis; sterile-site infections have a geographic variation of 21-33/10⁵ RISK GROUPS Elderly, children < age 2, African Americans, Native Americans, day care center inmates, and persons with underlying medical conditions including HIV infection and sickle-cell disease PROPHYLAXIS 88% of clinical isolates of *S pneumoniae* are serotypes in the 23-valent polysaccharide vaccine. See Meningitis

streptokinase Kabikinase®, Streptase® CARDIOLOGY A fibrinolytic enzyme that converts plasminogen to plasmin, and used as a thrombolytic for reperfusing acutely occluded coronary arteries. See Thrombolytic therapy. Cf APSAC, tPA

streptomycin INFECTIOUS DISEASE An antibiotic used for TB ADVERSE EFFECTS N&V, dizziness, rash, fever, ototoxicity, nephrotoxicity

streptozotocin ONCOLOGY An alkylating nitrosurea derivative used to treat lymphomas and other CA ADVERSE EFFECTS Renal tubular dysfunction, glycosuria, hypoglycemia

stress A force that causes a change in physical or mental health. See Biotic stress, Oxidative stress, Physical stress, Stressor PSYCHOLOGY A noxious physical or mental stimulus that may cause a loss of self-control CLINICAL Depression, over/undereating, too tired for sex, anger, crying, physical Sx fatigue, headache, backache, insomnia, anxiety, palpitations, ↑ colds/flu, nervous stomach, skin complaints; feeling of disorganization, loss of concentration. See Chronic stress, Job stress, Mental stress, Physician stress, Shear stress, Social stress, Workplace stress

STRESS CARDIOLOGY A clinical trial–Stent Restenosis Study comparing outcomes of coronary stent placement to balloon angioplasty in treating CAD. See Balloon angioplasty, Coronary angioplasty, Coronary artery disease, Stenting

stress echocardiography CARDIOLOGY An echocardiogram performed when the Pt is exercising. See Radionuclide perfusion imaging

stress fracture Fatigue fracture, insufficiency fracture ORTHOPEDICS A fracture due to repeated, unidirectional stress or strain on a particular musculoskeletal zone, or due to repeated relatively trivial bone trauma PEOPLE & PLACES Runners, ballet dancers–fibula, tibia; soldiers–metatarsal bones; jackhammer/pneumatic drill operators–metacarpal bones, office workers–coccyx. See March fracture

stress incontinence Dribbling bladder, fallen bladder; loss of pelvic support; urinary incontinence UROLOGY An involuntary loss of urine linked to ↑ internal abdominal pressure–eg, coughing, sneezing, laughing, physical activity; SI is a storage problem in which the urethral sphincter's strength is ↓, and sphincter reacts to ↑ abdominal pressure by releasing urine ETIOLOGY Weak pelvic muscles that support the bladder, or urethral sphincter defect, due to prior trauma to the urethral area, neurologic injury, drugs, prostate or pelvic surgery, multiple pregnancies, pelvic prolapse–protrusion of bladder, urethra, rectal wall into vaginal space, with a cystocele, cystourethrocele, rectocele, ↓ estrogen; SI affects 20% of ♀ ≥ age 75 RISK FACTORS Age, obesity, chronic bronchitis, asthma, childbearing

Stress Index PSYCHOLOGIC TESTING An analogue visual rating scale ranging from no stress–1 to unbearable stress–7

stress-related condition PSYCHOLOGY Any medical condition caused by physical or mental stress SRCs Bruxism, gastric ulcers, HTN, insomnia, irritable bowel syndrome, migraines, tachyarrhythmias, tension headaches, tics MANAGEMENT Possibly biofeedback training. See Biofeedback training, Stress management, Stress therapy

stress response Flight-or-fight response, see there

stress shutdown PSYCHIATRY A loss of cognitive abilities due to excess anxiety, which may involve psychologic mechanisms or a neurotransmitter defect

stress test A closely monitored clinical maneuver used to evaluate the body's ability to respond to a cardiac, mental, pharmacologic, physical, or supraphysiologic stressant; in practice, ST is usually understood to mean cardiac stress test. See Exercise tolerance test, Treadmill exercise stress test

stress ulcer Stress ulceration GI DISEASE An erosion of the gastric mucosa, attributed to physical or mental stress RISK FACTORS Respiratory failure, coag-

ulopathy MANAGEMENT Ranitidine. See Executive monkey, Ranitidine, 'Toxic core,' Type A personality. Cf Sucralfate

STRESS ULCER TYPES*

ACTIVITY ULCER A type of gastric erosion produced when rats are placed in a running wheel with access to food for only 1 hr/day

EXERTION ULCER Gastric ulceration that is associated with excessive and unexpected forced activity, eg a rotating cage keeps the rodents constantly running and the gastric juices flowing to the maximum

RESTRAINT ULCER An ulcer that appears in rats hours after placement in a very confined spaces, especially when the ambient temperature is lowered

SHOCK ULCER Gastric ulcer in humans related to burns, eg Curling's ulcer, ischemia, neurologic injury, eg Cushing's ulcer, sepsis or trauma

*Given the association of gastric ulcers with *H pylori*; many respond to antibiotics; because bleeding from SUs is relatively rare, but has a high mortality–in one report, 49% vs 9% without hemorrhage, the use of prophylactic measures, eg neutralization of gastric acid, ↓ gastric acid secretion and cytoprotection is commonly recommended; prophylaxis is best administered to those at highest risk–respiratory failure-odds ratio 15.6, coagulopathy-odds ratio 4.3) for GI bleeding and can be withheld from other Pts–NEJM 1994; 330:377ᴜᴀ

stressor A thing that ↑ mental or physical stress–eg, environment, housing, personal, sexual, social, work related. See Stress management PSYCHOLOGY A noxious physical or mental trigger that compromises mental function and control. See Stress

STRETCH CARDIOLOGY A clinical trial–Symptom, Tolerability, Response to Exercise Trial of Candesartan Cilexetil Atacand™ in heart failure

'stretch mark' A purplish vertical 'stripe' on the lower abdomen, thighs, iliac crests and breasts in pregnancy and after corticosteroid excess; such marks whiten after birth, but remain as lasting reminders of an expanded abdomen

stretch reflex Myotactic reflex NEUROPHYSIOLOGY Reflex contraction of a muscle when its tendon is stretched/pulled, especially abruptly; the SR is critical for maintaining an erect posture. See Reflex

stretching SPORTS MEDICINE An activity that ideally precedes exercise and is intended to keep the muscles supple and improve the joints' range of motion; in some exercises–eg, yoga, stretching *is* the exercise; a therapeutic maneuver designed to elongate shortened soft tissue structures and thereby ↑ flexibility. See Ballistic stretching. Cf Aerobic exercise

strict liability FORENSIC MEDICINE Liability without fault–eg, when one is responsible for all of the consequences of the actions of one's employees regardless of whether a result was intended or not. See Liability

stricture The closing of a luminal structure. See Biliary stricture, Esophageal stricture, Stenosis, Urethral stricture

stride length BIOMECHANICS The distance between 2 successive placements of the same foot, consisting of 2 step lengths; SL measured between successive positions of the left foot is always the same as that measured by the right foot, unless the subject is walking in a curve

stridor ENT A harsh medium- to high-pitched crowing heard when breathing, especially on inspiration, due to an airway obstruction in the larynx or trachea; in children, stridor may occur in a background of congenital laryngeal stridor–laryngomalacia, which usually improves with age, or persists or recurs due to allergies, URIs, papillomas, foreign bodies, mediastinal masses, cysts of lung parenchyma

string of beads sign String of pearls sign GI IMAGING A descriptor for the radiologic appearance of small intestine obstruction–the 'beads' correspond to pockets of gas oriented in an oblique line, a function of the amount of fluid and the intensity of peristalsis; characteristic of mechanical obstruction, the SOB sign may also be seen in adynamic ileus due to inflammation PULMONARY IMAGING A descriptor for the distribution of sarcoid granulomas along the pulmonary septae VASCULAR DISEASE A descriptor for the multiple arterial dilatations and strictures seen in fibromuscular dysplasia, a condition affecting small and medium-sized arteries including the renal and extracranial cephalic vessels

string sign GI IMAGING–COLON A linear fraying of the column of a barium study of the lower GI tract with luminal stenosis, spasm, ulceration and scarring, seen in the terminal ileum in Crohn's disease; grossly, affected intestine is thickened, rigid and likened to a garden hose STOMACH An elongated, narrowed and straight single, occasionally dual channel of contrast as seen in well-developed hypertrophic pyloric stenosis

string test OBSTETRICS See Spinnbarkeit RHEUMATOLOGY See Mucin clot test UROLOGY A macroscopic method for determining active spermatogenesis; the testicle is bisected, forceps are used to grasp the parenchyma and visually note the separation of the strands of fibers

strip See Filter strip, Urine test strip

stripping COSMETIC SURGERY See Lip stripping SURGERY Removal of the renal capsule from the kidney and pedicles in order to interrupt the lymphaticorenal fistulas–eg, to manage Pts with intractable chyluria induced by *Wuchereria bancrofti*

stroke Cerebrovascular accident NEUROLOGY A sudden focal neurologic defect lasting > 24 hrs, which is characterized by abrupt loss of consciousness due to either hemorrhage or vascular occlusion of cerebral blood vessels, leading to immediate paralysis, weakness, speech defects; a sudden onset of neurologic deficit of vascular origin; strokes are a leading cause of disability in developed countries–500,000 new victims/yr, US, 20-30% of whom are left with severe residua; strokes are the 3rd leading cause of death–20-30% early mortality; the incidence of stroke rises dramatically with age; the risk doubles every decade after age 35 STATISTICS, MORTALITY < 80 deaths/10^5: Whites in US–especially in the midwest, Australia, New Zealand, northern Europe, Egypt; >130 deaths/10^5: Black US–especially in south, Russia, mainland China, former Eastern Blocks, Argentina CLINICAL Paralysis, weakness, sensory loss, speech defects ETIOLOGY ASHD, dissection or stenosis of carotid artery, cocaine, embolism, HTN, fibromuscular dysplasia, syphilis TREATMENT Warfarin ↓ risk of stroke in Pts with A Fib or previous MI; in poor candidates for warfarin therapy, aspirin–which is less protective ± ticlopidine; carotid endarterectomy–useful if 70+% stenosis; CE's role in asymptomatic Pts is uncertain; dipyridamole and sulfinpyrazone are useless. See Completed stroke, Delayed stroke, Embolic stroke, Hemorrhagic stroke, Recent completed small stroke, Sunstroke, Working stroke

stroke belt EPIDEMIOLOGY A popular term for a region of southeastern US with a ± 2-fold ↑ in stroke mortality rate. See Stroke, Stroke buckle

stroke buckle EPIDEMIOLOGY A popular term for 3 southern states–the Carolinas and Georgia, with an extremely high stroke mortality rate. See Stroke belt

strokes Kudos PSYCHOLOGY A popular term for ego-gratifying verbalizations–eg, compliments, bestowed on a person by others. See Social interaction SPORTS MEDICINE The number of times the ball is whacked from tee-off to cup

Stromectol® Ivermectin, see there

strontium-89 RADIATION ONCOLOGY A beta-emitting bone-seeking radioisotope–$T_{1/2}$ 50.4 days, used to manage pain in Pts with diffuse bone metastases–eg, with end-stage breast or prostate CA for whom pain medication may be ineffective

Stroop test PSYCHOLOGY A test used to measure a person's sustained attention–eg, for word reading and color naming–with/without interference. See Psychological testing

structural imaging Anatomical imaging IMAGING An imaging philosophy–exemplified by film, CT, MRI, that examines the anatomic basis of changes caused by disease. See CT, MRI. Cf Molecular imaging

structure VOX POPULI The appearance of a thing. See Age structure, Bile duct structure, Cloverleaf structure, Pharyngeal structure, Sleep structure, Wolffian duct structure

struma ovarii The presence of mature thyroid tissue as the predominant tissue in an ovarian teratoma, which may cause clinical hyperthyroidism; 5-10% of struma ovarii become malignant, which is indicated by metastasis

strychnine TOXICOLOGY A highly toxic alkaloid from *Strychnos nux-vomica*, commonly used as a rodenticide, that elicits CNS hyperactivity, causing painful, recurrent tonic seizures, muscle tightness, cramping, risus sardonicus, marked flaccidity, decorticate posturing and death; Sx appear at 15 mg, death occurs with doses > 60 mg MANAGEMENT Control seizures with diazepam and phenobarbital; for muscle relaxation, curare, succinylcholine

Stryker saw ORTHOPEDICS A hand-held electric saw with an oscillating blade, used to cut the cranium, ribs and other bones. See Saw

STS 1. Serologic tests for syphilis, see there 2. Soft tissue sarcoma, see there

stubborn VOX POPULI → MEDTALK Refractory; unresponsive to therapy

stucco keratosis DERMATOLOGY A form of seborrheic keratosis; it is gray-white, symmetrical, measures 1-3 mm in diameter, on distal extremities, commonly on elderly ♂ feet TREATMENT Remove by scraping

student loan Medical student debt, see there

student's elbow Bursitis of the olecranon related to prolonged resting of the elbows on table tops, classically seen in students

study The formal examination of a phenomenon or the relationship between two or more factors in the pathogenesis or management of a disease. See ABC study, Analysis, BEIR study, Biological Indicators of Exposure study, Blinded study, CAESAR AIDS study, Cache County study, Case study, Case-control study, Cheese Study, Clinical research, Clinical study, Cohort study, Combination study, Concorde study, Contract study, Cross-sectional study, Descriptive study, Disease and symptom prevalence study, Dose-ranging study, Double blinded study, Double contrast study, Double labeling study, Electrophysiologic study, Enteroclysis study, Epidemiologic study, Experimental study, Feasibility study, FRIC study, Harvard-Hsiao study, Health outcomes study, Intracardiac electrophysiologic study, LETS study, Level II study, Longitudinal study, Massachusetts Male Aging study, Mechanistic study, Mixing study, MONICA, National Polyp study, Nerve conduction study, Nun study, Nurses' Health study, Observational study, Parametric study, pH study, phase I/II/II/IV study, Pilot study, Platelet aggregation study, Port Pirie Cohort study, Prevalence study, Primary study, Prospective study, Quasi-experimental study, Red cell survival study, Replication study, Retrospective study, Rochester study, Safari study, Schecter study, Seven-yr study, Six Cities study, Sixty Plus study, SMART surveillance study, Synthetic/integrative study, Trial, Triple-blinded study, Tuskegee study, Twin study, Understudy, Upper GI study, Viral study, Women's Health Initiative Observational Study

STUMP Smooth muscle tumor of undetermined malignant potential. See Borderline tumors

stump SURGERY That part of an extremity, or organ–eg, stomach, that remains after partial resection. See Amputation, Gastric stump

stump blowout SURGERY Leakage of a blind end–'stump' of a duodenum partially resected for an ulcer; the leak may be due to technical error or suture line failure, especially in a previously scarred or edematous duodenum; complications of leakage include peritonitis, hepatic bed abscess, pancreatitis, external fistula formation with electrolyte derangement

stump carcinoma ONCOLOGY A CA arising in the gastric 'stump' remaining after a subtotal gastrectomy. Cf Scar cancer

'stunned' myocardium Transient–hrs to days in duration postischemic contraction defects that follow myocardial reperfusion in acute MI. Cf Hibernating myocardium

Sturge-Weber syndrome Encephalotrigeminal angiomatosis NEUROLOGY A rare disorder characterized by mucocutaneous angiomatosis with port wine stains that also affects the meninges. See Neurocutaneous syndrome

stuttering CLINICAL MEDICINE *adjective* Referring to an intermittent progression of a disease state, characterized by a staccato pattern of deterioration, as classically occurs in multiple sclerosis or in a stroke in evolution SPEECH PATHOLOGY *noun* A defective speech pattern more common in ♂ which affects ± 1/100 characterized by irregular repetition of syllables, words, or phrases, hesitation and interruption of fluent speech, commonly, a staccato repetition of the first phoneme of a spoken phrase; it may be accompanied by facial grimacing, postural gestures, involuntary grunts or loss of airway control MANAGEMENT Bethanechol. See Cluttering

sty External hordeolum, eyelash follicle infection OPHTHALMOLOGY An acute infection of the eyelid's sebaceous glands or hair follicles/eyelashes due to blocked glands in eyelid; once glands are blocked, bacterial infection–often staphylococcal–ensues, resulting in a gob of pus/pimple; the infection resolves when the pus drains from the sty CAUTION Don't pop pimple, it could worsen; it usually resolves on its own

Stypven time A one-stage prothrombin time assay performed with Russell viper venom–Stypven, prolonged in defects of coagulation factors V, X, thrombocytopenia. See Prothrombin time

subacute *adjective* Relating to a disease process that develops more slowly than an acute process–often in 1 to 3 days, but more rapidly than a chronic process which, for some workers, is more than 2 wks

subacute bacterial endocarditis Subacute endocarditis CARDIOLOGY A chronic bacterial infection of the endocardium and heart valves eventually leading to valve destruction and deformity PROPHYLAXIS Antibiotics to prevent SBE are recommended for Pts undergoing heart surgery (1) Prosthetic cardiac valve of all types; (2) Previous bacterial endocarditis; (3) Most congenital cardiac malformations; (4) Rheumatic and other acquired valvular dysfunction; (5) Hypertrophic cardiomyopathy; (6) Mitral valve prolapse with valvular regurgitation; antibiotics are also recommended for (1) Dental procedures known to induce gingival bleeding–this includes cleaning; (2) Uretheral catheterization in a Pt with a UTI; (3) Incision/drainage of infected tissue–antibiotics are pathogen-targeted; (4) Vaginal delivery in the presence of infection

subacute combined degeneration Subacute combined degeneration of the spinal cord NEUROLOGY A disorder caused by vitamin B$_{12}$ deficiency, characterized by weakness, sensory defects, mental changes, vision defects, often associated with megaloblastic anemia CLINICAL SCD causes demyelination, affecting the spinal cord, brain, optic and peripheral nerves, with axonal destruction. See Vitamin B$_{12}$ deficiency

subacute necrotizing encephalomyelopathy Leigh's disease An AR condition of neonatal onset CLINICAL Swallowing and feeding difficulties, hypotonia, hyporeflexia, weakness, ataxia, peripheral neuropathy, external ophthalmoplegia, impaired hearing and vision, seizures, convulsions. See Wernicke's encephalopathy

subacute sclerosing panencephalitis Dawson's encephalitis, SSPE NEUROLOGY A slow virus-induced inflammation evoked by the measles virus or its vaccines CLINICAL Onset in childhood, with mental dysfunction, dyskinesia, myoclonus, hypotonia, emotional lability; it is usually fatal in 1-3 yrs LAB Paretic gold curve with ↑ CSF Igs MANAGEMENT None. Cf Prion

subacute thyroiditis De Quervain's, giant cell, granulomatous thyroiditis ENDOCRINOLOGY Inflammation of thyroid post infection–eg, mumps, influenza, coxsackievirus, adenovirus CLINICAL Persistent neck pain, hyperthyroidism RISK FACTORS Recent viral infection, ♀, young. See Hyperthyroidism

subarachnoid hemorrhage NEUROLOGY A severe intracerebral hemorrhage in the subarachnoid space, often due to a ruptured intracranial aneurysm CLINICAL Abrupt severe headache, loss of consciousness, vomiting ETIOLOGY Ruptured cerebral aneurysm, AV malformation, idiopathic RISK FACTORS Aneurysms, polycystic kidney disease; fibromuscular dysplasia, HTN MANAGEMENT Ventilation, oxgenation, fluid, tissue dehydration–eg, mannitol; surgical evacuation; bed rest, sedation, analgesia, anti-seizures. See Warning leak. Cf Intracerebral hemorrhage

subclavian steal syndrome Aortic arch syndrome, carotid artery occlusive syndrome, subclavian artery occlusive syndrome, vertebral-basilar artery disease NEUROLOGY Cerebrovascular insufficiency due to stenosis or occlusion of the left subclavian artery proximal to the origin of vertebral artery ETIOLOGY ASHD, blood clots, trauma, congenital defects and vascular malformations, after neurosurgery, Takayasu's arteritis, thrombosis, trauma, and tumors, exacerbated by exercise of the upper extremity CLINICAL Neurologic Sx, ↓ pulse, altered BP. See Steal. Cf 'Robin Hood syndrome'

subclinical hyperthyroidism A low serum TSH concentration in an asymptomatic person with normal serum thyroid hormone concentration; SH is more common in older–> age 60 Pts, and detected by measuring TSH ETIOLOGY Solitary thyroid adenoma, multinodular goiter, subclinical Graves' disease CLINICAL A Fib, atrial premature contractions, ↑ pulse rate, ↑ left ventricular mass and contractility, osteoporosis MANAGEMENT Treatment may be unnecessary

subclinical hypothyroidism An ↑ TSH before or after administration of TRH in the face of normal T$_3$ and T$_4$; SH affects 6-7% of ♀ and 2-3 of ♂, with 5-10% annual rate of progression to overt hypothyroidism in children and adolescents, SH is a benign remitting process, and may not require therapy

subclinical infection An infection in which Sx are mild or inapparent, and may not be diagnosed other than by positive confirmation of the ability to transmit the infection or serologically

subconjunctival hemorrhage OPHTHALMOLOGY A bright hemorrhagic patch on the bulbar conjunctiva caused by rupture and bleeding of a superficial small capillary, due to ↑ pressure–eg, violent sneezing or coughing; SHs occur in newborns as a bright red sickle-shaped hemorrhage at the margin of the cornea and conjunctiva, attributed to abrupt pressure changes over the infant's body during delivery

subconscious NEUROLOGY Obtunded, see there

subcutaneous insulin-resistance syndrome Insulin-resistance in type 1 DM attributed to an insulin-specific protease present in the subcutaneous tissue; the existence of same is increasingly controversial; if SIRS *does* exist, it is very rare. See Dawn phenomenon, Somogyi effect

subcutaneous mastectomy BREAST SURGERY Surgery in which up to 95% of the breast is removed without disturbing the skin and nipple; once the native tissue is removed, an implant is placed in the defect. See Radical mastectomy

subcutaneous port NURSING A tube inserted in a blood vessel and attached to a disk placed under the skin, to administer IV fluids/drugs and sample blood

subdural effusion NEUROLOGY A protein-rich CSF excess under the dura, seen in ± 50% of children with *H influenzae* meningitis CLINICAL, INFANTS Bulging fontanelle, widened sutures, ↑ head circumference, persistent fever, ↑ intracranial pressure → vomiting, seizures MANAGEMENT Antibiotics; drainage if no response

subdural empyema

subdural empyema NEUROLOGY An acute bacterial infection of the meninges, most common in infants with GNR or group B streptococcus meningitis CLINICAL Refractory seizures, motor defects MANAGEMENT Surgical evacuation

subdural hematoma NEUROLOGY Hemorrhage into the subdural space, which occurs when veins located between the meninges leak blood after a head injury; pressure from the "mass" damages brain tissue and causes loss of function which worsens as the hematoma enlarges and ↑ intracranial pressure, leading to cerebral edema, further ↑ of intracranial pressure, ±accompanied by herniation of the uncal gyri CLINICAL ↓ consciousness; acute SH progresses rapidly–Sx usually appear within 24 hrs of injury, followed by rapid deterioration; subacute SH usually develops Sx 2 to 10 days post injury because of slower leaking of blood RISKS FACTORS Head trauma in extremes of age, overuse of aspirin, anticoagulants, alcoholism. See Subdural hematoma. Cf Extradural hemorrhage

subfertility REPRODUCTION MEDICINE A ♂ condition in which semen parameters are below the lower limits of normal on ≥ 2 occasions CRITERIA Volume < 1.5 mL, sperm concentration < 20 million/mL, sperm viability < 60%, motility < 2–on a scale of 1 to 4, and > 60% abnormal forms; subfertile semen may also demonstrate hyperviscosity, sperm agglutination, polyspermia and/or hematospermia. Cf Anti-sperm antibodies, Infertility

subinternship Acting internship A training period for a senior medical student who acts as an intern during rotation through various services–eg, internal medicine, surgery, pediatrics; the subintern is closely supervised by a resident or an attending physician. Cf Extern, Intern, Scut monkey

subject CLINICAL RESEARCH A person being studied. See Human subject MEDTALK → VOX POPULI Person

subjective probability estimate INTENSIVE CARE A probability, estimated for an individual Pt, of dying or of having another defined outcome, based on the personal knowledge and experiences of the prognosticator–eg, a physician with previous Pts. Cf Objective probability estimate

subjective vertigo The sensation that a person is revolving in space. See Vertigo

sublethal *adjective* Referring to that which does not kill a cell or organism, but usually forces adaptation for survival

sublimation PSYCHIATRY An ego defense mechanism, operating unconsciously, by which instinctual drives, consciously unacceptable, are diverted into personally and socially acceptable channels

sublingual testosterone Androtest-SL® UROLOGY A format for delivering physiologic testosterone, where a molecule of testosterone forms an inclusion complex with hydroxypropyl-β-cyclodextrin. See Infertility

subluxation CHIROPRACTIC A motion segment in which the spine's alignment, movement or physiologic function is altered although contact between joint surfaces remains intact. See Dysfunction ORTHOPEDICS An incomplete joint dislocation with parts of the articular surfaces remaining in contact, with either a gradual displacement or partial dislocation within a joint; subluxations are contrary to a joint's plane or motion, or exceeds its ROM. See Range of motion. Cf Dislocation, Luxation

submammary abscess Breast abscess, see there

submarining PUBLIC HEALTH A popular term for a mechanism of trama consisting of the sliding of a child under the lap strap of a 3-point safety belt during an MVA EFFECT Face, neck injuries SOLUTION Restrain younger children in a child safety seat See Air bag, Seat belt

submersion syndrome Near drowning A Sx complex due to prolonged submersion without death CLINICAL Tachypnea, mild hyperthermia, restlessness, vertigo, confusion, N&V, shock, with pulmonary congestion, edema and, depending on the water temperature, hypothermia. See Drowning. Cf Muddy lung

submission hold SPORTS MEDICINE A hold–eg, a choke-hold, or joint-lock used in wrestling or ultimate fighting in which the receiver is virtually incapacitated. See Ultimate fighting

subphrenic interposition syndrome Chilaiditi syndrome A condition caused by interposition of the colon between the liver and diaphragm, most often symptomatic in children CLINICAL Abdominal pain, vomiting, anorexia, constipation, abdominal distention–bloating MANAGEMENT Avoid 'gassy' foods; improves with age

subspecialty MEDTALK A field of health care expertise requiring 1–2 yrs of post-residency training or fellowship in a recognized program, which often has a closure examination. See Fellowship. Cf Specialty

substance abuse Drug addiction PSYCHIATRY Use of any substance for nontherapeutic purposes; or use of medication for purposes other than those for which it is prescribed; SA includes: 1. Use of illicit, potentially addicting drugs–eg, cocaine; 2. Misuse of prescribed drugs that stimulate or depress the CNS–eg, amphetamines or barbiturates; 3. Habitual use of commercially-available substances with known desired and deleterious effects–eg, alcohol, tobacco. See Addiction, Alcohol, Cocaine, Crack, Ice, Marijuana

substance dependence PSYCHIATRY A maladaptive pattern of substance abuse, leading to clinically significant impairment or distress; DSD is formally defined by the DSM-IV as the presence of 3 or more clinical criteria. See Substance abuse

---

SUBSTANCE DEPENDENCE–3+ OF FOLLOWING

1. Tolerance, either
   a. A need for ↑ amounts, or
   b. ↓ Effect with continued use of same amount of substance
2. Withdrawal symptoms
3. The substance is taken in larger amounts than intended
4. A persistent but unsuccessful desire to ↓ substance intake
5. Much time is spent in activities needed to obtaining the substance or recovering from its effects
6. Important occupational, social, or recreational activities are sacrificed because of substance use
7. Continued substance use despite user's knowledge of its adverse physical and/or psychological effects

Modified from *Diagnostic & Statistical Manual of Mental Disorders, 4th ed, Wash, DC, Am Psychiatric Assn, 1994

---

substance use Substance abuse The *appropriate* use or ingestion of a substance or drug. Cf Substance abuse, Substance dependence

substantial factor test FORENSIC MEDICINE A test used to prove proximate cause in alleged negligence, when independent events are linked to harm ISSUE Was defendant's negligent act a substantial factor in causing the alleged harm. See 'But for' test, Negligence, Proximate cause

substantial pain Pain of a severity sufficient to impair function. See Brief Pain Inventory, Pain

substitution PSYCHIATRY An unconscious defense mechanism through which an unattainable or unacceptable goal, emotion, or object is replaced by one that is more attainable or acceptable. See Ego defense mechanism

substrate PSYCHIATRY The mental and/or emotional basis on which a particular response occurs. See Suicide substrate

subsyndromal depression Mild depression, see there

subtalar joint neutral Subtalar neutral ORTHOPEDICS The position in which the forefoot is locked on the rearfoot with maximum pronation of the midtarsal joint

subtotal laryngectomy SURGICAL ONCOLOGY Partial excision of the larynx for invasive CA of the epiglottis and false vocal cords. See Laryngectomy

succimer A water-soluble chelator administered per os for heavy metal poisoning–eg, lead poisoning in children > 2.17 μmol/L–US: 45 µg/dl or adults with lead poisoning due to gunshot wounds. See Lead, Saturnine gout

succinylcholine ANESTHESIOLOGY A potent, rapid depolarizing muscle relaxant that provides short-term paralysis for tracheal intubation during surgery, attenuating motor activity of seizures evoked by electroconvulsive therapy COMPLICATIONS Postoperative myalgia, transient ↑ in serum K+, ↑ in intraocular, intracranial, and intragastric pressure; malignant hyperthermia. See Cholinesterase, Dibucaine number, Depolarizing agent, Muscle relaxant

succussion splash PHYSICAL EXAM A splashing heard when rocking a Pt with a fluid collection in the pleural space, abdominal cavity, stomach or elsewhere

sucralfate A complex of sucrose sulfate and AlOH, used to prevent acute gastric injury, ↓ inflammation, heal peptic ulcers SIDE EFFECTS Constipation, dry mouth, N&V, headache, urticaria, rashes

suction The removal of a fluid or semifluid substance with a negative pressure device. See Liposuction

suction curettage Vacuum curettage GYNECOLOGY The removal of uterine contents postcervical dilation, applying suction through a hollow-tipped device, a common method for performing first trimester abortion

suction irrigator SURGERY Suction irrigator gravity flow–SIGF device for open general surgery, laparoscopic and open obstetric surgery, laparoscopic and open urologic surgery, endoscopic and open nasal surgery, otolaryngologic and plastic surgery

SUD 1. Substance use disorder 2. Sudden unexpected or unexplained death. See Sudden unexplained nocturnal death

sudden cardiac death CARDIOLOGY Death due to cardiac disease ≤ 1 hr after onset of Sx PUBLIC HEALTH A nontraumatic, nonviolent, unexpected event due to sudden cardiac arrest within 6 hrs of a previously witnessed usual state of normal health; SCD is the most common COD in the US–±400,000 deaths/yr, primarily of ♂ age 20-64; 75% are due to sustained ventricular arrhythmias ETIOLOGY Anomalous aortic origin–LCA or circumflex from right sinus of Valsalva–SOV; RCA from left SOV; CA origin from pulmonary artery, myocardial bridging, ostial stenosis, coronary arteritis or AV fistula, coronary artery dissection, embolism, hypoplasia, or spasm, idiopathic arterial calcification of infancy, fibromuscular dysplasia. See Sudden unexplained nocturnal death

sudden deafness AUDIOLOGY An abrupt hearing loss that follows a known cause of deafness–eg, an explosion, viral infection, or use of certain drugs

sudden death FORENSIC MEDICINE Precipitous demise, most commonly due to cardiovascular disease ETIOLOGY Ischemia, arrhythmia, shock, aortic dissection, CHF, accompanied by hypoxia, polycystic disease of heart, familial endocardial fibroelastosis, Kawasaki's disease, anaphylaxis, 'cafe coronary,' carbon monoxide, hydrogen sulfide, cyanide, nicotine, organophosphate pesticides, gastric rupture due to Mallory-Weiss syndrome, ulcers, septicemia, obstruction, bezoars, cerebrovascular lesions; SD is more common in alcoholics, smokers, nulliparous ♀, Pts with major psychiatric disease MOST COMMON AUTOPSY FINDING Pulmonary edema. See Atrial fibrillation, Cafe coronary. Cf Sudden unexplained nocturnal death

sudden infant death syndrome See SIDS

sudden unexplained nocturnal death Idiopathic death often in young, previously healthy, Southeastern Asian ♂, attributed to a conduction

system anomaly coupled with 'culture shock' and relocation-related stress in emigrants. Cf Sudden death

SUDS Sudden unexplained death syndrome. See Sudden unexplained nocturnal death

sufentanil ANESTHESIOLOGY An opioid anesthetic that is 1000-fold more potent than morphine used in cardiac anesthesia PROS Hemodynamic stability, no histamine release, no adverse SNS stimulation. See Opioid anesthetic. Cf Muscle relaxant

sugar A water-soluble, crystallizable carbohydrate that is the primary source of energy and structural components. See Amino sugar, Non-reducing sugar, Reducing sugar

sugar cube DRUG SLANG A popular street term for LSD, named for a common delivery "device", a sugar cube

sugar hypothesis POPULAR NUTRITION A controversial belief by many parents and some physicians, that refined sugars–eg, sucrose, further stimulate hyperactive children. See Diet, Feingold diet

sugar substitute NUTRITION Any of a group of carbohydrates–eg, fructose, sorbitol and xylitol, which are of potential use as replacements for the usual dietary sugars–glucose and sucrose in diabetics, as these molecules do not require insulin for certain steps in their metabolism; the efficacy of SSs in DM is less than optimal, since the diabetic liver converts a significant portion of fructose and its metabolites into glucose. See Artificial sweeteners, Aspartame, Cyclamates

suggestion PSYCHIATRY The influencing of a Pt to accept an idea, belief, or attitude suggested by a therapist. See Countersuggestion, Hypnosis, Hypnotic suggestion

suggestive of DECISION MAKING *adjective* Referring to a pattern by LM or imaging, that the interpreter associates with a particular–usually malignant lesion. See Aunt Millie approach, Defensive medicine

suicidal gesture Any behavior or action that might be–or might have been, in the case of successful completion thereof–interpreted as indicating a person's desire or intent to commit suicide. See Cry for help

suicidal ideation Suicidality PSYCHIATRY Mental thoughts and images which hinge around committing suicide. See Suicide

suicidal patient PSYCHIATRY A Pt at ↑ risk of committing suicide in the near future RISK FACTORS–♂: ≥ age 60, widowed, divorced, white, Native American, living alone, unemployed or having financial difficulties, substance abuse RISK FACTORS–♀: panic attacks, severe anhedonia, anxiety, hopelessness; 90% of Pts who commit suicide have a psychiatric disease, usually depression and/or alcohol abuse HIGH-RISK CONDITIONS Bipolar disorder, schizophrenia. See Suicide

suicide PUBLIC HEALTH The killing of oneself; at least 1% of any population consider suicide annually; many consult a non-psychiatric physician in the 6 months before suicide AGE Peaks in adolescence and college–age 15-25; more frequent in adolescent drug abusers MANNER–♂ Firearms 46%, hanging 22%, gas 16%, poison 10%; ♀ Poison 41%, strangulation 17%, gas 15%, drowning 10%, firearms 8.5%–suicide by firearms is increasingly popular in ♀ PHYSICIAN RATES Highest of all professionals; ♀ physicians are up to 3 times more likely to autodestruct than other ♀ professionals RISK FACTORS Mental illness, especially depression, schizophrenia; 15% of those with affective disorders die by suicide; 10-15% of alcoholics kill themselves, accounting for $1/4$ of all suicides; other 'at-risk' conditions include AIDS, cancer, spinal cord injuries, seizure disorders and Huntington's disease; $1/2$ are unmarried, whites are 2-fold more common than blacks INCIDENCE 28,000/yr–US, where it is the 8th leading COD, 12/10⁵; from 1950 to 1980, ♂ rate ↑ 305%; ♀ ↑ 67%; from 1955 to 1977, suicides jumped 230% in the

15-24 age group; suicide is attempted more often in ♀, but more often successful in ♂–♂ : ♀ ratio, 4:1 SUCCESS RATE Suicide attempt:success ratio, 5:1; North America has seasonal peaks in March, September; most occur at home; bodies are often discovered by family or friends. See Assisted suicide, Cluster suicide, Multishot suicide

suitcase sign Samsonite™ sign MEDTALK A popular term for a Pt's expectation that he/she will be admitted to the hospital, indicated by arrival with a packed suitcase

Sular® Nisoldipine, see there

sulcus sign ORTHOPEDICS A joint laxity test used clinically to diagnose shoulder instability. See Laxity test, Shoulder instability. Cf Provocative test

sulfa drug Any sulfur-based antibiotic, in particular sulfonamides

sulfadiazine PHARMACOLOGY A sulfonamide used with pyrimethamine to treat toxoplasmosis in AIDS and meningitis ADVERSE EFFECTS BM suppression, crystallization in urinary tract, leading to kidney failure. See Sulfonamide

sulfamethoxazole PHARMACOLOGY A sulfonamide used for UTIs or, with trimethoprim for PCP. See *Pneumocystis carinii* pneumonia, Trimethoprim-sulfamethoxazole

sulfamethoxazole/trimethoprim Bactrim, see there

sulfasalazine THERAPEUTICS An agent used for inflammatory conditions such as ulcerative colitis and arthritis ADVERSE EFFECTS Headaches, N&V, abdominal pain. See AIDS

sulfisoxazole Gantrisin® PHARMACOLOGY A sulfonamide used primarily to treat UTIs

sulfite Sulfiting agent FOOD INDUSTRY An agent used as a food preservative; up to 5% of asthmatics are sensitive to sulfites–possibly due to low levels of sulfite oxidase, and respond to sulfites with nausea, diarrhea, bronchospasm, pruritus, edema, hives, potentially anaphylactic shock and death; some drugs used for asthma may contain sulfiting agents. See Food allergies, Pseudoallergies

sulfone syndrome See Dapsone syndrome

sulforaphane Sulforaphane glucosinolate, sulphoraphane ONCOLOGY An isothiocyanate phytochemical or 'nutriceutical' in cruciferous vegetables–eg, broccoli, which may prevent CA. See Cruciferous vegetables, Nutriceutical

sulindac Clinoril® An NSAID that binds to plasma proteins. See NSAIDs

sulphur BRITISH Sulfur

sulprostone OBSTETRICS A PG analogue used with mifepristone–RU 486 as an abortifacient. See Abortion, Gemeprost, Mifepristone, Misoprostol, Morning-after pill

sumatriptan succinate Imitrex® NEUROLOGY An agent used for migraines PROS ↓ Headaches, clinical disability, nausea, photophobia CONTRAINDICATIONS CAD, uncontrolled HTN, Prinzmetal's angina, concurrent MAOI therapy ADVERSE EVENTS Angina, stroke, TIAs, HTN, ischemic bowel disease. See Migraine. Cf Losartan

summary suspension HOSPITAL PRACTICE The immediate termination of a physician's medical staff membership over concerns for the safety of Pts, employees, or any other persons in the hospital. See Request for corrective action. Cf Automatic suspension

summation gallop Triple gallop CARDIOLOGY An extra 'galloping' heart sound caused by tachycardia, best heard over the apex with the Pt in supine or left lateral position; it ↑ with inspiration, exercise, elevation of legs, ↑ venous return. Cf Fourth heart sound

sunblock PUBLIC HEALTH An opaque substance, usually formulated from zinc or titanium oxides, designed to completely prevent solar radiation from reaching the skin. See SPF rating. Cf Sunscreen

sunburst pattern OPHTHALMOLOGY A pattern seen on funduscopy of the retina in stage III proliferative retinopathy of sickle cell disease, characterized by areas of retinal pigment epithelial hyperplasia, which develop after intra– and subneural retinal hemorrhage, more commonly in Hb SS disease than in SC disease. See Proliferative sickle retinopathy. Cf Seafan pattern RADIOLOGY A descriptor for the appearance of a periosteal reaction, in which dense filiform spiculations are perpendicular to the periosteum, classically in bone infiltration by typical and parosteal osteosarcomas, as these usually evoke a minimal periosteal reaction; 'sunburst' also refers to irradiating spiculation of the ileal bone around the acetabulum in Voorhoeve syndrome–osteopathia striata

SUND Sudden unexplained nocturnal death, see there

Sunday neurosis PSYCHIATRY A popular term for a constellation of Sx in persons who function well in a planned and organized work setting, as scheduled activities tend to 'bind' their anxieties; when presented with an unstructured weekend or holiday, the usual defense mechanisms are 'offline,' and they may suffer anxiety, conversion reactions, dissociative states, obsessions, compulsions, phobias

sundown syndrome Sundowning PSYCHIATRY A term for disorientation , agitation, or general worsening of mental Sx affecting some elderly persons at dusk or nightfall CLINICAL Picking at bed clothing, banging on bedrails, shouting ETIOLOGY Sedatives, analgesics, hypnotics RISK FACTORS Dementia, organic brain syndrome, vision or hearing defects, dehydration, fatigue

sunglasses A tinted pair of glasses used to ↓ light arriving at the eye, which are labeled according to the amount of UV light blocked; nonprescription glasses are classified according to use and amount of UV radiation blocked

---

SUNGLASSES

COSMETIC SUNGLASSES Block 70% of UVB, 20% of UVA; designed for 'around town' use, 'looking cool'

GENERAL PURPOSE SUNGLASSES block 95% of UVB, 60% of UVA; designed for most outdoor activities

SPECIAL PURPOSE SUNGLASSES block 99% of UVB, 60% of UVA; designed for very bright environments–eg, snow, open boat yachting, laser work; prescription sunglasses are constructed of plastic or glass–eg, CR-39, polycarbonate, polychromic glass; the solar radiation-protecting chemical is allyl diglycol carbonate NY Times 16/11/93; C5

---

sun protection factor Skin protection factor, SPF DERMATOLOGY A scale for rating sunscreens; sunscreens with an SPF of ≥ 15 provide the best protection from the sun's harmful rays. See SPF rating. Cf Sunscreen

sunscreen PUBLIC HEALTH A transparent substance lotion or cream containing oxybenzone and dioxybenzone, which absorbs or scatters UVB light, to ↓ the risk of actinic-related CA. See SPF rating Cf Melanoma, Sunblock, Tanning salon, Ultraviolet light

sunstroke SPORTS MEDICINE *'The outdated term "sunstroke:" should not be used for heat intolerance conditions because these conditions may occur in absence of sun and have been known to occur indoors… with high heat and excessive humidity'.* See Heat intolerance

super aspirin CARDIOLOGY A popular term for any agent–eg, antiplatelet GP IIb/IIIa receptor antagonist–orbofiban, sibrafiban, xemilofiban; SA relieves angina and ↓ risks of acute MI, but may be less effective than aspirin in preventing blood clotting post coronary angioplasty. See Aspirin, Lamifiban. Cf Baby aspirin

Superbowl Sunday abuse ANIMAL PSYCHOLOGY A phenomenon unique to the US, affecting wives/girlfriends of ♂ watching the Superbowl football game, which closes the National Football League's season. See Spousal abuse. Cf Domestic violence

supercritical result LAB MEDICINE Any result for an analyte–eg, glucose, Ca²⁺, K⁺–that is so markedly out of range that it warrants a paramedic or police search to locate a person with these values. See Critical results

superfecundation OBSTETRICS Fertilization of a 2ⁿᵈ ovum by sperm from a different act of coitus, after a first ovum has been fertilized; SF explains the rare cases in which twins have different fathers–ouch! See Higher multiples, Twin. Cf Superfetation

superfetation OBSTETRICS Fertilization and subsequent development of a 2ⁿᵈ ovum after the 1ˢᵗ has already implanted in the uterus; superfetation provides the theoretic explanation for the difference in times of delivery of fraternal 'twins'. Cf Superfecundation

Super Glue WOUND CARE A proprietary adhesive used for nonsuture closure of simple skin lacerations. See Laceration

superinfection INFECTIOUS DISEASE An infection that follows a prior infection, which occurs when native regional flora are substantially reduced, often by antibiotics, allowing invasion by opportunistic organisms, as in pseudomembranous colitis or esophageal candidiasis

superovulation REPRODUCTION MEDICINE Hyperstimulation of the ovaries with gonadotropins to drive development of dominant follicles; SO is commonly used in couples with unexplained or male-factor infertility. See Artificial reproductive technology

SuperSkin® WOUND CARE A protective skin barrier for preventing ulcers in bedridden Pts. See Pressure ulcer

Superstat hemostatic pad WOUND CARE A biocompatible sponge-like wafer which, when applied to wounds and surgical sites, dissolves in blood and forms a hemostatic gel that slows blood flow

super stress test CARDIOLOGY A type of exercise stress test, used to detect T-wave alternans in Pts at high risk for sudden cardiac death

superwarfarin TOXICOLOGY A warfarin–eg, brodifacoum, bromadiolone, difenacou, which is up to 100-fold more toxic, and has a $T_{1/2}$ 60 times longer than warfarin; ingestion causes severe and prolonged coagulopathy

super woman syndrome(s) A condition that may affect any ♀ with ≥ 2 X chromosomes, which tends to be more pronounced with more X chromosomes CLINICAL ♀ with 3 X chromosomes may be asymptomatic or have mental retardation, menstrual dysfunction, microcephaly, dental defects, strabismus, and hypertelorism; with 4 X, mental retardation is the norm, which may be accompanied by midfacial hypoplasia, micrognathia, radial synostosis, 5ᵗʰ finger clinodactyly, narrow shoulders, web neck; with 5 X, defects include growth retardation, FTT, mongoloid slant, saddle nose, PDA, colobomas, limb defects. See XXX, XXXX, and XXXXX syndromes

superficial *adjective* Toward or near a surface; confined to or pertaining to a surface

superficial spreading carcinoma ONCOLOGY A malignancy which, in the esophagus, is defined as one with an intramucosal extension of ≥ 20 mm from the primary lesion; while an SSC tends to respect the mucosa-submucosal interface, locoregional metastases are not uncommon

superficial spreading melanoma DERMATOLOGY A melanoma, 70% of which affect Pts from age 30 to 60, especially ♀ in lower legs or trunk, as a flat lesion–radial growth phase that may be present for months to yrs, average 5-yr survival 75% ETIOLOGY Recreational suntanning. See Melanoma

superficial thrombophlebitis VASCULAR DISEASE Venous inflammation due to a blood clot, which may be linked IV access sites, local venous trauma, or clots RISK FACTORS ↑ clotting tendency, infection, current or recent pregnancy, varicose veins, chemical or other irritation, prolonged sitting, standing, or immobilization, abdominal CAs–eg, carcinoma of pancreas, DVT, thromboangiitis obliterans, PE

superior mesenteric artery syndrome An uncommon condition caused by compression of the superior mesenteric artery, with obstruction of the 3ʳᵈ portion of the duodenum ETIOLOGY Loss of cushioning regional adipose tissues that maintain an appropriate arterial angle, seen in excess weight loss, rapid growth in children without corresponding gain of weight, in those with an asthenic habitus, or in Pts fixed in a hyperextended position by spinal injury or surgery CLINICAL Postprandial epigastric pain, distension, nausea, abdominal cramps, weight loss DIAGNOSIS Distension of proximal duodenum by barium studies and narrowing of angle between the aorta and SMA by aortography or sonography MANAGEMENT SMAS is a diagnosis of exclusion that may respond to conservative therapy–eg, adoption of a postprandial knee-chest position, smaller meals, elemental diet

superior vena cava syndrome Superior vena cava obstruction, SVC obstruction VASCULAR DISEASE The narrowing or obstruction of the SVC, a rare condition often linked to mediastinal inflammation, primary or metastatic CA, lymphoma, mediastinal tumors, fibrosis, inflammation–eg, TB, histoplasmosis, thrombophlebitis, aortic aneurysm, constrictive pericarditis CLINICAL Swelling of arms, head MANAGEMENT Treat 1° cause

superiority complex PSYCHIATRY A popular term for a constellation of behaviors–eg, aggressiveness, assertiveness, self-aggrandization, etc, which may represent overcompensation for a deep-rooted sense of inadequacy. Cf Inferiority complex

supervising physician MEDICAL PRACTICE A licensed physician in good standing who, pursuant to state regulations, engages in direct supervision of physician assistants whose duties are encompassed by the supervising physician's scope of practice

supervision GRADUATE EDUCATION The oversight of physicians-in-training by an experienced physician as part of the residency training; supervision is a fine art that requires 'alert hovering' so that '…errors of thought don't become errors of action'. See Libby Zion, Medical supervision

supine IMAGING *adjective* Pertaining to a posture in which the anterior portion of the body faces upward, the torso is aligned parallel to the reference surface, and hips and knees extended MEDTALK Lying on the back. See Position

supple PHYSICAL EXAM *adjective* Referring to free movement of a body part

supplementary groove Accessory anatomy DENTISTRY A furrow placed on the surface of molar crown–'cap' to facilitate grinding of food

supplementation NUTRITION The adding of nutrients–minerals, vitamins, to a diet ORTHOPEDICS See Viscosupplementation

SUPPORT TERMINAL CARE A study–Study to Understand Prognoses & Preferences for Outcomes & Risks of Treatments intended to evaluate decision-making processes and outcomes of seriously ill, hospitalized adult Pts regarding quality of care–vis-á-vis pain management, prolongation of life, provided to 4301 Pts with advanced stages of 1+ of 9 life-threatening illnesses

support CRITICAL CARE *verb* To maintain all necessary vital structures and functions that might be compromised–eg, blocked airways, heart in asystole, and monitor those physiologic parameters–eg, GI tract, renal function, that may not represent immediate dangers to life. See Advanced life support, Ancillary support, Basic life support, Life support, Single support PSYCHOLOGY Any form of interpersonal assistance in the form of listening or suggesting alternative solutions for an individual suffering mental stress. See Psychosocial support, Spousal support, Support group RESEARCH The providing of funding and resources to an individual or group of researchers. See Recommended levels of future support

support group PSYCHOLOGY A group of people with a similar disease–eg, CA, AIDS, who share encouragement, consolation, information regarding recovery, who meet regularly to help each other cope with the disease and/or therapy SOCIAL MEDICINE Those persons in an individual's 'circle,' who can be called in times of personal crisis EXAMPLES Children, spouses, siblings, friends, etc, who may help the person through the crisis, often by merely being 'good listeners'. See Companionship, Marriage bonus, Most significant other, Psychoneuroimmunology, Twelve step program; Cf Social isolation

support services PSYCHOLOGY Non-health care-related ancillary services–eg, transportation, financial aid, support groups, homemaker services, respite services, and other services

supportive *adjective* Pertaining to a Pt management philosophy in which only the Sx of a particular condition are treated; supportive measures are often taken when no specific and/or effective therapy is available or accessible–eg, viral meningitis, or exotic infections

supportive therapy Secondary therapy MEDTALK Any therapy given in addition to a primary therapy–eg, fluids for sepsis. See Primary therapy

suppository PHARMACOLOGY A solid form of medication of various shapes which, after insertion in the rectum, vagina, or urethra, dissolves and is absorbed into the blood VEHICLES Cocoa butter, polyethylene glycol DRUGS Aspirin, barbital, chloral hydrate, phenobarbital, procaine, quinine, resorcinol

suppression Slowing down, restraint, inhibition PSYCHIATRY The conscious effort to control and conceal unacceptable impulses, thoughts, feelings, acts

suppression test A test or assay–eg, dexamethasone suppression test, used to determine whether a substance–hormone or protein being produced in excess is under the control of regulating or releasing factor(s), and therefore responsive to a feedback loop, or whether the excess production is autonomous, and not under feedback control. Cf Stimulation test

suppuration INFECTIOUS DISEASE 1. The formation of pus 2. The discharge of pus. See Pus

supracricoid laryngectomy SURGICAL ONCOLOGY A procedure for managing laryngeal CA, in which the entire larynx is removed except arytenoids and cricoid cartilage which, with an intact base of tongue, preserves the voice, albeit with a breathy texture, retains the airway and, for most Pts, allows safe deglutition. See Head and neck cancer

suprahepatic inferior vena cava CLINICAL ANATOMY A functional component of the hepatic venous outflow system defined as a segment of the inferior vena cava, which extends from the entry of the right, middle, and left hepatic veins to the junction between the inferior vena cava and the right atrium

supraphysiologic *adjective* ENDOCRINOLOGY Referring to administration of hormones or other substances that are normally present in the circulation, in doses in excess of those normally produced by the body. Cf Homeopathic, Pharmacologic, Physiologic

supravalvular aortic stenosis CARDIOLOGY A congenital condition which may be localized or diffuse, and accompanied by intimal thickening and medial hypertrophy CLINICAL Angina, syncope, right-sided HTN TREATMENT Surgical

supraventricular tachycardia Paroxysmal atrial tachycardia CARDIOLOGY An arrhythmia initiated by a premature atrial beat in the AV node, SA node, or bundle to the point of bifurcation; once the ventricle con-

tracts, an echo atrial beat is stimulated via a retrograde tract, resulting in a reverberating re-entry phenomenon, resulting in an atrial rate of 180-300 beats/min MANAGEMENT Simple vagal stimulation–eg, carotid sinus massage, ice bag, breath-holding; if intense, cardioversion or digoxin therapy

surdomutism PSYCHIATRY An anxiety-induced inability to speak. See Hysterical conversion

SurePress bandage WOUND CARE A high-compression absorbent padding used to treat venous ulcers. See Pressure ulcer, Wound care

surface *adjective* Superficial *verb* VOX POPULI The outer part of a solid structure. See Working surface

surface anatomy 1. Anatomic structures that can be identified on the outside of the body 2. The study of anatomic structures that can be identified on the outside of the body

surfactant PHYSIOLOGY A mixture of phospholipids and proteins lining the alveoli that allows optimal gas–CO_2, O_2 exchange, by minimizing surface tension; surfactant is produced by type II pneumocytes and secreted as lamellar bodies into the amniotic fluid; fetal surfactant production begins by the 20th wk of gestation; it effectively prevents respiratory distress syndrome after wk 35. See L/S ratio, Respiratory distress syndrome, SP-A, SP-B, SP-C, Surfactant replacement therapy

surfactant replacement therapy NEONATOLOGY Intratracheally administered bronchoalveolar fluid derived from calves–98% lipids, comprised of 90% phospholipid, especially dipalmitoyl-phosphatidylcholine and 2% apoproteins, which markedly improves gas exchange in premature infants; prophylactic bovine surfactant administered intratracheally ↓ need for neonatal respiratory support and ↑ survival of premature infants, especially < 30 wks. See Hyaline membrane disease

surge Any marked ↑ in flow of a substance above a relatively constant baseline REPRODUCTIVE PHYSIOLOGY An abrupt ↑ in LH secretion BACKGROUND LHRL–luteinizing hormone-releasing hormone is normally secreted in episodic bursts, resulting in cyclical peaks of LH; LHRH bursts are ↑ by estrogens and ↓ by progesterone and testosterone and ↑ in frequency until the end of the follicular phase, at which time a surge of LH signals the onset of endometrial secretion, in preparation for a fertilized egg

surgeon A physician trained in surgical arts. See Breast surgeon, Colorectal surgeon, General surgeon, Maxillofacial surgeon, Neurosurgeon surgeon, Oral surgeon, Plastic surgeon, Vascular surgeon

surgerize *pronounced*, sur-jur-ize A popular verb for surgical therapy, as in '...the Pt was surgerized'

surgery 1. That branch of 'procedural' medicine which addresses physical defects and/or acquired lesions by operative design 2. Any procedure to remove or repair damaged tissues or diagnose disease. See Abdominal surgery, Band-Aid™ surgery, Beating heart surgery, Billboard surgery, Brain-graft surgery, Cancer surgery, Cataract surgery, Cardiothoracic surgery, Cardiovascular surgery, Chemosurgery, Conservative surgery, Cosmetic surgery, Cranial base surgery, Craniofacial surgery, Debulking surgery, Dermatologic surgery, Dry run surgery, Disfiguring surgery, Elective surgery, Emergency surgery, Esthetic surgery, Facial plastic & reconstructive surgery, Functional (endonasal) endoscopic sinus surgery, Ghost surgery, Hand surgery, Hand-assisted laparoscopic surgery, Head & neck surgery, Heart port surgery, Heroic surgery, Image-directed surgery, Keyhole surgery, Kiss of death surgery, Laparoscopic surgery, Laser surgery, Love surgery, Lung-reduction surgery, Major surgery, Mastoid surgery, MIDCAB surgery, Minimally invasive surgery, Minimally invasive cardiac surgery, Minimally invasive valve surgery, Minor surgery, Mohs surgery, Mutilating surgery, Neurosurgery surgery, Nintendo® surgery, No problem surgery, Open heart surgery, Optional surgery, Outpatient surgery, Palliative surgery, Perineal surgery, Phonosurgery, Port access surgery, Psychosurgery, Radiation surgery, Radical surgery, Radioimmunoguided surgery, Reconstructive surgery, Re-do vascular surgery, Refractive surgery, Required surgery,

Robotic surgery, Roller ball surgery, Same-day surgery, Second-look surgery, Stereotactic radiosurgery, Thoracic surgery, Tommy John elbow surgery, Unnecessary surgery, Urgent surgery, Videotaped surgery

surgical abdomen Acute abdomen–AA that requires surgical intervention–eg, acute appendicitis, acute cholecystitis, acute diverticulitis with bowel obstruction, CA, acute vascular disease–eg, infarction, abdominal aneurysm. See Acute abdomen. Cf NASA

surgical anatomy The study of anatomic structures and their relationships as required to obtain optimal access to a particular surgical field. See Anatomy

surgical blood management TRANSFUSION MEDICINE The manner in which blood or volume expanders are used in surgery, with the intent of minimizing the risk of infection and immunoreactivity, while reducing or eliminating allogeneic transfusions. See Autologous transfusion, Colloids, Crystalloids, Hemodilution, Intraoperative blood salvage, Postoperative reinfusion, Preoperative autologous donation

SURGICAL BLOOD MANAGEMENT TECHNIQUES

INTRAOPERATIVE BLOOD SALVAGE

POSTOPERATIVE REINFUSION

TRANSFUSION

Allogeneic donation

Autologous (self) donation

HEMODILUTION

Colloids

Crystalloids

surgical castration UROLOGY Surgical removal of testes–orchiectomy or ovaries–oophorectomy to stop sex hormone production, which slows growth of certain cancers

surgical closure Those devices or artifices used to close an open surgical field. See Absorbable sutures, Catgut, Nonabsorbable sutures, Silk, Staples, Tape

surgical history A history of the surgical procedures that a particular person has had, and complications therefrom, if any

surgical isolation bubble system SURGERY A proprietary system designed to maintain sterile operative field

surgical menopause GYNECOLOGY Cessation of native estrogenic activity after bilateral oophorectomy in a premenopausal woman

surgical oncology Oncological surgery The field of surgery dedicated to the operative ablation of neoplasia, generally, 'solid' tumors

surgical pathology The field of pathology dedicated to the analysis of tissues removed during surgery. See Anatomic pathology; Cf Clinical pathology

surgical review committee Tissue review committee, see there

surgical robot SURGERY A device used to perform some surgical procedures PROS Smaller incisions, ↓ pain, correction of surgeon error and tremors, ↓ infections and other post-surgical complications, faster recovery, better outcomes CONS High–±$1,000,000 price, approved for only a few procedures. See DaVinci, Robotic surgery

surgical scrubs Cotton or cotton/polyester wearing apparel consisting of a short-sleeved shirt and drawstring pants, is the universal uniform of those daring men and women of action, the surgeons, often faded Kelly green. Cf Whites

surgical service HOSPITAL PRACTICE Any service that is invasive or supports surgical specialties SSs Anesthesiology, cardiovascular surgery, ENT, general surgery, Ob/Gyn, orthopedics, pathology, podiatry, radiology. See Service. Cf Medical service

surgical site infection SURGERY Any superficial infection that occurs at the site of a surgical incision

657

SURGICAL SITE INFECTION CLASSES

I Clean wound–75% of all surgery; nontraumatic wound; noninflamed; no break in technique; no entry into the GI, GU, respiratory tracts, or oropharynx; infection rate < 5%, usually ±1%

II Clean-contaminated wound; minimal break in surgical technique; infection rate < 10%

III Contaminated wound: open, fresh, traumatic wound from relatively clean source, or major break in surgical technique; infection rate < 20%

IV Dirty and/or infected wound with devitalized and/or necrotic tissue; infection rate 30–40%

surgical specialty A specialty of health care in which interventions constitute a significant component of Pt management EXAMPLES OB/GYN, ophthalmology, ENT, surgery–cardiothoracic, colorectal, general, neurologic, orthopedic, plastic, urology. Cf RAPERs, Hospital-based medicine, 'Medical' specialty, Primary care

surgical sterilization Mechanical sterilization GYNECOLOGY Sterilization that prevents passage of a fertilized egg to the uterus, or of sperm meeting egg; the more common form of SS is tubal ligation, but vasectomy is not uncommon. See Tubal ligation, Vasectomy

surgicenter MEDICAL PRACTICE A place where outpatient–minor or 'same day' surgical procedures are performed–eg, removal of cysts or skin lesions; most are physician owned and operated businesses

SurgiLav Plus SURGERY A hydro-debridement system for cleaning debris and necrotic tissue from chronic wounds. See Debridement

Surmontil® Trimipramine, see there

Surodex® OPHTHALMOLOGY A controlled-release, intraocular drug-delivery system containing dexamethasone instilled into the anterior chamber of the eye during cataract surgery to ↓ pain and swelling postoperatively

Surplix Imipramine, see there

surrogacy See Gestational surrogacy

surrogate marker LAB MEDICINE A parameter or measured to detect a pathologic condition when a more specific test doesn't exist, is impractical or not cost-effective; surrogate testing has been used for non-A, non-B hepatitis, measuring ALT and antibodies to HBV core antigen–anti-HB$_c$ and HC

surrogate mother Contract mother, surrogacy SOCIAL MEDICINE A ♀ who agrees to be (artificially) inseminated with the sperm of the partner of an infertile ♀; the SM carries the baby to term, at which point it is adopted by the biological father and his partner. See Gestational surrogacy

surrogate motherhood True surrogacy Motherhood in which a ♀ carries a gestational product in her uterus that is not her own genetically–one haploid set of genes is contributed by the genetic or natural father and the 2nd haploid is contributed by the genetic mother who, for various reasons–eg, hysterectomy, uterus didelphys, cannot carry fertilized ova. See Artificial reproduction, Baby M, Gestational surrogacy

surrogate parenting Artificial reproduction, see there

surveillance EPIDEMIOLOGY 1. The monitoring of diseases that have a certain prevalence in a population 2. The ongoing systematic collection, analysis, interpretation, and reporting of health data. See Epidemiologic surveillance, Fluoride surveillance, Health surveillance, HIV surveillance, Immunosurveillance, Medical surveillance, Public health surveillance, Sentinel surveillance, Site-specific surveillance

surveillance scanning IMAGING The use of various imaging modalities–eg, CT and MRI, to detect and follow Pts with certain diseases–eg, brain tumors

survey CLINICAL RESEARCH A nonexperimental observational or descriptive evaluation of a group of individuals for the presence or absence of characteristics of interest. See AIDS belief survey, AUTS survey, CAP survey, HANES survey, Horizontal survey, National Nosocomial Infection Survery, SF-12 survey, Satisfaction survey, SF-36 survey, Skeletal survey

survival MEDTALK The length of time that a person lives after being diagnosed with a particular disease. See Disease-free survival, Event-free survival, Five-yr survival, Median survival

survival analysis STATISTICS Assessment of the amount of time that a person or population lives after a particular intervention or condition

survival curve EPIDEMIOLOGY A curve that starts at 100% of the study population and shows the percentage of the population still surviving at successive times for as long as information is available. See Survival

survivorship effect GERIATRICS A finding among persons in different age groups, where those who survive longer may be more robust ab initio, and more likely to survive, even were they to live in countries with higher rates of mortality in younger Pts

susceptibility The likelihood of suffering from an adverse effect or disorder when exposed to a noxious stimulus or pathogen. See Cancer susceptibility

susceptibility test Antimicrobial susceptibility test, see there

susceptible *adjective* 1. At risk of infection by a pathogen 2. Able to be killed by a particular drug

sushi A Japanese raw fish delicacy that may be a vector for parasites–eg, *Anasakis*, which often affect sushi made from mackerel caught in early spring ENDOSCOPY Edema, gastritis, erosion CLINICAL Myalgia, abdominal pain RADIOLOGY Thread-like larvae may be seen in radiocontrast studies TREATMENT Endoscopic removal

sushi syncope A transient condition caused by ingesting a bolus of wasabi, a 'hot' mustard–active ingredient: isothiocyanate used to flavor sushi. See Horseradish, Seder syncope, Spicy food, Sushi; Cf Hot peppers

suspected child abuse or neglect SCAN FORENSIC MEDICINE A potential case of child abuse. See Battered child syndrome, Child abuse, Infanticide

suspended heart 'syndrome' A radiologic finding in which the heart appears as if suspended in the mid-thorax–ie, cardio-thoracic 'separation', when viewed in a left oblique and occasionally in the right oblique position; the 'syndrome' is accompanied by low T waves in the II lead and a prominent S-T depression in the III lead and has no known clinical significance

Suspend™ sling UROGYNECOLOGY A surgical implant for treating stress incontinence in ♀. See Stress incontinence

suspension 1. The termination of an activity. See Pregnancy suspension, Summary suspension 2. A fluid solute in a solvent. See Gadolite® oral suspension, Jones suspension

suspicion See Clinical suspicion, Index of suspicion

suspicious *adjective* Referring to the consideration of a particular disorder–eg, cancer, as a diagnostic possibility, as in 'suspicious for malignancy'

Sustiva™ Efavirenz AIDS An non-nucleoside reverse transcriptase inhibitor NNRTI antiretroviral drug for treating HIV-1 infection approved for once-daily dosing in combination with other antiretroviral agents. See AIDS

Sutton's law DECISION-MAKING A guideline evoked to temper the enthusiasm of externs–US medical students in their 3rd and 4th yrs of school–and other novices in clinical medicine, who want to 'work up' a condition–eg, an acute abdomen for porphyria, metastatic medulloblastoma or other esoteri-ca, while ignoring a particular symptom's more common causes; the 'law' is attributed to the noted bank robber, Willie Sutton who, when asked why he robbed banks, reportedly replied, '...*that's where the money is*'; to apply Sutton's law then, is to search for the most likely cause of a symptom–ie, to go where the 'money' is. See Hoofbeats. Cf Red herring, 'Zebras'

suture *noun* SURGERY A material–eg, wire, thread–used to hold tissues in apposition. See Biosyn™ suture, Knotless anchor suture, Linatrix suture, Safil® synthetic absorbable surgical suture, Synthetic absorbable suture, Synthetic nonabsorbable suture. Cf Stable *verb* To join tissues by sewing

Sutureloop SURGERY A colposuspension needle suture used in bladder neck suspension and other procedures. See Colposuspension

SVC obstruction Superior vena cava syndrome, see there

SVGLAD Saphenous vein graft to the left anterior descending artery

SVGRCA Saphenous vein graft to right coronary artery

SVS Apex Plus OPHTHALMOLOGY An excimer laser used with a mask to treat hyperopia–farsightedness. See Excimer laser

SVT Supraventricular tachycardia, see there

swallowing disorder 1. Dysphagia, see there 2. Any of a group of problems that interfere with the transfer of food from the mouth to the stomach

swamp fever Synonym for 1. Equine infectious anemia, a viral infection of horses, transmitted by hematophagous arthopods, causing weakness, recurrent fever, marked anemia and muscular atrophy 2. Malaria 3. Marsh fever A waterborne infection–vector is *Rattus rattus*–by the spirochete, *Leptospira grippotyphosa*, which causes fever, general malaise and aseptic meningitis

Swan-Ganz catheter CARDIOLOGY A flexible balloon-tipped device placed in the pulmonary artery to monitor hemodynamics of critically ill Pts

Swan-Ganz catheterization INTERVENTIONAL CARDIOLOGY Insertion of a SG catheter into the pulmonary artery for hemodynamic monitoring of a critically ill Pt INDICATIONS Acute MI with hemodynamic instability, ARDS, suspected cardiac tamponade, CHF that responds poorly to diuretics, rupture of papillary muscle of mitral valves, severe hypotension, septic shock, monitoring of hemodynamics during cardiovascular surgery–eg, CABG CONTRAINDICATIONS Severe coagulopathy, left BBB, hypothermia, inadequate monitoring equipment COMPLICATIONS Thrombosis, PE, infection, infarction, arrhythmias, perforation of pulmonary artery, hemothorax, pneumothorax, heart valve trauma. See Swan-Ganz catheter. Cf Left heart catheterization

swan neck deformity RHEUMATOLOGY A descriptor for the hyperextended proximal interphalangeal joint and compensatory flexion of the distal interphalangeal joints caused by shortening of the extensor tendon, classically seen in rheumatoid arthritis, but also in SLE, post-rheumatic fever arthritis, psoriatic arthritis, scleroderma, camptodactyly. Cf Boutonniere deformity

sweat test PEDIATRICS A test used to diagnose cystic fibrosis–CF, which is characterized by defects in secretion–especially of sodium and chloride–by exocrine glands. See Cystic fibrosis. Cf Sweat testing

sweat testing NEUROLOGY A test used to identify defects of the autonomic nervous system; in contrast to ST for cystic fibrosis–which evaluates a specific component of sweat–ie, Cl, Na+, ST determines whether sweating is occurring at all; failure to sweat indicates a defect in the efferent sympathetic pathway between the hypothalamus and the skin or, far less commonly, lack of sweat glands, which occurs in ectodermal dysplasia. Cf Sweat test

sweet protein A protein–eg, monellin or thaumatin, that binds specifically to taste receptors, and evokes a sensation of sweetness 100,000 times > sugar on a molar basis. See Monellin, Thaumatin; Cf Artificial sweeteners, Aspartame, Cyclamates, Sweet herb

Sweet syndrome Febrile neutrophilic dermatosis A rare condition characterized by fever, neutrophilia, and erythematous non-ulcerating papules on the face, neck, upper thorax, extremities

sweetener See Artificial sweetener

swelling of eyes See Angioedema

swelling of optic disk VOX POPULI Papilledema, see there

SWIFT CARDIOLOGY A clinical trial–Should We Intervene Following Thrombolysis–comparing delayed elective intervention and conservative strategies after tPA/antistreplase thrombolysis in Pts with acute MI. See Acute myocardial infarction, Coronary angioplasty, tPA

swimmer's ear Acute diffuse external otitis SPORTS MEDICINE The most common form of external otitis, occurring primarily in summer swimmers CLINICAL Pain, serous, then seropurulent discharge TREATMENT Hospitalization, long-term high-dose antibiotics, targeting the usual pathogen *Pseudomonas aeruginosa*, using aminoglycosides and semisynthetic penicillin for 6 wks

swimming pool granuloma Fish bowl granuloma, Fish tank granuloma DERMATOLOGY An indolent opportunistic skin infection CLINICAL Mimics spirotrichosis with cellulitis, lymphadenitis and joint infections ETIOLOGY Atypical mycobacteria, *M marinum* or *M kansasii*, after percutaneous inoculation with contaminated fresh or salt water; the lesion develops into a solitary nodule at sites of abrasion–fingers, hands, elbows, knees and feet, and later becomes indurated and ulcerated, resembling cutaneous TB, occasionally with satellite lesions, without evidence of systemic disease; the lesions generally disappear with time–up to a yr MANAGEMENT Doxycycline, minocycline, clarithromycin, rifampin, TS. See Atypical mycobacterium, *Mycobacterium kansasii*, *Mycobacterium marinum*. Cf Fish tank granuloma

SWING CARDIOLOGY A clinical trial–Sound Waves Inhibit Neointimal Growth

swinging flashlight test NEUROLOGY A test for comparing direct and consensual reactions of each pupil, to ID afferent pupillary defects. See Marcus Gunn pupil

'swinging heart' CARDIOLOGY A fanciful synonym for the EKG findings in electrical alternans, characterized by a regular alteration in the direction and/or amplitude of one or more components of the EKG reading–eg, simultaneous oscillation of the P waves, QRS complexes, and T waves–total electrical alternans; these features are highly characteristic of cardiac tamponade

swisher SEXOLOGY A regional term for a really queer queer, not that there's anything wrong with that

Swiss agammaglobulinemia IMMUNOLOGY An AR form of SCID, with a high mortality in early infancy due to a severe diarrhea, malabsorption with disaccharidase deficiency, villar atrophy; defective cellular and humoral immunity, infection by a menagerie of opportunistic pathogens–eg, *Candida albicans*, CMV, measles, PCP, varicella, GVHD LAB Lymphocytopenia, anemia, ↑ liver enzymes, electrolyte imbalance 2° to chronic diarrhea TREATMENT Antibiotics, gammaglobulins; HLA-matched BMT. See Adenosine deaminase deficiency, Severe combined immunodeficiency

switch ERGONOMICS See Foot switch INFORMATICS An input/output device with several ports, which allows a user to choose where data is to be sent–eg, to a fax machine, printer, network, etc

SWOG Southwestern Oncology Group

swollen lymph glands VOX POPULI Lymphadenopathy, see there

SWORD CARDIOLOGY A trial–Survival With Oral D-Sotalol–that evaluated the efficacy of D-sotalol therapy on the M&M in Pts with CHF. See Heart failure, D-Sotalol

Sx Symptom(s). See Symptom

Sydenham's chorea St. Vitus dance NEUROLOGY Rapid, involuntary movement of face–grins, grimaces, contortions, tics; extremities–erratic flailing, if unilateral, hemichorea; hands–repeated partial fisting of the hands or a 'milkmaid's grip'; tongue–muscular fasciculation is likened to a 'bag of worms'; SC is most common in children and adolescents, and a major criterion for diagnosing rheumatic fever–status/post untreated streptococal infection by the Jones' criteria; SC develops after a latency of wks to months, ± accompanied by emotional lability and rheumatic heart disease. See Rheumatic fever

Sydney System GASTROENTEROLOGY A system for classifying and grading gastritis by combining topographic, morphologic, and etiologic data into a common format, intended to generate reproducible and clinically useful diagnoses

symbiotic psychosis PSYCHIATRY A condition seen in young children with an abnormal relationship to a mother figure after a precipitating event–eg, birth of a sibling CLINICAL Intense separation anxiety, marked regression of social behavior and intellectual development, speech regression may occur and become garbled or jargonistic, leading to 2° autism, which may respond to therapy. See Folie à deux–shared psychotic delusion

Symmetra™ RADIATION ONCOLOGY A ¹²⁵I radiation seed used for brachytherapy for localized prostate CA. See Prostate cancer

sympathy PSYCHIATRY A feeling or capacity for sharing in the interests or concerns of another, often without emotional attachment to the sympathy's recipient. Cf Empathy

SYMPHONY CARDIOLOGY A clinical trial–Sibrafiban Versus Aspirin to yield Maximum Protection from Ischemic Heart Events Post-Acute Coronary Syndrome. See Xubix™

Symphony stent SURGERY A self-expanding, nitinol stent used in the legs to maintain arterial patency after balloon angioplasty, and in the bile ducts in Pts with obstruction. See Stenting

symphorophilia SEXOLOGY A paraphilia in which sexuoerotism hinges on stage-managing and watching a disaster–eg, fire or MVA

symposium MEDICAL COMMUNICATION A meeting or collection of views on one topic. See Independent symposium, Sponsored symposium

symptom A subjective manifestation–eg, nausea, light-headedness, itching, of a morbid condition reported by a person; often used loosely for signs or other evidence used of a particular condition. See B symptom, Cancer symptom, Cognitive symptom, Concomitant symptom, First rank symptom, Homeopathic symptom, Negative symptom, Positive symptom, Shake & bake symptom. Cf Sign

Symptom Checklist SCL-90R PSYCHOLOGY An instrument that assess 9 domains of psychiatric Sx–anxiety, depression, hostility, interpersonal sensitivity, obsessive-compulsiveness, paranoid ideation, phobic anxiety, pychoticism, somatization

symptom magnifier MEDICAL PRACTICE A 'problem Pt' who feigns or exaggerates medical problems to prolong disability payments. See Hypochondriasis, Munchhausen syndrome

symptom-oriented construct The tenet, held by mainstream medical practice, that an illness is most effectively treated by suppressing its Sx. See Alternative medicine

symptom picture See Homeopathic symptom

sympto-thermic method Combined method FAMILY PLANNING A method of natural family planning, based on bone changes in basal body temperature and in the normal changes in the cervical mucus at the time of ovulation. See Cervix mucus method, Natural family planning

Synagis™

Synagis™ IMMUNOLOGY A humanized monoclonal antibody for preventing winter RSV in children. See RSV

synaptic plasticity PHYSIOLOGY Malleability present in synapses in various forms–eg, presynaptic inhibition, homosynaptic depression, presynaptic facilitation and modulation of transmitter release by tonic depolarization of sensory neuron. See Neuroplasticity, Nitric oxide

synchronous pacing See Artificial pacemaker

synchrony CARDIOLOGY A-V synchrony, see there

syncope NEUROLOGY A transient loss of consciousness not explained by other altered states of consciousness in the history of the Pt, often linked to cerebral ischemia; fainting, loss of conciousness or vertigo due to a transient arrhythmia, cardiac conduction–heart block, or neurovascular tone. See Carotid sinus syncope, Deglutition syncope, Neurocardiogenic syncope, Seder syncope, Seinfeld syncope, Sushi syncope

syncytial knot OBSTETRICS A multinucleated protrusion from the trophoblastic surface which is commonly found in later pregnancy, excessively so in post-term gestation. See Chorionic villi. Cf Syncytial bud, Syncytial sprout

syndactyly NEONATOLOGY Fusion of fingers, which may be cutaneous, due to bridging soft tissues, or osseous, due to bone fusion of varying severity; in general, only soft tissue syndactylism is treated, without which ostosis develops at the articulations with loss of function. See Rosebud hands

syndesmosis sprain Syndesmosis ankle sprain SPORTS MEDICINE A severe high-energy leg injury that damages the interosseous membrane between the tibia and fibula; severe SSs may extend to the anterior tibiofibular ligament SPORTS Football, hockey MECHANISM Forced external rotation of foot or hyperdorsiflexion of ankle. See Sprain

syndrome MEDTALK An aggregate of clinical Sx and/or lab findings that are characteristic of a particular condition or morbid process or occur together and constitute a recognizable condition. See A&V syndrome, Aarskog syndrome, Aase-Smith syndrome, Abortion trauma syndrome, Achard-Thiers syndrome, Achoo syndrome, Acute coronary syndrome, Acute HIV syndrome, Acute radiation injury syndrome, Acute urethral syndrome, Adult respiratory distress syndrome, Afferent loop syndrome, A/G syndrome, Aging face syndrome, Ahumanda del Castillo syndrome, Aicardi syndrome, AIDS syndrome, Albatross syndrome, Alcoholic flush syndrome, Alcoholic rose gardener syndrome, Aldrich syndrome, Alice in Wonderland syndrome, Alport syndrome, Alstrom syndrome, Alveolar-capillary block syndrome, Amenorrhea-galactorrhea syndrome, Amniotic band syndrome, Amotivational syndrome, Angelman syndrome, Angry back syndrome, Anterior cleavage syndrome, Anterior compartment syndrome, Anterior cord syndrome, Antimongoloid syndrome, Antiphospholipid antibody syndrome, Apert syndrome, Arthritis-dermatitis syndrome, Austrian syndrome, Baby bottle syndrome, BADS syndrome, Bardet-Biedl syndrome, Bare lymphocyte syndrome, Basal cell nevus syndrome, Battered buttock syndrome, Battered wife syndrome, Beauty parlor stroke syndrome, Beckwith-Wiedemann syndrome, Behavioral syndrome, Beradinelli syndrome, Bernard-Soulier syndrome, Bianchi syndrome, BIDS syndrome, Bile plug syndrome, Bile reflux syndrome, Black thyroid syndrome, Blind loop syndrome, Blind spot syndrome, Blue rubber bleb nevus syndrome, Blue sac syndrome, Blue toe syndrome, Blue valve syndrome, Bobble-headed doll syndrome, Boerhaave syndrome, Bone pointing syndrome, Bonnet syndrome, Bonnevie-Ullrich syndrome, Böök syndrome, Bradykinetic syndrome, Brain-bone-fat syndrome, Branching snowflake syndrome, Branchio-oto-renal syndrome, Breakage syndrome, Brittle cornea syndrome, Brittle hair syndrome, Bronze baby syndrome, Brown-Sequard syndrome, Brueghel syndrome, Brugada syndrome, Budd-Chiari syndrome, Bulldog syndrome, Burning feet syndrome, Burning tongue syndrome, Burnout syndrome, Caffeine dependence syndrome, Caffeine withdrawal syndrome, Call syndrome, Candidiasis hypersensitivity syndrome, Capillary leakage syndrome, Caplan syndrome, CAR syndrome, Carcinoid syndrome, Cardiovocal syndrome, Carpal tunnel syndrome, Cartilage-hair hypoplasia syndrome, Cat eye syndrome, CATCH-22 syndrome, Cauda equina syndrome, Cavernous sinus syndrome, Central cord syndrome, Central spinal cord syndrome, Cervical disk syndrome, Chediak-Higashi syndrome, Chérambault-

Kandinsky syndrome, Cheshire cat syndrome, Chiari-Frommel syndrome, 'Chief,' CHILD syndrome, China syndrome, China paralytic syndrome, Chinese restaurant syndrome, Christian syndrome, Christ-Siemens-Touraine syndrome, Chromosome breakage syndrome, Chronic exertional compartment syndrome, Chronic fatigue syndrome, Chubby puffer syndrome, Churg-Strauss syndrome, Chronic prostatitis/chronic pelvic pain syndrome, Chronic vascular syndrome, Chylomicronemia syndrome, Clenched fist syndrome, Clinical syndrome, Clumsy child syndrome, Cocaine rhabdomyolysis syndrome, Cocaine withdrawal syndrome, Cockayne syndrome, Cocktail party syndrome, Cogan syndrome, Compartment syndrome, Complex, Complex regional pain syndrome, Computer vision syndrome, Cornelia de Lange syndrome, Concentration camp syndrome, Condition, Continual late syndrome, Continuous gene syndrome, Conn syndrome, Cracked tooth syndrome, Cri-du-chat syndrome, Crigler-Najjar syndrome, Crocodile tear syndrome, Crossed syndrome, Crush syndrome, Culture-bound syndrome, Curtis Fitz Hugh syndrome, Cushing syndrome, Cytogenetic syndrome, Damocles syndrome, Dancing eyes-dancing feet syndrome, Dapsone syndrome, Dawson syndrome, Dead fetus syndrome, Dead hand syndrome, De Clérambault syndrome, Delayed sleep phase syndrome, Deletion syndrome, Delilah syndrome, Dengue shock syndrome, Denial syndrome, De Sanctis-Cacchione syndrome, Descending perineum syndrome, DeVaal syndrome, Diabetes-dermatitis syndrome, Diamond-Blackfan syndrome, DIDMOAD syndrome, Diencephalic syndrome of infancy syndrome, DiGeorge syndrome, Diogenes syndrome, Disaster syndrome, Disconnection syndrome, Disequilibrium syndrome, Disialotransferrin developmental syndrome, Disease, Disk syndrome, Dissociative syndrome, Distress syndrome, Distributed computing syndrome, DNA repair syndrome, Don Juan syndrome, Dorsal root ganglion syndrome, Dorito™ syndrome, Double crush syndrome, Down syndrome, Drash syndrome, Duane syndrome, Dubin-Johnson syndrome, Dubowitz syndrome, Dueling PhDs syndrome, Dumping syndrome, D0 syndrome, Duncan syndrome, Duplication 9p syndrome, Dyggve-Melchior-Claussen syndrome, Dysarthria-clumsy hand syndrome, Dysmenic syndrome, Dysmyelopoietic syndrome, Dysplastic nevus syndrome, Economy class syndrome, Ectopic ACTH syndrome, Ectopic Cushing syndrome, Ectopic hormonal syndrome, EEC syndrome, Ehlers-Danlos syndrome, Eisenmenger syndrome, Ellis-van Creveld syndrome, Emperor's New Clothes syndrome, Empty nest syndrome, Empty sella syndrome, Empty scrotum syndrome, Eosinophilic-myalgia syndrome, Epstein-Barr immunodeficiency syndrome, Epstein-Barr virus-associated hemophagocytic syndrome, Erysichthon syndrome, ETF syndrome, Euthyroid syndrome, Excision repair syndrome, Expanded rubella syndrome, Exploding head syndrome, Explosive syndrome, Extrapyramidal syndrome, Facet syndrome, Fat pad impingement syndrome, Family cancer syndrome, Family melanoma syndrome, Fat overload syndrome, Felty syndrome, Femoral facial syndrome, Fertile Eunuch syndrome, Fetal alcohol syndrome, Fetal distress syndrome, Fetal hydantoin syndrome, Fetal tobacco syndrome, Fetal trimethadione syndrome, Fetal varicella syndrome, Fetal warfarin syndrome, FG syndrome, Fibromyalgia syndrome, Filippi syndrome, Finnish congenital nephrotic syndrome, First arch syndrome, First use syndrome, Fisch-Renwick syndrome, Fish malodor syndrome, Fisher syndrome, Fisher-Volavsek syndrome, 5p- syndrome, Floating Harbor syndrome, Floppy baby syndrome, Floppy head syndrome, Folded lung syndrome, Foot & mouth syndrome, Foregut pain syndrome, Fourth venticle syndrome, Fragile X syndrome, Fragile X-E syndrome, Fraser syndrome, French congenital nephrotic syndrome, Fructose intolerance syndrome, Full house syndrome, Functional somatic syndrome, G syndrome, G deletion syndrome, GALOP syndrome, GAPO, Gardner syndrome, Gas-bloat syndrome, Gasping syndrome, Gay bowel syndrome, Gilbert syndrome, Ginseng abuse syndrome, Globoside dysfunction syndrome, Glucagonoma syndrome, Goldberg syndrome, Goldenhar syndrome, Golden-Kantor syndrome, Goldscheider syndrome, Golf war syndrome, Good child syndrome, Goodpasture syndrome, Gorlin-Moss-Chaudhry syndrome, Gourmand syndrome, Graham Little syndrome, GRANDDAD syndrome, Gray syndrome, Gray platelet syndrome, Green nail syndrome, Green stool syndrome, Growth hormone insensitivity syndrome, Guillain-Barré syndrome, Gulf War syndrome, Hallervorden-Spatz syndrome, HAM syndrome, Hamman-Rich syndrome, Hand-arm vibration syndrome, Hand, foot & mouth syndrome, Hand-foot syndrome, Hand-foot-flat face syndrome, Hand-foot-genital syndrome, Hantavirus pulmonary syndrome, Heel fat pad syndrome, HELLP syndrome, Hemi 3 syndrome, Hemispheric disconnection syndrome, Hemolytic-uremic syndrome, Hemophagocytic syndrome, Hepatopulmonary syndrome, Hepatorenal syndrome, Hereditary cancer syndrome, Hermansky-Pudlak syndrome, HFG syndrome, HGPRT deficiency syndrome, HHH syndrome, HHHH syndrome, HHHO syndrome, HIV wasting syndrome, HOHD syndrome, Holt-Oram syndrome, H_2O syndrome, HOOD syndrome, Hormone resistance syndrome, Horner syndrome,

Housebound syndrome, Hungry bone syndrome, Hunter syndrome, Hutchinson-Gilford syndrome, Hydrolethalus syndrome, Hypereosinophilic syndrome, Hyper-IgD syndrome, Hyper-IgE syndrome, Hyper-IgM immunodeficiency syndrome, Hyperinfection syndrome, Hyperinsulinism-hyperammonemia syndrome, Hypermobile joint syndrome, Hypersensitivity syndrome, Hyperventilation syndrome, Hyperviscosity syndrome, Hypoperfusion syndrome, Hypoplastic left heart syndrome, Idiopathic postprandial syndrome, Iliotibial band syndrome, Immotile cilia syndrome, Immunoglobulin M deficiency syndrome, Impingement syndrome, Inactive residual extremity syndrome, Infant nephrotic syndrome, Influenza syndrome, Inspissated bile syndrome, Inspissated milk syndrome, Intermediate syndrome, Intersex syndrome, Intestinal knot syndrome, Irregular sleep-wake syndrome, Irritable bowel syndrome, IVIC syndrome, Ivory tower syndrome, J syndrome, Jackpot syndrome, Jadassohn-Lewandowsky syndrome, Japan syndrome, Jarcho-Levin syndrome, Jekyll & Hyde syndrome, Jo-1 syndrome, Job syndrome, Joubert syndrome, Jumper syndrome, Jumping Frenchman syndrome, Kabuki mask syndrome, Kallmann syndrome, Kartagener syndrome, Kassabach-Merritt syndrome, Kearns-Sayre syndrome, KID syndrome, Kiesselbach syndrome, King syndrome, Kitamura syndrome, Kleine-Levin syndrome, Klinefelter syndrome, Klippel Feil syndrome, Klippel-Trenaunay-Weber syndrome, Kluver-Bucy syndrome, Kneist syndrome, Koala bear syndrome, Laband syndrome, Lady Godiva syndrome, LAMB syndrome, Lambert-Eaton syndrome, Landau-Kleffner syndrome, Langer-Giedion syndrome, Larsen syndrome, Late whiplash syndrome, Latent tetany syndrome, Laurence-Moon syndrome, Lazy bowel syndrome, Lazy leukocyte syndrome, Leaky gut syndrome, Lean cuisine syndrome, Lecithin-cholesterol acyl transferase deficiency syndrome, Leigh syndrome, Lennox-Gastaut syndrome, LEOPARD syndrome, Lesch-Nyhan syndrome, Leukocyte adhesion deficiency syndrome, Li-Fraumeni syndrome, Little omen syndrome, Locked in syndrome, Loin pain-hematuria syndrome, Long leg syndrome, Long Q-T syndrome, Lot syndrome, Low back syndrome, Low cardiac output syndrome, Low T$_4$ syndrome, Lowe syndrome, syndrome, Lown-Ganong-Levine syndrome, LUF syndrome, Luxury perfusion syndrome, Lymphangiomyomatosis syndrome, Maine syndrome, Mad Hatter syndrome, Magic pill syndrome, Malabsorption syndrome, Malignant hyperthermia syndrome, Malignant neuroleptic syndrome, Malpractice stress syndrome, Man-in-the-barrel syndrome, Marcus Welby syndrome, Marfan syndrome, Maternal deprivation syndrome, Mayo Clinic syndrome, McCune-Albright syndrome, Meconium aspiration syndrome, MEDAC syndrome, Medial tibial stress syndrome, Medical school syndrome, Meige syndrome, Meigs' syndrome, Melkerson-Rosenthal syndrome, Mendenhall syndrome, Menopausal syndrome, Mesenteric artery syndrome, Metabolic syndrome, Methionine malabsorption syndrome, Microdeletion syndrome, Middle lobe syndrome, Milk-alkali syndrome, Milkman syndrome, Mirror syndrome, Mitral valve prolapse syndrome, Mirizzi syndrome, Modigliani syndrome, Montgomery syndrome, Montreal platelet syndrome, Morning glory syndrome, Morquio syndrome, Moses syndrome, Moth ball syndrome, Mournier-Kuhn syndrome, Muir Torre syndrome, Multiple chemical sensitivities syndrome, Multiple cholesterol emboli syndrome, Multiple evanescent white dot syndrome, Multiple hamartoma syndrome, Multiple malformation syndrome, Multiple primary malignancy syndrome, Munchausen syndrome, Munchausenby-proxy syndrome, Murine acquired immunodeficiency syndrome, Myelodysplastic syndrome, Myofascial pain syndrome, Nail-patella syndrome, NASA syndrome, National Disaster Medical syndrome, 'Near miss' sudden infant death syndrome, Neck-face syndrome, Neck-tongue syndrome, Negative love syndrome, Neonatal withdrawal syndrome, Nephritic syndrome, Nephrotic syndrome, Nervous bowel syndrome, Neuroblastoma/IVS syndrome, Neurocutaneous syndrome, Neuroleptic malignant syndrome, Neutrophic dysfunction syndrome, Nonsense syndrome, Obstructive sleep apnea syndrome, Oculocerebrorenal syndrome, ODD syndrome, Olfactory reference syndrome, Omenn syndrome, One & one-half syndrome, Orange person syndrome, Organic brain syndrome, Organic mood syndrome, Ovarian vein syndrome, Overgrowth syndrome, Overlap syndrome, Overuse syndrome, Pacemaker syndrome, PAD syndrome, Padded dashboard syndrome, Page syndrome, Paget-Schroetter syndrome, Painful bruising syndrome, Painful fat syndrome, Painful leg & moving toes syndrome, Painful red leg syndrome, Pallister-Hall syndrome, Parana hard skin syndrome, Parinaud syndrome, Parkinson plus syndrome, Parrot syndrome, Parting of the Red Sea syndrome, Patellofemoral pain syndrome, Pearson syndrome, Peeling skin syndrome, Pelvic congestion syndrome, Pepper syndrome, Periodic fever syndrome, Perlman syndrome, Persistent müllerian duct syndrome, Peter Pan syndrome, Peutz-Jegher syndrome, Phantom limb syndrome, Physician invulnerability syndrome, Pickwick syndrome, PIE syndrome, Pigment dispersion syndrome, Piriformis syndrome, Plummer-Vinson syndrome, POEMS syndrome, Poland syndrome, Polonowski syndrome, Polyglandular autoimmune syn-

drome, Popeye syndrome, Popliteal pterygium syndrome, Postconcussive syndrome, Post-gastrectomy syndrome, Postmyocardial infarction syndrome, Postperfusion syndrome, Postpericardiotomy syndrome, Post-polio syndrome, Postpump syndrome, Post-resuscitation syndrome, Post-splenectomy syndrome, Postal worker syndrome, Posterior cord syndrome, Potter syndrome, Power Breakfast syndrome, Prader-Willi syndrome, Premenstrual syndrome, Prisoner of war syndrome, Proteus syndrome, Prostatitis syndrome, Prune belly syndrome, pseudo-Bartter syndrome, Pseudo-Cushing syndrome, Pseudo-Hurler syndrome, Pseudonym syndrome, Pseudo-Turner syndrome, PSL syndrome, Psychogenic syndrome, Psycho-organic syndrome, PUGH syndrome, Pulmonary hemorrhagic syndrome, Pulmonary renal syndrome, Punch drunk syndrome, Purple glove syndrome, Purple people syndrome, Purple urine bag syndrome, Quadruple syndrome, Quasimodo syndrome, Rabbit syndrome, Rabson-Mendenhall syndrome, RAPADILINO syndrome, Rape trauma syndrome, Rapunzel syndrome, Reactive hemophagocytic syndrome, Red cell fragmentation syndrome, Red diaper syndrome, Red man syndrome, Reiter's syndrome, Respiratory distress syndrome, Respiratory syndrome, Restless legs syndrome, Retained acid syndrome, Retained antrum syndrome, Retirement syndrome, Reversible posterior leukoencephalopathy syndrome, Revolving door syndrome, Reye syndrome, Rh syndrome, Rib-tip syndrome, Rieger syndrome, Right ovarian vein syndrome, Rigid spine syndrome, Riley-Smith syndrome, Rodney Dangerfield syndrome, Rosewater syndrome, RS$_3$PE syndrome, Rubinstein-Taybi syndrome, Rumination syndrome, Russell-Silver syndrome, Sanfilippo syndrome, Santos syndrome, SAPHO syndrome, Savannah syndrome, Scarlet O'Hara syndrome, Scheie syndrome, Schnitzler syndrome, Scimitar syndrome, Sclerosteosis syndrome, Seat belt syndrome, Sebright bantam syndrome, Second impact syndrome, Sepsis syndrome, Sequence, Serotonin syndrome, Sexing-stealing-lying syndrome, Sézary syndrome, Shaken baby syndrome, Sheehan syndrome, Shift syndrome, Shift work maladaptation syndrome, Short bowel syndrome, Short PR syndrome, Short rib polydactyly syndrome, Shoulder-girdle syndrome, Shoulder-hand syndrome, Shoulder pointer syndrome, Shprintzen syndrome, Shy-Drager syndrome, Shwachman syndrome, SIADH syndrome, Sick building syndrome, Sick cell syndrome, Sick Santa syndrome, Sick sinus syndrome, Sickle chest syndrome, SIDS syndrome, Sisyphus syndrome, Sjögren syndrome, Sleep apnea syndrome, slow channel syndrome, SLUDGE syndrome, Small syndrome, Small capacity syndrome, Small cuff syndrome, Small left colon syndrome, Small meal syndrome, Smith-Lemli-Opitz syndrome, Sneddon syndrome, Social breakdown syndrome, Solitary rectal ulcer syndrome, Somatostatinoma syndrome, Sorsby syndrome, Sotos syndrome, Space adaptation syndrome, Specificity spillover syndrome, Spinal vasculature steal syndrome, Splenic flexure syndrome, Spotted leg syndrome, Squash drinking syndrome, Squat jump syndrome, Staphylococcal scalded skin syndrome, Steakhouse syndrome, Steering wheel syndrome, Stevens-Johnson syndrome, Stewart-Treves syndrome, Stick-man syndrome, Stiff baby syndrome, Stiff heart syndrome, Stiff man syndrome, Stiff skin syndrome, Still syndrome, Stockholm syndrome, Stokes-Adams syndrome, Stone heart syndrome, Straight back syndrome, Straw Peter syndrome, Streptococcal sex syndrome, Subclavian steal syndrome, Subcutaneous insulin-resistance syndrome, Submersion syndrome, Subphrenic interposition syndrome, Subwakefulness syndrome, Suits & suites syndrome, Sump syndrome, Sundown syndrome, Superior mesenteric artery syndrome, Superior vena cava syndrome, Superwoman syndrome, Suspended heart syndrome, Sweet syndrome, Systemic inflammatory response syndrome, Taffy candy syndrome, Talk and die syndrome, TAR syndrome, Tarsal tunnel syndrome, Television intoxication syndrome, Tenosynovitis-dermatitis syndrome, Tendon sheath syndrome, Terminal illness syndrome, Terminal reservoir syndrome, Testicular feminization syndrome, Tethered (spinal) cord syndrome, Thalamic syndrome, Third diabetic syndrome, Thoracic outlet syndrome, 3/B translocation syndrome, Thucydides syndrome, TMJ syndrome, Tn syndrome, Tobacco withdrawal syndrome, Tolosa-Hunt syndrome, Top of the basilar syndrome, Torture syndrome, Tourette syndrome, Toxic oil syndrome, Toxic shock-like syndrome, Toxic shock syndrome, Toxic sock syndrome, Trapped egg syndrome, Treacher-Collins syndrome, Trichodentoosseous syndrome, Tropical immersion foot syndrome, Tropical splenomegaly syndrome, Trotter syndrome, Tryptophan malabsorption syndrome, Tumor lysis syndrome, Turcot syndrome, Turner syndrome, Twin-to-twin transfusion syndrome, Two hand-one foot syndrome, Ulnar tunnel syndrome, Ulysses syndrome, Uncombable hair syndrome, Urethral syndrome, Usher syndrome, VACTERYL syndrome, Van Gogh syndrome, Vanishing bile duct syndrome, Vanishing diabetes mellitus syndrome, Vanishing lung syndrome, VATER syndrome, Venus syndrome, Vibration white finger syndrome, Vietnam syndrome, VIPoma syndrome, V.I.P. syndrome, Viral hemorrhagic fever syndrome, Virus-associated hemophagocytic syndrome, Vitamin B$_6$-dependency syndrome, Vogt-Koyanagi(-Harada) syn-

syndrome of inappropriate antidiuretic

drome, Vulnerable child syndrome, W syndrome, Waardenburg syndrome, WAGR syndrome, Warm shock syndrome, Wasting syndrome, Waterhouse-Friderichsen syndrome, WDHA syndrome, Weber-Christian syndrome, Weill-Marchesani syndrome, Werner syndrome, Westphal-Leyden syndrome, West syndrome, Whiplash shaken infant syndrome, Whistling face syndrome, White dot chorioretinal inflammatory syndrome, Williams syndrome, Williams-Campbell syndrome, Wiskott-Aldrich syndrome, Withdrawal syndrome, Withdrawal emergent syndrome, Wolf-Hirschhorn syndrome, Wolff-Parkinson-White syndrome, Wrinkly skin syndrome, WT syndrome, XTE syndrome, XX male syndrome, XXX syndrome, XXXX syndrome, XXXXX syndrome, XXXY syndrome, XYY syndrome, Yellow nail syndrome, Yentl syndrome, Zellweger syndrome, Zollinger-Ellison syndrome

syndrome of inappropriate antidiuretic hormone secretion SIADH, aka dilutional hyponatremia

syndrome X CARDIOLOGY Microvascular angina Angina without detectable CAD, which may affect 1% to 15% of Pts with anginal pain ENDOCRINOLOGY A condition of older adults with truncal–upper body or central obesity, characterized by glucose intolerance, insulin resistance, 1° HTN, dyslipidemia, type 2 DM, ovarian androgen hyperproduction LAB ↓ HDL-C, ↑ TGs. See Coronary artery disease. Cf Diabesity

synechia An adhesion, usually of the anterior pole of the eye, which may be anterior–iris to cornea, or posterior–iris to lens capsule

Synercid® Quinupristin/dalfopristin THERAPEUTICS An antibiotic effective against gram-positive multiresistant organism–eg, vancomycin-resistant and multiresistant *Enterococcus faecium*–VREF, nosocomial pneumonia, MRSA, methicillin-resistant *S epidermidis*–MRSE. See Antibiotic resistance, Methicillin-resistant *Staphylococcus aureus*

synergism Cooperative interaction between 2+ components in a system, such that the combined effect is greater than the sum of each part ANATOMY The combined action of muscle groups, resulting in a force greater than that which could be generated by the individual muscles PHARMACOLOGY Pharmacologic synergism An approach to recalcitrant bacterial infections or virulent malignancies in which the therapeutic agents each affect different pathways or steps in a metabolic pathway, making the treatment more efficient–eg, penicillin and an aminoglycoside. See Chemical synergism, Combination chemotherapy

synergist 1. Any agent, component, factor, or structure that facilitates the activity of another. See Synergism 2. A muscle that assists the action of the prime mover

synergistic necrotizing cellulitis A form of necrotizing fasciitis characterized by involvement of skin, subcutaneous tissue, fascia, muscle; lesions are usually located on legs or perineum, arising in a perirectal abscess RISK FACTORS DM, obesity, advancing age, cardiorenal disease CLINICAL Small skin ulcers that ooze a red-brown fetid liquid fancifully termed 'dishwater pus', surrounded by gangrenous patches punctuated by preserved islands of normal-appearing skin, pain, tenderness, tissue gas, systemic toxicity, bacteremia

syngeneic transplant Isograft, syngeneic transplantation NEPHROLOGY A tissue graft between 2 genetically identical individuals–eg, identical twins

SynMesh ORTHOPEDICS A titanium mesh used to reinforce weak bone in orthopedics

synovial effusion An excess of synovial fluid in a joint, often due to inflammation. See Synovial fluid analysis

synovial fluid analysis LAB MEDICINE The evaluation of SF obtained by aspiration from the knee, shoulder, hip, elbow, less commonly from another joint; SFA is commonly performed on younger Pts to detect infection–eg, with staphylococcus or TB and in older Pts to categorize type of inflammation–eg, rheumatoid arthritis, pseudogout, exclude gout

SYNOVIAL FLUID ANALYSIS*

TYPE	APPEARANCE	GLUCOSE‡	WBC/mL	PMNs	VISCOSITY
NORMAL	Clear yellow	< 10 mg/dL	< 200/μL	< 25%	↑
NONINFLAMMATORY	Clear yellow	< 10 mg/dL	< 2000/μL	< 25%	↑
INFLAMMATORY	Slightly turbid	< 20 mg/dL	<5000/μL	< 50%	↓
SEPTIC-MILD	Turbid	< 40 mg/dL	<50,000/μL	50-90%	↓
SEPTIC-SEVERE	Turbid/purulent	> 20 mg/dL	>50,000/μL	> 90%	↓

*after JB Henry, Ed, Clinical Diagnosis & Management by Laboratory Methods, 18th edition, WB Saunders, 1991
‡Synovial fluid and blood difference

synovial sarcoma SURGERY A mesenchymal malignancy that comprises ±10% of all soft tissue tumors; it is most common in young–age 20-40 ♂, knee, ankle, foot, etc PROGNOSIS 50% 5-yr survival; extensively calcified SSs have a higher–84% 5-yr survival. See Sarcoma, Soft tissue tumors

synovitis ORTHOPEDICS Inflammation of a synovial joint. See Dendritic synovitis, Florid synovitis, Pigmented villonodular synovitis, Toxic synovitis

synthesis The creation of a whole from simpler parts or components. See Biosynthesis, Cell-free synthesis, Combinatorial biosynthesis, Coordinated enzyme synthesis, Narcosynthesis, Parallel synthesis, Psychosynthesis, Split synthesis

synthetic absorbable suture SURGERY A suture material with a predictable loss of tensile strength, that evokes minimal inflammation in tissue; SASs are most useful in GI, urologic, and gynecologic surgery. See Surgical closure, Suture. Cf Synthetic nonabsorbable suture

synthetic nonabsorbable suture SURGERY An inert suture material that is stronger than wire, that minimizes 'spitting' but does not 'handle' as well as silk; NSs are required for cardiovascular surgery as they must function indefinitely; vascular anastomoses using prothetic grafts and NSs may lead to aneurysm formation. See Surgical closure. Cf Absorbable suture

Synthofil® SURGERY A nonabsorbable polyester braided/coated suture for skin closure and cardiovascular and arterial surgery. See Suture

Synvisc® RHEUMATOLOGY A viscosupplementation that lubricates and protects the knee in Pts with osteoarthritis. See Osteoarthritis

syphilis STD A multisystem STD caused by a spirochete, *Treponema pallidum*, when it penetrates broken or abraded mucotaneous tissue through sexual contact; ≥ 100,000 new cases/yr, US; it is the third most commonly reported infectious disease–gonorrhea is the first; it is common in urban areas, especially in the US South and affects young adults CLINICAL Primary stage or chancre stage causes a nasty looking, but painless rounded ulcer, which may heal spontaneously. See Benign late syphilis, Congenital syphilis, Latent syphilis, Neurosyphilis, Primary syphilis, Secondary syphilis, Tertiary syphilis

syphilitic aortitis INFECTIOUS DISEASE The most common systemic change of late syphilis, most prominent in the ascending aorta and transverse arch; the vasa vasorum is obliterated, vasa media is necrosed and fibrotic. See Syphilis

syringe A calibrated disposable plastic–or less commonly, a nondisposable glass tube with a rubber sealed plunger at one end and a tapered tip for the insertion of a needle at the other. See Electronic syringe, SofDraw™ safety syringe

syringe exchange program Needle exchange program, see there

syringomyelia NEUROLOGY A disorder characterized by damage to the spinal cord, caused by formation of a fluid-filled cavity within the spinal cord, due to trauma, tumors, or congenital defects; the cavity begins in the cervical region and slowly expands, causing progressive pressure-related damage to the spinal cord. Cf Spinal cord trauma

system A defined collection of related structures and processes and the components required for their function. See ABO system, Admission-discharge-transfer system, Adrenergic system, Aldrete Recovery Room Scoring system, All-payer system, Allocation system, Alternate delivery system, Auditory system, Autonomic nervous system, Barotypic system, Bartenieff fundamentals system, Beta-Cath system, Bethesda system, BioLogic-HT™ system, BioZ system, Bonus system, Buddy system, Central nervous system, CHARS, CHESS system, Circulatory system, Clinical laboratory information management system, Closed system, Closed loop system, Community support system, Community water system, Computer-assisted diagnostic system, Computerized thermal imaging system, Conduction system, Consecutive water system, Cotswolds Staging system, DBx diagnostic system, Decision support system, Digestive system, Disease simulation system, Distribution system, Dual review system, Dynamic system, Early warning system, Eclipse system, Enterprise liability system, Er:YAG laser system, Executive information system, Expert system, Expression system, Extrapyramidal system, Feedback system, FIGO staging system, Fluoridated water system, Fountain™ system, Freehand system, Genitourinary system, Geographic information system, Gleason grading system, Global positioning system, Graham system, Groundwater system, Hamilton score system, Hepatic portal system, Hospital information system, HCPCS system, Humoral immune system, Immune, system, Incardia CABG system, Incardia valve system, Individual water system, Injury Surveillance system, Integrated delivery system, Kallikrein-kinin system, Laboratory information system, Lane system, Legacy system, Lewis system, Limbic system, Linear system, Lymphatic system, Lyophilized liposomal delivery system, Magnocellular system, Management information system, Medical anthrotonic system, Medication Event Monitoring system, Medipatch™ system, MICRO21 automated microscope system, MIDCAB™ system, Mirizzi system, Mission-critical system, MLR system, Model system, Mononuclear phagocytic system, Morter HealthSystem, Mossy fiber system, Mountain staging system, Multiaxial system, Multipayer system, Multiple chemical sensitivities system, NADPH oxidase system, National Health system, Natural-Hip system, Natural-Knee system, Naturally fluoridated water system, Nervous system, Neuroendocrine system, Noncommunity water system, Nonlinear system, Open protocol system, Nontransient system, Operating system, Opioid-mediated analgesia system, OSCAR system, PACS system, Papile system, Parvocellular system, Patient accounting system, Pedicle screw fixation system, Pluralistic system, Pneumatic tube system, Prognostic scoring system, Prospective payment system, Prostar system, Public water system, Pyramid system, Quartet™ system, Quinton® Synergy™ cardiac information management system, RAI staging system, Read classification system, Red system, r/LS system, Remote Automated Laboratory system, Renin-angiotensin-aldosterone system, Reproductive system, Respiratory system, Revelation™ hip system, Rh system, Rosenkranz™ pediatric retractor system, Rye Staging system, Sabolich socket system, SalEst™ system, School water system, Second messenger system, Sentinel™ 2010 implantable cardioverter-defibrillator system, Sharplan™ SilkTouch laser system, Side Branch Occlusion system, Silhouette laser system, Simal cervical stabilization system, Single-payer system, SkinLaser™ system, Skinlight erbium-YAG laser system, Somnoplasty℠ system, Sonic accelerated fracture healing system, Static system, Station system, Stress system, Subsystem, Svennerholm system, Sydney system, Sympathetic nervous system, Television rating system, Thermochemic HT™ system, Thermoflex™ system, Thoratec® VAD system, Tort system, Total hip system, Total knee system, Trusted system, Two-tiered system, Unified health care system, Vacuum system, Vestibular system, Voluntary Resident Tracking system, Walter Reed Staging system, Withhold system

systematic desensitization Desensitization, exposure therapy PSYCHOLOGY A type of behavioral intervention for managing anxiety, phobic disorders and anticipatory side effects–eg, nausea associated with chemotherapy. See Anticipatory side effects, Aversion therapy, Behavioral therapy, Encounter group therapy, Flooding, Imaging aversion therapy, Relaxing imagery. Cf Attentional distraction, Reciprocal inhibition, Relaxation training

systemic circulation CARDIOLOGY The part of the circulatory system concerned with blood flow from the left ventricle of the heart to the entire body and back to the heart via the right atrium. Cf Pulmonary system

systemic idiopathic fibrosis A condition characterized by idiopathic retroperitoneal fibrosis, which may extend to the anterior chest wall CLINICAL Backache, fever, N&V, constipation, anemia, oliguria, peripheral vascular insufficiency; fibrosis-induced ureteral compression with urinary retention and renal failure

systemic immunoblastic proliferation A condition caused by proliferation of immature lymphocytes CLINICAL Dyspnea, rash, hepatosplenomegaly, lymphadenopathy, ↑ risk of immunoblastic lymphoma

systemic inflammatory response syndrome A term that *'was developed to imply a clinical response arising from a nonspecific insult and includes two or more of the following.'* See Sepsis, Septic shock, Severe sepsis

SYSTEMIC INFLAMMATORY RESPONSES

TEMPERATURE < 36ºC **OR** > 38ºC

HEART RATE > 90 beats/min

RESPIRATORY RATE pCO_2 < 32 mm Hg **OR** > 20 breaths/min

WBC COUNT < 4 **x** 10^9 **OR** 12 **x** 10^9 or , or the presence of > 0.10 immature neutrophils

systemic lupus erythematosus Disseminated LE, lupus, lupus erythematosus RHEUMATOLOGY An idiopathic multisystem collagen vascular (connective tissue) disorder EPIDEMIOLOGY Affects ±40/10⁵–North America, Europe, blacks/hispanics > whites, ♀:♂ ratio = 3:1; 80% onset during childbearing yrs CLINICAL Vasculitis, serositis, synovitis, cerebral, renal, skin involvement CAUSE OF DEATH Sepsis 40%, CVS 20%, CNS 10%, renal 10%, etc 20%. See Anticardiolipin antibody, Antinuclear antibody, Butterfly rash

SYSTEMIC LUPUS ERYTHEMATOSUS–1982 REVISED CRITERIA, REREVISED IN 1997

1. MALAR RASH Fixed erythema, in particular over the malar eminences

2. DISCOID RASH Raised erythematous patches with adherent hyperkeratotic scaling; atrophic scarring in some old lesions

3. PHOTOSENSITIVITY Unusual skin rashes in response to sunlight

4. ORAL ULCERS Oral or nasopharyngeal ulcers

5. ARTHRITIS Nonerosive arthritis of 2+ joints, accompanied by tenderness, swelling, or effusions

6. SEROSITIS

a. Pleuritis OR

b. Pericarditis

7. RENAL DISEASE

a. Persistent proteinuria > 0.5 g/day OR

b. Cellular casts

8. NEUROLOGIC DISORDER

a. Seizures without substance use or medical disease OR

b. Psychosis in absence of substance use or medical disease

9. HEMATOLOGIC DISORDER

a. Hemolytic anemia OR

b. Leukopenia OR

c. Lymphocytopenia

d. Thrombocytopenia

10. IMMUNOLOGIC DISORDER

a. Positive LE cell prep DELETED IN 1997

b. Anti-DNA antibody

c. Anti-Sm antibody

d. False positive serological test for syphilis BECOMES

Positive finding of antiphospholipid antibodies based on

1) an abnormal serum level of IgG or IgM anticardiolipin antibodies,

2) a positive test result for lupus anticoagulant using a standard method, or

3) a false-positive serologic test for syphilis known to be positive for at least 6 months and confirmed by *Treponema pallidum* immobilization or fluorescent treponemal antibody absorption test. Standard methods should be used in testing for the presence of antiphospholipid

11. ANTINUCLEAR ANTIBODY Abnormal titers of ANA in absence of drugs known to be associated with drug-induced LE www.rheumatology.org/research/classification/1982SLEupdate.html

systemic-onset juvenile rheumatoid arthritis Systemic-onset juvenile chronic arthritis A form of joint disease, arthritis, that presents with systemic signs and Sx–eg, high intermittent fever, a salmon-colored skin rash,

664 swollen lymph nodes, hepatosplenomegaly, and pleuritis and around the heart–pericarditis; the arthritis itself may not be immediately apparent but in time it surfaces and may persist after the systemic Sx are gone

systemic sclerosis Scleroderma RHEUMATOLOGY An idiopathic connective tissue disease with clinical and pathologic features of GVHD, characterized by subcutaneous and visceral fibrosis, anticentromere and antitopoisomerase antibodies and prominent microvascular changes with endothelial cell damage and proliferation of subendothelial connective tissue; ♀:♂ ratio, 3-8:1-after child-bearing yrs **Peak** Age 45 to 55. See Connective tissue disease, Graft-versus-host disease

systemic therapy THERAPEUTICS Any therapy that reaches target tissues via the systemic circulation

systemic vasculitis Noninfectious vasculitis VASCULAR DISEASE Any of a number of conditions characterized by inflammation of vessels in multiple sites in the body, which share many clinical and lab features CLINICAL Fever, fatigue, anorexia, weight loss, polymyalgia rheumatica, nondestructive oligoarthritis LAB ↑ ESR, anemia, thrombocytosis, ↓ albumin PATHOGENESIS Unknown MANAGEMENT Corticosteroids, cytotoxic agents; the various forms of SV have been formally classified by the Chapel Hill Consensus Conference according to size–small, medium-sized, large, and the major forms defined. See Vasculitis

SYSTEMIC VASCULITIS

LARGE VESSEL VASCULITIS

Giant cell (temporal) arteritis

Takayasu's arteritis

MEDIUM-SIZED VESSEL VASCULITIS

Polyarteritis nodosa

Kawasaki's disease

Primary granulomatous CNS vasculitis

SMALL VESSEL VASCULITIS

ANCA-associated small vessel vasculitis

Churg-Strauss syndrome

Drug-induced ANCA-associated vasculitis

Microscopic polyangiitis

Wegener's granulomatosis

Immune complex small vessel vasculitis

Behçet's disease

Cryoglobulinemic vasculitis

Drug-induced immune complex vasculitis

Goodpasture syndrome

Henoch-Schönlein purpura

Hypocomplementemic urticarial vasculitis

Infection-induced immune complex vasculitis

Lupus vasculitis

Rheumatoid vasculitis

Serum sickness vasculitis

Sjögren syndrome vasculitis

Paraneoplastic small vessel vasculitis

  Carcinoma-induced vasculitis

  Lymphoproliferative neoplasm-induced vasculitis

  Myeloproliferative neoplasm-induced vasculitis

Inflammatory bowel disease vasculitis

Syst-Eur CARDIOLOGY A trial designed to evaluate the effectiveness of isolated antihypertensive therapy in preventing cardiovascular complications in elderly Pts with isolated systolic HTN. See Antihypertensive, Calcium channel blockers, Hypertension

systole CARDIOLOGY Contraction of the heart, generally understood to be ventricular, with ejection of blood from the right ventricle into the pulmonary arteries and from the left ventricle into the aorta SIGNS OF LEFT VENTRICULAR SYSTOLE 1st heart sound, apical beat, arterial pulse. See Ejection fraction. Cf Diastole

systolic heart failure CARDIOLOGY Heart failure with a severely reduced systolic function–LV ejection fraction of ≤35%. Cf Diastolic heart failure

systolic time interval A noninvasive technique that combines the data provided by EKG and phonocardiography in order to evaluate ventricular function during systole

T 1. *Taenia* 2. Tamoxifen 3. Temperature 4. Term–obstetrics 5. Tetracycline 6. Theophylline 7. Threonine 8. Thoracic 9. Thymidine 10. Thymine 11. Tidal–lung physiology 12. Time 13. Tocopherol 14. Toxicity 15. Transferrin 16. Transfusion 17. Treatment 18. *Treponema* 19. Trimethoprim 20. Troponin 21. Tumor–TNM Classification 22. Typhoid MILITARY MEDICINE An organic compound developed as a vesicant type chemical weapon, which causes chemical burns or blisters of skin and mucosa including the conjunctiva. See Chemical weapons, Vesicant/blistering agent

t 1. Telocentric–cytogenetics 2. Tertiary–chemistry, commonly tert 3. Trans–chemistry 4. Transfer–genetics–eg, tRNA 5. Variable–statistics

T&A Tonsillectomy and adenoidectomy, see there

T-cell immunodeficiency syndrome Any of a group of immunodeficiency states arising from partial or absolute defects in T-cell function; T-cell defects are generally more severe than B-cell defects, have no effective therapy and are characterized by recurrent opportunistic infections–eg, by *Pneumocystis carinii*, cutaneous anergy, growth retardation, ↓ life span, wasting/'runting', diarrhea, ↑ susceptibility to GVHD, potentially fatal reactions to live viral or BCG vaccinations and an ↑ risk of malignancy; TCISs include DiGeorge syndrome–thymic hypoplasia, Nezelof syndrome–cellular immunodeficiency with immunoglobulins, and T-cell defects–eg, absence of inosine phosphorylase or purine nucleoside phosphorylase

T-cell lymphoma A malignant proliferation of T cells arising in the skin, diagnosed by detecting rearrangement of the T-cell receptor's β chain; TCLs are often 'driven' by EBV and other viral infections; 90% of all Pts with TCL have extracutaneous involvement when diagnosed. See Mycosis fungoides

T-CELL LYMPHOMAS

Small lymphocytic (well-differentiated lymphocyte-like) lymphoma–13%

Convoluted cell lymphoma or poorly-differentiated lymphocytic lymphoma–52%

Immunoblastic sarcoma or 'histiocytic' lymphoma–19%

Mycosis fungoides/Sézary syndrome–11%

Lymphoepithelial cell (Lennert's) lymphoma–5%

Cutaneous T-cell lymphoma

HTLV (I, II)–induced lymphoma

T1 line TELEMEDICINE A data line that uses copper wires, transferring data at 1.5 Mb/sec. See ATM, Bandwidth, Bit, Byte, Codec, Ethernet, Modem, Telemedicine

T-max CLINICAL MEDICINE A popular term for the maximum body temperature recorded in a 24-hr period

T-on-P phenomenon CARDIOLOGY An EKG finding consisting of sinus tachycardia with prolongation of Q-T and a delayed T wave, followed or overlapped by the succeeding P wave; a 'T on P' is suggestive of alkalosis

T&S Type and screen, see there

T-shaped fracture Y-shaped fracture ORTHOPEDICS A type of intercondylar fracture of the distal femur that occurs in falls from a height with the feet extended, resulting in a violent impact of the femur on the tibial plateau; fall-related fractures also occur as the distal tibia impacts on the ankle or as an intercondylar fracture to the distal humerus. Cf Lover's heels

t-test Student's t-test, see there

T-wave alternans CARDIOLOGY A subtle every-other-beat variation in T waves that is prognostic of Pts at high risk for life-threatening cardiac arrhythmias and sudden cardiac death. See Alternans test

T zone Transformation zone, see there

T$_3$ Triiodothyronine, see there

T$_3$ suppression test See Thyroid suppression test

T$_3$ thyrotoxicosis Hyperthyroidism in which T$_3$ but not T$_4$ is elevated; TT Pts are clinically heterogeneous and lack distinctive signs and Sx, comprising about 4% of those with hyperthyroidism due to Graves' disease, toxic nodular goiter and thyroid adenomas and a higher percentage of hyperthyroidism in regions with lower levels of iodine

T$_3$ uptake test Triiodothyronine uptake assay, T$_3$ resin uptake A test that indirectly measures free T$_4$ as a reflection of the amount of T$_3$ that can be bound to carrier proteins, in particular to thyroxin-binding globulin. See Thyroxine, Triiodothyronine

T$_4$ Thyroxine, see there

T$_4$ thyrotoxicosis Hyperthyroidism in which T$_4$ but not T$_3$ is ↑, which occurs in Pts with iodine-induced thyrotoxicosis and in euthyroid Pts who are sick for other reasons. See Euthyroid sick syndrome

T7 assay Free thyroxine index, see there

T12 assay Free thyroxine index, see there

T-ACE OBSTETRICS, SUBSTANCE ABUSE An abbreviated 4-question test that '…*circumvents the problems of denial and underreporting that historically make self-reporting–of alcohol abuse during pregnancy…of limited value…(it has the further advantage of not seeming to pry into current drinking habits, which might prompt untruthful answers. The key question concerns* T*olerance, one of the best predictors of continued drinking during pregnancy: 'How many drinks does it take to make you feel high?'…more than two …is…enough alcohol to bear a child with ARBD–alcohol-related birth defects. That risk is amplified by positive responses to one of T-ACE's other queries about whether she has been* A*nnoyed by criticism of her drinking, has felt she should* C*ut down, and has ever had an* E*ye-opener–a drink first thing in the morning to steady herself or get rid of a hangover*

TA-GVHD Transfusion-associated-graft-versus-host disease, see there

TAB 1. Tablet, see there 2. Therapeutic abortion, see there

666 **tabes dorsalis** Tabes syphilis, tabetic neurosyphilis NEUROLOGY Slowly progressive degeneration of the dorsal/posterior columns of the spinal cord typical of late or tertiary phase of syphilis, which occurs a decade or more after infection, when it is improperly treated CLINICAL Sharp–lightning pain, ataxia, optic nerve degeneration leading to blindness, urinary incontinence, loss of position sense, joint degeneration, aka Charcot's joints. See Charcot's joints, Syphilis. Cf Friedreich's tabes, Pseudotabes

table EPIDEMIOLOGY A set of data arranged in rows and columns. See Contingency table, Evidence table, Increment-decrement life table, Life table, Metropolitan Life table SURGERY The slab on which a Pt is placed to perform an operation

taboo SOCIOLOGY A culture-specific ban on certain actions–eg, adultery among intimate friends, behaviors–eg, incest and thoughts, the abrogation of which results in reproof, persecution, or exile by members of the group

TAC Tetracaine, adrenaline, cocaine AMBULATORY SURGERY A cocktail of topical anesthetics used for lacerations of highly vascular regions of the head and face ADVERSE EFFECTS Seizures and/or death. Cf LAT

tachophobia PSYCHOLOGY Fear of speed. See Phobia

tachyarrhythmia Tachycardia accompanied by arrhythmia; Herzmuskel schnell pumpfen wird; Herzleitungsbahnen nicht normale sein

tachyarrhythmia-mediated cardiomyopathy CARDIOLOGY A condition characterized by Sx of CHF and compromised left ventricular function, which improve with successful therapy of tachyarrhythmias and/or sinus rhythm

tachy-aye pronounced, tackyeye MEDTALK Jargon for a Pt who responds to pain by rapidly and repeatedly yelling aye-aye-aye

tachycardia CARDIOLOGY A rapid heart rate, usually defined as a rate over 100 beats/min. See Atrioventricular nodal reentrant tachycardia, Multifocal atrial tachycardia, Narrow QRS complex tachycardia, Pacemaker-mediated tachycardia, Supraventricular tachycardia. Cf Bradycardia

tachycardia-bradycardia syndrome Sick sinus syndrome, see there

tachypnea MEDTALK Abnormally fast breathing. Cf Dyspnea

tackle SPORTS MEDICINE 1. A maneuver in football and rugby in which a player on team A brings down another player, ideally on the opposite team, who is carrying the ball 2. The equipment used in certain sports–eg, fishing

tacrine Cognex® NEUROLOGY An aminoacridine-type cholinesterase inhibitor reported to improve–slightly–the cognitive status of Pts with Alzheimer's disease ADVERSE EFFECTS Hepatotoxicity

tacrolimus FK506, Prograf IMMUNOLOGY An immunosuppressant that inhibits IL-2 synthesis and binding; it is similar to, and synergistic with, cyclosporine, up to 50-fold more immunosuppressive than cyclosporine; it is used in BM, kidney, liver, lung, and other transplants ADVERSE EFFECTS Neurotoxicity–tremor, seizures, white matter disease, headache, nausea, paresthesias of hands, feet, insomnia; nephrotoxicity, hyperglycemia, hirsutism, paresthesia, ↑ lipids, ↑ K^+, HUS, ↓ Mg^{2+}. Cf Cyclosporine

TACT Tuned aperature computed tomography IMAGING A technology used to study ambiguous breast images, display 2-D images in 3-D and ↓ false negative mammograms. See Mammography

TACTICS Clinical trials 1. Thrombolysis And Counterpulsation To Improve Cardiogenic Shock Survival 2. Treat Angina with Aggrastat® and Determine Costs of Therapy with Invasive or Conservative Strategies–TIMI-18

tactile agnosia NEUROLOGY Impaired tactile sensation, with inability to recognize objects by touch, a common finding in parietal lobe tumors; TA is a subtle, nondisabling disorder caused by unilateral damage to parietotemporal cortices that may be severe in left brain infarcts DIFFDX Astereognosis, a complex somatosensory disorder, tactile aphasia. See Agnosia

tactile aphasia NEUROLOGY Impaired tactile sensation, in which the Pt cannot name the object being handled

tactile device NEUROLOGY A touch sensing and/or driven mechanical device that helps a person with certain disabilities–eg, deaf-blindness, to communicate. See Assistive technology

tadpole sign IMAGING A comma-shaped ultrasound 'shadow' seen under a tumor, which is of lesser density directly beneath the center of a mass than at the edges; 'tadpoles' appear in malignant masses and have ragged margins, due to necrotic tissue, which is a poor conductor of US waves. See Ultrasonography. Cf Tennis racquet sign

Taenia Latin, taenia, ribbon, tape PARASITOLOGY A genus of large segmented, with each segment or proglottid capable of producing eggs tapeworms–cestodes-that parasitize mammalian GI tracts EPIDEMIOLOGY Infestation is linked to consumption of raw pork–*T solium* or beef–*T saginata;* eggs are dispersed by individual or groups of proglottids, which detach and pass out with the stool. See *Taenia solium*

Taenia solium Armed tapeworm, measly tapeworm PARASITOLOGY The pork tapeworm, contracted from undercooked or measly pork–pork infected with larval forms of *T' solium*, which can grow 0.9-1.8 m long in the intestine EPIDEMIOLOGY Self-infected by ingesting eggs from own hands after scratching or wiping his/her anus; eggs hatch in GI tract; larvae migrate through tissues and encyst CLINICAL Larvae that migrate to the brain and encyst–formally, cysticercosis–and cause seizures and other neurologic problems

taffy candy 'syndrome' CARDIOLOGY A fanciful descriptor for the clinical Sx caused by elongation of the anterior leaflet of the mitral valve ETIOLOGY Idiopathic or linked to rheumatic fever and acute MI. See Mitral valve prolapse

taffy pulling effect Marked elongation of a structure, likened to stretched taffy candy CARDIOLOGY Marked elongation and thinning of the anterior mitral leaflet's chordae associated with mitral valve regurgitation, which may later rupture; Sx cause the so-called taffy candy syndrome SURGICAL PATHOLOGY Spindled, darkly basophilic streaks and strands due to the extreme delicacy of the nuclear chromatin; although this 'crush artefact' is highly characteristic of pulmonary small oat cell carcinomas, it may also be seen in lymphomas and chronic inflammation

Tafil® Alprazolam, see there

tag DERMATOLOGY A redundancy of mucocutaneous or other tissue. See Anal tag, Ear tag, Pleural tag, Skin tag VOX POPULI A label. See Blue tag, Pink tag, Yellow tag

Tagamet® Cimetidine, see there

TAH 1. Total abdominal hysterectomy 2. Transfusion-associated hepatitis, see there

TAH-BSO Total abdominal hysterectomy-bilateral salpingo-oophorectomy, see there

tail *adjective* Referring to an elongated terminal tapering of an organism, cell, molecule, statistic, or other component in a system that slowly arrives to a baseline or disappears *noun* SURGERY An elongated mass of tissue. See Axillary tail

tail coverage MALPRACTICE A malpractice insurance rider or supplement to a claims-made policy that provides coverage for an incident that occurred while the insurance was in effect but was not filed by the time the insurer-policyholder relationship terminated. See Malpractice, Statutes of limitations. Cf 'Going bare', Nose coverage

TAIM CARDIOLOGY A clinical trial–Trial of Antihypertensive Interventions & Management that evaluated the efficacy of various dietary interventions plus antihypertensive therapy with chlorthalidone vs atenolol on mild diastolic HTN, lifestyle intervention vs drugs, and effects of antihypertensives on sexual function. See Antihypertensive, Hypertension

Takayasu's arteritis Pulseless disease NEUROLOGY Idiopathic segmental inflammation of the aorta and major branches, that affects young ♀, especially in Africa and Asia LAB ↑ ESR, ↑ Igs

take IMMUNOLOGY *noun* A popular term for a vaccine's efficacy; it is said to 'have taken' if there is a ≥ 4-fold ↑ in antibody titers TRANSPLANT IMMUNOLOGY The adherence of a free skin graft occurring between days 3 and 5 of the transfer of skin VOX POPULI Opinion, as in, '…what's your 'take' on this…'

talc A dry lubricant used in products that may contact mucocutaneous surfaces–eg, condoms, dental dams, and non-surgical latex gloves; talc was banned from surgical gloves in 1991 by the FDA

talc granuloma A foreign-body giant cell reaction of three types.

TALC GRANULOMA OCCURRENCES

POST-SURGICAL The peritoneum, body cavities, or tissues, due to contamination by surgical glove lubricants, eg talc, lycopodium, mineral oil rice, or corn starch, or by cellulose fibers from disposable gauze pads, drapes, gowns and other paper products

IV EXPOSURE Various organs, commonly the lungs of IV drug abusers, where the substance of abuse, usually a white powder–eg cocaine, has been 'cut' with starch or talcum powder; the granulomas measure 14-50 μm in diameter, and are located in eccentric patches of connective tissue and fibrous septae; the lungs have mild medial hypertrophy of the pulmonary arteries, but are not associated with pulmonary hypertension

TALCUM POWDER Penetrates to subepithelial tissues of external genitalia, causing TGs of the vagina, cervix, uterus, fallopian tubes, urethra and bladder

talc insufflation Topical aerosolized administration of talc as a sclerosing agent, effecting pleurodesis; TI may be useful in controlling pleural effusions due to breast CA. See Pleurodesis

talcosis PUBLIC HEALTH Intravascular, perivascular, and alveolar granulomas in which talc and sometimes starch can be identified, a finding typical of IV drug abusers, which may cause defects in pulmonary function

Talent™ endoluminal stent-graft HEART SURGERY A device delivery system indicated for minimally invasive repair of abdominal aortic aneurysms, in which the stent/graft is delivered collapsed under fluoroscopy through a catheter sheath previously positioned in the femoral artery, then released and expanded, sealing the vessel. See Stent

talipes Latin, talipes = talus–ankle + pes–foot A general term for clubfoot–a congenital foot deformity involving the talus. See Clubfoot

talipes calcaneus ORTHOPEDICS Clubfoot characterized by heel walking. See Clubfoot

talipes equinus ORTHOPEDICS Clubfoot in which the heel is raised and weight thrown onto the front of the foot. See Clubfoot

talipes valgus ORTHOPEDICS Clubfoot in which the outer border of the foot is everted, with inward rotation of the tarsus and flattening of the plantar arch. See Clubfoot

talipes varus ORTHOPEDICS A type of clubfoot–talipes in which the foot is inverted with the load bearing falling on the lateral border. See Clubfoot

talismanic paraphilia Fetishism in the usual sexual sense; shoes, stockings, feet, you know… goofy stuff. See Fetish

talk & die 'syndrome' TRAUMATOLOGY A clinical presentation in acceleration-deceleration brain injury, which may cause massive cerebral edema, that may have a latency period–eg, 48-72 hrs, until death. Cf Subarachnoid hemorrhage

talking therapy Talking cure, verbal therapy PSYCHIATRY A popular term for psychotherapy patterned after Freudian psychoanalysis. See Humanistic psychology, Psychoanalysis

talon noire Black heel SPORTS MEDICINE A condition affecting adolescents engaging in sports requiring major 'footwork'–eg, lacrosse, tennis, football, basketball; TN is caused by shear-stress rupture of papillary capillaries due to sudden stops and twists on the heels, resulting in skin punctuated by black spots corresponding to petechiae. See Black palm, Sports dermatology.

talotibial joint Ankle joint, see there

Talwin® Pentazocine, see there

TAMI CARDIOLOGY A series of clinical trials–Thrombolysis & Angioplasty in Myocardial Infarction designed to examine the role of angioplasty, urokinase, heparin, and prostacyclin in managing acute MI. See Acute myocardial infarction, tPA, Urokinase. Cf TIMI

Tamiflu® Oseltamivir INFECTIOUS DISEASE An oral neuraminidase inhibitor used for influenza A and B

tamoxifen Adjuvant tamoxifen, Novaldex® ONCOLOGY A nonsteroidal antiestrogenic used to treat early estrogen receptor–ER-positive breast CA; prophylactic tamoxifen may used in postmenopausal ♀ at high risk for breast CA OTHER BENEFITS ↓ Serum lipids, ↓ risk of CAD; it maintains bone mass and ↓ osteoporosis CONS ↑ risk of endometrial CA . See Breast cancer, Chemokine, STAR. Cf Raloxifene

tampering The adulteration of a thing. See Drug tampering

tampon GYNECOLOGY A device inserted per vagina to absorb menses. See Toxic shock syndrome

tamponade CARDIOLOGY See Balloon tamponade MEDTALK A pathologic plugging of an organ–eg, cardiac tamponade, caused by massive accumulation of pericardial fluid, resulting in a mechanical limitation of organ function. See Cardiac tamponade

Tandearil® Oxyphenbutazone, see there

tangential speech NEUROLOGY Speech characterized by tightly linked associations that miss the goal of the communication by veering off in tangents ETIOLOGY Organic brain disease, bipolar disorder, schizophrenia

tangentialty NEUROLOGY A 'kink' in associative thought, in which the flow of thought veers off in tangents, rather than following a main topic. See Tangential speech. Cf Circumstantiality

Tangier disease Analphalipoproteinemia A rare AR condition caused by a deficiency in α-lipoprotein CLINICAL Deposits of cholesteryl esters in tonsils, other lymphoid tissues, lymphadenopathy, hepatosplenomegaly, peripheral neuropathy, intermittent diarrhea, corneal opacification LAB Absent HDL, ↓ cholesterol < 120 mg/dl, ↓ phospholipids, ↓ apoA-I and apoA-II, ↑ TGs PROGNOSIS Usually benign, rarely CAD. See Triglyceride

Tanner Developmental Scale Sexual maturity rating, Tanner staging PEDIATRICS A system for objectively determining sexual maturity, which correlates chronologic age with a group of anatomic parameters, determining the degree of adolescent maturation; the most commonly used system was delineated by Tanner; in ♀, 5 stages of maturation are recorded for pubic hair and breast development; in ♂, 5 stages are recorded for pubic hair, growth of penis and testicles

tanning DERMATOLOGY A sedentary 'activity' consisting of autobasting, beached-whale-like under a UVA lamp in 15-30 min dollops to achieve a natural look; heuristically 'logical' evidence that a sleek well-tanned jet-set wannabe' enviable 'look to die for' might be a bad thing confirms an ↑ risk of benign, premalignant, or frankly malignant–eg, SCC–RR of 2.5 and BCC-RR of 1.5–lesions linked to 'recreational' tanning. See Melanoma. See Tanning device

tanning device PUBLIC HEALTH A bed or booth fitted with UV lights that emit UV-A, and lesser amounts of UV-B radiation, homogeneously delivering maximal light in the minimum time. See Tanning

tanning pill Any of a number of OTC agents–BronzGlo, Darker Tan, Orobronze–that contain concentrated canthaxanthin, a beta-carotene-like substance which imparts a 'healthy' red-brown tint to skin TOXICOLOGY Canthaxanthin may evoke headaches, fatigue, weight loss, bruisability, allergic skin reactions, blurred nocturnal vision, hepatitis, and irreversible–ie, potentially fatal aplastic anemia. See Tanning, UV-A

the Tan Sheet PHARMACEUTICAL INDUSTRY A specialized weekly publication that provides business and federal regulatory information on non-prescription pharmaceuticals and nutritionals

tantrum VOX POPULI An inappropriate display of anger. See Adult temper tantrum, Temper tantrum

tap *noun* The fluid obtained when a body cavity is tapped *verb* To obtain a fluid or liquefied material from a body cavity or tissue by inserting a needle or catheter. See Abdominal tap, Dry tap, Pericardiocentesis

tap-out SPORTS MEDICINE A hand signal from a combatant in ultimate fighting that signals surrender, usually evoked by a submission hold–eg, a choke hold or joint-lock. See Ultimate fighting. Cf Marquis of Queensbury rules

Tapazol® Methimazole–an antithyroid agent

tape SURGERY An adhesive paper or cloth strip or ribbon used to close incisions; tape ↓ infection risk by eliminating a foreign body–eg, sutures that communicate with outside of the body. Cf Catgut, Staples VOX POPULI A magnetic recording medium. See Digital tape

tape booger MEDTALK A popular term for remnants on the skin of old tape or monitoring electrode/EKG goo

taper *verb* To gradually ↓ a dose, usually of a therapeutic agent–eg, corticosteroids, with potentially significant adverse effects, which cannot be abruptly halted, often due to rebound effects

tapering IMAGING The gradual narrowing of a lumen, often due to external compression. See Rat tail tapering SPORTS MEDICINE A weight training term for a slow 'weaning' from steroid use. See Anabolic steroids, Weight training THERAPEUTICS The gradual lowering of drug dose. See Drug tapering

tapering off SPORTS MEDICINE A format for competition training, where a world-class athlete ↓ frequency and intensity of training in the wks before an Olympic or other sport event of importance, with the hope that performance in the key event will be medal-worthy

tapeworm infection Teniasis INFECTIOUS DISEASE A general term for infestation by either *Hymenolepis* spp–*H diminuta*–rat tapeworm or *H nana*–dwarf tapeworm or *Taenia* spp–*T saginata*–cattle tapeworm, *T solium*–pork tapeworm. See *Hymenolepis, Taenia*

taphophobia PSYCHOLOGY Fear of being buried alive. See Phobia

tapir mouth NEUROLOGY A fanciful descriptor for the 'pouting' expression seen in protruding lips due to weakening of the orbicularis oris muscles, seen facioscapulohumeral muscular dystrophy of Landouzy-Déjerine

tapotement Percussion, see there

TAR syndrome Trombocytopenia-absent radius syndrome An AR condition of perinatal onset characterized by Thrombocytopenia with Absent Radius–thumbs are almost always present CLINICAL Profound thrombocytopenia, purpura with amegakaryocytosis in BM and bilateral aplasia of the radii, phocomelia, the fibula is often absent; up to ⅔ of Pts have leukemoid reactions, occasionally anemia, eosinophilia, cardiovascular disease–eg, ASD, Fallot's tetralogy, cutaneous and renal anomalies PROGNOSIS 50% die in the 1ˢᵗ yr of life due to intracranial hemorrhage. See Thrombocytopenia

Tarantula *Lycosa tarantula*, wolf spider ENTOMOLOGY A popular, much maligned and relatively harmless Grade B Movie prop. See Arachnid injuries

Tarasoff decision PSYCHIATRY A California decision that imposes a duty on a therapist to warn appropriate person(s) when she becomes aware that a Pt may present a risk of harm to a specific person or persons. See *Tarasoff v. Regents of the University of California*

tardive dyskinesia NEUROLOGY Slow involuntary sinuous rhythmic movements usually caused as a neurologic side effect of certain drugs–eg, tricyclic tranquilizers–phenothiazine and other tricyclics. See Abnormal Involuntary Movement Scale, 'Piano playing', Rabbit syndrome, Tricyclic antidepressant

target *adjective* Pertaining to a lesion or radiologic finding in which there are ≥ 3 relatively well-circumscribed, concentrically arranged annular patterns or radiodensities *noun* IMAGING The molecular defect that is examined in molecular imaging. See CT/PET, Molecular imaging. Cf Probe

target heart rate CARDIOLOGY A pre-determined pulse rate to target during aerobic exercise based on age when the cardiovascular system is functioning optimally.

target lesion DERMATOLOGY A lesion typical of erythema multiforme–EM in which a vesicle is surrounded by an often hemorrhagic maculopapule; EM is often self-limited, of acute onset, resolves in 3-6 wks, and has a cyclical pattern; EM lesions are 'multiform' and include macules, papules, vesicles, bullae ETIOLOGY Idiopathic or follow infections, drug therapy or occur in immunocompromised hosts

target organ ENDOCRINOLOGY A specific organ that a particular hormone affects

target sign GI IMAGING A smoothly-contoured radiopacity with both central and peripheral radiolucency, which may be seen in pedunculated colonic polyps, when viewed *en face* by double contrast–air-contrast barium studies; an 'eccentric' target sign is seen in GI diverticuli where a small amount of radiocontrast enters the pouch and is surrounded by the radiopaque body of a diverticulum LUNG IMAGING A circumscribed pulmonary aspergilloma or focus of necrotizing bronchopneumonia seen in a plain CXR. Cf Coin lesion

targeting MEDTALK The focusing of one's efforts on a particular goal. Gene targeting

Targis™ T3® system UROLOGY A microwave, catheter-based system for non-surgical treatment of BPH. See Benign prostatic hypertrophy

Tarka® VASCULAR DISEASE A combination drug for treating HTN containing an ACE inhibitor–trandolapril and CCB–verapamil. See ACE inhibitor, Calcium channel blocker

tarsal coalition ORTHOPEDICS A block between 2 bones, which may be osseous–synostosis, cartilaginous–synchondrosis, or fibrous–syndesmosis; TCs are often congenital, and first identified in adolescence when the person is being examined for other reasons; TCs may also be acquired due to fractures, tumors, infections, arthritis. See Calcaneonavicular coalition, Lover's heel, Talocalcaneal coalition

tarsal cyst Chalazian, see there

tarsal tunnel syndrome A carpal tunnel syndrome-like complex caused by post-traumatic fibrosis, abductor hallucis hypertrophy, tenosynovitis or fascial band entrapment by the posterior tibial nerve CLINICAL Pronounced plantar surface and toe causalgia that may irradiate to the calf, resulting in paresthesias, cyanosis, sensation of coldness, numbness TREATMENT Massage, steroid injection, weight reduction, surgical decompression of compartment

TART OSTEOPATHY A mnemonic–tenderness, asymmetry, restricted motion, tissue texture changes–for the most common findings in somatic dysfunction. See Osteopathy, Somatic dysfunction

tartar Calculus DENTISTRY Hardened gray-white preplaque goo composed of hydroxyapatite, food bacteria, which adheres to teeth after a meal; tartar and plaque cause bone inflammation around teeth known as periodontia. See Caries

tartrazine FD&C Yellow No. 5, see there

Tarui disease Glycogen storage disease VII, see there

task analysis ERGONOMICS A systematic breakdown of a task into its elements, specifically including a detailed task description of both manual and mental activities, task and element durations, task frequency, task allocation, task complexity, environmental conditions, necessary clothing and equipment, and any other unique factors involved in or required for one or more humans to perform a given task

taste disorder NEUROLOGY An inability to perceive different flavors ETIOLOGY Neurologic defects, poor oral hygiene, gum disease, hepatitis, drugs, chemotherapy

taste fatigue CRITICAL CARE A taste-driven loss of desire to ingest a particular food–eg, oral supplements, in a CA Pt undergoing oral feeding, because the Pt can no longer tolerate the taste. See Cancer cachexia

TAT INFECTIOUS DISEASE Tetanus antitoxin, see there LAB MEDICINE Turn-around time, see there PSYCHOLOGY Thematic Apperception Test A projection-type psychological test that evaluates a child's sense of reality and personality traits and gives insight into his fantasies. See Psychological testing

Tatlockia micdadei An alternative term for *Legionella micdadei*, aka Pittsburgh pneumonia agent and TATLOCK strain

tattoo DERMATOLOGY A permanent form of cutaneous decoration that may range from simple, often small dark-colored insignias, messages or symbols performed by amateurs in prison, to elaborate multi-colored animals, objects or scenes performed by more skilled workers under relatively sterile conditions; up to 25% of college-aged persons have tattoos; persons with tattoos have a 7-fold ↑ risk of HCV infection. See Laser surgery. Cf Tattooing FORENSIC PATHOLOGY An abnormal mark etched into tissue. See Powder tattoo

tattooing TRAUMATOLOGY Complex skin abrasions and wounds filled with debris, glass and dirt, resulting from being dragged along a road, common in pedestrian victims of MVAs MANAGEMENT Adequate debridement and often wound healing by second intention

TAUSA CARDIOLOGY A clinical trial–Thrombolysis & Angioplasty in Unstable Angina that evaluated the role of prophylactic intracoronary thrombolytics during angioplasty in Pts with unstable angina. See Acute myocardial infarction, Angioplasty, Urokinase

Tavor Lorazepam, see there

tax VOX POPULI Monies paid to a governing body. See Dean's tax, Fat tax, Provider tax, Sin tax

Taxagon Trazodone, see there

taxanes ONCOLOGY A family of antimitotic/antimicrotubule agents that inhibit cancer cell growth by stopping cell division

Taxol® Paclitaxel ONCOLOGY A chemotherapeutic used to manage Pts with ovarian, breast CAs, non-small cell lung CA, melanoma, KS who are not candidates for potentially curative surgery and/or RT MECHANISM Taxol blocks mitosis by stabilizing microtubules and promoting tubulin polymerization, resulting in non-functional microtubules. See Ovarian cancer

Taxotere® Docetaxol, see there

Tay-Sachs disease GM2-gangliosidosis PEDIATRIC NEUROLOGY A rare AR lipid storage disease, most common in Ashkenazi Jews–carrier frequency 1:30, due to hexosaminidase A deficiency, resulting in accumulation of gangliosides in neurons, cerebellum, axons CLINICAL Onset at 4-6 months with arrest and decline of psychomotor activities, irritability, hyperacusis, convulsions, chorioathetosis, spasticity, decerebrate rigidity, death by age 3. See Gangliosidosis

tazarotene Allergan®, Tazorac®, Zorac® DERMATOLOGY A topical, receptor-selective retinoid–vitamin A analog–used to manage stable plaque psoriasis, acne vulgaris, BCC. See Basal cell carcinoma, Retinoid

TB Tuberculosis, see there

TB screening program PUBLIC HEALTH A program in which employees and residents in a facility are periodically given tuberculin skin tests to identify active TB. See Tuberculosis

TBF Total body failure MEDTALK A term referring to a Pt, usually elderly, suffering from preterminal failure to thrive

TBG Thyroxine-binding globulin, see there

TBI 1. Thyroxine-binding index 2. Total body irradiation

TBT Transcervical balloon tuboplasty, see there

^{99m}Tc-labeled RBC scintigraphy IMAGING A noninvasive method used to localize slow or intermittent GI hemorrhage, in which RBCs are labeled with ^{99m}Tc; images are obtained from the Pt with a gamma camera, which detect RBCs extravasated from the large or small intestine. See Scintigraphy

^{99m}Tc sulfur colloid IMAGING A radiolabel used to identify sites of tumor development

TCA Tricyclic antidepressant, see there

TCE ENVIRONMENT A volatile chlorinated hydrocarbon that boils at 88°C and is highly soluble–1000 ppm in water, with various industrial uses TOXICITY Peripheral neuropathy, carcinogenic. See Bioremediation, EPA, Plumes, Superfund, Toxic dumps, 'White-out'. Cf Dioxin, PCBs

TCP/IP Transmission Control Protocol/Internet Protocol INFORMATICS The suite of computer protocols and rules for exchanging packets of information over networks, including the Internet. See IP Number, Internet, UNIX

Td Adult diphtheria and tetanus toxoids

Td vaccination A vaccine given to children over age 6 and adults as boosters for immunity to diphtheria and tetanus

TD$_{50}$ Toxic dose, see there

TDM Therapeutic drug monitoring, see there

TDMS™–Trex mammography IMAGING A full-field digital mammography system for early detection of breast CA. See Mammography

TE 1. Thromboembolic 2. Total estrogen 3. Tracheoesophageal

670

tea MAINSTREAM MEDICINE A 'clear liquid' prepared as an infusion from various leaves, used for rehydration, or to 'bind' Pts with diarrhea

TEA Thromboendarterectomy

tea-drinker's disease Theism NEUROLOGY A caffeine-induced condition characterized by nervousness, cerebrovascular congestion, excitement, depression, pallor, arrhythmias, hallucinations, insomnia. See Caffeine, Coffee, Green tea

teaching hospital A hospital, health care center, or institute that has 2 or more residency training programs approved by the Accreditation Council for Graduate Medical Education

teaching patient Standardized patient GRADUATE EDUCATION A Pt with–or an actor trained to portray–a particular condition, who can be used as a teaching "tool" for medical students

teaching surgeon ACADEMIC MEDICINE A surgeon who supervises and instructs a physician in training–resident, registrar, fellow. See Mentor

TEAM Techniques for Effective Alcohol Management PUBLIC HEALTH A partnership by the US government, professional sports, and US businesses, which fights the use of alcohol in and around pubic assembly facilities

team VOX POPULI A group of persons with common interests or goals. See Care team, Charge master team, Medical team, Operating team, Self-manage team, Sexual assault response team

team approach See Medical team

TEAM-2 CARDIOLOGY A randomized, double blind, multicenter trial that compared coronary artery patency after IV streptokinase versus APSAC in Pts with acute MI. See Acute myocardial infarction, APSAC, Streptokinase

tear MEDTALK Pronounced, TARE A rent or disruption of a flattened tissue or surface. See Job's tear, Mallory-Weiss tear, Meniscal tear, Re-entry tear, Skier's tear Pronounced, TEER The watery product of the lacrimal glands

teardrop bladder UROLOGY A descriptor for a markedly distended urinary bladder in which a trauma-induced hematoma surrounds the bladder base, lifting it out of the pelvis; external pressure narrows the bladder neck into an attenuated 'stem,' while a broad base at the bladder's apex imparts a piriform configuration by excretory urography ETIOLOGY Trauma, pelvic lipomatosis, inferior vena cava occlusion, psoas muscle hypertrophy, rarely, pelvic lymphadenopathy

teardrop fracture ORTHOPEDICS A lower cervical spine fracture-dislocation compression fracture of the anteroinferior corner of a cervical vertebra, caused by flexion with axial loading or hyperflexive compressive forces which may cause posterior subluxation of the vertebrae or burst the vertebral body, separating and displacing a wedge-shaped fragment of bone from the antero-inferior margin; such fractures are not usually associated with permanent neurologic sequelae IMAGING Lateral cervical film–wedging of vertebral body and an anterior-inferior fragment fractured from vertebral body; involved vertebrae are usually subluxed posteriorly; most Pts are rendered quadriplegic MANAGEMENT Surgical fusion to stabilize the injury and prevent kyphosis, pain, and further neurologic deterioration POTENTIAL DANGER Posterior displacement into spinal canal with cord compression. Cf Axial-load teardrop fracture, Wedge fracture

teardrop sign TRAUMATOLOGY A clinical finding characterized as an elongated soft-tissue mass that prolapses into the maxillary antrum; the TS may be seen in a plain film of the face in blunt trauma to the anterior rim of the orbit, which may cause a 'blowout' fracture of the orbital floor. See LeFort fracture

teaser advertising Advertising for a product that has not yet come to market. See Advertising

technical component MANAGED CARE Those charges made for the facility component of billing and claims for health care services

technetium pyrophosphate myocardial scanning CARDIOLOGY A technique used to evaluate the right and left ventricular volumes, systolic and diastolic function, and regional or global myocardial performance–eg, after an MI. Cf First-pass radionuclide ventriculography

technological disaster Man-made disaster, see there

technology The application of scientific or other organized knowledge--including any tool, technique, product, process, method, organization or system--to practical tasks. See Antibody technology, Biotechnology, Enabling technology, Green technology, Halfway technology, High tech, Mature technology, Microarray technology, Nanotechnology, Object-oriented technology, Pen-based technology, Platform technology, Push technology, Reference technology, SKY technology, Slam-bang technology, Store & forward technology, Web technology, Wireless technology

Techstar® SURGERY A percutaneous system with suture-based arterial access site closure post diagnostic and therapeutic catheterization. See Catheterization

Teczem® CARDIOLOGY A slow-release combination of enalapril maleate and diltiazem mealate, used for HTN. See ACE inhibitor

TEE Transesophageal echocardiography, see there

teenage pregnancy Adolescent pregnancy, teen pregnancy SOCIAL MEDICINE Pregnancy by a ♀, age 13 to 19; TP is usually understood to occur in a ♀ who has not completed her core education–secondary school, has few or no marketable skills, is financially dependent upon her parents and/or continues to live at home and is often mentally immature. See Adolescence

teeth grinding Bruxism, see there

TEF Tracheoesophageal fistula, see there

Teflon™ Polytetrafluoroethylene A proprietary name for a biologically inert polymeric molecule resistant to organic solvents, with a melting temperature of 225ºC, which is widely used in medicine MEDICAL DEVICES Teflon has diverse medical applications including use in annuloplasty rings, batteries for defibrillators and pacemakers, cardiac patches, vena caval clips, vascular grafts, implantable pumps, implants for impotence, implants for incontinence, prosthetic joints, bone replacements, pledgets, various types of sheeting, dialysis shunts, sutures, myringotomy tubes, valved conduits, and in low-temperature chemical reactions. Cf Gore-Tex™

Teflon™ injection therapy ENT The injection of a Teflon-based paste lateral to the conus elasticus to 'medialize' a unilaterally paralyzed vocal cord APPLICATIONS Unilateral recurrent laryngeal nerve paralysis, minimal vocal cord atrophy, mobile cricoarytemoid, abductor spasmodic dysphonia, abnormal patency of eustachian tube, cricoarytenoid joint ankylosis

tegafur Uftoral® ONCOLOGY An antimetabolite prodrug absorbed from the small intestine and metabolized in vivo to 5-FU. See 5-FU, Prodrug

Tegison Etretinate DERMATOLOGY A retinoid for severe psoriasis that is linked to major teratogenicity

Tegretol® Carbamazepine, see there

tegumentary *adjective* Relating to cutaneous or mucocutaneous surfaces

teicoplanin INFECTIOUS DISEASE An antibiotic that may be effective against teichomycin and vancomycin-resistant–multiply-resistant enterococci. See Antibiotic resistance

TEIS Endocrinology A clinical trial–Troglitazone and Exogenous Insulin Study comparing troglitazone with a placebo in Pts with poorly controlled type 2 DM. See Troglitazone

telangiectasia Dermatology A capillary with complex and tortuous dilation seen on mucocutaneous surfaces. See Hereditary hemorrhagic telangiectasia, Rendu-Osler-Weber syndrome

teleconsultation Virtual consultation Any medical consultation by either a non-medical consumer, or by a health care professional from a colleague on an electronic network–eg, intranet, Internet. See Telemedicine

telediagnosis Mainstream medicine Any method of transferring diagnostic data, including consultations and diagnostic images–telecytology, telepathology, and teleradiology. See Teleconsultation, Telemedicine

telegraphic speech Psychology A general term for speech characterized by marked 'abridgement'

telehealth Health informatics The effects of telecommunication and information technology on the efficiency and quality of health care, services, education, public health surveillance, research, administration. See Telemedicine

TeleMed forceps Surgery Oval-cup forceps used for endoscopic biopsies and retrieval of foreign bodies and stones from GI and GU tracts, lungs. See Biopsy

telemedicine Informatics Any form of medical practice in which diagnostic information–eg, telecytology, telemetry, telemicroscopy, telepathology, or teleradiology, is transmitted from a distance to a physician for analysis, who performs teleconsultation; telemedicine focuses on provider aspects of healthcare telecommunications, especially medical imaging. See Telemetry

telemetry Cardiology The monitoring of the cardiac rhythm and transmission of signals or data from one electronic unit to another by radio waves using a device that provides real-time measurement of a Pt's EKG for various lengths of time See Multiple parameter telemetry, Real-time telemetry

---

Telemetry–indications

Cardiac dysrhythmia

• Atrial fibrillation or flutter

• SVT–supraventricular tachycardia

• PVC–premature ventricular contraction under treatment

• Ventricular tachycardia or fibrillation

• Bradycardia < 50/min

Heart block–2nd or 3rd degree

Angina

Acute MI–transferred from ICU

Respiratory failure

Hypotension or HTN

Electrolyte imbalance

• Hypokalemia

• Hyperkalemia

• Hypocalcemia

• Hypercalcemia

Syncope

---

telepathology Telemedicine An evolving field of pathology which uses high-resolution video cameras and robotic microscopes to transmit images of cells and tissues via phone or fiberoptic lines to a distant center. See Telemedicine

telephone counseling The provision of advice and verbalized moral support to a person with a particular need by a group of either volunteers or a paid staff with some level of experience and/or expertise in the area of interest; TC may include crisis intervention–eg, for suicide, HIV infection, or acute or chronic addiction disorders. See Hotline, Support group

telephone receiver deformity ENT A popular term for an ear deformity, likened to a telephone handset, an undesired result of poorly healed trauma. See Cauliflower ear deformity

telephone scatologia Telephonicophilia Sexology A paraphilia in which sexuoeroticism hinges on exchanging sexually explicit, usually uninvited conversation with the listener; typically, the caller is not dangerous, only a nuisance, and usually ♂; TS is classified in the DSM-IV as a paraphilia, not otherwise specified. See Paraphilia, Sexual deviancy. Cf Phone sex

telephysician Online A physician who practices medicine on the Internet, by consulting, diagnosing and managing Pts. See Teleconsultation

telepresence surgery Telesurgery An evolving format for laparoscopic procedures, in which the surgeon is located at a distance from the Pt and operates using a joystick held in place by a 'confederate' at the Pt's bedside. See Robotic surgery

teleradiology Imaging A format for delivering imaging services by transmitting digitalized images from angiography, CT, MRI, PET scanning, ultrasonography, and other imaging devices by satellite or telephone cabling to radiologists who may be located at a distance from the imaging site; conducting radiology image exchange and/or image interpretations electronically, usually via videoconferencing or messaging. See Telemedicine

telescoped finger Doigt-en-lorgnette Orthopedics A finger in which there is concentric osteolysis, bone collapse Seen in Psoriasis, rheumatoid arthritis. See Psoriasis, Rheumatoid arthritis

Telescope™ catheter Interventional cardiology An OTW balloon catheter with telescoping segments allowing shaft shortening/lengthening during PTCA. See Percutaneous transluminal coronary angioplasty

telescoping The 'compression' or overlapping of clinical or pathologic features of a disease or lesion that is normally subdivided into chronologic stages of progression

telescoping fracture Orthopedics A fracture seen in osteogenesis imperfecta where marked osteoporosis facilitates an axial compaction fracture with collapse, shortening and thickening of long bones. See Compaction, Compression fracture, Osteogenesis imperfecta

telesurgery Telepresence surgery, see there

television theory of allergy Allergy medicine A popular term for the posit that asthma in children is partially related to excessive exposure to household antigens, especially cockroach antigen–CA and dust. See Cockroach antigen

TeliCam Dentistry An intraoral camera that captures and displays video images, enabling Pts to view their mouth as the dentist describes problem areas and proposed treatment

telmisartan Micardis® Cardiology An angiotensin II receptor antagonist for treating HTN . See Angiotensin II receptor antagonist

temafloxacin Omniflox® Antibiotics An advanced fluoroquinolone active against streptococci and anaerobes; it was withdrawn from the market. See Fluoroquinolone, Temafloxacin syndrome

temafloxacin syndrome Therapeutics A condition caused by temafloxacin characterized by vasculitis with microangiopathic hemolytic anemia, DIC, thrombocytopenia, renal dysfunction, hepatotoxicity, and CNS changes, including strokes FREQUENCY 1/2000 Pts treated; it was withdrawn from the market. See Fluoroquinolone, Temafloxacin

temazepam Temaze PHARMACOLOGY A benzodiazepine used for insomnia. See Benzodiazepine

temozolomide Temodal® ONCOLOGY An alkylating agent used for recurrent gliomas–eg, anaplastic astrocytoma, and possibly melanoma and other solid tumors. See Astrocytoma, Glioblastoma multiforme

temper *adjective* Relating to temper or temperament *noun* PSYCHOLOGY Temperament, see there VOX POPULI A term referring to a current state or display of temperament, commonly used in a negative context–eg, *he/she has a real bad temper today*. See Temper tantrum, temperament

temper tantrum PEDIATRICS A prolonged anger reaction in an infant or child, characterized by screaming, kicking, noisy and noisome behavior, or throwing him/her self on the ground to get his/her way from a parent/caretaker/warden. Cf Adult temper tantrum

temperament PSYCHOLOGY An inborn pattern of behavior that tends to remain constant throughout life; a constitutional predisposition to react in a particular way to stimuli. See Artistic temperament, Temper. Cf Personality

temperature method See Basal body temperature

temperature spike MEDTALK An abrupt rise in temperature of > 38ºC/101ºF

template COMPUTERS A 'boilerplate' document saved as a permanent file, which contains specifications and formatting details, to be used repeatedly and modified slightly for individual reports PSYCHOLOGY A pattern that regulates the shape or appearance of a construction or idea. See Lovemap

temporal *adjective* Relating to 1. Time, transient 2. Temporal bone or the temple, the cranial region lateral to the frontal region

temporal arteritis Cranial arteritis, giant cell arteritis, granulomatous arteritis NEUROLOGY A self-limited disease of middle-aged ♀ characterized by vasculitis of the carotid artery which evolves to systemic arteritis in 10-15% of Pts; blindness, strokes possible as late complications DIAGNOSIS Temporal artery biopsy MANAGEMENT High-dose prednisone, tapered to low-dose

temporal lobe seizure Psychomotor seizure NEUROLOGY A simple or complex seizure caused by abnormal electrical activity in the temporal lobes, which may result in transient changes in movement, sensation, autonomic function, alertness and awareness CLINICAL Temporary paralysis, fear sensation, hallucinations, sleep paralysis; TLSs may occur in any person at any age, as a single episode, or as a chronic seizure disorder DIAGNOSIS Abnormal electrical activity on EEG ETIOLOGY Temporal lobe damage–trauma, hypoxia–ischemia and/or infarction, tumors, infection or any other discrete lesion. See Seizure

temporal lobe syndrome The functional loss of major portions of the temporal lobes and rhinencephalon–amygdala, hippocampus, uncus and hippocampal gyrus CLINICAL Visual agnosia, tendency to examine all objects orally, loss of emotion, hypersexuality–heterosexual, autosexual and homosexual activity, and ↑ consumption of meat. See Hypersexuality

temporal lobe trauma NEUROLOGY Injury to the temporal lobe of the brain, of either external or internal origin, with ↑ risk of temporal lobe seizures–psychomotor epilepsy, complex partial seizure and, rarely, associated paraphilic attacks. See Psychomotor epilepsy

temporary insanity FORENSIC PSYCHIATRY A transient loss of control over one's judgement and sense of reason, a common cause of marriage–Author's note. See M'Naghten rule, Insanity defense

temporary lead CARDIOLOGY A pacing lead intended for short term placement, often used with an external pacemaker. See Lead. Cf Permanent lead

temporary privileges HOSPITAL PRACTICE Limited clinical privileges given by a hospital's medical staff to a practitioner–eg, a locum tenens physician, to practice medicine for a defined period of time. See Privileges. Cf Emergency privileges

temporomandibular joint syndrome TMJ syndrome, see there

TEN 1. Total enteral nutrition 2. Toxic epidermal necrolysis, see there

ten-fingered osteopath Lesion osteopath OSTEOPATHY A popular term for an osteopath who adheres to the original precepts delineated by AT Still, in which 10 fingers are used for osteopathic manipulation; Cf Three-fingered osteopath

ten percent tumor A mnemonic for pheochromocytomas: 10% are malignant, 10% bilateral, 10% extra-adrenal, 10% in children, 10% associated with other systemic disease–eg, von Recklinghausen's disease, von Hippel-Lindau syndrome, Sturge-Weber disease, MEN IIa and IIb

tenaculum SURGICAL ANATOMY A fibrotendinous component of a flexor or extensor retinaculum SURGERY A thin hooked clamp used to retract tissues away from an operative field

tender MEDTALK Sensitive to pressure, specifically by palpation. Cf Nontender

tender years doctrine CHILD REARING A gender-specific philosophy that once prevailed in the US court system, favoring placement of young children with the mother in a divorce or separation. See Father factor

tenderness MEDTALK Mild pain. See Rebound tenderness

tendinitis Tendonitis ORTHOPEDICS Inflammation, tearing, tightness, or weakness in the tendons at the point of attachment of muscle to bone–eg, of the elbow, knee, which may restrict or inhibit adduction, abduction, supination MANAGEMENT RICE–rest, ice, compression, elevation, NSAIDs. See Achilles tendinitis, Bicipital tendinitis, Crossover tendinitis, Golf elbow, Tennis elbow

tendinosis PODIATRY Noninflamed degeneration of the Achilles tendon, most commonly of the midthird of the tendon, a relatively avascular zone

tendon sheath syndrome Tendon sheath adherence syndrome OPHTHALMOLOGY A condition characterized by limited elevation of the eye in adduction, due to fibrosis, and shortening of the superior oblique ocular muscle

tenecteplase TNKase® CARDIOLOGY A single-bolus thrombolytic clinically similar to tPA ADVERSE EFFECTS Intracranial bleeding, stroke, major or minor bleeding, hematomas. See Thrombolytic, tPA

tenesmus Painful spasm and productionless straining at stool

teniasis Tapeworm infection PARASITOLOGY Tapeworm infection acquired by eating raw or undercooked beef–*Taenia saginata* or pork–*T solium*; larvae from infected meat develop in the human intestine into the adult tapeworm, attaining lengths up to 4 m; each segment/proglottid can produce eggs, which are dispersed by proglottids as they detach and pass with the stool; *T saginata* proglottids are motile and actively exit via the anus; in absence of appropriate hygiene, Pts may become self-infected by ingesting eggs from their own tapeworms, which they pick up as they pick on their hindend; eggs hatch in the GI tract, larvae migrate through tissue and encyst; CNS larvae–cysticercosis–cause seizures, neurologic defects. See *Diphyllobothrium latum*

teniposide Vumon® ONCOLOGY An epipodophyllotoxin-type chemotherapeutic used in children with relapsed ALL and neuroblastomas, and adults with lymphoproliferative disorders ADVERSE EFFECTS Myelosuppression, N&V. See ALL

tennis elbow Lateral epicondylitis SPORTS MEDICINE A condition associated with tennis playing–duh! with no limitation of movement, swelling or pain when the joint is moved passively, but painful when actively moved; there may be a partial tear of the tendon fibers at or near their point of insertion on the humerus CAUSE Repetitive twisting of the wrist against resistance or frequent forearm rotation DIAGNOSIS Active dorsiflexion of the wrist against resistance or firm fingertip pressure over the lateral humeral epicondyle produces sharp pain RISK FACTORS Forceful repetitive wrist or forearm movement MANAGEMENT Rest, splinting and, if necessary, steroid injection

tennis leg SPORTS MEDICINE Exercise-induced rupture of calf muscles that may occur following any violent exercise in which the rapidly moving body abruptly changes direction–eg, tennis, soccer, downhill–alpine skiing CLINICAL An audible snap may be heard in the popliteal space, accompanied by severe calf pain and hematoma, due to a rupture of the gastrocnemius TREATMENT Immobilization in plantar flexion and physical therapy

tennis wrist ORTHOPEDICS A wrist affected by tenovaginitis which may occur in tennis players. See Wrist

Tenormin® Atenolol, see there

tenosynovitis ORTHOPEDICS Inflammation of the tendon sheath and tendon, which may be caused by injury, overuse, strain, or, rarely, by infection. See Trigger finger

tenosynovitis-dermatitis syndrome INFECTIOUS DISEASES A clinical complex seen in gonococcal bacteremia characterized by painful and swollen tendons, monoarthralgias or, more commonly, asymmetric polyarthralgias and a skin rash which is macular at onset, and later pustular. See Gonococcal arthritis, Gonorrhea

Tensilon® Edrophonium chloride PHARMACOLOGY A short-acting anticholinesterase used to diagnose myasthenia gravis and reverse the effects of nondepolarizing neuromuscular blockers. See Myasthenia gravis, Reversal agent

Tensilon® test NEUROLOGY A clinical test used in Pts with known myasthenia gravis to distinguish between a myasthenic and cholinergic crisis. See Myasthenia gravis, Myasthenic crisis

Tensinyl Chlordiazepoxide, see there

tension VOX POPULI A general term for any form of actual or perceived pressure. See Tension headache

tension band wiring ORTHOPEDICS A format for orthopedic wiring of fracture fragments either alone or with a screw or Kirschner wire to force fragments together in compression

tension headache Benign headache, muscle contraction headache NEUROLOGY A common headache related to prolonged muscle contraction, which begins as an occipital non-pulsatile, vise-like pain extending fronto-temporally, with 'tight' posterior cervical, temporalis or masseter muscles; THs are most common in ♀, occur at any age but are more common in adults and adolescents, who also have migraines, and are related to postures requiring sustained contraction of neck and scalp muscles, especially in response to stress, depression, or anxiety, or malposition of head and neck, linked to typing, computer use, fine work with hands, use of a microscope; sleeping in cold rooms or with neck in an abnormal position can trigger THs; other causes include eye strain, fatigue, excess alcohol, smoking, or caffeine use, sinusitis, nasal congestion, overexertion, colds, influenza PREVENTION Exercise and stretch muscles of head & neck. See Headache

tension myalgia RHEUMATOLOGY A term that encompasses the largely incorrect, or inadequate terms of fibrositis, fibromyalgia, and myofascial pain syndrome. See Fibromyalgia

tension pneumothorax CRITICAL CARE A life-threatening emergency consisting of air under pressure in the pleural space, due to a one-way valve type mechanism, allowing ↑ entry of air and eventually complete lung collapse on the affected side, which is acompanied by mediastinal shift of thoracic organs–heart, trachea, esophagus, and great vessels towards the unaffected side of the chest, and compression of the opposite lung with compromise in the return flow of blood to the heart ETIOLOGY Penetrating trauma to the chest, infection, mechanical ventilation with high pressures, and as a complication of CPR MANAGEMENT Chest tube drainage, or pleurodesis. See Pneumothorax

tenting CARDIOLOGY A term for the symmetrical 'peaking' of the 'T' wave on the EKG, associated with a lengthening of P-R, typically seen in early hyperkalemia; with further ↑ of K+, the P wave disappears and a sine wave appears INTERNAL MEDICINE A sign consisting of light pinching of a Pt's skin which, under usual conditions, springs back to a flattened position; a delay in flattening–'tenting' is characteristic of relatively severe dehydration and in the elderly whose dermal collagen and elastin have undergone age-related cross-linking

tenuous INTENSIVE CARE *adjective* Referring to a 'touch-and-go,' uncertain, or otherwise 'iffy' clinical situation

tenure ACADEMIA A status granted to a person with a 'terminal' degree–eg, doctor of medicine–MD or doctor of philosophy–PhD, after a trial period, which protects him/her from summary dismissal; tenured academicians are expected to assume major duties in research, teaching and, if applicable, Pt care fostering, through their activities, the academic 'agenda' of their respective departments or institutions. See Endowed chair, Lecturer, Professor. Cf Chair

Teperin Amitriptyline, see there

Tequin® Gatofloxacin INFECTIOUS DISEASE A broad-spectrum quinolone used for UTIs and RTIs. See Fluoroquinolone

terabyte 1000 gigabytes. See Byte, Kilobyte

teratogen GENETICS Any agent, chemical, or factor that causes a physical defect in a developing embryo or fetus; maternal medications with known teratogenic effects include aminopterin–spontaneous abortion, malformations; anticoagulants; anticonvulsants; cytotoxic drugs; mepivacaine–bradycardia, death; methimazole & propylthiouracil-goiter; ^{131}I–destruction of fetal thyroid; ♂ sex hormones–methyltestosterone, 17-α-ethinyl-testosterone, 17-α-ethinyl-19-nortestosterone–causes masculinization of ♀, tetracycline–hypoplasia and pigmentation of tooth enamel; trimethadione–abortion, multiple malformations, mental retardation; ♀ sex hormones cause virilization with defective external genitalia, transplacental carcinogenesis by DES. See Fetal warfarin syndrome, Fetal hydantoin syndrome, Thalidomide. Cf Litogen

teratophobia PSYCHOLOGY Fear of bearing a deformed child; fear of monsters or deformed people. See Phobia

terazosin Hytrin® An antihypertensive quinazoline that selectively blocks post-synaptic α₁-adrenergic receptors CLINICAL EFFECTS ↓ Peripheral vascular resistance due to arterial and venous dilation, ↑ HDL-C; may improve glucose tolerance and CHF; it provides symptomatic relief in BPH, ↑urine flow, ↓ muscle tone in bladder ADVERSE EFFECTS Hypotension, vertigo, headache, drowsiness, weakness, palpitations. Cf Proscar

terbinafine Lamisil® An antifungal that is better than griseofulvin for treating onychomycosis

terbutaline Brethine® THERAPEUTICS A β₂-agonist used to ↑ airway patency in asthmatics. Cf Budesonide

terfenadine Seldane* A non-sedating antihistamine used for allergy and viral URIs ADVERSE EFFECTS Arrhythmias–eg, torsades de pointes; should not be combined with ketoconazole, an antifungal that alters its metabolism, resulting in accumulation of the parent compound, a K⁺ channel blocker that prolongs QT. See Allegra

term of art MEDTALK A word or phrase with a precise definition in a particular field or specialty–eg, 'ugly' in oncology, 'satellite' in molecular biology or 'significant' in statistics

terminal *adjective* Near death or the natural end of a process; relating to an end-stage of a particular process. See Terminal cancer

terminal cancer A malignancy expected to cause the Pt's death in a short period of time–ie, wks to several months; Pts with TCs have 1+ of the following: no response to any form of therapy, tumor-related cachexia/marked weight loss, florid metastases to multiple sites or 'secondary' metastases–ie, those arising from an already metastatic focus, marked jaundice–due to liver replacement by CA, a need for constant pain medication, and compression of vital stuctures of 'impossible' surgical access. See 'Heroic' surgery, Hospice, Most significant other. Cf Spontaneous remission of cancer, Unproven therapies for cancer

terminal condition CHOICE IN DYING A condition caused by injury, disease or illness from which, to a reasonable degree of medical probability, a Pt cannot recover, and 1. The Pt's death is imminent or 2. The Pt is in a persistent vegetative state

terminal event FORENSIC PATHOLOGY A complication of an underlying cause of death, which is uniformly fatal if not immediately reversed. See Mechanism of death

terminal ileitis Crohn's disease, see there

terminal illness syndrome A condition which includes one or more of the following: a causative illness with a progressive evolution–eg, AIDS or CA, survival defined in terms of days or wks, Karnovsky score of < 40%, single or multiorgan failure, failure of conventional or proven treatment measures, absence of other potential proven or experimental therapy, irreversible progressive complications. See Karnovsky scale

terminal reservoir syndrome GI TRACT A potentially massive dilation of the descending and sigmoid colon, particularly common in the elderly, which is initially caused by voluntary suppression of defecatory urge; as the rectal stretch receptors degenerate, a vicious cycle of overextension and fecal impaction develops

terminally ill MANAGED CARE The status of a person expected to die within 6 months from a specific condition, and thus may need hospice care

termination of pregnancy Induced abortion. See Abortion

Terminologia Anatomica International Anatomical Terminology ANATOMY The official body and grand poobah of anatomic nomenclature created by the Federative Committee on Anatomical Terminology and the 56 members of the Intl Associations of Anatomists–a chucklefest of brainy long-faced coves with serious intent NOTE The author, at the risk of completely alienating himself from all medicos, calls attention to the fact that English–*not Latin*–is the lingua franca of medicine; to stubbornly cling to the belief that anatomic terminology can be written in anything but English is to risk being labelled obsolete and irrelevant. See Nomina Anatomica

terminology VOX POPULI A body of names assigned to or used for a particular type of thing. See Current procedural terminology, Dictionary, Lexicon, Nomina Anatomica, Terminologia Anatomica

terodiline Micturin™ THERAPEUTICS A spasmolytic used to treat GI and GU colic

TERP Totally endoscopic radical prostatectomy. See Prostatectomy.

terrace SURGERY A layer of sutures used to close thick tissue from the most internal to most external layers. See Sutures

terrain therapy Dosed walking REHAB MEDICINE An incremental ↑ in the distance and the grade–steepness of the terrain, a person should walk when recuperating from heart disease or other health problems; TT may be part of the regimen at a heath spa

tertian fever Tertian malaria INFECTIOUS DISEASE A fever characterized by febrile paroxysms occurring every 3ʳᵈ day, as in the 48-hr febrile peaks in *P vivax* malaria–benign tertian malaria; malignant TF is caused by the virulent *P falciparum* which, in its most intense form, may be fatal within days. See Malaria. Cf Quartan fever

tertiary *adjective* Referring to a third level

tertiary care MANAGED CARE The most specialized health care, administered to Pts with complex diseases who may require high-risk pharmacologic regimens, surgical procedures, or high-cost high-tech resources; TC is provided in 'tertiary care centers', often university hospitals, as it requires sophisticated technology, multiple specialists and subspecialists, a diagnostic support group, and intensive care facilities. See Tertiary care center. Cf Primary care, Secondary care

tertiary care center HOSPITAL CARE A hospital or medical center for Pts often referred from secondary care centers, which provides subspecialty expertise

---

TERTIARY CARE CENTER

SURGERY Organ transplantation, pediatric cardiovascular surgery, stereotactic neurosurgery and others

INTERNAL MEDICINE Genetics, hepatology, adolescent psychiatry and others

DIAGNOSTIC MODALITIES PET–positron emission tomography and SQUID–superconducting quantum interface device scanning, color Doppler electrocardiography, electron microscopy, gene rearrangement, and molecular analysis

THERAPEUTIC MODALITIES Experimental protocols for treating advanced and/or potentially fatal disease–eg, AIDS, cancer, and inborn errors of metabolism

---

tertiary Lyme disease Late persistent Lyme disease INFECTIOUS DISEASE Late, persistent inflammatory response to *Borrelia burgdorferi*, which occurs months to yrs after infection CLINICAL Skin, neurologic, musculoskeletal manifestations. See Lyme disease

tertiary prevention MEDTALK Treatment that alters the course of clinical disease--eg, with CABG or PCTA. See Percutaneous transluminal coronary angioplasty PSYCHIATRY Measures to reduce impairment or disability following a disorder–eg, through rehabilitation

tertiary syphilis Late syphilis STD A late phase of syphilis preceded, often by many yrs, by primary and secondary syphilis, caused by the spirochetes of *Treponema pallidum* which continued to reproduce for up to 15 years; focal damage in the form of highly destructive gummas occurs in bone, skin, nervous tissue, heart, arteries; neurosyphilis has many clinical forms–tabes dorsalis, general paresis and optic atrophy; cardiovascular involvement leads to aortic aneurysms, valvular heart disease, aortitis. See Benign late syphilis, Congenital syphilis, Latent syphilis, Neurosyphilis, Primary syphilis, Secondary syphilis, Syphilis

Teslascan® Mangafodipir trisodium IMAGING A hepatocyte-specific contrast medium used with MRI to detect and characterize liver lesions. See Magnetic resonance imagingf

test Assay MEDTALK The quantitative or qualitative measurement of a substance or process. See Acid loading test, Acid hemolysis test, Acid perfusion test, Agar-diffusion test, Agglutination test, AIDS test, Anterior drawer test, Antimicrobial sensitivity test, Anti-DNase B test, Antiglobulin test, Anti-glomerular basement membrane test, Apprehension test, Arginine test, Arterial blood gas test, Ascorbate-cyanide test, Aspergillosis precipitin test, Aspergillus antigen skin test, Autohemolysis test, Back bleeding test, Bar-reading test, Barium burger test, BarOn test, Baseline test, Bedside test, Bentiromide test, Benton VFT test, Bethanechol sensivity test, Bile solubility test,

Billable test, Biomedical test, Bleeding time test, Block sorting test, Breath test, But for test, C-peptide suppression test, Calcitonin suppression test, Caloric test, Cancer screening test, Capillary fragility test, Carbon dioxide challenge test, Carrier test, Chi squared test, Chromogenic (enzyme) substrate test, *Cis-trans* test, Citric acid urine test, Clinical laboratory test, Clomiphene stimulation test, Clonidine suppression test, Cochran-Mantel-Haenzel test, Colorimetric test, Combined anterior pituitary test, Commodity test, Complement fixation test, Complex test, Contraction stress test, Coombs' test, Copper sulfate test, Cover test, CRH stimulation test, Definitive test, Denver developmental screening test, Deoxyuridine suppression test, Developmental screening test, Dexamethasone suppression test, Diagnostic test, Differential renal function test, D-dimer screening test, Direct antiglobulin test, Doerfler-Stewart test, Donsbach's nutrient deficiency test, Durham test, Draw-a-person test, d-xylose absorption test, EDAC test, ELM test, Esoteric test, Exercise test, F test, F-max test, Fagan test, Farr test, Fecal occult blood test, Feminist test, Filippi test, Finger-to-nose test, Fisher's exact test, 40 millimeter test, Free PSA test, Full-Range Picture-Vocabulary test, Function-way-result test, Functional test, Functional gene test, Functional intact fibrinogen test, Fungal test, Germ tube test, Glucose tolerance test, GnRH stimulation test, Goldstein-Scheerer test, Goodenough-Harris Drawing test, Ham test, Heel-to-knee-to-shin test, HER-2/neu test, Highly complex test, HIV test, Hollander test, Home test, Hormone receptor test, House-tree-person test, HRV test, Hybrid Capture® CMV DNA test, Hyperemia test, Illinois Test of Psycholinguistic Abilities, Indican test, Indirect antiglobulin (Coombs) test, Indole spot test, Irresistible impulse test, Kahn Test of Symbol Arrangement, Kappa-lambda test, Kiss of death test, Kleihauer-Betke test, Kveim test, Lab test, Labyrinthine test, Lachman test, ²H-lactose test, Lactose intolerance test, Latex agglutination test, Laxity test, Leishmanin skin test, Lepromin skin test, Leukocyte bactericidal test, Level 1 test, Level 2 test, Liebermann-Burchard test, *Limulus* amebocyte lysate test, Liver function test, Load & shift test, Lombard (voice-reflex) test, Lupus band test, Lysine decarboxylase test, Malonate test, Management test, Mann-Whitney U test, Mantoux (tuberculin skin) test, Marshmallow test, Match test, Mercaptoethanol agglutination inhibition test, Methyl red test, Milk ring test, Mills' test, Mini-mental test, Mirror test, M'Naughten test, Moderately complex test, Monospot test, Morton's test, Motility test, Mouse footpad test, Mucin clot test, Multiple puncture test, Multiple sleep latency test, Mumps skin test, Myoglobin cardiac diagnostic test, Nafarelin test, Naffziger's test, NAP test, Navarro urine test, NBT test, Neostigmine test, Neuraminidase inhibition test, Neutralization test, Neutrophil hypersegmentation test, Niacin test, Ninhydrin test, Nitrate reduction test, Nonparametric test, Nonstress test, Occult blood test, O-F test, One-tail test, Optochin (disk) test, Ordered test test, Orientation test, Ornithine decarboxylase test, Osmotic fragility test, Oxidase test, Oxygen test, Oxytocin stress test, Pancreolauryl test, Panel test, Parametric test of significance test, Parathyroid squeeze test, Patrick's test, Paul-Bunnell-Davidsohn test, Penile stress test, Pentagastrin test, Performance test, Personality test, Perthes' test, Phenolphthalein test, Phenolsulfonphthalein test, Phentolamine test, Physician test, Pituitary function test, Plantar ischemia test, Platelet adhesion test, Postage stamp test, PPD test, Prausnitz-Küstner test, Predictive gene test, Pregnancy test, Prenatal test, Prick test, PRIST test, Projective test, Potein-bound iodine test, PROVERBS test, Provocative test, Psychologic test, Psychomotor test, Pulmonary function test, Pulp vitality test, Q-tip test, Quellung test, QuickScreen™ at-home drug test, Rabbit ileal loop test, Radiosensitivity test, Radioimmunosorbent test, Radiosensitivity test, RAST test, Reinsch's test, Reverse CAMP test, Rey test, RF-rheumatoid factor test, Rickettsia antibody test, Rinne test, Romberg test, Rose-Waaler test, Rothera's (nitroprusside) test, RPR test, Rubin test, Sabin-Feldman dye test, Saccharin test, Scheffé's test, Schick test, Schiller's test, Schilling test, Schirmer's test, Schwabach's test, Schwarz test, Scotchtape™ test, Screening test, Seashore test, Secretin injection test, Seleny test, Seliwanoff's test, Sentinel™ urine HIV-1 test, Serial sevens test, Serologic test, Serologic test for syphilis, Serum neutralization test, Sherlock Holmes test, Sickle cell test, Simple test, Sims-Huhner test, Sink-or-float test, SISI test, Skin test, 'Sort of' test, Special test, Spot test, Standard acid reflux test, Standard of care test, Statistical test, Stenger test, Stimulation test, Stratified log-rank test, Stress test, String test, Stroop test, Student's 't' test, Substantial capacity test, Substantial factor test, Sucrose hemolysis test, Super stress test, Suppression test, Sweat test, Swinging flashlight test, Tensilon test, Thallium stress test, Thematic apperception—See TAT test, Thenar weakness test, Thörmahlen test, Three glass test, Thrombin time test, Thromboplastin generation test, Thyroid suppression test, Thyroid function test, Tilt test, Time manual performance test, Tolerance test, Tone decay test, Total catecholamine test, Tourniquet test, Trail-making test, TRAx CD4 test, *Treponema pallidum* hemagglutination test, *Treponema pallidum* immobilization test, TRH stimulation test, Triple test, TRUE test, Trypan blue dye exclusion test, Tryptophan loading test, TSH stim-

ulation test, Tubeless test, Tuberculin skin test, Tuberculosis test, Tukey test, Turing test, Turbidity test, TWEAK test, Twin-to-twin transfusion test, Two-tailed test, Tzanck test, Uni-Gold HIV test, Urea breath test, Urease test, Urecholine supersensitivity test, Urine concentration test, Urecholine sensitivity test, Van der Bergh test, VDRL test, Vestibular test, Viral load test, Vitamin K test, Vocabulary test, Voges-Proskauer test, Volhard's test, Wada test, Waived test, Washout test, Wassermann test, Water deprivation test, Watson-Schwartz test, Weber test, Wechsler test, Weil-Felix test, Whiff test, Whiskey test, Widal's test, Wilcoxon's rank-sum test, Wipe test, Wisconsin Card-Sorting test, Wonderlic Personnel test, Word association test, WRAT test, Xylose absorption test, ¹⁴C-Xylose test, Y chromosome test **VOX POPULI** The process of putting a product through exercises that simulate its anticipated use, to identify differences between its actual and expected behavior. See Beta testing

test of cure STD The re-culturing of a site of initial infection—eg, gonorrhea, to determine whether the Pt is cured. See *Neisseria gonorrhoeae*

test of healing GI DISEASE A therapeutic trial of H₂-blockers—eg, cimetidine, ranitidine, which are used in Pts with a gastric ulcer, in whom a ↓ in ulcer pain is equated with therapeutic success; non-resolution of Sx after 3-6 wks of H₂-blockers mandates endoscopy and endoscopic biopsy as gastric CA must be excluded. See H₂-blockers

test menu Menu LAB MEDICINE 1. The list of all the tests performed by a particular lab, exclusive of those tests it sends to reference laboratories 2. A list of tests that a lab offers to its 'consumers', or a list of tests that a particular analytical instrument is capable of performing

test panel Laboratory diagnosis-related group, multiple test panel, organ panel, panel LAB MEDICINE A battery of diagnostic tests that have been found to be the most cost-effective, sensitive and specific means for evaluating a particular organ, organ system, or identifying a morbid process. See Anemia panel, Bone/joint panel, Cardiac injury panel, Cardiac risk evaluation panel, Collagen disease and arthritis panel, Collagen disease/lupus erythematosus panel, Coma panel, Diabetic panel, Electrolyte/fluid balance panel, General health panel, Hepatitis–immunopathology panel, Hypertension panel, Kidney panel, Liver panel, Metastatic disease panel, Neoplasm panel, Pancreatic panel, Parathyroid panel, Pulmonary panel, Thyroid panel, TORCH panel

test-retest reliability PSYCHOLOGY A measure of the ability of a psychologic testing instrument to yield the same result for a single Pt at 2 different test periods, which are closely spaced so that any variation detected reflects reliability of the instrument rather than changes in the Pt's status. Cf Interrater reliability STASTISTICS The correlation between the 1st and 2nd test of a number of subjects. See Reliability

test-tube baby A term gestational product resulting from in vitro fertilization of an egg implanted in a uterus and carried to term by either the genetic mother or by a surrogate—gestational mother. See Artificial reproduction, Baby M, in vitro reproduction, Surrogate motherhood

testicle See Bell clapper testicle, Migrating testicle

testicular cancer Lance Armstrong tumor, testicular malignancy, testicular neoplasia ONCOLOGY The most common cancer in ♂ age 20 to 35, which is curable if treated early; often first identified by its owner as a lump in the testicle, usually noted after an episode of trauma RISK FACTORS Hx of cryptorchidism, mumps orchitis, inguinal hernia during childhood, or previous TC on other side MANAGEMENT Orchiectomy, RT. See Embryonal carcinoma, Germ cell tumor, Seminoma, Testicular lymphoma

testicular cancer panel Germ cell profile ONCOLOGY A battery of serum tests—AFP and β-hCG—used to identify metastases in testicular CA. See Germ cell tumor, Seminoma

testicular failure ENDOCRINOLOGY The inability of the testes to produce sperm or hormones ETIOLOGY Chromosome defects, testicular torsion, direct trauma, mumps orchitis, testicular CA, various drugs RISK FACTORS

Cryptorchidism, constant low-level scrotal trauma–eg, motorcycle use, frequent use of drugs affecting testicular function–eg, marijuana, some prescription drugs. See Cryptorchidism

testicular feminization Androgen-insensitivity syndrome, feminization; Morris syndrome ENDOCRINOLOGY An X-linked pseudohermaphroditic state occurring in a genotypic XY ♂ with a ♀ phenotype; there is normal secretion of and response to müllerian inhibiting hormone, resulting in a phenotypic ♀ CLINICAL ♀ habitus, sexual behavior and 2° sex characteristics, associated with a blind vaginal pouch without uterine tissue, scanty pubic and axillary hair, and internally, undescended testicles; although the gonads often harbor Sertoli cell adenomas, malignancy occurs in 4% of TF LAB Normal testosterone levels. See Male pseudohermaphroditism

testicular infection See 1. Epididymitis 2. Orchitis

testicular lymphoma An intermediate- to high-grade lymphoma arising in the lymphocytes of testicular stroma; TLs comprise 5% of testicular tumors, but are the most common ones in ♂ > age 60

testicular regression syndrome Vanishing testes syndrome, embryonic testicular regression syndrome NEONATOLOGY A heterogeneous group of ♂ pseudohermaphrodites resulting from a cessation of testicular hormone activity during ♂ sex differentiation–wks 8-14, resulting in an absence of gonads in a genotypic–XY ♂

testicular torsion Torsion of testes UROLOGY Twisting of the spermatic cord, with ↓ blood supply to the testes and scrotum; it is the most common cause of scrotal pain in boys, most of whom are < age 6, ±linked to ↓ connective tissue in the scrotum, trauma, or physical activity

testing Computer Assisted Testing–CAT NIHSPEAK A testing format which assesses concrete knowledge based on true/false or multiple-choice questions that are objectively scored–Ex. MMPI; subjective tests require learners to formulate a response VOX POPULI The objective evaluation of a thing or process. See Alpha testing, Alternate site testing, Animal testing, Athlete drug testing, Bedside testing, Beta testing, Cascade testing, Cytotoxic testing, Cytotoxicity testing, Direct genetic testing, Duplicate testing, Dynamic testing, Fluorescence lifetime testing, Fructosamine testing, Gene testing, Genetic testing, Histocompatibility testing, HIV drug resistance testing, Home testing, Home access HIV testing, Hypothesis testing, Lipid testing, Needle-free blood testing, Occult blood testing, Mobility testing, Molecular wellness testing, Muscle response testing, p24 antigen testing, Paternity testing, Patient-initiated lab testing, Pharmacologic testing, Point-of-care testing, Polarity testing, Preadmission testing, Predisposition testing, Preimplantation genetic testing, Presymptomatic testing, Proficiency testing, Psychological testing, Reality testing, Reflex diagnostic testing, Remedial proficiency testing, Reverse paternity testing, Round-robin testing, Silicone testing, Sink testing, Statistical testing, Sweat testing, Systematic nutritional muscle testing, Total viable organism testing, Toxicity testing, Water testing, Wellness testing, Zone testing

testis-determining factor A protein encoded by a gene in the SYR region on the short arm of the Y chromosome, which is responsible for development of 1° ♂ organs. See X-chromosome inactivation

Testoderm® TTS ENDOCRINOLOGY A patch that delivers low-dose testosterone transdermally for 24 hrs INDICATION Testosterone replacement in ♂. See Androtest

testosterone ENDOCRINOLOGY The principal and most potent of the C-19 androgenic steroids, produced from its precursor hormone, progesterone, by the Leydig cells of testis, and, in far lesser amounts, in the ovary and adrenal cortex, and drives the development and maintenance of ♂ sex characteristics ACTIVITY Nitrogen retention, buildup of protein, induction and maintenance of 2° ♂ characteristics–eg, facial hair; testosterone secretion is in turn regulated by LH, which is produced by the anterior pituitary INDICATIONS Treatment of sexual dysfunction, weight loss, depression REF RANGE ♂–total, 300–1000 ng/dL; free, 5.1–41.0 ng/dL; ♀–total, 20-90 ng/dL; free, 0.1–2.3 ng/dL ↑ IN ♂ sexual precocity, hyperplasia of adrenal cortex, adrenogenital

syndrome, polycystic ovary syndrome ↓ IN Alcoholism, anterior pituitary gland hypofunction, estrogen therapy, Klinefelter syndrome–aka 1° hypogonadism, testicular hypoactivity. See Sublingual testosterone Cf Progesterone

testosterone 17β-dehydrogenase–NADP+ deficiency 17-Ketosteroid reductase deficiency A rare cause of ♂ pseudohermaphroditism, which may also result in micropenis and, in ♀, polycystic ovary disease; partial 17-TDD may be of late onset and associated with gynecomastia, hydrocele, hypogonadotropic hypogonadism

testosterone replacement therapy Androgen replacement therapy, see there

testotoxicosis A state of ♂ precocious puberty characterized by autonomous testicular production of androgen, requiring antiandrogens and aromatase inhibitors to slow skeletal growth to a prepubertal rate

tetanospasmin TOXICOLOGY A 150 kD neurotoxin produced by *Clostridium tetani*, which is one of the most toxic substances known to man; it selectively blocks inhibitory nerve transmission from the spinal cord to the muscles, causing severe muscle spasms, which may be intense enough to tear muscle or cause compression fractures of the vertebrae; it evokes unrestrained muscle firing and sustained muscular contraction, causing lockjaw, dysphagia, or acute respiratory failure by tetany of the diaphragm. See Poisons

tetanus Lockjaw INFECTIOUS DISEASE Acute infection by the anaerobic spore-forming bacillus *Clostridium tetani*, manifest by uncontrolled muscle spasms due to tetanospasmin; it is often fatal, especially at the extremes of age, and preventable by immunization EPIDEMIOLOGY *C tetani* is ubiquitous, and may infect virtually any open wound; no longer a major health problem in socioeconomically advanced countries or the US–incidence ± 0.035/10⁵, in developing nations, it is a 'top 10' killer, causing ±1 million deaths/yr SUBSTRATE Tetanus develops in a menagerie of mishaps from minor mayhem to sloppy abortions, ♀ circumcision, and so on, rolling rural romanticism into a reality sandwich CLINICAL ±2 wk incubation, followed by localized or generalized weakness, cramping, dysphagia, trismus–lockjaw, ↑ muscle rigidity–eg, risus sardonicus, opisthotonus, laryngospasm, and ± death MANAGEMENT Benzodiazepines–↓ anxiety, sedation, anticonvulsant, muscle relaxant facilitating GABA-inhibitory transmission in the brain stem and spinal cord, ventilatory support, tetanus immune globulin VACCINE The tetanus vaccine is 96% effective; in the US rates of immunity to tetanus ranges from 80% in white ♂ to < 20% in Mexican-American ♀; immunity is ↑ in those with a ↑ education and income. See *Clostridium tetani*. Cf Tetany NEUROLOGY Tonic muscle contraction, see there

tetanus toxoid A small peptide fragment that selectively elicits helper immune response but not immune suppression

tetany NEUROLOGY Neuromuscular hyperexcitability evoked by an extreme drop in extracellular calcium ETIOLOGY Alkalosis, parathyroid hypofunction, vitamin D deficiency CLINICAL Carpal, laryngeal, pedal spasms, cramping, stridor, hyperreflexia, choreiform movement. Cf Tetanus

tethered cord syndrome NEUROLOGY An occult spinal dysraphism due to a defect in dorsal induction, the earliest major embryologic process in forming the brain and spinal cord; during 2° neurulation, the neural tube atrophies and the ventral remnant persists as the filum terminale; as the vertebral column grows, the filum terminale and nerve roots lengthen; tethering results if the primitive sacral cord does not degenerate or the filum terminale and nerve roots do not lengthen properly CLINICAL-CHILDREN Static sensorimotor defects, deformities of spine and feet, skin changes in lumbosacral region–eg, hypertrichosis, hemangiomas, dimples; pain is rarely prominent CLINICAL-ADULTS Gait defects, upper and lower motor neuron signs, and either a spastic small-capacity bladder or less commonly, a hypotonic large-capacity bladder DIAGNOSIS MRI; CT with intrathecal contrast, plain films, ultrasonography TREATMENT Laminectomy at appropriate lumbosacral level. Cf Neural tube defect

tetra– Greek root, four

tetra-X syndrome XXXX syndrome, see there

Tetracel™ vaccine DPT & HIB VACCINOLOGY A combination vaccine against diphtheria, pertussis, tetanus–DPT, *H influenzae* type B. See DPT, HIB

tetracycline Tet DERMATOLOGY An antibiotic widely used for acne vulgaris INFECTIOUS DISEASE Tetracycline is used to manage Rickettsiae–RMSF, typhus fever, typhus group, Q fever, rickettsialpox, tick fevers, *Mycoplasma pneumoniae*, *Chlamydia*–psittacosis/ornithosis, LGV, granuloma inguinale; *Borrelia recurrentis*–relapsing fever; gram-negative bugs–*Haemophilus ducreyi*–chancroid agent, *Pasteurella pestis*, *P tularensis*, *Bartonella bacilliformis*, *Bacteroides* spp, *Vibrio comma*, *V fetus*, *Brucella* spp ADVERSE EFFECTS GI–anorexia, epigastric distress, N&V, diarrhea, bulky loose stools, stomatitis, sore throat, glossitis, black hairy tongue, dysphagia, hoarseness, enterocolitis, anogenital lesions with monilial overgrowth MOLECULAR BIOLOGY A broad-spectrum bacteriostatic that inhibits protein synthesis by preventing aminoacyl tRNA from binding to ribosomes. See Tetracycline labeling

tetrahydrobiopterin METABOLIC DISEASE A natural cofactor of aromatic amino acid hydroxylases and nitric oxide synthase, which may be used as an alternative to low-phenylalanine diets in Pts with mild phenylketonuria. See Phenylketonuria

tetraiodothyronine Thyroxin, see T4

tetralogy of Fallot PEDIATRIC CARDIOLOGY A cyanotic congenital heart disease, defined by obstruction of the right ventricular outflow, VSD, right ventricular hypertrophy, and an overriding aorta RISK FACTORS Maternal rubella, viral illness during pregnancy, poor prenatal nutrition, maternal alcoholism, mother > age 40, gestational DM, trisomy 21 CLINICAL At birth, infants are not cyanotic, but later may develop episodic cyanosis from crying or feeding–called "Tet spells"

tetrodotoxin TOXICOLOGY A potent heat-stable neurotoxin concentrated in fishes of order Tetraodontoidea; tetrodotoxin blocks conduction of sodium across membranes, and neural transmission in skeletal muscle, and kills up to 60% of those who ingest it, causing 50 deaths/yr in Japan CLINICAL Paresthesias begin 10-45 mins after ingestion, commonly as intraoral and tongue tingling, often associated with N&V, lightheadedness, vertigo, feelings of doom, weakness, hypersalivation, muscle twitching, diaphoresis, pleuritic chest pain, dysphagia, aphonia, convulsions, hypotension, bradycardia, depressed corneal reflexes, and fixed dilated pupils MANAGEMENT IV hydration, gastric lavage, activated charcoal

Tetwrist REHAB MEDICINE A repetitive strain injury acquired after extended play of "addictive" computer games–eg, Tetris

teutophobia PSYCHOLOGY Fear of German things. See Phobia

Teveten® Eprosartan mesylate, see there

Texas Primary Reading Inventory NEUROLOGY An abbreviated test for dyslexia developed at the U Tex-Houston. See Dyslexia

TexCAPS CARDIOLOGY A clinical trial–Texas Coronary Atherosclerosis Prevention Study

text bite JOURNALISM A pithy 'one-liner' carved from text written in a journal article, editorial, review, or textbook, that summarizes a salient point

textbook INFORMATICS A treatise on a particular subject. See Bible

TFL syndrome Tensor fasciae latae syndrome. See Iliotibial band syndrome

TFR Total fertility rate, see there

THA Total hip arthroplasty. See Total hip replacement

thalamic syndrome Dejerine-Roussy syndrome, thalamic hyperesthetic anesthesia A condition caused by infarction of the posteroinferior thalamus CLINICAL Transient hemiparesis, loss of deep and superficial nociception, with preservation of positional sense; with time, trophic changes occur

thalassemia Beta thalassemia, clinical thalassemia, Cooley's anemia, Mediterranean anemia, thalassemia major HEMATOLOGY A group of genetic diseases by underproduction of hemoglobin due to mutations in the beta globin gene, which is more common in Mediterraneans HEREDITY Parents are carriers–heterozygotes; one in 4 children is homozygous for the mutation and thus has full-blown disease CLINICAL See Anemia. Cf Sickle cell anemia

α-thalassemia Hemoglobin Barts HEMATOLOGY An inherited condition caused by a defect in the synthesis of the Hb α chain; Hb Barts hemoglobinopathy is characterized by the presence of 4 gamma chains; it is more common in southeast Asians; the most severe form of alpha thalassemia causes stillbirth due to hydrops fetalis HEREDITY Parents are carriers–heterozygotes; one in 4 children is homozygous for the mutation and thus has full-blown disease CLINICAL Pallor, fatiguability, FTT, fever, infections, diarrhea MANAGEMENT Transfusions

β-thalassemia Thalassemia major HEMATOLOGY A hemoglobinopathy caused by a defect in the synthesis of Hb β chain CLINICAL Pallor, fatigability, FTT, fever due to infections, diarrhea, bone deformities, hepatosplenomegaly MANAGEMENT Transfusions, but iron overload can damage the heart, liver, and endocrine systems, ergo iron chelation–early use of deferiprone, deferoxamine ↓ transfusion-related iron overload and may protect against DM, cardiac disease, early death

δ-thalassemia HEMATOLOGY A condition characterized by a defect of Hb A_2–$α_2δ_2$; because Hb A_2 comprises only 3% of the circulating Hb, even its complete absence; δ-thalassemia has little clinical or hematologic impact

γ-thalassemia HEMATOLOGY A condition characterized by a defect of gamma–γ Hb chains found in Hb F–$α_2γ_2$; because Hb F is present primarily in the fetus and newborns, it is rarely seen outside of the neonatal period, but may cause transient neonatal hemolytic anemia

thalassemia minor Thalassemia trait HEMATOLOGY The carrier state for beta thalassemia–heterozygosity, possession of one thalassemia gene; the person is essentially normal

thalidomide THERAPEUTICS A drug first marketed as a sedative and sleeping aid, which caused 12,000-15,000 cases of embryopathy, often in the form of phocomelia or 'flipper' extremities NEW INDICATIONS Thalidomide may be effective in combating wasting in AIDS Pts–especially with oral ulcers, as well as rheumatoid arthritis, photodermatitis, Behçet's disease, SLE, GVHD, asbestosis, IBD, lepromatous leprosy–erythema nodosum leprosum, possibly TB; it is antiangiogenic, stops menstruation and may be useful for neoangiogenesis in macular degeneration and diabetic retinopathy. See Angiogenesis inhibitors GI DISEASE Benefits Pts with refractory Crohn's disease, due to TNF-α inhibition ONCOLOGY Thalidomide has orphan drug status for treating refractory myeloma, possibly due to thalidomide's inhibition of angiogenesis and modulation of cytokines; it may be used for Pts with head & neck SCC , and as an ancillary therapy in KS, melanoma, brain, breast, prostate CAs

thalidomide neuropathy NEUROLOGY A severe polyneuropathy seen in up 25% of those treated with thalidomide as a sedative, or higher in GVHD; in half, sensory loss is permanent

thallium stress test Pharmacologic stress imaging CARDIOLOGY A myocardial perfusion technique in which the radionuclide thallium-201–^{201}Tl, is injected as a diagnostic adjunct to cardiac stress tests, to detect regional ischemia or infarction; TST is an increasingly popular alternative to the exercise stress test–EST; it is used to evaluate Pts with CAD who cannot perform an adequate EST, and allows risk stratification. Cf Treadmill stress/exercise test

THAM Tris(hydroxymethyl)aminomethane, Tris buffer EMERGENCY MEDICINE An amine proton donor administered IV during early CPR to treat lactic acidosis

thanatophoric dwarfism *thanatos*, Greek, death, *phoric*, bearing, Chondrodysplasia punctata dwarfism of Conradi-Hünermann An AD form of dwarfism with a 2:1 ♂:♀ ratio, in which the infants are stillborn or die in early infancy CLINICAL Hydrocephaly, megalocephaly with frontal bossing, chondrodystrophy, narrow thorax with respiratory difficulties, congenital heart disease, hypotonia and hyporeflexia–floppy infant, hypertelorism, 'cloverleaf' skull, saddle nose, marked skeletal abnormalities with shortened deformed 'telephone receiver' long bones, affecting the epiphysis, causing micromelia, altered feet, lenticular opacity, shortened extremities with curved fingers, H or U-shaped vertebrae and pulmonary hypoplasia result in short postnatal survival

THBR Thyroid hormone binding ratio, see there

THC Tetrahydrocannabinol Any of a family of compounds present in *Cannabis sativa* var *indica*, the major constituent of which is the Δ^1-3,4-*trans* isomer, $^9\Delta$-THC. See Marijuana

theca cell tumor Thecoma GYNECOLOGIC PATHOLOGY A sex cord-stromal tumor, usually of the post-menopausal ovary CLINICAL TCTs may produce copious estrogens and, like thecomas, may evoke precocious puberty in the young or adenomatous hyperplasia of endometrium or WD endometrial CA in 3-20% of older Pts. See Sex cord-stromal tumors

thelarche GYNECOLOGY Peripubertal enlargement of the breasts

Thematic Apperception Test TAT PSYCHOLOGY A projective diagnostic technique which seeks to uncover personality dynamics TARGETED AREAS Main theme, hero/heroine, solving or blame. See Psychological testing

thenar muscles ANATOMY The intrinsic muscles of the thumb: adductor pollicis brevis, adductor pollicis brevis, flexor pollicis brevis, opponens pollicis See Hand, Thumb. Cf Hypothenar muscles

Theo-dur® Theophylline, see there

theobromine An alkaloid obtained from *Theobroma cacao*, the chocolate plant, which is similar to caffeine

theophylline THERAPEUTICS A xanthine derivative used in asthmatics, which relaxes smooth muscle; its effect is ↓ in smokers and in barbiturates and phenytoin therapy ACTION Relaxes smooth muscle of bronchial airways and pulmonary blood vessels, resulting in broncho- and vasodilation; it is also a diuretic, coronary vasodilator, cardiac and cerebral stimulant ADVERSE EFFECTS GI irritation-anorexia, N&V, epigastric pain, restlessness, insominia, headache CONTRAINDICATIONS Acute peptic ulcer disease, untreated seizure disorder; theophylline accumulates in organ failure–eg, heart–CHF, liver, lungs, and kidney, by interfering with drug metabolism–eg, allopurinol, cimetidine, erythromycin, propranolol; theophylline may cause seizures and arrhythmias. See Aminophylline

theory A hypothesis or explanation of a phenomenon based on available data STATISTICS A general statement predicting, explaining, or describing the relationships among a number of constructs

Theospan® Theophylline, see there

therapeutic *adjective* Referring to therapy or treatment

therapeutic alternative HEALTH INSURANCE A drug which differs chemically from one prescribed, but provides the same effect when administered to Pts. See Generic drug

therapeutic abortion GYNECOLOGY The termination of pregnancy before fetal viability in order to preserve maternal health. In its broadest definition, therapeutic abortion can be performed to (1) save the life of the mother, (2) preserve the health of the mother, (3) terminate a pregnancy that would result in the birth of a child with defects incompatible with life or associated with significant morbidity, (4) terminate a nonviable pregnancy, or (5) selectively reduce a multifetal pregnancy. Cf Elective abortion

therapeutic apheresis A form of exchange transfusion in which blood is removed from a Pt and fluids returned to the general circulation: 1. Plasma is replaced–plasmapheresis with a volume of albumin or a crystalloid solution. 2. Cells are removed–apheresis, to either prevent or reduce leukostasis (leukapheresis) in the brain and renal arteries, or to reduce thrombotic phenomena (platelet apheresis) Note: In the working parlance, the adjective *therapeutic* is usually deleted

therapeutic cloning MOLECULAR MEDICINE The cloning of human cell lines to replace nonfunctional tissue. See Cloning

therapeutic community ADDICTION DISORDERS A structured treatment environment with emphasis on group process, often with former addicts on staff. See Halfway house

therapeutic crisis PSYCHIATRY An abrupt and extreme positive or negative change in a response to therapy

therapeutic drug monitoring CLINICAL PHARMACOLOGY The regular measurement of serum levels of drugs requiring close 'titration' of doses in order to ensure that there are sufficient levels in the blood to be therapeutically effective, while avoiding potentially toxic excess; drug concentration in vivo is a function of multiple factors COMMON TDM DRUGS Carbamazepine, digoxin, gentamycin, procainamide, phenobarbital, phenytoin, theophylline, tobramycin, valproic acid, vancomycin

---

THERAPEUTIC DRUG LEVELS IN VIVO–FACTORS INVOLVED

PATIENT COMPLIANCE Ingestion of drug in the doses prescribed

BIOAVAILABILITY Access to circulation, interaction with cognate receptor(s); ionized and 'free', or bound to a carrier molecule, often albumin

PHARMACOKINETICS Drug equilibrium requires 4-6 half-lives of drug clearance (a period of time for $^1/_2$ of the drug to 'clear', either through metabolism or excretion, multiplied by 4-6); the drug is affected by

- **INTERACTION WITH FOODS OR OTHER DRUGS** at the site of absorption, eg tetracycline binding to cations or chelation with binding resins, eg bile acid-binding cholestyramine that also sequesters warfarin, thyroxine and digitoxin or interactions of various drugs with each other, eg digitalis with quinidine resulting in a 3-fold ↓ in digitalis clearance

- **ABSORPTION** may be changed by GI hypermotility or large molecule size

- **LIPID SOLUBILITY**, which affects the volume of distribution; highly lipid-soluble substances have high affinity for adipose tissue and a low tendency to remain in the vascular compartment, see Volume of distribution

- **BIOTRANSFORMATION**, with 'first pass' elimination by hepatic metabolism, in which polar groups are introduced into relatively insoluble molecules by oxidation, reduction or hydrolysis; for elimination, lipid-soluble drugs require the 'solubility' steps of glucuronidation or sulfatation in the liver; water-soluble molecules are eliminated directly via the kidneys, weak acidic drugs are eliminated by active tubular secretion that may be altered by therapy with methotrexate, penicillin, probenecid, salicylates, phenylbutazone and thiazide diuretics

FIRST ORDER KINETICS Drug elimination is proportional to its concentration

ZERO ORDER KINETICS Drug elimination is independent of the drug's concentration

PHYSIOLOGICAL FACTORS

- **AGE** Lower doses are required in both infants and the elderly, in the former because the metabolic machinery is not fully operational, in the latter because the machinery is decaying, with ↓ cardiac and renal function, enzyme activity, density of receptors on the cell surfaces and ↓ albumin, the major drug transporting molecule

- **ENZYME INDUCTION**, which is involved in a drug's metabolism may reduce the drug's activity; enzyme-inducing drugs include barbiturates, carbamazepine, glutethimide, phenytoin, primidone, rifampicin

- **ENZYME INHIBITION**, which is involved in drug metabolism, resulting in ↑ drug activity, prolonging the action of various drugs, including chloramphenicol, cimetidine, disulfiram (Antabuse), isoniazid, methyldopa, metronidazole, phenylbutazone and sulfonamides

GENETIC FACTORS play an as yet poorly defined role in therapeutic drug monitoring, as is the case of the poor ability of some racial groups to acetylate drugs

CONCOMITANT DISEASE, ie whether there are underlying conditions that may affect drug distribution or metabolism, eg renal disease with ↓ clearance and ↑ drug levels, or hepatic disease, in which there is ↓ albumin production and ↓ enzyme activity resulting in a functional ↑ in drug levels, due to ↓ availability of drug-carrying proteins

therapeutic equivalent HEALTH INSURANCE A drug that controls a symptom or condition in the exact same way as another. See Generic drug

therapeutic gain Placebo-subtracted response RESEARCH The response in a treatment group minus the response in a control/placebo group. See Placebo effect

therapeutic immune paralysis Chemical 'splenectomy' A method that inhibits splenic endocytosis of opsonized–ie, Ig- or complement-coated cells or microorganisms by blocking Fc receptors; the targets are thus bound, but not endocytosed; TIP is inducible by high-dose–corticosteroids or IVIG; the effect lasts as long as the therapy and is an alternative to surgery in immune-related hypersplenism–eg, autoimmune hemolytic anemia, autoimmune neutropenia, Felty syndrome

therapeutic index PHARMACOLOGY The ratio of a drug's toxic level to its therapeutic level, which is calculated as the toxic concentration–TC of a drug, divided by the effective concentration, expressed as TC_{50}/EC_{50}, a point at which 50% of Pts have a toxic reaction to the drug being monitored; the lower the therapeutic index, the more difficult it is to titrate a drug's dose in a Pt and the more critical it is that the drug be monitored. See Apparent volume of distribution, First-order kinetics, Peak levels, Trough levels, Volume of distribution, Zero-order kinetics

therapeutic intent NIHSPEAK A specific endpoint sought in a line of therapeutic research–eg, prolongation of life, shrinkage of tumor, or improved quality of life, even in absence of cure or dramatic improvement of a condition

therapeutic misadventure Boo-boo An unintentional–or 'functional' overdose of a therapeutic agent due to unanticipated effects of extraneous factors–eg, acetaminophen hepatotoxicity in alcoholics, due to cytochrome P-450 induction. See Opportunity for improvement

therapeutic phelobotomy TRANSFUSION MEDICINE Blood withdrawn for a therapeutic purpose with the possiblity of future allogeneic–homologous transfusion; TPs are done only at the request of the Pt's physician; use of said blood is determined by the prospective recipient's physician. See Phlebotomy

therapeutic privilege Therapeutic exception MEDICAL ETHICS A paternalistic principle under which truth is withheld from a Pt out of concern that if the details of a procedure are fully delineated, the Pt may choose to forego an operation that the physician believes is in the Pt's best interest or only option for improved quality of life and/or survival. See *Arato v Avedon*, Information overload, Doctor-patient interaction, Paternalism, Therapeutic privilege doctrine

therapeutic privilege doctrine FORENSIC MEDICINE A doctrine that protects the physician faced with a Pt who may be too emotional or apprehensive to fully and logically assess his needs for a therapeutic intervention; such situations may arise in emergencies, advanced age, dementia; when possible, permission should be obtained from the nearest relative. See Good Samaritan laws. Cf Doctor-patient interaction, Informed consent, Paternalism

therapeutic range THERAPEUTICS The range of concentrations at which a drug or other therapeutic agent is effective with minimal toxicity to most Pts. See Clearance

therapeutic rating PHARMACOLOGY A rating assigned by the FDA to drugs based on therapeutic efficacy; A-rated drugs present an important therapeutic gain, B-rated agents have a moderate therapeutic gain; C-rated agents have little or no therapeutic value. Cf Therapeutic index

therapeutic threshold INTERNAL MEDICINE The level of certainty that a Pt has a particular condition warranting treatment, as opposed to the Pt having another condition. See Benefit:risk ratio, Threshold value

therapeutic vaccine IMMUNOLOGY A vaccine–eg, Salk's Remune™ intended to treat a viral infection by stimulating the immune system. See Vaccine therapy

therapeutic window The well-defined range of a drug's serum concentration at which a desired effect occurs, below which there is little effect, above which toxicity occurs; the TW differs among Pts and may be determined empirically. See Therapeutic drug monitoring. Cf Window

therapist A person who provides therapy. See Enterostomal therapist, Licensed Massage therapist, Psychotherapist, Physical therapist

therapist-patient privacy PSYCHIATRY A privileged relationship between a Pt and a mental health professional–eg, psychiatrist, psychologist, or other, that is protected from public domain, assuming that the Pt does not pose a threat to him/herself or others. See Anne Sexton, Bennett-Leahy bill, Health information trustee, Medical privacy, Nondisclosure ceiling, *Tarasoff v. Regents of the University of California*

therapy A general term for any form of management of a particular condition; treatment intended and expected to alleviate a disease or disorder; any technique of recovery, which may be medical, psychiatric, or psychological. See Ablation therapy, 'Abracadabra therapy,' Adjunctive therapy, Adjuvant therapy, Air ionization therapy, Alternative therapy, Amino acid therapy, Androgen deprivation therapy, Androgen replacement therapy, Angiogenic gene therapy, Antiadhesive therapy, Antiarrhythmic therapy, Antibiotherapy, Antibody directed enzyme prodrug therapy, Anticoagulant therapy, Antioxidant therapy, Antiplatelet therapy, Antisense therapy, Antistenotic therapy, Autohemotherapy, Autolymphocyte therapy, Autosuggestion therapy, Aversion therapy, B chain therapy, B-1 therapy, Baggie™ therapy, Balance therapy, BCS therapy, Bee venom therapy, Behavior therapy, Benjamin system of muscular therapy, Biodynamic therapy, Biotherapy, Bladder-conserving therapy, Blood component therapy, Body therapy, Body-oriented psychotherapy, Boron therapy, Brachytherapy, Breast conservation therapy, Breath therapy, Bright light therapy, Burst therapy, Cardiovascular gene therapy, Castration therapy, Catheter ablation therapy, Cell therapy, Chelation therapy, Chemotherapy, Chiropractic therapy, Chronotherapy, Cognitive therapy, Cognitive behavioral therapy, Cold therapy, Colon therapy, Combination therapy, Combination chemotherapy, Combined modality therapy, Concentration therapy, Concept therapy, Condom therapy, Confrontation therapy, Conservative therapy, Continuous sleep therapy, Conventional therapy, Convergent therapy, Creative arts therapy, Cryotherapy, Cymatic therapy, Cytokine gene therapy, Dance (movement) therapy, Desensitization therapy, Detoxification therapy, Differentiation therapy, Directly-observed therapy, Double whammy therapy, Drama therapy, Dream therapy, E5 therapy, Electric therapy, Electroconvulsive therapy, Encapsulated cell therapy, Electroporation therapy, Encounter group therapy, Enzyme therapy, Enzyme replacement therapy, Estrogen replacement therapy, Exercise/movement therapy, Ex vivo therapy, Ex vivo gene therapy, Extracorporeal shock wave therapy, Fab therapy, Family therapy, Fever therapy, Fourth therapy, Germ cell gene therapy, Gene therapy, Gestalt therapy, Group therapy, HAART therapy, Heat therapy, Hematogenic oxidation therapy, Heroic therapy, High-dose chemotherapy, High pH therapy, Hippotherapy, Home infusion therapy, Hormonal therapy, Hormone-replacement therapy, Hydrogen peroxide therapy, Hypnotherapy, Hypnotic psychotherapy, Image aversion therapy, Immunoaugmentive therapy, Immunosuppressive therapy, Immunotherapy, IMRT, Induction therapy, Information therapy, Injection sclerotherapy, Integrative therapy, Interstim® continence-control therapy, Interstitial therapy, Interventional therapy, Intracavitary therapy, Intraoperative radiation therapy, Intrathecal baclofen therapy, Intravesicular therapy, Iron-chelation therapy, Laser therapy, Laughter therapy, Leukemia therapy, Life-extending therapy, Light therapy, Lipid therapy, Local therapy, Low-flow oxygen therapy, Magnet therapy, Magnetotherapy, Maintenance drug therapy, Maintenance therapy, Malariotherapy, Management, Massage therapy, Maternal blood clot patch therapy, Mechanotherapy, Medical fetal therapy, Megavitamin therapy, Mobilization therapy, Monopolar/electrohydrothermal coagulation therapy, Monotherapy, Movement therapy, Multimodal therapy, Myeloablative therapy, Myelosuppressive therapy, Myofascial release therapy, Neoadjuvant therapy, Nicotine replacement therapy, Nutritional therapy, Ocular photodynamic therapy, Open therapy, Oral chelation therapy, Oral rehydration therapy, Oxidative therapy, Ozone therapy, Palliative therapy, Patch therapy, Pet therapy, Photochemotherapy, Photodynamic therapy, Phototherapy, Physical therapy, Physiotherapy, Play therapy, Polarity therapy, Postural therapy, Preventive therapy, Pranic therapy, Psychotherapy, Preemptive therapy, Preemptive chemotherapy, Pressure point therapy, Prophylactic antibiotic therapy, Psychotherapy, PUVA therapy, Quality of life therapy, Radiation therapy, Radioimmunotherapy, Radiotherapy, Reconstructive therapy, Recreational therapy, Remission induction therapy, Reperfusion therapy, Respiratory therapy, REST, RIGS®/ACT therapy, Rogerian therapy, St John's neuromuscular therapy, Sclerotherapy, Sensory therapy, Shot therapy, Single-drug therapy, Steam inhalation therapy, Stem-cell therapy, Step-down therapy, Strain-counterstrain therapy, Stress therapy, Subliminal therapy, Suicide gene therapy, Surfactant replacement therapy, Superovulation therapy, Supplemental therapy, Supportive therapy, Synvisc® injection therapy, Systemic therapy, Third-line therapy, Thrombolytic therapy, TIL therapy, Topical chemotherapy, Topical immunotherapy, Treatment, Tremor control therapy, Trigger point injection therapy, Triple therapy, Ultrasound therapy, Vaccine therapy, Virtual reality exposure therapy, Vision therapy, Water-based therapy, Water-induced thermotherapy, Water therapy, Zone therapy

therapy addict PSYCHIATRY A popular term for a Pt mentally dependent on 'talk therapy' and often the therapist, who uses the therapy as a substitute for solving life problems

Theratope® vaccine ONCOLOGY A synthetic vaccine designed to stimulate the immune system to combat metastatic breast, colorectal and other CAs. See Therapeutic vaccine

ThermaChoice™ UBT system GYNECOLOGY Balloon thermal-ablation for treating menorrhagia instead of surgical D&C or electrosurgical ablation

thermal capsular shift ORTHOPEDICS A surgical procedure for managing multidirectional shoulder instability in which a probe is deployed arthroscopically to shrink the glenohumeral ligaments. See Multidirectional shoulder instability. Cf Inferior capsular shift

thermatic sterilization REPRODUCTIVE MEDICINE A method of inducing temporary sterility in ♂, by sitting in a hot bath for 45 mins/day for 3 wks, which allegedly renders a person sterile for 6 months. See Briefs

thermescent skin treatment POPULAR HEALTH A noninvasive procedure in which heat is applied to the skin to promote new collagen growth and allegedly combat aging. See Aging

ThermoChem HT™ system Biologic-HT™ system ONCOLOGY A system for administering intraperitoneal hyperthermic chemotherapy–IPHC, whole-body extracorporeal treatment of CA, AIDS. See Hyperthermic chemotherapy

Thermoflex™ system UROLOGY A water-induced thermotherapy system for treating BPH and associated urinary obstruction. See Benign prostatic hypertrophy

thermography IMAGING An abandoned method for diagnosing breast CA based on an ↑ warmth of skin overlying malignancy; heat is a relatively nonspecific finding that also occurs in mastitis; the technique was abandoned due to the unacceptably high rates of false positivity and false negativity; it has no role in breast CA diagnostics. Cf Mammography, Xeroradiography

thermolysis Heat-induced interruption of cells and tissue. See Selective photothermolysis

thermophobia PSYCHOLOGY Fear of heat. See Phobia

thermoreversible gel AIDS A temperature-sensitive gel which, alone or in concert with anti-HIV agents may block entry of HIV in vaginal, cervical, or anal mucosa. Cf Contraceptive gel

Thermo-STAT™ system A device for noninvasive treatment of hypothermia by rewarming the body core with negative pressure and heating the limb. See Hypothermia

theta waves NEUROLOGY 'Deep' brain waves that oscillate at 4–8 cycles/sec–Hz, normally present in minimal amounts in the temporal lobes; TWs are relatively more common in persons > age 60, and are intentionally induced in yoga and transcendental meditation. See Mantra, Meditation, Yoga; Cf Alpha waves, Autosuggestion therapy, Hypnosis, Silva mind control

thiamin deficiency Beri-beri NUTRITION A vitamin deficiency characterized by malnutrition, softened bones, depression, due to an absence of dietary thiamin; the clinical disease is divided into 'wet' beri-beri, which causes CHF and 'dry' beri-beri, which causes neurologic disease in the form of peripheral neuropathy and Wernicke-Korsakoff syndrome. See Thiamin

thiazolidinedione Glitazone, TZD ENDOCRINOLOGY A class of oral antidiabetic agents that inhibit tyrosine kinase, ↓ insulin resistance and correct hyperglycemia, hyperinsulinemia, hypertriglyceridemia ADVERSE EFFECTS Hepatic dysfunction. See Troglitazone, Tyrosine kinase inhibitor

thick conception of risk RISK ANALYSIS A broad assessment of risk that encompasses all conceivable risks associated with an intervention–eg, injury, death, social loss, costs. See Informed consent, Risk, Risk assessment. Cf Thin conception of risk

thick prep HEMATOLOGY A regionally popular term for a peripheral blood smear, which has been layered with multiple drops of blood, allowing the glass slide to dry between layerings; TPs are used to identify certain bloodborne infections–eg, babesiosis, borreliosis, malaria. Cf Thin prep, ThinPrep™

thin conception of risk RISK ANALYSIS A narrow assessment of risk which addresses only the common physical harms associated with an intervention–eg, injuries, death. See Informed consent, Risk, Risk assessment. Cf Thick conception of risk

thin melanoma Stage I cutaneous melanoma A melanoma measuring < 1 cm in diameter, characterized by virtually 100% survival; the prognosis is less favorable if the lesion is > 0.73 mm in thickness–73% 5-yr survival or > 1.5 mm–63% 5-yr survival. See Melanoma

think-drink effect PSYCHOLOGY The observation that behavior is more closely related to perceived alcohol consumption than it is to actual consumption

thinking VOX POPULI Cognition; mental evaluation, weighing or consideration. See Concrete thinking, Distorted thinking, Holistic thinking, Magical thinking, Sequential thinking

thinning of blood A popular term for use of anticoagulant to ↓ blood coagulability

ThinPrep™ CYTOLOGY A proprietary system–Cytec, Inc, in which specimens are obtained with a special fluid collection system and the slides for cytologic examination are spread in one-cell-thick layers. See thin prep

thin prep HEMATOLOGY A regionally popular term for a 'standard' peripheral blood smear, which is relatively thick at one end and thin and 'feathered' at the other. Cf Thick prep, ThinPrep™

thiocyanate S=C=N TOXICOLOGY A salt of thiocyanic acid that is a breakdown product of cyanide, which is ↑ in smokers, Pts receiving nitroprusside, and victims of smoke inhalation. See Cyanide

thioguanine Lanvis® ONCOLOGY An antimitotic purine analogue, used against AML. See Purine

thiopental Pentothal® ANESTHESIOLOGY An ultrashort IV barbiturate used to induce anesthesia

thioridazine Mellaril® PHARMACOLOGY A chlorpromazine-like antipsychotic, metabolized to the therapeutically active mesoridazine

thiotepa Triethylenethiophosphoramide, Thioplex® ONCOLOGY A relatively toxic alkylating agent used topically to manage transitional cell cancer of the urinary bladder

thiothixine Navane® An antipsychotic

third & fourth pharyngeal pouch syndrome DiGeorge syndrome, see there

third-degree block See Heart Block

third-degree burn See Burn

third degree collateral ligament injury ORTHOPEDICS A complete tear, dislocation, or stretching of the lateral collateral ligament from the bone attachments in the knee, or elsewhere over the ligament, caused by medial–varus or lateral–valgus force. Cf First degree MCL injury, Second degree MCL injury

third diabetic syndrome A form of DM described in young black Pts which may differ from type 1 DM, as 1. 30-40% have HLA-DR3 or HLA-DR4 antigen, in contrast to whites, 95% of whom express HLA-DR3 or -DR4 2. 40% have islet cell antibodies, versus 70-80% of whites 3. Most Pts are well controlled with diet or oral anti-diabetics, 4. AD inheritance pattern

the third ear PSYCHIATRY A popular term for the use of intuition, sensitivity, and awareness– a 'third ear' of subliminal cues to interpret clinical observations of Pts in therapy. See Intuition

third generation oral contraceptive GYNECOLOGY An OC containing desogestrol, gestodene OCs which skirt the adverse effects of progestins, and have a lower risk of acute MI relative to 2nd-generation OCs, at the risk of ↑ clot formation. See Oral contraceptive. Cf Second-generation oral contraceptive

third generation cephalosporin INFECTIOUS DISEASE A group of broad-spectrum antibiotics–eg, cefatoxime, ceftazidime, ceftriaxone and moxalactam that are structurally related to penicillins and used against penicillinase-producing bacteria; TGCs are more active against enteric bacteria, are stable against the β-lactamases of *H influenzae* and *N gonorrhoeae*, have a longer serum $T_{1/2}$ than 1st generation cephalosporins– administered twice/day, cross the blood-brain barrier and effective against gram-negative CNS infections

third heart sound S_3 gallop, ventricular gallop CARDIOLOGY An extra heart sound traditionally regarded as a sign of left ventricular systolic dysfunction–heart failure; its significance differs according to the valve involved and the type of underlying defect; it is caused by rapid ventricular filling, and best heard over the apex with the Pt in supine or left lateral position; it ↑ with inspiration, exercise, elevation of legs, ↑ venous return. Cf Fourth heart sound

third-line therapy Any alternative to conventional–surgery, chemotherapy or RT for malignancy. Cf Unproven forms of cancer therapy

third party MANAGED CARE A person or organization ancillary to the doctor-patient 'dyad', that participates in financing the services rendered–eg, a health insurance carrier, or administrates processing and paying claims for health services provided–eg, Blue Cross/Blue Shield, Medicare

third spacing MEDTALK A popular verb for hemorrhage or other accumulation of fluids into the so-called 3rd space or physiologic compartment, the interstitial space between the skin and fascia

30-day mortality EPIDEMIOLOGY A statistic defined as death occurring within 30 days of a hospital admission

36-item Medical Outcomes Study Short-Form General Health Survey See SF-36

Thomas spint ORTHOPEDICS A bone length-reducing cast splint held in place with plaster of Paris, used for provisional stabilization of a spiral fracture of the femoral shaft–eg, produced by a fall in which the foot was anchored while a twisting motion transmits to the femur–in such fractures, the bone fragments separate. See Traction

Thombran® Trazodone, see there

Thomsen's disease Myotonia congenita, see there

thoracentesis Pleural fluid tap, pleurocentesis The drainage of fluid for therapeutic or diagnostic purposes from the pleural space; the fluid is obtained with a long needle, which is then analyzed for chemical composition and cell types. Cf Pericardiocentesis

thoracic *adjective* Pertaining to the chest

thoracic inlet injury Traumatic injury to the base of the neck involving the superior mediastinal vessels–innominate, subclavian, proximal common carotid arteries and veins; damaged vessels in TII are surgically problematic as facile access is blocked by the claviosternal 'shield'; although correlation of anatomic defects with clinical Sx is poor, adequate regional exploration and hemostasis in this region results in ↓ mortality

thoracic outlet syndrome Scalenus syndrome, thoracic outlet compression syndrome THORACIC SURGERY Any of the neurovascular disorders characterized by compression of the inner branches of the brachial plexus and/or subclavian artery, related among other factors, to kinking of vessels over a cervical rib, fibrous bands passing from a prominent transverse process of the 7th cervical vertebra to the 1st rib, or edge of the scalenus anterior or medius muscle(s) CLINICAL Pain, unilateral paresthesiae along the medial border of the arm, forearm, and little finger, ischemia, atrophy of the small hand muscles, especially of the thenar eminence, myalgia, myasthenia, with vasomotor disorders, edema and thromboses; TOS most commonly affects ♀ with osteoporosis, and may also occur in pregnancy, trauma or overstretching; compromise of this space causes the cervical rib syndrome, scalenus anticus syndrome, costoclavicular syndrome, pectoralis minor syndrome, and first rib syndrome; vascular compromise occurs in 90% of TOSs, but Sx are predominantly neurogenic

thoracotomy SURGERY An operation to open the chest and to directly access the mediastinum to visualize, biopsy, or treat INDICATIONS Lesions inaccessible by, or too large for, mediastinoscopy or mediastinotomy; open cardiac massage. See Mediastinoscopy, Mediastinotomy, Open lung biopsy

Thoratec® system Thoratec® ventricular assist device system CARDIOLOGY A circulatory support device that provides left, right and biventricular total circulatory support for Pts with weakened hearts as a bridge while waiting for a donor heart. See Ventricular assist device

Thorazine® Chlorpromazine. see there

Thorburn position FORENSIC PATHOLOGY A reflex position assumed by the elbows immediately after injury to the spinal cord in the lower cervical region. See 'Magic bullet' theory

thorn sign IMAGING A vaguely defined spicular radiologic shadow that tapers medially from the lateral chest wall, which corresponds to a thickening of the minor fissure, most commonly seen in a right-sided pleural effusion; since the finding appears in most positions, the TS should prompt a lateral decubitus film and ultrasonography to confirm the presence of pleural fluid

thought disorder PSYCHIATRY A disturbance of speech, communication, or content of thought–eg, delusions, ideas of reference, poverty of thought, flight of ideas, perseveration, loosening of associations, etc; TDs can be functional emotional disorders or organic

thought insertion PSYCHIATRY A schizophrenic delusion in which the Pt believes in or acts on hallucinated external voices. See Schizophrenia

thought leader Influential physician MEDICAL PRACTICE A physican or other person in a specialty who has a big 'name,' is respected by his peers, or is capable of influencing the way in which his peers/disciples practice medicine and the medications they prescribe. See Authorship, Detailing, Ghostwriting, Independent symposium, Name, Pharmaceutical sales representative, Project house, Sponsored symposium

threadworm 1. Pinworm, see there 2. *Strongyloides stercoralis*

threat VOX POPULI A real or perceived danger. See Risk threat

threat level PUBLIC HEALTH A standard set by the National Institute of Justice, which determines the amount of ballistric protection afforded by soft body armor–bullet-proof vests, etc

threatened abortion Threatened miscarriage, threatened spontaneous abortion OBSTETRICS Vaginal bleeding at any time within the first 20 wks of pregnancy, accompanied by colicky pain, backache and a bright red to brownish discharge; TAs occur in up to 20% of early pregnancies of which $\frac{1}{2}$ progress to inevitable abortion; no therapy is consistently effective; bed rest, analgesics and sedatives are advised. See Abortion. Cf Spontaneous abortion

3D conformal external beam radiation therapy RADIATION ONCOLOGY A format for delivering RT designed to deliver an exact dose of radiation to a target volume. See Radiation oncology

3D radiotherapy RADIATION ONCOLOGY The administration of high dose radiation with minimal damage to surrounding tissues, made possible with stereotactic griding and beam focusing; 3-D RT is used for tumors of prostate, lungs, and nasopharynx, as these tissues are often surrounded by 'important' radiosensitive structures

3D ultrasound reconstruction IMAGING A computerized technique that transforms 2D ultrasound images into 3D images for more precise localization of lesions in cerebral vasculature

3D ultrasound reconstruction imaging IMAGING A computerized technique that transforms 2D ultrasound images into 3D images for more precise localization of lesions

3D volume reconstruction RADIATION ONCOLOGY A reconstruction of tumor volume and surrounding critical normal tissues from direct CT or MRI data in preparation for noncoplanar or coplanar RT. See Radiation oncology simulation, Simulator, Therapeutic radiology treatment planning, Treatment field, Treatment Port

three-finger rule CLINICAL ANATOMY A rule regarding the anatomic relationships of the frontal nerve: by placement of 3 fingers vertically over the forehead, and centered above the glabella, both major nerve branches emerge immediately lateral to the outer 3 fingers

three-glass test Three-glass urinalysis UROLOGY A crude clinical test in which the urinary stream in collected in succession in 3 different 3 ounce–± 75 mL receptacles

Three Main Complaints questionnaire PSYCHOLOGY An instrument that focuses on the 3 mental problems–eg, a paraphilia that most affects a Pt, determined by psychiatric evaluation, each of which is rated on a 13-point scale, from least to most distressing or severe. See Paraphilia

threshold MEDTALK The point, stage, or degree of intensity at which a particular effect occurs or action is taken. See Therapeutic threshold, Transfusion threshold

thrill CARDIOLOGY A palpable murmur that correlates with zones of maximum intensity of auscultated sounds; rough lower sternal border thrills occur in ventricular septal defect, apical systolic thrills are associated with mitral valve insufficiency; diastolic thrills may be palpated in AV valve stenosis VOX POPULI Whee, whoopie, yippie, yeeeehaaa, etc

thrill of victory, agony of defeat COSMETIC SURGERY A popular phrase used in the context of hair restoration, for the pros of transplanting a frontal forelock, coupled to the cons of leaving a Pt 'fur-less' on the vertex

throat culture The placing of a sample swabbed from the throat–posterior pharynx, tonsils and tonsillar fossa, and any area grossly suggestive of inflammation in the appropriate culture medium to detect the growth of a pathogen of interest–eg, group A streptococci

thromb- *prefix*, Greek, pertaining to a blood clot

thrombasthenia Glanzmann's disease, see there

thrombectomy The removal of a clot/thrombus

thrombin HEMATOLOGY A key clot promoting enzyme that converts fibrinogen to fibrin and protects against fibrinolysis by activating thrombin-activatable fibrinolysis inhibitor. See Fibrin, Fibrinolysis, TAFI

thrombin time Thrombin clotting time LAB MEDICINE A test used to evaluate the final stage of coagulation–the conversion of fibrinogen to fibrin; it is ↑ in deficiency or defects of fibrinogen, or in the presence of FDPs or heparin-coagulation products; it is used clinically to determine adequacy of thrombolytic therapy. See Reperfusion therapy, Streptokinase, tPA, Urokinase

thrombinogen Prothrombin The molecular precursor to thrombin. See Prothrombin

thromboangitis obliterans Buerger's disease VASCULAR SURGERY A disorder characterized by inflammation-induced obstruction of small arteries and veins; usually affects young ♂ with a Hx of smoking or chewing tobacco; it may accompany Raynaud's disease or autoimmune disease CLINICAL Pain, tissue damage or destruction, ↑ risk infections, gangrene. See Vasculitis

thrombocytopenia Thrombocythemia HEMATOLOGY An absolute ↓ platelet count, with ↑ bruising and bleeding from wounds, mucosae and other tissues EXAMPLES ITP, TTP, drug-induced immune and nonimmune thrombocytopenia, acquired platelet function defect, congenital platelet function defects, cyclic neutropenia. See Essential thrombocytopenia, Fetal alloimmune thrombocytopenia, Gestational thrombocytopenia, Heparin-induced thrombocytopenia, Neonatal thrombocytopenia, Platelet

thrombocytosis An excess of platelets. See Absolute thrombocytosis, Extreme thrombocytosis, Reactive thrombocytosis

thromboembolic state HEMATOLOGY A state that may induce coagulation–prolonged bedrest without ambulation or physical therapy, dehydration, poor positioning–eg, crossing legs, and prolonged use of an IV catheter. See Migratory thromboembolism, Thromboembolism

thrombokinase Factor X, see there

thrombolysis Dissolution of a blood clot/thrombus. See Laser thrombolysis

thrombolytic agent Clot-dissolving drug, thrombolytic An agent–eg, tPA, streptokinase, that effects thrombolysis and restores vascular patency–eg, in managing acute MIs. See Thrombolytic therapy, tPA

thrombolytic therapy CARDIOLOGY Any therapy, especially, enzymatic, that dissolves intravascular blood clots and/or fibrin thrombi, digests fibrinogen and other proteins, and recanalizes occluded vessels–usually arteries, to improve circulation; in acute MI, TT is most effective if given at time of Sx onset, if the Pt is stabilized and diagnosis of Q-wave MI is strongly suspected AGENTS Urokinase, streptokinase, acylated plasminogen streptokinase complex–APSAC, pro-urokinase, tPA INDICATIONS PTE, DVT, AMI, peripheral artery disease with intraarterial thrombi

thrombophlebitis Phlebitis, phlebothrombosis HEMATOLOGY A term that arose when DVT was thought to occur either without inflammation–phlebothrombosis, or with inflammation–thrombophlebitis, a concept that has long since been abandoned; the term continues to be used for inflamed or infected thrombi, which give rise to septic infarcts. See Deep vein thrombosis

thrombosis The formation of one or more blood clots or thrombi. See Coronary thrombosis, Deep vein thrombosis, Effort thrombosis, Late-stent thrombosis, Protein C deficiency, Pulmonary thromboembolism

thrombotic occlusion MEDTALK Any vascular blockage caused by a thrombus or by thromboembolism

thrombotic stroke NEUROLOGY A stroke caused by thrombosis or thromboembolism. See Stroke

thrombotic thrombocytopenic purpura Moschcowitz's disease; TTP HEMATOLOGY A rare–1:10⁶/yr disorder of the microcirculation, most common in ♀ age 20-50 CLINICAL Moschcowitz's pentad: Thrombocytopenia, splenomegaly, DIC, fever, neurologic signs–eg, headache, confusion, aphasia, transient paresis, ataxia, sensory defects and coma, as well as pallor, jaundice, fibrin thrombi in renal vessels, heart, liver and spleen, epistaxis, cerebral, retinal, vaginal hemorrhage, heart dysfunction, hepatomegaly, pancreatitis ASSOCIATIONS SLE, rheumatoid arthritis, Sjögren syndrome LAB ↓ Platelets, usually < 50,000/mm³, normal coagulation factors, ↓ complement proteins, Coombs-negative microangiopathic hemolytic anemia–often severe, 30% have ↓ Hb < 55g/L, US: < 5.5 g/dl and reticulocytosis, schistocytes, burr cells, helmet-shaped RBCs, normoblasts, reticulocytosis, ↑ unconjugated BR, ↑ Hb, ↑ hemosiderin, ↓ haptoglobin, proteinuria MANAGEMENT Corticosteroids, aspirin, dipyridamole, Igs; if no response, plasmapheresis. See Hemolytic-uremic syndrome

thrombus HEMATOLOGY An intravascular blood clot formed in vivo from fibrin thread accumulation around a platelet plug. See Deep vein thrombosis, Pulmonary thromboembolism

throw-away journal A medical journal or other magazine intended for health professionals sent free of charge or by non-paid subscription, which often contains non-peer-reviewed articles–eg, reviews, editorials, etc. Cf Peer-reviewed journal

thrush Pseudomembranous candidiasis A popular term for oral and mucocutaneous candidiasis, characterized by erythematous intraoral lesion overlaid by white, creamy patches, which correspond to necrotic debris, squames, fibrin, inflammatory cells, fungal hyphae and bacteria AT RISK GROUPS Infants, immunocompromised Pts, malnourished Pts in poor health, or post-antibiotic therapy

thrust VOX POPULI *noun* Pressure in a particular direction. See Recoil thrust, Rotational thrust

thud drill Thump practice SPORTS MEDICINE A format of football practice in which the players, garbed in helmet and shoulder pads, hit each other, but remain on their feet. See Spring practice

thumb See Bowler's thumb, Gamekeeper's thumb, Harp player's thumb, Hitch-hiker's thumb, Skier's thumb, Trigger thumb

thumb & little finger signs IMAGING A pair of findings on plain lateral films of the neck; the 'thumb' occurs in acute epiglottitis and corresponds to an edematous aryepiglottic fold–epiglottic shadow with near-complete obliteration of the valleculae and pyriform sinuses, likened to an adult's thumb; the 'little finger' is used for comparison, where the epiglottic shadow is svelte, resembling an adult's little finger

thumbprinting IMAGING A finding in a barium study of the colon, which consists of multiple broad, sharply defined short and rounded often symmetric indentations in the contrast column; it is described in 1. Mesenteric artery ischemia or ischemic 'colitis' with infarction and intramural hematoma formation–often at splenic flexure, descending colon and 2. Ulcerative colitis. Cf Collar button lesion

thumpversion Precordial thump CARDIOLOGY The administration of a blow(s) to the anterior chest of a person who has undergone cardiac arrest. Cf Heimlich maneuver

thunder & lightning, fear of PSYCHOLOGY Astraphobia, astrapophobia, brontophobia, keraunophobia. See Phobia

thymoma HEMATOLOGY A thymic epithelial neoplasm of the anterosuperior mediastinum, often first an incidental finding on a plain AP film CLINICAL Thymomas may be associated with connective tissue disease–giant cell polymyositis, rheumatoid arthritis, SLE, Sjögren syndrome, dermatomyositis, polymyositis, scleroderma, skin disease–pemphigus vulgaris, lichen planus, alopecia areata, hematologic disease–erythroid hypoplasia, pernicious anemia, aplastic anemia, P vera, myeloma, angioimmunoblastic lymphadenopathy, inflammation–eg, granulomatous myocarditis, meningoencephalitis, thyroiditis, immune disorders–eg, hypogammaglobulinemia, IgA deficiency, monoclonal gammopathy, mucocutaneous candidiasis, CA–seen in up to 17% of Pts, myasthenia gravis, adrenal atrophy, Cushing's disease, ulcerative colitis, Crohn's disease DIFFDX Lymphoma MANAGEMENT Excision PROGNOSIS Reflects capsular invasion, mitotic activity, cytologic atypia, nuclear hyperchromasia

thymus & parathyroid hypoplasia See DiGeorge syndrome

thyroglobulin A 669 kD iodine-rich glycoprotein secreted by thyroid follicular cells into thyroid colloid where it is iodinated; once resorbed by follicular cells, it is cleaved into multiple units of T₃–iodothyronine and T₄–thyroxine. See T₃, T₄, Thyroid gland

thyroglossal cyst Thyrolingual cyst A fluid-filled space present at birth at the midline of the neck, resulting from an incomplete closure of the thyroglossal duct which normally closes before birth

thyroid *adjective* Relating to the thyroid gland *noun* 1. Thyroid gland extract 2. Thyroid gland, see there

thyroid antibodies See Antimyeloperoxidase antibody, Antithyroglobulin antibody, Antithyroid peroxidase antibody

thyroid cancer Thyroid carcinoma ONCOLOGY A malignant epithelial lesion of the thyroid that affects 14,000/yr–US, 1100 deaths/yr; ♀ account for 77% of new cases and 61% of deaths; current overall 5-year survival with treatment is 95% RISK FACTORS Persons exposed to the upper body–especially head & neck radiation during childhood; persons with a family Hx of TC or MEN 2 syndrome; the risk of radiation-induced thyroid nodularity and CA ↑ with radiation dose and ↓ the older the person was at the time of irradiation; medullary TC, which comprises about 10% of all TCs, is inherited in 25% of cases as part of MEN 2 syndrome CLINICAL Palpable neck mass NOTE 0.14% of palpable neck masses ultimately prove to be TC MANAGEMENT Surgery, radioiodine PROGNOSIS Uncertain; recurrences may occur decades later; administration of TSH stimulates remaining thyroid tissue, and prevents symptomatic hypothyroidism, which occurs when thyroid hormone is discontinued, and stimulates radioiodine uptake by residual thyroid and thyroid CA tissue. See Thyrotropin

thyroid cascade Thyroid function testing cascade ENDOCRINOLOGY A format for reflex testing of thyroid function: serum TSH followed by measuring free T₄, if TSH is abnormal and measurement of T₃ and/or microsomal antibody, if indicated. See Free thyroxine index, T₃, T₄

thyroid disease

thyroid disease Thyroid disorder ENDOCRINOLOGY Any benign or malignant condition that affects the structure or function of the thyroid gland. See Anaplastic carcinoma of thyroid, Chronic thyroiditis–Hashimoto's disease, Hyperthyroidism, Hypoparathyroidism, Hypopituitarism, Hypothyroidism, Medullary thyroid carcinoma, Silent thyroiditis, Papillary carcinoma of thyroid, Primary hyperparathyroidism, Subacute thyroiditis, Thyroid cancer, Congenital goiter

thyroid function test ENDOCRINOLOGY Any test–eg, TSH, free & total thryoxine, T3 resin uptake–used to detect thyroid disease–eg, hyperthyroidism, hypothyroidism

thyroid scan A image obtained from the thyroid gland after oral administration of radioiodine. See Radioactive iodine uptake

thyroid-stimulating hormone Thyrotropin, TSH, thyrotropic hormone, thyrotrophin A 28 kD glycopeptide hormone produced by the anterior pituitary–hypophysis in response to TRH, released by the hypothalamus, which controls thyroid growth, development, and secretion; TSH is controlled by feedback loops in the hypothalamus, the central loop, which hinges on TRH and the peripheral loop located in the thyroid gland, which produces T3 and T4 ↑ IN 1° hypothyroidism, chronic–Hashimoto's thyroiditis, antithyroid therapy for hyperthyroidism, therapy with lithium or potassium iodide ↓ IN 2° hypothyroidism, hyperthyroidism–Graves' disease, malnutrition, vigorous exercise, hypofunction of anterior pituitary, or therapy with malnutrition, L-dopa, vigorous exercise, renal failure, therapy with corticosteroids, chlorpromazine, heparin, lithium, phenytoin, propranolol, reserpine, salicylates, sulfonamides, testosterone, tolbutamide. See Thyroid-stimulating hormone, Thyroxine-binding globulin, Triiodothyronine–T₃

thyroidectomy SURGERY The partial or total surgical removal of the thyroid gland INDICATIONS Hyperthyroidism or goiter COMPLICATIONS Hypothyroidism, vocal cord paralysis, accidental removal of parathyroid glands, resulting in low calcium levels–parathyroid glands regulate calcium

thyroiditis ENDOCRINOLOGY Inflammation of the thyroid, which releases hormones into the circulation, resulting in transient hyperthyroidism; after thyroid depletion, Pt is often hypothyroid for 3-6 months until thyroid recuperates DIAGNOSIS Thyroid scan with radioactive iodine. See Granulomatous thyroiditis, Hashimoto's thyroiditis, Silent thyroiditis, Subacute thyroiditis

thyromegaly ENDOCRINOLOGY Thyroid enlargement. See Goiter

thyroplasty Laryngeal framework surgery SURGERY A procedure for altering the structure of the larynx to change vocal quality–eg, the soft breathy quality typical of unilateral vocal cord paralysis, by altering laryngeal cartilage or changing vocal cord position or length. See Medialization thyroplasty

thyrotoxic crisis Thyroid crisis, thyrotoxic storm ENDOCRINOLOGY A hypermetabolic state superimposed on hyperthyroidism, often occurring in Grave's disease–especially if untreated, less common in toxic multinodular goiter, and rarely in Hashimoto's disease/Hashitoxicosis ETIOLOGY/'TRIGGERS' Infection–eg, streptococcal pharyngitis, thromboembolism, surgery, physical or mental stress, parturition, withdrawal from thyroid-blocking drugs CLINICAL Sx of hyperthyroidism–eg, weight loss, heat intolerance, myasthenia, ↓ concentration, CHF, PE, restlessness, diarrhea, diaphoresis, palpitations which, when accompanied by 'triggers,' become a life-threatening hypermetabolism, characterized by hyperthermia, tachycardia, cardiac, hepatic and cerebral dysfunction, death LAB ↑ T₄, ↑ T₃, ↓ cholesterol MANAGEMENT ↓ Hypermetabolism, inhibit T₃/T₄ release with iodine, block T₃/T₄ synthesis with propylthiouracil and other anti-thyroid agents, block adrenergic neurotransmission with beta blockers and glucocorticoids to 'cover' for functional hypoadrenalism. See Euthyroid sick syndrome, Hashimoto's disease, Hashitoxicosis, Thyrotoxicosis

thyrotoxic periodic paralysis ENDOCRINOLOGY A disorder characterized by intermittent episodes of muscle weakness that occur in thyrotoxicosis, a disorder associated with ↑ thyroid hormones. See Hypokalemic periodic paralysis

thyrotoxicosis CARDIOVASCULAR DISEASE High-output heart failure due to 1. Circulatory factors–↑ total blood volume, ↓ systemic vascular resistance, 2. Cardiac factors–↑ contractility, ↑ heart rate, ↑ diastolic relaxation, which translate into volume overload, ↓ myocardial contractile reserve, ↓ diastolic filling time, tachyarrhythmia MANAGEMENT Inorganic iodine abruptly inhibits T₃ and T₄ secretion, but should be preceded by methimazole or propyluracil to prevent oxidation and organification of iodine, followed by definitive treatment of 1. Thyrotoxicosis–radioactive iodine, surgical resection, 2. Heart disease–eg, furosemide–diuretic for volume overload, beta blockers–eg, propranolol, in absence of CHF, and anticoagulants for possible arterial thromboembolism due to A Fib. See Amiodarone-induced thyrotoxicosis, Hamburger thyrotoxicosis, Hyperthyroidism, Postpartum thyrotoxicosis, T₃ thyrotoxicosis, T₄ thyrotoxicosis. Cf Thyroid crisis ENDOCRINOLOGY See Thyrotoxic crisis

thyrotropin Thyroid stimulating hormone. See TSH

thyrotropin blocking antibody An autoantibody produced in thyroid autoimmune disease that blocks TSH activity; it is bioassayed based on its ability to block TSH on cAMP ↑ IN Atrophic thyroiditis, Hashimoto's thyroiditis, transient congenital hypothyroidism. See Thyroid autoimmune disease

thyrotropin releasing hormone stimulation test see TRH stimulation test

thyroxine T₄, 3,5,3',5'-Tetraiodothyronine A hormone that stimulates metabolism and O₂ consumption, which is secreted by the thyroid gland in response to TSH–thyrotropin produced in the adenohypophysis–anterior pituitary gland ↑ IN Hyperthyroidism, acute thyroiditis, myasthenia gravis, preeclampsia, pregnancy, viral hepatitis, therapy with clofibrate, OCs, estrogens, perphenazine ↓ IN Hypothyroidism, malnutrition, vigorous exercise, hypofunction of adenohypophysis–anterior pituitary gland, renal failure, therapy with corticosteroids, chlorpromazine, heparin, lithium, phenytoin, propranolol, reserpine, salicylates, sulfonamides, testosterone, tolbutamide. See Triiodothyronine–T₃, Thyroxine-binding globulin

thyroxine-binding globulin TBG An α₂-migrating protein that is the main–70% carrrier protein for thyroxine/T₄ and triiodothryonine/T₃ ↑ IN Acute intermittent porphyria, estrogens, hereditary defects, hepatic disease, hypothyroidism, neonates, OCs, perphenazine, pregnancy ↓ IN Acromegaly, androgens, hereditary defects, hepatic disease, nephrotic syndrome, phenytoin, prednisone, underlying illness, post-surgery, thyrotoxicosis

TI Therapeutic index, see there

Tiazac® Diltiazem, see there

TIBBS CARDIOLOGY A clinical trial–Total Ischemic Burden Bisoprolol Study that evaluated medical management in ↓ ischemia in Pts with stable angina. See Bisoprolol, Nifedipine, Stable angina

TIBC Total iron-binding capacity LAB MEDICINE A quantitative measurement of transferrin's ability to transport iron; normally, ±33% of transferrin's binding sites–BS are occupied by iron; in iron deficiency, pregnancy and viral hepatitis, 15% of transferrin's BS are occupied, therefore transferrin's capacity to bind iron or TIBC is ↑; in iron-overload syndromes–eg, hemochromatosis, transferrin has few sites available to bind iron and thus the TIBC is ↓

tibial nerve dysfunction Tibial nerve neuropathy NEUROLOGY A peripheral neuropathy characterized by impaired movement or sensation in the leg, caused by damage to the tibial nerve, which innervates the calf muscles or hamstrings and foot flexors, and supplies sensation to the sole of the foot

tibial valgum ORTHOPEDICS A frontal plane deformity where the distal ⅓ of the leg is angled away from the midsagittal plane more than the proximal end

tibial varum ORTHOPEDICS A frontal plane deformity where the distal $\frac{1}{3}$ of the leg is angled closer to the midsagittal plane than the proximal end

TIBO A class of benzodiazepine-related compounds with anti-HIV activity which, at low doses, ↓ in vitro viral replication by inhibiting reverse transcriptase; TIBO's inhibition of reverse transcriptase is 5 times more potent than zidovudine's. See Nonnucleoside reverse transcriptase inhibitor

tic Habit spasm A complex of multiple abrupt, coordinated involuntary and/or compulsive spasms, including eye blinking, facial gestures, vocalizations, shoulder shrugging, etc which, when controlled, may be followed by more intense and frequent 'rebound' contractions; tics may be exacerbated by stress and ameliorated by psychotherapy. See Dystonic tic, Motor tic, Tourette syndrome, Transient tic disorder, Verbal tic, Vocal tic. Cf Jumping Frenchmen of Maine syndrome MEDTALK A popular synonym for diverticulosis

ticarcillin® Trimentin® A semisynthetic broad-spectrum penicillin used for GNRs and gram-positive cocci

tic doloreaux Trigeminal neuralgia, see there

TICE® BCG BCG vaccine, TICE strain IMMUNOLOGY A BCG vaccine for treating bladder CIS, stage Ta or T1 papillary tumors of bladder, interstitial cystitis. See Bladder cancer

tick MEDICAL ENTOMOLOGY A hematophagous ectoparasitic arthropod of the superfamily Ixodoidea, which is either a hard tick–family Ixodidae or a soft tick–family Argasidae; ticks may be vectors of bacterial and viral infections. See Colorado tick fever, Deer tick, Lone Star tick, Lyme disease, Rocky Mountain spotted fever

TICKS OF INTEREST

Dermacentor andersoni, the North America vector, Rocky Mountain spotted fever–RMSF, Colorado tick fever–CTF, tularemia and tick paralysis

D marginatus Asian vector–Russian spring-summer fever virus, tick-borne encephalitis virus, possibly also Congo-Crimean hemorrhagic fever virus and *Babesia* reservoir for Omsk hemorrhagic fever virus

D occidentalis West coast North America–presumed vector for RMSF, CTF

D parumapertus Southwestern US–vector for RMSF, CTF

D variabilis Eastern US–vector for RMSF, tularemia, CTF, and tick paralysis; Asian and African ticks, vectors of Rickettsialpox

Ixodes dammini Northern deer tick–vector, *Babesia microti*, Lyme disease agent

tick-borne fever Relapsing fever, see there

tick-borne rickettsiosis A tick-borne rickettsial infection–eg, north Asian tick-borne rickettsiosis, Queensland tick typhus, and African tick typhus–fiévre boutonneuse, etc, which is similar to RMSF, but less severe, with fever, a small ulcer at the site of the tick bite, satellite lymphadenopathy, and a maculopapular rash. See Rocky Mountain spotted fever

tick fever See Rocky Mountain spotted fever

tick paralysis Ascending tick paralysis, tick toxicosis A flaccid ascending quadriplegia that resembles Guillain-Barré syndrome–GBS, due to the continued presence of certain ticks attached to the occipital or upper neck region in humans, attributed to an unidentified neurotoxin produced by the bite of certain pregnant ticks CLINICAL Onset several days after the tick begins feeding–unsteady gait, extremity and truncal ataxia, loss of deep tendon reflexes, drooling, tachypnea, affected children develop an unsteady gait–ataxia followed several days later by lower extremity weakness that gradually moves up to involve the upper limbs; paralysis may affect respiration, mandating ventilatory assistance DIFFDX Botulism, GBS, myasthenia gravis, postinfectious polyradiculopathy MANAGEMENT Tick removal, supportive care PROGNOSIS Tick removal results in complete recuperation; 10% mortality if not removed, due to respiratory paralysis

ticket VOX POPULI A chit or voucher that permits access to a particular activity; a price tag. See High ticket

tickler Clinical alert, see there

ticlopidine THERAPEUTICS A platelet aggregation inhibitor that alters the platelet membrane, interfering with membrane-fibrinogen interaction by blocking platelet gpIIb/IIIa receptor INDICATIONS ↓ Stroke in high risk Pts, minimize stroke progression, cerebral ischemia, progression of diabetic retinopathy SIDE EFFECTS BM suppression with severe reversible neutropenia, diarrhea, rash, gastritis or gastric ulcers–nausea, dyspepsia, hemorrhage, thrombocytopenia, ↑ cholesterol. See TIA

tid 3 times a day

tidal volume LUNG PHYSIOLOGY The volume of air drawn into the lungs during inspiration from the end-expiratory position, which leaves the lungs passively during expiration in the course of quiet breathing. See Lung volumes. Cf Residual volume

Tietze syndrome Costosternal syndrome, costochondritis A benign bone lesion involving ≥ 1 costosacral junction in Pts age 20 to 40 CLINICAL Pain, tenderness, swelling DIFFDX Enchondroma, chondrosarcoma RADIOLOGY Sclerotic bone; may be radiologically invisible, as it is poorly calcified COMPLICATIONS Rare; only if the lesion is so large that it compromises bone strength. Cf Enchondroma

Tiger Woods syndrome SPORTS MEDICINE A facetious coinage for skull fractures caused by young novice golfers with poor aim and/or wild swings, who began playing after TW's win in the 1997 Masters Tournament

Tikosyn® Dofetilide CARDIOLOGY A selective potassium channel blocker for treating A Fib, to maintain normal heart rhythm and ↓ fatigue, SOB, and other Sx. See DIAMOND CHF trial

TIL therapy ONCOLOGY An experimental CA therapy in which antigen-specific tumor-infiltrating T lymphocytes–TILs are isolated from biopsies of Pts with CA and co-administered with IL-2. Cf LAK/IL-2 therapy

Tilade® Nedocromil nebulizer RESPIRATORY MEDICINE An inhalable NSAID used as maintenance therapy in Pts with asthma See Asthma

tilt test Upright tilt test CARDIOLOGY A clinical maneuver in which a person is placed in a head-up position on a tilt table at a 40° to 80° from horizontal and maintained in a motionless upright position for ≥ 10-15 mins, the intent being to provoke syncope, bradycardia, or hypotension; the resulting gravitational pooling of blood volume evokes a decline in CVP, stroke volume, and bp; the TT may be used to detect neurocardiogenic syncope

tiludronate disodium Skelid® METABOLISM A biphosphonate used to manage osteoporosis and Paget's disease of bone; tiludronate controls abnormal bone growth without inhibiting normal bone formation. See Biphosphonate, Osteoporosis

time VOX POPULI The so-called fourth dimension, which corresponds to the duration of a particular event. See Activated partial thromboplastin time, Collision time, Contact time, Delay time, Doubling time, Dwell time, Emergence time, Euglobulin clot lysis time, Expiratory time, Forced expiratory time, Gastric emptying time, Interpulse time, Ivy bleeding time, Just in time, Lead time, Lethal time, Mean time between failure, Movement time, Overtime, Partial thromboplastin time, Pit recovery time, Plasma recalcification time, Quality time, Relaxation time, Prothrombin time, Real time, Relaxation time, Reptilase time, Retention time, Stypven time, Therapeutic turnaround time, Total lead time, Total sleep time, Transportation time, Tumor doubling time, Turnaround time, Turnover time, Wake time, Zero time

time bomb PSYCHIATRY A popular term for a person who is under such stress that a certain dread event–eg, suicide or homicide is inevitable

686 **time/dose** RADIATION ONCOLOGY The ratio between the dose delivered and the time over which the dose is distributed

time manual performance test CLINICAL MEDICINE An interviewer-based observation instrument used to evaluate physical functions through timed assessment of performance of structured manual tests PROS Assesses a Pt's actual performance; sensitive to small changes CONS Difficult to use in seriously ill or mentally impaired Pts. See ADL scale

time-out FORENSIC PSYCHIATRY A strategy for managing violent Pts in psychiatric units, consisting of temporary separation from a rewarding environment, as part of a planned and recorded therapeutic program to modify behavior

time-sampling method PEDIATRIC PSYCHOLOGY A method for evaluating children, based on multiple, ≤ 1 min, observation periods

time to thrombolysis See Door to drug

time trade-off metric CLINICAL DECISION-MAKING A tool used in outcomes management that quantifies a person's preferences for a specific health state, '...*by assessing how much time a Pt would be willing to give up to be freed from a reduced health state; the TTOU is the number of Sx-free yrs divided by the number of yrs with Sx, at the point of indifference'.* See Outcomes, Utility. Cf Standard gamble metric

timed collection The obtention of lab specimens with a certain periodicity

Timentin AIDS A parenteral antibacterial used to manage skin infections. See AIDS

TIMI classification CARDIOLOGY A system developed during the TIMI–Thrombolysis in Myocardial Infarction trials for grading the severity of stenosis and extent of blood flow through the coronary arteries

TIMI studies CARDIOLOGY A series of long-term multi-center, multi-national, multi-agent studies–Thrombolysis In Myocardial Infarction designed to evaluate various early interventions for improving the survival in Pts with AMIs; the TSs have examined the effects of early thrombolytic therapy in recanalizing occluded coronary arteries, in limiting the size of the MI and residual cardiac dysfunction, and ↓ mortality, analyzing various combinations of tPA, heparin, aspirin and coronary arteriography followed by prophylactic percutaneous transluminal angioplasty

timidity PSYCHOLOGY Shyness; a stance of nonaggression in social situations

timolol maleate CARDIOLOGY An antihypertensive and antiarrhythmic β-blocker, used to treat glaucoma ADVERSE EFFECTS Worsening of lipid profiles–↑ TGs, ↑ LDL, ↓ HDL. See β-blocker

tincture PHARMACOLOGY A medicinal preparation, often of herbal origin in which the ground substrate–eg, bark, root, nuts, or seeds is soaked in alcohol to extract oils or other substances of interest

tinea barbae DERMATOLOGY A group of superficial infections of the hair follicle, caused by *Staphylococcus* or a fungus, due to injury or damage to the hair follicle caused by friction from clothing, by blockage of follicle, or shaving. See Barber's itch, pseudofolliculitis barbae

tinea of body Tinea corporis, see there

tinea capitis Fungal scalp infection, tinea of scalp DERMATOLOGY A dermatophytosis, primarily of the pediatric scalp; it can be persistent and contagious, but often disappears spontaneously at puberty; the fungi of TC thrive in warm, moist areas, aided by poor hygiene, minor skin or scalp injuries; passed by direct contact with combs, hats, clothing, fomites or pets.

tinea corporis Fungal skin infection, ringworm of body, tinea circinata, tinea of body DERMATOLOGY An infection of the skin surface with dermatophytes. See Tinea capitis, Tinea cruris, Tinea pedis

tinea cruris Jock itch, see there

tinea of groin Tinea cruris, see there

tinea pedis Athlete's foot, see there

tinea of scalp Tinea capitis, see there

tinea versicolor DERMATOLOGY A chronic skin infection common in ♂ adolescents, caused by a fungus, *Pityrosporum orbiculare*, which is part of the normal human skin flora AFFECTED SITES Underarm, upper arms, chest, neck; the typical lesion is a flat discoloration with a sharp border and fine scales; the lesions are typically dark tan with a reddish cast; in blacks, pigmentary changes are common with hypo– or hyperpigmentation–increase in skin color; itching may or may not be present; TV is more common in hot climates and is associated with ↑ sweating

tingling throat 'syndrome' PARASITOLOGY The sensation of a worm–*Phocanema–Pseudoterranova decipiens*, a nematode occasionally found in sushi–slithering through the oropharynx or proximal esophagus. See Sushi

tinidazole PARASITOLOGY An antiprotozoal used to treat amebiasis, giardiasis, trichomoniasis, and *H pylori* infections in Pts with low-grade gastric lymphoma

Tinnel sign ORTHOPEDICS Pain in the fingers induced by percussion of the median nerve at the palmar wrist, classically linked to carpal tunnel syndrome

tinnitus NEUROLOGY Nonhallucinatory ringing, buzzing, clicking, clanging, roaring, etc, in the ears linked to loss of hearing ETIOLOGY Aspirin, NSAIDs, aging, auditory neuroma, acoustic trauma

tinted spectacles sign PSYCHIATRY The wearing of dark-colored glasses under normal lighting conditions, in absence of photophobia or photosensitivity; the TSS may be associated with underlying psychoneuroses

tinzaparin Innohep® VASCULAR DISEASE An LMW heparin for preventing and treating DVT, PTE. See Low-molecular weight heparin

tioconazole Vagistat® GYNECOLOGY An oral agent for vaginal candidiasis

tip electrode CARDIAC PACING The distal–usually, stimulating–electrode in a bipolar, transvenous pacing lead system. Cf Ring electrode

tip migration CARDIOLOGY A forward-into right atrium or backward movement of a diagnositc or therapeutic catheter. See Cardiac catheterization

tip-of-the-tongue error NEUROLOGY A specific type of word-finding deficit in which the individual can 'feel' a word, but has difficulty in retreiving, or may not be able to retrieve, the word from memory. See Dementia

tipped uterus Retroversion of uterus, see there

Tipramine Imipramine, see there

TIPS HEPATOLOGY Transjugular intrahepatic portosystemic stent-shunt procedure, see there PUBLIC HEALTH Training and Intervention Procedures for Alcohol Servers An organization that educates mixologists and alcohol servers on the most effective means of managing inebriated customers

tip-toe position ORTHOPEDICS A position of ankle joint plantiflexion at the time of injury, as occurs in dancers or falls from a height. See Cavalry fracture, Lover's fracture

tirofiban Aggrastat®, MK-383 CARDIOLOGY A potent short-acting, nonpeptide inhibitor of platelet glycoprotein IIb/IIIa or integrin $\alpha_{IIb}\beta_3$ receptor which, when administered with heparin, ↓ ischemic events in Pts with acute coronary syndrome. See PRISM, PRISM-PLUS

TISS Therapeutic Intervention Scoring System INTENSIVE CARE A system used to assess intensity of care administered in the ICU, based on the number of interventions the Pt needs. See Intensive care unit. Cf APACHE

Tisseel® Hemaseel HMN SURGERY A hemostatic/sealing surgical adhesive/biologic tissue glue comprised of fibrinogen, thrombin, aprotinin INDICATIONS Adjunct to hemostasis in procedures involving cardiopulmonary bypass, repair of spleen; sealant for closing colostomies. See Fibrin glue

tissue A group of similar cells and stroma, which perform a specific function. See Adipose tissue, Bronchiole-associated lymphoid tissue, Chromaffin tissue, Connective tissue, Granulation tissue, Gut-associated lymphoid tissue, Mucosa-related lymphoid tissue, Skin-associated lymphoid tissue, Soft tissue

tissue bank A repository of cadaveric tissues destined for transplantation; banked tissues include arteries, BM, cartilage, cornea, dura mater, fascia, heart valves, pericardium, semen, skin, tendons, veins. See LifeNet case

tissue factor Coagulation factor III, factor III, thromboplastin HEMATOLOGY A complex of phosphatides, lipoproteins, and cholesterol, that accelerates coagulation, which is obtained from homogenates of human tissue

tissue factor pathway inhibitor (extrinsic factor) lipoprotein-associated coagulation inhibitor HEMATOLOGY A coagulation factor X-dependent inhibitor of the factor VIIa/tissue factor complex; it is a plasma lipoprotein that regulates procoagulant effects of tissue factor; with Ca^{2+} and phospholipid, TFPI binds to and inhibits either free factor Xa or factors Xa and VIIa when associated with tissue factor-factor VIIa-Xa complex

tissue pathology Histopathology SURGICAL PATHOLOGY A general term for the evaluation of tissues obtained by biopsy or other surgical procedure

tissue plasminogen activator A thrombolytic protease, the natural form of which is the physiologic activator of the fibrinolytic system; TPA is released from vascular endothelium by epinephrine, exertion, adherent thrombi, or vascular compression; tPA is commercially available in a recombinant form, r-tPA; tPA ↓ mortality of MI in the immediate post-ischemic period; thrombolytic therapy given within the first post-MI hr ↓ mortality by 47%, ↓ mortality in PTE. See Thrombolytic therapy

tissue review committee Tissue committee QUALITY ASSURANCE A hospital committee that reviews the appropriateness of all surgical procedures performed in the institution, correlating pre- and post-operative surgical diagnoses with pathological findings See Appropriateness

tissue tracking CARDIOLOGY A component of cardiac ultrasonography that improves visualization of the heart's regional wall motion. See Ultrasonography

titer INFECTIOUS DISEASE Somewhat incorrectly, the term titer refers to the serum concentration of an antibiotic at which it is (bacteri)–cidal or static. See Serum bactericidal titer, Serum inhibitory titer

title VOX POPULI The official designation of a thing. See Job title

titration MEDTALK The serial dilution of a substance of interest. See Checkerboard titration

tit sign Pyloric tit sign IMAGING A finding in an upper GI series of films of a child with hypertrophic pyloric stenosis, in which the completely obstructed antrum and adjoining pyloric canal simulate the voluptuous curves of a breast and nipple; the nipple becomes pronounced as the pyloric muscle unsuccessfully attempts to contract and push the barium beyond the markedly stenosed canal; rarely, normal peristaltic pouches may transiently ↑ in size, simulating the tit sign

Titus Lorazepam, see there

tixocortol pivalate A steroid analogue with enhanced receptor binding activity, but more rapid presystemic metabolism

TLC 1. Thin-layer chromatography, see there 2. Total lung capacity, see there

TLD Tumor lethal dose

TLV Total lung volume, see there

T_m 1. Melting temperature 2. Tubular maximum

TM 1. Temporomandibular 2. Thalassemia major, see there 3. Transitional mucosa 4. Transmetatarsal 5. Tympanic membrane, see there

T_{max} PHARMACOLOGY The amount of time that a drug is present at the maximum concentration in serum. See C_{max}, Pharmacokinetics

TMJ Temporomandibular joint, see there

TMJ syndrome Dental occlusive disease, temporomandibular joint syndrome, temporomandibular pain, TMJ disorder, myofacial–pain dysfunction syndrome NEUROLOGY A complex neuromuscular disorder caused by dental malocclusion, possibly exacerbated by trauma, mental stress, bruxism CLINICAL Nonspecific unilateral preauricular and facial pain, masseter muscle spasms MANAGEMENT No specific therapy is available; nonspecific therapies include physical modalities—moist heat, analgesics, soft diet, surgery—eg, high intracapsular condylectomy. See Soft diet

TMP-SMX Trimethoprim-sulfamethoxazole, Bactrim, Septra AIDS A drug combination used for PCP pneumonia, toxoplasmosis as well ADVERSE EFFECTS Allergies, fever, itchy rash, N&V, neutropenia. See AIDS

TMR 1. Trainable mentally retarded 2. Transmyocardial revascularization, see there

TMT 1. Tarsometatarsal 2. Thermomechanical treatment 3. Treatment, see there

Tn syndrome HEMATOLOGY A chronic myelodysplasia associated with polyagglutination of RBCs CLINICAL Severe thrombocytopenia, hemolytic anemia, leukopenia; RBCs and platelets lack T-transferase—UDPGal:GalNAc-β-3-D-galactosyltransferase, resulting in inability to express GPIb glycoprotein and cryptantigen Tn, which reacts with naturally-occurring antibodies, causing global hemolysis and clotting

TNF Tumor necrosis factor, see there

TNKase® Tenecteplase, see there

TNK-tPA CARDIOLOGY A proprietary bioengineered agent administered as an IV bolus, which dissolves coronary clots without interfering with other clotting factors. See ASSENT I, Bolus thrombolytic

TNM classification ONCOLOGY An international system for staging malignancy which measures 3 major parameters of a cancer: T–size or extent of the primary tumor, as determined by clinical exam, endoscopy, laparoscopy, biopsy or resective procedures, categorized as being 1-4 and a-d depending upon site, size, and spread; N–number of involved lymph nodes, categorized as X–unknown status, 0–none identified, 1–one to several, 2–a whole bunch of positive LNs; M-X, unknown presence or absence of metastases. See Staging

---

TNM CLASSIFICATION FOR STAGING MALIGNANCY

T TUMOR

Tis   Carcinoma in situ

Ta   Non-invasive

Tx   Cannot be evaluated for nonspecified reasons

T0   Localized tumor

T1   Lesion extends to muscle–bladder, colon, breast

  T1a < 0.5 cm in greatest dimension

T1b < 1.0 cm in greatest dimension

T1c < 2.0 cm in greatest dimension

T2 Invasion into muscle

T3 Persistent induration of organ after resection

T3a Invasion to deep muscle

T3b Invasion through the organ

T4 Tumor invasion or fixation

T4a Adjacent organ invasion

T4b Fixation to bladder or colonic wall; in breast, edema

N NODES

N0 No lymph node metastasis

N1 One regional lymph node metastasis

N2 Multiple, mobile regional lymph node metastases

N3 Fixed regional lymph node metastases

N4 Beyond regional lymph node involvement

NX Lymph nodes, not evaluable

M METASTASIS

M0 No evidence of metastases

M1 Distant metastases are present

MX Distant metastases not evaluable

TNT 1. Treating to New Targets 2. Tumor necrosis therapy, see there

TNTC Too numerous to count LAB MEDICINE A popular abbreviation for a 'lawn' of bacteria on a culture plate that may be seen in UTIs–confluent growth is equal to $\pm 10^5$ colonies MEDTALK An abbreviation facetiously used for a hypochondriac Pt's list of complaints

TOAD Tuberculous old alcoholic derelict MEDTALK A deprecative acronym for an archetypal Pt seen in inner-city hospital wards. Cf Gomer

tobacco PUBLIC HEALTH Any product prepared from the dried leaves of *Nicotiana tabacum*, rich in the addictive alkaloid, nicotine TOBACCO MORTALITY–US ±425K/yr; cardiovascular deaths ±180K/yr; lung CA deaths ±120K/yr; 2nd-hand smoke deaths 9K/yr. See Black tobacco, Blonde tobacco, Environmental tobacco smoke, Nicotine, Smokeless tobacco, Smoking

tobacco control PUBLIC HEALTH Any measure–advertising bans in media–eg, billboards, television, newspapers, bans on smoking in public places–aimed at ↓ tobacco use SUBSTANCE ABUSE Smoking cessation, see there

tobacco dependence The psychologic and physiologic components of addiction to nicotine, which has 3 addictive components: habit–smoking is cued by daily activities, pleasure–nicotine produces euphoria similar to other addictive psychotropic stimulants, and self-medication–nicotine use ↓ negative affect and physical Sx

Tobacco Withdrawal Symptoms Inventory SUBSTANCE ABUSE An instrument designed to determine the Sx that smokers experience during a withdrawal period, and rate the Sx according to how upsetting they are or were, based on the Schneider–Smokers Complaint Scale and Shiffman-Jarvik withdrawal scale

tobacco withdrawal syndrome SUBSTANCE ABUSE A condition characterized by irritability, sleep disorders, GI disturbances, ↑ appetite, weight gain; the use of nicotine replacement therapy–eg, nicotine patches ameliorate some of the components of TWS, but the impact is modest and selective, and obviously does not address the psychological component of smoking. See Nicotine replacement therapy

TOBI® INFECTIOUS DISEASE An aerosolized tobramycin used to treat chronic lung infection in Pts with cystic fibrosis infected with *Pseudomonas aeruginosa* and other bacteria. See Cystic fibrosis

tobramycin Nebcin® A broad-spectrum aminoglycoside antibiotic effective against many gram-negative bacteria, in particular, *Pseudomonas* infections, unresponsiveness to penicillins or cephalosporins ADVERSE EFFECTS Ototoxicity, neurotoxicity, nephrotoxicity. See Therapeutic drug monitoring

tocainide Tonocard® CARDIOLOGY A negative inotropic antiarrhythmic similar to lidocaine and mexiletine, used to ↓ ventricular arrhythmias, tachycardia, PVCs ADVERSE EFFECTS Myelotoxicity with agranulocytosis–in 0.2% of Pts, limiting its usefulness, neurotoxicity–tremor, memory loss, vertigo, anxiety, tinnitus, N&V, anorexia, rashes, pulmonary fibrosis

tocol Any of a group of 8 naturally occurring fat soluble compounds with vitamin E activity; the most biologically active is D-α-tocopherol. See Vitamin E

tocolytic OBSTETRICS *adjective* Referring to an agent with tocolytic activity *noun* Any of a number of beta agonists–eg isoxsuprine, terbulatine, ritodrine, used to halt the progression of preterm labor. See Ritodrine

tocopherol Vitamin E, see there

tocophobia PSYCHOLOGY Fear of childbirth. See Phobia

toddlers' diarrhea PEDIATRICS A condition defined as the presence of unresolved diarrhea with mild malabsorption that persists after the resolution of acute gastroenteritis; TD may be a prelude to irritable bowel syndrome CLINICAL Abdominal pain, vomiting, loose, malodorous stool, highly-irritating rash of buttocks, dysuria and urinary urgency

todeserwartung German, awaiting death SOCIAL MEDICINE A condition affecting elderly persons relegated to nursing homes by society or progeny CLINICAL Hypochondriasis, hysteria, impulsiveness, loss of self-esteem, withdrawal from reality, obsession with death THERAPY Displays of affection, pets, involvement in child care; cortical atrophy and mental lassitude may be controlled, video games for mental stimulation. See Geriatrics. Cf Elderly abuse, Melanocholia, 'Shelterization'

toe See Claw toe, Curly toe, Hammer toe, Mallet toe, Overlapping toe, Pheasant hunter's toe, Pigeon toe, Sausage toe, Tennis toe, Turf toe, Underlapping toe

toe walking ORTHOPEDICS A defective gait, in which the Pts walk on 'tiptoes' due to force of habit, congenital tight heel cords or cerebral palsy with mild spasticity

Tofnil Imipramine, see there

Tofranil® See Imipramine and desimipramine

tofu A cheese-like soy product used as a low-fat meat surrogate. See Soy

TOHP CARDIOLOGY A series of clinical trials–Trial of Hypertension Prevention–that evaluated the efficacy of nonpharmacologic interventions vs drugs on HTN, effect of potassium supplementation on persons with normal high BP and effects of Ca^{2+} or Mg^{2+} supplements. See Antihypertensive, Hypertension

toilet MEDTALK 'Cleansing, as of an accidental wound and the surrounding skin, or of an obstetrical Pt after childbirth'. See Bronchial toilet VOX POPULI Commode, hopper, john, lazybowl, porcelain convenience. See Space toilet

toke *verb* SUBSTANCE ABUSE To inhale a large air volume while smoking a substance of abuse–eg, marijuana, less commonly cocaine or crack cocaine, maintaining the lungs expanded with a slight Valsalva maneuver, to maximize the substance's absorption. Cf 'Snort'

tolcapone Tasmar® NEUROLOGY A COMT inhibitor used with levodopa/carbidopa to treat Parkinson's disease ADVERSE EFFECTS Liver toxicity

Tolectin® Tolmetin, see there

tolerance IMMUNOLOGY Immune unresponsiveness to an antigenic challenge. See Immune tolerance, Self-tolerance PHARMACOLOGY An ↑ in dose of a drug required to achieve the same effect in a particular Pt, which is a function of ↑ metabolism–eg, by hypertrophy of the endoplasmic reticulum or ↑ expulsion of the drug from a cell–eg, amplification of the multidrug resistant gene by malignant cells. See Oral tolerance, MDR PSYCHIATRY Resistance to the effects of a sedative SUBSTANCE ABUSE 1. A state caused by regular use of opioids, where an increased dose is needed to produce the desired effect; tolerance may be a predictable sequelae of opioid use and does not imply addiction. See Drug tolerance, Physical dependence 2. The ability to 'hold liquor'–consume alcohol without overt signs of inebriation VOX POPULI A general term for a person's general 'mellowness,' which encompasses the ability to cope with stress, acceptance of others, complete with bumps and flaws, and other facets of social intelligence

tolerance test 1. Exercise tolerance test, see there 2. A maneuver in which the ability to metabolize a drug is tested by administration of a small dose thereof

tolmetin Tolectin® THERAPEUTICS An NSAID that binds albumin. See NSAIDs

Tolosa-Hunt syndrome NEUROLOGY A cavernous sinus syndrome, characterized by idiopathic local inflammation, and retro-orbital pain; THS peaks in the 4th–6th decades, ±accompanied by neurologic or systemic findings; it occurs in ±3% of painful ophthalmoplegia; it is a diagnosis of exclusion. See Cavernous sinus syndrome

tolterodine Detrol® UROLOGY An agent used to manage overactive bladder CONTRAINDICATIONS Urinary retention, gastric retention, narrow-angle glaucoma

toluidine TOXICOLOGY An aniline analogue used today to dye, and to manufacture chemicals

tomato effect CLINICAL DECISION-MAKING Rejection of an effective treatment for a disease for illogical reasons, as may occur when conventional logic dictates that a drug should have no therapeutic value or is toxic, as once occurred with colchicine, aspirin, and currently with thalidomide

tomboy PSYCHOLOGY A popular term for a girl whose developmental gender-identity/role is discordant with her genotype. Cf Sissy

tombstone advertisement MEDIA & MEDICINE An advertising layout format in which the 'copy' is a white field with text surrounded by a black border, often carrying a message with a potentially negative impact. See Cigarette advertising

TOMHS CARDIOLOGY A clinical trial–Treatment of Mild Hypertension Study that evaluated the efficacy of nonpharmacologic therapy and various monotherapies in mild HTN. See Antihypertensive, Calcium channel blockers, Hypertension

tomography IMAGING The creation of images at planes located at specific distances from an x-ray beam. See Brain imaging, Computed tomography, Contrast-enhanced electron-beam tomography, Electron beam computed tomography, Focused appendix computed tomography, High-resolution computed tomography, Nephrotomography, Optical coherence tomography, Positron emission tomography, SPECT tomography, Spiral computed tomography

tone MUSIC THERAPY A musical sound NEUROLOGY The degree of tension in a muscle PSYCHOLOGY The nuance of a spoken phrase SPORTS MEDICINE The baseline muscle tension, which usually reflects the amount of training. See Muscle tone

tone decay test AUDIOLOGY A hearing test that evaluates defects in adaptation to sounds, formally known as auditory fatigue–tone decay, which indicates a defect in retrocochlear transmission of nerve impulses from the ear, possibly caused by pressure or 8th cranial nerve damage. See Acoustic nerve

tongue *lingua* ANATOMY A complex, highly mobile muscular organ anchored in the floor of the mouth, which is central to speaking, chewing, swallowing, is covered by a mucosae invested with tastebuds, is the main organ of taste, assists in forming speech sounds and, when used indiscriminately, a major source of interpersonal problems. See Black hairy tongue, Coated tongue, Flycatcher tongue, Geographic tongue, Golden tongue, Hairy tongue, Liver tongue, Magenta tongue, Raspberry tongue, Scrotal tongue, Smart tongue, Strawberry tongue, White strawberry tongue

tongue inflammation Glossitis, see there

tongue-retaining device A device that keeps the tongue in a forward position while a person is asleep, as a means of preventing obstructive sleep apnea. See Sleep apnea syndrome

tongue scraping DENTISTRY The brushing of the tongue to ↓ intraoral bacteria, a maneuver that may ↓ caries

tongue tie Ankyloglossia ENT A condition in which the tongue's free movement is restricted, as the lingual frenum is attached too far forward on the tongue, restricting tongue motion. See Tongue

tongue worm A non-forensic blood-sucking parasite that infests the nasal cavity of carnivores

tonic ALTERNATIVE MEDICINE A medicinal preparation, usually of herbal origin–eg, ginseng, used in traditional Chinese and in ayurvedic medicine; tonics are said to be help build vital energy–*qi*. See Hoxsey tonic. Cf Bitter

tonic-clonic seizure Generalized tonic-clonic seizure, see there, aka grand mal seizure. See Grand mal seizure

tonic contraction Tetanus NEUROLOGY Smooth muscle contraction of a muscle. Cf Muscle twitching

tonometry BLOOD GASES See Blood gas analysis OPHTHALMOLOGY A technique that measures intraocular pressure–normal is ≤ 20 mm Hg by contact–indentation of, or applanation on or noncontact–by a puff of air on the outer part of the eye INDICATIONS Diagnose and manage glaucoma, ocular HTN, and in routine ocular examination

tonsillectomy & adenoidectomy ENT Excision of the tonsils and adenoids; in the current environment, the 2 procedures are often performed as separate procedures for different indications–tonsillectomy for recurrent pharyngitis or peritonsillar abscesses and adenoidectomy for chronic or recurrent otitis media MICROBIOLOGY Pathogens are found in up to 80% of pediatric T&As EXAMPLES α-hemolytic streptococci, *Haemophilus* spp, *Staphylococcus aureus* and *S pneumoniae*

tonsillitis ENT Inflammation of the tonsils, often due to bacterial or viral infection, which may expand locally, causing pharyngitis. See Tonsillectomy

toot *noun* Cocaine, see there *verb* To inhale cocaine

tooth abscess Dental abscess, periapical abscess, tooth infection DENTISTRY A complication of dental caries or trauma–chipping or breaking–to a tooth; an interruption in the enamel allows bacteria to infect the pulp, which may then infect the bone supporting the tooth. See Periodontal disease

tooth decay See Caries, Tooth abscess

tooth root hypersensitivity DENTISTRY Marked sensitivity to cold, hot, or sour foods due to exposure of the teeth roots 2° to periodontal disease and gum recession, which may be accompanied by bacterial infection

tooth whitener COSMETIC DENTISTRY Any agent–eg, carbamide peroxide, that removes yellowish stains on teeth See Cosmetic dentistry

tooth whitening DENTISTRY Any process is intended to whiten yellowed or stained teeth COMPLICATIONS Improperly applied, whiteners have been associated with tooth disintegration. See Cosmetic dentistry

toothache Dental pain, dentalgia VOX POPULI A potentially incapacitating pain of a tooth in the mandible CAUSES Caries/tooth decay, periodontal disease, pain of reference from other body regions

toothbrush DENTISTRY A device invented by an English prisoner, Wm Addis who saved a bone from his dinner, bored holes in it and wedged hard bristles, probably from a broom into the holes; once released, he became the Rockefeller of dental care

toothpaste sign A descriptor for a column of dense radiocontrast that has passed through a narrowed ureteral opening at the trigone and lies in a toothpaste-like fashion on the bladder floor; this finding may be caused by postoperative contraction of the vesical neck, as may occur in TURPs, as well as in suprapubic and perineal surgery

top of the basilar syndrome NEUROLOGY A condition caused by a lesion of the rostral brainstem, characterized by confusion, hallucinations–peduncular hallucinosis, prominent pupillary abnormalities, ptosis, and disorder ocular movements; the TotBS may be accompanied by somnolence, and inattention, due to infarction of the rostral mesencephalic reticular formation

Topamax® Topiramate NEUROLOGY An agent used to manage partial seizures. See Epilepsy

TOPAS VASCULAR SURGERY A clinical trial–Thrombolysis Or Peripheral Arterial Surgery, which compared the benefits of thrombolysis by catheter-directed intraarterial recombinant urokinase with vascular surgery–eg, thrombectomy or bypass surgery in Pts with acute, < 14 days duration, arterial obstruction of the legs. See Thrombolytic therapy. Cf Rochester study, STILE

tophus The pathognomonic lesion of gout, which appears grossly when preserved in alcohol or other non-aqueous solution as a nodular mass of white chalky, pasty material composed of crystalline and amorphous urates–eg, monosodium urate monohydrates, surrounded by mononuclear cells, fibroblasts and a foreign body-type giant cell reaction with epithelioid histiocytes. See Gouty toe

TOPIC A clinical trial–Tobramycin Once-daily Prescribing In Cystic fibrosis

topical *adjective* Local, focal, superficial; referring to a body surface

topical chemotherapy Treatment with anticancer drugs in a lotion or cream

topical microbicide INFECTIOUS DISEASE A chemical that can be applied to the surface of the body to kill microorganisms

Topiglan® UROLOGY A formulation of alprostadil–synthetic PGE_1 in the SEPA transdermal-delivery technology for treating ED. See Erectile dysfunction

topotecan Hycamtin® ONCOLOGY A topoisomerase I inhibitor used for refractory ovarian CA, small cell CA, and for colorectal and breast CAs ADVERSE EFFECTS Dose-limiting–never if baseline neutrophils are < 1500 cells/mm³–neutropenia, anemia, thrombocytopenia; N&V, diarrhea, alopecia CONTRAINDICATIONS Pregnancy, breastfeeding, prior BM depression. See Topoisomerase I

Toprol® Metoprolol succinate, see there

TOPS CARDIOLOGY A clinical trial which evaluated the effect of delayed vs no PCTA up to 14 days after IV thrombolytic therapy. See Percutaneous coronary angioplasty, Thrombolytic therapy NUTRITION A group therapy diet plan–Take Off Pounds Sensibly created by Esther Manz in the 1950s. See Diet. Cf Weight-watchers

TORCH panel TORCH antibody panel PEDIATRICS A serologic screen for diagnosing prenatal infection; the finding of ↑ IgM in the neonate implies in utero infection by one of the TORCH agents–toxoplasma, rubella, CMV, herpes simplex, which is then characterized by measuring specific IgM levels.

TORCH AGENTS

TOXOPLASMOSIS may cause periventricular microglial nodules, thrombosis and necrosis; obstruction of cerebral foramina causes hydrocephalus; with prolonged survival, there is intracranial calcification, hepatocellular, adrenal, pulmonary, cardiac necrosis and extramedullary hematopoiesis

RUBELLA may cause LBW, hepatosplenomegaly, petechiae and purpura, congenital heart disease, cataracts, microophthalmia and microcephaly; CNS symptoms include lethargy, irritability, dystonia, bulging fontanelles and seizures. See Congenital rubella syndrome

CYTOMEGALOVIRUS may cause hepatosplenomegaly, hyperbilirubinemia, neonatal thrombocytopenia, microcephaly and a mortality of 20-30%; later manifestations include mental retardation, deafness, psychomotor delays, dysodontogenesis, chorioretinitis, learning disabilities; ± 33 000 congenital cases/year–US, of which 10% are symptomatic

HERPES SIMPLEX may cause prematurity, and becomes symptomatic after the first week of life; CNS symptoms include irritability, seizures, chorioretinitis, hydrocephalus, flaccid or spastic paralysis, opisthotonos, decerebrate rigidity and coma; in neonatal HSV infection, no deaths occur in those with localized disease, 15% die if encephalitis is present and 57% die if HSV is disseminated, potentially evoking DIC NEJM 1991; 324:450

SYPHILIS–an optional 'TORCH' Congenital syphilis has ↑ to epidemic rates in the urban US since the mid-1980s; the clinical findings are nonspecific and include fever, lethargy, failure to thrive, and irritability

TORCH(e)S Toxoplasma, other, rubella, CMV, HS, syphilis PEDIATRICS An acronym for a group of in utero infections that may induce major fetal malformations and cause prominent and permanent neurologic defects–eg, seizures, hydrocephalus or microcephaly

toremifene ONCOLOGY An antiestrogen that may control certain CAs, ↓ recurrence. Cf Tamoxifen

tori palati Benign osseous 'tumors' of oral cavity that may be associated with SCC of overlying epithelium. Cf Pseudoepitheliomatous hyperplasia

torsade de pointes French, *torsade*–twist, CARDIOLOGY A form of polymorphic ventricular tachycardia with prolonged Q-T intervals initiated by a premature ventricular depolarization striking near the apex of a delayed T wave; torsades have irregular rates of 200-250/min with marked variability in amplitude and direction of a QRS wave that seems to twist around an isoelectric baseline; torsades may spontaneously resolve or evolve to ventricular tachyarrhythmia and may be nonspecific or due to drugs–eg, adrenergics, antihistamine, phenothiazine, procainamide, quinines, sotalol, and tricyclic antidepressants, electrolyte imbalance–eg, hypokalemia, hypomagnesemia, CNS hemorrhage or trauma, long Q-T wave syndrome, liquid diet, and underlying heart disease MANAGEMENT Isoproterenol

torsion dystonia NEUROLOGY An AD, possibly also AR condition most common in Jews, onset age 5 to 16 CLINICAL Gait defects, involuntary contractions and distortion of spine, hands, feet, hips, and eventually neck; lesions typically start in one body region, usually in an arm or leg, and spread to the rest of the body in ±5 yrs; most Pts are wheelchair bound by adulthood; muscles are hypertonic, under voluntary control, hypotonic at rest; IQ unaffected

torsion of testis Testicular torsion, see there

tort LAW & MEDICINE An act deemed unlawful and capable of triggering a civil action; the wrongdoer–tortfeasor may be held liable in damages. See Malpractice, Negligence OPTHALMOLOGY *verb* To rotate an eye on its antero-posterior axis

torticollis 1. Congenital torticollis 2. Torticollis Loxia, spasmodic torticollis, wry-neck NEUROLOGY The most common of the focal dystonias, in which the neck muscles cause the head to twist to one side, and possibly be shifted forward or backward, often linked to prolonged muscle contractions ETIOLOGY Idiopathic, congenital, acquired due to neuromuscular damage; congenital torticollis may be caused by malpositioning of head in uterus or by prenatal injury of muscles or blood supply in neck. See Febrile torticollis

tortious *adjective* Referring to an action that inflicts a civil wrong. Cf Tortuous

tortuous *adjective* Referring to complexly twisted thing. Cf Tortious

torture syndrome PSYCHOLOGY A post-torture complex characterized by a range of residua–eg, extreme anxiety, insomnia, nightmares, phobias and suspicion of others; the end-stage is termed 'post-traumatic cerebral syndrome. See Post-traumatic cerebral syndrome

Torulopsis glabrata A yeast-like saprobic fungus, with features of *Candida* and *Cryptococcus*, which may rarely cause opportunistic infections in immunocompromised hosts CLINICAL Spiking fever, hypotension, UTI, fungemia

torus fracture ORTHOPEDICS An incomplete fracture of the diaphysis of long bones with buckling of the cortex on the side opposite the fracture. See Greenstick fracture

toss-up DRUG SLANG A ♀ who trades sex for crack or money to buy crack. See Pill whore MEDICAL ETHICS A medical decision in which the difference between the outcomes following one strategy–eg, screening, or treating vs another–not screening, or not treating is negligible and the caring physician is faced with a 'heads you win,' 'tails you lose' decision, popularly termed a 'tossup'. See Clinical decision-making VOX POPULI Vomitus

total abdominal hysterectomy GYNECOLOGY The surgical removal of the uterus through an incision in abdomen. Cf Vaginal hysterectomy

total aphasia Global aphasia, complete aphasia Aphasia caused by lesions that destroy significant amounts of brain–eg, occlusion of middle cerebral or left internal carotid arteries, or tumors, hemorrhage, or other lesions CLINICAL Virtually complete impairment of speech and recognition thereof; afflicted Pts cannot read, write, or repeat what is said to them; although they may understand simple words or phrases, rapid fatigue and verbal and motor perseverence, they fail to carry out simple commands; TA of vascular origin is usually accompanied by right hemiplegia, hemianesthesia, homonymous hemianopia. See Aphasia, Motor aphasia

total-body clearance THERAPEUTICS The rate at which a drug is removed from the body, considered as a single unit, the sum of renal clearance and metabolic (hepatic) clearance, expressed as volume per unit time

total-body irradiation RADIATION ONCOLOGY RT to the entire body, usually followed by BMT or peripheral stem cell transplantation

total catecholamine test A test used to identify catecholamine-producing tumors–eg, pheochromocytomas, and neuroblastoma; catecholamines are excreted in the urine in both free–unconjugated and bound–conjugated to glucuronide and sulfate forms. See Urinary free catecholamine test

total drug concentration THERAPEUTICS The sum of unbound/free and protein-bound drug in the serum or plasma

total energy expenditure PHYSIOLOGY A metabolic 'unit' that is the sum of a number of energy 'units' RESTING ENERGY EXPENDITURE is 60% of TEE; THERMIC ENERGY OF FEEDING is 10% of TEE; NONRESTING ENERGY EXPENDITURE is 30% of TEE.

total estrogen block ONCOLOGY A therapy used to eliminate estrogen, by surgery, RT, chemotherapy, or combination. See Estrogen receptor

total hip replacement ORTHOPEDICS Surgery that replaces the femoral head and its articular surface with a mechanical surrogate INDICATIONS Advanced osteoarthritis and rheumatoid arthritis with disabling pain COMPLICATIONS Loosening of 1 or more of the synthetic components, dislocation, femoral head fracture, DVT, nerve damage and, rarely, infection

total hysterectomy Complete hysterectomy, see there

total ischemic burden CARDIOLOGY The sum of all episodes of symptomatic and asymptomatic/silent myocardial ischemia

total knee replacement ORTHOPEDICS A procedure that substitutes a painful arthritic knee and articular surface with a synthetic device INDICATIONS Advanced osteoarthritis, rheumatoid arthritis with disabling pain COMPLICATIONS Loosening of 1+ synthetic components, dislocation, femoral head fracture, infection, DVT, nerve damage COMPLICATIONS In elderly Pts, TKR carries ↑ risk of cardiovascular–CHF, arrhythmias, ischemia. AMI, syncope, neurologic–acute delirium, CVAs, hypoxic encephalopathy, new-onset dementia, depression, radiculopathy, peroneal nerve palsy. See Clinical decision-making

total knee system ORTHOPEDICS A prosthetic system used in TK replacement COMPONENTS Femoral, tibial, patellar implants

total laryngectomy SURGICAL ONCOLOGY The complete excision of the larynx for invasive CA, which is performed when the lesions cannot be removed by a more conservative–hemilaryngectomy, subtotal laryngectomy procedure. See Laryngectomy

total lung capacity LUNG PHYSIOLOGY The total volume of air contained in the lungs at the end of maximum respiration. See Lung volumes. Cf Functional residual capacity, Vital capacity

total lymphoid irradiation ONCOLOGY Sequential radiation therapy to the 'mantle' and 'inverted Y' lymphoid regions, a combination of fields that may be used in extensive stage IV Hodgkin's disease and NHLs

total parenteral nutrition TPN, see there

total pelvic exenteration Bruschwig procedure GYNECOLOGIC SURGERY A 'heroic' surgery for extensive cervical CA that persists after regional RT and/or prior total hysterectomy; because of its high M&M, TPE is reserved for biopsy-proven recurrences of tumor confined to the central pelvis, in a Pt believed capable of psychologically and physically coping with the stomas necessitated by the operation

total quality management MANAGED CARE The application of the management theory of quality, a concept that focuses on product quality as a means of ↑ sales and saving money. See Continuous quality improvement, Quality control

total range of motion ORTHOPEDICS The total available motion of one part to another sharing a common joint or joints

touchy-feely MEDTALK *adjective* Referring to the 'art' aspects of Pt management–ie, bedside manner. See The Match

tough love POP PSYCHOLOGY A parent's "active nonintervention" in the societal consequences of a child's or adolescent's misbehavior–eg, allowing the police to arrest her for drug possession or use, for DUI, etc

Toughlove® PSYCHOLOGY A network of self-help support groups for parents, care-takers, and/or guardians of 'out-of-control' children with mental and/or emotional problems

toughman fighting SPORTS MEDICINE A modern, largely unsupervised blood sport, in which 2 usually untrained contestants battle each other, virtually bare-fisted RULES 3 one-min rounds, rest periods, use of light gloves; brawling is encouraged. See Blood sport. Cf Boxing, Extreme fighting, Ultimate fighting

692 **Tourette syndrome** Gilles de la Tourette syndrome NEUROLOGY A neuropsychiatric condition characterized by motor and behavioral abnormalities, which begins in the first 2 decades of life, and has a 4:1 ♂ : ♀ ratio CLINICAL Motor Sx include multiple brief muscular spasms–convulsive tics of the face, neck, and shoulders, vocal tics–grunting, snorting, sniffing, barking, throat clearing, other unusual sounds, involuntary profanity–coprolalia, echolalia, self-mutilation MANAGEMENT Neuroleptics–eg, haloperidol, phenothiazines ↓ severity and frequency of tics in 75–90% of Pts, regardless of the severity of disease; other agents used include clonazepam, clonidine

Touretter A person with Tourette syndrome, see there

tourniquet A cord or constrictive band used to ↓ blood flow to 1+ extremity; tourniquets have clinical currency in ↓ the centripetal flow of toxins in snake and scorpion bites, and in ↓ the cardiac load in acute CHF, as may occur in an acute MI, where the tourniquets are rotated, simultaneously with other emergency measures–eg, O₂, lasix, nitroprusside, nitroglycerin; when used, a tourniquet should be confined to the proximal part of the extremity. See Rotating tourniquet

tourniquet paralysis A condition caused by prolonged tourniquet compression of an extremity, resulting in loss of touch, light pressure, vibration, position sensation

towel clamp SURGERY A hemostatic-like hand-held locking device with sharp prong, used to hold surgical drapes in place

towns ACADEMIA Popular for the clinical faculty of a medical school, who are in close contact with the community–ie, work in the 'town,' actively practice medicine, see Pts and are less–if at all–involved in research. Cf 'Gowns'

toxemia GYNECOLOGY A condition in which HTN and fluid retention occur late in pregnancy. See Preeclampsia

toxemia with seizures Eclampsia, see there

toxic *adjective* Referring to a potentially dangerous chemical or substance. See Highly toxic, Toxic chemical

toxic core PSYCHIATRY A 'factor' hypothesized to ↑ cardiovascular mortality–CM in type As, who are angry, cynical, distrustful, and repress marked hostility towards others. See Negative emotions, Toxic emotion, Type A personality

toxic dose TD₅₀ TOXICOLOGY The calculated dose of a chemical introduced by a route other than inhalation, that would cause a specific toxic effect in 50% of a defined experimental animal population Cf Lethal concentration, Lethal dose

toxic emotion PSYCHIATRY A negative emotion–eg, stress and anger, with an adverse effect on a person's mental, and possibly also, physical health. Cf Toxic core

toxic epidermal necrolysis Lyell syndrome DERMATOLOGY An acute life-threatening mucocutaneous reaction often to drugs–80% of cases, characterized by widespread and/or confluent erythema, necrosis, formation of multiple bullae that coalesce, followed by mucocutaneous sloughing ETIOLOGY Drug hypersensitivity–eg, allopurinol, anticonvulsants, barbiturates, carbamazepine, NSAIDs–eg, phenylbutazone, oxicam derivatives, sulfonamides–eg, T-S, infections, vaccination, RT, CA CLINICAL Prodrome–fever, malaise, erythema, followed by subepidermal bullae, epidermal sloughing COMPLICATIONS Dehydration, electrolyte imbalance or 'third space phenomenon', abscess formation, sepsis, renal failure, CHF, GI hemorrhage, shock, skin failure, oral–erosive stomatitis, anogenital, conjunctival–keratitis, corneal erosions, respiratory tract lesions; ±30% are fatal TREATMENT Symptomatic, as with 2ⁿᵈ degree burns, stop offending drug. See Erythema multiforme, Stevens-Johnson syndrome. Cf Staphylococcus scalded skin syndrome

toxic facelift ESTHETIC MEDICINE The injection of botulin toxin/botox in skin of the face with wrinkles–eg, crow's feet, frown lines, furrowed browns, which erases the lines for up to 15 wks before needing to repeat same

toxic megacolon Acute megacolon, toxic dilation of colon GASTROENTEROLOGY A life-threatening complication of GI disease, resulting in transmural inflammation and toxic Sx due to colonic dilation ETIOLOGY ulcerative colitis, Crohn's disease, amebiasis, pseudomembranous colitis, typhoid, bacterial dysentery, Hirschsprung's disease CLINICAL Abdominal pain, distension, fever, tachycardia, fatigue, dehydration, ↓ bowel sounds, tympany, rebound tenderness, hypotension CONTRIBUTING FACTORS Laxatives, opiate use, anticholinergics, hypopotassemia LAB Leukocytosis–> 20,000/mm³, anemia, hypoalbuminemia, ↓ K⁺ COMPLICATIONS Perforation, septicemia MANAGEMENT Resuscitation, metabolic support, correct fluid and electrolyte derangements, corticosteroids, subtotal colectomy and ileostomy, salvage rectal sphincter if possible. Cf Megacolon

toxic metal ENVIRONMENT Any metal known to be toxic to humans–eg, antimony, arsenic, beryllium, bismuth, cadmium, lead, mercury, nickel. Cf Nontoxic metal

toxic multinodular goiter ENDOCRINOLOGY A hyperthyroid state characterized by innumerable functionally active nodules producing excess thyroid hormone CLINICAL Hyperthyroidism without ophthalmoplegia RISK GROUPS ♀ > age 60. See Goiter

toxic shock syndrome INFECTIOUS DISEASE A disease caused by *Staphylococcus aureus* strains that produce a superantigen toxin, TSST-1; the early cases of TSS occurred in tampon users; most began as vaginal lesions; TSS are also linked to foreign bodies–eg, sutures CLINICAL Abrupt onset of high fever–> 40ºC, N&V, watery diarrhea, which may occur during menstruation, followed by an intense blanching mucocutaneous erythema, desquamative palmoplantar rash and cleavage of the basal layer of the epidermis; without therapy, the Pts deteriorate, become lethargic and confused, develop capillary leakage, hypotension, ARDS, renal, multiorgan failure and shock; with appropriate therapy–non-β-lactam antibiotics, mortality is 5-10%. See Superantigen. Cf Toxic shock-like syndrome, Toxic Sock syndrome

toxic shock-like syndrome 'Jim Henson's' disease An epidemic infection caused by a highly virulent, antibiotic-resistant strain of group A streptococcus, which begins as a mild skin infection or 'strep throat' and rapidly progresses to high fever, hypotension, focal vasodilatation and soft-tissue cellulitis of an intensity that may require amputation; most cases occur in Pts without predisposing factors. Cf Toxic shock syndrome

toxic sock syndrome SPORTS MEDICINE A popular term for pitted keratolysis of plantar skin of runners, tennis or basketball players, caused by *Corynebacterium* spp, so named because of its strong odor; TSS is characterized by rounded 1–3 mm punched-out lesions; it develops in a moist environment and is aggrevated by occlusive footwear MANAGEMENT Drying–eg, 20% aluminum chloride, 5% benzoyl peroxide solution, erythromycin

toxic staring NEUROLOGY Fixed staring, abnormal behavior, altered mental status, delirium, aphonia and coma, classically described in typhoid fever. See Rose spots

toxicity The sum of adverse effects 2º to exposure to a toxic substance, by mouth, through the skin or respiratory tract. See Amalgam toxicity, Botanical toxicity, Developmental toxicity, Digitalis/digoxin toxicity, Diphtheria toxicity, Excitotoxicity, Hashitoxicosis, Immunologic toxicity, Neurotoxicity, Nickel toxicity, Oxygen toxicity, Reproductive toxicity, Vitamin A toxicity

toxin A poison or noxious thing produced by animals, plants, or bacteria. See Amatoxin, Anaphylatoxin, Bacterial toxin, Batrachotoxin, Biotoxin, Botulinum toxin, Bungarotoxin, Coley's toxin, Endotoxin, Exotoxin, Heat-stable toxin, Immunotoxin, Lethal toxin, Middle molecule toxin, Neurotoxin, Phallotoxin, Picrotoxin, Recombinant toxin, Rhizotoxin, Shiga neurotoxin, Tetanospasmin

toxocariasis Visceral larva migrans PARASITOLOGY A disease primarily of children due to dog–*Toxocara canis* and cat–*T cati* parasites that inhabit the GI tract and release eggs in the feces; when children ingest eggs from contaminated plants, dirt, stool, eggs hatch into larvae in GI tract, burrow through the wall and migrate elsewhere, primarily to liver and lung, but also brain, eye, etc, causing inflammation and tissue damage in transit CLINICAL Fever, pulmonary complaints–eg, cough, wheezing, seizures, rash, ↓ visual acuity due to migration through ocular structures–periorbital edema, strabismus PROGNOSIS Generally self-limited, Sx eventually disappear; there is no specific therapy

toxoid IMMUNOLOGY A bacterial toxin or other antigen treated with formaldehyde to ↓ toxicity while preserving antigenicity; toxoids are used to prepare diphtheria and tetanus vaccines. Cf Freund's adjuvant

Toxoplasma gondii PARASITOLOGY A species of obligate intracellular coccidian protozoans that has its sexual cycle in the GI tract of its definitive host, *Felis catus*; infection–toxoplasmosis–is usually the result of ingestion of oocytes shed in cat feces or, very rarely, due to ingestion of meat contaminated by pseudocysts. See Toxoplasmosis

toxoplasmosis INFECTIOUS DISEASE An infection by *Toxoplasma gondii* which is either 1. Congenital–acquired transplacentally, often with major neurologic residua or 2. Acquired EPIDEMIOLOGY Ingestion of inadequately cooked meats with *T gondii*-filled cysts, or due to exposure to feces of infected cats, which are the definitive hosts CLINICAL In immunocompetent Pts, infection is benign with transient lymphadenopathy–80% of primary infections are asymptomatic; in immunocompromised Pts, *T gondii* infection may be accompanied by necrotizing encephalitis, myocarditis, pneumonitis, with CNS involvement in ≥ 50% TREATMENT Clindamycin, pyrimethamine; in AIDS, azithromycin. See Congenital toxoplasmosis

toy VOX POPULI A plaything. See Germ-fighting toy, Sex toy

t-PA Activase®, Alteplase®, tissue-plasminogen activator CARDIOLOGY A thrombolytic that upregulates fibrinolysis, activating conversion of plasminogen to plasmin, an enzyme that degrades fibrin clots, fibrinogen, and other clotting factors, ergo reperfusing Pts with acutely occluded coronary arteries; SK has a half-life of 6-8 minutes, an expected 60-80% rate of reperfusion, is nonallergenic and nonantigenic, evokes little hypotension, and is theoretically, more clot specific. See Thrombolytic therapy. Cf APSAC, Streptokinase

TPH Transplacental hemorrhage

TPHA test see *Treponema pallidum* hemagglutination test

TPI 1. A 10-site clinical trial–Thrombolytic Predictive Instrument funded by AHCPR 2. *Treponema pallidum* immobilization

TPN Total parenteral nutrition IV hyperalimentation CRITICAL CARE A modality that attempts to provide all the body's need for nutrition without using the GI tract INDICATIONS 1. Correction of nutritional depletion in the face of inadequate oral intake and/or intestinal absorption, as in Crohn's disease, malignancy, pseudo-obstruction, radiation enteritis, short bowel syndrome, sprue and 2. Conditions requiring bowel rest and nutritional restitution–eg, nonspecific colitides and associated growth retardation, enterocutaneous fistulas and pancreatitis TPN is used for children with diaphragmatic hernia, malrotation, esophageal atresia, tracheoesophageal fistula, gastroschisis, volvulus, meconium ileus and omphalocele; severely malnourished See Cancer cachexia. Cf Enteral nutrition

COMPLICATIONS OF TPN

HEPATIC DYSFUNCTION Cholestasis, cholelithiasis, hepatic dysfunction, jaundice, hepatomegaly, micronodular cirrhosis, lipofuscinosis and steatosis–most common in premature infants

RELATED TO INDWELLING IV LINE Misplacement of line, infections, eg *Candida* spp, aspergillosis

METABOLIC DEFECTS Hyperglycemia–osmotic diuresis, hyperosmolarity, post-infusion hypoglycemia, hyperosmolar coma, ketoacidosis and other metabolic derangements, excess or deficiency of electrolytes, including Na^+, K^+, Cl^-, eg hyperchloremic acidosis and mineral imbalances, affecting Mg^{2+}, PO_4, and Ca^{2+} with hypercalcemia and accompanying pancreatitis, hypercalciuria and metabolic bone disease

NUTRITIONAL IMBALANCES Generalized ↓ in essential fatty acids, trace minerals–copper, chromium, molybdenum, tin, zinc, and vitamins and ↑ TGs, cholesterol

TPR 1. Temperature, pulse, respiration 2. Third-party reimbursement, see there 3. Total pulmonary resistance

TQM Total quality managment, see there

TR 1. Repetition time–MRI 2. Temperature range 3. Timed release 4. Transfusion reaction 5. Treatment 6. Tremor 7. Tubular reabsorption

trabeculoplasty OPHTHALMOLOGY Surgery of the trabecular network of the eye, the site of egress of the aqueous humor, used to treat glaucoma

TRACE CARDIOLOGY A clinical trial–Trandolapril Cardiac Evaluation that evaluated the effect of an ACE inhibitor on the M&M of Pts with acute MI. See Acute myocardial infarction, Trandolapril MOLECULAR MEDICINE Trial of Genetic Assessment in Breast Cancer

traceback EPIDEMIOLOGY Any maneuver designed to follow a public health issue–eg, outbreak of infection or exposure to a toxic substance–to its source. See Lookback, Outbreak

trace evidence FORENSIC MEDICINE Hair and fibers, body fluids and other substances, usually collected by Police Forensic Unit, Sexual Assault Team or Medical Examiner Investigator

tracer IMAGING A substance, such as a radioisotope, used in imaging procedures

tracheal/bronchial rupture Torn tracheal mucosa EMERGENCY MEDICINE A tear of the trachea or bronchus ETIOLOGY Infection, ulcerations due to foreign objects, trauma

tracheitis Acute bacterial tracheitis, bacterial tracheitis PEDIATRICS A bacterial infection of the trachea capable of producing airway obstruction, most often caused by *Staphylococcus aureus* often following a viral URI, most commonly affecting young children, possibly because of the decreased diameter of the trachea relative to the swelling CLINICAL High pitched, crowing stridor–usually on inhalation, aka inspiratory stridor, dyspnea, high fever, toxic appearance MANAGEMENT Hospitalization, endotracheal tube, penicillin, cephalosporin. Cf Croup

tracheobronchial lavage Bronchial lavage, see there

tracheoesophageal fistula NEONATOLOGY A congenital communication between the lower esophagus and trachea, often complicated by aspiration into the lungs, causing pneumonia, choking and possibly death; TEFs are usually detected after birth when feeding is attempted and the infant coughs, chokes and becomes cyanotic; they are a surgical emergency, requiring immediate repair to prevent permanent damage. See Esophageal atresia

tracheomalacia PEDIATRICS Congenital weakness of the tracheal wall which occurs when the cartilage in the trachea fails to develop or mature in a timely manner, with ↓ rigidity CLINICAL Stridor, rattling breath that worsens with URIs; as the tracheal cartilage grows, noisy respirations and breathing difficulties recede. See Stridor

tracheopathia chondro-osteoplastica PULMONARY MEDICINE A rare squamous metaplasia of the trachea, associated with an accumulation of calcium phosphate and ingrowth of cartilage and bone in the subepithelial connective tissue; it is usually asymptomatic; large chondro-osseous masses may cause dyspnea

tracheostomy SURGERY The incision in the anterior wall of the trachea to establish an airway INDICATIONS Upper airway obstruction–due to congenital lesions or acute events-eg foreign body, diphtheria, bilateral voal cord paralysis, laryngeal neoplasms, regional trauma, edema, or anaphylactic reactions, or inability to handle upper or lower respiratory secretions PROS

Relieves obstruction, ↓ dead air space, therefore the work required for effective ventilation; facilitates lavage Cons Loss of effective cough; ↑ susceptibility to infection, especially with *Pseudomonas* spp.

tracheostomy button SURGERY A 1.5 to 4 cm-long plastic tube placed in a surgical tracheostomy to maintain patency

tracheostomy tube "Trake" tube SURGERY A curved metal or plastic tube placed in a surgical tracheostomy to maintain patency

tracheotomy A cutting into the trachea. See Tracheostomy

Trachipleistophora hominis INFECTIOUS DISEASE A microsporidium that most commonly infects immunocompromised hosts–eg, with AIDS. See Microsporidiasis

trachoma OPHTHALMOLOGY An ocular infection by *Chlamydia trachomatis* which, if untreated, leads to blindness INCUBATION PERIOD 5 to 7 days, begins as a mild conjunctivitis that develops into a fulminant infection producing large amounts of discharge, and swollen eyelids; the initial stage lasts several wks, followed by a chronic stage in which the lids remain very swollen, the cornea becomes eroded, scarred, and vascularized; the lids develop contractures and may turn outward, pulling away from the eye; 2° bacterial infections may cause blindness. See *Chlamydia trachomatis*

track *noun* SUBSTANCE ABUSE A punctate, erythematous linear scar on the skin of the extremities, neck, and groin, and on mucocutaneous surfaces, which may be accompanied by intense venous sclerosis and edema of the extremities, a typical finding in heroin addicts. Cf Skin 'popping'

tracking CARDIAC PACING Pacemaker behavior in which ventricular pacing is synchronized to sensed atria activity. See Atrial tracking LAB MEDICINE The following of a specimen from the point at which it is received in the lab until a report is generated. See Tissue tracking MANAGED CARE The evaluation of Pt flow in, through and outside of a healthcare network. See Referral tracking ONLINE The analysis of an Internet user's 'trail' on the Web. See Clickstream tracking PUBLIC HEALTH The monitoring of a person's status in terms of health-related activities–eg, return to primary care-giver for followup visits. See Parallel tracking, Surveillance

tract ANATOMY 1. A bundle of nerve fibers in the CNS. See Spinothalamic tract 2. A tube through which a substance or gas flows. See Aerodigestive tract, Biliary tract, Gastrointestinal tract, Olfactory tract, Respiratory tract, Serpiginous tract, Urogenital

traction ORTHOPEDICS The use of a pulling force to treat muscle and skeleton disorders and Fx steady pulling of a body part by manual or mechanical means; once applied, the force must remain constant until the fracture heals or sets PURPOSE Relieve pain, ↓ muscle spasms, align fracture ends; ↓ pressure on fracture ends. See Buck's traction, Cervical traction, Gallows traction, Hamilton Russell traction, Pelvic traction, Simple traction, Skeletal traction, Skin traction

trade name The trademark name or commercial trade name for a material or product

trading VOX POPULI The exchange of one thing of value for another. See Insider trading

traditional European massage MASSAGE THERAPY A massage therapy that focuses on 'soft tissue' manipulation–eg, effleurage, friction, percussion–tapotement, petrissage, vibration; the prototypic TEM is Swedish massage. See Massage therapy, Swedish massage

traditional medicine 1. Any system of health care that has ancient roots, cultural bonds, trained healers, and a theoretical construct; traditional systems include ayurvedic medicine, ethnomedicine, shamanism, traditional Chinese medicine 2. Mainstream medicine, see there

traffic 1. See Motor vehicles 2. Sexual traffic, see there

traffic accident Motor vehicle accident, see there

trail-making test Reitan's test A two-part test for assessing motor speed and integration, in which multiple dots are connected to form various objects; like the Bender-Gestalt test, the 'Trail-maker' screens for gross organic defects. See Psychological testing

train MEDTALK A series of a similar thing. See Spike train

train wreck MEDTALK A popular term for a multiproblem Pt in critical condition

trainer VOX POPULI One who trains or instructs a person in improving that person's performance in a particular activity. See Athletic trainer, Personal trainer

training A generic term for deliberate goal-oriented practice, in a mental or physical activity, with the intent of bettering one's performance MEDTALK Undergoing postgraduate education, as in, "in training" PHYSIOLOGY A program of regular exercise that results in physiologic muscle hypertrophy, especially of the heart, ↑ skeletal muscle blood supply due to ↑ capillaries, and change in the proportion of slow- or fast-twitch muscle, depending on the type of training activity. See Assertiveness training, Athletic training, Bates vision training, Biofeedback training, Bladder training, Cross-training, Eccentric training, Eye training, Endurance training, Exercise training, Parent training, Relaxation training, Resistance training, Spiritual training, Strength training, Weight training

trait GENETICS An attribute or characteristic of an individual in a species for which heritable differences can be defined. See Bipolar trait, Complex trait, Heritability, High-sensation seeking trait, Input trait, Output trait, Sickle cell trait, Species

TRAM flap Transverse rectus abdominus musculocutaneous flap GENERAL SURGERY A rotated piece of tissue used as an alternative to a prosthesis in post-radical mastectomy reconstructive breast surgery; the operating time and the recovery period is longer, but the cosmetic result is often better

tramadol Ultram® PHARMACOLOGY An analgesic ADVERSE EFFECTS Constipation, N&V, vertigo, headache, somnolence, anaphylactoid reactions, drug abuse, seizures

tramtrack appearance Tramline calcification RADIOLOGY A descriptive term for parallel, curved lines, radiopacities or radiolucencies of varying length BONE A descriptor for the split cortical thickening with parallel neo-osteogenesis or endosteal splitting, seen in infarctions of long bones in sickle cell anemia LUNG A descriptor for the parallel, thickened bronchial walls seen on a plain CXR in bronchiectasis. Cf Railroad track, Ring shadow

trance PSYCHIATRY A state of focused attention and diminished sensory and motor activity seen in hypnosis, hysterical neurosis, dissociative types. See Ecstatic religious state, Neurosis

trance logic POPULAR PSYCHOLOGY An altered mental state in which a person's normal capacity for critical analysis is suspended, and inconsistencies in logic are better tolerated; TL opens a hypnotized person to suggestion. See Hypnosis, Post-hypnotic suggestion, Trance state

Trandate® Labetalol, see there

trandolapril Mavik® An ACE inhibitor indicated as maintenance of Pts with post-MI left ventricular dysfunction, post-MI heart failure. See TRACE

Trankimazin Alprazolam, see there

tranquilizer A popular term for a sedative or sedative/hypnotic

***trans* fatty acid** An unsaturated fatty acid–present in minimal amounts in animal fat–prepared by hydrogenation, which ↑ serum cholesterol CARDIOVASCULAR DISEASE ↑ TFAs have a relative risk of 1.4 for CAD in

♂ in the upper quintile for intake of TFAs, and 1.4 for breast CA in ♀ in the highest quartile for TFA consumption. See Fatty acids CLINICAL NUTRITION TFAs are abundant in margarines, frying fats and shortenings; TFAs comprise 6–8% of ± 120 g/day/person fat consumption in developed nations; the most abundant TFA is elaidic acid; ↓ dietary TFA result in ↓ total cholesterol; the ratio of total cholesterol to HDL-C is lowest after consumption of soybean oil; Cf Cis fatty acid, Fatty acid, Fish, HDL-cholesterol, LDL-cholesterol, Olive oil, Polyunsaturated fatty acid, Tropical oils, Unsaturated fatty acid

transactional analysis PSYCHIATRY A psychodynamic psychotherapy based on role theory that attempts to understand the interplay between therapist and Pt and ultimately between Pt and external reality. See Humanistic psychology, Psychoanalysis, Psycho-therapy. Cf Gestalt therapy

transaminitis MEDTALK A regionally popular term for ↑ transaminases–AST, ALT, coupled with non-specific hepatitis, which may occur in a Pt undergoing the early stages of multiorgan failure. See Multiorgan failure

transbronchial lung biopsy A biopsy from the lung by an endoscopically-guided forceps, used to diagnose benign–eg, interstitial fibrosis, sarcoidosis and malignant–eg, cancer, lymphoma–lesions. See Transbronchial needle aspiration biopsy

transbronchial needle–aspiration biopsy An endoscopic technique used to obtain a specimen from a submucosal endobronchial lesion or an accessible extrabronchial mass, if there is evidence of extrinsic compression; TNAB is used to diagnose bronchogenic carcinoma, carcinoid, bronchogenic cyst, lymphoma, sarcoid, pneumonia, interstitial fibrosis, TB, abscesses COMPLICATIONS Pneumothorax, hemomediastinum, hemorrhage, bacteremia. See Bronchoalveolar lavage, 'Skinny needle' biopsy

TRANSCEND CARDIOLOGY A clinical trial–Telmisartan Randomized Assessment Study in ACE-1 Intolerant Patients with Cardiovascular Disease–designed to evaluate the benefits of telmisartan in Pts with HTN who cannot tolerate ACE inhibitors. See Renin-angiotensin system

transcervical balloon tuboplasty GYNECOLOGY The insertion of a balloon catheter via the cervix to dilate and re-establish patency of fallopian tubes stenosed by the vicissitudes of salpingitis; TBT is analogous to balloon angioplasty. See Pelvic inflammatory disease

transcranial Doppler ultrasonography A non-invasive modality of bedside imaging of the intracranial cerebral circulation in critically ill hospitalized Pts and outpatients, used to diagnose vasospasm, assess collateral circulation and stenoses, to confirm brain death and monitor circulation in neurosurgical Pts

transcranial electrotherapy Cranial electrotherapy stimulation, see there

transcranial magnetic stimulation NEUROLOGY A technique in which a high-intensity, 1 msec magnetic pulse is administered over the skull, disrupting normal brain activity, causing neurons to misfire

transcript GRADUATE EDUCATION A list of a person's academic record MANAGED CARE A document containing the details of a meeting

transcutaneous cardiac pacemaker A device used to generate electrical stimuli to pace the heart via external electrodes adherent to the chest wall

transcutaneous drug delivery Transdermal therapy THERAPEUTICS Use of topical prolonged-release forms of drugs; transcutaneous penetration of a drug requires that it traverse the intercellular lipid layer surrounding the cells of the stratum corneum–rather than through the cells themselves, but also the aqueous environment of the basal cells of the epidermis and the dermis MOST ABSORPTIVE SITES–in descending order–scrotum, jaw, forehead, scalp, axilla, arm, leg DRUGS/AGENTS APPROVED FOR TDD Clonidine HCl, Estradiol, fentanyl citrate, nicotine, nitroglycerin, scopolamine, testosterone replacement therapy

transcutaneous electrical nerve stimulation Electrotherapy A modality in which low electrical current is sent through a pad at an injury site, stimulating the brain to release endorphins REHABILITATION MEDICINE A modality for controlling pain by delivering low-level electric shocks to the skin; TENS effect is explained by the 'gate' theory of pain and is used to relieve pain of the lower back and neck, 'phantom' limb syndrome, amputation stump pain. See Electrical acupuncture; Cf Biofeedback training

TransCyte® SURGERY Biosynthetic temporary skin substitute wound covering for treating severe burn, protecting the wound bed, promoting healing and ↓ side effects of allografts. See Artificcial skin

transdiaphragmatic recurrence ONCOLOGY The appearance of signs and Sx of malignancy at a site on the opposite side of the diaphragm as one that had been previously treated and responded to therapy. See Relapse

transducer INSTRUMENTATION A device that transforms one form of energy to another–eg, a photocell that converts light into electrical energy; it is the major component in ultrasonographic devices, and contains an emitting and receiving piezoelectric crystal

transesophageal echocardiography Two-dimensional transesophageal color-flow Doppler echocardiography CARDIOLOGY An ultrasonographic imaging modality used to examine cardiac structures–valves, chambers and inflow and outflow tracts and function in which a transducer is placed immediately behind the heart in the esophagus and stomach; because there are no interfering air spaces or bone, the image is superior to that obtained with transthoracic echocardiography and is of particular use in evaluating the status of the endocardium–eg, to identify vegetations on the cardiac valves INDICATIONS for TEE ID cardiac source of embolism–35%, prosthetic heart valve malfunction–20%, endocarditis–16%, aortic dissection, cardiac tumor, valvular disease, and others, based on a series of 5000 Pts. See Biplane–intraoperative transesophageal echocardiography, Echocardiography

transfer MEDTALK *noun* 1. A popular term for a Pt whose care has been passed from one service to another 2. The changing of a thing's position with relationship to others. See Blastocyst transfer, Egg transfer, Electron transfer, Electronic funds transfer, Embryo transfer, Gamete intrafallopian transfer, Gene transfer, Linear transfer, Microvascular free toe transfer, Somatic-cell nuclear transfer, Zygote intrafallopian transfer *verb* To pass the care of a Pt from one service or ward to another

transference 1. The projection of attitudes, wishes, desires, libidinous and aggressive thoughts to another party, usually understood to mean to the psychoanalyst 2. An unconscious responsiveness that contributes to the Pt's confidence in a therapist and willingness to work cooperatively. See Countertransference, Parataxic distortion

transferrin Siderophilin An 80-90 kD iron-transporting β-globulin that binds up to 2 atoms of Fe(III) with bicarbonate REF RANGE 2–4 g/L, US: 250-300 mg/dL. See Carbohydrate-deficient transferrin

transformation ONCOLOGY Malignant transformation The change that a normal cell undergoes as it becomes malignant, either due to infection with an oncogenic virus or environmental factors. See Malignant transformation, Progressive transformation of germinal centers, Linear transformation, Sol-gel transformation

transformation zone GYNECOLOGY The area of actively maturing epithelium between the current squamocolumnar junction–SCJ and the original squamous epithelium COMPOSITION Intermingled squamous and columnar epithelium–squamous metaplasia, islands of columnar epithelium, gland openings, Nabothian cyst; > 90% of neoplasia arises from the TZ; cervical colposcopy is inadequate unless the TZ is fully visualized; its location varies in relation to the ectocervix and endocervix, related to Pt age and degree of squamous metaplasia. See CIN

transforming growth factor Any of a group of distinct polypeptides that have been isolated from virus-transformed rodent cells, which are capable of altering a cell's phenotype, causing fibroblasts to lose anchorage-dependence and stimulating angiogenesis

transforming growth factor–β1, –β2 MOLECULAR BIOLOGY Factors responsible for positive and negative autocrine growth regulation

transfusion TRANSFUSION MEDICINE The administration of a blood product to a recipient–with a relative deficit of the product being transfused. See Transfusion guideline STATISTICS–TRANSFUSION-RELATED INFECTIONS HBV 1:63,000; HCV 1:100,000; HIV 1:675,000; Chagas' disease, Brazil 10–20,000/yr. See Autologous transfusion, Blood transfusion, Exchange transfusion, Fetofetal transfusion, Fetomaternal transfusion, Intrauterine transfusion, Massive transfusion, Out-of-hospital transfusion, Platelet transfusion, Single-donor transfusion, Single-unit transfusion

transfusion-associated graft-versus-host disease IMMUNOLOGY A type of GVHD in which immunocompetent T cells administered during a transfusion to a recipient and attack the recipient's immune system RISK FACTORS Poorly defined; may include BM graft recipients and Pts immunocompromised by chemotherapy or malignancy CLINICAL Fever, skin rash, diarrhea, liver dysfunction–abnormal LFTs, severe pancytopenia PREVENTION Gamma irradiation of blood products with 15-25 Gy–1500–2500 rad; up to 50 Gy can be used for most blood products; the onset of TAGVHD is early–10-12 days, often accompanied by BM aplasia, poor response to immunosuppression, postoperative erythroderma, 90% mortality

transfusion criteria TRANSFUSION MEDICINE Any objective criterion–eg, Hct, Hb, that would indicate a need for transfusing a blood product

transfusion guidelines TRANSFUSION MEDICINE Guidlelines for use of blood components, which are usually written in a hospital's policy manual. See Transfusion criteria, Transfusion medicine

TRANSFUSION GUIDELINES, GENERAL CRITERIA
HEMOGLOBIN

< 8g/dL if healthy and stable

< 11g/dL if Pt is at risk of ischemia

ACUTE BLOOD LOSS ≥ 15% (est) blood volume, tachycardia, oliguria

SYMPTOMATIC ANEMIA resulting in tachycardia, change in mental status, cardiac ischemia, or SOB Transfusion 1996; 36:144. See Transfusion guidelines

PACKED RBCs
- **HEMORRHAGE**
- **ACTIVE** Physiologic instability, including tachycardia, ↓ in systolic BP > 30 mm Hg below baseline, orthostatic hypotension, angina, mental confusion, agitation
- **CHRONIC** Physiologic instability–see above, refractory state
- **SICKLE CELL ANEMIA**

    Refractory crisis, acute lung syndrome, CVA, priapism, hepatic infarct, acute papillary necrosis, general anesthesia, contrast studies

PLATELETS
- **PLATELET COUNT** < 30,000/μL
- **FUNCTIONAL PLATELET DEFICIT**
- **SURGICAL PROPHYLAXIS**
- **MASSIVE BLEEDING**

CRYOPRECIPITATE
- **ACTIVE BLEEDING**, fibrinogen < 100 mg/dL
- **MASSIVE BLEEDING**
- **DIC W/ BLEEDING**
- **1⁰ FIBRINOLYSIS**

DYSFIBRINOGENEMIA

MAJOR ↓ IN FIBRINOGEN, FACTOR **VIII**, VON WILLEBRAND FACTOR

± IN REVERSIBLE LIVER DISEASE

Queens Hospital Medical Center, 1990

transfusion medicine Blood banking A subspecialty of clinical pathology or internal medicine which is involved in Pt management through administration of blood cells and blood products including fresh-frozen plasma and cryoprecipitate; TM specialists are versant in relevant areas of hematology, immunology, infectious disease, both clinically and from a lab standpoint, participate in establishing standards for use of these products in a health care facility, and address the legal aspects of transfusions; in the US, the FDA and Am Assn of Blood Banks–AABB–set standards and provide guidelines in TM. See AABB, Transfusion criteria

transfusion reaction Blood transfusion reaction, incompatibility reaction TRANSFUSION MEDICINE Any untoward response to the transfusion of non-self blood products, in particular RBCs, which evokes febrile reactions that are either minor–occurring in 1:40 transfusions and attributed to nonspecific leukocyte-derived pyrogens, or major–occurring in 1:3000 transfusions and caused by a true immune reaction, which is graded according to the presence of urticaria, itching, chills, fever and, if the reaction is intense, collapse, cyanosis, chest and/or back pain and diffuse hemorrhage **NOTE: IF ANY OF ABOVE SIGNS APPEAR IN A TRANSFUSION REACTION, OR IF THE TEMPERATURE RISES 1°C, THE TRANSFUSION MUST BE STOPPED**; most Pts survive if < 200 ml has been transfused in cases of red cell incompatibility-induced transfusion reaction; over 50% die when 500 ml or more has been transfused; TF mortality is ± 1.13/10⁵ transfusions CLINICAL Flank pain, fever, chills, bloody urine, rash, hypotension, vertigo, fainting

TRANSFUSION REACTIONS
IMMUNE, NON-INFECTIOUS TRANSFUSION REACTIONS

- **ALLERGIC** Urticaria with immediate hypersensitivity
- **ANAPHYLAXIS** Spontaneous anti-IgA antibody formation, occurs in ± 1:30 of Pts with immunoglobulin A deficiency, which affects 1:600 of the general population–total frequency: 1/30 X 1/600 = 1/18,000
- **ANTIBODIES TO RED CELL ANTIGENS**, eg antibodies to ABH, Ii, MNSs, P1, HLA
- **SERUM SICKNESS** Antibodies to donor's immunoglobulins and proteins

NON-IMMUNE, NON-INFECTIOUS TRANSFUSION REACTIONS

- **AIR EMBOLISM** A problem of historic interest that occurred when air vents were included in transfusion sets
- **ANTICOAGULANT** Citrate anticoagulant may cause tremors and EKG changes
- **COAGULATION DEFECTS** Depletion of factors VIII and V; this 'dilutional' effect requires massive transfusion of 10 + units before becoming significant
- **COLD BLOOD** In ultra-emergent situations, blood stored at 4° C may be transfused prior to reaching body temperature at 37° C; warming a unit of blood from 4 to 37° C requires 30 kcal/L of energy, consumed as glucose; cold blood slows metabolism, exacerbates lactic acidosis, ↓ available calcium, ↑ hemoglobin's affinity for O₂ and causes K⁺ leakage, a major concern in cold hemoglobinuria
- **HEMOLYSIS** A phenomenon due to blood collection trauma, a clinically insignificant problem
- **HYPERAMMONEMIA AND LACTIC ACID** Both molecules accumulate during packed red cell storage and when transfused, require hepatorenal clearance, of concern in Pts with hepatic or renal dysfunction, who should receive the freshest units possible
- **HYPERKALEMIA** Hemolysis causes an ↑ of 1 mmol/L/day of potassium in a unit of stored blood, of concern in Pts with poor renal function, potentially causing arrhythmia
- **IRON OVERLOAD** Each unit of packed RBCs has 250 mg iron, potentially causing hemosiderosis in multi-transfused Pts

MICROAGGREGATES Sludged debris in the pulmonary vasculature causing ARDS may be removed with micropore filters

PSEUDOREACTION Transfusion reaction mimics, eg anxiety, anaphylaxis related to a drug being administered at the same time as the transfusion

INFECTIONS TRANSMITTED BY BLOOD TRANSFUSION

- **VIRUSES** B19, CMV, EBV, HAV, HBV, HCV, HDV, HEV, Creutzfeldt-Jakob disease, Colorado tick fever, tropical viruses–eg Rift Valley fever, Ebola, Lassa, dengue, HHV 6, HIV-1, HIV-2, HTLV-I, HTLV-II
- **BACTERIA** Transmission of bacterial infections from an infected donor is uncommon and includes brucellosis and syphilis in older reports; more recent reports include Lyme disease and *Yersinia enterocolitica* Note: Although virtually any bacteria could in theory be transmitted in blood, the usual cause is contamination during processing rather than transmission from an infected donor

• **PARASITES** Babesiosis, *Leishmania donovani*, *L tropica*, malaria, microfilariasis–*Brugia malayi, Loa loa, Mansonella perstans, Mansonella ozzardi, Toxoplasma gondii, Trypanosoma cruzi*

transfusion-related acute lung injury TRANSFUSION MEDICINE 1. A type of ARDS characterized by acute noncardiogenic pulmonary edema–pulmonary white-out–and hypoxia within 4 hrs of transfusing a blood product from a multiparous donor containing plasma with potent antigranulocytic antibodies. See Adult respiratory distress syndrome 2. A nonimmune response to transfusion of packed RBCs, attributed to platelet-specific antibodies CLINICAL Severe thrombocytopenia–eg, < 10 x 10^9/L–5–10 days after packed RBCs in Pt sensitized by prior transfusion or pregnancy MANAGEMENT Steroids, IVIG PROGNOSIS Spontaneous recovery. See Transfusion

transfusion siderosis METABOLIC DISEASE Iron accumulation due to multiple transfusions for intractable anemia linked to thalassemia, BM failure, aggressive CA treatment, often with cardiomyopathy MANAGEMENT Chelation with deferoxamine. See Transfusion

transfusion trigger Transfusion threshold TRANSFUSION MEDICINE The Hct and Hb values at or below which packed red cells are usually ordered for transfusion by a clinician; the most widely used current TT is a Hb of 70 g/L, after an NIH consensus = conference; lower TTs–eg, transfusion at 7g Hb, have better outcomes–eg, ↓ MIs and ↓ pulmonary edema 'PROPHYLACTIC' PLATELET TT Stable Pts < 5 **x** 10^9, Pts with recent hemorrhage and/or temperature ≥ 38.5° C < 10 **x** 10^9, Pts with significant coagulopathy, anatomic lesion(s), or heparin therapy < 20 **x** 10^9 'THERAPEUTIC' PLATELET TT Active hemorrhage or scheduled invasive procedure < 50 **x** 10^9, hemorrhage and/or thrombocytopenia, prn

transgendered *adjective* Relating to a person who has undergone genital/sexual reassignment surgery TRANSGENDER HEALTH ISSUES Hormonal therapy, cosmetic surgery, fertility options–eg, egg and sperm banking. See Sexual reassignment. Cf Transsexual

transglottic carcinoma SURGICAL ONCOLOGY An ad hoc term for a tumor that crosses the laryngeal ventricles vertically, to involve the supraglottis, the glottis and often the subglottis; most TCs are supraglottic extensions of glottic CA. See Glottis, Larynx

transient global amnesia NEUROLOGY The loss of memory largely attributable to ischemia, possibly vasospasms. See Amnesia

transient hypogammaglobulinemia of infancy PEDIATRICS A prolongation or accentuation of the normal physiologic decline in serum Igs that normally occurs in the first 3-7 months of life; in THOI, Ig levels may not normalize for 2-3 yrs; antibody production is normal

transient hypothyroxinemia A common finding in preterm infants, which, when severe and present at birth is associated with a ↑ cerebral palsy and ↓ mental development. See Cretinism

transient ischemic attack Little stroke, ministroke NEUROLOGY A focal, abrupt ischemia-induced loss of brain function, caused by ↓ blood flow, especially by microemboli from ASHD plaques and embolic showers, accompanied by disability of < 24 hrs, usually < 1 hr in duration; symptomatic TIAs often precede cerebral infarcts or strokes, more common in older Pts with marked cerebrovascular ASHD, and may affect any cerebral vessel; despite clinical resolution, CT demonstrates residual anatomic lesions in 15% with ischemic neurologic disability and TIAs ETIOLOGY ASHD, polycythemia, sickle cell anemia, hyperviscosity syndromes, spasm of small arteries in the brain, blood vessels defects caused by disorders such as fibromuscular dysplasia, vasculitis, polyarteritis, granulomatous angiitis, SLE, syphilis IMAGING MRI, carotid artery ultrasonography, MR angiography RISK FACTORS HTN, heart disease, migraines, smoking, DM, ↑ age ♂, African-Americans PREVENTION Low dose–eg, 30 mg aspirin, a level at which inhibition of aggregation due to ↓ production of platelet thromboxane A$_2$ is still

complete, but the production of prostacyclin, which has an antiaggregation effect is unaffected in endothelial cells; prostacyclin production is inhibited by higher doses of aspirin. See Reversible ischemic neurological disability

transient response imaging ECHOCARDIOGRAPHY A type of contrast medium-based ultrasonographic imaging, in which cardiac images are acquired intermittently by ECG-gating. See Contrast echocardiography. Cf Harmonic ultrasound imaging, Sonicated albumin

transient retinal ischemia Amaurosis fugax OPHTHALMOLOGY A transient unilateral loss of vision due to ↓ retinal perfusion, related to ASHD, embolism, or vasospasm MANAGEMENT CCBs–eg, nifedipine

transient tic disorder Simple tic NEUROLOGY A childhood disorder characterized by single or multiple motor tics–face, arms, legs, or other areas, often occurring 1+/day for 4 wks, which is either organic or psychogenic; TTD worsens with emotional stress and is absent while sleeping; tics may be precipitated in predisposed children with ADHD when given methylphenidate–Ritalin. See ADD

transillumination CLINICAL PRACTICE The shining of a light through a tissue or body region to detect masses or other lesions

transitional cell carcinoma Bladder cancer, see there

transitional-dyspnea index PULMONOLOGY A 3-part system used to quantify subjective improvement or deterioration of a Pt's functional lung capacity based on interview. See Elastic recoil

transitional year program GRADUATE EDUCATION A flexible internship in which an intern–a physician in the first year of graduate medical education is exposed to many fields–eg, OB/GYN, ER, orthopedics, anesthesia, medicine, surgery. See Internship, Purple book

transjugular access set IMAGING A series of coaxial catheters and needles designed to access the liver through the jugular vein and establish a pathway between the hepatic and portal veins preceding TIPS. See Transjugular intrahepatic portosystemic stent-shunt

transjugular intrahepatic portosystemic stent-shunt TIPS HEPATOLOGY A minimally invasive portosystemic shunt for portal HTN and bleeding complications due to liver disease–eg, esophageal varices INDICATIONS Esophageal varices that bleed after endoscopic banding, sclerotherapy, or surgical portosystemic shunt; refractory ascites, hepatic venoocclusive disease or Budd-Chiari syndrome COMPLICATIONS Hemorrhage, stenosis, occlusion–intraabdominal, biliary, subcapsular; stent migration, contrast-induced renal failure, pulmonary edema, hepatic infarction, hepatic arterial injury, gallbladder puncture, stent malpositions, hemobilia, hemolysis, fever, sepsis, hemoperitoneum, radiation injury, entry site hematoma, subcapsular hematoma, encephalopathy, liver failure OUTCOME Hepatic encephalopathy 25%; 18% rebleed at 1 yr; 85% one-yr survival

translational medicine MOLECULAR MEDICINE The constellation of activities which seek to translate the science of gene discovery, gene transfer, and functional genomics into gene-targeted therapies

translumbar amputation SURGICAL ONCOLOGY The amputation of most of a person's body–essentially everything below the lumbar vertebrae; TA is a draconian surgical procedure which may be the only therapeutic option in certain pelvic and lower extremity malignancies. See Heroic surgery

transluminal angiography Balloon angioplasty A general term for any procedure in which a device is introduced into an artery in order to dilate it and ↑ its internal diameter; TA is used both to establish the diagnosis of vascular occlusion–narrowing and to treat the condition

transmission INFECTIOUS DISEASE The process by which a pathogen passes from a source of infection to a new host MAJOR TYPES Horizontal transmission, which constitutes the majority, and consists of the spread from one person to another by direct contact, aerosol, fecal contamination, etc, and vertical transmission–mother to infant in the birth canal. See Aerosol transmission, Cyclopropagative transmission, Direct transmission, Droplet spread transmission, Hemo-oral transmission, Hereditary transmission, Horizontal transmission, Indirect transmission, Line-of-sight transmission, Maternal-infant transmission, Nondirect transmission, Reverse transmission, Vehicle-borne transmission, Vertical transmission

transmural *adjective* Through a wall

transmyocardial revascularization Laser heart surgery CARDIOVASCULAR SURGERY A technique used for Pts with incapacitating heart disease, in which 15 to 30 1-mm in diameter holes are 'drilled' by laser into the myocardium, in an operation that takes 60–90 mins; indications for TMR are similar to those for repeat CABGs–ie, severe heart disease PROS No need for heart-lung machine; lower costs–CABG is $15,000 to $20,000, recuperation is shorter MORTALITY ±8% OUTCOME 1 yr post-surgery, 90% of Pts have improved to NYHA class 1 to 2 angina Cf CABG

transparent TECHNOLOGY *adjective* Referring to a process that can be incorporated into one's lifestyle or profession in a seamless fashion, such that the user requires virtually no special expertise to use the technology. See User-friendly VOX POPULI Invisible

transparent dressing Transparent film WOUND CARE A waterproof dressing that is permeable to O_2 and moisture. See Wound care

transpersonal psychology Transpersonal counseling POP PSYCHOLOGY Any philosophy on development of mystical, spiritual, and psychic experiences, that is the springboard for maximizing the human potential; TP therapies are intended to remove a person's 'mask,' which may prevent a person from achieving a maximum potential. See Psychosynthesis, Zen therapy. Cf Humanistic psychology, Near-death experience, Past life/lives therapy, Rebirthing therapy

transplacental *adjective* Relating to the passage from the mother into a fetus via the placental circulation

transplant *noun* 1. Any organ or tissue that has been transplanted. See Corneal transplant, Domino liver transplant, Fetal mesencephalic tissue transplant, Laryngeal nerve transplant, Solid organ transplant 2. The process of transplanting an organ or tissue; transplantation *verb* To transplant an organ or tissue

Transplant Games SPORTS MEDICINE An Olympic competition open to transplant recipients. Cf Paralympics, Special Olympics

transplant rejection Graft rejection, organ rejection, tissue rejection IMMUNOLOGY The constellation of host immune responses evoked when an allograft tissue is transplanted into a recipient; rejection phenomena may be minimized by optimal matching of MHC antigens and ABO blood groups and ameliorated with immunosuppressants–eg, cyclosporin, tacrolimus, rapamycin EXCEPTIONS OF TR Corneal transplants, identical twins. See Graft-versus-host disease, HLA, MHC, Tissue typing

transplantation The moving of a tissue or organ from one person–the donor or, less commonly, from a different site on the same person, to another person–the recipient, to replace a malfunctioning organ or organ system; solid organ and hematopoietic precursor transplantations are performed with increasing immunologic impunity in BM, bone matrix, heart valves, heart, heart-lung, kidney, liver, pancreas, skin and intestine, largely due to the availability of agents–eg, cyclosporine and tacrolimus–FK 506, which minimize the otherwise limiting complications of GVHD COMPLICATIONS Transplanted tumors STATISTICS Kidney 14,800; liver 5,350; heart 2155; lung 1042; kidney/pancreas 905; pancreas; 349; intestine 83; heart/lung 22; in Nov, 2003, 83,200 were on waiting lists at the 255 US medical centers that perform transplantations. See Allogeneic transplantation, Allogeneic bone marrow transplantation, Autologous bone marrow transplantation, Autologous chondrocyte transplantation, Bone marrow transplantation, Death row transplantation, Fetal tissue transplantation, Graft-versus-host disease, Hair transplantation, Half-side transplantation, Hand transplantation, Heart transplantation, Heart-lung transplantation, Hematopoietic stem cell transplantation, Hepatocyte transplantation, Islet cell transplantation, Laser hair transplantation, Laser-assisted transplantation, Liver transplantation, Lung transplantation, Multiorgan transplantation, Organ transplantation, Organ cluster transplantation, Orthotopic transplantation, Pancreatic islet cell transplantation, Pancreatic transplantation, Procurement, Renal transplantation, Skin graft, Small intestine transplantation, Stem cell transplantation, Syngeneic transplantation, Transpecies transplantation, UNOS

transposition PEDIATRICS A malposition of an organ or tissues that occurs during embryogenesis PSYCHIATRY See Gender identity transposition SURGERY Plastic surgery in which a flap of tissue is moved from one site to another and allowed sufficient time to establish a new blood supply before severing the vascular connection with the donor site. See Sensory nerve transposition

transposition of gender-identity/role PSYCHOLOGY The interchange of masculine and feminine expectancies and stereotypes: mentally, behaviorly, physically

transposition of great vessels Transposition of great arteries PEDIATRIC SURGERY A congenital cyanotic heart defect in which the position of the aorta and pulmonary artery is transposed; absence of a communication between the pulmonary circulation and the systemic circulation is fatal; usually there is an associated defect that permits mixing of the systemic and pulmonary circulation to provide oxygenated blood to the body PRENATAL RISK FACTORS Maternal rubella or other viral illnesses during pregnancy, poor prenatal nutrition, maternal alcoholism, maternal age > 40, DM CLINICAL Early onset of cyanosis, SOB. See Congenital heart disease

transpulmonary pressure PHYSIOLOGY The difference between airway pressure and pleural pressure–$P_{AW} - P_{PL}$, a clinically important respiratory measure in ICU Pts; it is also derived by multiplying the airway pressure by the ratio of lung parenchyma elastance and total lung elastance; for a given P_{AW}, rise of the P_{PL} has effects on hemodynamics, lung distention and recruitment, and interstitial lung fluid–pulmonary edema

transrectal ultrasonography Transrectal ultrasound An imaging procedure used to screen the prostate for cancer

transseptal catheterization CARDIOLOGY A method used to measure the left atrial and left ventricular pressure and for percutaneous mitral balloon valvoplasty, catheter ablation of accessory–electrical pathways on the left side of the heart, and for evaluation of complex hemodynamics. See Cardiac catheterization

transsexual *adjective* Referring to transsexual behavior *noun* A person manifesting transsexualism, see there

transsexualism Gender identity disorder, see there

transspecies transplantation Xenotransplantation The transplantation of an organ from a lower mammal–eg, baboon, pig to a higher mammal–eg, human

transtelephonic transmitter/receiver CARDIAC PACING An instrument that converts signals for transmission over the telephone wires–the transmitter and then reconverts them into a representation of the original signal–the receiver–eg, EKGs, and telemetry information is transmitted from a pacemaker Pt and is received at the physician's office

transtentorial herniation Cerebellar herniation NEUROLOGY Brain herniation that occurs when part of the cerebellum is displaced through the foramen magnum, compressing the brainstem, causing death by destroying the respiratory center. Cf Uncal herniation

transthoracic echocardiography Two-dimensional transthoracic color-flow Doppler echocardiography TTE CARDIOLOGY A noninvasive imaging technique used as a screening method for analyzing defects of locoregional fluid distribution or blood flow patterns–eg, pericardial effusion or aortic regurgitation; TTE is less sensitive than transesophagel color-flow Doppler echocardiography or spin-echo MRI in detecting thoracic aneurysms because of anatomic and technical drawbacks–eg, limited field of view INDICATIONS Assessment of: acute endocarditis, acute MI, aortic disease, cardiotoxic agents–adriamycin, corticosteroids, opiates, carbon monoxide, radiation, cardiomyopathy, CAD, congenital heart disease, hypertensive heart disease-left ventricular hypertrophy, intracardiac lesions–thrombi, embolism, masses–eg, myxoma, pericardial disease, post-heart transplant, prosthetic valves, valve disease, ventricular dysfunction

transthyretin Prealbumin A 55 kD homotetrameric protein that binds thyroxine and retinol; transthyretin is ↓ in malnutrition and in acute and chronic inflammation, and thus is considered a 'negative acute phase protein', which may be defective in AD amyloidosis. Cf Proalbumin

transurethral laser-induced prostatectomy TULIP UROLOGY A proprietary–Intra-Sonix In–prostatectomy using an intraurethral laser. See Transurethral resection of prostate

transurethral microwave therapy TUMT UROLOGY A minimally invasive procedure using a microwave device–Prostatron™ for outpatient BPH management. Cf Transurethral needle ablation, Transurethral resection of prostate

transurethral needle ablation TUNA UROLOGY A proprietary–VidaMed procedure for managing BPH, which can be performed in the office as an alternative to TURP. See Benign prostatic hypertrophy. Cf Transurethral microwave therapy, Transurethral resection of prostate

transurethral resection of bladder tumor Resection of small low-grade noninvasive papillary transitional cell carcinomas of the urinary bladder with a cystoscope

transurethral resection of prostate TURP UROLOGY The standard method for managing prostate disease–BPH and CA, in which a curette is inserted transurethrally and crescent-shaped 'chips' are removed COMPLICATIONS Higher rates of postoperative cystoscopy, reoperation, postoperative urethral stricture, possibly higher 5-yr mortality. Cf Finasteride–Proscar®, Microwave prostatectomy, Transurethral microwave therapy, Transurethral needle ablation

transvaginal ultrasound IMAGING Ultrasonography of vagina, uterus, fallopian tubes, bladder

transvenous pacing CARDIOLOGY Pacing in which a lead is passed through a vein to reach the right atrium or right ventricle

transverse facial fracture LeFort III fracture, see there

transverse lie Shoulder presentation OBSTETRICS A non-cephalic, non-breech position, in which the fetus's long axis is perpendicular to that of the mother's; TLs occur in 1:300 births, due to lower uterine obstruction–eg, placenta previa, intrauterine leiomyomas or an ovarian tumor in the cul-de-sac, or in a multiparous uterus with a lax wall MANAGEMENT C-section, less commonly, gentle external version, if the membranes have not ruptured; risks of an internal version are unacceptably high and rarely performed

transverse myelitis NEUROLOGY An acute spinal cord disorder causing sudden low back pain and muscle weakness and abnormal sensory sensations in the legs; TM often remits spontaneously; severe or long-lasting cases may lead to permanent disability

transvestic fetishism A paraphilia of cross-dressing; clinical question is whether the TF is accompanied by gender dysphoria–persistent discomfort with present gender role or identity, which if extreme may eventuate in sexual reassignment. See Sexual reassignment

transvestism Cross-dressing PSYCHIATRY Sexuoeroticism from dressing or masquerading as one of the opposite sex, a possibly unconscious lifestyle choice, more common in ♂. See Gynemimesis, Gynemimetophila, Gender identity disorder

transvestite SEXOLOGY A person with a compulsion to dress as a member of the other sex, which may be essential to maintaining an erection and achieving orgasm. See Transsexual

Tranxene® Clorazepate, see there

trapdoor scar PLASTIC SURGERY A scar that puckers above the skin surface in a large healing 'horseshoe' avulsion flap, most common in MVA windshield injuries

trapedoidal rule THERAPEUTICS A formula used to calculate the area under a serum drug concentration-vs-time curve, see there

trapped in the wrong body See Transsexual

trapped lung A sequestered segment of lung in empyema, where part of a bacterially infected lobe and visceral pleura 'fix' the affected lung in a partially collapsed position. Cf Folded lung, Scimitar syndrome, Sequestration complex

trapping NEUROSURGERY A therapy for Pts with intracranial aneurysms who are poor candidates for clipping–ie, where the aneurysmal neck is ill-defined; trapping of the aneurysm may fail if collateral circulation is present

trastuzumab Herceptin® ONCOLOGY A bioengineered monoclonal antibody used to manage Pts with breast CA who overexpress HER2, which transmits growth signals to breast CA cells. See HER2

trauma A physical or emotional wound or injury. See Alternobaric trauma, Atmospheric inner ear barotrauma, Barotrauma, Birth trauma, Childhood trauma, Implantation trauma, Penetrating trauma, Phonotrauma

trauma center MEDTALK A hospital with personnel and equipment, on-site or available on short notice, to manage Pts suffering major injuries, usually–depending on the 'level'–I, II, III of care–without the need to send the Pt to another center. See Level I, II, III trauma centers

trauma score EMERGENCY MEDICINE An index measuring systolic BP, respiratory rate, lung expansion, capillary refill, eye opening, verbal and motor responses, placing them on a scale of 2 to 16. Cf Injury Severity Score

trauma X PEDIATRICS A regional euphemism for the physical signs of child abuse. See Battered child syndrome, Child abuse. Cf SIDS

TraumaSeal™ WOUND CARE A topical nonabsorbable product used in skin closure of lacerations and incisions as an alternative to surgical sutures and staples. See Wound care

traumatic aortic rupture TRAUMA SURGERY A trauma-induced tear of the aorta, a common, often fatal injury due to sudden deceleration in MVAs or falls from a height DIAGNOSIS Transesophageal echocardiography, arch aortography, CT, MRI, films. See Dissecting aneurysm

traumatic arthropathy ORTHOPEDICS A joint affected by trauma, characterized by a fracture line through the joint, resulting in hemorrhage, capsular swelling and distension, followed by adhesions between the pannus and synovia, granulation tissue covering the articular cartilage and fibrous ankylosis which may become ossified. Cf Shenton's line

traumatic disk rupture ORTHOPEDICS A lower cervical spine injury common in Pts with nerve root injury or incomplete spinal cord syndromes after cervical spine injury; most herniations occur with flexion/dislocation or flexion/compression injuries DIAGNOSIS MRI; anterior cervical diskectomy and fusion may enhance neurologic recovery

traumatic grief PSYCHOLOGY A severe form of bereavement–eg, loss of spouse, which may be similar to post-traumatic stress disorder CLINICAL Sleep disturbance, anguish, suicidal ideation, ↑ alcohol, ↑ food, ↑ tobacco consumption, HTN, heart disease, possibly, CA. See Bilateral mastectomy

traumatic injury of bladder and urethra UROLOGY A rare clinical event, given the bladder's location in the bony pelvis; TIBU may occur if the blow to the pelvis is severe enough to break bones, and bone fragments penetrate the bladder wall OTHER ASSOCIATIONS Surgery of pelvis or groin–eg, hernia repair and abdominal hysterectomy, injury to urethra, especially in ♂–cuts, tears, bruises, etc; injury to the bladder or urethra may cause urine to leak into the abdomen, leading to peritonitis, severe bleeding and fluid loss, scarring–stricture or obstruction of bladder or urethra from edema may develop, leading to urine retention, possibly vesicoureteric reflux or bilateral obstructive nephropathy; there is an ↑ risk of UTIs post urethral or bladder trauma due to stasis of retained urine

traumatic pneumothorax EMERGENCY MEDICINE Air or gas in the pleural cavity which causes the lung(s) to collapse, usually caused by trauma–eg, gunshot or knife wounds to chest, MVAs, scuba diving accidents, medical procedures–transbronchial Bx, pleural Bx, thoracentesis, chest tube placement, intracostal needle anesthesia, esophagoscopy–which introduce air in the pleural space. See Pneumothorax

traumatic tap A diagnostic lumbar puncture in which there is incidental hemorrhage due to violent Pt movement or tearing of vessels, a risk inherent in the procedure; TTs are differentiated from subarachnoid hemorrhage by absence of xanthochromia, ↓ RBCs in serial tubes, rapid coagulation

travel medicine Traveler's medicine MEDTALK The field of medicine dedicated to studying and treating health issues, vaccination, preventing infectious disease, injuries, ophthalmalogic conditions related to international travel PUBLIC HEALTH A subspecialty of infectious disease, which focuses primarily on prevention

travelers' diarrhea Montezuma's revenge, Aztec two-step, *E coli* enteritis, Turkey trot, Dehli belly INTERNAL MEDICINE A condition defined as '… the passage of at least 3 unformed stools in a 24-hr period … with N&V, abdominal pain or cramps, fecal urgency, tenesmus, passage of bloody or mucoid stools …in a person who normally resides in an industrialized region and who travels to a developing or semitropical country…Diarrhea > 1 wk in 10% of Pts, and > 1 month in 2%; ±20% of Pts are confined to bed for 1-2 days; most diarrhea in travelers is acquired orally and caused by the heat-stable and heat-labile toxins of *E coli* and *Shigella*; the intensity of infection depends on water supply quality, and previous host exposure and susceptibility PATHOGENS ON CRUISE SHIPS *Shigella*, *Salmonella*, *Vibrio parahemolyticus*, *Aeromonas hydrophila*, *Campylobacter jejuni*, *Plesiomonas shigelloides*, *V cholerae*–non-01, *V fluvialis*, *Yersinia enterocolitica* PARASITES *Giardia lamblia*, *Entamoeba histolytica*, *Balantidium coli*, *Cryptosporidium* spp, *Dientamoeba fragilis*, *Isospora belli*, *Strongyloides stercoralis* VIRUSES Norwalk-like agents, rotavirus MANAGEMENT Rehydration, bismuth subsalicylate, narcotic analogs to slow the motility and T-S if antibiotics are required PREVENTION Boil it, cook it, peel it, or forget it. See Airline food, Cruise ship-related diarrhea

traveling nurse A nurse who travels to find employment because of the relative lack of opportunity in his/her local area

Traverso-Longmire technique SURGERY A procedure for resecting pancreatic CA. See Pancreas cancer

Trazalon Trazodone, see there

Trazepam Diazepam, see there

trazodone Desyrel® PHARMACOLOGY An SSRI-antidepressant that blocks postsynaptic serotonin-5HT and alpha-adrenergic receptors; it is therapeutically similar to desimipramine, et al CONTRAINDICATIONS Pregnancy, arrhythmias, post-MI ADVERSE EFFECTS Sedation, postural hypotension, ventricular arrhythmias. See Selective serotonin reuptake inhibitor

treadmill test Exercise stress test, see there

treasure hunting MEDICAL MALPRACTICE A popular term for a search for the 'needle in a haystack' by a plaintiff's pathologist-expert in a lawsuit for a 'missed'–ie, false negative pap smear that subsequently proved to have cancer

TREAT CARDIOLOGY A clinical trial–Tranilast Restenosis following Angioplasty Trial

treated wood TOXICOLOGY Wood impregnated with preservatives–eg, chromium-copper-arsenate, creosote, inorganic arsenicals, pentachlorophenol, to ↑ its useful life, thwarting insects, fungi, etc; chronic exposure to the fumes of burning wood or skin contact therewith may produce heavy metal intoxication syndrome

treatment A therapy intended to stabilize or reverse a morbid process. Cf Management MEDTALK Therapy. See Cadillac treatment, Experimental treatment, Early treatment, Extraordinary treatment, Foregoing of treatment, Heavy ion treatment, Interim methadone treatment, Investigational treatment, Life-sustaining treatment, Local treatment, Metrazol shock treatment PUBLIC HEALTH The improvement of a water supply. See Aerobic waste treatment, Fluoride treatment, Water treatment

treatment arm Active arm CLINICAL TRIALS In a placebo- controlled trial, the arm that receives a particular therapy

treatment device RADIATION ONCOLOGY Any device or material–port block, stent, wedge, mold, bolus, collimator used to modify dose delivered to a particular site in a Pt undergoing RT. See Radiation oncology, Treatment device

treatment field RADIATION ONCOLOGY An area of interest on the surface or deep in the body that encompasses the tumor volume being treated, which may include vital structures. See Radiation oncology simulation, Simulation, Simulator, Therapeutic radiology treatment planning, Treatment port

treatment group STATISTICS A group formed by making all possible combinations of 2 factors

treatment modality MEDTALK The method used to treat a Pt for a particular condition

treatment pathway Clinical pathway, see there

treatment port Treatment portal RADIATION ONCOLOGY The mechanical opening through which a beam of therapeutic radiation is delivered–the beam impinges on a single surface area of the skin–defined by the collimated radiation beam of the device used to deliver RT; one RT region, area, or field may have > one TP converging on it. See Radiation oncology simulation, Simulation, Simulator, Therapeutic radiology treatment planning, Treatment field

treatment recommendation DECISION-MAKING A narrative in reviews, original articles, and metanalyses, that suggests the best therapeutic option for Pts with a particular condition. See Therapy

treatment window HEMATOLOGY A 4-6 day period during which a Pt with factor VIII inhibitor-producing hemophilia A may respond to a bolus of factor VIII which, by complexing with the inhibitors to ↓ levels enough to allow hemostasis

Trelex mesh SURGERY A mesh used to reinforce repair during inguinal herniorrhaphy

tremor NEUROLOGY Involuntary, rhythmic oscillations of a body part, commonly extremities, but also tongue, jaw, head, eyes, voice; tremors are a Sx and not a disease per se; they occur in primarily extrapyramidal, conditions–eg, advanced hepatic encephalopathy, Parkinson's disease, Wilson's disease, myoclonias TREMORS AS PRIMARY Sx Drug-induced tremor, essential tremor, familial tremor MANAGEMENT Beta-blockers–eg, propranolol, metoprolol, ethanol. See Drug-induced tremor, Essential tremor, Familial tremor, Flapping tremor, Intention tremor, Parkinson's disease, Vocal tremor

tremor at rest NEUROLOGY A tremor typical of Parkinson's disease and parkinsonism, which may be evoked by neuroleptics or other dopamine-blockers–eg, prochlorperazine and metoclopramide; TARs can affect any body part, and may be markedly asymmetrical; they are classically seen with flexion-extension movement of elbow, pronation-supination of forearm, and movement of thumbs across the fingers–'pill rolling' FREQUENCY 3–7 Hz; TARs disappear with movement, return at rest MANAGEMENT Dopaminergics, anticholinergics

trench fever A rickettsia-like disease caused by *Bartonella quintana* transmitted by the feces of the body louse–*Pediculus humanus* under crowded conditions of poor hygiene, seen in endemic form in developing nations CLINICAL Abrupt onset of paroxysmal fever, asthenia, chills, vertigo, headache, backache, characteristic shin pain, truncal rash, transient maculopapules and moderate leukocytosis; febrile relapse TREATMENT Tetracycline, broad-spectrum antibiotics, eradicate. See Saddleback curve

trench foot A condition described in World War I in soldiers in the trenches, whose feet were damp and exposed to near-freezing temperatures for prolonged periods, which caused acral vasoconstriction and heat loss; the prolonged cold is followed by vasodilation, burning pain and paresthesiae with formation of hemorrhagic blebs or gangrene, accompanied by cellulitis, lymphangitis, swelling, thrombophlebitis, and persistent hypersensitivity to cold with 2° Raynaud's phenomenon MANAGEMENT Slow warming of foot. See Immersion foot

trench mouth Acute necrotizing ulcerative gingivitis, stomatitis, Vincent's stomatitis ENT A form of gingivitis ETIOLOGY Opportunistic overgrowth of oral flora, resulting in painful ulcers, most common in young adults RISK FACTORS Poor oral hygiene, poor nutrition, oral infections, smoking, emotional stress. See Herpes

Trendelenburg position Trendelenburg ORTHOPEDICS A position in which the Pt is on an elevated and inclined plane, usually about 45°. with the pelvis higher than the head, and the feet over the edge of the table; the TP is used in abdominal surgery to scoot the abdominal organs toward the chest, and to help manage non head-trauma-related shock. See Gait, Position

Trental Pentoxifylline, see there

trephine biopsy HEMATOLOGY A BM Bx collected using a Jamshidi needle

Trepiline Amitryptiline, see there

Treponema pallidum INFECTIOUS DISEASE The spirochete that causes syphilis EPIDEMIOLOGY 9000 cases/yrs–US, primarily in the SE US. See FTA-ABS, Syphilis, TORCH. Cf Pinta, Yaws

Treponema pallidum hemagglutination test TPHA test LAB MEDICINE A nonspecific serologic test for syphilis, in which the Pt's serum is incubated with tanned sheep RBCs previously coated with *T pallidum* antigens; if syphilis is present, tanned RBCs agglutinate

Treponema pallidum immobilization test TPI test A test once the 'gold standard' for the serologic diagnosis of syphilis. Cf Rapid plasma reagin test, VDRL

tretinoin all-*trans*-retinoic acid DERMATOLOGY A synthetic derivative of vitamin A used topically to reverse some of the effects of photoaging, both clinically–↓ skin wrinkling, improved skin texture and color and microscopically–↑ epidermal thickness, ↑ collagen and dermal vessels and 'erasing' epithelial atypia and dysplasia; tretinoin restores production of collagen I in photodamaged skin and lightens postinflammatory hyperpigmentation, it is also used for acne, keratinization, dermatitis, as a cancer preventive agent, and to induce terminal differentiation of acute promyelocytic leukemia, driving it into a mature nonproliferative state of remission MECHANISM Unknown, possibly related to tretinoin's inhibition of collagenase, which degrades anchoring fibril collagen; tretinoin doubles the number of anchoring fibrils

at the dermoepidermal junction ADVERSE EFFECTS Skin blistering, dry skin, bone pain, headache, N&V, vertigo, ↑ transferases, hyperhistaminemia. See Retinal, Retinoic acid, Vitamin A

TRH Thyrotropin-releasing hormone A tripeptide–pyroglutamic acid-histidine-proline of hypothalamic origin that releases TSH after receptor attachment and activation of cAMPase. See TSH

TRH stimulation test Thyrotropin releasing hormone stimulation test A clinical test used to determine the level in the endocrine system that is responsible for ↓ secretion of TSH by the hypophysis–pituitary gland

triad A trilogy of clinical or pathologic findings, first described as typical for a particular disease, which often prove to be nonspecific. See Asthma triad, Autonomic triad, Behçet's triad, Carney's triad, Christian's triad, Charcot's triad, Epidemiologic triad, Female athlete triad, Hemochromatosis triad, Lennox's triad, Negative triad, Petit's triad, Renal cell carcinoma triad, Saint's triad, Somatostatinoma triad, Toxoplasmosis triad, Trotter's triad, Virchow's triad, Waterhouse-Friderichsen triad, Whipple's triad, Wilson's triad

triage *triage*, French, sorting EMERGENCY MEDICINE A method of ranking sick or injured people according to the severity of their sickness or injury in order to ensure that medical and nursing staff facilities are used most efficiently; assessment of injury intensity and the immediacy or urgency for medical attention. See Streamlined review

---

TRIAGE PRIORITIES

HIGHEST PRIORITY Respiratory, facial, neck, chest, cardiovascular, hemorrhage, neck injuries

VERY HIGH PRIORITY Shock, retroperitoneal or intraperitoneal hemorrhage

HIGH PRIORITY Cranial, cerebral, spinal cord, burns

LOW PRIORITY Lower genitourinary tract, peripheral nerves and vessels, splinted fractures, soft tissue lesions

---

trial Any formal study of a method or therapy. See AAASPS, ABC, ABCD, ACADEMIC, ACAS, ACCT, ACES, ACIP, ACME, ACRE, ACT, ACTION, ACUTE, ADAM, ADEPT, ADMIRAL, ADMIT, ADOPT, ADOPT-A, AEGIS, AFASAK, AFCAPS, AFFIRM, AFIB, AGENT, AGIS, AIMS, AIRE, AIREX, ALE\RT, ALIVE, ALLHAT, AMISTAD, APASS, APRES, APRICOT, ARCH, AREDS, ARMS, ARREST, ARTISTIC, ASAP, ASCENT, ASCOT, ASIST, ASPECT, ASSET, ASSENT, ASSIST, ASTRID, ASTRONAUT, ATACS, ATLANTIC, ATLANTIS, ATLAS, AT LAST, ATS, ATTRACT, AVERT, AVID, BAATAF, BAATAF, BAMI, BARI, BASIS, BBB, BEAT, Before-after trial, BEIR, BENESTENT, BERT, BESMART, BEST, BETTER, BHAT, BLAT, BLOSS, BOAT, BRAINS, BRAVO, BRITE, CABADAS, CABG Patch, CABRI, CADILLAC, CAESAR AIDS, CAFA, CALM-PD, CALYPSO, CAMELOT, CAMEO, CAMIAT, CANDLE, CAPARES, CAPE, CAPPP, CAPRI, CAPRIE, CAPS, CAPTEN, CAPTIN, CARDIA, CARE, CARET, CARISA, CARPORT, CART, CASCADE, CASCADE, CASES, CASH, CASS, CAST, CAST-2, CASTLE, CAT, CATS, CAVEAT, CAVEAT-1, CAVEAT-2, CBT-CD, CCCCCC, CEDARS, CHAMP, CHAOS, CHARM, CHEESE, CHS, CIBIS, CIGTS, CLAS, CLASP, CLASS, CLEERE, Clinical trial, COBALT, COMPANION, COMS, CONSENSUS, CONSENSUS-2, CONVINCE, COPERNICUS, CORE, COURAGE, COURT, CRASH, CREDO, CREST, Crossover trial, CRUISE, CURE, DAIS, DASH, DASH2, DAVIT, DAVIT II, DEBATE, DEBATE II, DECODE, DEFINITE, DESTINI-CFR, Diabetes Control & Complications, Diagnostic, DIADS, DIAMOND-CHF, DIG, DIGAMI, DINAMIT, DIRECT, DISC, Dose control, DouBLE, DPT-1, DRASTIC, DYSBOT, EAFT, EARS, EAST, EBMT, ECASS, ECCO 2000, EDGE™, ELLDOPA, ELITE, EMERAS, EMIAT, EMIP, ENABLE, ENCORE, ENRICHD, ENTIRE, EPHESUS, EPIC, EPICORE, EPILOG, EPISTENT, Equivalence trial, ERA, ERACI, ERASE, ESPRIM, ESPRIT, ESPS2, ESSENCE, ESVEM, EXCEL, EXCITE, FACET, FASTER, FATS, FIRST, FIT, Five-City project, FRESCO, FRIC, FRISC, GABI, GART, GESICA, GISG, GISSI-1, GISSI-2, GISSI-3, GREAT, GUSTO-I, GUSTO II, GUSTO-III, GUSTO-SPEED, GUSTO V, Half-side, HALT-C, HART, HASI, HeADDFIRST, EART, HEP, HERO, HERS, HIPS, HIT, HOPE, HOT, HPYLORI, IMAGE, IMAGES, IMPACT II, IMPRESS, InDDEx, INHIBIT, INJECT, INOS, INSIGHT, INTACT, INTERSALT, InTIME, INTRO-AMI, IRAS, ISAM, ISIS-2, ISIS-3, ISIS-4, IVAT, LACI, LAMP, Large simple, LARS, LATE, LIDS, LIFE, LIFE STUDY, LIMIT-2, LIPID, LONG WRIST, MAAS, MADIT, MAGIC, MARCATOR, MARISA, MDC, MDPIT, MEDENOX, MERCATOR, MERIT-HF, MICRO-HOPE, MIDAS, MILIS, MIRA, MIRACL, MIRACLE, MIRACLE ICD, MIRAGE, MITI-1, MITI-2, MONARCS,

MORE, MOXCON, MRC, MR IMAGES, MRFIT, MSMI, MUCOSA, MUST, MUST EECP, MUSST, MUSTIC, MUSTT, MUTT, N of 1, NASCET, NASCIS, NETT, NICE, NINDS, NIÑOS, Nonrandomized, NORDIL, OASIS, OASIS 2, OAT, OBJECT, OCTAVE, OHTS, OMNIUM, ONTARGET, Open-label, OPERA, OPTIMAAL, OPTIME-CHF, OPUS, ORBIT, OVERTURE, PAC-A-TACH, PACT, PAIR, Parallel groups, PAMI-1, PAMI-2, PARADIGM, PARAGON, PARIS, PARK, PCDD, PEACE, PEPI, PETHEMA, PHADE, PHAROS, PIMI, PIOPED, PIVOT, PLAC I, PLAC II, PLESS, POEM, POSCH, PRAISE, PRECEDENT, PREPIC, PRESTO, PREVENT, PRIMI, PRISM-PLUS, PROACT, PROBE evaluation, PROMISE, PROSPER, PROVE IT, PROVED, PROWESS, PURSUIT, Putative placebo, QUIET, QoLITY, RADIANCE, RADIUS, RALES, Randomized clinical, RAPID, RAPPORT, RAVEL, REACH, REACT, ReALIZe, RECIFE, REFLECT, REGRESS, REMATCH, RENAAL, RENAISSANCE, RESOLVD, RESTORE, RHYTHMS, RID--HD, RISC, RITA, RUTH, SADHAT, SAFE, SARECCO, SAVE, SAVED, SCD/HCT, Scopes monkey, SCRIP, SCRIPPS, SEAL, SECURE, Seeding, SENIC, SEROCO, SHARP, SHEP, SHOCK, SIAM, SIESTA, SILCAAT, SILENT, SMART, (S)MASH, SMILE, SOLVD-prevention, SOLVD-treatment, SPAF, SPAF-2, SPARCL, SPEED, SPICE, SPINAF, SPORT, SPRINT, SPS3, SSITT, STAR, START, STARS, STICH, STILE, STOP-Hypertension, STOPIT, STRESS, STRETCH, Superiority, SUPPORT, SWIFT, SWING, SWORD, SYMPHONY, Syst-Eur, TAIM, TAIST, TAMI-1, TAMI-5, TAMI-7, TAMI-9, TASS, TAUSA, TEAM-2, TEIS, TeqCES, TexCAPS, TIBBS, TIMI-2, TIMI-3, TIMI-4, TIMI-5, TIMI-7, TIMI-9, TOHP-1, TOMHS, TOPAS, TRACE, TRANSCEND, TREAT, TROPHY, TURBO, UK-HEART, UKPDS, UNASEM, V-HeFT-2, VA-HIT, VAL-HEFT, VALIANT, VALUE, VANILA, VANQWISH, VAST, VERT, VIGOR, Viral, VIRADAPT, VISP, VITATOPS, VMAC, WARIS, WARSS, WASID, WATCH, WIHS, WISE, WHIP, WHIMS, WINS, WOSCOPS, WRIST, XISHF, ZEUS

trial & error method Stochastic method, see there

Trial of Antihypertensive Interventions & Management See TAIM

trial of therapy The application of a particular therapy–eg, edrophonium–Tensilon to a person suspected of having a particular disease–eg, myasthenia gravis; response of the condition to the TOT confirms the diagnosis

Trialam Triazolam, see there

triamterene A K⁺-sparing diuretic that blocks resorption of sodium in the distal convoluted tubule of the kidney; it may be used alone, with a loop–eg, furosemide or thiazide diuretic–eg, hydrochlorothiazide, for managing HTN. See Diuretic, Potassium-sparing diuretic

triangulation ONCOLOGY A method for determining the exact position of a breast lesion localized by a guide wire of a lesion of the breast, based on 3 parameters: position, depth of penetration, wire angle. See Guided wire open biopsy. Cf Stereotactic biopsy

triazolam Halcion® PHARMACOLOGY A sedative-hypnotic benzodiazepine analogue. ADVERSE EFFECTS Drowsiness, ↓ concentration See Benzodiazepine

TRIC Trachoma-induced interstitial conjunctivitis

tricarboxylic acid cycle Citric acid cycle, see there

triceps reflex Elbow jerk, elbow reflex NEUROLOGY In-Fast METHOD With the elbow in flexion, sharply tap the triceps tendon, just proximal to the elbow, with a reflex hammer; the arm can also be abducted at the shoulder NORMAL RESPONSE Reflex contraction of triceps–ie, elbow extension. Cf Achilles reflex, Biceps reflex, Patellar reflex

triceps skin-fold thickness A value used to estimate body fat, which is measured on the right arm halfway between the olecranon process of the elbow and the acromial process of the scapula; normal: ♂ 12 mm; ♀ 23 mm; Cf Body-mass index, Mid-arm muscle mass, Obesity

triceps surae reflex Ankle (Achilles tendon) reflex, see there

Trichinella spiralis PARASITOLOGY A globally distributed nematode that causes trichinosis, which is found in carnivore and omnivore muscle after ingestion of larvae in undercooked meat–especially bear and pig meats MORPHOLOGY 1.5 mm, thinner in the front; the anus is subterminal. See Trichinosis

trichinosis PARASITOLOGY Infection by *Trichinella spiralis* due to ingestion of inadequately cooked pork, polar bear, wild carnivores TRANSMISSION Ingestion of inadequately cooked pork with encysted larvae in the muscle; once the cyst-laden meat passes into the lower small intestine, the larvae decloak and develop into viviparous adults which temporarily pitch tent in the ileum and reproduce; the resulting larvae penetrate the capillaries of the small intestinal mucosa, disseminate, eventually encyst in the muscle of the new host CLINICAL The 2 'waves' of Sx correspond to the stages of the life cycle of *T spiralis* in a host–the 1ˢᵗ wave occurs in the small intestine when larvae are liberated from cysts, evoking colicky pain and diarrhea; the 2ⁿᵈ occurs when the larvae penetrate capillaries and encyst, causing hypersensitivity reactions–facial and periorbital edema, eosinophilia, itching, urticaria, low-grade fever, inflammation–eg, myocarditis, myositis, muscle pain; early Sx of infestation are usually GI–cramping, diarrhea, larvae migrate to muscle, causing inflammation and myalgia, especially on movement; muscles in constant motion–eg, of respiration–diaphragm and intercostal muscles are painful; during migration of larvae through tissue, infected Pts often develops swelling in the face and around the eyes; larvae can invade myocardium, causing arrhythmias, myocardial damage MANAGEMENT Thiabendazole, Mebendazole. See *Trichinella spiralis*

trichloroacetic acid DERMATOLOGY An astringent antiseptic used as a exfoliant for Pts with extensive actinic keratosis

trichobezoar Bezoar, hairball PSYCHOLOGY A bolus of hair, usually found in the stomach of a person who may have "issues". See Bezoar

trichomoniasis STD A urogenital infection by *Trichomonas vaginalis* transmitted through sexual contact but also contact with contaminated surfaces–eg, wet wash cloths, toilet seats; ♂ only contract it from infected ♀ CLINICAL In ♂, infection is usually banal, and clears spontaneously; some ♂ experience urethral itching or discharge, burning after urination or ejaculation, or develop prostatitis or epididymitis; ♀ have voluminous fizzy, stinky, purulent, and generally nasty vaginal discharge, itching on labia and inner thighs and labial swelling; over half ♀ with gonorrhea also have trichomonas DIAGNOSIS Pap smear, culture MANAGEMENT Metronidazole, econazole, tinidazole. See Sexually transmitted disease

trichophytin test IMMUNOLOGY A skin sensitivity test for allergies to extract of dermatophytes–*Trichophyton* spp, used to measure cell-mediated immunity, which is compromised or lost in immunodeficiency states–eg, AIDS

Trichophyton MYCOLOGY A genus of Imperfect Fungi, family Moniliaceae, consisting of branched filaments and spores, which infects the skin surface, nails, hair

trichorrhexis nodosa DERMATOLOGY A defect in the hair shaft characterized by thickened weak points–nodes that cause hair breakage; TN may have a genetic basis but is precipitated by environmental factors–eg, "perming," blow drying, aggressive brushing, chemical exposure

Trichosporon beigelii An organism widely distributed in soil, stagnant and fresh water and animal excreta isolated from normal skin, urinary, respiratory, GI tract; although rarely pathogenic, it is linked to white piedra and a summer-type hypersensitivity pneumonia; disseminated infection is more common in immunosuppression or CA TREATMENT Amphotericin B

trichotillomania Hair-pulling, trichomania PSYCHIATRY A chronic traction alopecia of childhood onset affecting adolescent and adult ♀, affecting ± 8 million–US; it is a disorder of impulse control, linked to tics, and habit disorders–eg, thumb-sucking, attributed to psychodynamic conflicts MANAGEMENT Hypnosis, behavior mod, psychotherapy, IMAOs, amitriptyline, etc have been tried with varying degrees of failure; TCAs–eg, clomipramine, may be effective short-term. See Obsessive compulsive disorder

trichuriasis Whipworm infection INFECTIOUS DISEASE Infection by *Trichuris trichiura*, which occurs by oral contact with whipworm ova in contaminated soil; eggs hatch, worm embeds in GI mucosa, primarily in cecum and appendix CLINICAL Heavy infestation causes bloody, mucus-like diarrhea, rectal prolapse

Trichuris trichiura Whipworm PARASITOLOGY A common pathogen believed to infect 750 million people worldwide. See Trichiuriasis

trick movement HAND SURGERY A movement that an active and highly-motivated person performs to circumvent limitations of musculoskeletal paralysis; TMs are never normal and often bizarre and uncoordinated CONS Prolonged 'trick'-type compensation may stretch various hand structures and may persist as a habit after tendons have been surgically rerouted

Tricomin® DERMATOLOGY Hair care products based on the copper peptide technology for maintaining thinning hair . See Hair replacement

TriCor® Fenofibrate CARDIOLOGY An adjunct to diet for managing stratospheric serum TG levels in Pts unresponsive to diet or at risk of pancreatitis

tricuspid atresia CARDIOLOGY A rare congenital heart disease characterized by obstruction of blood flow from the right atrium to the right ventricle, by an absent or abnormally developed tricuspid valve, which compromises pulmonary blood flow; to maintain pulmonary blood flow, blood from the right atrium flows through the foramen ovale into the left atrium, then the left ventricle, then part of the blood flows directly into the right ventricle through a patent VSD or out the aorta via PDA allowing part of the aortic blood flow to flow into the pulmonary artery; this supplies the lungs with some, less than optimal, blood flow; it also puts a strain on the left ventricle which must pump both systemic and pulmonary blood CLINICAL Cyanosis, limited exercise tolerance, SOB. See Patent ductus arteriosus, Ventricular septal defect

tricuspid regurgitation Tricuspid insufficiency CARDIOLOGY Backflow of blood from the right ventricle to the right atrium during right ventricular contraction, due to damage to the tricuspid valve, right ventricular enlargement, rheumatic fever

tricyclic antidepressant PHARMACOLOGY Any psychoactive dibenzozepine derivative–eg, which have 3 central rings and a short linear chain attached to the terminal nitrogen and are thus tertiary amines LAB TA levels correlate poorly with the clinical status, as circulating levels of the highly lipid-soluble TCA represent a minute portion of the body load, and TCA metabolites with similar clinical effects may not be measured CLINICAL, OVERDOSE Parasympathetic disease with anticholinergic effects–eg, mydriasis, xerostomia, urinary retention, ↓ peristalsis, cardiac disease–intractable myocardial depression, hypotension, ventricular tachycardia, fibrillation, or heart block and CNS disease–confusion, agitation, hallucinations, myoclonus, seizures, lethargy that may progress to coma and respiratory arrest ADVERSE EFFECTS Dry mouth, constipation, blurred vision, hyperthermia, weakness, fatigue, lethargy, insomnia, confusion, seizures, ↓ peristalsis, hypotension, coma. See Desipramine, Imipramine, Nortriptyline, Therapeutic drug monitoring, Tricyclic antidepressant suicide. Cf Fluoxetine, Monoamine oxidase inhibitors

tricyclic antidepressant suicide A term of uncertain validity for a suicide linked to tricyclic antidepressant–TA therapy; TAs are widely prescribed and are the most common drug involved in suicide attempts by single, young ♀ without previous Hx of autodestructive thoughts

trident sign PEDIATRICS A finding in Turner syndrome, consisting of a low hairline that extends in 3 vaguely defined vertical bands down the characteristically webbed neck

Tridep Amitriptyline, see there

Tridione® Trimethadione and dimethadione

trifluoperazine Stelazine®, Etalazine® PSYCHIATRY An oral antipsychotic and anxiolytic phenothiazine analogue ADVERSE EFFECTS Extrapyramidal effects, sedation, hypotension

triflupromazine Vesprin® A phenothiazine analogue used as an antiemetic ADVERSE EFFECTS Extrapyramidal effects; sedation, hypotension

trifluralin PARASITOLOGY A dinitroaniline herbicide, which at micromolar concentrations selectively inhibits the proliferation and differentiation of *Leishmania mexicana amazonensis*, by binding to leishmania tubulin, suggesting potential use of trifluralin as an economical and safe antiparasitic agent. See Leishmaniasis

trigeminal ganglion balloon microcompression PAIN MANAGEMENT Compression of the trigeminal region with a miniature balloon for medically intractable pain COMPLICATIONS Anesthesia dolorosa, keratitis. Cf Microvascular decompression

trigeminal neuralgia NEUROLOGY A condition characterized by stabbing paroxysmal unilateral neuropathic pain in absence of sensory or motor paralysis, of the 2nd and 3rd divisions of the 5th cranial–trigeminal nerve; TN is most common in middle-aged ♀, evoked by touching trigger points, yawning, smiling, chewing, etc, or oral pathology, or regional tumors–eg, acoustic neuroma MANAGEMENT–MEDICAL Carbamazepine, phenytoin, alcohol injection MANAGEMENT–SURGICAL If pain is intractable, microvascular decompression may be the most effective OTHER METHODS Extracranial neurectomy of trigeminal nerve branches, percutaneous heat ablation to produce trigeminal nerve or ganglion lesions–radiofrequency thermal rhizotomy, injection of glycerol into trigeminal cistern–retrogasserian glycerol rhizotomy, or physical compression–trigeminal ganglion balloon microcompression

trigeminy CARDIOLOGY 1. An arrhythmia in which 2 normal QRS complexes are followed by a PVC; this rhythm usually does not progress to dangerous forms of fast ventricular rhythms; in this rhythm, two VPCs never occur one right after the other 2. A group of 3 beats, usually a sinus beat followed by 2 extrasystoles ETIOLOGY Digitalis toxicity, dizziness, caffeine. See Premature ventricular contraction

trigger PULMONOLOGY A factor that may exacerbate asthma; a stimulus that causes an ↑ in asthma Sx and/or airflow limitation

trigger finger Tenosynovitis RHEUMATOLOGY A digit in which the flexor tendon passes through a fibro-osseous tunnel, in which there is a fusiform swelling–congenital, edema or tenosynovitis of the tendon or tendon sheath causing a painful lock-snap sensation, leaving the finger or thumb in flexion or extension; TF is most common in ♀ in the 6th decade, and associated with de Quervain's disease, carpal tunnel syndrome, rheumatoid arthritis, and collagen vascular disease; TFs in children may be idiopathic or linked to chromosome defects; trigger/locked fingers may be caused by various fractures, tendinous or ligamentous lesions

trigger point NEUROLOGY Trigger zone An area of low neurologic activity which, when stimulated or stressed, can become an area of high neural activity with referred sensations to other parts of the body served by the excited nerves RHEUMATOLOGY A local region of ↑ tenderness that may occur in fibrositis, often located around the vertebrae medial to the scapula

trigger point injection therapy A pain therapy that attenuates muscle spasms by locoregional injection of a procaine solution into painful muscles. Cf Bonnie Prudden myotherapy

trigger zone NEUROLOGY A circumscribed region near a nerve, often in the head & neck which, when stimulated, even with light touch, may elicit marked neuralgia accompanied by lightning pain EXAMPLES Lips and buccal cavity–evoking trigeminal neuralgia and tic douloureux, tonsillar or posterior pharynx–glossopharyngeal neuralgia and muscles involved in myofascial pain syndrome

704 **triggered pacemaker** CARDIAC PACING A pacemaker that detects a spontaneous depolarization or other signal, and delivers an electrical pulse to the heart; in contrast, an inhibited pacemaker, withholds the stimulus upon sensing a spontaneous depolarization

triglitazone Rezulin® DIABETOLOGY An agent used as a monotherapy for Pts with type 2 DM with diet and exercise, or combined with sulfonylurea ADVERSE EFFECTS Liver injury or failure

triglyceride Triacylglycerol A long chain fatty acid ester of glycerol; TGs constitute 95% of fat by weight, and are the major form of stored lipids ↑ IN Acute MI, alcoholic cirrhosis, untreated DM, high carbohydrate diet, certain forms of hyperlipoproteinemia, HTN, hypothyroidism, nephrotic syndrome, pregnancy, OCs, estrogens ↓ IN Congenital β-lipoproteinemia, hyperthyroidism, malnutrition, vigorous exercise, therapy with ascorbic acid, clofibrate, metformin, phenformin. See Medium chain triglyceride. Cf Cholesterol

trihexylphenidyl Artane® NEUROPHARMACOLOGY An acetylcholine inhibitor that acts on central muscarinic receptors INDICATIONS Parkinson's disease as an adjunct to balance cholinergic activity in central synapses which controls drug-induced extrapyramidal disorders ADVERSE EFFECTS Dose related; dry mouth, blurred vision, N&V, dizziness, constipation, nervousness, urinary retention

trihydroxycoprostanic acid syndrome THCA syndrome An AR condition of neonatal onset characterized by hepatosplenomegaly, growth retardation, rickets, ↑ THCA in serum and bile

triiodothyronine T$_3$, 3,5,3´-triiodothyronine A hormone formed by removing an iodine ion from the chemical parent thyroxine–T$_4$, which occurs in the liver and kidney; in general, the T$_3$ and T$_4$ serum levels rise and fall together with certain exceptions–eg, T$_3$ thyrotoxicosis, in which T$_4$ and free T$_4$ values are in the normal range; most T$_3$ is *not* bound to a carrier protein in the circulation ↑ IN Hyperthyroidism–T$_3$ thyrotoxicosis, pregnancy, and therapy with clofibrate,–oral contraceptives–progestins, estrogens, methadone, perphenazine ↓ IN Euthyroid sick syndrome, ↑ free fatty acids, hypothyroidism, malnutrition, or therapy with corticosteroids, ethionamide, heparin, iodides, lithium, methimazole, phenylbutazone, phenytoin, propranolol, propylthiouracil, reserpine, salicylates, sulfonamides, testosterone, tolbutamide; Pts with T$_3$ thyrotoxicosis are clinically heterogeneous and lack distinctive signs and Sx, comprising 4% of hyperthyroid Pts due to Graves' disease, toxic nodular goiter and thyroid adenomas and a higher percentage of hyperthyroidism in regions with lower levels of iodine REF VALUE 0.92-2.46 nmol/L; 60-160 ng/dL. See Thyroxine–T$_4$, Thyroxine-binding globulin, TSH

triiodothyronine–resin uptake test T$_3$ uptake test, see there

Trilafon® Perphenazine, see there

trimester OBSTETRICS A period of 3 months, understood to be during pregnancy; the 1st trimester is 1-12 wks; the 2nd trimester is 13-26 wks; the 3rd trimester is 27 wks to delivery

trimethadione Tridione® PHARMACOLOGY An oxazolidinedione used for absence seizures ADVERSE EFFECTS Sedation, blurring of vision, exfoliative dermatitis, rashes, hepatitis, nephrosis, death

trimethoprim-sulfamethoxazole Bactrim®. Septra® A broad-spectrum combination antibacterial, formulated as a 1:20 ratio of T to S, which is effective in genitourinary, GI, and respiratory tract infections; it is the antibiotic of choice in PCP for which the failure rate of 5-20%; T-S is also effective against *Salmonella, Shigella, Nocardia* spp, *H influenzae, Listeria monocytogenes, S pneumoniae, Isospora belli*, and possibly *Toxoplasma gondii*; T-S is non-toxic in non-immunocompromised Pts; up to 60% of AIDS Pts have adverse effects–eg, ↑ LFTs, neutropenia, thrombocytopenia, erythematous maculo-papular rash, Stevens-Johnson syndrome, exfoliative dermatitis, N&V

trimetrexate glucuronate NeuTrexin® An antimetabolic lipophilic dihydrofolate reductase inhibitor–ie, antifolate structurally related to MTX, which is antineoplastic and antimicrobial; TG is approved for treating PCP when Bactrim/Septra cannot be used in immunocompromised Pts, and in Pts with colorectal, gastric, and pancreatic CA, acute leukemia, osteosarcoma; it may enhance 5-FU effectiveness in 5-FU and leucovorin-based therapy, improving survival. See *Pneumocystis carinii* pneumonia

trimipramine Surmontil® NEUROPHARMACOLOGY An oral TCA ADVERSE EFFECTS Anticholinergic effects, sedation, orthostatic hypotension. See Tricyclic antidepressant

trinucleotide repeat disease DNA triplet disease, triplet disease MOLECULAR MEDICINE A class of clinically heterogenous diseases, defined by the presence of an abnormal unstable expansion of DNA–triplet repeats in the mutant gene, with up to 200 copies of the repeated trinucleotide EXAMPLES Hereditary–CAG repeats–Huntington's disease, Kennedy's disease, CGG–fragile X syndrome, GCT–myotonic dystrophy, spinal and bulbar muscle atrophy; acquired–some forms of colorectal CA. See Fragile X syndrome, Friedreich's ataxia, Huntington's disease, Machado-Joseph syndrome, X-linked spinal and bulbar muscular atrophy

triosephosphate isomerase deficiency MOLECULAR MEDICINE An AD error of metabolism caused by a deficit of triose phosphate isomerase which leads to a block in glycolytic pathway and a generalized defect in energy metabolism within cells CLINICAL Hemolytic anemia, progressive neuromuscular impairment, spasticity

trip DRUG SLANG *noun* A popular term for a transient mind state evoked by hallucinogens–eg, by LSD. See Bad trip

TriPak™ A product containing 3 Zithromax pills used to manage URIs–eg, acute bacterial exacerbation of COPD due to *H influenzae, S pneumoniae, Moraxella catarrhalis*

Tripedia® PEDIATRICS A trivalent vaccine–diphtheria and tetanus toxoids, acellular pertussis

triphasic oral contraceptive Triphasic pill GYNECOLOGY An OC in which 3 different pills are taken during the cycle. See Oral contraceptive

triple airway maneuver EMERGENCY MEDICINE A procedure used to clear the air passages of Pts with upper airway obstruction, where the mandible is moved forward and rescue breathing is performed through the mouth and the nose

triple apical pulse CARDIOLOGY A double systolic pulse coupled to presystolic distension, an uncommon but characteristic finding by precordial palpation in Pts with hypertrophic cardiomyopathy

triple-blinded study CLINICAL THERAPEUTICS A study in which the Pts and researchers are unaware of whether a treatment–experimental drug or placebo is being administered–ie, 'double-blinded'; in addition, the team analyzing the data is unaware of which group's data they are evaluating–ie, from the protocol's placebo or treatment arm. See Blinding, Double blinding

triple bypass surgery Coronary arterial bypass graft, see there

triple combination therapy AIDS See HAART

triple flexion movement SLEEP DISORDERS A spasm in which the thigh flexes toward the pelvis, the calf toward the thigh, the foot toward the calf; TFMs are characteristic of the restless legs syndrome. See Restless legs syndrome, Sleep disorders

triple-marker screen Triple screen profile A popular term for a diagnosis based on measurement of 3 substances–eg, AFP, hCG, unconjugated estriol, which are ↑ in Down syndrome–DS, trisomy 21. See Alpha-fetoprotein, Estriol

triple phosphate calculus Struvite stone, triple stone A kidney stone composed of ammonium magnesium phosphate, formed in the renal pelvis-giving rise to a staghorn calculus, and urinary bladder; TPCs are caused by urea-splitting bacteria–eg, *Proteus, Pseudomonas, Klebsiella* and *Staphyloccus* spp; TPCs appear as coffin-lid crystals in alkaline urine

triple test The use of 3 diagnostic modalities–eg, clinical, radiographic, and cytopathologic data to arrive at a diagnosis; a positive TT is critical in areas where each method being used has a low specificity ONCOLOGY The use of 3 diagnostic modalities, physical exam, mammography, needle aspiration cytology to diagnose breast CA

triple therapy INFECTIOUS DISEASE The use of 3 antibiotics–ampicillin, gentamicin, metronidazole-to 'cover' for acute diverticulitis. See Acute diverticulitis MEDTALK A therapy that requires 3 modalities/agents–eg, TT for *H pylori* includes a proton-pump inhibitor–eg, omeprazole and 2 antibiotics–eg, amoxicillin, clarithromycin

triple threat physician A popular term for a rare and possibly vanishing breed of physician who is a 1. Researcher of a caliber enough to obtain self-supporting grants 2. Teacher with the skill of Socrates, 3. Clinician with active Pt contact. See Professor, Socratic method

triple X syndrome XXX syndrome, see there

triplet OBSTETRICS One of 3 children carried by one mother during the same gestational period

triplicate prescription form PHARMACOLOGY A 3 copy prescription form required in certain jurisdictions–eg, NY State, to document and monitor presciption of controlled drugs. See Controlled drug substances

triploidy syndrome XXX and XXY syndromes, see there

tripod fracture TRAUMATOLOGY A complex maxillofacial fracture that affects the zygoma, orbit and maxilla

tripod sign NEUROLOGY A nuchal-spinal sign in which the sitting position requires a rigid spine and both arms extended towards the back for support, typically seen in children with non-paralytic poliomyelitis

Triptizol Amitriptyline, see there

triptorelin THERAPEUTICS A long-acting gonadotropin-releasing hormone analogue, that induces intense reversible hypogonadism by inhibiting LH and, to a lesser degree, FSH release, resulting in ↓ testosterone INDICATIONS Advanced prostate and ovarian CA; it may have a role in Rx-refractory sexual deviancy. See Paraphilia

triradius DERMATOGLYPHICS A pattern of whorls seen on the palms of children with trisomy 13 and 21, a finding of itself without pathologic significance, which supports the diagnosis of these trisomies

triscadecaphobia Phobia of number 13. See Phobia

trisomy 13 Patau syndrome, D₁ trisomy syndrome GENETICS A congenital defect due to an extra copy of chromosome 13 CLINICAL Findings in ≥ 50% of Pts: Holoprosencephaly with incomplete development of forebrain and olfactory and optic nerves, accompanied by minor motor seizures, apneic spells, and profound mental defect, microcephaly, deafness, microphthalmia, retinal dysplasia, cleft lip and palate, abnormal helices ± low-set ears, polydactyly, simian crease, pelvic hypoplasia, cardiac defects present in 80% of Pts, including ventricular and/or atrial septal defect, dextroposition of great vessels, cryporchidism, bicornuate uterus, inguinal or umbilical hernias, persistence of embryonic or fetal Hbs INCIDENCE 1:5–15,000 births PROGNOSIS Death usually by age 2

trisomy 18 Edwards syndrome A chromosome defect due to duplication of chromosome 18 CLINICAL Mental retardation, hypotonia, polyhydramnios, FTT, small placenta, low-birth-weight, micrognathia, ASD, VSD, PDA, horseshoe kidney, unilateral or double kidney, double ureter, inguinal or umbilical hernia, nail hypoplasia, cleft lip and palate, deformed skull, low-set malformed ears, short sternum, cryptorchidism INCIDENCE 1:3-11,000 PROGNOSIS Death usually by age 2

trisomy 21 Down syndrome, mongolism PEDIATRICS A chromosomal dysgenesis caused by partial or complete duplication of all or part of chromosome 21 appears 3 times; in some only some cells contain the chromosomal defect–ie, mosaicism CLINICAL MR, microcephaly, short stature, a characteristic facial dysmorphia–flat hypoplastic facies, short/flat nose, upward slanting eyes–Mongolian slant; inner corner of the eyes often has a rounded fold of skin–epicanthal fold; small, low-set ears, thickened, protruding and fissured tongue, joint laxity, pelvic dysplasia, broad hands and feet with shortened digits, a single transverse palmar or simian crease, lenticular opacification, cardiac defects–eg, PDA, endocardial cushion defects, Fallot's tetralogy, esophageal atresia, duodenal atresia, muscle spasms, ↑ risk of ALL, retardation of growth and development is typical; most never reach average adult height RISK OF DS ↑ with maternal age–0.7/1000 births ≤ age 30; 34.6/1000 ≥ age 45 PROGNOSIS Compatible with long-term survival INCIDENCE 1:900. See Maternal age

Trittico® Trazodone, see there

trivial complaint MEDTALK A Sx, sign or lesion, usually identified by the Pt which, after appropriate evaluation, is deemed to have little or no impact on a Pt's well-being or management of an unrelated medical condition. See Hypochondriac, Organ recital

trivial name A popular, working, or common name for a thing or process that has a formal name. See CD, DSM-IV, EC, SI

TRIVIAL NAME

DISEASE–eg, Lou Gehrig's disease for amyotrophic lateral sclerosis

MOLECULE–eg, Teflon for polytetrafluoroethylene

ORGAN–eg, anterior pituitary for adenohypophysis

STRUCTURE–eg, vocal cord for vocal fold–*plica vocalis,* or fallopian tube for *tuba uterina,* which is not standard nomenclature or based on 'official' rules delineated by international agencies or organizations–eg, American Psychiatric Association, Enzyme Commission, the International System, Terminologia Anatomica, etc

Trivora® GYNECOLOGY A combination OC composed of ethinyl estradiol/levonorgestrel. See Oral contraceptive

trocar LAPAROSCOPIC SURGERY A blunt needle-shaped instrument used to withdraw fluid from a body cavity, or perform a 'centesis'. See Saber BT™ blunt-tip surgical

trocar site metastasis SURGICAL ONCOLOGY The development of a tumor at the site of laparoscopic resection of a gallbladder that later proves to have cancer MANAGEMENT Resect entire trocar site. See Laparoscopic surgery

troche Lozenge A form of oral medication formulated as a discoid solid containing a therapeutic agent in a flavored base; a troche is placed in the mouth and allowed to slowly dissolve, releasing its active ingredient–eg, analgesic, antibiotic, antihistaminic, antiseptic, antitussive, decongestant, local anesthetic. Cf Pill

trogliatazone Rezulin® DIABETOLOGY An agent that potentiates insulin in DM ADVERSE EFFECTS Potentially fatal hepatic dysfunction, cardiac arrhythmias *WITHDRAWN FROM MARKET* See Diabetes mellitus, Obesity, TEIS, Thiazolidinedione. Cf 15d-PGJ₂, Metformin

Trojan™ A proprietary condom–the term is used in the US like Kleenex

Trojan horse effect Any disastrous result of an anticipated gain; or, the masking of a dangerous agent within an innocent garb EPIDEMIOLOGY An unanticipated vector of an organism or potential route of disease transmission–eg, Hagnaya wreathes, which transport ectoparasites or used rubber tires that provide ideal breeding sites for the northern Asian mosquito, *Aedes albopictus,* a potential vector for Bunyaviridae and LaCrosse viruses

706

troleandromycin An erythromycin-like macrolide antibiotic used to treat severe asthma, by an unknown mechanism unrelated to antimicrobial activity ADVERSE EFFECTS Severe liver dysfunction–jaundice develops in 4%; inhibits carbamazepine, cisapramide, terfenadine, theophylline, etc, metabolism by CYP3A4 INDICATIONS Reserved for Pts requiring corticosteroids. See Asthma

Tropheryma whippelii The causative bacillus of Whipple's disease, and some cases of uveitis DIAGNOSIS Suspected is based on evidence with EM; confirmed by PCR to detect 16S ribosomal RNA gene–rDNA sequences of *T whippelii*; *T whippelii* has been isolated and cultivated by inoculation in a human fibroblast cell line, using a shell-vial assay. See Whipple's disease

trophic ulceration OPHTHALMOLOGY A noninfectious corneal ulcer caused by repeated trauma to the corneal epithelium and Bowman's membrane

trophoblastic tumor Choriocarcinoma, see there

TROPHY CARDIOLOGY A clinical trial–TRial for Preventing HYpertension

tropical immersion foot syndrome An infection caused by *Pseudomonas aeruginosa*, which commonly affects the 2nd, 3rd, and 4th toe webs PREDISPOSING FACTORS High temperature, high humidity, physical stress, tight interdigital spaces, and pre-existing tinea pedis CLINICAL scaling and maceration of skin, serous or seropurulent discharge, greenish discoloration if pyocyanin is produced MANAGEMENT Local measures; systemic antibiotics

tropical medicine The branch of internal medicine dedicated to the study and management of diseases found primarily in the tropics–in particular parasitic infections, but also 'exotic' viral, bacterial and fungal infections

tropical oil NUTRITION A cooking oil from palm and coconut trees, which differs from other vegetable oils in that, like animal fats, it is high in saturated fatty acids, and thus may have atherogenic potential. See *Cis* fatty acids, Fish oil, Olive oil, *Trans* fatty acids

tropical pulmonary eosinophilia Tropical eosinophilia A hypersensitivity response of the lungs to filarial worms ETIOLOGY *Ancylostoma duodenale*, *Brugia malayi*, *Dirofilaria immitis*, *Strongyloides stercoralis*, *Toxocara canis*, *Wuchereria bancrofti* CLINICAL Malaise, wasting, wheezing, bronchospasm, chronic productive cough with bilateral rales MANAGEMENT Diethylcarbamazine, Ivermectin

tropical splenomegaly syndrome An idiopathic splenomegaly affecting malnourished children and adult ♀ in malaria-endemic regions–eg, New Guinea, Africa, which may be a defective immune response to *P malariae* CLINICAL Massive splenomegaly, asthenia, fatigue LAB ↑ IgM antibodies against *P vivax*, ↓ T-helper cells ↓ CD4:CD8 ratio TREATMENT Chloroquine

tropical sprue TROPICAL MEDICINE An idiopathic malabsorption complex, described in the tropics, occurring either in miniepidemics, or in recent arrivals to a region; it has been linked to either subclinical defects of certain nutrients–protein, folate, vitamin B₁₂, fats and sugars or an as yet unidentified pathogen, resulting in diarrhea-induced weakness that favors the overgrowth of coliform bacteria indigenous to the tropics CLINICAL Malaise, fever, anorexia, intermittent diarrhea, chronic malabsorption, which in the epidemic form first affects adults; prolonged malabsorption causes vitamin deficiencies, muscle wasting, mucocutaneous pigmentation, edema TREATMENT Folic acid, vitamin B₁₂, broad-spectrum antibiotics–eg, tetracycline, intraluminal sulfonamides. Cf Celiac sprue

tropicamide Mydriacyl® OPHTHALMOLOGY An antichoinergic that blocks the action of acetylcholine, dilating the pupils in Alzheimer's disease–AD Pts in a highly diluted concentrations, a finding that some authors believe may be of use as a clinical test for AD; probably not

Tropium® Chlordiazepoxide, see there

troponin I CARDIOLOGY A contractile protein that ↑ in serum after myocardial necrosis; it is a sensitive and specific marker of acute MI, and better than CK-MB as a cardiac injury marker INDICATIONS Acute, recent, or perioperative MI, cardiac contusion, heart allograft rejection, assessment of side-branch occlusion during coronary revascularization procedures or thrombolysis. See Myocardial infarction. Cf Surrogate marker

Trotter syndrome Sinus of Morgagni syndrome A complex characterized by homolateral deafness, pain in the sensory zone of the mandibular branch of the trigeminal nerve, immobility of the ipsilateral palate and trismus, due to invasion by malignancy, often a SCC arising in the lateral nasopharynx, specifically near the sinus of Morgagni

troubleshoot LAB MEDICINE *verb* To determine the source of a systematic error and correct it

trough C₍min₎, trough serum concentration THERAPEUTIC DRUG MONITORING The point of minimum concentration of a drug or therapeutic agent on the SDC-vs-time curve; TSCs occur immediately before administering a drug's next dose. See MBC, MIC, Therapeutic drug monitoring, Therapeutic index. Cf Peak serum concentration

Trousseau sign INTERNAL MEDICINE Compression of the upper arm by a tourniquet or blood-pressure cuff causes carpal spasm and paresthesia, an indicator of ↓ Ca^{2+}

trovafloxacin mesylate Trovan® ANTIBIOTICS An advanced/'fourth' generation fluoroquinolone EFFICACY Available in IV and oral formulations, it is approved for treating bacterial infections with improved activity against streptococci and anaerobes INDICATIONS Pneumonia, sinusitis, surgical infections, gonorrhea, and prophylaxis before hysterectomy or colorectal surgery,. See Fluoroquinolone

Tru-cut biopsy See Core needle biopsy

true cholinesterase Cholinesterase, see there

true hermaphroditism ENDOCRINOLOGY A condition in which one person has ovarian *and* testicular tissue, genotypically either 46, XX or 46, XY; 50% to 75% are raised as boys; testicular tissue is dysgenic, doesn't produce sperm and may undergo malignant degeneration–requiring prophylactic removal; ovarian function in those raised as girls may be adequate to produce term pregnancy. See Hermaphroditism, Intersex disorder

true knot True umbilical cord knot NEONATOLOGY A knot of the umbilical cord at various locations and degrees of tightness; edema, grooving, narrowing, tightness, and differences of color and/or diameter on the opposite sides of the knot may indicate vascular obstruction; the placental disk is congested PROGNOSIS Consequences hinge on the tightness of the knot; tight knots cause fetal demise. Cf False knot

true memory NEUROLOGY Memory of an event that actually occurred; TMs appear to be stored in the temporal parietal region, where sounds are processed. See Memory. Cf False memory

true negative STATISTICS A negative test result, that accurately reflects the tested-for activity of an analyte

true positive STATISTICS A positive test result, that accurately reflects the tested-for activity of an analyte

true rib *costa vera*, vertebrosternal rib ANATOMY Any of the upper 7 ribs that connect the vertebral column to the sternum via costal cartilage. Cf Vertebral rib

TRUE test ALLERGY MEDICINE A format–Thin-layer, Rapid-Use Epicutaneous for testing skin sensitivity to allergens

true vocal cord ANATOMY A fold of laryngeal mucous membrane that produces sound when taut and vibrating

Truelove-Witts criteria GI DISEASE A scheme for stratifying disease severity in ulcerative colitis, see there

truncal obesity Abdominal obesity, see there

truncated *adjective* Shortened

truncus arteriosus CARDIOLOGY A congenital heart defect characterized by a single arterial trunk arising in the ventricle, which supplies the pulmonary, coronary, and systemic circulation; a large VSD always accompanies TAs, essentially converting the right and left ventricles into a single chamber; blood flows at a higher pressure through the arteries–the pressure in the pulmonary circulation is normally low; systemic pressure in the pulmonary circulation ↑ blood flow through the lungs, ↑ cardiac load, heart failure; ↑ pulmonary pressure eventually damages the pulmonary vessels, causing ↑ pulmonary resistance, ergo ↓ pulmonary blood flow and cyanosis CLINICAL SOB, fatigue, heart failure, poor growth, physical development and, untreated, death at a young age. See Cyanotic heart disease

TRUS Transrectal ultrasonography, see there

Trusopt OPHTHALMOLOGY A topical that ↓ intraocular pressure in Pts with ocular HTN or open-angle glaucoma. See Glaucoma

trustee VOX POPULI A person entrusted to perform a particular task. See Health information

trustworthiness ETHICS A principle in which a person both deserves the trust of others and does not violate that trust

truth-telling ETHICS A component of ethical behavior required of physicians–by the dictums of the Hippocratic Oath and integral to a high standard of professional behavior. See *Arato v Avedon*, Doctor-patient interaction, Hippocratic Oath, Information overload, Paternalism, Therapeutic privilege

Trynol Amitriptiline, see there

trypanosomiasis See American trypanosomiasis

tryptophan malabsorption syndrome Blue diaper syndrome An AR condition characterized by ↓ growth, hypercalcemia, nephrocalcinosis, recurrent infections, renal defects, and indicanuria; the blue color is due to oxidation of indican to indican blue upon exposure to air. Cf Black urine disease

TSH Thyroid-stimulating hormone, see there

TSLS Toxic shock-like syndrome, see there

TSP 1. Thrombospondin 2. Total serum protein 3. Tropical spastic paraparesis, see there

TSS Toxic shock syndrome, see there

TST 1. Toxic shock toxin 2. Treadmill stress test, see there

Tsukamurella INFECTIOUS DISEASE A genus of facultatively pathogenic actinomycetes that infect immunocompromised hosts, sites with foreign body or chronic infections–eg, TB. See Actinomycete

tsunami Seismic sea waves GEOMEDICINE A superwave that can circle an island, and pound the otherwise protected leeward side with ↑ velocity–up to 365 km/hr or 225 miles/hr; waves can reach a height of 25+ meters; the Boxing Day (2004) tsunami centered in Indonesia claimed 300,000 lives. See Geological disaster. Cf Tidal waves

T-TAC Transcervical tubal access catheter system GYNECOLOGY A system used to diagnose proximal tubal occlusion, for hysterosalpingography, laparoscopic chromopertubation, and intrauterine insemination. See Ectopic pregnancy

TTP-HUS The combination of thrombotic thrombocytopenic purpura–TTP and hemolytic-uremic syndrome–HUS, polar expressions of the same disease defined by a pentad of TTP features: Thrombocytopenia, microangiopathic hemolytic anemia, neurologic defects, fever, renal disease CLINICAL Abrupt onset in children after a viral URI; may be associated with a verotoxin in *E coli*-induced gastroenteritis; it may occur in pregnancy, at parturition, or during chemotherapy; spontaneously resolving renal failure–RF occurs in 60%–10% progress to chronic RF; RF is more common in *E coli* 0157:H7 infections, which are linked to 16% of cases with HUS LAB Reticulocytosis, ↑ BR, ↑ FDPs without DIC, ↓ haptoglobin MANAGEMENT Most resolve spontaneously, others require high-dose corticosteroids PROGNOSIS with prednisone and plasma exchange, 91% survival

tubal factor REPRODUCTION MEDICINE Structural or functional damage of one/both fallopian tubes that ↓ fertility

tubal ligation Tubal interruption GYNECOLOGY Any sterilization procedure in which the lumen of the fallopian tube is intentionally sealed to prevent future pregnancy. See Ectopic pregnancy

tubal pregnancy Ectopic pregnancy, see there

tube An elongated cylindrical structure with an opened central bore. See Blue top tube, Chest tube, Corvac tube, CTAD tube, Fallopian tube, Gray top tube, Green top tube, Lavender top tube, Minnesota tube, Myringotomy tube, Non-additive tube, Partial draw tube, Photomultiplier tube, Sengstaken-Blakemore tube, Tiger top tube, Tympanostomy tube, Yellow top tube

tube insertion Tympanostomy, see there

tubercle MEDTALK Bump, lump, nodule, protuberance, especially on a bone

tuberculid A non-infectious skin lesion due to hypersensitivity to *Mycobacterium* spp TYPES Papulonecrotic tuberculid, characterized by symmetrical waves of sterile papules with central ulceration and obliterative vasculitis; lichen scrofulosorum–groups of tiny, sarcoid-like red papules

tuberculin skin test Any intracutaneous test used to diagnose TB based on hypersensitivity to tuberculin, a concentrated preparation of TB antigen; the standard preparation of which is PPD–purified protein derivative; the most commonly TST is the Mantoux test; ≥ 15 mm is regarded as positive. See Mantoux test; Multiple-puncture test, PPD test

tuberculosis INFECTIOUS DISEASE A disease first known to the ancients; there are one million new cases of *Mycobacterium tuberculosis*/yr worldwide, of which ±10% of those in developing nations eventually die; 'smear'-positive cases in Africa–165/10⁵, are more often clinically inactive than those in Asia where the rate is 110/10⁵ US incidence: 9.3 cases/10⁵–white/Hispanic 5.7/10⁵, black 26.7/10⁵, Asian 49.6/10⁵; the previous trend of ↓ TB in the US reversed itself in the mid-1980s, due to ↑ of *M tuberculosis* and *M avium* complex in AIDS; up to 10 million in the US have latent TB–many of whom are poor, aged, malnourished; homeless or IVDAs CLINICAL Coughing, chest pain, hemoptysis, weight loss, fatigue, malaise, fever, night sweats DIAGNOSIS Ziehl-Neelsen or Kinyoun AFB stains, viewed by LM; auramine-rhodamine stain with fluorescent microscopy; NAP test, nucleic acid probes, PCR TREATMENT-1° DRUGS Isoniazid, ethambutol, rifampicin, streptomycin 2° DRUGS Ethionamide, capreomycin, kanamycin, cycloserine, pyrazinamide, para-aminosalicylic acid. See Latent tuberculosis, MOTT, Mycobacterial infection, Multidrug resistant tuberculosis, Runyon classification. Cf Pseudotuberculosis

tuberculosis test Any of a number of tests used to detect past exposure to, or current infection by *M tuberculosis*. See Mantoux test, Tuberculin skin test

TUBERCULOSIS TEST

1. Tests that document current infection, either by a direct smear of sputum, or by culturing a specimen, eg sputum, lymph node, lung biopsy; *M tuberculosis*, is notoriously slow to

grow, requiring 3+ weeks for definitive identification when cultured in the traditional fashion; more recently, molecular techniques, eg PCR-have been used to shorten the turn-around time for test results

2. Tests that evaluate the immune response, see Mantoux test, Tuberculin test

Obsolete TTs include colorimetric tests–once believed to measure substance(s) seen in ↑ amounts in Pts with TB, the mirror test and theurorosein test; other TTs, eg the tine and Sterneedle tests, that measure sensitiv-ity to tuberculin, reflect the Pt's immune status, and have been abandoned

tuberculous arthritis TB-induced articular inflammation, affecting ±1% of TB Pts JOINTS Vertebrae, hips, knees, wrists, ankles; TA can be very destruc-tive and lead to vertebral collapse, muscle atrophy, spasms. See Pott's disease

tuberculous meningitis NEUROLOGY *M tuberculosis* meningitis caused by spread from elsewhere in the body RISK FACTORS Hx pulmonary tuberculosis, alcoholism, AIDS. See Meningitis *Mycobacterium tuberculosis*

tuberculous mycobacteria Mycobacteria that can cause TB or TB-like clinical disease: *M tuberculosis, M bovis, M africanum*

tubular necrosis See Acute tubular necrosis

tuberculous pleuritis Tuberculous pleurisy INFECTIOUS DISEASE Pleural infection by *Mycobacterium tuberculosis*, which is characterized by pro-tein-rich effusions and abundant *M tuberculosis*. See Tuberculosis

tuberous sclerosis INTERNAL MEDICINE An AD neurocutaneous syn-drome characterized by skin lesions, variable mental retardation and seizures; TS is named after the typical brain lesion–a tuber. See Neurocutaneous disorder

TUBEROUS SCLEROSIS–GOMEZ CRITERIA

PRIMARY†	SECONDARY‡
Facial angiofibroma	1° relative with TS
Fibrous plaque on forehead	Hypomelanic macules
Ungual fibroma	Shagreen patch
Cortical tuber	Infantile spasms
Multiple retinal hamartomas	Single retinal hamartoma
Subependymal hamartoma	Cardiac rhabdomyoma
Bilateral renal angiomyolipomas or cysts	

†One required for diagnosis ‡Two required for diagnosis

tuberous sclerosis triad Mental retardation, seizures, facial angiofibro-mas

tubo-ovarian *adjective* Referring to fallopian tube and ovary

tuboplasty Tuberoplasty GYNECOLOGY Surgery of the fallopian tube to remove obstruction due to scarring or prior tubal ligation. See Pelvic inflam-matory disease, Reversal operation

TUBS ORTHOPEDICS Acronym for a type of shoulder joint instability, char-acterized as traumatic, unidirectional, Bankart lesion, best treated by sur-gery. See AMBRI, Shoulder instability

tubulovillous polyp Villotubular adenoma, see there

tularemia Deerfly fever, rabbit fever INFECTIOUS DISEASE An infection of wild small animals–eg, rabbits and rats caused by *Francisella tularensis*, a small, gram-negative aerobic bacillus, transmitted to man by bites or via arthropod vectors–eg, ticks, tabanids CLINICAL FORMS Oculoglandular, pneu-monic–atypical pneumonia, typhoidal, ulceroglandular

Tulio's phenomenon NEUROLOGY Vertigo after loud noise, a sign of inner ear fistula. See Hennebert's phenomenon, Inner ear fistula

TULIP Transurethral laser-induced prostatectomy, see there

Tumarkin crisis Otolithic crisis of Tumarkin ENT A drop attack–without

warning or loss of consciousness experienced by 1–2% of Ménière's disease Pts, attributed to mechanical disturbance in the otolithic organs of the ear which sense gravity. See Drop attack

tumescent anesthesia PAIN CONTROL Local anesthesia, in which large quantities of very dilute anesthetic with epinephrine are injected into the scalp, to cause it to balloon. See Local anesthesia

tumescent liposuction COSMETIC SURGERY A type of liposuction in which pressurized fluid–Klein/Hunstad formula is infiltrated into fat, distend-ing, anesthetizing and exsanguinating the region; the TT allows for almost bloodless and painless excision of excess fat during liposuction COMPLICATIONS Thrombosis, hypotension, perforation of organs by wand. See Liposuction

tummy tuck Abdominoplasty COSMETIC SURGERY A procedure in which a large horizontal ellipse of skin and fat is excised from the anterior lower abdomen, and an upper abdominal flap is stretched to the suprapubic inci-sion and sewn in place; the umbilicus is exteriorized through an incision in the flap at the proper level RAISON D'ETRE Vanity COST $4100 NUMBER PROCE-DURES 2000 58,000. See Cosmetic surgery

tumor 1. A neoplasm, benign or malignant. See Benign tumor, Blue cell tumor, Bone tumor, Borderline tumor, Brain tumor, Carcinoid tumor, Carcinoma ex-mixed tumor, Carotid body tumor, Desmoid tumor, Desmoplastic round cell tumor, Ewing fam-ily of tumors, Fibrous tumor of childhood, Gastrointestinal autonomic nerve tumor, Gastrointestinal stromal tumor, Giant cell tumor of tendon sheath, Glial tumor, Glomus tumor, Granular cell tumor, Heart tumor, Hilus tumor, Hürthle cell tumor, Internists' tumor, Intracranial tumor, Krükenberg tumor, Kulchitsky cell tumor, Leydig cell tumor, Malignant mesodermal tumor, Medullary tumor, Metastatic tumor, Metastatic brain tumor, Mixed tumor, Neuroectodermal tumor, Neuroendocrine tumor, Pancreatic endocrine tumor, Peripheral nerve sheath tumor, Phantom tumor, Pigmented neuroecto-dermal tumor of infancy, Potato tumor, Pott's puffy tumor, Primary tumor, Pseudotumor, Pseudotumor cerebri, Round cell tumor, Sclerosing stromal tumor, Secondary tumor, Sex cord-stromal tumor, Skin adnexal tumor, Small round cell tumor, Solid tumor, Solid tumor of childhood, Spaghetti tumor, Spontaneous tumor, Steroid cell tumor, Sugar tumor, Theca cell tumor, Triton tumor, Turban tumor, Warthin's tumor, Wilms tumor, Yolk sac tumor 2. A term of waning popularity for a local swelling

tumor-associated antigen IMMUNOLOGY, ONCOLOGY A molecule–eg, CA 15.3, 19-9, CA-125, that may be associated with specific tumors–eg, lymphomas, carcinomas, sarcomas, melanomas; TAAs may elic-it cellular and/or humoral immune responses against the tumor, but rarely defend the host against the tumor

TUMOR-ASSOCIATED ANTIGEN TYPES

CLASS 1 Highly specific for a particular tumor; they are present in one or only a few indi-viduals and not found in normal cells, eg tumor-specific transplantation antigen

CLASS 2 Present in a number of related tumors from different Pts

CLASS 3 Present on normal and malignant cells, but are expressed in ↑ amounts in malig-nant cells

Note: Class 2 antigens have the greatest potential for clinically useful assays, as they are present in many tumors and are rarely observed in normal subjects

tumor board Tumor conference ONCOLOGY A multidisciplinary conference held in larger US hospitals in which mechanisms, diagnostic modalities, and therapeutic options for Pts with CA are discussed. See MDT

tumor debulking SURGICAL ONCOLOGY The excision of as much of a CA as possible during a first procedure. See Second-look operation

tumor development A multistep process that occurs over yrs in which a tis-sue accumulates genetic hits that eventually translate into a neoplasm with metastatic potential. See One-hit, two-hit model

tumor implant ONCOLOGY A nodule of a malignant tumor that develops at a site in a body cavity at a distance from the initial site of cancer, a find-ing typical of ovarian CA

tumor lysis syndrome Acute tumor lysis syndrome A constellation of biochemical derangements seen early in chemotherapy–rapid cell death causes hyperkalemia, hyperphosphatemia, hyperuricemia, hypocalcemia; ATLS occurs in breast and small cell carcinoma of lungs, metastatic adeno CA, lymphosarcoma, NHL–eg, lymphoblastic and Burkitt's lymphomas, CML, ALL, myeloma, metastatic medulloblastoma

TUMOR LYSIS SYNDROME–METABOLIC DERANGEMENTS

HYPERURICEMIA Release of uric and nucleic acid precursors by massive cytolysis, aggravated by acid urine, renal dysfunction, dehydration and edema; may cause renal shutdown 2° to massive tubular precipitation of uric acid crystals

HYPERKALEMIA Release of intracellular K+, worsened by acidosis, renal dysfunction, potassium-sparing diuretics, adrenal insufficiency, DM, hypocalcemia or hyponatremia appearing as muscular weakness and cardiac arrhythmias, including AV block, tachyarrhythmia or cardiac arrest

HYPERPHOSPHATEMIA/HYPOCALCEMIA Minerals released by lysing cells cause soft tissue mineralization, exacerbating renal failure by intratubular crystallization, and may cause neuromuscular irritability, tetany, convulsions

tumor marker Biomarker ONCOLOGY Any substance present in ↑ amounts in serum that may be used for early CA detection; most TMs are nonspecific and little used to screen for CA; they may be used to detect recurrence–eg, CEA in Pts with known colorectal and other CAs EXAMPLES Oncofetal proteins–α-fetoprotein, CA-125–ovarian CA, CA-15-3–breast CA, CEA–ovarian, lung, breast, pancreas, GI tract CAs, PSA–prostate CA

tumor necrosis Death of tumor tissue, a common event in aggressive CAs in which the tumor rapidly outgrows its blood supply, resulting in tumor cell death. Cf Apoptosis

tumor necrosis factor Either of 2 molecules–TNF-α and TNF-β that mediate shock and tumor-related cachexia; TNF-α–aka cachexin production is ↑ in inflammation, sepsis, ↑ lipid and protein metabolism, hematopoiesis, angiogenesis, collagen vascular disease, terminal CHF–with activated renin-angiotensin system, in children with falciparum malaria, ↓ host resistance to viruses and parasites, in children with severe infectious purpura, CA SIDE EFFECTS Fever, headache, hypotension, cachexia, chills, fatigue, anorexia, thrombocytopenia; endotoxemia is a TNF-induced event with a shock-like clinical picture with interstitial pneumonia, vascular plugging by PMNs, acute tubular necrosis, GI tract, adrenal and pancreatic ischemia, hemorrhage LAB Hemoconcentration, metabolic acidosis, hypoglycemia and hyperkalemia; the levels of TNF-a in the CSF appear to correlate with the severity of multiple sclerosis, and may be of use to monitor response to therapy TOXICITY Fever, headache, hypotension, cachexia, chills, fatigue, anorexia, thrombocytopenia. See Biological response modifier

tumor promoter Cocarcinogen A substance, often lipid-soluble, that has no intrinsic carcinogenic potential, but which, when applied repeatedly, amplifies cancer-inducing effects of other (initiator) substances. See Antipromoter. Cf Tumor initiator

tumor seeding The spillage of tumor cell clusters and their subsequent growth as malignant implants at a site adjacent to an original tumor. See Tumor implant

tumor-to-tumor metastasis A rare event in which one CA metastasizes into another in a Pt with 2 or more CAs

tumoral calcinosis ONCOLOGY A condition characterized by single or multiple well-circumscribed calcium deposits in periarticular connective tissue, which may cause spinal cord compression and neurologic defects in mammals. See Calcinosis

tumorigenesis The process of tumor development. See Solid-state tumorigenesis

tumour Queen's English for tumor, see there

TUMT Transurethral microwave therapy, see there

TUNA Transurethral needle ablation of prostate UROLOGY A method for treating BPH, in which selective thermal energy is delivered to the prostate, while preserving the urethra and adjacent tissues CONTRAINDICATIONS Prostate CA, neurogenic bladder, acute UTI, active cholelithiasis, gross hematuria, urethral stricture, bladder neck contracture, active prostatitis, or DM affecting bladder function. See Benign prostatic hypertrophy

tunnel effect GI DISEASE Visualization of only the open end of the stainless steel tube in rigid sigmoidoscopy. See Sigmoidoscopy

tunnel infection INFECTIOUS DISEASE A catheter-related infection seen in tunneled central venous catheters CLINICAL Erythema, tenderness, induration of skin and subcutaneous tissue extending > 2 cm from the skin exit site. See Central venous catheter

tunnel vision NEUROLOGY A functional concentric constriction of both visual fields; the absolute size of the visual fields is the same regardless of distance from object being viewed ETIOLOGY Degeneration of calcarine cortex, seen in methylmercury poisoning, often accompanied by multiple scotomata, ↓ auditory acuity, changed mental status; infarction of sensory relay nuclei of the thalamus with relative sparing of the occipital lobes; functional cause of TV is hysterical reaction or hypersuggestibility. See Mercury. Cf 'Tunnel vision' VOX POPULI Popular for a myopic perspective adopted by a person, group, or organization

TURBO A clinical trial–The Ultrasound Removal of Blood Clots in Vein Grafts

TURBT Transurethral resection of bladder tumor

Turcot syndrome ONCOLOGY A syndrome characterized by multiple colon polyps and brain tumors–eg, ependymoma, glioblastoma, medulloblastoma

turf *noun* HOSPITAL PRACTICE A popular term for a 'power base' in a hospital, which may hinge on number of Pts and procedures under the control of a particular physician or department. See Turf war MEDICAL PRACTICE A popular term for a personal range or type of practice *verb* GRADUATE EDUCATION To place a Pt in another service's–eg, from a medical to a surgical–ward

turf toe SPORTS MEDICINE A ligamentous sprain of the metatarsophalangeal–MTP joint of the great toe affecting athletes in various sports–baseball, football, soccer played on artificial turf CLINICAL Commonly presents as an intensely painful, swollen, red, accompanied by acute flexor and extensor tendinitis TREATMENT Ice, NSAIDS, immobilization, rest, footwear designed for artificial turf, ↑ stiffness of shoe plus taping help to prevent hyperextension or hyperflexion of the MTP of great toe. See Sports dermatology

turgor PHYSICAL EXAMINATION Swelling of skin and subcutaneous tissue

turnaround time LAB MEDICINE A parameter of a clinical lab's efficiency, defined as the time between ordering a test or submitting a specimen to the lab etc. and reporting results. Cf Turnover time

Turner syndrome Gonadal dysgenesis, monosomy X, XO female GENETICS A condition characterized by an XO chromosome pattern in a phenotypic ♀ with pathognomonic Sx CLINICAL Absent ovaries–gonadal agenesis or dysgenesis, short stature, webbed neck, rarely, mental retardation VARIANTS A second X may be present, but partially deleted; in mosaics, some cells are 45,X and others 46,XX TREATMENT Female sex hormone at puberty to induce adult appearance and menses; most XOs are sterile

turnkey *adjective* Referring to a system–eg, information system or device–eg, PET-CT for which the time and effort required for it to become fully operational is minimal after installation and setup

TURP Transurethral resection of prostate, see there

TUSS A clinical trial–Tuberculosis Ultraviolet Shelter Study

Tussi-12™

Tussi-12™ Phenylephrine, chlorpheniramine, carbetapentane suspension POPULAR HEALTH An OTC agent for symptomatic relief of cough in common cold, asthma, bronchitis. See Common cold

Tutoplast® process SURGERY A tissue preservation and viral inactivation process for treating bioimplants used in neurosurgery, general, orthopedic, and reconstructive surgery

tutorial Dialogue EDUCATION An exercise which attempts to emulate a teacher/student interaction; tutorials typically arrange and present information in a meaningful fashion, foster interaction through appropriate objective and subjective questions, process various responses and provide feedback

TV test PSYCHOLOGICAL TESTING A simple test for detecting learning disorders–eg, Down syndrome, pure language disorders, and autism, which is based on the ability of a child to identify TV images of cats, dogs, or other babies

TVH Total vaginal hysterectomy. See Vaginal hysterectomy

TWA Time-weighted average, see there

TWAR agent MICROBIOLOGY A fastidious strain of *Chlamydia psittaci* that causes outbreaks of community-acquired pneumonia, and RTIs–the TWAR antibody occurs in 20-45% of normal adults, transmitted person-to-person and may be fatal in the elderly CLINICAL Pneumonia in young adults, pharyngitis, laryngitis TREATMENT Tetracycline

tweaking VOX POPULI Fine-tuning to produce optimal results

Tweedles disease Acute alveolitis linked to use of humidifier harboring bacterial pathogens

tweezers An instrument with pincers used to grasp or extract. See Optical tweezers

twelve lead EKG CARDIOLOGY An EKG generated from a 12-lead EKG–leads I, II, III, aVR, aVL, aVF, V1–V6. See Electrocardiogram

Twelve-Step Program ADDICTION DISORDERS Any program modeled after the 12-step self-help-group programs used for rehabilitating alcoholics, Alcoholics Anonymous; central to all 12-SPs is the belief in a God, transpersonal spiritual form of energy, or superhuman power

12 to 24 pH study See pH study

24 hour sleep-wake cycle SLEEP DISORDERS The clock hour relationships of the major sleep and wake phases in the 24 hr cycle. See Circadian rhythm, Phase transition

24-hour urine LAB MEDICINE Urine collected in an exact period of 24 hrs, which may be indicated for arsenic, chloride and others analytes; 24HUs are collected in a large, lab-supplied container with preservatives

24-hour virus VOX POPULI A popular term for a a gastroenteric or flu-like illness, presumed by its victim to be viral in nature, which generally resolves in about a day

23-hour stay MANAGED CARE A period of hospitalization of < 1 day in duration PROS 23HSs allow the management team to observe a Pt with signs of a condition–eg, stroke, AMI, hemorrhage that would require hospitalization for a prolonged period of time; because the Pt is admitted for < one day, all the services are billed at higher rates than would be allowed by the DRGs, were the Pt hospitalized. See TQM

twilight sleep ANESTHESIOLOGY A dream-like state of 'conscious sedation' induced by Versed, an agent used for minimally invasive surgery–eg, colonoscopy or minor oral procedures, without general anesthesia; Versed is associated with sudden deaths, possibly related to 'overshooting' therapeutic levels. See Versed. Cf Dauerschlaff, Continuous sleep therapy

twin One of 2 gestational products that develop during a single intrauterine gestational period. See Dizygotic twin, Monozygous twin, Higher multiples, Partial twin, Siamese twin, Vanishing twin. Cf Siamese twins

twin study EPIDEMIOLOGY A study using twins to determine the interplay of nature–genetic and nurture–environment and disease and its treatment; the interpretation of twin study data is often easier as there are exact genetic matches–ie, identical–monozygotic and partial genetic matches–ie, fraternal–dizygotic twins which are used as controls

twin-to-twin transfusion syndrome Fetal transfusion syndrome, fetofetal transfusion syndrome, stuck twin syndrome OBSTETRICS Intrauterine growth retardation in 1 twin due to an artery-to-artery vascular shunting, which may occur in a diamnionic-dichorionic placenta and be accompanied by hydramnios; the 'donor' twin is anemic, pale, lighter, smaller with organ hypoplasia; the recipient twin is plethoric, polycythemic, macrosomic, has ↑ vascular resistance and may develop heart failure. Cf Twin arterial reversed perfusion sequence

Twinkie® defense FORENSIC PSYCHIATRY A legal tack in which a defendant claims that a criminal act resulted from chemical imbalances induced by 'junk food,' and not criminal intent. Note: when the defense was first used, the jury, not surprisingly in California, "bought it" and reduced the conviction. See Black rage defense, Insanity defense, Television intoxication syndrome

twinkling star sign IMAGING A fanciful descriptor for the short linear radiations that extend from pulmonary vessels and parallel the long axis of the body on CT scans, most prominently seen at a 100-200HU window width; 'twinkling stars' help distinguish normal vessels from tumor masses, which may have the same CT density

TWiST Time without Sx of disease or toxicity AIDS A quality of life indicator, corresponding to the number of months preceding the development of a Sx with ≥ grade 3 HIV disease, whichever occurred first. See Q-TWiST

twisted ovarian cyst GYNECOLOGY A mucinous cyst or dermoid of the ovary that rotates around its vascular pedicle, compromising the blood supply and erupting into clinical disease manifest by generalized hypogastric tenderness, of often abrupt onset, abdominal colic and distension, fever, possibly a palpable mass DIAGNOSIS Ultrasound MANAGEMENT Resection

twitch Body twitch SLEEP DISORDERS A small body movement–eg, a facial grimace or finger jerk, not usually associated with arousal. See REM twitch

two b/three a inhibitor See GpIIb/IIIa inhibitor

two-dimensional echocardiography Cross-sectional echocardiography CARDIOLOGY A common ultrasound-based diagnostic method in cardiology, which provides high-resolution, 'real time' images of the heart and great vessels; it is the noninvasive method of choice for diagnosing and managing congenital, pericardial, myocardial, cardiac valve disease; it is used to evaluate end-diastolic intraventricular dimensions, septal and free wall thickness and depressed ventricular function. See Transesophageal echocardiography

two-parent advantage PSYCHOLOGY A psychosocial 'edge' that a child from a 2-parent household has when compared to children raised in a single-parent household. See Father factor, Nuclear family

2 plus 2 curriculum GRADUATE EDUCATION The traditonal format for medical school education in which the first 2 yrs of medical school are dedicated to learning basic sciences–eg, anatomy, physiology, pathology, and the last 2 yrs to clinical sciences–eg, surgery, pediatrics, gynecology

two-point discrimination NEUROLOGY The ability to discriminate 1 stimulus from 2 stimuli, which may be compromised in hand injuries

two-tiered system SOCIAL MEDICINE The existence of 2 levels of health benefits and care, depending on whether the Pt can afford to pay or not

TX 1. Thromboxane 2. Treatment 3. Traction

Tylciprine® Tranylcypromine, see there

Tylenol® Acetaminophen, see there

tympanic membrane perforation Perforated, punctured, ruptured ear drum ENT A disruption of the tympanic membrane due to acoustic trauma, direct injury, barotrauma, introduction of Q-tips or small objects, or infection with fluid buildup in the middle ear. See Tympanoplasty

tympanoplasty ENT A technique of middle ear reconstruction intended to restore hearing, which consists of 2 components 1. Tympanic membrane engraftment, using various materials including canal skin, fascia, and homografts–eg, dura, periosteum, knee cartilage, ossicles and ossicular replacement with hydroxyapatite prostheses 2. Ossicular chain reconstruction, see there

tympanoplasty failure ENT An unsuccessful tympanoplasty due to recurrent perforation, persistent drainage, conductive hearing loss, malleus handle pull-away, anterior sulcus blunting, cholesteatoma and graft material breakdown; most 'problem ears' can be repaired or prevented, requiring removal of offending residual mastoid and middle ear disease and reconstruction

tympanostomy ENT A surgical procedure in which small tube–popularly, a grommet–is inserted in the tympanic membrane; tympanostomy follows myringotomy–an incision in the in tympanic membrane for draining middle ear effusions typical of chronic otitis media; tympanostomy offers more disease-free time and improved hearing, at a 'cost' of otorrhea and persistent perforation of the tympanic membrane

tympanostomy tube OTORHINOLARYNGOLOGY A small metal or Teflon tube or grommet inserted in the tympanic membrane of Pts–often young children with recurring, antibiotic-refractory otitis media; while the ventilating tube is in place, there is a significant ↓ in episodes of otitis media; once removed, the treated ear has worse auditory discrimination, more otitis, tympanosclerosis, retraction and atrophy

Tyndall effect DERMATOLOGY The change that light undergoes as it passes through a turbid medium–eg, skin, causing the colors of the spectrum to scatter; colors with a longer wavelength–red, orange and yellow tend to continue traveling forward while those with a shorter wavelength–blue, indigo and violet scatter to the side and backward; the TE explains why a subcutaneous lesion, which should have a red-brown hue due to hemorrhage or melanin deposition, has a blue tinge

type A simple way of classifying practically anything is to divide it into 2 or more 'types'; in general, typing keeps the number of subgroups to a minimum, while satisfying those with obsessive-compulsive neuroses who are driven to classify diseases, objects, people and mechanisms

type A personality PSYCHOLOGY A relatively distinct set of character traits, commonly observed in aggressive, hard-driving, 'workaholics'; a temperament characterized by excessive drive, competitiveness, a sense of time urgency, impatience, unrealistic ambition, and need for control. See 'Toxic core'; Cf Type B personality, Type C personality, Type D personality

type B personality PSYCHOLOGY A personality typical of those persons who tend to be relaxed and inclined to do things 'mañana'; a temperament characterized by an easy-going demeanor; less time-bound and competitive than the type A personality. Cf Type A personality

type 1 diabetes mellitus Brittle DM, insulin-dependent DM, juvenile-onset DM ENDOCRINOLOGY A severe form of DM caused by ↓ endogenous insulin production by the pancreas, which comprises ± 10% of DM CLINICAL Extreme hyperglycemia, lability of glucose control and ketosis; type 1 DM of recent onset may have IgG autoantibodies against glucose transport proteins. See Honeymoon period. See Diabetes mellitus. Cf type 2 DM

type 2 diabetes mellitus Adult-onset diabetes, diabetes mellitus type 2, NIDDM, non-insulin-dependent diabetes mellitus ENDOCRINOLOGY A mild form of DM with an onset > age 40, ↓ incidence of DKA, accompanied by microvascular complications, which comprises 90% of DM; 80% of type 2 DM Pts are obese–an association known as 'diabesity', insulin-deficient, insulin-resistant DIAGNOSIS 1. Fasting glucose is 7.8 mmol/L–US > 140 mg/dL on ≥ 2 occasions or 2. When in a 75g GTT, the 2-hr and one other value–drawn at the 30, 60, or 90 min intervals are > 11.2 mmol/L–US > 200 mg/dL CLINICAL Blurred vision, poorly healing cuts, paresthesias in hands/feet, recurring skin, mouth, or bladder infections, any type 1 Sx–thirst, ↑ appetite, rapid weight loss, fatigue TREATMENT type 2 DM does not usually require exogenous insulin; insulin may be required during 'crises' PROGNOSIS Relatively good, especially if controllable by lifestyle modifications. See Glucose tolerance curve, MODY. Cf type 1 DM

type I error α error, false-positive error STATISTICS Rejection of the null hypothesis when it is true, or the error of falsely stating that 2 proportions are significantly different when they are the same

type II error β error, false-negative error STATISTICS Acceptance of the null hypothesis when it is false or incorrect, or the error of falsely stating that two proportions are not significantly different when they actually are. See Null hypothesis Cf Type 1 error

type & screen Group & screen TRANSFUSION MEDICINE A protocol used to determine the ABO and Rh groups of RBCs, and screen serum for the presence of potentially hemolyzing antibodies. See Cross-match, Immediate spin cross-match, Major cross-match

type T personality PSYCHOLOGY A personality type that takes risks; type Ts tend to be extroverted and creative, and crave novel experiences and excitement. See Extreme sports, Novelty seeking behavior

typhoid fever Enteric fever INFECTIOUS DISEASE A bacterial infection characterized by diarrhea, systemic disease, rash; most commonly caused by *Salmonella typhi* EPIDEMIOLOGY Oral, via contaminated food, drink, or water; rare in developed countries, ±16 million non-US new cases of TF/yr with 600 K deaths; after ingestion, *S typhi* spreads from the intestine to regional lymph nodes, liver, and spleen via the blood where they multiply, and may directly infect the gallbladder via the hepatic duct or spread to other areas of the body through the bloodstream CLINICAL Early–fever, malaise, abdominal pain, followed by higher fever, ≥ 40°C, weakness, fatigue, delirium, confusion, characteristic–"rose spots," which are small–2 mm, dark red, flat on the abdomen and chest; children usually have milder disease COMPLICATIONS GI bleeding, intestinal perforation, typhoid encephalopathy; some Pts become carriers of typhoid and shed bacteria in feces for yrs MANAGEMENT Chloramphenicol. See Typhoid Mary

Typhoid Mary EPIDEMIOLOGY A popular term or a human carrier of a virulent strain of bacteria–eg, salmonella, diphtheria or virus–eg, hepatitis or HIV-1. See High disseminator

typhus Brill-Zinsser disease; epidemic typhus; murine typhus, typhus fever INFECTIOUS DISEASE A rickettsiosis transmitted by a louse or a flea CLINICAL Fever, a transient rash, and hypotension caused by either *R prowazekii*–epidemic typhus, Brill disease–rare in US or *R typhi*–murine typhus, a mild summer-fall infection most common in southeastern and southern US; < 2% mortality. See Epidemic typhus, Queensland tick typhus, *Rickettsia* São Paulo typhus, Scrub typhus

typing MEDTALK The process of determining a particular type. See Back typing, Front typing, Immunotyping, Lancefield typing, Lymphocyte typing, Metabolic typing, Phage typing, Ribotyping

tyramine hypertension Cheese disease NUTRITION A complication of MAOI therapy, used to treat depression and panic disorders; MAOIs inhibit the metabolism of tyramines and catecholamines; ingestion of tyramine-rich food and/or beverages–eg, Chianti wine, cheddar cheese, beer, chicken liver or drugs–eg, ephedrine, amphetamines, evokes an acute hypertensive crisis due to the release of tissue catecholamines, which may be accompanied by sweating, tachycardia, or arrhythmias

tyrosine kinase An enzyme intimately linked to signal transduction–ST, either as a receptor-type TK, which participates in transmembrane signaling, or as an intracellular TK, participating in ST to the nucleus; ↑ or ↓ TK activity is associated with various diseases, and alteration of TK activity at various points in its signaling pathway is of potential therapeutic interest; ↑ TK activity implicated in many CAs and other malignant and nonmalignant diseases, ASHD, psoriasis, inflammation; ↓ TK activity is linked to DM, X-linked agammaglobulinemia

tyrosinemia type I METABOLIC DISEASE A rare AR condition due to fumarylacetoacetate hydrolase CLINICAL Progressive liver dysfunction, cirrhosis, and hepatocellular carcinoma of childhood onset, renal tubular damage, acute porphyria-like neurologic crises DIAGNOSIS ↑ Succinylacetone in prenatal testing, allele-specific oligonucleotide hybridization MANAGEMENT Liver transplant, NTBC– may help 'clear' toxic metabolites

characteristic deformity of the lumbar vertebrae in thanatophoric dysplasia. **713**
See Cloverleaf skull, 'Telephone receiver' deformity

U wave CARDIOLOGY An EKG wave that follows the T wave; it is upright and has 5% to 50% of the T wave amplitude; it is most prominent in V_2 and V_3, where it may reach 0.2 mV; the genesis of the U wave is uncertain, but has been attributed to either Purkinje fiber repolarization or to ventricular relaxation; the U wave fuses with the T wave in the face of ↓ ventricular filling and ejection, or when the Q-T interval is prolonged as in hypocalcemia or with quinidine therapy; the U wave is exaggerated in hypokalemia, digitalis therapy or with some antiarrhythmics; inverted in 1% of the normal population, or in Pts with HTN, aortic or mitral valve disease, right ventricular hypertrophy, and myocardial ischemia

UGH syndrome See PUGH 'syndrome'

ugliness, fear of PSYCHOLOGY Cacophobia. See Phobia

ugly A popular adjective for an aggressive disease or lesion INFECTIOUS DISEASE An 'ugly' infection responds poorly to aggressive antibiotic therapy, which most commonly occurs in immunocompromised Pts or those with terminal CA ONCOLOGY An 'ugly' malignancy is characterized by fulminant deterioration, multiorgan failure, infection, poor response to therapy SURGICAL PATHOLOGY 'Ugly' usually refers to a malignant lesion with bizarre, poorly differentiated cells characterized by marked nuclear atypia, pleomorphism and florid mitotic activity–eg, giant cell tumor of the lung. See Anaplasia

UGP Urinary gonadotropin fragment, see there

UICC Union International Contre le Cancer International Union against Cancer

UIP Usual interstitial pneumonia, see there

UK-HEART CARDIOLOGY A clinical trial–United Kingdom heart failure evaluation and assessment of risk trial

ulcer DERMATOLOGY A defect in a mucocutaneous surface. See Bairnesdale ulcer, Buruli ulcer, Esophageal ulcer, Kissing ulcer, Pressure ulcer, Rodent ulcer ENT Mouth ulcer, see there. See Aphthous ulcer GI DISEASE Duodenal ulcer, see there. See Cushing's ulcer, Dieulafoy ulcer, Peptic ulcer, Stercoral ulcer, Stress ulcer OPHTHALMOLOGY A defect on the epithelium of the eye. See Corneal ulcer, Corneal neurotrophic ulcer, Geographic ulcer, Peptic ulcer, Serpiginous ulcer

ulcerating tumor A malignancy which spreads along mucosal surfaces, causing ulceration

ulcerative colitis GASTROENTEROLOGY A chronic inflammatory bowel disease that primarily affects the colorectum, characterized by bloody diarrhea and abdominal pain EPIDEMIOLOGY High incidence in US, UK, Australia, Northern Europe, 3-15/10⁵/yr in whites; Jews, 13/10⁵; non-Jews, 3.8/10⁵; age of onset 20-40; smaller peak in elderly CLINICAL Diarrhea, intermittent rectal bleeding, passage of mucus, abdominal colic; if severe, fever, tachycardia, anemia, ↑ ESR ASSOCIATED LESIONS Inflammation in joints, spine, skin, eyes, liver, bile ducts DIAGNOSIS Colonoscopy, radiology–barium enema–distention and dilation, collar-stud ulcers; if long-standing, loss of haustration MANAGEMENT IV corticosteroids; in non-responders, sulfasalazine, mesalazine, IV cyclosporine; if refractory to cyclosporine, colectomy; transdermal nicotine may induce complete remission in acute UC, improving the clinical status, as well as histologic grade, ↓ the stool frequency and sense of urgency, at a cost of nausea, lightheadedness, headache and sleep disturbances; nicotine is of no use in maintaining remission

ulcerative stomatitis 1. Mouth ulcer,s, see there 2. Herpetic stomatitis, see there

ulnar nerve dysfunction Ulnar nerve neuropathy NEUROLOGY A peripheral mononeuropathy characterized by impaired movement or sensation in the

U Symbol for 1. International Unit of enzyme activity 2. Uracil, see there 3. Uranium, see there

u Atomic mass unit

U90 Delavirdine, see there

UA Unstable angina, see there

UAC Umbilical artery catheter

UB-92 HCFA 1450, Uniform/Universal Billing form 92 MANAGED CARE The official HCFA/CMS form used by hospitals and health care centers when submitting bills to Medicare and 3ʳᵈ-party payors for reimbursement for health services provided to Pts covered. See Compliance. Cf HCFA 1500

UBF Uterine blood flow

UBT™ breath test GI DISEASE A noninvasive breath test for detecting *H pylori*. See *Helicobacter pylori*

U curve phenomenon U shaped curve EPIDEMIOLOGY A curve that describes the relationship between alcohol intake and the risk of MI, when viewed vis-á-vis CAD-related mortality; the **U** curve phenomenon also occurs in weight loss/gain, where those subjects with the most stable weight have the lowest mortality for all–nonmalignant causes and for CAD; the mortality is said to ↑ in weight cycling. Cf J curve phenomenon

uE3 Unconjugated estriol. See Triple-marker screen

U-P junction obstruction Ureteropelvic junction obstruction, see there

U-shaped scar A deep fibrotic depression on the cortical surface of the kidney, which is associated with ASHD and chronic pyelonephritis

U-shaped vertebra IMAGING An inverted U-shaped vertebra–on an AP film, with marked flattening–and central narrowing–lateral film, which is a

wrist and hand due to ulnar nerve damage ETIOLOGY Trauma, prolonged external pressure on nerve, compression from adjacent structures, entrapment

ulnar tunnel syndrome ORTHOPEDICS Any of a group of clinical complexes caused by ulnar nerve compression, which is most often due to a ganglion 'cyst,' but which may also be occupation- or hobby-related or due to laceration, arteritis, fractures and inflammation CLINICAL Sx differ according to site of nerve compression, and include varying degrees of motor defects and sensory loss of the medial aspect of hand TREATMENT Physical therapy, occasionally surgery. Cf Carpal tunnel syndrome, Tarsal tunnel

ultimate fighting SPORTS MEDICINE A modern blood sport, in which 2 combatants battle each other without rounds or rest periods, to the finish, be it death, incapacitation, or surrender, in which one opponent is battered into submission, and signals abdication by a so-called 'tap-out' signal to the referee INJURIES Brain concussion and/or hemorrhage, skin and scalp lacerations, bone fractures, and death from hypoxia related to choke holds or laryngeal fracture, damage to the spinal cord or brainstem, eye damage or blindness. See Blood sport. Cf Boxing, Extreme fighting, Toughman fighting

Ultimate Xphoria SUBSTANCE ABUSE A proprietary form of ma huang which, given its high content of ephedrine-like alkaloids, may be self-administered to produce euphoria and, if taken in excess, cause death. See Herbal medicine, Natural drug, Rave party. Cf Peyote

ultradian rhythm PHYSIOLOGY A biological rhythm with an ultrashort period and much higher frequency—eg, heartbeat, breath. See Biologic rhythm, Circadian rhythm. Cf Infradian rhythm

ultraertia SEXOLOGY Excess intensity or multiple responses or arousals at the proceptive phase, which may continue into the acceptive phase. Cf Inertia

ultrafast CT Imatron™ IMAGING A CT imager that captures images in microsecs vs secs for conventional CT; the ↑ in speed eliminates the artifacts created by organ movement, making UCT ideal for evaluating coronary artery disease and other forms of heart disease. See Computer tomography

ultrafast Papanicolaou stain CYTOLOGY A stain for cytology specimens which improves the quality of preparation, and facilitates interpretation of specimens obtained by fine-needle aspiration

Ultraject® IMAGING A contrast medium in a syringe prefilled for accurate dosing and protection against accidental needlesticks. See Contrast medium

ultramarathon SPORTS MEDICINE A footrace that is longer—eg, > 50 miles/80 km—than a marathon—26.2 miles/42 km. See Marathon

ultrarapid detoxification Ultra-rapid opiate detoxification procedure SUBSTANCE ABUSE An expensive and seemingly effective method for effecting withdrawal from abuse substances, in which a drug addict is given an opiate antagonist, naltrexone or naloxone, which displaces opiates—heroin, cocaine, etc, while under general anesthesia, and thus does not suffer from the 'cold turkey' Sx of drug withdrawal. See Withdrawal syndrome

ultrasensitive C-reactive protein LAB MEDICINE A C-reactive protein which is more sensitive for acute MIs. See C-reactive protein

ultrasonic *adjective* Referring to an electromagnetic frequency faster than sound waves—eg, ≥ 30,000 Hz

ultrasonic osteogenic stimulation ORTHOPEDICS The use of a non-invasive ultrasonic device which emits low-intensity ultrasonic pulses across the skin to augment bone repair INDICATIONS Cast immobilization. See Osteogenic stimulation

ultrasonic pulse A mechanical reverberation of the transducer in a pulse-echo sonographic device after electrical stimulation. See Axial resolution

ultrasonography Ultrasound IMAGING The generation of diagnostic images—sonograms—based on differences in the acoustic impedance of tissues. See Aortic ultrasonography—US, Breast US, Carotid US, Doppler US, Duplex Doppler US, Endoscopic US, Hydrocolonic US, HIFU, Ocular US, Pancreatic US, Pelvic US, Prostate US, Transcranial US OBSTETRICS A noninvasive technique for visualizing the gestational sac or fetus in utero

ULTRASONOGRAPHY TYPES

A-MODE ULTRASONOGRAPHY An ultrasonographic modality that provides simple displays that are plotted as a series of peaks, the height of which represents the depth of the echoing structure from the transducer

B-MODE ULTRASONOGRAPHY Brightness-modulated display Ultrasonography with a wide range of applications including imaging of a fetus, kidneys, liver, gallbladder, uterus, cardiovascular structures, breast, prostate, early ovarian CA, liver transplant recipients preoperatively—a narrow or thrombosed portal vein precludes transplant, *and* postoperatively to assess various complications—rejection, infection, thrombosis and patency of biliary tracts and identifying gallbladder calculi; the most common clinical use of BMU is to evaluate fetal status, providing real-time 2-D evaluation of the fetus, presenting the images in rapid succession on a monitor, and likened to a motion picture; the 'biophysical profile' has a B-mode display, and measures the head—cephalometry, thorax, abdomen, estimates fetal maturation and identifies growth retardation and major congenital anomalies, including anencephaly, hydrocephaly, meningocele, congenital heart disease, dextrocardia, fetal tumors, diaphragmatic hernia, gastroschisis, omphalocele, polycystic kidneys, hydrops fetalis, GI obstruction and death; BMU helps localize the amniocentesis needle and is of use in identifying placental anomalies including hydatidiform mole or anomalous implantation, eg placenta previa. SIDE EFFECTS Minimal—the energy levels for diagnostic imaging are too low to produce tissue destruction; WBCs subjected to ultrasound may mutate, ? significance

DUPLEX ULTRASONOGRAPHY Ultrasonography that combines the standard real-time B-mode display with pulsed Doppler signals, allowing analysis of frequency shifts in an ultrasonographic signal, reflecting motion within a tissue, eg blood flow; DU is thus useful in evaluating ASHD of the carotid arteries, AV malformations and circulatory disturbances in the neonatal brain

M-MODE ULTRASONOGRAPHY Time-motion display A modality in which the echo signal is recorded on a continuously moving strip of paper, with the transducer is held in a fixed position over the aortic or mitral valves; each dot corresponding to a moving structure has a sinewy path, while stationary structures are represented as straight lines; M-mode was the first display used and continues to be useful for precise timing of cardiac valve opening and correlating valve motion with EKG, phonocardiography and Doppler echocardiography

ultrasonography angiographic catheterization A technique that uses a unique probe—1 mm diameter, 20-30 MHz energy emission v. the 10 cm diameter probe operating at 5 MHz used for fetal ultrasonography to analyze the intensity of atherosclerotic plaque deposition

ultrasound densitometry IMAGING The use of ultrasound to assess the mineral content of bone of peripheral sites; although it is less accurate than X-ray absorptiometry, it may provide useful information. See Osteoporosis

ultrasound therapy MAINSTREAM MEDICINE The application of ultrasound waves to soft tissue to heat and relax injured tissue and disperse edema

ultrastructural *adjective* Referring to structures of a size below the resolution of conventional LM, usually electron microscopy

ultraviolet blood irradiation TRANSFUSION MEDICINE A technique for ↓ blood-borne infections by zapping packed RBCs with UV light. See Blood filter, Leukoreduction

ultraviolet germicidal irradiation PUBLIC HEALTH The use of UV light to kill *Mycobacterium* spp contained in droplet nuclei

ultraviolet light Ultraviolet radiation PHYSICS The part of the invisible electromagnetic spectrum below violet, including photons emitted during electronic transition states WAVELENGTH 100-400 nm PUBLIC HEALTH The electromagnetic spectrum between 200 nm and 400 nm which is emitted by the sun. See SPF, Sunscreen

UV LIGHT TYPES

UV-A Long wave, 'near' UV, black light; short wave UV-A–320–340 nm; long wave UV-A–340–400 nm; no UV-A absorbed by ozone layer

UV-B Middle UV, sunburn radiation UV-B–290–320 nm; most UV-B is absorbed by ozone layer

UV-C Short wave, far UV, germicidal radiation, ≤ 290 nm; absorbed by ozone layer; has no role in photobiology of natural sunlight

Ulysses syndrome Ulysses sequence DECISION-MAKING A complication of false-positive diagnostic tests or clinical observations that trigger a complete and aggressive diagnostic work-up to elucidate the nature of what is, in fact, a non-disease, before the Pt can return to an original state of health

ULYSSES SYNDROME–TRIGGERING EVENTS

MISCHIEVOUS/UNNECESSARY INVESTIGATION That which is motivated by mass screening, eg 'blanket coverage' to pay for testing by an insurance company, house-staff 'overkill' to avoid criticism, requisition forms which have the laboratory's entire menu

UNCRITICAL EXAMINATION Lack of familiarity with a body region may mislead the examiner, especially if he encounters trivial anatomic variations of normal structures

SERPENTINO 'COMPLEX' Two snakes consuming each other tail first. A neurotic Pt may succeed in making himself ill when there is unexpected interest in an otherwise trivial complaint

INVERTED SERENDIPITY While Marie Curie's serendipitous dropping of a key on a pile of photographic film near radium was the founding event of radiology, 'discoveries' made while using an unfamiliar technique are usually 'red herrings'

NON-INVESTIGATIONAL INVESTIGATION When a laboratory request form has a new test on it, the new box is checked off with disproportionate frequency

A review of statistical principles in lab medicine makes it surprising that the Ulysses syndrome doesn't occur *more* often, since results of certain lab tests are placed on a standard Gaussian curve of distribution and any value > 2 standard deviations–SD above or below a mean is considered statistically abnormal (not biologically abnormal); this verification process is a function of daily fluctuations of machinery and other non-disease factors; thus 5%, ie, 1 in 20 of any normal population will be > 2 SD from the mean of a value, and therefore, abnormal; 1 in 400 normal subjects will be statistically abnormal in 2 tests and so on

Ulysses, who fought in the Trojan war, required 20 years for the return leg of the journey; all of the harrowing detours were unnecessary

umbilical artery catheter NEONATOLOGY A catheter inserted in the neonate's umbilical artery and used to monitor bp, draw blood, administer fluids

umbilical cord blood TRANSPLANTATION A source of primitive and stem cells that can be used to reconstitute BM destroyed by aplastic anemia or by RT or chemotherapy for CA, lymphoproliferative malignancies. See Bone marrow transplantation, Stem cell therapy

umbilical hernia PEDIATRIC SURGERY A painless protrusion of intraabdominal structure(s) through the umbilical ring–through which umbilical blood vessels pass in a developing fetus–due to incomplete closure. See Hernia

umbilical vein catheter NEONATOLOGY A support line for drawing and administering fluids which is placed in an infant in the first few hrs of life and can be left in place for up to 14 days with few complications INDICATIONS Resuscitation–eg, epinephrine, vasoactive agents, hypertonic fluids, parenteral nutrition, blood and blood products, exchange transfusion, measurement of central venous pressure CONTRAINDICATIONS Omphalitis, omphalocele, necrotizing enterocolitis, peritonitis COMPLICATIONS Sepsis–occurs in up to 16%, air embolism, thrombosis, catheter malposition, causing pericardial effusion or tamponade, arrhythmia, hydrothorax, liver necrosis, portal vein thrombosis, extravasation into the liver, sepsis. See Umbilical artery catheter

umbrella Mobin-Uddin umbrella HEART SURGERY A stainless steel sieve placed IV below the renal arteries via the jugular vein, designed to trap deep vein thrombi in the inferior vena cava. See Deep vein thrombosis, Pulmonary thromboembolism

UNASEM CARDIOLOGY A clinical trial–Unstable Angina Study using Eminase–that evaluated the outcome of thrombolytic therapy in Pts with unstable angina. See Acute myocardial infarction, tPA

unattached patient PATIENT CARE A Pt not known to have a regular attending physician when registered in an ER or other 'neutral' health care setting. See Flipping. Cf Attached patient

unbound drug Free drug THERAPEUTICS The fraction of drug in serum that is not bound to a carrier protein or other molecule, which generally is pharmacologically active

unbundling Exploding MEDICARE REIMBURSEMENT A fraudulent practice in which provider services–eg, blood or chemistry panels are broken down to their individual components, resulting in a higher payment by Medicare. See Fraud, HIPA Act, Rebundling, Upcoding

uncal herniation Brain herniation, herniation syndrome, transtentorial herniation NEUROLOGY A condition in which the uncus–the anterior hooked end of the parahippocampal gyrus on the basomedial surface of the temporal lobe of the brain is displaced due to ↑ intracranial causing progressive brain and brainstem damage ETIOLOGY Cerebral edema due to head injury, space-occupying lesions–primary and metastatic brain tumors or other lesions within the skull, bacterial meningitis; UH compresses cranial nerve III, the midbrain, and the posterior cerebral artery, leading to coma and respiratory arrest. Cf Cerebellar herniation

uncircumcised UROLOGY Referring to a ♂ or penis which has not been circumcised. See Circumcision

uncompensated care Medical treatment of a Pt provided in the US by a physician or other health care professional that is not paid by the Pt, the government, or an insurance carrier TYPES Charity care, bad debt, discounted Medicaid care. See Medicaid, 'Service' patient. Cf Pro bono

unconscious *adjective* 1. Not conscious, referring to a reflex movement 2. The psychic structure(s), per the psychoanalytic construct, of which a person is unaware PSYCHIATRY That part of the mind or mental functioning of which the content is only rarely subject to awareness; it is a repository for data that have never been conscious–primary repression or that may have been conscious briefly and later repressed–secondary repression

uncontrolled pain NEUROLOGY Pain that doesn't respond to medications at doses that usually provide appropriate analgesia

unconventional 'virus' An infectious agent, formerly regarded as slow viruses, now known as prions, a small infectious particle, consisting entirely of subverted cell protein; prions cause degenerative encephalopathies in sheep and goats–scrapie, cows–bovine spongiform encephalopathy, humans–kuru, Creutzfeldt-Jakob disease. See Kuru, Prions, Slow viruses

'under general' A popular term for a surgical procedure performed under general anesthesia

under the influence VOX POPULI A popular phrase, the long form of which is 'under the influence of alcohol' or, less commonly, other substances

'under local' A popular term for a surgical procedure performed under local anesthesia

undercall Underdiagnosis, underread DECISON-MAKING A noun and a verb for an error in which a benign diagnosis was rendered on a lesion that later proves malignant. See Misadventure, Overread

underdrive pacing CARDIAC PACING Pacing at a rate below the tachycardia rate, for the purpose of interrupting the heart's tachy circuit with randomly timed stimuli to gain control of the heart and restore natural rhythm. See Pacemaker

undergraduate education MEDTALK In the US, a 4+ yr college or university education leading to a baccalaureate degree, the minimum education level required for medical school admission; undergraduate medical education refers to the 4 yrs of medical school. Cf CME

underimmunization

underimmunization PUBLIC HEALTH A level of immunization which is suboptimal for a person or population CAUSES Financial factors, mother's age, race, education, socioeconomic status, birth order, prenatal care. See Racial inequality

underinsurance MANAGED CARE A generic term for insurance policies that require large out-of-pocket payments, and provide suboptimal coverage for common conditions EXAMPLES Lack of coverage for catastrophic medical expenses, pre-exisiting condition clauses, benefits-not-provided clauses, deductibles, coinsurance. See Underinsured

underlying cause of death Primary cause of death, proximate cause of death FORENSIC MEDICINE '…*the disease or injury that initiated the train of morbid events leading directly to death, or the circumstances or violence that produced the fatal injury.*' Cf Immediate cause of death

underrepresented minority SOCIAL MEDICINE Any ethnic group–African American, Hispanic, Native American–whose representation among professionals in biomedical sciences is disproportionately less than their proportion in the general population. See Affirmative action, *Bakke* decision, Project 3000 by 2000, Reverse discrimination

underride crash PUBLIC HEALTH An MVA in which a car slides under the container of an 18-wheeler/tractor trailer, often resulting in serious injury or death by decapitation of the driver and front passengers. See Motor vehicle accident

undersensing CARDIOLOGY Failure of a pacemaker to sense the P or R waves, which may cause the pacemaker to emit inappropriately timed impulses. See Pacemaker

underuse HEALTH CARE The failure to provide a medical intervention when it is likely to produce a favorable outcome for a Pt–eg, failure to give influenza vaccine to an elderly Pt with DM. Cf Misuse, Overuse

underwriting HEALTH INSURANCE The process of determining a person's insurability in terms of life, liability, home, and other insurance policies and whether it will accept an application for insurance. See Field underwriting, Medical malpractice underwriting

undescended testicle Cryptorchidism, see there

undifferentiated carcinoma An epithelial malignancy that lacks morphologic–by LM or functional indicators of its embryonic origin, which may arise in the lungs, larynx, bronchus, esophagus, colon, urinary bladder, uterine cervix, salivary glands EXAMPLES Anaplastic carcinoma, 'monstrocellular' carcinoma, oat/small cell carcinoma. See Anaplastic carcinoma

undifferentiated connective tissue disease An early stage of a connective tissue disease, in which the predominant organ of involvement or clinical form is not yet manifest. Cf Overlap syndrome, Palindromic rheumatism

undue hardship SOCIAL MEDICINE A term used in the context of the ADA, in which an employer may claim that the accommodations required to comply with the ADA are financially unviable and represent an undue hardship. See Americans with Disabilities Act, Environmental tobacco smoke. Cf Universal design

unethical medical research BIOMEDICAL ETHICS The performing of medical experiments on human subjects against their will or knowledge of therapeutic options. See Helsinki Declaration, Nazi 'science,' Nuremberg Code of Ethics

unexplained cause of infertility GYNECOLOGY Infertility for which no cause has been determined despite a comprehensive evaluation. See infertility

'unhappy gut' Functional colitis A term introduced because 'irritable gut' had lost its specificity; 'UG' refers to dysfunctional GI smooth muscle that is not 'in the mood' to function properly, while 'irritable gut' refers to colonic changes attributed to unidentified intraluminal irritants. See Inflammatory bowel disease, Irritable bowel syndrome.

unhealthful day PUBLIC HEALTH Any day in which the air quality for any of the 5 ambient air pollutants–carbon monoxide, nitrogen dioxide, ozone, particulate matter, sulfur dioxide exceeds an established air quality standard. See Pollution Standards Index. Cf Unhealthy day

unhealthy day SOCIAL MEDICINE A day in which a person experiences problems with physical–illness or injury, or mental–depression, stress, emotional illness–health. Cf Unhealthful day

unhealthy food Any food that is not regarded as being conducive to maintaining health; UFs include fats, in particular of animal origin, 'fast' foods–low in fiber and vitamins; 'junk food'–eg, potato and corn chips, pretzels, crackers–high in salt and tropical oils; 'white sauces'–northern Italian cuisine, high in fat. See Junk food. Cf Healthy foods

unhealthy lifestyle PUBLIC HEALTH A dissipated personal modus operandum, which may be characterized by one or more of the following: substance abuse–eg, alcohol, drug and/or tobacco use, debauchery, sexual promiscuity and/or teenage pregnancy, poor sleep hygiene, domestic violence, and other unhealthy habits

UNICEF United Nations International Children's Emergency Fund, now, United Nation's Children's Fund GLOBAL VILLAGE A part of the United Nations system which has as its primary raison d'etre, helping children living in poverty in developing nations

unicompartmental osteoarthritis ORTHOPEDICS Osteoarthritis which affects only part of a joint, classically seen in the knee, where only one of the 2 condyles is affected

unidimensional echocardiography M-mode echocardiography, see there

Unified Parkinson's Disease Rating Scale NEUROLOGY A measure of severity of Parkinson's disease, based on a scale from 0 to 160 total scale and 0 to 44 motor section. See Parkinson's disease

Uniform Anatomical Gift Act Legislation that allows a person to make an anatomic gift at the time of death–all or part of the body for medical education, scientific research, organ transplantation, by a signed document–eg, in a will or driver's license. See *Brotherton v. Cleveland*, Cadaver organ, Organ procurement, UNOS

uniform deceleration OBSTETRICS A fetal heart rate response to uterine contractions that is symmetrical and has a uniform temporal relationship therewith EARLY DECELERATION TYPE I DIP Due to vagal stimulation elicited in the 1st stage of labor by fetal head compression and LATE DECELERATION TYPE II DIP Due to uteroplacental insufficiency, potentially associated with a less favorable outcome and may signal early vasomotor lability. Cf Variable deceleration

unilateral hydronephrosis NEPHROLOGY Distention of the pelvis and calyces of a single kidney because urine can't drain into the bladder ETIOLOGY Unilateral ureteral obstruction, vesicoureteric reflux, kidney stones. See Hydronephrosis

unilateral locked facet ORTHOPEDICS A lower cervical spine subluxation and locking at the facet joint caused by falls from steps or stairs or motor vehicle crashes, and due to flexion/rotation forces IMAGING Subluxation of superior vertebral body–seen in ±25% on the lateral cervical spine films or is malignment of spinous processes on AP MANAGEMENT Reduction with cervical traction; if facet is fractured, halo; if not, posterior fusion. See Halo. Cf Bilateral locked facets

unilateral obstructive uropathy UROLOGY A disorder of blocked urine flow from the ureter of 1 kidney, resulting in backup of urine, distention of renal pelvis and calyces–hydronephrosis, and renal injury; abrupt blockage of a ureter causes acute UOU; progressive blockage causes chronic UOU ETIOLOGY Acute UOU–stones, trauma, congenital anatomic defects; chronic UOU–stones, local tumors or tumors from adjacent structures

unilateral vocal cord paralysis ENT A complication of intrathoracic malignancy–eg, CA of lung, esophagus, etc, which consists of significant dysfunction of speech, swallowing, ventilation, and effective coughing, due to lack of compensation of the nonparalyzed vocal cord CLINICAL Hoarseness, dyspnea, aspiration, weight loss, dysphagia, pneumonia TREATMENT Vocal cord medialization. See Vocal cord medialization

uninsurable HEALTH INSURANCE A high-risk person without health care coverage through private insurance who falls outside the parameters of risks of standard health underwriting practices. See Underwriting

unintentional injury Accidental injury PUBLIC HEALTH Any injury caused by an accident. See Injury

uninvolved bystander phenomenon PSYCHOLOGY The adoption by a group of persons of a 'herd mentality' of inaction in the face of a dread situation–homicides, sexual assaults, muggings that have occurred in view of others. See Kitty Genovese

unipolar electrogenerator LAPAROSCOPIC SURGERY A device used to coagulate tissue during laparoscopic surgery. Cf Bipolar electrogenerator

unique physician identifier number HEALTH INFORMATICS A unique ID number for each physician billing for services under Medicare; HCFA will replace the UPIN system with a National Provider Identifier system

unit 1. A general term for a single discrete entity 2. A quantity which is a standard of measurement. 3. A place where something is performed. See Accreditation unit, Allergy unit, Autologous unit, Bioequivalent allergy unit, British Thermal unit, Burst-forming unit, Central processing unit, Chest pain unit, Collective bargaining unit, Colony-forming unit, Critical care unit, Crossover unit, CSU/DSU unit, Electron unit, Fetoplacental unit, Hounsfield unit, ICU unit, International unit, Map unit, Mediational unit, MET unit, MICU, Montevideo unit, Neonatal intensive care unit, Nephelometric turbidity unit, Output unit, Relative value unit, Revive Ring/Vacuum Therapy unit, Section unit, Terminal duct lobular unit, Transciption unit

United Network for Organ Sharing See UNOS

United States Adopted Names See USAN

univariate *adjective* Determined, produced, or caused by only one variable

universal coverage Universal access HEALTH CARE REFORM The provision of health care for entire group–including preventive care–eg, vaccines, screening, outpatient visits to a generalist or specialist, hospitalization for basic and catastrophic needs. See Canadian plan, Clinton plan, Socialized medicine. Cf Underinsured, Uninsured

universal design PUBLIC HEALTH An industrial, mechanical or architectural design of a device, physical plant or workplace environment which is intended to be used by all–'universal'–ie, does not represent an impediment for individuals with disabilities. See Americans with Disabilities Act, Architectural barriers, Readily achievable. Cf Undue hardship

universal donor TRANSFUSION MEDICINE A person with blood group O, whose RBCs lack A and B antigens and thus do not elicit a hemolytic transfusion reaction when the blood is transfused to a person with blood group A, B, AB or O. Cf Universal recipient

universal precautions INFECTIOUS DISEASE A method of infection control–recommendations issued by CDC–in which all human blood, certain body fluids, as well as fresh tissues and cells of human origin, are treated as if known to be infected with HIV, HBV, and/or other blood-borne pathogens. See Precautions, Reverse precautions. Cf Body substance isolation

universal recipient TRANSFUSION MEDICINE A person with blood group AB, whose serum has no anti-A or anti-B antibodies and thus can receive any ABO blood without suffering an ABO-related hemolytic transfusion reaction when blood is transfused from a person with blood group A, B, AB or O. Cf Universal donor

university An institution of higher learning that can grant a certificate of academic achievement

univessel disease Single-vessel disease, see there

UNIX INFORMATICS A versatile, non-proprietary open operating system, which is the industry standard for mainframes, work stations and servers on the Internet, written in C programming language. See OS X. Cf MacOS, Windows

unknown primary cancer See Occult primary malignancy

unknown risk A new and/or unfamiliar risk with delayed consequences. See Negotiated safety. Cf Dread risk

unload MEDICAL PRACTICE A type of dumping, in which a physician opts not to treat a terminally ill Pt. See Dumping

Unna boot WOUND CARE A compressive dressing for varicose veins or ulcers consisting of a paste of zinc oxide, gelatin, glycerin and water–applied to the lower leg, covered with a bandage, and applied outside the bandage

unnatural death FORENSIC MEDICINE A death that is '...*caused by external causes–injury or poisoning... which includes death... due to intentional injury such as homicide or suicide, and death caused by unintentional injury in an accidental manner.*' See Depraved heart murder, Murder. Cf Natural death

unnecessary surgery SURGERY An instrumental procedure that is not needed EXAMPLES Some hysterectomies, C-sections, tonsillectomies

UNOS United Network for Organ Sharing TRANSPLANT SURGERY A database dedicated to optimizing the use of transplantable organs; according to UNOS statistics–1995, ± 20,000 major organs and tissues are transplanted/yr; since successful survival of transplanted tissue is a function of the degree of histocompatibility matches, the larger the pool of donor and recipient haplotypes, the better the long-term survival of these organs. See Transplantation

unproven Dubious, nonscientific, not proven, quack, questionable, unscientific *adjective* Relating to that which has not been validated by reproducible experiments or other scientific methods for determining effect or efficacy

unproven care Any therapy with no proven efficacy–eg, bioelectromagnetic applications, mind/body control, biological therapies, energy therapies, and ethnomedicine or traditional medicine. See Health fraud, Quackery, Unproven methods for cancer management

unproven method for cancer management FRINGE ONCOLOGY Any of the unorthodox therapies of uncertain benefit offered by self-proclaimed cancer specialists–who may have little formal medical training; in a comparison study at similar stages of terminal malignancy between Pts treated with conventional and unproven cancer therapies, those receiving conventional therapy had better survival and quality-of-life scores than those receiving an unproven regimen, which consisted of a combination of autogenous immune-enhancing vaccines, bacille Calmette-Guérin–bCG, vegetarian diets and coffee enemas–the difference does not reach statistical significance

METHODS OF UNPROVEN CARE

BIOELECTROMAGNETIC APPLICATIONS Electromagnetic field manipulation, electrostimulation and neurostimulation devices, magnetoresonance spectroscopy, blue and other colored light treatment, artificial light therapy

LIFESTYLE CHANGES Changes in habits and lifestyle, diet, nutrition, megavitamins, macrobiotics, nutritional supplements

MIND/BODY CONTROL Various external or internal therapies—art, humor, hypnosis, music, and prayer, relaxation techniques—biofeedback training, yoga, meditation, counseling

PHARMACOLOGICAL AND BIOLOGICAL TREATMENTS Antioxidizing agents—eg vitamin A and E, oxidizing agents,[1] cell treatment, chelation therapy,[2] herbal medicine, naturopathy, metabolic therapy

STRUCTURAL AND ENERGETIC THERAPIES Acupressure, aromatherapy, ayurvedic medicine, bodywork, chiropractic, herbal medicine, massage therapy, reflexology, shiatsu

TRADITIONAL AND ETHNOMEDICINE Acupuncture, ayurveda, herbs, homeopathy, Native American medicine, traditional Chinese medicine, folk medicine

[1]Eg hydrogen peroxide—H_2O_2, ozone, both identified by the FDA as fraudulent [2]Other than for accepted medical indications, eg desferroxamine for hemochromatosis AMN 21/11/94 p13

unresectable cancer SURGICAL ONCOLOGY A malignancy which can't be surgically removed, due either to the number of metastatic foci, or because it is in a surgical danger zone. See Heroic surgery

unresectable gallbladder cancer SURGICAL ONCOLOGY Gallbladder CA that can't be excised as it has spread to tissues around the gallbladder—liver, stomach, pancreas, intestine and/or lymph nodes

unresponsive NEUROLOGY *adjective* Referring to a total lack of response to neurologic stimuli

unscientific Unproven, see there

unstable *adjective* Referring to the structural or physical lability or instability of a chemical or other substance during production, transportation, or storage. See Unstable reactive substance

unstable angina Accelerating angina, new-onset angina, progressive angina CARDIOLOGY A spectrum of Sx of ischemic heart disease, intermediate in severity between stable angina pectoris—intense chest pain and acute MI—crushing chest pain; UA is a subacute < 6 months in duration—state with ↑ risk of MI and sudden death RISK FACTORS ♂, cigarette smoking, ↑ cholesterol—in particular, ↑ LDL-C and/or ↓ HDL-C, HTN, DM, family Hx of CAD < age 55, sedentary lifestyle, > 30% ideal body weight, sudden overwhelming stress CLINICAL Chest pain at rest, or ↑ in severity, frequency, or duration of chest pain at lower levels of activity DIFFDX MI, DM, HTN, myxedema, peripheral vascular disease, heart valve disease, cardiomyopathy, ASHD MANAGEMENT Antianginals—eg, nitrates, beta-blockers, CCBs, aspirin, IV heparin. See Angina pectoris, ATACS, FRISC, HASI, Myocardial infarction, RISC, Silent ischemia, TAUSA, TIMI-3A, TIMI-3B, TIMI-7, UNASEM. Cf Prinzmetal's angina

unstable bladder UROLOGY A bladder characterized by severe inappropriate contraction. See Urge incontinence

unusual life event PSYCHOLOGY A life event capable of evoking emotional stress ULEs Death of family or friends, disease, loss or change of job or emotional stress at work, change of interpersonal relationship(s), divorce

unvoluntary *adjective* Referring to a movement—eg, a motor tic, that relieves an unpleasant sensation; such movement may be perceived by a Pt as voluntary, and is often, but not always, suppressible which, unlike other hyperkinetic movement disorders, can be suppressed for only very short durations, if at all

unwanted pregnancy OBSTETRICS A pregnancy that is not desired by one or both biologic parents. See Teen pregnancy

unwed parent SOCIAL MEDICINE A person who is not married at the time of delivering a child

UOAC Uterine ostial access catheter GYNECOLOGY A device used in selective diagnostic salpingography via the uterine ostia

Upan Lorazepam, see there

upcoding MEDICARE A fraudulent practice in which provider services are billed for higher CPT procedure codes than were actually performed, resulting in a higher payment by Medicare or 3rd-party payors. See Coding, CPT coding, Health care fraud, HIPPA Act, Prompt payment law, Rebundling, Unbundling. Cf Downcoding

upgrade TECHNOLOGY Replacement of older equipment, software, services

UPJ obstruction Ureteropelvic junction obstruction, see there

upper DRUG SLANG A popular term for any amphetamine or neurostimulant. See Amphetamine, Ecstasy, Rave

upper airways A term that encompasses the nasal passages, nasopharynx, oropharynx, larynx. Cf Lower airways

upper body fat obesity Android obesity, male obesity, 'beer gut' obesity Obesity characterized by ↑ abdomen-to-thigh skin fold ratio, waist-to-hip circumference ratio, suprailiac-to-thigh skin fold ratio, waist circumference, thigh skin fold. See Obesity

upper cervical spine injury ORTHOPEDICS Any osteoarticular injury affecting the upper cervical spine. See Atlanto-axial displacement, C1 fracture, C2 fracture, Hangman's fracture, Jefferson fracture, Odontoid fracture, Os odontoideum

upper GI endoscopy A procedure, in which a fiberoptic endoscope—esophagogastroduodenoscope is inserted by mouth and the mucosa of the esophagus, stomach, duodenum, and proximal jejunum are examined for ulceration, polyps, bleeding sites, strictures, and other changes; at the time of the procedure, suspicious lesions may be sampled and the specimens submitted for culturing or microscopic analysis of cells, tissues, or fluids INDICATIONS UGE is used to evaluate acute upper GI bleeding—eg, of the esophagus in cirrhosis, dysphagia—which may be due to strictures, dyspepsia, esophageal pain, and evaluate changes seen by an upper GI study, to identify causes of the gastric outlet syndrome, and to monitor premalignant conditions—eg, Barrett's esophagus, lye-induced strictures; 'low-yield' indications for upper GI endoscopy include atypical chest pain, abdominal pain of unknown origin. Cf Upper GI study

upper GI study Upper GI, upper GI series IMAGING A radiologic study in which a radiocontrast—eg, a barium-based 'milkshake' is administered in the upper GI tract—esophagus, stomach, duodenum, and images obtained. Cf Upper GI endoscopy

upper rate CARDIAC PACING In atrial tracking dual-chamber pacemakers—VDD and DDD, a programmed limit to the rate at which the ventricles are paced in response to atrial activity. See Pacemaker

upper respiratory tract flora MICROBIOLOGY Bacteria that are normally located in the upper respiratory tract

upper respiratory tract infection URI INFECTIOUS DISEASE A nonspecific term used to describe acute infections involving the nose, paranasal sinuses, pharynx, and larynx, the prototypic URI is the common cold; flu/influenza is a systemic illness involving the URT and is differentiated from other URIs CLINICAL 1–3 days after exposure to pathogen, nasal congestion, sneezing, sore throat, conjunctivitis—with adenovirus infections; sore throat with pain on swallowing, fever, absence of cough, and exposure to a person with streptococcal pharyngitis in the prior 2 wks support Dx of GABHS-related pharyngitis; Pts with acute sinusitis often have fever for > 1 wk, facial pain—especially unilateral, maxillary toothache, headache, and excessive purulent nasal discharge; hoarseness suggests laryngitis; difficulty in swallowing oral secretions and stridor should raise suspicion for epiglottitis or pharyngeal abscess; influenza presents as a sudden illness characterized by high fever, severe headache, myalgia, dry cough, with lingering fatigue and malaise; elderly patients may also present with confusion and somnolence PHYSICAL EXAM Common cold–nasal voice, macerated skin over the nostrils, inflamed nasal mucosa; GABHS-related pharyngitis–pharyngeal erythema/exudate, palatal petechiae–popularly, "doughnut lesions," tender anterior cervical lymphadenopathy, and occasionally a scarlatiniform rash; pharyngeal or palatal vesicles and ulcers (herpangina) suggest enteroviral or herpetic pharyngitis; pharyngeal exudates are most common in GABHS-related pharyngitis, but can be seen with infectious mononucleosis due to EBV, acute retroviral syndrome, candidal infections, diphtheria; swelling, redness, and tenderness overlying affected sinuses and abnormal transillumination are specific for, but not commonly seen in acute sinusitis; generalized lymphadenopathy with sore throat, fever, and rash should raise the possibility of a systemic viral infection–eg, EBV, CMV, HIV; Pts with influenza appear toxic and may have pulmonary rhonchi and diffuse myalgias TYPES Pharyngitis, sinusitis, laryngitis/epiglottitis, otitis DIAGNOSIS Because viruses cause most URIs, the diagnostic role of lab and radiologic studies is limited; rapid antigen detection of influenza virus on a nasopharyngeal swab is indicated in cases where specific antiviral therapy is recommended; a rapid antigen detection test is also available for adenovirus, RSV, and parainfluenza virus; serologic tests for viruses that can cause a mononucleosis-type illness should be done in the correct clinical setting; influenza serologies have only epidemiologic value and should not be used for clinical care; pharyngeal swab for rapid antigen detection of GABHS is 80% to 95% sensitive and should be considered in all patients in whom GABHS-related pharyngitis is suspected; pharyngeal culture remains the gold standard for diagnosis and should be done if GABHS-related pharyngitis is highly likely on clinical grounds but in which the rapid antigen detection test is negative; cultures obtained by paranasal sinus puncture should be reserved for only severely ill patients with acute sinusitis and intracranial or orbital complications; blood cultures should be done in severely ill Pts or in those with epiglottitis or a pharyngeal abscess MANAGEMENT Symptomatic to relieve the most prominent Sx; rest, ↑ fluid intake are measures recommended for all URIs. Cf Acute laryngotracheobronchitis (croup), Common cold, Epiglottitis, Otitis media. Viral URI

UPPP Uvulopalatopharyngoplasty, see there

upright tilt test Tilt test, see there

Uprima® UROLOGY A dopamine receptor agonist, apomorphine–used for erectile dysfunction, which may improve blood flow to the penis. See Erectile dysfunction, Orgasm

upset stomach A popular term for transient gastralgia often accompanied by N&V, attributed to something its victim ingested. Cf Stomach virus

upside-down films Wangensteen-Rice technique IMAGING A series of radiologic studies performed on a child with a suspected imperforate anus, in which plain films are taken with the child's anus in the most superior position–ie, bottom up–in order to identify the level of atresia

upstream procedure MANAGED CARE A diagnostic procedure in which a defect may result in a therapeutic, or downstream, procedure

urate crystal A needle-shaped water-soluble crystal of precipitated uric acid, which measures 2–8 μm long, present in macrophages or freely floating in synovial fluid in Pts with gout. See Gout

urea breath test GI DISEASE A noninvasive test for the presence of *H pylori* in the stomach, based on *H pylori*'s urease activity. See *Helicobacter pylori*

urea cycle disorder Any disorder in which the body is unable to excrete waste nitrogen–ammonia–resulting in mental and behavioral dysfunction, coma, death

urea nitrogen See BUN

urea reduction ratio NEPHROLOGY The percent reduction in serum urea due to dialysis, an indicator of dialysis efficacy. Cf Kt/V

Ureaplasma urealyticum T strain mycoplasma MICROBIOLOGY A species of small gram-negative bacteria of the family Mycoplasmataceae that lack a cell wall and catabolize urea–to ammonia; *U urealyticum* resides in the genital tract and nasopharynx, is sexually transmitted, and causes nongonococcal urethritis, urethroprostatitis, epididymitis in ♂; UTIs, fetal wastage, chorioamnionitis, and ↑ infertility in ♀, URIs; CNS infections in neonates MANAGEMENT Tetracycline–eg, doxycycline; if resistant, erythromycin. See *Mycoplasma hominis*. Cf *Mycoplasma pneumoniae*

urecholine supersensitivity test Bethanechol sensitivity test UROLOGY A test used to evaluate bladder voiding activity, in which urecholine, a cholinergic agent, is administered while the bladder pressure is monitored; in a 'neurogenic' bladder, the intravesical pressure ↑ by > 15 cm above that of a control Pt. See Neurogenic bladder

uremia Prerenal azotemia, renal underperfusion NEPHROLOGY A constellation of Sx caused by the retention of urea and other products of protein catabolism due to inadequate kidney function in advanced renal failure CLINICAL N&V, pruritus, uremic frost, mental clouding, peripheral neuropathies, osteodystrophy, HTN, CHF, pericarditis, pulmonary edema LAB Acidosis, anemia, azotemia, ↓ Ca²⁺, ↑ PO₄, coagulopathy PATHOGENESIS 1° glomerular and/or tubular disease MANAGEMENT General support–restriction of protein, Na⁺, K⁺, and water; dialysis–hemodialysis, peritoneal dialysis; kidney transplant

uremic frost A finding in severe chronic renal failure, in which the concentration of urea is ↑↑↑ in the sweat, causing precipitation of crystallized urea in the skin. See Uremia

uremic heart disease Uremic cardiopathy CHF due to fluid retention by the kidney in advanced renal failure–uremia, which may be accompanied by moderate to severe HTN. See Uremia

ureteral obstruction Acute unilateral obstructive uropathy UROLOGY A unilateral block of urine flow through the ureter of 1 kidney, resulting in a backup of urine, distension of the renal pelvis and calyces, and hydronephrosis ETIOLOGY Kidney stone, trauma, stricture in children–of one ureter causes acute unilateral obstructive uropathy; slow, progressive blockage causes chronic unilateral obstructive uropathy; either may result in permanent damage and failure of one kidney and HTN, resulting in the so-called Goldblatt kidney, which is usually asymptomic–the remaining kidney usually has functional reserve. See Goldblatt kidney, Reflux nephropathy

ureteral vesical reflux Reflux nephropathy, see there

ureterocele UROLOGY A congenital disorder in which a ureter develops an outpouching as it enters the bladder

ureteropelvic junction obstruction Obstruction of ureteropelvic junction; UP junction obstruction UROLOGY A congenital obstruction at the junction between the ureter and renal pelvis, caused by ureteral stenosis, resulting in accumulation of urine in the pelvis under pressure, ergo hydronephrosis; UPJ obstruction is the most common cause of urinary obstruction in children CLINICAL Abdominal mass in infant with UTI, fever, back pain, hematuria MANAGEMENT Kidney decompression before or after delivery. See Hydronephrosis

urethral bulking agent UROGYNECOLOGY An injectable hydrogel composed of a biocompatible recombinant protein polymer used to bulk tissue about the intrinsic sphincter to provide complete closure and urinary control for ♀ with incontinence due to intrinsic sphincter deficiency. See Urinary incontinence

urethral injury Traumatic injury of bladder & urethra, see there

urethral meatal stenosis Meatal stenosis UROLOGY Narrowing of the bore of the meatus, which occurs exclusively in males--most of whom are not circumcised, due to irritation of the urethral opening at the end of the penis, leading to reactive hyperplasia and scarring across the opening CLINICAL Extremely narrow, needle-fine spray of urine rather than a full stream; buildup of pressure may result in discomfort with urination and occasionally there may be slight bleeding at the end of urination

urethral obstruction Acute bilateral obstructive uropathy, see there

urethral pressure profile UROLOGY A dataset used to determine the functional length, resting pressure, and maximal pressure of the urethral sphincter mechanisms

urethral stricture UROLOGY An abnormal narrowing of the urethra due to inflammation or scarring from prior urethral or prostate surgery, disease or injury, and rarely, due to external pressure from an enlarging periurethral tumor RISK FACTORS STDs, repeated urethritis, BPH, pelvic injury, trauma or surgery to pelvic region; congenital strictures are rare, as are true strictures in ♀. See Urethra

urethral syndrome Chronic urethritis UROLOGY Chronic inflammation and irritation of the urethra, usually caused by bacterial infection or structures, may be associated with emotional distress CLINICAL Urethritis, painful urination and urinary frequency, Sx which are common and account for 5 to 10% of all outpatient health care visits

urethralism SEXOLOGY Sexuoerotic arousal by urethral stimulation–insertion of rubber catethers, rods, objects, fluids, ballbearings, long flexible cathether-like electrodes–'sparklers', etc. See Catheterophilia; electrocutophilia

urethritis UROLOGY Inflammation of the urethra, ETIOLOGY Infection–*E coli*, *Klebsiella*, STDs–eg, *Chlamydia*, *N gonorrhoeae*, *Ureaplasma urealyticum*; chemical irritants–spermatocides in condoms, contraceptive jelly, cream, or foam; mechanical–insertion of stuff into urethra RISK FACTORS ♂ age 20–35, multiple sexual partners, high-risk sexual behavior–eg, anal intercourse. See Nongonococcal urethritis

urethrocele UROLOGY The prolapse of the ♀ urethra into the vaginal wall

urethroplasty UROLOGY Reconstructive surgery to create a urethra in a penis–eg, to correct injury or a birth defect

urge incontinence Detrusor instability; irritable bladder; spasmodic bladder; unstable bladder; urgency incontinence UROLOGY A condition characterized by a sudden and urgent need to urinate (urinary urgency), frequent urination; abdominal distention or discomfort immediately before an involuntary bladder contraction with a loss of a large amount of urine; a urine storage defect in which the bladder muscle contracts inappropriately regardless of the amount of urine in the bladder ETIOLOGY Idiopathic (most), neurologic disease–spinal cord injury, stroke, multiple sclerosis, infection, bladder CA, bladder stones, cystitis, bladder outlet obstruction–eg, BPH; UI is more common in elderly ♀; UI affects ±2% of older ♀ DIAGNOSIS Post void residual–PVR, to measure amount of urine left after urination, urinalysis to exclude UTI, urinary stress test–Pt stands with full bladder, then coughs, pad test–after placing a pre-weighed sanitary pad, Pt exercises after which pad is re-weighed to ID urine loss/spillage, pelvic ultrasound, X-rays with contrast dye, cystoscopy, urodynamic studies–to measure pressure and urine flow, rarely, EMG MANAGEMENT Bladder retraining therapy, medications to relax bladder contractions, surgery; antispasmodics–eg, oxybutynin, Kegel exercises– which strengthen the muscles of the pelvic floor and thus improves the Sx of UI, antibiotics for UTIs, bladder retraining exercises, biofeedback; anticholinergics–propantheline and dicyclomine–which is also a bladder muscle relaxant ADVERSE EFFECTS Dry mouth, dizziness, drowsiness, ↑ heart rate, difficulty urinating, terbutaline–a beta-adrenergic that ↑ bladder capaci-

ty–CONS Palpitations, insomnia, HTN, distribution of fluid intake throughout the day; surgery aimed to ↑ bladder storage capacity, while ↓ the pressure in the bladder. See Kegel exercises, Oxybutynin. Cf Dribbling

urgency The need to do a thing–eg, urinate, now dad blast it!

urgent care MANAGED CARE Immediate, nonemergent primary health care. See Urgent care center. Cf Emergent care

urgent surgery SURGERY Surgery required within < 48 hrs EXAMPLES Kidney stone, stomach obstruction or ulcer, bleeding hemorrhoids, ectopic pregnancy

URI Upper respiratory infection, see there

uric acid A small purine metabolite excreted primarily by the kidneys, less by the GI tract; ↑ UA occurs in gout, which primarily affects acral joints, associated with deposition of UA crystals in various tissues; ↑ UA occurs in rapid cell turnover–eg, cancer–leukemia, metastases, myeloma, as well as in alcoholism, dehydration due to diuretics, DM, hyperlipoproteinemia, lead poisoning, renal failure, rarely, idiopathic REF RANGE Serum, ♂, 3.6-8.3 mg/dL; ♀, 2.2-6.8 mg/dL. See Gout. Cf Synovial fluid analysis

uric acid 'infarct' A popular term for the yellowish streaks seen in young children at the tips of the renal papillae, which corresponds to normal collections of uric acid crystals in terminal collecting tubules CLINICAL Asymptomatic

urinalysis UA LAB MEDICINE A screening test in which a random urine specimen is examined grossly, by the microscopy and/or tested by reagent strips; UA is used to detect renal disease–eg, glomerulonephritis, urinary tract disease–eg, bladder infection, or metabolic diseases–eg, DM; routine chemical analysis of the urine is based on the use of reagent strips–dipsticks, which are coated with substances that change colors depending on the concentration of a substance of interest in the urine

urinary fistula UROLOGY A defective urinary passageway, usually due to a failed surgical repair, with leaking of urine into surrounding tissue

urinary free catecholamines The free/nonbound and nonmetabolized catecholamines, epinephrine and norepinephrine, excreted in urine in 24 hrs which, normally, is 2-4% of the total catecholamine production REF RANGE < 600 nmol/24 hrs–US: 100 μg/24 hrs, of which 473 nmol/24 hrs–US: 80 μg/24 hrs is norepinephrine, and 109 nmol/24 hrs–US: 20 μg/24 hrs is epinephrine. Cf Total catecholamine test

urinary retention UROLOGY The inability to completely void the urinary bladder

urinary tract infection Infection of the kidney, ureter, bladder, or urethra EPIDEMIOLOGY Affects ± 7 million/yr, often young ♀ with acute uncomplicated pyelonephritis or cystitis, or recurrent cystitis, and adults with asymptomatic bacteriuria or with complicated UTI EPIDEMIOLOGY Common in young, sexually active ♀, often associated with sexual intercourse, use of a diaphragm with a spermicide, Hx of recurrent UTIs CLINICAL Dysuria, burning, ↑ frequency, urgency LAB Leukocyte esterase + by dipstick, 10^2–10^5 colony-forming U/mL MICROBIOLOGY *E coli*–± 80%, *Staphylococcus saprophyticus*, *Proteus mirabilis*, *K pneumoniae* TREATMENT Oral T-S, norfloxacin, ciprofloxacin

urinary tract obstruction UROLOGY A block in urine flow, often caused by a stone. See Kidney stone

urine concentration test Concentrating ability test UROLOGY Any test in which the osmolality of the urine is measured after fluid restriction; the ability to concentrate urine indicates adequacy of both renal and endocrine function; the responsible hormone is vasopressin–ADH. See Water deprivation test

urine culture The placing of urine in a culture medium to detect the growth of a pathogen–eg, gram-negative rods; UCs require a clean-catch, midstream specimen to minimize contamination by commensals on the external genitalia

urinoma PEDIATRIC UROLOGY A tumor-like urine-filled cyst seen in the renal capsule of children, which is 2° to congenital urethral obstruction, in which urine extravasates into subcapsular or perirenal spaces via ruptured calyceal fornices

URL LAB MEDICINE Upper reference limit ONLINE Uniform Resource Locator An address on the web MEDICAL URLs Virtual Library of Medicine, Visible Man, Mayo Clinic. See Bookmark, Browser, Web

urn The container where cremated ashes are stored

uro- A root form for urine, urinary organs or tract

urodynamic evaluation UROLOGY A battery of clinical tests used to assess neuromuscular responses of the bladder to filling and emptying. See Cystometrogram, Urethral pressure profile, Urinary flow rate

urogynecology A subspecialty of urology or gynecology dedicated to the study and management of structural and functional changes of the urologic system in ♀. See Female circumcision, Gynecology, Urology

urokinase A fibrinolytic enzyme involved in extracellular proteolysis; it is more active during cell migration and tissue remodeling, and cleaves fibronectin. See Kringle domain. Cf Streptokinase, Reteplase, tPA

urolagnia Urophilia PSYCHIATRY A paraphilia in which sexuoeroticism is linked to urinary functions–eg, urinating on partner, drinking urine, etc. See Golden shower, Paraphilia, Water sports

Urolase™ UROLOGY A side-firing laser used to treat BPH. See Benign prostatic hypertrophy

urolithiasis See Kidney stones

urologic disorder Any condition affecting the kidneys, often understood to be of the collecting tubules and southward. Cf Kidney disease

urologist A physician specialized in diseases of ♀ urinary organs and ♂ urinary and sex organs MEAT & POTATOES DISEASES Kidney stones, BPH, impotence SALARY $222K. See Urogynecology

urology The subspecialty of medicine dedicated to the urogenital tract, primarily of men. See Urologist

urostomy Urinary ostomy GYNECOLOGY The creation of an artificial opening for elimination of urine due to loss of bladder or bladder function, necessitated by birth defects, bladder CA, injury or nerve damage. See Ostomy UROLOGY 1. An operation to create a new conduit for the passage of urine 2. The conduit itself. See Ileal stoma

ursodiol Actigall®, ursodeoxycholic acid HEPATOLOGY A pharmacologic preparation of ursodeoxylchoic acid, a 2° bile acid, used to dissolve bile gallstones and manage primary biliary cirrhosis. See Primary biliary cirrhosis

urticaria Hives DERMATOLOGY A condition characterized by pruritic raised red welts on the skin, associated with allergic reactions and histamine release or defects in the complement or kinin systems RISK FACTORS Prior allergic reactions–eg, hay fever and angioedema TRIGGERS Medications; foods–eg, berries, shellfish, fish, nuts, eggs, milk; pollen; animal dander–especially cats; insect bites; mechanical stimulants–eg, water, sunlight, cold or heat; emotional stress; post-infection; linked to other disease–eg, autoimmune diseases SLE, leukemia, etc; may be partially hereditary, dermographism, cold urticaria, echinococcus infection–dog tapeworm, hereditary angioedema, Henoch-Schönlein purpura, mononucleosis, hepatitis, mastocytosis

urticaria pigmentosa DERMATOLOGY An idiopathic self-limiting benign condition characterized by light brown skin lesions, intense itching, and hives at the site of rubbed lesions; affected children develop by age 2 lesions containing mast cells which produce histamine and cause the typical skin response–rubbed lesions rapidly develop a wheal and systemic Sx–flushing, headache, diarrhea, tachycardia, fainting; younger children may develop a fluid-filled blister over a traumatized lesion; UP often disappears by puberty. See Histamine, Mastocytosis

U/S Ultrasound

USAN United States Adopted Names PHARMACOLOGY A nonproprietary or generic drug name adopted by the USAN Council. See Generic drug

USB Universal Serial Bus INFORMATICS An external bus with transfer rates up to 12 Mbps, which accommodates up to 127 peripherals, replacing serial and parallel ports. Cf Firewire

use VOX POPULI Utilize. See Illegal use of controlled dangerous substances, Intravenous drug use, Misuse, Off-label use, Overuse, Substance use, Underuse

user *noun* A person who uses a thing. See Authorized user, Individual user

USFMG United States foreign medical graduate A North American unable to obtain a seat in an US or Canadian medical school, and graduated from a foreign medical school. See International medical graduate, Off-shore medical school

Usher syndrome An AR condition characterized by retinitis pigmentosa–RP and sensorineural deafness

USMLE United States Medical Licensing Examination GRADUATE EDUCATION A 3-step examination required for medical licensure in the US, sanctioned by the Natl Bd of Medical Examiners and Federation of State Medical Bds. See Off-shore medical school, USFMG. Cf FLEX

USOH Usual state of health

USP United States Pharmacopeia PHARMACOLOGY A compendium of drug standards that includes assays and tests for determining drug strength, purity and quality; USP is recognized US federal law, published by the authorities of the US Pharmacopeia convention. See National Formulary

usual interstitial pneumonia PULMONOLOGY A condition of middle-aged individuals, often associated with connective tissue disease, characterized by insidious deterioration of respiratory function with dyspnea, tachypnea, right-sided heart failure, ↓ lung capacity, ↓ residual volume IMAGING Early, 'ground-glass', linear or nodular markings, followed by coarsened shadows and cyst formation; diaphragmatic elevation in end-stage UIP reflects tissue loss MANAGEMENT None effective, possibly corticosteroids

uterine apoplexy Active uterine bleeding where myometrial vessels have been stripped of endometrium EXAMPLES Abruptio placenta, Couvelaire uterus, uteroplacental apoplexy, or in placenta acreta or in massive cardiovascular collapse–apoplexia uteri

uterine balloon ablation GYNECOLOGY A technique for managing hypermenorrhea unrelated to CA, fibroids/leiomyomas, or endometriosis, which require surgery; in UBA

uterine border GYNECOLOGY The lateral aspects of the uterus palpated through the abdomen

uterine cancer GYNECOLOGY Any cancer that arises in the uterus–eg, cervical CA, choriocarcinoma, endometrial CA, leiomyosarcoma, mesodermal mixed tumor. See Endometrial CA

uterine fibroid Fibroid, fibromyoma, leiomyoma, myoma GYNECOLOGY A benign smooth muscle tumor that arises in the myometrium, possibly linked to endogenous estrogen production; UFs are rare before age 20, involute after menopause, weigh up to a 1+ kg–average < 15 cm in diameter; they are the most common pelvic tumor, occur in up to 30% of reproductive-age ♀, and are 3 to 9 times more common in African-Americans than Caucasians; fibroids with a stalk are termed pedunculated

uterine monitoring Home uterine monitoring, pregnancy monitoring OBSTETRICS The use of a belt-like device with pressure sensor, to detect premature labor; some sensors measure pressure changes over time, which can be "ported" by telephone to a clinician INDICATIONS High-risk Pts–eg, with Hx of early labor or early contractions; it is unclear whether UM prevents premature labor

uterine polyp Endometrial polyp, see there

uterine prolapse Pelvic floor hernia; pudendal hernia GYNECOLOGY Falling or sliding of the uterus from its normal position in the pelvic cavity into the vaginal canal; the uterus is normally supported by pelvic connective tissue, pubococcygeus muscle, and uterine ligaments which, when weakened allow the uterus to descend into the vaginal canal ETIOLOGY Childbirth trauma, especially from large babies or abrupt L&D, loss of muscle tone due to aging and ↓ estrogens RISK FACTORS Multiparity, white, obesity, excess coughing due to chronic bronchitis and asthma, constipation and abdominal muscle strain, large uterine fibroids

uterine sarcoma See Leiomyosarcoma, Mixed müllerian tumor, Smooth muscle of uncertain malignant potential

UTI Urinary tract infection, see there

utilization Use MANAGED CARE The use or amount of usage per unit population, of health care services; the pattern of use of a service or type of service in a specified time, usually expressed in rate per unit of population-at-risk for a given period–eg, number of hospital admissions/yr/1,000 persons enrolled in an HMO. See Hospital utilization, Overutilization, Underutilization

utilization review Utilization control, utilization management MANAGED CARE Evaluation of the necessity–eg, for surgery or other therapy, appropriateness–of admissions or of services ordered, and efficiency of the use of health care services, procedures, and facilities. See Utilization

Uvadex® Methoxsalen DERMATOLOGY An agent used for extracorporeal administration with the UVAR® photopheresis system in palliating treatment-refractory changes of cutaneous T-cell lymphoma. See Cutaneous T-cell lymphoma

UVB phototherapy An artificial light treatment used for mild psoriasis. See Phototherapy

UVC Umbilical vein catheter, see there

uveitis Choroidoretinitis OPHTHALMOLOGY Inflammation of the uveal tract–iris, ciliary body, choroid, uvea–the vascular layer between sclera and retina TYPES Anterior uveitis–aka iritis, linked to autoimmune disease; posterior uveitis–granulomatous affects choroid and/or retina, and may follow systemic infection ETIOLOGY Allergy, infection–eg, *Toxoplasma gondii*–congenital uveitis, histoplasmosis, TB, CMV, trauma, chemicals, idiopathic, ulcerative colitis, pauciarticular rheumatoid arthritis, Kawasaki disease, HSV, ankylosing spondylitis, Behçet disease, sarcoidosis, psoriasis, Reiter syndrome, syphilis; posterior uveitis causes patchy scars that correspond to vision defects; if the macula is involved, central vision is impaired. See Granulomatous uveitis, Salt & pepper choroidoretinitis

uvulopalatopharyngoplasty UPPP ENT A surgical procedure for treating obstructive sleep apnea that consists of resection of the uvula, the distal margin of the soft palate, palatine tonsils and any excessive lateral pharyngeal tissue; UPPP is successful in ⅔ of selected cases with obstructive sleep apnea in which there is focal airway collapse, which may cause the Pts to snore. See Laser-assisted uvulopalatoplasty, Snore

Uxen Amitriptyline, see there

the microbe, thereby preventing disease. See Anthrax vaccination, Booster, Chickenpox vaccination, Childhood vaccination, DNA vaccine, DPT vaccination, Hepatitis A vaccination, Hepatitis B vaccination, HIB vaccination, HIV vaccination, MMR vaccination, Rubella vaccination, Supervaccine, Vaccine

723

vaccine IMMUNOLOGY A mixture of live, live-attenuated, killed, complete or incomplete microorganisms or their products, that contains antigens capable of stimulating production of specific protective antibodies against a pathogen; vaccines may be biochemically synthesized or made by recombinant DNA techniques EFFECTIVE VACCINES Diphtheria, HAV, HBV, influenza, measles, pertussis, *S pneumoniae*, tetanus

vaccine reaction Vaccine injury IMMUNOLOGY Any injury or condition caused by vaccination with certain childhood vaccines–eg, seizures, DPT–anaphylaxis, acute encephalopathy, shock, atypical shock, polio, MMR, possibly HBV. See Vaccine

vaccine therapy A therapeutic strategy to boost the immune response against a particular pathogen to such degree that the vaccine alone is enough to treat the Pt

vaccinology A nascent field of expertise related to the creation and deployment of vaccines; the field 'borrows' from epidemiology, immunology, infectious disease, pediatrics, preventive medicine, public health, virology

VACTERUL VATER, VACTER NEONATOLOGY A condition characterized by a multisystem association of Vertebral defects–eg, hemivertebra and scoliosis, Anorectal atresia, Cardiac malformations–eg, VSD, TracheoEsophageal fistula, Renal defects–agenesis, ureteropelvic junction obstruction, vesicoureteral reflux and crossed fused ectopia and radial Upper Limb anomalies–eg, polydactyly

vacuolar myelopathy NEUROLOGY A heterogeneous group of conditions characterized by spinal cord degeneration; VM occurs in up to 20% of AIDS Pts; it is most extensive in the lateral columns of the middle and lower thoracic segments. Cf Spongy degeneration

vacuolar myopathy Vacuolar degeneration A structural change–multiple membrane-bound vacuoles–seen in skeletal muscle in glycogen storage disease types II, III, V, VII, AIDS or nutritional deficiency. See Glycogen storage disease

Vacutainer™ A proprietary blood collection tube with a vacuum to facilitate blood collection

vacuum abortion Early abortion, early uterine evacuation, first trimester abortion, suction abortion OBSTETRICS Suction removal of endometrial tissue from the uterus via the cervix to terminate a pregnancy of ≤ 12 wks. Cf Vacuum curettage

vacuum curettage Endometrial aspiration, vabra, vacuum aspiration GYNECOLOGY Suction removal of endometrial tissue from the uterus via the cervix to either either correct menstrual defects or establish a diagnosis–eg, endometrial hyperplasia, CA, dysfunctional uterine bleeding. Cf Vacuum abortion

vacuum extraction OBSTETRICS Operator-assisted delivery in which suction is applied to the skull and the fetus delivered vaginally COMPLICATIONS Brachial plexus injury due to shoulder dystocia, scalp injuries, intracranial–especially, subgaleal–hemorrhage due to tentorial tearing and skull fractures, CNS depression, convulsions, mechanical ventilation. See Operative vaginal delivery. Cf Forceps

vacuum phenomenon IMAGING A linear or oval radiolucency that corresponds to gas in the intervertebral space, most often seen in degenerative disk disease DIFFDX Vertebral osteomyelitis, Schmorl's nodes, spondylosis deformans, necrotic vertebral collapse, often with loss of height and reactive osteosclerosis

V Symbol for: 1. Electric potential 2. Specific volume 3 Vaccinated 4. Valine 5. Valve 6. Vanadium 7. Variable region–immunology 8. Vein 9. Velocity 10. Venous 11. Version 12. Vertex 13. *Vibrio* 14. Vinblastine 15. Vincristine 16. Virus 17. Viscosity 18. Vision 19. Visual acuity 20. Volt 21. Volume 22. Vomiting

v Symbol for: 1.Specific volume 2. Velocity 3. Vitamin

V cholera *Vibrio cholerae*, see there

V fracture T fracture, see there

V-HEFT CARDIOLOGY A trial–Veterans Heart Failure Trial that evaluated the effect of combination vasodilator–hydralazine and isosorbide dinitrate therapy in Pts with severe heart failure. See ACE inhibitor, Congestive heart failure, Enalapril, Hydralazine-isosorbide dinitrate, NY Heart Association classification

V-shaped scar A sharply angled fibrotic scar on the renal cortical surface, often due to bacterial infection or cortical abscesses. Cf Rat-bitten kidneys

V tach Ventricular tachycardia, see there

v wave CARDIOLOGY The wave in the normal jugular phlebogram which corresponds to the atrial diastole

V-Y advance PLASTIC SURGERY A surgical advancement incision which lengthens a contracted scar–a 'Y' is sewn into a 'V' incision

V&P SURGERY Vagotomy and pyloroplasty

VA 1. Veterans Administration, Department of Veteran Affairs The hospital system sponsored by the federal government for providing medical care directly to civilians, most of whom are related to military veterans 2. Valproic acid, see there 3. Ventricular aneurysm, see there 4. Ventricular arrhythmia, see there 5. Vertebral artery, see there 6. Viral antigen

VA-HIT CARDIOLOGY A clinical trial–Veterans Affairs HDL Intervention Trial

V-A interval Physiologic retrograde conduction time CARDIAC PACING DVI, DDD The period of time elapsing in a dual-chamber pacemaker from a ventricular event–sensed or paced–to the next scheduled atrial pace. See Pacemaker

vaccination Immunization The injection of a killed bacteria or virus, or antigen therefrom, to stimulate the immune system to produce antibodies against

vacuum sign IMAGING A normal radiologic finding; traction applied to a joint causes coalescence of gas within a joint; the VC disappears in effusions

vacuum system UROLOGY A mechanical system used to facilitate and maintain an erection; an erection erector. Cf Penile implant

Vagifem® GYNECOLOGY A vaginal suppository with 17β-estradiol, used to manage estrogen-related postmenopausal atrophic vaginitis. See Atrophic vaginitis

vaginal atrophy Atrophic vaginitis, see there

vaginal birth after cesarean VBAC OBSTETRICS Vagina delivery of an infant after a cesarean section COMPLICATIONS Uterine apoplexy

vaginal cancer GYNECOLOGY Any malignancy of the vagina, including nonepithelial lesions–eg, *Sarcoma botryoides*; vaginal adenoCA is linked to maternal use of DES during pregnancy. See Diethylstilbestrol

vaginal candidiasis Vaginal mycosis, vaginal thrush, vaginal yeast infection GYNECOLOGY Infection of the lower ♀ genital tract with *Candida* spp, usually *C albicans*, a fungus normally found in the flora of the vagina, mouth, GI tract, skin–±25% of ♀ have *Candida* spp; Sx appear when the balance with other saprobes is changed, favoring candidal growth–eg, after antibiotics, especially tetracycline; related to estrogen-based OCs, DM, immunocompromise–eg, AIDS. See Candidiasis

vaginal cone UROGYNECOLOGY A weighted device inserted into the vagina to help perform Kegel exercises in ♀ with postpartum stress incontinence. See Kegel exercises. Cf Cone biopsy

vaginal cuff GYNECOLOGY The part of the vagina remaining after hysterectomy and cervicectomy

vaginal discharge OBSTETRICS A whitish fluid arising in the vagina, a normal finding in pregnancy

vaginal dryness GYNECOLOGY 1. Atrophic vaginitis, see there 2. ↓ vaginal lubrication or premature loss of same

vaginal hysterectomy Total vaginal hysterectomy GYNECOLOGY The excision of a uterus with benign disease or cervical CIS via an upper vaginal incision PROS Simple, easy, ↓ OR time, no visible scar. Cf Abdominal hysterectomy

vaginal pouch Female condom, see there

vaginal ring GYNECOLOGY An annular contraceptive device inserted in the vagina before coitus, which slowly releases levonorgestrel or progesterone. See Norplant, Pearl index. Cf Female condom

vaginal speculum OBSTETRICS An L-shaped device used to help visualize the upper vagina and uterine cervix. See Colposcopy

vaginal spotting OBSTETRICS A popular term for scant, dark brown, often intermittent bleeding, which may occur in Pts with ectopic bleeding, or in the 1st or 2nd trimester in Pts with placenta previa

vaginismus Unconsummated marriage PSYCHIATRY A sexual dysfunction characterized by '...*recurrent or persistent involuntary contraction of the perineal muscles surrounding the outer third of the vagina when penetration with penis, finger, tampon, or speculum is attempted*...(This causes)...*marked distress or interpersonal difficulty*...(and) *is not better accounted for by another...disorder.*' See Dyspareunia. Cf Vaginosis

vaginitis GYNECOLOGY Inflammation of the vagina, which may be nonspecific or induced by a specific organism. See Atrophic vaginitis, Hormone replacement therapy, Vaginal candidiasis

vaginoplasty GYNECOLOGY Reconstructive surgery of the vagina for birth defects, for conversion in ♂→♀ transsexualism, surgical or other trauma to the vagina. See Female circumcision, Love surgery

vaginosis GYNECOLOGY A vaginal infection without PMN infiltration; bacterial vaginosis is the most common vaginal infection of reproductive-aged ♀ ETIOLOGY *Mobiluncus* spp CLINICAL Range from asymptomatic to a copious and malodorous milky discharge which, when mixed with 10% KOH, has a fishy odor. Cf Vaginitis

Vagistat® Tioconazole, see there

vagotomy NEUROLOGY Surgical disruption of the vagal nerve

vagus nerve stimulation PSYCHIATRY Electroconvulsive therapy in which a pacemaker-like device stimulates the vagus nerve. See Electroconvulsive therapy

VAHS Virus-associated hemophagocytic syndrome, see there

Val-HeFT Valsartan heart failure trial CARDIOLOGY A multinational M&M study investigating valsartan use with standard therapy–eg, ACE inhibitors, in Pts with previously treated and untreated CHF. See Valsartan

valacyclovir Valtrex® INFECTIOUS DISEASE An antiviral used for 1st episode/recurrent genital herpes and acute herpes zoster, which may ↓ risk of CMV infection after renal transplantation. See Dofetilide, Herpes simplex

Valaxona Diazepam, see there

valdecoxib See Bextra®

Valeans Alprazolam, see there

valgus ORTHOPEDICS Fixation of an extremity in the position it would assume if everted; if in the frontal plane, the plantar surface is directed away from the midline. Cf Varus

VALIANT Valsartan in Acute Myocardial Infarction Trial CARDIOLOGY A series of multinational M&M trials to determine the effects of valsartan–Diovan® combined with, and compared to, an ACE inhibitor, captopril in Pts with recent MI See Myocardial infarction

validation set DECISION-MAKING A group of Pts with a clinical finding of interest–eg, chest pain, who are studied prospectively in order to verify facets of their disease that had been previously identified as possible predictors of outcome. See Derivation set

Valitran Diazepam, see there

Valium® Diazepam, see there

valley fever A form of 1° coccidioidomycosis, first described in San Joaquin Valley, California, most common in white ♀ CLINICAL Fever, cough, pleuritic pain, arthralgia, erythema multiforme or nodosum DIAGNOSIS IgM antibody to mycelial phase antigen, coccidioidin, ↑ CF antibody, ↑ latex agglutination MANAGEMENT Ketoconazole, amphotericin B, miconazole

valproic acid Depakote®. Divalproex, Valproate NEUROLOGY An anticonvulsant monotherapy for absence seizures, equivalent to carbamazepine for generalized tonic-clonic seizures ADVERSE EFFECTS Weight gain, hair loss, tremors COMPLICATIONS Pancreatitis, cholecystitis. See Seizures. Cf Carbamazepine

Valsalva maneuver Forced expiration against a closed glottis after full inspiration, described by Valsalva in 1704 to expulse pus from the middle ear; the 'Valsalva' ↑ intrathoracic pressure for ±10 secs, eliciting a complex series of changes in pulse rate and BP involving both vagal and sympathetic responses

Valstar® Valrubicin ONCOLOGY An anthracycline for intravesical use in Pts with biopsy-proven carcinoma in situ of bladder who are refractory to BCG immunotherapy, and in whom cystectomy is contraindicated. See BCG therapy, Bladder cancer

valuation MEDICAL PRACTICE The determination of the value of a thing–eg, a physician's private practice. See Practice valuation

value 1. The worth of a thing 2. A quantity. See Added value, Biological value, Ceiling value, Comparison value, Critical value, CT value, Daily value, Ethical value, Expected value, Fair market value, K$_{nuc}$ value, Log-transformed value, Negative predictive value, Panic value, Positive predictive value, Predictive value, Q value, R value, Reference value, Supplementary assigned value, Z value

VALUE CARDIOLOGY A series of multinational M&M trials–Valsartan Antihypertensive Long-term Use Evaluation comparing valsartan–Diovan® to a CCB, amlodipine in treating high-risk Pts with HTN. See Valsartan

valuing PSYCHOLOGY A process in which persons assign worth to a particular activity or object

valve 1. A flapped or flap-like structure. See Aortic valve, Bicuspid valve, Bioengineered heart valve, Bioprosthetic valve, Ileocecal valve, Mitral valve, Pulmonary valve, Ross pulmonary porcine valve, Safe-Connect™ valve, Shiley heart valve, Tricuspid valve 2. A device intended to limit the flow, or prevent the reversal of flow of a liquid or gas from point A to point B

valve dysplasia CARDIOLOGY A congenital defect in which the heart's semilunar valves–pulmonary, aortic–are myomatous, thickened, immobile; $^2/_3$ of Pts with Noonan syndrome have pulmonary stenosis due to VD. See Noonan syndrome, Pulmonary stenosis

valve stenosis Valvular stenosis CARDIOLOGY Any narrowing of a heart valve, the most common of which is stenosis of the mitral valve. See Mitral valve stenosis

valve vegetation Vegetative endocarditis Any variably sized excrescence that may be present on heart valves

valvoplasty Valvuloplasty HEART SURGERY Any procedure affecting a heart valve–eg, replacement, dilatation–for stenosis or tightening–for regurgitation or incompetence. See Balloon valvoplasty, Percutaneous balloon valvoplasty

valvular *adjective* Referring to a valve, see there

valvular heart disease CARDIOLOGY Any morbid condition caused by pathology of the cardiac, primarily mitral and tricuspid, valves

valvuloplasty Valvoplasty, see there

the Vampire project See Diversity project

vampirism The practice of drinking blood CLINICAL MEDICINE A quasi-facetious term for excessive blood tests, which causes iatrogenic anemia. See Anemia of investigation PSYCHIATRY A deviant behavior in which blood is ingested, variably accompanied by necrophilia, often in a background of schizophrenia, psychosis, sadomasochism, cult–eg, voodoo rituals, cannibalism, fetishism or drug intoxication. See Necrophilia

Vanatrip Amitriptyline, see there

Vanceril® Beclomethasone dipropionate CRITICAL CARE An aerosolized steroid for inhalation treatment of asthma. See Nebulizer

vancomycin A glycopeptide antibiotic effective against coccal bacteria, especially in recalcitrant staphylococcal infections–eg, MRSA, unresponsive to penicillins or cephalosporins ADVERSE EFFECTS Ototoxicity, nephrotoxicity

vancomycin-resistant enterococcus INFECTIOUS DISEASE An enterococcus, primarily *Enterococcus faecium*, resistant to most antibiotics, including aminoglycosides and vancomycin, once a 'last-resort' agent; VRE is primarily nosocomial, in long hospitalizations–especially in the ICU, enteral feeding, liver transplants, antibiotic-associated colitis; it may be community acquired MANAGEMENT Ampicillin, penicillin G, teicoplanin–not

US, quinupristin/dalfopristin or linuzolid, both investigational. See Vancomycin. Cf Methicillin-resistance *Staphylococcus aureus*

vancomycin-resistant *Staphylococcus aureus* VRSA INFECTIOUS DISEASE A long anticipated bacterium first identified in a clinical specimen in mid-2002; the isolate was susceptible to chloramphenicol, linezolid, quinupristin-dalfopristin, T-S. See Vancomycin-resistant *Enterococcus*

Vanconin Diazepam, see there

VANILA A clinical trial–Ventricular Arrhythmia Needing Intravenous Lidocaine/Amiodarone

vanishing lung syndrome Giant bullous emphysema PULMONOLOGY A disease affecting 1:1000 young usually ♂ subjects, characterized by multiple bullae of the apical portions of one or both lungs CLINICAL ↑ Resonance to percussion with radiologic disappearance of lung markings, hence, the term 'vanishing'; in absence of concomitant emphysema, the vital capacity is relatively normal; with extensive lesions, severe impairment of pulmonary function may occur TREATMENT Surgical excision of afunctional bullae results in expansion of normally functioning pulmonary tissue and clinical improvement

vanishing twin The spontaneous regression of a 2nd gestational product, which occurs in 20% of twin gestations DIAGNOSIS Ultrasound in 1st trimester. Cf Twins

VANQWISH CARDIOLOGY A clinical trial–Veterans Affairs Non-Q-Wave Infarction Strategies In-Hospital–which compared early conservative vs invasive strategies for acute MI; both have equivalent outcomes; superiority vis-á-vis cost or readmission is controversial

Vantin® Cefpodoxime proxetil INFECTIOUS DISEASE An antibiotic used for acute maxillary sinusitis and otitis media. See Cephalosporin

vapor lock CRITICAL CARE Obstruction of blood flow due to a trapped bolus of gas in a tube or vessel. See Cardiac index

vapor pressure OCCUPATIONAL HEALTH The pressure exerted by a saturated vapor–the gaseous phase of a liquid, a function of the amount of vapor given off by a chemical. See Materials Safety Data Sheets, OSHA. Cf Vapor density

vaporizer RESPIRATORY MEDICINE A device that adds moisture to ambient air to ↓ dryness or loosen upper airway secretions

vapors Vapours MEDICAL HISTORY An 18th century belief that nervous illness in ♀ resulted from vapors produced by the uterus which affect brain. See Hysteria MEDTALK A vague physical, perhaps hypochondriacal, somatization complaints, often flutteringly or histrionically displayed by a Pt. Cf Vapor

VaporTrode® UROLOGY An electrode that delivers high-density electrical current for removing prostate tissue

Vapr™ system Mitek Vapr™ system ORTHOPEDICS A bipolar radiofrequency device for controlling bleeding during arthroscopy of the knee, shoulder, wrist, ankle, and elbow. See Arthroscopy

V$_A$/Q mismatch PULMONOLOGY A defect in which ventilation–V and perfusion–Q are unevenly matched, a finding typical of COPD. See COPD, Right-to-left shunt

Vaqta A vaccine against HAV made of killed hepatitis A virus to stimulate the body's immune system to produce antibodies against HAV. See Hepatitis A

variable *noun* EPIDEMIOLOGY Any characteristic or attribute that can be measured. See Confounding variable, Continuous variable, Dependent variable, Independent variable, Instrumental variable, Intervening variable, Lurking variable, Natural variable, Predictor variable, Qualitative variable, Quantitative variable, Random variable

variable deceleration OBSTETRICS A fetal heart response that is asynchronous with respect to uterine contractions; the curves of VDs on the fetal heart monitor are angled and saw-toothed and may be caused by compromised placental blood flow–eg, umbilical cord compression and, like late decelerations, may signify delivery-related difficulties. See Fetal heart monitor. Cf Uniform deceleration

variable resistance SPORTS MEDICINE A resistance that changes over the ROM when an isotomic contraction is used to move a load. See Range of motion. Cf Static resistance

variant angina Coronary artery spasm, Prinzmetal's angina CARDIOLOGY A syndrome primarily affecting ♀ < age 50 characterized by muscle spasm of a normal or atherosclerotic coronary artery, resulting in focal ischemia; VA may be asymptomatic, evoke angina pain or, if prolonged, cause acute MI TRIGGERS Spontaneous or linked to exposure to cold, emotional stress, or vasoconstrictors–eg, cocaine use. See Angina

variceal bleeding Hemorrhage from dilated or variceal veins, usually understood to mean esophageal varices 2° to end-stage liver failure MANAGEMENT-SURGICAL Surgical shunting, endoscopic sclerotherapy, esophageal variceal ligation, transjugular intrahepatic portosystemic stent-shunt–TIPS procedure MANAGEMENT-MEDICAL Vasopressin, nitroglycerin, somatostatin, β-blockers, long-acting nitrates. See Transjugular intrahepatic portosystemic stent-shunt–TIPS procedure

varicella Chickenpox, see there

varicella embryopathy A constellation of malformations occurring sporadically in a fetus when the mother contracts varicella before the 20th wk of gestation CLINICAL Intrauterine growth retardation, motor and sensory defects, chorioretinitis, cerebral dysgenesis, zosteriform skin lesions, limb defects. See Chickenpox

varicella-zoster virus Chickenpox virus, human herpesvirus, HH-3 VIROLOGY A virus which causes chickenpox in children, becomes latent in cranial nerve and dorsal root ganglia and often reactivates decades later to produce shingles and postherpetic neuralgia CLINICAL The 1°–childhood form, chickenpox, presents commonly as a wave of pruritic vesicles that spread over the body, healing within 3–5 days. See Chickenpox, Herpes zoster

varices Plural of varix Any widely dilated vein, commonly, the saphenous veins, dilated through the viscissitudes of pregnancy, childbirth, life in general and esophagus. See Esophageal varices, Varicose veins

varicocele Varicose veins in scrotum UROLOGY Elongated and dilated veins of the pampiniform plexus–the network of veins leaving the testis which join to form the testicular vein; varicoceles appear bluish through the scrotum and impart a bag of worms-like sensation to palpation CLINICAL Varicoceles are caused by incompetent valves in spermatic cord veins; abnormal valves obstruct normal blood flow causing a backup of blood, resulting in venous dilation; varicoceles usually develop slowly and may be asymptomatic; incidence is higher in ♂ age 15 and 25; varicoceles are linked to infertility in 40% of ♂ treated for infertility; abrupt appearance of a varicocele in older ♂ may be caused by a renal tumor affecting the renal vein and altering the blood flow through the spermatic vein

varicose veins Varicosis; varicosity SURGERY Enlarged, twisted veins with nonfunctioning valves, resulting in IV pooling of blood and venous enlargement, most commonly in leg veins; VVs affect ±10% of the population, most commonly ♀, age 30 to 60 ETIOLOGY Congenital valve defects, thrombophlebitis, pregnancy, prolonged standing or sitting, poor posture, ↑ intraabdominal pressure. See Esophageal varices, Vein stripping. Cf Varicocele

variegate porphyria South African porphyria METABOLIC DISEASE An AD condition caused by protoporphyrinogen oxidase deficiency CLINICAL Abdominal pain, neuropsychiatric Sx, photosensitivity, increased sensitivity of skin to mechanical trauma LAB ↑ Protoporphyrin, coproporphyrin in stools; ↑ δ-aminolevulinic acid and porphobilinogen in urine

variegated *adjective* Multifaceted; with many colors, aspects, features, etc

Varigrip® system ORTHOPEDICS A spine fixation system that attaches to lamina without pedicle screws, providing multidirectional support, allowing extension of attachment to the thoracic spine. See Laminectomy

varix SURGERY An enlarged and convoluted vein, artery or lymphatic vessel. See Varicose veins

varus ORTHOPEDICS A fixation of a part in the position assumed if it were inverted–in a frontal plane, fixation where the plantar surface is directed toward the midline

vas- Latin, vessel

vascular access CLINICAL MEDICINE The ability to enter the vascular system; the ease with which the vascular system can be entered for administering therapy or obtaining blood for testing

vascular access specialist MEDTALK Phlebotomist, see there

vascular access thrombosis NURSING A thrombus which forms at the site of vascular access

vascular dementia NEUROLOGY A potentially preventable form of dementia, in which cerebral atrophy is due to various types of CVAs, resulting in variably-sized infarcts. See Multi-infarct dementia, Stroke. Cf Alzheimer's disease

vascular headache NEUROLOGY A headache–eg, migraine–attributed to arterial hypersensitivity to various triggers that cause vasospasm or vasoconstriction or vasodilation, which evokes throbbing pain. See Migraine

vascular/lymphatic invasion ONCOLOGY The penetration of vascular or other transport conduit by a malignancy. See Metastasis

vascular phase Prevascular phase, see there

vascular purpura Henoch-Schönlein purpura, see there

vascular skin changes See Capillary hemangioma, Cavernous hemangioma

vascular sling A rare congenital malformation in which the left pulmonary artery arises from the right pulmonary artery, crossing to the left side, and insinuates itself between the trachea and esophagus, forming a 'sling' around the trachea CLINICAL Tracheal stenosis–stridor, wheezing, choking TREATMENT Surgery

vascular spasm Vasospasm MEDTALK A sudden, brief constriction of a blood vessel, which may temporarily ↓ blood flow to 'downstream' tissue

vascular-ventricular coupling CARDIOLOGY A combining of cardiovascular benefits of dobutamine therapy, where there is moderate vasodilating activity, minimal ↑ in myocardial O₂ consumption–MOC, ↑ cardiac output with little change in MOC; VVC ↓ as dose ↑, due to positive chronotropic effects. See Dobutamine

vascularity A tissue blood supply

vasculitis Inflammation of blood vessel(s). See Essential cryoglobulinemic vasculitis, Hypersensitivity vasculitis, Large vessel vasculitis, Leukocytoclastic vasculitis, Medium vessel vasculitis, Necrotizing vasculitis, Rheumatoid vasculitis, Small vessel vasculitis, Systemic vasculitis

vasculitis profile RHEUMATIC DISEASE A battery of assays intended to ID and measure serum levels of antibodies–eg, anti-glomerular basement membrane, anti-myeloperoxidase, and anti-neutrophil cytoplasmic antibodies, often associated with systemic vasculitis. See Systemic–necrotizing vasculitis

vasculoma A popular term for severe vascular disease

vasculopath A Pt with vascular disease. See 'Mayo-ism'

vasectomy UROLOGY The surgical induction of infertility, denying sperm egress from the testes, by ligating or disrupting both vasa deferentia, a form of permanent contraception chosen by up to 500,000 US ♂/yr; 10-20 ejaculates are required post-procedure before the vasectomized are 'shooting blanks' and have no sperm; the vasectomized may experience long-term scrotal pain, possibly related to sperm granuloma formation; ⅓ develop circulating antisperm antibodies. See In vitro fertilization, Post-vasectomy sperm count, Vasectomy reversal

vasectomy reversal UROLOGY Surgical reanastomosis of disrupted vasa deferentia. See Antisperm antibodies

vasoactive substance Any of a group of circulating substances that regulate vascular tone, causing either vasodilation–ANP, kinins, and VIP, or vasoconstriction–angiotensin II, epinephrine, norepinephrine, vasopressin

vasoconstrictor A vasoactive agent that constricts blood vessels–eg, epinephrine, norepinephrine, which is produced endogenously to compensate for hypotension and maintain vascular tone in hemorrhagic and other forms of shock. Cf Vasodilator

vasodilator An agent that dilates blood vessels CARDIOLOGY A vasoactive agent used in CHF–eg, nitrates, nitroprusside, hydralazine, ACE inhibitors, prazocin, captopril, to ↓ preload and/or afterload

Vasofem® Phentolamine mesylate, see there

vasomotor center NEUROANATOMY A cluster of nerve cell bodies in the medulla oblongata that controls the diameter of blood vessels and regulates blood pressure

vasomotor rhinitis ENT Rhinitis characterized by intermittent episodic sneezing, rhinorrhea, and congestion of nasal mucosa, attributed to hypersensitivity to dry air, air pollutants, spicy food, alcohol, emotion, drugs. See Rhinitis

vaso-occlusive crisis Painful crisis HEMATOLOGY A 'crisis' common in sickle cell anemia, in which 'sludged' sickled RBCs cause capillary stasis and infarction, and incapacitating musculoskeletal pain or 'referral'-type organ pain, hemoptysis, hematuria, melena, CNS Sx; VOCs occur at a rate of 0.8 episodes/yr in sickle cell disease, 1.0/yr in sickle-β-thalassemia, and 0.4/yr in HbSC-β thalassemia; this frequency is translated into a 'pain rate', a measure of disease severity that correlates with early death in Ps with sickling anemias MANAGEMENT High-dose methylprednisolone ↓ duration of painful crises; rebound attacks ↑ with discontinuation of steroids. See Hemolytic crisis, Sickle cell anemia

vasopeptidase inhibitor CARDIOLOGY An agent–eg, omapatrilat, that inhibits neural endopeptidase and angiotensin-converting enzyme, modulating cardiac function and bp in Pts with HTN. See Isolated systolic hypertension, Omapatrilat, OPERA

vasopermeability reaction PHYSIOLOGY An inflammatory response to local ↑ in vasomotor substances, which evoke egress of fluids and cells from the vascular compartment during inflamation. See Slow-reactive substances of anaphylaxis, Triple response, Vasculitis

VASOPERMEABILITY REACTION–PHASES OF

IMMEDIATE-TRANSIENT RESPONSE mediated by histamine, as well as leukotriene E$_4$, serotonin, bradykinin and others, a reaction affecting small (< 100 μm in diameter) venules, but not capillaries; during this phase, the endothelial cells contract, widening the interendothelial cell gaps

IMMEDIATE-SUSTAINED RESPONSE follows severe injury, eg burns and is associated with endothelial cell necrosis, affecting the small arterioles, capillaries and venules and

DELAYED-PROLONGED LEAKAGE begins after a delay of hours to days, representing a response to a vast array of environmental 'toxins', eg burns, bacterial toxins, UV light, and X-rays and delayed hypersensitivity reactions, which affect venules and capillaries

vasopermeation enhancement ONCOLOGY An investigational therapy that ↑ up to 3-fold CA cell uptake of chemotherapeutics

vasopressin Antidiuretic hormone, see there

vasopressin resistance A therapeutic artefact seen in diabetes insipidus managed with pitressin, caused by inadequate mixing of the drug and injection only of an oily vehicle

VasoSeal® THERAPEUTICS A family of arterial puncture sealing devices, designed to effect hemostasis by delivering purified collagen directly to the arterial surface INDICATIONS Hemostasis during coronary angiography, angioplasty, radiologic procedures. See Cath lab

vasospasm An abrupt and/or excess local or regional contraction of vascular smooth muscle; vasospasm may underlying amaurosis fugax, thrombosis-negative MI, migraine, Prinzmetal's angina, Raynaud's phenomenon, subarachnoid hematoma-related CVAs

Vasotec® Enalapril maleate, see there

vasovagal syncope Neurocardiogenic syncope, see there

VAST PEDIATRICS A clinical trial–Vitamin A Supplementation Trial performed in Ghana, which evaluated the effect of vitamin A supplements on infant mortality. See Vitamin A

VATER complex VACTERUL, see there

VATER syndrome NEONATOLOGY A clinical syndrome characterized by Vertebral defects, imperforate anus, tracheoesophageal fistula, renal defects. See VACTERUL

VATK technique Fernald multisensory technique PSYCHOLOGY A method that integrates visual, auditory, kinesthetic, and tactile imagery as a tool for remediating children with a block in one of the sensory channels

Vatran Diazepam, see there

VATS Video-assisted thoracic surgery, see there

Vaughan Williams classification CARDIOLOGY The system by EM Vaughan Williams, used to categorize effects of antiarrhythmics by class. See Antiarrhythmic drugs. Cf Sicilian Gambit

VAUGHAN WILLIAMS CLASSIFICATION OF ANTIARRHYTHMIC DRUGS

CLASS I Sodium-channel blockers

IA Moderately slow conduction, moderately prolonged duration of action potential–active in atria, ventricles; most cardiotoxic group

EXAMPLES Quinidine, procainamide, diisopyramide

IB Minimally slowed conduction, shortened duration of action potential–active in atria; greatest potential for proarrhythmias

EXAMPLES Lidocaine, mexiletine, tocainamide, phenytoin

IC Markedly slowed conduction, minimally duration of action potential–active in atria, ventricles; most effective group

EXAMPLES Flecainide, encainide, propafenone, moricizine

CLASS II Beta blockers–active in AV nodes, ventricles; virtually no proarrhythmic effect

CLASS III Potassium-channel blockers; prolonged duration of action potential

EXAMPLES Amiodarone, bretylium, sotalol, ibutilide

CLASS IV Calcium-channel blockers–active in AV nodes; virtually no proarrhythmic effect

vaulting BIOMECHANICS A high energy stance phase modification where a person positions him/herself on his/her toes, which requires ↑ ground clearance of the swinging leg; the exaggerated vertical movement of the trunk imparts an ungainly appearance

Vaxid™ VACCINOLOGY A DNA vaccine for treating B-cell lymphoma which immunizes Pts after chemotherapy; it may eliminate residual disease and prevent relapse. See DNA vaccine

Vazen Diazepam, see there

VBAC Vaginal birth after cesarean section, see there

VBPAC Vaccines & Biological Products Advisory Committee

VC Abbreviation for: 1. Vascular catheterization 2. Vasoconstrictor 3. Vena cava 6. Venous capacitance 7. Ventilatory capacity 8. Ventricular complex 9. Ventricular coupling 10. Vital capacity 11. Vocal cord

VCUG Voiding cystourethrography, see there

VD Abbreviation for: 1. Vapor density 2. Vasodilator 3. Venereal disease–STD is widely preferred 4. Verbal discrimination–psychology 5. Volume of distribution–Vd is more commonly used

Vd Volume of distribution, see there

VDAC Vaginal delivery after cesarean section, see there

VDRL test Venereal Disease Research Laboratory test A reaginic test–RT for syphilis, used to screen for early syphilis; it is virtually always positive in 2° syphilis; it is variable in 3° syphilis and is negative in half of neurosyphilis DIFFDX–BIOLOGICAL FALSE POSITIVE–BFP Malaria–±90%, acute infection–10-30%, SLE–10-20%–classic cause of BFP VDRL, viral hepatitis–10%, infectious mononucleosis–20%, rheumatoid arthritis–5-10%, pneumococcal pneumonia, drug addiction, pregnancy, etc

VDT Video/visual display terminal, see there

vector EPIDEMIOLOGY 1. An 'inactive' vehicle of transport of an agent of disease; an intermediate host of parasites with indirect life cycles 2. A thing that transmits a pathogen–eg, an arthropod transporting viruses and parasites, or an inanimate intermediary in indirect transmission of an agent from a reservoir to a susceptible host; a carrier that transmits a pathogen from one host to another

vector-borne disease INFECTIOUS DISEASES Any infection, usually transmitted by insects–eg, ticks–eg, Lyme disease, Rocky Mountain spotted fever, ehrlichiosis, Colorado tick fever; mosquitos–eg, California-or La Crosse, St Louis, Eastern, Western encephalitides– See Vector

vectorcardiography Spatial electrocardiography, 3-D electrocardiography CARDIOLOGY A sophisticated type of EKG, which simultaneously demonstrates myocardial activity and damage; in VC, the strength and direction of electric currents passing through the heart are represented as vector loops APPLICATION Diagnose and monitor left-sided hypertrophy and hyperfunction, bundle branch blocks and MIs; VC informs on the direction of heart muscle activity as well as its force vis-à-vis contraction and relaxation.

vecuronium Norcuron® ANESTHESIOLOGY An nondepolarizing muscle relaxing anesthetic derived from pancuronium PROS Vecuronium has minimal hemodynamic effects–it does not evoke histamine release. See Muscle relaxant. Nondepolarizing agent. Cf Depolarizing agent

vegan Pure vegetarian, strict vegetarian NUTRITION A vegetarian who consumes only plant foods–vegetables, fruits, grains, beans, nuts–ie, no animal products, meat, fish, dairy products, and eggs; vegans are at ↑ risk for vitamin B_{12} deficiency; vegan adolescents may not meet energy requirements during the growth spurt, and become deficient in vitamin B_6, riboflavin, calcium, zinc, iron, trace minerals. See Diet, Pareve, Vegetarian diet, Vegetarianism; Cf Lacto-ovo vegetarian

vegetarianism CLINICAL NUTRITION The pursuing of a primarily, but not exclusively, vegetarian diet, with minimal amounts of animal proteins; a vegetarian diet consists mainly of whole-grain cereals, pulses–lentils, beans, peas, dried fruits, nuts, dairy products. See Diet, Kosher, Macrobiotic diet, Pareve. Cf Veganism

vegetation See Valve vegetations

vegetative state CLINICAL MEDICINE A state characterized by unresponsiveness to external stimuli TYPES Permanent VS, persistent VS

vehicle EPIDEMIOLOGY An inanimate intermediate in the indirect transmission of a pathogen from a reservoir or infected host to a susceptible host; vehicles include foods, clothing, instruments. Cf Vector PHARMACOLOGY An inert carrier or excipient for a therapeutic agent–eg, water, alcohol-containing elixirs or a sweetened syrup, which provides bulk or solubilizes a drug, facilitating deglutition. Cf Carrier, Schlepper, Vector

vein stripping SURGERY The surgical removal of a varicose vein of the leg by ligating it–eg, the great saphenous vein and its tributaries at the junction with the common femoral vein in the groin, then yanking it, which pops off the collaterals, 'stripping' it out in its entirety INDICATIONS VS is indicated for 2° varicosities with deep venous insufficiency; 90% of Pts treated by VS have satisfactory functional and cosmetic results; in 10%, varicosities recur. See Varicose veins

Velcro-like crackles PULMONOLOGY A popular term for dry or fine crackles–fancifully likened to the sound of a Velcro closure being opened–heard posterolaterally at the lung bases and associated with interstitial lung disease

vellus hair DERMATOLOGY Short, fine, usually nonpigmented which covers much of the body. See Hair. Cf Terminal hair

velocardiofacial syndrome Shprintzen syndrome CLINICAL GENETICS An AD condition characterized by cleft palate, cardiac defects, typical facies–prominent tubular nose, narrow palpebral fissures, slightly retrocessed mandible, learning disabilities, as well as microcephaly, mental and growth retardation, short stature, speech and feeding defects, minor ear anomalies, slender hands and digits, inguinal hernia. See FISH

velocity The rate of a body's motion in a given direction per unit of time; speed BALLISTICS The speed that a projectile/bullet attains while in flight; the difference in tissue destruction between high- or low-velocity bullets is caused by the fragmentation of the bullet–as occurs in an M-16 semiautomatic weapon, rather than the speed of the bullet. See Ballistics, Critical velocity

Velosef® Cephradine, see there

velvet ants ENTOMOLOGY A parasitic wasp indigenous to the Southern/Southwestern US, the stings of which evoke allergic reactions MANAGEMENT Avoid, anti-allergy injections; injection therapy. Cf Fire ant

vena cava filter CARDIOVASCULAR DISEASE A device implanted within veins, usually leg veins, that are at high risk of developing DVT and thromboembolism. See Deep vein thrombosis, Greenfield filter, Pulmonary thromboembolism

Venefon Imipramine, see there

venereal *adjective* Referring to sexual contact, after Venus, Roman goddess of love. See Sexually-transmitted disease

venereal disease Sexually transmitted disease, see there

venereal wart 1. Genital wart, see there 2. Condyloma acuminatum, see there

venesection Phlebotomy, see there

Venezuelan equine encephalitis An alphavirus infection first identified in a sick horse in Venezuela in 1938, which occurs as an epizootic infection in central and northern South America; most exposed humans develop flu-like Sx; ±4%, especially adolescents, develop encephalitis MORTALITY Up to 35% in the young; < 10% in adults MANAGEMENT Supportive. Cf Eastern equine encephalitis, Western equine encephalitis

veni-, veno- *prefix*, Latin, pertaining to veins

venipuncture The puncture of a vein with a needle with the intent to either obtain blood or administer a therapeutic substance. See Phlebotomy. Cf Cut-down

venlafaxine Effexor® NEUROPHARMACOLOGY A second-generation bicyclic member of the serotonin-norepinephrine reuptake inhibitor class of antidepressants. See Paxil, Prozac, Zoloft

venography Phlebography IMAGING A technique in which radiocontrast is injected to obtain radiographic or fluoroscopic images of veins, and detect DVT of legs. See CT venography, Percutaneous transcutaneous portal venography, Radionuclide venography. Cf Varicose veins

venom TOXICOLOGY A poisonous substance produced by an insect or animal, stored in specific sacs and sundry sites, and released by biting or stinging; venoms, the original biological weapons, are used for defense and to capture prey. See Snake venom, Yellow jacket venom

venom immunotherapy A type of allergic desensitization therapy for Pts who are highly susceptible to hymenopteran venom

veno-occlusive disease LIVER Obliteration of the small hepatic venules, which may lead to portal HTN and cirrhosis; VOD has been linked to ingestion of bush tea. See Jamaican vomiting sickness LUNG See Pulmonary veno-occlusive disease

venous hum A low-pitched hum in the neck, caused by jugular vein turbulence, with an altered or more intense flow pattern, that mimics the machinery murmur of patent ductus arteriosus; VH may be abolished by light lateral neck pressure; it may be innocent or seen in hyperthyroidism. See Machinery murmur

venous insufficiency INTERNAL MEDICINE A state in which one or more veins do not allow normal blood flow due damage of internal valves, resulting in leakage and pooling of blood in legs and feet CLINICAL Swelling and dull aching, heaviness, cramping of legs, discoloration of overlying skin. See Deep venous thrombosis, Stasis dermatitis, varicose veins

venous introducer MEDTALK A device used to facilitate insertion of a venous catheter. See Catheterization

venous lake GERIATRICS A dark blue-purple compressible papular vascular ectasia of older ♂ caused by dilation of venules, exacerbated by sun exposure CLINICAL Usually asymptomatic; pain, tenderness, excess bleeding may follow trauma DiffDx Melanoma, pigmented BCC

venous O₂ saturation CARDIOLOGY The O₂ saturation percentage of venous blood, which is influenced by metabolic demand; ↓ VOS mandates an ↑ in pacing rate. See Pacing

venous patterning FORENSIC PATHOLOGY Marbling, see there

venous port A surgically implanted indwelling–eg, Port-A-Cath, subcutaneous device, consisting of a reservoir attached to plastic tubing in a large vein used to provide venous access for a period of wks

venous stasis MEDTALK The pooling of venous blood in a particular region which, in the legs results in edema, hyperpigmentation and possibly ulceration

venous thrombus See Deep vein thrombosis, Thromboembolism

venovenous hemodialysis See Hemodialysis

vent Ventilation, ventilator

VentCheck™ A proprietary respiratory mechanics device for monitoring and verifying ventilator parameters and Pt respiratory mechanics. See Respirator

ventilation PULMONOLOGY The exchange of air between the lungs and the outside air. See Dead space, High-frequency ventilation, Jet ventilation, Maximum voluntary ventilation, Mechanical ventilation, Noninvasive positive pressure ventilation, Partial ventilation, Staircase ventilation PUBLIC HEALTH The circulation of air from one space to another, usually understood to mean the replacement of ambient air with fresh air from another source. See General exhaust, Local exhaust, Mechanical exhaust

ventilation/perfusion ratio mismatch V/Q mismatch INTENSIVE CARE An alteration in the normal relationship between ventilation and perfusion; normal ventilation with a defect in perfusion is typical of PTE

ventilation-perfusion scan Radionuclide scan of lung, V/Q scan CARDIOLOGY, PULMONOLOGY A noninvasive radionuclide test that provides functional information as a ratio of pulmonary ventilation–V to blood flow–perfusion, Q through the lungs; in one common protocol, the scan is performed after inhaling ¹³³Xe –20 mCi, 148 MBq and IV injection of ⁹⁹ᵐTc microacroaggregated albumin, for assessing inspiratory airflow, lung volume, air trapping and adequacy of perfusion; 'indeterminant' studies are not uncommon, and are more frequent in those with underlying lung disease. See Lung volumes, Pulmonary panel

ventilation scan A radionuclide study of pulmonary ventilation; in a commonly used protocol, the scan is performed after inhalation of ¹³³Xe. See Ventilation-perfusion scan

ventilation studies See Pulmonary function tests, Ventilation scan, Ventilation-perfusion scan

ventilation system PUBLIC HEALTH An air system designed to maintain negative pressure and exhaust air properly, to minimize the spread of TB and other respiratory pathogens in a health care facility

ventilator Artificial respirator A device that mechanically helps Pts exchange O₂ and CO₂. See High-frequency oscillating ventilator

ventilator-assistance pneumonia Ventilator pneumonia ICU Pneumonia linked to prolonged use of a ventilator, seen in 25% of Pts after 48 hrs of mechanical ventilation RISK FACTORS Age > 70, COPD, NG tubes, prior chest surgery, H-2 or antacid therapy CLINICAL New, persistent, progressive infiltrate, fever, leukocytosis, purulent tracheobronchial secretions–only half of Pts meeting these criteria have VAP COMMON PATHOGENS *Acinetobacter* spp, *S aureus, Pseudomonas aeruginosa, Enterobacter* spp, *Serratia marscescens* MANAGEMENT Vancomycin, amikacin, imipenem

ventilator-induced lung injury Volutrauma Those changes related to ventilatory support of Pts with acute respiratory failure and/or ARDS, which may exacerbate already compromised pulmonary function MANAGEMENT Permissive hypercapnia, see there

venting Ventilation PSYCHOLOGY The verbalization* of one's 'emotional baggage' to another person; qvetching

VenTrak™ system PULMONARY MEDICINE A proprietary system for monitoring and providing real-time measurement of pulmonary dead space, CO₂ production, volume of CO₂, and impact of ventilator changes on each parameter. See Pulmonary dead space

ventral *adjective* 1. Pertaining to the venter–abdomen; toward the front or facing surface; the opposite of dorsal 2. Anterior 3. Inferior

ventral hernia Abdominal hernia, see there

ventricular *adjective* Referring to a ventricle of the brain or heart

ventricular arrhythmia An abnormal, usually rapid, heart rhythm that arises in a ventricle; VAs are often life threatening and 2° to myocardial infarction EXAMPLES V tach, V fib

ventricular assist device HEART SURGERY A portable, battery-powered device that assists the flow of blood while a Pt is awaiting heart transplant; the VAD is connected at the apex of the left ventricle and pumps the blood past an effete ventricle and aortic valve directly into the aortic arch. Cf Artificial heart, Jarvik-7

ventricular fibrillation V fib CARDIOLOGY An abnormal, life-threatening irregular heart rhythm characterized by rapid uncoordinated fluttering contractions of the ventricles of the heart; VF is often life threatening and occur 2° to an acute MI or a healed infarction PHYSICAL EXAMINATION Loss of synchrony between the heartbeat and pulse beat. Cf Ventricular tachycardia

ventricular gallop Third heart sound, see there

ventricular late potential CARDIOLOGY An abnormal high-frequency, low-amplitude electrical signal which occurs in the terminal QRS complex or on the ST segment of the EKG during sinus rhythm; VLPs may represent delayed activation of regions of diseased myocardium, and indicate an arrhythmogenic substrate, especially in post-MI Pts. See Ventricular premature contractions

ventricular lead CARDIOLOGY A pacing lead placed in or on the ventricle. See Lead. Cf Atrial lead

ventricular pacemaker Single chamber pacemaker CARDIOLOGY A pacemaker with formal mode designations of VVIR–ventricular pacing, ventricular sensing, inhibition response, rate-adaptive. See PSE. Cf Dual chamber

ventricular premature beat Premature ventricular contraction, see there

ventricular remodeling Left ventricular diameter reduction CARDIOVASCULAR SURGERY An operative technique for CHF, which consists of excising the flabbiest portion of the dilated ventricle followed by side-to-side anastomosis; VR ↑ the pumping efficiency of the remaining tissue. See Transmyocardial revascularization

ventricular septal defect Interventricular septal defect; ventriculoseptal defect, VSD CARDIOLOGY A congenital heart defect–CHD in which there is direct communication between the right and left ventricles through a defect in the muscular or fibromembranous septum; VSDs result when there is a lack of growth or failure of the components to align or fuse; simple VSD accounts for 25% of all CHDs; 25% to 40% of VSDs close spontaneously by age 2, 90% by age 10; small VSDs are usually asymptomatic; if large, blood flow direction is determined by systemic vs pulmonary vascular resistance; with time, pulmonary vascular resistance ↑, resulting in pulmonary HTN–PH and R→L shunting ANATOMIC DEFECTS Most VSDs–70% located in the membranous portion of the interventricular septum; the remainder are located in the muscular portion of the septum, below the aortic valve, and at the atrioventricular canal RISK FACTORS FOR VSD Maternal rubella, other viral illness during pregnancy, poor nutrition, alcoholism, age ≥ 40, DM SEVERITY With major L→R shunting, LV impulse is laterally dsplaced and the RV impulse is weak in absence of PH HEART SOUNDS Holosystolic murmur loudest at the left sternal border, often accompanied by a palpable thrill; mid-diastolic apical rumble due to ↑ flow across the mitral valve; decrescendo diastolic murmur of aortic regurgitation if the VSD undermines the valve annulus; if PH is present, an RV heave and pulsation over the pulmonary trunk is palpable; with PH, the holosystolic thrill and murmur ↓ and a Graham Steel murmur–pulmonary regurgitation–appears EKG Large defects have evidence of left atrial and LV enlargement; with pH!RS axis shift appears CLINICAL Cyanosis, finger clubbing IMAGING CXR–LV enlargement and shunt vasculopathy; with PH, there is marked enlargement of the proximal pulmonary arteries, rapid

tapering of the peripheral pulmonary arteries, and oligemic lung fields; other techniques used include 2-D echocardiography with Doppler flow, catheterization and angiography MANAGEMENT Adults with small defects are generally asymptomatic, but have an ↑ risk of infective endocarditis; for adults with large defects surgical closure is indicated if the pulmonary obstructive vascular disease will allow treatment. See Atrial fibrillation, Congenital heart disease. Cf Atrial septal defect

---

VENTRICULAR SEPTAL DEFECT-FUNCTIONAL TYPES

SMALL-MEDIUM LEFT → RIGHT SHUNTS W/O PULMONARY HTN Acyanotic, asymptomatic, grade II-IV/VI pansystolic murmur MANAGEMENT May resolve spontaneously, or be too small to justify repair

LARGE LEFT → RIGHT SHUNTS W/O OR W/ PULMONARY HTN Acyanotic, easy fatigability, CHF in infancy, grade II-V/VI pansystolic murmur; hyperactive heart; biventricular enlargement, diastolic murmur at apex; P_2 usually accentuated MANAGEMENT Repair required; in infancy for CHF; before age 2 if progressive pulmonary HTN; otherwise, 'electively' between ages 2 and 5

---

ventricular tachycardia V tach, wide-complex tachycardia CARDIOLOGY A common, potentially life threatening abnormal rapid–160–240 beats/min–heart beat initiated in the ventricles, characterized by 3 or more consecutive premature ventricular beats; VT may compromise systemic pumping of blood TRIGGERS Spontaneous, post-acute MI, cardiomyopathy, mitral valve prolapse, myocarditis, post-heart surgery, antiarrhythmics, ↓ potassium, pH–acid-base changes, ↓ O_2 TYPES Nonsustained–eg, lasting < 30 secs, sustained, > 30 secs. See Torsade de pointes

ventriculitis INFECTIOUS DISEASE Inflammation of cerebral ventricles, most common in infants, caused by gram-negative bacilli or group B streptococci, and often an extension of purulent meningitis MANAGEMENT Ventriculostomy, intraventricular antibiotics. See Purulent meningitis

ventriculography IMAGING Imaging of cerebral or cardiac ventricles. See Radionuclide ventriculography

ventriculoperitoneal shunt VP shunt NEONATOLOGY A tube implanted in the cerebral ventricle in neonates with noncommunicating hydrocephalus which empties into the abdominal cavity

ventriculotomy SURGERY An incision into a ventricle, either of the brain or heart

verapamil CARDIOLOGY A CCB antiarrhythmic that slows transmission of impulses across the AV node and supresses the sinus node's intrinsic rhythm INDICATIONS Tachycardial supraventricular arrhythmias–eg, paroxysmal supraventricular tachycardia, A Fib with tachyarrhythmias, atrial flutter with rapid conduction, and is the parenteral emergency drug of choice ADVERSE EFFECTS Constipation, slowed AV conduction and sinus node automaticity, peripheral edema; it is a standard agent for managing HTN, especially in Pts with angina; it improves pain linked to obstructive cardiomyopathy, and prevents migraines CONTRAINDICATIONS Severe left ventricular dysfunction, hypotension, sick sinus syndrome, 2° or 3° AV block, WPW, heart failure, or in Pts with accessory bypass tracts, WPW syndrome, peripheral edema; caution in renal failure. See Calcium channel blocker, Therapeutic drug monitoring. Cf Beta blocker

verbal abuse PSYCHOLOGY A form of emotional abuse consisting of the use of abusive and demeaning language with a spouse, child, or elder, often by a caregiver or other person in a position of power. See Child abuse, Emotional abuse, Spousal abuse

verbal and motor perseverence NEUROLOGY A manifestation of total–or global, aphasia, characterized by obligate repetition of a word or motor act just after it has been used in another context. See Aphasia, Motor aphasia

verbal intelligence quotient PSYCHOLOGY A parameter of cognitive ability, measurable by the WAIS-R–Wechsler Adult Intelligence Scale-Revised. See Intelligence quotient

verbal overflow NEUROLOGY A manifestation of ideomotor apraxia, characterized by an inability to perform a task without the person first stating what he was going to do

verbal stereotypy Automatism, see there

verbal therapy 1. Imagineering, see there 2. Talking therapy, see there

verbigeration NEUROLOGY Stereotyped and seemingly meaningless repetition of words, sentences, or associations, often at the end of a thought CAUSES Bipolar disorder, organic brain disease, schizophrenia. See Perseveration

Verdia® Tasosartan CARDIOLOGY An antihypertensive angiotensin II receptor antagonist, withdrawn from the market by the manufacturer. See Angiotensin II receptor antagonist

Verdict THC LAB MEDICINE A one-step test for detecting marijuana. See THC

Verelan® Verapamil, see there

Veridien umbilical clamp OBSTETRICS A device that automatically clamps, cuts and seals the umbilical cord near the newborn

verification MANAGED CARE The process of evaluating a system, component or other product at the end of its development cycle to determine whether it meets projected performance goals. See Beta testing SPORTS MEDICINE The determination of validity of a structure or function. See Gender verification

verified longest lived individual POPULATION BIOLOGY A member of a given species whose maximum length of life has been observed and verified, which is an operational definition of the species' maximum life-span potential. See Lifespan, Life expectancy

Verluma™ Nofetumomab IMAGING An agent that detects a protein on the surface of most small cell lung CAs; it may be used to diagnose biopsy-confirmed tumors. See Small cell lung cancer

vermicide PARASITOLOGY An agent that kills intestinal parasites

Vermox® Mebendazole, see there

vernal conjunctivitis OPHTHALMOLOGY A seasonal–spring and summer–inflammation of the conjunctivae, largely attributed to allergies CLINICAL Itchy, watery eyes, photophobia, cobblestone-like changes under the eyelid, scarring around the limbus which, if it extends onto the cornea, may compromise vision. See Conjunctivitis

vernix caseosa NEONATOLOGY A white fatty goo composed of squames, lanugo, and sebaceous secretions, which covers and protects neonatal skin; the closer to term the infant, the less the VC; post-term infants have very little VC

vernix caseosa peritonitis NEONATOLOGY A rare complication of a C-section, caused by a granulomatous response to amniotic fluid spilled into the abdominal cavity

vero toxin Shiga neurotoxin, see there

verocytotoxin VTEC, see there

Veronal Barbital, see there

VerreScope™ SURGERY A microlaparoscope. See Laparoscopy

verruca 1. Condyloma, see there 2. Verruca vulgaris, see there 3. Wart, see there

verruca acuminatum Condyloma acuminatum, see there

verruca pedis Plantar wart, see there

verruca plana juvenilis DERMATOLOGY A flat wart most common on sun-exposed skin of the face, neck, hands, wrists, and knees. See Verruca vulgaris

verruca vulgaris Wart DERMATOLOGY A benign skin tumor induced by HPV and histologically characterized by hyperkeratosis, parakeratosis, acanthosis. See Human papillomavirus, Wart

verrucae Plural of verruca, see there

verruciform xanthoma A benign solitary lesion of the oral cavity–occasionally extraoral, described as a red, asymptomatic lesion occurring at age 40 TREATMENT Complete excision

verrucosis A condition characterized by multiple warty excrescences. See Lymphostatic verrucosis

verrucous carcinoma A WD SCC of the oral cavity, larynx, nasal cavity, esophagus, penis, anorectal region, vulva, vagina, uterine cervix and skin, especially on the sole of the foot; intraoral VCs occur in elderly ♂ smokeless tobacco users TREATMENT Resection. See Squamous cell carcinoma

verrucous hyperplasia Acanthosis in which epithelium–eg, of the foot or oral cavity is thickened, but not neoplastic; chronic VH–eg, of the foot with lymphedema may evolve to verrucous carcinoma

VersaPoint™ GYNECOLOGY Minimally-invasive electrosurgical device used with hysteroscopy for vaporizing, cutting or dissecting tissue, specifically for safe removal of uterine fibroids and as an alternative to hysterectomy. See Leiomyoma

Versaport SURGERY A trocar requiring a smaller incision site ↓ risk of herniation and improving cosmetic results. See Minimally invasive surgery

VersaPulse® UROLOGY A holmium laser used to resect a prostate and/or urethra with BPH, CA, strictures, stones. See HoLRP

Versed® Midazolam PHARMACOLOGY A preoperative sedative

version OBSTETRICS The manual changing of the axis of the fetus with respect to the mother. See External version, Inactive version, Internal version INFORMATICS The maximum number of different backup copies of files retained for files

VERT OSTEOPOROSIS A clinical trial–Vertebral Efficacy with Risedronate Actonel® Therapy. See Osteoporosis

vertebral-basilar artery occlusive syndrome Subclavian steal syndrome, see there

vertebral compression fracture Compression fracture of back ORTHOPEDICS A traumatic fracture of a vertebral body which may occur in a background of osteoporosis or malignancy and cause kyphosis and spinal cord pressure. See Herniated disk

vertebrobasilar circulatory disorder NEUROLOGY Any of several disorders involving the basilar and vertebral arteries which disrupts blood flow to the brain stem and cerebellum CLINICAL Episodic dizziness, double vision, weakness, and dysarthria. See Subclavian steal syndrome

verteporfin–benzoporphyrin BPD verteporfin THERAPEUTICS A photodynamic–light-activated agent in clinical trials for treating age-related macular degeneration–see Visudyne™–and psoriasis. See Photodynamic therapy

vertex compression NEUROLOGY Axial compression at the top of the skull

vertex presentation Anterior presentation, crown presentation, occiput presentation OBSTETRICS A head position at the time of delivery, where the crown of the baby is the presenting part; VP is the easiest presentation to deliver. See Cephalic presentation, Cesarean section

vertical transmission EPIDEMIOLOGY Mother-to-child transmission The transmission of an infection through the placenta to the fetus, as occurs in the 'TORCH' infections–toxoplasmosis, rubella, CMV, herpes, syphilis, and HIV. Cf Hereditary transmission, Horizontal transmission

vertiginous *adjective* Related to vertigo, dizzy

vertigo Dizziness NEUROLOGY A distortion of perception characterized by a sensation of rotational movement or loss of equilibrium, a finding typical of vestibular dysfunction CLINICAL Often accompanied by nystagmus and, if severe, N&V ETIOLOGY Benign positional vertigo, Ménière's disease, labyrinthitis, acoustic neuroma TREATMENT–MEDICAL If acute, diazepam; if recurrent, scopolamine; if nausea, antiemetic; if severe, bed rest; if recurrent, exercise TREATMENT–INTERVENTIONAL Transmastoid labyrinthectomy, vestibular nerve section, middle ear endoscopy, semicircular canal ablation, streptomycin infusion. See Benign paroxysmal positional vertigo, Objective vertigo, Positional vertigo, Subjective vertigo. Cf Dizziness, Dizzy spell, Pseudovertigo

---

VERTIGO–DURATION

SECONDS Benign paroxysmal positional vertigo

MINUTES TO HOURS

a. Idiopathic endolymphatic hydrops–Ménière's disease

b. Secondary endolymphatic hydrops

  1. Otic syphilis

  2. Delayed endolymphatic hydrops

  3. Cogan's disease

  4. Recurrent vestibulopathy

DAYS Vestibular neuronitis

VARIABLE DURATION

a. Inner ear fistula

b. Inner ear trauma

  1. Nonpenetrating trauma

  2. Penetrating trauma

  3. Barotrauma

---

very high-density lipoprotein VHDL A plasma lipoprotein with a density of > 1.210 kg/L–US: 1.210 g/dl; VHDL is 57% protein–predominantly apoA-I and apoA-II, 21% phospholipid, 17% cholesterol and 5% TGs and transports cholesterol from the intestine to the liver; the larger the HDL molecule, the more efficient the lipid transport and by extension, lipolysis. See HDL

very low birth weight VLBW NEONATOLOGY Referring to an infant weighing between 1000 g and 1500 g at birth; these children are at high risk for neurobehavioral dysfunction, poor school performance. See Low birth weight . Cf Extremely low birth weight

VERY LOW BIRTH WEIGHT INFANTS–OUTCOMES

Birthweight	≤ 750 g	750-1.5 g	≥ 1.5 Kg
Sample number	68	65	61
MPC score*	87	93	100
Mental retardation (IQ < 70)	21%	8%	2%
Poor cognitive function	22%	9%	2%
Poor academic skills	27%	9%	2%
Poor gross motor function	27%	9%	0%
Poor adaptive function	25%	14%	2%
Cerebral palsy	9%	6%	0%
Severe visual disability	25%	5%	2%
Hearing disability	24%	13%	3%
↓ Weight/height/head size	22/25/35%	11/5/14%	0/0/2%%

*Mental Processing Composite score

very low-density lipoprotein See VLDL

vesical *adjective* Referring to the urinary bladder

vesicle DERMATOLOGY A small skin blister

vesico-ureteral reflux Reflux nephropathy, ureteral reflux, vesicoureteric reflux UROLOGY The retrograde flow of urine from the bladder into the ureters which occurs when the unidirectional valve-like flow between the ureters and bladder fails; bladder infection may cause pyelonephritis, and expose the kidney to high pressures, which over time damages the kidney and causes scarring ETIOLOGY Bladder infection, stones, bladder outlet obstruction, neurogenic bladder, and abnormal ureters or number of ureters

vesicular *adjective* Referring to a vesicle or blisters

vesicular stomatitis virus A rhabdovirus which replicates in the cytoplasm of infected cells; most VSV victims were in direct contact with oral secretions of infected livestock CLINICAL Fever, chills, malaise, myalgia, N&V, pharyngitis. See Foot and mouth disease, Rabies

vesnarinone CARDIOLOGY A quinolinone with a narrow therapeutic range; withdrawn from the market. See Flosequinan

vessel See Blood vessel, Corkscrew vessels, Double-barreled blood vessel, Great vessel, Hairpin vessel PUBLIC HEALTH An enclosable structure. See Containment vessel

Vessel Sanitation Program See Sanitation score

vest CPR EMERGENCY MEDICINE CPR using a proprietary vest to ↑ chest pressure; the vest is pneumatically cycled and applies pressure in a circumferential fashion, vs point–sternal pressure of manual CPR. See Cardiopulmonary resuscitation

vest-over-pants repair SURGERY A method used in surgical correction of inguinal hernias, where the fascia above the hernia is brought down over the fascia from below, effecting a 2-layer closure. See Inguinal hernia

vestibular neuronitis NEUROLOGY A condition that presents with dramatic, abrupt onset of vertigo and vegetative Sx; vertigo for days, gradual improvement; slow phase of nystagmus is toward affected side and hypofunction is observed on caloric responses; auditory Sx are absent ETIOLOGY VN invariably follows a viral URI PROGNOSIS VN may recur and be bilateral; postural and motion instability with certain head movements occurs for months subsequently; 15% of Pts develop benign paroxysmal positional vertigo TREATMENT Supportive, symptomatic, early ambulation. See Benign paroxysmal positional vertigo

vestibular testing NEUROLOGY A battery of clinical tests for evaluating the neural component of the vestibular system in Pts with dysequilibrium, dizziness, loss of balance, nystagmus; VTs evaluate both the 'mechanical'–ie, the vestibule per se, and the 'neural processing' components–cerebellum of the vestibular system; alcohol, sedatives, tranquilizers are withheld for 2–3 days before VTs. See Nystagmus, Vertigo, Vestibule

vestibular window Oval window, see there

vestibulo-ocular reflex NEUROLOGY A reflex in which eye movement is equal and opposite to the head movement; loss of the VOR implies vestibular disease that may accompany aminoglycoside toxicity

veto A legislative nyet

Vexol® Rimexolone OPHTHALMOLOGY A topical for treating inflammation after ocular surgery and for anterior uveitis. See Uveitis

viability Survival capability; capable of extrauterine survival–eg, a fetus. See Limit of viability

viable infant NEONATOLOGY An infant who is likely to survive to the point of sustaining life independently, given the benefit of available medical therapy. Cf Nonviable fetus

Viadur® Duros® (leuprolide acetate) implant ONCOLOGY A device designed to provide once-yearly dosing of leuprolide for palliating advanced prostate CA. See Prostate cancer

Viagra® Sildenafil citrate UROLOGY An agent used to manage erectile dysfunction–male impotence that enhances the effects of endogenous nitric acid, which relaxes penile smooth muscle allowing an inflow of blood, ergo erection; it is ± 70% effective INDICATIONS ED due to DM, HTN, hypercholesterolemia, prostate surgery, spinal cord injuries and psychologic factors COMPLICATIONS Flushed skin, headaches, upset stomach, blue-tinted vision, pregnancy ADVERSE EFFECTS Stroke, MI, priapism CONTRAINDICATIONS Nitrate therapy–eg, nitroglycerin. See Erectile dysfunction, Impotence.

ViaScint® IMAGING A formulation of I¹²³-iodophenyl pentadecanoic acid used to evaluate post-CABG myocardium. See CABG

ViaSpan® NEPHROLOGY A preservation fluid that is the standard medium for maintaining organs and tissues–eg, kidneys, liver, pancreas, small bowel from time of procurement to transplantation. See Kidney transplant, Procurement

ViaStem™ TRANSPLANTATION A solution used to improve storage of stem cells from BM or peripheral blood while a Pt receives RT and/or chemotherapy for CA or immune disorders; clinical trials are planned for use of umbilical cord blood and platelets. See Bone marrow transplantation

Vibra-Surge™ SURGERY An ultrasonic surgical instrument used for excising soft tissue

vibration VOX POPULI Jittering, oscillation, grinding

vibrator Robopenis GYNECOLOGY A cylindrical battery powered device with an internal motor that generates periodic oscillations, traditionally used as an autostimulator/sexual aid. See Sex toy

Vibrio cholerae INFECTIOUS DISEASE The *Vibrio* that produces the heat-tolerant exotoxin which causes cholera EPIDEMIOLOGY Transmitted through poorly treated water supplies CLINICAL Abdominal cramping, diarrhea MANAGEMENT Rehydration is more important than antibiotics. See Broad Street pump, Cholera toxin

Vibrio vulnificus CDC group EF-3 BACTERIOLOGY A bacterium of brackish or salt water, which may contaminate oysters, and be part of the normal marine flora; may cause wound infections and septicemia, possibly also gastroenteritis 2° to exposure to contaminated water or seafood CLINICAL *V vulnificus* is a virulent noncholera vibrio; it may rarely cause acute, self-limited gastroenteritis in those receiving antibiotics; major clinical forms: 1. compromised hosts–eg, Pts with cirrhosis, *V vulnificus* crosses the GI mucosa, passes into the circulation and causes fever, chills, hypotension and, in most Pts, metastatic skin lesions within 36 hrs, by erythema, hemorrhagic vesicles and bullae, necrotic ulcers; the condition is fatal in V_2; 2. in otherwise healthy persons, *V vulnificus* may cause intense cellulitis, necrotizing vasculitis and ulceration, which requires debridement MANAGEMENT Tetracycline; penicillin, chloramphenicol. Cf *Vibrio cholerae*

vibrotactile aid AUDIOLOGY A mechanical device that help deaf persons detect and interpret sound through touch. See Deafness

vicarious menstruation Cyclical bleeding from various sites–eg, the lungs, which occurs in ♀ of reproductive age, usually from endometriosis. See Endometriosis

vicious viscous hypothesis Kinetic hypothesis HEMATOLOGY A hypothesis that attempts to explain the pathogenesis of the microvascular occlusions typical of sickle cell anemia; according to VVH, the ↑ in blood viscosity caused by the relatively inelastic sickled RBCs, slows blood flow in the microcirculation; this slowing results in relative hypoxia, which favors further sickling of upstream RBCs, resulting in vascular occlusion and infarction

Vicoprofen® PAIN MANAGEMENT An opioid/ibuprofen combination for treating acute pain in Pts with physical or psychologic opiate dependence. See Opiate dependence

victim One who suffers an injustice at another's hand. See Fashion victim

victimization SOCIAL MEDICINE The abuse of the disenfranchised–eg, those underage, elderly, ♀, mentally retarded, illegal aliens, or other, by coercing them into illegal activities–eg, drug trade, pornography, prostitution. See Homelessness, Pornography, Runaway

victimizer PSYCHOLOGY A victim who, having been physically, sexually, emotionally abused, reverses the role and abuses others

vidarabine Ara-A VIROLOGY An IV or topical antiviral active against DNA viruses; it is a competitive inhibitor of DNA-dependent DNA polymerase, and ↓ mortality of neonatal HSV, HSV encephalitis TOXICITY Thrombocytopenia, leukopenia, N&V, diarrhea, irritability, tremors, ataxia, paresthesias, hallucinations, ↓ renal, liver function

video-assisted surgery Video-assisted resection Surgery aided by use of a video camera that projects and enlarges the image on a television screen

video-assisted thoracic surgery SURGERY A format of minimally invasive thoracic surgery used to treat a number of specific lesions INDICATIONS Pulmonary nodules, effusions, infiltrates, pneumothorax, mediastinal mass, pleural mass, etc PROCEDURES Wedge resection–excision, examination, pleural biopsy, talc pleurodesis, decortication, excision of masses, application of fibrin glue to air leak COMPLICATIONS Persistent air leak, A Fib, respiratory failure CONVERSION TO OPEN PROCEDURE 33% require conversion, due to obscure nature of lesion, malignancy, obliterated pleural space MORTALITY 2%

video consultation TELEMEDICINE The transmission of video images to an expert/consultant located far from a Pt. See Telemedicine

videodefecography GI DISEASE A technique for evaluating rectoanal function, in which barium is instilled per rectum and the changes that occur in response to relaxing, coughing, or straining are evaluated in 'real time' by videofluoroscopy. Cf Defecography

video electroencephalography NEUROLOGY A method for evaluating seizure disorders, in which the EEG signal is amplified, encoded by an analog system, and transmitted to a central station, as an analog or digital video image of the Pt. See Seizure disorder

videolaryngoscope ENT A video camera attached to a flexible fibroendoscope, which is used to record and document, for medicolegal purposes, laryngeal lesions

video thoracoscopy ENDOSCOPY A minimally invasive technique in which a flexible endoscope with a video-camera interface and instrumentation is introduced through an intercostal incision

Videx® Didanosine–ddI, see there

Vietnam syndrome PSYCHIATRY A popular term for the psychosocial consequences of active participation in the Vietnam conflict–eg, substance abuse, depression. See Burned-out syndrome, Post-traumatic stress disorder. Cf Gulf War syndrome

view *noun* IMAGING The direction from which a radiologic image is obtained. See Jughandle view, Swimmer's view

Viewing Wand® IMAGING An intraoperative neurosurgical imaging device for head & neck procedures, and for spinal surgery

vigabatrin Gamma-vinyl GABA, Sabril® PHARMACOLOGY A 2nd-line antiepileptic used for convulsions or seizures refractory to other agents ADVERSE EFFECTS Somnolence, stupor, hyperkinesia, insomnia, ↑ weight, facial edema

vigilance NEUROLOGY The conscious and semiconscious focusing and sustained attention to subtle sensory signals within a determined modality–eg, auditory or visual, coupled to filtering out of distracting internal and external stimuli; PET studies of humans localize the 'attention center' to the prefrontal and superior parietal cortex primarily in the right hemisphere, regardless of the modality or laterality of the stimulus

VIGOR INTERNAL MEDICINE A clinical study–Vioxx GI Outcomes Report comparing a proprietary COX-2 inhibitor to standard NSAIDs

vigorous exercise A form of exercise that is intense enough to cause sweating and/or heavy breathing/ and/or ↑ heart rate to near maximum; VE is formally defined as that which requires > 6 METs; there is a graded inverse relationship between total physical activity–in particular vigorous exercise and mortality. See Exercise, MET. Cf the 'Zone'

villi Plural of villus A small vascular protrusion, especially from the free surface of a membrane

VIN Vulvar intraepithelial neoplasm, see there

vinblastine ONCOLOGY An IV vinca alkaloid used with other agents to treat Hodgkin's disease, leukemia, and other lymphoproliferative disorders, KS, CAs ADVERSE EFFECTS BM toxicity, neuropathy

vinca alkaloid ONCOLOGY Any agent that inhibits CA growth by stopping cell division, aka antimitotic or antimicrotubule agents, or mitotic inhibitors

Vincent's stomatitis Acute necrotizing ulcerative gingivitis, see there, aka Trench mouth

vincristine Oncovin® AIDS A dimeric vinca alkaloid used to treat KS, thrombocytopenia, leukemias, lymphomas, solid tumors and other CAs which, like vinblastine, interrupts the mitotic spindle, causing lysis of proliferating cells ADVERSE EFFECTS Reversible peripheral neuropathy, constipation. Cf Vinblastine

vindesine Eldisine® ONCOLOGY An experimental vinca alkaloid used to manage CA, Hodgkin's disease, leukemia, other lymphoproliferative disorders ADVERSE EFFECTS BM toxicity, neuropathy

vinorelbine ONCOLOGY An anticancer vinca alkaloid. See Vinblastine, Vincristine

vinyl chloride TOXICOLOGY A monomer that polymerizes to polyvinyl chloride–PVC, which is used in organic chemistry and manufacturing plastics ADVERSE EFFECTS VC is carcinogenic and carries a risk of liver, brain, lung CA, lymphoma, leukemia. See PVC–polyvinyl chloride

violent household PUBLIC HEALTH A household that fosters child, elder, or spousal abuse. See Domestic violence

violin band mark ENTOMOLOGY A distinct stradivariusoid macule on the dorsal cephalothorax of *Loxosceles reclusa*, the brown recluse spider of the southern US. See Brown recluse spider

violin string adhesions A descriptor for the taut fibrous bands between the anterior parietal peritoneum and the anterior face of the liver, typical of *N gonorrhoeae*-induced perihepatitis, which may be accompanied by RUQ pain, fever and a hepatic friction rub; VSAs also occur in the uterine adnexae in PID due to *Chlamydia trachomatis* and in fibrinous mesothelial inflammation of the pericardium, pleura, peritoneum

Vioxx® Rofecoxib PAIN MANAGEMENT A COX-2 inhibitor used to manage osteoarthritic pain, acute pain in adults, primary dysmenorrhea.

See COX-2 inhibitor

VIP receptor scan IMAGING A radioisotope-based diagnostic modality based on the finding that certain tumors–GI adenoCAs and endocrine tumors have high-affinity VIP receptors; after an IV bolus injection of radiolabeled–[123]I VIP, its distribution is evaluated using a gamma camera.

V.I.P. 'syndrome' A "condition" caused when a very important person–V.I.P. by virtue of fame, position or claim on public interest–disrupts the normal course of Pt care in a hospital. See Chief 'syndrome,' Code Purple

VIPoma ONCOLOGY A VIP–vasoactive intestinal peptide-producing tumor of the pancreas, which is associated with a cholera-like syndrome in absence of gastric hypersecretion. See VIPoma syndrome

VIPoma syndrome Pancreatic cholera syndrome, Verner-Morrison syndrome, WDHA syndrome ONCOLOGY Any of a number of clinical complexes characterized by watery diarrhea, ↓ K+, hypotension and dehydration caused by neuroendocrine tumors of the pancreas, most commonly VIPoma. See WDHA

Viprinex™ Ancrod, formerly Arvin™ CRITICAL CARE An orphan drug from Malayan pit viper venom, used in Pts with acute ischemic stroke to anticoagulate heparin-intolerant Pts undergoing CABG. See CABG

Viracept® Nelfinavir mesylate AIDS An antiretroviral HIV protease inhibitor for treating AIDS, prescribed as a monotherapy or in a triple-combination therapy with Retrovir®–AZT, Epivir®–3TC . See Protease inhibitor

VIRADAPT AIDS A French study of AIDS Pts unresponsive to protease inhibitors. See Genotype testing

viral arthritis RHEUMATOLOGY A transient and nonspecific inflammatory response of one or more joints to a viral infection–eg, mumps, rubella, human parvovirus, and HBV, which may also follow rubella vaccination. See Arthritis

viral culture A test in which a specimen–eg, throat swab, sputum, stool, CSF, urine, from a Pt is placed in live cells; various viruses–eg, adenovirus, enterovirus, herpes simplex, measles, mumps, myxovirus, paramyxovirus, rhinovirus, rubella, varicella-zoster, etc can be cultured from clinical specimens, but are not routinely performed as effective therapies are limited. See Shell vial assay

viral encephalitis Viral meningoencephalitis NEUROLOGY, INFECTIOUS DISEASE A general term for nonpurulent–'aseptic' viral infection of the CNS ETIOLOGY Coxsackie A and B–eg, A7, enterovirus 71, herpes simplex, etc CLINICAL If the viral load is extreme, asymmetric flaccid paralysis may ensue LAB CSF pleocytosis, ↑ lymphocytes. See Aseptic meningitis. Cf Cowdry type B inclusion

viral gastroenteritis Intestinal flu INFECTIOUS DISEASE A generic term for GE induced by viruses CLINICAL PRESENTATIONS 1. Epidemic VGE, most often caused by the Norwalk agent or Norwalk-like viruses CLINICAL N&V, diarrhea, abdominal pain, anorexia, headache, malaise, low-grade fever; it rarely requires hospitalization, but may be contribute to death in the elderly 2. Sporadic VGE is often caused by rotaviruses and differs from the epidemic form of VGE as it most commonly affects children < age 2, and evokes a wide range of responses, from subclinical infection to a mild form of the GE, as with the epidemic form, to a severe, potentially life-threatening dehydrating disease

viral hemorrhagic fever syndrome A condition caused by various viruses–eg, Filoviridae–eg, Ebola and Marburg agents, Arenaviridae–eg, Junin and Lassa agents, Togaviridae including Flaviviridae-eg, Dengue, Omsk and yellow fever agents, and Bunyaviridae–eg, Congo-Crimean, Hantan virus and Rift Valley fever agents VECTORS Arthropods–ticks or mosquitoes–eg, Dengue and Rift Valley and yellow fevers and environmental contaminants–eg, Argentine, Bolivian, Ebola, Lassa and Marburg HFs and

HF with renal syndrome CLINICAL Incubation 3 days to 3 wks, then headache, myalgias, dysphagia, vomiting, diarrhea, abdominal and/or chest pain, pharyngitis, conjunctivitis, cervical lymphadenopathy, macular rashes, shock LAB Leukopenia and/or thrombocytopenia, proteinuria, DIC

viral hepatitis GI DISEASE Liver inflammation caused by viruses—eg, hepatitis A, B, C, D, E, F, GB, and other viruses. See Hepatitis A, Hepatitis B, Hepatitis C, Hepatitis delta, Hepatitis E

viral load The amount of a specific virus—eg, HIV, in a person—eg, a Pt with AIDS. See bDNA signal amplification, HIV RNA

viral load test LAB MEDICINE A test that measures HIV RNA for prognosis and/or to monitor the efficacy of anti-HIV drug regimens

virally-induced cancer Any malignancy induced by either DNA or RNA viruses, which include HBV–liver cell carcinoma, HPV–squamous cell carcinoma of the cervix, HTLV-I–adult T-cell leukemia, and EBV–Burkitt's lymphoma-Africa, nasopharyngeal carcinoma-China; vaccines *may* prevent VICs

viral meningoencephalitis Viral encephalitis, see there

viral pharyngitis INFECTIOUS DISEASE Oropharyngeal inflammation caused by viruses; VP is the most common cause of a sore throat and diagnosis of exclusion, which is often linked to a recent URI. See Strep throat

viral pneumonia PULMONOLOGY Pneumonia of viral origin, which is more severe in the very young and very old COMMON PATHOGENS Adenovirus, influenza virus, parainfluenza virus, RSV, rhinovirus, HS, CMV. See Influenza, Pneumonia, Respiratory syncytial virus

viral profile A battery of immunoassays performed on the serum of Pts suspected of being infected by, or having been recently exposed to viruses

viral study VIROLOGY Any test or battery of tests used to detect or confirm past or present exposure to a particular virus TYPES Indirect–viral effects on the host are assessed by measuring antibody levels; direct–the virus itself is cultured. See Viral culture

viral transduction GENE THERAPY A method of gene insertion in which a modified virus infects a cell of interest, introducing viral genome containing the foreign genes. Cf Physical transfection

Viramune® Nevirapine, see there

Virchow's node Left supraclavicular lymph node SURGERY A lymph node that is a relatively common site of metastases from abdominal or pelvic CA. Cf Sister Mary Joseph node

viremia Presence of viral particles in the circulation

virgin birth SURROGACY A popular term for a child born out of wedlock to a ♀ through artificial insemination from an anonymous donor. See Artificial reproduction

virginal hypertrophy Gigantomastia A marked ↑ in breast size in adolescent ♀, which may be uni– or bilateral and characterized by ductal and stromal proliferation ETIOLOGY Uncertain MANAGEMENT Reduction mammoplasty

virginity Sexual naïveté Never having engaged in sexual congress, intercourse; the classic finding in ♀ virgins is an intact hymen. See Abstinence, Hymen

virilism ENDOCRINOLOGY The presence of ♂ physical and sexual characteristics in a ♀–hirsutism, receding frontal hairline, masculine body habitus, deepening of voice, ↑ sebaceous gland secretion, clitoral enlargement. See Hirsutism, Virilization

virilization Physical changes in a ♀ 2° to androgen excess CLINICAL Hirsutism, temporal balding, deepened voice, acne, enlarged clitoris and clitoral index–ratio between the sagittal and transverse diameters of the clitoris;

the androgen excess may be 1. Adrenal, due to Cushing's syndrome, congenital enzyme deficiency–eg, 11– or 21–hydroxylase deficiency or neoplasia, either benign or malignant or 2. Ovarian, due to hyperplasia–eg, polycystic ovaries or tumors–eg, arrhenoblastoma, steroid cell tumors or gynandroblastoma. Cf Hermaphroditism, Intersex, Precocious puberty, Virilism

virologic failure Antiretroviral therapy failure, see there

Viroptic® Trifluridine OPHTHALMOLOGY A topical solution used manage primary keratoconjunctivitis and recurrent HSV-1 and HSV-2 keratitis. See Keratoconjunctivitis

virosome vaccine A 'designer' vaccine for influenza in which hemagglutinin is extracted from the influenza virus and incorporated into the membrane of liposomes composed of phosphatidylcholine and phosphatidylethanol-amine

viroxan AIDS FRAUD An unsterilized mixture of uncertain nature once claimed to have anti-HIV activity. See AIDS fraud; Cf AIDS therapy

virtual animal Virtual critter COMPUTERS A small mobile piece of computer software that can send itself across a computer network and perform a task on a remote machine. Cf Agent

virtual colonoscopy IMAGING A diagnostic procedure in which 2-D helical CT images are converted into 3-D images of the colonic mucosa that are similar and an alternative to those obtained by conventional colonoscopy. See Colonoscopy, CT colography, Double contrast barium enema

virtual consultation ONLINE An Internet physician consultation for which, depending on the diagnosis, medical therapy may be prescribed

virtual endoscopy IMAGING The use of an imaging modality–eg, ultrafast CT, to obtain 2-D images, which are reconstructed to form a 3-D image that is similar to endoscopy; VE does not obtain a biopsy. See Position

virtual hospital 21ST CENTURY MEDICINE An evolving format for providing health care in a central location that relies on the components of telemedicine and high-tech communications devices to connect it to a regional hospital, where a greater range of expertise is available. See Telemedicine

virtual physician A hypothetical inhabitant of the world of 'cybermedicine', who 'lives' in the Internet as an entity providing information '24/7' on a person's disease, types, cost-effectiveness, and outcomes of various diagnostic modalities, therapeutic options. See Telemedicine, Video consultation

virtual reality exposure therapy PSYCHIATRY An exposure therapy using artificial or computer-generated sensory experiences, which may be effective for treating phobic disorders. See Phobic disorder

virtual visit ONLINE An Internet-based episode of interaction between a physician and Pt. See eMedicine. Cf Teleconsultation

virulence EPIDEMIOLOGY The proportion of persons with clinical disease, who after infection, become severely ill or die. See Neurovirulence

virus INFECTIOUS DISEASE A small, obligatorily intracellular agent ranging from 10^6 daltons–eg, Parvoviridae to 200×10^6–eg, Poxviridae; viral nucleic acid is single- or double-stranded, either DNA or RNA, and is a closed circle or opened and linear; viral nucleic acid is packaged within a protein coat–capsid composed of a few distinct types of protein; most have a helical or icosahedral symmetry; once inside the infected cell, the virus uses the host's synthetic capabilities to produce progeny virus; some viruses–eg, influenza virus, are 'studded' with external proteins–eg, hemagglutinins, neuraminidases

virus-associated hemophagocytic syndrome A hemophagocytotic condition seen in immunocompromised–♂:♀ ratio, 1:1 and in nonimmunocompromised–♂:♀ ratio, 4:1 subjects, often in a background of DNA virus–eg, CMV, EBV, herpesviruses, and adenovirus and rubella infections; VAHS may accompany brucellosis, candidiasis, leishmaniasis, TB, salmonellosis, GVHD, hemolytic anemia CLINICAL Hepatosplenomegaly, lymphadenopathy, pulmonary infiltrates, pancytopenia–with a depleted BM, skin rash DIFFDX Lymphoma, malignant histiocytosis, sinus histiocytosis with massive lymphadenopathy, lymphomatoid granulomatosis

virus scan INFORMATICS A computer program that can ID code–geek speak for computer program subroutines–often found in computer viruses. See Computer virus

vis-a-tergo CARDIOLOGY A term referring to the force driving the venous return of peripheral blood, which is supplied by the left ventricle; by the time blood has passed through the capillaries, the blood pressure, or vis-a-tergo, is 15 mm Hg

visceral *adjective* 1. Referring to viscera or organs 2. Referring to the serosa of viscera or organs

visceral larva migrans PARASITOLOGY A condition of children infected by GI tract parasites of dogs–*Toxocara canis* or cats–*T cati*; after ingestion, eggs hatch into larvae in the intestine, burrow through the GI wall and migrate elsewhere–eg, liver, lung, brain, eye, and other organs, causing inflammation and tissue damage as they migrate CLINICAL Self-limited Sx–eg, fever, lung complaints–cough and wheezing, seizures, hive-like rashes, ↓ visual acuity, periorbital edema, strabismus. See *Toxocara* spp

visceral manipulation OSTEOPATHY A general term for stretching, kneading, and gentle poking of deep tissues–eg, gallbladder and other internal organs, to enhance flow of fluids and energy, release restrictions. See Osteopathy

visceromegaly Organomegaly, see there

Viscoat® OPHTHALMOLOGY A proprietary viscoelastic material used in ophthalmic surgery

viscosity The tendency of a fluid to resist flow or the quality of resistance to flow; viscosity is measured with a viscometer to assess hyperviscosity syndromes associated with monoclonal gammopathies, rheumatoid arthritis, SLE, hyperfibrinogenemia REF RANGE 1.4-1.8 relative to water. See Apparent viscosity. Cf Specific gravity

viscosupplementation ORTHOPEDICS The injection of hyaluronan or derivatives–hylans into joints to manage osteoathritis, improve synovial fluid elasticity and viscosity and relieve pain. See Osteoarthritis

visible light transmittance OCCUPATIONAL SAFETY The amount of light that passes though an eye protection device–EPD to the retina

the Visible Man A 3-D computer-generated 'cadaver,' created from thousands of images of a human body donated to science, collected with state-of-the-art photography of cryosections and imaging–CT, MRI techniques. Cf VOXEL-MAN

Visiflo™ SURGERY A device that enhances visibility during MIDCAB procedures by blowing filtered, moist air on the suture line, maintaining a blood-free field, and providing additional light. See MIDCAB

Visilex® mesh SURGERY A polypropylene mesh used for laparoscopic hernia repair. See Hernia

vision VOX POPULI 1. The act of seeing 2. Visual acuity. See Binocular vision, Computer vision, Tunnel vision

Visipaque® Iodixanol IMAGING A nonionic radiocontrast INDICATIONS Intraarterial–eg, visceral intraarterial digital subtraction angiography, angiocardiography–eg, left ventriculography and selective coronary arteri-

ography, peripheral arteriography, visceral arteriography, cerebral arteriography; IV injection for CECT–contrast-enhanced CT imaging of head and body, excretory urography, peripheral venography. See Radiocontrast

visit See Patient encounter. See Site visit

Visken® Pindolol, see there

Vistaril® Hydroxyzine, see there

Vistide® Cidofovir, see there

visual acuity Central vision OPHTHALMOLOGY The ability to distinguish details and shapes of objects; a measure of the ability to resolve distinct objects or fine details with the eye

visual agnosia NEUROLOGY An inability to recognize objects and people, a common finding in parietal lobe tumors. See Agnosia

visual association PSYCHOLOGY The ability to relate visual stimuli to other, previously learned, visual images in a meaningful way

visual evoked-pattern shift response Visual evoked potential NEUROLOGY A clinical test in which the retina and optic nerve are stimulated with external stimuli in the form of a shifting checkerboard pattern; the response to the stimulus is recorded by EEG electrodes over the occipital scalp, appears 100 msec after the stimulus as a single positive peak on a graph–the so-called P-100; P-100 is slowed and/or its waveform altered in various eye lesions, classically in multiple sclerosis, but also in glaucoma, tumors, ischemic, nutritional, toxic neuropathies, and pseudotumor cerebri. See P-100

visual laser ablation of the trigone UROLOGY A laser ablation procedure for managing chronic bladder syndrome due to chronic trigone inflammation

visual memory PSYCHOLOGY The ability to recall visual images in the form of objects, events, or words

visual reception Visual decoding PSYCHOLOGY The ability to comprehend the meaning of symbols, written 'sight words', pictures, and other concrete objects; visual reception can be compromised because of difficulties in spacial perception or figure-ground

Visudyne® Verteporfin OPHTHALMOLOGY A photodynamic–light-activated agent used for age-related macular degeneration. See Macular degeneration

vital *adjective* Necessary to maintain life. See Vital signs

vital capacity LUNG PHYSIOLOGY The volume of air exhaled by a maxium expiration after a maxium inspiration. See Lung volumes. Cf Total lung capacity

vital signs Vitals CLINICAL MEDICINE Any objective parameter used to assess basic life functions–eg, bp, pulse, respiratory rate, and temperature

vital statistics EPIDEMIOLOGY Systematically tabulated information about births, deaths, marriages, and divorces, based on registration of vital events

vitamin Any of a number of organic accessory factors present in food–in addition to the basic components of carbohydrates, fats, proteins, minerals, water and fiber–which are necessary in minimal or trace amounts, often acting as coenzymes–daily requirements of individual vitamins are measured in mg to μg quantities, as the body either does not produce them or does so in minute quantities; water-soluble vitamins–B_1, B_2, B_6, B_{12}, C, are reasonably well-tolerated as they are easily excreted, while the lipid soluble–A, D, E, K vitamins accumulate in fat, have significant hepatotoxic potential. See Antioxidant vitamin, B complex vitamin, Multivitamin. Cf Chemoprevention, Pseudovitamin

vitamin A deficiency A condition characterized by night blindness, keratomalacia, ↑ urogenital and nasopharyngeal infections, dry eyes; VAD is rare, as many foods contain or are supplemented with vitamin A. toxicity

vitamin A embryopathy A condition induced by in utero exposure to vitamin A analogues, ±associated with cardiovascular–VSD, aortic arch, conotruncal defects, external ear deformities, cleft palate, micrognathia, CNS defects. See Category X drugs, Vitamin A, Vitamin A toxicity, Tretinoin

vitamin A toxicity Hypervitaminosis A, vitamin A intoxication NUTRITION A potentially fatal condition evoked by an acute or chronic excess of vitamin A CLINICAL Bone pain, dry skin, GI complaints–N&V, constipation, diarrhea; ↑ intracranial pressure, poor growth in children, affected infants may develop bulging fontanelles, craniotabes–softening of skull bones, pseudotumor cerebri, papilledema, drowsiness, severe headaches, insomnia, jaundice, menstrual disorders, stress, weight loss, as well as irritability, ↓ appetite, pruritus, hair loss, seborrhea, cracking at corners of mouth. See Vitamin A

vitamin B₁ Thiamin, see there

vitamin B₁ deficiency Beri-beri, see there

vitamin B₂ Riboflavin, see there

vitamin B₆ deficiency NUTRITION A deficiency state that may occur in malabsorption, DM, pregnancy, the elderly, and in those taking oral contraceptives CLINICAL Dermatitis, oral inflammation depression, insomnia, irritability, muscle fatigue, dizziness, weakness, anemia, N&V. See Vitamin B₆

vitamin B₆-dependency syndromes A group of functional or structural enzyme defects that respond to a megadoses–50-100-fold > minimum daily requirements of pyridoxine EXAMPLES Vitamin B₆-dependent convulsions, vitamin B₆-responsive anemia, xanthurenic aciduria, cystathioninuria, homocystinuria, possibly due to a defective structure of the apoenzyme, its coenzyme binding site or coenzyme synthesis CLINICAL Predominantly neurologic–eg, mental retardation, psychiatric Sx, seizures, convulsions, ataxia, spasticity, peripheral neuropathy

vitamin B₆-dependent streptococci MICROBIOLOGY Thiol-dependent, satelliting or nutritionally variant streptococci, which comprise 5% of all streptococci and may cause 'culture-negative' bacterial endocarditis

vitamin B₁₂ Cyanocobalamin, extrinsic factor, methylcobalamin CLINICAL NUTRITION A water-soluble vitamin of animal origin, required for DNA synthesis; absorbed from the GI tract bound to intrinsic factor; it is a glycoprotein produced and secreted by the gastic parietal cells; the body stores up to one yr's worth of vitamin B₁₂ in the liver, kidneys, heart; rapid cell turnover–eg, growth spurts in children, CA require ↑ amounts of vitamin B₁₂. ↑CML, COPD, CHF, liver disease, obesity, polycythemia vera, renal failure. ↓Atrophic gastritis, drugs–antibiotics, anticonvulsants, antimalarials, antituberculous agents, chemotherapy, contraceptives, diuretics, oral hypoglycemics, sedatives; IBD–eg, Crohn's disease, ulcerative colitis; intrinsic factor deficiency–causing megaloblastic anemia, malabsorption, malnutrition, parasites–eg, *Diphyllobotrium latum*, veganism. Note: A strict vegetarian diet's victim ingests no proteins of animal origin, including meat, fish and dairy products; all vegans are at risk for vitamin B₁₂ deficiency; in addition, vegan adolescents may not meet energy requirements during the growth spurt and may become deficient in vitamin B₆ and riboflavin; the high-fiber diet may chelate calcium, zinc and iron, and ↓ absorption of essential cations and trace minerals See Intrinsic factor, Pareve, Vegan, Vegetariansim; Cf Lacto-ovo vegetarian

vitamin B₁₂ deficiency Megalobalstic anemia, see there

vitamin C deficiency Scurvy A condition caused by inadequate intake of vitamin C, characterized by fatigue, bleeding gums, poor wound healing, ↓ resistance to infections, weight loss. See Rebound scurvy, Vitamin C

vitamin C intoxication A condition caused by megadoses of vitamin C CLINICAL Diarrhea, dental erosion, kidney stones, rebound scurvy when the megadoses are ↓; possibly death in Pts with sickle cell anemia LAB ↑ Iron absorption, ↓ copper, selenium, and vitamin B₁₂ absorption, ↑ estrogens in those receiving exogenous estrogens, uricosuria. See Rebound scurvy, Vitamin C

vitamin D₃ Cholecalciferol A vitamin is synthesized in the skin from 7-dehydrocholesterol on exposure to UV light; vitamin D₂ is obtained only from the diet; vitamins D₂ and D₃ are metabolized to 25-hydroxy vitamin D in the liver and then to the active 1,25 dihydroxy form in the kidney. See Rickets, Vitamin D deficiency, Vitamin D intoxication

vitamin D deficiency A condition caused by insufficient activated vitamin D and characterized by irritability, muscle hyperactivity, diarrhea, osteoporosis, osteomalacia in adults and rickets in children

vitamin D-dependent rickets Pseudovitamin D resistant rickets An AR disorder of bone and calcium metabolism characterized by signs and Sx of rickets–hypocalcemia, low-to-normal plasma phosphate, ↑ PTH

vitamin D-dependent rickets type I ENDOCRINOLOGY An AR form of rickets CLINICAL Hypotonia, weakness, growth failure, hypocalcemic seizures in infancy LAB ↓ Ca²⁺, ↑ PTH, aminoaciduria. See Rickets

vitamin D intoxication Hypervitaminosis D, vitamin D toxicity METABOLISM A condition that follows megadoses of vitamin D, or excess dairy product consumption CLINICAL Anorexia, headaches, muscle weakness, nausea, thirst, organ damage–heart, liver, kidney due to calcium deposition; infants given excess vitamin D may develop ASHD, mental retardation, facial dysmorphia, kidney damage, infections, FTT, death. See Vitamin D

vitamin E deficiency A condition that may result from chronic fat malabsorption, as occurs in cystic fibrosis, but rarely causes clinical disease in humans CLINICAL Spinocerebellar degeneration, gait ataxia, loss of proprioception, incoordination, dysarthria, ophthalmoplegia, pigmentary retinopathy, generalized muscle weakness, superficial sensory loss, chronic liver disease, possibly anemia. See Vitamin E

vitamin E intoxication A condition that may result from chronic excess–often self-prescribed of vitamin E CLINICAL GI complaints–eg, nausea, vertigo, blurred vision, fatigue, oral inflammation, ↓ glucose, thyroxine, wound healing, muscle weakness, increased lipids, cholesterol, bleeding. See Vitamin E

'vitamin H' EMERGENCY MEDICINE A popular term for haloperidol, which may be injected in a Pt undergoing acute psychiatric decompensation

vitamin K deficiency A rare condition–most vitamin K is produced in the large intestine by bacterial flora; it may occur in infants with inadequate intestinal bacteria and result in clotting defects. See Vitamin K

vitamin K-dependent proteins A group of coagulation factor proenzymes–factors II, VII, IX and X produced in the liver, which contain multiple residues of γ-carboxyglutamic acid, an amino acid produced by the posttranslational action of a vitamin K-dependent γ-carboxylase on certain glutamyl residues.

VITATOPS VASCULAR DISEASE A study–Vitamins To Prevent Stroke designed to determine whether vitamins prevent stroke

vitiligo DERMATOLOGY An acquired condition characterized by patchy depigmentation in the face, elbows, knees, hands, feet, and genitalia, often at sites of trauma and pressure, which may appear at any age; there is an ↑ incidence in some families; autoimmunity may be a factor, as there is selective loss of melanocytes in the involved area. See Melanocyte

Vitrasert® AIDS A surgical implant that delivers ganciclovir directly to an eye with CMV retinitis; lasts 6-8 months before replacement. See AIDS

Vitrase®

Vitrase® Hyaluronidase OPHTHALMOLOGY An agent injected directly into the vitreous cavity to treat hemorrhage. See Intraocular hemorrhage

Vitrax® OPHTHALMOLOGY A viscoelastic material used in ophthalmic surgery

vitrectomy OPHTHALMOLOGY A procedure in which the vitreous is removed to operate on the retina. See Diabetic retinopathy

Vivelle-Dot™ system GYNECOLOGY A patch that delivers 17β-estradiol through the skin into the bloodstream, to manage menopause. See Estrogen replacement therapy

Vividyl Nortriptyline, see there

Vivol® Diazepam, see there

VLBW Very low birth weight, see there

VLCD Very-low-calorie diet. See Diet

VLDL Very-low-density lipoprotein A plasma lipoprotein with a density of 0.950-1.006 kg/L–US: 0.950-1.006 g/dl; VLDL is 6-10% protein–apoB-100 and apoC, with some apoE, 15-20% phospholipid, 20-30% cholesterol and 45-65% TGs; VLDL is composed of endogenous TGs of hepatic origin

VLP Virus-like particles, see there

VMAC CARDIOLOGY A clinical trial–Vasodilation in Management of Acute CHF. See Congestive heart failure

vocabulary The body of words used in a particular language. See Controlled vocabulary

vocabulary test A component of IQ tests in which a person is asked to define words of varying level of difficulty, and use them in context, which provides the examiner with a measure of the person's intellectual achievement and aptitude. See IQ test

vocal anesthesia ANESTHESIOLOGY A popular term for the use of verbal reassurances–instead of more anesthetics–when local anesthesia begins to wear off, and the operative field is not completely closed

vocal cord cancer Laryngeal cancer, see there, popularly, throat cancer

vocal cord medialization ENT A technique used to treat unilateral vocal cord paralysis, which can acutely compromise pulmonary function, often caused by intrathoracic malignancy. See Unilateral vocal cord paralysis

vocal cord paralysis AUDIOLOGY Inability of one or both vocal folds (cords) to move because of brain or nerve damage. See Phonation

vocal tic Phonic tic NEUROLOGY An involuntary sound produced by moving air through the nose, mouth, or throat; VTs include throat-clearing sounds and sniffing to grunts to verbalizations of syllables and words, utterances of inappropriate, undesired statements or obscenities or coprolalia TYPES Simple–single sounds–eg, throat clearing, barking, sniffing; complex–verbalizations–expression of words–eg, coprolalia, echolalia, palilalia–repeating of person's own words; VTs frequently change and vary in severity over time; remissions and exacerbations are common. See Tourette syndrome. Cf Motor tic

vocal tremor NEUROLOGY Trembling of one or more muscles of phonation, resulting in an unsteady voice. See Tremor

Vocare™ bladder system NEUROLOGY A neuroprosthetic device which provides electrical stimulation to promote bladder control in Pts with spinal cord injuries. See Urinary incontinence

Vogt-Koyanagi syndrome Uveomeningoencephalitic syndrome OPHTHALMOLOGY A multisystem autoimmune disease characterized by acute, often bilateral anterior uveitis with leukodermia, canities, and dysacousia, accompanied by Harada syndrome–posterior granulomatous uveitis with a choroidal infiltrate, exudative, bilateral serous retinal detachment–retinopathy, with fluctuating central and peripheral signs of meningeal irritation, cranial neuropathy, tinnitus, deafness LAB ↑ protein and pleocytosis in CSF

voice AUDIOLOGY A series of sounds generally under voluntary control which are produced by air passing out through the larynx and upper respiratory tract. See Professional voice, Vocal cords

voice box VOX POPULI Larynx, see there

void *verb* To urinate

voiding cystourethrography VCUG, vesicoureterogram IMAGING A study in which a catheter is inserted into the bladder, radiocontrast injected and images obtained to determine whether urine is flowing normally–through the urethra to the outside or has partially reversed its flow into the ureters, as occurs in vesicoureteral reflux

volatile organic compound ENVIRONMENT Any toxic cabon-based (organic) substance that easily become vapors or gases–eg, solvents–paint thinners, lacquer thinner, degreasers, dry cleaning fluids

volatility A measure of the ease with which a substance forms a vapor at ordinary temperatures. See Vapor pressure

volenti non fit injuria See Assumption-of-risk doctrine

Volhard's test A test of a Pt's renal function based on the ability to produce dilute urine after ingesting 1500 mL of water on an empty stomach

volition test Irresistible impulse test, see there

volitional collapse Loss of the 'will to live' or volition. Cf Todeserwartung

Volkmann's ischemic contracture Ischemic contracture ORTHOPEDICS A fixed claw-like flexion contracture of the wrist and hand caused by ischemia due to pressure or crushing injury at the elbow

volume VOX POPULI A measure of the capacity or quantity of a thing. See Dwell volume, End-diastolic volume, End-systolic volume, Expiratory reserve volume, Extracellular fluid volume, Forced expiratory volume, Lung volume, Postvoid residual urine volume, Residual volume, Stroke volume, Void volume

volume depletion INTERNAL MEDICINE A state of vascular instability characterized by ↓ sodium in the extracellular space–intravascular and interstitial fluid after GI hemorrhage, vomiting, diarrhea, diuresis MANAGEMENT 0.9% saline ASAP. Cf Dehydration, Fluid overload

volume of distribution V_d PHARMACOLOGY A hypothetical volume of body fluid that would be required to dissolve the total amount of drug needed to achieve the same concentration as that found in the blood. See Apparent volume of distribution, Therapeutic drug monitoring

volume fraction ONCOLOGY The percentage of a resected tissue involved by cancer

volume-outcome analysis The analysis of a procedure's success rate based on the number of procedures performed. See Outcomes

volume overload PATHOPHYSIOLOGY A state of actual–eg, due to excess administration or ingestion, or functional–eg, due to CHF–fluid excess. Cf Dehydration

voluntary abortion Elective abortion, see there

voluntary admission See Commitment

voluntary formulary permitted CLINICAL PHARMACOLOGY A label that follows an 'order box' on a prescription form in the US. See Dispense as written

voluntary hospital Community hospital, see there

voluntary sector The sum of all organizations and agencies that are neither private–for profit, nor public–government-sponsored sectors; this 'third sector' is comprised of nonprofit and charitable organizations and philanthropies. See Howard Hughes Medical Institute, Imperial Cancer Research Fund

voluntary sterilization GYNECOLOGY The surgical deletion of reproductive capacity, by personal choice. See Sterilization. Cf Involuntary sterilization

volunteer An unpaid do-gooder. See Normal volunteer

volvulus PEDIATRIC SURGERY A condition characterized by torsion of the large intestine, resulting in obstruction and variable loss of blood supply; malrotation of the intestine during fetal development *may* predispose infants to volvulus, often early in life CLINICAL Abrupt onset of bowel obstruction Sx–eg, N&V, bloody stools, abdominal pain, constipation, shock MANAGEMENT Surgical fixation. See Childhood volvulus, Malrotation

vomit Vomitus Ejected matter from the stomach and upper GI tract which often follows nausea; reddish or coffee-ground colored vomitus may represent serious internal bleeding. See Black vomit

vomiting The act of ejecting vomitus. See Projectile vomiting

von Gierke's disease Glycogen storage disease–type 1a, glucose-6-phosphatase deficiency PEDIATRICS A rare AR metabolic disorder of glycogen storage, due to a defect in glucose-6-phosphatase, resulting in glycogen accumulation primarily in liver and kidney CLINICAL Hypoglycemia, lipidemia, xanthoma formation, ↑ uric acid, ↑ lactic acid, liver adenomas, which may become malignant, hepatomegaly, bleeding diathesis, vasoconstrictive pulmonary HTN, convulsions, failure to thrive, lordosis. See Glycogen storage disease

von Hippel-Lindau disease MOLECULAR MEDICINE An AD condition characterized by retinal angioma, CNS hemangioblastoma, renal cysts and CA, pheochromocytoma, pancreatic cysts, polycythemia 2° to ↑ erythropoietin production, epididymal cystadenoma.

von Recklinghausen's disease Neurofibromatosis, type 2 NEUROLOGY An AD condition characterized by cafe-au-lait skin spotting and pendulous fibrous tumors. See Neurofibromatosis

von Willebrand disease HEMATOLOGY A hereditary bleeding disorder caused by moderate-to-severe factor VIII deficiency and low-levels of factor VIII-related antigen–substances needed for coagulation CLINICAL Most are mild; bleeding may follow surgery or a tooth extraction; it is exacerbated by aspirin and NSAIDs; coagulation parameters may improve during pregnancy.

von Willebrand factor HEMATOLOGY A large–> 20,000 kD multimeric molecule composed of ± 200 kD monomers, synthesized by vascular endothelium, megakaryocytes and platelets; vWF's hemostatic efficiency is related to multimer size. See Rocket electrophoresis

VOR Vestibulo-ocular reflex, see there

voriconazole VFEND® INFECTIOUS DISEASE A potent antifungal ADVERSE EFFECTS Visual disturbance: enhanced brightness or blurry vision, ↑ LFTs–BR, alk phos–may indicate cholestasis, erythema, rash, photosensitivity. Cf Amphotericin B

vorozole ENDOCRINOLOGY An aromatase inhibiting antiestrogen used as 2nd-line therapy for advanced breast CA in postmenopausal ♀ with disease progression after tamoxifen therapy. See Antiestrogen therapy, Tamoxifen

vox populi Voice of the people SOCIOLOGY A language, as spoken, which includes slang and jargon. See Jargon, Slang

VOXEL-MAN A 3-D computer-generated 'cadaver' created from the thousands of images of the human body which constitute the US National Library of Medicine's Visible Man project. Cf Visible Man

voyeurism Scoptophilia PSYCHIATRY Sexuoeroticism and sexual fantasies involving the observing of unsuspecting person(s) who are naked or engaged in sexual activity. See Paraphilia, Peeping Tom, Sexual deviancy, Voyeurism. Cf Compulsion, Exhibitionism

VP 1. Vapor pressure 2. Vasopressin 3. Venous pressure 4. Ventral posterior 5. Ventriculoperitoneal 6.Volume pressure

VP shunt Ventriculoperitoneal shunt, see there

VPAP Variable positive airway pressure

V/Q scan Ventilation/perfusion scan, see there

VRE Vancomycin-resistent enterococcus, see there

VRSA Vancomycin-resistant *Staphylococcus aureus*. Cf Vancomycin-resistant enterococcus

VSD Ventricular septal defect, see there; also virtually safe dose

Vulcan nerve pinch OCCUPATIONAL MEDICINE Any ill-conceived keyboard command, the execution of which requires the user to contort the hands in an uncomfortable position

vulgarism A word or phrase not in good usage, coarse, unrefined

vulnerable child syndrome A 'condition' that affects the family of an infant or child who has suffered what the parents believe is a 'close call' with death and thereafter perceived as vulnerable to serious injury or accidents

vulvar anesthesia SEXOLOGY Attenuated vulvovaginal eroticism or sexual sensation; numbness. See Dyspareunia, Sexual adhedonia.

vulvar intraepithelial neoplasia GYNECOLOGY An umbrella term for a precancerous state characterized by the presence of dysplastic cells within the vulvar epithelium, which ranges from low-grade to carcinoma in situ. See Carcinoma in situ, CIN, Intraepithelial neoplasia

vulvitis GYNECOLOGY Vulvar inflammation ETIOLOGY Chronic dermatitis, allergies–eg, soaps, bubble bath, fragrances, fungal infections, pediculosis, scabies; can occur at all ages, in prepubescent and postmenopausal ♀ attributable to ↓ estrogen

vulvovaginitis Inflamed vulvae GYNECOLOGY Inflammation of vulva/vagina, most common in prepubertal ♀ ETIOLOGY Bubble baths, soaps, perfumes, tight or non-absorbent clothing, poor hygiene, STDs, allergens, bacteria–*N gonorrheae*, *Gardnerella vaginalis*, fecal or skin contaminants–eg, staphylococci, streptococci, yeasts–eg, *Candida albicans*, viruses, parasites–eg, pinworms

VUR Vesicoureteral reflux, see there

VVIR CARDIOLOGY A designation–ventricular pacing, ventricular sensing, inhibition response and rate-adaptive, used for ventricular pacemakers. See Dual-chamber pacemaker

VZV Varicella-zoster virus, see there

W Symbol for: 1. Tryptophan 2. Warfarin 3. Water 4. Watt 5. Web–as in WWW 6. Weber fraction–psychological testing 7. Weight 8. White–as in WBCs 9. Width 10. Withdrawal 11. Woman 12. Work

w Symbol for 1. Load/unit of length 2. Workshop–see there

W2 form MEDICAL WORK A form given to an employee at the end of a year, which reports to the US federal government the employee's total earnings, and any earnings withheld for various purposes, including federal, state, local, and FICA taxes

Waardenburg syndrome Klein-Waardenberg syndrome AUDIOLOGY An AD condition characterized by sensorineural deafness and partial albinism, a wide nasal bridge due to the lateral displacement of inner canthi, pigment defects–eg, white forelock, heterochromic iris–distinctive pale blue color of one or both eyes, leukoderma and defects in balance

Wada test NEUROSURGERY A test used to determine which side of the brain contains higher language functions. See Left brain

waddling gait Myopathic gait NEUROLOGY A gait in which the subject sways from side to side, due to a lack of hip stabilization; with ambulation, side A rolls up while side B rolls down in the opposite direction, accompanied by lateral trunk contortions; WG is typical of muscular dystrophy. See Gait

waffle DECISION-MAKING A popular synonym for indecision; can have either nominative–eg, the diagnosis is a complete waffle, or verbal–eg, he/she decided to waffle–status

WAGR syndrome An AD complex characterized by Wilms' tumor–WT, aniridia, genitourinary abnormalities–renal agenesis, gonadoblastoma, mental retardation

Wagstaffe fracture ORTHOPEDICS A fracture characterized by separation of the internal malleolus

WAIS PSYCHIATRY Wechsler Adult Intelligence Scale, see there

WAIS-R Wechsler Adult Intelligence Scale-Revised, see there

waist The relatively narrower part of the body between the thorax and the abdomen.

waist-to-hip ratio NUTRITION The circumference of the waist, divided by that of the hips, which is a measure of the obesity. See Obesity

waiting period MANAGED CARE The time a person must wait from the date of acceptance into an eligible class of insurance or from application, to when the insurance becomes effective. See Elimination period

waived test LAB MEDICINE A test regarded as being so simple (i.e., "idiot proof") that it would require a special talent NOT to perform it correctly. See CLIA-88, Physician office lab

waiver of liability MANAGED CARE A process in which a beneficiary signs an agreement with a hospital and/or health care provider to pay for certain medical services if the services being provided are found not to be medically necessary. See Advance beneficiary notice

waiver of premium HEALTH INSURANCE A provision that, under certain conditions, insurance will be kept in full force without further payment of premiums WHEN APPLICABLE Permanent and total disability. See COBRA, Disability

wake boarding SPORTS MEDICINE An extreme sport in which the participant waterskiis with a snowboard, allowing performance of various gyrations INJURIES Knee joints, ligaments, water skiers' enema. See Extreme sports

wake time SLEEP DISORDERS The total time that a Pt is scored awake in a polysomnogram between sleep onset and final wake-up. Cf Sleep time

Waldenström's macroglobulinemia Waldenstrom's disease ONCOLOGY A rare CA characterized by proliferation of B-cells that produce an IgM monoclonal protein there is no rationale for separating Pts based on monoclonal IgM levels; Pts may present with IgM > 3 g/dL and have a significant BM infiltration with lymphocytes and plasma cells, without major Sx, hepatosplenomegaly, lymphadenopathy or anemia; biologically, WM is an MGUS with IgM and BM involvement much greater than MGUS; asymptomatic Pts are classified as smoldering or asymptomatic macroglobulinemia CLINICAL Constitutional Sx–recurrent fever, night sweats, fatigue due to anemia, weight loss require therapy; anemia with a Hb ≤ 10 g/dL or platelets <100x10⁹/L due to BM infiltration justifies treatment; hyperviscosity syndrome–HS, symptomatic sensorimotor peripheral neuropathy, systemic amyloidosis, renal failure–rare, or symptomatic cryoglobulinemia may also be indications for therapy THERAPY Treatment should not be based on IgM levels because they correlate poorly with WM's clinical findings–ie, don't treat asymptomatic Pts; chlorambucil, cyclophosphamide, either alone or with other drugs; rituximab; symptomatic HS may warrant immediate plasmapheresis. See Bence-Jones protein, Hyperviscosity syndrome, Monoclonal gammopathy of undetermined significance

Waldeyer field Waldeyer port RADIATION ONCOLOGY An RT field encompassing preauricular lymph nodes, lymphoid tissue of Waldeyer's ring, high cervical lymph nodes. See Radiation therapy

walk *noun* Gait *verb* To ambulate afoot

walker A light-weight 3-sided support structure used by Pts with ambulation defects to help self-mobilization

walk-in clinic Ambulatory clinic, see there

walking POPULAR HEALTH A low-impact exercise; regular walking benefits arthritis, back pain, cardiovascular disease, type 2 DM, HTN, osteoporosis, upper body injuries. See Low-impact exercise, Low-impact sport. Cf High-impact exercise, High-impact sport

walking cast ORTHOPEDICS A leg cast that approximates and immobilizes bone fragments without compromising ambulation. See Cast

walking economy VASCULAR DISEASE The rate of O_2 consumption for distance covered during walking, related to walking biomechanics and cardiac burden; WE ↓ in Pts with claudication

walking pneumonia Mycoplasma pneumonia INFECTIOUS DISEASE A popular term for the Sx of atypical pneumonia caused by *Mycoplasma pneumoniae* CLINICAL Usually mild; insidious onset of malaise, headache, ± fever followed by an intense, usually nonproductive cough, with scant mucopurulent or hemorrhagic sputum, accompanied by chest tenderness on inspiration; auscultation may reveal wheezes, rhonchi and occasional moist râles; clinical and radiologic improvement of WP with pleural effusions may require 2–6 wks TREATMENT Tetracycline, erythromycin; a vaccine is now available for pigs/hogs. See *Mycoplasma pneumoniae*

walking-through phenomenon CARDIOLOGY Angina that disappears as the Pt 'walks through' the pain MECHANISM Unknown

walkway REHABILITATION MEDICINE An instrument used to measure the timing of foot contact and or position of the foot on the ground

wall sign Halo sign GYNECOLOGY A vague radiolucency imparted by the dense fibrous capsule of an ovarian dermoid cyst–mature teratoma, separating different sharply circumscribed tissue densities, often recognized by plain films of the lower abdomen

wall of silence Don't ask, don't tell ETHICS A popular phrase for the general unwillingness by physicians to report misconduct, incompetence, substance abuse, and other breaches of professionalism committed by colleagues; those who do bring unethical colleagues to public attention are ostracized and/or lose their jobs–Author's note

walled abscess An abscess surrounded by a thickened 'capsule' of fibrosis

walling-off reaction INFECTIOUS DISEASE A tissue response to acute and subacute inflammation, in which a focus of infection or inflammation is 'isolated,' and surrounded by fibrosis, which seals off the lesion, as occurs in the 'maturation' of an abscess

Walter Reed Staging System AIDS An abandoned system for staging AIDS, that failed to gain widespread support, largely due to heavy reliance on repeated delayed hypersensitivity tests, which is cumbersome, relatively expensive, and poorly reproducible. See WHO system

wanderer A person, often with a mental disorder, admitted to one hospital after another, defined as ≥ 4/yr. DESIGNATIONS WANDERING PATIENT WP So named if wanderer has a treatable illness. HABITUALLY WANDERING PATIENT HWP So named if behavior extends over 5 + years. Note: Wandering may be a phenomenon unique to the US Veterans Affairs hospitals, which offer free medical care and/or hospital admission for prolonged periods of time to any person who is a veteran of military service. While such behavior has features overlapping those of factitious disorders, Munchausen syndrome, and drug seeking behaviors, it is considered to be a 'pathology' a sui generis; in a 1991 report, 810 WPs were identified who averaged 8 admissions/year and 100 days of inpatient care, for substance abuse and/or mental disease for a total cost of $26.5 million; 35 HWPs were identified who totaled 2268 admissions and 7832 outpatient visits for a total cost of $6.5 million NEJM 1994; 331:1752OA, 1771ED

wandering pacemaker CARDIOLOGY An intermittent shift–or suppression–of one cardiac pacemaker, usually from the sinus node to another, often the AV node; the contraction gradually ↑ in length of the cycle, and is usually an innocuous finding in infants; WP may reflect fluctuating vagal tone; physical findings in WPs include bradycardia, variable 1st heart sounds, variable size and shape of P wave

wanton negligence MALPRACTICE The provision of health care without regard for potential injury to the Pt without an actual intent to cause injury–ie, 'reckless'. See Malpractice, Negligence

ward Unit HOSPITAL CARE A unit of in patient beds–usually from 10-40 designated for a particular type of service or care–eg, pediatric ward, OB ward–aka labor & delivery, CCU–coronary care unit, ICU–intensive care unit, oncology unit

ward X A euphemism for the morgue

Ward's triangle BONE DISEASE A radiolucent and fracture-prone triangular zone in the femoral head which can be imaged with plain films in Pts with osteoporosis See Osteoporosis

ware See Groupware, Hardware, Shareware, Software

warfarin Coumadin HEMATOLOGY An anticoagulant that inhibits synthesis of liver-dependent coagulation factors–the prothrombin complex, factors II, VII, IX and X, which are formed by γ-carboxylation of precursor proteins INDICATIONS Prevention of uncomplicated distal DVT, prophylaxis and prevention of thromboembolism, post-acute MI; warfarin therapy is monitored by serial evaluation of PT–2-3-fold > normal 12–16 secs; its activity is ↑ by phenylbutazone, clofibrate–by outcompeting with warfarin for plasma protein binding sites and ↓ by barbiturates, which stimulate hepatic metabolism. See Superwarfarin. Cf Heparin

WARIS CARDIOLOGY A Norwegian trial–Warfarin Reinfarction Study–in which the effects of coumadin was evaluated in elderly post-MI Pts. See Myocardial infarction, Warfarin

warm agglutinin disease IMMUNOLOGY An autoimmune hemolytic syndrome caused by IgG antibodies; 40% are 2° to underlying conditions–eg, neoplasia–eg, CLL, ovarian teratoma, connective tissue disease–eg, SLE, progressive systemic sclerosis, rheumatoid arthritis, ulcerative colitis, etc CLINICAL Brisk hemolysis, Sx of anemia–ie, pallor, fatigue, exertional dyspnea, vertigo, palpitations, jaundice, splenomegaly LAB Anemia, positive antiglobin–Coombs' test, spherocytes, schistocytes, erythrophagocytosis; BM–erythroid hyperplasia, ±underlying lymphoproliferative disorder TREATMENT Transfusions; steroids ↓ hemolysis in ⅔ of Pts; 20% achieve complete remission at >20 mg/day; ⅔ respond to splenectomy, but may relapse; modalities with varying degrees of failure include immunosuppressants and plasmapheresis PROGNOSIS 73% 10-yr survival. Cf Cold agglutinin disease

warm antibody Warm reactive antibody TRANSFUSION MEDICINE An antibody–usually IgG that reacts optimally at 37°C and has an affinity for certain RBC antigens–eg, Duffy, Kell, Kidd, MNSs and Rh and, if produced by a blood recipient, may cause immune hemolysis. Cf Cold agglutinin disease

warm nodule NUCLEAR MEDICINE A circumscribed ↑ in radioisotope concentration seen by radionuclide imaging of the thyroid–'thyroid scan' in which functional thyroid lesions suppress TSH synthesis, but don't cause hyperthyroidism–the latter of which evokes 'hot' nodules. Cf Cold nodule

warm reactive antibody Warm antibody, see there

warm shock syndrome A clinical complex seen in early septic shock CLINICAL Normal cardiac activity, no fluid losses, enhanced peripheral perfusion, minimal catecholamine effect

warm-up rotation GRADUATE EDUCATION A clinical rotation undertaken in a specific area of medicine–eg, cardiology, infectious disease, emergency medicine by a 4th-yr medical student, as a prelude to an acting internship. See Acting internship; Cf Internship

warning leak NEUROSURGERY A minor early hemorrhage in an evolving subarachnoid hematoma or hemorrhage, followed by an abrupt, often severe headache, variably accompanied by nuchal stiffness; significant subarachnoid hemorrhage is fatal if untreated; WLs must be recognized early, treated aggressively, and underlying pathologies–angiomatous malformations, neoplasia, trauma–addressed

WARSS A clinical trial–Warfarin aspirin recurrent stroke study

wart Verruca DERMATOLOGY A typically rough round or oval raised bump on mucocutaneous surfaces that may be lighter or darker than the surrounding normal skin, skin colored or rarely black induced by papoviruses, and single most common reason for dermatologic consultation; warts are most common in children and adolescents, and rarely develop de novo in adults TYPES Common wart–verruca vulgaris, filiform wart, plantar wart, juvenile flat wart LOCATION Anyplace, most common on hands, feet–plantar wart, around and under the fingernails or toenails–periungual or subungual warts–very difficult to treat, face; numerous very small smooth flat warts–pinhead size often in large numbers on children's faces, foreheads, arms and legs are called verrucae planae juvenili CLINICAL Ranges from spontaneous involution, common in flat warts to extreme recalcitrance, typical of periungual and moist plantar warts; plantar warts are identical to common warts but, because of their location on the soles of the feet, they can become extremely painful, especially if they are numerous, compromising running and walking; dermatologic consult is usually triggered by cosmetic considerations; genital/venereal warts are located on the genitals and are sexually transmitted MANAGEMENT 'Benign neglect' and 'abracadabra therapy' are most effective in young children–implying a component of biofeedback control of the immune system, chemocautery–5-20% formalin, phenol-nitric acid-salicylic acid, podophyllin, electrodissection, X-ray–narrow field, low dose, rarely used; DCNB immunotherapy PROGNOSIS Recurrence is common, as is spontaneous involution within 2 years. See Genital wart, HPV, Mosaic wart, Musician's wart, Prosector's wart

Warthin's tumor Papillary cystadenoma lymphomatosum A benign salivary gland tumor–SGT, which often affects the parotid gland; WTs represent up to 10% of SGTs; ♂ : ♀ ratio, 5:1 TREATMENT Excision

washed red cells TRANSFUSION MEDICINE RBCs that have been washed in sterile saline to remove WBCs, lytic mediators, non-self antigens; WRCs are most useful in IgA-deficient Pts who have circulating anti-IgA antibodies, used to ↓ febrile, urticarial and anaphylactic reactions. See Blood filters

washer ORTHOPEDICS A flattened disk of metal with a central hole used to distribute stress under a screw head to prevent thin cortical bone from splitting; serrated washers are used to affix avulsed ligaments, small avulsion fractures or comminuted fractures to the remainder of the bone.

washerwoman skin PEDIATRICS A term of art referring to 1. The loose, dry skin of post-term infants, likened to parchment 2. Loss of skin turgor due to prolonged diarrhea in cholera may cause a similar appearance

washout test NEPHROLOGY A method for estimating renal obstruction, based on the time needed for a radioactive substance to be completely cleared–'washed out' from the kidneys

WASID A clinical trial–Warfarin-Asprin Symptomatic Intracranial Disease

wasp sting A sting from wasps, bees, hornets and yellow jackets, which may trigger allergic reactions varying greatly in severity; avoidance and prompt treatment are essential MANAGEMENT Allergen injection therapy

wastage Fetal wastage, see there

waste ENVIRONMENT *adjective* Relating to materials that are discarded or disposed of–eg, waste water *noun* Materials that are discarded or disposed–eg, biohazardous, hazardous, and regulated waste MEDICINE *verb* To become emaciated, to lose body mass, as occurs in terminal stages of progressive disease–eg, AIDS or cancer

wastebasket diagnosis MEDTALK *noun* A diagnosis which may be faddish, not proven to exist, or physiologically explained, or described with specificity, and thus so broadly inclusive to the point of being scientifically useless

wasted SUBSTANCE ABUSE A popular usage–under the influence of a substance of abuse

wasting disease 1. Kwashiorkor, see there 2. Wasting syndrome, see there

wasting syndrome Any clinical complex associated with chronic renal insufficiency, attributed to a combination of poor nutrition, endocrine dysfunction, catabolic stresses–eg, infection, uremia, dialysis; starvation causes death at 66% of ideal body weight. See Starvation

WATCH CARDIOLOGY A clinical trial–Women Atorvastatin Trial on Cholesterol

watchful waiting Expectant management, observation, surveillance-only management CLINICAL DECISION-MAKING A stance in which a condition is closely monitored, but treatment withheld until Sx appear or change; WW is appropriate when there is a short–eg, ≤ 10 yrs–life expectancy, and/or the lesion being watched has minimal aggressiveness. See Benign neglect

water H_2O A colorless liquid critical for biologic reactions See Bottled water, Drinking water, Finished water, Fluoridated water, Hard water, Heavy water, Hydrotherapy, Individual water, Mineral water, Musket shot water, Raw water, Reagent grade water, Soft water, Source water, Spring water, Surface water

water in abdomen VOX POPULI Ascites, see there

water balance A state of equilibrium in which the fluid intake from water and other beverages, and foods equals fluid lost in the urine, GI tract, sweat, and other secretions. See Water

water-based therapy Aquatic therapy, hydrokinesic therapy PHYSICAL THERAPY Active exercise under water to improve strength in Pts with spastic paresis, quadriplegia, multiple sclerosis and other musculoskeletal disorders

waterbed A bed with a water-filled mattress that may have therapeutic currency NEONATOLOGY Oscillating waterbeds in preterm infants provide compensatory movement stimulation, ↓ uncomplicated apnea of prematurity, with ↑ quiet sleep, ↓ crying, fussiness and, ↑ growth; non-oscillating WBs are used for narcotic-exposed neonates. See Hydrotherapy

water-borne diarrhea outbreak PUBLIC HEALTH Diarrhea due to contaminated water; drinking water systems linked to WBDOs have one+ of following: untreated surface or ground water; treatment deficiency–eg, temporary interruption of disinfection, chronically inadequate disinfection, and inadequate/no filtration; distribution system deficiency–eg, cross-connection, contamination of water mains during construction/repair, and contamination of a storage facility; unknown or miscellaneous deficiency–eg, contaminated bottled water

water-borne pathogen A pathogen, usually bacterial, that infects via contaminated water; WBPs cause gastroenteritis when ingested via the GI tract, or URIs, when the microorganisms are aerosolized, as in legionellosis; virulence of WBPs ranges widely. from enterotoxigenic *E coli*, *Campylobacter jejuni*, and nontyphoid *Salmonella* which are rarely fatal; others are highly virulent–eg, *V cholerae*–15% mortality, or *Salmonella typhi* or *Shigella dysenteriae*–5-10% mortality

water bottle heart A descriptor for the globose cardiac shadow typical of large pericardial effusions–PE, which if < 250 ml are often radiologically 'silent'; PEs may be serous–due to CHF or ↓ proteins in nephrotic syndrome, hepatic failure, malnutrition, serosanguineous–blunt chest trauma, CPR, chylous–lymphatic obstruction, ½ of which are due to malignancy and cholesterol–'pseudochylous' effusion, either idiopathic or due to myxedema

water brash Foam at the mouth Hypersalivation characteristic of reflux esophagitis–variably accompanied by chronic blood loss, anemia, aspiration, regurgitation and recurrent pneumonitis. Cf Shaving cream appearance

water deprivation test ENDOCRINOLOGY A test of the ability to ↑ the concentration–osmolality of urine in response to withholding water. See ADH, Dehydration

water filter PUBLIC HEALTH A device that removes impurities from the water–eg, heavy metals, chlorine, microorganisms, pesticides. See Bottled water

water hammer pulse Collapsing pulse A booming, bounding, or pistol-shot-like sound auscultated in aortic regurgitation or in a large patent ductus arteriosus–PDA; the so-called 'small water hammer pulse' is due to a brisk rise in systemic arterial pressure in the face of normal pulse pressure

water in lung VOX POPULI Pulmonary edema, see there

water-induced thermotherapy UROLOGY An outpatient procedure for managing BPH by introducing heated–60°C water into the prostatic urethra via a heat-transmitting balloon catheter, destroying a predictable amount of tissue, which is reabsorbed, and the obstructed urethra reopened. See Thermoflex™ system

water intoxication Hyperhydration due to excess ingestion of water, resulting in dilutional hyponatremia; WI is most common in Pts with psychiatric or neurologic disease, and may be accompanied by impaired renal fluid excretion and ↑ secretion of ADH, altered mental status, irritability, seizures, somnolence, hypothermia, edema; it is common in infants living in poverty, whose parents 'stretch' powdered formula by adding water SPORTS MEDICINE Cerebral hyponatremia, hyponatremic encephalopathy A specific form of WI affecting the senses, which occurs in otherwise healthy long-distance runners–eg, ultramarathoners. See Marathon

water on brain VOX POPULI Hydrocephalus, see there

water pill Diuretic, see there

water retention Edema, see there

water softening LAB MEDICINE The conversion of 'hard'–mineral-laden to soft water by ion exchange chromatography. Cf Water treatment

water sports Urophilia, see there

water treatment PUBLIC HEALTH The processing of water to make it safely potable or otherwise suitable for consumption. Cf Water softening

Waterhouse-Friderichsen syndrome INFECTIOUS DISEASE A clinical crisis characterized by acute adrenal insufficiency, often accompanied by septicemia; WFS is most often caused by *N meningitidis*, but also *H influenzae*, *S pneumoniae*, staphylococci, less commonly, linked to hypoxia during a difficult L&D, acute adrenal insufficiency EPIDEMIOLOGY Most common in hot, dry climates, especially in West Africa CLINICAL Generalized purpuric rash, shock, hypotension MANAGEMENT Treat underlying infection ASAP

watering-pot perineum UROLOGY A fanciful descriptor for a complication of trauma to the urethra and covering structures–as seen in urethral surgery, in which multiple fistulas associated with inflammatory strictures and diverticuli develop parallel to the urethra and penetrate the perineum and scrotum, forming draining sinuses; as the Pt voids, infected urine is forced into these tracts, and the urine dribbles from multiple, watering-pot-like pores

Waters' position RADIOLOGY A position used to visualize the faciomaxillary bones and maxillary sinuses and determine maxillary sinus patency; in the WP, the Pt is placed at a 37° angle with the orbito-meatal line, perpendicular to the mid-sagittal plane

watershed infarct NEUROLOGY Infarction of a region peripheral to 2 arteries and susceptible to ischemia; WIs are often hemorrhagic, as restoration of the circulation allows blood to flow into damaged capillaries and 'leak' into the ischemic tissue

WATERSHED INFARCTIONS–LOCATIONS

BRAIN After internal carotid artery occlusion, causing vascular 'steal' phenomena, or between the anterior and middle cerebral arteries, which may be compromised in circle of Willis occlusions, often in a background of generalized atherosclerosis and as a possible complication of directed therapeutic embolization; cerebral perfusion may be impaired by cardiac arrest, pericardial tamponade and ex-sanguination

LARGE INTESTINE At either the splenic flexure, the site of anastomosis between the inferior and superior mesenteric arteries, or at the rectum, a region supplied by peripheral irrigation from the inferior mesenteric artery and the hypogastric artery

watery diarrhea-hypokalemia-achlorhydria syndrome WDHA syndrome, see there

watery runny nose VOX POPULI Rhinorrhea

wave MEDTALK A continuous, uniformly advancing oscillation about a "zero" point; a wavelike pattern. See A wave, Alpha wave, Blood pressure wave, Body wave, Brain wave, C wave, Cannon 'a' wave,' Compression wave, Delta wave, F wave, Fluid wave, H wave, Heat wave, Herald wave, J wave, Lambda wave, M wave, P wave, Pontine-geniculate-occipital wave, Q wave, R wave, S wave, Sine wave, Slow wave, Square wave, T wave, Theta wave, U wave, V wave, Zigzag QRS wave

wavefront phenomenon CARDIOLOGY The finding that prolonged coronary artery occlusion results in the expansion of small subendocardial infarcts into a larger transmural MI, seemingly spreading in waves

wavy change Contraction band CARDIOLOGY An undulent kinking of cardiac muscle seen by LM in early acute MIs, often within 1 hr of myocardial ischemia. See Myocardial infarction

wax blockage Cerumen impaction, ear blockage, ear impaction, ear wax ENT Blockage of the ear canal with cerumen produced by the hair follicles and glands; excess wax may harden in the canal and block the ear, resulting in acquired hearing loss

wax epilation Waxing DERMATOLOGY The removal of undesired hair, in particular on the legs, by pouring hot paraffin-like material on the skin and, once cooled, removing the 'cast' en bloc (ouch!). Cf Electrolysis

waxy casts NEPHROLOGY Homogeneous cylindrical structures seen in the urine, and correspond to the degenerated cellular casts, and are typical of long-term oliguria and tubule obstruction; WCs are found in chronic renal failure, kidney rejection, amyloidosis; very broad waxy casts or 'renal failure' casts are typical of ESRD. See End-stage renal disease

waxy exudates Tapetoretinal degeneration OPHTHALMOLOGY A descriptor for the 'hard', yellow-white macular aggregates of fatty and proteinaceous material that leak into the retina via thin, atrophic capillaries; WEs are seen in DM, and ↓ visual acuity

waxy flexibility Cerea flexibilitas, flexibilitas cerea NEUROLOGY Slow resistance to movement as the Pt allows the examiner to place his extremities in unusual positions; a flexibility often present in Pts with catatonic schizophrenia whose limbs remain in the position in which they are placed

WBC 1. White blood cell. See Leukocyte 2. White blood cell count

WDHA syndrome Watery diarrhea-hypokalemia-achlorhydria syndrome, Verner-Morrison syndrome, pancreatic 'cholera' A condition caused by ↑ serum VIP, caused by a VIPoma, a pancreatic islet cell tumor–½ are malignant CLINICAL Profuse watery diarrhea, dehydration, hypotension, shock, episodic flushing LAB Achlorhydria, ↑ or ↓ glucose, ↓ K⁺, ↑ Ca²⁺; WDHA in children is associated with ganglioneuromas

weak D Dᵘ TRANSFUSION MEDICINE A weakly expressed Rh D antigen

wean Old English, to accustom *verb* CRITICAL CARE To transition a Pt from dependence on mechanical ventilation–MV to spontaneous auto-regulated breathing. See Pulmonary function tests, V/Q ratio PEDIATRICS To transition an infant from dependence on maternal–ie, breast milk to milk from other sources or other forms of sustenance

weaning diarrhea PEDIATRICS A condition occurring in a background of poor sanitation, affecting infants 6–24 months of age, and a major cause of infant mortality in developing nations; weaning from maternal milk results in exposure of the infant to new organisms, deterioration of nutrition–a mechanism similar to kwashiorkor and loss of passively transferred IgA CLINICAL Acute, sporadic, watery diarrhea, low-grade fever, vomiting, most common in the summer; in a well-nourished child, the process resolves in 2-3 days with adequate hydration, in malnourished children, diarrhea persists, and may be associated with comorbidity AGENTS Enterotoxic *E coli*, rotavirus, *Shigella* spp DIAGNOSIS CIE, ELISA, EM of stool

weapon VOX POPULI A man-made or -modified artifice designed to kill or maim. See Biological weapon, Chemical weapon, Safe weapon, Special weapon

weapons-related injury FORENSIC MEDICINE Any injury due to guns, sharp instruments, blunt instruments

wear-and-tear arthritis Osteoarthritis, see there

weasel VOX POPULI An unethical ne'er-do-well

weather VOX POPULI That which people talk about when everything else is taboo

WEATHER US RECORD-HOLDERS–NATL WEATHER SERV	
COLDEST Average temperature–36.4°F–Intl Falls, MN	
WARMEST–77.7°F–Key West, FL	
DRIEST–2.65 inches–Yuma, AZ	
WETTEST–104.5 inches rainfall–Quillayute, WA	
CLOUDIEST–242 cloudy days–Quillayute, WA	
SNOWIEST–243.2 inches of snow–Blue Canyon, CA	
SUNNIEST–348 sunny days–Yuma, AZ	
WINDIEST–15.4 mph–Blue Hill, MA	

Note: For international figures, visit Worldatlas.com/geoquiz/the list.htm

web INFORMATICS the Web, World Wide Web ONLINE The Internet's worldwide, HTML-based, hypertext linked information system; a group of databases within the Internet that uses hypertext technology to access text, pictures and other multimedia with a mouse click ; the Web is an 'entity' on the Internet that ties it together so that all data is accessible simply and consistently. See Browser, FTP, HTTP, Internet, URL, WAIS, Web browser MEDTALK A flattened tissue or membrane. See Esophageal web, Laryngeal web

web browser Browser ONLINE A software tool–eg, Internet Explorer, Foxfire, Opera that requests and displays HTML documents and other Internet or intranet resources using HTTP; WBs enable Internet navigation to Web sites, by typing in key words or phrases on the subject of interest; once a Web address–or URL–is typed the WB sends the user to a 'home page'. See Surf, URL, Web, Web site

web server HTTP server INFORMATICS A server that stores and retrieves HTML documents and other Internet or intranet resources, using HTTP. See HyperText Transfer Protocol, Server

website ONLINE A group of related files, including text, graphics and hypertext links, located on the 'Web, and accessed with a web browser or by typing a unique address; a WS usually includes a home page and layers of supporting pages. See URL, Web, Web browser

webbed neck Pterygium colli, Sphinx neck A sphinx-like neck characterized by a thick web or fold of skin that extends from behind the ears to the distal clavicle and to acromial process; WN is typical of Turner gonadal dysgenesis

syndrome, but also occurs in fetal hydantoin, Noonan, trisomy 18 syndromes. See Turner syndrome

webEBM CYBERMEDICINE A website, which promoted web-accessible peer-reviewed evidence-based medicine clinical pathways; may have disappeared from the internet–Author's note. See Clinical pathway, Evidence-based medicine

Weber-Christian syndrome Relapsing nodular nonsuppurative panniculitis A rare idiopathic condition, which primarily affects adult ♀ CLINICAL Crops of subcutaneous nodules on the thighs, abdomen, arms, and breasts, accompanied by fever, local inflammation, rheumatic complaints–arthritis, arthralgia, myalgia, hepatosplenomegaly, abdominal pain, & episcleritis LAB Nonspecific; leukopenia, ↑ ESR, rheumatoid factor, cryoglobulinemia DIFFDX Erythema induratum, sarcoidosis, postinjection fat necrosis MANAGEMENT Symptomatic relief with corticosteroids, chloroquine, phenylbutazone. See Panniculitis

Weber test NEUROLOGY A hearing test in which a vibrating tuning fork is used to evaluate hearing by placing it on the middle of the head–the vertex. See Hearing test

Wechsler Adult Intelligence Scale-Revised WAIS-R PSYCHOLOGY A measure of a person's cognitive abilities. See Psychological tests

Wechsler tests CHILD PSYCHIATRY A series of verbal and performance IQ tests used in schools

wedding ring appearance CARDIOLOGY A fanciful term for a regurgitating mitral valve deformity, in which the valve ring and commissures are calcified, and 'frozen' in an opened position

wedge biopsy SURGICAL PATHOLOGY An excisional biopsy in which a lesion identified at the time of a surgical procedure is removed, with a wedge of normal surrounding tissue. See Biopsy, Excisional biopsy, Wedge resection

wedge excision Lumpectomy, see there

wedge fracture A sharply angled post-traumatic spinal fracture, most common in the thoracic vertebrae in a background of osteoporosis, and often accompanied by anterior wedging. Cf Teardrop fracture

wedge pressure CARDIOLOGY An intravascular pressure recording obtained when a recording catheter is wedged in–ie, occludes–a vessel, which allows assessment of cardiopulmonary pressure gradients

wedge resection A triangular piece of tissue removed in surgery, most commonly obtained in 2 distinct contexts GYNECOLOGY A cuneiform section from an ovary which, by an unknown mechanism, may induce ovulation in polycystic ovaries, usually performed after clomiphene and gonadotropic therapy have failed to correct infertility in ♀ with polycystic ovaries–Stein-Leventhal syndrome SURGICAL ONCOLOGY A wedge of lung, often subpleural, which contains a small radiologically identified lesion, which may harbor a single focus of malignancy

wedge shadow IMAGING A vague radiopacity extending from the lung's hilum, which is characteristic of pulmonary edema

WEE Western equine encephalitis, see there

weed DRUG SLANG Marijuana, see there

weekend call MEDTALK The providing of consultative or non-urgent health care to a defined group–eg, on a particular floor, or type–eg, OB–of Pt, who may or may not be hospitalized. See Call, Coverage

weekend facelift Lunchtime facelift COSMETIC SURGERY A popular term for a facelift performed with advanced techniques that are less invasive and allow rapid recovery–ie, during lunchtime or over a weekend. See Facelift

Wegener's granulomatosis Midline granulomatosis, Wegener syndrome INTERNAL MEDICINE A rare condition characterized by small vessel vasculitis and by granulomatous inflammation of the respiratory tract, formation of ANCA and necrotizing vasculitis of small-to-medium-sized vessels–eg, capillaries, arteries, arterioles, and venules, which is often accompanied by necrotizing glomerulonephritis ; ♂:♀ ratio 2:1, middle aged population, hemoptysis, fever, rash, prostration, arthritis, neuropathy, splenomegaly, progressive glomerulonephritis ending in terminal renal failure ETIOLOGY Possibly hypersensitivity or autoimmunity DIAGNOSIS Renal biopsy determines extent of renal involvement; open lung Bx of a solid or cavitating lesion DIFFDX polyarteritis, vascular renal phase of SBE, rapidly or slowly progressive glomerulonephritis, SLE, and lethal midline granuloma–ie, lymphoma CLINICAL–EARLY Sx anorexia, weight loss, fatigue, malaise, persistent FUO, night sweats, migratory polyarthritis, granulomatous skin lesions, proptosis, ocular manifestations with nasolacrimal duct obstruction, episcleritis, chondritis of ear, acute MI from vasculitis, aseptic meningitis and nonhealing granulomas of CNS may occur; upper respiratory tract complaints–ulcers, pneumonia, granulomas of oral cavity and periorbital regions, skin lesions; eye problems–eg, conjunctivitis; renal disease may be asymptomatic, necrotizing hemorrhagic rhinorrhea, paranasal sinusitis, nasal ulcerations–with 2° bacterial infection, serous or purulent otitis media with hearing loss, cough, hemoptysis, pleuritis, and progressive dyspnea–later requiring O₂; Pts usually present with a granuloma of nose that simulates chronic sinusitis; nasal mucosa is red, raised, granular, friable, bleeds easily, and may perforate; eventually, a disseminated vascular phase may develop and is associated with necrotizing inflammatory skin lesions, pulmonary lesions with cavitation, diffuse leukocytoclastic vasculitis, focal glomerulitis that may progress to generalized crescentic glomerulonephritis with HTN and uremia LAB Anemia may be profound; normal/↑ complement, ↑ ESR, leukocytosis, ↑ antineutrophilic cytoplasmic antibodies URINALYSIS Proteinuria, hematuria, RBC casts MANAGEMENT SIDE EFFECTS Cyclophosphamide–leukopenia, ergo infections; hemorrhagic cystitis; gonadal dysfunction; and some hair loss, reversible on discontinuing the drug MANAGEMENT Corticosteroids–eg, prednisone, MTX JAMA 1995; 273:1288GR, azathioprine, cyclophosphamide induces remission in 75%; long-term prophylactic T-S may be effective for upper respiratory tract lesions; kidney transplant if renal failure PROGNOSIS Dramatically improved by immunosuppressants; ↑ solid tumors post high-dose cyclophosphamide; complete WG usually progresses rapidly to renal failure once diffuse vascular phase begins; Pts with the limited form of the disease may have nasal and pulmonary lesions, with little or no systemic involvement. See Small vessel vasculitis, Systemic vasculitis

weight PHYSICS The mass of a body multiplied by the force of gravity. See Atomic weight, Molecular weight PHYSIOLOGY The mass of a person, a measure of health. See Critical weight, Desirable weight, Extremely low birth weight, Healthy weight, Ideal body weight, Lean body weight, Low birth weight, Obesity, Overweight, Very low birth weight STATISTICS The relative importance of a datum

weight-bearing *adjective* Referring to the ability of a part of the body to resist or support weight.

weight discrimination Size discrimination SOCIAL MEDICINE Any restriction of individual rights, employment or academic opportunities, or biases against overweight persons. Cf Chubby chaser

weight-lifter's headache SPORTS MEDICINE Severe headache and neck pain due to wrenching or tearing of cervical ligaments by the exertion of weight-lifting. See Weight training

weight lifting belt Back belt, see there

weight training A health-promoting exercise effective in any age group, particularly the elderly; WT ↑ muscle mass, walking speed, climbing ability, and sense of well-being; it may be beneficial for arthritis, cardiovascular disease, depression, type 2 DM, ↑ GI transit time, obesity, osteoporosis

Weight Watchers® NUTRITION A weight control program based on sound principles of nutrition–ie, eating 'healthy foods', spurning of diet fads, exercise, and group support. See Diet

Weill-Marchesani syndrome Spheroplakia-brachymorphia An AR connective tissue disorder characterized by short stature, joint stiffness, myopia, subluxation of small, rounded lens–microspherophakia, ectopia lentis, glaucoma, brachydactyly, brachycephaly

welfare Pubic funds provided to persons and families whose income falls below certain federal poverty levels. See Aid to Families with Dependent Children, Child welfare SOCIAL MEDICINE The constellation of public assistance programs for indigent families

well MEDTALK Not sick, not mentally disturbed, healthy

well baby visit PEDIATRICS An outpatient visit that monitors the status of a presumably healthy child, and IDs changes in status in a timely fashion. See Screening

Well-Being at Work or in Daily Routine Scale PSYCHOLOGIC TESTING An 11-item testing instrument relating to a Pt's daily work responsibilities or performance of daily role or routine, if retired, a student, or not employed outside the home. See Overall Quality of Life Scale

well-developed, well-nourished PHYSICAL EXAM *adjective* Referring to a general state of wellness

Wellbutrin® Bupropion, see there

Wellferon® Lymphoblastoid IFN alfa-n1 IMMUNOLOGY A drug used for KS, hairy cell leukemia, HCV in Pts w/o active liver disease. See Interferon

wellness PUBLIC HEALTH A state of well-being. See Health

wellness testing PREVENTIVE MEDICINE A diagnostic procedure or risk assessment strategy applied to a population presumed to not have a particular illness. See Molecular wellness testing

Wenckebach block Möbitz I block, secondary AV heart block, Wenckebach AV block, Wenckebach phenomenon CARDIOLOGY A heart block in which there is progressive prolongation of the P-R interval until a beat is dropped, usually attributable to a block at the AV node. See Möbitz II block

Wendy dilemma PSYCHIATRY A marital situation in which a wife is trapped into acting as her husband's surrogate mother, although it reinforces his immature behavior. See Peter Pan & Wendy

Werdnig-Hoffmann disease Spinal muscle atrophy, see there

Werner-His disease His-Werner disease, trench fever, Wolhynia fever A louse-borne disease caused by *Rochalimaea quintana*, a rickettsia that multiplies in the gut of body lice EPIDEMIOLOGY Mexico, northern Africa, eastern Europe, etc; transmission by rubbing infected louse feces into abraded–scuffed skin or conjunctiva CLINICAL Abrupt onset of high fever, headache, back and leg pain, fleeting rash; recovery takes a month or more; relapses common MANAGEMENT If Sx are systemic or prolonged, ciprofloxacine, doxycycline

Wernicke-Korsakoff syndrome Alcoholic encephalopathy, Korsakoff psychosis, Wernicke's disease NEUROLOGY Organic mental disease linked to chronic alcohol abuse, usually accompanied by thiamin deficiency CLINICAL Memory loss, confabulation

Wernicke's aphasia NEUROLOGY Loss of ability to comprehend language, and production of inappropriate language

Werther effect PUBLIC HEALTH An ↑ suicide rate linked to media coverage of suicide(s), or which occurs in persons 'inspired' by reading about or having had a close relationship with a 'successful' suicide

WEST Weinstein enhanced sensory test NEUROLOGY A nerve testing format used to evaluate tactile sensation, using a pocket-sized device with calibrated nylon filaments which deliver specified force to the skin

West Nile fever West Nile meningoencephalitis INFECTIOUS DISEASE An acute, mosquito-borne flaviviral infection endemic–rarely, epidemic–in the Near East, Africa, former Soviet Union, India CLINICAL After a 3-6 day incubation, children present with a nonspecific febrile illness; older Pts develop a mild dengue-like disease with fever, rash that clears without desquamation, frontal headaches, orbital pain, backaches, myalgia, anorexia, lymphadenopathy, leukopenia, sore throat and possibly meningoencephalitis in the elderly; usually it is self-limited, and resolves in a wk. See Dengue fever

West syndrome Massive myoclonus NEUROLOGY An occasionally X-linked condition characterized by infantile spasms–seizures and 2° generalized epilepsy, hypsarrhythmia, encephalopathy with mental retardation and arrested psychomotor development, ± with immune dysfunction and death at an early age due to bronchopneumonia. See Salaam convulsions

Westergren method A method for determining ESR. See Zetacrit

Westermark sign An abrupt radiologic 'cutoff' in pulmonary vessels seen in a standard AP CXR in PE without infarction, often accompanied by dilation of vessels before the cutoff–eg, the right descending pulmonary artery. Cf Hampton sign

Western blot Immunoblot MOLECULAR DIAGNOSTICS An assay that IDs antibodies to proteins of specific molecular weights, used to confirm HIV infection–after a positive ELISA screening assay, Lyme disease. See Blot, Electrophoresis, Protein. Cf Northern blot, Southern blot

Western diet PUBLIC HEALTH A diet loosely defined as one high in saturated fats, red meats, 'empty' carbohydrates–junk food, and low in fresh fruits and vegetables, whole grains, seafoods, poultry. See Junk food. Cf Food pyramid, Mediterranean diet

Westphal-Leyden syndrome Acute ataxia An AD disease of childhood that progesses to death within 10 yrs of onset CLINICAL Vertigo, vomiting, proximal muscular rigidity, convulsive seizures, and mental defects

wet dream Nocturnal emission, see there

wet drowning FORENSIC PATHOLOGY The usual form of drowning or near-drowning, in which the victim aspirates water, seen at the time of resuscition or at post-mortem examination. See Drowning. Cf Dry drowning

wet gangrene A condition caused by relatively acute vascular occlusion–eg, burns, freezing, crush injuries and thromboembolism, resulting in liquefactive necrosis, bleb and bullae formation with violaceous discoloration. See Gangrene

wheal-and-flare reaction Triple response of Lewis IMMUNOLOGY A series of 3 sequential responses that occur in skin subjected to minor trauma, caused by tissue release of histamine; WFR tests immediate hypersensitivity–'reaginic' or allergic reaction to an antigen, and can be transferred to a nonallergic person. See P-K test

wheeze Sibilant rhonchus PULMONARY MEDICINE A type of continuous–> 250 msec, high-pitched, hissing lung sound, with a frequency of ≥ 400 Hz . See End-expiratory wheeze, Expiratory wheeze. Cf Rhonchus

wheezing PULMONARY MEDICINE A hissing breath sound associated with asthma; most wheezing in infants is transient–eg, due to viral respiratory infection, which ↓ airway diameter at birth but rarely evolves to asthma

whiff test GYNECOLOGY A test in which vaginal secretions are mixed with 10% KOH, resulting in a fishy odor typical of bacterial vaginosis. See Vaginosis, Wet prep

WHIP CARDIOLOGY A trial–Weight-based Heparin for Interventional Pts–designed to determine whether weight-based vs physician-choice heparin dosing would achieve optimal and timely anticoagulation and sheath removal in Pts undergoing coronary intervention. See Interventional cardiology, Low-weight heparin

whiplash An abrupt to-and-fro movement likened to the cracking of a whip, which almost invariably refers to whiplash injury, see there

whiplash injury EMERGENCY MEDICINE Hyperextension injury to the neck, most often in passengers from vehicle A, struck from behind by a fast-moving vehicle B; WIs have been ↓↓↓ by the high-backed headrests now standard in automobiles; nonetheless, WIs affect from 120,000 to 1 million people/yr–US; although most become asymptomatic within a few wks, 20-40% develop the so-called late whiplash syndrome; cervical zygapophyseal joint pain is common in WI, and was once treated with corticosteroids; 90% of WIs are self-limited, requiring neither radiology nor treatment MANAGEMENT Per Quebec Task Force on Whiplash-Associated Disorders. See Late whiplash syndrome, Quebec classification, Steering wheel syndrome

whiplash shaken infant syndrome Shaken baby syndrome FORENSIC MEDICINE A form of child abuse that may be either fatal or leave permanent neurologic sequelae; severe shaking of an infant who has virtually no neck muscle tone may cause bilateral subdural hematomas, due to laceration of veins bridging the dura mater and cerebral cortex caused by rapid acceleration and deceleration as the chin strikes the chest and the occipital bone strikes the back, subarachnoid and retinal hemorrhage, cerebral edema and cortical contusions without signs of external cranial trauma. See Battered child syndrome, Bucket handle fracture. Cf Infanticide

Whipple's disease Intestinal lipodystrophy, lipodystrophy GI DISEASE A systemic malabsorption syndrome characterized by diffuse tissue infiltration with glycoprotein-rich macrophages and abundant bacillary forms CLINICAL ♂:♀ ratio, 10:1, accompanied by acropachyia digestiva, cutaneous hyperpigmentation, malabsorption, diarrhea, steatorrhea, arthritis, lymphadenopathy, endocarditis, fever, pleuritis CNS alterations-eg demyelination of posterior columns; bacilliform inclusions occur in the skin, nervous system, joints, heart, vessels, kidney, lung, serosal membranes, lymph nodes, spleen, liver TREATMENT Penicillin + streptomycin for 2 wks, then long-term tetracycline. Cf Pseudo-Whipple's disease

Whipple procedure Pancreaticoduodenectomy SURGERY The removal of part or all of the pancreas, duodenum, proximal jejunum, distal stomach, common bile duct; it is the surgery of choice for periampullary pancreas CA. See Pancreas cancer

Whipple's triad ENDOCRINOLOGY A triad of features of an insulin-producing tumor: hypoglycemia < 50 mg/dL, CNS and/or vasomotor Sx, relief of Sx by administration of glucose. See Insulinoma

whipworm infection Trichuriasis INFECTIOUS DISEASE Infestation by the nematode, *Trichuris trichiura*, which causes the world's most common parasitic infection EPIDEMIOLOGY After oral-fecal contact with contaminated soil in a background of poor hygiene; ingested eggs hatch and the worm embeds in the large intestine, primarily in the cecum, appendix CLINICAL Light worm loads are virtually asymptomatic; heavy loads cause abdominal malaise, colic, and bloating and, when massive, mild anemia, bloody diarrhea, rectal prolapse TREATMENT Mebendazol. See O&P

whiskering appearance IMAGING Short, linear spiculation seen by plain films of bone at the sites of muscle insertion and osseous stress, ± accompanied by osteosclerosis, cystic changes and dystrophic calcification; it is most common in the iliac, ischial and calcaneus bones, due to ankylosing spondylitis, DISH in renal osteodystrophy

whistle-blowing The act of informing authorities of another person's alleged wrong-doings; WB in science is used in the context of research fraud, alleged fraud or blatant misinterpretation of data. See Baltimore affair, 'Dingellization', Fraud in science, Qui tam suit

white DRUG SLANG A street term for amphetamine, cocaine, heroin MULTICULTURE Caucasian–a person with ancestry in Europe, North Africa, or the Middle East. Cf Hispanic VOX POPULI Absence of color. See Calcofluor White

white blood cell Leukocyte HEMATOLOGY A cell in the general circulation which is whitish to the naked eye when centrifuged TYPES Nonspecific immune response cells–eg, monocytes and granulocytes–neutrophils, basophils and eosinophils and specific immune response cells, B and T lymphocytes. See Granulocyte, Lymphocyte, Monocyte

white blood cell differential count See 'Diff'

white cell count Leukocyte count, WBC count The number of WBCs per volume (microliter–µL or cubic millimeter–mm³) of blood NORMAL VALUE ♂: 4.1-5.4 x 10¹²/L: 4.1-5.4 x 10⁶/µL; ♀: 3.8-5.2 x 10¹²/L: 3.8-5.2 x 10⁶/µL. See Differential white cell count

white coat 1. A popular term for any health care professional who, in performing his/her duties, dons a white coat 2. The physician's robe

white coat hyperglycemia A spurious ↑ in serum glucose attributed to the mental stress of being examined by those with 'white coats' in a clinical setting

white coat hypertension Office hypertension A transient ↑ in blood pressure that occurs in apprehensive Pts on seeing a 'white coat', especially if the Pt is ♀ and the doctor ♂, possibly resulting in inappropriate anti-hypertensive therapy. Cf Pseudohypertension, Small cuff syndrome

white graft An anemic–ie, 'white', transplant tissue–eg, skin or renal graft undergoing hyperacute rejection and anoxia; vascularization doesn't occur as, despite successful host and graft vessel anastomosis, the arteries are occluded by preformed antibodies, causing infarction, mandating tissue removal. See Rejection

white head Closed comedo, milium A blocked sebaceous gland covered by skin and characterized by a small cyst. Cf Blackhead

white noise A sound with equal energy at all frequencies, which can be eliminated by raising the 'cut-off' of the detection device. Cf Chaos

white of eye VOX POPULI → MEDTALK Bulbar conjunctiva

white-out DRUG SLANG A street term for isobutyl nitrite SUBSTANCE ABUSE A generic term for commercial products used to correct typed or written errors on paper documents by painting over them with an opaque white, rapid-drying liquid. See 'Gateway' drugs, Proposition 65, TCE. Cf Glue-sniffing WILDNESS MEDICINE An arctic condition characterized by loss of object visualization caused by bright sunlight reflecting off snow, resulting in snow blindness. See Snow blindness

white plague SUBSTANCE ABUSE A popular term for the epidemic of cocaine abuse

white pulp disease A lymphoproliferative disorder primarily of the spleen's white pulp EXAMPLES Lymphocytic leukemia, histiocytic lymphoma, Hodgkin's disease. Cf Benign lymphadenopathy patterns

white sponge nevus DERMATOLOGY An AD lesion characterized by variably-sized, painless, albescent plaques which, by LM, display a 'basket-weave' epithelium, hyperparakeratosis and acanthosis; WSNs occur on the mucosa of the nasopharynx, mouth, esophagus, larynx, vagina, and anus, WSNs may respond to penicillin

white spot Any hypopigmented cutaneous macule DIFFDX Addison's disease,

alopecia areata, DM, halo nevus, hypomelanosis of Ito, amelanotic melanoma, piebaldism, thyroid disease, tuberous sclerosis, vitamin B₁₂ deficiency, vitiligo, Vogt-Koyanagi-Harada syndrome

white strawberry tongue PEDIATRICS Red, congested and edematous fungiform papillae of the tongue, in a whitish background of filiform papillae; the WST is found between day 2 and 5 of scarlet fever; it may be seen in measles and in other febrile viral enanthemas. See Scarlet fever

whites A popular term for the white 'uniform' worn by medicos the world over, which changes in length according to rank. Cf Scrubs, Greens

White's classification A system for categorizing gestational DM, designed to predict the maternal and fetal M&M

WHITE'S CLASSIFICATION—MODIFIED

CLASS	DURATION	LESIONS	THERAPY
A	Any	None	Diet only
B	< 10 Yrs	None	Insulin
C	10-19 Yrs	None	Insulin
D	> 20 Yrs	Benign retinopathy	Insulin
F	Any	Nephropathy	Insulin
H	Any	Heart disease	Insulin
R	Any	Proliferative retinopathy	Insulin

whitlow A painful bacterial dermatitis of the finger, seeded by contact. See Herpetic whitlow, Melanotic whitlow

WHO classification SURGICAL PATHOLOGY A classification of non-Hodgkins lymphomas based upon an updated version of the REAL classification system; the WHO classification system is currently accepted as the authoritative standard for diagnosing lymphomas; it does not group lymphomas into prognostic categories, but, rather, considers each lymphoma as a separate disease; the WHO system uses morphology, immunophenotyping, and genotyping to classify the malignancy, which is often graded, typically from I to IV. See Non-Hodgkin's lymphoma, REAL classification. Cf Hodgkin's disease, Working formulation

WHO staging system AIDS A simplified AIDS staging system proposed by the WHO Global Programme on AIDS that is flexible enough to be used in different regions, based on 4 groups of clinical conditions that have prognostic significance and therefore constitute stages, plus an assessment of physical activity performance expressed as a 4-point score; Pts are classified according to the highest stage recorded for either clinical condition or physical activity CATEGORIES Asymptomatic; mildly symptomatic; moderately symptomatic; severely symptomatic. See AIDS, HIV-1

WHO STAGING SYSTEM FOR HIV INFECTION

CLINICAL STAGE 1

1. Asymptomatic infection
2. Persistent generalized lymphadenopathy
3. Acute retroviral infection

Performance Stage 1: asymptomatic, normal activity

CLINICAL STAGE 2

4. Unintentional weight loss < 10% body weight
5. Minor mucocutaneous manifestations (e.g., dermatitis, prurigo, fungal nail infections, angular cheilitis)
6. Herpes zoster within previous 5 years
7. Recurrent upper respiratory tract infections

Performance Stage 2: symptoms, but nearly fully ambulatory

CLINICAL STAGE 3

8. Unintentional weight loss > 10% body weight
9. Chronic diarrhea > 1 month
10. Prolonged fever > 1 month (constant or intermittent)

11. Oral candidiasis
12. Oral hairy leukoplakia
13. Pulmonary tuberculosis within the previous year
14. Severe bacterial infections
15. Vulvovaginal candidiasis

Performance Stage 3: in bed > normal but < 50% of normal daytime during the previous month

CLINICAL STAGE 4

16. HIV wasting syndrome
17. *Pneumocystis carinii* pneumonia
18. Toxoplasmosis of the brain
19. Crytosporidiosis with diarrhea > 1 month
20. Isosporiasis with diarrhea > 1 month
21. Cryptococcosis, extrapulmonary
22. Cytomegalovirus disease of an organ other than liver, spleen or lymph node
23. Herpes simplex virus infection, mucocutaneous
24. Progressive multifocal leukoencephalopathy
25. Any disseminated endemic mycosis (e.g., histoplasmosis)
26. Candidiasis of the esophagus, trachea, bronchi, or lung
27. Atypical mycobacteriosis, disseminated
28. Non-typhoid Salmonella septicemia
29. Extrapulmonary tuberculosis
30. Lymphoma
31. Kaposi's sarcoma
32. HIV encephalopathy

http://hivinsite.ucsf.edu/InSite?page=kb-01-01#S4X

whole blood TRANSFUSION MEDICINE Blood containing all the components–RBCs, WBCs, platelets, plasma proteins and fluid–of blood. See Packed red cells

whole-body dose RADIATION ONCOLOGY Radiation exposure to gamma rays from outside the body, which irradiates the entire body; each organ receives about the same dose. See Dosimetry

whole-body hyperthermia ONCOLOGY The heating of the body as an adjunct to managing various CAs–eg, non-small cell lung CA. See Hyperthermia

whole-gut scintigraphy IMAGING A technique in which 2 radiolabeled markers are used to evaluate colonic transit–$^{111}InCl_3$-labeled capsules that disintegrate in the terminal ileum and proximal colon and gastric emptying and small bowel transit–^{99m}Tc mixed in a test meal; constipation ↓ GI transit time detected by WGS; diarrhea ↑ transit time

whooping cough PEDIATRICS The characteristic cough of pertussis, which infects ± 60 million and kills 1 million/year worldwide, caused by the gram-negative nonmotile *Bordetella pertussis*; the 'whoop' occurs during the paroxysmal stage of infection–1-2 wks after onset, lasting 2-6 wks; a 'paroxysm' consists of 10-20 coughs of ↑ intensity, a deep inspiration–the 'whoop' after which a thick, viscid plug of mucus is expelled, occasionally with vomiting; the paroxysms may occur every $\frac{1}{2}$ hr, often accompanied by ↑ venous pressure, plethoric conjunctivae, periorbital edema, petechiae and epistaxis, infants may be cyanotic until relieved of the obstructing plug of mucus MANAGEMENT Supportive; young infants may need hospitalization if coughing is severe PREVENTION Immunization with DPT vaccine. See DTP

'whore' 'Hired gun', see there

wicking INFECTIOUS DISEASE Enhanced penetration of liquids, and small pathogens, through minute holes in latex membranes–eg, surgical gloves, which may develop when washed with surfactants, an effect that militates against the re-use of certain materials

wide-complex tachycardia Ventricular tachycardia, see there

Wide Range Achievement Test PSYCHOMETRIC TESTING A test that measures ability to read, write, and use arithmetic; results are matched with criteria for intelligence rating. See Intelligence quotient

Widmark formula FORENSIC MEDICINE A formula used to calculate a person's blood alcohol concentration, based on the alcohol dose, alcohol distribution in body water, body weight, and rate of elimination. See Blood alcohol concentration

'widow-maker' CARDIOLOGY Severe stenosis of the left anterior descending coronary artery, so named as occlusion thereof is classically associated with sudden death post acute MI

wife abuser PUBLIC HEALTH A ♂ who abuses his spouse or common law partner. See Domestic violence, Spousal abuse

WIHS AIDS A clinical trial–Women's Interagency HIV Study

wild *adjective* Referring to that which has been neither intentionally inbred or genetically manipulated

wilderness medicine A specialty of sports medicine that studies the effects of environmental extremes–eg, high mountain, glacier, desert, etc conditions–on health and disease. See Extreme medicine

wilding FORENSIC PSYCHIATRY An unprovoked, motiveless attack by a group, usually of young ♂, on another, often a stranger, merely for violence's sake EXAMPLES Gang rape, stabbing, shooting, homicide. See Evil, Gang

will 1. Desire, volition, as in the 'will to live', see there 2. *The legal expression or declaration of a person's mind or wishes as to the disposition of his property, to be performed or take effect after his death'*. See Advance directive, Living will

Will Rogers phenomenon Stage migration EPIDEMIOLOGY The improved survival of Pts with CA or other disease by either reclassifying Pts into different prognostic groups, recognizing more subtle disease manifestations, or by using diagnostic modalities that allow a disease to be diagnosed at an earlier stage; the WRP results in a 'zero-time shift' and improved prognosis, without affecting survival

will to live PSYCHOLOGY The sense of self-preservation, usually coupled to a 'future sense'–ie, dreams, and aspirations, and expectations for future improvement in one's state in life

willful negligence MALPRACTICE Provision of health care in an intentionally substandard fashion, the most serious form of negligence, which may carry with it criminal charges. See Malpractice, Negligence

Wilms' tumor Nephroblastoma ONCOLOGY An embryonal renal tumor that arises in aberrant mesenchymal renal stem cell lines, coupled with loss of functioning tumor suppressor genes; WT comprises 92% of 1° childhood renal CAs–frequency: 1:10,000–the rest are sarcomas CLINICAL WT presents as a mass, often with HTN 60%, hematuria 24%, nephritis–Wilms' nephritis, serosal effusions–ascites, pleural effusions; 5% are bilateral; anaplasia connotes a poor prognosis. See WAGR syndrome

WILMS' TUMOR STAGING

I CA in kidney only; can be completely removed by surgery

II CA spread to tissue near kidney, to blood vessels, or to the renal sinus–a part of the kidney through which blood and fluid enter and exit; CA can be completely removed by surgery

III CA has spread to tissues near the kidney; cannot be completely removed by surgery. CA may have spread to blood vessels or organs near the kidney or throughout the abdomen and/or to regional lymph nodes

IV CA spread to organs distant from kidney–eg, lungs, liver, bone, brain

V CA in both kidneys

Wilson's disease NEUROLOGY An AR disorder of copper metabolism characterized by accumulation of copper in the brain, liver, and other organs CLINICAL Cirrhosis, degeneration of basal ganglia and neurologic deterioration with involuntary movement, tremors, muscular rigidity, spastic contractures, psychiatric defects, dysphagia, Kayser-Fleischer ring. See Ceruloplasmin, Copper

wind *pronounced* WIN'd VOX POPULI The rushing of air from one point to another, generally induced by differences in land temperature. See Fire wind

wind chill factor WILDERNESS MEDICINE An index used to adjust the actual air temperature to express the intensity of cooling expected from a cold environment as a function of the ambient temperature and wind speed; the WCF is a measure of the effect of air temperature and wind speed on human comfort and safety. The correct term is windchill index but is unlikely to prevail in the forseeable future. Humidity and radiant heat energy tend to be less important in cold environments. See Hypothermia

WIND CHILL EQUIVALENTS IN ºC/ºF FOR REFERENCE
WIND SPEEDS OF 2.5 KPH/4MPH

TEMP	2.5/4	6/10	12/20	19/30	25/40	30/50
4.5ºC/+40ºF	4.5ºC/+40ºF	-2ºC/+28ºF	-8ºC/+18ºF	-10.5ºC/+13ºF	-12ºC/+10ºF	-13ºC/+9ºF
-1ºC/30ºF	-1ºC/30ºF	-9ºC/+18ºF	-18ºC/+4ºF	-19ºC/-2ºF	-21ºC/-8ºF	-22ºC/-7ºF
-7ºC/20ºF	-7ºC/20ºF	-18ºC/+4ºF	-23ºC/-10ºF	-28ºC/-18ºF	-30ºC/-22ºF	-30ºC/-23ºF
-12ºC/10ºF	-12ºC/10ºF	-23ºC/-9ºF	-32ºC/-25ºF	-36ºC/-33ºF	-38ºC/-37ºF	-39ºC/-39ºF
-18ºC/0ºF	-18ºC/0ºF	-29ºC/-21ºF	-39ºC/-39ºF	-44ºC/-48ºF	-47ºC/-53ºF	-49ºC/-55ºF
-23ºC/-10ºF	-23ºC/-10ºF	-38ºC/-33ºF	-47ºC/-53ºF	-53ºC/-64ºF	-58ºC/-69ºF	-57ºC/-71ºF
-29ºC/-20ºF	-29ºC/-20ºF	-43ºC/-46ºF	-55ºC/-67ºF	-62ºC/-79ºF	-65ºC/-85ºF	-66ºC/-87ºF

Last, Wallace, Eds, Public Health and Preventive Medicine, 13ᵗʰ ed, Appleton & Lange, Norwalk, 1992

window *adjective* Referring to an interruption in time or space. See Core window, Fertilization window, Round window, Square window, Therapeutic window RADIOLOGY An interval of photon energies used in a scintillation counter–gamma-ray detector; the so-called 'pulse height analyzer' rejects any photon energy falling outside of the window–and is thus not counted SURGERY A region of an abscess in closest contact with the abdominal wall–or any accessible skin surface without an intervening visceral organ, which can be opened for relatively safe drainage

window of opportunity MALPRACTICE The time during which treatment has the best chance of preventing permanent injury. See Golden hour, Point of irreversible injury

window period AIDS An interval between initial infection of HIV and development of anti-HIV-1 antibodies, usually 3–6 months IMMUNOLOGY An interval between the time of inoculation or exposure to a microorganism, usually viral, and the ability to detect its presence by serologic assays which detect the production of antibodies by the host, or the presence of the pathogen's antigens

windpipe VOX POPULI Trachea

windsock deformity CARDIOLOGY A descriptor for an aneurysm of the sinus of Valsalva, as seen by aortography, in which there is a wide base at the aortic origin and a nipple-like apex projecting into the right atrium, the usual site of rupture; the aneurysm is characterized by attenuation and separation of the aortic media within the sinus from the common fibrous junction of the aorta, the left ventricle and the aortic valve PEDIATRIC SURGERY A descriptor for the characteristic balloon-like dilatation seen on a barium enema in neonates with membranous atresia of the colon, ileum or jejunum; when the deformity is colonic, the barium column reaches the membrane but is forced back by the fecal flow in the transverse colon, forming a cup-shaped or windsock-like radiolucency

windup Central sensitization, neuroplasticity ANESTHESIOLOGY A physiologic ↑ sensitization of excitable neurons, coupled to a ↓ threshold for peripheral afferent pain terminals; windup may be mediated by excitatory amino acid neurotransmitters and neuropeptides; NMDA receptor inhibitors prevent 'windup' and may play a central role in managing chronic pain. See NMDA receptor antagonist, Preemptive analgesia. Cf Multimodality pain control

winging of scapula NEUROLOGY A clinical finding characterized by scapular lifting, likened to an angel's wing; it is seen when the arm is abducted, is typical of limb-girdle dystrophy, and due to paralysis of the long thoracic nerve–5th, 6th, 7th nerve roots; it also occurs in wrinkly skin syndrome and in heavy manual labor, especially in those who carry heavy, sharp objects on the shoulders, which causes the so-called 'hod carrier's palsy'

WINS PUBLIC HEALTH A 10-yr prospective randomized clinical trial–Women's Intervention Nutrition Study funded by the NCI which began in 1995. See Women's health

winter depression Seasonal affective disorder, see there

winter sport SPORTS MEDICINE A sport–eg, skating, skiing, mountaineering, practiced in the cold–usually < 0ºC/32ºF temperatures, often at higher altitudes. See Extreme medicine, Ice climbing, Wilderness medicine

winter vomiting disease Epidemic vomiting A 1-3 day, often parvovirus-induced intestinal 'flu' most common in the winter in temperate climates CLINICAL Either mild, afebrile watery diarrhea or more severe, febrile with vomiting, headache, systemic complaints

Wintin Lorazepam, see there

wire ORTHOPEDICS Any wire used in orthopedic surgery to anchor or stabilize bone fragments, tendons, or other soft tissues. See Cerclage wire, Commander guide wire, Figure-of-eight wire, Guide wire, Kirschner (K) wire, Tension band wiring

wire-guided excisional biopsy SURGERY A biopsy, usually of a breast mass deemed suspicious by mammography, which is guided to the site of excision by a localization wire placed at the time of mammography. See Breast biopsy

WISC Wechsler Intelligence Scale for Children PSYCHOLOGY A 10-category test that measures both verbal and performance IQ. See Psychological testing

wisdom teeth DENTISTRY A popular term for the 3rd molars that usually erupt during late adolescence and which, given the physical confines of the jaw, often become impacted. See Permanent teeth

WISE CARDIOLOGY A clinical trial–Women's Ischemia Syndrome Evaluation

wishful thinking PSYCHOLOGY Dereitic thought that a thing or event should have a specified outcome

Wiskott-Aldrich syndrome HEMATOLOGY An XR immunodeficiency CLINICAL Eczema, proneness to infection due to recurrent pyogenic infections, especially with encapsulated bacteria–eg, *H influenzae, S pneumoniae, N meningitidis*, thrombocytopenia, bloody diarrhea, ↑ susceptibility for lymphoproliferative disorders LAB ↓ IgM, ↑ IgA, ↑ IgE, ↓ platelets MANAGEMENT Splenectomy, kidney transplant, BMT PROGNOSIS Death usually before age 10

witch-hunt 'Whistleblowing', see there

witch's milk PEDIATRICS A clear-to-lactescent discharge from the nipples of ♂ and ♀ neonates, which is accompanied by breast hypertrophy, and caused by transplacental hormonal effects; it is most common in term infants. Cf Breast milk, Certified milk, Maternal milk, Raw milk, Unpasteurized milk

withdrawal PSYCHOLOGY A retreat from interpersonal contact, which may be a normal reaction–eg, to uncomfortable social situations or unemployment, or a sign of mental disorders–eg, schizophrenia, depression, bipolar disorder SUBSTANCE ABUSE A specific constellation of signs and Sx due to the abrupt cessation of, or reduction in, regularly administered opioids;

opioid withdrawal is characterized by 3 or more of the following Sx that develop within hrs to several days after abrupt cessation of the substance: 1. Dysphoric mood, 2. N&V, 3. muscle aches & abdominal cramps, 4. lacrimation or rhinorrhea, 5. pupillary dilation, piloerection or sweating, 6. diarrhea, 7. yawning, 8. fever, 9. insomnia. See Alcohol withdrawal syndrome, Physical dependence

withdrawal emergent syndrome NEUROLOGY A tardive dyskinesia-like clinical complex that affects children and adolescents upon abrupt withdrawal of neuroleptic–antipsychotic drugs; the WES is attributed to hypersensitivity of dopamine receptors, characterized by choreoathetosis, myoclonus MANAGEMENT Reintroduce drug, taper dose slowly or await spontaneous resolution

withdrawal syndrome CARDIOLOGY A constellation of findings, including angina and acute MI, that may follow abrupt cessation of β-blockers in Pts with HTN PSYCHOLOGY See Withdrawal SUBSTANCE ABUSE A constellation of Sx that follow the abrupt cessation of psychoactive agents, which is largely a function of the withdrawn agent

withdrawing CHILD PSYCHIATRY Behavior characterized by ↓ interest in or contact with other people; WBs include ↓ speech, regression to babyhood, exhibition of many fears, depression, refusing contact with other people

withhold system MANAGED CARE Any financial incentive by an HMO which ↓ compensation to its doctors for 'inefficient' use of services–ie, by excess diagnostic testing, hospitalization and referrals to specialists; withholds have fallen into disfavor as they tend to encourage physicians to undertreat Pts. See Covert financial incentive, ERISA. Cf Bonus system

within normal limits MEDTALK A term for a CT, MRI or other image or specimen, which the examiner regards as having no significant changes

witness CHOICE IN DYING A person who is not a spouse or blood relative of a dying Pt; employees of health care facilities, who act in good faith, can act as witnesses with regard to end-of-life decisions FORENSIC MEDICINE 1. A person who has seen an act 2. A person qualified by education and/or experience to testify as to a thing. See Expert witness, Physician expert witness

witzelsucht German, joke addiction, moria NEUROLOGY A manifestation of organic brain disease–eg, tumors or atherosclerosis-induced lesions of the frontal cortex, in which a Pt compulsively tells puns, or relates silly jokes, often accompanied by childish behavior, as a defense mechanism to deflect attention from memory defects

WIZARD CARDIOLOGY A clinical trial–Weekly Intervention with Zithromax® against Atherosclerosis and Related Disorders–examining the effect of reducing the risk of ASHD with antibiotics that treat *Chlamydia pneumoniae*

WNL Within Normal Limits–see there

wobbling VOX POPULI Ataxia, see there

Wohlfahrtia A genus of flesh-eating flies (family Sarcophagidae) linked to myiasis

Wolff-Chaikoff effect ENDOCRINOLOGY An acute adaptive response to high doses of iodine–↑ intracellular iodide blocks the organic-binding and coupling reactions in the thyroid, functionally turning off the thyroid; the WCE used to prepare the thyroid for surgery. See Hyperthyroidism

Wolff-Parkinson-White syndrome Preexcitation syndrome CARDIOLOGY A conduction disorder with ↑ susceptibility to supraventricular paroxysmal tachyarrhythmias, in which the sinoatrial impulse travels directly to the AV node or anomalously via Kent's bundle, which is more rapid, resulting in PVCs EKG Short PR interval–< 0.1 sec, a slur on the upstroke of the R or on the downstroke of an S wave or delta wave, prolonged QRS complex, often ST-T changes MANAGEMENT Cardioversion, procainamide, lidocaine or radiofrequency ablation of accessory AV pathway

Wolman's disease Cholesterol ester storage disease GENETICS An AR storage disease seen in consanguineous cohorts, due to a defect in lysosomal sterol esterase with accumulation of TGs, and cholesterol esters CLINICAL Hepatosplenomegaly, bloating, wasting, adrenal calcification, xanthomatosis, steatorrhea LAB Anemia, ↑ TGs, cholesterol esters, foam cells in BM, vacuolated lymphocytes in circulation PROGNOSIS Onset in early infancy, death by age 1

WOMAC Western Ontario McMaster University Osteoarthritis Index RHEUMATOLOGY An arthritic pain scoring system ranging from 0–no pain/disability to 100–most severe pain/disability

woman VOX POPULI A female person. See Abdominal woman, Declared pregnant woman

women's health PUBLIC HEALTH Disease is unfair; in general, men die more often through acts of war, homicide, suicide–and general stupidity, and earlier from AIDS, alcohol, drugs of abuse, tobacco, most cancers; ♀ live longer, better, more disease-free; CA mortality before age 20 is 22% lower in ♀; in certain tumors–eg, sarcomas and carcinomas, ♀ linger longer COD IN ♀ 233,000 MIs; 87,000 strokes; 55,000 lung CA; 43,000 breast CA–US, 1996. See Violence Against Women Act

Women's Health Initiative A 15-yr, $628 million project involving 1. An observational study of the health habits and medical Hx of ±100,000 ♀ 2. A randomized controlled clinical trial of HRT in ±65,500 postmenopausal ♀ between age 50–79 3. A prevention study to improve healthy behaviors in ♀ of all ethnic and socioeconomic backgrounds ☎ 1.800.54-WOMEN

wonder drug Any therapeutic–eg, penicillin, tricyclic antidepressants, etc that rapidly cures or clinically improves the status of a particular condition

Wonderlic Personnel Test PSYCHOLOGY A brief psychological test which assesses a person's ability to learn and grasp principles and apply them in new settings

wooden sensation NEUROLOGY Tactile numbness of long–eg, 10± years duration, associated with limited cutaneous sclerosis, 2° to occupational use of vibrating machines. See Vibration white finger syndrome

woolly appearance RADIOLOGY A finding in Paget's disease of the skull bones, characterized by multiple irregular dense nodules and small areas of rarefication. See Paget's disease of the bone

woolly hair disease PEDIATRICS Tight curly hair seen at birth in Caucasians that is 1. Sporadic and associated with woolly hair nevus 2. AR, in which the scalp hair is ash-white and body hair short and kinky or 3. AD, in which the scalp hair is also ash-white, but with normal body hair. Cf Kinky hair disease, Uncombable hair syndrome

woolsorter's disease Inhalation anthrax, see there

word approximation NEUROLOGY The use of a word similar in meaning–eg, heat for fever–to another; a feature of positive thought disorders

word association test PSYCHOLOGY A test used to ID disorders of thought processes, based on a person's response to a stimulus word, which may be either as a free association or examiner-controlled association. See Free association, Psychological testing

word salad PSYCHIATRY A mixture of neologisms, words, and phrases that lack comprehensive meaning or logical coherence; WSs are characterized by loosening of associations, shifting of topics that may progress to near incoherence, and a lack of logical connection; WSs are typical of disordered thought processes–eg, schizophrenia. Cf Witzelsucht

work That which occurs when a forces moves an object over a distance W=Fd PHYSICS Force applied to an object times the distance the object is moved, defined by the SI unit, joule. See Joule VOX POPULI Labor. See Light work, Rootwork, Scut work, Shift work, Social work, Statement of work, Substantive programmatic work

work adjuster MEDTALK A number used to calculate physicians' reimbursement from Medicare, based on the RBRVS formula. See Geographic practice cost index, RBRVS

work conditioning Work hardening OCCUPATIONAL MEDICINE A rehabilitation program that prepares a client for return to work through conditioning to improve biomechanical, neuromuscular, cardiovascular and metabolic functions of a worker, with real or simulated work activities. See Occupational therapy

workout SPORTS MEDICINE *verb* To undertake a 30+ minute-session of usually isometric exercise–eg, weight-lifting, treadmill 'work,' rowing, other muscle training, often in a gym or health club. See Culture VOX POPULI *verb* To resolve a problem

work practice controls OCCUPATIONAL SAFETY An OSHA term for controls that ↓ the risk of exposure–to blood-borne pathogens–by altering the manner in which a task is performed–eg, prohibition of 2-handed needle recapping, mouth pipetting, other high-risk practices

work-up *noun* The constellation of procedures–eg, taking a medical Hx, performing a physical exam, and ordering and evaluating lab tests and imaging procedures, on which a Pt's diagnosis and therapy is based. See Diagnostic 'overkill', Ulysses syndrome *verb* To evaluate a Pt–ie, to work him/her up. See Fehldiagnose

workers' compensation Workman's compensation OCCUPATIONAL MEDICINE The benefits provided to employees for injuries suffered in the workplace EPIDEMIOLOGY Work-related injuries affect 8 million employees/yr–US resulting in the loss of 73 million work days and 10,000 lives

workflow INFORMATICS The process by which tasks are done, by whom, in what order and how quickly

working diagnosis Preliminary diagnosis, provisional diagnosis DECISION-MAKING A diagnosis based on experience, clinical epidemiology, and early confirmatory evidence provided by ancillary studies–eg, radiologic findings; WDs allow early disease management, while awaiting special or more definitive studies–eg, immunoperoxidase stains or results from a reference lab

working memory Short-term memory, see there

working through PSYCHIATRY Exploration of a problem by a Pt and therapist until a satisfactory solution has been found or a symptom traced to its unconscious roots

workplace stress OCCUPATIONAL MEDICINE Work-related mental tension, which causes exhaustion, insomnia, muscle pain, depression, and eventually job burnout. See Burnout

workplace violence OCCUPATIONAL MEDICINE Any act of violence that occurs in a work environment, which may be committed by one worker against another, by outsiders, or by former employees. See Postal worker syndrome. Cf Domestic violence

works DRUG SLANG A street term for the paraphenalia for injecting drugs. See Needle exchange program

workshop ACADEMIA A specialized conference in which experts in a particular area of science or medicine, convene in order to compare data and establish or standardize criteria for a substance, disease, or phenomenon. Cf 'Wetshop'

World Medical Games Medical Olympics SPORTS MEDICINE An Olympic event for health care professionals–physicians, veterinarians, dentists, physiotherapists, or last-yr students in above groups; Cf Paralympics, Special Olympics

World Trade Center Twin Towers GLOBAL VILLAGE A pair of really tall buildings in NYC demolished by terrorists who flew fuel-laden airliners into each tower on September 11, 2001. See Post-traumatic stress disorder

World Wide Web See the Web

world's most expensive drug Alglucerase, formally, Ceredase®, see there

worm PARASITOLOGY A soft-bodied, elongate invertebrate of certain phyla: Annelida–class Hirudinea, Aschelminthes–class Nematodes, Platyhelminthes–eg, class Trematoda. See Hookworm, Ringworm, Screwworm

worn tennis ball appearance IMAGING A fanciful descriptor for the 'fuzzy' margins surrounding a well-circumscribed peripheral adeno CA of the lung seen in a plain CXR, an appearance that contrasts with a smooth margin more typical of a peripheral SCC. See Squamous cell carcinoma

the worried well A person in a low-risk population who is concerned that a past picadillo might trigger a dread disease

worst headache of my life MEDTALK A statement made during the chief complaint of a Pt that may result in a million-dollar work-up, including physical exam, lab work, IV fluids, CT scan, lumbar puncture, neurology

worth VOX POPULI The value of a thing. See Social worth

WOSCOPS CARDIOLOGY A trial–West of Scotland Coronary Prevention Study–of the effect of pravastatin on M&M–risk of CHD, malignancy and other outcome data–in men with hypercholesterolemia. See Lipid-lowering therapy, Pravastatin. Cf MAAS, PLAC I, PLAC II, 4S–Scandanavian Simvastatin Survival Study

Wotherspoon criteria GI TRACT A scoring system used to grade the proliferation of lymphocytes in gastric biopsies infected with *H pylori*. See Helicobacter pylori, MALT lymphoma

wound MEDTALK An injury caused by physical means. See Clean wound, Defense wound, Dirty wound, Entrance wound, Execution wound, Exit wound, Hesitation wound, Problem wound

wound care product NURSING A product, ointment, antibacterial, absorbant, H_2O_2, betadine, packing dressings used to manage wounds and decubitus ulcers. See Decubitus ulcer

wound culture MICROBIOLOGY The placing of material from a wound–eg, pus, tissue, fluids in growth media, under aeorobic and anaerobic conditions to optimize the proliferation and identification of pathogens. See Culture

wound healing PHYSIOLOGY The repair of a wound STEPS Inflammation, repair and closure, remodeling, final healing; repair of incisions may be either simple–'clean' wounds with little loss of tissue heal by 'primary intention', or 'dirty' wounds heal by 'secondary intention'

WPW syndrome Wolff-Parkinson-White syndrome, see there

wraparound policy MANAGED CARE A high cost 'major medical' health insurance policy that provides full payment for any fees charged by a hospital and/or physician, which exceed the monies reimbursed by the 'basic carriers,' eg Blue Cross–for hospitals, and Blue Shield–for physicians. Cf Blue Cross/Blue Shield, 'Major medical'

wraparound procedure SURGERY An 'anti-reflux' operation–eg, Nissen, Belsey, Hill procedures that restore sphincter function to the lower esophagus; the Nissen procedure may afford the greatest long-term success. See 'Inkwell'

WRAT Wide Range Achievement Test PSYCHOLOGY A test that evaluates a child's basic skills of spelling, mathematics and reading–ie, educational achievement. See Psychological testing. Cf Psychiatric testing

wrecking ball effect CARDIOLOGY A tethered ball-like effect of a pedunculated tumor in the atrial or ventricular chambers, often due to a calcified myxoma, which may damage the mitral valve, rupture chordae tendineae or cause severe mitral regurgitation, as it pounds against endocardial structures

wringer injury A lesion once seen in children feeding clothes into or playing with an electrically driven laundry wringer, who literally got 'caught up' in the work CLINICAL Crushing of hand & upper extremity, friction burns at points of ↑ diameters–eg, metacarpophalangeal joints, thenar and base of hand, elbow and shoulder, neurapraxia, edema, lacerations, hematomas, fractures and potentially complete avulsion of a hand or arm; similar injuries may occur in industrial accidents–eg, in hot rollers in a printing press or in rotating farm equipment; WIs may be accompanied by thermal injury, extensive crushing and soil-related infection–eg, *Clostridium* spp contamination

wrinkles, fear of getting PSYCHOLOGY Rhytiphobia. See Phobia

WRIST CARDIOLOGY A clinical trial–Washington Radiation for In-Stent restenosis Trial

wrist See CLIP wrist, Golfer's wrist, SLAC wrist, Tennis wrist

wristband An identifying bracelet attached to a Pt's wrist at the time of admission to a health care facility, which may be the only identifier used during a person's stay in a hospital

wristdrop NEUROLOGY A condition caused by partial or complete paralysis of wrist extensors, a manifestation of peripheral neuropathy accompanied by a loss of acral motor activity ETIOLOGY Amyloidosis, Charcot-Marie-Tooth syndrome, collagen vascular diseases, diabetic mononeuropathy, lead poisoning, severe vitamin B_{12} deficiency. See Footdrop

wrist sign RHEUMATOLOGY A finding in Pts with Marfan and other heritable connective tissue disorders characterized by thinned distal extremities; the WS is positive if the distal phalanges of the thumb and 5th finger overlap when encircling the opposite wrist. See Marfan syndrome

write-up *noun* ADMINISTRATION A document detailing a variance in Pt management–a euphemism for an error or, expressed more optimistically, an "opportunity for improvement", which may have resulted in injury or death to the Pt MEDTALK A document written about a Pt, either by a medical student or physician-in-training, or practicing physician referring the Pt to another physician for a consultation *verb* To write a Pt's Hx, as in, "... to write the Pt up."

writer's block PSYCHIATRY An occupational neurosis of authors, in whom creative juices are temporarily or permanently inspissated

writer's cramp Conscious immobility NEUROLOGY A transient focal dystonia, often occupational neurosis of the hand and sometimes the forearm, characterized by a loss of skilled motor sequences–eg, handwriting, which may be a psychogenic conversion reaction

written expression disorder NEUROLOGY A disorder seen in 3% to 10% of school children, characterized by poor spelling, punctuation, grammar, and handwriting RELATED CONDITIONS Reading disorder, expressive language disorder, mathematics disorder, developmental coordination disorder. See Dyslexia

wrongful FORENSIC MEDICINE An adjective with considerable medicolegal currency, used in several contexts. See Negligence

WRONGFUL

WRONGFUL DEATH An event that is usually regarded as negligent. See Negligence

WRONGFUL BIRTH An event resulting from the failure of a contraceptive or sterilization procedure, eg fallopian tube liagation, failure to diagnose pregnancy, or an unsuccessful attempt to abort a conceptus.

WRONGFUL LIFE An event in which legal action may be taken by–or on behalf of the baby suffering from a hereditary or congenital defect, eg Down syndrome or other disease, eg rubella, who would not have been born had the parents had the knowledge to opt for an abortion; WL represents either the failure to diagnose in utero a condition that would lead to a major life-long handicap or recognize such a condition in a sibling, allowing a 2°, similarly afflicted, child to be born; the child is the defendant named in a lawsuit initiated to defray the incurred and anticipated medical, nursing and related health expenses; in both WB and WL, the defendant may be liable for support and care of the infant from 'cradle to grave'

wryneck Congenital torticollis, torticollis A focal dystonia consisting of one-sided contracture awith palpable induration of the sternocleidomastoid muscle, causing the chin to turn towards the opposite side and the head to rotate towards the lesion; WN is accompanied by facial muscle dysplasia ETIOLOGY Congenital form– unclear–possibly due to in utero or peripartum trauma to venous drainage, causing asymmetric development of the face and skull; the later WN is recognized, the more likely surgical intervention is required

Wt Weight, see there

WTC World Trade Center, see there

WVN West Nile virus, see there

WWW World Wide Web. See the Web

Wypax Lorazepam, see there

WYSIWYG What you see is what you get *pronounced*, wissy-wig COMPUTERS A 'hard' or paper copy that corresponds exactly to what is viewed on a computer monitor or display

x Symbol for: 1. Abscissa 2. Mean–statistics 3. Xylose

X Symbol for: 1. Any unknown amino acid 2. Any unknown quantity 3. Arithmetic mean–statistics 4. The term for the axis on a graph 5. Ecstasy–a designer drug, see there 6. Exphoria distance–ophthalmology 7. Female chromosome 8. X-ray 9. Xanthine 10. Xanthosine

X chromosome A sex chromosome with 2 copies in normal ♀ and one copy in normal ♂. Cf Y chromosome

X factor HAIR REPLACEMENT A popular term for an unidentified cause of follicular destruction, which occurs in up to 1% of Pts undergoing hair transplantation

X Games SPORTS MEDICINE The official Olympics of 'extreme sports' sponsored by ESPN, held annually during the summer. See Extreme sports

X-linkage The linking of a trait or gene to an X chromosome

X-linked Sex linked GENETICS *adjective* Referring to a mode of transmission in which a trait or gene is linked to the X chromosome and generally expressed only in males. See Gene, Mutation, Sex chromosome

X-linked adrenoleukodystrophy An X-R peroxisomal disease due to a defective gene on chromosome Xq28, characterized by impaired degradation of saturated very long chain fatty acids–VLCFAs that accumulate with the cholesterol ester and gangliosides in the CNS, adrenal glands, plasma cells, leukocytes, causing progressive multifocal demyelination CLINICAL Variable; often first seen in childhood with major neurologic deterioration and death in a few yrs MANAGEMENT Modifications in lipid intake may normalize VLCFA levels, the long-term effects of which are unknown, BMT. See Lorenzo's oil

X-linked agammaglobulinemia An X-linked disease with defective humoral immunity due to an intrinsic B-cell gene defect that 'maps' to chromosome Xq21.3-22 CLINICAL Recurrent pyogenic infections, affecting boys by 5-6 months of life, coincident with fall in IgG acquired in utero LAB ↓ B cells, ↓ Igs, intact T-cell function MANAGEMENT Antibiotics, monthly injections of gammaglobulins PROGNOSIS Death in early childhood, often due to fulminant lung infection

X-linked centronuclear–myotubular myopathy An often fatal, X-R myopathy, which causes severe hypotonia; adult survivors have minimal residual functional incapacity; the associated proliferation of sarcotubular organelles may be caused by impaired innervation

X-linked disorder A condition in which the defective gene is on the X chromosome; because ♂ have one X chromosome, they are usually affected if the defective gene is structural–ie, necessary for function. See Autosomal dominant, Autosomal recessive

X-linked hypophosphatemic rickets A condition characterized by impaired proximal renal tubule resorption of phosphate, coupled with a relative ↓ of 1,25-dihydroxyvitamin D production CLINICAL Growth failure, bowing of legs, hypophosphatemia, radiologic changes of rickets MANAGEMENT Calcitriol, phosphate COMPLICATIONS Kidney stone formation reflects phosphate levels

X-linked lymphoproliferative disorder Duncan disease, Epstein-Barr immunodeficiency syndrome, Epstein-Barr virus-induced lymphoproliferative disorder IMMUNOLOGY An X-linked or far less commonly AR–primary immunodeficiency disorder characterized by a defective immune response to EBV infection, resulting in severe, life-threatening infectious mononucleosis–IM, severe hypogammaglobulinemia–ergo ↑ susceptibility to various infections, B-cell lymphomas CLINICAL Onset between 6 months and 10 yrs of age; ± half of Pts have IM characterized by fever, pharyngitis, lymphadenopathy, splenomegaly, hepatomegaly, abnormal LFTs, jaundice; severe cases may be accompanied by lymphocytic and histiocytic infiltrates with severe liver damage and/or failure, BM replacement, causing aplastic anemia, leukocytopenia, thrombocytopenia; XLP may be accompanied by congenital cardiovascular and CNS defects, CAUSE OF DEATH Hepatitis, immune suppression, B-cell lymphomas

X-linked recessive GENETICS *adjective* Referring to a mode of inheritance, in which a gene on the X chromosome requires one copy for phenotypic expression in ♂, but 2 copies for expression in ♀; with the gene only on the X chromosome, ♀ are carriers; ♂ get the disease. Cf Autosomal dominant, Autosomal recessive

X-ray High-energy radiation A range of the electromagnetic spectrum used in low doses to diagnose disease and in high doses to treat CA. See Soft X-rays

X-RAY EXPOSURE

DIAGNOSTIC X-RAYS Impart 30-150 keV of energy; rare reports vaguely suggest a relationship between exposure to low- level X-rays and a slight ↑ in myeloproliferative disorders and a minimal ↑ risk for developing myeloma

THERAPEUTIC X-RAYS

• Low level radiation, eg 5-10 keV or 'grenz' radiation–may be used to treat recalcitrant skin conditions–eg, psoriasis

• High level radiation, eg megaelectron-volt–MeV) radiation–may be used to treat internal malignancy

X-ray pelvimetry See Radiologic pelvimetry

Xalatan® Latanoprost OPHTHALMOLOGY A topical agent for managing open-angle glaucoma and ocular HTN. See Open-angle galucoma

Xanax® Alprazolam, see there

xanthelasma Fatty skin growth DERMATOLOGY A condition characterized by multiple 1-2 mm yellowish plaques–lipid-laden histiocytes surrounding blood vessels, formally, xanthomas of the inner eyelid, commonly seen in normocholesterolemic elderly, or hypercholesterolemic younger persons. See Xanthoma

xanthogranulomatous pyelonephritis NEPHROLOGY A malakoplakia-like kidney lesion that lacks Michaelis-Gutmann bodies, arising in an *E coli* or *Proteus* spp infection COMPLICATIONS Hydronephrosis, nephrolithiasis

xanthoma Fatty skin growth DERMATOLOGY A yellow-orange lipid-filled papule or plaque located on the skin or tendons, which is common in the general population and Pts with DM, and may be ↑ in various 'lipid disorders–eg, lipoprotein lipase deficiency, abetalipoproteinemia, hypercholesterolemia, hypertriglyceridemia

754

xanthomatosis A generalized, nonspecific ↑ in xanthomas that may occur in malignancy–eg, lymphoma, multiple myeloma, or occur in other disease–eg, familial hypercholesterolemia syndrome, Langerhans' cell histiocytosis-Hand-Schüller-Christian type, Wolman's disease

Xeloda® Capecitabine ONCOLOGY An anticancer agent used for therapy-resistant metastatic breast CA, which targets and is activated by tumor cell enzymes; once activated, the drug is converted to 5-FU. See Breast cancer, 5-FU

xemilofiban A GP IIb/IIIa receptor antagonist or 'superaspirin' oral antiplatelet agent designed to prevent blood clot formation after coronary angioplasty; it was assessed during EXCITE trials to evaluate whether daily postprocedure xemilofiban for 6 months would ↓ clot-related cardiac events in Pts undergoing PCTA and stent placement; xemilofiban was less effective than aspirin and has been abandoned

Xenical® Orlistat OBESITY A lipase inhibitor that blocks absorption of ±30% of the fat eaten in food; is the first non-CNS, nonsystemic drugs for treating obesity INDICATIONS Obesity management (weight loss and maintenance, reduced-calorie diet)

xenophobia PSYCHIATRY An abnormal/morbid fear of strangers or foreigners. See Phobia

Xenopsylla Rat flea A genus of fleas, family Pulicidae, which is a vector for dwarf and rat tapeworms (*Xenopsylla cheopis*), murine typhus–*X cheopis* and plague–*X astia, X brasiliense, X cheopis*

xenotransplantation Xenogeneic transplantation TRANSPLANT BIOLOGY The transplantation of cells or tissues from one species to another; the use of live, nonhuman animal cells, tissues, and organs in humans. See Xenograft

Xepin Doxepin, see there

Xerecept® hCRF NEUROLOGY A synthetic corticotropin-releasing factor–CRF–with orphan drug status for managing neurologic Sx caused by peritumoral brain edema to ↓ post-brain injury cerebral edema. See Cerebral edema

xeroderma Dry skin, see there

xeroderma pigmentosa An AR genetic condition characterized by a sensitivity to all sources of UV radiation; XP Pts cannot repair UV light-induced damage to pyrimidine nucleotide–cytosine, thymidine dimers due to lack of one or more multigene products or complementation groups CLINICAL Incidence 2/10⁶; normal at birth, freckling, xeroderma by age 3, accompanied by telangiectasia, keratoacanthomas, keratoses, scarring, skin cancer–eg, BCC, SCC, melanoma, internal malignancy–fibrosarcoma, angiosarcoma TREATMENT High-dose isotretinoin may be effective in chemoprophylaxis of skin CA in Pts with XP. See Photoactivation

xerophthalmia Dry eye syndrome, see there

xerostomia Dry mouth syndrome, see there

Xigris® Drotrecogin INFECTIOUS DISEASE An agent used to manage severe sepsis

XO syndrome Turner's syndrome, see there

Xopenex® RESPIRATORY CARE A nebulized levalbuterol solution used manage bronchospasm in adults with reversible obstructive airway disease. See Bronchospasm

XTC See Ecstasy, MDMA

Xubix™ Sibrafiban, see there

XX male syndrome An AR sex chromosome anomaly that clinically mimics Klinefelter syndrome CLINICAL ♂ psychosexual orientation, masculine appearance, weak 2° sexual characteristics, azoospermia, ↓ androgen levels, small testes

XXX syndrome 47, XXX The most common–1:1000 chromosomal abnormality of ♀ CLINICAL Clinodactyly, epicanthal folds, ocular hypertelorism, an ↑ incidence of congenital defects in Pts' children; most have normal IQs, but tend to be emotionally labile, with speech and learning defects

XXXX syndrome 48, XXXX, superwoman A ♀ chromosome defect associated with moderate mental retardation, behavioral defects, midfacial hypoplasia, clinodactyly, radial synostosis, hypertelorism, micrognathia, webneck, dysmenorrhea

XXXXX syndrome 49, XXXXX, penta-X syndrome A ♀ presenting in infancy with PDA, mental and growth retardation, upward slanting of eyes, carpal hypoplasia, clinodactyly of 5th finger

XXXY syndrome(s) XXXXY syndrome A group of chromosome defects with multiple X chromosomes and one Y chromosome CLINICAL Somatic defects overlap Klinefelter syndrome–small undescended testes, hypoplastic penis, gynecomastia, mental retardation, wide-set eyes, ulnar and radial abnormalities; > ½ of those with the 49, XXXXY syndrome present with low birth weight, muscle hypotonicity, profound mental and growth retardation, malformed ears, a short neck, hypertelorism with a mongoloid slant, flattened nose, hypoplastic external male genitalia, clinodactyly, radioulnar synostoses, coxa vara, genu varum, pes planus TREATMENT None, testosterone may improve 2° sexual characteristics

XXY syndrome Klinefelter syndrome, see there

XY gonadal agenesis syndrome Embryonic testicular regression syndrome, see there

Xylocaine® Lidocaine, see there

xylophonist Flateur; farter; one with noisy flatus. See Flatulence

XYY hypothesis See Crime genes

XYY syndrome YY syndrome A condition characterized by an extra Y chromosome, an array of typical clinical features–facial asymmetry, long ears, teeth and fingers, poor musculoskeletal development, cranial synostosis, prolonged P-R on EKG, and a possible tendency toward antisocial behavior

Y Symbol for: 1. Luminance 2. Male chromosome 3. A pyrimidine nucleotide 4. Tyrosine 5. Year 6. Yttrium, see there

Y chromosome The sex chromosome of normal ♂. Cf X chromosome

Y construct A general term for any pathogenic sequelae in which there are two or more results arising from a common pathway

Y fracture T fracture, see there

Y-linked *adjective* Referring to a gene of sequence of DNA which is located on the Y chromosome and, by necessity, is passed from father to son. Cf X-linked

Y-linked inheritance Holandric inheritance GENETICS Inheritance by genes on the Y chromosome. See X-linked inheritance

Y protein Ligandin, see there

Y-shaped fracture T-shaped fracture, see there

YAG laser Yttrium-aluminum-garnet laser, Nd:YAG–neodymium:yttrium-aluminum-garnet–laser. See Laser

yaw FORENSIC PATHOLOGY The angle between the line of flight and a bullet's long axis; bullets enter tissue and tumble once inside, rotating 180°, and exit with the base forward, which explains the large size of some exit wounds. See Ballistics

yawn The involuntary opening of mouth, often caused by suggestion–"contagious" and accompanied by breathing inward then outward; repeated yawning may indicate drowsiness, depression, or boredom

yearly checkup MEDICAL PRACTICE An annual evaluation of a person's health status, which includes a physical exam, and routine screening tests to insure continued health, or identify early, and often treatable stages of a disease. See Private physician

years of potential life lost PUBLIC HEALTH A measure of the impact of premature mortality on a population, calculated as the sum of the differences between a predetermined minimum or desired life span usually set at

years of potential life lost rate VITAL STATISTICS The rate of yrs of potential life lost/10^5 persons due to premature death from a particular high-risk activity–eg, cigarette smoking. See YPLL

yeast A unicellular spherical-to-oval 3–5 μm budding fungus that reproduces both sexually and asexually, primarily by budding–some by binary fission which, when adherent in end-to-end rows are termed pseudohyphae; most fungi are saprobes; many are used in commercial fermentation of foods and beverages; 7 genera–class Deuteromycetes–Imperfect Fungi–are human pathogens: *Candida, Crytococcus, Geotrichum, Pityrosporum, Rhodotorula, Torulopsis, Trichosporon.* See *Candida*, YAC cloning

yeast infection of esophagus Candida esophagitis, see there

yeast infection of skin Candida dermatitis, see there

yeast infection of vagina Candida vaginitis, see there

yellow fever TROPICAL MEDICINE An acute mosquito-borne infection–*Aedes aegyptii, A africanus, Haemagogus* spp of tropical Africa, Central and South America, by a flavivirus of the Togavirus family CLINICAL From benign nonspecific 'flu' Sx–mild subclinical infection is more common in children–to severe disease characterized by abrupt high fever with bradycardia–Faget sign, headache, anorexia, myalgia, lumbosacral pain, N&V–vomitus becomes blood-tinged up to 7–10 days post-onset, photophobia, flushing of face, red eyes, red 'strawberry' tongue, insomnia, constipation, hypotension, jaundice ± day 4, generalized petechial hemorrhages and hematemesis, hepatic and renal necrosis, dilated cardiomyopathy, stomach pain, delirium and seizures followed by coma; this is followed by a period of 'remission' of hrs to several days in duration, after which the Pt deteriorates and dies–up to 50% mortality LAB Leukopenia, ± pronounced, coagulation factor depletion, DIC, ↑ bilirubin, proteinuria, albuminuria with renal failure, ↑ neutralizing antibodies, ↑ transaminases

yellow fever vaccination A live attenuated–weakened viral vaccine recommended for people traveling to or living in tropical areas in the Americas and Africa where yellow fever occurs

yellow jacket venom A substance produced by hymenopteran insects–eg, bees, wasps, which may be fatal via an anaphylactic reaction with release of vasoactive substances–leukotrienes, ECF-A, histamine, causing generalized urticaria, throat and chest tightness, stridor, fever, chills and cardiovascular collapse; toxic overload of hymenopteran venom may also be fatal to victims of 'swarmings'. See Swarming

yellow nail syndrome A clinical complex characterized by slow-growing indurated flavescence of all or the distal part of the nail with the loss of the lunula, onycholysis, primary lymphedema of the face and extremities, idiopathic pleural effusions, chronic bronchitis and sinusitis; the nails are thickened without visible lunulae and have an exaggerated lateral curvature, and slowed growth, attributed to primary stromal sclerosis, leading to lymphatic obstruction. See Harlequin nail

yellow pill See Little Yellow Pill

yellow urine A yellow-tinged urine which, in acidic pH urine, may be due to excretion of dinitrophenol, phenacetin or chrysarobin or in alkaline pH urine, due to ↑ secretion of anthocyanin, or associated with ingestion of beets or blackberries; pure YU is associated with ↑ acriflavine and has a green fluorescence; yellow-orange urine may be highly concentrated or contain ↑ bilirubin-biliverdin

Yentl 'syndrome' EPIDEMIOLOGY A phenomenon in which once a ♀ shows herself to be equal to a ♂–in terms of risk factors, she receives equal medical treatment. See Women and disease, Women's Health Initiative

756 *Yersinia* A genus of pathogenic gram-negative, facultatively anaerobic, coccobacillary bacteria, which cause bubonic plague–*Y pestis*, intestinal infections–*Y enterocolitica*, mesenteric lymphadenitis–*Y pseudotuberculosis*, which mimics appendicitis

yo-yo dieting NUTRITION Undesirable dietary cycling characterized by a rapid weight loss then regain. See Starvation diet

yoga Hatha yoga AYURVEDIC MEDICINE A holistic system of health care and maintenance, widely practiced throughout the world; the purpose of yoga is to join the mind, body, and breath as one unit; if the mind is disturbed, the breath and body are affected; as the body's activity increases, the mind is altered and the rate and depth of breath changes; yoga attempts to join the 3 units through proper breathing and by assuming *asanas*–yogic poses; regular practice of yoga may ↓ stress, heart rate, BP and possibly retard aging

yogurt NUTRITION A smooth semisolid dairy product produced by fermenting milk with *Lactobacillus acidophilus* and other bacteria that convert lactose into lactic acid; it is regarded as a healthy food

yohimbine PHARMACOGNOSY An alkaloid used to treat erectile dysfunction, which is chemically similar to reserpine; it may be of use for orthostatic hypotension and ischemic vascular disease caused by autonomonic dysfunction in DM, narcotic overdose, possibly acting to ↑ dopamine release, and for psychogenic and organic impotence, possibly ↑ inflow to and/or ↓ outflow of blood from the penis, with ↑ erection without ↑ sexual desire

Yokohama asthma Tokyo asthma ENVIRONMENT A smog-induced asthmatiform complex seen in Yokohama, Tokyo's port city, where nocturnal atmospheric conditions favor a dense buildup of industrial pollutants CLINICAL Wheezing, dyspnea in Pts with hyperirritable airways; worse in Pts with COPD. See Bhopal, Smog

Yom Kippur effect Spontaneous premature onset of labor, which occurs in pregnant Jewish ♀ who observe Yom Kippur, Judaism's annual 24-hr religious fast, with a total abstinence from food and water. See Judaism–practice of

Young syndrome A condition seen in men in UK and Australia between 1960 and 1980, accompanied by bronchiectasis and azoospermia due to epididymal obstruction ETIOLOGY Uncertain; some authors have postulated that YS was caused by mercury intoxication. See Mercury intoxication

yoursurgery.com ONLINE A website that provides basic information on commonly performed surgical procedures. See Internet

youth Adolescence VOX POPULI 1. A nebulously defined period of development between puberty and maturity 2. A young person

YPLL Years of potential life lost, see there

yttrium An ultrarare metalic element–atomic number, 39; atomic weight, 88.9–named after Ytterby in southern Sweden, which may be used in nuclear medicine scans

yuppie disease Chronic fatigue syndrome, see there

yusho Kanemi oil intoxication TOXICOLOGY A major epidemic of food intoxication that occurred in 1968 in Japan, affecting ±13,000 CLINICAL N&V, persistent ARDS, meibomian gland hyperplasia, chloroacne, cardiovascular and CNS disease, petechial and erythematous rashes with pulmonary HTN. See PCBs, Toxic oil syndrome, Yucheng disease

Yuzpe protocol Yuzpe method GYNECOLOGY A hormone protocol for emergency postcoital contraception. See Contraception, RU-486

YY syndrome See XYY syndrome

Z Symbol for: 1. Atomic number 2. Aza–a nucleoside substitite 3. Glutamic acid and glutamine 4. Impedance 5. DRUG SLANG A street term for 1 ounce– 28 g. of heroin 6. Net charge 7. Proton number 8. Zone

Z-Pak® INFECTIOUS DISEASE Zithromax–azithromycin, that contains 6 capsules–2 on day 1; the rest are taken once daily until consumed. See Azithromycin

Z-plasty SURGERY A surgical incision that lengthens a zone of skin or a muscle, 'breaking up' a linear scar or repositioning an incision to a line of least tension. See Y plasty

Zacetin Alprazolam, see there

Zadaxin® Thymosin alpha 1, see there

zafirkulast Accolate® ALLERGY MEDICINE An antileukotriene that ↓ inflammation and bronchoconstriction of asthma ADVERSE EFFECTS ↑ in PT in Pts receiving warfarin. See Antileukotrienes, Asthma

Zagam® Sparfloxacin, see there

zalcitabine ddC, Hivid AIDS A purine analogue that inhibits HIV reverse transcriptase; it is similar didanosine vis-á-vis survival, progression, CD4 count, adverse effects, efficacy. See Didanosine, Reverse transcriptase inhibitor. Cf Protease inhibitor

zaleplon Sonata®, see there

Zanaflex® Tizanidine NEUROLOGY An agent used to manage ↑ muscle tone in spasticity–eg, in spinal cord injury, multiple sclerosis. See Spasticity

zanamivir Relenza® INFECTIOUS DISEASE An inhaled viral neuraminidase inhibitor used to treat and prevent influenza A and B infection. See Influenza. Cf Amantadine, Rimantidine

Zanapam Alprazolam, see there

Zantac® Ranitidine, see there

ZDV Zidovudine, see there

zebra MEDTALK A popular 'short form' of the aphorism often quoted to medical students during clinical rotations–*'when you hear hoofbeats, don't think of zebras'*. See Sutton's law. Cf 'Red herring'

zeitgeber *Zeit*, German, time, *geber*, keeper A factor in the environment with a periodicity, capable of synchronizing the endogenous circadian rhythm into a 24-hr cycle; without zeitgebers, the free-running human clock is 25.3 hrs. See Circadian rhythm, Jet lag, Shift work, Sleep disorders. Cf REM sleep

Zeldox® Ziprasidone PSYCHIATRY A neuropharmacologic used for schizophrenia and to prevent psychotic relapses. See Schizophrenia

Zellweger syndrome Cerebrohepatorenal syndrome An AR disease characterized by defective peroxisomes, due to inability to import matrix proteins CLINICAL Profound neurologic impairment–seizures, flaccidity, metabolic dysfunction, hyoid bone and thyroid cartilage calcification, dolichocephaly, persistent wormian bones, hand and foot deformities and bone immaturity, facial dysmorphia, cataracts, contractures, renal cortical cysts, liver fibrosis, death in early infancy

Zemplar® Paricalcitol ENDOCRINOLOGY A synthetic vitamin D for preventing and treating 2° hyperparathyroidism due to chronic renal failure. See Hyperparathyroidism

Zen therapy FRINGE PSYCHOLOGY A therapy that encourages self understanding through prolonged meditation. See Meditation, Zen garden therapy

Zenapax® Daclizumab, see there

Zenax Alprazolam, see there

Zenecin Chlordiazepoxide, see there

Zepaxid Diazepam, see there

Zerit® D4T, stavudine AIDS An oral stavudine for treating HIV-infected infants. See AIDS, Stavudine

zero PHYSICS Null, naught, nada. See Audiometric zero VOX POPULI A popular term for a person with no personality and/or life

zero error standard MEDICAL PRACTICE A popular term for the unreasonable expectation on the part of the public that physicians can make no mistakes

zero-order kinetics THERAPEUTICS The in vivo dynamics of drug elimination, which is linear with time, proportional to the concentration of the enzyme responsible for catabolism, and independent of substrate concentration. See Michaelis-Menten equation, Therapeutic drug monitoring. Cf First order kinetics

zero time EPIDEMIOLOGY The time at which a particular condition being studied–eg, AIDS-is first identified or diagnosed TYPES Delayed ZT–the condition is diagnosed by others and referred to a center where a study is being performed, and prompt ZT–if the diagnosis is established for the first time

zero tolerance policy SUBSTANCE ABUSE A stance taken by US government, that any type of drug abuse is punishable by incarceration. See Correctional facility, War on Drugs

zero work pool MEDICAL PRACTICE An allocation of Medicare practice expense payment money shared by all physician fee schedule services that do not have a physician work component

Zetia® Ezetimibe, see there

Zetran Diazepam, see there

ZEUS NEUROLOGY A clinical trial–Zomaril™–iloperidone Efficacy, Utility & Safety–in treating schizophrenia. See Zomaril SURGERY A robotic system for cardiothoracic surgery and E-CABG via computerized maneuvering of instruments and endoscope. See Robotic surgery

Ziagen® Abacavir, see there

zidovudine AZT 3'-azido-3'-deoxythimidine AIDS A thymidine analogue and HIV-1 antimetabolite, which slows progression of AIDS and terminating HIV's DNA chain growth INDICATIONS AIDS and *Pneumocystis carinii* pneumonia, CD4+ T-helper cells < 200/mm³, zidovudine therapy–ZT ↓ vertical maternal-infant transmission of HIV, HIV-1 infected children with neurodevelopmental abnormalities EFFECT Improved sense of well-being, ↑ CD4+ T cells, ↓ AIDS-related complications SIDE EFFECTS Dose-limiting myelosuppression with granulocytopenia–which may respond to lithium, anemia, headache, insomnia, mania, seizures, nausea, myalgia, lymphoma. See AIDS therapy

zidovudine failure Drug failure defined as the development of opportunistic infection despite adequate zidovudine therapy, due to infection with zidovudine-resistant HIV-1

ZIFT Zygote intrafallopian transfer, see there

zigzag QRS waves Sinusoidal QRS complexes seen by EKG in ventricular flutter–240-280/min

zileuton Zyflo® CLINICAL MEDICINE An inhibitor of 5-lipoxygenase, an enzyme central to leukotriene synthesis, which mediates inflammatory responses INDICATIONS Asthma. See Asthma

zinc deficiency A condition caused by inadequate ingestion of zinc CLINICAL Dermatitis–cheeks, elbows and knees and tissue about the mouth and anus, diarrhea, delayed hypersensitivity response, fatigue, hair loss–balding of the scalp, eyebrows and lashes, loss of taste, difficult pregnancy, anemia, hyperpigmentation, hepatosplenomegaly, hypogonadism, poor wound healing; in children, stunted growth and sexual development, recurrent infections due to immune deficiency. See Acrodermatitis enteropathica, Zinc

zinc intoxication Zinc poisoning A condition caused by industrial exposure or by excess ingestion of zinc supplements CLINICAL Headache, N&V, dehydration, stomach ache, incoordination, renal damage, immune deficiency

zinc oxide TOXICOLOGY A compound used in welding which, in excess, may cause zinc intoxication USES Topically, astringent, calamine lotion. See Zinc

Zinecard® Dexrazoxane, see there

ziprasidone See Zeldox™

Zithromax® Azithromycin, see there. See Z-Pak

Zocor® Simvastatin, see there

Zofran® Ondansetron THERAPEUTICS An orally disintegrating tablet–ODT indicated for preventing RT and chemoherapy-induced and postoperative N&V. See Ondansetron

Zoladex® Goserelin acetate implant, see there

Zoldac® Alprazolam, see there

zoldipem Ambiem® PHARMACOLOGY A nonbenzodiazepine sedative hypnotic used to manage insomnia in older adults PROS Rapid onset of action, short T$_{1/2}$, doesn't cross-react with drugs in standard urine drug screens

zoledronate Zoledronic acid, Zometa® OSTEOPOROSIS A potent bisphosphonate used to ↓ fractures, ↓ bone pain INDICATIONS Postmenopausal osteoporosis, CA metastases to bone ADVERSE EVENTS Myalgia, pyrexia. See Biphosphonate, Osteoporosis

Zollinger-Ellison syndrome A condition characterized by multiple duodenal ulcers and gastrin-secreting tumors that cause recurrent and refractory upper GI ulceration, diarrhea, steatorrhea, hypoglycemia–due to excess 'little' gastrin, pyrosis and dysphagia due to gastroesophageal reflux and ↑ gastric acid; 25% also have MEN-I; malignant gastric carcinoid may be related to chronic gastrin hypersecretion DIAGNOSIS CT, typical clinical findings LAB ↑ Gastrin, ↑ basal acid output–>15 mmol/hr, ↑ post-histamine stimulation acid production, ↑ ratio of basal acid output to maximum acid output–stimulation testing reveals fasting gastrin > 1000 pg/mL, ↑ Ca²⁺ DIFFDX-HYPERGASTRINEMIA Antral hyperplasia, retained antrum syndrome, chronic atrophic gastritis, pernicious anemia, short bowel syndrome, gastric outlet syndrome or obstruction, gastric CA, pheochromocytoma, renal failure MANAGEMENT H₂ receptor antagonists; at higher doses may be coadministered with antimuscarinics to inhibit gastric secretion by H₂ blockage; if unresponsive, total gastrectomy and/or vagotomy

zolmitriptan Zomig™ NEUROLOGY A selective 5-HT$_{1B/1D}$ agonist for acute migraine with/without aura, effective in ± ⅔ of Pts within 1 hr of therapy ADVERSE EFFECTS Paresthesia, anesthesia, head & neck pain/tightness/pressure, dizziness CONTRAINDICATIONS ASHD, CAD, HTN, MI

Zomaril™ Iloperidone PSYCHIATRY An antipsychotic for treating schizophrenia. See Schizophrenia, ZEUS

Zomax® Zomepirac, see there

zombie effect A descriptor for personality changes caused by haloperidol's extrapyramidal effects, which simulate Parkinsonism–hypomimia, bradykinesia, flat affect. See Extapyramidal effect, Parkinson's disease

zomepirac Zomax® THERAPEUTICS An oral NSAID that tightly binds to plasma proteins; it is excreted in the urine. See NSAIDs

Zomig® Zolmitriptan, see there

Zonalon® Doxepin, see there

the 'zone' SPORTS MEDICINE 1. A popular term used by some amateur athletes for heart rates–zones that reflect the intensity of a workout or exercise routine 2. A state of maximum physical, mental and psychological performance achieved by star athletes for fleeting moments in their careers. See Zone-favorable diet VOX POPULI A place, area or region with specified boundaries. See Body buffer zone, Convergence zone, Convergent zone, Epileptic zone, Forbidden zone, Gray zone, Health zone, Hot zone, Quiet zone, Safety zone, Transformation zone, Trigger zone, Twilight zone

zone electrophoresis Any electrophoretic technique in which components are separated into zones or bands in a buffer, and stabilized in solid, porous, or any other support medium–eg, filter paper, agar gel, or polyacrylamide gel

zone of hyperalgesia Pressure point, see there

Zonegran® Zonisamide NEUROLOGY An agent used to manage partial seizures in adults with epilepsy. See Seizures

zoonosis EPIDEMIOLOGY An infection in which the microbe's infectious cycle is completed between man and mammal

zoopharmacognosy HERBAL MEDICINE The formal study of plants used by animals in their native habitats, which they are believed to use to 'treat' their own illnesses; Cf Pharmacognosy

zoopherin Vitamin B₁₂, see there

zoophilia Bestiality SEXOLOGY A paraphilia in which sexuoerotism and/or orgasm depend on engaging in sexual activity with an animal. See Paraphilia

Zorac® Tazarotene, see there

zoster Herpes zoster, see there

zosteriform *adjective* Referring to a band-like unilateral skin lesion located along the cutaneous distribution of a spinal or a branch of the trigeminal nerves, usually seen in the recrudescence of herpes zoster but also–rarely seen in metastatic breast CA and hemangiomas of Sturge-Weber disease

Zostrix® Capsaicin, see there

Zosyn® INFECTIOUS DISEASE A broad-spectrum IV antibiotic containing piperacillin and tazobactam packaged in the ADD-Vantage® drug-delivery system. See Piperacillin

Zotran Alprazolam, see there

Zovirax® Acyclovir, see there

ZUMI Zinnanti Uterine Manipulator Injector SURGERY A device for manipulating the uterus during gynecologic procedures. See Hysterectomy

Zung depression scale PSYCHIATRY An objective rating instrument that evaluates depression, anxiety, hostility, phobias, paranoid ideation, obsessive compulsiveness and others

Zydone® NEUROLOGY A proprietary combination of hydrocodone bitartrate/acetaminophen used for mild-to-moderate pain. See Pain management

Zyflo® Zileuton, see there

zygoma 1. Zygomatic arch, see there 2. Zygomatic bone, see there 3. Zygomatic process of the temporal bone

zygoma guard ENT A hardened plastic arc worn by Pts with complex zygoma fractures. See LeFort fracture

zygomycetes Fungi–eg, *Absidia, Cunninghamella, Mortierella, Mucor, Rhizopus,* that are pathogenic to immunocompromised hosts; other fungi–eg, *Basiobolus, Conidiobolus,* and *Rhizomucor,* proliferate but are not usually pathogenic. See Fungi

zygote adoption Adoptive pregnancy, embryo adoption SOCIAL MEDICINE The transfer of a fertilized egg to a gestational mother who will carry the child to term, deliver and raise the child

zygote intrafallopian transfer REPRODUCTION MEDICINE An artificial reproduction method in which eggs are collected from the ovaries and fertilized outside the body; the resulting zygote is then placed by laparoscoy into the fallopian tube via abdominal incisions

Zyloprim® Allopurinol, see there

Zyprexa® Olanzapine PSYCHIATRY An antipsychotic. See Antipsychotic

Zyrtec® Cetirizine, see there

Zyvox™ Linezolid INFECTIOUS DISEASE An oxazolidinone class antibiotic active against gram-positive bacteria INDICATIONS Skin and soft-tissue infections, community-acquired and hospital-acquired pneumonia, gram-positive–eg, MRSA, VRE bacteremia. See Methicillin-resistant *Staphylococcus aureus,* Vancomycin-resistant enterococci

Abbreviations Used in Text

1° Primary

2° Secondary

2-D Two-dimensional

3° Tertiary

3-D Three-dimensional

aa Amino acid

ABG Arterial blood gases

ACE Angiotensin-converting enzyme

ACEI Angiotensin-converting enzyme inhibitor

AD Autosomal dominant

ADD Attention deficit (hyperactivity) disorder

ADH Antidiuretic hormone (vasopressin)

AF Atrial fibrillation

AFB Acid-fast bacillus

AFP Alpha-fetoprotein

AIDS Acquired immunodeficiency syndrome

aka Also known as

alk phos Alkaline phosphatase

ALL Acute lymphocytic (lymphoblastic) leukemia

ALS Amyotrophic lateral sclerosis

ALT Alanine aminotransferase (formerly GPT)

Am American

AMA American Medical Association

ama Against medical advice

AMI Acute myocardial infarction

AML Acute myelocytic (granulocytic, myeloid, myelogenous) leukemia

AMN American Medical News

ANA Antinuclear antibody

ANLL Acute nonlymphocytic leukemia

ANP Atrial natriuretic protein

AP Anteroposterior

APLM Archives of Pathology and Laboratory Medicine

apo Apolipoprotein

aPTT Activated partial thromboplastin time

AR Autosomal recessive

ARDS 1. Acute respiratory distress syndrome 2. Adult respiratory distress syndrome

ASAP As soon as possible

ASCUS Atypical squamous cells of undetermined significance

ASD Atrial septal defect

ASHD Atherosclerotic heart disease

Ass Association

AST Aspartate aminotransferase (formerly GPT)

ATP Adenosine triphosphate

AV 1. Arteriovenous 2. Atrioventricular

AW Atomic weight

BAL Bronchoalveolar lavage

BBB 1. Bundle branch block 2. Blood-brain barrier

BCC Basal cell carcinoma

β-blocker Beta adrenergic receptor blocking agent

BFP Biological false positive

BID Twice a day

BM 1. Bone marrow 2. Basement membrane 3. Bowel movement

BMJ British Medical Journal

BMT Bone marrow transplantation

bp Base pairs

BP Blood pressure

BPH Benign prostatic hypertrophy

BR Bilirubin

BUN Blood urea nitrogen

Bx Biopsy

CA Cancer, malignancy

CABG Coronary artery bypass graft

CAD Coronary artery disease

C albicans *Candida albicans*

cAMP Cyclic adenosine monophosphate

CAP College of American Pathologists

CBC Complete blood count

CCB Calcium channel blocking agent

CDC Centers for Disease Control & Prevention–based in Atlanta

cDNA Complementary DNA

CEA Carcinoembryonic antigen

CF Complement fixation

CHD Coronary heart disease

CHF Congestive heart failure

CIE Counter-immunoelectrophoresis

CIN Cervical intraepithelial neoplasia

CIS Carcinoma-in-situ

CK Creatinine (phospho)kinase

CLL Chronic lymphocytic (lymphoblastic, lymphoid) leukemia

CML Chronic myelocytic (granulocytic, myelogenous, myeloid) leukemia

CMV Cytomegalovirus

CNS Central nervous system

CO$_2$ Carbon dioxide

COD Cause of death

Coll College

Cons Disadvantages

COPD Chronic obstructive pulmonary disease

CPR Cardiopulmonary resuscitation

CPT Current Procedural Terminology, cocktail party trivia

CRH Corticotropin-releasing hormone

C-section Cesarean section

CSF Cerebrospinal fluid

CT Computed tomography

C trachomatis *Chlamydia trachomatis*

CVA Cerebrovascular accident

c/w Consistent with

CXR Chest X-ray

DAD Diffuse alveolar damage

DEA Drug Enforcement Administration, US Department of Justice

DIC Disseminated intravascular coagulation

DiffDx Differential diagnosis

DKA Diabetic ketoacidosis

DM Diabetes mellitus

D melanogaster *Drosophila melanogaster*

DNA Deoxyribonucleic acid

DOA Dead on arrival

DOE Dyspnea on exertion

DRG Diagnosis-related group

DSM-IV Diagnostic and Statistical Manual, 4th edition

DVT Deep vein thrombosis

DWI Driving while intoxicated

Dx Diagnosis

EBV Epstein-Barr virus

EC Enzyme Commission

E coli *Escherichia coli*

EDTA Ethylene diamine tetraacetic acid

EEG Electroencephalogram, electroencephalographic

eg *exempli gratia*, for example

EGF Epidermal growth factor

EIA Enzyme immunoassay

EKG Electrocardiography

ELISA Enzyme-linked immunosorbent assay

EM Electron microscopy, ultrastructure

EMG Electromyography

EMS Emergency medical services

EMT Emergency medical technician

ENT Ears, nose, and throat, otorhinolaryngology

EPA Environmental Protection Agency

ER 1. Emergency room, emergency ward 2. Endoplasmic reticulum

ERCP Endoscopic retrograde cholangiography

ERT Estrogen replacement therapy

ESR Erythrocyte sedimentation rate

ESRD End-stage renal disease

et al And others

etc Etcetera; yada, yada, yada

FDA United States Food and Drug Administration

FDP Fibrinogen degradation product(s)

FFA Free fatty acid

FISH Fluorescence in situ hybridization

FNA Fine-needle aspiration (biopsy or cytology)

FSH Follicle-stimulating hormone

FTT Failure to thrive

FUO Fever of unknown origin

GABA Gamma-aminobutyric acid

GC Gas chromatography

GC-MS Gas chromatography-mass spectroscopy

G-CSF Granulocyte-colony stimulating factor

GERD Gastroesophageal reflux disease

GFR Glomerular filtration rate

GGT Gamma-glutamyl transferase

GH-RH Growth hormone-releasing hormone

GI Gastrointestinal

GLC Gas-liquid chromatography

GM-CSF Granulocyte-macrophage colony-stimulating factor

GMS Gomori-methenamine-silver

GN Glomerulonephritis

GNP Gross National Product

GNR Gram-negative rods (bacilli)

G6PD Glucose-6-phosphate dehydrogenase

GRH Gonadotropin-releasing hormone

GTP Guanosine triphosphate

GTT Glucose tolerence test

GU Genitourinary

GVHD Graft-versus-host disease

H Hour

H⁺ Hydrogen ion

HAV Hepatitis A virus

Hb Hemoglobin

HBV Hepatitis B virus

hCG Human chorionic gonadotropin

HCFA Health Care & Financing Administration

HCl Hydrochloric acid

HCV Hepatitis C virus

HDL High-density lipoprotein

HDL-C High-density lipoprotein-cholesterol

HDN Hemolytic disease of the newborn

H&E Hematoxylin & eosin

hGH Human growth hormone

HHS US Department of Health and Human Services

HHV Human herpesvirus (HHV-1, HHV-etc)

H influenzae *Haemophilus influenzae*

HIV Human immunodeficiency virus

HLA Human leukocyte antigen (the major histocompatibility complex of humans)

HMO Health maintenance organization

H₂O₂ Hydrogen peroxide

HPLC High-performance liquid chromatography

HPV Human papillomavirus

H pylori *Helicobacter pylori*

HR Heart rate

HSV Herpes simplex virus

HTLV-I Human T cell leukemia/lymphoma virus

HUS Hemolytic-uremic syndrome

HVZ Herpes varicella-zoster

Hx History

IA Intraarterial

IBD Inflammatory bowel disease

ICP Intracranial pressure

ICU Intensive care unit

ID Identification, identify; infectious disease

IDDM Insulin-dependent diabetes mellitus

ie *id est*, that is to say

IFA Indirect fluorescent antibody

IFN Interferon

Ig Immunoglobulin

IGF-I Insulin-like growth factor I

IL Interleukin

IM Intramuscular

ImPx Immunoperoxidase

INR International normalized ratio

IQ Intelligence quotient

IR Infrared

IRMA Immunoradiometric assay

ISH In situ hybridization

ITP Idiopathic thrombocytopenic purpura

IU International Units

IUBMB International Union of Biochemistry and Molecular Biology

IUD Intrauterine (contraceptive) device

IV Intravenous

IVDU Intravenous drug use/user

IVP Intravenous pyelogram

JACC Journal of the American College of Cardiology

JAMA Journal of the American Medical Association

JCAHO Joint Commission of Accredited Hospitals Organization

JVP Jugular venous pulse

K⁺ Potassium

kb Kilobase

kbp Kilobasepair

kD Kilodalton

KS Kaposi sarcoma

LBW Low birth weight

LC Liquid chromatography

L&D Labor and delivery

LDH Lactate dehydrogenase

LDL Low-density lipoprotein

LDL-C Low-density lipoprotein-cholesterol

LES Lower esophageal sphincter

LGV Lymphogranuloma venereum

LH Luteinizing hormone

LH-RH Growth hormone-releasing hormone

LLQ Left lower quadrant

LM Light micoscopy

LMP Last menstrual period

LN Lymph node

LOC Loss of consciousness

LP Lumbar puncture

LT Leukotriene

LUQ Left upper quadrant

LVEF Left ventricular ejection fraction

MAb Monoclonal antibody

MAC *Mycobacterium avium* complex

MAOI Monoamine oxidase inhibitor

MCP Mayo Clinic Proceedings

MCTD Mixed connective tissue disease

MD Physician, medical doctor

MEN Multiple endocrine neoplasia

µg Microgram

mg Milligram

MHC Major histocompatibility complex

MI Myocardial infarction

MIM Mendelian Inheritance in Man

min Minute (time)

M&M Morbidity and mortality

mo/ma Monocyte/macrophage (tissue histiocyte)

MPS Mucopolysaccaride(s), mucopolysaccharidosis

MRI Magnetic resonance imaging

mRNA Messenger RNA (ribonucleic acid)

MRSA Methicillin-resistant *Staphylococcus aureus*

MS Multiple sclerosis

M tuberculosis *Mycobacterium tuberculosis*

MTX Methotrexate

MVA Motor vehicle accident

MW Molecular weight

NA Nomina Anatomica

Na⁺ Sodium

N/C ratio Nuclear/cytoplasmic ratio

N-CAM Neuronal-cell adhesion molecule

NCI National Cancer Institute (US)

NEJM New England Journal of Medicine

NG Nasogastric

NGF Nerve growth factor

N gonorrheae *Neisseria gonorrheae*

NYHA New York Hear Association

NHL Non-Hodgkin's lymphoma

NIDDM Non-insulin-dependent diabetes mellitus

NIH National Institutes of Health

NK cell Natural killer cell

NMDA *N*-methyl-D-aspartic acid, *N*-methyl-D-aspartate

NO Nitric oxide

NO₂ Nitrogen dioxide

NOS Nitric oxide synthase; not otherwise specified

NPO nothing by mouth

NSAID Nonsteroidal anti-inflammatory drug

NY New York

NYC New York City

O₂ Oxygen

Ob/Gyn Obstetrics and gynecology

OC Oral contraceptive(s)

OCD Obsessive-compulsive disorder

OD Overdose, optical density

O&P Ova and parasites

OR Operating room, operating suite

OSHA Occupational Safety and Health Administration

OTC Over-the-counter

PAF Platelet-activating factor

PAS Periodic acid-Schiff

PBC Primary biliary cirrhosis

PC Personal computer; politically correct

PCBs Polychlorinated biphenyls

PCP 1. Phencyclidine 2. *Pneumocystis carinii* pneumonia

PCR Polymerase chain reaction

PCTA Percutaneous transluminal angioplasty

PDA Patent ductus arteriosus

PDGF Platelet-derived growth factor

PEEP Positive end-expiratory pressure

PEF Peak expiratory flow

PET Positron emission tomography

PG Prostaglandin

PID Pelvic inflammatory disease

PKU Phenylketonuria

PMN Polymorphonuclear neutrophil or leukocyte, segmented neutrophil

PMS Premenstrual syndrome

PNS Parasympathetic nervous system, peripheral nervous system (see context)

PO By mouth

POC Products of conception

ppb Parts per billion

ppm Parts per million

PP&M Pharmacodynamics, pharmacokinetics, and metabolism

ppt Parts per trillion

PPID GL Mandell et al, Principles & Practice of Infectious Diseases, 4th ed, Churchill-Livingstone, 1995

ppm Parts per million

pron Pronounced

Pros Advantages

PRN *pro re nata*, as needed

PROM Premature rupture of membranes

PSA Prostate-specific antigen

PT Prothrombin time

Pt Patient

PTE Pulmonary thromboembolism

PTH Parathyroid hormone

aPTT (activated) Partial thromboplastin time

PUD Peptic ulcer disease

PVC Premature ventricular contraction

Px prognosis

Q Every

Q 4 H Every four hours

QA Quality assurance

QC Quality control

QD Every day

QID Four times a day

QOL Quality of life

r Recombinant

RAA Renin-aldosterone-angiotensin (system)

RA Rheumatoid arthritis

RAST Radioallergosorbent test

RBCs Red blood cells, erythrocytes

RBRVS Resource-based relative value scale

R&D Research & development

RDS Respiratory distress syndrome

re Regarding, about, concerning, in reference to

REM sleep Rapid eye movement sleep

RFLP Restriction fragment length polymorphism

rh Recombinant human

rH Rhesus factor

RIA Radioimmunoassay

RLQ Right lower quadrant

RN Registered nurse

R/O Rule out, exclude from diagnostic consideration

ROM Range of movement

RR Relative risk

rRNA Ribosomal RNA (ribonucleic acid)

RSV Respiratory syncytial virus

RT Radiation therapy, reverse transcriptase

RTI Respiratory tract infection

RUQ Right upper quadrant

Rx Therapy

S aureus Staphylococcus aureus

SCID Severe combined immunodeficiency

SD Standard deviation

sec Second (time)

SI International System (of units)

SIADH Syndrome of inappropriate secretion of ADH

SIDS Sudden infant death syndrome

SIL Squamous intraepithelial lesion

SIV Simian immunodeficiency virus

SLE Systemic lupus erythematosus

SNS Sympathetic nervous system

SO$_2$ Sulfur dioxide

SOB Shortness of breath

SOM School of medicine

SPECT Single photon emission-computed tomographic

SPEP Serum protein electrophoresis

S pneumoniae Streptococcus pneumoniae

spp Species

ssp Subspecies

STD Sexually transmitted disease

SVT Supraventricular tachycardia

Sx Symptoms

T$_3$ Triiodothyronine

T$_4$ Thyroxin

TAH-BSO Total abdominal hysterectomy with bilateral salpingo-oophorectomy

TB Tuberculosis

TDM Therapeutic drug monitoring

TdT Terminal deoxytransferase

TG Triglyceride

TGF-β Transforming growth factor-β

TIA Transient ischemic attack

TIBC Total iron-binding capacity

TID Three times a day

TLC Thin-layer chromatography

TNF Tumor necrosis factor

TNM Tumor, Node Metastasis classification

TMJ Temporomandibular joint

TOP Termination of pregnancy

tPA Tissue plasminogen activator

TPN Total parenteral nutrition

tRNA Transfer RNA (ribonucleic acid)

TRH Thyrotropin-releasing hormone

T-S Trimethoprim-sulfamethoxazole

TSH Thyroid-stimulating hormone, thyrotropin

TTP Thrombotic thrombocytopenic purpura

TURP Transurethral resection of the prostate

TX Thromboxane

Tx Tissue, pathology

U 1. Unit 2. University

UK United Kingdom

UN United Nations

URI Upper respiratory tract infection

US 1. Ultrasound, ultrasonographic 2. United States

USDA United States Department of Agriculture

usw *und so weiter*, etcetera

UTI Urinary tract infection

UV Ultraviolet

VDRL Venereal disease research laboratory (test) for syphilis

VEGF Vascular endothelial growth factor

VIP Vasoactive intestinal polypeptide

VLDL Very low density lipoportein

V/Q Ventilation/perfusion

VNTR Variable number of terminal repeats

VRE Vancomycin-resistant enterococcus

vs Versus, in contrast to, in comparison with

VSD Ventricular septal defect

VZV Varicella-zoster virus

WBCs White blood cells, leukocytes

WHO World Health Organization

WNL Within normal limits

WPW Wolff-Parkinson-White syndrome

WSJ Wall Street Journal

X-R X-linked recessive

SYMBOLS

- • Bullet point
- < Less than
- <<< Much less than
- ≤ Less than or equal to
- ? Uncertain
- > More than
- >>> Much more than

≥ More than or equal to

→ Leads to, results in

↔ Reversible

↓ Decrease, decreased, decreases, decreasing, reduces

↑ Increase, increased, increases, increasing

♀ Female, women

♂ Male, men

± About, approximately, circa

‡ See there

† Deceased, dead, mortality